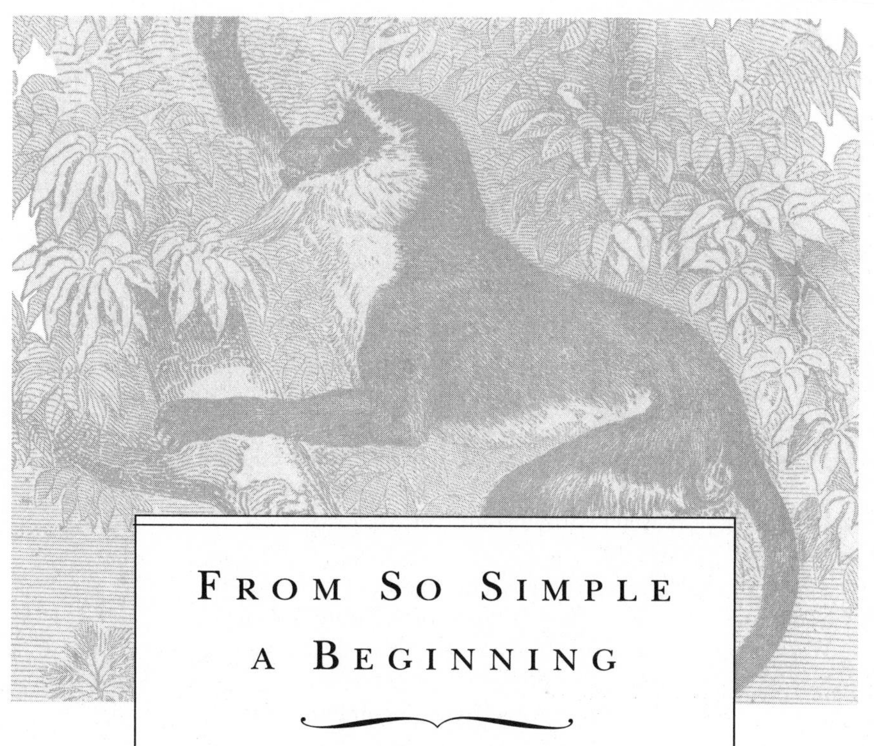

From So Simple A Beginning

The Four Great Books of Charles Darwin

Edited, with introductions by Edward O. Wilson

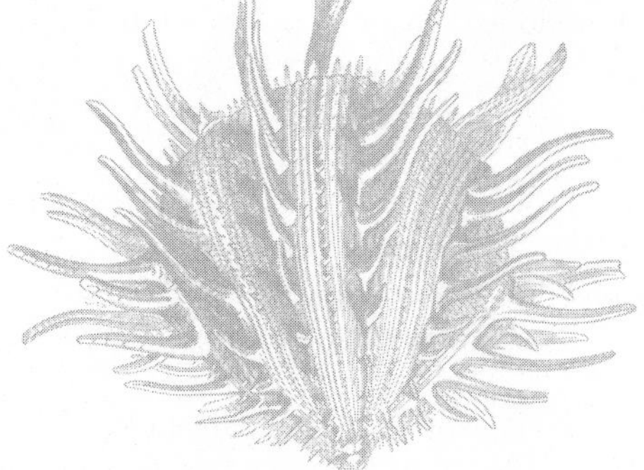

W. W. Norton & Company
New York London

Printed in the United States of America
First Edition

Manufacturing by The Haddon Craftsmen, Inc.
Book design by Judith Stagnitto Abbate
Production manager: Amanda Morrison

Library of Congress Cataloging-in-Publication Data

Darwin, Charles, 1809–1882.
From so simple a beginning : the four great books of Charles Darwin /
edited, with introductions by Edward O. Wilson.— 1st ed.
p. cm.
Includes bibliographical references and index.
ISBN 0-393-06134-5 (hardcover)
1. Evolution (Biology) 2. Natural selection.
3. Human beings–Origin. 4. Beagle Expedition (1831-1836)
5. Emotions. I. Wilson, Edward O. II. Title.
QH365.A1 2006
576.8'2—dc22
2005025825

W. W. Norton & Company, Inc.
500 Fifth Avenue, New York, N.Y. 10110
www.wwnorton.com

W. W. Norton & Company Ltd.
Castle House, 75/76 Wells Street, London W1T 3QT

1 2 3 4 5 6 7 8 9 0

From So Simple a Beginning

Ch. Darwin

Contents

We must acknowledge, as it seems to me, that man with all his noble qualities, with sympathy which feels for the most debased, with benevolence which extends not only to other men but to the humblest living creature, with his god-like intellect which has penetrated into the movements and constitution of the solar system—with all these exalted powers—Man still bears in his bodily frame the indelible stamp of his lowly origin.

—Charles Darwin
The Descent of Man, and Selection in Relation to Sex

From So Simple a Beginning

General Introduction

Great scientific discoveries are like sunrises. They illuminate first the steeples of the unknown, then its dark hollows. Such expansive influence has been enjoyed by the scientific writings of Charles Darwin. For over 150 years his books, the four most influential of which are reprinted here for the first time as a bound set, have spread light on the living world and the human condition. They have not lost their freshness: more than any other work in history's scientific canon, they are both timeless and persistently inspirational.

The four classics, flowing along one to the next like a well-wrought narrative, trace the development of Darwin's thought across almost all of his adult life. The first, *Voyage of the Beagle* (1845), one of literature's great travel books, is richly stocked with observations in natural history of the kind that were to guide the young Darwin toward his evolutionary worldview. Next comes the "one long argument," as he later put it, of *On the Origin of Species* (1859), arguably history's most influential book. In it the now middle-aged Darwin massively documents the evidences of organic evolution and introduces the theory of natural selection. *The Descent of Man* (1871) then addresses the burning topic foretold in *On the Origin of Species*: "Light will be thrown on the origin of man and his history." Finally, *The Expression of the Emotions in Man and Animals* (1872) draws close to the heart of the matter that concerns us all: the origin and nature of mind, the "citadel" that Darwin could see but knew that science at the time could not conquer.

The adventure that Darwin launched on all our behalf, and which continues into the twenty-first century, is driven by a deceptively simple idea, of which Darwin's friend and staunch supporter Thomas Henry Huxley said, and spoke for many to follow, "How extremely stupid of me not to have thought of that!" Evolution by natural selection is perhaps the only one true law unique to biological systems, as opposed to nonliving physical systems, and in recent decades it has taken on the solidity of a mathematical theorem. It states simply that if a population of organisms contains multiple hereditary variants in some trait (say, red versus blue eyes in a bird population), and if one of these variants succeeds in contributing more offspring to the next generation than the other variants, the overall composition of the population changes, and *evolution has occurred.* Further, if new genetic variants appear regularly in the population (by mutation or immigration), *evolution never ends.* Think of red-eyed and blue-eyed birds in a breeding population, and let the red-eyed birds be better adapted to the environment. The population will in time come to consist mostly or entirely of red-eyed birds. Now let green-eyed mutants appear that are even better adapted to the environment than the red-eyed form. As a consequence the species eventually becomes green-eyed. Evolution has thus taken two more small steps.

The full importance of Darwin's theory can be better understood by realizing that modern biology is guided by two overwhelmingly powerful and creative ideas. The first is that all biological processes are ultimately obedient to, even though far from fully explained by, the laws of physics and chemistry. The second is that all biological processes arose through evolution of these physicochemical systems through natural selection. The first principle is concerned with the *how* of biology. The second is concerned with the ways the systems adapted to the environment over periods of time long enough for evolution to occur—in other words the *why* of biology.

Knowledge addressing the first principle is called functional biology; that addressing the second is called evolutionary biology. If a moving automobile were an organism, functional biology would explain how it is constructed and operates, while evolutionary biology would reconstruct its origin and history—how it came to be made and its journey thus far.

The impact of the theory of evolution by natural selection, nowadays grown very sophisticated (and often referred to as the Modern Synthesis), has been profound. To the extent it can be upheld, and the evidence to date has done so compellingly, we must conclude that life has diversified on Earth autonomously without any kind of external guidance. Evolution in a pure Darwinian world has no goal or purpose: the exclusive driving force is random mutations sorted out by natural selection from one generation to the next.

What then are we to make of the purposes and goals obviously chosen by human beings? They are, in Darwinian interpretation, processes evolved as adaptive devices by an otherwise purposeless natural selection. Evolution by natural selection means, finally, that the essential qualities of the human mind also evolved autonomously. Humanity was thus born of Earth. However elevated in power over the rest of life, however exalted in self-image, we were descended from animals by the same blind force that created those animals, and we remain a member species of this planet's biosphere.

The revolution in astronomy begun by Nicolaus Copernicus in 1543 proved that Earth is not the center of the universe, nor even the center of the solar system. The revolution begun by Darwin was even more humbling: it showed that humanity is not the center of creation, and not its purpose either. But in freeing our minds from our imagined demigod bondage, even at the price of humility, Darwin turned our attention to the astounding power of the natural creative process and the magnificence of its products:

> There is grandeur in this view of life, with its several powers, having been originally breathed into a few forms or into one; and that, whilst this planet has gone cycling on according to the fixed law of gravity, from so simple a beginning endless forms most beautiful and most wonderful have been, and are being, evolved.
>
> DARWIN, *On the Origin of Species*
> (first edition, 1859)

Charles Darwin age 31 (1840)

The Voyage of the Beagle

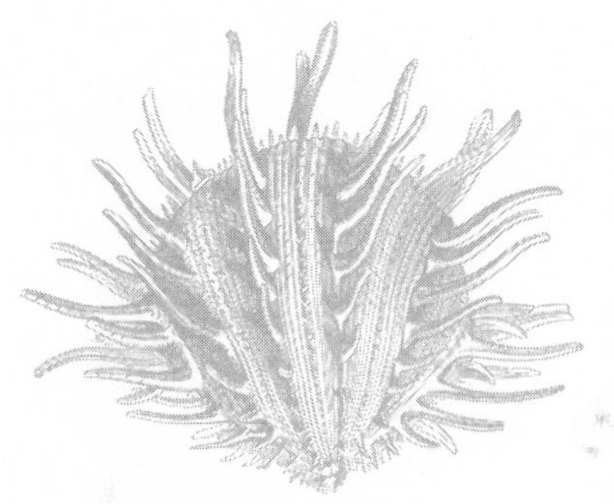

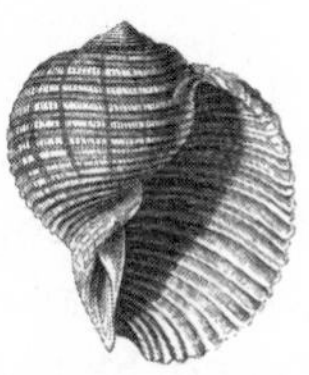

Introduction to *The Voyage of the Beagle*

IMAGINE: FOR FIVE YEARS no television, radio, telephone, newspapers, and no contact of any kind with home except months-old letters delivered at infrequent ports of call. Set before you, however, is something far better: an unexplored world to navigate and record for history, with unimaginable discoveries that await at every turn. For a young naturalist launched upon such a great voyage, with the vast majority of Earth's plants and animals yet unknown to science, all the specimens he could collect valuable to museums, and all the observations he made worth recording, and with no duties except gentleman companion to the captain and ship's naturalist, the world was Charles Darwin's to possess.

From such an epic trip, undertaken by Darwin at age twenty-two in 1831 and concluded in 1836, came *Journal of Researches into the Natural History and Geology of the Countries Visited During the Voyage of H. M. S. Beagle Round the World,* or, as now cited universally, *Voyage of the Beagle.* It is today regarded as intellectually the most important travel book of all time. Published in 1845, it also became the Victorian equivalent of a bestseller. Such *mémoire,* supplemented with an occasional personal lecture from the author on tour, were the vicarious and sole means most people in Western countries had of traveling beyond the confines of Europe and North America. *Voyage of the Beagle* is still an experience to enjoy, whether read from front to back in order to repeat the entire journey, or merely browsed for sporadic pleasure. The book, moreover, has great value, apart from the historical significance of the author, in its function as a kind of time machine. Its prose transports the reader not just in space but back in time across nearly two centuries. We enter the great Atlantic Forest of Brazil before it was logged; visit the Fuegian people at the tip of South America before they were extinguished; and tour places that even today seem almost impossibly remote: Ascension Island, St. Paul's Rocks, the Falkland Islands, Fernando de Noronha, the Cocos (Keeling) Islands. Darwin spares no detail; his obsessiveness is as fortunate for the general reader as it is for the historian. Most important, we read in *Voyage of the Beagle* of the experiences that were to be seminal in Darwin's later conception of evolution. When Darwin first spoke publicly of the theory three decades later, it was from an unprecedented experience of global natural history. The journey, he testified in later years, was the transformative experience of his life.

Charles Darwin was singularly primed to make the most of such a voyage. Born into a wealthy and famous family on February 12, 1809 (the same year and day as Abraham Lincoln), he was free to pursue any course of life that suited him. His choice was not reached easily, however. When he entered Cambridge University in the 1820s, he was already focused on science but found little formal education available except in mathematics and the classics. Moreover, he tried but could muster little enthusiasm for medicine and the ministry, the two callings most easily available to him out of university. Fortunately, he found mentors on the faculty, including John Stevens Henslow and Adam Sedgwick, who encouraged his study of science, and particularly his bent toward scientific natural history, the forerunner of the modern disciplines of ecology and evolutionary biology. Like so many young scientists down to the present time, Darwin drew inspiration from books by masters of his hoped-for calling. By his own reckoning, two stood out: *Preliminary Discourse on the Study of Natural Philosophy* (1831), by the famous astronomer John F. W. Herschel, which exposited the fundamental procedures and standards of science; and *Personal Narrative of Travels to the Equinoctial Regions of the New Continent* (1814–29), by the pioneering explorer-naturalist Alexander von Humboldt. To this, Darwin added his passion for collecting beetles, which provided him with a certain exactitude of attention: no creature was too small to excite his interest.

Darwin by temperament and circumstance was thus an ambitious young man preconditioned to be a scientist and explorer-naturalist. In 1831, with the help of Henslow, professor of botany at Cambridge, he obtained a berth on H.M.S. *Beagle.* The journey began, and the world opened before him.

Voyage of the Beagle can be read from several perspectives and interpreted according to taste. One very important but seldom noticed feature is its exemplification of the *Wanderjahre* (years of wandering) in the genesis of the scientific mind. No English term conveys the exact same meaning as the German. It refers originally to the medieval custom of sending young men to other villages or towns to learn a craft and more of the world in a different setting. History has shown that there is no more fruitful way to launch the career of a naturalist than by such an interlude, during which the adventurer travels alone, searching, freed from domestic ties, and energized by the visceral ambition of the young.

The pages of *Voyage of the Beagle* are the diary of Darwin's *Wanderjahre.* As he proceeds around the world (England–South America–Galapagos–South Pacific–South Africa–South America–England), the young naturalist unconsciously builds the foundation of what was to be his evolutionary view of life. He rejoices in the exuberant diversity of life of Brazil's Atlantic rainforest. He unearths fossils of extinct mammals in the beds of Argentina and notices that while they are distinct from living species in some respects, they resemble them in others. He is impressed by the similarity of different species of the contemporary faunas found at different parts of South America. Among his famous collections made in the Galapagos Islands are mockingbirds that vary consistently from one island of the archipelago to the next. When this fact is pointed out to him back in England by the ornithologist and artist John Gould, he at last becomes convinced of the general occurrence of evolution and begins his quest for the process that causes it.

The Voyage of the Beagle

Journal of Researches into the Natural History and Geology of the Countries Visited During the Voyage of H.M.S. Beagle Round the World

Charles Darwin

· 1845 ·

Contents

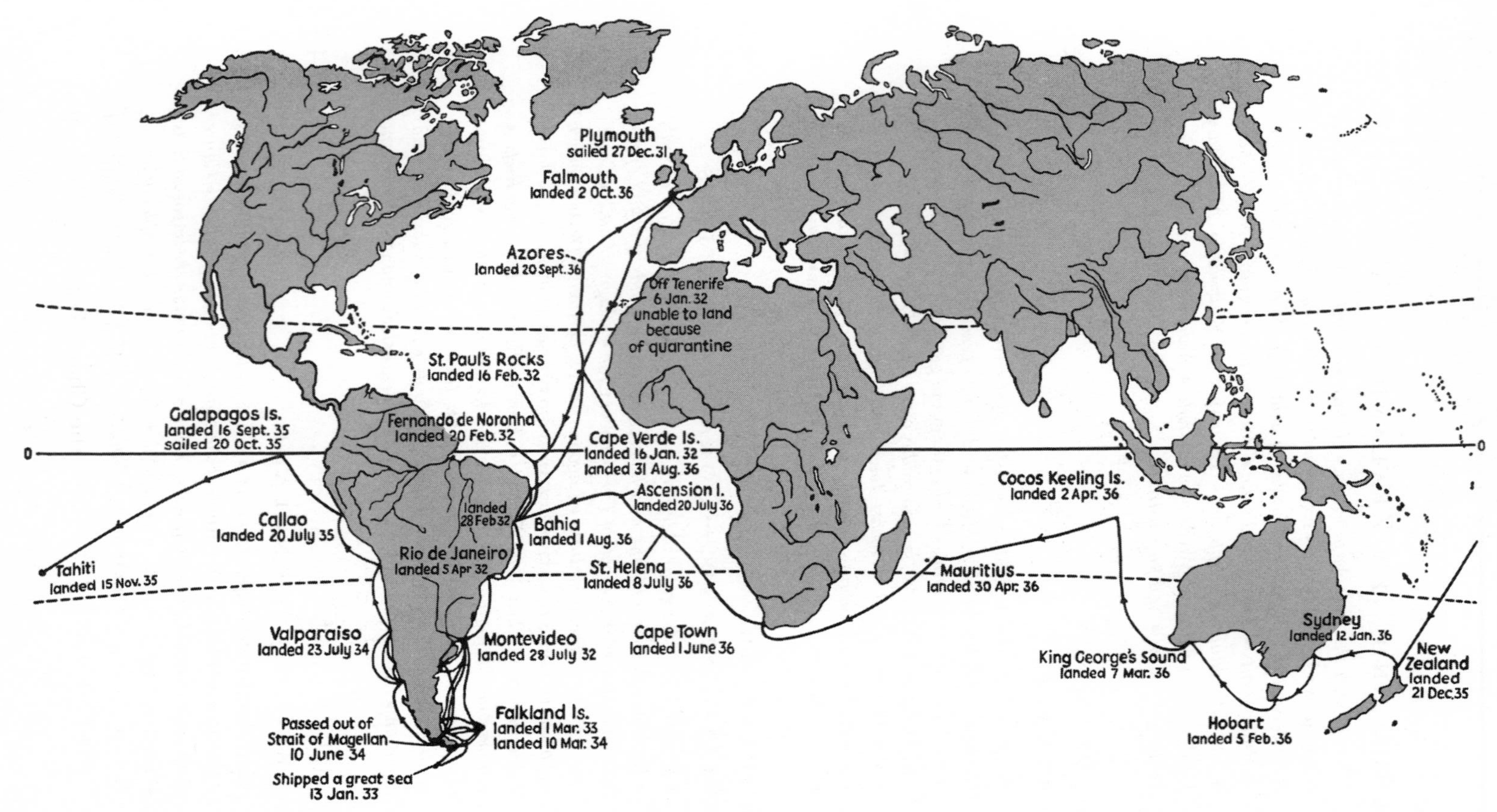

The Voyage of the Beagle, *1831–1836.*

I HAVE STATED in the preface to the first Edition of this work, and in the Zoology of the Voyage of the Beagle, that it was in consequence of a wish expressed by Captain Fitz Roy, of having some scientific person on board, accompanied by an offer from him of giving up part of his own accommodations, that I volunteered my services, which received, through the kindness of the hydrographer, Captain Beaufort, the sanction of the Lords of the Admiralty. As I feel that the opportunities which I enjoyed of studying the Natural History of the different countries we visited, have been wholly due to Captain Fitz Roy, I hope I may here be permitted to repeat my expression of gratitude to him; and to add that, during the five years we were together, I received from him the most cordial friendship and steady assistance. Both to Captain Fitz Roy and to all the Officers of the Beagle[1] I shall ever feel most thankful for the undeviating kindness with which I was treated during our long voyage.

This volume contains, in the form of a Journal, a history of our voyage, and a sketch of those observations in Natural History and Geology, which I think will possess some interest for the general reader. I have in this edition largely condensed and corrected some parts, and have added a little to others, in order to render the volume more fitted for popular reading; but I trust that naturalists will remember, that they must refer for details to the larger publications which comprise the scientific results of the Expedition. The Zoology of the Voyage of the Beagle includes an account of the Fossil Mammalia, by Professor Owen; of the Living Mammalia, by Mr. Waterhouse; of the Birds, by Mr. Gould; of the Fish, by the Rev. L. Jenyns; and of the Reptiles, by Mr. Bell. I have appended to the descriptions of each species an account of its habits and range. These works, which I owe to the high talents and disinterested zeal of the above distinguished authors, could not have been undertaken, had it not been for the liberality of the Lords Commissioners of Her Majesty's Treasury, who, through the representation of the Right Honourable the Chancellor of the Exchequer, have been pleased to grant a sum of one thousand pounds towards defraying part of the expenses of publication.

I have myself published separate volumes on the 'Structure and Distribution of Coral Reefs;' on the 'Volcanic Islands visited during the Voyage of the Beagle;' and on the 'Geology of South America.' The sixth volume of the 'Geological Transactions' contains two papers of mine on the Erratic Boulders and Volcanic Phenomena of South America. Messrs. Waterhouse, Walker, Newman, and White, have published several able papers on the Insects which were collected, and I trust that many others will hereafter follow. The plants from the southern parts of America will be given by Dr. J. Hooker, in his great work on the Botany of the Southern Hemisphere. The Flora of the Galapagos Archipelago is the subject of a separate memoir by him, in the 'Linnean Transactions.' The Reverend Professor Henslow has published a list of the plants collected by me at the Keeling Islands; and the Reverend J. M. Berkeley has described my cryptogamic plants.

I shall have the pleasure of acknowledging the great assistance which I have received from several other naturalists, in the course of this and my other works; but I must be here allowed to return my most sincere thanks to the Reverend Professor Henslow, who, when I was an undergraduate at Cambridge, was one chief means of giving me a taste for Natural History,—who, during my absence, took charge of the collections I sent home, and by his correspondence directed my endeavours,—and who, since my return, has constantly rendered me every assistance which the kindest friend could offer.

DOWN, BROMLEY, KENT,
June, 1845.

CHAPTER I

ST. JAGO—CAPE DE VERD ISLANDS

PORTO PRAYA • RIBEIRA GRANDE • ATMOSPHERIC DUST WITH INFUSORIA • HABITS OF A SEA-SLUG AND CUTTLE-FISH • ST. PAUL'S ROCKS, NON-VOLCANIC • SINGULAR INCRUSTATIONS • INSECTS THE FIRST COLONISTS OF ISLANDS • FERNANDO NORONHA • BAHIA • BURNISHED ROCKS • HABITS OF A DIODON • PELAGIC CONFERVÆ AND INFUSORIA • CAUSES OF DISCOLOURED SEA

AFTER HAVING BEEN twice driven back by heavy southwestern gales, Her Majesty's ship Beagle, a ten-gun brig, under the command of Captain Fitz Roy, R. N., sailed from Devonport on the 27th of December, 1831. The object of the expedition was to complete the survey of Patagonia and Tierra del Fuego, commenced under Captain King in 1826 to 1830—to survey the shores of Chile, Peru, and of some islands in the Pacific—and to carry a chain of chronometrical measurements round the World. On the 6th of January we reached Teneriffe, but were prevented landing, by fears of our bringing the cholera: the next morning we saw the sun rise behind the rugged outline of the Grand Canary island, and suddenly illuminate the Peak of Teneriffe, whilst the lower parts were veiled in fleecy clouds. This was the first of many delightful days never to be forgotten. On the 16th of January, 1832, we anchored at Porto Praya, in St. Jago, the chief island of the Cape de Verd archipelago.

The neighbourhood of Porto Praya, viewed from the sea, wears a desolate aspect. The volcanic fires of a past age, and the scorching heat of a tropical sun, have in most places rendered the soil unfit for vegetation. The country rises in successive steps of table-land, interspersed with some truncate conical hills, and the horizon is bounded by an irregular chain of more lofty mountains. The scene, as beheld through the hazy atmosphere of this climate, is one of great interest; if, indeed, a person, fresh from sea, and who has just walked, for the first time, in a grove of cocoa-nut trees, can be a judge of anything but his own happiness. The island would generally be considered as very uninteresting; but to anyone accustomed only to an English landscape, the novel aspect of an utterly sterile land possesses a grandeur which more vegetation might spoil. A single green leaf can scarcely be discovered over wide tracts of the lava plains; yet flocks of goats, together with a few cows, contrive to exist. It rains very seldom, but during a short portion of the year heavy torrents fall, and immediately afterwards a light vegetation springs out of every crevice. This soon withers; and upon such naturally formed hay the animals live. It had not now rained for an entire year. When the island was discovered, the immediate neighbourhood of Porto Praya was clothed with 'rees,[2] the reckless destruction of which has caused here, as at St. Helena, and at some of the Canary islands, almost entire sterility. The broad, flat-bottomed valleys, many of which serve during a few days only in the season as water-courses, are clothed with thickets of leafless bushes. Few living creatures inhabit these valleys. The commonest bird is a kingfisher (Dacelo Iagoensis), which tamely sits on the branches of the castor-oil plant, and thence darts on grasshoppers and lizards. It is brightly coloured, but not so beautiful as the European species: in its flight, manners, and place of habitation, which is generally in the driest valley, there is also a wide difference.

One day, two of the officers and myself rode to Ribeira Grande, a village a few miles eastward of Porto Praya. Until we reached the valley of St. Martin, the country presented its usual dull brown appearance; but here, a very small rill of water produces a most refreshing margin of luxuriant vegetation. In the course of an hour we arrived at Ribeira Grande, and were surprised at the sight of a large ruined fort and cathedral. This little town, before its harbour was filled up, was the principal place in the island: it now presents a melancholy, but very picturesque appearance. Having procured a black Padre for a guide, and a Spaniard who had served in the Peninsular war as an interpreter, we visited a collection of buildings, of which an ancient church formed the principal part. It is here the governors and captain-generals of the islands have been buried. Some of the tombstones recorded dates of the sixteenth century.[3] The heraldic ornaments were the only things in this retired place that reminded us of Europe. The church or chapel formed one side of a quadrangle, in the middle of which a large clump of bananas were growing. On another side was a hospital, containing about a dozen miserable-looking inmates.

We returned to the Vênda to eat our dinners. A considerable number of men, women, and children, all as black as jet, collected to watch us. Our companions were extremely merry; and everything we said or did was followed by their hearty laughter. Before leaving the town we visited the cathedral. It does not appear so rich as the smaller church, but boasts of a little organ, which sent forth singularly inharmonious cries. We presented the black priest with a few shillings, and the Spaniard, patting him on the head, said, with much candour, he thought his colour made no great difference. We then returned, as fast as the ponies would go, to Porto Praya.

Another day we rode to the village of St. Domingo, situated near the centre of the island. On a small plain which we crossed, a few stunted acacias were growing; their tops had been bent by the steady trade-wind, in a singular manner—some of them even at right angles to their trunks. The direction of the branches was exactly N. E. by N., and S. W. by S., and these natural vanes must indicate the prevailing direction of the force of the trade-wind. The travelling had made so little impression on the barren soil, that we here missed our track, and took that to Fuentes. This we did not find out till we arrived there; and we were afterwards glad of our mistake. Fuentes is a pretty village, with a small stream, and everything appeared to prosper well, excepting, indeed, that which ought to do so most—its inhabitants. The black children, completely naked, and looking very wretched, were carrying bundles of firewood half as big as their own bodies.

Near Fuentes we saw a large flock of guinea-fowl—probably fifty or sixty in number. They were extremely wary, and could not be approached. They avoided us, like partridges on a rainy day in September, running with their heads cocked up; and if pursued, they readily took to the wing.

The scenery of St. Domingo possesses a beauty totally unexpected, from the prevalent gloomy character of the rest of the island. The village is situated at the bottom of a valley, bounded by lofty and jagged walls of stratified lava. The black rocks afford a most striking contrast with the bright green vegetation, which follows the banks of a little stream of clear water. It happened to be a grand feast-day, and the village was full of people. On our return we overtook a party of about twenty young black girls, dressed in excellent taste; their black skins and snow-white linen being set off by coloured turbans and large shawls. As soon as we approached near, they suddenly all turned round, and covering the path with their shawls, sung with great energy a wild song, beating time with their hands upon their legs. We threw them some vintéms, which were received with screams of laughter, and we left them redoubling the noise of their song.

One morning the view was singularly clear; the distant mountains being projected with the sharpest outline on a heavy bank of dark blue clouds. Judging from the appearance, and from similar cases in England, I supposed that the air was saturated with moisture. The fact, however, turned out quite the contrary. The hygrometer gave a difference of 29.6 degrees, between the temperature of the air, and the point at which dew was precipitated. This difference was nearly double that which I had observed on the previous mornings. This unusual degree of atmospheric dryness was accompanied by continual flashes of lightning. Is it not an uncommon case, thus to find a remarkable degree of aerial transparency with such a state of weather?

Generally the atmosphere is hazy; and this is caused by the falling of impalpably fine dust, which was found to have slightly injured the astronomical instruments. The morning before we anchored at Porto Praya, I collected a little packet of this brown-coloured fine dust, which appeared to have been filtered from the wind by the gauze of the vane at the masthead. Mr. Lyell has also given me four packets of dust which fell on a vessel a few hundred miles northward of these islands. Professor Ehrenberg[4] finds that this dust consists in great part of infusoria with siliceous shields, and of the siliceous tissue of plants. In five little packets which I sent him, he has ascertained no less than sixty-seven different organic forms! The infusoria, with the exception of two marine species, are all inhabitants of fresh-water. I have found no less than fifteen different accounts of dust having fallen on vessels when far out in the Atlantic. From the direction of the wind whenever it has fallen, and from its having always fallen during those months when the harmattan is known to raise clouds of dust high into the atmosphere, we may feel sure that it all comes from Africa. It is, however, a very singular fact, that, although Professor Ehrenberg knows many species of infusoria peculiar to Africa, he finds none of these in the dust which I sent him. On the other hand, he finds in it two species which hitherto he knows as living only in South America. The dust falls in such quantities as to dirty everything on board, and to hurt people's eyes; vessels even have run on shore owing to the obscurity of the atmosphere. It has often fallen on ships when several hundred, and even more than a thousand miles from the coast of Africa, and at points sixteen hundred miles distant in a north and south direction. In some dust which was collected on a vessel three hundred miles from the land, I was much surprised to find particles of stone above the thousandth of an inch square, mixed with finer matter. After this fact one need not be surprised at the diffusion of the far lighter and smaller sporules of cryptogamic plants.

The geology of this island is the most interesting part of its natural history. On entering the harbour, a perfectly horizontal white band, in the face of the sea cliff, may be seen running for some miles along the coast, and at the height of about forty-five feet above the water. Upon examination, this white stratum is found to consist of calcareous matter, with numerous shells embedded, most or all of which now exist on the neighbouring coast. It rests on ancient volcanic rocks, and has been covered by a stream of basalt, which must have entered the sea when the white shelly bed was lying at the bottom. It is interesting to trace the changes, produced by the heat of the overlying lava, on the friable mass, which in parts has been converted into a crystalline limestone, and in other parts into a compact spotted stone. Where the lime has been caught up by the scoriaceous fragments of the lower surface of the stream, it is converted into groups of beautifully radiated fibres resembling arragonite. The beds of lava rise in successive gently-sloping plains, towards the interior, whence the deluges of melted stone have originally proceeded. Within historical times, no signs of volcanic activity have, I believe, been manifested in any part of St. Jago. Even the form of a crater can but rarely be discovered on the summits of the many red cindery hills; yet the more recent streams can be distinguished on the coast, forming lines of cliffs of less height, but stretching out in advance of those belonging to an older series: the height of the cliffs thus affording a rude measure of the age of the streams.

During our stay, I observed the habits of some marine animals. A large Aplysia is very common. This sea-slug is about five inches long; and is of a dirty yellowish colour, veined with purple. On each side of the lower surface, or foot, there is a broad membrane, which appears sometimes to act as a ventilator, in causing a current of water to flow over the dorsal branchiæ or lungs. It feeds on the delicate sea-weeds which grow among the stones in muddy and shallow water; and I found in its stomach several small pebbles, as in the gizzard of a bird. This slug, when disturbed, emits a very fine purplish-red fluid, which stains the water for the space of a foot around. Besides this means of defence, an acrid secretion, which is spread over its body, causes a sharp, stinging sensation, similar to that produced by the Physalia, or Portuguese man-of-war.

I was much interested, on several occasions, by watching the habits of an Octopus, or cuttle-fish. Although common in the pools of water left by the retiring tide, these animals were not easily caught. By means of their long arms and suckers, they could drag their bodies into very narrow crevices; and when thus fixed, it required great force to remove them. At other times they darted tail first, with the rapidity of an arrow, from one side of the pool to the other, at the same instant discolouring the water with a dark chestnut-brown ink. These animals also escape detection by a very extraordinary, chameleon-like power of changing their colour. They appear to vary their tints according to the nature of the ground over which they pass: when in deep water, their general shade was brownish purple, but when placed on the land, or in shallow water, this dark tint changed into one of a yellowish green. The colour, examined more carefully, was a French grey, with numerous minute spots of bright yellow: the former of these varied in intensity; the latter entirely disappeared and appeared again by turns. These changes were effected in such a manner, that clouds, varying in tint between a hyacinth red and a chestnut-brown,[5] were continually passing over the body. Any part, being subjected to a slight shock of galvanism, became almost black: a similar effect, but in a less degree, was produced by scratching the skin with a needle. These clouds, or blushes as they may be called, are said to be produced by the alternate expansion and contraction of minute vesicles containing variously coloured fluids.[6]

This cuttle-fish displayed its chameleon-like power both during the act of swimming and whilst remaining stationary at the bottom. I was much amused by the various arts to escape detection used by one individual, which seemed fully aware that I was watching it. Remaining for a time motionless, it would then stealthily advance an inch or two, like a cat after a mouse; sometimes changing its colour: it thus proceeded, till having gained a deeper part, it darted away, leaving a dusky train of ink to hide the hole into which it had crawled.

While looking for marine animals, with my head about two feet above the rocky shore, I was more than once saluted by a jet of water, accompanied by a slight grating noise. At first I could not think what it was, but afterwards I found out that it was this cuttle-fish, which, though concealed in a hole, thus often led me to its discovery. That it possesses the power of ejecting water there is no doubt, and it appeared to me that it could certainly take good aim by directing the tube or siphon on the under side of its body. From the difficulty which these animals have in carrying their heads, they cannot crawl with ease when placed on the ground. I observed that one which I kept in the cabin was slightly phosphorescent in the dark.

St. Paul's Rocks.—In crossing the Atlantic we hove-to, during the morning of February 16th, close to the island of St. Paul's. This cluster of rocks is situated in 0° 58' north latitude, and 29° 15' west longitude. It is 540 miles distant from the coast of America, and 350 from the island of Fernando Noronha. The highest point is only fifty feet above the level of the sea, and the entire circumference is under three-quarters of a mile. This small point rises abruptly out of the depths of the ocean. Its mineralogical constitution is not simple; in some parts the rock is of a cherry, in others of a felspathic nature, including thin veins of serpentine. It is a remarkable fact, that all the many small islands, lying far from any continent, in the Pacific, Indian, and Atlantic Oceans, with the exception of the Seychelles and this little point of rock, are, I believe, composed either of coral or of erupted matter. The volcanic nature of these oceanic islands is evidently an extension of that law, and the effect of those same causes, whether chemical or mechanical, from which it results that a vast majority of the volcanoes now in action stand either near sea-coasts or as islands in the midst of the sea.

The rocks of St. Paul appear from a distance of a brilliantly white colour. This is partly owing to the dung of a vast multitude of seafowl, and partly to a coating of a hard glossy substance with a pearly lustre, which is intimately united to the surface of the rocks. This, when examined with a lens, is found to consist of numerous exceedingly thin layers, its total thickness being about the tenth of an inch. It contains much animal matter, and its origin, no doubt, is due to the action of the rain or spray on the birds' dung. Below some small masses of guano at Ascension, and on the Abrolhos Islets, I found certain stalactitic branching bodies, formed apparently in the same manner as the thin white coating on these rocks. The branching bodies so closely resembled in general appearance certain nulliporæ (a family of hard calcareous sea-plants), that in lately looking hastily over my collection I did not perceive the difference. The globular extremities of the branches are of a pearly texture, like the enamel of teeth, but so hard as just to scratch plate-glass. I may here mention, that on a part of the coast of Ascension, where there is a vast accumulation of shelly sand, an incrustation is deposited on the tidal rocks by the water of the sea, resembling, as represented in the woodcut, certain cryptogamic plants (Marchantiæ) often seen on damp walls. The surface of the fronds is beautifully glossy; and those parts formed where fully exposed to the light are of a jet black colour, but those shaded under ledges are only grey. I have shown specimens of this incrustation to several geologists, and they all thought that they were of volcanic or igneous origin! In its hardness and translucency—in its polish, equal to that of the finest oliva-shell—in the bad smell given out, and loss of colour under the blowpipe—it shows a close similarity with living sea-shells. Moreover, in sea-shells, it is known that the parts habitually covered and shaded by the mantle of the animal, are of a paler colour than those fully exposed to the light, just as is the case with this incrustation. When we remember that lime, either as a phosphate or carbonate, enters into the composition of the hard parts, such as bones and shells, of all living animals, it is an interesting physiological fact[7] to find substances harder than the enamel of teeth, and coloured surfaces as well polished as those of a fresh shell, reformed through inorganic means from dead organic matter—mocking, also, in shape, some of the lower vegetable productions.

We found on St. Paul's only two kinds of birds—the booby and the noddy. The former is a species of gannet, and the latter a tern. Both are of a tame and stupid disposition, and are so unaccustomed to visitors, that I could have killed any number of them with my geological hammer. The booby lays her eggs on the bare rock; but the tern makes a very simple nest with sea-weed. By the side of many of these nests a small flying-fish was placed; which, I suppose, had been brought by the male bird for its partner. It was amusing to watch how quickly a large and active crab (Graspus), which inhabits the crevices of the rock, stole the fish from the side of the nest, as soon as we had disturbed the parent birds. Sir W. Symonds, one of the few persons who have landed here, informs me that he saw the crabs dragging even the young birds out of their nests, and devouring them. Not a single plant, not even a lichen, grows on this islet; yet it is inhabited by several insects and spiders. The following list completes, I believe, the terrestrial fauna: a fly (Olfersia) living on the booby, and a tick which must have come here as a parasite on the birds; a small brown moth, belonging to a genus that feeds on feathers; a beetle (Quedius) and a woodlouse from beneath the dung; and lastly, numerous spiders, which I suppose prey on these small attendants and scavengers of the water-fowl. The often repeated description of the stately palm and other noble tropical plants, then birds, and lastly man, taking possession of the coral islets as soon as formed, in the Pacific, is probably not correct; I fear it destroys the poetry of this story, that feather and dirt-feeding and parasitic insects and spiders should be the first inhabitants of newly formed oceanic land.

The smallest rock in the tropical seas, by giving a foundation for the growth of innumerable kinds of seaweed and compound animals, supports likewise a large number of fish. The sharks and the seamen in the boats maintained a constant struggle which should secure the greater share of the prey caught by the fishing-lines. I have heard that a rock near the Bermudas, lying many miles out at sea, and at a considerable depth, was first discovered by the circumstance of fish having been observed in the neighbourhood.

FERNANDO NORONHA, *Feb. 20th.*—As far as I was enabled to observe, during the few hours we stayed at this place, the constitution of the island is volcanic, but probably not of a recent date. The most remarkable feature is a conical hill, about one thousand feet high, the upper part of which is exceedingly steep, and on one side overhangs its base. The rock is phonolite, and is divided into irregular columns. On viewing one of these isolated masses, at first one is inclined to believe that it has been suddenly pushed up in a semi-fluid state. At St. Helena, however, I ascertained that some pinnacles, of a nearly similar figure and constitution, had been formed by the injection of melted rock into yielding strata, which thus had formed the moulds for these gigantic obelisks. The whole island is covered with wood; but from the dryness of the climate there is no appearance of luxuriance. Half-way up the mountain, some great masses of the columnar rock, shaded by laurel-like trees, and ornamented by others covered with fine pink flowers but without a single leaf, gave a pleasing effect to the nearer parts of the scenery.

BAHIA, OR SAN SALVADOR. BRAZIL, *Feb. 29th.*—The day has passed delightfully. Delight itself, however, is a weak term to express the feelings of a naturalist who, for the first time, has wandered by himself in a Brazilian forest. The elegance of the grasses, the novelty of the parasitical plants, the beauty of the flowers, the glossy green of the foliage, but above all the general luxuriance of the vegetation, filled me with admiration. A most paradoxical mixture of sound and silence pervades the shady parts of the wood. The noise from the insects is so loud, that it may be heard even in a vessel anchored several hundred yards from the shore; yet within the recesses of the forest a universal silence appears to reign. To a person fond of natural history, such a day as this brings with it a deeper pleasure than he can ever hope to experience again. After wandering about for some hours, I returned to the landing-place; but, before reaching it, I was overtaken by a tropical storm. I tried to find shelter under a tree, which was so thick that it would never have been penetrated by common English rain; but here, in a couple of minutes, a little torrent flowed down the trunk. It is to this violence of the rain that we must attribute the verdure at the bottom of the thickest woods: if the showers were like those of a colder climate, the greater part would be absorbed or evaporated before it reached the ground. I will not at present attempt to describe the gaudy scenery of this noble bay, because, in our homeward voyage, we called here a second time, and I shall then have occasion to remark on it.

Along the whole coast of Brazil, for a length of at least 2000 miles, and certainly for a considerable space inland, wherever solid rock occurs, it belongs to a granitic formation. The circumstance of this enormous area being constituted of materials which most geologists believe to have been crystallized when heated under pressure, gives rise to many curious reflections. Was this effect produced beneath the depths of a profound ocean? or did a covering of strata formerly extend over it, which has since been removed? Can we believe that any power, acting for a time short of infinity, could have denuded the granite over so many thousand square leagues?

On a point not far from the city, where a rivulet entered the sea, I observed a fact connected with a subject discussed by Humboldt.[8] At the cataracts of the great rivers Orinoco, Nile, and Congo, the syenitic rocks are coated by a black substance, appearing as if they had been polished with plumbago. The layer is of extreme thinness; and on analysis by Berzelius it was found to consist of the oxides of manganese and iron. In the Orinoco it occurs on the rocks periodically washed by the floods, and in those parts alone where the stream is rapid; or, as the Indians say, "the rocks are black where the waters are white." Here the coating is of a rich brown instead of a black colour, and seems to be composed of ferruginous matter alone. Hand specimens fail to give a just idea of these brown burnished stones which glitter in the sun's rays. They occur only within the limits of the tidal waves; and as the rivulet slowly trickles down, the surf must supply the polishing power of the cataracts in the great rivers. In like manner, the rise and fall of the tide probably answer to the periodical inundations; and thus the same effects are produced under apparently different but really similar circumstances. The origin, however, of these coatings of metallic oxides, which seem as if cemented to the rocks, is not understood; and no reason, I believe, can be assigned for their thickness remaining the same.

One day I was amused by watching the habits of the Diodon antennatus, which was caught swimming near the shore. This fish, with its flabby skin, is well known to possess the singular power of distending itself into a nearly spherical form. After having been taken out of water for a short time, and then again immersed in it, a considerable quantity both of water and air is absorbed by the mouth, and perhaps likewise by the branchial orifices. This process is effected by two methods: the air is swallowed, and is then forced into the cavity of the body, its return being prevented by a muscular contraction which is externally visible: but the water enters in a gentle stream through the mouth, which is kept wide open and motionless; this latter action must, therefore, depend on suction. The skin about the abdomen is much looser than that on the back; hence, during the inflation, the lower surface becomes far more distended than the upper; and the fish, in consequence, floats with its back downwards. Cuvier doubts whether the Diodon in this position is able to swim; but not only can it thus move forward in a straight line, but it can turn round to either side. This latter movement is effected solely by the aid of the pectoral fins; the tail being collapsed, and not used. From the body being buoyed up with so much air, the branchial openings are out of water, but a stream drawn in by the mouth constantly flows through them.

The fish, having remained in this distended state for a short time, generally expelled the air and water with considerable force from the branchial apertures and mouth. It could emit, at will, a certain portion of the water; and it appears, therefore, probable that this fluid is taken in partly for the sake of regulating its specific gravity. This Diodon possessed several means of defence. It could give a severe bite, and could eject water from its mouth to some distance, at the same time making a curious noise by the movement of its jaws. By the inflation of its body, the papillæ, with which the skin is covered, become erect and pointed. But the most curious circumstance is, that it secretes from the skin of its belly, when handled, a most beautiful carmine-red fibrous matter, which stains ivory and paper in so permanent a manner that the tint is retained with all its brightness to the present day: I am quite ignorant of the nature and use of this secretion. I have heard from Dr. Allan of Forres, that he has frequently found a Diodon, floating alive and distended, in the stomach of the shark; and that on several occasions he has known it eat its way, not only through the coats of the stomach, but through the sides of the monster, which has thus been killed. Who would ever have imagined that a little soft fish could have destroyed the great and savage shark?

March 18th.—We sailed from Bahia. A few days afterwards, when not far distant from the Abrolhos Islets, my attention was called to a reddish-brown appearance in the sea. The whole surface of the water, as it appeared under a weak lens, seemed as if covered by chopped bits of hay, with their ends jagged. These are minute cylindrical confervæ, in bundles or rafts of from twenty to sixty in each. Mr. Berkeley informs me that they are the same species (Trichodesmium erythræum) with that found over large spaces in the Red Sea, and whence its name of Red Sea is derived.[9] Their numbers must be infinite: the ship passed through several bands of them, one of which was about ten yards wide, and, judging from the mudlike colour of the water, at least two and a half miles long. In almost every long voyage some account is given of these confervæ. They appear especially common in the sea near Australia; and off Cape Leeuwin I found an allied but smaller and apparently different species. Captain Cook, in his third voyage, remarks, that the sailors gave to this appearance the name of sea-sawdust.

Near Keeling Atoll, in the Indian Ocean, I observed many little masses of confervæ a few inches square, consisting of long cylindrical threads of excessive thinness, so as to be barely visible to the naked eye, mingled with other rather larger bodies, finely conical at both ends.

Two of these are shown in the woodcut united together. They vary in length from .04 to .06, and even to .08 of an inch in length; and in diameter from .006 to .008 of an inch. Near one extremity of the cylindrical part, a green septum, formed of granular matter, and thickest in the middle, may generally be seen. This, I believe, is the bottom of a most delicate, colourless sac, composed of a pulpy substance, which lines the exterior case, but does not extend within the extreme conical points. In some specimens, small but perfect spheres of brownish granular matter supplied the places of the septa; and I observed the curious process by which they were produced. The pulpy matter of the internal coating suddenly grouped itself into lines, some of which assumed a form radiating from a common centre; it then continued, with an irregular and rapid movement, to contract itself, so that in the course of a second the whole was united into a perfect little sphere, which occupied the position of the septum at one end of the now quite hollow case. The formation of the granular sphere was hastened by any accidental injury. I may add, that frequently a pair of these bodies were attached to each other, as represented above, cone beside cone, at that end where the septum occurs.

I will add here a few other observations connected with the discoloration of the sea from organic causes. On the coast of Chile, a few leagues north of Concepcion, the Beagle one day passed through great bands of muddy water, exactly like that of a swollen river; and again, a degree south of Valparaiso, when fifty miles from the land, the same appearance was still more extensive. Some of the water placed in a glass was of a pale reddish tint; and, examined under a microscope, was seen to swarm with minute animalcula darting about, and often exploding. Their shape is oval, and contracted in the middle by a ring of vibrating curved ciliæ. It was, however, very difficult to examine them with care, for almost the instant motion ceased, even while crossing the field of vision, their bodies burst. Sometimes both ends burst at once, sometimes only one, and a quantity of coarse, brownish, granular matter was ejected. The animal an instant before bursting expanded to half again its natural size; and the explosion took place about fifteen seconds after the rapid progressive motion had ceased; in a few cases it was preceded for a short interval by a rotatory movement on the longer axis. About two minutes after any number were isolated in a drop of water, they thus perished. The animals move with the narrow apex forwards, by the aid of their vibratory ciliæ, and generally by rapid starts. They are exceedingly minute, and quite invisible to the naked eye, only covering a space equal to the square of the thousandth of an inch. Their numbers were infinite; for the smallest drop of water which I could remove contained very many. In one day we passed through two spaces of water thus stained, one of which alone must have extended over several square miles. What incalculable numbers of these microscopical animals! The colour of the water, as seen at some distance, was like that of a river which has flowed through a red clay district; but under the shade of the vessel's side it was quite as dark as chocolate. The line where the red and blue water joined was distinctly defined. The weather for some days previously had been calm, and the ocean abounded, to an unusual degree, with living creatures.[10]

In the sea around Tierra del Fuego, and at no great distance from the land, I have seen narrow lines of water of a bright red colour, from the number of crustacea, which somewhat resemble in form large prawns. The sealers call them whale-food. Whether whales feed on them I do not know; but terns, cormorants, and immense herds of great unwieldy seals derive, on some parts of the coast, their chief sustenance from these swimming crabs. Seamen invariably attribute the discoloration of the water to spawn; but I found this to be the case only on one occasion. At the distance of several leagues from the Archipelago of the Galapagos, the ship sailed through three strips of a dark yellowish, or mudlike water; these strips were some miles long, but only a few yards wide, and they were separated from the surrounding water by a sinuous yet distinct margin. The colour was caused by little gelatinous balls, about the fifth of an inch in diameter, in which numerous minute spherical ovules were imbedded: they were of two distinct kinds, one being of a reddish colour and of a different shape from the other. I cannot form a conjecture as to what two kinds of animals these belonged. Captain Colnett remarks, that this appearance is very common among the Galapagos Islands, and that the directions of the bands indicate that of the currents; in the described case, however, the line was caused by the wind. The only other appearance which I have to notice, is a thin oily coat on the water which displays iridescent colours. I saw a considerable tract of the ocean thus covered on the coast of Brazil; the seamen attributed it to the putrefying carcase of some whale, which probably was floating at no great distance. I do not here mention the minute gelatinous particles, hereafter to be referred to, which are frequently dispersed throughout the water, for they are not sufficiently abundant to create any change of colour.

There are two circumstances in the above accounts which appear remarkable: first, how do the various bodies which form the bands with defined edges keep together? In the case of the prawn-like crabs, their movements were as coinstantaneous as in a regiment of soldiers; but this cannot happen from anything like voluntary action with the ovules, or the confervæ, nor is it probable among the infusoria. Secondly, what causes the length and narrowness of the bands? The appearance so much resembles that which may be seen in every torrent, where the stream uncoils into long streaks the froth collected in the eddies, that I must attribute the effect to a similar action either of the currents of the air or sea. Under this supposition we must believe that the various organized bodies are produced in certain favourable places, and are thence removed by the set of either wind or water. I confess, however, there is a very great difficulty in imagining any one spot to be the birthplace of the millions of millions of animalcula and confervæ: for whence come the germs at such points?—the parent bodies having been distributed by the winds and waves over the immense ocean. But on no other hypothesis can I understand their linear grouping. I may add that Scoresby remarks that green water abounding with pelagic animals is invariably found in a certain part of the Arctic Sea.

CHAPTER II

Rio de Janeiro

RIO DE JANEIRO • EXCURSION NORTH OF CAPE FRIO • GREAT EVAPORATION • SLAVERY • BOTOFOGO BAY • TERRESTRIAL PLANARIAE • CLOUDS ON THE CORCOVADO • HEAVY RAIN • MUSICAL FROGS • PHOSPHORESCENT INSECTS • ELATER, SPRINGING POWERS OF • BLUE HAZE • NOISE MADE BY A BUTTERFLY • ENTOMOLOGY • ANTS • WASP KILLING A SPIDER • PARASITICAL SPIDER • ARTIFICES OF AN EPEIRA • GREGARIOUS SPIDER • SPIDER WITH AN UNSYMMETRICAL WEB

April 4th to July 5th, 1832.—A few days after our arrival I became acquainted with an Englishman who was going to visit his estate, situated rather more than a hundred miles from the capital, to the northward of Cape Frio. I gladly accepted his kind offer of allowing me to accompany him.

April 8th.—Our party amounted to seven. The first stage was very interesting. The day was powerfully hot, and as we passed through the woods, everything was motionless, excepting the large and brilliant butterflies, which lazily fluttered about. The view seen when crossing the hills behind Praia Grande was most beautiful; the colours were intense, and the prevailing tint a dark blue; the sky and the calm waters of the bay vied with each other in splendour. After passing through some cultivated country, we entered a forest, which in the grandeur of all its parts could not be exceeded. We arrived by midday at Ithacaia; this small village is situated on a plain, and round the central house are the huts of the negroes. These, from their regular form and position, reminded me of the drawings of the Hottentot habitations in Southern Africa. As the moon rose early, we determined to start the same evening for our sleeping-place at the Lagoa Marica. As it was growing dark we passed under one of the massive, bare, and steep hills of granite which are so common in this country. This spot is notorious from having been, for a long time, the residence of some run-away slaves, who, by cultivating a little ground near the top, contrived to eke out a subsistence. At length they were discovered, and a party of soldiers being sent, the whole were seized with the exception of one old woman, who, sooner than again be led into slavery, dashed herself to pieces from the summit of the mountain. In a Roman matron this would have been called the noble love of freedom: in a poor negress it is mere brutal obstinacy. We continued riding for some hours. For the few last miles the road was intricate, and it passed through a desert waste of marshes and lagoons. The scene by the dimmed light of the moon was most desolate. A few fireflies flitted by us; and the solitary snipe, as it rose, uttered its plaintive cry. The distant and sullen roar of the sea scarcely broke the stillness of the night.

April 9th. —We left our miserable sleeping-place before sunrise. The road passed through a narrow sandy plain, lying between the sea and the interior salt lagoons. The number of beautiful fishing birds, such as egrets and cranes, and the succulent plants assuming most fantastical forms, gave to the scene an interest which it would not otherwise have possessed. The few stunted trees were loaded with parasitical plants, among which the beauty and delicious fragrance of some of the orchideæ were most to be admired. As the sun rose, the day became extremely hot, and the reflection of the light and heat from the white sand was very distressing. We dined at Mandetiba; the thermometer in the shade being 84°. The beautiful view of the distant wooded hills, reflected in the perfectly calm water of an extensive lagoon, quite refreshed us. As the vênda[11] here was a very good one, and I have the pleasant, but rare remembrance, of an excellent dinner, I will be grateful and presently describe it, as the type of its class. These houses are often large, and are built of thick upright posts, with boughs interwoven, and afterwards plastered. They seldom have floors, and never glazed windows; but are generally pretty well roofed. Universally the front part is open, forming a kind of verandah, in which tables and benches are placed. The bed-rooms join on each side, and here the passenger may sleep as comfortably as he can, on a wooden platform, covered by a thin straw mat. The vênda stands in a courtyard, where the horses are fed. On first arriving it was our custom to unsaddle the horses and give them their Indian corn; then, with a low bow, to ask the senhôr to do us the favour to give us something to eat. "Anything you choose, sir," was his usual answer. For the few first times, vainly I thanked providence for having guided us to so good a man. The conversation proceeding, the case universally became deplorable. "Any fish can you do us the favour of giving?"—"Oh! no, sir."—"Any soup?"—"No, sir."—"Any bread?"—"Oh! no, sir."—"Any dried meat?"—"Oh! no, sir." If we were lucky, by waiting a couple of hours, we obtained fowls, rice, and farinha. It not unfrequently happened, that we were obliged to kill, with stones, the poultry for our own supper. When, thoroughly exhausted by fatigue and hunger, we timorously hinted that we should be glad of our meal, the pompous, and (though true) most unsatisfactory answer was, "It will be ready when it is ready." If we had dared to remonstrate any further, we should have been told to proceed on our journey, as being too impertinent. The hosts are most ungracious and disagreeable in their manners; their houses and their persons are often filthily dirty; the want of the accommodation of forks, knives, and spoons is common; and I am sure no cottage or hovel in England could be found in a state so utterly destitute of every comfort. At Campos Novos, however, we fared sumptuously; having rice and fowls, biscuit, wine, and spirits, for dinner; coffee in the evening, and fish with coffee for breakfast. All this, with good food for the horses, only cost 2*s.* 6*d.* per head. Yet the host of this vênda, being asked if he knew anything of a whip which one of the party had lost, gruffly answered, "How should I know? why did you not take care of it?—I suppose the dogs have eaten it."

Leaving Mandetiba, we continued to pass through an intricate wilderness of lakes; in some of which were fresh, in others salt water shells. Of the former kinds, I found a Limnæa in great numbers in a lake, into which, the inhabitants assured me that the sea enters once a year, and sometimes oftener, and makes the water quite salt. I have no doubt many interesting facts, in relation to marine and fresh water animals, might be observed in this chain of lagoons, which skirt the coast of Brazil. M. Gay[12] has stated that he found in the neighbourhood of Rio, shells of the marine genera solen and mytilus, and fresh water ampullariæ, living together in brackish water. I also frequently observed in the lagoon near the Botanic Garden, where the water is only a little less salt than in the sea, a species of hydrophilus, very similar to a water-beetle common in the ditches of England: in the same lake the only shell belonged to a genus generally found in estuaries.

Leaving the coast for a time, we again entered the forest. The trees were very lofty, and remarkable, compared with those of Europe, from the whiteness of their trunks. I see by my note-book, "wonderful and beautiful, flowering parasites," invariably struck me as the most novel object in these grand scenes. Travelling onwards we passed through tracts of pasturage, much injured by the enormous conical ants' nests, which were nearly twelve feet high. They gave to the plain exactly the appearance of the mud volcanos at Jorullo, as figured by Humboldt. We arrived at Engenhodo after it was dark, having been ten hours on horseback. I never ceased, during the whole journey, to be surprised at the amount of labour which the horses were capable of enduring; they appeared also to recover from any injury much sooner than those of our English breed. The Vampire bat is often the cause of much trouble, by biting the horses on their withers. The injury is generally not so much owing to the loss of blood, as to the inflammation which the pressure of the saddle afterwards produces. The whole circumstance has lately been doubted in England; I was therefore fortunate in being present when one (Desmodus d'orbignyi, Wat.) was actually caught on a horse's back. We were bivouacking late one evening near Coquimbo, in Chile, when my servant, noticing that one of the horses was very restive, went to see what was the matter, and fancying he could distinguish something, suddenly put his hand on the beast's withers, and secured the vampire. In the morning the spot where the bite had been inflicted was easily distinguished from being slightly swollen and bloody. The third day afterwards we rode the horse, without any ill effects.

April 13th.—After three days' travelling we arrived at Socêgo, the estate of Senhôr Manuel Figuireda, a relation of one of our party. The house was simple, and, though like a barn in form, was well suited to the climate. In the sitting-room gilded chairs and sofas were oddly contrasted with the white-washed walls, thatched roof, and windows without glass. The house, together with the granaries, the stables, and workshops for the blacks, who had been taught various trades, formed a rude kind of quadrangle; in the centre of which a large pile of coffee was drying. These buildings stand on a little hill, overlooking the cultivated ground, and surrounded on every side by a wall of dark green luxuriant forest. The chief produce of this part of the country is coffee. Each tree is supposed to yield annually, on an average, two pounds; but some give as much as eight. Mandioca or cassada is likewise cultivated in great quantity. Every part of this plant is useful; the leaves and stalks are eaten by the horses, and the roots are ground into a pulp, which, when pressed dry and baked, forms the farinha, the principal article of sustenance in the Brazils. It is a curious, though well-known fact, that the juice of this most nutritious plant is highly poisonous. A few years ago a cow died at this Fazênda, in consequence of having drunk some of it. Senhôr Figuireda told me that he had planted, the year before, one bag of feijaô or beans, and three of rice; the former of which produced eighty, and the latter three hundred and twenty fold. The pasturage supports a fine stock of cattle, and the woods are so full of game that a deer had been killed on each of the three previous days. This profusion of food showed itself at dinner, where, if the tables did not groan, the guests surely did; for each person is expected to eat of every dish. One day, having, as I thought, nicely calculated so that nothing should go away untasted, to my utter dismay a roast turkey and a pig appeared in all their substantial reality. During the meals, it was the employment of a man to drive out of the room sundry old hounds, and dozens of little black children, which crawled in together, at every opportunity. As long as the idea of slavery could be banished, there was something exceedingly fascinating in this simple and patriarchal style of living: it was such a perfect retirement and independence from the rest of the world.

As soon as any stranger is seen arriving, a large bell is set tolling, and generally some small cannon are fired. The event is thus announced to the rocks and woods, but to nothing else. One morning I walked out an hour before daylight to admire the solemn stillness of the scene; at last, the silence was broken by the morning hymn, raised on high by the whole body of the blacks; and in this manner their daily work is generally begun. On such fazêndas as these, I have no doubt the slaves pass happy and contented lives. On Saturday and Sunday they work for themselves, and in this fertile climate the labour of two days is sufficient to support a man and his family for the whole week.

April 14th.—Leaving Socêgo, we rode to another estate on the Rio Macâe, which was the last patch of cultivated ground in that direction. The estate was two and a half miles long, and the owner had forgotten how many broad. Only a very small piece had been cleared, yet almost every acre was capable of yielding all the various rich productions of a tropical land. Considering the enormous area of Brazil, the proportion of cultivated ground can scarcely be considered as anything, compared to that which is left in the state of nature: at some future age, how vast a population it will support! During the second day's journey we found the road so shut up, that it was necessary that a man should go ahead with a sword to cut away the creepers. The forest abounded with beautiful objects; among which the tree ferns, though not large, were, from their bright green foliage, and the elegant curvature of their fronds, most worthy of admiration. In the evening it rained very heavily, and although the thermometer stood at 65°, I felt very cold. As soon as the rain ceased, it was curious to observe the extraordinary evaporation which commenced over the whole extent of the forest. At the height of a hundred feet the hills were buried in a dense white vapour, which rose like columns of smoke from the most thickly wooded parts, and especially from the valleys. I observed this phenomenon on several occasions. I suppose it is owing to the large surface of foliage, previously heated by the sun's rays.

While staying at this estate, I was very nearly being an eye-witness to one of those atrocious acts which can only take place in a slave country. Owing to a quarrel and a lawsuit, the owner was on the point of taking all the women and children from the male slaves, and selling them separately at the public auction at Rio. Interest, and not any feeling of compassion, prevented this act. Indeed, I do not believe the inhumanity of separating thirty families, who had lived together for many years, even occurred to the owner. Yet I will pledge myself, that in humanity and good feeling he was superior to the common run of men. It may be said there exists no limit to the blindness of interest and selfish habit. I may mention one very trifling anecdote, which at the time struck me more forcibly than any story of cruelty. I was crossing a ferry with a negro, who was uncommonly stupid. In endeavouring to make him understand, I talked loud, and made signs, in doing which I passed my hand near his face. He, I suppose, thought I was in a passion, and was going to strike him; for instantly, with a frightened look and half-shut eyes, he dropped his hands. I shall never forget my feelings of surprise, disgust, and shame, at seeing a great powerful man afraid even to ward off a blow, directed, as he thought, at his face. This man had been trained to a degradation lower than the slavery of the most helpless animal.

April 18th.—In returning we spent two days at Socêgo, and I employed them in collecting insects in the forest. The greater number of trees, although so lofty, are not more than three or four feet in circumference. There are, of course, a few of much greater dimensions. Senhôr Manuel was then making a canoe 70 feet in length from a solid trunk, which had originally been 110 feet long, and of great thickness. The contrast of palm trees, growing amidst the common branching kinds, never fails to give the scene an intertropical character. Here the woods were ornamented by the Cabbage Palm—one of the most beautiful of its family. With a stem so narrow that it might be clasped with the two hands, it waves its elegant head at the height of forty or fifty feet above the ground. The woody creepers, themselves covered by other creepers, were of great thickness: some which I measured were two feet in circumference. Many of the older trees presented a very curious appearance from the tresses of a liana hanging from their boughs, and resembling bundles of hay. If the eye was turned from the world of foliage above, to the ground beneath, it was attracted by the extreme elegance of the leaves of the ferns and mimosæ. The latter, in some parts, covered the surface with a brushwood only a few inches high. In walking across these thick beds of mimosæ, a broad track was marked by the change of shade, produced by the drooping of their sensitive petioles. It is easy to specify the individual objects of admiration in these grand scenes; but it is not possible to give an adequate idea of the higher feelings of wonder, astonishment, and devotion, which fill and elevate the mind.

April 19th.—Leaving Socêgo, during the two first days, we retraced our steps. It was very wearisome work, as the road generally ran across a glaring hot sandy plain, not far from the coast. I noticed that each time the horse put its foot on the fine siliceous sand, a gentle chirping noise was produced. On the third day we took a different line, and passed through the gay little village of Madre de Deôs. This is one of the principal lines of road in Brazil; yet it was in so bad a state that no wheeled vehicle, excepting the clumsy bullock-wagon, could pass along. In our whole journey we did not cross a single bridge built of stone; and those made of logs of wood were frequently so much out of repair, that it was necessary to go on one side to avoid them. All distances are inaccurately known. The road is often marked by crosses, in the place of milestones, to signify where human blood has been spilled. On the evening of the 23rd we arrived at Rio, having finished our pleasant little excursion.

DURING THE REMAINDER of my stay at Rio, I resided in a cottage at Botofogo Bay. It was impossible to wish for anything more delightful than thus to spend some weeks in so magnificent a country. In England any person fond of natural history enjoys in his walks a great advantage, by always having something to attract his attention; but in these fertile climates, teeming with life, the attractions are so numerous, that he is scarcely able to walk at all.

The few observations which I was enabled to make were almost exclusively confined to the invertebrate animals. The existence of a division of the genus Planaria, which inhabits the dry land, interested me much. These animals are of so simple a structure, that Cuvier has arranged them with the intestinal worms, though never found within the bodies of other animals. Numerous species inhabit both salt and fresh water; but those to which I allude were found, even in the drier parts of the forest, beneath logs of rotten wood, on which I believe they feed. In general form they resemble little slugs, but are very much narrower in proportion, and several of the species are beautifully coloured with longitudinal stripes. Their structure is very simple: near the middle of the under or crawling surface there are two small transverse slits, from the anterior one of which a funnel-shaped and highly irritable mouth can be protruded. For some time after the rest of the animal was completely dead from the effects of salt water or any other cause, this organ still retained its vitality.

I found no less than twelve different species of terrestrial Planariæ in different parts of the southern hemisphere.[13] Some specimens which I obtained at Van Dieman's Land, I kept alive for nearly two months, feeding them on rotten wood. Having cut one of them transversely into two nearly equal parts, in the course of a fortnight both had the shape of perfect animals. I had, however, so divided the body, that one of the halves contained both the inferior orifices, and the other, in consequence, none. In the course of twenty-five days from the operation, the more perfect half could not have been distinguished from any other specimen. The other had increased much in size; and towards its posterior end, a clear space was formed in the parenchymatous mass, in which a rudimentary cup-shaped mouth could clearly be distinguished; on the under surface, however, no corresponding slit was yet open. If the increased heat of the weather, as we approached the equator, had not destroyed all the individuals, there can be no doubt that this last step would have completed its structure. Although so well-known an experiment, it was interesting to watch the gradual production of every essential organ, out of the simple extremity of another animal. It is extremely difficult to preserve these Planariæ; as soon as the cessation of life allows the ordinary laws of change to act, their entire bodies become soft and fluid, with a rapidity which I have never seen equalled.

I first visited the forest in which these Planariæ were found, in company with an old Portuguese priest who took me out to hunt with him. The sport consisted in turning into the cover a few dogs, and then patiently waiting to fire at any animal which might appear. We were accompanied by the son of a neighbouring farmer—a good specimen of a wild Brazilian youth. He was dressed in a tattered old shirt and trousers, and had his head uncovered: he carried an old-fashioned gun and a large knife. The habit of carrying the knife is universal; and in traversing a thick wood it is almost necessary, on account of the creeping plants. The frequent occurrence of murder may be partly attributed to this habit. The Brazilians are so dexterous with the knife, that they can throw it to some distance with precision, and with sufficient force to cause a fatal wound. I have seen a number of little boys practising this art as a game of play, and from their skill in hitting an upright stick, they promised well for more earnest attempts. My companion, the day before, had shot two large bearded monkeys. These animals have prehensile tails, the extremity of which, even after death, can support the whole weight of the body. One of them thus remained fast to a branch, and it was necessary to cut down a large tree to procure it. This was soon effected, and down came tree and monkey with an awful crash. Our day's sport, besides the monkey, was confined to sundry small green parrots and a few toucans. I profited, however, by my acquaintance with the Portuguese padre, for on another occasion he gave me a fine specimen of the Yagouaroundi cat.

Every one has heard of the beauty of the scenery near Botofogo. The house in which I lived was seated close beneath the well-known mountain of the Corcovado. It has been remarked, with much truth, that abruptly conical hills are characteristic of the formation which Humboldt designates as gneiss-granite. Nothing can be more striking than the effect of these huge rounded masses of naked rock rising out of the most luxuriant vegetation.

I was often interested by watching the clouds, which, rolling in from seaward, formed a bank just beneath the highest point of the Corcovado. This mountain, like most others, when thus partly veiled, appeared to rise to a far prouder elevation than its real height of 2300 feet. Mr. Daniell has observed, in his meteorological essays, that a cloud sometimes appears fixed on a mountain summit, while the wind continues to blow over it. The same phenomenon here presented a slightly different appearance. In this case the cloud was clearly seen to curl over, and rapidly pass by the summit, and yet was neither diminished nor increased in size. The sun was setting, and a gentle southerly breeze, striking against the southern side of the rock, mingled its current with the colder air above; and the vapour was thus condensed; but as the light wreaths of cloud passed over the ridge, and came within the influence of the warmer atmosphere of the northern sloping bank, they were immediately re-dissolved.

The climate, during the months of May and June, or the beginning of winter, was delightful. The mean temperature, from observations taken at nine o'clock, both morning and evening, was only 72°. It often rained heavily, but the drying southerly winds soon again rendered the walks pleasant. One morning, in the course of six hours, 1.6 inches of rain fell. As this storm passed over the forests which surround the Corcovado, the sound produced by the drops pattering on the countless multitude of leaves was very remarkable, it could be heard at the distance of a quarter of a mile, and was like the rushing of a great body of water. After the hotter days, it was delicious to sit quietly in the garden and watch the evening pass into night. Nature, in these climes, chooses her vocalists from more humble performers than in Europe. A small frog, of the genus Hyla, sits on a blade of grass about an inch above the surface of the water, and sends forth a pleasing chirp: when several are together they sing in harmony on different notes. I had some difficulty in catching a specimen of this frog. The genus Hyla has its toes terminated by small suckers; and I found this animal could crawl up a pane of glass, when placed absolutely perpendicular. Various cicidæ and crickets, at the same time, keep up a ceaseless shrill cry, but which, softened by the distance, is not unpleasant. Every evening after dark this great concert commenced; and often have I sat listening to it, until my attention has been drawn away by some curious passing insect.

At these times the fireflies are seen flitting about from hedge to hedge. On a dark night the light can be seen at about two hundred paces distant. It is remarkable that in all the different kinds of glowworms, shining elaters, and various marine animals (such as the crustacea, medusæ, nereidæ, a coralline of the genus Clytia, and Pyrosma), which I have observed, the light has been of a well-marked green colour. All the fireflies, which I caught here, belonged to the Lampyridæ (in which family the English glowworm is included), and the greater number of specimens were of Lampyris occidentalis.[14] I found that this insect emitted the most brilliant flashes when irritated: in the intervals, the abdominal rings were obscured. The flash was almost co-instantaneous in the two rings, but it was just perceptible first in the anterior one. The shining matter was fluid and very adhesive: little spots, where the skin had been torn, continued bright with a slight scintillation, whilst the uninjured parts were obscured. When the insect was decapitated the rings remained uninterruptedly bright, but not so brilliant as before: local irritation with a needle always increased the vividness of the light. The rings in one instance retained their luminous property nearly twenty-four hours after the death of the insect. From these facts it would appear probable, that the animal has only the power of concealing or extinguishing the light for short intervals, and that at other times the display is involuntary. On the muddy and wet gravel-walks I found the larvæ of this lampyris in great numbers: they resembled in general form the female of the English glowworm. These larvæ possessed but feeble luminous powers; very differently from their parents, on the slightest touch they feigned death and ceased to shine; nor did irritation excite any fresh display. I kept several of them alive for some time: their tails are very singular organs, for they act, by a well-fitted contrivance, as suckers or organs of attachment, and likewise as reservoirs for saliva, or some such fluid. I repeatedly fed them on raw meat; and I invariably observed, that every now and then the extremity of the tail was applied to the mouth, and a drop of fluid exuded on the meat, which was then in the act of being consumed. The tail, notwithstanding so much practice, does not seem to be able to find its way to the mouth; at least the neck was always touched first, and apparently as a guide.

When we were at Bahia, an elater or beetle (Pyrophorus luminosus, Illig.) seemed the most common luminous insect. The light in this case was also rendered more brilliant by irritation. I amused myself one day by observing the springing powers of this insect, which have not, as it appears to me, been properly described.[15] The elater, when placed on its back and preparing to spring, moved its head and thorax backwards, so that the pectoral spine was drawn out, and rested on the edge of its sheath. The same backward movement being continued, the spine, by the full action of the muscles, was bent like a spring; and the insect at this moment rested on the extremity of its head and wing-cases. The effort being suddenly relaxed, the head and thorax flew up, and in consequence, the base of the wing-cases struck the supporting surface with such force, that the insect by the reaction was jerked upwards to the height of one or two inches. The projecting points of the thorax, and the sheath of the spine, served to steady the whole body during the spring. In the descriptions which I have read, sufficient stress does not appear to have been laid on the elasticity of the spine: so sudden a spring could not be the result of simple muscular contraction, without the aid of some mechanical contrivance.

On several occasions I enjoyed some short but most pleasant excursions in the neighbouring country. One day I went to the Botanic Garden, where many plants, well known for their great utility, might be seen growing. The leaves of the camphor, pepper, cinnamon, and clove trees were delightfully aromatic; and the bread-fruit, the jaca, and the mango, vied with each other in the magnificence of their foliage. The landscape in the neighbourhood of Bahia almost takes its character from the two latter trees. Before seeing them, I had no idea that any trees could cast so black a shade on the ground. Both of them bear to the evergreen vegetation of these climates the same kind of relation which laurels and hollies in England do to the lighter green of the deciduous trees. It may be observed, that the houses within the tropics are surrounded by the most beautiful forms of vegetation, because many of them are at the same time most useful to man. Who can doubt that these qualities are united in the banana, the cocoa-nut, the many kinds of palm, the orange, and the bread-fruit tree?

During this day I was particularly struck with a remark of Humboldt's, who often alludes to "the thin vapour which, without changing the transparency of the air, renders its tints more harmonious, and softens its effects." This is an appearance which I have never observed in the temperate zones. The atmosphere, seen through a short space of half or three-quarters of a mile, was perfectly lucid, but at a greater distance all colours were blended into a most beautiful haze, of a pale French grey, mingled with a little blue. The condition of the atmosphere between the morning and about noon, when the effect was most evident, had undergone little change, excepting in its dryness. In the interval, the difference between the dew point and temperature had increased from 7°.5 to 17°.

On another occasion I started early and walked to the Gavia, or topsail mountain. The air was delightfully cool and fragrant; and the drops of dew still glittered on the leaves of the large liliaceous plants, which shaded the streamlets of clear water. Sitting down on a block of granite, it was delightful to watch the various insects and birds as they flew past. The humming-bird seems particularly fond of such shady retired spots. Whenever I saw these little creatures buzzing round a flower, with their wings vibrating so rapidly as to be scarcely visible, I was reminded of the sphinx moths: their movements and habits are indeed in many respects very similar.

Following a pathway, I entered a noble forest, and from a height of five or six hundred feet, one of those splendid views was presented, which are so common on every side of Rio. At this elevation the landscape attains its most brilliant tint; and every form, every shade, so completely surpasses in magnificence all that the European has ever beheld in his own country, that he knows not how to express his feelings. The general effect frequently recalled to my mind the gayest scenery of the Opera-house or the great theatres. I never returned from these excursions empty-handed. This day I found a specimen of a curious fungus, called Hymenophallus. Most people know the English Phallus, which in autumn taints the air with its odious smell: this, however, as the entomologist is aware, is, to some of our beetles a delightful fragrance. So was it here; for a Strongylus, attracted by the odour, alighted on the fungus as I carried it in my hand. We here see in two distant countries a similar relation between plants and insects of the same families, though the species of both are different. When man is the agent in introducing into a country a new species, this relation is often broken: as one instance of this I may mention, that the leaves of the cabbages and lettuces, which in England afford food to such a multitude of slugs and caterpillars, in the gardens near Rio are untouched.

During our stay at Brazil I made a large collection of insects. A few general observations on the comparative importance of the different orders may be interesting to the English entomologist. The large and brilliantly coloured Lepidoptera bespeak the zone they inhabit, far more plainly than any other race of animals. I allude only to the butterflies; for the moths, contrary to what might have been expected from the rankness of the vegetation, certainly appeared in much fewer numbers than in our own temperate regions. I was much surprised at the habits of Papilio feronia. This butterfly is not uncommon, and generally frequents the orange-groves. Although a high flier, yet it very frequently alights on the trunks of trees. On these occasions its head is invariably placed downwards; and its wings are expanded in a horizontal plane, instead of being folded vertically, as is commonly the case. This is the only butterfly which I have ever seen, that uses its legs for running. Not being aware of this fact, the insect, more than once, as I cautiously approached with my forceps, shuffled on one side just as the instrument was on the point of closing, and thus escaped. But a far more singular fact is the power which this species possesses of making a noise.[16] Several times when a pair, probably male and female, were chasing each other in an irregular course, they passed within a few yards of me; and I distinctly heard a clicking noise, similar to that produced by a toothed wheel passing under a spring catch. The noise was continued at short intervals, and could be distinguished at about twenty yards' distance: I am certain there is no error in the observation.

I was disappointed in the general aspect of the Coleoptera. The number of minute and obscurely coloured beetles is exceedingly great.[17] The cabinets of Europe can, as yet, boast only of the larger species from tropical climates. It is sufficient to disturb the composure of an entomologist's mind, to look forward to the future dimensions of a complete catalogue.

The carnivorous beetles, or Carabidæ, appear in extremely few numbers within the tropics: this is the more remarkable when compared to the case of the carnivorous quadrupeds, which are so abundant in hot countries. I was struck with this observation both on entering Brazil, and when I saw the many elegant and active forms of the Harpalidæ re-appearing on the temperate plains of La Plata. Do the very numerous spiders and rapacious Hymenoptera supply the place of the carnivorous beetles? The carrion-feeders and Brachelytra are very uncommon; on the other hand, the Rhyncophora and Chrysomelidæ, all of which depend on the vegetable world for subsistence, are present in astonishing numbers. I do not here refer to the number of different species, but to that of the individual insects; for on this it is that the most striking character in the entomology of different countries depends. The orders Orthoptera and Hemiptera are particularly numerous; as likewise is the stinging division of the Hymenoptera; the bees, perhaps, being excepted. A person, on first entering a tropical forest, is astonished at the labours of the ants: well-beaten paths branch off in every direction, on which an army of never-failing foragers may be seen, some going forth, and others returning, burdened with pieces of green leaf, often larger than their own bodies.

A small dark-coloured ant sometimes migrates in countless numbers. One day, at Bahia, my attention was drawn by observing many spiders, cockroaches, and other insects, and some lizards, rushing in the greatest agitation across a bare piece of ground. A little way behind, every stalk and leaf was blackened by a small ant. The swarm having crossed the bare space, divided itself, and descended an old wall. By this means many insects were fairly enclosed; and the efforts which the poor little creatures made to extricate themselves from such a death were wonderful. When the ants came to the road they changed their course, and in narrow files reascended the wall. Having placed a small stone so as to intercept one of the lines, the whole body attacked it, and then immediately retired. Shortly afterwards another body came to the charge, and again having failed to make any impression, this line of march was entirely given up. By going an inch round, the file might have avoided the stone, and this doubtless would have happened, if it had been originally there: but having been attacked, the lion-hearted little warriors scorned the idea of yielding.

Certain wasp-like insects, which construct in the corners of the verandahs clay cells for their larvæ, are very numerous in the neighbourhood of Rio. These cells they stuff full of half-dead spiders and caterpillars, which they seem wonderfully to know how to sting to that degree as to leave them paralysed but alive, until their eggs are hatched; and the larvæ feed on the horrid mass of powerless, half-killed victims—a sight which has been described by an enthusiastic naturalist[18] as curious and pleasing! I was much interested one day by watching a deadly contest between a Pepsis and a large spider of the genus Lycosa. The wasp made a sudden dash at its prey, and then flew away: the spider was evidently wounded, for, trying to escape, it rolled down a little slope, but had still strength sufficient to crawl into a thick tuft of grass. The wasp soon returned, and seemed surprised at not immediately finding its victim. It then commenced as regular a hunt as ever hound did after fox; making short semicircular casts, and all the time rapidly vibrating its wings and antennæ. The spider, though well concealed, was soon discovered; and the wasp, evidently still afraid of its adversary's jaws, after much manœuvring, inflicted two stings on the under side of its thorax. At last, carefully examining with its antennæ the now motionless spider, it proceeded to drag away the body. But I stopped both tyrant and prey.[19]

The number of spiders, in proportion to other insects, is here compared with England very much larger; perhaps more so than with any other division of the articulate animals. The variety of species among the jumping spiders appears almost infinite. The genus, or rather family, of Epeira, is here characterized by many singular forms; some species have pointed coriaceous shells, others enlarged and spiny tibiæ. Every path in the forest is barricaded with the strong yellow web of a species, belonging to the same division with the Epeira clavipes of Fabricius, which was formerly said by Sloane to make, in the West Indies, webs so strong as to catch birds. A small and pretty kind of spider, with very long fore-legs, and which appears to belong to an undescribed genus, lives as a parasite on almost every one of these webs. I suppose it is too insignificant to be noticed by the great Epeira, and is therefore allowed to prey on the minute insects, which, adhering to the lines, would otherwise be wasted. When frightened, this little spider either feigns death by extending its front legs, or suddenly drops from the web. A large Epeira of the same division with Epeira tuberculata and conica is extremely common, especially in dry situations. Its web, which is generally placed among the great leaves of the common agave, is sometimes strengthened near the centre by a pair or even four zigzag ribbons, which connect two adjoining rays. When any large insect, as a grasshopper or wasp, is caught, the spider, by a dexterous movement, makes it revolve very rapidly, and at the same time emitting a band of threads from its spinners, soon envelops its prey in a case like the cocoon of a silkworm. The spider now examines the powerless victim, and gives the fatal bite on the hinder part of its thorax; then retreating, patiently waits till the poison has taken effect. The virulence of this poison may be judged of from the fact that in half a minute I opened the mesh, and found a large wasp quite lifeless. This Epeira always stands with its head downwards near the centre of the web. When disturbed, it acts differently according to circumstances: if there is a thicket below, it suddenly falls down; and I have distinctly seen the thread from the spinners lengthened by the animal while yet stationary, as preparatory to its fall. If the ground is clear beneath, the Epeira seldom falls, but moves quickly through a central passage from one to the other side. When still further disturbed, it practises a most curious manœuvre: standing in the middle, it violently jerks the web, which is attached to elastic twigs, till at last the whole acquires such a rapid vibratory movement, that even the outline of the spider's body becomes indistinct.

It is well known that most of the British spiders, when a large insect is caught in their webs, endeavour to cut the lines and liberate their prey, to save their nets from being entirely spoiled. I once, however, saw in a hothouse in Shropshire a large female wasp caught in the irregular web of a quite small spider; and this spider, instead of cutting the web, most perseveringly continued to entangle the body, and especially the wings, of its prey. The wasp at first aimed in vain repeated thrusts with its sting at its little antagonist. Pitying the wasp, after allowing it to struggle for more than an hour, I killed it and put it back into the web. The spider soon returned; and an hour afterwards I was much surprised to find it with its jaws buried in the orifice, through which the sting is protruded by the living wasp. I drove the spider away two or three times, but for the next twenty-four hours I always found it again sucking at the same place. The spider became much distended by the juices of its prey, which was many times larger than itself.

I may here just mention, that I found, near St. Fé Bajada, many large black spiders, with ruby-coloured marks on their backs, having gregarious habits. The webs were placed vertically, as is invariably the case with the genus Epeira: they were separated from each other by a space of about two feet, but were all attached to certain common lines, which were of great length, and extended to all parts of the community. In this manner the tops of some large bushes were encompassed by the united nets. Azara[20] has described a gregarious spider in Paraguay, which Walckenaer thinks must be a Theridion, but probably it is an Epeira, and perhaps even the same species with mine. I cannot, however, recollect seeing a central nest as large as a hat, in which, during autumn, when the spiders die, Azara says the eggs are deposited. As all the spiders which I saw were of the same size, they must have been nearly of the same age. This gregarious habit, in so typical a genus as Epeira, among insects, which are so bloodthirsty and solitary that even the two sexes attack each other, is a very singular fact.

In a lofty valley of the Cordillera, near Mendoza, I found another spider with a singularly-formed web. Strong lines radiated in a vertical plane from a common centre, where the insect had its station; but only two of the rays were connected by a symmetrical mesh-work; so that the net, instead of being, as is generally the case, circular, consisted of a wedge-shaped segment. All the webs were similarly constructed.

CHAPTER III

Maldonado

MONTE VIDEO • MALDONADO • EXCURSION TO R. POLANCO • LAZO AND BOLAS • PARTRIDGES • ABSENCE OF TREES • DEER • CAPYBARA, OR RIVER HOG • TUCUTUCO • MOLOTHRUS, CUCKOO-LIKE HABITS • TYRANT-FLYCATCHER • MOCKING-BIRD • CARRION HAWKS • TUBES FORMED BY LIGHTNING • HOUSE STRUCK

July 5th, 1832.—In the morning we got under way, and stood out of the splendid harbour of Rio de Janeiro. In our passage to the Plata, we saw nothing particular, excepting on one day a great shoal of porpoises, many hundreds in number. The whole sea was in places furrowed by them; and a most extraordinary spectacle was presented, as hundreds, proceeding together by jumps, in which their whole bodies were exposed, thus cut the water. When the ship was running nine knots an hour, these animals could cross and recross the bows with the greatest ease, and then dash away right ahead. As soon as we entered the estuary of the Plata, the weather was very unsettled. One dark night we were surrounded by numerous seals and penguins, which made such strange noises, that the officer on watch reported he could hear the cattle bellowing on shore. On a second night we witnessed a splendid scene of natural fireworks; the mast-head and yard-arm-ends shone with St. Elmo's light; and the form of the vane could almost be traced, as if it had been rubbed with phosphorus. The sea was so highly luminous, that the tracks of the penguins were marked by a fiery wake, and the darkness of the sky was momentarily illuminated by the most vivid lightning.

When within the mouth of the river, I was interested by observing how slowly the waters of the sea and river mixed. The latter, muddy and discoloured, from its less specific gravity, floated on the surface of the salt water. This was curiously exhibited in the wake of the vessel, where a line of blue water was seen mingling in little eddies, with the adjoining fluid.

July 26th.—We anchored at Monte Video. The Beagle was employed in surveying the extreme southern and eastern coasts of America, south of the Plata, during the two succeeding years. To prevent useless repetitions, I will extract those parts of my journal which refer to the same districts, without always attending to the order in which we visited them.

MALDONADO is situated on the northern bank of the Plata, and not very far from the mouth of the estuary. It is a most quiet, forlorn, little town; built, as is universally the case in these countries, with the streets running at right angles to each other, and having in the middle a large plaza or square, which, from its size, renders the scantiness of the population more evident. It possesses scarcely any trade; the exports being confined to a few hides and living cattle. The inhabitants are chiefly landowners, together with a few shopkeepers and the necessary tradesmen, such as blacksmiths and carpenters, who do nearly all the business for a circuit of fifty miles round. The town is separated from the river by a band of sand-hillocks, about a mile broad: it is surrounded, on all other sides, by an open slightly-undulating country, covered by one uniform layer of fine green turf, on which countless herds of cattle, sheep, and horses graze. There is very little land cultivated even close to the town. A few hedges, made of cacti and agave, mark out where some wheat or Indian corn has been planted. The features of the country are very similar along the whole northern bank of the Plata. The only difference is, that here the granitic hills are a little bolder. The scenery is very uninteresting; there is scarcely a house, an enclosed piece of ground, or even a tree, to give it an air of cheerfulness. Yet, after being imprisoned for some time in a ship, there is a charm in the unconfined feeling of walking over boundless plains of turf. Moreover, if your view is limited to a small space, many objects possess beauty. Some of the smaller birds are brilliantly coloured; and the bright green sward, browsed short by the cattle, is ornamented by dwarf flowers, among which a plant, looking like the daisy, claimed the place of an old friend. What would a florist say to whole tracts, so thickly covered by the Verbena melindres, as, even at a distance, to appear of the most gaudy scarlet?

I stayed ten weeks at Maldonado, in which time a nearly perfect collection of the animals, birds, and reptiles, was procured. Before making any observations respecting them, I will give an account of a little excursion I made as far as the river Polanco, which is about seventy miles distant, in a northerly direction. I may mention, as a proof how cheap everything is in this country, that I paid only two dollars a day, or eight shillings, for two men, together with a troop of about a dozen riding-horses. My companions were well armed with pistols and sabres; a precaution which I thought rather unnecessary: but the first piece of news we heard was, that, the day before, a traveller from Monte Video had been found dead on the road, with his throat cut. This happened close to a cross, the record of a former murder.

On the first night we slept at a retired little country-house; and there I soon found out that I possessed two or three articles, especially a pocket compass, which created unbounded astonishment. In every house I was asked to show the compass, and by its aid, together with a map, to point out the direction of various places. It excited the liveliest admiration that I, a perfect stranger, should know the road (for direction and road are synonymous in this open country) to places where I had never been. At one house a young woman, who was ill in bed, sent to entreat me to come and show her the compass. If their surprise was great, mine was greater, to find such ignorance among people who possessed their thousands of cattle, and "estancias" of great extent. It can only be accounted for by the circumstance that this retired part of the country is seldom visited by foreigners. I was asked whether the earth or sun moved; whether it was hotter or colder to the north; where Spain was, and many other such questions. The greater number of the inhabitants had an indistinct idea that England, London, and North America, were different names for the same place; but the better informed well knew that London and North America were separate countries close together, and that England was a large town in London! I carried with me some promethean matches, which I ignited by biting; it was thought so wonderful that a man should strike fire with his teeth, that it was usual to collect the whole family to see it: I was once offered a dollar for a single one. Washing my face in the morning caused much speculation at the village of Las Minas; a superior tradesman closely cross-questioned me about so singular a practice; and likewise why on board we wore our beards; for he had heard from my guide that we did so. He eyed me with much suspicion; perhaps he had heard of ablutions in the Mahomedan religion, and knowing me to be a heretick, probably he came to the conclusion that all hereticks were Turks. It is the general custom in this country to ask for a night's lodging at the first convenient house. The astonishment at the compass, and my other feats of jugglery, was to a certain degree advantageous, as with that, and the long stories my guides told of my breaking stones, knowing venomous from harmless snakes, collecting insects, etc., I repaid them for their hospitality. I am writing as if I had been among the inhabitants of central Africa: Banda Oriental would not be flattered by the comparison; but such were my feelings at the time.

The next day we rode to the village of Las Minas. The country was rather more hilly, but otherwise continued the same; an inhabitant of the Pampas no doubt would have considered it as truly Alpine. The country is so thinly inhabited, that during the whole day we scarcely met a single person. Las Minas is much smaller than Maldonado. It is seated on a little plain, and is surrounded by low rocky mountains. It is of the usual symmetrical form; and with its whitewashed church standing in the centre, had rather a pretty appearance. The outskirting houses rose out of the plain like isolated beings, without the accompaniment of gardens or courtyards. This is generally the case in the country, and all the houses have, in consequence, an uncomfortable aspect. At night we stopped at a pulperia, or drinking-shop. During the evening a great number of Gauchos came in to drink spirits and smoke cigars: their appearance is very striking; they are generally tall and handsome, but with a proud and dissolute expression of countenance. They frequently wear their moustaches and long black hair curling down their backs. With their brightly coloured garments, great spurs clanking about their heels, and knives stuck as daggers (and often so used) at their waists, they look a very different race of men from what might be expected from their name of Gauchos, or simple countrymen. Their politeness is excessive; they never drink their spirits without expecting you to taste it; but whilst making their exceedingly graceful bow, they seem quite as ready, if occasion offered, to cut your throat.

On the third day we pursued rather an irregular course, as I was employed in examining some beds of marble. On the fine plains of turf we saw many ostriches (Struthio rhea). Some of the flocks contained as many as twenty or thirty birds. These, when standing on any little eminence, and seen against the clear sky, presented a very noble appearance. I never met with such tame ostriches in any other part of the country: it was easy to gallop up within a short distance of them; but then, expanding their wings, they made all sail right before the wind, and soon left the horse astern.

At night we came to the house of Don Juan Fuentes, a rich landed proprietor, but not personally known to either of my companions. On approaching the house of a stranger, it is usual to follow several little points of etiquette: riding up slowly to the door, the salutation of Ave Maria is given, and until somebody comes out and asks you to alight, it is not customary even to get off your horse: the formal answer of the owner is, "sin pecado concebida"—that is, conceived without sin. Having entered the house, some general conversation is kept up for a few minutes, till permission is asked to pass the night there. This is granted as a matter of course. The stranger then takes his meals with the family, and a room is assigned him, where with the horsecloths belonging to his recado (or saddle of the Pampas) he makes his bed. It is curious how similar circumstances produce such similar results in manners. At the Cape of Good Hope the same hospitality, and very nearly the same points of etiquette, are universally observed. The difference, however, between the character of the Spaniard and that of the Dutch boer is shown, by the former never asking his guest a single question beyond the strictest rule of politeness, whilst the honest Dutchman demands where he has been, where he is going, what is his business, and even how many brothers, sisters, or children he may happen to have.

Shortly after our arrival at Don Juan's, one of the largest herds of cattle was driven in towards the house, and three beasts were picked out to be slaughtered for the supply of the establishment. These half-wild cattle are very active; and knowing full well the fatal lazo, they led the horses a long and laborious chase. After witnessing the rude wealth displayed in the number of cattle, men, and horses, Don Juan's miserable house was quite curious. The floor consisted of hardened mud, and the windows were without glass; the sitting-room boasted only of a few of the roughest chairs and stools, with a couple of tables. The supper, although several strangers were present, consisted of two huge piles, one of roast beef, the other of boiled, with some pieces of pumpkin: besides this latter there was no other vegetable, and not even a morsel of bread. For drinking, a large earthenware jug of water served the whole party. Yet this man was the owner of several square miles of land, of which nearly every acre would produce corn, and, with a little trouble, all the common vegetables. The evening was spent in smoking, with a little impromptu singing, accompanied by the guitar. The signoritas all sat together in one corner of the room, and did not sup with the men.

So many works have been written about these countries, that it is almost superfluous to describe either the lazo or the bolas. The lazo consists of a very strong, but thin, well-plaited rope, made of raw hide. One end is attached to the broad surcingle, which fastens together the complicated gear of the recado, or saddle used in the Pampas; the other is terminated by a small ring of iron or brass, by which a noose can be formed. The Gaucho, when he is going to use the lazo, keeps a small coil in his bridle-hand, and in the other holds the running noose, which is made very large, generally having a diameter of about eight feet. This he whirls round his head, and by the dexterous movement of his wrist keeps the noose open; then, throwing it, he causes it to fall on any particular spot he chooses. The lazo, when not used, is tied up in a small coil to the after part of the recado. The bolas, or balls, are of two kinds: the simplest, which is chiefly used for catching ostriches, consists of two round stones, covered with leather, and united by a thin plaited thong, about eight feet long. The other kind differs only in having three balls united by the thongs to a common centre. The Gaucho holds the smallest of the three in his hand, and whirls the other two round and round his head; then, taking aim, sends them like chain shot revolving through the air. The balls no sooner strike any object, than, winding round it, they cross each other, and become firmly hitched. The size and weight of the balls vary, according to the purpose for which they are made: when of stone, although not larger than an apple, they are sent with such force as sometimes to break the leg even of a horse. I have seen the balls made of wood, and as large as a turnip, for the sake of catching these animals without injuring them. The balls are sometimes made of iron, and these can be hurled to the greatest distance. The main difficulty in using either lazo or bolas is to ride so well as to be able at full speed, and while suddenly turning about, to whirl them so steadily round the head, as to take aim: on foot any person would soon learn the art. One day, as I was amusing myself by galloping and whirling the balls round my head, by accident the free one struck a bush; and its revolving motion being thus destroyed, it immediately fell to the ground, and, like magic, caught one hind leg of my horse; the other ball was then jerked out of my hand, and the horse fairly secured. Luckily he was an old practised ainimal, and knew what it meant; otherwise he would probably have kicked till he had thrown himself down. The Gauchos roared with laugher; they cried out that they had seen every sort of animal caught, but had never before seen a man caught by himself.

During the two succeeding days, I reached the furthest point which I was anxious to examine. The country wore the same aspect, till at last the fine green turf became more wearisome than a dusty turnpike road. We everywhere saw great numbers of partridges (Nothura major). These birds do not go in coveys, nor do they conceal themselves like the English kind. It appears a very silly bird. A man on horseback by riding round and round in a circle, or rather in a spire, so as to approach closer each time, may knock on the head as many as he pleases. The more common method is to catch them with a running noose, or little lazo, made of the stem of an ostrich's feather, fastened to the end of a long stick. A boy on a quiet old horse will frequently thus catch thirty or forty in a day. In Arctic North America[21] the Indians catch the Varying Hare by walking spirally round and round it, when on its form: the middle of the day is reckoned the best time, when the sun is high, and the shadow of the hunter not very long.

On our return to Maldonado, we followed rather a different line of road. Near Pan de Azucar, a landmark well known to all those who have sailed up the Plata, I stayed a day at the house of a most hospitable old Spaniard. Early in the morning we ascended the Sierra de las Animas. By the aid of the rising sun the scenery was almost picturesque. To the westward the view extended over an immense level plain as far as the Mount, at Monte Video, and to the eastward, over the mammillated country of

Maldonado. On the summit of the mountain there were several small heaps of stones, which evidently had lain there for many years. My companion assured me that they were the work of the Indians in the old time. The heaps were similar, but on a much smaller scale, to those so commonly found on the mountains of Wales. The desire to signalize any event, on the highest point of the neighbouring land, seems an universal passion with mankind. At the present day, not a single Indian, either civilized or wild, exists in this part of the province; nor am I aware that the former inhabitants have left behind them any more permanent records than these insignificant piles on the summit of the Sierra de las Animas.

THE GENERAL, and almost entire absence of trees in Banda Oriental is remarkable. Some of the rocky hills are partly covered by thickets, and on the banks of the larger streams, especially to the north of Las Minas, willow-trees are not uncommon. Near the Arroyo Tapes I heard of a wood of palms; and one of these trees, of considerable size, I saw near the Pan de Azucar, in lat. 35°. These, and the trees planted by the Spaniards, offer the only exceptions to the general scarcity of wood. Among the introduced kinds may be enumerated poplars, olives, peach, and other fruit trees: the peaches succeed so well, that they afford the main supply of firewood to the city of Buenos Ayres. Extremely level countries, such as the Pampas, seldom appear favourable to the growth of trees. This may possibly be attributed either to the force of the winds, or the kind of drainage. In the nature of the land, however, around Maldonado, no such reason is apparent; the rocky mountains afford protected situations, enjoying various kinds of soil; streamlets of water are common at the bottoms of nearly every valley; and the clayey nature of the earth seems adapted to retain moisture. It has been inferred with much probability, that the presence of woodland is generally determined[22] by the annual amount of moisture; yet in this province abundant and heavy rain falls during the winter; and the summer, though dry, is not so in any excessive degree.[23] We see nearly the whole of Australia covered by lofty trees, yet that country possesses a far more arid climate. Hence we must look to some other and unknown cause.

Confining our view to South America, we should certainly be tempted to believe that trees flourished only under a very humid climate; for the limit of the forest-land follows, in a most remarkable manner, that of the damp winds. In the southern part of the continent, where the western gales, charged with moisture from the Pacific, prevail, every island on the broken west coast, from lat. 38° to the extreme point of Tierra del Fuego, is densely covered by impenetrable forests. On the eastern side of the Cordillera, over the same extent of latitude, where a blue sky and a fine climate prove that the atmosphere has been deprived of its moisture by passing over the mountains, the arid plains of Patagonia support a most scanty vegetation. In the more northern parts of the continent, within the limits of the constant south-eastern trade-wind, the eastern side is ornamented by magnificent forests; whilst the western coast, from lat. 4° S. to lat. 32° S., may be described as a desert; on this western coast, northward of lat. 4° S., where the trade-wind loses its regularity, and heavy torrents of rain fall periodically, the shores of the Pacific, so utterly desert in Peru, assume near Cape Blanco the character of luxuriance so celebrated at Guyaquil and Panama. Hence in the southern and northern parts of the continent, the forest and desert lands occupy reversed positions with respect to the Cordillera, and these positions are apparently determined by the direction of the prevalent winds. In the middle of the continent there is a broad intermediate band, including central Chile and the provinces of La Plata, where the rain-bringing winds have not to pass over lofty mountains, and where the land is neither a desert nor covered by forests. But even the rule, if confined to South America, of trees flourishing only in a climate rendered humid by rain-bearing winds, has a strongly marked exception in the case of the Falkland Islands. These islands, situated in the same latitude with Tierra del Fuego and only between two and three hundred miles distant from it, having a nearly similar climate, with a geological formation almost identical, with favourable situations and the same kind of peaty soil, yet can boast of few plants deserving even the title of bushes; whilst in Tierra del Fuego it is impossible to find an acre of land not covered by the densest forest. In this case, both the direction of the heavy gales of wind and of the currents of the sea are favourable to the transport of seeds from Tierra del Fuego, as is shown by the canoes and trunks of trees drifted from that country, and frequently thrown on the shores of the Western Falkland. Hence perhaps it is, that there are many plants in common to the two countries: but with respect to the trees of Tierra del Fuego, even attempts made to transplant them have failed.

During our stay at Maldonado I collected several quadrupeds, eighty kinds of birds, and many reptiles, including nine species of snakes. Of the indigenous mammalia, the only one now left of any size, which is common, is the Cervus campestris. This deer is exceedingly abundant, often in small herds, throughout the countries bordering the Plata and in Northern Patagonia. If a person crawling close along the ground, slowly advances towards a herd, the deer frequently, out of curiosity, approach to reconnoitre him. I have by this means, killed from one spot, three out of the same herd. Although so tame and inquisitive, yet when approached on horseback, they are exceedingly wary. In this country nobody goes on foot, and the deer knows man as its enemy only when he is mounted and armed with the bolas. At Bahia Blanca, a recent establishment in Northern Patagonia, I was surprised to find how little the deer cared for the noise of a gun: one day I fired ten times from within eighty yards at one animal; and it was much more startled at the ball cutting up the ground than at the report of the rifle. My powder being exhausted, I was obliged to get up (to my shame as a sportsman be it spoken, though well able to kill birds on the wing) and halloo till the deer ran away.

The most curious fact with respect to this animal, is the overpoweringly strong and offensive odour which proceeds from the buck. It is quite indescribable: several times whilst skinning the specimen which is now mounted at the Zoological Museum, I was almost overcome by nausea. I tied up the skin in a silk pocket-handkerchief, and so carried it home: this handkerchief, after being well washed, I continually used, and it was of course as repeatedly washed; yet every time, for a space of one year and seven months, when first unfolded, I distinctly perceived the odour. This appears an astonishing instance of the permanence of some matter, which nevertheless in its nature must be most subtile and volatile. Frequently, when passing at the distance of half a mile to leeward of a herd, I have perceived the whole air tainted with the effluvium. I believe the smell from the buck is most powerful at the period when its horns are perfect, or free from the hairy skin. When in this state the meat is, of course, quite uneatable; but the Gauchos assert, that if buried for some time in fresh earth, the taint is removed. I have somewhere read that the islanders in the north of Scotland treat the rank carcasses of the fish-eating birds in the same manner.

The order Rodentia is here very numerous in species: of mice alone I obtained no less than eight kinds.[24] The largest gnawing animal in the world, the Hydrochærus capybara (the water-hog), is here also common. One which I shot at Monte Video weighed ninety-eight pounds: its length, from the end of the snout to the stump-like tail, was three feet two inches; and its girth three feet eight. These great Rodents occasionally frequent the islands in the mouth of the Plata, where the water is quite salt, but are far more abundant on the borders of fresh-water lakes and rivers. Near Maldonado three or four generally live together. In the daytime they either lie among the aquatic plants, or openly feed on the turf plain.[25] When viewed at a distance, from their manner of walking and colour they resemble pigs: but when seated on their haunches, and attentively watching any object with one eye, they reassume the appearance of their congeners, cavies and rabbits. Both the front and side view of their head has quite a ludicrous aspect, from the great depth of their jaw. These animals, at Maldonado, were very tame; by cautiously walking, I approached within three yards of four old ones. This tameness may probably be accounted for, by the Jaguar having been banished for some years, and by the Gaucho not thinking it worth his while to hunt them. As I approached nearer and nearer they frequently made their peculiar noise, which is a low abrupt grunt, not having much actual sound, but rather arising from the sudden expulsion of air: the only noise I know at all like it, is the first hoarse bark of a large dog. Having watched the four from almost within arm's length (and they me) for several minutes, they rushed into the water at full gallop with the greatest impetuosity, and emitted at the same time their bark. After diving a short distance they came again to the surface, but only just showed the upper part of their heads. When the female is swimming in the water, and has young ones, they are said to sit on her back. These animals are easily killed in numbers; but their skins are of trifling value, and the meat is very indifferent. On the islands in the Rio Parana they are exceedingly abundant, and afford the ordinary prey to the Jaguar.

The Tucutuco (Ctenomys Brasiliensis) is a curious small animal, which may be briefly described as a Gnawer, with the habits of a mole. It is extremely numerous in some parts of the country, but it is difficult to be procured, and never, I believe, comes out of the ground. It throws up at the mouth of its burrows hillocks of earth like those of the mole, but smaller. Considerable tracts of country are so completely undermined by these animals, that horses in passing over, sink above their fetlocks. The tucutucos appear, to a certain degree, to be gregarious: the man who procured the specimens for me had caught six together, and he said this was a common occurrence. They are nocturnal in their habits; and their principal food is the roots of plants, which are the object of their extensive and superficial burrows. This animal is universally known by a very peculiar noise which it makes when beneath the ground. A person, the first time he hears it, is much surprised; for it is not easy to tell whence it comes, nor is it possible to guess what kind of creature utters it. The noise consists in a short, but not rough, nasal grunt, which is monotonously repeated about four times in quick succession:[26] the name Tucutuco is given in imitation of the sound. Where this animal is abundant, it may be heard at all times of the day, and sometimes directly beneath one's feet. When kept in a room, the tucutucos move both slowly and clumsily, which appears owing to the outward action of their hind legs; and they are quite incapable, from the socket of the thigh-bone not having a certain ligament, of jumping even the smallest vertical height. They are very stupid in making any attempt to escape; when angry or frightened they utter the tucutuco. Of those I kept alive several, even the first day, became quite tame, not attempting to bite or to run away; others were a little wilder.

The man who caught them asserted that very many are invariably found blind. A specimen which I preserved in spirits was in this state; Mr. Reid considers it to be the effect of inflammation in the nictitating membrane.

When the animal was alive I placed my finger within half an inch of its head, and not the slightest notice was taken: it made its way, however, about the room nearly as well as the others. Considering the strictly subterranean habits of the tucutuco, the blindness, though so common, cannot be a very serious evil; yet it appears strange that any animal should possess an organ frequently subject to be injured. Lamarck would have been delighted with this fact, had he known it, when speculating[27] (probably with more truth than usual with him) on the gradually *acquired* blindness of the Asphalax, a Gnawer living under ground, and of the Proteus, a reptile living in dark caverns filled with water; in both of which animals the eye is in an almost rudimentary state, and is covered by a tendinous membrane and skin. In the common mole the eye is extraordinarily small but perfect, though many anatomists doubt whether it is connected with the true optic nerve; its vision must certainly be imperfect, though probably useful to the animal when it leaves its burrow. In the tucutuco, which I believe never comes to the surface of the ground, the eye is rather larger, but often rendered blind and useless, though without apparently causing any inconvenience to the animal; no doubt Lamarck would have said that the tucutuco is now passing into the state of the Asphalax and Proteus.

Birds of many kinds are extremely abundant on the undulating grassy plains around Maldonado. There are several species of a family allied in structure and manners to our Starling: one of these (Molothrus niger) is remarkable from its habits. Several may often be seen standing together on the back of a cow or horse; and while perched on a hedge, pluming themselves in the sun, they sometimes attempt to sing, or rather to hiss; the noise being very peculiar, resembling that of bubbles of air passing rapidly from a small orifice under water, so as to produce an acute sound. According to Azara, this bird, like the cuckoo, deposits its eggs in other birds' nests. I was several times told by the country people that there certainly is some bird having this habit; and my assistant in collecting, who is a very accurate person, found a nest of the sparrow of this country (Zonotrichia matutina), with one egg in it larger than the others, and of a different colour and shape. In North America there is another species of Molothrus (M. pecoris), which has a similar cuckoo-like habit, and which is most closely allied in every respect to the species from the Plata, even in such trifling peculiarities as standing on the backs of cattle; it differs only in being a little smaller, and in its plumage and eggs being of a slightly different shade of colour. This close agreement in structure and habits, in representative species coming from opposite quarters of a great continent, always strikes one as interesting, though of common occurrence.

Mr. Swainson has well remarked,[28] that with the exception of the Molothrus pecoris, to which must be added the M. niger, the cuckoos are the only birds which can be called truly parasitical; namely, such as "fasten themselves, as it were, on another living animal, whose animal heat brings their young into life, whose food they live upon, and whose death would cause theirs during the period of infancy." It is remarkable that some of the species, but not all, both of the Cuckoo and Molothrus, should agree in this one strange habit of their parasitical propagation, whilst opposed to each other in almost every other habit: the molothrus, like our starling, is eminently sociable, and lives on the open plains without art or disguise: the cuckoo, as every one knows, is a singularly shy bird; it frequents the most retired thickets, and feeds on fruit and caterpillars. In structure also these two genera are widely removed from each other. Many theories, even phrenological theories, have been advanced to explain the origin of the cuckoo laying its eggs in other birds' nests. M. Prévost alone, I think, has thrown light by his observations[29] on this puzzle: he finds that the female cuckoo, which, according to most observers, lays at least from four to six eggs, must pair with the male each time after laying only one or two eggs. Now, if the cuckoo was obliged to sit on her own eggs, she would either have to sit on all together, and therefore leave those first laid so long, that they probably would become addled; or she would have to hatch separately each egg, or two eggs, as soon as laid: but as the cuckoo stays a shorter time in this country than any other migratory bird, she certainly would not have time enough for the successive hatchings. Hence we can perceive in the fact of the cuckoo pairing several times, and laying her eggs at intervals, the cause of her depositing her eggs in other birds' nests, and leaving them to the care of foster-parents. I am strongly inclined to believe that this view is correct, from having been independently led (as we shall hereafter see) to an analogous conclusion with regard to the South American ostrich, the females of which are parasitical, if I may so express it, on each other; each female laying several eggs in the nests of several other females, and the male ostrich undertaking all the cares of incubation, like the strange foster-parents with the cuckoo.

I will mention only two other birds, which are very common, and render themselves prominent from their habits. The Saurophagus sulphuratus is typical of the great American tribe of tyrant-flycatchers. In its structure it closely approaches the true shrikes, but in its habits may be compared to many birds. I have frequently observed it, hunting a field, hovering over one spot like a hawk, and then proceeding on to another. When seen thus suspended in the air, it might very readily at a short distance be mistaken for one of the Rapacious order; its stoop, however, is very inferior in force and rapidity to that of a hawk. At other times the Saurophagus haunts the neighbourhood of water, and there, like a kingfisher, remaining stationary, it catches any small fish which may come near the margin. These birds are not unfrequently kept either in cages or in courtyards, with their wings cut. They soon become tame, and are very amusing from their cunning odd manners, which were described to me as being similar to those of the common magpie. Their flight is undulatory, for the weight of the head and bill appears too great for the body. In the evening the Saurophagus takes its stand on a bush, often by the roadside, and continually repeats without a change a shrill and rather agreeable cry, which somewhat resembles articulate words: the Spaniards say it is like the words "Bien te veo" (I see you well), and accordingly have given it this name.

A mocking-bird (Mimus orpheus), called by the inhabitants Calandria, is remarkable, from possessing a song far superior to that of any other bird in the country: indeed, it is nearly the only bird in South America which I have observed to take its stand for the purpose of singing. The song may be compared to that of the Sedge warbler, but is more powerful; some harsh notes and some very high ones, being mingled with a pleasant warbling. It is heard only during the spring. At other times its cry is harsh and far from harmonious. Near Maldonado these birds were tame and bold; they constantly attended the country houses in numbers, to pick the meat which was hung up on the posts or walls: if any other small bird joined the feast, the Calandria soon chased it away. On the wide uninhabited plains of Patagonia another closely allied species, O. Patagonica of d'Orbigny, which frequents the valleys clothed with spiny bushes, is a wilder bird, and has a slightly different tone of voice. It appears to me a curious circumstance, as showing the fine shades of difference in habits, that judging from this latter respect alone, when I first saw this second species, I thought it was different from the Maldonado kind. Having afterwards procured a specimen, and comparing the two without particular care, they appeared so very similar, that I changed my opinion; but now Mr. Gould says that they are certainly distinct; a conclusion in conformity with the trifling difference of habit, of which, of course, he was not aware.

The number, tameness, and disgusting habits of the carrion-feeding hawks of South America make them pre-eminently striking to any one accustomed only to the birds of Northern Europe. In this list may be included four species of the Caracara or Polyborus, the Turkey buzzard, the Gallinazo, and the Condor. The Caracaras are, from their structure, placed among the eagles: we shall soon see how ill they become so high a rank. In their habits they well supply the place of our carrion-crows, magpies, and ravens; a tribe of birds widely distributed over the rest of the world, but entirely absent in South America. To begin with the Polyborus Brasiliensis: this is a common bird, and has a wide geographical range; it is most numerous on the grassy savannahs of La Plata (where it goes by the name of Carrancha), and is far from unfrequent throughout the sterile plains of

Patagonia. In the desert between the rivers Negro and Colorado, numbers constantly attend the line of road to devour the carcasses of the exhausted animals which chance to perish from fatigue and thirst. Although thus common in these dry and open countries, and likewise on the arid shores of the Pacific, it is nevertheless found inhabiting the damp impervious forests of West Patagonia and Tierra del Fuego. The Carranchas, together with the Chimango, constantly attend in numbers the estancias and slaughtering-houses. If an animal dies on the plain the Gallinazo commences the feast, and then the two species of Polyborus pick the bones clean. These birds, although thus commonly feeding together, are far from being friends. When the Carrancha is quietly seated on the branch of a tree or on the ground, the Chimango often continues for a long time flying backwards and forwards, up and down, in a semicircle, trying each time at the bottom of the curve to strike its larger relative. The Carrancha takes little notice, except by bobbing its head. Although the Carranchas frequently assemble in numbers, they are not gregarious; for in desert places they may be seen solitary, or more commonly by pairs.

The Carranchas are said to be very crafty, and to steal great numbers of eggs. They attempt, also, together with the Chimango, to pick off the scabs from the sore backs of horses and mules. The poor animal, on the one hand, with its ears down and its back arched; and, on the other, the hovering bird, eyeing at the distance of a yard the disgusting morsel, form a picture, which has been described by Captain Head with his own peculiar spirit and accuracy. These false eagles most rarely kill any living bird or animal; and their vulture-like, necrophagous habits are very evident to any one who has fallen asleep on the desolate plains of Patagonia, for when he wakes, he will see, on each surrounding hillock, one of these birds patiently watching him with an evil eye: it is a feature in the landscape of these countries, which will be recognised by every one who has wandered over them. If a party of men go out hunting with dogs and horses, they will be accompanied, during the day, by several of these attendants. After feeding, the uncovered craw protrudes; at such times, and indeed generally, the Carrancha is an inactive, tame, and cowardly bird. Its flight is heavy and slow, like that of an English rook. It seldom soars; but I have twice seen one at a great height gliding through the air with much ease. It runs (in contradistinction to hopping), but not quite so quickly as some of its congeners. At times the Carrancha is noisy, but is not generally so: its cry is loud, very harsh and peculiar, and may be likened to the sound of the Spanish guttural *g*, followed by a rough double *r r;* when uttering this cry it elevates its head higher and higher, till at last, with its beak wide open, the crown almost touches the lower part of the back. This fact, which has been doubted, is quite true; I have seen them several times with their heads backwards in a completely inverted position. To these observations I may add, on the high authority of Azara, that the Carrancha feeds on worms, shells, slugs, grasshoppers, and frogs; that it destroys young lambs by tearing the umbilical cord; and that it pursues the Gallinazo, till that bird is compelled to vomit up the carrion it may have recently gorged. Lastly, Azara states that several Carranchas, five or six together, will unite in chase of large birds, even such as herons. All these facts show that it is a bird of very versatile habits and considerable ingenuity.

The Polyborus Chimango is considerably smaller than the last species. It is truly omnivorous, and will eat even bread; and I was assured that it materially injures the potato-crops in Chiloe, by stocking up the roots when first planted. Of all the carrion-feeders it is generally the last which leaves the skeleton of a dead animal, and may often be seen within the ribs of a cow or horse, like a bird in a cage. Another species is the Polyborus Novæ Zelandiæ, which is exceedingly common in the Falkland Islands. These birds in many respects resemble in their habits the Carranchas. They live on the flesh of dead animals and on marine productions; and on the Ramirez rocks their whole sustenance must depend on the sea. They are extraordinarily tame and fearless, and haunt the neighborhood of houses for offal. If a hunting party kills an animal, a number soon collect and patiently await, standing on the ground on all sides. After eating, their uncovered craws are largely protruded, giving them a disgusting appearance. They readily attack wounded birds: a cormorant in this state having taken to the shore, was immediately seized on by several, and its death hastened by their blows. The Beagle was at the Falklands only during the summer, but the officers of the Adventure, who were there in the winter, mention many extraordinary instances of the boldness and rapacity of these birds. They actually pounced on a dog that was lying fast asleep close by one of the party; and the sportsmen had difficulty in preventing the wounded geese from being seized before their eyes. It is said that several together (in this respect resembling the Carranchas) wait at the mouth of a rabbit-hole, and together seize on the animal when it comes out. They were constantly flying on board the vessel when in the harbour; and it was necessary to keep a good look out to prevent the leather being torn from the rigging, and the meat or game from the stern. These birds are very mischievous and inquisitive; they will pick up almost anything from the ground; a large black glazed hat was carried nearly a mile, as was a pair of the heavy balls used in catching cattle. Mr. Usborne experienced during the survey a more severe loss, in their stealing a small Kater's compass in a red morocco leather case, which was never recovered. These birds are, moreover, quarrelsome and very passionate; tearing up the grass with their bills from rage. They are not truly gregarious; they do not soar, and their flight is heavy and clumsy; on the ground they run extremely fast, very much like pheasants. They are noisy, uttering several harsh cries, one of which is like that of the English rook; hence the sealers always call them rooks. It is a curious circumstance that, when crying out, they throw their heads upwards and backwards, after the same manner as the Carrancha. They build in the rocky cliffs of the sea-coast, but only on the small adjoining islets, and not on the two main islands: this is a singular precaution in so tame and fearless a bird. The sealers say that the flesh of these birds, when cooked, is quite white, and very good eating; but bold must the man be who attempts such a meal.

We have now only to mention the turkey-buzzard (Vultur aura), and the Gallinazo. The former is found wherever the country is moderately damp, from Cape Horn to North America. Differently from the Polyborus Brasiliensis and Chimango, it has found its way to the Falkland Islands. The turkey-buzzard is a solitary bird, or at most goes in pairs. It may at once be recognised from a long distance, by its lofty, soaring, and most elegant flight. It is well known to be a true carrion-feeder. On the west coast of Patagonia, among the thickly-wooded islets and broken land, it lives exclusively on what the sea throws up, and on the carcasses of dead seals. Wherever these animals are congregated on the rocks, there the vultures may be seen. The Gallinazo (Cathartes atratus) has a different range from the last species, as it never occurs southward of lat. 41°. Azara states that there exists a tradition that these birds, at the time of the conquest, were not found near Monte Video, but that they subsequently followed the inhabitants from more northern districts. At the present day they are numerous in the valley of the Colorado, which is three hundred miles due south of Monte Video. It seems probable that this additional migration has happened since the time of Azara. The Gallinazo generally prefers a humid climate, or rather the neighbourhood of fresh water; hence it is extremely abundant in Brazil and La Plata, while it is never found on the desert and arid plains of Northern Patagonia, excepting near some stream. These birds frequent the whole Pampas to the foot of the Cordillera, but I never saw or heard of one in Chile; in Peru they are preserved as scavengers. These vultures certainly may be called gregarious, for they seem to have pleasure in society, and are not solely brought together by the attraction of a common prey. On a fine day a flock may often be observed at a great height, each bird wheeling round and round without closing its wings, in the most graceful evolutions. This is clearly performed for the mere pleasure of the exercise, or perhaps is connected with their matrimonial alliances.

I have now mentioned all the carrion-feeders, excepting the condor, an account of which will be more appropriately introduced when we visit a country more congenial to its habits than the plains of La Plata.

In a broad band of sand-hillocks which separate the Laguna del Potrero from the shores of the Plata, at the distance of a few miles from Maldonado, I found a group of those vitrified, siliceous tubes, which are formed by lightning entering loose sand. These tubes resemble in every particular those from Drigg in Cumberland, described in the Geological Transactions.[30] The sand-hillocks of Maldonado, not being protected by vegetation, are constantly changing their position. From this cause the tubes projected above the surface, and numerous fragments lying near, showed that they had formerly been buried to a greater depth. Four sets entered the sand perpendicularly: by working with my hands I traced one of them two feet deep; and some fragments which evidently had belonged to the same tube, when added to the other part, measured five feet three inches. The diameter of the whole tube was nearly equal, and therefore we must suppose that originally it extended to a much greater depth. These dimensions are however small, compared to those of the tubes from Drigg, one of which was traced to a depth of not less than thirty feet.

The internal surface is completely vitrified, glossy, and smooth. A small fragment examined under the microscope appeared, from the number of minute entangled air or perhaps steam bubbles, like an assay fused before the blowpipe. The sand is entirely, or in greater part, siliceous; but some points are of a black colour, and from their glossy surface possess a metallic lustre. The thickness of the wall of the tube varies from a thirtieth to a twentieth of an inch, and occasionally even equals a tenth. On the outside the grains of sand are rounded, and have a slightly glazed appearance: I could not distinguish any signs of crystallization. In a similar manner to that described in the Geological Transactions, the tubes are generally compressed, and have deep longitudinal furrows, so as closely to resemble a shrivelled vegetable stalk, or the bark of the elm or cork tree. Their circumference is about two inches, but in some fragments, which are cylindrical and without any furrows, it is as much as four inches. The compression from the surrounding loose sand, acting while the tube was still softened from the effects of the intense heat, has evidently caused the creases or furrows. Judging from the uncompressed fragments, the measure or bore of the lightning (if such a term may be used) must have been about one inch and a quarter. At Paris, M. Hachette and M. Beudant[31] succeeded in making tubes, in most respects similar to these fulgurites, by passing very strong shocks of galvanism through finely-powdered glass: when salt was added, so as to increase its fusibility, the tubes were larger in every dimension. They failed both with powdered felspar and quartz. One tube, formed with pounded glass, was very nearly an inch long, namely .982, and had an internal diameter of .019 of an inch. When we hear that the strongest battery in Paris was used, and that its power on a substance of such easy fusibility as glass was to form tubes so diminutive, we must feel greatly astonished at the force of a shock of lightning, which, striking the sand in several places, has formed cylinders, in one instance of at least thirty feet long, and having an internal bore, where not compressed, of full an inch and a half; and this in a material so extraordinarily refractory as quartz!

The tubes, as I have already remarked, enter the sand nearly in a vertical direction. One, however, which was less regular than the others, deviated from a right line, at the most considerable bend, to the amount of thirty-three degrees. From this same tube, two small branches, about a foot apart, were sent off; one pointed downwards, and the other upwards. This latter case is remarkable, as the electric fluid must have turned back at the acute angle of 26°, to the line of its main course. Besides the four tubes which I found vertical, and traced beneath the surface, there were several other groups of fragments, the original sites of which without doubt were near. All occurred in a level area of shifting sand, sixty yards by twenty, situated among some high sand-hillocks, and at the distance of about half a mile from a chain of hills four or five hundred feet in height. The most remarkable circumstance, as it appears to me, in this case as well as in that of Drigg, and in one described by M. Ribbentrop in Germany, is the number of tubes found within such limited spaces. At Drigg, within an area of fifteen yards, three were observed, and the same number occurred in Germany. In the case which I have described, certainly more than four existed within the space of the sixty by twenty yards. As it does not appear probable that the tubes are produced by successive distinct shocks, we must believe that the lightning, shortly before entering the ground, divides itself into separate branches.

The neighbourhood of the Rio Plata seems peculiarly subject to electric phenomena. In the year 1793,[32] one of the most destructive thunderstorms perhaps on record happened at Buenos Ayres: thirty-seven places within the city were struck by lightning, and nineteen people killed. From facts stated in several books of travels, I am inclined to suspect that thunderstorms are very common near the mouths of great rivers. Is it not possible that the mixture of large bodies of fresh and salt water may disturb the electrical equilibrium? Even during our occasional visits to this part of South America, we heard of a ship, two churches, and a house having been struck. Both the church and the house I saw shortly afterwards: the house belonged to Mr. Hood, the consul-general at Monte Video. Some of the effects were curious: the paper, for nearly a foot on each side of the line where the bell-wires had run, was blackened. The metal had been fused, and although the room was about fifteen feet high, the globules, dropping on the chairs and furniture, had drilled in them a chain of minute holes. A part of the wall was shattered, as if by gunpowder, and the fragments had been blown off with force sufficient to dent the wall on the opposite side of the room. The frame of a looking-glass was blackened, and the gilding must have been volatilized, for a smelling-bottle, which stood on the chimney-piece, was coated with bright metallic particles, which adhered as firmly as if they had been enamelled.

CHAPTER IV

Rio Negro to Bahia Blanca

RIO NEGRO • ESTANCIAS ATTACKED BY THE INDIANS • SALT-LAKES • FLAMINGOES • R. NEGRO TO R. COLORADO • SACRED TREE • PATAGONIAN HARE • INDIAN FAMILIES • GENERAL ROSAS • PROCEED TO BAHIA BLANCA • SAND DUNES • NEGRO LIEUTENANT • BAHIA BLANCA • SALINE INCRUSTATIONS • PUNTA ALTA • ZORILLO

JULY 24TH, 1833.—THE BEAGLE sailed from Maldonado, and on August the 3rd she arrived off the mouth of the Rio Negro. This is the principal river on the whole line of coast between the Strait of Magellan and the Plata. It enters the sea about three hundred miles south of the estuary of the Plata. About fifty years ago, under the old Spanish government, a small colony was established here; and it is still the most southern position (lat. 41°) on this eastern coast of America inhabited by civilized man.

The country near the mouth of the river is wretched in the extreme: on the south side a long line of perpendicular cliffs commences, which exposes a section of the geological nature of the country. The strata are of sand-stone, and one layer was remarkable from being composed of a firmly-cemented conglomerate of pumice pebbles, which must have travelled more than four hundred miles, from the Andes. The surface is everywhere covered up by a thick bed of gravel, which extends far and wide over the open plain. Water is extremely scarce, and, where found, is almost invariably brackish. The vegetation is scanty; and although there are bushes of many kinds, all are armed with formidable thorns, which seem to warn the stranger not to enter on these inhospitable regions.

The settlement is situated eighteen miles up the river. The road follows the foot of the sloping cliff, which forms the northern boundary of the great valley, in which the Rio Negro flows. On the way we passed the ruins of some fine "estancias," which a few years since had been destroyed by the Indians. They withstood several attacks. A man present at one gave me a very lively description of what took place. The inhabitants had sufficient notice to drive all the cattle and horses into the "corral"[33] which surrounded the house, and likewise to mount some small cannon. The Indians were Araucanians from the south of Chile; several hundreds in number, and highly disciplined. They first appeared in two bodies on a neighbouring hill; having there dismounted, and taken off their fur mantles, they advanced naked to the charge. The only weapon of an Indian is a very long bamboo or chuzo, ornamented with ostrich feathers, and pointed by a sharp spearhead. My informer seemed to remember with the greatest horror the quivering of these chuzos as they approached near. When close, the cacique Pincheira hailed the besieged to give up their arms, or he would cut all their throats. As this would probably have been the result of their entrance under any circumstances, the answer was given by a volley of musketry. The Indians, with great steadiness, came to the very fence of the corral: but to their surprise they found the posts fastened together by iron nails instead of leather thongs, and, of course, in vain attempted to cut them with their knives. This saved the lives of the Christians: many of the wounded Indians were carried away by their companions, and at last, one of the under caciques being wounded, the bugle sounded a retreat. They retired to their horses, and seemed to hold a council of war. This was an awful pause for the Spaniards, as all their ammunition, with the exception of a few cartridges, was expended. In an instant the Indians mounted their horses, and galloped out of sight. Another attack was still more quickly repulsed. A cool Frenchman managed the gun; he stopped till the Indians approached close, and then raked their line with grape-shot: he thus laid thirty-nine of them on the ground; and, of course, such a blow immediately routed the whole party.

The town is indifferently called El Carmen or Patagones. It is built on the face of a cliff which fronts the river, and many of the houses are excavated even in the sandstone. The river is about two or three hundred yards wide, and is deep and rapid. The many islands, with their willow-trees, and the flat headlands, seen one behind the other on the northern boundary of the broad green valley, form, by the aid of a bright sun, a view almost picturesque. The number of inhabitants does not exceed a few hundreds. These Spanish colonies do not, like our British ones, carry within themselves the elements of growth. Many Indians of pure blood reside here: the tribe of the Cacique Lucanee constantly have their Toldos[34] on the outskirts of the town. The local government partly supplies them with provisions, by giving them all the old worn-out horses, and they earn a little by making horse-rugs and other articles of riding-gear. These Indians are considered civilized; but what their character may have gained by a lesser degree of ferocity, is almost counterbalanced by their entire immorality. Some of the younger men are, however, improving; they are willing to labour, and a short time since a party went on a sealing-voyage, and behaved very well. They were now enjoying the fruits of their labour, by being dressed in very gay, clean clothes, and by being very idle. The taste they showed in their dress was admirable; if you could have turned one of these young Indians into a statue of bronze, his drapery would have been perfectly graceful.

One day I rode to a large salt-lake, or Salina, which is distant fifteen miles from the town. During the winter it consists of a shallow lake of brine, which in summer is converted into a field of snow-white salt. The layer near the margin is from four to five inches thick, but towards the centre its thickness increases. This lake was two and a half miles long, and one broad. Others occur in the neighbourhood many times larger, and with a floor of salt, two and three feet in thickness, even when under water during the winter. One of these brilliantly white and level expanses, in the midst of the brown and desolate plain, offers an extraordinary spectacle. A large quantity of salt is annually drawn from the salina: and great piles, some hundred tons in weight, were lying ready for exportation. The season for working the salinas forms the harvest of Patagones; for on it the prosperity of the place depends. Nearly the whole population encamps on the bank of the river, and the people are employed in drawing out the salt in bullock-waggons. This salt is crystallized in great cubes, and is remarkably pure: Mr. Trenham Reeks has kindly analyzed some for me, and he finds in it only 0.26 of gypsum and 0.22 of earthy matter. It is a singular fact, that it does not serve so well for preserving meat as sea-salt from the Cape de Verd islands; and a merchant at Buenos Ayres told me that he considered it as fifty per cent less valuable. Hence the Cape de Verd salt is constantly imported, and is mixed with that from these salinas. The purity of the Patagonian salt, or absence from it of those other saline bodies found in all sea-water, is the only assignable cause for this inferiority: a conclusion which no one, I think, would have suspected, but which is supported by the fact lately ascertained,[35] that those salts answer best for preserving cheese which contain most of the deliquescent chlorides.

The border of this lake is formed of mud: and in this numerous large crystals of gypsum, some of which are three inches long, lie embedded; whilst on the surface others of sulphate of soda lie scattered about. The Gauchos call the former the "Padre del sal," and the latter the "Madre;" they state that these progenitive salts always occur on the borders of the salinas, when the water begins to evaporate. The mud is black, and has a fetid odour. I could not at first imagine the cause of this, but I afterwards perceived that the froth which the wind drifted on shore was coloured green, as if by confervæ; I attempted to carry home some of this green matter, but from an accident failed. Parts of the lake seen from a short distance appeared of a reddish colour, and this perhaps was owing to some infusorial animalcula. The mud in many places was thrown up by numbers of some kind of worm, or annelidous animal. How surprising it is that any creatures should be able to exist in brine, and that they should be crawling among crystals of sulphate of soda and lime! And what becomes of these worms when, during the long summer, the surface is hardened into a solid layer of salt? Flamingoes in considerable numbers inhabit this lake, and breed here; throughout Patagonia, in Northern Chile, and at the Galapagos Islands, I met with these birds wherever there were lakes of brine. I saw them here wading about in search of food—probably for the worms which burrow in the mud; and these latter probably feed on infusoria or confervæ. Thus we have a little living world within itself, adapted to these inland lakes of brine. A minute crustaceous animal (Cancer salinus) is said[36] to live in countless numbers in the brine-pans at Lymington: but only in those in which the fluid has attained, from evaporation, considerable strength—namely, about a quarter of a pound of salt to a pint of water. Well may we affirm that every part of the world is habitable! Whether lakes of brine, or those subterranean ones hidden beneath volcanic mountains—warm mineral springs—the wide expanse and depths of the ocean—the upper regions of the atmosphere, and even the surface of perpetual snow—all support organic beings.

To the northward of the Rio Negro, between it and the inhabited country near Buenos Ayres, the Spaniards have only one small settlement, recently established at Bahia Blanca. The distance in a straight line to Buenos Ayres is very nearly five hundred British miles. The wandering tribes of horse Indians, which have always occupied the greater part of this country, having of late much harassed the outlying estancias, the government at Buenos Ayres equipped some time since an army under the command of General Rosas for the purpose of exterminating them. The troops were now encamped on the banks of the Colorado; a river lying about eighty miles northward of the Rio Negro. When General Rosas left Buenos Ayres he struck in a direct line across the unexplored plains: and as the country was thus pretty well cleared of Indians, he left behind him, at wide intervals, a small party of soldiers with a troop of horses *(a posta)*, so as to be enabled to keep up a communication with the capital. As the Beagle intended to call at Bahia Blanca, I determined to proceed there by land; and ultimately I extended my plan to travel the whole way by the postas to Buenos Ayres.

August 11th.—Mr. Harris, an Englishman residing at Patagones, a guide, and five Gauchos who were proceeding to the army on business, were my companions on the journey. The Colorado, as I have already said, is nearly eighty miles distant: and as we travelled slowly, we were two days and a half on the road. The whole line of country deserves scarcely a better name than that of a desert. Water is found only in two small wells; it is called fresh; but even at this time of the year, during the rainy season, it was quite brackish. In the summer this must be a distressing passage; for now it was sufficiently desolate. The valley of the Rio Negro, broad as it is, has merely been excavated out of the sandstone plain; for immediately above the bank on which the town stands, a level country commences, which is interrupted only by a few trifling valleys and depressions. Everywhere the landscape wears the same sterile aspect; a dry gravelly soil supports tufts of brown withered grass, and low scattered bushes, armed with thorns.

Shortly after passing the first spring we came in sight of a famous tree, which the Indians reverence as the altar of Walleechu. It is situated on a high part of the plain; and hence is a landmark visible at a great distance. As soon as a tribe of Indians come in sight of it, they offer their adorations by loud shouts. The tree itself is low, much branched, and thorny: just above the root it has a diameter of about three feet. It stands by itself without any neighbour, and was indeed the first tree we saw; afterwards we met with a few others of the same kind, but they were far from common. Being winter the tree had no leaves, but in their place numberless threads, by which the various offerings, such as cigars, bread, meat, pieces of cloth, etc., had been suspended. Poor Indians, not having anything better, only pull a thread out of their ponchos, and fasten it to the tree. Richer Indians are accustomed to pour spirits and maté into a certain hole, and likewise to smoke upwards, thinking thus to afford all possible gratification to Walleechu. To complete the scene, the tree was surrounded by the bleached bones of horses which had been slaughtered as sacrifices. All Indians of every age and sex make their offerings; they then think that their horses will not tire, and that they themselves shall be prosperous. The Gaucho who told me this, said that in the time of peace he had witnessed this scene, and that he and others used to wait till the Indians had passed by, for the sake of stealing from Walleechu the offerings.

The Gauchos think that the Indians consider the tree as the god itself, but it seems far more probable that they regard it as the altar. The only cause which I can imagine for this choice, is its being a landmark in a dangerous passage. The Sierra de la Ventana is visible at an immense distance; and a Gaucho told me that he was once riding with an Indian a few miles to the north of the Rio Colorado, when the Indian commenced making the same loud noise, which is usual at the first sight of the distant tree; putting his hand to his head, and then pointing in the direction of the Sierra. Upon being asked the reason of this, the Indian said in broken Spanish, "First see the Sierra." About two leagues beyond this curious tree we halted for the night: at this instant an unfortunate cow was spied by the lynx-eyed Gauchos, who set off in full chase, and in a few minutes dragged her in with their lazos, and slaughtered her. We here had the four necessaries of life "en el campo,"—pasture for the horses, water (only a muddy puddle), meat and firewood. The Gauchos were in high spirits at finding all these luxuries; and we soon set to work at the poor cow. This was the first night which I passed under the open sky, with the gear of the recado for my bed. There is high enjoyment in the independence of the Gaucho life—to be able at any moment to pull up your horse, and say, "Here we will pass the night." The death-like stillness of the plain, the dogs keeping watch, the gipsy-group of Gauchos making their beds round the fire, have left in my mind a strongly-marked picture of this first night, which will never be forgotten.

The next day the country continued similar to that above described. It is inhabited by few birds or animals of any kind. Occasionally a deer, or a Guanaco (wild Llama) may be seen; but the Agouti (Cavia Patagonica) is the commonest quadruped. This animal here represents our hares. It differs, however, from that genus in many essential respects; for instance, it has only three toes behind. It is also nearly twice the size, weighing from twenty to twenty-five pounds. The Agouti is a true friend of the desert; it is a common feature of the landscape to see two or three hopping quickly one after the other in a straight line across these wild plains. They are found as far north as the Sierra Tapalguen (lat. 37° 30'), where the plain rather suddenly becomes greener and more humid; and their southern limit is between Port Desire and St. Julian, where there is no change in the nature of the country. It is a singular fact, that although the Agouti is not now found as far south as Port St. Julian, yet that Captain Wood, in his voyage in 1670, talks of them as being numerous there. What cause can have altered, in a wide, uninhabited, and rarely-visited country, the range of an animal like this? It appears also, from the number shot by Captain Wood in one day at Port Desire, that they must have been considerably more abundant there formerly than at present. Where the Bizcacha lives and makes its burrows, the Agouti uses them; but where, as at Bahia Blanca, the Bizcacha is not found, the Agouti burrows for itself. The same thing occurs with the little owl of the Pampas (Athene cunicularia), which has so often been described as standing like a sentinel at the mouth of the burrows; for in Banda Oriental, owing to the absence of the Bizcacha, it is obliged to hollow out its own habitation.

The next morning, as we approached the Rio Colorado, the appearance of the country changed; we soon came on a plain covered with turf, which, from its flowers, tall clover, and little owls, resembled the Pampas. We passed also a muddy swamp of considerable extent, which in summer dries, and becomes incrusted with various salts; and hence is called a salitral. It was covered by low succulent plants, of the same kind with those growing on the sea-shore. The Colorado, at the pass where we crossed it, is only about sixty yards wide; generally it must be nearly double that width. Its course is very tortuous, being marked by willow-trees and beds of reeds: in a direct line the distance to the mouth of the river is said to be nine leagues, but by water twenty-five. We were delayed crossing in the canoe by some immense troops of mares, which were swimming the river in order to follow a division of troops into the interior. A more ludicrous spectacle I never beheld than the hundreds and hundreds of heads, all directed one way, with pointed ears and distended snorting nostrils, appearing just above the water like a great shoal of some amphibious animal. Mare's flesh is the only food which the soldiers have when on an expedition. This gives them a great facility of movement; for the distance to which horses can be driven over these plains is quite surprising: I have been assured that an unloaded horse can travel a hundred miles a day for many days successively.

The encampment of General Rosas was close to the river. It consisted of a square formed by waggons, artillery, straw huts, etc. The soldiers were nearly all cavalry; and I should think such a villainous, banditti-like army was never before collected together. The greater number of men were of a mixed breed, between Negro, Indian, and Spaniard. I know not the reason, but men of such origin seldom have a good expression of countenance. I called on the Secretary to show my passport. He began to cross-question me in the most dignified and mysterious manner. By good luck I had a letter of recommendation from the government of Buenos Ayres[37] to the commandant of Patagones. This was taken to General Rosas, who sent me a very obliging message; and the Secretary returned all smiles and graciousness. We took up our residence in the *rancho,* or hovel, of a curious old Spaniard, who had served with Napoleon in the expedition against Russia.

We stayed two days at the Colorado; I had little to do, for the surrounding country was a swamp, which in summer (December), when the snow melts on the Cordillera, is overflowed by the river. My chief amusement was watching the Indian families as they came to buy little articles at the rancho where we stayed. It was supposed that General Rosas had about six hundred Indian allies. The men were a tall, fine race, yet it was afterwards easy to see in the Fuegian savage the same countenance rendered hideous by cold, want of food, and less civilization. Some authors, in defining the primary races of mankind, have separated these Indians into two classes; but this is certainly incorrect. Among the young women or chinas, some deserve to be called even beautiful. Their hair was coarse, but bright and black; and they wore it in two plaits hanging down to the waist. They had a high colour, and eyes that glistened with brilliancy; their legs, feet, and arms were small and elegantly formed; their ankles, and sometimes their wrists, were ornamented by broad bracelets of blue beads. Nothing could be more interesting than some of the family groups. A mother with one or two daughters would often come to our rancho, mounted on the same horse. They ride like men, but with their knees tucked up much higher. This habit, perhaps, arises from their being accustomed, when travelling, to ride the loaded horses. The duty of the women is to load and unload the horses; to make the tents for the night; in short to be, like the wives of all savages, useful slaves. The men fight, hunt, take care of the horses, and make the riding gear. One of their chief indoor occupations is to knock two stones together till they become round, in order to make the bolas. With this important weapon the Indian catches his game, and also his horse, which roams free over the plain. In fighting, his first attempt is to throw down the horse of his adversary with the bolas, and when entangled by the fall to kill him with the chuzo. If the balls only catch the neck or body of an animal, they are often carried away and lost. As the making the stones round is the labour of two days, the manufacture of the balls is a very common employment. Several of the men and women had their faces painted red, but I never saw the horizontal bands which are so common among the Fuegians. Their chief pride consists in having everything made of silver; I have seen a cacique with his spurs, stirrups, handle of his knife, and bridle made of this metal: the head-stall and reins being of wire, were not thicker than whipcord; and to see a fiery steed wheeling about under the command of so light a chain, gave to the horsemanship a remarkable character of elegance.

General Rosas intimated a wish to see me; a circumstance which I was afterwards very glad of. He is a man of an extraordinary character, and has a most predominant influence in the country, which it seems he will use to its prosperity and advancement.[38] He is said to be the owner of seventy-four square leagues of land, and to have about three hundred thousand head of cattle. His estates are admirably managed, and are far more productive of corn than those of others. He first gained his celebrity by his laws for his own estancias, and by disciplining several hundred men, so as to resist with success the attacks of the Indians. There are many stories current about the rigid manner in which his laws were enforced. One of these was, that no man, on penalty of being put into the stocks, should carry his knife on a Sunday: this being the principal day for gambling and drinking, many quarrels arose, which from the general manner of fighting with the knife often proved fatal. One Sunday the Governor came in great form to pay the estancia a visit, and General Rosas, in his hurry, walked out to receive him with his knife, as usual, stuck in his belt. The steward touched his arm, and reminded him of the law; upon which, turning to the Governor, he said he was extremely sorry, but that he must go into the stocks, and that till let out, he possessed no power even in his own house. After a little time the steward was persuaded to open the stocks, and to let him out, but no sooner was this done, than he turned to the steward and said, "You now have broken the laws, so you must take my place in the stocks." Such actions as these delighted the Gauchos, who all possess high notions of their own equality and dignity.

General Rosas is also a perfect horseman—an accomplishment of no small consequence in a country where an assembled army elected its general by the following trial: A troop of unbroken horses being driven into a corral, were let out through a gateway, above which was a cross-bar: it was agreed whoever should drop from the bar on one of these wild animals, as it rushed out, and should be able, without saddle or bridle, not only to ride it, but also to bring it back to the door of the corral, should be their general. The person who succeeded was accordingly elected; and doubtless made a fit general for such an army. This extraordinary feat has also been performed by Rosas.

By these means, and by conforming to the dress and habits of the Gauchos, he has obtained an unbounded popularity in the country, and in consequence a despotic power. I was assured by an English merchant, that a man who had murdered another, when arrested and questioned concerning his motive, answered, "He spoke disrespectfully of General Rosas, so I killed him." At the end of a week the murderer was at liberty. This doubtless was the act of the general's party, and not of the general himself.

In conversation he is enthusiastic, sensible, and very grave. His gravity is carried to a high pitch: I heard one of his mad buffoons (for he keeps two, like the barons of old) relate the following anecdote. "I wanted very much to hear a certain piece of music, so I went to the general two or three times to ask him; he said to me, 'Go about your business, for I am engaged.' I went a second time; he said, 'If you come again I will punish you.' A third time I asked, and he laughed. I rushed out of the tent, but it was too late; he ordered two soldiers to catch and stake me. I begged by all the saints in heaven he would let me off; but it would not do;—when the general laughs he spares neither mad man nor sound." The poor flighty gentleman looked quite dolorous, at the very recollection of the staking. This is a very severe punishment; four posts are driven into the ground, and the man is extended by his arms and legs horizontally, and there left to stretch for several hours. The idea is evidently taken from the usual method of drying hides. My interview passed away, without a smile, and I obtained a passport and order for the government post-horses, and this he gave me in the most obliging and ready manner.

In the morning we started for Bahia Blanca, which we reached in two days. Leaving the regular encampment, we passed by the toldos of the Indians. These are round like ovens, and covered with hides; by the mouth of each, a tapering chuzo was stuck in the ground. The toldos were divided into separate groups, which belong to the different caciques' tribes, and the groups were again divided into smaller ones, according to the relationship of the owners. For several miles we travelled along the valley of the Colorado. The alluvial plains on the side appeared fertile, and it is supposed that they are well adapted to the growth of corn. Turning northward from the river, we soon entered on a country, differing from the plains south of the river. The land still continued dry and sterile: but it supported many different kinds of plants, and the grass, though brown and withered, was more abundant, as the thorny bushes were less so. These latter in a short space entirely disappeared, and the plains were left without a thicket to cover their nakedness. This change in the vegetation marks the commencement of the grand calcareo argillaceous deposit, which forms the wide extent of the Pampas, and covers the granitic rocks of Banda Oriental. From the Strait of Magellan to the Colorado, a distance of about eight hundred miles, the face of the country is everywhere composed of shingle: the pebbles are chiefly of porphyry, and probably owe their origin to the rocks of the Cordillera. North of the Colorado this bed thins out, and the pebbles become exceedingly small, and here the characteristic vegetation of Patagonia ceases.

Having ridden about twenty-five miles, we came to a broad belt of sand-dunes, which stretches, as far as the eye can reach, to the east and west. The sand-hillocks resting on the clay, allow small pools of water to collect, and thus afford in this dry country an invaluable supply of fresh water. The great advantage arising from depressions and elevations of the soil, is not often brought home to the mind. The two miserable springs in the long passage between the Rio Negro and Colorado were caused by trifling inequalities in the plain; without them not a drop of water would have been found. The belt of sand-dunes is about eight miles wide; at some former period, it probably formed the margin of a grand estuary, where the Colorado now flows. In this district, where absolute proofs of the recent elevation of the land occur, such speculations can hardly be neglected by any one, although merely considering the physical geography of the country. Having crossed the sandy tract, we arrived in the evening at one of the post-houses; and, as the fresh horses were grazing at a distance we determined to pass the night there.

The house was situated at the base of a ridge between one and two hundred feet high—a most remarkable feature in this country. This posta was commanded by a negro lieutenant, born in Africa: to his credit be it said, there was not a rancho between the Colorado and Buenos Ayres in nearly such neat order as his. He had a little room for strangers, and a small corral for the horses, all made of sticks and reeds; he had also dug a ditch round his house as a defence in case of being attacked. This would, however, have been of little avail, if the Indians had come; but his chief comfort seemed to rest in the thought of selling his life dearly. A short time before, a body of Indians had travelled past in the night; if they had been aware of the posta, our black friend and his four soldiers would assuredly have been slaughtered. I did not anywhere meet a more civil and obliging man than this negro; it was therefore the more painful to see that he would not sit down and eat with us.

In the morning we sent for the horses very early, and started for another exhilarating gallop. We passed the Cabeza del Buey, an old name given to the head of a large marsh, which extends from Bahia Blanca. Here we changed horses, and passed through some leagues of swamps and saline marshes. Changing horses for the last time, we again began wading through the mud. My animal fell and I was well soused in black mire—a very disagreeable accident when one does not possess a change of clothes. Some miles from the fort we met a man, who told us that a great gun had been fired, which is a signal that Indians are near. We immediately left the road, and followed the edge of a marsh, which when chased offers the best mode of escape. We were glad to arrive within the walls, when we found all the alarm was about nothing, for the Indians turned out to be friendly ones, who wished to join General Rosas.

Bahia Blanca scarcely deserves the name of a village. A few houses and the barracks for the troops are enclosed by a deep ditch and fortified wall. The settlement is only of recent standing (since 1828); and its growth has been one of trouble. The government of Buenos Ayres unjustly occupied it by force, instead of following the wise example of the Spanish Viceroys, who purchased the land near the older settlement of the Rio Negro, from the Indians. Hence the need of the fortifications; hence the few houses and little cultivated land without the limits of the walls; even the cattle are not safe from the attacks of the Indians beyond the boundaries of the plain, on which the fortress stands.

The part of the harbour where the Beagle intended to anchor being distant twenty-five miles, I obtained from the Commandant a guide and horses, to take me to see whether she had arrived. Leaving the plain of green turf, which extended along the course of a little brook, we soon entered on a wide level waste consisting either of sand, saline marshes, or bare mud. Some parts were clothed by low thickets, and others with those succulent plants, which luxuriate only where salt abounds. Bad as the country was, ostriches, deer, agoutis, and armadilloes, were abundant. My guide told me, that two months before he had a most narrow escape of his life: he was out hunting with two other men, at no great distance from this part of the country, when they were suddenly met by a party of Indians, who giving chase, soon overtook and killed his two friends. His own horse's legs were also caught by the bolas; but he jumped off, and with his knife cut them free: while doing this he was obliged to dodge round his horse, and received two severe wounds from their chuzos. Springing on the saddle, he managed, by a most wonderful exertion, just to keep ahead of the long spears of his pursuers, who followed him to within sight of the fort. From that time there was an order that no one should stray far from the settlement. I did not know of this when I started, and was surprised to observe how earnestly my guide watched a deer, which appeared to have been frightened from a distant quarter.

We found the Beagle had not arrived, and consequently set out on our return, but the horses soon tiring, we were obliged to bivouac on the plain. In the morning we had caught an armadillo, which, although a most excellent dish when roasted in its shell, did not make a very substantial breakfast and dinner for two hungry men. The ground at the place where we stopped for the night, was incrusted with a layer of sulphate of soda, and hence, of course, was without water. Yet many of the smaller rodents managed to exist even here, and the tucutuco was making its odd little grunt beneath my head, during half the night. Our horses were very poor ones, and in the morning they were soon exhausted from not having had anything to drink, so that we were obliged to walk. About noon the dogs killed a kid, which we roasted. I ate some of it, but it made me intolerably thirsty. This was the more distressing as the road, from some recent rain, was full of little puddles of clear water, yet not a drop was drinkable. I had scarcely been twenty hours without water, and only part of the time under a hot sun, yet the thirst rendered me very weak. How people survive two or three days under such circumstances, I cannot imagine: at the same time, I must confess that my guide did not suffer at all, and was astonished that one day's deprivation should be so troublesome to me.

I have several times alluded to the surface of the ground being incrusted with salt. This phenomenon is quite different from that of the salinas, and more extraordinary. In many parts of South America, wherever the climate is moderately dry, these incrustations occur; but I have nowhere seen them so abundant as near Bahia Blanca. The salt here, and in other parts of Patagonia, consists chiefly of sulphate of soda with some common salt. As long as the ground remains moist in the salitrales (as the Spaniards improperly call them, mistaking this substance for saltpetre), nothing is to be seen but an extensive plain composed of a black, muddy soil, supporting scattered tufts of succulent plants. On returning through one of these tracts, after a week's hot weather, one is surprised to see square miles of the plain white, as if from a slight fall of snow, here and there heaped up by the wind into little drifts. This latter appearance is chiefly caused by the salts being drawn up, during the slow evaporation of the moisture, round blades of dead grass, stumps of wood, and pieces of broken earth, instead of being crystallized at the bottoms of the puddles of water. The salitrales occur either on level tracts elevated only a few feet above the level of the sea, or on alluvial land bordering rivers. M. Parchappe[39] found that the saline incrustation on the plain, at the distance of some miles from the sea, consisted chiefly of sulphate of soda, with only seven per cent of common salt; whilst nearer to the coast, the common salt increased to 37 parts in a hundred. This circumstance would tempt one to believe that the sulphate of soda is generated in the soil, from the muriate, left on the surface during the slow and recent elevation of this dry country. The whole phenomenon is well worthy the attention of naturalists. Have the succulent, salt-loving plants, which are well known to contain much soda, the power of decomposing the muriate? Does the black fetid mud, abounding with organic matter, yield the sulphur and ultimately the sulphuric acid?

Two days afterwards I again rode to the harbour: when not far from our destination, my companion, the same man as before, spied three people hunting on horseback. He immediately dismounted, and watching them intently, said, "They don't ride like Christians, and nobody can leave the fort." The three hunters joined company, and likewise dismounted from their horses. At last one mounted again and rode over the hill out of sight. My companion said, "We must now get on our horses: load your pistol;" and he looked to his own sword. I asked, "Are they Indians?"—"Quien sabe? (who knows?) if there are no more than three, it does not signify." It then struck me, that the one man had gone over the hill to fetch the rest of his tribe. I suggested this; but all the answer I could extort was, "Quien sabe?" His head and eye never for a minute ceased scanning slowly the distant horizon. I thought his uncommon coolness too good a joke, and asked him why he did not return home. I was startled when he answered, "We are returning, but in a line so as to pass near a swamp, into which we can gallop the horses as far as they can go, and then trust to our own legs; so that there is no danger." I did not feel quite so confident of this, and wanted to increase our pace. He said, "No, not until they do." When any little inequality concealed us, we galloped; but when in sight, continued walking. At last we reached a valley, and turning to the left, galloped quickly to the foot of a hill; he gave me his horse to hold, made the dogs lie down, and then crawled on his hands and knees to reconnoitre. He remained in this position for some time, and at last, bursting out in laughter, exclaimed, "Mugeres!" (women!). He knew them to be the wife and sister-in-law of the major's son, hunting for ostrich's eggs. I have described this man's conduct, because he acted under the full impression that they were Indians. As soon, however, as the absurd mistake was found out, he gave me a hundred reasons why they could not have been Indians; but all these were forgotten at the time. We then rode on in peace and quietness to a low point called Punta Alta, whence we could see nearly the whole of the great harbour of Bahia Blanca.

The wide expanse of water is choked up by numerous great mud-banks, which the inhabitants call Cangrejales, or *crabberies,* from the number of small crabs. The mud is so soft that it is impossible to walk over them, even for the shortest distance. Many of the banks have their surfaces covered with long rushes, the tops of which alone are visible at high water. On one occasion, when in a boat, we were so entangled by these shallows that we could hardly find our way. Nothing was visible but the flat beds of mud; the day was not very clear, and there was much refraction, or as the sailors expressed it, "things loomed high." The only object within our view which was not level was the horizon; rushes looked like bushes unsupported in the air, and water like mud-banks, and mud-banks like water.

We passed the night in Punta Alta, and I employed myself in searching for fossil bones; this point being a perfect catacomb for monsters of extinct races. The evening was perfectly calm and clear; the extreme monotony of the view gave it an interest even in the midst of mud-banks and gulls, sand-hillocks and solitary vultures. In riding back in the morning we came across a very fresh track of a Puma, but did not succeed in finding it. We saw also a couple of Zorillos, or skunks,—odious animals, which are far from uncommon. In general appearance, the Zorillo resembles a polecat, but it is rather larger, and much thicker in proportion. Conscious of its power, it roams by day about the open plain, and fears neither dog nor man. If a dog is urged to the attack, its courage is instantly checked by a few drops of the fetid oil, which brings on violent sickness and running at the nose. Whatever is once polluted by it, is for ever useless. Azara says the smell can be perceived at a league distant; more than once, when entering the harbour of Monte Video, the wind being off shore, we have perceived the odour on board the Beagle. Certain it is, that every animal most willingly makes room for the Zorillo.

CHAPTER V

BAHIA BLANCA

BAHIA BLANCA • GEOLOGY • NUMEROUS GIGANTIC EXTINCT QUADRUPEDS • RECENT EXTINCTION • LONGEVITY OF SPECIES • LARGE ANIMALS DO NOT REQUIRE A LUXURIANT VEGETATION • SOUTHERN AFRICA • SIBERIAN FOSSILS • TWO SPECIES OF OSTRICH • HABITS OF OVEN-BIRD • ARMADILLOES • VENOMOUS SNAKE, TOAD, LIZARD • HYBERNATION OF ANIMALS • HABITS OF SEA-PEN • INDIAN WARS AND MASSACRES • ARROW-HEAD, ANTIQUARIAN RELIC

THE BEAGLE ARRIVED HERE on the 24th of August, and a week afterwards sailed for the Plata. With Captain Fitz Roy's consent I was left behind, to travel by land to Buenos Ayres. I will here add some observations, which were made during this visit and on a previous occasion, when the Beagle was employed in surveying the harbour.

The plain, at the distance of a few miles from the coast, belongs to the great Pampean formation, which consists in part of a reddish clay, and in part of a highly calcareous marly rock. Nearer the coast there are some plains formed from the wreck of the upper plain, and from mud, gravel, and sand thrown up by the sea during the slow elevation of the land, of which elevation we have evidence in upraised beds of recent shells, and in rounded pebbles of pumice scattered over the country. At Punta Alta we have a section of one of these later-formed little plains, which is highly interesting from the number and extraordinary character of the remains of gigantic land-animals embedded in it. These have been fully described by Professor Owen, in the Zoology of the voyage of the Beagle, and are deposited in the College of Surgeons. I will here give only a brief outline of their nature.

First, parts of three heads and other bones of the Megatherium, the huge dimensions of which are expressed by its name. Secondly, the Megalonyx, a great allied animal. Thirdly, the Scelidotherium, also an allied animal, of which I obtained a nearly perfect skeleton. It must have been as large as a rhinoceros: in the structure of its head it comes, according to Mr. Owen, nearest to the Cape Anteater, but in some other respects it approaches to the armadilloes. Fourthly, the Mylodon Darwinii, a closely related genus of little inferior size. Fifthly, another gigantic edental quadruped. Sixthly, a large animal, with an osseous coat in compartments, very like that of an armadillo. Seventhly, an extinct kind of horse, to which I shall have again to refer. Eighthly, a tooth of a Pachydermatous animal, probably the same with the Macrauchenia, a huge beast with a long neck like a camel, which I shall also refer to again. Lastly, the Toxodon, perhaps one of the strangest animals ever discovered: in size it equalled an elephant or megatherium, but the structure of its teeth, as Mr. Owen states, proves indisputably that it was intimately related to the Gnawers, the order which, at the present day, includes most of the smallest quadrupeds: in many details it is allied to the Pachydermata: judging from the position of its eyes, ears, and nostrils, it was probably aquatic, like the Dugong and Manatee, to which it is also allied. How wonderfully are the different Orders, at the present time so well separated, blended together in different points of the structure of the Toxodon!

The remains of these nine great quadrupeds, and many detached bones, were found embedded on the beach, within the space of about 200 yards square. It is a remarkable circumstance that so many different species should be found together; and it proves how numerous in kind the ancient inhabitants of this country must have been. At the distance of about thirty miles from Punta Alta, in a cliff of red earth, I found several fragments of bones, some of large size. Among them were the teeth of a gnawer, equalling in size and closely resembling those of the Capybara, whose habits have been described; and therefore, probably, an aquatic animal. There was also part of the head of a Ctenomys; the species being different from the Tucutuco, but with a close general resemblance. The red earth, like that of the Pampas, in which these remains were embedded, contains, according to Professor Ehrenberg, eight fresh-water and one salt-water infusorial animalcule; therefore, probably, it was an estuary deposit.

The remains at Punta Alta were embedded in stratified gravel and reddish mud, just such as the sea might now wash up on a shallow bank. They were associated with twenty-three species of shells, of which thirteen are recent and four others very closely related to recent forms.[40] From the bones of the Scelidotherium, including even the knee-cap, being intombed in their proper relative positions, and from the osseous armour of the great armadillo-like animal being so well preserved, together with the bones of one of its legs, we may feel assured that these remains were fresh and united by their ligaments, when deposited in the gravel together with the shells.[41] Hence we have good evidence that the above enumerated gigantic quadrupeds, more different from those of the present day than the oldest of the tertiary quadrupeds of Europe, lived whilst the sea was peopled with most of its present inhabitants; and we have confirmed that remarkable law so often insisted on by Mr. Lyell, namely, that the "longevity of the species in the mammalia is upon the whole inferior to that of the testacea."[42]

The great size of the bones of the Megatheroid animals, including the Megatherium, Megalonyx, Scelidotherium, and Mylodon, is truly wonderful. The habits of life of these animals were a complete puzzle to naturalists, until Professor Owen[43] solved the problem with remarkable ingenuity. The teeth indicate, by their simple structure, that these Megatheroid animals lived on vegetable food, and probably on the leaves and small twigs of trees; their ponderous forms and great strong curved claws seem so little adapted for locomotion, that some eminent naturalists have actually believed, that, like the sloths, to which they are intimately related, they subsisted by climbing back downwards on trees, and feeding on the leaves. It was a bold, not to say preposterous, idea to conceive even antediluvian trees, with branches strong enough to bear animals as large as elephants. Professor Owen, with far more probability, believes that, instead of climbing on the trees, they pulled the branches down to them, and tore up the smaller ones by the roots, and so fed on the leaves. The colossal breadth and weight of their hinder quarters, which can hardly be imagined without having been seen, become on this view, of obvious service, instead of being an incumbrance: their apparent clumsiness disappears. With their great tails and their huge heels firmly fixed like a tripod on the ground, they could freely exert the full force of their most powerful arms and great claws. Strongly rooted, indeed, must that tree have been, which could have resisted such force! The Mylodon, moreover, was furnished with a long extensile tongue like that of the giraffe, which, by one of those beautiful provisions of nature, thus reaches with the aid of its long neck its leafy food. I may remark, that in Abyssinia the elephant, according to Bruce, when it cannot reach with its proboscis the branches, deeply scores with its tusks the trunk of the tree, up and down and all round, till it is sufficiently weakened to be broken down.

The beds including the above fossil remains, stand only from fifteen to twenty feet above the level of high-water; and hence the elevation of the land has been small (without there has been an intercalated period of subsidence, of which we have no evidence) since the great quadrupeds wandered over the surrounding plains; and the external features of the country must then have been very nearly the same as now. What, it may naturally be asked, was the character of the vegetation at that period; was the country as wretchedly sterile as it now is? As so many of the co-embedded shells are the same with those now living in the bay, I was at first inclined to think that the former vegetation was probably similar to the existing one; but this would have been an erroneous inference, for some of these same shells live on the luxuriant coast of Brazil; and generally, the character of the inhabitants of the sea are useless as guides to judge of those on the land. Nevertheless, from the following considerations, I do not believe that the simple fact of many gigantic quadrupeds having lived on the plains round Bahia Blanca, is any sure guide that they formerly were clothed with a luxuriant vegetation: I have no doubt that the sterile country a little southward, near the Rio Negro, with its scattered thorny trees, would support many and large quadrupeds.

THAT LARGE ANIMALS require a luxuriant vegetation, has been a general assumption which has passed from one work to another; but I do not hesitate to say that it is completely false, and that it has vitiated the reasoning of geologists on some points of great interest in the ancient history of the world. The prejudice has probably been derived from India, and the Indian islands, where troops of elephants, noble forests, and impenetrable jungles, are associated together in every one's mind. If, however, we refer to any work of travels through the southern parts of Africa, we shall find allusions in almost every page either to the desert character of the country, or to the numbers of large animals inhabiting it. The same thing is rendered evident by the many engravings which have been published of various parts of the interior. When the Beagle was at Cape Town, I made an excursion of some days' length into the country, which at least was sufficient to render that which I had read more fully intelligible.

Dr. Andrew Smith, who, at the head of his adventurous party, has lately succeeded in passing the Tropic of Capricorn, informs me that, taking into consideration the whole of the southern part of Africa, there can be no doubt of its being a sterile country. On the southern and south-eastern coasts there are some fine forests, but with these exceptions, the traveller may pass for days together through open plains, covered by a poor and scanty vegetation. It is difficult to convey any accurate idea of degrees of comparative fertility; but it may be safely said that the amount of vegetation supported at any one time[44] by Great Britain, exceeds, perhaps even tenfold, the quantity on an equal area, in the interior parts of Southern Africa. The fact that bullock-waggons can travel in any direction, excepting near the coast, without more than occasionally half an hour's delay in cutting down bushes, gives, perhaps, a more definite notion of the scantiness of the vegetation. Now, if we look to the animals inhabiting these wide plains, we shall find their numbers extraordinarily great, and their bulk immense. We must enumerate the elephant, three species of rhinoceros, and probably, according to Dr. Smith, two others, the hippopotamus, the giraffe, the bos caffer—as large as a full-grown bull, and the elan—but little less, two zebras, and the quaccha, two gnus, and several antelopes even larger than these latter animals. It may be supposed that although the species are numerous, the individuals of each kind are few. By the kindness of Dr. Smith, I am enabled to show that the case is very different. He informs me, that in lat. 24°, in one day's march with the bullock-waggons, he saw, without wandering to any great distance on either side, between one hundred and one hundred and fifty rhinoceroses, which belonged to three species: the same day he saw several herds of giraffes, amounting together to nearly a hundred; and that although no elephant was observed, yet they are found in this district. At the distance of a little more than one hour's march from their place of encampment on the previous night, his party actually killed at one spot eight hippopotamuses, and saw many more. In this same river there were likewise crocodiles. Of course it was a case quite extraordinary, to see so many great animals crowded together, but it evidently proves that they must exist in great numbers. Dr. Smith describes the country passed through that day, as "being thinly covered with grass, and bushes about four feet high, and still more thinly with mimosa-trees." The waggons were not prevented travelling in a nearly straight line.

Besides these large animals, every one the least acquainted with the natural history of the Cape, has read of the herds of antelopes, which can be compared only with the flocks of migratory birds. The numbers indeed of the lion, panther, and hyæna, and the multitude of birds of prey, plainly speak of the abundance of the smaller quadrupeds: one evening seven lions were counted at the same time prowling round Dr. Smith's encampment. As this able naturalist remarked to me, the carnage each day in Southern Africa must indeed be terrific! I confess it is truly surprising how such a number of animals can find support in a country producing so little food. The larger quadrupeds no doubt roam over wide tracts in search of it; and their food chiefly consists of underwood, which probably contains much nutriment in a small bulk. Dr. Smith also informs me that the vegetation has a rapid growth; no sooner is a part consumed, than its place is supplied by a fresh stock. There can be no doubt, however, that our ideas respecting the apparent amount of food necessary for the support of large quadrupeds are much exaggerated: it should have been remembered that the camel, an animal of no mean bulk, has always been considered as the emblem of the desert.

The belief that where large quadrupeds exist, the vegetation must necessarily be luxuriant, is the more remarkable, because the converse is far from true. Mr. Burchell observed to me that when entering Brazil, nothing struck him more forcibly than the splendour of the South American vegetation contrasted with that of South Africa, together with the absence of all large quadrupeds. In his Travels,[45] he has suggested that the comparison of the respective weights (if there were sufficient data) of an equal number of the largest herbivorous quadrupeds of each country would be extremely curious. If we take on the one side, the elephant,[46] hippopotamus, giraffe, bos caffer, elan, certainly three, and probably five species of rhinoceros; and on the American side, two tapirs, the guanaco, three deer, the vicuna, peccari, capybara (after which we must choose from the monkeys to complete the number), and then place these two groups alongside each other, it is not easy to conceive ranks more disproportionate in size. After the above facts, we are compelled to conclude, against anterior probability,[47] that among the mammalia there exists no close relation between the *bulk* of the species, and the *quantity* of the vegetation, in the countries which they inhabit.

With regard to the number of large quadrupeds, there certainly exists no quarter of the globe which will bear comparison with Southern Africa. After the different statements which have been given, the extremely desert character of that region will not be disputed. In the European division of the world, we must look back to the tertiary epochs, to find a condition of things among the mammalia, resembling that now existing at the Cape of Good Hope. Those tertiary epochs, which we are apt to consider as abounding to an astonishing degree with large animals, because we find the remains of many ages accumulated at certain spots, could hardly boast of more large quadrupeds than Southern Africa does at present. If we speculate on the condition of the vegetation during these epochs, we are at least bound so far to consider existing analogies, as not to urge as absolutely necessary a luxuriant vegetation, when we see a state of things so totally different at the Cape of Good Hope.

We know[48] that the extreme regions of North America, many degrees beyond the limit where the ground at the depth of a few feet remains perpetually congealed, are covered by forests of large and tall trees. In a like manner, in Siberia, we have woods of birch, fir, aspen, and larch, growing in a latitude[49] (64°) where the mean temperature of the air falls below the freezing point, and where the earth is so completely frozen, that the carcass of an animal embedded in it is perfectly preserved. With these facts we must grant, as far as *quantity alone* of vegetation is concerned, that the great quadrupeds of the later tertiary epochs might, in most parts of Northern Europe and Asia, have lived on the spots where their remains are now found. I do not here speak of the *kind* of vegetation necessary for their support; because, as there is evidence of physical changes, and as the animals have become extinct, so may we suppose that the species of plants have likewise been changed.

These remarks, I may be permitted to add, directly bear on the case of the Siberian animals preserved in ice. The firm conviction of the necessity of a vegetation possessing a character of tropical luxuriance, to support such large animals, and the impossibility of reconciling this with the proximity of perpetual congelation, was one chief cause of the several theories of sudden revolutions of climate, and of overwhelming catastrophes, which were invented to account for their entombment. I am far from supposing that the climate has not changed since the period when those animals lived, which now lie buried in the ice. At present I only wish to show, that as far as *quantity* of food *alone* is concerned, the ancient rhinoceroses might have roamed over the *steppes* of central Siberia (the northern parts probably being under water) even in their present condition, as well as the living rhinoceroses and elephants over the *Karros* of Southern Africa.

I WILL NOW give an account of the habits of some of the more interesting birds which are common on the wild plains of Northern Patagonia; and first for the largest, or South American ostrich. The ordinary habits of the ostrich are familiar to every one. They live on vegetable matter, such as roots and grass; but at Bahia Blanca I have repeatedly seen three or four come down at low water to the extensive mud-banks which are then dry, for the sake, as the Gauchos say, of feeding on small fish. Although the ostrich in its habits is so shy, wary, and solitary, and although so fleet in its pace, it is caught without much difficulty by the Indian or Gaucho armed with the bolas. When several horsemen appear in a semicircle, it becomes confounded, and does not know which way to escape. They generally prefer running against the wind; yet at the first start they expand their wings, and like a vessel make all sail. On one fine hot day I saw several ostriches enter a bed of tall rushes, where they squatted concealed, till quite closely approached. It is not generally known that ostriches readily take to the water. Mr. King informs me that at the Bay of San Blas, and at Port Valdes in Patagonia, he saw these birds swimming several times from island to island. They ran into the water both when driven down to a point, and likewise of their own accord when not frightened: the distance crossed was about two hundred yards. When swimming, very little of their bodies appear above water; their necks are extended a little forward, and their progress is slow. On two occasions I saw some ostriches swimming across the Santa Cruz river, where its course was about four hundred yards wide, and the stream rapid. Captain Sturt,[50] when descending the Murrumbidgee, in Australia, saw two emus in the act of swimming.

The inhabitants of the country readily distinguish, even at a distance, the cock bird from the hen. The former is larger and darker-coloured,[51] and has a bigger head. The ostrich, I believe the cock, emits a singular, deep-toned, hissing note: when first I heard it, standing in the midst of some sand-hillocks, I thought it was made by some wild beast, for it is a sound that one cannot tell whence it comes, or from how far distant. When we were at Bahia Blanca in the months of September and October, the eggs, in extraordinary numbers, were found all over the country. They lie either scattered and single, in which case they are never hatched, and are called by the Spaniards huachos; or they are collected together into a shallow excavation, which forms the nest. Out of the four nests which I saw, three contained twenty-two eggs each, and the fourth twenty-seven. In one day's hunting on horseback sixty-four eggs were found; forty-four of these were in two nests, and the remaining twenty, scattered huachos. The Gauchos unanimously affirm, and there is no reason to doubt their statement, that the male bird alone hatches the eggs, and for some time afterwards accompanies the young. The cock when on the nest lies very close; I have myself almost ridden over one. It is asserted that at such times they are occasionally fierce, and even dangerous, and that they have been known to attack a man on horseback, trying to kick and leap on him. My informer pointed out to me an old man, whom he had seen much terrified by one chasing him. I observe in Burchell's travels in South Africa, that he remarks, "Having killed a male ostrich, and the feathers being dirty, it was said by the Hottentots to be a nest bird." I understand that the male emu in the Zoological Gardens takes charge of the nest: this habit, therefore, is common to the family.

The Gauchos unanimously affirm that several females lay in one nest. I have been positively told that four or five hen birds have been watched to go in the middle of the day, one after the other, to the same nest. I may add, also, that it is believed in Africa, that two or more females lay in one nest.[52] Although this habit at first appears very strange, I think the cause may be explained in a simple manner. The number of eggs in the nest varies from twenty to forty, and even to fifty; and according to Azara, sometimes to seventy or eighty. Now, although it is most probable, from the number of eggs found in one district being so extraordinarily great in proportion to the parent birds, and likewise from the state of the ovarium of the hen, that she may in the course of the season lay a large number, yet the time required must be very long. Azara states,[53] that a female in a state of domestication laid seventeen eggs, each at the interval of three days one from another. If the hen was obliged to hatch her own eggs, before the last was laid the first probably would be addled; but if each laid a few eggs at successive periods, in different nests, and several hens, as is stated to be the case, combined together, then the eggs in one collection would be nearly of the same age. If the number of eggs in one of these nests is, as I believe, not greater on an average than the number laid by one female in the season, then there must be as many nests as females, and each cock bird will have its fair share of the labour of incubation; and that during a period when the females probably could not sit, from not having finished laying.[54] I have before mentioned the great numbers of huachos, or deserted eggs; so that in one day's hunting twenty were found in this state. It appears odd that so many should be wasted. Does it not arise from the difficulty of several females associating together, and finding a male ready to undertake the office of incubation? It is evident that there must at first be some degree of association between at least two females; otherwise the eggs would remain scattered over the wide plain, at distances far too great to allow of the male collecting them into one nest: some authors have believed that the scattered eggs were deposited for the young birds to feed on. This can hardly be the case in America, because the huachos, although often found addled and putrid, are generally whole.

When at the Rio Negro in Northern Patagonia, I repeatedly heard the Gauchos talking of a very rare bird which they called Avestruz Petise. They described it as being less than the common ostrich (which is there abundant), but with a very close general resemblance. They said its colour was dark and mottled, and that its legs were shorter, and feathered lower down than those of the common ostrich. It is more easily caught by the bolas than the other species. The few inhabitants who had seen both kinds, affirmed they could distinguish them apart from a long distance. The eggs of the small species appeared, however, more generally known; and it was remarked, with surprise, that they were very little less than those of the Rhea, but of a slightly different form, and with a tinge of pale blue. This species occurs most rarely on the plains bordering the Rio Negro; but about a degree and a half further south they are tolerably abundant. When at Port Desire, in Patagonia (lat. 48°), Mr. Martens shot an ostrich; and I looked at it, forgetting at the moment, in the most unaccountable manner, the whole subject of the Petises, and thought it was a not full-grown bird of the common sort. It was cooked and eaten before my memory returned. Fortunately the head, neck, legs, wings, many of the larger feathers, and a large part of the skin, had been preserved; and from these a very nearly perfect specimen has been put together, and is now exhibited in the museum of the Zoological Society. Mr. Gould, in describing this new species, has done me the honour of calling it after my name.

Among the Patagonian Indians in the Strait of Magellan, we found a half Indian, who had lived some years with the tribe, but had been born in the northern provinces. I asked him if he had ever heard of the Avestruz Petise? He answered by saying, "Why, there are none others in these southern countries." He informed me that the number of eggs in the nest of the petise is considerably less than in that of the other kind, namely, not more than fifteen on an average; but he asserted that more than one female deposited them. At Santa Cruz we saw several of these birds. They were excessively wary: I think they could see a person approaching when too far off to be distinguished themselves. In ascending the river few were seen; but in our quiet and rapid descent, many, in pairs and by fours or fives, were observed. It was remarked that this bird did not expand its wings, when first starting at full speed, after the manner of the northern kind. In conclusion I may observe, that the Struthio rhea inhabits the country of La Plata as far as a little south of the Rio Negro in lat. 41°, and that the Struthio Darwinii takes its place in Southern Patagonia; the part about the Rio Negro being neutral territory. M. A. d'Orbigny,[55] when at the Rio Negro, made great exertions to procure this bird, but never had the good fortune to succeed. Dobrizhoffer[56] long ago was aware of there being two kinds of ostriches; he says, "You must know, moreover, that Emus differ in size and habits in different tracts of land; for those that inhabit the plains of Buenos Ayres and Tucuman are larger, and have black, white and grey, feathers; those near to the Strait of Magellan are smaller and more beautiful, for their white feathers are tipped with black at the extremity, and their black ones in like manner terminate in white."

A VERY SINGULAR little bird, Tinochorus rumicivorus, is here common: in its habits and general appearance, it nearly equally partakes of the characters, different as they are, of the quail and snipe. The Tinochorus is found in the whole of southern South America, wherever there are sterile plains, or open dry pasture land. It frequents in pairs or small flocks the most desolate places, where scarcely another living creature can exist. Upon being approached they squat close, and then are very difficult to be distinguished from the ground. When feeding they walk rather slowly, with their legs wide apart. They dust themselves in roads and sandy places, and frequent particular spots, where they may be found day after day: like partridges, they take wing in a flock. In all these respects, in the muscular gizzard adapted for vegetable food, in the arched beak and fleshy nostrils, short legs and form of foot, the Tinochorus has a close affinity with quails. But as soon as the bird is seen flying, its whole appearance changes; the long pointed wings, so different from those in the gallinaceous order, the irregular manner of flight, and plaintive cry uttered at the moment of rising, recall the idea of a snipe. The sportsmen of the Beagle unanimously called it the short-billed snipe. To this genus, or rather to the family of the Waders, its skeleton shows that it is really related.

The Tinochorus is closely related to some other South American birds. Two species of the genus Attagis are in almost every respect ptarmigans in their habits; one lives in Tierra del Fuego, above the limits of the forest land; and the other just beneath the snow-line on the Cordillera of Central Chile. A bird of another closely allied genus, Chionis alba, is an inhabitant of the antarctic regions; it feeds on sea-weed and shells on the tidal rocks. Although not web-footed, from some unaccountable habit, it is frequently met with far out at sea. This small family of birds is one of those which, from its varied relations to other families, although at present offering only difficulties to the systematic naturalist, ultimately may assist in revealing the grand scheme, common to the present and past ages, on which organized beings have been created.

The genus Furnarius contains several species, all small birds, living on the ground, and inhabiting open dry countries. In structure they cannot be compared to any European form. Ornithologists have generally included them among the creepers, although opposed to that family in every habit. The best known species is the common oven-bird of La Plata, the Casara or housemaker of the Spaniards. The nest, whence it takes its name, is placed in the most exposed situations, as on the top of a post, a bare rock, or on a cactus. It is composed of mud and bits of straw, and has strong thick walls: in shape it precisely resembles an oven, or depressed beehive. The opening is large and arched, and directly in front, within the nest, there is a partition, which reaches nearly to the roof, thus forming a passage or antechamber to the true nest.

Another and smaller species of Furnarius (F. cunicularius), resembles the oven-bird in the general reddish tint of its plumage, in a peculiar shrill reiterated cry, and in an odd manner of running by starts. From its affinity, the Spaniards call it Casarita (or little housebuilder), although its nidification is quite different. The Casarita builds its nest at the bottom of a narrow cylindrical hole, which is said to extend horizontally to nearly six feet under ground. Several of the country people told me, that when boys, they had attempted to dig out the nest, but had scarcely ever succeeded in getting to the end of the passage. The bird chooses any low bank of firm sandy soil by the side of a road or stream. Here (at Bahia Blanca) the walls round the houses are built of hardened mud; and I noticed that one, which enclosed a courtyard where I lodged, was bored through by round holes in a score of places. On asking the owner the cause of this, he bitterly complained of the little casarita, several of which I afterwards observed at work. It is rather curious to find how incapable these birds must be of acquiring any notion of thickness, for although they were constantly flitting over the low wall, they continued vainly to bore through it, thinking it an excellent bank for their nests. I do not doubt that each bird, as often as it came to daylight on the opposite side, was greatly surprised at the marvellous fact.

I have already mentioned nearly all the mammalia common in this country. Of armadilloes three species occur, namely, the Dasypus minutus or *pichy,* the D. villosus or *peludo,* and the *apar.* The first extends ten degrees further south than any other kind; a fourth species, the *Mulita,* does not come as far south as Bahia Blanca. The four species have nearly similar habits; the *peludo,* however, is nocturnal, while the others wander by day over the open plains, feeding on beetles, larvæ, roots, and even small snakes. The *apar,* commonly called *mataco,* is remarkable by having only three moveable bands; the rest of its tesselated covering being nearly inflexible. It has the power of rolling itself into a perfect sphere, like one kind of English woodlouse. In this state it is safe from the attack of dogs; for the dog not being able to take the whole in its mouth, tries to bite one side, and the ball slips away. The smooth hard covering of the *mataco* offers a better defence than the sharp spines of the hedgehog. The *pichy* prefers a very dry soil; and the sand-dunes near the coast, where for many months it can never taste water, is its favourite resort: it often tries to escape notice, by squatting close to the ground. In the course of a day's ride, near Bahia Blanca, several were generally met with. The instant one was perceived, it was necessary, in order to catch it, almost to tumble off one's horse; for in soft soil the animal burrowed so quickly, that its hinder quarters would almost disappear before one could alight. It seems almost a pity to kill such nice little animals, for as a Gaucho said, while sharpening his knife on the back of one, "Son tan mansos" (they are so quiet).

Of reptiles there are many kinds: one snake (a Trigonocephalus, or Cophias[57]), from the size of the poison channel in its fangs, must be very deadly. Cuvier, in opposition to some other naturalists, makes this a subgenus of the rattlesnake, and intermediate between it and the viper. In confirmation of this opinion, I observed a fact, which appears to me very curious and instructive, as showing how every character, even though it may be in some degree independent of structure, has a tendency to vary by slow degrees. The extremity of the tail of this snake is terminated by a point, which is very slightly enlarged; and as the animal glides along, it constantly vibrates the last inch; and this part striking against the dry grass and brushwood, produces a rattling noise, which can be distinctly heard at the distance of six feet. As often as the animal was irritated or surprised, its tail was shaken; and the vibrations were extremely rapid. Even as long as the body retained its irritability, a tendency to this habitual movement was evident. This Trigonocephalus has, therefore, in some respects the structure of a viper, with the habits of a rattlesnake: the noise, however, being produced by a simpler device. The expression of this snake's face was hideous and fierce; the pupil consisted of a vertical slit in a mottled and coppery iris; the jaws were broad at the base, and the nose terminated in a triangular projection. I do not think I ever saw anything more ugly, excepting, perhaps, some of the vampire bats. I imagine this repulsive aspect originates from the features being placed in positions, with respect to each other, somewhat proportional to those of the human face; and thus we obtain a scale of hideousness.

Amongst the Batrachian reptiles, I found only one little toad (Phryniscus nigricans), which was most singular from its colour. If we imagine, first, that it had been steeped in the blackest ink, and then, when dry, allowed to crawl over a board, freshly painted with the brightest vermilion, so as to colour the soles of its feet and parts of its stomach, a good idea of its appearance will be gained. If it had been an unnamed species, surely it ought to have been called *Diabolicus*, for it is a fit toad to preach in the ear of Eve. Instead of being nocturnal in its habits, as other toads are, and living in damp obscure recesses, it crawls during the heat of the day about the dry sand-hillocks and arid plains, where not a single drop of water can be found. It must necessarily depend on the dew for its moisture; and this probably is absorbed by the skin, for it is known, that these reptiles possess great powers of cutaneous absorption. At Maldonado, I found one in a situation nearly as dry as at Bahia Blanca, and thinking to give it a great treat, carried it to a pool of water; not only was the little animal unable to swim, but I think without help it would soon have been drowned.

Of lizards there were many kinds, but only one (Proctotretus multimaculatus) remarkable from its habits. It lives on the bare sand near the sea-coast, and from its mottled colour, the brownish scales being speckled with white, yellowish red, and dirty blue, can hardly be distinguished from the surrounding surface. When frightened, it attempts to avoid discovery by feigning death, with outstretched legs, depressed body, and closed eyes: if further molested, it buries itself with great quickness in the loose sand. This lizard, from its flattened body and short legs, cannot run quickly.

I will here add a few remarks on the hybernation of animals in this part of South America. When we first arrived at Bahia Blanca, September 7th, 1832, we thought nature had granted scarcely a living creature to this sandy and dry country. By digging, however, in the ground, several insects, large spiders, and lizards were found in a half-torpid state. On the 15th, a few animals began to appear, and by the 18th (three days from the equinox), everything announced the commencement of spring. The plains were ornamented by the flowers of a pink wood-sorrel, wild peas, œnotheræ, and geraniums; and the birds began to lay their eggs. Numerous Lamellicorn and Heteromerous insects, the latter remarkable for their deeply sculptured bodies, were slowly crawling about; while the lizard tribe, the constant inhabitants of a sandy soil, darted about in every direction. During the first eleven days, whilst nature was dormant, the mean temperature taken from observations made every two hours on board the Beagle, was 51°; and in the middle of the day the thermometer seldom ranged above 55°. On the eleven succeeding days, in which all living things became so animated, the mean was 58°, and the range in the middle of the day between 60° and 70°. Here, then, an increase of seven degrees in mean temperature, but a greater one of extreme heat, was sufficient to awake the functions of life. At Monte Video, from which we had just before sailed, in the twenty-three days included between the 26th of July and the 19th of August, the mean temperature from 276 observations was 58°.4; the mean hottest day being 65°.5, and the coldest 46°. The lowest point to which the thermometer fell was 41°.5, and occasionally in the middle of the day it rose to 69° or 70°. Yet with this high temperature, almost every beetle, several genera of spiders, snails, and land-shells, toads and lizards were all lying torpid beneath stones. But we have seen that at Bahia Blanca, which is four degrees southward, and therefore with a climate only a very little colder, this same temperature with a rather less extreme heat, was sufficient to awake all orders of animated beings. This shows how nicely the stimulus required to arouse hybernating animals is governed by the usual climate of the district, and not by the absolute heat. It is well known that within the tropics, the hybernation, or more properly æstivation, of animals is determined not by the temperature, but by the times of drought. Near Rio de Janeiro, I was at first surprised to observe, that, a few days after some little depressions had been filled with water, they were peopled by numerous full-grown shells and beetles, which must have been lying dormant. Humboldt has related the strange accident of a hovel having been erected over a spot where a young crocodile lay buried in the hardened mud. He adds, "The Indians often find enormous boas, which they call Uji, or water serpents, in the same lethargic state. To reanimate them, they must be irritated or wetted with water."

I will only mention one other animal, a zoophyte (I believe Virgularia Patagonica), a kind of sea-pen. It consists of a thin, straight, fleshy stem, with alternate rows of polypi on each side, and surrounding an elastic stony axis, varying in length from eight inches to two feet. The stem at one extremity is truncate, but at the other is terminated by a vermiform fleshy appendage. The stony axis which gives strength to the stem may be traced at this extremity into a mere vessel filled with granular matter. At low water hundreds of these zoophytes might be seen, projecting like stubble, with the truncate end upwards, a few inches above the surface of the muddy sand. When touched or pulled they suddenly drew themselves in with force, so as nearly or quite to disappear. By this action, the highly elastic axis must be bent at the lower extremity, where it is naturally slightly curved; and I imagine it is by this elasticity alone that the zoophyte is enabled to rise again through the mud. Each polypus, though closely united to its brethren, has a distinct mouth, body, and tentacula. Of these polypi, in a large specimen, there must be many thousands; yet we see that they act by one movement: they have also one central axis connected with a system of obscure circulation, and the ova are produced in an organ distinct from the separate individuals.[58] Well may one be allowed to ask, what is an individual? It is always interesting to discover the foundation of the strange tales of the old voyagers; and I have no doubt but that the habits of this Virgularia explain one such case. Captain Lancaster, in his voyage[59] in 1601, narrates that on the sea-sands of the Island of Sombrero, in the East Indies, he "found a small twig growing up like a young tree, and on offering to pluck it up it shrinks down to the ground, and sinks, unless held very hard. On being plucked up, a great worm is found to be its root, and as the tree groweth in greatness, so doth the worm diminish; and as soon as the worm is entirely turned into a tree it rooteth in the earth, and so becomes great. This transformation is one of the strangest wonders that I saw in all my travels: for if this tree is plucked up, while young, and the leaves and bark stripped off, it becomes a hard stone when dry, much like white coral: thus is this worm twice transformed into different natures. Of these we gathered and brought home many."

DURING MY STAY at Bahia Blanca, while waiting for the Beagle, the place was in a constant state of excitement, from rumours of wars and victories, between the troops of Rosas and the wild Indians. One day an account came that a small party forming one of the postas on the line to Buenos Ayres, had been found all murdered. The next day three hundred men arrived from the Colorado, under the command of Commandant Miranda. A large portion of these men were Indians (*mansos*, or tame), belonging to the tribe of the Cacique Bernantio. They passed the night here; and it was impossible to conceive anything more wild and savage than the scene of their bivouac. Some drank till they were intoxicated; others swallowed the steaming blood of the cattle slaughtered for their suppers, and then, being sick from drunkenness, they cast it up again, and were besmeared with filth and gore.

> Nam simul expletus dapibus, vinoque sepultus
> Cervicem inflexam posuit, jacuitque per antrum
> Immensus, saniem eructans, ac frusta cruenta
> Per somnum commixta mero.

In the morning they started for the scene of the murder, with orders to follow the "rastro," or track, even if it led them to Chile. We subsequently heard that the wild Indians had escaped into the great Pampas, and from some cause the track had been missed. One glance at the rastro tells these people a whole history. Supposing they examine the track of a thousand horses, they will soon guess the number of mounted ones by seeing how many have cantered; by the depth of the other impressions, whether any horses were loaded with cargoes; by the irregularity of the footsteps, how far tired; by the manner in which the food has been cooked, whether the pursued travelled in haste; by the general appearance, how long it has been since they passed. They consider a rastro of ten days or a fortnight, quite recent enough to be hunted out. We also heard that Miranda struck from the west end of the Sierra Ventana, in a direct line to the island of Cholechel, situated seventy leagues up the Rio Negro. This is a distance of between two and three hundred miles, through a country completely unknown. What other troops in the world are so independent? With the sun for their guide, mare's flesh for food, their saddle-cloths for beds,—as long as there is a little water, these men would penetrate to the end of the world.

A few days afterwards I saw another troop of these banditti-like soldiers start on an expedition against a tribe of Indians at the small Salinas, who had been betrayed by a prisoner cacique. The Spaniard who brought the orders for this expedition was a very intelligent man. He gave me an account of the last engagement at which he was present. Some Indians, who had been taken prisoners, gave information of a tribe living north of the Colorado. Two hundred soldiers were sent; and they first discovered the Indians by a cloud of dust from their horses' feet, as they chanced to be travelling. The country was mountainous and wild, and it must have been far in the interior, for the Cordillera were in sight. The Indians, men, women, and children, were about one hundred and ten in number, and they were nearly all taken or killed, for the soldiers sabre every man. The Indians are now so terrified that they offer no resistance in a body, but each flies, neglecting even his wife and children; but when overtaken, like wild animals, they fight against any number to the last moment. One dying Indian seized with his teeth the thumb of his adversary, and allowed his own eye to be forced out sooner than relinquish his hold. Another, who was wounded, feigned death, keeping a knife ready to strike one more fatal blow. My informer said, when he was pursuing an Indian, the man cried out for mercy, at the same time that he was covertly loosing the bolas from his waist, meaning to whirl it round his head and so strike his pursuer. "I however struck him with my sabre to the ground, and then got off my horse, and cut his throat with my knife." This is a dark picture; but how much more shocking is the unquestionable fact, that all the women who appear above twenty years old are massacred in cold blood! When I exclaimed that this appeared rather inhuman, he answered, "Why, what can be done? they breed so!"

Every one here is fully convinced that this is the most just war, because it is against barbarians. Who would believe in this age that such atrocities could be committed in a Christian civilized country? The children of the Indians are saved, to be sold or given away as servants, or rather slaves for as long a time as the owners can make them believe themselves slaves; but I believe in their treatment there is little to complain of.

In the battle four men ran away together. They were pursued, one was killed, and the other three were taken alive. They turned out to be messengers or ambassadors from a large body of Indians, united in the common cause of defence, near the Cordillera. The tribe to which they had been sent was on the point of holding a grand council; the feast of mare's flesh was ready, and the dance prepared: in the morning the ambassadors were to have returned to the Cordillera. They were remarkably fine men, very fair, above six feet high, and all under thirty years of age. The three survivors of course possessed very valuable information; and to extort this they were placed in a line. The two first being questioned, answered, "No sé" (I do not know), and were one after the other shot. The third also said "No sé;" adding, "Fire, I am a man, and can die!" Not one syllable would they breathe to injure the united cause of their country! The conduct of the above-mentioned cacique was very different; he saved his life by betraying the intended plan of warfare, and the point of union in the Andes. It was believed that there were already six or seven hundred Indians together, and that in summer their numbers would be doubled. Ambassadors were to have been sent to the Indians at the small Salinas, near Bahia Blanca, whom I have mentioned that this same cacique had betrayed. The communication, therefore, between the Indians, extends from the Cordillera to the coast of the Atlantic.

General Rosas's plan is to kill all stragglers, and having driven the remainder to a common point, to attack them in a body, in the summer, with the assistance of the Chilenos. This operation is to be repeated for three successive years. I imagine the summer is chosen as the time for the main attack, because the plains are then without water, and the Indians can only travel in particular directions. The escape of the Indians to the south of the Rio Negro, where in such a vast unknown country they would be safe, is prevented by a treaty with the Tehuelches to this effect;—that Rosas pays them so much to slaughter every Indian who passes to the south of the river, but if they fail in so doing, they themselves are to be exterminated. The war is waged chiefly against the Indians near the Cordillera; for many of the tribes on this eastern side are fighting with Rosas. The general, however, like Lord Chesterfield, thinking that his friends may in a future day become his enemies, always places them in the front ranks, so that their numbers may be thinned. Since leaving South America we have heard that this war of extermination completely failed.

Among the captive girls taken in the same engagement, there were two very pretty Spanish ones, who had been carried away by the Indians when young, and could now only speak the Indian tongue. From their account they must have come from Salta, a distance in a straight line of nearly one thousand miles. This gives one a grand idea of the immense territory over which the Indians roam: yet, great as it is, I think there will not, in another half-century, be a wild Indian northward of the Rio Negro. The warfare is too bloody to last; the Christians killing every Indian, and the Indians doing the same by the Christians. It is melancholy to trace how the Indians have given way before the Spanish invaders. Schirdel[60] says that in 1535, when Buenos Ayres was founded, there were villages containing two and three thousand inhabitants. Even in Falconer's time (1750) the Indians made inroads as far as Luxan, Areco, and Arrecife, but now they are driven beyond the Salado. Not only have whole tribes been exterminated, but the remaining Indians have become more barbarous: instead of living in large villages, and being employed in the arts of fishing, as well as of the chase, they now wander about the open plains, without home or fixed occupation.

I heard also some account of an engagement which took place, a few weeks previously to the one mentioned, at Cholechel. This is a very important station on account of being a pass for horses; and it was, in consequence, for some time the head-quarters of a division of the army. When the troops first arrived there they found a tribe of Indians, of whom they killed twenty or thirty. The cacique escaped in a manner which astonished every one. The chief Indians always have one or two picked horses, which they keep ready for any urgent occasion. On one of these, an old white horse, the cacique sprung, taking with him his little son. The horse had neither saddle nor bridle. To avoid the shots, the Indian rode in the peculiar method of his nation; namely, with an arm round the horse's neck, and one leg only on its back. Thus hanging on one side, he was seen patting the horse's head, and talking to him. The pursuers urged every effort in the chase; the Commandant three times changed his horse, but all in vain. The old Indian father and his son escaped, and were free. What a fine picture one can form in one's mind,—the naked, bronze-like figure of the old man with his little boy, riding like a Mazeppa on the white horse, thus leaving far behind him the host of his pursuers!

I saw one day a soldier striking fire with a piece of flint, which I immediately recognised as having been a part of the head of an arrow. He told me it was found near the island of Cholechel, and that they are frequently picked up there. It was between two and three inches long, and therefore twice as large as those now used in Tierra del Fuego: it was made of opaque cream-coloured flint, but the point and barbs had been intentionally broken off. It is well known that no Pampas Indians now use bows and arrows. I believe a small tribe in Banda Oriental must be excepted; but they are widely separated from the Pampas Indians, and border close on those tribes that inhabit the forest, and live on foot. It appears, therefore, that these arrow-heads are antiquarian[61] relics of the Indians, before the great change in habits consequent on the introduction of the horse into South America.

CHAPTER VI

Bahia Blanca to Buenos Ayres

SET OUT FOR BUENOS AYRES • RIO SAUCE • SIERRA VENTANA • THIRD POSTA • DRIVING HORSES • BOLAS • PARTRIDGES AND FOXES • FEATURES OF THE COUNTRY • LONG-LEGGED PLOVER • TERU-TERO • HAIL-STORM • NATURAL ENCLOSURES IN THE SIERRA TAPALGUEN • FLESH OF PUMA • MEAT DIET • GUARDIA DEL MONTE • EFFECTS OF CATTLE ON THE VEGETATION • CARDOON • BUENOS AYRES • CORRAL WHERE CATTLE ARE SLAUGHTERED

SEPTEMBER 18TH.—I HIRED a Gaucho to accompany me on my ride to Buenos Ayres, though with some difficulty, as the father of one man was afraid to let him go, and another, who seemed willing, was described to me as so fearful, that I was afraid to take him, for I was told that even if he saw an ostrich at a distance, he would mistake it for an Indian, and would fly like the wind away. The distance to Buenos Ayres is about four hundred miles, and nearly the whole way through an uninhabited country. We started early in the morning; ascending a few hundred feet from the basin of green turf on which Bahia Blanca stands, we entered on a wide desolate plain. It consists of a crumbling argillaceo-calcareous rock, which, from the dry nature of the climate, supports only scattered tufts of withered grass, without a single bush or tree to break the monotonous uniformity. The weather was fine, but the atmosphere remarkably hazy; I thought the appearance foreboded a gale, but the Gauchos said it was owing to the plain, at some great distance in the interior, being on fire. After a long gallop, having changed horses twice, we reached the Rio Sauce: it is a deep, rapid, little stream, not above twenty-five feet wide. The second posta on the road to Buenos Ayres stands on its banks; a little above there is a ford for horses, where the water does not reach to the horses' belly; but from that point, in its course to the sea, it is quite impassable, and hence makes a most useful barrier against the Indians.

Insignificant as this stream is, the Jesuit Falconer, whose information is generally so very correct, figures it as a considerable river, rising at the foot of the Cordillera. With respect to its source, I do not doubt that this is the case; for the Gauchos assured me, that in the middle of the dry summer, this stream, at the same time with the Colorado, has periodical floods; which can only originate in the snow melting on the Andes. It is extremely improbable that a stream so small as the Sauce then was, should traverse the entire width of the continent; and indeed, if it were the residue of a large river, its waters, as in other ascertained cases, would be saline. During the winter we must look to the springs round the Sierra Ventana as the source of its pure and limpid stream. I suspect the plains of Patagonia, like those of Australia, are traversed by many water-courses, which only perform their proper parts at certain periods. Probably this is the case with the water which flows into the head of Port Desire, and likewise with the Rio Chupat, on the banks of which masses of highly cellular scoriæ were found by the officers employed in the survey.

As it was early in the afternoon when we arrived, we took fresh horses, and a soldier for a guide, and started for the Sierra de la Ventana. This mountain is visible from the anchorage at Bahia Blanca; and Capt. Fitz Roy calculates its height to be 3340 feet—an altitude very remarkable on this eastern side of the continent. I am not aware that any foreigner, previous to my visit, had ascended this mountain; and indeed very few of the soldiers at Bahia Blanca knew anything about it. Hence we heard of beds of coal, of gold and silver, of caves, and of forests, all of which inflamed my curiosity, only to disappoint it. The distance from the posta was about six leagues, over a level plain of the same character as before. The ride was, however, interesting, as the mountain began to show its true form. When we reached the foot of the main ridge, we had much difficulty in finding any water, and we thought we should have been obliged to have passed the night without any. At last we discovered some by looking close to the mountain, for at the distance even of a few hundred yards, the streamlets were buried and entirely lost in the friable calcareous stone and loose detritus. I do not think Nature ever made a more solitary, desolate pile of rock;—it well deserves its name of *Hurtado,* or separated. The mountain is steep, extremely rugged, and broken, and so entirely destitute of trees, and even bushes, that we actually could not make a skewer to stretch out our meat over the fire of thistle-stalks.[62] The strange aspect of this mountain is contrasted by the sea-like plain, which not only abuts against its steep sides, but likewise separates the parallel ranges. The uniformity of the colouring gives an extreme quietness to the view;—

the whitish grey of the quartz rock, and the light brown of the withered grass of the plain, being unrelieved by any brighter tint. From custom, one expects to see in the neighbourhood of a lofty and bold mountain, a broken country strewed over with huge fragments. Here nature shows that the last movement before the bed of the sea is changed into dry land may sometimes be one of tranquillity. Under these circumstances I was curious to observe how far from the parent rock any pebbles could be found. On the shores of Bahia Blanca, and near the settlement, there were some of quartz, which certainly must have come from this source: the distance is forty-five miles.

The dew, which in the early part of the night wetted the saddle-cloths under which we slept, was in the morning frozen. The plain, though appearing horizontal, had insensibly sloped up to a height of between 800 and 900 feet above the sea. In the morning (9th of September) the guide told me to ascend the nearest ridge, which he thought would lead me to the four peaks that crown the summit. The climbing up such rough rocks was very fatiguing; the sides were so indented, that what was gained in one five minutes was often lost in the next. At last, when I reached the ridge, my disappointment was extreme in finding a precipitous valley as deep as the plain, which cut the chain transversely in two, and separated me from the four points. This valley is very narrow, but flat-bottomed, and it forms a fine horse-pass for the Indians, as it connects the plains on the northern and southern sides of the range. Having descended, and while crossing it, I saw two horses grazing: I immediately hid myself in the long grass, and began to reconnoitre; but as I could see no signs of Indians I proceeded cautiously on my second ascent. It was late in the day, and this part of the mountain, like the other, was steep and rugged. I was on the top of the second peak by two o'clock, but got there with extreme difficulty; every twenty yards I had the cramp in the upper part of both thighs, so that I was afraid I should not have been able to have got down again. It was also necessary to return by another road, as it was out of the question to pass over the saddle-back. I was therefore obliged to give up the two higher peaks. Their altitude was but little greater, and every purpose of geology had been answered; so that the attempt was not worth the hazard of any further exertion. I presume the cause of the cramp was the great change in the kind of muscular action, from that of hard riding to that of still harder climbing. It is a lesson worth remembering, as in some cases it might cause much difficulty.

I have already said the mountain is composed of white quartz rock, and with it a little glossy clay-slate is associated. At the height of a few hundred feet above the plain, patches of conglomerate adhered in several places to the solid rock. They resembled in hardness, and in the nature of the cement, the masses which may be seen daily forming on some coasts. I do not doubt these pebbles were in a similar manner aggregated, at a period when the great calcareous formation was depositing beneath the surrounding sea. We may believe that the jagged and battered forms of the hard quartz yet show the effects of the waves of an open ocean.

I was, on the whole, disappointed with this ascent. Even the view was insignificant;—a plain like the sea, but without its beautiful colour and defined outline. The scene, however, was novel, and a little danger, like salt to meat, gave it a relish. That the danger was very little was certain, for my two companions made a good fire—a thing which is never done when it is suspected that Indians are near. I reached the place of our bivouac by sunset, and drinking much maté, and smoking several cigaritos, soon made up my bed for the night. The wind was very strong and cold, but I never slept more comfortably.

September 10th.—In the morning, having fairly scudded before the gale, we arrived by the middle of the day at the Sauce posta. On the road we saw great numbers of deer, and near the mountain a guanaco. The plain, which abuts against the Sierra, is traversed by some curious gullies, of which one was about twenty feet wide, and at least thirty deep; we were obliged in consequence to make a considerable circuit before we could find a pass. We stayed the night at the posta, the conversation, as was generally the case, being about the Indians. The Sierra Ventana was formerly a great place of resort; and three or four years ago there was much fighting there. My guide had been present when many Indians were killed: the women escaped to the top of the ridge, and fought most desperately with great stones; many thus saving themselves.

September 11th.—Proceeded to the third posta in company with the lieutenant who commanded it. The distance is called fifteen leagues; but it is only guess-work, and is generally overstated. The road was uninteresting, over a dry grassy plain; and on our left hand at a greater or less distance there were some low hills; a continuation of which we crossed close to the posta. Before our arrival we met a large herd of cattle and horses, guarded by fifteen soldiers; but we were told many had been lost. It is very difficult to drive animals across the plains; for if in the night a puma, or even a fox, approaches, nothing can prevent the horses dispersing in every direction; and a storm will have the same effect. A short time since, an officer left Buenos Ayres with five hundred horses, and when he arrived at the army he had under twenty.

Soon afterwards we perceived by the cloud of dust, that a party of horsemen were coming towards us; when far distant my companions knew them to be Indians, by their long hair streaming behind their backs. The Indians generally have a fillet round their heads, but never any covering; and their black hair blowing across their swarthy faces, heightens to an uncommon degree the wildness of their appearance. They turned out to be a party of Bernantio's friendly tribe, going to a salina for salt. The Indians eat much salt, their children sucking it like sugar. This habit is very different from that of the Spanish Gauchos, who, leading the same kind of life, eat scarcely any; according to Mungo Park,[63] it is people who live on vegetable food who have an unconquerable desire for salt. The Indians gave us good-humoured nods as they passed at full gallop, driving before them a troop of horses, and followed by a train of lanky dogs.

September 12th and 13th.—I stayed at this posta two days, waiting for a troop of soldiers, which General Rosas had the kindness to send to inform me, would shortly travel to Buenos Ayres; and he advised me to take the opportunity of the escort. In the morning we rode to some neighbouring hills to view the country, and to examine the geology. After dinner the soldiers divided themselves into two parties for a trial of skill with the bolas. Two spears were stuck in the ground twenty-five yards apart, but they were struck and entangled only once in four or five times. The balls can be thrown fifty or sixty yards, but with little certainty. This, however, does not apply to a man on horseback; for when the speed of the horse is added to the force of the arm, it is said, that they can be whirled with effect to the distance of eighty yards. As a proof of their force, I may mention, that at the Falkland Islands, when the Spaniards murdered some of their own countrymen and all the Englishmen, a young friendly Spaniard was running away, when a great tall man, by name Luciano, came at full gallop after him, shouting to him to stop, and saying that he only wanted to speak to him. Just as the Spaniard was on the point of reaching the boat, Luciano threw the balls: they struck him on the legs with such a jerk, as to throw him down and to render him for some time insensible. The man, after Luciano had had his talk, was allowed to escape. He told us that his legs were marked by great weals, where the thong had wound round, as if he had been flogged with a whip. In the middle of the day two men arrived who brought a parcel from the next posta to be forwarded to the general: so that besides these two, our party consisted this evening of my guide and self, the lieutenant, and his four soldiers. The latter were strange beings; the first a fine young negro; the second half Indian and negro; and the two others nondescripts; namely, an old Chilian miner, the colour of mahogany, and another partly a mulatto; but two such mongrels, with such detestable expressions, I never saw before. At night, when they were sitting round the fire, and playing at cards, I retired to view such a Salvator Rosa scene. They were seated under a low cliff, so that I could look down upon them; around the party were lying dogs, arms, remnants of deer and ostriches; and their long spears were stuck in the turf. Further in the dark background, their horses were tied up, ready for any sudden danger. If the stillness of the desolate plain was broken by one of the dogs barking, a soldier, leaving the fire, would place his head close to the ground, and thus slowly scan the horizon. Even if the noisy teru-tero uttered its scream, there would be a pause in the conversation, and every head, for a moment, a little inclined.

What a life of misery these men appear to us to lead! They were at least ten leagues from the Sauce posta, and since the murder committed by the Indians, twenty from another. The Indians are supposed to have made their attack in the middle of the night; for very early in the morning after the murder, they were luckily seen approaching this posta. The whole party here, however, escaped, together with the troop of horses; each one taking a line for himself, and driving with him as many animals as he was able to manage.

The little hovel, built of thistle-stalks, in which they slept, neither kept out the wind nor rain; indeed in the latter case the only effect the roof had, was to condense it into larger drops. They had nothing to eat excepting what they could catch, such as ostriches, deer, armadilloes, etc., and their only fuel was the dry stalks of a small plant, somewhat resembling an aloe. The sole luxury which these men enjoyed was smoking the little paper cigars, and sucking maté. I used to think that the carrion vultures, man's constant attendants on these dreary plains, while seated on the little neighbouring cliffs seemed by their very patience to say, "Ah! when the Indians come we shall have a feast."

In the morning we all sallied forth to hunt, and although we had not much success, there were some animated chases. Soon after starting the party separated, and so arranged their plans, that at a certain time of the day (in guessing which they show much skill) they should all meet from different points of the compass on a plain piece of ground, and thus drive together the wild animals. One day I went out hunting at Bahia Blanca, but the men there merely rode in a crescent, each being about a quarter of a mile apart from the other. A fine male ostrich being turned by the headmost riders, tried to escape on one side. The Gauchos pursued at a reckless pace, twisting their horses about with the most admirable command, and each man whirling the balls round his head. At length the foremost threw them, revolving through the air: in an instant the ostrich rolled over and over, its legs fairly lashed together by the thong.

The plains abound with three kinds of partridge,[64] two of which are as large as hen pheasants. Their destroyer, a small and pretty fox, was also singularly numerous; in the course of the day we could not have seen less than forty or fifty. They were generally near their earths, but the dogs killed one. When we returned to the posta, we found two of the party returned who had been hunting by themselves. They had killed a puma, and had found an ostrich's nest with twenty-seven eggs in it. Each of these is said to equal in weight eleven hen's eggs; so that we obtained from this one nest as much food as 297 hen's eggs would have given.

September 14th. —As the soldiers belonging to the next posta meant to return, and we should together make a party of five, and all armed, I determined not to wait for the expected troops. My host, the lieutenant, pressed me much to stop. As he had been very obliging—not only providing me with food, but lending me his private horses—I wanted to make him some remuneration. I asked my guide whether I might do so, but he told me certainly not; that the only answer I should receive, probably would be, "We have meat for the dogs in our country, and therefore do not grudge it to a Christian." It must not be supposed that the rank of lieutenant in such an army would at all prevent the acceptance of payment: it was only the high sense of hospitality, which every traveller is bound to acknowledge as nearly universal throughout these provinces. After galloping some leagues, we came to a low swampy country, which extends for nearly eighty miles northward, as far as the Sierra Tapalguen. In some parts there were fine damp plains, covered with grass, while others had a soft, black, and peaty soil. There were also many extensive but shallow lakes, and large beds of reeds. The country on the whole resembled the better parts of the Cambridgeshire fens. At night we had some difficulty in finding amidst the swamps, a dry place for our bivouac.

September 15th. —Rose very early in the morning and shortly after passed the posta where the Indians had murdered the five soldiers. The officer had eighteen chuzo wounds in his body. By the middle of the day, after a hard gallop, we reached the fifth posta: on account of some difficulty in procuring horses we stayed there the night. As this point was the most exposed on the whole line, twenty-one soldiers were stationed here; at sunset they returned from hunting, bringing with them seven deer, three ostriches, and many armadilloes and partridges. When riding through the country, it is a common practice to set fire to the plain; and hence at night, as on this occasion, the horizon was illuminated in several places by brilliant conflagrations. This is done partly for the sake of puzzling any stray Indians, but chiefly for improving the pasture. In grassy plains unoccupied by the larger ruminating quadrupeds, it seems necessary to remove the superfluous vegetation by fire, so as to render the new year's growth serviceable.

The rancho at this place did not boast even of a roof, but merely consisted of a ring of thistle-stalks, to break the force of the wind. It was situated on the borders of an extensive but shallow lake, swarming with wild fowl, among which the black-necked swan was conspicuous.

The kind of plover, which appears as if mounted on stilts (Himantopus nigricollis), is here common in flocks of considerable size. It has been wrongfully accused of inelegance; when wading about in shallow water, which is its favourite resort, its gait is far from awkward. These birds in a flock utter a noise, that singularly resembles the cry of a pack of small dogs in full chase: waking in the night, I have more than once been for a moment startled at the distant sound. The teru-tero (Vanellus cayanus) is another bird, which often disturbs the stillness of the night. In appearance and habits it resembles in many respects our peewits; its wings, however, are armed with sharp spurs, like those on the legs of the common cock. As our peewit takes its name from the sound of its voice, so does the teru-tero. While riding over the grassy plains, one is constantly pursued by these birds, which appear to hate mankind, and I am sure deserve to be hated for their never-ceasing, unvaried, harsh screams. To the sportsman they are most annoying, by telling every other bird and animal of his approach: to the traveller in the country, they may possibly, as Molina says, do good, by warning him of the midnight robber. During the breeding season, they attempt, like our peewits, by feigning to be wounded, to draw away from their nests dogs and other enemies. The eggs of this bird are esteemed a great delicacy.

September 16th.—To the seventh posta at the foot of the Sierra Tapalguen. The country was quite level, with a coarse herbage and a soft peaty soil. The hovel was here remarkably neat, the posts and rafters being made of about a dozen dry thistle-stalks bound together with thongs of hide; and by the support of these Ionic-like columns, the roof and sides were thatched with reeds. We were here told a fact, which I would not have credited, if I had not had partly ocular proof of it; namely, that, during the previous night hail as large as small apples, and extremely hard, had fallen with such violence, as to kill the greater number of the wild animals. One of the men had already found thirteen deer (Cervus campestris) lying dead, and I saw their *fresh* hides; another of the party, a few minutes after my arrival, brought in seven more. Now I well know, that one man without dogs could hardly have killed seven deer in a week. The men believed they had seen about fifteen ostriches (part of one of which we had for dinner); and they said that several were running about evidently blind in one eye. Numbers of smaller birds, as ducks, hawks, and partridges, were killed. I saw one of the latter with a black mark on its back, as if it had been struck with a paving-stone. A fence of thistle-stalks round the hovel was nearly broken down, and my informer, putting his head out to see what was the matter, received a severe cut, and now wore a bandage. The storm was said to have been of limited extent: we certainly saw from our last night's bivouac a dense cloud and lightning in this direction. It is marvellous how such strong animals as deer could thus have been killed; but I have no doubt, from the evidence I have given, that the story is not in the least exaggerated. I am glad, however, to have its credibility supported by the Jesuit Dobrizhoffen,[65] who, speaking of a country much to the northward, says, hail fell of an enormous size and killed vast numbers of cattle: the Indians hence called the place *Lalegraicavalca,* meaning "the little white things." Dr. Malcolmson, also, informs me that he witnessed in 1831 in India, a hail-storm, which killed numbers of large birds and much injured the cattle. These hailstones were flat, and one was ten inches in circumference, and another weighed two ounces. They ploughed up a gravel-walk like musket-balls, and passed through glass-windows, making round holes, but not cracking them.

Having finished our dinner, of hail-stricken meat, we crossed the Sierra Tapalguen; a low range of hills, a few hundred feet in height, which commences at Cape Corrientes. The rock in this part is pure quartz; further eastward I understand it is granitic. The hills are of a remarkable form; they consist of flat patches of table-land, surrounded by low perpendicular cliffs, like the outliers of a sedimentary deposit. The hill which I ascended was very small, not above a couple of hundred yards in diameter; but I saw others larger. One which goes by the name of the "Corral," is said to be two or three miles in diameter, and encompassed by perpendicular cliffs between thirty and forty feet high, excepting at one spot, where the entrance lies. Falconer[66] gives a curious account of the Indians driving troops of wild horses into it, and then by guarding the entrance, keeping them secure. I have never heard of any other instance of table-land in a formation of quartz, and which, in the hill I examined, had neither cleavage nor stratification. I was told that the rock of the "Corral" was white, and would strike fire.

We did not reach the posta on the Rio Tapalguen till after it was dark. At supper, from something which was said, I was suddenly struck with horror at thinking that I was eating one of the favourite dishes of the country, namely, a half-formed calf, long before its proper time of birth. It turned out to be Puma; the meat is very white, and remarkably like veal in taste. Dr. Shaw was laughed at for stating that "the flesh of the lion is in great esteem, having no small affinity with veal, both in colour, taste, and flavour." Such certainly is the case with the Puma. The Gauchos differ in their opinion, whether the Jaguar is good eating, but are unanimous in saying that cat is excellent.

September 17th.—We followed the course of the Rio Tapalguen, through a very fertile country, to the ninth posta. Tapalguen, itself, or the town of Tapalguen, if it may be so called, consists of a perfectly level plain, studded over, as far as the eye can reach, with the toldos or oven-shaped huts of the

Indians. The families of the friendly Indians, who were fighting on the side of Rosas, resided here. We met and passed many young Indian women, riding by two or three together on the same horse: they, as well as many of the young men, were strikingly handsome,—their fine ruddy complexions being the picture of health. Besides the toldos, there were three ranchos; one inhabited by the Commandant, and the two others by Spaniards with small shops.

We were here able to buy some biscuit. I had now been several days without tasting anything besides meat: I did not at all dislike this new regimen; but I felt as if it would only have agreed with me with hard exercise. I have heard that patients in England, when desired to confine themselves exclusively to an animal diet, even with the hope of life before their eyes, have hardly been able to endure it. Yet the Gaucho in the Pampas, for months together, touches nothing but beef. But they eat, I observe, a very large proportion of fat, which is of a less animalized nature; and they particularly dislike dry meat, such as that of the Agouti. Dr. Richardson,[67] also, has remarked, "that when people have fed for a long time solely upon lean animal food, the desire for fat becomes so insatiable, that they can consume a large quantity of unmixed and even oily fat without nausea:" this appears to me a curious physiological fact. It is, perhaps, from their meat regimen that the Gauchos, like other carnivorous animals, can abstain long from food. I was told that at Tandeel, some troops voluntarily pursued a party of Indians for three days, without eating or drinking.

We saw in the shops many articles, such as horsecloths, belts, and garters, woven by the Indian women. The patterns were very pretty, and the colours brilliant; the workmanship of the garters was so good that an English merchant at Buenos Ayres maintained they must have been manufactured in England, till he found the tassels had been fastened by split sinew.

September 18th.—We had a very long ride this day. At the twelfth posta, which is seven leagues south of the Rio Salado, we came to the first estancia with cattle and white women. Afterwards we had to ride for many miles through a country flooded with water above our horses' knees. By crossing the stirrups, and riding Arab-like with our legs bent up, we contrived to keep tolerably dry. It was nearly dark when we arrived at the Salado; the stream was deep, and about forty yards wide; in summer, however, its bed becomes almost dry, and the little remaining water nearly as salt as that of the sea. We slept at one of the great estancias of General Rosas. It was fortified, and of such an extent, that arriving in the dark I thought it was a town and fortress. In the morning we saw immense herds of cattle, the general here having seventy-four square leagues of land. Formerly nearly three hundred men were employed about this estate, and they defied all the attacks of the Indians.

September 19th.—Passed the Guardia del Monte. This is a nice scattered little town, with many gardens, full of peach and quince trees. The plain here looked like that around Buenos Ayres; the turf being short and bright green, with beds of clover and thistles, and with bizcacha holes. I was very much struck with the marked change in the aspect of the country after having crossed the Salado. From a coarse herbage we passed on to a carpet of fine green verdure. I at first attributed this to some change in the nature of the soil, but the inhabitants assured me that here, as well as in Banda Oriental, where there is as great a difference between the country round Monte Video and the thinly-inhabited savannahs of Colonia, the whole was to be attributed to the manuring and grazing of the cattle. Exactly the same fact has been observed in the prairies[68] of North America, where coarse grass, between five and six feet high, when grazed by cattle, changes into common pasture land. I am not botanist enough to say whether the change here is owing to the introduction of new species, to the altered growth of the same, or to a difference in their proportional numbers. Azara has also observed with astonishment this change: he is likewise much perplexed by the immediate appearance of plants not occurring in the neighbourhood, on the borders of any track that leads to a newly-constructed hovel. In another part he says,[69] "ces chevaux (sauvages) ont la manie de préférer les chemins, et le bord des routes pour déposer leurs excrémens, dont on trouve des monceaux dans ces endroits." Does this not partly explain the circumstance? We thus have lines of richly manured land serving as channels of communication across wide districts.

Near the Guardia we find the southern limit of two European plants, now become extraordinarily common. The fennel in great profusion covers the ditch-banks in the neighbourhood of Buenos Ayres, Monte Video, and other towns. But the cardoon (Cynara cardunculus) has a far wider range:[70] it occurs in these latitudes on both sides of the Cordillera, across the continent. I saw it in unfrequented spots in Chile, Entre Rios, and Banda Oriental. In the latter country alone, very many (probably several hundred) square miles are covered by one mass of these prickly plants, and are impenetrable by man or beast. Over the undulating plains, where these great beds occur, nothing else can now live. Before their introduction, however, the surface must have supported, as in other parts, a rank herbage. I doubt whether any case is on record of an invasion on so grand a scale of one plant over the aborigines. As I have already said, I nowhere saw the cardoon south of the Salado; but it is probable that in proportion as that country becomes inhabited, the cardoon will extend its limits. The case is different with the giant thistle (with variegated leaves) of the Pampas, for I met with it in the valley of the Sauce. According to the principles so well laid down by Mr. Lyell, few countries have undergone more remarkable changes, since the year 1535, when the first colonist of La Plata landed with seventy-two horses. The countless herds of horses, cattle, and sheep, not only have altered the whole aspect of the vegetation, but they have almost banished the guanaco, deer and ostrich. Numberless other changes must likewise have taken place; the wild pig in some parts probably replaces the peccari; packs of wild dogs may be heard howling on the wooded banks of the less-frequented streams; and the common cat, altered into a large and fierce animal, inhabits rocky hills. As M. d'Orbigny has remarked, the increase in numbers of the carrion-vulture, since the introduction of the domestic animals, must have been infinitely great; and we have given reasons for believing that they have extended their southern range. No doubt many plants, besides the cardoon and fennel, are naturalized; thus the islands near the mouth of the Parana, are thickly clothed with peach and orange trees, springing from seeds carried there by the waters of the river.

While changing horses at the Guardia several people questioned us much about the army,—I never saw anything like the enthusiasm for Rosas, and for the success of the "most just of all wars, because against barbarians." This expression, it must be confessed, is very natural, for till lately, neither man, woman nor horse, was safe from the attacks of the Indians. We had a long day's ride over the same rich green plain, abounding with various flocks, and with here and there a solitary estancia, and its one *ombu* tree. In the evening it rained heavily: on arriving at a post-house we were told by the owner, that if we had not a regular passport we must pass on, for there were so many robbers he would trust no one. When he read, however, my passport, which began with "El Naturalista Don Carlos," his respect and civility were as unbounded as his suspicions had been before. What a naturalist might be, neither he nor his countrymen, I suspect, had any idea; but probably my title lost nothing of its value from that cause.

September 20th.—We arrived by the middle of the day at Buenos Ayres. The outskirts of the city looked quite pretty, with the agave hedges, and groves of olive, peach and willow trees, all just throwing out their fresh green leaves. I rode to the house of Mr. Lumb, an English merchant, to whose kindness and hospitality, during my stay in the country, I was greatly indebted.

The city of Buenos Ayres is large;[71] and I should think one of the most regular in the world. Every street is at right angles to the one it crosses, and the parallel ones being equi-distant, the houses are collected into solid squares of equal dimensions, which are called quadras. On the other hand, the houses themselves are hollow squares; all the rooms opening into a neat little courtyard. They are generally only one story high, with flat roofs, which are fitted with seats, and are much frequented by the inhabitants in summer. In the centre of the town is the Plaza, where the public offices, fortress, cathedral, etc., stand. Here also, the old viceroys, before the revolution, had their palaces. The general assemblage of buildings possesses considerable architectural beauty, although none individually can boast of any.

The great *corral,* where the animals are kept for slaughter to supply food to this beef-eating population, is one of the spectacles best worth seeing. The strength of the horse as compared to that of the bullock is quite astonishing: a man on horseback having thrown his lazo round the horns of a beast, can drag it anywhere he chooses. The animal ploughing up the ground with outstretched legs, in vain efforts to resist the force, generally dashes at full speed to one side; but the horse immediately turning to receive the shock, stands so firmly that the bullock is almost thrown down, and it is surprising that their necks are not broken. The struggle is not, however, one of fair strength; the horse's girth being matched against the bullock's extended neck. In a similar manner a man can hold the wildest horse, if caught with the lazo, just behind the ears. When the bullock has been dragged to the spot where it is to be slaughtered, the *matador* with great caution cuts the hamstrings. Then is given the death bellow; a noise more expressive of fierce agony than any I know. I have often distinguished it from a long distance, and have always known that the struggle was then drawing to a close. The whole sight is horrible and revolting: the ground is almost made of bones; and the horses and riders are drenched with gore.

CHAPTER VII

Buenos Ayres and St. Fé

EXCURSION TO ST. FÉ • THISTLE BEDS • HABITS OF THE BIZCACHA • LITTLE OWL • SALINE STREAMS • LEVEL PLAINS • MASTODON • ST. FÉ • CHANGE IN LANDSCAPE • GEOLOGY • TOOTH OF EXTINCT HORSE • RELATION OF THE FOSSIL AND RECENT QUADRUPEDS OF NORTH AND SOUTH AMERICA • EFFECTS OF A GREAT DROUGHT • PARANA • HABITS OF THE JAGUAR • SCISSOR-BEAK • KINGFISHER, PARROT, AND SCISSOR-TAIL • REVOLUTION • BUENOS AYRES • STATE OF GOVERNMENT

September 27th.—In the evening I set out on an excursion to St. Fé, which is situated nearly three hundred English miles from Buenos Ayres, on the banks of the Parana. The roads in the neighbourhood of the city after the rainy weather, were extraordinarily bad. I should never have thought it possible for a bullock-waggon to have crawled along: as it was, they scarcely went at the rate of a mile an hour, and a man was kept ahead, to survey the best line for making the attempt. The bullocks were terribly jaded: it is a great mistake to suppose that with improved roads, and an accelerated rate of travelling, the sufferings of the animals increase in the same proportion. We passed a train of waggons and a troop of beasts on their road to Mendoza. The distance is about 580 geographical miles, and the journey is generally performed in fifty days. These waggons are very long, narrow, and thatched with reeds; they have only two wheels, the diameter of which in some cases is as much as ten feet. Each is drawn by six bullocks, which are urged on by a goad at least twenty feet long: this is suspended from within the roof; for the wheel bullocks a smaller one is kept; and for the intermediate pair, a point projects at right angles from the middle of the long one.

The whole apparatus looked like some implement of war.

September 28th.—We passed the small town of Luxan, where there is a wooden bridge over the river—a most unusual convenience in this country. We passed also Areco. The plains appeared level, but were not so in fact; for in various places the horizon was distant. The estancias are here wide apart; for there is little good pasture, owing to the land being covered by beds either of an acrid clover, or of the great thistle. The latter, well known from the animated description given by Sir F. Head, were at this time of the year two-thirds grown; in some parts they were as high as the horse's back, but in others they had not yet sprung up, and the ground was bare and dusty as on a turnpike-road. The clumps were of the most brilliant green, and they made a pleasing miniature-likeness of broken forest land. When the thistles are full grown, the great beds are impenetrable, except by a few tracts, as intricate as those in a labyrinth. These are only known to the robbers, who at this season inhabit them, and sally forth at night to rob and cut throats with impunity. Upon asking at a house whether robbers were numerous, I was answered, "The thistles are not up yet;"—the meaning of which reply was not at first very obvious. There is little interest in passing over these tracts, for they are inhabited by few animals or birds, excepting the bizcacha and its friend the little owl.

The bizcacha[72] is well known to form a prominent feature in the zoology of the Pampas. It is found as far south as the Rio Negro, in lat. 41°, but not beyond. It cannot, like the agouti, subsist on the gravelly and desert plains of Patagonia, but prefers a clayey or sandy soil, which produces a different and more abundant vegetation. Near Mendoza, at the foot of the Cordillera, it occurs in close neighbourhood with the allied alpine species. It is a very curious circumstance in its geographical distribution, that it has never been seen, fortunately for the inhabitants of Banda Oriental, to the eastward of the river Uruguay: yet in this province there are plains which appear admirably adapted to its habits. The Uruguay has formed an insuperable obstacle to its migration: although the broader barrier of the Parana has been passed, and the bizcacha is common in Entre Rios, the province between these two great rivers. Near Buenos Ayres these animals are exceedingly common. Their most favourite resort appears to be those parts of the plain which during one-half of the year are covered with giant thistles, to the exclusion of other plants. The Gauchos affirm that it lives on roots; which, from the great strength of its gnawing teeth, and the kind of places frequented by it, seems probable. In the evening the bizcachas come out in numbers, and quietly sit at the mouths of their burrows on their haunches. At such times they are very tame, and a man on horseback passing by seems only to present an object for their grave contemplation. They run very awkwardly, and when running out of danger, from their elevated tails and short front legs, much resemble great rats. Their flesh, when cooked, is very white and good, but it is seldom used.

The bizcacha has one very singular habit; namely, dragging every hard object to the mouth of its burrow: around each group of holes many bones of cattle, stones, thistle-stalks, hard lumps of earth, dry dung, etc., are collected into an irregular heap, which frequently amounts to as much as a wheelbarrow would contain. I was credibly informed that a gentleman, when riding on a dark night, dropped his watch; he returned in the morning, and by searching the neighbourhood of every bizcacha hole on the line of road, as he expected, he soon found it. This habit of picking up whatever may be lying on the ground anywhere near its habitation, must cost much trouble. For what purpose it is done, I am quite unable to form even the most remote conjecture: it cannot be for defence, because the rubbish is chiefly placed above the mouth of the burrow, which enters the ground at a very small inclination. No doubt there must exist some good reason; but the inhabitants of the country are quite ignorant of it. The only fact which I know analogous to it, is the habit of that extraordinary Australian bird, the Calodera maculata, which makes an elegant vaulted passage of twigs for playing in, and which collects near the spot, land and sea-shells, bones, and the feathers of birds, especially brightly coloured ones. Mr. Gould, who has described these facts, informs me, that the natives, when they lose any hard object, search the playing passages, and he has known a tobacco-pipe thus recovered.

The little owl (Athene cunicularia), which has been so often mentioned, on the plains of Buenos Ayres exclusively inhabits the holes of the bizcacha; but in Banda Oriental it is its own workman. During the open day, but more especially in the evening, these birds may be seen in every direction standing frequently by pairs on the hillock near their burrows. If disturbed they either enter the hole, or, uttering a shrill harsh cry, move with a remarkably undulatory flight to a short distance, and then turning round, steadily gaze at their pursuer. Occasionally in the evening they may be heard hooting. I found in the stomachs of two which I opened the remains of mice, and I one day saw a small snake killed and carried away. It is said that snakes are their common prey during the daytime. I may here mention, as showing on what various kinds of food owls subsist, that a species killed among the islets of the Chonos Archipelago, had its stomach full of good-sized crabs. In India[73] there is a fishing genus of owls, which likewise catches crabs.

In the evening we crossed the Rio Arrecife on a simple raft made of barrels lashed together, and slept at the post-house on the other side. I this day paid horse-hire for thirty-one leagues; and although the sun was glaring hot

I was but little fatigued. When Captain Head talks of riding fifty leagues a day, I do not imagine the distance is equal to 150 English miles. At all events, the thirty-one leagues was only 76 miles in a straight line, and in an open country I should think four additional miles for turnings would be a sufficient allowance.

29th and 30th.—We continued to ride over plains of the same character. At San Nicolas I first saw the noble river of the Parana. At the foot of the cliff on which the town stands, some large vessels were at anchor. Before arriving at Rozario, we crossed the Saladillo, a stream of fine clear running water, but too saline to drink. Rozario is a large town built on a dead level plain, which forms a cliff about sixty feet high over the Parana. The river here is very broad, with many islands, which are low and wooded, as is also the opposite shore. The view would resemble that of a great lake, if it were not for the linear-shaped islets, which alone give the idea of running water. The cliffs are the most picturesque part; sometimes they are absolutely perpendicular, and of a red colour; at other times in large broken masses, covered with cacti and mimosa-trees. The real grandeur, however, of an immense river like this, is derived from reflecting how important a means of communication and commerce it forms between one nation and another; to what a distance it travels; and from how vast a territory it drains the great body of fresh water which flows past your feet.

For many leagues north and south of San Nicolas and Rozario, the country is really level. Scarcely anything which travellers have written about its extreme flatness, can be considered as exaggeration. Yet I could never find a spot where, by slowly turning round, objects were not seen at greater distances in some directions than in others; and this manifestly proves inequality in the plain. At sea, a person's eye being six feet above the surface of the water, his horizon is two miles and four-fifths distant. In like manner, the more level the plain, the more nearly does the horizon approach within these narrow limits; and this, in my opinion, entirely destroys that grandeur which one would have imagined that a vast level plain would have possessed.

October 1st.—We started by moonlight and arrived at the Rio Tercero by sunrise. The river is also called the Saladillo, and it deserves the name, for the water is brackish. I stayed here the greater part of the day, searching for fossil bones. Besides a perfect tooth of the Toxodon, and many scattered bones, I found two immense skeletons near each other, projecting in bold relief from the perpendicular cliff of the Parana. They were, however, so completely decayed, that I could only bring away small fragments of one of the great molar teeth; but these are sufficient to show that the remains belonged to a Mastodon, probably to the same species with that, which formerly must have inhabited the Cordillera in Upper Peru in such great numbers. The men who took me in the canoe, said they had long known of these skeletons, and had often wondered how they had got there: the necessity of a theory being felt, they came to the conclusion that, like the bizcacha, the mastodon was formerly a burrowing animal! In the evening we rode another stage, and crossed the Monge, another brackish stream, bearing the dregs of the washings of the Pampas.

October 2nd.—We passed through Corunda, which, from the luxuriance of its gardens, was one of the prettiest villages I saw. From this point to St. Fé the road is not very safe. The western side of the Parana northward, ceases to be inhabited; and hence the Indians sometimes come down thus far, and waylay travellers. The nature of the country also favours this, for instead of a grassy plain, there is an open woodland, composed of low prickly mimosas. We passed some houses that had been ransacked and since deserted; we saw also a spectacle, which my guides viewed with high satisfaction; it was the skeleton of an Indian with the dried skin hanging on the bones, suspended to the branch of a tree.

In the morning we arrived at St. Fé. I was surprised to observe how great a change of climate a difference of only three degrees of latitude between this place and Buenos Ayres had caused. This was evident from the dress and complexion of the men—from the increased size of the ombu-trees—the number of new cacti and other plants—and especially from the birds. In the course of an hour I remarked half-a-dozen birds, which I had never seen at Buenos Ayres. Considering that there is no natural boundary between the two places, and that the character of the country is nearly similar, the difference was much greater than I should have expected.

October 3rd and 4th.—I was confined for these two days to my bed by a headache. A good-natured old woman, who attended me, wished me to try many odd remedies. A common practice is, to bind an orange-leaf or a bit of black plaster to each temple: and a still more general plan is, to split a bean into halves, moisten them, and place one on each temple, where they will easily adhere. It is not thought proper ever to remove the beans or plaster, but to allow them to drop off; and sometimes, if a man, with patches on his head, is asked, what is the matter? he will answer, "I had a headache the day before yesterday." Many of the remedies used by the people of the country are ludicrously strange, but too disgusting to be mentioned. One of the least nasty is to kill and cut open two puppies and bind them on each side of a broken limb. Little hairless dogs are in great request to sleep at the feet of invalids.

St. Fé is a quiet little town, and is kept clean and in good order. The governor, Lopez, was a common soldier at the time of the revolution; but has now been seventeen years in power. This stability of government is owing to his tyrannical habits; for tyranny seems as yet better adapted to these countries than republicanism. The governor's favourite occupation is hunting Indians: a short time since he slaughtered forty-eight, and sold the children at the rate of three or four pounds apiece.

October 5th.—We crossed the Parana to St. Fé Bajada, a town on the opposite shore. The passage took some hours, as the river here consisted of a labyrinth of small streams, separated by low wooded islands. I had a letter of introduction to an old Catalonian Spaniard, who treated me with the most uncommon hospitality. The Bajada is the capital of Entre Rios. In 1825 the town contained 6000 inhabitants, and the province 30,000; yet, few as the inhabitants are, no province has suffered more from bloody and desperate revolutions. They boast here of representatives, ministers, a standing army, and governors: so it is no wonder that they have their revolutions. At some future day this must be one of the richest countries of La Plata. The soil is varied and productive; and its almost insular form gives it two grand lines of communication by the rivers Parana and Uruguay.

I WAS DELAYED here five days, and employed myself in examining the geology of the surrounding country, which was very interesting. We here see at the bottom of the cliffs, beds containing sharks' teeth and sea-shells of extinct species, passing above into an indurated marl, and from that into the red clayey earth of the Pampas, with its calcareous concretions and the bones of terrestrial quadrupeds. This vertical section clearly tells us of a large bay of pure salt-water, gradually encroached on, and at last converted into the bed of a muddy estuary, into which floating carcasses were swept. At Punta Gorda, in Banda Oriental, I found an alternation of the Pampæan estuary deposit, with a limestone containing some of the same extinct sea-shells; and this shows either a change in the former currents, or more probably an oscillation of level in the bottom of the ancient estuary. Until lately, my reasons for considering the Pampæan formation to be an estuary deposit were, its general appearance, its position at the mouth of the existing great river the Plata, and the presence of so many bones of terrestrial quadrupeds: but now Professor Ehrenberg has had the kindness to examine for me a little of the red earth, taken from low down in the deposit, close to the skeletons of the mastodon, and he finds in it many infusoria, partly salt-water and partly fresh-water forms, with the latter rather preponderating; and therefore, as he remarks, the water must have been brackish.

M. A. d'Orbigny found on the banks of the Parana, at the height of a hundred feet, great beds of an estuary shell, now living a hundred miles lower down nearer the sea; and I found similar shells at a less height on the banks of the Uruguay; this shows that just before the Pampas was slowly elevated into dry land, the water covering it was brackish. Below Buenos Ayres there are upraised beds of sea-shells of existing species, which also proves that the period of elevation of the Pampas was within the recent period.

In the Pampæan deposit at the Bajada I found the osseous armour of a gigantic armadillo-like animal, the inside of which, when the earth was removed, was like a great cauldron; I found also teeth of the Toxodon and Mastodon, and one tooth of a Horse, in the same stained and decayed state. This latter tooth greatly interested me,[74] and I took scrupulous care in ascertaining that it had been embedded contemporaneously with the other remains; for I was not then aware that amongst the fossils from Bahia Blanca there was a horse's tooth hidden in the matrix: nor was it then known with certainty that the remains of horses are common in North America. Mr. Lyell has lately brought from the United States a tooth of a horse; and it is an interesting fact, that Professor Owen could find in no species, either fossil or recent, a slight but peculiar curvature characterizing it, until he thought of comparing it with my specimen found here: he has named this American horse Equus curvidens. Certainly it is a marvellous fact in the history of the Mammalia, that in South America a native horse should have lived and disappeared, to be succeeded in after ages by the countless herds descended from the few introduced with the Spanish colonists!

The existence in South America of a fossil horse, of the mastodon, possibly of an elephant,[75] and of a hollow-horned ruminant, discovered by MM. Lund and Clausen in the caves of Brazil, are highly interesting facts with respect to the geographical distribution of animals. At the present time, if we divide America, not by the Isthmus of Panama, but by the southern part of Mexico[76] in lat. 20°, where the great table-land presents an obstacle to the migration of species, by affecting the climate, and by forming, with the exception of some valleys and of a fringe of low land on the coast, a broad barrier; we shall then have the two zoological provinces of North and South America strongly contrasted with each other. Some few species alone have passed the barrier, and may be considered as wanderers from the south, such as the puma, opossum, kinkajou, and peccari. South America is characterized by possessing many peculiar gnawers, a family of monkeys, the llama, peccari, tapir, opossums, and, especially, several genera of Edentata, the order which includes the sloths, ant-eaters, and armadilloes. North America, on the other hand, is characterized (putting on one side a few wandering species) by numerous peculiar gnawers, and by four genera (the ox, sheep, goat, and antelope) of hollow-horned ruminants, of which great division South America is not known to possess a single species. Formerly, but within the period when most of the now existing shells were living, North America possessed, besides hollow-horned ruminants, the elephant, mastodon, horse, and three genera of Edentata, namely, the Megatherium, Megalonyx, and Mylodon. Within nearly this same period (as proved by the shells at Bahia Blanca) South America possessed, as we have just seen, a mastodon, horse, hollow-horned ruminant, and the same three genera (as well as several others) of the Edentata. Hence it is evident that North and South America, in having within a late geological period these several genera in common, were much more closely related in the character of their terrestrial inhabitants than they now are. The more I reflect on this case, the more interesting it appears: I know of no other instance where we can almost mark the period and manner of the splitting up of one great region into two well-characterized zoological provinces. The geologist, who is fully impressed with the vast oscillations of level which have affected the earth's crust within late periods, will not fear to speculate on the recent elevation of the Mexican platform, or, more probably, on the recent submergence of land in the West Indian Archipelago, as the cause of the present zoological separation of North and South America. The South American character of the West Indian mammals[77] seems to indicate that this archipelago was formerly united to the southern continent, and that it has subsequently been an area of subsidence.

When America, and especially North America, possessed its elephants, mastodons, horse, and hollow-horned ruminants, it was much more closely related in its zoological characters to the temperate parts of Europe and Asia than it now is. As the remains of these genera are found on both sides of Behring's Straits[78] and on the plains of Siberia, we are led to look to the north-western side of North America as the former point of communication between the Old and so-called New World. And as so many species, both living and extinct, of these same genera inhabit and have inhabited the Old World, it seems most probable that the North American elephants, mastodons, horse, and hollow-horned ruminants migrated, on land since submerged near Behring's Straits, from Siberia into North America, and thence, on land since submerged in the West Indies, into South America, where for a time they mingled with the forms characteristic of that southern continent, and have since become extinct.

While travelling through the country, I received several vivid descriptions of the effects of a late great drought; and the account of this may throw some light on the cases where vast numbers of animals of all kinds have been embedded together. The period included between the years 1827 and 1830 is called the "gran seco," or the great drought. During this time so little rain fell, that the vegetation, even to the thistles, failed; the brooks were dried up, and the whole country assumed the appearance of a dusty high road. This was especially the case in the northern part of the province of Buenos Ayres and the southern part of St. Fé. Very great numbers of birds, wild animals, cattle, and horses perished from the want of food and water. A man told me that the deer[79] used to come into his court-yard to the well, which he had been obliged to dig to supply his own family with water; and that the partridges had hardly strength to fly away when pursued. The lowest estimation of the loss of cattle in the province of Buenos Ayres alone, was taken at one million head. A proprietor at San Pedro had previously to these years 20,000 cattle; at the end not one remained. San Pedro is situated in the middle of the finest country; and even now abounds again with animals; yet, during the latter part of the "gran seco," live cattle were brought in vessels for the consumption of the inhabitants. The animals roamed from their estancias, and, wandering far southward, were mingled together in such multitudes, that a government commission was sent from Buenos Ayres to settle the disputes of the owners. Sir Woodbine Parish informed me of another and very curious source of dispute; the ground being so long dry, such quantities of dust were blown about, that in this open country the landmarks became obliterated, and people could not tell the limits of their estates.

I was informed by an eye-witness that the cattle in herds of thousands rushed into the Parana, and being exhausted by hunger they were unable to crawl up the muddy banks, and thus were drowned. The arm of the river which runs by San Pedro was so full of putrid carcasses, that the master of a vessel told me that the smell rendered it quite impassable. Without doubt several hundred thousand animals thus perished in the river: their bodies when putrid were seen floating down the stream; and many in all probability were deposited in the estuary of the Plata. All the small rivers became highly saline, and this caused the death of vast numbers in particular spots; for when an animal drinks of such water it does not recover. Azara describes[80] the fury of the wild horses on a similar occasion rushing into the marshes, those which arrived first being overwhelmed and crushed by those which followed. He adds that more than once he has seen the carcasses of upwards of a thousand wild horses thus destroyed. I noticed that the smaller streams in the Pampas were paved with a breccia of bones, but this probably is the effect of a gradual increase, rather than of the destruction at any one period. Subsequently to the drought of 1827 to 1832, a very rainy season followed, which caused great floods. Hence it is almost certain that some thousands of the skeletons were buried by the deposits of the very next year. What would be the opinion of a geologist, viewing such an enormous collection of bones, of all kinds of animals and of all ages, thus embedded in one thick earthy mass? Would he not attribute it to a flood having swept over the surface of the land, rather than to the common order of things?[81]

October 12th.—I had intended to push my excursion further, but not being quite well, I was compelled to return by a balandra, or one-masted vessel of about a hundred tons' burden, which was bound to Buenos Ayres. As the weather was not fair, we moored early in the day to a branch of a tree on one of the islands. The Parana is full of islands, which undergo a constant round of decay and renovation. In the memory of the master several large ones had disappeared, and others again had been formed and protected by vegetation. They are composed of muddy sand, without even the smallest pebble, and were then about four feet above the level of the river; but during the periodical floods they are inundated. They all present one character; numerous willows and a few other trees are bound together by a great variety of creeping plants, thus forming a thick jungle. These thickets afford a retreat for capybaras and jaguars. The fear of the latter animal quite destroyed all pleasure in scrambling through the woods. This evening I had not proceeded a hundred yards, before finding indubitable signs of the recent presence of the tiger, I was obliged to come back. On every island there were tracks; and as on the former excursion "el rastro de los Indios" had been the subject of conversation, so in this was "el rastro del tigre."

The wooded banks of the great rivers appear to be the favourite haunts of the jaguar; but south of the Plata, I was told that they frequented the reeds bordering lakes: wherever they are, they seem to require water. Their common prey is the capybara, so that it is generally said, where capybaras are numerous there is little danger from the jaguar. Falconer states that near the southern side of the mouth of the Plata there are many jaguars, and that they chiefly live on fish; this account I have heard repeated. On the Parana they have killed many wood-cutters, and have even entered vessels at night. There is a man now living in the Bajada, who, coming up from below when it was dark, was seized on the deck; he escaped, however, with the loss of the use of one arm. When the floods drive these animals from the islands, they are most dangerous. I was told that a few years since a very large one found its way into a church at St. Fé: two padres entering one after the other were killed, and a third, who came to see what was the matter, escaped with difficulty. The beast was destroyed by being shot from a corner of the building which was unroofed. They commit also at these times great ravages among cattle and horses. It is said that they kill their prey by breaking their necks. If driven from the carcass, they seldom return to it. The Gauchos say that the jaguar, when wandering about at night, is much tormented by the foxes yelping as they follow him. This is a curious coincidence with the fact which is generally affirmed of the jackals accompanying, in a similarly officious manner, the East Indian tiger. The jaguar is a noisy animal, roaring much by night, and especially before bad weather.

One day, when hunting on the banks of the Uruguay, I was shown certain trees, to which these animals constantly recur for the purpose, as it is said, of sharpening their claws. I saw three well-known trees; in front, the bark was worn smooth, as if by the breast of the animal, and on each side there were deep scratches, or rather grooves, extending in an oblique line, nearly a yard in length. The scars were of different ages. A common method of ascertaining whether a jaguar is in the neighbourhood is to examine these trees. I imagine this habit of the jaguar is exactly similar to one which may any day be seen in the common cat, as with outstretched legs and exserted claws it scrapes the leg of a chair; and I have heard of young fruit-trees in an orchard in England having been thus much injured. Some such habit must also be common to the puma, for on the bare hard soil of Patagonia I have frequently seen scores so deep that no other animal could have made them. The object of this practice is, I believe, to tear off the ragged points of their claws, and not, as the Gauchos think, to sharpen them. The jaguar is killed without much difficulty, by the aid of dogs baying and driving him up a tree where he is despatched with bullets.

Owing to bad weather we remained two days at our moorings. Our only amusement was catching fish for our dinner: there were several kinds, and all good eating. A fish called the "armado" (a Silurus) is remarkable from a harsh grating noise which it makes when caught by hook and line, and which can be distinctly heard when the fish is beneath the water. This same fish has the power of firmly catching hold of any object, such as the blade of an oar or the fishing-line, with the strong spine both of its pectoral and dorsal fin. In the evening the weather was quite tropical, the thermometer standing at 79°. Numbers of fireflies were hovering about, and the musquitoes were very troublesome. I exposed my hand for five minutes, and it was soon black with them; I do not suppose there could have been less than fifty, all busy sucking.

October 15th.—We got under way and passed Punta Gorda, where there is a colony of tame Indians from the province of Missiones. We sailed rapidly down the current, but before sunset, from a silly fear of bad weather, we brought-to in a narrow arm of the river. I took the boat and rowed some distance up this creek. It was very narrow, winding, and deep; on each side a wall thirty or forty feet high, formed by trees intwined with creepers, gave to the canal a singularly gloomy appearance. I here saw a very extraordinary bird, called the Scissor-beak (Rhynchops nigra). It has short legs, web feet, extremely long-pointed wings, and is of about the size of a tern. The beak is flattened laterally, that is, in a plane at right angles to that of a spoonbill or duck. It is as flat and elastic as an ivory paper-cutter, and the lower mandible, differing from every other bird, is an inch and a half longer than the upper. In a lake near Maldonado, from which the water had been nearly drained, and which, in consequence, swarmed with small fry, I saw several of these birds, generally in small flocks, flying rapidly backwards and forwards close to the surface of the lake. They kept their bills wide open, and the lower mandible half buried in the water. Thus skimming the surface, they ploughed it in their course: the water was quite smooth, and it formed a most curious spectacle to behold a flock, each bird leaving its narrow wake on the mirror-like surface. In their flight they frequently twist about with extreme quickness, and dexterously manage with their projecting lower mandible to plough up small fish, which are secured by the upper and shorter half of their scissor-like bills. This fact I repeatedly saw, as, like swallows, they continued to fly backwards and forwards close before me. Occasionally when leaving the surface of the water their flight was wild, irregular, and rapid; they then uttered loud harsh cries. When these birds are fishing, the advantage of the long primary feathers of their wings, in keeping them dry, is very evident. When thus employed, their forms resemble the symbol by which many artists represent marine birds. Their tails are much used in steering their irregular course.

These birds are common far inland along the course of the Rio Parana; it is said that they remain here during the whole year, and breed in the marshes. During the day they rest in flocks on the grassy plains at some distance from the water. Being at anchor, as I have said, in one of the deep creeks between the islands of the Parana, as the evening drew to a close, one of these scissor-beaks suddenly appeared. The water was quite still, and many little fish were rising. The bird continued for a long time to skim the surface, flying in its wild and irregular manner up and down the narrow canal, now dark with the growing night and the shadows of the overhanging trees. At Monte Video, I observed that some large flocks during the day remained

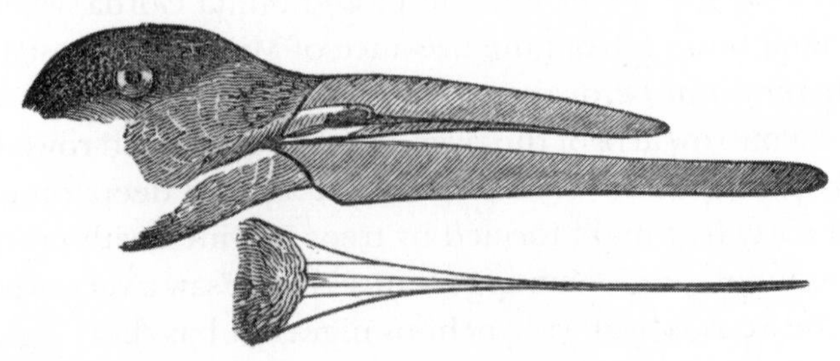

on the mud-banks at the head of the harbour, in the same manner as on the grassy plains near the Parana; and every evening they took flight seaward. From these facts I suspect that the Rhynchops generally fishes by night, at which time many of the lower animals come most abundantly to the surface. M. Lesson states that he has seen these birds opening the shells of the mactræ buried in the sand-banks on the coast of Chile: from their weak bills, with the lower mandible so much projecting, their short legs and long wings, it is very improbable that this can be a general habit.

In our course down the Parana, I observed only three other birds, whose habits are worth mentioning. One is a small kingfisher (Ceryle Americana); it has a longer tail than the European species, and hence does not sit in so stiff and upright a position. Its flight also, instead of being direct and rapid, like the course of an arrow, is weak and undulatory, as among the soft-billed birds. It utters a low note, like the clicking together of two small stones. A small green parrot (Conurus murinus), with a grey breast, appears to prefer the tall trees on the islands to any other situation for its building-place. A number of nests are placed so close together as to form one great mass of sticks. These parrots always live in flocks, and commit great ravages on the corn-fields. I was told, that near Colonia 2500 were killed in the course of one year. A bird with a forked tail, terminated by two long feathers (Tyrannus savana), and named by the Spaniards scissor-tail is very common near Buenos Ayres: it commonly sits on a branch of the *ombu* tree, near a house, and thence takes a short flight in pursuit of insects, and returns to the same spot. When on the wing it presents in its manner of flight and general appearance a caricature-likeness of the common swallow. It has the power of turning very shortly in the air, and in so doing opens and shuts its tail, sometimes in a horizontal or lateral and sometimes in a vertical direction, just like a pair of scissors.

October 16th.—Some leagues below Rozario, the western shore of the Parana is bounded by perpendicular cliffs, which extend in a long line to below San Nicolas; hence it more resembles a sea-coast than that of a freshwater river. It is a great drawback to the scenery of the Parana, that, from the soft nature of its banks, the water is very muddy. The Uruguay, flowing through a granitic country, is much clearer; and where the two channels unite at the head of the Plata, the waters may for a long distance be distinguished by their black and red colours. In the evening, the wind being not quite fair, as usual we immediately moored, and the next day, as it blew rather freshly, though with a favouring current, the master was much too indolent to think of starting. At Bajada, he was described to me as "hombre muy aflicto"—a man always miserable to get on; but certainly he bore all delays with admirable resignation. He was an old Spaniard, and had been many years in this country. He professed a great liking to the English, but stoutly maintained that the battle of Trafalgar was merely won by the Spanish captains having been all bought over; and that the only really gallant action on either side was performed by the Spanish admiral. It struck me as rather characteristic, that this man should prefer his countrymen being thought the worst of traitors, rather than unskilful or cowardly.

18th and 19th.—We continued slowly to sail down the noble stream: the current helped us but little. We met, during our descent, very few vessels. One of the best gifts of nature, in so grand a channel of communication, seems here wilfully thrown away—a river in which ships might navigate from a temperate country, as surprisingly abundant in certain productions as destitute of others, to another possessing a tropical climate, and a soil which, according to the best of judges, M. Bonpland, is perhaps unequalled in fertility in any part of the world. How different would have been the aspect of this river if English colonists had by good fortune first sailed up the Plata! What noble towns would now have occupied its shores! Till the death of Francia, the Dictator of Paraguay, these two countries must remain distinct, as if placed on opposite sides of the globe. And when the old bloody-minded tyrant is gone to his long account, Paraguay will be torn by revolutions, violent in proportion to the previous unnatural calm. That country will have to learn, like every other South American state, that a republic cannot succeed till it contains a certain body of men imbued with the principles of justice and honour.

October 20th.—Being arrived at the mouth of the Parana, and as I was very anxious to reach Buenos Ayres, I went on shore at Las Conchas, with the intention of riding there. Upon landing, I found to my great surprise that I was to a certain degree a prisoner. A violent revolution having broken out, all the ports were laid under an embargo. I could not return to my vessel, and as for going by land to the city, it was out of the question. After a long conversation with the commandant, I obtained permission to go the next day to General Rolor, who commanded a division of the rebels on this side the capital. In the morning I rode to the encampment. The general, officers, and soldiers, all appeared, and I believe really were, great villains. The general, the very evening before he left the city, voluntarily went to the Governor, and with his hand to his heart, pledged his word of honour that he at least would remain faithful to the last. The general told me that the city was in a state of close blockade, and that all he could do was to give me a passport to the commander-in-chief of the rebels at Quilmes. We had therefore to take a great sweep round the city, and it was with much difficulty that we procured horses. My reception at the encampment was quite civil, but I was told it was impossible that I could be allowed to enter the city. I was very anxious about this, as I anticipated the Beagle's departure from the Rio Plata earlier than it took place. Having mentioned, however, General Rosas's obliging kindness to me when at the Colorado, magic itself could not have altered circumstances quicker than did this conversation. I was instantly told that though they could not give me a passport, if I chose to leave my guide and horses, I might pass their sentinels. I was too glad to accept of this, and an officer was sent with me to give directions that I should not be stopped at the bridge. The road for the space of a league was quite deserted. I met one party of soldiers, who were satisfied by gravely looking at an old passport: and at length I was not a little pleased to find myself within the city.

This revolution was supported by scarcely any pretext of grievances: but in a state which, in the course of nine months (from February to October, 1820), underwent fifteen changes in its government—each governor, according to the constitution, being elected for three years—it would be very unreasonable to ask for pretexts. In this case, a party of men—who, being attached to Rosas, were disgusted with the governor Balcarce—to the number of seventy left the city, and with the cry of Rosas the whole country took arms. The city was then blockaded, no provisions, cattle or horses, were allowed to enter; besides this, there was only a little skirmishing, and a few men daily killed. The outside party well knew that by stopping the supply of meat they would certainly be victorious. General Rosas could not have known of this rising; but it appears to be quite consonant with the plans of his party. A year ago he was elected governor, but he refused it, unless the Sala would also confer on him extraordinary powers. This was refused, and since then his party have shown that no other governor can keep his place. The warfare on both sides was avowedly protracted till it was possible to hear from Rosas. A note arrived a few days after I left Buenos Ayres, which stated that the General disapproved of peace having been broken, but that he thought the outside party had justice on their side. On the bare reception of this, the Governor, ministers, and part of the military, to the number of some hundreds, fled from the city. The rebels entered, elected a new governor, and were paid for their services to the number of 5500 men. From these proceedings, it was clear that Rosas ultimately would become the dictator: to the term king, the people in this, as in other republics, have a particular dislike. Since leaving South America, we have heard that Rosas has been elected, with powers and for a time altogether opposed to the constitutional principles of the republic.

CHAPTER VIII

Banda Oriental and Patagonia

EXCURSION TO COLONIA DEL SACRAMIENTO • VALUE OF AN ESTANCIA • CATTLE, HOW COUNTED • SINGULAR BREED OF OXEN • PERFORATED PEBBLES • SHEPHERD DOGS • HORSES BROKEN-IN, GAUCHOS RIDING • CHARACTER OF INHABITANTS • RIO PLATA • FLOCKS OF BUTTERFLIES • AËRONAUT SPIDERS • PHOSPHORESCENCE OF THE SEA • PORT DESIRE • GUANACO • PORT ST. JULIAN • GEOLOGY OF PATAGONIA • FOSSIL GIGANTIC ANIMAL • TYPES OF ORGANIZATION CONSTANT • CHANGE IN THE ZOOLOGY OF AMERICA • CAUSES OF EXTINCTION

Having been delayed for nearly a fortnight in the city, I was glad to escape on board a packet bound for Monte Video. A town in a state of blockade must always be a disagreeable place of residence; in this case moreover there were constant apprehensions from robbers within. The sentinels were the worst of all; for, from their office and from having arms in their hands, they robbed with a degree of authority which other men could not imitate.

Our passage was a very long and tedious one. The Plata looks like a noble estuary on the map; but is in truth a poor affair. A wide expanse of muddy water has neither grandeur nor beauty. At one time of the day, the two shores, both of which are extremely low, could just be distinguished from the deck. On arriving at Monte Video I found that the Beagle would not sail for some time, so I prepared for a short excursion in this part of Banda Oriental. Everything which I have said about the country near Maldonado is applicable to Monte Video; but the land, with the one exception of the Green Mount

450 feet high, from which it takes its name, is far more level. Very little of the undulating grassy plain is enclosed; but near the town there are a few hedge-banks, covered with agaves, cacti, and fennel.

November 14th.—We left Monte Video in the afternoon. I intended to proceed to Colonia del Sacramiento, situated on the northern bank of the Plata and opposite to Buenos Ayres, and thence, following up the Uruguay, to the village of Mercedes on the Rio Negro (one of the many rivers of this name in South America), and from this point to return direct to Monte Video. We slept at the house of my guide at Canelones. In the morning we rose early, in the hopes of being able to ride a good distance; but it was a vain attempt, for all the rivers were flooded. We passed in boats the streams of Canelones, St. Lucia, and San José, and thus lost much time. On a former excursion I crossed the Lucia near its mouth, and I was surprised to observe how easily our horses, although not used to swim, passed over a width of at least six hundred yards. On mentioning this at Monte Video, I was told that a vessel containing some mountebanks and their horses, being wrecked in the Plata, one horse swam seven miles to the shore. In the course of the day I was amused by the dexterity with which a Gaucho forced a restive horse to swim a river. He stripped off his clothes, and jumping on its back, rode into the water till it was out of its depth; then slipping off over the crupper, he caught hold of the tail, and as often as the horse turned round, the man frightened it back by splashing water in its face. As soon as the horse touched the bottom on the other side, the man pulled himself on, and was firmly seated, bridle in hand, before the horse gained the bank. A naked man on a naked horse is a fine spectacle; I had no idea how well the two animals suited each other. The tail of a horse is a very useful appendage; I have passed a river in a boat with four people in it, which was ferried across in the same way as the Gaucho. If a man and horse have to cross a broad river, the best plan is for the man to catch hold of the pommel or mane, and help himself with the other arm.

We slept and stayed the following day at the post of Cufre. In the evening the postman or letter-carrier arrived. He was a day after his time, owing to the Rio Rozario being flooded. It would not, however, be of much consequence; for, although he had passed through some of the principal towns in Banda Oriental, his luggage consisted of two letters! The view from the house was pleasing; an undulating green surface, with distant glimpses of the Plata. I find that I look at this province with very different eyes from what I did upon my first arrival. I recollect I then thought it singularly level; but now, after galloping over the Pampas, my only surprise is, what could have induced me ever to call it level. The country is a series of undulations, in themselves perhaps not absolutely great, but, as compared to the plains of St. Fé, real mountains. From these inequalities there is an abundance of small rivulets, and the turf is green and luxuriant.

November 17th.—We crossed the Rozario, which was deep and rapid, and passing the village of Colla, arrived at midday at Colonia del Sacramiento. The distance is twenty leagues, through a country covered with fine grass, but poorly stocked with cattle or inhabitants. I was invited to sleep at Colonia, and to accompany on the following day a gentleman to his estancia, where there were some limestone rocks. The town is built on a stony promontory something in the same manner as at Monte Video. It is strongly fortified, but both fortifications and town suffered much in the Brazilian war. It is very ancient; and the irregularity of the streets, and the surrounding groves of old orange and peach trees, gave it a pretty appearance. The church is a curious ruin; it was used as a powder-magazine, and was struck by lightning in one of the ten thousand thunder-storms of the Rio Plata. Two-thirds of the building were blown away to the very foundation; and the rest stands a shattered and curious monument of the united powers of lightning and gunpowder. In the evening I wandered about the half-demolished walls of the town. It was the chief seat of the Brazilian war;—a war most injurious to this country, not so much in its immediate effects, as in being the origin of a multitude of generals and all other grades of officers. More generals are numbered (but not paid) in the United Provinces of La Plata than in the United Kingdom of Great Britain. These gentlemen have learned to like power, and do not object to a little skirmishing. Hence there are many always on the watch to create disturbance and to overturn a government which as yet has never rested on any stable foundation. I noticed, however, both here and in other places, a very general interest in the ensuing election for the President; and this appears a good sign for the prosperity of this little country. The inhabitants do not require much education in their representatives; I heard some men discussing the merits of those for Colonia; and it was said that, "although they were not men of business, they could all sign their names:" with this they seemed to think every reasonable man ought to be satisfied.

18th.—Rode with my host to his estancia, at the Arroyo de San Juan. In the evening we took a ride round the estate: it contained two square leagues and a half, and was situated in what is called a rincon; that is, one side was fronted by the Plata, and the two others guarded by impassable brooks. There was an excellent port for little vessels, and an abundance of small wood, which is valuable as supplying fuel to Buenos Ayres. I was curious to know the value of so complete an estancia. Of cattle there were 3000, and it would well support three or four times that number; of mares 800, together with 150 broken-in horses, and 600 sheep. There was plenty of water and limestone, a rough house, excellent corrals, and a peach orchard. For all this he had been offered £2000, and he only wanted £500 additional, and probably would sell it for less. The chief trouble with an estancia is driving the cattle twice a week to a central spot, in order to make them tame, and to count them. This latter operation would be thought difficult, where there are ten or fifteen thousand head together. It is managed on the principle that the cattle invariably divide themselves into little troops of from forty to one hundred. Each troop is recognised by a few peculiarly marked animals, and its number is known: so that, one being lost out of ten thousand, it is perceived by its absence from one of the tropillas. During a stormy night the cattle all mingle together; but the next morning the tropillas separate as before; so that each animal must know its fellow out of ten thousand others.

On two occasions I met with in this province some oxen of a very curious breed, called nata or niata. They appear externally to hold nearly the same relation to other cattle, which bull or pug dogs do to other dogs. Their forehead is very short and broad, with the nasal end turned up, and the upper lip much drawn back; their lower jaws project beyond the upper, and have a corresponding upward curve; hence their teeth are always exposed. Their nostrils are seated high up and are very open; their eyes project outwards. When walking they carry their heads low, on a short neck; and their hinder legs are rather longer compared with the front legs than is usual. Their bare teeth, their short heads, and upturned nostrils give them the most ludicrous self-confident air of defiance imaginable.

Since my return, I have procured a skeleton head, through the kindness of my friend Captain Sulivan, R. N., which is now deposited in the College of Surgeons.[82] Don F. Muniz, of Luxan, has kindly collected for me all the information which he could respecting this breed. From his account it seems that about eighty or ninety years ago, they were rare and kept as curiosities at Buenos Ayres. The breed is universally believed to have originated amongst the Indians southward of the Plata; and that it was with them the commonest kind. Even to this day, those reared in the provinces near the Plata show their less civilized origin, in being fiercer than common cattle, and in the cow easily deserting her first calf, if visited too often or molested. It is a singular fact that an almost similar structure to the abnormal[83] one of the niata breed, characterizes, as I am informed by Dr. Falconer, that great extinct ruminant of India, the Sivatherium. The breed is very *true;* and a niata bull and cow invariably produce niata calves. A niata bull with a common cow, or the reverse cross, produces offspring having an intermediate character, but with the niata characters strongly displayed:

according to Señor Muniz, there is the clearest evidence, contrary to the common belief of agriculturists in analogous cases, that the niata cow when crossed with a common bull transmits her peculiarities more strongly than the niata bull when crossed with a common cow. When the pasture is tolerably long, the niata cattle feed with the tongue and palate as well as common cattle; but during the great droughts, when so many animals perish, the niata breed is under a great disadvantage, and would be exterminated if not attended to; for the common cattle, like horses, are able just to keep alive, by browsing with their lips on twigs of trees and reeds; this the niatas cannot so well do, as their lips do not join, and hence they are found to perish before the common cattle. This strikes me as a good illustration of how little we are able to judge from the ordinary habits of life, on what circumstances, occurring only at long intervals, the rarity or extinction of a species may be determined.

November 19th.—Passing the valley of Las Vacas, we slept at a house of a North American, who worked a lime-kiln on the Arroyo de las Vivoras. In the morning we rode to a projecting headland on the banks of the river, called Punta Gorda. On the way we tried to find a jaguar. There were plenty of fresh tracks, and we visited the trees, on which they are said to sharpen their claws; but we did not succeed in disturbing one. From this point the Rio Uruguay presented to our view a noble volume of water. From the clearness and rapidity of the stream, its appearance was far superior to that of its neighbour the Parana. On the opposite coast, several branches from the latter river entered the Uruguay. As the sun was shining, the two colours of the waters could be seen quite distinct.

In the evening we proceeded on our road towards Mercedes on the Rio Negro. At night we asked permission to sleep at an estancia at which we happened to arrive. It was a very large estate, being ten leagues square, and the owner is one of the greatest landowners in the country. His nephew had charge of it, and with him there was a captain in the army, who the other day ran away from Buenos Ayres. Considering their station, their conversation was rather amusing. They expressed, as was usual, unbounded astonishment at the globe being round, and could scarcely credit that a hole would, if deep enough, come out on the other side. They had, however, heard of a country where there were six months of light and six of darkness, and where the inhabitants were very tall and thin! They were curious about the price and condition of horses and cattle in England. Upon finding out we did not catch our animals with the lazo, they cried out, "Ah, then, you use nothing but the bolas:" the idea of an enclosed country was quite new to them. The captain at last said, he had one question to ask me, which he should be very much obliged if I would answer with all truth. I trembled to think how deeply scientific it would be: it was, "Whether the ladies of Buenos Ayres were not the handsomest in the world." I replied, like a renegade, "Charmingly so." He added, "I have one other question: Do ladies in any other part of the world wear such large combs?" I solemnly assured him that they did not. They were absolutely delighted. The captain exclaimed, "Look there! a man who has seen half the world says it is the case; we always thought so, but now we know it." My excellent judgment in combs and beauty procured me a most hospitable reception; the captain forced me to take his bed, and he would sleep on his recado.

21st.—Started at sunrise, and rode slowly during the whole day. The geological nature of this part of the province was different from the rest, and closely resembled that of the Pampas. In consequence, there were immense beds of the thistle, as well as of the cardoon: the whole country, indeed, may be called one great bed of these plants. The two sorts grow separate, each plant in company with its own kind. The cardoon is as high as a horse's back, but the Pampas thistle is often higher than the crown of the rider's head. To leave the road for a yard is out of the question; and the road itself is partly, and in some cases entirely closed. Pasture, of course there is none; if cattle or horses once enter the bed, they are for the time completely lost. Hence it is very hazardous to attempt to drive cattle at this season of the year; for when jaded enough to face the thistles, they rush among them, and are seen no more. In these districts there are very few estancias, and these few are situated in the neighbourhood of damp valleys, where fortunately neither of these overwhelming plants can exist. As night came on before we arrived at our journey's end, we slept at a miserable little hovel inhabited by the poorest people. The extreme though rather formal courtesy of our host and hostess, considering their grade of life, was quite delightful.

November 22nd.—Arrived at an estancia on the Berquelo belonging to a very hospitable Englishman, to whom I had a letter of introduction from my friend Mr. Lumb. I stayed here three days. One morning I rode with my host to the Sierra del Pedro Flaco, about twenty miles up the Rio Negro. Nearly the whole country was covered with good though coarse grass, which was as high as a horse's belly; yet there were square leagues without a single head of cattle. The province of Banda Oriental, if well stocked, would support an astonishing number of animals; at present the annual export of hides from Monte Video amounts to three hundred thousand; and the home consumption, from waste, is very considerable. An "estanciero" told me that he often had to send large herds of cattle a long journey to a salting establishment, and that the tired beasts were frequently obliged to be killed and skinned; but that he could never persuade the Gauchos to eat of them, and every evening a fresh beast was slaughtered for their suppers! The view of the Rio Negro from the Sierra was more picturesque than any other which I saw in this province. The river, broad, deep, and rapid, wound at the foot of a rocky precipitous cliff: a belt of wood followed its course, and the horizon terminated in the distant undulations of the turf-plain.

When in this neighbourhood, I several times heard of the Sierra de las Cuentas: a hill distant many miles to the northward. The name signifies hill of beads. I was assured that vast numbers of little round stones, of various colours, each with a small cylindrical hole, are found there. Formerly the Indians used to collect them, for the purpose of making necklaces and bracelets—a taste, I may observe, which is common to all savage nations, as well as to the most polished. I did not know what to understand from this story, but upon mentioning it at the Cape of Good Hope to Dr. Andrew Smith, he told me that he recollected finding on the south-eastern coast of Africa, about one hundred miles to the eastward of St. John's river, some quartz crystals with their edges blunted from attrition, and mixed with gravel on the sea-beach. Each crystal was about five lines in diameter, and from an inch to an inch and a half in length. Many of them had a small canal extending from one extremity to the other, perfectly cylindrical, and of a size that readily admitted a coarse thread or a piece of fine catgut. Their colour was red or dull white. The natives were acquainted with this structure in crystals. I have mentioned these circumstances because, although no crystallized body is at present known to assume this form, it may lead some future traveller to investigate the real nature of such stones.

WHILE STAYING AT THIS ESTANCIA, I was amused with what I saw and heard of the shepherd-dogs of the country.[84] When riding, it is a common thing to meet a large flock of sheep guarded by one or two dogs, at the distance of some miles from any house or man. I often wondered how so firm a friendship had been established. The method of education consists in separating the puppy, while very young, from the bitch, and in accustoming it to its future companions. An ewe is held three or four times a day for the little thing to suck, and a nest of wool is made for it in the sheep-pen; at no time is it allowed to associate with other dogs, or with the children of the family. The puppy is, moreover, generally castrated; so that, when grown up, it can scarcely have any feelings in common with the rest of its kind. From this education it has no wish to leave the flock, and just as another dog will defend its master, man, so will these the sheep. It is amusing to observe, when approaching a flock, how the dog immediately advances barking, and the sheep all close in his rear, as if round the oldest ram. These dogs are also easily taught to bring home the flock, at a certain hour in the evening. Their most troublesome fault, when young, is their desire of playing with the sheep; for in their sport they sometimes gallop their poor subjects most unmercifully.

The shepherd-dog comes to the house every day for some meat, and as soon as it is given him, he skulks away as if ashamed of himself. On these occasions the house-dogs are very tyrannical, and the least of them will attack and pursue the stranger. The minute, however, the latter has reached the flock, he turns round and begins to bark, and then all the house-dogs take very quickly to their heels. In a similar manner a whole pack of the hungry wild dogs will scarcely ever (and I was told by some never) venture to attack a flock guarded by even one of these faithful shepherds. The whole account appears to me a curious instance of the pliability of the affections in the dog; and yet, whether wild or however educated, he has a feeling of respect or fear for those that are fulfilling their instinct of association. For we can understand on no principle the wild dogs being driven away by the single one with its flock, except that they consider, from some confused notion, that the one thus associated gains power, as if in company with its own kind. F. Cuvier has observed, that all animals that readily enter into domestication, consider man as a member of their own society, and thus fulfil their instinct of association. In the above case the shepherd-dog ranks the sheep as its fellow-brethren, and thus gains confidence; and the wild dogs, though knowing that the individual sheep are not dogs, but are good to eat, yet partly consent to this view when seeing them in a flock with a shepherd-dog at their head.

One evening a "domidor" (a subduer of horses) came for the purpose of breaking-in some colts. I will describe the preparatory steps, for I believe they have not been mentioned by other travellers. A troop of wild young horses is driven into the corral, or large enclosure of stakes, and the door is shut. We will suppose that one man alone has to catch and mount a horse, which as yet had never felt bridle or saddle. I conceive, except by a Gaucho, such a feat would be utterly impracticable. The Gaucho picks out a full-grown colt; and as the beast rushes round the circus, he throws his lazo so as to catch both the front legs. Instantly the horse rolls over with a heavy shock, and whilst struggling on the ground, the Gaucho, holding the lazo tight, makes a circle, so as to catch one of the hind legs, just beneath the fetlock, and draws it close to the two front legs: he then hitches the lazo, so that the three are bound together. Then sitting on the horse's neck, he fixes a strong bridle, without a bit, to the lower jaw: this he does by passing a narrow thong through the eye-holes at the end of the reins, and several times round both jaw and tongue. The two front legs are now tied closely together with a strong leathern thong, fastened by a slip-knot. The lazo, which bound the three together, being then loosed, the horse rises with difficulty. The Gaucho now holding fast the bridle fixed to the lower jaw, leads the horse outside the corral. If a second man is present (otherwise the trouble is much greater) he holds the animal's head, whilst the first puts on the horsecloths and saddle, and girths the whole together. During this operation, the horse, from dread and astonishment at thus being bound round the waist, throws himself over and over again on the ground, and, till beaten, is unwilling to rise. At last, when the saddling is finished, the poor animal can hardly breathe from fear, and is white with foam and sweat. The man now prepares to mount by pressing heavily on the stirrup, so that the horse may not lose its balance; and at the moment that he throws his leg over the animal's back, he pulls the slip-knot binding the front legs, and the beast is free. Some "domidors" pull the knot while the animal is lying on the ground, and, standing over the saddle, allow him to rise beneath them. The horse, wild with dread, gives a few most violent bounds, and then starts off at full gallop: when quite exhausted, the man, by patience, brings him back to the corral, where, reeking hot and scarcely alive, the poor beast is let free. Those animals which will not gallop away, but obstinately throw themselves on the ground, are by far the most troublesome. This process is tremendously severe, but in two or three trials the horse is tamed. It is not, however, for some weeks that the animal is ridden with the iron bit and solid ring, for it must learn to associate the will of its rider with the feel of the rein, before the most powerful bridle can be of any service.

Animals are so abundant in these countries, that humanity and self-interest are not closely united; therefore I fear it is that the former is here scarcely known. One day, riding in the Pampas with a very respectable "estanciero," my horse, being tired, lagged behind. The man often shouted to me to spur him. When I remonstrated that it was a pity, for the horse was quite exhausted, he cried out, "Why not?—never mind—spur him—it is *my* horse." I had then some difficulty in making him comprehend that it was for the horse's sake, and not on his account, that I did not choose to use my spurs. He exclaimed, with a look of great surprise, "Ah, Don Carlos, que cosa!" It was clear that such an idea had never before entered his head.

The Gauchos are well known to be perfect riders. The idea of being thrown, let the horse do what it likes, never enters their head. Their criterion of a good rider is, a man who can manage an untamed colt, or who, if his horse falls, alights on his own feet, or can perform other such exploits. I have heard of a man betting that he would throw his horse down twenty times, and that nineteen times he would not fall himself. I recollect seeing a Gaucho riding a very stubborn horse, which three times successively reared so high as to fall backwards with great violence. The man judged with uncommon coolness the proper moment for slipping off, not an instant before or after the right time; and as soon as the horse got up, the man jumped on his back, and at last they started at a gallop. The Gaucho never appears to exert any muscular force. I was one day watching a good rider, as we were galloping along at a rapid pace, and thought to myself, "Surely if the horse starts, you appear so careless on your seat, you must fall." At this moment, a male ostrich sprang from its nest right beneath the horse's nose: the young colt bounded on one side like a stag; but as for the man, all that could be said was, that he started and took fright with his horse.

In Chile and Peru more pains are taken with the mouth of the horse than in La Plata, and this is evidently a consequence of the more intricate nature of the country. In Chile a horse is not considered perfectly broken, till he can be brought up standing, in the midst of his full speed, on any particular spot,—for instance, on a cloak thrown on the ground: or, again, he will charge a wall, and rearing, scrape the surface with his hoofs. I have seen an animal bounding with spirit, yet merely reined by a fore-finger and thumb, taken at full gallop across a court-yard, and then made to wheel round the post of a veranda with great speed, but at so equal a distance, that the rider, with outstretched arm, all the while kept one finger rubbing the post. Then making a demi-volte in the air, with the other arm outstretched in a like manner, he wheeled round, with astonishing force, in an opposite direction.

Such a horse is well broken; and although this at first may appear useless, it is far otherwise. It is only carrying that which is daily necessary into perfection. When a bullock is checked and caught by the lazo, it will sometimes gallop round and round in a circle, and the horse being alarmed at the great strain, if not well broken, will not readily turn like the pivot of a wheel. In consequence many men have been killed; for if the lazo once takes a twist round a man's body, it will instantly, from the power of the two opposed animals, almost cut him in twain. On the same principle the races are managed; the course is only two or three hundred yards long, the wish being to have horses that can make a rapid dash. The racehorses are trained not only to stand with their hoofs touching a line, but to draw all four feet together, so as at the first spring to bring into play the full action of the hind-quarters. In Chile I was told an anecdote, which I believe was true; and it offers a good illustration of the use of a well-broken animal. A respectable man riding one day met two others, one of whom was mounted on a horse, which he knew to have been stolen from himself. He challenged them; they answered him by drawing their sabres and giving chase. The man, on his good and fleet beast, kept just ahead: as he passed a thick bush he wheeled round it, and brought up his horse to a dead check. The pursuers were obliged to shoot on one side and ahead. Then instantly dashing on, right behind them, he buried his knife in the back of one, wounded the other, recovered his horse from the dying robber, and rode home. For these feats of horsemanship two things are necessary: a most severe bit, like the Mameluke, the power of which, though seldom used, the horse knows full well; and large blunt spurs, that can be applied either as a mere touch, or as an instrument of extreme pain. I conceive that with English spurs, the slightest touch of which pricks the skin, it would be impossible to break in a horse after the South American fashion.

At an estancia near Las Vacas large numbers of mares are weekly slaughtered for the sake of their hides, although worth only five paper dollars, or about half a crown apiece. It seems at first strange that it can answer to kill mares for such a trifle; but as it is thought ridiculous in this country ever to break in or ride a mare, they are of no value except for breeding. The only thing for which I ever saw mares used, was to tread out wheat from the ear; for which purpose they were driven round a circular enclosure, where the wheat-sheaves were strewed. The man employed for slaughtering the mares happened to be celebrated for his dexterity with the lazo. Standing at the distance of twelve yards from the mouth of the corral, he has laid a wager that he would catch by the legs every animal, without missing one, as it rushed past him. There was another man who said he would enter the corral on foot, catch a mare, fasten her front legs together, drive her out, throw her down, kill, skin, and stake the hide for drying (which latter is a tedious job); and he engaged that he would perform this whole operation on twenty-two animals in one day. Or he would kill and take the skin off fifty in the same time. This would have been a prodigious task, for it is considered a good day's work to skin and stake the hides of fifteen or sixteen animals.

November 26th.—I set out on my return in a direct line for Monte Video. Having heard of some giant's bones at a neighbouring farm-house on the Sarandis, a small stream entering the Rio Negro, I rode there accompanied by my host, and purchased for the value of eighteen pence the head of the Toxodon.[85] When found it was quite perfect; but the boys knocked out some of the teeth with stones, and then set up the head as a mark to throw at. By a most fortunate chance I found a perfect tooth, which exactly fitted one of the sockets in this skull, embedded by itself on the banks of the Rio Tercero, at the distance of about 180 miles from this place. I found remains of this extraordinary animal at two other places, so that it must formerly have been common. I found here, also, some large portions of the armour of a gigantic armadillo-like animal, and part of the great head of a Mylodon. The bones of this head are so fresh, that they contain, according to the analysis by Mr. T. Reeks, seven per cent of animal matter; and when placed in a spirit-lamp, they burn with a small flame. The number of the remains embedded in the grand estuary deposit which forms the Pampas and covers the granitic rocks of Banda Oriental, must be extraordinarily great. I believe a straight line drawn in any direction through the Pampas would cut through some skeleton or bones. Besides those which I found during my short excursions, I heard of many others, and the origin of such names as "the stream of the animal," "the hill of the giant," is obvious. At other times I heard of the marvellous property of certain rivers, which had the power of changing small bones into large; or, as some maintained, the bones themselves grew. As far as I am aware, not one of these animals perished, as was formerly supposed, in the marshes or muddy river-beds of the present land, but their bones have been exposed by the streams intersecting the subaqueous deposit in which they were originally embedded. We may conclude that the whole area of the Pampas is one wide sepulchre of these extinct gigantic quadrupeds.

By the middle of the day, on the 28th, we arrived at Monte Video, having been two days and a half on the road. The country for the whole way was of a very uniform character, some parts being rather more rocky and hilly than near the Plata. Not far from Monte Video we passed through the village of Las Pietras, so named from some large rounded masses of syenite. Its appearance was rather pretty. In this country a few fig-trees round a group of houses, and a site elevated a hundred feet above the general level, ought always to be called picturesque.

DURING THE LAST six months I have had an opportunity of seeing a little of the character of the inhabitants of these provinces. The Gauchos, or countrymen, are very superior to those who reside in the towns. The Gaucho is invariably most obliging, polite, and hospitable: I did not meet with even one instance of rudeness or inhospitality. He is modest, both respecting himself and country, but at the same time a spirited, bold fellow. On the other hand, many robberies are committed, and there is much bloodshed: the habit of constantly wearing the knife is the chief cause of the latter. It is lamentable to hear how many lives are lost in trifling quarrels. In fighting, each party tries to mark the face of his adversary by slashing his nose or eyes; as is often attested by deep and horrid-looking scars. Robberies are a natural consequence of universal gambling, much drinking, and extreme indolence. At Mercedes I asked two men why they did not work. One gravely said the days were too long; the other that he was too poor. The number of horses and the profusion of food are the destruction of all industry. Moreover, there are so many feast-days; and again, nothing can succeed without it be begun when the moon is on the increase; so that half the month is lost from these two causes.

Police and justice are quite inefficient. If a man who is poor commits murder and is taken, he will be imprisoned, and perhaps even shot; but if he is rich and has friends, he may rely on it no very severe consequence will ensue. It is curious that the most respectable inhabitants of the country invariably assist a murderer to escape: they seem to think that the individual sins against the government, and not against the people. A traveller has no protection besides his fire-arms; and the constant habit of carrying them is the main check to more frequent robberies.

The character of the higher and more educated classes who reside in the towns, partakes, but perhaps in a lesser degree, of the good parts of the Gaucho, but is, I fear, stained by many vices of which he is free. Sensuality, mockery of all religion, and the grossest corruption, are far from uncommon. Nearly every public officer can be bribed. The head man in the post-office sold forged government franks. The governor and prime minister openly combined to plunder the state. Justice, where gold came into play, was hardly expected by any one. I knew an Englishman, who went to the Chief Justice (he told me, that not then understanding the ways of the place, he trembled as he entered the room), and said, "Sir, I have come to offer you two hundred (paper) dollars (value about five pounds sterling) if you will arrest before a certain time a man who has cheated me. I know it is against the law, but my lawyer (naming him) recommended me to take this step." The Chief Justice smiled acquiescence, thanked him, and the man before night was safe in prison. With this entire want of principle in many of the leading men, with the country full of ill-paid turbulent officers, the people yet hope that a democratic form of government can succeed!

On first entering society in these countries, two or three features strike one as particularly remarkable. The polite and dignified manners pervading every rank of life, the excellent taste displayed by the women in their dresses, and the equality amongst all ranks. At the Rio Colorado some men who kept the humblest shops used to dine with General Rosas. A son of a major at Bahia Blanca gained his livelihood by making paper cigars, and he wished to accompany me, as guide or servant, to Buenos Ayres, but his father objected on the score of the danger alone. Many officers in the army can neither read nor write, yet all meet in society as equals. In Entre Rios, the Sala consisted of only six representatives. One of them kept a common shop, and evidently was not degraded by the office. All this is what would be expected in a new country; nevertheless the absence of gentlemen by profession appears to an Englishman something strange.

When speaking of these countries, the manner in which they have been brought up by their unnatural parent, Spain, should always be borne in mind. On the whole, perhaps, more credit is due for what has been done, than blame for that which may be deficient. It is impossible to doubt but that the extreme liberalism of these countries must ultimately lead to good results. The very general toleration of foreign religions, the regard paid to the means of education, the freedom of the press, the facilities offered to all foreigners, and especially, as I am bound to add, to every one professing the humblest pretensions to science, should be recollected with gratitude by those who have visited Spanish South America.

December 6th.—The Beagle sailed from the Rio Plata, never again to enter its muddy stream. Our course was directed to Port Desire, on the coast of Patagonia. Before proceeding any further, I will here put together a few observations made at sea.

Several times when the ship has been some miles off the mouth of the Plata, and at other times when off the shores of Northern Patagonia, we have been surrounded by insects. One evening, when we were about ten miles from the Bay of San Blas, vast numbers of butterflies, in bands or flocks of countless myriads, extended as far as the eye could range. Even by the aid of a telescope it was not possible to see a space free from butterflies. The seamen cried out "it was snowing butterflies," and such in fact was the appearance. More species than one were present, but the main part belonged to a kind very similar to, but not identical with, the common English Colias edusa. Some moths and hymenoptera accompanied the butterflies; and a fine beetle (Calosoma) flew on board. Other instances are known of this beetle having been caught far out at sea; and this is the more remarkable, as the greater number of the Carabidæ seldom or never take wing. The day had been fine and calm, and the one previous to it equally so, with light and variable airs. Hence we cannot suppose that the insects were blown off the land, but we must conclude that they voluntarily took flight. The great bands of the Colias seem at first to afford an instance like those on record of the migrations of another butterfly, Vanessa cardui;[86] but the presence of other insects makes the case distinct, and even less intelligible. Before sunset a strong breeze sprung up from the north, and this must have caused tens of thousands of the butterflies and other insects to have perished.

On another occasion, when seventeen miles off Cape Corrientes, I had a net overboard to catch pelagic animals. Upon drawing it up, to my surprise, I found a considerable number of beetles in it, and although in the open sea, they did not appear much injured by the salt water. I lost some of the specimens, but those which I preserved belonged to the genera Colymbetes, Hydroporus, Hydrobius (two species), Notaphus, Cynucus, Adimonia, and Scarabæus. At first I thought that these insects had been blown from the shore; but upon reflecting that out of the eight species four were aquatic, and two others partly so in their habits, it appeared to me most probable that they were floated into the sea by a small stream which drains a lake near Cape Corrientes. On any supposition it is an interesting circumstance to find live insects swimming in the open ocean seventeen miles from the nearest point of land. There are several accounts of insects having been blown off the Patagonian shore. Captain Cook observed it, as did more lately Captain King of the Adventure. The cause probably is due to the want of shelter, both of trees and hills, so that an insect on the wing with an off-shore breeze, would be very apt to be blown out to sea. The most remarkable instance I have known of an insect being caught far from the land, was that of a large grasshopper (Acrydium), which flew on board, when the Beagle was to windward of the Cape de Verd Islands, and when the nearest point of land, not directly opposed to the trade-wind, was Cape Blanco on the coast of Africa, 370 miles distant.[87]

On several occasions, when the Beagle has been within the mouth of the Plata, the rigging has been coated with the web of the Gossamer Spider. One day (November 1st, 1832) I paid particular attention to this subject. The weather had been fine and clear, and in the morning the air was full of patches of the flocculent web, as on an autumnal day in England. The ship was sixty miles distant from the land, in the direction of a steady though light breeze. Vast numbers of a small spider, about one-tenth of an inch in length, and of a dusky red colour, were attached to the webs. There must have been, I should suppose, some thousands on the ship. The little spider, when first coming in contact with the rigging, was always seated on a single thread, and not on the flocculent mass. This latter seems merely to be produced by the entanglement of the single threads. The spiders were all of one species, but of both sexes, together with young ones. These latter were distinguished by their smaller size and more dusky colour. I will not give the description of this spider, but merely state that it does not appear to me to be included in any of Latreille's genera. The little aëronaut as soon as it arrived on board was very active, running about, sometimes letting itself fall, and then reascending the same thread; sometimes employing itself in making a small and very irregular mesh in the corners between the ropes. It could run with facility on the surface of the water. When disturbed it lifted up its front legs, in the attitude of attention. On its first arrival it appeared very thirsty, and with exserted maxillæ drank eagerly of drops of water; this same circumstance has been observed by Strack: may it not be in consequence of the little insect having passed through a dry and rarefied atmosphere? Its stock of web seemed inexhaustible. While watching some that were suspended by a single thread, I several times observed that the slightest breath of air bore them away out of sight, in a horizontal line.

On another occasion (25th) under similar circumstances, I repeatedly observed the same kind of small spider, either when placed or having crawled on some little eminence, elevate its abdomen, send forth a thread, and then sail away horizontally, but with a rapidity which was quite unaccountable. I thought I could perceive that the spider, before performing the above preparatory steps, connected its legs together with the most delicate threads, but I am not sure whether this observation was correct.

One day, at St. Fé, I had a better opportunity of observing some similar facts. A spider which was about three-tenths of an inch in length, and which in its general appearance resembled a Citigrade (therefore quite different from the gossamer), while standing on the summit of a post, darted forth four or five threads from its spinners. These, glittering in the sunshine, might be compared to diverging rays of light; they were not, however, straight, but in undulations like films of silk blown by the wind. They were more than a yard in length, and diverged in an ascending direction from the orifices. The spider then suddenly let go its hold of the post, and was quickly borne out of sight. The day was hot and apparently calm; yet under such circumstances, the atmosphere can never be so tranquil as not to affect a vane so delicate as the thread of a spider's web. If during a warm day we look either at the shadow of any object cast on a bank, or over a level plain at a distant landmark, the effect of an ascending current of heated air is almost always evident: such upward currents, it has been remarked, are also shown by the ascent of soap-bubbles, which will not rise in an in-doors room. Hence I think there is not much difficulty in understanding the ascent of the fine lines projected from a spider's spinners, and afterwards of the spider itself; the divergence of the lines has been attempted to be explained, I believe by Mr. Murray, by their similar electrical condition. The circumstance of spiders of the same species, but of different sexes and ages, being found on several occasions at the distance of many leagues from the land, attached in vast numbers to the lines, renders it probable that the habit of sailing through the air is as characteristic of this tribe, as that of diving is of the Argyroneta. We may then reject Latreille's supposition, that the gossamer owes its origin indifferently to the young of several genera of spiders: although, as we have seen, the young of other spiders do possess the power of performing aërial voyages.[88]

During our different passages south of the Plata, I often towed astern a net made of bunting, and thus caught many curious animals. Of Crustacea there were many strange and undescribed genera. One, which in some respects is allied to the Notopods (or those crabs which have their posterior legs placed almost on their backs, for the purpose of adhering to the under side of rocks), is very remarkable from the structure of its hind pair of legs. The penultimate joint, instead of terminating in a simple claw, ends in three bristle-like appendages of dissimilar lengths—the longest equalling that of the entire leg. These claws are very thin, and are serrated with the finest teeth, directed backwards: their curved extremities are flattened, and on this part five most minute cups are placed which seem to act in the same manner as the suckers on the arms of the cuttle-fish. As the animal lives in the open sea, and probably wants a place of rest, I suppose this beautiful and most anomalous structure is adapted to take hold of floating marine animals.

In deep water, far from the land, the number of living creatures is extremely small: south of the latitude 35°, I never succeeded in catching anything besides some beroe, and a few species of minute entomostracous crustacea. In shoaler water, at the distance of a few miles from the coast, very many kinds of crustacea and some other animals are numerous, but only during the night. Between latitudes 56° and 57° south of Cape Horn, the net was put astern several times; it never, however, brought up anything besides a few of two extremely minute species of Entomostraca. Yet whales and seals, petrels and albatross, are exceedingly abundant throughout this part of the ocean. It has always been a mystery to me on what the albatross, which lives far from the shore, can subsist; I presume that, like the condor, it is able to fast long; and that one good feast on the carcass of a putrid whale lasts for a long time. The central and intertropical parts of the Atlantic swarm with Pteropoda, Crustacea, and Radiata, and with their devourers the flying-fish, and again with their devourers the bonitos and albicores; I presume that the numerous lower pelagic animals feed on the Infusoria, which are now known, from the researches of Ehrenberg, to abound in the open ocean: but on what, in the clear blue water, do these Infusoria subsist?

While sailing a little south of the Plata on one very dark night, the sea presented a wonderful and most beautiful spectacle. There was a fresh breeze, and every part of the surface, which during the day is seen as foam, now glowed with a pale light. The vessel drove before her bows two billows of liquid phosphorus, and in her wake she was followed by a milky train. As far as the eye reached, the crest of every wave was bright, and the sky above the horizon, from the reflected glare of these livid flames, was not so utterly obscure as over the vault of the heavens.

As we proceed further southward the sea is seldom phosphorescent; and off Cape Horn I do not recollect more than once having seen it so, and then it was far from being brilliant. This circumstance probably has a close connection with the scarcity of organic beings in that part of the ocean. After the elaborate paper,[89] by Ehrenberg, on the phosphorescence of the sea, it is almost superfluous on my part to make any observations on the subject. I may however add, that the same torn and irregular particles of gelatinous matter, described by Ehrenberg, seem in the southern as well as in the northern hemisphere, to be the common cause of this phenomenon. The particles were so minute as easily to pass through fine gauze; yet many were distinctly visible by the naked eye. The water when placed in a tumbler and agitated, gave out sparks, but a small portion in a watch-glass scarcely ever was luminous. Ehrenberg states that these particles all retain a certain degree of irritability. My observations, some of which were made directly after taking up the water, gave a different result. I may also mention, that having used the net during one night, I allowed it to become partially dry, and having occasion twelve hours afterwards to employ it again, I found the whole surface sparkled as brightly as when first taken out of the water. It does not appear probable in this case, that the particles could have remained so long alive. On one occasion having kept a jelly-fish of the genus Dianæa till it was dead, the water in which it was placed became luminous. When the waves scintillate with bright green sparks, I believe it is generally owing to minute crustacea. But there can be no doubt that very many other pelagic animals, when alive, are phosphorescent.

On two occasions I have observed the sea luminous at considerable depths beneath the surface. Near the mouth of the Plata some circular and oval patches, from two to four yards in diameter, and with defined outlines, shone with a steady but pale light; while the surrounding water only gave out a few sparks. The appearance resembled the reflection of the moon, or some luminous body; for the edges were sinuous from the undulations of the surface. The ship, which drew thirteen feet of water, passed over, without disturbing these patches. Therefore we must suppose that some animals were congregated together at a greater depth than the bottom of the vessel.

Near Fernando Noronha the sea gave out light in flashes. The appearance was very similar to that which might be expected from a large fish moving rapidly through a luminous fluid. To this cause the sailors attributed it; at the time, however, I entertained some doubts, on account of the frequency and rapidity of the flashes. I have already remarked that the phenomenon is very much more common in warm than in cold countries; and I have sometimes imagined that a disturbed electrical condition of the atmosphere was most favourable to its production. Certainly I think the sea is most luminous after a few days of more calm weather than ordinary, during which time it has swarmed with various animals. Observing that the water charged with gelatinous particles is in an impure state, and that the luminous appearance in all common cases is produced by the agitation of the fluid in contact with the atmosphere, I am inclined to consider that the phosphorescence is the result of the decomposition of the organic particles, by which process (one is tempted almost to call it a kind of respiration) the ocean becomes purified.

December 23rd.—We arrived at Port Desire, situated in lat. 47°, on the coast of Patagonia. The creek runs for about twenty miles inland, with an irregular width. The Beagle anchored a few miles within the entrance, in front of the ruins of an old Spanish settlement.

The same evening I went on shore. The first landing in any new country is very interesting, and especially when, as in this case, the whole aspect bears the stamp of a marked and individual character. At the height of between two and three hundred feet above some masses of porphyry a wide plain extends, which is truly characteristic of Patagonia. The surface is quite level, and is composed of well-rounded shingle mixed with a whitish earth. Here and there scattered tufts of brown wiry grass are supported, and still more rarely, some low thorny bushes. The weather is dry and pleasant, and the fine blue sky is but seldom obscured. When standing in the middle of one of these desert plains and looking towards the interior, the view is generally bounded by the escarpment of another plain, rather higher, but equally level and desolate; and in every other direction the horizon is indistinct from the trembling mirage which seems to rise from the heated surface.

In such a country the fate of the Spanish settlement was soon decided; the dryness of the climate during the greater part of the year, and the occasional hostile attacks of the wandering Indians, compelled the colonists to desert their half-finished buildings. The style, however, in which they were commenced shows the strong and liberal hand of Spain in the old time. The result of all the attempts to colonize this side of America south of 41°, has been miserable. Port Famine expresses by its name the lingering and extreme sufferings of several hundred wretched people, of whom one alone survived to relate their misfortunes. At St. Joseph's Bay, on the coast of Patagonia, a small settlement was made; but during one Sunday the Indians made an attack and massacred the whole party, excepting two men, who remained captives during many years. At the Rio Negro I conversed with one of these men, now in extreme old age.

The zoology of Patagonia is as limited as its flora.[90] On the arid plains a few black beetles (Heteromera) might be seen slowly crawling about, and occasionally a lizard darted from side to side. Of birds we have three carrion hawks, and in the valleys a few finches and insect-feeders. An ibis (Theristicus melanops—a species said to be found in central Africa) is not uncommon on the most desert parts: in their stomachs I found grasshoppers, cicadæ, small lizards, and even scorpions.[91] At one time of the year these birds go in flocks, at another in pairs; their cry is very loud and singular, like the neighing of the guanaco.

The guanaco, or wild llama, is the characteristic quadruped of the plains of Patagonia; it is the South American representative of the camel of the East. It is an elegant animal in a state of nature, with a long slender neck and fine legs. It is very common over the whole of the temperate parts of the continent, as far south as the islands near Cape Horn. It generally lives in small herds of from half a dozen to thirty in each; but on the banks of the St. Cruz we saw one herd which must have contained at least five hundred.

They are generally wild and extremely wary. Mr. Stokes told me, that he one day saw through a glass a herd of these animals which evidently had been frightened, and were running away at full speed, although their distance was so great that he could not distinguish them with his naked eye. The sportsman frequently receives the first notice of their presence, by hearing from a long distance their peculiar shrill neighing note of alarm. If he then looks attentively, he will probably see the herd standing in a line on the side of some distant hill. On approaching nearer, a few more squeals are given, and off they set at an apparently slow, but really quick canter, along some narrow beaten track to a neighbouring hill. If, however, by chance he abruptly meets a single animal, or several together, they will generally stand motionless and intently gaze at him; then perhaps move on a few yards, turn round, and look again. What is the cause of this difference in their shyness? Do they mistake a man in the distance for their chief enemy the puma? Or does curiosity overcome their timidity? That they are curious is certain; for if a person lies on the ground, and plays strange antics, such as throwing up his feet in the air, they will almost always approach by degrees to reconnoitre him. It was an artifice that was repeatedly practised by our sportsmen with success, and it had moreover the advantage of allowing several shots to be fired, which were all taken as parts of the performance. On the mountains of Tierra del Fuego, I have more than once seen a guanaco, on being approached, not only neigh and squeal, but prance and leap about in the most ridiculous manner, apparently in defiance as a challenge. These animals are very easily domesticated, and I have seen some thus kept in northern Patagonia near a house, though not under any restraint. They are in this state very bold, and readily attack a man by striking him from behind with both knees. It is asserted that the motive for these attacks is jealousy on account of their females. The wild guanacos, however, have no idea of defence; even a single dog will secure one of these large animals, till the huntsman can come up. In many of their habits they are like sheep in a flock. Thus when they see men approaching in several directions on horseback, they soon become bewildered, and know not which way to run. This greatly facilitates the Indian method of hunting, for they are thus easily driven to a central point, and are encompassed.

The guanacos readily take to the water: several times at Port Valdes they were seen swimming from island to island. Byron, in his voyage, says he saw them drinking salt water. Some of our officers likewise saw a herd apparently drinking the briny fluid from a salina near Cape Blanco. I imagine in several parts of the country, if they do not drink salt water, they drink none at all. In the middle of the day they frequently roll in the dust, in saucer-shaped hollows. The males fight together; two one day passed quite close to me, squealing and trying to bite each other; and several were shot with their hides deeply scored. Herds sometimes appear to set out on exploring parties: at Bahia Blanca, where, within thirty miles of the coast, these animals are extremely unfrequent, I one day saw the tracks of thirty or forty, which had come in a direct line to a muddy salt-water creek. They then must have perceived that they were approaching the sea, for they had wheeled with the regularity of cavalry, and had returned back in as straight a line as they had advanced. The guanacos have one singular habit, which is to me quite inexplicable; namely, that on successive days they drop their dung in the same defined heap. I saw one of these heaps which was eight feet in diameter, and was composed of a large quantity. This habit, according to M. A. d'Orbigny, is common to all the species of the genus; it is very useful to the Peruvian Indians, who use the dung for fuel, and are thus saved the trouble of collecting it.

The guanacos appear to have favourite spots for lying down to die. On the banks of the St. Cruz, in certain circumscribed spaces, which were generally bushy and all near the river, the ground was actually white with bones. On one such spot I counted between ten and twenty heads. I particularly examined the bones; they did not appear, as some scattered ones which I had seen, gnawed or broken, as if dragged together by beasts of prey. The animals in most cases must have crawled, before dying, beneath and amongst the bushes. Mr. Bynoe informs me that during a former voyage he observed the same circumstance on the banks of the Rio Gallegos. I do not at all understand the reason of this, but I may observe, that the wounded guanacos at the St. Cruz invariably walked towards the river. At St. Jago in the Cape de Verd Islands, I remember having seen in a ravine a retired corner covered with bones of the goat; we at the time exclaimed that it was the burial ground of all the goats in the island. I mention these trifling circumstances, because in certain cases they might explain the occurrence of a number of uninjured bones in a cave, or buried under alluvial accumulations; and likewise the cause why certain animals are more commonly embedded than others in sedimentary deposits.

One day the yawl was sent under the command of Mr. Chaffers with three days' provisions to survey the upper part of the harbour. In the morning we searched for some watering-places mentioned in an old Spanish chart. We found one creek, at the head of which there was a trickling rill (the first we had seen) of brackish water. Here the tide compelled us to wait several hours; and in the interval I walked some miles into the interior. The plain as usual consisted of gravel, mingled with soil resembling chalk in appearance, but very different from it in nature. From the softness of these materials it was worn into many gulleys. There was not a tree, and, excepting the guanaco, which stood on the hill-top a watchful sentinel over its herd, scarcely an animal or a bird. All was stillness and desolation. Yet in passing over these scenes, without one bright object near, an ill-defined but strong sense of pleasure is vividly excited. One asked how many ages the plain had thus lasted, and how many more it was doomed thus to continue.

> "None can reply—all seems eternal now.
> The wilderness has a mysterious tongue,
> Which teaches awful doubt."[92]

In the evening we sailed a few miles further up, and then pitched the tents for the night. By the middle of the next day the yawl was aground, and from the shoalness of the water could not proceed any higher. The water being found partly fresh, Mr.Chaffers took the dingey and went up two or three miles further, where she also grounded, but in a fresh-water river. The water was muddy, and though the stream was most insignificant in size, it would be difficult to account for its origin, except from the melting snow on the Cordillera. At the spot where we bivouacked, we were surrounded by bold cliffs and steep pinnacles of porphyry. I do not think I ever saw a spot which appeared more secluded from the rest of the world, than this rocky crevice in the wide plain.

The second day after our return to the anchorage, a party of officers and myself went to ransack an old Indian grave, which I had found on the summit of a neighbouring hill. Two immense stones, each probably weighing at least a couple of tons, had been placed in front of a ledge of rock about six feet high. At the bottom of the grave on the hard rock there was a layer of earth about a foot deep, which must have been brought up from the plain below. Above it a pavement of flat stones was placed, on which others were piled, so as to fill up the space between the ledge and the two great blocks. To complete the grave, the Indians had contrived to detach from the ledge a huge fragment, and to throw it over the pile so as to rest on the two blocks. We undermined the grave on both sides, but could not find any relics, or even bones. The latter probably had decayed long since (in which case the grave must have been of extreme antiquity), for I found in another place some smaller heaps, beneath which a very few crumbling fragments could yet be distinguished as having belonged to a man. Falconer states, that where an Indian dies he is buried, but that subsequently his bones are carefully taken up and carried, let the distance be ever so great, to be deposited near the sea-coast. This custom, I think, may be accounted for by recollecting, that before the introduction of horses, these Indians must have led nearly the same life as the Fuegians now do, and therefore generally have resided in the neighbourhood of the sea. The common prejudice of lying where one's ancestors have lain, would make the now roaming Indians bring the less perishable part of their dead to their ancient burial-ground on the coast.

January 9th, 1834.—Before it was dark the Beagle anchored in the fine spacious harbour of Port St. Julian, situated about one hundred and ten miles to the south of Port Desire. We remained here eight days. The country is nearly similar to that of Port Desire, but perhaps rather more sterile. One day a party accompanied Captain Fitz Roy on a long walk round the head of the harbour. We were eleven hours without tasting any water, and some of the party were quite exhausted. From the summit of a hill (since well named Thirsty Hill) a fine lake was spied, and two of the party proceeded with concerted signals to show whether it was fresh water. What was our disappointment to find a snow-white expanse of salt, crystallized in great cubes! We attributed our extreme thirst to the dryness of the atmosphere; but whatever the cause might be, we were exceedingly glad late in the evening to get back to the boats. Although we could nowhere find, during our whole visit, a single drop of fresh water, yet some must exist; for by an odd chance I found on the surface of the salt water, near the head of the bay, a Colymbetes not quite dead, which must have lived in some not far distant pool. Three other insects (a Cincindela, like *hybrida,* a Cymindis, and a Harpalus, which all live on muddy flats occasionally overflowed by the sea), and one other found dead on the plain, complete the list of the beetles. A good-sized fly (Tabanus) was extremely numerous, and tormented us by its painful bite. The common horsefly, which is so troublesome in the shady lanes of England, belongs to this same genus. We here have the puzzle that so frequently occurs in the case of musquitoes—on the blood of what animals do these insects commonly feed? The guanaco is nearly the only warm-blooded quadruped, and it is found in quite inconsiderable numbers compared with the multitude of flies.

THE GEOLOGY OF Patagonia is interesting. Differently from Europe, where the tertiary formations appear to have accumulated in bays, here along hundreds of miles of coast we have one great deposit, including many tertiary shells, all apparently extinct. The most common shell is a massive gigantic oyster, sometimes even a foot in diameter. These beds are covered by others of a peculiar soft white stone, including much gypsum, and resembling chalk, but really of a pumiceous nature. It is highly remarkable, from being composed, to at least one-tenth of its bulk, of Infusoria. Professor Ehrenberg has already ascertained in it thirty oceanic forms. This bed extends for 500 miles along the coast, and probably for a considerably greater distance. At Port St. Julian its thickness is more than 800 feet! These white beds are everywhere capped by a mass of gravel, forming probably one of the largest beds of shingle in the world: it certainly extends from near the Rio Colorado to between 600 and 700 nautical miles southward; at Santa Cruz (a river a little south of St. Julian), it reaches to the foot of the Cordillera; half way up the river, its thickness is more than 200 feet; it probably everywhere extends to this great chain, whence the well-rounded pebbles of porphyry have been derived: we may consider its average breadth as 200 miles, and its average thickness as about 50 feet. If this great bed of pebbles, without including the mud necessarily derived from their attrition, was piled into a mound, it would form a great mountain chain! When we consider that all these pebbles, countless as the grains of sand in the desert, have been derived from the slow falling of masses of rock on the old coast-lines and banks of rivers; and that these fragments have been dashed into smaller pieces, and that each of them has since been slowly rolled, rounded, and far transported, the mind is stupefied in thinking over the long, absolutely necessary, lapse of years. Yet all this gravel has been transported, and probably rounded, subsequently to the deposition of the white beds, and long subsequently to the underlying beds with the tertiary shells.

Everything in this southern continent has been effected on a grand scale: the land, from the Rio Plata to Tierra del Fuego, a distance of 1200 miles, has been raised in mass (and in Patagonia to a height of between 300 and 400 feet), within the period of the now existing sea-shells. The old and weathered shells left on the surface of the upraised plain still partially retain their colours. The uprising movement has been interrupted by at least eight long periods of rest, during which the sea ate deeply back into the land, forming at successive levels the long lines of cliffs, or escarpments, which separate the different plains as they rise like steps one behind the other. The elevatory movement, and the eating-back power of the sea during the periods of rest, have been equable over long lines of coast; for I was astonished to find that the step-like plains stand at nearly corresponding heights at far distant points. The lowest plain is 90 feet high; and the highest, which I ascended near the coast, is 950 feet; and of this, only relics are left in the form of flat gravel-capped hills. The upper plain of Santa Cruz slopes up to a height of 3000 feet at the foot of the Cordillera. I have said that within the period of existing sea-shells, Patagonia has been upraised 300 to 400 feet: I may add, that within the period when icebergs transported boulders over the upper plain of Santa Cruz, the elevation has been at least 1500 feet. Nor has Patagonia been affected only by upward movements: the extinct tertiary shells from Port St. Julian and Santa Cruz cannot have lived, according to Professor E. Forbes, in a greater depth of water than from 40 to 250 feet; but they are now covered with sea-deposited strata from 800 to 1000 feet in thickness: hence the bed of the sea, on which these shells once lived, must have sunk downwards several hundred feet, to allow of the accumulation of the superincumbent strata. What a history of geological changes does the simply-constructed coast of Patagonia reveal!

At Port St. Julian,[93] in some red mud capping the gravel on the 90-feet plain, I found half the skeleton of the Macrauchenia Patachonica, a remarkable quadruped, full as large as a camel. It belongs to the same division of the Pachydermata with the rhinoceros, tapir, and palæotherium; but in the structure of the bones of its long neck it shows a clear relation to the camel, or rather to the guanaco and llama. From recent sea-shells being found on two of the higher step-formed plains, which must have been modelled and upraised before the mud was deposited in which the Macrauchenia was entombed, it is certain that this curious quadruped lived long after the sea was inhabited by its present shells. I was at first much surprised how a large quadruped could so lately have subsisted, in lat. 49° 15', on these wretched gravel plains, with their stunted vegetation; but the relationship of the Macrauchenia to the Guanaco, now an inhabitant of the most sterile parts, partly explains this difficulty.

The relationship, though distant, between the Macrauchenia and the Guanaco, between the Toxodon and the Capybara,—the closer relationship between the many extinct Edentata and the living sloths, ant-eaters, and armadillos, now so eminently characteristic of South American zoology,—and the still closer relationship between the fossil and living species of Ctenomys and Hydrochærus, are most interesting facts. This relationship is shown wonderfully—as wonderfully as between the fossil and extinct Marsupial animals of Australia—by the great collection lately brought to Europe from the caves of Brazil by MM. Lund and Clausen. In this collection there are extinct species of all the thirty-two genera, excepting four, of the terrestrial quadrupeds now inhabiting the provinces in which the caves occur; and the extinct species are much more numerous than those now living: there are fossil ant-eaters, armadillos, tapirs, peccaries, guanacos, opossums, and numerous South American gnawers and monkeys, and other animals. This wonderful relationship in the same continent between the dead and the living, will, I do not doubt, hereafter throw more light on the appearance of organic beings on our earth, and their disappearance from it, than any other class of facts.

It is impossible to reflect on the changed state of the American continent without the deepest astonishment. Formerly it must have swarmed with great monsters: now we find mere pigmies, compared with the antecedent, allied races. If Buffon had known of the gigantic sloth and armadillo-like animals, and of the lost Pachydermata, he might have said with a greater semblance of truth that the creative force in America had lost its power, rather than that it had never possessed great vigour. The greater number, if not all, of these extinct quadrupeds lived at a late period, and were the contemporaries of most of the existing sea-shells. Since they lived, no very great change in the form of the land can have taken place. What, then, has exterminated so many species and whole genera? The mind at first is irresistibly hurried into the belief of some great catastrophe; but thus to destroy animals, both large and small, in Southern Patagonia, in Brazil, on the Cordillera of Peru, in North America up to Behring's Straits, we must shake the entire framework of the globe. An examination, moreover, of the geology of La Plata and Patagonia, leads to the belief that all the features of the land result from slow and gradual changes. It appears from the character of the fossils in Europe, Asia, Australia, and in North and South America, that those conditions which favour the life of the *larger* quadrupeds were lately co-extensive with the world: what those conditions were, no one has yet even conjectured. It could hardly have been a change of temperature, which at about the same time destroyed the inhabitants of tropical, temperate, and arctic latitudes on both sides of the globe. In North America we positively know from Mr. Lyell, that the large quadrupeds lived subsequently to that period, when boulders were brought into latitudes at which icebergs now never arrive: from conclusive but indirect reasons we may feel sure, that in the southern hemisphere the Macrauchenia, also, lived long subsequently to the ice-transporting boulder-period. Did man, after his first inroad into South America, destroy, as has been suggested, the unwieldy Megatherium and the other Edentata? We must at least look to some other cause for the destruction of the little tucutuco at Bahia Blanca, and of the many fossil mice and other small quadrupeds in Brazil. No one will imagine that a drought, even far severer than those which cause such losses in the provinces of La Plata, could destroy every individual of every species from Southern Patagonia to Behring's Straits. What shall we say of the extinction of the horse? Did those plains fail of pasture, which have since been overrun by thousands and hundreds of thousands of the descendants of the stock introduced by the Spaniards? Have the subsequently introduced species consumed the food of the great antecedent races? Can we believe that the Capybara has taken the food of the Toxodon, the Guanaco of the Macrauchenia, the existing small Edentata of their numerous gigantic prototypes? Certainly, no fact in the long history of the world is so startling as the wide and repeated exterminations of its inhabitants.

Nevertheless, if we consider the subject under another point of view, it will appear less perplexing. We do not steadily bear in mind, how profoundly ignorant we are of the conditions of existence of every animal; nor do we always remember, that some check is constantly preventing the too rapid increase of every organized being left in a state of nature. The supply of food, on an average, remains constant; yet the tendency in every animal to increase by propagation is geometrical; and its surprising effects have nowhere been more astonishingly shown, than in the case of the European animals run wild during the last few centuries in America. Every animal in a state of nature regularly breeds; yet in a species long established, any *great* increase in numbers is obviously impossible, and must be checked by some means. We are, nevertheless, seldom able with certainty to tell in any given species, at what period of life, or at what period of the year, or whether only at long intervals, the check falls; or, again, what is the precise nature of the check. Hence probably it is, that we feel so little surprise at one, of two species closely allied in habits, being rare and the other abundant in the same district; or, again, that one should be abundant in one district, and another, filling the same place in the economy of nature, should be abundant in a neighbouring district, differing very little in its conditions. If asked how this is, one immediately replies that it is determined by some slight difference, in climate, food, or the number of enemies: yet how rarely, if ever, we can point out the precise cause and manner of action of the check! We are, therefore, driven to the conclusion, that causes generally quite inappreciable by us, determine whether a given species shall be abundant or scanty in numbers.

In the cases where we can trace the extinction of a species through man, either wholly or in one limited district, we know that it becomes rarer and rarer, and is then lost: it would be difficult to point out any just distinction[94] between a species destroyed by man or by the increase of its natural enemies. The evidence of rarity preceding extinction, is more striking in the successive tertiary strata, as remarked by several able observers; it has often been found that a shell very common in a tertiary stratum is now most rare, and has even long been thought to be extinct. If then, as appears probable, species first become rare and then extinct—if the too rapid increase of every species, even the most favoured, is steadily checked, as we must admit, though how and when it is hard to say—and if we see, without the smallest surprise, though unable to assign the precise reason, one species abundant and another closely allied species rare in the same district—why should we feel such great astonishment at the rarity being carried a step further to extinction? An action going on, on every side of us, and yet barely appreciable, might surely be carried a little further, without exciting our observation. Who would feel any great surprise at hearing that the Magalonyx was formerly rare compared with the Megatherium, or that one of the fossil monkeys was few in number compared with one of the now living monkeys? and yet in this comparative rarity, we should have the plainest evidence of less favourable conditions for their existence. To admit that species generally become rare before they become extinct—to feel no surprise at the comparative rarity of one species with another, and yet to call in some extraordinary agent and to marvel greatly when a species ceases to exist, appears to me much the same as to admit that sickness in the individual is the prelude to death—to feel no surprise at sickness—but when the sick man dies to wonder, and to believe that he died through violence.

CHAPTER IX

Santa Cruz, Patagonia, and the Falkland Islands

SANTA CRUZ • EXPEDITION UP THE RIVER • INDIANS • IMMENSE STREAMS OF BASALTIC LAVA • FRAGMENTS NOT TRANSPORTED BY THE RIVER • EXCAVATION OF THE VALLEY • CONDOR, HABITS OF • CORDILLERA • ERRATIC BOULDERS OF GREAT SIZE • INDIAN RELICS • RETURN TO THE SHIP • FALKLAND ISLANDS • WILD HORSES, CATTLE, RABBITS • WOLF-LIKE FOX • FIRE MADE OF BONES • MANNER OF HUNTING WILD CATTLE • GEOLOGY • STREAMS OF STONES • SCENES OF VIOLENCE • PENGUIN • GEESE • EGGS OF DORIS • COMPOUND ANIMALS

April 13th, 1834.—The Beagle anchored within the mouth of the Santa Cruz. This river is situated about sixty miles south of Port St. Julian. During the last voyage Captain Stokes proceeded thirty miles up it, but then, from the want of provisions, was obliged to return. Excepting what was discovered at that time, scarcely anything was known about this large river. Captain Fitz Roy now determined to follow its course as far as time would allow. On the 18th three whale-boats started, carrying three weeks' provisions; and the party consisted of twenty-five souls—a force which would have been sufficient to have defied a host of Indians. With a strong flood-tide and a fine day we made a good run, soon drank some of the fresh water, and were at night nearly above the tidal influence.

The river here assumed a size and appearance which, even at the highest point we ultimately reached, was scarcely diminished. It was generally from three to four hundred yards broad, and in the middle about seventeen feet deep. The rapidity of the current, which in its whole course runs at the rate of from four to six knots an hour, is perhaps its most remarkable feature. The water is of a fine blue colour, but with a slight milky tinge, and not so transparent as at first sight would have been expected. It flows over a bed of pebbles, like those which compose the beach and the surrounding plains. It runs in a winding course through a valley, which extends in a direct line westward. This valley varies from five to ten miles in breadth; it is bounded by step-formed terraces, which rise in most parts, one above the other, to the height of five hundred feet, and have on the opposite sides a remarkable correspondence.

April 19th.—Against so strong a current it was, of course, quite impossible to row or sail: consequently the three boats were fastened together head and stern, two hands left in each, and the rest came on shore to track. As the general arrangements made by Captain Fitz Roy were very good for facilitating the work of all, and as all had a share in it, I will describe the system. The party including every one, was divided into two spells, each of which hauled at the tracking line alternately for an hour and a half. The officers of each boat lived with, ate the same food, and slept in the same tent with their crew, so that each boat was quite independent of the others. After sunset the first level spot where any bushes were growing, was chosen for our night's lodging. Each of the crew took it in turns to be cook. Immediately the boat was hauled up, the cook made his fire; two others pitched the tent; the coxswain handed the things out of the boat; the rest carried them up to the tents and collected firewood. By this order, in half an hour everything was ready for the night. A watch of two men and an officer was always kept, whose duty it was to look after the boats, keep up the fire, and guard against Indians. Each in the party had his one hour every night.

During this day we tracked but a short distance, for there were many islets, covered by thorny bushes, and the channels between them were shallow.

April 20th.—We passed the islands and set to work. Our regular day's march, although it was hard enough, carried us on an average only ten miles in a straight line, and perhaps fifteen or twenty altogether. Beyond the place where we slept last night, the country is completely *terra incognita,* for it was there that Captain Stokes turned back. We saw in the distance a great smoke, and found the skeleton of a horse, so we knew that Indians were in the neighbourhood. On the next morning (21st) tracks of a party of horse, and marks left by the trailing of the chuzos, or long spears, were observed on the ground. It was generally thought that the Indians had reconnoitred us during the night. Shortly afterwards we came to a spot where, from the fresh footsteps of men, children, and horses, it was evident that the party had crossed the river.

April 22nd.—The country remained the same, and was extremely uninteresting. The complete similarity of the productions throughout Patagonia is one of its most striking characters. The level plains of arid shingle support the same stunted and dwarf plants; and in the valleys the same thorn-bearing bushes grow. Everywhere we see the same birds and insects. Even the very banks of the river and of the clear streamlets which entered it, were scarcely enlivened by a brighter tint of green. The curse of sterility is on the land, and the water flowing over a bed of pebbles partakes of the same curse. Hence the number of waterfowl is very scanty; for there is nothing to support life in the stream of this barren river.

Patagonia, poor as she is in some respects, can however boast of a greater stock of small rodents[95] than perhaps any other country in the world. Several species of mice are externally characterized by large thin ears and a very fine fur. These little animals swarm amongst the thickets in the valleys, where they cannot for months together taste a drop of water excepting the dew. They all seem to be cannibals; for no sooner was a mouse caught in one of my traps than it was devoured by others. A small and delicately shaped fox, which is likewise very abundant, probably derives its entire support from these small animals. The guanaco is also in his proper district; herds of fifty or a hundred were common; and, as I have stated, we saw one which must have contained at least five hundred. The puma, with the condor and other carrion-hawks in its train, follows and preys upon these animals. The footsteps of the puma were to be seen almost everywhere on the banks of the river; and the remains of several guanacos, with their necks dislocated and bones broken, showed how they had met their death.

April 24th.—Like the navigators of old when approaching an unknown land, we examined and watched for the most trivial sign of a change. The drifted trunk of a tree, or a boulder of primitive rock, was hailed with joy, as if we had seen a forest growing on the flanks of the Cordillera. The top, however, of a heavy bank of clouds, which remained almost constantly in one position, was the most promising sign, and eventually turned out a true harbinger. At first the clouds were mistaken for the mountains themselves, instead of the masses of vapour condensed by their icy summits.

April 26th.—We this day met with a marked change in the geological structure of the plains. From the first starting I had carefully examined the gravel in the river, and for the two last days had noticed the presence of a few small pebbles of a very cellular basalt. These gradually increased in number and in size, but none were as large as a man's head. This morning, however, pebbles of the same rock, but more compact, suddenly became abundant, and in the course of half an hour we saw, at the distance of five or six miles, the angular edge of a great basaltic platform. When we arrived at its base we found the stream bubbling among the fallen blocks. For the next twenty-eight miles the river-course was encumbered with these basaltic masses. Above that limit immense fragments of primitive rocks, derived from its surrounding boulder-formation, were equally numerous. None of the fragments of any considerable size had been washed more than three or four miles down the river below their parent-source: considering the singular rapidity of the great body of water in the Santa Cruz, and that no still reaches occur in any part, this example is a most striking one, of the inefficiency of rivers in transporting even moderately-sized fragments.

The basalt is only lava, which has flowed beneath the sea; but the eruptions must have been on the grandest scale. At the point where we first met this formation it was 120 feet in thickness; following up the river course, the surface imperceptibly rose and the mass became thicker, so that at forty miles above the first station it was 320 feet thick. What the thickness may be close to the Cordillera, I have no means of knowing, but the platform there attains a height of about three thousand feet above the level of the sea: we must therefore look to the mountains of that great chain for its source; and worthy of such a source are streams that have flowed over the gently inclined bed of the sea to a distance of one hundred miles. At the first glance of the basaltic cliffs on the opposite sides of the valley, it was evident that the strata once were united. What power, then, has removed along a whole line of country, a solid mass of very hard rock, which had an average thickness of nearly three hundred feet, and a breadth varying from rather less than two miles to four miles? The river, though it has so little power in transporting even inconsiderable fragments, yet in the lapse of ages might produce by its gradual erosion an effect of which it is difficult to judge the amount. But in this case, independently of the insignificance of such an agency, good reasons can be assigned for believing that this valley was formerly occupied by an arm of the sea. It is needless in this work to detail the arguments leading to this conclusion, derived from the form and the nature of the step-formed terraces on both sides of the valley, from the manner in which the bottom of the valley near the Andes expands into a great estuary-like plain with sand-hillocks on it, and from the occurrence of a few sea-shells lying in the bed of the river. If I had space I could prove that South America was formerly here cut off by a strait, joining the Atlantic and Pacific oceans, like that of Magellan. But it may yet be asked, how has the solid basalt been moved? Geologists formerly would have brought into play, the violent action of some overwhelming debacle; but in this case such a supposition would have been quite inadmissible; because, the same step-like plains with existing sea-shells lying on their surface, which front the long line of the Patagonian coast, sweep up on each side of the valley of Santa Cruz. No possible action of any flood could thus have modelled the land, either within the valley or along the open coast; and by the formation of such step-like plains or terraces the valley itself had been hollowed out. Although we know that there are tides, which run within the Narrows of the Strait of Magellan at the rate of eight knots an hour, yet we must confess that it makes the head almost giddy to reflect on the number of years, century after century, which the tides, unaided by a heavy surf, must have required to have corroded so vast an area and thickness of solid basaltic lava. Nevertheless, we must believe that the strata undermined by the waters of this ancient strait, were broken up into huge fragments, and these lying scattered on the beach, were reduced first to smaller blocks, then to pebbles and lastly to the most impalpable mud, which the tides drifted far into the Eastern or Western Ocean.

With the change in the geological structure of the plains the character of the landscape likewise altered. While rambling up some of the narrow and rocky defiles, I could almost have fancied myself transported back again to the barren valleys of the island of St. Jago. Among the basaltic cliffs, I found some plants which I had seen nowhere else, but others I recognised as being wanderers from Tierra del Fuego. These porous rocks serve as a reservoir for the scanty rain-water; and consequently on the line where the igneous and sedimentary formations unite, some small springs (most rare occurrences in Patagonia) burst forth; and they could be distinguished at a distance by the circumscribed patches of bright green herbage.

April 27th.—The bed of the river became rather narrower, and hence the stream more rapid. It here ran at the rate of six knots an hour. From this cause, and from the many great angular fragments, tracking the boats became both dangerous and laborious.

THIS DAY I shot a condor. It measured from tip to tip of the wings, eight and a half feet, and from beak to tail, four feet. This bird is known to have a wide geographical range, being found on the west coast of South America, from the Strait of Magellan along the Cordillera as far as eight degrees north of the equator. The steep cliff near the mouth of the Rio Negro is its northern limit on the Patagonian coast; and they have there wandered about four hundred miles from the great central line of their habitation in the Andes. Further south, among the bold precipices at the head of Port Desire, the condor is not uncommon; yet only a few stragglers occasionally visit the sea-coast. A line of cliff near the mouth of the Santa Cruz is frequented by these birds, and about eighty miles up the river, where the sides of the valley are formed by steep basaltic precipices, the condor reappears. From these facts, it seems that the condors require perpendicular cliffs. In Chile, they haunt, during the greater part of the year, the lower country near the shores of the Pacific, and at night several roost together in one tree; but in the early part of summer, they retire to the most inaccessible parts of the inner Cordillera, there to breed in peace.

With respect to their propagation, I was told by the country people in Chile, that the condor makes no sort of nest, but in the months of November and December lays two large white eggs on a shelf of bare rock. It is said that the young condors cannot fly for an entire year; and long after they are able, they continue to roost by night, and hunt by day with their parents. The old birds generally live in pairs; but among the inland basaltic cliffs of the Santa Cruz, I found a spot, where scores must usually haunt. On coming suddenly to the brow of the precipice, it was a grand spectacle to see between twenty and thirty of these great birds start heavily from their resting-place, and wheel away in majestic circles. From the quantity of dung on the rocks, they must long have frequented this cliff for roosting and breeding. Having gorged themselves with carrion on the plains below, they retire to these favourite ledges to digest their food. From these facts, the condor, like the gallinazo, must to a certain degree be considered as a gregarious bird. In this part of the country they live altogether on the guanacos which have died a natural death, or as more commonly happens, have been killed by the pumas. I believe, from what I saw in Patagonia, that they do not on ordinary occasions extend their daily excursions to any great distance from their regular sleeping-places.

The condors may oftentimes be seen at a great height, soaring over a certain spot in the most graceful circles. On some occasions I am sure that they do this only for pleasure, but on others, the Chileno countryman tells you that they are watching a dying animal, or the puma devouring its prey. If the condors glide down, and then suddenly all rise together, the Chileno knows that it is the puma which, watching the carcass, has sprung out to drive away the robbers. Besides feeding on carrion, the condors frequently attack young goats and lambs; and the shepherd-dogs are trained, whenever they pass over, to run out, and looking upwards to bark violently. The Chilenos destroy and catch numbers. Two methods are used; one is to place a carcass on a level piece of ground within an enclosure of sticks with an opening, and when the condors are gorged, to gallop up on horseback to the entrance, and thus enclose them: for when this bird has not space to run, it cannot give its body sufficient momentum to rise from the ground. The second method is to mark the trees in which, frequently to the number of five or six together, they roost, and then at night to climb up and noose them. They are such heavy sleepers, as I have myself witnessed, that this is not a difficult task. At Valparaiso, I have seen a living condor sold for sixpence, but the common price is eight or ten shillings. One which I saw brought in, had been tied with rope, and was much injured; yet, the moment the line was cut by which its bill was secured, although surrounded by people, it began ravenously to tear a piece of carrion. In a garden at the same place, between twenty and thirty were kept alive. They were fed only once a week, but they appeared in pretty good health.[96] The Chileno countrymen assert that the condor will live, and retain its vigour, between five and six weeks without eating: I cannot answer for the truth of this, but it is a cruel experiment, which very likely has been tried.

When an animal is killed in the country, it is well known that the condors, like other carrion-vultures, soon gain intelligence of it, and congregate in an inexplicable manner. In most cases it must not be overlooked, that the birds have discovered their prey, and have picked the skeleton clean, before the flesh is in the least degree tainted. Remembering the experiments of M. Audubon, on the little smelling powers of carrion-hawks, I tried in the above-mentioned garden the following experiment: the condors were tied, each by a rope, in a long row at the bottom of a wall; and having folded up a piece of meat in white paper, I walked backwards and forwards, carrying it in my hand at the distance of about three yards from them, but no notice whatever was taken. I then threw it on the ground, within one yard of an old male bird; he looked at it for a moment with attention, but then regarded it no more. With a stick I pushed it closer and closer, until at last he touched it with his beak; the paper was then instantly torn off with fury, and at the same moment, every bird in the long row began struggling and flapping its wings. Under the same circumstances, it would have been quite impossible to have deceived a dog. The evidence in favour of and against the acute smelling powers of carrion-vultures is singularly balanced. Professor Owen has demonstrated that the olfactory nerves of the turkey-buzzard (Cathartes aura) are highly developed; and on the evening when Mr. Owen's paper was read at the Zoological Society, it was mentioned by a gentleman that he had seen the carrion-hawks in the West Indies on two occasions collect on the roof of a house, when a corpse had become offensive from not having been buried; in this case, the intelligence could hardly have been acquired by sight. On the other hand, besides the experiments of Audubon and that one by myself, Mr. Bachman has tried in the United States many varied plans, showing that neither the turkey-buzzard (the species dissected by Professor Owen) nor the gallinazo find their food by smell. He covered portions of highly-offensive offal with a thin canvas cloth, and strewed pieces of meat on it: these the carrion-vultures ate up, and then remained quietly standing, with their beaks within the eighth of an inch of the putrid mass, without discovering it. A small rent was made in the canvas, and the offal was immediately discovered; the canvas was replaced by a fresh piece, and meat again put on it, and was again devoured by the vultures without their discovering the hidden mass on which they were trampling. These facts are attested by the signatures of six gentlemen, besides that of Mr. Bachman.[97]

Often when lying down to rest on the open plains, on looking upwards, I have seen carrion-hawks sailing through the air at a great height. Where the country is level I do not believe a space of the heavens, of more than fifteen degrees above the horizon, is commonly viewed with any attention by a person either walking or on horseback. If such be the case, and the vulture is on the wing at a height of between three and four thousand feet, before it could come within the range of vision, its distance in a straight line from the beholder's eye, would be rather more than two British miles. Might it not thus readily be overlooked? When an animal is killed by the sportsman in a lonely valley, may he not all the while be watched from above by the sharp-sighted bird? And will not the manner of its descent proclaim throughout the district to the whole family of carrion-feeders, that their prey is at hand?

When the condors are wheeling in a flock round and round any spot, their flight is beautiful. Except when rising from the ground, I do not recollect ever having seen one of these birds flap its wings. Near Lima, I watched several for nearly half an hour, without once taking off my eyes: they moved in large curves, sweeping in circles, descending and ascending without giving a single flap. As they glided close over my head, I intently watched from an oblique position, the outlines of the separate and great terminal feathers of each wing; and these separate feathers, if there had been the least vibratory movement, would have appeared as if blended together; but they were seen distinct against the blue sky. The head and neck were moved frequently, and apparently with force; and the extended wings seemed to form the fulcrum on which the movements of the neck, body, and tail acted. If the bird wished to descend, the wings were for a moment collapsed; and when again expanded with an altered inclination, the momentum gained by the rapid descent seemed to urge the bird upwards with the even and steady movement of a paper kite. In the case of any bird *soaring*, its motion must be sufficiently rapid so that the action of the inclined surface of its body on the atmosphere may counterbalance its gravity. The force to keep up the momentum of a body moving in a horizontal plane in the air (in which there is so little friction) cannot be great, and this force is all that is wanted. The movement of the neck and body of the condor, we must suppose, is sufficient for this. However this may be, it is truly wonderful and beautiful to see so great a bird, hour after hour, without any apparent exertion, wheeling and gliding over mountain and river.

April 29th.—From some high land we hailed with joy the white summits of the Cordillera, as they were seen occasionally peeping through their dusky envelope of clouds. During the few succeeding days we continued to get on slowly, for we found the river-course very tortuous, and strewed with immense fragments of various ancient slaty rocks, and of granite. The plain bordering the valley had here attained an elevation of about 1100 feet above the river, and its character was much altered. The well-rounded pebbles of porphyry were mingled with many immense angular fragments of basalt and of primary rocks. The first of these erratic boulders which I noticed, was sixty-seven miles distant from the nearest mountain; another which I measured was five yards square, and projected five feet above the gravel. Its edges were so angular, and its size so great, that I at first mistook it for a rock *in situ,* and took out my compass to observe the direction of its cleavage. The plain here was not quite so level as that nearer the coast, but yet it betrayed no signs of any great violence. Under these circumstances it is, I believe, quite impossible to explain the transportal of these gigantic masses of rock so many miles from their parent-source, on any theory except by that of floating icebergs.

During the two last days we met with signs of horses, and with several small articles which had belonged to the Indians—such as parts of a mantle and a bunch of ostrich feathers—but they appeared to have been lying long on the ground. Between the place where the Indians had so lately crossed the river and this neighbourhood, though so many miles apart, the country appears to be quite unfrequented. At first, considering the abundance of the guanacos, I was surprised at this; but it is explained by the stony nature of the plains, which would soon disable an unshod horse from taking part in the chase. Nevertheless, in two places in this very central region, I found small heaps of stones, which I do not think could have been accidentally thrown together. They were placed on points, projecting over the edge of the highest lava cliff, and they resembled, but on a small scale, those near Port Desire.

May 4th.—Captain Fitz Roy determined to take the boats no higher. The river had a winding course, and was very rapid; and the appearance of the country offered no temptation to proceed any further. Everywhere we met with the same productions, and the same dreary landscape. We were now one hundred and forty miles distant from the Atlantic, and about sixty from the nearest arm of the Pacific. The valley in this upper part expanded into a wide basin, bounded on the north and south by the basaltic platforms, and fronted by the long range of the snow-clad Cordillera. But we viewed these grand mountains with regret, for we were obliged to imagine their nature and productions, instead of standing, as we had hoped, on their summits.

Besides the useless loss of time which an attempt to ascend the river any higher would have cost us, we had already been for some days on half allowance of bread. This, although really enough for reasonable men, was, after a hard day's march, rather scanty food: a light stomach and an easy digestion are good things to talk about, but very unpleasant in practice.

5th.—Before sunrise we commenced our descent. We shot down the stream with great rapidity, generally at the rate of ten knots an hour. In this one day we effected what had cost us five-and-a-half hard days' labour in ascending. On the 8th, we reached the Beagle after our twenty-one days' expedition. Every one, excepting myself, had cause to be dissatisfied; but to me the ascent afforded a most interesting section of the great tertiary formation of Patagonia.

ON MARCH 1ST, 1833, and again on March 16th, 1834, the Beagle anchored in Berkeley Sound, in East Falkland Island. This archipelago is situated in nearly the same latitude with the mouth of the Strait of Magellan; it covers a space of one hundred and twenty by sixty geographical miles, and is a little more than half the size of Ireland. After the possession of these miserable islands had been contested by France, Spain, and England, they were left uninhabited. The government of Buenos Ayres then sold them to a private individual, but likewise used them, as old Spain had done before, for a penal settlement. England claimed her right and seized them. The Englishman who was left in charge of the flag was consequently murdered. A British officer was next sent, unsupported by any power: and when we arrived, we found him in charge of a population, of which rather more than half were runaway rebels and murderers.

The theatre is worthy of the scenes acted on it. An undulating land, with a desolate and wretched aspect, is everywhere covered by a peaty soil and wiry grass, of one monotonous brown colour. Here and there a peak or ridge of grey quartz rock breaks through the smooth surface. Every one has heard of the climate of these regions; it may be compared to that which is experienced at the height of between one and two thousand feet, on the mountains of North Wales; having however less sunshine and less frost, but more wind and rain.[98]

16th.—I will now describe a short excursion which I made round a part of this island. In the morning I started with six horses and two Gauchos: the latter were capital men for the purpose, and well accustomed to living on their own resources. The weather was very boisterous and cold, with heavy hail-storms. We got on, however, pretty well, but, except the geology, nothing could be less interesting than our day's ride. The country is uniformly the same undulating moorland; the surface being covered by light brown withered grass and a few very small shrubs, all springing out of an elastic peaty soil. In the valleys here and there might be seen a small flock of wild geese, and everywhere the ground was so soft that the snipe were able to feed. Besides these two birds there were few others. There is one main range of hills, nearly two thousand feet in height, and composed of quartz rock, the rugged and barren crests of which gave us some trouble to cross. On the south side we came to the best country for wild cattle; we met, however, no great number, for they had been lately much harassed.

In the evening we came across a small herd. One of my companions, St. Jago by name, soon separated a fat cow; he threw the bolas, and it struck her legs, but failed in becoming entangled. Then dropping his hat to mark the spot where the balls were left, while at full gallop, he uncoiled his lazo, and after a most severe chase, again came up to the cow, and caught her round the horns. The other Gaucho had gone on ahead with the spare horses, so that St. Jago had some difficulty in killing the furious beast. He managed to get her on a level piece of ground, by taking advantage of her as often as she rushed at him; and when she would not move, my horse, from having been trained, would canter up, and with his chest give her a violent push. But when on level ground it does not appear an easy job for one man to kill a beast mad with terror. Nor would it be so, if the horse, when left to itself without its rider, did not soon learn, for its own safety, to keep the lazo tight; so that, if the cow or ox moves forward, the horse moves just as quickly forward; otherwise, it stands motionless leaning on one side. This horse, however, was a young one, and would not stand still, but gave in to the cow as she struggled. It was admirable to see with what dexterity St. Jago dodged behind the beast, till at last he contrived to give the fatal touch to the main tendon of the hind leg; after which, without much difficulty, he drove his knife into the head of the spinal marrow, and the cow dropped as if struck by lightning. He cut off pieces of flesh with the skin to it, but without any bones, sufficient for our expedition. We then rode on to our sleeping-place, and had for supper "carne con cuero," or meat roasted with the skin on it. This is as superior to common beef as venison is to mutton. A large circular piece taken from the back is roasted on the embers with the hide downwards and in the form of a saucer, so that none of the gravy is lost. If any worthy alderman had supped with us that evening, "carne con cuero," without doubt, would soon have been celebrated in London.

During the night it rained, and the next day (17th) was very stormy, with much hail and snow. We rode across the island to the neck of land which joins the Rincon del Toro (the great peninsula at the S. W. extremity) to the rest of the island. From the great number of cows which have been killed, there is a large proportion of bulls. These wander about single, or two and three together, and are very savage. I never saw such magnificent beasts; they equalled in the size of their huge heads and necks the Grecian marble sculptures. Capt. Sulivan informs me that the hide of an average-sized bull weighs forty-seven pounds, whereas a hide of this weight, less thoroughly dried, is considered as a very heavy one at Monte Video. The young bulls generally run away, for a short distance; but the old ones do not stir a step, except to rush at man and horse; and many horses have been thus killed. An old bull crossed a boggy stream, and took his stand on the opposite side to us; we in vain tried to drive him away, and failing, were obliged to make a large circuit. The Gauchos in revenge determined to emasculate him and render him for the future harmless. It was very interesting to see how art completely mastered force. One lazo was thrown over his horns as he rushed at the horse, and another round his hind legs: in a minute the monster was stretched powerless on the ground. After the lazo has once been drawn tightly round the horns of a furious animal, it does not at first appear an easy thing to disengage it again without killing the beast: nor, I apprehend, would it be so if the man was by himself. By the aid, however, of a second person throwing his lazo so as to catch both hind legs, it is quickly managed: for the animal, as long as its hind legs are kept outstretched, is quite helpless, and the first man can with his hands loosen his lazo from the horns, and then quietly mount his horse; but the moment the second man, by backing ever so little, relaxes the strain, the lazo slips off the legs of the struggling beast, which then rises free, shakes himself, and vainly rushes at his antagonist.

During our whole ride we saw only one troop of wild horses. These animals, as well as the cattle, were introduced by the French in 1764, since which time both have greatly increased. It is a curious fact, that the horses have never left the eastern end of the island, although there is no natural boundary to prevent them from roaming, and that part of the island is not more tempting than the rest. The Gauchos whom I asked, though asserting this to be the case, were unable to account for it, except from the strong attachment which horses have to any locality to which they are accustomed. Considering that the island does not appear fully stocked, and that there are no beasts of prey, I was particularly curious to know what has checked their originally rapid increase. That in a limited island some check would sooner or later supervene, is inevitable; but why has the increase of the horse been checked sooner than that of the cattle? Capt. Sulivan has taken much pains for me in this inquiry. The Gauchos employed here attribute it chiefly to the stallions constantly roaming from place to place, and compelling the mares to accompany them, whether or not the young foals are able to follow. One Gaucho told Capt. Sulivan that he had watched a stallion for a whole hour, violently kicking and biting a mare till he forced her to leave her foal to its fate. Capt. Sulivan can so far corroborate this curious account, that he has several times found young foals dead, whereas he has never found a dead calf. Moreover, the dead bodies of full-grown horses are more frequently found, as if more subject to disease or accidents, than those of the cattle. From the softness of the ground their hoofs often grow irregularly to a great length, and this causes lameness. The predominant colours are roan and iron-grey. All the horses bred here, both tame and wild, are rather small-sized, though generally in good condition; and they have lost so much strength, that they are unfit to be used in taking wild cattle with the lazo: in consequence, it is necessary to go to the great expense of importing fresh horses from the Plata. At some future period the southern hemisphere probably will have its breed of Falkland ponies, as the northern has its Shetland breed.

The cattle, instead of having degenerated like the horse, seem, as before remarked, to have increased in size; and they are much more numerous than the horses. Capt. Sulivan informs me that they vary much less in the general form of their bodies and in the shape of their horns than English cattle. In colour they differ much; and it is a remarkable circumstance, that in different parts of this one small island, different colours predominate. Round Mount Usborne, at a height of from 1000 to 1500 feet above the sea, about half of some of the herds are mouse or lead-coloured, a tint which is not common in other parts of the island. Near Port Pleasant dark brown prevails, whereas south of Choiseul Sound (which almost divides the island into two parts), white beasts with black heads and feet are the most common: in all parts black, and some spotted animals may be observed. Capt. Sulivan remarks, that the difference in the prevailing colours was so obvious, that in looking for the herds near Port Pleasant, they appeared from a long distance like black spots, whilst south of Choiseul Sound they appeared like white spots on the hill-sides. Capt. Sulivan thinks that the herds do not mingle; and it is a singular fact, that the mouse-coloured cattle, though living on the high land, calve about a month earlier in the season than the other coloured beasts on the lower land. It is interesting thus to find the once domesticated cattle breaking into three colours, of which some one colour would in all probability ultimately prevail over the others, if the herds were left undisturbed for the next several centuries.

The rabbit is another animal which has been introduced, and has succeeded very well; so that they abound over large parts of the island. Yet, like the horses, they are confined within certain limits; for they have not crossed the central chain of hills, nor would they have extended even so far as its base, if, as the Gauchos informed me, small colonies had not been carried there. I should not have supposed that these animals, natives of northern

Africa, could have existed in a climate so humid as this, and which enjoys so little sunshine that even wheat ripens only occasionally. It is asserted that in Sweden, which any one would have thought a more favourable climate, the rabbit cannot live out of doors. The first few pairs, moreover, had here to contend against pre-existing enemies, in the fox and some large hawks. The French naturalists have considered the black variety a distinct species, and called it Lepus Magellanicus.[99] They imagined that Magellan, when talking of an animal under the name of "conejos" in the Strait of Magellan, referred to this species; but he was alluding to a small cavy, which to this day is thus called by the Spaniards. The Gauchos laughed at the idea of the black kind being different from the grey, and they said that at all events it had not extended its range any further than the grey kind; that the two were never found separate; and that they readily bred together, and produced piebald offspring. Of the latter I now possess a specimen, and it is marked about the head differently from the French specific description. This circumstance shows how cautious naturalists should be in making species; for even Cuvier, on looking at the skull of one of these rabbits, thought it was probably distinct!

The only quadruped native to the island[100] is a large wolf-like fox (Canis antarcticus), which is common to both East and West Falkland. I have no doubt it is a peculiar species, and confined to this archipelago; because many sealers, Gauchos, and Indians, who have visited these islands, all maintain that no such animal is found in any part of South America.

Molina, from a similarity in habits, thought that this was the same with his "culpeu;"[101] but I have seen both, and they are quite distinct. These wolves are well known, from Byron's account of their tameness and curiosity, which the sailors, who ran into the water to avoid them, mistook for fierceness. To this day their manners remain the same. They have been observed to enter a tent, and actually pull some meat from beneath the head of a sleeping seaman. The Gauchos also have frequently in the evening killed them, by holding out a piece of meat in one hand, and in the other a knife ready to stick them. As far as I am aware, there is no other instance in any part of the world, of so small a mass of broken land, distant from a continent, possessing so large an aboriginal quadruped peculiar to itself. Their numbers have rapidly decreased; they are already banished from that half of the island which lies to the eastward of the neck of land between St. Salvador Bay and Berkeley Sound. Within a very few years after these islands shall have become regularly settled, in all probability this fox will be classed with the dodo, as an animal which has perished from the face of the earth.

At night (17th) we slept on the neck of land at the head of Choiseul Sound, which forms the south-west peninsula. The valley was pretty well sheltered from the cold wind; but there was very little brushwood for fuel. The Gauchos, however, soon found what, to my great surprise, made nearly as hot a fire as coals; this was the skeleton of a bullock lately killed, from which the flesh had been picked by the carrion-hawks. They told me that in winter they often killed a beast, cleaned the flesh from the bones with their knives, and then with these same bones roasted the meat for their suppers.

18th.—It rained during nearly the whole day. At night we managed, however, with our saddle-cloths to keep ourselves pretty well dry and warm; but the ground on which we slept was on each occasion nearly in the state of a bog, and there was not a dry spot to sit down on after our day's ride. I have in another part stated how singular it is that there should be absolutely no trees on these islands, although Tierra del Fuego is covered by one large forest. The largest bush in the island (belonging to the family of Compositæ) is scarcely so tall as our gorse. The best fuel is afforded by a green little bush about the size of common heath, which has the useful property of burning while fresh and green. It was very surprising to see the Gauchos, in the midst of rain and everything soaking wet, with nothing more than a tinder-box and a piece of rag, immediately make a fire. They sought beneath the tufts of grass and bushes for a few dry twigs, and these they rubbed into fibres; then surrounding them with coarser twigs, something like a bird's nest, they put the rag with its spark of fire in the middle and covered it up. The nest being then held up to the wind, by degrees it smoked more and more, and at last burst out in flames. I do not think any other method would have had a chance of succeeding with such damp materials.

19th.—Each morning, from not having ridden for some time previously, I was very stiff. I was surprised to hear the Gauchos, who have from infancy almost lived on horseback, say that, under similar circumstances, they always suffer. St. Jago told me, that having been confined for three months by illness, he went out hunting wild cattle, and in consequence, for the next two days, his thighs were so stiff that he was obliged to lie in bed. This shows that the Gauchos, although they do not appear to do so, yet really must exert much muscular effort in riding. The hunting wild cattle, in a country so difficult to pass as this is on account of the swampy ground, must be very hard work. The Gauchos say they often pass at full speed over ground which would be impassable at a slower pace; in the same manner as a man is able to skate over thin ice. When hunting, the party endeavours to get as close as possible to the herd without being discovered. Each man carries four or five pair of the bolas; these he throws one after the other at as many cattle, which, when once entangled, are left for some days, till they become a little exhausted by hunger and struggling. They are then let free and driven towards a small herd of tame animals, which have been brought to the spot on purpose. From their previous treatment, being too much terrified to leave the herd, they are easily driven, if their strength last out, to the settlement.

The weather continued so very bad that we determined to make a push, and try to reach the vessel before night. From the quantity of rain which had fallen, the surface of the whole country was swampy. I suppose my horse fell at least a dozen times, and sometimes the whole six horses were floundering in the mud together. All the little streams are bordered by soft peat, which makes it very difficult for the horses to leap them without falling. To complete our discomforts we were obliged to cross the head of a creek of the sea, in which the water was as high as our horses' backs; and the little waves, owing to the violence of the wind, broke over us, and made us very wet and cold. Even the iron-framed Gauchos professed themselves glad when they reached the settlement, after our little excursion.

THE GEOLOGICAL STRUCTURE of these islands is in most respects simple. The lower country consists of clay-slate and sandstone, containing fossils, very closely related to, but not identical with, those found in the Silurian formations of Europe; the hills are formed of white granular quartz rock. The strata of the latter are frequently arched with perfect symmetry, and the appearance of some of the masses is in consequence most singular. Pernety[102] has devoted several pages to the description of a Hill of Ruins, the successive strata of which he has justly compared to the seats of an amphitheatre. The quartz rock must have been quite pasty when it underwent such remarkable flexures without being shattered into fragments. As the quartz insensibly passes into the sandstone, it seems probable that the former owes its origin to the sandstone having been heated to such a degree that it became viscid, and upon cooling crystallized. While in the soft state it must have been pushed up through the overlying beds.

In many parts of the island the bottoms of the valleys are covered in an extraordinary manner by myriads of great loose angular fragments of the quartz rock, forming "streams of stones." These have been mentioned with surprise by every voyager since the time of Pernety. The blocks are not waterworn, their angles being only a little blunted; they vary in size from one or two feet in diameter to ten, or even more than twenty times as much. They are not thrown together into irregular piles, but are spread out into level sheets or great streams. It is not possible to ascertain their thickness, but the water of small streamlets can be heard trickling through the stones many feet below the surface. The actual depth is probably great, because the crevices between the lower fragments must long ago have been filled up with sand. The width of these sheets of stones varies from a few hundred feet to a mile; but the peaty soil daily encroaches on the borders, and even forms islets wherever a few fragments happen to lie close together. In a valley south of Berkeley Sound, which some of our party called the "great valley of fragments," it was necessary to cross an uninterrupted band half a mile wide, by jumping from one pointed stone to another. So large were the fragments, that being overtaken by a shower of rain, I readily found shelter beneath one of them.

Their little inclination is the most remarkable circumstance in these "streams of stones." On the hill-sides I have seen them sloping at an angle of ten degrees with the horizon; but in some of the level, broad-bottomed valleys, the inclination is only just sufficient to be clearly perceived. On so rugged a surface there was no means of measuring the angle; but to give a common illustration, I may say that the slope would not have checked the speed of an English mail-coach. In some places, a continuous stream of these fragments followed up the course of a valley, and even extended to the very crest of the hill. On these crests huge masses, exceeding in dimensions any small building, seemed to stand arrested in their headlong course: there, also, the curved strata of the archways lay piled on each other, like the ruins of some vast and ancient cathedral. In endeavouring to describe these scenes of violence one is tempted to pass from one simile to another. We may imagine that streams of white lava had flowed from many parts of the mountains into the lower country, and that when solidified they had been rent by some enormous convulsion into myriads of fragments. The expression "streams of stones," which immediately occurred to every one, conveys the same idea. These scenes are on the spot rendered more striking by the contrast of the low rounded forms of the neighbouring hills.

I was interested by finding on the highest peak of one range (about 700 feet above the sea) a great arched fragment, lying on its convex side, or back downwards. Must we believe that it was fairly pitched up in the air, and thus turned? Or, with more probability, that there existed formerly a part of the same range more elevated than the point on which this monument of a great convulsion of nature now lies. As the fragments in the valleys are neither rounded nor the crevices filled up with sand, we must infer that the period of violence was subsequent to the land having been raised above the waters of the sea. In a transverse section within these valleys, the bottom is nearly level, or rises but very little towards either side. Hence the fragments appear to have travelled from the head of the valley; but in reality it seems more probable that they have been hurled down from the nearest slopes; and that since, by a vibratory movement of overwhelming force,[103] the fragments have been levelled into one continuous sheet. If during the earthquake[104] which in 1835 overthrew Concepcion, in Chile, it was thought wonderful that small bodies should have been pitched a few inches from the ground, what must we say to a movement which has caused fragments many tons in weight, to move onwards like so much sand on a vibrating board, and find their level? I have seen, in the Cordillera of the Andes, the evident marks where stupendous mountains have been broken into pieces like so much thin crust, and the strata thrown on their vertical edges; but never did any scene, like these "streams of stones," so forcibly convey to my mind the idea of a convulsion, of which in historical records we might in vain seek for any counterpart: yet the progress of knowledge will probably some day give a simple explanation of this phenomenon, as it already has of the so long-thought inexplicable transportal of the erratic boulders, which are strewed over the plains of Europe.

I have little to remark on the zoology of these islands. I have before described the carrion-vulture of Polyborus. There are some other hawks, owls, and a few small land-birds. The water-fowl are particularly numerous, and they must formerly, from the accounts of the old navigators, have been much more so. One day I observed a cormorant playing with a fish which it had caught. Eight times successively the bird let its prey go, then dived after it, and although in deep water, brought it each time to the surface. In the Zoological Gardens I have seen the otter treat a fish in the same manner, much as a cat does a mouse: I do not know of any other instance where dame Nature appears so wilfully cruel. Another day, having placed myself between a penguin (Aptenodytes demersa) and the water, I was much amused by watching its habits. It was a brave bird; and till reaching the sea, it regularly fought and drove me backwards. Nothing less than heavy blows would have stopped him; every inch he gained he firmly kept, standing close before me erect and determined. When thus opposed he continually rolled his head from side to side, in a very odd manner, as if the power of distinct vision lay only in the anterior and basal part of each eye. This bird is commonly called the jackass penguin, from its habit, while on shore, of throwing its head backwards, and making a loud strange noise, very like the braying of an ass; but while at sea, and undisturbed, its note is very deep and solemn, and is often heard in the night-time. In diving, its little wings are used as fins; but on the land, as front legs. When crawling, it may be said on four legs, through the tussocks or on the side of a grassy cliff, it moves so very quickly that it might easily be mistaken for a quadruped. When at sea and fishing, it comes to the surface for the purpose of breathing with such a spring, and dives again so instantaneously, that I defy any one at first sight to be sure that it was not a fish leaping for sport.

Two kinds of geese frequent the Falklands. The upland species (Anas Magellanica) is common, in pairs and in small flocks, throughout the island. They do not migrate, but build on the small outlying islets. This is supposed to be from fear of the foxes: and it is perhaps from the same cause that these birds, though very tame by day, are shy and wild in the dusk of the evening. They live entirely on vegetable matter.

The rock-goose, so called from living exclusively on the sea-beach (Anas antarctica), is common both here and on the west coast of America, as far north as Chile. In the deep and retired channels of Tierra del Fuego, the snow-white gander, invariably accompanied by his darker consort, and standing close by each other on some distant rocky point, is a common feature in the landscape.

In these islands a great loggerheaded duck or goose (Anas brachyptera), which sometimes weighs twenty-two pounds, is very abundant. These birds were in former days called, from their extraordinary manner of paddling and splashing upon the water, race-horses; but now they are named, much more appropriately, steamers. Their wings are too small and weak to allow of flight, but by their aid, partly swimming and partly flapping the surface of the water, they move very quickly. The manner is something like that by which the common house-duck escapes when pursued by a dog; but I am nearly sure that the steamer moves its wings alternately, instead of both together, as in other birds. These clumsy, loggerheaded ducks make such a noise and splashing, that the effect is exceedingly curious.

Thus we find in South America three birds which use their wings for other purposes besides flight; the penguins as fins, the steamer as paddles, and the ostrich as sails: and the Apteryz of New Zealand, as well as its gigantic extinct prototype the Deinornis, possess only rudimentary representatives of wings. The steamer is able to dive only to a very short distance. It feeds entirely on shell-fish from the kelp and tidal rocks: hence the beak and head, for the purpose of breaking them, are surprisingly heavy and strong: the head is so strong that I have scarcely been able to fracture it with my geological hammer; and all our sportsmen soon discovered how tenacious these birds were of life. When in the evening pluming themselves in a flock, they make the same odd mixture of sounds which bull-frogs do within the tropics.

IN TIERRA DEL FUEGO, as well as in the Falkland Islands, I made many observations on the lower marine animals,[105] but they are of little general interest. I will mention only one class of facts, relating to certain zoophytes in the more highly organized division of that class. Several genera (Flustra, Eschara, Cellaria, Crisia, and others) agree in having singular moveable organs (like those of Flustra avicularia, found in the European seas) attached to their cells. The organ, in the greater number of cases, very closely resembles the head of a vulture; but the lower mandible can be opened much wider than in a real bird's beak. The head itself possesses considerable powers of movement, by means of a short neck. In one zoophyte the head itself was fixed, but the lower jaw free: in another it was replaced by a triangular hood, with a beautifully-fitted trap-door, which evidently answered to the lower mandible. In the greater number of species, each cell was provided with one head, but in others each cell had two.

The young cells at the end of the branches of these corallines contain quite immature polypi, yet the vulture-heads attached to them, though small, are in every respect perfect. When the polypus was removed by a needle from any of the cells, these organs did not appear in the least affected. When one of the vulture-like heads was cut off from the cell, the lower mandible retained its power of opening and closing. Perhaps the most singular part of their structure is, that when there were more than two rows of cells on a branch, the central cells were furnished with these appendages, of only one-fourth the size of the outside ones. Their movements varied according to the species; but in some I never saw the least motion; while others, with the lower mandible generally wide open, oscillated backwards and forwards at the rate of about five seconds each turn; others moved rapidly and by starts. When touched with a needle, the beak generally seized the point so firmly, that the whole branch might be shaken.

These bodies have no relation whatever with the production of the eggs or gemmules, as they are formed before the young polypi appear in the cells at the end of the growing branches; as they move independently of the polypi, and do not appear to be in any way connected with them; and as they differ in size on the outer and inner rows of cells, I have little doubt, that in their functions, they are related rather to the horny axis of the branches than to the polypi in the cells. The fleshy appendage at the lower extremity of the sea-pen (described at Bahia Blanca) also forms part of the zoophyte, as a whole, in the same manner as the roots of a tree form part of the whole tree, and not of the individual leaf or flower-buds.

In another elegant little coralline (Crisia?), each cell was furnished with a long-toothed bristle, which had the power of moving quickly. Each of these bristles and each of the vulture-like heads generally moved quite independently of the others, but sometimes all on both sides of a branch, sometimes only those on one side, moved together coinstantaneously; sometimes each moved in regular order one after another. In these actions we apparently behold as perfect a transmission of will in the zoophyte, though composed of thousands of distinct polypi, as in any single animal. The case, indeed, is not different from that of the sea-pens, which, when touched, drew themselves into the sand on the coast of Bahia Blanca. I will state one other instance of uniform action, though of a very different nature, in a zoophyte closely allied to Clytia, and therefore very simply organized. Having kept a large tuft of it in a basin of salt-water, when it was dark I found that as often as I rubbed any part of a branch, the whole became strongly phosphorescent with a green light: I do not think I ever saw any object more beautifully so. But the remarkable circumstance was, that the flashes of light always proceeded up the branches, from the base towards the extremities.

The examination of these compound animals was always very interesting to me. What can be more remarkable than to see a plant-like body producing an egg, capable of swimming about and of choosing a proper place to adhere to, which then sprouts into branches, each crowded with innumerable distinct animals, often of complicated organizations? The branches, moreover, as we have just seen, sometimes possess organs capable of movement and independent of the polypi. Surprising as this union of separate individuals in a common stock must always appear, every tree displays the same fact, for buds must be considered as individual plants. It is, however, natural to consider a polypus, furnished with a mouth, intestines, and other organs, as a distinct individual, whereas the individuality of a leaf-bud is not easily realised; so that the union of separate individuals in a common body is more striking in a coralline than in a tree. Our conception of a compound animal, where in some respects the individuality of each is not completed, may be aided, by reflecting on the production of two distinct creatures by bisecting a single one with a knife, or where Nature herself performs the task of bisection. We may consider the polypi in a zoophyte, or the buds in a tree, as cases where the division of the individual has not been completely effected. Certainly in the case of trees, and judging from analogy in that of corallines, the individuals propagated by buds seem more intimately related to each other, than eggs or seeds are to their parents. It seems now pretty well established that plants propagated by buds all partake of a common duration of life; and it is familiar to every one, what singular and numerous peculiarities are transmitted with certainty, by buds, layers, and grafts, which by seminal propagation never or only casually reappear.

CHAPTER X

Tierra del Fuego

TIERRA DEL FUEGO, FIRST ARRIVAL • GOOD SUCCESS BAY • AN ACCOUNT OF THE FUEGIANS ON BOARD • INTERVIEW WITH THE SAVAGES • SCENERY OF THE FORESTS • CAPE HORN • WIGWAM COVE • MISERABLE CONDITION OF THE SAVAGES • FAMINES • CANNIBALS • MATRICIDE • RELIGIOUS FEELINGS • GREAT GALE • BEAGLE CHANNEL • PONSONBY SOUND • BUILD WIGWAMS AND SETTLE THE FUEGIANS • BIFURCATION OF THE BEAGLE CHANNEL • GLACIERS • RETURN TO THE SHIP • SECOND VISIT IN THE SHIP TO THE SETTLEMENT • EQUALITY OF CONDITION AMONGST THE NATIVES

December 17th, 1832.—Having now finished with Patagonia and the Falkland Islands, I will describe our first arrival in Tierra del Fuego. A little after noon we doubled Cape St. Diego, and entered the famous strait of Le Maire. We kept close to the Fuegian shore, but the outline of the rugged, inhospitable Statenland was visible amidst the clouds. In the afternoon we anchored in the Bay of Good Success. While entering we were saluted in a manner becoming the inhabitants of this savage land. A group of Fuegians partly concealed by the entangled forest, were perched on a wild point overhanging the sea; and as we passed by, they sprang up and waving their tattered cloaks sent forth a loud and sonorous shout. The savages followed the ship, and just before dark we saw their fire, and again heard their wild cry. The harbour consists of a fine piece of water half surrounded by low rounded mountains of clay-slate, which are covered to the water's edge by one dense gloomy forest. A single glance at the landscape was sufficient to show me how widely different it was from anything I had ever beheld. At night it blew a gale of wind, and heavy squalls from the mountains swept past us. It would have been a bad time out at sea, and we, as well as others, may call this Good Success Bay.

In the morning the Captain sent a party to communicate with the Fuegians. When we came within hail, one of the four natives who were present advanced to receive us, and began to shout most vehemently, wishing to direct us where to land. When we were on shore the party looked rather alarmed, but continued talking and making gestures with great rapidity. It was without exception the most curious and interesting spectacle I ever beheld: I could not have believed how wide was the difference between savage and civilized man: it is greater than between a wild and domesticated animal, inasmuch as in man there is a greater power of improvement. The chief spokesman was old, and appeared to be the head of the family; the three others were powerful young men, about six feet high. The women and children had been sent away. These Fuegians are a very different race from the stunted, miserable wretches farther westward; and they seem closely allied to the famous Patagonians of the Strait of Magellan. Their only garment consists of a mantle made of guanaco skin, with the wool outside: this they wear just thrown over their shoulders, leaving their persons as often exposed as covered. Their skin is of a dirty coppery-red colour.

The old man had a fillet of white feathers tied round his head, which partly confined his black, coarse, and entangled hair. His face was crossed by two broad transverse bars; one, painted bright red, reached from ear to ear and included the upper lip; the other, white like chalk, extended above and parallel to the first, so that even his eyelids were thus coloured. The other two men were ornamented by streaks of black powder, made of charcoal. The party altogether closely resembled the devils which come on the stage in plays like Der Freischutz.

Their very attitudes were abject, and the expression of their countenances distrustful, surprised, and startled. After we had presented them with some scarlet cloth, which they immediately tied round their necks, they became good friends. This was shown by the old man patting our breasts, and making a chuckling kind of noise, as people do when feeding chickens. I walked with the old man, and this demonstration of friendship was repeated several times; it was concluded by three hard slaps, which were given me on the breast and back at the same time. He then bared his bosom for me to return the compliment, which being done, he seemed highly pleased. The language of these people, according to our notions, scarcely deserves to be called articulate. Captain Cook has compared it to a man clearing his throat, but certainly no European ever cleared his throat with so many hoarse, guttural, and clicking sounds.

They are excellent mimics: as often as we coughed or yawned, or made any odd motion, they immediately imitated us. Some of our party began to squint and look awry; but one of the young Fuegians (whose whole face was painted black, excepting a white band across his eyes) succeeded in making far more hideous grimaces. They could repeat with perfect correctness each word in any sentence we addressed them, and they remembered such words for some time. Yet we Europeans all know how difficult it is to distinguish apart the sounds in a foreign language. Which of us, for instance, could follow an American Indian through a sentence of more than three words? All savages appear to possess, to an uncommon degree, this power of mimicry. I was told, almost in the same words, of the same ludicrous habit among the Caffres; the Australians, likewise, have long been notorious for being able to imitate and describe the gait of any man, so that he may be recognised. How can this faculty be explained? is it a consequence of the more practised habits of perception and keener senses, common to all men in a savage state, as compared with those long civilized?

When a song was struck up by our party, I thought the Fuegians would have fallen down with astonishment. With equal surprise they viewed our dancing; but one of the young men, when asked, had no objection to a little waltzing. Little accustomed to Europeans as they appeared to be, yet they knew and dreaded our fire-arms; nothing would tempt them to take a gun in their hands. They begged for knives, calling them by the Spanish word "cuchilla." They explained also what they wanted, by acting as if they had a piece of blubber in their mouth, and then pretending to cut instead of tear it.

I have not as yet noticed the Fuegians whom we had on board. During the former voyage of the Adventure and Beagle in 1826 to 1830, Captain Fitz Roy seized on a party of natives, as hostages for the loss of a boat, which had been stolen, to the great jeopardy of a party employed on the survey; and some of these natives, as well as a child whom he bought for a pearl-button, he took with him to England, determining to educate them and instruct them in religion at his own expense. To settle these natives in their own country, was one chief inducement to Captain Fitz Roy to undertake our present voyage; and before the Admiralty had resolved to send out this expedition, Captain Fitz Roy had generously chartered a vessel, and would himself have taken them back. The natives were accompanied by a missionary, R. Matthews; of whom and of the natives, Captain Fitz Roy has published a full and excellent account. Two men, one of whom died in England of the small-pox, a boy and a little girl, were originally taken; and we had now on board, York Minster, Jemmy Button (whose name expresses his purchase-money), and Fuegia Basket. York Minster was a full-grown, short, thick, powerful man: his disposition was reserved, taciturn, morose, and when excited violently passionate; his affections were very strong towards a few friends on board; his intellect good. Jemmy Button was a universal favourite, but likewise passionate; the expression of his face at once showed his nice disposition. He was merry and often laughed, and was remarkably sympathetic with any one in pain: when the water was rough, I was often a little sea-sick, and he used to come to me and say in a plaintive voice, "Poor, poor fellow!" but the notion, after his aquatic life, of a man being sea-sick, was too ludicrous, and he was generally obliged to turn on one side to hide a smile or laugh, and then he would repeat his "Poor, poor fellow!" He was of a patriotic disposition; and he liked to praise his own tribe and country, in which he truly said there were "plenty of trees," and he abused all the other tribes: he stoutly declared that there was no Devil in his land. Jemmy was short, thick, and fat, but vain of his personal appearance; he used always to wear gloves, his hair was neatly cut, and he was distressed if his well-polished shoes were dirtied. He was fond of admiring himself in a looking glass; and a merry-faced little Indian boy from the Rio Negro, whom we had for some months on board, soon perceived this, and used to mock him: Jemmy, who was always rather jealous of the attention paid to this little boy, did not at all like this, and used to say, with rather a contemptuous twist of his head, "Too much skylark." It seems yet wonderful to me, when I think over all his many good qualities, that he should have been of the same race, and doubtless partaken of the same character, with the miserable, degraded savages whom we first met here. Lastly, Fuegia Basket was a nice, modest, reserved young girl, with a rather pleasing but sometimes sullen expression, and very quick in learning anything, especially languages. This she showed in picking up some Portuguese and Spanish, when left on shore for only a short time at Rio de Janeiro and Monte Video, and in her knowledge of English. York Minster was very jealous of any attention paid to her; for it was clear he determined to marry her as soon as they were settled on shore.

Although all three could both speak and understand a good deal of English, it was singularly difficult to obtain much information from them, concerning the habits of their countrymen; this was partly owing to their apparent difficulty in understanding the simplest alternative. Every one accustomed to very young children, knows how seldom one can get an answer even to so simple a question as whether a thing is black *or* white; the idea of black or white seems alternately to fill their minds. So it was with these Fuegians, and hence it was generally impossible to find out, by cross-questioning, whether one had rightly understood anything which they had asserted. Their sight was remarkably acute; it is well known that sailors, from long practice, can make out a distant object much better than a landsman; but both York and Jemmy were much superior to any sailor on board: several times they have declared what some distant object has been, and though doubted by every one, they have proved right, when it has been examined through a telescope. They were quite conscious of this power; and Jemmy, when he had any little quarrel with the officer on watch, would say, "Me see ship, me no tell."

It was interesting to watch the conduct of the savages, when we landed, towards Jemmy Button: they immediately perceived the difference between him and ourselves, and held much conversation one with another on the subject. The old man addressed a long harangue to Jemmy, which it seems was to invite him to stay with them. But Jemmy understood very little of their language, and was, moreover, thoroughly ashamed of his countrymen. When York Minster afterwards came on shore, they noticed him in the same way, and told him he ought to shave; yet he had not twenty dwarf hairs on his face, whilst we all wore our untrimmed beards. They examined the colour of his skin, and compared it with ours. One of our arms being bared, they expressed the liveliest surprise and admiration at its whiteness, just in the same way in which I have seen the ourangoutang do at the Zoological Gardens. We thought that they mistook two or three of the officers, who were rather shorter and fairer, though adorned with large beards, for the ladies of our party. The tallest amongst the Fuegians was evidently much pleased at his height being noticed. When placed back to back with the tallest of the boat's crew, he tried his best to edge on higher ground, and to stand on tiptoe. He opened his mouth to show his teeth, and turned his face for a side view; and all this was done with such alacrity, that I dare say he thought himself the handsomest man in Tierra del Fuego. After our first feeling of grave astonishment was over, nothing could be more ludicrous than the odd mixture of surprise and imitation which these savages every moment exhibited.

THE NEXT DAY I attempted to penetrate some way into the country. Tierra del Fuego may be described as a mountainous land, partly submerged in the sea, so that deep inlets and bays occupy the place where valleys should exist. The mountain sides, except on the exposed western coast, are covered from the water's edge upwards by one great forest. The trees reach to an elevation of between 1000 and 1500 feet, and are succeeded by a band of peat, with minute alpine plants; and this again is succeeded by the line of perpetual snow, which, according to Captain King, in the Strait of Magellan descends to between 3000 and 4000 feet. To find an acre of level land in any part of the country is most rare. I recollect only one little flat piece near Port Famine, and another of rather larger extent near Goeree Road. In both places, and everywhere else, the surface is covered by a thick bed of swampy peat. Even within the forest, the ground is concealed by a mass of slowly putrefying vegetable matter, which, from being soaked with water, yields to the foot.

Finding it nearly hopeless to push my way through the wood, I followed the course of a mountain torrent. At first, from the waterfalls and number of dead trees, I could hardly crawl along; but the bed of the stream soon became a little more open, from the floods having swept the sides. I continued slowly to advance for an hour along the broken and rocky banks, and was amply repaid by the grandeur of the scene. The gloomy depth of the ravine well accorded with the universal signs of violence. On every side were lying irregular masses of rock and torn-up trees; other trees, though still erect, were decayed to the heart and ready to fall. The entangled mass of the thriving and the fallen reminded me of the forests within the tropics—yet there was a difference: for in these still solitudes, Death, instead of Life, seemed the predominant spirit. I followed the water-course till I came to a spot where a great slip had cleared a straight space down the mountain side. By this road I ascended to a considerable elevation, and obtained a good view of the surrounding woods. The trees all belong to one kind, the Fagus betuloides; for the number of the other species of Fagus and of the Winter's Bark, is quite inconsiderable. This beech keeps its leaves throughout the year; but its foliage is of a peculiar brownish-green colour, with a tinge of yellow. As the whole landscape is thus coloured, it has a sombre, dull appearance; nor is it often enlivened by the rays of the sun.

December 20th.—One side of the harbour is formed by a hill about 1500 feet high, which Captain Fitz Roy has called after Sir J. Banks, in commemoration of his disastrous excursion, which proved fatal to two men of his party, and nearly so to Dr. Solander. The snowstorm, which was the cause of their misfortune, happened in the middle of January, corresponding to our July, and in the latitude of Durham! I was anxious to reach the summit of this mountain to collect alpine plants; for flowers of any kind in the lower parts are few in number. We followed the same water-course as on the previous day, till it dwindled away, and we were then compelled to crawl blindly among the trees. These, from the effects of the elevation and of the impetuous winds, were low, thick and crooked. At length we reached that which from a distance appeared like a carpet of fine green turf, but which, to our vexation, turned out to be a compact mass of little beech-trees about four or five feet high. They were as thick together as box in the border of a garden, and we were obliged to struggle over the flat but treacherous surface. After a little more trouble we gained the peat, and then the bare slate rock.

A ridge connected this hill with another, distant some miles, and more lofty, so that patches of snow were lying on it. As the day was not far advanced, I determined to walk there and collect plants along the road. It would have been very hard work, had it not been for a well-beaten and straight path made by the guanacos; for these animals, like sheep, always follow the same line. When we reached the hill we found it the highest in the immediate neighbourhood, and the waters flowed to the sea in opposite directions. We obtained a wide view over the surrounding country: to the north a swampy moorland extended, but to the south we had a scene of savage magnificence, well becoming Tierra del Fuego. There was a degree of mysterious grandeur in mountain behind mountain, with the deep intervening valleys, all covered by one thick, dusky mass of forest. The atmosphere, likewise, in this climate, where gale succeeds gale, with rain, hail, and sleet, seems blacker than anywhere else. In the Strait of Magellan looking due southward from Port Famine, the distant channels between the mountains appeared from their gloominess to lead beyond the confines of this world.

December 21st.—The Beagle got under way: and on the succeeding day, favoured to an uncommon degree by a fine easterly breeze, we closed in with the Barnevelts, and running past Cape Deceit with its stony peaks, about three o'clock doubled the weather-beaten Cape Horn. The evening was calm and bright, and we enjoyed a fine view of the surrounding isles. Cape Horn, however, demanded his tribute, and before night sent us a gale of wind directly in our teeth. We stood out to sea, and on the second day again made the land, when we saw on our weather-bow this notorious promontory in its proper form—veiled in a mist, and its dim outline surrounded by a storm of wind and water. Great black clouds were rolling across the heavens, and squalls of rain, with hail, swept by us with such extreme violence, that the Captain determined to run into Wigwam Cove. This is a snug little harbour, not far from Cape Horn; and here, at Christmas-eve, we anchored in smooth water. The only thing which reminded us of the gale outside, was every now and then a puff from the mountains, which made the ship surge at her anchors.

December 25th.—Close by the Cove, a pointed hill, called Kater's Peak, rises to the height of 1700 feet. The surrounding islands all consist of conical masses of greenstone, associated sometimes with less regular hills of baked and altered clay-slate. This part of Tierra del Fuego may be considered as the extremity of the submerged chain of mountains already alluded to. The cove takes its name of "Wigwam" from some of the Fuegian habitations; but every bay in the neighbourhood might be so called with equal propriety. The inhabitants, living chiefly upon shell-fish, are obliged constantly to change their place of residence; but they return at intervals to the same spots, as is evident from the piles of old shells, which must often amount to many tons in weight. These heaps can be distinguished at a long distance by the bright green colour of certain plants, which invariably grow on them. Among these may be enumerated the wild celery and scurvy grass, two very serviceable plants, the use of which has not been discovered by the natives.

The Fuegian wigwam resembles, in size and dimensions, a haycock. It merely consists of a few broken branches stuck in the ground, and very imperfectly thatched on one side with a few tufts of grass and rushes. The whole cannot be the work of an hour, and it is only used for a few days. At Goeree Roads I saw a place where one of these naked men had slept, which absolutely offered no more cover than the form of a hare. The man was evidently living by himself, and York Minster said he was "very bad man," and that probably he had stolen something. On the west coast, however, the wigwams are rather better, for they are covered with seal-skins. We were detained here several days by the bad weather. The climate is certainly wretched: the summer solstice was now passed, yet every day snow fell on the hills, and in the valleys there was rain, accompanied by sleet. The thermometer generally stood about 45°, but in the night fell to 38° or 40°. From the damp and boisterous state of the atmosphere, not cheered by a gleam of sunshine, one fancied the climate even worse than it really was.

While going one day on shore near Wollaston Island, we pulled alongside a canoe with six Fuegians. These were the most abject and miserable creatures I anywhere beheld. On the east coast the natives, as we have seen, have guanaco cloaks, and on the west they possess seal-skins. Amongst these central tribes the men generally have an otter-skin, or some small scrap about as large as a pocket-handkerchief, which is barely sufficient to cover their backs as low down as their loins. It is laced across the breast by strings, and according as the wind blows, it is shifted from side to side. But these Fuegians in the canoe were quite naked, and even one full-grown woman was absolutely so. It was raining heavily, and the fresh water, together with the spray, trickled down her body. In another harbour not far distant, a woman, who was suckling a recently-born child, came one day alongside the vessel, and remained there out of mere curiosity, whilst the sleet fell and thawed on her naked bosom, and on the skin of her naked baby! These poor wretches were stunted in their growth, their hideous faces bedaubed with white paint, their skins filthy and greasy, their hair entangled, their voices discordant, and their gestures violent. Viewing such men, one can hardly make one's self believe that they are fellow-creatures, and inhabitants of the same world. It is a common subject of conjecture what pleasure in life some of the lower animals can enjoy: how much more reasonably the same question may be asked with respect to these barbarians! At night, five or six human beings, naked and scarcely protected from the wind and rain of this tempestuous climate, sleep on the wet ground coiled up like animals. Whenever it is low water, winter or summer, night or day, they must rise to pick shell-fish from the rocks; and the women either dive to collect sea-eggs, or sit patiently in their canoes, and with a baited hair-line without any hook, jerk out little fish. If a seal is killed, or the floating carcass of a putrid whale is discovered, it is a feast; and such miserable food is assisted by a few tasteless berries and fungi.

They often suffer from famine: I heard Mr. Low, a sealing-master intimately acquainted with the natives of this country, give a curious account of the state of a party of one hundred and fifty natives on the west coast, who were very thin and in great distress. A succession of gales prevented the women from getting shell-fish on the rocks, and they could not go out in their canoes to catch seal. A small party of these men one morning set out, and the other Indians explained to him, that they were going a four days' journey for food: on their return, Low went to meet them, and he found them excessively tired, each man carrying a great square piece of putrid whale's-blubber with a hole in the middle, through which they put their heads, like the Gauchos do through their ponchos or cloaks. As soon as the blubber was brought into a wigwam, an old man cut off thin slices, and muttering over them, broiled them for a minute, and distributed them to the famished party, who during this time preserved a profound silence. Mr. Low believes that whenever a whale is cast on shore, the natives bury large pieces of it in the sand, as a resource in time of famine; and a native boy, whom he had on board, once found a stock thus buried. The different tribes when at war are cannibals. From the concurrent, but quite independent evidence of the boy taken by Mr. Low, and of Jemmy Button, it is certainly true, that when pressed in winter by hunger, they kill and devour their old women before they kill their dogs: the boy, being asked by Mr. Low why they did this, answered, "Doggies catch otters, old women no." This boy described the manner in which they are killed by being held over smoke and thus choked; he imitated their screams as a joke, and described the parts of their bodies which are considered best to eat. Horrid as such a death by the hands of their friends and relatives must be, the fears of the old women, when hunger begins to press, are more painful to think of; we are told that they then often run away into the mountains, but that they are pursued by the men and brought back to the slaughter-house at their own firesides!

Captain Fitz Roy could never ascertain that the Fuegians have any distinct belief in a future life. They sometimes bury their dead in caves, and sometimes in the mountain forests; we do not know what ceremonies they perform. Jemmy Button would not eat land-birds, because "eat dead men": they are unwilling even to mention their dead friends. We have no reason to believe that they perform any sort of religious worship; though perhaps the muttering of the old man before he distributed the putrid blubber to his famished party, may be of this nature. Each family or tribe has a wizard or conjuring doctor, whose office we could never clearly ascertain. Jemmy believed in dreams, though not, as I have said, in the devil: I do not think that our Fuegians were much more superstitious than some of the sailors; for an old quartermaster firmly believed that the successive heavy gales, which we encountered off Cape Horn, were caused by our having the Fuegians on board. The nearest approach to a religious feeling which I heard of, was shown by York Minster, who, when Mr. Bynoe shot some very young ducklings as specimens, declared in the most solemn manner, "Oh, Mr. Bynoe, much rain, snow, blow much." This was evidently a retributive punishment for wasting human food. In a wild and excited manner he also related, that his brother, one day whilst returning to pick up some dead birds which he had left on the coast, observed some feathers blown by the wind. His brother said (York imitating his manner), "What that?" and crawling onwards, he peeped over the cliff, and saw "wild man" picking his birds; he crawled a little nearer, and then hurled down a great stone and killed him. York declared for a long time afterwards storms raged, and much rain and snow fell. As far as we could make out, he seemed to consider the elements themselves as the avenging agents: it is evident in this case, how naturally, in a race a little more advanced in culture, the elements would become personified. What the "bad wild men" were, has always appeared to me most mysterious: from what York said, when we found the place like the form of a hare, where a single man had slept the night before, I should have thought that they were thieves who had been driven from their tribes; but other obscure speeches made me doubt this; I have sometimes imagined that the most probable explanation was that they were insane.

The different tribes have no government or chief; yet each is surrounded by other hostile tribes, speaking different dialects, and separated from each other only by a deserted border or neutral territory: the cause of their warfare appears to be the means of subsistence. Their country is a broken mass of wild rocks, lofty hills, and useless forests: and these are viewed through mists and endless storms. The habitable land is reduced to the stones on the beach; in search of food they are compelled unceasingly to wander from spot to spot, and so steep is the coast, that they can only move about in their wretched canoes. They cannot know the feeling of having a home, and still less that of domestic affection; for the husband is to the wife a brutal master to a laborious slave. Was a more horrid deed ever perpetrated, than that witnessed on the west coast by Byron, who saw a wretched mother pick up her bleeding dying infant-boy, whom her husband had mercilessly dashed on the stones for dropping a basket of sea-eggs! How little can the higher powers of the mind be brought into play: what is there for imagination to picture, for reason to compare, for judgment to decide upon? to knock a limpet from the rock does not require even cunning, that lowest power of the mind. Their skill in some respects may be compared to the instinct of animals; for it is not improved by experience: the canoe, their most ingenious work, poor as it is, has remained the same, as we know from Drake, for the last two hundred and fifty years.

Whilst beholding these savages, one asks, whence have they come? What could have tempted, or what change compelled a tribe of men, to leave the fine regions of the north, to travel down the Cordillera or backbone of America, to invent and build canoes, which are not used by the tribes of Chile, Peru, and Brazil, and then to enter on one of the most inhospitable countries within the limits of the globe? Although such reflections must at first seize on the mind, yet we may feel sure that they are partly erroneous. There is no reason to believe that the Fuegians decrease in number; therefore we must suppose that they enjoy a sufficient share of happiness, of whatever kind it may be, to render life worth having. Nature by making habit omnipotent, and its effects hereditary, has fitted the Fuegian to the climate and the productions of his miserable country.

AFTER HAVING BEEN DETAINED six days in Wigwam Cove by very bad weather, we put to sea on the 30th of December. Captain Fitz Roy wished to get westward to land York and Fuegia in their own country. When at sea we had a constant succession of gales, and the current was against us: we drifted to 57° 23' south. On the 11th of January, 1833, by carrying a press of sail, we fetched within a few miles of the great rugged mountain of York Minster (so called by Captain Cook, and the origin of the name of the elder Fuegian), when a violent squall compelled us to shorten sail and stand out to sea. The surf was breaking fearfully on the coast, and the spray was carried over a cliff estimated to 200 feet in height. On the 12th the gale was very heavy, and we did not know exactly where we were: it was a most unpleasant sound to hear constantly repeated, "keep a good look-out to leeward." On the 13th the storm raged with its full fury: our horizon was narrowly limited by the sheets of spray borne by the wind. The sea looked ominous, like a dreary waving plain with patches of drifted snow: whilst the ship laboured heavily, the albatross glided with its expanded wings right up the wind. At noon a great sea broke over us, and filled one of the whale boats, which was obliged to be instantly cut away. The poor Beagle trembled at the shock, and for a few minutes would not obey her helm; but soon, like a good ship that she was, she righted and came up to the wind again. Had another sea followed the first, our fate would have been decided soon, and for ever. We had now been twenty-four days trying in vain to get westward; the men were worn out with fatigue, and they had not had for many nights or days a dry thing to put on. Captain Fitz Roy gave up the attempt to get westward by the outside coast. In the evening we ran in behind False Cape Horn, and dropped our anchor in forty-seven fathoms, fire flashing from the windlass as the chain rushed round it. How delightful was that still night, after having been so long involved in the din of the warring elements!

January 15th, 1833.—The Beagle anchored in Goeree Roads. Captain Fitz Roy having resolved to settle the Fuegians, according to their wishes, in Ponsonby Sound, four boats were equipped to carry them there through the Beagle Channel. This channel, which was discovered by Captain Fitz Roy during the last voyage, is a most remarkable feature in the geography of this, or indeed of any other country: it may be compared to the valley of Lochness in Scotland, with its chain of lakes and friths. It is about one hundred and twenty miles long, with an average breadth, not subject to any very great variation, of about two miles; and is throughout the greater part so perfectly straight, that the view, bounded on each side by a line of mountains, gradually becomes indistinct in the long distance. It crosses the southern part of Tierra del Fuego in an east and west line, and in the middle is joined at right angles on the south side by an irregular channel, which has been called Ponsonby Sound. This is the residence of Jemmy Button's tribe and family.

19th.—Three whale-boats and the yawl, with a party of twenty-eight, started under the command of Captain Fitz Roy. In the afternoon we entered the eastern mouth of the channel, and shortly afterwards found a snug little cove concealed by some surrounding islets. Here we pitched our tents and lighted our fires. Nothing could look more comfortable than this scene. The glassy water of the little harbour, with the branches of the trees hanging over the rocky beach, the boats at anchor, the tents supported by the crossed oars, and the smoke curling up the wooded valley, formed a picture of quiet retirement. The next day (20th) we smoothly glided onwards in our little fleet, and came to a more inhabited district. Few if any of these natives could ever have seen a white man; certainly nothing could exceed their astonishment at the apparition of the four boats. Fires were lighted on every point (hence the name of Tierra del Fuego, or the land of fire), both to attract our attention and to spread far and wide the news. Some of the men ran for miles along the shore. I shall never forget how wild and savage one group appeared: suddenly four or five men came to the edge of an over-hanging cliff; they were absolutely naked, and their long hair streamed about their faces; they held rugged staffs in their hands, and, springing from the ground, they waved their arms round their heads, and sent forth the most hideous yells.

At dinner-time we landed among a party of Fuegians. At first they were not inclined to be friendly; for until the Captain pulled in ahead of the other boats, they kept their slings in their hands. We soon, however, delighted them by trifling presents, such as tying red tape round their heads. They liked our biscuit: but one of the savages touched with his finger some of the meat preserved in tin cases which I was eating, and feeling it soft and cold, showed as much disgust at it, as I should have done at putrid blubber. Jemmy was thoroughly ashamed of his countrymen, and declared his own tribe were quite different, in which he was wofully mistaken. It was as easy to please as it was difficult to satisfy these savages. Young and old, men and children, never ceased repeating the word "yammerschooner," which means "give me." After pointing to almost every object, one after the other, even to the buttons on our coats, and saying their favourite word in as many intonations as possible, they would then use it in a neuter sense, and vacantly repeat "yammerschooner." After yammerschoonering for any article very eagerly, they would by a simple artifice point to their young women or little children, as much as to say, "If you will not give it me, surely you will to such as these."

At night we endeavoured in vain to find an uninhabited cove; and at last were obliged to bivouac not far from a party of natives. They were very inoffensive as long as they were few in numbers, but in the morning (21st) being joined by others they showed symptoms of hostility, and we thought that we should have come to a skirmish. A European labours under great disadvantages when treating with savages like these, who have not the least idea of the power of fire-arms. In the very act of levelling his musket he appears to the savage far inferior to a man armed with a bow and arrow, a spear, or even a sling. Nor is it easy to teach them our superiority except by striking a fatal blow. Like wild beasts, they do not appear to compare numbers; for each individual, if attacked, instead of retiring, will endeavour to dash your brains out with a stone, as certainly as a tiger under similar circumstances would tear you. Captain Fitz Roy on one occasion being very anxious, from good reasons, to frighten away a small party, first flourished a cutlass near them, at which they only laughed; he then twice fired his pistol close to a native. The man both times looked astounded, and carefully but quickly rubbed his head; he then stared awhile, and gabbled to his companions, but he never seemed to think of running away. We can hardly put ourselves in the position of these savages, and understand their actions. In the case of this Fuegian, the possibility of such a sound as the report of a gun close to his ear could never have entered his mind. He perhaps literally did not for a second know whether it was a sound or a blow, and therefore very naturally rubbed his head. In a similar manner, when a savage sees a mark struck by a bullet, it may be some time before he is able at all to understand how it is effected; for the fact of a body being invisible from its velocity would perhaps be to him an idea totally inconceivable. Moreover, the extreme force of a bullet, that penetrates a hard substance without tearing it, may convince the savage that it has no force at all. Certainly I believe that many savages of the lowest grade, such as these of Tierra del Fuego, have seen objects struck, and even small animals killed by the musket, without being in the least aware how deadly an instrument it is.

22nd.—After having passed an unmolested night, in what would appear to be neutral territory between Jemmy's tribe and the people whom we saw yesterday, we sailed pleasantly along. I do not know anything which shows more clearly the hostile state of the different tribes, than these wide border or neutral tracts. Although Jemmy Button well knew the force of our party, he was, at first, unwilling to land amidst the hostile tribe nearest to his own. He often told us how the savage Oens men "when the leaf red," crossed the mountains from the eastern coast of Tierra del Fuego, and made inroads on the natives of this part of the country. It was most curious to watch him when thus talking, and see his eyes gleaming and his whole face assume a new and wild expression. As we proceeded along the Beagle Channel, the scenery assumed a peculiar and very magnificent character; but the effect was much lessened from the lowness of the point of view in a boat, and from looking along the valley, and thus losing all the beauty of a succession of ridges. The mountains were here about three thousand feet high, and terminated in sharp and jagged points. They rose in one unbroken sweep from the water's edge, and were covered to the height of fourteen or fifteen hundred feet by the dusky-coloured forest. It was most curious to observe, as far as the eye could range, how level and truly horizontal the line on the mountain side was, at which trees ceased to grow: it precisely resembled the high-water mark of drift-weed on a sea-beach.

At night we slept close to the junction of Ponsonby Sound with the Beagle Channel. A small family of Fuegians, who were living in the cove, were quiet and inoffensive, and soon joined our party round a blazing fire. We were well clothed, and though sitting close to the fire were far from too warm; yet these naked savages, though further off, were observed, to our great surprise, to be streaming with perspiration at undergoing such a roasting. They seemed, however, very well pleased, and all joined in the chorus of the seamen's songs: but the manner in which they were invariably a little behindhand was quite ludicrous.

During the night the news had spread, and early in the morning (23rd) a fresh party arrived, belonging to the Tekenika, or Jemmy's tribe. Several of them had run so fast that their noses were bleeding, and their mouths frothed from the rapidity with which they talked; and with their naked bodies all bedaubed with black, white,[106] and red, they looked like so many demoniacs who had been fighting. We then proceeded (accompanied by twelve canoes, each holding four or five people) down Ponsonby Sound to the spot where poor Jemmy expected to find his mother and relatives. He had already heard that his father was dead; but as he had had a "dream in his head" to that effect, he did not seem to care much about it, and repeatedly comforted himself with the very natural reflection—"Me no help it." He was not able to learn any particulars regarding his father's death, as his relations would not speak about it.

Jemmy was now in a district well known to him, and guided the boats to a quiet pretty cove named Woollya, surrounded by islets, every one of which and every point had its proper native name. We found here a family of Jemmy's tribe, but not his relations: we made friends with them; and in the evening they sent a canoe to inform Jemmy's mother and brothers. The cove was bordered by some acres of good sloping land, not covered (as elsewhere) either by peat or by forest-trees. Captain Fitz Roy originally intended, as before stated, to have taken York Minster and Fuegia to their own tribe on the west coast; but as they expressed a wish to remain here, and as the spot was singularly favourable, Captain Fitz Roy determined to settle here the whole party, including Matthews, the missionary. Five days were spent in building for them three large wigwams, in landing their goods, in digging two gardens, and sowing seeds.

The next morning after our arrival (the 24th) the Fuegians began to pour in, and Jemmy's mother and brothers arrived. Jemmy recognised the stentorian voice of one of his brothers at a prodigious distance. The meeting was less interesting than that between a horse, turned out into a field, when he joins an old companion. There was no demonstration of affection; they simply stared for a short time at each other; and the mother immediately went to look after her canoe. We heard, however, through York that the mother has been inconsolable for the loss of Jemmy, and had searched everywhere for him, thinking that he might have been left after having been taken in the boat. The women took much notice of and were very kind to Fuegia. We had already perceived that Jemmy had almost forgotten his own language. I should think there was scarcely another human being with so small a stock of language, for his English was very imperfect. It was laughable, but almost pitiable, to hear him speak to his wild brother in English, and then ask him in Spanish ("no sabe?") whether he did not understand him.

Everything went on peaceably during the three next days, whilst the gardens were digging and wigwams building. We estimated the number of natives at about one hundred and twenty. The women worked hard, whilst the men lounged about all day long, watching us. They asked for everything they saw, and stole what they could. They were delighted at our dancing and singing, and were particularly interested at seeing us wash in a neighbouring brook; they did not pay much attention to anything else, not even to our boats. Of all the things which York saw, during his absence from his country, nothing seems more to have astonished him than an ostrich, near Maldonado: breathless with astonishment, he came running to Mr. Bynoe, with whom he was out walking—"Oh, Mr. Bynoe, oh, bird all same horse!" Much as our white skins surprised the natives, by Mr. Low's account a negro-cook to a sealing vessel, did so more effectually; and the poor fellow was so mobbed and shouted at that he would never go on shore again. Everything went on so quietly, that some of the officers and myself took long walks in the surrounding hills and woods. Suddenly, however, on the 27th, every woman and child disappeared. We were all uneasy at this, as neither York nor Jemmy could make out the cause. It was thought by some that they had been frightened by our cleaning and firing off our muskets on the previous evening; by others, that it was owing to offence taken by an old savage, who, when told to keep further off, had coolly spit in the sentry's face, and had then, by gestures acted over a sleeping Fuegian, plainly showed, as it was said, that he should like to cut up and eat our man. Captain Fitz Roy, to avoid the chance of an encounter, which would have been fatal to so many of the Fuegians, thought it advisable for us to sleep at a cove a few miles distant. Matthews, with his usual quiet fortitude (remarkable in a man apparently possessing little energy of character), determined to stay with the Fuegians, who evinced no alarm for themselves; and so we left them to pass their first awful night.

On our return in the morning (28th) we were delighted to find all quiet, and the men employed in their canoes spearing fish. Captain Fitz Roy determined to send the yawl and one whale-boat back to the ship; and to proceed with the two other boats, one under his own command (in which he most kindly allowed me to accompany him), and one under Mr. Hammond, to survey the western parts of the Beagle Channel, and afterwards to return and visit the settlement. The day to our astonishment was overpoweringly hot, so that our skins were scorched: with this beautiful weather, the view in the middle of the Beagle Channel was very remarkable. Looking towards either hand, no object intercepted the vanishing points of this long canal between the mountains. The circumstance of its being an arm of the sea was rendered very evident by several huge whales[107] spouting in different directions. On one occasion I saw two of these monsters, probably male and female, slowly swimming one after the other, within less than a stone's throw of the shore, over which the beech-tree extended its branches.

We sailed on till it was dark, and then pitched our tents in a quiet creek. The greatest luxury was to find for our beds a beach of pebbles, for they were dry and yielded to the body. Peaty soil is damp; rock is uneven and hard; sand gets into one's meat, when cooked and eaten boat-fashion; but when lying in our blanket-bags, on a good bed of smooth pebbles, we passed most comfortable nights.

It was my watch till one o'clock. There is something very solemn in these scenes. At no time does the consciousness in what a remote corner of the world you are then standing, come so strongly before the mind. Everything tends to this effect; the stillness of the night is interrupted only by the heavy breathing of the seamen beneath the tents, and sometimes by the cry of a night-bird. The occasional barking of a dog, heard in the distance, reminds one that it is the land of the savage.

January 29th.—Early in the morning we arrived at the point where the Beagle Channel divides into two arms; and we entered the northern one. The scenery here becomes even grander than before. The lofty mountains on the north side compose the granitic axis, or backbone of the country, and boldly rise to a height of between three and four thousand feet, with one peak above six thousand feet. They are covered by a wide mantle of perpetual snow, and numerous cascades pour their waters, through the woods, into the narrow channel below. In many parts, magnificent glaciers extend from the mountain side to the water's edge. It is scarcely possible to imagine anything more beautiful than the beryl-like blue of these glaciers, and especially as contrasted with the dead white of the upper expanse of snow. The fragments which had fallen from the glacier into the water were floating away, and the channel with its icebergs presented, for the space of a mile, a miniature likeness of the Polar Sea. The boats being hauled on shore at our dinner-hour, we were admiring from the distance of half a mile a perpendicular cliff of ice, and were wishing that some more fragments would fall. At last, down came a mass with a roaring noise, and immediately we saw the smooth outline of a wave travelling towards us. The men ran down as quickly as they could to the boats; for the chance of their being dashed to pieces was evident. One of the seamen just caught hold of the bows, as the curling breaker reached it: he was knocked over and over, but not hurt; and the boats, though thrice lifted on high and let fall again, received no damage. This was most fortunate for us, for we were a hundred miles distant from the ship, and we should have been left without provisions or fire-arms. I had previously observed that some large fragments of rock on the beach had been lately displaced; but until seeing this wave, I did not understand the cause. One side of the creek was formed by a spur of mica-slate; the head by a cliff of ice about forty feet high; and the other side by a promontory fifty feet high, built up of huge rounded fragments of granite and mica-slate, out of which old trees were growing. This promontory was evidently a moraine, heaped up at a period when the glacier had greater dimensions.

When we reached the western mouth of this northern branch of the Beagle Channel, we sailed amongst many unknown desolate islands, and the weather was wretchedly bad. We met with no natives. The coast was almost everywhere so steep, that we had several times to pull many miles before we could find space enough to pitch our two tents: one night we slept on large round boulders, with putrefying sea-weed between them; and when the tide rose, we had to get up and move our blanket-bags. The farthest point westward which we reached was Stewart Island, a distance of about one hundred and fifty miles from our ship. We returned into the Beagle Channel by the southern arm, and thence proceeded, with no adventure, back to Ponsonby Sound.

February 6th.—We arrived at Woollya. Matthews gave so bad an account of the conduct of the Fuegians, that Captain Fitz Roy determined to take him back to the Beagle; and ultimately he was left at New Zealand, where his brother was a missionary. From the time of our leaving, a regular system of plunder commenced; fresh parties of the natives kept arriving: York and Jemmy lost many things, and Matthews almost everything which had not been concealed underground. Every article seemed to have been torn up and divided by the natives. Matthews described the watch he was obliged always to keep as most harassing; night and day he was surrounded by the natives, who tried to tire him out by making an incessant noise close to his head. One day an old man, whom Matthews asked to leave his wigwam, immediately returned with a large stone in his hand: another day a whole party came armed with stones and stakes, and some of the younger men and Jemmy's brother were crying: Matthews met them with presents. Another party showed by signs that they wished to strip him naked and pluck all the hairs out of his face and body. I think we arrived just in time to save his life. Jemmy's relatives had been so vain and foolish, that they had showed to strangers their plunder, and their manner of obtaining it. It was quite melancholy leaving the three Fuegians with their savage countrymen; but it was a great comfort that they had no personal fears. York, being a powerful resolute man, was pretty sure to get on well, together with his wife Fuegia. Poor Jemmy looked rather disconsolate, and would then, I have little doubt, have been glad to have returned with us. His own brother had stolen many things from him; and as he remarked, "What fashion call that:" he abused his countrymen, "all bad men, no sabe (know) nothing," and, though I never heard him swear before, "damned fools." Our three

Fuegians, though they had been only three years with civilized men, would, I am sure, have been glad to have retained their new habits; but this was obviously impossible. I fear it is more than doubtful, whether their visit will have been of any use to them.

In the evening, with Matthews on board, we made sail back to the ship, not by the Beagle Channel, but by the southern coast. The boats were heavily laden and the sea rough, and we had a dangerous passage. By the evening of the 7th we were on board the Beagle after an absence of twenty days, during which time we had gone three hundred miles in the open boats. On the 11th, Captain Fitz Roy paid a visit by himself to the Fuegians and found them going on well; and that they had lost very few more things.

ON THE LAST DAY of February in the succeeding year (1834), the Beagle anchored in a beautiful little cove at the eastern entrance of the Beagle Channel. Captain Fitz Roy determined on the bold, and as it proved successful, attempt to beat against the westerly winds by the same route, which we had followed in the boats to the settlement at Woollya. We did not see many natives until we were near Ponsonby Sound, where we were followed by ten or twelve canoes. The natives did not at all understand the reason of our tacking, and, instead of meeting us at each tack, vainly strove to follow us in our zigzag course. I was amused at finding what a difference the circumstance of being quite superior in force made, in the interest of beholding these savages. While in the boats I got to hate the very sound of their voices, so much trouble did they give us. The first and last word was "yammerschooner." When, entering some quiet little cove, we have looked round and thought to pass a quiet night, the odious word "yammerschooner" has shrilly sounded from some gloomy nook, and then the little signal-smoke has curled up to spread the news far and wide. On leaving some place we have said to each other, "Thank Heaven, we have at last fairly left these wretches!" when one more faint hallo from an all-powerful voice, heard at a prodigious distance, would reach our ears, and clearly could we distinguish—"yammerschooner." But now, the more Fuegians the merrier; and very merry work it was. Both parties laughing, wondering, gaping at each other; we pitying them, for giving us good fish and crabs for rags, etc.; they grasping at the chance of finding people so foolish as to exchange such splendid ornaments for a good supper. It was most amusing to see the undisguised smile of satisfaction with which one young woman with her face painted black, tied several bits of scarlet cloth round her head with rushes. Her husband, who enjoyed the very universal privilege in this country of possessing two wives, evidently became jealous of all the attention paid to his young wife; and, after a consultation with his naked beauties, was paddled away by them.

Some of the Fuegians plainly showed that they had a fair notion of barter. I gave one man a large nail (a most valuable present) without making any signs for a return; but he immediately picked out two fish, and handed them up on the point of his spear. If any present was designed for one canoe, and it fell near another, it was invariably given to the right owner. The Fuegian boy, whom Mr. Low had on board, showed, by going into the most violent passion, that he quite understood the reproach of being called a liar, which in truth he was. We were this time, as on all former occasions, much surprised at the little notice, or rather none whatever, which was taken of many things, the use of which must have been evident to the natives. Simple circumstances—such as the beauty of scarlet cloth or blue beads, the absence of women, our care in washing ourselves,—excited their admiration far more than any grand or complicated object, such as our ship. Bougainville has well remarked concerning these people, that they treat the "chefs-d'oeuvre de l'industrie humaine, comme ils traitent les loix de la nature et ses phénomènes."

On the 5th of March, we anchored in a cove at Woollya, but we saw not a soul there. We were alarmed at this, for the natives in Ponsonby Sound showed by gestures, that there had been fighting; and we afterwards heard that the dreaded Oens men had made a descent. Soon a canoe, with a little flag flying, was seen approaching, with one of the men in it washing the paint off his face. This man was poor Jemmy,—now a thin, haggard savage, with long disordered hair, and naked, except a bit of blanket round his waist. We did not recognise him till he was close to us, for he was ashamed of himself, and turned his back to the ship. We had left him plump, fat, clean, and well-dressed;—I never saw so complete and grievous a change. As soon, however, as he was clothed, and the first flurry was over, things wore a good appearance. He dined with Captain Fitz Roy, and ate his dinner as tidily as formerly. He told us that he had "too much" (meaning enough) to eat, that he was not cold, that his relations were very good people, and that he did not wish to go back to England: in the evening we found out the cause of this great change in Jemmy's feelings, in the arrival of his young and nice-looking wife. With his usual good feeling, he brought two beautiful otter-skins for two of his best friends, and some spear-heads and arrows made with his own hands for the Captain. He said he had built a canoe for himself, and he boasted that he could talk a little of his own language! But it is a most singular fact, that he appears to have taught all his tribe some English: an old man spontaneously announced "Jemmy Button's wife." Jemmy had lost all his property. He told us that York Minster had built a large canoe, and with his wife Fuegia,[108] had several months since gone to his own country, and had taken farewell by an act of consummate villainy; he persuaded Jemmy and his mother to come with him, and then on the way deserted them by night, stealing every article of their property.

Jemmy went to sleep on shore, and in the morning returned, and remained on board till the ship got under way, which frightened his wife, who continued crying violently till he got into his canoe. He returned loaded with valuable property. Every soul on board was heartily sorry to shake hands with him for the last time. I do not now doubt that he will be as happy as, perhaps happier than, if he had never left his own country. Every one must sincerely hope that Captain Fitz Roy's noble hope may be fulfilled, of being rewarded for the many generous sacrifices which he made for these Fuegians, by some shipwrecked sailor being protected by the descendants of Jemmy Button and his tribe! When Jemmy reached the shore, he lighted a signal fire, and the smoke curled up, bidding us a last and long farewell, as the ship stood on her course into the open sea.

THE PERFECT EQUALITY among the individuals composing the Fuegian tribes must for a long time retard their civilization. As we see those animals, whose instinct compels them to live in society and obey a chief, are most capable of improvement, so is it with the races of mankind. Whether we look at it as a cause or a consequence, the more civilized always have the most artificial governments. For instance, the inhabitants of Otaheite, who, when first discovered, were governed by hereditary kings, had arrived at a far higher grade than another branch of the same people, the New Zealanders,—who, although benefited by being compelled to turn their attention to agriculture, were republicans in the most absolute sense. In Tierra del Fuego, until some chief shall arise with power sufficient to secure any acquired advantage, such as the domesticated animals, it seems scarcely possible that the political state of the country can be improved. At present, even a piece of cloth given to one is torn into shreds and distributed; and no one individual becomes richer than another. On the other hand, it is difficult to understand how a chief can arise till there is property of some sort by which he might manifest his superiority and increase his power.

I believe, in this extreme part of South America, man exists in a lower state of improvement than in any other part of the world. The South Sea Islanders, of the two races inhabiting the Pacific, are comparatively civilized. The Esquimau in his subterranean hut, enjoys some of the comforts of life, and in his canoe, when fully equipped, manifests much skill. Some of the tribes of Southern Africa, prowling about in search of roots, and living concealed on the wild and arid plains, are sufficiently wretched. The Australian, in the simplicity of the arts of life, comes nearest the Fuegian:

he can, however, boast of his boomerang, his spear and throwing-stick, his method of climbing trees, of tracking animals, and of hunting. Although the Australian may be superior in acquirements, it by no means follows that he is likewise superior in mental capacity: indeed, from what I saw of the Fuegians when on board, and from what I have read of the Australians, I should think the case was exactly the reverse.

CHAPTER XI

Strait of Magellan—Climate of the Southern Coasts

STRAIT OF MAGELLAN • PORT FAMINE • ASCENT OF MOUNT TARN • FORESTS • EDIBLE FUNGUS • ZOOLOGY • GREAT SEA-WEED • LEAVE TIERRA DEL FUEGO • CLIMATE • FRUIT-TREES AND PRODUCTIONS OF THE SOUTHERN COASTS • HEIGHT OF SNOW-LINE ON THE CORDILLERA • DESCENT OF GLACIERS TO THE SEA • ICEBERGS FORMED • TRANSPORTAL OF BOULDERS • CLIMATE AND PRODUCTIONS OF THE ANTARCTIC ISLANDS • PRESERVATION OF FROZEN CARCASSES • RECAPITULATION

In the end of May, 1834, we entered for a second time the eastern mouth of the Strait of Magellan. The country on both sides of this part of the Strait consists of nearly level plains, like those of Patagonia. Cape Negro, a little within the second Narrows, may be considered as the point where the land begins to assume the marked features of Tierra del Fuego. On the east coast, south of the Strait, broken park-like scenery in a like manner connects these two countries, which are opposed to each other in almost every feature. It is truly surprising to find in a space of twenty miles such a change in the landscape. If we take a rather greater distance, as between Port Famine and Gregory Bay, that is about sixty miles, the difference is still more wonderful. At the former place, we have rounded mountains concealed by impervious forests, which are drenched with the rain, brought by an endless succession of gales; while at Cape Gregory, there is a clear and bright blue sky over the dry and sterile plains. The atmospheric currents,[109] although rapid, turbulent, and unconfined by any apparent limits, yet seem to follow, like a river in its bed, a regularly determined course.

During our previous visit (in January), we had an interview at Cape Gregory with the famous so-called gigantic Patagonians, who gave us a cordial reception. Their height appears greater than it really is, from their large guanaco mantles, their long flowing hair, and general figure: on an average, their height is about six feet, with some men taller and only a few shorter; and the women are also tall; altogether they are certainly the tallest race which we anywhere saw. In features they strikingly resemble the more northern Indians whom I saw with Rosas, but they have a wilder and more formidable appearance: their faces were much painted with red and black, and one man was ringed and dotted with white like a Fuegian. Captain Fitz Roy offered to take any three of them on board, and all seemed determined to be of the three. It was long before we could clear the boat; at last we got on board with our three giants, who dined with the Captain, and behaved quite like gentlemen, helping themselves with knives, forks, and spoons: nothing was so much relished as sugar. This tribe has had so much communication with sealers and whalers that most of the men can speak a little English and Spanish; and they are half civilized, and proportionally demoralized.

The next morning a large party went on shore, to barter for skins and ostrich-feathers; fire-arms being refused, tobacco was in greatest request, far more so than axes or tools. The whole population of the toldos, men, women, and children, were arranged on a bank. It was an amusing scene, and it was impossible not to like the so-called giants, they were so thoroughly good-humoured and unsuspecting: they asked us to come again. They seem to like to have Europeans to live with them; and old Maria, an important woman in the tribe, once begged Mr. Low to leave any one of his sailors with them. They spend the greater part of the year here; but in summer they hunt along the foot of the Cordillera: sometimes they travel as far as the Rio Negro, 750 miles to the north. They are well stocked with horses, each man having, according to Mr. Low, six or seven, and all the women, and even children, their one own horse. In the time of Sarmiento (1580), these Indians had bows and arrows, now long since disused; they then also possessed some horses. This is a very curious fact, showing the extraordinarily rapid multiplication of horses in South America. The horse was first landed at Buenos Ayres in 1537, and the colony being then for a time deserted, the horse ran wild;[110] in 1580, only forty-three years afterwards, we hear of them at the Strait of Magellan! Mr. Low informs me, that a neighbouring tribe of foot-Indians is now changing into horse-Indians: the tribe at Gregory Bay giving them their worn-out horses, and sending in winter a few of their best skilled men to hunt for them.

June 1st.—We anchored in the fine bay of Port Famine. It was now the beginning of winter, and I never saw a more cheerless prospect; the dusky woods, piebald with snow, could be only seen indistinctly, through a drizzling hazy atmosphere. We were, however, lucky in getting two fine days. On one of these, Mount Sarmiento, a distant mountain 6800 feet high, presented a very noble spectacle. I was frequently surprised in the scenery of Tierra del Fuego, at the little apparent elevation of mountains really lofty. I suspect it is owing to a cause which would not at first be imagined, namely, that the whole mass, from the summit to the water's edge, is generally in full view. I remember having seen a mountain, first from the Beagle Channel, where the whole sweep from the summit to the base was full in view, and then from Ponsonby Sound across several successive ridges; and it was curious to observe in the latter case, as each fresh ridge afforded fresh means of judging of the distance, how the mountain rose in height.

Before reaching Port Famine, two men were seen running along the shore and hailing the ship. A boat was sent for them. They turned out to be two sailors who had run away from a sealing-vessel, and had joined the Patagonians. These Indians had treated them with their usual disinterested hospitality. They had parted company through accident, and were then proceeding to Port Famine in hopes of finding some ship. I dare say they were worthless vagabonds, but I never saw more miserable-looking ones. They had been living for some days on mussel-shells and berries, and their tattered clothes had been burnt by sleeping so near their fires. They had been exposed night and day, without any shelter, to the late incessant gales, with rain, sleet, and snow, and yet they were in good health.

During our stay at Port Famine, the Fuegians twice came and plagued us. As there were many instruments, clothes, and men on shore, it was thought necessary to frighten them away. The first time a few great guns were fired, when they were far distant. It was most ludicrous to watch through a glass the Indians, as often as the shot struck the water, take up stones, and, as a bold defiance, throw them towards the ship, though about a mile and a half distant! A boat was sent with orders to fire a few musket-shots wide of them. The Fuegians hid themselves behind the trees, and for every discharge of the muskets they fired their arrows; all, however, fell short of the boat, and the officer as he pointed at them laughed. This made the Fuegians frantic with passion, and they shook their mantles in vain rage. At last, seeing the balls cut and strike the trees, they ran away, and we were left in peace and quietness. During the former voyage the Fuegians were here very troublesome, and to frighten them a rocket was fired at night over their wigwams; it answered effectually, and one of the officers told me that the clamour first raised, and the barking of the dogs, was quite ludicrous in contrast with the profound silence which in a minute or two afterwards prevailed. The next morning not a single Fuegian was in the neighbourhood.

When the Beagle was here in the month of February, I started one morning at four o'clock to ascend Mount Tarn, which is 2600 feet high, and is the most elevated point in this immediate district. We went in a boat to the foot of the mountain (but unluckily not to the best part), and then began our ascent. The forest commences at the line of high-water mark, and during the first two hours I gave over all hopes of reaching the summit. So thick was the wood, that it was necessary to have constant recourse to the compass; for every landmark, though in a mountainous country, was completely shut out. In the deep ravines, the death-like scene of desolation exceeded all description; outside it was blowing a gale, but in these hollows, not even a breath of wind stirred the leaves of the tallest trees. So gloomy, cold, and wet was every part, that not even the fungi, mosses, or ferns could flourish. In the valleys it was scarcely possible to crawl along, they were so completely barricaded by great mouldering trunks, which had fallen down in every direction. When passing over these natural bridges, one's course was often arrested by sinking knee deep into the rotten wood; at other times, when attempting to lean against a firm tree, one was startled by finding a mass of decayed matter ready to fall at the slightest touch. We at last found ourselves among the stunted trees, and then soon reached the bare ridge, which conducted us to the summit. Here was a view characteristic of Tierra del Fuego; irregular chains of hills, mottled with patches of snow, deep yellowish-green valleys, and arms of the sea intersecting the land in many directions. The strong wind was piercingly cold, and the atmosphere rather hazy, so that we did not stay long on the top of the mountain. Our descent was not quite so laborious as our ascent; for the weight of the body forced a passage, and all the slips and falls were in the right direction.

I have already mentioned the sombre and dull character of the evergreen forests,[111] in which two or three species of trees grow, to the exclusion of all others. Above the forest land, there are many dwarf alpine plants, which all spring from the mass of peat, and help to compose it: these plants are very remarkable from their close alliance with the species growing on the mountains of Europe, though so many thousand miles distant. The central part of Tierra del Fuego, where the clay-slate formation occurs, is most favourable to the growth of trees; on the outer coast the poorer granitic soil, and a situation more exposed to the violent winds, do not allow of their attaining any great size. Near Port Famine I have seen more large trees than anywhere else: I measured a Winter's Bark which was four feet six inches in girth, and several of the beech were as much as thirteen feet. Captain King also mentions a beech which was seven feet in diameter, seventeen feet above the roots.

There is one vegetable production deserving notice from its importance as an article of food to the Fuegians. It is a globular, bright-yellow fungus, which grows in vast numbers on the beech-trees. When young it is elastic and turgid, with a smooth surface; but when mature it shrinks, becomes tougher, and has its entire surface deeply pitted or honey-combed, as represented in the accompanying wood-cut. This fungus belongs to a new and curious genus;[112] I found a second species on another species of beech in Chile; and Dr. Hooker informs me, that just lately a third species has been discovered on a third species of beech in Van Dieman's Land. How singular is this relationship between parasitical fungi and the trees on which they grow, in distant parts of the world! In Tierra del Fuego the fungus in its tough and mature state is collected in large quantities by the women and children, and is eaten uncooked. It has a mucilaginous, slightly sweet taste, with a faint smell like that of a mushroom. With the exception of a few berries, chiefly of a dwarf arbutus, the natives eat no vegetable food besides this fungus. In New Zealand, before the introduction of the potato, the roots of the fern were largely consumed; at the present time, I believe, Tierra del Fuego is the only country in the world where a cryptogamic plant affords a staple article of food.

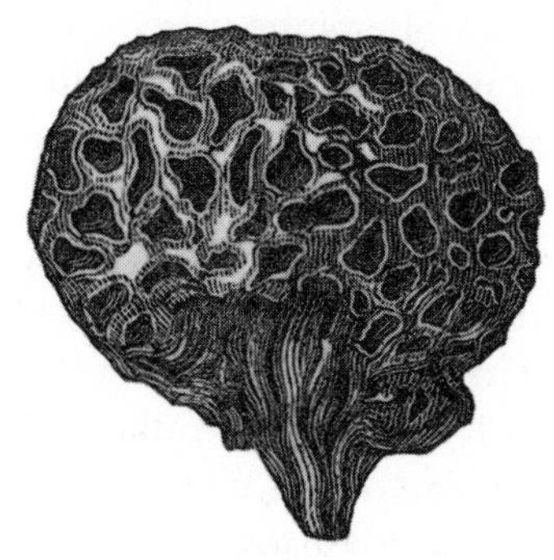

The zoology of Tierra del Fuego, as might have been expected from the nature of its climate and vegetation, is very poor. Of mammalia, besides whales and seals, there is one bat, a kind of mouse (Reithrodon chinchilloides), two true mice, a ctenomys allied to or identical with the tucutuco, two foxes (Canis Magellanicus and C. Azaræ), a sea-otter, the guanaco, and a deer. Most of these animals inhabit only the drier eastern parts of the country; and the deer has never been seen south of the Strait of Magellan. Observing the general correspondence of the cliffs of soft sandstone, mud, and shingle, on the opposite sides of the Strait, and on some intervening islands, one is strongly tempted to believe that the land was once joined, and thus allowed animals so delicate and helpless as the tucutuco and Reithrodon to pass over. The correspondence of the cliffs is far from proving any junction; because such cliffs generally are formed by the intersection of sloping deposits, which, before the elevation of the land, had been accumulated near the then existing shores. It is, however, a remarkable coincidence, that in the two large islands cut off by the Beagle Channel from the rest of Tierra del Fuego, one has cliffs composed of matter that may be called stratified alluvium, which front similar ones on the opposite side of the channel,—while the other is exclusively bordered by old crystalline rocks: in the former, called Navarin Island, both foxes and guanacos occur; but in the latter, Hoste Island, although similar in every respect, and only separated by a channel a little more than half a mile wide, I have the word of Jemmy Button for saying that neither of these animals are found.

The gloomy woods are inhabited by few birds: occasionally the plaintive note of a white-tufted tyrant-flycatcher (Myiobius albiceps) may be heard, concealed near the summit of the most lofty trees; and more rarely the loud strange cry of a black wood-pecker, with a fine scarlet crest on its head. A little, dusky-coloured wren (Scytalopus Magellanicus) hops in a skulking manner among the entangled mass of the fallen and decaying trunks. But the creeper (Oxyurus tupinieri) is the commonest bird in the country. Throughout the beech forests, high up and low down, in the most gloomy, wet, and impenetrable ravines, it may be met with. This little bird no doubt appears more numerous than it really is, from its habit of following with seeming curiosity any person who enters these silent woods: continually uttering a harsh twitter, it flutters from tree to tree, within a few feet of the intruder's face. It is far from wishing for the modest concealment of the true creeper (Certhia familiaris); nor does it, like that bird, run up the trunks of trees, but industriously, after the manner of a willow-wren, hops about, and searches for insects on every twig and branch. In the more open parts, three or four species of finches, a thrush, a starling (or Icterus), two Opetiorhynchi, and several hawks and owls occur.

The absence of any species whatever in the whole class of Reptiles, is a marked feature in the zoology of this country, as well as in that of the Falkland Islands. I do not ground this statement merely on my own observation, but I heard it from the Spanish inhabitants of the latter place, and from Jemmy Button with regard to Tierra del Fuego. On the banks of the Santa Cruz, in 50° south, I saw a frog; and it is not improbable that these animals, as well as lizards, may be found as far south as the Strait of Magellan, where the country retains the character of Patagonia; but within the damp and cold limit of Tierra del Fuego not one occurs. That the climate would not have suited some of the orders, such as lizards, might have been foreseen; but with respect to frogs, this was not so obvious.

Beetles occur in very small numbers: it was long before I could believe that a country as large as Scotland, covered with vegetable productions and with a variety of stations, could be so unproductive. The few which I found were alpine species (Harpalidæ and Heteromidæ) living under stones. The vegetable-feeding Chrysomelidæ, so eminently characteristic of the Tropics, are here almost entirely absent;[113] I saw very few flies, butterflies, or bees, and no crickets or Orthoptera. In the pools of water I found but a few aquatic beetles, and not any fresh-water shells: Succinea at first appears an exception; but here it must be called a terrestrial shell, for it lives on the damp herbage far from the water. Land-snails could be procured only in the same alpine situations with the beetles. I have already contrasted the climate as well as the general appearance of Tierra del Fuego with that of Patagonia; and the difference is strongly exemplified in the entomology. I do not believe they have one species in common; certainly the general character of the insects is widely different.

If we turn from the land to the sea, we shall find the latter as abundantly stocked with living creatures as the former is poorly so. In all parts of the world a rocky and partially protected shore perhaps supports, in a given space, a greater number of individual animals than any other station. There is one marine production which, from its importance, is worthy of a particular history. It is the kelp, or Macrocystis pyrifera. This plant grows on every rock from low-water mark to a great depth, both on the outer coast and within the channels.[114] I believe, during the voyages of the Adventure and Beagle, not one rock near the surface was discovered which was not buoyed by this floating weed. The good service it thus affords to vessels navigating near this stormy land is evident; and it certainly has saved many a one from being wrecked. I know few things more surprising than to see this plant growing and flourishing amidst those great breakers of the western ocean, which no mass of rock, let it be ever so hard, can long resist. The stem is round, slimy, and smooth, and seldom has a diameter of so much as an inch. A few taken together are sufficiently strong to support the weight of the large loose stones, to which in the inland channels they grow attached; and yet some of these stones were so heavy that when drawn to the surface, they could scarcely be lifted into a boat by one person. Captain Cook, in his second voyage, says, that this plant at Kerguelen Land rises from a greater depth than twenty-four fathoms; "and as it does not grow in a perpendicular direction, but makes a very acute angle with the bottom, and much of it afterwards spreads many fathoms on the surface of the sea, I am well warranted to say that some of it grows to the length of sixty fathoms and upwards." I do not suppose the stem of any other plant attains so great a length as three hundred and sixty feet, as stated by Captain Cook. Captain Fitz Roy, moreover, found it growing[115] up from the greater depth of forty-five fathoms. The beds of this sea-weed, even when of not great breadth, make excellent natural floating breakwaters. It is quite curious to see, in an exposed harbour, how soon the waves from the open sea, as they travel through the straggling stems, sink in height, and pass into smooth water.

The number of living creatures of all Orders, whose existence intimately depends on the kelp, is wonderful. A great volume might be written, describing the inhabitants of one of these beds of sea-weed. Almost all the leaves, excepting those that float on the surface, are so thickly incrusted with

	Latitude	Summer Temp.	Winter Temp.	Mean of Summer and Winter
Tierra del Fuego	53° 38' S.	50°	33°.08	41°.54
Falkland Islands	51 38 S.	51	—	—
Dublin	53 21 N.	59.54	39.2	49.37

Hence we see that the central part of Tierra del Fuego is colder in winter, and no less than 9½° less hot in summer, than Dublin. According to von Buch, the mean temperature of July (not the hottest month in the year) at Saltenfiord in Norway, is as high as 57°.8, and this place is actually 13° nearer the pole than Port Famine![116] Inhospitable as this climate appears to our feelings, evergreen trees flourish luxuriantly under it. Humming-birds may be seen sucking the flowers, and parrots feeding on the seeds of the Winter's Bark, in lat. 55° S. I have already remarked to what a degree the sea swarms with living creatures; and the shells (such as the Patellæ, Fissurellæ, Chitons, and Barnacles), according to Mr. G. B. Sowerby, are of a much larger size and of a more vigorous growth, than the analogous species in the northern hemisphere. A large Voluta is abundant in southern Tierra del Fuego and the Falkland Islands. At Bahia Blanca, in lat. 39° S., the most abundant shells were three species of Oliva (one of large size), one or two Volutas, and a Terebra. Now, these are amongst the best characterized tropical forms. It is doubtful whether even one small species of Oliva exists on the southern shores of Europe, and there are no species of the two other genera. If a geologist were to find in lat. 39° on the coast of Portugal a bed containing numerous shells belonging to three species of Oliva, to a Voluta and Terebra, he would probably assert that the climate at the period of their existence must have been tropical; but judging from South America, such an inference might be erroneous.

The equable, humid, and windy climate of Tierra del Fuego extends, with only a small increase of heat, for many degrees along the west coast of the continent. The forests for 600 miles northward of Cape Horn, have a very similar aspect. As a proof of the equable climate, even for 300 or 400 miles still further northward, I may mention that in Chiloe (corresponding in latitude with the northern parts of Spain) the peach seldom produces fruit, whilst strawberries and apples thrive to perfection. Even the crops of barley and wheat[117] are often brought into the houses to be dried and ripened. At Valdivia (in the same latitude of 40°, with Madrid) grapes and figs ripen, but are not common; olives seldom ripen even partially, and oranges not at all. These fruits, in corresponding latitudes in Europe, are well known to succeed to perfection; and even in this continent, at the Rio Negro, under nearly the same parallel with Valdivia, sweet potatoes (convolvulus) are cultivated; and grapes, figs, olives, oranges, water and musk melons, produce abundant fruit. Although the humid and equable climate of Chiloe, and of the coast northward and southward of it, is so unfavourable to our fruits, yet the native forests, from lat. 45° to 38°, almost rival in luxuriance those of the glowing intertropical regions. Stately trees of many kinds, with smooth and highly coloured barks, are loaded by parasitical monocotyledonous plants; large and elegant ferns are numerous, and arborescent grasses entwine the trees into one entangled mass to the height of thirty or forty feet above the ground. Palm-trees grow in lat 37°; an arborescent grass, very like a bamboo, in 40°; and another closely allied kind, of great length, but not erect, flourishes even as far south as 45° S.

An equable climate, evidently due to the large area of sea compared with the land, seems to extend over the greater part of the southern hemisphere; and, as a consequence, the vegetation partakes of a semi-tropical character. Tree-ferns thrive luxuriantly in Van Diemen's Land (lat. 45°), and I measured one trunk no less than six feet in circumference. An arborescent fern was found by Forster in New Zealand in 46°, where orchideous plants are parasitical on the trees. In the Auckland Islands, ferns, according to Dr. Dieffenbach,[118] have trunks so thick and high that they may be almost called tree-ferns; and in these islands, and even as far south as lat. 55° in the Macquarrie Islands, parrots abound.

On the Height of the Snow-line, and on the Descent of the Glaciers in South America.—For the detailed authorities for the following table, I must refer to the former edition:—

Latitude	Height in feet of Snow-line	Observer
Equatorial region; mean result	15,748	Humboldt.
Bolivia, lat. 16° to 18° S.	17,000	Pentland.
Central Chile, lat. 33° S.	14,500 to 15,000	Gillies, and the Author.
Chiloe, lat. 41° to 43° S.	6,000	Officers of the Beagle, and the Author.
Tierra del Fuego, 54° S.	3,500 to 4,000	King.

As the height of the plane of perpetual snow seems chiefly to be determined by the extreme heat of the summer, rather than by the mean temperature of the year, we ought not to be surprised at its descent in the Strait of Magellan, where the summer is so cool, to only 3500 or 4000 feet above the level of the sea; although in Norway, we must travel to between lat. 67° and 70° N., that is, about 14° nearer the pole, to meet with perpetual snow at this low level. The difference in height, namely, about 9000 feet, between the snow-line on the Cordillera behind Chiloe (with its highest points ranging from only 5600 to 7500 feet) and in central Chile[119] (a distance of only 9° of latitude), is truly wonderful. The land from the southward of Chiloe to near Concepcion (lat. 37°) is hidden by one dense forest dripping with moisture. The sky is cloudy, and we have seen how badly the fruits of southern Europe succeed. In central Chile, on the other hand, a little northward of Concepcion, the sky is generally clear, rain does not fall for the seven summer months, and southern European fruits succeed admirably; and even the sugar-cane has been cultivated.[120] No doubt the plane of perpetual snow undergoes the above remarkable flexure of 9000 feet, unparalleled in other parts of the world, not far from the latitude of Concepcion, where the land ceases to be covered with forest-trees; for trees in South America indicate a rainy climate, and rain a clouded sky and little heat in summer.

The descent of glaciers to the sea must, I conceive, mainly depend (subject, of course, to a proper supply of snow in the upper region) on the lowness of the line of perpetual snow on steep mountains near the coast. As the snow-line is so low in Tierra del Fuego, we might have expected that many of the glaciers would have reached the sea. Nevertheless, I was astonished when I first saw a range, only from 3000 to 4000 feet in height, in the latitude of Cumberland, with every valley filled with streams of ice descending to the sea-coast. Almost every arm of the sea, which penetrates to the interior higher chain, not only in Tierra del Fuego, but on the coast for 650 miles northwards, is terminated by "tremendous and astonishing glaciers," as described by one of the officers on the survey. Great masses of ice frequently fall from these icy cliffs, and the crash reverberates like the broadside of a man-of-war through the lonely channels. These falls, as noticed in the last chapter, produce great waves which break on the adjoining coasts. It is known that earthquakes frequently cause masses of earth to fall from sea-cliffs: how terrific, then, would be the effect of a severe shock (and such occur here[121]) on a body like a glacier, already in motion, and traversed by fissures! I can readily believe that the water would be fairly beaten back out of the deepest channel, and then, returning with an overwhelming force, would whirl about huge masses of rock like so much chaff. In Eyre's Sound, in the latitude of Paris, there are immense glaciers, and yet the loftiest neighbouring mountain is only 6200 feet high. In this Sound, about fifty icebergs were seen at one time floating outwards, and one of them must have been *at least* 168 feet in total height. Some of the icebergs were loaded with blocks of no inconsiderable size, of granite and other rocks, different from the clay-slate of the surrounding mountains. The glacier furthest from the pole, surveyed during the voyages of the Adventure and Beagle, is in lat. 46° 50´, in the Gulf of Penas. It is 15 miles long, and in one part 7 broad and descends to the sea-coast. But even a few miles northward of this glacier, in

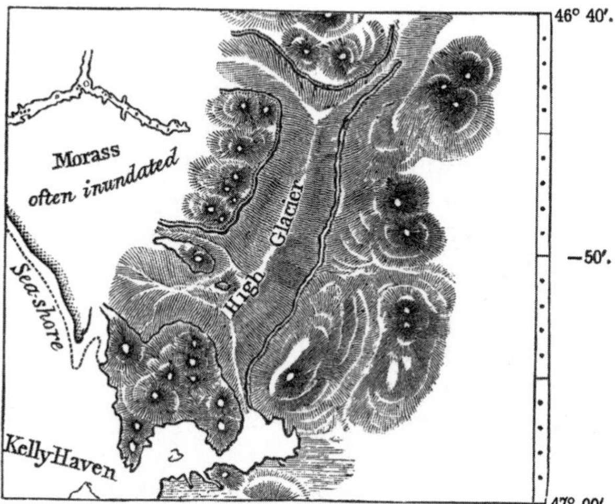

the Laguna de San Rafael, some Spanish missionaries[122] encountered "many icebergs, some great, some small, and others middle-sized," in a narrow arm of the sea, on the 22nd of the month corresponding with our June, and in a latitude corresponding with that of the Lake of Geneva!

In Europe, the most southern glacier which comes down to the sea is met with, according to Von Buch, on the coast of Norway, in lat. 67°. Now, this is more than 20° of latitude, or 1230 miles, nearer the pole than the Laguna de San Rafael. The position of the glaciers at this place and in the Gulf of Penas may be put even in a more striking point of view, for they descend to the sea-coast within 7½ ° of latitude, or 450 miles, of a harbour, where three species of Oliva, a Voluta, and a Terebra, are the commonest shells, within less than 9° from where palms grow, within 4½° of a region where the jaguar and puma range over the plains, less than 2½° from arborescent grasses, and (looking to the westward in the same hemisphere) less than 2° from orchideous parasites, and within a single degree of tree-ferns!

These facts are of high geological interest with respect to the climate of the northern hemisphere at the period when boulders were transported. I will not here detail how simply the theory of icebergs being charged with fragments of rock, explain the origin and position of the gigantic boulders of eastern Tierra del Fuego, on the high plain of Santa Cruz, and on the island of Chiloe. In Tierra del Fuego, the greater number of boulders lie on the lines of old sea-channels, now converted into dry valleys by the elevation of the land. They are associated with a great unstratified formation of mud and sand, containing rounded and angular fragments of all sizes, which has originated[123] in the repeated ploughing up of the sea-bottom by the stranding of icebergs, and by the matter transported on them. Few geologists now doubt that those erratic boulders which lie near lofty mountains have been pushed forward by the glaciers themselves, and that those distant from mountains, and embedded in subaqueous deposits, have been conveyed thither either on icebergs or frozen in coast-ice. The connection between the transportal of boulders and the presence of ice in some form, is strikingly shown by their geographical distribution over the earth. In South America they are not found further than 48° of latitude, measured from the southern pole; in North America it appears that the limit of their transportal extends to 53½° from the northern pole; but in Europe to not more than 40° of latitude, measured from the same point. On the other hand, in the intertropical parts of America, Asia, and Africa, they have never been observed; nor at the Cape of Good Hope, nor in Australia.[124]

On the Climate and Productions of the Antarctic Islands.—Considering the rankness of the vegetation in Tierra del Fuego, and on the coast northward of it, the condition of the islands south and south-west of America is truly surprising. Sandwich Land, in the latitude of the north part of Scotland, was found by Cook, during the hottest month of the year, "covered many fathoms thick with everlasting snow;" and there seems to be scarcely any vegetation. Georgia, an island 96 miles long and 10 broad, in the latitude of Yorkshire, "in the very height of summer, is in a manner wholly covered with frozen snow." It can boast only of moss, some tufts of grass, and wild burnet; it has only one land-bird (Anthus correndera), yet Iceland, which is 10° nearer the pole, has, according to Mackenzie, fifteen land-birds. The South Shetland Islands, in the same latitude as the southern half of Norway, possess only some lichens, moss, and a little grass; and Lieut. Kendall[125] found the bay, in which he was at anchor, beginning to freeze at a period corresponding with our 8th of September. The soil here consists of ice and volcanic ashes interstratified; and at a little depth beneath the surface it must remain perpetually congealed, for Lieut. Kendall found the body of a foreign sailor which had long been buried, with the flesh and all the features perfectly preserved. It is a singular fact, that on the two great continents in the northern hemisphere (but not in the broken land of Europe between them), we have the zone of perpetually frozen undersoil in a low latitude—namely, in 56° in North America at the depth of three feet,[126] and in 62° in Siberia at the depth of twelve to fifteen feet—as the result of a directly opposite condition of things to those of the southern hemisphere. On the northern continents, the winter is rendered excessively cold by the radiation from a large area of land into a clear sky, nor is it moderated by the warmth-bringing currents of the sea; the short summer, on the other hand, is hot. In the Southern Ocean the winter is not so excessively cold, but the summer is far less hot, for the clouded sky seldom allows the sun to warm the ocean, itself a bad absorbent of heat; and hence the mean temperature of the year, which regulates the zone of perpetually congealed under-soil, is low. It is evident that a rank vegetation, which does not so much require heat as it does protection from intense cold, would approach much nearer to this zone of perpetual congelation under the equable climate of the southern hemisphere, than under the extreme climate of the northern continents.

The case of the sailor's body perfectly preserved in the icy soil of the South Shetland Islands (lat. 62° to 63° S.), in a rather lower latitude than that (lat. 64° N.) under which Pallas found the frozen rhinoceros in Siberia, is very interesting. Although it is a fallacy, as I have endeavoured to show in a former chapter, to suppose that the larger quadrupeds require a luxuriant vegetation for their support, nevertheless it is important to find in the South Shetland Islands, a frozen under-soil within 360 miles of the forest-clad islands near Cape Horn, where, as far as the *bulk* of vegetation is concerned, any number of great quadrupeds might be supported. The perfect preservation of the carcasses of the Siberian elephants and rhinoceroses is certainly one of the most wonderful facts in geology; but independently of the imagined difficulty of supplying them with food from the adjoining countries, the whole case is not, I think, so perplexing as it has generally been considered. The plains of Siberia, like those of the Pampas, appear to have been formed under the sea, into which rivers brought down the bodies of many animals; of the greater number of these, only the skeletons have been preserved, but of others the perfect carcass. Now, it is known that in the shallow sea on the Arctic coast of America the bottom freezes,[127] and does not thaw in spring so soon as the surface of the land; moreover at greater depths, where the bottom of the sea does not freeze, the mud a few feet beneath the top layer might remain even in summer below 32°, as in the case on the land with the soil at the depth of a few feet. At still greater depths, the temperature of the mud and water would probably not be low enough to preserve the flesh; and hence, carcasses drifted beyond the shallow parts near an Arctic coast, would have only their skeletons preserved: now in the extreme northern parts of Siberia bones are infinitely numerous, so that even islets are said to be almost composed of them;[128] and those islets lie no less than ten degrees of latitude north of the place where Pallas found the frozen rhinoceros. On the other hand, a carcass washed by a flood into a shallow part of the Arctic Sea, would be preserved for an indefinite period, if it were soon afterwards covered with mud sufficiently thick to prevent the heat of the summer-water penetrating to it; and if, when the sea-bottom was upraised into land, the covering was sufficiently thick to prevent the heat of the summer air and sun thawing and corrupting it.

Recapitulation.—I will recapitulate the principal facts with regard to the climate, ice-action, and organic productions of the southern hemisphere, transposing the places in imagination to Europe, with which we are so much better acquainted. Then, near Lisbon, the commonest sea-shells, namely, three species of Oliva, a Voluta, and a Terebra, would have a tropical character. In the southern provinces of France, magnificent forests, intwined by arborescent grasses and with the trees loaded with parasitical plants, would hide the face of the land. The puma and the jaguar would haunt the Pyrenees. In the latitude of Mont Blanc, but on an island as far westward as Central North America, tree-ferns and parasitical Orchideæ would thrive amidst the thick woods. Even as far north as central Denmark, humming-birds would be seen fluttering about delicate flowers, and parrots feeding amidst the evergreen woods; and in the sea there, we should have a Voluta, and all the shells of large size and vigorous growth. Nevertheless, on some islands only 360 miles northward of our new Cape Horn in Denmark, a carcass buried in the soil (or if washed into a shallow sea, and covered up with mud) would be preserved perpetually frozen. If some bold navigator attempted to penetrate northward of these islands, he would run a thousand dangers amidst gigantic icebergs, on some of which he would see great blocks of rock borne far away from their original site. Another island of large size in the latitude of southern Scotland, but twice as far to the west, would be "almost wholly covered with everlasting snow," and would have each bay terminated by ice-cliffs, whence great masses would be yearly detached: this island would boast only of a little moss, grass, and burnet, and a titlark would be its only land inhabitant. From our new Cape Horn in Denmark, a chain of mountains, scarcely half the height of the Alps, would run in a straight line due southward; and on its western flank every deep creek of the sea, or fiord, would end in "bold and astonishing glaciers." These lonely channels would frequently reverberate with the falls of ice, and so often would great waves rush along their coasts; numerous icebergs, some as tall as cathedrals, and occasionally loaded with "no inconsiderable blocks of rock," would be stranded on the outlying islets; at intervals violent earthquakes would shoot prodigious masses of ice into the waters below. Lastly, some missionaries attempting to penetrate a long arm of the sea, would behold the not lofty surrounding mountains, sending down their many grand icy streams to the sea-coast, and their progress in the boats would be checked by the innumerable floating icebergs, some small and some great; and this would have occurred on our twenty-second of June, and where the Lake of Geneva is now spread out![129]

CHAPTER XII

CENTRAL CHILE

VALPARAISO • EXCURSION TO THE FOOT OF THE ANDES • STRUCTURE OF THE LAND • ASCEND THE BELL OF QUILLOTA • SHATTERED MASSES OF GREENSTONE • IMMENSE VALLEYS • MINES • STATE OF MINERS • SANTIAGO • HOT-BATHS OF CAUQUENES • GOLD-MINES • GRINDING-MILLS • PERFORATED STONES • HABITS OF THE PUMA • EL TURCO AND TAPACOLO • HUMMING-BIRDS

JULY 23RD.—THE BEAGLE anchored late at night in the bay of Valparaiso, the chief seaport of Chile. When morning came, everything appeared delightful. After Tierra del Fuego, the climate felt quite delicious—the atmosphere so dry, and the heavens so clear and blue with the sun shining brightly, that all nature seemed sparkling with life. The view from the anchorage is very pretty. The town is built at the very foot of a range of hills, about 1600 feet high, and rather steep. From its position, it consists of one long, straggling street, which runs parallel to the beach, and wherever a ravine comes down, the houses are piled up on each side of it. The rounded hills, being only partially protected by a very scanty vegetation, are worn into numberless little gullies, which expose a singularly bright red soil. From this cause, and from the low whitewashed houses with tile roofs, the view reminded me of St. Cruz in Teneriffe. In a north-westerly direction there are some fine glimpses of the Andes: but these mountains appear much grander when viewed from the neighbouring hills; the great distance at which they are situated can then more readily be perceived. The volcano of Aconcagua is particularly magnificent. This huge and irregularly conical mass has an elevation greater than that of Chimborazo; for, from measurements made by the officers in the Beagle, its height is no less than 23,000 feet. The Cordillera, however, viewed from this point, owe the greater part of their beauty to the atmosphere through which they are seen. When the sun was setting in the Pacific, it was admirable to watch how clearly their rugged outlines could be distinguished, yet how varied and how delicate were the shades of their colour.

I had the good fortune to find living here Mr. Richard Corfield, an old schoolfellow and friend, to whose hospitality and kindness I was greatly indebted, in having afforded me a most pleasant residence during the Beagle's stay in Chile. The immediate neighbourhood of Valparaiso is not very productive to the naturalist. During the long summer the wind blows steadily from the southward, and a little off shore, so that rain never falls; during the three winter months, however, it is sufficiently abundant. The vegetation in consequence is very scanty: except in some deep valleys, there are no trees, and only a little grass and a few low bushes are scattered over the less steep parts of the hills. When we reflect, that at the distance of 350 miles to the south, this side of the Andes is completely hidden by one impenetrable forest, the contrast is very remarkable. I took several long walks while collecting objects of natural history. The country is pleasant for exercise. There are many very beautiful flowers; and, as in most other dry climates, the plants and shrubs possess strong and peculiar odours—even one's clothes by brushing through them became scented. I did not cease from wonder at finding each succeeding day as fine as the foregoing. What a difference does climate make in the enjoyment of life! How opposite are the sensations when viewing black mountains half enveloped in clouds, and seeing another range through the light blue haze of a fine day! The one for a time may be very sublime; the other is all gaiety and happy life.

August 14th.—I set out on a riding excursion, for the purpose of geologizing the basal parts of the Andes, which alone at this time of the year are not shut up by the winter snow. Our first day's ride was northward along the sea-coast. After dark we reached the Hacienda of Quintero, the estate which formerly belonged to Lord Cochrane. My object in coming here was to see the great beds of shells, which stand some yards above the level of the sea, and are burnt for lime. The proofs of the elevation of this whole line of coast are unequivocal: at the height of a few hundred feet old-looking shells are numerous, and I found some at 1300 feet. These shells either lie loose on the surface, or are embedded in a reddish-black vegetable mould. I was much surprised to find under the microscope that this vegetable mould is really marine mud, full of minute particles of organic bodies.

15th.—We returned towards the valley of Quillota. The country was exceedingly pleasant; just such as poets would call pastoral: green open lawns, separated by small valleys with rivulets, and the cottages, we may suppose of the shepherds, scattered on the hill-sides. We were obliged to cross the ridge of the Chilicauquen. At its base there were many fine evergreen forest-trees, but these flourished only in the ravines, where there was running water. Any person who had seen only the country near Valparaiso, would never have imagined that there had been such picturesque spots in Chile. As soon as we reached the brow of the Sierra, the valley of Quillota was immediately under our feet. The prospect was one of remarkable artificial luxuriance. The valley is very broad and quite flat, and is thus easily irrigated in all parts. The little square gardens are crowded with orange and olive trees, and every sort of vegetable. On each side huge bare mountains rise, and this from the contrast renders the patchwork valley the more pleasing. Whoever called "Valparaiso" the "Valley of Paradise," must have been thinking of Quillota. We crossed over to the Hacienda de San Isidro, situated at the very foot of the Bell Mountain.

Chile, as may be seen in the maps, is a narrow strip of land between the Cordillera and the Pacific; and this strip is itself traversed by several mountain-lines, which in this part run parallel to the great range. Between these outer lines and the main Cordillera, a succession of level basins, generally opening into each other by narrow passages, extend far to the southward: in these, the principal towns are situated, as San Felipe, Santiago, San Fernando. These basins or plains, together with the transverse flat valleys (like that of Quillota) which connect them with the coast, I have no doubt are the bottoms of ancient inlets and deep bays, such as at the present day intersect every part of Tierra del Fuego and the western coast. Chile must formerly have resembled the latter country in the configuration of its land and water. The resemblance was occasionally shown strikingly when a level fog-bank covered, as with a mantle, all the lower parts of the country: the white vapour curling into the ravines, beautifully represented little coves and bays; and here and there a solitary hillock peeping up, showed that it had formerly stood there as an islet. The contrast of these flat valleys and basins with the irregular mountains, gave the scenery a character which to me was new and very interesting.

From the natural slope to seaward of these plains, they are very easily irrigated, and in consequence singularly fertile. Without this process the land would produce scarcely anything, for during the whole summer the sky is cloudless. The mountains and hills are dotted over with bushes and low trees, and excepting these the vegetation is very scanty. Each landowner in the valley possesses a certain portion of hill-country, where his half-wild cattle, in considerable numbers, manage to find sufficient pasture. Once every year there is a grand "rodeo," when all the cattle are driven down, counted, and marked, and a certain number separated to be fattened in the irrigated fields. Wheat is extensively cultivated, and a good deal of

Indian corn: a kind of bean is, however, the staple article of food for the common labourers. The orchards produce an overflowing abundance of peaches, figs, and grapes. With all these advantages, the inhabitants of the country ought to be much more prosperous than they are.

16th.—The mayor-domo of the Hacienda was good enough to give me a guide and fresh horses; and in the morning we set out to ascend the Campana, or Bell Mountain, which is 6400 feet high. The paths were very bad, but both the geology and scenery amply repaid the trouble. We reached, by the evening, a spring called the Agua del Guanaco, which is situated at a great height. This must be an old name, for it is very many years since a guanaco drank its waters. During the ascent I noticed that nothing but bushes grew on the northern slope, whilst on the southern slope there was a bamboo about fifteen feet high. In a few places there were palms, and I was surprised to see one at an elevation of at least 4500 feet. These palms are, for their family, ugly trees. Their stem is very large, and of a curious form, being thicker in the middle than at the base or top. They are excessively numerous in some parts of Chile, and valuable on account of a sort of treacle made from the sap. On one estate near Petorca they tried to count them, but failed, after having numbered several hundred thousand. Every year in the early spring, in August, very many are cut down, and when the trunk is lying on the ground, the crown of leaves is lopped off. The sap then immediately begins to flow from the upper end, and continues so doing for some months: it is, however, necessary that a thin slice should be shaved off from that end every morning, so as to expose a fresh surface. A good tree will give ninety gallons, and all this must have been contained in the vessels of the apparently dry trunk. It is said that the sap flows much more quickly on those days when the sun is powerful; and likewise, that it is absolutely necessary to take care, in cutting down the tree, that it should fall with its head upwards on the side of the hill; for if it falls down the slope, scarcely any sap will flow; although in that case one would have thought that the action would have been aided, instead of checked, by the force of gravity. The sap is concentrated by boiling, and is then called treacle, which it very much resembles in taste.

We unsaddled our horses near the spring, and prepared to pass the night. The evening was fine, and the atmosphere so clear, that the masts of the vessels at anchor in the bay of Valparaiso, although no less than twenty-six geographical miles distant, could be distinguished clearly as little black streaks. A ship doubling the point under sail, appeared as a bright white speck. Anson expresses much surprise, in his voyage, at the distance at which his vessels were discovered from the coast; but he did not sufficiently allow for the height of the land, and the great transparency of the air.

The setting of the sun was glorious; the valleys being black whilst the snowy peaks of the Andes yet retained a ruby tint. When it was dark, we made a fire beneath a little arbour of bamboos, fried our charqui (or dried slips of beef), took our maté, and were quite comfortable. There is an inexpressible charm in thus living in the open air. The evening was calm and still;—the shrill noise of the mountain bizcacha, and the faint cry of a goatsucker, were occasionally to be heard. Besides these, few birds, or even insects, frequent these dry, parched mountains.

August 17th.—In the morning we climbed up the rough mass of greenstone which crowns the summit. This rock, as frequently happens, was much shattered and broken into huge angular fragments. I observed, however, one remarkable circumstance, namely, that many of the surfaces presented every degree of freshness—some appearing as if broken the day before, whilst on others lichens had either just become, or had long grown, attached. I so fully believed that this was owing to the frequent earthquakes, that I felt inclined to hurry from below each loose pile. As one might very easily be deceived in a fact of this kind, I doubted its accuracy, until ascending Mount Wellington, in Van Diemen's Land, where earthquakes do not occur; and there I saw the summit of the mountain similarly composed and similarly shattered, but all the blocks appeared as if they had been hurled into their present position thousands of years ago.

We spent the day on the summit, and I never enjoyed one more thoroughly. Chile, bounded by the Andes and the Pacific, was seen as in a map. The pleasure from the scenery, in itself beautiful, was heightened by the many reflections which arose from the mere view of the Campana range with its lesser parallel ones, and of the broad valley of Quillota directly intersecting them. Who can avoid wondering at the force which has upheaved these mountains, and even more so at the countless ages which it must have required to have broken through, removed, and levelled whole masses of them? It is well in this case to call to mind the vast shingle and sedimentary beds of Patagonia, which, if heaped on the Cordillera, would increase its height by so many thousand feet. When in that country, I wondered how any mountain-chain could have supplied such masses, and not have been utterly obliterated. We must not now reverse the wonder, and doubt whether all-powerful time can grind down mountains—even the gigantic Cordillera—into gravel and mud.

The appearance of the Andes was different from that which I had expected. The lower line of the snow was of course horizontal, and to this line the even summits of the range seemed quite parallel. Only at long intervals, a group of points or a single cone showed where a volcano had existed, or does now exist. Hence the range resembled a great solid wall, surmounted here and there by a tower, and making a most perfect barrier to the country.

Almost every part of the hill had been drilled by attempts to open gold-mines: the rage for mining has left scarcely a spot in Chile unexamined. I spent the evening as before, talking round the fire with my two companions. The Guasos of Chile, who correspond to the Gauchos of the Pampas, are, however, a very different set of beings. Chile is the more civilized of the two countries, and the inhabitants, in consequence, have lost much individual character. Gradations in rank are much more strongly marked: the Guaso does not by any means consider every man his equal; and I was quite surprised to find that my companions did not like to eat at the same time with myself. This feeling of inequality is a necessary consequence of the existence of an aristocracy of wealth. It is said that some few of the greater landowners possess from five to ten thousand pounds sterling per annum: an inequality of riches which I believe is not met with in any of the cattle-breeding countries eastward of the Andes. A traveller does not here meet that unbounded hospitality which refuses all payment, but yet is so kindly offered that no scruples can be raised in accepting it. Almost every house in Chile will receive you for the night, but a trifle is expected to be given in the morning; even a rich man will accept two or three shillings. The Gaucho, although he may be a cutthroat, is a gentleman; the Guaso is in few respects better, but at the same time a vulgar, ordinary fellow. The two men, although employed much in the same manner, are different in their habits and attire; and the peculiarities of each are universal in their respective countries. The Gaucho seems part of his horse, and scorns to exert himself except when on his back: the Guaso may be hired to work as a labourer in the fields. The former lives entirely on animal food; the latter almost wholly on vegetable. We do not here see the white boots, the broad drawers and scarlet chilipa; the picturesque costume of the Pampas. Here, common trousers are protected by black and green worsted leggings. The poncho, however, is common to both. The chief pride of the Guaso lies in his spurs, which are absurdly large. I measured one which was six inches in the *diameter* of the rowel, and the rowel itself contained upwards of thirty points. The stirrups are on the same scale, each consisting of a square, carved block of wood, hollowed out, yet weighing three or four pounds. The Guaso is perhaps more expert with the lazo than the Gaucho; but, from the nature of the country, he does not know the use of the bolas.

August 18th. —We descended the mountain, and passed some beautiful little spots, with rivulets and fine trees. Having slept at the same hacienda as before, we rode during the two succeeding days up the valley, and passed through Quillota, which is more like a collection of nursery-gardens than a town. The orchards were beautiful, presenting one mass of peach-blossoms. I saw, also, in one or two places the date-palm; it is a most stately tree; and I should think a group of them in their native Asiatic or African deserts must be superb. We passed likewise San Felipe, a pretty straggling town like Quillota. The valley in this part expands into one of those great bays or plains, reaching to the foot of the Cordillera, which have been mentioned as forming so curious a part of the scenery of Chile. In the evening we reached the mines of Jajuel, situated in a ravine at the flank of the great chain. I stayed here five days. My host, the superintendent of the mine, was a shrewd but rather ignorant Cornish miner. He had married a Spanish woman, and did not mean to return home; but his admiration for the mines of Cornwall remained unbounded. Amongst many other questions, he asked me, "Now that George Rex is dead, how many more of the family of Rexes are yet alive?" This Rex certainly must be a relation of the great author Finis, who wrote all books!

These mines are of copper, and the ore is all shipped to Swansea, to be smelted. Hence the mines have an aspect singularly quiet, as compared to those in England: here no smoke, furnaces, or great steam-engines, disturb the solitude of the surrounding mountains.

The Chilian government, or rather the old Spanish law, encourages by every method the searching for mines. The discoverer may work a mine on any ground, by paying five shillings; and before paying this he may try, even in the garden of another man, for twenty days.

It is now well known that the Chilian method of mining is the cheapest. My host says that the two principal improvements introduced by foreigners have been, first, reducing by previous roasting the copper pyrites—which, being the common ore in Cornwall, the English miners were astounded on their arrival to find thrown away as useless: secondly, stamping and washing the scoriæ from the old furnaces—by which process particles of metal are recovered in abundance. I have actually seen mules carrying to the coast, for transportation to England, a cargo of such cinders. But the first case is much the most curious. The Chilian miners were so convinced that copper pyrites contained not a particle of copper, that they laughed at the Englishmen for their ignorance, who laughed in turn, and bought their richest veins for a few dollars. It is very odd that, in a country where mining had been extensively carried on for many years, so simple a process as gently roasting the ore to expel the sulphur previous to smelting it, had never been discovered. A few improvements have likewise been introduced in some of the simple machinery; but even to the present day, water is removed from some mines by men carrying it up the shaft in leathern bags!

The labouring men work very hard. They have little time allowed for their meals, and during summer and winter they begin when it is light, and leave off at dark. They are paid one pound sterling a month, and their food is given them: this for breakfast consists of sixteen figs and two small loaves of bread; for dinner, boiled beans; for supper, broken roasted wheat grain. They scarcely ever taste meat; as, with the twelve pounds per annum, they have to clothe themselves, and support their families. The miners who work in the mine itself have twenty-five shillings per month, and are allowed a little charqui. But these men come down from their bleak habitations only once in every fortnight or three weeks.

During my stay here I thoroughly enjoyed scrambling about these huge mountains. The geology, as might have been expected, was very interesting. The shattered and baked rocks, traversed by innumerable dykes of greenstone, showed what commotions had formerly taken place. The scenery was much the same as that near the Bell of Quillota—dry barren mountains, dotted at intervals by bushes with a scanty foliage. The cactuses, or rather opuntias, were here very numerous. I measured one of a spherical figure, which, including the spines, was six feet and four inches in circumference. The height of the common cylindrical, branching kind, is from twelve to fifteen feet, and the girth (with spines) of the branches between three and four feet.

A heavy fall of snow on the mountains prevented me, during the last two days, from making some interesting excursions. I attempted to reach a lake which the inhabitants, from some unaccountable reason, believe to be an arm of the sea. During a very dry season, it was proposed to attempt cutting a channel from it for the sake of the water, but the padre, after a consultation, declared it was too dangerous, as all Chile would be inundated, if, as generally supposed, the lake was connected with the Pacific. We ascended to a great height, but becoming involved in the snow-drifts failed in reaching this wonderful lake, and had some difficulty in returning. I thought we should have lost our horses; for there was no means of guessing how deep the drifts were, and the animals, when led, could only move by jumping. The black sky showed that a fresh snow-storm was gathering, and we therefore were not a little glad when we escaped. By the time we reached the base the storm commenced, and it was lucky for us that this did not happen three hours earlier in the day.

August 26th.—We left Jajuel and again crossed the basin of San Felipe. The day was truly Chilian: glaringly bright, and the atmosphere quite clear. The thick and uniform covering of newly fallen snow rendered the view of the volcano of Aconcagua and the main chain quite glorious. We were now on the road to Santiago, the capital of Chile. We crossed the Cerro del Talguen, and slept at a little rancho. The host, talking about the state of Chile as com-

circumstance,—which, if true, certainly is very curious: for we must suppose that the snow-water, being conducted through porous strata to the regions of heat, is again thrown up to the surface by the line of dislocated and injected rocks at Cauquenes; and the regularity of the phenomenon would seem to indicate that in this district heated rock occurred at a depth not very great.

One day I rode up the valley to the farthest inhabited spot. Shortly above that point, the Cachapual divides into two deep tremendous ravines, which penetrate directly into the great range. I scrambled up a peaked mountain, probably more than six thousand feet high. Here, as indeed everywhere else, scenes of the highest interest presented themselves. It was by one of these ravines, that Pincheira entered Chile and ravaged the neighbouring country. This is the same man whose attack on an estancia at the Rio Negro I have described. He was a renegade half-caste Spaniard, who collected a great body of Indians together and established himself by a stream in the Pampas, which place none of the forces sent after him could ever discover. From this point he used to sally forth, and crossing the Cordillera by passes hitherto unattempted, he ravaged the farm-houses and drove the cattle to his secret rendezvous. Pincheira was a capital horseman, and he made all around him equally good, for he invariably shot any one who hesitated to follow him. It was against this man, and other wandering Indian tribes, that Rosas waged the war of extermination.

September 13th.—We left the baths of Cauquenes, and, rejoining the main road, slept at the Rio Clara. From this place we rode to the town of San Fernando. Before arriving there, the last land-locked basin had expanded into a great plain, which extended so far to the south, that the snowy summits of the more distant Andes were seen as if above the horizon of the sea. San Fernando is forty leagues from Santiago; and it was my farthest point southward; for we here turned at right angles towards the coast. We slept at the gold-mines of Yaquil, which are worked by Mr. Nixon, an American gentleman, to whose kindness I was much indebted during the four days I stayed at his house. The next morning we rode to the mines, which are situated at the distance of some leagues, near the summit of a lofty hill. On the way we had a glimpse of the lake Tagua-tagua, celebrated for its floating islands, which have been described by M. Gay.[131] They are composed of the stalks of various dead plants intertwined together, and on the surface of which other living ones take root. Their form is generally circular, and their thickness from four to six feet, of which the greater part is immersed in the water. As the wind blows, they pass from one side of the lake to the other, and often carry cattle and horses as passengers.

When we arrived at the mine, I was struck by the pale appearance of many of the men, and inquired from Mr. Nixon respecting their condition. The mine is 450 feet deep, and each man brings up about 200 pounds weight of stone. With this load they have to climb up the alternate notches cut in the trunks of trees, placed in a zigzag line up the shaft. Even beardless young men, eighteen and twenty years old, with little muscular development of their bodies (they are quite naked excepting drawers) ascend with this great load from nearly the same depth. A strong man, who is not accustomed to this labour, perspires most profusely, with merely carrying up his own body. With this very severe labour, they live entirely on boiled beans and bread. They would prefer having bread alone; but their masters, finding that they cannot work so hard upon this, treat them like horses, and make them eat the beans. Their pay is here rather more than at the mines of Jajuel, being from 24 to 28 shillings per month. They leave the mine only once in three weeks; when they stay with their families for two days. One of the rules of this mine sounds very harsh, but answers pretty well for the master. The only method of stealing gold is to secrete pieces of the ore, and take them out as occasion may offer. Whenever the major-domo finds a lump thus hidden, its full value is stopped out of the wages of all the men; who thus, without they all combine, are obliged to keep watch over each other.

When the ore is brought to the mill, it is ground into an impalpable powder; the process of washing removes all the lighter particles, and amalgamation finally secures the gold-dust. The washing, when described, sounds a very simple process; but it is beautiful to see how the exact adaptation of the current of water to the specific gravity of the gold, so easily separates the powdered matrix from the metal. The mud which passes from the mills is collected into pools, where it subsides, and every now and then is cleared out, and thrown into a common heap. A great deal of chemical action then commences, salts of various kinds effloresce on the surface, and the mass becomes hard. After having been left for a year or two, and then rewashed, it yields gold; and this process may be repeated even six or seven times; but the gold each time becomes less in quantity, and the intervals required (as the inhabitants say, to generate the metal) are longer. There can be no doubt that the chemical action, already mentioned, each time liberates fresh gold from some combination. The discovery of a method to effect this before the first grinding, would without doubt raise the value of gold-ores many fold. It is curious to find how the minute particles of gold, being scattered about and not corroding, at last accumulate in some quantity. A short time since a few miners, being out of work, obtained permission to scrape the ground round the house and mills; they washed the earth thus got together, and so procured thirty dollars' worth of gold. This is an exact counterpart of what takes place in nature. Mountains suffer degradation and wear away, and with them the metallic veins which they contain. The hardest rock is worn into impalpable mud, the ordinary metals oxidate, and both are removed; but gold, platina, and a few others are nearly indestructible, and from their weight, sinking to the bottom, are left behind. After whole mountains have passed through this grinding mill, and have been washed by the hand of nature, the residue becomes metalliferous, and man finds it worth his while to complete the task of separation.

Bad as the above treatment of the miners appears, it is gladly accepted of by them; for the condition of the labouring agriculturists is much worse. Their wages are lower, and they live almost exclusively on beans. This poverty must be chiefly owing to the feudal-like system on which the land is tilled: the landowner gives a small plot of ground to the labourer, for building on and cultivating, and in return has his services (or those of a proxy) for every day of his life, without any wages. Until a father has a grown-up son, who can by his labour pay the rent, there is no one, except on occasional days, to take care of his own patch of ground. Hence extreme poverty is very common among the labouring classes in this country.

There are some old Indian ruins in this neighbourhood, and I was shown one of the perforated stones, which Molina mentions as being found in many places in considerable numbers. They are of a circular flattened form, from five to six inches in diameter, with a hole passing quite through the centre. It has generally been supposed that they were used as heads to clubs, although their form does not appear at all well adapted for that purpose. Burchell[132] states that some of the tribes in Southern Africa dig up roots by the aid of a stick pointed at one end, the force and weight of which are increased by a round stone with a hole in it, into which the other end is firmly wedged. It appears probable that the Indians of Chile formerly used some such rude agricultural instrument.

One day, a German collector in natural history, of the name of Renous, called, and nearly at the same time an old Spanish lawyer. I was amused at being told the conversation which took place between them. Renous speaks Spanish so well, that the old lawyer mistook him for a Chilian. Renous, alluding to me, asked him what he thought of the King of England sending out a collector to their country, to pick up lizards and beetles, and to break stones? The old gentleman thought seriously for some time, and then said, "It is not well,—*hay un gato encerrado aqui* (there is a cat shut up here). No man is so rich as to send out people to pick up such rubbish. I do not like it: if one of us were to go and do such things in England, do not you think the King of England would very soon send us out of his country?" And this old gentleman, from his profession, belongs to the better informed and more intelligent classes! Renous himself, two or three years before, left in a house at San Fernando some caterpillars, under charge of a girl to feed, that they might turn into butterflies. This was rumoured through the town, and at last the padres and governor consulted together, and agreed it must be some heresy. Accordingly, when Renous returned, he was arrested.

September 19th.—We left Yaquil, and followed the flat valley, formed like that of Quillota, in which the Rio Tinderidica flows. Even at these few miles south of Santiago the climate is much damper; in consequence there are fine tracts of pasturage, which are not irrigated. (20th.) We followed this valley till it expanded into a great plain, which reaches from the sea to the mountains west of Rancagua. We shortly lost all trees and even bushes; so that the inhabitants are nearly as badly off for firewood as those in the Pampas. Never having heard of these plains, I was much surprised at meeting with such scenery in Chile. The plains belong to more than one series of different elevations, and they are traversed by broad flat-bottomed valleys; both of which circumstances, as in Patagonia, bespeak the action of the sea on gently rising land. In the steep cliffs bordering these valleys, there are some large caves, which no doubt were originally formed by the waves: one of these is celebrated under the name of Cueva del Obispo; having formerly been consecrated. During the day I felt very unwell, and from that time till the end of October did not recover.

September 22nd.—We continued to pass over green plains without a tree. The next day we arrived at a house near Navedad, on the sea-coast, where a rich Haciendero gave us lodgings. I stayed here the two ensuing days, and although very unwell, managed to collect from the tertiary formation some marine shells.

24th.—Our course was now directed towards Valparaiso, which with great difficulty I reached on the 27th, and was there confined to my bed till the end of October. During this time I was an inmate in Mr. Corfield's house, whose kindness to me I do not know how to express.

I WILL HERE ADD a few observations on some of the animals and birds of Chile. The Puma, or South American Lion, is not uncommon. This animal has a wide geographical range; being found from the equatorial forests, throughout the deserts of Patagonia, as far south as the damp and cold latitudes (53° to 54°) of Tierra del Fuego. I have seen its footsteps in the Cordillera of central Chile, at an elevation of at least 10,000 feet. In La Plata the puma preys chiefly on deer, ostriches, bizcacha, and other small quadrupeds; it there seldom attacks cattle or horses, and most rarely man. In Chile, however, it destroys many young horses and cattle, owing probably to the scarcity of other quadrupeds: I heard, likewise, of two men and a woman who had been thus killed. It is asserted that the puma always kills its prey by springing on the shoulders, and then drawing back the head with one of its paws, until the vertebræ break: I have seen in Patagonia the skeletons of guanacos, with their necks thus dislocated.

The puma, after eating its fill, covers the carcass with many large bushes, and lies down to watch it. This habit is often the cause of its being discovered; for the condors wheeling in the air every now and then descend to partake of the feast, and being angrily driven away, rise all together on the wing. The Chileno Guaso then knows there is a lion watching his prey—the word is given—and men and dogs hurry to the chase. Sir F. Head says that a Gaucho in the Pampas, upon merely seeing some condors wheeling in the air, cried "A lion!" I could never myself meet with any one who pretended to such powers of discrimination. It is asserted that, if a puma has once been betrayed by thus watching the carcass, and has then been hunted, it never resumes this habit; but that, having gorged itself, it wanders far away. The puma is easily killed. In an open country, it is first entangled with the bolas, then lazoed, and dragged along the ground till rendered insensible. At Tandeel (south of the Plata), I was told that within three months one hundred were thus destroyed. In Chile they are generally driven up bushes or trees, and are then either shot, or baited to death by dogs. The dogs employed in this chase belong to a particular breed, called Leoneros: they are weak, slight animals, like long-legged terriers, but are born with a particular instinct for this sport. The puma is described as being very crafty: when pursued, it often returns on its former track, and then suddenly making a spring on one side, waits there till the dogs have passed by. It is a very silent animal, uttering no cry even when wounded, and only rarely during the breeding season.

Of birds, two species of the genus Pteroptochos (megapodius and albicollis of Kittlitz) are perhaps the most conspicuous. The former, called by the Chilenos "el Turco," is as large as a fieldfare, to which bird it has some alliance; but its legs are much longer, tail shorter, and beak stronger: its colour is a reddish brown. The Turco is not uncommon. It lives on the ground, sheltered among the thickets which are scattered over the dry and sterile hills. With its tail erect, and stilt-like legs, it may be seen every now and then popping from one bush to another with uncommon quickness. It really requires little imagination to believe that the bird is ashamed of itself, and is aware of its most ridiculous figure. On first seeing it, one is tempted to exclaim, "A vilely stuffed specimen has escaped from some museum, and has come to life again!" It cannot be made to take flight without the greatest trouble, nor does it run, but only hops. The various loud cries which it utters when concealed amongst the bushes, are as strange as its appearance. It is said to build its nest in a deep hole beneath the ground. I dissected several specimens: the gizzard, which was very muscular, contained beetles, vegetable fibres, and pebbles. From this character, from the length of its legs, scratching feet, membranous covering to the nostrils, short and arched wings, this bird seems in a certain degree to connect the thrushes with the gallinaceous order.

The second species (or P. albicollis) is allied to the first in its general form. It is called Tapacolo, or "cover your posterior;" and well does the shameless little bird deserve its name; for it carries its tail more than erect, that is, inclined backwards towards its head. It is very common, and frequents the bottoms of hedge-rows, and the bushes scattered over the barren hills, where scarcely another bird can exist. In its general manner of feeding, of quickly hopping out of the thickets and back again, in its desire of concealment, unwillingness to take flight, and nidification, it bears a close resemblance to the Turco; but its appearance is not quite so ridiculous. The Tapacolo is very crafty: when frightened by any person, it will remain motionless at the bottom of a bush, and will then, after a little while, try with much address to crawl away on the opposite side. It is also an active bird, and continually making a noise: these noises are various and strangely odd; some are like the cooing of doves, others like the bubbling of water, and many defy all similes. The country people say it changes its cry five times in the year—according to some change of season, I suppose.[133]

Two species of humming-birds are common; Trochilus forficatus is found over a space of 2500 miles on the west coast, from the hot dry country of Lima, to the forests of Tierra del Fuego—where it may be seen flitting about in snow-storms. In the wooded island of Chiloe, which has an extremely humid climate, this little bird, skipping from side to side amidst the dripping foliage, is perhaps more abundant than almost any other kind. I opened the stomachs of several specimens, shot in different parts of the continent, and in all, remains of insects were as numerous as in the stomach of a creeper. When this species migrates in the summer southward, it is replaced by the arrival of another species coming from the north. This second kind (Trochilus gigas) is a very large bird for the delicate family to which it belongs: when on the wing its appearance is singular. Like others of the genus, it moves from place to place with a rapidity which may be compared to that of Syrphus amongst flies, and Sphinx among moths; but whilst hovering over a flower, it flaps its wings with a very slow and powerful movement, totally different from that vibratory one common to most of the species, which produces the humming noise. I never saw any other bird where the force of its wings appeared (as in a butterfly) so powerful in proportion to the weight of its body. When hovering by a flower, its tail is constantly expanded and shut like a fan, the body being kept in a nearly vertical position. This action appears to steady and support the bird, between the slow movements of its wings. Although flying from flower to flower in search of food, its stomach generally contained abundant remains of insects, which I suspect are much more the object of its search than honey. The note of this species, like that of nearly the whole family, is extremely shrill.

CHAPTER XIII

Chiloe and Chonos Islands

CHILOE • GENERAL ASPECT • BOAT EXCURSION • NATIVE INDIANS • CASTRO • TAME FOX • ASCEND SAN PEDRO • CHONOS ARCHIPELAGO • PENINSULA OF TRES MONTES • GRANITIC RANGE • BOAT-WRECKED SAILORS • LOW'S HARBOUR • WILD POTATO • FORMATION OF PEAT • MYOPOTAMUS, OTTER AND MICE • CHEUCAU AND BARKING-BIRD • OPETIORHYNCHUS • SINGULAR CHARACTER OF ORNITHOLOGY • PETRELS

November 10th.—The Beagle sailed from Valparaiso to the south, for the purpose of surveying the southern part of Chile, the island of Chiloe, and the broken land called the Chonos Archipelago, as far south as the Peninsula of Tres Montes. On the 21st we anchored in the bay of S. Carlos, the capital of Chiloe.

This island is about ninety miles long, with a breadth of rather less than thirty. The land is hilly, but not mountainous, and is covered by one great forest, except where a few green patches have been cleared round the thatched cottages. From a distance the view somewhat resembles that of Tierra del Fuego; but the woods, when seen nearer, are incomparably more beautiful. Many kinds of fine evergreen trees, and plants with a tropical character, here take the place of the gloomy beech of the southern shores. In winter the climate is detestable, and in summer it is only a little better. I should think there are few parts of the world, within the temperate regions, where so much rain falls. The winds are very boisterous, and the sky almost always clouded: to have a week of fine weather is something wonderful. It is even difficult to get a single glimpse of the Cordillera: during our first visit, once only the volcano of Osorno stood out in bold relief, and that was before sunrise; it was curious to watch, as the sun rose, the outline gradually fading away in the glare of the eastern sky.

The inhabitants, from their complexion and low stature, appear to have three-fourths of Indian blood in their veins. They are an humble, quiet, industrious set of men. Although the fertile soil, resulting from the decomposition of the volcanic rocks, supports a rank vegetation, yet the climate is not favourable to any production which requires much sunshine to ripen it. There is very little pasture for the larger quadrupeds; and in consequence, the staple articles of food are pigs, potatoes, and fish. The people all dress in strong woollen garments, which each family makes for itself, and dyes with indigo of a dark blue colour. The arts, however, are in the rudest state;—as may be seen in their strange fashion of ploughing, their method of spinning, grinding corn, and in the construction of their boats. The forests are so impenetrable, that the land is nowhere cultivated except near the coast and on the adjoining islets. Even where paths exist, they are scarcely passable from the soft and swampy state of the soil. The inhabitants, like those of Tierra del Fuego, move about chiefly on the beach or in boats. Although with plenty to eat, the people are very poor: there is no demand for labour, and consequently the lower orders cannot scrape together money sufficient to purchase even the smallest luxuries. There is also a great deficiency of a circulating medium. I have seen a man bringing on his back a bag of charcoal, with which to buy some trifle, and another carrying a plank to exchange for a bottle of wine. Hence every tradesman must also be a merchant, and again sell the goods which he takes in exchange.

November 24th.—The yawl and whale-boat were sent under the command of Mr. (now Captain) Sulivan, to survey the eastern or inland coast of Chiloe; and with orders to meet the Beagle at the southern extremity of the island; to which point she would proceed by the outside, so as thus to circumnavigate the whole. I accompanied this expedition, but instead of going in the boats the first day, I hired horses to take me to Chacao, at the northern extremity of the island. The road followed the coast; every now and then crossing promontories covered by fine forests. In these shaded paths it is absolutely necessary that the whole road should be made of logs of wood, which are squared and placed by the side of each other. From the rays of the sun never penetrating the evergreen foliage, the ground is so damp and soft, that except by this means neither man nor horse would be able to pass along. I arrived at the village of Chacao shortly after the tents belonging to the boats were pitched for the night.

The land in this neighbourhood has been extensively cleared, and there were many quiet and most picturesque nooks in the forest. Chacao was formerly the principal port in the island; but many vessels having been lost, owing to the dangerous currents and rocks in the straits, the Spanish government burnt the church, and thus arbitrarily compelled the greater number of inhabitants to migrate to S. Carlos. We had not long bivouacked, before the barefooted son of the governor came down to reconnoitre us. Seeing the English flag hoisted at the yawl's mast-head, he asked with the utmost indifference, whether it was always to fly at Chacao. In several places the inhabitants were much astonished at the appearance of men-of-war's boats, and hoped and believed it was the forerunner of a Spanish fleet, coming to recover the island from the patriot government of Chile. All the men in power, however, had been informed of our intended visit, and were exceedingly civil. While we were eating our supper, the governor paid us a visit. He had been a lieutenant-colonel in the Spanish service, but now was miserably poor. He gave us two sheep, and accepted in return two cotton handkerchiefs, some brass trinkets, and a little tobacco.

25th.—Torrents of rain: we managed, however, to run down the coast as far as Huapi-lenou. The whole of this eastern side of Chiloe has one aspect; it is a plain, broken by valleys and divided into little islands, and the whole thickly covered with one impervious blackish-green forest. On the margins there are some cleared spaces, surrounding the high-roofed cottages.

26th.—The day rose splendidly clear. The volcano of Orsono was spouting out volumes of smoke. This most beautiful mountain, formed like a perfect cone, and white with snow, stands out in front of the Cordillera. Another great volcano, with a saddle-shaped summit, also emitted from its immense crater little jets of steam. Subsequently we saw the lofty-peaked Corcovado—well deserving the name of "el famoso Corcovado." Thus we beheld, from one point of view, three great active volcanoes, each about seven thousand feet high. In addition to this, far to the south, there were other lofty cones covered with snow, which, although not known to be active, must be in their origin volcanic. The line of the Andes is not, in this neighbourhood, nearly so elevated as in Chile; neither does it appear to form so perfect a barrier between the regions of the earth. This great range, although running in a straight north and south line, owing to an optical deception, always appeared more or less curved; for the lines drawn from each peak to the beholder's eye, necessarily converged like the radii of a semicircle, and as it was not possible (owing to the clearness of the atmosphere and the absence of all intermediate objects) to judge how far distant the farthest peaks were off, they appeared to stand in a flattish semicircle.

Landing at midday, we saw a family of pure Indian extraction. The father was singularly like York Minster; and some of the younger boys, with their ruddy complexions, might have been mistaken for Pampas Indians.

Everything I have seen, convinces me of the close connection of the different American tribes, who nevertheless speak distinct languages. This party could muster but little Spanish, and talked to each other in their own tongue. It is a pleasant thing to see the aborigines advanced to the same degree of civilization, however low that may be, which their white conquerors have attained. More to the south we saw many pure Indians: indeed, all the inhabitants of some of the islets retain their Indian surnames. In the census of 1832, there were in Chiloe and its dependencies forty-two thousand souls; the greater number of these appear to be of mixed blood. Eleven thousand retain their Indian surnames, but it is probable that not nearly all of these are of a pure breed. Their manner of life is the same with that of the other poor inhabitants, and they are all Christians; but it is said that they yet retain some strange superstitious ceremonies, and that they pretend to hold communication with the devil in certain caves. Formerly, every one convicted of this offence was sent to the Inquisition at Lima. Many of the inhabitants who are not included in the eleven thousand with Indian surnames, cannot be distinguished by their appearance from Indians. Gomez, the governor of Lemuy, is descended from noblemen of Spain on both sides; but by constant inter-marriages with the natives the present man is an Indian. On the other hand, the governor of Quinchao boasts much of his purely kept Spanish blood.

We reached at night a beautiful little cove, north of the island of Caucahue. The people here complained of want of land. This is partly owing to their own negligence in not clearing the woods, and partly to restrictions by the government, which makes it necessary, before buying ever so small a piece, to pay two shillings to the surveyor for measuring each quadra (150 yards square), together with whatever price he fixes for the value of the land. After his valuation, the land must be put up three times to auction, and if no one bids more, the purchaser can have it at that rate. All these exactions must be a serious check to clearing the ground, where the inhabitants are so extremely poor. In most countries, forests are removed without much difficulty by the aid of fire; but in Chiloe, from the damp nature of the climate, and the sort of trees, it is necessary first to cut them down. This is a heavy drawback to the prosperity of Chiloe. In the time of the Spaniards the Indians could not hold land; and a family, after having cleared a piece of ground, might be driven away, and the property seized by the government. The Chilian authorities are now performing an act of justice by making retribution to these poor Indians, giving to each man, according to his grade of life, a certain portion of land. The value of uncleared ground is very little. The government gave Mr. Douglas (the present surveyor, who informed me of these circumstances) eight and a half square miles of forest near S. Carlos, in lieu of a debt; and this he sold for 350 dollars, or about 70*l.* sterling.

The two succeeding days were fine, and at night we reached the island of Quinchao. This neighbourhood is the most cultivated part of the Archipelago; for a broad strip of land on the coast of the main island, as well as on many of the smaller adjoining ones, is almost completely cleared. Some of the farm-houses seemed very comfortable. I was curious to ascertain how rich any of these people might be, but Mr. Douglas says that no one can be considered as possessing a regular income. One of the richest land-owners might possibly accumulate, in a long industrious life, as much as 1000*l.* sterling; but should this happen, it would all be stowed away in some secret corner, for it is the custom of almost every family to have a jar or treasure-chest buried in the ground.

November 30th.—Early on Sunday morning we reached Castro, the ancient capital of Chiloe, but now a most forlorn and deserted place. The usual quadrangular arrangement of Spanish towns could be traced, but the streets and plaza were coated with fine green turf, on which sheep were browsing. The church, which stands in the middle, is entirely built of plank, and has a picturesque and venerable appearance. The poverty of the place may be conceived from the fact, that although containing some hundreds of inhabitants, one of our party was unable anywhere to purchase either a pound of sugar or an ordinary knife. No individual possessed either a watch or a clock; and an old man, who was supposed to have a good idea of time, was employed to strike the church bell by guess. The arrival of our boats was a rare event in this quiet retired corner of the world; and nearly all the inhabitants came down to the beach to see us pitch our tents. They were very civil, and offered us a house; and one man even sent us a cask of cider as a present. In the afternoon we paid our respects to the governor—a quiet old man, who, in his appearance and manner of life, was scarcely superior to an English cottager. At night heavy rain set in, which was hardly sufficient to drive away from our tents the large circle of lookers-on. An Indian family, who had come to trade in a canoe from Caylen, bivouacked near us. They had no shelter during the rain. In the morning I asked a young Indian, who was wet to the skin, how he had passed the night. He seemed perfectly content, and answered, "Muy bien, señor."

December 1st.—We steered for the island of Lemuy. I was anxious to examine a reported coal-mine which turned out to be lignite of little value, in the sandstone (probably of an ancient tertiary epoch) of which these islands are composed. When we reached Lemuy we had much difficulty in finding any place to pitch our tents, for it was spring-tide, and the land was wooded down to the water's edge. In a short time we were surrounded by a large group of the nearly pure Indian inhabitants. They were much surprised at our arrival, and said one to the other, "This is the reason we have seen so many parrots lately; the cheucau (an odd red-breasted little bird, which inhabits the thick forest, and utters very peculiar noises) has not cried 'beware' for nothing." They were soon anxious for barter. Money was scarcely worth anything, but their eagerness for tobacco was something quite extraordinary. After tobacco, indigo came next in value; then capsicum, old clothes, and gunpowder. The latter article was required for a very innocent purpose: each parish has a public musket, and the gunpowder was wanted for making a noise on their saint or feast days.

The people here live chiefly on shell-fish and potatoes. At certain seasons they catch also, in "corrales," or hedges under water, many fish which are left on the mud-banks as the tide falls. They occasionally possess fowls, sheep, goats, pigs, horses, and cattle; the order in which they are here mentioned, expressing their respective numbers. I never saw anything more obliging and humble than the manners of these people. They generally began with stating that they were poor natives of the place, and not Spaniards, and that they were in sad want of tobacco and other comforts. At Caylen, the most southern island, the sailors bought with a stick of tobacco, of the value of three-half-pence, two fowls, one of which, the Indian stated, had skin between its toes, and turned out to be a fine duck; and with some cotton handkerchiefs, worth three shillings, three sheep and a large bunch of onions were procured. The yawl at this place was anchored some way from the shore, and we had fears for her safety from robbers during the night. Our pilot, Mr. Douglas, accordingly told the constable of the district that we always placed sentinels with loaded arms, and not understanding Spanish, if we saw any person in the dark, we should assuredly shoot him. The constable, with much humility, agreed to the perfect propriety of this arrangement, and promised us that no one should stir out of his house during that night.

During the four succeeding days we continued sailing southward. The general features of the country remained the same, but it was much less thickly inhabited. On the large island of Tanqui there was scarcely one cleared spot, the trees on every side extending their branches over the sea-beach. I one day noticed, growing on the sandstone cliffs, some very fine plants of the panke (Gunnera scabra), which somewhat resembles the rhubarb on a gigantic scale. The inhabitants eat the stalks, which are subacid, and tan leather with the roots, and prepare a black dye from them. The leaf is nearly circular, but deeply indented on its margin. I measured one which was nearly eight feet in diameter, and therefore no less than twenty-four in circumference! The stalk is rather more than a yard high, and each plant sends out four or five of these enormous leaves, presenting together a very noble appearance.

December 6th.—We reached Caylen, called "el fin del Cristiandad." In the morning we stopped for a few minutes at a house on the northern end of Laylec, which was the extreme point of South American Christendom, and a miserable hovel it was. The latitude is 43° 10´, which is two degrees farther south than the Rio Negro on the Atlantic coast. These extreme Christians were very poor, and, under the plea of their situation, begged for some tobacco. As a proof of the poverty of these Indians, I may mention that shortly before this, we had met a man, who had travelled three days and a half on foot, and had as many to return, for the sake of recovering the value of a small axe and a few fish. How very difficult it must be to buy the smallest article, when such trouble is taken to recover so small a debt.

In the evening we reached the island of San Pedro, where we found the Beagle at anchor. In doubling the point, two of the officers landed to take a round of angles with the theodolite. A fox (Canis fulvipes), of a kind said to be peculiar to the island, and very rare in it, and which is a new species, was sitting on the rocks. He was so intently absorbed in watching the work of the officers, that I was able, by quietly walking up behind, to knock him on the head with my geological hammer. This fox, more curious or more scientific, but less wise, than the generality of his brethren, is now mounted in the museum of the Zoological Society.

We stayed three days in this harbour, on one of which Captain Fitz Roy, with a party, attempted to ascend to the summit of San Pedro. The woods here had rather a different appearance from those on the northern part of the island. The rock, also, being micaceous slate, there was no beach, but the steep sides dipped directly beneath the water. The general aspect in consequence was more like that of Tierra del Fuego than of Chiloe. In vain we tried to gain the summit: the forest was so impenetrable, that no one who has not beheld it can imagine so entangled a mass of dying and dead trunks. I am sure that often, for more than ten minutes together, our feet never touched the ground, and we were frequently ten or fifteen feet above it, so that the seamen as a joke called out the soundings. At other times we crept one after another on our hands and knees, under the rotten trunks. In the lower part of the mountain, noble trees of the Winter's Bark, and a laurel like the sassafras with fragrant leaves, and others, the names of which I do not know, were matted together by a trailing bamboo or cane. Here we were more like fishes struggling in a net than any other animal. On the higher parts, brushwood takes the place of larger trees, with here and there a red cedar or an alerce pine. I was also pleased to see, at an elevation of a little less than 1000 feet, our old friend the southern beech. They were, however, poor stunted trees; and I should think that this must be nearly their northern limit. We ultimately gave up the attempt in despair.

December 10th.—The yawl and whale-boat, with Mr. Sulivan, proceeded on their survey, but I remained on board the Beagle, which the next day left San Pedro for the southward. On the 13th we ran into an opening in the southern part of Guayatecas, or the Chonos Archipelago; and it was fortunate we did so, for on the following day a storm, worthy of Tierra del Fuego, raged with great fury. White massive clouds were piled up against a dark blue sky, and across them black ragged sheets of vapour were rapidly driven. The successive mountain ranges appeared like dim shadows; and the setting sun cast on the woodland a yellow gleam, much like that produced by the flame of spirits of wine. The water was white with the flying spray, and the wind lulled and roared again through the rigging: it was an ominous, sublime scene. During a few minutes there was a bright rainbow, and it was curious to observe the effect of the spray, which being carried along the surface of the water, changed the ordinary semicircle into a circle—a band of prismatic colours being continued, from both feet of the common arch across the bay, close to the vessel's side: thus forming a distorted, but very nearly entire ring.

We stayed here three days. The weather continued bad: but this did not much signify, for the surface of the land in all these islands is all but impassable. The coast is so very rugged that to attempt to walk in that direction requires continued scrambling up and down over the sharp rocks of mica-slate; and as for the woods, our faces, hands, and shin-bones all bore witness to the maltreatment we received, in merely attempting to penetrate their forbidden recesses.

December 18th.—We stood out to sea. On the 20th we bade farewell to the south, and with a fair wind turned the ship's head northward. From Cape Tres Montes we sailed pleasantly along the lofty weather-beaten coast, which is remarkable for the bold outline of its hills, and the thick covering of forest even on the almost precipitous flanks. The next day a harbour was discovered, which on this dangerous coast might be of great service to a distressed vessel. It can easily be recognised by a hill 1600 feet high, which is even more perfectly conical than the famous sugar-loaf at Rio de Janeiro. The next day, after anchoring, I succeeded in reaching the summit of this hill. It was a laborious undertaking, for the sides were so steep that in some parts it was necessary to use the trees as ladders. There were also several extensive brakes of the Fuchsia, covered with its beautiful drooping flowers, but very difficult to crawl through. In these wild countries it gives much delight to gain the summit of any mountain. There is an indefinite expectation of seeing something very strange, which, however often it may be balked, never failed with me to recur on each successive attempt. Every one must know the feeling of triumph and pride which a grand view from a height communicates to the mind. In these little frequented countries there is also joined to it some vanity, that you perhaps are the first man who ever stood on this pinnacle or admired this view.

A strong desire is always felt to ascertain whether any human being has previously visited an unfrequented spot. A bit of wood with a nail in it, is picked up and studied as if it were covered with hieroglyphics. Possessed with this feeling, I was much interested by finding, on a wild part of the coast, a bed made of grass beneath a ledge of rock. Close by it there had been a fire, and the man had used an axe. The fire, bed, and situation showed the dexterity of an Indian; but he could scarcely have been an Indian, for the race is in this part extinct, owing to the Catholic desire of making at one blow Christians and Slaves. I had at the time some misgivings that the solitary man who had made his bed on this wild spot, must have been some poor shipwrecked sailor, who, in trying to travel up the coast, had here laid himself down for his dreary night.

December 28th.—The weather continued very bad, but it at last permitted us to proceed with the survey. The time hung heavy on our hands, as it always did when we were delayed from day to day by successive gales of wind. In the evening another harbour was discovered, where we anchored. Directly afterwards a man was seen waving a shirt, and a boat was sent which brought back two seamen. A party of six had run away from an American whaling vessel, and had landed a little to the southward in a boat, which was shortly afterwards knocked to pieces by the surf. They had now been wandering up and down the coast for fifteen months, without knowing which way to go, or where they were. What a singular piece of good fortune it was that this harbour was now discovered! Had it not been for this one chance, they might have wandered till they had grown old men, and at last have perished on this wild coast. Their sufferings had been very great, and one of their party had lost his life by falling from the cliffs. They were sometimes obliged to separate in search of food, and this explained the bed of the solitary man. Considering what they had undergone, I think they had kept a very good reckoning of time, for they had lost only four days.

December 30th.—We anchored in a snug little cove at the foot of some high hills, near the northern extremity of Tres Montes. After breakfast the next morning, a party ascended one of these mountains, which was 2400 feet high. The scenery was remarkable. The chief part of the range was composed of grand, solid, abrupt masses of granite, which appeared as if they had been coeval with the beginning of the world. The granite was capped with mica-slate, and this in the lapse of ages had been worn into strange finger-shaped points. These two formations, thus differing in their outlines, agree in being almost destitute of vegetation. This barrenness had to our eyes a strange appearance, from having been so long accustomed to the sight of an almost universal forest of dark-green trees. I took much delight in examining the structure of these mountains. The complicated and lofty ranges bore a noble aspect of durability—equally profitless, however, to man and to all other animals. Granite to the geologist is classic ground: from its widespread limits, and its beautiful and compact texture, few rocks have been more anciently recognised. Granite has given rise, perhaps, to more discussion concerning its origin than any other formation. We generally see it constituting the fundamental rock, and, however formed, we know it is the deepest layer in the crust of this globe to which man has penetrated. The limit of man's knowledge in any subject possesses a high interest, which is perhaps increased by its close neighbourhood to the realms of imagination.

January 1st, 1835.—The new year is ushered in with the ceremonies proper to it in these regions. She lays out no false hopes: a heavy north-western gale, with steady rain, bespeaks the rising year. Thank God, we are not destined here to see the end of it, but hope then to be in the Pacific Ocean, where a blue sky tells one there is a heaven,—a something beyond the clouds above our heads.

The north-west winds prevailing for the next four days, we only managed to cross a great bay, and then anchored in another secure harbour. I accompanied the Captain in a boat to the head of a deep creek. On the way the number of seals which we saw was quite astonishing: every bit of flat rock, and parts of the beach, were covered with them. They appeared to be of a loving disposition, and lay huddled together, fast asleep, like so many pigs; but even pigs would have been ashamed of their dirt, and of the foul smell which came from them. Each herd was watched by the patient but inauspicious eyes of the turkey-buzzard. This disgusting bird, with its bald scarlet head, formed to wallow in putridity, is very common on the west coast, and their attendance on the seals shows on what they rely for their food. We found the water (probably only that of the surface) nearly fresh: this was caused by the number of torrents which, in the form of cascades, came tumbling over the bold granite mountains into the sea. The fresh water attracts the fish, and these bring many terns, gulls, and two kinds of cormorant. We saw also a pair of the beautiful black-necked swans, and several small sea-otters, the fur of which is held in such high estimation. In returning, we were again amused by the impetuous manner in which the heap of seals, old and young, tumbled into the water as the boat passed. They did not remain long under water, but rising, followed us with outstretched necks, expressing great wonder and curiosity.

7th.—Having run up the coast, we anchored near the northern end of the Chonos Archipelago, in Low's Harbour, where we remained a week. The islands were here, as in Chiloe, composed of a stratified, soft, littoral deposit; and the vegetation in consequence was beautifully luxuriant. The woods came down to the sea-beach, just in the manner of an evergreen shrubbery over a gravel walk. We also enjoyed from the anchorage a splendid view of four great snowy cones of the Cordillera, including "el famoso Corcovado;" the range itself had in this latitude so little height, that few parts of it appeared above the tops of the neighbouring islets. We found here a party of five men from Caylen, "el fin del Cristiandad," who had most adventurously crossed in their miserable boat-canoe, for the purpose of fishing, the open space of sea which separates Chonos from Chiloe. These islands will, in all probability, in a short time become peopled like those adjoining the coast of Chiloe.

THE WILD POTATO grows on these islands in great abundance, on the sandy, shelly soil near the sea-beach. The tallest plant was four feet in height. The tubers were generally small, but I found one, of an oval shape, two inches in diameter: they resembled in every respect, and had the same smell as English potatoes; but when boiled they shrunk much, and were watery and insipid, without any bitter taste. They are undoubtedly here indigenous: they grow as far south, according to Mr. Low, as lat. 50°, and are called Aquinas by the wild Indians of that part: the Chilotan Indians have a different name for them. Professor Henslow, who has examined the dried specimens which I brought home, says that they are the same with those described by Mr. Sabine[134] from Valparaiso, but that they form a variety which by some botanists has been considered as specifically distinct. It is remarkable that the same plant should be found on the sterile mountains of central Chile, where a drop of rain does not fall for more than six months, and within the damp forests of these southern islands.

In the central parts of the Chonos Archipelago (lat. 45°), the forest has very much the same character with that along the whole west coast, for 600 miles southward to Cape Horn. The arborescent grass of Chiloe is not found here; while the beech of Tierra del Fuego grows to a good size, and forms a considerable proportion of the wood; not, however, in the same exclusive manner as it does farther southward. Cryptogamic plants here find a most congenial climate. In the Strait of Magellan, as I have before remarked, the country appears too cold and wet to allow of their arriving at perfection; but in these islands, within the forest, the number of species and great abundance of mosses, lichens, and small ferns, is quite extraordinary.[135] In Tierra del Fuego trees grow only on the hillsides; every level piece of land being invariably covered by a thick bed of peat; but in Chiloe flat land supports the most luxuriant forests. Here, within the Chonos Archipelago, the nature of the climate more closely approaches that of Tierra del Fuego than that of northern Chiloe; for every patch of level ground is covered by two species of plants (Astelia pumila and Donatia magellanica), which by their joint decay compose a thick bed of elastic peat.

In Tierra del Fuego, above the region of woodland, the former of these eminently sociable plants is the chief agent in the production of peat. Fresh leaves are always succeeding one to the other round the central tap-root; the lower ones soon decay, and in tracing a root downwards in the peat, the leaves, yet holding their place, can be observed passing through every stage of decomposition, till the whole becomes blended in one confused mass. The Astelia is assisted by a few other plants,—here and there a small creeping Myrtus (M. nummularia), with a woody stem like our cranberry and with a sweet berry,—an Empetrum (E. rubrum), like our health,—a rush (Juncus grandiflorus), are nearly the only ones that grow on the swampy surface. These plants, though possessing a very close general resemblance to the English species of the same genera, are different. In the more level parts of the country, the surface of the peat is broken up into little pools of water, which stand at different heights, and appear as if artificially excavated. Small streams of water, flowing underground, complete the disorganization of the vegetable matter, and consolidate the whole.

The climate of the southern part of America appears particularly favourable to the production of peat. In the Falkland Islands almost every kind of plant, even the coarse grass which covers the whole surface of the land, becomes converted into this substance: scarcely any situation checks its growth; some of the beds are as much as twelve feet thick, and the lower part becomes so solid when dry, that it will hardly burn. Although every plant lends its aid, yet in most parts the Astelia is the most efficient. It is rather a singular circumstance, as being so very different from what occurs in Europe, that I nowhere saw moss forming by its decay any portion of the peat in South America. With respect to the northern limit, at which the climate allows of that peculiar kind of slow decomposition which is necessary for its production, I believe that in Chiloe (lat. 41° to 42°), although there is much swampy ground, no well-characterized peat occurs: but in the Chonos Islands, three degrees farther southward, we have seen that it is abundant. On the eastern coast in La Plata (lat. 35°) I was told by a Spanish resident who had visited Ireland, that he had often sought for this substance, but had never been able to find any. He showed me, as the nearest approach to it which he had discovered, a black peaty soil, so penetrated with roots as to allow of an extremely slow and imperfect combustion.

THE ZOOLOGY OF these broken islets of the Chonos Archipelago is, as might have been expected, very poor. Of quadrupeds two aquatic kinds are common. The Myopotamus Coypus (like a beaver, but with a round tail) is well known from its fine fur, which is an object of trade throughout the tributaries of La Plata. It here, however, exclusively frequents salt water; which same circumstance has been mentioned as sometimes occurring with the great rodent, the Capybara. A small sea-otter is very numerous; this animal does not feed exclusively on fish, but, like the seals, draws a large supply from a small red crab, which swims in shoals near the surface of the water. Mr. Bynoe saw one in Tierra del Fuego eating a cuttle-fish; and at Low's Harbour, another was killed in the act of carrying to its hole a large volute shell. At one place I caught in a trap a singular little mouse (M. brachiotis); it appeared common on several of the islets, but the Chilotans at Low's Harbour said that it was not found in all. What a succession of chances,[136] or what changes of level must have been brought into play, thus to spread these small animals throughout this broken archipelago!

In all parts of Chiloe and Chonos, two very strange birds occur, which are allied to, and replace, the Turco and Tapacolo of central Chile. One is called by the inhabitants "Cheucau" (Pteroptochos rubecula): it frequents the most gloomy and retired spots within the damp forests. Sometimes, although its cry may be heard close at hand, let a person watch ever so attentively he will not see the cheucau; at other times, let him stand motionless and the red-breasted little bird will approach within a few feet in the most familiar manner. It then busily hops about the entangled mass of rotting canes and branches, with its little tail cocked upwards. The cheucau is held in superstitious fear by the Chilotans, on account of its strange and varied cries. There are three very distinct cries: one is called "chiduco," and is an omen of good; another, "huitreu," which is extremely unfavourable; and a third, which I have forgotten. These words are given in imitation of the noises; and the natives are in some things absolutely governed by them. The Chilotans assuredly have chosen a most comical little creature for their prophet. An allied species, but rather larger, is called by the natives "Guidguid" (Pteroptochos Tarnii), and by the English the barking-bird. This latter name is well given; for I defy any one at first to feel certain that a small dog is not yelping somewhere in the forest. Just as with the cheucau, a person will sometimes hear the bark close by, but in vain many endeavour by watching, and with still less chance by beating the bushes, to see the bird;

yet at other times the guid-guid fearlessly comes near. Its manner of feeding and its general habits are very similar to those of the cheucau.

On the coast[137] a small dusky-coloured bird (Opetiorhynchus Patagonicus) is very common. It is remarkable from its quiet habits; it lives entirely on the sea-beach, like a sandpiper. Besides these birds only few others inhabit this broken land. In my rough notes I describe the strange noises, which, although frequently heard within these gloomy forests, yet scarcely disturb the general silence. The yelping of the guid-guid, and the sudden whew-whew of the cheucau, sometimes come from afar off, and sometimes from close at hand; the little black wren of Tierra del Fuego occasionally adds its cry; the creeper (Oxyurus) follows the intruder screaming and twittering; the humming-bird may be seen every now and then darting from side to side, and emitting, like an insect, its shrill chirp; lastly, from the top of some lofty tree the indistinct but plaintive note of the white-tufted tyrant-flycatcher (Myiobius) may be noticed. From the great preponderance in most countries of certain common genera of birds, such as the finches, one feels at first surprised at meeting with the peculiar forms above enumerated, as the commonest birds in any district. In central Chile two of them, namely, the Oxyurus and Scytalopus, occur, although most rarely. When finding, as in this case, animals which seem to play so insignificant a part in the great scheme of nature, one is apt to wonder why they were created.

But it should always be recollected, that in some other country perhaps they are essential members of society, or at some former period may have been so. If America south of 37° were sunk beneath the waters of the ocean, these two birds might continue to exist in central Chile for a long period, but it is very improbable that their numbers would increase. We should then see a case which must inevitably have happened with very many animals.

These southern seas are frequented by several species of Petrels: the largest kind, Procellaria gigantea, or nelly (quebrantahuesos, or break-bones, of the Spaniards), is a common bird, both in the inland channels and on the open sea. In its habits and manner of flight, there is a very close resemblance with the albatross; and as with the albatross, a person may watch it for hours together without seeing on what it feeds. The "break-bones" is, however, a rapacious bird, for it was observed by some of the officers at Port St. Antonio chasing a diver, which tried to escape by diving and flying, but was continually struck down, and at last killed by a blow on its head. At Port St. Julian these great petrels were seen killing and devouring young gulls. A second species (Puffinus cinereus), which is common to Europe, Cape Horn, and the coast of Peru, is of much smaller size than the P. gigantea, but, like it, of a dirty black colour. It generally frequents the inland sounds in very large flocks: I do not think I ever saw so many birds of

S. Carlos, which is no less than ninety-three miles from the Corcovado. In the morning the volcano became tranquil.

I was surprised at hearing afterwards that Aconcagua in Chile, 480 miles northwards, was in action on the same night; and still more surprised to hear that the great eruption of Coseguina (2700 miles north of Aconcagua), accompanied by an earthquake felt over a 1000 miles, also occurred within six hours of this same time. This coincidence is the more remarkable, as Coseguina had been dormant for twenty-six years; and Aconcagua most rarely shows any signs of action. It is difficult even to conjecture whether this coincidence was accidental, or shows some subterranean connection. If Vesuvius, Etna, and Hecla in Iceland (all three relatively nearer each other than the corresponding points in South America), suddenly burst forth in eruption on the same night, the coincidence would be thought remarkable; but it is far more remarkable in this case, where the three vents fall on the same great mountain-chain, and where the vast plains along the entire eastern coast, and the upraised recent shells along more than 2000 miles on the western coast, show in how equable and connected a manner the elevatory forces have acted.

Captain Fitz Roy being anxious that some bearings should be taken on the outer coast of Chiloe, it was planned that Mr. King and myself should ride to Castro, and thence across the island to the Capella de Cucao, situated on the west coast. Having hired horses and a guide, we set out on the morning of the 22nd. We had not proceeded far, before we were joined by a woman and two boys, who were bent on the same journey. Every one on this road acts on a "hail fellow well met" fashion; and one may here enjoy the privilege, so rare in South America, of travelling without firearms. At first, the country consisted of a succession of hills and valleys: nearer to Castro it became very level. The road itself is a curious affair; it consists in its whole length, with the exception of very few parts, of great logs of wood, which are either broad and laid longitudinally, or narrow and placed transversely. In summer the road is not very bad; but in winter, when the wood is rendered slippery from rain, travelling is exceedingly difficult. At that time of the year, the ground on each side becomes a morass, and is often overflowed: hence it is necessary that the longitudinal logs should be fastened down by transverse poles, which are pegged on each side into the earth. These pegs render a fall from a horse dangerous, as the chance of alighting on one of them is not small. It is remarkable, however, how active custom has made the Chilotan horses. In crossing bad parts, where the logs had been displaced, they skipped from one to the other, almost with the quickness and certainty of a dog. On both hands the road is bordered by the lofty forest-trees, with their bases matted together by canes. When occasionally a long reach of this avenue could be beheld, it presented a curious scene of uniformity: the white line of logs, narrowing in perspective, became hidden by the gloomy forest, or terminated in a zigzag which ascended some steep hill.

Although the distance from S. Carlos to Castro is only twelve leagues in a straight line, the formation of the road must have been a great labour. I was told that several people had formerly lost their lives in attempting to cross the forest. The first who succeeded was an Indian, who cut his way through the canes in eight days, and reached S. Carlos: he was rewarded by the Spanish government with a grant of land. During the summer, many of the Indians wander about the forests (but chiefly in the higher parts, where the woods are not quite so thick) in search of the half-wild cattle which live on the leaves of the cane and certain trees. It was one of these huntsmen who by chance discovered, a few years since, an English vessel, which had been wrecked on the outer coast. The crew were beginning to fail in provisions, and it is not probable that, without the aid of this man, they would ever have extricated themselves from these scarcely penetrable woods. As it was, one seaman died on the march, from fatigue. The Indians in these excursions steer by the sun; so that if there is a continuance of cloudy weather, they cannot travel.

The day was beautiful, and the number of trees which were in full flower perfumed the air; yet even this could hardly dissipate the effects of the gloomy dampness of the forest. Moreover, the many dead trunks that stand like skeletons, never fail to give to these primeval woods a character of solemnity, absent in those of countries long civilized. Shortly after sunset we bivouacked for the night. Our female companion, who was rather good-looking, belonged to one of the most respectable families in Castro: she rode, however, astride, and without shoes or stockings. I was surprised at the total want of pride shown by her and her brother. They brought food with them, but at all our meals sat watching Mr. King and myself whilst eating, till we were fairly shamed into feeding the whole party. The night was cloudless; and while lying in our beds, we enjoyed the sight (and it is a high enjoyment) of the multitude of stars which illumined the darkness of the forest.

January 23rd.—We rose early in the morning, and reached the pretty quiet town of Castro by two o'clock. The old governor had died since our last visit, and a Chileno was acting in his place. We had a letter of introduction to Don Pedro, whom we found exceedingly hospitable and kind, and more disinterested than is usual on this side of the continent. The next day Don Pedro procured us fresh horses, and offered to accompany us himself. We proceeded to the south—generally following the coast, and passing through several hamlets, each with its large barn-like chapel built of wood. At Vilipilli, Don Pedro asked the commandant to give us a guide to Cucao.

The old gentleman offered to come himself; but for a long time nothing would persuade him that two Englishmen really wished to go to such an out-of-the-way place as Cucao. We were thus accompanied by the two greatest aristocrats in the country, as was plainly to be seen in the manner of all the poorer Indians towards them. At Chonchi we struck across the island, following intricate winding paths, sometimes passing through magnificent forests, and sometimes through pretty cleared spots, abounding with corn and potato crops. This undulating woody country, partially cultivated, reminded me of the wilder parts of England, and therefore had to my eye a most fascinating aspect. At Vilinco, which is situated on the borders of the lake of Cucao, only a few fields were cleared; and all the inhabitants appeared to be Indians. This lake is twelve miles long, and runs in an east and west direction. From local circumstances, the sea-breeze blows very regularly during the day, and during the night it falls calm: this has given rise to strange exaggerations, for the phenomenon, as described to us at S. Carlos, was quite a prodigy.

The road to Cucao was so very bad that we determined to embark in a *periagua.* The commandant, in the most authoritative manner, ordered six Indians to get ready to pull us over, without deigning to tell them whether they would be paid. The periagua is a strange rough boat, but the crew were still stranger: I doubt if six uglier little men ever got into a boat together. They pulled, however, very well and cheerfully. The stroke-oarsman gabbled Indian, and uttered strange cries, much after the fashion of a pig-driver driving his pigs. We started with a light breeze against us, but yet reached the Capella de Cucao before it was late. The country on each side of the lake was one unbroken forest. In the same periagua with us, a cow was embarked. To get so large an animal into a small boat appears at first a difficulty, but the Indians managed it in a minute. They brought the cow alongside the boat, which was heeled towards her; then placing two oars under her belly, with their ends resting on the gunwale, by the aid of these levers they fairly tumbled the poor beast heels over head into the bottom of the boat, and then lashed her down with ropes. At Cucao we found an uninhabited hovel (which is the residence of the padre when he pays this Capella a visit), where, lighting a fire, we cooked our supper, and were very comfortable.

The district of Cucao is the only inhabited part on the whole west coast of Chiloe. It contains about thirty or forty Indian families, who are scattered along four or five miles of the shore. They are very much secluded from the rest of Chiloe, and have scarcely any sort of commerce, except sometimes in a little oil, which they get from seal-blubber. They are tolerably dressed in clothes of their own manufacture, and they have plenty to eat. They seemed, however, discontented, yet humble to a degree which it was quire painful to witness. These feelings are, I think, chiefly to be attributed to the harsh and authoritative manner in which they are treated by their rulers. Our companions, although so very civil to us, behaved to the poor Indians as if they had been slaves, rather than free men. They ordered provisions and the use of their horses, without ever condescending to say how much, or indeed whether the owners should be paid at all. In the morning, being left alone with these poor people, we soon ingratiated ourselves by presents of cigars and maté. A lump of white sugar was divided between all present, and tasted with the greatest curiosity. The Indians ended all their complaints by saying, "And it is only because we are poor Indians, and know nothing; but it was not so when we had a King."

The next day after breakfast, we rode a few miles northward to Punta Huantamó. The road lay along a very broad beach, on which, even after so many fine days, a terrible surf was breaking. I was assured that after a heavy gale, the roar can be heard at night even at Castro, a distance of no less than twenty-one sea-miles across a hilly and wooded country. We had some difficulty in reaching the point, owing to the intolerably bad paths; for everywhere in the shade the ground soon becomes a perfect quagmire. The point itself is a bold rocky hill. It is covered by a plant allied, I believe, to Bromelia, and called by the inhabitants Chepones. In scrambling through the beds, our hands were very much scratched. I was amused by observing the precaution our Indian guide took, in turning up his trousers, thinking that they were more delicate than his own hard skin. This plant bears a fruit, in shape like an artichoke, in which a number of seed-vessels are packed: these contain a pleasant sweet pulp, here much esteemed. I saw at Low's Harbour the Chilotans making chichi, or cider, with this fruit: so true is it, as Humboldt remarks, that almost everywhere man finds means of preparing some kind of beverage from the vegetable kingdom. The savages, however, of Tierra del Fuego, and I believe of Australia, have not advanced thus far in the arts.

The coast to the north of Punta Huantamó is exceedingly rugged and broken, and is fronted by many breakers, on which the sea is eternally roaring. Mr. King and myself were anxious to return, if it had been possible, on foot along this coast; but even the Indians said it was quite impracticable. We were told that men have crossed by striking directly through the woods from Cucao to S. Carlos, but never by the coast. On these expeditions, the Indians carry with them only roasted corn, and of this they eat sparingly twice a day.

26th.—Re-embarking in the periagua, we returned across the lake, and then mounted our horses. The whole of Chiloe took advantage of this week of unusually fine weather, to clear the ground by burning. In every direction volumes of smoke were curling upwards. Although the inhabitants were so assiduous in setting fire to every part of the wood, yet I did not see a single fire which they had succeeded in making extensive. We dined with our friend the commandant, and did not reach Castro till after dark. The next morning we started very early. After having ridden for some time, we obtained from the brow of a steep hill an extensive view (and it is a rare thing on this road) of the great forest. Over the horizon of trees, the volcano of Corcovado, and the great flat-topped one to the north, stood out in proud pre-eminence: scarcely another peak in the long range showed its snowy summit. I hope it will be long before I forget this farewell view of the magnificent Cordillera fronting Chiloe. At night we bivouacked under a cloudless sky, and the next morning reached S. Carlos. We arrived on the right day, for before evening heavy rain commenced.

February 4th.—Sailed from Chiloe. During the last week I made several short excursions. One was to examine a great bed of now-existing shells, elevated 350 feet above the level of the sea: from among these shells, large forest-trees were growing. Another ride was to P. Huechucucuy. I had with me a guide who knew the country far too well; for he would pertinaciously tell me endless Indian names for every little point, rivulet, and creek. In the same manner as in Tierra del Fuego, the Indian language appears singularly well adapted for attaching names to the most trivial features of the land. I believe every one was glad to say farewell to Chiloe; yet if we could forget the gloom and ceaseless rain of winter, Chiloe might pass for a charming island. There is also something very attractive in the simplicity and humble politeness of the poor inhabitants.

We steered northward along shore, but owing to thick weather did not reach Valdivia till the night of the 8th. The next morning the boat proceeded to the town, which is distant about ten miles. We followed the course of the river, occasionally passing a few hovels, and patches of ground cleared out of the otherwise unbroken forest; and sometimes meeting a canoe with an Indian family. The town is situated on the low banks of the stream, and is so completely buried in a wood of apple-trees that the streets are merely paths in an orchard. I have never seen any country, where apple-trees appeared to thrive so well as in this damp part of South America: on the borders of the roads there were many young trees evidently self-grown. In Chiloe the inhabitants possess a marvellously short method of making an orchard. At the lower part of almost every branch, small, conical, brown, wrinkled points project: these are always ready to change into roots, as may sometimes be seen, where any mud has been accidentally splashed against the tree. A branch as thick as a man's thigh is chosen in the early spring, and is cut off just beneath a group of these points, all the smaller branches are lopped off, and it is then placed about two feet deep in the ground. During the ensuing summer the stump throws out long shoots, and sometimes even bears fruit: I was shown one which had produced as many as twenty-three apples, but this was thought very unusual. In the third season the stump is changed (as I have myself seen) into a well-wooded tree, loaded with fruit. An old man near Valdivia illustrated his motto, "Necesidad es la madre del invención," by giving an account of the several useful things he manufactured from his apples. After making cider, and likewise wine, he extracted from the refuse a white and finely flavoured spirit; by another process he procured a sweet treacle, or, as he called it, honey. His children and pigs seemed almost to live, during this season of the year, in his orchard.

February 11th.—I set out with a guide on a short ride, in which, however, I managed to see singularly little, either of the geology of the country or of its inhabitants. There is not much cleared land near Valdivia: after crossing a river at the distance of a few miles, we entered the forest, and then passed only one miserable hovel, before reaching our sleeping-place for the night. The short difference in latitude, of 150 miles, has given a new aspect to the forest, compared with that of Chiloe. This is owing to a slightly different proportion in the kinds of trees. The evergreens do not appear to be quite so numerous, and the forest in consequence has a brighter tint. As in Chiloe, the lower parts are matted together by canes: here also another kind (resembling the bamboo of Brazil and about twenty feet in height) grows in clusters, and ornaments the banks of some of the streams in a very pretty manner. It is with this plant that the Indians make their chuzos, or long tapering spears. Our resting-house was so dirty that I preferred sleeping outside: on these journeys the first night is generally very uncomfortable, because one is not accustomed to the tickling and biting of the fleas. I am sure, in the morning, there was not a space on my legs the size of a shilling which had not its little red mark where the flea had feasted.

12th.—We continued to ride through the uncleared forest; only occasionally meeting an Indian on horseback, or a troop of fine mules bringing alerce-planks and corn from the southern plains. In the afternoon one of the horses knocked up: we were then on a brow of a hill, which commanded a fine view of the Llanos. The view of these open plains was very refreshing, after being hemmed in and buried in the wilderness of trees. The uniformity of a forest soon becomes very wearisome. This west coast makes me remember with pleasure the free, unbounded plains of Patagonia; yet, with the true spirit of contradiction, I cannot forget how sublime is the silence of the forest. The Llanos are the most fertile and thickly peopled parts of the country, as

pened to be on shore, and was lying down in the wood to rest myself. It came on suddenly, and lasted two minutes, but the time appeared much longer. The rocking of the ground was very sensible. The undulations appeared to my companion and myself to come from due east, whilst others thought they proceeded from south-west: this shows how difficult it sometimes is to perceive the directions of the vibrations. There was no difficulty in standing upright, but the motion made me almost giddy: it was something like the movement of a vessel in a little cross-ripple, or still more like that felt by a person skating over thin ice, which bends under the weight of his body.

A bad earthquake at once destroys our oldest associations: the earth, the very emblem of solidity, has moved beneath our feet like a thin crust over a fluid;—one second of time has created in the mind a strange idea of insecurity, which hours of reflection would not have produced. In the forest, as a breeze moved the trees, I felt only the earth tremble, but saw no other effect. Captain Fitz Roy and some officers were at the town during the shock, and there the scene was more striking; for although the houses, from being built of wood, did not fall, they were violently shaken, and the boards creaked and rattled together. The people rushed out of doors in the greatest alarm. It is these accompaniments that create that perfect horror of earthquakes, experienced by all who have thus seen, as well as felt, their effects. Within the forest it was a deeply interesting, but by no means an awe-exciting phenomenon. The tides were very curiously affected. The great shock took place at the time of low water; and an old woman who was on the beach told me that the water flowed very quickly, but not in great waves, to high-water mark, and then as quickly returned to its proper level; this was also evident by the line of wet sand. The same kind of quick but quiet movement in the tide happened a few years since at Chiloe, during a slight earthquake, and created much causeless alarm. In the course of the evening there were many weaker shocks, which seemed to produce in the harbour the most complicated currents, and some of great strength.

March 4th.—We entered the harbour of Concepcion. While the ship was beating up to the anchorage, I landed on the island of Quiriquina. The mayor-domo of the estate quickly rode down to tell me the terrible news of the great earthquake of the 20th:—"That not a house in Concepcion or Talcahuano (the port) was standing; that seventy villages were destroyed; and that a great wave had almost washed away the ruins of Talcahuano." Of this latter statement I soon saw abundant proofs—the whole coast being strewed over with timber and furniture as if a thousand ships had been wrecked. Besides chairs, tables, book-shelves, etc., in great numbers, there were several roofs of cottages, which had been transported almost whole.

The storehouses at Talcahuano had been burst open, and great bags of cotton, yerba, and other valuable merchandise were scattered on the shore. During my walk round the island, I observed that numerous fragments of rock, which, from the marine productions adhering to them, must recently have been lying in deep water, had been cast up high on the beach; one of these was six feet long, three broad, and two thick.

The island itself as plainly showed the overwhelming power of the earthquake, as the beach did that of the consequent great wave. The ground in many parts was fissured in north and south lines, perhaps caused by the yielding of the parallel and steep sides of this narrow island. Some of the fissures near the cliffs were a yard wide. Many enormous masses had already fallen on the beach; and the inhabitants thought that when the rains commenced far greater slips would happen. The effect of the vibration on the hard primary slate, which composes the foundation of the island, was still more curious: the superficial parts of some narrow ridges were as completely shivered as if they had been blasted by gun-powder. This effect, which was rendered conspicuous by the fresh fractures and displaced soil, must be confined to near the surface, for otherwise there would not exist a block of solid rock throughout Chile; nor is this improbable, as it is known that the surface of a vibrating body is affected differently from the central part. It is, perhaps, owing to this same reason, that earthquakes do not cause quite such terrific havoc within deep mines as would be expected. I believe this convulsion has been more effectual in lessening the size of the island of Quiriquina, than the ordinary wear-and-tear of the sea and weather during the course of a whole century.

The next day I landed at Talcahuano, and afterwards rode to Concepcion. Both towns presented the most awful yet interesting spectacle I ever beheld. To a person who had formerly known them, it possibly might have been still more impressive; for the ruins were so mingled together, and the whole scene possessed so little the air of a habitable place, that it was scarcely possible to imagine its former condition. The earthquake commenced at half-past eleven o'clock in the forenoon. If it had happened in the middle of the night, the greater number of the inhabitants (which in this one province must amount to many thousands) must have perished, instead of less than a hundred: as it was, the invariable practice of running out of doors at the first trembling of the ground, alone saved them. In Concepcion each house, or row of houses, stood by itself, a heap or line of ruins; but in Talcahuano, owing to the great wave, little more than one layer of bricks, tiles, and timber, with here and there part of a wall left standing, could be distinguished. From this circumstance Concepcion, although not so completely desolated, was a more terrible, and if I may so call it, picturesque sight. The first shock was very sudden. The mayor-domo at

Quiriquina told me, that the first notice he received of it, was finding both the horse he rode and himself, rolling together on the ground. Rising up, he was again thrown down. He also told me that some cows which were standing on the steep side of the island were rolled into the sea. The great wave caused the destruction of many cattle; on one low island, near the head of the bay, seventy animals were washed off and drowned. It is generally thought that this has been the worst earthquake ever recorded in Chile; but as the very severe ones occur only after long intervals, this cannot easily be known; nor indeed would a much worse shock have made any difference, for the ruin was now complete. Innumerable small tremblings followed the great earthquake, and within the first twelve days no less than three hundred were counted.

After viewing Concepcion, I cannot understand how the greater number of inhabitants escaped unhurt. The houses in many parts fell outwards; thus forming in the middle of the streets little hillocks of brickwork and rubbish. Mr. Rouse, the English consul, told us that he was at breakfast when the first movement warned him to run out. He had scarcely reached the middle of the court-yard, when one side of his house came thundering down. He retained presence of mind to remember, that if he once got on the top of that part which had already fallen, he would be safe. Not being able from the motion of the ground to stand, he crawled up on his hands and knees; and no sooner had he ascended this little eminence, than the other side of the house fell in, the great beams sweeping close in front of his head. With his eyes blinded, and his mouth choked with the cloud of dust which darkened the sky, at last he gained the street. As shock succeeded shock, at the interval of a few minutes, no one dared approach the shattered ruins; and no one knew whether his dearest friends and relations were not perishing from the want of help. Those who had saved any property were obliged to keep a constant watch, for thieves prowled about, and at each little trembling of the ground, with one hand they beat their breasts and cried "Misericordia!" and then with the other filched what they could from the ruins. The thatched roofs fell over the fires, and flames burst forth in all parts. Hundreds knew themselves ruined, and few had the means of providing food for the day.

Earthquakes alone are sufficient to destroy the prosperity of any country. If beneath England the now inert subterranean forces should exert those powers, which most assuredly in former geological ages they have exerted, how completely would the entire condition of the country be changed! What would become of the lofty houses, thickly packed cities, great manufactories, the beautiful public and private edifices? If the new period of disturbance were first to commence by some great earthquake in the dead of the night, how terrific would be the carnage! England would at once be bankrupt; all papers, records, and accounts would from that moment be lost. Government being unable to collect the taxes, and failing to maintain its authority, the hand of violence and rapine would remain uncontrolled. In every large town famine would go forth, pestilence and death following in its train.

Shortly after the shock, a great wave was seen from the distance of three or four miles, approaching in the middle of the bay with a smooth outline; but along the shore it tore up cottages and trees, as it swept onwards with irresistible force. At the head of the bay it broke in a fearful line of white breakers, which rushed up to a height of 23 vertical feet above the highest spring-tides. Their force must have been prodigious; for at the Fort a cannon with its carriage, estimated at four tons in weight, was moved 15 feet inwards. A schooner was left in the midst of the ruins, 200 yards from the beach. The first wave was followed by two others, which in their retreat carried away a vast wreck of floating objects. In one part of the bay, a ship was pitched high and dry on shore, was carried off, again driven on shore, and again carried off. In another part, two large vessels anchored near together were whirled about, and their cables were thrice wound round each other; though anchored at a depth of 36 feet, they were for some minutes aground. The great wave must have travelled slowly, for the inhabitants of Talcahuano had time to run up the hills behind the town; and some sailors pulled out seaward, trusting successfully to their boat riding securely over the swell, if they could reach it before it broke. One old woman with a little boy, four or five years old, ran into a boat, but there was nobody to row it out: the boat was consequently dashed against an anchor and cut in twain; the old woman was drowned, but the child was picked up some hours afterwards clinging to the wreck. Pools of salt-water were still standing amidst the ruins of the houses, and children, making boats with old tables and chairs, appeared as happy as their parents were miserable. It was, however, exceedingly interesting to observe, how much more active and cheerful all appeared than could have been expected. It was remarked with much truth, that from the destruction being universal, no one individual was humbled more than another, or could suspect his friends of coldness—that most grievous result of the loss of wealth. Mr. Rouse, and a large party whom he kindly took under his protection, lived for the first week in a garden beneath some apple-trees. At first they were as merry as if it had been a picnic; but soon afterwards heavy rain caused much discomfort, for they were absolutely without shelter.

In Captain Fitz Roy's excellent account of the earthquake, it is said that two explosions, one like a column of smoke and another like the blowing of a great whale, were seen in the bay. The water also appeared everywhere to be boiling; and it "became black, and exhaled a most disagreeable sulphureous smell." These latter circumstances were observed in the Bay of Valparaiso during the earthquake of 1822; they may, I think, be accounted for, by the disturbance of the mud at the bottom of the sea containing organic matter in decay. In the Bay of Callao, during a calm day, I noticed, that as the ship dragged her cable over the bottom, its course was marked by a line of bubbles. The lower orders in Talcahuano thought that the earthquake was caused by some old Indian women, who two years ago, being offended, stopped the volcano of Antuco. This silly belief is curious, because it shows that experience has taught them to observe, that there exists a relation between the suppressed action of the volcanos, and the trembling of the ground. It was necessary to apply the witchcraft to the point where their perception of cause and effect failed; and this was the closing of the volcanic vent. This belief is the more singular in this particular instance, because, according to Captain Fitz Roy, there is reason to believe that Antuco was noways affected.

The town of Concepcion was built in the usual Spanish fashion, with all the streets running at right angles to each other; one set ranging S.W. by W., and the other set N.W. by N. The walls in the former direction certainly stood better than those in the latter; the greater number of the masses of brickwork were thrown down towards the N.E. Both these circumstances perfectly agree with the general idea, of the undulations having come from the S.W., in which quarter subterranean noises were also heard; for it is evident that the walls running S.W. and N.E. which presented their ends to the point whence the undulations came, would be much less likely to fall than those walls which, running N.W. and S.E., must in their whole lengths have been at the same instant thrown out of the perpendicular; for the undulations, coming from the S.W., must have extended in N.W. and S.E. waves, as they passed under the foundations. This may be illustrated by placing books edgeways on a carpet, and then, after the manner suggested by Michell, imitating the undulations of an earthquake: it will be found that they fall with more or less readiness, according as their direction more or less nearly coincides with the line of the waves. The fissures in the ground generally, though not uniformly, extended in a S.E. and N.W. direction, and therefore corresponded to the lines of undulation or of principal flexure. Bearing in mind all these circumstances, which so clearly point to the S.W. as the chief focus of disturbance, it is a very interesting fact that the island of S. Maria, situated in that quarter, was, during the general uplifting of the land, raised to nearly three times the height of any other part of the coast.

The different resistance offered by the walls, according to their direction, was well exemplified in the case of the Cathedral. The side which fronted the N.E. presented a grand pile of ruins, in the midst of which door-cases and masses of timber stood up, as if floating in a stream. Some of the angular blocks of brickwork were of great dimensions; and they were rolled to a distance on the level plaza, like fragments of rock at the base of some high mountain. The side walls (running S.W. and N.E.), though exceedingly fractured, yet remained standing; but the vast buttresses (at right angles to them, and therefore parallel to the walls that fell) were in many cases cut clean off, as if by a chisel, and hurled to the ground. Some square ornaments on the coping of these same walls, were moved by the earthquake into a diagonal position. A similar circumstance was observed after an earthquake at Valparaiso, Calabria, and other places, including some of the ancient Greek temples.[138] This twisting displacement, at first appears to indicate a vorticose movement beneath each point thus affected; but this is highly improbable. May it not be caused by a tendency in each stone to arrange itself in some particular position, with respect to the lines of vibration,—in a manner somewhat similar to pins on a sheet of paper when shaken? Generally speaking, arched doorways or windows stood much better than any other part of the buildings. Nevertheless, a poor lame old man, who had been in the habit, during trifling shocks, of crawling to a certain doorway, was this time crushed to pieces.

I have not attempted to give any detailed description of the appearance of Concepcion, for I feel that it is quite impossible to convey the mingled feelings which I experienced. Several of the officers visited it before me, but their strongest language failed to give a just idea of the scene of desolation. It is a bitter and humiliating thing to see works, which have cost man so much time and labour, overthrown in one minute; yet compassion for the inhabitants was almost instantly banished, by the surprise in seeing a state of things produced in a moment of time, which one was accustomed to attribute to a succession of ages. In my opinion, we have scarcely beheld, since leaving England, any sight so deeply interesting.

In almost every severe earthquake, the neighbouring waters of the sea are said to have been greatly agitated. The disturbance seems generally, as in the case of Concepcion, to have been of two kinds: first, at the instant of the shock, the water swells high up on the beach with a gentle motion, and then as quietly retreats; secondly, some time afterwards, the whole body of the sea retires from the coast, and then returns in waves of overwhelming force. The first movement seems to be an immediate consequence of the earthquake affecting differently a fluid and a solid, so that their respective levels are slightly deranged: but the second case is a far more important phenomenon. During most earthquakes, and especially during those on the west coast of America, it is certain that the first great movement of the waters has been a retirement. Some authors have attempted to explain this, by supposing that the water retains its level, whilst the land oscillates upwards; but surely the water close to the land, even on a rather steep coast,

confidently come to the conclusion, that the forces which slowly and by little starts uplift continents, and those which at successive periods pour forth volcanic matter from open orifices, are identical. From many reasons, I believe that the frequent quakings of the earth on this line of coast are caused by the rending of the strata, necessarily consequent on the tension of the land when upraised, and their injection by fluidified rock. This rending and injection would, if repeated often enough (and we know that earthquakes repeatedly affect the same areas in the same manner), form a chain of hills;—and the linear island of S. Mary, which was upraised thrice the height of the neighbouring country, seems to be undergoing this process. I believe that the solid axis of a mountain, differs in its manner of formation from a volcanic hill, only in the molten stone having been repeatedly injected, instead of having been repeatedly ejected. Moreover, I believe that it is impossible to explain the structure of great mountain-chains, such as that of the Cordillera, where the strata, capping the injected axis of plutonic rock, have been thrown on their edges along several parallel and neighbouring lines of elevation, except on this view of the rock of the axis having been repeatedly injected, after intervals sufficiently long to allow the upper parts or wedges to cool and become solid;—for if the strata had been thrown into their present highly inclined, vertical, and even inverted positions, by a single blow, the very bowels of the earth would have gushed out; and instead of beholding abrupt mountain-axes of rock solidified under great pressure, deluges of lava would have flowed out at innumerable points on every line of elevation.[139]

CHAPTER XV

Passage of the Cordillera

VALPARAISO • PORTILLO PASS • SAGACITY OF MULES • MOUNTAIN-TORRENTS • MINES, HOW DISCOVERED • PROOFS OF THE GRADUAL ELEVATION OF THE CORDILLERA • EFFECT OF SNOW ON ROCKS • GEOLOGICAL STRUCTURE OF THE TWO MAIN RANGES, THEIR DISTINCT ORIGIN AND UPHEAVAL • GREAT SUBSIDENCE • RED SNOW • WINDS • PINNACLES OF SNOW • DRY AND CLEAR ATMOSPHERE • ELECTRICITY • PAMPAS • ZOOLOGY OF THE OPPOSITE SIDES OF THE ANDES • LOCUSTS • GREAT BUGS • MENDOZA • USPALLATA PASS • SILICIFIED TREES BURIED AS THEY GREW • INCAS BRIDGE • BADNESS OF THE PASSES EXAGGERATED • CUMBRE • CASUCHAS • VALPARAISO

MARCH 7TH, 1835.—WE STAYED three days at Concepcion, and then sailed for Valparaiso. The wind being northerly, we only reached the mouth of the harbour of Concepcion before it was dark. Being very near the land, and a fog coming on, the anchor was dropped. Presently a large American whaler appeared alongside of us; and we heard the Yankee swearing at his men to keep quiet, whilst he listened for the breakers. Captain Fitz Roy hailed him, in a loud clear voice, to anchor where he then was. The poor man must have thought the voice came from the shore: such a Babel of cries issued at once from the ship—every one hallooing out, "Let go the anchor! veer cable! shorten sail!" It was the most laughable thing I ever heard. If the ship's crew had been all captains, and no men, there could not have been a greater uproar of orders. We afterwards found that the mate stuttered: I suppose all hands were assisting him in giving his orders.

On the 11th we anchored at Valparaiso, and two days afterwards I set out to cross the Cordillera. I proceeded to Santiago, where Mr. Caldcleugh most kindly assisted me in every possible way in making the little preparations which were necessary. In this part of Chile there are two passes across the Andes to Mendoza: the one most commonly used—namely, that of Aconcagua or Uspallata—is situated some way to the north; the other, called the Portillo, is to the south, and nearer, but more lofty and dangerous.

March 18th.—We set out for the Portillo pass. Leaving Santiago we crossed the wide burnt-up plain on which that city stands, and in the afternoon arrived at the Maypu, one of the principal rivers in Chile. The valley, at the point where it enters the first Cordillera, is bounded on each side by lofty barren mountains; and although not broad, it is very fertile. Numerous cottages were surrounded by vines, and by orchards of apple, nectarine, and peach-trees—their boughs breaking with the weight of the beautiful ripe fruit. In the evening we passed the custom-house, where our luggage was examined. The frontier of Chile is better guarded by the Cordillera, than by the waters of the sea. There are very few valleys which lead to the central ranges, and the mountains are quite impassable in other parts by beasts of burden. The custom-house officers were very civil, which was perhaps partly owing to the passport which the President of the Republic had given me; but I must express my admiration at the natural politeness of almost every Chileno. In this instance, the contrast with the same class of men in most other countries was strongly marked. I may mention an anecdote with which I was at the time much pleased: we met near Mendoza a little and very fat negress, riding astride on a mule. She had a *goître* so enormous that it was scarcely possible to avoid gazing at her for a moment; but my two companions almost instantly, by way of apology, made the common salute of the country by taking off their hats. Where would one of the lower or higher classes in Europe, have shown such feeling politeness to a poor and miserable object of a degraded race?

At night we slept at a cottage. Our manner of travelling was delightfully independent. In the inhabited parts we bought a little firewood, hired pasture for the animals, and bivouacked in the corner of the same field with them. Carrying an iron pot, we cooked and ate our supper under a cloudless sky, and knew no trouble. My companions were Mariano Gonzales, who had formerly accompanied me in Chile, and an "arriero," with his ten mules and a "madrina." The madrina (or god-mother) is a most important personage: she is an old steady mare, with a little bell round her neck; and wherever she goes, the mules, like good children, follow her. The affection of these animals for their madrinas saves infinite trouble. If several large troops are turned into one field to graze, in the morning the muleteers have only to lead the madrinas a little apart, and tinkle their bells; although there may be two or three hundred together, each mule immediately knows the bell of its own madrina, and comes to her. It is nearly impossible to lose an old mule; for if detained for several hours by force, she will, by the power of smell, like a dog, track out her companions, or rather the madrina, for, according to the muleteer, she is the chief object of affection. The feeling, however, is not of an individual nature; for I believe I am right in saying that any animal with a bell will serve as a madrina. In a troop each animal carries on a level road, a cargo weighing 416 pounds (more than 29 stone), but in a mountainous country 100 pounds less; yet with what delicate slim limbs, without any proportional bulk of muscle, these animals support so great a burden! The mule always appears to me a most surprising animal. That a hybrid should possess more reason, memory, obstinacy, social affection, powers of muscular endurance, and length of life, than either of its parents, seems to indicate that art has here outdone nature. Of our ten animals, six were intended for riding, and four for carrying cargoes, each taking turn about. We carried a good deal of food, in case we should be snowed up, as the season was rather late for passing the Portillo.

March 19th.—We rode during this day to the last, and therefore most elevated, house in the valley. The number of inhabitants became scanty; but wherever water could be brought on the land, it was very fertile. All the main valleys in the Cordillera are characterized by having, on both sides, a fringe or terrace of shingle and sand, rudely stratified, and generally of considerable thickness. These fringes evidently once extended across the valleys and were united; and the bottoms of the valleys in northern Chile, where there are no streams, are thus smoothly filled up. On these fringes the roads are generally carried, for their surfaces are even, and they rise with a very gentle slope up the valleys: hence, also, they are easily cultivated by irrigation. They may be traced up to a height of between 7000 and 9000 feet, where they become hidden by the irregular piles of debris. At the lower end or mouths of the valleys, they are continuously united to those land-locked plains (also formed of shingle) at the foot of the main Cordillera, which I have described in a former chapter as characteristic of the scenery of Chile, and which were undoubtedly deposited when the sea penetrated Chile, as it now does the more southern coasts. No one fact in the geology of South America, interested me more than these terraces of rudely-stratified shingle. They precisely resemble in composition the matter which the torrents in each valley would deposit, if they were checked in their course by any cause, such as entering a lake or arm of the sea; but the torrents, instead of depositing matter, are now steadily at work wearing away both the solid rock and these alluvial deposits, along the whole line of every main valley and side valley. It is impossible here to give the reasons, but I am convinced that the shingle terraces were accumulated, during the gradual elevation of the Cordillera, by the torrents delivering, at successive levels, their detritus on the beachheads of long narrow arms of the sea, first high up the valleys, then lower and lower down as the land slowly rose. If this be so, and I cannot doubt it, the grand and broken chain of the Cordillera, instead of having been suddenly thrown up, as was till lately the universal, and still is the common opinion of geologists, has been slowly upheaved in mass, in the same gradual manner as the coasts of the Atlantic and Pacific have risen within the recent period. A multitude of facts in the structure of the Cordillera, on this view receive a simple explanation.

The rivers which flow in these valleys ought rather to be called mountain-torrents. Their inclination is very great, and their water the colour of mud. The roar which the Maypu made, as it rushed over the great rounded fragments, was like that of the sea. Amidst the din of rushing waters, the noise from the stones, as they rattled one over another, was most distinctly audible even from a distance. This rattling noise, night and day, may be heard along the whole course of the torrent. The sound spoke eloquently to the geologist; the thousands and thousands of stones, which, striking against each other, made the one dull uniform sound, were all hurrying in one direction. It was like thinking on time, where the minute that now glides past is irrevocable. So was it with these stones; the ocean is their eternity, and each note of that wild music told of one more step towards their destiny.

It is not possible for the mind to comprehend, except by a slow process, any effect which is produced by a cause repeated so often, that the multiplier itself conveys an idea, not more definite than the savage implies when he points to the hairs of his head. As often as I have seen beds of mud, sand, and shingle, accumulated to the thickness of many thousand feet, I have felt inclined to exclaim that causes, such as the present rivers and the present beaches, could never have ground down and produced such masses. But, on the other hand, when listening to the rattling noise of these torrents, and calling to mind that whole races of animals have passed away from the face of the earth, and that during this whole period, night and day, these stones have gone rattling onwards in their course, I have thought to myself, can any mountains, any continent, withstand such waste?

In this part of the valley, the mountains on each side were from 3000 to 6000 or 8000 feet high, with rounded outlines and steep bare flanks. The general colour of the rock was dullish purple, and the stratification very distinct. If the scenery was not beautiful, it was remarkable and grand. We met during the day several herds of cattle, which men were driving down from the higher valleys in the Cordillera. This sign of the approaching winter hurried our steps, more than was convenient for geologizing. The house where we slept was situated at the foot of a mountain, on the summit of which are the mines of S. Pedro de Nolasko. Sir F. Head marvels how mines have been discovered in such extraordinary situations, as the bleak summit of the mountain of S. Pedro de Nolasko. In the first place, metallic veins in this country are generally harder than the surrounding strata: hence, during the gradual wear of the hills, they project above the surface of the ground. Secondly, almost every labourer, especially in the northern parts of Chile, understands something about the appearance of ores. In the great mining provinces of Coquimbo and Copiapó, firewood is very scarce, and men search for it over every hill and dale; and by this means nearly all the richest mines have there been discovered. Chanuncillo, from which silver to the value of many hundred thousand pounds has been raised in the course of a few years, was discovered by a man who threw a stone at his loaded donkey, and thinking that it was very heavy, he picked it up, and found it full of pure silver: the vein occurred at no great distance, standing up like a wedge of metal. The miners, also, taking a crowbar with them, often wander on Sundays over the mountains. In this south part of Chile, the men who drive cattle into the Cordillera, and who frequent every ravine where there is a little pasture, are the usual discoverers.

20th.—As we ascended the valley, the vegetation, with the exception of a few pretty alpine flowers, became exceedingly scanty; and of quadrupeds, birds, or insects, scarcely one could be seen. The lofty mountains, their summits marked with a few patches of snow, stood well separated from each other, the valleys being filled up with an immense thickness of stratified alluvium. The features in the scenery of the Andes which struck me most, as contrasted with the other mountain chains with which I am acquainted, were,—the flat fringes sometimes expanding into narrow plains on each side of the valleys,—the bright colours, chiefly red and purple, of the utterly bare and precipitous hills of porphyry, the grand and continuous wall-like dykes,—the plainly-divided strata which, where nearly vertical, formed the picturesque and wild central pinnacles, but where less inclined, composed the great massive mountains on the outskirts of the range,—and lastly, the smooth conical piles of fine and brightly coloured detritus, which sloped up at a high angle from the base of the mountains, sometimes to a height of more than 2000 feet.

I frequently observed, both in Tierra del Fuego and within the Andes, that where the rock was covered during the greater part of the year with snow, it was shivered in a very extraordinary manner into small angular fragments. Scoresby[140] has observed the same fact in Spitzbergen. The case appears to me rather obscure: for that part of the mountain which is protected by a mantle of snow, must be less subject to repeated and great changes of temperature than any other part. I have sometimes thought, that the earth and fragments of stone on the surface, were perhaps less effectually removed by slowly percolating snow-water[141] than by rain, and therefore that the appearance of a quicker disintegration of the solid rock under the snow, was deceptive. Whatever the cause may be, the quantity of crumbling stone on the Cordillera is very great. Occasionally in the spring, great masses of this detritus slide down the mountains, and cover the snow-drifts in the valleys, thus forming natural ice-houses. We rode over one, the height of which was far below the limit of perpetual snow.

As the evening drew to a close, we reached a singular basin-like plain, called the Valle del Yeso. It was covered by a little dry pasture, and we had the pleasant sight of a herd of cattle amidst the surrounding rocky deserts. The valley takes its name of Yeso from a great bed, I should think at least 2000 feet thick, of white, and in some parts quite pure, gypsum. We slept with a party of men, who were employed in loading mules with this substance, which is used in the manufacture of wine. We set out early in the morning (21st), and continued to follow the course of the river, which had become very small, till we arrived at the foot of the ridge, that separates the waters flowing into the Pacific and Atlantic Oceans. The road, which as yet had been good with a steady but very gradual ascent, now changed into a steep zigzag track up the great range, dividing the republics of Chile and Mendoza.

I will here give a very brief sketch of the geology of the several parallel lines forming the Cordillera. Of these lines, there are two considerably higher than the others; namely, on the Chilian side, the Peuquenes ridge, which, where the road crosses it, is 13,210 feet above the sea; and the Portillo ridge, on the Mendoza side, which is 14,305 feet. The lower beds of the Peuquenes ridge, and of the several great lines to the westward of it, are composed of a vast pile, many thousand feet in thickness, of porphyries which have flowed as submarine lavas, alternating with angular and rounded fragments of the same rocks, thrown out of the submarine craters. These alternating masses are covered in the central parts, by a great thickness of red sandstone, conglomerate, and calcareous clay-slate, associated with, and passing into, prodigious beds of gypsum. In these upper beds shells are tolerably frequent; and they belong to about the period of the lower chalk of Europe. It is an old story, but not the less wonderful, to hear of shells which were once crawling on the bottom of the sea, now standing nearly 14,000 feet above its level. The lower beds in this great pile of strata, have been dislocated, baked, crystallized and almost blended together, through the agency of mountain masses of a peculiar white soda-granitic rock.

The other main line, namely, that of the Portillo, is of a totally different formation: it consists chiefly of grand bare pinnacles of a red potash-granite, which low down on the western flank are covered by a sandstone, converted by the former heat into a quartz-rock. On the quartz, there rest beds of a conglomerate several thousand feet in thickness, which have been upheaved by the red granite, and dip at an angle of 45° towards the Peuquenes line. I was astonished to find that this conglomerate was partly composed of pebbles, derived from the rocks, with their fossil shells, of the Peuquenes range; and partly of red potash-granite, like that of the Portillo. Hence we must conclude, that both the Peuquenes and Portillo ranges were partially upheaved and exposed to wear and tear, when the conglomerate was forming; but as the beds of the conglomerate have been thrown off at an angle of 45° by the red Portillo granite (with the underlying sandstone baked by it), we may feel sure, that the greater part of the injection and upheaval of the already partially formed Portillo line, took place after the accumulation of the conglomerate, and long after the elevation of the Peuquenes ridge. So that the Portillo, the loftiest line in this part of the Cordillera, is not so old as the less lofty line of the Peuquenes. Evidence derived from an inclined stream of lava at the eastern base of the Portillo, might be adduced to show, that it owes part of its great height to elevations of a still later date. Looking to its earliest origin, the red granite seems to have been injected on an ancient pre-existing line of white granite and mica-slate. In most parts, perhaps in all parts, of the Cordillera, it may be concluded that each line has been formed by repeated upheavals and injections; and that the several parallel lines are of different ages. Only thus can we gain time, at all sufficient to explain the truly astonishing amount of denudation, which these great, though comparatively with most other ranges recent, mountains have suffered.

Finally, the shells in the Peuquenes or oldest ridge, prove, as before remarked, that it has been upraised 14,000 feet since a Secondary period, which in Europe we are accustomed to consider as far from ancient; but since these shells lived in a moderately deep sea, it can be shown that the area now occupied by the Cordillera, must have subsided several thousand feet—in northern Chile as much as 6000 feet—so as to have allowed that amount of submarine strata to have been heaped on the bed on which the shells lived. The proof is the same with that by which it was shown, that at a much later period, since the tertiary shells of Patagonia lived, there must have been there a subsidence of several hundred feet, as well as an ensuing elevation. Daily it is forced home on the mind of the geologist, that nothing, not even the wind that blows, is so unstable as the level of the crust of this earth.

I will make only one other geological remark: although the Portillo chain is here higher than the Peuquenes, the waters draining the intermediate valleys, have burst through it. The same fact, on a grander scale, has been remarked in the eastern and loftiest line of the Bolivian Cordillera, through which the rivers pass: analogous facts have also been observed in other quarters of the world. On the supposition of the subsequent and gradual elevation of the Portillo line, this can be understood; for a chain of islets would at first appear, and, as these were lifted up, the tides would be always wearing deeper and broader channels between them. At the present day, even in the most retired Sounds on the coast of Tierra del Fuego, the currents in the transverse breaks which connect the longitudinal channels, are very strong, so that in one transverse channel even a small vessel under sail was whirled round and round.

About noon we began the tedious ascent of the Peuquenes ridge, and then for the first time experienced some little difficulty in our respiration. The mules would halt every fifty yards, and after resting for a few seconds the poor willing animals started of their own accord again. The short breathing from the rarefied atmosphere is called by the Chilenos "puna;" and they have most ridiculous notions concerning its origin. Some say "all the waters here have puna;" others that "where there is snow there is puna;"—and this no doubt is true. The only sensation I experienced was a slight tightness across the head and chest, like that felt on leaving a warm room and running quickly in frosty weather. There was some imagination even in this; for upon finding fossil shells on the highest ridge, I entirely forgot the puna in my delight. Certainly the exertion of walking was extremely great, and the respiration became deep and laborious: I am told that in Potosi (about 13,000 feet above the sea) strangers do not become thoroughly accustomed to the atmosphere for an entire year. The inhabitants all recommend onions for the puna; as this vegetable has sometimes been given in Europe for pectoral complaints, it may possibly be of real service:—for my part I found nothing so good as the fossil shells!

When about half-way up we met a large party with seventy loaded mules. It was interesting to hear the wild cries of the muleteers, and to watch the long descending string of the animals; they appeared so diminutive, there being nothing but the black mountains with which they could be compared. When near the summit, the wind, as generally happens, was impetuous and extremely cold. On each side of the ridge, we had to pass over broad bands of perpetual snow, which were now soon to be covered by a fresh layer. When we reached the crest and looked backwards, a glorious view was presented. The atmosphere resplendently clear; the sky an intense blue; the profound valleys; the wild broken forms; the heaps of ruins, piled up during the lapse of ages; the bright-coloured rocks, contrasted with the quiet mountains of snow; all these together produced a scene no one could have imagined. Neither plant nor bird, excepting a few condors wheeling around the higher pinnacles, distracted my attention from the inanimate mass. I felt glad that I was alone: it was like watching a thunderstorm, or hearing in full orchestra a chorus of the Messiah.

On several patches of the snow I found the Protococcus nivalis, or red snow, so well known from the accounts of Arctic navigators. My attention was called to it, by observing the footsteps of the mules stained a pale red, as if their hoofs had been slightly bloody. I at first thought that it was owing to dust blown from the surrounding mountains of red porphyry; for from the magnifying power of the crystals of snow, the groups of these microscopical plants appeared like coarse particles. The snow was coloured only where it had thawed very rapidly, or had been accidentally crushed. A little rubbed on paper gave it a faint rose tinge mingled with a little brick-red. I afterwards scraped some off the paper, and found that it consisted of groups of little spheres in colourless cases, each of the thousandth part of an inch in diameter.

The wind on the crest of the Peuquenes, as just remarked, is generally impetuous and very cold: it is said[142] to blow steadily from the westward or Pacific side. As the observations have been chiefly made in summer, this wind must be an upper and return current. The Peak of Teneriffe, with a less elevation, and situated in lat. 28°, in like manner falls within an upper return stream. At first it appears rather surprising, that the trade-wind along the northern parts of Chile and on the coast of Peru, should blow in so very southerly a direction as it does; but when we reflect that the Cordillera, running in a north and south line, intercepts, like a great wall, the entire depth of the lower atmospheric current, we can easily see that the trade-wind must be drawn northward, following the line of mountains, towards the equatorial regions, and thus lose part of that easterly movement which it otherwise would have gained from the earth's rotation. At Mendoza, on the eastern foot of the Andes, the climate is said to be subject to long calms, and to frequent though false appearances of gathering rainstorms: we may imagine that the wind, which coming from the eastward is thus banked up by the line of mountains, would become stagnant and irregular in its movements.

Having crossed the Peuquenes, we descended into a mountainous country, intermediate between the two main ranges, and then took up our quarters for the night. We were now in the republic of Mendoza. The elevation was probably not under 11,000 feet, and the vegetation in consequence exceedingly scanty. The root of a small scrubby plant served as fuel, but it made a miserable fire, and the wind was piercingly cold. Being quite tired with my day's work, I made up my bed as quickly as I could, and went to sleep. About midnight I observed the sky became suddenly clouded: I awakened the arriero to know if there was any danger of bad weather; but he said that without thunder and lightning there was no risk of a heavy snow-storm. The peril is imminent, and the difficulty of subsequent escape great, to any one overtaken by bad weather between the two ranges. A certain cave offers the only place of refuge: Mr. Caldcleugh, who crossed on this same day of the month, was detained there for some time by a heavy fall of snow. Casuchas, or houses of refuge, have not been built in this pass as in that of Uspallata, and, therefore, during the autumn, the Portillo is little frequented. I may here remark that within the main Cordillera rain never falls, for during the summer the sky is cloudless, and in winter snow-storms alone occur.

At the place where we slept water necessarily boiled, from the diminished pressure of the atmosphere, at a lower temperature than it does in a less lofty country; the case being the converse of that of a Papin's digester. Hence the potatoes, after remaining for some hours in the boiling water, were nearly as hard as ever. The pot was left on the fire all night, and next morning it was boiled again, but yet the potatoes were not cooked. I found out this, by overhearing my two companions discussing the cause; they had come to the simple conclusion, "that the cursed pot [which was a new one] did not choose to boil potatoes."

March 22nd.—After eating our potato-less breakfast, we travelled across the intermediate tract to the foot of the Portillo range. In the middle of summer cattle are brought up here to graze; but they had now all been removed: even the greater number of the Guanacos had decamped, knowing well that if overtaken here by a snow-storm, they would be caught in a trap. We had a fine view of a mass of mountains called Tupungato, the whole clothed with unbroken snow, in the midst of which there was a blue patch, no doubt a glacier;—a circumstance of rare occurrence in these mountains. Now commenced a heavy and long climb, similar to that of the Peuquenes. Bold conical hills of red granite rose on each hand; in the valleys there were several broad fields of perpetual snow. These frozen masses, during the process of thawing, had in some parts been converted into pinnacles or columns,[143] which, as they were high and close together, made it difficult for the cargo mules to pass. On one of these columns of ice, a frozen horse was sticking as on a pedestal, but with its hind legs straight up in the air. The animal, I suppose, must have fallen with its head downward into a hole, when the snow was continuous, and afterwards the surrounding parts must have been removed by the thaw.

When nearly on the crest of the Portillo, we were enveloped in a falling cloud of minute frozen spicula. This was very unfortunate, as it continued the whole day, and quite intercepted our view. The pass takes its name of

Portillo, from a narrow cleft or doorway on the highest ridge, through which the road passes. From this point, on a clear day, those vast plains which uninterruptedly extend to the Atlantic Ocean can be seen. We descended to the upper limit of vegetation, and found good quarters for the night under the shelter of some large fragments of rock. We met here some passengers, who made anxious inquiries about the state of the road. Shortly after it was dark the clouds suddenly cleared away, and the effect was quite magical. The great mountains, bright with the full moon, seemed impending over us on all sides, as over a deep crevice: one morning, very early, I witnessed the same striking effect. As soon as the clouds were dispersed it froze severely; but as there was no wind, we slept very comfortably.

The increased brilliancy of the moon and stars at this elevation, owing to the perfect transparency of the atmosphere, was very remarkable. Travellers having observed the difficulty of judging heights and distances amidst lofty mountains, have generally attributed it to the absence of objects of comparison. It appears to me, that it is fully as much owing to the transparency of the air confounding objects at different distances, and likewise partly to the novelty of an unusual degree of fatigue arising from a little exertion,—habit being thus opposed to the evidence of the senses. I am sure that this extreme clearness of the air gives a peculiar character to the landscape, all objects appearing to be brought nearly into one plane, as in a drawing or panorama. The transparency is, I presume, owing to the equable and high state of atmospheric dryness. This dryness was shown by the manner in which woodwork shrank (as I soon found by the trouble my geological hammer gave me); by articles of food, such as bread and sugar, becoming extremely hard; and by the preservation of the skin and parts of the flesh of the beasts, which had perished on the road. To the same cause we must attribute the singular facility with which electricity is excited. My flannel waistcoat, when rubbed in the dark, appeared as if it had been washed with phosphorus;—every hair on a dog's back crackled;—even the linen sheets, and leathern straps of the saddle, when handled, emitted sparks.

March 23rd.—The descent on the eastern side of the Cordillera is much shorter or steeper than on the Pacific side; in other words, the mountains rise more abruptly from the plains than from the alpine country of Chile. A level and brilliantly white sea of clouds was stretched out beneath our feet, shutting out the view of the equally level Pampas. We soon entered the band of clouds, and did not again emerge from it that day. About noon, finding pasture for the animals and bushes for firewood at Los Arenales, we stopped for the night. This was near the uppermost limit of bushes, and the elevation, I suppose, was between seven and eight thousand feet.

I was much struck with the marked difference between the vegetation of these eastern valleys and those on the Chilian side: yet the climate, as well as the kind of soil, is nearly the same, and the difference of longitude very trifling. The same remark holds good with the quadrupeds, and in a lesser degree with the birds and insects. I may instance the mice, of which I obtained thirteen species on the shores of the Atlantic, and five on the Pacific, and not one of them is identical. We must except all those species, which habitually or occasionally frequent elevated mountains; and certain birds, which range as far south as the Strait of Magellan. This fact is in perfect accordance with the geological history of the Andes; for these mountains have existed as a great barrier since the present races of animals have appeared; and therefore, unless we suppose the same species to have been created in two different places, we ought not to expect any closer similarity between the organic beings on the opposite sides of the Andes than on the opposite shores of the ocean. In both cases, we must leave out of the question those kinds which have been able to cross the barrier, whether of solid rock or salt-water.[144]

A great number of the plants and animals were absolutely the same as, or most closely allied to, those of Patagonia. We here have the agouti, bizcacha, three species of armadillo, the ostrich, certain kinds of partridges and other birds, none of which are ever seen in Chile, but are the characteristic animals of the desert plains of Patagonia. We have likewise many of the same (to the eyes of a person who is not a botanist) thorny stunted bushes, withered grass, and dwarf plants. Even the black slowly crawling beetles are closely similar, and some, I believe, on rigorous examination, absolutely identical. It had always been to me a subject of regret, that we were unavoidably compelled to give up the ascent of the S. Cruz river before reaching the mountains: I always had a latent hope of meeting with some great change in the features of the country; but I now feel sure, that it would only have been following the plains of Patagonia up a mountainous ascent.

March 24th.—Early in the morning I climbed up a mountain on one side of the valley, and enjoyed a far extended view over the Pampas. This was a spectacle to which I had always looked forward with interest, but I was disappointed: at the first glance it much resembled a distant view of the ocean, but in the northern parts many irregularities were soon distinguishable. The most striking feature consisted in the rivers, which, facing the rising sun, glittered like silver threads, till lost in the immensity of the distance. At midday we descended the valley, and reached a hovel, where an officer and three soldiers were posted to examine passports. One of these men was a thoroughbred Pampas Indian: he was kept much for the same purpose as a bloodhound, to track out any person who might pass by secretly, either on foot or horseback. Some years ago, a passenger endeavoured to escape detection, by making a long circuit over a neighbouring mountain; but this Indian, having by chance crossed his track, followed it for the whole day over dry and very stony hills, till at last he came on his prey hidden in a gully. We here heard that the silvery clouds, which we had admired from the bright region above, had poured down torrents of rain. The valley from this point gradually opened, and the hills became mere water-worn hillocks compared to the giants behind: it then expanded into a gently sloping plain of shingle, covered with low trees and bushes. This talus, although appearing narrow, must be nearly ten miles wide before it blends into the apparently dead level Pampas. We passed the only house in this neighbourhood, the Estancia of Chaquaio: and at sunset we pulled up in the first snug corner, and there bivouacked.

March 25th.—I was reminded of the Pampas of Buenos Ayres, by seeing the disk of the rising sun, intersected by an horizon level as that of the ocean. During the night a heavy dew fell, a circumstance which we did not experience within the Cordillera. The road proceeded for some distance due east across a low swamp; then meeting the dry plain, it turned to the north towards Mendoza. The distance is two very long days' journey. Our first day's journey was called fourteen leagues to Estacado, and the second seventeen to Luxan, near Mendoza. The whole distance is over a level desert plain, with not more than two or three houses. The sun was exceedingly powerful, and the ride devoid of all interest. There is very little water in this "traversia," and in our second day's journey we found only one little pool. Little water flows from the mountains, and it soon becomes absorbed by the dry and porous soil; so that, although we travelled at the distance of only ten or fifteen miles from the outer range of the Cordillera, we did not cross a single stream. In many parts the ground was incrusted with a saline efflorescence; hence we had the same salt-loving plants which are common near Bahia Blanca. The landscape has a uniform character from the Strait of Magellan, along the whole eastern coast of Patagonia, to the Rio Colorado; and it appears that the same kind of country extends inland from this river, in a sweeping line as far as San Luis, and perhaps even further north. To the eastward of this curved line lies the basin of the comparatively damp and green plains of Buenos Ayres. The sterile plains of Mendoza and Patagonia consist of a bed of shingle, worn smooth and accumulated by the waves of the sea; while the Pampas, covered by thistles, clover, and grass, have been formed by the ancient estuary mud of the Plata.

After our two days' tedious journey, it was refreshing to see in the distance the rows of poplars and willows growing round the village and river of Luxan. Shortly before we arrived at this place, we observed to the south a ragged cloud of dark reddish-brown colour. At first we thought that it was smoke from some great fire on the plains; but we soon found that it was a swarm of locusts. They were flying northward; and with the aid of a light breeze, they overtook us at a rate of ten or fifteen miles an hour. The main body filled the air from a height of twenty feet, to that, as it appeared, of two or three thousand above the ground; "and the sound of their wings was as the sound of chariots of many horses running to battle:" or rather, I should say, like a strong breeze passing through the rigging of a ship. The sky, seen through the advanced guard, appeared like a mezzo-tinto engraving, but the main body was impervious to sight; they were not, however, so thick together, but that they could escape a stick waved backwards and forwards. When they alighted, they were more numerous than the leaves in the field, and the surface became reddish instead of being green: the swarm having once alighted, the individuals flew from side to side in all directions. Locusts are not an uncommon pest in this country: already during the season, several smaller swarms had come up from the south, where, as apparently in all other parts of the world, they are bred in the deserts. The poor cottagers in vain attempted by lighting fires, by shouts, and by waving branches to avert the attack. This species of locust closely resembles, and perhaps is identical with, the famous Gryllus migratorius of the East.

We crossed the Luxan, which is a river of considerable size, though its course towards the sea-coast is very imperfectly known: it is even doubtful whether, in passing over the plains, it is not evaporated and lost. We slept in the village of Luxan, which is a small place surrounded by gardens, and forms the most southern cultivated district in the Province of Mendoza; it is five leagues south of the capital. At night I experienced an attack (for it deserves no less a name) of the *Benchuca,* a species of Reduvius, the great black bug of the Pampas. It is most disgusting to feel soft wingless insects, about an inch long, crawling over one's body. Before sucking they are quite thin, but afterwards they become round and bloated with blood, and in this state are easily crushed. One which I caught at Iquique, (for they are found in Chile and Peru,) was very empty. When placed on a table, and though surrounded by people, if a finger was presented, the bold insect would immediately protrude its sucker, make a charge, and if allowed, draw blood. No pain was caused by the wound. It was curious to watch its body during the act of sucking, as in less than ten minutes it changed from being as flat as a wafer to a globular form. This one feast, for which the benchuca was indebted to one of the officers, kept it fat during four whole months; but, after the first fortnight, it was quite ready to have another suck.

March 27th.—We rode on to Mendoza. The country was beautifully cultivated, and resembled Chile. This neighbourhood is celebrated for its fruit;

and certainly nothing could appear more flourishing than the vineyards and the orchards of figs, peaches, and olives. We bought water-melons nearly twice as large as a man's head, most deliciously cool and well-flavoured, for a halfpenny apiece; and for the value of threepence, half a wheelbarrowful of peaches. The cultivated and enclosed part of this province is very small; there is little more than that which we passed through between Luxan and the capital. The land, as in Chile, owes its fertility entirely to artificial irrigation; and it is really wonderful to observe how extraordinarily productive a barren traversia is thus rendered.

We stayed the ensuing day in Mendoza. The prosperity of the place has much declined of late years. The inhabitants say "it is good to live in, but very bad to grow rich in." The lower orders have the lounging, reckless manners of the Gauchos of the Pampas; and their dress, riding-gear, and habits of life, are nearly the same. To my mind the town had a stupid, forlorn aspect. Neither the boasted alameda, nor the scenery, is at all comparable with that of Santiago; but to those who, coming from Buenos Ayres, have just crossed the unvaried Pampas, the gardens and orchards must appear delightful. Sir F. Head, speaking of the inhabitants, says, "They eat their dinners, and it is so very hot, they go to sleep—and could they do better?" I quite agree with Sir F. Head: the happy doom of the Mendozinos is to eat, sleep and be idle.

March 29th.—We set out on our return to Chile, by the Uspallata pass situated north of Mendoza. We had to cross a long and most sterile traversia of fifteen leagues. The soil in parts was absolutely bare, in others covered by numberless dwarf cacti, armed with formidable spines, and called by the inhabitants "little lions." There were, also, a few low bushes. Although the plain is nearly three thousand feet above the sea, the sun was very powerful; and the heat, as well as the clouds of impalpable dust, rendered the travelling extremely irksome. Our course during the day lay nearly parallel to the Cordillera, but gradually approaching them. Before sunset we entered one of the wide valleys, or rather bays, which open on the plain: this soon narrowed into a ravine, where a little higher up the house of Villa Vicencio is situated. As we had ridden all day without a drop of water, both our mules and selves were very thirsty, and we looked out anxiously for the stream which flows down this valley. It was curious to observe how gradually the water made its appearance: on the plain the course was quite dry; by degrees it became a little damper; then puddles of water appeared; these soon became connected; and at Villa Vicencio there was a nice little rivulet.

30th.—The solitary hovel which bears the imposing name of Villa Vicencio, has been mentioned by every traveller who has crossed the Andes. I stayed here and at some neighbouring mines during the two succeeding days. The geology of the surrounding country is very curious. The Uspallata range is separated from the main Cordillera by a long narrow plain or basin, like those so often mentioned in Chile, but higher, being six thousand feet above the sea. This range has nearly the same geographical position with respect to the Cordillera, which the gigantic Portillo line has, but it is of a totally different origin: it consists of various kinds of submarine lava, alternating with volcanic sandstones and other remarkable sedimentary deposits; the whole having a very close resemblance to some of the tertiary beds on the shores of the Pacific. From this resemblance I expected to find silicified wood, which is generally characteristic of those formations. I was gratified in a very extraordinary manner. In the central part of the range, at an elevation of about seven thousand feet, I observed on a bare slope some snow-white projecting columns. These were petrified trees, eleven being silicified, and from thirty to forty converted into coarsely-crystallized white calcareous spar. They were abruptly broken off, the upright stumps projecting a few feet above the ground. The trunks measured from three to five feet each in circumference. They stood a little way apart from each other, but the whole formed one group. Mr. Robert Brown has been kind enough to examine the wood: he says it belongs to the fir tribe, partaking of the character of the Araucarian family, but with some curious points of affinity with the yew. The volcanic sandstone in which the trees were embedded, and from the lower part of which they must have sprung, had accumulated in successive thin layers around their trunks; and the stone yet retained the impression of the bark.

It required little geological practice to interpret the marvellous story which this scene at once unfolded; though I confess I was at first so much astonished that I could scarcely believe the plainest evidence. I saw the spot where a cluster of fine trees once waved their branches on the shores of the Atlantic, when that ocean (now driven back 700 miles) came to the foot of the Andes. I saw that they had sprung from a volcanic soil which had been raised above the level of the sea, and that subsequently this dry land, with its upright trees, had been let down into the depths of the ocean. In these depths, the formerly dry land was covered by sedimentary beds, and these again by enormous streams of submarine lava—one such mass attaining the thickness of a thousand feet; and these deluges of molten stone and aqueous deposits five times alternately had been spread out. The ocean which received such thick masses, must have been profoundly deep; but again the subterranean forces exerted themselves, and I now beheld the bed of that ocean, forming a chain of mountains more than seven thousand feet in height. Nor had those antagonistic forces been dormant, which are always at work wearing down the surface of the land; the great piles of strata had been intersected by many wide valleys, and the trees now changed into silex, were exposed projecting from the volcanic soil, now changed into rock, whence formerly, in a green and budding state, they had raised their lofty heads. Now, all is utterly irreclaimable and desert; even the lichen cannot adhere to the stony casts of former trees. Vast, and scarcely comprehensible as such changes must ever appear, yet they have all occurred within a period, recent when compared with the history of the Cordillera; and the Cordillera itself is absolutely modern as compared with many of the fossiliferous strata of Europe and America.

April 1st.—We crossed the Upsallata range, and at night slept at the custom-house—the only inhabited spot on the plain. Shortly before leaving the mountains, there was a very extraordinary view; red, purple, green, and quite white sedimentary rocks, alternating with black lavas, were broken up and thrown into all kinds of disorder by masses of porphyry of every shade of colour, from dark brown to the brightest lilac. It was the first view I ever saw, which really resembled those pretty sections which geologists make of the inside of the earth.

The next day we crossed the plain, and followed the course of the same great mountain stream which flows by Luxan. Here it was a furious torrent, quite impassable, and appeared larger than in the low country, as was the case with the rivulet of Villa Vicencio. On the evening of the succeeding day, we reached the Rio de las Vacas, which is considered the worst stream in the Cordillera to cross. As all these rivers have a rapid and short course, and are formed by the melting of the snow, the hour of the day makes a considerable difference in their volume. In the evening the stream is muddy and full, but about daybreak it becomes clearer, and much less impetuous. This we found to be the case with the Rio Vacas, and in the morning we crossed it with little difficulty.

The scenery thus far was very uninteresting, compared with that of the Portillo pass. Little can be seen beyond the bare walls of the one grand flat-bottomed valley, which the road follows up to the highest crest. The valley and the huge rocky mountains are extremely barren: during the two previous nights the poor mules had absolutely nothing to eat, for excepting a few low resinous bushes, scarcely a plant can be seen. In the course of this day we crossed some of the worst passes in the Cordillera, but their danger has been much exaggerated. I was told that if I attempted to pass on foot, my head would turn giddy, and that there was no room to dismount; but I did not see a place where any one might not have walked over backwards, or got off his mule on either side. One of the bad passes, called *las Animas* (the souls), I had crossed, and did not find out till a day afterwards, that it was one of the awful dangers. No doubt there are many parts in which, if

CHAPTER XVI

Northern Chile and Peru

COAST-ROAD TO COQUIMBO • GREAT LOADS CARRIED BY THE MINERS • COQUIMBO • EARTHQUAKE • STEP-FORMED TERRACES • ABSENCE OF RECENT DEPOSITS • CONTEMPORANEOUSNESS OF THE TERTIARY FORMATIONS • EXCURSION UP THE VALLEY • ROAD TO GUASCO • DESERTS • VALLEY OF COPIAPÓ • RAIN AND EARTHQUAKES • HYDROPHOBIA • THE DESPOBLADO • INDIAN RUINS • PROBABLE CHANGE OF CLIMATE • RIVER-BED ARCHED BY AN EARTHQUAKE • COLD GALES OF WIND • NOISES FROM A HILL • IQUIQUE • SALT ALLUVIUM • NITRATE OF SODA • LIMA • UNHEALTHY COUNTRY • RUINS OF CALLAO, OVERTHROWN BY AN EARTHQUAKE • RECENT SUBSIDENCE • ELEVATED SHELLS ON SAN LORENZO, THEIR DECOMPOSITION • PLAIN WITH EMBEDDED SHELLS AND FRAGMENTS OF POTTERY • ANTIQUITY OF THE INDIAN RACE

APRIL 27TH.—I SET OUT on a journey to Coquimbo, and thence through Guasco to Copiapó, where Captain Fitz Roy kindly offered to pick me up in the Beagle. The distance in a straight line along the shore northward is only 420 miles; but my mode of travelling made it a very long journey. I bought four horses and two mules, the latter carrying the luggage on alternate days. The six animals together only cost the value of twenty-five pounds sterling, and at Copiapó I sold them again for twenty-three. We travelled in the same independent manner as before, cooking our own meals, and sleeping in the open air. As we rode towards the Viño del Mar, I took a farewell view of Valparaiso, and admired its picturesque appearance. For geological purposes I made a detour from the high road to the foot of the

Bell of Quillota. We passed through an alluvial district rich in gold, to the neighbourhood of Limache, where we slept. Washing for gold supports the inhabitants of numerous hovels, scattered along the sides of each little rivulet; but, like all those whose gains are uncertain, they are unthrifty in all their habits, and consequently poor.

28th.—In the afternoon we arrived at a cottage at the foot of the Bell mountain. The inhabitants were freeholders, which is not very usual in Chile. They supported themselves on the produce of a garden and a little field, but were very poor. Capital is here so deficient, that the people are obliged to sell their green corn while standing in the field, in order to buy necessaries for the ensuing year. Wheat in consequence was dearer in the very district of its production than at Valparaiso, where the contractors live. The next day we joined the main road to Coquimbo. At night there was a very light shower of rain: this was the first drop that had fallen since the heavy rain of September 11th and 12th, which detained me a prisoner at the Baths of Cauquenes. The interval was seven and a half months; but the rain this year in Chile was rather later than usual. The distant Andes were now covered by a thick mass of snow, and were a glorious sight.

May 2nd.—The road continued to follow the coast, at no great distance from the sea. The few trees and bushes which are common in central Chile decreased rapidly in numbers, and were replaced by a tall plant, something like a yucca in appearance. The surface of the country, on a small scale, was singularly broken and irregular; abrupt little peaks of rock rising out of small plains or basins. The indented coast and the bottom of the neighbouring sea, studded with breakers, would, if converted into dry land, present similar forms; and such a conversion without doubt has taken place in the part over which we rode.

3rd.—Quilimari to Conchalee. The country became more and more barren. In the valleys there was scarcely sufficient water for any irrigation; and the intermediate land was quite bare, not supporting even goats. In the spring, after the winter showers, a thin pasture rapidly springs up, and cattle are then driven down from the Cordillera to graze for a short time. It is curious to observe how the seeds of the grass and other plants seem to accommodate themselves, as if by an acquired habit, to the quantity of rain which falls upon different parts of this coast. One shower far northward at Copiapó produces as great an effect on the vegetation, as two at Guasco, and three or four in this district. At Valparaiso a winter so dry as greatly to injure the pasture, would at Guasco produce the most unusual abundance. Proceeding northward, the quantity of rain does not appear to decrease in strict proportion to the latitude. At Conchalee, which is only 67 miles north of Valparaiso, rain is not expected till the end of May; whereas at Valparaiso some generally falls early in April: the annual quantity is likewise small in proportion to the lateness of the season at which it commences.

4th.—Finding the coast-road devoid of interest of any kind, we turned inland towards the mining district and valley of Illapel. This valley, like every other in Chile, is level, broad, and very fertile: it is bordered on each side, either by cliffs of stratified shingle, or by bare rocky mountains. Above the straight line of the uppermost irrigating ditch, all is brown as on a high road; while all below is of as bright a green as verdigris, from the beds of alfalfa, a kind of clover. We proceeded to Los Hornos, another mining district, where the principal hill was drilled with holes, like a great ants'-nest. The Chilian miners are a peculiar race of men in their habits. Living for weeks together in the most desolate spots, when they descend to the villages on feast-days, there is no excess of extravagance into which they do not run. They sometimes gain a considerable sum, and then, like sailors with prize-money, they try how soon they can contrive to squander it. They drink excessively, buy quantities of clothes, and in a few days return penniless to their miserable abodes, there to work harder than beasts of burden. This thoughtlessness, as with sailors, is evidently the result of a similar manner of life. Their daily food is found them, and they acquire no habits of carefulness: moreover, temptation and the means of yielding to it are placed in their power at the same time. On the other hand, in Cornwall, and some other parts of England, where the system of selling part of the vein is followed, the miners, from being obliged to act and think for themselves, are a singularly intelligent and well-conducted set of men.

The dress of the Chilian miner is peculiar and rather picturesque. He wears a very long shirt of some dark-coloured baize, with a leathern apron; the whole being fastened round his waist by a bright-coloured sash. His trousers are very broad, and his small cap of scarlet cloth is made to fit the head closely. We met a party of these miners in full costume, carrying the body of one of their companions to be buried. They marched at a very quick trot, four men supporting the corpse. One set having run as hard as they could for about two hundred yards, were relieved by four others, who had previously dashed on ahead on horseback. Thus they proceeded, encouraging each other by wild cries: altogether the scene formed a most strange funeral.

We continued travelling northward, in a zigzag line; sometimes stopping a day to geologize. The country was so thinly inhabited, and the track so obscure, that we often had difficulty in finding our way. On the 12th I stayed at some mines. The ore in this case was not considered particularly good, but from being abundant it was supposed the mine would sell for about thirty or forty thousand dollars (that is, 6000 or 8000 pounds sterling); yet it had been bought by one of the English Associations for an ounce of gold (*3l.* 8*s.*). The ore is yellow pyrites, which, as I have already remarked, before the arrival of the English, was not supposed to contain a particle of copper. On a scale of profits nearly as great as in the above instance, piles of cinders, abounding with minute globules of metallic copper, were purchased; yet with these advantages, the mining associations, as is well known, contrived to lose immense sums of money. The folly of the greater number of the commissioners and shareholders amounted to infatuation;—a thousand pounds per annum given in some cases to entertain the Chilian authorities; libraries of well-bound geological books; miners brought out for particular metals, as tin, which are not found in Chile; contracts to supply the miners with milk, in parts where there are no cows; machinery, where it could not possibly be used; and a hundred similar arrangements, bore witness to our absurdity, and to this day afford amusement to the natives. Yet there can be no doubt, that the same capital well employed in these mines would have yielded an immense return, a confidential man of business, a practical miner and assayer, would have been all that was required.

Captain Head has described the wonderful load which the "Apires," truly beasts of burden, carry up from the deepest mines. I confess I thought the account exaggerated; so that I was glad to take an opportunity of weighing one of the loads, which I picked out by hazard. It required considerable exertion on my part, when standing directly over it, to lift it from the ground. The load was considered under weight when found to be 197 pounds. The apire had carried this up eighty perpendicular yards,—part of the way by a steep passage, but the greater part up notched poles, placed in a zigzag line up the shaft. According to the general regulation, the apire is not allowed to halt for breath, except the mine is six hundred feet deep. The average load is considered as rather more than 200 pounds, and I have been assured that one of 300 pounds (twenty-two stone and a half) by way of a trial has been brought up from the deepest mine! At this time the apires were bringing up the usual load twelve times in the day; that is 2400 pounds from eighty yards deep; and they were employed in the intervals in breaking and picking ore.

These men, excepting from accidents, are healthy, and appear cheerful. Their bodies are not very muscular. They rarely eat meat once a week, and never oftener, and then only the hard dry charqui. Although with a knowledge that the labour was voluntary, it was nevertheless quite revolting to see the state in which they reached the mouth of the mine; their bodies bent forward, leaning with their arms on the steps, their legs bowed, their muscles quivering, the perspiration streaming from their faces over their breasts, their nostrils distended, the corners of their mouth forcibly drawn back, and the expulsion of their breath most laborious. Each time they draw their breath, they utter an articulate cry of "ay-ay," which ends in a sound rising from deep in the chest, but shrill like the note of a fife. After staggering to the pile of ore, they emptied the "carpacho;" in two or three seconds recovering their breath, they wiped the sweat from their brows, and apparently quite fresh descended the mine again at a quick pace. This appears to me a wonderful instance of the amount of labour which habit, for it can be nothing else, will enable a man to endure.

In the evening, talking with the mayor-domo of these mines about the number of foreigners now scattered over the whole country, he told me that, though quite a young man, he remembers when he was a boy at school at Coquimbo, a holiday being given to see the captain of an English ship, who was brought to the city to speak to the governor. He believes that nothing would have induced any boy in the school, himself included, to have gone close to the Englishman; so deeply had they been impressed with an idea of the heresy, contamination, and evil to be derived from contact with such a person. To this day they relate the atrocious actions of the bucaniers; and especially of one man, who took away the figure of the Virgin Mary, and returned the year after for that of St. Joseph, saying it was a pity the lady should not have a husband. I heard also of an old lady who, at a dinner at Coquimbo, remarked how wonderfully strange it was that she should have lived to dine in the same room with an Englishman; for she remembered as a girl, that twice, at the mere cry of "Los Ingleses," every soul, carrying what valuables they could, had taken to the mountains.

14th.—We reached Coquimbo, where we stayed a few days. The town is remarkable for nothing but its extreme quietness. It is said to contain from 6000 to 8000 inhabitants. On the morning of the 17th it rained lightly, the first time this year, for about five hours. The farmers, who plant corn near the sea-coast where the atmosphere is most humid, taking advantage of this shower, would break up the ground; after a second they would put the seed in; and if a third shower should fall, they would reap a good harvest in the spring. It was interesting to watch the effect of this trifling amount of moisture. Twelve hours afterwards the ground appeared as dry as ever; yet after an interval of ten days, all the hills were faintly tinged with green patches; the grass being sparingly scattered in hair-like fibres a full inch in length. Before this shower every part of the surface was bare as on a high road.

In the evening, Captain Fitz Roy and myself were dining with Mr. Edwards, an English resident well known for his hospitality by all who have visited Coquimbo, when a sharp earthquake happened. I heard the fore-coming rumble, but from the screams of the ladies, the running of the servants, and the rush of several of the gentlemen to the doorway, I could not distinguish the motion. Some of the women afterwards were crying with terror, and one gentleman said he should not be able to sleep all night, or if he did, it would only be to dream of falling houses. The father of this person had lately lost all his property at Talcahuano, and he himself had only just escaped a falling roof at Valparaiso, in 1822. He mentioned a curious coincidence which then happened: he was playing at cards, when a German, one of the party, got up, and said he would never sit in a room in these countries with the door shut, as owing to his having done so, he had nearly lost his life at Copiapó. Accordingly he opened the door; and no sooner had he done this, than he cried out, "Here it comes again!" and the famous shock commenced. The whole party escaped. The danger in an earthquake is not from the time lost in opening the door, but from the chance of its becoming jammed by the movement of the walls.

It is impossible to be much surprised at the fear which natives and old residents, though some of them known to be men of great command of mind, so generally experience during earthquakes. I think, however, this excess of panic may be partly attributed to a want of habit in governing their fear, as it is not a feeling they are ashamed of. Indeed, the natives do not like to see a person indifferent. I heard of two Englishmen who, sleeping in the open air during a smart shock, knowing that there was no danger, did not rise. The natives cried out indignantly, "Look at those heretics, they do not even get out of their beds!"

I SPENT SOME DAYS in examining the step-formed terraces of shingle, first noticed by Captain B. Hall, and believed by Mr. Lyell to have been formed by the sea, during the gradual rising of the land. This certainly is the true explanation, for I found numerous shells of existing species on these terraces. Five narrow, gently sloping, fringe-like terraces rise one behind the other, and where best developed are formed of shingle: they front the bay, and sweep up both sides of the valley. At Guasco, north of Coquimbo, the phenomenon is displayed on a much grander scale, so as to strike with surprise even some of the inhabitants. The terraces are there much broader, and may be called plains; in some parts there are six of them, but generally only five: they run up the valley for thirty-seven miles from the coast. These step-formed terraces or fringes closely resemble those in the valley of S. Cruz, and, except in being on a smaller scale, those great ones along the whole coast-line of Patagonia. They have undoubtedly been formed by the denuding power of the sea, during long periods of rest in the gradual elevation of the continent.

Shells of many existing species not only lie on the surface of the terraces at Coquimbo (to a height of 250 feet), but are embedded in a friable calcareous rock, which in some places is as much as between twenty and thirty feet in thickness, but is of little extent. These modern beds rest on an ancient tertiary formation containing shells, apparently all extinct. Although I examined so many hundred miles of coast on the Pacific, as well as Atlantic side of the continent, I found no regular strata containing sea-shells of recent species, excepting at this place, and at a few points northward on the road to Guasco. This fact appears to me highly remarkable; for the explanation generally given by geologists, of the absence in any district of stratified fossiliferous deposits of a given period, namely, that the surface then existed as dry land, is not here applicable; for we know from the shells strewed on the surface and embedded in loose sand or mould, that the land for thousands of miles along both coasts has lately been submerged. The explanation, no doubt, must be sought in the fact; that the whole southern part of the continent has been for a long time slowly rising; and therefore that all matter deposited along shore in shallow water, must have been soon brought up and slowly exposed to the wearing action of the sea-beach; and it is only in comparatively shallow water that the greater number of marine organic beings can flourish, and in such water it is obviously impossible that strata of any great thickness can accumulate. To show the vast power of the wearing action of sea-beaches, we need only appeal to the great cliffs along the present coast of Patagonia, and to the escarpments or ancient sea-cliffs at different levels, one above another, on that same line of coast.

The old underlying tertiary formation at Coquimbo, appears to be of about the same age with several deposits on the coast of Chile (of which that of Navedad is the principal one), and with the great formation of Patagonia. Both at Navedad and in Patagonia there is evidence, that since the shells (a list of which has been seen by Professor E. Forbes) there entombed were living, there has been a subsidence of several hundred feet, as well as an ensuing elevation. It may naturally be asked, how it comes that, although no extensive fossiliferous deposits of the recent period, nor of any period intermediate between it and the ancient tertiary epoch, have been preserved on either side of the continent, yet that at this ancient tertiary epoch, sedimentary matter containing fossil remains, should have been deposited and preserved at different points in north and south lines, over a space of 1100 miles on the shores of the Pacific, and of at least 1350 miles on the shores of the Atlantic, and in an east and west line of 700 miles across the widest part of the continent? I believe the explanation is not difficult, and that it is perhaps applicable to nearly analogous facts observed in other quarters of the world. Considering the enormous power of denudation which the sea possesses, as shown by numberless facts, it is not probable that a sedimentary deposit, when being upraised, could pass through the ordeal of the beach, so as to be preserved in sufficient masses to last to a distant period, without it were originally of wide extent and of considerable thickness: now it is impossible on a moderately shallow bottom, which alone is favourable to most living creatures, that a thick and widely extended covering of sediment could be spread out, without the bottom sank down to receive the successive layers. This seems to have actually taken place at about the same period in southern Patagonia and Chile, though these places are a thousand miles apart. Hence, if prolonged movements of approximately contemporaneous subsidence are generally widely extensive, as I am strongly inclined to believe from my examination of the Coral Reefs of the great oceans—or if, confining our view to South America, the subsiding movements have been coextensive with those of elevation, by which, within the same period of existing shells, the shores of Peru, Chile, Tierra del Fuego, Patagonia, and La Plata have been upraised—then we can see that at the same time, at far distant points, circumstances would have been favourable to the formation of fossiliferous deposits of wide extent and of considerable thickness; and such deposits, consequently, would have a good chance of resisting the wear and tear of successive beach-lines, and of lasting to a future epoch.

May 21st.—I set out in company with Don Jose Edwards to the silver-mine of Arqueros, and thence up the valley of Coquimbo. Passing through a mountainous country, we reached by nightfall the mines belonging to Mr. Edwards. I enjoyed my night's rest here from a reason which will not be fully appreciated in England, namely, the absence of fleas! The rooms in Coquimbo swarm with them; but they will not live here at the height of only three or four thousand feet: it can scarcely be the trifling diminution of temperature, but some other cause which destroys these troublesome insects at this place. The mines are now in a bad state, though they formerly yielded about 2000 pounds in weight of silver a year. It has been said that "a person with a copper-mine will gain; with silver he may gain; but with gold he is sure to lose." This is not true: all the large Chilian fortunes have been made by mines of the more precious metals. A short time since an English physician returned to England from Copiapó, taking with him the profits of one share of a silver-mine, which amounted to about 24,000 pounds sterling. No doubt a copper-mine with care is a sure game, whereas the other is gambling, or rather taking a ticket in a lottery. The owners lose great quantities of rich ores; for no precautions can prevent robberies. I heard of a gentleman laying a bet with another, that one of his men should rob him before his face. The ore when brought out of the mine is broken into pieces, and the useless stone thrown on one side. A couple of the miners who were thus employed, pitched, as if by accident, two fragments away at the same moment, and then cried out for a joke "Let us see which rolls furthest." The owner, who was standing by, bet a cigar with his friend on the race. The miner by this means watched the very point amongst the rubbish where the stone lay. In the evening he picked it up and carried it to his master, showing him a rich mass of silver-ore, and saying, "This was the stone on which you won a cigar by its rolling so far."

May 23rd.—We descended into the fertile valley of Coquimbo, and followed it till we reached an Hacienda belonging to a relation of Don Jose, where we stayed the next day. I then rode one day's journey further, to see what were declared to be some petrified shells and beans, which latter turned out to be small quartz pebbles. We passed through several small villages; and the valley was beautifully cultivated, and the whole scenery very grand. We were here near the main Cordillera, and the surrounding hills were lofty. In all parts of northern Chile, fruit trees produce much more abundantly at a considerable height near the Andes than in the lower country. The figs and grapes of this district are famous for their excellence, and are cultivated to a great extent. This valley is, perhaps, the most productive one north of Quillota. I believe it contains, including Coquimbo, 25,000 inhabitants. The next day I returned to the Hacienda, and thence, together with Don Jose, to Coquimbo.

June 2nd.—We set out for the valley of Guasco, following the coast-road, which was considered rather less desert than the other. Our first day's ride was to a solitary house, called Yerba Buena, where there was pasture for our horses. The shower mentioned as having fallen, a fortnight ago, only reached about half-way to Guasco; we had, therefore, in the first part of our journey a most faint tinge of green, which soon faded quite away. Even where brightest, it was scarcely sufficient to remind one of the fresh turf and budding flowers of the spring of other countries. While travelling through these deserts one feels like a prisoner shut up in a gloomy court, who longs to see something green and to smell a moist atmosphere.

June 3rd.—Yerba Buena to Carizal. During the first part of the day we crossed a mountainous rocky desert, and afterwards a long deep sandy plain, strewed with broken sea-shells. There was very little water, and that little saline: the whole country, from the coast to the Cordillera, is an uninhabited desert. I saw traces only of one living animal in abundance, namely, the shells of a Bulimus, which were collected together in extraordinary numbers on the driest spots. In the spring one humble little plant sends out a few leaves, and on these the snails feed. As they are seen only very early in the morning, when the ground is slightly damp with dew, the Guascos believe that they are bred from it. I have observed in other places that extremely dry and sterile districts, where the soil is calcareous, are extraordinarily favourable to land-shells. At Carizal there were a few cottages, some brackish water, and a trace of cultivation: but it was with difficulty that we purchased a little corn and straw for our horses.

4th.—Carizal to Sauce. We continued to ride over desert plains, tenanted by large herds of guanaco. We crossed also the valley of Chaneral; which, although the most fertile one between Guasco and Coquimbo, is very narrow, and produces so little pasture, that we could not purchase any for our horses. At Sauce we found a very civil old gentleman, superintendent of a copper-smelting furnace. As an especial favour, he allowed me to purchase at a high price an armful of dirty straw, which was all the poor horses had for supper after their long day's journey. Few smelting-furnaces are now at work in any part of Chile; it is found more profitable, on account of the extreme scarcity of firewood, and from the Chilian method of reduction being so unskilful, to ship the ore for Swansea. The next day we crossed some mountains to Freyrina, in the valley of Guasco. During each day's ride further northward, the vegetation became more and more scanty; even the great chandelier-like cactus was here replaced by a different and much smaller species. During the winter months, both in northern Chile and in Peru, a uniform bank of clouds hangs, at no great height, over the Pacific. From the mountains we had a very striking view of this white and brilliant aërial-field, which sent arms up the valleys, leaving islands and promontories in the same manner, as the sea does in the Chonos archipelago and in Tierra del Fuego.

We stayed two days at Freyrina. In the valley of Guasco there are four small towns. At the mouth there is the port, a spot entirely desert, and without any water in the immediate neighbourhood. Five leagues higher up stands Freyrina, a long straggling village, with decent whitewashed houses. Again, ten leagues further up Ballenar is situated, and above this Guasco Alto, a horticultural village, famous for its dried fruit. On a clear day the view up the valley is very fine; the straight opening terminates in the far-distant snowy Cordillera; on each side an infinity of crossing-lines are blended together in a beautiful haze. The foreground is singular from the number of parallel and step-formed terraces; and the included strip of green valley, with its willow-bushes, is contrasted on both hands with the naked hills. That the surrounding country was most barren will be readily believed, when it is known that a shower of rain had not fallen during the last thirteen months. The inhabitants heard with the greatest envy of the rain at Coquimbo; from the appearance of the sky they had hopes of equally good fortune, which, a fortnight afterwards, were realized. I was at Copiapó at the time; and there the people, with equal envy, talked of the abundant rain at Guasco. After two or three very dry years, perhaps with not more than one shower during the whole time, a rainy year generally follows; and this does more harm than even the drought. The rivers swell, and cover with gravel and sand the narrow strips of ground, which alone are fit for cultivation. The floods also injure the irrigating ditches. Great devastation had thus been caused three years ago.

June 8th.—We rode on to Ballenar, which takes its name from Ballenagh in Ireland, the birthplace of the family of O'Higgins, who, under the Spanish government, were presidents and generals in Chile. As the rocky mountains on each hand were concealed by clouds, the terrace-like plains gave to the valley an appearance like that of Santa Cruz in Patagonia. After spending one day at Ballenar I set out, on the 10th, for the upper part of the valley of Copiapó. We rode all day over an uninteresting country. I am tired of repeating the epithets barren and sterile. These words, however, as commonly used, are comparative; I have always applied them to the plains of Patagonia, which can boast of spiny bushes and some tufts of grass; and this is absolute fertility, as compared with northern Chile. Here again, there are not many spaces of two hundred yards square, where some little bush, cactus or lichen, may not be discovered by careful examination; and in the soil seeds lie dormant ready to spring up during the first rainy winter. In Peru real deserts occur over wide tracts of country. In the evening we arrived at a valley, in which the bed of the streamlet was damp: following it up, we came to tolerably good water. During the night, the stream, from not being evaporated and absorbed so quickly, flows a league lower down than during the day. Sticks were plentiful for firewood, so that it was a good place to bivouac for us; but for the poor animals there was not a mouthful to eat.

June 11th.—We rode without stopping for twelve hours, till we reached an old smelting-furnace, where there was water and firewood; but our horses again had nothing to eat, being shut up in an old courtyard. The line of road was hilly, and the distant views interesting, from the varied colours of the bare mountains. It was almost a pity to see the sun shining constantly over so useless a country; such splendid weather ought to have brightened fields and pretty gardens. The next day we reached the valley of Copiapó. I was heartily glad of it; for the whole journey was a continued source of anxiety; it was most disagreeable to hear, whilst eating our own suppers, our horses gnawing the posts to which they were tied, and to have no means of relieving their hunger. To all appearance, however, the animals were quite fresh; and no one could have told that they had eaten nothing for the last fifty-five hours.

I had a letter of introduction to Mr. Bingley, who received me very kindly at the Hacienda of Potrero Seco. This estate is between twenty and thirty miles long, but very narrow, being generally only two fields wide, one on each side the river. In some parts the estate is of no width, that is to say, the land cannot be irrigated, and therefore is valueless, like the surrounding rocky desert. The small quantity of cultivated land in the whole line of valley, does not so much depend on inequalities of level, and consequent unfitness for irrigation, as on the small supply of water. The river this year was remarkably full: here, high up the valley, it reached to the horse's belly, and was about fifteen yards wide, and rapid; lower down it becomes smaller and smaller, and is generally quite lost, as happened during one period of thirty years, so that not a drop entered the sea. The inhabitants watch a storm over the Cordillera with great interest; as one good fall of snow provides them with water for the ensuing year. This is of infinitely more consequence than rain in the lower country. Rain, as often as it falls, which is about once in every two or three years, is a great advantage, because the cattle and mules can for some time afterwards find a little pasture in the mountains. But without snow on the Andes, desolation extends throughout the valley. It is on record that three times nearly all the inhabitants have been obliged to emigrate to the south. This year there was plenty of water, and every man irrigated his ground as much as he chose; but it has frequently been necessary to post soldiers at the sluices, to see that each estate took only its proper allowance during so many hours in the week. The valley is said to contain 12,000 souls, but its produce is sufficient only for three months in the year; the rest of the supply being drawn from Valparaiso and the south. Before the discovery of the famous silver-mines of Chanuncillo, Copiapó was in a rapid state of decay; but now it is in a very thriving condition; and the town, which was completely overthrown by an earthquake, has been rebuilt.

The valley of Copiapó, forming a mere ribbon of green in a desert, runs in a very southerly direction; so that it is of considerable length to its source in the Cordillera. The valleys of Guasco and Copiapó may both be considered as long narrow islands, separated from the rest of Chile by deserts of rock instead of by salt water. Northward of these, there is one other very miserable valley, called Paposo, which contains about two hundred souls; and then there extends the real desert of Atacama—a barrier far worse than the most turbulent ocean. After staying a few days at Potrero Seco, I proceeded up the valley to the house of Don Benito Cruz, to whom I had a letter of introduction. I found him most hospitable; indeed it is impossible to bear too strong testimony to the kindness with which travellers are received in almost every part of South America. The next day I hired some mules to take me by the ravine of Jolquera into the central Cordillera. On the second night the weather seemed to foretell a storm of snow or rain, and whilst lying in our beds we felt a trifling shock of an earthquake.

The connection between earthquakes and the weather has been often disputed: it appears to me to be a point of great interest, which is little understood. Humboldt has remarked in one part of the Personal Narrative,[145] that it would be difficult for any person who had long resided in New Andalusia, or in Lower Peru, to deny that there exists some connection between these phenomena: in another part, however, he seems to think the connection fanciful. At Guayaquil, it is said that a heavy shower in the dry season is invariably followed by an earthquake. In Northern Chile, from the extreme infrequency of rain, or even of weather foreboding rain, the probability of accidental coincidences becomes very small; yet the inhabitants are here most firmly convinced of some connection between the state of the atmosphere and of the trembling of the ground: I was much struck by this, when mentioning to some people at Copiapó that there had been a sharp shock at Coquimbo: they immediately cried out, "How fortunate! there will be plenty of pasture there this year." To their minds an earthquake foretold rain, as surely as rain foretold abundant pasture. Certainly it did so happen that on the very day of the earthquake, that shower of rain fell, which I have described as in ten days' time producing a thin sprinkling of grass. At other times rain has followed earthquakes at a period of the year when it is a far greater prodigy than the earthquake itself: this happened after the shock of November, 1822, and again in 1829, at Valparaiso; also after that of September, 1833, at Tacna. A person must be somewhat habituated to the climate of these countries to perceive the extreme improbability of rain falling at such seasons, except as a consequence of some law quite unconnected with the ordinary course of the weather. In the cases of great volcanic eruptions, as that of Coseguina, where torrents of rain fell at a time of the year most unusual for it, and "almost unprecedented in Central America," it is not difficult to understand that the volumes of vapour and clouds of ashes might have disturbed the atmospheric equilibrium. Humboldt extends this view to the case of earthquakes unaccompanied by eruptions; but I can hardly conceive it possible, that the small quantity of aëriform fluids which then escape from the fissured ground, can produce such remarkable effects. There appears much probability in the view first proposed by Mr. P. Scrope, that when the barometer is low, and when rain might naturally be expected to fall, the diminished pressure of the atmosphere over a wide extent of country, might well determine the precise day on which the earth, already stretched to the utmost by the subterranean forces, should yield, crack, and consequently tremble. It is, however, doubtful how far this idea will explain the circumstances of torrents of rain falling in the dry season during several days, after an earthquake unaccompanied by an eruption; such cases seem to bespeak some more intimate connection between the atmospheric and subterranean regions.

Finding little of interest in this part of the ravine, we retraced our steps to the house of Don Benito, where I stayed two days collecting fossil shells and wood. Great prostrate silicified trunks of trees, embedded in a conglomerate, were extraordinarily numerous. I measured one, which was fifteen feet in circumference: how surprising it is that every atom of the woody matter in this great cylinder should have been removed and replaced by silex so perfectly, that each vessel and pore is preserved! These trees flourished at about the period of our lower chalk; they all belonged to the fir-tribe. It was amusing to hear the inhabitants discussing the nature of the fossil shells which I collected, almost in the same terms as were used a century ago in Europe,—namely, whether or not they had been thus "born by nature." My geological examination of the country generally created a good deal of surprise amongst the Chilenos: it was long before they could be convinced that I was not hunting for mines. This was sometimes troublesome: I found the most ready way of explaining my employment, was to ask them how it was that they themselves were not curious concerning earthquakes and volcanos?—why some springs were hot and others cold?—why there were mountains in Chile, and not a hill in La Plata? These bare questions at once satisfied and silenced the greater number; some, however (like a few in England who are a century behind-hand), thought that all such inquiries were useless and impious; and that it was quite sufficient that God had thus made the mountains.

An order had recently been issued that all stray dogs should be killed, and we saw many lying dead on the road. A great number had lately gone mad, and several men had been bitten and had died in consequence. On several occasions hydrophobia has prevailed in this valley. It is remarkable thus to find so strange and dreadful a disease, appearing time after time in the same isolated spot. It has been remarked that certain villages in England are in like manner much more subject to this visitation than others. Dr. Unanùe states that hydrophobia was first known in South America in 1803: this statement is corroborated by Azara and Ulloa having never heard of it in their time. Dr. Unanùe says that it broke out in Central America, and slowly travelled southward. It reached Arequipa in 1807; and it is said that some men there, who had not been bitten, were affected, as were some negroes, who had eaten a bullock which had died of hydrophobia. At Ica forty-two people thus miserably perished. The disease came on between twelve and ninety days after the bite; and in those cases where it did come on, death ensued invariably within five days. After 1808, a long interval ensued without any cases. On inquiry, I did not hear of hydrophobia in Van Diemen's Land, or in Australia; and Burchell says, that during the five years he was at the Cape of Good Hope, he never heard of an instance of it. Webster asserts that at the Azores hydrophobia has never occurred; and the same assertion has been made with respect to Mauritius and St. Helena.[146] In so strange a disease some information might possibly be gained by considering the circumstances under which it originates in distant climates; for it is improbable that a dog already bitten, should have been brought to these distant countries.

At night, a stranger arrived at the house of Don Benito, and asked permission to sleep there. He said he had been wandering about the mountains for seventeen days, having lost his way. He started from Guasco, and being accustomed to travelling in the Cordillera, did not expect any difficulty in following the track to Copiapó; but he soon became involved in a labyrinth of mountains, whence he could not escape. Some of his mules had fallen over precipices, and he had been in great distress. His chief difficulty arose from not knowing where to find water in the lower country, so that he was obliged to keep bordering the central ranges.

We returned down the valley, and on the 22nd reached the town of Copiapó. The lower part of the valley is broad, forming a fine plain like that of Quillota. The town covers a considerable space of ground, each house possessing a garden: but it is an uncomfortable place, and the dwellings are poorly furnished. Every one seems bent on the one object of making money, and then migrating as quickly as possible. All the inhabitants are more or less directly concerned with mines; and mines and ores are the sole subjects of conversation. Necessaries of all sorts are extremely dear; as the distance from the town to the port is eighteen leagues, and the land carriage very expensive. A fowl costs five or six shillings; meat is nearly as dear as in England; firewood, or rather sticks, are brought on donkeys from a distance of two and three days' journey within the Cordillera; and pasturage for animals is a shilling a day: all this for South America is wonderfully exorbitant.

June 26th.—I hired a guide and eight mules to take me into the Cordillera by a different line from my last excursion. As the country was utterly desert, we took a cargo and a half of barley mixed with chopped straw. About two leagues above the town a broad valley called the "Despoblado," or uninhabited, branches off from that one by which we had arrived. Although a valley of the grandest dimensions, and leading to a pass across the Cordillera, yet it is completely dry, excepting perhaps for a few days during some very rainy winter. The sides of the crumbling mountains were furrowed by scarcely any ravines; and the bottom of the main valley, filled with shingle, was smooth and nearly level. No considerable torrent could ever have flowed down this bed of shingle; for if it had, a great cliff-bounded channel, as in all the southern valleys, would assuredly have been formed. I feel little doubt that this valley, as well as those mentioned by travellers in Peru, were left in the state we now see them by the waves of the sea, as the land slowly rose. I observed in one place, where the Despoblado was joined by a ravine (which in almost any other chain would have been called a grand valley), that its bed, though composed merely of sand and gravel, was higher than that of its tributary. A mere rivulet of water, in the course of an hour, would have cut a channel for itself; but it was evident that ages had passed away, and no such rivulet had drained this great tributary. It was curious to behold the machinery, if such a term may be used, for the drainage, all, with the last trifling exception, perfect, yet without any signs of action. Every one must have remarked how mud-banks, left by the retiring tide, imitate in miniature a country with hill and dale; and here we have the original model in rock, formed as the continent rose during the secular retirement of the ocean, instead of during the ebbing and flowing of the tides. If a shower of rain falls on the mud-bank, when left dry, it deepens the already-formed shallow lines of excavation; and so it is with the rain of successive centuries on the bank of rock and soil, which we call a continent.

We rode on after it was dark, till we reached a side ravine with a small well, called "Agua amarga." The water deserved its name, for besides being saline it was most offensively putrid and bitter; so that we could not force ourselves to drink either tea or maté. I suppose the distance from the river of Copiapó to this spot was at least twenty-five or thirty English miles; in the whole space there was not a single drop of water, the country deserving the name of desert in the strictest sense. Yet about half way we passed some old Indian ruins near Punta Gorda: I noticed also in front of some of the valleys, which branch off from the Despoblado, two piles of stones placed a little way apart, and directed so as to point up the mouths of these small valleys. My companions knew nothing about them, and only answered my queries by their imperturbable "quién sabe?"

I observed Indian ruins in several parts of the Cordillera: the most perfect which I saw, were the Ruinas de Tambillos, in the Uspallata Pass. Small square rooms were there huddled together in separate groups: some of the doorways were yet standing; they were formed by a cross slab of stone only about three feet high. Ulloa has remarked on the lowness of the doors in the ancient Peruvian dwellings. These houses, when perfect, must have been capable of containing a considerable number of persons. Tradition says, that they were used as halting-places for the Incas, when they crossed the mountains. Traces of Indian habitations have been discovered in many other parts, where it does not appear probable that they were used as mere resting-places, but yet where the land is as utterly unfit for any kind of cultivation, as it is near the Tambillos or at the Incas Bridge, or in the Portillo Pass, at all which places I saw ruins. In the ravine of Jajuel, near Aconcagua, where there is no pass, I heard of remains of houses situated at a great height, where it is extremely cold and sterile. At first I imagined that these buildings had been places of refuge, built by the Indians on the first arrival of the Spaniards; but I have since been inclined to speculate on the probability of a small change of climate.

In this northern part of Chile, within the Cordillera, old Indian houses are said to be especially numerous: by digging amongst the ruins, bits of woollen articles, instruments of precious metals, and heads of Indian corn, are not unfrequently discovered: an arrow-head made of agate, and of precisely the same form with those now used in Tierra del Fuego, was given me. I am aware that the Peruvian Indians now frequently inhabit most lofty and bleak situations; but at Copiapó I was assured by men who had spent their lives in travelling through the Andes, that there were very many *(muchisimas)* buildings at heights so great as almost to border upon the perpetual snow, and in parts where there exist no passes, and where the land produces absolutely nothing, and what is still more extraordinary, where there is no water. Nevertheless it is the opinion of the people of the country (although they are much puzzled by the circumstance), that, from the appearance of the houses, the Indians must have used them as places of residence. In this valley, at Punta Gorda, the remains consisted of seven or eight square little rooms, which were of a similar form with those at Tambillos, but built chiefly of mud, which the present inhabitants cannot, either here or, according to Ulloa, in Peru, imitate in durability. They were situated in the most conspicuous and defenceless position, at the bottom of the flat broad valley. There was no water nearer than three or four leagues, and that only in very small quantity, and bad: the soil was absolutely sterile; I looked in vain even for a lichen adhering to the rocks. At the present day, with the advantage of beasts of burden, a mine, unless it were very rich, could scarcely be worked here with profit. Yet the Indians formerly chose it as a place of residence! If at the present time two or three showers of rain were to fall annually, instead of one, as now is the case during as many years, a small rill of water would probably be formed in this great valley; and then, by irrigation (which was formerly so well understood by the Indians), the soil would easily be rendered sufficiently productive to support a few families.

I have convincing proofs that this part of the continent of South America has been elevated near the coast at least from 400 to 500, and in some parts from 1000 to 1300 feet, since the epoch of existing shells; and further inland the rise possibly may have been greater. As the peculiarly arid character of the climate is evidently a consequence of the height of the Cordillera, we may feel almost sure that before the later elevations, the atmosphere could not have been so completely drained of its moisture as it now is; and as the rise has been gradual, so would have been the change in climate. On this notion of a change of climate since the buildings were inhabited, the ruins must be of extreme antiquity, but I do not think their preservation under the Chilian climate any great difficulty. We must also admit on this notion (and this perhaps is a greater difficulty) that man has inhabited South America for an immensely long period, inasmuch as any change of climate effected by the elevation of the land must have been extremely gradual. At Valparaiso, within the last 220 years, the rise has been somewhat less than 19 feet: at Lima a sea-beach has certainly been upheaved from 80 to 90 feet, within the Indo-human period: but such small elevations could have had little power in deflecting the moisture-bringing atmospheric currents. Dr. Lund, however, found human skeletons in the caves of Brazil, the appearance of which induced him to believe that the Indian race has existed during a vast lapse of time in South America.

When at Lima, I conversed on these subjects[147] with Mr. Gill, a civil engineer, who had seen much of the interior country. He told me that a conjecture of a change of climate had sometimes crossed his mind; but that he thought that the greater portion of land, now incapable of cultivation, but covered with Indian ruins, had been reduced to this state by the water-conduits, which the Indians formerly constructed on so wonderful a scale, having been injured by neglect and by subterranean movements. I may here mention, that the Peruvians actually carried their irrigating streams in tunnels through hills of solid rock. Mr. Gill told me, he had been employed professionally to examine one: he found the passage low, narrow, crooked, and not of uniform breadth, but of very considerable length. Is it not most wonderful that men should have attempted such operations, without the use of iron or gunpowder? Mr. Gill also mentioned to me a most interesting, and, as far as I am aware, quite unparalleled case, of a subterranean disturbance having changed the drainage of a country. Travelling from Casma to Huaraz (not very far distant from Lima), he found a plain covered with ruins and marks of ancient cultivation, but now quite barren. Near it was the dry course of a considerable river, whence the water for irrigation had formerly been conducted. There was nothing in the appearance of the water-course to indicate that the river had not flowed there a few years previously; in some parts, beds of sand and gravel were spread out; in others, the solid rock had been worn into a broad channel, which in one spot was about 40 yards in breadth and 8 feet deep. It is self-evident that a person following up the course of a stream, will always ascend at a greater or less inclination: Mr. Gill, therefore, was much astonished, when walking up the bed of this ancient river, to find himself suddenly going down hill. He imagined that the downward slope had a fall of about 40 or 50 feet perpendicular. We here have unequivocal evidence that a ridge had been uplifted right across the old bed of a stream. From the moment the river-course was thus arched, the water must necessarily have been thrown back, and a new channel formed. From that moment, also, the neighbouring plain must have lost its fertilizing stream, and become a desert.

June 27th.—We set out early in the morning, and by mid-day reached the ravine of Paypote, where there is a tiny rill of water, with a little vegetation, and even a few algarroba trees, a kind of mimosa. From having fire-wood, a smelting-furnace had formerly been built here: we found a solitary man in charge of it, whose sole employment was hunting guanacos. At night it froze sharply; but having plenty of wood for our fire, we kept ourselves warm.

28th.—We continued gradually ascending, and the valley now changed into a ravine. During the day we saw several guanacos, and the track of the closely-allied species, the Vicuña: this latter animal is pre-eminently alpine in its habits; it seldom descends much below the limit of perpetual snow, and therefore haunts even a more lofty and sterile situation than the guanaco. The only other animal which we saw in any number was a small fox: I suppose this animal preys on the mice and other small rodents, which, as long as there is the least vegetation, subsist in considerable numbers in very desert places. In Patagonia, even on the borders of the salinas, where a drop of fresh water can never be found, excepting dew, these little animals swarm. Next to lizards, mice appear to be able to support existence on the smallest and driest portions of the earth—even on islets in the midst of great oceans.

The scene on all sides showed desolation, brightened and made palpable by a clear, unclouded sky. For a time such scenery is sublime, but this feeling cannot last, and then it becomes uninteresting. We bivouacked at the foot of the "primera linea," or the first line of the partition of waters. The streams, however, on the east side do not flow to the Atlantic, but into an elevated district, in the middle of which there is a large saline, or salt lake; thus forming a little Caspian Sea at the height, perhaps, of ten thousand feet. Where we slept, there were some considerable patches of snow, but they do not remain throughout the year. The winds in these lofty regions obey very regular laws: every day a fresh breeze blows up the valley, and at night, an hour or two after sunset, the air from the cold regions above descends as through a funnel. This night it blew a gale of wind, and the temperature must have been considerably below the freezing-point, for water in a vessel soon became a block of ice. No clothes seemed to oppose any obstacle to the air; I suffered very much from the cold, so that I could not sleep, and in the morning rose with my body quite dull and benumbed.

In the Cordillera further southward, people lose their lives from snow-storms; here, it sometimes happens from another cause. My guide, when a boy of fourteen years old, was passing the Cordillera with a party in the month of May; and while in the central parts, a furious gale of wind arose, so that the men could hardly cling on their mules, and stones were flying along the ground. The day was cloudless, and not a speck of snow fell, but the temperature was low. It is probable that the thermometer could not have stood very many degrees below the freezing-point, but the effect on their bodies, ill protected by clothing, must have been in proportion to the rapidity of the current of cold air. The gale lasted for more than a day; the men began to lose their strength, and the mules would not move onwards. My guide's brother tried to return, but he perished, and his body was found two years afterwards, lying by the side of his mule near the road, with the bridle still in his hand. Two other men in the party lost their fingers and toes; and out of two hundred mules and thirty cows, only fourteen mules escaped alive. Many years ago the whole of a large party are supposed to have perished from a similar cause, but their bodies to this day have never been discovered. The union of a cloudless sky, low temperature, and a furious gale of wind, must be, I should think, in all parts of the world an unusual occurrence.

June 29th.—We gladly travelled down the valley to our former night's lodging, and thence to near the Agua amarga. On July 1st we reached the valley of Copiapó. The smell of the fresh clover was quite delightful, after the scentless air of the dry, sterile Despoblado. Whilst staying in the town I heard an account from several of the inhabitants, of a hill in the neighbourhood which they called "El Bramador,"—the roarer or bellower. I did not at the time pay sufficient attention to the account; but, as far as I understood, the hill was covered by sand, and the noise was produced only when people, by ascending it, put the sand in motion. The same circumstances are described in detail on the authority of Seetzen and Ehrenberg,[148] as the cause of the sounds which have been heard by many travellers on Mount Sinai near the Red Sea. One person with whom I conversed had himself heard the noise: he described it as very surprising; and he distinctly stated that, although he could not understand how it was caused, yet it was necessary to set the sand rolling down the acclivity. A horse walking over dry coarse sand, causes a peculiar chirping noise from the friction of the particles; a circumstance which I several times noticed on the coast of Brazil.

Three days afterwards I heard of the Beagle's arrival at the Port, distant eighteen leagues from the town. There is very little land cultivated down the valley; its wide expanse supports a wretched wiry grass, which even the donkeys can hardly eat. This poorness of the vegetation is owing to the quantity of saline matter with which the soil is impregnated. The Port consists of an assemblage of miserable little hovels, situated at the foot of a sterile plain. At present, as the river contains water enough to reach the sea, the inhabitants enjoy the advantage of having fresh water within a mile and a half. On the beach there were large piles of merchandise, and the little place had an air of activity. In the evening I gave my adios, with a hearty good-will, to my companion Mariano Gonzales, with whom I had ridden so many leagues in Chile. The next morning the Beagle sailed for Iquique.

July 12th.—We anchored in the port of Iquique, in lat. 20° 12', on the coast of Peru. The town contains about a thousand inhabitants, and stands on a little plain of sand at the foot of a great wall of rock, 2000 feet in height, here forming the coast. The whole is utterly desert. A light shower of rain falls only once in very many years; and the ravines consequently are filled with detritus, and the mountain-sides covered by piles of fine white sand, even to a height of a thousand feet. During this season of the year a heavy bank of clouds, stretched over the ocean, seldom rises above the wall of rocks on the coast. The aspect of the place was most gloomy; the little port, with its few vessels, and small group of wretched houses, seemed overwhelmed and out of all proportion with the rest of the scene.

The inhabitants live like persons on board a ship: every necessary comes from a distance: water is brought in boats from Pisagua, about forty miles northward, and is sold at the rate of nine reals *(4s. 6d.)* an eighteen-gallon cask: I bought a wine-bottle full for threepence. In like manner firewood, and of course every article of food, is imported. Very few animals can be maintained in such a place: on the ensuing morning I hired with difficulty, at the price of four pounds sterling, two mules and a guide to take me to the nitrate of soda works. These are at present the support of Iquique. This salt was first exported in 1830: in one year an amount in value of one hundred thousand pounds sterling, was sent to France and England. It is principally used as a manure and in the manufacture of nitric acid: owing to its deliquescent property it will not serve for gun-powder. Formerly there were two exceedingly rich silver-mines in this neighbourhood, but their produce is now very small.

Our arrival in the offing caused some little apprehension. Peru was in a state of anarchy; and each party having demanded a contribution, the poor town of Iquique was in tribulation, thinking the evil hour was come. The people had also their domestic troubles; a short time before, three

French carpenters had broken open, during the same night, the two churches, and stolen all the plate: one of the robbers, however, subsequently confessed, and the plate was recovered. The convicts were sent to Arequipa, which though the capital of this province, is two hundred leagues distant; the government there thought it a pity to punish such useful workmen, who could make all sorts of furniture; and accordingly liberated them. Things being in this state, the churches were again broken open, but this time the plate was not recovered. The inhabitants became dreadfully enraged, and declaring that none but heretics would thus "eat God Almighty," proceeded to torture some Englishmen, with the intention of afterwards shooting them. At last the authorities interfered, and peace was established.

13th.—In the morning I started for the saltpetre-works, a distance of fourteen leagues. Having ascended the steep coast-mountains by a zigzag sandy track, we soon came in view of the mines of Guantajaya and St. Rosa. These two small villages are placed at the very mouths of the mines; and being perched up on hills, they had a still more unnatural and desolate appearance than the town of Iquique. We did not reach the saltpetre-works till after sunset, having ridden all day across an undulating country, a complete and utter desert. The road was strewed with the bones and dried skins of many beasts of burden which had perished on it from fatigue. Excepting the Vultur aura, which preys on the carcasses, I saw neither bird, quadruped, reptile, nor insect. On the coast-mountains, at the height of about 2000 feet, where during this season the clouds generally hang, a very few cacti were growing in the clefts of rock; and the loose sand was strewed over with a lichen, which lies on the surface quite unattached. This plant belongs to the genus Cladonia, and somewhat resembles the reindeer lichen. In some parts it was in sufficient quantity to tinge the sand, as seen from a distance, of a pale yellowish colour. Further inland, during the whole ride of fourteen leagues, I saw only one other vegetable production, and that was a most minute yellow lichen, growing on the bones of the dead mules. This was the first true desert which I had seen: the effect on me was not impressive; but I believe this was owing to my having become gradually accustomed to such scenes, as I rode northward from Valparaiso, through Coquimbo, to Copiapó. The appearance of the country was remarkable, from being covered by a thick crust of common salt, and of a stratified saliferous alluvium, which seems to have been deposited as the land slowly rose above the level of the sea. The salt is white, very hard, and compact: it occurs in water-worn nodules projecting from the agglutinated sand, and is associated with much gypsum. The appearance of this superficial mass very closely resembled that of a country after snow, before the last dirty patches are thawed. The existence of this crust of a soluble substance over the whole face of the country, shows how extraordinarily dry the climate must have been for a long period.

At night I slept at the house of the owner of one of the saltpetre mines. The country is here as unproductive as near the coast; but water, having rather a bitter and brackish taste, can be procured by digging wells. The well at this house was thirty-six yards deep: as scarcely any rain falls, it is evident the water is not thus derived; indeed if it were, it could not fail to be as salt as brine, for the whole surrounding country is incrusted with various saline substances. We must therefore conclude that it percolates under ground from the Cordillera, though distant many leagues. In that direction there are a few small villages, where the inhabitants, having more water, are enabled to irrigate a little land, and raise hay, on which the mules and asses, employed in carrying the saltpetre, are fed. The nitrate of soda was now selling at the ship's side at fourteen shillings per hundred pounds: the chief expense is its transport to the sea-coast. The mine consists of a hard stratum, between two and three feet thick, of the nitrate mingled with a little of the sulphate of soda and a good deal of common salt. It lies close beneath the surface, and follows for a length of one hundred and fifty miles the margin of a grand basin or plain; this, from its outline, manifestly must once have been a lake, or more probably an inland arm of the sea, as may be inferred from the presence of iodic salts in the saline stratum. The surface of the plain is 3300 feet above the Pacific.

19th.—We anchored in the Bay of Callao, the seaport of Lima, the capital of Peru. We stayed here six weeks, but from the troubled state of public affairs, I saw very little of the country. During our whole visit the climate was far from being so delightful, as it is generally represented. A dull heavy bank of clouds constantly hung over the land, so that during the first sixteen days I had only one view of the Cordillera behind Lima. These mountains, seen in stages, one above the other, through openings in the clouds, had a very grand appearance. It is almost become a proverb, that rain never falls in the lower part of Peru. Yet this can hardly be considered correct; for during almost every day of our visit there was a thick drizzling mist, which was sufficient to make the streets muddy and one's clothes damp: this the people are pleased to call Peruvian dew. That much rain does not fall is very certain, for the houses are covered only with flat roofs made of hardened mud; and on the mole shiploads of wheat were piled up, being thus left for weeks together without any shelter.

I cannot say I liked the very little I saw of Peru: in summer, however, it is said that the climate is much pleasanter. In all seasons, both inhabitants and foreigners suffer from severe attacks of ague. This disease is common on the whole coast of Peru, but is unknown in the interior. The attacks of illness which arise from miasma never fail to appear most mysterious. So difficult is it to judge from the aspect of a country, whether or not it is healthy, that if a person had been told to choose within the tropics a situation appearing favourable for health, very probably he would have named this coast. The plain round the outskirts of Callao is sparingly covered with a coarse grass, and in some parts there are a few stagnant, though very small, pools of water. The miasma, in all probability, arises from these: for the town of Arica was similarly circumstanced, and its healthiness was much improved by the drainage of some little pools. Miasma is not always produced by a luxuriant vegetation with an ardent climate; for many parts of Brazil, even where there are marshes and a rank vegetation, are much more healthy than this sterile coast of Peru. The densest forests in a temperate climate, as in Chiloe, do not seem in the slightest degree to affect the healthy condition of the atmosphere.

The island of St. Jago, at the Cape de Verds, offers another strongly marked instance of a country, which any one would have expected to find most healthy, being very much the contrary. I have described the bare and open plains as supporting, during a few weeks after the rainy season, a thin vegetation, which directly withers away and dries up: at this period the air appears to become quite poisonous; both natives and foreigners often being affected with violent fevers. On the other hand, the Galapagos Archipelago, in the Pacific, with a similar soil, and periodically subject to the same process of vegetation, is perfectly healthy. Humboldt has observed, that, "under the torrid zone, the smallest marshes are the most dangerous, being surrounded, as at Vera Cruz and Carthagena, with an arid and sandy soil, which raises the temperature of the ambient air."[149] On the coast of Peru, however, the temperature is not hot to any excessive degree; and perhaps in consequence, the intermittent fevers are not of the most malignant order. In all unhealthy countries the greatest risk is run by sleeping on shore. Is this owing to the state of the body during sleep, or to a greater abundance of miasma at such times? It appears certain that those who stay on board a vessel, though anchored at only a short distance from the coast, generally suffer less than those actually on shore. On the other hand, I have heard of one remarkable case where a fever broke out among the crew of a man-of-war some hundred miles off the coast of Africa, and at the same time one of those fearful periods[150] of death commenced at Sierra Leone.

No state in South America, since the declaration of independence, has suffered more from anarchy than Peru. At the time of our visit, there were four chiefs in arms contending for supremacy in the government: if one succeeded in becoming for a time very powerful, the others coalesced against him; but no sooner were they victorious, than they were again hostile to each other. The other day, at the Anniversary of the Independence, high mass was performed, the President partaking of the sacrament: during the *Te Deum laudamus*, instead of each regiment displaying the Peruvian flag, a black one with death's head was unfurled. Imagine a government under which such a scene could be ordered, on such an occasion, to be typical of their determination of fighting to death! This state of affairs happened at a time very unfortunately for me, as I was precluded from taking any excursions much beyond the limits of the town. The barren island of St. Lorenzo, which forms the harbour, was nearly the only place where one could walk securely. The upper part, which is upwards of 1000 feet in height, during this season of the year (winter), comes within the lower limit of the clouds; and in consequence, an abundant cryptogamic vegetation, and a few flowers cover the summit. On the hills near Lima, at a height but little greater, the ground is carpeted with moss, and beds of beautiful yellow lilies, called Amancaes. This indicates a very much greater degree of humidity, than at a corresponding height at Iquique. Proceeding northward of Lima, the climate becomes damper, till on the banks of the Guayaquil, nearly under the equator, we find the most luxuriant forests. The change, however, from the sterile coast of Peru to that fertile land is described as taking place rather abruptly in the latitude of Cape Blanco, two degrees south of Guayaquil.

Callao is a filthy, ill-built, small seaport. The inhabitants, both here and at Lima, present every imaginable shade of mixture, between European, Negro, and Indian blood. They appear a depraved, drunken set of people. The atmosphere is loaded with foul smells, and that peculiar one, which may be perceived in almost every town within the tropics, was here very strong. The fortress, which withstood Lord Cochrane's long siege, has an imposing appearance. But the President, during our stay, sold the brass guns, and proceeded to dismantle parts of it. The reason assigned was, that he had not an officer to whom he could trust so important a charge. He himself had good reason for thinking so, as he had obtained the presidentship by rebelling while in charge of this same fortress. After we left South America, he paid the penalty in the usual manner, by being conquered, taken prisoner, and shot.

Lima stands on a plain in a valley, formed during the gradual retreat of the sea. It is seven miles from Callao, and is elevated 500 feet above it; but from the slope being very gradual, the road appears absolutely level; so that when at Lima it is difficult to believe one has ascended even one hundred feet: Humboldt has remarked on this singularly deceptive case. Steep, barren hills rise like islands from the plain, which is divided, by straight mud-walls, into large green fields. In these scarcely a tree grows excepting a few willows, and an occasional clump of bananas and of oranges. The city of Lima is now in a wretched state of decay: the streets are nearly unpaved; and heaps of filth are piled up in all directions, where the black gallinazos, tame as poultry, pick up bits of carrion. The houses have generally an upper story, built on account of the earthquakes, of plastered wood-work; but some of the old ones, which are now used by several families, are immensely large, and would rival in suites of apartments the most magnificent in any place. Lima, the City of the Kings, must formerly have been a splendid town. The extraordinary number of churches gives it, even at the present day, a peculiar and striking character, especially when viewed from a short distance.

One day I went out with some merchants to hunt in the immediate vicinity of the city. Our sport was very poor; but I had an opportunity of seeing the ruins of one of the ancient Indian villages, with its mound like a natural hill in the centre. The remains of houses, enclosures, irrigating streams, and burial mounds, scattered over this plain, cannot fail to give one a high idea of the condition and number of the ancient population. When their earthenware, woollen clothes, utensils of elegant forms cut out of the hardest rocks, tools of copper, ornaments of precious stones, palaces, and hydraulic works, are considered, it is impossible not to respect the considerable advance made by them in the arts of civilization. The burial mounds, called Huacas, are really stupendous; although in some places they appear to be natural hills incased and modelled.

There is also another and very different class of ruins, which possesses some interest, namely, those of old Callao, overwhelmed by the great earthquake of 1746, and its accompanying wave. The destruction must have been more complete even than at Talcahuano. Quantities of shingle almost conceal the foundations of the walls, and vast masses of brickwork appear to have been whirled about like pebbles by the retiring waves. It has been stated that the land subsided during this memorable shock: I could not discover any proof of this; yet it seems far from improbable, for the form of the coast must certainly have undergone some change since the foundation of the old town; as no people in their senses would willingly have chosen for their building place, the narrow spit of shingle on which the ruins now stand. Since our voyage, M. Tschudi has come to the conclusion, by the comparison of old and modern maps, that the coast both north and south of Lima has certainly subsided.

On the island of San Lorenzo, there are very satisfactory proofs of elevation within the recent period; this of course is not opposed to the belief, of a small sinking of the ground having subsequently taken place. The side of this island fronting the Bay of Callao, is worn into three obscure terraces, the lower one of which is covered by a bed a mile in length, almost wholly composed of shells of eighteen species, now living in the adjoining sea. The height of this bed is eighty-five feet. Many of the shells are deeply corroded, and have a much older and more decayed appearance than those at the height of 500 or 600 feet on the coast of Chile. These shells are associated with much common salt, a little sulphate of lime (both probably left by the evaporation of the spray, as the land slowly rose), together with sulphate of soda and muriate of lime. They rest on fragments of the underlying sandstone, and are covered by a few inches thick of detritus. The shells, higher up on this terrace, could be traced scaling off in flakes, and falling into an impalpable powder; and on an upper terrace, at the height of 170 feet, and likewise at some considerably higher points, I found a layer of saline powder of exactly similar appearance, and lying in the same relative position. I have no doubt that this upper layer originally existed as a bed of shells, like that on the eighty-five-feet ledge; but it does not now contain even a trace of organic structure. The powder has been analyzed for me by Mr. T. Reeks; it consists of sulphates and muriates both of lime and soda, with very little carbonate of lime. It is known that common salt and carbonate of lime left in a mass for some time together, partly decompose each other; though this does not happen with small quantities in solution. As the half-decomposed shells in the lower parts are associated with much common salt, together with some of the saline substances composing the upper saline layer, and as these shells are corroded and decayed in a remarkable manner, I strongly suspect that this double decomposition has here taken place. The resultant salts, however, ought to be carbonate of soda and muriate of lime; the latter is present, but not the carbonate of soda. Hence I am led to imagine that by some unexplained means, the carbonate of soda becomes changed into the sulphate. It is obvious that the saline layer could not have been preserved in any country in which abundant rain occasionally fell: on the other hand, this very circumstance, which at first sight appears so highly favourable to the long preservation of exposed shells, has probably been the indirect means, through the common salt not having been washed away, of their decomposition and early decay.

I was much interested by finding on the terrace, at the height of eighty-five feet, *embedded* amidst the shells and much sea-drifted rubbish, some bits of cotton thread, plaited rush, and the head of a stalk of Indian corn: I compared these relics with similar ones taken out of the Huacas, or old Peruvian tombs, and found them identical in appearance. On the mainland in front of San Lorenzo, near Bellavista, there is an extensive and level plain about a hundred feet high, of which the lower part is formed of alternating layers of sand and impure clay, together with some gravel, and the surface, to the depth of from three to six feet, of a reddish loam, containing a few scattered sea-shells and numerous small fragments of coarse red earthenware, more abundant at certain spots than at others. At first I was inclined to believe that this superficial bed, from its wide extent and smoothness, must have been deposited beneath the sea; but I afterwards found in one spot, that it lay on an artificial floor of round stones. It seems, therefore, most probable that at a period when the land stood at a lower level there was a plain very similar to that now surrounding Callao, which being protected by a shingle beach, is raised but very little above the level of the sea. On this plain, with its underlying red-clay beds, I imagine that the Indians manufactured their earthen vessels; and that, during some violent earthquake, the sea broke over the beach, and converted the plain into a temporary lake, as happened round Callao in 1713 and 1746. The water would then have deposited mud, containing fragments of pottery from the kilns, more abundant at some spots than at others, and shells from the sea. This bed, with fossil earthenware, stands at about the same height with the shells on the lower terrace of San Lorenzo, in which the cotton-thread and other relics were embedded.

Hence we may safely conclude, that within the Indo-human period there has been an elevation, as before alluded to, of more than eighty-five feet; for some little elevation must have been lost by the coast having subsided since the old maps were engraved. At Valparaiso, although in the 220 years before our visit, the elevation cannot have exceeded nineteen feet, yet subsequently to 1817, there has been a rise, partly insensible and partly by a start during the shock of 1822, of ten or eleven feet. The antiquity of the Indo-human race here, judging by the eighty-five feet rise of the land since the relics were embedded, is the more remarkable, as on the coast of Patagonia, when the land stood about the same number of feet lower, the Macrauchenia was a living beast; but as the Patagonian coast is some way distant from the Cordillera, the rising there may have been slower than here. At Bahia Blanca, the elevation has been only a few feet since the numerous gigantic quadrupeds were there entombed; and, according to the generally received opinion, when these extinct animals were living, man did not exist. But the rising of that part of the coast of Patagonia, is perhaps no way connected with the Cordillera, but rather with a line of old volcanic rocks in Banda Oriental, so that it may have been infinitely slower than on the shores of Peru. All these speculations, however, must be vague; for who will pretend to say that there may not have been several periods of subsidence, intercalated between the movements of elevation; for we know that along the whole coast of Patagonia, there have certainly been many and long pauses in the upward action of the elevatory forces.

CHAPTER XVII

Galapagos Archipelago

THE WHOLE GROUP VOLCANIC • NUMBERS OF CRATERS • LEAFLESS BUSHES • COLONY AT CHARLES ISLAND • JAMES ISLAND • SALT-LAKE IN CRATER • NATURAL HISTORY OF THE GROUP • ORNITHOLOGY, CURIOUS FINCHES • REPTILES • GREAT TORTOISES, HABITS OF • MARINE LIZARD, FEEDS ON SEAWEED • TERRESTRIAL LIZARD, BURROWING HABITS, HERBIVOROUS • IMPORTANCE OF REPTILES IN THE ARCHIPELAGO • FISH, SHELLS, INSECTS • BOTANY • AMERICAN TYPE OF ORGANIZATION • DIFFERENCES IN THE SPECIES OR RACES ON DIFFERENT ISLANDS • TAMENESS OF THE BIRDS • FEAR OF MAN, AN ACQUIRED INSTINCT

September 15th.—This archipelago consists of ten principal islands, of which five exceed the others in size. They are situated under the Equator, and between five and six hundred miles westward of the coast of America. They are all formed of volcanic rocks; a few fragments of granite curiously glazed and altered by the heat, can hardly be considered as an exception. Some of the craters, surmounting the larger islands, are of immense size, and they rise to a height of between three and four thousand feet. Their flanks are studded by innumerable smaller orifices. I scarcely hesitate to affirm, that there must be in the whole archipelago at least two thousand craters. These consist either of lava or scoriæ, or of finely-stratified, sandstone-like tuff. Most of the latter are beautifully symmetrical; they owe their origin to eruptions of volcanic mud without any lava: it is a remarkable circumstance that every one of the twenty-eight tuff-craters which were examined, had their southern sides either much lower than the other sides, or quite broken down and removed. As all these craters appar-

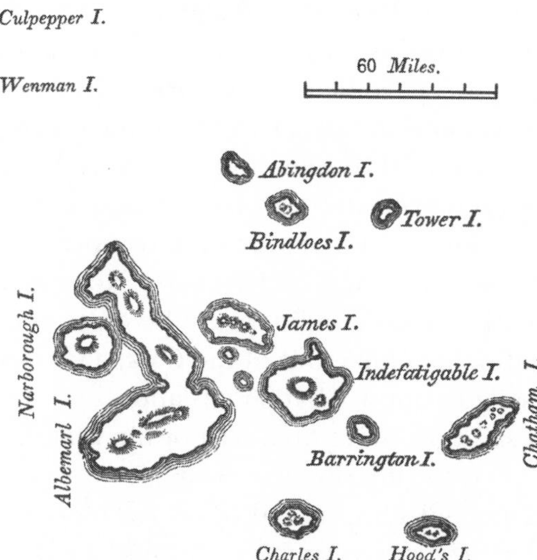

ently have been formed when standing in the sea, and as the waves from the trade-wind and the swell from the open Pacific here unite their forces on the southern coasts of all the islands, this singular uniformity in the broken state of the craters, composed of the soft and yielding tuff, is easily explained.

Considering that these islands are placed directly under the equator, the climate is far from being excessively hot; this seems chiefly caused by the singularly low temperature of the surrounding water, brought here by the great southern Polar current. Excepting during one short season, very little rain falls, and even then it is irregular; but the clouds generally hang low. Hence, whilst the lower parts of the islands are very sterile, the upper parts, at a height of a thousand feet and upwards, possess a damp climate and a tolerably luxuriant vegetation. This is especially the case on the windward sides of the islands, which first receive and condense the moisture from the atmosphere.

In the morning (17th) we landed on Chatham Island, which, like the others, rises with a tame and rounded outline, broken here and there by scattered hillocks, the remains of former craters. Nothing could be less inviting than the first appearance. A broken field of black basaltic lava, thrown into the most rugged waves, and crossed by great fissures, is everywhere covered by stunted, sun-burnt brushwood, which shows little signs of life. The dry and parched surface, being heated by the noon-day sun, gave to the air a close and sultry feeling, like that from a stove: we fancied even that the bushes smelt unpleasantly. Although I diligently tried to collect as many plants as possible, I succeeded in getting very few; and such wretched-looking little weeds would have better become an arctic than an equatorial Flora. The brushwood appears, from a short distance, as leafless as our trees during winter; and it was some time before I discovered that not only almost every plant was now in full leaf, but that the greater number were in flower. The commonest bush is one of the Euphorbiaceæ: an acacia and a great odd-looking cactus are the only trees which afford any shade. After the season of heavy rains, the islands are said to appear for a short time partially green. The volcanic island of Fernando Noronha, placed in many respects under nearly similar conditions, is the only other country where I have seen a vegetation at all like this of the Galapagos Islands.

The Beagle sailed round Chatham Island, and anchored in several bays. One night I slept on shore on a part of the island, where black truncated cones were extraordinarily numerous: from one small eminence I counted sixty of them, all surmounted by craters more or less perfect. The greater number consisted merely of a ring of red scoriæ or slags, cemented together: and their height above the plain of lava was not more than from fifty to a hundred feet; none had been very lately active. The entire surface of this part of the island seems to have been permeated, like a sieve, by the subterranean vapours: here and there the lava, whilst soft, has been blown into great bubbles; and in other parts, the tops of caverns similarly formed have fallen in, leaving circular pits with steep sides. From the regular form of the many craters, they gave to the country an artificial appearance, which vividly reminded me of those parts of Staffordshire, where the great iron-foundries are most numerous. The day was glowing hot, and the scrambling over the rough surface and through the intricate thickets, was very fatiguing; but I was well repaid by the strange Cyclopean scene. As I was walking along I met two large tortoises, each of which must have weighed at least two hundred pounds: one was eating a piece of cactus, and as I approached, it stared at me and slowly walked away; the other gave a deep hiss, and drew in its head. These huge reptiles, surrounded by the black lava, the leafless shrubs, and large cacti, seemed to my fancy like some antediluvian animals. The few dull-coloured birds cared no more for me than they did for the great tortoises.

23rd.—The Beagle proceeded to Charles Island. This archipelago has long been frequented, first by the bucaniers, and latterly by whalers, but it is only within the last six years, that a small colony has been established here. The inhabitants are between two and three hundred in number; they are nearly all people of colour, who have been banished for political crimes from the Republic of the Equator, of which Quito is the capital. The settlement is placed about four and a half miles inland, and at a height probably of a thousand feet. In the first part of the road we passed through leafless thickets, as in Chatham Island. Higher up, the woods gradually became greener; and as soon as we crossed the ridge of the island, we were cooled by a fine southerly breeze, and our sight refreshed by a green and thriving vegetation. In this upper region coarse grasses and ferns abound; but there are no tree-ferns: I saw nowhere any member of the palm family, which is the more singular, as 360 miles northward, Cocos Island takes its name from the number of cocoa-nuts. The houses are irregularly scattered over a flat space of ground, which is cultivated with sweet potatoes and bananas. It will not easily be imagined how pleasant the sight of black mud was to us, after having been so long accustomed to the parched soil of Peru and northern Chile. The inhabitants, although complaining of poverty, obtain, without much trouble, the means of subsistence. In the woods there are many wild pigs and goats; but the staple article of animal food is supplied by the tortoises. Their numbers have of course been greatly reduced in this island, but the people yet count on two days' hunting giving them food for the rest of the week. It is said that formerly single vessels have taken away as many as seven hundred, and that the ship's company of a frigate some years since brought down in one day two hundred tortoises to the beach.

September 29th.—We doubled the south-west extremity of Albemarle Island, and the next day were nearly becalmed between it and Narborough Island. Both are covered with immense deluges of black naked lava, which have flowed either over the rims of the great caldrons, like pitch over the rim of a pot in which it has been boiled, or have burst forth from smaller orifices on the flanks; in their descent they have spread over miles of the sea-coast. On both of these islands, eruptions are known to have taken place; and in Albemarle, we saw a small jet of smoke curling from the summit of one of the great craters. In the evening we anchored in Bank's Cove, in Albemarle Island. The next morning I went out walking. To the south of the broken tuff-crater, in which the Beagle was anchored, there was another beautifully symmetrical one of an elliptic form; its longer axis was a little less than a mile, and its depth about 500 feet. At its bottom there was a shallow lake, in the middle of which a tiny crater formed an islet. The day was overpoweringly hot, and the lake looked clear and blue: I hurried down the cindery slope, and, choked with dust, eagerly tasted the water—but, to my sorrow, I found it salt as brine.

The rocks on the coast abounded with great black lizards, between three and four feet long; and on the hills, an ugly yellowish-brown species was equally common. We saw many of this latter kind, some clumsily running out of the way, and others shuffling into their burrows. I shall presently describe in more detail the habits of both these reptiles. The whole of this northern part of Albemarle Island is miserably sterile.

October 8th.—We arrived at James Island: this island, as well as Charles Island, were long since thus named after our kings of the Stuart line. Mr. Bynoe, myself, and our servants were left here for a week, with provisions and a tent, whilst the Beagle went for water. We found here a party of Spaniards, who had been sent from Charles Island to dry fish, and to salt tortoise-meat. About six miles inland, and at the height of nearly 2000 feet, a hovel had been built in which two men lived, who were employed in catching tortoises, whilst the others were fishing on the coast. I paid this party two visits, and slept there one night. As in the other islands, the lower region was covered by nearly leafless bushes, but the trees were here of a larger growth than elsewhere, several being two feet and some even two feet nine inches in diameter. The upper region being kept damp by the clouds, supports a green and flourishing vegetation. So damp was the ground, that there were large beds of a coarse cyperus, in which great numbers of a very small water-rail lived and bred. While staying in this upper region, we lived entirely upon tortoise-meat: the breast-plate roasted (as the Gauchos do *carne con cuero),* with the flesh on it, is very good; and the young tortoises make excellent soup; but otherwise the meat to my taste is indifferent.

One day we accompanied a party of the Spaniards in their whale-boat to a salina, or lake from which salt is procured. After landing, we had a very rough walk over a rugged field of recent lava, which has almost surrounded a tuff-crater, at the bottom of which the salt-lake lies. The water is only three or four inches deep, and rests on a layer of beautifully crystallized, white salt. The lake is quite circular, and is fringed with a border of bright green succulent plants; the almost precipitous walls of the crater are clothed with wood, so that the scene was altogether both picturesque and curious. A few years since, the sailors belonging to a sealing-vessel murdered their captain in this quiet spot; and we saw his skull lying among the bushes.

During the greater part of our stay of a week, the sky was cloudless, and if the trade-wind failed for an hour, the heat became very oppressive. On two days, the thermometer within the tent stood for some hours at 93°; but in the open air, in the wind and sun, at only 85°. The sand was extremely hot; the thermometer placed in some of a brown colour immediately rose to 137°, and how much above that it would have risen, I do not know, for it was not graduated any higher. The black sand felt much hotter, so that even in thick boots it was quite disagreeable to walk over it.

THE NATURAL HISTORY of these islands is eminently curious, and well deserves attention. Most of the organic productions are aboriginal creations, found nowhere else; there is even a difference between the inhabitants of the different islands; yet all show a marked relationship with those of America, though separated from that continent by an open space of ocean, between 500 and 600 miles in width. The archipelago is a little world within itself, or rather a satellite attached to America, whence it has derived a few stray colonists, and has received the general character of its indigenous productions. Considering the small size of the islands, we feel the more astonished at the number of their aboriginal beings, and at their confined range. Seeing every height crowned with its crater, and the boundaries of most of the lava-streams still distinct, we are led to believe that within a period geologically recent the unbroken ocean was here spread out. Hence, both in space and time, we seem to be brought somewhat near to that great fact—that mystery of mysteries—the first appearance of new beings on this earth.

Of terrestrial mammals, there is only one which must be considered as indigenous, namely, a mouse (Mus Galapagoensis), and this is confined, as far as I could ascertain, to Chatham Island, the most easterly island of the group. It belongs, as I am informed by Mr. Waterhouse, to a division of the family of mice characteristic of America. At James Island, there is a rat sufficiently distinct from the common kind to have been named and described by Mr. Waterhouse; but as it belongs to the old-world division of the family, and as this island has been frequented by ships for the last hundred and fifty years, I can hardly doubt that this rat is merely a variety produced by the new and peculiar climate, food, and soil, to which it has been subjected. Although no one has a right to speculate without distinct facts, yet even with respect to the Chatham Island mouse, it should be borne in mind, that it may possibly be an American species imported here; for I have seen, in a most unfrequented part of the Pampas, a native mouse living in the roof of a newly built hovel, and therefore its transportation in a vessel is not improbable: analogous facts have been observed by Dr. Richardson in North America.

Of land-birds I obtained twenty-six kinds, all peculiar to the group and found nowhere else, with the exception of one lark-like finch from North America (Dolichonyx oryzivorus), which ranges on that continent as far north as 54°, and generally frequents marshes. The other twenty-five birds consist, firstly, of a hawk, curiously intermediate in structure between a buzzard and the American group of carrion-feeding Polybori; and with these latter birds it agrees most closely in every habit and even tone of voice. Secondly, there are two owls, representing the short-eared and white barn-owls of Europe. Thirdly, a wren, three tyrant-flycatchers (two of them species of Pyrocephalus, one or both of which would be ranked by some ornithologists as only varieties), and a dove—all analogous to, but distinct from, American species. Fourthly, a swallow, which though differing from the Progne purpurea of both Americas, only in being rather duller colored, smaller, and slenderer, is considered by Mr. Gould as specifically distinct. Fifthly, there are three species of mocking thrush—a form highly characteristic of America. The remaining land-birds form a most singular group of finches, related to each other in the structure of their beaks, short tails, form of body and plumage: there are thirteen species, which Mr. Gould has divided into four subgroups. All these species are peculiar to this archipelago; and so is the whole group, with the exception of one species of the sub-group Cactornis, lately brought from Bow Island, in the Low Archipelago. Of Cactornis, the two species may be often seen climbing about the flowers of the great cactus-trees; but all the other species of this group of finches, mingled together in flocks, feed on the dry and sterile ground of the lower districts. The males of all, or certainly of the greater number, are jet black; and the females (with perhaps one or two exceptions) are brown. The most curious fact is the perfect gradation in the size of the beaks in the different species of Geospiza, from one as large as that of a hawfinch to that of a chaffinch, and (if Mr. Gould is right in including his sub-group, Certhidea, in the main group) even to that of a warbler. The largest beak in the genus Geospiza is shown in Fig. 1, and the smallest in Fig. 3; but instead of there being only one intermediate species, with a beak of the size shown in Fig. 2, there are no less than six species with insensibly graduated beaks. The beak of the sub-group Certhidea, is shown in Fig. 4. The beak of Cactornis is somewhat like that of a starling; and that of the fourth sub-group, Camarhynchus, is slightly parrot-shaped. Seeing this gradation and diversity of structure in one small, intimately related group of birds, one might really fancy that from an original paucity of birds in this archipelago, one species had been taken

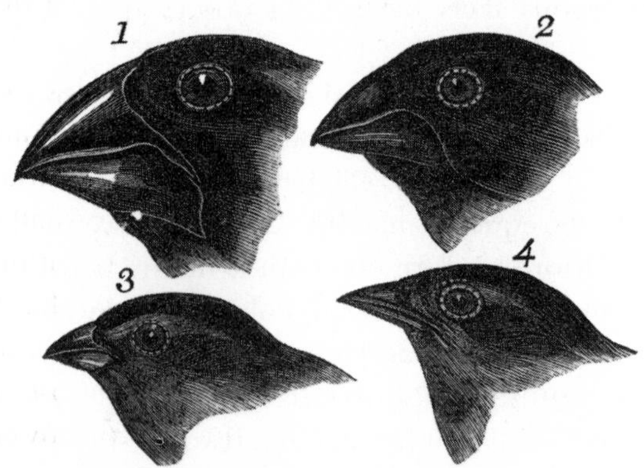

1. Geospiza magnirostris. *2. Geospiza fortis.*
3. Geospiza parvula. *4. Certhidea olivasea.*

and modified for different ends. In a like manner it might be fancied that a bird originally a buzzard, had been induced here to undertake the office of the carrion-feeding Polybori of the American continent.

Of waders and water-birds I was able to get only eleven kinds, and of these only three (including a rail confined to the damp summits of the islands) are new species. Considering the wandering habits of the gulls, I was surprised to find that the species inhabiting these islands is peculiar, but allied to one from the southern parts of South America. The far greater peculiarity of the land-birds, namely, twenty-five out of twenty-six, being new species, or at least new races, compared with the waders and web-footed birds, is in accordance with the greater range which these latter orders have in all parts of the world. We shall hereafter see this law of aquatic forms, whether marine or fresh-water, being less peculiar at any given point of the earth's surface than the terrestrial forms of the same classes, strikingly illustrated in the shells, and in a lesser degree in the insects of this archipelago.

Two of the waders are rather smaller than the same species brought from other places: the swallow is also smaller, though it is doubtful whether or not it is distinct from its analogue. The two owls, the two tyrant-flycatchers (Pyrocephalus) and the dove, are also smaller than the analogous but distinct species, to which they are most nearly related; on the other hand, the gull is rather larger. The two owls, the swallow, all three species of mocking-thrush, the dove in its separate colours though not in its whole plumage, the Totanus, and the gull, are likewise duskier coloured than their analogous species; and in the case of the mocking-thrush and Totanus, than any other species of the two genera. With the exception of a wren with a fine yellow breast, and of a tyrant-flycatcher with a scarlet tuft and breast, none of the birds are brilliantly coloured, as might have been expected in an equatorial district. Hence it would appear probable, that the same causes which here make the immigrants of some peculiar species smaller, make most of the peculiar Galapageian species also smaller, as well as very generally more dusky coloured. All the plants have a wretched, weedy appearance, and I did not see one beautiful flower. The insects, again, are small-sized and dull-coloured, and, as Mr. Waterhouse informs me, there is nothing in their general appearance which would have led him to imagine that they had come from under the equator.[151] The birds, plants, and insects have a desert character, and are not more brilliantly coloured than those from southern Patagonia; we may, therefore, conclude that the usual gaudy colouring of the inter-tropical productions, is not related either to the heat or light of those zones, but to some other cause, perhaps to the conditions of existence being generally favourable to life.

We will now turn to the order of reptiles, which gives the most striking character to the zoology of these islands. The species are not numerous, but the numbers of individuals of each species are extraordinarily great. There is one small lizard belonging to a South American genus, and two species (and probably more) of the Amblyrhynchus—a genus confined to the Galapagos Islands. There is one snake which is numerous; it is identical, as I am informed by M. Bibron, with the Psammophis Temminckii from Chile.[152] Of sea-turtle I believe there are more than one species; and of tortoises there are, as we shall presently show, two or three species or races. Of toads and frogs there are none: I was surprised at this, considering how well suited for them the temperate and damp upper woods appeared to be. It recalled to my mind the remark made by Bory St. Vincent,[153] namely, that none of this family are found on any of the volcanic islands in the great oceans. As far as I can ascertain from various works, this seems to hold good throughout the Pacific, and even in the large islands of the Sandwich archipelago. Mauritius offers an apparent exception, where I saw the Rana Mascariensis in abundance: this frog is said now to inhabit the Seychelles, Madagascar, and Bourbon; but on the other hand, Du Bois, in his voyage in 1669, states that there were no reptiles in Bourbon except tortoises; and the Officier du Roi asserts that before 1768 it had been attempted, without success, to introduce frogs into Mauritius—I presume for the purpose of eating: hence it may be well doubted whether this frog is an aboriginal of these islands. The absence of the frog family in the oceanic islands is the more remarkable, when contrasted with the case of lizards, which swarm on most of the smallest islands. May this difference not be caused, by the greater facility with which the eggs of lizards, protected by calcareous shells, might be transported through salt-water, than could the slimy spawn of frogs?

I will first describe the habits of the tortoise (Testudo nigra, formerly called Indica), which has been so frequently alluded to. These animals are found, I believe, on all the islands of the archipelago; certainly on the greater number. They frequent in preference the high damp parts, but they likewise live in the lower and arid districts. I have already shown, from the numbers which have been caught in a single day, how very numerous they must be. Some grow to an immense size: Mr. Lawson, an Englishman, and vice-governor of the colony, told us that he had seen several so large, that it required six or eight men to lift them from the ground; and that some had afforded as much as two hundred pounds of meat. The old males are the largest, the females rarely growing to so great a size: the male can readily be distinguished from the female by the greater length of its tail. The tortoises which live on those islands where there is no water, or in the lower and arid parts of the others, feed chiefly on the succulent cactus. Those which frequent the higher and damp regions, eat the leaves of various trees, a kind of berry (called guayavita) which is acid and austere, and likewise a pale green filamentous lichen (Usnera plicata), that hangs from the boughs of the trees.

The tortoise is very fond of water, drinking large quantities, and wallowing in the mud. The larger islands alone possess springs, and these are always situated towards the central parts, and at a considerable height. The tortoises, therefore, which frequent the lower districts, when thirsty, are obliged to travel from a long distance. Hence broad and well-beaten paths branch off in every direction from the wells down to the sea-coast; and the Spaniards by following them up, first discovered the watering-places. When I landed at Chatham Island, I could not imagine what animal travelled so methodically along well-chosen tracks. Near the springs it was a curious spectacle to behold many of these huge creatures, one set eagerly travelling onwards with outstretched necks, and another set returning, after having drunk their fill. When the tortoise arrives at the spring, quite regardless of any spectator, he buries his head in the water above his eyes, and greedily swallows great mouthfuls, at the rate of about ten in a minute. The inhabitants say each animal stays three or four days in the neighbourhood of the water, and then returns to the lower country; but they differed respecting the frequency of these visits. The animal probably regulates them according to the nature of the food on which it has lived. It is, however, certain, that tortoises can subsist even on these islands where there is no other water than what falls during a few rainy days in the year.

I believe it is well ascertained, that the bladder of the frog acts as a reservoir for the moisture necessary to its existence: such seems to be the case with the tortoise. For some time after a visit to the springs, their urinary bladders are distended with fluid, which is said gradually to decrease in volume, and to become less pure. The inhabitants, when walking in the lower district, and overcome with thirst, often take advantage of this circumstance, and drink the contents of the bladder if full: in one I saw killed, the fluid was quite limpid, and had only a very slightly bitter taste. The inhabitants, however, always first drink the water in the pericardium, which is described as being best.

The tortoises, when purposely moving towards any point, travel by night and day, and arrive at their journey's end much sooner than would be expected. The inhabitants, from observing marked individuals, consider that they travel a distance of about eight miles in two or three days. One large tortoise, which I watched, walked at the rate of sixty yards in ten minutes, that is 360 yards in the hour, or four miles a day,—allowing a little time for it to eat on the road. During the breeding season, when the male and female are together, the male utters a hoarse roar or bellowing, which, it is said, can be heard at the distance of more than a hundred yards. The female never uses her voice, and the male only at these times; so that when the people hear this noise, they know that the two are together. They were at this time (October) laying their eggs. The female, where the soil is sandy, deposits them together, and covers them up with sand; but where the ground is rocky she drops them indiscriminately in any hole: Mr. Bynoe found seven placed in a fissure. The egg is white and spherical; one which I measured was seven inches and three-eighths in circumference, and therefore larger than a hen's egg. The young tortoises, as soon as they are hatched, fall a prey in great numbers to the carrion-feeding buzzard. The old ones seem generally to die from accidents, as from falling down precipices: at least, several of the inhabitants told me, that they never found one dead without some evident cause.

The inhabitants believe that these animals are absolutely deaf; certainly they do not overhear a person walking close behind them. I was always amused when overtaking one of these great monsters, as it was quietly pacing along, to see how suddenly, the instant I passed, it would draw in its head and legs, and uttering a deep hiss fall to the ground with a heavy sound, as if struck dead. I frequently got on their backs, and then giving a few raps on the hinder part of their shells, they would rise up and walk away;—but I found it very difficult to keep my balance. The flesh of this animal is largely employed, both fresh and salted; and a beautifully clear oil is prepared from the fat. When a tortoise is caught, the man makes a slit in the skin near its tail, so as to see inside its body, whether the fat under the dorsal plate is thick. If it is not, the animal is liberated and it is said to recover soon from this strange operation. In order to secure the tortoise, it is not sufficient to turn them like turtle, for they are often able to get on their legs again.

There can be little doubt that this tortoise is an aboriginal inhabitant of the Galapagoes; for it is found on all, or nearly all, the islands, even on some of the smaller ones where there is no water; had it been an imported species, this would hardly have been the case in a group which has been so little frequented. Moreover, the old Bucaniers found this tortoise in greater numbers even than at present: Wood and Rogers also, in 1708, say that it is the opinion of the Spaniards, that it is found nowhere else in this quarter of the world. It is now widely distributed; but it may be questioned whether it is in any other place an aboriginal. The bones of a tortoise at Mauritius, associated with those of the extinct Dodo, have generally been considered as belonging to this tortoise; if this had been so, undoubtedly it must have been there indigenous; but M. Bibron informs me that he believes that it was distinct, as the species now living there certainly is.

The Amblyrhynchus, a remarkable genus of lizards, is confined to this

Amblyrhynchus cristatus
a. Tooth of, natural size, and likewise magnified.

archipelago; there are two species, resembling each other in general form, one being terrestrial and the other aquatic. This latter species (A. cristatus) was first characterized by Mr. Bell, who well foresaw, from its short, broad head, and strong claws of equal length, that its habits of life would turn out very peculiar, and different from those of its nearest ally, the Iguana. It is extremely common on all the islands throughout the group, and lives exclusively on the rocky sea-beaches, being never found, at least I never saw one, even ten yards in-shore. It is a hideous-looking creature, of a dirty black colour, stupid, and sluggish in its movements. The usual length of a full-grown one is about a yard, but there are some even four feet long; a large one weighed twenty pounds: on the island of Albemarle they seem to grow to a greater size than elsewhere. Their tails are flattened sideways, and all four feet partially webbed. They are occasionally seen some hundred yards from the shore, swimming about; and Captain Collnett, in his Voyage says, "They go to sea in herds a-fishing, and sun themselves on the rocks; and may be called alligators in miniature." It must not, however, be supposed that they live on fish. When in the water this lizard swims with perfect ease and quickness, by a serpentine movement of its body and flattened tail—the legs being motionless and closely collapsed on its sides. A seaman on board sank one, with a heavy weight attached to it, thinking thus to kill it directly; but when, an hour afterwards, he drew up the line, it was quite active. Their limbs and strong claws are admirably adapted for crawling over the rugged and fissured masses of lava, which everywhere form the coast. In such situations, a group of six or seven of these hideous reptiles may oftentimes be seen on the black rocks, a few feet above the surf, basking in the sun with outstretched legs.

I opened the stomachs of several, and found them largely distended with minced sea-weed (Ulvæ), which grows in thin foliaceous expansions of a bright green or a dull red colour. I do not recollect having observed this sea-weed in any quantity on the tidal rocks; and I have reason to believe it grows at the bottom of the sea, at some little distance from the coast. If such be the case, the object of these animals occasionally going out to sea is explained. The stomach contained nothing but the sea-weed. Mr. Baynoe, however, found a piece of crab in one; but this might have got in accidentally, in the same manner as I have seen a caterpillar, in the midst of some lichen, in the paunch of a tortoise. The intestines were large, as in other herbivorous animals. The nature of this lizard's food, as well as the structure of its tail and feet, and the fact of its having been seen voluntarily swimming out at sea, absolutely prove its aquatic habits; yet there is in this respect one strange anomaly, namely, that when frightened it will not enter the water. Hence it is easy to drive these lizards down to any little point overhanging the sea, where they will sooner allow a person to catch hold of their tails than jump into the water. They do not seem to have any notion of biting; but when much frightened they squirt a drop of fluid from each nostril. I threw one several times as far as I could, into a deep pool left by the retiring tide; but it invariably returned in a direct line to the spot where I stood. It swam near the bottom, with a very graceful and rapid movement, and occasionally aided itself over the uneven ground with its feet. As soon as it arrived near the edge, but still being under water, it tried to conceal itself in the tufts of sea-weed, or it entered some crevice. As soon as it thought the danger was past, it crawled out on the dry rocks, and shuffled away as quickly as it could. I several times caught this same lizard, by driving it down to a point, and though possessed of such perfect powers of diving and swimming, nothing would induce it to enter the water; and as often as I threw it in, it returned in the manner above described. Perhaps this singular piece of apparent stupidity may be accounted for by the circumstance, that this reptile has no enemy whatever on shore, whereas at sea it must often fall a prey to the numerous sharks. Hence, probably, urged by a fixed and hereditary instinct that the shore is its place of safety, whatever the emergency may be, it there takes refuge.

During our visit (in October), I saw extremely few small individuals of this species, and none I should think under a year old. From this circumstance it seems probable that the breeding season had not then commenced. I asked several of the inhabitants if they knew where it laid its eggs: they said that they knew nothing of its propagation, although well acquainted with the eggs of the land kind—a fact, considering how very common this lizard is, not a little extraordinary.

We will now turn to the terrestrial species (A. Demarlii), with a round tail, and toes without webs. This lizard, instead of being found like the other on all the islands, is confined to the central part of the archipelago, namely to Albemarle, James, Barrington, and Indefatigable islands. To the southward, in Charles, Hood, and Chatham islands, and to the northward, in Towers, Bindloes, and Abingdon, I neither saw nor heard of any. It would appear as if it had been created in the centre of the archipelago, and thence had been dispersed only to a certain distance. Some of these lizards inhabit the high and damp parts of the islands, but they are much more numerous in the lower and sterile districts near the coast. I cannot give a more forcible proof of their numbers, than by stating that when we were left at James Island, we could not for some time find a spot free from their burrows on which to pitch our single tent. Like their brothers the sea-kind, they are ugly animals, of a yellowish orange beneath, and of a brownish red colour above: from their low facial angle they have a singularly stupid appearance. They are, perhaps, of a rather less size than the marine species; but several of them weighed between ten and fifteen pounds. In their movements they are lazy and half torpid. When not frightened, they slowly crawl along with their tails and bellies dragging on the ground. They often stop, and doze for a minute or two, with closed eyes and hind legs spread out on the parched soil.

They inhabit burrows, which they sometimes make between fragments of lava, but more generally on level patches of the soft sandstone-like tuff. The holes do not appear to be very deep, and they enter the ground at a small angle; so that when walking over these lizard-warrens, the soil is constantly giving way, much to the annoyance of the tired walker. This animal, when making its burrow, works alternately the opposite sides of its body. One front leg for a short time scratches up the soil, and throws it towards the hind foot, which is well placed so as to heave it beyond the mouth of the hole. That side of the body being tired, the other takes up the task, and so on alternately. I watched one for a long time, till half its body was buried; I then walked up and pulled it by the tail; at this it was greatly astonished, and soon shuffled up to see what was the matter; and then stared me in the face, as much as to say, "What made you pull my tail?"

They feed by day, and do not wander far from their burrows; if frightened, they rush to them with a most awkward gait. Except when running down hill, they cannot move very fast, apparently from the lateral position of their legs. They are not at all timorous: when attentively watching any one, they curl their tails, and, raising themselves on their front legs, nod their heads vertically, with a quick movement, and try to look very fierce; but in reality they are not at all so: if one just stamps on the ground, down go their tails, and off they shuffle as quickly as they can. I have frequently observed small fly-eating lizards, when watching anything, nod their heads in precisely the same manner; but I do not at all know for what purpose. If this Amblyrhynchus is held and plagued with a stick, it will bite it very severely; but I caught many by the tail, and they never tried to bite me. If two are placed on the ground and held together, they will fight, and bite each other till blood is drawn.

The individuals, and they are the greater number, which inhabit the lower country, can scarcely taste a drop of water throughout the year; but they consume much of the succulent cactus, the branches of which are occasionally broken off by the wind. I several times threw a piece to two or three of them when together; and it was amusing enough to see them trying to seize and carry it away in their mouths, like so many hungry dogs with a bone. They eat very deliberately, but do not chew their food. The little birds are aware how harmless these creatures are: I have seen one of the thick-billed finches picking at one end of a piece of cactus (which is much relished by all the animals of the lower region), whilst a lizard was eating at the other end; and afterwards the little bird with the utmost indifference hopped on the back of the reptile.

I opened the stomachs of several, and found them full of vegetable fibres and leaves of different trees, especially of an acacia. In the upper region they live chiefly on the acid and astringent berries of the guayavita, under which trees I have seen these lizards and the huge tortoises feeding together. To obtain the acacia-leaves they crawl up the low stunted trees; and it is not uncommon to see a pair quietly browsing, whilst seated on a branch several feet above the ground. These lizards, when cooked, yield a white meat, which is liked by those whose stomachs soar above all prejudices.

Humboldt has remarked that in intertropical South America, all lizards which inhabit dry regions are esteemed delicacies for the table. The inhabitants state that those which inhabit the upper damp parts drink water, but that the others do not, like the tortoises, travel up for it from the lower sterile country. At the time of our visit, the females had within their bodies numerous, large, elongated eggs, which they lay in their burrows: the inhabitants seek them for food.

These two species of Amblyrhynchus agree, as I have already stated, in their general structure, and in many of their habits. Neither have that rapid movement, so characteristic of the genera Lacerta and Iguana. They are both herbivorous, although the kind of vegetation on which they feed is so very different. Mr. Bell has given the name to the genus from the shortness of the snout: indeed, the form of the mouth may almost be compared to that of the tortoise: one is led to suppose that this is an adaptation to their herbivorous appetites. It is very interesting thus to find a well-characterized genus, having its marine and terrestrial species, belonging to so confined a portion of the world. The aquatic species is by far the most remarkable, because it is the only existing lizard which lives on marine vegetable productions. As I at first observed, these islands are not so remarkable for the number of the species of reptiles, as for that of the individuals; when we remember the well-beaten paths made by the thousands of huge tortoises—the many turtles—the great warrens of the terrestrial Amblyrhynchus—and the groups of the marine species basking on the coast-rocks of every island—we must admit that there is no other quarter of the world where this Order replaces the herbivorous mammalia in so extraordinary a manner. The geologist on hearing this will probably refer back in his mind to the Secondary epochs, when lizards, some herbivorous, some carnivorous, and of dimensions comparable only with our existing whales, swarmed on the land and in the sea. It is, therefore, worthy of his observation, that this archipelago, instead of possessing a humid climate and rank vegetation, cannot be considered otherwise than extremely arid, and, for an equatorial region, remarkably temperate.

To finish with the zoology: the fifteen kinds of sea-fish which I procured here are all new species; they belong to twelve genera, all widely distributed, with the exception of Prionotus, of which the four previously known species live on the eastern side of America. Of land-shells I collected sixteen kinds (and two marked varieties), of which, with the exception of one Helix found at Tahiti, all are peculiar to this archipelago: a single fresh-water shell (Paludina) is common to Tahiti and Van Diemen's Land. Mr. Cuming, before our voyage, procured here ninety species of sea-shells, and this does not include several species not yet specifically examined, of Trochus, Turbo, Monodonta, and Nassa. He has been kind enough to give me the following interesting results: Of the ninety shells, no less than forty-seven are unknown elsewhere—a wonderful fact, considering how widely distributed sea-shells generally are. Of the forty-three shells found in other parts of the world, twenty-five inhabit the western coast of America, and of these eight are distinguishable as varieties; the remaining eighteen (including one variety) were found by Mr. Cuming in the Low Archipelago, and some of them also at the Philippines. This fact of shells from islands in the central parts of the Pacific occurring here, deserves notice, for not one single sea-shell is known to be common to the islands of that ocean and to the west coast of America. The space of open sea running north and south off the west coast, separates two quite distinct conchological provinces; but at the Galapagos Archipelago we have a halting-place, where many new forms have been created, and whither these two great conchological provinces have each sent up several colonists. The American province has also sent here representative species; for there is a Galapageian species of Monoceros, a genus only found on the west coast of America; and there are Galapageian species of Fissurella and Cancellaria, genera common on the west coast, but not found (as I am informed by Mr. Cuming) in the central islands of the Pacific. On the other hand, there are Galapageian species of Oniscia and Stylifer, genera common to the West Indies and to the Chinese and Indian seas, but not found either on the west coast of America or in the central Pacific. I may here add, that after the comparison by Messrs. Cuming and Hinds of about 2000 shells from the eastern and western coasts of America, only one single shell was found in common, namely, the Purpura patula, which inhabits the West Indies, the coast of Panama, and the Galapagos. We have, therefore, in this quarter of the world, three great conchological sea-provinces, quite distinct, though surprisingly near each other, being separated by long north and south spaces either of land or of open sea.

I took great pains in collecting the insects, but excepting Tierra del Fuego, I never saw in this respect so poor a country. Even in the upper and damp region I procured very few, excepting some minute Diptera and Hymenoptera, mostly of common mundane forms. As before remarked, the insects, for a tropical region, are of very small size and dull colours. Of beetles I collected twenty-five species (excluding a Dermestes and Corynetes imported, wherever a ship touches); of these, two belong to the Harpalidæ, two to the Hydrophilidæ, nine to three families of the Heteromera, and the remaining twelve to as many different families. This circumstance of insects (and I may add plants), where few in number, belonging to many different families, is, I believe, very general. Mr. Waterhouse, who has published[154] an account of the insects of this archipelago, and to whom I am indebted for the above details, informs me that there are several new genera: and that of the genera not new, one or two are American, and the rest of mundane distribution. With the exception of a wood-feeding Apate, and of one or probably two water-beetles from the American continent, all the species appear to be new.

The botany of this group is fully as interesting as the zoology. Dr. J. Hooker will soon publish in the "Linnean Transactions" a full account of the Flora, and I am much indebted to him for the following details. Of flowering plants there are, as far as at present is known, 185 species, and 40 cryptogamic species, making altogether 225; of this number I was fortunate enough to bring home 193. Of the flowering plants, 100 are new species, and are probably confined to this archipelago. Dr. Hooker conceives that, of the plants not so confined, at least 10 species found near the cultivated ground at Charles Island, have been imported. It is, I think, surprising that more American species have not been introduced naturally, considering that the distance is only between 500 and 600 miles from the continent; and that (according to Collnet, p. 58) drift-wood, bamboos, canes, and the nuts of a palm, are often washed on the south-eastern shores. The proportion of 100 flowering plants out of 185 (or 175 excluding the imported weeds) being new, is sufficient, I conceive, to make the Galapagos Archipelago a distinct botanical province; but this Flora is not nearly so peculiar as that of St. Helena, nor, as I am informed by Dr. Hooker, of Juan Fernandez. The peculiarity of the Galapageian Flora is best shown in certain families;—thus there are 21 species of Compositæ, of which 20 are peculiar to this archipelago; these belong to twelve genera, and of these genera no less than ten are confined to the archipelago! Dr. Hooker informs me that the Flora has an undoubtedly Western American character; nor can he detect in it any affinity with that of the Pacific. If, therefore, we except the eighteen marine, the one fresh-water, and one land-shell, which have apparently come here as colonists from the central islands of the Pacific, and likewise the one distinct Pacific species of the Galapageian group of finches, we see that this archipelago, though standing in the Pacific Ocean, is zoologically part of America.

If this character were owing merely to immigrants from America, there would be little remarkable in it; but we see that a vast majority of all the land animals, and that more than half of the flowering plants, are aboriginal productions. It was most striking to be surrounded by new birds, new reptiles, new shells, new insects, new plants, and yet by innumerable trifling details of structure, and even by the tones of voice and plumage of the birds, to have the temperate plains of Patagonia, or rather the hot dry deserts of Northern Chile, vividly brought before my eyes. Why, on these small points of land, which within a late geological period must have been covered by the ocean, which are formed by basaltic lava, and therefore differ in geological character from the American continent, and which are placed under a peculiar climate,—why were their aboriginal inhabitants, associated, I may add, in different proportions both in kind and number from those on the continent, and therefore acting on each other in a different manner—why were they created on American types of organization? It is probable that the islands of the Cape de Verd group resemble, in all their physical conditions, far more closely the Galapagos Islands, than these latter physically resemble the coast of America, yet the aboriginal inhabitants of the two groups are totally unlike; those of the Cape de Verd Islands bearing the impress of Africa, as the inhabitants of the Galapagos Archipelago are stamped with that of America.

I HAVE NOT AS YET noticed by far the most remarkable feature in the natural history of this archipelago; it is, that the different islands to a considerable extent are inhabited by a different set of beings. My attention was first called to this fact by the Vice-Governor, Mr. Lawson, declaring that the tortoises differed from the different islands, and that he could with certainty tell from which island any one was brought. I did not for some time pay sufficient attention to this statement, and I had already partially mingled together the collections from two of the islands. I never dreamed that islands, about 50 or 60 miles apart, and most of them in sight of each other, formed of precisely the same rocks, placed under a quite similar climate, rising to a nearly equal height, would have been differently tenanted; but we shall soon see that this is the case. It is the fate of most voyagers, no sooner to discover what is most interesting in any locality, than they are hurried from it; but I ought, perhaps, to be thankful that I obtained sufficient materials to establish this most remarkable fact in the distribution of organic beings.

The inhabitants, as I have said, state that they can distinguish the tortoises from the different islands; and that they differ not only in size, but in other characters. Captain Porter has described[155] those from Charles and from the nearest island to it, namely, Hood Island, as having their shells in front thick and turned up like a Spanish saddle, whilst the tortoises from James Island are rounder, blacker, and have a better taste when cooked. M. Bibron, moreover, informs me that he has seen what he considers two distinct species of tortoise from the Galapagos, but he does not know from which islands. The specimens that I brought from three islands were young ones: and probably owing to this cause neither Mr. Gray nor myself could find in them any specific differences. I have remarked that the marine Amblyrhynchus was larger at Albemarle Island than elsewhere; and M. Bibron informs me that he has seen two distinct aquatic species of this genus; so that the different islands probably have their representative species or races of the Amblyrhynchus, as well as of the tortoise. My attention was first thoroughly aroused, by comparing together the numerous specimens, shot by myself and several other parties on board, of the mocking-thrushes, when, to my astonishment, I discovered that all those from Charles Island belonged to one species (Mimus trifasciatus); all from Albemarle Island to M. parvulus; and all from James and Chatham Islands (between which two other islands are situated, as connecting links) belonged to M. melanotis. These two latter species are closely allied, and would by some ornithologists be considered as only well-marked races or varieties; but the Mimus trifasciatus is very distinct. Unfortunately most of the specimens of the finch tribe were mingled together; but I have strong reasons to suspect that some of the species of the sub-group Geospiza are confined to separate islands. If the different islands have their representatives of Geospiza, it may help to explain the singularly large number of the species of this sub-group in this one small archipelago, and as a probable consequence of their numbers, the perfectly graduated series in the size of their beaks. Two species of the sub-group Cactornis, and two of the

Camarhynchus, were procured in the archipelago; and of the numerous specimens of these two sub-groups shot by four collectors at James Island, all were found to belong to one species of each; whereas the numerous specimens shot either on Chatham or Charles Island (for the two sets were mingled together) all belonged to the two other species: hence we may feel almost sure that these islands possess their respective species of these two sub-groups. In land-shells this law of distribution does not appear to hold good. In my very small collection of insects, Mr. Waterhouse remarks, that of those which were ticketed with their locality, not one was common to any two of the islands.

If we now turn to the Flora, we shall find the aboriginal plants of the different islands wonderfully different. I give all the following results on the high authority of my friend Dr. J. Hooker. I may premise that I indiscriminately collected everything in flower on the different islands, and fortunately kept my collections separate. Too much confidence, however, must not be placed in the proportional results, as the small collections brought home by some other naturalists, though in some respects confirming the results, plainly show that much remains to be done in the botany of this group: the Leguminosæ, moreover, has as yet been only approximately worked out:—

Name of Island.	Total No. of Species.	No. of Species found in other parts of the world.	No. of Species confined to the Galapagos Archipelago.	No. confined to the one Island.	No. of Species confined to the Galapagos Archipelago, but found on more than the one Island.
James	71	33	38	30	8
Albemarle	4	18	26	22	4
Chatham	32	16	16	12	4
Charles	68*	39	29	21	8

*or 29, if the probably imported plants be subtracted

Hence, we have the truly wonderful fact, that in James Island, of the thirty-eight Galapageian plants, or those found in no other part of the world, thirty are exclusively confined to this one island; and in Albemarle Island, of the twenty-six aboriginal Galapageian plants, twenty-two are confined to this one island, that is, only four are at present known to grow in the other islands of the archipelago; and so on, as shown in the above table, with the plants from Chatham and Charles Islands. This fact will, perhaps, be rendered even more striking, by giving a few illustrations:—thus, Scalesia, a remarkable arborescent genus of the Compositæ, is confined to the archipelago: it has six species: one from Chatham, one from Albemarle, one from Charles Island, two from James Island, and the sixth from one of the three latter islands, but it is not known from which: not one of these six species grows on any two islands. Again, Euphorbia, a mundane or widely distributed genus, has here eight species, of which seven are confined to the archipelago, and not one found on any two islands: Acalypha and Borreria, both mundane genera, have respectively six and seven species, none of which have the same species on two islands, with the exception of one Borreria, which does occur on two islands. The species of the Compositæ are particularly local; and Dr. Hooker has furnished me with several other most striking illustrations of the difference of the species on the different islands. He remarks that this law of distribution holds good both with those genera confined to the archipelago, and those distributed in other quarters of the world: in like manner we have seen that the different islands have their proper species of the mundane genus of tortoise, and of the widely distributed American genus of the mocking-thrush, as well as of two of the Galapageian sub-groups of finches, and almost certainly of the Galapageian genus Amblyrhynchus.

The distribution of the tenants of this archipelago would not be nearly so wonderful, if, for instance, one island had a mocking-thrush, and a second island some other quite distinct genus;—if one island had its genus of lizard, and a second island another distinct genus, or none whatever;—or if the different islands were inhabited, not by representative species of the same genera of plants, but by totally different genera, as does to a certain extent hold good: for, to give one instance, a large berry-bearing tree at James Island has no representative species in Charles Island. But it is the circumstance, that several of the islands possess their own species of the tortoise, mocking-thrush, finches, and numerous plants, these species having the same general habits, occupying analogous situations, and obviously filling the same place in the natural economy of this archipelago, that strikes me with wonder. It may be suspected that some of these representative species, at least in the case of the tortoise and of some of the birds, may hereafter prove to be only well-marked races; but this would be of equally great interest to the philosophical naturalist. I have said that most of the islands are in sight of each other: I may specify that Charles Island is fifty miles from the nearest part of Chatham Island, and thirty-three miles from the nearest part of Albemarle Island. Chatham Island is sixty miles from the nearest part of James Island, but there are two intermediate islands between them which were not visited by me. James Island is only ten miles from the nearest part of Albemarle Island, but the two points where the collections were made are thirty-two miles apart. I must repeat, that neither the nature of the soil, nor height of the land, nor the climate, nor the general character of the associated beings, and therefore their action one on another, can differ much in the different islands. If there be any sensible difference in their climates, it must be between the Windward group (namely, Charles and Chatham Islands), and that to leeward; but there seems to be no corresponding difference in the productions of these two halves of the archipelago.

The only light which I can throw on this remarkable difference in the inhabitants of the different islands, is, that very strong currents of the sea running in a westerly and W.N.W. direction must separate, as far as transportal by the sea is concerned, the southern islands from the northern ones; and between these northern islands a strong N.W. current was observed, which must effectually separate James and Albemarle Islands. As the archipelago is free to a most remarkable degree from gales of wind, neither the birds, insects, nor lighter seeds, would be blown from island to island. And lastly, the profound depth of the ocean between the islands, and their apparently recent (in a geological sense) volcanic origin, render it highly unlikely that they were ever united; and this, probably, is a far more important consideration than any other, with respect to the geographical distribution of their inhabitants. Reviewing the facts here given, one is astonished at the amount of creative force, if such an expression may be used, displayed on these small, barren, and rocky islands; and still more so, at its diverse yet analogous action on points so near each other. I have said that the Galapagos Archipelago might be called a satellite attached to America, but it should rather be called a group of satellites, physically similar, organically distinct, yet intimately related to each other, and all related in a marked, though much lesser degree, to the great American continent.

I will conclude my description of the natural history of these islands, by giving an account of the extreme tameness of the birds.

This disposition is common to all the terrestrial species; namely, to the mocking-thrushes, the finches, wrens, tyrant-flycatchers, the dove, and carrion-buzzard. All of them are often approached sufficiently near to be killed with a switch, and sometimes, as I myself tried, with a cap or hat. A gun is here almost superfluous; for with the muzzle I pushed a hawk off the branch of a tree. One day, whilst lying down, a mocking-thrush alighted on the edge of a pitcher, made of the shell of a tortoise, which I held in my hand, and began very quietly to sip the water; it allowed me to lift it from the ground whilst seated on the vessel: I often tried, and very nearly succeeded, in catching these birds by their legs. Formerly the birds appear to have been even tamer than at present. Cowley (in the year 1684) says that the "Turtledoves were so tame, that they would often alight on our hats and arms, so as that we could take them alive; they not fearing man, until such time as some of our company did fire at them, whereby they were rendered more shy." Dampier also, in the same year, says that a man in a morning's walk might kill six or seven dozen of these doves. At present, although certainly very tame, they do not alight on people's arms, nor do they suffer themselves to be killed in such large numbers. It is surprising that they have not become wilder; for these islands during the last hundred and fifty years have been frequently visited by bucaniers and whalers; and the sailors, wandering through the wood in search of tortoises, always take cruel delight in knocking down the little birds.

These birds, although now still more persecuted, do not readily become wild. In Charles Island, which had then been colonized about six years, I saw a boy sitting by a well with a switch in his hand, with which he killed the doves and finches as they came to drink. He had already procured a little heap of them for his dinner; and he said that he had constantly been in the habit of waiting by this well for the same purpose. It would appear that the birds of this archipelago, not having as yet learnt that man is a more dangerous animal than the tortoise or the Amblyrhynchus, disregard him, in the same manner as in England shy birds, such as magpies, disregard the cows and horses grazing in our fields.

The Falkland Islands offer a second instance of birds with a similar disposition. The extraordinary tameness of the little Opetiorhynchus has been remarked by Pernety, Lesson, and other voyagers. It is not, however, peculiar to that bird: the Polyborus, snipe, upland and lowland goose, thrush, bunting, and even some true hawks, are all more or less tame. As the birds are so tame there, where foxes, hawks, and owls occur, we may infer that the absence of all rapacious animals at the Galapagos, is not the cause of their tameness here. The upland geese at the Falklands show, by the precaution they take in building on the islets, that they are aware of their danger from the foxes; but they are not by this rendered wild towards man. This tameness of the birds, especially of the waterfowl, is strongly contrasted with the habits of the same species in Tierra del Fuego, where for ages past they have been persecuted by the wild inhabitants. In the Falklands, the sportsman may sometimes kill more of the upland geese in one day than he can carry home; whereas in Tierra del Fuego it is nearly as difficult to kill one, as it is in England to shoot the common wild goose.

In the time of Pernety (1763), all the birds there appear to have been much tamer than at present; he states that the Opetiorhynchus would almost perch on his finger; and that with a wand he killed ten in half an hour. At that period the birds must have been about as tame as they now are at the Galapagos. They appear to have learnt caution more slowly at these latter islands than at the Falklands, where they have had proportionate means of experience; for besides frequent visits from vessels, those islands have been at intervals colonized during the entire period. Even formerly, when all the birds were so tame, it was impossible by Pernety's account to kill the black-necked swan—a bird of passage, which probably brought with it the wisdom learnt in foreign countries.

I may add that, according to Du Bois, all the birds at Bourbon in 1571–72, with the exception of the flamingoes and geese, were so extremely tame, that they could be caught by the hand, or killed in any number with a stick. Again, at Tristan d'Acunha in the Atlantic, Carmichael[156] states that the only two land-birds, a thrush and a bunting, were "so tame as to suffer themselves to be caught with a hand-net." From these several facts we may, I think, conclude, first, that the wildness of birds with regard to man, is a particular instinct directed against *him*, and not dependent upon any general degree of caution arising from other sources of danger; secondly, that it is not acquired by individual birds in a short time, even when much persecuted; but that in the course of successive generations it becomes hereditary. With domesticated animals we are accustomed to see new mental habits or instincts acquired or rendered hereditary; but with animals in a state of nature, it must always be most difficult to discover instances of acquired hereditary knowledge. In regard to the wildness of birds towards man, there is no way of accounting for it, except as an inherited habit: comparatively few young birds, in any one year, have been injured by man in England, yet almost all, even nestlings, are afraid of him; many individuals, on the other hand, both at the Galapagos and at the Falklands, have been pursued and injured by man, yet have not learned a salutary dread of him. We may infer from these facts, what havoc the introduction of any new beast of prey must cause in a country, before the instincts of the indigenous inhabitants have become adapted to the stranger's craft or power.

CHAPTER XVIII

TAHITI AND NEW ZEALAND

PASS THROUGH THE LOW ARCHIPELAGO • TAHITI • ASPECT • VEGETATION ON THE MOUNTAINS • VIEW OF EIMEO • EXCURSION INTO THE INTERIOR • PROFOUND RAVINES • SUCCESSION OF WATERFALLS • NUMBER OF WILD USEFUL PLANTS • TEMPERANCE OF THE INHABITANTS • THEIR MORAL STATE • PARLIAMENT CONVENED • NEW ZEALAND • BAY OF ISLANDS • HIPPAHS • EXCURSION TO WAIMATE • MISSIONARY ESTABLISHMENT • ENGLISH WEEDS NOW RUN WILD • WAIOMIO • FUNERAL OF A NEW ZEALAND WOMAN • SAIL FOR AUSTRALIA

OCTOBER 20TH.—THE SURVEY of the Galapagos Archipelago being concluded, we steered towards Tahiti and commenced our long passage of 3200 miles. In the course of a few days we sailed out of the gloomy and clouded ocean-district which extends during the winter far from the coast of South America. We then enjoyed bright and clear weather, while running pleasantly along at the rate of 150 or 160 miles a day before the steady trade-wind. The temperature in this more central part of the Pacific is higher than near the American shore. The thermometer in the poop cabin, by night and day, ranged between 80° and 83°, which feels very pleasant; but with one degree or two higher, the heat becomes oppressive. We passed through the Low or Dangerous Archipelago, and saw several of those most curious rings of coral land, just rising above the water's edge, which have been called Lagoon Islands. A long and brilliantly white beach is capped by a margin of green vegetation; and the strip, looking either way, rapidly narrows away in the distance, and sinks beneath the horizon. From the mast-head a wide expanse of smooth water can be seen within the ring. These low hollow coral islands bear no proportion to the vast ocean out of which they abruptly rise; and it seems wonderful, that such weak invaders are not overwhelmed, by the all-powerful and never-tiring waves of that great sea, miscalled the Pacific.

November 15th.—At daylight, Tahiti, an island which must for ever remain classical to the voyager in the South Sea, was in view. At a distance the appearance was not attractive. The luxuriant vegetation of the lower part could not yet be seen, and as the clouds rolled past, the wildest and most precipitous peaks showed themselves towards the centre of the island. As soon as we anchored in Matavai Bay, we were surrounded by canoes. This was our Sunday, but the Monday of Tahiti: if the case had been reversed, we should not have received a single visit; for the injunction not to launch a canoe on the sabbath is rigidly obeyed. After dinner we landed to enjoy all the delights produced by the first impressions of a new country, and that country the charming Tahiti. A crowd of men, women, and children, was collected on the memorable Point Venus, ready to receive us with laughing, merry faces. They marshalled us towards the house of Mr. Wilson, the missionary of the district, who met us on the road, and gave us a very friendly reception. After sitting a very short time in his house, we separated to walk about, but returned there in the evening.

The land capable of cultivation, is scarcely in any part more than a fringe of low alluvial soil, accumulated round the base of the mountains, and protected from the waves of the sea by a coral reef, which encircles the entire line of coast. Within the reef there is an expanse of smooth water, like that of a lake, where the canoes of the natives can ply with safety and where ships anchor. The low land which comes down to the beach of coral-sand, is covered by the most beautiful productions of the intertropical regions. In the midst of bananas, orange, cocoa-nut, and bread-fruit trees, spots are cleared where yams, sweet potatoes, and sugar-cane, and pine-apples are cultivated. Even the brushwood is an imported fruit-tree, namely, the guava, which from its abundance has become as noxious as a weed. In Brazil I have often admired the varied beauty of the bananas, palms, and orange-trees contrasted together; and here we also have the bread-fruit, conspicuous from its large, glossy, and deeply digitated leaf. It is admirable to behold groves of a tree, sending forth its branches with the vigour of an English oak, loaded with large and most nutritious fruit. However seldom the usefulness of an object can account for the pleasure of beholding it, in the case of these beautiful woods, the knowledge of their high productiveness no doubt enters largely into the feeling of admiration. The little winding paths, cool from the surrounding shade, led to the scattered houses; the owners of which everywhere gave us a cheerful and most hospitable reception.

I was pleased with nothing so much as with the inhabitants. There is a mildness in the expression of their countenances which at once banishes the idea of a savage; and an intelligence which shows that they are advancing in civilization. The common people, when working, keep the upper part of their bodies quite naked; and it is then that the Tahitians are seen to advantage. They are very tall, broad-shouldered, athletic, and well-proportioned. It has been remarked, that it requires little habit to make a dark skin more pleasing and natural to the eye of an European than his own colour. A white man bathing by the side of a Tahitian, was like a plant bleached by the gardener's art compared with a fine dark green one growing vigorously in the open fields. Most of the men are tattooed, and the ornaments follow the curvature of the body so gracefully, that they have a very elegant effect. One common pattern, varying in its details, is somewhat like the crown of a palm-tree. It springs from the central line of the back, and gracefully curls round both sides. The simile may be a fanciful one, but I thought the body of a man thus ornamented was like the trunk of a noble tree embraced by a delicate creeper.

Many of the elder people had their feet covered with small figures, so placed as to resemble a sock. This fashion, however, is partly gone by, and has been succeeded by others. Here, although fashion is far from immutable, every one must abide by that prevailing in his youth. An old man has thus his age for ever stamped on his body, and he cannot assume the airs of a young dandy. The women are tattooed in the same manner as the men, and very commonly on their fingers. One unbecoming fashion is now almost universal: namely, shaving the hair from the upper part of the head, in a circular form, so as to leave only an outer ring. The missionaries have tried to persuade the people to change this habit; but it is the fashion, and that is a sufficient answer at Tahiti, as well as at Paris. I was much disappointed in the personal appearance of the women: they are far inferior in every respect to the men. The custom of wearing a white or scarlet flower in the back of the head, or through a small hole in each ear, is pretty. A crown of woven cocoa-nut leaves is also worn as a shade for the eyes. The women appear to be in greater want of some becoming costume even than the men.

Nearly all the natives understand a little English—that is, they know the names of common things; and by the aid of this, together with signs, a lame sort of conversation could be carried on. In returning in the evening to the boat, we stopped to witness a very pretty scene. Numbers of children were playing on the beach, and had lighted bonfires which illumined the placid sea and surrounding trees; others, in circles, were singing Tahitian verses. We seated ourselves on the sand, and joined their party. The songs were impromptu, and I believe related to our arrival: one little girl sang a line, which the rest took up in parts, forming a very pretty chorus. The whole scene made us unequivocally aware that we were seated on the shores of an island in the far-famed South Sea.

17th.—This day is reckoned in the log-book as Tuesday the 17th, instead of Monday the 16th, owing to our, so far, successful chase of the sun. Before breakfast the ship was hemmed in by a flotilla of canoes; and when the natives were allowed to come on board, I suppose there could not have been less than two hundred. It was the opinion of every one that it would have been difficult to have picked out an equal number from any other nation, who would have given so little trouble. Everybody brought something for sale: shells were the main articles of trade. The Tahitians now fully understand the value of money, and prefer it to old clothes or other articles. The various coins, however, of English and Spanish denomination puzzle them, and they never seemed to think the small silver quite secure until changed into dollars. Some of the chiefs have accumulated considerable sums of money. One chief, not long since, offered 800 dollars (about 160*l.* sterling) for a small vessel; and frequently they purchase whale-boats and horses at the rate of from 50 to 100 dollars.

After breakfast I went on shore, and ascended the nearest slope to a height of between two and three thousand feet. The outer mountains are smooth and conical, but steep; and the old volcanic rocks, of which they are formed, have been cut through by many profound ravines, diverging from the central broken parts of the island to the coast. Having crossed the narrow low girt of inhabited and fertile land, I followed a smooth steep ridge between two of the deep ravines. The vegetation was singular, consisting almost exclusively of small dwarf ferns, mingled higher up, with coarse grass; it was not very dissimilar from that on some of the Welsh hills, and this so close above the orchard of tropical plants on the coast was very surprising. At the highest point, which I reached, trees again appeared. Of the three zones of comparative luxuriance, the lower one owes its moisture, and therefore fertility, to its flatness; for, being scarcely raised above the level of the sea, the water from the higher land drains away slowly. The intermediate zone does not, like the upper one, reach into a damp and cloudy atmosphere, and therefore remains sterile. The woods in the upper zone are very pretty, tree-ferns replacing the cocoa-nuts on the coast. It must not, however, be supposed that these woods at all equal in splendour the forests of Brazil. The vast numbers of productions, which characterize a continent, cannot be expected to occur in an island.

From the highest point which I attained, there was a good view of the distant island of Eimeo, dependent on the same sovereign with Tahiti. On the lofty and broken pinnacles, white massive clouds were piled up, which formed an island in the blue sky, as Eimeo itself did in the blue ocean. The island, with the exception of one small gateway, is completely encircled by a reef. At this distance, a narrow but well-defined brilliantly white line was alone visible, where the waves first encountered the wall of coral. The mountains rose abruptly out of the glassy expanse of the lagoon, included within this narrow white line, outside which the heaving waters of the ocean were dark-coloured. The view was striking: it may aptly be compared to a framed engraving, where the frame represents the breakers, the marginal paper the smooth lagoon, and the drawing the island itself. When in the evening I descended from the mountain, a man, whom I had pleased with a trifling gift, met me, bringing with him hot roasted bananas, a pine-apple, and cocoa-nuts. After walking under a burning sun, I do not know anything more delicious than the milk of a young cocoa-nut. Pine-apples are here so abundant that the people eat them in the same wasteful manner as we might turnips. They are of an excellent flavor—perhaps even better than those cultivated in England; and this I believe is the highest compliment which can be paid to any fruit. Before going on board, Mr. Wilson interpreted for me to the Tahitian who had paid me so adroit an attention, that I wanted him and another man to accompany me on a short excursion into the mountains.

18th.—In the morning I came on shore early, bringing with me some provisions in a bag, and two blankets for myself and servant. These were lashed to each end of a long pole, which was alternately carried by my Tahitian companions on their shoulders. These men are accustomed thus to carry, for a whole day, as much as fifty pounds at each end of their poles. I told my guides to provide themselves with food and clothing; but they said that there was plenty of food in the mountains, and for clothing, that their skins were sufficient. Our line of march was the valley of Tiaauru, down which a river flows into the sea by Point Venus. This is one of the principal streams in the island, and its source lies at the base of the loftiest central pinnacles, which rise to a height of about 7000 feet. The whole island is so mountainous that the only way to penetrate into the interior is to follow up the valleys. Our road, at first, lay through woods which bordered each side of the river; and the glimpses of the lofty central peaks, seen as through an avenue, with here and there a waving cocoa-nut tree on one side, were extremely picturesque. The valley soon began to narrow, and the sides to grow lofty and more precipitous. After having walked between three and four hours, we found the width of the ravine scarcely exceeded that of the bed of the stream. On each hand the walls were nearly vertical; yet from the soft nature of the volcanic strata, trees and a rank vegetation sprung from every projecting ledge. These precipices must have been some thousand feet high; and the whole formed a mountain gorge far more magnificent than anything which I had ever before beheld. Until the midday sun stood vertically over the ravine, the air felt cool and damp, but now it became very sultry. Shaded by a ledge of rock, beneath a façade of columnar lava, we ate our dinner. My guides had already procured a dish of small fish and fresh-water prawns. They carried with them a small net stretched on a hoop; and where the water was deep and in eddies, they dived, and like otters, with their eyes open followed the fish into holes and corners, and thus caught them.

The Tahitians have the dexterity of amphibious animals in the water. An anecdote mentioned by Ellis shows how much they feel at home in this element. When a horse was landing for Pomarre in 1817, the slings broke, and it fell into the water; immediately the natives jumped overboard, and by their cries and vain efforts at assistance almost drowned it. As soon, however, as it reached the shore, the whole population took to flight, and tried to hide themselves from the man-carrying pig, as they christened the horse.

A little higher up, the river divided itself into three little streams. The two northern ones were impracticable, owing to a succession of waterfalls which descended from the jagged summit of the highest mountain; the other to all appearance was equally inaccessible, but we managed to ascend it by a most extraordinary road. The sides of the valley were here nearly precipitous; but, as frequently happens with stratified rocks, small ledges projected, which were thickly covered by wild bananas, lilaceous plants, and other luxuriant productions of the tropics. The Tahitians, by climbing amongst these ledges, searching for fruit, had discovered a track by which the whole precipice could be scaled. The first ascent from the valley was very dangerous; for it was necessary to pass a steeply inclined face of naked rock, by the aid of ropes which we brought with us. How any person discovered that this formidable spot was the only point where the side of the mountain was practicable, I cannot imagine. We then cautiously walked along one of the ledges till we came to one of the three streams. This ledge formed a flat spot, above which a beautiful cascade, some hundred feet in height, poured down its waters, and beneath, another high cascade fell into the main stream in the valley below. From this cool and shady recess we made a circuit to avoid the overhanging waterfall. As before, we followed little projecting ledges, the danger being partly concealed by the thickness of the vegetation. In passing from one of the ledges to another, there was a vertical wall of rock. One of the Tahitians, a fine active man, placed the trunk of a tree against this, climbed up it, and then by the aid of crevices reached the summit. He fixed the ropes to a projecting point, and lowered them for our dog and luggage, and then we clambered up ourselves. Beneath the ledge on which the dead tree was placed, the precipice must have been five or six hundred feet deep; and if the abyss had not been partly concealed by the overhanging ferns and lilies my head would have turned giddy, and nothing should have induced me to have attempted it. We continued to ascend, sometimes along ledges, and sometimes along knife-edged ridges, having on each hand profound ravines. In the Cordillera I have seen mountains on a far grander scale, but for abruptness, nothing at all comparable with this. In the evening we reached a flat little spot on the banks of the same stream, which we had continued to follow, and which descends in a chain of waterfalls: here we bivouacked for the night. On each side of the ravine there were great beds of the mountain-banana, covered with ripe fruit. Many of these plants were from twenty to twenty-five feet high, and from three to four in circumference. By the aid of strips of bark for rope, the stems of bamboos for rafters, and the large leaf of the banana for a thatch, the Tahitians in a few minutes built us an excellent house; and with withered leaves made a soft bed.

They then proceeded to make a fire, and cook our evening meal. A light was procured, by rubbing a blunt pointed stick in a groove made in another, as if with intention of deepening it, until by the friction the dust became ignited. A peculiarly white and very light wood (the Hibiscus tiliaceus) is alone used for this purpose: it is the same which serves for poles to carry any burden, and for the floating out-riggers to their canoes. The fire was produced in a few seconds: but to a person who does not understand the art, it requires, as I found, the greatest exertion; but at last, to my great pride, I succeeded in igniting the dust. The Gaucho in the Pampas uses a different method: taking an elastic stick about eighteen inches long, he presses one end on his breast, and the other pointed end into a hole in a piece of wood, and then rapidly turns the curved part, like a carpenter's centre-bit. The Tahitians having made a small fire of sticks, placed a score of stones, of about the size of cricket-balls, on the burning wood. In about ten minutes the sticks were consumed, and the stones hot. They had previously folded up in small parcels of leaves, pieces of beef, fish, ripe and unripe bananas, and the tops of the wild arum. These green parcels were laid in a layer between two layers of the hot stones, and the whole then covered up with earth, so that no smoke or steam could escape. In about a quarter of an hour, the whole was most deliciously cooked. The choice green parcels were now laid on a cloth of banana leaves, and with a cocoa-nut shell we drank the cool water of the running stream; and thus we enjoyed our rustic meal.

I could not look on the surrounding plants without admiration. On every side were forests of banana; the fruit of which, though serving for food in various ways, lay in heaps decaying on the ground. In front of us there was an extensive brake of wild sugar-cane; and the stream was shaded by the dark green knotted stem of the Ava,—so famous in former days for its powerful intoxicating effects. I chewed a piece, and found that it had an acrid and unpleasant taste, which would have induced any one at once to have pronounced it poisonous. Thanks to the missionaries, this plant now thrives only in these deep ravines, innocuous to every one. Close by I saw the wild arum, the roots of which, when well baked, are good to eat, and the young leaves better than spinach. There was the wild yam, and a liliaceous plant called Ti, which grows in abundance, and has a soft brown root, in shape and size like a huge log of wood: this served us for dessert, for it is as sweet as treacle, and with a pleasant taste. There were, moreover, several other wild fruits, and useful vegetables. The little stream, besides its cool water, produced eels, and cray-fish. I did indeed admire this scene, when I compared it with an uncultivated one in the temperate zones. I felt the force of the remark, that man, at least savage man, with his reasoning powers only partly developed, is the child of the tropics.

As the evening drew to a close, I strolled beneath the gloomy shade of the bananas up the course of the stream. My walk was soon brought to a close, by coming to a waterfall between two and three hundred feet high; and again above this there was another. I mention all these waterfalls in this one brook, to give a general idea of the inclination of the land. In the little recess where the water fell, it did not appear that a breath of wind had ever blown. The thin edges of the great leaves of the banana, damp with spray, were unbroken, instead of being, as is so generally the case, split into a thousand shreds. From our position, almost suspended on the mountain-side, there were glimpses into the depths of the neighbouring valleys; and the lofty points of the central mountains, towering up within sixty degrees of the zenith, hid half the evening sky. Thus seated, it was a sublime spectacle to watch the shades of night gradually obscuring the last and highest pinnacles.

Before we laid ourselves down to sleep, the elder Tahitian fell on his knees, and with closed eyes repeated a long prayer in his native tongue. He prayed as a Christian should do, with fitting reverence, and without the fear of ridicule or any ostentation of piety. At our meals neither of the men would taste food, without saying beforehand a short grace. Those travellers who think that a Tahitian prays only when the eyes of the missionary are fixed on him, should have slept with us that night on the mountain-side. Before morning it rained very heavily; but the good thatch of banana-leaves kept us dry.

November 19th.—At daylight my friends, after their morning prayer, prepared an excellent breakfast in the same manner as in the evening. They themselves certainly partook of it largely; indeed I never saw any men eat near so much. I suppose such enormously capacious stomachs must be the effect of a large part of their diet consisting of fruit and vegetables, which contain, in a given bulk, a comparatively small portion of nutriment. Unwittingly, I was the means of my companions breaking, as I afterwards learned, one of their own laws, and resolutions: I took with me a flask of spirits, which they could not refuse to partake of; but as often as they drank a little, they put their fingers before their mouths, and uttered the word "Missionary." About two years ago, although the use of the ava was prevented, drunkenness from the introduction of spirits became very prevalent. The missionaries prevailed on a few good men, who saw that their country was rapidly going to ruin, to join with them in a Temperance Society. From good sense or shame, all the chiefs and the queen were at last persuaded to join. Immediately a law was passed, that no spirits should be allowed to be introduced into the island, and that he who sold and he who bought the forbidden article should be punished by a fine. With remarkable justice, a certain period was allowed for stock in hand to be sold, before the law came into effect. But when it did, a general search was made, in which even the houses of the missionaries were not exempted, and all the ava (as the natives call all ardent spirits) was poured on the ground. When one reflects on the effect of intemperance on the aborigines of the two Americas, I think it will be acknowledged that every well-wisher of Tahiti owes no common debt of gratitude to the missionaries. As long as the little island of St. Helena remained under the government of the East India Company, spirits, owing to the great injury they had produced, were not allowed to be imported; but wine was supplied from the Cape of Good Hope. It is rather a striking and not very gratifying fact, that in the same year that spirits were allowed to be sold in Helena, their use was banished from Tahiti by the free will of the people.

After breakfast we proceeded on our journey. As my object was merely to see a little of the interior scenery, we returned by another track, which descended into the main valley lower down. For some distance we wound, by a most intricate path, along the side of the mountain which formed the valley. In the less precipitous parts we passed through extensive groves of the wild banana. The Tahitians, with their naked, tattooed bodies, their heads ornamented with flowers, and seen in the dark shade of these groves, would have formed a fine picture of man inhabiting some primeval land. In our descent we followed the line of ridges; these were exceedingly narrow, and for considerable lengths steep as a ladder; but all clothed with vegetation. The extreme care necessary in poising each step rendered the walk fatiguing. I did not cease to wonder at these ravines and precipices: when viewing the country from one of the knife-edged ridges, the point of support was so small, that the effect was nearly the same as it must be from a balloon. In this descent we had occasion to use the ropes only once, at the point where we entered the main valley. We slept under the same ledge of rock where we had dined the day before: the night was fine, but from the depth and narrowness of the gorge, profoundly dark.

Before actually seeing this country, I found it difficult to understand two facts mentioned by Ellis; namely, that after the murderous battles of former times, the survivors on the conquered side retired into the mountains, where a handful of men could resist a multitude. Certainly half a dozen men, at the spot where the Tahitian reared the old tree, could easily have repulsed thousands. Secondly, that after the introduction of Christianity, there were wild men who lived in the mountains, and whose retreats were unknown to the more civilized inhabitants.

November 20th.—In the morning we started early, and reached Matavai at noon. On the road we met a large party of noble athletic men, going for wild bananas. I found that the ship, on account of the difficulty in watering, had moved to the harbour of Papawa, to which place I immediately walked. This is a very pretty spot. The cove is surrounded by reefs, and the water as smooth as in a lake. The cultivated ground, with its beautiful productions, interspersed with cottages, comes close down to the water's edge.

From the varying accounts which I had read before reaching these islands, I was very anxious to form, from my own observation, a judgment of their moral state,—although such judgment would necessarily be very imperfect. First impressions at all times very much depend on one's previously acquired ideas. My notions were drawn from Ellis's "Polynesian Researches"—an admirable and most interesting work, but naturally looking at everything under a favourable point of view; from Beechey's Voyage; and from that of Kotzebue, which is strongly adverse to the whole missionary system. He who compares these three accounts will, I think, form a tolerably accurate conception of the present state of Tahiti. One of my impressions, which I took from the two last authorities, was decidedly incorrect; viz., that the Tahitians had become a gloomy race, and lived in fear of the missionaries. Of the latter feeling I saw no trace, unless, indeed, fear and respect be confounded under one name. Instead of discontent being a common feeling, it would be difficult in Europe to pick out of a crowd half so many merry and happy faces. The prohibition of the flute and dancing is inveighed against as wrong and foolish;—the more than presbyterian manner of keeping the sabbath is looked at in a similar light. On these points I will not pretend to offer any opinion to men who have resided as many years as I was days on the island.

On the whole, it appears to me that the morality and religion of the inhabitants are highly creditable. There are many who attack, even more acrimoniously than Kotzebue, both the missionaries, their system, and the effects produced by it. Such reasoners never compare the present state with

that of the island only twenty years ago; nor
day; but they compare it with the high stand
expect the missionaries to effect that which
to do. Inasmuch as the condition of the peo
dard, blame is attached to the missionary, i
he has effected. They forget, or will not re
and the power of an idolatrous priesthood—
leled in any other part of the world—infanti
tem—bloody wars, where the conquerors
children—that all these have been abolishe
perance, and licentiousness have been greatl
of Christianity. In a voyager to forget these
should he chance to be at the point of shipw
he will most devoutly pray that the lesson
extended thus far.

In point of morality, the virtue of the wo
most open to exception. But before they are
well distinctly to call to mind the scenes descr
Banks, in which the grandmothers and moth
a part. Those who are most severe, should con
ity of the women in Europe is owing to the sys
ers on their daughters, and how much in
precepts of religion. But it is useless to argu
believe that, disappointed in not finding the fi
open as formerly, they will not give credit to a
wish to practise, or to a religion which they un

Sunday, 22nd.—The harbour of Papiéte, wher
considered as the capital of the island: it is also
the chief resort of shipping. Captain Fitz Roy t
hear divine service, first in the Tahitian langu
own. Mr. Pritchard, the leading missionary in
service. The chapel consisted of a large airy fra
filled to excess by tidy, clean people, of all ages
disappointed in the apparent degree of attenti
tations were raised too high. At all events the ap
that in a country church in England. The singi
edly very pleasing, but the language from th
delivered, did not sound well: a constant repeti
mata mai," rendered it monotonous. After Engli
on foot to Matavai. It was a pleasant walk, som
and sometimes under the shade of the many be

all vessel un
ants of the I
een of Tahiti.
act by some
nent demand
arly three thou
ember. The C
ncerning this
Fitz Roy acco
ce famous fr
a parliament
al chiefs of th
to describe wh
in Fitz Roy. The
ed reasons were
ress our general
moderation, can
sides. I believe
Tahitians, from
ple resolved to
tain Fitz Roy urg
sacrificed for the
re grateful for
nd that they wer
tion and its pron
ning, made a pe
and good feeling.
on was ended, se
in Fitz Roy many i
relating to the tre
n as the decision
ahitian parliament
z Roy invited Quee

ening four boats we
and the yards mann
ost of the chiefs. T
thing, and seemed
queen is a large aw
e has only one royal
r all circumstances, a

The roc
shore, al
also mu
boistero
return o

26th.—I
New Zea
Tahiti—
admirati

Decembe
may nov
sail ove
onwards
foundly
specks,
drawn
togethe
land is t
wise bee
league
tions of
this airy
it, and a
man m
has late
voyage,

Decemb
being b
anchora
line, an
the bay.
ture, bu
as in pa
of the l
distanc
tle villa
water's
now an
extrem

came alongside. This, and the aspect of the whole scene, afforded a remarkable, and not very pleasing contrast, with our joyful and boisterous welcome at Tahiti.

In the afternoon we went on shore to one of the larger groups of houses, which yet hardly deserves the title of a village. Its name is Pahia: it is the residence of the missionaries; and there are no native residents except servants and labourers. In the vicinity of the Bay of Islands, the number of Englishmen, including their families, amounts to between two and three hundred. All the cottages, many of which are white-washed and look very neat, are the property of the English. The hovels of the natives are so diminutive and paltry, that they can scarcely be perceived from a distance. At Pahia, it was quite pleasing to behold the English flowers in the gardens before the houses; there were roses of several kinds, honey-suckle, jasmine, stocks, and whole hedges of sweetbrier.

December 22nd. —In the morning I went out walking; but I soon found that the country was very impracticable. All the hills are thickly covered with tall fern, together with a low bush which grows like a cypress; and very little ground has been cleared or cultivated. I then tried the sea-beach; but proceeding towards either hand, my walk was soon stopped by salt-water creeks and deep brooks. The communication between the inhabitants of the different parts of the bay, is (as in Chiloe) almost entirely kept up by boats. I was surprised to find that almost every hill which I ascended, had been at some former time more or less fortified. The summits were cut into steps or successive terraces, and frequently they had been protected by deep trenches. I afterwards observed that the principal hills inland in like manner showed an artificial outline. These are the Pas, so frequently mentioned by Captain Cook under the name of "hippah;" the difference of sound being owing to the prefixed article.

That the Pas had formerly been much used, was evident from the piles of shells, and the pits in which, as I was informed, sweet potatoes used to be kept as a reserve. As there was no water on these hills, the defenders could never have anticipated a long siege, but only a hurried attack for plunder, against which the successive terraces would have afforded good protection. The general introduction of fire-arms has changed the whole system of warfare; and an exposed situation on the top of a hill is now worse than useless. The Pas in consequence are, at the present day, always built on a level piece of ground. They consist of a double stockade of thick and tall posts, placed in a zigzag line, so that every part can be flanked. Within the stockade a mound of earth is thrown up, behind which the defenders can rest in safety, or use their fire-arms over it. On the level of the ground little archways sometimes pass through this breastwork, by which means the defenders can me to pay him a visit there. Mr. Bushby, the British resident, offered to take me in his boat by a creek, where I should see a pretty waterfall, and by which means my walk would be shortened. He likewise procured for me a guide.

Upon asking a neighbouring chief to recommend a man, the chief himself offered to go; but his ignorance of the value of money was so complete, that at first he asked how many pounds I would give him, but afterwards was well contented with two dollars. When I showed the chief a very small bundle, which I wanted carried, it became absolutely necessary for him to take a slave. These feelings of pride are beginning to wear away; but formerly a leading man would sooner have died, than undergone the indignity of carrying the smallest burden. My companion was a light active man, dressed in a dirty blanket, and with his face completely tattooed. He had formerly been a great warrior. He appeared to be on very cordial terms with Mr. Bushby; but at various times they had quarrelled violently. Mr. Bushby remarked that a little quiet irony would frequently silence any one of these natives in their most blustering moments. This chief has come and harangued Mr. Bushby in a hectoring manner, saying, "great chief, a great man, a friend of mine, has come to pay me a visit—you must give him something good to eat, some fine presents, etc." Mr. Bushby has allowed him to finish his discourse, and then has quietly replied by some answer such as, "What else shall your slave do for you?" The man would then instantly, with a very comical expression, cease his braggadocio.

Some time ago, Mr. Bushby suffered a far more serious attack. A chief and a party of men tried to break into his house in the middle of the night, and not finding this so easy, commenced a brisk firing with their muskets. Mr. Bushby was slightly wounded, but the party was at length driven away. Shortly afterwards it was discovered who was the aggressor; and a general meeting of the chiefs was convened to consider the case. It was considered by the New Zealanders as very atrocious, inasmuch as it was a night attack, and that Mrs. Bushby was lying ill in the house: this latter circumstance, much to their honour, being considered in all cases as a protection. The chiefs agreed to confiscate the land of the aggressor to the King of England. The whole proceeding, however, in thus trying and punishing a chief was entirely without precedent. The aggressor, moreover, lost caste in the estimation of his equals and this was considered by the British as of more consequence than the confiscation of his land.

As the boat was shoving off, a second chief stepped into her, who only wanted the amusement of the passage up and down the creek. I never saw a more horrid and ferocious expression than this man had. It immediately struck me I had somewhere seen his likeness: it will be found in Retzch's outlines to Schiller's ballad of Fridolin, where two men are pushing Robert into the burning iron furnace. It is the man who has his arm on Robert's breast. Physiognomy here spoke the truth; this chief had been a notorious murderer, and was an arrant coward to boot. At the point where the boat landed, Mr. Bushby accompanied me a few hundred yards on the road: I could not help admiring the cool impudence of the hoary old villain, whom we left lying in the boat, when he shouted to Mr. Bushby, "Do not you stay long, I shall be tired of waiting here."

We now commenced our walk. The road lay along a well beaten path, bordered on each side by the tall fern, which covers the whole country. After travelling some miles, we came to a little country village, where a few hovels were collected together, and some patches of ground cultivated with potatoes. The introduction of the potato has been the most essential benefit to the island; it is now much more used than any native vegetable. New Zealand is favoured by one great natural advantage; namely, that the inhabitants can never perish from famine. The whole country abounds with fern: and the roots of this plant, if not very palatable, yet contain much nutriment. A native can always subsist on these, and on the shell-fish, which are abundant on all parts of the sea-coast. The villages are chiefly conspicuous by the platforms which are raised on four posts ten or twelve feet above the ground, and on which the produce of the fields is kept secure from all accidents.

On coming near one of the huts I was much amused by seeing in due form the ceremony of rubbing, or, as it ought to be called, pressing noses. The women, on our first approach, began uttering something in a most dolorous voice; they then squatted themselves down and held up their faces; my companion standing over them, one after another, placed the bridge of his nose at right angles to theirs, and commenced pressing. This lasted rather longer than a cordial shake of the hand with us; and as we vary the force of the grasp of the hand in shaking, so do they in pressing. During the process they uttered comfortable little grunts, very much in the same manner as two pigs do, when rubbing against each other. I noticed that the slave would press noses with any one he met, indifferently either before or after his master the chief. Although among the savages, the chief has absolute power of life and death over his slave, yet there is an entire absence of ceremony between them. Mr. Burchell has remarked the same thing in Southern Africa, with the rude Bachapins. Where civilization has arrived at a certain point, complex formalities soon arise between the different grades of society: thus at Tahiti all were formerly obliged to uncover themselves as low as the waist in presence of the king.

The ceremony of pressing noses having been duly completed with all present, we seated ourselves in a circle in the front of one of the hovels, and rested there half-an-hour. All the hovels have nearly the same form and dimensions, and all agree in being filthily dirty. They resemble a cowshed with one end open, but having a partition a little way within, with a square hole in it, making a small gloomy chamber. In this the inhabitants keep all their property, and when the weather is cold they sleep there. They eat, however, and pass their time in the open part in front. My guides having finished their pipes, we continued our walk. The path led through the same undulating country, the whole uniformly clothed as before with fern. On our right hand we had a serpentine river, the banks of which were fringed with trees, and here and there on the hill-sides there was a clump of wood. The whole scene, in spite of its green colour, had rather a desolate aspect. The sight of so much fern impresses the mind with an idea of sterility: this, however, is not correct; for wherever the fern grows thick and breast-high, the land by tillage becomes productive. Some of the residents think that all this extensive open country originally was covered with forests, and that it has been cleared by fire. It is said, that by digging in the barest spots, lumps of the kind of resin which flows from the kauri pine are frequently found. The natives had an evident motive in clearing the country; for the fern, formerly a staple article of food, flourishes only in the open cleared tracks. The almost entire absence of associated grasses, which forms so remarkable a feature in the vegetation of this island, may perhaps be accounted for by the land having been aboriginally covered with forest-trees.

The soil is volcanic; in several parts we passed over shaggy lavas, and craters could clearly be distinguished on several of the neighbouring hills. Although the scenery is nowhere beautiful, and only occasionally pretty, I enjoyed my walk. I should have enjoyed it more, if my companion, the chief, had not possessed extraordinary conversational powers. I knew only three words: "good," "bad," and "yes:" and with these I answered all his remarks, without of course having understood one word he said. This, however, was quite sufficient: I was a good listener, an agreeable person, and he never ceased talking to me.

At length we reached Waimate. After having passed over so many miles of an uninhabited useless country, the sudden appearance of an English farm-house, and its well-dressed fields, placed there as if by an enchanter's wand, was exceedingly pleasant. Mr. Williams not being at home, I received in Mr. Davies's house a cordial welcome. After drinking tea with his family party, we took a stroll about the farm. At Waimate there are three large houses, where the missionary gentlemen, Messrs. Williams, Davies, and Clarke, reside; and near them are the huts of the native labourers. On an adjoining slope, fine crops of barley and wheat were standing in full ear; and in another part, fields of potatoes and clover. But I cannot attempt to describe all I saw; there were large gardens, with every fruit and vegetable which England produces; and many belonging to a warmer clime. I may instance asparagus, kidney beans, cucumbers, rhubarb, apples, pears, figs, peaches, apricots, grapes, olives, gooseberries, currants, hops, gorse for fences, and English oaks; also many kinds of flowers. Around the farm-yard there were stables, a thrashing-barn with its winnowing machine, a blacksmith's forge, and on the ground ploughshares and other tools: in the middle was that happy mixture of pigs and poultry, lying comfortably together, as in every English farm-yard. At the distance of a few hundred yards, where the water of a little rill had been dammed up into a pool, there was a large and substantial water-mill.

All this is very surprising, when it is considered that five years ago nothing but the fern flourished here. Moreover, native workmanship, taught by the missionaries, has effected this change;—the lesson of the missionary is the enchanter's wand. The house had been built, the windows framed, the fields ploughed, and even the trees grafted, by a New Zealander. At the mill, a New Zealander was seen powdered white with flour, like his brother miller in England. When I looked at this whole scene, I thought it admirable. It was not merely that England was brought vividly before my mind; yet, as the evening drew to a close, the domestic sounds, the fields of corn, the distant undulating country with its trees might well have been mistaken for our fatherland: nor was it the triumphant feeling at seeing what Englishmen could effect; but rather the high hopes thus inspired for the future progress of this fine island.

Several young men, redeemed by the missionaries from slavery, were employed on the farm. They were dressed in a shirt, jacket, and trousers, and had a respectable appearance. Judging from one trifling anecdote, I should think they must be honest. When walking in the fields, a young labourer came up to Mr. Davies, and gave him a knife and gimlet, saying that he had found them on the road, and did not know to whom they belonged! These young men and boys appeared very merry and good-humoured. In the evening I saw a party of them at cricket: when I thought of the austerity of which the missionaries have been accused, I was amused by observing one of their own sons taking an active part in the game. A more decided and pleasing change was manifested in the young women, who acted as servants within the houses. Their clean, tidy, and healthy appearance, like that of the dairy-maids in England, formed a wonderful contrast with the women of the filthy hovels in Kororadika. The wives of the missionaries tried to persuade them not to be tattooed; but a famous operator having arrived from the south, they said, "We really must just have a few lines on our lips; else when we grow old, our lips will shrivel, and we shall be so very ugly." There is not nearly so much tattooing as formerly; but as it is a badge of distinction between the chief and the slave, it will probably long be practised. So soon does any train of ideas become habitual, that the missionaries told me that even in their eyes a plain face looked mean, and not like that of a New Zealand gentleman.

Late in the evening I went to Mr. Williams's house, where I passed the night. I found there a large party of children, collected together for Christmas Day, and all sitting round a table at tea. I never saw a nicer or more merry group; and to think that this was in the centre of the land of cannibalism, murder, and all atrocious crimes! The cordiality and happiness so plainly pictured in the faces of the little circle, appeared equally felt by the older persons of the mission.

December 24th.—In the morning, prayers were read in the native tongue to the whole family. After breakfast I rambled about the gardens and farm. This was a market-day, when the natives of the surrounding hamlets bring their potatoes, Indian corn, or pigs, to exchange for blankets, tobacco, and sometimes, through the persuasions of the missionaries, for soap. Mr. Davies's eldest son, who manages a farm of his own, is the man of business in the market. The children of the missionaries, who came while young to the island, understand the language better than their parents, and can get anything more readily done by the natives.

A little before noon Messrs. Williams and Davies walked with me to a part of a neighbouring forest, to show me the famous kauri pine. I measured one of the noble trees, and found it thirty-one feet in circumference above the roots. There was another close by, which I did not see, thirty-three feet; and I heard of one no less than forty feet. These trees are remarkable for their smooth cylindrical boles, which run up to a height of sixty, and even ninety feet, with a nearly equal diameter, and without a single branch. The crown of branches at the summit is out of all proportion small to the trunk; and the leaves are likewise small compared with the branches. The forest was here almost composed of the kauri; and the largest trees, from the parallelism of their sides, stood up like gigantic columns of wood. The timber of the kauri is the most valuable production of the island; moreover, a quantity of resin oozes from the bark, which is sold at a penny a pound to the Americans, but its use was then unknown. Some of the New Zealand forest must be impenetrable to an extraordinary degree. Mr. Matthews informed me that one forest only thirty-four miles in width, and separating two inhabited districts, had only lately, for the first time, been crossed. He and another missionary, each with a party of about fifty men, undertook to open a road; but it cost more than a fortnight's labour! In the woods I saw very few birds. With regard to animals, it is a most remarkable fact, that so large an island, extending over more than 700 miles in latitude, and in many parts ninety broad, with varied stations, a fine climate, and land of all heights, from 14,000 feet downwards, with the exception of a small rat, did not possess one indigenous animal. The several species of that gigantic genus of birds, the Deinornis seem here to have replaced mammiferous quadrupeds, in the same manner as the reptiles still do at the Galapagos archipelago. It is said that the common Norway rat, in the short space of two years, annihilated in this northern end of the island, the New Zealand species. In many places I noticed several sorts of weeds, which, like the rats, I was forced to own as countrymen. A leek has overrun whole districts, and will prove very troublesome, but it was imported as a favour by a French vessel. The common dock is also widely disseminated, and will, I fear, for ever remain a proof of the rascality of an Englishman, who sold the seeds for those of the tobacco plant.

On returning from our pleasant walk to the house, I dined with Mr. Williams; and then, a horse being lent me, I returned to the Bay of Islands. I took leave of the missionaries with thankfulness for their kind welcome, and with feelings of high respect for their gentlemanlike, useful, and upright characters. I think it would be difficult to find a body of men better adapted for the high office which they fulfil.

Christmas Day.—In a few more days the fourth year of our absence from England will be completed. Our first Christmas Day was spent at Plymouth; the second at St. Martin's Cove, near Cape Horn; the third at Port Desire, in Patagonia; the fourth at anchor in a wild harbour in the peninsula of Tres Montes; this fifth here; and the next, I trust in Providence, will be in England. We attended divine service in the chapel of Pahia; part of the service being read in English, and part in the native language. Whilst at New Zealand we did not hear of any recent acts of cannibalism; but Mr. Stokes found burnt human bones strewed round a fire-place on a small island near the anchorage; but these remains of a comfortable banquet might have been lying there for several years. It is probable that the moral state of the people will rapidly improve. Mr. Bushby mentioned one pleasing anecdote as a proof of the sincerity of some, at least, of those who profess Christianity. One of his young men left him, who had been accustomed to read prayers to the rest of the servants. Some weeks afterwards, happening to pass late in the evening by an outhouse, he saw and heard one of his men reading the Bible with difficulty by the light of the fire, to the others. After this the party knelt and prayed: in their prayers they mentioned Mr. Bushby and his family, and the missionaries, each separately in his respective district.

December 26th.—Mr. Bushby offered to take Mr. Sulivan and myself in his boat some miles up the river to Cawa-Cawa; and proposed afterwards to walk on to the village of Waiomio, where there are some curious rocks. Following one of the arms of the bay, we enjoyed a pleasant row, and passed through pretty scenery, until we came to a village, beyond which the boat could not pass. From this place a chief and a party of men volunteered to

CHAPTER XIX

AUSTRALIA

SYDNEY • EXCURSION TO BATHURST • ASPECT OF THE WOODS • PARTY OF NATIVES • GRADUAL EXTINCTION OF THE ABORIGINES • INFECTION GENERATED BY ASSOCIATED MEN IN HEALTH • BLUE MOUNTAINS • VIEW OF THE GRAND GULF-LIKE VALLEYS • THEIR ORIGIN AND FORMATION • BATHURST, GENERAL CIVILITY OF THE LOWER ORDERS • STATE OF SOCIETY • VAN DIEMEN'S LAND • HOBART TOWN • ABORIGINES ALL BANISHED • MOUNT WELLINGTON • KING GEORGE'S SOUND • CHEERLESS ASPECT OF THE COUNTRY • BALD HEAD, CALCAREOUS CASTS OF BRANCHES OF TREES • PARTY OF NATIVES • LEAVE AUSTRALIA

JANUARY 12TH, 1836.—EARLY IN THE MORNING a light air carried us towards the entrance of Port Jackson. Instead of beholding a verdant country, interspersed with fine houses, a straight line of yellowish cliff brought to our minds the coast of Patagonia. A solitary lighthouse, built of white stone, alone told us that we were near a great and populous city. Having entered the harbour, it appears fine and spacious, with cliff-formed shores of horizontally stratified sandstone. The nearly level country is covered with thin scrubby trees, bespeaking the curse of sterility. Proceeding further inland, the country improves: beautiful villas and nice cottages are here and there scattered along the beach. In the distance stone houses, two and three stories high, and windmills standing on the edge of a bank, pointed out to us the neighbourhood of the capital of Australia.

At last we anchored within Sydney Cove. We found the little basin occupied by many large ships, and surrounded by warehouses. In the evening I walked through the town, and returned full of admiration at the whole scene. It is a most magnificent testimony to the power of the British nation. Here, in a less promising country, scores of years have done many more times more than an equal number of centuries have effected in South America. My first feeling was to congratulate myself that I was born an Englishman. Upon seeing more of the town afterwards, perhaps my admiration fell a little; but yet it is a fine town. The streets are regular, broad, clean, and kept in excellent order; the houses are of a good size, and the shops well furnished. It may be faithfully compared to the large suburbs which stretch out from London and a few other great towns in England; but not even near London or Birmingham is there an appearance of such rapid growth. The number of large houses and other buildings just finished was truly surprising; nevertheless, every one complained of the high rents and difficulty in procuring a house. Coming from South America, where in the towns every man of property is known, no one thing surprised me more than not being able to ascertain at once to whom this or that carriage belonged.

I hired a man and two horses to take me to Bathurst, a village about one hundred and twenty miles in the interior, and the centre of a great pastoral district. By this means I hoped to gain a general idea of the appearance of the country. On the morning of the 16th (January) I set out on my excursion. The first stage took us to Paramatta, a small country town, next to Sydney in importance. The roads were excellent, and made upon the MacAdam principle, whinstone having been brought for the purpose from the distance of several miles. In all respects there was a close resemblance to England: perhaps the alehouses here were more numerous. The iron gangs, or parties of convicts who have committed here some offence, appeared the least like England: they were working in chains, under the charge of sentries with loaded arms.

The power which the government possesses, by means of forced labour, of at once opening good roads throughout the country, has been, I believe, one main cause of the early prosperity of this colony. I slept at night at a very comfortable inn at Emu ferry, thirty-five miles from Sydney, and near the ascent of the Blue Mountains. This line of road is the most frequented, and has been the longest inhabited of any in the colony. The whole land is enclosed with high railings, for the farmers have not succeeded in rearing hedges. There are many substantial houses and good cottages scattered about; but although considerable pieces of land are under cultivation, the greater part yet remains as when first discovered.

The extreme uniformity of the vegetation is the most remarkable feature in the landscape of the greater part of New South Wales. Everywhere we have an open woodland, the ground being partially covered with a very thin pasture, with little appearance of verdure. The trees nearly all belong to one family, and mostly have their leaves placed in a vertical, instead of, as in Europe, in a nearly horizontal position: the foliage is scanty, and of a peculiar pale green tint; without any gloss. Hence the woods appear light and shadowless: this, although a loss of comfort to the traveller under the scorching rays of summer, is of importance to the farmer, as it allows grass to grow where it otherwise would not. The leaves are not shed periodically: this character appears common to the entire southern hemisphere, namely, South America, Australia, and the Cape of Good Hope. The inhabitants of this hemisphere, and of the intertropical regions, thus lose perhaps one of the most glorious, though to our eyes common, spectacles in the world—the first bursting into full foliage of the leafless tree. They may, however, say that we pay dearly for this by having the land covered with mere naked skeletons for so many months. This is too true; but our senses thus acquire a keen relish for the exquisite green of the spring, which the eyes of those living within the tropics, sated during the long year with the gorgeous productions of those glowing climates, can never experience. The greater number of the trees, with the exception of some of the Blue-gums, do not attain a large size; but they grow tall and tolerably straight, and stand well apart. The bark of some of the Eucalypti falls annually, or hangs dead in long shreds which swing about with the wind, and give to the woods a desolate and untidy appearance. I cannot imagine a more complete contrast, in every respect, than between the forests of Valdivia or Chiloe, and the woods of Australia.

At sunset, a party of a score of the black aborigines passed by, each carrying, in their accustomed manner, a bundle of spears and other weapons. By giving a leading young man a shilling, they were easily detained, and threw their spears for my amusement. They were all partly clothed, and several could speak a little English: their countenances were good-humoured and pleasant, and they appeared far from being such utterly degraded beings as they have usually been represented. In their own arts they are admirable. A cap being fixed at thirty yards distance, they transfixed it with a spear, delivered by the throwing-stick with the rapidity of an arrow from the bow of a practised archer. In tracking animals or men they show most wonderful sagacity; and I heard of several of their remarks which manifested considerable acuteness. They will not, however, cultivate the ground, or build houses and remain stationary, or even take the trouble of tending a flock of sheep when given to them. On the whole they appear to me to stand some few degrees higher in the scale of civilization than the Fuegians.

It is very curious thus to see in the midst of a civilized people, a set of harmless savages wandering about without knowing where they shall sleep at night, and gaining their livelihood by hunting in the woods. As the white man has travelled onwards, he has spread over the country belonging to several tribes. These, although thus enclosed by one common people, keep up their ancient distinctions, and sometimes go to war with each other. In an engagement which took place lately, the two parties most singularly chose the centre of the village of Bathurst for the field of battle. This was of service to the defeated side, for the runaway warriors took refuge in the barracks.

The number of aborigines is rapidly decreasing. In my whole ride, with the exception of some boys brought up by Englishmen, I saw only one other party. This decrease, no doubt, must be partly owing to the introduction of spirits, to European diseases (even the milder ones of which, such as the measles,[157] prove very destructive), and to the gradual extinction of the wild animals. It is said that numbers of their children invariably perish in very early infancy from the effects of their wandering life; and as the difficulty of procuring food increases, so must their wandering habits increase; and hence the population, without any apparent deaths from famine, is repressed in a manner extremely sudden compared to what happens in civilized countries, where the father, though in adding to his labour he may injure himself, does not destroy his offspring.

Besides the several evident causes of destruction, there appears to be some more mysterious agency generally at work. Wherever the European has trod, death seems to pursue the aboriginal. We may look to the wide extent of the Americas, Polynesia, the Cape of Good Hope, and Australia, and we find the same result. Nor is it the white man alone that thus acts the destroyer; the Polynesian of Malay extraction has in parts of the East Indian archipelago, thus driven before him the dark-coloured native. The varieties of man seem to act on each other in the same way as different species of animals—the stronger always extirpating the weaker. It was melancholy at New Zealand to hear the fine energetic natives saying, that they knew the land was doomed to pass from their children. Every one has heard of the inexplicable reduction of the population in the beautiful and healthy island of Tahiti since the date of Captain Cook's voyages: although in that case we might have expected that it would have been increased; for infanticide, which formerly prevailed to so extraordinary a degree, has ceased; profligacy has greatly diminished, and the murderous wars become less frequent.

The Rev. J. Williams, in his interesting work[158], says, that the first intercourse between natives and Europeans, "is invariably attended with the introduction of fever, dysentery, or some other disease, which carries off numbers of the people." Again he affirms, "It is certainly a fact, which cannot be controverted, that most of the diseases which have raged in the islands during my residence there, have been introduced by ships;[159] and what renders this fact remarkable is, that there might be no appearance of disease among the crew of the ship which conveyed this destructive importation." This statement is not quite so extraordinary as it at first appears; for several cases are on record of the most malignant fevers having broken out, although the parties themselves, who were the cause, were not affected. In the early part of the reign of George III., a prisoner who had been confined in a dungeon, was taken in a coach with four constables before a magistrate; and although the man himself was not ill, the four constables died from a short putrid fever; but the contagion extended to no others. From these facts it would almost appear as if the effluvium of one set of men shut up for some time together was poisonous when inhaled by others; and possibly more so, if the men be of different races. Mysterious as this circumstance appears to be, it is not more surprising than that the body of one's fellow-creature, directly after death, and before putrefaction has commenced, should often be of so deleterious a quality, that the mere puncture from an instrument used in its dissection, should prove fatal.

17th.—Early in the morning we passed the Nepean in a ferry-boat. The river, although at this spot both broad and deep, had a very small body of running water. Having crossed a low piece of land on the opposite side, we reached the slope of the Blue Mountains. The ascent is not steep, the road having been cut with much care on the side of a sandstone cliff. On the summit an almost level plain extends, which, rising imperceptibly to the westward, at last attains a height of more than 3000 feet. From so grand a title as Blue Mountains, and from their absolute altitude, I expected to have seen a bold chain of mountains crossing the country; but instead of this, a sloping plain presents merely an inconsiderable front to the low land near the coast. From this first slope, the view of the extensive woodland to the east was striking, and the surrounding trees grew bold and lofty. But when once on the sandstone platform, the scenery becomes exceedingly monotonous; each side of the road is bordered by scrubby trees of the never-failing Eucalyptus family; and with the exception of two or three small inns, there are no houses or cultivated land: the road, moreover, is solitary; the most frequent object being a bullock-waggon, piled up with bales of wool.

In the middle of the day we baited our horses at a little inn, called the Weatherboard. The country here is elevated 2800 feet above the sea. About a mile and a half from this place there is a view exceedingly well worth visiting. Following down a little valley and its tiny rill of water, an immense gulf unexpectedly opens through the trees which border the pathway, at the depth of perhaps 1500 feet. Walking on a few yards, one stands on the brink of a vast precipice, and below one sees a grand bay or gulf, for I know not what other name to give it, thickly covered with forest. The point of view is situated as if at the head of a bay, the line of cliff diverging on each side, and showing headland behind headland, as on a bold sea-coast. These cliffs are composed of horizontal strata of whitish sandstone; and are so absolutely vertical, that in many places a person standing on the edge and throwing down a stone, can see it strike the trees in the abyss below. So unbroken is the line of cliff, that in order to reach the foot of the waterfall, formed by this little stream, it is said to be necessary to go sixteen miles round. About five miles distant in front, another line of cliff extends, which thus appears completely to encircle the valley; and hence the name of bay is justified, as applied to this grand amphitheatrical depression. If we imagine a winding harbour, with its deep water surrounded by bold cliff-like shores, to be laid dry, and a forest to spring up on its sandy bottom, we should then have the appearance and structure here exhibited. This kind of view was to me quite novel, and extremely magnificent.

In the evening we reached the Blackheath. The sandstone plateau has here attained the height of 3400 feet; and is covered, as before, with the same scrubby woods. From the road, there were occasional glimpses into a profound valley, of the same character as the one described; but from the steepness and depth of its sides, the bottom was scarcely ever to be seen. The Blackheath is a very comfortable inn, kept by an old soldier; and it reminded me of the small inns in North Wales.

18th.—Very early in the morning, I walked about three miles to see Govett's Leap; a view of a similar character with that near the Weatherboard, but perhaps even more stupendous. So early in the day the gulf was filled with a thin blue haze, which, although destroying the general effect of the view added to the apparent depth at which the forest was stretched out beneath our feet. These valleys, which so long presented an insuperable barrier to the attempts of the most enterprising of the colonists to reach the interior, are most remarkable. Great arm-like bays, expanding at their upper ends, often branch from the main valleys and penetrate the sandstone platform; on the other hand, the platform often sends promontories into the valleys, and even leaves in them great, almost insulated, masses. To descend into some of these valleys, it is necessary to go round twenty miles; and into others, the surveyors have only lately penetrated, and the colonists have not yet been able to drive in their cattle. But the most remarkable feature in their structure is, that although several miles wide at their heads, they generally contract towards their mouths to such a degree as to become impassable. The Surveyor-General, Sir T. Mitchell,[160] endeavoured in vain, first walking and then by crawling between the great fallen fragments of sandstone, to ascend through the gorge by which the river Grose joins the Nepean; yet the valley of the Grose in its upper part, as I saw, forms a magnificent level basin some miles in width, and is on all sides surrounded by cliffs, the summits of which are believed to be nowhere less than 3000 feet above the level of the sea. When cattle are driven into the valley of the Wolgan by a path (which I descended), partly natural and partly made by the owner of the land, they cannot escape; for this valley is in every other part surrounded by perpendicular cliffs, and eight miles lower down, it contracts from an average width of half a mile, to a mere chasm, impassable to man or beast. Sir T. Mitchell states that the great valley of the Cox river with all its branches, contracts, where it unites with the Nepean, into a gorge 2200 yards in width, and about 1000 feet in depth. Other similar cases might have been added.

The first impression, on seeing the correspondence of the horizontal strata on each side of these valleys and great amphitheatrical depressions, is that they have been hollowed out, like other valleys, by the action of water; but when one reflects on the enormous amount of stone, which on this view must have been removed through mere gorges or chasms, one is led to ask whether these spaces may not have subsided. But considering the form of the irregularly branching valleys, and of the narrow promontories projecting into them from the platforms, we are compelled to abandon this notion. To attribute these hollows to the present alluvial action would be preposterous; nor does the drainage from the summit-level always fall, as I remarked near the Weatherboard, into the head of these valleys, but into one side of their bay-like recesses. Some of the inhabitants remarked to me that they never viewed one of those bay-like recesses, with the headlands receding on both hands, without being struck with their resemblance to a bold sea-coast. This is certainly the case; moreover, on the present coast of New South Wales, the numerous, fine, widely-branching harbours, which are generally connected with the sea by a narrow mouth worn through the sandstone coast-cliffs, varying from one mile in width to a quarter of a mile, present a likeness, though on a miniature scale, to the great valleys of the interior. But then immediately occurs the startling difficulty, why has the sea worn out these great, though circumscribed depressions on a wide platform, and left mere gorges at the openings, through which the whole vast amount of triturated matter must have been carried away? The only light I can throw upon this enigma, is by remarking that banks of the most irregular forms appear to be now forming in some seas, as in parts of the West Indies and in the Red Sea, and that their sides are exceedingly steep. Such banks, I have been led to suppose, have been formed by sediment heaped by strong currents on an irregular bottom. That in some cases the sea, instead of spreading out sediment in a uniform sheet, heaps it round submarine rocks and islands, it is hardly possible to doubt, after examining the charts of the West Indies; and that the waves have power to form high and precipitous cliffs, even in land-locked harbours, I have noticed in many parts of South America. To apply these ideas to the sandstone platforms of

New South Wales, I imagine that the strata were heaped by the action of strong currents, and of the undulations of an open sea, on an irregular bottom; and that the valley-like spaces thus left unfilled had their steeply sloping flanks worn into cliffs, during a slow elevation of the land; the worn-down sandstone being removed, either at the time when the narrow gorges were cut by the retreating sea, or subsequently by alluvial action.

SOON AFTER LEAVING the Blackheath, we descended from the sandstone platform by the pass of Mount Victoria. To effect this pass, an enormous quantity of stone has been cut through; the design, and its manner of execution, being worthy of any line of road in England. We now entered upon a country less elevated by nearly a thousand feet, and consisting of granite. With the change of rock, the vegetation improved; the trees were both finer and stood farther apart; and the pasture between them was a little greener and more plentiful. At Hassan's Walls, I left the high road, and made a short detour to a farm called Walerawang; to the superintendent of which I had a letter of introduction from the owner in Sydney. Mr. Browne had the kindness to ask me to stay the ensuing day, which I had much pleasure in doing. This place offers an example of one of the large farming, or rather sheep-grazing establishments of the colony. Cattle and horses are, however, in this case rather more numerous than usual, owing to some of the valleys being swampy and producing a coarser pasture. Two or three flat pieces of ground near the house were cleared and cultivated with corn, which the harvestmen were now reaping: but no more wheat is sown than sufficient for the annual support of the labourers employed on the establishment. The usual number of assigned convict-servants here is about forty, but at the present time there were rather more. Although the farm was well stocked with every necessary, there was an apparent absence of comfort; and not one single woman resided here. The sunset of a fine day will generally cast an air of happy contentment on any scene; but here, at this retired farm-house, the brightest tints on the surrounding woods could not make me forget that forty hardened, profligate men were ceasing from their daily labours, like the slaves from Africa, yet without their holy claim for compassion.

Early on the next morning, Mr. Archer, the joint superintendent, had the kindness to take me out kangaroo-hunting. We continued riding the greater part of the day, but had very bad sport, not seeing a kangaroo, or even a wild dog. The greyhounds pursued a kangaroo rat into a hollow tree, out of which we dragged it: it is an animal as large as a rabbit, but with the figure of a kangaroo. A few years since this country abounded with wild animals; but now the emu is banished to a long distance, and the kangaroo is become scarce; to both the English greyhound has been highly destructive. It may be long before these animals are altogether exterminated, but their doom is fixed. The aborigines are always anxious to borrow the dogs from the farm-houses: the use of them, the offal when an animal is killed, and some milk from the cows, are the peace-offerings of the settlers, who push farther and farther towards the interior. The thoughtless aboriginal, blinded by these trifling advantages, is delighted at the approach of the white man, who seems predestined to inherit the country of his children.

Although having poor sport, we enjoyed a pleasant ride. The woodland is generally so open that a person on horseback can gallop through it. It is traversed by a few flat-bottomed valleys, which are green and free from trees: in such spots the scenery was pretty like that of a park. In the whole country I scarcely saw a place without the marks of a fire; whether these had been more or less recent—whether the stumps were more or less black, was the greatest change which varied the uniformity, so wearisome to the traveller's eye. In these woods there are not many birds; I saw, however, some large flocks of the white cockatoo feeding in a corn-field, and a few most beautiful parrots; crows, like our jackdaws were not uncommon, and another bird something like the magpie. In the dusk of the evening I took a stroll along a chain of ponds, which in this dry country represented the course of a river, and had the good fortune to see several of the famous Ornithorhynchus paradoxus. They were diving and playing about the surface of the water, but showed so little of their bodies, that they might easily have been mistaken for water-rats. Mr. Browne shot one: certainly it is a most extraordinary animal; a stuffed specimen does not at all give a good idea of the appearance of the head and beak when fresh; the latter becoming hard and contracted.[161]

20th.—A long day's ride to Bathurst. Before joining the highroad we followed a mere path through the forest; and the country, with the exception of a few squatters' huts, was very solitary. We experienced this day the sirocco-like wind of Australia, which comes from the parched deserts of the interior. Clouds of dust were travelling in every direction; and the wind felt as if it had passed over a fire. I afterwards heard that the thermometer out of doors had stood at 119°, and in a closed room at 96°. In the afternoon we came in view of the downs of Bathurst. These undulating but nearly smooth plains are very remarkable in this country, from being absolutely destitute of trees. They support only a thin brown pasture. We rode some miles over this country, and then reached the township of Bathurst, seated in the middle of what may be called either a very broad valley, or narrow plain. I was told at Sydney not to form too bad an opinion of Australia by judging of the country from the roadside, nor too good a one from Bathurst; in this latter respect, I did not feel myself in the least danger of being prejudiced. The season, it must be owned, had been one of great drought, and the country did not wear a favourable aspect; although I understand it was incomparably worse two or three months before. The secret of the rapidly growing prosperity of Bathurst is, that the brown pasture which appears to the stranger's eye so wretched, is excellent for sheep-grazing. The town stands, at the height of 2200 feet above the sea, on the banks of the Macquarie. This is one of the rivers flowing into the vast and scarcely known interior. The line of watershed, which divides the inland streams from those on the coast, has a height of about 3000 feet, and runs in a north and south direction at the distance of from eighty to a hundred miles from the sea-side. The Macquarie figures in the map as a respectable river, and it is the largest of those draining this part of the water-shed; yet to my surprise I found it a mere chain of ponds, separated from each other by spaces almost dry. Generally a small stream is running; and sometimes there are high and impetuous floods. Scanty as the supply of the water is throughout this district, it becomes still scantier further inland.

22nd.—I commenced my return, and followed a new road called Lockyer's Line, along which the country is rather more hilly and picturesque. This was a long day's ride; and the house where I wished to sleep was some way off the road, and not easily found. I met on this occasion, and indeed on all others, a very general and ready civility among the lower orders, which, when one considers what they are, and what they have been, would scarcely have been expected. The farm where I passed the night, was owned by two young men who had only lately come out, and were beginning a settler's life. The total want of almost every comfort was not attractive; but future and certain prosperity was before their eyes, and that not far distant.

The next day we passed through large tracts of country in flames, volumes of smoke sweeping across the road. Before noon we joined our former road, and ascended Mount Victoria. I slept at the Weatherboard, and before dark took another walk to the amphitheatre. On the road to Sydney I spent a very pleasant evening with Captain King at Dunheved; and thus ended my little excursion in the colony of New South Wales.

Before arriving here the three things which interested me most were—the state of society amongst the higher classes, the condition of the convicts, and the degree of attraction sufficient to induce persons to emigrate. Of course, after so very short a visit, one's opinion is worth scarcely anything; but it is as difficult not to form some opinion, as it is to form a correct judgment. On the whole, from what I heard, more than from what I saw, I was disappointed in the state of society. The whole community is rancorously divided into parties on almost every subject. Among those who, from their station in life, ought to be the best, many live in such open profligacy that respectable people cannot associate with them. There is much jealousy between the children of the rich emancipist and the free settlers, the former being pleased to consider honest men as interlopers. The whole population, poor and rich, are bent on acquiring wealth: amongst the higher orders, wool and sheep-grazing form the constant subject of conversation. There are many serious drawbacks to the comforts of a family, the chief of which, perhaps, is being surrounded by convict servants. How thoroughly odious to every feeling, to be waited on by a man who the day before, perhaps, was flogged, from your representation, for some trifling misdemeanor. The female servants are of course, much worse: hence children learn the vilest expressions, and it is fortunate, if not equally vile ideas.

On the other hand, the capital of a person, without any trouble on his part, produces him treble interest to what it will in England; and with care he is sure to grow rich. The luxuries of life are in abundance, and very little dearer than in England, and most articles of food are cheaper. The climate is splendid, and perfectly healthy; but to my mind its charms are lost by the uninviting aspect of the country. Settlers possess a great advantage in finding their sons of service when very young. At the age of from sixteen to twenty, they frequently take charge of distant farming stations. This, however, must happen at the expense of their boys associating entirely with convict servants. I am not aware that the tone of society has assumed any peculiar character; but with such habits, and without intellectual pursuits, it can hardly fail to deteriorate. My opinion is such, that nothing but rather sharp necessity should compel me to emigrate.

The rapid prosperity and future prospects of this colony are to me, not understanding these subjects, very puzzling. The two main exports are wool and whale-oil, and to both of these productions there is a limit. The country is totally unfit for canals, therefore there is a not very distant point, beyond which the land-carriage of wool will not repay the expense of shearing and tending sheep. Pasture everywhere is so thin that settlers have already pushed far into the interior: moreover, the country further inland becomes extremely poor. Agriculture, on account of the droughts, can never succeed on an extended scale: therefore, so far as I can see, Australia must ultimately depend upon being the centre of commerce for the southern hemisphere, and perhaps on her future manufactories. Possessing coal, she always has the moving power at hand. From the habitable country extending along the coast, and from her English extraction, she is sure to be a maritime nation. I formerly imagined that Australia would rise to be as grand and powerful a country as North America, but now it appears to me that such future grandeur is rather problematical.

With respect to the state of the convicts, I had still fewer opportunities of judging than on other points. The first question is, whether their condition is at all one of punishment: no one will maintain that it is a very severe one. This, however, I suppose, is of little consequence as long as it continues to be an object of dread to criminals at home. The corporeal wants of the convicts are tolerably well supplied: their prospect of future liberty and comfort is not distant, and, after good conduct, certain. A "ticket of leave," which, as long as a man keeps clear of suspicion as well as of crime, makes him free within a certain district, is given upon good conduct, after years proportional to the length of the sentence; yet with all this, and overlooking the previous imprisonment and wretched passage out, I believe the years of assignment are passed away with discontent and unhappiness. As an intelligent man remarked to me, the convicts know no pleasure beyond sensuality, and in this they are not gratified. The enormous bribe which Government possesses in offering free pardons, together with the deep horror of the secluded penal settlements, destroys confidence between the convicts, and so prevents crime. As to a sense of shame, such a feeling does not appear to be known, and of this I witnessed some very singular proofs. Though it is a curious fact, I was universally told that the character of the convict population is one of arrant cowardice: not unfrequently some become desperate, and quite indifferent as to life, yet a plan requiring cool or continued courage is seldom put into execution. The worst feature in the whole case is, that although there exists what may be called a legal reform, and comparatively little is committed which the law can touch, yet that any moral reform should take place appears to be quite out of the question. I was assured by well-informed people, that a man who should try to improve, could not while living with other assigned servants;—his life would be one of intolerable misery and persecution. Nor must the contamination of the convict-ships and prisons, both here and in England, be forgotten. On the whole, as a place of punishment, the object is scarcely gained; as a real system of reform it has failed, as perhaps would every other plan; but as a means of making men outwardly honest,—of converting vagabonds, most useless in one hemisphere, into active citizens of another, and thus giving birth to a new and splendid country—a grand centre of civilization—it has succeeded to a degree perhaps unparalleled in history.

30th.—The Beagle sailed for Hobart Town in Van Diemen's Land. On the 5th of February, after a six days' passage, of which the first part was fine, and the latter very cold and squally, we entered the mouth of Storm Bay: the weather justified this awful name. The bay should rather be called an estuary, for it receives at its head the waters of the Derwent. Near the mouth, there are some extensive basaltic platforms; but higher up the land becomes mountainous, and is covered by a light wood. The lower parts of the hills which skirt the bay are cleared; and the bright yellow fields of corn, and dark green ones of potatoes, appear very luxuriant. Late in the evening we anchored in the snug cove, on the shores of which stands the capital of Tasmania. The first aspect of the place was very inferior to that of Sydney; the latter might be called a city, this is only a town. It stands at the base of Mount Wellington, a mountain 3100 feet high, but of little picturesque beauty; from this source, however, it receives a good supply of water. Round the cove there are some fine warehouses, and on one side a small fort. Coming from the Spanish settlements, where such magnificent care has generally been paid to the fortifications, the means of defence in these colonies appeared very contemptible. Comparing the town with Sydney, I was chiefly struck with the comparative fewness of the large houses, either built or building. Hobart Town, from the census of 1835, contained 13,826 inhabitants, and the whole of Tasmania 36,505.

All the aborigines have been removed to an island in Bass's Straits, so that Van Diemen's Land enjoys the great advantage of being free from a native population. This most cruel step seems to have been quite unavoidable, as the only means of stopping a fearful succession of robberies, burnings, and murders, committed by the blacks; and which sooner or later would have ended in their utter destruction. I fear there is no doubt, that this train of evil and its consequences, originated in the infamous conduct of some of our countrymen. Thirty years is a short period, in which to have banished the last aboriginal from his native island,—and that island nearly as large as Ireland. The correspondence on this subject, which took place between the government at home and that of Van Diemen's Land, is very interesting. Although numbers of natives were shot and taken prisoners in the skirmishing, which was going on at intervals for several years; nothing seems fully to have impressed them with the idea of our overwhelming power, until the whole island, in 1830, was put under martial law, and by proclamation the whole population commanded to assist in one great attempt to secure the entire race. The plan adopted was nearly similar to that of the great hunting-matches in India: a line was formed reaching across the island, with the intention of driving the natives into a *cul-de-sac* on Tasman's peninsula. The attempt failed; the natives, having tied up their dogs, stole during one night through the lines. This is far from surprising, when their practised senses, and usual manner of crawling after wild animals is considered. I have been assured that they can conceal themselves on almost bare ground, in a manner which until witnessed is scarcely credible; their dusky bodies being easily mistaken for the blackened stumps which are scattered all over the country. I was told of a trial between a party of

Englishmen and a native, who was to stand in full view on the side of a bare hill; if the Englishmen closed their eyes for less than a minute, he would squat down, and then they were never able to distinguish him from the surrounding stumps. But to return to the hunting-match; the natives understanding this kind of warfare, were terribly alarmed, for they at once perceived the power and numbers of the whites. Shortly afterwards a party of thirteen belonging to two tribes came in; and, conscious of their unprotected condition, delivered themselves up in despair. Subsequently by the intrepid exertions of Mr. Robinson, an active and benevolent man, who fearlessly visited by himself the most hostile of the natives, the whole were induced to act in a similar manner. They were then removed to an island, where food and clothes were provided them. Count Strzelecki states,[162] that "at the epoch of their deportation in 1835, the number of natives amounted to 210. In 1842, that is, after the interval of seven years, they mustered only fifty-four individuals; and, while each family of the interior of New South Wales, uncontaminated by contact with the whites, swarms with children, those of Flinders' Island had during eight years an accession of only fourteen in number!"

The Beagle stayed here ten days, and in this time I made several pleasant little excursions, chiefly with the object of examining the geological structure of the immediate neighbourhood. The main points of interest consist, first in some highly fossiliferous strata, belonging to the Devonian or Carboniferous period; secondly, in proofs of a late small rise of the land; and lastly, in a solitary and superficial patch of yellowish limestone or travertin, which contains numerous impressions of leaves of trees, together with land-shells, not now existing. It is not improbable that this one small quarry includes the only remaining record of the vegetation of Van Diemen's Land during one former epoch.

The climate here is damper than in New South Wales, and hence the land is more fertile. Agriculture flourishes: the cultivated fields look well, and the gardens abound with thriving vegetables and fruit-trees. Some of the farm-houses, situated in retired spots, had a very attractive appearance. The general aspect of the vegetation is similar to that of Australia; perhaps it is a little more green and cheerful; and the pasture between the trees rather more abundant. One day I took a long walk on the side of the bay opposite to the town: I crossed in a steamboat, two of which are constantly plying backwards and forwards. The machinery of one of these vessels was entirely manufactured in this colony, which, from its very foundation, then numbered only three and thirty years! Another day I ascended Mount Wellington; I took with me a guide, for I failed in a first attempt, from the thickness of the wood. Our guide, however, was a stupid fellow, and conducted us to the southern and damp side of the mountain, where the vegetation was very luxuriant; and where the labour of the ascent, from the number of rotten trunks, was almost as great as on a mountain in Tierra del Fuego or in Chiloe. It cost us five and a half hours of hard climbing before we reached the summit. In many parts the Eucalypti grew to a great size, and composed a noble forest. In some of the dampest ravines, tree-ferns flourished in an extraordinary manner; I saw one which must have been at least twenty feet high to the base of the fronds, and was in girth exactly six feet. The fronds forming the most elegant parasols, produced a gloomy shade, like that of the first hour of the night. The summit of the mountain is broad and flat, and is composed of huge angular masses of naked greenstone. Its elevation is 3100 feet above the level of the sea. The day was splendidly clear, and we enjoyed a most extensive view; to the north, the country appeared a mass of wooded mountains, of about the same height with that on which we were standing, and with an equally tame outline: to the south the broken land and water, forming many intricate bays, was mapped with clearness before us. After staying some hours on the summit, we found a better way to descend, but did not reach the Beagle till eight o'clock, after a severe day's work.

February 7th.—The Beagle sailed from Tasmania, and, on the 6th of the ensuing month, reached King George's Sound, situated close to the S. W. corner of Australia. We stayed there eight days; and we did not during our voyage pass a more dull and uninteresting time. The country, viewed from an eminence, appears a woody plain, with here and there rounded and partly bare hills of granite protruding. One day I went out with a party, in hopes of seeing a kangaroo hunt, and walked over a good many miles of country. Everywhere we found the soil sandy, and very poor; it supported either a coarse vegetation of thin, low brushwood and wiry grass, or a forest of stunted trees. The scenery resembled that of the high sandstone platform of the Blue Mountains; the Casuarina (a tree somewhat resembling a Scotch fir) is, however, here in greater number, and the Eucalyptus in rather less. In the open parts there were many grass-trees,—a plant which, in appearance, has some affinity with the palm; but, instead of being surmounted by a crown of noble fronds, it can boast merely of a tuft of very coarse grass-like leaves. The general bright green colour of the brushwood and other plants, viewed from a distance, seemed to promise fertility. A single walk, however, was enough to dispel such an illusion; and he who thinks with me will never wish to walk again in so uninviting a country.

One day I accompanied Captain Fitz Roy to Bald Head; the place mentioned by so many navigators, where some imagined that they saw corals, and others that they saw petrified trees, standing in the position in which they had grown. According to our view, the beds have been formed by the wind having heaped up fine sand, composed of minute rounded particles of shells and corals, during which process branches and roots of trees, together with many land-shells, became enclosed. The whole then became consolidated by the percolation of calcareous matter; and the cylindrical cavities left by the decaying of the wood, were thus also filled up with a hard pseudo-stalactical stone. The weather is now wearing away the softer parts, and in consequence the hard casts of the roots and branches of the trees project above the surface, and, in a singularly deceptive manner, resemble the stumps of a dead thicket.

A large tribe of natives, called the White Cockatoo men, happened to pay the settlement a visit while we were there. These men, as well as those of the tribe belonging to King George's Sound, being tempted by the offer of some tubs of rice and sugar, were persuaded to hold a "corrobery," or great dancing-party. As soon as it grew dark, small fires were lighted, and the men commenced their toilet, which consisted in painting themselves white in spots and lines. As soon as all was ready, large fires were kept blazing, round which the women and children were collected as spectators; the Cockatoo and King George's men formed two distinct parties, and generally danced in answer to each other. The dancing consisted in their running either sideways or in Indian file into an open space, and stamping the ground with great force as they marched together. Their heavy footsteps were accompanied by a kind of grunt, by beating their clubs and spears together, and by various other gesticulations, such as extending their arms and wriggling their bodies. It was a most rude, barbarous scene, and, to our ideas, without any sort of meaning; but we observed that the black women and children watched it with the greatest pleasure. Perhaps these dances originally represented actions, such as wars and victories; there was one called the Emu dance, in which each man extended his arm in a bent manner, like the neck of that bird. In another dance, one man imitated the movements of a kangaroo grazing in the woods, whilst a second crawled up, and pretended to spear him. When both tribes mingled in the dance, the ground trembled with the heaviness of their steps, and the air resounded with their wild cries. Every one appeared in high spirits, and the group of nearly naked figures, viewed by the light of the blazing fires, all moving in hideous harmony, formed a perfect display of a festival amongst the lowest barbarians. In Tierra del Fuego, we have beheld many curious scenes in savage life, but never, I think, one where the natives were in such high spirits, and so perfectly at their ease. After the dancing was over, the whole party formed a great circle on the ground, and the boiled rice and sugar was distributed, to the delight of all.

After several tedious delays from clouded weather, on the 14th of March, we gladly stood out of King George's Sound on our course to Keeling Island. Farewell, Australia! you are a rising child, and doubtless some day will reign a great princess in the South: but you are too great and ambitious for affection, yet not great enough for respect. I leave your shores without sorrow or regret.

Keeling Island:—Coral Formations

KEELING ISLAND • SINGULAR APPEARANCE • SCANTY FLORA • TRANSPORT OF SEEDS • BIRDS AND INSECTS • EBBING AND FLOWING SPRINGS • FIELDS OF DEAD CORAL • STONES TRANSPORTED IN THE ROOTS OF TREES • GREAT CRAB • STINGING CORALS • CORAL-EATING FISH • CORAL FORMATIONS • LAGOON ISLANDS, OR ATOLLS • DEPTH AT WHICH REEF-BUILDING CORALS CAN LIVE • VAST AREAS INTERSPERSED WITH LOW CORAL ISLANDS • SUBSIDENCE OF THEIR FOUNDATIONS • BARRIER REEFS • FRINGING REEFS • CONVERSION OF FRINGING REEFS INTO BARRIER REEFS, AND INTO ATOLLS • EVIDENCE OF CHANGES IN LEVEL • BREACHES IN BARRIER REEFS • MALDIVA ATOLLS; THEIR PECULIAR STRUCTURE • DEAD AND SUBMERGED REEFS • AREAS OF SUBSIDENCE AND ELEVATION • DISTRIBUTION OF VOLCANOES • SUBSIDENCE SLOW, AND VAST IN AMOUNT

April 1st.—We arrived in view of the Keeling or Cocos Islands, situated in the Indian Ocean, and about six hundred miles distant from the coast of Sumatra. This is one of the lagoon-islands (or atolls) of coral formation, similar to those in the Low Archipelago which we passed near. When the ship was in the channel at the entrance, Mr. Liesk, an English resident, came off in his boat. The history of the inhabitants of this place, in as few words as possible, is as follows. About nine years ago, Mr. Hare, a worthless character, brought from the East Indian archipelago a number of Malay slaves, which now, including children, amount to more than a hundred. Shortly afterwards, Captain Ross, who had before visited these islands in his merchant-ship, arrived from England, bringing with him his family and goods for settlement: along with him came Mr. Liesk, who had been a mate in his vessel. The Malay slaves soon ran away from the islet on which Mr. Hare was settled, and joined Captain Ross's party. Mr. Hare upon this was ultimately obliged to leave the place.

The Malays are now nominally in a state of freedom, and certainly are so, as far as regards their personal treatment; but in most other points they are considered as slaves. From their discontented state, from the repeated removals from islet to islet, and perhaps also from a little mismanagement, things are not very prosperous. The island has no domestic quadruped, excepting the pig, and the main vegetable production is the cocoa-nut. The whole prosperity of the place depends on this tree: the only exports being oil from the nut, and the nuts themselves, which are taken to Singapore and Mauritius, where they are chiefly used, when grated, in making curries. On the cocoa-nut, also, the pigs, which are loaded with fat, almost entirely subsist, as do the ducks and poultry. Even a huge land-crab is furnished by nature with the means to open and feed on this most useful production.

The ring-formed reef of the lagoon-island is surmounted in the greater part of its length by linear islets. On the northern or leeward side, there is an opening through which vessels can pass to the anchorage within. On entering, the scene was very curious and rather pretty; its beauty, however, entirely depended on the brilliancy of the surrounding colours. The shallow, clear, and still water of the lagoon, resting in its greater part on white sand, is, when illumined by a vertical sun, of the most vivid green. This brilliant expanse, several miles in width, is on all sides divided, either by a line of snow-white breakers from the dark heaving waters of the ocean, or from the blue vault of heaven by the strips of land, crowned by the level tops of the cocoa-nut trees. As a white cloud here and there affords a pleasing contrast with the azure sky, so in the lagoon, bands of living coral darken the emerald green water.

The next morning after anchoring, I went on shore on Direction Island. The strip of dry land is only a few hundred yards in width; on the lagoon side there is a white calcareous beach, the radiation from which under this sultry climate was very oppressive; and on the outer coast, a solid broad flat of coral-rock served to break the violence of the open sea. Excepting near the lagoon, where there is some sand, the land is entirely composed of rounded fragments of coral. In such a loose, dry, stony soil, the climate of the intertropical regions alone could produce a vigorous vegetation. On some of the smaller islets, nothing could be more elegant than the manner in which the young and full-grown cocoa-nut trees, without destroying each other's symmetry, were mingled into one wood. A beach of glittering white sand formed a border to these fairy spots.

I will now give a sketch of the natural history of these islands, which, from its very paucity, possesses a peculiar interest. The cocoa-nut tree, at the first glance, seems to compose the whole wood; there are, however, five or six other trees. One of these grows to a very large size, but from the extreme softness of its wood, is useless; another sort affords excellent timber for ship-building. Besides the trees, the number of plants is exceedingly limited, and consists of insignificant weeds. In my collection, which includes, I believe, nearly the perfect Flora, there are twenty species, without reckoning a moss, lichen, and fungus. To this number two trees must be added; one of which was not in flower, and the other I only heard of. The latter is a solitary tree of its kind, and grows near the beach, where, without doubt, the one seed was thrown up by the waves. A Guilandina also grows on only one of the islets. I do not include in the above list the sugar-cane, banana, some other vegetables, fruit-trees, and imported grasses. As the islands consist entirely of coral, and at one time must have existed as mere water-washed reefs, all their terrestrial productions must have been transported here by the waves of the sea. In accordance with this, the Florula has quite the character of a refuge for the destitute: Professor Henslow informs me that of the twenty species nineteen belong to different genera, and these again to no less than sixteen families![163]

In Holman's[164] Travels an account is given, on the authority of Mr. A. S. Keating, who resided twelve months on these islands, of the various seeds and other bodies which have been known to have been washed on shore. "Seeds and plants from Sumatra and Java have been driven up by the surf on the windward side of the islands. Among them have been found the Kimiri, native of Sumatra and the peninsula of Malacca; the cocoa-nut of Balci, known by its shape and size; the Dadass, which is planted by the Malays with the pepper-vine, the latter intwining round its trunk, and supporting itself by the prickles on its stem; the soap-tree; the castor-oil plant; trunks of the sago palm; and various kinds of seeds unknown to the Malays settled on the islands. These are all supposed to have been driven by the N. W. monsoon to the coast of New Holland, and thence to these islands by the S. E. trade-wind. Large masses of Java teak and Yellow wood have also been found, besides immense trees of red and white cedar, and the blue gumwood of New Holland, in a perfectly sound condition. All the hardy seeds, such as creepers, retain their germinating power, but the softer kinds, among which is the mangostin, are destroyed in the passage. Fishing-canoes, apparently from Java, have at times been washed on shore." It is interesting thus to discover how numerous the seeds are, which, coming from several countries, are drifted over the wide ocean. Professor Henslow tells me, he believes that nearly all the plants which I brought from these islands, are common littoral species in the East Indian archipelago. From the direction, however, of the winds and currents, it seems scarcely possible that they could have come here in a direct line. If, as suggested with much probability by Mr. Keating, they were first carried towards the coast of New Holland, and thence drifted back together with the productions of that country, the seeds, before germinating, must have travelled between 1800 and 2400 miles.

Chamisso,[165] when describing the Radack Archipelago, situated in the western part of the Pacific, states that "the sea brings to these islands the seeds and fruits of many trees, most of which have yet not grown here. The greater part of these seeds appear to have not yet lost the capability of growing."

It is also said that palms and bamboos from somewhere in the torrid zone, and trunks of northern firs, are washed on shore: these firs must have come from an immense distance. These facts are highly interesting. It cannot be doubted that if there were land-birds to pick up the seeds when first cast on shore, and a soil better adapted for their growth than the loose blocks of coral, that the most isolated of the lagoon-islands would in time possess a far more abundant Flora than they now have.

The list of land animals is even poorer than that of the plants. Some of the islets are inhabited by rats, which were brought in a ship from the Mauritius, wrecked here. These rats are considered by Mr. Waterhouse as identical with the English kind, but they are smaller, and more brightly coloured. There are no true land-birds, for a snipe and a rail (Rallus Phillippensis), though living entirely in the dry herbage, belong to the order of Waders. Birds of this order are said to occur on several of the small low islands in the Pacific. At Ascension, where there is no land-bird, a rail (Porphyrio simplex) was shot near the summit of the mountain, and it was evidently a solitary straggler. At Tristan d'Acunha, where, according to Carmichael, there are only two land-birds, there is a coot. From these facts I believe that the waders, after the innumerable web-footed species, are generally the first colonists of small isolated islands. I may add, that whenever I noticed birds, not of oceanic species, very far out at sea, they always belonged to this order; and hence they would naturally become the earliest colonists of any remote point of land.

Of reptiles I saw only one small lizard. Of insects I took pains to collect every kind. Exclusive of spiders, which were numerous, there were thirteen species.[166] Of these, one only was a beetle. A small ant swarmed by thousands under the loose dry blocks of coral, and was the only true insect which was abundant. Although the productions of the land are thus scanty, if we look to the waters of the surrounding sea, the number of organic beings is indeed infinite. Chamisso has described[167] the natural history of a lagoon-island in the Radack Archipelago; and it is remarkable how closely its inhabitants, in number and kind, resemble those of Keeling Island.

There is one lizard and two waders, namely, a snipe and curlew. Of plants there are nineteen species, including a fern; and some of these are the same with those growing here, though on a spot so immensely remote, and in a different ocean.

The long strips of land, forming the linear islets, have been raised only to that height to which the surf can throw fragments of coral, and the wind heap up calcareous sand. The solid flat of coral rock on the outside, by its breadth, breaks the first violence of the waves, which otherwise, in a day, would sweep away these islets and all their productions. The ocean and the land seem here struggling for mastery: although terra firma has obtained a footing, the denizens of the water think their claim at least equally good. In every part one meets hermit crabs of more than one species,[168] carrying on their backs the shells which they have stolen from the neighbouring beach. Overhead, numerous gannets, frigate-birds, and terns, rest on the trees; and the wood, from the many nests and from the smell of the atmosphere, might be called a sea-rookery. The gannets, sitting on their rude nests, gaze at one with a stupid yet angry air. The noddies, as their name expresses, are silly little creatures. But there is one charming bird: it is a small, snow-white tern, which smoothly hovers at the distance of a few feet above one's head, its large black eye scanning, with quiet curiosity, your expression. Little imagination is required to fancy that so light and delicate a body must be tenanted by some wandering fairy spirit.

Sunday, April 3rd.—After service I accompanied Captain Fitz Roy to the settlement, situated at the distance of some miles, on the point of an islet thickly covered with tall cocoa-nut trees. Captain Ross and Mr. Liesk live in a large barn-like house open at both ends, and lined with mats made of woven bark. The houses of the Malays are arranged along the shore of the lagoon. The whole place had rather a desolate aspect, for there were no gardens to show the signs of care and cultivation. The natives belong to different islands in the East Indian archipelago, but all speak the same language: we saw the inhabitants of Borneo, Celebes, Java, and Sumatra. In colour they resemble the Tahitians, from whom they do not widely differ in features. Some of the women, however, show a good deal of the Chinese character. I liked both their general expressions and the sound of their voices. They appeared poor, and their houses were destitute of furniture; but it was evident, from the plumpness of the little children, that cocoa-nuts and turtle afford no bad sustenance.

On this island the wells are situated, from which ships obtain water. At first sight it appears not a little remarkable that the fresh water should regularly ebb and flow with the tides; and it has even been imagined, that sand has the power of filtering the salt from the sea-water. These ebbing wells are common on some of the low islands in the West Indies. The compressed sand, or porous coral rock, is permeated like a sponge with the salt water; but the rain which falls on the surface must sink to the level of the surrounding sea, and must accumulate there, displacing an equal bulk of the salt water. As the water in the lower part of the great sponge-like coral mass rises and falls with the tides, so will the water near the surface; and this will keep fresh, if the mass be sufficiently compact to prevent much mechanical admixture; but where the land consists of great loose blocks of coral with open interstices, if a well be dug, the water, as I have seen, is brackish.

After dinner we stayed to see a curious half superstitious scene acted by the Malay women. A large wooden spoon dressed in garments, and which had been carried to the grave of a dead man, they pretend becomes inspired at the full of the moon, and will dance and jump about. After the proper preparations, the spoon, held by two women, became convulsed, and danced in good time to the song of the surrounding children and women. It was a most foolish spectacle; but Mr. Liesk maintained that many of the Malays believed in its spiritual movements. The dance did not commence till the moon had risen, and it was well worth remaining to behold her bright orb so quietly shining through the long arms of the cocoa-nut trees as they waved in the evening breeze. These scenes of the tropics are in themselves so delicious, that they almost equal those dearer ones at home, to which we are bound by each best feeling of the mind.

The next day I employed myself in examining the very interesting, yet simple structure and origin of these islands. The water being unusually smooth, I waded over the outer flat of dead rock as far as the living mounds of coral, on which the swell of the open sea breaks. In some of the gullies and hollows there were beautiful green and other coloured fishes, and the form and tints of many of the zoophytes were admirable. It is excusable to grow enthusiastic over the infinite numbers of organic beings with which the sea of the tropics, so prodigal of life, teems; yet I must confess I think those naturalists who have described, in well-known words, the submarine grottoes decked with a thousand beauties, have indulged in rather exuberant language.

April 6th.—I accompanied Captain Fitz Roy to an island at the head of the lagoon: the channel was exceedingly intricate, winding through fields of delicately branched corals. We saw several turtle and two boats were then employed in catching them. The water was so clear and shallow, that although at first a turtle quickly dives out of sight, yet in a canoe or boat under sail, the pursuers after no very long chase come up to it. A man standing ready in the bow, at this moment dashes through the water upon the turtle's back; then clinging with both hands by the shell of its neck, he is carried away till the animal becomes exhausted and is secured. It was quite an interesting chase to see the two boats thus doubling about, and the men dashing head foremost into the water trying to seize their prey. Captain Moresby informs me that in the Chagos archipelago in this same ocean, the natives, by a horrible process, take the shell from the back of the living turtle. "It is covered with burning charcoal, which causes the outer shell to curl upwards; it is then forced off with a knife, and before it becomes cold flattened between boards. After this barbarous process the animal is suffered to regain its native element, where, after a certain time, a new shell is formed; it is, however, too thin to be of any service, and the animal always appears languishing and sickly."

When we arrived at the head of the lagoon, we crossed a narrow islet, and found a great surf breaking on the windward coast. I can hardly explain the reason, but there is to my mind much grandeur in the view of the outer shores of these lagoon-islands. There is a simplicity in the barrier-like beach; the margin of green bushes and tall cocoa-nuts, the solid flat of dead coral-rock, strewed here and there with great loose fragments, and the line of furious breakers, all rounding away towards either hand. The ocean throwing its waters over the broad reef appears an invincible, all-powerful enemy; yet we see it resisted, and even conquered, by means which at first seem most weak and inefficient. It is not that the ocean spares the rock of coral; the great fragments scattered over the reef, and heaped on the beach, whence the tall cocoa-nut springs, plainly bespeak the unrelenting power of the waves. Nor are any periods of repose granted. The long swell caused by the gentle but steady action of the trade-wind, always blowing in one direction over a wide area, causes breakers, almost equalling in force those during a gale of wind in the temperate regions, and which never cease to rage. It is impossible to behold these waves without feeling a conviction that an island, though built of the hardest rock, let it be porphyry, granite, or quartz, would ultimately yield and be demolished by such an irresistible power. Yet these low, insignificant coral-islets stand and are victorious: for here another power, as an antagonist, takes part in the contest. The organic forces separate the atoms of carbonate of lime, one by one, from the foaming breakers, and unite them into a symmetrical structure. Let the hurricane tear up its thousand huge fragments; yet what will that tell against the accumulated labour of myriads of architects at work night and day, month after month? Thus do we see the soft and gelatinous body of a polypus, through the agency of the vital laws, conquering the great mechanical power of the waves of an ocean which neither the art of man nor the inanimate works of nature could successfully resist.

We did not return on board till late in the evening, for we stayed a long time in the lagoon, examining the fields of coral and the gigantic shells of the chama, into which, if a man were to put his hand, he would not, as long as the animal lived, be able to withdraw it. Near the head of the lagoon I was much surprised to find a wide area, considerably more than a mile square, covered with a forest of delicately branching corals, which, though standing upright, were all dead and rotten. At first I was quite at a loss to understand the cause; afterwards it occurred to me that it was owing to the following rather curious combination of circumstances. It should, however, first be stated, that corals are not able to survive even a short exposure in the air to the sun's rays, so that their upward limit of growth is determined by that of lowest water at spring tides. It appears, from some old charts, that the long island to windward was formerly separated by wide channels into several islets; this fact is likewise indicated by the trees being younger on these portions. Under the former condition of the reef, a strong breeze, by throwing more water over the barrier, would tend to raise the level of the lagoon. Now it acts in a directly contrary manner; for the water within the lagoon not only is not increased by currents from the outside, but is itself blown outwards by the force of the wind. Hence it is observed, that the tide near the head of the lagoon does not rise so high during a strong breeze as it does when it is calm. This difference of level, although no doubt very small, has, I believe, caused the death of those coral-groves, which under the former and more open condition of the outer reef has attained the utmost possible limit of upward growth.

A few miles north of Keeling there is another small atoll, the lagoon of which is nearly filled up with coral-mud. Captain Ross found embedded in the conglomerate on the outer coast, a well-rounded fragment of greenstone, rather larger than a man's head: he and the men with him were so much surprised at this, that they brought it away and preserved it as a curiosity. The occurrence of this one stone, where every other particle of matter is calcareous, certainly is very puzzling. The island has scarcely ever been visited, nor is it probable that a ship had been wrecked there. From the absence of any better explanation, I came to the conclusion that it must have come entangled in the roots of some large tree: when, however, I considered the great distance from the nearest land, the combination of chances against a stone thus being entangled, the tree washed into the sea, floated so far, then landed safely, and the stone finally so embedded as to allow of its discovery, I was almost afraid of imagining a means of transport apparently so improbable. It was therefore with great interest that I found Chamisso, the justly distinguished naturalist who accompanied Kotzebue, stating that the inhabitants of the Radack archipelago, a group of lagoon-islands in the midst of the Pacific, obtained stones for sharpening their instruments by searching the roots of trees which are cast upon the beach. It will be evident that this must have happened several times, since laws have been established that such stones belong to the chief, and a punishment is inflicted on any one who attempts to steal them. When the isolated position of these small islands in the midst of a vast ocean—their great distance from any land excepting that of coral formation, attested by the value which the inhabitants, who are such bold navigators, attach to a stone of any kind,[169]—and the slowness of the currents of the open sea, are all considered, the occurrence of pebbles thus transported does appear wonderful. Stones may often be thus carried; and if the island on which they are stranded is constructed of any other substance besides coral, they would scarcely attract attention, and their origin at least would never be guessed. Moreover, this agency may long escape discovery from the probability of trees, especially those loaded with stones, floating beneath the surface. In the channels of Tierra del Fuego large quantities of drift timber are cast upon the beach, yet it is extremely rare to meet a tree swimming on the water. These facts may possibly throw light on single stones, whether angular or rounded, occasionally found embedded in fine sedimentary masses.

During another day I visited West Islet, on which the vegetation was perhaps more luxuriant than on any other. The cocoa-nut trees generally grow separate, but here the young ones flourished beneath their tall parents, and formed with their long and curved fronds the most shady arbours. Those alone who have tried it, know how delicious it is to be seated in such shade, and drink the cool pleasant fluid of the cocoa-nut. In this island there is a large bay-like space, composed of the finest white sand: it is quite level and is only covered by the tide at high water; from this large bay smaller creeks penetrate the surrounding woods. To see a field of glittering white sand, representing water, with the cocoa-nut trees extending their tall and waving trunks around the margin, formed a singular and very pretty view.

I have before alluded to a crab which lives on the cocoa-nuts; it is very common on all parts of the dry land, and grows to a monstrous size: it is closely allied or identical with the Birgus latro. The front pair of legs terminate in very strong and heavy pincers, and the last pair are fitted with others weaker and much narrower. It would at first be thought quite impossible for a crab to open a strong cocoa-nut covered with the husk; but Mr. Liesk assures me that he has repeatedly seen this effected. The crab begins by tearing the husk, fibre by fibre, and always from that end under which the three eye-holes are situated; when this is completed, the crab commences hammering with its heavy claws on one of the eye-holes till an opening is made. Then turning round its body, by the aid of its posterior and narrow pair of pincers, it extracts the white albuminous substance. I think this is as curious a case of instinct as ever I heard of, and likewise of adaptation in structure between two objects apparently so remote from each other in the scheme of nature, as a crab and a cocoa-nut tree. The Birgus is diurnal in its habits; but every night it is said to pay a visit to the sea, no doubt for the purpose of moistening its branchiæ. The young are likewise hatched, and live for some time, on the coast. These crabs inhabit deep burrows, which they hollow out beneath the roots of trees; and where they accumulate surprising quantities of the picked fibres of the cocoa-nut husk, on which they rest as on a bed. The Malays sometimes take advantage of this, and collect the fibrous mass to use as junk. These crabs are very good to eat; moreover, under the tail of the larger ones there is a mass of fat, which, when melted, sometimes yields as much as a quart bottle full of limpid oil. It has been stated by some authors that the Birgus crawls up the cocoa-nut trees for the purpose of stealing the nuts: I very much doubt the possibility of this; but with the Pandanus[170] the task would be very much easier. I was told by Mr. Liesk that on these islands the Birgus lives only on the nuts which have fallen to the ground.

Captain Moresby informs me that this crab inhabits the Chagos and Seychelle groups, but not the neighbouring Maldiva archipelago. It formerly abounded at Mauritius, but only a few small ones are now found there. In the Pacific, this species, or one with closely allied habits, is said[171] to inhabit a single coral island, north of the Society group. To show the wonderful strength of the front pair of pincers, I may mention, that Captain Moresby confined one in a strong tin-box, which had held biscuits, the lid being secured with wire; but the crab turned down the edges and escaped. In turning down the edges, it actually punched many small holes quite through the tin!

I was a good deal surprised by finding two species of coral of the genus Millepora (M. complanata and alcicornis), possessed of the power of stinging. The stony branches or plates, when taken fresh from the water, have a harsh feel and are not slimy, although possessing a strong and disagreeable smell. The stinging property seems to vary in different specimens: when a piece was pressed or rubbed on the tender skin of the face or arm, a pricking sensation was usually caused, which came on after the interval of a second, and lasted only for a few minutes. One day, however, by merely touching my face with one of the branches, pain was instantaneously caused; it increased as usual after a few seconds, and remaining sharp for some minutes, was perceptible for half an hour afterwards. The sensation was as bad as that from a nettle, but more like that caused by the Physalia or Portuguese man-of-war. Little red spots were produced on the tender skin of the arm, which appeared as if they would have formed watery pustules, but did not. M. Quoy mentions this case of the Millepora; and I have heard of stinging corals in the West Indies. Many marine animals seem to have this power of stinging: besides the Portuguese man-of-war, many jelly-fish, and the Aplysia or sea-slug of the Cape de Verd Islands, it is stated in the voyage of the Astrolabe, that an Actinia or sea-anemone, as well as a flexible coralline allied to Sertularia, both possess this means of offence or defence. In the East Indian sea, a stinging sea-weed is said to be found.

Two species of fish, of the genus Scarus, which are common here, exclusively feed on coral: both are coloured of a splendid bluish-green, one living invariably in the lagoon, and the other amongst the outer breakers. Mr. Liesk assured us, that he had repeatedly seen whole shoals grazing with their strong bony jaws on the tops of the coral branches: I opened the intestines of several, and found them distended with yellowish calcareous sandy mud. The slimy disgusting Holuthuriæ (allied to our starfish), which the Chinese gourmands are so fond of, also feed largely, as I am informed by Dr. Allan, on corals; and the bony apparatus within their bodies seems well adapted for this end. These Holuthuriæ, the fish, the numerous burrowing shells, and nereidous worms, which perforate every block of dead coral, must be very efficient agents in producing the fine white mud which lies at the bottom and on the shores of the lagoon. A portion, however, of this mud, which when wet resembled pounded chalk, was found by Professor Ehrenberg to be partly composed of siliceous-shielded infusoria.

April 12th.—In the morning we stood out of the lagoon on our passage to the Isle of France. I am glad we have visited these islands: such formations surely rank high amongst the wonderful objects of this world. Captain Fitz Roy found no bottom with a line 7200 feet in length, at the distance of only 2200 yards from the shore; hence this island forms a lofty submarine mountain, with sides steeper even than those of the most abrupt volcanic cone. The saucer-shaped summit is nearly ten miles across; and every single atom,[172] from the least particle to the largest fragment of rock, in this great pile, which however is small compared with very many other lagoon-islands, bears the stamp of having been subjected to organic arrangement. We feel surprise when travellers tell us of the vast dimensions of the Pyramids and other great ruins, but how utterly insignificant are the greatest of these, when compared to these mountains of stone accumulated by the agency of various minute and tender animals! This is a wonder which does not at first strike the eye of the body, but, after reflection, the eye of reason.

I WILL NOW GIVE a very brief account of the three great classes of coral-reefs; namely, Atolls, Barrier, and Fringing-reefs, and will explain my views[173] on their formation. Almost every voyager who has crossed the Pacific has expressed his unbounded astonishment at the lagoon-islands, or as I shall for the future call them by their Indian name of atolls, and has attempted some explanation. Even as long ago as the year 1605, Pyrard de

Laval well exclaimed, "C'est une merveille de voir chacun de ces atollons, environné d'un grand banc de pierre tout autour, n'y ayant point d'artifice humain." The accompanying sketch of Whitsunday Island in the Pacific, copied from Capt. Beechey's admirable Voyage, gives but a faint idea of the singular aspect of an atoll: it is one of the smallest size, and has its narrow islets united together in a ring. The immensity of the ocean, the fury of the breakers, contrasted with the lowness of the land and the smoothness of the bright green water within the lagoon, can hardly be imagined without having been seen.

The earlier voyagers fancied that the coral-building animals instinctively built up their great circles to afford themselves protection in the inner parts; but so far is this from the truth, that those massive kinds, to whose growth on the exposed outer shores the very existence of the reef depends, cannot live within the lagoon, where other delicately-branching kinds flourish. Moreover, on this view, many species of distinct genera and families are supposed to combine for one end; and of such a combination, not a single instance can be found in the whole of nature. The theory that has been most generally received is, that atolls are based on submarine craters; but when we consider the form and size of some, the number, proximity, and relative positions of others, this idea loses its plausible character: thus Suadiva atoll is 44 geographical miles in diameter in one line, by 34 miles in another line; Rimsky is 54 by 20 miles across, and it has a strangely sinuous margin; Bow atoll is 30 miles long, and on an average only 6 in width; Menchicoff atoll consists of three atolls united or tied together. This theory, moreover, is totally inapplicable to the northern Maldiva atolls in the Indian Ocean (one of which is 88 miles in length, and between 10 and 20 in breadth), for they are not bounded like ordinary atolls by narrow reefs, but by a vast number of separate little atolls; other little atolls rising out of the great central lagoon-like spaces. A third and better theory was advanced by Chamisso, who thought that from the corals growing more vigorously where exposed to the open sea, as undoubtedly is the case, the outer edges would grow up from the general foundation before any other part, and that this would account for the ring or cup-shaped structure. But we shall immediately see, that in this, as well as in the crater-theory, a most important consideration has been overlooked, namely, on what have the reef-building corals, which cannot live at a great depth, based their massive structures?

Numerous soundings were carefully taken by Captain Fitz Roy on the steep outside of Keeling atoll, and it was found that within ten fathoms, the prepared tallow at the bottom of the lead, invariably came up marked with the impression of living corals, but as perfectly clean as if it had been dropped on a carpet of turf; as the depth increased, the impressions became less numerous, but the adhering particles of sand more and more numerous, until at last it was evident that the bottom consisted of a smooth sandy layer: to carry on the analogy of the turf, the blades of grass grew thinner and thinner, till at last the soil was so sterile, that nothing sprang from it. From these observations, confirmed by many others, it may be safely inferred that the utmost depth at which corals can construct reefs is between 20 and 30 fathoms. Now there are enormous areas in the Pacific and Indian Ocean, in which every single island is of coral formation, and is raised only to that height to which the waves can throw up fragments, and the winds pile up sand. Thus Radack group of atolls is an irregular square, 520 miles long and 240 broad; the Low Archipelago is elliptic-formed, 840 miles in its longer, and 420 in its shorter axis: there are other small groups and single low islands between these two archipelagoes, making a linear space of ocean actually more than 4000 miles in length, in which not one single island rises above the specified height. Again, in the Indian Ocean there is a space of ocean 1500 miles in length, including three archipelagoes, in which every island is low and of coral formation. From the fact of the reef-building corals not living at great depths, it is absolutely certain that throughout these vast areas, wherever there is now an atoll, a foundation must have originally existed within a depth of from 20 to 30 fathoms from the surface. It is improbable in the highest degree that broad, lofty, isolated, steep-sided banks of sediment, arranged in groups and lines hundreds of leagues in length, could have been deposited in the central and profoundest parts of the Pacific and Indian Oceans, at an immense distance from any continent, and where the water is perfectly limpid. It is equally improbable that the elevatory forces should have uplifted throughout the above vast areas, innumerable great rocky banks within 20 to 30 fathoms, or 120 to 180 feet, of the surface of the sea, and not one single point above that level; for where on the whole surface of the globe can we find a single chain of mountains, even a few hundred miles in length, with their many summits rising within a few feet of a given level, and not one pinnacle above it? If then the foundations, whence the atoll-building corals sprang, were not formed of sediment, and if they were not lifted up to the required level, they must of necessity have subsided into it; and this at once solves the difficulty. For as mountain after mountain, and island after island, slowly sank beneath the water, fresh bases would be successively afforded for the growth of the corals. It is impossible here to enter into all the necessary details, but I venture to defy[174] any one to explain in any other manner how it is possible that numerous islands should be distributed throughout vast areas—all the islands being low—all being built of corals, absolutely requiring a foundation within a limited depth from the surface.

Before explaining how atoll-formed reefs acquire their peculiar structure, we must turn to the second great class, namely, Barrier-reefs. These either extend in straight lines in front of the shores of a continent or of a large island, or they encircle smaller islands; in both cases, being separated from the land by a broad and rather deep channel of water, analogous to the lagoon within an atoll. It is remarkable how little attention has been paid to encircling barrier-reefs; yet they are truly wonderful structures. The following sketch represents part of the barrier encircling the island of Bolabola in the Pacific, as seen from one of the central peaks. In this instance the whole line of reef has been converted into land; but usually a snow-white line of great breakers, with only here and there a single low islet crowned with cocoa-nut trees, divides the dark heaving waters of the ocean from the light-green expanse of the lagoon-channel. And the quiet waters of this channel generally bathe a fringe of low alluvial soil, loaded with the most beautiful productions of the tropics, and lying at the foot of the wild, abrupt, central mountains.

Encircling barrier-reefs are of all sizes, from three miles to no less than forty-four miles in diameter; and that which fronts one side, and encircles both ends, of New Caledonia, is 400 miles long. Each reef includes one, two, or several rocky islands of various heights; and in one instance, even as many as twelve separate islands. The reef runs at a greater or less distance

from the included land; in the Society archipelago generally from one to three or four miles; but at Hogoleu the reef is 20 miles on the southern side, and 14 miles on the opposite or northern side, from the included islands. The depth within the lagoon-channel also varies much; from 10 to 30 fathoms may be taken as an average; but at Vanikoro there are spaces no less than 56 fathoms or 363 feet deep. Internally the reef either slopes gently into the lagoon-channel, or ends in a perpendicular wall sometimes between two and three hundred feet under water in height: externally the reef rises, like an atoll, with extreme abruptness out of the profound depths of the ocean. What can be more singular than these structures? We see an island, which may be compared to a castle situated on the summit of a lofty submarine mountain, protected by a great wall of coral-rock, always steep externally and sometimes internally, with a broad level summit, here and there breached by a narrow gateway, through which the largest ships can enter the wide and deep encircling moat.

As far as the actual reef of coral is concerned, there is not the smallest difference, in general size, outline, grouping, and even in quite trifling details of structure, between a barrier and an atoll. The geographer Balbi has well remarked, that an encircled island is an atoll with high land rising out of its lagoon; remove the land from within, and a perfect atoll is left.

But what has caused these reefs to spring up at such great distances from the shores of the included islands? It cannot be that the corals will not grow close to the land; for the shores within the lagoon-channel, when not surrounded by alluvial soil, are often fringed by living reefs; and we shall presently see that there is a whole class, which I have called Fringing Reefs from their close attachment to the shores both of continents and of islands. Again, on what have the reef-building corals, which cannot live at great depths, based their encircling structures? This is a great apparent difficulty, analogous to that in the case of atolls, which has generally been overlooked. It will be perceived more clearly by inspecting the following sections, which are real ones, taken in north and south lines, through the islands with their barrier-reefs, of Vanikoro, Gambier, and Maurua; and they are laid down, both vertically and horizontally, on the same scale of a quarter of an inch to a mile.

It should be observed that the sections might have been taken in any direction through these islands, or through many other encircled islands, and the general features would have been the same. Now, bearing in mind that reef-building coral cannot live at a greater depth than from 20 to 30 fathoms, and that the scale is so small that the plummets on the right hand show a depth of 200 fathoms, on what are these barrier-reefs based? Are we to suppose that each island is surrounded by a collar-like submarine ledge of rock, or by a great bank of sediment, ending abruptly where the reef ends?

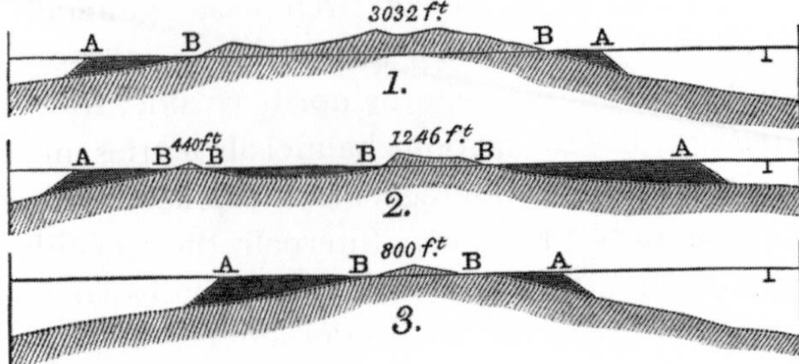

1. Vanikoro. 2. Gambier Islands. 3. Maurua.

The horizontal shading shows the barrier-reefs and lagoon-channels. The inclined shading above the level of the sea (AA) shows the actual form of the land; the inclined shading below this line, shows its probable prolongation under water.

If the sea had formerly eaten deeply into the islands, before they were protected by the reefs, thus having left a shallow ledge round them under water, the present shores would have been invariably bounded by great precipices; but this is most rarely the case. Moreover, on this notion, it is not possible to explain why the corals should have sprung up, like a wall, from the extreme outer margin of the ledge, often leaving a broad space of water within, too deep for the growth of corals. The accumulation of a wide bank of sediment all round these islands, and generally widest where the included islands are smallest, is highly improbable, considering their exposed positions in the central and deepest parts of the ocean. In the case of the barrier-reef of New Caledonia, which extends for 150 miles beyond the northern point of the islands, in the same straight line with which it fronts the west coast, it is hardly possible to believe that a bank of sediment could thus have been straightly deposited in front of a lofty island, and so far beyond its termination in the open sea. Finally, if we look to other oceanic islands of about the same height and of similar geological constitution, but not encircled by coral-reefs, we may in vain search for so trifling a circumambient depth as 30 fathoms, except quite near to their shores; for usually land that rises abruptly out of water, as do most of the encircled and non-encircled oceanic islands, plunges abruptly under it. On what then, I repeat, are these barrier-reefs based? Why, with their wide and deep moat-like channels, do they stand so far from the included land? We shall soon see how easily these difficulties disappear.

We come now to our third class of Fringing-reefs, which will require a very short notice. Where the land slopes abruptly under water, these reefs are only a few yards in width, forming a mere ribbon or fringe round the shores: where the land slopes gently under the water the reef extends further, sometimes even as much as a mile from the land; but in such cases the soundings outside the reef always show that the submarine prolongation of the land is gently inclined. In fact, the reefs extend only to that distance from the shore, at which a foundation within the requisite depth from 20 to 30 fathoms is found. As far as the actual reef is concerned, there is no essential difference between it and that forming a barrier or an atoll: it is, however, generally of less width, and consequently few islets have been formed on it. From the corals growing more vigorously on the outside, and from the noxious effect of the sediment washed inwards, the outer edge of the reef is the highest part, and between it and the land there is generally a shallow sandy channel a few feet in depth. Where banks or sediments have accumulated near to the surface, as in parts of the West Indies, they sometimes become fringed with corals, and hence in some degree resemble lagoon-islands or atolls, in the same manner as fringing-reefs, surrounding gently sloping islands, in some degree resemble barrier-reefs.

NO THEORY on the formation of coral-reefs can be considered satisfactory which does not include the three great classes. We have seen that we are driven to believe in the subsidence of those vast areas, interspersed with low islands, of which not one rises above the height to which the wind and waves can throw up matter, and yet are constructed by animals requiring a foundation, and that foundation to lie at no great depth. Let us then take an island surrounded by fringing-reefs, which offer no difficulty in their structure; and let this island with its reefs, represented by the unbroken lines in the woodcut, slowly subside. Now, as the island sinks down, either a few feet at a time or quite insensibly, we may safely infer, from what is known of the conditions favourable to the growth of coral, that the living masses, bathed by the surf on the margin of the reef, will soon regain the surface. The water, however, will encroach little by little on the shore, the island becoming lower and smaller, and the space between the inner edge of the reef and the beach proportionately broader. A section of the reef and island in this state, after a subsidence of several hundred feet, is given by the dotted lines. Coral islets are supposed to have been formed on the reef; and a ship is anchored in the lagoon-channel. This channel will be more or less deep, according to the rate of subsidence, to the amount of sediment accumulated in it, and to the growth of the delicately branched corals which can live there. The section in this state resembles in every respect

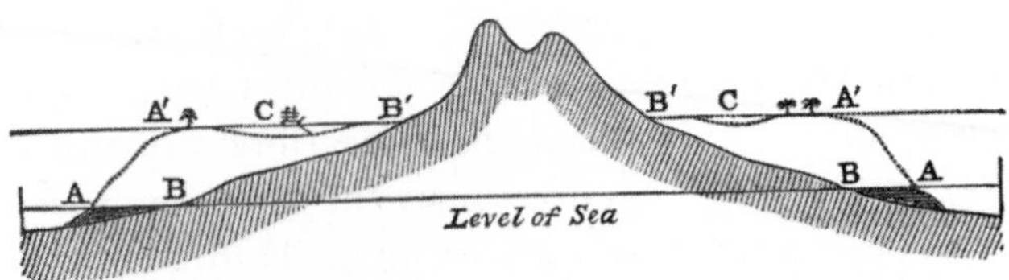

AA. Outer edges of the fringing-reef, at the level of the sea.
BB. The shores of the fringed island.
A'A'. Outer edges of the reef, after its upward growth during a period of subsidence, now converted into a barrier, with islets on it.
B'B'. The shores of the now encircled island.
CC. Lagoon-channel.
N. B.—In this and the following woodcut, the subsidence of the land could be represented only by an apparent rise in the level of sea.

one drawn through an encircled island: in fact, it is a real section (on the scale of .517 of an inch to a mile) through Bolabola in the Pacific. We can now at once see why encircling barrier-reefs stand so far from the shores which they front. We can also perceive, that a line drawn perpendicularly down from the outer edge of the new reef, to the foundation of solid rock beneath the old fringing-reef, will exceed by as many feet as there have been feet of subsidence, that small limit of depth at which the effective corals can live:—the little architects having built up their great wall-like mass, as the whole sank down, upon a basis formed of other corals and their consolidated fragments. Thus the difficulty on this head, which appeared so great, disappears.

If, instead of an island, we had taken the shore of a continent fringed with reefs, and had imagined it to have subsided, a great straight barrier, like that of Australia or New Caledonia, separated from the land by a wide and deep channel, would evidently have been the result.

Let us take our new encircling barrier-reef, of which the section is now represented by unbroken lines, and which, as I have said, is a real section through Bolabola, and let it go on subsiding. As the barrier-reef slowly sinks down, the corals will go on vigorously growing upwards; but as the island sinks, the water will gain inch by inch on the shore—the separate mountains first forming separate islands within one great reef—and finally, the last and highest pinnacle disappearing. The instant this takes place, a perfect atoll is formed: I have said, remove the high land from within an encircling barrier-reef, and an atoll is left, and the land has been removed. We

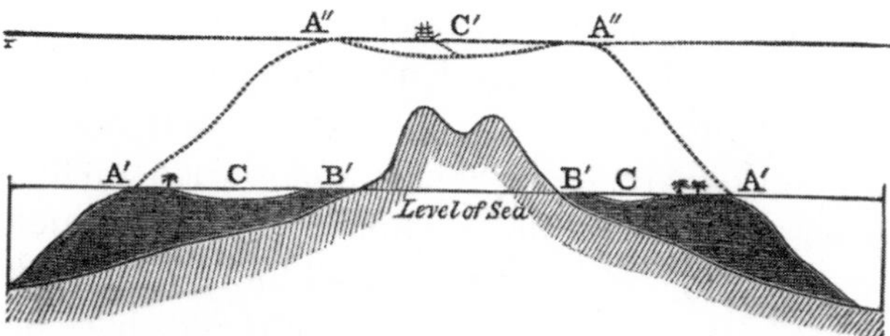

A′A′. Outer edges of the barrier-reef at the level of the sea, with islets on it.
B′B′. The shores of the included island. CC. The lagoon-channel.
A″A″. Outer edges of the reef, now converted into an atoll.
C′ The lagoon of the new atoll.
N. B.—According to the true scale, the depths of the lagoon-channel and lagoon are much exaggerated.

can now perceive how it comes that atolls, having sprung from encircling barrier-reefs, resemble them in general size, form, in the manner in which they are grouped together, and in their arrangement in single or double lines; for they may be called rude outline charts of the sunken islands over which they stand. We can further see how it arises that the atolls in the Pacific and Indian Oceans extend in lines parallel to the generally prevailing strike of the high islands and great coast-lines of those oceans. I venture, therefore, to affirm, that on the theory of the upward growth of the corals during the sinking of the land,[175] all the leading features in those wonderful structures, the lagoon-islands or atolls, which have so long excited the attention of voyagers, as well as in the no less wonderful barrier-reefs, whether encircling small islands or stretching for hundreds of miles along the shores of a continent, are simply explained.

It may be asked, whether I can offer any direct evidence of the subsidence of barrier-reefs or atolls; but it must be borne in mind how difficult it must ever be to detect a movement, the tendency of which is to hide under water the part affected. Nevertheless, at Keeling atoll I observed on all sides of the lagoon old cocoa-nut trees undermined and falling; and in one place the foundation-posts of a shed, which the inhabitants asserted had stood seven years before just above high-water mark, but now was daily washed by every tide: on inquiry I found that three earthquakes, one of them severe, had been felt here during the last ten years. At Vanikoro, the lagoon-channel is remarkably deep, scarcely any alluvial soil has accumulated at the foot of the lofty included mountains, and remarkably few islets have been formed by the heaping of fragments and sand on the wall-like barrier reef; these facts, and some analogous ones, led me to believe that this island must lately have subsided and the reef grown upwards: here again earthquakes are frequent and very severe. In the Society archipelago, on the other hand, where the lagoon-channels are almost choked up, where much low alluvial land has accumulated, and where in some cases long islets have been formed on the barrier-reefs—facts all showing that the islands have not very lately subsided—only feeble shocks are most rarely felt. In these coral formations, where the land and water seem struggling for mastery, it must be ever difficult to decide between the effects of a change in the set of the tides and of a slight subsidence: that many of these reefs and atolls are subject to changes of some kind is certain; on some atolls the islets appear to have increased greatly within a late period; on others they have been partially or wholly washed away. The inhabitants of parts of the Maldiva archipelago know the date of the first formation of some islets; in other parts, the corals are now flourishing on water-washed reefs, where holes made for graves attest the former existence of inhabited land. It is difficult to believe in frequent changes in the tidal currents of an open ocean; whereas, we have in the earthquakes recorded by the natives on some atolls, and in the great fissures observed on other atolls, plain evidence of changes and disturbances in progress in the subterranean regions.

It is evident, on our theory, that coasts merely fringed by reefs cannot have subsided to any perceptible amount; and therefore they must, since the growth of their corals, either have remained stationary or have been upheaved. Now, it is remarkable how generally it can be shown, by the presence of upraised organic remains, that the fringed islands have been elevated: and so far, this is indirect evidence in favour of our theory. I was particularly struck with this fact, when I found, to my surprise, that the descriptions given by MM. Quoy and Gaimard were applicable, not to reefs in general as implied by them, but only to those of the fringing class; my surprise, however, ceased when I afterwards found that, by a strange chance, all the several islands visited by these eminent naturalists, could be shown by their own statements to have been elevated within a recent geological era.

Not only the grand features in the structure of barrier-reefs and of atolls, and to their likeness to each other in form, size, and other characters, are explained on the theory of subsidence—which theory we are independently forced to admit in the very areas in question, from the necessity of finding bases for the corals within the requisite depth—but many details in structure and exceptional cases can thus also be simply explained. I will give only a few instances. In barrier-reefs it has long been remarked with surprise, that the passages through the reef exactly face valleys in the included land, even in cases where the reef is separated from the land by a lagoon-channel so wide and so much deeper than the actual passage itself, that it seems hardly possible that the very small quantity of water or sediment brought down could injure the corals on the reef. Now, every reef of the fringing class is breached by a narrow gateway in front of the smallest rivulet, even if dry during the greater part of the year, for the mud, sand, or gravel, occasionally washed down kills the corals on which it is deposited. Consequently, when an island thus fringed subsides, though most of the narrow gateways will probably become closed by the outward and upward growth of the corals, yet any that are not closed (and some must always be kept open by the sediment and impure water flowing out of the lagoon-channel) will still continue to front exactly the upper parts of those valleys, at the mouths of which the original basal fringing-reef was breached.

We can easily see how an island fronted only on one side, or on one side with one end or both ends encircled by barrier-reefs, might after long-continued subsidence be converted either into a single wall-like reef, or into an atoll with a great straight spur projecting from it, or into two or three atolls tied together by straight reefs—all of which exceptional cases actually occur. As the reef-building corals require food, are preyed upon by other animals, are killed by sediment, cannot adhere to a loose bottom, and may be easily carried down to a depth whence they cannot spring up again, we need feel no surprise at the reefs both of atolls and barriers becoming in parts imperfect. The great barrier of New Caledonia is thus imperfect and broken in many parts; hence, after long subsidence, this great reef would not produce one great atoll 400 miles in length, but a chain or archipelago of atolls, of very nearly the same dimension with those in the Maldiva archipelago. Moreover, in an atoll once breached on opposite sides, from the likelihood of the oceanic and tidal currents passing straight through the breaches, it is extremely improbable that the corals, especially during continued subsidence, would ever be able again to unite the rim; if they did not, as the whole sank downwards, one atoll would be divided into two or more. In the Maldiva archipelago there are distinct atolls so related to each other in position, and separated by channels either unfathomable or very deep (the channel between Ross and Ari atolls is 150 fathoms, and that between the north and south Nillandoo atolls is 200 fathoms in depth), that it is impossible to look at a map of them without believing that they were once more intimately related. And in this same archipelago, Mahlos-Mahdoo atoll is divided by a bifurcating channel from 100 to 132 fathoms in depth, in such a manner, that it is scarcely possible to say whether it ought strictly to be called three separate atolls, or one great atoll not yet finally divided.

I will not enter on many more details; but I must remark that the curious structure of the northern Maldiva atolls receives (taking into consideration the free entrance of the sea through their broken margins) a simple explanation in the upward and outward growth of the corals, originally based both on small detached reefs in their lagoons, such as occur in common atolls, and on broken portions of the linear marginal reef, such as bounds every atoll of the ordinary form. I cannot refrain from once again remarking on the singularity of these complex structures—a great sandy and generally concave disk rises abruptly from the unfathomable ocean, with its central expanse studded, and its edge symmetrically bordered with oval basins of coral-rock just lipping the surface of the sea, sometimes clothed with vegetation, and each containing a lake of clear water!

One more point in detail: as in the two neighbouring archipelagoes corals flourish in one and not in the other, and as so many conditions before enumerated must affect their existence, it would be an inexplicable fact if, during the changes to which earth, air, and water are subjected, the reef-building corals were to keep alive for perpetuity on any one spot or area. And as by our theory the areas including atolls and barrier-reefs are subsiding, we ought occasionally to find reefs both dead and submerged. In all reefs, owing to the sediment being washed out of the lagoon-channel to leeward, that side is least favourable to the long-continued vigorous growth of the corals; hence dead portions of reef not unfrequently occur on the leeward side; and these, though still retaining their proper wall-like form, are now in several instances sunk several fathoms beneath the surface. The Chagos group appears from some cause, possibly from the subsidence having been too rapid, at present to be much less favourably circumstanced for the growth of reefs than formerly: one atoll has a portion of its marginal reef, nine miles in length, dead and submerged; a second has only a few quite small living points which rise to the surface; a third and fourth are entirely dead and submerged; a fifth is a mere wreck, with its structure almost obliterated. It is remarkable that in all these cases, the dead reefs and portions of reef lie at nearly the same depth, namely, from six to eight fathoms beneath the surface, as if they had been carried down by one uniform movement. One of these "half-drowned atolls," so called by Capt. Moresby (to whom I am indebted for much invaluable information), is of vast size, namely, ninety nautical miles across in one direction, and seventy miles in another line; and is in many respects eminently curious. As by our theory it follows that new atolls will generally be formed in each new area of subsidence, two weighty objections might have been raised, namely, that atolls must be increasing indefinitely in number; and secondly, that in old areas of subsidence each separate atoll must be increasing indefinitely in thickness, if proofs of their occasional destruction could not have been adduced. Thus have we traced the history of these great rings of coral-rock, from their first origin through their normal changes, and through the occasional accidents of their existence, to their death and final obliteration.

In my volume on "Coral Formations" I have published a map, in which I have coloured all the atolls dark-blue, the barrier-reefs pale-blue, and the fringing reefs red. These latter reefs have been formed whilst the land has been stationary, or, as appears from the frequent presence of upraised organic remains, whilst it has been slowly rising: atolls and barrier-reefs, on the other hand, have grown up during the directly opposite movement of subsidence, which movement must have been very gradual, and in the case of atolls so vast in amount as to have buried every mountain-summit over wide ocean-spaces. Now in this map we see that the reefs tinted pale and dark-blue, which have been produced by the same order of movement, as a general rule manifestly stand near each other. Again we see, that the areas with the two blue tints are of wide extent; and that they lie separate from extensive lines of coast coloured red, both of which circumstances might naturally have been inferred, on the theory of the nature of the reefs having been governed by the nature of the earth's movement. It deserves notice, that in more than one instance where single red and blue circles approach near each other, I can show that there have been oscillations of level; for in such cases the red or fringed circles consist of atolls, originally by our theory formed during subsidence, but subsequently upheaved; and on the other hand, some of the pale-blue or encircled islands are composed of coral-rock, which must have been uplifted to its present height before that subsidence took place, during which the existing barrier-reefs grew upwards.

Authors have noticed with surprise, that although atolls are the commonest coral-structures throughout some enormous oceanic tracts, they are entirely absent in other seas, as in the West Indies: we can now at once perceive the cause, for where there has not been subsidence, atolls cannot have been formed; and in the case of the West Indies and parts of the East Indies, these tracts are known to have been rising within the recent period. The larger areas, coloured red and blue, are all elongated; and between the two colours there is a degree of rude alternation, as if the rising of one had balanced the sinking of the other. Taking into consideration the proofs of recent elevation both on the fringed coasts and on some others (for instance, in South America) where there are no reefs, we are led to conclude that the great continents are for the most part rising areas: and from the nature of the coral-reefs, that the central parts of the great oceans are sinking areas. The East Indian archipelago, the most broken land in the world, is in most parts an area of elevation, but surrounded and penetrated, probably in more lines than one, by narrow areas of subsidence.

I have marked with vermilion spots all the many known active volcanos within the limits of this same map. Their entire absence from every one of the great subsiding areas, coloured either pale or dark blue, is most striking and not less so is the coincidence of the chief volcanic chains with the parts coloured red, which we are led to conclude have either long remained stationary, or more generally have been recently upraised. Although a few of the vermilion spots occur within no great distance of single circles tinted blue, yet not one single active volcano is situated within several hundred miles of an archipelago, or even small group of atolls. It is, therefore, a striking fact that in the Friendly archipelago, which consists of a group of atolls upheaved and since partially worn down, two volcanos, and perhaps more, are historically known to have been in action. On the other hand, although most of the islands in the Pacific which are encircled by barrier-reefs, are of volcanic origin, often with the remnants of craters still distinguishable, not one of them is known to have ever been in eruption. Hence in these cases it would appear, that volcanos burst forth into action and become extinguished on the same spots, accordingly as elevatory or subsiding movements prevail there. Numberless facts could be adduced to prove that upraised organic remains are common wherever there are active volcanos; but until it could be shown that in areas of subsidence, volcanos were either absent or inactive, the inference, however probable in itself, that their distribution depended on the rising or falling of the earth's surface, would have been hazardous. But now, I think, we may freely admit this important deduction.

Taking a final view of the map, and bearing in mind the statements made with respect to the upraised organic remains, we must feel astonished at the vastness of the areas, which have suffered changes in level either downwards or upwards, within a period not geologically remote. It would appear also, that the elevatory and subsiding movements follow nearly the same laws. Throughout the spaces interspersed with atolls, where not a single peak of high land has been left above the level of the sea, the sinking must have been immense in amount. The sinking, moreover, whether continuous, or recurrent with intervals sufficiently long for the corals again to bring up their living edifices to the surface, must necessarily have been extremely slow. This conclusion is probably the most important one which can be deduced from the study of coral formations;—and it is one which it is difficult to imagine how otherwise could ever have been arrived at. Nor can I quite pass over the probability of the former existence of large archipelagoes of lofty islands, where now only rings of coral-rock scarcely break the open expanse of the sea, throwing some light on the distribution of the inhabitants of the other high islands, now left standing so immensely remote from each other in the midst of the great oceans. The reef-constructing corals have indeed reared and preserved wonderful memorials of the subterranean oscillations of level; we see in each barrier-reef a proof that the land has there subsided, and in each atoll a monument over an island now lost. We may thus, like unto a geologist who had lived his ten thousand years and kept a record of the passing changes, gain some insight into the great system by which the surface of this globe has been broken up, and land and water interchanged.

CHAPTER XXI

Mauritius to England

MAURITIUS, BEAUTIFUL APPEARANCE OF • GREAT CRATERIFORM RING OF MOUNTAINS • HINDOOS • ST. HELENA • HISTORY OF THE CHANGES IN THE VEGETATION • CAUSE OF THE EXTINCTION OF LANDSNAILS • ASCENSION • VARIATION IN THE IMPORTED RATS • VOLCANIC BOMBS • BEDS OF INFUSORIA • BAHIA • BRAZIL • SPLENDOUR OF TROPICAL SCENERY • PERNAMBUCO • SINGULAR REEF • SLAVERY • RETURN TO ENGLAND • RETROSPECT ON OUR VOYAGE

April 29th.—In the morning we passed round the northern end of Mauritius, or the Isle of France. From this point of view the aspect of the island equalled the expectations raised by the many well-known descriptions of its beautiful scenery. The sloping plain of the Pamplemousses, interspersed with houses, and coloured by the large fields of sugar-cane of a bright green, composed the foreground. The brilliancy of the green was the more remarkable because it is a colour which generally is conspicuous only from a very short distance. Towards the centre of the island groups of wooded mountains rose out of this highly cultivated plain; their summits, as so commonly happens with ancient volcanic rocks, being jagged into the sharpest points. Masses of white clouds were collected around these pinnacles, as if for the sake of pleasing the stranger's eye. The whole island, with its sloping border and central mountains, was adorned with an air of perfect elegance: the scenery, if I may use such an expression, appeared to the sight harmonious.

I spent the greater part of the next day in walking about the town and visiting different people. The town is of considerable size, and is said to contain 20,000 inhabitants; the streets are very clean and regular. Although the island has been so many years under the English Government, the general character of the place is quite French: Englishmen speak to their servants in French, and the shops are all French; indeed, I should think that Calais or Boulogne was much more Anglified. There is a very pretty little theatre, in which operas are excellently performed. We were also surprised at seeing large booksellers' shops, with well-stored shelves;—music and reading bespeak our approach to the old world of civilization; for in truth both Australia and America are new worlds.

The various races of men walking in the streets afford the most interesting spectacle in Port Louis. Convicts from India are banished here for life; at present there are about 800, and they are employed in various public works. Before seeing these people, I had no idea that the inhabitants of India were such noble-looking figures. Their skin is extremely dark, and many of the older men had large mustaches and beards of a snow-white colour; this, together with the fire of their expression, gave them quite an imposing aspect. The greater number had been banished for murder and the worst crimes; others for causes which can scarcely be considered as moral faults, such as for not obeying, from superstitious motives, the English laws. These men are generally quiet and well-conducted; from their outward conduct, their cleanliness, and faithful observance of their strange religious rites, it was impossible to look at them with the same eyes as on our wretched convicts in New South Wales.

May 1st.—Sunday. I took a quiet walk along the sea-coast to the north of the town. The plain in this part is quite uncultivated; it consists of a field of black lava, smoothed over with coarse grass and bushes, the latter being chiefly Mimosas. The scenery may be described as intermediate in character between that of the Galapagos and of Tahiti; but this will convey a definite idea to very few persons. It is a very pleasant country, but it has not the charms of Tahiti, or the grandeur of Brazil. The next day I ascended La Pouce, a mountain so called from a thumb-like projection, which rises close behind the town to a height of 2600 feet. The centre of the island consists of a great platform, surrounded by old broken basaltic mountains, with their strata dipping seawards. The central platform, formed of comparatively recent streams of lava, is of an oval shape, thirteen geographical miles across, in the line of its shorter axis. The exterior bounding mountains come into that class of structures called Craters of Elevation, which are supposed to have been formed not like ordinary craters, but by a great and sudden upheaval. There appears to me to be insuperable objections to this view: on the other hand, I can hardly believe, in this and in some other cases, that these marginal crateriform mountains are merely the basal remnants of immense volcanos, of which the summits either have been blown off, or swallowed up in subterranean abysses.

From our elevated position we enjoyed an excellent view over the island. The country on this side appears pretty well cultivated, being divided into fields and studded with farm-houses. I was, however, assured that of the whole land, not more than half is yet in a productive state; if such be the case, considering the present large export of sugar, this island, at some future period when thickly peopled, will be of great value. Since England has taken possession of it, a period of only twenty-five years, the export of sugar is said to have increased seventy-five fold. One great cause of its prosperity is the excellent state of the roads. In the neighbouring Isle of Bourbon, which remains under the French government, the roads are still in the same miserable state as they were here only a few years ago. Although the French residents must have largely profited by the increased prosperity of their island, yet the English government is far from popular.

3rd.—In the evening Captain Lloyd, the Surveyor-general, so well known from his examination of the Isthmus of Panama, invited Mr. Stokes and myself to his country-house, which is situated on the edge of Wilheim Plains, and about six miles from the Port. We stayed at this delightful place two days; standing nearly 800 feet above the sea, the air was cool and fresh, and on every side there were delightful walks. Close by, a grand ravine has been worn to a depth of about 500 feet through the slightly inclined streams of lava, which have flowed from the central platform.

5th.—Captain Lloyd took us to the Rivière Noire, which is several miles to the southward, that I might examine some rocks of elevated coral. We passed through pleasant gardens, and fine fields of sugar-cane growing amidst huge blocks of lava. The roads were bordered by hedges of Mimosa, and near many of the houses there were avenues of the mango. Some of the views, where the peaked hills and the cultivated farms were seen together, were exceedingly picturesque; and we were constantly tempted to exclaim, "How pleasant it would be to pass one's life in such quiet abodes!" Captain Lloyd possessed an elephant, and he sent it half way with us, that we might enjoy a ride in true Indian fashion. The circumstance which surprised me most was its quite noiseless step. This elephant is the only one at present on the island; but it is said others will be sent for.

May 9th.—We sailed from Port Louis, and, calling at the Cape of Good Hope, on the 8th of July, we arrived off St. Helena. This island, the forbidding aspect of which has been so often described, rises abruptly like a huge black castle from the ocean. Near the town, as if to complete nature's defence, small forts and guns fill up every gap in the rugged rocks. The town runs up a flat and narrow valley; the houses look respectable, and are interspersed with a very few green trees. When approaching the anchorage there was one striking view: an irregular castle perched on the summit of a lofty hill, and surrounded by a few scattered fir-trees, boldly projected against the sky.

The next day I obtained lodgings within a stone's throw of Napoleon's tomb;[176] it was a capital central situation, whence I could make excursions in every direction. During the four days I stayed here, I wandered over the island from morning to night, and examined its geological history. My lodgings were situated at a height of about 2000 feet; here the weather was cold and boisterous, with constant showers of rain; and every now and then the whole scene was veiled in thick clouds.

Near the coast the rough lava is quite bare: in the central and higher parts, feldspathic rocks by their decomposition have produced a clayey soil, which, where not covered by vegetation, is stained in broad bands of many bright colours. At this season, the land moistened by constant showers, produces a singularly bright green pasture, which lower and lower down, gradually fades away and at last disappears. In latitude 16°, and at the trifling elevation of 1500 feet, it is surprising to behold a vegetation possessing a character decidedly British. The hills are crowned with irregular plantations of Scotch firs; and the sloping banks are thickly scattered over with thickets of gorse, covered with its bright yellow flowers. Weeping-willows are common on the banks of the rivulets, and the hedges are made of the blackberry, producing its well-known fruit. When we consider that the number of plants now found on the island is 746, and that out of these fifty-two alone are indigenous species, the rest having been imported, and most of them from England, we see the reason of the British character of the vegetation. Many of these English plants appear to flourish better than in their native country; some also from the opposite quarter of Australia succeed remarkably well. The many imported species must have destroyed some of the native kinds; and it is only on the highest and steepest ridges that the indigenous Flora is now predominant.

The English, or rather Welsh character of the scenery, is kept up by the numerous cottages and small white houses; some buried at the bottom of the deepest valleys, and others mounted on the crests of the lofty hills. Some of the views are striking, for instance that from near Sir W. Doveton's house, where the bold peak called Lot is seen over a dark wood of firs, the whole being backed by the red water-worn mountains of the southern coast. On viewing the island from an eminence, the first circumstance which strikes one, is the number of the roads and forts: the labour bestowed on the public works, if one forgets its character as a prison, seems out of all proportion to its extent or value. There is so little level or useful land, that it seems surprising how so many people, about 5000, can subsist here. The lower orders, or the emancipated slaves, are I believe extremely poor: they complain of the want of work. From the reduction in the number of public servants, owing to the island having been given up by the East Indian Company, and the consequent emigration of many of the richer people, the poverty probably will increase. The chief food of the working class is rice with a little salt meat; as neither of these articles are the products of the island, but must be purchased with money, the low wages tell heavily on the poor people. Now that the people are blessed with freedom, a right which I believe they value fully, it seems probable that their numbers will quickly increase: if so, what is to become of the little state of St. Helena?

My guide was an elderly man, who had been a goatherd when a boy, and knew every step amongst the rocks. He was of a race many times crossed, and although with a dusky skin, he had not the disagreeable expression of a mulatto. He was a very civil, quiet old man, and such appears the character of the greater number of the lower classes. It was strange to my ears to hear a man, nearly white and respectably dressed, talking with indifference of the times when he was a slave. With my companion, who carried our dinners and a horn of water, which is quite necessary, as all the water in the lower valleys is saline, I every day took long walks.

Beneath the upper and central green circle, the wild valleys are quite desolate and untenanted. Here, to the geologist, there were scenes of high interest, showing successive changes and complicated disturbances. According to my views, St. Helena has existed as an island from a very remote epoch: some obscure proofs, however, of the elevation of the land are still extant. I believe that the central and highest peaks form parts of the rim of a great crater, the southern half of which has been entirely removed by the waves of the sea: there is, moreover, an external wall of black basaltic rocks, like the coast-mountains of Mauritius, which are older than the central volcanic streams. On the higher parts of the island, considerable numbers of a shell, long thought to be a marine species, occur imbedded in the soil.

It proved to be a Cochlogena, or land-snail of a very peculiar form;[177] with it I found six other kinds; and in another spot an eighth species. It is remarkable that none of them are now found living. Their extinction has probably been caused by the entire destruction of the woods, and the consequent loss of food and shelter, which occurred during the early part of the last century.

The history of the changes, which the elevated plains of Longwood and Deadwood have undergone, as given in General Beatson's account of the island, is extremely curious. Both plains, it is said in former times were covered with wood, and were therefore called the Great Wood. So late as the year 1716 there were many trees, but in 1724 the old trees had mostly fallen; and as goats and hogs had been suffered to range about, all the young trees had been killed. It appears also from the official records, that the trees were unexpectedly, some years afterwards, succeeded by a wire grass which spread over the whole surface.[178] General Beatson adds that now this plain "is covered with fine sward, and is become the finest piece of pasture on the island." The extent of surface, probably covered by wood at a former period, is estimated at no less than two thousand acres; at the present day scarcely a single tree can be found there. It is also said that in 1709 there were quantities of dead trees in Sandy Bay; this place is now so utterly desert, that nothing but so well attested an account could have made me believe that they could ever have grown there. The fact, that the goats and hogs destroyed all the young trees as they sprang up, and that in the course of time the old ones, which were safe from their attacks, perished from age, seems clearly made out. Goats were introduced in the year 1502; eighty-six years afterwards, in the time of Cavendish, it is known that they were exceedingly numerous. More than a century afterwards, in 1731, when the evil was complete and irretrievable, an order was issued that all stray animals should be destroyed. It is very interesting thus to find, that the arrival of animals at St. Helena in 1501, did not change the whole aspect of the island, until a period of two hundred and twenty years had elapsed: for the goats were introduced in 1502, and in 1724 it is said "the old trees had mostly fallen." There can be little doubt that this great change in the vegetation affected not only the land-snails, causing eight species to become extinct, but likewise a multitude of insects.

St. Helena, situated so remote from any continent, in the midst of a great ocean, and possessing a unique Flora, excites our curiosity. The eight land-snails, though now extinct, and one living Succinea, are peculiar species found nowhere else. Mr. Cuming, however, informs me that an English Helix is common here, its eggs no doubt having been imported in some of the many introduced plants. Mr. Cuming collected on the coast sixteen species of sea-shells, of which seven, as far as he knows, are confined to this island. Birds and insects,[179] as might have been expected, are very few in number; indeed I believe all the birds have been introduced within late years. Partridges and pheasants are tolerably abundant; the island is much too English not to be subject to strict game-laws. I was told of a more unjust sacrifice to such ordinances than I ever heard of even in England. The poor people formerly used to burn a plant, which grows on the coast-rocks, and export the soda from its ashes; but a peremptory order came out prohibiting this practice, and giving as a reason that the partridges would have nowhere to build.

In my walks I passed more than once over the grassy plain bounded by deep valleys, on which Longwood stands. Viewed from a short distance, it appears like a respectable gentleman's country-seat. In front there are a few cultivated fields, and beyond them the smooth hill of coloured rocks called the Flagstaff, and the rugged square black mass of the Barn. On the whole the view was rather bleak and uninteresting. The only inconvenience I suffered during my walks was from the impetuous winds. One day I noticed a curious circumstance; standing on the edge of a plain, terminated by a great cliff of about a thousand feet in depth, I saw at the distance of a few yards right to windward, some tern, struggling against a very strong breeze, whilst, where I stood, the air was quite calm. Approaching close to the brink, where the current seemed to be deflected upwards from the face of the cliff, I stretched out my arm, and immediately felt the full force of the wind: an invisible barrier, two yards in width, separated perfectly calm air from a strong blast.

I so much enjoyed my rambles among the rocks and mountains of St. Helena, that I felt almost sorry on the morning of the 14th to descend to the town. Before noon I was on board, and the Beagle made sail.

On the 19th of July we reached Ascension. Those who have beheld a volcanic island, situated under an arid climate, will at once be able to picture to themselves the appearance of Ascension. They will imagine smooth conical hills of a bright red colour, with their summits generally truncated, rising separately out of a level surface of black rugged lava. A principal mound in the centre of the island, seems the father of the lesser cones. It is called Green Hill: its name being taken from the faintest tinge of that colour, which at this time of the year is barely perceptible from the anchorage. To complete the desolate scene, the black rocks on the coast are lashed by a wild and turbulent sea.

The settlement is near the beach; it consists of several houses and barracks placed irregularly, but well built of white freestone. The only inhabitants are marines, and some negroes liberated from slave-ships, who are paid and victualled by government. There is not a private person on the island. Many of the marines appeared well contented with their situation; they think it better to serve their one-and-twenty years on shore, let it be what it may, than in a ship; in this choice, if I were a marine, I should most heartily agree.

The next morning I ascended Green Hill, 2840 feet high, and thence walked across the island to the windward point. A good cart-road leads from the coast-settlement to the houses, gardens, and fields, placed near the summit of the central mountain. On the roadside there are milestones, and likewise cisterns, where each thirsty passer-by can drink some good water. Similar care is displayed in each part of the establishment, and especially in the management of the springs, so that a single drop of water may not be lost: indeed the whole island may be compared to a huge ship kept in first-rate order. I could not help, when admiring the active industry, which had created such effects out of such means, at the same time regretting that it had been wasted on so poor and trifling an end. M. Lesson has remarked with justice, that the English nation would have thought of making the island of Ascension a productive spot; any other people would have held it as a mere fortress in the ocean.

Near this coast nothing grows; further inland, an occasional green castor-oil plant, and a few grasshoppers, true friends of the desert, may be met with. Some grass is scattered over the surface of the central elevated region, and the whole much resembles the worse parts of the Welsh mountains. But scanty as the pasture appears, about six hundred sheep, many goats, a few cows and horses, all thrive well on it. Of native animals, land-crabs and rats swarm in numbers. Whether the rat is really indigenous, may well be doubted; there are two varieties as described by Mr. Waterhouse; one is of a black colour, with fine glossy fur, and lives on the grassy summit; the other is brown-coloured and less glossy, with longer hairs, and lives near the settlement on the coast. Both these varieties are one-third smaller than the common black rat (M. rattus); and they differ from it both in the colour and character of their fur, but in no other essential respect. I can hardly doubt that these rats (like the common mouse, which has also run wild) have been imported, and, as at the Galapagos, have varied from the effect of the new conditions to which they have been exposed: hence the variety on the summit of the island differs from that on the coast. Of native birds there are none; but the guinea-fowl, imported from the Cape de Verd Islands, is abundant, and the common fowl has likewise run wild. Some cats, which were originally turned out to destroy the rats and mice, have increased, so as to become a great plague. The island is entirely without trees, in which, and in every other respect, it is very far inferior to St. Helena.

One of my excursions took me towards the S. W. extremity of the island. The day was clear and hot, and I saw the island, not smiling with beauty, but staring with naked hideousness. The lava streams are covered with hummocks, and are rugged to a degree which, geologically speaking, is not of easy explanation. The intervening spaces are concealed with layers of pumice, ashes and volcanic tuff. Whilst passing this end of the island at sea, I could not imagine what the white patches were with which the whole plain was mottled; I now found that they were seafowl, sleeping in such full confidence, that even in midday a man could walk up and seize hold of them. These birds were the only living creatures I saw during the whole day. On the beach a great surf, although the breeze was light, came tumbling over the broken lava rocks.

The geology of this island is in many respects interesting. In several places I noticed volcanic bombs, that is, masses of lava which have been shot through the air whilst fluid, and have consequently assumed a spherical or pear-shape. Not only their external form, but, in several cases, their internal structure shows in a very curious manner that they have revolved in their aërial course. The internal structure of one of these bombs, when broken, is represented very accurately in the woodcut. The central part is coarsely cellular, the cells decreasing in size towards the exterior; where there is a shell-like case about the third of an inch in thickness, of compact stone, which again is overlaid by the outside crust of finely cellular lava. I think there can be little doubt, first that the external crust cooled rapidly in the state in which we now see it; secondly, that the still fluid lava within, was packed by the centrifugal force, generated by the revolving of the bomb, against the external cooled crust, and so produced the solid shell of stone; and lastly, that the centrifugal force, by relieving the pressure in the more central parts of the bomb, allowed the heated vapours to expand their cells, thus forming the coarse cellular mass of the centre.

A hill, formed of the older series of volcanic rocks, and which has been incorrectly considered as the crater of a volcano, is remarkable from its broad, slightly hollowed, and circular summit having been filled up with many successive layers of ashes and fine scoriæ. These saucer-shaped layers crop out on the margin, forming perfect rings of many different colours, giving to the summit a most fantastic appearance; one of these rings is white and broad, and resembles a course round which horses have been exercised; hence the hill has been called the Devil's Riding School. I brought away specimens of one of the tufaceous layers of a pinkish colour and it is a most extraordinary fact, that Professor Ehrenberg[180] finds it almost wholly composed of matter which has been organized: he detects in it some siliceous-shielded fresh-water infusoria, and no less than twenty-five different kinds of the siliceous tissue of plants, chiefly of grasses. From the absence of all carbonaceous matter, Professor Ehrenberg believes that these organic bodies have passed through the volcanic fire, and have been erupted in the state in which we now see them. The appearance of the layers induced me to believe that they had been deposited under water, though from the extreme dryness of the climate I was forced to imagine, that torrents of rain had probably fallen during some great eruption, and that thus a temporary lake had been formed into which the ashes fell. But it may now be suspected that the lake was not a temporary one. Anyhow, we may feel sure, that at some former epoch the climate and productions of Ascension were very different from what they now are. Where on the face of the earth can we find a spot, on which close investigation will not discover signs of that endless cycle of change, to which this earth has been, is, and will be subjected?

On leaving Ascension, we sailed for Bahia, on the coast of Brazil, in order to complete the chronometrical measurement of the world. We arrived there on August 1st, and stayed four days, during which I took several long walks. I was glad to find my enjoyment in tropical scenery had not decreased from the want of novelty, even in the slightest degree. The elements of the scenery are so simple, that they are worth mentioning, as a proof on what trifling circumstances exquisite natural beauty depends.

The country may be described as a level plain of about three hundred feet in elevation, which in all parts has been worn into flat-bottomed valleys. This structure is remarkable in a granitic land, but is nearly universal in all those softer formations of which plains are usually composed. The whole surface is covered by various kinds of stately trees, interspersed with patches of cultivated ground, out of which houses, convents, and chapels arise. It must be remembered that within the tropics, the wild luxuriance of nature is not lost even in the vicinity of large cities: for the natural vegetation of the hedges and hill-sides overpowers in picturesque effect the artificial labour of man. Hence, there are only a few spots where the bright red soil affords a strong contrast with the universal clothing of green. From the edges of the plain there are distant views either of the ocean, or of the great Bay with its low-wooded shores, and on which numerous boats and canoes show their white sails. Excepting from these points, the scene is extremely limited; following the level pathways, on each hand, only glimpses into the wooded valleys below can be obtained. The houses I may add, and especially the sacred edifices, are built in a peculiar and rather fantastic style of architecture. They are all whitewashed; so that when illumined by the brilliant sun of midday, and as seen against the pale blue sky of the horizon, they stand out more like shadows than real buildings.

Such are the elements of the scenery, but it is a hopeless attempt to paint the general effect. Learned naturalists describe these scenes of the tropics by naming a multitude of objects, and mentioning some characteristic feature of each. To a learned traveller this possibly may communicate some definite ideas: but who else from seeing a plant in an herbarium can imagine its appearance when growing in its native soil? Who from seeing choice plants in a hothouse, can magnify some into the dimensions of forest trees, and crowd others into an entangled jungle? Who when examining in the cabinet of the entomologist the gay exotic butterflies, and singular cicadas, will associate with these lifeless objects, the ceaseless harsh music of the latter, and the lazy flight of the former,—the sure accompaniments of the still, glowing noonday of the tropics? It is when the sun has attained its greatest height, that such scenes should be viewed: then the dense splendid foliage of the mango hides the ground with its darkest shade, whilst the upper branches are rendered from the profusion of light of the most brilliant green. In the temperate zones the case is different—the vegetation there is not so dark or so rich, and hence the rays of the declining sun, tinged of a red, purple, or bright yellow color, add most to the beauties of those climes.

When quietly walking along the shady pathways, and admiring each successive view, I wished to find language to express my ideas. Epithet after epithet was found too weak to convey to those who have not visited the intertropical regions, the sensation of delight which the mind experiences. I have said that the plants in a hothouse fail to communicate a just idea of the vegetation, yet I must recur to it. The land is one great wild, untidy, luxuriant hothouse, made by Nature for herself, but taken possession of by man, who has studded it with gay houses and formal gardens. How great would be the desire in every admirer of nature to behold, if such were possible, the scenery of another planet! yet to every person in Europe, it may be truly said, that at the distance of only a few degrees from his native soil, the glories of another world are opened to him. In my last walk I stopped again and again to gaze on these beauties, and endeavoured to fix in my mind forever, an impression which at the time I knew sooner or later must fail. The form of the orange-tree, the cocoa-nut, the palm, the mango, the tree-fern, the banana, will remain clear and separate; but the thousand beauties which unite these into one perfect scene must fade away: yet they will leave, like a tale heard in childhood, a picture full of indistinct, but most beautiful figures.

August 6th.—In the afternoon we stood out to sea, with the intention of making a direct course to the Cape de Verd Islands. Unfavourable winds, however, delayed us, and on the 12th we ran into Pernambuco,—a large city on the coast of Brazil, in latitude 8° south. We anchored outside the reef; but in a short time a pilot came on board and took us into the inner harbour, where we lay close to the town.

Pernambuco is built on some narrow and low sand-banks, which are separated from each other by shoal channels of salt-water. The three parts of the town are connected together by two long bridges built on wooden piles. The town is in all parts disgusting, the streets being narrow, ill-paved, and filthy; the houses, tall and gloomy. The season of heavy rains had hardly come to an end, and hence the surrounding country, which is scarcely raised above the level of the sea, was flooded with water; and I failed in all my attempts to take walks.

The flat swampy land on which Pernambuco stands is surrounded, at the distance of a few miles, by a semicircle of low hills, or rather by the edge of a country elevated perhaps two hundred feet above the sea. The old city of Olinda stands on one extremity of this range. One day I took a canoe, and proceeded up one of the channels to visit it; I found the old town from its situation both sweeter and cleaner than that of Pernambuco. I must here commemorate what happened for the first time during our nearly five years' wandering, namely, having met with a want of politeness. I was refused in a sullen manner at two different houses, and obtained with difficulty from a third, permission to pass through their gardens to an uncultivated hill, for the purpose of viewing the country. I feel glad that this happened in the land of the Brazilians, for I bear them no good-will—a land also of slavery, and therefore of moral debasement. A Spaniard would have felt ashamed at the very thought of refusing such a request, or of behaving to a stranger with rudeness. The channel by which we went to and returned from Olinda, was bordered on each side by mangroves, which sprang like a miniature forest out of the greasy mud-banks. The bright green colour of these bushes always reminded me of the rank grass in a churchyard: both are nourished by putrid exhalations; the one speaks of death past, and the other too often of death to come.

The most curious object which I saw in this neighbourhood, was the reef that forms the harbour. I doubt whether in the whole world any other natural structure has so artificial an appearance.[181] It runs for a length of several miles in an absolutely straight line, parallel to, and not far distant from, the shore. It varies in width from thirty to sixty yards, and its surface is level and smooth; it is composed of obscurely stratified hard sandstone. At high water the waves break over it; at low water its summit is left dry, and it might then be mistaken for a breakwater erected by Cyclopean workmen. On this coast the currents of the sea tend to throw up in front of the land, long spits and bars of loose sand, and on one of these, part of the town of Pernambuco stands. In former times a long spit of this nature seems to have become consolidated by the percolation of calcareous matter, and afterwards to have been gradually upheaved; the outer and loose parts during this process having been worn away by the action of the sea, and the solid nucleus left as we now see it. Although night and day the waves of the open Atlantic, turbid with sediment, are driven against the steep outside edges of this wall of stone, yet the oldest pilots know of no tradition of any change in its appearance. This durability is much the most curious fact in its history: it is due to a tough layer, a few inches thick, of calcareous matter, wholly formed by the successive growth and death of the small shells of Serpulæ, together with some few barnacles and nulliporæ. These nulliporæ, which are hard, very simply-organized sea-plants, play an analogous and important part in protecting the upper surfaces of coral-reefs, behind and within the breakers, where the true corals, during the outward growth of the mass, become killed by exposure to the sun and air. These insignificant organic beings, especially the Serpulæ, have done good service to the people of Pernambuco; for without their protective aid the bar of sandstone would inevitably have been long ago worn away and without the bar, there would have been no harbour.

On the 19th of August we finally left the shores of Brazil. I thank God, I shall never again visit a slave-country. To this day, if I hear a distant scream, it recalls with painful vividness my feelings, when passing a house near Pernambuco, I heard the most pitiable moans, and could not but suspect that some poor slave was being tortured, yet knew that I was as powerless as a child even to remonstrate. I suspected that these moans were from a tortured slave, for I was told that this was the case in another instance. Near Rio de Janeiro I lived opposite to an old lady, who kept screws to crush the fingers of her female slaves. I have stayed in a house where a young household mulatto, daily and hourly, was reviled, beaten, and persecuted enough to break the spirit of the lowest animal. I have seen a little boy, six or seven years old, struck thrice with a horse-whip (before I could interfere) on his naked head, for having handed me a glass of water not quite clean; I saw his father tremble at a mere glance from his master's eye. These latter cruelties were witnessed by me in a Spanish colony, in which it has always been said, that slaves are better treated than by the Portuguese, English, or other European nations. I have seen at Rio de Janeiro a powerful negro afraid to ward off a blow directed, as he thought, at his face. I was present when a kind-hearted man was on the point of separating forever the men, women, and little children of a large number of families who had long lived together. I will not even allude to the many heart-sickening atrocities which I authentically heard of;—nor would I have mentioned the above revolting details, had I not met with several people, so blinded by the constitutional gaiety of the negro as to speak of slavery as a tolerable evil. Such people have generally visited at the houses of the upper classes, where the domestic slaves are usually well treated; and they have not, like myself, lived amongst the lower classes. Such inquirers will ask slaves about their condition; they forget that the slave must indeed be dull, who does not calculate on the chance of his answer reaching his master's ears.

It is argued that self-interest will prevent excessive cruelty; as if self-interest protected our domestic animals, which are far less likely than degraded slaves, to stir up the rage of their savage masters. It is an argument long since protested against with noble feeling, and strikingly exemplified, by the ever-illustrious Humboldt. It is often attempted to palliate slavery by comparing the state of slaves with our poorer countrymen: if the misery of our poor be caused not by the laws of nature, but by our institutions, great is our sin; but how this bears on slavery, I cannot see; as well might the use of the thumb-screw be defended in one land, by showing that men in another land suffered from some dreadful disease. Those who look tenderly at the slave owner, and with a cold heart at the slave, never seem to put themselves into the position of the latter; what a cheerless prospect, with not even a hope of change! picture to yourself the chance, ever hanging over you, of your wife and your little children—those objects which nature urges even the slave to call his own—being torn from you and sold like beasts to the first bidder! And these deeds are done and palliated by men, who profess to love their neighbours as themselves, who believe in God, and pray that his Will be done on earth! It makes one's blood boil, yet heart tremble, to think that we Englishmen and our American descendants, with their boastful cry of liberty, have been and are so guilty: but it is a consolation to reflect, that we at least have made a greater sacrifice, than ever made by any nation, to expiate our sin.

On the last day of August we anchored for the second time at Porto Praya in the Cape de Verd archipelago; thence we proceeded to the Azores, where we stayed six days. On the 2nd of October we made the shores of England; and at Falmouth I left the Beagle, having lived on board the good little vessel nearly five years.

Our Voyage having come to an end, I will take a short retrospect of the advantages and disadvantages, the pains and pleasures, of our circumnavigation of the world. If a person asked my advice, before undertaking a long voyage, my answer would depend upon his possessing a decided taste for some branch of knowledge, which could by this means be advanced. No doubt it is a high satisfaction to behold various countries and the many races of mankind, but the pleasures gained at the time do not counterbalance the evils. It is necessary to look forward to a harvest, however distant that may be, when some fruit will be reaped, some good effected.

Many of the losses which must be experienced are obvious; such as that of the society of every old friend, and of the sight of those places with which every dearest remembrance is so intimately connected. These losses, however, are at the time partly relieved by the exhaustless delight of anticipating the long wished-for day of return. If, as poets say, life is a dream, I am sure in a voyage these are the visions which best serve to pass away the long night. Other losses, although not at first felt, tell heavily after a period: these are the want of room, of seclusion, of rest; the jading feeling of constant hurry; the privation of small luxuries, the loss of domestic society and even of music and the other pleasures of imagination. When such trifles are mentioned, it is evident that the real grievances, excepting from accidents, of a sea-life are at an end. The short space of sixty years has made an astonishing difference in the facility of distant navigation. Even in the time of Cook, a man who left his fireside for such expeditions underwent severe privations. A yacht now, with every luxury of life, can circumnavigate the globe. Besides the vast improvements in ships and naval resources, the whole western shores of America are thrown open, and Australia has become the capital of a rising continent. How different are the circumstances to a man shipwrecked at the present day in the Pacific, to what they were in the time of Cook! Since his voyage a hemisphere has been added to the civilized world.

If a person suffer much from sea-sickness, let him weigh it heavily in the balance. I speak from experience: it is no trifling evil, cured in a week. If, on the other hand, he take pleasure in naval tactics, he will assuredly have full scope for his taste. But it must be borne in mind, how large a proportion of the time, during a long voyage, is spent on the water, as compared with the days in harbour. And what are the boasted glories of the illimitable ocean. A tedious waste, a desert of water, as the Arabian calls it. No doubt there are some delightful scenes. A moonlight night, with the clear heavens and the dark glittering sea, and the white sails filled by the soft air of a gently blowing trade-wind, a dead calm, with the heaving surface polished like a mirror, and all still except the occasional flapping of the canvas. It is well once to behold a squall with its rising arch and coming fury, or the heavy gale of wind and mountainous waves. I confess, however, my imagination had painted something more grand, more terrific in the full-grown storm. It is an incomparably finer spectacle when beheld on shore, where the waving trees, the wild flight of the birds, the dark shadows and bright lights, the rushing of the torrents all proclaim the strife of the unloosed elements. At sea the albatross and little petrel fly as if the storm were their proper sphere, the water rises and sinks as if fulfilling its usual task, the ship alone and its inhabitants seem the objects of wrath. On a forlorn and weather-beaten coast, the scene is indeed different, but the feelings partake more of horror than of wild delight.

Let us now look at the brighter side of the past time. The pleasure derived from beholding the scenery and the general aspect of the various countries we have visited, has decidedly been the most constant and highest source of enjoyment. It is probable that the picturesque beauty of many parts of Europe exceeds anything which we beheld. But there is a growing pleasure in comparing the character of the scenery in different countries, which to a certain degree is distinct from merely admiring its beauty. It depends chiefly on an acquaintance with the individual parts of each view. I am strongly induced to believe that as in music, the person who understands every note will, if he also possesses a proper taste, more thoroughly enjoy the whole, so he who examines each part of a fine view, may also thoroughly comprehend the full and combined effect. Hence, a traveller should be a botanist, for in all views plants form the chief embellishment. Group masses of naked rock, even in the wildest forms, and they may for a time afford a sublime spectacle, but they will soon grow monotonous. Paint them with bright and varied colours, as in Northern Chile, they will become fantastic; clothe them with vegetation, they must form a decent, if not a beautiful picture.

When I say that the scenery of parts of Europe is probably superior to anything which we beheld, I except, as a class by itself, that of the intertropical zones. The two classes cannot be compared together; but I have already often enlarged on the grandeur of those regions. As the force of impressions generally depends on preconceived ideas, I may add, that mine were taken from the vivid descriptions in the Personal Narrative of Humboldt, which far exceed in merit anything else which I have read. Yet with these high-wrought ideas, my feelings were far from partaking of a tinge of disappointment on my first and final landing on the shores of Brazil.

Among the scenes which are deeply impressed on my mind, none exceed in sublimity the primeval forests undefaced by the hand of man; whether those of Brazil, where the powers of Life are predominant, or those of Tierra del Fuego, where Death and Decay prevail. Both are temples filled with the varied productions of the God of Nature:—no one can stand in these solitudes unmoved, and not feel that there is more in man than the mere breath of his body. In calling up images of the past, I find that the plains of Patagonia frequently cross before my eyes; yet these plains are pronounced by all wretched and useless. They can be described only by negative characters; without habitations, without water, without trees, without mountains, they support merely a few dwarf plants. Why, then, and the case is not peculiar to myself, have these arid wastes taken so firm a hold on my memory? Why have not the still more level, the greener and more fertile Pampas, which are serviceable to mankind, produced an equal impression? I can scarcely analyze these feelings: but it must be partly owing to the free scope given to the imagination. The plains of Patagonia are boundless, for they are scarcely passable, and hence unknown: they bear the stamp of having lasted, as they are now, for ages, and there appears no limit to their duration through future time. If, as the ancients supposed, the flat earth was surrounded by an impassable breadth of water, or by deserts heated to himself, and of making the best of every occurrence. In short, he ought to partake of the characteristic qualities of most sailors. Travelling ought also to teach him distrust; but at the same time he will discover, how many truly kind-hearted people there are, with whom he never before had, or ever again will have any further communication, who yet are ready to offer him the most disinterested assistance.

Charles Darwin age 51 (1860)

On The Origin of Species

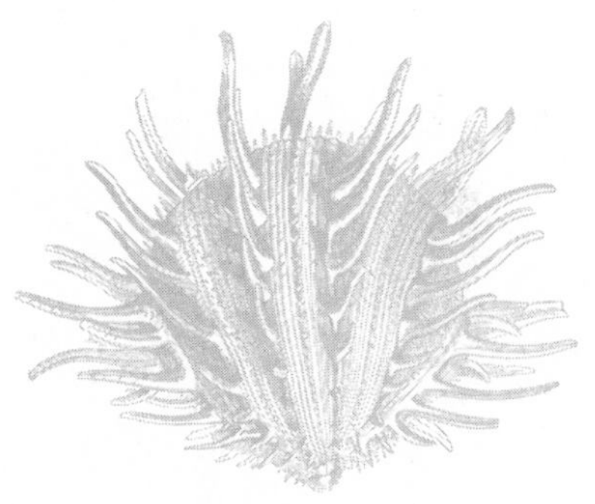

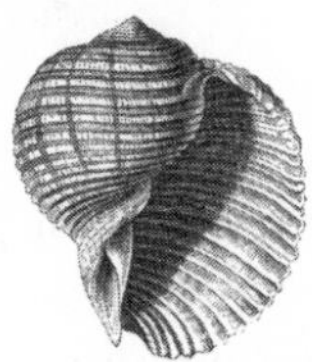

Introduction to *On The Origin of Species*

Darwinism is the word often used to denote evolution by natural selection and the many consequences that process has for nature and humanity. The concept was introduced by Charles Darwin to a broad public in what many scholars rightly consider the greatest scientific book of all time: *On the Origin of Species by Means of Natural Selection, Or the Preservation of Favoured Races in the Struggle for Life,* usually cited as *On the Origin of Species* or just *The Origin of Species.* The first edition (1859) of the six published during Darwin's lifetime is the one reprinted here.

The concepts most popularly used to characterize natural selection are "struggle for existence" and "survival of the fittest," first explicitly and fully explained by Darwin here. The process, despite its almost universal dissolving power in biology, is supremely simple in conception. To recapitulate the description given in my introductory essay, we start with a population whose individuals differ one from another in some hereditary trait, say running speed or the ability to digest certain food. Individuals with more of the trait (or, in some instances that can be imagined, less of it) survive longer or breed faster—in any case, leave more offspring. Thus in the next generation there are more individuals with the favored trait. On average, members of the population now run faster or digest the food better. Evolution by natural selection has occurred. If there exists a second process by which new hereditary forms (mutations in the modern parlance) are continuously created, evolution continues indefinitely.

The Darwinian calculus of genetic change, pitting variant against variant, often inspires the image of nature red in tooth and claw. Biologists are blunt about this stereotype: for the most part nature *is* red in tooth and claw. But that is not the whole story. A great deal of evidence has accumulated to show that under many circumstances natural selection can also generate hereditary cooperation and even self-sacrifice. This ameliorative process can be easily explained as follows. If groups of cooperating individuals are better than solitary animals at obtaining food, building shelters, repelling enemies, or (red tooth and claw again, I grant) exploiting others, then the species will evolve toward the formation of social groups. Such has been the case, for example, in corals, ants, and human beings.

There is a further twist in this theory. If instead of nature, human beings are the agent that decides which varieties in the population best survive and reproduce, then the result is still evolution, but evolution by artificial selection. The result has been the amazingly diverse domestic breeds of plants and animals created by human beings during the past 10,000 years. In *On the Origin of Species* Darwin stressed the close parallel between natural and artificial selection, making it a cornerstone of his argument for the occurrence of evolution in nature.

In hindsight, evolution by natural selection may seem not only simple but obvious. Such is not the case, however: the theory could easily have been wrong! Before Darwin the best available theory of evolution had been Lamarckism, or the inheritance of acquired characters. Proposed by the Chevalier de Lamarck in the early 1800s, it held that when plants or animals strive to achieve some adaptation to the environment in a way that alters their personal ability to accomplish the adaptation—say again, run faster, digest the new food better, or, in Lamarck's famous example of the giraffe, acquire a longer neck to feed on higher foliage—some of this change is passed on to the offspring. Evolution has occurred, but in a fundamentally different way from natural selection.

Another rival theory to explain evolution was created after Mendel's laws of heredity were rediscovered in 1900, when geneticists recognized that the units of heredity are particulate (the particles are now called *genes*). The first mutations found were large and conspicuous in effect—otherwise they would likely not have been discovered—leading to the proposal that evolution occurs by large jumps, not the small, gradual changes envisioned by Darwin's theory of natural selection. One more theory to add to the competition of ideas was concocted by some early paleontologists, who, working with a still limited number of fossils, decided that evolution is somehow guided in a straight line toward a predetermined goal. This theory of "orthogenesis," at least the predetermined-goal part, has been revived by creationists under the rubric of "evolution by design." In their view, of course, the designer is most likely the Creator of their personal religious belief.

Any one of these theories of evolution might have proved to be true to the exclusion of the others. But none did. A century and a half of research on all fronts in biology has shown that Darwin was correct in his reasoning: natural selection is the general driving force of evolution.

In *On the Origin of Species* Darwin took the very original and, as it turns out, necessary step of conceiving evolution as a phenomenon of populations. It is the ensemble of individuals, exchanging hereditary material each generation by recombination, that evolves as a whole in the crucible of environmental pressure. Evolution is not a change in a line of individuals within a population, as conceived by Lamarck, nor is it a sequence of hereditary jumps within such lineages. No evidence has ever been adduced that evolution is guided by any ultimate purpose or goal.

Darwin also realized that evolution has two dimensions. The first is descent with modification of entire populations. The second is the multiplication of species by the splitting off and divergence of subpopulations during the course of the overall descent with modification.

The triumph of *On the Origin of Species* did not spring solely from the soundness of its logic. It owed at least as much to the massive documentation that Darwin assembled, piece by piece, upon that theoretical foundation. His masterwork is impressive enough for the quality of his empirical evidence alone. Much of what he wrote contains valuable information for biological researchers and general readers who open this book today.

On the Origin of Species by Means of Natural Selection, or the Preservation of Favoured Races in the Struggle for Life.

Charles Darwin

· 1859 ·

"But with regard to the material world, we can at least go so far as this—we can perceive that events are brought about not by insulated interpositions of Divine power, exerted in each particular case, but by the establishment of general laws."

W. WHEWELL: *Bridgewater Treatise.*

"To conclude, therefore, let no man out of a weak conceit of sobriety, or an ill-applied moderation, think or maintain, that a man can search too far or be too well studied in the book of God's word, or in the book of God's works; divinity or philosophy; but rather let men endeavour an endless progress or proficience in both."

BACON: *Advancement of Learning.*

DOWN, BROMLEY, KENT,
October 1st, 1859.

When on board H.M.S. 'Beagle,' as naturalist, I was much struck with certain facts in the distribution of the inhabitants of South America, and in the geological relations of the present to the past inhabitants of that continent. These facts seemed to me to throw some light on the origin of species—that mystery of mysteries, as it has been called by one of our greatest philosophers. On my return home, it occurred to me, in 1837, that something might perhaps be made out on this question by patiently accumulating and reflecting on all sorts of facts which could possibly have any bearing on it. After five years' work I allowed myself to speculate on the subject, and drew up some short notes; these I enlarged in 1844 into a sketch of the conclusions, which then seemed to me probable: from that period to the present day I have steadily pursued the same object. I hope that I may be excused for entering on these personal details, as I give them to show that I have not been hasty in coming to a decision.

My work is now nearly finished; but as it will take me two or three more years to complete it, and as my health is far from strong, I have been urged to publish this Abstract. I have more especially been induced to do this, as Mr. Wallace, who is now studying the natural history of the Malay archipelago, has arrived at almost exactly the same general conclusions that I have on the origin of species. Last year he sent to me a memoir on this subject, with a request that I would forward it to Sir Charles Lyell, who sent it to the Linnean Society, and it is published in the third volume of the Journal of that Society. Sir C. Lyell and Dr. Hooker, who both knew of my work—the latter having read my sketch of 1844—honoured me by thinking it advisable to publish, with Mr. Wallace's excellent memoir, some brief extracts from my manuscripts.

This Abstract, which I now publish, must necessarily be imperfect. I cannot here give references and authorities for my several statements; and I must trust to the reader reposing some confidence in my accuracy. No doubt errors will have crept in, though I hope I have always been cautious in trusting to good authorities alone. I can here give only the general conclusions at which I have arrived, with a few facts in illustration, but which, I hope, in most cases will suffice. No one can feel more sensible than I do of the necessity of hereafter publishing in detail all the facts, with references, on which my conclusions have been grounded; and I hope in a future work to do this. For I am well aware that scarcely a single point is discussed in this volume on which facts cannot be adduced, often apparently leading to conclusions directly opposite to those at which I have arrived. A fair result can be obtained only by fully stating and balancing the facts and arguments on both sides of each question; and this cannot possibly be here done.

I much regret that want of space prevents my having the satisfaction of acknowledging the generous assistance which I have received from very many naturalists, some of them personally unknown to me. I cannot, however, let this opportunity pass without expressing my deep obligations to Dr. Hooker, who for the last fifteen years has aided me in every possible way by his large stores of knowledge and his excellent judgment.

In considering the Origin of Species, it is quite conceivable that a naturalist, reflecting on the mutual affinities of organic beings, on their embryological relations, their geographical distribution, geological succession, and other such facts, might come to the conclusion that each species had not been independently created, but had descended, like varieties, from other species. Nevertheless, such a conclusion, even if well founded, would be unsatisfactory, until it could be shown how the innumerable species inhabiting this world have been modified, so as to acquire that perfection of structure and coadaptation which most justly excites our admiration. Naturalists continually refer to external conditions, such as climate, food, &c., as the only possible cause of variation. In one very limited sense, as we shall hereafter see, this may be true; but it is preposterous to attribute to mere external conditions, the structure, for instance, of the woodpecker, with its feet, tail, beak, and tongue, so admirably adapted to catch insects under the bark of trees. In the case of the mistletoe, which draws its nourishment from certain trees, which has seeds that must be transported by certain birds, and which has flowers with separate sexes absolutely requiring the agency of certain insects to bring pollen from one flower to the other, it is equally preposterous to account for the structure of this parasite, with its relations to several distinct organic beings, by the effects of external conditions, or of habit, or of the volition of the plant itself.

The author of the 'Vestiges of Creation' would, I presume, say that, after a certain unknown number of generations, some bird had given birth to a woodpecker, and some plant to the mistletoe, and that these had been produced perfect as we now see them; but this assumption seems to me to be no explanation, for it leaves the case of the coadaptations of organic beings to each other and to their physical conditions of life, untouched and unexplained.

It is, therefore, of the highest importance to gain a clear insight into the means of modification and coadaptation. At the commencement of my observations it seemed to me probable that a careful study of domesticated animals and of cultivated plants would offer the best chance of making out this obscure problem. Nor have I been disappointed; in this and in all other perplexing cases I have invariably found that our knowledge, imperfect though it be, of variation under domestication, afforded the best and safest clue. I may venture to express my conviction of the high value of such studies, although they have been very commonly neglected by naturalists.

From these considerations, I shall devote the first chapter of this Abstract to Variation under Domestication. We shall thus see that a large amount of hereditary modification is at least possible; and, what is equally or more important, we shall see how great is the power of man in accumulating by his Selection successive slight variations. I will then pass on to the variability of species in a state of nature; but I shall, unfortunately, be compelled to treat this subject far too briefly, as it can be treated properly only by giving long catalogues of facts. We shall, however, be enabled to discuss what circumstances are most favourable to variation. In the next chapter the Struggle for Existence amongst all organic beings throughout the world, which inevitably follows from their high geometrical powers of increase, will be treated of. This is the doctrine of Malthus, applied to the whole animal and vegetable kingdoms. As many more individuals of each species are born than can possibly survive; and as, consequently, there is a frequently recurring struggle for existence, it follows that any being, if it vary however slightly in any manner profitable to itself, under the complex and sometimes varying conditions of life, will have a better chance of surviving, and thus be *naturally selected.* From the strong principle of inheritance, any selected variety will tend to propagate its new and modified form.

This fundamental subject of Natural Selection will be treated at some length in the fourth chapter; and we shall then see how Natural Selection almost inevitably causes much Extinction of the less improved forms of life, and induces what I have called Divergence of Character. In the next chapter I shall discuss the complex and little known laws of variation and of correlation of growth. In the four succeeding chapters, the most apparent and gravest difficulties on the theory will be given: namely, first, the difficulties of transitions, or in understanding how a simple being or a simple organ can be changed and perfected into a highly developed being or elaborately constructed organ; secondly, the subject of Instinct, or the mental powers of animals; thirdly, Hybridism, or the infertility of species and the fertility of varieties when intercrossed; and fourthly, the imperfection of the Geological Record. In the next chapter I shall consider the geological succession of organic beings throughout time; in the eleventh and twelfth, their geographical distribution throughout space; in the thirteenth, their classification or mutual affinities, both when mature and in an embryonic condition. In the last chapter I shall give a brief recapitulation of the whole work, and a few concluding remarks.

No one ought to feel surprise at much remaining as yet unexplained in regard to the origin of species and varieties, if he makes due allowance for our profound ignorance in regard to the mutual relations of all the beings which live around us. Who can explain why one species ranges widely and is very numerous, and why another allied species has a narrow range and is rare? Yet these relations are of the highest importance, for they determine the present welfare, and, as I believe, the future success and modification of every inhabitant of this world. Still less do we know of the mutual relations of the innumerable inhabitants of the world during the many past geological epochs in its history. Although much remains obscure, and will long remain obscure, I can entertain no doubt, after the most deliberate study and dispassionate judgment of which I am capable, that the view which most naturalists entertain, and which I formerly entertained—namely, that each species has been independently created—is erroneous. I am fully convinced that species are not immutable; but that those belonging to what are called the same genera are lineal descendants of some other and generally extinct species, in the same manner as the acknowledged varieties of any one species are the descendants of that species. Furthermore, I am convinced that Natural Selection has been the main but not exclusive means of modification.

CHAPTER I

Variation under Domestication

CAUSES OF VARIABILITY • EFFECTS OF HABIT • CORRELATION OF GROWTH • INHERITANCE • CHARACTER OF DOMESTIC VARIETIES • DIFFICULTY OF DISTINGUISHING BETWEEN VARIETIES AND SPECIES • ORIGIN OF DOMESTIC VARIETIES FROM ONE OR MORE SPECIES • DOMESTIC PIGEONS, THEIR DIFFERENCES AND ORIGIN • PRINCIPLE OF SELECTION ANCIENTLY FOLLOWED, ITS EFFECTS • METHODICAL AND UNCONSCIOUS SELECTION • UNKNOWN ORIGIN OF OUR DOMESTIC PRODUCTIONS • CIRCUMSTANCES FAVOURABLE TO MAN'S POWER OF SELECTION

When we look to the individuals of the same variety or sub-variety of our older cultivated plants and animals, one of the first points which strikes us, is, that they generally differ much more from each other, than do the individuals of any one species or variety in a state of nature. When we reflect on the vast diversity of the plants and animals which have been cultivated, and which have varied during all ages under the most different climates and treatment, I think we are driven to conclude that this greater variability is simply due to our domestic productions having been raised under conditions of life not so uniform as, and somewhat different from, those to which the parent-species have been exposed under nature. There is, also, I think, some probability in the view propounded by Andrew Knight, that this variability may be partly connected with excess of food. It seems pretty clear that organic beings must be exposed during several generations to the new conditions of life to cause any appreciable amount of variation; and that when the organisation has once begun to vary, it generally continues to vary for many generations. No case is on record of a variable being ceasing to be variable under cultivation. Our oldest cultivated plants, such as wheat, still often yield new varieties: our oldest domesticated animals are still capable of rapid improvement or modification.

It has been disputed at what period of life the causes of variability, whatever they may be, generally act; whether during the early or late period of development of the embryo, or at the instant of conception. Geoffroy St. Hilaire's experiments show that unnatural treatment of the embryo causes monstrosities; and monstrosities cannot be separated by any clear line of distinction from mere variations. But I am strongly inclined to suspect that the most frequent cause of variability may be attributed to the male and female reproductive elements having been affected prior to the act of conception. Several reasons make me believe in this; but the chief one is the remarkable effect which confinement or cultivation has on the functions of the reproductive system; this system appearing to be far more susceptible than any other part of the organisation, to the action of any change in the conditions of life. Nothing is more easy than to tame an animal, and few things more difficult than to get it to breed freely under confinement, even in the many cases when the male and female unite. How many animals there are which will not breed, though living long under not very close confinement in their native country! This is generally attributed to vitiated instincts; but how many cultivated plants display the utmost vigour, and yet rarely or never seed! In some few such cases it has been found out that very trifling changes, such as a little more or less water at some particular period of growth, will determine whether or not the plant sets a seed. I cannot here enter on the copious details which I have collected on this curious subject; but to show how singular the laws are which determine the reproduction of animals under confinement, I may just mention that carnivorous animals, even from the tropics, breed in this country pretty freely under confinement, with the exception of the plantigrades or bear family; whereas, carnivorous birds, with the rarest exceptions, hardly ever lay fertile eggs. Many exotic plants have pollen utterly worthless, in the same exact condition as in the most sterile hybrids. When, on the one hand, we see domesticated animals and plants, though often weak and sickly, yet breeding quite freely under confinement; and when, on the other hand, we see individuals, though taken young from a state of nature, perfectly tamed, long-lived, and healthy (of which I could give numerous instances), yet having their reproductive system so seriously affected by unperceived causes as to fail in acting, we need not be surprised at this system, when it does act under confinement, acting not quite regularly, and producing offspring not perfectly like their parents or variable.

Sterility has been said to be the bane of horticulture; but on this view we owe variability to the same cause which produces sterility; and variability is the source of all the choicest productions of the garden. I may add, that as some organisms will breed most freely under the most unnatural conditions (for instance, the rabbit and ferret kept in hutches), showing that their reproductive system has not been thus affected; so will some animals and plants withstand domestication or cultivation, and vary very slightly—perhaps hardly more than in a state of nature.

A long list could easily be given of "sporting plants;" by this term gardeners mean a single bud or offset, which suddenly assumes a new and sometimes very different character from that of the rest of the plant. Such buds can be propagated by grafting, &c., and sometimes by seed. These "sports" are extremely rare under nature, but far from rare under cultivation; and in this case we see that the treatment of the parent has affected a bud or offset, and not the ovules or pollen. But it is the opinion of most physiologists that there is no essential difference between a bud and an ovule in their earliest stages of formation; so that, in fact, "sports" support my view, that variability may be largely attributed to the ovules or pollen, or to both, having been affected by the treatment of the parent prior to the act of conception. These cases anyhow show that variation is not necessarily connected, as some authors have supposed, with the act of generation.

Seedlings from the same fruit, and the young of the same litter, sometimes differ considerably from each other, though both the young and the parents, as Müller has remarked, have apparently been exposed to exactly the same conditions of life; and this shows how unimportant the direct effects of the conditions of life are in comparison with the laws of reproduction, and of growth, and of inheritance; for had the action of the conditions been direct, if any of the young had varied, all would probably have varied in the same manner. To judge how much, in the case of any variation, we should attribute to the direct action of heat, moisture, light, food, &c., is most difficult: my impression is, that with animals such agencies have produced very little direct effect, though apparently more in the case of plants. Under this point of view, Mr. Buckman's recent experiments on plants seem extremely valuable. When all or nearly all the individuals exposed to certain conditions are affected in the same way, the change at first appears to be directly due to such conditions; but in some cases it can be shown that quite opposite conditions produce similar changes of structure. Nevertheless some slight amount of change may, I think, be attributed to the direct action of the conditions of life—as, in some cases, increased size from amount of food, colour from particular kinds of food and from light, and perhaps the thickness of fur from climate.

Habit also has a decided influence, as in the period of flowering with plants when transported from one climate to another. In animals it has a more marked effect; for instance, I find in the domestic duck that the bones of the wing weigh less and the bones of the leg more, in proportion to the whole skeleton, than do the same bones in the wild-duck; and I presume that this change may be safely attributed to the domestic duck flying much less, and walking more, than its wild parent. The great and inherited development of the udders in cows and goats in countries where they are habitually milked, in comparison with the state of these organs in other countries, is another instance of the effect of use. Not a single domestic animal can be named which has not in some country drooping ears; and the view suggested by some authors, that the drooping is due to the disuse of the muscles of the ear, from the animals not being much alarmed by danger, seems probable.

There are many laws regulating variation, some few of which can be dimly seen, and will be hereafter briefly mentioned. I will here only allude to what may be called correlation of growth. Any change in the embryo or larva will almost certainly entail changes in the mature animal. In monstrosities, the correlations between quite distinct parts are very curious; and many instances are given in Isidore Geoffroy St. Hilaire's great work on this subject. Breeders believe that long limbs are almost always accompanied by an elongated head. Some instances of correlation are quite whimsical: thus cats with blue eyes are invariably deaf; colour and constitutional peculiarities go together, of which many remarkable cases could be given amongst animals and plants. From the facts collected by Heusinger, it appears that white sheep and pigs are differently affected from coloured individuals by certain vegetable poisons. Hairless dogs have imperfect teeth; long-haired and coarse-haired animals are apt to have, as is asserted, long or many horns; pigeons with feathered feet have skin between their outer toes; pigeons with short beaks have small feet, and those with long beaks large feet. Hence, if man goes on selecting, and thus augmenting, any peculiarity, he will almost certainly unconsciously modify other parts of the structure, owing to the mysterious laws of the correlation of growth.

The result of the various, quite unknown, or dimly seen laws of variation is infinitely complex and diversified. It is well worth while carefully to study the several treatises published on some of our old cultivated plants, as on the hyacinth, potato, even the dahlia, &c.; and it is really surprising to note the endless points in structure and constitution in which the varieties and sub-varieties differ slightly from each other. The whole organisation seems to have become plastic, and tends to depart in some small degree from that of the parental type.

Any variation which is not inherited is unimportant for us. But the number and diversity of inheritable deviations of structure, both those of slight and those of considerable physiological importance, is endless. Dr. Prosper Lucas's treatise, in two large volumes, is the fullest and the best on this subject. No breeder doubts how strong is the tendency to inheritance: like produces like is his fundamental belief: doubts have been thrown on this principle by theoretical writers alone. When a deviation appears not unfrequently, and we see it in the father and child, we cannot tell whether it may not be due to the same original cause acting on both; but when amongst individuals, apparently exposed to the same conditions, any very rare deviation, due to some extraordinary combination of circumstances, appears in the parent—say, once amongst several million individuals—and it reappears in the child, the mere doctrine of chances almost compels us to attribute its reappearance to inheritance. Every one must have heard of cases of albinism, prickly skin, hairy bodies, &c., appearing in several members of the same family. If strange and rare deviations of structure are truly inherited, less strange and commoner deviations may be freely admitted to be inheritable. Perhaps the correct way of viewing the whole subject, would be, to look at the inheritance of every character whatever as the rule, and non-inheritance as the anomaly.

The laws governing inheritance are quite unknown; no one can say why the same peculiarity in different individuals of the same species, and in individuals of different species, is sometimes inherited and sometimes not so; why the child often reverts in certain characters to its grandfather or grandmother or other much more remote ancestor; why a peculiarity is often transmitted from one sex to both sexes, or to one sex alone, more commonly but not exclusively to the like sex. It is a fact of some little importance to us, that peculiarities appearing in the males of our domestic breeds are often transmitted either exclusively, or in a much greater degree, to males alone. A much more important rule, which I think may be trusted, is that, at whatever period of life a peculiarity first appears, it tends to appear in the offspring at a corresponding age, though sometimes earlier. In many cases this could not be otherwise: thus the inherited peculiarities in the horns of cattle could appear only in the offspring when nearly mature; peculiarities in the silkworm are known to appear at the corresponding caterpillar or cocoon stage. But hereditary diseases and some other facts make me believe that the rule has a wider extension, and that when there is no apparent reason why a peculiarity should appear at any particular age, yet that it does tend to appear in the offspring at the same period at which it first appeared in the parent. I believe this rule to be of the highest importance in explaining the laws of embryology. These remarks are of course confined to the first *appearance* of the peculiarity, and not to its primary cause, which may have acted on the ovules or male element; in nearly the same manner as in the crossed offspring from a short-horned cow by a long-horned bull, the greater length of horn, though appearing late in life, is clearly due to the male element.

Having alluded to the subject of reversion, I may here refer to a statement often made by naturalists—namely, that our domestic varieties, when run wild, gradually but certainly revert in character to their aboriginal stocks. Hence it has been argued that no deductions can be drawn from domestic races to species in a state of nature. I have in vain endeavoured to discover on what decisive facts the above statement has so often and so boldly been made. There would be great difficulty in proving its truth: we may safely conclude that very many of the most strongly-marked domestic varieties could not possibly live in a wild state. In many cases we do not know what the aboriginal stock was, and so could not tell whether or not nearly perfect reversion had ensued. It would be quite necessary, in order to prevent the effects of intercrossing, that only a single variety should be turned loose in its new home. Nevertheless, as our varieties certainly do occasionally revert in some of their characters to ancestral forms, it seems to me not improbable, that if we could succeed in naturalising, or were to cultivate, during many generations, the several races, for instance, of the cabbage, in very poor soil (in which case, however, some effect would have to be attributed to the direct action of the poor soil), that they would to a large extent, or even wholly, revert to the wild aboriginal stock. Whether or not the experiment would succeed, is not of great importance for our line of argument; for by the experiment itself the conditions of life are changed. If it could be shown that our domestic varieties manifested a strong tendency to reversion,—that is, to lose their acquired characters, whilst kept under unchanged conditions, and whilst kept in a considerable body, so that free intercrossing might check, by blending together, any slight deviations of structure, in such case, I grant that we could deduce nothing from domestic varieties in regard to species. But there is not a shadow of evidence in favour of this view: to assert that we could not breed our cart and race-horses, long and short-horned cattle, and poultry of various breeds, and esculent vegetables, for an almost infinite number of generations, would be opposed to all experience. I may add, that when under nature the conditions of life do change, variations and reversions of character probably do occur; but natural selection, as will hereafter be explained, will determine how far the new characters thus arising shall be preserved.

When we look to the hereditary varieties or races of our domestic animals and plants, and compare them with species closely allied together, we generally perceive in each domestic race, as already remarked, less uniformity of character than in true species. Domestic races of the same species, also, often have a somewhat monstrous character; by which I mean, that, although differing from each other, and from the other species of the same genus, in several trifling respects, they often differ in an extreme degree in some one part, both when compared one with another, and more especially when compared with all the species in nature to which they are nearest allied. With these exceptions (and with that of the perfect fertility of varieties when crossed,—a subject hereafter to be discussed), domestic races of the same species differ from each other in the same manner as, only in most cases in a lesser degree than, do closely-allied species of the same genus in a state of nature. I think this must be admitted, when we find that there are hardly any domestic races, either amongst animals or plants, which have not been ranked by some competent judges as mere varieties, and by other competent judges as the descendants of aboriginally distinct species. If any marked distinction existed between domestic races and species, this source of doubt could not so perpetually recur. It has often been stated that domestic races do not differ from each other in characters of generic value. I think it could be shown that this statement is hardly correct; but naturalists differ most widely in determining what characters are of generic value; all such valuations being at present empirical. Moreover, on the view of the origin of genera which I shall presently give, we have no right to expect often to meet with generic differences in our domesticated productions.

When we attempt to estimate the amount of structural difference between the domestic races of the same species, we are soon involved in doubt, from not knowing whether they have descended from one or several parent-species. This point, if it could be cleared up, would be interesting; if, for instance, it could be shown that the greyhound, bloodhound, terrier, spaniel, and bull-dog, which we all know propagate their kind so truly, were the offspring of any single species, then such facts would have great weight in making us doubt about the immutability of the many very closely allied and natural species—for instance, of the many foxes—inhabiting different quarters of the world. I do not believe, as we shall presently see, that all our dogs have descended from any one wild species; but, in the case of some other domestic races, there is presumptive, or even strong, evidence in favour of this view.

It has often been assumed that man has chosen for domestication animals and plants having an extraordinary inherent tendency to vary, and likewise to withstand diverse climates. I do not dispute that these capacities have added largely to the value of most of our domesticated productions; but how could a savage possibly know, when he first tamed an animal, whether it would vary in succeeding generations, and whether it would endure other climates? Has the little variability of the ass or guinea-fowl, or the small power of endurance of warmth by the reindeer, or of cold by the common camel, prevented their domestication? I cannot doubt that if other animals and plants, equal in number to our domesticated productions, and belonging to equally diverse classes and countries, were taken from a state of nature, and could be made to breed for an equal number of generations under domestication, they would vary on an average as largely as the parent species of our existing domesticated productions have varied.

In the case of most of our anciently domesticated animals and plants, I do not think it is possible to come to any definite conclusion, whether they have descended from one or several species. The argument mainly relied on by those who believe in the multiple origin of our domestic animals is, that we find in the most ancient records, more especially on the monuments of Egypt, much diversity in the breeds; and that some of the breeds closely resemble, perhaps are identical with, those still existing. Even if this latter fact were found more strictly and generally true than seems to me to be the case, what does it show, but that some of our breeds originated there, four or five thousand years ago? But Mr. Horner's researches have rendered it in some degree probable that man sufficiently civilized to have manufactured pottery existed in the valley of the Nile thirteen or fourteen thousand years ago; and who will pretend to say how long before these ancient periods, savages, like those of Tierra del Fuego or Australia, who possess a semi-domestic dog, may not have existed in Egypt?

The whole subject must, I think, remain vague; nevertheless, I may, without here entering on any details, state that, from geographical and other considerations, I think it highly probable that our domestic dogs have descended from several wild species. In regard to sheep and goats I can form no opinion. I should think, from facts communicated to me by Mr. Blyth, on the habits, voice, and constitution, &c., of the humped Indian cattle, that these had descended from a different aboriginal stock from our European cattle; and several competent judges believe that these latter have had more than one wild parent. With respect to horses, from reasons which I cannot give here, I am doubtfully inclined to believe, in opposition to several authors, that all the races have descended from one wild stock. Mr. Blyth, whose opinion, from his large and varied stores of knowledge, I should value more than that of almost any one, thinks that all the breeds of poultry have proceeded from the common wild Indian fowl (Gallus bankiva). In regard to ducks and rabbits, the breeds of which differ considerably from each other in structure, I do not doubt that they all have descended from the common wild duck and rabbit.

The doctrine of the origin of our several domestic races from several aboriginal stocks, has been carried to an absurd extreme by some authors. They believe that every race which breeds true, let the distinctive characters be ever so slight, has had its wild prototype. At this rate there must have existed at least a score of species of wild cattle, as many sheep, and several goats in Europe alone, and several even within Great Britain. One author believes that there formerly existed in Great Britain eleven wild species of sheep peculiar to it! When we bear in mind that Britain has now hardly one peculiar mammal, and France but few distinct from those of Germany and conversely, and so with Hungary, Spain, &c., but that each of these kingdoms possesses several peculiar breeds of cattle, sheep, &c., we must admit that many domestic breeds have originated in Europe; for whence could they have been derived, as these several countries do not possess a number of peculiar species as distinct parent-stocks? So it is in India. Even in the case of the domestic dogs of the whole world, which I fully admit have probably descended from several wild species, I cannot doubt that there has been an immense amount of inherited variation. Who can believe that animals closely resembling the Italian greyhound, the bloodhound, the bull-dog, or Blenheim spaniel, &c.—so unlike all wild Canidæ—ever existed freely in a state of nature? It has often been loosely said that all our races of dogs have been produced by the crossing of a few aboriginal species; but by crossing we can get only forms in some degree intermediate between their parents; and if we account for our several domestic races by this process, we must admit the former existence of the most extreme forms, as the Italian greyhound, bloodhound, bull-dog, &c., in the wild state. Moreover, the possibility of making distinct races by crossing has been greatly exaggerated. There can be no doubt that a race may be modified by occasional crosses, if aided by the careful selection of those individual mongrels, which present any desired character; but that a race could be obtained nearly intermediate between two extremely different races or species, I can hardly believe. Sir J. Sebright expressly experimentised for this object, and failed. The offspring from the first cross between two pure breeds is tolerably and sometimes (as I have found with pigeons) extremely uniform, and everything seems simple enough; but when these mongrels are crossed one with another for several generations, hardly two of them will be alike, and then the extreme difficulty, or rather utter hopelessness, of the task becomes apparent. Certainly, a breed intermediate between *two very distinct* breeds could not be got without extreme care and long-continued selection; nor can I find a single case on record of a permanent race having been thus formed.

On the Breeds of the Domestic Pigeon.—Believing that it is always best to study some special group, I have, after deliberation, taken up domestic pigeons. I have kept every breed which I could purchase or obtain, and have been most kindly favoured with skins from several quarters of the world, more especially by the Hon. W. Elliot from India, and by the Hon. C. Murray from Persia. Many treatises in different languages have been published on pigeons, and some of them are very important, as being of considerable antiquity. I have associated with several eminent fanciers, and have been permitted to join two of the London Pigeon Clubs. The diversity of the breeds is something astonishing. Compare the English carrier and the short-faced tumbler, and see the wonderful difference in their beaks, entailing corresponding differences in their skulls. The carrier, more especially the male bird, is also remarkable from the wonderful development of the carunculated skin about the head, and this is accompanied by greatly elongated eyelids, very large external orifices to the nostrils, and a wide gape of mouth. The short-faced tumbler has a beak in outline almost like that of a finch; and the common tumbler has the singular and strictly inherited habit of flying at a great height in a compact flock, and tumbling in the air head over heels. The runt is a bird of great size, with long, massive beak and large feet; some of the sub-breeds of runts have very long necks, others very long wings and tails, others singularly short tails. The barb is allied to the carrier, but, instead of a very long beak, has a very short and very broad one. The pouter has a much elongated body, wings, and legs; and its enormously developed crop, which it glories in inflating, may well excite astonishment and even laughter. The turbit has a very short and conical beak, with a line of reversed feathers down the breast; and it has the habit of continually expanding slightly the upper part of the œsophagus. The Jacobin has the feathers so much reversed along the back of the neck that they form a hood, and it has, proportionally to its size, much elongated wing and tail feathers. The trumpeter and laugher, as their names express, utter a very different coo from the other breeds. The fantail has thirty or even forty tail-feathers, instead of twelve or fourteen, the normal number in all members of the great pigeon family; and these feathers are kept expanded, and are carried so erect that in good birds the head and tail touch; the oil-gland is quite aborted. Several other less distinct breeds might have been specified.

In the skeletons of the several breeds, the development of the bones of the face in length and breadth and curvature differs enormously. The shape, as well as the breadth and length of the ramus of the lower jaw, varies in a highly remarkable manner. The number of the caudal and sacral vertebræ vary; as does the number of the ribs, together with their relative breadth and the presence of processes. The size and shape of the apertures in the sternum are highly variable; so is the degree of divergence and relative size of the two arms of the furcula. The proportional width of the gape of mouth, the proportional length of the eyelids, of the orifice of the nostrils, of the tongue (not always in strict correlation with the length of beak), the size of the crop and of the upper part of the œsophagus; the development and abortion of the oil-gland; the number of the primary wing and caudal feathers; the relative length of wing and tail to each other and to the body; the relative length of leg and of the feet; the number of scutellæ on the toes, the development of skin between the toes, are all points of structure which are variable. The period at which the perfect plumage is acquired varies, as does the state of the down with which the nestling birds are clothed when hatched. The shape and size of the eggs vary. The manner of flight differs remarkably; as does in some breeds the voice and disposition. Lastly, in certain breeds, the males and females have come to differ to a slight degree from each other.

Altogether at least a score of pigeons might be chosen, which if shown to an ornithologist, and he were told that they were wild birds, would certainly, I think, be ranked by him as well-defined species. Moreover, I do not believe that any ornithologist would place the English carrier, the short-faced tumbler, the runt, the barb, pouter, and fantail in the same genus; more especially as in each of these breeds several truly-inherited sub-breeds, or species as he might have called them, could be shown him.

Great as the differences are between the breeds of pigeons, I am fully convinced that the common opinion of naturalists is correct, namely, that all have descended from the rock-pigeon (Columba livia), including under this term several geographical races or sub-species, which differ from each other in the most trifling respects. As several of the reasons which have led me to this belief are in some degree applicable in other cases, I will here briefly give them. If the several breeds are not varieties, and have not proceeded from the rock-pigeon, they must have descended from at least seven or eight aboriginal stocks; for it is impossible to make the present domestic breeds by the crossing of any lesser number: how, for instance, could a pouter be produced by crossing two breeds unless one of the parent-stocks possessed the characteristic enormous crop? The supposed aboriginal stocks must all have been rock-pigeons, that is, not breeding or willingly perching on trees. But besides C. livia, with its geographical sub-species, only two or three other species of rock-pigeons are known; and these have not any of the characters of the domestic breeds. Hence the supposed aboriginal stocks must either still exist in the countries where they were originally domesticated, and yet be unknown to ornithologists; and this, considering their size, habits, and remarkable characters, seems very improbable; or they must have become extinct in the wild state. But birds breeding on precipices, and good fliers, are unlikely to be exterminated; and the common rock-pigeon, which has the same habits with the domestic breeds, has not been exterminated even on several of the smaller British islets, or on the shores of the Mediterranean. Hence the supposed extermination of so many species having similar habits with the rock-pigeon seems to me a very rash assumption. Moreover, the several above-named domesticated breeds have been transported to all parts of the world, and, therefore, some of them must have been carried back again into their native country; but not one has ever become wild or feral, though the dovecot-pigeon, which is the rock-pigeon in a very slightly altered state, has become feral in several places. Again, all recent experience shows that it is most difficult to get any wild animal to breed freely under domestication; yet on the hypothesis of the multiple origin of our pigeons, it must be assumed that at least seven or eight species were so thoroughly domesticated in ancient times by half-civilized man, as to be quite prolific under confinement.

An argument, as it seems to me, of great weight, and applicable in several other cases, is, that the above-specified breeds, though agreeing generally in constitution, habits, voice, colouring, and in most parts of their structure, with the wild rock-pigeon, yet are certainly highly abnormal in other parts of their structure: we may look in vain throughout the whole great family of Columbidæ for a beak like that of the English carrier, or that of the short-faced tumbler, or barb; for reversed feathers like those of the jacobin; for a crop like that of the pouter; for tail-feathers like those of the fantail. Hence it must be assumed not only that half-civilized man succeeded in thoroughly domesticating several species, but that he intentionally or by chance picked out extraordinarily abnormal species; and further, that these very species have since all become extinct or unknown. So many strange contingencies seem to me improbable in the highest degree.

Some facts in regard to the colouring of pigeons well deserve consideration. The rock-pigeon is of a slaty-blue, and has a white rump (the Indian sub-species, C. intermedia of Strickland, having it bluish); the tail has a terminal dark bar, with the bases of the outer feathers externally edged with white; the wings have two black bars; some semi-domestic breeds and some apparently truly wild breeds have, besides the two black bars, the wings chequered with black. These several marks do not occur together in any other species of the whole family. Now, in every one of the domestic breeds, taking thoroughly well-bred birds, all the above marks, even to the white edging of the outer tail-feathers, sometimes concur perfectly developed. Moreover, when two birds belonging to two distinct breeds are crossed, neither of which is blue or has any of the above-specified marks, the mongrel offspring are very apt suddenly to acquire these characters; for instance, I crossed some uniformly white fantails with some uniformly black barbs, and they produced mottled brown and black birds; these I again crossed together, and one grandchild of the pure white fantail and pure black barb was of as beautiful a blue colour, with the white rump, double black wing-bar, and barred and white-edged tail-feathers, as any wild rock-pigeon! We can understand these facts, on the well-known principle of reversion to ancestral characters, if all the domestic breeds have descended from the rock-pigeon. But if we deny this, we must make one of the two following highly improbable suppositions. Either, firstly, that all the several imagined aboriginal stocks were coloured and marked like the rock-pigeon, although no other existing species is thus coloured and marked, so that in each separate breed there might be a tendency to revert to the very same colours and markings. Or, secondly, that each breed, even the purest, has within a dozen or, at most, within a score of generations, been crossed by the rock-pigeon: I say within a dozen or twenty generations, for we know of no fact countenancing the belief that the child ever reverts to some one ancestor, removed by a greater number of generations. In a breed which has been crossed only once with some distinct breed, the tendency to reversion to any character derived from such cross will naturally become less and less, as in each succeeding generation there will be less of the foreign blood; but when there has been no cross with a distinct breed, and there is a tendency in both parents to revert to a character, which has been lost during some former generation, this tendency, for all that we can see to the contrary, may be transmitted undiminished for an indefinite number of generations. These two distinct cases are often confounded in treatises on inheritance.

Lastly, the hybrids or mongrels from between all the domestic breeds of pigeons are perfectly fertile. I can state this from my own observations, purposely made on the most distinct breeds. Now, it is difficult, perhaps impossible, to bring forward one case of the hybrid offspring of two animals *clearly distinct* being themselves perfectly fertile. Some authors believe that long-continued domestication eliminates this strong tendency to sterility: from the history of the dog I think there is some probability in this hypothesis, if applied to species closely related together, though it is unsupported by a single experiment. But to extend the hypothesis so far as to suppose that species, aboriginally as distinct as carriers, tumblers, pouters, and fantails now are, should yield offspring perfectly fertile, *inter se,* seems to me rash in the extreme.

From these several reasons, namely, the improbability of man having formerly got seven or eight supposed species of pigeons to breed freely under domestication; these supposed species being quite unknown in a wild state, and their becoming nowhere feral; these species having very abnormal characters in certain respects, as compared with all other Columbidæ, though so like in most other respects to the rock-pigeon; the blue colour and various marks occasionally appearing in all the breeds, both when kept pure and when crossed; the mongrel offspring being perfectly fertile;—from these several reasons, taken together, I can feel no doubt that all our domestic breeds have descended from the Columba livia with its geographical sub-species.

In favour of this view, I may add, firstly, that C. livia, or the rock-pigeon, has been found capable of domestication in Europe and in India; and that it agrees in habits and in a great number of points of structure with all the domestic breeds. Secondly, although an English carrier or short-faced tumbler differs immensely in certain characters from the rock-pigeon, yet by comparing the several sub-breeds of these breeds, more especially those brought from distant countries, we can make an almost perfect series between the extremes of structure. Thirdly, those characters which are mainly distinctive of each breed, for instance the wattle and length of beak of the carrier, the shortness of that of the tumbler, and the number of tail-feathers in the fantail, are in each breed eminently variable; and the explanation of this fact will be obvious when we come to treat of selection. Fourthly, pigeons have been watched, and tended with the utmost care, and loved by many people. They have been domesticated for thousands of years in several quarters of the world; the earliest known record of pigeons is in the fifth Ægyptian dynasty, about 3000 B.C., as was pointed out to me by Professor Lepsius; but Mr. Birch informs me that pigeons are given in a bill of fare in the previous dynasty. In the time of the Romans, as we hear from Pliny, immense prices were given for pigeons; "nay, they are come to this pass, that they can reckon up their pedigree and race." Pigeons were much valued by Akbar Khan in India, about the year 1600; never less than 20,000 pigeons were taken with the court. "The monarchs of Iran and Turan sent him some very rare birds;" and, continues the courtly historian, "His Majesty by crossing the breeds, which method was never practised before, has improved them astonishingly." About this same period the Dutch were as eager about pigeons as were the old Romans. The paramount importance of these considerations in explaining the immense amount of variation which pigeons have undergone, will be obvious when we treat of Selection. We shall then, also, see how it is that the breeds so often have a somewhat monstrous character. It is also a most favourable circumstance for the production of distinct breeds, that male and female pigeons can be easily mated for life; and thus different breeds can be kept together in the same aviary.

I have discussed the probable origin of domestic pigeons at some, yet quite insufficient, length; because when I first kept pigeons and watched the several kinds, knowing well how true they bred, I felt fully as much difficulty in believing that they could ever have descended from a common parent, as any naturalist could in coming to a similar conclusion in regard to the many species of finches, or other large groups of birds, in nature. One circumstance has struck me much; namely, that all the breeders of the various domestic animals and the cultivators of plants, with whom I have ever conversed, or whose treatises I have read, are firmly convinced that the several breeds to which each has attended, are descended from so many aboriginally distinct species. Ask, as I have asked, a celebrated raiser of Hereford cattle, whether his cattle might not have descended from long-horns, and he will laugh you to scorn. I have never met a pigeon, or poultry, or duck, or rabbit fancier, who was not fully convinced that each main breed was descended from a distinct species. Van Mons, in his treatise on pears and apples, shows how utterly he disbelieves that the several sorts, for instance a Ribston-pippin or Codlin-apple, could ever have proceeded from the seeds of the same tree. Innumerable other examples could be given. The explanation, I think, is simple: from long-continued study they are strongly impressed with the differences between the several races; and though they well know that each race varies slightly, for they win their prizes by selecting such slight differences, yet they ignore all general arguments, and refuse to sum up in their minds slight differences accumulated during many successive generations. May not those naturalists who, knowing far less of the laws of inheritance than does the breeder, and knowing no more than he does of the intermediate links in the long lines of descent, yet admit that many of our domestic races have descended from the same parents—may they not learn a lesson of caution, when they deride the idea of species in a state of nature being lineal descendants of other species?

Selection.—Let us now briefly consider the steps by which domestic races have been produced, either from one or from several allied species. Some little effect may, perhaps, be attributed to the direct action of the external conditions of life, and some little to habit; but he would be a bold man who would account by such agencies for the differences of a dray and race horse, a greyhound and bloodhound, a carrier and tumbler pigeon. One of the most remarkable features in our domesticated races is that we see in them adaptation, not indeed to the animal's or plant's own good, but to man's use or fancy. Some variations useful to him have probably arisen suddenly, or by one step; many botanists, for instance, believe that the fuller's teazle, with its hooks, which cannot be rivalled by any mechanical contrivance, is only a variety of the wild Dipsacus; and this amount of change may have suddenly arisen in a seedling. So it has probably been with the turnspit dog; and this is known to have been the case with the ancon sheep. But when we compare the dray-horse and race-horse, the dromedary and camel, the various breeds of sheep fitted either for cultivated land or mountain pasture, with the wool of one breed good for one purpose, and that of another breed for another purpose; when we compare the many breeds of dogs, each good for man in very different ways; when we compare the game-cock, so pertinacious in battle, with other breeds so little quarrelsome, with "everlasting layers" which never desire to sit, and with the bantam so small and elegant; when we compare the host of agricultural, culinary, orchard, and flower-garden races of plants, most useful to man at different seasons and for different purposes, or so beautiful in his eyes, we must, I think, look further than to mere variability. We cannot suppose that all the breeds were suddenly produced as perfect and as useful as we now see them; indeed, in several cases, we know that this has not been their history. The key is man's power of accumulative selection: nature gives successive variations; man adds them up in certain directions useful to him. In this sense he may be said to make for himself useful breeds.

The great power of this principle of selection is not hypothetical. It is certain that several of our eminent breeders have, even within a single lifetime, modified to a large extent some breeds of cattle and sheep. In order fully to realise what they have done, it is almost necessary to read several of the many treatises devoted to this subject, and to inspect the animals. Breeders habitually speak of an animal's organisation as something quite plastic, which they can model almost as they please. If I had space I could quote numerous passages to this effect from highly competent authorities. Youatt, who was probably better acquainted with the works of agriculturists than almost any other individual, and who was himself a very good judge of an animal, speaks of the principle of selection as "that which enables the agriculturist, not only to modify the character of his flock, but to change it altogether. It is the magician's wand, by means of which he may summon into life whatever form and mould he pleases." Lord Somerville, speaking of what breeders have done for sheep, says:— "It would seem as if they had chalked out upon a wall a form perfect in itself, and then had given it existence." That most skilful breeder, Sir John Sebright, used to say, with respect to pigeons, that "he would produce any given feather in three years, but it would take him six years to obtain head and beak." In Saxony the importance of the principle of selection in regard to merino sheep is so fully recognised, that men follow it as a trade: the sheep are placed on a table and are studied, like a picture by a connoisseur; this is done three times at intervals of months, and the sheep are each time marked and classed, so that the very best may ultimately be selected for breeding.

What English breeders have actually effected is proved by the enormous prices given for animals with a good pedigree; and these have now been exported to almost every quarter of the world. The improvement is by no means generally due to crossing different breeds; all the best breeders are strongly opposed to this practice, except sometimes amongst closely allied sub-breeds. And when a cross has been made, the closest selection is far more indispensable even than in ordinary cases. If selection consisted merely in separating some very distinct variety, and breeding from it, the principle would be so obvious as hardly to be worth notice; but its importance consists in the great effect produced by the accumulation in one direction, during successive generations, of differences absolutely inappreciable by an uneducated eye—differences which I for one have vainly attempted to appreciate. Not one man in a thousand has accuracy of eye and judgment sufficient to become an eminent breeder. If gifted with these qualities, and he studies his subject for years, and devotes his lifetime to it with indomitable perseverance, he will succeed, and may make great improvements; if he wants any of these qualities, he will assuredly fail. Few would readily believe in the natural capacity and years of practice requisite to become even a skilful pigeon-fancier.

The same principles are followed by horticulturists; but the variations are here often more abrupt. No one supposes that our choicest productions have been produced by a single variation from the aboriginal stock. We have proofs that this is not so in some cases, in which exact records have been kept; thus, to give a very trifling instance, the steadily-increasing size of the common gooseberry may be quoted. We see an astonishing improvement in many florists' flowers, when the flowers of the present day are compared with drawings made only twenty or thirty years ago. When a race of plants is once pretty well established, the seed-raisers do not pick out the best plants, but merely go over their seed-beds, and pull up the "rogues," as they call the plants that deviate from the proper standard. With animals this kind of selection is, in fact, also followed; for hardly any one is so careless as to allow his worst animals to breed.

In regard to plants, there is another means of observing the accumulated effects of selection—namely, by comparing the diversity of flowers in the different varieties of the same species in the flower-garden; the diversity of leaves, pods, or tubers, or whatever part is valued, in the kitchen-garden, in comparison with the flowers of the same varieties; and the diversity of fruit of the same species in the orchard, in comparison with the leaves and flowers of the same set of varieties. See how different the leaves of the cabbage are, and how extremely alike the flowers; how unlike the flowers of the heartsease are, and how alike the leaves; how much the fruit of the different kinds of gooseberries differ in size, colour, shape, and hairiness, and yet the flowers present very slight differences. It is not that the varieties which differ largely in some one point do not differ at all in other points; this is hardly ever, perhaps never, the case. The laws of correlation of growth, the importance of which should never be overlooked, will ensure some differences; but, as a general rule, I cannot doubt that the continued selection of slight variations, either in the leaves, the flowers, or the fruit, will produce races differing from each other chiefly in these characters.

It may be objected that the principle of selection has been reduced to methodical practice for scarcely more than three-quarters of a century; it has certainly been more attended to of late years, and many treatises have been published on the subject; and the result, I may add, has been, in a corresponding degree, rapid and important. But it is very far from true that the principle is a modern discovery. I could give several references to the full acknowledgment of the importance of the principle in works of high antiquity. In rude and barbarous periods of English history choice animals were often imported, and laws were passed to prevent their exportation: the destruction of horses under a certain size was ordered, and this may be compared to the "roguing" of plants by nurserymen. The principle of selection I find distinctly given in an ancient Chinese encyclopædia. Explicit rules are laid down by some of the Roman classical writers. From passages in Genesis, it is clear that the colour of domestic animals was at that early period attended to. Savages now sometimes cross their dogs with wild canine animals, to improve the breed, and they formerly did so, as is attested by passages in Pliny. The savages in South Africa match their draught cattle by colour, as do some of the Esquimaux their teams of dogs. Livingstone shows how much good domestic breeds are valued by the negroes of the interior of Africa who have not associated with Europeans. Some of these facts do not show actual selection, but they show that the breeding of domestic animals was carefully attended to in ancient times, and is now attended to by the lowest savages. It would, indeed, have been a strange fact, had attention not been paid to breeding, for the inheritance of good and bad qualities is so obvious.

At the present time, eminent breeders try by methodical selection, with a distinct object in view, to make a new strain or sub-breed, superior to anything existing in the country. But, for our purpose, a kind of Selection, which may be called Unconscious, and which results from every one trying to possess and breed from the best individual animals, is more important. Thus, a man who intends keeping pointers naturally tries to get as good dogs as he can, and afterwards breeds from his own best dogs, but he has no wish or expectation of permanently altering the breed. Nevertheless I cannot doubt that this process, continued during centuries, would improve and modify any breed, in the same way as Bakewell, Collins, &c., by this very same process, only carried on more methodically, did greatly modify, even during their own lifetimes, the forms and qualities of their cattle. Slow and insensible changes of this kind could never be recognised unless actual measurements or careful drawings of the breeds in question had been made long ago, which might serve for comparison. In some cases, however, unchanged or but little changed individuals of the same breed may be found in less civilised districts, where the breed has been less improved. There is reason to believe that King Charles's spaniel has been unconsciously modified to a large extent since the time of that monarch. Some highly competent authorities are convinced that the setter is directly derived from the spaniel, and has probably been slowly altered from it. It is known that the English pointer has been greatly changed within the last century, and in this case the change has, it is believed, been chiefly effected by crosses with the fox-hound; but what concerns us is, that the change has been effected unconsciously and gradually, and yet so effectually, that, though the old Spanish pointer certainly came from Spain, Mr. Borrow has not seen, as I am informed by him, any native dog in Spain like our pointer.

By a similar process of selection, and by careful training, the whole body of English racehorses have come to surpass in fleetness and size the parent Arab stock, so that the latter, by the regulations for the Goodwood Races, are favoured in the weights they carry. Lord Spencer and others have shown how the cattle of England have increased in weight and in early maturity, compared with the stock formerly kept in this country. By comparing the accounts given in old pigeon treatises of carriers and tumblers with these breeds as now existing in Britain, India, and Persia, we can, I think, clearly trace the stages through which they have insensibly passed, and come to differ so greatly from the rock-pigeon.

Youatt gives an excellent illustration of the effects of a course of selection, which may be considered as unconsciously followed, in so far that the breeders could never have expected or even have wished to have produced the result which ensued—namely, the production of two distinct strains. The two flocks of Leicester sheep kept by Mr. Buckley and Mr. Burgess, as Mr. Youatt remarks, "have been purely bred from the original stock of Mr. Bakewell for upwards of fifty years. There is not a suspicion existing in the mind of any one at all acquainted with the subject that the owner of either of them has deviated in any one instance from the pure blood of Mr. Bakewell's flock, and yet the difference between the sheep possessed by these two gentlemen is so great that they have the appearance of being quite different varieties."

If there exist savages so barbarous as never to think of the inherited character of the offspring of their domestic animals, yet any one animal particularly useful to them, for any special purpose, would be carefully preserved during famines and other accidents, to which savages are so liable, and such choice animals would thus generally leave more offspring than the inferior ones; so that in this case there would be a kind of unconscious selection going on. We see the value set on animals even by the barbarians of Tierra del Fuego, by their killing and devouring their old women, in times of dearth, as of less value than their dogs.

In plants the same gradual process of improvement, through the occasional preservation of the best individuals, whether or not sufficiently distinct to be ranked at their first appearance as distinct varieties, and whether or not two or more species or races have become blended together by crossing, may plainly be recognised in the increased size and beauty which we now see in the varieties of the heartsease, rose, pelargonium, dahlia, and other plants, when compared with the older varieties or with their parent-stocks. No one would ever expect to get a first-rate heartsease or dahlia from the seed of a wild plant. No one would expect to raise a first-rate melting pear from the seed of the wild pear, though he might succeed from a poor seedling growing wild, if it had come from a garden-stock. The pear, though cultivated in classical times, appears, from Pliny's description, to have been a fruit of very inferior quality. I have seen great surprise expressed in horticultural works at the wonderful skill of gardeners, in having produced such splendid results from such poor materials; but the art, I cannot doubt, has been simple, and, as far as the final result is concerned, has been followed almost unconsciously. It has consisted in always cultivating the best known variety, sowing its seeds, and, when a slightly better variety has chanced to appear, selecting it, and so onwards. But the gardeners of the classical period, who cultivated the best pear they could procure, never thought what splendid fruit we should eat; though we owe our excellent fruit, in some small degree, to their having naturally chosen and preserved the best varieties they could anywhere find.

A large amount of change in our cultivated plants, thus slowly and unconsciously accumulated, explains, as I believe, the well-known fact, that in a vast number of cases we cannot recognise, and therefore do not know, the wild parent-stocks of the plants which have been longest cultivated in our flower and kitchen gardens. If it has taken centuries or thousands of years to improve or modify most of our plants up to their present standard of usefulness to man, we can understand how it is that neither Australia, the Cape of Good Hope, nor any other region inhabited by quite uncivilised man, has afforded us a single plant worth culture. It is not that these countries, so rich in species, do not by a strange chance possess the aboriginal stocks of any useful plants, but that the native plants have not been improved by continued selection up to a standard of perfection comparable with that given to the plants in countries anciently civilised.

In regard to the domestic animals kept by uncivilised man, it should not be overlooked that they almost always have to struggle for their own food, at least during certain seasons. And in two countries very differently circumstanced, individuals of the same species, having slightly different constitutions or structure, would often succeed better in the one country than in the other, and thus by a process of "natural selection," as will hereafter be more fully explained, two sub-breeds might be formed. This, perhaps, partly explains what has been remarked by some authors, namely, that the varieties kept by savages have more of the character of species than the varieties kept in civilised countries.

On the view here given of the all-important part which selection by man has played, it becomes at once obvious, how it is that our domestic races show adaptation in their structure or in their habits to man's wants or fancies. We can, I think, further understand the frequently abnormal character of our domestic races, and likewise their differences being so great in external characters and relatively so slight in internal parts or organs. Man can hardly select, or only with much difficulty, any deviation of structure excepting such as is externally visible; and indeed he rarely cares for what is internal. He can never act by selection, excepting on variations which are first given to him in some slight degree by nature. No man would ever try to make a fantail, till he saw a pigeon with a tail developed in some slight degree in an unusual manner, or a pouter till he saw a pigeon with a crop of somewhat unusual size; and the more abnormal or unusual any character was when it first appeared, the more likely it would be to catch his attention. But to use such an expression as trying to make a fantail, is, I have no doubt, in most cases, utterly incorrect. The man who first selected a pigeon with a slightly larger tail, never dreamed what the descendants of that pigeon would become through long-continued, partly unconscious and partly methodical selection. Perhaps the parent bird of all fantails had only fourteen tail-feathers somewhat expanded, like the present Java fantail, or like individuals of other and distinct breeds, in which as many as seventeen tail-feathers have been counted. Perhaps the first pouter-pigeon did not inflate its crop much more than the turbit now does the upper part of its œsophagus,—a habit which is disregarded by all fanciers, as it is not one of the points of the breed.

Nor let it be thought that some great deviation of structure would be necessary to catch the fancier's eye: he perceives extremely small differences, and it is in human nature to value any novelty, however slight, in one's own possession. Nor must the value which would formerly be set on any slight differences in the individuals of the same species, be judged of by the value which would now be set on them, after several breeds have once fairly been established. Many slight differences might, and indeed do now, arise amongst pigeons, which are rejected as faults or deviations from the standard of perfection of each breed. The common goose has not given rise to any marked varieties; hence the Thoulouse and the common breed, which differ only in colour, that most fleeting of characters, have lately been exhibited as distinct at our poultry-shows.

I think these views further explain what has sometimes been noticed—namely, that we know nothing about the origin or history of any of our domestic breeds. But, in fact, a breed, like a dialect of a language, can hardly be said to have had a definite origin. A man preserves and breeds from an individual with some slight deviation of structure, or takes more care than usual in matching his best animals and thus improves them, and the improved individuals slowly spread in the immediate neighbourhood. But as yet they will hardly have a distinct name, and from being only slightly valued, their history will be disregarded. When further improved by the same slow and gradual process, they will spread more widely, and will get recognised as something distinct and valuable, and will then probably first receive a provincial name. In semi-civilised countries, with little free communication, the spreading and knowledge of any new sub-breed will be a slow process. As soon as the points of value of the new sub-breed are once fully acknowledged, the principle, as I have called it, of unconscious selection will always tend,—perhaps more at one period than at another, as the breed rises or falls in fashion,—perhaps more in one district than in another, according to the state of civilisation of the inhabitants,—slowly to add to the characteristic features of the breed, whatever they may be. But the chance will be infinitely small of any record having been preserved of such slow, varying, and insensible changes.

I must now say a few words on the circumstances, favourable, or the reverse, to man's power of selection. A high degree of variability is obviously favourable, as freely giving the materials for selection to work on; not that mere individual differences are not amply sufficient, with extreme care, to allow of the accumulation of a large amount of modification in almost any desired direction. But as variations manifestly useful or pleasing to man appear only occasionally, the chance of their appearance will be much increased by a large number of individuals being kept; and hence this comes to be of the highest importance to success. On this principle Marshall has remarked, with respect to the sheep of parts of Yorkshire, that "as they generally belong to poor people, and are mostly *in small lots*, they never can be improved." On the other hand, nurserymen, from raising large stocks of the same plants, are generally far more successful than amateurs in getting new and valuable varieties. The keeping of a large number of individuals of a species in any country requires that the species should be placed under favourable conditions of life, so as to breed freely in that country. When the individuals of any species are scanty, all the individuals, whatever their quality may be, will generally be allowed to breed, and this will effectually prevent selection. But probably the most important point of all, is, that the animal or plant should be so highly useful to man, or so much valued by him, that the closest attention should be paid to even the slightest deviation in the qualities or structure of each individual. Unless such attention be paid nothing can be effected. I have seen it gravely remarked, that it was most fortunate that the strawberry began to vary just when gardeners began to attend closely to this plant. No doubt the strawberry had always varied since it was cultivated, but the slight varieties had been neglected. As soon, however, as gardeners picked out individual plants with slightly larger, earlier, or better fruit, and raised seedlings from them, and again picked out the best seedlings and bred from them, then, there appeared (aided by some crossing with distinct species) those many admirable varieties of the strawberry which have been raised during the last thirty or forty years.

In the case of animals with separate sexes, facility in preventing crosses is an important element of success in the formation of new races,—at least, in a country which is already stocked with other races. In this respect enclosure of the land plays a part. Wandering savages or the inhabitants of open plains rarely possess more than one breed of the same species. Pigeons can be mated for life, and this is a great convenience to the fancier, for thus many races may be kept true, though mingled in the same aviary; and this circumstance must have largely favoured the improvement and formation of new breeds. Pigeons, I may add, can be propagated in great numbers and at a very quick rate, and inferior birds may be freely rejected, as when killed they serve for food. On the other hand, cats, from their nocturnal rambling habits, cannot be matched, and, although so much valued by women and children, we hardly ever see a distinct breed kept up; such breeds as we do sometimes see are almost always imported from some other country, often from islands. Although I do not doubt that some domestic animals vary less than others, yet the rarity or absence of distinct breeds of the cat, the donkey, peacock, goose, &c., may be attributed in main part to selection not having been brought into play: in cats, from the difficulty in pairing them; in donkeys, from only a few being kept by poor people, and little attention paid to their breeding; in peacocks, from not being very easily reared and a large stock not kept; in geese, from being valuable only for two purposes, food and feathers, and more especially from no pleasure having been felt in the display of distinct breeds.

To sum up on the origin of our Domestic Races of animals and plants. I believe that the conditions of life, from their action on the reproductive system, are so far of the highest importance as causing variability. I do not believe that variability is an inherent and necessary contingency, under all circumstances, with all organic beings, as some authors have thought. The effects of variability are modified by various degrees of inheritance and of reversion. Variability is governed by many unknown laws, more especially by that of correlation of growth. Something may be attributed to the direct action of the conditions of life. Something must be attributed to use and disuse. The final result is thus rendered infinitely complex. In some cases, I do not doubt that the intercrossing of species, aboriginally distinct, has played an important part in the origin of our domestic productions. When in any country several domestic breeds have once been established, their occasional intercrossing, with the aid of selection, has, no doubt, largely aided in the formation of new sub-breeds; but the importance of the crossing of varieties has, I believe, been greatly exaggerated, both in regard to animals and to those plants which are propagated by seed. In plants which are temporarily propagated by cuttings, buds, &c., the importance of the crossing both of distinct species and of varieties is immense; for the cultivator here quite disregards the extreme variability both of hybrids and mongrels, and the frequent sterility of hybrids; but the cases of plants not propagated by seed are of little importance to us, for their endurance is only temporary. Over all these causes of Change I am convinced that the accumulative action of Selection, whether applied methodically and more quickly, or unconsciously and more slowly, but more efficiently, is by far the predominant Power.

CHAPTER II

Variation under Nature

VARIABILITY • INDIVIDUAL DIFFERENCES • DOUBTFUL SPECIES • WIDE RANGING, MUCH DIFFUSED, AND COMMON SPECIES VARY MOST • SPECIES OF THE LARGER GENERA IN ANY COUNTRY VARY MORE THAN THE SPECIES OF THE SMALLER GENERA • MANY OF THE SPECIES OF THE LARGER GENERA RESEMBLE VARIETIES IN BEING VERY CLOSELY, BUT UNEQUALLY, RELATED TO EACH OTHER, AND IN HAVING RESTRICTED RANGES

BEFORE APPLYING THE PRINCIPLES arrived at in the last chapter to organic beings in a state of nature, we must briefly discuss whether these latter are subject to any variation. To treat this subject at all properly, a long catalogue of dry facts should be given; but these I shall reserve for my future work. Nor shall I here discuss the various definitions which have been given of the term species. No one definition has as yet satisfied all naturalists; yet every naturalist knows vaguely what he means when he speaks of a species. Generally the term includes the unknown element of a distinct act of creation. The term "variety" is almost equally difficult to define; but here community of descent is almost universally implied, though it can rarely be proved. We have also what are called monstrosities; but they graduate into varieties. By a monstrosity I presume is meant some considerable deviation of structure in one part, either injurious to or not useful to the species, and not generally propagated. Some authors use the term "variation" in a technical sense, as implying a modification directly due to the physical conditions of life; and "variations" in this sense are supposed not to be inherited: but who can say that the dwarfed condition of shells in the brackish waters of the Baltic, or dwarfed plants on Alpine summits, or the thicker fur of an animal from far northwards, would not in some cases be inherited for at least some few generations? and in this case I presume that the form would be called a variety.

Again, we have many slight differences which may be called individual differences, such as are known frequently to appear in the offspring from the same parents, or which may be presumed to have thus arisen, from being frequently observed in the individuals of the same species inhabiting the same confined locality. No one supposes that all the individuals of the same species are cast in the very same mould. These individual differences are highly important for us, as they afford materials for natural selection to accumulate, in the same manner as man can accumulate in any given direction individual differences in his domesticated productions. These individual differences generally affect what naturalists consider unimportant parts; but I could show by a long catalogue of facts, that parts which must be called important, whether viewed under a physiological or classificatory point of view, sometimes vary in the individuals of the same species. I am convinced that the most experienced naturalist would be surprised at the number of the cases of variability, even in important parts of structure, which he could collect on good authority, as I have collected, during a course of years. It should be remembered that systematists are far from pleased at finding variability in important characters, and that there are not many men who will laboriously examine internal and important organs, and compare them in many specimens of the same species. I should never have expected that the branching of the main nerves close to the great central ganglion of an insect would have been variable in the same species; I should have expected that changes of this nature could have been effected only by slow degrees: yet quite recently Mr. Lubbock has shown a degree of variability in these main nerves in Coccus, which may almost be compared to the irregular branching of the stem of a tree. This philosophical naturalist, I may add, has also quite recently shown that the muscles in the larvæ of certain insects are very far from uniform. Authors sometimes argue in a circle when they state that important organs never vary; for these same authors practically rank that character as important (as some few naturalists have honestly confessed) which does not vary; and, under this point of view, no instance of an important part varying will ever be found: but under any other point of view many instances assuredly can be given.

There is one point connected with individual differences, which seems to me extremely perplexing: I refer to those genera which have sometimes been called "protean" or "polymorphic," in which the species present an inordinate amount of variation; and hardly two naturalists can agree which forms to rank as species and which as varieties. We may instance Rubus, Rosa, and Hieracium amongst plants, several genera of insects, and several

ascend mountains to different heights; they have different geographical ranges; and lastly, according to very numerous experiments made during several years by that most careful observer Gärtner, they can be crossed only with much difficulty. We could hardly wish for better evidence of the two forms being specifically distinct. On the other hand, they are united by many intermediate links, and it is very doubtful whether these links are hybrids; and there is, as it seems to me, an overwhelming amount of experimental evidence, showing that they descend from common parents, and consequently must be ranked as varieties.

Close investigation, in most cases, will bring naturalists to an agreement how to rank doubtful forms. Yet it must be confessed, that it is in the best-known countries that we find the greatest number of forms of doubtful value. I have been struck with the fact, that if any animal or plant in a state of nature be highly useful to man, or from any cause closely attract his attention, varieties of it will almost universally be found recorded. These varieties, moreover, will be often ranked by some authors as species. Look at the common oak, how closely it has been studied; yet a German author makes more than a dozen species out of forms, which are very generally considered as varieties; and in this country the highest botanical authorities and practical men can be quoted to show that the sessile and pedunculated oaks are either good and distinct species or mere varieties.

When a young naturalist commences the study of a group of organisms quite unknown to him, he is at first much perplexed to determine what differences to consider as specific, and what as varieties; for he knows nothing of the amount and kind of variation to which the group is subject; and this shows, at least, how very generally there is some variation. But if he confine his attention to one class within one country, he will soon make up his mind how to rank most of the doubtful forms. His general tendency will be to make many species, for he will become impressed, just like the pigeon or poultry-fancier before alluded to, with the amount of difference in the forms which he is continually studying; and he has little general knowledge of analogical variation in other groups and in other countries, by which to correct his first impressions. As he extends the range of his observations, he will meet with more cases of difficulty; for he will encounter a greater number of closely-allied forms. But if his observations be widely extended, he will in the end generally be enabled to make up his own mind which to call varieties and which species; but he will succeed in this at the expense of admitting much variation,—and the truth of this admission will often be disputed by other naturalists. When, moreover, he comes to study allied forms brought from countries not now continuous, in which case he can hardly hope to find the intermediate links between his doubtful forms, he will have to trust almost entirely to analogy, and his difficulties will rise to a climax.

Certainly no clear line of demarcation has as yet been drawn between species and sub-species—that is, the forms which in the opinion of some naturalists come very near to, but do not quite arrive at the rank of species; or, again, between sub-species and well-marked varieties, or between lesser varieties and individual differences. These differences blend into each other in an insensible series; and a series impresses the mind with the idea of an actual passage.

Hence I look at individual differences, though of small interest to the systematist, as of high importance for us, as being the first step towards such slight varieties as are barely thought worth recording in works on natural history. And I look at varieties which are in any degree more distinct and permanent, as steps leading to more strongly marked and more permanent varieties; and at these latter, as leading to sub-species, and to species. The passage from one stage of difference to another and higher stage may be, in some cases, due merely to the long-continued action of different physical conditions in two different regions; but I have not much faith in this view; and I attribute the passage of a variety, from a state in which it differs very slightly from its parent to one in which it differs more, to the action of natural selection in accumulating (as will hereafter be more fully explained) differences of structure in certain definite directions. Hence I believe a well-marked variety may be justly called an incipient species; but whether this belief be justifiable must be judged of by the general weight of the several facts and views given throughout this work.

It need not be supposed that all varieties or incipient species necessarily attain the rank of species. They may whilst in this incipient state become extinct, or they may endure as varieties for very long periods, as has been shown to be the case by Mr. Wollaston with the varieties of certain fossil land-shells in Madeira. If a variety were to flourish so as to exceed in numbers the parent species, it would then rank as the species, and the species as the variety; or it might come to supplant and exterminate the parent species; or both might co-exist, and both rank as independent species. But we shall hereafter have to return to this subject.

From these remarks it will be seen that I look at the term species, as one arbitrarily given for the sake of convenience to a set of individuals closely resembling each other, and that it does not essentially differ from the term variety, which is given to less distinct and more fluctuating forms. The term variety, again, in comparison with mere individual differences, is also applied arbitrarily, and for mere convenience sake.

Guided by theoretical considerations, I thought that some interesting results might be obtained in regard to the nature and relations of the species which vary most, by tabulating all the varieties in several well-worked floras. At first this seemed a simple task; but Mr. H. C. Watson, to whom I am much indebted for valuable advice and assistance on this subject, soon convinced me that there were many difficulties, as did subsequently Dr. Hooker, even in stronger terms. I shall reserve for my future work the discussion of these difficulties, and the tables themselves of the proportional numbers of the varying species. Dr. Hooker permits me to add, that after having carefully read my manuscript, and examined the tables, he thinks that the following statements are fairly well established. The whole subject, however, treated as it necessarily here is with much brevity, is rather perplexing, and allusions cannot be avoided to the "struggle for existence," "divergence of character," and other questions, hereafter to be discussed.

Alph. De Candolle and others have shown that plants which have very wide ranges generally present varieties; and this might have been expected, as they become exposed to diverse physical conditions, and as they come into competition (which, as we shall hereafter see, is a far more important circumstance) with different sets of organic beings. But my tables further show that, in any limited country, the species which are most common, that is abound most in individuals, and the species which are most widely diffused within their own country (and this is a different consideration from wide range, and to a certain extent from commonness), often give rise to varieties sufficiently well-marked to have been recorded in botanical works. Hence it is the most flourishing, or, as they may be called, the dominant species,—those which range widely over the world, are the most diffused in their own country, and are the most numerous in individuals,—which oftenest produce well-marked varieties, or, as I consider them, incipient species. And this, perhaps, might have been anticipated; for, as varieties, in order to become in any degree permanent, necessarily have to struggle with the other inhabitants of the country, the species which are already dominant will be the most likely to yield offspring which, though in some slight degree modified, will still inherit those advantages that enabled their parents to become dominant over their compatriots.

If the plants inhabiting a country and described in any Flora be divided into two equal masses, all those in the larger genera being placed on one side, and all those in the smaller genera on the other side, a somewhat larger number of the very common and much diffused or dominant species will be found on the side of the larger genera. This, again, might have been anticipated; for the mere fact of many species of the same genus inhabiting any country, shows that there is something in the organic or inorganic conditions of that country favourable to the genus; and, consequently, we might have expected to have found in the larger genera, or those including many species, a large proportional number of dominant species. But so many causes tend to obscure this result, that I am surprised that my tables show even a small majority on the side of the larger genera. I will here allude to only two causes of obscurity. Fresh-water and salt-loving plants have generally very wide ranges and are much diffused, but this seems to be connected with the nature of the stations inhabited by them, and has little or no relation to the size of the genera to which the species belong. Again, plants low in the scale of organisation are generally much more widely diffused than plants higher in the scale; and here again there is no close relation to the size of the genera. The cause of lowly-organised plants ranging widely will be discussed in our chapter on geographical distribution.

From looking at species as only strongly-marked and well-defined varieties, I was led to anticipate that the species of the larger genera in each country would oftener present varieties, than the species of the smaller genera; for wherever many closely related species (*i. e.* species of the same genus) have been formed, many varieties or incipient species ought, as a general rule, to be now forming. Where many large trees grow, we expect to find saplings. Where many species of a genus have been formed through variation, circumstances have been favourable for variation; and hence we might expect that the circumstances would generally be still favourable to variation. On the other hand, if we look at each species as a special act of creation, there is no apparent reason why more varieties should occur in a group having many species, than in one having few.

To test the truth of this anticipation I have arranged the plants of twelve countries, and the coleopterous insects of two districts, into two nearly equal masses, the species of the larger genera on one side, and those of the smaller genera on the other side, and it has invariably proved to be the case that a larger proportion of the species on the side of the larger genera present varieties, than on the side of the smaller genera. Moreover, the species of the large genera which present any varieties, invariably present a larger average number of varieties than do the species of the small genera. Both these results follow when another division is made, and when all the smallest genera, with from only one to four species, are absolutely excluded from the tables. These facts are of plain signification on the view that species are only strongly marked and permanent varieties; for wherever many species of the same genus have been formed, or where, if we may use the expression, the manufactory of species has been active, we ought generally to find the manufactory still in action, more especially as we have every reason to believe the process of manufacturing new species to be a slow one. And this certainly is the case, if varieties be looked at as incipient species; for my tables clearly show as a general rule that, wherever many species of a genus have been formed, the species of that genus present a number of varieties, that is of incipient species, beyond the average. It is not that all large genera are now varying much, and are thus increasing in the number of their species, or that no small genera are now varying and increasing; for if this had been so, it would have been fatal to my theory; inasmuch as geology plainly tells us that small genera have in the lapse of time often increased greatly in size; and that large genera have often come to their maxima, declined, and disappeared. All that we want to show is, that where many species of a genus have been formed, on an average many are still forming; and this holds good.

There are other relations between the species of large genera and their recorded varieties which deserve notice. We have seen that there is no infallible criterion by which to distinguish species and well-marked varieties; and in those cases in which intermediate links have not been found between doubtful forms, naturalists are compelled to come to a determination by the amount of difference between them, judging by analogy whether or not the amount suffices to raise one or both to the rank of species. Hence the amount of difference is one very important criterion in settling whether two forms should be ranked as species or varieties. Now Fries has remarked in regard to plants, and Westwood in regard to insects, that in large genera the amount of difference between the species is often exceedingly small. I have endeavoured to test this numerically by averages, and, as far as my imperfect results go, they always confirm the view. I have also consulted some sagacious and most experienced observers, and, after deliberation, they concur in this view. In this respect, therefore, the species of the larger genera resemble varieties, more than do the species of the smaller genera. Or the case may be put in another way, and it may be said, that in the larger genera, in which a number of varieties or incipient species greater than the average are now manufacturing, many of the species already manufactured still to a certain extent resemble varieties, for they differ from each other by a less than usual amount of difference.

Moreover, the species of the large genera are related to each other, in the same manner as the varieties of any one species are related to each other. No naturalist pretends that all the species of a genus are equally distinct from each other; they may generally be divided into sub-genera, or sections, or lesser groups. As Fries has well remarked, little groups of species are generally clustered like satellites around certain other species. And what are varieties but groups of forms, unequally related to each other, and clustered round certain forms—that is, round their parent-species? Undoubtedly there is one most important point of difference between varieties and species; namely, that the amount of difference between varieties, when compared with each other or with their parent-species, is much less than that between the species of the same genus. But when we come to discuss the principle, as I call it, of Divergence of Character, we shall see how this may be explained, and how the lesser differences between varieties will tend to increase into the greater differences between species.

There is one other point which seems to me worth notice. Varieties generally have much restricted ranges: this statement is indeed scarcely more than a truism, for if a variety were found to have a wider range than that of its supposed parent-species, their denominations ought to be reversed. But there is also reason to believe, that those species which are very closely allied to other species, and in so far resemble varieties, often have much restricted ranges. For instance, Mr. H. C. Watson has marked for me in the well-sifted London Catalogue of plants (4th edition) 63 plants which are therein ranked as species, but which he considers as so closely allied to other species as to be of doubtful value: these 63 reputed species range on an average over 6·9 of the provinces into which Mr. Watson has divided Great Britain. Now, in this same catalogue, 53 acknowledged varieties are recorded, and these range over 7·7 provinces; whereas, the species to which these varieties belong range over 14·3 provinces. So that the acknowledged varieties have very nearly the same restricted average range, as have those very closely allied forms, marked for me by Mr. Watson as doubtful species, but which are almost universally ranked by British botanists as good and true species.

FINALLY, THEN, varieties have the same general characters as species, for they cannot be distinguished from species,—except, firstly, by the discovery of intermediate linking forms, and the occurrence of such links cannot affect the actual characters of the forms which they connect; and except, secondly, by a certain amount of difference, for two forms, if differing very little, are generally ranked as varieties, notwithstanding that intermediate linking forms have not been discovered; but the amount of difference considered necessary to give to two forms the rank of species is quite indefinite. In genera having more than the average number of species in any country, the species of these genera have more than the average number of varieties. In large genera the species are apt to be closely, but unequally, allied together, forming little clusters round certain species. Species very closely allied to other species apparently have restricted ranges. In all these several respects the species of large genera present a strong analogy with varieties. And we can clearly understand these analogies, if species have once existed as varieties, and have thus originated: whereas, these analogies are utterly inexplicable if each species has been independently created.

We have, also, seen that it is the most flourishing and dominant species of the larger genera which on an average vary most; and varieties, as we shall hereafter see, tend to become converted into new and distinct species.

The larger genera thus tend to become larger; and throughout nature the forms of life which are now dominant tend to become still more dominant by leaving many modified and dominant descendants. But by steps hereafter to be explained, the larger genera also tend to break up into smaller genera. And thus, the forms of life throughout the universe become divided into groups subordinate to groups.

CHAPTER III

Struggle for Existence

BEARS ON NATURAL SELECTION • THE TERM USED IN A WIDE SENSE • GEOMETRICAL POWERS OF INCREASE • RAPID INCREASE OF NATURALISED ANIMALS AND PLANTS • NATURE OF THE CHECKS TO INCREASE • COMPETITION UNIVERSAL • EFFECTS OF CLIMATE • PROTECTION FROM THE NUMBER OF INDIVIDUALS • COMPLEX RELATIONS OF ALL ANIMALS AND PLANTS THROUGHOUT NATURE • STRUGGLE FOR LIFE MOST SEVERE BETWEEN INDIVIDUALS AND VARIETIES OF THE SAME SPECIES; OFTEN SEVERE BETWEEN SPECIES OF THE SAME GENUS • THE RELATION OF ORGANISM TO ORGANISM THE MOST IMPORTANT OF ALL RELATIONS

Before entering on the subject of this chapter, I must make a few preliminary remarks, to show how the struggle for existence bears on Natural Selection. It has been seen in the last chapter that amongst organic beings in a state of nature there is some individual variability; indeed I am not aware that this has ever been disputed. It is immaterial for us whether a multitude of doubtful forms be called species or sub-species or varieties; what rank, for instance, the two or three hundred doubtful forms of British plants are entitled to hold, if the existence of any well-marked varieties be admitted. But the mere existence of individual variability and of some few well-marked varieties, though necessary as the foundation for the work, helps us but little in understanding how species arise in nature. How have all those exquisite adaptations of one part of the organisation to another part, and to the conditions of life, and of one distinct organic being to another being, been perfected? We see these beautiful co-adaptations most plainly in the woodpecker and mistletoe; and only a little less plainly in the humblest parasite which clings to the hairs of a quadruped or feathers of a bird; in the structure of the beetle which dives through the water; in the plumed seed which is wafted by the gentlest breeze; in short, we see beautiful adaptations everywhere and in every part of the organic world.

Again, it may be asked, how is it that varieties, which I have called incipient species, become ultimately converted into good and distinct species, which in most cases obviously differ from each other far more than do the varieties of the same species? How do those groups of species, which constitute what are called distinct genera, and which differ from each other more than do the species of the same genus, arise? All these results, as we shall more fully see in the next chapter, follow inevitably from the struggle for life. Owing to this struggle for life, any variation, however slight and from whatever cause proceeding, if it be in any degree profitable to an individual of any species, in its infinitely complex relations to other organic beings and to external nature, will tend to the preservation of that individual, and will generally be inherited by its offspring. The offspring, also, will thus have a better chance of surviving, for, of the many individuals of any species which are periodically born, but a small number can survive. I have called this principle, by which each slight variation, if useful, is preserved, by the term of Natural Selection, in order to mark its relation to man's power of selection. We have seen that man by selection can certainly produce great results, and can adapt organic beings to his own uses, through the accumulation of slight but useful variations, given to him by the hand of Nature. But Natural Selection, as we shall hereafter see, is a power incessantly ready for action, and is as immeasurably superior to man's feeble efforts, as the works of Nature are to those of Art.

We will now discuss in a little more detail the struggle for existence. In my future work this subject shall be treated, as it well deserves, at much greater length. The elder De Candolle and Lyell have largely and philosophically shown that all organic beings are exposed to severe competition. In regard to plants, no one has treated this subject with more spirit and ability than W. Herbert, Dean of Manchester, evidently the result of his great horticultural knowledge. Nothing is easier than to admit in words the truth of the universal struggle for life, or more difficult—at least I have found it so—than constantly to bear this conclusion in mind. Yet unless it be thoroughly engrained in the mind, I am convinced that the whole economy of nature, with every fact on distribution, rarity, abundance, extinction, and variation, will be dimly seen or quite misunderstood. We behold the face of nature bright with gladness, we often see superabundance of food; we do not see, or we forget, that the birds which are idly singing round us mostly live on insects or seeds, and are thus constantly destroying life; or we forget how largely these songsters, or their eggs, or their

slaughtered for food, and that in a state of nature an equal number would have somehow to be disposed of.

The only difference between organisms which annually produce eggs or seeds by the thousand, and those which produce extremely few, is, that the slow-breeders would require a few more years to people, under favourable conditions, a whole district, let it be ever so large. The condor lays a couple of eggs and the ostrich a score, and yet in the same country the condor may be the more numerous of the two: the Fulmar petrel lays but one egg, yet it is believed to be the most numerous bird in the world. One fly deposits hundreds of eggs, and another, like the hippobosca, a single one; but this difference does not determine how many individuals of the two species can be supported in a district. A large number of eggs is of some importance to those species, which depend on a rapidly fluctuating amount of food, for it allows them rapidly to increase in number. But the real importance of a large number of eggs or seeds is to make up for much destruction at some period of life; and this period in the great majority of cases is an early one. If an animal can in any way protect its own eggs or young, a small number may be produced, and yet the average stock be fully kept up; but if many eggs or young are destroyed, many must be produced, or the species will become extinct. It would suffice to keep up the full number of a tree, which lived on an average for a thousand years, if a single seed were produced once in a thousand years, supposing that this seed were never destroyed, and could be ensured to germinate in a fitting place. So that in all cases, the average number of any animal or plant depends only indirectly on the number of its eggs or seeds.

In looking at Nature, it is most necessary to keep the foregoing considerations always in mind—never to forget that every single organic being around us may be said to be striving to the utmost to increase in numbers; that each lives by a struggle at some period of its life; that heavy destruction inevitably falls either on the young or old, during each generation or at recurrent intervals. Lighten any check, mitigate the destruction ever so little, and the number of the species will almost instantaneously increase to any amount. The face of Nature may be compared to a yielding surface, with ten thousand sharp wedges packed close together and driven inwards by incessant blows, sometimes one wedge being struck, and then another with greater force.

What checks the natural tendency of each species to increase in number is most obscure. Look at the most vigorous species; by as much as it swarms in numbers, by so much will its tendency to increase be still further increased. We know not exactly what the checks are in even one single instance. Nor will this surprise any one who reflects how ignorant we are on this head, even in regard to mankind, so incomparably better known than any other animal. This subject has been ably treated by several authors, and I shall, in my future work, discuss some of the checks at considerable length, more especially in regard to the feral animals of South America. Here I will make only a few remarks, just to recall to the reader's mind some of the chief points. Eggs or very young animals seem generally to suffer most, but this is not invariably the case. With plants there is a vast destruction of seeds, but, from some observations which I have made, I believe that it is the seedlings which suffer most from germinating in ground already thickly stocked with other plants. Seedlings, also, are destroyed in vast numbers by various enemies; for instance, on a piece of ground three feet long and two wide, dug and cleared, and where there could be no choking from other plants, I marked all the seedlings of our native weeds as they came up, and out of the 357 no less than 295 were destroyed, chiefly by slugs and insects. If turf which has long been mown, and the case would be the same with turf closely browsed by quadrupeds, be let to grow, the more vigorous plants gradually kill the less vigorous, though fully grown, plants: thus out of twenty species growing on a little plot of turf (three feet by four) nine species perished from the other species being allowed to grow up freely.

The amount of food for each species of course gives the extreme limit to which each can increase; but very frequently it is not the obtaining food, but the serving as prey to other animals, which determines the average numbers of a species. Thus, there seems to be little doubt that the stock of partridges, grouse, and hares on any large estate depends chiefly on the destruction of vermin. If not one head of game were shot during the next twenty years in England, and, at the same time, if no vermin were destroyed, there would, in all probability, be less game than at present, although hundreds of thousands of game animals are now annually killed. On the other hand, in some cases, as with the elephant and rhinoceros, none are destroyed by beasts of prey: even the tiger in India most rarely dares to attack a young elephant protected by its dam.

Climate plays an important part in determining the average numbers of a species, and periodical seasons of extreme cold or drought, I believe to be the most effective of all checks. I estimated that the winter of 1854–55 destroyed four-fifths of the birds in my own grounds; and this is a tremendous destruction, when we remember that ten per cent. is an extraordinarily severe mortality from epidemics with man. The action of climate seems at first sight to be quite independent of the struggle for existence; but in so far as climate chiefly acts in reducing food, it brings on the most severe struggle between the individuals, whether of the same or of distinct species, which subsist on the same kind of food. Even when climate, for instance extreme cold, acts directly, it will be the least vigorous, or those which have got least food through the advancing winter, which will suffer most. When we travel from south to north, or from a damp region to a dry, we invariably see some species gradually getting rarer and rarer, and finally disappearing; and the change of climate being conspicuous, we are tempted to attribute the whole effect to its direct action. But this is a very false view: we forget that each species, even where it most abounds, is constantly suffering enormous destruction at some period of its life, from enemies or from competitors for the same place and food; and if these enemies or competitors be in the least degree favoured by any slight change of climate, they will increase in numbers, and, as each area is already fully stocked with inhabitants, the other species will decrease. When we travel southward and see a species decreasing in numbers, we may feel sure that the cause lies quite as much in other species being favoured, as in this one being hurt. So it is when we travel northward, but in a somewhat lesser degree, for the number of species of all kinds, and therefore of competitors, decreases northwards; hence in going northward, or in ascending a mountain, we far oftener meet with stunted forms, due to the *directly* injurious action of climate, than we do in proceeding southwards or in descending a mountain. When we reach the Arctic regions, or snow-capped summits, or absolute deserts, the struggle for life is almost exclusively with the elements.

That climate acts in main part indirectly by favouring other species, we may clearly see in the prodigious number of plants in our gardens which can perfectly well endure our climate, but which never become naturalised, for they cannot compete with our native plants, nor resist destruction by our native animals.

When a species, owing to highly favourable circumstances, increases inordinately in numbers in a small tract, epidemics—at least, this seems generally to occur with our game animals—often ensue: and here we have a limiting check independent of the struggle for life. But even some of these so-called epidemics appear to be due to parasitic worms, which have from some cause, possibly in part through facility of diffusion amongst the crowded animals, been disproportionably favoured: and here comes in a sort of struggle between the parasite and its prey.

On the other hand, in many cases, a large stock of individuals of the same species, relatively to the numbers of its enemies, is absolutely necessary for its preservation. Thus we can easily raise plenty of corn and rapeseed, &c., in our fields, because the seeds are in great excess compared with the number of birds which feed on them; nor can the birds, though having a superabundance of food at this one season, increase in number proportionally to the supply of seed, as their numbers are checked during winter: but any one who has tried, knows how troublesome it is to get seed from a few wheat or other such plants in a garden; I have in this case lost every single seed. This view of the necessity of a large stock of the same species for its preservation, explains, I believe, some singular facts in nature, such as that of very rare plants being sometimes extremely abundant in the few spots where they do occur; and that of some social plants being social, that is, abounding in individuals, even on the extreme confines of their range. For in such cases, we may believe, that a plant could exist only where the conditions of its life were so favourable that many could exist together, and thus save each other from utter destruction. I should add that the good effects of frequent intercrossing, and the ill effects of close interbreeding, probably come into play in some of these cases; but on this intricate subject I will not here enlarge.

Many cases are on record showing how complex and unexpected are the checks and relations between organic beings, which have to struggle together in the same country. I will give only a single instance, which, though a simple one, has interested me. In Staffordshire, on the estate of a relation where I had ample means of investigation, there was a large and extremely barren heath, which had never been touched by the hand of man; but several hundred acres of exactly the same nature had been enclosed twenty-five years previously and planted with Scotch fir. The change in the native vegetation of the planted part of the heath was most remarkable, more than is generally seen in passing from one quite different soil to another: not only the proportional numbers of the heath-plants were wholly changed, but twelve species of plants (not counting grasses and carices) flourished in the plantations, which could not be found on the heath. The effect on the insects must have been still greater, for six insectivorous birds were very common in the plantations, which were not to be seen on the heath; and the heath was frequented by two or three distinct insectivorous birds. Here we see how potent has been the effect of the introduction of a single tree, nothing whatever else having been done, with the exception that the land had been enclosed, so that cattle could not enter. But how important an element enclosure is, I plainly saw near Farnham, in Surrey. Here there are extensive heaths, with a few clumps of old Scotch firs on the distant hilltops: within the last ten years large spaces have been enclosed, and self-sown firs are now springing up in multitudes, so close together that all cannot live. When I ascertained that these young trees had not been sown or planted, I was so much surprised at their numbers that I went to several points of view, whence I could examine hundreds of acres of the unenclosed heath, and literally I could not see a single Scotch fir, except the old planted clumps. But on looking closely between the stems of the heath, I found a multitude of seedlings and little trees, which had been perpetually browsed down by the cattle. In one square yard, at a point some hundred yards distant from one of the old clumps, I counted thirty-two little trees; and one of them, judging from the rings of growth, had during twenty-six years tried to raise its head above the stems of the heath, and had failed. No wonder that, as soon as the land was enclosed, it became thickly clothed with vigorously growing young firs. Yet the heath was so extremely barren and so extensive that no one would ever have imagined that cattle would have so closely and effectually searched it for food.

Here we see that cattle absolutely determine the existence of the Scotch fir; but in several parts of the world insects determine the existence of cattle. Perhaps Paraguay offers the most curious instance of this; for here neither cattle nor horses nor dogs have ever run wild, though they swarm southward and northward in a feral state; and Azara and Rengger have shown that this is caused by the greater number in Paraguay of a certain fly, which lays its eggs in the navels of these animals when first born. The increase of these flies, numerous as they are, must be habitually checked by some means, probably by birds. Hence, if certain insectivorous birds (whose numbers are probably regulated by hawks or beasts of prey) were to increase in Paraguay, the flies would decrease—then cattle and horses would become feral, and this would certainly greatly alter (as indeed I have observed in parts of South America) the vegetation: this again would largely affect the insects; and this, as we just have seen in Staffordshire, the insectivorous birds, and so onwards in ever-increasing circles of complexity. We began this series by insectivorous birds, and we have ended with them. Not that in nature the relations can ever be as simple as this. Battle within battle must ever be recurring with varying success; and yet in the long-run the forces are so nicely balanced, that the face of nature remains uniform for long periods of time, though assuredly the merest trifle would often give the victory to one organic being over another. Nevertheless so profound is our ignorance, and so high our presumption, that we marvel when we hear of the extinction of an organic being; and as we do not see the cause, we invoke cataclysms to desolate the world, or invent laws on the duration of the forms of life!

I am tempted to give one more instance showing how plants and animals, most remote in the scale of nature, are bound together by a web of complex relations. I shall hereafter have occasion to show that the exotic Lobelia fulgens, in this part of England, is never visited by insects, and consequently, from its peculiar structure, never can set a seed. Many of our orchidaceous plants absolutely require the visits of moths to remove their pollen-masses and thus to fertilise them. I have, also, reason to believe that humble-bees are indispensable to the fertilisation of the heartsease (Viola tricolor), for other bees do not visit this flower. From experiments which I have tried, I have found that the visits of bees, if not indispensable, are at least highly beneficial to the fertilisation of our clovers; but humble-bees alone visit the common red clover (Trifolium pratense), as other bees cannot reach the nectar. Hence I have very little doubt, that if the whole genus of humble-bees became extinct or very rare in England, the heartsease and red clover would become very rare, or wholly disappear. The number of humble-bees in any district depends in a great degree on the number of field-mice, which destroy their combs and nests; and Mr. H. Newman, who has long attended to the habits of humble-bees, believes that "more than two-thirds of them are thus destroyed all over England." Now the number of mice is largely dependent, as every one knows, on the number of cats; and Mr. Newman says, "Near villages and small towns I have found the nests of humble-bees more numerous than elsewhere, which I attribute to the number of cats that destroy the mice." Hence it is quite credible that the presence of a feline animal in large numbers in a district might determine, through the intervention first of mice and then of bees, the frequency of certain flowers in that district!

In the case of every species, many different checks, acting at different periods of life, and during different seasons or years, probably come into play; some one check or some few being generally the most potent, but all concurring in determining the average number or even the existence of the species. In some cases it can be shown that widely-different checks act on the same species in different districts. When we look at the plants and bushes clothing an entangled bank, we are tempted to attribute their proportional numbers and kinds to what we call chance. But how false a view is this! Every one has heard that when an American forest is cut down, a very different vegetation springs up; but it has been observed that the trees now growing on the ancient Indian mounds, in the Southern United States, display the same beautiful diversity and proportion of kinds as in the surrounding virgin forests. What a struggle between the several kinds of trees must here have gone on during long centuries, each annually scattering its seeds by the thousand; what war between insect and insect—between insects, snails, and other animals with birds and beasts of prey—all striving to increase, and all feeding on each other or on the trees or their seeds and seedlings, or on the other plants which first clothed the ground and thus checked the growth of the trees! Throw up a handful of feathers, and all must fall to the ground according to definite laws; but how simple is this problem compared to the action and reaction of the innumerable plants and animals which have determined, in the course of centuries, the proportional numbers and kinds of trees now growing on the old Indian ruins!

The dependency of one organic being on another, as of a parasite on its prey, lies generally between beings remote in the scale of nature. This is often the case with those which may strictly be said to struggle with each other for existence, as in the case of locusts and grass-feeding quadrupeds.

But the struggle almost invariably will be most severe between the individuals of the same species, for they frequent the same districts, require the same food, and are exposed to the same dangers. In the case of varieties of the same species, the struggle will generally be almost equally severe, and we sometimes see the contest soon decided: for instance, if several varieties of wheat be sown together, and the mixed seed be resown, some of the varieties which best suit the soil or climate, or are naturally the most fertile, will beat the others and so yield more seed, and will consequently in a few years quite supplant the other varieties. To keep up a mixed stock of even such extremely close varieties as the variously coloured sweet-peas, they must be each year harvested separately, and the seed then mixed in due proportion, otherwise the weaker kinds will steadily decrease in numbers and disappear. So again with the varieties of sheep: it has been asserted that certain mountain-varieties will starve out other mountain-varieties, so that they cannot be kept together. The same result has followed from keeping together different varieties of the medicinal leech. It may even be doubted whether the varieties of any one of our domestic plants or animals have so exactly the same strength, habits, and constitution, that the original proportions of a mixed stock could be kept up for half a dozen generations, if they were allowed to struggle together, like beings in a state of nature, and if the seed or young were not annually sorted.

As species of the same genus have usually, though by no means invariably, some similarity in habits and constitution, and always in structure, the struggle will generally be more severe between species of the same genus, when they come into competition with each other, than between species of distinct genera. We see this in the recent extension over parts of the United States of one species of swallow having caused the decrease of another species. The recent increase of the missel-thrush in parts of Scotland has caused the decrease of the song-thrush. How frequently we hear of one species of rat taking the place of another species under the most different climates! In Russia the small Asiatic cockroach has everywhere driven before it its great congener. One species of charlock will supplant another, and so in other cases. We can dimly see why the competition should be most severe between allied forms, which fill nearly the same place in the economy of nature; but probably in no one case could we precisely say why one species has been victorious over another in the great battle of life.

A corollary of the highest importance may be deduced from the foregoing remarks, namely, that the structure of every organic being is related, in the most essential yet often hidden manner, to that of all other organic beings, with which it comes into competition for food or residence, or from which it has to escape, or on which it preys. This is obvious in the structure of the teeth and talons of the tiger; and in that of the legs and claws of the parasite which clings to the hair on the tiger's body. But in the beautifully plumed seed of the dandelion, and in the flattened and fringed legs of the water-beetle, the relation seems at first confined to the elements of air and water. Yet the advantage of plumed seeds no doubt stands in the closest relation to the land being already thickly clothed by other plants; so that the seeds may be widely distributed and fall on unoccupied ground. In the water-beetle, the structure of its legs, so well adapted for diving, allows it to compete with other aquatic insects, to hunt for its own prey, and to escape serving as prey to other animals.

The store of nutriment laid up within the seeds of many plants seems at first sight to have no sort of relation to other plants. But from the strong growth of young plants produced from such seeds (as peas and beans), when sown in the midst of long grass, I suspect that the chief use of the nutriment in the seed is to favour the growth of the young seedling, whilst struggling with other plants growing vigorously all around.

Look at a plant in the midst of its range, why does it not double or quadruple its numbers? We know that it can perfectly well withstand a little more heat or cold, dampness or dryness, for elsewhere it ranges into slightly hotter or colder, damper or drier districts. In this case we can clearly see that if we wished in imagination to give the plant the power of increasing in number, we should have to give it some advantage over its competitors, or over the animals which preyed on it. On the confines of its geographical range, a change of constitution with respect to climate would clearly be an advantage to our plant; but we have reason to believe that only a few plants or animals range so far, that they are destroyed by the rigour of the climate alone. Not until we reach the extreme confines of life, in the arctic regions or on the borders of an utter desert, will competition cease. The land may be extremely cold or dry, yet there will be competition between some few species, or between the individuals of the same species, for the warmest or dampest spots.

Hence, also, we can see that when a plant or animal is placed in a new country amongst new competitors, though the climate may be exactly the same as in its former home, yet the conditions of its life will generally be changed in an essential manner. If we wished to increase its average numbers in its new home, we should have to modify it in a different way to what we should have done in its native country; for we should have to give it some advantage over a different set of competitors or enemies.

It is good thus to try in our imagination to give any form some advantage over another. Probably in no single instance should we know what to do, so as to succeed. It will convince us of our ignorance on the mutual relations of all organic beings; a conviction as necessary, as it seems to be difficult to acquire. All that we can do, is to keep steadily in mind that each organic being is striving to increase at a geometrical ratio; that each at some period of its life, during some season of the year, during each generation or at intervals, has to struggle for life, and to suffer great destruction. When we reflect on this struggle, we may console ourselves with the full belief, that the war of nature is not incessant, that no fear is felt, that death is generally prompt, and that the vigorous, the healthy, and the happy survive and multiply.

CHAPTER IV

Natural Selection

NATURAL SELECTION • ITS POWER COMPARED WITH MAN'S SELECTION • ITS POWER ON CHARACTERS OF TRIFLING IMPORTANCE • ITS POWER AT ALL AGES AND ON BOTH SEXES • SEXUAL SELECTION • ON THE GENERALITY OF INTERCROSSES BETWEEN INDIVIDUALS OF THE SAME SPECIES • CIRCUMSTANCES FAVOURABLE AND UNFAVOURABLE TO NATURAL SELECTION, NAMELY, INTERCROSSING, ISOLATION, NUMBER OF INDIVIDUALS • SLOW ACTION • EXTINCTION CAUSED BY NATURAL SELECTION • DIVERGENCE OF CHARACTER, RELATED TO THE DIVERSITY OF INHABITANTS OF ANY SMALL AREA, AND TO NATURALISATION • ACTION OF NATURAL SELECTION, THROUGH DIVERGENCE OF CHARACTER AND EXTINCTION, ON THE DESCENDANTS FROM A COMMON PARENT • EXPLAINS THE GROUPING OF ALL ORGANIC BEINGS

How will the struggle for existence, discussed too briefly in the last chapter, act in regard to variation? Can the principle of selection, which we have seen is so potent in the hands of man, apply in nature? I think we shall see that it can act most effectually. Let it be borne in mind in what an endless number of strange peculiarities our domestic productions, and, in a lesser degree, those under nature, vary; and how strong the hereditary tendency is. Under domestication, it may be truly said that the whole organisation becomes in some degree plastic. Let it be borne in mind how infinitely complex and close-fitting are the mutual relations of all organic beings to each other and to their physical conditions of life. Can it, then, be thought improbable, seeing that variations useful to man have undoubtedly occurred, that other variations useful in some way to each being in the great and complex battle of life, should sometimes occur in the course of thousands of generations? If such do occur, can we doubt (remembering that many more individuals are born than can possibly survive) that individuals having any advantage, however slight, over others, would have the best chance of surviving and of procreating their kind? On the other hand, we may feel sure that any variation in the least degree injurious would be rigidly destroyed. This preservation of favourable variations and the rejection of injurious variations, I call Natural Selection. Variations neither useful nor injurious would not be affected by natural selection, and would be left a fluctuating element, as perhaps we see in the species called polymorphic.

We shall best understand the probable course of natural selection by taking the case of a country undergoing some physical change, for instance, of climate. The proportional numbers of its inhabitants would almost immediately undergo a change, and some species might become extinct. We may conclude, from what we have seen of the intimate and complex manner in which the inhabitants of each country are bound together, that any change in the numerical proportions of some of the inhabitants, independently of the change of climate itself, would most seriously affect many of the others. If the country were open on its borders, new forms would certainly immigrate, and this also would seriously disturb the relations of some of the former inhabitants. Let it be remembered how powerful the influence of a single introduced tree or mammal has been shown to be. But in the case of an island, or of a country partly surrounded by barriers, into which new and better adapted forms could not freely enter, we should then have places in the economy of nature which would assuredly be better filled up, if some of the original inhabitants were in some manner modified; for, had the area been open to immigration, these same places would have been seized on by intruders. In such case, every slight modification, which in the course of ages chanced to arise, and which in any way favoured the individuals of any of the species, by better adapting them to their altered conditions, would tend to be preserved; and natural selection would thus have free scope for the work of improvement.

We have reason to believe, as stated in the first chapter, that a change in the conditions of life, by specially acting on the reproductive system, causes or increases variability; and in the foregoing case the conditions of life are supposed to have undergone a change, and this would manifestly be favourable to natural selection, by giving a better chance of profitable variations occurring; and unless profitable variations do occur, natural selection can do nothing. Not that, as I believe, any extreme amount of variability is necessary; as man can certainly produce great results by adding up in any given direction mere individual differences, so could Nature, but far more easily, from having incomparably longer time at her disposal. Nor do I believe that any great physical change, as of climate, or any unusual degree of isolation to check immigration, is actually necessary to produce new and unoccupied places for natural selection to fill up by modifying and improving some of the varying inhabitants. For as all the inhabitants of each country are struggling together with nicely balanced forces, extremely slight modifications in the structure or habits of one inhabitant would often give it an advantage over others; and still further modifications of the same kind would often still further increase the advantage. No country can be named in which all the native inhabitants are now so perfectly adapted to each other and to the physical conditions under which they live, that none of them could anyhow be improved; for in all countries, the natives have been so far conquered by naturalised productions, that they have allowed foreigners to take firm possession of the land. And as foreigners have thus everywhere beaten some of the natives, we may safely conclude that the natives might have been modified with advantage, so as to have better resisted such intruders.

As man can produce and certainly has produced a great result by his methodical and unconscious means of selection, what may not nature effect? Man can act only on external and visible characters: nature cares nothing for appearances, except in so far as they may be useful to any being. She can act on every internal organ, on every shade of constitutional difference, on the whole machinery of life. Man selects only for his own good; Nature only for that of the being which she tends. Every selected character is fully exercised by her; and the being is placed under well-suited conditions of life. Man keeps the natives of many climates in the same country; he seldom exercises each selected character in some peculiar and fitting manner; he feeds a long and a short beaked pigeon on the same food; he does not exercise a long-backed or long-legged quadruped in any peculiar manner; he exposes sheep with long and short wool to the same climate. He does not allow the most vigorous males to struggle for the females. He does not rigidly destroy all inferior animals, but protects during each varying season, as far as lies in his power, all his productions. He often begins his selection by some half-monstrous form; or at least by some modification prominent enough to catch his eye, or to be plainly useful to him. Under nature, the slightest difference of structure or constitution may well turn the nicely-balanced scale in the struggle for life, and so be preserved. How fleeting are the wishes and efforts of man! how short his time! and consequently how poor will his products be, compared with those accumulated by nature during whole geological periods. Can we wonder, then, that nature's productions should be far "truer" in character than man's productions; that they should be infinitely better adapted to the most complex conditions of life, and should plainly bear the stamp of far higher workmanship?

It may be said that natural selection is daily and hourly scrutinising, throughout the world, every variation, even the slightest; rejecting that which is bad, preserving and adding up all that is good; silently and insensibly working, whenever and wherever opportunity offers, at the improvement of each organic being in relation to its organic and inorganic conditions of life. We see nothing of these slow changes in progress, until the hand of time has marked the long lapse of ages, and then so imperfect is our view into long past geological ages, that we only see that the forms of life are now different from what they formerly were.

Although natural selection can act only through and for the good of each being, yet characters and structures, which we are apt to consider as of very trifling importance, may thus be acted on. When we see leaf-eating insects green, and bark-feeders mottled-grey; the alpine ptarmigan white in winter, the red-grouse the colour of heather, and the black-grouse that of peaty earth, we must believe that these tints are of service to these birds and insects in preserving them from danger. Grouse, if not destroyed at some period of their lives, would increase in countless numbers; they are known to suffer largely from birds of prey; and hawks are guided by eyesight to their prey,—so much so, that on parts of the Continent persons are warned not to keep white pigeons, as being the most liable to destruction. Hence I can see no reason to doubt that natural selection might be most effective in giving the proper colour to each kind of grouse, and in keeping that colour, when once acquired, true and constant. Nor ought we to think that the occasional destruction of an animal of any particular colour would produce little effect: we should remember how essential it is in a flock of white sheep to destroy every lamb with the faintest trace of black. In plants the down on the fruit and the colour of the flesh are considered by botanists as characters of the most trifling importance: yet we hear from an excellent horticulturist, Downing, that in the United States smooth-skinned fruits suffer far more from a beetle, a curculio, than those with down; that purple plums suffer far more from a certain disease than yellow plums; whereas another disease attacks yellow-fleshed peaches far more than those with other coloured flesh. If, with all the aids of art, these slight differences make a great difference in cultivating the several varieties, assuredly, in a state of nature, where the trees would have to struggle with other trees and with a host of enemies, such differences would effectually settle which variety, whether a smooth or downy, a yellow or purple fleshed fruit, should succeed.

In looking at many small points of difference between species, which, as far as our ignorance permits us to judge, seem to be quite unimportant, we must not forget that climate, food, &c., probably produce some slight and direct effect. It is, however, far more necessary to bear in mind that there are many unknown laws of correlation of growth, which, when one part of the organisation is modified through variation, and the modifications are accumulated by natural selection for the good of the being, will cause other modifications, often of the most unexpected nature.

As we see that those variations which under domestication appear at any particular period of life, tend to reappear in the offspring at the same period;—for instance, in the seeds of the many varieties of our culinary and agricultural plants; in the caterpillar and cocoon stages of the varieties of the silkworm; in the eggs of poultry, and in the colour of the down of their chickens; in the horns of our sheep and cattle when nearly adult;—so in a state of nature, natural selection will be enabled to act on and modify organic beings at any age, by the accumulation of profitable variations at that age, and by their inheritance at a corresponding age. If it profit a plant to have its seeds more and more widely disseminated by the wind, I can see no greater difficulty in this being effected through natural selection, than in the cotton-planter increasing and improving by selection the down in the pods on his cotton-trees. Natural selection may modify and adapt the larva of an insect to a score of contingencies, wholly different from those which concern the mature insect. These modifications will no doubt affect, through the laws of correlation, the structure of the adult; and probably in the case of those insects which live only for a few hours, and which never feed, a large part of their structure is merely the correlated result of successive changes in the structure of their larvæ. So, conversely, modifications in the adult will probably often affect the structure of the larva; but in all cases natural selection will ensure that modifications consequent on other modifications at a different period of life, shall not be in the least degree injurious: for if they became so, they would cause the extinction of the species.

Natural selection will modify the structure of the young in relation to the parent, and of the parent in relation to the young. In social animals it will adapt the structure of each individual for the benefit of the community; if each in consequence profits by the selected change. What natural selection cannot do, is to modify the structure of one species, without giving it any advantage, for the good of another species; and though statements to this effect may be found in works of natural history, I cannot find one case which will bear investigation. A structure used only once in an animal's whole life, if of high importance to it, might be modified to any extent by natural selection; for instance, the great jaws possessed by certain insects, and used exclusively for opening the cocoon—or the hard tip to the beak of nestling birds, used for breaking the egg. It has been asserted, that of the best short-beaked tumbler-pigeons more perish in the egg than are able to get out of it; so that fanciers assist in the act of hatching. Now, if nature had to make the beak of a full-grown pigeon very short for the bird's own advantage, the process of modification would be very slow, and there would be simultaneously the most rigorous selection of the young birds within the egg, which had the most powerful and hardest beaks, for all with weak beaks would inevitably perish: or, more delicate and more easily broken shells might be selected, the thickness of the shell being known to vary like every other structure.

Sexual Selection.—Inasmuch as peculiarities often appear under domestication in one sex and become hereditarily attached to that sex, the same fact probably occurs under nature, and if so, natural selection will be able to modify one sex in its functional relations to the other sex, or in relation to wholly different habits of life in the two sexes, as is sometimes the case with insects. And this leads me to say a few words on what I call Sexual Selection. This depends, not on a struggle for existence, but on a struggle between the males for possession of the females; the result is not death to the unsuccessful competitor, but few or no offspring. Sexual selection is, therefore, less rigorous than natural selection. Generally, the most vigorous males, those which are best fitted for their places in nature, will leave most progeny. But in many cases, victory will depend not on general vigour, but on having special weapons, confined to the male sex. A hornless stag or spurless cock would have a poor chance of leaving offspring. Sexual selection by always allowing the victor to breed might surely give indomitable courage, length to the spur, and strength to the wing to strike in the spurred leg, as well as the brutal cock-fighter, who knows well that he can improve his breed by careful selection of the best cocks. How low in the scale of nature this law of battle descends, I know not; male alligators have been described as fighting, bellowing, and whirling round, like Indians in a war-dance, for the possession of the females; male salmons have been seen fighting all day long; male stag-beetles often bear wounds from the huge mandibles of other males. The war is, perhaps, severest between the males of polygamous animals, and these seem oftenest provided with special weapons. The males of carnivorous animals are already well armed; though to them and to others, special means of defence may be given through means of sexual selection, as the mane to the lion, the shoulder-pad to the boar, and the hooked jaw to the male salmon; for the shield may be as important for victory, as the sword or spear.

Amongst birds, the contest is often of a more peaceful character. All those who have attended to the subject, believe that there is the severest rivalry between the males of many species to attract by singing the females. The rock-thrush of Guiana, birds of Paradise, and some others, congregate; and successive males display their gorgeous plumage and perform strange antics before the females, which standing by as spectators, at last choose the most attractive partner. Those who have closely attended to birds in confinement well know that they often take individual preferences and dislikes: thus Sir R. Heron has described how one pied peacock was eminently attractive to all his hen birds. It may appear childish to attribute any effect to such apparently weak means: I cannot here enter on the details necessary to support this view; but if man can in a short time give elegant carriage and beauty to his bantams, according to his standard of beauty, I can see no good reason to doubt that female birds, by selecting, during thousands of generations, the most melodious or beautiful males, according to their standard of beauty, might produce a marked effect. I strongly suspect that some well-known laws with respect to the plumage of male and female birds, in comparison with the plumage of the young, can be explained on the view of plumage having been chiefly modified by sexual selection, acting when the birds have come to the breeding age or during the breeding season; the modifications thus produced being inherited at corresponding ages or seasons, either by the males alone, or by the males and females; but I have not space here to enter on this subject.

Thus it is, as I believe, that when the males and females of any animal have the same general habits of life, but differ in structure, colour, or ornament, such differences have been mainly caused by sexual selection; that is, individual males have had, in successive generations, some slight advantage over other males, in their weapons, means of defence, or charms; and have transmitted these advantages to their male offspring. Yet, I would not wish to attribute all such sexual differences to this agency: for we see peculiarities arising and becoming attached to the male sex in our domestic animals (as the wattle in male carriers, horn-like protuberances in the cocks of certain fowls, &c.), which we cannot believe to be either useful to the males in battle, or attractive to the females. We see analogous cases under nature, for instance, the tuft of hair on the breast of the turkey-cock, which can hardly be either useful or ornamental to this bird;—indeed, had the tuft appeared under domestication, it would have been called a monstrosity.

Illustrations of the action of Natural Selection.—In order to make it clear how, as I believe, natural selection acts, I must beg permission to give one or two imaginary illustrations. Let us take the case of a wolf, which preys on various animals, securing some by craft, some by strength, and some by fleetness; and let us suppose that the fleetest prey, a deer for instance, had from any change in the country increased in numbers, or that other prey had decreased in numbers, during that season of the year when the wolf is hardest pressed for food. I can under such circumstances see no reason to doubt that the swiftest and slimmest wolves would have the best chance of surviving, and so be preserved or selected,—provided always that they retained strength to master their prey at this or at some other period of the year, when they might be compelled to prey on other animals. I can see no more reason to doubt this, than that man can improve the fleetness of his greyhounds by careful and methodical selection, or by that unconscious selection which results from each man trying to keep the best dogs without any thought of modifying the breed.

Even without any change in the proportional numbers of the animals on which our wolf preyed, a cub might be born with an innate tendency to pursue certain kinds of prey. Nor can this be thought very improbable; for we often observe great differences in the natural tendencies of our domestic animals; one cat, for instance, taking to catch rats, another mice; one cat, according to Mr. St. John, bringing home winged game, another hares or rabbits, and another hunting on marshy ground and almost nightly catching woodcocks or snipes. The tendency to catch rats rather than mice is known to be inherited. Now, if any slight innate change of habit or of structure benefited an individual wolf, it would have the best chance of surviving and of leaving offspring. Some of its young would probably inherit the same habits or structure, and by the repetition of this process, a new variety might be formed which would either supplant or coexist with the parent-form of wolf. Or, again, the wolves inhabiting a mountainous district, and those frequenting the lowlands, would naturally be forced to hunt different prey; and from the continued preservation of the individuals best fitted for the two sites, two varieties might slowly be formed. These varieties would cross and blend where they met; but to this subject of intercrossing we shall soon have to return. I may add, that, according to Mr. Pierce, there are two varieties of the wolf inhabiting the Catskill Mountains in the United States, one with a light greyhound-like form, which pursues deer, and the other more bulky, with shorter legs, which more frequently attacks the shepherd's flocks.

Let us now take a more complex case. Certain plants excrete a sweet juice, apparently for the sake of eliminating something injurious from their sap: this is effected by glands at the base of the stipules in some Leguminosæ, and at the back of the leaf of the common laurel. This juice, though small in quantity, is greedily sought by insects. Let us now suppose a little sweet juice or nectar to be excreted by the inner bases of the petals of a flower. In this case insects in seeking the nectar would get dusted with pollen, and would certainly often transport the pollen from one flower to the stigma of another flower. The flowers of two distinct individuals of the same species would thus get crossed; and the act of crossing, we have good reason to believe (as will hereafter be more fully alluded to), would produce very vigorous seedlings, which consequently would have the best chance of flourishing and surviving. Some of these seedlings would probably inherit the nectar-excreting power. Those individual flowers which had the largest glands or nectaries, and which excreted most nectar, would be oftenest visited by insects, and would be oftenest crossed; and so in the long-run would gain the upper hand. Those flowers, also, which had their stamens and pistils placed, in relation to the size and habits of the particular insects which visited them, so as to favour in any degree the transportal of their pollen from flower to flower, would likewise be favoured or selected. We might have taken the case of insects visiting flowers for the sake of collecting pollen instead of nectar; and as pollen is formed for the sole object of fertilisation, its destruction appears a simple loss to the plant; yet if a little pollen were carried, at first occasionally and then habitually, by the pollen-devouring insects from flower to flower, and a cross thus effected, although nine-tenths of the pollen were destroyed, it might still be a great gain to the plant; and those individuals which produced more and more pollen, and had larger and larger anthers, would be selected.

When our plant, by this process of the continued preservation or natural selection of more and more attractive flowers, had been rendered highly attractive to insects, they would, unintentionally on their part, regularly carry pollen from flower to flower; and that they can most effectually do this, I could easily show by many striking instances. I will give only one—not as a very striking case, but as likewise illustrating one step in the separation of the sexes of plants, presently to be alluded to. Some holly-trees bear only male flowers, which have four stamens producing rather a small quantity of pollen, and a rudimentary pistil; other holly-trees bear only female flowers; these have a full-sized pistil, and four stamens with shrivelled anthers, in which not a grain of pollen can be detected. Having found a female tree exactly sixty yards from a male tree, I put the stigmas of twenty flowers, taken from different branches, under the microscope, and on all, without exception, there were pollen-grains, and on some a profusion of pollen. As the wind had set for several days from the female to the male tree, the pollen could not thus have been carried. The weather had been cold and boisterous, and therefore not favourable to bees, nevertheless every female flower which I examined had been effectually fertilised by the bees, accidentally dusted with pollen, having flown from tree to tree in search of nectar. But to return to our imaginary case: as soon as the plant had been rendered so highly attractive to insects that pollen was regularly carried from flower to flower, another process might commence. No naturalist doubts the advantage of what has been called the "physiological division of labour;" hence we may believe that it would be advantageous to a plant to produce stamens alone in one flower or on one whole plant, and pistils alone in another flower or on another plant. In plants under culture and placed under new conditions of life, sometimes the male organs and sometimes the female organs become more or less impotent; now if we suppose this to occur in ever so slight a degree under nature, then as pollen is already carried regularly from flower to flower, and as a more complete separation of the sexes of our plant would be advantageous on the principle of the division of labour, individuals with this tendency more and more increased, would be continually favoured or selected, until at last a complete separation of the sexes would be effected.

Let us now turn to the nectar-feeding insects in our imaginary case: we may suppose the plant of which we have been slowly increasing the nectar by continued selection, to be a common plant; and that certain insects depended in main part on its nectar for food. I could give many facts, showing how anxious bees are to save time; for instance, their habit of cutting holes and sucking the nectar at the bases of certain flowers, which they can, with a very little more trouble, enter by the mouth. Bearing such facts in mind, I can see no reason to doubt that an accidental deviation in the size and form of the body, or in the curvature and length of the proboscis, &c., far too slight to be appreciated by us, might profit a bee or other insect, so that an individual so characterised would be able to obtain its food more quickly, and so have a better chance of living and leaving descendants. Its descendants would probably inherit a tendency to a similar slight deviation of structure. The tubes of the corollas of the common red and incarnate clovers (Trifolium pratense and incarnatum) do not on a hasty glance appear to differ in length; yet the hive-bee can easily suck the nectar out of the incarnate clover, but not out of the common red clover, which is visited by humble-bees alone; so that whole fields of the red clover offer in vain an abundant supply of precious nectar to the hive-bee. Thus it might be a great advantage to the hive-bee to have a slightly longer or differently constructed proboscis. On the other hand, I have found by experiment that the fertility of clover greatly depends on bees visiting and moving parts of the corolla, so as to push the pollen on to the stigmatic surface. Hence, again, if humble-bees were to become rare in any country, it might be a great advantage to the red clover to have a shorter or more deeply divided tube to its corolla, so that the hive-bee could visit its flowers. Thus I can understand how a flower and a bee might slowly become, either simultaneously or one after the other, modified and adapted in the most perfect manner to each other, by the continued preservation of individuals presenting mutual and slightly favourable deviations of structure.

I am well aware that this doctrine of natural selection, exemplified in the above imaginary instances, is open to the same objections which were at first urged against Sir Charles Lyell's noble views on "the modern changes of the earth, as illustrative of geology;" but we now very seldom hear the action, for instance, of the coast-waves, called a trifling and insignificant cause, when applied to the excavation of gigantic valleys or to the formation of the longest lines of inland cliffs. Natural selection can act only by the preservation and accumulation of infinitesimally small inherited modifications, each profitable to the preserved being; and as modern geology has almost banished such views as the excavation of a great valley by a single diluvial wave, so will natural selection, if it be a true principle, banish the belief of the continued creation of new organic beings, or of any great and sudden modification in their structure.

On the Intercrossing of Individuals.—I must here introduce a short digression. In the case of animals and plants with separated sexes, it is of course obvious that two individuals must always unite for each birth; but in the case of hermaphrodites this is far from obvious. Nevertheless I am strongly inclined to believe that with all hermaphrodites two individuals, either occasionally or habitually, concur for the reproduction of their kind. This view, I may add, was first suggested by Andrew Knight. We shall presently see its importance; but I must here treat the subject with extreme brevity, though I have the materials prepared for an ample discussion. All vertebrate animals, all insects, and some other large groups of animals, pair for each birth. Modern research has much diminished the number of supposed hermaphrodites, and of real hermaphrodites a large number pair; that is, two individuals regularly unite for reproduction, which is all that concerns us. But still there are many hermaphrodite animals which certainly do not habitually pair, and a vast majority of plants are hermaphrodites. What reason, it may be asked, is there for supposing in these cases that two individuals ever concur in reproduction? As it is impossible here to enter on details, I must trust to some general considerations alone.

In the first place, I have collected so large a body of facts, showing, in accordance with the almost universal belief of breeders, that with animals and plants a cross between different varieties, or between individuals of the same variety but of another strain, gives vigour and fertility to the offspring; and on the other hand, that *close* interbreeding diminishes vigour and fertility; that these facts alone incline me to believe that it is a general law of nature (utterly ignorant though we be of the meaning of the law) that no organic being self-fertilises itself for an eternity of generations; but that a cross with another individual is occasionally—perhaps at very long intervals—indispensable.

On the belief that this is a law of nature, we can, I think, understand several large classes of facts, such as the following, which on any other view are inexplicable. Every hybridizer knows how unfavourable exposure to wet is to the fertilisation of a flower, yet what a multitude of flowers have their anthers and stigmas fully exposed to the weather! but if an occasional cross be indispensable, the fullest freedom for the entrance of pollen from another individual will explain this state of exposure, more especially as the plant's own anthers and pistil generally stand so close together that self-fertilisation seems almost inevitable. Many flowers, on the other hand, have their organs of fructification closely enclosed, as in the great papilionaceous or pea-family; but in several, perhaps in all, such flowers, there is a very curious adaptation between the structure of the flower and the manner in which bees suck the nectar; for, in doing this, they either push the flower's own pollen on the stigma, or bring pollen from another flower. So necessary are the visits of bees to papilionaceous flowers, that I have found, by experiments published elsewhere, that their fertility is greatly diminished if these visits be prevented. Now, it is scarcely possible that bees should fly from flower to flower, and not carry pollen from one to the other, to the great good, as I believe, of the plant. Bees will act like a camel-hair pencil, and it is quite sufficient just to touch the anthers of one flower and then the stigma of another with the same brush to ensure fertilisation; but it must not be supposed that bees would thus produce a multitude of hybrids between distinct species; for if you bring on the same brush a plant's own pollen and pollen from another species, the former will have such a prepotent effect, that it will invariably and completely destroy, as has been shown by Gärtner, any influence from the foreign pollen.

When the stamens of a flower suddenly spring towards the pistil, or slowly move one after the other towards it, the contrivance seems adapted solely to ensure self-fertilisation; and no doubt it is useful for this end: but, the agency of insects is often required to cause the stamens to spring forward, as Kölreuter has shown to be the case with the barberry; and curiously in this very genus, which seems to have a special contrivance for self-fertilisation, it is well known that if very closely-allied forms or varieties are planted near each other, it is hardly possible to raise pure seedlings, so largely do they naturally cross. In many other cases, far from there being any aids for self-fertilisation, there are special contrivances, as I could show from the writings of C. C. Sprengel and from my own observations, which effectually prevent the stigma receiving pollen from its own flower: for instance, in Lobelia fulgens, there is a really beautiful and elaborate contrivance by which every one of the infinitely numerous pollen-granules are swept out of the conjoined anthers of each flower, before the stigma of that individual flower is ready to receive them; and as this flower is never visited, at least in my garden, by insects, it never sets a seed, though by placing pollen from one flower on the stigma of another, I raised plenty of seedlings, and whilst another species of Lobelia growing close by, which is visited by bees, seeds freely. In very many other cases, though there be no special mechanical contrivance to prevent the stigma of a flower receiving its own pollen, yet, as C. C. Sprengel has shown, and as I can confirm, either the anthers burst before the stigma is ready for fertilisation, or the stigma is ready before the pollen of that flower is ready, so that these plants have in fact separated sexes, and must habitually be crossed. How strange are these facts! How strange that the pollen and stigmatic surface of the same flower, though placed so close together, as if for the very purpose of self-fertilisation, should in so many cases be mutually useless to each other! How simply are these facts explained on the view of an occasional cross with a distinct individual being advantageous or indispensable!

If several varieties of the cabbage, radish, onion, and of some other plants, be allowed to seed near each other, a large majority, as I have found, of the seedlings thus raised will turn out mongrels: for instance, I raised 233 seedling cabbages from some plants of different varieties growing near each other, and of these only 78 were true to their kind, and some even of these were not perfectly true. Yet the pistil of each cabbage-flower is surrounded not only by its own six stamens, but by those of the many other flowers on the same plant. How, then, comes it that such a vast number of the seedlings are mongrelized? I suspect that it must arise from the pollen of a distinct *variety* having a prepotent effect over a flower's own pollen; and that this is part of the general law of good being derived from the intercrossing of distinct individuals of the same species. When distinct *species* are crossed the case is directly the reverse, for a plant's own pollen is always prepotent over foreign pollen; but to this subject we shall return in a future chapter.

In the case of a gigantic tree covered with innumerable flowers, it may be objected that pollen could seldom be carried from tree to tree, and at most only from flower to flower on the same tree, and that flowers on the same tree can be considered as distinct individuals only in a limited sense. I believe this objection to be valid, but that nature has largely provided against it by giving to trees a strong tendency to bear flowers with separated sexes. When the sexes are separated, although the male and female flowers may be produced on the same tree, we can see that pollen must be regularly carried from flower to flower; and this will give a better chance of pollen being occasionally carried from tree to tree. That trees belonging to all Orders have their sexes more often separated than other plants, I find to be the case in this country; and at my request Dr. Hooker tabulated the trees of New Zealand, and Dr. Asa Gray those of the United States, and the result was as I anticipated. On the other hand, Dr. Hooker has recently informed me that he finds that the rule does not hold in Australia; and I have made these few remarks on the sexes of trees simply to call attention to the subject.

Turning for a very brief space to animals: on the land there are some hermaphrodites, as land-mollusca and earth-worms; but these all pair. As yet I have not found a single case of a terrestrial animal which fertilises itself. We can understand this remarkable fact, which offers so strong a contrast with terrestrial plants, on the view of an occasional cross being indispensable, by considering the medium in which terrestrial animals live, and the nature of the fertilising element; for we know of no means, analogous to the action of insects and of the wind in the case of plants, by which an occasional cross could be effected with terrestrial animals without the concurrence of two individuals. Of aquatic animals, there are many self-fertilising hermaphrodites; but here currents in the water offer an obvious means for an occasional cross. And, as in the case of flowers, I have as yet failed, after consultation with one of the highest authorities, namely, Professor Huxley, to discover a single case of an hermaphrodite animal with the organs of reproduction so perfectly enclosed within the body, that access from without and the occasional influence of a distinct individual can be shown to be physically impossible. Cirripedes long appeared to me to present a case of very great difficulty under this point of view; but I have been enabled, by a fortunate chance, elsewhere to prove that two individuals, though both are self-fertilising hermaphrodites, do sometimes cross.

It must have struck most naturalists as a strange anomaly that, in the case of both animals and plants, species of the same family and even of the same genus, though agreeing closely with each other in almost their whole organisation, yet are not rarely, some of them hermaphrodites, and some of them unisexual. But if, in fact, all hermaphrodites do occasionally intercross with other individuals, the difference between hemaphrodites and unisexual species, as far as function is concerned, becomes very small.

From these several considerations and from the many special facts which I have collected, but which I am not here able to give, I am strongly inclined to suspect that, both in the vegetable and animal kingdoms, an occasional intercross with a distinct individual is a law of nature. I am well aware that there are, on this view, many cases of difficulty, some of which I am trying to investigate. Finally then, we may conclude that in many organic beings, a cross between two individuals is an obvious necessity for each birth; in many others it occurs perhaps only at long intervals; but in none, as I suspect, can self-fertilisation go on for perpetuity.

Circumstances favourable to Natural Selection.—This is an extremely intricate subject. A large amount of inheritable and diversified variability is favourable, but I believe mere individual differences suffice for the work. A large number of individuals, by giving a better chance for the appearance within any given period of profitable variations, will compensate for a lesser amount of variability in each individual, and is, I believe, an extremely important element of success. Though nature grants vast periods of time for the work of natural selection, she does not grant an indefinite period; for as all organic beings are striving, it may be said, to seize on each place in the economy of nature, if any one species does not become modified and improved in a corresponding degree with its competitors, it will soon be exterminated.

In man's methodical selection, a breeder selects for some definite object, and free intercrossing will wholly stop his work. But when many men, without intending to alter the breed, have a nearly common standard of perfection, and all try to get and breed from the best animals, much improvement and modification surely but slowly follow from this unconscious process of selection, notwithstanding a large amount of crossing with inferior animals. Thus it will be in nature; for within a confined area, with some place in its polity not so perfectly occupied as might be, natural selection will always tend to preserve all the individuals varying in the right direction, though in different degrees, so as better to fill up the unoccupied place. But if the area be large, its several districts will almost certainly present different conditions of life; and then if natural selection be modifying and improving a species in the several districts, there will be intercrossing with the other individuals of the same species on the confines of each. And in this case the effects of intercrossing can hardly be counterbalanced by natural selection always tending to modify all the individuals in each district in exactly the same manner to the conditions of each; for in a continuous area, the conditions will generally graduate away insensibly from one district to another. The intercrossing will most affect those animals which unite for each birth, which wander much, and which do not breed at a very quick rate. Hence in animals of this nature, for instance in birds, varieties will generally be confined to separated countries; and this I believe to be the case. In hermaphrodite organisms which cross only occasionally, and likewise in animals which unite for each birth, but which wander little and which can increase at a very rapid rate, a new and improved variety might be quickly formed on any one spot, and might there maintain itself in a body, so that whatever intercrossing took place would be chiefly between the individuals of the same new variety. A local variety when once thus formed might subsequently slowly spread to other districts. On the above principle, nurserymen always prefer getting seed from a large body of plants of the same variety, as the chance of intercrossing with other varieties is thus lessened.

Even in the case of slow-breeding animals, which unite for each birth, we must not overrate the effects of intercrosses in retarding natural selection; for I can bring a considerable catalogue of facts, showing that within the same area, varieties of the same animal can long remain distinct, from haunting different stations, from breeding at slightly different seasons, or from varieties of the same kind preferring to pair together.

Intercrossing plays a very important part in nature in keeping the individuals of the same species, or of the same variety, true and uniform in character. It will obviously thus act far more efficiently with those animals which unite for each birth; but I have already attempted to show that we have reason to believe that occasional intercrosses take place with all animals and with all plants. Even if these take place only at long intervals, I am convinced that the young thus produced will gain so much in vigour and fertility over the offspring from long-continued self-fertilisation, that they will have a better chance of surviving and propagating their kind; and thus, in the long run, the influence of intercrosses, even at rare intervals, will be great. If there exist organic beings which never intercross, uniformity of character can be retained amongst them, as long as their conditions of life remain the same, only through the principle of inheritance, and through natural selection destroying any which depart from the proper type; but if their conditions of life change and they undergo modification, uniformity of character can be given to their modified offspring, solely by natural selection preserving the same favourable variations.

Isolation, also, is an important element in the process of natural selection. In a confined or isolated area, if not very large, the organic and inorganic conditions of life will generally be in a great degree uniform; so that natural selection will tend to modify all the individuals of a varying species throughout the area in the same manner in relation to the same conditions. Intercrosses, also, with the individuals of the same species, which otherwise would have inhabited the surrounding and differently circumstanced districts, will be prevented. But isolation probably acts more efficiently in checking the immigration of better adapted organisms, after any physical change, such as of climate or elevation of the land, &c.; and thus new places in the natural economy of the country are left open for the old inhabitants to struggle for, and become adapted to, through modifications in their structure and constitution. Lastly, isolation, by checking immigration and consequently competition, will give time for any new variety to be slowly improved; and this may sometimes be of importance in the production of new species. If, however, an isolated area be very small, either from being surrounded by barriers, or from having very peculiar physical conditions, the total number of the individuals supported on it will necessarily be very small; and fewness of individuals will greatly retard the production of new species through natural selection, by decreasing the chance of the appearance of favourable variations.

If we turn to nature to test the truth of these remarks, and look at any small isolated area, such as an oceanic island, although the total number of the species inhabiting it, will be found to be small, as we shall see in our chapter on geographical distribution; yet of these species a very large proportion are endemic,—that is, have been produced there, and nowhere else. Hence an oceanic island at first sight seems to have been highly favourable for the production of new species. But we may thus greatly deceive ourselves, for to ascertain whether a small isolated area, or a large open area like a continent, has been most favourable for the production of new organic forms, we ought to make the comparison within equal times; and this we are incapable of doing.

Although I do not doubt that isolation is of considerable importance in the production of new species, on the whole I am inclined to believe that largeness of area is of more importance, more especially in the production of species, which will prove capable of enduring for a long period, and of spreading widely. Throughout a great and open area, not only will there be a better chance of favourable variations arising from the large number of individuals of the same species there supported, but the conditions of life are infinitely complex from the large number of already existing species; and if some of these many species become modified and improved, others will have to be improved in a corresponding degree or they will be exterminated. Each new form, also, as soon as it has been much improved, will be able to spread over the open and continuous area, and will thus come into competition with many others. Hence more new places will be formed, and the competition to fill them will be more severe, on a large than on a small and isolated area. Moreover, great areas, though now continuous, owing to oscillations of level, will often have recently existed in a broken condition, so that the good effects of isolation will generally, to a certain extent, have concurred. Finally, I conclude that, although small isolated areas probably have been in some respects highly favourable for the production of new species, yet that the course of modification will generally have been more rapid on large areas; and what is more important, that the new forms produced on large areas, which already have been victorious over many competitors, will be those that will spread most widely, will give rise to most new varieties and species, and will thus play an important part in the changing history of the organic world.

We can, perhaps, on these views, understand some facts which will be again alluded to in our chapter on geographical distribution; for instance, that the productions of the smaller continent of Australia have formerly yielded, and apparently are now yielding, before those of the larger Europæo-Asiatic area. Thus, also, it is that continental productions have everywhere become so largely naturalised on islands. On a small island, the race for life will have been less severe, and there will have been less modification and less extermination. Hence, perhaps, it comes that the flora of Madeira, according to Oswald Heer, resembles the extinct tertiary flora of Europe. All fresh-water basins, taken together, make a small area compared with that of the sea or of the land; and, consequently, the competition between fresh-water productions will have been less severe than elsewhere; new forms will have been more slowly formed, and old forms more slowly exterminated. And it is in fresh water that we find seven genera of Ganoid fishes, remnants of a once preponderant order: and in fresh water we find some of the most anomalous forms now known in the world, as the Ornithorhynchus and Lepidosiren, which, like fossils, connect to a certain extent orders now widely separated in the natural scale. These anomalous forms may almost be called living fossils; they have endured to the present day, from having inhabited a confined area, and from having thus been exposed to less severe competition.

To sum up the circumstances favourable and unfavourable to natural selection, as far as the extreme intricacy of the subject permits. I conclude, looking to the future, that for terrestrial productions a large continental area, which will probably undergo many oscillations of level, and which consequently will exist for long periods in a broken condition, will be the most favourable for the production of many new forms of life, likely to endure long and to spread widely. For the area will first have existed as a continent, and the inhabitants, at this period numerous ini ndividuals and kinds, will have been subjected to very severe competition. When converted by subsidence into large separate islands, there will still exist many individuals of the same species on each island: intercrossing on the confines of the range of each species will thus be checked: after physical changes of any kind, immigration will be prevented, so that new places in the polity of each island will have to be filled up by modifications of the old inhabitants; and time will be allowed for the varieties in each to become well modified and perfected. When, by renewed elevation, the islands shall be re-converted into a continental area, there will again be severe competition: the most favoured or improved varieties will be enabled to spread: there will be much extinction of the less improved forms, and the relative proportional numbers of the various inhabitants of the renewed continent will again be changed; and again there will be a fair field for natural selection to improve still further the inhabitants, and thus produce new species.

That natural selection will always act with extreme slowness, I fully admit. Its action depends on there being places in the polity of nature, which can be better occupied by some of the inhabitants of the country undergoing modification of some kind. The existence of such places will often depend on physical changes, which are generally very slow, and on the immigration of better adapted forms having been checked. But the action of natural selection will probably still oftener depend on some of the inhabitants becoming slowly modified; the mutual relations of many of the other inhabitants being thus disturbed. Nothing can be effected, unless favourable variations occur, and variation itself is apparently always a very slow process. The process will often be greatly retarded by free intercrossing. Many will exclaim that these several causes are amply sufficient wholly to stop the action of natural selection. I do not believe so. On the other hand, I do believe that natural selection will always act very slowly, often only at long intervals of time, and generally on only a very few of the inhabitants of the same region at the same time. I further believe, that this very slow, intermittent action of natural selection accords perfectly well with what geology tells us of the rate and manner at which the inhabitants of this world have changed.

Slow though the process of selection may be, if feeble man can do much by his powers of artificial selection, I can see no limit to the amount of change, to the beauty and infinite complexity of the co-adaptations between all organic beings, one with another and with their physical conditions of life, which may be effected in the long course of time by nature's power of selection.

Extinction.—This subject will be more fully discussed in our chapter on Geology; but it must be here alluded to from being intimately connected with natural selection. Natural selection acts solely through the preservation of variations in some way advantageous, which consequently endure. But as from the high geometrical powers of increase of all organic beings, each area is already fully stocked with inhabitants, it follows that as each selected and favoured form increases in number, so will the less favoured forms decrease and become rare. Rarity, as geology tells us, is the precursor to extinction. We can, also, see that any form represented by few individuals will, during fluctuations in the seasons or in the number of its enemies, run a good chance of utter extinction. But we may go further than this; for as new forms are continually and slowly being produced, unless we believe that the number of specific forms goes on perpetually and almost indefinitely increasing, numbers inevitably must become extinct. That the number of specific forms has not indefinitely increased, geology shows us plainly; and indeed we can see reason why they should not have thus increased, for the number of places in the polity of nature is not indefinitely great,—not that we have any means of knowing that any one region has as yet got its maximum of species. Probably no region is as yet fully stocked, for at the Cape of Good Hope, where more species of plants are crowded together than in any other quarter of the world, some foreign plants have become naturalised, without causing, as far as we know, the extinction of any natives.

Furthermore, the species which are most numerous in individuals will have the best chance of producing within any given period favourable variations. We have evidence of this, in the facts given in the second chapter, showing that it is the common species which afford the greatest number of recorded varieties, or incipient species. Hence, rare species will be less quickly modified or improved within any given period, and they will consequently be beaten in the race for life by the modified descendants of the commoner species.

From these several considerations I think it inevitably follows, that as new species in the course of time are formed through natural selection, others will become rarer and rarer, and finally extinct. The forms which stand in closest competition with those undergoing modification and improvement, will naturally suffer most. And we have seen in the chapter on the Struggle for Existence that it is the most closely-allied forms,—varieties of the same species, and species of the same genus or of related genera,—which, from having nearly the same structure, constitution, and habits, generally come into the severest competition with each other. Consequently, each new variety or species, during the progress of its formation, will generally press hardest on its nearest kindred, and tend to exterminate them. We see the same process of extermination amongst our domesticated productions, through the selection of improved forms by man. Many curious instances could be given showing how quickly new breeds of cattle, sheep, and other animals, and varieties of flowers, take the place of older and inferior kinds. In Yorkshire, it is historically known that the ancient black cattle were displaced by the long-horns, and that these "were swept away by the short-horns" (I quote the words of an agricultural writer) "as if by some murderous pestilence."

Divergence of Character.—The principle, which I have designated by this term, is of high importance on my theory, and explains, as I believe, several important facts. In the first place, varieties, even strongly-marked ones, though having somewhat of the character of species—as is shown by the hopeless doubts in many cases how to rank them—yet certainly differ from each other far less than do good and distinct species. Nevertheless, according to my view, varieties are species in the process of formation, or are, as I have called them, incipient species. How, then, does the lesser difference between varieties become augmented into the greater difference between species? That this does habitually happen, we must infer from most of the innumerable species throughout nature presenting well-marked differences; whereas varieties, the supposed prototypes and parents of future well-marked species, present slight and ill-defined differences. Mere chance, as we may call it, might cause one variety to differ in some character from its parents, and the offspring of this variety again to differ from its parent in the very same character and in a greater degree; but this alone would never account for so habitual and large an amount of difference as that between varieties of the same species and species of the same genus.

As has always been my practice, let us seek light on this head from our domestic productions. We shall here find something analogous. A fancier is struck by a pigeon having a slightly shorter beak; another fancier is struck by a pigeon having a rather longer beak; and on the acknowledged principle that "fanciers do not and will not admire a medium standard, but like extremes," they both go on (as has actually occurred with tumbler-pigeons) choosing and breeding from birds with longer and longer beaks, or with shorter and shorter beaks. Again, we may suppose that at an early period one man preferred swifter horses; another stronger and more bulky horses. The early differences would be very slight; in the course of time, from the continued selection of swifter horses by some breeders, and of stronger ones by others, the differences would become greater, and would be noted as forming two sub-breeds; finally, after the lapse of centuries, the sub-breeds would become converted into two well-established and distinct breeds. As the differences slowly become greater, the inferior animals with intermediate characters, being neither very swift nor very strong, will have been neglected, and will have tended to disappear. Here, then, we see in man's productions the action of what may be called the principle of divergence, causing differences, at first barely appreciable, steadily to increase, and the breeds to diverge in character both from each other and from their common parent.

But how, it may be asked, can any analogous principle apply in nature? I believe it can and does apply most efficiently, from the simple circumstance that the more diversified the descendants from any one species become in structure, constitution, and habits, by so much will they be better enabled to seize on many and widely diversified places in the polity of nature, and so be enabled to increase in numbers.

We can clearly see this in the case of animals with simple habits. Take the case of a carnivorous quadruped, of which the number that can be supported in any country has long ago arrived at its full average. If its natural powers of increase be allowed to act, it can succeed in increasing (the country not undergoing any change in its conditions) only by its varying descendants seizing on places at present occupied by other animals: some of them, for instance, being enabled to feed on new kinds of prey, either dead or alive; some inhabiting new stations, climbing trees, frequenting water, and some perhaps becoming less carnivorous. The more diversified in habits and structure the descendants of our carnivorous animal became, the more places they would be enabled to occupy. What applies to one animal will apply throughout all time to all animals—that is, if they vary—for otherwise natural selection can do nothing. So it will be with plants. It has been experimentally proved, that if a plot of ground be sown with one species of grass, and a similar plot be sown with several distinct genera of grasses, a greater number of plants and a greater weight of dry herbage can thus be raised. The same has been found to hold good when first one variety and then several mixed varieties of wheat have been sown on equal spaces of ground. Hence, if any one species of grass were to go on varying, and those varieties were continually selected which differed from each other in at all the same manner as distinct species and genera of grasses differ from each other, a greater number of individual plants of this species of grass, including its modified descendants, would succeed in living on the same piece of ground. And we well know that each species and each variety of grass is annually sowing almost countless seeds; and thus, as it may be said, is striving its utmost to increase its numbers. Consequently, I cannot doubt that in the course of many thousands of generations, the most distinct varieties of any one species of grass would always have the best chance of succeeding and of increasing in numbers, and thus of supplanting the less distinct varieties; and varieties, when rendered very distinct from each other, take the rank of species.

The truth of the principle, that the greatest amount of life can be supported by great diversification of structure, is seen under many natural circumstances. In an extremely small area, especially if freely open to immigration, and where the contest between individual and individual must be severe, we always find great diversity in its inhabitants. For instance, I found that a piece of turf, three feet by four in size, which had been exposed for many years to exactly the same conditions, supported twenty species of plants, and these belonged to eighteen genera and to eight orders, which shows how much these plants differed from each other. So it is with the plants and insects on small and uniform islets; and so in small ponds of fresh water. Farmers find that they can raise most food by a rotation of plants belonging to the most different orders: nature follows what may be called a simultaneous rotation. Most of the animals and plants which live close round any small piece of ground, could live on it (supposing it not to be in any way peculiar in its nature), and may be said to be striving to the utmost to live there; but, it is seen, that where they come into the closest competition with each other, the advantages of diversification of structure, with the accompanying differences of habit and constitution, determine that the inhabitants, which thus jostle each other most closely, shall, as a general rule, belong to what we call different genera and orders.

The same principle is seen in the naturalisation of plants through man's agency in foreign lands. It might have been expected that the plants which have succeeded in becoming naturalised in any land would generally have been closely allied to the indigenes; for these are commonly looked at as specially created and adapted for their own country. It might, also, perhaps have been expected that naturalised plants would have belonged to a few groups more especially adapted to certain stations in their new homes. But the case is very different; and Alph. De Candolle has well remarked in his great and admirable work, that floras gain by naturalisation, proportionally with the number of the native genera and species, far more in new genera than in new species. To give a single instance: in the last edition of Dr. Asa Gray's 'Manual of the Flora of the Northern United States,' 260 naturalised plants are enumerated, and these belong to 162 genera. We thus see that these naturalised plants are of a highly diversified nature. They differ, moreover, to a large extent from the indigenes, for out of the 162 genera, no less than 100 genera are not there indigenous, and thus a large proportional addition is made to the genera of these States.

By considering the nature of the plants or animals which have struggled successfully with the indigenes of any country, and have there become naturalised, we can gain some crude idea in what manner some of the natives would have had to be modified, in order to have gained an advantage over the other natives; and we may, I think, at least safely infer that diversification of structure, amounting to new generic differences, would have been profitable to them.

The advantage of diversification in the inhabitants of the same region is, in fact, the same as that of the physiological division of labour in the organs of the same individual body—a subject so well elucidated by Milne Edwards. No physiologist doubts that a stomach by being adapted to digest vegetable matter alone, or flesh alone, draws most nutriment from these substances. So in the general economy of any land, the more widely and perfectly the animals and plants are diversified for different habits of life, so will a greater number of individuals be capable of there supporting themselves. A set of animals, with their organisation but little diversified, could hardly compete with a set more perfectly diversified in structure. It may be doubted, for instance, whether the Australian marsupials, which are divided into groups differing but little from each other, and feebly representing, as Mr. Waterhouse and others have remarked, our carnivorous, ruminant, and rodent mammals, could successfully compete with these well-pronounced orders. In the Australian mammals, we see the process of diversification in an early and incomplete stage of development.

After the foregoing discussion, which ought to have been much amplified, we may, I think, assume that the modified descendants of any one species will succeed by so much the better as they become more diversified in structure, and are thus enabled to encroach on places occupied by other beings. Now let us see how this principle of great benefit being derived from divergence of character, combined with the principles of natural selection and of extinction, will tend to act.

The accompanying diagram will aid us in understanding this rather perplexing subject. Let A to L represent the species of a genus large in its own country; these species are supposed to resemble each other in unequal degrees, as is so generally the case in nature, and as is represented in the diagram by the letters standing at unequal distances. I have said a large genus, because we have seen in the second chapter, that on an average more of the species of large genera vary than of small genera; and the varying species of the large genera present a greater number of varieties. We have, also, seen that the species, which are the commonest and the most widely-diffused, vary more than rare species with restricted ranges. Let (A) be a common, widely-diffused, and varying species, belonging to a genus large in its own country. The little fan of diverging dotted lines of unequal lengths proceeding from (A), may represent its varying offspring. The variations are supposed to be extremely slight, but of the most diversified nature; they are not supposed all to appear simultaneously, but often after long intervals of time; nor are they all supposed to endure for equal periods. Only those variations which are in some way profitable will be preserved or naturally selected. And here the importance of the principle of benefit being derived from divergence of character comes in; for this will generally lead to the most different or divergent variations (represented by the outer dotted lines) being preserved and accumulated by natural selection. When a dotted line reaches one of the horizontal lines, and is there marked by a small numbered letter, a sufficient amount of variation is supposed to have been accumulated to have formed a fairly well-marked variety, such as would be thought worthy of record in a systematic work.

The intervals between the horizontal lines in the diagram, may represent each a thousand generations; but it would have been better if each had represented ten thousand generations. After a thousand generations, species (A) is supposed to have produced two fairly well-marked varieties, namely a^1 and m^1. These two varieties will generally continue to be exposed to the same conditions which made their parents variable, and the tendency to variability is in itself hereditary, consequently they will tend to vary, and generally to vary

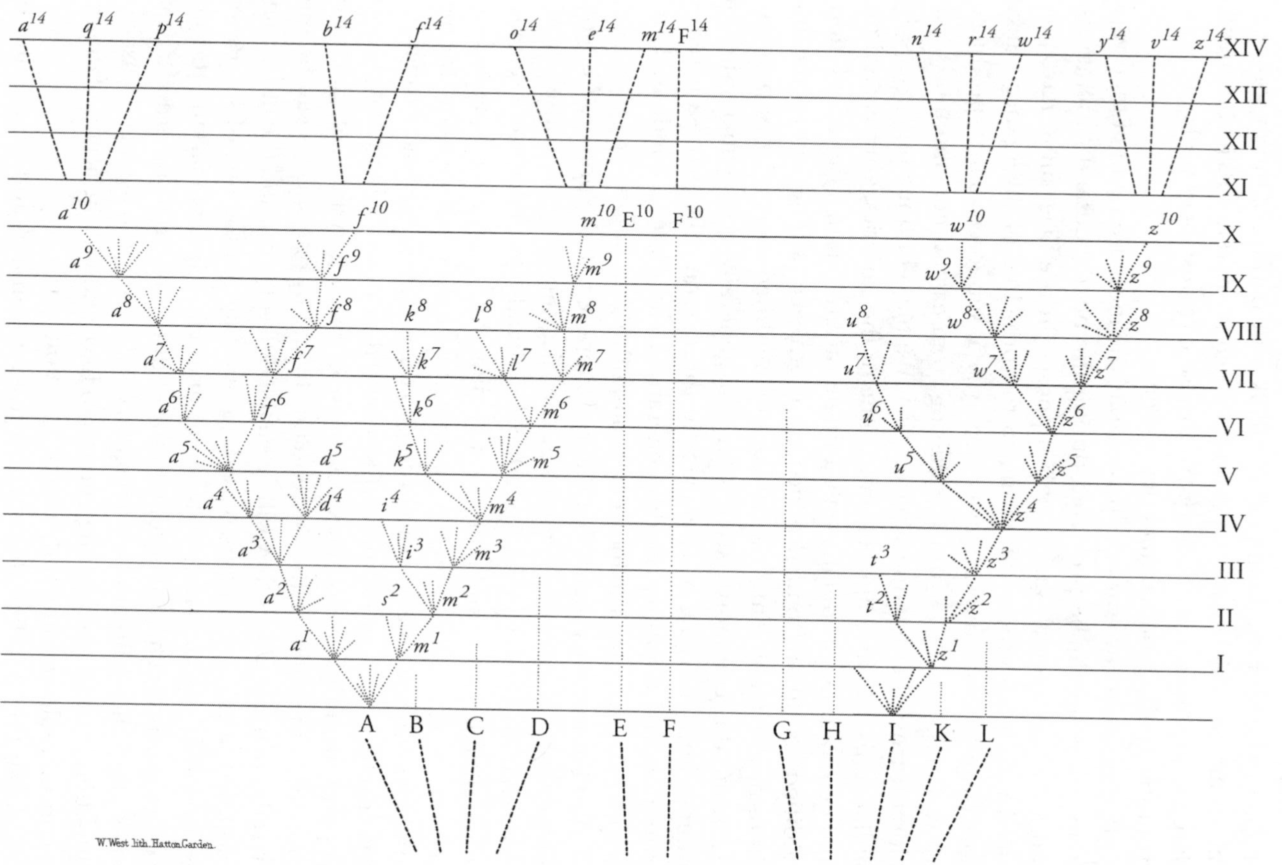
a14 q14 p14 b14 f14 o14 e14 m14 F14 n14 r14 w14 y14 v14 z14
XIV
XIII
XII
XI
a10 f10 m10 E10 F10 w10 z10
X
a9 f9 m9 w9 z9
IX
a8 f8 k8 l8 m8 u8 w8 z8
VIII
a7 f7 k7 l7 m7 u7 w7 z7
VII
a6 f6 k6 m6 u6 z6
VI
a5 d5 k5 m5 u5 z5
V
a4 d4 i4 m4 z4
IV
a3 i3 m3 t3 z3
III
a2 s2 m2 t2 z2
II
a1 m1 z1
I
A B C D E F G H I K L
W. West lith. Hatton Garden.

in nearly the same manner as their parents varied. Moreover, these two varieties, being only slightly modified forms, will tend to inherit those advantages which made their common parent (A) more numerous than most of the other inhabitants of the same country; they will likewise partake of those more general advantages which made the genus to which the parent-species belonged, a large genus in its own country. And these circumstances we know to be favourable to the production of new varieties.

If, then, these two varieties be variable, the most divergent of their variations will generally be preserved during the next thousand generations. And after this interval, variety a^1 is supposed in the diagram to have produced variety a^2, which will, owing to the principle of divergence, differ more from (A) than did variety a^1. Variety m^1 is supposed to have produced two varieties, namely m^2 and s^2, differing from each other, and more considerably from their common parent (A). We may continue the process by similar steps for any length of time; some of the varieties, after each thousand generations, producing only a single variety, but in a more and more modified condition, some producing two or three varieties, and some failing to produce any. Thus the varieties or modified descendants, proceeding from the common parent (A), will generally go on increasing in number and diverging in character. In the diagram the process is represented up to the ten-thousandth generation, and under a condensed and simplified form up to the fourteen-thousandth generation.

But I must here remark that I do not suppose that the process ever goes on so regularly as is represented in the diagram, though in itself made somewhat irregular. I am far from thinking that the most divergent varieties will invariably prevail and multiply: a medium form may often long endure, and may or may not produce more than one modified descendant; for natural selection will always act according to the nature of the places which are either unoccupied or not perfectly occupied by other beings; and this will depend on infinitely complex relations. But as a general rule, the more diversified in structure the descendants from any one species can be rendered, the more places they will be enabled to seize on, and the more their modified progeny will be increased. In our diagram the line of succession is broken at regular intervals by small numbered letters marking the successive forms which have become sufficiently distinct to be recorded as varieties. But these breaks are imaginary, and might have been inserted anywhere, after intervals long enough to have allowed the accumulation of a considerable amount of divergent variation.

As all the modified descendants from a common and widely-diffused species, belonging to a large genus, will tend to partake of the same advantages which made their parent successful in life, they will generally go on multiplying in number as well as diverging in character: this is represented in the diagram by the several divergent branches proceeding from (A). The modified offspring from the later and more highly improved branches in the lines of descent, will, it is probable, often take the place of, and so destroy, the earlier and less improved branches: this is represented in the diagram by some of the lower branches not reaching to the upper horizontal lines. In some cases I do not doubt that the process of modification will be confined to a single line of descent, and the number of the descendants will not be increased; although the amount of divergent modification may have been increased in the successive generations. This case would be represented in the diagram, if all the lines proceeding from (A) were removed, excepting that from a^1 to a^{10}. In the same way, for instance, the English race-horse and English pointer have apparently both gone on slowly diverging in character from their original stocks, without either having given off any fresh branches or races.

After ten thousand generations, species (A) is supposed to have produced three forms, a^{10}, f^{10}, and m^{10}, which, from having diverged in character during the successive generations, will have come to differ largely, but perhaps unequally, from each other and from their common parent. If we suppose the amount of change between each horizontal line in our diagram to be excessively small, these three forms may still be only well-marked varieties; or they may have arrived at the doubtful category of sub-species; but we have only to suppose the steps in the process of modification to be more numerous or greater in amount, to convert these three forms into well-defined species: thus the diagram illustrates the steps by which the small differences distinguishing varieties are increased into the larger differences distinguishing species. By continuing the same process for a greater number of generations (as shown in the diagram in a condensed and simplified manner), we get eight species, marked by the letters between a^{14} and m^{14}, all descended from (A). Thus, as I believe, species are multiplied and genera are formed.

In a large genus it is probable that more than one species would vary. In the diagram I have assumed that a second species (I) has produced, by analogous steps, after ten thousand generations, either two well-marked varieties (w^{10} and z^{10}) or two species, according to the amount of change supposed to be represented between the horizontal lines. After fourteen thousand generations, six new species, marked by the letters n^{14} to z^{14}, are supposed to have been produced. In each genus, the species, which are already extremely different in character, will generally tend to produce the greatest number of modified descendants; for these will have the best chance of filling new and widely different places in the polity of nature: hence in the diagram I have chosen the extreme species (A), and the nearly extreme species (I), as those which have largely varied, and have given rise to new varieties and species. The other nine species (marked by capital letters) of our original genus, may for a long period continue transmitting unaltered descendants; and this is shown in the diagram by the dotted lines not prolonged far upwards from want of space.

But during the process of modification, represented in the diagram, another of our principles, namely that of extinction, will have played an important part. As in each fully stocked country natural selection necessarily acts by the selected form having some advantage in the struggle for life over other forms, there will be a constant tendency in the improved descendants of any one species to supplant and exterminate in each stage of descent their predecessors and their original parent. For it should be remembered that the competition will generally be most severe between those forms which are most nearly related to each other in habits, constitution, and structure. Hence all the intermediate forms between the earlier and later states, that is between the less and more improved state of a species, as well as the original parent-species itself, will generally tend to become extinct. So it probably will be with many whole collateral lines of descent, which will be conquered by later and improved lines of descent. If, however, the modified offspring of a species get into some distinct country, or become quickly adapted to some quite new station, in which child and parent do not come into competition, both may continue to exist.

If then our diagram be assumed to represent a considerable amount of modification, species (A) and all the earlier varieties will have become extinct, having been replaced by eight new species (a^{14} to m^{14}); and (I) will have been replaced by six (n^{14} to z^{14}) new species.

But we may go further than this. The original species of our genus were supposed to resemble each other in unequal degrees, as is so generally the case in nature; species (A) being more nearly related to B, C, and D, than to the other species; and species (I) more to G, H, K, L, than to the others. These two species (A) and (I), were also supposed to be very common and widely diffused species, so that they must originally have had some advantage over most of the other species of the genus. Their modified descendants, fourteen in number at the fourteen-thousandth generation, will probably have inherited some of the same advantages: they have also been modified and improved in a diversified manner at each stage of descent, so as to have become adapted to many related places in the natural economy of their country. It seems, therefore, to me extremely probable that they will have taken the places of, and thus exterminated, not only their parents (A) and (I), but likewise some of the original species which were most nearly related to their parents. Hence very few of the original species will have transmitted offspring to the fourteen-thousandth generation. We may suppose that only one (F), of the two species which were least closely related to the other nine original species, has transmitted descendants to this late stage of descent.

The new species in our diagram descended from the original eleven species, will now be fifteen in number. Owing to the divergent tendency of natural selection, the extreme amount of difference in character between species a^{14} and z^{14} will be much greater than that between the most different of the original eleven species. The new species, moreover, will be allied to each other in a widely different manner. Of the eight descendants from (A) the three marked a^{14}, q^{14}, p^{14}, will be nearly related from having recently branched off from a^{10}; b^{14} and f^{14}, from having diverged at an earlier period from a^5, will be in some degree distinct from the three first-named species; and lastly, o^{14}, e^{14}, and m^{14}, will be nearly related one to the other, but from having diverged at the first commencement of the process of modification, will be widely different from the other five species, and may constitute a sub-genus or even a distinct genus.

The six descendants from (I) will form two sub-genera or even genera. But as the original species (I) differed largely from (A), standing nearly at the extreme points of the original genus, the six descendants from (I) will, owing to inheritance, differ considerably from the eight descendants from (A); the two groups, moreover, are supposed to have gone on diverging in different directions. The intermediate species, also (and this is a very important consideration), which connected the original species (A) and (I), have all become, excepting (F), extinct, and have left no descendants. Hence the six new species descended from (I), and the eight descended from (A), will have to be ranked as very distinct genera, or even as distinct sub-families.

Thus it is, as I believe, that two or more genera are produced by descent, with modification, from two or more species of the same genus. And the two or more parent-species are supposed to have descended from some one species of an earlier genus. In our diagram, this is indicated by the broken lines, beneath the capital letters, converging in sub-branches downwards towards a single point; this point representing a single species, the supposed single parent of our several new sub-genera and genera.

It is worth while to reflect for a moment on the character of the new species F^{14}, which is supposed not to have diverged much in character, but to have retained the form of (F), either unaltered or altered only in a slight degree. In this case, its affinities to the other fourteen new species will be of a curious and circuitous nature. Having descended from a form which stood between the two parent-species (A) and (I), now supposed to be extinct and unknown, it will be in some degree intermediate in character between the two groups descended from these species. But as these two groups have gone on diverging in character from the type of their parents, the new species (F^{14}) will not be directly intermediate between them, but rather between types of the two groups; and every naturalist will be able to bring some such case before his mind.

In the diagram, each horizontal line has hitherto been supposed to represent a thousand generations, but each may represent a million or hundred million generations, and likewise a section of the successive strata of the earth's crust including extinct remains. We shall, when we come to our chapter on Geology, have to refer again to this subject, and I think we shall then see that the diagram throws light on the affinities of extinct beings, which, though generally belonging to the same orders, or families, or genera, with those now living, yet are often, in some degree, intermediate in character between existing groups; and we can understand this fact, for the extinct species lived at very ancient epochs when the branching lines of descent had diverged less.

I see no reason to limit the process of modification, as now explained, to the formation of genera alone. If, in our diagram, we suppose the amount of change represented by each successive group of diverging dotted lines to be very great, the forms marked a^{14} to p^{14}, those marked b^{14} and f^{14}, and those marked o^{14} to m^{14}, will form three very distinct genera. We shall also have two very distinct genera descended from (I); and as these latter two genera, both from continued divergence of character and from inheritance from a different parent, will differ widely from the three genera descended from (A), the two little groups of genera will form two distinct families, or even orders, according to the amount of divergent modification supposed to be represented in the diagram. And the two new families, or orders, will have descended from two species of the original genus; and these two species are supposed to have descended from one species of a still more ancient and unknown genus.

We have seen that in each country it is the species of the larger genera which oftenest present varieties or incipient species. This, indeed, might have been expected; for as natural selection acts through one form having some advantage over other forms in the struggle for existence, it will chiefly act on those which already have some advantage; and the largeness of any group shows that its species have inherited from a common ancestor some advantage in common. Hence, the struggle for the production of new and modified descendants, will mainly lie between the larger groups, which are all trying to increase in number. One large group will slowly conquer another large group, reduce its numbers, and thus lessen its chance of further variation and improvement. Within the same large group, the later and more highly perfected sub-groups, from branching out and seizing on many new places in the polity of Nature, will constantly tend to supplant and destroy the earlier and less improved sub-groups. Small and broken groups and sub-groups will finally tend to disappear. Looking to the future, we can predict that the groups of organic beings which are now large and triumphant, and which are least broken up, that is, which as yet have suffered least extinction, will for a long period continue to increase. But which groups will ultimately prevail, no man can predict; for we well know that many groups, formerly most extensively developed, have now become extinct. Looking still more remotely to the future, we may predict that, owing to the continued and steady increase of the larger groups, a multitude of smaller groups will become utterly extinct, and leave no modified descendants; and consequently that of the species living at any one period, extremely few will transmit descendants to a remote futurity. I shall have to return to this subject in the chapter on Classification, but I may add that on this view of extremely few of the more ancient species having transmitted descendants, and on the view of all the descendants of the same species making a class, we can understand how it is that there exist but very few classes in each main division of the animal and vegetable kingdoms. Although extremely few of the most ancient species may now have living and modified descendants, yet at the most remote geological period, the earth may have been as well peopled with many species of many genera, families, orders, and classes, as at the present day.

Summary of Chapter.—If during the long course of ages and under varying conditions of life, organic beings vary at all in the several parts of their organisation, and I think this cannot be disputed; if there be, owing to the high geometrical powers of increase of each species, at some age, season, or year, a severe struggle for life, and this certainly cannot be disputed; then, considering the infinite complexity of the relations of all organic beings to each other and to their conditions of existence, causing an infinite diversity in structure, constitution, and habits, to be advantageous to them, I think it would be a most extraordinary fact if no variation ever had occurred useful to each being's own welfare, in the same way as so many variations have occurred useful to man. But if variations useful to any organic being do occur, assuredly individuals thus characterised will have the best chance of being preserved in the struggle for life; and from the strong principle of inheritance they will tend to produce offspring similarly characterised. This principle of preservation, I have called, for the sake of brevity, Natural Selection. Natural selection, on the principle of qualities being inherited at corresponding ages, can modify the egg, seed, or young, as easily as the adult. Amongst many animals, sexual selection will give its aid to ordinary selection, by assuring to the most vigorous and best adapted males the greatest number of offspring. Sexual selection will also give characters useful to the males alone, in their struggles with other males.

Whether natural selection has really thus acted in nature, in modifying and adapting the various forms of life to their several conditions and stations, must be judged of by the general tenour and balance of evidence given in the following chapters. But we already see how it entails extinction; and how largely extinction has acted in the world's history, geology plainly declares. Natural selection, also, leads to divergence of character; for more living beings can be supported on the same area the more they diverge in structure, habits, and constitution, of which we see proof by looking at the inhabitants of any small spot or at naturalised productions. Therefore during the modification of the descendants of any one species, and during the incessant struggle of all species to increase in numbers, the more diversified these descendants become, the better will be their chance of succeeding in the battle of life. Thus the small differences distinguishing varieties of the same species, will steadily tend to increase till they come to equal the greater differences between species of the same genus, or even of distinct genera.

We have seen that it is the common, the widely-diffused, and widely-ranging species, belonging to the larger genera, which vary most; and these will tend to transmit to their modified offspring that superiority which now makes them dominant in their own countries. Natural selection, as has just been remarked, leads to divergence of character and to much extinction of the less improved and intermediate forms of life. On these principles, I believe, the nature of the affinities of all organic beings may be explained. It is a truly wonderful fact—the wonder of which we are apt to overlook from familiarity—that all animals and all plants throughout all time and space should be related to each other in group subordinate to group, in the manner which we everywhere behold—namely, varieties of the same species most closely related together, species of the same genus less closely and unequally related together, forming sections and sub-genera, species of distinct genera much less closely related, and genera related in different degrees, forming sub-families, families, orders, sub-classes, and classes. The several subordinate groups in any class cannot be ranked in a single file, but seem rather to be clustered round points, and these round other points, and so on in almost endless cycles. On the view that each species has been independently created, I can see no explanation of this great fact in the classification of all organic beings; but, to the best of my judgment, it is explained through inheritance and the complex action of natural selection, entailing extinction and divergence of character, as we have seen illustrated in the diagram.

The affinities of all the beings of the same class have sometimes been represented by a great tree. I believe this simile largely speaks the truth. The green and budding twigs may represent existing species; and those produced during each former year may represent the long succession of extinct species. At each period of growth all the growing twigs have tried to branch out on all sides, and to overtop and kill the surrounding twigs and branches, in the same manner as species and groups of species have tried to overmaster other species in the great battle for life. The limbs divided into great branches, and these into lesser and lesser branches, were themselves once, when the tree was small, budding twigs; and this connexion of the former and present buds by ramifying branches may well represent the classification of all extinct and living species in groups subordinate to groups. Of the many twigs which flourished when the tree was a mere bush, only two or three, now grown into great branches, yet survive and bear all the other branches; so with the species which lived during long-past geological periods, very few now have living and modified descendants. From the first growth of the tree, many a limb and branch has decayed and dropped off; and these lost branches of various sizes may represent those whole orders, families, and genera which have now no living representatives, and which are known to us only from having been found in a fossil state. As we here and there see a thin straggling branch springing from a fork low down in a tree, and which by some chance has been favoured and is still alive on its summit, so we occasionally see an animal like the Ornithorhynchus or Lepidosiren, which in some small degree connects by its affinities two large branches of life, and which has apparently been saved from fatal competition by having inhabited a protected station. As buds give rise by growth to fresh buds, and these, if vigorous, branch out and overtop on all sides many a feebler branch, so by generation I believe it has been with the great Tree of Life, which fills with its dead and broken branches the crust of the earth, and covers the surface with its ever branching and beautiful ramifications.

CHAPTER V

Laws of Variation

EFFECTS OF EXTERNAL CONDITIONS • USE AND DISUSE, COMBINED WITH NATURAL SELECTION; ORGANS OF FLIGHT AND OF VISION • ACCLIMATISATION • CORRELATION OF GROWTH • COMPENSATION AND ECONOMY OF GROWTH • FALSE CORRELATIONS • MULTIPLE, RUDIMENTARY, AND LOWLY ORGANISED STRUCTURES VARIABLE • PARTS DEVELOPED IN AN UNUSUAL MANNER ARE HIGHLY VARIABLE: SPECIFIC CHARACTERS MORE VARIABLE THAN GENERIC: SECONDARY SEXUAL CHARACTERS VARIABLE • SPECIES OF THE SAME GENUS VARY IN AN ANALOGOUS MANNER • REVERSIONS TO LONG LOST CHARACTERS • SUMMARY

I HAVE HITHERTO sometimes spoken as if the variations—so common and multiform in organic beings under domestication, and in a lesser degree in those in a state of nature—had been due to chance. This, of course, is a wholly incorrect expression, but it serves to acknowledge plainly our ignorance of the cause of each particular variation. Some authors believe it to be as much the function of the reproductive system to produce individual differences, or very slight deviations of structure, as to make the child like its parents. But the much greater variability, as well as the greater frequency of monstrosities, under domestication or cultivation, than under nature, leads me to believe that deviations of structure are in some way due to the nature of the conditions of life, to which the parents and their more remote ancestors have been exposed during several generations. I have remarked in the first chapter—but a long catalogue of facts which cannot be here given would be necessary to show the truth of the remark—that the reproductive system is eminently susceptible to changes in the conditions of life; and to this system being functionally disturbed in the parents, I chiefly attribute the varying or plastic condition of the offspring. The male and female sexual elements seem to be affected before that union takes place which is to form a new being. In the case of "sporting" plants, the bud, which in its earliest condition does not apparently differ essentially from an ovule, is alone affected. But why, because the reproductive system is disturbed, this or that part should vary more or less, we are profoundly ignorant. Nevertheless, we can here and there dimly catch a faint ray of light, and we may feel sure that there must be some cause for each deviation of structure, however slight.

How much direct effect difference of climate, food, &c., produces on any being is extremely doubtful. My impression is, that the effect is extremely small in the case of animals, but perhaps rather more in that of plants. We may, at least, safely conclude that such influences cannot have produced the many striking and complex co-adaptations of structure between one organic being and another, which we see everywhere throughout nature. Some little influence may be attributed to climate, food, &c.: thus, E. Forbes speaks confidently that shells at their southern limit, and when living in shallow water, are more brightly coloured than those of the same species further north or from greater depths. Gould believes that birds of the same species are more brightly coloured under a clear atmosphere, than when living on islands or near the coast. So with insects, Wollaston is convinced that residence near the sea affects their colours. Moquin-Tandon gives a list of plants which when growing near the seashore have their leaves in some degree fleshy, though not elsewhere fleshy. Several other such cases could be given.

The fact of varieties of one species, when they range into the zone of habitation of other species, often acquiring in a very slight degree some of the characters of such species, accords with our view that species of all kinds are only well-marked and permanent varieties. Thus the species of shells which are confined to tropical and shallow seas are generally brighter-coloured than those confined to cold and deeper seas. The birds which are confined to continents are, according to Mr. Gould, brighter-coloured than those of islands. The insect-species confined to sea-coasts, as every collector knows, are often brassy or lurid. Plants which live exclusively on the sea-side are very apt to have fleshy leaves. He who believes in the creation of each species, will have to say that this shell, for instance, was created with bright colours for a warm sea; but that this other shell became bright-coloured by variation when it ranged into warmer or shallower waters.

When a variation is of the slightest use to a being, we cannot tell how much of it to attribute to the accumulative action of natural selection, and how much to the conditions of life. Thus, it is well known to furriers that animals of the same species have thicker and better fur the more severe the climate is under which they have lived; but who can tell how much of this difference may be due to the warmest-clad individuals having been favoured and preserved during many generations, and how much to the direct action of the severe climate? for it would appear that climate has some direct action on the hair of our domestic quadrupeds.

Instances could be given of the same variety being produced under conditions of life as different as can well be conceived; and, on the other hand, of different varieties being produced from the same species under the same conditions. Such facts show how indirectly the conditions of life must act. Again, innumerable instances are known to every naturalist of species keeping true, or not varying at all, although living under the most opposite climates. Such considerations as these incline me to lay very little weight on the direct action of the conditions of life. Indirectly, as already remarked, they seem to play an important part in affecting the reproductive system, and in thus inducing variability; and natural selection will then accumulate all profitable variations, however slight, until they become plainly developed and appreciable by us.

Effects of Use and Disuse.—From the facts alluded to in the first chapter, I think there can be little doubt that use in our domestic animals strengthens and enlarges certain parts, and disuse diminishes them; and that such modifications are inherited. Under free nature, we can have no standard of comparison, by which to judge of the effects of long-continued use or disuse, for we know not the parent-forms; but many animals have structures which can be explained by the effects of disuse. As Professor Owen has remarked, there is no greater anomaly in nature than a bird that cannot fly; yet there are several in this state. The logger-headed duck of South America can only flap along the surface of the water, and has its wings in nearly the same condition as the domestic Aylesbury duck. As the larger ground-feeding birds seldom take flight except to escape danger, I believe that the nearly wingless condition of several birds, which now inhabit or have lately inhabited several oceanic islands, tenanted by no beast of prey, has been caused by disuse. The ostrich indeed inhabits continents and is exposed to danger from which it cannot escape by flight, but by kicking it can defend itself from enemies, as well as any of the smaller quadrupeds. We may imagine that the early progenitor of the ostrich had habits like those of a bustard, and that as natural selection increased in successive generations the size and weight of its body, its legs were used more, and its wings less, until they became incapable of flight.

Kirby has remarked (and I have observed the same fact) that the anterior tarsi, or feet, of many male dung-feeding beetles are very often broken off; he examined seventeen specimens in his own collection, and not one had even a relic left. In the Onites apelles the tarsi are so habitually lost, that the insect has been described as not having them. In some other genera they are present, but in a rudimentary condition. In the Ateuchus or sacred beetle of the Egyptians, they are totally deficient. There is not sufficient evidence to induce us to believe that mutilations are ever inherited; and I should prefer explaining the entire absence of the anterior tarsi in Ateuchus, and their rudimentary condition in some other genera, by the long-continued effects of disuse in their progenitors; for as the tarsi are almost always lost in many dung-feeding beetles, they must be lost early in life, and therefore cannot be much used by these insects.

In some cases we might easily put down to disuse modifications of structure which are wholly, or mainly, due to natural selection. Mr. Wollaston has discovered the remarkable fact that 200 beetles, out of the 550 species inhabiting Madeira, are so far deficient in wings that they cannot fly; and that of the twenty-nine endemic genera, no less than twenty-three genera have all their species in this condition! Several facts, namely, that beetles in many parts of the world are very frequently blown to sea and perish; that the beetles in Madeira, as observed by Mr. Wollaston, lie much concealed, until the wind lulls and the sun shines; that the proportion of wingless beetles is larger on the exposed Dezertas than in Madeira itself; and especially the extraordinary fact, so strongly insisted on by Mr. Wollaston, of the almost entire absence of certain large groups of beetles, elsewhere excessively numerous, and which groups have habits of life almost necessitating frequent flight;—these several considerations have made me believe that the wingless condition of so many Madeira beetles is mainly due to the action of natural selection, but combined probably with disuse. For during thousands of successive generations each individual beetle which flew least, either from its wings having been ever so little less perfectly developed or from indolent habit, will have had the best chance of surviving from not being blown out to sea; and, on the other hand, those beetles which most readily took to flight will oftenest have been blown to sea and thus have been destroyed.

The insects in Madeira which are not ground-feeders, and which, as the flower-feeding coleoptera and lepidoptera, must habitually use their wings to gain their subsistence, have, as Mr. Wollaston suspects, their wings not at all reduced, but even enlarged. This is quite compatible with the action of natural selection. For when a new insect first arrived on the island, the tendency of natural selection to enlarge or to reduce the wings, would depend on whether a greater number of individuals were saved by successfully battling with the winds, or by giving up the attempt and rarely or never flying. As with mariners shipwrecked near a coast, it would have been better for the good swimmers if they had been able to swim still further, whereas it would have been better for the bad swimmers if they had not been able to swim at all and had stuck to the wreck.

The eyes of moles and of some burrowing rodents are rudimentary in size, and in some cases are quite covered up by skin and fur. This state of the eyes is probably due to gradual reduction from disuse, but aided perhaps by natural selection. In South America, a burrowing rodent, the tuco-tuco, or Ctenomys, is even more subterranean in its habits than the mole; and I was assured by a Spaniard, who had often caught them, that they were frequently blind; one which I kept alive was certainly in this condition, the cause, as appeared on dissection, having been inflammation of the nictitating membrane. As frequent inflammation of the eyes must be injurious to any animal, and as eyes are certainly not indispensable to animals with subterranean habits, a reduction in their size with the adhesion of the eyelids and growth of fur over them, might in such case be an advantage; and if so, natural selection would constantly aid the effects of disuse.

It is well known that several animals, belonging to the most different classes, which inhabit the caves of Styria and of Kentucky, are blind. In some of the crabs the foot-stalk for the eye remains, though the eye is gone; the stand for the telescope is there, though the telescope with its glasses has been lost. As it is difficult to imagine that eyes, though useless, could be in any way injurious to animals living in darkness, I attribute their loss wholly to disuse. In one of the blind animals, namely, the cave-rat, the eyes are of immense size; and Professor Silliman thought that it regained, after living some days in the light, some slight power of vision. In the same manner as in Madeira the wings of some of the insects have been enlarged, and the wings of others have been reduced by natural selection aided by use and disuse, so in the case of the cave-rat natural selection seems to have struggled with the loss of light and to have increased the size of the eyes; whereas with all the other inhabitants of the caves, disuse by itself seems to have done its work.

It is difficult to imagine conditions of life more similar than deep limestone caverns under a nearly similar climate: so that on the common view of the blind animals having been separately created for the American and European caverns, close similarity in their organisation and affinities might have been expected; but, as Schiödte and others have remarked, this is not the case, and the cave-insects of the two continents are not more closely allied than might have been anticipated from the general resemblance of the other inhabitants of North America and Europe. On my view we must suppose that American animals, having ordinary powers of vision, slowly migrated by successive generations from the outer world into the deeper and deeper recesses of the Kentucky caves, as did European animals into the caves of Europe. We have some evidence of this gradation of habit; for, as Schiödte remarks, "animals not far remote from ordinary forms, prepare the transition from light to darkness. Next follow those that are constructed for twilight; and, last of all, those destined for total darkness." By the time that an animal had reached, after numberless generations, the deepest recesses, disuse will on this view have more or less perfectly obliterated its eyes, and natural selection will often have effected other changes, such as an increase in the length of the antennæ or palpi, as a compensation for blindness. Notwithstanding such modifications, we might expect still to see in the cave-animals of America, affinities to the other inhabitants of that continent, and in those of Europe, to the inhabitants of the European continent. And this is the case with some of the American cave-animals, as I hear from Professor Dana; and some of the European cave-insects are very closely allied to those of the surrounding country. It would be most difficult to give any rational explanation of the affinities of the blind cave-animals to the other inhabitants of the two continents on the ordinary view of their independent creation. That several of the inhabitants of the caves of the Old and New Worlds should be closely related, we might expect from the well-known relationship of most of their other productions. Far from feeling any surprise that some of the cave-animals should be very anomalous, as Agassiz has remarked in regard to the blind fish, the Amblyopsis, and as is the case with the blind Proteus with reference to the reptiles of Europe, I am only surprised that more wrecks of ancient life have not been preserved, owing to the less severe competition to which the inhabitants of these dark abodes will probably have been exposed.

Acclimatisation.—Habit is hereditary with plants, as in the period of flowering, in the amount of rain requisite for seeds to germinate, in the time of sleep, &c., and this leads me to say a few words on acclimatisation. As it is extremely common for species of the same genus to inhabit very hot and very cold countries, and as I believe that all the species of the same genus have descended from a single parent, if this view be correct, acclimatisation must be readily effected during long-continued descent. It is notorious that each species is adapted to the climate of its own home: species from an arctic or even from a temperate region cannot endure a tropical climate, or conversely. So again, many succulent plants cannot endure a damp climate. But the degree of adaptation of species to the climates under which they live is often overrated. We may infer this from our frequent inability to predict whether or not an imported plant will endure our climate, and from the number of plants and animals brought from warmer countries which here enjoy good health. We have reason to believe that species in a state of nature are limited in their ranges by the competition of other organic beings quite as much as, or more than, by adaptation to particular climates. But whether or not the adaptation be generally very close, we have evidence, in the case of some few plants, of their becoming, to a certain extent, naturally habituated to different temperatures, or becoming acclimatised: thus the pines and rhododendrons, raised from seed collected by Dr. Hooker from trees growing at different heights on the Himalaya, were found in this country to possess different constitutional powers of resisting cold. Mr. Thwaites informs me that he has observed similar facts in Ceylon, and analogous observations have been made by Mr. H. C. Watson on European species of plants brought from the Azores to England. In regard to animals, several authentic cases could be given of species within historical times having largely extended their range from warmer to cooler latitudes, and conversely; but we do not positively know that these animals were strictly adapted to their native climate, but in all ordinary cases we assume such to be the case; nor do we know that they have subsequently become acclimatised to their new homes.

As I believe that our domestic animals were originally chosen by uncivilised man because they were useful and bred readily under confinement, and not because they were subsequently found capable of far-extended transportation, I think the common and extraordinary capacity in our domestic animals of not only withstanding the most different climates but of being perfectly fertile (a far severer test) under them, may be used as an argument that a large proportion of other animals, now in a state of nature, could easily be brought to bear widely different climates. We must not, however, push the foregoing argument too far, on account of the probable origin of some of our domestic animals from several wild stocks: the blood, for instance, of a tropical and arctic wolf or wild dog may perhaps be mingled in our domestic breeds. The rat and mouse cannot be considered as domestic animals, but they have been transported by man to many parts of the world, and now have a far wider range than any other rodent, living free under the cold climate of Faroe in the north and of the Falklands in the south, and on many islands in the torrid zones. Hence I am inclined to look at adaptation to any special climate as a quality readily grafted on an innate wide flexibility of constitution, which is common to most animals. On this view, the capacity of enduring the most different climates by man himself and by his domestic animals, and such facts as that former species of the elephant and rhinoceros were capable of enduring a glacial climate, whereas the living species are now all tropical or sub-tropical in their habits, ought not to be looked at as anomalies, but merely as examples of a very common flexibility of constitution, brought, under peculiar circumstances, into play.

How much of the acclimatisation of species to any peculiar climate is due to mere habit, and how much to the natural selection of varieties having different innate constitutions, and how much to both means combined, is a very obscure question. That habit or custom has some influence I must believe, both from analogy, and from the incessant advice given in agricultural works, even in the ancient Encyclopædias of China, to be very cautious in transposing animals from one district to another; for it is not likely that man should have succeeded in selecting so many breeds and sub-breeds with constitutions specially fitted for their own districts: the result must, I think, be due to habit. On the other hand, I can see no reason to doubt that natural selection will continually tend to preserve those individuals which are born with constitutions best adapted to their native countries. In treatises on many kinds of cultivated plants, certain varieties are said to withstand certain climates better than others: this is very strikingly shown in works on fruit trees published in the United States, in which certain varieties are habitually recommended for the northern, and others for the southern States; and as most of these varieties are of recent origin, they cannot owe their constitutional differences to habit. The case of the Jerusalem artichoke, which is never propagated by seed, and of which consequently new varieties have not been produced, has even been advanced—for it is now as tender as ever it was—as proving that acclimatisation cannot be effected! The case, also, of the kidney-bean has been often cited for a similar purpose, and with much greater weight; but until some one will sow, during a score of generations, his kidney-beans so early that a very large proportion are destroyed by frost, and then collect seed from the few survivors, with care to prevent accidental crosses, and then again get seed from these seedlings, with the same precautions, the experiment cannot be said to have been even tried. Nor let it be supposed that no differences in the constitution of seedling kidney-beans ever appear, for an account has been published how much more hardy some seedlings appeared to be than others.

On the whole, I think we may conclude that habit, use, and disuse, have, in some cases, played a considerable part in the modification of the constitution, and of the structure of various organs; but that the effects of use and disuse have often been largely combined with, and sometimes overmastered by, the natural selection of innate differences.

Correlation of Growth.—I mean by this expression that the whole organisation is so tied together during its growth and development, that when slight variations in any one part occur, and are accumulated through natural selection, other parts become modified. This is a very important subject, most imperfectly understood. The most obvious case is, that modifications accumulated solely for the good of the young or larva, will, it may safely be concluded, affect the structure of the adult; in the same manner as any malconformation affecting the early embryo, seriously affects the whole organisation of the adult. The several parts of the body which are homologous, and which, at an early embryonic period, are alike, seem liable to vary in an allied manner: we see this in the right and left sides of the body varying in the same manner; in the front and hind legs, and even in the jaws and limbs, varying together, for the lower jaw is believed to be homologous with the limbs. These tendencies, I do not doubt, may be mastered more or less completely by natural selection: thus a family of stags once existed with an antler only on one side; and if this had been of any great use to the breed it might probably have been rendered permanent by natural selection.

Homologous parts, as has been remarked by some authors, tend to cohere; this is often seen in monstrous plants; and nothing is more common than the union of homologous parts in normal structures, as the union of the petals of the corolla into a tube. Hard parts seem to affect the form of adjoining soft parts; it is believed by some authors that the diversity in the shape of the pelvis in birds causes the remarkable diversity in the shape of their kidneys. Others believe that the shape of the pelvis in the human mother influences by pressure the shape of the head of the child. In snakes, according to Schlegel, the shape of the body and the manner of swallowing determine the position of several of the most important viscera.

The nature of the bond of correlation is very frequently quite obscure. M. Is. Geoffroy St. Hilaire has forcibly remarked, that certain malconformations very frequently, and that others rarely coexist, without our being able to assign any reason. What can be more singular than the relation between blue eyes and deafness in cats, and the tortoise-shell colour with the female sex; the feathered feet and skin between the outer toes in pigeons, and the presence of more or less down on the young birds when first hatched, with the future colour of their plumage; or, again, the relation between the hair and teeth in the naked Turkish dog, though here probably homology comes into play? With respect to this latter case of correlation, I think it can hardly be accidental, that if we pick out the two orders of mammalia which are most abnormal in their dermal covering, viz. Cetacea (whales) and Edentata (armadilloes, scaly ant-eaters, &c.), that these are likewise the most abnormal in their teeth.

I know of no case better adapted to show the importance of the laws of correlation in modifying important structures, independently of utility and, therefore, of natural selection, than that of the difference between the outer and inner flowers in some Compositous and Umbelliferous plants. Every one knows the difference in the ray and central florets of, for instance, the daisy, and this difference is often accompanied with the abortion of parts of the flower. But, in some Compositous plants, the seeds also differ in shape and sculpture; and even the ovary itself, with its accessory parts, differs, as has been described by Cassini. These differences have been attributed by some authors to pressure, and the shape of the seeds in the ray-florets in some Compositæ countenances this idea; but, in the case of the corolla of the Umbelliferæ, it is by no means, as Dr. Hooker informs me, in species with the densest heads that the inner and outer flowers most frequently differ. It might have been thought that the development of the ray-petals by drawing nourishment from certain other parts of the flower had caused their abortion; but in some Compositæ there is a difference in the seeds of the outer and inner florets without any difference in the corolla. Possibly, these several differences may be connected with some difference in the flow of nutriment towards the central and external flowers: we know, at least, that in irregular flowers, those nearest to the axis are oftenest subject to peloria, and become regular. I may add, as an instance of this, and of a striking case of correlation, that I have recently observed in some garden pelargoniums, that the central flower of the truss often loses the patches of darker colour in the two upper petals; and that when this occurs, the adherent nectary is quite aborted; when the colour is absent from only one of the two upper petals, the nectary is only much shortened.

With respect to the difference in the corolla of the central and exterior flowers of a head or umbel, I do not feel at all sure that C. C. Sprengel's idea that the ray-florets serve to attract insects, whose agency is highly advantageous in the fertilisation of plants of these two orders, is so far-fetched, as it may at first appear: and if it be advantageous, natural selection may have come into play. But in regard to the differences both in the internal and external structure of the seeds, which are not always correlated with any differences in the flowers, it seems impossible that they can be in any way advantageous to the plant: yet in the Umbelliferæ these differences are of such apparent importance—the seeds being in some cases, according to Tausch, orthospermous in the exterior flowers and cœlospermous in the central flowers,—that the elder De Candolle founded his main divisions of the order on analogous differences. Hence we see that modifications of structure, viewed by systematists as of high value, may be wholly due to unknown laws of correlated growth, and without being, as far as we can see, of the slightest service to the species.

We may often falsely attribute to correlation of growth, structures which are common to whole groups of species, and which in truth are simply due to inheritance; for an ancient progenitor may have acquired through natural selection some one modification in structure, and, after thousands of generations, some other and independent modification; and these two modifications, having been transmitted to a whole group of descendants with diverse habits, would naturally be thought to be correlated in some necessary manner. So, again, I do not doubt that some apparent correlations, occurring throughout whole orders, are entirely due to the manner alone in which natural selection can act. For instance, Alph. De Candolle has remarked that winged seeds are never found in fruits which do not open: I should explain the rule by the fact that seeds could not gradually become winged through natural selection, except in fruits which opened; so that the individual plants producing seeds which were a little better fitted to be wafted further, might get an advantage over those producing seed less fitted for dispersal; and this process could not possibly go on in fruit which did not open.

The elder Geoffroy and Goethe propounded, at about the same period, their law of compensation or balancement of growth; or, as Goethe expressed it, "in order to spend on one side, nature is forced to economise on the other side." I think this holds true to a certain extent with our domestic productions: if nourishment flows to one part or organ in excess, it rarely flows, at least in excess, to another part; thus it is difficult to get a cow to give much milk and to fatten readily. The same varieties of the cabbage do not yield abundant and nutritious foliage and a copious supply of oil-bearing seeds. When the seeds in our fruits become atrophied, the fruit itself gains largely in size and quality. In our poultry, a large tuft of feathers on the head is generally accompanied by a diminished comb, and a large beard by diminished wattles. With species in a state of nature it can hardly be maintained that the law is of universal application; but many good observers, more especially botanists, believe in its truth. I will not, however, here give any instances, for I see hardly any way of distinguishing between the effects, on the one hand, of a part being largely developed through natural selection and another and adjoining part being reduced by this same process or by disuse, and, on the other hand, the actual withdrawal of nutriment from one part owing to the excess of growth in another and adjoining part.

I suspect, also, that some of the cases of compensation which have been advanced, and likewise some other facts, may be merged under a more general principle, namely, that natural selection is continually trying to economise in every part of the organisation. If under changed conditions of life a structure before useful becomes less useful, any diminution, however slight, in its development, will be seized on by natural selection, for it will profit the individual not to have its nutriment wasted in building up an useless structure. I can thus only understand a fact with which I was much struck when examining cirripedes, and of which many other instances could be given: namely, that when a cirripede is parasitic within another and is thus protected, it loses more or less completely its own shell or carapace.

This is the case with the male Ibla, and in a truly extraordinary manner with the Proteolepas: for the carapace in all other cirripedes consists of the three highly-important anterior segments of the head enormously developed, and furnished with great nerves and muscles; but in the parasitic and protected Proteolepas, the whole anterior part of the head is reduced to the merest rudiment attached to the bases of the prehensile antennæ. Now the saving of a large and complex structure, when rendered superfluous by the parasitic habits of the Proteolepas, though effected by slow steps, would be a decided advantage to each successive individual of the species; for in the struggle for life to which every animal is exposed, each individual Proteolepas would have a better chance of supporting itself, by less nutriment being wasted in developing a structure now become useless.

Thus, as I believe, natural selection will always succeed in the long run in reducing and saving every part of the organisation, as soon as it is rendered superfluous, without by any means causing some other part to be largely developed in a corresponding degree. And, conversely, that natural selection may perfectly well succeed in largely developing any organ, without requiring as a necessary compensation the reduction of some adjoining part.

It seems to be a rule, as remarked by Is. Geoffroy St. Hilaire, both in varieties and in species, that when any part or organ is repeated many times in the structure of the same individual (as the vertebræ in snakes, and the stamens in polyandrous flowers) the number is variable; whereas the number of the same part or organ, when it occurs in lesser numbers, is constant. The same author and some botanists have further remarked that multiple parts are also very liable to variation in structure. Inasmuch as this "vegetative repetition," to use Prof. Owen's expression, seems to be a sign of low organisation; the foregoing remark seems connected with the very general opinion of naturalists, that beings low in the scale of nature are more variable than those which are higher. I presume that lowness in this case means that the several parts of the organisation have been but little specialised for particular functions; and as long as the same part has to perform diversified work, we can perhaps see why it should remain variable, that is, why natural selection should have preserved or rejected each little deviation of form less carefully than when the part has to serve for one special purpose alone. In the same way that a knife which has to cut all sorts of things may be of almost any shape; whilst a tool for some particular object had better be of some particular shape. Natural selection, it should never be forgotten, can act on each part of each being, solely through and for its advantage.

Rudimentary parts, it has been stated by some authors, and I believe with truth, are apt to be highly variable. We shall have to recur to the general subject of rudimentary and aborted organs; and I will here only add that their variability seems to be owing to their uselessness, and therefore to natural selection having no power to check deviations in their structure. Thus rudimentary parts are left to the free play of the various laws of growth, to the effects of long-continued disuse, and to the tendency to reversion.

A part developed in any species in an extraordinary degree or manner, in comparison with the same part in allied species, tends to be highly variable.—Several years ago I was much struck with a remark, nearly to the above effect, published by Mr. Waterhouse. I infer also from an observation made by Professor Owen, with respect to the length of the arms of the ourang-outang, that he has come to a nearly similar conclusion. It is hopeless to attempt to convince any one of the truth of this proposition without giving the long array of facts which I have collected, and which cannot possibly be here introduced. I can only state my conviction that it is a rule of high generality. I am aware of several causes of error, but I hope that I have made due allowance for them. It should be understood that the rule by no means applies to any part, however unusually developed, unless it be unusually developed in comparison with the same part in closely allied species. Thus, the bat's wing is a most abnormal structure in the class mammalia; but the rule would not here apply, because there is a whole group of bats having wings; it would apply only if some one species of bat had its wings developed in some remarkable manner in comparison with the other species of the same genus. The rule applies very strongly in the case of secondary sexual characters, when displayed in any unusual manner. The term, secondary sexual characters, used by Hunter, applies to characters which are attached to one sex, but are not directly connected with the act of reproduction. The rule applies to males and females; but as females more rarely offer remarkable secondary sexual characters, it applies more rarely to them. The rule being so plainly applicable in the case of secondary sexual characters, may be due to the great variability of these characters, whether or not displayed in any unusual manner—of which fact I think there can be little doubt. But that our rule is not confined to secondary sexual characters is clearly shown in the case of hermaphrodite cirripedes; and I may here add, that I particularly attended to Mr. Waterhouse's remark, whilst investigating this Order, and I am fully convinced that the rule almost invariably holds good with cirripedes. I shall, in my future work, give a list of the more remarkable cases; I will here only briefly give one, as it illustrates the rule in its largest application. The opercular valves of sessile cirripedes (rock barnacles) are, in every sense of the word, very important structures, and they differ extremely little even in different genera; but in the several species of one genus, Pyrgoma, these valves present a marvellous amount of diversification: the homologous valves in the different species being sometimes wholly unlike in shape; and the amount of variation in the individuals of several of the species is so great, that it is no exaggeration to state that the varieties differ more from each other in the characters of these important valves than do other species of distinct genera.

As birds within the same country vary in a remarkably small degree, I have particularly attended to them, and the rule seems to me certainly to hold good in this class. I cannot make out that it applies to plants, and this would seriously have shaken my belief in its truth, had not the great variability in plants made it particularly difficult to compare their relative degrees of variability.

When we see any part or organ developed in a remarkable degree or manner in any species, the fair presumption is that it is of high importance to that species; nevertheless the part in this case is eminently liable to variation. Why should this be so? On the view that each species has been independently created, with all its parts as we now see them, I can see no explanation. But on the view that groups of species have descended from other species, and have been modified through natural selection, I think we can obtain some light. In our domestic animals, if any part, or the whole animal, be neglected and no selection be applied, that part (for instance, the comb in the Dorking fowl) or the whole breed will cease to have a nearly uniform character. The breed will then be said to have degenerated. In rudimentary organs, and in those which have been but little specialised for any particular purpose, and perhaps in polymorphic groups, we see a nearly parallel natural case; for in such cases natural selection either has not or cannot come into full play, and thus the organisation is left in a fluctuating condition. But what here more especially concerns us is, that in our domestic animals those points, which at the present time are undergoing rapid change by continued selection, are also eminently liable to variation. Look at the breeds of the pigeon; see what a prodigious amount of difference there is in the beak of the different tumblers, in the beak and wattle of the different carriers, in the carriage and tail of our fantails, &c., these being the points now mainly attended to by English fanciers. Even in the sub-breeds, as in the short-faced tumbler, it is notoriously difficult to breed them nearly to perfection, and frequently individuals are born which depart widely from the standard. There may be truly said to be a constant struggle going on between, on the one hand, the tendency to reversion to a less modified state, as well as an innate tendency to further variability of all kinds, and, on the other hand, the power of steady selection to keep the breed true. In the long run selection gains the day, and we do not expect to fail so far as to breed a bird as coarse as a common tumbler from a good short-faced strain. But as long as selection is rapidly going on, there may always be expected to be much variability in the structure undergoing modification. It further deserves notice that these variable characters, produced by man's selection, sometimes become attached, from causes quite unknown to us, more to one sex than to the other, generally to the male sex, as with the wattle of carriers and the enlarged crop of pouters.

Now let us turn to nature. When a part has been developed in an extraordinary manner in any one species, compared with the other species of the same genus, we may conclude that this part has undergone an extraordinary amount of modification, since the period when the species branched off from the common progenitor of the genus. This period will seldom be remote in any extreme degree, as species very rarely endure for more than one geological period. An extraordinary amount of modification implies an unusually large and long-continued amount of variability, which has continually been accumulated by natural selection for the benefit of the species. But as the variability of the extraordinarily-developed part or organ has been so great and long-continued within a period not excessively remote, we might, as a general rule, expect still to find more variability in such parts than in other parts of the organisation, which have remained for a much longer period nearly constant. And this, I am convinced, is the case. That the struggle between natural selection on the one hand, and the tendency to reversion and variability on the other hand, will in the course of time cease; and that the most abnormally developed organs may be made constant, I can see no reason to doubt. Hence when an organ, however abnormal it may be, has been transmitted in approximately the same condition to many modified descendants, as in the case of the wing of the bat, it must have existed, according to my theory, for an immense period in nearly the same state; and thus it comes to be no more variable than any other structure. It is only in those cases in which the modification has been comparatively recent and extraordinarily great that we ought to find the *generative variability,* as it may be called, still present in a high degree. For in this case the variability will seldom as yet have been fixed by the continued selection of the individuals varying in the required manner and degree, and by the continued rejection of those tending to revert to a former and less modified condition.

The principle included in these remarks may be extended. It is notorious that specific characters are more variable than generic. To explain by a simple example what is meant. If some species in a large genus of plants had blue flowers and some had red, the colour would be only a specific character, and no one would be surprised at one of the blue species varying into red, or conversely; but if all the species had blue flowers, the colour would become a generic character, and its variation would be a more unusual circumstance. I have chosen this example because an explanation is not in this case applicable, which most naturalists would advance, namely, that specific characters are more variable than generic, because they are taken from parts of less physiological importance than those commonly used for classing genera. I believe this explanation is partly, yet only indirectly, true; I shall, however, have to return to this subject in our chapter on Classification. It would be almost superfluous to adduce evidence in support of the above statement, that specific characters are more variable than generic; but I have repeatedly noticed in works on natural history, that when an author has remarked with surprise that some *important* organ or part, which is generally very constant throughout large groups of species, has *differed* considerably in closely-allied species, that it has, also, been *variable* in the individuals of some of the species. And this fact shows that a character, which is generally of generic value, when it sinks in value and becomes only of specific value, often becomes variable, though its physiological importance may remain the same. Something of the same kind applies to monstrosities: at least Is. Geoffroy St. Hilaire seems to entertain no doubt, that the more an organ normally differs in the different species of the same group, the more subject it is to individual anomalies.

On the ordinary view of each species having been independently created, why should that part of the structure, which differs from the same part in other independently-created species of the same genus, be more variable than those parts which are closely alike in the several species? I do not see that any explanation can be given. But on the view of species being only strongly marked and fixed varieties, we might surely expect to find them still often continuing to vary in those parts of their structure which have varied within a moderately recent period, and which have thus come to differ. Or to state the case in another manner:—the points in which all the species of a genus resemble each other, and in which they differ from the species of some other genus, are called generic characters; and these characters in common I attribute to inheritance from a common progenitor, for it can rarely have happened that natural selection will have modified several species, fitted to more or less widely-different habits, in exactly the same manner: and as these so-called generic characters have been inherited from a remote period, since that period when the species first branched off from their common progenitor, and subsequently have not varied or come to differ in any degree, or only in a slight degree, it is not probable that they should vary at the present day. On the other hand, the points in which species differ from other species of the same genus, are called specific characters; and as these specific characters have varied and come to differ within the period of the branching off of the species from a common progenitor, it is probable that they should still often be in some degree variable,—at least more variable than those parts of the organisation which have for a very long period remained constant.

In connexion with the present subject, I will make only two other remarks. I think it will be admitted, without my entering on details, that secondary sexual characters are very variable; I think it also will be admitted that species of the same group differ from each other more widely in their secondary sexual characters, than in other parts of their organisation; compare, for instance, the amount of difference between the males of gallinaceous birds, in which secondary sexual characters are strongly displayed, with the amount of difference between their females; and the truth of this proposition will be granted. The cause of the original variability of secondary sexual characters is not manifest; but we can see why these characters should not have been rendered as constant and uniform as other parts of the organisation; for secondary sexual characters have been accumulated by sexual selection, which is less rigid in its action than ordinary selection, as it does not entail death, but only gives fewer offspring to the less favoured males. Whatever the cause may be of the variability of secondary sexual characters, as they are highly variable, sexual selection will have had a wide scope for action, and may thus readily have succeeded in giving to the species of the same group a greater amount of difference in their sexual characters, than in other parts of their structure.

It is a remarkable fact, that the secondary sexual differences between the two sexes of the same species are generally displayed in the very same parts of the organisation in which the different species of the same genus differ from each other. Of this fact I will give in illustration two instances, the first which happen to stand on my list; and as the differences in these cases are of a very unusual nature, the relation can hardly be accidental. The same number of joints in the tarsi is a character generally common to very large groups of beetles, but in the Engidæ, as Westwood has remarked, the number varies greatly; and the number likewise differs in the two sexes of the same species: again in fossorial hymenoptera, the manner of neuration of the wings is a character of the highest importance, because common to large groups; but in certain genera the neuration differs in the different species, and likewise in the two sexes of the same species. This relation has a clear meaning on my view of the subject: I look at all the species of the same genus as having as certainly descended from the same progenitor, as have the two sexes of any one of the species. Consequently, whatever part of the structure of the common progenitor, or of its early descendants, became variable; variations of this part would, it is highly probable, be taken advantage of by natural and sexual selection, in order to fit the several species to their several places in the economy of nature, and likewise to fit the two sexes of the same species to each other, or to fit the males and females to different habits of life, or the males to struggle with other males for the possession of the females.

Finally, then, I conclude that the greater variability of specific characters, or those which distinguish species from species, than of generic characters, or those which the species possess in common;—that the frequent extreme variability of any part which is developed in a species in an extraordinary manner in comparison with the same part in its congeners; and the not great degree of variability in a part, however extraordinarily it may be developed, if it be common to a whole group of species;—that the great variability of secondary sexual characters, and the great amount of difference in these same characters between closely allied species;—that secondary sexual and ordinary specific differences are generally displayed in the same parts of the organisation,—are all principles closely connected together. All being mainly due to the species of the same group having descended from a common progenitor, from whom they have inherited much in common,—to parts which have recently and largely varied being more likely still to go on varying than parts which have long been inherited and have not varied,—to natural selection having more or less completely, according to the lapse of time, overmastered the tendency to reversion and to further variability,—to sexual selection being less rigid than ordinary selection,—and to variations in the same parts having been accumulated by natural and sexual selection, and thus adapted for secondary sexual, and for ordinary specific purposes.

Distinct species present analogous variations; and a variety of one species often assumes some of the characters of an allied species, or reverts to some of the characters of an early progenitor.—These propositions will be most readily understood by looking to our domestic races. The most distinct breeds of pigeons, in countries most widely apart, present sub-varieties with reversed feathers on the head and feathers on the feet,—characters not possessed by the aboriginal rock-pigeon; these then are analogous variations in two or more distinct races. The frequent presence of fourteen or even sixteen tail-feathers in the pouter, may be considered as a variation representing the normal structure of another race, the fantail. I presume that no one will doubt that all such analogous variations are due to the several races of the pigeon having inherited from a common parent the same constitution and tendency to variation, when acted on by similar unknown influences. In the vegetable kingdom we have a case of analogous variation, in the enlarged stems, or roots as commonly called, of the Swedish turnip and Ruta baga, plants which several botanists rank as varieties produced by cultivation from a common parent: if this be not so, the case will then be one of analogous variation in two so-called distinct species; and to these a third may be added, namely, the common turnip. According to the ordinary view of each species having been independently created, we should have to attribute this similarity in the enlarged stems of these three plants, not to the *vera causa* of community of descent, and a consequent tendency to vary in a like manner, but to three separate yet closely related acts of creation.

With pigeons, however, we have another case, namely, the occasional appearance in all the breeds, of slaty-blue birds with two black bars on the wings, a white rump, a bar at the end of the tail, with the outer feathers externally edged near their bases with white. As all these marks are characteristic of the parent rock-pigeon, I presume that no one will doubt that this is a case of reversion, and not of a new yet analogous variation appearing in the several breeds. We may I think confidently come to this conclusion, because, as we have seen, these coloured marks are eminently liable to appear in the crossed offspring of two distinct and differently coloured breeds; and in this case there is nothing in the external conditions of life to cause the reappearance of the slaty-blue, with the several marks, beyond the influence of the mere act of crossing on the laws of inheritance.

No doubt it is a very surprising fact that characters should reappear after having been lost for many, perhaps for hundreds of generations. But when a breed has been crossed only once by some other breed, the offspring occasionally show a tendency to revert in character to the foreign breed for many generations—some say, for a dozen or even a score of generations. After twelve generations, the proportion of blood, to use a common expression, of any one ancestor, is only 1 in 2048; and yet, as we see, it is generally believed that a tendency to reversion is retained by this very small proportion of foreign blood. In a breed which has not been crossed, but in which *both* parents have lost some character which their progenitor possessed, the tendency, whether strong or weak, to reproduce the lost character might be, as was formerly remarked, for all that we can see to the contrary, transmitted for almost any number of generations. When a character which has been lost in a breed, reappears after a great number of generations, the most probable hypothesis is, not that the offspring suddenly takes after an ancestor some hundred generations distant, but that in each successive generation there has been a tendency to reproduce the character in question, which at last, under unknown favourable conditions, gains an ascendancy. For instance, it is probable that in each generation of the barb-pigeon, which produces most rarely a blue and black-barred bird, there has been a tendency in each generation in the plumage to assume this colour. This view is hypothetical, but could be supported by some facts; and I can see no more abstract improbability in a tendency to produce any character being inherited for an endless number of generations, than in quite useless or rudimentary organs being, as we all know them to be, thus inherited. Indeed, we may sometimes observe a mere tendency to produce a rudiment inherited: for instance, in the common snap-dragon (Antirrhinum) a rudiment of a fifth stamen so often appears, that this plant must have an inherited tendency to produce it.

As all the species of the same genus are supposed, on my theory, to have descended from a common parent, it might be expected that they would occasionally vary in an analogous manner; so that a variety of one species would resemble in some of its characters another species; this other species being on my view only a well-marked and permanent variety. But characters thus gained would probably be of an unimportant nature, for the presence of all important characters will be governed by natural selection, in accordance with the diverse habits of the species, and will not be left to the mutual action of the conditions of life and of a similar inherited constitution. It might further be expected that the species of the same genus would occasionally exhibit reversions to lost ancestral characters. As, however, we never know the exact character of the common ancestor of a group, we could not distinguish these two cases: if, for instance, we did not know that the rock-pigeon was not feather-footed or turn-crowned, we could not have told, whether these characters in our domestic breeds were reversions or only analogous variations; but we might have inferred that the blueness was a case of reversion, from the number of the markings, which are correlated with the blue tint, and which it does not appear probable would all appear together from simple variation. More especially we might have inferred this, from the blue colour and marks so often appearing when distinct breeds of diverse colours are crossed. Hence, though under nature it must generally be left doubtful, what cases are reversions to an anciently existing character, and what are new but analogous variations, yet we ought, on my theory, sometimes to find the varying offspring of a species assuming characters (either from reversion or from analogous variation) which already occur in some other members of the same group. And this undoubtedly is the case in nature.

A considerable part of the difficulty in recognising a variable species in our systematic works, is due to its varieties mocking, as it were, some of the other species of the same genus. A considerable catalogue, also, could be given of forms intermediate between two other forms, which themselves must be doubtfully ranked as either varieties or species; and this shows, unless all these forms be considered as independently created species, that the one in varying has assumed some of the characters of the other, so as to produce the intermediate form. But the best evidence is afforded by parts or organs of an important and uniform nature occasionally varying so as to acquire, in some degree, the character of the same part or organ in an allied species. I have collected a long list of such cases; but here, as before, I lie under a great disadvantage in not being able to give them. I can only repeat that such cases certainly do occur, and seem to me very remarkable.

I will, however, give one curious and complex case, not indeed as affecting any important character, but from occurring in several species of the same genus, partly under domestication and partly under nature. It is a case apparently of reversion. The ass not rarely has very distinct transverse bars on its legs, like those on the legs of the zebra: it has been asserted that these are plainest in the foal, and from inquiries which I have made, I believe this to be true. It has also been asserted that the stripe on each shoulder is sometimes double. The shoulder-stripe is certainly very variable in length and outline. A white ass, but *not* an albino, has been described without either spinal or shoulder stripe; and these stripes are sometimes very obscure, or actually quite lost, in dark-coloured asses. The koulan of Pallas is said to have been seen with a double shoulder-stripe. The hemionus has no shoulder-stripe; but traces of it, as stated by Mr. Blyth and others, occasionally appear: and I have been informed by Colonel Poole that the foals of this species are generally striped on the legs, and faintly on the shoulder. The quagga, though so plainly barred like a zebra over the body, is without bars on the legs; but Dr. Gray has figured one specimen with very distinct zebra-like bars on the hocks.

With respect to the horse, I have collected cases in England of the spinal stripe in horses of the most distinct breeds, and of *all* colours; transverse bars on the legs are not rare in duns, mouse-duns, and in one instance in a chestnut: a faint shoulder-stripe may sometimes be seen in duns, and I have seen a trace in a bay horse. My son made a careful examination and sketch for me of a dun Belgian cart-horse with a double stripe on each shoulder and with leg-stripes; and a man, whom I can implicitly trust, has examined for me a small dun Welch pony with *three* short parallel stripes on each shoulder.

In the north-west part of India the Kattywar breed of horses is so generally striped, that, as I hear from Colonel Poole, who examined the breed for the Indian Government, a horse without stripes is not considered as purely-bred. The spine is always striped; the legs are generally barred; and the shoulder-stripe, which is sometimes double and sometimes treble, is common; the side of the face, moreover, is sometimes striped. The stripes are plainest in the foal; and sometimes quite disappear in old horses. Colonel Poole has seen both gray and bay Kattywar horses striped when first foaled. I have, also, reason to suspect, from information given me by Mr. W. W. Edwards, that with the English race-horse the spinal stripe is much commoner in the foal than in the full-grown animal. Without here entering on further details, I may state that I have collected cases of leg and shoulder stripes in horses of very different breeds, in various countries from Britain to Eastern China; and from Norway in the north to the Malay Archipelago in the south. In all parts of the world these stripes occur far oftenest in duns and mouse-duns; by the term dun a large range of colour is included, from one between brown and black to a close approach to cream-colour.

I am aware that Colonel Hamilton Smith, who has written on this subject, believes that the several breeds of the horse have descended from several aboriginal species—one of which, the dun, was striped; and that the above-described appearances are all due to ancient crosses with the dun stock. But I am not at all satisfied with this theory, and should be loth to apply it to breeds so distinct as the heavy Belgian cart-horse, Welch ponies, cobs, the lanky Kattywar race, &c., inhabiting the most distant parts of the world.

Now let us turn to the effects of crossing the several species of the horse-genus. Rollin asserts, that the common mule from the ass and horse is particularly apt to have bars on its legs. I once saw a mule with its legs so much striped that any one at first would have thought that it must have been the product of a zebra; and Mr. W. C. Martin, in his excellent treatise on the horse, has given a figure of a similar mule. In four coloured drawings, which I have seen, of hybrids between the ass and zebra, the legs were much more plainly barred than the rest of the body; and in one of them there was a double shoulder-stripe. In Lord Moreton's famous hybrid from a chestnut mare and male quagga, the hybrid, and even the pure offspring subsequently produced from the mare by a black Arabian sire, were much more plainly barred across the legs than is even the pure quagga. Lastly, and this is another most remarkable case, a hybrid has been figured by Dr. Gray (and he informs me that he knows of a second case) from the ass and the hemionus; and this hybrid, though the ass seldom has stripes on its legs and the hemionus has none and has not even a shoulder-stripe, nevertheless had all four legs barred, and had three short shoulder-stripes, like those on the dun Welch pony, and even had some zebra-like stripes on the sides of its face. With respect to this last fact, I was so convinced that not even a stripe of colour appears from what would commonly be called an accident, that I was led solely from the occurrence of the face-stripes on this hybrid from the ass and hemionus, to ask Colonel Poole whether such face-stripes ever occur in the eminently striped Kattywar breed of horses, and was, as we have seen, answered in the affirmative.

What now are we to say to these several facts? We see several very distinct species of the horse-genus becoming, by simple variation, striped on the legs like a zebra, or striped on the shoulders like an ass. In the horse we see this tendency strong whenever a dun tint appears—a tint which approaches to that of the general colouring of the other species of the genus. The appearance of the stripes is not accompanied by any change of form or by any other new character. We see this tendency to become striped most strongly displayed in hybrids from between several of the most distinct species. Now observe the case of the several breeds of pigeons: they are descended from a pigeon (including two or three sub-species or geographical races) of a bluish colour, with certain bars and other marks; and when any breed assumes by simple variation a bluish tint, these bars and other marks invariably reappear; but without any other change of form or character. When the oldest and truest breeds of various colours are crossed, we see a strong tendency for the blue tint and bars and marks to reappear in the mongrels. I have stated that the most probable hypothesis to account for the reappearance of very ancient characters, is—that there is a *tendency* in the young of each successive generation to produce the long-lost character, and that this tendency, from unknown causes, sometimes prevails. And we have just seen that in several species of the horse-genus the stripes are either plainer or appear more commonly in the young than in the old. Call the breeds of pigeons, some of which have bred true for centuries, species; and how exactly parallel is the case with that of the species of the horse-genus! For myself, I venture confidently to look back thousands on thousands of generations, and I see an animal striped like a zebra, but perhaps otherwise very differently constructed, the common parent of our domestic horse, whether or not it be descended from one or more wild stocks, of the ass, the hemionus, quagga, and zebra.

He who believes that each equine species was independently created, will, I presume, assert that each species has been created with a tendency to vary, both under nature and under domestication, in this particular manner, so as often to become striped like other species of the genus; and that each has been created with a strong tendency, when crossed with species inhabiting distant quarters of the world, to produce hybrids resembling in their stripes, not their own parents, but other species of the genus. To admit this view is, as it seems to me, to reject a real for an unreal, or at least for an unknown, cause. It makes the works of God a mere mockery and deception; I would almost as soon believe with the old and ignorant cosmogonists, that fossil shells had never lived, but had been created in stone so as to mock the shells now living on the sea-shore.

Summary.—Our ignorance of the laws of variation is profound. Not in one case out of a hundred can we pretend to assign any reason why this or that part differs, more or less, from the same part in the parents. But whenever we have the means of instituting a comparison, the same laws appear to have acted in producing the lesser differences between varieties of the same species, and the greater differences between species of the same genus. The external conditions of life, as climate and food, &c., seem to have induced some slight modifications. Habit in producing constitutional differences, and use in strengthening, and disuse in weakening and diminishing organs, seem to have been more potent in their effects. Homologous parts tend to vary in the same way, and homologous parts tend to cohere. Modifications in hard parts and in external parts sometimes affect softer and internal parts. When one part is largely developed, perhaps it tends to draw nourishment from the adjoining parts; and every part of the structure which can be saved without detriment to the individual, will be saved. Changes of structure at an early age will generally affect parts subsequently developed; and there are very many other correlations of growth, the nature of which we are utterly unable to understand. Multiple parts are variable in number and in structure, perhaps arising from such parts not having been closely specialised to any particular function, so that their modifications have not been closely checked by natural selection. It is probably from this same cause that organic beings low in the scale of nature are more variable than those which have their whole organisation more specialised, and are higher in the scale. Rudimentary organs, from being useless, will be disregarded by natural selection, and hence probably are variable. Specific characters—that is, the characters which have come to differ since the several species of the same genus branched off from a common parent—are more variable than generic characters, or those which have long been inherited, and have not differed within this same period. In these remarks we have referred to special parts or organs being still variable, because they have recently varied and thus come to differ; but we have also seen in the second Chapter that the same principle applies to the whole individual; for in a district where many species of any genus are found—that is, where there has been much former variation and differentiation, or where the manufactory of new specific forms has been actively at work—there, on an average, we now find most varieties or incipient species. Secondary sexual characters are highly variable, and such characters differ much in the species of the same group. Variability in the same parts of the organisation has generally been taken advantage of in giving secondary sexual differences to the sexes of the same species, and specific differences to the several species of the same genus. Any part or organ developed to an extraordinary size or in an extraordinary manner, in comparison with the same part or organ in the allied species, must have gone through an extraordinary amount of modification since the genus arose; and thus we can understand why it should often still be variable in a much higher degree than other parts; for variation is a long-continued and slow process, and natural selection will in such cases not as yet have had time to overcome the tendency to further variability and to reversion to a less modified state. But when a species with any extraordinarily-developed organ has become the parent of many modified descendants—which on my view must be a very slow process, requiring a long lapse of time—in this case, natural selection may readily have succeeded in giving a fixed character to the organ, in however extraordinary a manner it may be developed. Species inheriting nearly the same constitution from a common parent and exposed to similar influences will naturally tend to present analogous variations, and these same species may occasionally revert to some of the characters of their ancient progenitors. Although new and important modifications may not arise from reversion and analogous variation, such modifications will add to the beautiful and harmonious diversity of nature.

Whatever the cause may be of each slight difference in the offspring from their parents—and a cause for each must exist—it is the steady accumulation, through natural selection, of such differences, when beneficial to the individual, that gives rise to all the more important modifications of structure, by which the innumerable beings on the face of this earth are enabled to struggle with each other, and the best adapted to survive.

CHAPTER VI

DIFFICULTIES ON THEORY

DIFFICULTIES ON THE THEORY OF DESCENT WITH MODIFICATION • TRANSITIONS • ABSENCE OR RARITY OF TRANSITIONAL VARIETIES • TRANSITIONS IN HABITS OF LIFE • DIVERSIFIED HABITS IN THE SAME SPECIES • SPECIES WITH HABITS WIDELY DIFFERENT FROM THOSE OF THEIR ALLIES • ORGANS OF EXTREME PERFECTION • MEANS OF TRANSITION • CASES OF DIFFICULTY • NATURA NON FACIT SALTUM • ORGANS OF SMALL IMPORTANCE • ORGANS NOT IN ALL CASES ABSOLUTELY PERFECT • THE LAW OF UNITY OF TYPE AND OF THE CONDITIONS OF EXISTENCE EMBRACED BY THE THEORY OF NATURAL SELECTION

LONG BEFORE HAVING ARRIVED at this part of my work, a crowd of difficulties will have occurred to the reader. Some of them are so grave that to this day I can never reflect on them without being staggered; but, to the best of my judgment, the greater number are only apparent, and those that are real are not, I think, fatal to my theory.

These difficulties and objections may be classed under the following heads:—Firstly, why, if species have descended from other species by insensibly fine gradations, do we not everywhere see innumerable transitional forms? Why is not all nature in confusion instead of the species being, as we see them, well defined?

Secondly, is it possible that an animal having, for instance, the structure and habits of a bat, could have been formed by the modification of some animal with wholly different habits? Can we believe that natural selection could produce, on the one hand, organs of trifling importance, such as the tail of a giraffe, which serves as a fly-flapper, and, on the other hand, organs of such wonderful structure, as the eye, of which we hardly as yet fully understand the inimitable perfection?

Thirdly, can instincts be acquired and modified through natural selection? What shall we say to so marvellous an instinct as that which leads the bee to make cells, which have practically anticipated the discoveries of profound mathematicians?

Fourthly, how can we account for species, when crossed, being sterile and producing sterile offspring, whereas, when varieties are crossed, their fertility is unimpaired?

The two first heads shall be here discussed—Instinct and Hybridism in separate chapters.

On the absence or rarity of transitional varieties.—As natural selection acts solely by the preservation of profitable modifications, each new form will tend in a fully-stocked country to take the place of, and finally to exterminate, its own less improved parent or other less-favoured forms with which it comes into competition. Thus extinction and natural selection will, as we have seen, go hand in hand. Hence, if we look at each species as descended from some other unknown form, both the parent and all the transitional varieties will generally have been exterminated by the very process of formation and perfection of the new form.

But, as by this theory innumerable transitional forms must have existed, why do we not find them embedded in countless numbers in the crust of the earth? It will be much more convenient to discuss this question in the chapter on the Imperfection of the geological record; and I will here only state that I believe the answer mainly lies in the record being incomparably less perfect than is generally supposed; the imperfection of the record being chiefly due to organic beings not inhabiting profound depths of the sea, and to their remains being embedded and preserved to a future age only in masses of sediment sufficiently thick and extensive to withstand an enormous amount of future degradation; and such fossiliferous masses can be accumulated only where much sediment is deposited on the shallow bed of the sea, whilst it slowly subsides. These contingencies will concur only rarely, and after enormously long intervals. Whilst the bed of the sea is stationary or is rising, or when very little sediment is being deposited, there will be blanks in our geological history. The crust of the earth is a vast museum; but the natural collections have been made only at intervals of time immensely remote.

But it may be urged that when several closely-allied species inhabit the same territory we surely ought to find at the present time many transitional forms. Let us take a simple case: in travelling from north to south over a continent, we generally meet at successive intervals with closely allied or representative species, evidently filling nearly the same place in the natural economy of the land. These representative species often meet and interlock; and as the one becomes rarer and rarer, the other becomes more and more frequent, till the one replaces the other. But if we compare these species where they intermingle, they are generally as absolutely distinct from each other in every detail of structure as are specimens taken from the metropolis inhabited by each. By my theory these allied species have descended from a common parent; and during the process of modification, each has become adapted to the conditions of life of its own region, and has supplanted and exterminated its original parent and all the transitional varieties between its past and present states. Hence we ought not to expect at the present time to meet with numerous transitional varieties in each region, though they must have existed there, and may be embedded there in a fossil condition. But in the intermediate region, having intermediate conditions of life, why do we not now find closely-linking intermediate varieties? This difficulty for a long time quite confounded me. But I think it can be in large part explained.

In the first place we should be extremely cautious in inferring, because an area is now continuous, that it has been continuous during a long period. Geology would lead us to believe that almost every continent has been broken up into islands even during the later tertiary periods; and in such islands distinct species might have been separately formed without the possibility of intermediate varieties existing in the intermediate zones. By changes in the form of the land and of climate, marine areas now continuous must often have existed within recent times in a far less continuous and uniform condition than at present. But I will pass over this way of escaping from the difficulty; for I believe that many perfectly defined species have been formed on strictly continuous areas; though I do not doubt that the formerly broken condition of areas now continuous has played an important part in the formation of new species, more especially with freely-crossing and wandering animals.

In looking at species as they are now distributed over a wide area, we generally find them tolerably numerous over a large territory, then becoming somewhat abruptly rarer and rarer on the confines, and finally disappearing. Hence the neutral territory between two representative species is generally narrow in comparison with the territory proper to each. We see the same fact in ascending mountains, and sometimes it is quite remarkable how abruptly, as Alph. De Candolle has observed, a common alpine species disappears. The same fact has been noticed by Forbes in sounding the depths of the sea with the dredge. To those who look at climate and the physical conditions of life as the all-important elements of distribution, these facts ought to cause surprise, as climate and height or depth graduate away insensibly. But when we bear in mind that almost every species, even in its metropolis, would increase immensely in numbers, were it not for other competing species; that nearly all either prey on or serve as prey for others; in short, that each organic being is either directly or indirectly related in the most important manner to other organic beings, we must see that the range of the inhabitants of any country by no means exclusively depends on insensibly changing physical conditions, but in large part on the presence of other species, on which it depends, or by which it is destroyed, or with which it comes into competition; and as these species are already defined objects (however they may have become so), not blending one into another by insensible gradations, the range of any one species, depending as it does on the range of others, will tend to be sharply defined. Moreover, each species on the confines of its range, where it exists in lessened numbers, will, during fluctuations in the number of its enemies or of its prey, or in the seasons, be extremely liable to utter extermination; and thus its geographical range will come to be still more sharply defined.

If I am right in believing that allied or representative species, when inhabiting a continuous area, are generally so distributed that each has a wide range, with a comparatively narrow neutral territory between them, in which they become rather suddenly rarer and rarer; then, as varieties do not essentially differ from species, the same rule will probably apply to both; and if we in imagination adapt a varying species to a very large area, we shall have to adapt two varieties to two large areas, and a third variety to a narrow intermediate zone. The intermediate variety, consequently, will exist in lesser numbers from inhabiting a narrow and lesser area; and practically, as far as I can make out, this rule holds good with varieties in a state of nature. I have met with striking instances of the rule in the case of varieties intermediate between well-marked varieties in the genus Balanus. And it would appear from information given me by Mr. Watson, Dr. Asa Gray, and Mr. Wollaston, that generally when varieties intermediate between two other forms occur, they are much rarer numerically than the forms which they connect. Now, if we may trust these facts and inferences, and therefore conclude that varieties linking two other varieties together have generally existed in lesser numbers than the forms which they connect, then, I think, we can understand why intermediate varieties should not endure for very long periods;—why as a general rule they should be exterminated and disappear, sooner than the forms which they originally linked together.

For any form existing in lesser numbers would, as already remarked, run a greater chance of being exterminated than one existing in large numbers; and in this particular case the intermediate form would be eminently liable to the inroads of closely allied forms existing on both sides of it. But a far more important consideration, as I believe, is that, during the process of further modification, by which two varieties are supposed on my theory to be converted and perfected into two distinct species, the two which exist in larger numbers from inhabiting larger areas, will have a great advantage over the intermediate variety, which exists in smaller numbers in a narrow and intermediate zone. For forms existing in larger numbers will always have a better chance, within any given period, of presenting further favourable variations for natural selection to seize on, than will the rarer forms which exist in lesser numbers. Hence, the more common forms, in the race for life, will tend to beat and supplant the less common forms, for these will be more slowly modified and improved. It is the same principle which, as I believe, accounts for the common species in each country, as shown in the second chapter, presenting on an average a greater number of well-marked varieties than do the rarer species. I may illustrate what I mean by supposing three varieties of sheep to be kept, one adapted to an extensive mountainous region; a second to a comparatively narrow, hilly tract; and a third to wide plains at the base; and that the inhabitants are all trying with equal steadiness and skill to improve their stocks by selection; the chances in this case will be strongly in favour of the great holders on the mountains or on the plains improving their breeds more quickly than the small holders on the intermediate narrow, hilly tract; and consequently the improved mountain or plain breed will soon take the place of the less improved hill breed; and thus the two breeds, which originally existed in greater numbers, will come into close contact with each other, without the interposition of the supplanted, intermediate hill-variety.

To sum up, I believe that species come to be tolerably well-defined objects, and do not at any one period present an inextricable chaos of varying and intermediate links: firstly, because new varieties are very slowly formed, for variation is a very slow process, and natural selection can do nothing until favourable variations chance to occur, and until a place in the natural polity of the country can be better filled by some modification of some one or more of its inhabitants. And such new places will depend on slow changes of climate, or on the occasional immigration of new inhabitants, and, probably, in a still more important degree, on some of the old inhabitants becoming slowly modified, with the new forms thus produced and the old ones acting and reacting on each other. So that, in any one region and at any one time, we ought only to see a few species presenting slight modifications of structure in some degree permanent; and this assuredly we do see.

Secondly, areas now continuous must often have existed within the recent period in isolated portions, in which many forms, more especially amongst the classes which unite for each birth and wander much, may have separately been rendered sufficiently distinct to rank as representative species. In this case, intermediate varieties between the several representative species and their common parent, must formerly have existed in each broken portion of the land, but these links will have been supplanted and exterminated during the process of natural selection, so that they will no longer exist in a living state.

Thirdly, when two or more varieties have been formed in different portions of a strictly continuous area, intermediate varieties will, it is probable, at first have been formed in the intermediate zones, but they will generally have had a short duration. For these intermediate varieties will, from reasons already assigned (namely from what we know of the actual distribution of closely allied or representative species, and likewise of acknowledged varieties), exist in the intermediate zones in lesser numbers than the varieties which they tend to connect. From this cause alone the intermediate varieties will be liable to accidental extermination; and during the process of further modification through natural selection, they will almost certainly be beaten and supplanted by the forms which they connect; for these from existing in greater numbers will, in the aggregate, present more variation, and thus be further improved through natural selection and gain further advantages.

Lastly, looking not to any one time, but to all time, if my theory be true, numberless intermediate varieties, linking most closely all the species of the same group together, must assuredly have existed; but the very process of natural selection constantly tends, as has been so often remarked, to exterminate the parent-forms and the intermediate links. Consequently evidence of their former existence could be found only amongst fossil remains, which are preserved, as we shall in a future chapter attempt to show, in an extremely imperfect and intermittent record.

On the origin and transitions of organic beings with peculiar habits and structure.—It has been asked by the opponents of such views as I hold, how, for instance, a land carnivorous animal could have been converted into one with aquatic habits; for how could the animal in its transitional state have subsisted? It would be easy to show that within the same group carnivorous animals exist having every intermediate grade between truly aquatic and strictly terrestrial habits; and as each exists by a struggle for life, it is clear that each is well adapted in its habits to its place in nature. Look at the Mustela vison of North America, which has webbed feet and which resembles an otter in its fur, short legs, and form of tail; during summer this animal dives for and preys on fish, but during the long winter it leaves the frozen waters, and preys like other pole-cats on mice and land animals. If a different case had been taken, and it had been asked how an insectivorous quadruped could possibly have been converted into a flying bat, the question would have been far more difficult, and I could have given no answer. Yet I think such difficulties have very little weight.

Here, as on other occasions, I lie under a heavy disadvantage, for out of the many striking cases which I have collected, I can give only one or two instances of transitional habits and structures in closely allied species of the same genus; and of diversified habits, either constant or occasional, in the same species. And it seems to me that nothing less than a long list of such cases is sufficient to lessen the difficulty in any particular case like that of the bat.

Look at the family of squirrels; here we have the finest gradation from animals with their tails only slightly flattened, and from others, as Sir J. Richardson has remarked, with the posterior part of their bodies rather wide and with the skin on their flanks rather full, to the so-called flying squirrels; and flying squirrels have their limbs and even the base of the tail united by a broad expanse of skin, which serves as a parachute and allows them to glide through the air to an astonishing distance from tree to tree. We cannot doubt that each structure is of use to each kind of squirrel in its own country, by enabling it to escape birds or beasts of prey, or to collect food more quickly, or, as there is reason to believe, by lessening the danger from occasional falls. But it does not follow from this fact that the structure of each squirrel is the best that it is possible to conceive under all natural conditions. Let the climate and vegetation change, let other competing rodents or new beasts of prey immigrate, or old ones become modified, and all analogy would lead us to believe that some at least of the squirrels would decrease in numbers or become exterminated, unless they also became modified and improved in structure in a corresponding manner. Therefore, I can see no difficulty, more especially under changing conditions of life, in the continued preservation of individuals with fuller and fuller flank-membranes, each modification being useful, each being propagated, until by the accumulated effects of this process of natural selection, a perfect so-called flying squirrel was produced.

Now look at the Galeopithecus or flying lemur, which formerly was falsely ranked amongst bats. It has an extremely wide flank-membrane, stretching from the corners of the jaw to the tail, and including the limbs and the elongated fingers: the flank-membrane is, also, furnished with an extensor muscle. Although no graduated links of structure, fitted for gliding through the air, now connect the Galeopithecus with the other Lemuridæ, yet I can see no difficulty in supposing that such links formerly existed, and that each had been formed by the same steps as in the case of the less perfectly gliding squirrels; and that each grade of structure had been useful to its possessor. Nor can I see any insuperable difficulty in further believing it possible that the membrane-connected fingers and fore-arm of the Galeopithecus might be greatly lengthened by natural selection; and this, as far as the organs of flight are concerned, would convert it into a bat. In bats which have the wing-membrane extended from the top of the shoulder to the tail, including the hind-legs, we perhaps see traces of an apparatus originally constructed for gliding through the air rather than for flight.

If about a dozen genera of birds had become extinct or were unknown, who would have ventured to have surmised that birds might have existed which used their wings solely as flappers, like the logger-headed duck (Micropterus of Eyton); as fins in the water and front legs on the land, like the penguin; as sails, like the ostrich; and functionally for no purpose, like the Apteryx. Yet the structure of each of these birds is good for it, under the conditions of life to which it is exposed, for each has to live by a struggle; but it is not necessarily the best possible under all possible conditions. It must not be inferred from these remarks that any of the grades of wing-structure here alluded to, which perhaps may all have resulted from disuse, indicate the natural steps by which birds have acquired their perfect power of flight; but they serve, at least, to show what diversified means of transition are possible.

Seeing that a few members of such water-breathing classes as the Crustacea and Mollusca are adapted to live on the land, and seeing that we have flying birds and mammals, flying insects of the most diversified types, and formerly had flying reptiles, it is conceivable that flying-fish, which now glide far through the air, slightly rising and turning by the aid of their fluttering fins, might have been modified into perfectly winged animals. If this had been effected, who would have ever imagined that in an early transitional state they had been inhabitants of the open ocean, and had used their incipient organs of flight exclusively, as far as we know, to escape being devoured by other fish?

When we see any structure highly perfected for any particular habit, as the wings of a bird for flight, we should bear in mind that animals displaying early transitional grades of the structure will seldom continue to exist to the present day, for they will have been supplanted by the very process of perfection through natural selection. Furthermore, we may conclude that transitional grades between structures fitted for very different habits of life will rarely have been developed at an early period in great numbers and under many subordinate forms. Thus, to return to our imaginary illustration of the flying-fish, it does not seem probable that fishes capable of true flight would have been developed under many subordinate forms, for taking prey of many kinds in many ways, on the land and in the water, until their organs of flight had come to a high stage of perfection, so as to have given them a decided advantage over other animals in the battle for life. Hence the chance of discovering species with transitional grades of structure in a fossil condition will always be less, from their having existed in lesser numbers, than in the case of species with fully developed structures.

I will now give two or three instances of diversified and of changed habits in the individuals of the same species. When either case occurs, it would be easy for natural selection to fit the animal, by some modification of its structure, for its changed habits, or exclusively for one of its several different habits. But it is difficult to tell, and immaterial for us, whether habits generally change first and structure afterwards; or whether slight modifications of structure lead to changed habits; both probably often change almost simultaneously. Of cases of changed habits it will suffice merely to allude to that of the many British insects which now feed on exotic plants, or exclusively on artificial substances. Of diversified habits innumerable instances could be given: I have often watched a tyrant flycatcher (Saurophagus sulphuratus) in South America, hovering over one spot and then proceeding to another, like a kestrel, and at other times standing stationary on the margin of water, and then dashing like a kingfisher at a fish. In our own country the larger titmouse (Parus major) may be seen climbing branches, almost like a creeper; it often, like a shrike, kills small birds by blows on the head; and I have many times seen and heard it hammering the seeds of the yew on a branch, and thus breaking them like a nuthatch. In North America the black bear was seen by Hearne swimming for hours with widely open mouth, thus catching, like a whale, insects in the water. Even in so extreme a case as this, if the supply of insects were constant, and if better adapted competitors did not already exist in the country, I can see no difficulty in a race of bears being rendered, by natural selection, more and more aquatic in their structure and habits, with larger and larger mouths, till a creature was produced as monstrous as a whale.

As we sometimes see individuals of a species following habits widely different from those both of their own species and of the other species of the same genus, we might expect, on my theory, that such individuals would occasionally have given rise to new species, having anomalous habits, and with their structure either slightly or considerably modified from that of their proper type. And such instances do occur in nature. Can a more striking instance of adaptation be given than that of a woodpecker for climbing trees and for seizing insects in the chinks of the bark? Yet in North America there are woodpeckers which feed largely on fruit, and others with elongated wings which chase insects on the wing; and on the plains of La Plata, where not a tree grows, there is a woodpecker, which in every essential part of its organisation, even in its colouring, in the harsh tone of its voice, and undulatory flight, told me plainly of its close blood-relationship to our common species; yet it is a woodpecker which never climbs a tree!

Petrels are the most aërial and oceanic of birds, yet in the quiet Sounds of Tierra del Fuego, the Puffinuria berardi, in its general habits, in its astonishing power of diving, its manner of swimming, and of flying when unwillingly it takes flight, would be mistaken by any one for an auk or grebe; nevertheless, it is essentially a petrel, but with many parts of its organisation profoundly modified. On the other hand, the acutest observer by examining the dead body of the water-ouzel would never have suspected its sub-aquatic habits; yet this anomalous member of the strictly terrestrial thrush family wholly subsists by diving,—grasping the stones with its feet and using its wings under water.

He who believes that each being has been created as we now see it, must occasionally have felt surprise when he has met with an animal having habits and structure not at all in agreement. What can be plainer than that the webbed feet of ducks and geese are formed for swimming? yet there are upland geese with webbed feet which rarely or never go near the water; and no one except Audubon has seen the frigate-bird, which has all its four toes webbed, alight on the surface of the sea. On the other hand, grebes and coots are eminently aquatic, although their toes are only bordered by membrane. What seems plainer than that the long toes of grallatores are formed for walking over swamps and floating plants, yet the water-hen is nearly as aquatic as the coot; and the landrail nearly as terrestrial as the quail or partridge. In such cases, and many others could be given, habits have changed without a corresponding change of structure. The webbed feet of the upland goose may be said to have become rudimentary in function, though not in structure. In the frigate-bird, the deeply-scooped membrane between the toes shows that structure has begun to change.

He who believes in separate and innumerable acts of creation will say, that in these cases it has pleased the Creator to cause a being of one type to take the place of one of another type; but this seems to me only restating the fact in dignified language. He who believes in the struggle for existence and in the principle of natural selection, will acknowledge that every organic being is constantly endeavouring to increase in numbers; and that if any one being vary ever so little, either in habits or structure, and thus gain an advantage over some other inhabitant of the country, it will seize on the place of that inhabitant, however different it may be from its own place. Hence it will cause him no surprise that there should be geese and frigate-birds with webbed feet, either living on the dry land or most rarely alighting on the water; that there should be long-toed corncrakes living in meadows instead of in swamps; that there should be woodpeckers where not a tree grows; that there should be diving thrushes, and petrels with the habits of auks.

Organs of extreme perfection and complication.—To suppose that the eye, with all its inimitable contrivances for adjusting the focus to different distances, for admitting different amounts of light, and for the correction of spherical and chromatic aberration, could have been formed by natural selection, seems, I freely confess, absurd in the highest possible degree. Yet reason tells me, that if numerous gradations from a perfect and complex eye to one very imperfect and simple, each grade being useful to its possessor, can be shown to exist; if further, the eye does vary ever so slightly, and the variations be inherited, which is certainly the case; and if any variation or modification in the organ be ever useful to an animal under changing conditions of life, then the difficulty of believing that a perfect and complex eye could be formed by natural selection, though insuperable by our imagination, can hardly be considered real. How a nerve comes to be sensitive to light, hardly concerns us more than how life itself first originated; but I may remark that several facts make me suspect that any sensitive nerve may be rendered sensitive to light, and likewise to those coarser vibrations of the air which produce sound.

In looking for the gradations by which an organ in any species has been perfected, we ought to look exclusively to its lineal ancestors; but this is scarcely ever possible, and we are forced in each case to look to species of the same group, that is to the collateral descendants from the same original parent-form, in order to see what gradations are possible, and for the chance of some gradations having been transmitted from the earlier stages of descent, in an unaltered or little altered condition. Amongst existing Vertebrata, we find but a small amount of gradation in the structure of the eye, and from fossil species we can learn nothing on this head. In this great class we should probably have to descend far beneath the lowest known fossiliferous stratum to discover the earlier stages, by which the eye has been perfected.

In the Articulata we can commence a series with an optic nerve merely coated with pigment, and without any other mechanism; and from this low stage, numerous gradations of structure, branching off in two fundamentally different lines, can be shown to exist, until we reach a moderately high stage of perfection. In certain crustaceans, for instance, there is a double cornea, the inner one divided into facets, within each of which there is a lens shaped swelling. In other crustaceans the transparent cones which are coated by pigment, and which properly act only by excluding lateral pencils of light, are convex at their upper ends and must act by convergence; and at their lower ends there seems to be an imperfect vitreous substance. With these facts, here far too briefly and imperfectly given, which show that there is much graduated diversity in the eyes of living crustaceans, and bearing in mind how small the number of living animals is in proportion to those which have become extinct, I can see no very great difficulty (not more than in the case of many other structures) in believing that natural selection has converted the simple apparatus of an optic nerve merely coated with pigment and invested by transparent membrane, into an optical instrument as perfect as is possessed by any member of the great Articulate class.

He who will go thus far, if he find on finishing this treatise that large bodies of facts, otherwise inexplicable, can be explained by the theory of descent, ought not to hesitate to go further, and to admit that a structure even as perfect as the eye of an eagle might be formed by natural selection, although in this case he does not know any of the transitional grades. His reason ought to conquer his imagination; though I have felt the difficulty far too keenly to be surprised at any degree of hesitation in extending the principle of natural selection to such startling lengths.

It is scarcely possible to avoid comparing the eye to a telescope. We know that this instrument has been perfected by the long-continued efforts of the highest human intellects; and we naturally infer that the eye has been formed by a somewhat analogous process. But may not this inference be presumptuous? Have we any right to assume that the Creator works by intellectual powers like those of man? If we must compare the eye to an optical instrument, we ought in imagination to take a thick layer of transparent tissue, with a nerve sensitive to light beneath, and then suppose every part of this layer to be continually changing slowly in density, so as to separate into layers of different densities and thicknesses, placed at different distances from each other, and with the surfaces of each layer slowly changing in form. Further we must suppose that there is a power always intently watching each slight accidental alteration in the transparent layers; and carefully selecting each alteration which, under varied circumstances, may in any way, or in any degree, tend to produce a distincter image. We must suppose each new state of the instrument to be multiplied by the million; and each to be preserved till a better be produced, and then the old ones to be destroyed. In living bodies, variation will cause the slight alterations, generation will multiply them almost infinitely, and natural selection will pick out with unerring skill each improvement. Let this process go on for millions on millions of years; and during each year on millions of individuals of many kinds; and may we not believe that a living optical instrument might thus be formed as superior to one of glass, as the works of the Creator are to those of man?

If it could be demonstrated that any complex organ existed, which could not possibly have been formed by numerous, successive, slight modifications, my theory would absolutely break down. But I can find out no such case. No doubt many organs exist of which we do not know the transitional grades, more especially if we look to much-isolated species, round which, according to my theory, there has been much extinction. Or again, if we look to an organ common to all the members of a large class, for in this latter case the organ must have been first formed at an extremely remote period, since which all the many members of the class have been developed; and in order to discover the early transitional grades through which the organ has passed, we should have to look to very ancient ancestral forms, long since become extinct.

We should be extremely cautious in concluding that an organ could not have been formed by transitional gradations of some kind. Numerous cases could be given amongst the lower animals of the same organ performing at the same time wholly distinct functions; thus the alimentary canal respires, digests, and excretes in the larva of the dragon-fly and in the fish Cobites. In the Hydra, the animal may be turned inside out, and the exterior surface will then digest and the stomach respire. In such cases natural selection might easily specialise, if any advantage were thus gained, a part or organ, which had performed two functions, for one function alone, and thus wholly change its nature by insensible steps. Two distinct organs sometimes perform simultaneously the same function in the same individual; to give one instance, there are fish with gills or branchiæ that breathe the air dissolved in the water, at the same time that they breathe free air in their swimbladders, this latter organ having a ductus pneumaticus for its supply, and being divided by highly vascular partitions. In these cases, one of the two organs might with ease be modified and perfected so as to perform all the work by itself, being aided during the process of modification by the other organ; and then this other organ might be modified for some other and quite distinct purpose, or be quite obliterated.

The illustration of the swimbladder in fishes is a good one, because it shows us clearly the highly important fact that an organ originally constructed for one purpose, namely flotation, may be converted into one for a wholly different purpose, namely respiration. The swimbladder has, also, been worked in as an accessory to the auditory organs of certain fish, or, for I do not know which view is now generally held, a part of the auditory apparatus has been worked in as a complement to the swimbladder. All physiologists admit that the swimbladder is homologous, or "ideally similar," in position and structure with the lungs of the higher vertebrate animals:

hence there seems to me to be no great difficulty in believing that natural selection has actually converted a swimbladder into a lung, or organ used exclusively for respiration.

I can, indeed, hardly doubt that all vertebrate animals having true lungs have descended by ordinary generation from an ancient prototype, of which we know nothing, furnished with a floating apparatus or swimbladder. We can thus, as I infer from Professor Owen's interesting description of these parts, understand the strange fact that every particle of food and drink which we swallow has to pass over the orifice of the trachea, with some risk of falling into the lungs, notwithstanding the beautiful contrivance by which the glottis is closed. In the higher Vertebrata the branchiæ have wholly disappeared—the slits on the sides of the neck and the loop-like course of the arteries still marking in the embryo their former position. But it is conceivable that the now utterly lost branchiæ might have been gradually worked in by natural selection for some quite distinct purpose: in the same manner as, on the view entertained by some naturalists that the branchiæ and dorsal scales of Annelids are homologous with the wings and wing-covers of insects, it is probable that organs which at a very ancient period served for respiration have been actually converted into organs of flight.

In considering transitions of organs, it is so important to bear in mind the probability of conversion from one function to another, that I will give one more instance. Pedunculated cirripedes have two minute folds of skin, called by me the ovigerous frena, which serve, through the means of a sticky secretion, to retain the eggs until they are hatched within the sack. These cirripedes have no branchiæ, the whole surface of the body and sack, including the small frena, serving for respiration. The Balanidæ or sessile cirripedes, on the other hand, have no ovigerous frena, the eggs lying loose at the bottom of the sack, in the well-enclosed shell; but they have large folded branchiæ. Now I think no one will dispute that the ovigerous frena in the one family are strictly homologous with the branchiæ of the other family; indeed, they graduate into each other. Therefore I do not doubt that little folds of skin, which originally served as ovigerous frena, but which, likewise, very slightly aided the act of respiration, have been gradually converted by natural selection into branchiæ, simply through an increase in their size and the obliteration of their adhesive glands. If all pedunculated cirripedes had become extinct, and they have already suffered far more extinction than have sessile cirripedes, who would ever have imagined that the branchiæ in this latter family had originally existed as organs for preventing the ova from being washed out of the sack?

Although we must be extremely cautious in concluding that any organ could not possibly have been produced by successive transitional gradations, yet, undoubtedly, grave cases of difficulty occur, some of which will be discussed in my future work.

One of the gravest is that of neuter insects, which are often very differently constructed from either the males or fertile females; but this case will be treated of in the next chapter. The electric organs of fishes offer another case of special difficulty; it is impossible to conceive by what steps these wondrous organs have been produced; but, as Owen and others have remarked, their intimate structure closely resembles that of common muscle; and as it has lately been shown that Rays have an organ closely analogous to the electric apparatus, and yet do not, as Matteuchi asserts, discharge any electricity, we must own that we are far too ignorant to argue that no transition of any kind is possible.

The electric organs offer another and even more serious difficulty; for they occur in only about a dozen fishes, of which several are widely remote in their affinities. Generally when the same organ appears in several members of the same class, especially if in members having very different habits of life, we may attribute its presence to inheritance from a common ancestor; and its absence in some of the members to its loss through disuse or natural selection. But if the electric organs had been inherited from one ancient progenitor thus provided, we might have expected that all electric fishes would have been specially related to each other. Nor does geology at all lead to the belief that formerly most fishes had electric organs, which most of their modified descendants have lost. The presence of luminous organs in a few insects, belonging to different families and orders, offers a parallel case of difficulty. Other cases could be given; for instance in plants, the very curious contrivance of a mass of pollen-grains, borne on a footstalk with a sticky gland at the end, is the same in Orchis and Asclepias,—genera almost as remote as possible amongst flowering plants. In all these cases of two very distinct species furnished with apparently the same anomalous organ, it should be observed that, although the general appearance and function of the organ may be the same, yet some fundamental difference can generally be detected. I am inclined to believe that in nearly the same way as two men have sometimes independently hit on the very same invention, so natural selection, working for the good of each being and taking advantage of analogous variations, has sometimes modified in very nearly the same manner two parts in two organic beings, which owe but little of their structure in common to inheritance from the same ancestor.

Although in many cases it is most difficult to conjecture by what transitions an organ could have arrived at its present state; yet, considering that the proportion of living and known forms to the extinct and unknown is very small, I have been astonished how rarely an organ can be named, towards which no transitional grade is known to lead. The truth of this remark is indeed shown by that old canon in natural history of "Natura non facit saltum." We meet with this admission in the writings of almost every experienced naturalist; or, as Milne Edwards has well expressed it, nature is prodigal in variety, but niggard in innovation. Why, on the theory of Creation, should this be so? Why should all the parts and organs of many independent beings, each supposed to have been separately created for its proper place in nature, be so invariably linked together by graduated steps? Why should not Nature have taken a leap from structure to structure? On the theory of natural selection, we can clearly understand why she should not; for natural selection can act only by taking advantage of slight successive variations; she can never take a leap, but must advance by the shortest and slowest steps.

Organs of little apparent importance.—As natural selection acts by life and death,—by the preservation of individuals with any favourable variation, and by the destruction of those with any unfavourable deviation of structure,—I have sometimes felt much difficulty in understanding the origin of simple parts, of which the importance does not seem sufficient to cause the preservation of successively varying individuals. I have sometimes felt as much difficulty, though of a very different kind, on this head, as in the case of an organ as perfect and complex as the eye.

In the first place, we are much too ignorant in regard to the whole economy of any one organic being, to say what slight modifications would be of importance or not. In a former chapter I have given instances of most trifling characters, such as the down on fruit and the colour of the flesh, which, from determining the attacks of insects or from being correlated with constitutional differences, might assuredly be acted on by natural selection. The tail of the giraffe looks like an artificially constructed fly-flapper; and it seems at first incredible that this could have been adapted for its present purpose by successive slight modifications, each better and better, for so trifling an object as driving away flies; yet we should pause before being too positive even in this case, for we know that the distribution and existence of cattle and other animals in South America absolutely depends on their power of resisting the attacks of insects: so that individuals which could by any means defend themselves from these small enemies, would be able to range into new pastures and thus gain a great advantage. It is not that the larger quadrupeds are actually destroyed (except in some rare cases) by the flies, but they are incessantly harassed and their strength reduced, so that they are more subject to disease, or not so well enabled in a coming dearth to search for food, or to escape from beasts of prey.

Organs now of trifling importance have probably in some cases been of high importance to an early progenitor, and, after having been slowly perfected at a former period, have been transmitted in nearly the same state, although now become of very slight use; and any actually injurious deviations in their structure will always have been checked by natural selection. Seeing how important an organ of locomotion the tail is in most aquatic animals, its general presence and use for many purposes in so many land animals, which in their lungs or modified swimbladders betray their aquatic origin, may perhaps be thus accounted for. A well-developed tail having been formed in an aquatic animal, it might subsequently come to be worked in for all sorts of purposes, as a fly-flapper, an organ of prehension, or as an aid in turning, as with the dog, though the aid must be slight, for the hare, with hardly any tail, can double quickly enough.

In the second place, we may sometimes attribute importance to characters which are really of very little importance, and which have originated from quite secondary causes, independently of natural selection. We should remember that climate, food, &c., probably have some little direct influence on the organisation; that characters reappear from the law of reversion; that correlation of growth will have had a most important influence in modifying various structures; and finally, that sexual selection will often have largely modified the external characters of animals having a will, to give one male an advantage in fighting with another or in charming the females. Moreover when a modification of structure has primarily arisen from the above or other unknown causes, it may at first have been of no advantage to the species, but may subsequently have been taken advantage of by the descendants of the species under new conditions of life and with newly acquired habits.

To give a few instances to illustrate these latter remarks. If green woodpeckers alone had existed, and we did not know that there were many black and pied kinds, I dare say that we should have thought that the green colour was a beautiful adaptation to hide this tree-frequenting bird from its enemies; and consequently that it was a character of importance and might have been acquired through natural selection; as it is, I have no doubt that the colour is due to some quite distinct cause, probably to sexual selection. A trailing bamboo in the Malay Archipelego climbs the loftiest trees by the aid of exquisitely constructed hooks clustered around the ends of the branches, and this contrivance, no doubt, is of the highest service to the plant; but as we see nearly similar hooks on many trees which are not climbers, the hooks on the bamboo may have arisen from unknown laws of growth, and have been subsequently taken advantage of by the plant undergoing further modification and becoming a climber. The naked skin on the head of a vulture is generally looked at as a direct adaptation for wallowing in putridity; and so it may be, or it may possibly be due to the direct action of putrid matter; but we should be very cautious in drawing any such inference, when we see that the skin on the head of the clean-feeding male turkey is likewise naked. The sutures in the skulls of young mammals have been advanced as a beautiful adaptation for aiding parturition, and no doubt they facilitate, or may be indispensable for this act; but as sutures occur in the skulls of young birds and reptiles, which have only to escape from a broken egg, we may infer that this structure has arisen from the laws of growth, and has been taken advantage of in the parturition of the higher animals.

We are profoundly ignorant of the causes producing slight and unimportant variations; and we are immediately made conscious of this by reflecting on the differences in the breeds of our domesticated animals in different countries,—more especially in the less civilized countries where there has been but little artificial selection. Careful observers are convinced that a damp climate affects the growth of the hair, and that with the hair the horns are correlated. Mountain breeds always differ from lowland breeds; and a mountainous country would probably affect the hind limbs from exercising them more, and possibly even the form of the pelvis; and then by the law of homologous variation, the front limbs and even the head would probably be affected. The shape, also, of the pelvis might affect by pressure the shape of the head of the young in the womb. The laborious breathing necessary in high regions would, we have some reason to believe, increase the size of the chest; and again correlation would come into play. Animals kept by savages in different countries often have to struggle for their own subsistence, and would be exposed to a certain extent to natural selection, and individuals with slightly different constitutions would succeed best under different climates; and there is reason to believe that constitution and colour are correlated. A good observer, also, states that in cattle susceptibility to the attacks of flies is correlated with colour, as is the liability to be poisoned by certain plants; so that colour would be thus subjected to the action of natural selection. But we are far too ignorant to speculate on the relative importance of the several known and unknown laws of variation; and I have here alluded to them only to show that, if we are unable to account for the characteristic differences of our domestic breeds, which nevertheless we generally admit to have arisen through ordinary generation, we ought not to lay too much stress on our ignorance of the precise cause of the slight analogous differences between species. I might have adduced for this same purpose the differences between the races of man, which are so strongly marked; I may add that some little light can apparently be thrown on the origin of these differences, chiefly through sexual selection of a particular kind, but without here entering on copious details my reasoning would appear frivolous.

The foregoing remarks lead me to say a few words on the protest lately made by some naturalists, against the utilitarian doctrine that every detail of structure has been produced for the good of its possessor. They believe that very many structures have been created for beauty in the eyes of man, or for mere variety. This doctrine, if true, would be absolutely fatal to my theory. Yet I fully admit that many structures are of no direct use to their possessors. Physical conditions probably have had some little effect on structure, quite independently of any good thus gained. Correlation of growth has no doubt played a most important part, and a useful modification of one part will often have entailed on other parts diversified changes of no direct use. So again characters which formerly were useful, or which formerly had arisen from correlation of growth, or from other unknown cause, may reappear from the law of reversion, though now of no direct use. The effects of sexual selection, when displayed in beauty to charm the females, can be called useful only in rather a forced sense. But by far the most important consideration is that the chief part of the organisation of every being is simply due to inheritance; and consequently, though each being assuredly is well fitted for its place in nature, many structures now have no direct relation to the habits of life of each species. Thus, we can hardly believe that the webbed feet of the upland goose or of the frigate-bird are of special use to these birds; we cannot believe that the same bones in the arm of the monkey, in the fore leg of the horse, in the wing of the bat, and in the flipper of the seal, are of special use to these animals. We may safely attribute these structures to inheritance. But to the progenitor of the upland goose and of the frigate-bird, webbed feet no doubt were as useful as they now are to the most aquatic of existing birds. So we may believe that the progenitor of the seal had not a flipper, but a foot with five toes fitted for walking or grasping; and we may further venture to believe that the several bones in the limbs of the monkey, horse, and bat, which have been inherited from a common progenitor, were formerly of more special use to that progenitor, or its progenitors, than they now are to these animals having such widely diversified habits. Therefore we may infer that these several bones might have been acquired through natural selection, subjected formerly, as now, to the several laws of inheritance, reversion, correlation of growth, &c. Hence every detail of structure in every living creature (making some little allowance for the direct action of physical conditions) may be viewed, either as having been of special use to some ancestral form, or as being now of special use to the descendants of this form—either directly, or indirectly through the complex laws of growth.

Natural selection cannot possibly produce any modification in any one species exclusively for the good of another species; though throughout nature one species incessantly takes advantage of, and profits by, the structure of another. But natural selection can and does often produce structures for the direct injury of other species, as we see in the fang of the adder, and in the ovipositor of the ichneumon, by which its eggs are deposited in the living bodies of other insects. If it could be proved that any part of the structure of any one species had been formed for the exclusive good of another species, it would annihilate my theory, for such could not have been produced through natural selection. Although many statements may be found in works on natural history to this effect, I cannot find even one which seems to me of any weight. It is admitted that the rattlesnake has a poison-fang for its own defence and for the destruction of its prey; but some authors suppose that at the same time this snake is furnished with a rattle for its own injury, namely, to warn its prey to escape. I would almost as soon believe that the cat curls the end of its tail when preparing to spring, in order to warn the doomed mouse. But I have not space here to enter on this and other such cases.

Natural selection will never produce in a being anything injurious to itself, for natural selection acts solely by and for the good of each. No organ will be formed, as Paley has remarked, for the purpose of causing pain or for doing an injury to its possessor. If a fair balance be struck between the good and evil caused by each part, each will be found on the whole advantageous. After the lapse of time, under changing conditions of life, if any part comes to be injurious, it will be modified; or if it be not so, the being will become extinct, as myriads have become extinct.

Natural selection tends only to make each organic being as perfect as, or slightly more perfect than, the other inhabitants of the same country with which it has to struggle for existence. And we see that this is the degree of perfection attained under nature. The endemic productions of New Zealand, for instance, are perfect one compared with another; but they are now rapidly yielding before the advancing legions of plants and animals introduced from Europe. Natural selection will not produce absolute perfection, nor do we always meet, as far as we can judge, with this high standard under nature. The correction for the aberration of light is said, on high authority, not to be perfect even in that most perfect organ, the eye. If our reason leads us to admire with enthusiasm a multitude of inimitable contrivances in nature, this same reason tells us, though we may easily err on both sides, that some other contrivances are less perfect. Can we consider the sting of the wasp or of the bee as perfect, which, when used against many attacking animals, cannot be withdrawn, owing to the backward serratures, and so inevitably causes the death of the insect by tearing out its viscera?

If we look at the sting of the bee, as having originally existed in a remote progenitor as a boring and serrated instrument, like that in so many members of the same great order, and which has been modified but not perfected for its present purpose, with the poison originally adapted to cause galls subsequently intensified, we can perhaps understand how it is that the use of the sting should so often cause the insect's own death: for if on the whole the power of stinging be useful to the community, it will fulfil all the requirements of natural selection, though it may cause the death of some few members. If we admire the truly wonderful power of scent by which the males of many insects find their females, can we admire the production for this single purpose of thousands of drones, which are utterly useless to the community for any other end, and which are ultimately slaughtered by their industrious and sterile sisters? It may be difficult, but we ought to admire the savage instinctive hatred of the queen-bee, which urges her instantly to destroy the young queens her daughters as soon as born, or to perish herself in the combat; for undoubtedly this is for the good of the community; and maternal love or maternal hatred, though the latter fortunately is most rare, is all the same to the inexorable principle of natural selection. If we admire the several ingenious contrivances, by which the flowers of the orchis and of many other plants are fertilised through insect agency, can we consider as equally perfect the elaboration by our fir-trees of dense clouds of pollen, in order that a few granules may be wafted by a chance breeze on to the ovules?

Summary of Chapter.—We have in this chapter discussed some of the difficulties and objections which may be urged against my theory. Many of them are very grave; but I think that in the discussion light has been thrown on several facts, which on the theory of independent acts of creation are utterly obscure. We have seen that species at any one period are not indefinitely variable, and are not linked together by a multitude of intermediate gradations, partly because the process of natural selection will always be very slow, and will act, at any one time, only on a very few forms; and partly because the very process of natural selection almost implies the continual supplanting and extinction of preceding and intermediate gradations. Closely allied species, now living on a continuous area, must often have been formed when the area was not continuous, and when the conditions of life did not insensibly graduate away from one part to another. When two varieties are formed in two districts of a continuous area, an intermediate variety will often be formed, fitted for an intermediate zone; but from reasons assigned, the intermediate variety will usually exist in lesser numbers than the two forms which it connects; consequently the two latter, during the course of further modification, from existing in greater numbers, will have a great advantage over the less numerous intermediate variety, and will thus generally succeed in supplanting and exterminating it.

We have seen in this chapter how cautious we should be in concluding that the most different habits of life could not graduate into each other; that a bat, for instance, could not have been formed by natural selection from an animal which at first could only glide through the air.

We have seen that a species may under new conditions of life change its habits, or have diversified habits, with some habits very unlike those of its nearest congeners. Hence we can understand, bearing in mind that each organic being is trying to live wherever it can live, how it has arisen that there are upland geese with webbed feet, ground woodpeckers, diving thrushes, and petrels with the habits of auks.

Although the belief that an organ so perfect as the eye could have been formed by natural selection, is more than enough to stagger any one; yet in the case of any organ, if we know of a long series of gradations in complexity, each good for its possessor, then, under changing conditions of life, there is no logical impossibility in the acquirement of any conceivable degree of perfection through natural selection. In the cases in which we know of no intermediate or transitional states, we should be very cautious in concluding that none could have existed, for the homologies of many organs and their intermediate states show that wonderful metamorphoses in function are at least possible. For instance, a swim-bladder has apparently been converted into an air-breathing lung. The same organ having performed simultaneously very different functions, and then having been specialised for one function; and two very distinct organs having performed at the same time the same function, the one having been perfected whilst aided by the other, must often have largely facilitated transitions.

We are far too ignorant, in almost every case, to be enabled to assert that any part or organ is so unimportant for the welfare of a species, that modifications in its structure could not have been slowly accumulated by means of natural selection. But we may confidently believe that many modifications, wholly due to the laws of growth, and at first in no way advantageous to a species, have been subsequently taken advantage of by the still further modified descendants of this species. We may, also, believe that a part formerly of high importance has often been retained (as the tail of an aquatic animal by its terrestrial descendants), though it has become of such small importance that it could not, in its present state, have been acquired by natural selection,—a power which acts solely by the preservation of profitable variations in the struggle for life.

Natural selection will produce nothing in one species for the exclusive good or injury of another; though it may well produce parts, organs, and excretions highly useful or even indispensable, or highly injurious to another species, but in all cases at the same time useful to the owner. Natural selection in each well-stocked country, must act chiefly through the competition of the inhabitants one with another, and consequently will produce perfection, or strength in the battle for life, only according to the standard of that country. Hence the inhabitants of one country, generally the smaller one, will often yield, as we see they do yield, to the inhabitants of another and generally larger country. For in the larger country there will have existed more individuals, and more diversified forms, and the competition will have been severer, and thus the standard of perfection will have been rendered higher. Natural selection will not necessarily produce absolute perfection; nor, as far as we can judge by our limited faculties, can absolute perfection be everywhere found.

On the theory of natural selection we can clearly understand the full meaning of that old canon in natural history, "Natura non facit saltum." This canon, if we look only to the present inhabitants of the world, is not strictly correct, but if we include all those of past times, it must by my theory be strictly true.

It is generally acknowledged that all organic beings have been formed on two great laws—Unity of Type, and the Conditions of Existence. By unity of type is meant that fundamental agreement in structure, which we see in organic beings of the same class, and which is quite independent of their habits of life. On my theory, unity of type is explained by unity of descent. The expression of conditions of existence, so often insisted on by the illustrious Cuvier, is fully embraced by the principle of natural selection. For natural selection acts by either now adapting the varying parts of each being to its organic and inorganic conditions of life; or by having adapted them during long-past periods of time: the adaptations being aided in some cases by use and disuse, being slightly affected by the direct action of the external conditions of life, and being in all cases subjected to the several laws of growth. Hence, in fact, the law of the Conditions of Existence is the higher law; as it includes, through the inheritance of former adaptations, that of Unity of Type.

CHAPTER VII

INSTINCT

INSTINCTS COMPARABLE WITH HABITS, BUT DIFFERENT IN THEIR ORIGIN • INSTINCTS GRADUATED • APHIDES AND ANTS • INSTINCTS VARIABLE • DOMESTIC INSTINCTS, THEIR ORIGIN • NATURAL INSTINCTS OF THE CUCKOO, OSTRICH, AND PARASITIC BEES • SLAVE-MAKING ANTS • HIVE-BEE, ITS CELL-MAKING INSTINCT • DIFFICULTIES ON THE THEORY OF THE NATURAL SELECTION OF INSTINCTS • NEUTER OR STERILE INSECTS • SUMMARY

THE SUBJECT OF INSTINCT might have been worked into the previous chapters; but I have thought that it would be more convenient to treat the subject separately, especially as so wonderful an instinct as that of the hive-bee making its cells will probably have occurred to many readers, as a difficulty sufficient to overthrow my whole theory. I must premise, that I have nothing to do with the origin of the primary mental powers, any more than I have with that of life itself. We are concerned only with the diversities of instinct and of the other mental qualities of animals within the same class.

I will not attempt any definition of instinct. It would be easy to show that several distinct mental actions are commonly embraced by this term; but every one understands what is meant, when it is said that instinct impels the cuckoo to migrate and to lay her eggs in other birds' nests. An action, which we ourselves should require experience to enable us to perform, when performed by an animal, more especially by a very young one, without any experience, and when performed by many individuals in the same way, without their knowing for what purpose it is performed, is usually said to be instinctive. But I could show that none of these characters of instinct are universal. A little dose, as Pierre Huber expresses it, of judgment or reason, often comes into play, even in animals very low in the scale of nature.

Frederick Cuvier and several of the older metaphysicians have compared instinct with habit. This comparison gives, I think, a remarkably accurate notion of the frame of mind under which an instinctive action is performed, but not of its origin. How unconsciously many habitual actions are performed, indeed not rarely in direct opposition to our conscious will! yet they may be modified by the will or reason. Habits easily become associated with other habits, and with certain periods of time and states of the body. When once acquired, they often remain constant throughout life. Several other points of resemblance between instincts and habits could be pointed out. As in repeating a well-known song, so in instincts, one action follows another by a sort of rhythm; if a person be interrupted in a song, or in repeating anything by rote, he is generally forced to go back to recover the habitual train of thought: so P. Huber found it was with a caterpillar, which makes a very complicated hammock; for if he took a caterpillar which had completed its hammock up to, say, the sixth stage of construction, and put it into a hammock completed up only to the third stage, the caterpillar simply re-performed the fourth, fifth, and sixth stages of construction. If, however, a caterpillar were taken out of a hammock made up, for instance, to the third stage, and were put into one finished up to the sixth stage, so that much of its work was already done for it, far from feeling the benefit of this, it was much embarrassed, and, in order to complete its hammock, seemed forced to start from the third stage, where it had left off, and thus tried to complete the already finished work.

If we suppose any habitual action to become inherited—and I think it can be shown that this does sometimes happen—then the resemblance between what originally was a habit and an instinct becomes so close as not to be distinguished. If Mozart, instead of playing the pianoforte at three years old with wonderfully little practice, had played a tune with no practice at all, he might truly be said to have done so instinctively. But it would be the most serious error to suppose that the greater number of instincts have been acquired by habit in one generation, and then transmitted by inheritance to succeeding generations. It can be clearly shown that the most wonderful instincts with which we are acquainted, namely, those of the hive-bee and of many ants, could not possibly have been thus acquired.

It will be universally admitted that instincts are as important as corporeal structure for the welfare of each species, under its present conditions of life. Under changed conditions of life, it is at least possible that slight modifications of instinct might be profitable to a species; and if it can be shown that instincts do vary ever so little, then I can see no difficulty in natural selection preserving and continually accumulating variations of instinct to any extent that may be profitable. It is thus, as I believe, that all the most complex and wonderful instincts have originated. As modifications of corporeal structure arise from, and are increased by, use or habit, and are diminished or lost by disuse, so I do not doubt it has been with instincts. But I believe that the effects of habit are of quite subordinate importance to the effects of the natural selection of what may be called accidental variations of instincts;—that is of variations produced by the same unknown causes which produce slight deviations of bodily structure.

No complex instinct can possibly be produced through natural selection, except by the slow and gradual accumulation of numerous, slight, yet profitable, variations. Hence, as in the case of corporeal structures, we ought to find in nature, not the actual transitional gradations by which each complex instinct has been acquired—for these could be found only in the lineal ancestors of each species—but we ought to find in the collateral lines of descent some evidence of such gradations; or we ought at least to be able to show that gradations of some kind are possible; and this we certainly can do. I have been surprised to find, making allowance for the instincts of animals having been but little observed except in Europe and North America, and for no instinct being known amongst extinct species, how very generally gradations, leading to the most complex instincts, can be discovered. The canon of "Natura non facit saltum" applies with almost equal force to instincts as to bodily organs. Changes of instinct may sometimes be facilitated by the same species having different instincts at different periods of life, or at different seasons of the year, or when placed under different circumstances, &c.; in which case either one or the other instinct might be preserved by natural selection. And such instances of diversity of instinct in the same species can be shown to occur in nature.

Again as in the case of corporeal structure, and conformably with my theory, the instinct of each species is good for itself, but has never, as far as we can judge, been produced for the exclusive good of others. One of the strongest instances of an animal apparently performing an action for the sole good of another, with which I am acquainted, is that of aphides voluntarily yielding their sweet excretion to ants: that they do so voluntarily, the following facts show. I removed all the ants from a group of about a dozen aphides on a dock-plant, and prevented their attendance during several hours. After this interval, I felt sure that the aphides would want to excrete. I watched them for some time through a lens, but not one excreted; I then tickled and stroked them with a hair in the same manner, as well as I could, as the ants do with their antennæ; but not one excreted. Afterwards I allowed an ant to visit them, and it immediately seemed, by its eager way of running about, to be well aware what a rich flock it had discovered; it then began to play with its antennæ on the abdomen first of one aphis and then of another; and each aphis, as soon as it felt the antennæ, immediately lifted up its abdomen and excreted a limpid drop of sweet juice, which was eagerly devoured by the ant. Even the quite young aphides behaved in this manner, showing that the action was instinctive, and not the result of experience. But as the excretion is extremely viscid, it is probably a convenience to the aphides to have it removed; and therefore probably the aphides do not instinctively excrete for the sole good of the ants. Although I do not believe that any animal in the world performs an action for the exclusive good of another of a distinct species, yet each species tries to take advantage of the instincts of others, as each takes advantage of the weaker bodily structure of others. So again, in some few cases, certain instincts cannot be considered as absolutely perfect; but as details on this and other such points are not indispensable, they may be here passed over.

As some degree of variation in instincts under a state of nature, and the inheritance of such variations, are indispensable for the action of natural selection, as many instances as possible ought to have been here given; but want of space prevents me. I can only assert, that instincts certainly do vary—for instance, the migratory instinct, both in extent and direction, and in its total loss. So it is with the nests of birds, which vary partly in dependence on the situations chosen, and on the nature and temperature of the country inhabited, but often from causes wholly unknown to us: Audubon has given several remarkable cases of differences in nests of the same species in the northern and southern United States. Fear of any particular enemy is certainly an instinctive quality, as may be seen in nestling birds, though it is strengthened by experience, and by the sight of fear of the same enemy in other animals. But fear of man is slowly acquired, as I have elsewhere shown, by various animals inhabiting desert islands; and we may see an instance of this, even in England, in the greater wildness of all our large birds than of our small birds; for the large birds have been most persecuted by man. We may safely attribute the greater wildness of our large birds to this cause; for in uninhabited islands large birds are not more fearful than small; and the magpie, so wary in England, is tame in Norway, as is the hooded crow in Egypt.

That the general disposition of individuals of the same species, born in a state of nature, is extremely diversified, can be shown by a multitude of facts. Several cases also, could be given, of occasional and strange habits in certain species, which might, if advantageous to the species, give rise, through natural selection, to quite new instincts. But I am well aware that these general statements, without facts given in detail, can produce but a feeble effect on the reader's mind. I can only repeat my assurance, that I do not speak without good evidence.

The possibility, or even probability, of inherited variations of instinct in a state of nature will be strengthened by briefly considering a few cases under domestication. We shall thus also be enabled to see the respective parts which habit and the selection of so-called accidental variations have played in modifying the mental qualities of our domestic animals. A number of curious and authentic instances could be given of the inheritance of all shades of disposition and tastes, and likewise of the oddest tricks, associated with certain frames of mind or periods of time. But let us look to the familiar case of the several breeds of dogs: it cannot be doubted that young pointers (I have myself seen a striking instance) will sometimes point and even back other dogs the very first time that they are taken out; retrieving is certainly in some degree inherited by retrievers; and a tendency to run round, instead of at, a flock of sheep, by shepherd-dogs. I cannot see that these actions, performed without experience by the young, and in nearly the same manner by each individual, performed with eager delight by each breed, and without the end being known,—for the young pointer can no more know that he points to aid his master, than the white butterfly knows why she lays her eggs on the leaf of the cabbage,—I cannot see that these actions differ essentially from true instincts. If we were to see one kind of wolf, when young and without any training, as soon as it scented its prey, stand motionless like a statue, and then slowly crawl forward with a peculiar gait; and another kind of wolf rushing round, instead of at, a herd of deer, and driving them to a distant point, we should assuredly call these actions instinctive. Domestic instincts, as they may be called, are certainly far less fixed or invariable than natural instincts; but they have been acted on by far less rigorous selection, and have been transmitted for an incomparably shorter period, under less fixed conditions of life.

How strongly these domestic instincts, habits, and dispositions are inherited, and how curiously they become mingled, is well shown when different breeds of dogs are crossed. Thus it is known that a cross with a bulldog has affected for many generations the courage and obstinacy of greyhounds; and a cross with a greyhound has given to a whole family of shepherd-dogs a tendency to hunt hares. These domestic instincts, when thus tested by crossing, resemble natural instincts, which in a like manner become curiously blended together, and for a long period exhibit traces of the instincts of either parent: for example, Le Roy describes a dog, whose great-grandfather was a wolf, and this dog showed a trace of its wild parentage only in one way, by not coming in a straight line to his master when called.

Domestic instincts are sometimes spoken of as actions which have become inherited solely from long-continued and compulsory habit, but this, I think, is not true. No one would ever have thought of teaching, or probably could have taught, the tumbler-pigeon to tumble,—an action which, as I have witnessed, is performed by young birds, that have never seen a pigeon tumble. We may believe that some one pigeon showed a slight tendency to this strange habit, and that the long-continued selection of the best individuals in successive generations made tumblers what they now are; and near Glasgow there are house-tumblers, as I hear from Mr. Brent, which cannot fly eighteen inches high without going head over heels. It may be doubted whether any one would have thought of training a dog to point, had not some one dog naturally shown a tendency in this line; and this is known occasionally to happen, as I once saw in a pure terrier. When the first tendency was once displayed, methodical selection and the inherited effects of compulsory training in each successive generation would soon complete the work; and unconscious selection is still at work, as each man tries to procure, without intending to improve the breed, dogs which will stand and hunt best. On the other hand, habit alone in some cases has sufficed; no animal is more difficult to tame than the young of the wild rabbit; scarcely any animal is tamer than the young of the tame rabbit; but I do not suppose that domestic rabbits have ever been selected for tameness; and I presume that we must attribute the whole of the inherited change from extreme wildness to extreme tameness, simply to habit and long-continued close confinement.

Natural instincts are lost under domestication: a remarkable instance of this is seen in those breeds of fowls which very rarely or never become "broody," that is, never wish to sit on their eggs. Familiarity alone prevents our seeing how universally and largely the minds of our domestic animals have been modified by domestication. It is scarcely possible to doubt that the love of man has become instinctive in the dog. All wolves, foxes, jackals, and species of the cat genus, when kept tame, are most eager to attack poultry, sheep, and pigs; and this tendency has been found incurable in dogs which have been brought home as puppies from countries, such as Tierra del Fuego and Australia, where the savages do not keep these domestic animals. How rarely, on the other hand, do our civilised dogs, even when quite young, require to be taught not to attack poultry, sheep, and pigs! No doubt they occasionally do make an attack, and are then beaten; and if not cured, they are destroyed; so that habit, with some degree of selection, has probably concurred in civilising by inheritance our dogs. On the other hand, young chickens have lost, wholly by habit, that fear of the dog and cat which no doubt was originally instinctive in them, in the same way as it is so plainly instinctive in young pheasants, though reared under a hen. It is not that chickens have lost all fear, but fear only of dogs and cats, for if the hen gives the danger-chuckle, they will run (more especially young turkeys) from under her, and conceal themselves in the surrounding grass or thickets; and this is evidently done for the instinctive purpose of allowing, as we see in wild ground-birds, their mother to fly away. But this instinct retained by our chickens has become useless under domestication, for the mother-hen has almost lost by disuse the power of flight.

Hence, we may conclude, that domestic instincts have been acquired and natural instincts have been lost partly by habit, and partly by man selecting and accumulating during successive generations, peculiar mental habits and actions, which at first appeared from what we must in our ignorance call an accident. In some cases compulsory habit alone has sufficed to produce such inherited mental changes; in other cases compulsory habit has done nothing, and all has been the result of selection, pursued both methodically and unconsciously; but in most cases, probably, habit and selection have acted together.

We shall, perhaps, best understand how instincts in a state of nature have become modified by selection, by considering a few cases. I will select only three, out of the several which I shall have to discuss in my future work,—namely, the instinct which leads the cuckoo to lay her eggs in other birds' nests; the slave-making instinct of certain ants; and the comb-making power of the hive-bee: these two latter instincts have generally, and most justly, been ranked by naturalists as the most wonderful of all known instincts.

It is now commonly admitted that the more immediate and final cause of the cuckoo's instinct is, that she lays her eggs, not daily, but at intervals of two or three days; so that, if she were to make her own nest and sit on her own eggs, those first laid would have to be left for some time unincubated, or there would be eggs and young birds of different ages in the same nest. If this were the case, the process of laying and hatching might be inconveniently long, more especially as she has to migrate at a very early period; and the first hatched young would probably have to be fed by the male alone. But the American cuckoo is in this predicament; for she makes her own nest and has eggs and young successively hatched, all at the same time. It has been asserted that the American cuckoo occasionally lays her eggs in other birds' nests; but I hear on the high authority of Dr. Brewer, that this is a mistake. Nevertheless, I could give several instances of various birds which have been known occasionally to lay their eggs in other birds' nests. Now let us suppose that the ancient progenitor of our European cuckoo had the habits of the American cuckoo; but that occasionally she laid an egg in another bird's nest. If the old bird profited by this occasional habit, or if the young were made more vigorous by advantage having been taken of the mistaken maternal instinct of another bird, than by their own mother's care, encumbered as she can hardly fail to be by having eggs and young of different ages at the same time; then the old birds or the fostered young would gain an advantage. And analogy would lead me to believe, that the young thus reared would be apt to follow by inheritance the occasional and aberrant habit of their mother, and in their turn would be apt to lay their eggs in other birds' nests, and thus be successful in rearing their young. By a continued process of this nature, I believe that the strange instinct of our cuckoo could be, and has been, generated. I may add that, according to Dr. Gray and to some other observers, the European cuckoo has not utterly lost all maternal love and care for her own offspring.

The occasional habit of birds laying their eggs in other birds' nests, either of the same or of a distinct species, is not very uncommon with the Gallinaceæ; and this perhaps explains the origin of a singular instinct in the allied group of ostriches. For several hen ostriches, at least in the case of the American species, unite and lay first a few eggs in one nest and then in another; and these are hatched by the males. This instinct may probably be accounted for by the fact of the hens laying a large number of eggs; but, as in the case of the cuckoo, at intervals of two or three days. This instinct, however, of the American ostrich has not as yet been perfected; for a surprising number of eggs lie strewed over the plains, so that in one day's hunting I picked up no less than twenty lost and wasted eggs.

Many bees are parasitic, and always lay their eggs in the nests of bees of other kinds. This case is more remarkable than that of the cuckoo; for these bees have not only their instincts but their structure modified in accordance with their parasitic habits; for they do not possess the pollen-collecting apparatus which would be necessary if they had to store food for their own young. Some species, likewise, of Sphegidæ (wasp-like insects) are parasitic on other species; and M. Fabre has lately shown good reason for believing that although the Tachytes nigra generally makes its own burrow and stores it with paralysed prey for its own larvæ to feed on, yet that when this insect finds a burrow already made and stored by another sphex, it takes advantage of the prize, and becomes for the occasion parasitic. In this case, as with the supposed case of the cuckoo, I can see no difficulty in natural selection making an occasional habit permanent, if of advantage to the species, and if the insect whose nest and stored food are thus feloniously appropriated, be not thus exterminated.

Slave-making instinct.—This remarkable instinct was first discovered in the Formica (Polyerges) rufescens by Pierre Huber, a better observer even than his celebrated father. This ant is absolutely dependent on its slaves; without their aid, the species would certainly become extinct in a single year. The males and fertile females do no work. The workers or sterile females, though most energetic and courageous in capturing slaves, do no other work. They are incapable of making their own nests, or of feeding their own larvæ. When the old nest is found inconvenient, and they have to migrate, it is the slaves which determine the migration, and actually carry their masters in their jaws. So utterly helpless are the masters, that when Huber shut up thirty of them without a slave, but with plenty of the food which they like best, and with their larvæ and pupæ to stimulate them to work, they did nothing; they could not even feed themselves, and many perished of hunger. Huber then introduced a single slave (F. fusca), and she instantly set to work, fed and saved the survivors; made some cells and tended the larvæ, and put all to rights. What can be more extraordinary than these well-ascertained facts? If we had not known of any other slave-making ant, it would have been hopeless to have speculated how so wonderful an instinct could have been perfected.

Formica sanguinea was likewise first discovered by P. Huber to be a slave-making ant. This species is found in the southern parts of England, and its habits have been attended to by Mr. F. Smith, of the British Museum, to whom I am much indebted for information on this and other subjects. Although fully trusting to the statements of Huber and Mr. Smith, I tried to approach the subject in a sceptical frame of mind, as any one may well be excused for doubting the truth of so extraordinary and odious an instinct as that of making slaves. Hence I will give the observations which I have myself made, in some little detail. I opened fourteen nests of F. sanguinea, and found a few slaves in all. Males and fertile females of the slave-species are found only in their own proper communities, and have never been observed in the nests of F. sanguinea. The slaves are black and not above half the size of their red masters, so that the contrast in their appearance is very great. When the nest is slightly disturbed, the slaves occasionally come out, and like their masters are much agitated and defend the nest: when the nest is much disturbed and the larvæ and pupæ are exposed, the slaves work energetically with their masters in carrying them away to a place of safety. Hence, it is clear, that the slaves feel quite at home. During the months of June and July, on three successive years, I have watched for many hours several nests in Surrey and Sussex, and never saw a slave either leave or enter a nest. As, during these months, the slaves are very few in number, I thought that they might behave differently when more numerous; but Mr. Smith informs me that he has watched the nests at various hours during May, June and August, both in Surrey and Hampshire, and has never seen the slaves, though present in large numbers in August, either leave or enter the nest. Hence he considers them as strictly household slaves. The masters, on the other hand, may be constantly seen bringing in materials for the nest, and food of all kinds. During the present year, however, in the month of July, I came across a community with an unusually large stock of slaves, and I observed a few slaves mingled with their masters leaving the nest, and marching along the same road to a tall Scotch-fir-tree, twenty-five yards distant, which they ascended together, probably in search of aphides or cocci. According to Huber, who had ample opportunities for observation, in Switzerland the slaves habitually work with their masters in making the nest, and they alone open and close the doors in the morning and evening; and, as Huber expressly states, their principal office is to search for aphides. This difference in the usual habits of the masters and slaves in the two countries, probably depends merely on the slaves being captured in greater numbers in Switzerland than in England.

One day I fortunately chanced to witness a migration from one nest to another, and it was a most interesting spectacle to behold the masters carefully carrying, as Huber has described, their slaves in their jaws. Another day my attention was struck by about a score of the slave-makers haunting the same spot, and evidently not in search of food; they approached and were vigorously repulsed by an independent community of the slave species (F. fusca); sometimes as many as three of these ants clinging to the legs of the slave-making F. sanguinea. The latter ruthlessly killed their small opponents, and carried their dead bodies as food to their nest, twenty-nine yards distant; but they were prevented from getting any pupæ to rear as slaves. I then dug up a small parcel of the pupæ of F. fusca from another nest, and put them down on a bare spot near the place of combat; they were eagerly seized, and carried off by the tyrants, who perhaps fancied that, after all, they had been victorious in their late combat.

At the same time I laid on the same place a small parcel of the pupæ of another species, F. flava, with a few of these little yellow ants still clinging to the fragments of the nest. This species is sometimes, though rarely, made into slaves, as has been described by Mr. Smith. Although so small a species, it is very courageous, and I have seen it ferociously attack other ants. In one instance I found to my surprise an independent community of F. flava under a stone beneath a nest of the slave-making F. sanguinea; and when I had accidentally disturbed both nests, the little ants attacked their big neighbours with surprising courage. Now I was curious to ascertain whether F. sanguinea could distinguish the pupæ of F. fusca, which they habitually make into slaves, from those of the little and furious F. flava, which they rarely capture, and it was evident that they did at once distinguish them: for we have seen that they eagerly and instantly seized the pupæ of F. fusca, whereas they were much terrified when they came across the pupæ, or even the earth from the nest of F. flava, and quickly ran away; but in about a quarter of an hour, shortly after all the little yellow ants had crawled away, they took heart and carried off the pupæ.

One evening I visited another community of F. sanguinea, and found a number of these ants entering their nest, carrying the dead bodies of F. fusca (showing that it was not a migration) and numerous pupæ. I traced the returning file burthened with booty, for about forty yards, to a very thick clump of heath, whence I saw the last individual of F. sanguinea emerge, carrying a pupa; but I was not able to find the desolated nest in the thick heath. The nest, however, must have been close at hand, for two or three individuals of F. fusca were rushing about in the greatest agitation, and one was perched motionless with its own pupa in its mouth on the top of a spray of heath over its ravaged home.

Such are the facts, though they did not need confirmation by me, in regard to the wonderful instinct of making slaves. Let it be observed what a contrast the instinctive habits of F. sanguinea present with those of the F. rufescens. The latter does not build its own nest, does not determine its own migrations, does not collect food for itself or its young, and cannot even feed itself: it is absolutely dependent on its numerous slaves. Formica sanguinea, on the other hand, possesses much fewer slaves, and in the early part of the summer extremely few. The masters determine when and where a new nest shall be formed, and when they migrate, the masters carry the slaves. Both in Switzerland and England the slaves seem to have the exclusive care of the larvæ, and the masters alone go on slave-making expeditions. In Switzerland the slaves and masters work together, making and bringing materials for the nest: both, but chiefly the slaves, tend, and milk as it may be called, their aphides; and thus both collect food for the community. In England the masters alone usually leave the nest to collect building materials and food for themselves, their slaves and larvæ. So that the masters in this country receive much less service from their slaves than they do in Switzerland.

By what steps the instinct of F. sanguinea originated I will not pretend to conjecture. But as ants, which are not slave-makers, will, as I have seen, carry off pupæ of other species, if scattered near their nests, it is possible that pupæ originally stored as food might become developed; and the ants thus unintentionally reared would then follow their proper instincts, and do what work they could. If their presence proved useful to the species which had seized them—if it were more advantageous to this species to capture workers than to procreate them—the habit of collecting pupæ originally for food might by natural selection be strengthened and rendered permanent for the very different purpose of raising slaves. When the instinct was once acquired, if carried out to a much less extent even than in our British F. sanguinea, which, as we have seen, is less aided by its slaves than the same species in Switzerland, I can see no difficulty in natural selection increasing and modifying the instinct—always supposing each modification to be of use to the species—until an ant was formed as abjectly dependent on its slaves as is the Formica rufescens.

Cell-making instinct of the Hive-Bee.—I will not here enter on minute details on this subject, but will merely give an outline of the conclusions at which I have arrived. He must be a dull man who can examine the exquisite structure of a comb, so beautifully adapted to its end, without enthusiastic admiration. We hear from mathematicians that bees have practically solved a recondite problem, and have made their cells of the proper shape to hold the greatest possible amount of honey, with the least possible consumption of precious wax in their construction. It has been remarked that a skilful workman, with fitting tools and measures, would find it very difficult to make cells of wax of the true form, though this is perfectly effected by a crowd of bees working in a dark hive. Grant whatever instincts you please, and it seems at first quite inconceivable how they can make all the necessary angles and planes, or even perceive when they are correctly made. But the difficulty is not nearly so great as it at first appears: all this beautiful work can be shown, I think, to follow from a few very simple instincts.

I was led to investigate this subject by Mr. Waterhouse, who has shown that the form of the cell stands in close relation to the presence of adjoining cells; and the following view may, perhaps, be considered only as a modification of his theory. Let us look to the great principle of gradation, and see whether Nature does not reveal to us her method of work. At one end of a short series we have humble-bees, which use their old cocoons to hold honey, sometimes adding to them short tubes of wax, and likewise making separate and very irregular rounded cells of wax. At the other end of the series we have the cells of the hive-bee, placed in a double layer: each cell, as is well known, is an hexagonal prism, with the basal edges of its six sides bevelled so as to join on to a pyramid, formed of three rhombs. These rhombs have certain angles, and the three which form the pyramidal base of a single cell on one side of the comb, enter into the composition of the bases of three adjoining cells on the opposite side. In the series between the extreme perfection of the cells of the hive-bee and the simplicity of those of the humble-bee, we have the cells of the Mexican Melipona domestica, carefully described and figured by Pierre Huber. The Melipona itself is intermediate in structure between the hive and humble bee, but more nearly related to the latter: it forms a nearly regular waxen comb of cylindrical cells, in which the young are hatched, and, in addition, some large cells of wax for holding honey. These latter cells are nearly spherical and of nearly equal sizes, and are aggregated into an irregular mass. But the important point to notice, is that these cells are always made at that degree of nearness to each other, that they would have intersected or broken into each other, if the spheres had been completed; but this is never permitted, the bees building perfectly flat walls of wax between the spheres which thus tend to intersect. Hence each cell consists of an outer spherical portion and of two, three, or more perfectly flat surfaces, according as the cell adjoins two, three, or more other cells. When one cell comes into contact with three other cells, which, from the spheres being nearly of the same size, is very frequently and necessarily the case, the three flat surfaces are united into a pyramid; and this pyramid, as Huber has remarked, is manifestly a gross imitation of the three-sided pyramidal basis of the cell of the hive-bee. As in the cells of the hive-bee, so here, the three plane surfaces in any one cell necessarily enter into the construction of three adjoining cells. It is obvious that the Melipona saves wax by this manner of building; for the flat walls between the adjoining cells are not double, but are of the same thickness as the outer spherical portions, and yet each flat portion forms a part of two cells.

Reflecting on this case, it occurred to me that if the Melipona had made its spheres at some given distance from each other, and had made them of equal sizes and had arranged them symmetrically in a double layer, the resulting structure would probably have been as perfect as the comb of the hive-bee. Accordingly I wrote to Professor Miller, of Cambridge, and this geometer has kindly read over the following statement, drawn up from his information, and tells me that it is strictly correct:—

If a number of equal spheres be described with their centres placed in two parallel layers; with the centre of each sphere at the distance of radius $\times \sqrt{2}$, or radius $\times$ 1.41421 (or at some lesser distance), from the centres of the six surrounding spheres in the same layer; and at the same distance from the centres of the adjoining spheres in the other and parallel layer; then, if planes of intersection between the several spheres in both layers be formed, there will result a double layer of hexagonal prisms united together by pyramidal bases formed of three rhombs; and the rhombs and the sides of the hexagonal prisms will have every angle identically the same with the best measurements which have been made of the cells of the hive-bee.

Hence we may safely conclude that if we could slightly modify the instincts already possessed by the Melipona, and in themselves not very wonderful, this bee would make a structure as wonderfully perfect as that of the hive-bee. We must suppose the Melipona to make her cells truly spherical, and of equal sizes; and this would not be very surprising, seeing that she already does so to a certain extent, and seeing what perfectly cylindrical burrows in wood many insects can make, apparently by turning round on a fixed point. We must suppose the Melipona to arrange her cells in level layers, as she already does her cylindrical cells; and we must further suppose, and this is the greatest difficulty, that she can somehow judge accurately at what distance to stand from her fellow-labourers when several are making their spheres; but she is already so far enabled to judge of distance, that she always describes her spheres so as to intersect largely; and then she unites the points of intersection by perfectly flat surfaces. We have further to suppose, but this is no difficulty, that after hexagonal prisms have been formed by the intersection of adjoining spheres in the same layer, she can prolong the hexagon to any length requisite to hold the stock of honey; in the same way as the rude humble-bee adds cylinders of wax to the circular mouths of her old cocoons. By such modifications of instincts in themselves not very wonderful,—hardly more wonderful than those which guide a bird to make its nest,—I believe that the hive-bee has acquired, through natural selection, her inimitable architectural powers.

But this theory can be tested by experiment. Following the example of Mr. Tegetmeier, I separated two combs, and put between them a long, thick, square strip of wax: the bees instantly began to excavate minute circular pits in it; and as they deepened these little pits, they made them wider and wider until they were converted into shallow basins, appearing to the eye perfectly true or parts of a sphere, and of about the diameter of a cell. It was most interesting to me to observe that wherever several bees had begun to excavate these basins near together, they had begun their work at such a distance from each other, that by the time the basins had acquired the above stated width (*i. e.* about the width of an ordinary cell), and were in depth about one sixth of the diameter of the sphere of which they formed a part, the rims of the basins intersected or broke into each other. As soon as this occurred, the bees ceased to excavate, and began to build up flat walls of wax on the lines of intersection between the basins, so that each hexagonal prism was built upon the festooned edge of a smooth basin, instead of on the straight edges of a three-sided pyramid as in the case of ordinary cells.

I then put into the hive, instead of a thick, square piece of wax, a thin and narrow, knife-edged ridge, coloured with vermilion. The bees instantly began on both sides to excavate little basins near to each other, in the same way as before; but the ridge of wax was so thin, that the bottoms of the basins, if they had been excavated to the same depth as in the former experiment, would have broken into each other from the opposite sides. The bees, however, did not suffer this to happen, and they stopped their excavations in due time; so that the basins, as soon as they had been a little deepened, came to have flat bottoms; and these flat bottoms, formed by thin little plates of the vermilion wax having been left ungnawed, were situated, as far as the eye could judge, exactly along the planes of imaginary intersection between the basins on the opposite sides of the ridge of wax. In parts, only little bits, in other parts, large portions of a rhombic plate had been left between the opposed basins, but the work, from the unnatural state of things, had not been neatly performed. The bees must have worked at very nearly the same rate on the opposite sides of the ridge of vermilion wax, as they circularly gnawed away and deepened the basins on both sides, in order to have succeeded in thus leaving flat plates between the basins, by stopping work along the intermediate planes or planes of intersection.

Considering how flexible thin wax is, I do not see that there is any difficulty in the bees, whilst at work on the two sides of a strip of wax, perceiving when they have gnawed the wax away to the proper thinness, and then stopping their work. In ordinary combs it has appeared to me that the bees do not always succeed in working at exactly the same rate from the opposite sides; for I have noticed half-completed rhombs at the base of a just-commenced cell, which were slightly concave on one side, where I suppose that the bees had excavated too quickly, and convex on the opposed side, where the bees had worked less quickly. In one well-marked instance, I put the comb back into the hive, and allowed the bees to go on working for a short time, and again examined the cell, and I found that the rhombic plate had been completed, and had become *perfectly flat:* it was absolutely impossible, from the extreme thinness of the little rhombic plate, that they could have effected this by gnawing away the convex side; and I suspect that the bees in such cases stand in the opposed cells and push and bend the ductile and warm wax (which as I have tried is easily done) into its proper intermediate plane, and thus flatten it.

From the experiment of the ridge of vermilion wax, we can clearly see that if the bees were to build for themselves a thin wall of wax, they could make their cells of the proper shape, by standing at the proper distance from each other, by excavating at the same rate, and by endeavouring to make equal spherical hollows, but never allowing the spheres to break into each other. Now bees, as may be clearly seen by examining the edge of a growing comb, do make a rough, circumferential wall or rim all round the comb; and they gnaw into this from the opposite sides, always working circularly as they deepen each cell. They do not make the whole three-sided pyramidal base of any one cell at the same time, but only the one rhombic plate which stands on the extreme growing margin, or the two plates, as the case may be; and they never complete the upper edges of the rhombic plates, until the hexagonal walls are commenced. Some of these statements differ from those made by the justly celebrated elder Huber, but I am convinced of their accuracy; and if I had space, I could show that they are conformable with my theory.

Huber's statement that the very first cell is excavated out of a little parallel-sided wall of wax, is not, as far as I have seen, strictly correct; the first commencement having always been a little hood of wax; but I will not here enter on these details. We see how important a part excavation plays in the construction of the cells; but it would be a great error to suppose that the bees cannot build up a rough wall of wax in the proper position—that is, along the plane of intersection between two adjoining spheres. I have several specimens showing clearly that they can do this. Even in the rude circumferential rim or wall of wax round a growing comb, flexures may sometimes be observed, corresponding in position to the planes of the rhombic basal plates of future cells. But the rough wall of wax has in every case to be finished off, by being largely gnawed away on both sides. The manner in which the bees build is curious; they always make the first rough wall from ten to twenty times thicker than the excessively thin finished wall of the cell, which will ultimately be left. We shall understand how they work, by supposing masons first to pile up a broad ridge of cement, and then to begin cutting it away equally on both sides near the ground, till a smooth, very thin wall is left in the middle; the masons always piling up the cut-away cement, and adding fresh cement, on the summit of the ridge. We shall thus have a thin wall steadily growing upward; but always crowned by a gigantic coping. From all the cells, both those just commenced and those completed, being thus crowned by a strong coping of wax, the bees can cluster and crawl over the comb without injuring the delicate hexagonal walls, which are only about one four-hundredth of an inch in thickness; the plates of the pyramidal basis being about twice as thick. By this singular manner of building, strength is continually given to the comb, with the utmost ultimate economy of wax.

It seems at first to add to the difficulty of understanding how the cells are made, that a multitude of bees all work together; one bee after working a short time at one cell going to another, so that, as Huber has stated, a score of individuals work even at the commencement of the first cell. I was able practically to show this fact by covering the edges of the hexagonal walls of a single cell, or the extreme margin of the circumferential rim of a growing comb, with an extremely thin layer of melted vermilion wax; and I invariably found that the colour was most delicately diffused by the bees—as delicately as a painter could have done with his brush—by atoms of the coloured wax having been taken from the spot on which it had been placed, and worked into the growing edges of the cells all round. The work of construction seems to be a sort of balance struck between many bees, all instinctively standing at the same relative distance from each other, all trying to sweep equal spheres, and then building up, or leaving ungnawed, the planes of intersection between these spheres. It was really curious to note in cases of difficulty, as when two pieces of comb met at an angle, how often the bees would entirely pull down and rebuild in different ways the same cell, sometimes recurring to a shape which they had at first rejected.

When bees have a place on which they can stand in their proper positions for working,—for instance, on a slip of wood, placed directly under the middle of a comb growing downwards so that the comb has to be built over one face of the slip—in this case the bees can lay the foundations of one wall of a new hexagon, in its strictly proper place, projecting beyond the other completed cells. It suffices that the bees should be enabled to stand at their proper relative distances from each other and from the walls of the last completed cells, and then, by striking imaginary spheres, they can build up a wall intermediate between two adjoining spheres; but, as far as I have seen, they never gnaw away and finish off the angles of a cell till a large part both of that cell and of the adjoining cells has been built. This capacity in bees of laying down under certain circumstances a rough wall in its proper place between two just-commenced cells, is important, as it bears on a fact, which seems at first quite subversive of the foregoing theory; namely, that the cells on the extreme margin of wasp-combs are sometimes strictly hexagonal; but I have not space here to enter on this subject. Nor does there seem to me any great difficulty in a single insect (as in the case of a queen-wasp) making hexagonal cells, if she work alternately on the inside and outside of two or three cells commenced at the same time, always standing at the proper relative distance from the parts of the cells just begun, sweeping spheres or cylinders, and building up intermediate planes. It is even conceivable that an insect might, by fixing on a point at which to commence a cell, and then moving outside, first to one point, and then to five other points, at the proper relative distances from the central point and from each other, strike the planes of intersection, and so make an isolated hexagon: but I am not aware that any such case has been observed; nor would any good be derived from a single hexagon being built, as in its construction more materials would be required than for a cylinder.

As natural selection acts only by the accumulation of slight modifications of structure or instinct, each profitable to the individual under its conditions of life, it may reasonably be asked, how a long and graduated succession of modified architectural instincts, all tending towards the present perfect plan of construction, could have profited the progenitors of the hive-bee? I think the answer is not difficult: it is known that bees are often hard pressed to get sufficient nectar; and I am informed by Mr. Tegetmeier that it has been experimentally found that no less than from twelve to fifteen pounds of dry sugar are consumed by a hive of bees for the secretion of each pound of wax; so that a prodigious quantity of fluid nectar must be collected and consumed by the bees in a hive for the secretion of the wax necessary for the construction of their combs. Moreover, many bees have to remain idle for many days during the process of secretion. A large store of honey is indispensable to support a large stock of bees during the winter; and the security of the hive is known mainly to depend on a large number of bees being supported. Hence the saving of wax by largely saving honey must be a most important element of success in any family of bees. Of course the success of any species of bee may be dependent on the number of its parasites or other enemies, or on quite distinct causes, and so be altogether independent of the quantity of honey which the bees could collect. But let us suppose that this latter circumstance determined, as it probably often does determine, the numbers of a humble-bee which could exist in a country; and let us further suppose that the community lived throughout the winter, and consequently required a store of honey: there can in this case be no doubt that it would be an advantage to our humble-bee, if a slight modification of her instinct led her to make her waxen cells near together, so as to intersect a little; for a wall in common even to two adjoining cells, would save some little wax. Hence it would continually be more and more advantageous to our humble-bee, if she were to make her cells more and more regular, nearer together, and aggregated into a mass, like the cells of the Melipona; for in this case a large part of the bounding surface of each cell would serve to bound other cells, and much wax would be saved. Again, from the same cause, it would be advantageous to the Melipona, if she were to make her cells closer together, and more regular in every way than at present; for then, as we have seen, the spherical surfaces would wholly disappear, and would all be replaced by plane surfaces; and the Melipona would make a comb as perfect as that of the hive-bee. Beyond this stage of perfection in architecture, natural selection could not lead; for the comb of the hive-bee, as far as we can see, is absolutely perfect in economising wax.

Thus, as I believe, the most wonderful of all known instincts, that of the hive-bee, can be explained by natural selection having taken advantage of numerous, successive, slight modifications of simpler instincts; natural selection having by slow degrees, more and more perfectly, led the bees to sweep equal spheres at a given distance from each other in a double layer, and to build up and excavate the wax along the planes of intersection. The bees, of course, no more knowing that they swept their spheres at one particular distance from each other, than they know what are the several angles of the hexagonal prisms and of the basal rhombic plates. The motive power of the process of natural selection having been economy of wax; that individual swarm which wasted least honey in the secretion of wax, having succeeded best, and having transmitted by inheritance its newly acquired economical instinct to new swarms, which in their turn will have had the best chance of succeeding in the struggle for existence.

No doubt many instincts of very difficult explanation could be opposed to the theory of natural selection,—cases, in which we cannot see how an instinct could possibly have originated; cases, in which no intermediate gradations are known to exist; cases of instinct of apparently such trifling importance, that they could hardly have been acted on by natural selection; cases of instincts almost identically the same in animals so remote in the scale of nature, that we cannot account for their similarity by inheritance from a common parent, and must therefore believe that they have been acquired by independent acts of natural selection. I will not here enter on these several cases, but will confine myself to one special difficulty, which at first appeared to me insuperable, and actually fatal to my whole theory. I allude to the neuters or sterile females in insect-communities: for these neuters often differ widely in instinct and in structure from both the males and fertile females, and yet, from being sterile, they cannot propagate their kind.

The subject well deserves to be discussed at great length, but I will here take only a single case, that of working or sterile ants. How the workers have been rendered sterile is a difficulty; but not much greater than that of any other striking modification of structure; for it can be shown that some insects and other articulate animals in a state of nature occasionally become sterile; and if such insects had been social, and it had been profitable to the community that a number should have been annually born capable of work, but incapable of procreation, I can see no very great difficulty in this being effected by natural selection. But I must pass over this preliminary difficulty. The great difficulty lies in the working ants differing widely from both the males and the fertile females in structure, as in the shape of the thorax and in being destitute of wings and sometimes of eyes, and in instinct. As far as instinct alone is concerned, the prodigious difference in this respect between the workers and the perfect females, would have been far better exemplified by the hive-bee. If a working ant or other neuter insect had been an animal in the ordinary state, I should have unhesitatingly assumed that all its characters had been slowly acquired through natural selection; namely, by an individual having been born with some slight profitable modification of structure, this being inherited by its offspring, which again varied and were again selected, and so onwards. But with the working ant we have an insect differing greatly from its parents, yet absolutely sterile; so that it could never have transmitted successively acquired modifications of structure or instinct to its progeny. It may well be asked how is it possible to reconcile this case with the theory of natural selection?

First, let it be remembered that we have innumerable instances, both in our domestic productions and in those in a state of nature, of all sorts of differences of structure which have become correlated to certain ages, and to either sex. We have differences correlated not only to one sex, but to that short period alone when the reproductive system is active, as in the nuptial plumage of many birds, and in the hooked jaws of the male salmon. We have even slight differences in the horns of different breeds of cattle in relation to an artificially imperfect state of the male sex; for oxen of certain breeds have longer horns than in other breeds, in comparison with the horns of the bulls or cows of these same breeds. Hence I can see no real difficulty in any character having become correlated with the sterile condition of certain members of insect-communities: the difficulty lies in understanding how such correlated modifications of structure could have been slowly accumulated by natural selection.

This difficulty, though appearing insuperable, is lessened, or, as I believe, disappears, when it is remembered that selection may be applied to the family, as well as to the individual, and may thus gain the desired end. Thus, a well-flavoured vegetable is cooked, and the individual is destroyed; but the horticulturist sows seeds of the same stock, and confidently expects to get nearly the same variety; breeders of cattle wish the flesh and fat to be well marbled together; the animal has been slaughtered, but the breeder goes with confidence to the same family. I have such faith in the powers of selection, that I do not doubt that a breed of cattle, always yielding oxen with extraordinarily long horns, could be slowly formed by carefully watching which individual bulls and cows, when matched, produced oxen with the longest horns; and yet no one ox could ever have propagated its kind. Thus I believe it has been with social insects: a slight modification of structure, or instinct, correlated with the sterile condition of certain members of the community, has been advantageous to the community: consequently the fertile males and females of the same community flourished, and transmitted to their fertile offspring a tendency to produce sterile members having the same modification. And I believe that this process has been repeated, until that prodigious amount of difference between the fertile and sterile females of the same species has been produced, which we see in many social insects.

But we have not as yet touched on the climax of the difficulty; namely, the fact that the neuters of several ants differ, not only from the fertile females and males, but from each other, sometimes to an almost incredible degree, and are thus divided into two or even three castes. The castes, moreover, do not generally graduate into each other, but are perfectly well defined; being as distinct from each other, as are any two species of the same genus, or rather as any two genera of the same family. Thus in Eciton, there are working and soldier neuters, with jaws and instincts extraordinarily different: in Cryptocerus, the workers of one caste alone carry a wonderful sort of shield on their heads, the use of which is quite unknown: in the Mexican Myrmecocystus, the workers of one caste never leave the nest; they are fed by the workers of another caste, and they have an enormously developed abdomen which secretes a sort of honey, supplying the place of that excreted by the aphides, or the domestic cattle as they may be called, which our European ants guard or imprison.

It will indeed be thought that I have an overweening confidence in the principle of natural selection, when I do not admit that such wonderful and well-established facts at once annihilate my theory. In the simpler case of neuter insects all of one caste or of the same kind, which have been rendered by natural selection, as I believe to be quite possible, different from the fertile males and females,—in this case, we may safely conclude from the analogy of ordinary variations, that each successive, slight, profitable modification did not probably at first appear in all the individual neuters in the same nest, but in a few alone; and that by the long-continued selection of the fertile parents which produced most neuters with the profitable modification, all the neuters ultimately came to have the desired character. On this view we ought occasionally to find neuter-insects of the same species, in the same nest, presenting gradations of structure; and this we do find, even often, considering how few neuter-insects out of Europe have been carefully examined. Mr. F. Smith has shown how surprisingly the neuters of several British ants differ from each other in size and sometimes in colour; and that the extreme forms can sometimes be perfectly linked together by individuals taken out of the same nest: I have myself compared perfect gradations of this kind. It often happens that the larger or the smaller sized workers are the most numerous; or that both large and small are numerous, with those of an intermediate size scanty in numbers. Formica flava has larger and smaller workers, with some of intermediate size; and, in this species, as Mr. F. Smith has observed, the larger workers have simple eyes (ocelli), which though small can be plainly distinguished, whereas the smaller workers have their ocelli rudimentary. Having carefully dissected several specimens of these workers, I can affirm that the eyes are far more rudimentary in the smaller workers than can be accounted for merely by their proportionally lesser size; and I fully believe, though I dare not assert so positively, that the workers of intermediate size have their ocelli in an exactly intermediate condition. So that we here have two bodies of sterile workers in the same nest, differing not only in size, but in their organs of vision, yet connected by some few members in an intermediate condition. I may digress by adding, that if the smaller workers had been the most useful to the community, and those males and females had been continually selected, which produced more and more of the smaller workers, until all the workers had come to be in this condition; we should then have had a species of ant with neuters very nearly in the same condition with those of Myrmica. For the workers of Myrmica have not even rudiments of ocelli, though the male and female ants of this genus have well-developed ocelli.

I may give one other case: so confidently did I expect to find gradations in important points of structure between the different castes of neuters in the same species, that I gladly availed myself of Mr. F. Smith's offer of numerous specimens from the same nest of the driver ant (Anomma) of West Africa. The reader will perhaps best appreciate the amount of difference in these workers, by my giving not the actual measurements, but a strictly accurate illustration: the difference was the same as if we were to see a set of workmen building a house of whom many were five feet four inches high, and many sixteen feet high; but we must suppose that the larger workmen had heads four instead of three times as big as those of the smaller men, and jaws nearly five times as big. The jaws, moreover, of the working ants of the several sizes differed wonderfully in shape, and in the form and number of the teeth. But the important fact for us is, that though the workers can be grouped into castes of different sizes, yet they graduate insensibly into each other, as does the widely-different structure of their jaws. I speak confidently on this latter point, as Mr. Lubbock made drawings for me with the camera lucida of the jaws which I had dissected from the workers of the several sizes.

With these facts before me, I believe that natural selection, by acting on the fertile parents, could form a species which should regularly produce neuters, either all of large size with one form of jaw, or all of small size with jaws having a widely different structure; or lastly, and this is our climax of difficulty, one set of workers of one size and structure, and simultaneously another set of workers of a different size and structure; —a graduated series having been first formed, as in the case of the driver ant, and then the extreme forms, from being the most useful to the community, having been produced in greater and greater numbers through the natural selection of the parents which generated them; until none with an intermediate structure were produced.

Thus, as I believe, the wonderful fact of two distinctly defined castes of sterile workers existing in the same nest, both widely different from each other and from their parents, has originated. We can see how useful their production may have been to a social community of insects, on the same principle that the division of labour is useful to civilised man. As ants work by inherited instincts and by inherited tools or weapons, and not by acquired knowledge and manufactured instruments, a perfect division of labour could be effected with them only by the workers being sterile; for had they been fertile, they would have intercrossed, and their instincts and structure would have become blended. And nature has, as I believe, effected this admirable division of labour in the communities of ants, by the means of natural selection. But I am bound to confess, that, with all my faith in this principle, I should never have anticipated that natural selection could have been efficient in so high a degree, had not the case of these neuter insects convinced me of the fact. I have, therefore, discussed this case, at some little but wholly insufficient length, in order to show the power of natural selection, and likewise because this is by far the most serious special difficulty, which my theory has encountered. The case, also, is very interesting, as it proves that with animals, as with plants, any amount of modification in structure can be effected by the accumulation of numerous, slight, and as we must call them accidental, variations, which are in any manner profitable, without exercise or habit having come into play. For no amount of exercise, or habit, or volition, in the utterly sterile members of a community could possibly have affected the structure or instincts of the fertile members, which alone leave descendants. I am surprised that no one has advanced this demonstrative case of neuter insects, against the well-known doctrine of Lamarck.

Summary.—I have endeavoured briefly in this chapter to show that the mental qualities of our domestic animals vary, and that the variations are inherited. Still more briefly I have attempted to show that instincts vary slightly in a state of nature. No one will dispute that instincts are of the highest importance to each animal. Therefore I can see no difficulty, under changing conditions of life, in natural selection accumulating slight modifications of instinct to any extent, in any useful direction. In some cases habit or use and disuse have probably come into play. I do not pretend that the facts given in this chapter strengthen in any great degree my theory; but none of the cases of difficulty, to the best of my judgment, annihilate it. On the other hand, the fact that instincts are not always absolutely perfect and are liable to mistakes;—that no instinct has been produced for the exclusive good of other animals, but that each animal takes advantage of the instincts of others;—that the canon in natural history, of "natura non facit saltum" is applicable to instincts as well as to corporeal structure, and is plainly explicable on the foregoing views, but is otherwise inexplicable,—all tend to corroborate the theory of natural selection.

This theory is, also, strengthened by some few other facts in regard to instincts; as by that common case of closely allied, but certainly distinct, species, when inhabiting distant parts of the world and living under considerably different conditions of life, yet often retaining nearly the same instincts. For instance, we can understand on the principle of inheritance, how it is that the thrush of South America lines its nest with mud, in the same peculiar manner as does our British thrush: how it is that the male wrens (Troglodytes) of North America, build "cock-nests," to roost in, like the males of our distinct Kitty-wrens,—a habit wholly unlike that of any other known bird. Finally, it may not be a logical deduction, but to my imagination it is far more satisfactory to look at such instincts as the young cuckoo ejecting its foster-brothers,—ants making slaves,—the larvæ of ichneumonidæ feeding within the live bodies of caterpillars,—not as specially endowed or created instincts, but as small consequences of one general law, leading to the advancement of all organic beings, namely, multiply, vary, let the strongest live and the weakest die.

CHAPTER VIII

Hybridism

DISTINCTION BETWEEN THE STERILITY OF FIRST CROSSES AND OF HYBRIDS • STERILITY VARIOUS IN DEGREE, NOT UNIVERSAL, AFFECTED BY CLOSE INTERBREEDING, REMOVED BY DOMESTICATION • LAWS GOVERNING THE STERILITY OF HYBRIDS • STERILITY NOT A SPECIAL ENDOWMENT, BUT INCIDENTAL ON OTHER DIFFERENCES • CAUSES OF THE STERILITY OF FIRST CROSSES AND OF HYBRIDS • PARALLELISM BETWEEN THE EFFECTS OF CHANGED CONDITIONS OF LIFE AND CROSSING • FERTILITY OF VARIETIES WHEN CROSSED AND OF THEIR MONGREL OFFSPRING NOT UNIVERSAL • HYBRIDS AND MONGRELS COMPARED INDEPENDENTLY OF THEIR FERTILITY • SUMMARY

THE VIEW GENERALLY ENTERTAINED by naturalists is that species, when intercrossed, have been specially endowed with the quality of sterility, in order to prevent the confusion of all organic forms. This view certainly seems at first probable, for species within the same country could hardly have kept distinct had they been capable of crossing freely. The importance of the fact that hybrids are very generally sterile, has, I think, been much underrated by some late writers. On the theory of natural selection the case is especially important, inasmuch as the sterility of hybrids could not possibly be of any advantage to them, and therefore could not have been acquired by the continued preservation of successive profitable degrees of sterility. I hope, however, to be able to show that sterility is not a specially acquired or endowed quality, but is incidental on other acquired differences.

In treating this subject, two classes of facts, to a large extent fundamentally different, have generally been confounded together; namely, the sterility of two species when first crossed, and the sterility of the hybrids produced from them.

Pure species have of course their organs of reproduction in a perfect condition, yet when intercrossed they produce either few or no offspring. Hybrids, on the other hand, have their reproductive organs functionally impotent, as may be clearly seen in the state of the male element in both plants and animals; though the organs themselves are perfect in structure, as far as the microscope reveals. In the first case the two sexual elements which go to form the embryo are perfect; in the second case they are either not at all developed, or are imperfectly developed. This distinction is important, when the cause of the sterility, which is common to the two cases, has to be considered. The distinction has probably been slurred over, owing to the sterility in both cases being looked on as a special endowment, beyond the province of our reasoning powers.

The fertility of varieties, that is of the forms known or believed to have descended from common parents, when intercrossed, and likewise the fertility of their mongrel offspring, is, on my theory, of equal importance with the sterility of species; for it seems to make a broad and clear distinction between varieties and species.

First, for the sterility of species when crossed and of their hybrid offspring. It is impossible to study the several memoirs and works of those two conscientious and admirable observers, Kölreuter and Gärtner, who almost devoted their lives to this subject, without being deeply impressed with the high generality of some degree of sterility. Kölreuter makes the rule universal; but then he cuts the knot, for in ten cases in which he found two forms, considered by most authors as distinct species, quite fertile together, he unhesitatingly ranks them as varieties. Gärtner, also, makes the rule equally universal; and he disputes the entire fertility of Kölreuter's ten cases. But in these and in many other cases, Gärtner is obliged carefully to count the seeds, in order to show that there is any degree of sterility. He always compares the maximum number of seeds produced by two species when crossed and by their hybrid offspring, with the average number produced by both pure parent-species in a state of nature. But a serious cause of error seems to me to be here introduced: a plant to be hybridised must be castrated, and, what is often more important, must be secluded in order to prevent pollen being brought to it by insects from other plants. Nearly all the plants experimentised on by Gärtner were potted, and apparently were kept in a chamber in his house. That these processes are often injurious to the fertility of a plant cannot be doubted; for Gärtner gives in his table about a score of cases of plants which he castrated, and artificially fertilised with their own pollen, and (excluding all cases such as the Leguminosæ, in which there is an acknowledged difficulty in the manipulation) half of these twenty plants had their fertility in some degree impaired. Moreover, as Gärtner during several years repeatedly crossed the primrose and cowslip, which we have such good reason to believe to be varieties, and only once or twice succeeded in getting fertile seed; as he found the common red and blue pimpernels (Anagallis arvensis and cœrulea), which the best botanists rank as varieties, absolutely sterile together; and as he came to the same concluson in several other analogous cases; it seems to me that we may well be permitted to doubt whether many other species are really so sterile, when intercrossed, as Gärtner believes.

It is certain, on the one hand, that the sterility of various species when crossed is so different in degree and graduates away so insensibly, and, on the other hand, that the fertility of pure species is so easily affected by various circumstances, that for all practical purposes it is most difficult to say where perfect fertility ends and sterility begins. I think no better evidence of this can be required than that the two most experienced observers who have ever lived, namely, Kölreuter and Gärtner, should have arrived at diametrically opposite conclusions in regard to the very same species. It is also most instructive to compare—but I have not space here to enter on details—the evidence advanced by our best botanists on the question whether certain doubtful forms should be ranked as species or varieties, with the evidence from fertility adduced by different hybridisers, or by the same author, from experiments made during different years. It can thus be shown that neither sterility nor fertility affords any clear distinction between species and varieties; but that the evidence from this source graduates away, and is doubtful in the same degree as is the evidence derived from other constitutional and structural differences.

In regard to the sterility of hybrids in successive generations; though Gärtner was enabled to rear some hybrids, carefully guarding them from a cross with either pure parent, for six or seven, and in one case for ten generations, yet he asserts positively that their fertility never increased, but generally greatly decreased. I do not doubt that this is usually the case, and that the fertility often suddenly decreases in the first few generations. Nevertheless I believe that in all these experiments the fertility has been diminished by an independent cause, namely, from close interbreeding. I have collected so large a body of facts, showing that close interbreeding lessens fertility, and, on the other hand, that an occasional cross with a distinct individual or variety increases fertility, that I cannot doubt the correctness of this almost universal belief amongst breeders. Hybrids are seldom raised by experimentalists in great numbers; and as the parent-species, or other allied hybrids, generally grow in the same garden, the visits of insects must be carefully prevented during the flowering season: hence hybrids will generally be fertilised during each generation by their own individual pollen; and I am convinced that this would be injurious to their fertility, already lessened by their hybrid origin. I am strengthened in this conviction by a remarkable statement repeatedly made by Gärtner, namely, that if even the less fertile hybrids be artificially fertilised with hybrid pollen of the same kind, their fertility, notwithstanding the frequent ill effects of manipulation, sometimes decidedly increases, and goes on increasing. Now, in artificial fertilisation pollen is as often taken by chance (as I know from my own experience) from the anthers of another flower, as from the anthers of the flower itself which is to be fertilised; so that a cross between two flowers, though probably on the same plant, would be thus effected. Moreover, whenever complicated experiments are in progress, so careful an observer as Gärtner would have castrated his hybrids, and this would have insured in each generation a cross with the pollen from a distinct flower, either from the same plant or from another plant of the same hybrid nature. And thus, the strange fact of the increase of fertility in the successive generations of *artificially fertilised* hybrids may, I believe, be accounted for by close interbreeding having been avoided.

Now let us turn to the results arrived at by the third most experienced hybridiser, namely, the Hon. and Rev. W. Herbert. He is as emphatic in his conclusion that some hybrids are perfectly fertile—as fertile as the pure parent-species—as are Kölreuter and Gärtner that some degree of sterility between distinct species is a universal law of nature. He experimentised on some of the very same species as did Gärtner. The difference in their results may, I think, be in part accounted for by Herbert's great horticultural skill, and by his having hothouses at his command. Of his many important statements I will here give only a single one as an example, namely, that "every ovule in a pod of Crinum capense fertilised by C. revolutum produced a plant, which (he says) I never saw to occur in a case of its natural fecundation." So that we here have perfect, or even more than commonly perfect, fertility in a first cross between two distinct species.

This case of the Crinum leads me to refer to a most singular fact, namely, that there are individual plants, as with certain species of Lobelia, and with all the species of the genus Hippeastrum, which can be far more easily fertilised by the pollen of another and distinct species, than by their own pollen. For these plants have been found to yield seed to the pollen of a distinct species, though quite sterile with their own pollen, notwithstanding that their own pollen was found to be perfectly good, for it fertilised distinct species. So that certain individual plants and all the individuals of certain species can actually be hybridised much more readily than they can be self-fertilised! For instance, a bulb of Hippeastrum aulicum produced four flowers; three were fertilised by Herbert with their own pollen, and the fourth was subsequently fertilised by the pollen of a compound hybrid descended from three other and distinct species: the result was that "the ovaries of the three first flowers soon ceased to grow, and after a few days perished entirely, whereas the pod impregnated by the pollen of the hybrid made vigorous growth and rapid progress to maturity, and bore good seed, which vegetated freely." In a letter to me, in 1839, Mr. Herbert told me that he had then tried the experiment during five years, and he continued to try it during several subsequent years, and always with the same result. This result has, also, been confirmed by other observers in the case of Hippeastrum with its sub-genera, and in the case of some other genera, as Lobelia, Passiflora and Verbascum. Although the plants in these experiments appeared perfectly healthy, and although both the ovules and pollen of the same flower were perfectly good with respect to other species, yet as they were functionally imperfect in their mutual self-action, we must infer that the plants were in an unnatural state. Nevertheless these facts show on what slight and mysterious causes the lesser or greater fertility of species when crossed, in comparison with the same species when self-fertilised, sometimes depends.

The practical experiments of horticulturists, though not made with scientific precision, deserve some notice. It is notorious in how complicated a manner the species of Pelargonium, Fuchsia, Calceolaria, Petunia, Rhododendron, &c., have been crossed, yet many of these hybrids seed freely. For instance, Herbert asserts that a hybrid from Calceolaria integrifolia and plantaginea, species most widely dissimilar in general habit, "reproduced itself as perfectly as if it had been a natural species from the mountains of Chile." I have taken some pains to ascertain the degree of fertility of some of the complex crosses of Rhododendrons, and I am assured that many of them are perfectly fertile. Mr. C. Noble, for instance, informs me that he raises stocks for grafting from a hybrid between Rhod. Ponticum and Catawbiense, and that this hybrid "seeds as freely as it is possible to imagine." Had hybrids, when fairly treated, gone on decreasing in fertility in each successive generation, as Gärtner believes to be the case, the fact would have been notorious to nurserymen. Horticulturists raise large beds of the same hybrids, and such alone are fairly treated, for by insect agency the several individuals of the same hybrid variety are allowed to freely cross with each other, and the injurious influence of close interbreeding is thus prevented. Any one may readily convince himself of the efficiency of insect-agency by examining the flowers of the more sterile kinds of hybrid rhododendrons, which produce no pollen, for he will find on their stigmas plenty of pollen brought from other flowers.

In regard to animals, much fewer experiments have been carefully tried than with plants. If our systematic arrangements can be trusted, that is if the genera of animals are as distinct from each other, as are the genera of plants, then we may infer that animals more widely separated in the scale of nature can be more easily crossed than in the case of plants; but the hybrids themselves are, I think, more sterile. I doubt whether any case of a perfectly fertile hybrid animal can be considered as thoroughly well authenticated. It should, however, be borne in mind that, owing to few animals breeding freely under confinement, few experiments have been fairly tried: for instance, the canary-bird has been crossed with nine other finches, but as not one of these nine species breeds freely in confinement, we have no right to expect that the first crosses between them and the canary, or that their hybrids, should be perfectly fertile. Again, with respect to the fertility in successive generations of the more fertile hybrid animals, I hardly know of an instance in which two families of the same hybrid have been raised at the same time from different parents, so as to avoid the ill effects of close interbreeding. On the contrary, brothers and sisters have usually been crossed in each successive generation, in opposition to the constantly repeated admonition of every breeder. And in this case, it is not at all surprising that the inherent sterility in the hybrids should have gone on increasing. If we were to act thus, and pair brothers and sisters in the case of any pure animal, which from any cause had the least tendency to sterility, the breed would assuredly be lost in a very few generations.

Although I do not know of any thoroughly well-authenticated cases of perfectly fertile hybrid animals, I have some reason to believe that the hybrids from Cervulus vaginalis and Reevesii, and from Phasianus colchicus with P. torquatus and with P. versicolor are perfectly fertile. The hybrids from the common and Chinese geese (A. cygnoides), species which are so different that they are generally ranked in distinct genera, have often bred in this country with either pure parent, and in one single instance they have bred *inter se.* This was effected by Mr. Eyton, who raised two hybrids from the same parents but from different hatches; and from these two birds he raised no less than eight hybrids (grandchildren of the pure geese) from one nest. In India, however, these cross-bred geese must be far more fertile; for I am assured by two eminently capable judges, namely Mr. Blyth and Capt. Hutton, that whole flocks of these crossed geese are kept in various parts of the country; and as they are kept for profit, where neither pure parent-species exists, they must certainly be highly fertile.

A doctrine which originated with Pallas, has been largely accepted by modern naturalists; namely, that most of our domestic animals have descended from two or more aboriginal species, since commingled by intercrossing. On this view, the aboriginal species must either at first have produced quite fertile hybrids, or the hybrids must have become in subsequent generations quite fertile under domestication. This latter alternative seems to me the most probable, and I am inclined to believe in its truth, although it rests on no direct evidence. I believe, for instance, that our dogs have descended from several wild stocks; yet, with perhaps the exception of certain indigenous domestic dogs of South America, all are quite fertile together; and analogy makes me greatly doubt, whether the several aboriginal species would at first have freely bred together and have produced quite fertile hybrids. So again there is reason to believe that our European and the humped Indian cattle are quite fertile together; but from facts communicated to me by Mr. Blyth, I think they must be considered as distinct species. On this view of the origin of many of our domestic animals, we must either give up the belief of the almost universal sterility of distinct species of animals when crossed; or we must look at sterility, not as an indelible characteristic, but as one capable of being removed by domestication.

Finally, looking to all the ascertained facts on the intercrossing of plants and animals, it may be concluded that some degree of sterility, both in first crosses and in hybrids, is an extremely general result; but that it cannot, under our present state of knowledge, be considered as absolutely universal.

Laws governing the Sterility of first Crosses and of Hybrids.—We will now consider a little more in detail the circumstances and rules governing the sterility of first crosses and of hybrids. Our chief object will be to see whether or not the rules indicate that species have specially been endowed with this quality, in order to prevent their crossing and blending together in utter confusion. The following rules and conclusions are chiefly drawn up from Gärtner's admirable work on the hybridisation of plants. I have taken much pains to ascertain how far the rules apply to animals, and considering how scanty our knowledge is in regard to hybrid animals, I have been surprised to find how generally the same rules apply to both kingdoms.

It has been already remarked, that the degree of fertility, both of first crosses and of hybrids, graduates from zero to perfect fertility. It is surprising in how many curious ways this gradation can be shown to exist; but only the barest outline of the facts can here be given. When pollen from a plant of one family is placed on the stigma of a plant of a distinct family, it exerts no more influence than so much inorganic dust. From this absolute zero of fertility, the pollen of different species of the same genus applied to the stigma of some one species, yields a perfect gradation in the number of seeds produced, up to nearly complete or even quite complete fertility; and, as we have seen, in certain abnormal cases, even to an excess of fertility, beyond that which the plant's own pollen will produce. So in hybrids themselves, there are some which never have produced, and probably never would produce, even with the pollen of either pure parent, a single fertile seed: but in some of these cases a first trace of fertility may be detected, by the pollen of one of the pure parent-species causing the flower of the hybrid to wither earlier than it otherwise would have done; and the early withering of the flower is well known to be a sign of incipient fertilisation. From this extreme degree of sterility we have self-fertilised hybrids producing a greater and greater number of seeds up to perfect fertility.

Hybrids from two species which are very difficult to cross, and which rarely produce any offspring, are generally very sterile; but the parallelism between the difficulty of making a first cross, and the sterility of the hybrids thus produced—two classes of facts which are generally confounded together—is by no means strict. There are many cases, in which two pure species can be united with unusual facility, and produce numerous hybrid-offspring, yet these hybrids are remarkably sterile. On the other hand, there are species which can be crossed very rarely, or with extreme difficulty, but the hybrids, when at last produced, are very fertile. Even within the limits of the same genus, for instance in Dianthus, these two opposite cases occur.

The fertility, both of first crosses and of hybrids, is more easily affected by unfavourable conditions, than is the fertility of pure species. But the degree of fertility is likewise innately variable; for it is not always the same when the same two species are crossed under the same circumstances, but depends in part upon the constitution of the individuals which happen to have been chosen for the experiment. So it is with hybrids, for their degree of fertility is often found to differ greatly in the several individuals raised from seed out of the same capsule and exposed to exactly the same conditions.

By the term systematic affinity is meant, the resemblance between species in structure and in constitution, more especially in the structure of parts which are of high physiological importance and which differ little in the allied species. Now the fertility of first crosses between species, and of the hybrids produced from them, is largely governed by their systematic affinity. This is clearly shown by hybrids never having been raised between species ranked by systematists in distinct families; and on the other hand, by very closely allied species generally uniting with facility. But the correspondence between systematic affinity and the facility of crossing is by no means strict. A multitude of cases could be given of very closely allied species which will not unite, or only with extreme difficulty; and on the other hand of very distinct species which unite with the utmost facility. In the same family there may be a genus, as Dianthus, in which very many species can most readily be crossed; and another genus, as Silene, in which the most persevering efforts have failed to produce between extremely close species a single hybrid. Even within the limits of the same genus, we meet with this same difference; for instance, the many species of Nicotiana have been more largely crossed than the species of almost any other genus; but Gärtner found that N. acuminata, which is not a particularly distinct species, obstinately failed to fertilise, or to be fertilised by, no less than eight other species of Nicotiana. Very many analogous facts could be given.

No one has been able to point out what kind, or what amount, of difference in any recognisable character is sufficient to prevent two species crossing. It can be shown that plants most widely different in habit and general appearance, and having strongly marked differences in every part of the flower, even in the pollen, in the fruit, and in the cotyledons, can be crossed. Annual and perennial plants, deciduous and evergreen trees, plants inhabiting different stations and fitted for extremely different climates, can often be crossed with ease.

By a reciprocal cross between two species, I mean the case, for instance, of a stallion-horse being first crossed with a female-ass, and then a male-ass with a mare: these two species may then be said to have been reciprocally crossed. There is often the widest possible difference in the facility of making reciprocal crosses. Such cases are highly important, for they prove that the capacity in any two species to cross is often completely independent of their systematic affinity, or of any recognisable difference in their whole organisation. On the other hand, these cases clearly show that the capacity for crossing is connected with constitutional differences imperceptible by us, and confined to the reproductive system. This difference in the result of reciprocal crosses between the same two species was long ago observed by Kölreuter. To give an instance: Mirabilis jalappa can easily be fertilised by the pollen of M. longiflora, and the hybrids thus produced are sufficiently fertile; but Kölreuter tried more than two hundred times, during eight following years, to fertilise reciprocally M. longiflora with the pollen of M. jalappa, and utterly failed. Several other equally striking cases could be given. Thuret has observed the same fact with certain sea-weeds or Fuci. Gärtner, moreover, found that this difference of facility in making reciprocal crosses is extremely common in a lesser degree. He has observed it even between forms so closely related (as Matthiola annua and glabra) that many botanists rank them only as varieties. It is also a remarkable fact, that hybrids raised from reciprocal crosses, though of course compounded of the very same two species, the one species having first been used as the father and then as the mother, generally differ in fertility in a small, and occasionally in a high degree.

Several other singular rules could be given from Gärtner: for instance, some species have a remarkable power of crossing with other species; other species of the same genus have a remarkable power of impressing their likeness on their hybrid offspring; but these two powers do not at all necessarily go together. There are certain hybrids which instead of having, as is usual, an intermediate character between their two parents, always closely resemble one of them; and such hybrids, though externally so like one of their pure parent-species, are with rare exceptions extremely sterile. So again amongst hybrids which are usually intermediate in structure between their parents, exceptional and abnormal individuals sometimes are born, which closely resemble one of their pure parents; and these hybrids are almost always utterly sterile, even when the other hybrids raised from seed from the same capsule have a considerable degree of fertility. These facts show how completely fertility in the hybrid is independent of its external resemblance to either pure parent.

Considering the several rules now given, which govern the fertility of first crosses and of hybrids, we see that when forms, which must be considered as good and distinct species, are united, their fertility graduates from zero to perfect fertility, or even to fertility under certain conditions in excess. That their fertility, besides being eminently susceptible to favourable and unfavourable conditions, is innately variable. That it is by no means always the same in degree in the first cross and in the hybrids produced from this cross. That the fertility of hybrids is not related to the degree in which they resemble in external appearance either parent. And lastly, that the facility of making a first cross between any two species is not always governed by their systematic affinity or degree of resemblance to each other. This latter statement is clearly proved by reciprocal crosses between the same two species, for according as the one species or the other is used as the father or the mother, there is generally some difference, and occasionally the widest possible difference, in the facility of effecting an union. The hybrids, moreover, produced from reciprocal crosses often differ in fertility.

Now do these complex and singular rules indicate that species have been endowed with sterility simply to prevent their becoming confounded in nature? I think not. For why should the sterility be so extremely different in degree, when various species are crossed, all of which we must suppose it would be equally important to keep from blending together? Why should the degree of sterility be innately variable in the individuals of the same species? Why should some species cross with facility, and yet produce very sterile hybrids; and other species cross with extreme difficulty, and yet produce fairly fertile hybrids? Why should there often be so great a difference in the result of a reciprocal cross between the same two species? Why, it may even be asked, has the production of hybrids been permitted? to grant to species the special power of producing hybrids, and then to stop their further propagation by different degrees of sterility, not strictly related to the facility of the first union between their parents, seems to be a strange arrangement.

The foregoing rules and facts, on the other hand, appear to me clearly to indicate that the sterility both of first crosses and of hybrids is simply incidental or dependent on unknown differences, chiefly in the reproductive systems, of the species which are crossed. The differences being of so peculiar and limited a nature, that, in reciprocal crosses between two species the male sexual element of the one will often freely act on the female sexual element of the other, but not in a reversed direction. It will be advisable to explain a little more fully by an example what I mean by sterility being incidental on other differences, and not a specially endowed quality. As the capacity of one plant to be grafted or budded on another is so entirely unimportant for its welfare in a state of nature, I presume that no one will suppose that this capacity is a *specially* endowed quality, but will admit that it is incidental on differences in the laws of growth of the two plants. We can sometimes see the reason why one tree will not take on another, from differences in their rate of growth, in the hardness of their wood, in the period of the flow or nature of their sap, &c.; but in a multitude of cases we can assign no reason whatever. Great diversity in the size of two plants, one being woody and the other herbaceous, one being evergreen and the other deciduous, and adaptation to widely different climates, does not always prevent the two grafting together. As in hybridisation, so with grafting, the capacity is limited by systematic affinity, for no one has been able to graft trees together belonging to quite distinct families; and, on the other hand, closely allied species, and varieties of the same species, can usually, but not invariably, be grafted with ease. But this capacity, as in hybridisation, is by no means absolutely governed by systematic affinity. Although many distinct genera within the same family have been grafted together, in other cases species of the same genus will not take on each other. The pear can be grafted far more readily on the quince, which is ranked as a distinct genus, than on the apple, which is a member of the same genus. Even different varieties of the pear take with different degrees of facility on the quince; so do different varieties of the apricot and peach on certain varieties of the plum.

As Gärtner found that there was sometimes an innate difference in different *individuals* of the same two species in crossing; so Sagaret believes this to be the case with different individuals of the same two species in being grafted together. As in reciprocal crosses, the facility of effecting an union is often very far from equal, so it sometimes is in grafting; the common gooseberry, for instance, cannot be grafted on the currant, whereas the currant will take, though with difficulty, on the gooseberry.

We have seen that the sterility of hybrids, which have their reproductive organs in an imperfect condition, is a very different case from the difficulty of uniting two pure species, which have their reproductive organs perfect;

yet these two distinct cases run to a certain extent parallel. Something analogous occurs in grafting; for Thouin found that three species of Robinia, which seeded freely on their own roots, and which could be grafted with no great difficulty on another species, when thus grafted were rendered barren. On the other hand, certain species of Sorbus, when grafted on other species, yielded twice as much fruit as when on their own roots. We are reminded by this latter fact of the extraordinary case of Hippeastrum, Lobelia, &c., which seeded much more freely when fertilised with the pollen of distinct species, than when self-fertilised with their own pollen.

We thus see, that although there is a clear and fundamental difference between the mere adhesion of grafted stocks, and the union of the male and female elements in the act of reproduction, yet that there is a rude degree of parallelism in the results of grafting and of crossing distinct species. And as we must look at the curious and complex laws governing the facility with which trees can be grafted on each other as incidental on unknown differences in their vegetative systems, so I believe that the still more complex laws governing the facility of first crosses, are incidental on unknown differences, chiefly in their reproductive systems. These differences, in both cases, follow to a certain extent, as might have been expected, systematic affinity, by which every kind of resemblance and dissimilarity between organic beings is attempted to be expressed. The facts by no means seem to me to indicate that the greater or lesser difficulty of either grafting or crossing together various species has been a special endowment; although in the case of crossing, the difficulty is as important for the endurance and stability of specific forms, as in the case of grafting it is unimportant for their welfare.

Causes of the Sterility of first Crosses and of Hybrids.—We may now look a little closer at the probable causes of the sterility of first crosses and of hybrids. These two cases are fundamentally different, for, as just remarked, in the union of two pure species the male and female sexual elements are perfect, whereas in hybrids they are imperfect. Even in first crosses, the greater or lesser difficulty in effecting a union apparently depends on several distinct causes. There must sometimes be a physical impossibility in the male element reaching the ovule, as would be the case with a plant having a pistil too long for the pollen-tubes to reach the ovarium. It has also been observed that when pollen of one species is placed on the stigma of a distantly allied species, though the pollen-tubes protrude, they do not penetrate the stigmatic surface. Again, the male element may reach the female element, but be incapable of causing an embryo to be developed, as seems to have been the case with some of

Thuret's experiments on Fuci. No explanation can be given of these facts, any more than why certain trees cannot be grafted on others. Lastly, an embryo may be developed, and then perish at an early period. This latter alternative has not been sufficiently attended to; but I believe, from observations communicated to me by Mr. Hewitt, who has had great experience in hybridising gallinaceous birds, that the early death of the embryo is a very frequent cause of sterility in first crosses. I was at first very unwilling to believe in this view; as hybrids, when once born, are generally healthy and long-lived, as we see in the case of the common mule. Hybrids, however, are differently circumstanced before and after birth: when born and living in a country where their two parents can live, they are generally placed under suitable conditions of life. But a hybrid partakes of only half of the nature and constitution of its mother, and therefore before birth, as long as it is nourished within its mother's womb or within the egg or seed produced by the mother, it may be exposed to conditions in some degree unsuitable, and consequently be liable to perish at an early period; more especially as all very young beings seem eminently sensitive to injurious or unnatural conditions of life.

In regard to the sterility of hybrids, in which the sexual elements are imperfectly developed, the case is very different. I have more than once alluded to a large body of facts, which I have collected, showing that when animals and plants are removed from their natural conditions, they are extremely liable to have their reproductive systems seriously affected. This, in fact, is the great bar to the domestication of animals. Between the sterility thus superinduced and that of hybrids, there are many points of similarity. In both cases the sterility is independent of general health, and is often accompanied by excess of size or great luxuriance. In both cases, the sterility occurs in various degrees; in both, the male element is the most liable to be affected; but sometimes the female more than the male. In both, the tendency goes to a certain extent with systematic affinity, for whole groups of animals and plants are rendered impotent by the same unnatural conditions; and whole groups of species tend to produce sterile hybrids. On the other hand, one species in a group will sometimes resist great changes of conditions with unimpaired fertility; and certain species in a group will produce unusually fertile hybrids. No one can tell, till he tries, whether any particular animal will breed under confinement or any plant seed freely under culture; nor can he tell, till he tries, whether any two species of a genus will produce more or less sterile hybrids. Lastly, when organic beings are placed during several generations under conditions not natural to them, they are extremely liable to vary, which is due, as I believe, to their reproductive systems having been specially affected, though in a lesser degree than when sterility ensues. So it is with hybrids, for hybrids in successive generations are eminently liable to vary, as every experimentalist has observed.

Thus we see that when organic beings are placed under new and unnatural conditions, and when hybrids are produced by the unnatural crossing of two species, the reproductive system, independently of the general state of health, is affected by sterility in a very similar manner. In the one case, the conditions of life have been disturbed, though often in so slight a degree as to be inappreciable by us; in the other case, or that of hybrids, the external conditions have remained the same, but the organisation has been disturbed by two different structures and constitutions having been blended into one. For it is scarcely possible that two organisations should be compounded into one, without some disturbance occurring in the development, or periodical action, or mutual relation of the different parts and organs one to another, or to the conditions of life. When hybrids are able to breed *inter se,* they transmit to their offspring from generation to generation the same compounded organisation, and hence we need not be surprised that their sterility, though in some degree variable, rarely diminishes.

It must, however, be confessed that we cannot understand, excepting on vague hypotheses, several facts with respect to the sterility of hybrids; for instance, the unequal fertility of hybrids produced from reciprocal crosses; or the increased sterility in those hybrids which occasionally and exceptionally resemble closely either pure parent. Nor do I pretend that the foregoing remarks go to the root of the matter: no explanation is offered why an organism, when placed under unnatural conditions, is rendered sterile. All that I have attempted to show, is that in two cases, in some respects allied, sterility is the common result,—in the one case from the conditions of life having been disturbed, in the other case from the organisation having been disturbed by two organisations having been compounded into one.

It may seem fanciful, but I suspect that a similar parallelism extends to an allied yet very different class of facts. It is an old and almost universal belief, founded, I think, on a considerable body of evidence, that slight changes in the conditions of life are beneficial to all living things. We see this acted on by farmers and gardeners in their frequent exchanges of seed, tubers, &c., from one soil or climate to another, and back again. During the convalescence of animals, we plainly see that great benefit is derived from almost any change in the habits of life. Again, both with plants and animals, there is abundant evidence, that a cross between very distinct individuals of the same species, that is between members of different strains or sub-breeds, gives vigour and fertility to the offspring. I believe, indeed, from the facts alluded to in our fourth chapter, that a certain amount of crossing is indispensable even with hermaphrodites; and that close interbreeding continued during several generations between the nearest relations, especially if these be kept under the same conditions of life, always induces weakness and sterility in the progeny.

Hence it seems that, on the one hand, slight changes in the conditions of life benefit all organic beings, and on the other hand, that slight crosses, that is crosses between the males and females of the same species which have varied and become slightly different, give vigour and fertility to the offspring. But we have seen that greater changes, or changes of a particular nature, often render organic beings in some degree sterile; and that greater crosses, that is crosses between males and females which have become widely or specifically different, produce hybrids which are generally sterile in some degree. I cannot persuade myself that this parallelism is an accident or an illusion. Both series of facts seem to be connected together by some common but unknown bond, which is essentially related to the principle of life.

Fertility of Varieties when crossed, and of their Mongrel offspring.—It may be urged, as a most forcible argument, that there must be some essential distinction between species and varieties, and that there must be some error in all the foregoing remarks, inasmuch as varieties, however much they may differ from each other in external appearance, cross with perfect facility, and yield perfectly fertile offspring. I fully admit that this is almost invariably the case. But if we look to varieties produced under nature, we are immediately involved in hopeless difficulties; for if two hitherto reputed varieties be found in any degree sterile together, they are at once ranked by most naturalists as species. For instance, the blue and red pimpernel, the primrose and cowslip, which are considered by many of our best botanists as varieties, are said by Gärtner not to be quite fertile when crossed, and he consequently ranks them as undoubted species. If we thus argue in a circle, the fertility of all varieties produced under nature will assuredly have to be granted.

If we turn to varieties, produced, or supposed to have been produced, under domestication, we are still involved in doubt. For when it is stated, for instance, that the German Spitz dog unites more easily than other dogs with foxes, or that certain South American indigenous domestic dogs do not readily cross with European dogs, the explanation which will occur to every one, and probably the true one, is that these dogs have descended from several aboriginally distinct species. Nevertheless the perfect fertility of so many domestic varieties, differing widely from each other in appearance, for instance of the pigeon or of the cabbage, is a remarkable fact; more especially when we reflect how many species there are, which, though resembling each other most closely, are utterly sterile when intercrossed. Several considerations, however, render the fertility of domestic varieties less remarkable than at first appears. It can, in the first place, be clearly shown that mere external dissimilarity between two species does not determine their greater or lesser degree of sterility when crossed; and we may apply the same rule to domestic varieties. In the second place, some eminent naturalists believe that a long course of domestication tends to eliminate sterility in the successive generations of hybrids, which were at first only slightly sterile; and if this be so, we surely ought not to expect to find sterility both appearing and disappearing under nearly the same conditions of life. Lastly, and this seems to me by far the most important consideration, new races of animals and plants are produced under domestication by man's methodical and unconscious power of selection, for his own use and pleasure: he neither wishes to select, nor could select, slight differences in the reproductive system, or other constitutional differences correlated with the reproductive system. He supplies his several varieties with the same food; treats them in nearly the same manner, and does not wish to alter their general habits of life. Nature acts uniformly and slowly during vast periods of time on the whole organisation, in any way which may be for each creature's own good; and thus she may, either directly, or more probably indirectly, through correlation, modify the reproductive system in the several descendants from any one species. Seeing this difference in the process of selection, as carried on by man and nature, we need not be surprised at some difference in the result.

I have as yet spoken as if the varieties of the same species were invariably fertile when intercrossed. But it seems to me impossible to resist the evidence of the existence of a certain amount of sterility in the few following cases, which I will briefly abstract. The evidence is at least as good as that from which we believe in the sterility of a multitude of species. The evidence is, also, derived from hostile witnesses, who in all other cases consider fertility and sterility as safe criterions of specific distinction. Gärtner kept during several years a dwarf kind of maize with yellow seeds, and a tall variety with red seeds, growing near each other in his garden; and although these plants have separated sexes, they never naturally crossed. He then fertilised thirteen flowers of the one with the pollen of the other; but only a single head produced any seed, and this one head produced only five grains. Manipulation in this case could not have been injurious, as the plants have separated sexes. No one, I believe, has suspected that these varieties of maize are distinct species; and it is important to notice that the hybrid plants thus raised were themselves *perfectly* fertile; so that even Gärtner did not venture to consider the two varieties as specifically distinct.

Girou de Buzareingues crossed three varieties of gourd, which like the maize has separated sexes, and he asserts that their mutual fertilisation is by so much the less easy as their differences are greater. How far these experiments may be trusted, I know not; but the forms experimentised on, are ranked by Sagaret, who mainly founds his classification by the test of infertility, as varieties.

The following case is far more remarkable, and seems at first quite incredible; but it is the result of an astonishing number of experiments made during many years on nine species of Verbascum, by so good an observer and so hostile a witness, as Gärtner: namely, that yellow and white varieties of the same species of Verbascum when intercrossed produce less seed, than do either coloured varieties when fertilised with pollen from their own coloured flowers. Moreover, he asserts that when yellow and white varieties of one species are crossed with yellow and white varieties of a *distinct* species, more seed is produced by the crosses between the same coloured flowers, than between those which are differently coloured. Yet these varieties of Verbascum present no other difference besides the mere colour of the flower; and one variety can sometimes be raised from the seed of the other.

From observations which I have made on certain varieties of hollyhock, I am inclined to suspect that they present analogous facts.

Kölreuter, whose accuracy has been confirmed by every subsequent observer, has proved the remarkable fact, that one variety of the common tobacco is more fertile, when crossed with a widely distinct species, than are the other varieties. He experimentised on five forms, which are commonly reputed to be varieties, and which he tested by the severest trial, namely, by reciprocal crosses, and he found their mongrel offspring perfectly fertile. But one of these five varieties, when used either as father or mother, and crossed with the Nicotiana glutinosa, always yielded hybrids not so sterile as those which were produced from the four other varieties when crossed with N. glutinosa. Hence the reproductive system of this one variety must have been in some manner and in some degree modified.

From these facts; from the great difficulty of ascertaining the infertility of varieties in a state of nature, for a supposed variety if infertile in any degree would generally be ranked as species; from man selecting only external characters in the production of the most distinct domestic varieties, and from not wishing or being able to produce recondite and functional differences in the reproductive system; from these several considerations and facts, I do not think that the very general fertility of varieties can be proved to be of universal occurrence, or to form a fundamental distinction between varieties and species. The general fertility of varieties does not seem to me sufficient to overthrow the view which I have taken with respect to the very general, but not invariable, sterility of first crosses and of hybrids, namely, that it is not a special endowment, but is incidental on slowly acquired modifications, more especially in the reproductive systems of the forms which are crossed.

Hybrids and Mongrels compared, independently of their fertility.—Independently of the question of fertility, the offspring of species when crossed and of varieties when crossed may be compared in several other respects. Gärtner, whose strong wish was to draw a marked line of distinction between species and varieties, could find very few and, as it seems to me, quite unimportant differences between the so-called hybrid offspring of species, and the so-called mongrel offspring of varieties. And, on the other hand, they agree most closely in very many important respects.

I shall here discuss this subject with extreme brevity. The most important distinction is, that in the first generation mongrels are more variable than hybrids; but Gärtner admits that hybrids from species which have long been cultivated are often variable in the first generation; and I have myself seen striking instances of this fact. Gärtner further admits that hybrids between very closely allied species are more variable than those from very distinct species; and this shows that the difference in the degree of variability graduates away. When mongrels and the more fertile hybrids are propagated for several generations an extreme amount of variability in their offspring is notorious; but some few cases both of hybrids and mongrels long retaining uniformity of character could be given. The variability, however, in the successive generations of mongrels is, perhaps, greater than in hybrids.

This greater variability of mongrels than of hybrids does not seem to me at all surprising. For the parents of mongrels are varieties, and mostly domestic varieties (very few experiments having been tried on natural varieties), and this implies in most cases that there has been recent variability; and therefore we might expect that such variability would often continue and be super-added to that arising from the mere act of crossing. The slight degree of variability in hybrids from the first cross or in the first generation, in contrast with their extreme variability in the succeeding generations, is a curious fact and deserves attention. For it bears on and corroborates the view which I have taken on the cause of ordinary variability; namely, that it is due to the reproductive system being eminently sensitive to any change in the conditions of life, being thus often rendered either impotent or at least incapable of its proper function of producing offspring identical with the parent-form. Now hybrids in the first generation are descended from species (excluding those long cultivated) which have not had their reproductive systems in any way affected, and they are not variable; but hybrids themselves have their reproductive systems seriously affected, and their descendants are highly variable.

But to return to our comparison of mongrels and hybrids: Gärtner states that mongrels are more liable than hybrids to revert to either parent-form; but this, if it be true, is certainly only a difference in degree. Gärtner further insists that when any two species, although most closely allied to each other, are crossed with a third species, the hybrids are widely different from each other; whereas if two very distinct varieties of one species are crossed with another species, the hybrids do not differ much. But this conclusion, as far as I can make out, is founded on a single experiment; and seems directly opposed to the results of several experiments made by Kölreuter.

These alone are the unimportant differences, which Gärtner is able to point out, between hybrid and mongrel plants. On the other hand, the resemblance in mongrels and in hybrids to their respective parents, more especially in hybrids produced from nearly related species, follows according to Gärtner the same laws. When two species are crossed, one has sometimes a prepotent power of impressing its likeness on the hybrid; and so I believe it to be with varieties of plants. With animals one variety certainly often has this prepotent power over another variety. Hybrid plants produced from a reciprocal cross, generally resemble each other closely; and so it is with mongrels from a reciprocal cross. Both hybrids and mongrels can be reduced to either pure parent-form, by repeated crosses in successive generations with either parent.

These several remarks are apparently applicable to animals; but the subject is here excessively complicated, partly owing to the existence of secondary sexual characters; but more especially owing to prepotency in transmitting likeness running more strongly in one sex than in the other, both when one species is crossed with another, and when one variety is crossed with another variety. For instance, I think those authors are right, who maintain that the ass has a prepotent power over the horse, so that both the mule and the hinny more resemble the ass than the horse; but that the prepotency runs more strongly in the male-ass than in the female, so that the mule, which is the offspring of the male-ass and mare, is more like an ass, than is the hinny, which is the offspring of the female-ass and stallion.

Much stress has been laid by some authors on the supposed fact, that mongrel animals alone are born closely like one of their parents; but it can be shown that this does sometimes occur with hybrids; yet I grant much less frequently with hybrids than with mongrels. Looking to the cases which I have collected of cross-bred animals closely resembling one parent, the resemblances seem chiefly confined to characters almost monstrous in their nature, and which have suddenly appeared—such as albinism, melanism, deficiency of tail or horns, or additional fingers and toes; and do not relate to characters which have been slowly acquired by selection. Consequently, sudden reversions to the perfect character of either parent would be more likely to occur with mongrels, which are descended from varieties often suddenly produced and semi-monstrous in character, than with hybrids, which are descended from species slowly and naturally produced. On the whole I entirely agree with Dr. Prosper Lucas, who, after arranging an enormous body of facts with respect to animals, comes to the conclusion, that the laws of resemblance of the child to its parents are the same, whether the two parents differ much or little from each other, namely in the union of individuals of the same variety, or of different varieties, or of distinct species.

Laying aside the question of fertility and sterility, in all other respects there seems to be a general and close similarity in the offspring of crossed species, and of crossed varieties. If we look at species as having been specially created, and at varieties as having been produced by secondary laws, this similarity would be an astonishing fact. But it harmonises perfectly with the view that there is no essential distinction between species and varieties.

Summary of Chapter.—First crosses between forms sufficiently distinct to be ranked as species, and their hybrids, are very generally, but not universally, sterile. The sterility is of all degrees, and is often so slight that the two most careful experimentalists who have ever lived, have come to diametrically opposite conclusions in ranking forms by this test. The sterility is innately variable in individuals of the same species, and is eminently susceptible of favourable and unfavourable conditions. The degree of sterility does not strictly follow systematic affinity, but is governed by several curious and complex laws. It is generally different, and sometimes widely different, in reciprocal crosses between the same two species. It is not always equal in degree in a first cross and in the hybrid produced from this cross.

In the same manner as in grafting trees, the capacity of one species or variety to take on another, is incidental on generally unknown differences in their vegetative systems, so in crossing, the greater or less facility of one species to unite with another, is incidental on unknown differences in their reproductive systems. There is no more reason to think that species have been specially endowed with various degrees of sterility to prevent them crossing and blending in nature, than to think that trees have been specially endowed with various and somewhat analogous degrees of difficulty in being grafted together in order to prevent them becoming inarched in our forests.

The sterility of first crosses between pure species, which have their reproductive systems perfect, seems to depend on several circumstances; in some cases largely on the early death of the embryo. The sterility of hybrids, which have their reproductive systems imperfect, and which have had this system and their whole organisation disturbed by being compounded of two distinct species, seems closely allied to that sterility which so frequently affects pure species, when their natural conditions of life have been disturbed. This view is supported by a parallelism of another kind;—namely, that the crossing of forms only slightly different is favourable to the vigour and fertility of their offspring; and that slight changes in the conditions of life are apparently favourable to the vigour and fertility of all organic beings. It is not surprising that the degree of difficulty in uniting two species, and the degree of sterility of their hybrid-offspring should generally correspond, though due to distinct causes; for both depend on the amount of difference of some kind between the species which are crossed. Nor is it surprising that the facility of effecting a first cross, the fertility of the hybrids produced, and the capacity of being grafted together—though this latter capacity evidently depends on widely different circumstances—should all run, to a certain extent, parallel with the systematic affinity of the forms which are subjected to experiment; for systematic affinity attempts to express all kinds of resemblance between all species.

First crosses between forms known to be varieties, or sufficiently alike to be considered as varieties, and their mongrel offspring, are very generally, but not quite universally, fertile. Nor is this nearly general and perfect fertility surprising, when we remember how liable we are to argue in a circle with respect to varieties in a state of nature; and when we remember that the greater number of varieties have been produced under domestication by the selection of mere external differences, and not of differences in the reproductive system. In all other respects, excluding fertility, there is a close general resemblance between hybrids and mongrels. Finally, then, the facts briefly given in this chapter do not seem to me opposed to, but even rather to support the view, that there is no fundamental distinction between species and varieties.

CHAPTER IX

On the Imperfection of the Geological Record

ON THE ABSENCE OF INTERMEDIATE VARIETIES AT THE PRESENT DAY • ON THE NATURE OF EXTINCT INTERMEDIATE VARIETIES; ON THEIR NUMBER • ON THE VAST LAPSE OF TIME, AS INFERRED FROM THE RATE OF DEPOSITION AND OF DENUDATION • ON THE POORNESS OF OUR PALAEONTOLOGICAL COLLECTIONS • ON THE INTERMITTENCE OF GEOLOGICAL FORMATIONS • ON THE ABSENCE OF INTERMEDIATE VARIETIES IN ANY ONE FORMATION • ON THE SUDDEN APPEARANCE OF GROUPS OF SPECIES • ON THEIR SUDDEN APPEARANCE IN THE LOWEST KNOWN FOSSILIFEROUS STRATA

In the sixth chapter I enumerated the chief objections which might be justly urged against the views maintained in this volume. Most of them have now been discussed. One, namely the distinctness of specific forms, and their not being blended together by innumerable transitional links, is a very obvious difficulty. I assigned reasons why such links do not commonly occur at the present day, under the circumstances apparently most favourable for their presence, namely on an extensive and continuous area with graduated physical conditions. I endeavoured to show, that the life of each species depends in a more important manner on the presence of other already defined organic forms, than on climate; and, therefore, that the really governing conditions of life do not graduate away quite insensibly like heat or moisture. I endeavoured, also, to show that intermediate varieties, from existing in lesser numbers than the forms which they connect, will generally be beaten out and exterminated during the course of further modification and improvement. The main cause, however, of innumerable intermediate links not now occurring everywhere throughout nature depends on the very process of natural selection, through which new varieties continually take the places of and exterminate their parent-forms. But just in proportion as this process of extermination has acted on an enormous scale, so must the number of intermediate varieties, which have formerly existed on the earth, be truly enormous. Why then is not every geological formation and every stratum full of such intermediate links? Geology assuredly does not reveal any such finely graduated organic chain; and this, perhaps, is the most obvious and gravest objection which can be urged against my theory. The explanation lies, as I believe, in the extreme imperfection of the geological record.

In the first place it should always be borne in mind what sort of intermediate forms must, on my theory, have formerly existed. I have found it difficult, when looking at any two species, to avoid picturing to myself, forms *directly* intermediate between them. But this is a wholly false view; we should always look for forms intermediate between each species and a common but unknown progenitor; and the progenitor will generally have differed in some respects from all its modified descendants. To give a simple illustration: the fantail and pouter pigeons have both descended from the rock-pigeon; if we possessed all the intermediate varieties which have ever existed, we should have an extremely close series between both and the rock-pigeon; but we should have no varieties directly intermediate between the fantail and pouter; none, for instance, combining a tail somewhat expanded with a crop somewhat enlarged, the characteristic features of these two breeds. These two breeds, moreover, have become so much modified, that if we had no historical or indirect evidence regarding their origin, it would not have been possible to have determined from a mere comparison of their structure with that of the rock-pigeon, whether they had descended from this species or from some other allied species, such as C. oenas.

So with natural species, if we look to forms very distinct, for instance to the horse and tapir, we have no reason to suppose that links ever existed directly intermediate between them, but between each and an unknown common parent. The common parent will have had in its whole organisation much general resemblance to the tapir and to the horse; but in some points of structure may have differed considerably from both, even perhaps more than they differ from each other. Hence in all such cases, we should be unable to recognise the parent-form of any two or more species, even if we closely compared the structure of the parent with that of its modified descendants, unless at the same time we had a nearly perfect chain of the intermediate links.

It is just possible by my theory, that one of two living forms might have descended from the other; for instance, a horse from a tapir; and in this case *direct* intermediate links will have existed between them. But such a case would imply that one form had remained for a very long period unaltered, whilst its descendants had undergone a vast amount of change; and the principle of competition between organism and organism, between child and parent, will render this a very rare event; for in all cases the new and improved forms of life will tend to supplant the old and unimproved forms.

By the theory of natural selection all living species have been connected with the parent-species of each genus, by differences not greater than we see between the varieties of the same species at the present day; and these parent-species, now generally extinct, have in their turn been similarly connected with more ancient species; and so on backwards, always converging to the common ancestor of each great class. So that the number of intermediate and transitional links, between all living and extinct species, must have been inconceivably great. But assuredly, if this theory be true, such have lived upon this earth.

On the lapse of Time.—Independently of our not finding fossil remains of such infinitely numerous connecting links, it may be objected, that time will not have sufficed for so great an amount of organic change, all changes having been effected very slowly through natural selection. It is hardly possible for me even to recall to the reader, who may not be a practical geologist, the facts leading the mind feebly to comprehend the lapse of time. He who can read Sir Charles Lyell's grand work on the Principles of Geology, which the future historian will recognise as having produced a revolution in natural science, yet does not admit how incomprehensibly vast have been the past periods of time, may at once close this volume. Not that it suffices to study the Principles of Geology, or to read special treatises by different observers on separate formations, and to mark how each author attempts to give an inadequate idea of the duration of each formation or even each stratum. A man must for years examine for himself great piles of superimposed strata, and watch the sea at work grinding down old rocks and making fresh sediment, before he can hope to comprehend anything of the lapse of time, the monuments of which we see around us.

It is good to wander along lines of sea-coast, when formed of moderately hard rocks, and mark the process of degradation. The tides in most cases reach the cliffs only for a short time twice a day, and the waves eat into them only when they are charged with sand or pebbles; for there is reason to believe that pure water can effect little or nothing in wearing away rock. At last the base of the cliff is undermined, huge fragments fall down, and these remaining fixed, have to be worn away, atom by atom, until reduced in size they can be rolled about by the waves, and then are more quickly ground into pebbles, sand, or mud. But how often do we see along the bases of retreating cliffs rounded boulders, all thickly clothed by marine productions, showing how little they are abraded and how seldom they are rolled about! Moreover, if we follow for a few miles any line of rocky cliff, which is undergoing degradation, we find that it is only here and there, along a short length or round a promontory, that the cliffs are at the present time suffering. The appearance of the surface and the vegetation show that elsewhere years have elapsed since the waters washed their base.

He who most closely studies the action of the sea on our shores, will, I believe, be most deeply impressed with the slowness with which rocky coasts are worn away. The observations on this head by Hugh Miller, and by that excellent observer Mr. Smith of Jordan Hill, are most impressive. With the mind thus impressed, let any one examine beds of conglomerate many thousand feet in thickness, which, though probably formed at a quicker rate than many other deposits, yet, from being formed of worn and rounded pebbles, each of which bears the stamp of time, are good to show how slowly the mass has been accumulated. Let him remember Lyell's profound remark, that the thickness and extent of sedimentary formations are the result and measure of the degradation which the earth's crust has elsewhere suffered. And what an amount of degradation is implied by the sedimentary deposits of many countries! Professor Ramsay has given me the maximum thickness, in most cases from actual measurement, in a few cases from estimate, of each formation in different parts of Great Britain; and this is the result:—

	Feet.
Palæozoic strata (not including igneous beds)	57,154
Secondary strata	13,190
Tertiary strata	2,240

—making altogether 72,584 feet; that is, very nearly thirteen and three-quarters British miles. Some of these formations, which are represented in England by thin beds, are thousands of feet in thickness on the Continent. Moreover, between each successive formation, we have, in the opinion of most geologists, enormously long blank periods. So that the lofty pile of sedimentary rocks in Britain, gives but an inadequate idea of the time which has elapsed during their accumulation; yet what time this must have consumed! Good observers have estimated that sediment is deposited by the great Mississippi river at the rate of only 600 feet in a hundred thousand years. This estimate may be quite erroneous; yet, considering over what wide spaces very fine sediment is transported by the currents of the sea, the process of accumulation in any one area must be extremely slow.

But the amount of denudation which the strata have in many places suffered, independently of the rate of accumulation of the degraded matter, probably offers the best evidence of the lapse of time. I remember having been much struck with the evidence of denudation, when viewing volcanic islands, which have been worn by the waves and pared all round into perpendicular cliffs of one or two thousand feet in height; for the gentle slope of the lava-streams, due to their formerly liquid state, showed at a glance how far the hard, rocky beds had once extended into the open ocean. The same story is still more plainly told by faults,—those great cracks along which the strata have been upheaved on one side, or thrown down on the other, to the height or depth of thousands of feet; for since the crust cracked, the surface of the land has been so completely planed down by the action of the sea, that no trace of these vast dislocations is externally visible.

The Craven fault, for instance, extends for upwards of 30 miles, and along this line the vertical displacement of the strata has varied from 600 to 3000 feet. Prof. Ramsay has published an account of a downthrow in Anglesea of 2300 feet; and he informs me that he fully believes there is one in Merionethshire of 12,000 feet; yet in these cases there is nothing on the surface to show such prodigious movements; the pile of rocks on the one or other side having been smoothly swept away. The consideration of these facts impresses my mind almost in the same manner as does the vain endeavour to grapple with the idea of eternity.

I am tempted to give one other case, the well-known one of the denudation of the Weald. Though it must be admitted that the denudation of the Weald has been a mere trifle, in comparison with that which has removed masses of our palæozoic strata, in parts ten thousand feet in thickness, as shown in Prof. Ramsay's masterly memoir on this subject. Yet it is an admirable lesson to stand on the North Downs and to look at the distant South Downs; for, remembering that at no great distance to the west the northern and southern escarpments meet and close, one can safely picture to oneself the great dome of rocks which must have covered up the Weald within so limited a period as since the latter part of the Chalk formation. The distance from the northern to the southern Downs is about 22 miles, and the thickness of the several formations is on an average about 1100 feet, as I am informed by Prof. Ramsay. But if, as some geologists suppose, a range of older rocks underlies the Weald, on the flanks of which the overlying sedimentary deposits might have accumulated in thinner masses than elsewhere, the above estimate would be erroneous; but this source of doubt probably would not greatly affect the estimate as applied to the western extremity of the district. If, then, we knew the rate at which the sea commonly wears away a line of cliff of any given height, we could measure the time requisite to have denuded the Weald. This, of course, cannot be done; but we may, in order to form some crude notion on the subject, assume that the sea would eat into cliffs 500 feet in height at the rate of one inch in a century. This will at first appear much too small an allowance; but it is the same as if we were to assume a cliff one yard in height to be eaten back along a whole line of coast at the rate of one yard in nearly every twenty-two years. I doubt whether any rock, even as soft as chalk, would yield at this rate excepting on the most exposed coasts; though no doubt the degradation of a lofty cliff would be more rapid from the breakage of the fallen fragments. On the other hand, I do not believe that any line of coast, ten or twenty miles in length, ever suffers degradation at the same time along its whole indented length; and we must remember that almost all strata contain harder layers or nodules, which from long resisting attrition form a breakwater at the base. Hence, under ordinary circumstances, I conclude that for a cliff 500 feet in height, a denudation of one inch per century for the whole length would be an ample allowance. At this rate, on the above data, the denudation of the Weald must have required 306,662,400 years; or say three hundred million years.

The action of fresh water on the gently inclined Wealden district, when upraised, could hardly have been great, but it would somewhat reduce the above estimate. On the other hand, during oscillations of level, which we know this area has undergone, the surface may have existed for millions of years as land, and thus have escaped the action of the sea: when deeply submerged for perhaps equally long periods, it would, likewise, have escaped the action of the coast-waves. So that in all probability a far longer period than 300 million years has elapsed since the latter part of the Secondary period.

I have made these few remarks because it is highly important for us to gain some notion, however imperfect, of the lapse of years. During each of these years, over the whole world, the land and the water has been peopled by hosts of living forms. What an infinite number of generations, which the mind cannot grasp, must have succeeded each other in the long roll of years! Now turn to our richest geological museums, and what a paltry display we behold!

On the poorness of our Palæontological collections.—That our palæontological collections are very imperfect, is admitted by every one. The remark of that admirable palæontologist, the late Edward Forbes, should not be forgotten, namely, that numbers of our fossil species are known and named from single and often broken specimens, or from a few specimens collected on some one spot. Only a small portion of the surface of the earth has been geologically explored, and no part with sufficient care, as the important discoveries made every year in Europe prove. No organism wholly soft can be preserved. Shells and bones will decay and disappear when left on the bottom of the sea, where sediment is not accumulating. I believe we are continually taking a most erroneous view, when we tacitly admit to ourselves that sediment is being deposited over nearly the whole bed of the sea, at a rate sufficiently quick to embed and preserve fossil remains. Throughout an enormously large proportion of the ocean, the bright blue tint of the water bespeaks its purity. The many cases on record of a formation conformably covered, after an enormous interval of time, by another and later formation, without the underlying bed having suffered in the interval any wear and tear, seem explicable only on the view of the bottom of the sea not rarely lying for ages in an unaltered condition. The remains which do become embedded, if in sand or gravel, will when the beds are upraised generally be dissolved by the percolation of rain-water. I suspect that but few of the very many animals which live on the beach between high and low watermark are preserved. For instance, the several species of the Chthamalinæ (a sub-family of sessile cirripedes) coat the rocks all over the world in infinite numbers: they are all strictly littoral, with the exception of a single Mediterranean species, which inhabits deep water and has been found fossil in Sicily, whereas not one other species has hitherto been found in any tertiary formation: yet it is now known that the genus Chthamalus existed during the chalk period. The molluscan genus Chiton offers a partially analogous case.

With respect to the terrestrial productions which lived during the Secondary and Palæozoic periods, it is superfluous to state that our evidence from fossil remains is fragmentary in an extreme degree. For instance, not a land shell is known belonging to either of these vast periods, with one exception discovered by Sir C. Lyell in the carboniferous strata of North America. In regard to mammiferous remains, a single glance at the historical table published in the Supplement to Lyell's Manual, will bring home the truth, how accidental and rare is their preservation, far better than pages of detail. Nor is their rarity surprising, when we remember how large a proportion of the bones of tertiary mammals have been discovered either in caves or in lacustrine deposits; and that not a cave or true lacustrine bed is known belonging to the age of our secondary or palæozoic formations.

But the imperfection in the geological record mainly results from another and more important cause than any of the foregoing; namely, from the several formations being separated from each other by wide intervals of time. When we see the formations tabulated in written works, or when we follow them in nature, it is difficult to avoid believing that they are closely consecutive. But we know, for instance, from Sir R. Murchison's great work on Russia, what wide gaps there are in that country between the superimposed formations; so it is in North America, and in many other parts of the world. The most skilful geologist, if his attention had been exclusively confined to these large territories, would never have suspected that during the periods which were blank and barren in his own country, great piles of sediment, charged with new and peculiar forms of life, had elsewhere been accumulated. And if in each separate territory, hardly any idea can be formed of the length of time which has elapsed between the consecutive formations, we may infer that this could nowhere be ascertained. The frequent and great changes in the mineralogical composition of consecutive formations, generally implying great changes in the geography of the surrounding lands, whence the sediment has been derived, accords with the belief of vast intervals of time having elapsed between each formation.

But we can, I think, see why the geological formations of each region are almost invariably intermittent; that is, have not followed each other in close sequence. Scarcely any fact struck me more when examining many hundred miles of the South American coasts, which have been upraised several hundred feet within the recent period, than the absence of any recent deposits sufficiently extensive to last for even a short geological period. Along the whole west coast, which is inhabited by a peculiar marine fauna, tertiary beds are so scantily developed, that no record of several successive and peculiar marine faunas will probably be preserved to a distant age. A little reflection will explain why along the rising coast of the western side of South America, no extensive formations with recent or tertiary remains can anywhere be found, though the supply of sediment must for ages have been great, from the enormous degradation of the coast-rocks and from muddy streams entering the sea. The explanation, no doubt, is, that the littoral and sub-littoral deposits are continually worn away, as soon as they are brought up by the slow and gradual rising of the land within the grinding action of the coast-waves.

We may, I think, safely conclude that sediment must be accumulated in extremely thick, solid, or extensive masses, in order to withstand the incessant action of the waves, when first upraised and during subsequent oscillations of level. Such thick and extensive accumulations of sediment may be formed in two ways; either, in profound depths of the sea, in which case, judging from the researches of E. Forbes, we may conclude that the bottom will be inhabited by extremely few animals, and the mass when upraised will give a most imperfect record of the forms of life which then existed; or, sediment may be accumulated to any thickness and extent over a shallow bottom, if it continue slowly to subside. In this latter case, as long as the rate of subsidence and supply of sediment nearly balance each other, the sea will remain shallow and favourable for life, and thus a fossiliferous formation thick enough, when upraised, to resist any amount of degradation, may be formed.

I am convinced that all our ancient formations, which are rich in fossils, have thus been formed during subsidence. Since publishing my views on this subject in 1845, I have watched the progress of Geology, and have been surprised to note how author after author, in treating of this or that great formation, has come to the conclusion that it was accumulated during subsidence. I may add, that the only ancient tertiary formation on the west coast of South America, which has been bulky enough to resist such degradation as it has as yet suffered, but which will hardly last to a distant geological age, was certainly deposited during a downward oscillation of level, and thus gained considerable thickness.

All geological facts tell us plainly that each area has undergone numerous slow oscillations of level, and apparently these oscillations have affected wide spaces. Consequently formations rich in fossils and sufficiently thick and extensive to resist subsequent degradation, may have been formed over wide spaces during periods of subsidence, but only where the supply of sediment was sufficient to keep the sea shallow and to embed and preserve the remains before they had time to decay. On the other hand, as long as the bed of the sea remained stationary, *thick* deposits could not have been accumulated in the shallow parts, which are the most favourable to life. Still less could this have happened during the alternate periods of elevation; or, to speak more accurately, the beds which were then accumulated will have been destroyed by being upraised and brought within the limits of the coast-action.

Thus the geological record will almost necessarily be rendered intermittent. I feel much confidence in the truth of these views, for they are in strict accordance with the general principles inculcated by Sir C. Lyell; and E. Forbes independently arrived at a similar conclusion.

One remark is here worth a passing notice. During periods of elevation the area of the land and of the adjoining shoal parts of the sea will be increased, and new stations will often be formed;—all circumstances most favourable, as previously explained, for the formation of new varieties and species; but during such periods there will generally be a blank in the geological record. On the other hand, during subsidence, the inhabited area and number of inhabitants will decrease (excepting the productions on the shores of a continent when first broken up into an archipelago), and consequently during subsidence, though there will be much extinction, fewer new varieties or species will be formed; and it is during these very periods of subsidence, that our great deposits rich in fossils have been accumulated. Nature may almost be said to have guarded against the frequent discovery of her transitional or linking forms.

From the foregoing considerations it cannot be doubted that the geological record, viewed as a whole, is extremely imperfect; but if we confine our attention to any one formation, it becomes more difficult to understand, why we do not therein find closely graduated varieties between the allied species which lived at its commencement and at its close. Some cases are on record of the same species presenting distinct varieties in the upper and lower parts of the same formation, but, as they are rare, they may be here passed over. Although each formation has indisputably required a vast number of years for its deposition, I can see several reasons why each should not include a graduated series of links between the species which then lived; but I can by no means pretend to assign due proportional weight to the following considerations.

Although each formation may mark a very long lapse of years, each perhaps is short compared with the period requisite to change one species into another. I am aware that two palæontologists, whose opinions are worthy of much deference, namely Bronn and Woodward, have concluded that the average duration of each formation is twice or thrice as long as the average duration of specific forms. But insuperable difficulties, as it seems to me, prevent us coming to any just conclusion on this head. When we see a species first appearing in the middle of any formation, it would be rash in the extreme to infer that it had not elsewhere previously existed. So again when we find a species disappearing before the uppermost layers have been deposited, it would be equally rash to suppose that it then became wholly extinct. We forget how small the area of Europe is compared with the rest of the world; nor have the several stages of the same formation throughout Europe been correlated with perfect accuracy.

With marine animals of all kinds, we may safely infer a large amount of migration during climatal and other changes; and when we see a species first appearing in any formation, the probability is that it only then first immigrated into that area. It is well known, for instance, that several species appeared somewhat earlier in the palæozoic beds of North America than in those of Europe; time having apparently been required for their migration from the American to the European seas. In examining the latest deposits of various quarters of the world, it has everywhere been noted, that some few still existing species are common in the deposit, but have become extinct in the immediately surrounding sea; or, conversely, that some are now abundant in the neighbouring sea, but are rare or absent in this particular deposit. It is an excellent lesson to reflect on the ascertained amount of migration of the inhabitants of Europe during the Glacial period, which forms only a part of one whole geological period; and likewise to reflect on the great changes of level, on the inordinately great change of climate, on the prodigious lapse of time, all included within this same glacial period. Yet it may be doubted whether in any quarter of the world, sedimentary deposits, *including fossil remains,* have gone on accumulating within the same area during the whole of this period. It is not, for instance, probable that sediment was deposited during the whole of the glacial period near the mouth of the Mississippi, within that limit of depth at which marine animals can flourish; for we know what vast geographical changes occurred in other parts of America during this space of time. When such beds as were deposited in shallow water near the mouth of the Mississippi during some part of the glacial period shall have been upraised, organic remains will probably first appear and disappear at different levels, owing to the migration of species and to geographical changes. And in the distant future, a geologist examining these beds, might be tempted to conclude that the average duration of life of the embedded fossils had been less than that of the glacial period, instead of having been really far greater, that is extending from before the glacial epoch to the present day.

In order to get a perfect gradation between two forms in the upper and lower parts of the same formation, the deposit must have gone on accumulating for a very long period, in order to have given sufficient time for the slow process of variation; hence the deposit will generally have to be a very thick one; and the species undergoing modification will have had to live on the same area throughout this whole time. But we have seen that a thick fossiliferous formation can only be accumulated during a period of subsidence; and to keep the depth approximately the same, which is necessary in order to enable the same species to live on the same space, the supply of sediment must nearly have counter-balanced the amount of subsidence. But this same movement of subsidence will often tend to sink the area whence the sediment is derived, and thus diminish the supply whilst the downward movement continues. In fact, this nearly exact balancing between the supply of sediment and the amount of subsidence is probably a rare contingency; for it has been observed by more than one palæontologist, that very thick deposits are usually barren of organic remains, except near their upper or lower limits.

It would seem that each separate formation, like the whole pile of formations in any country, has generally been intermittent in its accumulation. When we see, as is so often the case, a formation composed of beds of different mineralogical composition, we may reasonably suspect that the process of deposition has been much interrupted, as a change in the currents of the sea and a supply of sediment of a different nature will generally have been due to geographical changes requiring much time. Nor will the closest inspection of a formation give any idea of the time which its deposition has consumed. Many instances could be given of beds only a few feet in thickness, representing formations, elsewhere thousands of feet in thickness, and which must have required an enormous period for their accumulation; yet no one ignorant of this fact would have suspected the vast lapse of time represented by the thinner formation. Many cases could be given of the lower beds of a formation having been upraised, denuded, submerged, and then re-covered by the upper beds of the same formation,—facts, showing what wide, yet easily overlooked, intervals have occurred in its accumulation. In other cases we have the plainest evidence in great fossilised trees, still standing upright as they grew, of many long intervals of time and changes of level during the process of deposition, which would never even have been suspected, had not the trees chanced to have been preserved: thus, Messrs. Lyell and Dawson found carboniferous beds 1400 feet thick in Nova Scotia, with ancient root-bearing strata, one above the other, at no less than sixty-eight different levels. Hence, when the same species occur at the bottom, middle, and top of a formation, the probability is that they have not lived on the same spot during the whole period of deposition, but have disappeared and reappeared, perhaps many times, during the same geological period. So that if such species were to undergo a considerable amount of modification during any one geological period, a section would not probably include all the fine intermediate gradations which must on my theory have existed between them, but abrupt, though perhaps very slight, changes of form.

It is all-important to remember that naturalists have no golden rule by which to distinguish species and varieties; they grant some little variability to each species, but when they meet with a somewhat greater amount of difference between any two forms, they rank both as species, unless they are enabled to connect them together by close intermediate gradations. And this from the reasons just assigned we can seldom hope to effect in any one geological section. Supposing B and C to be two species, and a third, A, to be found in an underlying bed; even if A were strictly intermediate between B and C, it would simply be ranked as a third and distinct species, unless at the same time it could be most closely connected with either one or both forms by intermediate varieties. Nor should it be forgotten, as before explained, that A might be the actual progenitor of B and C, and yet might not at all necessarily be strictly intermediate between them in all points of structure. So that we might obtain the parent-species and its several modified descendants from the lower and upper beds of a formation, and unless we obtained numerous transitional gradations, we should not recognise their relationship, and should consequently be compelled to rank them all as distinct species.

It is notorious on what excessively slight differences many palæontologists have founded their species; and they do this the more readily if the specimens come from different sub-stages of the same formation. Some experienced conchologists are now sinking many of the very fine species of D'Orbigny and others into the rank of varieties; and on this view we do find the kind of evidence of change which on my theory we ought to find. Moreover, if we look to rather wider intervals, namely, to distinct but consecutive stages of the same great formation, we find that the embedded fossils, though almost universally ranked as specifically different, yet are far more closely allied to each other than are the species found in more widely separated formations; but to this subject I shall have to return in the following chapter.

One other consideration is worth notice: with animals and plants that can propagate rapidly and are not highly locomotive, there is reason to suspect, as we have formerly seen, that their varieties are generally at first local; and that such local varieties do not spread widely and supplant their parent-forms until they have been modified and perfected in some considerable degree. According to this view, the chance of discovering in a formation in any one country all the early stages of transition between any two forms, is small, for the successive changes are supposed to have been local or confined to some one spot. Most marine animals have a wide range; and we have seen that with plants it is those which have the widest range, that oftenest present varieties; so that with shells and other marine animals, it is probably those which have had the widest range, far exceeding the limits of the known geological formations of Europe, which have oftenest given rise, first to local varieties and ultimately to new species; and this again would greatly lessen the chance of our being able to trace the stages of transition in any one geological formation.

It should not be forgotten, that at the present day, with perfect specimens for examination, two forms can seldom be connected by intermediate varieties and thus proved to be the same species, until many specimens have been collected from many places; and in the case of fossil species this could rarely be effected by palæontologists. We shall, perhaps, best perceive the improbability of our being enabled to connect species by numerous, fine, intermediate, fossil links, by asking ourselves whether, for instance, geologists at some future period will be able to prove, that our different breeds of cattle, sheep, horses, and dogs have descended from a single stock or from several aboriginal stocks; or, again, whether certain sea-shells inhabiting the shores of North America, which are ranked by some conchologists as distinct species from their European representatives, and by other conchologists as only varieties, are really varieties or are, as it is called, specifically distinct. This could be effected only by the future geologist discovering in a fossil state numerous intermediate gradations; and such success seems to me improbable in the highest degree.

Geological research, though it has added numerous species to existing and extinct genera, and has made the intervals between some few groups less wide than they otherwise would have been, yet has done scarcely anything in breaking down the distinction between species, by connecting them together by numerous, fine, intermediate varieties; and this not having been effected, is probably the gravest and most obvious of all the many objections which may be urged against my views. Hence it will be worth while to sum up the foregoing remarks, under an imaginary illustration. The Malay Archipelago is of about the size of Europe from the North Cape to the Mediterranean, and from Britain to Russia; and therefore equals all the geological formations which have been examined with any accuracy, excepting those of the United States of America. I fully agree with Mr. Godwin-Austen, that the present condition of the Malay Archipelago, with its numerous large islands separated by wide and shallow seas, probably represents the former state of Europe, when most of our formations were accumulating. The Malay Archipelago is one of the richest regions of the whole world in organic beings; yet if all the species were to be collected which have ever lived there, how imperfectly would they represent the natural history of the world!

But we have every reason to believe that the terrestrial productions of the archipelago would be preserved in an excessively imperfect manner in the formations which we suppose to be there accumulating. I suspect that not many of the strictly littoral animals, or of those which lived on naked submarine rocks, would be embedded; and those embedded in gravel or sand, would not endure to a distant epoch. Wherever sediment did not accumulate on the bed of the sea, or where it did not accumulate at a sufficient rate to protect organic bodies from decay, no remains could be preserved.

In our archipelago, I believe that fossiliferous formations could be formed of sufficient thickness to last to an age, as distant in futurity as the secondary formations lie in the past, only during periods of subsidence. These periods of subsidence would be separated from each other by enormous intervals, during which the area would be either stationary or rising; whilst rising, each fossiliferous formation would be destroyed, almost as soon as accumulated, by the incessant coast-action, as we now see on the shores of South America. During the periods of subsidence there would probably be much extinction of life; during the periods of elevation, there would be much variation, but the geological record would then be least perfect.

It may be doubted whether the duration of any one great period of subsidence over the whole or part of the archipelago, together with a contemporaneous accumulation of sediment, would *exceed* the average duration of the same specific forms; and these contingencies are indispensable for the preservation of all the transitional gradations between any two or more species. If such gradations were not fully preserved, transitional varieties would merely appear as so many distinct species. It is, also, probable that each great period of subsidence would be interrupted by oscillations of level, and that slight climatal changes would intervene during such lengthy periods; and in these cases the inhabitants of the archipelago would have to migrate, and no closely consecutive record of their modifications could be preserved in any one formation.

Very many of the marine inhabitants of the archipelago now range thousands of miles beyond its confines; and analogy leads me to believe that it would be chiefly these far-ranging species which would oftenest produce new varieties; and the varieties would at first generally be local or confined to one place, but if possessed of any decided advantage, or when further modified and improved, they would slowly spread and supplant their parent-forms. When such varieties returned to their ancient homes, as they would differ from their former state, in a nearly uniform, though perhaps extremely slight degree, they would, according to the principles followed by many palæontologists, be ranked as new and distinct species.

If then, there be some degree of truth in these remarks, we have no right to expect to find in our geological formations, an infinite number of those fine transitional forms, which on my theory assuredly have connected all the past and present species of the same group into one long and branching chain of life. We ought only to look for a few links, some more closely, some more distantly related to each other; and these links, let them be ever so close, if found in different stages of the same formation, would, by most palæontologists, be ranked as distinct species. But I do not pretend that I should ever have suspected how poor a record of the mutations of life, the best preserved geological section presented, had not the difficulty of our not discovering innumerable transitional links between the species which appeared at the commencement and close of each formation, pressed so hardly on my theory.

On the sudden appearance of whole groups of Allied Species.—The abrupt manner in which whole groups of species suddenly appear in certain formations, has been urged by several palæontologists, for instance, by Agassiz, Pictet, and by none more forcibly than by Professor Sedgwick, as a fatal objection to the belief in the transmutation of species. If numerous species, belonging to the same genera or families, have really started into life all at once, the fact would be fatal to the theory of descent with slow modification through natural selection. For the development of a group of forms, all of which have descended from some one progenitor, must have been an extremely slow process; and the progenitors must have lived long ages before their modified descendants. But we continually over-rate the perfection of the geological record, and falsely infer, because certain genera or families have not been found beneath a certain stage, that they did not exist before that stage. We continually forget how large the world is, compared with the area over which our geological formations have been carefully examined; we forget that groups of species may elsewhere have long existed and have slowly multiplied before they invaded the ancient archipelagoes of Europe and of the United States. We do not make due allowance for the enormous intervals of time, which have probably elapsed between our consecutive formations,—longer perhaps in some cases than the time required for the accumulation of each formation. These intervals will have given time for the multiplication of species from some one or some few parent-forms; and in the succeeding formation such species will appear as if suddenly created.

I may here recall a remark formerly made, namely that it might require a long succession of ages to adapt an organism to some new and peculiar line of life, for instance to fly through the air; but that when this had been effected, and a few species had thus acquired a great advantage over other organisms, a comparatively short time would be necessary to produce many divergent forms, which would be able to spread rapidly and widely throughout the world.

I will now give a few examples to illustrate these remarks; and to show how liable we are to error in supposing that whole groups of species have suddenly been produced. I may recall the well-known fact that in geological treatises, published not many years ago, the great class of mammals was always spoken of as having abruptly come in at the commencement of the tertiary series. And now one of the richest known accumulations of fossil mammals belongs to the middle of the secondary series; and one true mammal has been discovered in the new red sandstone at nearly the commencement of this great series. Cuvier used to urge that no monkey occurred in any tertiary stratum; but now extinct species have been discovered in India, South America, and in Europe even as far back as the eocene stage. The most striking case, however, is that of the Whale family; as these animals have huge bones, are marine, and range over the world, the fact of not a single bone of a whale having been discovered in any secondary formation, seemed fully to justify the belief that this great and distinct order had been suddenly produced in the interval between the latest secondary and earliest tertiary formation. But now we may read in the Supplement to Lyell's 'Manual,' published in 1858, clear evidence of the existence of whales in the upper greensand, some time before the close of the secondary period.

I may give another instance, which from having passed under my own eyes has much struck me. In a memoir on Fossil Sessile Cirripedes, I have stated that, from the number of existing and extinct tertiary species; from the extraordinary abundance of the individuals of many species all over the world, from the Arctic regions to the equator, inhabiting various zones of depths from the upper tidal limits to 50 fathoms; from the perfect manner in which specimens are preserved in the oldest tertiary beds; from the ease with which even a fragment of a valve can be recognised; from all these circumstances, I inferred that had sessile cirripedes existed during the secondary periods, they would certainly have been preserved and discovered; and as not one species had been discovered in beds of this age, I concluded that this great group had been suddenly developed at the commencement of the tertiary series. This was a sore trouble to me, adding as I thought one more instance of the abrupt appearance of a great group of species. But my work had hardly been published, when a skilful palæontologist, M. Bosquet, sent me a drawing of a perfect specimen of an unmistakeable sessile cirripede, which he had himself extracted from the chalk of Belgium. And, as if to make the case as striking as possible, this sessile cirripede was a Chthamalus, a very common, large, and ubiquitous genus, of which not one specimen has as yet been found even in any tertiary stratum. Hence we now positively know that sessile cirripedes existed during the secondary period; and these cirripedes might have been the progenitors of our many tertiary and existing species.

The case most frequently insisted on by palæontologists of the apparently sudden appearance of a whole group of species, is that of the teleostean fishes, low down in the Chalk period. This group includes the large majority of existing species. Lately, Professor Pictet has carried their existence one sub-stage further back; and some palæontologists believe that certain much older fishes, of which the affinities are as yet imperfectly known, are really teleostean. Assuming, however, that the whole of them did appear, as Agassiz believes, at the commencement of the chalk formation, the fact would certainly be highly remarkable; but I cannot see that it would be an insuperable difficulty on my theory, unless it could likewise be shown that the species of this group appeared suddenly and simultaneously throughout the world at this same period. It is almost superfluous to remark that hardly any fossil-fish are known from south of the equator; and by running through Pictet's Palæontology it will be seen that very few species are known from several formations in Europe. Some few families of fish now have a confined range; the teleostean fish might formerly have had a similarly confined range, and after having been largely developed in some one sea, might have spread widely. Nor have we any right to suppose that the seas of the world have always been so freely open from south to north as they are at present. Even at this day, if the Malay Archipelago were converted into land, the tropical parts of the Indian Ocean would form a large and perfectly enclosed basin, in which any great group of marine animals might be multiplied; and here they would remain confined, until some of the species became adapted to a cooler climate, and were enabled to double the southern capes of Africa or Australia, and thus reach other and distant seas.

From these and similar considerations, but chiefly from our ignorance of the geology of other countries beyond the confines of Europe and the United States; and from the revolution in our palæontological ideas on many points, which the discoveries of even the last dozen years have effected, it seems to me to be about as rash in us to dogmatize on the succession of organic beings throughout the world, as it would be for a naturalist to land for five minutes on some one barren point in Australia, and then to discuss the number and range of its productions.

On the sudden appearance of groups of Allied Species in the lowest known fossiliferous strata.—There is another and allied difficulty, which is much graver. I allude to the manner in which numbers of species of the same group, suddenly appear in the lowest known fossiliferous rocks. Most of the arguments which have convinced me that all the existing species of the same group have descended from one progenitor, apply with nearly equal force to the earliest known species. For instance, I cannot doubt that all the Silurian trilobites have descended from some one crustacean, which must have lived long before the Silurian age, and which probably differed greatly from any known animal. Some of the most ancient Silurian animals, as the Nautilus, Lingula, &c., do not differ much from living species; and it cannot on my theory be supposed, that these old species were the progenitors of all the species of the orders to which they belong, for they do not present characters in any degree intermediate between them. If, moreover, they had been the progenitors of these orders, they would almost certainly have been long ago supplanted and exterminated by their numerous and improved descendants.

Consequently, if my theory be true, it is indisputable that before the lowest Silurian stratum was deposited, long periods elapsed, as long as, or probably far longer than, the whole interval from the Silurian age to the present day; and that during these vast, yet quite unknown, periods of time, the world swarmed with living creatures.

To the question why we do not find records of these vast primordial periods, I can give no satisfactory answer. Several of the most eminent geologists, with Sir R. Murchison at their head, are convinced that we see in the organic remains of the lowest Silurian stratum the dawn of life on this planet. Other highly competent judges, as Lyell and the late E. Forbes, dispute this conclusion. We should not forget that only a small portion of the world is known with accuracy. M. Barrande has lately added another and lower stage to the Silurian system, abounding with new and peculiar species. Traces of life have been detected in the Longmynd beds beneath Barrande's so-called primordial zone. The presence of phosphatic nodules and bituminous matter in some of the lowest azoic rocks, probably indicates the former existence of life at these periods. But the difficulty of understanding the absence of vast piles of fossiliferous strata, which on my theory no doubt were somewhere accumulated before the Silurian epoch, is very great. If these most ancient beds had been wholly worn away by denudation, or obliterated by metamorphic action, we ought to find only small remnants of the formations next succeeding them in age, and these ought to be very generally in a metamorphosed condition. But the descriptions which we now possess of the Silurian deposits over immense territories in Russia and in North America, do not support the view, that the older a formation is, the more it has suffered the extremity of denudation and metamorphism.

The case at present must remain inexplicable; and may be truly urged as a valid argument against the views here entertained. To show that it may hereafter receive some explanation, I will give the following hypothesis. From the nature of the organic remains, which do not appear to have inhabited profound depths, in the several formations of Europe and of the United States; and from the amount of sediment, miles in thickness, of which the formations are composed, we may infer that from first to last large islands or tracts of land, whence the sediment was derived, occurred in the neighbourhood of the existing continents of Europe and North America. But we do not know what was the state of things in the intervals between the successive formations; whether Europe and the United States during these intervals existed as dry land, or as a submarine surface near land, on which sediment was not deposited, or again as the bed of an open and unfathomable sea.

Looking to the existing oceans, which are thrice as extensive as the land, we see them studded with many islands; but not one oceanic island is as yet known to afford even a remnant of any palæozoic or secondary formation. Hence we may perhaps infer, that during the palæozoic and secondary periods, neither continents nor continental islands existed where our oceans now extend; for had they existed there, palæozoic and secondary formations would in all probability have been accumulated from sediment derived from their wear and tear; and would have been at least partially upheaved by the oscillations of level, which we may fairly conclude must have intervened during these enormously long periods. If then we may infer anything from these facts, we may infer that where our oceans now extend, oceans have extended from the remotest period of which we have any record; and on the other hand, that where continents now exist, large tracts of land have existed, subjected no doubt to great oscillations of level, since the earliest silurian period. The coloured map appended to my volume on Coral Reefs, led me to conclude that the great oceans are still mainly areas of subsidence, the great archipelagoes still areas of oscillations of level, and the continents areas of elevation. But have we any right to assume that things have thus remained from eternity? Our continents seem to have been formed by a preponderance, during many oscillations of level, of the force of elevation; but may not the areas of preponderant movement have changed in the lapse of ages? At a period immeasurably antecedent to the silurian epoch, continents may have existed where oceans are now spread out; and clear and open oceans may have existed where our continents now stand. Nor should we be justified in assuming that if, for instance, the bed of the Pacific Ocean were now converted into a continent, we should there find formations older than the silurian strata, supposing such to have been formerly deposited; for it might well happen that strata which had subsided some miles nearer to the centre of the earth, and which had been pressed on by an enormous weight of superincumbent water, might have undergone far more metamorphic action than strata which have always remained nearer to the surface. The immense areas in some parts of the world, for instance in South America, of bare metamorphic rocks, which must have been heated under great pressure, have always seemed to me to require some special explanation; and we may perhaps believe that we see in these large areas, the many formations long anterior to the silurian epoch in a completely metamorphosed condition.

THE SEVERAL DIFFICULTIES here discussed, namely our not finding in the successive formations infinitely numerous transitional links between the many species which now exist or have existed; the sudden manner in which whole groups of species appear in our European formations; the almost entire absence, as at present known, of fossiliferous formations beneath the Silurian strata, are all undoubtedly of the gravest nature. We see this in the plainest manner by the fact that all the most eminent palæontologists, namely Cuvier, Owen, Agassiz, Barrande, Falconer, E. Forbes, &c., and all our greatest geologists, as Lyell, Murchison, Sedgwick,

&c., have unanimously, often vehemently, maintained the immutability of species. But I have reason to believe that one great authority, Sir Charles Lyell, from further reflexion entertains grave doubts on this subject. I feel how rash it is to differ from these great authorities, to whom, with others, we owe all our knowledge. Those who think the natural geological record in any degree perfect, and who do not attach much weight to the facts and arguments of other kinds given in this volume, will undoubtedly at once reject my theory. For my part, following out Lyell's metaphor, I look at the natural geological record, as a history of the world imperfectly kept, and written in a changing dialect; of this history we possess the last volume alone, relating only to two or three countries. Of this volume, only here and there a short chapter has been preserved; and of each page, only here and there a few lines. Each word of the slowly-changing language, in which the history is supposed to be written, being more or less different in the interrupted succession of chapters, may represent the apparently abruptly changed forms of life, entombed in our consecutive, but widely separated, formations. On this view, the difficulties above discussed are greatly diminished, or even disappear.

CHAPTER X

On the Geological Succession of Organic Beings

ON THE SLOW AND SUCCESSIVE APPEARANCE OF NEW SPECIES • ON THEIR DIFFERENT RATES OF CHANGE • SPECIES ONCE LOST DO NOT REAPPEAR • GROUPS OF SPECIES FOLLOW THE SAME GENERAL RULES IN THEIR APPEARANCE AND DISAPPEARANCE AS DO SINGLE SPECIES • ON EXTINCTION • ON SIMULTANEOUS CHANGES IN THE FORMS OF LIFE THROUGHOUT THE WORLD • ON THE AFFINITIES OF EXTINCT SPECIES TO EACH OTHER AND TO LIVING SPECIES • ON THE STATE OF DEVELOPMENT OF ANCIENT FORMS • ON THE SUCCESSION OF THE SAME TYPES WITHIN THE SAME AREAS • SUMMARY OF PRECEDING AND PRESENT CHAPTERS

LET US NOW SEE whether the several facts and rules relating to the geological succession of organic beings, better accord with the common view of the immutability of species, or with that of their slow and gradual modification, through descent and natural selection.

New species have appeared very slowly, one after another, both on the land and in the waters. Lyell has shown that it is hardly possible to resist the evidence on this head in the case of the several tertiary stages; and every year tends to fill up the blanks between them, and to make the percentage system of lost and new forms more gradual. In some of the most recent beds, though undoubtedly of high antiquity if measured by years, only one or two species are lost forms, and only one or two are new forms, having here appeared for the first time, either locally, or, as far as we know, on the face of the earth. If we may trust the observations of Philippi in Sicily, the successive changes in the marine inhabitants of that island have been many and most gradual. The secondary formations are more broken; but, as Bronn has remarked, neither the appearance nor disappearance of their many now extinct species has been simultaneous in each separate formation.

Species of different genera and classes have not changed at the same rate, or in the same degree. In the oldest tertiary beds a few living shells may still be found in the midst of a multitude of extinct forms. Falconer has given a striking instance of a similar fact, in an existing crocodile associated with many strange and lost mammals and reptiles in the sub-Himalayan deposits. The Silurian Lingula differs but little from the living species of this genus; whereas most of the other Silurian Molluscs and all the Crustaceans have changed greatly. The productions of the land seem to change at a quicker rate than those of the sea, of which a striking instance has lately been observed in Switzerland. There is some reason to believe that organisms, considered high in the scale of nature, change more quickly than those that are low: though there are exceptions to this rule. The amount of organic change, as Pictet has remarked, does not strictly correspond with the succession of our geological formations; so that between each two consecutive formations, the forms of life have seldom changed in exactly the same degree. Yet if we compare any but the most closely related formations, all the species will be found to have undergone some change. When a species has once disappeared from the face of the earth, we have reason to believe that the same identical form never reappears. The strongest apparent exception to this latter rule, is that of the so-called "colonies" of M. Barrande, which intrude for a period in the midst of an older formation, and then allow the pre-existing fauna to reappear; but Lyell's explanation, namely, that it is a case of temporary migration from a distinct geographical province, seems to me satisfactory.

These several facts accord well with my theory. I believe in no fixed law of development, causing all the inhabitants of a country to change abruptly, or simultaneously, or to an equal degree. The process of modification must be extremely slow. The variability of each species is quite independent of that of all others. Whether such variability be taken advantage of by natural selection, and whether the variations be accumulated to a greater or lesser amount, thus causing a greater or lesser amount of modification in the varying species, depends on many complex contingencies,—on the variability being of a beneficial nature, on the power of intercrossing, on the rate of breeding, on the slowly changing physical conditions of the country, and more especially on the nature of the other inhabitants with which the varying species comes into competition. Hence it is by no means surprising that one species should retain the same identical form much longer than others; or, if changing, that it should change less. We see the same fact in geographical distribution; for instance, in the land-shells and coleopterous insects of Madeira having come to differ considerably from their nearest allies on the continent of Europe, whereas the marine shells and birds have remained unaltered. We can perhaps understand the apparently quicker rate of change in terrestrial and in more highly organised productions compared with marine and lower productions, by the more complex relations of the higher beings to their organic and inorganic conditions of life, as explained in a former chapter. When many of the inhabitants of a country have become modified and improved, we can understand, on the principle of competition, and on that of the many all-important relations of organism to organism, that any form which does not become in some degree modified and improved, will be liable to be exterminated. Hence we can see why all the species in the same region do at last, if we look to wide enough intervals of time, become modified; for those which do not change will become extinct.

In members of the same class the average amount of change, during long and equal periods of time, may, perhaps, be nearly the same; but as the accumulation of long-enduring fossiliferous formations depends on great masses of sediment having been deposited on areas whilst subsiding, our formations have been almost necessarily accumulated at wide and irregularly intermittent intervals; consequently the amount of organic change exhibited by the fossils embedded in consecutive formations is not equal. Each formation, on this view, does not mark a new and complete act of creation, but only an occasional scene, taken almost at hazard, in a slowly changing drama.

We can clearly understand why a species when once lost should never reappear, even if the very same conditions of life, organic and inorganic, should recur. For though the offspring of one species might be adapted (and no doubt this has occurred in innumerable instances) to fill the exact place of another species in the economy of nature, and thus supplant it; yet the two forms—the old and the new—would not be identically the same; for both would almost certainly inherit different characters from their distinct progenitors. For instance, it is just possible, if our fantail-pigeons were all destroyed, that fanciers, by striving during long ages for the same object, might make a new breed hardly distinguishable from our present fantail; but if the parent rock-pigeon were also destroyed, and in nature we have every reason to believe that the parent-form will generally be supplanted and exterminated by its improved offspring, it is quite incredible that a fantail, identical with the existing breed, could be raised from any other species of pigeon, or even from the other well-established races of the domestic pigeon, for the newly-formed fantail would be almost sure to inherit from its new progenitor some slight characteristic differences.

Groups of species, that is, genera and families, follow the same general rules in their appearance and disappearance as do single species, changing more or less quickly, and in a greater or lesser degree. A group does not reappear after it has once disappeared; or its existence, as long as it lasts, is continuous. I am aware that there are some apparent exceptions to this rule, but the exceptions are surprisingly few, so few, that E. Forbes, Pictet, and Woodward (though all strongly opposed to such views as I maintain) admit its truth; and the rule strictly accords with my theory. For as all the species of the same group have descended from some one species, it is clear that as long as any species of the group have appeared in the long succession of ages, so long must its members have continuously existed, in order to have generated either new and modified or the same old and unmodified forms. Species of the genus Lingula, for instance, must have continuously existed by an unbroken succession of generations, from the lowest Silurian stratum to the present day.

We have seen in the last chapter that the species of a group sometimes falsely appear to have come in abruptly; and I have attempted to give an explanation of this fact, which if true would have been fatal to my views. But such cases are certainly exceptional; the general rule being a gradual increase in number, till the group reaches its maximum, and then, sooner or later, it gradually decreases. If the number of the species of a genus, or the number of the genera of a family, be represented by a vertical line of varying thickness, crossing the successive geological formations in which the species are found, the line will sometimes falsely appear to begin at its lower end, not in a sharp point, but abruptly; it then gradually thickens upwards, sometimes keeping for a space of equal thickness, and ultimately thins out in the upper beds, marking the decrease and final extinction of the species. This gradual increase in number of the species of a group is strictly conformable with my theory; as the species of the same genus, and the genera of the same family, can increase only slowly and progressively; for the process of modification and the production of a number of allied forms must be slow and gradual,—one species giving rise first to two or three varieties, these being slowly converted into species, which in their turn produce by equally slow steps other species, and so on, like the branching of a great tree from a single stem, till the group becomes large.

On Extinction.—We have as yet spoken only incidentally of the disappearance of species and of groups of species. On the theory of natural selection the extinction of old forms and the production of new and improved forms are intimately connected together. The old notion of all the inhabitants of the earth having been swept away at successive periods by catastrophes, is very generally given up, even by those geologists, as Elie de

Beaumont, Murchison, Barrande, &c., whose general views would naturally lead them to this conclusion. On the contrary, we have every reason to believe, from the study of the tertiary formations, that species and groups of species gradually disappear, one after another, first from one spot, then from another, and finally from the world. Both single species and whole groups of species last for very unequal periods; some groups, as we have seen, having endured from the earliest known dawn of life to the present day; some having disappeared before the close of the palæozoic period. No fixed law seems to determine the length of time during which any single species or any single genus endures. There is reason to believe that the complete extinction of the species of a group is generally a slower process than their production: if the appearance and disappearance of a group of species be represented, as before, by a vertical line of varying thickness, the line is found to taper more gradually at its upper end, which marks the progress of extermination, than at its lower end, which marks the first appearance and increase in numbers of the species. In some cases, however, the extermination of whole groups of beings, as of ammonites towards the close of the secondary period, has been wonderfully sudden.

The whole subject of the extinction of species has been involved in the most gratuitous mystery. Some authors have even supposed that as the individual has a definite length of life, so have species a definite duration. No one I think can have marvelled more at the extinction of species, than I have done. When I found in La Plata the tooth of a horse embedded with the remains of Mastodon, Megatherium, Toxodon, and other extinct monsters, which all co-existed with still living shells at a very late geological period, I was filled with astonishment; for seeing that the horse, since its introduction by the Spaniards into South America, has run wild over the whole country and has increased in numbers at an unparalleled rate, I asked myself what could so recently have exterminated the former horse under conditions of life apparently so favourable. But how utterly groundless was my astonishment! Professor Owen soon perceived that the tooth, though so like that of the existing horse, belonged to an extinct species. Had this horse been still living, but in some degree rare, no naturalist would have felt the least surprise at its rarity; for rarity is the attribute of a vast number of species of all classes, in all countries. If we ask ourselves why this or that species is rare, we answer that something is unfavourable in its conditions of life; but what that something is, we can hardly ever tell. On the supposition of the fossil horse still existing as a rare species, we might have felt certain from the analogy of all other mammals, even of the slow-breeding elephant, and from the history of the naturalisation of the domestic horse in South America, that under more favourable conditions it would in a very few years have stocked the whole continent. But we could not have told what the unfavourable conditions were which checked its increase, whether some one or several contingencies, and at what period of the horse's life, and in what degree, they severally acted. If the conditions had gone on, however slowly, becoming less and less favourable, we assuredly should not have perceived the fact, yet the fossil horse would certainly have become rarer and rarer, and finally extinct;—its place being seized on by some more successful competitor.

It is most difficult always to remember that the increase of every living being is constantly being checked by unperceived injurious agencies; and that these same unperceived agencies are amply sufficient to cause rarity, and finally extinction. We see in many cases in the more recent tertiary formations, that rarity precedes extinction; and we know that this has been the progress of events with those animals which have been exterminated, either locally or wholly, through man's agency. I may repeat what I published in 1845, namely, that to admit that species generally become rare before they become extinct—to feel no surprise at the rarity of a species, and yet to marvel greatly when it ceases to exist, is much the same as to admit that sickness in the individual is the forerunner of death—to feel no surprise at sickness, but when the sick man dies, to wonder and to suspect that he died by some unknown deed of violence.

The theory of natural selection is grounded on the belief that each new variety, and ultimately each new species, is produced and maintained by having some advantage over those with which it comes into competition; and the consequent extinction of less-favoured forms almost inevitably follows. It is the same with our domestic productions: when a new and slightly improved variety has been raised, it at first supplants the less improved varieties in the same neighbourhood; when much improved it is transported far and near, like our short-horn cattle, and takes the place of other breeds in other countries. Thus the appearance of new forms and the disappearance of old forms, both natural and artificial, are bound together. In certain flourishing groups, the number of new specific forms which have been produced within a given time is probably greater than that of the old forms which have been exterminated; but we know that the number of species has not gone on indefinitely increasing, at least during the later geological periods, so that looking to later times we may believe that the production of new forms has caused the extinction of about the same number of old forms.

The competition will generally be most severe, as formerly explained and illustrated by examples, between the forms which are most like each other in all respects. Hence the improved and modified descendants of a species will generally cause the extermination of the parent-species; and if many new forms have been developed from any one species, the nearest allies of that species, *i.e.* the species of the same genus, will be the most liable to extermination. Thus, as I believe, a number of new species descended from one species, that is a new genus, comes to supplant an old genus, belonging to the same family. But it must often have happened that a new species belonging to some one group will have seized on the place occupied by a species belonging to a distinct group, and thus caused its extermination; and if many allied forms be developed from the successful intruder, many will have to yield their places; and it will generally be allied forms, which will suffer from some inherited inferiority in common. But whether it be species belonging to the same or to a distinct class, which yield their places to other species which have been modified and improved, a few of the sufferers may often long be preserved, from being fitted to some peculiar line of life, or from inhabiting some distant and isolated station, where they have escaped severe competition. For instance, a single species of Trigonia, a great genus of shells in the secondary formations, survives in the Australian seas; and a few members of the great and almost extinct group of Ganoid fishes still inhabit our fresh waters. Therefore the utter extinction of a group is generally, as we have seen, a slower process than its production.

With respect to the apparently sudden extermination of whole families or orders, as of Trilobites at the close of the palæozoic period and of Ammonites at the close of the secondary period, we must remember what has been already said on the probable wide intervals of time between our consecutive formations; and in these intervals there may have been much slow extermination. Moreover, when by sudden immigration or by unusually rapid development, many species of a new group have taken possession of a new area, they will have exterminated in a correspondingly rapid manner many of the old inhabitants; and the forms which thus yield their places will commonly be allied, for they will partake of some inferiority in common.

Thus, as it seems to me, the manner in which single species and whole groups of species become extinct, accords well with the theory of natural selection. We need not marvel at extinction; if we must marvel, let it be at our presumption in imagining for a moment that we understand the many complex contingencies, on which the existence of each species depends. If we forget for an instant, that each species tends to increase inordinately, and that some check is always in action, yet seldom perceived by us, the whole economy of nature will be utterly obscured. Whenever we can precisely say why this species is more abundant in individuals than that; why this species and not another can be naturalised in a given country; then, and not till then, we may justly feel surprise why we cannot account for the extinction of this particular species or group of species.

On the Forms of Life changing almost simultaneously throughout the World.— Scarcely any palæontological discovery is more striking than the fact, that the forms of life change almost simultaneously throughout the world. Thus our European Chalk formation can be recognised in many distant parts of the world, under the most different climates, where not a fragment of the mineral chalk itself can be found; namely, in North America, in equatorial South America, in Tierra del Fuego, at the Cape of Good Hope, and in the peninsula of India. For at these distant points, the organic remains in certain beds present an unmistakeable degree of resemblance to those of the Chalk. It is not that the same species are met with; for in some cases not one species is identically the same, but they belong to the same families, genera, and sections of genera, and sometimes are similarly characterised in such trifling points as mere superficial sculpture. Moreover other forms, which are not found in the Chalk of Europe, but which occur in the formations either above or below, are similarly absent at these distant points of the world. In the several successive palæozoic formations of Russia, Western Europe and North America, a similar parallelism in the forms of life has been observed by several authors: so it is, according to Lyell, with the several European and North American tertiary deposits. Even if the few fossil species which are common to the Old and New Worlds be kept wholly out of view, the general parallelism in the successive forms of life, in the stages of the widely separated palæozoic and tertiary periods, would still be manifest, and the several formations could be easily correlated.

These observations, however, relate to the marine inhabitants of distant parts of the world: we have not sufficient data to judge whether the productions of the land and of fresh water change at distant points in the same parallel manner. We may doubt whether they have thus changed: if the Megatherium, Mylodon, Macrauchenia, and Toxodon had been brought to Europe from La Plata, without any information in regard to their geological position, no one would have suspected that they had coexisted with still living sea-shells; but as these anomalous monsters coexisted with the Mastodon and Horse, it might at least have been inferred that they had lived during one of the later tertiary stages.

When the marine forms of life are spoken of as having changed simultaneously throughout the world, it must not be supposed that this expression relates to the same thousandth or hundred-thousandth year, or even that it has a very strict geological sense; for if all the marine animals which live at the present day in Europe, and all those that lived in Europe during the pleistocene period (an enormously remote period as measured by years, including the whole glacial epoch), were to be compared with those now living in South America or in Australia, the most skilful naturalist would hardly be able to say whether the existing or the pleistocene inhabitants of Europe resembled most closely those of the southern hemisphere. So, again, several highly competent observers believe that the existing productions of the United States are more closely related to those which lived in Europe during certain later tertiary stages, than to those which now live here; and if this be so, it is evident that fossiliferous beds deposited at the present day on the shores of North America would hereafter be liable to be classed with somewhat older European beds. Nevertheless, looking to a remotely future epoch, there can, I think, be little doubt that all the more modern *marine* formations, namely, the upper pliocene, the pleistocene and strictly modern beds, of Europe, North and South America, and Australia, from containing fossil remains in some degree allied, and from not including those forms which are only found in the older underlying deposits, would be correctly ranked as simultaneous in a geological sense.

The fact of the forms of life changing simultaneously, in the above large sense, at distant parts of the world, has greatly struck those admirable observers, MM. De Verneuil and d'Archiac. After referring to the parallelism of the palæozoic forms of life in various parts of Europe, they add, "If struck by this strange sequence, we turn our attention to North America, and there discover a series of analogous phenomena, it will appear certain that all these modifications of species, their extinction, and the introduction of new ones, cannot be owing to mere changes in marine currents or other causes more or less local and temporary, but depend on general laws which govern the whole animal kingdom." M. Barrande has made forcible remarks to precisely the same effect. It is, indeed, quite futile to look to changes of currents, climate, or other physical conditions, as the cause of these great mutations in the forms of life throughout the world, under the most different climates. We must, as Barrande has remarked, look to some special law. We shall see this more clearly when we treat of the present distribution of organic beings, and find how slight is the relation between the physical conditions of various countries, and the nature of their inhabitants.

This great fact of the parallel succession of the forms of life throughout the world, is explicable on the theory of natural selection. New species are formed by new varieties arising, which have some advantage over older forms; and those forms, which are already dominant, or have some advantage over the other forms in their own country, would naturally oftenest give rise to new varieties or incipient species; for these latter must be victorious in a still higher degree in order to be preserved and to survive. We have distinct evidence on this head, in the plants which are dominant, that is, which are commonest in their own homes, and are most widely diffused, having produced the greatest number of new varieties. It is also natural that the dominant, varying, and far-spreading species, which already have invaded to a certain extent the territories of other species, should be those which would have the best chance of spreading still further, and of giving rise in new countries to new varieties and species. The process of diffusion may often be very slow, being dependent on climatal and geographical changes, or on strange accidents, but in the long run the dominant forms will generally succeed in spreading. The diffusion would, it is probable, be slower with the terrestrial inhabitants of distinct continents than with the marine inhabitants of the continuous sea. We might therefore expect to find, as we apparently do find, a less strict degree of parallel succession in the productions of the land than of the sea.

Dominant species spreading from any region might encounter still more dominant species, and then their triumphant course, or even their existence, would cease. We know not at all precisely what are all the conditions most favourable for the multiplication of new and dominant species; but we can, I think, clearly see that a number of individuals, from giving a better chance of the appearance of favourable variations, and that severe competition with many already existing forms, would be highly favourable, as would be the power of spreading into new territories. A certain amount of isolation, recurring at long intervals of time, would probably be also favourable, as before explained. One quarter of the world may have been most favourable for the production of new and dominant species on the land, and another for those in the waters of the sea. If two great regions had been for a long period favourably circumstanced in an equal degree, whenever their inhabitants met, the battle would be prolonged and severe; and some from one birthplace and some from the other might be victorious. But in the course of time, the forms dominant in the highest degree, wherever produced, would tend everywhere to prevail. As they prevailed, they would cause the extinction of other and inferior forms; and as these inferior forms would be allied in groups by inheritance, whole groups would tend slowly to disappear; though here and there a single member might long be enabled to survive.

Thus, as it seems to me, the parallel, and, taken in a large sense, simultaneous, succession of the same forms of life throughout the world, accords well with the principle of new species having been formed by dominant species spreading widely and varying; the new species thus produced being themselves dominant owing to inheritance, and to having already had some advantage over their parents or over other species; these again spreading, varying, and producing new species. The forms which are beaten and which yield their places to the new and victorious forms, will generally be allied in groups, from inheriting some inferiority in common;

and therefore as new and improved groups spread throughout the world, old groups will disappear from the world; and the succession of forms in both ways will everywhere tend to correspond.

There is one other remark connected with this subject worth making. I have given my reasons for believing that all our greater fossiliferous formations were deposited during periods of subsidence; and that blank intervals of vast duration occurred during the periods when the bed of the sea was either stationary or rising, and likewise when sediment was not thrown down quickly enough to embed and preserve organic remains. During these long and blank intervals I suppose that the inhabitants of each region underwent a considerable amount of modification and extinction, and that there was much migration from other parts of the world. As we have reason to believe that large areas are affected by the same movement, it is probable that strictly contemporaneous formations have often been accumulated over very wide spaces in the same quarter of the world; but we are far from having any right to conclude that this has invariably been the case, and that large areas have invariably been affected by the same movements. When two formations have been deposited in two regions during nearly, but not exactly the same period, we should find in both, from the causes explained in the foregoing paragraphs, the same general succession in the forms of life; but the species would not exactly correspond; for there will have been a little more time in the one region than in the other for modification, extinction, and immigration.

I suspect that cases of this nature have occurred in Europe. Mr. Prestwich, in his admirable Memoirs on the eocene deposits of England and France, is able to draw a close general parallelism between the successive stages in the two countries; but when he compares certain stages in England with those in France, although he finds in both a curious accordance in the numbers of the species belonging to the same genera, yet the species themselves differ in a manner very difficult to account for, considering the proximity of the two areas,—unless, indeed, it be assumed that an isthmus separated two seas inhabited by distinct, but contemporaneous, faunas. Lyell has made similar observations on some of the later tertiary formations. Barrande, also, shows that there is a striking general parallelism in the successive Silurian deposits of Bohemia and Scandinavia; nevertheless he finds a surprising amount of difference in the species. If the several formations in these regions have not been deposited during the same exact periods,—a formation in one region often corresponding with a blank interval in the other,—and if in both regions the species have gone on slowly changing during the accumulation of the several formations and during the long intervals of time between them; in this case, the several formations in the two regions could be arranged in the same order, in accordance with the general succession of the form of life, and the order would falsely appear to be strictly parallel; nevertheless the species would not all be the same in the apparently corresponding stages in the two regions.

On the Affinities of extinct Species to each other, and to living forms.—Let us now look to the mutual affinities of extinct and living species. They all fall into one grand natural system; and this fact is at once explained on the principle of descent. The more ancient any form is, the more, as a general rule, it differs from living forms. But, as Buckland long ago remarked, all fossils can be classed either in still existing groups, or between them. That the extinct forms of life help to fill up the wide intervals between existing genera, families, and orders, cannot be disputed. For if we confine our attention either to the living or to the extinct alone, the series is far less perfect than if we combine both into one general system. With respect to the Vertebrata, whole pages could be filled with striking illustrations from our great palæontologist, Owen, showing how extinct animals fall in between existing groups. Cuvier ranked the Ruminants and Pachyderms, as the two most distinct orders of mammals; but Owen has discovered so many fossil links, that he has had to alter the whole classification of these two orders; and has placed certain pachyderms in the same sub-order with ruminants: for example, he dissolves by fine gradations the apparently wide difference between the pig and the camel. In regard to the Invertebrata, Barrande, and a higher authority could not be named, asserts that he is every day taught that palæozoic animals, though belonging to the same orders, families, or genera with those living at the present day, were not at this early epoch limited in such distinct groups as they now are.

Some writers have objected to any extinct species or group of species being considered as intermediate between living species or groups. If by this term it is meant that an extinct form is directly intermediate in all its characters between two living forms, the objection is probably valid. But I apprehend that in a perfectly natural classification many fossil species would have to stand between living species, and some extinct genera between living genera, even between genera belonging to distinct families. The most common case, especially with respect to very distinct groups, such as fish and reptiles, seems to be, that supposing them to be distinguished at the present day from each other by a dozen characters, the ancient members of the same two groups would be distinguished by a somewhat lesser number of characters, so that the two groups, though formerly quite distinct, at that period made some small approach to each other.

It is a common belief that the more ancient a form is, by so much the more it tends to connect by some of its characters groups now widely separated from each other. This remark no doubt must be restricted to those groups which have undergone much change in the course of geological ages; and it would be difficult to prove the truth of the proposition, for every now and then even a living animal, as the Lepidosiren, is discovered having affinities directed towards very distinct groups. Yet if we compare the older Reptiles and Batrachians, the older Fish, the older Cephalopods, and the eocene Mammals, with the more recent members of the same classes, we must admit that there is some truth in the remark.

Let us see how far these several facts and inferences accord with the theory of descent with modification. As the subject is somewhat complex, I must request the reader to turn to the diagram in the fourth chapter. We may suppose that the numbered letters represent genera, and the dotted lines diverging from them the species in each genus. The diagram is much too simple, too few genera and too few species being given, but this is unimportant for us. The horizontal lines may represent successive geological formations, and all the forms beneath the uppermost line may be considered as extinct. The three existing genera, a^{14}, q^{14}, p^{14}, will form a small family; b^{14} and f^{14} a closely allied family or sub-family; and o^{14}, e^{14}, m^{14}, a third family. These three families, together with the many extinct genera on the several lines of descent diverging from the parent-form A, will form an order; for all will have inherited something in common from their ancient and common progenitor. On the principle of the continued tendency to divergence of character, which was formerly illustrated by this diagram, the more recent any form is, the more it will generally differ from its ancient progenitor. Hence we can understand the rule that the most ancient fossils differ most from existing forms. We must not, however, assume that divergence of character is a necessary contingency; it depends solely on the descendants from a species being thus enabled to seize on many and different places in the economy of nature. Therefore it is quite possible, as we have seen in the case of some Silurian forms, that a species might go on being slightly modified in relation to its slightly altered conditions of life, and yet retain throughout a vast period the same general characteristics. This is represented in the diagram by the letter F^{14}.

All the many forms, extinct and recent, descended from A, make, as before remarked, one order; and this order, from the continued effects of extinction and divergence of character, has become divided into several sub-families and families, some of which are supposed to have perished at different periods, and some to have endured to the present day.

By looking at the diagram we can see that if many of the extinct forms, supposed to be embedded in the successive formations, were discovered at several points low down in the series, the three existing families on the uppermost line would be rendered less distinct from each other. If, for instance, the genera a^1, a^5, a^{10}, f^8, m^3, m^6, m^9, were disinterred, these three families would be so closely linked together that they probably would have to be united into one great family, in nearly the same manner as has occurred with ruminants and pachyderms. Yet he who objected to call the extinct genera, which thus linked the living genera of three families together, intermediate in character, would be justified, as they are intermediate, not directly, but only by a long and circuitous course through many widely different forms. If many extinct forms were to be discovered above one of the middle horizontal lines or geological formations—for instance, above No. VI.—but none from beneath this line, then only the two families on the left hand (namely, a^{14}, &c., and b^{14}, &c.) would have to be united into one family; and the two other families (namely, a^{14} to f^{14} now including five genera, and o^{14} to m^{14}) would yet remain distinct. These two families, however, would be less distinct from each other than they were before the discovery of the fossils. If, for instance, we suppose the existing genera of the two families to differ from each other by a dozen characters, in this case the genera, at the early period marked VI., would differ by a lesser number of characters; for at this early stage of descent they have not diverged in character from the common progenitor of the order, nearly so much as they subsequently diverged. Thus it comes that ancient and extinct genera are often in some slight degree intermediate in character between their modified descendants, or between their collateral relations.

In nature the case will be far more complicated than is represented in the diagram; for the groups will have been more numerous, they will have endured for extremely unequal lengths of time, and will have been modified in various degrees. As we possess only the last volume of the geological record, and that in a very broken condition, we have no right to expect, except in very rare cases, to fill up wide intervals in the natural system, and thus unite distinct families or orders. All that we have a right to expect, is that those groups, which have within known geological periods undergone much modification, should in the older formations make some slight approach to each other; so that the older members should differ less from each other in some of their characters than do the existing members of the same groups; and this by the concurrent evidence of our best palæontologists seems frequently to be the case.

Thus, on the theory of descent with modification, the main facts with respect to the mutual affinities of the extinct forms of life to each other and to living forms, seem to me explained in a satisfactory manner. And they are wholly inexplicable on any other view.

On this same theory, it is evident that the fauna of any great period in the earth's history will be intermediate in general character between that which preceded and that which succeeded it. Thus, the species which lived at the sixth great stage of descent in the diagram are the modified offspring of those which lived at the fifth stage, and are the parents of those which became still more modified at the seventh stage; hence they could hardly fail to be nearly intermediate in character between the forms of life above and below. We must, however, allow for the entire extinction of some preceding forms, and for the coming in of quite new forms by immigration, and for a large amount of modification, during the long and blank intervals between the successive formations. Subject to these allowances, the fauna of each geological period undoubtedly is intermediate in character, between the preceding and succeeding faunas. I need give only one instance, namely, the manner in which the fossils of the Devonian system, when this system was first discovered, were at once recognised by palæontologists as intermediate in character between those of the overlying carboniferous, and underlying Silurian system. But each fauna is not necessarily exactly intermediate, as unequal intervals of time have elapsed between consecutive formations.

It is no real objection to the truth of the statement, that the fauna of each period as a whole is nearly intermediate in character between the preceding and succeeding faunas, that certain genera offer exceptions to the rule. For instance, mastodons and elephants, when arranged by Dr. Falconer in two series, first according to their mutual affinities and then according to their periods of existence, do not accord in arrangement. The species extreme in character are not the oldest, or the most recent; nor are those which are intermediate in character, intermediate in age. But supposing for an instant, in this and other such cases, that the record of the first appearance and disappearance of the species was perfect, we have no reason to believe that forms successively produced necessarily endure for corresponding lengths of time: a very ancient form might occasionally last much longer than a form elsewhere subsequently produced, especially in the case of terrestrial productions inhabiting separated districts. To compare small things with great: if the principal living and extinct races of the domestic pigeon were arranged as well as they could be in serial affinity, this arrangement would not closely accord with the order in time of their production, and still less with the order of their disappearance; for the parent rock-pigeon now lives; and many varieties between the rock-pigeon and the carrier have become extinct; and carriers which are extreme in the important character of length of beak originated earlier than short-beaked tumblers, which are at the opposite end of the series in this same respect.

Closely connected with the statement, that the organic remains from an intermediate formation are in some degree intermediate in character, is the fact, insisted on by all palæontologists, that fossils from two consecutive formations are far more closely related to each other, than are the fossils from two remote formations. Pictet gives as a well-known instance, the general resemblance of the organic remains from the several stages of the chalk formation, though the species are distinct in each stage. This fact alone, from its generality, seems to have shaken Professor Pictet in his firm belief in the immutability of species. He who is acquainted with the distribution of existing species over the globe, will not attempt to account for the close resemblance of the distinct species in closely consecutive formations, by the physical conditions of the ancient areas having remained nearly the same. Let it be remembered that the forms of life, at least those inhabiting the sea, have changed almost simultaneously throughout the world, and therefore under the most different climates and conditions. Consider the prodigious vicissitudes of climate during the pleistocene period, which includes the whole glacial period, and note how little the specific forms of the inhabitants of the sea have been affected.

On the theory of descent, the full meaning of the fact of fossil remains from closely consecutive formations, though ranked as distinct species, being closely related, is obvious. As the accumulation of each formation has often been interrupted, and as long blank intervals have intervened between successive formations, we ought not to expect to find, as I attempted to show in the last chapter, in any one or two formations all the intermediate varieties between the species which appeared at the commencement and close of these periods; but we ought to find after intervals, very long as measured by years, but only moderately long as measured geologically, closely allied forms, or, as they have been called by some authors, representative species; and these we assuredly do find. We find, in short, such evidence of the slow and scarcely sensible mutation of specific forms, as we have a just right to expect to find.

On the state of Development of Ancient Forms.—There has been much discussion whether recent forms are more highly developed than ancient. I will not here enter on this subject, for naturalists have not as yet defined to each other's satisfaction what is meant by high and low forms. But in one particular sense the more recent forms must, on my theory, be higher than the more ancient; for each new species is formed by having had some advantage in the struggle for life over other and preceding forms. If under a nearly similar climate, the eocene inhabitants of one quarter of the world were put into competition with the existing inhabitants of the same or some other quarter, the eocene fauna or flora would certainly be beaten and exterminated; as would a secondary fauna by an eocene, and a palæozoic fauna by a secondary fauna. I do not doubt that this process of improvement has affected in a marked and sensible manner the organisation of the more recent and victorious forms of life, in comparison with the ancient and beaten forms; but I can see no way of testing this sort of progress. Crustaceans, for instance, not the highest in their own class, may have beaten the highest molluscs. From the extraordinary manner in which European productions have recently spread over New Zealand, and have seized on places which must have been previously occupied, we may believe, if all the animals and plants of Great Britain were set free in New Zealand, that in the course of time a multitude of British forms would become thoroughly naturalized there, and would exterminate many of the natives. On the other hand, from what we see now occurring in New Zealand, and from hardly a single inhabitant of the southern hemisphere having become wild in any part of Europe, we may doubt, if all the productions of New Zealand were set free in Great Britain, whether any considerable number would be enabled to seize on places now occupied by our native plants and animals. Under this point of view, the productions of Great Britain may be said to be higher than those of New Zealand. Yet the most skilful naturalist from an examination of the species of the two countries could not have foreseen this result.

Agassiz insists that ancient animals resemble to a certain extent the embryos of recent animals of the same classes; or that the geological succession of extinct forms is in some degree parallel to the embryological development of recent forms. I must follow Pictet and Huxley in thinking that the truth of this doctrine is very far from proved. Yet I fully expect to see it hereafter confirmed, at least in regard to subordinate groups, which have branched off from each other within comparatively recent times. For this doctrine of Agassiz accords well with the theory of natural selection. In a future chapter I shall attempt to show that the adult differs from its embryo, owing to variations supervening at a not early age, and being inherited at a corresponding age. This process, whilst it leaves the embryo almost unaltered, continually adds, in the course of successive generations, more and more difference to the adult.

Thus the embryo comes to be left as a sort of picture, preserved by nature, of the ancient and less modified condition of each animal. This view may be true, and yet it may never be capable of full proof. Seeing, for instance, that the oldest known mammals, reptiles, and fish strictly belong to their own proper classes, though some of these old forms are in a slight degree less distinct from each other than are the typical members of the same groups at the present day, it would be vain to look for animals having the common embryological character of the Vertebrata, until beds far beneath the lowest Silurian strata are discovered—a discovery of which the chance is very small.

On the Succession of the same Types within the same areas, during the later tertiary periods.—Mr. Clift many years ago showed that the fossil mammals from the Australian caves were closely allied to the living marsupials of that continent. In South America, a similar relationship is manifest, even to an uneducated eye, in the gigantic pieces of armour like those of the armadillo, found in several parts of La Plata; and Professor Owen has shown in the most striking manner that most of the fossil mammals, buried there in such numbers, are related to South American types. This relationship is even more clearly seen in the wonderful collection of fossil bones made by MM. Lund and Clausen in the caves of Brazil. I was so much impressed with these facts that I strongly insisted, in 1839 and 1845, on this "law of the succession of types,"—on "this wonderful relationship in the same continent between the dead and the living." Professor Owen has subsequently extended the same generalisation to the mammals of the Old World. We see the same law in this author's restorations of the extinct and gigantic birds of New Zealand. We see it also in the birds of the caves of Brazil. Mr. Woodward has shown that the same law holds good with sea-shells, but from the wide distribution of most genera of molluscs, it is not well displayed by them. Other cases could be added, as the relation between the extinct and living land-shells of Madeira; and between the extinct and living brackish-water shells of the Aralo-Caspian Sea.

Now what does this remarkable law of the succession of the same types within the same areas mean? He would be a bold man, who after comparing the present climate of Australia and of parts of South America under the same latitude, would attempt to account, on the one hand, by dissimilar physical conditions for the dissimilarity of the inhabitants of these two continents, and, on the other hand, by similarity of conditions, for the uniformity of the same types in each during the later tertiary periods. Nor can it be pretended that it is an immutable law that marsupials should have been chiefly or solely produced in Australia; or that Edentata and other American types should have been solely produced in South America. For we know that Europe in ancient times was peopled by numerous marsupials; and I have shown in the publications above alluded to, that in America the law of distribution of terrestrial mammals was formerly different from what it now is. North America formerly partook strongly of the present character of the southern half of the continent; and the southern half was formerly more closely allied, than it is at present, to the northern half. In a similar manner we know from Falconer and Cautley's discoveries, that northern India was formerly more closely related in its mammals to Africa than it is at the present time. Analogous facts could be given in relation to the distribution of marine animals.

On the theory of descent with modification, the great law of the long enduring, but not immutable, succession of the same types within the same areas, is at once explained; for the inhabitants of each quarter of the world will obviously tend to leave in that quarter, during the next succeeding period of time, closely allied though in some degree modified descendants. If the inhabitants of one continent formerly differed greatly from those of another continent, so will their modified descendants still differ in nearly the same manner and degree. But after very long intervals of time and after great geographical changes, permitting much inter-migration, the feebler will yield to the more dominant forms, and there will be nothing immutable in the laws of past and present distribution.

It may be asked in ridicule, whether I suppose that the megatherium and other allied huge monsters have left behind them in South America the sloth, armadillo, and anteater, as their degenerate descendants. This cannot for an instant be admitted. These huge animals have become wholly extinct, and have left no progeny. But in the caves of Brazil, there are many extinct species which are closely allied in size and in other characters to the species still living in South America; and some of these fossils may be the actual progenitors of living species. It must not be forgotten that, on my theory, all the species of the same genus have descended from some one species; so that if six genera, each having eight species, be found in one geological formation, and in the next succeeding formation there be six other allied or representative genera with the same number of species, then we may conclude that only one species of each of the six older genera has left modified descendants, constituting the six new genera. The other seven species of the old genera have all died out and have left no progeny. Or, which would probably be a far commoner case, two or three species of two or three alone of the six older genera will have been the parents of the six new genera; the other old species and the other whole genera having become utterly extinct. In failing orders, with the genera and species decreasing in numbers, as apparently is the case of the Edentata of South America, still fewer genera and species will have left modified blood-descendants.

Summary of the preceding and present Chapters.—I have attempted to show that the geological record is extremely imperfect; that only a small portion of the globe has been geologically explored with care; that only certain classes of organic beings have been largely preserved in a fossil state; that the number both of specimens and of species, preserved in our museums, is absolutely as nothing compared with the incalculable number of generations which must have passed away even during a single formation; that, owing to subsidence being necessary for the accumulation of fossiliferous deposits thick enough to resist future degradation, enormous intervals of time have elapsed between the successive formations; that there has probably been more extinction during the periods of subsidence, and more variation during the periods of elevation, and during the latter the record will have been least perfectly kept; that each single formation has not been continuously deposited; that the duration of each formation is, perhaps, short compared with the average duration of specific forms; that migration has played an important part in the first appearance of new forms in any one area and formation; that widely ranging species are those which have varied most, and have oftenest given rise to new species; and that varieties have at first often been local. All these causes taken conjointly, must have tended to make the geological record extremely imperfect, and will to a large extent explain why we do not find interminable varieties, connecting together all the extinct and existing forms of life by the finest graduated steps.

He who rejects these views on the nature of the geological record, will rightly reject my whole theory. For he may ask in vain where are the numberless transitional links which must formerly have connected the closely allied or representative species, found in the several stages of the same great formation. He may disbelieve in the enormous intervals of time which have elapsed between our consecutive formations; he may overlook how important a part migration must have played, when the formations of any one great region alone, as that of Europe, are considered; he may urge the apparent, but often falsely apparent, sudden coming in of whole groups of species. He may ask where are the remains of those infinitely numerous organisms which must have existed long before the first bed of the Silurian system was deposited: I can answer this latter question only hypothetically, by saying that as far as we can see, where our oceans now extend they have for an enormous period extended, and where our oscillating continents now stand they have stood ever since the Silurian epoch; but that long before that period, the world may have presented a wholly different aspect; and that the older continents, formed of formations older than any known to us, may now all be in a metamorphosed condition, or may lie buried under the ocean.

Passing from these difficulties, all the other great leading facts in palæontology seem to me simply to follow on the theory of descent with modification through natural selection. We can thus understand how it is that new species come in slowly and successively; how species of different classes do not necessarily change together, or at the same rate, or in the same degree; yet in the long run that all undergo modification to some extent. The extinction of old forms is the almost inevitable consequence of the production of new forms. We can understand why when a species has once disappeared it never reappears. Groups of species increase in numbers slowly, and endure for unequal periods of time; for the process of modification is necessarily slow, and depends on many complex contingencies. The dominant species of the larger dominant groups tend to leave many modified descendants, and thus new sub-groups and groups are formed. As these are formed, the species of the less vigorous groups, from their inferiority inherited from a common progenitor, tend to become extinct together, and to leave no modified offspring on the face of the earth. But the utter extinction of a whole group of species may often be a very slow process, from the survival of a few descendants, lingering in protected and isolated situations. When a group has once wholly disappeared, it does not reappear; for the link of generation has been broken.

We can understand how the spreading of the dominant forms of life, which are those that oftenest vary, will in the long run tend to people the world with allied, but modified, descendants; and these will generally succeed in taking the places of those groups of species which are their inferiors in the struggle for existence. Hence, after long intervals of time, the productions of the world will appear to have changed simultaneously.

We can understand how it is that all the forms of life, ancient and recent, make together one grand system; for all are connected by generation. We can understand, from the continued tendency to divergence of character, why the more ancient a form is, the more it generally differs from those now living. Why ancient and extinct forms often tend to fill up gaps between existing forms, sometimes blending two groups previously classed as distinct into one; but more commonly only bringing them a little closer together. The more ancient a form is, the more often, apparently, it displays characters in some degree intermediate between groups now distinct; for the more ancient a form is, the more nearly it will be related to, and consequently resemble, the common progenitor of groups, since become widely divergent. Extinct forms are seldom directly intermediate between existing forms; but are intermediate only by a long and circuitous course through many extinct and very different forms. We can clearly see why the organic remains of closely consecutive formations are more closely allied to each other, than are those of remote formations; for the forms are more closely linked together by generation: we can clearly see why the remains of an intermediate formation are intermediate in character.

The inhabitants of each successive period in the world's history have beaten their predecessors in the race for life, and are, in so far, higher in the scale of nature; and this may account for that vague yet ill-defined sentiment, felt by many palæontologists, that organisation on the whole has progressed. If it should hereafter be proved that ancient animals resemble to a certain extent the embryos of more recent animals of the same class, the fact will be intelligible. The succession of the same types of structure within the same areas during the later geological periods ceases to be mysterious, and is simply explained by inheritance.

If then the geological record be as imperfect as I believe it to be, and it may at least be asserted that the record cannot be proved to be much more perfect, the main objections to the theory of natural selection are greatly diminished or disappear. On the other hand, all the chief laws of palæontology plainly proclaim, as it seems to me, that species have been produced by ordinary generation: old forms having been supplanted by new and improved forms of life, produced by the laws of variation still acting round us, and preserved by Natural Selection.

CHAPTER XI

Geographical Distribution

PRESENT DISTRIBUTION CANNOT BE ACCOUNTED FOR BY DIFFERENCES IN PHYSICAL CONDITIONS • IMPORTANCE OF BARRIERS • AFFINITY OF THE PRODUCTIONS OF THE SAME CONTINENT • CENTRES OF CREATION • MEANS OF DISPERSAL, BY CHANGES OF CLIMATE AND OF THE LEVEL OF THE LAND, AND BY OCCASIONAL MEANS • DISPERSAL DURING THE GLACIAL PERIOD CO-EXTENSIVE WITH THE WORLD

IN CONSIDERING THE DISTRIBUTION of organic beings over the face of the globe, the first great fact which strikes us is, that neither the similarity nor the dissimilarity of the inhabitants of various regions can be accounted for by their climatal and other physical conditions. Of late, almost every author who has studied the subject has come to this conclusion. The case of America alone would almost suffice to prove its truth: for if we exclude the northern parts where the circumpolar land is almost continuous, all authors agree that one of the most fundamental divisions in geographical distribution is that between the New and Old Worlds; yet if we travel over the vast American continent, from the central parts of the United States to its extreme southern point, we meet with the most diversified conditions; the most humid districts, arid deserts, lofty mountains, grassy plains, forests, marshes, lakes, and great rivers, under almost every temperature. There is hardly a climate or condition in the Old World which cannot be paralleled in the New—at least as closely as the same species generally require; for it is a most rare case to find a group of organisms confined to any small spot, having conditions peculiar in only a slight degree; for instance, small areas in the Old World could be pointed out hotter than any in the New World, yet these are not inhabited by a peculiar fauna or flora. Notwithstanding this parallelism in the conditions of the Old and New Worlds, how widely different are their living productions!

In the southern hemisphere, if we compare large tracts of land in Australia, South Africa, and western South America, between latitudes 25° and 35°, we shall find parts extremely similar in all their conditions, yet it would not be possible to point out three faunas and floras more utterly dissimilar. Or again we may compare the productions of South America south of lat. 35° with those north of 25°, which consequently inhabit a considerably different climate, and they will be found incomparably more closely related to each other, than they are to the productions of Australia or Africa under nearly the same climate. Analogous facts could be given with respect to the inhabitants of the sea.

A second great fact which strikes us in our general review is, that barriers of any kind, or obstacles to free migration, are related in a close and important manner to the differences between the productions of various regions. We see this in the great difference of nearly all the terrestrial productions of the New and Old Worlds, excepting in the northern parts, where the land almost joins, and where, under a slightly different climate, there might have been free migration for the northern temperate forms, as there now is for the strictly arctic productions. We see the same fact in the great difference between the inhabitants of Australia, Africa, and South America under the same latitude: for these countries are almost as much isolated from each other as is possible. On each continent, also, we see the same fact; for on the opposite sides of lofty and continuous mountain-ranges, and of great deserts, and sometimes even of large rivers, we find different productions; though as mountain-chains, deserts, &c., are not as impassable, or likely to have endured so long as the oceans separating continents, the differences are very inferior in degree to those characteristic of distinct continents.

Turning to the sea, we find the same law. No two marine faunas are more distinct, with hardly a fish, shell, or crab in common, than those of the eastern and western shores of South and Central America; yet these great faunas are separated only by the narrow, but impassable, isthmus of Panama. Westward of the shores of America, a wide space of open ocean extends, with not an island as a halting-place for emigrants; here we have a barrier of another kind, and as soon as this is passed we meet in the eastern islands of the Pacific, with another and totally distinct fauna. So that here three marine faunas range far northward and southward, in parallel lines not far from each other, under corresponding climates; but from being separated from each other by impassable barriers, either of land or open sea, they are wholly distinct. On the other hand, proceeding still further westward from the eastern islands of the tropical parts of the Pacific, we encounter no impassable barriers, and we have innumerable islands as halting-places, until after travelling over a hemisphere we come to the shores of Africa; and over this vast space we meet with no well-defined and distinct marine faunas. Although hardly one shell, crab or fish is common to the above-named three approximate faunas of Eastern and Western America and the eastern Pacific islands, yet many fish range from the Pacific into the Indian Ocean, and many shells are common to the eastern islands of the Pacific and the eastern shores of Africa, on almost exactly opposite meridians of longitude.

A third great fact, partly included in the foregoing statements, is the affinity of the productions of the same continent or sea, though the species themselves are distinct at different points and stations. It is a law of the widest generality, and every continent offers innumerable instances. Nevertheless the naturalist in travelling, for instance, from north to south never fails to be struck by the manner in which successive groups of beings, specifically distinct, yet clearly related, replace each other. He hears from closely allied, yet distinct kinds of birds, notes nearly similar, and sees their nests similarly constructed, but not quite alike, with eggs coloured in nearly the same manner. The plains near the Straits of Magellan are inhabited by one species of Rhea (American ostrich), and northward the plains of La Plata by another species of the same genus; and not by a true ostrich or emeu, like those found in Africa and Australia under the same latitude. On these same plains of La Plata, we see the agouti and bizcacha, animals having nearly the same habits as our hares and rabbits and belonging to the same order of Rodents, but they plainly display an American type of structure. We ascend the lofty peaks of the Cordillera and we find an alpine species of bizcacha; we look to the waters, and we do not find the beaver or musk-rat, but the coypu and capybara, rodents of the American type. Innumerable other instances could be given. If we look to the islands off the American shore, however much they may differ in geological structure, the inhabitants, though they may be all peculiar species, are essentially American. We may look back to past ages, as shown in the last chapter, and we find American types then prevalent on the American continent and in the American seas. We see in these facts some deep organic bond, prevailing throughout space and time, over the same areas of land and water, and independent of their physical conditions. The naturalist must feel little curiosity, who is not led to inquire what this bond is.

This bond, on my theory, is simply inheritance, that cause which alone, as far as we positively know, produces organisms quite like, or, as we see in the case of varieties nearly like each other. The dissimilarity of the inhabitants of different regions may be attributed to modification through natural selection, and in a quite subordinate degree to the direct influence of different physical conditions. The degree of dissimilarity will depend on the migration of the more dominant forms of life from one region into another having been effected with more or less ease, at periods more or less remote;—on the nature and number of the former immigrants;—and on their action and reaction, in their mutual struggles for life;—the relation of organism to organism being, as I have already often remarked, the most important of all relations. Thus the high importance of barriers comes into play by checking migration; as does time for the slow process of modification through natural selection. Widely-ranging species, abounding in individuals, which have already triumphed over many competitors in their own widely-extended homes will have the best chance of seizing on new places, when they spread into new countries. In their new homes they will be exposed to new conditions, and will frequently undergo further modification and improvement; and thus they will become still further victorious, and will produce groups of modified descendants. On this principle of inheritance with modification, we can understand how it is that sections of genera, whole genera, and even families are confined to the same areas, as is so commonly and notoriously the case.

I believe, as was remarked in the last chapter, in no law of necessary development. As the variability of each species is an independent property, and will be taken advantage of by natural selection, only so far as it profits the individual in its complex struggle for life, so the degree of modification in different species will be no uniform quantity. If, for instance, a number of species, which stand in direct competition with each other, migrate in a body into a new and afterwards isolated country, they will be little liable to modification; for neither migration nor isolation in themselves can do anything. These principles come into play only by bringing organisms into new relations with each other, and in a lesser degree with the surrounding physical conditions. As we have seen in the last chapter that some forms have retained nearly the same character from an enormously remote geological period, so certain species have migrated over vast spaces, and have not become greatly modified.

On these views, it is obvious, that the several species of the same genus, though inhabiting the most distant quarters of the world, must originally have proceeded from the same source, as they have descended from the same progenitor. In the case of those species, which have undergone during whole geological periods but little modification, there is not much difficulty in believing that they may have migrated from the same region; for during the vast geographical and climatal changes which will have supervened since ancient times, almost any amount of migration is possible. But in many other cases, in which we have reason to believe that the species of a genus have been produced within comparatively recent times, there is great difficulty on this head. It is also obvious that the individuals of the same species, though now inhabiting distant and isolated regions, must have proceeded from one spot, where their parents were first produced: for, as explained in the last chapter, it is incredible that individuals identically the same should ever have been produced through natural selection from parents specifically distinct.

We are thus brought to the question which has been largely discussed by naturalists, namely, whether species have been created at one or more points of the earth's surface. Undoubtedly there are very many cases of extreme difficulty, in understanding how the same species could possibly have migrated from some one point to the several distant and isolated points, where now found. Nevertheless the simplicity of the view that each species was first produced within a single region captivates the mind. He who rejects it, rejects the *vera causa* of ordinary generation with subsequent migration, and calls in the agency of a miracle. It is universally admitted, that in most cases the area inhabited by a species is continuous; and when a plant or animal inhabits two points so distant from each other, or with an interval of such a nature, that the space could not be easily passed over by migration, the fact is given as something remarkable and exceptional. The capacity of migrating across the sea is more distinctly limited in terrestrial mammals, than perhaps in any other organic beings; and, accordingly, we find no inexplicable cases of the same mammal inhabiting distant points of the world. No geologist will feel any difficulty in such cases as Great Britain having been formerly united to Europe, and consequently possessing the same quadrupeds. But if the same species can be produced at two separate points, why do we not find a single mammal common to Europe and Australia or South America? The conditions of life are nearly the same, so that a multitude of European animals and plants have become naturalised in America and Australia; and some of the aboriginal plants are identically the same at these distant points of the northern and southern hemispheres? The answer, as I believe, is, that mammals have not been able to migrate, whereas some plants, from their varied means of dispersal, have migrated across the vast and broken interspace. The great and striking influence which barriers of every kind have had on distribution, is intelligible only on the view that the great majority of species have been produced on one side alone, and have not been able to migrate to the other side. Some few families, many sub-families, very many genera, and a still greater number of sections of genera are confined to a single region; and it has been observed by several naturalists, that the most natural genera, or those genera in which the species are most closely related to each other, are generally local, or confined to one area. What a strange anomaly it would be, if, when coming one step lower in the series, to the individuals of the same species, a directly opposite rule prevailed; and species were not local, but had been produced in two or more distinct areas!

Hence it seems to me, as it has to many other naturalists, that the view of each species having been produced in one area alone, and having subsequently migrated from that area as far as its powers of migration and subsistence under past and present conditions permitted, is the most probable. Undoubtedly many cases occur, in which we cannot explain how the same species could have passed from one point to the other. But the geographical and climatal changes, which have certainly occurred within recent geological times, must have interrupted or rendered discontinuous the formerly continuous range of many species. So that we are reduced to consider whether the exceptions to continuity of range are so numerous and of so grave a nature, that we ought to give up the belief, rendered probable by general considerations, that each species has been produced within one area, and has migrated thence as far as it could. It would be hopelessly tedious to discuss all the exceptional cases of the same species, now living at distant and separated points; nor do I for a moment pretend that any explanation could be offered of many such cases. But after some preliminary remarks, I will discuss a few of the most striking classes of facts; namely, the existence of the same species on the summits of distant mountain-ranges, and at distant points in the arctic and antarctic regions; and secondly (in the following chapter), the wide distribution of fresh-water productions; and thirdly, the occurrence of the same terrestrial species on islands and on the mainland, though separated by hundreds of miles of open sea. If the existence of the same species at distant and isolated points of the earth's surface, can in many instances be explained on the view of each species having migrated from a single birthplace; then, considering our ignorance with respect to former climatal and geographical changes and various occasional means of transport, the belief that this has been the universal law, seems to me incomparably the safest.

In discussing this subject, we shall be enabled at the same time to consider a point equally important for us, namely, whether the several distinct species of a genus, which on my theory have all descended from a common progenitor, can have migrated (undergoing modification during some part of their migration) from the area inhabited by their progenitor. If it can be shown to be almost invariably the case, that a region, of which most of its inhabitants are closely related to, or belong to the same genera with the species of a second region, has probably received at some former period immigrants from this other region, my theory will be strengthened; for we can clearly understand, on the principle of modification, why the inhabitants of a region should be related to those of another region, whence it has been stocked. A volcanic island, for instance, upheaved and formed at the distance of a few hundreds of miles from a continent, would probably receive from it in the course of time a few colonists, and their descendants, though modified, would still be plainly related by inheritance to the inhabitants of the continent. Cases of this nature are common, and are, as we shall hereafter more fully see, inexplicable on the theory of independent creation. This view of the relation of species in one region to those in another, does not differ much (by substituting the word variety for species) from that lately advanced in an ingenious paper by Mr. Wallace, in which he concludes, that "every species has come into existence coincident both in space and time with a pre-existing closely allied species." And I now know from correspondence, that this coincidence he attributes to generation with modification.

The previous remarks on "single and multiple centres of creation" do not directly bear on another allied question,—namely whether all the individuals of the same species have descended from a single pair, or single hermaphrodite, or whether, as some authors suppose, from many individuals simultaneously created. With those organic beings which never intercross (if such exist), the species, on my theory, must have descended from a succession of improved varieties, which will never have blended with other individuals or varieties, but will have supplanted each other; so that, at each successive stage of modification and improvement, all the individuals of each variety will have descended from a single parent. But in the majority of cases, namely, with all organisms which habitually unite for each birth, or which often intercross, I believe that during the slow process of modification the individuals of the species will have been kept nearly uniform by intercrossing; so that many individuals will have gone on simultaneously changing, and the whole amount of modification will not have been due, at each stage, to descent from a single parent. To illustrate what I mean: our English race-horses differ slightly from the horses of every other breed; but they do not owe their difference and superiority to descent from any single pair, but to continued care in selecting and training many individuals during many generations.

Before discussing the three classes of facts, which I have selected as presenting the greatest amount of difficulty on the theory of "single centres of creation," I must say a few words on the means of dispersal.

Means of Dispersal.—Sir C. Lyell and other authors have ably treated this subject. I can give here only the briefest abstract of the more important facts. Change of climate must have had a powerful influence on migration: a region when its climate was different may have been a high road for migration, but now be impassable; I shall, however, presently have to discuss this branch of the subject in some detail. Changes of level in the land must also have been highly influential: a narrow isthmus now separates two marine faunas; submerge it, or let it formerly have been sub-merged, and the two faunas will now blend or may formerly have blended: where the sea now extends, land may at a former period have connected islands or possibly even continents together, and thus have allowed terrestrial productions to pass from one to the other. No geologist will dispute that great mutations of level, have occurred within the period of existing organisms. Edward Forbes insisted that all the islands in the Atlantic must recently have been connected with Europe or Africa, and Europe likewise with America. Other authors have thus hypothetically bridged over every ocean, and have united almost every island to some mainland. If indeed the arguments used by Forbes are to be trusted, it must be admitted that scarcely a single island exists which has not recently been united to some continent. This view cuts the Gordian knot of the dispersal of the same species to the most distant points, and removes many a difficulty: but to the best of my judgment we are not authorized in admitting such enormous geographical changes within the period of existing species. It seems to me that we have abundant evidence of great oscillations of level in our continents; but not of such vast changes in their position and extension, as to have united them within the recent period to each other and to the several intervening oceanic islands. I freely admit the former existence of many islands, now buried beneath the sea, which may have served as halting places for plants and for many animals during their migration. In the coral-producing oceans such sunken islands are now marked, as I believe, by rings of coral or atolls standing over them. Whenever it is fully admitted, as I believe it will some day be, that each species has proceeded from a single birthplace, and when in the course of time we know something definite about the means of distribution, we shall be enabled to speculate with security on the former extension of the land. But I do not believe that it will ever be proved that within the recent period continents which are now quite separate, have been continuously, or almost continuously, united with each other, and with the many existing oceanic islands. Several facts in distribution,—such as the great difference in the marine faunas on the opposite sides of almost every continent,—the close relation of the tertiary inhabitants of several lands and even seas to their present inhabitants,—a certain degree of relation (as we shall hereafter see) between the distribution of mammals and the depth of the sea,—these and other such facts seem to me opposed to the admission of such prodigious geographical revolutions within the recent period, as are necessitated on the view advanced by Forbes and admitted by his many followers. The nature and relative proportions of the inhabitants of oceanic islands likewise seem to me opposed to the belief of their former continuity with continents. Nor does their almost universally volcanic composition favour the admission that they are the wrecks of sunken continents;—if they had originally existed as mountain-ranges on the land, some at least of the islands would have been formed, like other mountain-summits, of granite, metamorphic schists, old fossiliferous or other such rocks, instead of consisting of mere piles of volcanic matter.

I must now say a few words on what are called accidental means, but which more properly might be called occasional means of distribution. I shall here confine myself to plants. In botanical works, this or that plant is stated to be ill adapted for wide dissemination; but for transport across the sea, the greater or less facilities may be said to be almost wholly unknown. Until I tried, with Mr. Berkeley's aid, a few experiments, it was not even known how far seeds could resist the injurious action of sea-water. To my surprise I found that out of 87 kinds, 64 germinated after an immersion of 28 days, and a few survived an immersion of 137 days. For convenience sake I chiefly tried small seeds, without the capsule or fruit; and as all of these sank in a few days, they could not be floated across wide spaces of the sea, whether or not they were injured by the salt-water. Afterwards I tried some larger fruits, capsules, &c., and some of these floated for a long time. It is well known what a difference there is in the buoyancy of green and seasoned timber; and it occurred to me that floods might wash down plants or branches, and that these might be dried on the banks, and then by a fresh rise in the stream be washed into the sea. Hence I was led to dry stems and branches of 94 plants with ripe fruit, and to place them on sea water. The majority sank quickly, but some which whilst green floated for a very short time, when dried floated much longer; for instance, ripe hazel-nuts sank immediately, but when dried, they floated for 90 days and afterwards when planted they germinated; an asparagus plant with ripe berries floated for 23 days, when dried it floated for 85 days, and the seeds afterwards germinated: the ripe seeds of Helosciadium sank in two days, when dried they floated for above 90 days, and afterwards germinated. Altogether out of the 94 dried plants, 18 floated for above 28 days, and some of the 18 floated for a very much longer period. So that as 64/87 seeds germinated after an immersion of 28 days; and as 18/94 plants with ripe fruit (but not all the same species as in the foregoing experiment) floated, after being dried, for above 28 days, as far as we may infer anything from these scanty facts, we may conclude that the seeds of 14/100 plants of any country might be floated by sea-currents during 28 days, and would retain their power of germination. In Johnston's Physical Atlas, the average rate of the several Atlantic currents is 33 miles per diem (some currents running at the rate of 60 miles per diem); on this average, the seeds of 14/100 plants belonging to one country might be floated across 924 miles of sea to another country; and when stranded, if blown to a favourable spot by an inland gale, they would germinate.

Subsequently to my experiments, M. Martens tried similar ones, but in a much better manner, for he placed the seeds in a box in the actual sea, so that they were alternately wet and exposed to the air like really floating plants. He tried 98 seeds, mostly different from mine; but he chose many large fruits and likewise seeds from plants which live near the sea; and this would have favoured the average length of their flotation and of their resistance to the injurious action of the salt-water. On the other hand he did not previously dry the plants or branches with the fruit; and this, as we have seen, would have caused some of them to have floated much longer. The result was that 18/98 of his seeds floated for 42 days, and were then capable of germination. But I do not doubt that plants exposed to the waves would float for a less time than those protected from violent movement as in our experiments. Therefore it would perhaps be safer to assume that the seeds of about 10/100 plants of a flora, after having been dried, could be floated across a space of sea 900 miles in width, and would then germinate. The fact of the larger fruits often floating longer than the small, is interesting; as plants with large seeds or fruit could hardly be transported by any other means; and Alph. de Candolle has shown that such plants generally have restricted ranges.

But seeds may be occasionally transported in another manner. Drift timber is thrown up on most islands, even on those in the midst of the widest oceans; and the natives of the coral-islands in the Pacific, procure stones for their tools, solely from the roots of drifted trees, these stones being a valuable royal tax. I find on examination, that when irregularly shaped stones are embedded in the roots of trees, small parcels of earth are very frequently enclosed in their interstices and behind them,—so perfectly that not a particle could be washed away in the longest transport: out of one small portion of earth thus *completely* enclosed by wood in an oak about 50 years old, three dicotyledonous plants germinated: I am certain of the accuracy of this observation. Again, I can show that the carcasses of birds, when floating on the sea, sometimes escape being immediately devoured; and seeds of many kinds in the crops of floating birds long retain their vitality: peas and vetches, for instance, are killed by even a few days' immersion in sea-water; but some taken out of the crop of a pigeon, which had floated on artificial salt-water for 30 days, to my surprise nearly all germinated.

Living birds can hardly fail to be highly effective agents in the transportation of seeds. I could give many facts showing how frequently birds of many kinds are blown by gales to vast distances across the ocean. We may I think safely assume that under such circumstances their rate of flight would often be 35 miles an hour; and some authors have given a far higher estimate. I have never seen an instance of nutritious seeds passing through the intestines of a bird; but hard seeds of fruit will pass uninjured through even the digestive organs of a turkey. In the course of two months, I picked up in my garden 12 kinds of seeds, out of the excrement of small birds, and these seemed perfect, and some of them, which I tried, germinated. But the following fact is more important: the crops of birds do not secrete gastric juice, and do not in the least injure, as I know by trial, the germination of seeds; now after a bird has found and devoured a large supply of food, it is positively asserted that all the grains do not pass into the gizzard for 12 or even 18 hours. A bird in this interval might easily be blown to the distance of 500 miles, and hawks are known to look out for tired birds, and the contents of their torn crops might thus readily get scattered. Mr. Brent informs me that a friend of his had to give up flying carrier-pigeons from France to England, as the hawks on the English coast destroyed so many on their arrival. Some hawks and owls bolt their prey whole, and after an interval of from twelve to twenty hours, disgorge pellets, which, as I know from experiments made in the Zoological Gardens, include seeds capable of germination. Some seeds of the oat, wheat, millet, canary, hemp, clover, and beet germinated after having been from twelve to twenty-one hours in the stomachs of different birds of prey; and two seeds of beet grew after having been thus retained for two days and fourteen hours. Fresh-water fish, I find, eat seeds of many land and water plants: fish are frequently devoured by birds, and thus the seeds might be transported from place to place. I forced many kinds of seeds into the stomachs of dead fish, and then gave their bodies to fishing-eagles, storks, and pelicans; these birds after an interval of many hours, either rejected the seeds in pellets or passed them in their excrement; and several of these seeds retained their power of germination. Certain seeds, however, were always killed by this process.

Although the beaks and feet of birds are generally quite clean, I can show that earth sometimes adheres to them: in one instance I removed twenty-two grains of dry argillaceous earth from one foot of a partridge, and in this earth there was a pebble quite as large as the seed of a vetch. Thus seeds might occasionally be transported to great distances; for many facts could be given showing that soil almost everywhere is charged with seeds. Reflect for a moment on the millions of quails which annually cross the Mediterranean; and can we doubt that the earth adhering to their feet would sometimes include a few minute seeds? But I shall presently have to recur to this subject.

As icebergs are known to be sometimes loaded with earth and stones, and have even carried brushwood, bones, and the nest of a land-bird, I can hardly doubt that they must occasionally have transported seeds from one part to another of the arctic and antarctic regions, as suggested by Lyell; and during the Glacial period from one part of the now temperate regions to another. In the Azores, from the large number of the species of plants common to Europe, in comparison with the plants of other oceanic islands nearer to the mainland, and (as remarked by Mr. H. C. Watson) from the somewhat northern character of the flora in comparison with the latitude, I suspected that these islands had been partly stocked by ice-borne seeds, during the Glacial epoch. At my request Sir C. Lyell wrote to M. Hartung to inquire whether he had observed erratic boulders on these islands, and he answered that he had found large fragments of granite and other rocks, which do not occur in the archipelago. Hence we may safely infer that icebergs formerly landed their rocky burthens on the shores of these mid-ocean islands, and it is at least possible that they may have brought thither the seeds of northern plants.

Considering that the several above means of transport, and that several other means, which without doubt remain to be discovered, have been in action year after year, for centuries and tens of thousands of years, it would I think be a marvellous fact if many plants had not thus become widely transported. These means of transport are sometimes called accidental, but this is not strictly correct: the currents of the sea are not accidental, nor is the direction of prevalent gales of wind. It should be observed that scarcely any means of transport would carry seeds for very great distances; for seeds do not retain their vitality when exposed for a great length of time to the action of sea-water; nor could they be long carried in the crops or intestines of birds. These means, however, would suffice for occasional transport across tracts of sea some hundred miles in breadth, or from island to island, or from a continent to a neighbouring island, but not from one distant continent to another. The floras of distant continents would not by such means become mingled in any great degree; but would remain as distinct as we now see them to be. The currents, from their course, would never bring seeds from North America to Britain, though they might and do bring seeds from the West Indies to our western shores, where, if not killed by so long an immersion in salt-water, they could not endure our climate. Almost every year, one or two land-birds are blown across the whole Atlantic Ocean, from North America to the western shores of Ireland and England; but seeds could be transported by these wanderers only by one means, namely, in dirt sticking to their feet, which is in itself a rare accident. Even in this case, how small would the chance be of a seed falling on favourable soil, and coming to maturity! But it would be a great error to argue that because a well-stocked island, like Great Britain, has not, as far as is known (and it would be very difficult to prove this), received within the last few centuries, through occasional means of transport, immigrants from Europe or any other continent, that a poorly-stocked island, though standing more remote from the mainland, would not receive colonists by similar means. I do not doubt that out of twenty seeds or animals transported to an island, even if far less well-stocked than Britain, scarcely more than one would be so well fitted to its new home, as to become naturalised. But this, as it seems to me, is no valid argument against what would be effected by occasional means of transport, during the long lapse of geological time, whilst an island was being upheaved and formed, and before it had become fully stocked with inhabitants. On almost bare land, with few or no destructive insects or birds living there, nearly every seed, which chanced to arrive, would be sure to germinate and survive.

Dispersal during the Glacial period.—The identity of many plants and animals, on mountain-summits, separated from each other by hundreds of miles of lowlands, where the Alpine species could not possibly exist, is one of the most striking cases known of the same species living at distant points, without the apparent possibility of their having migrated from one to the other. It is indeed a remarkable fact to see so many of the same plants living on the snowy regions of the Alps or Pyrenees, and in the extreme northern parts of Europe; but it is far more remarkable, that the plants on the White Mountains, in the United States of America, are all the same with those of Labrador, and nearly all the same, as we hear from Asa Gray, with those on the loftiest mountains of Europe. Even as long ago as 1747, such facts led Gmelin to conclude that the same species must have been independently created at several distinct points; and we might have remained in this same belief, had not Agassiz and others called vivid attention to the Glacial period, which, as we shall immediately see, affords a simple explanation of these facts. We have evidence of almost every conceivable kind, organic and inorganic, that within a very recent geological period, central Europe and North America suffered under an Arctic climate. The ruins of a house burnt by fire do not tell their tale more plainly, than do the mountains of Scotland and Wales, with their scored flanks, polished surfaces, and perched boulders, of the icy streams with which their valleys were lately filled. So greatly has the climate of Europe changed, that in Northern Italy, gigantic moraines, left by old glaciers, are now clothed by the vine and maize. Throughout a large part of the United States, erratic boulders, and rocks scored by drifted icebergs and coast-ice, plainly reveal a former cold period.

The former influence of the glacial climate on the distribution of the inhabitants of Europe, as explained with remarkable clearness by Edward

Forbes, is substantially as follows. But we shall follow the changes more readily, by supposing a new glacial period to come slowly on, and then pass away, as formerly occurred. As the cold came on, and as each more southern zone became fitted for arctic beings and ill-fitted for their former more temperate inhabitants, the latter would be supplanted and arctic productions would take their places. The inhabitants of the more temperate regions would at the same time travel southward, unless they were stopped by barriers, in which case they would perish. The mountains would become covered with snow and ice, and their former Alpine inhabitants would descend to the plains. By the time that the cold had reached its maximum, we should have a uniform arctic fauna and flora, covering the central parts of Europe, as far south as the Alps and Pyrenees, and even stretching into Spain. The now temperate regions of the United States would likewise be covered by arctic plants and animals, and these would be nearly the same with those of Europe; for the present circumpolar inhabitants, which we suppose to have everywhere travelled southward, are remarkably uniform round the world. We may suppose that the Glacial period came on a little earlier or later in North America than in Europe, so will the southern migration there have been a little earlier or later; but this will make no difference in the final result.

As the warmth returned, the arctic forms would retreat northward, closely followed up in their retreat by the productions of the more temperate regions. And as the snow melted from the bases of the mountains, the arctic forms would seize on the cleared and thawed ground, always ascending higher and higher, as the warmth increased, whilst their brethren were pursuing their northern journey. Hence, when the warmth had fully returned, the same arctic species, which had lately lived in a body together on the lowlands of the Old and New Worlds, would be left isolated on distant mountain-summits (having been exterminated on all lesser heights) and in the arctic regions of both hemispheres.

Thus we can understand the identity of many plants at points so immensely remote as on the mountains of the United States and of Europe. We can thus also understand the fact that the Alpine plants of each mountain-range are more especially related to the arctic forms living due north or nearly due north of them: for the migration as the cold came on, and the re-migration on the returning warmth, will generally have been due south and north. The Alpine plants, for example, of Scotland, as remarked by Mr. H. C. Watson, and those of the Pyrenees, as remarked by Ramond, are more especially allied to the plants of northern Scandinavia; those of the United States to Labrador; those of the mountains of Siberia to the arctic regions of that country. These views, grounded as they are on the perfectly well-ascertained occurrence of a former Glacial period, seem to me to explain in so satisfactory a manner the present distribution of the Alpine and Arctic productions of Europe and America, that when in other regions we find the same species on distant mountain-summits, we may almost conclude without other evidence, that a colder climate permitted their former migration across the low intervening tracts, since become too warm for their existence.

If the climate, since the Glacial period, has ever been in any degree warmer than at present (as some geologists in the United States believe to have been the case, chiefly from the distribution of the fossil Gnathodon), then the arctic and temperate productions will at a very late period have marched a little further north, and subsequently have retreated to their present homes; but I have met with no satisfactory evidence with respect to this intercalated slightly warmer period, since the Glacial period.

The arctic forms, during their long southern migration and re-migration northward, will have been exposed to nearly the same climate, and, as is especially to be noticed, they will have kept in a body together; consequently their mutual relations will not have been much disturbed, and, in accordance with the principles inculcated in this volume, they will not have been liable to much modification. But with our Alpine productions, left isolated from the moment of the returning warmth, first at the bases and ultimately on the summits of the mountains, the case will have been somewhat different; for it is not likely that all the same arctic species will have been left on mountain ranges distant from each other, and have survived there ever since; they will, also, in all probability have become mingled with ancient Alpine species, which must have existed on the mountains before the commencement of the Glacial epoch, and which during its coldest period will have been temporarily driven down to the plains; they will, also, have been exposed to somewhat different climatal influences. Their mutual relations will thus have been in some degree disturbed; consequently they will have been liable to modification; and this we find has been the case; for if we compare the present Alpine plants and animals of the several great European mountain-ranges, though very many of the species are identically the same, some present varieties, some are ranked as doubtful forms, and some few are distinct yet closely allied or representative species.

In illustrating what, as I believe, actually took place during the Glacial period, I assumed that at its commencement the arctic productions were as uniform round the polar regions as they are at the present day. But the foregoing remarks on distribution apply not only to strictly arctic forms, but also to many sub-arctic and to some few northern temperate forms, for some of these are the same on the lower mountains and on the plains of North America and Europe; and it may be reasonably asked how I account for the necessary degree of uniformity of the sub-arctic and northern temperate forms round the world, at the commencement of the Glacial period. At the present day, the sub-arctic and northern temperate productions of the Old and New Worlds are separated from each other by the Atlantic Ocean and by the extreme northern part of the Pacific. During the Glacial period, when the inhabitants of the Old and New Worlds lived further southwards than at present, they must have been still more completely separated by wider spaces of ocean. I believe the above difficulty may be surmounted by looking to still earlier changes of climate of an opposite nature. We have good reason to believe that during the newer Pliocene period, before the Glacial epoch, and whilst the majority of the inhabitants of the world were specifically the same as now, the climate was warmer than at the present day. Hence we may suppose that the organisms now living under the climate of latitude 60°, during the Pliocene period lived further north under the Polar Circle, in latitude 66°–67°; and that the strictly arctic productions then lived on the broken land still nearer to the pole. Now if we look at a globe, we shall see that under the Polar Circle there is almost continuous land from western Europe, through Siberia, to eastern America. And to this continuity of the circumpolar land, and to the consequent freedom for intermigration under a more favourable climate, I attribute the necessary amount of uniformity in the sub-arctic and northern temperate productions of the Old and New Worlds, at a period anterior to the Glacial epoch.

Believing, from reasons before alluded to, that our continents have long remained in nearly the same relative position, though subjected to large, but partial oscillations of level, I am strongly inclined to extend the above view, and to infer that during some earlier and still warmer period, such as the older Pliocene period, a large number of the same plants and animals inhabited the almost continuous circumpolar land; and that these plants and animals, both in the Old and New Worlds, began slowly to migrate southwards as the climate became less warm, long before the commencement of the Glacial period. We now see, as I believe, their descendants, mostly in a modified condition, in the central parts of Europe and the United States. On this view we can understand the relationship, with very little identity, between the productions of North America and Europe,—a relationship which is most remarkable, considering the distance of the two areas, and their separation by the Atlantic Ocean. We can further understand the singular fact remarked on by several observers, that the productions of Europe and America during the later tertiary stages were more closely related to each other than they are at the present time; for during these warmer periods the northern parts of the Old and New Worlds will have been almost continuously united by land, serving as a bridge, since rendered impassable by cold, for the inter-migration of their inhabitants.

During the slowly decreasing warmth of the Pliocene period, as soon as the species in common, which inhabited the New and Old Worlds, migrated south of the Polar Circle, they must have been completely cut off from each other. This separation, as far as the more temperate productions are concerned, took place long ages ago. And as the plants and animals migrated southward, they will have become mingled in the one great region with the native American productions, and have had to compete with them; and in the other great region, with those of the Old World. Consequently we have here everything favourable for much modification,—for far more modification than with the Alpine productions, left isolated, within a much more recent period, on the several mountain-ranges and on the arctic lands of the two Worlds. Hence it has come, that when we compare the now living productions of the temperate regions of the New and Old Worlds, we find very few identical species (though Asa Gray has lately shown that more plants are identical than was formerly supposed), but we find in every great class many forms, which some naturalists rank as geographical races, and others as distinct species; and a host of closely allied or representative forms which are ranked by all naturalists as specifically distinct.

As on the land, so in the waters of the sea, a slow southern migration of a marine fauna, which during the Pliocene or even a somewhat earlier period, was nearly uniform along the continuous shores of the Polar Circle, will account, on the theory of modification, for many closely allied forms now living in areas completely sundered. Thus, I think, we can understand the presence of many existing and tertiary representative forms on the eastern and western shores of temperate North America; and the still more striking case of many closely allied crustaceans (as described in Dana's admirable work), of some fish and other marine animals, in the Mediterranean and in the seas of Japan,—areas now separated by a continent and by nearly a hemisphere of equatorial ocean.

These cases of relationship, without identity, of the inhabitants of seas now disjoined, and likewise of the past and present inhabitants of the temperate lands of North America and Europe, are inexplicable on the theory of creation. We cannot say that they have been created alike, in correspondence with the nearly similar physical conditions of the areas; for if we compare, for instance, certain parts of South America with the southern continents of the Old World, we see countries closely corresponding in all their physical conditions, but with their inhabitants utterly dissimilar.

But we must return to our more immediate subject, the Glacial period. I am convinced that Forbes's view may be largely extended. In Europe we have the plainest evidence of the cold period, from the western shores of Britain to the Oural range, and southward to the Pyrenees. We may infer, from the frozen mammals and nature of the mountain vegetation, that Siberia was similarly affected. Along the Himalaya, at points 900 miles apart, glaciers have left the marks of their former low descent; and in Sikkim, Dr. Hooker saw maize growing on gigantic ancient moraines. South of the equator, we have some direct evidence of former glacial action in New Zealand; and the same plants, found on widely separated mountains in this island, tell the same story. If one account which has been published can be trusted, we have direct evidence of glacial action in the south-eastern corner of Australia.

Looking to America; in the northern half, ice-borne fragments of rock have been observed on the eastern side as far south as lat. 36°–37°, and on the shores of the Pacific, where the climate is now so different, as far south as lat. 46°; erratic boulders have, also, been noticed on the Rocky Mountains. In the Cordillera of Equatorial South America, glaciers once extended far below their present level. In central Chile I was astonished at the structure of a vast mound of detritus, about 800 feet in height, crossing a valley of the Andes; and this I now feel convinced was a gigantic moraine, left far below any existing glacier. Further south on both sides of the continent, from lat. 41° to the southernmost extremity, we have the clearest evidence of former glacial action, in huge boulders transported far from their parent source.

We do not know that the Glacial epoch was strictly simultaneous at these several far distant points on opposite sides of the world. But we have good evidence in almost every case, that the epoch was included within the latest geological period. We have, also, excellent evidence, that it endured for an enormous time, as measured by years, at each point. The cold may have come on, or have ceased, earlier at one point of the globe than at another, but seeing that it endured for long at each, and that it was contemporaneous in a geological sense, it seems to me probable that it was, during a part at least of the period, actually simultaneous throughout the world. Without some distinct evidence to the contrary, we may at least admit as probable that the glacial action was simultaneous on the eastern and western sides of North America, in the Cordillera under the equator and under the warmer temperate zones, and on both sides of the southern extremity of the continent. If this be admitted, it is difficult to avoid believing that the temperature of the whole world was at this period simultaneously cooler. But it would suffice for my purpose, if the temperature was at the same time lower along certain broad belts of longitude.

On this view of the whole world, or at least of broad longitudinal belts, having been simultaneously colder from pole to pole, much light can be thrown on the present distribution of identical and allied species. In America, Dr. Hooker has shown that between forty and fifty of the flowering plants of Tierra del Fuego, forming no inconsiderable part of its scanty flora, are common to Europe, enormously remote as these two points are; and there are many closely allied species. On the lofty mountains of equatorial America a host of peculiar species belonging to European genera occur. On the highest mountains of Brazil, some few European genera were found by Gardner, which do not exist in the wide intervening hot countries. So on the Silla of Caraccas the illustrious Humboldt long ago found species belonging to genera characteristic of the Cordillera. On the mountains of Abyssinia, several European forms and some few representatives of the peculiar flora of the Cape of Good Hope occur. At the Cape of Good Hope a very few European species, believed not to have been introduced by man, and on the mountains, some few representative European forms are found, which have not been discovered in the intertropical parts of Africa. On the Himalaya, and on the isolated mountain-ranges of the peninsula of India, on the heights of Ceylon, and on the volcanic cones of Java, many plants occur, either identically the same or representing each other, and at the same time representing plants of Europe, not found in the intervening hot lowlands. A list of the genera collected on the loftier peaks of Java raises a picture of a collection made on a hill in Europe! Still more striking is the fact that southern Australian forms are clearly represented by plants growing on the summits of the mountains of Borneo. Some of these Australian forms, as I hear from Dr. Hooker, extend along the heights of the peninsula of Malacca, and are thinly scattered, on the one hand over India and on the other as far north as Japan.

On the southern mountains of Australia, Dr. F. Muller has discovered several European species; other species, not introduced by man, occur on the lowlands; and a long list can be given, as I am informed by Dr. Hooker, of European genera, found in Australia, but not in the intermediate torrid regions. In the admirable 'Introduction to the Flora of New Zealand,' by Dr. Hooker, analogous and striking facts are given in regard to the plants of that large island. Hence we see that throughout the world, the plants growing on the more lofty mountains, and on the temperate lowlands of the northern and southern hemispheres, are sometimes identically the same; but they are much oftener specifically distinct, though related to each other in a most remarkable manner.

This brief abstract applies to plants alone: some strictly analogous facts could be given on the distribution of terrestrial animals. In marine productions, similar cases occur; as an example, I may quote a remark by the highest authority, Prof. Dana, that "it is certainly a wonderful fact that New Zealand should have a closer resemblance in its crustacea to Great Britain, its antipode, than to any other part of the world." Sir J. Richardson, also, speaks of the reappearance on the shores of New Zealand, Tasmania, &c.,

of northern forms of fish. Dr. Hooker informs me that twenty-five species of Algæ are common to New Zealand and to Europe, but have not been found in the intermediate tropical seas.

It should be observed that the northern species and forms found in the southern parts of the southern hemisphere, and on the mountain-ranges of the intertropical regions, are not arctic, but belong to the northern temperate zones. As Mr. H. C. Watson has recently remarked, "In receding from polar towards equatorial latitudes, the Alpine or mountain floras really become less and less arctic." Many of the forms living on the mountains of the warmer regions of the earth and in the southern hemisphere are of doubtful value, being ranked by some naturalists as specifically distinct, by others as varieties; but some are certainly identical, and many, though closely related to northern forms, must be ranked as distinct species.

Now let us see what light can be thrown on the foregoing facts, on the belief, supported as it is by a large body of geological evidence, that the whole world, or a large part of it, was during the Glacial period simultaneously much colder than at present. The Glacial period, as measured by years, must have been very long; and when we remember over what vast spaces some naturalised plants and animals have spread within a few centuries, this period will have been ample for any amount of migration. As the cold came slowly on, all the tropical plants and other productions will have retreated from both sides towards the equator, followed in the rear by the temperate productions, and these by the arctic; but with the latter we are not now concerned. The tropical plants probably suffered much extinction; how much no one can say; perhaps formerly the tropics supported as many species as we see at the present day crowded together at the Cape of Good Hope, and in parts of temperate Australia. As we know that many tropical plants and animals can withstand a considerable amount of cold, many might have escaped extermination during a moderate fall of temperature, more especially by escaping into the warmest spots. But the great fact to bear in mind is, that all tropical productions will have suffered to a certain extent. On the other hand, the temperate productions, after migrating nearer to the equator, though they will have been placed under somewhat new conditions, will have suffered less. And it is certain that many temperate plants, if protected from the inroads of competitors, can withstand a much warmer climate than their own. Hence, it seems to me possible, bearing in mind that the tropical productions were in a suffering state and could not have presented a firm front against intruders, that a certain number of the more vigorous and dominant temperate forms might have penetrated the native ranks and have reached or even crossed the equator. The invasion would, of course, have been greatly favoured by high land, and perhaps by a dry climate; for Dr. Falconer informs me that it is the damp with the heat of the tropics which is so destructive to perennial plants from a temperate climate. On the other hand, the most humid and hottest districts will have afforded an asylum to the tropical natives. The mountain-ranges north-west of the Himalaya, and the long line of the Cordillera, seem to have afforded two great lines of invasion: and it is a striking fact, lately communicated to me by Dr. Hooker, that all the flowering plants, about forty-six in number, common to Tierra del Fuego and to Europe still exist in North America, which must have lain on the line of march. But I do not doubt that some temperate productions entered and crossed even the *lowlands* of the tropics at the period when the cold was most intense,—when arctic forms had migrated some twenty-five degrees of latitude from their native country and covered the land at the foot of the Pyrenees. At this period of extreme cold, I believe that the climate under the equator at the level of the sea was about the same with that now felt there at the height of six or seven thousand feet. During this the coldest period, I suppose that large spaces of the tropical lowlands were clothed with a mingled tropical and temperate vegetation, like that now growing with strange luxuriance at the base of the Himalaya, as graphically described by Hooker.

Thus, as I believe, a considerable number of plants, a few terrestrial animals, and some marine productions, migrated during the Glacial period from the northern and southern temperate zones into the intertropical regions, and some even crossed the equator. As the warmth returned, these temperate forms would naturally ascend the higher mountains, being exterminated on the lowlands; those which had not reached the equator, would re-migrate northward or southward towards their former homes; but the forms, chiefly northern, which had crossed the equator, would travel still further from their homes into the more temperate latitudes of the opposite hemisphere. Although we have reason to believe from geological evidence that the whole body of arctic shells underwent scarcely any modification during their long southern migration and re-migration northward, the case may have been wholly different with those intruding forms which settled themselves on the intertropical mountains, and in the southern hemisphere. These being surrounded by strangers will have had to compete with many new forms of life; and it is probable that selected modifications in their structure, habits, and constitutions will have profited them. Thus many of these wanderers, though still plainly related by inheritance to their brethren of the northern or southern hemispheres, now exist in their new homes as well-marked varieties or as distinct species.

It is a remarkable fact, strongly insisted on by Hooker in regard to America, and by Alph. de Candolle in regard to Australia, that many more identical plants and allied forms have apparently migrated from the north to the south, than in a reversed direction. We see, however, a few southern vegetable forms on the mountains of Borneo and Abyssinia. I suspect that this preponderant migration from north to south is due to the greater extent of land in the north, and to the northern forms having existed in their own homes in greater numbers, and having consequently been advanced through natural selection and competition to a higher stage of perfection or dominating power, than the southern forms. And thus, when they became commingled during the Glacial period, the northern forms were enabled to beat the less powerful southern forms. Just in the same manner as we see at the present day, that very many European productions cover the ground in La Plata, and in a lesser degree in Australia, and have to a certain extent beaten the natives; whereas extremely few southern forms have become naturalised in any part of Europe, though hides, wool, and other objects likely to carry seeds have been largely imported into Europe during the last two or three centuries from La Plata, and during the last thirty or forty years from Australia. Something of the same kind must have occurred on the intertropical mountains: no doubt before the Glacial period they were stocked with endemic Alpine forms; but these have almost everywhere largely yielded to the more dominant forms, generated in the larger areas and more efficient workshops of the north. In many islands the native productions are nearly equalled or even outnumbered by the naturalised; and if the natives have not been actually exterminated, their numbers have been greatly reduced, and this is the first stage towards extinction. A mountain is an island on the land; and the intertropical mountains before the Glacial period must have been completely isolated; and I believe that the productions of these islands on the land yielded to those produced within the larger areas of the north, just in the same way as the productions of real islands have everywhere lately yielded to continental forms, naturalised by man's agency.

I am far from supposing that all difficulties are removed on the view here given in regard to the range and affinities of the allied species which live in the northern and southern temperate zones and on the mountains of the intertropical regions. Very many difficulties remain to be solved. I do not pretend to indicate the exact lines and means of migration, or the reason why certain species and not others have migrated; why certain species have been modified and have given rise to new groups of forms, and others have remained unaltered. We cannot hope to explain such facts, until we can say why one species and not another becomes naturalised by man's agency in a foreign land; why one ranges twice or thrice as far, and is twice or thrice as common, as another species within their own homes.

I have said that many difficulties remain to be solved: some of the most remarkable are stated with admirable clearness by Dr. Hooker in his botanical works on the antarctic regions. These cannot be here discussed. I will only say that as far as regards the occurrence of identical species at points so enormously remote as Kerguelen Land, New Zealand, and Fuegia, I believe that towards the close of the Glacial period, icebergs, as suggested by Lyell, have been largely concerned in their dispersal. But the existence of several quite distinct species, belonging to genera exclusively confined to the south, at these and other distant points of the southern hemisphere, is, on my theory of descent with modification, a far more remarkable case of difficulty. For some of these species are so distinct, that we cannot suppose that there has been time since the commencement of the Glacial period for their migration, and for their subsequent modification to the necessary degree. The facts seem to me to indicate that peculiar and very distinct species have migrated in radiating lines from some common centre; and I am inclined to look in the southern, as in the northern hemisphere, to a former and warmer period, before the commencement of the Glacial period, when the antarctic lands, now covered with ice, supported a highly peculiar and isolated flora. I suspect that before this flora was exterminated by the Glacial epoch, a few forms were widely dispersed to various points of the southern hemisphere by occasional means of transport, and by the aid, as halting-places, of existing and now sunken islands, and perhaps at the commencement of the Glacial period, by icebergs. By these means, as I believe, the southern shores of America, Australia, New Zealand have become slightly tinted by the same peculiar forms of vegetable life.

Sir C. Lyell in a striking passage has speculated, in language almost identical with mine, on the effects of great alternations of climate on geographical distribution. I believe that the world has recently felt one of his great cycles of change; and that on this view, combined with modification through natural selection, a multitude of facts in the present distribution both of the same and of allied forms of life can be explained. The living waters may be said to have flowed during one short period from the north and from the south, and to have crossed at the equator; but to have flowed with greater force from the north so as to have freely inundated the south. As the tide leaves its drift in horizontal lines, though rising higher on the shores where the tide rises highest, so have the living waters left their living drift on our mountain-summits, in a line gently rising from the arctic lowlands to a great height under the equator. The various beings thus left stranded may be compared with savage races of man, driven up and surviving in the mountain-fastnesses of almost every land, which serve as a record, full of interest to us, of the former inhabitants of the surrounding lowlands.

CHAPTER XII

Geographical Distribution—continued

DISTRIBUTION OF FRESH-WATER PRODUCTIONS • ON THE INHABITANTS OF OCEANIC ISLANDS • ABSENCE OF BATRACHIANS AND OF TERRESTRIAL MAMMALS • ON THE RELATION OF THE INHABITANTS OF ISLANDS TO THOSE OF THE NEAREST MAINLAND • ON COLONISATION FROM THE NEAREST SOURCE WITH SUBSEQUENT MODIFICATION • SUMMARY OF THE LAST AND PRESENT CHAPTERS

As lakes and river-systems are separated from each other by barriers of land, it might have been thought that fresh-water productions would not have ranged widely within the same country, and as the sea is apparently a still more impassable barrier, that they never would have extended to distant countries. But the case is exactly the reverse. Not only have many fresh-water species, belonging to quite different classes, an enormous range, but allied species prevail in a remarkable manner throughout the world. I well remember, when first collecting in the fresh waters of Brazil, feeling much surprise at the similarity of the fresh-water insects, shells, &c., and at the dissimilarity of the surrounding terrestrial beings, compared with those of Britain.

But this power in fresh-water productions of ranging widely, though so unexpected, can, I think, in most cases be explained by their having become fitted, in a manner highly useful to them, for short and frequent migrations from pond to pond, or from stream to stream; and liability to wide dispersal would follow from this capacity as an almost necessary consequence. We can here consider only a few cases. In regard to fish, I believe that the same species never occur in the fresh waters of distant continents.

But on the same continent the species often range widely and almost capriciously; for two river-systems will have some fish in common and some different. A few facts seem to favour the possibility of their occasional transport by accidental means; like that of the live fish not rarely dropped by whirlwinds in India, and the vitality of their ova when removed from the water. But I am inclined to attribute the dispersal of fresh-water fish mainly to slight changes within the recent period in the level of the land, having caused rivers to flow into each other. Instances, also, could be given of this having occurred during floods, without any change of level. We have evidence in the loess of the Rhine of considerable changes of level in the land within a very recent geological period, and when the surface was peopled by existing land and fresh-water shells. The wide difference of the fish on opposite sides of continuous mountain-ranges, which from an early period must have parted river-systems and completely prevented their inosculation, seems to lead to this same conclusion. With respect to allied fresh-water fish occurring at very distant points of the world, no doubt there are many cases which cannot at present be explained: but some fresh-water fish belong to very ancient forms, and in such cases there will have been ample time for great geographical changes, and consequently time and means for much migration. In the second place, salt-water fish can with care be slowly accustomed to live in fresh water; and, according to Valenciennes, there is hardly a single group of fishes confined exclusively to fresh water, so that we may imagine that a marine member of a fresh-water group might travel far along the shores of the sea, and subsequently become modified and adapted to the fresh waters of a distant land.

Some species of fresh-water shells have a very wide range, and allied species, which, on my theory, are descended from a common parent and must have proceeded from a single source, prevail throughout the world. Their distribution at first perplexed me much, as their ova are not likely to be transported by birds, and they are immediately killed by sea water, as are the adults. I could not even understand how some naturalised species have rapidly spread throughout the same country. But two facts, which I have observed—and no doubt many others remain to be observed—throw some light on this subject. When a duck suddenly emerges from a pond covered with duck-weed, I have twice seen these little plants adhering to its back; and it has happened to me, in removing a little duck-weed from one aquarium to another, that I have quite unintentionally stocked the one with fresh-water shells from the other. But another agency is perhaps more effectual: I suspended a duck's feet, which might represent those of a bird sleeping in a natural pond, in an aquarium, where many ova of fresh-water shells were hatching; and I found that numbers of the extremely minute and just hatched shells crawled on the feet, and clung to them so firmly that when taken out of the water they could not be jarred off, though at a somewhat more advanced age they would voluntarily drop off. These just hatched molluscs, though aquatic in their nature, survived on the duck's feet, in damp air, from twelve to twenty hours; and in this length of time a duck or heron might fly at least six or seven hundred miles, and would be sure to alight on a pool or rivulet, if blown across sea to an oceanic island or to any other distant point. Sir Charles Lyell also informs me that a Dyticus has been caught with an Ancylus (a fresh-water shell like a limpet) firmly adhering to it; and a water-beetle of the same family, a Colymbetes, once flew on board the 'Beagle,' when forty-five miles distant from the nearest land: how much farther it might have flown with a favouring gale no one can tell.

With respect to plants, it has long been known what enormous ranges many fresh-water and even marsh-species have, both over continents and to the most remote oceanic islands. This is strikingly shown, as remarked by Alph. de Candolle, in large groups of terrestrial plants, which have only a very few aquatic members; for these latter seem immediately to acquire, as if in consequence, a very wide range. I think favourable means of dispersal explain this fact. I have before mentioned that earth occasionally, though rarely, adheres in some quantity to the feet and beaks of birds. Wading birds, which frequent the muddy edges of ponds, if suddenly flushed, would be the most likely to have muddy feet. Birds of this order I can show are the greatest wanderers, and are occasionally found on the most remote and barren islands in the open ocean; they would not be likely to alight on the surface of the sea, so that the dirt would not be washed off their feet; when making land, they would be sure to fly to their natural fresh-water haunts. I do not believe that botanists are aware how charged the mud of ponds is with seeds: I have tried several little experiments, but will here give only the most striking case: I took in February three table-spoonfuls of mud from three different points, beneath water, on the edge of a little pond; this mud when dry weighed only 6 3/4 ounces; I kept it covered up in my study for six months, pulling up and counting each plant as it grew; the plants were of many kinds, and were altogether 537 in number; and yet the viscid mud was all contained in a breakfast cup! Considering these facts, I think it would be an inexplicable circumstance if water-birds did not transport the seeds of fresh-water plants to vast distances, and if consequently the range of these plants was not very great. The same agency may have come into play with the eggs of some of the smaller fresh-water animals.

Other and unknown agencies probably have also played a part. I have stated that fresh-water fish eat some kinds of seeds, though they reject many other kinds after having swallowed them; even small fish swallow seeds of moderate size, as of the yellow water-lily and Potamogeton. Herons and other birds, century after century, have gone on daily devouring fish; they then take flight and go to other waters, or are blown across the sea; and we have seen that seeds retain their power of germination, when rejected in pellets or in excrement, many hours afterwards. When I saw the great size of the seeds of that fine water-lily, the Nelumbium, and remembered Alph. de Candolle's remarks on this plant, I thought that its distribution must remain quite inexplicable; but Audubon states that he found the seeds of the great southern water-lily (probably, according to Dr. Hooker, the Nelumbium luteum) in a heron's stomach; although I do not know the fact, yet analogy makes me believe that a heron flying to another pond and getting a hearty meal of fish, would probably reject from its stomach a pellet containing the seeds of the Nelumbium undigested; or the seeds might be dropped by the bird whilst feeding its young, in the same way as fish are known sometimes to be dropped.

In considering these several means of distribution, it should be remembered that when a pond or stream is first formed, for instance, on a rising islet, it will be unoccupied; and a single seed or egg will have a good chance of succeeding. Although there will always be a struggle for life between the individuals of the species, however few, already occupying any pond, yet as the number of kinds is small, compared with those on the land, the competition will probably be less severe between aquatic than between terrestrial species; consequently an intruder from the waters of a foreign country, would have a better chance of seizing on a place, than in the case of terrestrial colonists. We should, also, remember that some, perhaps many, fresh-water productions are low in the scale of nature, and that we have reason to believe that such low beings change or become modified less quickly than the high; and this will give longer time than the average for the migration of the same aquatic species. We should not forget the probability of many species having formerly ranged as continuously as fresh-water productions ever can range, over immense areas, and having subsequently become extinct in intermediate regions. But the wide distribution of fresh-water plants and of the lower animals, whether retaining the same identical form or in some degree modified, I believe mainly depends on the wide dispersal of their seeds and eggs by animals, more especially by fresh-water birds, which have large powers of flight, and naturally travel from one to another and often distant piece of water. Nature, like a careful gardener, thus takes her seeds from a bed of a particular nature, and drops them in another equally well fitted for them.

On the Inhabitants of Oceanic Islands.—We now come to the last of the three classes of facts, which I have selected as presenting the greatest amount of difficulty, on the view that all the individuals both of the same and of allied species have descended from a single parent; and therefore have all proceeded from a common birthplace, notwithstanding that in the course of time they have come to inhabit distant points of the globe. I have already stated that I cannot honestly admit Forbes's view on continental extensions, which, if legitimately followed out, would lead to the belief that within the recent period all existing islands have been nearly or quite joined to some continent. This view would remove many difficulties, but it would not, I think, explain all the facts in regard to insular productions. In the following remarks I shall not confine myself to the mere question of dispersal; but shall consider some other facts, which bear on the truth of the two theories of independent creation and of descent with modification.

The species of all kinds which inhabit oceanic islands are few in number compared with those on equal continental areas: Alph. de Candolle admits this for plants, and Wollaston for insects. If we look to the large size and varied stations of New Zealand, extending over 780 miles of latitude, and compare its flowering plants, only 750 in number, with those on an equal area at the Cape of Good Hope or in Australia, we must, I think, admit that something quite independently of any difference in physical conditions has caused so great a difference in number. Even the uniform county of Cambridge has 847 plants, and the little island of Anglesea 764, but a few ferns and a few introduced plants are included in these numbers, and the comparison in some other respects is not quite fair. We have evidence that the barren island of Ascension aboriginally possessed under half-a-dozen flowering plants; yet many have become naturalised on it, as they have on New Zealand and on every other oceanic island which can be named. In St. Helena there is reason to believe that the naturalised plants and animals have nearly or quite exterminated many native productions. He who admits the doctrine of the creation of each separate species, will have to admit, that a sufficient number of the best adapted plants and animals have not been created on oceanic islands; for man has unintentionally stocked them from various sources far more fully and perfectly than has nature.

Although in oceanic islands the number of kinds of inhabitants is scanty, the proportion of endemic species (*i.e.* those found nowhere else in the world) is often extremely large. If we compare, for instance, the number of the endemic land-shells in Madeira, or of the endemic birds in the Galapagos Archipelago, with the number found on any continent, and then compare the area of the islands with that of the continent, we shall see that this is true. This fact might have been expected on my theory, for, as already explained, species occasionally arriving after long intervals in a new and isolated district, and having to compete with new associates, will be eminently liable to modification, and will often produce groups of modified descendants. But it by no means follows, that, because in an island nearly all the species of one class are peculiar, those of another class, or of another section of the same class, are peculiar; and this difference seems to depend on the species which do not become modified having immigrated with facility and in a body, so that their mutual relations have not been much disturbed. Thus in the Galapagos Islands nearly every land-bird, but only two out of the eleven marine birds, are peculiar; and it is obvious that marine birds could arrive at these islands more easily than land-birds. Bermuda, on the other hand, which lies at about the same distance from North America as the Galapagos Islands do from South America, and which has a very peculiar soil, does not possess one endemic land bird; and we know from Mr. J. M. Jones's admirable account of Bermuda, that very many North American birds, during their great annual migrations, visit either periodically or occasionally this island. Madeira does not possess one peculiar bird, and many European and African birds are almost every year blown there, as I am informed by Mr. E. V. Harcourt. So that these two islands of Bermuda and Madeira have been stocked by birds, which for long ages have struggled together in their former homes, and have become mutually adapted to each other; and when settled in their new homes, each kind will have been kept by the others to their proper places and habits, and will consequently have been little liable to modification. Madeira, again, is inhabited by a wonderful number of peculiar land-shells, whereas not one species of sea-shell is confined to its shores: now, though we do not know how sea-shells are dispersed, yet we can see that their eggs or larvæ, perhaps attached to seaweed or floating timber, or to the feet of wading-birds, might be transported far more easily than land-shells, across three or four hundred miles of open sea. The different orders of insects in Madeira apparently present analogous facts.

Oceanic islands are sometimes deficient in certain classes, and their places are apparently occupied by the other inhabitants; in the Galapagos Islands reptiles, and in New Zealand gigantic wingless birds, take the place of mammals. In the plants of the Galapagos Islands, Dr. Hooker has shown that the proportional numbers of the different orders are very different from what they are elsewhere. Such cases are generally accounted for by the physical conditions of the islands; but this explanation seems to me not a little doubtful. Facility of immigration, I believe, has been at least as important as the nature of the conditions.

Many remarkable little facts could be given with respect to the inhabitants of remote islands. For instance, in certain islands not tenanted by mammals, some of the endemic plants have beautifully hooked seeds; yet few relations are more striking than the adaptation of hooked seeds for transportal by the wool and fur of quadrupeds. This case presents no difficulty on my view, for a hooked seed might be transported to an island by some other means; and the plant then becoming slightly modified, but still retaining its hooked seeds, would form an endemic species, having as useless an appendage as any rudimentary organ,—for instance, as the shrivelled wings under the soldered elytra of many insular beetles. Again, islands often possess trees or bushes belonging to orders which elsewhere include only herbaceous species; now trees, as Alph. de Candolle has shown, generally have, whatever the cause may be, confined ranges. Hence trees would be little likely to reach distant oceanic islands; and an herbaceous plant, though it would have no chance of successfully competing in stature with a fully developed tree, when established on an island and having to compete with herbaceous plants alone, might readily gain an advantage by growing taller and taller and overtopping the other plants. If so, natural selection would often tend to add to the stature of herbaceous plants when growing on an island, to whatever order they belonged, and thus convert them first into bushes and ultimately into trees.

With respect to the absence of whole orders on oceanic islands, Bory St. Vincent long ago remarked that Batrachians (frogs, toads, newts) have never been found on any of the many islands with which the great oceans are studded. I have taken pains to verify this assertion, and I have found it strictly true. I have, however, been assured that a frog exists on the mountains of the great island of New Zealand; but I suspect that this exception (if the information be correct) may be explained through glacial agency. This general absence of frogs, toads, and newts on so many oceanic islands cannot be accounted for by their physical conditions; indeed it seems that islands are peculiarly well fitted for these animals; for frogs have been introduced into Madeira, the Azores, and Mauritius, and have multiplied so as to become a nuisance. But as these animals and their spawn are known to be immediately killed by sea-water, on my view we can see that there would be great difficulty in their transportal across the sea, and therefore why they do not exist on any oceanic island. But why, on the theory of creation, they should not have been created there, it would be very difficult to explain.

Mammals offer another and similar case. I have carefully searched the oldest voyages, but have not finished my search; as yet I have not found a single instance, free from doubt, of a terrestrial mammal (excluding domesticated animals kept by the natives) inhabiting an island situated above 300 miles from a continent or great continental island; and many islands situated at a much less distance are equally barren. The Falkland Islands, which are inhabited by a wolf-like fox, come nearest to an exception; but this group cannot be considered as oceanic, as it lies on a bank connected with the mainland; moreover, icebergs formerly brought boulders to its western shores, and they may have formerly transported foxes, as so frequently now happens in the arctic regions. Yet it cannot be said that small islands will not support small mammals, for they occur in many parts of the world on very small islands, if close to a continent; and hardly an island can be named on which our smaller quadrupeds have not become naturalised and greatly multiplied. It cannot be said, on the ordinary view of creation, that there has not been time for the creation of mammals; many volcanic islands are sufficiently ancient, as shown by the stupendous degradation which they have suffered and by their tertiary strata: there has also been time for the production of endemic species belonging to other classes; and on continents it is thought that mammals appear and disappear at a quicker rate than other and lower animals. Though terrestrial mammals do not occur on oceanic islands, aërial mammals do occur on almost every island. New Zealand possesses two bats found nowhere else in the world: Norfolk Island, the Viti Archipelago, the Bonin Islands, the Caroline and Marianne Archipelagoes, and Mauritius, all possess their peculiar bats. Why, it may be asked, has the supposed creative force produced bats and no other mammals on remote islands? On my view this question can easily be answered; for no terrestrial mammal can be transported across a wide space of sea, but bats can fly across. Bats have been seen wandering by day far over the Atlantic Ocean; and two North American species either regularly or occasionally visit Bermuda, at the distance of 600 miles from the mainland. I hear from Mr. Tomes, who has specially studied this family, that many of the same species have enormous ranges, and are found on continents and on far distant islands. Hence we have only to suppose that such wandering species have been modified through natural selection in their new homes in relation to their new position, and we can understand the presence of endemic bats on islands, with the absence of all terrestrial mammals.

Besides the absence of terrestrial mammals in relation to the remoteness of islands from continents, there is also a relation, to a certain extent independent of distance, between the depth of the sea separating an island from the neighbouring mainland, and the presence in both of the same mammiferous species or of allied species in a more or less modified condition. Mr. Windsor Earl has made some striking observations on this head in regard to the great Malay Archipelago, which is traversed near Celebes by a space of deep ocean; and this space separates two widely distinct mammalian faunas. On either side the islands are situated on moderately deep submarine banks, and they are inhabited by closely allied or identical quadrupeds. No doubt some few anomalies occur in this great archipelago, and there is much difficulty in forming a judgment in some cases owing to the probable naturalisation of certain mammals through man's agency; but we shall soon have much light thrown on the natural history of this archipelago by the admirable zeal and researches of Mr. Wallace. I have not as yet had time to follow up this subject in all other quarters of the world; but as far as I have gone, the relation generally holds good. We see Britain separated by a shallow channel from Europe, and the mammals are the same on both sides; we meet with analogous facts on many islands separated by similar channels from Australia. The West Indian Islands stand on a deeply submerged bank, nearly 1000 fathoms in depth, and here we find American forms, but the species and even the genera are distinct. As the amount of modification in all cases depends to a certain degree on the lapse of time, and as during changes of level it is obvious that islands separated by shallow channels are more likely to have been continuously united within a recent period to the mainland than islands separated by deeper channels, we can understand the frequent relation between the depth of the sea and the degree of affinity of the mammalian inhabitants of islands with those of a neighbouring continent,—an inexplicable relation on the view of independent acts of creation.

All the foregoing remarks on the inhabitants of oceanic islands,—namely, the scarcity of kinds—the richness in endemic forms in particular classes or sections of classes,—the absence of whole groups, as of batrachians, and of terrestrial mammals notwithstanding the presence of aërial bats,—the singular proportions of certain orders of plants,—herbaceous forms having been developed into trees, &c.,—seem to me to accord better with the view of occasional means of transport having been largely efficient in the long course of time, than with the view of all our oceanic islands having been formerly connected by continuous land with the nearest continent; for on this latter view the migration would probably have been more complete; and if modification be admitted, all the forms of life would have been more equally modified, in accordance with the paramount importance of the relation of organism to organism.

I do not deny that there are many and grave difficulties in understanding how several of the inhabitants of the more remote islands, whether still retaining the same specific form or modified since their arrival, could have reached their present homes. But the probability of many islands having existed as halting-places, of which not a wreck now remains, must not be overlooked. I will here give a single instance of one of the cases of difficulty. Almost all oceanic islands, even the most isolated and smallest, are inhabited by land-shells, generally by endemic species, but sometimes by species found elsewhere. Dr. Aug. A. Gould has given several interesting cases in regard to the land-shells of the islands of the Pacific. Now it is notorious that land-shells are very easily killed by salt; their eggs, at least such as I have tried, sink in sea-water and are killed by it. Yet there must be, on my view, some unknown, but highly efficient means for their transportal. Would the just-hatched young occasionally crawl on and adhere to the feet of birds roosting on the ground, and thus get transported? It occurred to me that land-shells, when hybernating and having a membranous diaphragm over the mouth of the shell, might be floated in chinks of drifted timber across moderately wide arms of the sea. And I found that several species did in this state withstand uninjured an immersion in sea-water during seven days: one of these shells was the Helix pomatia, and after it had again hybernated I put it in sea-water for twenty days, and it perfectly recovered. As this species has a thick calcareous operculum, I removed it, and when it had formed a new membranous one, I immersed it for fourteen days in sea-water, and it recovered and crawled away: but more experiments are wanted on this head.

The most striking and important fact for us in regard to the inhabitants of islands, is their affinity to those of the nearest mainland, without being actually the same species. Numerous instances could be given of this fact. I will give only one, that of the Galapagos Archipelago, situated under the equator, between 500 and 600 miles from the shores of South America. Here almost every product of the land and water bears the unmistakeable stamp of the American continent. There are twenty-six land birds, and twenty-five of these are ranked by Mr. Gould as distinct species, supposed to have been created here; yet the close affinity of most of these birds to American species in every character, in their habits, gestures, and tones of voice, was manifest. So it is with the other animals, and with nearly all the plants, as shown by Dr. Hooker in his admirable memoir on the Flora of this archipelago. The naturalist, looking at the inhabitants of these volcanic islands in the Pacific, distant several hundred miles from the continent, yet feels that he is standing on American land. Why should this be so? why should the species which are supposed to have been created in the Galapagos Archipelago, and nowhere else, bear so plain a stamp of affinity to those created in America? There is nothing in the conditions of life, in the geological nature of the islands, in their height or climate, or in the proportions in which the several classes are associated together, which resembles closely the conditions of the South American coast: in fact there is a considerable dissimilarity in all these respects. On the other hand, there is a considerable degree of resemblance in the volcanic nature of the soil, in climate, height, and size of the islands, between the Galapagos and Cape de Verde Archipelagos: but what an entire and absolute difference in their inhabitants! The inhabitants of the Cape de Verde Islands are related to those of Africa, like those of the Galapagos to America. I believe this grand fact can receive no sort of explanation on the ordinary view of independent creation; whereas on the view here maintained, it is obvious that the Galapagos Islands would be likely to receive colonists, whether by occasional means of transport or by formerly continuous land, from America;

and the Cape de Verde Islands from Africa; and that such colonists would be liable to modification;—the principle of inheritance still betraying their original birthplace.

Many analogous facts could be given: indeed it is an almost universal rule that the endemic productions of islands are related to those of the nearest continent, or of other near islands. The exceptions are few, and most of them can be explained. Thus the plants of Kerguelen Land, though standing nearer to Africa than to America, are related, and that very closely, as we know from Dr. Hooker's account, to those of America: but on the view that this island has been mainly stocked by seeds brought with earth and stones on icebergs, drifted by the prevailing currents, this anomaly disappears. New Zealand in its endemic plants is much more closely related to Australia, the nearest mainland, than to any other region: and this is what might have been expected; but it is also plainly related to South America, which, although the next nearest continent, is so enormously remote, that the fact becomes an anomaly. But this difficulty almost disappears on the view that both New Zealand, South America, and other southern lands were long ago partially stocked from a nearly intermediate though distant point, namely from the antarctic islands, when they were clothed with vegetation, before the commencement of the Glacial period. The affinity, which, though feeble, I am assured by Dr. Hooker is real, between the flora of the south-western corner of Australia and of the Cape of Good Hope, is a far more remarkable case, and is at present inexplicable: but this affinity is confined to the plants, and will, I do not doubt, be some day explained.

The law which causes the inhabitants of an archipelago, though specifically distinct, to be closely allied to those of the nearest continent, we sometimes see displayed on a small scale, yet in a most interesting manner, within the limits of the same archipelago. Thus the several islands of the Galapagos Archipelago are tenanted, as I have elsewhere shown, in a quite marvellous manner, by very closely related species; so that the inhabitants of each separate island, though mostly distinct, are related in an incomparably closer degree to each other than to the inhabitants of any other part of the world. And this is just what might have been expected on my view, for the islands are situated so near each other that they would almost certainly receive immigrants from the same original source, or from each other. But this dissimilarity between the endemic inhabitants of the islands may be used as an argument against my views; for it may be asked, how has it happened in the several islands situated within sight of each other, having the same geological nature, the same height, climate, &c., that many of the immigrants should have been differently modified, though only in a small degree. This long appeared to me a great difficulty: but it arises in chief part from the deeply-seated error of considering the physical conditions of a country as the most important for its inhabitants; whereas it cannot, I think, be disputed that the nature of the other inhabitants, with which each has to compete, is at least as important, and generally a far more important element of success. Now if we look to those inhabitants of the Galapagos Archipelago which are found in other parts of the world (laying on one side for the moment the endemic species, which cannot be here fairly included, as we are considering how they have come to be modified since their arrival), we find a considerable amount of difference in the several islands. This difference might indeed have been expected on the view of the islands having been stocked by occasional means of transport—a seed, for instance, of one plant having been brought to one island, and that of another plant to another island. Hence when in former times an immigrant settled on any one or more of the islands, or when it subsequently spread from one island to another, it would undoubtedly be exposed to different conditions of life in the different islands, for it would have to compete with different sets of organisms: a plant, for instance, would find the best-fitted ground more perfectly occupied by distinct plants in one island than in another, and it would be exposed to the attacks of somewhat different enemies. If then it varied, natural selection would probably favour different varieties in the different islands. Some species, however, might spread and yet retain the same character throughout the group, just as we see on continents some species spreading widely and remaining the same.

The really surprising fact in this case of the Galapagos Archipelago, and in a lesser degree in some analogous instances, is that the new species formed in the separate islands have not quickly spread to the other islands. But the islands, though in sight of each other, are separated by deep arms of the sea, in most cases wider than the British Channel, and there is no reason to suppose that they have at any former period been continuously united. The currents of the sea are rapid and sweep across the archipelago, and gales of wind are extraordinarily rare; so that the islands are far more effectually separated from each other than they appear to be on a map. Nevertheless a good many species, both those found in other parts of the world and those confined to the archipelago, are common to the several islands, and we may infer from certain facts that these have probably spread from some one island to the others. But we often take, I think, an erroneous view of the probability of closely allied species invading each other's territory, when put into free intercommunication. Undoubtedly if one species has any advantage whatever over another, it will in a very brief time wholly or in part supplant it; but if both are equally well fitted for their own places in nature, both probably will hold their own places and keep separate for almost any length of time. Being familiar with the fact that many species, naturalised through man's agency, have spread with astonishing rapidity over new countries, we are apt to infer that most species would thus spread; but we should remember that the forms which become naturalised in new countries are not generally closely allied to the aboriginal inhabitants, but are very distinct species, belonging in a large proportion of cases, as shown by Alph. de Candolle, to distinct genera. In the Galapagos Archipelago, many even of the birds, though so well adapted for flying from island to island, are distinct on each; thus there are three closely-allied species of mocking-thrush, each confined to its own island. Now let us suppose the mocking-thrush of Chatham Island to be blown to Charles Island, which has its own mocking-thrush: why should it succeed in establishing itself there? We may safely infer that Charles Island is well stocked with its own species, for annually more eggs are laid there than can possibly be reared; and we may infer that the mocking-thrush peculiar to Charles Island is at least as well fitted for its home as is the species peculiar to Chatham Island. Sir C. Lyell and Mr. Wollaston have communicated to me a remarkable fact bearing on this subject; namely, that Madeira and the adjoining islet of Porto Santo possess many distinct but representative land-shells, some of which live in crevices of stone; and although large quantities of stone are annually transported from Porto Santo to Madeira, yet this latter island has not become colonised by the Porto Santo species: nevertheless both islands have been colonised by some European land-shells, which no doubt had some advantage over the indigenous species. From these considerations I think we need not greatly marvel at the endemic and representative species, which inhabit the several islands of the Galapagos Archipelago, not having universally spread from island to island. In many other instances, as in the several districts of the same continent, pre-occupation has probably played an important part in checking the commingling of species under the same conditions of life. Thus, the south-east and south-west corners of Australia have nearly the same physical conditions, and are united by continuous land, yet they are inhabited by a vast number of distinct mammals, birds, and plants.

The principle which determines the general character of the fauna and flora of oceanic islands, namely, that the inhabitants, when not identically the same, yet are plainly related to the inhabitants of that region whence colonists could most readily have been derived,—the colonists having been subsequently modified and better fitted to their new homes,—is of the widest application throughout nature. We see this on every mountain, in every lake and marsh. For Alpine species, excepting in so far as the same forms, chiefly of plants, have spread widely throughout the world during the recent Glacial epoch, are related to those of the surrounding lowlands;—thus we have in South America, Alpine humming-birds, Alpine rodents, Alpine plants, &c., all of strictly American forms, and it is obvious that a mountain, as it became slowly upheaved, would naturally be colonised from the surrounding lowlands. So it is with the inhabitants of lakes and marshes, excepting in so far as great facility of transport has given the same general forms to the whole world. We see this same principle in the blind animals inhabiting the caves of America and of Europe. Other analogous facts could be given. And it will, I believe, be universally found to be true, that wherever in two regions, let them be ever so distant, many closely allied or representative species occur, there will likewise be found some identical species, showing, in accordance with the foregoing view, that at some former period there has been intercommunication or migration between the two regions. And wherever many closely-allied species occur, there will be found many forms which some naturalists rank as distinct species, and some as varieties; these doubtful forms showing us the steps in the process of modification.

This relation between the power and extent of migration of a species, either at the present time or at some former period under different physical conditions, and the existence at remote points of the world of other species allied to it, is shown in another and more general way. Mr. Gould remarked to me long ago, that in those genera of birds which range over the world, many of the species have very wide ranges. I can hardly doubt that this rule is generally true, though it would be difficult to prove it. Amongst mammals, we see it strikingly displayed in Bats, and in a lesser degree in the Felidæ and Canidæ. We see it, if we compare the distribution of butterflies and beetles. So it is with most fresh-water productions, in which so many genera range over the world, and many individual species have enormous ranges. It is not meant that in world-ranging genera all the species have a wide range, or even that they have on an *average* a wide range; but only that some of the species range very widely; for the facility with which widely-ranging species vary and give rise to new forms will largely determine their average range. For instance, two varieties of the same species inhabit America and Europe, and the species thus has an immense range; but, if the variation had been a little greater, the two varieties would have been ranked as distinct species, and the common range would have been greatly reduced. Still less is it meant, that a species which apparently has the capacity of crossing barriers and ranging widely, as in the case of certain powerfully-winged birds, will necessarily range widely; for we should never forget that to range widely implies not only the power of crossing barriers, but the more important power of being victorious in distant lands in the struggle for life with foreign associates. But on the view of all the species of a genus having descended from a single parent, though now distributed to the most remote points of the world, we ought to find, and I believe as a general rule we do find, that some at least of the species range very widely; for it is necessary that the unmodified parent should range widely, undergoing modification during its diffusion, and should place itself under diverse conditions favourable for the conversion of its offspring, firstly into new varieties and ultimately into new species.

In considering the wide distribution of certain genera, we should bear in mind that some are extremely ancient, and must have branched off from a common parent at a remote epoch; so that in such cases there will have been ample time for great climatal and geographical changes and for accidents of transport; and consequently for the migration of some of the species into all quarters of the world, where they may have become slightly modified in relation to their new conditions. There is, also, some reason to believe from geological evidence that organisms low in the scale within each great class, generally change at a slower rate than the higher forms; and consequently the lower forms will have had a better chance of ranging widely and of still retaining the same specific character. This fact, together with the seeds and eggs of many low forms being very minute and better fitted for distant transportation, probably accounts for a law which has long been observed, and which has lately been admirably discussed by Alph. de Candolle in regard to plants, namely, that the lower any group of organisms is, the more widely it is apt to range.

The relations just discussed,—namely, low and slowly-changing organisms ranging more widely than the high,—some of the species of widely-ranging genera themselves ranging widely,—such facts, as alpine, lacustrine, and marsh productions being related (with the exceptions before specified) to those on the surrounding low lands and dry lands, though these stations are so different—the very close relation of the distinct species which inhabit the islets of the same archipelago,—and especially the striking relation of the inhabitants of each whole archipelago or island to those of the nearest mainland,—are, I think, utterly inexplicable on the ordinary view of the independent creation of each species, but are explicable on the view of colonisation from the nearest and readiest source, together with the subsequent modification and better adaptation of the colonists to their new homes.

Summary of last and present Chapters.—In these chapters I have endeavoured to show, that if we make due allowance for our ignorance of the full effects of all the changes of climate and of the level of the land, which have certainly occurred within the recent period, and of other similar changes which may have occurred within the same period; if we remember how profoundly ignorant we are with respect to the many and curious means of occasional transport,—a subject which has hardly ever been properly experimentised on; if we bear in mind how often a species may have ranged continuously over a wide area, and then have become extinct in the intermediate tracts, I think the difficulties in believing that all the individuals of the same species, wherever located, have descended from the same parents, are not insuperable. And we are led to this conclusion, which has been arrived at by many naturalists under the designation of single centres of creation, by some general considerations, more especially from the importance of barriers and from the analogical distribution of sub-genera, genera, and families.

With respect to the distinct species of the same genus, which on my theory must have spread from one parent-source; if we make the same allowances as before for our ignorance, and remember that some forms of life change most slowly, enormous periods of time being thus granted for their migration, I do not think that the difficulties are insuperable; though they often are in this case, and in that of the individuals of the same species, extremely grave.

As exemplifying the effects of climatal changes on distribution, I have attempted to show how important has been the influence of the modern Glacial period, which I am fully convinced simultaneously affected the whole world, or at least great meridional belts. As showing how diversified are the means of occasional transport, I have discussed at some little length the means of dispersal of fresh-water productions.

If the difficulties be not insuperable in admitting that in the long course of time the individuals of the same species, and likewise of allied species, have proceeded from some one source; then I think all the grand leading facts of geographical distribution are explicable on the theory of migration (generally of the more dominant forms of life), together with subsequent modification and the multiplication of new forms. We can thus understand the high importance of barriers, whether of land or water, which separate our several zoological and botanical provinces. We can thus understand the localisation of sub-genera, genera, and families; and how it is that under different latitudes, for instance in South America, the inhabitants of the plains and mountains, of the forests, marshes, and deserts, are in so mysterious a manner linked together by affinity, and are likewise linked to the extinct beings which formerly inhabited the same continent. Bearing in mind that the mutual relations of organism to organism are of the highest importance, we can see why two areas having nearly the same physical conditions should often be inhabited by very different forms of life; for according to the length of time which has elapsed since new inhabitants entered one region; according to the nature of the communication which allowed certain forms and not others to enter, either in greater or lesser numbers; according or not, as those which entered happened to come in more or less direct competition with each other and with the aborigines; and according as the immigrants were capable of varying more or less rapidly, there would ensue in different regions, independently of their physical conditions, infinitely diversified conditions of life,—there would be an almost endless amount of organic action and reaction,—and we should find, as we do find, some groups of beings greatly, and some only slightly modified,—some developed in great force, some existing in scanty numbers—in the different great geographical provinces of the world.

On these same principles, we can understand, as I have endeavoured to show, why oceanic islands should have few inhabitants, but of these a great number should be endemic or peculiar; and why, in relation to the means of migration, one group of beings, even within the same class, should have all its species endemic, and another group should have all its species common to other quarters of the world. We can see why whole groups of organisms, as batrachians and terrestrial mammals, should be absent from oceanic islands, whilst the most isolated islands possess their own peculiar species of aërial mammals or bats. We can see why there should be some relation between the presence of mammals, in a more or less modified condition, and the depth of the sea between an island and the mainland. We can clearly see why all the inhabitants of an archipelago, though specifically distinct on the several islets, should be closely related to each other, and likewise be related, but less closely, to those of the nearest continent or other source whence immigrants were probably derived. We can see why in two areas, however distant from each other, there should be a correlation, in the presence of identical species, of varieties, of doubtful species, and of distinct but representative species.

As the late Edward Forbes often insisted, there is a striking parallelism in the laws of life throughout time and space: the laws governing the succession of forms in past times being nearly the same with those governing at the present time the differences in different areas. We see this in many facts. The endurance of each species and group of species is continuous in time; for the exceptions to the rule are so few, that they may fairly be attributed to our not having as yet discovered in an intermediate deposit the forms which are therein absent, but which occur above and below: so in space, it certainly is the general rule that the area inhabited by a single species, or by a group of species, is continuous; and the exceptions, which are not rare, may, as I have attempted to show, be accounted for by migration at some former period under different conditions or by occasional means of transport, and by the species having become extinct in the intermediate tracts. Both in time and space, species and groups of species have their points of maximum development. Groups of species, belonging either to a certain period of time, or to a certain area, are often characterised by trifling characters in common, as of sculpture or colour. In looking to the long succession of ages, as in now looking to distant provinces throughout the world, we find that some organisms differ little, whilst others belonging to a different class, or to a different order, or even only to a different family of the same order, differ greatly. In both time and space the lower members of each class generally change less than the higher; but there are in both cases marked exceptions to the rule. On my theory these several relations throughout time and space are intelligible; for whether we look to the forms of life which have changed during successive ages within the same quarter of the world, or to those which have changed after having migrated into distant quarters, in both cases the forms within each class have been connected by the same bond of ordinary generation; and the more nearly any two forms are related in blood, the nearer they will generally stand to each other in time and space; in both cases the laws of variation have been the same, and modifications have been accumulated by the same power of natural selection.

CHAPTER XIII

Mutual Affinities of Organic Beings: Morphology: Embryology: Rudimentary Organs

CLASSIFICATION, GROUPS SUBORDINATE TO GROUPS • NATURAL SYSTEM • RULES AND DIFFICULTIES IN CLASSIFICATION, EXPLAINED ON THE THEORY OF DESCENT WITH MODIFICATION • CLASSIFICATION OF VARIETIES • DESCENT ALWAYS USED IN CLASSIFICATION • ANALOGICAL OR ADAPTIVE CHARACTERS • AFFINITIES, GENERAL, COMPLEX AND RADIATING • EXTINCTION SEPARATES AND DEFINES GROUPS • MORPHOLOGY, BETWEEN MEMBERS OF THE SAME CLASS, BETWEEN PARTS OF THE SAME INDIVIDUAL • EMBRYOLOGY, LAWS OF, EXPLAINED BY VARIATIONS NOT SUPERVENING AT AN EARLY AGE, AND BEING INHERITED AT A CORRESPONDING AGE • RUDIMENTARY ORGANS; THEIR ORIGIN EXPLAINED • SUMMARY

FROM THE FIRST DAWN of life, all organic beings are found to resemble each other in descending degrees, so that they can be classed in groups under groups. This classification is evidently not arbitrary like the grouping of the stars in constellations. The existence of groups would have been of simple signification, if one group had been exclusively fitted to inhabit the land, and another the water; one to feed on flesh, another on vegetable matter, and so on; but the case is widely different in nature; for it is notorious how commonly members of even the same sub-group have different habits. In our second and fourth chapters, on Variation and on Natural Selection, I have attempted to show that it is the widely ranging, the much diffused and common, that is the dominant species belonging to the larger genera, which vary most. The varieties, or incipient species, thus produced ultimately become converted, as I believe, into new and distinct species; and these, on the principle of inheritance, tend to produce other new and dominant species. Consequently the groups which are now large, and which generally include many dominant species, tend to go on increasing indefinitely in size. I further attempted to show that from the varying descendants of each species trying to occupy as many and as different places as possible in the economy of nature, there is a constant tendency in their characters to diverge. This conclusion was supported by looking at the great diversity of the forms of life which, in any small area, come into the closest competition, and by looking to certain facts in naturalisation.

I attempted also to show that there is a constant tendency in the forms which are increasing in number and diverging in character, to supplant and exterminate the less divergent, the less improved, and preceding forms. I request the reader to turn to the diagram illustrating the action, as formerly explained, of these several principles; and he will see that the inevitable result is that the modified descendants proceeding from one progenitor become broken up into groups subordinate to groups. In the diagram each letter on the uppermost line may represent a genus including several species; and all the genera on this line form together one class, for all have descended from one ancient but unseen parent, and, consequently, have inherited something in common. But the three genera on the left hand have, on this same principle, much in common, and form a sub-family, distinct from that including the next two genera on the right hand, which diverged from a common parent at the fifth stage of descent. These five genera have also much, though less, in common; and they form a family distinct from that including the three genera still further to the right hand, which diverged at a still earlier period. And all these genera, descended from (A), form an order distinct from the genera descended from (I). So that we here have many species descended from a single progenitor grouped into genera; and the genera are included in, or subordinate to, sub-families, families, and orders, all united into one class. Thus, the grand fact in natural history of the subordination of group under group, which, from its familiarity, does not always sufficiently strike us, is in my judgment fully explained.

Naturalists try to arrange the species, genera, and families in each class, on what is called the Natural System. But what is meant by this system? Some authors look at it merely as a scheme for arranging together those living objects which are most alike, and for separating those which are most unlike; or as an artificial means for enunciating, as briefly as possible, general propositions,—that is, by one sentence to give the characters common, for instance, to all mammals, by another those common to all carnivora, by another those common to the dog-genus, and then by adding a single sentence, a full description is given of each kind of dog. The ingenuity and utility of this system are indisputable. But many naturalists think that something more is meant by the Natural System; they believe that it reveals the plan of the Creator; but unless it be specified whether order in time or space, or what else is meant by the plan of the Creator, it seems to me that nothing is thus added to our knowledge. Such expressions as that famous one of Linnæus, and which we often meet with in a more or less concealed form, that the characters do not make the genus, but that the genus gives the characters, seem to imply that something more is included in our classification, than mere resemblance. I believe that something more is included; and that propinquity of descent,—the only known cause of the similarity of organic beings,—is the bond, hidden as it is by various degrees of modification, which is partially revealed to us by our classifications.

Let us now consider the rules followed in classification, and the difficulties which are encountered on the view that classification either gives some unknown plan of creation, or is simply a scheme for enunciating general propositions and of placing together the forms most like each other. It might have been thought (and was in ancient times thought) that those parts of the structure which determined the habits of life, and the general place of each being in the economy of nature, would be of very high importance in classification. Nothing can be more false. No one regards the external similarity of a mouse to a shrew, of a dugong to a whale, of a whale to a fish, as of any importance. These resemblances, though so intimately connected with the whole life of the being, are ranked as merely "adaptive or analogical characters;" but to the consideration of these resemblances we shall have to recur. It may even be given as a general rule, that the less any part of the organisation is concerned with special habits, the more important it becomes for classification. As an instance: Owen, in speaking of the dugong, says, "The generative organs being those which are most remotely related to the habits and food of an animal, I have always regarded as affording very clear indications of its true affinities. We are least likely in the modifications of these organs to mistake a merely adaptive for an essential character." So with plants, how remarkable it is that the organs of vegetation, on which their whole life depends, are of little signification, excepting in the first main divisions; whereas the organs of reproduction, with their product the seed, are of paramount importance!

We must not, therefore, in classifying, trust to resemblances in parts of the organisation, however important they may be for the welfare of the being in relation to the outer world. Perhaps from this cause it has partly arisen, that almost all naturalists lay the greatest stress on resemblances in organs of high vital or physiological importance. No doubt this view of the classificatory importance of organs which are important is generally, but by no means always, true. But their importance for classification, I believe, depends on their greater constancy throughout large groups of species; and this constancy depends on such organs having generally been subjected to less change in the adaptation of the species to their conditions of life. That the mere physiological importance of an organ does not determine its classificatory value, is almost shown by the one fact, that in allied groups, in which the same organ, as we have every reason to suppose, has nearly the same physiological value, its classificatory value is widely different. No naturalist can have worked at any group without being struck with this fact; and it has been most fully acknowledged in the writings of almost every author. It will suffice to quote the highest authority, Robert Brown, who in speaking of certain organs in the Proteaceæ, says their generic importance, "like that of all their parts, not only in this but, as I apprehend, in every natural family, is very unequal, and in some cases seems to be entirely lost." Again in another work he says, the genera of the Connaraceæ "differ in having one or more ovaria, in the existence or absence of albumen, in the imbricate or valvular æstivation. Any one of these characters singly is frequently of more than generic importance, though here even when all taken together they appear insufficient to separate Cnestis from Connarus." To give an example amongst insects, in one great division of the Hymenoptera, the antennæ, as Westwood has remarked, are most constant in structure; in another division they differ much, and the differences are of quite subordinate value in classification; yet no one probably will say that the antennæ in these two divisions of the same order are of unequal physiological importance. Any number of instances could be given of the varying importance for classification of the same important organ within the same group of beings.

Again, no one will say that rudimentary or atrophied organs are of high physiological or vital importance; yet, undoubtedly, organs in this condition are often of high value in classification. No one will dispute that the rudimentary teeth in the upper jaws of young ruminants, and certain rudimentary bones of the leg, are highly serviceable in exhibiting the close affinity between Ruminants and Pachyderms. Robert Brown has strongly insisted on the fact that the rudimentary florets are of the highest importance in the classification of the Grasses.

Numerous instances could be given of characters derived from parts which must be considered of very trifling physiological importance, but which are universally admitted as highly serviceable in the definition of whole groups. For instance, whether or not there is an open passage from the nostrils to the mouth, the only character, according to Owen, which absolutely distinguishes fishes and reptiles—the inflection of the angle of the jaws in Marsupials—the manner in which the wings of insects are folded—mere colour in certain Algæ—mere pubescence on parts of the flower in grasses—the nature of the dermal covering, as hair or feathers, in the Vertebrata. If the Ornithorhynchus had been covered with feathers instead of hair, this external and trifling character would, I think, have been considered by naturalists as important an aid in determining the degree of affinity of this strange creature to birds and reptiles, as an approach in structure in any one internal and important organ.

The importance, for classification, of trifling characters, mainly depends on their being correlated with several other characters of more or less importance. The value indeed of an aggregate of characters is very evident in natural history. Hence, as has often been remarked, a species may depart from its allies in several characters, both of high physiological importance and of almost universal prevalence, and yet leave us in no doubt where it should be ranked. Hence, also, it has been found, that a classification founded on any single character, however important that may be, has always failed; for no part of the organisation is universally constant. The importance of an aggregate of characters, even when none are important, alone explains, I think, that saying of Linnæus, that the characters do not give the genus, but the genus gives the characters; for this saying seems founded on an appreciation of many trifling points of resemblance, too slight to be defined. Certain plants, belonging to the Malpighiaceæ, bear perfect and degraded flowers; in the latter, as A. de Jussieu has remarked, "the greater number of the characters proper to the species, to the genus, to the family, to the class, disappear, and thus laugh at our classification." But when Aspicarpa produced in France, during several years, only degraded flowers, departing so wonderfully in a number of the most important points of structure from the proper type of the order, yet M. Richard sagaciously saw, as Jussieu observes, that this genus should still be retained amongst the Malpighiaceæ. This case seems to me well to illustrate the spirit with which our classifications are sometimes necessarily founded.

Practically when naturalists are at work, they do not trouble themselves about the physiological value of the characters which they use in defining a group, or in allocating any particular species. If they find a character nearly uniform, and common to a great number of forms, and not common to others, they use it as one of high value; if common to some lesser number, they use it as of subordinate value. This principle has been broadly confessed by some naturalists to be the true one; and by none more clearly than by that excellent botanist, Aug. St. Hilaire. If certain characters are always found correlated with others, though no apparent bond of connexion can be discovered between them, especial value is set on them. As in most groups of animals, important organs, such as those for propelling the blood, or for aërating it, or those for propagating the race, are found nearly uniform, they are considered as highly serviceable in classification; but in some groups of animals all these, the most important vital organs, are found to offer characters of quite subordinate value.

We can see why characters derived from the embryo should be of equal importance with those derived from the adult, for our classifications of course include all ages of each species. But it is by no means obvious, on the ordinary view, why the structure of the embryo should be more important for this purpose than that of the adult, which alone plays its full part in the economy of nature. Yet it has been strongly urged by those great naturalists, Milne Edwards and Agassiz, that embryonic characters are the most important of any in the classification of animals; and this doctrine has very generally been admitted as true. The same fact holds good with flowering plants, of which the two main divisions have been founded on characters derived from the embryo,—on the number and position of the embryonic leaves or cotyledons, and on the mode of development of the plumule and radicle. In our discussion on embryology, we shall see why such characters are so valuable, on the view of classification tacitly including the idea of descent.

Our classifications are often plainly influenced by chains of affinities. Nothing can be easier than to define a number of characters common to all birds; but in the case of crustaceans, such definition has hitherto been found impossible. There are crustaceans at the opposite ends of the series, which have hardly a character in common; yet the species at both ends, from being plainly allied to others, and these to others, and so onwards, can be recognised as unequivocally belonging to this, and to no other class of the Articulata.

Geographical distribution has often been used, though perhaps not quite logically, in classification, more especially in very large groups of closely allied forms. Temminck insists on the utility or even necessity of this practice in certain groups of birds; and it has been followed by several entomologists and botanists.

Finally, with respect to the comparative value of the various groups of species, such as orders, sub-orders, families, sub-families, and genera, they seem to be, at least at present, almost arbitrary. Several of the best botanists, such as Mr. Bentham and others, have strongly insisted on their arbitrary value. Instances could be given amongst plants and insects, of a group of forms, first ranked by practised naturalists as only a genus, and then raised to the rank of a sub-family or family; and this has been done, not because further research has detected important structural differences, at first overlooked, but because numerous allied species, with slightly different grades of difference, have been subsequently discovered.

All the foregoing rules and aids and difficulties in classification are explained, if I do not greatly deceive myself, on the view that the natural system is founded on descent with modification; that the characters which naturalists consider as showing true affinity between any two or more species, are those which have been inherited from a common parent, and, in so far, all true classification is genealogical; that community of descent is the hidden bond which naturalists have been unconsciously seeking, and not some unknown plan of creation, or the enunciation of general propositions, and the mere putting together and separating objects more or less alike.

But I must explain my meaning more fully. I believe that the *arrangement* of the groups within each class, in due subordination and relation to the other groups, must be strictly genealogical in order to be natural; but that the *amount* of difference in the several branches or groups, though allied in the same degree in blood to their common progenitor, may differ greatly, being due to the different degrees of modification which they have undergone; and this is expressed by the forms being ranked under different genera, families, sections, or orders. The reader will best understand what is meant, if he will take the trouble of referring to the diagram in the fourth chapter. We will suppose the letters A to L to represent allied genera, which lived during the Silurian epoch, and these have descended from a species which existed at an unknown anterior period. Species of three of these genera (A, F, and I) have transmitted modified descendants to the present day, represented by the fifteen genera (a^{14} to z^{14}) on the uppermost horizontal line. Now all these modified descendants from a single species, are represented as related in blood or descent to the same degree; they may metaphorically be called cousins to the same millionth degree; yet they differ widely and in different degrees from each other. The forms descended from A, now broken up into two or three families, constitute a distinct order from those descended from I, also broken up into two families. Nor can the existing species, descended from A, be ranked in the same genus with the parent A; or those from I, with the parent I. But the existing genus F^{14} may be supposed to have been but slightly modified; and it will then rank with the parent-genus F; just as some few still living organic beings belong to Silurian genera. So that the amount or value of the differences between organic beings all related to each other in the same degree in blood, has come to be widely different. Nevertheless their genealogical *arrangement* remains strictly true, not only at the present time, but at each successive period of descent. All the modified descendants from A will have inherited something in common from their common parent, as will all the descendants from I; so will it be with each subordinate branch of descendants, at each successive period. If, however, we choose to suppose that any of the descendants of A or of I have been so much modified as to have more or less completely lost traces of their parentage, in this case, their places in a natural classification will have been more or less completely lost,—as sometimes seems to have occurred with existing organisms. All the descendants of the genus F, along its whole line of descent, are supposed to have been but little modified, and they yet form a single genus. But this genus, though much isolated, will still occupy its proper intermediate position; for F originally was intermediate in character between A and I, and the several genera descended from these two genera will have inherited to a certain extent their characters. This natural arrangement is shown, as far as is possible on paper, in the diagram, but in much too simple a manner. If a branching diagram had not been used, and only the names of the groups had been written in a linear series, it would have been still less possible to have given a natural arrangement; and it is notoriously not possible to represent in a series, on a flat surface, the affinities which we discover in nature amongst the beings of the same group. Thus, on the view which I hold, the natural system is genealogical in its arrangement, like a pedigree; but the degrees of modification which the different groups have undergone, have to be expressed by ranking them under different so-called genera, sub-families, families, sections, orders, and classes.

It may be worth while to illustrate this view of classification, by taking the case of languages. If we possessed a perfect pedigree of mankind, a genealogical arrangement of the races of man would afford the best classification of the various languages now spoken throughout the world; and if all extinct languages, and all intermediate and slowly changing dialects, had to be included, such an arrangement would, I think, be the only possible one. Yet it might be that some very ancient language had altered little, and had given rise to few new languages, whilst others (owing to the spreading and subsequent isolation and states of civilisation of the several races, descended from a common race) had altered much, and had given rise to many new languages and dialects. The various degrees of difference in the languages from the same stock, would have to be expressed by groups subordinate to groups; but the proper or even only possible arrangement would still be genealogical; and this would be strictly natural, as it would connect together all languages, extinct and modern, by the closest affinities, and would give the filiation and origin of each tongue.

In confirmation of this view, let us glance at the classification of varieties, which are believed or known to have descended from one species. These are grouped under species, with sub-varieties under varieties; and with our domestic productions, several other grades of difference are requisite, as we have seen with pigeons. The origin of the existence of groups subordinate to groups, is the same with varieties as with species, namely, closeness of descent with various degrees of modification. Nearly the same rules are followed in classifying varieties, as with species. Authors have insisted on the necessity of classing varieties on a natural instead of an artificial system; we are cautioned, for instance, not to class two varieties of the pine-apple together, merely because their fruit, though the most important part, happens to be nearly identical; no one puts the Swedish and common turnips together, though the esculent and thickened stems are so similar. Whatever part is found to be most constant, is used in classing varieties: thus the great agriculturist Marshall says the horns are very useful for this purpose with cattle, because they are less variable than the shape or colour of the body, &c.; whereas with sheep the horns are much less serviceable, because less constant. In classing varieties, I apprehend if we had a real pedigree, a genealogical classification would be universally preferred; and it has been attempted by some authors. For we might feel sure, whether there had been more or less modification, the principle of inheritance would keep the forms together which were allied in the greatest number of points. In tumbler pigeons, though some sub-varieties differ from the others in the important character of having a longer beak, yet all are kept together from having the common habit of tumbling; but the short-faced breed has nearly or quite lost this habit; nevertheless, without any reasoning or thinking on the subject, these tumblers are kept in the same group, because allied in blood and alike in some other respects. If it could be proved that the Hottentot had descended from the Negro, I think he would be classed under the Negro group, however much he might differ in colour and other important characters from negroes.

With species in a state of nature, every naturalist has in fact brought descent into his classification; for he includes in his lowest grade, or that of a species, the two sexes; and how enormously these sometimes differ in the most important characters, is known to every naturalist: scarcely a single fact can be predicated in common of the males and hermaphrodites of certain cirripedes, when adult, and yet no one dreams of separating them. The naturalist includes as one species the several larval stages of the same individual, however much they may differ from each other and from the adult; as he likewise includes the so-called alternate generations of Steenstrup, which can only in a technical sense be considered as the same individual. He includes monsters; he includes varieties, not solely because they closely resemble the parent-form, but because they are descended from it. He who believes that the cowslip is descended from the primrose, or conversely, ranks them together as a single species, and gives a single definition. As soon as three Orchidean forms (Monochanthus, Myanthus, and Catasetum), which had previously been ranked as three distinct genera, were known to be sometimes produced on the same spike, they were immediately included as a single species. But it may be asked, what ought we to do, if it could be proved that one species of kangaroo had been produced, by a long course of modification, from a bear? Ought we to rank this one species with bears, and what should we do with the other species? The supposition is of course preposterous; and I might answer by the *argumentum ad hominem,* and ask what should be done if a perfect kangaroo were seen to come out of the womb of a bear? According to all analogy, it would be ranked with bears; but then assuredly all the other species of the kangaroo family would have to be classed under the bear genus. The whole case is preposterous; for where there has been close descent in common, there will certainly be close resemblance or affinity.

As descent has universally been used in classing together the individuals of the same species, though the males and females and larvæ are sometimes extremely different; and as it has been used in classing varieties which have undergone a certain, and sometimes a considerable amount of modification, may not this same element of descent have been unconsciously used in grouping species under genera, and genera under higher groups, though in these cases the modification has been greater in degree, and has taken a longer time to complete? I believe it has thus been unconsciously used; and only thus can I understand the several rules and guides which have been followed by our best systematists. We have no written pedigrees; we have to make out community of descent by resemblances of any kind. Therefore we choose those characters which, as far as we can judge, are the least likely to have been modified in relation to the conditions of life to which each species has been recently exposed. Rudimentary structures on this view are as good as, or even sometimes better than, other parts of the organisation. We care not how trifling a character may be—let it be the mere inflection of the angle of the jaw, the manner in which an insect's wing is folded, whether the skin be covered by hair or feathers—if it prevail throughout many and different species, especially those having very different habits of life, it assumes high value; for we can account for its presence in so many forms with such different habits, only by its inheritance from a common parent. We may err in this respect in regard to single points of structure, but when several characters, let them be ever so trifling, occur together throughout a large group of beings having different habits, we may feel almost sure, on the theory of descent, that these characters have been inherited from a common ancestor. And we know that such correlated or aggregated characters have especial value in classification.

We can understand why a species or a group of species may depart, in several of its most important characteristics, from its allies, and yet be safely classed with them. This may be safely done, and is often done, as long as a sufficient number of characters, let them be ever so unimportant, betrays the hidden bond of community of descent. Let two forms have not a single character in common, yet if these extreme forms are connected together by a chain of intermediate groups, we may at once infer their community of descent, and we put them all into the same class. As we find organs of high physiological importance—those which serve to preserve life under the most diverse conditions of existence—are generally the most constant, we attach especial value to them; but if these same organs, in another group or section of a group, are found to differ much, we at once value them less in our classification. We shall hereafter, I think, clearly see why embryological characters are of such high classificatory importance. Geographical distribution may sometimes be brought usefully into play in classing large and widely-distributed genera, because all the species of the same genus, inhabiting any distinct and isolated region, have in all probability descended from the same parents.

We can understand, on these views, the very important distinction between real affinities and analogical or adaptive resemblances. Lamarck first called attention to this distinction, and he has been ably followed by Macleay and others. The resemblance, in the shape of the body and in the fin-like anterior limbs, between the dugong, which is a pachydermatous animal, and the whale, and between both these mammals and fishes, is analogical. Amongst insects there are innumerable instances: thus Linnæus, misled by external appearances, actually classed an homopterous insect as a moth. We see something of the same kind even in our domestic varieties, as in the thickened stems of the common and Swedish turnip. The resemblance of the greyhound and racehorse is hardly more fanciful than the analogies which have been drawn by some authors between very distinct animals. On my view of characters being of real importance for classification, only in so far as they reveal descent, we can clearly understand why analogical or adaptive character, although of the utmost importance to the welfare of the being, are almost valueless to the systematist. For animals, belonging to two most distinct lines of descent, may readily become adapted to similar conditions, and thus assume a close external resemblance; but such resemblances will not reveal—will rather tend to conceal their blood-relationship to their proper lines of descent. We can also understand the apparent paradox, that the very same characters are analogical when one class, or order is compared with another, but give true affinities when the members of the same class or order are compared one with another: thus the shape of the body and fin-like limbs are only analogical when whales are compared with fishes, being adaptations in both classes for swimming through the water; but the shape of the body and fin-like limbs serve as characters exhibiting true affinity between the several members of the whale family; for these cetaceans agree in so many characters, great and small, that we cannot doubt that they have inherited their general shape of body and structure of limbs from a common ancestor. So it is with fishes.

As members of distinct classes have often been adapted by successive slight modifications to live under nearly similar circumstances,—to inhabit for instance the three elements of land, air, and water,—we can perhaps understand how it is that a numerical parallelism has sometimes been observed between the sub-groups in distinct classes. A naturalist, struck by a parallelism of this nature in any one class, by arbitrarily raising or sinking the value of the groups in other classes (and all our experience shows that this valuation has hitherto been arbitrary), could easily extend the parallelism over a wide range; and thus the septenary, quinary, quaternary, and ternary classifications have probably arisen.

As the modified descendants of dominant species, belonging to the larger genera, tend to inherit the advantages, which made the groups to which they belong large and their parents dominant, they are almost sure to spread widely, and to seize on more and more places in the economy of nature. The larger and more dominant groups thus tend to go on increasing in size; and they consequently supplant many smaller and feebler groups. Thus we can account for the fact that all organisms, recent and extinct, are included under a few great orders, under still fewer classes, and all in one great natural system. As showing how few the higher groups are in number, and how widely spread they are throughout the world, the fact is striking, that the discovery of Australia has not added a single insect belonging to a new order; and that in the vegetable kingdom, as I learn from Dr. Hooker, it has added only two or three orders of small size.

In the chapter on geological succession I attempted to show, on the principle of each group having generally diverged much in character during the long-continued process of modification, how it is that the more ancient forms of life often present characters in some slight degree intermediate between existing groups. A few old and intermediate parent-forms having occasionally transmitted to the present day descendants but little modified, will give to us our so-called osculant or aberrant groups. The more aberrant any form is, the greater must be the number of connecting forms which on my theory have been exterminated and utterly lost. And we have some evidence of aberrant forms having suffered severely from extinction, for they are generally represented by extremely few species; and such species as do occur are generally very distinct from each other, which again implies extinction. The genera Ornithorhynchus and Lepidosiren, for example, would not have been less aberrant had each been represented by a dozen species instead of by a single one; but such richness in species, as I find after some investigation, does not commonly fall to the lot of aberrant genera. We can, I think, account for this fact only by looking at aberrant forms as failing groups conquered by more successful competitors, with a few members preserved by some unusual coincidence of favourable circumstances.

Mr. Waterhouse has remarked that, when a member belonging to one group of animals exhibits an affinity to a quite distinct group, this affinity in most cases is general and not special: thus, according to Mr. Waterhouse, of all Rodents, the bizcacha is most nearly related to Marsupials; but in the points in which it approaches this order, its relations are general, and not to any one marsupial species more than to another. As the points of affinity of the bizcacha to Marsupials are believed to be real and not merely adaptive, they are due on my theory to inheritance in common. Therefore we must suppose either that all Rodents, including the bizcacha, branched off from some very ancient Marsupial, which will have had a character in some degree intermediate with respect to all existing Marsupials; or that both Rodents and Marsupials branched off from a common progenitor, and that both groups have since undergone much modification in divergent directions. On either view we may suppose that the bizcacha has retained, by inheritance, more of the character of its ancient progenitor than have other Rodents; and therefore it will not be specially related to any one existing Marsupial, but indirectly to all or nearly all Marsupials, from having partially retained the character of their common progenitor, or of an early member of the group. On the other hand, of all Marsupials, as Mr. Waterhouse has remarked, the phascolomys resembles most nearly, not any one species, but the general order of Rodents. In this case, however, it may be strongly suspected that the resemblance is only analogical, owing to the phascolomys having become adapted to habits like those of a Rodent. The elder De Candolle has made nearly similar observations on the general nature of the affinities of distinct orders of plants.

On the principle of the multiplication and gradual divergence in character of the species descended from a common parent, together with their retention by inheritance of some characters in common, we can understand the excessively complex and radiating affinities by which all the members of the same family or higher group are connected together. For the common parent of a whole family of species, now broken up by extinction into distinct groups and sub-groups, will have transmitted some of its characters, modified in various ways and degrees, to all; and the several species will consequently be related to each other by circuitous lines of affinity of various lengths (as may be seen in the diagram so often referred to), mounting up through many predecessors. As it is difficult to show the blood-relationship between the numerous kindred of any ancient and noble family, even by the aid of a genealogical tree, and almost impossible to do this without this aid, we can understand the extraordinary difficulty which naturalists have experienced in describing, without the aid of a diagram, the various affinities which they perceive between the many living and extinct members of the same great natural class.

Extinction, as we have seen in the fourth chapter, has played an important part in defining and widening the intervals between the several groups in each class. We may thus account even for the distinctness of whole classes from each other—for instance, of birds from all other vertebrate animals—by the belief that many ancient forms of life have been utterly lost, through which the early progenitors of birds were formerly connected with the early progenitors of the other vertebrate classes. There has been less entire extinction of the forms of life which once connected fishes with batrachians. There has been still less in some other classes, as in that of the Crustacea, for here the most wonderfully diverse forms are still tied together by a long, but broken, chain of affinities. Extinction has only separated groups: it has by no means made them; for if every form which has ever lived on this earth were suddenly to reappear, though it would be quite impossible to give definitions by which each group could be distinguished from other groups, as all would blend together by steps as fine as those between the finest existing varieties, nevertheless a natural classification, or at least a natural arrangement, would be possible. We shall see this by turning to the diagram: the letters, A to L, may represent eleven Silurian genera, some of which have produced large groups of modified descendants. Every intermediate link between these eleven genera and their primordial parent, and every intermediate link in each branch and sub-branch of their descendants, may be supposed to be still alive; and the links to be as fine as those between the finest varieties. In this case it would be quite impossible to give any definition by which the several members of the several groups could be distinguished from their more immediate parents; or these parents from their ancient and unknown progenitor. Yet the natural arrangement in the diagram would still hold good; and, on the principle of inheritance, all the forms descended from A, or from I, would have something in common. In a tree we can specify this or that branch, though at the actual fork the two unite and blend together. We could not, as I have said, define the several groups; but we could pick out types, or forms, representing most of the characters of each group, whether large or small, and thus give a general idea of the value of the differences between them. This is what we should be driven to, if we were ever to succeed in collecting all the forms in any class which have lived throughout all time and space. We shall certainly never succeed in making so perfect a collection: nevertheless, in certain classes, we are tending in this direction; and Milne Edwards has lately insisted, in an able paper, on the high importance of looking to types, whether or not we can separate and define the groups to which such types belong.

Finally, we have seen that natural selection, which results from the struggle for existence, and which almost inevitably induces extinction and divergence of character in the many descendants from one dominant parent-species, explains that great and universal feature in the affinities of all organic beings, namely, their subordination in group under group. We use the element of descent in classing the individuals of both sexes and of all ages, although having few characters in common, under one species; we use descent in classing acknowledged varieties, however different they may be from their parent; and I believe this element of descent is the hidden bond of connexion which naturalists have sought under the term of the Natural System. On this idea of the natural system being, in so far as it has been perfected, genealogical in its arrangement, with the grades of difference between the descendants from a common parent, expressed by the terms genera, families, orders, &c., we can understand the rules which we are compelled to follow in our classification. We can understand why we value certain resemblances far more than others; why we are permitted to use rudimentary and useless organs, or others of trifling physiological importance; why, in comparing one group with a distinct group, we summarily reject analogical or adaptive characters, and yet use these same characters within the limits of the same group. We can clearly see how it is that all living and extinct forms can be grouped together in one great system; and how the several members of each class are connected together by the most complex and radiating lines of affinities. We shall never, probably, disentangle the inextricable web of affinities between the members of any one class; but when we have a distinct object in view, and do not look to some unknown plan of creation, we may hope to make sure but slow progress.

Morphology.—We have seen that the members of the same class, independently of their habits of life, resemble each other in the general plan of their organisation. This resemblance is often expressed by the term "unity of type;" or by saying that the several parts and organs in the different species of the class are homologous. The whole subject is included under the general name of Morphology. This is the most interesting department of natural history, and may be said to be its very soul. What can be more curious than that the hand of a man, formed for grasping, that of a mole for digging, the leg of the horse, the paddle of the porpoise, and the wing of the bat, should all be constructed on the same pattern, and should include the same bones, in the same relative positions? Geoffroy St. Hilaire has insisted strongly on the high importance of relative connexion in homologous organs: the parts may change to almost any extent in form and size, and yet they always remain connected together in the same order. We never find, for instance, the bones of the arm and forearm, or of the thigh and leg, transposed. Hence the same names can be given to the homologous bones in widely different animals. We see the same great law in the construction of the mouths of insects: what can be more different than the immensely long spiral proboscis of a sphinx-moth, the curious folded one of a bee or bug, and the great jaws of a beetle?—yet all these organs, serving for such different purposes, are formed by infinitely numerous modifications of an upper lip, mandibles, and two pairs of maxillæ. Analogous laws govern the construction of the mouths and limbs of crustaceans. So it is with the flowers of plants.

Nothing can be more hopeless than to attempt to explain this similarity of pattern in members of the same class, by utility or by the doctrine of final causes. The hopelessness of the attempt has been expressly admitted by Owen in his most interesting work on the 'Nature of Limbs.' On the ordinary view of the independent creation of each being, we can only say that so it is;—that it has so pleased the Creator to construct each animal and plant.

The explanation is manifest on the theory of the natural selection of successive slight modifications,—each modification being profitable in some way to the modified form, but often affecting by correlation of growth other parts of the organisation. In changes of this nature, there will be little or no tendency to modify the original pattern, or to transpose parts. The bones of a limb might be shortened and widened to any extent, and become gradually enveloped in thick membrane, so as to serve as a fin; or a webbed foot might have all its bones, or certain bones, lengthened to any extent, and the membrane connecting them increased to any extent, so as to serve as a wing: yet in all this great amount of modification there will be no tendency to alter the framework of bones or the relative connexion of the several parts. If we suppose that the ancient progenitor, the archetype as it may be called, of all mammals, had its limbs constructed on the existing general pattern, for whatever purpose they served, we can at once perceive the plain signification of the homologous construction of the limbs throughout the whole class. So with the mouths of insects, we have only to suppose that their common progenitor had an upper lip, mandibles, and two pair of maxillæ, these parts being perhaps very simple in form; and then natural selection will account for the infinite diversity in structure and function of the mouths of insects. Nevertheless, it is conceivable that the general pattern of an organ might become so much obscured as to be finally lost, by the atrophy and ultimately by the complete abortion of certain parts, by the soldering together of other parts, and by the doubling or multiplication of others,—variations which we know to be within the limits of possibility. In the paddles of the extinct gigantic sea-lizards, and in the mouths of certain suctorial crustaceans, the general pattern seems to have been thus to a certain extent obscured.

There is another and equally curious branch of the present subject; namely, the comparison not of the same part in different members of a class, but of the different parts or organs in the same individual. Most physiologists believe that the bones of the skull are homologous with—that is correspond in number and in relative connexion with—the elemental parts of a certain number of vertebræ. The anterior and posterior limbs in each member of the vertebrate and articulate classes are plainly homologous. We see the same law in comparing the wonderfully complex jaws and legs in crustaceans. It is familiar to almost every one, that in a flower the relative position of the sepals, petals, stamens, and pistils, as well as their intimate structure, are intelligible on the view that they consist of metamorphosed leaves, arranged in a spire. In monstrous plants, we often get direct evidence of the possibility of one organ being transformed into another; and we can actually see in embryonic crustaceans and in many other animals, and in flowers, that organs, which when mature become extremely different, are at an early stage of growth exactly alike.

How inexplicable are these facts on the ordinary view of creation! Why should the brain be enclosed in a box composed of such numerous and such extraordinarily shaped pieces of bone? As Owen has remarked, the benefit derived from the yielding of the separate pieces in the act of parturition of mammals, will by no means explain the same construction in the skulls of birds. Why should similar bones have been created in the formation of the wing and leg of a bat, used as they are for such totally different purposes? Why should one crustacean, which has an extremely complex mouth formed of many parts, consequently always have fewer legs; or conversely, those with many legs have simpler mouths? Why should the sepals, petals, stamens, and pistils in any individual flower, though fitted for such widely different purposes, be all constructed on the same pattern?

On the theory of natural selection, we can satisfactorily answer these questions. In the vertebrata, we see a series of internal vertebræ bearing certain processes and appendages; in the articulata, we see the body divided into a series of segments, bearing external appendages; and in flowering plants, we see a series of successive spiral whorls of leaves. An indefinite repetition of the same part or organ is the common characteristic (as Owen has observed) of all low or little-modified forms; therefore we may readily believe that the unknown progenitor of the vertebrata possessed many vertebræ; the unknown progenitor of the articulata, many segments; and the unknown progenitor of flowering plants, many spiral whorls of leaves. We have formerly seen that parts many times repeated are eminently liable to vary in number and structure; consequently it is quite probable that natural selection, during a long-continued course of modification, should have seized on a certain number of the primordially similar elements, many times repeated, and have adapted them to the most diverse purposes. And as the whole amount of modification will have been effected by slight successive steps, we need not wonder at discovering in such parts or organs, a certain degree of fundamental resemblance, retained by the strong principle of inheritance.

In the great class of molluscs, though we can homologise the parts of one species with those of another and distinct species, we can indicate but few serial homologies; that is, we are seldom enabled to say that one part or organ is homologous with another in the same individual. And we can understand this fact; for in molluscs, even in the lowest members of the class, we do not find nearly so much indefinite repetition of any one part, as we find in the other great classes of the animal and vegetable kingdoms.

Naturalists frequently speak of the skull as formed of metamorphosed vertebræ: the jaws of crabs as metamorphosed legs; the stamens and pistils of flowers as metamorphosed leaves; but it would in these cases probably be more correct, as Professor Huxley has remarked, to speak of both skull and vertebræ, both jaws and legs, &c.,—as having been metamorphosed, not one from the other, but from some common element. Naturalists, however, use such language only in a metaphorical sense: they are far from meaning that during a long course of descent, primordial organs of any kind—vertebræ in the one case and legs in the other—have actually been modified into skulls or jaws. Yet so strong is the appearance of a modification of this nature having occurred, that naturalists can hardly avoid employing language having this plain signification. On my view these terms may be used literally; and the wonderful fact of the jaws, for instance, of a crab retaining numerous characters, which they would probably have retained through inheritance, if they had really been metamorphosed during a long course of descent from true legs, or from some simple appendage, is explained.

Embryology.—It has already been casually remarked that certain organs in the individual, which when mature become widely different and serve for different purposes, are in the embryo exactly alike. The embryos, also, of distinct animals within the same class are often strikingly similar: a better proof of this cannot be given, than a circumstance mentioned by Agassiz, namely, that having forgotten to ticket the embryo of some vertebrate animal, he cannot now tell whether it be that of a mammal, bird, or reptile. The vermiform larvae of moths, flies, beetles, &c., resemble each other much more closely than do the mature insects; but in the case of larvæ, the embryos are active, and have been adapted for special lines of life. A trace of the law of embryonic resemblance, sometimes lasts till a rather late age: thus birds of the same genus, and of closely allied genera, often resemble each other in their first and second plumage; as we see in the spotted feathers in the thrush group. In the cat tribe, most of the species are striped or spotted in lines; and stripes can be plainly distinguished in the whelp of the lion. We occasionally though rarely see something of this kind in plants: thus the embryonic leaves of the ulex or furze, and the first leaves of the phyllodineous acaceas, are pinnate or divided like the ordinary leaves of the leguminosæ.

The points of structure, in which the embryos of widely different animals of the same class resemble each other, often have no direct relation to their conditions of existence. We cannot, for instance, suppose that in the embryos of the vertebrata the peculiar loop-like course of the arteries near the branchial slits are related to similar conditions,—in the young mammal which is nourished in the womb of its mother, in the egg of the bird which is hatched in a nest, and in the spawn of a frog under water. We have no more reason to believe in such a relation, than we have to believe that the same bones in the hand of a man, wing of a bat, and fin of a porpoise, are related to similar conditions of life. No one will suppose that the stripes on the whelp of a lion, or the spots on the young blackbird, are of any use to these animals, or are related to the conditions to which they are exposed.

The case, however, is different when an animal during any part of its embryonic career is active, and has to provide for itself. The period of activity may come on earlier or later in life; but whenever it comes on, the adaptation of the larva to its conditions of life is just as perfect and as beautiful as in the adult animal. From such special adaptations, the similarity of the larvæ or active embryos of allied animals is sometimes much obscured; and cases could be given of the larvæ of two species, or of two groups of species, differing quite as much, or even more, from each other than do their adult parents. In most cases, however, the larvæ, though active, still obey more or less closely the law of common embryonic resemblance. Cirripedes afford a good instance of this: even the illustrious Cuvier did not perceive that a barnacle was, as it certainly is, a crustacean; but a glance at the larva shows this to be the case in an unmistakeable manner. So again the two main divisions of cirripedes, the pedunculated and sessile, which differ widely in external appearance, have larvæ in all their several stages barely distinguishable.

The embryo in the course of development generally rises in organisation: I use this expression, though I am aware that it is hardly possible to define clearly what is meant by the organisation being higher or lower. But no one probably will dispute that the butterfly is higher than the caterpillar. In some cases, however, the mature animal is generally considered as lower in the scale than the larva, as with certain parasitic crustaceans. To refer once again to cirripedes: the larvæ in the first stage have three pairs of legs, a very simple single eye, and a probosciformed mouth, with which they feed largely, for they increase much in size. In the second stage, answering to the chrysalis stage of butterflies, they have six pairs of beautifully constructed natatory legs, a pair of magnificent compound eyes, and extremely complex antennæ; but they have a closed and imperfect mouth, and cannot feed: their function at this stage is, to search by their well-developed organs of sense, and to reach by their active powers of swimming, a proper place on which to become attached and to undergo their final metamorphosis. When this is completed they are fixed for life: their legs are now converted into prehensile organs; they again obtain a well-constructed mouth; but they have no antennæ, and their two eyes are now reconverted into a minute, single, and very simple eye-spot. In this last and complete state, cirripedes may be considered as either more highly or more lowly organised than they were in the larval condition. But in some genera the larvæ become developed either into hermaphrodites having the ordinary structure, or into what I have called complemental males: and in the latter, the development has assuredly been retrograde; for the male is a mere sack, which lives for a short time, and is destitute of mouth, stomach, or other organ of importance, excepting for reproduction.

We are so much accustomed to see differences in structure between the embryo and the adult, and likewise a close similarity in the embryos of widely different animals within the same class, that we might be led to look at these facts as necessarily contingent in some manner on growth. But there is no obvious reason why, for instance, the wing of a bat, or the fin of a porpoise, should not have been sketched out with all the parts in proper proportion, as soon as any structure became visible in the embryo. And in some whole groups of animals and in certain members of other groups, the embryo does not at any period differ widely from the adult: thus Owen has remarked in regard to cuttle-fish, "there is no metamorphosis; the cephalopodic character is manifested long before the parts of the embryo are completed;" and again in spiders, "there is nothing worthy to be called a metamorphosis." The larvæ of insects, whether adapted to the most diverse and active habits, or quite inactive, being fed by their parents or placed in the midst of proper nutriment, yet nearly all pass through a similar worm-like stage of development; but in some few cases, as in that of Aphis, if we look to the admirable drawings by Professor Huxley of the development of this insect, we see no trace of the vermiform stage.

How, then, can we explain these several facts in embryology,—namely the very general, but not universal difference in structure between the embryo and the adult;—of parts in the same individual embryo, which ultimately become very unlike and serve for diverse purposes, being at this early period of growth alike;—of embryos of different species within the same class, generally, but not universally, resembling each other;—of the structure of the embryo not being closely related to its conditions of existence, except when the embryo becomes at any period of life active and has to provide for itself;—of the embryo apparently having sometimes a higher organisation than the mature animal, into which it is developed. I believe that all these facts can be explained, as follows, on the view of descent with modification.

It is commonly assumed, perhaps from monstrosities often affecting the embryo at a very early period, that slight variations necessarily appear at an equally early period. But we have little evidence on this head—indeed the evidence rather points the other way; for it is notorious that breeders of cattle, horses, and various fancy animals, cannot positively tell, until some time after the animal has been born, what its merits or form will ultimately turn out. We see this plainly in our own children; we cannot always tell whether the child will be tall or short, or what its precise features will be. The question is not, at what period of life any variation has been caused, but at what period it is fully displayed. The cause may have acted, and I believe generally has acted, even before the embryo is formed; and the variation may be due to the male and female sexual elements having been affected by the conditions to which either parent, or their ancestors, have been exposed. Nevertheless an effect thus caused at a very early period, even before the formation of the embryo, may appear late in life; as when an hereditary disease, which appears in old age alone, has been communicated to the offspring from the reproductive element of one parent. Or again, as when the horns of cross-bred cattle have been affected by the shape of the horns of either parent. For the welfare of a very young animal, as long as it remains in its mother's womb, or in the egg, or as long as it is nourished and protected by its parent, it must be quite unimportant whether most of its characters are fully acquired a little earlier or later in life. It would not signify, for instance, to a bird which obtained its food best by having a long beak, whether or not it assumed a beak of this particular length, as long as it was fed by its parents. Hence, I conclude, that it is quite possible, that each of the many successive modifications, by which each species has acquired its present structure, may have supervened at a not very early period of life; and some direct evidence from our domestic animals supports this view. But in other cases it is quite possible that each successive modification, or most of them, may have appeared at an extremely early period.

I have stated in the first chapter, that there is some evidence to render it probable, that at whatever age any variation first appears in the parent, it tends to reappear at a corresponding age in the offspring. Certain variations can only appear at corresponding ages, for instance, peculiarities in the caterpillar, cocoon, or imago states of the silk-moth; or, again, in the horns of almost full-grown cattle. But further than this, variations which, for all that we can see, might have appeared earlier or later in life, tend to appear at a corresponding age in the offspring and parent. I am far from meaning that this is invariably the case; and I could give a good many cases of variations (taking the word in the largest sense) which have supervened at an earlier age in the child than in the parent.

These two principles, if their truth be admitted, will, I believe, explain all the above specified leading facts in embryology. But first let us look at a few analogous cases in domestic varieties. Some authors who have written on Dogs, maintain that the greyhound and bull-dog, though appearing so different, are really varieties most closely allied, and have probably descended from the same wild stock; hence I was curious to see how far their puppies differed from each other: I was told by breeders that they differed just as much as their parents, and this, judging by the eye, seemed almost to be the case; but on actually measuring the old dogs and their six-days old puppies, I found that the puppies had not nearly acquired their full amount of proportional difference. So, again, I was told that the foals of cart and race-horses differed as much as the full-grown animals; and this surprised me greatly, as I think it probable that the difference between these two breeds has been wholly caused by selection under domestication; but having had careful measurements made of the dam and of a three-days old colt of a race and heavy cart-horse, I find that the colts have by no means acquired their full amount of proportional difference.

As the evidence appears to me conclusive, that the several domestic breeds of Pigeon have descended from one wild species, I compared young pigeons of various breeds, within twelve hours after being hatched; I carefully measured the proportions (but will not here give details) of the beak, width of mouth, length of nostril and of eyelid, size of feet and length of leg, in the wild stock, in pouters, fantails, runts, barbs, dragons, carriers, and tumblers. Now some of these birds, when mature, differ so extraordinarily in length and form of beak, that they would, I cannot doubt, be ranked in distinct genera, had they been natural productions. But when the nestling birds of these several breeds were placed in a row, though most of them could be distinguished from each other, yet their proportional differences in the above specified several points were incomparably less than in the full-grown birds. Some characteristic points of difference—for instance, that of the width of mouth—could hardly be detected in the young. But there was one remarkable exception to this rule, for the young of the short-faced tumbler differed from the young of the wild rock-pigeon and of the other breeds, in all its proportions, almost exactly as much as in the adult state.

The two principles above given seem to me to explain these facts in regard to the later embryonic stages of our domestic varieties. Fanciers select their horses, dogs, and pigeons, for breeding, when they are nearly grown up: they are indifferent whether the desired qualities and structures have been acquired earlier or later in life, if the full-grown animal possesses them. And the cases just given, more especially that of pigeons, seem to show that the characteristic differences which give value to each breed, and which have been accumulated by man's selection, have not generally first appeared at an early period of life, and have been inherited by the offspring at a corresponding not early period. But the case of the short-faced tumbler, which when twelve hours old had acquired its proper proportions, proves that this is not the universal rule; for here the characteristic differences must either have appeared at an earlier period than usual, or, if not so, the differences must have been inherited, not at the corresponding, but at an earlier age.

Now let us apply these facts and the above two principles—which latter, though not proved true, can be shown to be in some degree probable—to species in a state of nature. Let us take a genus of birds, descended on my theory from some one parent-species, and of which the several new species have become modified through natural selection in accordance with their diverse habits. Then, from the many slight successive steps of variation having supervened at a rather late age, and having been inherited at a corresponding age, the young of the new species of our supposed genus will manifestly tend to resemble each other much more closely than do the adults, just as we have seen in the case of pigeons. We may extend this view to whole families or even classes. The fore-limbs, for instance, which served as legs in the parent-species, may become, by a long course of modification, adapted in one descendant to act as hands, in another as paddles, in another as wings; and on the above two principles—namely of each successive modification supervening at a rather late age, and being inherited at a corresponding late age—the fore-limbs in the embryos of the several descendants of the parent-species will still resemble each other closely, for they will not have been modified. But in each individual new species, the embryonic fore-limbs will differ greatly from the fore-limbs in the mature animal; the limbs in the latter having undergone much modification at a rather late period of life, and having thus been converted into hands, or paddles, or wings. Whatever influence long-continued exercise or use on the one hand, and disuse on the other, may have in modifying an organ, such influence will mainly affect the mature animal, which has come to its full powers of activity and has to gain its own living; and the effects thus produced will be inherited at a corresponding mature age. Whereas the young will remain unmodified, or be modified in a lesser degree, by the effects of use and disuse.

In certain cases the successive steps of variation might supervene, from causes of which we are wholly ignorant, at a very early period of life, or each step might be inherited at an earlier period than that at which it first appeared. In either case (as with the short-faced tumbler) the young or embryo would closely resemble the mature parent-form. We have seen that this is the rule of development in certain whole groups of animals, as with cuttle-fish and spiders, and with a few members of the great class of insects, as with Aphis. With respect to the final cause of the young in these cases not undergoing any metamorphosis, or closely resembling their parents from their earliest age, we can see that this would result from the two following contingencies; firstly, from the young, during a course of modification carried on for many generations, having to provide for their own wants at a very early stage of development, and secondly, from their following exactly the same habits of life with their parents; for in this case, it would be indispensable for the existence of the species, that the child should be modified at a very early age in the same manner with its parents, in accordance with their similar habits. Some further explanation, however, of the embryo not undergoing any metamorphosis is perhaps requisite. If, on the other hand, it profited the young to follow habits of life in any degree different from those of their parent, and consequently to be constructed in a slightly different manner, then, on the principle of inheritance at corresponding ages, the active young or larvæ might easily be rendered by natural selection different to any conceivable extent from their parents. Such differences might, also, become correlated with successive stages of development; so that the larvæ, in the first stage, might differ greatly from the larvæ in the second stage, as we have seen to be the case with cirripedes. The adult might become fitted for sites or habits, in which organs of locomotion or of the senses, &c., would be useless; and in this case the final metamorphosis would be said to be retrograde.

As all the organic beings, extinct and recent, which have ever lived on this earth have to be classed together, and as all have been connected by the finest gradations, the best, or indeed, if our collections were nearly perfect, the only possible arrangement, would be genealogical. Descent being on my view the hidden bond of connexion which naturalists have been seeking under the term of the natural system. On this view we can understand how it is that, in the eyes of most naturalists, the structure of the embryo is even more important for classification than that of the adult. For the embryo is the animal in its less modified state; and in so far it reveals the structure of its progenitor. In two groups of animal, however much they may at present differ from each other in structure and habits, if they pass through the same or similar embryonic stages, we may feel assured that they have both descended from the same or nearly similar parents, and are therefore in that degree closely related. Thus, community in embryonic structure reveals community of descent. It will reveal this community of descent, however much the structure of the adult may have been modified and obscured; we have seen, for instance, that cirripedes can at once be recognised by their larvæ as belonging to the great class of crustaceans. As the embryonic state of each species and group of species partially shows us the structure of their less modified ancient progenitors, we can clearly see why ancient and extinct forms of life should resemble the embryos of their descendants,—our existing species. Agassiz believes this to be a law of nature; but I am bound to confess that I only hope to see the law hereafter proved true. It can be proved true in those cases alone in which the ancient state, now supposed to be represented in many embryos, has not been obliterated, either by the successive variations in a long course of modification having supervened at a very early age, or by the variations having been inherited at an earlier period than that at which they first appeared. It should also be borne in mind, that the supposed law of resemblance of ancient forms of life to the embryonic stages of recent forms, may be true, but yet, owing to the geological record not extending far enough back in time, may remain for a long period, or for ever, incapable of demonstration.

Thus, as it seems to me, the leading facts in embryology, which are second in importance to none in natural history, are explained on the principle of slight modifications not appearing, in the many descendants from some one ancient progenitor, at a very early period in the life of each, though perhaps caused at the earliest, and being inherited at a corresponding not early period. Embryology rises greatly in interest, when we thus look at the embryo as a picture, more or less obscured, of the common parent-form of each great class of animals.

Rudimentary, atrophied, or aborted organs.—Organs or parts in this strange condition, bearing the stamp of inutility, are extremely common throughout nature. For instance, rudimentary mammæ are very general in the males of mammals: I presume that the "bastard-wing" in birds may be safely considered as a digit in a rudimentary state: in very many snakes one lobe of the lungs is rudimentary; in other snakes there are rudiments of the pelvis and hind limbs. Some of the cases of rudimentary organs are extremely curious; for instance, the presence of teeth in fœtal whales, which when grown up have not a tooth in their heads; and the presence of teeth, which never cut through the gums, in the upper jaws of our unborn calves. It has even been stated on good authority that rudiments of teeth can be detected in the beaks of certain embryonic birds. Nothing can be plainer than that wings are formed for flight, yet in how many insects do we see wings so reduced in size as to be utterly incapable of flight, and not rarely lying under wing-cases, firmly soldered together!

The meaning of rudimentary organs is often quite unmistakeable: for instance there are beetles of the same genus (and even of the same species) resembling each other most closely in all respects, one of which will have full-sized wings, and another mere rudiments of membrane; and here it is impossible to doubt, that the rudiments represent wings. Rudimentary organs sometimes retain their potentiality, and are merely not developed: this seems to be the case with the mammæ of male mammals, for many instances are on record of these organs having become well developed in full-grown males, and having secreted milk. So again there are normally four developed and two rudimentary teats in the udders of the genus Bos, but in our domestic cows the two sometimes become developed and give milk. In individual plants of the same species the petals sometimes occur as mere rudiments, and sometimes in a well-developed state. In plants with separated sexes, the male flowers often have a rudiment of a pistil; and Kölreuter found that by crossing such male plants with an hermaphrodite species, the rudiment of the pistil in the hybrid offspring was much increased in size; and this shows that the rudiment and the perfect pistil are essentially alike in nature.

An organ serving for two purposes, may become rudimentary or utterly aborted for one, even the more important purpose; and remain perfectly efficient for the other. Thus in plants, the office of the pistil is to allow the pollen-tubes to reach the ovules protected in the ovarium at its base. The pistil consists of a stigma supported on the style; but in some Compositæ, the male florets, which of course cannot be fecundated, have a pistil, which is in a rudimentary state, for it is not crowned with a stigma; but the style remains well developed, and is clothed with hairs as in other compositæ, for the purpose of brushing the pollen out of the surrounding anthers. Again, an organ may become rudimentary for its proper purpose, and be used for a distinct object: in certain fish the swim-bladder seems to be rudimentary for its proper function of giving buoyancy, but has become converted into a nascent breathing organ or lung. Other similar instances could be given.

Rudimentary organs in the individuals of the same species are very liable to vary in degree of development and in other respects. Moreover, in closely allied species, the degree to which the same organ has been rendered rudimentary occasionally differs much. This latter fact is well exemplified in the state of the wings of the female moths in certain groups. Rudimentary organs may be utterly aborted; and this implies, that we find in an animal or plant no trace of an organ, which analogy would lead us to expect to find, and which is occasionally found in monstrous individuals of the species. Thus in the snapdragon (antirrhinum) we generally do not find a rudiment of a fifth stamen; but this may sometimes be seen. In tracing the homologies of the same part in different members of a class, nothing is more common, or more necessary, than the use and discovery of rudiments. This is well shown in the drawings given by Owen of the bones of the leg of the horse, ox, and rhinoceros.

It is an important fact that rudimentary organs, such as teeth in the upper jaws of whales and ruminants, can often be detected in the embryo, but afterwards wholly disappear. It is also, I believe, a universal rule, that a rudimentary part or organ is of greater size relatively to the adjoining parts in the embryo, than in the adult; so that the organ at this early age is less rudimentary, or even cannot be said to be in any degree rudimentary. Hence, also, a rudimentary organ in the adult, is often said to have retained its embryonic condition.

I have now given the leading facts with respect to rudimentary organs. In reflecting on them, every one must be struck with astonishment: for the same reasoning power which tells us plainly that most parts and organs are exquisitely adapted for certain purposes, tells us with equal plainness that these rudimentary or atrophied organs, are imperfect and useless. In works on natural history rudimentary organs are generally said to have been created "for the sake of symmetry," or in order "to complete the scheme of nature;" but this seems to me no explanation, merely a restatement of the fact. Would it be thought sufficient to say that because planets revolve in elliptic courses round the sun, satellites follow the same course round the planets, for the sake of symmetry, and to complete the scheme of nature? An eminent physiologist accounts for the presence of rudimentary organs, by supposing that they serve to excrete matter in excess, or injurious to the system; but can we suppose that the minute papilla, which often represents the pistil in male flowers, and which is formed merely of cellular tissue, can thus act? Can we suppose that the formation of rudimentary teeth which are subsequently absorbed, can be of any service to the rapidly growing embryonic calf by the excretion of precious phosphate of lime? When a man's fingers have been amputated, imperfect nails sometimes appear on the stumps: I could as soon believe that these vestiges of nails have appeared, not from unknown laws of growth, but in order to excrete horny matter, as that the rudimentary nails on the fin of the manatee were formed for this purpose.

On my view of descent with modification, the origin of rudimentary organs is simple. We have plenty of cases of rudimentary organs in our domestic productions,—as the stump of a tail in tailless breeds,—the vestige of an ear in earless breeds,—the reappearance of minute dangling horns in hornless breeds of cattle, more especially, according to Youatt, in young animals,—and the state of the whole flower in the cauliflower. We often see rudiments of various parts in monsters. But I doubt whether any of these cases throw light on the origin of rudimentary organs in a state of nature, further than by showing that rudiments can be produced; for I doubt whether species under nature ever undergo abrupt changes. I believe that disuse has been the main agency; that it has led in successive generations to the gradual reduction of various organs, until they have become rudimentary,—as in the case of the eyes of animals inhabiting dark caverns, and of the wings of birds inhabiting oceanic islands, which have seldom been forced to take flight, and have ultimately lost the power of flying. Again, an organ useful under certain conditions, might become injurious under others, as with the wings of beetles living on small and exposed islands; and in this case natural selection would continue slowly to reduce the organ, until it was rendered harmless and rudimentary.

Any change in function, which can be effected by insensibly small steps, is within the power of natural selection; so that an organ rendered, during changed habits of life, useless or injurious for one purpose, might easily be modified and used for another purpose. Or an organ might be retained for one alone of its former functions. An organ, when rendered useless, may well be variable, for its variations cannot be checked by natural selection. At whatever period of life disuse or selection reduces an organ, and this will generally be when the being has come to maturity and to its full powers of action, the principle of inheritance at corresponding ages will reproduce the organ in its reduced state at the same age, and consequently will seldom affect or reduce it in the embryo. Thus we can understand the greater relative size of rudimentary organs in the embryo, and their lesser relative size in the adult. But if each step of the process of reduction were to be inherited, not at the corresponding age, but at an extremely early period of life (as we have good reason to believe to be possible) the rudimentary part would tend to be wholly lost, and we should have a case of complete abortion. The principle, also, of economy, explained in a former chapter, by which the materials forming any part or structure, if not useful to the possessor, will be saved as far as is possible, will probably often come into play; and this will tend to cause the entire obliteration of a rudimentary organ.

As the presence of rudimentary organs is thus due to the tendency in every part of the organisation, which has long existed, to be inherited—we can understand, on the genealogical view of classification, how it is that systematists have found rudimentary parts as useful as, or even sometimes more useful than, parts of high physiological importance. Rudimentary organs may be compared with the letters in a word, still retained in the spelling, but become useless in the pronunciation, but which serve as a clue in seeking for its derivation. On the view of descent with modification, we may conclude that the existence of organs in a rudimentary, imperfect, and useless condition, or quite aborted, far from presenting a strange difficulty, as they assuredly do on the ordinary doctrine of creation, might even have been anticipated, and can be accounted for by the laws of inheritance.

Summary.—In this chapter I have attempted to show, that the subordination of group to group in all organisms throughout all time; that the nature of the relationship, by which all living and extinct beings are united by complex, radiating, and circuitous lines of affinities into one grand system; the rules followed and the difficulties encountered by naturalists in their classifications; the value set upon characters, if constant and prevalent, whether of high vital importance, or of the most trifling importance, or, as in rudimentary organs, of no importance; the wide opposition in value between analogical or adaptive characters, and characters of true affinity; and other such rules;—all naturally follow on the view of the common parentage of those forms which are considered by naturalists as allied, together with their modification through natural selection, with its contingencies of extinction and divergence of character. In considering this view of classification, it should be borne in mind that the element of descent has been universally used in ranking together the sexes, ages, and acknowledged varieties of the same species, however different they may be in structure. If we extend the use of this element of descent,—the only certainly known cause of similarity in organic beings,—we shall understand what is meant by the natural system: it is genealogical in its attempted arrangement, with the grades of acquired difference marked by the terms varieties, species, genera, families, orders, and classes.

On this same view of descent with modification, all the great facts in Morphology become intelligible,—whether we look to the same pattern displayed in the homologous organs, to whatever purpose applied, of the different species of a class; or to the homologous parts constructed on the same pattern in each individual animal and plant.

On the principle of successive slight variations, not necessarily or generally supervening at a very early period of life, and being inherited at a corresponding period, we can understand the great leading facts in Embryology; namely, the resemblance in an individual embryo of the homologous parts, which when matured will become widely different from each other in structure and function; and the resemblance in different species of a class of the homologous parts or organs, though fitted in the adult members for purposes as different as possible. Larvæ are active embryos, which have become specially modified in relation to their habits of life, through the principle of modifications being inherited at corresponding ages. On this same principle—and bearing in mind, that when organs are reduced in size, either from disuse or selection, it will generally be at that period of life when the being has to provide for its own wants, and bearing in mind how strong is the principle of inheritance—the occurrence of rudimentary organs and their final abortion, present to us no inexplicable difficulties; on the contrary, their presence might have been even anticipated. The importance of embryological characters and of rudimentary organs in classification is intelligible, on the view that an arrangement is only so far natural as it is genealogical.

Finally, the several classes of facts which have been considered in this chapter, seem to me to proclaim so plainly, that the innumerable species, genera, and families of organic beings, with which this world is peopled, have all descended, each within its own class or group, from common parents, and have all been modified in the course of descent, that I should without hesitation adopt this view, even if it were unsupported by other facts or arguments.

CHAPTER XIV

Recapitulation and Conclusion

RECAPITULATION OF THE DIFFICULTIES ON THE THEORY OF NATURAL SELECTION • RECAPITULATION OF THE GENERAL AND SPECIAL CIRCUMSTANCES IN ITS FAVOUR • CAUSES OF THE GENERAL BELIEF IN THE IMMUTABILITY OF SPECIES • HOW FAR THE THEORY OF NATURAL SELECTION MAY BE EXTENDED • EFFECTS OF ITS ADOPTION ON THE STUDY OF NATURAL HISTORY • CONCLUDING REMARKS

As this whole volume is one long argument, it may be convenient to the reader to have the leading facts and inferences briefly recapitulated.

That many and grave objections may be advanced against the theory of descent with modification through natural selection, I do not deny. I have endeavoured to give to them their full force. Nothing at first can appear more difficult to believe than that the more complex organs and instincts should have been perfected, not by means superior to, though analogous with, human reason, but by the accumulation of innumerable slight variations, each good for the individual possessor. Nevertheless, this difficulty, though appearing to our imagination insuperably great, cannot be considered real if we admit the following propositions, namely,—that gradations in the perfection of any organ or instinct, which we may consider, either do now exist or could have existed, each good of its kind,—that all organs and instincts are, in ever so slight a degree, variable,—and, lastly, that there is a struggle for existence leading to the preservation of each profitable deviation of structure or instinct. The truth of these propositions cannot, I think, be disputed.

It is, no doubt, extremely difficult even to conjecture by what gradations many structures have been perfected, more especially amongst broken and failing groups of organic beings; but we see so many strange gradations in nature, as is proclaimed by the canon, "Natura non facit saltum," that we ought to be extremely cautious in saying that any organ or instinct, or any whole being, could not have arrived at its present state by many graduated steps. There are, it must be admitted, cases of special difficulty on the theory of natural selection; and one of the most curious of these is the existence of two or three defined castes of workers or sterile females in the same community of ants; but I have attempted to show how this difficulty can be mastered.

With respect to the almost universal sterility of species when first crossed, which forms so remarkable a contrast with the almost universal fertility of varieties when crossed, I must refer the reader to the recapitulation of the facts given at the end of the eighth chapter, which seem to me conclusively to show that this sterility is no more a special endowment than is the incapacity of two trees to be grafted together; but that it is incidental on constitutional differences in the reproductive systems of the intercrossed species. We see the truth of this conclusion in the vast difference in the result, when the same two species are crossed reciprocally; that is, when one species is first used as the father and then as the mother.

The fertility of varieties when intercrossed and of their mongrel offspring cannot be considered as universal; nor is their very general fertility surprising when we remember that it is not likely that either their constitutions or their reproductive systems should have been profoundly modified. Moreover, most of the varieties which have been experimentised on have been produced under domestication; and as domestication apparently tends to eliminate sterility, we ought not to expect it also to produce sterility.

The sterility of hybrids is a very different case from that of first crosses, for their reproductive organs are more or less functionally impotent; whereas in first crosses the organs on both sides are in a perfect condition. As we continually see that organisms of all kinds are rendered in some degree sterile from their constitutions having been disturbed by slightly different and new conditions of life, we need not feel surprise at hybrids being in some degree sterile, for their constitutions can hardly fail to have been disturbed from being compounded of two distinct organisations. This parallelism is supported by another parallel, but directly opposite, class of facts; namely, that the vigour and fertility of all organic beings are increased by slight changes in their conditions of life, and that the offspring of slightly modified forms or varieties acquire from being crossed increased vigour and fertility. So that, on the one hand, considerable changes in the conditions of life and crosses between greatly modified forms, lessen fertility; and on the other hand, lesser changes in the conditions of life and crosses between less modified forms, increase fertility.

Turning to geographical distribution, the difficulties encountered on the theory of descent with modification are grave enough. All the individuals of the same species, and all the species of the same genus, or even higher group, must have descended from common parents; and therefore, in however distant and isolated parts of the world they are now found, they must in the course of successive generations have passed from some one part to the others. We are often wholly unable even to conjecture how this could have been effected. Yet, as we have reason to believe that some species have retained the same specific form for very long periods, enormously long as measured by years, too much stress ought not to be laid on the occasional wide diffusion of the same species; for during very long periods of time there will always be a good chance for wide migration by many means. A broken or interrupted range may often be accounted for by the extinction of the species in the intermediate regions. It cannot be denied that we are as yet very ignorant of the full extent of the various climatal and geographical changes which have affected the earth during modern periods; and such changes will obviously have greatly facilitated migration. As an example, I have attempted to show how potent has been the influence of the Glacial period on the distribution both of the same and of representative species throughout the world. We are as yet profoundly ignorant of the many occasional means of transport. With respect to distinct species of the same genus inhabiting very distant and isolated regions, as the process of modification has necessarily been slow, all the means of migration will have been possible during a very long period; and consequently the difficulty of the wide diffusion of species of the same genus is in some degree lessened.

As on the theory of natural selection an interminable number of intermediate forms must have existed, linking together all the species in each group by gradations as fine as our present varieties, it may be asked, Why do we not see these linking forms all around us? Why are not all organic beings blended together in an inextricable chaos? With respect to existing forms, we should remember that we have no right to expect (excepting in rare cases) to discover *directly* connecting links between them, but only between each and some extinct and supplanted form. Even on a wide area, which has during a long period remained continuous, and of which the climate and other conditions of life change insensibly in going from a district occupied by one species into another district occupied by a closely allied species, we have no just right to expect often to find intermediate varieties in the intermediate zone. For we have reason to believe that only a few species are undergoing change at any one period; and all changes are slowly effected. I have also shown that the intermediate varieties which will at first probably exist in the intermediate zones, will be liable to be supplanted by the allied forms on either hand; and the latter, from existing in greater numbers, will generally be modified and improved at a quicker rate than the intermediate varieties, which exist in lesser numbers; so that the intermediate varieties will, in the long run, be supplanted and exterminated.

On this doctrine of the extermination of an infinitude of connecting links, between the living and extinct inhabitants of the world, and at each successive period between the extinct and still older species, why is not every geological formation charged with such links? Why does not every collection of fossil remains afford plain evidence of the gradation and mutation of the forms of life? We meet with no such evidence, and this is the most obvious and forcible of the many objections which may be urged against my theory. Why, again, do whole groups of allied species appear, though certainly they often falsely appear, to have come in suddenly on the several geological stages? Why do we not find great piles of strata beneath the Silurian system, stored with the remains of the progenitors of the Silurian groups of fossils? For certainly on my theory such strata must somewhere have been deposited at these ancient and utterly unknown epochs in the world's history.

I can answer these questions and grave objections only on the supposition that the geological record is far more imperfect than most geologists believe. It cannot be objected that there has not been time sufficient for any amount of organic change; for the lapse of time has been so great as to be utterly inappreciable by the human intellect. The number of specimens in all our museums is absolutely as nothing compared with the countless generations of countless species which certainly have existed. We should not be able to recognise a species as the parent of any one or more species if we were to examine them ever so closely, unless we likewise possessed many of the intermediate links between their past or parent and present states; and these many links we could hardly ever expect to discover, owing to the imperfection of the geological record. Numerous existing doubtful forms could be named which are probably varieties; but who will pretend that in future ages so many fossil links will be discovered, that naturalists will be able to decide, on the common view, whether or not these doubtful forms are varieties? As long as most of the links between any two species are unknown, if any one link or intermediate variety be discovered, it will simply be classed as another and distinct species. Only a small portion of the world has been geologically explored. Only organic beings of certain classes can be preserved in a fossil condition, at least in any great number. Widely ranging species vary most, and varieties are often at first local,—both causes rendering the discovery of intermediate links less likely. Local varieties will not spread into other and distant regions until they are considerably modified and improved; and when they do spread, if discovered in a geological formation, they will appear as if suddenly created there, and will be simply classed as new species. Most formations have been intermittent in their accumulation; and their duration, I am inclined to believe, has been shorter than the average duration of specific forms. Successive formations are separated from each other by enormous blank intervals of time; for fossiliferous formations, thick enough to resist future degradation, can be accumulated only where much sediment is deposited on the subsiding bed of the sea. During the alternate periods of elevation and of stationary level the record will be blank. During these latter periods there will probably be more variability in the forms of life; during periods of subsidence, more extinction.

With respect to the absence of fossiliferous formations beneath the lowest Silurian strata, I can only recur to the hypothesis given in the ninth chapter. That the geological record is imperfect all will admit; but that it is imperfect to the degree which I require, few will be inclined to admit. If we look to long enough intervals of time, geology plainly declares that all species have changed; and they have changed in the manner which my theory requires, for they have changed slowly and in a graduated manner. We clearly see this in the fossil remains from consecutive formations invariably being much more closely related to each other, than are the fossils from formations distant from each other in time.

Such is the sum of the several chief objections and difficulties which may justly be urged against my theory; and I have now briefly recapitulated the answers and explanations which can be given to them. I have felt these difficulties far too heavily during many years to doubt their weight. But it deserves especial notice that the more important objections relate to questions on which we are confessedly ignorant; nor do we know how ignorant we are. We do not know all the possible transitional gradations between the simplest and the most perfect organs; it cannot be pretended that we know all the varied means of Distribution during the long lapse of years, or that we know how imperfect the Geological Record is. Grave as these several difficulties are, in my judgment they do not overthrow the theory of descent with modification.

NOW LET US TURN to the other side of the argument. Under domestication we see much variability. This seems to be mainly due to the reproductive system being eminently susceptible to changes in the conditions of life; so that this system, when not rendered impotent, fails to reproduce offspring exactly like the parent-form. Variability is governed by many complex laws,—by correlation of growth, by use and disuse, and by the direct action of the physical conditions of life. There is much difficulty in ascertaining how much modification our domestic productions have undergone; but we may safely infer that the amount has been large, and that modifications can be inherited for long periods. As long as the conditions of life remain the same, we have reason to believe that a modification, which has already been inherited for many generations, may continue to be inherited for an almost infinite number of generations. On the other hand we have evidence that variability, when it has once come into play, does not wholly cease; for new varieties are still occasionally produced by our most anciently domesticated productions.

Man does not actually produce variability; he only unintentionally exposes organic beings to new conditions of life, and then nature acts on the organisation, and causes variability. But man can and does select the variations given to him by nature, and thus accumulate them in any desired manner. He thus adapts animals and plants for his own benefit or pleasure. He may do this methodically, or he may do it unconsciously by preserving the individuals most useful to him at the time, without any thought of altering the breed. It is certain that he can largely influence the character of a breed by selecting, in each successive generation, individual differences so slight as to be quite inappreciable by an uneducated eye. This process of selection has been the great agency in the production of the most distinct and useful domestic breeds. That many of the breeds produced by man have to a large extent the character of natural species, is shown by the inextricable doubts whether very many of them are varieties or aboriginal species.

There is no obvious reason why the principles which have acted so efficiently under domestication should not have acted under nature. In the preservation of favoured individuals and races, during the constantly-recurrent Struggle for Existence, we see the most powerful and ever-acting means of selection. The struggle for existence inevitably follows from the high geometrical ratio of increase which is common to all organic beings. This high rate of increase is proved by calculation, by the effects of a succession of peculiar seasons, and by the results of naturalisation, as explained in the third chapter. More individuals are born than can possibly survive. A grain in the balance will determine which individual shall live and which shall die,—which variety or species shall increase in number, and which shall decrease, or finally become extinct. As the individuals of the same species come in all respects into the closest competition with each other, the struggle will generally be most severe between them; it will be almost equally severe between the varieties of the same species, and next in severity between the species of the same genus. But the struggle will often be very severe between beings most remote in the scale of nature. The slightest advantage in one being, at any age or during any season, over those with which it comes into competition, or better adaptation in however slight a degree to the surrounding physical conditions, will turn the balance.

With animals having separated sexes there will in most cases be a struggle between the males for possession of the females. The most vigorous individuals, or those which have most successfully struggled with their conditions of life, will generally leave most progeny. But success will often depend on having special weapons or means of defence, or on the charms of the males; and the slightest advantage will lead to victory.

As geology plainly proclaims that each land has undergone great physical changes, we might have expected that organic beings would have varied under nature, in the same way as they generally have varied under the changed conditions of domestication. And if there be any variability under nature, it would be an unaccountable fact if natural selection had not come into play. It has often been asserted, but the assertion is quite incapable of proof, that the amount of variation under nature is a strictly limited quantity. Man, though acting on external characters alone and often capriciously, can produce within a short period a great result by adding up mere individual differences in his domestic productions; and every one admits that there are at least individual differences in species under nature. But, besides such differences, all naturalists have admitted the existence of varieties, which they think sufficiently distinct to be worthy of record in systematic works. No one can draw any clear distinction between individual differences and slight varieties; or between more plainly marked varieties and sub-species, and species. Let it be observed how naturalists differ in the rank which they assign to the many representative forms in Europe and North America.

If then we have under nature variability and a powerful agent always ready to act and select, why should we doubt that variations in any way useful to beings, under their excessively complex relations of life, would be preserved, accumulated, and inherited? Why, if man can by patience select variations most useful to himself, should nature fail in selecting variations useful, under changing conditions of life, to her living products? What limit can be put to this power, acting during long ages and rigidly scrutinising the whole constitution, structure, and habits of each creature,—favouring the good and rejecting the bad? I can see no limit to this power, in slowly and beautifully adapting each form to the most complex relations of life. The theory of natural selection, even if we looked no further than this, seems to me to be in itself probable. I have already recapitulated, as fairly as I could, the opposed difficulties and objections: now let us turn to the special facts and arguments in favour of the theory.

On the view that species are only strongly marked and permanent varieties, and that each species first existed as a variety, we can see why it is that no line of demarcation can be drawn between species, commonly supposed to have been produced by special acts of creation, and varieties which are acknowledged to have been produced by secondary laws. On this same view we can understand how it is that in each region where many species of a genus have been produced, and where they now flourish, these same species should present many varieties; for where the manufactory of species has been active, we might expect, as a general rule, to find it still in action; and this is the case if varieties be incipient species. Moreover, the species of the larger genera, which afford the greater number of varieties or incipient species, retain to a certain degree the character of varieties; for they differ from each other by a less amount of difference than do the species of smaller genera. The closely allied species also of the larger genera apparently have restricted ranges, and they are clustered in little groups round other species—in which respects they resemble varieties. These are strange relations on the view of each species having been independently created, but are intelligible if all species first existed as varieties.

As each species tends by its geometrical ratio of reproduction to increase inordinately in number; and as the modified descendants of each species will be enabled to increase by so much the more as they become more diversified in habits and structure, so as to be enabled to seize on many and widely different places in the economy of nature, there will be a constant tendency in natural selection to preserve the most divergent offspring of any one species. Hence during a long-continued course of modification, the slight differences, characteristic of varieties of the same species, tend to be augmented into the greater differences characteristic of species of the same genus. New and improved varieties will inevitably supplant and exterminate the older, less improved and intermediate varieties; and thus species are rendered to a large extent defined and distinct objects. Dominant species belonging to the larger groups tend to give birth to new and dominant forms; so that each large group tends to become still larger, and at the same time more divergent in character. But as all groups cannot thus succeed in increasing in size, for the world would not hold them, the more dominant groups beat the less dominant. This tendency in the large groups to go on increasing in size and diverging in character, together with the almost inevitable contingency of much extinction, explains the arrangement of all the forms of life, in groups subordinate to groups, all within a few great classes, which we now see everywhere around us, and which has prevailed throughout all time. This grand fact of the grouping of all organic beings seems to me utterly inexplicable on the theory of creation.

As natural selection acts solely by accumulating slight, successive, favourable variations, it can produce no great or sudden modification; it can act only by very short and slow steps. Hence the canon of "Natura non facit saltum," which every fresh addition to our knowledge tends to make more strictly correct, is on this theory simply intelligible. We can plainly see why nature is prodigal in variety, though niggard in innovation. But why this should be a law of nature if each species has been independently created, no man can explain.

Many other facts are, as it seems to me, explicable on this theory. How strange it is that a bird, under the form of woodpecker, should have been created to prey on insects on the ground; that upland geese, which never or rarely swim, should have been created with webbed feet; that a thrush should have been created to dive and feed on sub-aquatic insects; and that a petrel should have been created with habits and structure fitting it for the life of an auk or grebe! and so on in endless other cases. But on the view of each species constantly trying to increase in number, with natural selection always ready to adapt the slowly varying descendants of each to any unoccupied or ill-occupied place in nature, these facts cease to be strange, or perhaps might even have been anticipated.

As natural selection acts by competition, it adapts the inhabitants of each country only in relation to the degree of perfection of their associates; so that we need feel no surprise at the inhabitants of any one country, although on the ordinary view supposed to have been specially created and adapted for that country, being beaten and supplanted by the naturalised productions from another land. Nor ought we to marvel if all the contrivances in nature be not, as far as we can judge, absolutely perfect; and if some of them be abhorrent to our ideas of fitness. We need not marvel at the sting of the bee causing the bee's own death; at drones being produced in such vast numbers for one single act, and being then slaughtered by their sterile sisters; at the astonishing waste of pollen by our fir-trees; at the instinctive hatred of the queen bee for her own fertile daughters; at ichneumonidæ feeding within the live bodies of caterpillars; and at other such cases. The wonder indeed is, on the theory of natural selection, that more cases of the want of absolute perfection have not been observed.

The complex and little known laws governing variation are the same, as far as we can see, with the laws which have governed the production of so-called specific forms. In both cases physical conditions seem to have produced but little direct effect; yet when varieties enter any zone, they occasionally assume some of the characters of the species proper to that zone. In both varieties and species, use and disuse seem to have produced some effect; for it is difficult to resist this conclusion when we look, for instance, at the logger-headed duck, which has wings incapable of flight, in nearly the same condition as in the domestic duck; or when we look at the burrowing tucutucu, which is occasionally blind, and then at certain moles, which are habitually blind and have their eyes covered with skin; or when we look at the blind animals inhabiting the dark caves of America and

Europe. In both varieties and species correlation of growth seems to have played a most important part, so that when one part has been modified other parts are necessarily modified. In both varieties and species reversions to long-lost characters occur. How inexplicable on the theory of creation is the occasional appearance of stripes on the shoulder and legs of the several species of the horse-genus and in their hybrids! How simply is this fact explained if we believe that these species have descended from a striped progenitor, in the same manner as the several domestic breeds of pigeon have descended from the blue and barred rock-pigeon!

On the ordinary view of each species having been independently created, why should the specific characters, or those by which the species of the same genus differ from each other, be more variable than the generic characters in which they all agree? Why, for instance, should the colour of a flower be more likely to vary in any one species of a genus, if the other species, supposed to have been created independently, have differently coloured flowers, than if all the species of the genus have the same coloured flowers? If species are only well-marked varieties, of which the characters have become in a high degree permanent, we can understand this fact; for they have already varied since they branched off from a common progenitor in certain characters, by which they have come to be specifically distinct from each other; and therefore these same characters would be more likely still to be variable than the generic characters which have been inherited without change for an enormous period. It is inexplicable on the theory of creation why a part developed in a very unusual manner in any one species of a genus, and therefore, as we may naturally infer, of great importance to the species, should be eminently liable to variation; but, on my view, this part has undergone, since the several species branched off from a common progenitor, an unusual amount of variability and modification, and therefore we might expect this part generally to be still variable. But a part may be developed in the most unusual manner, like the wing of a bat, and yet not be more variable than any other structure, if the part be common to many subordinate forms, that is, if it has been inherited for a very long period; for in this case it will have been rendered constant by long-continued natural selection.

Glancing at instincts, marvellous as some are, they offer no greater difficulty than does corporeal structure on the theory of the natural selection of successive, slight, but profitable modifications. We can thus understand why nature moves by graduated steps in endowing different animals of the same class with their several instincts. I have attempted to show how much light the principle of gradation throws on the admirable architectural powers of the hive-bee. Habit no doubt sometimes comes into play in modifying instincts; but it certainly is not indispensable, as we see, in the case of neuter insects, which leave no progeny to inherit the effects of long-continued habit. On the view of all the species of the same genus having descended from a common parent, and having inherited much in common, we can understand how it is that allied species, when placed under considerably different conditions of life, yet should follow nearly the same instincts; why the thrush of South America, for instance, lines her nest with mud like our British species. On the view of instincts having been slowly acquired through natural selection we need not marvel at some instincts being apparently not perfect and liable to mistakes, and at many instincts causing other animals to suffer.

If species be only well-marked and permanent varieties, we can at once see why their crossed offspring should follow the same complex laws in their degrees and kinds of resemblance to their parents,—in being absorbed into each other by successive crosses, and in other such points,—as do the crossed offspring of acknowledged varieties. On the other hand, these would be strange facts if species have been independently created, and varieties have been produced by secondary laws.

If we admit that the geological record is imperfect in an extreme degree, then such facts as the record gives, support the theory of descent with modification. New species have come on the stage slowly and at successive intervals; and the amount of change, after equal intervals of time, is widely different in different groups. The extinction of species and of whole groups of species, which has played so conspicuous a part in the history of the organic world, almost inevitably follows on the principle of natural selection; for old forms will be supplanted by new and improved forms. Neither single species nor groups of species reappear when the chain of ordinary generation has once been broken. The gradual diffusion of dominant forms, with the slow modification of their descendants, causes the forms of life, after long intervals of time, to appear as if they had changed simultaneously throughout the world. The fact of the fossil remains of each formation being in some degree intermediate in character between the fossils in the formations above and below, is simply explained by their intermediate position in the chain of descent. The grand fact that all extinct organic beings belong to the same system with recent beings, falling either into the same or into intermediate groups, follows from the living and the extinct being the offspring of common parents. As the groups which have descended from an ancient progenitor have generally diverged in character, the progenitor with its early descendants will often be intermediate in character in comparison with its later descendants; and thus we can see why the more ancient a fossil is, the oftener it stands in some degree intermediate between existing and allied groups. Recent forms are generally looked at as being, in some vague sense, higher than ancient and extinct forms;

and they are in so far higher as the later and more improved forms have conquered the older and less improved organic beings in the struggle for life. Lastly, the law of the long endurance of allied forms on the same continent,—of marsupials in Australia, of edentata in America, and other such cases,—is intelligible, for within a confined country, the recent and the extinct will naturally be allied by descent.

Looking to geographical distribution, if we admit that there has been during the long course of ages much migration from one part of the world to another, owing to former climatal and geographical changes and to the many occasional and unknown means of dispersal, then we can understand, on the theory of descent with modification, most of the great leading facts in Distribution. We can see why there should be so striking a parallelism in the distribution of organic beings throughout space, and in their geological succession throughout time; for in both cases the beings have been connected by the bond of ordinary generation, and the means of modification have been the same. We see the full meaning of the wonderful fact, which must have struck every traveller, namely, that on the same continent, under the most diverse conditions, under heat and cold, on mountain and lowland, on deserts and marshes, most of the inhabitants within each great class are plainly related; for they will generally be descendants of the same progenitors and early colonists. On this same principle of former migration, combined in most cases with modification, we can understand, by the aid of the Glacial period, the identity of some few plants, and the close alliance of many others, on the most distant mountains, under the most different climates; and likewise the close alliance of some of the inhabitants of the sea in the northern and southern temperate zones, though separated by the whole intertropical ocean. Although two areas may present the same physical conditions of life, we need feel no surprise at their inhabitants being widely different, if they have been for a long period completely separated from each other; for as the relation of organism to organism is the most important of all relations, and as the two areas will have received colonists from some third source or from each other, at various periods and in different proportions, the course of modification in the two areas will inevitably be different.

On this view of migration, with subsequent modification, we can see why oceanic islands should be inhabited by few species, but of these, that many should be peculiar. We can clearly see why those animals which cannot cross wide spaces of ocean, as frogs and terrestrial mammals, should not inhabit oceanic islands; and why, on the other hand, new and peculiar species of bats, which can traverse the ocean, should so often be found on islands far distant from any continent. Such facts as the presence of peculiar species of bats, and the absence of all other mammals, on oceanic islands, are utterly inexplicable on the theory of independent acts of creation.

The existence of closely allied or representative species in any two areas, implies, on the theory of descent with modification, that the same parents formerly inhabited both areas; and we almost invariably find that wherever many closely allied species inhabit two areas, some identical species common to both still exist. Wherever many closely allied yet distinct species occur, many doubtful forms and varieties of the same species likewise occur. It is a rule of high generality that the inhabitants of each area are related to the inhabitants of the nearest source whence immigrants might have been derived. We see this in nearly all the plants and animals of the Galapagos archipelago, of Juan Fernandez, and of the other American islands being related in the most striking manner to the plants and animals of the neighbouring American mainland; and those of the Cape de Verde archipelago and other African islands to the African mainland. It must be admitted that these facts receive no explanation on the theory of creation.

The fact, as we have seen, that all past and present organic beings constitute one grand natural system, with group subordinate to group, and with extinct groups often falling in between recent groups, is intelligible on the theory of natural selection with its contingencies of extinction and divergence of character. On these same principles we see how it is, that the mutual affinities of the species and genera within each class are so complex and circuitous. We see why certain characters are far more serviceable than others for classification;—why adaptive characters, though of paramount importance to the being, are of hardly any importance in classification; why characters derived from rudimentary parts, though of no service to the being, are often of high classificatory value; and why embryological characters are the most valuable of all. The real affinities of all organic beings are due to inheritance or community of descent. The natural system is a genealogical arrangement, in which we have to discover the lines of descent by the most permanent characters, however slight their vital importance may be.

The framework of bones being the same in the hand of a man, wing of a bat, fin of the porpoise, and leg of the horse,—the same number of vertebræ forming the neck of the giraffe and of the elephant,—and innumerable other such facts, at once explain themselves on the theory of descent with slow and slight successive modifications. The similarity of pattern in the wing and leg of a bat, though used for such different purpose,—in the jaws and legs of a crab,—in the petals, stamens, and pistils of a flower, is likewise intelligible on the view of the gradual modification of parts or organs, which were alike in the early progenitor of each class. On the principle of successive variations not always supervening at an early age, and being inherited at a corresponding not early period of life, we can clearly see why the embryos of mammals, birds, reptiles, and fishes should be so closely alike, and should be so unlike the adult forms. We may cease marvelling at the embryo of an air-breathing mammal or bird having branchial slits and arteries running in loops, like those in a fish which has to breathe the air dissolved in water, by the aid of well-developed branchiæ.

Disuse, aided sometimes by natural selection, will often tend to reduce an organ, when it has become useless by changed habits or under changed conditions of life; and we can clearly understand on this view the meaning of rudimentary organs. But disuse and selection will generally act on each creature, when it has come to maturity and has to play its full part in the struggle for existence, and will thus have little power of acting on an organ during early life; hence the organ will not be much reduced or rendered rudimentary at this early age. The calf, for instance, has inherited teeth, which never cut through the gums of the upper jaw, from an early progenitor having well-developed teeth; and we may believe, that the teeth in the mature animal were reduced, during successive generations, by disuse or by the tongue and palate having been fitted by natural selection to browse without their aid; whereas in the calf, the teeth have been left untouched by selection or disuse, and on the principle of inheritance at corresponding ages have been inherited from a remote period to the present day. On the view of each organic being and each separate organ having been specially created, how utterly inexplicable it is that parts, like the teeth in the embryonic calf or like the shrivelled wings under the soldered wing-covers of some beetles, should thus so frequently bear the plain stamp of inutility! Nature may be said to have taken pains to reveal, by rudimentary organs and by homologous structures, her scheme of modification, which it seems that we wilfully will not understand.

I HAVE NOW recapitulated the chief facts and considerations which have thoroughly convinced me that species have changed, and are still slowly changing by the preservation and accumulation of successive slight favourable variations. Why, it may be asked, have all the most eminent living naturalists and geologists rejected this view of the mutability of species? It cannot be asserted that organic beings in a state of nature are subject to no variation; it cannot be proved that the amount of variation in the course of long ages is a limited quantity; no clear distinction has been, or can be, drawn between species and well-marked varieties. It cannot be maintained that species when intercrossed are invariably sterile, and varieties invariably fertile; or that sterility is a special endowment and sign of creation. The belief that species were immutable productions was almost unavoidable as long as the history of the world was thought to be of short duration; and now that we have acquired some idea of the lapse of time, we are too apt to assume, without proof, that the geological record is so perfect that it would have afforded us plain evidence of the mutation of species, if they had undergone mutation.

But the chief cause of our natural unwillingness to admit that one species has given birth to other and distinct species, is that we are always slow in admitting any great change of which we do not see the intermediate steps. The difficulty is the same as that felt by so many geologists, when Lyell first insisted that long lines of inland cliffs had been formed, and great valleys excavated, by the slow action of the coast-waves. The mind cannot possibly grasp the full meaning of the term of a hundred million years; it cannot add up and perceive the full effects of many slight variations, accumulated during an almost infinite number of generations.

Although I am fully convinced of the truth of the views given in this volume under the form of an abstract, I by no means expect to convince experienced naturalists whose minds are stocked with a multitude of facts all viewed, during a long course of years, from a point of view directly opposite to mine. It is so easy to hide our ignorance under such expressions as the "plan of creation," "unity of design," &c., and to think that we give an explanation when we only restate a fact. Any one whose disposition leads him to attach more weight to unexplained difficulties than to the explanation of a certain number of facts will certainly reject my theory. A few naturalists, endowed with much flexibility of mind, and who have already begun to doubt on the immutability of species, may be influenced by this volume; but I look with confidence to the future, to young and rising naturalists, who will be able to view both sides of the question with impartiality. Whoever is led to believe that species are mutable will do good service by conscientiously expressing his conviction; for only thus can the load of prejudice by which this subject is overwhelmed be removed.

Several eminent naturalists have of late published their belief that a multitude of reputed species in each genus are not real species; but that other species are real, that is, have been independently created. This seems to me a strange conclusion to arrive at. They admit that a multitude of forms, which till lately they themselves thought were special creations, and which are still thus looked at by the majority of naturalists, and which consequently have every external characteristic feature of true species,—they admit that these have been produced by variation, but they refuse to extend the same view to other and very slightly different forms. Nevertheless they do not pretend that they can define, or even conjecture, which are the created forms of life, and which are those produced by secondary laws. They admit variation as a *vera causa* in one case, they arbitrarily reject it in another, without assigning any distinction in the two cases. The day will come when this will be given as a curious illustration of the blindness of preconceived opinion. These authors seem no more startled at a miraculous act of creation than at an ordinary birth. But do they really believe that at innumerable periods in the earth's history certain elemental atoms have been commanded suddenly to flash into living tissues? Do they believe that at each supposed act of creation one individual or many were produced? Were all the infinitely numerous kinds of animals and plants created as eggs or seed, or as full grown? and in the case of mammals, were they created bearing the false marks of nourishment from the mother's womb? Although naturalists very properly demand a full explanation of every difficulty from those who believe in the mutability of species, on their own side they ignore the whole subject of the first appearance of species in what they consider reverent silence.

It may be asked how far I extend the doctrine of the modification of species. The question is difficult to answer, because the more distinct the forms are which we may consider, by so much the arguments fall away in force. But some arguments of the greatest weight extend very far. All the members of whole classes can be connected together by chains of affinities, and all can be classified on the same principle, in groups subordinate to groups. Fossil remains sometimes tend to fill up very wide intervals between existing orders. Organs in a rudimentary condition plainly show that an early progenitor had the organ in a fully developed state; and this in some instances necessarily implies an enormous amount of modification in the descendants. Throughout whole classes various structures are formed on the same pattern, and at an embryonic age the species closely resemble each other. Therefore I cannot doubt that the theory of descent with modification embraces all the members of the same class. I believe that animals have descended from at most only four or five progenitors, and plants from an equal or lesser number.

Analogy would lead me one step further, namely, to the belief that all animals and plants have descended from some one prototype. But analogy may be a deceitful guide. Nevertheless all living things have much in common, in their chemical composition, their germinal vesicles, their cellular structure, and their laws of growth and reproduction. We see this even in so trifling a circumstance as that the same poison often similarly affects plants and animals; or that the poison secreted by the gall-fly produces monstrous growths on the wild rose or oak-tree. Therefore I should infer from analogy that probably all the organic beings which have ever lived on this earth have descended from some one primordial form, into which life was first breathed.

WHEN THE VIEWS entertained in this volume on the origin of species, or when analogous views are generally admitted, we can dimly foresee that there will be a considerable revolution in natural history. Systematists will be able to pursue their labours as at present; but they will not be incessantly haunted by the shadowy doubt whether this or that form be in essence a species. This I feel sure, and I speak after experience, will be no slight relief. The endless disputes whether or not some fifty species of British brambles are true species will cease. Systematists will have only to decide (not that this will be easy) whether any form be sufficiently constant and distinct from other forms, to be capable of definition; and if definable, whether the differences be sufficiently important to deserve a specific name. This latter point will become a far more essential consideration than it is at present; for differences, however slight, between any two forms, if not blended by intermediate gradations, are looked at by most naturalists as sufficient to raise both forms to the rank of species. Hereafter we shall be compelled to acknowledge that the only distinction between species and well-marked varieties is, that the latter are known, or believed, to be connected at the present day by intermediate gradations, whereas species were formerly thus connected. Hence, without quite rejecting the consideration of the present existence of intermediate gradations between any two forms, we shall be led to weigh more carefully and to value higher the actual amount of difference between them. It is quite possible that forms now generally acknowledged to be merely varieties may hereafter be thought worthy of specific names, as with the primrose and cowslip; and in this case scientific and common language will come into accordance. In short, we shall have to treat species in the same manner as those naturalists treat genera, who admit that genera are merely artificial combinations made for convenience. This may not be a cheering prospect; but we shall at least be freed from the vain search for the undiscovered and undiscoverable essence of the term species.

The other and more general departments of natural history will rise greatly in interest. The terms used by naturalists of affinity, relationship, community of type, paternity, morphology, adaptive characters, rudimentary and aborted organs, &c., will cease to be metaphorical, and will have a plain signification. When we no longer look at an organic being as a savage looks at a ship, as at something wholly beyond his comprehension; when we regard every production of nature as one which has had a history; when we contemplate every complex structure and instinct as the summing up of many contrivances, each useful to the possessor, nearly in the same way as when we look at any great mechanical invention as the summing up of the labour, the experience, the reason, and even the blunders of numerous workmen; when we thus view each organic being, how far more interesting, I speak from experience, will the study of natural history become!

A grand and almost untrodden field of inquiry will be opened, on the causes and laws of variation, on correlation of growth, on the effects of use and disuse, on the direct action of external conditions, and so forth. The study of domestic productions will rise immensely in value. A new variety raised by man will be a far more important and interesting subject for study than one more species added to the infinitude of already recorded species. Our classifications will come to be, as far as they can be so made, genealogies; and will then truly give what may be called the plan of creation. The rules for classifying will no doubt become simpler when we have a definite object in view. We possess no pedigrees or armorial bearings; and we have to discover and trace the many diverging lines of descent in our natural genealogies, by characters of any kind which have long been inherited. Rudimentary organs will speak infallibly with respect to the nature of long-lost structures. Species and groups of species, which are called aberrant, and which may fancifully be called living fossils, will aid us in forming a picture of the ancient forms of life. Embryology will reveal to us the structure, in some degree obscured, of the prototypes of each great class.

When we can feel assured that all the individuals of the same species, and all the closely allied species of most genera, have within a not very remote period descended from one parent, and have migrated from some one birthplace; and when we better know the many means of migration, then, by the light which geology now throws, and will continue to throw, on former changes of climate and of the level of the land, we shall surely be enabled to trace in an admirable manner the former migrations of the inhabitants of the whole world. Even at present, by comparing the differences of the inhabitants of the sea on the opposite sides of a continent, and the nature of the various inhabitants of that continent in relation to their apparent means of immigration, some light can be thrown on ancient geography.

The noble science of Geology loses glory from the extreme imperfection of the record. The crust of the earth with its embedded remains must not be looked at as a well-filled museum, but as a poor collection made at hazard and at rare intervals. The accumulation of each great fossiliferous formation will be recognised as having depended on an unusual concurrence of circumstances, and the blank intervals between the successive stages as having been of vast duration. But we shall be able to gauge with some security the duration of these intervals by a comparison of the preceding and succeeding organic forms. We must be cautious in attempting to correlate as strictly contemporaneous two formations, which include few identical species, by the general succession of their forms of life. As species are produced and exterminated by slowly acting and still existing causes, and not by miraculous acts of creation and by catastrophes; and as the most important of all causes of organic change is one which is almost independent of altered and perhaps suddenly altered physical conditions, namely, the mutual relation of organism to organism,—the improvement of one being entailing the improvement or the extermination of others; it follows, that the amount of organic change in the fossils of consecutive formations probably serves as a fair measure of the lapse of actual time. A number of species, however, keeping in a body might remain for a long period unchanged, whilst within this same period, several of these species, by migrating into new countries and coming into competition with foreign associates, might become modified; so that we must not overrate the accuracy of organic change as a measure of time. During early periods of the earth's history, when the forms of life were probably fewer and simpler, the rate of change was probably slower; and at the first dawn of life, when very few forms of the simplest structure existed, the rate of change may have been slow in an extreme degree. The whole history of the world, as at present known, although of a length quite incomprehensible by us, will hereafter be recognised as a mere fragment of time, compared with the ages which have elapsed since the first creature, the progenitor of innumerable extinct and living descendants, was created.

In the distant future I see open fields for far more important researches. Psychology will be based on a new foundation, that of the necessary acquirement of each mental power and capacity by gradation. Light will be thrown on the origin of man and his history,

Authors of the highest eminence seem to be fully satisfied with the view that each species has been independently created. To my mind it accords better with what we know of the laws impressed on matter by the Creator, that the production and extinction of the past and present inhabitants of the world should have been due to secondary causes, like those determining the birth and death of the individual. When I view all beings not as special creations, but as the lineal descendants of some few beings which lived long before the first bed of the Silurian system was deposited, they seem to me to become ennobled. Judging from the past, we may safely infer that not one living species will transmit its unaltered likeness to a distant futurity. And of the species now living very few will transmit progeny of any kind to a far distant futurity; for the manner in which all organic beings are grouped, shows that the greater number of species of each genus, and all the species of many genera, have left no descendants, but have become utterly extinct. We can so far take a prophetic glance into futurity as to foretel that it will be the common and widely-spread species, belonging to the larger and dominant groups, which will ultimately prevail and procreate new and dominant species. As all the living forms of life are the lineal descendants of those which lived long before the Silurian epoch, we may feel certain that the ordinary succession by generation has never once been broken, and that no cataclysm has desolated the whole world. Hence we may look with some confidence to a secure future of equally inappreciable length. And as natural selection works solely by and for the good of each being, all corporeal and mental endowments will tend to progress towards perfection.

It is interesting to contemplate an entangled bank, clothed with many plants of many kinds, with birds singing on the bushes, with various insects flitting about, and with worms crawling through the damp earth, and to reflect that these elaborately constructed forms, so different from each other, and dependent on each other in so complex a manner, have all been produced by laws acting around us. These laws, taken in the largest sense, being Growth with Reproduction; Inheritance which is almost implied by reproduction; Variability from the indirect and direct action of the external conditions of life, and from use and disuse; a Ratio of Increase so high as to lead to a Struggle for Life, and as a consequence to Natural Selection, entailing Divergence of Character and the Extinction of less-improved forms. Thus, from the war of nature, from famine and death, the most exalted object which we are capable of conceiving, namely, the production of the higher animals, directly follows. There is grandeur in this view of life, with its several powers, having been originally breathed into a few forms or into one; and that, whilst this planet has gone cycling on according to the fixed law of gravity, from so simple a beginning endless forms most beautiful and most wonderful have been, and are being, evolved.

Charles Darwin age 65 (1874)

The Descent of Man

INTRODUCTION TO *THE DESCENT OF MAN*

THE DESCENT OF MAN, AND SELECTION IN RELATION TO SEX was the other shoe that Charles Darwin dropped after *On the Origin of Species.* Victorian society, whose wrath Charles Darwin so feared as to hold back on the publication of his theory of evolution, might not have been scandalized by the descent of plants and animals, but they were quite unprepared to accept the descent of man from "some pre-existing form," as Darwin delicately put it, and most certainly from apes, as the evidence implied from the very start.

Yet Darwin had to take that step. It was logical from the premise that evolution is universal. Further, the indelible stamp of evolution on humanity is everywhere to be seen. The first part of *The Descent of Man,* the book to be presented next, is relatively short. The reasons are clear. In *On the Origin of Species* Darwin had already laid, in "one long argument," in intricately detailed and ever cautious terms, the case for evolution by the process of natural selection. "The great principle of evolution," he now felt confident enough to proclaim in *The Descent of Man,* "stands up clear and firm, when these groups of facts are considered in connection with others, such as the mutual affinities of the members of the same group, their geographical distribution in past and present times, and their geological succession. It is incredible that all these facts should speak falsely." The multifarious evidences locked together in such orderly manner, Darwin believed, point straight to the origin of the human species from a lower form. "The grounds upon which this rests," Darwin concludes, "will never be shaken."

The evidence includes the close similarity of the embryonic development of humans and other mammals highlighted by Darwin in illustrations of early fetuses of a dog and human; unmistakable homologues in the adult anatomy of humans and primates; and vestiges of structures in humans that bespeak an animal ancestry. In the more than 130 years since the publication of *The Descent of Man,* the circumstantial case has been ratcheted up in fine detail in every aspect of human biology. The new evidence has arrived from the decoding of genomes, which places *Homo sapiens* closest to chimpanzees among living animals, from cell and tissue physiology, and even from comparative analyses of brain circuits.

Darwin's treatment of human origins is relatively short, however, for a second reason: he completely lacked fossil evidence, which in courtroom parlance would be called the smoking gun. Now that too is abundantly available. Over a century of worldwide searches have turned up such large numbers of prehuman and human fossils as to establish at least the broad outlines of the origins of *Homo sapiens.* The family of man, it turns out, originated in Africa and spread over the rest of the world in later waves. The line antecedent to modern humans evidently split between 4 and 7 million years ago from the separate line leading to our closest living relatives, chimpanzees and gorillas. By 4 million years before the present, the earliest direct human ancestor *(Australopithecus afarensis,* known best from the female fossil "Lucy") had acquired an erect posture and bipedal locomotion but still possessed a chimpanzee-sized brain. The first "truly human" species *(Homo habilis)* arose by 2.2 million years ago and possessed a brain capacity one-third larger. It gave rise within half a million years to *Homo erectus,* who possessed a brain nearly as large as that of modern *Homo sapiens.* This, our immediate ancestor, manufactured crude stone tools, evidently employed fire, and was the first of the hominids to spread to other continents.

At the time Darwin was piecing together the fragments of cooperative anatomy and behavior that made his case for human evolution, he was also creating the theory of sexual selection. As reported in the second and longer section of *The Descent of Man,* a great many of the most striking anatomical and behavioral traits of animals (and also, he believed, human beings) have arisen by competition for mates, chiefly through seductive displays but also and primarily among males through bluff and combat. The displays generate many of the most beautiful and dramatic sights and sounds of nature. They occur in enormous variety, from the fan of the peacock to the song of the nightingale, the horns of rhinoceros beetles, and the antler clashes of fighting deer. This principle is applied in the concluding chapters, and with considerable daring, to human gender differences and the instincts that guide much of human social behavior.

The Descent of Man,

and

Selection in Relation to Sex

Charles Darwin

· 1871 ·

Postscript

Chapter VIII, pp. 946–947.—I have fallen into a serious and unfortunate error, in relation to the sexual differences of animals, in attempting to explain what seemed to me a singular coincidence in the late period of life at which the necessary variations have arisen in many cases, and the late period at which sexual selection acts. The explanation given is wholly erroneous, as I have discovered by working out an illustration in figures. Moreover, the supposed coincidence of period is far from general, and is not remarkable; for, as I have elsewhere attempted to show, variations arising early in life have often been accumulated through sexual selection, being then commonly transmitted to both sexes. On the other hand, variations arising late in life cannot fail to coincide approximately in period with that of the process of sexual selection. Allusions to these erroneous views reappear in Chapter XV, p. 1112 and Chapter XVI, pp. 1154–1155.

THE NATURE OF THE FOLLOWING WORK will be best understood by a brief account of how it came to be written. During many years I collected notes on the origin or descent of man, without any intention of publishing on the subject, but rather with the determination not to publish, as I thought that I should thus only add to the prejudices against my views. It seemed to me sufficient to indicate, in the first edition of my 'Origin of Species,' that by this work "light would be thrown on the origin of man and his history;" and this implies that man must be included with other organic beings in any general conclusion respecting his manner of appearance on this earth. Now the case wears a wholly different aspect. When a naturalist like Carl Vogt ventures to say in his address as President of the National Institution of Geneva (1869), "personne, en Europe au moins, n'ose plus soutenir la création indépendante et de toutes pièces, des espèces," it is manifest that at least a large number of naturalists must admit that species are the modified descendants of other species; and this especially holds good with the younger and rising naturalists. The greater number accept the agency of natural selection; though some urge, whether with justice the future must decide, that I have greatly overrated its importance. Of the older and honoured chiefs in natural science, many unfortunately are still opposed to evolution in every form.

In consequence of the views now adopted by most naturalists, and which will ultimately, as in every other case, be followed by other men, I have been led to put together my notes, so as to see how far the general conclusions arrived at in my former works were applicable to man. This seemed all the more desirable as I had never deliberately applied these views to a species taken singly. When we confine our attention to any one form, we are deprived of the weighty arguments derived from the nature of the affinities which connect together whole groups of organisms—their geographical distribution in past and present times, and their geological succession. The homological structure, embryological development, and rudimentary organs of a species, whether it be man or any other animal, to which our attention may be directed, remain to be considered; but these great classes of facts afford, as it appears to me, ample and conclusive evidence in favour of the principle of gradual evolution. The strong support derived from the other arguments should, however, always be kept before the mind.

The sole object of this work is to consider, firstly, whether man, like every other species, is descended from some pre-existing form; secondly, the manner of his development; and thirdly, the value of the differences between the so-called races of man. As I shall confine myself to these points, it will not be necessary to describe in detail the differences between the several races—an enormous subject which has been fully discussed in many valuable works. The high antiquity of man has recently been demonstrated by the labours of a host of eminent men, beginning with M. Boucher de Perthes; and this is the indispensable basis for understanding his origin. I shall, therefore, take this conclusion for granted, and may refer my readers to the admirable treatises of Sir Charles Lyell, Sir John Lubbock, and others. Nor shall I have occasion to do more than to allude to the amount of difference between man and the anthropomorphous apes; for Prof. Huxley, in the opinion of most competent judges, has conclusively shewn that in every single visible character man differs less from the higher apes than these do from the lower members of the same order of Primates.

This work contains hardly any original facts in regard to man; but as the conclusions at which I arrived, after drawing up a rough draft, appeared to me interesting, I thought that they might interest others. It has often and confidently been asserted, that man's origin can never be known: but ignorance more frequently begets confidence than does knowledge: it is those who know little, and not those who know much, who so positively assert that this or that problem will never be solved by science. The conclusion that man is the co-descendant with other species of some ancient, lower, and extinct form, is not in any degree new. Lamarck long ago came to this conclusion, which has lately been maintained by several eminent naturalists and philosophers; for instance by Wallace, Huxley, Lyell, Vogt, Lubbock, Büchner, Rolle, &c.,[1] and especially by Häckel. This last naturalist, besides his great work, 'Generelle Morphologie' (1866), has recently (1868, with a second edit. in 1870), published his 'Natürliche Schöpfungsgeschichte,' in which he fully discusses the genealogy of man. If this work had appeared before my essay had been written, I should probably never have completed it. Almost all the conclusions at which I have arrived I find confirmed by this naturalist, whose knowledge on many points is much fuller than mine. Wherever I have added any fact or view from Prof. Häckel's writings, I give his authority in the text, other statements I leave as they originally stood in my manuscript, occasionally giving in the foot-notes references to his works, as a confirmation of the more doubtful or interesting points.

During many years it has seemed to me highly probable that sexual selection has played an important part in differentiating the races of man; but in my 'Origin of Species' (first edition, p. 199) I contented myself by merely alluding to this belief. When I came to apply this view to man, I found it indispensable to treat the whole subject in full detail.[2] Consequently the second part of the present work, treating of sexual selection, has extended to an inordinate length, compared with the first part; but this could not be avoided.

I had intended adding to the present volumes an essay on the expression of the various emotions by man and the lower animals. My attention was called to this subject many years ago by Sir Charles Bell's admirable work. This illustrious anatomist maintains that man is endowed with certain muscles solely for the sake of expressing his emotions. As this view is obviously opposed to the belief that man is descended from some other and lower form, it was necessary for me to consider it. I likewise wished to ascertain how far the emotions are expressed in the same manner by the different races of man. But owing to the length of the present work, I have thought it better to reserve my essay, which is partially completed, for separate publication.

Part I

The Descent or Origin of Man

The Evidence of the Descent of Man from some Lower Form

NATURE OF THE EVIDENCE BEARING ON THE ORIGIN OF MAN • HOMOLOGOUS STRUCTURES IN MAN AND THE LOWER ANIMALS • MISCELLANEOUS POINTS OF CORRESPONDENCE • DEVELOPMENT • RUDIMENTARY STRUCTURES, MUSCLES, SENSE-ORGANS, HAIR, BONES, REPRODUCTIVE ORGANS, &C. • THE BEARING OF THESE THREE GREAT CLASSES OF FACTS ON THE ORIGIN OF MAN

He who wishes to decide whether man is the modified descendant of some pre-existing form, would probably first enquire whether man varies, however slightly, in bodily structure and in mental faculties; and if so, whether the variations are transmitted to his offspring in accordance with the laws which prevail with the lower animals; such as that of the transmission of characters to the same age or sex. Again, are the variations the result, as far as our ignorance permits us to judge, of the same general causes, and are they governed by the same general laws, as in the case of other organisms; for instance by correlation, the inherited effects of use and disuse, &c.? Is man subject to similar malconformations, the result of arrested development, of reduplication of parts, &c., and does he display in any of his anomalies reversion to some former and ancient type of structure? It might also naturally be enquired whether man, like so many other animals, has given rise to varieties and sub-races, differing but slightly from each other, or to races differing so much that they must be classed as doubtful species? How are such races distributed over the world; and how, when crossed, do they react on each other, both in the first and succeeding generations? And so with many other points.

The enquirer would next come to the important point, whether man tends to increase at so rapid a rate, as to lead to occasional severe struggles for existence, and consequently to beneficial variations, whether in body or mind, being preserved, and injurious ones eliminated. Do the races or species of men, whichever term may be applied, encroach on and replace each other, so that some finally become extinct? We shall see that all these questions, as indeed is obvious in respect to most of them, must be answered in the affirmative, in the same manner as with the lower animals. But the several considerations just referred to may be conveniently deferred for a time; and we will first see how far the bodily structure of man shows traces, more or less plain, of his descent from some lower form. In the two succeeding chapters the mental powers of man, in comparison with those of the lower animals, will be considered.

The Bodily Structure of Man.—It is notorious that man is constructed on the same general type or model with other mammals. All the bones in his skeleton can be compared with corresponding bones in a monkey, bat, or seal. So it is with his muscles, nerves, blood-vessels and internal viscera. The brain, the most important of all the organs, follows the same law, as shewn by Huxley and other anatomists. Bischoff,[3] who is a hostile witness, admits that every chief fissure and fold in the brain of man has its analogy in that of the orang; but he adds that at no period of development do their brains perfectly agree; nor could this be expected, for otherwise their mental powers would have been the same. Vulpian[4] remarks: "Les différences réelles qui existent entre l'encéphale de l'homme et celui des singes supérieurs, sont bien minimes. Il ne faut pas se faire d'illusions à cet égard. L'homme est bien plus près des singes anthropomorphes par les caractères anatomiques de son cerveau que ceux-ci ne le sont non seulement des autres mammifères, mais mêmes de certains quadrumanes, des guenons et des macaques." But it would be superfluous here to give further details on the correspondence between man and the higher mammals in the structure of the brain and all other parts of the body.

It may, however, be worth while to specify a few points, not directly or obviously connected with structure, by which this correspondence or relationship is well shewn.

Man is liable to receive from the lower animals, and to communicate to them, certain diseases as hydrophobia, variola, the glanders, &c.; and this fact proves the close similarity of their tissues and blood, both in minute structure and composition, far more plainly than does their comparison under the best microscope, or by the aid of the best chemical analysis. Monkeys are liable to many of the same non-contagious diseases as we are;

thus Rengger,[5] who carefully observed for a long time the *Cebus Azaræ* in its native land, found it liable to catarrh, with the usual symptoms, and which when often recurrent led to consumption. These monkeys suffered also from apoplexy, inflammation of the bowels, and cataract in the eye. The younger ones when shedding their milk-teeth often died from fever. Medicines produced the same effect on them as on us. Many kinds of monkeys have a strong taste for tea, coffee, and spirituous liquors: they will also, as I have myself seen, smoke tobacco with pleasure. Brehm asserts that the natives of north-eastern Africa catch the wild baboons by exposing vessels with strong beer, by which they are made drunk. He has seen some of these animals, which he kept in confinement, in this state; and he gives a laughable account of their behaviour and strange grimaces. On the following morning they were very cross and dismal; they held their aching heads with both hands and wore a most pitiable expression: when beer or wine was offered them, they turned away with disgust, but relished the juice of lemons.[6] An American monkey, an Ateles, after getting drunk on brandy, would never touch it again, and thus was wiser than many men. These trifling facts prove how similar the nerves of taste must be in monkeys and man, and how similarly their whole nervous system is affected.

Man is infested with internal parasites, sometimes causing fatal effects, and is plagued by external parasites, all of which belong to the same genera or families with those infesting other mammals. Man is subject like other mammals, birds, and even insects, to that mysterious law, which causes certain normal processes, such as gestation, as well as the maturation and duration of various diseases, to follow lunar periods.[7] His wounds are repaired by the same process of healing; and the stumps left after the amputation of his limbs occasionally possess, especially during an early embryoni period, some power of regeneration, as in the lowest animals.[8]

The whole process of that most important function, the reproduction of the species, is strikingly the same in all mammals, from the first act of courtship by the male[9] to the birth and nurturing of the young. Monkeys are born in almost as helpless a condition as our own infants; and in certain genera the young differ fully as much in appearance from the adults, as do our children from their full-grown parents.[10] It has been urged by some writers as an important distinction, that with man the young arrive at maturity at a much later age than with any other animal: but if we look to the races of mankind which inhabit tropical countries the difference is not great, for the orang is believed not to be adult till the age of from ten to fifteen years.[11] Man differs from woman in size, bodily strength, hairyness, &c., as well as in mind, in the same manner as do the two sexes of many mammals. It is, in short, scarcely possible to exaggerate the close correspondence in general structure, in the minute structure of the tissues, in chemical composition and in constitution, between man and the higher animals, especially the anthropomorphous apes.

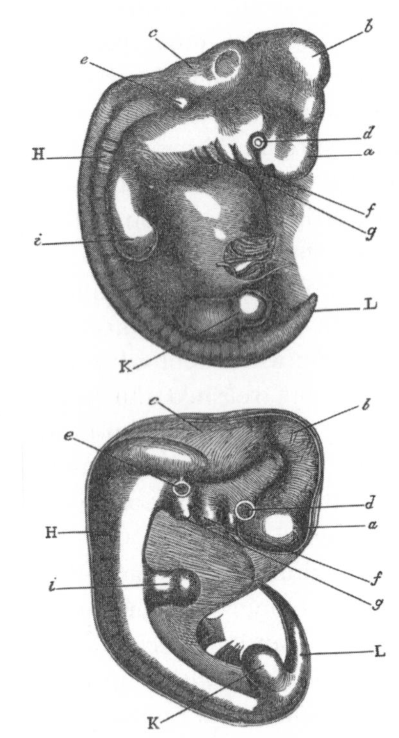

Fig. 1.
Upper figure human embryo, from Ecker. Lower figure that of a dog, from Bischoff.

a. *Fore-brain, cerebral hemispheres, &c.*
b. *Mid-brain, corpora quadrigemina.*
c. *Hind-brain, cerebellum, medulla oblongata.*
d. *Eye.*
e. *Ear.*
f. *First visceral arch.*
g. *Second visceral arch.*
H. *Vertebral columns and muscles in process of development.*
i. I*nterior.* } *extremities.*
K. Posterior. }
L. *Tail or os coccyx.*

Embryonic Development.—Man is developed from an ovule, about the 125th of an inch in diameter, which differs in no respect from the ovules of other animals. The embryo itself at a very early period can hardly be distinguished from that of other members of the vertebrate kingdom. At this period the arteries run in arch-like branches, as if to carry the blood to branchiae which are not present in the higher vertebrata, though the slits on the sides of the neck still remain (*f*, *g*, fig. 1), marking their former position. At a somewhat later period, when the extremities are developed, "the feet of lizards and mammals," as the illustrious Von Baer remarks, "the wings and feet of birds, no less than the hands and feet of man, all arise from the same fundamental form." It is, says Prof. Huxley,[12] "quite in the later stages of development that the young human being presents marked differences from the young ape, while the latter departs as much from the dog in its developments, as the man does. Startling as this last assertion may appear to be, it is demonstrably true."

As some of my readers may never have seen a drawing of an embryo, I have given one of man and another of a dog, at about the same early stage of development, carefully copied from two works of undoubted accuracy.[13]

After the foregoing statements made by such high authorities, it would be superfluous on my part to give a number of borrowed details, shewing that the embryo of man closely resembles that of other mammals. It may, however, be added that the human embryo likewise resembles in various points of structure certain low forms when adult. For instance, the heart at first exists as a simple pulsating vessel; the excreta are voided through a cloacal passage; and the os coccyx projects like a true tail, "extending considerably beyond the rudimentary legs."[14] In the embryos of all air-breathing vertebrates, certain glands called the corpora Wolffiana, correspond with and act like the kidneys of mature fishes.[15] Even at a later embryonic period, some striking resemblances between man and the lower animals may be observed. Bischoff says that the convolutions of the brain in a human fœtus at the end of the seventh month reach about the same stage of development as in a baboon when adult.[16] The great toe, as Prof. Owen remarks,[17] "which forms the fulcrum when standing or walking, is perhaps the most characteristic peculiarity in the human structure;" but in an embryo, about an inch in length, Prof. Wyman [18] found "that the great toe was shorter than the others, and, instead of being parallel to them, projected at an angle from the side of the foot, thus corresponding with the permanent condition of this part in the quadrumana." I will conclude with a quotation from Huxley,[19] who after asking, does man originate in a different way from a dog, bird, frog or fish? says, "the reply is not doubtful for a moment; without question, the mode of origin and the early stages of the development of man are identical with those of the animals immediately below him in the scale: without a doubt in these respects, he is far nearer to apes, than the apes are to the dog."

Rudiments.—This subject, though not intrinsically more important than the two last, will for several reasons be here treated with more fullness.[20] Not one of the higher animals can be named which does not bear some part in a rudimentary condition; and man forms no exception to the rule. Rudimentary organs must be distinguished from those that are nascent; though in some cases the distinction is not easy. The former are either absolutely useless, such as the mammæ of male quadrupeds, or the incisor teeth of ruminants which never cut through the gums; or they are of such slight service to their present possessors, that we cannot suppose that they were developed under the conditions which now exist. Organs in this latter state are not strictly rudimentary, but they are tending in this direction. Nascent organs, on the other hand, though not fully developed, are of high service to their possessors, and are capable of further development. Rudimentary organs are eminently variable; and this is partly intelligible, as they are useless or nearly useless, and consequently are no longer subjected to natural selection. They often become wholly suppressed. When this occurs, they are nevertheless liable to occasional reappearance through reversion; and this is a circumstance well worthy of attention.

Disuse at that period of life, when an organ is chiefly used, and this is generally during maturity, together with inheritance at a corresponding period of life, seem to have been the chief agents in causing organs to become rudimentary. The term "disuse" does not relate merely to the lessened action of muscles, but includes a diminished flow of blood to a part or organ, from being subjected to fewer alternations of pressure, or from becoming in any way less habitually active. Rudiments, however, may occur in one sex of parts normally present in the other sex; and such rudiments, as we shall hereafter see, have often originated in a distinct manner. In some cases organs have been reduced by means of natural selection, from having become injurious to the species under changed habits of life. The process of reduction is probably often aided through the two principles of compensation and economy of growth; but the later stages of reduction, after disuse has done all that can fairly be attributed to it, and when the saving to be effected by the economy of growth would be very small,[21] are difficult to understand. The final and complete suppression of a part, already useless and much reduced in size, in which case neither compensation nor economy can come into play, is perhaps intelligible by the aid of the hypothesis of pangenesis, and apparently in no other way. But as the whole subject of rudimentary organs has been fully discussed and illustrated in my former works,[22] I need here say no more on this head.

Rudiments of various muscles have been observed in many parts of the human body;[23] and not a few muscles, which are regularly present in some of the lower animals can occasionally be detected in man in a greatly reduced condition. Every one must have noticed the power which many animals, especially horses, possess of moving or twitching their skin; and this is effected by the panniculus carnosus. Remnants of this muscle in an efficient state are found in various parts of our bodies; for instance, on the forehead, by which the eyebrows are raised. The *platysma myoides,* which is well developed on the neck, belongs to this system, but cannot be voluntarily brought into action. Prof. Turner, of Edinburgh, has occasionally detected, as he informs me, muscular fasciculi in five different situations, namely in the axillæ, near the scapulæ, &c., all of which must be referred to the system of the panniculus. He has also shewn[24] that the *musculus sternalis* or *sternalis brutorum,* which is not an extension of the *rectus abdominalis,* but is closely allied to the panniculus, occurred in the proportion of about 3 per cent in upwards of 600 bodies: he adds, that this muscle affords "an excellent illustration of the statement that occasional and rudimentary structures are especially liable to variation in arrangement."

Some few persons have the power of contracting the superficial muscles on their scalps; and these muscles are in a variable and partially rudimentary condition. M. A. de Candolle has communicated to me a curious instance of the long-continued persistence or inheritance of this power, as well as of its unusual development. He knows a family, in which one member, the present head of a family, could, when a youth, pitch several heavy books from his head by the movement of the scalp alone; and he won wagers by performing this feat. His father, uncle, grandfather, and all his three children possess the same power to the same unusual degree. This family became divided eight generations ago into two branches; so that the head of the above-mentioned branch is cousin in the seventh degree to the head of the other branch. This distant cousin resides in another part of France, and on being asked whether he possessed the same faculty, immediately exhibited his power. This case offers a good illustration how persistently an absolutely useless faculty may be transmitted.

The extrinsic muscles which serve to move the whole external ear, and the intrinsic muscles which move the different parts, all of which belong to the system of the panniculus, are in a rudimentary condition in man; they are also variable in development, or at least in function. I have seen one man who could draw his ears forwards, and another who could draw them backwards;[25] and from what one of these persons told me, it is probable that most of us by often touching our ears and thus directing our attention towards them, could by repeated trials recover some power of movement. The faculty of erecting the ears and of directing them to different points of the compass, is no doubt of the highest service to many animals, as they thus perceive the point of danger; but I have never heard of a man who possessed the least power of erecting his ears,—the one movement which might be of use to him. The whole external shell of the ear may be considered a rudiment, together with the various folds and prominences (helix and anti-helix, tragus and anti-tragus, &c.) which in the lower animals strengthen and support the ear when erect, without adding much to its weight. Some authors, however, suppose that the cartilage of the shell serves to transmit vibrations to the acoustic nerve; but Mr. Toynbee,[26] after collecting all the known evidence on this head, concludes that the external shell is of no distinct use. The ears of the chimpanzee and orang are curiously like those of man, and I am assured by the keepers in the Zoological Gardens that these animals never move or erect them; so that they are in an equally rudimentary condition, as far as function is concerned, as in man. Why these animals, as well as the progenitors of man, should have lost the power of erecting their ears we cannot say. It may be, though I am not quite satisfied with this view, that owing to their arboreal habits and great strength they were but little exposed to danger, and so during a lengthened period moved their ears but little, and thus gradually lost the power of moving them. This would be a parallel case with that of those large and heavy birds, which from inhabiting oceanic islands have not been exposed to the attacks of beasts of prey, and have consequently lost the power of using their wings for flight.

The celebrated sculptor, Mr. Woolner, informs me of one little peculiarity in the external ear, which he has often observed both in men and women, and of which he perceived the full signification. His attention was first called to the subject whilst at work on his figure of Puck, to which he had given pointed ears. He was thus led to examine the ears of various monkeys, and subsequently more carefully those of man. The peculiarity consists in a little blunt point, projecting from the inwardly folded margin, or helix. Mr. Woolner made an exact model of one such case, and has sent me the accompanying drawing. (Fig. 2.) These points not only project inwards, but often a little outwards, so that they are visible when the head is viewed from directly in front or behind. They are variable in size and somewhat in position, standing either a little higher or lower; and they sometimes occur on one ear and not on the other. Now the meaning of these projections is not, I think, doubtful; but it may be thought that they offer too trifling a character to be worth notice. This thought, however, is as false as it is natural. Every character, however slight, must be the result of some definite cause; and if it occurs in many individuals deserves consideration. The helix obviously consists of the extreme margin of the ear folded inwards; and this folding appears to be in some manner connected with the whole external ear being permanently pressed backwards. In many monkeys, which do not stand high in the order, as baboons and some species of macacus,[27] the upper portion of the ear is slightly pointed, and the margin is not at all folded inwards; but if the margin were to be thus folded, a slight point would necessarily project inwards and probably a little outwards. This could actually be observed in a specimen of the *Ateles beelzebuth* in the Zoological Gardens; and we may safely conclude that it is a similar structure—a vestige of formerly pointed ears—which occasionally reappears in man.

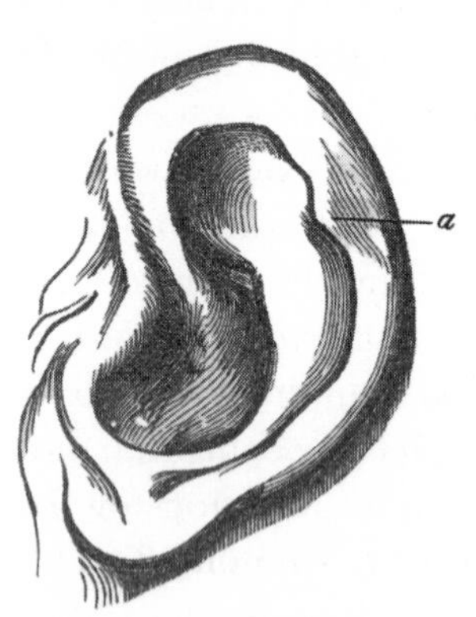

Fig. 2. *Human Ear, modelled and drawn by Mr. Woolner.*
a. *The projecting point.*

The nictitating membrane, or third eyelid, with its accessory muscles and other structures, is especially well developed in birds, and is of much functional importance to them, as it can be rapidly drawn across the whole eye-ball. It is found in some reptiles and amphibians, and in certain fishes, as in sharks. It is fairly well developed in the two lower divisions of the mammalian series, namely, in the monotremata and marsupials, and in some few of the higher mammals, as in the walrus. But in man, the quadrumana, and most other mammals, it exists, as is admitted by all anatomists, as a mere rudiment, called the semilunar fold.[28]

The sense of smell is of the highest importance to the greater number of mammals—to some, as the ruminants, in warning them of danger; to others, as the carnivora, in finding their prey; to others, as the wild boar, for both purposes combined. But the sense of smell is of extremely slight service, if any, even to savages, in whom it is generally more highly developed than in the civilised races. It does not warn them of danger, nor guide them to their food; nor does it prevent the Esquimaux from sleeping in the most fetid atmosphere, nor many savages from eating half-putrid meat. Those who believe in the principle of gradual evolution, will not readily admit that this sense in its present state was originally acquired by man, as he now exists. No doubt he inherits the power in an enfeebled and so far rudimentary condition, from some early progenitor, to whom it was highly serviceable and by whom it was continually used. We can thus perhaps understand how it is, as Dr. Maudsley has truly remarked,[29] that the sense of smell in man "is singularly effective in recalling vividly the ideas and images of forgotten scenes and places;" for we see in those animals, which have this sense highly developed, such as dogs and horses, that old recollections of persons and places are strongly associated with their odour.

Man differs conspicuously from all the other Primates in being almost naked. But a few short straggling hairs are found over the greater part of the body in the male sex, and fine down on that of the female sex. In individuals belonging to the same race these hairs are highly variable, not only in abundance, but likewise in position: thus the shoulders in some Europeans are quite naked, whilst in others they bear thick tufts of hair.[30] There can be little doubt that the hairs thus scattered over the body are the rudiments of the uniform hairy coat of the lower animals. This view is rendered all the more probable, as it is known that fine short, and pale-coloured hairs on the limbs and other parts of the body occasionally become developed into "thickset, long, and rather coarse dark hairs," when abnormally nourished near old-standing inflamed surfaces.[31]

I am informed by Mr. Paget that persons belonging to the same family often have a few hairs in their eyebrows much longer than the others; so that this slight peculiarity seems to be inherited. These hairs apparently represent the vibrissæ, which are used as organs of touch by many of the lower animals. In a young chimpanzee I observed that a few upright, rather long, hairs, projected above the eyes, where the true eyebrows, if present, would have stood.

The fine wool-like hair, or so-called lanugo, with which the human fœtus during the sixth month is thickly covered, offers a more curious case. It is first developed, during the fifth month, on the eyebrows and face, and especially round the mouth, where it is much longer than that on the head. A moustache of this kind was observed by Eschricht[32] on a female fœtus; but this is not so surprising a circumstance as it may at first appear, for the two sexes generally resemble each other in all external characters during an early period of growth. The direction and arrangement of the hairs on all parts of the fœtal body are the same as in the adult, but are subject to much variability. The whole surface, including even the forehead and ears, is thus thickly clothed; but it is a significant fact that the palms of the hands and the soles of the feet are quite naked, like the inferior surfaces of all four extremities in most of the lower animals. As this can hardly be an accidental coincidence, we must consider the woolly covering of the fœtus to be the rudimental representative of the first permanent coat of hair in those mammals which are born hairy. This representation is much more complete, in accordance with the usual law of embryological development, than that afforded by the straggling hairs on the body of the adult.

It appears as if the posterior molar or wisdom-teeth were tending to become rudimentary in the more civilised races of man. These teeth are rather smaller than the other molars, as is likewise the case with the corresponding teeth in the chimpanzee and orang; and they have only two separate fangs. They do not cut through the gums till about the seventeenth year, and I am assured by dentists that they are much more liable to decay, and are earlier lost, than the other teeth. It is also remarkable that they are much more liable to vary both in structure and in the period of their development than the other teeth.[33] In the Melanian races, on the other hand, the wisdom-teeth are usually furnished with three separate fangs, and are generally sound: they also differ from the other molars in size less than in the Caucasian races.[34] Prof. Schaaffhausen accounts for this difference between the races by "the posterior dental portion of the jaw being always shortened" in those that are civilised,[35] and this shortening may, I presume, be safely attributed to civilised men habitually feeding on soft, cooked food, and thus using their jaws less. I am informed by Mr. Brace that it is becoming quite a common practice in the United States to remove some of the molar teeth of children, as the jaw does not grow large enough for the perfect development of the normal number.

With respect to the alimentary canal I have met with an account of only a single rudiment, namely the vermiform appendage of the cæcum. The cæcum is a branch or diverticulum of the intestine, ending in a cul-de-sac, and it is extremely long in many of the lower vegetable-feeding mammals. In the marsupial koala it is actually more than thrice as long as the whole body.[36] It is sometimes produced into a long gradually-tapering point, and is sometimes constricted in parts. It appears as if, in consequence of changed diet or habits, the cæcum had become much shortened in various animals, the vermiform appendage being left as a rudiment of the shortened part. That this appendage is a rudiment, we may infer from its small size, and from the evidence which Prof. Canestrini [37] has collected of its variability in man. It is occasionally quite absent, or again is largely developed. The passage is sometimes completely closed for half or two-thirds of its length, with the terminal part consisting of a flattened solid expansion. In the orang this appendage is long and convoluted: in man it arises from the end of the short cæcum, and is commonly from four to five inches in length, being only about the third of an inch in diameter. Not only is it useless, but it is sometimes the cause of death, of which fact I have lately heard two instances: this is due to small hard bodies, such as seeds, entering the passage and causing inflammation.[38]

In the Quadrumana and some other orders of mammals, especially in the Carnivora, there is a passage near the lower end of the humerus, called the supra-condyloid foramen, through which the great nerve of the fore limb passes, and often the great artery. Now in the humerus of man, as Dr. Struthers[39] and others have shewn, there is generally a trace of this passage, and it is sometimes fairly well developed, being formed by a depending hook-like process of bone, completed by a band of ligament. When present the great nerve invariably passes through it, and this clearly indicates that it is the homologue and rudiment of the supra-condyloid foramen of the lower animals. Prof. Turner estimates, as he informs me, that it occurs in about one per cent. of recent skeletons; but during ancient times it appears to have been much more common. Mr. Busk[40] has collected the following evidence on this head: Prof. Broca "noticed the perforation in four and a half per cent of the arm-bones collected in the 'Cimetière du Sud' at Paris; and in the Grotto of Orrony, the contents of which are referred to the Bronze period, as many as eight humeri out of thirty-two were perforated; but this extraordinary proportion, he thinks, might be due to the cavern having been a sort of 'family vault.' Again, M. Dupont found 30 per cent of perforated bones in the caves of the Valley of the "Lesse, belonging to the Reindeer period; whilst M. Leguay, in a sort of *dolmen* at Argenteuil, observed twenty-five per cent to be perforated; and M. Pruner-Bey found twenty-six per cent in the same condition in bones from Vauréal. Nor should it be left unnoticed that M. Pruner-Bey states that this condition is common in Guanche skeletons." The fact that ancient races, in this and several other cases, more frequently present structures which resemble those of the lower animals than do the modern races, is interesting. One chief cause seems to be that ancient races stand somewhat nearer than modern races in the long line of descent to their remote animal-like progenitors.

The os coccyx in man, though functionless as a tail, plainly represents this part in other vertebrate animals. At an early embryonic period it is free, and, as we have seen, projects beyond the lower extremities. In certain rare and anomalous cases it has been known, according to Isidore Geoffroy St.-Hilaire and others,[41] to form a small external rudiment of a tail. The os coccyx is short, usually including only four vertebræ: and these are in a rudimental condition, for they consist, with the exception of the basal one, of the centrum alone.[42] They are furnished with some small muscles; one of which, as I am informed by Prof. Turner, has been expressly described by Theile as a rudimentary repetition of the extensor of the tail, which is so largely developed in many mammals.

The spinal cord in man extends only as far downwards as the last dorsal or first lumbar vertebra; but a thread-like structure (the *filum terminale)* runs down the axis of the sacral part of the spinal canal, and even along the back of the coccygeal bones. The upper part of this filament, as Prof. Turner informs me, is undoubtedly homologous with the spinal cord; but the lower part apparently consists merely of the *pia mater,* or vascular investing membrane. Even in this case the os coccyx may be said to possess a vestige of so important a structure as the spinal cord, though no longer enclosed within a bony canal. The following fact, for which I am also indebted to Prof. Turner, shews how closely the os coccyx corresponds with the true tail in the lower animals: Luschka has recently discovered at the extremity of the coccygeal bones a very peculiar convoluted body, which is continuous with the middle sacral artery; and this discovery led Krause and Meyer to examine the tail of a monkey (Macacus) and of a cat, in both of which they found, though not at the extremity, a similarly convoluted body.

The reproductive system offers various rudimentary structures; but these differ in one important respect from the foregoing cases. We are not here concerned with a vestige of a part which does not belong to the species in an efficient state; but with a part which is always present and efficient in the one sex, being represented in the other by a mere rudiment. Nevertheless, the occurrence of such rudiments is as difficult to explain on the belief of the separate creation of each species, as in the foregoing cases. Hereafter I shall have to recur to these rudiments, and shall shew that their presence generally depends merely on inheritance; namely, on parts acquired by one sex having been partially transmitted to the other. Here I will only give some instances of such rudiments. It is well known that in the males of all mammals, including man, rudimentary mammæ exist. These in several instances have become well developed, and have yielded a copious supply of milk. Their essential identity in the two sexes is likewise shewn by their occasional sympathetic enlargement in both during an attack of the measles. The *vesicula prostatica,* which has been observed in many male mammals, is now universally acknowledged to be the homologue of the female uterus, together with the connected passage. It is impossible to read Leuckart's able description of this organ, and his reasoning, without admitting the justness of his conclusion. This is especially clear in the case of those mammals in which the true female uterus bifurcates, for in the males of these the vesicula likewise bifurcates.[43] Some additional rudimentary structures belonging to the reproductive system might here have been adduced.[44]

THE BEARING OF THE THREE great classes of facts now given is unmistakeable. But it would be superfluous here fully to recapitulate the line of argument given in detail in my 'Origin of Species.' The homological construction of the whole frame in the members of the same class is intelligible, if we admit their descent from a common progenitor, together with their subsequent adaptation to diversified conditions. On any other view the similarity of pattern between the hand of a man or monkey, the foot of a horse, the flipper of a seal, the wing of a bat, &c., is utterly inexplicable. It is no scientific explanation to assert that they have all been formed on the same ideal plan. With respect to development, we can clearly understand, on the principle of variations supervening at a rather late embryonic period, and being inherited at a corresponding period, how it is that the embryos of wonderfully different forms should still retain, more or less perfectly, the structure of their common progenitor. No other explanation has ever been given of the marvellous fact that the embryo of a man, dog, seal, bat, reptile, &c., can at first hardly be distinguished from each other. In order to understand the existence of rudimentary organs, we have only to suppose that a former progenitor possessed the parts in question in a perfect state, and that under changed habits of life they became greatly reduced, either from simple disuse, or through the natural selection of those individuals which were least encumbered with a superfluous part, aided by the other means previously indicated.

Thus we can understand how it has come to pass that man and all other vertebrate animals have been constructed on the same general model, why they pass through the same early stages of development, and why they retain certain rudiments in common. Consequently we ought frankly to admit their community of descent: to take any other view, is to admit that our own structure and that of all the animals around us, is a mere snare laid to entrap our judgment. This conclusion is greatly strengthened, if we look to the members of the whole animal series, and consider the evidence derived from their affinities or classification, their geographical distribution and geological succession. It is only our natural prejudice, and that arrogance which made our forefathers declare that they were descended from demi-gods, which leads us to demur to this conclusion. But the time will before long come when it will be thought wonderful, that naturalists, who were well acquainted with the comparative structure and development of man and other mammals, should have believed that each was the work of a separate act of creation.

CHAPTER II

Comparison of the Mental Powers of Man and the Lower Animals

THE DIFFERENCE IN MENTAL POWER BETWEEN THE HIGHEST APE AND THE LOWEST SAVAGE, IMMENSE • CERTAIN INSTINCTS IN COMMON • THE EMOTIONS • CURIOSITY • IMITATION • ATTENTION • MEMORY • IMAGINATION • REASON • PROGRESSIVE IMPROVEMENT • TOOLS AND WEAPONS USED BY ANIMALS • LANGUAGE • SELF-CONSCIOUSNESS • SENSE OF BEAUTY • BELIEF IN GOD, SPIRITUAL AGENCIES, SUPERSTITIONS

We have seen in the last chapter that man bears in his bodily structure clear traces of his descent from some lower form; but it may be urged that, as man differs so greatly in his mental power from all other animals, there must be some error in this conclusion. No doubt the difference in this respect is enormous, even if we compare the mind of one of the lowest savages, who has no words to express any number higher than four, and who uses no abstract terms for the commonest objects or affections,[45] with that of the most highly organised ape. The difference would, no doubt, still remain immense, even if one of the higher apes had been improved or civilised as much as a dog has been in comparison with its parent-form, the wolf or jackal. The Fuegians rank amongst the lowest barbarians; but I was continually struck with surprise how closely the three natives on board H.M.S. "Beagle," who had lived some years in England and could talk a little English, resembled us in disposition and in most of our mental faculties. If no organic being excepting man had possessed any mental power, or if his powers had been of a wholly different nature from those of the lower animals, then we should never have been able to convince ourselves that our high faculties had been gradually developed. But it can be clearly shewn that there is no fundamental difference of this kind. We must also admit that there is a much wider interval in mental power between one of the lowest fishes, as a lamprey or lancelet, and one of the higher apes, than between an ape and man; yet this immense interval is filled up by numberless gradations.

Nor is the difference slight in moral disposition between a barbarian, such as the man described by the old navigator Byron, who dashed his child on the rocks for dropping a basket of sea-urchins, and a Howard or Clarkson; and in intellect, between a savage who does not use any abstract terms, and a Newton or Shakespeare. Differences of this kind between the highest men of the highest races and the lowest savages, are connected by the finest gradations. Therefore it is possible that they might pass and be developed into each other.

My object in this chapter is solely to shew that there is no fundamental difference between man and the higher mammals in their mental faculties. Each division of the subject might have been extended into a separate essay, but must here be treated briefly. As no classification of the mental powers has been universally accepted, I shall arrange my remarks in the order most convenient for my purpose; and will select those facts which have most struck me, with the hope that they may produce some effect on the reader.

With respect to animals very low in the scale, I shall have to give some additional facts under Sexual Selection, shewing that their mental powers are higher than might have been expected. The variability of the faculties in the individuals of the same species is an important point for us, and some few illustrations will here be given. But it would be superfluous to enter into many details on this head, for I have found on frequent enquiry, that it is the unanimous opinion of all those who have long attended to animals of many kinds, including birds, that the individuals differ greatly in every mental characteristic. In what manner the mental powers were first developed in the lowest organisms, is as hopeless an enquiry as how life first originated. These are problems for the distant future, if they are ever to be solved by man.

As man possesses the same senses with the lower animals, his fundamental intuitions must be the same. Man has also some few instincts in common, as that of self-preservation, sexual love, the love of the mother for her new-born offspring, the power possessed by the latter of sucking, and so forth. But man, perhaps, has somewhat fewer instincts than those possessed by the animals which come next to him in the series. The orang in the Eastern islands, and the chimpanzee in Africa, build platforms on which they sleep; and, as both species follow the same habit, it might be argued that this was due to instinct, but we cannot feel sure that it is not the result of both animals having similar wants and possessing similar powers of reasoning. These apes, as we may assume, avoid the many poisonous fruits of the tropics, and man has no such knowledge; but as our domestic animals, when taken to foreign lands and when first turned out in the spring, often eat poisonous herbs, which they afterwards avoid, we cannot feel sure that the apes do not learn from their own experience or from that of their parents what fruits to select. It is however certain, as we shall presently see, that apes have an instinctive dread of serpents, and probably of other dangerous animals.

The fewness and the comparative simplicity of the instincts in the higher animals are remarkable in contrast with those of the lower animals. Cuvier maintained that instinct and intelligence stand in an inverse ratio to each other; and some have thought that the intellectual faculties of the higher animals have been gradually developed from their instincts. But Pouchet, in an interesting essay,[46] has shewn that no such inverse ratio really exists. Those insects which possess the most wonderful instincts are certainly the most intelligent. In the vertebrate series, the least intelligent members, namely fishes and amphibians, do not possess complex instincts; and amongst mammals the animal most remarkable for its instincts, namely the beaver, is highly intelligent, as will be admitted by every one who has read Mr. Morgan's excellent account of this animal.[47]

Although the first dawnings of intelligence, according to Mr. Herbert Spencer,[48] have been developed through the multiplication and co-ordination of reflex actions, and although many of the simpler instincts graduate into actions of this kind and can hardly be distinguished from them, as in the case of young animals sucking, yet the more complex instincts seem to have originated independently of intelligence. I am, however, far from wishing to deny that instinctive actions may lose their fixed and untaught character, and be replaced by others performed by the aid of the free will. On the other hand, some intelligent actions—as when birds on oceanic islands first learn to avoid man—after being performed during many generations, become converted into instincts and are inherited. They may then be said to be degraded in character, for they are no longer performed through reason or from experience. But the greater number of the more complex instincts appear to have been gained in a wholly different manner, through the natural selection of variations of simpler instinctive actions. Such variations appear to arise from the same unknown causes acting on the cerebral organisation, which induce slight variations or individual differences in other parts of the body; and these variations, owing to our ignorance, are often said to arise spontaneously. We can, I think, come to no other conclusion with respect to the origin of the more complex instincts, when we reflect on the marvellous instincts of sterile worker-ants and bees, which leave no offspring to inherit the effects of experience and of modified habits.

Although a high degree of intelligence is certainly compatible with the existence of complex instincts, as we see in the insects just named and in the beaver, it is not improbable that they may to a certain extent interfere with each other's development. Little is known about the functions of the brain, but we can perceive that as the intellectual powers become highly developed, the various parts of the brain must be connected by the most intricate channels of intercommunication; and as a consequence each separate part would perhaps tend to become less well fitted to answer in a definite and uniform, that is instinctive, manner to particular sensations or associations.

I have thought this digression worth giving, because we may easily underrate the mental powers of the higher animals, and especially of man, when we compare their actions founded on the memory of past events, on foresight, reason, and imagination, with exactly similar actions instinctively performed by the lower animals; in this latter case the capacity of performing such actions having been gained, step by step, through the variability of the mental organs and natural selection, without any conscious intelligence on the part of the animal during each successive generation. No doubt, as Mr. Wallace has argued,[49] much of the intelligent work done by man is due to imitation and not to reason; but there is this great difference between his actions and many of those performed by the lower animals, namely, that man cannot, on his first trial, make, for instance, a stone hatchet or a canoe, through his power of imitation. He has to learn his work by practice; a beaver, on the other hand, can make its dam or canal, and a bird its nest, as well, or nearly as well, the first time it tries, as when old and experienced.

To return to our immediate subject: the lower animals, like man, manifestly feel pleasure and pain, happiness and misery. Happiness is never better exhibited than by young animals, such as puppies, kittens, lambs, &c., when playing together, like our own children. Even insects play together, as has been described by that excellent observer, P. Huber,[50] who saw ants chasing and pretending to bite each other, like so many puppies.

The fact that the lower animals are excited by the same emotions as ourselves is so well established, that it will not be necessary to weary the reader by many details. Terror acts in the same manner on them as on us, causing the muscles to tremble, the heart to palpitate, the sphincters to be relaxed, and the hair to stand on end. Suspicion, the offspring of fear, is eminently characteristic of most wild animals. Courage and timidity are extremely variable qualities in the individuals of the same species, as is plainly seen in our dogs. Some dogs and horses are ill-tempered and easily turn sulky; others are good-tempered; and these qualities are certainly inherited. Every one knows how liable animals are to furious rage, and how plainly they show it. Many anecdotes, probably true, have been published on the long-delayed and artful revenge of various animals. The accurate Rengger and Brehm[51] state that the American and African monkeys which they kept tame, certainly revenged themselves. The love of a dog for his master is notorious; in the agony of death he has been known to caress his master, and every one has heard of the dog suffering under vivisection, who licked the hand of the operator; this man, unless he had a heart of stone, must have felt remorse to the last hour of his life. As Whewell[52] has remarked, "who that reads the touching instances of maternal affection, related so often of the women of all nations, and of the females of all animals, can doubt that the principle of action is the same in the two cases?"

We see maternal affection exhibited in the most trifling details; thus Rengger observed an American monkey (a Cebus) carefully driving away the flies which plagued her infant; and Duvaucel saw a Hylobates washing the faces of her young ones in a stream. So intense is the grief of female monkeys for the loss of their young, that it invariably caused the death of certain kinds kept under confinement by Brehm in N. Africa. Orphan-monkeys were always adopted and carefully guarded by the other monkeys, both males and females. One female baboon had so capacious a heart that she not only adopted young monkeys of other species, but stole young dogs and cats, which she continually carried about. Her kindness, however, did not go so far as to share her food with her adopted offspring, at which Brehm was surprised, as his monkeys always divided everything quite fairly with their own young ones. An adopted kitten scratched the above-mentioned affectionate baboon, who certainly had a fine intellect, for she was much astonished at being scratched, and immediately examined the kitten's feet, and without more ado bit off the claws. In the Zoological Gardens, I heard from the keeper that an old baboon *(C. chacma)* had adopted a Rhesus monkey; but when a young drill and mandrill were placed in the cage, she seemed to perceive that these monkeys, though distinct species, were her nearer relatives, for she at once rejected the Rhesus and adopted both of them. The young Rhesus, as I saw, was greatly discontented at being thus rejected, and it would, like a naughty child, annoy and attack the young drill and mandrill whenever it could do so with safety; this conduct exciting great indignation in the old baboon. Monkeys will also, according to Brehm, defend their master when attacked by any one, as well as dogs to whom they are attached, from the attacks of other dogs. But we here trench on the subject of sympathy, to which I shall recur. Some of Brehm's monkeys took much delight in teasing, in various ingenious ways, a certain old dog whom they disliked, as well as other animals.

Most of the more complex emotions are common to the higher animals and ourselves. Every one has seen how jealous a dog is of his master's affection, if lavished on any other creature; and I have observed the same fact with monkeys. This shews that animals not only love, but have the desire to be loved. Animals manifestly feel emulation. They love approbation or praise; and a dog carrying a basket for his master exhibits in a high degree self-complacency or pride. There can, I think, be no doubt that a dog feels shame, as distinct from fear, and something very like modesty when begging too often for food. A great dog scorns the snarling of a little dog, and this may be called magnanimity. Several observers have stated that monkeys certainly dislike being laughed at; and they sometimes invent imaginary offences. In the Zoological Gardens I saw a baboon who always got into a furious rage when his keeper took out a letter or book and read it aloud to him; and his rage was so violent that, as I witnessed on one occasion, he bit his own leg till the blood flowed.

We will now turn to the more intellectual emotions and faculties, which are very important, as forming the basis for the development of the higher mental powers. Animals manifestly enjoy excitement and suffer from ennui, as may be seen with dogs, and, according to Rengger, with monkeys. All animals feel Wonder, and many exhibit Curiosity. They sometimes suffer from this latter quality, as when the hunter plays antics and thus attracts them; I have witnessed this with deer, and so it is with the wary chamois, and with some kinds of wild-ducks. Brehm gives a curious account of the instinctive dread which his monkeys exhibited towards snakes; but their curiosity was so great that they could not desist from occasionally satiating their horror in a most human fashion, by lifting up the lid of the box in which the snakes were kept. I was so much surprised at his account, that I took a stuffed and coiled-up snake into the monkey-house at the Zoological Gardens, and the excitement thus caused was one of the most curious spectacles which I ever beheld. Three species of Cercopithecus were the most alarmed; they dashed about their cages and uttered sharp signal-cries of danger, which were understood by the other monkeys. A few young monkeys and one old Anubis baboon alone took no notice of the snake. I then placed the stuffed specimen on the ground in one of the larger compartments. After a time all the monkeys collected round it in a large circle, and staring intently, presented a most ludicrous appearance. They became extremely nervous; so that when a wooden ball, with which they were familiar as a plaything, was accidently moved in the straw, under which it was partly hidden, they all instantly started away. These monkeys behaved very differently when a dead fish, a mouse, and some other new objects were placed in their cages; for though at first frightened, they soon approached, handled and examined them. I then placed a live snake in a paper bag, with the mouth loosely closed, in one of the larger compartments. One of the monkeys immediately approached, cautiously opened the bag a little, peeped in, and instantly dashed away. Then I witnessed what Brehm has described, for monkey after monkey, with head raised high and turned on one side, could not resist taking momentary peeps into the upright bag, at the dreadful object lying quiet at the bottom. It would almost appear as if monkeys had some notion of zoological affinities, for those kept by Brehm exhibited a strange, though mistaken, instinctive dread of innocent lizards and frogs. An orang, also, has been known to be much alarmed at the first sight of a turtle.[53]

The principle of *Imitation* is strong in man, and especially in man in a barbarous state. Desor[54] has remarked that no animal voluntarily imitates an action performed by man, until in the ascending scale we come to monkeys, which are well-known to be ridiculous mockers. Animals, however, sometimes imitate each others' actions: thus two species of wolves, which had been reared by dogs, learned to bark, as does sometimes the jackal,[55] but whether this can be called voluntary imitation is another question. From one account which I have read, there is reason to believe that puppies nursed by cats sometimes learn to lick their feet and thus to clean their faces: it is at least certain, as I hear from a perfectly trustworthy friend, that some dogs behave in this manner. Birds imitate the songs of their parents, and sometimes those of other birds; and parrots are notorious imitators of any sound which they often hear.

Hardly any faculty is more important for the intellectual progress of man than the power of *Attention*. Animals clearly manifest this power, as when a cat watches by a hole and prepares to spring on its prey. Wild animals sometimes become so absorbed when thus engaged, that they may be easily approached. Mr. Bartlett has given me a curious proof how variable this faculty is in monkeys. A man who trains monkeys to act used to purchase common kinds from the Zoological Society at the price of five pounds for each; but he offered to give double the price, if he might keep three or four of them for a few days, in order to select one. When asked how he could possibly so soon learn whether a particular monkey would turn out a good actor, he answered that it all depended on their power of attention. If when he was talking and explaining anything to a monkey, its attention was easily distracted, as by a fly on the wall or other trifling object, the case was hopeless. If he tried by punishment to make an inattentive monkey act, it turned sulky. On the other hand, a monkey which carefully attended to him could always be trained.

It is almost superfluous to state that animals have excellent *Memories* for persons and places. A baboon at the Cape of Good Hope, as I have been informed by Sir Andrew Smith, recognised him with joy after an absence of nine months. I had a dog who was savage and averse to all strangers, and I purposely tried his memory after an absence of five years and two days. I went near the stable where he lived, and shouted to him in my old manner; he showed no joy, but instantly followed me out walking and obeyed me, exactly as if I had parted with him only half-an-hour before. A train of old associations, dormant during five years, had thus been instantaneously awakened in his mind. Even ants, as P. Huber[56] has clearly shewn, recognised their fellow-ants belonging to the same community after a separation of four months. Animals can certainly by some means judge of the intervals of time between recurrent events.

The *Imagination* is one of the highest prerogatives of man. By this faculty he unites, independently of the will, former images and ideas, and thus creates brilliant and novel results. A poet, as Jean Paul Richter remarks,[57] "who must reflect whether he shall make a character say yes or no—to the devil with him; he is only a stupid corpse." Dreaming gives us the best notion of this power; as Jean Paul again says, "The dream is an involuntary art of poetry." The value of the products of our imagination depends of course on the number, accuracy, and clearness of our impressions; on our judgment and taste in selecting or rejecting the involuntary combinations, and to a certain extent on our power of voluntarily combining them. As dogs, cats, horses, and probably all the higher animals, even birds, as is stated on good authority,[58] have vivid dreams, and this is shewn by their movements and voice, we must admit that they possess some power of imagination.

Of all the faculties of the human mind, it will, I presume, be admitted that *Reason* stands at the summit. Few persons any longer dispute that animals possess some power of reasoning. Animals may constantly be seen to pause, deliberate, and resolve. It is a significant fact, that the more the habits of any particular animal are studied by a naturalist, the more he attributes to reason and the less to unlearnt instincts.[59] In future chapters we shall see that some animals extremely low in the scale apparently display a certain amount of reason. No doubt it is often difficult to distinguish between the power of reason and that of instinct. Thus Dr. Hayes, in his work on 'The Open Polar Sea,' repeatedly remarks that his dogs, instead of continuing to draw the sledges in a compact body, diverged and separated when they came to thin ice, so that their weight might be more evenly distributed. This was often the first warning and notice which the travellers received that the ice was becoming thin and dangerous. Now, did the dogs act thus from the experience of each individual, or from the example of the older and wiser dogs, or from an inherited habit, that is from an instinct? This instinct might possibly have arisen since the time, long ago, when dogs were first employed by the natives in drawing their sledges; or the Arctic wolves, the parent-stock of the Esquimaux dog, may have acquired this instinct, impelling them not to attack their prey in a close pack when on thin ice. Questions of this kind are most difficult to answer.

So many facts have been recorded in various works shewing that animals possess some degree of reason, that I will here give only two or three instances, authenticated by Rengger, and relating to American monkeys, which stand low in their order. He states that when he first gave eggs to his monkeys, they smashed them and thus lost much of their contents; afterwards they gently hit one end against some hard body, and picked off the bits of shell with their fingers. After cutting themselves only once with any sharp tool, they would not touch it again, or would handle it with the greatest care. Lumps of sugar were often given them wrapped up in paper; and Rengger sometimes put a live wasp in the paper, so that in hastily unfolding it they got stung; after this had once happened, they always first held the packet to their ears to detect any movement within. Any one who is not convinced by such facts as these, and by what he may observe with his own dogs, that animals can reason, would not be convinced by anything that I could add. Nevertheless I will give one case with respect to dogs, as it rests on two distinct observers, and can hardly depend on the modification of any instinct.

Mr. Colquhoun[60] winged two wild-ducks, which fell on the opposite side of a stream; his retriever tried to bring over both at once, but could not succeed; she then, though never before known to ruffle a feather, deliberately killed one, brought over the other, and returned for the dead bird. Col. Hutchinson relates that two partridges were shot at once, one being killed, the other wounded; the latter ran away, and was caught by the retriever, who on her return came across the dead bird; "she stopped, evidently greatly puzzled, and after one or two trials, finding she could not take it up without permitting the escape of the winged bird, she considered a moment, then deliberately murdered it by giving it a severe crunch, and afterwards brought away both together. This was the only known instance of her ever having wilfully injured any game." Here we have reason, though not quite perfect, for the retriever might have brought the wounded bird first and then returned for the dead one, as in the case of the two wild-ducks.

The muleteers in S. America say, "I will not give you the mule whose step is easiest, but *la mas racional*,—the one that reasons best;" and Humboldt[61] adds, "this popular expression, dictated by long experience, combats the system of animated machines, better perhaps than all the arguments of speculative philosophy."

IT HAS, I THINK, now been shewn that man and the higher animals, especially the Primates, have some few instincts in common. All have the same senses, intuitions and sensations—similar passions, affections, and emotions, even the more complex ones; they feel wonder and curiosity; they possess the same faculties of imitation, attention, memory, imagination, and reason, though in very different degrees. Nevertheless many authors have insisted that man is separated through his mental faculties by an impassable barrier from all the lower animals. I formerly made a collection of above a score of such aphorisms, but they are not worth giving, as their wide difference and number prove the difficulty, if not the impossibility, of the attempt. It has been asserted that man alone is capable of progressive improvement; that he alone makes use of tools or fire, domesticates other animals, possesses property, or employs language; that no other animal is self-conscious, comprehends itself, has the power of abstraction, or possesses general ideas; that man alone has a sense of beauty, is liable to caprice, has the feeling of gratitude, mystery, &c.; believes in God, or is endowed with a conscience. I will hazard a few remarks on the more important and interesting of these points.

Archbishop Sumner formerly maintained[62] that man alone is capable of progressive improvement. With animals, looking first to the individual, every one who has had any experience in setting traps knows that young animals can be caught much more easily than old ones; and they can be much more easily approached by an enemy. Even with respect to old animals, it is impossible to catch many in the same place and in the same kind of trap, or to destroy them by the same kind of poison; yet it is improbable that all should have partaken of the poison, and impossible that all should have been caught in the trap. They must learn caution by seeing their brethren caught or poisoned. In North America, where the fur-bearing animals have long been pursued, they exhibit, according to the unanimous testimony of all observers, an almost incredible amount of sagacity, caution, and cunning; but trapping has been there so long carried on that inheritance may have come into play.

If we look to successive generations, or to the race, there is no doubt that birds and other animals gradually both acquire and lose caution in relation to man or other enemies;[63] and this caution is certainly in chief part an inherited habit or instinct, but in part the result of individual experience. A good observer, Leroy,[64] states that in districts where foxes are much hunted, the young when they first leave their burrows are incontestably much more wary than the old ones in districts where they are not much disturbed.

Our domestic dogs are descended from wolves and jackals,[65] and though they may not have gained in cunning, and may have lost in waryness and suspicion, yet they have progressed in certain moral qualities, such as in affection, trust-worthiness, temper, and probably in general intelligence. The common rat has conquered and beaten several other species throughout Europe, in parts of North America, New Zealand, and recently in Formosa, as well as on the mainland of China. Mr. Swinhoe,[66] who describes these latter cases, attributes the victory of the common rat over the large *Mus coninga* to its superior cunning; and this latter quality may be attributed to the habitual exercise of all its faculties in avoiding extirpation by man, as well as to nearly all the less cunning or weak-minded rats having been successively destroyed by him. To maintain, independently of any direct evidence, that no animal during the course of ages has progressed in intellect or other mental faculties, is to beg the question of the evolution of species. Hereafter we shall see that, according to Lartet, existing mammals belonging to several orders have larger brains than their ancient tertiary prototypes.

It has often been said that no animal uses any tool; but the chimpanzee in a state of nature cracks a native fruit, somewhat like a walnut, with a stone.[67] Rengger[68] easily taught an American monkey thus to break open hard palm-nuts, and afterwards of its own accord it used stones to open other kinds of nuts, as well as boxes. It thus also removed the soft rind of fruit that had a disagreeable flavour. Another monkey was taught to open the lid of a large box with a stick, and afterwards it used the stick as a lever to move heavy bodies; and I have myself seen a young orang put a stick into a crevice, slip his hand to the other end, and use it in the proper manner as a lever. In the cases just mentioned stones and sticks were employed as implements; but they are likewise used as weapons. Brehm [69] states, on the authority of the well-known traveller Schimper, that in Abyssinia when the baboons belonging to one species *(C. gelada)* descend in troops from the mountains to plunder the fields, they sometimes encounter troops of another species *(C. hamadryas),* and then a fight ensues. The Geladas roll down great stones, which the Hamadryas try to avoid, and then both species, making a great uproar, rush furiously against each other. Brehm, when accompanying the Duke of Coburg-Gotha, aided in an attack with fire-arms on a troop of baboons in the pass of Mensa in Abyssinia. The baboons in return rolled so many stones down the mountain, some as large as a man's head, that the attackers had to beat a hasty retreat; and the pass was actually for a time closed against the caravan. It deserves notice that these baboons thus acted in concert. Mr. Wallace[70] on three occasions saw female orangs, accompanied by their young, "breaking off branches and the great spiny fruit of the Durian tree, with every appearance of rage; causing such a shower of missiles as effectually kept us from approaching too near the tree."

In the Zoological Gardens a monkey which had weak teeth used to break open nuts with a stone; and I was assured by the keepers that this animal, after using the stone, hid it in the straw, and would not let any other monkey touch it. Here, then, we have the idea of property; but this idea is common to every dog with a bone, and to most or all birds with their nests.

The Duke of Argyll[71] remarks, that the fashioning of an implement for a special purpose is absolutely peculiar to man; and he considers that this forms an immeasurable gulf between him and the brutes. It is no doubt a very important distinction, but there appears to me much truth in Sir J. Lubbock's suggestion,[72] that when primeval man first used flint-stones for any purpose, he would have accidentally splintered them, and would then have used the sharp fragments. From this step it would be a small one to intentionally break the flints, and not a very wide step to rudely fashion them. This latter advance, however, may have taken long ages, if we may judge by the immense interval of time which elapsed before the men of the neolithic period took to grinding and polishing their stone tools. In breaking the flints, as Sir J. Lubbock likewise remarks, sparks would have been emitted, and in grinding them heat would have been evolved: "thus the two usual methods of obtaining fire may have originated." The nature of fire would have been known in the many volcanic regions where lava occasionally flows through forests. The anthropomorphous apes, guided probably by instinct, build for themselves temporary platforms; but as many instincts are largely controlled by reason, the simpler ones, such as this of building a platform, might readily pass into a voluntary and conscious act. The orang is known to cover itself at night with the leaves of the Pandanus; and Brehm states that one of his baboons used to protect itself from the heat of the sun by throwing a straw-mat over its head. In these latter habits, we probably see the first steps towards some of the simpler arts; namely rude architecture and dress, as they arose amongst the early progenitors of man.

Language.—This faculty has justly been considered as one of the chief distinctions between man and the lower animals. But man, as a highly competent judge, Archbishop Whately remarks, "is not the only animal that can make use of language to express what is passing in his mind, and can understand, more or less, what is so expressed by another."[73] In Paraguay the *Cebus azaræ* when excited utters at least six distinct sounds, which excite in other monkeys similar emotions.[74] The movements of the features and gestures of monkeys are understood by us, and they partly understand ours, as Rengger and others declare. It is a more remarkable fact that the dog, since being domesticated, has learnt to bark[75] in at least four or five distinct tones. Although barking is a new art, no doubt the wild species, the parents of the dog, expressed their feelings by cries of various kinds. With the domesticated dog we have the bark of eagerness, as in the chase; that of anger; the yelping or howling bark of despair, as when shut up; that of joy, as when starting on a walk with his master; and the very distinct one of demand or supplication, as when wishing for a door or window to be opened.

Articulate language is, however, peculiar to man; but he uses in common with the lower animals inarticulate cries to express his meaning, aided by gestures and the movements of the muscles of the face.[76] This especially holds good with the more simple and vivid feelings, which are but little connected with our higher intelligence. Our cries of pain, fear, surprise, anger, together with their appropriate actions, and the murmur of a mother to her beloved child, are more expressive than any words. It is not the mere power of articulation that distinguishes man from other animals, for as every one knows, parrots can talk; but it is his large power of connecting definite sounds with definite ideas; and this obviously depends on the development of the mental faculties.

As Horne Tooke, one of the founders of the noble science of philology, observes, language is an art, like brewing or baking; but writing would have been a much more appropriate simile. It certainly is not a true instinct, as every language has to be learnt. It differs, however, widely from all ordinary arts, for man has an instinctive tendency to speak, as we see in the babble of our young children; whilst no child has an instinctive tendency to brew, bake, or write. Moreover, no philologist now supposes that any language has been deliberately invented; each has been slowly and unconsciously developed by many steps. The sounds uttered by birds offer in several respects the nearest analogy to language, for all the members of the same species utter the same instinctive cries expressive of their emotions; and all the kinds that have the power of singing exert this power instinctively; but the actual song, and even the call-notes, are learnt from their parents or foster-parents. These sounds, as Daines Barrington[77] has proved, "are no more innate than language is in man." The first attempts to sing "may be compared to the imperfect endeavour in a child to babble." The young males continue practising, or, as the bird-catchers say, recording, for ten or eleven months. Their first essays show hardly a rudiment of the future song; but as they grow older we can perceive what they are aiming at; and at last they are said "to sing their song round." Nestlings which have learnt the song of a distinct species, as with the canary-birds educated in the Tyrol, teach and transmit their new song to their offspring. The slight natural differences of song in the same species inhabiting different districts may be appositely compared, as Barrington remarks, "to provincial dialects;" and the songs of allied, though distinct species may be compared with the languages of distinct races of man. I have given the foregoing details to shew that an instinctive tendency to acquire an art is not a peculiarity confined to man.

With respect to the origin of articulate language, after having read on the one side the highly interesting works of Mr. Hensleigh Wedgwood, the Rev. F. Farrar, and Prof. Schleicher,[78] and the celebrated lectures of Prof. Max Müller on the other side, I cannot doubt that language owes its origin to the imitation and modification, aided by signs and gestures, of various natural sounds, the voices of other animals, and man's own instinctive cries. When we treat of sexual selection we shall see that primeval man, or rather some early progenitor of man, probably used his voice largely, as does one of the gibbon-apes at the present day, in producing true musical cadences, that is in singing; we may conclude from a widely-spread analogy that this power would have been especially exerted during the courtship of the sexes, serving to express various emotions, as love, jealousy, triumph, and serving as a challenge to their rivals. The imitation by articulate sounds of musical cries might have given rise to words expressive of various complex emotions. As bearing on the subject of imitation, the strong tendency in our nearest allies, the monkeys, in microcephalous idiots,[79] and in the barbarous races of mankind, to imitate whatever they hear deserves notice. As monkeys certainly understand much that is said to them by man, and as in a state of nature they utter signal-cries of danger to their fellows,[80] it does not appear altogether incredible, that some unusually wise ape-like animal should have thought of imitating the growl of a beast of prey, so as to indicate to his fellow monkeys the nature of the expected danger. And this would have been a first step in the formation of a language.

As the voice was used more and more, the vocal organs would have been strengthened and perfected through the principle of the inherited effects of use; and this would have reacted on the power of speech. But the relation between the continued use of language and the development of the brain has no doubt been far more important. The mental powers in some early progenitor of man must have been more highly developed than in any existing ape, before even the most imperfect form of speech could have come into use; but we may confidently believe that the continued use and advancement of this power would have reacted on the mind by enabling and encouraging it to carry on long trains of thought. A long and complex train of thought can no more be carried on without the aid of words, whether spoken or silent, than a long calculation without the use of figures or algebra. It appears, also, that even ordinary trains of thought almost require some form of language, for the dumb, deaf, and blind girl, Laura Bridgman, was observed to use her fingers whilst dreaming.[81] Nevertheless a long succession of vivid and connected ideas, may pass through the mind without the aid of any form of language, as we may infer from the prolonged dreams of dogs. We have, also, seen that retriever-dogs are able to reason to a certain extent; and this they manifestly do without the aid of language. The intimate connection between the brain, as it is now developed in us, and the faculty of speech, is well shewn by those curious cases of brain-disease, in which speech is specially affected, as when the power to remember substantives is lost, whilst other words can be correctly used.[82] There is no more improbability in the effects of the continued use of the vocal and mental organs being inherited, than in the case of hand-writing, which depends partly on the structure of the hand and partly on the disposition of the mind; and hand-writing is certainly inherited.[83]

Why the organs now used for speech should have been originally perfected for this purpose, rather than any other organs, it is not difficult to see. Ants have considerable powers of intercommunication by means of their antennæ, as shewn by Huber, who devotes a whole chapter to their language. We might have used our fingers as efficient instruments, for a person with practice can report to a deaf man every word of a speech rapidly delivered at a public meeting; but the loss of our hands, whilst thus employed, would have been a serious inconvenience. As all the higher mammals possess vocal organs constructed on the same general plan with ours, and which are used as a means of communication, it was obviously probable, if the power of communication had to be improved, that these same organs would have been still further developed; and this has been effected by the aid of adjoining and well-adapted parts, namely the tongue and lips.[84] The fact of the higher apes not using their vocal organs for speech, no doubt depends on their intelligence not having been sufficiently advanced. The possession by them of organs, which with long-continued practice might have been used for speech, although not thus used, is paralleled by the case of many birds which possess organs fitted for singing, though they never sing. Thus, the nightingale and crow have vocal organs similarly constructed, these being used by the former for diversified song, and by the latter merely for croaking.[85]

The formation of different languages and of distinct species, and the proofs that both have been developed through a gradual process, are curiously the same.[86] But we can trace the origin of many words further back than in the case of species, for we can perceive that they have arisen from the imitation of various sounds, as in alliterative poetry. We find in distinct languages striking homologies due to community of descent, and analogies due to a similar process of formation. The manner in which certain letters or sounds change when others change is very like correlated growth. We have in both cases the reduplication of parts, the effects of long-continued use, and so forth. The frequent presence of rudiments, both in languages and in species, is still more remarkable. The letter *m* in the word *am*, means *I;* so that in the expression *I am,* a superfluous and useless rudiment has been retained. In the spelling also of words, letters often remain as the rudiments of ancient forms of pronunciation. Languages, like organic beings, can be classed in groups under groups; and they can be classed either naturally according to descent, or artificially by other characters. Dominant languages and dialects spread widely and lead to the gradual extinction of other tongues. A language, like a species, when once extinct, never, as Sir C. Lyell remarks, reappears. The same language never has two birth-places. Distinct languages may be crossed or blended together.[87] We see variability in every tongue, and new words are continually cropping up; but as there is a limit to the powers of the memory, single words, like whole languages, gradually become extinct. As Max Müller[88] has well remarked:—"A struggle for life is constantly going on amongst the words and grammatical forms in each language. The better, the shorter, the easier forms are constantly gaining the upper hand, and they owe their success to their own inherent virtue." To these more important causes of the survival of certain words, mere novelty may, I think, be added; for there is in the mind of man a strong love for slight changes in all things. The survival or preservation of certain favoured words in the struggle for existence is natural selection.

The perfectly regular and wonderfully complex construction of the languages of many barbarous nations has often been advanced as a proof, either of the divine origin of these languages, or of the high art and former civilisation of their founders. Thus F. von Schlegel writes: "In those languages which appear to be at the lowest grade of intellectual culture, we frequently observe a very high and elaborate degree of art in their grammatical structure. This is especially the case with the Basque and the Lapponian, and many of the American languages."[89] But it is assuredly an error to speak of any language as an art in the sense of its having been elaborately and methodically formed. Philologists now admit that conjugations, declensions, &c., originally existed as distinct words, since joined together; and as such words express the most obvious relations between objects and persons, it is not surprising that they should have been used by the men of most races during the earliest ages. With respect to perfection, the following illustration will best shew how easily we may err: a Crinoid sometimes consists of no less than 150,000 pieces of shell,[90] all arranged with perfect symmetry in radiating lines; but a naturalist does not consider an animal of this kind as more perfect than a bilateral one with comparatively few parts, and with none of these alike, excepting on the opposite sides of the body. He justly considers the differentiation and specialisation of organs as the test of perfection. So with languages, the most symmetrical and complex ought not to be ranked above irregular, abbreviated, and bastardised languages, which have borrowed expressive words and useful forms of construction from various conquering, or conquered, or immigrant races.

From these few and imperfect remarks I conclude that the extremely complex and regular construction of many barbarous languages, is no proof that they owe their origin to a special act of creation.[91] Nor, as we have seen, does the faculty of articulate speech in itself offer any insuperable objection to the belief that man has been developed from some lower form.

Self-consciousness, Individuality, Abstraction, General Ideas, &c.—It would be useless to attempt discussing these high faculties, which, according to several recent writers, make the sole and complete distinction between man and the brutes, for hardly two authors agree in their definitions. Such faculties could not have been fully developed in man until his mental powers had advanced to a high standard, and this implies the use of a perfect language. No one supposes that one of the lower animals reflects whence he comes or whither he goes,—what is death or what is life, and so forth. But can we feel sure that an old dog with an excellent memory and some power of imagination, as shewn by his dreams, never reflects on his past pleasures in the chase? and this would be a form of self-consciousness. On the other hand, as Büchner[92] has remarked, how little can the hard-worked wife of a degraded Australian savage, who uses hardly any abstract words and cannot count above four, exert her self-consciousness, or reflect on the nature of her own existence.

That animals retain their mental individuality is unquestionable. When my voice awakened a train of old associations in the mind of the above-mentioned dog, he must have retained his mental individuality, although every atom of his brain had probably undergone change more than once during the interval of five years. This dog might have brought forward the argument lately advanced to crush all evolutionists, and said, "I abide amid all mental moods and all material changes. . . . The teaching that atoms leave their impressions as legacies to other atoms falling into the places they have vacated is contradictory of the utterance of consciousness, and is therefore false; but it is the teaching necessitated by evolutionism, consequently the hypothesis is a false one."[93]

Sense of Beauty.—This sense has been declared to be peculiar to man. But when we behold male birds elaborately displaying their plumes and splendid colours before the females, whilst other birds not thus decorated make no such display, it is impossible to doubt that the females admire the beauty of their male partners. As women everywhere deck themselves with these plumes, the beauty of such ornaments cannot be disputed. The Bower-birds by tastefully ornamenting their playing-passages with gaily-coloured objects, as do certain humming-birds their nests, offer additional evidence that they possess a sense of beauty. So with the song of birds, the sweet strains poured forth by the males during the season of love are certainly admired by the females, of which fact evidence will hereafter be given. If female birds had been incapable of appreciating the beautiful colours, the ornaments, and voices of their male partners, all the labour and anxiety exhibited by them in displaying their charms before the females would have been thrown away; and this it is impossible to admit. Why certain bright colours and certain sounds should excite pleasure, when in harmony, cannot, I presume, be explained any more than why certain flavours and scents are agreeable; but assuredly the same colours and the same sounds are admired by us and by many of the lower animals.

The taste for the beautiful, at least as far as female beauty is concerned, is not of a special nature in the human mind; for it differs widely in the different races of man, as will hereafter be shewn, and is not quite the same even in the different nations of the same race. Judging from the hideous ornaments and the equally hideous music admired by most savages, it might be urged that their æsthetic faculty was not so highly developed as in certain animals, for instance, in birds. Obviously no animal would be capable of admiring such scenes as the heavens at night, a beautiful landscape, or refined music; but such high tastes, depending as they do on culture and complex associations, are not enjoyed by barbarians or by uneducated persons.

Many of the faculties, which have been of inestimable service to man for his progressive advancement, such as the powers of the imagination, wonder, curiosity, an undefined sense of beauty, a tendency to imitation, and the love of excitement or novelty, could not fail to have led to the most capricious changes of customs and fashions. I have alluded to this point, because a recent writer[94] has oddly fixed on Caprice "as one of the most remarkable and typical differences between savages and brutes." But not only can we perceive how it is that man is capricious, but the lower animals are, as we shall hereafter see, capricious in their affections, aversions, and sense of beauty. There is also good reason to suspect that they love novelty, for its own sake.

Belief in God—Religion.—There is no evidence that man was aboriginally endowed with the ennobling belief in the existence of an Omnipotent God. On the contrary there is ample evidence, derived not from hasty travellers, but from men who have long resided with savages, that numerous races have existed and still exist, who have no idea of one or more gods, and who have no words in their languages to express such an idea.[95] The question is of course wholly distinct from that higher one, whether there exists a Creator and Ruler of the universe; and this has been answered in the affirmative by the highest intellects that have ever lived.

If, however, we include under the term "religion" the belief in unseen or spiritual agencies, the case is wholly different; for this belief seems to be almost universal with the less civilised races. Nor is it difficult to comprehend how it arose. As soon as the important faculties of the imagination, wonder, and curiosity, together with some power of reasoning, had become partially developed, man would naturally have craved to understand what was passing around him, and have vaguely speculated on his own existence. As Mr. M'Lennan[96] has remarked, "Some explanation of the phenomena of life, a man must feign for himself; and to judge from the universality of it, the simplest hypothesis, and the first to occur to men, seems to have been that natural phenomena are ascribable to the presence in animals, plants, and things, and in the forces of nature, of such spirits prompting to action as men are conscious they themselves possess." It is probable, as Mr. Tylor has clearly shewn, that dreams may have first given rise to the notion of spirits; for savages do not readily distinguish between subjective and objective impressions. When a savage dreams, the figures which appear before him are believed to have come from a distance and to stand over him; or "the soul of the dreamer goes out on its travels, and comes home with a remembrance of what it has seen."[97] But until the above-named faculties of imagination, curiosity, reason, &c., had been fairly well developed in the mind of man, his dreams would not have led him to believe in spirits, any more than in the case of a dog.

The tendency in savages to imagine that natural objects and agencies are animated by spiritual or living essences, is perhaps illustrated by a little fact which I once noticed: my dog, a full-grown and very sensible animal, was lying on the lawn during a hot and still day; but at a little distance a slight breeze occasionally moved an open parasol, which would have been wholly disregarded by the dog, had any one stood near it. As it was, every time that the parasol slightly moved, the dog growled fiercely and barked. He must, I think, have reasoned to himself in a rapid and unconscious manner, that movement without any apparent cause indicated the presence of some strange living agent, and no stranger had a right to be on his territory.

The belief in spiritual agencies would easily pass into the belief in the existence of one or more gods. For savages would naturally attribute to spirits the same passions, the same love of vengeance or simplest form of justice, and the same affections which they themselves experienced. The Fuegians appear to be in this respect in an intermediate condition, for when the surgeon on board the "Beagle" shot some young ducklings as specimens, York Minster declared in the most solemn manner, "Oh! Mr. Bynoe, much rain, much snow, blow much;" and this was evidently a retributive punishment for wasting human food. So again he related how, when his brother killed a "wild man," storms long raged, much rain and snow fell.

Yet we could never discover that the Fuegians believed in what we should call a God, or practised any religious rites; and Jemmy Button, with justifiable pride, stoutly maintained that there was no devil in his land. This latter assertion is the more remarkable, as with savages the belief in bad spirits is far more common than the belief in good spirits.

The feeling of religious devotion is a highly complex one, consisting of love, complete submission to an exalted and mysterious superior, a strong sense of dependence,[98] fear, reverence, gratitude, hope for the future, and perhaps other elements. No being could experience so complex an emotion until advanced in his intellectual and moral faculties to at least a moderately high level. Nevertheless we see some distant approach to this state of mind, in the deep love of a dog for his master, associated with complete submission, some fear, and perhaps other feelings. The behaviour of a dog when returning to his master after an absence, and, as I may add, of a monkey to his beloved keeper, is widely different from that towards their fellows. In the latter case the transports of joy appear to be somewhat less, and the sense of equality is shewn in every action. Professor Braubach[99] goes so far as to maintain that a dog looks on his master as on a god.

The same high mental faculties which first led man to believe in unseen spiritual agencies, then in fetishism, polytheism, and ultimately in monotheism, would infallibly lead him, as long as his reasoning powers remained poorly developed, to various strange superstitions and customs. Many of these are terrible to think of—such as the sacrifice of human beings to a blood-loving god; the trial of innocent persons by the ordeal of poison or fire; witchcraft, &c.—yet it is well occasionally to reflect on these superstitions, for they shew us what an infinite debt of gratitude we owe to the improvement of our reason, to science, and our accumulated knowledge.[100] As Sir J. Lubbock has well observed, "it is not too much to say that the horrible dread of unknown evil hangs like a thick cloud over savage life, and embitters every pleasure." These miserable and indirect consequences of our highest faculties may be compared with the incidental and occasional mistakes of the instincts of the lower animals.

CHAPTER III

Comparison of the Mental Powers of Man and the Lower Animals—continued

THE MORAL SENSE • FUNDAMENTAL PROPOSITION • THE QUALITIES OF SOCIAL ANIMALS • ORIGIN OF SOCIABILITY • STRUGGLE BETWEEN OPPOSED INSTINCTS • MAN A SOCIAL ANIMAL • THE MORE ENDURING SOCIAL INSTINCTS CONQUER OTHER LESS PERSISTENT INSTINCTS • THE SOCIAL VIRTUES ALONE REGARDED BY SAVAGES • THE SELF-REGARDING VIRTUES ACQUIRED AT A LATER STAGE OF DEVELOPMENT • THE IMPORTANCE OF THE JUDGMENT OF THE MEMBERS OF THE SAME COMMUNITY ON CONDUCT • TRANSMISSION OF MORAL TENDENCIES • SUMMARY

I FULLY SUBSCRIBE to the judgment of those writers[101] who maintain that of all the differences between man and the lower animals, the moral sense or conscience is by far the most important. This sense, as Mackintosh[102] remarks, "has a rightful supremacy over every other principle of human action;" it is summed up in that short but imperious word *ought,* so full of high significance. It is the most noble of all the attributes of man, leading him without a moment's hesitation to risk his life for that of a fellow-creature; or after due deliberation, impelled simply by the deep feeling of right or duty, to sacrifice it in some great cause. Immanuel Kant exclaims, "Duty! Wondrous thought, that workest neither by fond insinuation, flattery, nor by any threat, but merely by holding up thy naked law in the soul, and so extorting for thyself always reverence, if not always obedience; before whom all appetites are dumb, however secretly they rebel; whence thy original?"[103]

This great question has been discussed by many writers[104] of consummate ability; and my sole excuse for touching on it is the impossibility of here passing it over, and because, as far as I know, no one has approached it exclusively from the side of natural history. The investigation possesses, also, some independent interest, as an attempt to see how far the study of the lower animals can throw light on one of the highest psychical faculties of man.

The following proposition seems to me in a high degree probable—namely, that any animal whatever, endowed with well-marked social instincts,[105] would inevitably acquire a moral sense or conscience, as soon as its intellectual powers had become as well developed, or nearly as well developed, as in man. For, *firstly*, the social instincts lead an animal to take pleasure in the society of its fellows, to feel a certain amount of sympathy with them, and to perform various services for them. The services may be of a definite and evidently instinctive nature; or there may be only a wish and readiness, as with most of the higher social animals, to aid their fellows in certain general ways. But these feelings and services are by no means extended to all the individuals of the same species, only to those of the same association. *Secondly*, as soon as the mental faculties had become highly developed, images of all past actions and motives would be incessantly passing through the brain of each individual; and that feeling of dissatisfaction which invariably results, as we shall hereafter see, from any unsatisfied instinct, would arise, as often as it was perceived that the enduring and always present social instinct had yielded to some other instinct, at the time stronger, but neither enduring in its nature, nor leaving behind it a very vivid impression. It is clear that many instinctive desires, such as that of hunger, are in their nature of short duration; and after being satisfied are not readily or vividly recalled. *Thirdly*, after the power of language had been acquired and the wishes of the members of the same community could be distinctly expressed, the common opinion how each member ought to act for the public good, would naturally become to a large extent the guide to action. But the social instincts would still give the impulse to act for the good of the community, this impulse being strengthened, directed, and sometimes even deflected by public opinion, the power of which rests, as we shall presently see, on instinctive sympathy. *Lastly*, habit in the individual would ultimately play a very important part in guiding the conduct of each member; for the social instincts and impulses, like all other instincts, would be greatly strengthened by habit, as would obedience to the wishes and judgment of the community. These several subordinate propositions must now be discussed; and some of them at considerable length.

It may be well first to premise that I do not wish to maintain that any strictly social animal, if its intellectual faculties were to become as active and as highly developed as in man, would acquire exactly the same moral sense as ours. In the same manner as various animals have some sense of beauty, though they admire widely different objects, so they might have a sense of right and wrong, though led by it to follow widely different lines of conduct. If, for instance, to take an extreme case, men were reared under precisely the same conditions as hive-bees, there can hardly be a doubt that our unmarried females would, like the worker-bees, think it a sacred duty to kill their brothers, and mothers would strive to kill their fertile daughters; and no one would think of interfering. Nevertheless the bee, or any other social animal, would in our supposed case gain, as it appears to me, some feeling of right and wrong, or a conscience. For each individual would have an inward sense of possessing certain. stronger or more enduring instincts, and others less strong or enduring; so that there would often be a struggle which impulse should be followed; and satisfaction or dissatisfaction would be felt, as past impressions were compared during their incessant passage through the mind. In this case an inward monitor would tell the animal that it would have been better to have followed the one impulse rather than the other. The one course ought to have been followed: the one would have been right and the other wrong; but to these terms I shall have to recur.

Sociability.—Animals of many kinds are social; we find even distinct species living together, as with some American monkeys, and with the united flocks of rooks, jackdaws, and starlings. Man shows the same feeling in his strong love for the dog, which the dog returns with interest. Every one must have noticed how miserable horses, dogs, sheep, &c. are when separated from their companions; and what affection at least the two former kinds show on their reunion. It is curious to speculate on the feelings of a dog, who will rest peacefully for hours in a room with his master or any of the family, without the least notice being taken of him; but if left for a short time by himself, barks or howls dismally. We will confine our attention to the higher social animals, excluding insects, although these aid each other in many important ways. The most common service which the higher animals perform for each other, is the warning each other of danger by means of the united senses of all. Every sportsman knows, as Dr. Jaeger remarks,[106] how difficult it is to approach animals in a herd or troop. Wild horses and cattle do not, I believe, make any danger-signal; but the attitude of any one who first discovers an enemy, warns the others. Rabbits stamp loudly on the ground with their hind-feet as a signal: sheep and chamois do the same, but with their fore-feet, uttering likewise a whistle. Many birds and some mammals post sentinels, which in the case of seals are said[107] generally to be the females. The leader of a troop of monkeys acts as the sentinel, and utters cries expressive both of danger and of safety.[108] Social animals perform many little services for each other: horses nibble, and cows lick each other, on any spot which itches: monkeys search for each other's external parasites; and Brehm states that after a troop of the *Cercopithecus griseo-viridis* has rushed through a thorny brake, each monkey stretches itself on a branch, and another monkey sitting by "conscientiously" examines its fur and extracts every thorn or burr.

Animals also render more important services to each other: thus wolves and some other beasts of prey hunt in packs, and aid each other in attacking their victims. Pelicans fish in concert. The Hamadryas baboons turn over stones to find insects, &c.; and when they come to a large one, as many as can stand round, turn it over together and share the booty. Social animals mutually defend each other. The males of some ruminants come to the front when there is danger and defend the herd with their horns. I shall also in a future chapter give cases of two young wild bulls attacking an old one in concert, and of two stallions together trying to drive away a third stallion from a troop of mares. Brehm encountered in Abyssinia a great troop of baboons which were crossing a valley: some had already ascended the opposite mountain, and some were still in the valley: the latter were attacked by the dogs, but the old males immediately hurried down from the rocks, and with mouths widely opened roared so fearfully, that the dogs precipitately retreated. They were again encouraged to the attack; but by this time all the baboons had reascended the heights, excepting a young one, about six months old, who, loudly calling for aid, climbed on a block of rock and was surrounded. Now one of the largest males, a true hero, came down again from the mountain, slowly went to the young one, coaxed him, and triumphantly led him away—the dogs being too much astonished to make an attack. I cannot resist giving another scene which was witnessed by this same naturalist; an eagle seized a young Cercopithecus, which, by clinging to a branch, was not at once carried off; it cried loudly for assistance, upon which the other members of the troop with much uproar rushed to the rescue, surrounded the eagle, and pulled out so many feathers, that he no longer thought of his prey, but only how to escape. This eagle, as Brehm remarks, assuredly would never again attack a monkey in a troop.

It is certain that associated animals have a feeling of love for each other which is not felt by adult and non-social animals. How far in most cases they actually sympathise with each other's pains and pleasures is more doubtful, especially with respect to the latter. Mr. Buxton, however, who had excellent means of observation,[109] states that his macaws, which lived free in Norfolk, took "an extravagant interest" in a pair with a nest, and whenever the female left it, she was surrounded by a troop "screaming horrible acclamations in her honour." It is often difficult to judge whether animals have any feeling for each other's sufferings. Who can say what cows feel, when they surround and stare intently on a dying or dead companion? That animals sometimes are far from feeling any sympathy is too certain; for they will expel a wounded animal from the herd, or gore or worry it to death. This is almost the blackest fact in natural history unless indeed the explanation which has been suggested is true, that their instinct or reason leads them to expel an injured companion, lest beasts of prey, including man, should be tempted to follow the troop. In this case their conduct is not much worse than that of the North American Indians who leave their feeble comrades to perish on the plains, or the Fijians, who, when their parents get old or fall ill, bury them alive.[110]

Many animals, however, certainly sympathise with each other's distress or danger. This is the case even with birds; Capt. Stansbury[111] found on a salt lake in Utah an old and completely blind pelican, which was very fat, and must have been long and well fed by his companions. Mr. Blyth, as he informs me, saw Indian crows feeding two or three of their companions which were blind; and I have heard of an analogous case with the domestic cock. We may, if we choose, call these actions instinctive; but such cases are much too rare for the development of any special instinct.[112] I have myself seen a dog, who never passed a great friend of his, a cat which lay sick in a basket, without giving her a few licks with his tongue, the surest sign of kind feeling in a dog.

It must be called sympathy that leads a courageous dog to fly at any one who strikes his master, as he certainly will. I saw a person pretending to beat a lady who had a very timid little dog on her lap, and the trial had never before been made. The little creature instantly jumped away, but after the pretended beating was over, it was really pathetic to see how perseveringly he tried to lick his mistress's face and comfort her. Brehm[113] states that when a baboon in confinement was pursued to be punished, the others tried to protect him. It must have been sympathy in the cases above given which led the baboons and Cercopitheci to defend their young comrades from the dogs and the eagle. I will give only one other instance of sympathetic and heroic conduct in a little American monkey. Several years ago a keeper at the Zoological Gardens, showed me some deep and scarcely healed wounds on the nape of his neck, inflicted on him whilst kneeling on the floor by a fierce baboon. The little American monkey, who was a warm friend of this keeper, lived in the same large compartment, and was dreadfully afraid of the great baboon. Nevertheless, as soon as he saw his friend the keeper in peril, he rushed to the rescue, and by screams and bites so distracted the baboon that the man was able to escape, after running great risk, as the surgeon who attended him thought, of his life.

Besides love and sympathy, animals exhibit other qualities which in us would be called moral; and I agree with Agassiz[114] that dogs possess something very like a conscience. They certainly possess some power of self-command, and this does not appear to be wholly the result of fear. As Braubach[115] remarks, a dog will refrain from stealing food in the absence of his master. Dogs have long been accepted as the very type of fidelity and obedience. All animals living in a body which defend each other or attack their enemies in concert, must be in some degree faithful to each other; and those that follow a leader must be in some degree obedient. When the baboons in Abyssinia[116] plunder a garden, they silently follow their leader; and if an imprudent young animal makes a noise, he receives a slap from the others to teach him silence and obedience; but as soon as they are sure that there is no danger, all show their joy by much clamour.

With respect to the impulse which leads certain animals to associate together, and to aid each other in many ways, we may infer that in most cases they are impelled by the same sense of satisfaction or pleasure which they experience in performing other instinctive actions; or by the same sense of dissatisfaction, as in other cases of prevented instinctive actions. We see this in innumerable instances, and it is illustrated in a striking manner by the acquired instincts of our domesticated animals; thus a young shepherd-dog delights in driving and running round a flock of sheep, but not in worrying them; a young foxhound delights in hunting a fox, whilst some other kinds of dogs as I have witnessed, utterly disregard foxes. What a strong feeling of inward satisfaction must impel a bird, so full of activity, to brood day after day over her eggs. Migratory birds are miserable if prevented from migrating, and perhaps they enjoy starting on their long flight. Some few instincts are determined solely by painful feelings, as by fear, which leads to self-preservation, or is specially directed against certain enemies. No one, I presume, can analyse the sensations of pleasure or pain. In many cases, however, it is probable that instincts are persistently followed from the mere force of inheritance, without the stimulus of either pleasure or pain. A young pointer, when it first scents game, apparently cannot help pointing. A squirrel in a cage who pats the nuts which it cannot eat, as if to bury them in the ground, can hardly be thought to act thus either from pleasure or pain. Hence the common assumption that men must be impelled to every action by experiencing some pleasure or pain may be erroneous. Although a habit may be blindly and implicitly followed, independently of any pleasure or pain felt at the moment, yet if it be forcibly and abruptly checked, a vague sense of dissatisfaction is generally experienced; and this is especially true in regard to persons of feeble intellect.

It has often been assumed that animals were in the first place rendered social, and that they feel as a consequence uncomfortable when separated from each other, and comfortable whilst together; but it is a more probable view that these sensations were first developed, in order that those animals which would profit by living in society, should be induced to live together. In the same manner as the sense of hunger and the pleasure of eating were, no doubt, first acquired in order to induce animals to eat. The feeling of pleasure from society is probably an extension of the parental or filial affections; and this extension may be in chief part attributed to natural selection, but perhaps in part to mere habit. For with those animals which were benefited by living in close association, the individuals which took the greatest pleasure in society would best escape various dangers; whilst those that cared least for their comrades and lived solitary would perish in greater numbers. With respect to the origin of the parental and filial affections, which apparently lie at the basis of the social affections, it is hopeless to speculate; but we may infer that they have been to a large extent gained through natural selection. So it has almost certainly been with the unusual and opposite feeling of hatred between the nearest relations, as with the worker-bees which kill their brother-drones, and with the queen-bees which kill their daughter-queens; the desire to destroy, instead of loving, their nearest relations having been here of service to the community.

The all-important emotion of sympathy is distinct from that of love. A mother may passionately love her sleeping and passive infant, but she can then hardly be said to feel sympathy for it. The love of a man for his dog is distinct from sympathy, and so is that of a dog for his master. Adam Smith formerly argued, as has Mr. Bain recently, that the basis of sympathy lies in our strong retentiveness of former states of pain or pleasure. Hence, "the sight of another person enduring hunger, cold, fatigue, revives in us some recollection of these states, which are painful even in idea." We are thus impelled to relieve the sufferings of another, in order that our own painful feelings may be at the same time relieved. In like manner we are led to participate in the pleasures of others.[117] But I cannot see how this view explains the fact that sympathy is excited in an immeasurably stronger degree by a beloved than by an indifferent person. The mere sight of suffering, independently of love, would suffice to call up in us vivid recollections and associations. Sympathy may at first have originated in the manner above suggested; but it seems now to have become an instinct, which is especially directed towards beloved objects, in the same manner as fear with animals is especially directed against certain enemies. As sympathy is thus directed, the mutual love of the members of the same community will extend its limits. No doubt a tiger or lion feels sympathy for the sufferings of its own young, but not for any other animal. With strictly social animals the feeling will be more or less extended to all the associated members, as we know to be the case. With mankind selfishness, experience, and imitation probably add, as Mr. Bain has shewn, to the power of sympathy; for we are led by the hope of receiving good in return to perform acts of sympathetic kindness to others; and there can be no doubt that the feeling of sympathy is much strengthened by habit. In however complex a manner this feeling may have originated, as it is one of high importance to all those animals which aid and defend each other, it will have been increased, through natural selection; for those communities, which included the greatest number of the most sympathetic members, would flourish best and rear the greatest number of offspring.

In many cases it is impossible to decide whether certain social instincts have been acquired through natural selection, or are the indirect result of other instincts and faculties, such as sympathy, reason, experience, and a tendency to imitation; or again, whether they are simply the result of long-continued habit. So remarkable an instinct as the placing sentinels to warn the community of danger, can hardly have been the indirect result of any other faculty; it must therefore have been directly acquired. On the other hand, the habit followed by the males of some social animals, of defending the community and of attacking their enemies or their prey in concert, may perhaps have originated from mutual sympathy; but courage, and in most cases strength, must have been previously acquired, probably through natural selection.

Of the various instincts and habits, some are much stronger than others, that is, some either give more pleasure in their performance and more distress in their prevention than others; or, which is probably quite as important, they are more persistently followed through inheritance without exciting any special feeling of pleasure or pain. We are ourselves conscious that some habits are much more difficult to cure or change than others. Hence a struggle may often be observed in animals between different instincts, or between an instinct and some habitual disposition; as when a dog rushes after a hare, is rebuked, pauses, hesitates, pursues again or returns ashamed to his master; or as between the love of a female dog for her young puppies and for her master, for she may be seen to slink away to them, as if half ashamed of not accompanying her master. But the most curious instance known to me of one instinct conquering another, is the migratory instinct conquering the maternal instinct. The former is wonderfully strong; a confined bird will at the proper season beat her breast against the wires of her cage, until it is bare and bloody. It causes young salmon to leap out of the fresh water, where they could still continue to live, and thus unintentionally to commit suicide. Every one knows how strong the maternal instinct is, leading even timid birds to face great danger, though with hesitation and in opposition to the instinct of self-preservation. Nevertheless the migratory instinct is so powerful that late in the autumn swallows and house-martins frequently desert their tender young, leaving them to perish miserably in their nests.[118]

We can perceive that an instinctive impulse, if it be in any way more beneficial to a species than some other or opposed instinct, would be rendered the more potent of the two through natural selection; for the individuals which had it most strongly developed would survive in larger numbers. Whether this is the case with the migratory in comparison with the maternal instinct, may well be doubted. The great persistence or steady action of the former at certain seasons of the year during the whole day, may give it for a time paramount force.

Man a social animal.—Most persons admit that man is a social being. We see this in his dislike of solitude, and in his wish for society beyond that of his own family. Solitary confinement is one of the severest punishments which can be inflicted. Some authors suppose that man primevally lived in single families; but at the present day, though single families, or only two or three together, roam the solitudes of some savage lands, they are always, as far as I can discover, friendly with other families inhabiting the same district. Such families occasionally meet in council, and they unite for their common defence. It is no argument against savage man being a social animal, that the tribes inhabiting adjacent districts are almost always at war with each other; for the social instincts never extend to all the individuals of the same species. Judging from the analogy of the greater number of the Quadrumana, it is probable that the early ape-like progenitors of man were likewise social; but this is not of much importance for us. Although man, as he now exists, has few special instincts, having lost any which his early progenitors may have possessed, this is no reason why he should not have retained from an extremely remote period some degree of instinctive love and sympathy for his fellows. We are indeed all conscious that we do possess such sympathetic feelings;[119] but our consciousness does not tell us whether they are instinctive, having originated long ago in the same manner as with the lower animals, or whether they have been acquired by each of us during our early years. As man is a social animal, it is also probable that he would inherit a tendency to be faithful to his comrades, for this quality is common to most social animals. He would in like manner possess some capacity for self-command, and perhaps of obedience to the leader of the community. He would from an inherited tendency still be willing to defend, in concert with others, his fellow-men, and would be ready to aid them in any way which did not too greatly interfere with his own welfare or his own strong desires.

The social animals which stand at the bottom of the scale are guided almost exclusively, and those which stand higher in the scale are largely guided, in the aid which they give to the members of the same community, by special instincts; but they are likewise in part impelled by mutual love and sympathy, assisted apparently by some amount of reason. Although man, as just remarked, has no special instincts to tell him how to aid his fellow-men, he still has the impulse, and with his improved intellectual faculties would naturally be much guided in this respect by reason and experience. Instinctive sympathy would, also, cause him to value highly the approbation of his fellow-men; for, as Mr. Bain has clearly shewn,[120] the love of praise and the strong feeling of glory, and the still stronger horror of scorn and infamy, "are due to the workings of sympathy." Consequently man would be greatly influenced by the wishes, approbation, and blame of his fellow-men, as expressed by their gestures and language. Thus the social instincts, which must have been acquired by man in a very rude state, and probably even by his early ape-like progenitors, still give the impulse to many of his best actions; but his actions are largely determined by the expressed wishes and judgment of his fellow-men, and unfortunately still oftener by his own strong, selfish desires. But as the feelings of love and sympathy and the power of self-command become strengthened by habit, and as the power of reasoning becomes clearer so that man can appreciate the justice of the judgments of his fellow-men, he will feel himself impelled, independently of any pleasure or pain felt at the moment, to certain lines of conduct. He may then say, I am the supreme judge of my own conduct, and in the words of Kant, I will not in my own person violate the dignity of humanity.

The more enduring Social Instincts conquer the less Persistent Instincts.—We have, however, not as yet considered the main point, on which the whole question of the moral sense hinges. Why should a man feel that he ought to obey one instinctive desire rather than another? Why does he bitterly regret if he has yielded to the strong sense of self-preservation, and has not risked his life to save that of a fellow-creature; or why does he regret having stolen food from severe hunger?

It is evident in the first place, that with mankind the instinctive impulses have different degrees of strength; a young and timid mother urged by the maternal instinct will, without a moment's hesitation, run the greatest danger for her infant, but not for a mere fellow-creature. Many a man, or even boy, who never before risked his life for another, but in whom courage and sympathy were well developed, has, disregarding the instinct of self-preservation, instantaneously plunged into a torrent to save a drowning fellow-creature. In this case man is impelled by the same instinctive motive, which caused the heroic little American monkey, formerly described, to attack the great and dreaded baboon, to save his keeper. Such actions as the above appear to be the simple result of the greater strength of the social or maternal instincts than of any other instinct or motive; for they are performed too instantaneously for reflection, or for the sensation of pleasure or pain; though if prevented distress would be caused.

I am aware that some persons maintain that actions performed impulsively, as in the above cases, do not come under the dominion of the moral sense, and cannot be called moral. They confine this term to actions done deliberately, after a victory over opposing desires, or to actions prompted by some lofty motive. But it appears scarcely possible to draw any clear line of distinction of this kind; though the distinction may be real. As far as exalted motives are concerned, many instances have been recorded of barbarians, destitute of any feeling of general benevolence towards mankind, and not guided by any religious motive, who have deliberately as prisoners sacrificed their lives,[121] rather than betray their comrades; and surely their conduct ought to be considered as moral. As far as deliberation and the victory over opposing motives are concerned, animals may be seen doubting between opposed instincts, as in rescuing their offspring or comrades from danger; yet their actions, though done for the good of others, are not called moral. Moreover, an action repeatedly performed by us, will at last be done without deliberation or hesitation, and can then hardly be distinguished from an instinct; yet surely no one will pretend that an action thus done ceases to be moral. On the contrary, we all feel that an act cannot be considered as perfect, or as performed in the most noble manner, unless it be done impulsively, without deliberation or effort, in the same manner as by a man in whom the requisite qualities are innate. He who is forced to overcome his fear or want of sympathy before he acts, deserves, however, in one way higher credit than the man whose innate disposition leads him to a good act without effort. As we cannot distinguish between motives, we rank all actions of a certain class as moral, when they are performed by a moral being. A moral being is one who is capable of comparing his past and future actions or motives, and of approving or disapproving of them. We have no reason to suppose that any of the lower animals have this capacity; therefore when a monkey faces danger to rescue its comrade, or takes charge of an orphan-monkey, we do not call its conduct moral. But in the case of man, who alone can with certainty be ranked as a moral being, actions of a certain class are called moral, whether performed deliberately after a struggle with opposing motives, or from the effects of slowly-gained habit, or impulsively through instinct.

But to return to our more immediate subject; although some instincts are more powerful than others, thus leading to corresponding actions, yet it cannot be maintained that the social instincts are ordinarily stronger in man, or have become stronger through long-continued habit, than the instincts, for instance, of self-preservation, hunger, lust, vengeance, &c. Why then does man regret, even though he may endeavour to banish any such regret, that he has followed the one natural impulse, rather than the other; and why does he further feel that he ought to regret his conduct? Man in this respect differs profoundly from the lower animals. Nevertheless we can, I think, see with some degree of clearness the reason of this difference.

Man, from the activity of his mental faculties, cannot avoid reflection: past impressions and images are incessantly passing through his mind with distinctness. Now with those animals which live permanently in a body, the social instincts are ever present and persistent. Such animals are always ready to utter the danger-signal, to defend the community, and to give aid to their fellows in accordance with their habits; they feel at all times, without the stimulus of any special passion or desire, some degree of love and sympathy for them; they are unhappy if long separated from them, and always happy to be in their company. So it is with ourselves. A man who possessed no trace of such feelings would be an unnatural monster. On the other hand, the desire to satisfy hunger, or any passion, such as vengeance, is in its nature temporary, and can for a time be fully satisfied. Nor is it easy, perhaps hardly possible, to call up with complete vividness the feeling, for instance, of hunger; nor indeed, as has often been remarked, of any suffering. The instinct of self-preservation is not felt except in the presence of danger; and many a coward has thought himself brave until he has met his enemy face to face. The wish for another man's property is perhaps as persistent a desire as any that can be named; but even in this case the satisfaction of actual possession is generally a weaker feeling than the desire: many a thief, if not an habitual one, after success has wondered why he stole some article.

Thus, as man cannot prevent old impressions continually repassing through his mind, he will be compelled to compare the weaker impressions of, for instance, past hunger, or of vengeance satisfied or danger avoided at the cost of other men, with the instinct of sympathy and good-will to his fellows, which is still present and ever in some degree active in his mind. He will then feel in his imagination that a stronger instinct has yielded to one which now seems comparatively weak; and then that sense of dissatisfaction will inevitably be felt with which man is endowed, like every other animal, in order that his instincts may be obeyed. The case before given, of the swallow, affords an illustration, though of a reversed nature, of a temporary though for the time strongly persistent instinct conquering another instinct which is usually dominant over all others. At the proper season these birds seem all day long to be impressed with the desire to migrate; their habits change; they become restless, are noisy, and congregate in flocks. Whilst the mother-bird is feeding or brooding over her nestlings, the maternal instinct is probably stronger than the migratory; but the instinct which is more persistent gains the victory, and at last, at a moment when her young ones are not in sight, she takes flight and deserts them. When arrived at the end of her long journey, and the migratory instinct ceases to act, what an agony of remorse each bird would feel, if, from being endowed with great mental activity, she could not prevent the image continually passing before her mind of her young ones perishing in the bleak north from cold and hunger.

At the moment of action, man will no doubt be apt to follow the stronger impulse; and though this may occasionally prompt him to the noblest deeds, it will far more commonly lead him to gratify his own desires at the expense of other men. But after their gratification, when past and weaker impressions are contrasted with the ever-enduring social instincts, retribution will surely come. Man will then feel dissatisfied with himself, and will resolve with more or less force to act differently for the future. This is conscience; for conscience looks backwards and judges past actions, inducing that kind of dissatisfaction, which if weak we call regret, and if severe remorse.

These sensations are, no doubt, different from those experienced when other instincts or desires are left unsatisfied; but every unsatisfied instinct has its own proper prompting sensation, as we recognise with hunger, thirst, &c. Man thus prompted, will through long habit acquire such perfect self-command, that his desires and passions will at last instantly yield to his social sympathies, and there will no longer be a struggle between them. The still hungry, or the still revengeful man will not think of stealing food, or of wreaking his vengeance. It is possible, or, as we shall hereafter see, even probable, that the habit of self-command may, like other habits, be inherited. Thus at last man comes to feel, through acquired and perhaps inherited habit, that it is best for him to obey his more persistent instincts. The imperious word *ought* seems merely to imply the consciousness of the existence of a persistent instinct, either innate or partly acquired, serving him as a guide, though liable to be disobeyed. We hardly use the word *ought* in a metaphorical sense, when we say hounds ought to hunt, pointers to point, and retrievers to retrieve their game. If they fail thus to act, they fail in their duty and act wrongly.

If any desire or instinct, leading to an action opposed to the good of others, still appears to a man, when recalled to mind, as strong as, or stronger than, his social instinct, he will feel no keen regret at having followed it; but he will be conscious that if his conduct were known to his fellows, it would meet with their disapprobation; and few are so destitute of sympathy as not to feel discomfort when this is realised. If he has no such sympathy, and if his desires leading to bad actions are at the time strong, and when recalled are not overmastered by the persistent social instincts, then he is essentially a bad man;[122] and the sole restraining motive left is the fear of punishment, and the conviction that in the long run it would be best for his own selfish interests to regard the good of others rather than his own.

It is obvious that every one may with an easy conscience gratify his own desires, if they do not interfere with his social instincts, that is with the good of others; but in order to be quite free from self-reproach, or at least of anxiety, it is almost necessary for him to avoid the disapprobation, whether reasonable or not, of his fellow men. Nor must he break through the fixed habits of his life, especially if these are supported by reason; for if he does, he will assuredly feel dissatisfaction. He must likewise avoid the reprobation of the one God or gods, in whom according to his knowledge or superstition he may believe; but in this case the additional fear of divine punishment often supervenes.

The strictly Social Virtues at first alone regarded.—The above view of the first origin and nature of the moral sense, which tells us what we ought to do, and of the conscience which reproves us if we disobey it, accords well with what we see of the early and undeveloped condition of this faculty in mankind. The virtues which must be practised, at least generally, by rude men, so that they may associate in a body, are those which are still recognised as the most important. But they are practised almost exclusively in relation to the men of the same tribe; and their opposites are not regarded as crimes in relation to the men of other tribes. No tribe could hold together if murder, robbery, treachery, &c., were common; consequently such crimes within the limits of the same tribe "are branded with everlasting infamy;"[123] but excite no such sentiment beyond these limits. A North-American Indian is well pleased with himself, and is honoured by others, when he scalps a man of another tribe; and a Dyak cuts off the head of an unoffending person and dries it as a trophy. The murder of infants has prevailed on the largest scale throughout the world,[124] and has met with no reproach; but infanticide, especially of females, has been thought to be good for the tribe, or at least not injurious. Suicide during former times was not generally considered as a crime,[125] but rather from the courage displayed as an honourable act; and it is still largely practised by some semi-civilised nations without reproach, for the loss to a nation of a single individual is not felt: whatever the explanation may be, suicide, as I hear from Sir J. Lubbock, is rarely practised by the lowest barbarians. It has been recorded that an Indian Thug conscientiously regretted that he had not strangled and robbed as many travellers as did his father before him. In a rude state of civilisation the robbery of strangers is, indeed, generally considered as honourable.

The great sin of Slavery has been almost universal, and slaves have often been treated in an infamous manner. As barbarians do not regard the opinion of their women, wives are commonly treated like slaves. Most savages are utterly indifferent to the sufferings of strangers, or even delight in witnessing them. It is well known that the women and children of the North-American Indians aided in torturing their enemies. Some savages take a horrid pleasure in cruelty to animals,[126] and humanity with them is an unknown virtue. Nevertheless, feelings of sympathy and kindness are common, especially during sickness, between the members of the same tribe, and are sometimes extended beyond the limits of the tribe. Mungo Park's touching account of the kindness of the negro women of the interior to him is well known. Many instances could be given of the noble fidelity of savages towards each other, but not to strangers; common experience justifies the maxim of the Spaniard, "Never, never trust an Indian." There cannot be fidelity without truth; and this fundamental virtue is not rare between the members of the same tribe: thus Mungo Park heard the negro women teaching their young children to love the truth. This, again, is one of the virtues which becomes so deeply rooted in the mind that it is sometimes practised by savages even at a high cost, towards strangers; but to lie to your enemy has rarely been thought a sin, as the history of modern diplomacy too plainly shews. As soon as a tribe has a recognised leader, disobedience becomes a crime, and even abject submission is looked at as a sacred virtue.

As during rude times no man can be useful or faithful to his tribe without courage, this quality has universally been placed in the highest rank; and although, in civilised countries, a good, yet timid, man may be far more useful to the community than a brave one, we cannot help instinctively honouring the latter above a coward, however benevolent. Prudence, on the other hand, which does not concern the welfare of others, though a very useful virtue, has never been highly esteemed. As no man can practise the virtues necessary for the welfare of his tribe without self-sacrifice, self-command, and the power of endurance, these qualities have been at all times highly and most justly valued. The American savage voluntarily submits without a groan to the most horrid tortures to prove and strengthen his fortitude and courage; and we cannot help admiring him, or even an Indian Fakir, who, from a foolish religious motive, swings suspended by a hook buried in his flesh.

The other self-regarding virtues, which do not obviously, though they may really, affect the welfare of the tribe, have never been esteemed by savages, though now highly appreciated by civilised nations. The greatest intemperance with savages is no reproach. Their utter licentiousness, not to mention unnatural crimes, is something astounding.[127] As soon, however, as marriage, whether polygamous or monogamous, becomes common, jealousy will lead to the inculcation of female virtue; and this being honoured will tend to spread to the unmarried females. How slowly it spreads to the male sex we see at the present day. Chastity eminently requires self-command; therefore it has been honoured from a very early period in the moral history of civilised man. As a consequence of this, the senseless practice of celibacy has been ranked from a remote period as a virtue.[128] The hatred of indecency, which appears to us so natural as to be thought innate, and which is so valuable an aid to chastity, is a modern virtue, appertaining exclusively, as Sir G. Staunton remarks,[129] to civilised life. This is shewn by the ancient religious rites of various nations, by the drawings on the walls of Pompeii, and by the practices of many savages.

We have now seen that actions are regarded by savages, and were probably so regarded by primeval man, as good or bad, solely as they affect in an obvious manner the welfare of the tribe,—not that of the species, nor that of man as an individual member of the tribe. This conclusion agrees well with the belief that the so-called moral sense is aboriginally derived from the social instincts, for both relate at first exclusively to the community. The chief causes of the low morality of savages, as judged by our standard, are, firstly, the confinement of sympathy to the same tribe. Secondly, insufficient powers of reasoning, so that the bearing of many virtues, especially of the self-regarding virtues, on the general welfare of the tribe is not recognised. Savages, for instance, fail to trace the multiplied evils consequent on a want of temperance, chastity, &c. And, thirdly, weak power of self-command; for this power has not been strengthened through long-continued, perhaps inherited, habit, instruction and religion. I have entered into the above details on the immorality of savages,[130] because some authors have recently taken a high view of their moral nature, or have attributed most of their crimes to mistaken benevolence.[131] These authors appear to rest their conclusion on savages possessing, as they undoubtedly do possess, and often in a high degree, those virtues which are serviceable, or even necessary, for the existence of a tribal community.

Concluding Remarks.—Philosophers of the derivative[132] school of morals formerly assumed that the foundation of morality lay in a form of Selfishness; but more recently in the "Greatest Happiness principle." According to the view given above, the moral sense is fundamentally identical with the social instincts; and in the case of the lower animals it would be absurd to speak of these instincts as having been developed from selfishness, or for the happiness of the community. They have, however, certainly been developed for the general good of the community. The term, general good, may be defined as the means by which the greatest possible number of individuals can be reared in full vigour and health, with all their faculties perfect, under the conditions to which they are exposed. As the social instincts both of man and the lower animals have no doubt been developed by the same steps, it would be advisable, if found practicable, to use the same definition in both cases, and to take as the test of morality, the general good or welfare of the community, rather than the general happiness; but this definition would perhaps require some limitation on account of political ethics.

When a man risks his life to save that of a fellow-creature, it seems more appropriate to say that he acts for the general good or welfare, rather than for the general happiness of mankind. No doubt the welfare and the happiness of the individual usually coincide; and a contented, happy tribe will flourish better than one that is discontented and unhappy. We have seen that at an early period in the history of man, the expressed wishes of the community will have naturally influenced to a large extent the conduct of each member; and as all wish for happiness, the "greatest happiness principle" will have become a most important secondary guide and object; the social instincts, including sympathy, always serving as the primary impulse and guide. Thus the reproach of laying the foundation of the most noble part of our nature in the base principle of selfishness is removed; unless indeed the satisfaction which every animal feels when it follows its proper instincts, and the dissatisfaction felt when prevented, be called selfish.

The expression of the wishes and judgment of the members of the same community, at first by oral and afterwards by written language, serves, as just remarked, as a most important secondary guide of conduct, in aid of the social instincts, but sometimes in opposition to them. This latter fact is well exemplified by the *Law of Honour*, that is the law of the opinion of our equals, and not of all our countrymen. The breach of this law, even when the breach is known to be strictly accordant with true morality, has caused many a man more agony than a real crime. We recognise the same influence in the burning sense of shame which most of us have felt even after the interval of years, when calling to mind some accidental breach of a trifling though fixed rule of etiquette. The judgment of the community will generally be guided by some rude experience of what is best in the long run for all the members; but this judgment will not rarely err from ignorance and from weak powers of reasoning. Hence the strangest customs and superstitions, in complete opposition to the true welfare and happiness of mankind, have become all-powerful throughout the world. We see this in the horror felt by a Hindoo who breaks his caste, in the shame of a Mahometan woman who exposes her face, and in innumerable other instances. It would be difficult to distinguish between the remorse felt by a Hindoo who has eaten unclean food, from that felt after committing a theft; but the former would probably be the more severe.

How so many absurd rules of conduct, as well as so many absurd religious beliefs, have originated we do not know; nor how it is that they have become, in all quarters of the world, so deeply impressed on the mind of men; but it is worthy of remark that a belief constantly inculcated during the early years of life, whilst the brain is impressible, appears to acquire almost the nature of an instinct; and the very essence of an instinct is that it is followed independently of reason. Neither can we say why certain admirable virtues, such as the love of truth, are much more highly appreciated by some savage tribes than by others;[133] nor, again, why similar differences prevail even amongst civilised nations. Knowing how firmly fixed many strange customs and superstitions have become, we need feel no surprise that the self-regarding virtues should now appear to us so natural, supported as they are by reason, as to be thought innate, although they were not valued by man in his early condition.

Notwithstanding many sources of doubt, man can generally and readily distinguish between the higher and lower moral rules. The higher are founded on the social instincts, and relate to the welfare of others. They are supported by the approbation of our fellow-men and by reason. The lower rules, though some of them when implying self-sacrifice hardly deserve to be called lower, relate chiefly to self, and owe their origin to public opinion, when matured by experience and cultivated; for they are not practised by rude tribes.

As man advances in civilisation, and small tribes are united into larger communities, the simplest reason would tell each individual that he ought to extend his social instincts and sympathies to all the members of the same nation, though personally unknown to him. This point being once reached, there is only an artificial barrier to prevent his sympathies extending to the men of all nations and races. If, indeed, such men are separated from him by great differences in appearance or habits, experience unfortunately shews us how long it is before we look at them as our fellow-creatures. Sympathy beyond the confines of man, that is humanity to the lower animals, seems to be one of the latest moral acquisitions. It is apparently unfelt by savages, except towards their pets. How little the old Romans knew of it is shewn by their abhorrent gladiatorial exhibitions. The very idea of humanity, as far as I could observe, was new to most of the Gauchos of the Pampas. This virtue, one of the noblest with which man is endowed, seems to arise incidentally from our sympathies becoming more tender and more widely diffused, until they are extended to all sentient beings. As soon as this virtue is honoured and practised by some few men, it spreads through instruction and example to the young, and eventually through public opinion.

The highest stage in moral culture at which we can arrive, is when we recognise that we ought to control our thoughts, and "not even in inmost thought to think again the sins that made the past so pleasant to us."[134] Whatever makes any bad action familiar to the mind, renders its performance by so much the easier. As Marcus Aurelius long ago said, "Such as are thy habitual thoughts, such also will be the character of thy mind; for the soul is dyed by the thoughts."[135]

Our great philosopher, Herbert Spencer, has recently explained his views on the moral sense. He says,[136] "I believe that the experiences of utility organised and consolidated through all past generations of the human race, have been producing corresponding modifications, which, by continued transmission and accumulation, have become in us certain faculties of moral intuition—certain emotions responding to right and wrong conduct, which have no apparent basis in the individual experiences of utility." There is not the least inherent improbability, as it seems to me, in virtuous tendencies being more or less strongly inherited; for, not to mention the various dispositions and habits transmitted by many of our domestic animals, I have heard of cases in which a desire to steal and a tendency to lie appeared to run in families of the upper ranks; and as stealing is so rare a crime in the wealthy classes, we can hardly account by accidental coincidence for the tendency occurring in two or three members of the same family. If bad tendencies are transmitted, it is probable that good ones are likewise transmitted. Excepting through the principle of the transmission of moral tendencies, we cannot understand the differences believed to exist in this respect between the various races of mankind. We have, however, as yet, hardly sufficient evidence on this head.

Even the partial transmission of virtuous tendencies would be an immense assistance to the primary impulse derived directly from the social instincts, and indirectly from the approbation of our fellow-men. Admitting for the moment that virtuous tendencies are inherited, it appears probable, at least in such cases as chastity, temperance, humanity to animals, &c., that they become first impressed on the mental organisation through habit, instruction, and example, continued during several generations in the same family, and in a quite subordinate degree, or not at all, by the individuals possessing such virtues, having succeeded best in the struggle for life. My chief source of doubt with respect to any such inheritance, is that senseless customs, superstitions, and tastes, such as the horror of a Hindoo for unclean food, ought on the same principle to be transmitted. Although this in itself is perhaps not less probable than that animals should acquire inherited tastes for certain kinds of food or fear of certain foes, I have not met with any evidence in support of the transmission of superstitious customs or senseless habits.

FINALLY, THE SOCIAL INSTINCTS which no doubt were acquired by man, as by the lower animals, for the good of the community, will from the first have given to him some wish to aid his fellows, and some feeling of sympathy. Such impulses will have served him at a very early period as a rude rule of right and wrong. But as man gradually advanced in intellectual power and was enabled to trace the more remote consequences of his actions; as he acquired sufficient knowledge to reject baneful customs and superstitions; as he regarded more and more not only the welfare but the happiness of his fellow-men; as from habit, following on beneficial experience, instruction, and example, his sympathies became more tender and widely diffused, so as to extend to the men of all races, to the imbecile, the maimed, and other useless members of society, and finally to the lower animals,—so would the standard of his morality rise higher and higher. And it is admitted by moralists of the derivative school and by some intuitionists, that the standard of morality has risen since an early period in the history of man.[137]

As a struggle may sometimes be seen going on between the various instincts of the lower animals, it is not surprising that there should be a struggle in man between his social instincts, with their derived virtues, and his lower, though at the moment, stronger impulses or desires. This, as Mr. Galton[138] has remarked, is all the less surprising, as man has emerged from a state of barbarism within a comparatively recent period. After having yielded to some temptation we feel a sense of dissatisfaction, analogous to that felt from other unsatisfied instincts, called in this case conscience; for we cannot prevent past images and impressions continually passing through our minds, and these in their weakened state we compare with the ever-present social instincts, or with habits gained in early youth and strengthened during our whole lives, perhaps inherited, so that they are at last rendered almost as strong as instincts. Looking to future generations, there is no cause to fear that the social instincts will grow weaker, and we may expect that virtuous habits will grow stronger, becoming perhaps fixed by inheritance. In this case the struggle between our higher and lower impulses will be less severe, and virtue will be triumphant.

Summary of the two last Chapters.—There can be no doubt that the difference between the mind of the lowest man and that of the highest animal is immense. An anthropomorphous ape, if he could take a dispassionate view of his own case, would admit that though he could form an artful plan to plunder a garden—though he could use stones for fighting or for breaking open nuts, yet that the thought of fashioning a stone into a tool was quite beyond his scope. Still less, as he would admit, could he follow out a train of metaphysical reasoning, or solve a mathematical problem, or reflect on God, or admire a grand natural scene. Some apes, however, would probably declare that they could and did admire the beauty of the coloured skin and fur of their partners in marriage. They would admit, that though they could make other apes understand by cries some of their perceptions and simpler wants, the notion of expressing definite ideas by definite sounds had never crossed their minds. They might insist that they were ready to aid their fellow-apes of the same troop in many ways, to risk their lives for them, and to take charge of their orphans; but they would be forced to acknowledge that disinterested love for all living creatures, the most noble attribute of man, was quite beyond their comprehension.

Nevertheless the difference in mind between man and the higher animals, great as it is, is certainly one of degree and not of kind. We have seen that the senses and intuitions, the various emotions and faculties, such as love, memory, attention, curiosity, imitation, reason, &c., of which man boasts, may be found in an incipient, or even sometimes in a well-developed condition, in the lower animals. They are also capable of some inherited improvement, as we see in the domestic dog compared with the wolf or jackal. If it be maintained that certain powers, such as self-consciousness, abstraction, &c., are peculiar to man, it may well be that these are the incidental results of other highly-advanced intellectual faculties; and these again are mainly the result of the continued use of a highly developed language. At what age does the new-born infant possess the power of abstraction, or become self-conscious and reflect on its own existence? We cannot answer; nor can we answer in regard to the ascending organic scale. The half-art and half-instinct of language still bears the stamp of its gradual evolution. The ennobling belief in God is not universal with man; and the belief in active spiritual agencies naturally follows from his other mental powers. The moral sense perhaps affords the best and highest distinction between man and the lower animals; but I need not say anything on this head, as I have so lately endeavoured to shew that the social instincts,—the prime principle of man's moral constitution[139]—with the aid of active intellectual powers and the effects of habit, naturally lead to the golden rule, "As ye would that men should do to you, do ye to them likewise;" and this lies at the foundation of morality.

In a future chapter I shall make some few remarks on the probable steps and means by which the several mental and moral faculties of man have been gradually evolved. That this at least is possible ought not to be denied, when we daily see their development in every infant; and when we may trace a perfect gradation from the mind of an utter idiot, lower than that of the lowest animal, to the mind of a Newton.

CHAPTER IV

ON THE MANNER OF DEVELOPMENT OF MAN FROM SOME LOWER FORM

VARIABILITY OF BODY AND MIND IN MAN • INHERITANCE • CAUSES OF VARIABILITY • LAWS OF VARIATION THE SAME IN MAN AS IN THE LOWER ANIMALS • DIRECT ACTION OF THE CONDITIONS OF LIFE • EFFECTS OF THE INCREASED USE AND DISUSE OF PARTS • ARRESTED DEVELOPMENT • REVERSION • CORRELATED VARIATION • RATE OF INCREASE • CHECKS TO INCREASE • NATURAL SELECTION • MAN THE MOST DOMINANT ANIMAL IN THE WORLD • IMPORTANCE OF HIS CORPOREAL STRUCTURE • THE CAUSES WHICH HAVE LED TO HIS BECOMING ERECT • CONSEQUENT CHANGES OF STRUCTURE • DECREASE IN SIZE OF THE CANINE TEETH • INCREASED SIZE AND ALTERED SHAPE OF THE SKULL • NAKEDNESS • ABSENCE OF A TAIL • DEFENCELESS CONDITION OF MAN

WE HAVE SEEN in the first chapter that the homological structure of man, his embryological development and the rudiments which he still retains, all declare in the plainest manner that he is descended from some lower form. The possession of exalted mental powers is no insuperable objection to this conclusion. In order that an ape-like creature should have been transformed into man, it is necessary that this early form, as well as many successive links, should all have varied in mind and body. It is impossible to obtain direct evidence on this head; but if it can be shewn that man now varies—that his variations are induced by the same general causes, and obey the same general laws, as in the case of the lower animals—there can be little doubt that the preceding intermediate links varied in a like manner. The variations at each successive stage of descent must, also, have been in some manner accumulated and fixed.

The facts and conclusions to be given in this chapter relate almost exclusively to the probable means by which the transformation of man has been effected, as far as his bodily structure is concerned. The following chapter will be devoted to the development of his intellectual and moral faculties. But the present discussion likewise bears on the origin of the different races or species of mankind, whichever term may be preferred.

It is manifest that man is now subject to much variability. No two individuals of the same race are quite alike. We may compare millions of faces, and each will be distinct. There is an equally great amount of diversity in the proportions and dimensions of the various parts of the body; the length of the legs being one of the most variable points.[140] Although in some quarters of the world an elongated skull, and in other quarters a short skull prevails, yet there is great diversity of shape even within the limits of the same race, as with the aborigines of America and South Australia,—the latter a race "probably as pure and homogeneous in blood, customs, and language as any in existence"—and even with the inhabitants of so confined an area as the Sandwich Islands.[141] An eminent dentist assures me that there is nearly as much diversity in the teeth, as in the features. The chief arteries so frequently run in abnormal courses, that it has been found useful for surgical purposes to calculate from 12,000 corpses how often each course prevails.[142] The muscles are eminently variable: thus those of the foot were found by Prof. Turner[143] not to be strictly alike in any two out of fifty bodies; and in some the deviations were considerable. Prof. Turner adds that the power of performing the appropriate movements must have been modified in accordance with the several deviations. Mr. J. Wood has recorded[144] the occurrence of 295 muscular variations in thirty-six subjects, and in another set of the same number no less than 558 variations, reckoning both sides of the body as one. In the last set, not one body out of the thirty-six was "found totally wanting in departures from the standard descriptions of the muscular system given in anatomical text-books." A single body presented the extraordinary number of twenty-five distinct abnormalities. The same muscle sometimes varies in many ways: thus Prof. Macalister describes[145] no less than twenty distinct variations in the *palmaris accessorius.*

The famous old anatomist, Wolff,[146] insists that the internal viscera are more variable than the external parts: *Nulla particula est quæ non aliter et aliter in aliis se habeat hominibus.* He has even written a treatise on the choice of typical examples of the viscera for representation. A discussion on the beau-ideal of the liver, lungs, kidneys, &c., as of the human face divine, sounds strange in our ears.

The variability or diversity of the mental faculties in men of the same race, not to mention the greater differences between the men of distinct races, is so notorious that not a word need here be said. So it is with the lower animals, as has been illustrated by a few examples in the last chapter. All who have had charge of menageries admit this fact, and we see it plainly in our dogs and other domestic animals. Brehm especially insists that each individual monkey of those which he kept under confinement in Africa had its own peculiar disposition and temper: he mentions one baboon remarkable for its high intelligence; and the keepers in the Zoological Gardens pointed out to me a monkey, belonging to the New World division, equally remarkable for intelligence. Rengger, also, insists on the diversity in the various mental characters of the monkeys of the same species which he kept in Paraguay; and this diversity, as he adds, is partly innate, and partly the result of the manner in which they have been treated or educated.[147]

I have elsewhere[148] so fully discussed the subject of Inheritance that I need here add hardly anything. A greater number of facts have been collected with respect to the transmission of the most trifling, as well as of the most important characters in man than in any of the lower animals; though the facts are copious enough with respect to the latter. So in regard to mental qualities, their transmission is manifest in our dogs, horses, and other domestic animals. Besides special tastes and habits, general intelligence, courage, bad and good temper, &c., are certainly transmitted. With man we see similar facts in almost every family; and we now know through the admirable labours of Mr. Galton[149] that genius, which implies a wonderfully complex combination of high faculties, tends to be inherited; and, on the other hand, it is too certain that insanity and deteriorated mental powers likewise run in the same families.

With respect to the causes of variability we are in all cases very ignorant; but we can see that in man as in the lower animals, they stand in some relation with the conditions to which each species has been exposed during several generations. Domesticated animals vary more than those in a state of nature; and this is apparently due to the diversified and changing nature of their conditions. The different races of man resemble in this respect domesticated animals, and so do the individuals of the same race when inhabiting a very wide area, like that of America. We see the influence of diversified conditions in the more civilised nations, the members of which belong to different grades of rank and follow different occupations, presenting a greater range of character than the members of barbarous nations. But the uniformity of savages has often been exaggerated, and in some cases can hardly be said to exist.[150] It is nevertheless an error to speak of man, even if we look only to the conditions to which he has been subjected, as "far more domesticated"[151] than any other animal. Some savage races, such as the Australians, are not exposed to more diversified conditions than are many species which have very wide ranges. In another and much more important respect, man differs widely from any strictly domesticated animal; for his breeding has not been controlled, either through methodical or unconscious selection. No race or body of men has been so completely subjugated by other men, that certain individuals have been preserved and thus unconsciously selected, from being in some way more useful to their masters. Nor have certain male and female individuals been intentionally picked out and matched, except in the well-known case of the Prussian grenadiers; and in this case man obeyed, as might have been expected, the law of methodical selection; for it is asserted that many tall men were reared in the villages inhabited by the grenadiers with their tall wives.

If we consider all the races of man, as forming a single species, his range is enormous; but some separate races, as the Americans and Polynesians, have very wide ranges. It is a well-known law that widely-ranging species are much more variable than species with restricted ranges; and the variability of man may with more truth be compared with that of widely-ranging species, than with that of domesticated animals.

Not only does variability appear to be induced in man and the lower animals by the same general causes, but in both the same characters are affected in a closely analogous manner. This has been proved in such full detail by Godron and Quatrefages, that I need here only refer to their works.[152] Monstrosities, which graduate into slight variations, are likewise so similar in man and the lower animals, that the same classification and the same terms can be used for both, as may be seen in Isidore Geoffroy St.-Hilaire's great work.[153] This is a necessary consequence of the same laws of change prevailing throughout the animal kingdom. In my work on the variation of domestic animals, I have attempted to arrange in a rude fashion the laws of variation under the following heads:—The direct and definite action of changed conditions, as shewn by all or nearly all the individuals of the same species varying in the same manner under the same circumstances. The effects of the long-continued use or disuse of parts. The cohesion of homologous parts. The variability of multiple parts. Compensation of growth; but of this law I have found no good instances in the case of man. The effects of the mechanical pressure of one part on another; as of the pelvis on the cranium of the infant in the womb. Arrests of development, leading to the diminution or suppression of parts. The reappearance of long-lost characters through reversion. And lastly, correlated variation. All these so-called laws apply equally to man and the lower animals; and most of them even to plants. It would be superfluous here to discuss all of them;[154] but several are so important for us, that they must be treated at considerable length.

The direct and definite action of changed conditions.—This is a most perplexing subject. It cannot be denied that changed conditions produce some effect, and occasionally a considerable effect, on organisms of all kinds; and it seems at first probable that if sufficient time were allowed this would be the invariable result. But I have failed to obtain clear evidence in favour of this conclusion; and valid reasons may be urged on the other side, at least as far as the innumerable structures are concerned, which are adapted for special ends. There can, however, be no doubt that changed conditions induce an almost indefinite amount of fluctuating variability, by which the whole organisation is rendered in some degree plastic.

In the United States, above 1,000,000 soldiers, who served in the late war, were measured, and the States in which they were born and reared recorded.[155] From this astonishing number of observations it is proved that local influences of some kind act directly on stature; and we further learn that "the State where the physical growth has in great measure taken place, and the State of birth, which indicates the ancestry, seem to exert a marked influence on the stature." For instance it is established, "that residence in the Western States, during the years of growth, tends to produce increase of stature." On the other hand, it is certain that with sailors, their manner of life delays growth, as shewn "by the great difference between the statures of soldiers and sailors at the ages of 17 and 18 years." Mr. B. A. Gould endeavoured to ascertain the nature of the influences which thus act on stature; but he arrived only at negative results, namely, that they did not relate to climate, the elevation of the land, soil, nor even "in any controlling degree" to the abundance or need of the comforts of life. This latter conclusion is directly opposed to that arrived at by Villermé from the statistics of the height of the conscripts in different parts of France. When we compare the differences in stature between the Polynesian chiefs and the lower orders within the same islands, or between the inhabitants of the fertile volcanic and low barren coral islands of the same ocean,[156] or again between the Fuegians on the eastern and western shores of their country, where the means of subsistence are very different, it is scarcely possible to avoid the conclusion that better food and greater comfort do influence stature. But the preceding statements shew how difficult it is to arrive at any precise result. Dr. Beddoe has lately proved that, with the inhabitants of Britain, residence in towns and certain occupations have a deteriorating influence on height; and he infers that the result is to a certain extent inherited, as is likewise the case in the United States. Dr. Beddoe further believes that wherever a "race attains its maximum of physical development, it rises highest in energy and moral vigour."[157]

Whether external conditions produce any other direct effect on man is not known. It might have been expected that differences of climate would have had a marked influence, as the lungs and kidneys are brought into fuller activity under a low temperature, and the liver and skin under a high one.[158] It was formerly thought that the colour of the skin and the character of the hair were determined by light or heat; and although it can hardly be denied that some effect is thus produced, almost all observers now agree that the effect has been very small, even after exposure during many ages. But this subject will be more properly discussed when we treat of the different races of mankind. With our domestic animals there are grounds for believing that cold and damp directly affect the growth of the hair; but I have not met with any evidence on this head in the case of man.

Effects of the increased Use and Disuse of Parts.—It is well known that use strengthens the muscles in the individual, and complete disuse, or the destruction of the proper nerve, weakens them. When the eye is destroyed the optic nerve often becomes atrophied. When an artery is tied, the lateral channels increase not only in diameter, but in the thickness and strength of their coats. When one kidney ceases acting from disease, the other increases in size and does double work. Bones increase not only in thickness, but in length, from carrying a greater weight.[159] Different occupations habitually followed lead to changed proportions in various parts of the body. Thus it was clearly ascertained by the United States Commission[160] that the legs of the sailors employed in the late war were longer by 0·217 of an inch than those of the soldiers, though the sailors were on an average shorter men; whilst their arms were shorter by 1·09 of an inch, and therefore out of proportion shorter in relation to their lesser height. This shortness of the arms is apparently due to their greater use, and is an unexpected result; but sailors chiefly use their arms in pulling and not in supporting weights. The girth of the neck and the depth of the instep are greater, whilst the circumference of the chest, waist, and hips is less in sailors than in soldiers.

Whether the several foregoing modifications would become hereditary, if the same habits of life were followed during many generations, is not known, but is probable. Rengger[161] attributes the thin legs and thick arms of the Payaguas Indians to successive generations having passed nearly their whole lives in canoes, with their lower extremities motionless. Other writers have come to a similar conclusion in other analogous cases. According to Cranz,[162] who lived for a long time with the Esquimaux, "the natives believe that ingenuity and dexterity in seal-catching (their highest art and virtue) is hereditary; there is really something in it, for the son of a celebrated seal-catcher will distinguish himself though he lost his father in childhood." But in this case it is mental aptitude, quite as much as bodily structure, which appears to be inherited. It is asserted that the hands of English labourers are at birth larger than those of the gentry.[163] From the correlation which exists, at least in some cases,[164] between the development of the extremities and of the jaws, it is possible that in those classes which do not labour much with their hands and feet, the jaws would be reduced in size from this cause. That they are generally smaller in refined and civilised men than in hard-working men or savages, is certain. But with savages, as Mr. Herbert Spencer[165] has remarked, the greater use of the jaws in chewing coarse, uncooked food, would act in a direct manner on the masticatory muscles and on the bones to which they are attached. In infants long before birth, the skin on the soles of the feet is thicker than on any other part of the body;[166] and it can hardly be doubted that this is due to the inherited effects of pressure during a long series of generations.

It is familiar to every one that watchmakers and engravers are liable to become short-sighted, whilst sailors and especially savages are generally long-sighted. Short-sight and long-sight certainly tend to be inherited.[167] The inferiority of Europeans, in comparison with savages, in eye-sight and in the other senses, is no doubt the accumulated and transmitted effect of lessened use during many generations; for Rengger[168] states that he has repeatedly observed Europeans, who had been brought up and spent their whole lives with the wild Indians, who nevertheless did not equal them in the sharpness of their senses. The same naturalist observes that the cavities in the skull for the reception of the several sense-organs are larger in the American aborigines than in Europeans; and this no doubt indicates a corresponding difference in the dimensions of the organs themselves. Blumenbach has also remarked on the large size of the nasal cavities in the skulls of the American aborigines, and connects this fact with their remarkably acute power of smell. The Mongolians of the plains of Northern Asia, according to Pallas, have wonderfully perfect senses; and Prichard believes that the great breadth of their skulls across the zygomas follows from their highly-developed sense-organs.[169]

The Quechua Indians inhabit the lofty plateaux of Peru, and Alcide d'Orbigny states[170] that from continually breathing a highly rarefied atmosphere they have acquired chests and lungs of extraordinary dimensions. The cells, also, of the lungs are larger and more numerous than in Europeans. These observations have been doubted; but Mr. D. Forbes carefully measured many Aymaras, an allied race, living at the height of between ten and fifteen thousand feet; and he informs me[171] that they differ conspicuously from the men of all other races seen by him, in the circumference and length of their bodies. In his table of measurements, the stature of each man is taken at 1000, and the other measurements are reduced to this standard. It is here seen that the extended arms of the Aymaras are shorter than those of Europeans, and much shorter than those of Negroes. The legs are likewise shorter, and they present this remarkable peculiarity, that in every Aymara measured the femur is actually shorter than the tibia. On an average the length of the femur to that of the tibia is as 211 to 252; whilst in two Europeans measured at the same time, the femora to the tibiæ were as 244 to 230; and in three Negroes as 258 to 241. The humerus is likewise shorter relatively to the forearm. This shortening of that part of the limb which is nearest to the body, appears to be, as suggested to me by Mr. Forbes, a case of compensation in relation with the greatly increased length of the trunk. The Aymaras present some other singular points of structure, for instance, the very small projection of the heel.

These men are so thoroughly acclimatised to their cold and lofty abode, that when formerly carried down by the Spaniards to the low Eastern plains, and when now tempted down by high wages to the gold-washings, they suffer a frightful rate of mortality. Nevertheless Mr. Forbes found a few pure families which had survived during two generations; and he observed that they still inherited their characteristic peculiarities. But it was manifest, even without measurement, that these peculiarities had all decreased; and on measurement their bodies were found not to be so much elongated as those of the men on the high plateau; whilst their femora had become somewhat lengthened, as had their tibiæ but in a less degree. The actual measurements may be seen by consulting Mr. Forbes' memoir. From these valuable observations, there can, I think, be no doubt that residence during many generations at a great elevation tends, both directly and indirectly, to induce inherited modifications in the proportions of the body.[172]

Although man may not have been much modified during the latter stages of his existence through the increased or decreased use of parts, the facts now given shew that his liability in this respect has not been lost; and we positively know that the same law holds good with the lower animals. Consequently we may infer, that when at a remote epoch the progenitors of man were in a transitional state, and were changing from quadrupeds into bipeds, natural selection would probably have been greatly aided by the inherited effects of the increased or diminished use of the different parts of the body.

Arrests of Development.—Arrested development differs from arrested growth, as parts in the former state continue to grow whilst still retaining their early condition. Various monstrosities come under this head, and some are known to be occasionally inherited, as a cleft-palate. It will suffice for our purpose to refer to the arrested brain-development of microcephalous idiots, as described in Vogt's great memoir.[173] Their skulls are smaller, and the convolutions of the brain are less complex than in normal men. The frontal sinus, or the projection over the eyebrows, is largely developed, and the jaws are prognathous to an *"effrayant"* degree; so that these idiots somewhat resemble the lower types of mankind. Their intelligence and most of their mental faculties are extremely feeble. They cannot acquire the power of speech, and are wholly incapable of prolonged attention, but are much given to imitation. They are strong and remarkably active, continually gamboling and jumping about, and making grimaces. They often ascend stairs on all-fours; and are curiously fond of climbing up furniture or trees. We are thus reminded of the delight shewn by almost all boys in climbing trees; and this again reminds us how lambs and kids, originally alpine animals, delight to frisk on any hillock, however small.

Reversion.—Many of the cases to be here given might have been introduced under the last heading. Whenever a structure is arrested in its development, but still continues growing until it closely resembles a corresponding structure in some lower and adult member of the same group, we may in one sense consider it as a case of reversion. The lower members in a group give us some idea how the common progenitor of the group was probably constructed; and it is hardly credible that a part arrested at an early phase of embryonic development should be enabled to continue growing so as ultimately to perform its proper function, unless it had acquired this power of continued growth during some earlier state of existence, when the present exceptional or arrested structure was normal. The simple brain of a macrocephalous idiot, in as far as it resembles that of an ape, may in this sense be said to offer a case of reversion. There are other cases which come more strictly under our present heading of reversion. Certain structures, regularly occurring in the lower members of the group to which man belongs, occasionally make their appearance in him, though not found in the normal human embryo; or, if present in the normal human embryo, they become developed in an abnormal manner, though this manner of development is proper to the lower members of the same group. These remarks will be rendered clearer by the following illustrations.

In various mammals the uterus graduates from a double organ with two distinct orifices and two passages, as in the marsupials, into a single organ, showing no signs of doubleness except a slight internal fold, as in the higher apes and man. The rodents exhibit a perfect series of gradations between these two extreme states. In all mammals the uterus is developed from two simple primitive tubes, the inferior portions of which form the cornua; and it is in the words of Dr. Farre "by the coalescence of the two cornua at their lower extremities that the body of the uterus is formed in man; while in those animals in which no middle portion or body exists, the cornua remain un-united. As the development of the uterus proceeds, the two cornua become gradually shorter, until at length they are lost, or, as it were, absorbed into the body of the uterus." The angles of the uterus are still produced into cornua, even so high in the scale as in the lower apes, and their allies the lemurs.

Now in women anomalous cases are not very infrequent, in which the mature uterus is furnished with cornua, or is partially divided into two organs; and such cases, according to Owen, repeat "the grade of concentrative development," attained by certain rodents. Here perhaps we have an instance of a simple arrest of embryonic development, with subsequent growth and perfect functional development, for either side of the partially double uterus is capable of performing the proper office of gestation. In other and rarer cases, two distinct uterine cavities are formed, each having its proper orifice and passage.[174] No such stage is passed through during the ordinary development of the embryo, and it is difficult to believe, though perhaps not impossible, that the two simple, minute, primitive tubes could know how (if such an expression may be used) to grow into two distinct uteri, each with a well-constructed orifice and passage, and each furnished with numerous muscles, nerves, glands and vessels, if they had not formerly passed through a similar course of development, as in the case of existing marsupials. No one will pretend that so perfect a structure as the abnormal double uterus in woman could be the result of mere chance. But the principle of reversion, by which long-lost dormant structures are called back into existence, might serve as the guide for the full development of the organ, even after the lapse of an enormous interval of time.

Professor Canestrini,[175] after discussing the foregoing and various analogous cases, arrives at the same conclusion as that just given. He adduces, as another instance, the malar bone, which, in some of the Quadrumana and other mammals, normally consists of two portions. This is its condition in the two-months-old human fœtus; and thus it sometimes remains, through arrested development, in man when adult, more especially in the lower prognathous races. Hence Canestrini concludes that some ancient progenitor of man must have possessed this bone normally divided into two portions, which subsequently became fused together. In man the frontal bone consists of a single piece, but in the embryo and in children, and in almost all the lower mammals, it consists of two pieces separated by a distinct suture. This suture occasionally persists, more or less distinctly, in man after maturity, and more frequently in ancient than in recent crania, especially as Canestrini has observed in those exhumed from the Drift and belonging to the brachycephalic type. Here again he comes to the same conclusion as in the analogous case of the malar bones. In this and other instances presently to be given, the cause of ancient races approaching the lower animals in certain characters more frequently than do the modern races, appears to be that the latter stand at a somewhat greater distance in the long line of descent from their early semi-human progenitors.

Various other anomalies in man, more or less analogous with the foregoing, have been advanced by different authors[176] as cases of reversion; but these seem not a little doubtful, for we have to descend extremely low in the mammalian series before we find such structures normally present.[177]

In man the canine teeth are perfectly efficient instruments for mastication. But their true canine character, as Owen[178] remarks, "is indicated by the conical form of the crown, which terminates in an obtuse point, is convex outward and flat or sub-concave within, at the base of which surface there is a feeble prominence. The conical form is best expressed in the Melanian races, especially the Australian. The canine is more deeply implanted, and by a stronger fang than the incisors." Nevertheless this tooth no longer serves man as a special weapon for tearing his enemies or prey; it may, therefore, as far as its proper function is concerned, be considered as rudimentary. In every large collection of human skulls some may be found, as Häckel[179] observes, with the canine teeth projecting considerably beyond the others in the same manner, but in a less degree, as in the anthropomorphous apes. In these cases, open spaces between the teeth in the one jaw are left for the reception of the canines belonging to the opposite jaw. An interspace of this kind in a Kaffir skull, figured by Wagner, is surprisingly wide.[180] Considering how few ancient skulls have been examined in comparison with recent skulls, it is an interesting fact that in at least three cases the canines project largely; and in the Naulette jaw they are spoken of as enormous.[181]

The males alone of the anthropomorphous apes have their canines fully developed; but in the female gorilla, and in a less degree in the female orang, these teeth project considerably beyond the others; therefore the fact that women sometimes have, as I have been assured, considerably projecting canines, is no serious objection to the belief that their occasional great development in man is a case of reversion to an ape-like progenitor. He who rejects with scorn the belief that the shape of his own canines, and their occasional great development in other men, are due to our early progenitors having been provided with these formidable weapons, will probably reveal by sneering the line of his descent. For though he no longer intends, nor has the power, to use these teeth as weapons, he will unconsciously retract his "snarling muscles" (thus named by Sir C. Bell)[182] so as to expose them ready for action, like a dog prepared to fight.

Many muscles are occasionally developed in man, which are proper to the Quadrumana or other mammals. Professor Vlacovich[183] examined forty male subjects, and found a muscle, called by him the ischio-pubic, in nineteen of them; in three others there was a ligament which represented this muscle; and in the remaining eighteen no trace of it. Out of thirty female subjects this muscle was developed on both sides in only two, but in three others the rudimentary ligament was present. This muscle, therefore, appears to be much more common in the male than in the female sex; and on the principle of the descent of man from some lower form, its presence can be understood; for it has been detected in several of the lower animals, and in all of these it serves exclusively to aid the male in the act of reproduction.

Mr. J. Wood, in his valuable series of papers,[184] has minutely described a vast number of muscular variations in man, which resemble normal structures in the lower animals. Looking only to the muscles which closely resemble those regularly present in our nearest allies, the Quadrumana, they are too numerous to be here even specified. In a single male subject, having a strong bodily frame and well-formed skull, no less than seven muscular variations were observed, all of which plainly represented muscles proper to various kinds of apes. This man, for instance, had on both sides of his neck a true and powerful *"levator claviculæ,"* such as is found in all kinds of apes, and which is said to occur in about one out of sixty human subjects.[185] Again, this man had "a special abductor of the metatarsal bone of the fifth digit, such as Professor Huxley and Mr. Flower have shewn to exist uniformly in the higher and lower apes." The hands and arms of man are eminently characteristic structures, but their muscles are extremely liable to vary, so as to resemble the corresponding muscles in the lower animals.[186] Such resemblances are either complete and perfect or imperfect, yet in this latter case manifestly of a transitional nature. Certain variations are more common in man, and others in woman, without our being able to assign any reason. Mr. Wood, after describing numerous cases, makes the following pregnant remark: "Notable departures from the ordinary type of the muscular structures run in grooves or directions, which must be taken to indicate some unknown factor, of much importance to a comprehensive knowledge of general and scientific anatomy."[187]

That this unknown factor is reversion to a former state of existence may be admitted as in the highest degree probable. It is quite incredible that a man should through mere accident abnormally resemble, in no less than seven of his muscles, certain apes, if there had been no genetic connection between them. On the other hand, if man is descended from some ape-like creature, no valid reason can be assigned why certain muscles should not suddenly reappear after an interval of many thousand generations, in the same manner as with horses, asses, and mules, dark-coloured stripes suddenly reappear on the legs and shoulders, after an interval of hundreds, or more probably thousands, of generations.

These various cases of reversion are so closely related to those of rudimentary organs given in the first chapter, that many of them might have been indifferently introduced in either chapter. Thus a human uterus furnished with cornua may be said to represent in a rudimentary condition the same organ in its normal state in certain mammals. Some parts which are rudimental in man, as the os coccyx in both sexes and the mammæ in the male sex, are always present; whilst others, such as the supracondyloid foramen, only occasionally appear, and therefore might have been introduced under the head of reversion. These several reversionary, as well as the strictly rudimentary, structures reveal the descent of man from some lower form in an unmistakeable manner.

Correlated Variation.—In man, as in the lower animals, many structures are so intimately related, that when one part varies so does another, without our being able, in most cases, to assign any reason. We cannot say whether the one part governs the other, or whether both are governed by some earlier developed part. Various monstrosities, as I. Geoffroy repeatedly insists, are thus intimately connected. Homologous structures are particularly liable to change together, as we see on the opposite sides of the body, and in the upper and lower extremities. Meckel long ago remarked that when the muscles of the arm depart from their proper type, they almost always imitate those of the leg; and so conversely with the muscles of the legs. The organs of sight and hearing, the teeth and hair, the colour of the skin and hair, colour and constitution, are more or less correlated.[188] Professor Schaaffhausen first drew attention to the relation apparently existing between a muscular frame and strongly-pronounced supra-orbital ridges, which are so characteristic of the lower races of man.

Besides the variations which can be grouped with more or less probability under the foregoing heads, there is a large class of variations which may be provisionally called spontaneous, for they appear, owing to our ignorance, to arise without any exciting cause. It can, however, be shewn that such variations, whether consisting of slight individual differences, or of strongly-marked and abrupt deviations of structure, depend much more on the constitution of the organism than on the nature of the conditions to which it has been subjected.[189]

Rate of Increase.—Civilised populations have been known under favourable conditions, as in the United States, to double their number in twenty-five years; and according to a calculation by Euler, this might occur in a little over twelve years.[190] At the former rate the present population of the United

States, namely, thirty millions, would in 657 years cover the whole terraqueous globe so thickly, that four men would have to stand on each square yard of surface. The primary or fundamental check to the continued increase of man is the difficulty of gaining subsistence and of living in comfort. We may infer that this is the case from what we see, for instance, in the United States, where subsistence is easy and there is plenty of room. If such means were suddenly doubled in Great Britain, our number would be quickly doubled. With civilised nations the above primary check acts chiefly by restraining marriages. The greater death-rate of infants in the poorest classes is also very important; as well as the greater mortality at all ages, and from various diseases, of the inhabitants of crowded and miserable houses. The effects of severe epidemics and wars are soon counterbalanced, and more than counterbalanced, in nations placed under favourable conditions. Emigration also comes in aid as a temporary check, but not to any great extent with the extremely poor classes.

There is reason to suspect, as Malthus has remarked, that the reproductive power is actually less in barbarous than in civilised races. We know nothing positively on this head, for with savages no census has been taken; but from the concurrent testimony of missionaries, and of others who have long resided with such people, it appears that their families are usually small, and large ones rare. This may be partly accounted for, as it is believed, by the women suckling their infants for a prolonged period; but it is highly probable that savages, who often suffer much hardship, and who do not obtain so much nutritious food as civilised men, would be actually less prolific. I have shewn in a former work,[191] that all our domesticated quadrupeds and birds, and all our cultivated plants, are more fertile than the corresponding species in a state of nature. It is no valid objection to this conclusion that animals suddenly supplied with an excess of food, or when rendered very fat, and that most plants when suddenly removed from very poor to very rich soil, are rendered more or less sterile. We might, therefore, expect that civilised men, who in one sense are highly domesticated, would be more prolific than wild men. It is also probable that the increased fertility of civilised nations would become, as with our domestic animals, an inherited character: it is at least known that with mankind a tendency to produce twins runs in families.[192]

Notwithstanding that savages appear to be less prolific than civilised people, they would no doubt rapidly increase if their numbers were not by some means rigidly kept down. The Santali, or hill-tribes of India, have recently afforded a good illustration of this fact; for they have increased, as shewn by Mr. Hunter,[193] at an extraordinary rate since vaccination has been introduced, other pestilences mitigated, and war sternly repressed.

This increase, however, would not have been possible had not these rude people spread into the adjoining districts and worked for hire. Savages almost always marry; yet there is some prudential restraint, for they do not commonly marry at the earliest possible age. The young men are often required to show that they can support a wife, and they generally have first to earn the price with which to purchase her from her parents. With savages the difficulty of obtaining subsistence occasionally limits their number in a much more direct manner than with civilised people, for all tribes periodically suffer from severe famines. At such times savages are forced to devour much bad food, and their health can hardly fail to be injured. Many accounts have been published of their protruding stomachs and emaciated limbs after and during famines. They are then, also, compelled to wander much about, and their infants, as I was assured in Australia, perish in large numbers. As famines are periodical, depending chiefly on extreme seasons, all tribes must fluctuate in number. They cannot steadily and regularly increase, as there is no artificial increase in the supply of food. Savages when hardly pressed encroach on each other's territories, and war is the result; but they are indeed almost always at war with their neighbours. They are liable to many accidents on land and water in their search for food; and in some countries they must suffer much from the larger beasts of prey. Even in India, districts have been depopulated by the ravages of tigers.

Malthus has discussed these several checks, but he does not lay stress enough on what is probably the most important of all, namely infanticide, especially of female infants, and the habit of procuring abortion. These practices now prevail in many quarters of the world, and infanticide seems formerly to have prevailed, as Mr. M'Lennan[194] has shewn, on a still more extensive scale. These practices appear to have originated in savages recognising the difficulty, or rather the impossibility of supporting all the infants that are born. Licentiousness may also be added to the foregoing checks; but this does not follow from failing means of subsistence; though there is reason to believe that in some cases (as in Japan) it has been intentionally encouraged as a means of keeping down the population.

If we look back to an extremely remote epoch, before man had arrived at the dignity of manhood, he would have been guided more by instinct and less by reason than are savages at the present time. Our early semi-human progenitors would not have practised infanticide, for the instincts of the lower animals are never so perverted as to lead them regularly to destroy their own offspring. There would have been no prudential restraint from marriage, and the sexes would have freely united at an early age. Hence the progenitors of man would have tended to increase rapidly, but checks of some kind, either periodical or constant, must have kept down their numbers, even more severely than with existing savages. What the precise nature of these checks may have been, we cannot say, any more than with most other animals. We know that horses and cattle, which are not highly prolific animals, when first turned loose in South America, increased at an enormous rate. The slowest breeder of all known animals, namely the elephant, would in a few thousand years stock the whole world. The increase of every species of monkey must be checked by some means; but not, as Brehm remarks, by the attacks of beasts of prey. No one will assume that the actual power of reproduction in the wild horses and cattle of America, was at first in any sensible degree increased; or that, as each district became fully stocked, this same power was diminished. No doubt in this case and in all others, many checks concur, and different checks under different circumstances; periodical dearths, depending on unfavourable seasons, being probably the most important of all. So it will have been with the early progenitors of man.

Natural Selection.—We have now seen that man is variable in body and mind; and that the variations are induced, either directly or indirectly, by the same general causes, and obey the same general laws, as with the lower animals. Man has spread widely over the face of the earth, and must have been exposed, during his incessant migrations,[195] to the most diversified conditions. The inhabitants of Tierra del Fuego, the Cape of Good Hope, and Tasmania in the one hemisphere, and of the Arctic regions in the other, must have passed through many climates and changed their habits many times, before they reached their present homes.[196] The early progenitors of man must also have tended, like all other animals, to have increased beyond their means of subsistence; they must therefore occasionally have been exposed to a struggle for existence, and consequently to the rigid law of natural selection. Beneficial variations of all kinds will thus, either occasionally or habitually, have been preserved, and injurious ones eliminated. I do not refer to strongly-marked deviations of structure, which occur only at long intervals of time, but to mere individual differences. We know, for instance, that the muscles of our hands and feet, which determine our powers of movement, are liable, like those of the lower animals,[197] to incessant variability. If then the ape-like progenitors of man which inhabited any district, especially one undergoing some change in its conditions, were divided into two equal bodies, the one half which included all the individuals best adapted by their powers of movement for gaining subsistence or for defending themselves, would on an average survive in greater number and procreate more offspring than the other and less well endowed half.

Man in the rudest state in which he now exists is the most dominant animal that has ever appeared on the earth. He has spread more widely than any other highly organised form; and all others have yielded before him. He manifestly owes this immense superiority to his intellectual faculties, his social habits, which lead him to aid and defend his fellows, and to his corporeal structure. The supreme importance of these characters has been proved by the final arbitrament of the battle for life. Through his powers of intellect, articulate language has been evolved; and on this his wonderful advancement has mainly depended. He has invented and is able to use various weapons, tools, traps, &c., with which he defends himself, kills or catches prey, and otherwise obtains food. He has made rafts or canoes on which to fish or cross over to neighbouring fertile islands. He has discovered the art of making fire, by which hard and stringy roots can be rendered digestible, and poisonous roots or herbs innocuous. This last discovery, probably the greatest, excepting language, ever made by man, dates from before the dawn of history. These several inventions, by which man in the rudest state has become so pre-eminent, are the direct result of the development of his powers of observation, memory, curiosity, imagination, and reason. I cannot, therefore, understand how it is that Mr. Wallace[198] maintains, that "natural selection could only have endowed the savage with a brain a little superior to that of an ape."

Although the intellectual powers and social habits of man are of paramount importance to him, we must not underrate the importance of his bodily structure, to which subject the remainder of this chapter will be devoted. The development of the intellectual and social or moral faculties will be discussed in the following chapter.

Even to hammer with precision is no easy matter, as every one who has tried to learn carpentry will admit. To throw a stone with as true an aim as can a Fuegian in defending himself, or in killing birds, requires the most consummate perfection in the correlated action of the muscles of the hand, arm, and shoulder, not to mention a fine sense of touch. In throwing a stone or spear, and in many other actions, a man must stand firmly on his feet; and this again demands the perfect coadaptation of numerous muscles. To chip a flint into the rudest tool, or to form a barbed spear or hook from a bone, demands the use of a perfect hand; for, as a most capable judge, Mr. Schoolcraft,[199] remarks, the shaping fragments of stone into knives, lances, or arrow-heads, shews "extraordinary ability and long practice." We have evidence of this in primeval men having practised a division of labour; each man did not manufacture his own flint tools or rude pottery; but certain individuals appear to have devoted themselves to such work, no doubt receiving in exchange the produce of the chase. Archæologists are convinced that an enormous interval of time elapsed before our ancestors thought of grinding chipped flints into smooth tools. A man-like animal who possessed a hand and arm sufficiently perfect to throw a stone with precision or to form a flint into a rude tool, could, it can hardly be doubted, with sufficient practice make almost anything, as far as mechanical skill alone is concerned, which a civilised man can make. The structure of the hand in this respect may be compared with that of the vocal organs, which in the apes are used for uttering various signal-cries, or, as in one species, musical cadences; but in man closely similar vocal organs have become adapted through the inherited effects of use for the utterance of articulate language.

Turning now to the nearest allies of man, and therefore to the best representatives of our early progenitors, we find that the hands in the Quadrumana are constructed on the same general pattern as in us, but are far less perfectly adapted for diversified uses. Their hands do not serve so well as the feet of a dog for locomotion; as may be seen in those monkeys which walk on the outer margins of the palms, or on the backs of their bent fingers, as in the chimpanzee and orang.[200] Their hands, however, are admirably adapted for climbing trees. Monkeys seize thin branches or ropes, with the thumb on one side and the fingers and palm on the other side, in the same manner as we do. They can thus also carry rather large objects, such as the neck of a bottle, to their mouths. Baboons turn over stones and scratch up roots with their hands. They seize nuts, insects, or other small objects with the thumb in opposition to the fingers, and no doubt they thus extract eggs and the young from the nests of birds. American monkeys beat the wild oranges on the branches until the rind is cracked, and then tear it off with the fingers of the two hands. Other monkeys open mussel-shells with the two thumbs. With their fingers they pull out thorns and burs, and hunt for each other's parasites. In a state of nature they break open hard fruits with the aid of stones. They roll down stones or throw them at their enemies; nevertheless, they perform these various actions clumsily, and they are quite unable, as I have myself seen, to throw a stone with precision.

It seems to me far from true that because "objects are grasped clumsily" by monkeys, "a much less specialised organ of prehension" would have served them[201] as well as their present hands. On the contrary, I see no reason to doubt that a more perfectly constructed hand would have been an advantage to them, provided, and it is important to note this, that their hands had not thus been rendered less well adapted for climbing trees. We may suspect that a perfect hand would have been disadvantageous for climbing; as the most arboreal monkeys in the world, namely Ateles in America and Hylobates in Asia, either have their thumbs much reduced in size and even rudimentary, or their fingers partially coherent, so that their hands are converted into mere grasping-hooks.[202]

As soon as some ancient member in the great series of the Primates came, owing to a change in its manner of procuring subsistence, or to a change in the conditions of its native country, to live somewhat less on trees and more on the ground, its manner of progression would have been modified; and in this case it would have had to become either more strictly quadrupedal or bipedal. Baboons frequent hilly and rocky districts, and only from necessity climb up high trees;[203] and they have acquired almost the gait of a dog. Man alone has become a biped; and we can, I think, partly see how he has come to assume his erect attitude, which forms one of the most conspicuous differences between him and his nearest allies. Man could not have attained his present dominant position in the world without the use of his hands which are so admirably adapted to act in obedience to his will. As Sir C. Bell[204] insists "the hand supplies all instruments, and by its correspondence with the intellect gives him universal dominion." But the hands and arms could hardly have become perfect enough to have manufactured weapons, or to have hurled stones and spears with a true aim, as long as they were habitually used for locomotion and for supporting the whole weight of the body, or as long as they were especially well adapted, as previously remarked, for climbing trees. Such rough treatment would also have blunted the sense of touch, on which their delicate use largely depends. From these causes alone it would have been an advantage to man to have become a biped; but for many actions it is almost necessary that both arms and the whole upper part of the body should be free; and he must for this end stand firmly on his feet. To gain this great advantage, the feet have been rendered flat, and the great toe peculiarly modified, though this has entailed the loss of the power of prehension. It accords with the principle of the division of physiological labour, which prevails throughout the animal kingdom, that as the hands became perfected for prehension, the feet should have become perfected for support and locomotion. With some savages, however, the foot has not altogether lost its prehensile power, as shewn by their manner of climbing trees and of using them in other ways.[205]

If it be an advantage to man to have his hands and arms free and to stand firmly on his feet, of which there can be no doubt from his pre-eminent success in the battle of life, then I can see no reason why it should not have been advantageous to the progenitors of man to have become more and more erect or bipedal. They would thus have been better able to have defended themselves with stones or clubs, or to have attacked their prey, or otherwise obtained food. The best constructed individuals would in the long run have succeeded best, and have survived in larger numbers. If the gorilla and a few allied forms had become extinct, it might have been argued with great force and apparent truth, that an animal could not have been gradually converted from a quadruped into a biped; as all the individuals in an intermediate condition would have been miserably ill-fitted for progression. But we know (and this is well worthy of reflection) that several kinds of apes are now actually in this intermediate condition; and no one doubts that they are on the whole well adapted for their conditions of life. Thus the gorilla runs with a sidelong shambling gait, but more commonly progresses by resting on its bent hands. The long-armed apes occasionally use their arms like crutches, swinging their bodies forward between them, and some kinds of Hylobates, without having been taught, can walk or run upright with tolerable quickness; yet they move awkwardly, and much less securely than man. We see, in short, with existing monkeys various gradations between a form of progression strictly like that of a quadruped and that of a biped or man.

As the progenitors of man became more and more erect, with their hands and arms more and more modified for prehension and other purposes, with their feet and legs at the same time modified for firm support and progression, endless other changes of structure would have been necessary. The pelvis would have had to be made broader, the spine peculiarly curved and the head fixed in an altered position, and all these changes have been attained by man. Prof. Schaaffhausen[206] maintains that "the powerful mastoid processes of the human skull are the result of his erect position;" and these processes are absent in the orang, chimpanzee, &c., and are smaller in the gorilla than in man. Various other structures might here have been specified, which appear connected with man's erect position. It is very difficult to decide how far all these correlated modifications are the result of natural selection, and how far of the inherited effects of the increased use of certain parts, or of the action of one part on another. No doubt these means of change act and react on each other: thus when certain muscles, and the crests of bone to which they are attached, become enlarged by habitual use, this shews that certain actions are habitually performed and must be serviceable. Hence the individuals which performed them best, would tend to survive in greater numbers.

The free use of the arms and hands, partly the cause and partly the result of man's erect position, appears to have led in an indirect manner to other modifications of structure. The early male progenitors of man were, as previously stated, probably furnished with great canine teeth; but as they gradually acquired the habit of using stones, clubs, or other weapons, for fighting with their enemies, they would have used their jaws and teeth less and less. In this case, the jaws, together with the teeth, would have become reduced in size, as we may feel sure from innumerable analogous cases. In a future chapter we shall meet with a closely-parallel case, in the reduction or complete disappearance of the canine teeth in male ruminants, apparently in relation with the development of their horns; and in horses, in relation with their habit of fighting with their incisor teeth and hoofs.

In the adult male anthropomorphous apes, as Rütimeyer,[207] and others have insisted, it is precisely the effect which the jaw-muscles by their great development have produced on the skull, that causes it to differ so greatly in many respects from that of man, and has given to it "a truly frightful physiognomy." Therefore as the jaws and teeth in the progenitors of man gradually become reduced in size, the adult skull would have presented nearly the same characters which it offers in the young of the anthropomorphous apes, and would thus have come to resemble more nearly that of existing man. A great reduction of the canine teeth in the males would almost certainly, as we shall hereafter see, have affected through inheritance the teeth of the females.

As the various mental faculties were gradually developed, the brain would almost certainly have become larger. No one, I presume, doubts that the large size of the brain in man, relatively to his body, in comparison with that of the gorilla or orang, is closely connected with his higher mental powers. We meet with closely analogous facts with insects, in which the cerebral ganglia are of extraordinary dimensions in ants; these ganglia in all the Hymenoptera being many times larger than in the less intelligent orders, such as beetles.[208] On the other hand, no one supposes that the intellect of any two animals or of any two men can be accurately gauged by the cubic contents of their skulls. It is certain that there may be extraordinary mental activity with an extremely small absolute mass of nervous matter: thus the wonderfully diversified instincts, mental powers, and affections of ants are generally known, yet their cerebral ganglia are not so large as the quarter of a small pin's head. Under this latter point of view, the brain of an ant is one of the most marvellous atoms of matter in the world, perhaps more marvellous than the brain of man.

The belief that there exists in man some close relation between the size of the brain and the development of the intellectual faculties is supported by the comparison of the skulls of savage and civilised races, of ancient and modern people, and by the analogy of the whole vertebrate series. Dr. J. Barnard Davis has proved[209] by many careful measurements, that the mean internal capacity of the skull in Europeans is 92·3 cubic inches; in Americans 87·5; in Asiatics 87·1; and in Australians only 81·9 inches. Professor Broca[210] found that skulls, from graves in Paris of the nineteenth century, were larger than those from vaults of the twelfth century, in the proportion of 1484 to 1426; and Prichard is persuaded that the present inhabitants of Britain have "much more capacious brain-cases" than the ancient inhabitants. Nevertheless it must be admitted that some skulls of very high antiquity, such as the famous one of Neanderthal, are well developed and capacious. With respect to the lower animals, M. E. Lartet,[211] by comparing the crania of tertiary and recent mammals, belonging to the same groups, has come to the remarkable conclusion that the brain is generally larger and the convolutions more complex in the more recent form. On the other hand I have shewn[212] that the brains of domestic rabbits are considerably reduced in bulk, in comparison with those of the wild rabbit or hare; and this may be attributed to their having been closely confined during many generations, so that they have exerted but little their intellect, instincts, senses, and voluntary movements.

The gradually increasing weight of the brain and skull in man must have influenced the development of the supporting spinal column, more especially whilst he was becoming erect. As this change of position was being brought about, the internal pressure of the brain, will, also, have influenced the form of the skull; for many facts shew how easily the skull is thus affected. Ethnologists believe that it is modified by the kind of cradle in which infants sleep. Habitual spasms of the muscles and a cicatrix from a severe burn have permanently modified the facial bones. In young persons whose heads from disease have become fixed either sideways or backwards, one of the eyes has changed its position, and the bones of the skull have been modified; and this apparently results from the brain pressing in a new direction.[213] I have shewn that with long-eared rabbits, even so trifling a cause as the lopping forward of one ear drags forward on that side almost every bone of the skull; so that the bones on the opposite sides no longer strictly correspond. Lastly, if any animal were to increase or diminish much in general size, without any change in its mental powers; or if the mental powers were to be much increased or diminished without any great change in the size of the body; the shape of the skull would almost certainly be altered. I infer this from my observations on domestic rabbits, some kinds of which have become very much larger than the wild animal, whilst others have retained nearly the same size, but in both cases the brain has been much reduced relatively to the size of the body. Now I was at first much surprised by finding that in all these rabbits the skull had become elongated or dolichocephalic; for instance, of two skulls of nearly equal breadth, the one from a wild rabbit and the other from a large domestic kind, the former was only 3·15 and the latter 4·3 inches in length.[214] One of the most marked distinctions in different races of man is that the skull in some is elongated, and in others rounded; and here the explanation suggested by the case of the rabbits may partially hold good; for Welcker finds that short "men incline more to brachycephaly, and tall men to dolichocephaly;"[215] and tall men may be compared with the larger and longer-bodied rabbits, all of which have elongated skulls, or are dolichocephalic.

From these several facts we can to a certain extent understand the means through which the great size and more or less rounded form of the skull has been acquired by man; and these are characters eminently distinctive of him in comparison with the lower animals.

Another most conspicuous difference between man and the lower animals is the nakedness of his skin. Whales and dolphins (Cetacea), dugongs (Sirenia) and the hippopotamus are naked; and this may be advantageous to them for gliding through the water; nor would it be injurious to them from the loss of warmth, as the species which inhabit the colder regions are protected by a thick layer of blubber, serving the same purpose as the fur of seals and otters. Elephants and rhinoceroses are almost hairless; and as certain extinct species which formerly lived under an arctic climate were covered with long wool or hair, it would almost appear as if the existing species of both genera had lost their hairy covering from exposure to heat. This appears the more probable, as the elephants in India which live on elevated and cool districts are more hairy[216] than those on the lowlands. May we then infer that man became divested of hair from having aboriginally inhabited some tropical land? The fact of the hair being chiefly retained in the male sex on the chest and face, and in both sexes at the junction of all four limbs with the trunk, favours this inference, assuming that the hair was lost before man became erect; for the parts which now retain most hair would then have been most protected from the heat of the sun. The crown of the head, however, offers a curious exception, for at all times it must have been one of the most exposed parts, yet it is thickly clothed with hair. In this respect man agrees with the great majority of quadrupeds, which generally have their upper and exposed surfaces more thickly clothed than the lower surface. Nevertheless, the fact that the other members of the order of Primates, to which man belongs, although inhabiting various hot regions, are well clothed with hair, generally thickest on the upper surface,[217] is strongly opposed to the supposition that man became naked through the action of the sun. I am inclined to believe, as we shall see under sexual selection, that man, or rather primarily woman, became divested of hair for ornamental purposes; and according to this belief it is not surprising that man should differ so greatly in hairiness from all his lower brethren, for characters gained through sexual selection often differ in closely-related forms to an extraordinary degree.

According to a popular impression, the absence of a tail is eminently distinctive of man; but as those apes which come nearest to man are destitute of this organ, its disappearance does not especially concern us. Nevertheless it may be well to own that no explanation, as far as I am aware, has ever been given of the loss of the tail by certain apes and man. Its loss, however, is not surprising, for it sometimes differs remarkably in length in species of the same genera: thus in some species of Macacus the tail is longer than the whole body, consisting of twenty-four vertebræ; in others it consists of a scarcely visible stump, containing only three or four vertebræ. In some kinds of baboons there are twenty-five, whilst in the mandrill there are ten very small stunted caudal vertebræ, or, according to Cuvier,[218] sometimes only five. This great diversity in the structure and length of the tail in animals belonging to the same genera, and following nearly the same habits of life, renders it probable that the tail is not of much importance to them; and if so, we might have expected that it would sometimes have become more or less rudimentary, in accordance with what we incessantly see with other structures. The tail almost always tapers towards the end whether it be long or short; and this, I presume, results from the atrophy, through disuse, of the terminal muscles together with their arteries and nerves, leading to the atrophy of the terminal bones. With respect to the os coccyx, which in man and the higher apes manifestly consists of the few basal and tapering segments of an ordinary tail, I have heard it asked how could these have become completely embedded within the body; but there is no difficulty in this respect, for in many monkeys the basal segments of the true tail are thus embedded. For instance, Mr. Murie informs me that in the skeleton of a not full-grown *Macacus inornatus,* he counted nine or ten caudal vertebræ, which altogether were only 1·8 inch in length. Of these the three basal ones appeared to have been embedded; the remainder forming the free part of the tail, which was only one inch in length, and half an inch in diameter. Here, then, the three embedded caudal vertebræ plainly correspond with the four coalesced vertebræ of the human os coccyx.

I HAVE NOW ENDEAVOURED to shew that some of the most distinctive characters of man have in all probability been acquired, either directly, or more commonly indirectly, through natural selection. We should bear in mind that modifications in structure or constitution, which are of no service to an organism in adapting it to its habits of life, to the food which it consumes, or passively to the surrounding conditions, cannot have been thus acquired. We must not, however, be too confident in deciding what modifications are of service to each being: we should remember how little we know about the use of many parts, or what changes in the blood or tissues may serve to fit an organism for a new climate or some new kind of food. Nor must we forget the principle of correlation, by which, as Isidore Geoffroy has shewn in the case of man, many strange deviations of structure are tied together. Independently of correlation, a change in one part often leads through the increased or decreased use of other parts, to other changes of a quite unexpected nature. It is also well to reflect on such facts, as the wonderful growth of galls on plants caused by the poison of an insect, and on the remarkable changes of colour in the plumage of parrots when fed on certain fishes, or inoculated with the poison of toads;[219] for we can thus see that the fluids of the system, if altered for some special purpose, might induce other strange changes. We should especially bear in mind that modifications acquired and continually used during past ages for some useful purpose would probably become firmly fixed and might be long inherited.

Thus a very large yet undefined extension may safely be given to the direct and indirect results of natural selection; but I now admit, after reading the essay by Nägeli on plants, and the remarks by various authors with respect to animals, more especially those recently made by Professor Broca, that in the earlier editions of my 'Origin of Species' I probably attributed too much to the action of natural selection or the survival of the fittest. I have altered the fifth edition of the Origin so as to confine my remarks to adaptive changes of structure. I had not formerly sufficiently considered the existence of many structures which appear to be, as far as we can judge, neither beneficial nor injurious; and this I believe to be one of the greatest oversights as yet detected in my work. I may be permitted to say as some excuse, that I had two distinct objects in view, firstly, to shew that species had not been separately created, and secondly, that natural selection had been the chief agent of change, though largely aided by the inherited effects of habit, and slightly by the direct action of the surrounding conditions. Nevertheless I was not able to annul the influence of my former belief, then widely prevalent, that each species had been purposely created; and this led to my tacitly assuming that every detail of structure, excepting rudiments, was of some special, though unrecognised, service. Any one with this assumption in his mind would naturally extend the action of natural selection, either during past or present times, too far. Some of those who admit the principle of evolution, but reject natural selection, seem to forget, when criticising my book, that I had the above two objects in view; hence if I have erred in giving to natural selection great power, which I am far from admitting, or in having exaggerated its power, which is in itself probable, I have at least, as I hope, done good service in aiding to overthrow the dogma of separate creations.

That all organic beings, including man, present many modifications of structure which are of no service to them at present, nor have been formerly, is, as I can now see, probable. We know not what produces the numberless slight differences between the individuals of each species, for reversion only carries the problem a few steps backwards; but each peculiarity must have had its own efficient cause. If these causes, whatever they may be, were to act more uniformly and energetically during a lengthened period (and no reason can be assigned why this should not sometimes occur), the result would probably be not mere slight individual differences, but well-marked, constant modifications. Modifications which are in no way beneficial cannot have been kept uniform through natural selection, though any which were injurious would have been thus eliminated. Uniformity of character would, however, naturally follow from the assumed uniformity of the exciting causes, and likewise from the free intercrossing of many individuals. The same organism might acquire in this manner during successive periods successive modifications, and these would be transmitted in a nearly uniform state as long as the exciting causes remained the same and there was free intercrossing. With respect to the exciting causes we can only say, as when speaking of so-called spontaneous variations, that they relate much more closely to the constitution of the varying organism, than to the nature of the conditions to which it has been subjected.

Conclusion.—In this chapter we have seen that as man at the present day is liable, like every other animal, to multiform individual differences or slight variations, so no doubt were the early progenitors of man; the variations being then as now induced by the same general causes, and governed by the same general and complex laws. As all animals tend to multiply beyond their means of subsistence, so it must have been with the progenitors of man; and this will inevitably have led to a struggle for existence and to natural selection. This latter process will have been greatly aided by the inherited effects of the increased use of parts; these two processes incessantly reacting on each other. It appears, also, as we shall hereafter see, that various unimportant characters have been acquired by man through sexual selection. An unexplained residuum of change, perhaps a large one, must be left to the assumed uniform action of those unknown agencies, which occasionally induce strongly-marked and abrupt deviations of structure in our domestic productions.

Judging from the habits of savages and of the greater number of the Quadrumana, primeval men, and even the ape-like progenitors of man, probably lived in society. With strictly social animals, natural selection sometimes acts indirectly on the individual, through the preservation of variations which are beneficial only to the community. A community including a large number of well-endowed individuals increases in number and is victorious over other and less well-endowed communities; although each separate member may gain no advantage over the other members of the same community. With associated insects many remarkable structures, which are of little or no service to the individual or its own offspring, such as the pollen-collecting apparatus, or the sting of the worker-bee, or the great jaws of soldier-ants, have been thus acquired. With the higher social animals, I am not aware that any structure has been modified solely for the good of the community, though some are of secondary service to it. For instance, the horns of ruminants and the great canine teeth of baboons appear to have been acquired by the males as weapons for sexual strife, but they are used in defence of the herd or troop. In regard to certain mental faculties the case, as we shall see in the following chapter, is wholly different; for these faculties have been chiefly, or even exclusively, gained for the benefit of the community; the individuals composing the community being at the same time indirectly benefited.

It has often been objected to such views as the foregoing, that man is one of the most helpless and defenceless creatures in the world; and that during his early and less well-developed condition he would have been still more helpless. The Duke of Argyll, for instance, insists[220] that "the human frame has diverged from the structure of brutes, in the direction of greater physical helplessness and weakness. That is to say, it is a divergence which of all others it is most impossible to ascribe to mere natural selection." He adduces the naked and unprotected state of the body, the absence of great teeth or claws for defence, the little strength of man, his small speed in running, and his slight power of smell, by which to discover food or to avoid danger. To these deficiencies there might have been added the still more serious loss of the power of quickly climbing trees, so as to escape from enemies. Seeing that the unclothed Fuegians can exist under their wretched climate, the loss of hair would not have been a great injury to primeval man, if he inhabited a warm country. When we compare defenceless man with the apes, many of which are provided with formidable canine teeth, we must remember that these in their fully-developed condition are possessed by the males alone, being chiefly used by them for fighting with their rivals; yet the females which are not thus provided, are able to survive.

In regard to bodily size or strength, we do not know whether man is descended from some comparatively small species, like the chimpanzee, or from one as powerful as the gorilla; and, therefore, we cannot say whether man has become larger and stronger, or smaller and weaker, in comparison with his progenitors. We should, however, bear in mind that an animal possessing great size, strength, and ferocity, and which, like the gorilla, could defend itself from all enemies, would probably, though not necessarily, have failed to become social; and this would most effectually have checked the acquirement by man of his higher mental qualities, such as sympathy and the love of his fellow-creatures. Hence it might have been an immense advantage to man to have sprung from some comparatively weak creature.

The slight corporeal strength of man, his little speed, his want of natural weapons, &c., are more than counterbalanced, firstly by his intellectual powers, through which he has, whilst still remaining in a barbarous state, formed for himself weapons, tools, &c., and secondly by his social qualities which lead him to give aid to his fellow-men and to receive it in return. No country in the world abounds in a greater degree with dangerous beasts than Southern Africa; no country presents more fearful physical hardships than the Arctic regions; yet one of the puniest races, namely, the Bushmen, maintain themselves in Southern Africa, as do the dwarfed Esquimaux in the Arctic regions. The early progenitors of man were, no doubt, inferior in intellect, and probably in social disposition, to the lowest existing savages; but it is quite conceivable that they might have existed, or even flourished, if, whilst they gradually lost their brute-like powers, such as climbing trees, &c., they at the same time advanced in intellect. But granting that the progenitors of man were far more helpless and defenceless than any existing savages, if they had inhabited some warm continent or large island, such as Australia or New Guinea, or Borneo (the latter island being now tenanted by the orang), they would not have been exposed to any special danger. In an area as large as one of these islands, the competition between tribe and tribe would have been sufficient, under favourable conditions, to have raised man, through the survival of the fittest, combined with the inherited effects of habit, to his present high position in the organic scale.

On the Development of the Intellectual and Moral Faculties during Primeval and Civilised Times

THE ADVANCEMENT OF THE INTELLECTUAL POWERS THROUGH NATURAL SELECTION • IMPORTANCE OF IMITATION • SOCIAL AND MORAL FACULTIES • THEIR DEVELOPMENT WITHIN THE LIMITS OF THE SAME TRIBE • NATURAL SELECTION AS AFFECTING CIVILISED NATIONS • EVIDENCE THAT CIVILISED NATIONS WERE ONCE BARBAROUS

THE SUBJECTS TO BE DISCUSSED in this chapter are of the highest interest, but are treated by me in a most imperfect and fragmentary manner. Mr. Wallace, in an admirable paper before referred to,[221] argues that man after he had partially acquired those intellectual and moral faculties which distinguish him from the lower animals, would have been but little liable to have had his bodily structure modified through natural selection or any other means. For man is enabled through his mental faculties "to keep with an unchanged body in harmony with the changing universe." He has great power of adapting his habits to new conditions of life. He invents weapons, tools and various stratagems, by which he procures food and defends himself. When he migrates into a colder climate he uses clothes, builds sheds, and makes fires; and, by the aid of fire, cooks food otherwise indigestible. He aids his fellow-men in many ways, and anticipates future events. Even at a remote period he practised some subdivision of labour.

The lower animals, on the other hand, must have their bodily structure modified in order to survive under greatly changed conditions. They must be rendered stronger, or acquire more effective teeth or claws, in order to defend themselves from new enemies; or they must be reduced in size so as to escape detection and danger. When they migrate into a colder climate they must become clothed with thicker fur, or have their constitutions altered. If they fail to be thus modified, they will cease to exist.

The case, however, is widely different, as Mr. Wallace has with justice insisted, in relation to the intellectual and moral faculties of man. These faculties are variable; and we have every reason to believe that the variations tend to be inherited. Therefore, if they were formerly of high importance to primeval man and to his ape-like progenitors, they would have been perfected or advanced through natural selection. Of the high importance of the intellectual faculties there can be no doubt, for man mainly owes to them his pre-eminent position in the world. We can see that, in the rudest state of society, the individuals who were the most sagacious, who invented and used the best weapons or traps, and who were best able to defend themselves, would rear the greatest number of offspring. The tribes which included the largest number of men thus endowed would increase in number and supplant other tribes. Numbers depend primarily on the means of subsistence, and this, partly on the physical nature of the country, but in a much higher degree on the arts which are there practised. As a tribe increases and is victorious, it is often still further increased by the absorption of other tribes.[222] The stature and strength of the men of a tribe are likewise of some importance for its success, and these depend in part on the nature and amount of the food which can be obtained. In Europe the men of the Bronze period were supplanted by a more powerful and, judging from their sword-handles, larger-handed race;[223] but their success was probably due in a much higher degree to their superiority in the arts.

All that we know about savages, or may infer from their traditions and from old monuments, the history of which is quite forgotten by the present inhabitants, shew that from the remotest times successful tribes have supplanted other tribes. Relics of extinct or forgotten tribes have been discovered throughout the civilised regions of the earth, on the wild plains of America, and on the isolated islands in the Pacific Ocean. At the present day civilised nations are everywhere supplanting barbarous nations, excepting where the climate opposes a deadly barrier; and they succeed mainly, though not exclusively, through their arts, which are the products of the intellect. It is, therefore, highly probable that with mankind the intellectual faculties have been gradually perfected through natural selection; and this conclusion is sufficient for our purpose. Undoubtedly it would have been very interesting to have traced the development of each separate faculty from the state in which it exists in the lower animals to that in which it exists in man; but neither my ability nor knowledge permit the attempt.

It deserves notice that as soon as the progenitors of man became social (and this probably occurred at a very early period), the advancement of the intellectual faculties will have been aided and modified in an important manner, of which we see only traces in the lower animals, namely, through the principle of imitation, together with reason and experience. Apes are much given to imitation, as are the lowest savages; and the simple fact previously referred to, that after a time no animal can be caught in the same place by the same sort of trap, shews that animals learn by experience, and imitate each others' caution. Now, if some one man in a tribe, more sagacious than the others, invented a new snare or weapon, or other means of attack or defence, the plainest self-interest, without the assistance of much reasoning power, would prompt the other members to imitate him; and all would thus profit. The habitual practice of each new art must likewise in some slight degree strengthen the intellect. If the new invention were an important one, the tribe would increase in number, spread, and supplant other tribes. In a tribe thus rendered more numerous there would always be a rather better chance of the birth of other superior and inventive members. If such men left children to inherit their mental superiority, the chance of the birth of still more ingenious members would be somewhat better, and in a very small tribe decidedly better. Even if they left no children, the tribe would still include their blood-relations; and it has been ascertained by agriculturists[224] that by preserving and breeding from the family of an animal, which when slaughtered was found to be valuable, the desired character has been obtained.

TURNING NOW TO THE SOCIAL and moral faculties. In order that primeval men, or the ape-like progenitors of man, should have become social, they must have acquired the same instinctive feelings which impel other animals to live in a body; and they no doubt exhibited the same general disposition. They would have felt uneasy when separated from their comrades, for whom they would have felt some degree of love; they would have warned each other of danger, and have given mutual aid in attack or defence. All this implies some degree of sympathy, fidelity, and courage. Such social qualities, the paramount importance of which to the lower animals is disputed by no one, were no doubt acquired by the progenitors of man in a similar manner, namely, through natural selection, aided by inherited habit. When two tribes of primeval man, living in the same country, came into competition, if the one tribe included (other circumstances being equal) a greater number of courageous, sympathetic, and faithful members, who were always ready to warn each other of danger, to aid and defend each other, this tribe would without doubt succeed best and conquer the other. Let it be borne in mind how all-important, in the never-ceasing wars of savages, fidelity and courage must be. The advantage which disciplined soldiers have over undisciplined hordes follows chiefly from the confidence which each man feels in his comrades. Obedience, as Mr. Bagehot has well shewn,[225] is of the highest value, for any form of government is better than none. Selfish and contentious people will not cohere, and without coherence nothing can be effected. A tribe possessing the above qualities in a high degree would spread and be victorious over other tribes; but in the course of time it would, judging from all past history, be in its turn overcome by some other and still more highly endowed tribe. Thus the social and moral qualities would tend slowly to advance and be diffused throughout the world.

But it may be asked, how within the limits of the same tribe did a large number of members first become endowed with these social and moral qualities, and how was the standard of excellence raised? It is extremely doubtful whether the offspring of the more sympathetic and benevolent parents, or of those which were the most faithful to their comrades, would be reared in greater number than the children of selfish and treacherous parents of the same tribe. He who was ready to sacrifice his life, as many a savage has been, rather than betray his comrades, would often leave no offspring to inherit his noble nature. The bravest men, who were always willing to come to the front in war, and who freely risked their lives for others, would on an average perish in larger number than other men. Therefore it seems scarcely possible (bearing in mind that we are not here speaking of one tribe being victorious over another) that the number of men gifted with such virtues, or that the standard of their excellence, could be increased through natural selection, that is, by the survival of the fittest.

Although the circumstances which lead to an increase in the number of men thus endowed within the same tribe are too complex to be clearly followed out, we can trace some of the probable steps. In the first place, as the reasoning powers and foresight of the members became improved, each man would soon learn from experience that if he aided his fellow-men, he would commonly receive aid in return. From this low motive he might acquire the habit of aiding his fellows; and the habit of performing benevolent actions certainly strengthens the feeling of sympathy, which gives the first impulse to benevolent actions. Habits, moreover, followed during many generations probably tend to be inherited.

But there is another and much more powerful stimulus to the development of the social virtues, namely, the praise and the blame of our fellow-men. The love of approbation and the dread of infamy, as well as the bestowal of praise or blame, are primarily due, as we have seen in the third chapter, to the instinct of sympathy; and this instinct no doubt was originally acquired, like all the other social instincts, through natural selection. At how early a period the progenitors of man, in the course of their development, became capable of feeling and being impelled by the praise or blame of their fellow-creatures, we cannot, of course, say. But it appears that even dogs appreciate encouragement, praise, and blame. The rudest savages feel the sentiment of glory, as they clearly show by preserving the trophies of their prowess, by their habit of excessive boasting, and even by the extreme care which they take of their personal appearance and decorations; for unless they regarded the opinion of their comrades, such habits would be senseless.

They certainly feel shame at the breach of some of their lesser rules; but how far they experience remorse is doubtful. I was at first surprised that I could not recollect any recorded instances of this feeling in savages; and Sir J. Lubbock[226] states that he knows of none. But if we banish from our minds all cases given in novels and plays and in death-bed confessions made to priests, I doubt whether many of us have actually witnessed remorse; though we may have often seen shame and contrition for smaller offences. Remorse is a deeply hidden feeling. It is incredible that a savage, who will sacrifice his life rather than betray his tribe, or one who will deliver himself up as a prisoner rather than break his parole,[227] would not feel remorse in his inmost soul, though he might conceal it, if he had failed in a duty which he held sacred.

We may therefore conclude that primeval man, at a very remote period, would have been influenced by the praise and blame of his fellows. It is obvious, that the members of the same tribe would approve of conduct which appeared to them to be for the general good, and would reprobate that which appeared evil. To do good unto others—to do unto others as ye would they should do unto you,—is the foundation-stone of morality. It is, therefore, hardly possible to exaggerate the importance during rude times of the love of praise and the dread of blame. A man who was not impelled by any deep, instinctive feeling, to sacrifice his life for the good of others, yet was roused to such actions by a sense of glory, would by his example excite the same wish for glory in other men, and would strengthen by exercise the noble feeling of admiration. He might thus do far more good to his tribe than by begetting offspring with a tendency to inherit his own high character.

With increased experience and reason, man perceives the more remote consequences of his actions, and the self-regarding virtues, such as temperance, chastity, &c., which during early times are, as we have before seen, utterly disregarded, come to be highly esteemed or even held sacred. I need not, however, repeat what I have said on this head in the third chapter. Ultimately a highly complex sentiment, having its first origin in the social instincts, largely guided by the approbation of our fellow-men, ruled by reason, self-interest, and in later times by deep religious feelings, confirmed by instruction and habit, all combined, constitute our moral sense or conscience.

It must not be forgotten that although a high standard of morality gives but a slight or no advantage to each individual man and his children over the other men of the same tribe, yet that an advancement in the standard of morality and an increase in the number of well-endowed men will certainly give an immense advantage to one tribe over another. There can be no doubt that a tribe including many members who, from possessing in a high degree the spirit of patriotism, fidelity, obedience, courage, and sympathy, were always ready to give aid to each other and to sacrifice themselves for the common good, would be victorious over most other tribes; and this would be natural selection. At all times throughout the world tribes have supplanted other tribes; and as morality is one element in their success, the standard of morality and the number of well-endowed men will thus everywhere tend to rise and increase.

It is, however, very difficult to form any judgment why one particular tribe and not another has been successful and has risen in the scale of civilisation. Many savages are in the same condition as when first discovered several centuries ago. As Mr. Bagehot has remarked, we are apt to look at progress as the normal rule in human society; but history refutes this. The ancients did not even entertain the idea; nor do the oriental nations at the present day. According to another high authority, Mr. Maine,[228] "the greatest part of mankind has never shewn a particle of desire that its civil institutions should be improved." Progress seems to depend on many concurrent favourable conditions, far too complex to be followed out. But it has often been remarked, that a cool climate from leading to industry and the various arts has been highly favourable, or even indispensable for this end. The Esquimaux, pressed by hard necessity, have succeeded in many ingenious inventions, but their climate has been too severe for continued progress. Nomadic habits, whether over wide plains, or through the dense forests of the tropics, or along the shores of the sea, have in every case been highly detrimental. Whilst observing the barbarous inhabitants of Tierra del Fuego, it struck me that the possession of some property, a fixed abode, and the union of many families under a chief, were the indispensable requisites for civilisation. Such habits almost necessitate the cultivation of the ground; and the first steps in cultivation would probably result, as I have elsewhere shewn,[229] from some such accident as the seeds of a fruit-tree falling on a heap of refuse and producing an unusually fine variety. The problem, however, of the first advance of savages towards civilisation is at present much too difficult to be solved.

Natural Selection as affecting Civilised Nations.—In the last and present chapters I have considered the advancement of man from a former semi-human condition to his present state as a barbarian. But some remarks on the agency of natural selection on civilised nations may be here worth adding. This subject has been ably discussed by Mr. W. R. Greg,[230] and previously by Mr. Wallace and Mr. Galton.[231] Most of my remarks are taken from these three authors. With savages, the weak in body or mind are soon eliminated; and those that survive commonly exhibit a vigorous state of health. We civilised men, on the other hand, do our utmost to check the process of elimination; we build asylums for the imbecile, the maimed, and the sick; we institute poor-laws; and our medical men exert their utmost skill to save the life of every one to the last moment. There is reason to believe that vaccination has preserved thousands, who from a weak constitution would formerly have succumbed to small-pox. Thus the weak members of civilised societies propagate their kind. No one who has attended to the breeding of domestic animals will doubt that this must be highly injurious to the race of man. It is surprising how soon a want of care, or care wrongly directed, leads to the degeneration of a domestic race; but excepting in the case of man himself, hardly any one is so ignorant as to allow his worst animals to breed.

The aid which we feel impelled to give to the helpless is mainly an incidental result of the instinct of sympathy, which was originally acquired as part of the social instincts, but subsequently rendered, in the manner previously indicated, more tender and more widely diffused. Nor could we check our sympathy, if so urged by hard reason, without deterioration in the noblest part of our nature. The surgeon may harden himself whilst performing an operation, for he knows that he is acting for the good of his patient; but if we were intentionally to neglect the weak and helpless, it could only be for a contingent benefit, with a certain and great present evil. Hence we must bear without complaining the undoubtedly bad effects of the weak surviving and propagating their kind; but there appears to be at least one check in steady action, namely the weaker and inferior members of society not marrying so freely as the sound; and this check might be indefinitely increased, though this is more to be hoped for than expected, by the weak in body or mind refraining from marriage.

In all civilised countries man accumulates property and bequeaths it to his children. So that the children in the same country do not by any means start fair in the race for success. But this is far from an unmixed evil; for without the accumulation of capital the arts could not progress; and it is chiefly through their power that the civilised races have extended, and are now everywhere extending, their range, so as to take the place of the lower races. Nor does the moderate accumulation of wealth interfere with the process of selection. When a poor man becomes rich, his children enter trades or professions in which there is struggle enough, so that the able in body and mind succeed best. The presence of a body of well-instructed men, who have not to labour for their daily bread, is important to a degree which cannot be over-estimated; as all high intellectual work is carried on by them, and on such work material progress of all kinds mainly depends, not to mention other and higher advantages. No doubt wealth when very great tends to convert men into useless drones, but their number is never large; and some degree of elimination here occurs, as we daily see rich men, who happen to be fools or profligate, squandering away all their wealth.

Primogeniture with entailed estates is a more direct evil, though it may formerly have been a great advantage by the creation of a dominant class, and any government is better than anarchy. The eldest sons, though they may be weak in body or mind, generally marry, whilst the younger sons, however superior in these respects, do not so generally marry. Nor can worthless eldest sons with entailed estates squander their wealth. But here, as elsewhere, the relations of civilised life are so complex that some compensatory checks intervene. The men who are rich through primogeniture are able to select generation after generation the more beautiful and charming women; and these must generally be healthy in body and active in mind. The evil consequences, such as they may be, of the continued preservation of the same line of descent, without any selection, are checked by men of rank always wishing to increase their wealth and power; and this they effect by marrying heiresses. But the daughters of parents who have produced single children, are themselves, as Mr. Galton has shewn,[232] apt to be sterile; and thus noble families are continually cut off in the direct line, and their wealth flows into some side channel; but unfortunately this channel is not determined by superiority of any kind.

Although civilisation thus checks in many ways the action of natural selection, it apparently favours, by means of improved food and the freedom from occasional hardships, the better development of the body. This may be inferred from civilised men having been found, wherever compared, to be physically stronger than savages. They appear also to have equal powers of endurance, as has been proved in many adventurous expeditions. Even the great luxury of the rich can be but little detrimental; for the expectation of life of our aristocracy, at all ages and of both sexes, is very little inferior to that of healthy English lives in the lower classes.[233]

We will now look to the intellectual faculties alone. If in each grade of society the members were divided into two equal bodies, the one including the intellectually superior and the other the inferior, there can be little doubt that the former would succeed best in all occupations and rear a greater number of children. Even in the lowest walks of life, skill and ability must be of some advantage, though in many occupations, owing to the great division of labour, a very small one. Hence in civilised nations there will be some tendency to an increase both in the number and in the standard of the intellectually able. But I do not wish to assert that this tendency may not be more than counterbalanced in other ways, as by the multiplication of the reckless and improvident; but even to such as these, ability must be some advantage.

It has often been objected to views like the foregoing, that the most eminent men who have ever lived have left no offspring to inherit their great intellect. Mr. Galton says,[234] "I regret I am unable to solve the simple question whether, and how far, men and women who are prodigies of genius are infertile. I have, however, shewn that men of eminence are by no means so." Great lawgivers, the founders of beneficent religions, great philosophers and discoverers in science, aid the progress of mankind in a far higher degree by their works than by leaving a numerous progeny. In the case of corporeal structures, it is the selection of the slightly better-endowed and the elimination of the slightly less well-endowed individuals, and not the preservation of strongly-marked and rare anomalies, that leads to the advancement of a species.[235] So it will be with the intellectual faculties, namely from the some-what more able men in each grade of society succeeding rather better than the less able, and consequently increasing in number, if not otherwise prevented. When in any nation the standard of intellect and the number of intellectual men have increased, we may expect from the law of the deviation from an average, as shewn by Mr. Galton, that prodigies of genius will appear somewhat more frequently than before.

In regard to the moral qualities, some elimination of the worst dispositions is always in progress even in the most civilised nations. Malefactors are executed, or imprisoned for long periods, so that they cannot freely transmit their bad qualities. Melancholic and insane persons are confined, or commit suicide. Violent and quarrelsome men often come to a bloody end. Restless men who will not follow any steady occupation—and this relic of barbarism is a great check to civilisation[236]—emigrate to newly-settled countries, where they prove useful pioneers. Intemperance is so highly destructive, that the expectation of life of the intemperate, at the age, for instance, of thirty, is only 13·8 years; whilst for the rural labourers of England at the same age it is 40·59 years.[237] Profligate women bear few children, and profligate men rarely marry; both suffer from disease. In the breeding of domestic animals, the elimination of those individuals, though few in number, which are in any marked manner inferior, is by no means an unimportant element towards success. This especially holds good with injurious characters which tend to reappear through reversion, such as blackness in sheep; and with mankind some of the worst dispositions, which occasionally without any assignable cause make their appearance in families, may perhaps be reversions to a savage state, from which we are not removed by very many generations. This view seems indeed recognised in the common expression that such men are the black sheep of the family.

With civilised nations, as far as an advanced standard of morality, and an increased number of fairly well-endowed men are concerned, natural selection apparently effects but little; though the fundamental social instincts were originally thus gained. But I have already said enough, whilst treating of the lower races, on the causes which lead to the advance of morality, namely, the approbation of our fellow-men—the strengthening of our sympathies by habit—example and imitation—reason—experience and even self-interest—instruction during youth, and religious feelings.

A most important obstacle in civilised countries to an increase in the number of men of a superior class has been strongly urged by Mr. Greg and Mr. Galton,[238] namely, the fact that the very poor and reckless, who are often degraded by vice, almost invariably marry early, whilst the careful and frugal, who are generally otherwise virtuous, marry late in life, so that they may be able to support themselves and their children in comfort. Those who marry early produce within a given period not only a greater number of generations, but, as shewn by Dr. Duncan,[239] they produce many more children. The children, moreover, that are born by mothers during the prime of life are heavier and larger, and therefore probably more vigorous, than those born at other periods. Thus the reckless, degraded, and often vicious members of society, tend to increase at a quicker rate than the provident and generally virtuous members. Or as Mr. Greg puts the case: "The careless, squalid, unaspiring Irishman multiplies like rabbits: the frugal, foreseeing, self-respecting, ambitious Scot, stern in his morality, spiritual in his faith, sagacious and disciplined in his intelligence, passes his best years in struggle and in celibacy, marries late, and leaves few behind him. Given a land originally peopled by a thousand Saxons and a thousand Celts—and in a dozen generations five-sixths of the population would be Celts, but five-sixths of the property, of the power, of the intellect, would belong to the one-sixth of Saxons that remained. In the eternal 'struggle for existence,' it would be the inferior and *less* favoured race that had prevailed—and prevailed by virtue not of its good qualities but of its faults."

There are, however, some checks to this downward tendency. We have seen that the intemperate suffer from a high rate of mortality, and the extremely profligate leave few offspring. The poorest classes crowd into towns, and it has been proved by Dr. Stark from the statistics of ten years in

Scotland,[240] that at all ages the death-rate is higher in towns than in rural districts, "and during the first five years of life the town death-rate is almost exactly double that of the rural districts." As these returns include both the rich and the poor, no doubt more than double the number of births would be requisite to keep up the number of the very poor inhabitants in the towns, relatively to those in the country. With women, marriage at too early an age is highly injurious; for it has been found in France that, "twice as many wives under twenty die in the year, as died out of the same number of the unmarried." The mortality, also, of husbands under twenty is "excessively high,"[241] but what the cause of this may be seems doubtful. Lastly, if the men who prudently delay marrying until they can bring up their families in comfort, were to select, as they often do, women in the prime of life, the rate of increase in the better class would be only slightly lessened.

It was established from an enormous body of statistics, taken during 1853, that the unmarried men throughout France, between the ages of twenty and eighty, die in a much larger proportion than the married: for instance, out of every 1000 unmarried men, between the ages of twenty and thirty, 11·3 annually died, whilst of the married only 6·5 died.[242] A similar law was proved to hold good, during the years 1863 and 1864, with the entire population above the age of twenty in Scotland: for instance, out of every 1000 unmarried men, between the ages of twenty and thirty, 14·97 annually died, whilst of the married only 7·24 died, that is less than half.[243] Dr. Stark remarks on this, "Bachelorhood is more destructive to life than the most unwholesome trades, or than residence in an unwholesome house or district where there has never been the most distant attempt at sanitary improvement." He considers that the lessened mortality is the direct result of "marriage, and the more regular domestic habits which attend that state." He admits, however, that the intemperate, profligate, and criminal classes, whose duration of life is low, do not commonly marry; and it must likewise be admitted that men with a weak constitution, ill health, or any great infirmity in body or mind, will often not wish to marry, or will be rejected. Dr. Stark seems to have come to the conclusion that marriage in itself is a main cause of prolonged life, from finding that aged married men still have a considerable advantage in this respect over the unmarried of the same advanced age; but every one must have known instances of men, who with weak health during youth did not marry, and yet have survived to old age, though remaining weak and therefore always with a lessened chance of life. There is another remarkable circumstance which seems to support Dr. Stark's conclusion, namely, that widows and widowers in France suffer in comparison with the married a very heavy rate of mortality; but Dr. Farr attributes this to the poverty and evil habits consequent on the disruption of the family, and to grief. On the whole we may conclude with Dr. Farr that the lesser mortality of married than of unmarried men, which seems to be a general law, "is mainly due to the constant elimination of imperfect types, and to the skilful selection of the finest individuals out of each successive generation;" the selection relating only to the marriage state, and acting on all corporeal, intellectual, and moral qualities. We may, therefore, infer that sound and good men who out of prudence remain for a time unmarried do not suffer a high rate of mortality.

If the various checks specified in the two last paragraphs, and perhaps others as yet unknown, do not prevent the reckless, the vicious and otherwise inferior members of society from increasing at a quicker rate than the better class of men, the nation will retrograde, as has occurred too often in the history of the world. We must remember that progress is no invariable rule. It is most difficult to say why one civilised nation rises, becomes more powerful, and spreads more widely, than another; or why the same nation progresses more at one time than at another. We can only say that it depends on an increase in the actual number of the population, on the number of the men endowed with high intellectual and moral faculties, as well as on their standard of excellence. Corporeal structure, except so far as vigour of body leads to vigour of mind, appears to have little influence.

It has been urged by several writers that as high intellectual powers are advantageous to a nation, the old Greeks, who stood some grades higher in intellect than any race that has ever existed,[244] ought to have risen, if the power of natural selection were real, still higher in the scale, increased in number, and stocked the whole of Europe. Here we have the tacit assumption, so often made with respect to corporeal structures, that there is some innate tendency towards continued development in mind and body. But development of all kinds depends on many concurrent favourable circumstances. Natural selection acts only in a tentative manner. Individuals and races may have acquired certain indisputable advantages, and yet have perished from failing in other characters. The Greeks may have retrograded from a want of coherence between the many small states, from the small size of their whole country, from the practice of slavery, or from extreme sensuality; for they did not succumb until "they were enervated and corrupt to the very core."[245] The western nations of Europe, who now so immeasurably surpass their former savage progenitors and stand at the summit of civilisation, owe little or none of their superiority to direct inheritance from the old Greeks; though they owe much to the written works of this wonderful people.

Who can positively say why the Spanish nation, so dominant at one time, has been distanced in the race. The awakening of the nations of Europe from the dark ages is a still more perplexing problem. At this early period, as Mr. Galton[246] has remarked, almost all the men of a gentle nature, those given to meditation or culture of the mind, had no refuge except in the bosom of the Church which demanded celibacy; and this could hardly fail to have had a deteriorating influence on each successive generation. During this same period the Holy Inquisition selected with extreme care the freest and boldest men in order to burn or imprison them. In Spain alone some of the best men—those who doubted and questioned, and without doubting there can be no progress—were eliminated during three centuries at the rate of a thousand a year. The evil which the Catholic Church has thus effected, though no doubt counterbalanced to a certain, perhaps large extent in other ways, is incalculable; nevertheless, Europe has progressed at an unparalleled rate.

The remarkable success of the English as colonists over other European nations, which is well illustrated by comparing the progress of the Canadians of English and French extraction, has been ascribed to their "daring and persistent energy;" but who can say how the English gained their energy. There is apparently much truth in the belief that the wonderful progress of the United States, as well as the character of the people, are the results of natural selection; the more energetic, restless, and courageous men from all parts of Europe having emigrated during the last ten or twelve generations to that great country, and having there succeeded best.[247] Looking to the distant future, I do not think that the Rev. Mr. Zincke takes an exaggerated view when he says:[248] "All other series of events—as that which resulted in the culture of mind in Greece, and that which resulted in the empire of Rome—only appear to have purpose and value when viewed in connection with, or rather as subsidiary to . . . . the great stream of Anglo-Saxon emigration to the west." Obscure as is the problem of the advance of civilisation, we can at least see that a nation which produced during a lengthened period the greatest number of highly intellectual, energetic, brave, patriotic, and benevolent men, would generally prevail over less favoured nations.

Natural selection follows from the struggle for existence; and this from a rapid rate of increase. It is impossible not bitterly to regret, but whether wisely is another question, the rate at which man tends to increase; for this leads in barbarous tribes to infanticide and many other evils, and in civilised nations to abject poverty, celibacy, and to the late marriages of the prudent. But as man suffers from the same physical evils with the lower animals, he has no right to expect an immunity from the evils consequent on the struggle for existence. Had he not been subjected to natural selection, assuredly he would never have attained to the rank of manhood. When we see in many parts of the world enormous areas of the most fertile land peopled by a few wandering savages, but which are capable of supporting numerous happy homes, it might be argued that the struggle for existence had not been sufficiently severe to force man upwards to his highest standard. Judging from all that we know of man and the lower animals, there has always been sufficient variability in the intellectual and moral faculties, for their steady advancement through natural selection. No doubt such advancement demands many favourable concurrent circumstances; but it may well be doubted whether the most favourable would have sufficed, had not the rate of increase been rapid, and the consequent struggle for existence severe to an extreme degree.

On the evidence that all civilised nations were once barbarous.—As we have had to consider the steps by which some semi-human creature has been gradually raised to the rank of man in his most perfect state, the present subject cannot be quite passed over. But it has been treated in so full and admirable a manner by Sir J. Lubbock,[249] Mr. Tylor, Mr. M'Lennan, and others, that I need here give only the briefest summary of their results. The arguments recently advanced by the Duke of Argyll[250] and formerly by Archbishop Whately, in favour of the belief that man came into the world as a civilised being and that all savages have since undergone degradation, seem to me weak in comparison with those advanced on the other side. Many nations, no doubt, have fallen away in civilisation, and some may have lapsed into utter barbarism, though on this latter head I have not met with any evidence. The Fuegians were probably compelled by other conquering hordes to settle in their inhospitable country, and they may have become in consequence somewhat more degraded; but it would be difficult to prove that they have fallen much below the Botocudos who inhabit the finest parts of Brazil.

The evidence that all civilised nations are the descendants of barbarians, consists, on the one side, of clear traces of their former low condition in still-existing customs, beliefs, language, &c.; and on the other side, of proofs that savages are independently able to raise themselves a few steps in the scale of civilisation, and have actually thus risen. The evidence on the first head is extremely curious, but cannot be here given: I refer to such cases as that, for instance, of the art of enumeration, which, as Mr. Tylor clearly shews by the words still used in some places, originated in counting the fingers, first of one hand and then of the other, and lastly of the toes. We have traces of this in our own decimal system, and in the Roman numerals, which after reaching to the number V., change into VI, &c., when the other hand no doubt was used. So again, "when we speak of three-score and ten, we are counting by the vigesimal system, each score thus ideally made, standing for 20—for 'one man' as a Mexican or Carib would put it."[251] According to a large and increasing school of philologists, every language bears the marks of its slow and gradual evolution. So it is with the art of writing, as letters are rudiments of pictorial representations. It is hardly possible to read Mr. M'Lennan's work[252] and not admit that almost all civilised nations still retain some traces of such rude habits as the forcible capture of wives. What ancient nation, as the same author asks, can be named that was originally monogamous? The primitive idea of justice, as shewn by the law of battle and other customs of which traces still remain, was likewise most rude. Many existing superstitions are the remnants of former false religious beliefs. The highest form of religion—the grand idea of God hating sin and loving righteousness —was unknown during primeval times.

Turning to the other kind of evidence: Sir J. Lubbock has shewn that some savages have recently improved a little in some of their simpler arts. From the extremely curious account which he gives of the weapons, tools, and arts, used or practised by savages in various parts of the world, it cannot be doubted that these have nearly all been independent discoveries, excepting perhaps the art of making fire.[253] The Australian boomerang is a good instance of one such independent discovery. The Tahitians when first visited had advanced in many respects beyond the inhabitants of most of the other Polynesian islands. There are no just grounds for the belief that the high culture of the native Peruvians and Mexicans was derived from any foreign source;[254] many native plants were there cultivated, and a few native animals domesticated. We should bear in mind that a wandering crew from some semi-civilised land, if washed to the shores of America, would not, judging from the small influence of most missionaries, have produced any marked effect on the natives, unless they had already become somewhat advanced. Looking to a very remote period in the history of the world, we find, to use Sir J. Lubbock's well-known terms, a paleolithic and neolithic period; and no one will pretend that the art of grinding rough flint tools was a borrowed one. In all parts of Europe, as far east as Greece, in Palestine, India, Japan, New Zealand, and Africa, including Egypt, flint tools have been discovered in abundance; and of their use the existing inhabitants retain no tradition. There is also indirect evidence of their former use by the Chinese and ancient Jews. Hence there can hardly be a doubt that the inhabitants of these many countries, which include nearly the whole civilised world, were once in a barbarous condition. To believe that man was aboriginally civilised and then suffered utter degradation in so many regions, is to take a pitiably low view of human nature. It is apparently a truer and more cheerful view that progress has been much more general than retrogression; that man has risen, though by slow and interrupted steps, from a lowly condition to the highest standard as yet attained by him in knowledge, morals, and religion.

CHAPTER VI

On the Affinities and Genealogy of Man

POSITION OF MAN IN THE ANIMAL SERIES • THE NATURAL SYSTEM GENEALOGICAL • ADAPTIVE CHARACTERS OF SLIGHT VALUE • VARIOUS SMALL POINTS OF RESEMBLANCE BETWEEN MAN AND THE QUADRUMANA • RANK OF MAN IN THE NATURAL SYSTEM • BIRTHPLACE AND ANTIQUITY OF MAN • ABSENCE OF FOSSIL CONNECTING-LINKS • LOWER STAGES IN THE GENEALOGY OF MAN, AS INFERRED, FIRSTLY FROM HIS AFFINITIES AND SECONDLY FROM HIS STRUCTURE • EARLY ANDROGYNOUS CONDITION OF THE VERTEBRATA • CONCLUSION

EVEN IF IT BE GRANTED that the difference between man and his nearest allies is as great in corporeal structure as some naturalists maintain, and although we must grant that the difference between them is immense in mental power, yet the facts given in the previous chapters declare, as it appears to me, in the plainest manner, that man is descended from some lower form, notwithstanding that connecting-links have not hitherto been discovered.

Man is liable to numerous, slight, and diversified variations, which are induced by the same general causes, are governed and transmitted in accordance with the same general laws, as in the lower animals. Man tends to multiply at so rapid a rate that his offspring are necessarily exposed to a struggle for existence, and consequently to natural selection. He has given rise to many races, some of which are so different that they have often been ranked by naturalists as distinct species. His body is constructed on the same homological plan as that of other mammals, independently of the uses to which the several parts may be put. He passes through the same phases of embryological development. He retains many rudimentary and useless structures, which no doubt were once serviceable. Characters occasionally make their re-appearance in him, which we have every reason to believe were possessed by his early progenitors. If the origin of man had been wholly different from that of all other animals, these various appearances would be mere empty deceptions; but such an admission is incredible. These appearances, on the other hand, are intelligible, at least to a large extent, if man is the co-descendant with other mammals of some unknown and lower form.

Some naturalists, from being deeply impressed with the mental and spiritual powers of man, have divided the whole organic world into three kingdoms, the Human, the Animal, and the Vegetable, thus giving to man a separate kingdom.[255] Spiritual powers cannot be compared or classed by the naturalist; but he may endeavour to shew, as I have done, that the mental faculties of man and the lower animals do not differ in kind, although immensely in degree. A difference in degree, however great, does not justify us in placing man in a distinct kingdom, as will perhaps be best illustrated by comparing the mental powers of two insects, namely, a coccus or scale-insect and an ant, which undoubtedly belong to the same class. The difference is here greater, though of a somewhat different kind, than that between man and the highest mammal. The female coccus, whilst young, attaches itself by its proboscis to a plant; sucks the sap but never moves again; is fertilised and lays eggs; and this is its whole history. On the other hand, to describe the habits and mental powers of a female ant, would require, as Pierre Huber has shewn, a large volume; I may, however, briefly specify a few points. Ants communicate information to each other, and several unite for the same work, or games of play. They recognise their fellow-ants after months of absence. They build great edifices, keep them clean, close the doors in the evening, and post sentries. They make roads, and even tunnels under rivers. They collect food for the community, and when an object, too large for entrance, is brought to the nest, they enlarge the door, and afterwards build it up again.[256] They go out to battle in regular bands, and freely sacrifice their lives for the common weal. They emigrate in accordance with a preconcerted plan. They capture slaves. They keep Aphides as milch-cows. They move the eggs of their aphides, as well as their own eggs and cocoons, into warm parts of the nest, in order that they may be quickly hatched; and endless similar facts could be given. On the whole, the difference in mental power between an ant and a coccus is immense; yet no one has ever dreamed of placing them in distinct classes, much less in distinct kingdoms. No doubt this interval is bridged over by the intermediate mental powers of many other insects; and this is not the case with man and the higher apes. But we have every reason to believe that breaks in the series are simply the result of many forms having become extinct.

Professor Owen, relying chiefly on the structure of the brain, has divided the mammalian series into four sub-classes. One of these he devotes to man; in another he places both the marsupials and the monotremata; so that he makes man as distinct from all other mammals as are these two latter groups conjoined. This view has not been accepted, as far as I am aware, by any naturalist capable of forming an independent judgment, and therefore need not here be further considered.

We can understand why a classification founded on any single character or organ—even an organ so wonderfully complex and important as the brain—or on the high development of the mental faculties, is almost sure to prove unsatisfactory. This principle has indeed been tried with hymenopterous insects; but when thus classed by their habits or instincts, the arrangement proved thoroughly artificial.[257] Classifications may, of course, be based on any character whatever, as on size, colour, or the element inhabited; but naturalists have long felt a profound conviction that there is a natural system. This system, it is now generally admitted, must be, as far as possible, genealogical in arrangement,—that is, the co-descendants of the same form must be kept together in one group, separate from the co-descendants of any other form; but if the parent-forms are related, so will be their descendants, and the two groups together will form a larger group. The amount of difference between the several groups—that is the amount of modification which each has undergone—will be expressed by such terms as genera, families, orders, and classes. As we have no record of the lines of descent, these lines can be discovered only by observing the degrees of resemblance between the beings which are to be classed. For this object numerous points of resemblance are of much more importance than the amount of similarity or dissimilarity in a few points. If two languages were found to resemble each other in a multitude of words and points of construction, they would be universally recognised as having sprung from a common source, notwithstanding that they differed greatly in some few words or points of construction. But with organic beings the points of resemblance must not consist of adaptations to similar habits of life: two animals may, for instance, have had their whole frames modified for living in the water, and yet they will not be brought any nearer to each other in the natural system. Hence we can see how it is that resemblances in unimportant structures, in useless and rudimentary organs, and in parts not as yet fully developed or functionally active, are by far the most serviceable for classification; for they can hardly be due to adaptations within a late period; and thus they reveal the old lines of descent or of true affinity.

We can further see why a great amount of modification in some one character ought not to lead us to separate widely any two organisms. A part which already differs much from the same part in other allied forms has already, according to the theory of evolution, varied much; consequently it would (as long as the organism remained exposed to the same exciting conditions) be liable to further variations of the same kind; and these, if beneficial, would be preserved, and thus continually augmented. In many cases the continued development of a part, for instance, of the beak of a bird, or of the teeth of a mammal, would not be advantageous to the species for gaining its food, or for any other object; but with man we can see no definite limit, as far as advantage is concerned, to the continued development of the brain and mental faculties. Therefore in determining the position of man in the natural or genealogical system, the extreme development of his brain ought not to outweigh a multitude of resemblances in other less important or quite unimportant points.

The greater number of naturalists who have taken into consideration the whole structure of man, including his mental faculties, have followed Blumenbach and Cuvier, and have placed man in a separate Order, under the title of the Bimana, and therefore on an equality with the Orders of the Quadrumana, Carnivora, &c. Recently many of our best naturalists have recurred to the view first propounded by Linnæus, so remarkable for his sagacity, and have placed man in the same Order with the Quadrumana, under the title of the Primates. The justice of this conclusion will be admitted if, in the first place, we bear in mind the remarks just made on the comparatively small importance for classification of the great development of the brain in man; bearing, also, in mind that the strongly-marked differences between the skulls of man and the Quadrumana (lately insisted upon by Bischoff, Aeby, and others) apparently follow from their differently developed brains. In the second place, we must remember that nearly all the other and more important differences between man and the Quadrumana are manifestly adaptive in their nature, and relate chiefly to the erect position of man; such as the structure of his hand, foot, and pelvis, the curvature of his spine, and the position of his head. The family of seals offers a good illustration of the small importance of adaptive characters for classification. These animals differ from all other Carnivora in the form of their bodies and in the structure of their limbs, far more than does man from the higher apes; yet in every system, from that of Cuvier to the most recent one by Mr. Flower,[258] seals are ranked as a mere family in the Order of the Carnivora. If man had not been his own classifier, he would never have thought of founding a separate order for his own reception.

It would be beyond my limits, and quite beyond my knowledge, even to name the innumerable points of structure in which man agrees with the other Primates. Our great anatomist and philosopher, Prof. Huxley, has fully discussed this subject,[259] and has come to the conclusion that man in all parts of his organisation differs less from the higher apes, than these do from the lower members of the same group. Consequently there "is no justification for placing man in a distinct order."

In an early part of this volume I brought forward various facts, shewing how closely man agrees in constitution with the higher mammals; and this agreement, no doubt, depends on our close similarity in minute structure and chemical composition. I gave, as instances, our liability to the same diseases, and to the attacks of allied parasites; our tastes in common for the same stimulants, and the similar effects thus produced, as well as by various drugs; and other such facts.

As small unimportant points of resemblance between man and the higher apes are not commonly noticed in systematic works, and as, when numerous, they clearly reveal our relationship, I will specify a few such points. The relative position of the features are manifestly the same in man and the Quadrumana; and the various emotions are displayed by nearly similar movements of the muscles and skin, chiefly above the eyebrows and round the mouth. Some few expressions are, indeed, almost the same, as in the weeping of certain kinds of monkeys, and in the laughing noise made by others, during which the corners of the mouth are drawn backwards, and the lower eyelids wrinkled. The external ears are curiously alike. In man the nose is much more prominent than in most monkeys; but we may trace the commencement of an aquiline curvature in the nose of the Hoolock Gibbon; and this in the *Semnopithecus nasica* is carried to a ridiculous extreme.

The faces of many monkeys are ornamented with beards, whiskers, or moustaches. The hair on the head grows to a great length in some species of Semnopithecus;[260] and in the Bonnet monkey *(Macacus radiatus)* it radiates from a point on the crown, with a parting down the middle, as in man. It is commonly said that the forehead gives to man his noble and intellectual appearance; but the thick hair on the head of the Bonnet monkey terminates abruptly downwards, and is succeeded by such short and fine hair, or down, that at a little distance the forehead, with the exception of the eyebrows, appears quite naked. It has been erroneously asserted that eyebrows are not present in any monkey. In the species just named the degree of nakedness of the forehead differs in different individuals; and Eschricht states[261] that in our children the limit between the hairy scalp and the naked forehead is sometimes not well defined; so that here we seem to have a trifling case of reversion to a progenitor, in whom the forehead had not as yet become quite naked.

It is well known that the hair on our arms tends to converge from above and below to a point at the elbow. This curious arrangement, so unlike that in most of the lower mammals, is common to the gorilla, chimpanzee, orang, some species of Hylobates, and even to some few American monkeys. But in *Hylobates agilis* the hair on the fore-arm is directed downwards or towards the wrist in the ordinary manner; and in *H. lar* it is nearly erect, with only a very slight forward inclination; so that in this latter species it is in a transitional state. It can hardly be doubted that with most mammals the thickness of the hair and its direction on the back is adapted to throw off the rain; even the transverse hairs on the fore-legs of a dog may serve for this end when he is coiled up asleep. Mr. Wallace remarks that the convergence of the hair towards the elbow on the arms of the orang (whose habits he has so carefully studied) serves to throw off the rain, when, as is the custom of this animal, the arms are bent, with the hands clasped round a branch or over its own head. We should, however, bear in mind that the attitude of an animal may perhaps be in part determined by the direction of the hair; and not the direction of the hair by the attitude. If the above explanation is correct in the case of the orang, the hair on our fore-arms offers a curious record of our former state; for no one supposes that it is now of any use in throwing off the rain, nor in our present erect condition is it properly directed for this purpose.

It would, however, be rash to trust too much to the principle of adaptation in regard to the direction of the hair in man or his early progenitors; for it is impossible to study the figures given by Eschricht of the arrangement of the hair on the human fœtus (this being the same as in the adult) and not agree with this excellent observer that other and more complex causes have intervened. The points of convergence seem to stand in some relation to those points in the embryo which are last closed in during development. There appears, also, to exist some relation between the arrangement of the hair on the limbs, and the course of the medullary arteries.[262]

It must not be supposed that the resemblances between man and certain apes in the above and many other points—such as in having a naked forehead, long tresses on the head, &c.—are all necessarily the result of unbroken inheritance from a common progenitor thus characterised, or of subsequent reversion. Many of these resemblances are more probably due to analogous variation, which follows, as I have elsewhere attempted to shew,[263] from co-descended organisms having a similar constitution and having been acted on by similar causes inducing variability. With respect to the similar direction of the hair on the fore-arms of man and certain monkeys, as this character is common to almost all the anthropomorphous apes, it may probably be attributed to inheritance; but not certainly so, as some very distinct American monkeys are thus characterised. The same remark is applicable to the tailless condition of man; for the tail is absent in all the anthropomorphous apes. Nevertheless this character cannot with certainty be attributed to inheritance, as the tail, though not absent, is rudimentary in several other Old World and in some New World species, and is quite absent in several species belonging to the allied group of Lemurs.

Although, as we have now seen, man has no just right to form a separate Order for his own reception, he may perhaps claim a distinct Sub-order or Family. Prof. Huxley, in his last work,[264] divides the Primates into three Sub-orders; namely, the Anthropidæ with man alone, the Simiadæ including monkeys of all kinds, and the Lemuridæ with the diversified genera of lemurs. As far as differences in certain important points of structure are concerned, man may no doubt rightly claim the rank of a Sub-order; and this rank is too low, if we look chiefly to his mental faculties. Nevertheless, under a genealogical point of view it appears that this rank is too high, and that man ought to form merely a Family, or possibly even only a Sub-family. If we imagine three lines of descent proceeding from a common source, it is quite conceivable that two of them might after the lapse of ages be so slightly changed as still to remain as species of the same genus; whilst the third line might become so greatly modified as to deserve to rank as a distinct Sub-family, Family, or even Order. But in this case it is almost certain that the third line would still retain through inheritance numerous small points of resemblance with the other two lines. Here then would occur the difficulty, at present insoluble, how much weight we ought to assign in our classifications to strongly-marked differences in some few points,—that is to the amount of modification undergone; and how much to close resemblance in numerous unimportant points, as indicating the lines of descent or genealogy. The former alternative is the most obvious, and perhaps the safest, though the latter appears the most correct as giving a truly natural classification.

To form a judgment on this head, with reference to man we must glance at the classification of the Simiadæ. This family is divided by almost all naturalists into the Catarhine group, or Old World monkeys, all of which are characterised (as their name expresses) by the peculiar structure of their nostrils and by having four premolars in each jaw; and into the Platyrhine group or New World monkeys (including two very distinct sub-groups), all of which are characterised by differently-constructed nostrils and by having six pre-molars in each jaw. Some other small differences might be mentioned. Now man unquestionably belongs in his dentition, in the structure of his nostrils, and some other respects, to the Catarhine or Old World division; nor does he resemble the Platyrhines more closely than the Catarhines in any characters, excepting in a few of not much importance and apparently of an adaptive nature. Therefore it would be against all probability to suppose that some ancient New World species had varied, and had thus produced a man-like creature with all the distinctive characters proper to the Old World division; losing at the same time all its own distinctive characters. There can consequently hardly be a doubt that man is an offshoot from the Old World Simian stem; and that under a genealogical point of view, he must be classed with the Catarhine division.[265]

The anthropomorphous apes, namely the gorilla, chimpanzee, orang, and hylobates, are separated as a distinct sub-group from the other Old World monkeys by most naturalists. I am aware that Gratiolet, relying on the structure of the brain, does not admit the existence of this sub-group, and no doubt it is a broken one; thus the orang, as Mr. St. G. Mivart remarks,[266] "is one of the most peculiar and aberrant forms to be found in the Order." The remaining, non-anthropomorphous, Old World monkeys, are again divided by some naturalists into two or three smaller sub-groups; the genus Semnopithecus, with its peculiar sacculated stomach, being the type of one such sub-group. But it appears from M. Gaudry's wonderful discoveries in Attica, that during the Miocene period a form existed there, which connected Semnopithecus and Macacus; and this probably illustrates the manner in which the other and higher groups were once blended together.

If the anthropomorphous apes be admitted to form a natural sub-group, then as man agrees with them, not only in all those characters which he possesses in common with the whole Catarhine group, but in other peculiar characters, such as the absence of a tail and of callosities and in general appearance, we may infer that some ancient member of the anthropomorphous sub-group gave birth to man. It is not probable that a member of one of the other lower sub-groups should, through the law of analogous variation, have given rise to a man-like creature, resembling the higher anthropomorphous apes in so many respects. No doubt man, in comparison with most of his allies, has undergone an extraordinary amount of modification, chiefly in consequence of his greatly developed brain and erect position; nevertheless we should bear in mind that he "is but one of several exceptional forms of Primates."[267]

Every naturalist, who believes in the principle of evolution, will grant that the two main divisions of the Simiadæ, namely the Catarhine and Platyrhine monkeys, with their sub-groups, have all proceeded from some one extremely ancient progenitor. The early descendants of this progenitor, before they had diverged to any considerable extent from each other, would still have formed a single natural group; but some of the species or incipient genera would have already begun to indicate by their diverging characters the future distinctive marks of the Catarhine and Platyrhine divisions. Hence the members of this supposed ancient group would not have been so uniform in their dentition or in the structure of their nostrils, as are the existing Catarhine monkeys in one way and the Platyrhines in another way, but would have resembled in this respect the allied Lemuridæ which differ greatly from each other in the form of their muzzles,[268] and to an extraordinary degree in their dentition.

The Catarhine and Platyrhine monkeys agree in a multitude of characters, as is shewn by their unquestionably belonging to one and the same Order. The many characters which they possess in common can hardly have been independently acquired by so many distinct species; so that these characters must have been inherited. But an ancient form which possessed many characters common to the Catarhine and Platyrhine monkeys, and others in an intermediate condition, and some few perhaps distinct from those now present in either group, would undoubtedly have been ranked, if seen by a naturalist, as an ape or monkey. And as man under a genealogical point of view belongs to the Catarhine or Old World stock, we must conclude, however much the conclusion may revolt our pride, that our early progenitors would have been properly thus designated.[269] But we must not fall into the error of supposing that the early progenitor of the whole Simian stock, including man, was identical with, or even closely resembled, any existing ape or monkey.

On the Birthplace and Antiquity of Man.—We are naturally led to enquire where was the birthplace of man at that stage of descent when our progenitors diverged from the Catarhine stock. The fact that they belonged to this stock clearly shews that they inhabited the Old World; but not Australia nor any oceanic island, as we may infer from the laws of geographical distribution. In each great region of the world the living mammals are closely related to the extinct species of the same region. It is therefore probable that Africa was formerly inhabited by extinct apes closely allied to the gorilla and chimpanzee; and as these two species are now man's nearest allies, it is somewhat more probable that our early progenitors lived on the African continent than elsewhere. But it is useless to speculate on this subject, for an ape nearly as large as a man, namely the Dryopithecus of Lartet, which was closely allied to the anthropomorphous Hylobates, existed in Europe during the Upper Miocene period; and since so remote a period the earth has certainly undergone many great revolutions, and there has been ample time for migration on the largest scale.

At the period and place, whenever and wherever it may have been, when man first lost his hairy covering, he probably inhabited a hot country; and this would have been favourable for a frugiferous diet, on which, judging from analogy, he subsisted. We are far from knowing how long ago it was when man first diverged from the Catarhine stock; but this may have occurred at an epoch as remote as the Eocene period; for the higher apes had diverged from the lower apes as early as the Upper Miocene period, as shewn by the existence of the Dryopithecus. We are also quite ignorant at how rapid a rate organisms, whether high or low in the scale, may under favourable circumstances be modified: we know, however, that some have retained the same form during an enormous lapse of time. From what we see going on under domestication, we learn that within the same period some of the co-descendants of the same species may be not at all changed, some a little, and some greatly changed. Thus it may have been with man, who has undergone a great amount of modification in certain characters in comparison with the higher apes.

The great break in the organic chain between man and his nearest allies, which cannot be bridged over by any extinct or living species, has often been advanced as a grave objection to the belief that man is descended from some lower form; but this objection will not appear of much weight to those who, convinced by general reasons, believe in the general principle of evolution. Breaks incessantly occur in all parts of the series, some being wide, sharp and defined, others less so in various degrees; as between the orang and its nearest allies—between the Tarsius and the other Lemuridæ—between the elephant and in a more striking manner between the Ornithorhynchus or Echidna, and other mammals. But all these breaks depend merely on the number of related forms which have become extinct. At some future period, not very distant as measured by centuries, the civilised races of man will almost certainly exterminate and replace throughout the world the savage races. At the same time the anthropomorphous apes, as Professor Schaaffhausen has remarked,[270] will no doubt be exterminated. The break will then be rendered wider, for it will intervene between man in a more civilised state, as we may hope, than the Caucasian, and some ape as low as a baboon, instead of as at present between the negro or Australian and the gorilla.

With respect to the absence of fossil remains, serving to connect man with his ape-like progenitors, no one will lay much stress on this fact, who will read Sir C. Lyell's discussion,[271] in which he shews that in all the vertebrate classes the discovery of fossil remains has been an extremely slow and fortuitous process. Nor should it be forgotten that those regions which are the most likely to afford remains connecting man with some extinct ape-like creature, have not as yet been searched by geologists.

Lower Stages in the Genealogy of Man.—We have seen that man appears to have diverged from the Catarhine or Old World division of the Simiadæ, after these had diverged from the New World division. We will now endeavour to follow the more remote traces of his genealogy, trusting in the first place to the mutual affinities between the various classes and orders, with some slight aid from the periods, as far as ascertained, of their successive appearance on the earth. The Lemuridæ stand below and close to the Simiadæ, constituting a very distinct family of the Primates, or, according to Häckel, a distinct Order. This group is diversified and broken to an extraordinary degree, and includes many aberrant forms. It has, therefore, probably suffered much extinction. Most of the remnants survive on islands, namely in Madagascar and in the islands of the Malayan archipelago, where they have not been exposed to such severe competition as they would have been on well-stocked continents. This group likewise presents many gradations, leading, as Huxley remarks,[272] "insensibly from the crown and summit of the animal creation down to creatures from which there is but a step, as it seems, to the lowest, smallest, and least intelligent of the placental mammalia." From these various considerations it is probable that the Simiadæ were originally developed from the progenitors of the existing Lemuridæ; and these in their turn from forms standing very low in the mammalian series.

The Marsupials stand in many important characters below the placental mammals. They appeared at an earlier geological period, and their range was formerly much more extensive than what it now is. Hence the Placentata are generally supposed to have been derived from the Implacentata or Marsupials; not, however, from forms closely like the existing Marsupials, but from their early progenitors. The Monotremata are plainly allied to the Marsupials; forming a third and still lower division in the great mammalian series. They are represented at the present day solely by the Ornithorhynchus and Echidna; and these two forms may be safely considered as relics of a much larger group which have been preserved in Australia through some favourable concurrence of circumstances. The Monotremata are eminently interesting, as in several important points of structure they lead towards the class of reptiles.

In attempting to trace the genealogy of the Mammalia, and therefore of man, lower down in the series, we become involved in greater and greater obscurity. He who wishes to see what ingenuity and knowledge can effect, may consult Prof. Häckel's works.[273] I will content myself with a few general remarks. Every evolutionist will admit that the five great vertebrate classes, namely, mammals, birds, reptiles, amphibians, and fishes, are all descended from some one prototype; for they have much in common, especially during their embryonic state. As the class of fishes is the most lowly organised and appeared before the others, we may conclude that all the members of the vertebrate kingdom are derived from some fish-like animal, less highly organised than any as yet found in the lowest known formations. The belief that animals so distinct as a monkey or elephant and a humming-bird, a snake, frog, and fish, &c., could all have sprung from the same parents, will appear monstrous to those who have not attended to the recent progress of natural history. For this belief implies the former existence of links closely binding together all these forms, now so utterly unlike.

Nevertheless it is certain that groups of animals have existed, or do now exist, which serve to connect more or less closely the several great vertebrate classes. We have seen that the Ornithorhynchus graduates towards reptiles; and Prof. Huxley has made the remarkable discovery, confirmed by Mr. Cope and others, that the old Dinosaurians are intermediate in many important respects between certain reptiles and certain birds—the latter consisting of the ostrich-tribe (itself evidently a widely-diffused remnant of a larger group) and of the Archeopteryx, that strange Secondary bird having a long tail like that of the lizard. Again, according to Prof. Owen,[274] the Ichthyosaurians—great sea-lizards furnished with paddles—present many affinities with fishes, or rather, according to Huxley, with amphibians. This latter class (including in its highest division frogs and toads) is plainly allied to the Ganoid fishes. These latter fishes swarmed during the earlier geological periods, and were constructed on what is called a highly generalised type, that is they presented diversified affinities with other groups of organisms. The amphibians and fishes are also so closely united by the Lepidosiren, that naturalists long disputed in which of these two classes it ought to be placed. The Lepidosiren and some few Ganoid fishes have been preserved from utter extinction by inhabiting our rivers, which are harbours of refuge, bearing the same relation to the great waters of the ocean that islands bear to continents.

Lastly, one single member of the immense and diversified class of fishes, namely the lancelet or amphioxus, is so different from all other fishes, that Häckel maintains that it ought to form a distinct class in the vertebrate kingdom. This fish is remarkable for its negative characters; it can hardly be said to possess a brain, vertebral column, or heart, &c.; so that it was classed by the older naturalists amongst the worms. Many years ago Prof. Goodsir perceived that the lancelet presented some affinities with the Ascidians, which are invertebrate, hermaphrodite, marine creatures permanently attached to a support. They hardly appear like animals, and consist of a simple, tough, leathery sack, with two small projecting orifices. They belong to the Molluscoida of Huxley—a lower division of the great kingdom of the Mollusca; but they have recently been placed by some naturalists amongst the Vermes or worms. Their larvæ somewhat resemble tadpoles in shape,[275] and have the power of swimming freely about. Some observations lately made by M. Kowalevsky,[276] since confirmed by Prof. Kuppfer, will form a discovery of extraordinary interest, if still further extended, as I hear from M. Kowalevsky in Naples he has now effected. The discovery is that the larvæ of Ascidians are related to the Vertebrata, in their manner of development, in the relative position of the nervous system, and in possessing a structure closely like the *chorda dorsalis* of vertebrate animals. It thus appears, if we may rely on embryology, which has always proved the safest guide in classification, that we have at last gained a clue to the source whence the Vertebrata have been derived. We should thus be justified in believing that at an extremely remote period a group of animals existed, resembling in many respects the larvæ of our present Ascidians, which diverged into two great branches—the one retrograding in development and producing the present class of Ascidians, the other rising to the crown and summit of the animal kingdom by giving birth to the Vertebrata.

WE HAVE THUS FAR endeavoured rudely to trace the genealogy of the Vertebrata by the aid of their mutual affinities. We will now look to man as he exists; and we shall, I think, be able partially to restore during successive periods, but not in due order of time, the structure of our early progenitors. This can be effected by means of the rudiments which man still retains, by the characters which occasionally make their appearance in him through reversion, and by the aid of the principles of morphology and embryology. The various facts, to which I shall here allude, have been given in the previous chapters. The early progenitors of man were no doubt once covered with hair, both sexes having beards; their ears were pointed and capable of movement; and their bodies were provided with a tail, having the proper muscles. Their limbs and bodies were also acted on by many muscles which now only occasionally reappear, but are normally present in the Quadrumana. The great artery and nerve of the humerus ran through a supra-condyloid foramen. At this or some earlier period, the intestine gave forth a much larger diverticulum or cæcum than that now existing. The foot, judging from the condition of the great toe in the fœtus, was then prehensile; and our progenitors, no doubt, were arboreal in their habits, frequenting some warm, forest-clad land. The males were provided with great canine teeth, which served them as formidable weapons.

At a much earlier period the uterus was double; the excreta were voided through a cloaca; and the eye was protected by a third eyelid or nictitating membrane. At a still earlier period the progenitors of man must have been aquatic in their habits; for morphology plainly tells us that our lungs consist of a modified swim-bladder, which once served as a float. The clefts on the neck in the embryo of man show where the branchiæ once existed. At about this period the true kidneys were replaced by the corpora wolffiana. The heart existed as a simple pulsating vessel; and the chorda dorsalis took the place of a vertebral column. These early predecessors of man, thus seen in the dim recesses of time, must have been as lowly organised as the lancelet or amphioxus, or even still more lowly organised.

There is one other point deserving a fuller notice. It has long been known that in the vertebrate kingdom one sex bears rudiments of various accessory parts, appertaining to the reproductive system, which properly belong to the opposite sex; and it has now been ascertained that at a very early embryonic period both sexes possess true male and female glands. Hence some extremely remote progenitor of the whole vertebrate kingdom appears to have been hermaphrodite or androgynous.[277] But here we encounter a singular difficulty. In the mammalian class the males possess in their vesiculæ prostaticæ rudiments of a uterus with the adjacent passage; they bear also rudiments of mammæ, and some male marsupials have rudiments of a marsupial sack.[278] Other analogous facts could be added. Are we, then, to suppose that some extremely ancient mammal possessed organs proper to both sexes, that is, continued androgynous after it had acquired the chief distinctions of its proper class, and therefore after it had diverged from the lower classes of the vertebrate kingdom? This seems improbable in the highest degree; for had this been the case, we might have expected that some few members of the two lower classes, namely fishes[279] and amphibians, would still have remained androgynous. We must, on the contrary, believe that when the five vertebrate classes diverged from their common progenitor the sexes had already become separated. To account, however, for male mammals possessing rudiments of the accessory female organs, and for female mammals possessing rudiments of the masculine organs, we need not suppose that their early progenitors were still androgynous after they had assumed their chief mammalian characters. It is quite possible that as the one sex gradually acquired the accessory organs proper to it, some of the successive steps or modifications were transmitted to the opposite sex. When we treat of sexual selection, we shall meet with innumerable instances of this form of transmission,—as in the case of the spurs, plumes, and brilliant colours, acquired by male birds for battle or ornament, and transferred to the females in an imperfect or rudimentary condition.

The possession by male mammals of functionally imperfect mammary organs is, in some respects, especially curious. The Monotremata have the proper milk-secreting glands with orifices, but no nipples; and as these animals stand at the very base of the mammalian series, it is probable that the progenitors of the class possessed, in like manner, the milk-secreting glands, but no nipples. This conclusion is supported by what is known of their manner of development; for Professor Turner informs me, on the authority of Kölliker and Lauger, that in the embryo the mammary glands can be distinctly traced before the nipples are in the least visible; and it should be borne in mind that the development of successive parts in the individual generally seems to represent and accord with the development of successive beings in the same line of descent. The Marsupials differ from the Monotremata by possessing nipples; so that these organs were probably first acquired by the Marsupials after they had diverged from, and risen above, the Monotremata, and were then transmitted to the placental mammals. No one will suppose that after the Marsupials had approximately acquired their present structure, and therefore at a rather late period in the development of the mammalian series, any of its members still remained androgynous. We seem, therefore, compelled to recur to the foregoing view, and to conclude that the nipples were first developed in the females of some very early marsupial form, and were then, in accordance with a common law of inheritance, transferred in a functionally imperfect condition to the males.

Nevertheless a suspicion has sometimes crossed my mind that long after the progenitors of the whole mammalian class had ceased to be androgynous, both sexes might have yielded milk and thus nourished their young; and in the case of the Marsupials, that both sexes might have carried their young in marsupial sacks. This will not appear utterly incredible, if we reflect that the males of syngnathous fishes receive the eggs of the females in their abdominal pouches, hatch them, and afterwards, as some believe, nourish the young;[280]—that certain other male fishes hatch the eggs within their mouths or branchial cavities;—that certain male toads take the chaplets of eggs from the females and wind them round their own thighs, keeping them there until the tadpoles are born;—that certain male birds undertake the whole duty of incubation, and that male pigeons, as well as the females, feed their nestlings with a secretion from their crops. But the above suspicion first occurred to me from the mammary glands in male mammals being developed so much more perfectly than the rudiments of those other accessory reproductive parts, which are found in the one sex though proper to the other. The mammary glands and nipples, as they exist in male mammals, can indeed hardly be called rudimentary; they are simply not fully developed and not functionally active. They are sympathetically affected under the influence of certain diseases, like the same organs in the female. At birth they often secrete a few drops of milk; and they have been known occasionally in man and other mammals to become well developed, and to yield a fair supply of milk. Now if we suppose that during a former prolonged period male mammals aided the females in nursing their offspring, and that afterwards from some cause, as from a smaller number of young being produced, the males ceased giving this aid, disuse of the organs during maturity would lead to their becoming inactive; and from two well-known principles of inheritance this state of inactivity would probably be transmitted to the males at the corresponding age of maturity. But at all earlier ages these organs would be left unaffected, so that they would be equally well developed in the young of both sexes.

Conclusion.—The best definition of advancement or progress in the organic scale ever given, is that by Von Baer; and this rests on the amount of differentiation and specialisation of the several parts of the same being, when arrived, as I should be inclined to add, at maturity. Now as organisms have become slowly adapted by means of natural selection for diversified lines of life, their parts will have become, from the advantage gained by the division of physiological labour, more and more differentiated and specialised for various functions. The same part appears often to have been modified first for one purpose, and then long afterwards for some other and quite distinct purpose; and thus all the parts are rendered more and more complex. But each organism will still retain the general type of structure of the progenitor from which it was aboriginally derived. In accordance with this view it seems, if we turn to geological evidence, that organisation on the whole has advanced throughout the world by slow and interrupted steps. In the great kingdom of the Vertebrata it has culminated in man. It must not, however, be supposed that groups of organic beings are always supplanted and disappear as soon as they have given birth to other and more perfect groups. The latter, though victorious over their predecessors, may not have become better adapted for all places in the economy of nature. Some old forms appear to have survived from inhabiting protected sites, where they have not been exposed to very severe competition; and these often aid us in constructing our genealogies, by giving us a fair idea of former and lost populations. But we must not fall into the error of looking at the existing members of any lowly-organised group as perfect representatives of their ancient predecessors.

The most ancient progenitors in the kingdom of the Vertebrata, at which we are able to obtain an obscure glance, apparently consisted of a group of marine animals,[281] resembling the larvæ of existing Ascidians. These animals probably gave rise to a group of fishes, as lowly organised as the lancelet; and from these the Ganoids, and other fishes like the Lepidosiren, must have been developed. From such fish a very small advance would carry us on to the amphibians. We have seen that birds and reptiles were once intimately connected together; and the Monotremata now, in a slight degree, connect mammals with reptiles. But no one can at present say by what line of descent the three higher and related classes, namely, mammals, birds, and reptiles, were derived from either of the two lower vertebrate classes, namely amphibians and fishes. In the class of mammals the steps are not difficult to conceive which led from the ancient Monotremata to the ancient Marsupials; and from these to the early progenitors of the placental mammals. We may thus ascend to the Lemuridæ; and the interval is not wide from these to the Simiadæ. The Simiadæ then branched off into two great stems, the New World and Old World monkeys; and from the latter, at a remote period, Man, the wonder and glory of the Universe, proceeded.

Thus we have given to man a pedigree of prodigious length, but not, it may be said, of noble quality. The world, it has often been remarked, appears as if it had long been preparing for the advent of man; and this, in one sense is strictly true, for he owes his birth to a long line of progenitors. If any single link in this chain had never existed, man would not have been exactly what he now is. Unless we wilfully close our eyes, we may, with our present knowledge, approximately recognise our parentage; nor need we feel ashamed of it. The most humble organism is something much higher than the inorganic dust under our feet; and no one with an unbiassed mind can study any living creature, however humble, without being struck with enthusiasm at its marvellous structure and properties.

CHAPTER VII

On the Races of Man

THE NATURE AND VALUE OF SPECIFIC CHARACTERS • APPLICATION TO THE RACES OF MAN • ARGUMENTS IN FAVOUR OF, AND OPPOSED TO, RANKING THE SO-CALLED RACES OF MAN AS DISTINCT SPECIES • SUB-SPECIES • MONOGENISTS AND POLYGENISTS • CONVERGENCE OF CHARACTER • NUMEROUS POINTS OF RESEMBLANCE IN BODY AND MIND BETWEEN THE MOST DISTINCT RACES OF MAN • THE STATE OF MAN WHEN HE FIRST SPREAD OVER THE EARTH • EACH RACE NOT DESCENDED FROM A SINGLE PAIR • THE EXTINCTION OF RACES • THE FORMATION OF RACES • THE EFFECTS OF CROSSING • SLIGHT INFLUENCE OF THE DIRECT ACTION OF THE CONDITIONS OF LIFE • SLIGHT OR NO INFLUENCE OF NATURAL SELECTION • SEXUAL SELECTION

IT IS NOT MY INTENTION here to describe the several so-called races of men; but to inquire what is the value of the differences between them under a classificatory point of view, and how they have originated. In determining whether two or more allied forms ought to be ranked as species or varieties, naturalists are practically guided by the following considerations; namely, the amount of difference between them, and whether such differences relate to few or many points of structure, and whether they are of physiological importance; but more especially whether they are constant. Constancy of character is what is chiefly valued and sought for by naturalists. Whenever it can be shewn, or rendered probable, that the forms in question have remained distinct for a long period, this becomes an argument of much weight in favour of treating them as species. Even a slight degree of sterility between any two forms when first crossed, or in their offspring, is generally considered as a decisive test of their specific distinctness; and their continued persistence without blending within the same area, is usually accepted as sufficient evidence, either of some degree of mutual sterility, or in the case of animals of some repugnance to mutual pairing.

Independently of blending from intercrossing, the complete absence, in a well-investigated region, of varieties linking together any two closely-allied forms, is probably the most important of all the criterions of their specific distinctness; and this is a somewhat different consideration from mere constancy of character, for two forms may be highly variable and yet not yield intermediate varieties. Geographical distribution is often unconsciously and sometimes consciously brought into play; so that forms living in two widely separated areas, in which most of the other inhabitants are specifically distinct, are themselves usually looked at as distinct; but in truth this affords no aid in distinguishing geographical races from so-called good or true species.

Now let us apply these generally-admitted principles to the races of man, viewing him in the same spirit as a naturalist would any other animal. In regard to the amount of difference between the races, we must make some allowance for our nice powers of discrimination gained by the long habit of observing ourselves. In India, as Elphinstone remarks,[282] although a newly-arrived European cannot at first distinguish the various native races, yet they soon appear to him extremely dissimilar; and the Hindoo cannot at first perceive any difference between the several European nations. Even the most distinct races of man, with the exception of certain negro tribes, are much more like each other in form than would at first be supposed. This is well shewn by the French photographs in the Collection Anthropologique du Muséum of the men belonging to various races, the greater number of which, as many persons to whom I have shown them have remarked, might pass for Europeans. Nevertheless, these men if seen alive would undoubtedly appear very distinct, so that we are clearly much influenced in our judgment by the mere colour of the skin and hair, by slight differences in the features, and by expression.

There is, however, no doubt that the various races, when carefully compared and measured, differ much from each other,—as in the texture of the hair, the relative proportions of all parts of the body,[283] the capacity of the lungs, the form and capacity of the skull, and even in the convolutions of the brain.[284] But it would be an endless task to specify the numerous points of structural difference. The races differ also in constitution, in acclimatisation, and in liability to certain diseases. Their mental characteristics are likewise very distinct; chiefly as it would appear in their emotional, but partly in their intellectual, faculties. Every one who has had the opportunity of comparison, must have been struck with the contrast between the taciturn, even morose, aborigines of S. America and the light-hearted, talkative negroes. There is a nearly similar contrast between the Malays and the Papuans,[285] who live under the same physical conditions, and are separated from each other only by a narrow space of sea.

We will first consider the arguments which may be advanced in favour of classing the races of man as distinct species, and then those on the other side. If a naturalist, who had never before seen such beings, were to compare a Negro, Hottentot, Australian, or Mongolian, he would at once perceive that they differed in a multitude of characters, some of slight and some of considerable importance. On inquiry he would find that they were adapted to live under widely different climates, and that they differed somewhat in bodily constitution and mental disposition. If he were then told that hundreds of similar specimens could be brought from the same countries, he would assuredly declare that they were as good species as many to which he had been in the habit of affixing specific names. This conclusion would be greatly strengthened as soon as he had ascertained that these forms had all retained the same character for many centuries; and that negroes, apparently identical with existing negroes, had lived at least 4000 years ago.[286] He would also hear from an excellent observer, Dr. Lund,[287] that the human skulls found in the caves of Brazil, entombed with many extinct mammals, belonged to the same type as that now prevailing throughout the American Continent.

Our naturalist would then perhaps turn to geographical distribution, and he would probably declare that forms differing not only in appearance, but fitted for the hottest and dampest or driest countries, as well as for the arctic regions, must be distinct species. He might appeal to the fact that no one species in the group next to man, namely the Quadrumana, can resist a low temperature or any considerable change of climate; and that those species which come nearest to man have never been reared to maturity, even under the temperate climate of Europe. He would be deeply impressed with the fact, first noticed by Agassiz,[288] that the different races of man are distributed over the world in the same zoological provinces, as those inhabited by undoubtedly distinct species and genera of mammals. This is manifestly the case with the Australian, Mongolian, and Negro races of man; in a less well-marked manner with the Hottentots; but plainly with the Papuans and Malays, who are separated, as Mr. Wallace has shewn, by nearly the same line which divides the great Malayan and Australian zoological provinces. The aborigines of America range throughout the Continent; and this at first appears opposed to the above rule, for most of the productions of the Southern and Northern halves differ widely; yet some few living forms, as the opossum, range from the one into the other, as did formerly some of the gigantic Edentata. The Esquimaux, like other Arctic animals, extend round the whole polar regions. It should be observed that the mammalian forms which inhabit the several zoological provinces, do not differ from each other in the same degree; so that it can hardly be considered as an anomaly that the Negro differs more, and the American much less, from the other races of man than do the mammals of the same continents from those of the other provinces. Man, it may be added, does not appear to have aboriginally inhabited any oceanic island; and in this respect he resembles the other members of his class.

In determining whether the varieties of the same kind of domestic animal should be ranked as specifically distinct, that is, whether any of them are descended from distinct wild species, every naturalist would lay much stress on the fact, if established, of their external parasites being specifically distinct. All the more stress would be laid on this fact, as it would be an exceptional one, for I am informed by Mr. Denny that the most different kinds of dogs, fowls, and pigeons, in England, are infested by the same species of Pediculi or lice. Now Mr. A. Murray has carefully examined the Pediculi collected in different countries from the different races of man;[289]and he finds that they differ, not only in colour, but in the structure of their claws and limbs. In every case in which numerous specimens were obtained the differences were constant. The surgeon of a whaling ship in the Pacific assured me that when the Pediculi, with which some Sandwich Islanders on board swarmed, strayed on to the bodies of the English sailors, they died in the course of three or four days. These Pediculi were darker coloured and appeared different from those proper to the natives of Chiloe in South America, of which he gave me specimens. These, again, appeared larger and much softer than European lice. Mr. Murray procured four kinds from Africa, namely from the Negroes of the Eastern and Western coasts, from the Hottentots and Caffres; two kinds from the natives of Australia; two from North, and two from South America. In these latter cases it may be presumed that the Pediculi came from natives inhabiting different districts. With insects slight structural differences, if constant, are generally esteemed of specific value: and the fact of the races of man being infested by parasites, which appear to be specifically distinct, might fairly be urged as an argument that the races themselves ought to be classed as distinct species.

Our supposed naturalist having proceeded thus far in his investigation, would next inquire whether the races of men, when crossed, were in any degree sterile. He might consult the work[290] of a cautious and philosophical observer, Professor Broca; and in this he would find good evidence that some races were quite fertile together; but evidence of an opposite nature in regard to other races. Thus it has been asserted that the native women of

Australia and Tasmania rarely produce children to European men; the evidence, however, on this head has now been shewn to be almost valueless. The half-castes are killed by the pure blacks; and an account has lately been published of eleven half-caste youths murdered and burnt at the same time, whose remains were found by the police.[291] Again, it has often been said that when mulattoes intermarry they produce few children; on the other hand, Dr. Bachman of Charlestown[292] positively asserts that he has known mulatto families which have intermarried for several generations, and have continued on an average as fertile as either pure whites or pure blacks. Inquiries formerly made by Sir C. Lyell on this subject led him, as he informs me, to the same conclusion. In the United States the census for the year 1854 included, according to Dr. Bachman, 405,751 mulattoes; and this number, considering all the circumstances of the case, seems small; but it may partly be accounted for by the degraded and anomalous position of the class, and by the profligacy of the women. A certain amount of absorption of mulattoes into negroes must always be in progress; and this would lead to an apparent diminution of the former. The inferior vitality of mulattoes is spoken of in a trustworthy work[293] as a well-known phenomenon; but this is a different consideration from their lessened fertility; and can hardly be advanced as a proof of the specific distinctness of the parent races. No doubt both animal and vegetable hybrids, when produced from extremely distinct species, are liable to premature death; but the parents of mulattoes cannot be put under the category of extremely distinct species. The common Mule, so notorious for long life and vigour, and yet so sterile, shews how little necessary connection there is in hybrids between lessened fertility and vitality: other analogous cases could be added.

Even if it should hereafter be proved that all the races of men were perfectly fertile together, he who was inclined from other reasons to rank them as distinct species, might with justice argue that fertility and sterility are not safe criterions of specific distinctness. We know that these qualities are easily affected by changed conditions of life or by close inter-breeding, and that they are governed by highly complex laws, for instance that of the unequal fertility of reciprocal crosses between the same two species. With forms which must be ranked as undoubted species, a perfect series exists from those which are absolutely sterile when crossed, to those which are almost or quite fertile. The degrees of sterility do not coincide strictly with the degrees of difference in external structure or habits of life. Man in many respects may be compared with those animals which have long been domesticated, and a large body of evidence can be advanced in favour of the Pallasian doctrine[294] that domestication tends to eliminate the sterility which is so general a result of the crossing of species in a state of nature. From these several considerations, it may be justly urged that the perfect fertility of the intercrossed races of man, if established, would not absolutely preclude us from ranking them as distinct species.

Independently of fertility, the character of the offspring from a cross has sometimes been thought to afford evidence whether the parent-forms ought to be ranked as species or varieties; but after carefully studying the evidence, I have come to the conclusion that no general rules of this kind can be trusted. Thus with mankind the offspring of distinct races resemble in all respects the offspring of true species and of varieties. This is shewn, for instance, by the manner in which the characters of both parents are blended, and by one form absorbing another through repeated crosses. In this latter case the progeny both of crossed species and varieties retain for a long period a tendency to revert to their ancestors, especially to that one which is prepotent in transmission. When any character has suddenly appeared in a race or species as the result of a single act of variation, as is general with monstrosities,[295] and this race is crossed with another not thus characterised, the characters in question do not commonly appear in a blended condition in the young, but are transmitted to them either perfectly developed or not at all. As with the crossed races of man cases of this kind rarely or never occur, this may be used as an argument against the view suggested by some ethnologists, namely that certain characters, for instance the blackness of the negro, first appeared as a sudden variation or sport. Had this occurred, it is probable that mulattoes would often have been born, either completely black or completely white.

We have now seen that a naturalist might feel himself fully justified in ranking the races of man as distinct species; for he has found that they are distinguished by many differences in structure and constitution, some being of importance. These differences have, also, remained nearly constant for very long periods of time. He will have been in some degree influenced by the enormous range of man, which is a great anomaly in the class of mammals, if mankind be viewed as a single species. He will have been struck with the distribution of the several so-called races, in accordance with that of other undoubtedly distinct species of mammals. Finally he might urge that the mutual fertility of all the races has not as yet been fully proved; and even if proved would not be an absolute proof of their specific identity.

On the other side of the question, if our supposed naturalist were to enquire whether the forms of man kept distinct like ordinary species, when mingled together in large numbers in the same country, he would immediately discover that this was by no means the case. In Brazil he would behold an immense mongrel population of Negroes and Portuguese; in Chiloe and other parts of South America, he would behold the whole population consisting of Indians and Spaniards blended in various degrees.[296] In many parts of the same continent he would meet with the most complex crosses between Negroes, Indians, and Europeans; and such triple crosses afford the severest test, judging from the vegetable kingdom, of the mutual fertility of the parent-forms. In one island of the Pacific he would find a small population of mingled Polynesian and English blood; and in the Viti Archipelago a population of Polynesians and Negritos crossed in all degrees. Many analogous cases could be added, for instance, in South Africa. Hence the races of man are not sufficiently distinct to co-exist without fusion; and this it is, which in all ordinary cases affords the usual test of specific distinctness.

Our naturalist would likewise be much disturbed as soon as he perceived that the distinctive characters of every race of man were highly variable. This strikes every one when he first beholds the negro-slaves in Brazil, who have been imported from all parts of Africa. The same remark holds good with the Polynesians, and with many other races. It may be doubted whether any character can be named which is distinctive of a race and is constant. Savages, even within the limits of the same tribe, are not nearly so uniform in character, as has often been said. Hottentot women offer certain peculiarities, more strongly marked than those occurring in any other race, but these are known not to be of constant occurrence. In the several American tribes, colour and hairyness differ considerably; as does colour to a certain degree, and the shape of the features greatly, in the Negroes of Africa. The shape of the skull varies much in some races;[297] and so it is with every other character. Now all naturalists have learnt by dearly-bought experience, how rash it is to attempt to define species by the aid of inconstant characters.

But the most weighty of all the arguments against treating the races of man as distinct species, is that they graduate into each other, independently in many cases, as far as we can judge, of their having intercrossed. Man has been studied more carefully than any other organic being, and yet there is the greatest possible diversity amongst capable judges whether he should be classed as a single species or race, or as two (Virey), as three (Jacquinot), as four (Kant), five (Blumenbach), six (Buffon), seven (Hunter), eight (Agassiz), eleven (Pickering), fifteen (Bory St. Vincent), sixteen (Desmoulins), twenty-two (Morton), sixty (Crawfurd), or as sixty-three, according to Burke.[298] This diversity of judgment does not prove that the races ought not to be ranked as species, but it shews that they graduate into each other, and that it is hardly possible to discover clear distinctive characters between them.

Every naturalist who has had the misfortune to undertake the description of a group of highly varying organisms, has encountered cases (I speak after experience) precisely like that of man; and if of a cautious disposition, he will end by uniting all the forms which graduate into each other as a single species; for he will say to himself that he has no right to give names to objects which he cannot define. Cases of this kind occur in the Order which includes man, namely in certain genera of monkeys; whilst in other genera, as in Cercopithecus, most of the species can be determined with certainty. In the American genus Cebus, the various forms are ranked by some naturalists as species, by others as mere geographical races. Now if numerous specimens of Cebus were collected from all parts of South America, and those forms which at present appear to be specifically distinct, were found to graduate into each other by close steps, they would be ranked by most naturalists as mere varieties or races; and thus the greater number of naturalists have acted with respect to the races of man. Nevertheless it must be confessed that there are forms, at least in the vegetable kingdom,[299] which we cannot avoid naming as species, but which are connected together, independently of intercrossing, by numberless gradations.

Some naturalists have lately employed the term "sub-species" to designate forms which possess many of the characteristics of true species, but which hardly deserve so high a rank. Now if we reflect on the weighty arguments, above given, for raising the races of man to the dignity of species, and the insuperable difficulties on the other side in defining them, the term "sub-species" might here be used with much propriety. But from long habit the term "race" will perhaps always be employed. The choice of terms is only so far important as it is highly desirable to use, as far as that may be possible, the same terms for the same degrees of difference. Unfortunately this is rarely possible; for within the same family the larger genera generally include closely-allied forms, which can be distinguished only with much difficulty, whilst the smaller genera include forms that are perfectly distinct; yet all must equally be ranked as species. So again the species within the same large genus by no means resemble each other to the same degree: on the contrary, in most cases some of them can be arranged in little groups round other species, like satellites round planets.[300]

The question whether mankind consists of one or several species has of late years been much agitated by anthropologists, who are divided into two schools of monogenists and polygenists. Those who do not admit the principle of evolution, must look at species either as separate creations or as in some manner distinct entities; and they must decide what forms to rank as species by the analogy of other organic beings which are commonly thus received. But it is a hopeless endeavour to decide this point on sound grounds, until some definition of the term "species" is generally accepted; and the definition must not include an element which cannot possibly be ascertained, such as an act of creation. We might as well attempt without any definition to decide whether a certain number of houses should be called a village, or town, or city. We have a practical illustration of the difficulty in the never-ending doubts whether many closely-allied mammals, birds, insects, and plants, which represent each other in North America and Europe, should be ranked species or geographical races; and so it is with the productions of many islands situated at some little distance from the nearest continent.

Those naturalists, on the other hand, who admit the principle of evolution, and this is now admitted by the greater number of rising men, will feel no doubt that all the races of man are descended from a single primitive stock; whether or not they think fit to designate them as distinct species, for the sake of expressing their amount of difference.[301] With our domestic animals the question whether the various races have arisen from one or more species is different. Although all such races, as well as all the natural species within the same genus, have undoubtedly sprung from the same primitive stock, yet it is a fit subject for discussion, whether, for instance, all the domestic races of the dog have acquired their present differences since some one species was first domesticated and bred by man; or whether they owe some of their characters to inheritance from distinct species, which had already been modified in a state of nature. With mankind no such question can arise, for he cannot be said to have been domesticated at any particular period.

When the races of man diverged at an extremely remote epoch from their common progenitor, they will have differed but little from each other, and been few in number; consequently they will then, as far as their distinguishing characters are concerned, have had less claim to rank as distinct species, than the existing so-called races. Nevertheless such early races would perhaps have been ranked by some naturalists as distinct species, so arbitrary is the term, if their differences, although extremely slight, had been more constant than at present, and had not graduated into each other.

It is, however, possible, though far from probable, that the early progenitors of man might at first have diverged much in character, until they became more unlike each other than are any existing races; but that subsequently, as suggested by Vogt,[302] they converged in character. When man selects for the same object the offspring of two distinct species, he sometimes induces, as far as general appearance is concerned, a considerable amount of convergence. This is the case, as shewn by Von Nathusius,[303] with the improved breeds of pigs, which are descended from two distinct species; and in a less well-marked manner with the improved breeds of cattle. A great anatomist, Gratiolet, maintains that the anthropomorphous apes do not form a natural sub-group; but that the orang is a highly developed gibbon or semnopithecus; the chimpanzee a highly developed macacus; and the gorilla a highly developed mandrill. If this conclusion, which rests almost exclusively on brain-characters, be admitted, we should have a case of convergence at least in external characters, for the anthropomorphous apes are certainly more like each other in many points than they are to other apes. All analogical resemblances, as of a whale to a fish, may indeed be said to be cases of convergence; but this term has never been applied to superficial and adaptive resemblances. It would be extremely rash in most cases to attribute to convergence close similarity in many points of structure in beings which had once been widely different. The form of a crystal is determined solely by the molecular forces, and it is not surprising that dissimilar substances should sometimes assume the same form; but with organic beings we should bear in mind that the form of each depends on an infinitude of complex relations, namely on the variations which have arisen, these being due to causes far too intricate to be followed out,—on the nature of the variations which have been preserved, and this depends on the surrounding physical conditions, and in a still higher degree on the surrounding organisms with which each has come into competition,—and lastly, on inheritance (in itself a fluctuating element) from innumerable progenitors, all of which have had their forms determined through equally complex relations. It appears utterly incredible that two organisms, if differing in a marked manner, should ever afterwards converge so closely as to lead to a near approach to identity throughout their whole organisation. In the case of the convergent pigs above referred to, evidence of their descent from two primitive stocks is still plainly retained, according to Von Nathusius, in certain bones of their skulls. If the races of man were descended, as supposed by some naturalists, from two or more distinct species, which had differed as much, or nearly as much, from each other, as the orang differs from the gorilla, it can hardly be doubted that marked differences in the structure of certain bones would still have been discoverable in man as he now exists.

Although the existing races of man differ in many respects, as in colour, hair, shape of skull, proportions of the body, &c., yet if their whole organisation be taken into consideration they are found to resemble each other closely in a multitude of points. Many of these points are of so unimportant or of so singular a nature, that it is extremely improbable that they should have been independently acquired by aboriginally distinct species or races. The same remark holds good with equal or greater force with respect to the numerous points of mental similarity between the most distinct races of man. The American aborigines, Negroes and Europeans differ as much from each other in mind as any three races that can be named; yet I was incessantly struck, whilst living with the Fuegians on board the "Beagle," with the many little traits of character, shewing how similar their minds were to ours; and so it was with a full-blooded negro with whom I happened once to be intimate.

He who will carefully read Mr. Tylor's and Sir J. Lubbock's interesting works[304] can hardly fail to be deeply impressed with the close similarity between the men of all races in tastes, dispositions and habits. This is shewn by the pleasure which they all take in dancing, rude music, acting, painting, tattooing, and otherwise decorating themselves,—in their mutual comprehension of gesture-language—and, as I shall be able to shew in a future essay, by the same expression in their features, and by the same inarticulate cries, when they are excited by various emotions. This similarity, or rather identity, is striking, when contrasted with the different expressions which may be observed in distinct species of monkeys. There is good evidence that the art of shooting with bows and arrows has not been handed down from any common progenitor of mankind, yet the stone arrow-heads, brought from the most distant parts of the world and manufactured at the most remote periods, are, as Nilsson has shewn,[305] almost identical; and this fact can only be accounted for by the various races having similar inventive or mental powers. The same observation has been made by archæologists [306] with respect to certain widely-prevalent ornaments, such as zigzags, &c; and with respect to various simple beliefs and customs, such as the burying of the dead under megalithic structures. I remember observing in South America,[307] that there, as in so many other parts of the world, man has generally chosen the summits of lofty hills, on which to throw up piles of stones, either for the sake of recording some remarkable event, or for burying his dead.

Now when naturalists observe a close agreement in numerous small details of habits, tastes and dispositions between two or more domestic races, or between nearly-allied natural forms, they use this fact as an argument that all are descended from a common progenitor who was thus endowed; and consequently that all should be classed under the same species. The same argument may be applied with much force to the races of man.

As it is improbable that the numerous and unimportant points of resemblance between the several races of man in bodily structure and mental faculties (I do not here refer to similar customs) should all have been independently acquired, they must have been inherited from progenitors who were thus characterised. We thus gain some insight into the early state of man, before he had spread step by step over the face of the earth. The spreading of man to regions widely separated by the sea, no doubt, preceded any considerable amount of divergence of character in the several races; for otherwise we should sometimes meet with the same race in distinct continents; and this is never the case. Sir J. Lubbock, after comparing the arts now practised by savages in all parts of the world, specifies those which man could not have known, when he first wandered from his original birth-place; for if once learnt they would never have been forgotten.[308] He thus shews that "the spear, which is but a development of the knife-point, and the club, which is but a long hammer, are the only things left." He admits, however, that the art of making fire probably had already been discovered, for it is common to all the races now existing, and was known to the ancient cave-inhabitants of Europe. Perhaps the art of making rude canoes or rafts was likewise known; but as man existed at a remote epoch, when the land in many places stood at a very different level, he would have been able, without the aid of canoes, to have spread widely. Sir J. Lubbock further remarks how improbable it is that our earliest ancestors could have "counted as high as ten, considering that so many races now in existence cannot get beyond four." Nevertheless, at this early period, the intellectual and social faculties of man could hardly have been inferior in any extreme degree to those now possessed by the lowest savages; otherwise primeval man could not have been so eminently successful in the struggle for life, as proved by his early and wide diffusion.

From the fundamental differences between certain languages, some philologists have inferred that when man first became widely diffused he was not a speaking animal; but it may be suspected that languages, far less perfect than any now spoken, aided by gestures, might have been used, and yet have left no traces on subsequent and more highly-developed tongues. Without the use of some language, however imperfect, it appears doubtful whether man's intellect could have risen to the standard implied by his dominant position at an early period.

Whether primeval man, when he possessed very few arts of the rudest kind, and when his power of language was extremely imperfect, would have deserved to be called man, must depend on the definition which we employ. In a series of forms graduating insensibly from some ape-like creature to man as he now exists, it would be impossible to fix on any definite point when the term "man" ought to be used. But this is a matter of very little importance. So again it is almost a matter of indifference whether the so-called races of man are thus designated, or are ranked as species or sub-species; but the latter term appears the most appropriate. Finally, we may conclude that when the principles of evolution are generally accepted, as they surely will be before long, the dispute between the monogenists and the polygenists will die a silent and unobserved death.

ONE OTHER QUESTION ought not to be passed over without notice, namely, whether, as is sometimes assumed, each sub-species or race of man has sprung from a single pair of progenitors. With our domestic animals a new race can readily be formed from a single pair possessing some new character, or even from a single individual thus characterised, by carefully matching the varying offspring; but most of our races have been formed, not intentionally from a selected pair, but unconsciously by the preservation of many individuals which have varied, however slightly, in some useful or desired manner. If in one country stronger and heavier horses, and in another country lighter and fleeter horses, were habitually preferred, we may feel sure that two distinct sub-breeds would, in the course of time, be produced, without any particular pairs or individuals having been separated and bred from in either country. Many races have been thus formed, and their manner of formation is closely analogous with that of natural species. We know, also, that the horses which have been brought to the Falkland Islands have become, during successive generations, smaller and weaker, whilst those which have run wild on the Pampas have acquired larger and coarser heads; and such changes are manifestly due, not to any one pair, but to all the individuals having been subjected to the same conditions, aided, perhaps, by the principle of reversion. The new sub-breeds in none of these cases are descended from any single pair, but from many individuals which have varied in different degrees, but in the same general manner; and we may conclude that the races of man have been similarly produced, the modifications being either the direct result of exposure to different conditions, or the indirect result of some form of selection. But to this latter subject we shall presently return.

On the Extinction of the Races of Man.—The partial and complete extinction of many races and sub-races of man are historically known events. Humboldt saw in South America a parrot which was the sole living creature that could speak the language of a lost tribe. Ancient monuments and stone implements found in all parts of the world, of which no tradition is preserved by the present inhabitants, indicate much extinction. Some small and broken tribes, remnants of former races, still survive in isolated and generally mountainous districts. In Europe the ancient races were all, according to Schaaffhausen,[309] "lower in the scale than the rudest living savages;" they must therefore have differed, to a certain extent, from any existing race. The remains described by Professor Broca[310] from Les Eyzies, though they unfortunately appear to have belonged to a single family, indicate a race with a most singular combination of low or simious and high characteristics, and is "entirely different from any other race, ancient or modern, that we have ever heard of." It differed, therefore, from the quaternary race of the caverns of Belgium.

Unfavourable physical conditions appear to have had but little effect in the extinction of races.[311] Man has long lived in the extreme regions of the North, with no wood wherewith to make his canoes or other implements, and with blubber alone for burning and giving him warmth, but more especially for melting the snow. In the Southern extremity of America the Fuegians survive without the protection of clothes, or of any building worthy to be called a hovel. In South Africa the aborigines wander over the most arid plains, where dangerous beasts abound. Man can withstand the deadly influence of the Terai at the foot of the Himalaya, and the pestilential shores of tropical Africa.

Extinction follows chiefly from the competition of tribe with tribe, and race with race. Various checks are always in action, as specified in a former chapter, which serve to keep down the numbers of each savage tribe,—such as periodical famines, the wandering of the parents and the consequent deaths of infants, prolonged suckling, the stealing of women, wars, accidents, sickness, licentiousness, especially infanticide, and, perhaps, lessened fertility from less nutritious food, and many hardships. If from any cause any one of these checks is lessened, even in a slight degree, the tribe thus favoured will tend to increase; and when one of two adjoining tribes becomes more numerous and powerful than the other, the contest is soon settled by war, slaughter, cannibalism, slavery, and absorption. Even when a weaker tribe is not thus abruptly swept away, if it once begins to decrease, it generally goes on decreasing until it is extinct.[312]

When civilised nations come into contact with barbarians the struggle is short, except where a deadly climate gives its aid to the native race. Of the causes which lead to the victory of civilised nations, some are plain and some very obscure. We can see that the cultivation of the land will be fatal in many ways to savages, for they cannot, or will not, change their habits. New diseases and vices are highly destructive; and it appears that in every nation a new disease causes much death, until those who are most susceptible to its destructive influence are gradually weeded out;[313] and so it may be with the evil effects from spirituous liquors, as well as with the unconquerably strong taste for them shewn by so many savages. It further appears, mysterious as is the fact, that the first meeting of distinct and separated people generates disease.[314] Mr. Sproat, who in Vancouver Island closely attended to the subject of extinction, believes that changed habits of life, which always follow from the advent of Europeans, induces much ill-health. He lays, also, great stress on so trifling a cause as that the natives become "bewildered and dull by the new life around them; they lose the motives for exertion, and get no new ones in their place."[315]

The grade of civilisation seems a most important element in the success of nations which come in competition. A few centuries ago Europe feared the inroads of Eastern barbarians; now, any such fear would be ridiculous. It is a more curious fact, that savages did not formerly waste away, as Mr. Bagehot has remarked, before the classical nations, as they now do before modern civilised nations; had they done so, the old moralists would have mused over the event; but there is no lament in any writer of that period over the perishing barbarians.[316]

Although the gradual decrease and final extinction of the races of man is an obscure problem, we can see that it depends on many causes, differing in different places and at different times. It is the same difficult problem as that presented by the extinction of one of the higher animals—of the fossil horse, for instance, which disappeared from South America, soon afterwards to be replaced, within the same districts, by countless troops of the Spanish horse. The New Zealander seems conscious of this parallelism, for he compares his future fate with that of the native rat almost exterminated by the European rat. The difficulty, though great to our imagination, and really great if we wish to ascertain the precise causes, ought not to be so to our reason, as long as we keep steadily in mind that the increase of each species and each race is constantly hindered by various checks; so that if any new check, or cause of destruction, even a slight one, be superadded, the race will surely decrease in number; and as it has everywhere been observed that savages are much opposed to any change of habits, by which means injurious checks could be counterbalanced, decreasing numbers will sooner or later lead to extinction; the end, in most cases, being promptly determined by the inroads of increasing and conquering tribes.

On the Formation of the Races of Man.—It may be premised that when we find the same race, though broken up into distinct tribes, ranging over a great area, as over America, we may attribute their general resemblance to descent from a common stock. In some cases the crossing of races already distinct has led to the formation of new races. The singular fact that Europeans and Hindoos, who belong to the same Aryan stock and speak a language fundamentally the same, differ widely in appearance, whilst Europeans differ but little from Jews, who belong to the Semitic stock and speak quite another language, has been accounted for by Broca[317] through the Aryan branches having been largely crossed during their wide diffusion by various indigenous tribes. When two races in close contact cross, the first result is a heterogeneous mixture: thus Mr. Hunter, in describing the Santali or hill-tribes of India, says that hundreds of imperceptible gradations may be traced "from the black, squat tribes of the mountains to the tall olive-coloured Brahman, with his intellectual brow, calm eyes, and high but narrow head;" so that it is necessary in courts of justice to ask the witnesses whether they are Santalis or Hindoos.[318] Whether a heterogeneous people, such as the inhabitants of some of the Polynesian islands, formed by the crossing of two distinct races, with few or no pure members left, would ever become homogeneous, is not known from direct evidence. But as with our domesticated animals, a crossed breed can certainly, in the course of a few generations, be fixed and made uniform by careful selection,[319] we may infer that the free and prolonged intercrossing during many generations of a heterogeneous mixture would supply the place of selection, and overcome any tendency to reversion, so that a crossed race would ultimately become homogeneous, though it might not partake in an equal degree of the characters of the two parent-races.

Of all the differences between the races of man, the colour of the skin is the most conspicuous and one of the best marked. Differences of this kind, it was formerly thought, could be accounted for by long exposure under different climates; but Pallas first shewed that this view is not tenable, and he has been followed by almost all anthropologists.[320]The view has been rejected chiefly because the distribution of the variously coloured races, most of whom must have long inhabited their present homes, does not coincide with corresponding differences of climate. Weight must also be given to such cases as that of the Dutch families, who, as we hear on excellent authority,[321] have not undergone the least change of colour, after residing for three centuries in South Africa. The uniform appearance in various parts of the world of gypsies and Jews, though the uniformity of the latter has been somewhat exaggerated,[322] is likewise an argument on the same side. A very damp or a very dry atmosphere has been supposed to be more influential in modifying the colour of the skin than mere heat; but as D'Orbigny in South America, and Livingstone in Africa, arrived at diametrically opposite conclusions with respect to dampness and dryness, any conclusion on this head must be considered as very doubtful.[323]

Various facts, which I have elsewhere given, prove that the colour of the skin and hair is sometimes correlated in a surprising manner with a complete immunity from the action of certain vegetable poisons and from the attacks of certain parasites. Hence it occurred to me, that negroes and other dark races might have acquired their dark tints by the darker individuals escaping during a long series of generations from the deadly influence of the miasmas of their native countries.

I afterwards found that the same idea had long ago occurred to Dr. Wells.[324] That negroes, and even mulattoes, are almost completely exempt from the yellow-fever, which is so destructive in tropical America, has long been known.[325] They likewise escape to a large extent the fatal intermittent fevers that prevail along, at least, 2600 miles of the shores of Africa, and which annually cause one-fifth of the white settlers to die, and another fifth to return home invalided.[326] This immunity in the negro seems to be partly inherent, depending on some unknown peculiarity of constitution, and partly the result of acclimatisation. Pouchet[327] states that the negro regiments, borrowed from the Viceroy of Egypt for the Mexican war, which had been recruited near the Soudan, escaped the yellow-fever almost equally well with the negroes originally brought from various parts of Africa, and accustomed to the climate of the West Indies. That acclimatisation plays a part is shewn by the many cases in which negroes, after having resided for some time in a colder climate, have become to a certain extent liable to tropical fevers.[328] The nature of the climate under which the white races have long resided, likewise has some influence on them; for during the fearful epidemic of yellow-fever in Demerara during 1837, Dr. Blair found that the death-rate of the immigrants was proportional to the latitude of the country whence they had come. With the negro the immunity, as far as it is the result of acclimatisation, implies exposure during a prodigious length of time; for the aborigines of tropical America, who have resided there from time immemorial, are not exempt from yellow-fever; and the Rev. B. Tristram states, that there are districts in Northern Africa which the native inhabitants are compelled annually to leave, though the negroes can remain with safety.

That the immunity of the negro is in any degree correlated with the colour of his skin is a mere conjecture: it may be correlated with some difference in his blood, nervous system, or other tissues. Nevertheless, from the facts above alluded to, and from some connection apparently existing between complexion and a tendency to consumption, the conjecture seemed to me not improbable. Consequently I endeavoured, with but little success,[329] to ascertain how far it held good. The late Dr. Daniell, who had long lived on the West Coast of Africa, told me that he did not believe in any such relation. He was himself unusually fair, and had withstood the climate in a wonderful manner. When he first arrived as a boy on the coast, an old and experienced negro chief predicted from his appearance that this would prove the case. Dr. Nicholson, of Antigua, after having attended to this subject, wrote to me that he did not think that dark-coloured Europeans escaped the yellow-fever better than those that were light-coloured. Mr. J. M. Harris altogether denies[330] that Europeans with dark hair withstand a hot climate better than other men; on the contrary, experience has taught him in making a selection of men for service on the coast of Africa, to choose those with red hair. As far, therefore, as these slight indications serve, there seems no foundation for the hypothesis, which has been accepted by several writers, that the colour of the black races may have resulted from darker and darker individuals having survived in greater numbers, during their exposure to the fever-generating miasmas of their native countries.

Although with our present knowledge we cannot account for the strongly-marked differences in colour between the races of man, either through correlation with constitutional peculiarities, or through the direct action of climate; yet we must not quite ignore the latter agency, for there is good reason to believe that some inherited effect is thus produced.[331]

We have seen in our third chapter that the conditions of life, such as abundant food and general comfort, affect in a direct manner the development of the bodily frame, the effects being transmitted. Through the combined influences of climate and changed habits of life, European settlers in the United States undergo, as is generally admitted, a slight but extraordinarily rapid change of appearance. There is, also, a considerable body of evidence shewing that in the Southern States the house-slaves of the third generation present a markedly different appearance from the field-slaves.[332]

If, however, we look to the races of man, as distributed over the world, we must infer that their characteristic differences cannot be accounted for by the direct action of different conditions of life, even after exposure to them for an enormous period of time. The Esquimaux live exclusively on animal food; they are clothed in thick fur, and are exposed to intense cold and to prolonged darkness; yet they do not differ in any extreme degree from the inhabitants of Southern China, who live entirely on vegetable food and are exposed almost naked to a hot, glaring climate. The unclothed Fuegians live on the marine productions of their inhospitable shores; the Botocudos of Brazil wander about the hot forests of the interior and live chiefly on vegetable productions; yet these tribes resemble each other so closely that the Fuegians on board the "Beagle" were mistaken by some Brazilians for Botocudos. The Botocudos again, as well as the other inhabitants of tropical America, are wholly different from the Negroes who inhabit the opposite shores of the Atlantic, are exposed to a nearly similar climate, and follow nearly the same habits of life.

Nor can the differences between the races of man be accounted for, except to a quite insignificant degree, by the inherited effects of the increased or decreased use of parts. Men who habitually live in canoes, may have their legs somewhat stunted; those who inhabit lofty regions have their chests enlarged; and those who constantly use certain sense-organs have the cavities in which they are lodged somewhat increased in size, and their features consequently a little modified. With civilised nations, the reduced size of the jaws from lessened use, the habitual play of different muscles serving to express different emotions, and the increased size of the brain from greater intellectual activity, have together produced a considerable effect on their general appearance in comparison with savages.[333] It is also possible that increased bodily stature, with no corresponding increase in the size of the brain, may have given to some races (judging from the previously adduced cases of the rabbits) an elongated skull of the dolichocephalic type.

Lastly, the little-understood principle of correlation will almost certainly have come into action, as in the case of great muscular development and strongly projecting supra-orbital ridges. It is not improbable that the texture of the hair, which differs much in the different races, may stand in some kind of correlation with the structure of the skin; for the colour of the hair and skin are certainly correlated, as is its colour and texture with the Mandans.[334] The colour of the skin and the odour emitted by it are likewise in some manner connected. With the breeds of sheep the number of hairs within a given space and the number of the excretory pores stand in some relation to each other.[335] If we may judge from the analogy of our domesticated animals, many modifications of structure in man probably come under this principle of correlated growth.

We have now seen that the characteristic differences between the races of man cannot be accounted for in a satisfactory manner by the direct action of the conditions of life, nor by the effects of the continued use of parts, nor through the principle of correlation. We are therefore led to inquire whether slight individual differences, to which man is eminently liable, may not have been preserved and augmented during a long series of generations through natural selection. But here we are at once met by the objection that beneficial variations alone can be thus preserved; and as far as we are enabled to judge (although always liable to error on this head) not one of the external differences between the races of man are of any direct or special service to him. The intellectual and moral or social faculties must of course be excepted from this remark; but differences in these faculties can have had little or no influence on external characters. The variability of all the characteristic differences between the races, before referred to, likewise indicates that these differences cannot be of much importance; for, had they been important, they would long ago have been either fixed and preserved, or eliminated. In this respect man resembles those forms, called by naturalists protean or polymorphic, which have remained extremely variable, owing, as it seems, to their variations being of an indifferent nature, and consequently to their having escaped the action of natural selection.

We have thus far been baffled in all our attempts to account for the differences between the races of man; but there remains one important agency, namely Sexual Selection, which appears to have acted as powerfully on man, as on many other animals. I do not intend to assert that sexual selection will account for all the differences between the races. An unexplained residuum is left, about which we can in our ignorance only say, that as individuals are continually born with, for instance, heads a little rounder or narrower, and with noses a little longer or shorter, such slight differences might become fixed and uniform, if the unknown agencies which induced them were to act in a more constant manner, aided by long-continued intercrossing. Such modifications come under the provisional class, alluded to in our fourth chapter, which for the want of a better term have been called spontaneous variations. Nor do I pretend that the effects of sexual selection can be indicated with scientific precision; but it can be shewn that it would be an inexplicable fact if man had not been modified by this agency, which has acted so powerfully on innumerable animals, both high and low in the scale. It can further be shewn that the differences between the races of man, as in colour, hairyness, form of features, &c., are of the nature which it might have been expected would have been acted on by sexual selection. But in order to treat this subject in a fitting manner, I have found it necessary to pass the whole animal kingdom in review; I have therefore devoted to it the Second Part of this work. At the close I shall return to man, and, after attempting to shew how far he has been modified through sexual selection, will give a brief summary of the chapters in this First Part.

PART II

SEXUAL SELECTION

CHAPTER VIII

Principles of Sexual Selection

SECONDARY SEXUAL CHARACTERS • SEXUAL SELECTION • MANNER OF ACTION • EXCESS OF MALES • POLYGAMY • THE MALE ALONE GENERALLY MODIFIED THROUGH SEXUAL SELECTION • EAGERNESS OF THE MALE • VARIABILITY OF THE MALE • CHOICE EXERTED BY THE FEMALE • SEXUAL COMPARED WITH NATURAL SELECTION • INHERITANCE, AT CORRESPONDING PERIODS OF LIFE, AT CORRESPONDING SEASONS OF THE YEAR, AND AS LIMITED BY SEX • RELATIONS BETWEEN THE SEVERAL FORMS OF INHERITANCE • CAUSES WHY ONE SEX AND THE YOUNG ARE NOT MODIFIED THROUGH SEXUAL SELECTION • SUPPLEMENT ON THE PROPORTIONAL NUMBERS OF THE TWO SEXES THROUGHOUT THE ANIMAL KINGDOM • ON THE LIMITATION OF THE NUMBERS OF THE TWO SEXES THROUGH NATURAL SELECTION

With animals which have their sexes separated, the males necessarily differ from the females in their organs of reproduction; and these afford the primary sexual characters. But the sexes often differ in what Hunter has called secondary sexual characters, which are not directly connected with the act of reproduction; for instance, in the male possessing certain organs of sense or locomotion, of which the female is quite destitute, or in having them more highly-developed, in order that he may readily find or reach her; or again, in the male having special organs of prehension so as to hold her securely. These latter organs of infinitely diversified kinds graduate into, and in some cases can hardly be distinguished from, those which are commonly ranked as primary, such as the complex appendages at the apex of the abdomen in male insects. Unless indeed we confine the term "primary" to the reproductive glands, it is scarcely possible to decide, as far as the organs of prehension are concerned, which ought to be called primary and which secondary.

The female often differs from the male in having organs for the nourishment or protection of her young, as the mammary glands of mammals, and the abdominal sacks of the marsupials. The male, also, in some few cases differs from the female in possessing analogous organs, as the receptacles for the ova possessed by the males of certain fishes, and those temporarily developed in certain male frogs. Female bees have a special apparatus for collecting and carrying pollen, and their ovipositor is modified into a sting for the defence of their larvæ and the community. In the females of many insects the ovipositor is modified in the most complex manner for the safe placing of the eggs. Numerous similar cases could be given, but they do not here concern us. There are, however, other sexual differences quite disconnected with the primary organs with which we are more especially concerned—such as the greater size, strength, and pugnacity of the male, his weapons of offence or means of defence against rivals, his gaudy colouring and various ornaments, his power of song, and other such characters.

Besides the foregoing primary and secondary sexual differences, the male and female sometimes differ in structures connected with different habits of life, and not at all, or only indirectly, related to the reproductive functions. Thus the females of certain flies (Culicidæ and Tabanidæ) are blood-suckers, whilst the males live on flowers and have their mouths destitute of mandibles.[336] The males alone of certain moths and of some crustaceans (*e.g.* Tanais) have imperfect, closed mouths, and cannot feed. The Complemental males of certain cirripedes live like epiphytic plants either on the female or hermaphrodite form, and are destitute of a mouth and prehensile limbs. In these cases it is the male which has been modified and has lost certain important organs, which the other members of the same group possess. In other cases it is the female which has lost such parts; for instance, the female glowworm is destitute of wings, as are many female moths, some of which never leave their cocoons. Many female parasitic crustaceans have lost their natatory legs. In some weevil-beetles (Curculionidæ) there is a great difference between the male and female in the length of the rostrum or snout;[337] but the meaning of this and of many analogous differences, is not at all understood. Differences of structure between the two sexes in relation to different habits of life are generally confined to the lower animals; but with some few birds the beak of the male differs from that of the female. No doubt in most, but apparently not in all these cases, the differences are indirectly connected with the propagation of the species: thus a female which has to nourish a multitude of ova will require more food than the male, and consequently will require special means for procuring it. A male animal which lived for a very short time might without detriment lose through disuse its organs for procuring food; but he would retain his locomotive organs in a perfect state, so that he might reach the female. The female, on the other hand, might safely lose her organs for flying, swimming, or walking, if she gradually acquired habits which rendered such powers useless.

We are, however, here concerned only with that kind of selection, which I have called sexual selection. This depends on the advantage which certain individuals have over other individuals of the same sex and species, in exclusive relation to reproduction. When the two sexes differ in structure in relation to different habits of life, as in the cases above mentioned, they have no doubt been modified through natural selection, accompanied by inheritance limited to one and the same sex. So again the primary sexual organs, and those for nourishing or protecting the young, come under this same head; for those individuals which generated or nourished their offspring best, would leave, *cæteris paribus,* the greatest number to inherit their superiority; whilst those which generated or nourished their offspring badly, would leave but few to inherit their weaker powers. As the male has to search for the female, he requires for this purpose organs of sense and locomotion, but if these organs are necessary for the other purposes of life, as is generally the case, they will have been developed through natural selection. When the male has found the female he sometimes absolutely requires prehensile organs to hold her; thus Dr. Wallace informs me that the males of certain moths cannot unite with the females if their tarsi or feet are broken. The males of many oceanic crustaceans have their legs and antennæ modified in an extraordinary manner for the prehension of the female; hence we may suspect that owing to these animals being washed about by the waves of the open sea, they absolutely require these organs in order to propagate their kind, and if so their development will have been the result of ordinary or natural selection.

When the two sexes follow exactly the same habits of life, and the male has more highly developed sense or locomotive organs than the female, it may be that these in their perfected state are indispensable to the male for finding the female; but in the vast majority of cases, they serve only to give one male an advantage over another, for the less well-endowed males, if time were allowed them, would succeed in pairing with the females; and they would in all other respects, judging from the structure of the female, be equally well adapted for their ordinary habits of life. In such cases sexual selection must have come into action, for the males have acquired their present structure, not from being better fitted to survive in the struggle for existence, but from having gained an advantage over other males, and from having transmitted this advantage to their male offspring alone. It was the importance of this distinction which led me to designate this form of selection as sexual selection. So again, if the chief service rendered to the male by his prehensile organs is to prevent the escape of the female before the arrival of other males, or when assaulted by them, these organs will have been perfected through sexual selection, that is by the advantage acquired by certain males over their rivals. But in most cases it is scarcely possible to distinguish between the effects of natural and sexual selection. Whole chapters could easily be filled with details on the differences between the sexes in their sensory, locomotive, and prehensile organs. As, however, these structures are not more interesting than others adapted for the ordinary purposes of life, I shall almost pass them over, giving only a few instances under each class.

There are many other structures and instincts which must have been developed through sexual selection—such as the weapons of offence and the means of defence possessed by the males for fighting with and driving away their rivals—their courage and pugnacity—their ornaments of many kinds—their organs for producing vocal or instrumental music—and their glands for emitting odours; most of these latter structures serving only to allure or excite the female. That these characters are the result of sexual and not of ordinary selection is clear, as unarmed, unornamented, or unattractive males would succeed equally well in the battle for life and in leaving a numerous progeny, if better endowed males were not present. We may infer that this would be the case, for the females, which are unarmed and unornamented, are able to survive and procreate their kind. Secondary sexual characters of the kind just referred to, will be fully discussed in the following chapters, as they are in many respects interesting, but more especially as they depend on the will, choice, and rivalry of the individuals of either sex. When we behold two males fighting for the possession of the female, or several male birds displaying their gorgeous plumage, and performing the strangest antics before an assembled body of females, we cannot doubt that, though led by instinct, they know what they are about, and consciously exert their mental and bodily powers.

In the same manner as man can improve the breed of his game-cocks by the selection of those birds which are victorious in the cockpit, so it appears that the strongest and most vigorous males, or those provided with the best weapons, have prevailed under nature, and have led to the improvement of the natural breed or species. Through repeated deadly contests, a slight degree of variability, if it led to some advantage, however slight, would suffice for the work of sexual selection; and it is certain that secondary sexual characters are eminently variable. In the same manner as man can give beauty, according to his standard of taste, to his male poultry—can give to the Sebright bantam a new and elegant plumage, an erect and peculiar carriage—so it appears that in a state of nature female birds, by having long selected the more attractive males, have added to their beauty. No doubt this implies powers of discrimination and taste on the part of the female which will at first appear extremely improbable; but I hope hereafter to shew that this is not the case.

From our ignorance on several points, the precise manner in which sexual selection acts is to a certain extent uncertain. Nevertheless if those naturalists who already believe in the mutability of species, will read the following chapters, they will, I think, agree with me that sexual selection has played an important part in the history of the organic world. It is certain that with almost all animals there is a struggle between the males for the possession of the female. This fact is so notorious that it would be superfluous to give instances. Hence the females, supposing that their mental capacity sufficed for the exertion of a choice, could select one out of several males. But in numerous cases it appears as if it had been specially arranged that there should be a struggle between many males. Thus with migratory birds, the males generally arrive before the females at their place of breeding, so that many males are ready to contend for each female. The bird-catchers assert that this is invariably the case with the nightingale and blackcap, as I am informed by Mr. Jenner Weir, who confirms the statement with respect to the latter species.

Mr. Swaysland of Brighton, who has been in the habit, during the last forty years, of catching our migratory birds on their first arrival, writes to me that he has never known the females of any species to arrive before their males. During one spring he shot thirty-nine males of Ray's wagtail *(Budytes Raii)* before he saw a single female. Mr. Gould has ascertained by dissection, as he informs me, that male snipes arrive in this country before the females. In the case of fish, at the period when the salmon ascend our rivers, the males in large numbers are ready to breed before the females. So it apparently is with frogs and toads. Throughout the great class of insects the males almost always emerge from the pupal state before the other sex, so that they generally swarm for a time before any females can be seen.[338] The cause of this difference between the males and females in their periods of arrival and maturity is sufficiently obvious. Those males which annually first migrated into any country, or which in the spring were first ready to breed, or were the most eager, would leave the largest number of offspring; and these would tend to inherit similar instincts and constitutions. On the whole there can be no doubt that with almost all animals, in which the sexes are separate, there is a constantly recurrent struggle between the males for the possession of the females.

Our difficulty in regard to sexual selection lies in understanding how it is that the males which conquer other males, or those which prove the most attractive to the females, leave a greater number of offspring to inherit their superiority than the beaten and less attractive males. Unless this result followed, the characters which gave to certain males an advantage over others, could not be perfected and augmented through sexual selection. When the sexes exist in exactly equal numbers, the worst-endowed males will ultimately find females (excepting where polygamy prevails), and leave as many offspring, equally well fitted for their general habits of life, as the best-endowed males. From various facts and considerations, I formerly inferred that with most animals, in which secondary sexual characters were well developed, the males considerably exceeded the females in number; and this does hold good in some few cases. If the males were to the females as two to one, or as three to two, or even in a somewhat lower ratio, the whole affair would be simple; for the better-armed or more attractive males would leave the largest number of offspring. But after investigating, as far as possible, the numerical proportions of the sexes, I do not believe that any great inequality in number commonly exists. In most cases sexual selection appears to have been effective in the following manner.

Let us take any species, a bird for instance, and divide the females inhabiting a district into two equal bodies: the one consisting of the more vigorous and better-nourished individuals, and the other of the less vigorous and healthy. The former, there can be little doubt, would be ready to breed in the spring before the others; and this is the opinion of Mr. Jenner Weir, who has during many years carefully attended to the habits of birds. There can also be no doubt that the most vigorous, healthy, and best-nourished females would on an average succeed in rearing the largest number of offspring. The males, as we have seen, are generally ready to breed before the females; of the males the strongest, and with some species the best armed, drive away the weaker males; and the former would then unite with the more vigorous and best-nourished females, as these are the first to breed. Such vigorous pairs would surely rear a larger number of offspring than the retarded females, which would be compelled, supposing the sexes to be numerically equal, to unite with the conquered and less powerful males; and this is all that is wanted to add, in the course of successive generations, to the size, strength and courage of the males, or to improve their weapons.

But in a multitude of cases the males which conquer other males, do not obtain possession of the females, independently of choice on the part of the latter. The courtship of animals is by no means so simple and short an affair as might be thought. The females are most excited by, or prefer pairing with, the more ornamented males, or those which are the best songsters, or play the best antics; but it is obviously probable, as has been actually observed in some cases, that they would at the same time prefer the more vigorous and lively males.[339] Thus the more vigorous females, which are the first to breed, will have the choice of many males; and though they may not always select the strongest or best armed, they will select those which are vigorous and well armed, and in other respects the most attractive. Such early pairs would have the same advantage in rearing offspring on the female side as above explained, and nearly the same advantage on the male side. And this apparently has sufficed during a long course of generations to add not only to the strength and fighting-powers of the males, but likewise to their various ornaments or other attractions.

In the converse and much rarer case of the males selecting particular females, it is plain that those which were the most vigorous and had conquered others, would have the freest choice; and it is almost certain that they would select vigorous as well as attractive females. Such pairs would have an advantage in rearing offspring, more especially if the male had the power to defend the female during the pairing-season, as occurs with some of the higher animals, or aided in providing for the young. The same principles would apply if both sexes mutually preferred and selected certain individuals of the opposite sex; supposing that they selected not only the more attractive, but likewise the more vigorous individuals.

Numerical Proportion of the Two Sexes.—I have remarked that sexual selection would be a simple affair if the males considerably exceeded in number the females. Hence I was led to investigate, as far as I could, the proportions between the two sexes of as many animals as possible; but the materials are scanty. I will here give only a brief abstract of the results, retaining the details for a supplementary discussion, so as not to interfere with the course of my argument. Domesticated animals alone afford the opportunity of ascertaining the proportional numbers at birth; but no records have been specially kept for this purpose. By indirect means, however, I have collected a considerable body of statistical data, from which it appears that with most of our domestic animals the sexes are nearly equal at birth. Thus with race-horses, 25,560 births have been recorded during twenty-one years, and the male births have been to the female births as 99·7 to 100. With greyhounds the inequality is greater than with any other animal, for during twelve years, out of 6878 births, the male births have been as 110·1 to 100 female births. It is, however, in some degree doubtful whether it is safe to infer that the same proportional numbers would hold good under natural conditions as under domestication; for slight and unknown differences in the conditions affect to a certain extent the proportion of the sexes. Thus with mankind, the male births in England are as

104·5, in Russia as 108·9, and with the Jews of Livornia as 120 to 100 females. The proportion is also mysteriously affected by the circumstance of the births being legitimate or illegitimate.

For our present purpose we are concerned with the proportion of the sexes, not at birth, but at maturity, and this adds another element of doubt; for it is a well ascertained fact that with man a considerably larger proportion of males than of females die before or during birth, and during the first few years of infancy. So it almost certainly is with male lambs, and so it may be with the males of other animals. The males of some animals kill each other by fighting; or they drive each other about until they become greatly emaciated. They must, also, whilst wandering about in eager search for the females, be often exposed to various dangers. With many kinds of fish the males are much smaller than the females, and they are believed often to be devoured by the latter or by other fishes. With some birds the females appear to die in larger proportion than the males: they are also liable to be destroyed on their nests, or whilst in charge of their young. With insects the female larvæ are often larger than those of the males, and would consequently be more likely to be devoured: in some cases the mature females are less active and less rapid in their movements than the males, and would not be so well able to escape from danger. Hence, with animals in a state of nature, in order to judge of the proportions of the sexes at maturity, we must rely on mere estimation; and this, except perhaps when the inequality is strongly marked, is but little trustworthy. Nevertheless, as far as a judgment can be formed, we may conclude from the facts given in the supplement, that the males of some few mammals, of many birds, of some fish and insects, considerably exceed in number the females.

The proportion between the sexes fluctuates slightly during successive years: thus with race-horses, for every 100 females born, the males varied from 107·1 in one year to 92·6 in another year, and with greyhounds from 116·3 to 95·3. But had larger numbers been tabulated throughout a more extensive area than England, these fluctuations would probably have disappeared; and such as they are, they would hardly suffice to lead under a state of nature to the effective action of sexual selection. Nevertheless with some few wild animals, the proportions seem, as shewn in the supplement, to fluctuate either during different seasons or in different localities in a sufficient degree to lead to such action. For it should be observed that any advantage gained during certain years or in certain localities by those males which were able to conquer other males, or were the most attractive to the females, would probably be transmitted to the offspring and would not subsequently be eliminated. During the succeeding seasons, when from the equality of the sexes every male was everywhere able to procure a female, the stronger or more attractive males previously produced would still have at least as good a chance of leaving offspring as the less strong or less attractive.

Polygamy.—The practice of polygamy leads to the same results as would follow from an actual inequality in the number of the sexes; for if each male secures two or more females, many males will not be able to pair; and the latter assuredly will be the weaker or less attractive individuals. Many mammals and some few birds are polygamous, but with animals belonging to the lower classes I have found no evidence of this habit. The intellectual powers of such animals are, perhaps, not sufficient to lead them to collect and guard a harem of females. That some relation exists between polygamy and the development of secondary sexual characters, appears nearly certain; and this supports the view that a numerical preponderance of males would be eminently favourable to the action of sexual selection. Nevertheless many animals, especially birds, which are strictly monogamous, display strongly-marked secondary sexual characters; whilst some few animals, which are polygamous, are not thus characterised.

We will first briefly run through the class of mammals, and then turn to birds. The gorilla seems to be a polygamist, and the male differs considerably from the female; so it is with some baboons which live in herds containing twice as many adult females as males. In South America the *Mycetes caraya* presents well-marked sexual differences in colour, beard, and vocal organs, and the male generally lives with two or three wives: the male of the *Cebus capucinus* differs somewhat from the female, and appears to be polygamous.[340] Little is known on this head with respect to most other monkeys, but some species are strictly monogamous. The ruminants are eminently polygamous, and they more frequently present sexual differences than almost any other group of mammals, especially in their weapons, but likewise in other characters. Most deer, cattle, and sheep are polygamous; as are most antelopes, though some of the latter are monogamous. Sir Andrew Smith, in speaking of the antelopes of South Africa, says that in herds of about a dozen there was rarely more than one mature male. The Asiatic *Antilope saiga* appears to be the most inordinate polygamist in the world; for Pallas[341] states that the male drives away all rivals, and collects a herd of about a hundred, consisting of females and kids: the female is hornless and has softer hair, but does not otherwise differ much from the male. The horse is polygamous, but, except in his greater size and in the proportions of his body, differs but little from the mare. The wild boar, in his great tusks and some other characters, presents well-marked sexual characters; in Europe and in India he leads a solitary life, except during the breeding-season; but at this season he consorts in India with several females, as Sir W. Elliot, who has had large experience in observing this animal, believes: whether this holds good in Europe is doubtful, but is supported by some statements. The adult male Indian elephant, like the boar, passes much of his time in solitude; but when associating with others, "it is rare to find," as Dr. Campbell states, "more than one male with a whole herd of females." The larger males expel or kill the smaller and weaker ones. The male differs from the female by his immense tusks and greater size, strength, and endurance; so great is the difference in these latter respects, that the males when caught are valued at twenty per cent above the females.[342] With other pachydermatous animals the sexes differ very little or not at all, and they are not, as far as known, polygamists. Hardly a single species amongst the Cheiroptera and Edentata, or in the great Orders of the Rodents and Insectivora, presents well-developed secondary sexual differences; and I can find no account of any species being polygamous, excepting, perhaps, the common rat, the males of which, as some rat-catchers affirm, live with several females.

The lion in South Africa, as I hear from Sir Andrew Smith, sometimes lives with a single female, but generally with more than one, and, in one case, was found with as many as five females, so that he is polygamous. He is, as far as I can discover, the sole polygamist in the whole group of the terrestrial Carnivora, and he alone presents well-marked sexual characters. If, however, we turn to the marine Carnivora, the case is widely different; for many species of seals offer, as we shall hereafter see, extraordinary sexual differences, and they are eminently polygamous. Thus the male sea-elephant of the Southern Ocean, always possesses, according to Péron, several females, and the sea-lion of Forster is said to be surrounded by from twenty to thirty females. In the North, the male sea-bear of Steller is accompanied by even a greater number of females.

With respect to birds, many species, the sexes of which differ greatly from each other, are certainly monogamous. In Great Britain we see well-marked sexual differences in, for instance, the wild-duck which pairs with a single female, with the common blackbird, and with the bullfinch which is said to pair for life. So it is, as I am informed by Mr. Wallace, with the Chatterers or Cotingidæ of South America, and numerous other birds. In several groups I have not been able to discover whether the species are polygamous or monogamous. Lesson says that birds of paradise, so remarkable for their sexual differences, are polygamous, but Mr. Wallace doubts whether he had sufficient evidence. Mr. Salvin informs me that he has been led to believe that humming-birds are polygamous. The male widow-bird, remarkable for his caudal plumes, certainly seems to be a polygamist.[343] I have been assured by Mr. Jenner Weir and by others, that three starlings not rarely frequent the same nest; but whether this is a case of polygamy or polyandry has not been ascertained.

The Gallinaceæ present almost as strongly marked sexual differences as birds of paradise or humming-birds, and many of the species are, as is well known, polygamous; others being strictly monogamous. What a contrast is presented between the sexes by the polygamous peacock or pheasant, and the monogamous guinea-fowl or partridge! Many similar cases could be given, as in the grouse tribe, in which the males of the polygamous capercailzie and black-cock differ greatly from the females; whilst the sexes of the monogamous red grouse and ptarmigan differ very little. Amongst the Cursores, no great number of species offer strongly-marked sexual differences, except the bustards, and the great bustard *(Otis tarda)*, is said to be polygamous. With the Grallatores, extremely few species differ sexually, but the ruff *(Machetes pugnax)* affords a strong exception, and this species is believed by Montagu to be a polygamist. Hence it appears that with birds there often exists a close relation between polygamy and the development of strongly-marked sexual differences. On asking Mr. Bartlett, at the Zoological Gardens, who has had such large experience with birds, whether the male tragopan (one of the Gallinaceæ) was polygamous, I was struck by his answering, "I do not know, but should think so from his splendid colours."

It deserves notice that the instinct of pairing with a single female is easily lost under domestication. The wild-duck is strictly monogamous, the domestic-duck highly polygamous. The Rev. W. D. Fox informs me that with some half-tamed wild-ducks, kept on a large pond in his neighbourhood, so many mallards were shot by the gamekeeper that only one was left for every seven or eight females; yet unusually large broods were reared. The guinea-fowl is strictly monogamous; but Mr. Fox finds that his birds succeed best when he keeps one cock to two or three hens.[344] Canary-birds pair in a state of nature, but the breeders in England successfully put one male to four or five females; nevertheless the first female, as Mr. Fox has been assured, is alone treated as the wife, she and her young ones being fed by him; the others are treated as concubines. I have noticed these cases, as it renders it in some degree probable that monogamous species, in a state of nature, might readily become either temporarily or permanently polygamous.

With respect to reptiles and fishes, too little is known of their habits to enable us to speak of their marriage arrangements. The stickleback (Gasterosteus), however, is said to be a polygamist;[345] and the male during the breeding-season differs conspicuously from the female.

To sum up on the means through which, as far as we can judge, sexual selection has led to the development of secondary sexual characters. It has been shewn that the largest number of vigorous offspring will be reared from the pairing of the strongest and best-armed males, which have conquered other males, with the most vigorous and best-nourished females, which are the first to breed in the spring. Such females, if they select the more attractive, and at the same time vigorous, males, will rear a larger number of offspring than the retarded females, which must pair with the less vigorous and less attractive males. So it will be if the more vigorous males select the more attractive and at the same time healthy and vigorous females; and this will especially hold good if the male defends the female, and aids in providing food for the young. The advantage thus gained by the more vigorous pairs in rearing a larger number of offspring has apparently sufficed to render sexual selection efficient. But a large preponderance in number of the males over the females would be still more efficient; whether the preponderance was only occasional and local, or permanent; whether it occurred at birth, or subsequently from the greater destruction of the females; or whether it indirectly followed from the practice of polygamy.

The Male generally more modified than the Female.—Throughout the animal kingdom, when the sexes differ from each other in external appearance, it is the male which, with rare exceptions, has been chiefly modified; for the female still remains more like the young of her own species, and more like the other members of the same group. The cause of this seems to lie in the males of almost all animals having stronger passions than the females. Hence it is the males that fight together and sedulously display their charms before the females; and those which are victorious transmit their superiority to their male offspring. Why the males do not transmit their characters to both sexes will hereafter be considered. That the males of all mammals eagerly pursue the females is notorious to every one. So it is with birds; but many male birds do not so much pursue the female, as display their plumage, perform strange antics, and pour forth their song, in her presence. With the few fish which have been observed, the male seems much more eager than the female; and so it is with alligators, and apparently with Batrachians. Throughout the enormous class of insects, as Kirby remarks,[346] "the law is, that the male shall seek the female." With spiders and crustaceans, as I hear from two great authorities, Mr. Blackwall and Mr. C. Spence Bate, the males are more active and more erratic in their habits than the females. With insects and crustaceans, when the organs of sense or locomotion are present in the one sex and absent in the other, or when, as is frequently the case, they are more highly developed in the one than the other, it is almost invariably the male, as far as I can discover, which retains such organs, or has them most developed; and this shews that the male is the more active member in the courtship of the sexes.[347]

The female, on the other hand, with the rarest exception, is less eager than the male. As the illustrious Hunter[348] long ago observed, she generally "requires to be courted;" she is coy, and may often be seen endeavouring for a long time to escape from the male. Every one who has attended to the habits of animals will be able to call to mind instances of this kind. Judging from various facts, hereafter to be given, and from the results which may fairly be attributed to sexual selection, the female, though comparatively passive, generally exerts some choice and accepts one male in preference to others. Or she may accept, as appearances would sometimes lead us to believe, not the male which is the most attractive to her, but the one which is the least distasteful. The exertion of some choice on the part of the female seems almost as general a law as the eagerness of the male.

We are naturally led to enquire why the male in so many and such widely distinct classes has been rendered more eager than the female, so that he searches for her and plays the more active part in courtship. It would be no advantage and some loss of power if both sexes were mutually to search for each other; but why should the male almost always be the seeker? With plants, the ovules after fertilisation have to be nourished for a time; hence the pollen is necessarily brought to the female organs—being placed on the stigma, through the agency of insects or of the wind, or by the spontaneous movements of the stamens; and with the Algæ, &c., by the locomotive power of the antherozooids. With lowly-organised animals permanently affixed to the same spot and having their sexes separate, the male element is invariably brought to the female; and we can see the reason; for the ova, even if detached before being fertilised and not requiring subsequent nourishment or protection, would be, from their larger relative size, less easily transported than the male element. Hence plants[349] and many of the lower animals are, in this respect, analogous. In the case of animals not affixed to the same spot, but enclosed within a shell with no power of protruding any part of their bodies, and in the case of animals having little power of locomotion, the males must trust the fertilising element to the risk of at least a short transit through the waters of the sea. It would, therefore, be a great advantage to such animals, as their organisation became perfected, if the males when ready to emit the fertilising element, were to acquire the habit of approaching the female as closely as possible. The males of various lowly-organised animals having thus aboriginally acquired the habit of approaching and seeking the females, the same habit would naturally be transmitted to their more highly developed male descendants; and in order that they should become efficient seekers, they would have to be endowed with strong passions. The acquirement of such passions would naturally follow from the more eager males leaving a larger number of offspring than the less eager.

The great eagerness of the male has thus indirectly led to the much more frequent development of secondary sexual characters in the male than in the female. But the development of such characters will have been much aided, if the conclusion at which I arrived after studying domesticated animals, can be trusted, namely, that the male is more liable to vary than the female. I am aware how difficult it is to verify a conclusion of this kind. Some slight evidence, however, can be gained by comparing the two sexes in mankind, as man has been more carefully observed than any other animal. During the Novara Expedition[350] a vast number of measurements of various parts of the body in different races were made, and the men were found in almost every case to present a greater range of variation than the women; but I shall have to recur to this subject in a future chapter. Mr. J. Wood,[351] who has carefully attended to the variation of the muscles in man, puts in italics the conclusion that "the greatest number of abnormalities in each subject is found in the males." He had previously remarked that "altogether in 102 subjects the varieties of redundancy were found to be half as many again as in females, contrasting widely with the greater frequency of deficiency in females before described." Professor Macalister likewise remarks[352] that variations in the muscles "are probably more common in males than females." Certain muscles which are not normally present in mankind are also more frequently developed in the male than in the female sex, although exceptions to this rule are said to occur. Dr. Burt Wilder[353] has tabulated the cases of 152 individuals with supernumerary digits, of which 86 were males, and 39, or less than half, females; the remaining 27 being of unknown sex. It should not, however, be overlooked that women would more frequently endeavour to conceal a deformity of this kind than men. Whether the large proportional number of deaths of the male offspring of man and apparently of sheep, compared with the female offspring, before, during, and shortly after birth (see supplement), has any relation to a stronger tendency in the organs of the male to vary and thus to become abnormal in structure or function, I will not pretend to conjecture.

In various classes of animals a few exceptional cases occur, in which the female instead of the male has acquired well pronounced secondary sexual characters, such as brighter colours, greater size, strength, or pugnacity. With birds, as we shall hereafter see, there has sometimes been a complete transposition of the ordinary characters proper to each sex; the females having become the more eager in courtship, the males remaining comparatively passive, but apparently selecting, as we may infer from the results, the more attractive females. Certain female birds have thus been rendered more highly coloured or otherwise ornamented, as well as more powerful and pugnacious than the males, these characters being transmitted to the female offspring alone.

It may be suggested that in some cases a double process of selection has been carried on; the males having selected the more attractive females, and the latter the more attractive males. This process however, though it might lead to the modification of both sexes, would not make the one sex different from the other, unless indeed their taste for the beautiful differed; but this is a supposition too improbable in the case of any animal, excepting man, to be worth considering. There are, however, many animals, in which the sexes resemble each other, both being furnished with the same ornaments, which analogy would lead us to attribute to the agency of sexual selection. In such cases it may be suggested with more plausibility, that there has been a double or mutual process of sexual selection; the more vigorous and precocious females having selected the more attractive and vigorous males, the latter having rejected all except the more attractive females. But from what we know of the habits of animals, this view is hardly probable, the male being generally eager to pair with any female. It is more probable that the ornaments common to both sexes were acquired by one sex, generally the male, and then transmitted to the offspring of both sexes. If, indeed, during a lengthened period the males of any species were greatly to exceed the females in number, and then during another lengthened period under different conditions the reverse were to occur, a double, but not simultaneous, process of sexual selection might easily be carried on, by which the two sexes might be rendered widely different.

We shall hereafter see that many animals exist, of which neither sex is brilliantly coloured or provided with special ornaments, and yet the members of both sexes or of one alone have probably been modified through sexual selection. The absence of bright tints or other ornaments may be the result of variations of the right kind never having occurred, or of the animals themselves preferring simple colours, such as plain black or white. Obscure colours have often been acquired through natural selection for the sake of protection, and the acquirement through sexual selection of conspicuous colours, may have been checked from the danger thus incurred. But in other cases the males have probably struggled together during long ages, through brute force, or by the display of their charms, or by both means combined, and yet no effect will have been produced unless a larger number of offspring were left by the more successful males to inherit their superiority, than by the less successful males; and this, as previously shewn, depends on various complex contingencies.

Sexual selection acts in a less rigorous manner than natural selection. The latter produces its effects by the life or death at all ages of the more or less successful individuals. Death, indeed, not rarely ensues from the conflicts of rival males. But generally the less successful male merely fails to obtain a female, or obtains later in the season a retarded and less vigorous female, or, if polygamous, obtains fewer females; so that they leave fewer, or less vigorous, or no offspring. In regard to structures acquired through ordinary or natural selection, there is in most cases, as long as the conditions of life remain the same, a limit to the amount of advantageous modification in relation to certain special ends; but in regard to structures adapted to make one male victorious over another, either in fighting or in charming the female, there is no definite limit to the amount of advantageous modification; so that as long as the proper variations arise the work of sexual selection will go on. This circumstance may partly account for the frequent and extraordinary amount of variability presented by secondary sexual characters. Nevertheless, natural selection will determine that characters of this kind shall not be acquired by the victorious males, which would be injurious to them in any high degree, either by expending too much of their vital powers, or by exposing them to any great danger. The development, however, of certain structures—of the horns, for instance, in certain stags—has been carried to a wonderful extreme; and in some instances to an extreme which, as far as the general conditions of life are concerned, must be slightly injurious to the male. From this fact we learn that the advantages which favoured males have derived from conquering other males in battle or courtship, and thus leaving a numerous progeny, have been in the long run greater than those derived from rather more perfect adaptation to the external conditions of life. We shall further see, and this could never have been anticipated, that the power to charm the female has been in some few instances more important than the power to conquer other males in battle.

LAWS OF INHERITANCE

IN ORDER TO UNDERSTAND how sexual selection has acted, and in the course of ages has produced conspicuous results with many animals of many classes, it is necessary to bear in mind the laws of inheritance, as far as they are known. Two distinct elements are included under the term "inheritance," namely the transmission and the development of characters; but as these generally go together, the distinction is often overlooked. We see this distinction in those characters which are transmitted through the early years of life, but are developed only at maturity or during old age. We see the same distinction more clearly with secondary sexual characters, for these are transmitted through both sexes, though developed in one alone. That they are present in both sexes, is manifest when two species, having strongly-marked sexual characters, are crossed, for each transmits the characters proper to its own male and female sex to the hybrid offspring of both sexes. The same fact is likewise manifest, when characters proper to the male are occasionally developed in the female when she grows old or becomes diseased; and so conversely with the male. Again, characters occasionally appear, as if transferred from the male to the female, as when, in certain breeds of the fowl, spurs regularly appear in the young and healthy females; but in truth they are simply developed in the female; for in every breed each detail in the structure of the spur is transmitted through the female to her male offspring. In all cases of reversion, characters are transmitted through two, three, or many generations, and are then under certain unknown favourable conditions developed. This important distinction between transmission and development will be easiest kept in mind by the aid of the hypothesis of pangenesis, whether or not it be accepted as true. According to this hypothesis, every unit or cell of the body throws off gemmules or undeveloped atoms, which are transmitted to the offspring of both sexes, and are multiplied by self-division. They may remain undeveloped during the early years of life or during successive generations; their development into units or cells, like those from which they were derived, depending on their affinity for, and union with, other units or cells previously developed in the due order of growth.

Inheritance at Corresponding Periods of Life.—This tendency is well established. If a new character appears in an animal whilst young, whether it endures throughout life or lasts only for a time, it will reappear, as a general rule, at the same age and in the same manner in the offspring. If, on the other hand, a new character appears at maturity, or even during old age, it tends to reappear in the offspring at the same advanced age. When deviations from this rule occur, the transmitted characters much oftener appear before than after the corresponding age. As I have discussed this subject at sufficient length in another work,[354] I will here merely give two or three instances, for the sake of recalling the subject to the reader's mind. In several breeds of the Fowl, the chickens whilst covered with down, in their first true plumage, and in their adult plumage, differ greatly from each other, as well as from their common parent-form, the *Gallus bankiva*; and these characters are faithfully transmitted by each breed to their offspring at the corresponding period of life. For instance, the chickens of spangled Hamburghs, whilst covered with down, have a few dark spots on the head and rump, but are not longitudinally striped, as in many other breeds; in their first true plumage, "they are beautifully pencilled," that is each feather is transversely marked by numerous dark bars; but in their second plumage the feathers all become spangled or tipped with a dark round spot.[355] Hence in this breed variations have occurred and have been transmitted at three distinct periods of life. The Pigeon offers a more remarkable case, because the aboriginal parent-species does not undergo with advancing age any change of plumage, excepting that at maturity the breast becomes more iridescent; yet there are breeds which do not acquire their characteristic colours until they have moulted two, three, or four times; and these modifications of plumage are regularly transmitted.

Inheritance at Corresponding Seasons of the Year.—With animals in a state of nature innumerable instances occur of characters periodically appearing at different seasons. We see this with the horns of the stag, and with the fur of arctic animals which becomes thick and white during the winter. Numerous birds acquire bright colours and other decorations during the breeding-season alone. I can throw but little light on this form of inheritance from facts observed under domestication. Pallas states,[356] that in Siberia domestic cattle and horses periodically become lighter-coloured during the winter; and I have observed a similar marked change of colour in certain ponies in England. Although I do not know that this tendency to assume a differently coloured coat during different seasons of the year is transmitted, yet it probably is so, as all shades of colour are strongly inherited by the horse. Nor is this form of inheritance, as limited by season, more remarkable than inheritance as limited by age or sex.

Inheritance as Limited by Sex.—The equal transmission of characters to both sexes is the commonest form of inheritance, at least with those animals which do not present strongly-marked sexual differences, and indeed with many of these. But characters are not rarely transferred exclusively to that sex, in which they first appeared. Ample evidence on this head has been advanced in my work on Variation under Domestication; but a few instances may here be given. There are breeds of the sheep and goat, in which the horns of the male differ greatly in shape from those of the female; and these differences, acquired under domestication, are regularly transmitted to the same sex. With tortoise-shell cats the females alone, as a general rule, are thus coloured, the males being rusty-red. With most breeds of the fowl, the characters proper to each sex are transmitted to the same sex alone. So general is this form of transmission that it is an anomaly when we see in certain breeds variations transmitted equally to both sexes. There are also certain sub-breeds of the fowl in which the males can hardly be distinguished from each other, whilst the females differ considerably in colour. With the pigeon the sexes of the parent-species do not differ in any external character; nevertheless in certain domesticated breeds the male is differently coloured from the female.[357] The wattle in the English Carrier pigeon and the crop in the Pouter are more highly developed in the male than in the female; and although these characters have been gained through long-continued selection by man, the difference between the two sexes is wholly due to the form of inheritance which has prevailed; for it has arisen, not from, but rather in opposition to, the wishes of the breeder.

Most of our domestic races have been formed by the accumulation of many slight variations; and as some of the successive steps have been transmitted to one sex alone, and some to both sexes, we find in the different breeds of the same species all gradations between great sexual dissimilarity and complete similarity. Instances have already been given with the breeds of the fowl and pigeon; and under nature analogous cases are of frequent occurrence. With animals under domestication, but whether under nature I will not venture to say, one sex may lose characters proper to it, and may thus come to resemble to a certain extent the opposite sex; for instance, the males of some breeds of the fowl have lost their masculine plumes and hackles. On the other hand the differences between the sexes may be increased under domestication, as with merino sheep, in which the ewes have lost their horns. Again, characters proper to one sex may suddenly appear in the other sex; as with those sub-breeds of the fowl in which the hens whilst young acquire spurs; or, as in certain Polish sub-breeds, in which the females, as there is reason to believe, originally acquired a crest, and subsequently transferred it to the males. All these cases are intelligible on the hypothesis of pangenesis; for they depend on the gemmules of certain units of the body, although present in both sexes, becoming through the influence of domestication dormant in the one sex; or if naturally dormant, becoming developed.

There is one difficult question which it will be convenient to defer to a future chapter; namely, whether a character at first developed in both sexes, can be rendered through selection limited in its development to one sex alone. If, for instance, a breeder observed that some of his pigeons (in which species characters are usually transferred in an equal degree to both sexes) varied into pale blue; could he by long-continued selection make a breed, in which the males alone should be of this tint, whilst the females remained unchanged? I will here only say, that this, though perhaps not impossible, would be extremely difficult; for the natural result of breeding from the pale-blue males would be to change his whole stock, including both sexes, into this tint. If, however, variations of the desired tint appeared, which were from the first limited in their development to the male sex, there would not be the least difficulty in making a breed characterised by the two sexes being of a different colour, as indeed has been effected with a Belgian breed, in which the males alone are streaked with black. In a similar manner, if any variation appeared in a female pigeon, which was from the first sexually limited in its development, it would be easy to make a breed with the females alone thus characterised; but if the variation was not thus originally limited, the process would be extremely difficult, perhaps impossible.

On the Relation between the period of Development of a Character and its transmission to one sex or to both sexes.—Why certain characters should be inherited by both sexes, and other characters by one sex alone, namely by that sex in which the character first appeared, is in most cases quite unknown. We cannot even conjecture why with certain sub-breeds of the pigeon, black striæ, though transmitted through the female, should be developed in the male alone, whilst every other character is equally transferred to both sexes. Why, again, with cats, the tortoise-shell colour should, with rare exceptions, be developed in the female alone. The very same characters, such as deficient or supernumerary digits, colour-blindness, &c., may with mankind be inherited by the males alone of one family, and in another family by the females alone, though in both cases transmitted through the opposite as well as the same sex.[358] Although we are thus ignorant, two rules often hold good, namely that variations which first appear in either sex at a late period of life, tend to be developed in the same sex alone; whilst variations which first appear early in life in either sex tend to be developed in both sexes. I am, however, far from supposing that this is the sole determining cause. As I have not elsewhere discussed this subject, and as it has an important bearing on sexual selection, I must here enter into lengthy and somewhat intricate details.

It is in itself probable that any character appearing at an early age would tend to be inherited equally by both sexes, for the sexes do not differ much in constitution, before the power of reproduction is gained. On the other hand, after this power has been gained and the sexes have come to differ in constitution, the gemmules (if I may again use the language of pangenesis) which are cast off from each varying part in the one sex would be much more likely to possess the proper affinities for uniting with the tissues of the same sex, and thus becoming developed, than with those of the opposite sex.

I was first led to infer that a relation of this kind exists, from the fact that whenever and in whatever manner the adult male has come to differ from the adult female, he differs in the same manner from the young of both sexes. The generality of this fact is quite remarkable: it holds good with almost all mammals, birds, amphibians, and fishes; also with many crustaceans, spiders and some few insects, namely certain orthoptera and libellulæ. In all these cases the variations, through the accumulation of which the male acquired his proper masculine characters, must have occurred at a somewhat late period of life; otherwise the young males would have been similarly characterised; and conformably with our rule, they are transmitted to and developed in the adult males alone. When, on the other hand, the adult male closely resembles the young of both sexes (these, with rare exceptions, being alike), he generally resembles the adult female; and in most of these cases the variations through which the young and old acquired their present characters, probably occurred in conformity with our rule during youth. But there is here room for doubt, as characters are sometimes transferred to the offspring at an earlier age than that at which they first appeared in the parents, so that the parents may have varied when adult, and have transferred their characters to their offspring whilst young. There are, moreover, many animals, in which the two sexes closely resemble each other, and yet both differ from their young; and here the characters of the adults must have been acquired late in life; nevertheless, these characters in apparent contradiction to our rule, are transferred to both sexes. We must not, however, overlook the possibility or even probability of successive variations of the same nature sometimes occurring, under exposure to similar conditions, simultaneously in both sexes at a rather late period of life; and in this case the variations would be transferred to the offspring of both sexes at a corresponding late age; and there would be no real contradiction to our rule of the variations which occur late in life being transferred exclusively to the sex in which they first appeared. This latter rule seems to hold true more generally than the second rule, namely, that variations which occur in either sex early in life tend to be transferred to both sexes. As it was obviously impossible even to estimate in how large a number of cases throughout the animal kingdom these two propositions hold good, it occurred to me to investigate some striking or crucial instances, and to rely on the result.

An excellent case for investigation is afforded by the Deer Family. In all the species, excepting one, the horns are developed in the male alone, though certainly transmitted through the female, and capable of occasional abnormal development in her. In the reindeer, on the other hand, the female is provided with horns; so that in this species, the horns ought, according to our rule, to appear early in life, long before the two sexes had arrived at maturity and had come to differ much in constitution. In all the other species of deer the horns ought to appear later in life, leading to their development in that sex alone, in which they first appeared in the progenitor of the whole Family. Now in seven species, belonging to distinct sections of the family and inhabiting different regions, in which the stags alone bear horns, I find that the horns first appear at periods varying from nine months after birth in the roebuck to ten or twelve or even more months in the stags of the six other larger species.[359] But with the reindeer the case is widely different, for as I hear from Prof. Nilsson, who kindly made special enquiries for me in Lapland, the horns appear in the young animals within four or five weeks after birth, and at the same time in both sexes. So that here we have a structure, developed at a most unusually early age in one species of the family, and common to both sexes in this one species.

In several kinds of antelopes the males alone are provided with horns, whilst in the greater number both sexes have horns. With respect to the period of development, Mr. Blyth informs me that there lived at one time in the Zoological Gardens a young koodoo *(Ant. strepsiceros),* in which species the males alone are horned, and the young of a closely-allied species, viz. the eland *(Ant. oreas),* in which both sexes are horned. Now in strict conformity with our rule, in the young male koodoo, although arrived at the age of ten months, the horns were remarkably small considering the size ultimately attained by them: whilst in the young male eland, although only three months old, the horns were already very much larger than in the koodoo. It is also worth notice that in the prong-horned antelope,[360] in which species the horns, though present in both sexes, are almost rudimentary in the female, they do not appear until about five or six months after birth. With sheep, goats and cattle, in which the horns are well developed in both sexes, though not quite equal in size, they can be felt, or even seen, at birth or soon afterwards.[361] Our rule, however, fails in regard to some breeds of sheep, for instance merinos, in which the rams alone are horned; for I cannot find on enquiry,[362] that the horns are developed later in life in this breed than in ordinary sheep in which both sexes are horned. But with domesticated sheep the presence or absence of horns is not a firmly fixed character; a certain proportion of the merino ewes bearing small horns, and some of the rams being hornless; whilst with ordinary sheep hornless ewes are occasionally produced.

In most of the species of the splendid family of the Pheasants, the males differ conspicuously from the females, and they acquire their ornaments at a rather late period of life. The eared pheasant *(Crossoptilon auritum),* however, offers a remarkable exception, for both sexes possess the fine caudal plumes, the large ear-tufts and the crimson velvet about the head; and I find on enquiry in the Zoological Gardens that all these characters, in accordance with our rule, appear very early in life. The adult male can, however, be distinguished from the adult female by one character, namely by the presence of spurs; and conformably with our rule, these do not begin to be developed, as I am assured by Mr. Bartlett, before the age of six months, and even at this age, can hardly be distinguished in the two sexes.[363] The male and female

Peacock differ conspicuously from each other in almost every part of their plumage, except in the elegant head-crest, which is common to both sexes; and this is developed very early in life, long before the other ornaments which are confined to the male. The wild-duck offers an analogous case, for the beautiful green speculum on the wings is common to both sexes, though duller and somewhat smaller in the female, and it is developed early in life, whilst the curled tail-feathers and other ornaments peculiar to the male are developed later.[364] Between such extreme cases of close sexual resemblance and wide dissimilarity, as those of the Crossoptilon and peacock, many intermediate ones could be given, in which the characters follow in their order of development our two rules.

As most insects emerge from their pupal state in a mature condition, it is doubtful whether the period of development determines the transference of their characters to one or both sexes. But we do not know that the coloured scales, for instance, in two species of butterflies, in one of which the sexes differ in colour, whilst in the other they are alike, are developed at the same relative age in the cocoon. Nor do we know whether all the scales are simultaneously developed on the wings of the same species of butterfly, in which certain coloured marks are confined to one sex, whilst other marks are common to both sexes. A difference of this kind in the period of development is not so improbable as it may at first appear; for with the Orthoptera, which assume their adult state, not by a single metamorphosis, but by a succession of moults, the young males of some species at first resemble the females, and acquire their distinctive masculine characters only during a later moult. Strictly analogous cases occur during the successive moults of certain male crustaceans.

We have as yet only considered the transference of characters, relatively to their period of development, with species in a natural state; we will now turn to domesticated animals; first touching on monstrosities and diseases. The presence of supernumerary digits, and the absence of certain phalanges, must be determined at an early embryonic period—the tendency to profuse bleeding is at least congenital, as is probably colour-blindness—yet these peculiarities, and other similar ones, are often limited in their transmission to one sex; so that the rule that characters which are developed at an early period tend to be transmitted to both sexes, here wholly fails. But this rule, as before remarked, does not appear to be nearly so generally true as the converse proposition, namely, that characters which appear late in life in one sex are transmitted exclusively to the same sex. From the fact of the above abnormal peculiarities becoming attached to one sex, long before the sexual functions are active, we may infer that there must be a difference of some kind between the sexes at an extremely early age. With respect to sexually-limited diseases, we know too little of the period at which they originate, to draw any fair conclusion. Gout, however, seems to fall under our rule; for it is generally caused by intemperance after early youth, and is transmitted from the father to his sons in a much more marked manner than to his daughters.

In the various domestic breeds of sheep, goats, and cattle, the males differ from their respective females in the shape or development of their horns, forehead, mane, dewlap, tail, and hump on the shoulders; and these peculiarities, in accordance with our rule, are not fully developed until rather late in life. With dogs, the sexes do not differ, except that in certain breeds, especially in the Scotch deer-hound, the male is much larger and heavier than the female; and as we shall see in a future chapter, the male goes on increasing in size to an unusually late period of life, which will account, according to our rule, for his increased size being transmitted to his male offspring alone. On the other hand, the tortoise-shell colour of the hair, which is confined to female cats, is quite distinct at birth, and this case violates our rule. There is a breed of pigeons in which the males alone are streaked with black, and the streaks can be detected even in the nestlings; but they become more conspicuous at each successive moult, so that this case partly opposes and partly supports the rule. With the English Carrier and Pouter pigeon the full development of the wattle and the crop occurs rather late in life, and these characters, conformably with our rule, are transmitted in full perfection to the males alone. The following cases perhaps come within the class previously alluded to, in which the two sexes have varied in the same manner at a rather late period of life, and have consequently transferred their new characters to both sexes at a corresponding late period; and if so, such cases are not opposed to our rule. Thus there are sub-breeds of the pigeon, described by Neumeister,[365] both sexes of which change colour after moulting twice or thrice, as does likewise the Almond Tumbler; nevertheless these changes, though occurring rather late in life, are common to both sexes. One variety of the Canary-bird, namely the London Prize, offers a nearly analogous case.

With the breeds of the Fowl the inheritance of various characters by one sex or by both sexes, seems generally determined by the period at which such characters are developed. Thus in all the many breeds in which the adult male differs greatly in colour from the female and from the adult male parent-species, he differs from the young male, so that the newly acquired characters must have appeared at a rather late period of life. On the other hand with most of the breeds in which the two sexes resemble each other, the young are coloured in nearly the same manner as their parents, and this renders it probable that their colours first appeared early in life. We have instances of this fact in all black and white breeds, in which the young and old of both sexes are alike; nor can it be maintained that there is something peculiar in a black or white plumage, leading to its transference to both sexes; for the males alone of many natural species are either black or white, the females being very differently coloured. With the so-called Cuckoo sub-breeds of the fowl, in which the feathers are transversely pencilled with dark stripes, both sexes and the chickens are coloured in nearly the same manner. The laced plumage of the Sebright bantam is the same in both sexes, and in the chickens the feathers are tipped with black, which makes a near approach to lacing. Spangled Hamburghs, however, offer a partial exception, for the two sexes, though not quite alike, resemble each other more closely than do the sexes of the aboriginal parent-species, yet they acquire their characteristic plumage late in life, for the chickens are distinctly pencilled. Turning to other characters besides colour: the males alone of the wild parent-species and of most domestic breeds possess a fairly well developed comb, but in the young of the Spanish fowl it is largely developed at a very early age, and apparently in consequence of this it is of unusual size in the adult females. In the Game breeds pugnacity is developed at a wonderfully early age, of which curious proofs could be given; and this character is transmitted to both sexes, so that the hens, from their extreme pugnacity, are now generally exhibited in separate pens. With the Polish breeds the bony protuberance of the skull which supports the crest is partially developed even before the chickens are hatched, and the crest itself soon begins to grow, though at first feebly;[366] and in this breed a great bony protuberance and an immense crest characterise the adults of both sexes.

Finally, from what we have now seen of the relation which exists in many natural species and domesticated races, between the period of the development of their characters and the manner of their transmission—for example the striking fact of the early growth of the horns in the reindeer, in which both sexes have horns, in comparison with their much later growth in the other species in which the male alone bears horns—we may conclude that one cause, though not the sole cause, of characters being exclusively inherited by one sex, is their development at a late age. And secondly, that one, though apparently a less efficient, cause of characters being inherited by both sexes is their development at an early age, whilst the sexes differ but little in constitution. It appears, however, that some difference must exist between the sexes even during an early embryonic period, for characters developed at this age not rarely become attached to one sex.

Summary and concluding remarks.—From the foregoing discussion on the various laws of inheritance, we learn that characters often or even generally tend to become developed in the same sex, at the same age, and periodically at the same season of the year, in which they first appeared in the parents. But these laws, from unknown causes, are very liable to change. Hence the successive steps in the modification of a species might readily be transmitted in different ways; some of the steps being transmitted to one sex, and some to both; some to the offspring at one age, and some at all ages. Not only are the laws of inheritance extremely complex, but so are the causes which induce and govern variability. The variations thus caused are preserved and accumulated by sexual selection, which is in itself an extremely complex affair, depending, as it does, on ardour in love, courage, and the rivalry of the males, and on the powers of perception, taste, and will of the female. Sexual selection will also be dominated by natural selection for the general welfare of the species. Hence the manner in which the individuals of either sex or of both sexes are affected through sexual selection cannot fail to be complex in the highest degree.

When variations occur late in life in one sex, and are transmitted to the same sex at the same age, the other sex and the young are necessarily left unmodified. When they occur late in life, but are transmitted to both sexes at the same age, the young alone are left unmodified. Variations, however, may occur at any period of life in one sex or in both, and be transmitted to both sexes at all ages, and then all the individuals of the species will be similarly modified. In the following chapters it will be seen that all these cases frequently occur under nature.

Sexual selection can never act on any animal whilst young, before the age for reproduction has arrived. From the great eagerness of the male it has generally acted on this sex and not on the females. The males have thus become provided with weapons for fighting with their rivals, or with organs for discovering and securely holding the female, or for exciting and charming her. When the sexes differ in these respects, it is also, as we have seen, an extremely general law that the adult male differs more or less from the young male; and we may conclude from this fact that the successive variations, by which the adult male became modified, cannot have occurred much before the age for reproduction. How then are we to account for this general and remarkable coincidence between the period of variability and that of sexual selection,—principles which are quite independent of each other? I think we can see the cause: it is not that the males have never varied at an early age, but that such variations have commonly been lost, whilst those occurring at a later age have been preserved.

All animals produce more offspring than can survive to maturity; and we have every reason to believe that death falls heavily on the weak and inexperienced young. If then a certain proportion of the offspring were to vary at birth or soon afterwards, in some manner which at this age was of no service to them, the chance of the preservation of such variations would be small. We have good evidence under domestication how soon variations of all kinds are lost, if not selected. But variations which occurred at or near maturity, and which were of immediate service to either sex, would probably be preserved; as would similar variations occurring at an earlier period in any individuals which happened to survive. As this principle has an important bearing on sexual selection, it may be advisable to give an imaginary illustration. We will take a pair of animals, neither very fertile nor the reverse, and assume that after arriving at maturity they live on an average for five years, producing each year five young. They would thus produce 25 offspring; and it would not, I think, be an unfair estimate to assume that 18 or 20 out of the 25 would perish before maturity, whilst still young and inexperienced; the remaining seven or five sufficing to keep up the stock of mature individuals. If so, we can see that variations which occurred during youth, for instance in brightness, and which were not of the least service to the young, would run a good chance of being utterly lost. Whilst similar variations, which occurring at or near maturity in the comparatively few individuals surviving to this age, and which immediately gave an advantage to certain males, by rendering them more attractive to the females, would be likely to be preserved. No doubt some of the variations in brightness which occurred at an earlier age would by chance be preserved, and eventually give to the male the same advantage as those which appeared later; and this will account for the young males commonly partaking to a certain extent (as may be observed with many birds) of the bright colours of their adult male parents. If only a few of the successive variations in brightness were to occur at a late age, the adult male would be only a little brighter than the young male; and such cases are common.

In this illustration I have assumed that the young varied in a manner which was of no service to them; but many characters proper to the adult male would be actually injurious to the young,—as bright colours from making them conspicuous, or horns of large size from expending much vital force. Such variations in the young would promptly be eliminated through natural selection. With the adult and experienced males, on the other hand, the advantage thus derived in their rivalry with other males would often more than counterbalance exposure to some degree of danger. Thus we can understand how it is that variations which must originally have appeared rather late in life have alone or in chief part been preserved for the development of secondary sexual characters; and the remarkable coincidence between the periods of variability and of sexual selection is intelligible.

As variations which give to the male an advantage in fighting with other males, or in finding, securing, or charming the female, would be of no use to the female, they will not have been preserved in this sex either during youth or maturity. Consequently such variations would be extremely liable to be lost; and the female, as far as these characters are concerned, would be left unmodified, excepting in so far as she may have received them by transference from the male. No doubt if the female varied and transferred serviceable characters to her male offspring, these would be favoured through sexual selection; and then both sexes would thus far be modified in the same manner. But I shall hereafter have to recur to these more intricate contingencies.

In the following chapters, I shall treat of the secondary sexual characters in animals of all classes, and shall endeavour in each case to apply the principles explained in the present chapter. The lowest classes will detain us for a very short time, but the higher animals, especially birds, must be treated at considerable length. It should be borne in mind that for reasons already assigned, I intend to give only a few illustrative instances of the innumerable structures by the aid of which the male finds the female, or, when found, holds her. On the other hand, all structures and instincts by which the male conquers other males, and by which he allures or excites the female, will be fully discussed, as these are in many ways the most interesting.

SUPPLEMENT ON THE PROPORTIONAL NUMBERS OF THE TWO SEXES IN ANIMALS BELONGING TO VARIOUS CLASSES

As no one, as far as I can discover, has paid attention to the relative numbers of the two sexes throughout the animal kingdom, I will here give such materials as I have been able to collect, although they are extremely imperfect. They consist in only a few instances of actual enumeration, and the numbers are not very large. As the proportions are known with certainty on a large scale in the case of man alone, I will first give them, as a standard of comparison.

Man.—In England during ten years (from 1857 to 1866) 707,120 children on an annual average have been born alive, in the proportion of 104·5 males to 100 females. But in 1857 the male births throughout England were as 105·2, and in 1865 as 104·0 to 100. Looking to separate districts, in Buckinghamshire (where on an average 5000 children are annually born) the *mean* proportion of male to female births, during the whole period of the above ten years, was as 102·8 to 100; whilst in N. Wales (where the average annual births are 12,873) it was as high as 106·2 to 100. Taking a still smaller district, viz., Rutlandshire (where the annual births average only 739), in 1864 the male births were as 114·6, and in 1862 as 97·0 to 100; but even in this small district the average of the 7385 births during the whole ten years was as 104·5 to 100; that is in the same ratio as throughout England.[367] The proportions are sometimes slightly disturbed by unknown causes; thus Prof. Faye states "that in some districts of Norway there has been during a decennial period a steady deficiency of boys, whilst in others the opposite condition has existed." In France during forty-four years the male to the female births have been as 106·2 to 100; but during this period it has occurred five times in one department, and six times in another, that the female births have exceeded the males. In Russia the average proportion is as high as 108·9 to 100.[368] It is a singular fact that with Jews the proportion of male births is decidedly larger than with Christians: thus in Prussia the proportion is as 113, in Breslau as 114, and in Livonia as 120 to 100; the Christian births in these countries being the same as usual, for instance, in Livonia as 104 to 100.[369] It is a still more singular fact that in different nations, under different conditions and climates, in Naples, Prussia, Westphalia, France and England, the excess of male over female births is less when they are illegitimate than when legitimate.[370]

In various parts of Europe, according to Prof. Faye and other authors, "a still greater preponderance of males would be met with, if death struck both sexes in equal proportion in the womb and during birth. But the fact is, that for every 100 still-born females, we have in several countries from 134·6 to 144·9 still-born males." Moreover during the first four or five years of life more male children die than females; "for example in England, during the first year, 126 boys die for every 100 girls,—a proportion which in France is still more unfavourable."[371] As a consequence of this excess in the death-rate of male children, and of the exposure of men when adult to various dangers, and of their tendency to emigrate, the females in all old-settled countries, where statistical records have been kept,[372] are found to preponderate considerably over the males.

It has often been supposed that the relative ages of the parents determine the sex of the offspring; and Prof. Leuckart[373] has advanced what he considers sufficient evidence, with respect to man and certain domesticated animals, to shew that this is one important factor in the result. So again the period of impregnation has been thought to be the efficient cause; but recent observations discountenance this belief. Again, with mankind polygamy has been supposed to lead to the birth of a greater proportion of female infants; but Dr. J. Campbell[374] carefully attended to this subject in the harems of Siam, and he concludes that the proportion of male to female births is the same as from monogamous unions. Hardly any animal has been rendered so highly polygamous as our English race-horses, and we shall immediately see that their male and female offspring are almost exactly equal in number.

Horses.—Mr. Tegetmeier has been so kind as to tabulate for me from the 'Racing Calendar' the births of race-horses during a period of twenty-one years, viz. from 1846 to 1867; 1849 being omitted, as no returns were that year published. The total births have been 25,560,[375] consisting of 12,763 males and 12,797 females, or in the proportion of 99·7 males to 100 females. As these numbers are tolerably large, and as they are drawn from all parts of England, during several years, we may with much confidence conclude that with the domestic horse, or at least with the race-horse, the two sexes are produced in almost equal numbers. The fluctuations in the proportions during successive years are closely like those which occur with mankind, when a small and thinly-populated area is considered: thus in 1856 the male horses were as 107·1, and in 1867 as only 92·6 to 100 females. In the tabulated returns the proportions vary in cycles, for the males exceeded the females during six successive years; and the females exceeded the males during two periods each of four years: this, however, may be accidental; at least I can detect nothing of the kind with man in the decennial table in the Registrar's Report for 1866. I may add that certain mares, and this holds good with certain cows and with women, tend to produce more of one sex than of the other; Mr. Wright of Yeldersley House, informs me that one of his Arab mares, though put seven times to different horses, produced seven fillies.

Dogs.—During a period of twelve years, from 1857 to 1868, the births of a large number of greyhounds, throughout England, have been sent to the 'Field' newspaper; and I am again indebted to Mr. Tegetmeier for carefully tabulating the results. The recorded births have been 6878, consisting of 3605 males and 3273 females, that is, in the proportion of 110·1 males to 100 females. The greatest fluctuations occurred in 1864, when the proportion was as 95·3 males, and in 1867, as 116·3 males to 100 females. The above average proportion of 110·1 to 100 is probably nearly correct in the case of the greyhound, but whether it would hold with other domesticated breeds is in some degree doubtful. Mr. Cupples has enquired from several great breeders of dogs, and finds that all without exception believe that females are produced in excess; he suggests that this belief may have arisen from females being less valued and the consequent disappointment producing a stronger impression on the mind.

Sheep.—The sexes of sheep are not ascertained by agriculturists until several months after birth, at the period when the males are castrated; so that the following returns do not give the proportions at birth. Moreover, I find that several great breeders in Scotland, who annually raise some thousand sheep, are firmly convinced that a larger proportion of males than of females die during the first one or two years; therefore the proportion of males would be somewhat greater at birth than at the age of castration. This is a remarkable coincidence with what occurs, as we have seen, with mankind, and both cases probably depend on some common cause. I have received returns from four gentlemen in England who have bred lowland sheep, chiefly Leicesters, during the last ten or sixteen years; they amount altogether to 8965 births, consisting of 4407 males and 4558 females; that is in the proportion of 96·7 males to 100 females. With respect to Cheviot and black-faced sheep bred in Scotland, I have received returns from six breeders, two of them on a large scale, chiefly for the years 1867–1869, but some of the returns extending back to 1862. The total number recorded amounts to 50,685, consisting of 25,071 males and 25,614 females, or in the proportion of 97·9 males to 100 females. If we take the English and Scotch returns together, the total number amounts to 59,650, consisting of 29,478 males and 30,172 females, or as 97·7 to 100. So that with sheep at the age of castration the females are certainly in excess of the males; but whether this would hold good at birth is doubtful, owing to the greater liability in the males to early death.[376]

Of Cattle I have received returns from nine gentlemen of 982 births, too few to be trusted; these consisted of 477 bull-calves and 505 cow-calves; *i.e.* in the proportion of 94·4 males to 100 females. The Rev. W. D. Fox informs me that in 1867 out of 34 calves born on a farm in Derbyshire only one was a bull. Mr. Harrison Weir writes to me that he has enquired from several breeders of *Pigs,* and most of them estimate the male to the female births as about 7 to 6. This same gentleman has bred *Rabbits* for many years, and has noticed that a far greater number of bucks are produced than does.

Of Mammalia in a state of nature I have been able to learn very little. In regard to the common rat, I have received conflicting statements. Mr. R. Elliot of Laighwood, informs me that a rat-catcher assured him that he had always found the males in great excess, even with the young in the nest. In consequence of this, Mr. Elliot himself subsequently examined some hundred old ones, and found the statement true. Mr. F. Buckland has bred a large number of white rats, and he also believes that the males greatly exceed the females. In regard to Moles, it is said that "the males are much more numerous than the females;"[377] and as the catching of these animals is a special occupation, the statement may perhaps be trusted. Sir A. Smith, in describing an antelope of S. Africa[378] *(Kobus ellipsiprymnus),* remarks, that in the herds of this and other species, the males are few in number compared with the females: the natives believe that they are born in this proportion; others believe that the younger males are expelled from the herds, and Sir A. Smith says, that though he has himself never seen herds consisting of young males alone, others affirm that this does occur. It appears probable that the young males when expelled from the herd, would be likely to fall a prey to the many beasts of prey of the country.

Birds

With respect to the *Fowl*, I have received only one account, namely, that out of 1001 chickens of a highly-bred stock of Cochins, reared during eight years by Mr. Stretch, 487 proved males and 514 females: *i.e.* as 94.7 to 100. In regard to domestic pigeons there is good evidence that the males are produced in excess, or that their lives are longer; for these birds invariably pair, and single males, as Mr. Tegetmeier informs me, can always be purchased cheaper than females. Usually the two birds reared from the two eggs laid in the same nest consist of a male and female; but Mr. Harrison Weir, who has been so large a breeder, says that he has often bred two cocks from the same nest, and seldom two hens; moreover the hen is generally the weaker of the two, and more liable to perish.

With respect to birds in a state of nature, Mr. Gould and others[379]are convinced that the males are generally the more numerous; and as the young males of many species resemble the females, the latter would naturally appear to be the most numerous. Large numbers of pheasants are reared by Mr. Baker of Leadenhall from eggs laid by wild birds, and he informs Mr. Jenner Weir that four or five males to one female are generally produced. An experienced observer remarks[380] that in Scandinavia the broods of the capercailzie and black-cock contain more males than females; and that with the Dal-ripa (a kind of ptarmigan) more males than females attend the *leks* or places of courtship; but this latter circumstance is accounted for by some observers by a greater number of hen birds being killed by vermin. From various facts given by White of Selborne,[381] it seems clear that the males of the partridge must be in considerable excess in the south of England; and I have been assured that this is the case in Scotland. Mr. Weir on enquiring from the dealers who receive at certain seasons large numbers of ruffs *(Machetes pugnax)* was told that the males are much the most numerous. This same naturalist has also enquired for me from the bird-catchers, who annually catch an astonishing number of various small species alive for the London market, and he was unhesitatingly answered by an old and trustworthy man, that with the chaffinch the males are in large excess; he thought as high as 2 males to I female, or at least as high as 5 to 3.[382] The males of the blackbird, he likewise maintained, were by far the most numerous, whether caught by traps or by netting at night. These statements may apparently be trusted, because the same man said that the sexes are about equal with the lark, the twite *(Linaria montana)*, and goldfinch.

On the other hand he is certain that with the common linnet, the females preponderate greatly, but unequally during different years; during some years he has found the females to the males as four to one. It should, however, be borne in mind, that the chief season for catching birds does not begin till September, so that with some species partial migrations may have begun, and the flocks at this period often consist of hens alone. Mr. Salvin paid particular attention to the sexes of the humming-birds in Central America, and he is convinced that with most of the species the males are in excess; thus one year he procured 204 specimens belonging to ten species, and these consisted of 166 males and of 38 females. With two other species the females were in excess: but the proportions apparently vary either during different seasons or in different localities; for on one occasion the males of *Campylopterus hemileucurus* were to the females as five to two, and on another occasion[383] in exactly the reversed ratio. As bearing on this latter point, I may add, that Mr. Powys found in Corfu and Epirus the sexes of the chaffinch keeping apart, and "the females by far the most numerous;" whilst in Palestine Mr. Tristram found "the male flocks appearing greatly to exceed the female in number."[384] So again with the *Quiscalus major,* Mr. G. Taylor[385] says, that in Florida there were "very few females in proportion to the males," whilst in Honduras the proportion was the other way, the species there having the character of a polygamist.

Fish

With Fish the proportional numbers of the sexes can be ascertained only by catching them in the adult or nearly adult state; and there are many difficulties in arriving at any just conclusion.[386] Infertile females might readily be mistaken for males, as Dr. Günther has remarked to me in regard to trout. With some species the males are believed to die soon after fertilising the ova. With many species the males are of much smaller size than the females, so that a large number of males would escape from the same net by which the females were caught. M. Carbonnier,[387] who has especially attended to the natural history of the pike *(Esox lucius)* states that many males, owing to their small size, are devoured by the larger females; and he believes that the males of almost all fish are exposed from the same cause to greater danger than the females. Nevertheless in the few cases in which the proportional numbers have been actually observed, the males appear to be largely in excess. Thus Mr. R. Buist, the superintendent of the Stormontfield experiments, says that in 1865, out of 70 salmon first landed for the purpose of obtaining the ova, upwards of 60 were males. In 1867 he again "calls attention to the vast disproportion of the males to the females. We had at the outset at least ten males to one female." Afterwards sufficient females for obtaining ova were procured. He adds, "from the great proportion of the males, they are constantly fighting and tearing each other on the spawning-beds."[388] This disproportion, no doubt, can be accounted for in part, but whether wholly is very doubtful, by the males ascending the rivers before the females. Mr. F. Buckland remarks in regard to trout, that "it is a curious fact that the males preponderate very largely in number over the females. It *invariably* happens that when the first rush of fish is made to the net, there will be at least seven or eight males to one female found captive. I cannot quite account for this; either the males are more numerous than the females, or the latter seek safety by concealment rather than flight." He then adds, that by carefully searching the banks, sufficient females for obtaining ova can be found.[389] Mr. H. Lee informs me that out of 212 trout, taken for this purpose in Lord Portsmouth's park, 150 were males and 62 females.

With the Cyprinidæ the males likewise seem to be in excess; but several members of this Family, viz., the carp, tench, bream and minnow, appear regularly to follow the practice, rare in the animal kingdom, of polyandry; for the female whilst spawning is always attended by two males, one on each side, and in the case of the bream by three or four males. This fact is so well known, that it is always recommended to stock a pond with two male tenches to one female, or at least with three males to two females. With the minnow, an excellent observer states, that on the spawning-beds the males are ten times as numerous as the females; when a female comes amongst the males, "she is immediately pressed closely by a male on each side; and when they have been in that situation for a time, are superseded by other two males." [390]

Insects

In this class, the Lepidoptera alone afford the means of judging of the proportional numbers of the sexes; for they have been collected with special care by many good observers, and have been largely bred from the egg or caterpillar state. I had hoped that some breeders of silk-moths might have kept an exact record, but after writing to France and Italy, and consulting various treatises, I cannot find that this has ever been done. The general opinion appears to be that the sexes are nearly equal, but in Italy as I hear from Professor Canestrini, many breeders are convinced that the females are produced in excess. The same naturalist, however, informs me, that in the two yearly broods of the Ailanthus silk-moth *(Bombyx cynthia)*, the males greatly preponderate in the first, whilst in the second the two sexes are nearly equal, or the females rather in excess.

In regard to Butterflies in a state of nature, several observers have been much struck by the apparently enormous preponderance of the males.[391] Thus Mr. Bates,[392] in speaking of the species, no less than about a hundred in number, which inhabit the Upper Amazons, says that the males are much more numerous than the females, even in the proportion of a hundred to one. In North America, Edwards, who had great experience, estimates in the genus Papilio the males to the females as four to one; and Mr. Walsh, who informed me of this statement, says that with *P. turnus* this is certainly the case. In South Africa, Mr. R. Trimen found the males in excess in 19 species;[393] and in one of these, which swarms in open places, he estimated the number of males as fifty to one female. With another species, in which the males are numerous in certain localities, he collected during seven years only five females. In the island of Bourbon, M. Maillard states that the males of one species of Papilio are twenty times as numerous as the females.[394] Mr. Trimen informs me that as far as he has himself seen, or heard from others, it is rare for the females of any butterfly to exceed in number the males; but this is perhaps the case with three South African species. Mr. Wallace[395] states that the females of *Ornithoptera crœsus,* in the Malay archipelago, are more common and more easily caught than the males; but this is a rare butterfly. I may here add, that in Hyperythra, a genus of moths, Guenée says, that from four to five females are sent in collections from India for one male.

When this subject of the proportional numbers of the sexes of insects was brought before the Entomological Society,[396] it was generally admitted that the males of most Lepidoptera, in the adult or imago state, are caught in greater numbers than the females; but this fact was attributed by various observers to the more retiring habits of the females, and to the males emerging earlier from the cocoon. This latter circumstance is well known to occur with most Lepidoptera, as well as with other insects. So that, as M. Personnat remarks, the males of the domesticated *Bombyx Yamamai,* are lost at the beginning of the season, and the females at the end, from the want of mates.[397] I cannot however persuade myself that these causes suffice to explain the great excess of males in the cases, above given, of butterflies which are extremely common in their native countries. Mr. Stainton, who has paid such close attention during many years to the smaller moths, informs me that when he collected them in the imago state, he thought that the males were ten times as numerous as the females, but that since he has reared them on a large scale from the caterpillar state, he is convinced that the females are the most numerous. Several entomologists concur in this view. Mr. Doubleday, however, and some others, take an opposite view, and are convinced that they have reared from the egg and caterpillar states a larger proportion of males than of females.

Besides the more active habits of the males, their earlier emergence from the cocoon, and their frequenting in some cases more open stations, other causes may be assigned for an apparent or real difference in the proportional numbers of the sexes of Lepidoptera, when captured in the imago state, and when reared from the egg or caterpillar state. It is believed by many breeders in Italy, as I hear from Professor Canestrini, that the female caterpillar of the silk-moth suffers more from the recent disease than the male; and Dr. Staudinger informs me that in rearing Lepidoptera more females die in the cocoon than males. With many species the female caterpillar is larger than the male, and a collector would naturally choose the finest specimens, and thus unintentionally collect a larger number of females. Three collectors have told me that this was their practice; but Dr. Wallace is sure that most collectors take all the specimens which they can find of the rarer kinds, which alone are worth the trouble of rearing. Birds when surrounded by caterpillars would probably devour the largest; and Professor Canestrini informs me that in Italy some breeders believe, though on insufficient evidence, that in the first brood of the Ailanthus silk-moth, the wasps destroy a larger number of the female than of the male caterpillars. Dr. Wallace further remarks that female caterpillars, from being larger than the males, require more time for their development and consume more food and moisture; and thus they would be exposed during a longer time to danger from ichneumons, birds, &c., and in times of scarcity would perish in greater numbers. Hence it appears quite possible that, in a state of nature, fewer female Lepidoptera may reach maturity than males; and for our special object we are concerned with the numbers at maturity, when the sexes are ready to propagate their kind.

The manner in which the males of certain moths congregate in extraordinary numbers round a single female, apparently indicates a great excess of males, though this fact may perhaps be accounted for by the earlier emergence of the males from their cocoons. Mr. Stainton informs me that from twelve to twenty males may often be seen congregated round a female *Elachista rufocinerea.* It is well known that if a virgin *Lasiocampa quercus* or *Saturnia carpini* be exposed in a cage, vast numbers of males collect round her, and if confined in a room will even come down the chimney to her. Mr. Doubleday believes that he has seen from fifty to a hundred males of both these species attracted in the course of a single day by a female under confinement. Mr. Trimen exposed in the Isle of Wight a box in which a female of the Lasiocampa had been confined on the previous day, and five males soon endeavoured to gain admittance. M. Verreaux, in Australia, having placed the female of a small Bombyx in a box in his pocket, was followed by a crowd of males, so that about 200 entered the house with him.[398]

Mr. Doubleday has called my attention to Dr. Staudinger's[399] list of Lepidoptera, which gives the prices of the males and females of 300 species or well-marked varieties of (Rhopalocera) butterflies. The prices for both sexes of the very common species are of course the same; but with 114 of the rarer species they differ; the males being in all cases, excepting one, the cheapest. On an average of the prices of the 113 species, the price of the male to that of the female is as 100 to 149; and this apparently indicates that inversely the males exceed the females in number in the same proportion. About 2000 species or varieties of moths (Heterocera) are catalogued, those with wingless females being here excluded on account of the difference in habits of the two sexes: of these 2000 species, 141 differ in price according to sex, the males of 130 being cheaper, and the males of only 11 being dearer than the females. The average price of the males of the 130 species, to that of the females, is as 100 to 143. With respect to the butterflies in this priced list, Mr. Doubleday thinks (and no man in England has had more experience), that there is nothing in the habits of the species which can account for the difference in the prices of the two sexes, and that it can be accounted for only by an excess in the numbers of the males. But I am bound to add that Dr. Staudinger himself, as he informs me, is of a different opinion. He thinks that the less active habits of the females and the earlier emergence of the males will account for his collectors securing a larger number of males than of females, and consequently for the lower prices of the former. With respect to specimens reared from the caterpillar-state, Dr. Staudinger believes, as previously stated, that a greater number of females than of males die under confinement in the cocoons. He adds that with certain species one sex seems to preponderate over the other during certain years.

Of direct observations on the sexes of Lepidoptera, reared either from eggs or caterpillars, I have received only the few following cases:—

	Males	**Females**
The Rev. J. Hellins[400] of Exeter reared, during 1868, imagos of 73 species, which consisted of	153	137
Mr. Albert Jones of Eltham reared, during 1868, imagos of 9 species, which consisted of	159	126
During 1869 he reared imagos from 4 species, consisting of	114	112
Mr. Buckler of Emsworth, Hants, during 1869, reared imagos from 74 species, consisting of	180	169

	Males	Females
Dr. Wallace of Colchester reared from one brood of Bombyx cynthia	52	48
Dr. Wallace raised, from cocoons of Bombyx Pernyi sent from China, during 1869	224	123
Dr. Wallace raised, during 1868 and 1869, from two lots of cocoons of Bombyx yama-mai	52	46
Total	**934**	**761**

So that in these eight lots of cocoons and eggs, males were produced in excess. Taken together the proportion of males is as 122·7 to 100 females. But the numbers are hardly large enough to be trustworthy.

On the whole, from the above various sources of evidence, all pointing to the same direction, I infer that with most species of Lepidoptera, the males in the imago state generally exceed the females in number, whatever the proportions may be at their first emergence from the egg.

With reference to the other Orders of insects, I have been able to collect very little reliable information. With the stag-beetle *(Lucanus cervus)* "the males appear to be much more numerous than the females;" but when, as Cornelius remarked during 1867, an unusual number of these beetles appeared in one part of Germany, the females appeared to exceed the males as six to one. With one of the Elateridæ, the males are said to be much more numerous than the females, and "two or three are often found united with one female;[401] so that here polyandry seems to prevail. With Siagonium (Staphylinidæ), in which the males are furnished with horns, "the females are far more numerous than the opposite sex." Mr. Janson stated at the Entomological Society that the females of the bark-feeding *Tomicus villosus* are so common as to be a plague, whilst the males are so rare as to be hardly known. In other Orders, from unknown causes, but apparently in some instances owing to parthenogenesis, the males of certain species have never been discovered or are excessively rare, as with several of the Cynipidæ.[402] In all the gall-making Cynipidæ known to Mr. Walsh, the females are four or five times as numerous as the males; and so it is, as he informs me, with the gall-making Cecidomyiiæ (Diptera). With some common species of Saw-flies (Tenthredinæ) Mr. F. Smith has reared hundreds of specimens from larvæ of all sizes, but has never reared a single male: on the other hand Curtis says,[403] that with certain species (Athalia), bred by him, the males to the females were as six to one; whilst exactly the reverse occurred with the mature insects of the same species caught in the fields. With the Neuroptera, Mr. Walsh states that in many, but by no means in all, the species of the Odonatous groups (Ephemerina), there is a great overplus of males: in the genus Hetærina, also, the males are generally at least four times as numerous as the females. In certain species in the genus Gomphus the males are equally numerous, whilst in two other species, the females are twice or thrice as numerous as the males. In some European species of Psocus thousands of females may be collected without a single male, whilst with other species of the same genus both sexes are common.[404] In England, Mr. MacLachlan has captured hundreds of the female *Apatania muliebris,* but has never seen the male; and of *Boreus hyemalis* only four or five males have been here seen.[405] With most of these species (excepting, as I have heard, with the Tenthredinæ) there is no reason to suppose that the females are subject to parthenogenesis; and thus we see how ignorant we are on the causes of the apparent discrepancy in the proportional numbers of the two sexes.

In the other Classes of the Articulata I have been able to collect still less information. With Spiders, Mr. Blackwall, who has carefully attended to this class during many years, writes to me that the males from their more erratic habits are more commonly seen, and therefore appear to be the more numerous. This is actually the case with a few species; but he mentions several species in six genera, in which the females appear to be much more numerous than the males.[406] The small size of the males in comparison with the females, which is sometimes carried to an extreme degree, and their widely different appearance, may account in some instances for their rarity in collections.[407]

Some of the lower Crustaceans are able to propagate their kind asexually, and this will account for the extreme rarity of the males. "With some other forms (as with Tanais and Cypris) there is reason to believe, as Fritz Müller informs me, that the male is much shorter-lived than the female, which, supposing the two sexes to be at first equal in number, would explain the scarcity of the males. On the other hand this same naturalist has invariably taken, on the shores of Brazil, far more males than females of the Diastylidæ and of Cypridina; thus with a species in the latter genus, 63 specimens caught the same day, included 57 males; but he suggests that this preponderance may be due to some unknown difference in the habits of the two sexes. With one of the higher Brazilian crabs, namely a Gelasimus, Fritz Müller found the males to be more numerous than the females. The reverse seems to be the case, according to the large experience of Mr. C. Spence Bate, with six common British crabs, the names of which he has given me.

On the Power of Natural Selection to regulate the proportional Numbers of the Sexes, and General Fertility.—In some peculiar cases, an excess in the number of one sex over the other might be a great advantage to a species, as with the sterile females of social insects, or with those animals in which more than one male is requisite to fertilise the female, as with certain cirripedes and perhaps certain fishes. An inequality between the sexes in these cases might have been acquired through natural selection, but from their rarity they need not here be further considered. In all ordinary cases an inequality would be no advantage or disadvantage to certain individuals more than to others; and therefore it could hardly have resulted from natural selection. We must attribute the inequality to the direct action of those unknown conditions, which with mankind lead to the males being born in a somewhat larger excess in certain countries than in others, or which cause the proportion between the sexes to differ slightly in legitimate and illegitimate births.

Let us now take the case of a species producing from the unknown causes just alluded to, an excess of one sex—we will say of males—these being superfluous and useless, or nearly useless. Could the sexes be equalised through natural selection? We may feel sure, from all characters being variable, that certain pairs would produce a somewhat less excess of males over females than other pairs. The former, supposing the actual number of the offspring to remain constant, would necessarily produce more females, and would therefore be more productive. On the doctrine of chances a greater number of the offspring of the more productive pairs would survive; and these would inherit a tendency to procreate fewer males and more females. Thus a tendency towards the equalisation of the sexes would be brought about. But our supposed species would by this process be rendered, as just remarked, more productive; and this would in many cases be far from an advantage; for whenever the limit to the numbers which exist, depends, not on destruction by enemies, but on the amount of food, increased fertility will lead to severer competition and to most of the survivors being badly fed. In this case, if the sexes were equalised by an increase in the number of the females, a simultaneous decrease in the total number of the offspring would be beneficial, or even necessary, for the existence of the species; and this, I believe, could be effected through natural selection in the manner hereafter to be described. The same train of reasoning is applicable in the above, as well as in the following case, if we assume that females instead of males are produced in excess, for such females from not uniting with males would be superfluous and useless. So it would be with polygamous species, if we assume the excess of females to be inordinately great.

An excess of either sex, we will again say of the males, could, however, apparently be eliminated through natural selection in another and indirect manner, namely by an actual diminution of the males, without any increase of the females, and consequently without any increase in the productiveness of the species. From the variability of all characters, we may feel assured that some pairs, inhabiting any locality, would produce a rather smaller excess of superfluous males, but an equal number of productive females. When the offspring from the more and the less male-productive parents were all mingled together, none would have any direct advantage over the others; but those that produced few superfluous males would have one great indirect advantage, namely that their ova or embryos would probably be larger and finer, or their young better nurtured in the womb and afterwards. We see this principle illustrated with plants; as those which bear a vast number of seed produce small ones; whilst those which bear comparatively few seeds, often produce large ones well-stocked with nutriment for the use of the seedlings.[408] Hence the offspring of the parents which had wasted least force in producing superfluous males would be the most likely to survive, and would inherit the same tendency not to produce superfluous males, whilst retaining their full fertility in the production of females. So it would be with the converse case of the female sex. Any slight excess, however, of either sex could hardly be checked in so indirect a manner. Nor indeed has a considerable inequality between the sexes been always prevented, as we have seen in some of the cases given in the previous discussion. In these cases the unknown causes which determine the sex of the embryo, and which under certain conditions lead to the production of one sex in excess over the other, have not been mastered by the survival of those varieties which were subjected to the least waste of organised matter and force by the production of superfluous individuals of either sex. Nevertheless we may conclude that natural selection will always tend, though sometimes inefficiently, to equalise the relative numbers of the two sexes.

Having said this much on the equalisation of the sexes, it may be well to add a few remarks on the regulation through natural selection of the ordinary fertility of species. Mr. Herbert Spencer has shewn in an able discussion[409] that with all organisms a ratio exists between what he calls individuation and genesis; whence it follows that beings which consume much matter or force in their growth, complicated structure or activity, or which produce ova and embryos of large size, or which expend much energy in nurturing their young, cannot be so productive as beings of an opposite nature. Mr. Spencer further shews that minor differences in fertility will be regulated through natural selection. Thus the fertility of each species will tend to increase, from the more fertile pairs producing a larger number of offspring, and these from their mere number will have the best chance of surviving, and will transmit their tendency to greater fertility. The only check to a continued augmentation of fertility in each organism seems to be either the expenditure of more power and the greater risks run by the parents that produce a more numerous progeny, or the contingency of very numerous eggs and young being produced of smaller size, or less vigorous, or subsequently not so well nurtured. To strike a balance in any case between the disadvantages which follow from the production of a numerous progeny, and the advantages (such as the escape of at least some individuals from various dangers) is quite beyond our power of judgment.

When an organism has once been rendered extremely fertile, how its fertility can be reduced through natural selection is not so clear as how this capacity was first acquired. Yet it is obvious that if individuals of a species, from a decrease of their natural enemies, were habitually reared in larger numbers than could be supported, all the members would suffer. Nevertheless the offspring from the less fertile parents would have no direct advantage over the offspring from the more fertile parents, when all were mingled together in the same district. All the individuals would mutually tend to starve each other. The offspring indeed of the less fertile parents would lie under one great disadvantage, for from the simple fact of being produced in smaller numbers, they would be the most liable to extermination. Indirectly, however, they would partake of one great advantage; for under the supposed condition of severe competition, when all were pressed for food, it is extremely probable that those individuals which from some variation in their constitution produced fewer eggs or young, would produce them of greater size or vigour; and the adults reared from such eggs or young would manifestly have the best chance of surviving, and would inherit a tendency towards lessened fertility. The parents, moreover, which had to nourish or provide for fewer offspring would themselves be exposed to a less severe strain in the struggle for existence, and would have a better chance of surviving. By these steps, and by no others as far as I can see, natural selection under the above conditions of severe competition for food, would lead to the formation of a new race less fertile, but better adapted for survival, than the parent-race.

Secondary Sexual Characters in the Lower Classes of the Animal Kingdom

THESE CHARACTERS ABSENT IN THE LOWEST CLASSES • BRILLIANT COLOURS • MOLLUSCA • ANNELIDS • CRUSTACEA, SECONDARY SEXUAL CHARACTERS STRONGLY DEVELOPED; DIMORPHISM; COLOUR; CHARACTERS NOT ACQUIRED BEFORE MATURITY • SPIDERS, SEXUAL COLOURS OF; STRIDULATION BY THE MALES • MYRIAPODA

In the lowest classes the two sexes are not rarely united in the same individual, and therefore secondary sexual characters cannot be developed. In many cases in which the two sexes are separate, both are permanently attached to some support, and the one cannot search or struggle for the other. Moreover it is almost certain that these animals have too imperfect senses and much too low mental powers to feel mutual rivalry, or to appreciate each other's beauty or other attractions.

Hence in these classes, such as the Protozoa, Cœlenterata, Echinodermata, Scolecida, true secondary sexual characters do not occur; and this fact agrees with the belief that such characters in the higher classes have been acquired through sexual selection, which depends on the will, desires, and choice of either sex. Nevertheless some few apparent exceptions occur; thus, as I hear from Dr. Baird, the males of certain Entozoa, or internal parasitic worms, differ slightly in colour from the females; but we have no reason to suppose that such differences have been augmented through sexual selection.

Many of the lower animals, whether hermaphrodites or with the sexes separate, are ornamented with the most brilliant tints, or are shaded and striped in an elegant manner. This is the case with many corals and sea-anemonies (Actiniæ), with some jelly-fish (Medusæ, Porpita, &c.), with some Planariæ, Ascidians, numerous Star-fishes, Echini, &c.; but we may conclude from the reasons already indicated, namely the union of the two sexes in some of these animals, the permanently affixed condition of others, and the low mental powers of all, that such colours do not serve as a sexual attraction, and have not been acquired through sexual selection. With the higher animals the case is very different; for with them when one sex is much more brilliantly or conspicuously coloured than the other, and there is no difference in the habits of the two sexes which will account for this difference, we have reason to believe in the influence of sexual selection; and this belief is strongly confirmed when the more ornamented individuals, which are almost always the males, display their attractions before the other sex. We may also extend this conclusion to both sexes, when coloured alike, if their colours are plainly analogous to those of one sex alone in certain other species of the same group.

How, then, are we to account for the beautiful or even gorgeous colours of many animals in the lowest classes? It appears very doubtful whether such colours usually serve as a protection; but we are extremely liable to err in regard to characters of all kinds in relation to protection, as will be admitted by every one who has read Mr. Wallace's excellent essay on this subject. It would not, for instance, at first occur to any one that the perfect transparency of the Medusæ, or jelly-fishes, was of the highest service to them as a protection; but when we are reminded by Häckel that not only the medusæ but many floating mollusca, crustaceans, and even small oceanic fishes partake of this same glass-like structure, we can hardly doubt that they thus escape the notice of pelagic birds and other enemies.

Notwithstanding our ignorance how far colour in many cases serves as a protection, the most probable view in regard to the splendid tints of many of the lowest animals seems to be that their colours are the direct result either of the chemical nature or the minute structure of their tissues, independently of any benefit thus derived. Hardly any colour is finer than that of arterial blood; but there is no reason to suppose that the colour of the blood is in itself any advantage; and though it adds to the beauty of the maiden's cheek, no one will pretend that it has been acquired for this purpose. So again with many animals, especially the lower ones, the bile is richly coloured; thus the extreme beauty of the Eolidæ (naked sea-slugs) is chiefly due, as I am informed by Mr. Hancock, to the biliary glands seen through the translucent integuments; this beauty being probably of no service to these animals. The tints of the decaying leaves in an American forest, are described by every one as gorgeous; yet no one supposes that these tints are of the least advantage to the trees. Bearing in mind how many substances closely analogous to natural organic compounds have been recently formed by chemists, and which exhibit the most splendid colours, it would have been a strange fact if substances similarly coloured had not often originated, independently of any useful end being thus gained, in the complex laboratory of living organisms.

The sub-kingdom of the Mollusca.—Throughout this great division (taken in its largest acceptation) of the animal kingdom, secondary sexual characters, such as we are here considering, never, as far as I can discover, occur. Nor could they be expected in the three lowest classes, namely in the Ascidians, Polyzoa, and Brachiopods (constituting the Molluscoida of Huxley), for most of these animals are permanently affixed to a support or have their sexes united in the same individual. In the Lamellibranchiata, or bivalve shells, hermaphroditism is not rare. In the next higher class of the Gasteropoda, or marine univalve shells, the sexes are either united or separate. But in this latter case the males never possess special organs for finding, securing, or charming the females, or for fighting with other males. The sole external difference between the sexes consists, as I am informed by Mr. Gwyn Jeffreys, in the shell sometimes differing a little in form; for instance, the shell of the male periwinkle *(Littorina littorea)* is narrower and has a more elongated spire than that of the female. But differences of this nature, it may be presumed, are directly connected with the act of reproduction or with the development of the ova.

The Gasteropoda, though capable of locomotion and furnished with imperfect eyes, do not appear to be endowed with sufficient mental powers for the members of the same sex to struggle together in rivalry, and thus to acquire secondary sexual characters. Nevertheless with the pulmoniferous gasteropods, or land-snails, the pairing is preceded by courtship; for these animals, though hermaphrodites, are compelled by their structure to pair together. Agassiz remarks,[410] "Quiconque a eu l'occasion d'observer les amours des limaçons, ne saurait mettre en doute la séduction, déployée dans les mouvements et les allures qui préparent et accomplissent le double embrassement de ces hermaphrodites." These animals appear also susceptible of some degree of permanent attachment: an accurate observer, Mr. Lonsdale, informs me that he placed a pair of land-snails *(Helix pomatia)*, one of which was weakly, into a small and ill-provided garden. After a short time the strong and healthy individual disappeared, and was traced by its track of slime over a wall into an adjoining well-stocked garden. Mr. Lonsdale concluded that it had deserted its sickly mate; but after an absence of twenty-four hours it returned, and apparently communicated the result of its successful exploration, for both then started along the same track and disappeared over the wall.

Even in the highest class of the Mollusca, namely the Cephalopoda or cuttle-fishes, in which the sexes are separate, secondary sexual characters of the kind which we are here considering, do not, as far as I can discover, occur. This is a surprising circumstance, as these animals possess highly-developed sense-organs and have considerable mental powers, as will be admitted by every one who has watched their artful endeavours to escape from an enemy.[411] Certain Cephalopoda, however, are characterised by one extraordinary sexual character, namely, that the male element collects within one of the arms or tentacles, which is then cast off, and, clinging by its sucking-discs to the female, lives for a time an independent life. So completely does the cast-off arm resemble a separate animal, that it was described by Cuvier as a parasitic worm under the name of Hectocotyle. But this marvellous structure may be classed as a primary rather than as a secondary sexual character.

Although with the Mollusca sexual selection does not seem to have come into play; yet many univalve and bivalve shells, such as volutes, cones, scallops, &c., are beautifully coloured and shaped. The colours do not appear in most cases to be of any use as a protection; they are probably the direct result, as in the lowest classes, of the nature of the tissues; the patterns and the sculpture of the shell depending on its manner of growth. The amount of light seems to a certain extent to be influential; for although, as repeatedly stated by Mr. Gwyn Jeffreys, the shells of some species living at a profound depth are brightly coloured, yet we generally see the lower surfaces and the parts covered by the mantle less highly coloured than the upper and exposed surfaces.[412] In some cases, as with shells living amongst corals or brightly-tinted sea-weeds, the bright colours may serve as a protection. But many of the nudibranch mollusca, or sea-slugs, are as beautifully coloured as any shells, as may be seen in Messrs. Alder and Hancock's magnificent work; and from information kindly given me by Mr. Hancock, it is extremely doubtful whether these colours usually serve as a protection. With some species this may be the case, as with one which lives on the green leaves of algæ, and is itself bright-green. But many brightly-coloured, white or otherwise conspicuous species, do not seek concealment; whilst again some equally conspicuous species, as well as other dull-coloured kinds, live under stones and in dark recesses. So that with these nudibranch molluscs, colour apparently does not stand in any close relation to the nature of the places which they inhabit.

These naked sea-slugs are hermaphrodites, yet they pair together, as do land-snails, many of which have extremely pretty shells. It is conceivable that two hermaphrodites, attracted by each others' greater beauty, might unite and leave offspring which would inherit their parents' greater beauty. But with such lowly-organised creatures this is extremely improbable. Nor is it at all obvious how the offspring from the more beautiful pairs of hermaphrodites would have any advantage, so as to increase in numbers, over the offspring of the less beautiful, unless indeed vigour and beauty generally coincided. We have not here a number of males becoming mature before the females, and the more beautiful ones selected by the more vigorous females. If, indeed, brilliant colours were beneficial to an hermaphrodite animal in relation to its general habits of life, the more brightly-tinted individuals would succeed best and would increase in number; but this would be a case of natural and not of sexual selection.

Sub-kingdom of the Vermes or Annulosa: Class, Annelida (or Sea-worms).—In this class, although the sexes (when separate) sometimes differ from each other in characters of such importance that they have been placed under distinct genera or even families, yet the differences do not seem of the kind which can be safely attributed to sexual selection. These animals, like those in the preceding classes, apparently stand too low in the scale, for the individuals of either sex to exert any choice in selecting a partner, or for the individuals of the same sex to struggle together in rivalry.

Sub-kingdom of the Arthropoda: Class, Crustacea.—In this great class we first meet with undoubted secondary sexual characters, often developed in a remarkable manner. Unfortunately the habits of crustaceans are very imperfectly known, and we cannot explain the uses of many structures peculiar to one sex. With the lower parasitic species the males are of small size, and they alone are furnished with perfect swimming-legs, antennæ and sense-organs; the females being destitute of these organs, with their bodies often consisting of a mere distorted mass. But these extraordinary differences between the two sexes are no doubt related to their widely different habits of life, and consequently do not concern us. In various crustaceans, belonging to distinct families, the anterior antennas are furnished with peculiar thread-like bodies, which are believed to act as smelling-organs, and these are much more numerous in the males than in the females. As the males, without any unusual development of their olfactory organs, would almost certainly be able sooner or later to find the females, the increased number of the smelling-threads has probably been acquired through sexual selection, by the better provided males having been the most successful in finding partners and in leaving offspring.

Fritz Müller has described a remarkable dimorphic species of Tanais, in which the male is represented by two distinct forms, never graduating into each other. In the one form the male is furnished with more numerous smelling-threads, and in the other form with more powerful and more elongated chelæ or pincers which serve to hold the female. Fritz Müller suggests that these differences between the two male forms of the same species must have originated in certain individuals having varied in the number of the smelling-threads, whilst other individuals varied in the shape and size of their chelæ; so that of the former, those which were best able to find the female, and of the latter, those which were best able to hold her when found, have left the greater number of progeny to inherit their respective advantages.[413]

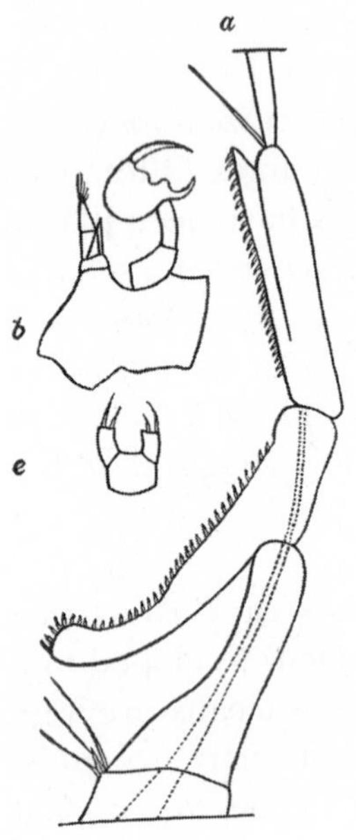

Fig. 3. *Labidocera Darwinii (from Lubbock).*
a. *Part of right-hand anterior antenna of male, forming a prehensile organ.*
b. *Posterior pair of thoracic legs of male.*
c. *Ditto of female.*

In some of the lower crustaceans, the right-hand anterior antenna of the male differs greatly in structure from the left-hand one, the latter resembling in its simple tapering joints the antennæ of the female. In the male the modified antenna is either swollen in the middle or angularly bent, or converted (fig. 3) into an elegant, and sometimes wonderfully complex, prehensile organ.[414] It serves, as I hear from Sir J. Lubbock, to hold the female, and for this same purpose one of the two posterior legs *(b)* on the same side of the body is converted into a forceps. In another family the inferior or posterior antennæ are "curiously zigzagged" in the males alone.

In the higher crustaceans the anterior legs form a pair of chelæ or pincers, and these are generally larger in the male than in the female. In many species the chelæ on the opposite sides of the body are of unequal size, the right-hand one being, as I am informed by Mr. C. Spence Bate, generally, though not invariably, the largest. This inequality is often much greater in the male than in the female. The two chelæ also often differ in structure (figs. 4, 5 and 6), the smaller one resembling those of the female. What advantage is gained by their inequality in size on the opposite sides of the body, and by the inequality being much greater in the male than in the female;

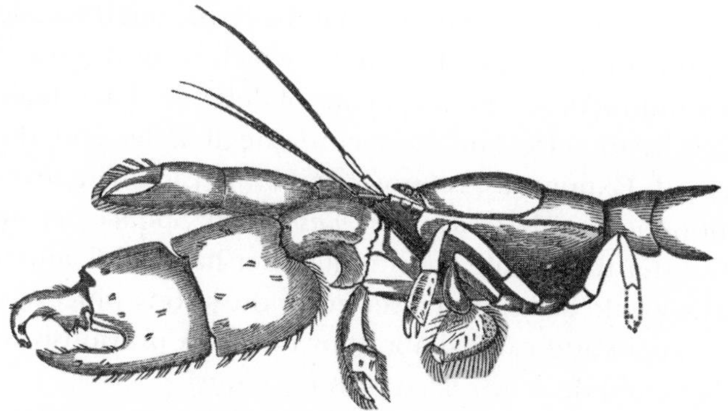

Fig. 4. *Anterior part of body of Callianassa (from Milne-Edwards), showing the unequal and differently-constructed right and left-hand chelæ of the male.*

N.B.—*The artist by mistake has reversed the drawing, and made the left-hand chela the largest.*

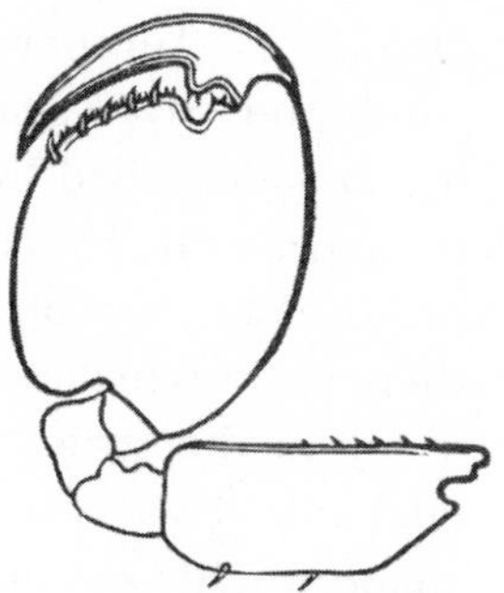

Fig. 5. *Second leg of male Orchestia Tucuratinga (from Fritz Müller).*

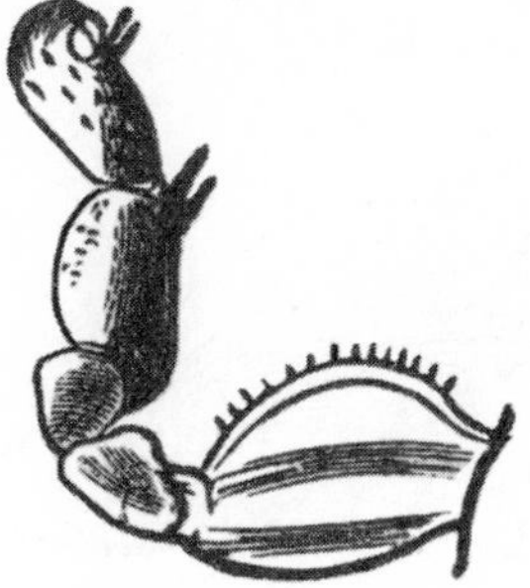

Fig. 6. *Ditto of female.*

and why, when they are of equal size, both are often much larger in the male than in the female, is not known. The chelæ are sometimes of such length and size that they cannot possibly be used, as I hear from Mr. Spence Bate, for carrying food to the mouth. In the males of certain fresh-water prawns (Palæmon) the right leg is actually longer than the whole body.[415] It is probable that the great size of one leg with its chelæ may aid the male in fighting with his rivals; but this use will not account for their inequality in the female on the opposite sides of the body. In Gelasimus, according to a statement quoted by Milne-Edwards,[416] the male and female live in the same burrow, which is worth notice, as shewing that they pair, and the male closes the mouth of the burrow with one of its chelæ, which is enormously developed; so that here it indirectly serves as a means of defence. Their main use, however, probably is to seize and to secure the female, and this in some instances, as with Gammarus, is known to be the case. The sexes, however, of the common shore-crab *(Carcinus mænas),* as Mr. Spence Bate informs me, unite directly after the female has moulted her hard shell, and when she is so soft that she would be injured if seized by the strong pincers of the male; but as she is caught and carried about by the male previously to the act of moulting, she could then be seized with impunity.

Fritz Müller states that certain species of Melita are distinguished from all other amphipods by the females having "the coxal lamellæ of the penultimate pair of feet produced into hook-like processes, of which the males lay hold with the hands of the first pair." The development of these hook-like processes probably resulted from those females which were the most securely held during the act of reproduction, having left the largest number of offspring. Another Brazilian amphipod *(Orchestia Darwinii,* fig. 7) is described by Fritz Müller, as presenting a case of dimorphism, like that of Tanais; for there are two male forms, which differ in the structure of their chelæ.[417] As chelæ of either shape would certainly have sufficed to hold the female, for both are now used for this purpose, the two male forms probably originated, by some having varied in one manner and some in another; both forms having derived certain special, but nearly equal advantages, from their differently shaped organs.

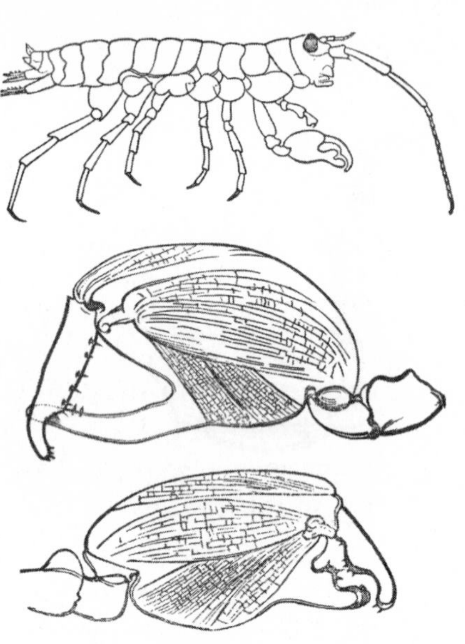

Fig. 7. *Orchestia Darwinii (from Fritz Müller), showing the differently-constructed chelæ of the two male forms.*

It is not known that male crustaceans fight together for the possession of the females, but this is probable; for with most animals when the male is larger than the female, he seems to have acquired his greater size by having conquered during many generations other males. Now Mr. Spence Bate informs me that in most of the crustacean orders, especially in the highest or the Brachyura, the male is larger than the female; the parasitic genera, however, in which the sexes follow different habits of life, and most of the Entomostraca must be excepted. The chelæ of many crustaceans are weapons well adapted for fighting. Thus a Devil-crab *(Portunus puber)* was seen by a son of Mr. Bate fighting with a *Carcinus mænas,* and the latter was soon thrown on its back, and had every limb torn from its body. When several males of a Brazilian Gelasimus, a species furnished with immense pincers, were placed together by Fritz Müller in a glass vessel, they mutilated and killed each other. Mr. Bate put a large male *Carcinus mænas* into a pan of water, inhabited by a female paired with a smaller male; the latter was soon dispossessed, but, as Mr. Bate adds, "if they fought, the victory was a bloodless one, for I saw no wounds." This same naturalist separated a male sand-skipper (so common on our sea-shores), *Gammarus marinus,* from its female, both of which were imprisoned in the same vessel with many individuals of the same species. The female being thus divorced joined her comrades. After an interval the male was again put into the same vessel and he then, after swimming about for a time, dashed into the crowd, and without any fighting at once took away his wife. This fact shews that in the Amphipoda, an order low in the scale, the males and females recognise each other, and are mutually attached.

The mental powers of the Crustacea are probably higher than might have been expected. Any one who has tried to catch one of the shore-crabs, so numerous on many tropical coasts, will have perceived how wary and alert they are. There is a large crab *(Birgus latro),* found on coral islands, which makes at the bottom of a deep burrow a thick bed of the picked fibres of the cocoa-nut. It feeds on the fallen fruit of this tree by tearing off the husk, fibre by fibre; and it always begins at that end where the three eye-like depressions are situated. It then breaks through one of these eyes by hammering with its heavy front pincers, and turning round, extracts the albuminous core with its narrow posterior pincers. But these actions are probably instinctive, so that they would be performed as well by a young as by an old animal. The following case, however, can hardly be so considered: a trustworthy naturalist, Mr. Gardner,[418] whilst watching a shore-crab (Gelasimus) making its burrow, threw some shells towards the hole. One rolled in, and three other shells remained within a few inches of the mouth. In about five minutes the crab brought out the shell which had fallen in, and carried it away to the distance of a foot; it then saw the three other shells lying near, and evidently thinking that they might likewise roll in, carried them to the spot where it had laid the first. It would, I think, be difficult to distinguish this act from one performed by man by the aid of reason.

With respect to colour which so often differs in the two sexes of animals belonging to the higher classes, Mr. Spence Bate does not know of any well-marked instances with our British crustaceans. In some cases, however, the male and female differ slightly in tint, but Mr. Bate thinks not more than may be accounted for by their different habits of life, such as by the male wandering more about and being thus more exposed to the light. In a curious Bornean crab, which inhabits sponges, Mr. Bate could always distinguish the sexes by the male not having the epidermis so much rubbed off. Dr. Power tried to distinguish by colour the sexes of the species which inhabit the Mauritius, but always failed, except with one species of Squilla, probably the *S. stylifera,* the male of which is described as being "of a beautiful blueish-green," with some of the appendages cherry-red, whilst the female is clouded with brown and grey, "with the red about her much less vivid than in the male."[419] In this case, we may suspect the agency of sexual selection. With Saphirina (an oceanic genus of Entomostraca, and therefore low in the scale) the males are furnished with minute shields or cell-like bodies, which exhibit beautiful changing colours; these being absent in the females, and in the case of one species in both sexes.[420] It would, however, be extremely rash to conclude that these curious organs serve merely to attract the females. In the female of a Brazilian species of Gelasimus, the whole body, as I am informed by Fritz Müller, is of a nearly uniform greyish-brown. In the male the posterior part of the cephalo-thorax is pure white, with the anterior part of a rich green, shading into dark brown; and it is remarkable that these colours are liable to change in the course of a few minutes—the white becoming dirty grey or even black, the green "losing much of its brilliancy." The males apparently are much more numerous than the females. It deserves especial notice that they do not acquire their bright colours until they become mature. They differ also from the females in the larger size of their chelæ. In some species of the genus, probably in all, the sexes pair and inhabit the same burrow. They are also, as we have seen, highly intelligent animals. From these various considerations it seems highly probable that the male in this species has become gaily ornamented in order to attract or excite the female.

It has just been stated that the male Gelasimus does not acquire his conspicuous colours until mature and nearly ready to breed. This seems the general rule in the whole class with the many remarkable differences in structure between the two sexes. We shall hereafter find the same law prevailing throughout the great sub-kingdom of the Vertebrata, and in all cases it is eminently distinctive of characters which have been acquired through sexual selection. Fritz Müller[421] gives some striking instances of this law; thus the male sand-hopper (Orchestia) does not acquire his large claspers, which are very differently constructed from those of the female, until nearly full-grown; whilst young his claspers resemble those of the female.

Thus, again, the male Brachyscelus possesses, like all other amphipods, a pair of posterior antennæ; the female, and this is a most extraordinary circumstance, is destitute of them, and so is the male as long as he remains immature.

Class, Arachnida (Spiders).—The males are often darker, but sometimes lighter than the females, as may be seen in Mr. Blackwall's magnificent work.[422] In some species the sexes differ conspicuously from each other in colour; thus the female of *Sparassus smaragdulus* is dullish-green; whilst the adult male has the abdomen of a fine yellow, with three longitudinal stripes of rich red. In some species of Thomisus the two sexes closely resemble each other; in others they differ much; thus in *T. citreus* the legs and body of the female are pale-yellow or green, whilst the front legs of the male are reddish-brown: in *T. floricolens*, the legs of the female are pale-green, those of the male being ringed in a conspicuous manner with various tints. Numerous analogous cases could be given in the genera Epeira, Nephila, Philodromus, Theridion, Linyphia, &c. It is often difficult to say which of the two sexes departs most from the ordinary coloration of the genus to which the species belong; but Mr. Blackwall thinks that, as a general rule, it is the male. Both sexes whilst young, as I am informed by the same author, usually resemble each other; and both often undergo great changes in colour during their successive moults before arriving at maturity. In other cases the male alone appears to change colour. Thus the male of the abovementioned brightly-coloured Sparassus at first resembles the female and acquires his peculiar tints only when nearly adult. Spiders are possessed of acute senses, and exhibit much intelligence. The females often shew, as is well known, the strongest affection for their eggs, which they carry about enveloped in a silken web. On the whole it appears probable that well-marked differences in colour between the sexes have generally resulted from sexual selection, either on the male or female side. But doubts may be entertained on this head from the extreme variability in colour of some species, for instance of *Theridion lineatum,* the sexes of which differ when adult; this great variability indicates that their colours have not been subjected to any form of selection.

Mr. Blackwall does not remember to have seen the males of any species fighting together for the possession of the female. Nor, judging from analogy, is this probable; for the males are generally much smaller than the females, sometimes to an extraordinary degree.[423] Had the males been in the habit of fighting together, they would, it is probable, have gradually acquired greater size and strength. Mr. Blackwall has sometimes seen two or more males on the same web with a single female; but their courtship is too tedious and prolonged an affair to be easily observed. The male is extremely cautious in making his advances, as the female carries her coyness to a dangerous pitch. De Geer saw a male that "in the midst of his preparatory caresses was seized by the object of his attentions, enveloped by her in a web and then devoured, a sight which, as he adds, filled him with horror and indignation."[424]

Westring has made the interesting discovery that the males of several species of Theridion[425] have the power of making a stridulating sound (like that made by many beetles and other insects, but feebler), whilst the females are quite mute. The apparatus consists of a serrated ridge at the base of the abdomen, against which the hard hinder part of the thorax is rubbed; and of this structure not a trace could be detected in the females. From the analogy of the Orthoptera and Homoptera, to be described in the next chapter, we may feel almost sure that the stridulation serves, as Westring remarks, either to call or to excite the female; and this is the first case in the ascending scale of the animal kingdom, known to me, of sounds emitted for this purpose.

Class, Myriapoda.—In neither of the two orders in this class, including the millipedes and centipedes, can I find any well-marked instances of sexual differences such as more particularly concern us. In *Glomeris limbata,* however, and perhaps in some few other species, the males differ slightly in colour from the females; but this Glomeris is a highly variable species. In the males of the Diplopoda, the legs belonging to one of the anterior segments of the body, or to the posterior segment, are modified into prehensile hooks which serve to secure the female. In some species of Iulus the tarsi of the male are furnished with membranous suckers for the same purpose. It is a much more unusual circumstance, as we shall see when we treat of Insects, that it is the female in Lithobius which is furnished with prehensile appendages at the extremity of the body for holding the male.[426]

CHAPTER X

Secondary Sexual Characters of Insects

DIVERSIFIED STRUCTURES POSSESSED BY THE MALES FOR SEIZING THE FEMALES • DIFFERENCES BETWEEN THE SEXES, OF WHICH THE MEANING IS NOT UNDERSTOOD • DIFFERENCE IN SIZE BETWEEN THE SEXES • THYSANURA • DIPTERA • HEMIPTERA • HOMOPTERA, MUSICAL POWERS POSSESSED BY THE MALES ALONE • ORTHOPTERA, MUSICAL INSTRUMENTS OF THE MALES, MUCH DIVERSIFIED IN STRUCTURE; PUGNACITY; COLOURS • NEUROPTERA, SEXUAL DIFFERENCES IN COLOUR • HYMENOPTERA, PUGNACITY AND COLOURS • COLEOPTERA, COLOURS; FURNISHED WITH GREAT HORNS, APPARENTLY AS AN ORNAMENT; BATTLES; STRIDULATING ORGANS GENERALLY COMMON TO BOTH SEXES

In the immense class of insects the sexes sometimes differ in their organs for locomotion, and often in their sense-organs, as in the pectinated and beautifully plumose antennæ of the males of many species. In one of the Ephemeræ, namely Chloëon, the male has great pillared eyes, of which the female is entirely destitute.[427] The ocelli are absent in the females of certain other insects, as in the Mutillidæ, which are likewise destitute of wings. But we are chiefly concerned with structures by which one male is enabled to conquer another, either in battle or courtship, through his strength, pugnacity, ornaments, or music. The innumerable contrivances, therefore, by which the male is able to seize the female, may be briefly passed over. Besides the complex structures at the apex of the abdomen, which ought perhaps to be ranked as primary organs,[428] "it is astonishing," as Mr. B. D. Walsh[429] has remarked, "how many different organs are worked in by nature, for the seemingly insignificant object of enabling the male to grasp the female firmly." The mandibles or jaws are sometimes used for this purpose; thus the male *Corydalis cornutus* (a neuropterous insect in some degree allied to the Dragon-flies, &c.) has immense curved jaws, many times longer than those of the female; and they are smooth instead of being toothed, by which means he is enabled to seize her without injury.[430] One of the stag-beetles of North America *(Lucanus elaphus)* uses his jaws, which are much larger than those of the female, for the same purpose, but probably likewise for fighting. In one of the sand-wasps *(Ammophila)* the jaws in the two sexes are closely alike, but are used for widely different purposes; the males, as Professor Westwood observes, "are exceedingly ardent, seizing their partners round the neck with their sickle-shaped jaws;"[431] whilst the females use these organs for burrowing in sand-banks and making their nests.

The tarsi of the front-legs are dilated in many male beetles, or are furnished with broad cushions of hairs; and in many genera of water-beetles they are armed with a round flat sucker, so that the male may adhere to the slippery body of the female. It is a much more unusual circumstance that the females of some water-beetles (Dytiscus) have their elytra deeply grooved, and in *Acilius sulcatus* thickly set with hairs, as an aid to the male. The females of some other water-beetles (Hydroporus) have their elytra punctured for the same object.[432] In the male of *Crabro cribrarius* (fig. 8), it is the tibia which is dilated into a broad horny plate, with minute membraneous dots, giving to it a singular appearance like that of a riddle.[433] In the male of Penthe (a genus of beetles) a few of the middle joints of the antennæ are dilated and furnished on the inferior surface with cushions of hair, exactly like those on the tarsi of the Carabidae, "and obviously for the same end." In male dragon-flies, "the appendages at the tip of the tail are modified in an almost infinite variety of curious patterns to enable them to embrace the neck of the female." Lastly in the males of many insects, the legs are furnished with peculiar spines, knobs or spurs; or the whole leg is bowed or thickened, but this is by no means invariably a sexual character; or one pair, or all three pairs are elongated, sometimes to an extravagant length.[434]

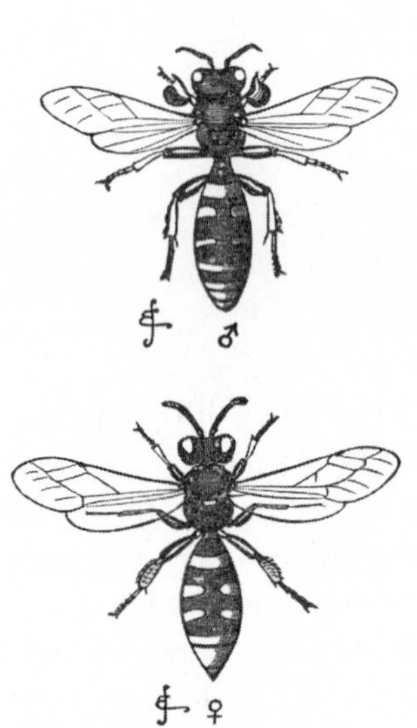

Fig. 8. *Crabro cribrarius. Upper figure, male; lower figure, female.*

In all the orders, the sexes of many species present differences, of which the meaning is not understood. One curious case is that of a beetle (fig. 9), the male of which has the left mandible much enlarged; so that the mouth is greatly distorted. In another Carabidous beetle, the Eurygnathus,[435] we have the unique case, as far as known to Mr. Wollaston, of the head of the female being much broader and larger, though in a variable degree, than that of the male. Any number of such cases could be given. They abound in the Lepidoptera: one of the most extraordinary is that certain male butterflies have their fore-legs more or less atrophied, with the tibiæ and tarsi reduced to mere rudimentary knobs. The wings, also, in the two sexes often differ in neuration,[436] and sometimes considerably in outline, as in the *Aricoris epitus,* which was shown to me in the British Museum by Mr. A. Butler. The males of certain South American butterflies have tufts of hair on the margins of the wings, and horny excrescences on the discs of the posterior pair.[437] In several British butterflies, the males alone, as shewn by Mr. Wonfor, are in parts clothed with peculiar scales.

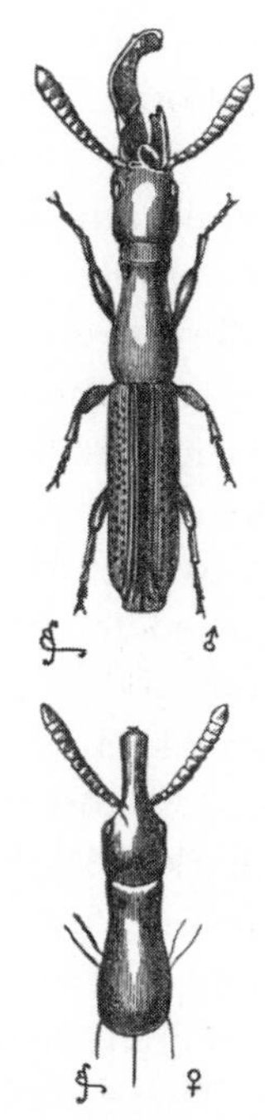

Fig. 9. *Taphroderes distortus (much enlarged). Upper figure, male; lower figure, female.*

The purpose of the luminosity in the female glow-worm is likewise not understood; for it is very doubtful whether the primary use of the light is to guide the male to the female. It is no serious objection to this latter belief that the males emit a feeble light; for secondary sexual characters proper to one sex are often developed in a slight degree in the other sex. It is a more valid objection that the larvæ shine, and in some species brilliantly: Fritz Müller informs me that the most luminous insect which he ever beheld in Brazil, was the larva of some beetle. Both sexes of certain luminous species of Elater emit light. Kirby and Spence suspect that the phosphorescence serves to frighten and drive away enemies.

Difference in Size between the Sexes.—With insects of all kinds the males are commonly smaller than the females;[438] and this difference can often be detected even in the larval state. So considerable is the difference between the male and female cocoons of the silk-moth *(Bombyx mori),* that in France they are separated by a particular mode of weighing.[439] In the lower classes of the animal kingdom, the greater size of the females seems generally to depend on their developing an enormous number of ova; and this may to a certain extent hold good with insects. But Dr. Wallace has suggested a much more probable explanation. He finds, after carefully attending to the development of the caterpillars of *Bombyx cynthia* and *yamamai,* and especially of some dwarfed caterpillars reared from a second brood on unnatural food, "that in proportion as the individual moth is finer, so is the time required for its metamorphosis longer; and for this reason the female, which is the larger and heavier insect, from having to carry her numerous eggs, will be preceded by the male, which is smaller and has less to mature."[440] Now as most insects are short-lived, and as they are exposed to many dangers, it would manifestly be advantageous to the female to be impregnated as soon as possible. This end would be gained by the males being first matured in large numbers ready for the advent of the females; and this again would naturally follow, as Mr. A. R. Wallace has remarked,[441] through natural selection; for the smaller males would be first matured, and thus would procreate a large number of offspring which would inherit the reduced size of their male parents, whilst the larger males from being matured later would leave fewer offspring.

There are, however, exceptions to the rule of male insects being smaller than the females; and some of these exceptions are intelligible. Size and strength would be an advantage to the males, which fight for the possession of the female; and in these cases the males, as with the stag-beetle (Lucanus), are larger than the females. There are, however, other beetles which are not known to fight together, of which the males exceed the females in size; and the meaning of this fact is not known; but in some of these cases, as with the huge Dynastes and Megasoma, we can at least see that there would be no necessity for the males to be smaller than the females, in order to be matured before them, for these beetles are not short-lived, and there would be ample time for the pairing of the sexes. So, again, male dragon-flies (Libellulidæ) are sometimes sensibly larger, and never smaller, than the females;[442] and they do not, as Mr. MacLachlan believes, generally pair with the females, until a week or fortnight has elapsed, and until they have assumed their proper masculine colours. But the most curious case, shewing on what complex and easily-overlooked relations, so trifling a character as a difference in size between the sexes may depend, is that of the aculeate Hymenoptera; for Mr. F. Smith informs me that throughout nearly the whole of this large group the males, in accordance with the general rule, are smaller than the females and emerge about a week before them; but amongst the Bees, the males of *Apis mellifica, Anthidium manicatum* and *Anthophora acervorum,* and amongst the Fossores, the males of the *Methoca ichneumonides,* are larger than the females. The explanation of this anomaly is that a marriage-flight is absolutely necessary with these species, and the males require great strength and size in order to carry the females through the air. Increased size has here been acquired in opposition to the usual relation between size and the period of development, for the males, though larger, emerge before the smaller females.

We will now review the several Orders, selecting such facts as more particularly concern us. The Lepidoptera (Butterflies and Moths) will be retained for a separate chapter.

Order, Thysanura.—The members of this Order are lowly organised for their class. They are wingless, dull-coloured, minute insects, with ugly, almost misshapen heads and bodies. The sexes do not differ; but they offer one interesting fact, by shewing that the males pay sedulous court to their females even low down in the animal scale. Sir J. Lubbock[443] in describing the *Smynthurus luteus,* says: "it is very amusing to see these little creatures coquetting together. The male, which is much smaller than the female, runs round her, and they butt one another, standing face to face, and moving backward and forward like two playful lambs. Then the female pretends to run away and the male runs after her with a queer appearance of anger, gets in front and stands facing her again; then she turns coyly round, but he, quicker and more active, scuttles round too, and seems to whip her with his antennæ; then for a bit they stand face to face, play with their antennæ, and seem to be all in all to one another."

Order, Diptera (Flies).—The sexes differ little in colour. The greatest difference, known to Mr. F. Walker, is in the genus Bibio, in which the males are blackish or quite black, and the females obscure brownish-orange. The genus Elaphomyia, discovered by Mr. Wallace[444] in New Guinea, is highly remarkable, as the males are furnished with horns, of which the females are quite destitute. The horns spring from beneath the eyes, and curiously resemble those of stags, being either branched or palmated. They equal in length the whole of the body in one of the species. They might be thought to serve for fighting, but as in one species they are of a beautiful pink colour, edged with black, with a pale central stripe, and as these insects have altogether a very elegant appearance, it is perhaps more probable that the horns serve as ornaments. That the males of some Diptera fight together is certain; for Prof. Westwood[445] has several times seen this with some species of Tipula or Harry-long-legs. Many observers believe that when gnats (Culicidæ) dance in the air in a body, alternately rising and falling, the males are courting the females. The mental faculties of the Diptera are probably fairly well developed, for their nervous system is more highly developed than in most other Orders of insects.[446]

Order, Hemiptera (Field-Bugs).—Mr. J. W. Douglas, who has particularly attended to the British species, has kindly given me an account of their sexual differences. The males of some species are furnished with wings, whilst the females are wingless; the sexes differ in the form of the body and elytra; in the second joints of their antennæ and in their tarsi; but as the signification of these differences is quite unknown, they may be here passed over. The females are generally larger and more robust than the males. With British, and, as far as Mr. Douglas knows, with exotic species, the sexes do not commonly differ much in colour; but in about six British species the male is considerably darker than the female, and in about four other species the female is darker than the male. Both sexes of some species are beautifully marked with vermilion and black. It is doubtful whether these colours serve as a protection. If in any species the males had differed from the females in an analogous manner, we might have been justified in attributing such conspicuous colours to sexual selection with transference to both sexes.

Some species of Reduvidæ make a stridulating noise; and, in the case of *Pirates stridulus,* this is said[447] to be effected by the movement of the neck within the pro-thoracic cavity. According to Westring, *Reduvius personatus* also stridulates. But I have not been able to learn any particulars about these insects; nor have I any reason to suppose that they differ sexually in this respect.

Order, Homoptera.—Every one who has wandered in a tropical forest must have been astonished at the din made by the male Cicadæ. The females are mute; as the Grecian poet Xenarchus says, "Happy the Cicadas live, since they all have voiceless wives." The noise thus made could be plainly heard on board the "Beagle," when anchored at a quarter of a mile from the shore of Brazil; and Captain Hancock says it can be heard at the distance of a mile. The Greeks formerly kept, and the Chinese now keep, these insects in cages for the sake of their song, so that it must be pleasing to the ears of some men.[448] The Cicadidæ usually sing during the day; whilst the Fulgoridæ appear to be night-songsters. The sound, according to Landois,[449] who has recently studied the subject, is produced by the vibration of the lips of the spiracles, which are set into motion by a current of air emitted from the tracheae. It is increased by a wonderfully complex resounding apparatus, consisting of two cavities covered by scales. Hence the sound may truly be called a voice. In the female the musical apparatus is present, but very much less developed than in the male, and is never used for producing sound.

With respect to the object of the music, Dr. Hartman in speaking of the *Cicada septemdecim* of the United States, says,[450] "the drums are now (June 6th and 7th, 1851) heard in all directions. This I believe to be the marital summons from the males. Standing in thick chestnut sprouts about as high as my head, where hundreds were around me, I observed the females coming around the drumming males." He adds, "this season (Aug. 1868) a dwarf pear-tree in my garden produced about fifty larvæ of *Cic. pruinosa;* and I several times noticed the females to alight near a male while he was uttering his clanging notes." Fritz Müller writes to me from S. Brazil that he has often listened to a musical contest between two or three males of a Cicada, having a particularly loud voice, and seated at a considerable distance from each other. As soon as the first had finished his song, a second immediately began; and after he had concluded, another began, and so on. As there is so much rivalry between the males, it is probable that the females not only discover them by the sounds emitted, but that, like female birds, they are excited or allured by the male with the most attractive voice.

I have not found any well-marked cases of ornamental differences between the sexes of the Homoptera. Mr. Douglas informs me that there are three British species, in which the male is black or marked with black bands, whilst the females are pale-coloured or obscure.

Order, Orthoptera.—The males in the three saltatorial families belonging to this Order are remarkable for their musical powers, namely the Achetidæ or crickets, the Locustidæ for which there is no exact equivalent name in English, and the Acridiidæ or grasshoppers. The stridulation produced by some of the Locustidæ is so loud that it can be heard during the night at the distance of a mile;[451] and that made by certain species is not unmusical even to the human ear, so that the Indians on the Amazons keep them in wicker cages. All observers agree that the sounds serve either to call or excite the mute females. But it has been noticed[452] that the male migratory locust of Russia (one of the Acridiidæ) whilst coupled with the female, stridulates from anger or jealousy when approached by another male. The house-cricket when surprised at night uses its voice to warn its fellows.[453] In North America the Katy-did *(Platyphyllum concavum,* one of the Locustidæ) is described[454] as mounting on the upper branches of a tree, and in the evening beginning "his noisy babble, while rival notes issue from the neighbouring trees, and the groves resound with the call of *Katy-did-she-did,* the live-long night." Mr. Bates, in speaking of the European field-cricket (one of the Achetidæ), says, "the male has been observed to place itself in the evening at the entrance of its burrow, and stridulate until a female approaches, when the louder notes are succeeded by a more subdued tone, whilst the successful musician caresses with his antennæ the mate he has won."[455] Dr. Scudder was able to excite one of these insects to answer him, by rubbing on a file with a quill.[456] In both sexes a remarkable auditory apparatus has been discovered by Von Siebold, situated in the front legs.[457]

In the three Families the sounds are differently produced. In the males of the Achetidæ both wing-covers have the same structure; and this in the field-

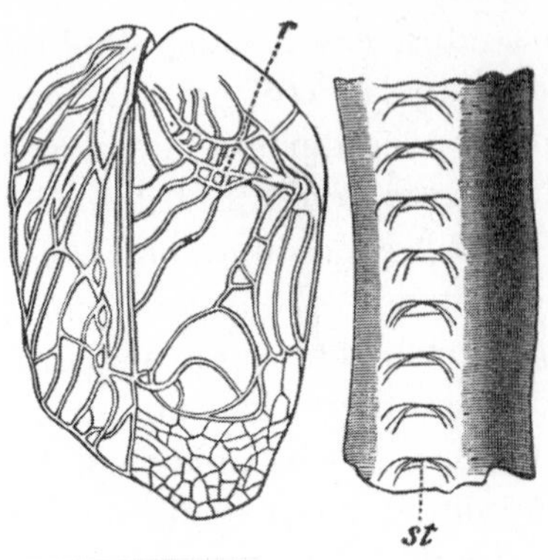

Fig.10. *Gryllus campestris (from Landois).*

Right-hand figure, under side of part of the wing-nervure, much magnified, showing the teeth, st.

Left-hand figure, upper surface of wing-cover, with the projecting, smooth nervure, r., *across which the teeth* (st) *are scraped.*

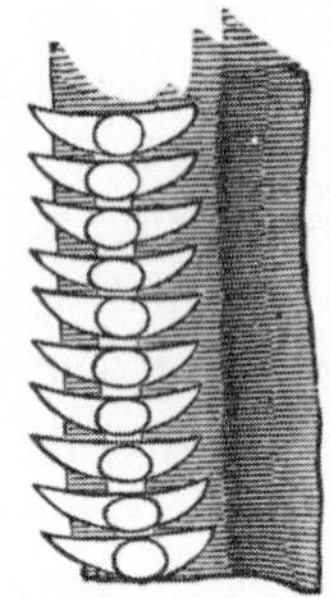

Fig.11. *Teeth of Nervure of Gryllus domesticus (from Landois).*

cricket *(Gryllus campestris,* fig. 10) consists, as described by Landois,[458] of from 131 to 138 sharp, transverse ridges or teeth *(st)* on the under side of one of the nervures of the wing-cover. This toothed nervure is rapidly scraped across a projecting, smooth, hard nervure *(r)* on the upper surface of the opposite wing. First one wing is rubbed over the other, and then the movement is reversed. Both wings are raised a little at the same time, so as to increase the resonance. In some species the wing-covers of the males are furnished at the base with a talc-like plate.[459] I have here given a drawing (fig. 11) of the teeth on the under side of the nervure of another species of Gryllus, viz. *G. domesticus.*

In the Locustidæ the opposite wing-covers differ in structure (fig. 12), and cannot, as in the last family, be indifferently used in a reversed manner. The left wing, which acts as the bow of the fiddle, lies over the right wing which serves as the fiddle itself. One of the nervures *(a)* on the under surface of the former is finely serrated, and is scraped across the prominent nervures on the upper surface of the opposite or right wing. In our British *Phasgonura viridissima* it appeared to me that the serrated nervure is rubbed against the rounded hind corner of the opposite wing, the edge of which is thickened, coloured brown, and very sharp. In the right wing, but not in the left, there is a little plate, as transparent as talc, surrounded by nervures, and called the speculum. In *Ephippiger vitium,* a member of this same family, we have a curious subordinate modification; for the wing-covers are greatly-reduced in size, but "the posterior part of the pro-thorax is elevated into a kind of dome over the wing-covers, and which has probably the effect of increasing the sound."[460]

We thus see that the musical apparatus is more differentiated or specialised in the Locustidæ, which includes I believe the most powerful performers in the Order, than in the Achetidæ, in which both wing-covers have the same structure and the same function.[461] Landois, however, detected in one of the Locustidæ, namely in Decticus, a short and narrow row of small teeth, mere rudiments, on the inferior surface of the right wing-cover, which underlies the other and is never used as the bow. I observed the same rudimentary structure on the under side of the right wing-cover in *Phasgonura viridissima*. Hence we may with confidence infer that the Locustidæ are descended from a form, in which, as in the existing Achetidæ, both wing-covers had serrated nervures on the under surface, and could be indifferently used as the bow; but that in the Locustidæ the two wing-covers gradually became differentiated and perfected, on the principle of the division of labour, the one to act exclusively as the bow and the other as the fiddle. By what steps the more simple apparatus in the Achetidæ originated, we do not know, but it is probable that the basal portions of the wing-covers overlapped each other formerly as at present, and that the friction of the nervures produced a grating sound, as I find is now the case with the wing-covers of the females.[462] A grating sound thus occasionally and accidentally made by the males, if it served them ever so little as a love-call to the females, might readily have been intensified through sexual selection by fitting variations in the roughness of the nervures having been continually preserved.

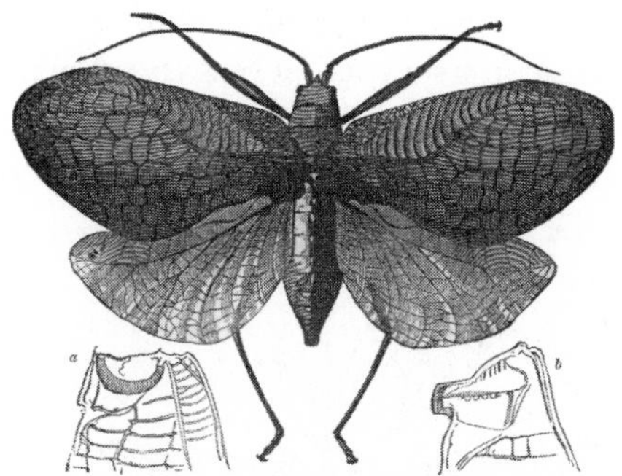

Fig. 12. *Chlorocœlus Tanana (from Bates).*

a, b. *Lobes of opposite wing-covers.*

In the last and third Family, namely the Acridiidæ or grasshoppers, the stridulation is produced in a very different manner, and is not so shrill, according to Dr. Scudder, as in the preceding Families. The inner surface of the femur (fig. 13, *r*) is furnished with a longitudinal row of

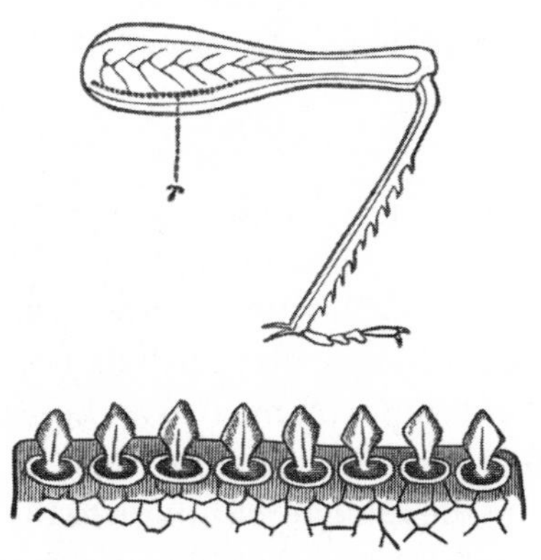

Fig. 13. *Hind-leg of Stenobothrus pratorum:* r, *the stridulating ridge; lower figure, the teeth forming the ridge, much magnified (from Landois).*

minute, elegant, lancet-shaped, elastic teeth, from 85 to 93 in number;[463] and these are scraped across the sharp, projecting nervures on the wing-covers, which are thus made to vibrate and resound. Harris[464] says that when one of the males begins to play, he first "bends the shank of the hind-leg beneath "the thigh, where it is lodged in a furrow designed to receive it, and then draws the leg briskly up and down. He does not play both fiddles together, but alternately first upon one and then on the other." In many species, the base of the abdomen is hollowed out into a great cavity which is believed to act as a resounding board. In Pneumora (fig. 14), a S. African genus belonging to this same family, we meet with a new and remarkable modification: in the males a small notched ridge projects obliquely from each side of the abdomen, against which the hind femora are rubbed.[465] As the male is furnished with wings, the female being wingless, it is remarkable that the thighs are not rubbed in the usual manner against the wing-covers; but this may perhaps be accounted for by the unusually small size of the hind-legs. I have not been able to examine the inner surface of the thighs, which, judging from analogy, would be finely serrated. The species of Pneumora have been more profoundly modified for the sake of stridulation than any other orthopterous insect; for in the male the whole body has been converted into a musical instrument, being distended with air, like a great pellucid bladder, so as to increase the resonance. Mr. Trimen informs me that at the Cape of Good Hope these insects make a wonderful noise during the night.

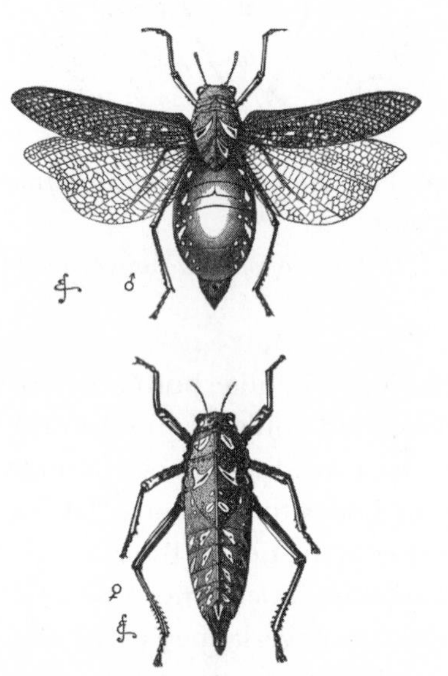

Fig. 14. *Pneumora (from specimens in the British Museum).*

Upper figure, male; lower figure, female.

There is one exception to the rule that the females in these three Families are destitute of an efficient musical apparatus; for both sexes of Ephippiger (Locustidæ) are said[466] to be thus provided. This case may be compared with that of the reindeer, in which species alone both sexes possess horns. Although the female orthoptera are thus almost invariably mute, yet Landois[467] found rudiments of the stridulating organs on the femora of the female Acridiidæ, and similar rudiments on the under surface of the wing-covers of the female Achetidæ; but he failed to find any rudiments in the females of Decticus, one of the Locustidæ. In the Homoptera the mute females of Cicada, have the proper musical apparatus in an undeveloped state; and we shall hereafter meet in other divisions of the animal kingdom with innumerable instances of structures proper to the male being present in a rudimentary condition in the female. Such cases appear at first sight to indicate that both sexes were primordially constructed in the same manner, but that certain organs were subsequently lost by the females. It is, however, a more probable view, as previously explained, that the organs in question were acquired by the males and partially transferred to the females.

Landois has observed another interesting fact, namely that in the females of the Acridiidæ, the stridulating teeth on the femora remain throughout life in the same condition in which they first appear in both sexes during the larval state. In the males, on the other hand, they become fully developed and acquire their perfect structure at the last moult, when the insect is mature and ready to breed.

From the facts now given, we see that the means by which the males produce their sounds are extremely diversified in the Orthoptera, and are altogether different from those employed by the Homoptera. But throughout the animal kingdom we incessantly find the same object gained by the most diversified means; this being due to the whole organisation undergoing in the course of ages multifarious changes; and as part after part varies, different variations are taken advantage of for the same general purpose. The diversification of the means for producing sound in the three families of the Orthoptera and in the Homoptera, impresses the mind with the high importance of these structures to the males, for the sake of calling or alluring the females. We need feel no surprise at the amount of modification which the Orthoptera have undergone in this respect, as we now know, from Dr. Scudder's remarkable discovery,[468] that there has been more than ample time. This naturalist has lately found a fossil insect in the Devonian formation of New Brunswick, which is furnished with "the well-known tympanum or stridulating apparatus of the male Locustidæ." This insect, though in most respects related to the Neuroptera, appears to connect, as is so often the case with very ancient forms, the two Orders of the Neuroptera and Orthoptera which are now generally ranked as quite distinct.

I have but little more to say on the Orthoptera. Some of the species are very pugnacious: when two male field-crickets *(Gryllus campestris)* are confined together, they fight till one kills the other; and the species of Mantis are described as manœuvring with their sword-like front-limbs, like hussars with their sabres. The Chinese keep these insects in little bamboo cages and match them like game-cocks.[469] With respect to colour, some exotic locusts are beautifully ornamented; the posterior wings being marked with red, blue, and black; but as throughout the Order the two sexes rarely differ much in colour, it is doubtful whether they owe these bright tints to sexual selection. Conspicuous colours may be of use to these insects as a protection, on the principle to be explained in the next chapter, by giving notice to their enemies that they are unpalatable. Thus it has been observed[470] that an Indian brightly-coloured locust was invariably rejected when offered to birds and lizards. Some cases, however, of sexual differences in colour in this Order are known. The male of an American cricket[471] is described as being as white as ivory, whilst the female varies from almost white to greenish-yellow or dusky. Mr. Walsh informs me that the adult male of *Spectrum femoratum* (one of the Phasmidæ) "is of a shining brownish-yellow colour; the adult female being of a dull, opaque, cinereous-brown; the young of both sexes being green." Lastly, I may mention that the male of one curious kind of cricket[472] is furnished with "a long membranous appendage, which falls over the face like a veil;" but whether this serves as an ornament is not known.

Order, Neuroptera.—Little need here be said, except in regard to colour. In the Ephemeridæ the sexes often differ slightly in their obscure tints;[473] but it is not probable that the males are thus rendered attractive to the females. The Libellulidæ or dragon-flies are ornamented with splendid green, blue, yellow, and vermilion metallic tints; and the sexes often differ. Thus, the males of some of the Agrionidæ, as Prof. Westwood remarks,[474] "are of a rich blue with black wings, whilst the females are fine green with colourless wings." But in *Agrion Ramburii* these colours are exactly reversed in the two sexes.[475] In the extensive N. American genus of Hetærina, the males alone have a beautiful carmine spot at the base of each wing. In *Anax junius* the basal part of the abdomen in the male is a vivid ultra-marine blue, and in the female grass-green. In the allied genus Gomphus, on the other hand, and in some other genera, the sexes differ but little in colour. Throughout the animal kingdom, similar cases of the sexes of closely-allied forms either differing greatly, or very little, or not at all, are of frequent occurrence. Although with many Libellulidæ there is so wide a difference in colour between the sexes, it is often difficult to say which is the most brilliant; and the ordinary coloration of the two sexes is exactly reversed, as we have just seen, in one species of Agrion. It is not probable that their colours in any case have been gained as a protection. As Mr. MacLachlan, who has closely attended to this family, writes to me, dragon-flies—the tyrants of the insect-world—are the least liable of any insect to be attacked by birds or other enemies. He believes that their bright colours serve as a sexual attraction. It deserves notice, as bearing on this subject, that certain dragon-flies appear to be attracted by particular colours: Mr. Patterson observed[476] that the species of Agrionidæ, of which the males are blue, settled in numbers on the blue float of a fishing line; whilst two other species were attracted by shining white colours.

It is an interesting fact, first observed by Schelver, that the males, in several genera belonging to two sub-families, when they first emerge from the pupal state are coloured exactly like the females; but that their bodies in a short time assume a conspicuous milky-blue tint, owing to the exudation of a kind of oil, soluble in ether and alcohol. Mr. MacLachlan believes that in the male of *Libellula depressa* this change of colour does not occur until nearly a fortnight after the metamorphosis, when the sexes are ready to pair.

Certain species of Neurothemis present, according to Brauer[477] a curious case of dimorphism, some of the females having their wings netted in the usual manner; whilst other females have them "very richly netted as in the males of the same species." Brauer "explains the phenomenon on Darwinian principles by the supposition that the close netting of the veins is a secondary sexual character in the males." This latter character is generally developed in the males alone, but being, like every other masculine character, latent in the female, is occasionally developed in them. We have here an illustration of the manner in which the two sexes of many animals have probably come to resemble each other, namely by variations first appearing in the males, being preserved in them, and then transmitted to and developed in the females; but in this particular genus a complete transference is occasionally and abruptly effected. Mr. MacLachlan informs me of another case of dimorphism occurring in several species of Agrion in which a certain number of individuals are found of an orange colour, and these are invariably females. This is probably a case of reversion, for in the true Libellulæ, when the sexes differ in colour, the females are always orange or yellow, so that supposing Agrion to be descended from some primordial form having the characteristic sexual colours of the typical Libellulæ, it would not be surprising that a tendency to vary in this manner should occur in the females alone.

Although many dragon-flies are such large, powerful, and fierce insects, the males have not been observed by Mr. MacLachlan to fight together, except, as he believes, in the case of some of the smaller species of Agrion. In another very distinct group in this Order, namely in the Termites or white ants, both sexes at the time of swarming may be seen running about, "the male after the female, sometimes two chasing one female, and contending with great eagerness who shall win the prize."[478]

Order, Hymenoptera.—That inimitable observer, M. Fabre,[479] in describing the habits of Cerceris, a wasp-like insect, remarks that "fights frequently ensue between the males for the possession of some particular female, who sits an apparently unconcerned beholder of the struggle for supremacy, and when the victory is decided, quietly flies away in company with the conqueror." Westwood[480] says that the males of one of the saw-flies (Tenthredinæ) "have been found fighting together, with their mandibles locked." As M. Fabre speaks of the males of Cerceris striving to obtain a particular female, it may be well to bear in mind that insects belonging to this Order have the power of recognising each other after long intervals of time, and are deeply attached. For instance, Pierre Huber, whose accuracy no one doubts, separated some ants, and when after an interval of four months they met others which had formerly belonged to the same community, they mutually recognised and caressed each other with their antennæ. Had they been strangers they would have fought together. Again, when two communities engage in a battle, the ants on the same side in the general confusion sometimes attack each other, but they soon perceive their mistake, and the one ant soothes the other.[481]

In this Order slight differences in colour, according to sex, are common, but conspicuous differences are rare except in the family of Bees; yet both sexes of certain groups are so brilliantly coloured—for instance in Chrysis, in which vermilion and metallic greens prevail—that we are tempted to attribute the result to sexual selection. In the Ichneumonidæ, according to Mr. Walsh,[482] the males are almost universally lighter coloured than the females. On the other hand, in the Tenthredinidæ the males are generally darker than the females. In the Siricidæ the sexes frequently differ; thus the male of *Sirex juvencus* is banded with orange, whilst the female is dark purple; but it is difficult to say which sex is the most ornamented. In *Tremex columbæ* the female is much brighter coloured than the male. With ants, as I am informed by Mr. F. Smith, the males of several species are black, the females being testaceous. In the family of Bees, especially in the solitary species, as I hear from the same distinguished entomologist, the sexes often differ in colour. The males are generally the brightest, and in Bombus as well as in Apathus, much more variable in colour than the females. In *Anthophora retusa* the male is of a rich fulvous-brown, whilst the female is quite black: so are the females of several species of Xylocopa, the males being bright yellow. In an Australian bee *(Lestis bombylans),* the female is of an extremely brilliant steel-blue, sometimes tinted with vivid green; the male being of a bright brassy colour clothed with rich fulvous pubescence. As in this group the females are provided with excellent defensive weapons in their stings, it is not probable that they have come to differ in colour from the males for the sake of protection.

Mutilla Europæa emits a stridulating noise; and according to Goureau[483] both sexes have this power. He attributes the sound to the friction of the third and preceding abdominal segments; and I find that these surfaces are marked with very fine concentric ridges, but so is the projecting thoracic collar, on which the head articulates; and this collar, when scratched with the point of a needle, emits the proper sound. It is rather surprising that both sexes should have the power of stridulating, as the male is winged and the female wingless. It is notorious that Bees express certain emotions, as of anger, by the tone of their humming, as do some dipterous insects; but I have not referred to these sounds, as they are not known to be in any way connected with the act of courtship.

Order, Coleoptera (Beetles).—Many beetles are coloured so as to resemble the surfaces which they habitually frequent. Other species are ornamented with gorgeous metallic tints,—for instance, many Carabidæ, which live on the ground and have the power of defending themselves by an intensely acrid secretion,—the splendid diamond-beetles which are protected by an extremely hard covering,—many species of Chrysomela, such as *C. cerealis,* a large species beautifully striped with various colours, and in Britain confined to the bare summit of Snowdon,—and a host of other species. These splendid colours, which are often arranged in stripes, spots, crosses and other elegant patterns, can hardly be beneficial, as a protection, except in the case of some flower-feeding species; and we cannot believe that they are purposeless. Hence the suspicion arises, that they serve as a sexual attraction; but we have no evidence on this head, for the sexes rarely differ in colour. Blind beetles, which cannot of course behold each other's beauty, never exhibit, as I hear from Mr. Waterhouse, jun., bright colours, though they often have polished coats: but the explanation of their obscurity may be that blind insects inhabit caves and other obscure stations.

Some Longicorns, however, especially certain Prionidæ, offer an exception to the common rule that the sexes of beetles do not differ in colour. Most of these insects are large and splendidly coloured. The males in the genus Pyrodes,[484] as I saw in Mr. Bates' collection, are generally redder but rather duller than the females, the latter being coloured of a more or less splendid golden green. On the other hand, in one species the male is golden-green, the female being richly tinted with red and purple. In the genus Esmeralda the sexes differ so greatly in colour that they have been ranked as distinct species: in one species both are of a beautiful shining green, but the male has a red thorax. On the whole, as far as I could judge, the females of those Prionidæ, in which the sexes differ, are coloured more richly than the males; and this does not accord with the common rule in regard to colour when acquired through sexual selection.

A most remarkable distinction between the sexes of many beetles is presented by the great horns which rise from the head, thorax, or clypeus of the males; and in some few cases from the under surface of the body. These horns, in the great family of the Lamellicorns, resemble those of various

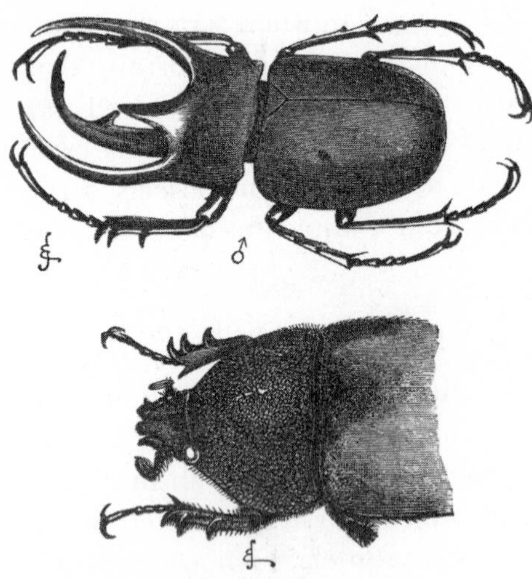

Fig. 15. *Chalcosoma atlas.* *Upper figure, male (reduced); lower figure, female (nat. size).*

quadrupeds, such as stags, rhinoceroses, &c., and are wonderful both from their size and diversified shapes. Instead of describing them, I have given figures of the males and females of some of the more remarkable forms. (Figs. 15 to 19.) The females generally exhibit rudiments of the horns in the form of small knobs or ridges; but some are destitute of even a rudiment. On the other hand, the horns are nearly as well developed in the female as in the male of *Phanæus lancifer*; and only a little less well developed in the females of some other species of the same genus and of Copris. In the several sub-divisions of the family, the differences in structure of the horns do not run parallel, as I am informed by Mr. Bates, with their more important and characteristic differences; thus within the same natural section of the genus Onthophagus, there are species which have either a single cephalic horn, or two distinct horns.

In almost all cases, the horns are remarkable from their excessive variability; so that a graduated series can be formed, from the most highly developed males to others so degenerate that they can barely be distinguished from the females. Mr. Walsh[485] found that in *Phanæus carnifex* the horns were thrice as long in some males as in others. Mr. Bates, after examining above a hundred males of *Onthophagus rangifer* (fig. 19), thought that he had at last discovered a species in which the horns did not vary; but further research proved the contrary.

The extraordinary size of the horns, and their widely different structure in closely-allied forms, indicate that they have been formed for some important purpose; but their excessive variability in the males of the same species leads to the inference that this purpose cannot be of a definite nature. The horns do not show marks of friction, as if used for any ordinary work. Some authors suppose[486] that as the males wander much more than the females, they require horns as a defence against their enemies; but in many cases the horns do not seem well adapted for defence, as they are not sharp. The most obvious conjecture is that they are used by the males for fighting together; but they have never been observed to fight; nor could Mr. Bates, after a careful examination of numerous species, find any suffi-

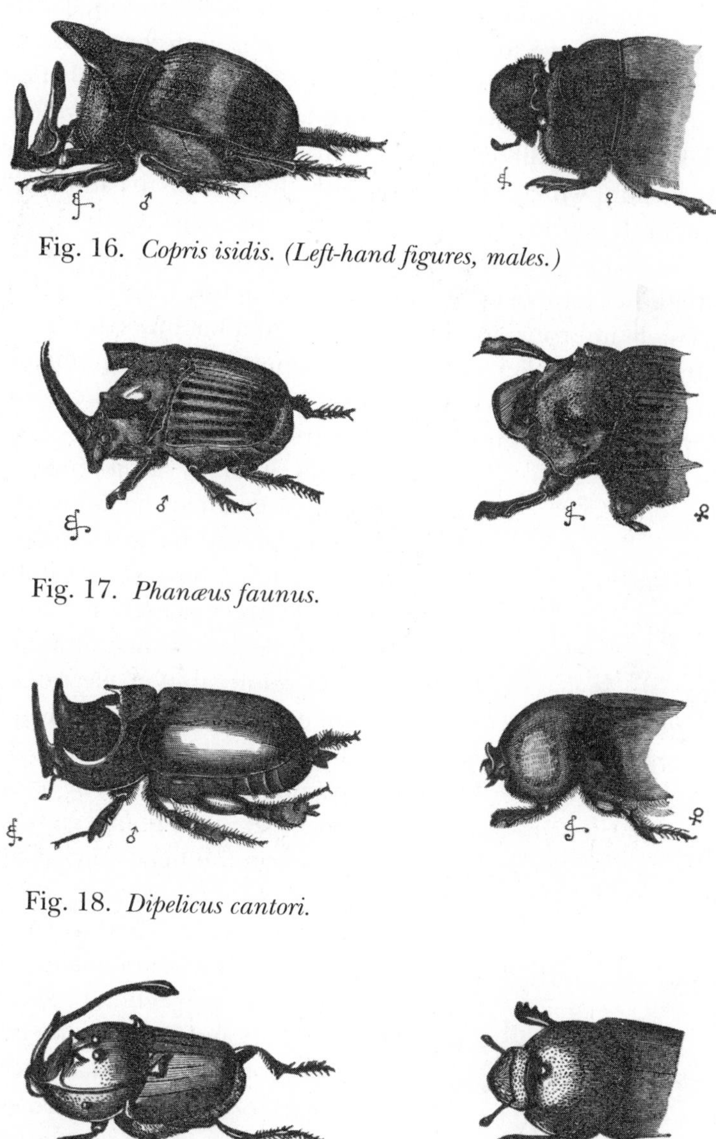

Fig. 16. *Copris isidis. (Left-hand figures, males.)*

Fig. 17. *Phanæus faunus.*

Fig. 18. *Dipelicus cantori.*

Fig. 19. *Onthophagus rangifer, enlarged.*

cient evidence in their mutilated or broken condition of their having been thus used. If the males had been habitual fighters, their size would probably have been increased through sexual selection, so as to have exceeded that of the female; but Mr. Bates, after comparing the two sexes in above a hundred species of the Copridæ, does not find in well-developed individuals any marked difference in this respect. There is, moreover, one beetle, belonging to the same great division of the Lamellicorns, namely Lethrus, the males of which are known to fight, but they are not provided with horns, though their mandibles are much larger than those of the female.

The conclusion, which best agrees with the fact of the horns having been so immensely yet not fixedly developed,—as shewn by their extreme variability in the same species and by their extreme diversity in closely-allied species—is that they have been acquired as ornaments. This view will at first appear extremely improbable; but we shall hereafter find with many animals, standing much higher in the scale, namely fishes, amphibians, reptiles and birds, that various kinds of crests, knobs, horns and combs have been developed apparently for this sole purpose.

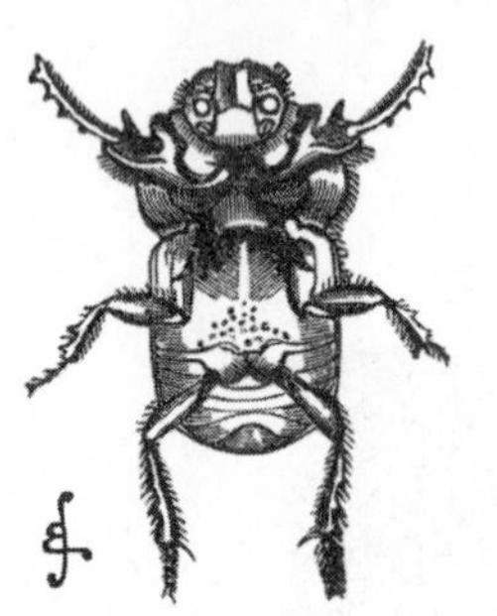

Fig. 20. *Onitis furcifer, male, viewed from beneath.*

The males of *Onitis furcifer* (fig. 20) are furnished with singular projections on their anterior femora, and with a great fork or pair of horns on the lower surface of the thorax. This situation seems extremely ill adapted for the display of these projections, and they may be of some real service; but no use can at present be assigned to them. It is a highly remarkable fact, that although the males do not exhibit even a trace of horns on the upper surface of the body, yet in the females a rudiment of a single horn on the head (fig. 21, *a*), and of a crest *(b)* on the thorax, are plainly visible. That the slight thoracic crest in the female is a rudiment of a projection proper to the male, though entirely absent in the male of this particular species, is clear: for the female of *Bubas bison* (a form which comes next to Onitis) has a similar slight crest on the thorax, and the male has in the same situation a great projection. So again there can be no doubt that the little point *(a)* on the head of the female *Onitis furcifer*, as well of the females of two or three allied species, is a rudimentary representative of the cephalic horn, which is common to the males of so many lamellicorn beetles, as in Phanæus, fig. 17. The

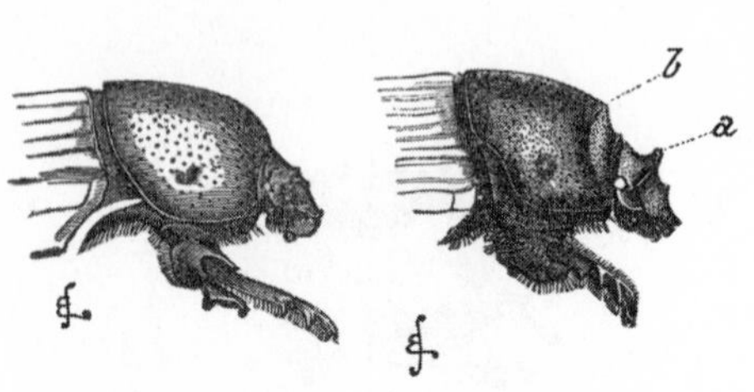

Fig. 21. *Left-hand figure, male of Onitis furcifer, viewed laterally. Right-hand figure, female.*
a. *Rudiment of cephalic horn.*
b. *Trace of thoracic horn or crest.*

males indeed of some unnamed beetles in the British Museum, which are believed actually to belong to the genus Onitis, are furnished with a similar horn. The remarkable nature of this case will be best perceived by an illustration: the Ruminant quadrupeds run parallel with the lamellicorn beetles, in some females possessing horns as large as those of the male, in others having them much smaller, or existing as mere rudiments (though this is as rare with ruminants as it is common with Lamellicorns), or in having none at all. Now if a new species of deer or sheep were discovered with the female bearing distinct rudiments of horns, whilst the head of the male was absolutely smooth, we should have a case like that of *Onitis furcifer.*

In this case the old belief of rudiments having been created to complete the scheme of nature is so far from holding good, that all ordinary rules are completely broken through. The view which seems the most probable is that some early progenitor of Onitis acquired, like other Lamellicorns, horns on the head and thorax, and then transferred them, in a rudimentary condition, as with so many existing species, to the female, by whom they have ever since been retained. The subsequent loss of the horns by the male may have resulted through the principle of compensation from the development of the projections on the lower surface, whilst the female has not been thus affected, as she is not furnished with these projections, and consequently has retained the rudiments of the horns on the upper surface. Although this view is supported by the case of Bledius immediately to be given, yet the projections on the lower surface differ greatly in structure and development in the males of the several species of Onitis, and are even rudimentary in some; nevertheless the upper surface in all these species is quite destitute of horns. As secondary sexual characters are so eminently variable, it is possible that the projections on the lower surface may have been first acquired by some progenitor of Onitis and produced their effect through compensation, and then have been in certain cases almost completely lost.

All the cases hitherto given refer to the Lamellicorns, but the males of some few other beetles, belonging to two widely distinct groups, namely, the Curculionidæ and Staphylinidæ, are furnished with horns,—in the former on the lower surface of the body,[487] in the latter on the upper surface of the head and thorax. In the Staphylinidæ the horns of the males in the same species are extraordinarily variable, just as we have seen with the Lamellicorns. In Siagonium we have a case of dimorphism, for the males can be divided into two sets, differing greatly in the size of their bodies, and in the development of their horns, without any intermediate gradations. In a species of Bledius (fig. 22), also belonging to the Staphylinidæ, male specimens can be found in the same locality, as Professor Westwood states, "in which the central horn of the thorax is very large, but the horns of the head quite rudimental; and others, in which the thoracic horn is much shorter, whilst

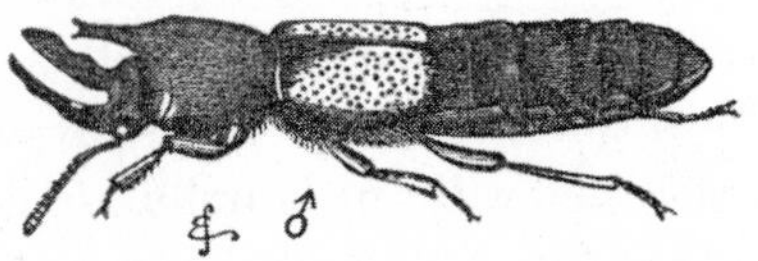

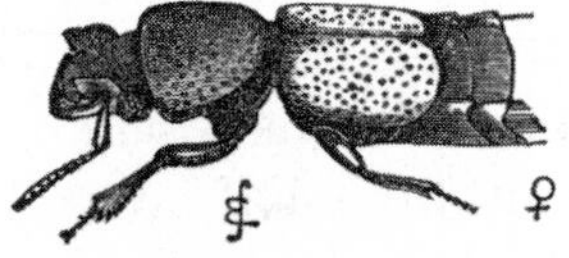

Fig. 22. *Bledius taurus, magnified.*
Left-hand figure, male; right-hand figure, female.

the protuberances on the head are long."[488] Here, then, we apparently have an instance of compensation of growth, which throws light on the curious case just given of the loss of the upper horns by the males of *Onitis furcifer.*

Law of Battle.—Some male beetles, which seem ill fitted for fighting, nevertheless engage in conflicts for the possession of the females. Mr. Wallace[489] saw two males of *Leptorhynchus angustatus,* a linear beetle with a much elongated rostrum, "fighting for a female, who stood close by busy at her boring. They pushed at each other with their rostra, and clawed and thumped, apparently in the greatest rage." The smaller male, however, "soon ran away, acknowledging himself vanquished." In some few cases the males are well adapted for fighting, by possessing great toothed mandibles, much larger than those of the females. This is the case with the common stag-beetle *(Lucanus cervus),* the males of which emerge from the pupal state about a week before the other sex, so that several may often be seen pursuing the same female. At this period they engage in fierce conflicts. When Mr. A. H. Davis[490] enclosed two males with one female in a box, the larger male severely pinched the smaller one, until he resigned his pretensions. A friend informs me that when a boy he often put the males together to see them fight, and he noticed that they were much bolder and fiercer than the females, as is well known to be the case with the higher animals. The males would seize hold of his finger, if held in front, but not so the females. With many of the Lucanidæ, as well as with the above-mentioned Leptorhynchus, the males are larger and more powerful insects than the females. The two sexes of *Lethrus cephalotes* (one of the Lamellicorns) inhabit the same burrow; and the male has larger mandibles than the female. If, during the breeding-season, a strange male attempts to enter the burrow, he is attacked; the female does not remain passive, but closes the mouth of the burrow, and encourages her mate by continually pushing him on from behind. The action does not cease until the aggressor is killed or runs away.[491] The two sexes of another lamellicorn beetle, the *Ateuchus cicatricosus* live in pairs, and seem much attached to each other; the male excites the female to roll the balls of dung in which the ova are deposited; and if she is removed, he becomes much agitated. If the male is removed, the female ceases all work, and as M. Brulerie[492] believes, would remain on the spot until she died.

The great mandibles of the male Lucanidæ are extremely variable both in size and structure, and in this respect resemble the horns on the head and thorax of many male Lamellicorns and Staphylinidæ. A perfect series can be formed from the best-provided to the worst-provided or degenerate males. Although the mandibles of the common stag-beetle, and probably of many other species, are used as efficient weapons for fighting, it is doubtful whether their great size can thus be accounted for. We have seen that with the *Lucanus elaphus* of N. America they are used for seizing the female. As they are so conspicuous and so elegantly branched, the suspicion has sometimes crossed my mind that they may be serviceable to the males as an ornament, in the same manner as the horns on the head and thorax of the various above described species. The male *Chiasognathus grantii* of S. Chile—a splendid beetle belonging to the same family—has enormously-developed mandibles (fig. 23); he is bold and pugnacious; when threatened on any side he faces round, opening his great jaws, and at the same time stridulating loudly; but the mandibles were not strong enough to pinch my finger so as to cause actual pain.

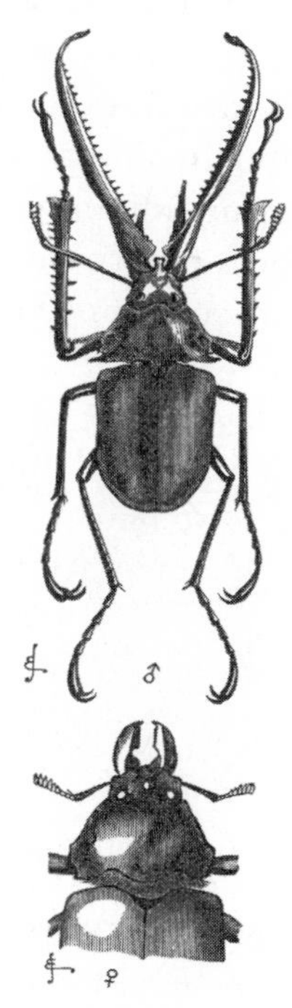

Fig. 23. *Chiasognathus grantii, reduced. Upper figure, male; lower figure, female.*

Sexual selection, which implies the possession of considerable perceptive powers and of strong passions, seems to have been more effective with the Lamellicorns than with any other family of the Coleoptera or beetles. With some species the males are provided with weapons for fighting; some live in pairs and show mutual affection; many have the power of stridulating when excited; many are furnished with the most extraordinary horns, apparently for the sake of ornament; some which are diurnal in their habits are gorgeously coloured; and, lastly, several of the largest beetles in the world belong to this family, which was placed by Linnæus and Fabricius at the head of the Order of the Coleoptera.[493]

Stridulating organs.—Beetles belonging to many and widely distinct families possess these organs. The sound can sometimes be heard at the distance of several feet or even yards,[494] but is not comparable with that produced by the Orthoptera. The part which may be called the rasp generally consists of a narrow slightly-raised surface, crossed by very fine, parallel ribs, sometimes so fine as to cause iridescent colours, and having a very elegant appearance under the microscope. In some cases, for instance, with Typhœus, it could be plainly seen that extremely minute, bristly, scale-like prominences, which cover the whole surrounding surface in approximately parallel lines, give rise to the ribs of the rasp by becoming confluent and straight, and at the same time more prominent and smooth. A hard ridge on any adjoining part of the body, which in some cases is specially modified for the purpose, serves as the scraper for the rasp. The scraper is rapidly moved across the rasp, or conversely the rasp across the scraper.

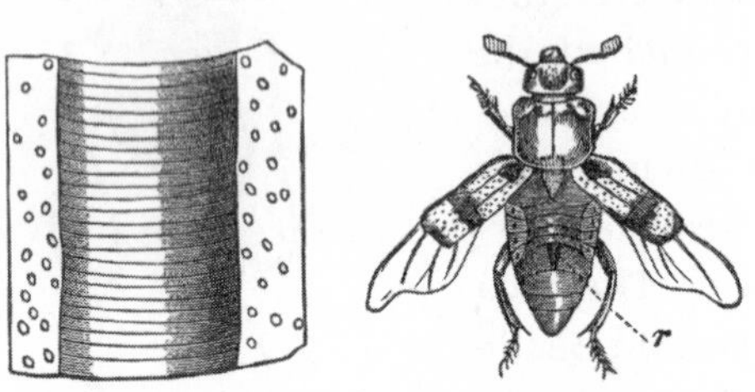

Fig. 24. *Necrophorus (from Landois).* r. *The two rasps. Left-hand figure, part of the rasp highly magnified.*

These organs are situated in widely different positions. In the carrion-beetles (Necrophorus) two parallel rasps (*r*, fig. 24) stand on the dorsal surface of the fifth abdominal segment, each rasp being crossed, as described by Landois,[495] by from 126 to 140 fine ribs. These ribs are scraped by the posterior margins of the elytra, a small portion of which projects beyond the general outline. In many Crioceridæ, and in *Clythra 4-punctata* (one of the Chrysomelidæ), and in some Tenebrionidæ, &c.,[496] the rasp is seated on the dorsal apex of the abdomen, on the pygidium or pro-pygidium, and is scraped as above by the elytra. In Heterocerus, which belongs to another family, the rasps are placed on the sides of the first abdominal segment, and are scraped by ridges on the femora.[497] In certain Curculionidæ and Carabidæ,[498] the parts are completely reversed in position, for the rasps are seated on the inferior surface of the elytra, near their apices, or along their outer margins, and the edges of the abdominal segments serve as the scrapers. In *Pelobius hermanni* (one of Dytiscidæ or water-beetles) a strong ridge runs parallel and near to the sutural margin of the elytra, and is crossed by ribs, coarse in the middle part, but becoming gradually finer at both ends, especially at the upper end; when this insect is held under water or in the air, a stridulating noise is produced by scraping the extreme horny margin of the abdomen against the rasp. In a great number of long-horned beetles (Longicornia) the organs are altogether differently situated, the rasp being on the mesothorax, which is rubbed against the pro-thorax; Landois counted 238 very fine ribs on the rasp of *Cerambyx heros.*

Many Lamellicorns have the power of stridulating, and the organs differ greatly in position. Some species stridulate very loudly, so that when Mr. F. Smith caught a *Trox sabulosus,* a gamekeeper who stood by thought that he had caught a mouse; but I failed to discover the proper organs in this beetle. In Geotrupes and Typhæus a narrow ridge runs obliquely across (*r,* fig. 25) the coxa of each hind-leg, having in *G. stercorarius* 84 ribs, which are scraped by a specially-projecting part of one of the abdominal segments. In the nearly allied *Copris lunaris,* an excessively narrow fine rasp runs along the sutural margin of the elytra, with another short rasp near the basal outer margin; but in some other Coprini the rasp is seated, according to Leconte,[499] on the dorsal surface of the abdomen. In Oryctes it is seated on the pro-pygidium, and in some other Dynastini, according to the same entomologist, on the under surface of the elytra. Lastly, Westring states that in *Omaloplia brunnea* the rasp is placed on the pro-sternum, and the scraper on the meta-sternum, the parts thus occupying the under surface of the body, instead of the upper surface as in the Longicorns.

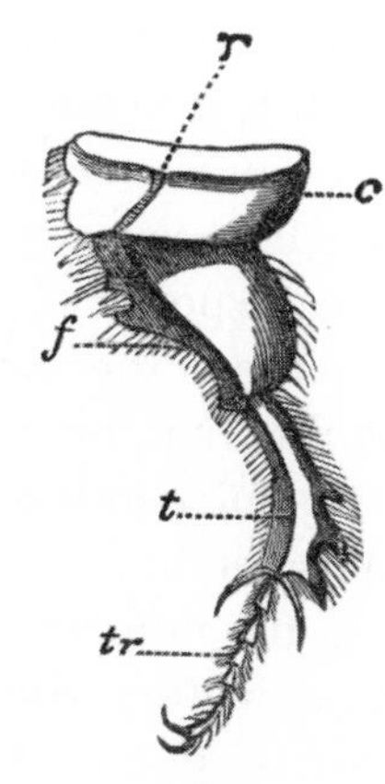

Fig. 25. *Hind-leg of Geotrupes stercorarius (from Landois).*
r. Rasp.
c. Coxa.
f. Femur
t. Tibia.
tr. Tarsi.

We thus see that the stridulating organs in the different coleopterous families are wonderfully diversified in position, but not much in structure. Within the same family some species are provided with these organs, and some are quite destitute of them. This diversity is intelligible, if we suppose that originally various species made a shuffling or hissing noise by the rubbing together of the hard and rough parts of their bodies which were in contact; and that from the noise thus produced being in some way useful, the rough surfaces were gradually developed into regular stridulating organs. Some beetles as they move, now produce, either intentionally or unintentionally, a shuffling noise, without possessing any proper organs for the purpose. Mr. Wallace informs me that the *Euchirus longimanus* (a Lamellicorn, with the anterior legs wonderfully elongated in the male) "makes, whilst moving, a low hissing sound by the protrusion and contraction of the abdomen; and when seized it produces a grating sound by rubbing its hind-legs against the edges of the elytra." The hissing sound is clearly due to a narrow rasp running along the sutural margin of each elytron; and I could likewise make the grating sound by rubbing the shagreened surface of the femur against the granulated margin of the corresponding elytron; but I could not here detect any proper rasp; nor is it likely that I could have overlooked it in so large an insect. After examining Cychrus and reading what Westring has written in his two papers about this beetle, it seems very doubtful whether it possesses any true rasp, though it has the power of emitting a sound.

From the analogy of the Orthoptera and Homoptera, I expected to find that the stridulating organs in the Coleoptera differed according to sex; but Landois, who has carefully examined several species, observed no such difference; nor did Westring; nor did Mr. G. R. Crotch in preparing the numerous specimens which he had the kindness to send me for examination. Any slight sexual difference, however, would be difficult to detect, on account of the great variability of these organs. Thus in the first pair of the *Necrophorus humator* and of the *Pelobius* which I examined, the rasp was considerably larger in the male than in the female; but not so with succeeding specimens. In *Geotrupes stercorarius* the rasp appeared to me thicker, opaquer, and more prominent in three males than in the same number of females; consequently my son, Mr. F. Darwin, in order to discover whether the sexes differed in their power of stridulating, collected 57 living specimens, which he separated into two lots, according as they made, when held in the same manner, a greater or lesser noise. He then examined their sexes, but found that the males were very nearly in the same proportion to the females in both lots. Mr. F. Smith has kept alive numerous specimens of *Mononychus pseudacori* (Curculionidæ), and is satisfied that both sexes stridulate, and apparently in an equal degree.

Nevertheless the power of stridulating is certainly a sexual character in some few Coleoptera. Mr. Crotch has discovered that the males alone of two species of Heliopathes (Tenebrionidæ) possess stridulating organs. I examined five males of *H. gibbus,* and in all these there was a well-developed rasp, partially divided into two, on the dorsal surface of the terminal abdominal segment; whilst in the same number of females there was not even a rudiment of the rasp, the membrane of this segment being transparent and much thinner than in the male. In *H. cribratostriatus* the male has a similar rasp, excepting that it is not partially divided into two portions, and the female is completely destitute of this organ; but in addition the male has on the apical margins of the elytra, on each side of the suture, three or four short longitudinal ridges, which are crossed by extremely fine ribs, parallel to and resembling those on the abdominal rasp; whether these ridges serve as an independent rasp, or as a scraper for the abdominal rasp, I could not decide: the female exhibits no trace of this latter structure.

Again, in three species of the Lamellicorn genus Oryctes, we have a nearly parallel case. In the females of *O. gryphus* and *nasicornis* the ribs on the rasp of the pro-pygidium are less continuous and less distinct than in the males; but the chief difference is that the whole upper surface of this segment, when held in the proper light, is seen to be clothed with hairs, which are absent or are represented by excessively fine down in the males. It should be noticed that in all Coleoptera the effective part of the rasp is destitute of hairs. In *O. senegalensis* the difference between the sexes is more strongly marked, and this is best seen when the proper segment is cleaned and viewed as a transparent object. In the female the whole surface is covered with little separate crests, bearing spines; whilst in the male these crests become, in proceeding towards the apex, more and more confluent, regular, and naked; so that three-fourths of the segment is covered with extremely fine parallel ribs, which are quite absent in the female. In the females, however, of all three species of Oryctes, when the abdomen of a softened specimen is pushed backwards and forwards, a slight grating or stridulating sound can be produced.

In the case of the Heliopathes and Oryctes there can hardly be a doubt that the males stridulate in order to call or to excite the females; but with most beetles the stridulation apparently serves both sexes as a mutual call. This view is not rendered improbable from beetles stridulating under various emotions; we know that birds use their voices for many purposes besides singing to their mates. The great Chiasognathus stridulates in anger or defiance; many species do the same from distress or fear, when held so that they cannot escape; Messrs. Wollaston and Crotch were able, by striking the hollow stems of trees in the Canary Islands, to discover the presence of beetles belonging to the genus Acalles by their stridulation. Lastly the male Ateuchus stridulates to encourage the female in her work, and from distress when she is removed.[500] Some naturalists believe that beetles make this noise to frighten away their enemies; but I cannot think that the quadrupeds and birds which are able to devour the larger beetles with their extremely hard coats, would be frightened by so slight a grating sound. The belief that the stridulation serves as a sexual call is supported by the fact that death-ticks *(Anobium tessellatum)* are well known to answer each other's ticking, or, as I have myself observed, a tapping noise artificially made; and Mr. Doubleday informs me that he has twice or thrice observed a female ticking,[501] and in the course of an hour or two has found her united with a male, and on one occasion surrounded by several males. Finally, it seems probable that the two sexes of many kinds of beetles were at first enabled to find each other by the slight shuffling noise produced by the rubbing together of the adjoining parts of their hard bodies; and that as the males or females which made the greatest noise succeeded best in finding partners, the rugosities on various parts of their bodies were gradually developed by means of sexual selection into true stridulating organs.

CHAPTER XI

Insects—continued. Order Lepidoptera

COURTSHIP OF BUTTERFLIES • BATTLES • TICKING NOISE • COLOURS COMMON TO BOTH SEXES, OR MORE BRILLIANT IN THE MALES • EXAMPLES • NOT DUE TO THE DIRECT ACTION OF THE CONDITIONS OF LIFE • COLOURS ADAPTED FOR PROTECTION • COLOURS OF MOTHS • DISPLAY • PERCEPTIVE POWERS OF THE LEPIDOPTERA • VARIABILITY • CAUSES OF THE DIFFERENCE IN COLOUR BETWEEN THE MALES AND FEMALES • MIMICKRY, FEMALE BUTTERFLIES MORE BRILLIANTLY COLOURED THAN THE MALES • BRIGHT COLOURS OF CATERPILLARS • SUMMARY AND CONCLUDING REMARKS ON THE SECONDARY SEXUAL CHARACTERS OF INSECTS • BIRDS AND INSECTS COMPARED

In this great Order the most interesting point for us is the difference in colour between the sexes of the same species, and between the distinct species of the same genus. Nearly the whole of the following chapter will be devoted to this subject; but I will first make a few remarks on one or two other points. Several males may often be seen pursuing and crowding round the same female. Their courtship appears to be a prolonged affair, for I have frequently watched one or more males pirouetting round a female until I became tired, without seeing the end of the courtship. Although butterflies are such weak and fragile creatures, they are pugnacious, and an Emperor butterfly[502] has been captured with the tips of its wings broken from a conflict with another male. Mr. Collingwood in speaking of the frequent battles between the butterflies of Borneo says, "They whirl round each other with the greatest rapidity, and appear to be incited by the greatest ferocity." One case is known of a butterfly, namely the *Ageronia feronia,* which makes a noise like that produced by a toothed wheel passing under a spring catch, and which could be heard at the distance of several yards. At Rio de Janeiro this sound was noticed by me, only when two were chasing each other in an irregular course, so that it is probably made during the courtship of the sexes; but I neglected to attend to this point.[503]

Every one has admired the extreme beauty of many butterflies and of some moths; and we are led to ask, how has this beauty been acquired? Have their colours and diversified patterns simply resulted from the direct action of the physical conditions to which these insects have been exposed, without any benefit being thus derived? Or have successive variations been accumulated and determined either as a protection or for some unknown purpose, or that one sex might be rendered attractive to the other? And, again, what is the meaning of the colours being widely different in the males and females of certain species, and alike in the two sexes of other species? Before attempting to answer these questions a body of facts must be given.

With most of our English butterflies, both those which are beautiful, such as the admiral, peacock, and painted lady (Vanessæ), and those which are plain-coloured, such as the meadow-browns (Hipparchiæ), the sexes are alike. This is also the case with the magnificent Heliconidæ and Danaidæ of the tropics. But in certain other tropical groups, and with some of our English butterflies, as the purple emperor, orange-tip, &c. *(Apatura Iris* and *Anthocharis cardamines),* the sexes differ either greatly or slightly in colour. No language suffices to describe the splendour of the males of some tropical species. Even within the same genus we often find species presenting an extraordinary difference between the sexes, whilst others have their sexes closely alike. Thus in the South American genus Epicalia, Mr. Bates, to whom I am much indebted for most of the following facts and for looking over this whole discussion, informs me that he knows twelve species, the two sexes of which haunt the same stations (and this is not always the case with butterflies), and therefore cannot have been differently affected by external conditions.[504] In nine of these species the males rank amongst the most brilliant of all butterflies, and differ so greatly from the comparatively plain females that they were formerly placed in distinct genera. The females of these nine species resemble each other in their general type of coloration, and likewise resemble both sexes in several allied genera, found in various parts of the world. Hence in accordance with the descent-theory we may infer that these nine species, and probably all the others of the genus, are descended from an ancestral form which was coloured in nearly the same manner. In the tenth species the female still retains the same general colouring, but the male resembles her, so that he is coloured in a much less gaudy and contrasted manner than the males of the previous species. In the eleventh and twelfth species, the females depart from the type of colouring which is usual with their sex in this genus, for they are gaily decorated in nearly the same manner as the males, but in a somewhat less degree. Hence in these two species the bright colours of the males seem to have been transferred to the females; whilst the male of the tenth species has either retained or recovered the plain colours of the female as well as of the parent-form of the genus; the two sexes being thus rendered in both cases, though in an opposite manner, nearly alike. In the allied genus Eubagis, both sexes of some of the species are plain-coloured and nearly alike; whilst with the greater number the males are decorated with beautiful metallic tints, in a diversified manner, and differ much from their females. The females throughout the genus retain the same general style of colouring, so that they commonly resemble each other much more closely than they resemble their own proper males.

In the genus Papilio, all the species of the Æneas group are remarkable for their conspicuous and strongly contrasted colours, and they illustrate the frequent tendency to gradation in the amount of difference between the sexes. In a few species, for instance in *P. ascanius,* the males and females are alike; in others the males are a little or very much more superbly coloured than the females. The genus Junonia allied to our Vanessæ offers a nearly parallel case, for although the sexes of most of the species resemble each other and are destitute of rich colours, yet in certain species, as in *J. œnone,* the male is rather more brightly coloured than the female, and in a few (for instance *J. andremiaja)* the male is so different from the female that he might be mistaken for an entirely distinct species.

Another striking case was pointed out to me in the British museum by Mr. A. Butler, namely one of the Tropical American Theclæ, in which both sexes are nearly alike and wonderfully splendid; in another, the male is coloured in a similarly gorgeous manner, whilst the whole upper surface of the female is of a dull uniform brown. Our common little English blue butterflies of the genus Lycæna, illustrate the various differences in colour between the sexes, almost as well, though not in so striking a manner, as the above exotic genera. In *Lycæna agestis* both sexes have wings of a brown colour, bordered with small ocellated orange spots, and are consequently alike. In *L. œgon* the wings of the male are of a fine blue, bordered with black; whilst the wings of the female are brown, with a similar border, and closely resemble those of *L. agestis.* Lastly, in *L. arion* both sexes are of a blue colour and nearly alike, though in the female the edges of the wings are rather duskier, with the black spots plainer; and in a bright blue Indian species both sexes are still more closely alike.

I have given the foregoing cases in some detail in order to shew, in the first place, that when the sexes of butterflies differ, the male as a general rule is the most beautiful, and departs most from the usual type of colouring of the group to which the species belongs. Hence in most groups the females of the several species resemble each other much more closely than do the males. In some exceptional cases, however, to which I shall hereafter allude, the females are coloured more splendidly than the males. In the second place these cases have been given to bring clearly before the mind that within the same genus, the two sexes frequently present every gradation from no difference in colour to so great a difference that it was long before the two were placed by entomologists in the same genus. In the third place, we have seen that when the sexes nearly resemble each other, this apparently may be due either to the male having transferred his colours to the female, or to the male having retained, or perhaps recovered, the primordial colours of the genus to which the species belongs. It also deserves notice that in those groups in which the sexes present any difference of colour, the females usually resemble the males to a certain extent, so that when the males are beautiful to an extraordinary degree, the females almost invariably exhibit some degree of beauty. From the numerous cases of gradation in the amount of difference between the sexes, and from the prevalence of the same general type of coloration throughout the whole of the same group, we may conclude that the causes, whatever they may be, which have determined the brilliant colouring of the males alone of some species, and of both sexes in a more or less equal degree of other species, have generally been the same.

As so many gorgeous butterflies inhabit the tropics, it has often been supposed that they owe their colours to the great heat and moisture of these zones; but Mr. Bates[505] has shewn by the comparison of various closely-allied groups of insects from the temperate and tropical regions, that this view cannot be maintained; and the evidence becomes conclusive when brilliantly-coloured males and plain-coloured females of the same species inhabit the same district, feed on the same food, and follow exactly the same habits of life. Even when the sexes resemble each other, we can hardly believe that their brilliant and beautifully-arranged colours are the purposeless result of the nature of the tissues, and the action of the surrounding conditions.

With animals of all kinds, whenever colour has been modified for some special purpose, this has been, as far as we can judge, either for protection or as an attraction between the sexes. With many species of butterflies the upper surfaces of the wings are obscurely coloured, and this in all probability leads to their escaping observation and danger. But butterflies when at rest would be particularly liable to be attacked by their enemies; and almost all the kinds when resting raise their wings vertically over their backs, so that the lower sides alone are exposed to view. Hence it is this side which in many cases is obviously coloured so as to imitate the surfaces on which these insects commonly rest. Dr. Rössler, I believe, first noticed the similarity of the closed wings of certain Vanessæ and other butterflies to the bark of trees. Many analogous and striking facts could be given. The most interesting one is that recorded by Mr. Wallace[506] of a common Indian and Sumatran butterfly (Kallima), which disappears like magic when it settles in a bush; for it hides its head and antennæ between its closed wings, and these in form, colour, and veining cannot be distinguished from a withered leaf together with the footstalk. In some other cases the lower surfaces of the wings are brilliantly coloured, and yet are protective; thus in *Thecla rubi* the wings when closed are of an emerald green and resemble the young leaves of the bramble, on which this butterfly in the spring may often be seen seated.

Although the obscure tints of the upper or under surface of many butterflies no doubt serve to conceal them, yet we cannot possibly extend this view to the brilliant and conspicuous colours of many kinds, such as our admiral and peacock Vanessæ, our white cabbage-butterflies (Pieris), or the great swallow-tail Papilio which haunts the open fens—for these butterflies are thus rendered visible to every living creature. With these species both sexes are alike; but in the common brimstone butterfly *(Gonepteryx rhamni),* the male is of an intense yellow, whilst the female is much paler; and in the orange-tip *(Anthocharis cardamines)* the males alone have the bright orange tips to their wings. In these cases the males and females are equally conspicuous, and it is not credible that their difference in colour stands in any relation to ordinary protection. Nevertheless it is possible that the conspicuous colours of many species may be in an indirect manner beneficial, as will hereafter be explained, by leading their enemies at once to recognise them as unpalatable. Even in this case it does not certainly follow that their bright colours and beautiful patterns were acquired for this special purpose. In some other remarkable cases, beauty has been gained for the sake of protection, through the imitation of other beautiful species, which inhabit the same district and enjoy an immunity from attack by being in some way offensive to their enemies.

The female of our orange-tip buttterfly, above referred to, and of an American species *(Anth. genutia)* probably shew us, as Mr. Walsh has remarked to me, the primordial colours of the parent-species of the genus; for both sexes of four or five widely-distributed species are coloured in nearly the same manner. We may infer here, as in several previous cases, that it is the males of *Anth. cardamines* and *genutia* which have departed from the usual type of colouring of their genus. In the *Anth. sara* from California, the orange-tips have become partially developed in the female; for her wings are tipped with reddish-orange, but paler than in the male, and slightly different in some other respects. In an allied Indian form, the *Iphias glaucippe,* the orange-tips are fully developed in both sexes. In this Iphias the under surface of the wings marvellously resembles, as pointed out to me by Mr. A. Butler, a pale-coloured leaf; and in our English orange-tip, the under surface resembles the flower-head of the wild parsley, on which it may be seen going to rest at night.[507] The same reasoning power which compels us to believe that the lower surfaces have here been coloured for the sake of protection, leads us to deny that the wings have been tipped, especially when this character is confined to the males, with bright orange for the same purpose.

TURNING NOW TO MOTHS: most of these rest motionless with their wings depressed during the whole or greater part of the day; and the upper surfaces of their wings are often shaded and coloured in an admirable manner, as Mr. Wallace has remarked, for escaping detection. With most of the Bombycidæ and Noctuidæ,[508] when at rest, the front-wings overlap and conceal the hind-wings; so that the latter might be brightly coloured without much risk; and they are thus coloured in many species of both families. During the act of flight, moths would often be able to escape from their enemies; nevertheless, as the hind-wings are then fully exposed to view, their bright colours must generally have been acquired at the cost of some little risk. But the following fact shews us how cautious we ought to be in drawing conclusions on this head. The common yellow under-wings (Triphaena) often fly about during the day or early evening, and are then conspicuous from the colour of their hind-wings. It would naturally be thought that this would be a source of danger; but Mr. J. Jenner Weir believes that it actually serves them as a means of escape, for birds strike at these brightly coloured and fragile surfaces, instead of at the body. For instance, Mr. Weir turned into his aviary a vigorous specimen of *Triphaena pronuba,* which was instantly pursued by a robin; but the bird's attention being caught by the coloured wings, the moth was not captured until after about fifty attempts, and small portions of the wings were repeatedly broken off. He tried the same experiment, in the open air, with a *T. fimbria* and swallow; but the large size of this moth probably interfered with its capture.[509] We are thus reminded of a statement made by Mr. Wallace,[510] namely, that in the Brazilian forests and Malayan islands, many common and highly-decorated butterflies are weak flyers, though furnished with a broad expanse of wings; and they "are often captured with pierced and broken wings, as if they had been seized by birds, from which they had escaped: if the wings had been much smaller in proportion to the body, it seems probable that the insect would more frequently have been struck or pierced in a vital part, and thus the increased expanse of the wings may have been indirectly beneficial."

Display.—The bright colours of butterflies and of some moths are specially arranged for display, whether or not they serve in addition as a protection. Bright colours would not be visible during the night; and there can be no doubt that moths, taken as a body, are much less gaily decorated than butterflies, all of which are diurnal in their habits. But the moths in certain families, such as the Zygænidæ, various Sphingidæ, Uraniidæ, some Arctiidæ and Saturniidæ, fly about during the day or early evening, and many of these are extremely beautiful, being far more brightly coloured than the strictly nocturnal kinds. A few exceptional cases, however, of brightly-coloured nocturnal species have been recorded.[511]

There is evidence of another kind in regard to display. Butterflies, as before remarked, elevate their wings when at rest, and whilst basking in the sunshine often alternately raise and depress them, thus exposing to full view both surfaces; and although the lower surface is often coloured in an obscure manner as a protection, yet in many species it is as highly coloured as the upper surface, and sometimes in a very different manner. In some tropical species the lower surface is even more brilliantly coloured than the upper.[512] In one English fritillary, the *Argynnis aglaia,* the lower surface alone is ornamented with shining silver discs. Nevertheless, as a general rule, the upper surface, which is probably the most fully exposed, is coloured more brightly and in a more diversified manner than the lower. Hence the lower surface generally affords to entomologists the most useful character for detecting the affinities of the various species.

Now if we turn to the enormous group of moths, which do not habitually expose to full view the under surface of their wings, this side is very rarely, as I hear from Mr. Stainton, coloured more brightly than the upper side, or even with equal brightness. Some exceptions to the rule, either real or apparent, must be noticed, as that of Hypopyra, specified by Mr. Wormald.[513] Mr. R. Trimen informs me that in Guenée's great work, three moths are figured, in which the under surface is much the most brilliant. For instance, in the Australian Gastrophora the upper surface of the forewing is pale greyish-ochreous, while the lower surface is magnificently ornamented by an ocellus of cobalt-blue, placed in the midst of a black mark, surrounded by orange-yellow, and this by bluish-white. But the habits of these three moths are unknown; so that no explanation can be given of their unusual style of colouring. Mr. Trimen also informs me that the lower surface of the wings in certain other Geometræ[514] and quadrifid Noctuæ are either more variegated or more brightly-coloured than the upper surface; but some of these species have the habit of "holding their wings quite erect over their backs, retaining them in this position for a considerable time," and thus exposing to view the under surface. Other species when settled on the ground or herbage have the habit of now and then suddenly and slightly lifting up their wings. Hence the lower surface of the wings being more brightly-coloured than the upper surface in certain moths is not so anomalous a circumstance as it at first appears. The Saturniidæ include some of the most beautiful of all moths, their wings being decorated, as in our British Emperor moth, with fine ocelli; and Mr. T. W. Wood[515] observes that they resemble butterflies in some of their movements; "for instance, in the gentle waving up and down of the wings, as if for display, which is more characteristic of diurnal than of nocturnal Lepidoptera."

It is a singular fact that no British moths, nor as far as I can discover hardly any foreign species, which are brilliantly coloured, differ much in colour according to sex; though this is the case with many brilliant butterflies. The male, however, of one American moth, the *Saturnia Io,* is described as having its fore-wings deep yellow, curiously marked with purplish-red spots; whilst the wings of the female are purple-brown, marked with grey lines.[516] The British moths which differ sexually in colour are all brown, or various tints of dull yellow, or nearly white. In several species the males are much darker than the females,[517] and these belong to groups which generally fly about during the afternoon. On the other hand, in many genera, as Mr. Stainton informs me, the males have the hind-wings whiter than those of the female—of which fact *Agrotis exclamationis* offers a good instance. The males are thus rendered more conspicuous than the females, whilst flying about in the dusk. In the Ghost Moth *(Hepialus humuli)* the difference is more strongly marked; the males being white and the females yellow with darker markings. It is difficult to conjecture what the meaning can be of these differences between the sexes in the shades of darkness or lightness; but we can hardly suppose that they are the result of mere variability with sexually-limited inheritance, independently of any benefit thus derived.

From the foregoing statements it is impossible to admit that the brilliant colours of butterflies and of some few moths, have commonly been acquired for the sake of protection. We have seen that their colours and elegant patterns are arranged and exhibited as if for display. Hence I am led to suppose that the females generally prefer, or are most excited by the more brilliant males; for on any other supposition the males would be ornamented, as far as we can see, for no purpose. We know that ants and certain lamellicorn beetles are capable of feeling an attachment for each other, and that ants recognise their fellows after an interval of several months. Hence there is no abstract improbability in the Lepidoptera, which probably stand nearly or quite as high in the scale as these insects, having sufficient mental capacity to admire bright colours. They certainly discover flowers by colour, and, as I have elsewhere shewn, the plants which are fertilised exclusively by the wind never have a conspicuously-coloured corolla. The Humming-bird Sphinx may often be seen to swoop down from a distance on a bunch of flowers in the midst of green foliage; and I have been assured by a friend, that these moths repeatedly visited flowers painted on the walls of a room in the South of France. The common white butterfly, as I hear from Mr. Doubleday, often flies down to a bit of paper on the ground, no doubt mistaking it for one of its own species. Mr. Collingwood[518] in speaking of the difficulty of collecting certain butterflies in the Malay Archipelago, states that "a dead specimen pinned upon a conspicuous twig will often arrest an insect of the same species in its headlong flight, and bring it down within easy reach of the net, especially if it be of the opposite sex."

The courtship of butterflies is a prolonged affair. The males sometimes fight together in rivalry; and many may be seen pursuing or crowding round the same female. If, then, the females do not prefer one male to another, the pairing must be left to mere chance, and this does not appear to me a probable event. If, on the other hand, the females habitually, or even occasionally, prefer the more beautiful males, the colours of the latter will have been rendered brighter by degrees, and will have been transmitted to both sexes or to one sex, according to which law of inheritance prevailed. The process of sexual selection will have been much facilitated, if the conclusions arrived at from various kinds of evidence in the supplement to the ninth chapter can be trusted; namely that the males of many Lepidoptera, at least in the imago state, greatly exceed in number the females.

Some facts, however, are opposed to the belief that female butterflies prefer the more beautiful males; thus, as I have been assured by several observers, fresh females may frequently be seen paired with battered, faded or dingy males; but this is a circumstance which could hardly fail often to follow from the males emerging from their cocoons earlier than the females. With moths of the family of the Bombycidæ, the sexes pair immediately after assuming the imago state; for they cannot feed, owing to the rudimentary condition of their mouths. The females, as several entomologists have remarked to me, lie in an almost torpid state, and appear not to evince the least choice in regard to their partners. This is the case with the common silk-moth *(B. mori),* as I have been told by some continental and English breeders. Dr. Wallace, who has had such immense experience in breeding *Bombyx cynthia,* is convinced that the females evince no choice or preference. He has kept above 300 of these moths living together, and has often found the most vigorous females mated with stunted males. The reverse apparently seldom occurs; for, as he believes, the more vigorous males pass over the weakly females, being attracted by those endowed with most vitality. Although we have been indirectly induced to believe that the females of many species prefer the more beautiful males, I have no reason to suspect, either with moths or butterflies, that the males are attracted by the beauty of the females. If the more beautiful females had been continually preferred, it is almost certain, from the colours of butterflies being so frequently transmitted to one sex alone, that the females would often have been rendered more beautiful than their male partners. But this does not occur except in a few instances; and these can be explained, as we shall presently see, on the principle of mimickry and protection.

As SEXUAL SELECTION primarily depends on variability, a few words must be added on this subject. In respect to colour there is no difficulty, as any number of highly variable Lepidoptera could be named. One good instance will suffice. Mr. Bates shewed me a whole series of specimens of *Papilio sesostris* and *childrenæ;* in the latter the males varied much in the extent of the beautifully enamelled green patch on the fore-wings, and in the size of the white mark, as well as of the splendid crimson stripe on the hind-wings; so that there was a great contrast between the most and least gaudy males. The male of *Papilio sesostris,* though a beautiful insect, is much less so than *P. childrenæ.* It likewise varies a little in the size of the green patch on the fore-wings, and in the occasional appearance of a small crimson stripe on the hind-wings, borrowed, as it would seem, from its own female; for the females of this and of many other species in the Æneas group possess this crimson stripe. Hence between the brightest specimens of *P. sesostris* and the least bright of *P. childrenæ,* there was but a small interval; and it was evident that as far as mere variability is concerned, there would be no difficulty in permanently increasing by means of selection the beauty of either species. The variability is here almost confined to the male sex; but Mr. Wallace and Mr. Bates have shewn[519] that the females of some other species are extremely variable, the males being nearly constant. As I have before mentioned the Ghost Moth *(Hepialus humuli)* as one of the best instances in Britain of a difference in colour between the sexes of moths, it may be worth adding[520] that in the Shetland Islands, males are frequently found which closely resemble the females. In a future chapter I shall have occasion to shew that the beautiful eye-like spots or ocelli, so common on the wings of many Lepidoptera, are eminently variable.

ON THE WHOLE, although many serious objections may be urged, it seems probable that most of the species of Lepidoptera which are brilliantly coloured, owe their colours to sexual selection, excepting in certain cases, presently to be mentioned, in which conspicuous colours are beneficial as a protection. From the ardour of the male throughout the animal kingdom, he is generally willing to accept any female; and it is the female which usually exerts a choice. Hence if sexual selection has here acted, the male, when the sexes differ, ought to be the most brilliantly coloured; and this undoubtedly is the ordinary rule. When the sexes are brilliantly coloured and resemble each other, the characters acquired by the males appear to have been transmitted to both sexes. But will this explanation of the similarity and dissimilarity in colour between the sexes suffice?

The males and females of the same species of butterfly are known[521] in several cases to inhabit different stations, the former commonly basking in the sunshine, the latter haunting gloomy forests. It is therefore possible that different conditions of life may have acted directly on the two sexes; but this is not probable,[522] as in the adult state they are exposed during a very short period to different conditions; and the larvæ of both are exposed to the same conditions. Mr. Wallace believes that the less brilliant colours of the female have been specially gained in all or almost all cases for the sake of protection. On the contrary it seems to me more probable that the males alone, in the large majority of cases, have acquired their bright colours through sexual selection, the females having been but little modified. Consequently the females of distinct but allied species ought to resemble each other much more closely than do the males of the same species; and this is the general rule. The females thus approximately show us the primordial colouring of the parent-species of the group to which they belong. They have, however, almost always been modified to a certain extent by some of the successive steps of variation, through the accumulation of which the males were rendered beautiful, having been transferred to them. The males and females of allied though distinct species will also generally have been exposed during their prolonged larval state to different conditions, and may have been thus indirectly affected; though with the males any slight change of colour thus caused will often have been completely masked by the brilliant tints gained through sexual selection. When we treat of Birds, I shall have to discuss the whole question whether the differences in colour between the males and females have been in part specially gained by the latter as a protection; so that I will here only give unavoidable details.

In all cases when the more common form of equal inheritance by both sexes has prevailed, the selection of bright-coloured males would tend to make the females bright-coloured; and the selection of dull-coloured females would tend to make the males dull. If both processes were carried on simultaneously, they would tend to neutralise each other. As far as I can see, it would be extremely difficult to change through selection the one form of inheritance into the other. But by the selection of successive variations, which were from the first sexually limited in their transmission, there would not be the slightest difficulty in giving bright colours to the males alone, and at the same time or subsequently, dull colours to the females alone. In this latter manner female butterflies and moths may, as I fully admit, have been rendered inconspicuous for the sake of protection, and widely different from their males.

Mr. Wallace[523] has argued with much force in favour of his view that when the sexes differ, the female has been specially modified for the sake of protection; and that this has been effected by one form of inheritance, namely, the transmission of characters to both sexes, having been changed through the agency of natural selection into the other form, namely, transmission to one sex. I was at first strongly inclined to accept this view; but the more I have studied the various classes throughout the animal kingdom, the less probable it has appeared. Mr. Wallace urges that both sexes of the *Heliconidæ, Danaidæ, Acræidæ* are equally brilliant because both are protected from the attacks of birds and other enemies, by their offensive odour; but that in other groups, which do not possess this immunity, the females have been rendered inconspicuous, from having more need of protection than the males. This supposed difference in the "need of protection by the two sexes" is rather deceptive, and requires some discussion. It is obvious that brightly-coloured individuals, whether males or females, would equally attract, and obscurely-coloured individuals equally escape, the attention of their enemies. But we are concerned with the effects of the destruction or preservation of certain individuals of either sex, on the character of the race. With insects, after the male has fertilised the female, and after the latter has laid her eggs, the greater or less immunity from danger of either sex could not possibly have any effect on the offspring. Before the sexes have performed their proper functions, if they existed in equal numbers and if they strictly paired (all other circumstances being the same), the preservation of the males and females would be equally important for the existence of the species and for the character of the offspring. But with most animals, as is known to be the case with the domestic silk-moth, the male can fertilise two or three females; so that the destruction of the males would not be so injurious to the species as that of the females. On the other hand, Dr. Wallace believes that with moths the progeny from a second or third fertilisation is apt to be weakly, and therefore would not have so good chance of surviving. When the males exist in much greater numbers than the females, no doubt many males might be destroyed with impunity to the species; but I cannot see that the results of ordinary selection for the sake of protection would be influenced by the sexes existing in unequal numbers; for the same proportion of the more conspicuous individuals, whether males or females, would probably be destroyed. If indeed the males presented a greater range of variation in colour, the result would be different; but we need not here follow out such complex details. On the whole I cannot perceive that an inequality in the numbers of the two sexes would influence in any marked manner the effects of ordinary selection on the character of the offspring.

Female Lepidoptera require, as Mr. Wallace insists, some days to deposit their fertilised ova and to search for a proper place; during this period (whilst the life of the male was of no importance) the brighter-coloured females would be exposed to danger and would be liable to be destroyed. The duller-coloured females on the other hand would survive, and thus would influence, it might be thought, in a marked manner the character of the species,—either of both sexes or of one sex, according to which form of inheritance prevailed. But it must not be forgotten that the males emerge from the cocoon-state some days before the females, and during this period, whilst the unborn females were safe, the brighter-coloured males would be exposed to danger; so that ultimately both sexes would probably be exposed during a nearly equal length of time to danger, and the elimination of conspicuous colours would not be much more effective in the one than the other sex.

It is a more important consideration that female Lepidoptera, as Mr. Wallace remarks, and as is known to every collector, are generally slower flyers than the males. Consequently the latter, if exposed to greater danger from being conspicuously coloured, might be able to escape from their enemies, whilst the similarly-coloured females would be destroyed; and thus the females would have the most influence in modifying the colour of their progeny.

There is one other consideration: bright colours, as far as sexual selection is concerned, are commonly of no service to the females; so that if the latter varied in brightness, and the variations were sexually limited in their transmission, it would depend on mere chance whether the females had their bright colours increased; and this would tend throughout the Order to diminish the number of species with brightly-coloured females in comparison with the species having brightly-coloured males. On the other hand, as bright colours are supposed to be highly serviceable to the males in their love-struggles, the brighter males (as we shall see in the chapter on Birds) although exposed to rather greater danger, would on an average procreate a greater number of offspring than the duller males. In this case, if the variations were limited in their transmission to the male sex, the males alone would be rendered more brilliantly coloured; but if the variations were not thus limited, the preservation and augmentation of such variations would depend on whether more evil was caused to the species by the females being rendered conspicuous, than good to the males by certain individuals being successful over their rivals.

As there can hardly be a doubt that both sexes of many butterflies and moths have been rendered dull-coloured for the sake of protection, so it may have been with the females alone of some species in which successive variations towards dullness first appeared in the female sex and were from the first limited in their transmission to the same sex. If not thus limited, both sexes would become dull-coloured. We shall immediately see, when we treat of mimickry, that the females alone of certain butterflies have been rendered extremely beautiful for the sake of protection, without any of the successive protective variations having been transferred to the male, to whom they could not possibly have been in the least degree injurious, and therefore could not have been eliminated through natural selection. Whether in each particular species, in which the sexes differ in colour, it is the female which has been specially modified for the sake of protection; or whether it is the male which has been specially modified for the sake of sexual attraction, the female having retained her primordial colouring only slightly changed through the agencies before alluded to; or whether again both sexes have been modified, the female for protection and the male for sexual attraction, can only be definitely decided when we know the life-history of each species.

Without distinct evidence, I am unwilling to admit that a double process of selection has long been going on with a multitude of species,—the males having been rendered more brilliant by beating their rivals; and the females more dull-coloured by having escaped from their enemies. We may take as an instance the common brimstone butterfly (Gonepteryx), which appears early in the spring before any other kind. The male of this species is of a far more intense yellow than the female, though she is almost equally conspicuous; and in this case it does not seem probable that she specially acquired her pale tints as a protection, though it is probable that the male acquired his bright colours as a sexual attraction. The female of *Anthocharis cardamines* does not possess the beautiful orange tips to her wings with which the male is ornamented; consequently she closely resembles the white butterflies (Pieris) so common in our gardens; but we have no evidence that this resemblance is beneficial. On the contrary, as she resembles both sexes of several species of the same genus inhabiting various quarters of the world, it is more probable that she has simply retained to a large extent her primordial colours.

Various facts support the conclusion that with the greater number of brilliantly-coloured Lepidoptera, it is the male which has been modified; the two sexes having come to differ from each other, or to resemble each other, according to which form of inheritance has prevailed. Inheritance is governed by so many unknown laws or conditions, that they seem to us to be most capricious in their action;[524] and we can so far understand how it is that with closely-allied species the sexes of some differ to an astonishing degree, whilst the sexes of others are identical in colour. As the successive steps in the process of variation are necessarily all transmitted through the female, a greater or less number of such steps might readily become developed in her; and thus we can understand the frequent gradations from an extreme difference to no difference at all between the sexes of the species within the same group. These cases of gradation are much too common to favour the supposition that we here see females actually undergoing the process of transition and losing their brightness for the sake of protection; for we have every reason to conclude that at any one time the greater number of species are in a fixed condition. With respect to the differences between the females of the species in the same genus or family, we can perceive that they depend, at least in part, on the females partaking of the colours of their respective males. This is well illustrated in those groups in which the males are ornamented to an extraordinary degree; for the females in these groups generally partake to a certain extent of the splendour of their male partners. Lastly, we continually find, as already remarked, that the females of almost all the species in the same genus, or even family, resemble each other much more closely in colour than do the males; and this indicates that the males have undergone a greater amount of modification than the females.

Mimickry.—This principle was first made clear in an admirable paper by Mr. Bates,[525] who thus threw a flood of light on many obscure problems. It had previously been observed that certain butterflies in S. America belonging to quite distinct families, resembled the Heliconidæ so closely in every stripe and shade of colour that they could not be distinguished except by an experienced entomologist. As the Heliconidæ are coloured in their usual manner, whilst the others depart from the usual colouring of the groups to which they belong, it is clear that the latter are the imitators, and the Heliconidæ the imitated. Mr. Bates further observed that the imitating species are comparatively rare, whilst the imitated swarm in large numbers; the two sets living mingled together. From the fact of the Heliconidæ being conspicuous and beautiful insects, yet so numerous in individuals and species, he concluded that they must be protected from the attacks of birds by some secretion or odour; and this hypothesis has now been confirmed by a considerable body of curious evidence.[526] From these considerations Mr. Bates inferred that the butterflies which imitate the protected species had acquired their present marvellously deceptive appearance, through variation and natural selection, in order to be mistaken for the protected kinds and thus to escape being devoured. No explanation is here attempted of the brilliant colours of the imitated, but only of the imitating butterflies. We must account for the colours of the former in the same general manner, as in the cases previously discussed in this chapter. Since the publication of Mr. Bates' paper, similar and equally striking facts have been observed by Mr. Wallace[527] in the Malayan region, and by Mr. Trimen in South Africa.

As some writers[528] have felt much difficulty in understanding how the first steps in the process of mimickry could have been effected through natural selection, it may be well to remark that the process probably has never commenced with forms widely dissimilar in colour. But with two species moderately like each other, the closest resemblance if beneficial to either form could readily be thus gained; and if the imitated form was subsequently and gradually modified through sexual selection or any other means, the imitating form would be led along the same track, and thus be modified to almost any extent, so that it might ultimately assume an appearance or colouring wholly unlike that of the other members of the group to which it belonged. As extremely slight variations in colour would not in many cases suffice to render a species so like another protected species as to lead to its preservation, it should be remembered that many species of Lepidoptera are liable to considerable and abrupt variations in colour. A few instances have been given in this chapter; but under this point of view Mr. Bates' original paper on mimickry, as well as Mr. Wallace's papers, should be consulted.

In the foregoing cases both sexes of the imitating species resemble the imitated; but occasionally the female alone mocks a brilliantly-coloured and protected species inhabiting the same district. Consequently the female differs in colour from her own male, and, which is a rare and anomalous circumstance, is the more brightly-coloured of the two. In all the few species of Pieridæ, in which the female is more conspicuously coloured than the male, she imitates, as I am informed by Mr. Wallace, some protected species inhabiting the same region. The female of *Diadema anomala* is rich purple-brown with almost the whole surface glossed with satiny blue, and she closely imitates the *Euplœa midamus*, "one of the commonest butterflies of the East;" whilst the male is bronzy or olive-brown, with only a slight blue gloss on the outer parts of the wings.[529] Both sexes of this Diadema and of *D. bolina* follow the same habits of life, so that the differences in colour between the sexes cannot be accounted for by exposure to different conditions;[530] even if this explanation were admissible in other instances.[531]

The above cases of female butterflies which are more brightly-coloured than the males, shew us, firstly, that variations have arisen in a state of nature in the female sex, and have been transmitted exclusively, or almost exclusively, to the same sex; and, secondly, that this form of inheritance has not been determined through natural selection. For if we assume that the females, before they became brightly coloured in imitation of some protected kind, were exposed during each season for a longer period to danger than the males; or if we assume that they could not escape so swiftly from their enemies, we can understand how they alone might originally have acquired through natural selection and sexually-limited inheritance their present protective colours. But except on the principle of these variations having been transmitted exclusively to the female offspring, we cannot understand why the males should have remained dull-coloured; for it would surely not have been in any way injurious to each individual male to have partaken by inheritance of the protective colours of the female, and thus to have had a better chance of escaping destruction. In a group in which brilliant colours are so common as with butterflies, it cannot be supposed that the males have been kept dull-coloured through sexual selection by the females rejecting the individuals which were rendered as beautiful as themselves. We may, therefore, conclude that in these cases inheritance by one sex is not due to the modification through natural selection of a tendency to equal inheritance by both sexes.

It may be well here to give an analogous case in another Order, of characters acquired only by the female, though not in the least injurious, as far as we can judge, to the male. Amongst the Phasmidæ, or spectre-insects, Mr. Wallace states that "it is often the females alone that so strikingly resemble leaves, while the males show only a rude approximation." Now, whatever may be the habits of these insects, it is highly improbable that it could be disadvantageous to the males to escape detection by resembling leaves.[532] Hence we may conclude that the females alone in this latter as in the previous cases originally varied in certain characters; these characters having been preserved and augmented through ordinary selection for the sake of protection and from the first transmitted to the female offspring alone.

Bright Colours of Caterpillars.—Whilst reflecting on the beauty of many butterflies, it occurred to me that some caterpillars were splendidly coloured, and as sexual selection could not possibly have here acted, it appeared rash to attribute the beauty of the mature insect to this agency, unless the bright colours of their larvæ could be in some manner explained. In the first place it may be observed that the colours of caterpillars do not stand in any close correlation with those of the mature insect. Secondly, their bright colours do not serve in any ordinary manner as a protection. As an instance of this, Mr. Bates informs me that the most conspicuous caterpillar which he ever beheld (that of a Sphinx) lived on the large green leaves of a tree on the open llanos of South America; it was about four inches in length, transversely banded with black and yellow, and with its head, legs, and tail of a bright red. Hence it caught the eye of any man who passed by at the distance of many yards, and no doubt of every passing bird.

I then applied to Mr. Wallace, who has an innate genius for solving difficulties. After some consideration he replied: "Most caterpillars require protection, as may be inferred from some kinds being furnished with spines or irritating hairs, and from many being coloured green like the leaves on which they feed, or curiously like the twigs of the trees on which they live." I may add as another instance of protection, that there is a caterpillar of a moth, as I am informed by Mr. J. Mansel Weale, which lives on the mimosas in South Africa, and fabricates for itself a case, quite undistinguishable from the surrounding thorns. From such considerations Mr. Wallace thought it probable that conspicuously-coloured caterpillars were protected by having a nauseous taste; but as their skin is extremely tender, and as their intestines readily protrude from a wound, a slight peck from the beak of a bird would be as fatal to them as if they had been devoured. Hence, as Mr. Wallace remarks, "distastefulness alone would be insufficient to protect a caterpillar unless some outward sign indicated to its would-be destroyer that its prey was a disgusting morsel." Under these circumstances it would be highly advantageous to a caterpillar to be instantaneously and certainly recognised as unpalatable by all birds and other animals. Thus the most gaudy colours would be serviceable, and might have been gained by variation and the survival of the most easily-recognised individuals.

This hypothesis appears at first sight very bold; but when it was brought before the Entomological Society[533] it was supported by various statements; and Mr. J. Jenner Weir, who keeps a large number of birds in an aviary, has made, as he informs me, numerous trials, and finds no exception to the rule, that all caterpillars of nocturnal and retiring habits with smooth skins, all of a green colour, and all which imitate twigs, are greedily devoured by his birds. The hairy and spinose kinds are invariably rejected, as were four conspicuously-coloured species. When the birds rejected a caterpillar, they plainly shewed, by shaking their heads and cleansing their beaks, that they were disgusted by the taste.[534] Three conspicuous kinds of caterpillars and moths were also given by Mr. A. Butler to some lizards and frogs, and were rejected; though other kinds were eagerly eaten. Thus the probable truth of Mr. Wallace's view is confirmed, namely, that certain caterpillars have been made conspicuous for their own good, so as to be easily recognised by their enemies, on nearly the same principle that certain poisons are coloured by druggists for the good of man. This view will, it is probable, be hereafter extended to many animals, which are coloured in a conspicuous manner.

Summary and Concluding Remarks on Insects.— Looking back to the several Orders, we have seen that the sexes often differ in various characters, the meaning of which is not understood. The sexes, also, often differ in their organs of sense or locomotion, so that the males may quickly discover or reach the females, and still oftener in the males possessing diversified contrivances for retaining the females when found. But we are not here much concerned with sexual differences of these kinds.

In almost all the Orders, the males of some species, even of weak and delicate kinds, are known to be highly pugnacious; and some few are furnished with special weapons for fighting with their rivals. But the law of battle does not prevail nearly so widely with insects as with the higher animals. Hence probably it is that the males have not often been rendered larger and stronger than the females. On the contrary they are usually smaller, in order that they may be developed within a shorter time, so as to be ready in large numbers for the emergence of the females.

In two families of the Homoptera the males alone possess, in an efficient state, organs which may be called vocal; and in three families of the Orthoptera the males alone possess stridulating organs. In both cases these organs are incessantly used during the breeding-season, not only for calling the females, but for charming or exciting them in rivalry with other males. No one who admits the agency of natural selection, will dispute that these musical instruments have been acquired through sexual selection. In four other Orders the members of one sex, or more commonly of both sexes, are provided with organs for producing various sounds, which apparently serve merely as call-notes. Even when both sexes are thus provided, the individuals which were able to make the loudest or most continuous noise would gain partners before those which were less noisy, so that their organs have probably been gained through sexual selection. It is instructive to reflect on the wonderful diversity of the means for producing sound, possessed by the males alone or by both sexes in no less than six Orders, and which were possessed by at least one insect at an extremely remote geological epoch. We thus learn how effectual sexual selection has been in leading to modifications of structure, which sometimes, as with the Homoptera, are of an important nature.

From the reasons assigned in the last chapter, it is probable that the great horns of the males of many lamellicorn, and some other beetles, have been acquired as ornaments. So perhaps it may be with certain other peculiarities confined to the male sex. From the small size of insects, we are apt to undervalue their appearance. If we could imagine a male Chalcosoma (fig. 15) with its polished, bronzed coat of mail, and vast complex horns, magnified to the size of a horse or even of a dog, it would be one of the most imposing animals in the world.

The colouring of insects is a complex and obscure subject. When the male differs slightly from the female, and neither are brilliantly coloured, it is probable that the two sexes have varied in a slightly different manner, with the variations transmitted to the same sex, without any benefit having been thus derived or evil suffered. When the male is brilliantly coloured and differs conspicuously from the female, as with some dragon-flies and many butterflies, it is probable that he alone has been modified, and that he owes his colours to sexual selection; whilst the female has retained a primordial or very ancient type of colouring, slightly modified by the agencies before explained, and has therefore not been rendered obscure, at least in most cases, for the sake of protection. But the female alone has sometimes been coloured brilliantly so as to imitate other protected species inhabiting the same district. When the sexes resemble each other and both are obscurely coloured, there is no doubt that they have been in a multitude of cases coloured for the sake of protection. So it is in some instances when both are brightly coloured, causing them to resemble surrounding objects such as flowers, or other protected species, or indirectly by giving notice to their enemies that they are of an unpalatable nature. In many other cases in which the sexes resemble each other and are brilliantly coloured, especially when the colours are arranged for display, we may conclude that they have been gained by the male sex as an attraction, and have been transferred to both sexes. We are more especially led to this conclusion whenever the same type of coloration prevails throughout a group, and we find that the males of some species differ widely in colour from the females, whilst both sexes of other species are quite alike, with intermediate gradations connecting these extreme states.

In the same manner as bright colours have often been partially transferred from the males to the females, so it has been with the extraordinary horns of many lamellicorn and some other beetles. So, again, the vocal or instrumental organs proper to the males of the Homoptera and Orthoptera have generally been transferred in a rudimentary, or even in a nearly perfect condition to the females; yet not sufficiently perfect to be used for producing sound. It is also an interesting fact, as bearing on sexual selection, that the stridulating organs of certain male Orthoptera are not fully developed until the last moult; and that the colours of certain male dragon-flies are not fully developed until some little time after their emergence from the pupal state, and when they are ready to breed.

Sexual selection implies that the more attractive individuals are preferred by the opposite sex; and as with insects, when the sexes differ, it is the male which, with rare exceptions, is the most ornamented and departs most from the type to which the species belongs;—and as it is the male which searches eagerly for the female, we must suppose that the females habitually or occasionally prefer the more beautiful males, and that these have thus acquired their beauty. That in most or all the orders the females have the power of rejecting any particular male, we may safely infer from the many singular contrivances possessed by the males, such as great jaws, adhesive cushions, spines, elongated legs, &c., for seizing the female; for these contrivances shew that there is some difficulty in the act. In the case of unions between distinct species, of which many instances have been recorded, the female must have been a consenting party. Judging from what we know of the perceptive powers and affections of various insects, there is no antecedent improbability in sexual selection having come largely into action; but we have as yet no direct evidence on this head, and some facts are opposed to the belief. Nevertheless, when we see many males pursuing the same female, we can hardly believe that the pairing is left to blind chance—that the female exerts no choice, and is not influenced by the gorgeous colours or other ornaments, with which the male alone is decorated.

If we admit that the females of the Homoptera and Orthoptera appreciate the musical tones emitted by their male partners, and that the various instruments for this purpose have been perfected through sexual selection, there is little improbability in the females of other insects appreciating beauty in form or colour, and consequently in such characters having been thus gained by the males. But from the circumstance of colour being so variable, and from its having been so often modified for the sake of protection, it is extremely difficult to decide in how large a proportion of cases sexual selection has come into play. This is more especially difficult in those Orders, such as the Orthoptera, Hymenoptera, and Coleoptera, in which the two sexes rarely differ much in colour; for we are thus cut off from our best evidence of some relation between the reproduction of the species and colour. With the Coleoptera, however, as before remarked, it is in the great lamellicorn group, placed by some authors at the head of the Order, and in which we sometimes see a mutual attachment between the sexes, that we find the males of some species possessing weapons for sexual strife, others furnished with wonderful horns, many with stridulating organs, and others ornamented with splendid metallic tints. Hence it seems probable that all these characters have been gained through the same means, namely sexual selection.

When we treat of Birds, we shall see that they present in their secondary sexual characters the closest analogy with insects, Thus, many male birds are highly pugnacious, and some are furnished with special weapons for fighting with their rivals. They possess organs which are used during the breeding-season for producing vocal and instrumental music. They are frequently ornamented with combs, horns, wattles and plumes of the most diversified kinds, and are decorated with beautiful colours, all evidently for the sake of display. We shall find that, as with insects, both sexes, in certain groups, are equally beautiful, and are equally provided with ornaments which are usually confined to the male sex. In other groups both sexes are equally plain-coloured and unornamented. Lastly, in some few anomalous cases, the females are more beautiful than the males. We shall often find, in the same group of birds, every gradation from no difference between the sexes, to an extreme difference. In the latter case we shall see that the females, like female insects, often possess more or less plain traces of the characters which properly belong to the males. The analogy, indeed, in all these respects between birds and insects, is curiously close. Whatever explanation applies to the one class probably applies to the other; and this explanation, as we shall hereafter attempt to shew, is almost certainly sexual selection.

CHAPTER XII

Secondary Sexual Characters of Fishes, Amphibians, and Reptiles

FISHES: COURTSHIP AND BATTLES OF THE MALES • LARGER SIZE OF THE FEMALES • MALES, BRIGHT COLOURS AND ORNAMENTAL APPENDAGES; OTHER STRANGE CHARACTERS • COLOURS AND APPENDAGES ACQUIRED BY THE MALES DURING THE BREEDING-SEASON ALONE • FISHES WITH BOTH SEXES BRILLIANTLY COLOURED • PROTECTIVE COLOURS • THE LESS CONSPICUOUS COLOURS OF THE FEMALE CANNOT BE ACCOUNTED FOR ON THE PRINCIPLE OF PROTECTION • MALE FISHES BUILDING NESTS, AND TAKING CHARGE OF THE OVA AND YOUNG. **AMPHIBIANS:** DIFFERENCES IN STRUCTURE AND COLOUR BETWEEN THE SEXES • VOCAL ORGANS. **REPTILES:** CHELONIANS • CROCODILES • SNAKES, COLOURS IN SOME CASES PROTECTIVE • LIZARDS, BATTLES OF • ORNAMENTAL APPENDAGES • STRANGE DIFFERENCES IN STRUCTURE BETWEEN THE SEXES • COLOURS • SEXUAL DIFFERENCES ALMOST AS GREAT AS WITH BIRDS

We have now arrived at the great sub-kingdom of the Vertebrata, and will commence with the lowest class, namely Fishes. The males of Plagiostomous fishes (sharks, rays) and of Chimæroid fishes are provided with claspers which serve to retain the female, like the various structures possessed by so many of the lower animals. Besides the claspers, the males of many rays have clusters of strong sharp spines on their heads, and several rows along "the upper outer surface of their pectoral fins." These are present in the males of some species, which have the other parts of their bodies smooth. They are only temporarily developed during the breeding-season; and Dr. Günther suspects that they are brought into action as prehensile organs by the doubling inwards and downwards of the two sides of the body. It is a remarkable fact that the females and not the males of some species, as of *Raia clavata,* have their backs studded with large hook-formed spines.[535]

Owing to the element which fishes inhabit, little is known about their courtship, and not much about their battles. The male stickleback *(Gasterosteus leiurus)* has been described as "mad with delight" when the female comes out of her hiding-place and surveys the nest which he has made for her. "He darts round her in every direction, then to his accumulated materials for the nest, then back again in an instant; and as she does not advance he endeavours to push her with his snout, and then tries to pull her by the tail and side-spine to the nest."[536] The males are said to be polygamists;[537] they are extraordinarily bold and pugnacious, whilst "the females are quite pacific." Their battles are at times desperate; "for these puny combatants fasten tight on each other for several seconds, tumbling over and over again, until their strength appears completely exhausted." With the rough-tailed stickleback *(G. trachurus)* the males whilst fighting swim round and round each other, biting and endeavouring to pierce each other with their raised lateral spines. The same writer adds,[538] "the bite of these little furies is very severe. They also use their lateral spines with such fatal effect, that I have seen one during a battle absolutely rip his opponent quite open, so that he sank to the bottom and died." When a fish is conquered, "his gallant bearing forsakes him; his gay colours fade away; and he hides his disgrace among his peaceable companions, but is for some time the constant object of his conqueror's persecution."

The male salmon is as pugnacious as the little stickleback; and so is the male trout, as I hear from Dr. Günther. Mr. Shaw saw a violent contest between two male salmons which lasted the whole day; and Mr. R. Buist, Superintendent of Fisheries, informs me that he has often watched from the bridge at Perth the males driving away their rivals whilst the females were spawning. The males "are constantly fighting and tearing each other on the spawning-beds, and many so injure each other as to cause the death of numbers, many being seen swimming near the banks of the river in a state of exhaustion, and apparently in a dying state."[539] The keeper of the Stormontfield breeding-ponds visited, as Mr. Buist informs me, in June, 1868, the northern Tyne, and found about 300 dead salmon, all of which with one exception were males; and he was convinced that they had lost their lives by fighting.

The most curious point about the male salmon is that during the breeding-season, besides a slight change in colour, "the lower jaw elongates, and a cartilaginous projection turns upwards from the point, which, when the jaws are closed, occupies a deep cavity between the intermaxillary bones of the upper jaw."[540] (Figs. 26 and 27.) In our salmon this change of structure lasts only during the breeding-season; but in the *Salmo lycaodon* of N.W. America the change, as Mr. J. K. Lord[541] believes, is permanent and best marked in the older males which have previously ascended the rivers. In these old males the jaws become developed into immense hook-like projections, and the teeth grow into regular fangs, often more than half an inch in length. With the European salmon, according to Mr. Lloyd,[542] the temporary hook-like structure serves to strengthen and protect the jaws, when one male charges another with wonderful violence; but the greatly developed teeth of the male American salmon may be compared with the tusks of many male mammals, and they indicate an offensive rather than a protective purpose.

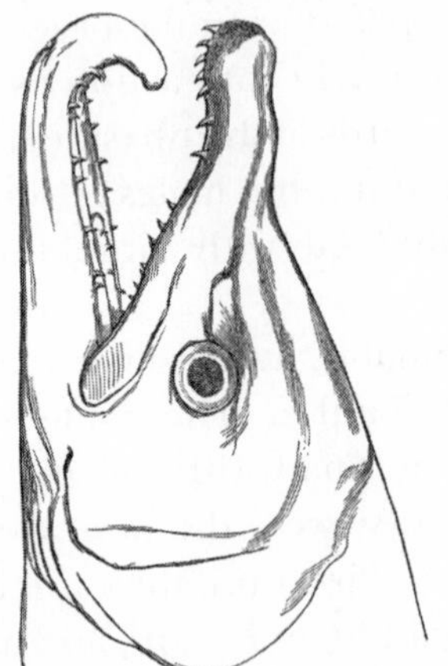

Fig. 26. *Head of male of common salmon (Salmo salar) during the breeding-season. [This drawing, as well as all the others in the present chapter, have been executed by the well-known artist, Mr. G. Ford, under the kind superintendence of Dr. Günther, from specimens in the British Museum.]*

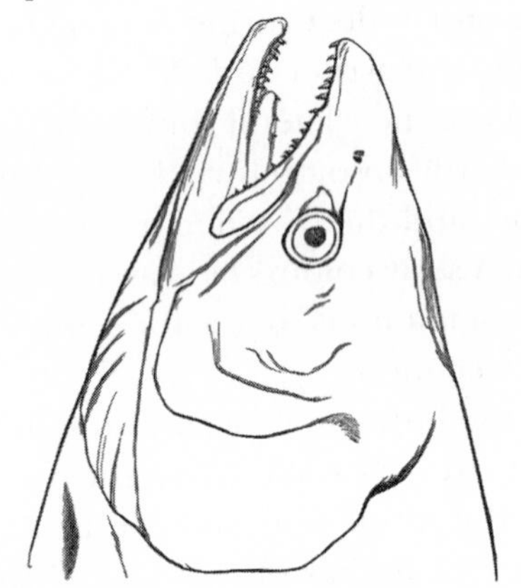

Fig. 27. *Head of female salmon.*

The salmon is not the only fish in which the teeth differ in the two sexes. This is the case with many rays. In the thornback *(Raia clavata)* the adult male has sharp, pointed teeth, directed backwards, whilst those of the female are broad and flat, forming a pavement; so that these teeth differ in the two sexes of the same species more than is usual in distinct genera of the same family. The teeth of the male become sharp only when he is adult: whilst young they are broad and flat like those of the female. As so frequently occurs with secondary sexual characters, both sexes of some species of rays, for instance *R. batis,* possess, when adult, sharp, pointed teeth; and here a character, proper to and primarily gained by the male, appears to have been transmitted to the offspring of both sexes. The teeth are likewise pointed in both sexes of *R. maculata,* but only when completely adult; the males acquiring them at an earlier age than the females. We shall hereafter meet with analogous cases with certain birds, in which the male acquires the plumage common to both adult sexes, at a somewhat earlier age than the female. With other species of rays the males even when old never possess sharp teeth, and consequently both sexes when adult are provided with broad, flat teeth like those of the young, and of the mature females of the above-mentioned species.[543] As the rays are bold, strong and voracious fishes, we may suspect that the males require their sharp teeth for fighting with their rivals; but as they possess many parts modified and adapted for the prehension of the female, it is possible that their teeth may be used for this purpose.

In regard to size, M. Carbonnier[544] maintains that with almost all fishes the female is larger than the male; and Dr. Günther does not know of a single instance in which the male is actually larger than the female. With some Cyprinodonts the male is not even half as large as the female. As with many kinds of fishes the males habitually fight together; it is surprising that they have not generally become through the effects of sexual selection larger and stronger than the females. The males suffer from their small size, for according to M. Carbonnier they are liable to be devoured by the females of their own species when carnivorous, and no doubt by other species. Increased size must be in some manner of more importance to the females, than strength and size are to the males for fighting with other males; and this perhaps is to allow of the production of a vast number of ova.

In many species the male alone is ornamented with bright colours; or these are much brighter in the male than the female. The male, also, is sometimes provided with appendages which appear to be of no more use to him for the ordinary purposes of life than are the tail-feathers to the peacock. I am indebted for most of the following facts to the great kindness of Dr. Günther. There is reason to suspect that many tropical fishes differ sexually in colour and structure; and there are some striking cases with our British fishes. The male *Callionymus lyra* has been called the g*emmeous dragonet* "from its brilliant gem-like colours." When freshly taken from the sea the body is yellow of various shades, striped and spotted with vivid blue on the head; the dorsal fins are pale brown with dark longitudinal bands; the ventral, caudal and anal fins being bluish-black. The female, or sordid dragonet, was considered by Linnæus and by many subsequent naturalists as a distinct species; it is of a dingy reddish-brown, with the dorsal fin brown and the other fins white. The sexes differ also in the proportional size of the head and mouth, and in the position of the eyes;[545] but the most striking differrence is the extraordinary elongation in the male (fig. 28) of the dorsal fin. The young males resemble in structure and colour the adult

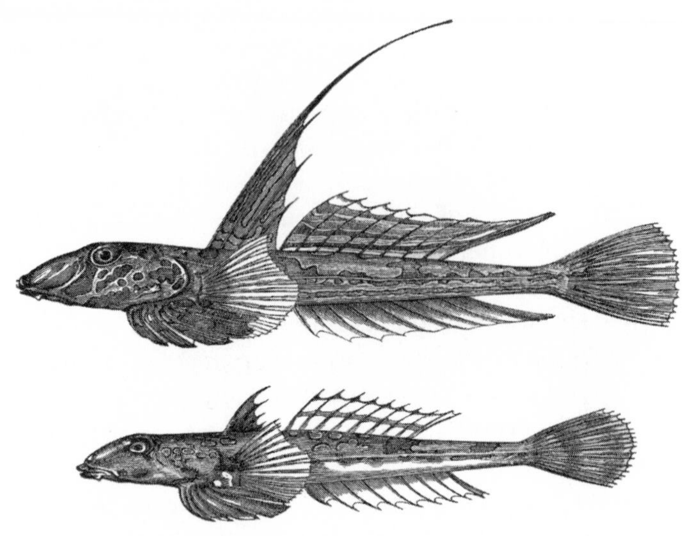

Fig. 28. *Callionymus lyra.*
Upper figure, male; lower figure, female.

females. Throughout the genus Callionymus,[546] the male is generally much more brightly spotted than the female, and in several species, not only the dorsal, but the anal fin of the male is much elongated.

The male of the *Cottus scorpius,* or sea-scorpion, is more slender and smaller than the female. There is also a great difference in colour between them. It is difficult, as Mr. Lloyd[547] remarks, "for any one, who has not seen this fish during the spawning-season, when its hues are brightest, to conceive the admixture of brilliant colours with which it, in other respects so ill-favoured, is at that time adorned." Both sexes of the *Labrus mixtus,* although very different in colour, are beautiful; the male being orange with bright-blue stripes, and the female bright-red with some black spots on the back.

In the very distinct family of the Cyprinodontidæ—inhabitants of the fresh waters of foreign lands—the sexes sometimes differ much in various characters. In the male of the *Mollienesia petenensis,*[548] the dorsal fin is greatly developed and is marked with a row of large, round, ocellated, bright-coloured spots; whilst the same fin in the female is smaller, of a different shape, and marked only with irregularly-curved brown spots. In the male the basal margin of the anal fin is also a little produced and dark-coloured. In the male of an allied form, the *Xiphophorus Hellerii* (fig. 29), the inferior margin of the anal fin is developed into a long filament, which is striped, as I hear from Dr. Günther, with bright colours. This filament does not contain any muscles and apparently cannot be of any direct use to

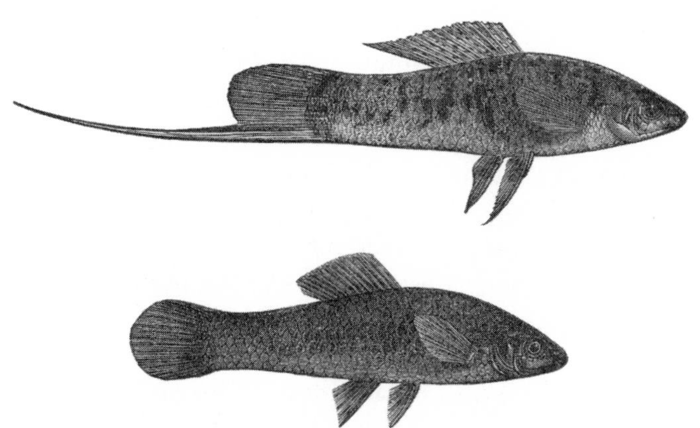

Fig. 29. *Xiphophorus Hellerii.*
Upper figure, male; lower figure, female.

the fish. As in the case of the Callionymus, the males whilst young resemble in colour and structure the adult females. Sexual differences such as these may be strictly compared with those which are so frequent with gallinaceous birds.[549]

Fig. 30. *Plecostomus barbatus.*
Upper figure, head of male; lower figure, female.

In a siluroid fish, inhabiting the fresh waters of South America, namely the *Plecostomus barbatus*[550] (fig. 30), the male has its mouth and interoperculum fringed with a beard of stiff hairs, of which the female shews hardly a trace. These hairs are of the nature of scales. In another species of the same genus, soft flexible tentacles project from the front part of the head of the male, which are absent in the female. These tentacles are prolongations of the true skin, and therefore are not homologous with the stiff hairs of the former species; but it can hardly be doubted that both serve the same purpose. What this purpose may be it is difficult to conjecture; ornament does not here seem probable, but we can hardly suppose that stiff hairs and flexible filaments can be useful in any ordinary way to the males alone. The *Monacanthus scopas,* which was shewn to me in the British Museum by Dr. Günther, presents a nearly analogous case. The male has a cluster of stiff, straight spines, like those of a comb, on the sides of the tail; and these in a specimen six inches long were nearly an inch and a half in length; the female has on the same place a cluster of bristles, which may be compared with those of a tooth-brush. In another species, the *M. peronii,* the male has a brush like that possessed by the female of the last species, whilst the sides of the tail in the female are smooth. In some other species the same part of the tail can be perceived to be a little roughened in the male and perfectly smooth in the female; and lastly in others, both sexes have smooth sides. In that strange monster, the *Chimæra monstrosa,* the male has a hook-shaped bone on the top of the head, directed forwards, with its rounded end covered with sharp spines; in the female "this crown is altogether absent," but what its use may be is utterly unknown.[551]

The structures as yet referred to are permanent in the male after he has arrived at maturity; but with some Blennies and in another allied genus[552] a crest is developed on the head of the male only during the breeding-season, and their bodies at the same time become more brightly-coloured. There can be little doubt that this crest serves as a temporary sexual ornament, for the female does not exhibit a trace of it. In other species of the same genus both sexes possess a crest, and in at least one species neither sex is thus provided. In this case and in that of the Monacanthus, we have good instances to how great an extent the sexual characters of closely-allied forms may differ. In many of the Chromidæ, for instance in Geophagus and especially in Cichla, the males, as I hear from Professor Agassiz,[553] have a conspicuous protuberance on the forehead, which is wholly wanting in the females and in the young males. Professor Agassiz adds, "I have often observed these fishes at the time of spawning when the protuberance is largest, and at other seasons when it is totally wanting and the two sexes shew no difference whatever in the outline of the profile of the head. I never could ascertain that it subserves any special function, and the Indians on the Amazon know nothing about its use." These protuberances in their periodical appearance resemble the fleshy caruncles on the heads of certain birds; but whether they serve as ornaments must remain at present doubtful.

The males of those fishes, which differ permanently in colour from the females, often become more brilliant, as I hear from Professor Agassiz and Dr. Günther, during the breeding-season. This is likewise the case with a multitude of fishes, the sexes of which at all other seasons of the year are identical in colour. The tench, roach, and perch may be given as instances. The male salmon at this season is "marked on the cheeks with orange-coloured stripes, which give it the appearance of a Labrus, and the body partakes of a golden-orange tinge. The females are dark in colour, and are commonly called black-fish."[554] An analogous and even greater change takes place with the *Salmo eriox* or bulltrout; the males of the char *(S. umbla)* are likewise at this season rather brighter in colour than the females.[555] The colours of the pike *(Esox reticulatus)* of the United States, especially of the male, become, during the breeding-season, exceedingly intense, brilliant, and iridescent.[556] Another striking instance out of many is afforded by the male stickleback *(Gasterosteus leiurus),* which is described by Mr. Warington,[557] as being then "beautiful beyond description." The back and eyes of the female are simply brown, and the belly white. The eyes of the male, on the other hand, are "of the most splendid green, having a metallic lustre like the green feathers of some humming-birds. The throat and belly are of a bright crimson, the back of an ashy-green, and the whole fish appears as though it were somewhat translucent and glowed with an internal incandescence." After the breeding-season these colours all change, the throat and belly become of a paler red, the back more green, and the glowing tints subside.

That with fishes there exists some close relation between their colours and their sexual functions we can clearly see;—firstly, from the adult males of certain species being differently coloured from the females, and often much more brilliantly;—secondly, from these same males, whilst immature, resembling the mature females;—and, lastly, from the males, even of those species which at all other times of the year are identical in colour with the females, often acquiring brilliant tints during the spawning-season. We know that the males are ardent in their courtship and sometimes fight desperately together. If we may assume that the females have the power of exerting a choice and of selecting the more highly-ornamented males, all the above facts become intelligible through the principle of sexual selection. On the other hand, if the females habitually deposited and left their ova to be fertilised by the first male which chanced to approach, this fact would be fatal to the efficiency of sexual selection; for there could be no choice of a partner. But, as far as is known, the female never willingly spawns except in the close presence of a male, and the male never fertilises the ova except in the close presence of a female. It is obviously difficult to obtain direct evidence with respect to female fishes selecting their partners. An excellent observer,[558] who carefully watched the spawning of minnows *(Cyprinus phoxinus),* remarks that owing to the males, which were ten times as numerous as the females, crowding closely round them, he could "speak only doubtfully on their operations. When a female came among a number of males they immediately pursued her; if she was not ready for shedding her spawn, she made a precipitate retreat; but if she was ready, she came boldly in among them, and was immediately pressed closely by a male on each side; and when they had been in that situation a short time, were superseded by other two, who wedged themselves in between them and the female, who appeared to treat all her lovers with the same kindness." Notwithstanding this last statement, I cannot, from the several previous considerations, give up the belief that the males which are the most attractive to the females, from their brighter colours or other ornaments, are commonly preferred by them; and that the males have thus been rendered more beautiful in the course of ages.

We have next to inquire whether this view can be extended, through the law of the equal transmission of characters to both sexes, to those groups in which the males and females are brilliant in the same or nearly the same degree and manner. In such a genus as Labrus, which includes some of the most splendid fishes in the world, for instance, the Peacock Labrus *(L. pavo),* described,[559] with pardonable exaggeration, as formed of polished scales of gold encrusting lapis-lazuli, rubies, sapphires, emeralds and amethysts, we may, with much probability, accept this belief; for we have seen that the sexes in at least one species differ greatly in colour. With some fishes, as with many of the lowest animals, splendid colours may be the direct result of the nature of their tissues and of the surrounding conditions, without any aid from selection. The gold-fish *(Cyprinus auratus),* judging from the analogy of the golden variety of the common carp, is, perhaps, a case in point, as it may owe its splendid colours to a single abrupt variation, due to the conditions to which this fish has been subjected under confinement. It is, however, more probable that these colours have been intensified through artificial selection, as this species has been carefully bred in China from a remote period.[560] Under natural conditions it does not seem probable that beings so highly organised as fishes, and which live under such complex relations, should become brilliantly coloured without suffering some evil or receiving some benefit from so great a change, and consequently without the intervention of natural selection.

What, then, must we conclude in regard to the many fishes, both sexes of which are splendidly coloured? Mr. Wallace[561] believes that the species which frequent reefs, where corals and other brightly-coloured organisms abound, are brightly coloured in order to escape detection by their enemies; but according to my recollection they were thus rendered highly conspicuous. In the fresh-waters of the Tropics there are no brilliantly-coloured corals or other organisms for the fishes to resemble; yet many species in the Amazons are beautifully coloured, and many of the carnivorous Cyprinidæ in India are ornamented with "bright longitudinal lines of various tints."[562] Mr. M'Clelland, in describing these fishes goes so far as to suppose that "the peculiar brilliancy of their colours" serves as "a better mark for king-fishers, terns, and other birds which are destined to keep the number of these fishes in check;" but at the present day few naturalists will admit that any animal has been made conspicuous as an aid to its own destruction. It is possible that certain fishes may have been rendered conspicuous in order to warn birds and beasts of prey (as explained when treating of caterpillars) that they were unpalatable; but it is not, I believe, known that any fish, at least any fresh-water fish, is rejected from being distasteful to fish-devouring animals. On the whole, the most probable view in regard to the fishes, of which both sexes are brilliantly coloured, is that their colours have been acquired by the males as a sexual ornament, and have been transferred in an equal or nearly equal degree to the other sex.

We have now to consider whether, when the male differs in a marked manner from the female in colour or in other ornaments, he alone has been modified, with the variations inherited only by his male offspring; or whether the female has been specially modified and rendered inconspicuous for the sake of protection, such modifications being inherited only by the females. It is impossible to doubt that colour has been acquired by many fishes as a protection: no one can behold the speckled upper surface of a flounder, and overlook its resemblance to the sandy bed of the sea on which it lives. One of the most striking instances ever recorded of an animal gaining protection by its colour (as far as can be judged in preserved specimens) and by its form, is that given by Dr. Günther[563] of a pipe-fish, which, with its reddish streaming filaments, is hardly distinguishable from the sea-weed to which it clings with its prehensile tail. But the question now under consideration is whether the females alone have been modified for this object. Fishes offer valuable evidence on this head. We can see that one sex will not be modified through natural selection for the sake of protection more than the other, supposing both to vary, unless one sex is exposed for a longer period to danger, or has less power of escaping from such danger than the other sex; and it does not appear that with fishes the sexes differ in these respects. As far as there is any difference, the males, from being generally of smaller size, and from wandering more about, are exposed to greater danger than the females; and yet, when the sexes differ, the males are almost always the most conspicuously coloured. The ova are fertilised immediately after being deposited, and when this process lasts for several days, as in the case of the salmon,[564] the female, during the whole time, is attended by the male. After the ova are fertilised they are, in most cases, left unprotected by both parents, so that the males and females, as far as oviposition is concerned, are equally exposed to danger, and both are equally important for the production of fertile ova; consequently the more or less brightly-coloured individuals of either sex would be equally liable to be destroyed or preserved, and both would have an equal influence on the colours of their offspring or the race.

Certain fishes, belonging to several families, make nests; and some of these fishes take care of their young when hatched. Both sexes of the brightly-coloured *Crenilabrus massa* and *melops* work together in building their nests with sea-weed, shells, &c.[565] But the males of certain fishes do all the work, and afterwards take exclusive charge of the young. This is the case with the dull-coloured gobies,[566] in which the sexes are not known to differ in colour, and likewise with the sticklebacks (Gasterosteus), in which the males become brilliantly coloured during the spawning-season. The male of the smooth-tailed stickleback *(G. leiurus)* performs during a long time the duties of a nurse with exemplary care and vigilance, and is continually employed in gently leading back the young to the nest when they stray too far. He courageously drives away all enemies, including the females of his own species. It would indeed be no small relief to the male if the female, after depositing her eggs, were immediately devoured by some enemy, for he is forced incessantly to drive her from the nest.[567]

The males of certain other fishes inhabiting South America and Ceylon, and belonging to two distinct orders, have the extraordinary habit of hatching the eggs laid by the females within their mouths or branchial cavities.[568] With the Amazonian species which follow this habit, the males, as I am informed by the kindness of Professor Agassiz, "not only are generally brighter than the females, but the difference is greater at the spawning-season than at any other time." The species of Geophagus act in the same manner; and in this genus, a conspicuous protuberance becomes developed on the forehead of the males during the breeding-season. With the various species of Chromids, as Professor Agassiz likewise informs me, sexual differences in colour may be observed, "whether they lay their eggs in the water among aquatic plants, or deposit them in holes, leaving them to come out without further care, or build shallow nests in the river-mud, over which they sit, as our Promotis does. It ought also to be observed that these sitters are among the brightest species in their respective families; for instance, Hygrogonus is bright green, with large black ocelli, encircled with the most brilliant red." Whether with all the species of Chromids it is the male alone which sits on the eggs is not known. It is, however, manifest that the fact of the eggs being protected or unprotected, has had little or no influence on the differences in colour between the sexes. It is further manifest, in all the cases in which the males take exclusive charge of the nests and young, that the destruction of the brighter-coloured males would be far more influential on the character of the race, than the destruction of the brighter-coloured females; for the death of the male during the period of incubation or nursing would entail the death of the young, so that these could not inherit his peculiarities; yet, in many of these very cases the males are more conspicuously coloured than the females.

In most of the Lophobranchii (Pipe-fish, Hippocampi, &c.) the males have either marsupial sacks or hemispherical depressions on the abdomen, in which the ova laid by the female are hatched. The males also shew great attachment to their young.[569] The sexes do not commonly differ much in colour; but Dr. Günther believes that the male Hippocampi are rather brighter than the females. The genus Solenostoma, however, offers a very curious exceptional case,[570] for the female is much more vividly coloured and spotted than the male, and she alone has a marsupial sack and hatches the eggs; so that the female of Solenostoma differs from all the other Lophobranchii in this latter respect, and from almost all other fishes, in being more brightly coloured than the male. It is improbable that this remarkable double inversion of character in the female should be an accidental coincidence. As the males of several fishes which take exclusive charge of the eggs and young are more brightly coloured than the females, and as here the female Solenostoma takes the same charge and is brighter than the male, it might be argued that the conspicuous colours of the sex which is the most important of the two for the welfare of the offspring must serve, in some manner, as a protection. But from the multitude of fishes, the males of which are either permanently or periodically brighter than the females, but whose life is not at all more important than that of the female for the welfare of the species, this view can hardly be maintained. When we treat of birds we shall meet with analogous cases in which there has been a complete inversion of the usual attributes of the two sexes, and we shall then give what appears to be the probable explanation, namely, that the males have selected the more attractive females, instead of the latter having selected, in accordance with the usual rule throughout the animal kingdom, the more attractive males.

On the whole we may conclude, that with most fishes, in which the sexes differ in colour or in other ornamental characters, the males originally varied, with their variations transmitted to the same sex, and accumulated through sexual selection by attracting or exciting the females. In many cases, however, such characters have been transferred, either partially or completely, to the females. In other cases, again, both sexes have been coloured alike for the sake of protection; but in no instance does it appear that the female alone has had her colours or other characters specially modified for this purpose.

The last point which need be noticed is that in many parts of the world fishes are known to make peculiar noises, which are described in some cases as being musical. Very little has been ascertained with respect to the means by which such sounds are produced, and even less about their purpose. The drumming of the Umbrinas in the European seas is said to be audible from a depth of twenty fathoms. The fishermen of Rochelle assert "that the males alone make the noise during the spawning-time; and that it is possible by imitating it, to take them without bait."[571] If this statement is trustworthy, we have an instance in this, the lowest class of the Vertebrata, of what we shall find prevailing throughout the other vertebrate classes, and which prevails, as we have already seen, with insects and spiders; namely, that vocal and instrumental sounds so commonly serve as a love-call or as a love-charm, that the power of producing them was probably first developed in connection with the propagation of the species.

Amphibians

Urodela.—First for the tailed amphibians. The sexes of salamanders or newts often differ much both in colour and structure. In some species prehensile claws are developed on the fore-legs of the males during the breeding-season; and at this season in the male *Triton palmipes* the hind-feet are provided with a swimming web, which is almost completely absorbed during the winter; so that their feet then resemble those of the female.[572] This structure no doubt aids the male in his eager search and pursuit of the female. With our common newts *(Triton punctatus* and *cristatus)* a deep, much-indented crest is developed along the back and tail of the male during the breeding-season, being absorbed during the winter. It is not furnished, as Mr. St. George Mivart informs me, with muscles, and therefore cannot be

Fig. 31. *Triton cristatus (half natural size, from Bell's 'British Reptiles').*

Upper figure, male during the breeding-season; lower figure, female.

used for locomotion. As during the season of courtship it becomes edged with bright colours, it serves, there can hardly be a doubt, as a masculine ornament. In many species the body presents strongly contrasted, though lurid tints; and these become more vivid during the breeding-season. The male, for instance, of our common little newt *(Triton punctatus)* is "brownish-grey above, passing into yellow beneath, which in the spring becomes a rich bright orange, marked everywhere with round dark spots." The edge of the crest also is then tipped with bright red or violet. The female is usually of a yellowish-brown colour with scattered brown dots; and the lower surface is often quite plain.[573] The young are obscurely tinted. The ova are fertilised during the act of deposition and are not subsequently tended by either parent. We may therefore conclude that the males acquired their strongly-marked colours and ornamental appendages through sexual selection; these being transmitted either to the male offspring alone or to both sexes.

Anura or Batrachia.—With many frogs and toads the colours evidently serve as a protection, such as the bright green tints of tree-frogs and the obscure mottled shades of many terrestrial species. The most conspicuously coloured toad which I ever saw, namely the *Phryniscus nigricans,*[574] had the whole upper surface of the body as black as ink, with the soles of the feet and parts of the abdomen spotted with the brightest vermilion. It crawled about the bare sandy or open grassy plains of La Plata under a scorching sun, and could not fail to catch the eye of every passing creature. These colours may be beneficial by making this toad known to all birds of prey as a nauseous mouthful; for it is familiar to every one that these animals emit a poisonous secretion, which causes the mouth of a dog to froth, as if attacked by hydrophobia. I was the more struck with the conspicuous colours of this toad, as close by I found a lizard *(Proctotretus multimaculatus)* which, when frightened, flattened its body, closed its eyes, and then from its mottled tints could hardly be distinguishable from the surrounding sand.

With respect to sexual differences of colour, Dr. Günther knows of no striking instance with frogs or toads; yet he can often distinguish the male from the female, by the tints of the former being a little more intense. Nor does Dr. Günther know of any striking difference in external structure between the sexes, excepting the prominences which become developed during the breeding-season on the front-legs of the male, by which he is enabled to hold the female. The *Megalophrys montana*[575] (fig. 32) offers the best case of a certain amount of structural difference between the sexes; for in the male the tip of the nose and the eyelids are produced into triangular flaps of skin, and there is a little black tubercle on the back—characters which are absent, or only feebly developed, in the females. It is surprising that frogs

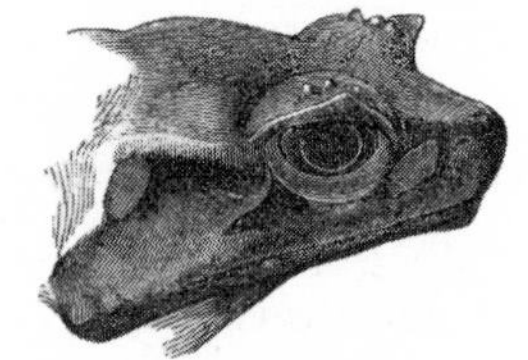
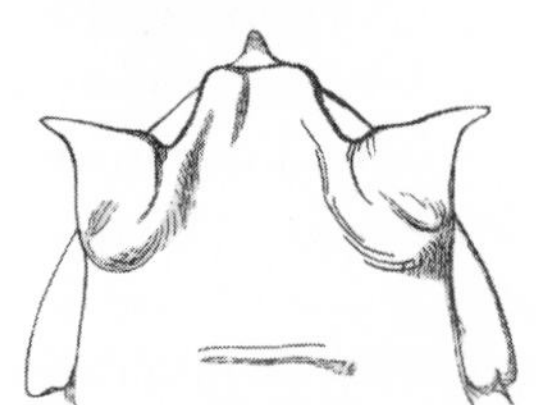
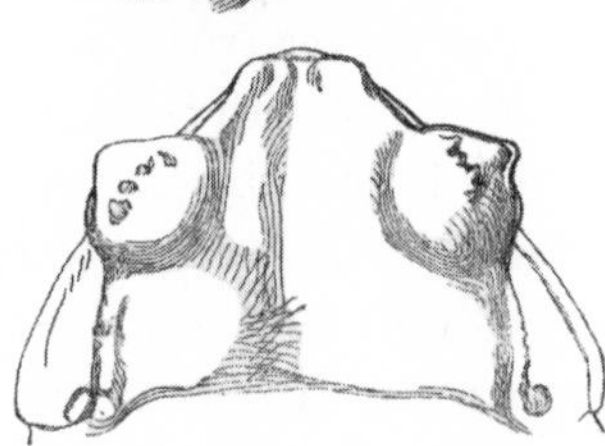

Fig. 32. *Megalophrys montana. The two left-hand figures, the male; the two right-hand figures, the female.*

and toads should not have acquired more strongly-marked sexual differences; for though cold-blooded, their passions are strong. Dr. Günther informs me that he has several times found an unfortunate female toad dead and smothered from having been so closely embraced by three or four males.

These animals, however, offer one interesting sexual difference, namely in the musical powers possessed by the males; but to speak of music, when applied to the discordant and overwhelming sounds emitted by male bull-frogs and some other species, seems, according to our taste, a singularly inappropriate expression. Nevertheless certain frogs sing in a decidedly pleasing manner. Near Rio de Janeiro I used often to sit in the evening to listen to a number of little Hylæ, which, perched on blades of grass close to the water, sent forth sweet chirping notes in harmony. The various sounds are emitted chiefly by the males during the breeding-season, as in the case of the croaking of our common frog.[576] In accordance with this fact the vocal organs of the males are more highly developed than those of the females. In some genera the males alone are provided with sacs which open into the larynx.[577] For instance, in the edible frog *(Rana esculenta)* "the sacs are peculiar to the males, and become, when filled with air in the act of croaking, large globular bladders, standing out one on each side of the head, near the corners of the mouth." The croak of the male is thus rendered exceedingly powerful; whilst that of the female is only a slight groaning noise.[578] The vocal organs differ considerably in structure in the several genera of the family; and their development in all cases may be attributed to sexual selection.

Reptiles

Chelonia.—Tortoises and turtles do not offer well-marked sexual differences. In some species, the tail of the male is longer than that of the female. In some, the plastron or lower surface of the shell of the male is slightly concave in relation to the back of the female. The male of the mud-turtle of the United States *(Chrysemys picta)* has claws on its front-feet twice as long as those of the female; and these are used when the sexes unite.[579] With the huge tortoise of the Galapagos Islands *(Testudo nigra)* the males are said to grow to a larger size than the females: during the pairing-season, and at no other time, the male utters a hoarse, bellowing noise, which can be heard at the distance of more than a hundred yards; the female, on the other hand, never uses her voice.[580]

Crocodilia.—The sexes apparently do not differ in colour; nor do I know that the males fight together, though this is probable, for some kinds make a prodigious display before the females. Bartram[581] describes the male alligator as striving to win the female by splashing and roaring in the midst of a lagoon, "swollen to an extent ready to burst, with his head and tail lifted up, he spins or twirls round on the surface of the water, like an Indian chief rehearsing his feats of war." During the season of love, a musky odour is emitted by the submaxillary glands of the crocodile, and pervades their haunts.[582]

Ophidia.—I have little to say about Snakes. Dr. Günther informs me that the males are always smaller than the females, and generally have longer and slenderer tails; but he knows of no other difference in external structure. In regard to colour, Dr. Günther can almost always distinguish the male from the female by his more strongly-pronounced tints; thus the black zigzag band on the back of the male English viper is more distinctly defined than in the female. The difference is much plainer in the Rattle-snakes of N. America, the male of which, as the keeper in the Zoological Gardens shewed me, can instantly be distinguished from the female by having more lurid yellow about its whole body. In S. Africa the *Bucephalus capensis* presents an analogous difference, for the female "is never so fully variegated with yellow on the sides, as the male."[583] The male of the Indian *Dipsas cynodon,* on the other hand, is blackish-brown, with the belly partly black, whilst the female is reddish or yellowish-olive with the belly either uniform yellowish or marbled with black. In the *Tragops dispar* of the same country, the male is bright green, and the female bronze-coloured.[584] No doubt the colours of some snakes serve as a protection, as the green tints of tree-snakes and the various mottled shades of the species which live in sandy places; but it is doubtful whether the colours of many kinds, for instance of the common English snake or viper, serve to conceal them; and this is still more doubtful with the many foreign species which are coloured with extreme elegance.

During the breeding-season their anal scent-glands are in active function;[585] and so it is with the same glands in lizards, and as we have seen with the sub-maxillary glands of crocodiles. As the males of most animals search for the females, these odoriferous glands probably serve to excite or charm the female, rather than to guide her to the spot where the male may be found.[586] Male snakes, though appearing so sluggish, are amorous; for many have been observed crowding round the same female, and even round the dead body of a female. They are not known to fight together from rivalry. Their intellectual powers are higher than might have been anticipated. An excellent observer in Ceylon, Mr. E. Layard,[587] saw a Cobra thrust its head through a narrow hole and swallow a toad. "With this incumbrance he could not withdraw himself; finding this, he reluctantly disgorged the precious morsel, which began to move off; this was too much for snake philosophy to bear, and the toad was again seized, and again was the snake, after violent efforts to escape, compelled to part with its prey. This time, however, a lesson had been learnt, and the toad was seized by one leg, withdrawn, and then swallowed in triumph."

It does not, however, follow because snakes have some reasoning power and strong passions, that they should likewise be endowed with sufficient taste to admire brilliant colours in their partners, so as to lead to the adornment of the species through sexual selection. Nevertheless it is difficult to account in any other manner for the extreme beauty of certain species; for instance, of the coral-snakes of S. America, which are of a rich red with black and yellow transverse bands. I well remember how much surprise I felt at the beauty of the first coral-snake which I saw gliding across a path in Brazil. Snakes coloured in this peculiar manner, as Mr. Wallace states on the authority of Dr. Günther,[588] are found nowhere else in the world except in S. America, and here no less than four genera occur. One of these, Elaps, is venomous; a second and widely-distinct genus is doubtfully venomous, and the two others are quite harmless. The species belonging to these distinct genera inhabit the same districts, and are so like each other, that no one "but a naturalist would distinguish the harmless from the poisonous kinds." Hence, as Mr. Wallace believes, the innocuous kinds have probably acquired their colours as a protection, on the principle of imitation; for they would naturally be thought dangerous by their enemies. The cause, however, of the bright colours of the venomous Elaps remains to be explained, and this may perhaps be sexual selection.

Lacertilia.—The males of some, probably of many kinds of lizards fight together from rivalry. Thus the arboreal *Anolis cristatellus* of S. America is extremely pugnacious: "During the spring and early part of the summer, two adult males rarely meet without a contest. On first seeing one another, they nod their heads up and down three or four times, at the same time expanding the frill or pouch beneath the throat; their eyes glisten with rage, and after waving their tails from side to side for a few seconds, as if to gather energy, they dart at each other furiously, rolling over and over, and holding firmly with their teeth. The conflict generally ends in one of the combatants losing his tail, which is often devoured by the victor." The male of this species is considerably larger than the female;[589] and this, as far as Dr. Günther has been able to ascertain, is the general rule with lizards of all kinds.

The sexes often differ greatly in various external characters. The male of the above-mentioned Anolis is furnished with a crest, which runs along the back and tail, and can be erected at pleasure; but of this crest the female does not exhibit a trace. In the Indian *Cophotis ceylanica,* the female possesses a dorsal crest, though much less developed than in the male; and so it is, as Dr. Günther informs me, with the females of many Iguanas, Chameleons and other lizards. In some species, however, the crest is equally developed in both sexes, as in the *Iguana tuberculata.* In the genus Sitana, the males alone are furnished with a large throat-pouch (fig. 33), which can be folded up like a fan, and is coloured blue, black, and red; but these splendid colours are exhibited only during the pairing-season. The female does not possess even a rudiment of this appendage. In the *Anolis cristatellus,* according to Mr. Austen, the throat-pouch, which is bright red marbled with yellow, is present, though in a rudimental condition, in the female. Again, in certain other lizards, both sexes are equally well provided with throat-pouches. Here, as in so many previous cases, we see with species belonging to the same group, the same character confined to the males, or more largely developed in the males than in the females, or equally developed in both sexes. The little lizards of the genus Draco, which glide through the air on their rib-supported parachutes, and which in the beauty of their colours baffle description, are furnished with skinny appendages to the throat, "like the wattles of gallinaceous birds." These

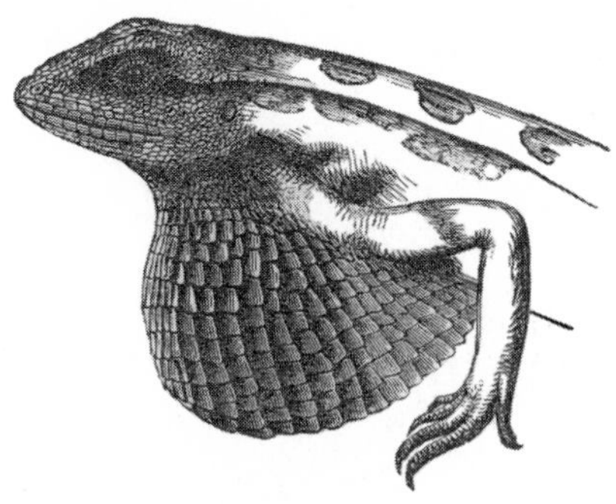

Fig. 33. *Sitana minor. Male, with the gular pouch expanded (from Günther's 'Reptiles of India').*

become erected when the animal is excited. They occur in both sexes, but are best developed in the male when arrived at maturity, at which age the middle appendage is sometimes twice as long as the head. Most of the species likewise have a low crest running along the neck; and this is much more developed in the full-grown males, than in the females or young males.[590]

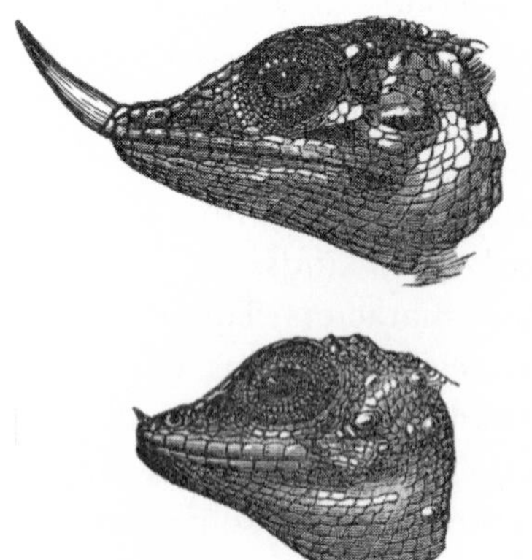

Fig. 34. *Ceratophora Stoddartii.*
Upper figure, male; lower figure, female.

There are other and much more remarkable differences between the sexes of certain lizards. The male of *Ceratophora aspera* bears on the extremity of his snout an appendage half as long as the head. It is cylindrical, covered with scales, flexible, and apparently capable of erection: in the female it is quite rudimental. In a second species of the same genus a terminal scale forms a minute horn on the summit of the flexible appendage; and in a third species *(C. Stoddartii,* fig. 34*)* the whole appendage is converted into a horn, which is usually of a white colour, but assumes a purplish tint when the animal is excited. In the adult male of this latter species the horn is half an inch in length, but is of quite minute size in the female and in the young. These appendages, as Dr. Günther has remarked to me, may be compared with the combs of gallinaceous birds, and apparently serve as ornaments.

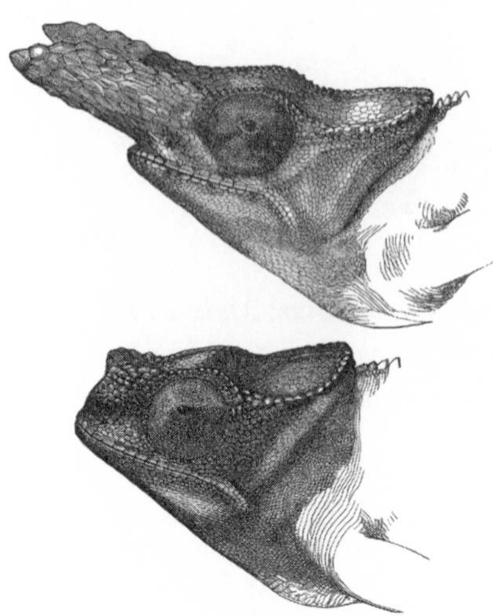

Fig. 35. *Chamæleon bifurcus.*
Upper figure, male; lower figure, female.

In the genus Chamæleon we come to the climax of difference between the sexes. The upper part of the skull of the male *C. bifurcus* (fig. 35), an inhabitant of Madagascar, is produced into two great, solid, bony projections, covered with scales like the rest of the head; and of this wonderful modification of structure the female exhibits only a rudiment. Again, in *Chamæleon Owenii* (fig. 36), from the West Coast of Africa, the male bears on his snout and forehead three curious horns, of which the female has not a trace. These horns consist of an excrescence of bone covered with a smooth sheath, forming part of the general integuments of the body, so that they are identical in structure with those of a bull, goat, or other sheath-horned ruminant. Although the three horns differ so much in appearance from the two great prolongations of the skull in *C. bifurcus,* we can hardly doubt that they serve the same general purpose in the economy of these two animals. The first conjecture which will occur to every one is that they are used by the males for fighting together; but Dr. Günther, to whom I am indebted for the foregoing details, does not believe that such peaceable creatures would ever become pugnacious. Hence we are driven to infer that these almost monstrous deviations of structure serve as masculine ornaments.

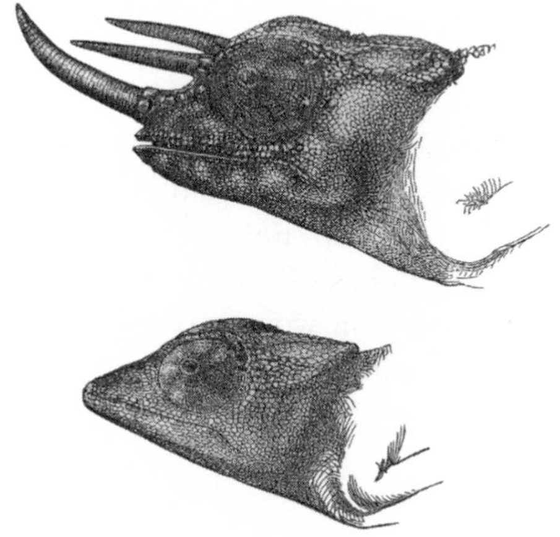

Fig. 36. *Chamæleon Owenii. Upper figure, male; lower figure, female.*

With many kinds of lizards, the sexes differ slightly in colour, the tints and stripes of the males being brighter and more distinctly defined than in the females. This, for instance, is the case with the previously-mentioned Cophotis and with the *Acanthodactylus capensis* of S. Africa. In a Cordylus of the latter country, the male is either much redder or greener than the female. In the Indian *Calotes nigrilabris* there is a greater difference in colour between the sexes; the lips also of the male are black, whilst those of the female are green. In our common little viviparous lizard *(Zootoca vivipara)* "the underside of the body and base of the tail in the male are bright orange, spotted with black; in the female these parts are pale greyish-green without spots."[591] We have seen that the males alone of Sitana possess a throat-pouch; and this is splendidly tinted with blue, black, and red. In the *Proctotretus tenuis* of Chile the male alone is marked with spots of blue, green, and coppery-red.[592] I collected in S. America fourteen species of this genus, and though I neglected to record the sexes, I observed that certain individuals alone were marked with emerald-like green spots, whilst others had orange-coloured gorges; and these in both cases no doubt were the males.

In the foregoing species, the males are more brightly coloured than the females, but with many lizards both sexes are coloured in the same elegant or even magnificent manner; and there is no reason to suppose that such conspicuous colours are protective. With some lizards, however, the green tints no doubt serve for concealment; and an instance has already been incidently given of one species of Proctotretus which closely resembles the sand on which it lives. On the whole we may conclude with tolerable safety that the beautiful colours of many lizards, as well as various appendages and other strange modifications of structure, have been gained by the males through sexual selection for the sake of ornament, and have been transmitted either to their male offspring alone or to both sexes. Sexual selection, indeed, seems to have played almost as important a part with reptiles as with birds. But the less conspicuous colours of the females in comparison with those of the males cannot be accounted for, as Mr. Wallace believes to be the case with birds, by the exposure of the females to danger during incubation.

CHAPTER XIII

Secondary Sexual Characters of Birds

SEXUAL DIFFERENCES • LAW OF BATTLE • SPECIAL WEAPONS • VOCAL ORGANS • INSTRUMENTAL MUSIC • LOVE-ANTICS AND DANCES • DECORATIONS, PERMANENT AND SEASONAL • DOUBLE AND SINGLE ANNUAL MOULTS • DISPLAY OF ORNAMENTS BY THE MALES

Secondary sexual characters are more diversified and conspicuous in birds, though not perhaps entailing more important changes of structure, than in any other class of animals. I shall, therefore, treat the subject at considerable length. Male birds sometimes, though rarely, possess special weapons for fighting with each other. They charm the females by vocal or instrumental music of the most varied kinds. They are ornamented by all sorts of combs, wattles, protuberances, horns, air-distended sacs, top-knots, naked shafts, plumes and lengthened feathers gracefully springing from all parts of the body. The beak and naked skin about the head, and the feathers are often gorgeously coloured. The males sometimes pay their court by dancing, or by fantastic antics performed either on the ground or in the air. In one instance, at least, the male emits a musky odour which we may suppose serves to charm or excite the female; for that excellent observer, Mr. Ramsay,[593] says of the Australian musk-duck *(Biziura lobata)* that "the smell which the male emits during the summer months is confined to that sex, and in some individuals is retained throughout the year; I have never even in the breeding-season, shot a female which had any smell of musk." So powerful is this odour during the pairing-season, that it can be detected long before the bird can be seen.[594] On the whole, birds appear to be the most æsthetic of all animals, excepting of course man, and they have nearly the same taste for the beautiful as we have. This is shewn by our enjoyment of the singing of birds, and by our women, both civilised and savage, decking their heads with borrowed plumes, and using gems which are hardly more brilliantly coloured than the naked skin and wattles of certain birds.

Before treating of the characters with which we are here more particularly concerned, I may just allude to certain differences between the sexes which apparently depend on differences in their habits of life; for such cases, though common in the lower, are rare in the higher classes. Two humming-birds belonging to the genus Eustephanus, which inhabit the island of Juan Fernandez, were long thought to be specifically distinct, but are now known, as Mr. Gould informs me, to be the sexes of the same species, and they differ slightly in the form of the beak. In another genus of humming-birds *(Grypus)*, the beak of the male is serrated along the margin and hooked at the extremity, thus differing much from that of the female. In the curious Neomorpha of New Zealand, there is a still wider difference in the form of the beak; and Mr. Gould has been informed that the male with his "straight and stout beak" tears off the bark of trees, in order that the female may feed on the uncovered larvæ with her weaker and more curved beak. Something of the same kind may be observed with our goldfinch *(Carduelis elegans)*, for I am assured by Mr. J. Jenner Weir that the bird-catchers can distinguish the males by their slightly longer beaks. The flocks of males, as an old and trustworthy bird-catcher asserted, are commonly found feeding on the seeds of the teazle (Dipsacus) which they can reach with their elongated beaks, whilst the females more commonly feed on the seeds of the betony or Scrophularia. With a slight difference of this nature as a foundation, we can see how the beaks of the two sexes might be made to differ greatly through natural selection. In all these cases, however, especially in that of the quarrelsome humming-birds, it is possible that the differences in the beaks may have been first acquired by the males in relation to their battles, and afterwards led to slightly changed habits of life.

Law of Battle.—Almost all male birds are extremely pugnacious, using their beaks, wings, and legs for fighting together. We see this every spring with our robins and sparrows. The smallest of all birds, namely the humming-bird, is one of the most quarrelsome. Mr. Gosse[595] describes a battle, in which a pair of humming-birds seized hold of each other's beaks, and whirled round and round, till they almost fell to the ground; and M. Montes de Oca, in speaking of another genus, says that two males rarely meet without a fierce aerial encounter: when kept in cages "their fighting has mostly ended in the splitting of the tongue of one of the two, which then surely dies from being unable to feed."[596] With Waders, the males of the common water-hen *(Gallinula chloropus)* "when pairing, fight violently for the females: they stand nearly upright in the water and strike with their feet." Two were seen to be thus engaged for half an hour, until one got hold of the head of the other which would have been killed, had not the observer interfered; the female all the time looking on as a quiet spectator.[597] The males of an allied bird *(Gallicrex cristatus)*, as Mr. Blyth informs me, are one third larger than the females, and are so pugnacious during the breeding-season, that they are kept by the natives of Eastern Bengal for the sake of fighting. Various other birds are kept in India for the same purpose, for instance the Bulbuls *(Pycnonotus hæmorrhous)* which fight with great spirit."[598]

The polygamous Ruff (*Machetes pugnax*, fig. 37) is notorious for his extreme pugnacity; and in the spring, the males, which are considerably larger than the females, congregate day after day at a particular spot, where the females propose to lay their eggs. The fowlers discover these spots by the turf being trampled somewhat bare. Here they fight very much like game-cocks, seizing each other with their beaks and striking with their wings. The great ruff of feathers round the neck is then erected, and according to Col. Montagu "sweeps the ground as a shield to defend the more tender parts;" and this is the only instance known to me in the case of birds, of any structure serving as a shield. The ruff of feathers, however, from its varied and rich colours probably serves in chief part as an ornament. Like most pugnacious birds, they seem always ready to fight, and when closely confined

Fig. 37. *The Ruff or Machetes pugnax (from Brehm's 'Thierleben').*

often kill each other; but Montagu observed that their pugnacity becomes greater during the spring, when the long feathers on their necks are fully developed; and at this period the least movement by any one bird provokes a general battle.[599] Of the pugnacity of web-footed birds, two instances will suffice: in Guiana "bloody fights occur during the breeding-season between the males of the wild musk-duck *(Cairina moschata);* and where these fights have occurred the river is covered for some distance with feathers."[600] Birds which seem ill-adapted for fighting engage in fierce conflicts; thus with the pelican the stronger males drive away the weaker ones, snapping with their huge beaks and giving heavy blows with their wings. Male snipes fight together, "tugging and pushing each other with their bills in the most curious manner imaginable." Some few species are believed never to fight; this is the case, according to Audubon, with one of the woodpeckers of the United States *(Picus auratus),* although "the hens are followed by even half a dozen of their gay suitors."[601]

The males of many birds are larger than the females, and this no doubt is an advantage to them in their battles with their rivals, and has been gained through sexual selection. The difference in size between the two sexes is carried to an extreme point in several Australian species; thus the male musk-duck (Biziura) and the male *Cincloramphus cruralis* (allied to our pipits) are by measurement actually twice as large as their respective females.[602] With many other birds the females are larger than the males; and as formerly remarked, the explanation often given, namely that the females have most of the work in feeding their young, will not suffice. In some few cases, as we shall hereafter see, the females apparently have acquired their greater size and strength for the sake of conquering other females and obtaining possession of the males.

The males of many gallinaceous birds, especially of the polygamous kinds, are furnished with special weapons for fighting with their rivals, namely spurs, which can be used with fearful effect. It has been recorded by a trustworthy writer[603] that in Derbyshire a kite struck at a game-hen accompanied by her chickens, when the cock rushed to the rescue and drove his spur right through the eye and skull of the aggressor. The spur was with difficulty drawn from the skull, and as the kite though dead retained his grasp, the two birds were firmly locked together; but the cock when disentangled was very little injured. The invincible courage of the game-cock is notorious: a gentleman who long ago witnessed the following brutal scene, told me that a bird had both its legs broken by some accident in the cock-pit, and the owner laid a wager that if the legs could be spliced so that the bird could stand upright, he would continue fighting. This was effected on the spot, and the bird fought with undaunted courage until he received his death-stroke. In Ceylon a closely-allied and wild species, the *Gallus Stanleyi,* is known to fight desperately "in defence of his seraglio," so that one of the combatants is frequently found dead.[604] An Indian partridge *(Ortygornis gularis)*, the male of which is furnished with strong and sharp spurs, is so quarrelsome, "that the scars of former fights disfigure the breast of almost every bird you kill."[605]

The males of almost all gallinaceous birds, even those which are not furnished with spurs, engage during the breeding-season in fierce conflicts. The Capercailzie and Black-cock *(Tetrao urogallus and T. tetrix)*, which are both polygamists, have regular appointed places, where during many weeks they congregate in numbers to fight together and to display their charms before the females. M. W. Kowalevsky informs me that in Russia he has seen the snow all bloody on the arenas where the Capercailzie have fought; and the Black-cocks "make the feathers fly in every direction," when several "engage in a battle royal." The elder Brehm gives a curious account of the Balz, as the love-dance and love-song of the Black-cock is called in Germany. The bird utters almost continuously the most strange noises: "he holds his tail up and spreads it out like a fan, he lifts up his head and neck with all the feathers erect, and stretches his wings from the body. Then he takes a few jumps in different directions, sometimes in a circle, and presses the under part of his beak so hard against the ground that the chin-feathers are rubbed off. During these movements he beats his wings and turns round and round. The more ardent he grows the more lively he becomes, until at last the bird appears like a frantic creature." At such times the black-cocks are so absorbed that they become almost blind and deaf, but less so than the capercailzie: hence bird after bird may be shot on the same spot, or even caught by the hand. After performing these antics the males begin to fight: and the same black-cock, in order to prove his strength over several antagonists, will visit in the course of one morning several Balz-places, which remain the same during successive years.[606]

The peacock with his long train appears more like a dandy than a warrior, but he sometimes engages in fierce contests: the Rev. W. Darwin Fox informs me that two peacocks became so excited whilst fighting at some little distance from Chester that they flew over the whole city, still fighting, until they alighted on the top of St. John's tower.

The spur, in those gallinaceous birds which are thus provided, is generally single; but Polyplectron (see fig. 51, p. 1071) has two or more on each leg; and one of the Blood-pheasants *(Ithaginis cruentus)* has been seen with five spurs. The spurs are generally confined to the male, being represented by mere knobs or rudiments in the female; but the females of the Java peacock *(Pavo muticus)* and, as I am informed by Mr. Blyth, of the small fire-backed pheasant *(Euplocamus erythropthalmus)* possess spurs. In Galloperdix it is usual for the males to have two spurs, and for the females to have only one on each leg.[607] Hence spurs may safely be considered as a masculine character, though occasionally transferred in a greater or less degree to the females. Like most other secondary sexual characters, the spurs are highly variable both in number and development in the same species.

Various birds have spurs on their wings. But the Egyptian goose *(Chenalopex ægyptiacus)* has only "bare obtuse knobs," and these probably shew us the first steps by which true spurs have been developed in other allied birds. In the spur-winged goose, *Plectropterus gambensis,* the males have much larger spurs than the females; and they use them, as I am informed by Mr. Bartlett, in fighting together, so that, in this case, the wing-spurs serve as sexual weapons; but according to Livingstone, they are chiefly used in the defence of the young. The Palamedea (fig. 38) is armed with a pair of spurs on each wing; and these are such formidable weapons that a single blow has driven a dog howling away. But it does not appear that the spurs in this case, or in that of some of the spur-winged rails, are larger in the male than in the female.[608] In certain plovers, however, the wing-spurs must be considered as a sexual character. Thus in the male of our common peewit *(Vanellus cristatus)* the tubercle on the shoulder of the wing becomes more prominent dur-

Fig. 38. *Palamedea cornuta (from Brehm), shewing the double-wing-spurs, and the filament on the head.*

ing the breeding-season, and the males are known to fight together. In some species of Lobivanellus a similar tubercle becomes developed during the breeding-season "into short horny spur." In the Australian *L. lobatus* both sexes have spurs, but these are much larger in the males than in the females. In an allied bird, the *Hoplopterus armatus,* the spurs do not increase in size during the breeding-season; but these birds have been seen in Egypt to fight together, in the same manner as our peewits, by turning suddenly in the air and striking sideways at each other, sometimes with a fatal result. Thus also they drive away other enemies.[609]

The season of love is that of battle; but the males of some birds, as of the game-fowl and ruff, and even the young males of the wild turkey and grouse,[610] are ready to fight whenever they meet. The presence of the female is the *teterrima belli causa.* The Bengali baboos make the pretty little males of the amadavat *(Estrelda amandava)* fight together by placing three small cages in a row, with a female in the middle; after a little time the two males are turned loose, and immediately a desperate battle ensues.[611] When many males congregate at the same appointed spot and fight together, as in the case of grouse and various other birds, they are generally attended by the females,[612] which afterwards pair with the victorious combatants. But in some cases the pairing precedes instead of succeeding the combat: thus, according to Audubon,[613] several males of the Virginian goat-sucker *(Caprimulgus Virginianus)* "court, in a highly entertaining manner, the female, and no sooner has she made her choice, than her approved gives chase to all intruders, and drives them beyond his dominions." Generally the males try with all their power to drive away or kill their rivals before they pair. It does not, however, appear that the females invariably prefer the victorious males. I have indeed been assured by M. W. Kowalevsky that the female capercailzie sometimes steals away with a young male who has not dared to enter the arena with the older cocks; in the same manner as occasionally happens with the does of the red-deer in Scotland. When two males contend in presence of a single female, the victor, no doubt, commonly gains his desire; but some of these battles are caused by wandering males trying to distract the peace of an already mated pair.[614]

Even with the most pugnacious species it is probable that the pairing does not depend exclusively on the mere strength and courage of the male: for such males are generally decorated with various ornaments which often become more brilliant during the breeding-season, and which are sedulously displayed before the females. The males also endeavour to charm or excite their mates by love-notes, songs, and antics; and the courtship is, in many instances, a prolonged affair. Hence it is not probable that the females are indifferent to the charms of the opposite sex, or that they are invariably compelled to yield to the victorious males. It is more probable that the females are excited, either before or after the conflict, by certain males, and thus unconsciously prefer them. In the case of *Tetrao umbellus,* a good observer[615] goes so far as to believe that the battles of the males "are all a sham, performed to show themselves to the greatest advantage before the admiring females who assemble around; for I have never been able to find a maimed hero, and seldom more than a broken feather." I shall have to recur to this subject, but I may here add that with the *Tetrao cupido* of the United States, about a score of males assemble at a particular spot, and strutting about make the whole air resound with their extraordinary noises. At the first answer from a female the males begin to fight furiously, and the weaker give way; but then, according to Audubon, both the victors and vanquished search for the female, so that the females must either then exert a choice, or the battle must be renewed. So, again, with one of the Field-starlings of the United States *(Sturnella ludoviciana)* the males engage in fierce conflicts, "but at the sight of a female they all fly after her, as if mad."[616]

Vocal and instrumental Music.—With birds the voice serves to express various emotions, such as distress, fear, anger, triumph, or mere happiness. It is apparently sometimes used to excite terror, as with the hissing noise made by some nestling-birds. Audubon[617] relates that a night-heron *(Ardea nycticorax,* Linn.) which he kept tame, used to hide itself when a cat approached, and then "suddenly start up uttering one of the most frightful cries, apparently enjoying the cat's alarm and flight." The common domestic cock clucks to the hen, and the hen to her chickens, when a dainty morsel is found. The hen, when she has laid an egg, "repeats the same note very often, and concludes with the sixth above, which she holds for a longer time;"[618] and thus she expresses her joy. Some social birds apparently call to each other for aid; and as they flit from tree to tree, the flock is kept together by chirp answering chirp. During the nocturnal migrations of geese and other water-fowl, sonorous clangs from the van may be heard in the darkness overhead, answered by clangs in the rear. Certain cries serve as danger-signals, which, as the sportsman knows to his cost, are well understood by the same species and by others. The domestic cock crows, and the humming-bird chirps, in triumph over a defeated rival. The true song, however, of most birds and various strange cries are chiefly uttered during the breeding-season, and serve as a charm, or merely as a call-note, to the other sex.

Naturalists are much divided with respect to the object of the singing of birds. Few more careful observers ever lived than Montagu, and he maintained that the "males of song-birds and of many others do not in general search for the female, but, on the contrary, their business in the spring is to perch on some conspicuous spot breathing out their full and amorous notes, which, by instinct, the female knows, and repairs to the spot to choose her mate."[619] Mr. Jenner Weir informs me that this is certainly the case with the nightingale. Bechstein, who kept birds during his whole life, asserts, "that the female canary always chooses the best singer, and that in a state of nature the female finch selects that male out of a hundred whose notes please her most."[620] There can be no doubt that birds closely attend to each other's song. Mr. Weir has told me of the case of a bullfinch which had been taught to pipe a German waltz, and who was so good a performer that he cost ten guineas; when this bird was first introduced into a room where other birds were kept and he began to sing, all the others, consisting of about twenty linnets and canaries, ranged themselves on the nearest side of their cages, and listened with the greatest interest to the new performer. Many naturalists believe that the singing of birds is almost exclusively "the effect of rivalry and emulation," and not for the sake of charming their mates. This was the opinion of Daines Barrington and White of Selborne, who both especially attended to this subject.[621] Barrington, however, admits that "superiority in song gives to birds an amazing ascendancy over others, as is well known to bird-catchers."

It is certain that there is an intense degree of rivalry between the males in their singing. Bird-fanciers match their birds to see which will sing longest; and I was told by Mr. Yarrell that a first-rate bird will sometimes sing till he drops down almost dead, or, according to Bechstein,[622] quite dead from rupturing a vessel in the lungs. Whatever the cause may be, male birds, as I hear from Mr. Weir, often die suddenly during the season of song. That the habit of singing is sometimes quite independent of love is clear, for a sterile hybrid canary-bird has been described[623] as singing whilst viewing itself in a mirror, and then dashing at its own image; it likewise attacked with fury a female canary when put into the same cage. The jealousy excited by the act of singing is constantly taken advantage of by bird-catchers; a male, in good song, is hidden and protected, whilst a stuffed bird, surrounded by limed twigs, is exposed to view. In this manner a man, as Mr. Weir informs me, has caught, in the course of a single day, fifty, and in one instance seventy, male chaffinches. The power and inclination to sing differ so greatly with birds that although the price of an ordinary male chaffinch is only sixpence, Mr. Weir saw one bird for which the bird-catcher asked three pounds; the test of a really good singer being that it will continue to sing whilst the cage is swung round the owner's head.

That birds should sing from emulation as well as for the sake of charming the female, is not at all incompatible; and, indeed, might have been expected to go together, like decoration and pugnacity. Some authors, however, argue that the song of the male cannot serve to charm the female, because the females of some few species, such as the canary, robin, lark, and bullfinch, especially, as Bechstein remarks, when in a state of widowhood, pour forth fairly melodious strains. In some of these cases the habit of singing may be in part attributed to the females having been highly fed and confined,[624] for this disturbs all the usual functions connected with the reproduction of the species. Many instances have already been given of the partial transference of secondary masculine characters to the female, so that it is not at all surprising that the females of some species should possess the power of song. It has also been argued, that the song of the male cannot serve as a charm, because the males of certain species, for instance, of the robin, sing during the autumn.[625] But nothing is more common than for animals to take pleasure in practising whatever instinct they follow at other times for some real good. How often do we see birds which fly easily, gliding and sailing through the air obviously for pleasure. The cat plays with the captured mouse, and the cormorant with the captured fish. The weaver-bird (Ploceus), when confined in a cage, amuses itself by neatly weaving blades of grass between the wires of its cage. Birds which habitually fight during the breeding-season are generally ready to fight at all times; and the males of the capercailzie sometimes hold their *balzens* or *leks* at the usual place of assemblage during the autumn.[626] Hence it is not at all surprising that male birds should continue singing for their own amusement after the season for courtship is over.

Singing is to a certain extent, as shewn in a previous chapter, an art, and is much improved by practice. Birds can be taught various tunes, and even the unmelodious sparrow has learnt to sing like a linnet. They acquire the song of their foster-parents,[627] and sometimes that of their neighbours.[628] All the common songsters belong to the Order of Insessores, and their vocal organs are much more complex than those of most other birds; yet it is a singular fact that some of the Insessores, such as ravens, crows, and magpies, possess the proper apparatus,[629] though they never sing, and do not naturally modulate their voices to any great extent. Hunter asserts[630] that with the true songsters the muscles of the larynx are stronger in the males than in the females; but with this slight exception there is no difference in the vocal organs of the two sexes, although the males of most species sing so much better and more continuously than the females.

It is remarkable that only small birds properly sing. The Australian genus Menura, however, must be excepted; for the *Menura Alberti*, which is about the size of a half-grown turkey, not only mocks other birds, but "its own whistle is exceedingly beautiful and varied." The males congregate and form "*corroborying* places," where they sing, raising and spreading their tails like peacocks and drooping their wings.[631] It is also remarkable that the birds which sing are rarely decorated with brilliant colours or other ornaments. Of our British birds, excepting the bullfinch and goldfinch, the best songsters are plain-coloured. The king-fisher, bee-eater, roller, hoopee, woodpeckers, &c., utter harsh cries; and the brilliant birds of the tropics are hardly ever songsters.[632] Hence bright colours and the power of song seem to replace each other. We can perceive that if the plumage did not vary in brightness, or if bright colours were dangerous to the species, other means would have to be employed to charm the females; and the voice being rendered melodious would offer one such means.

In some birds the vocal organs differ greatly in the two sexes. In the *Tetrao cupido* (fig. 39) the male has two bare, orange-coloured sacks, one on each side of the neck; and these are largely inflated when the male, during the breeding-season, makes a curious hollow sound, audible at a great distance. Audubon proved that the sound was intimately connected with this apparatus, which reminds us of the air-sacks on each side of the mouth of certain male frogs, for he found that the sound was much diminished when one of the sacks of a tame bird was pricked, and when both were pricked it was altogether stopped. The female has "a somewhat similar, though smaller, naked space of skin on the neck; but this is not capable of inflation."[633] The male of another kind of grouse *(Tetrao urophasianus),* whilst courting the female, has his "bare yellow œsophagus inflated to a prodigious size, fully half as large as the body;" and he then utters various grating, deep hollow tones. With his neck-feathers erect, his wings lowered and buzzing on the ground, and his long pointed tail spread out like a fan, he displays a variety of grotesque attitudes. The œsophagus of the female is not in any way remarkable.[634]

Fig 39. *Tetrao cupido: male (from Brehm).*

It seems now well made out that the great throat-pouch of the European male bustard *(Otis tarda),* and of at least four other species, does not serve, as was formerly supposed, to hold water, but is connected with the utterance during the breeding-season of a peculiar sound resembling "ock." The bird whilst uttering this sound throws himself into the most extraordinary attitudes. It is a singular fact that with the males of the same species the sack is not developed in all the individuals.[635] A crow-like bird inhabiting South America *(Cephalopterus ornatus,* fig. 40) is called the umbrella-bird, from its immense top-knot, formed of bare white quills surmounted by dark-blue plumes, which it can elevate into a great dome no less than five inches in diameter, covering the whole head. This bird has on its neck a long, thin, cylindrical, fleshy appendage, which is thickly clothed with scale-like blue feathers. It probably serves in part as an ornament, but likewise as a resounding apparatus, for Mr. Bates found that it is connected "with an unusual development of the trachea and vocal organs." It is dilated when the bird utters its singularly deep, loud, and long-sustained fluty note. The head-crest and neck-appendage are rudimentary in the female.[636]

The vocal organs of various web-footed and wading birds are extraordinarily complex, and differ to a certain extent in the two sexes. In some cases the trachea is convoluted, like a French horn, and is deeply embedded in the sternum. In the wild swan *(Cygnus ferus)* it is more deeply embedded

Fig. 40. *The Umbrella-Bird or Cephalopterus ornatus (male, from Brehm).*

in the adult male than in the female or young male. In the male Merganser the enlarged portion of the trachea is furnished with an additional pair of muscles.[637] But the meaning of these differences between the sexes of many Anatidæ is not at all understood; for the male is not always the more vociferous; thus with the common duck, the male hisses, whilst the female utters a loud quack.[638] In both sexes of one of the cranes *(Grus virgo)* the trachea penetrates the sternum, but presents "certain sexual modifications." In the male of the black stork there is also a well-marked sexual difference in the length and curvature of the bronchi.[639] So that highly important structures have in these cases been modified according to sex.

It is often difficult to conjecture whether the many strange cries and notes, uttered by male birds during the breeding-season, serve as a charm or merely as a call to the female. The soft cooing of the turtle-dove and of many pigeons, it may be presumed, pleases the female. When the female of the wild turkey utters her call in the morning, the male answers by a different note from the gobbling noise which he makes, when with erected feathers, rustling wings and distended wattles, he puffs and struts before her.[640] The *spel* of the black-cock certainly serves as a call to the female, for it has been known to bring four or five females from a distance to a male under confinement; but as the black-cock continues his *spel* for hours during successive days, and in the case of the capercailzie "with an agony of passion," we are led to suppose that the females which are already present are thus charmed.[641] The voice of the common rook is known to alter during the breeding-season, and is therefore in some way sexual.[642] But what shall we say about the harsh screams of, for instance, some kinds of macaws; have these birds as bad taste for musical sounds as they apparently have for colour, judging by the inharmonious contrast of their bright yellow and blue plumage? It is indeed possible that the loud voices of many male birds may be the result, without any advantage being thus gained, of the inherited effects of the continued use of their vocal organs, when they are excited by the strong passions of love, jealousy, and rage; but to this point we shall recur when we treat of quadrupeds.

We have as yet spoken only of the voice, but the males of various birds practise, during their courtship, what may be called instrumental music. Peacocks and Birds of Paradise rattle their quills together, and the vibratory movement apparently serves merely to make a noise, for it can hardly add to the beauty of their plumage. Turkey-cocks scrape their wings against the ground, and some kinds of grouse thus produce a buzzing sound. Another North American grouse, the *Tetrao umbellus,* when with his tail erect, his ruffs displayed, "he shows off his finery to the females, who lie hid in the neighbourhood," drums rapidly with his "lowered wings on the trunk of a fallen tree," or, according to Audubon, against his own body; the sound thus produced is compared by some to distant thunder, and by others to the quick roll of a drum. The female never drums, "but flies directly to the place where the male is thus engaged." In the Himalayas the male of the Kalij-pheasant "often makes a singular drumming noise with his wings, not unlike the sound produced by shaking a stiff piece of cloth." On the west coast of Africa the little black-weavers (Ploceus?) congregate in a small party on the bushes round a small open space, and sing and glide through the air with quivering wings, "which make a rapid whirring sound like a child's rattle." One bird after another thus performs for hours together, but only during the courting-season. At this same season the males of certain night-jars (Caprimulgus) make a most strange noise with their wings. The various species of woodpeckers strike a sonorous branch with their beaks, with so rapid a vibratory movement that "the head appears to be in two places at once." The sound thus produced is audible at a considerable distance, but cannot be described; and I feel sure that its cause would never be conjectured by any one who heard it for the first time. As this jarring sound is made chiefly during the breeding-season, it has been considered as a love-song; but it is perhaps more strictly a love-call. The female, when driven from her nest, has been observed thus to call her mate, who answered in the same manner and soon appeared. Lastly the male Hoopoe *(Upupa epops)* combines vocal and instrumental music; for during the breeding-season this bird, as Mr. Swinhoe saw, first draws in air and then taps the end of its beak perpendicularly down against a stone or the trunk of a tree, "when the breath being forced down the tubular bill produces the correct sound." When the male utters its cry without striking his beak the sound is quite different.[643]

In the foregoing cases sounds are made by the aid of structures already present and otherwise necessary; but in the following cases certain feathers have been specially modified for the express purpose of producing the sounds. The drumming, or bleating, or neighing, or thundering noise, as expressed by different observers, which is made by the common snipe *(Scolopax gallinago)* must have surprised every one who has ever heard it. This bird, during the pairing-season, flies to "perhaps a thousand feet in height," and after zig-zagging about for a time descends in a curved line, with outspread tail and quivering pinions, with surprising velocity to the earth. The sound is emitted only during this rapid descent. No one was able to explain the cause, until M. Meves observed that on each side of the tail the outer feathers are peculiarly formed (fig. 41), having a stiff sabre-shaped shaft, with the oblique barbs of unusual length, the outer webs being strongly bound together. He found that by blowing on these feathers, or by fastening them to a long thin stick and waving them rapidly through the air, he could exactly reproduce the drumming noise made by the living bird. Both sexes are furnished with these feathers, but they are generally larger in the male than in the female, and emit a deeper note. In some species, as in *S. frenata* (fig. 42), four feathers, and in *S. javensis* (fig. 43), no less than eight on each side of the tail are greatly modified. Different tones are emitted by the feathers of the different species when waved through the air; and the *Scolopax Wilsonii* of the United States makes a switching noise whilst descending rapidly to the earth.[644]

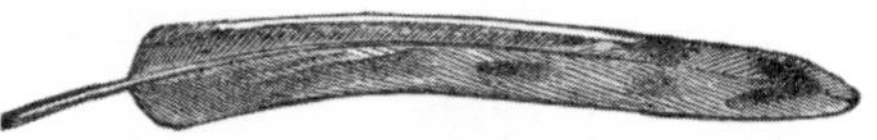

Fig. 41. *Outer tail-feather of Scolopax gallinago (from Proc. Zool. Soc. 1858).*

Fig. 42. *Outer tail-feather of Scolopax frenata.*

Fig. 43. *Outer tail-feather of Scolopax javensis.*

Fig. 44. *Primary wing-feather of a Humming-bird, the Selasphorus platycerus (from a sketch by Mr. Salvin).*

Upper figure, that of male; lower figure, corresponding feather of female.

In the male of the *Chamæpetes unicolor* (a large gallinaceous bird of America) the first primary wing-feather is arched towards the tip and is much more attenuated than in the female. In an allied bird, the *Penelope nigra,* Mr. Salvin observed a male, which, whilst it flew downwards "with outstretched wings, gave forth a kind of crashing, rushing noise," like the falling of a tree.[645] The male alone of one of the Indian bustards *(Sypheotides auritus)* has its primary wing-feathers greatly acuminated; and the male of an allied species is known to make a humming noise whilst courting the female.[646] In a widely different group of birds, namely the Humming-birds, the males alone of certain kinds have either the shafts of their primary wing-feathers broadly dilated, or the webs abruptly excised towards the extremity. The male, for instance, of *Selasphorus platycercus,* when adult, has the first primary wing-feather (fig. 44), excised in this manner. Whilst flying from flower to flower he makes "a shrill, almost whistling, noise;"[647] but it did not appear to Mr. Salvin that the noise was intentionally made.

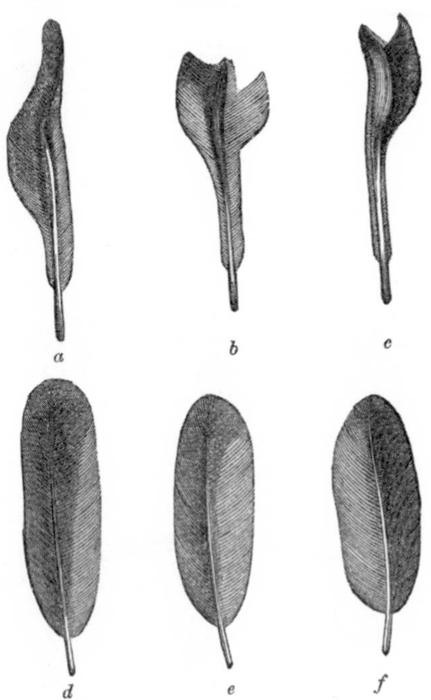

Fig. 45. *Secondary wing-feathers of Pipra deliciosa (from Mr. Sclater, in Proc. Zool. Soc. 1860).*

The three upper feathers, a, b, c, *from the male; the three lower corresponding feathers,* d, e, f, *from the female.*

a. *and* d. *Fifth secondary wing-feather of male and female, upper surface.*

b *and* e. *Sixth secondary, upper surface.*

c *and* f. *Seventh secondary, lower surface.*

Lastly, in several species of a sub-genus of Pipra or Manakin, the males have their *secondary* wing-feathers modified, as described by Mr. Sclater, in a still more remarkable manner. In the brilliantly-coloured *P. deliciosa* the first three secondaries are thick-stemmed and curved towards the body; in the fourth and fifth (fig. 45, *a)* the change is greater; and in the sixth and seventh *(b, c)* the shaft "is thickened to an extraordinary degree, forming a solid horny lump." The barbs also are greatly changed in shape, in comparison with the corresponding feathers *(d, e, f)* in the female. Even the bones of the wing which support these singular feathers in the male are said by Mr. Fraser to be much thickened. These little birds make an extraordinary noise, the first "sharp note being not unlike the crack of a whip."[648]

The diversity of the sounds, both vocal and instrumental, made by the males of many species during the breeding-season, and the diversity of the means for producing such sounds, are highly remarkable. We thus gain a high idea of their importance for sexual purposes, and are reminded of the same conclusion with respect to insects. It is not difficult to imagine the steps by which the notes of a bird, primarily used as a mere call or for some other purpose, might have been improved into a melodious love-song. This is somewhat more difficult in the case of the modified feathers, by which the drumming, whistling, or roaring noises are produced. But we have seen that some birds during their courtship flutter, shake, or rattle their unmodified feathers together; and if the females were led to select the best performers, the males which possessed the strongest or thickest, or most attenuated feathers, situated on any part of the body, would be the most successful; and thus by slow degrees the feathers might be modified to almost any extent. The females, of course, would not notice each slight successive alteration in shape, but only the sounds thus produced. It is a curious fact that in the same class of animals, sounds so different as the drumming of the snipe's tail, the tapping of the woodpecker's beak, the harsh trumpet-like cry of certain water-fowl, the cooing of the turtle-dove, and the song of the nightingale, should all be pleasing to the females of the several species. But we must not judge the tastes of distinct species by a uniform standard; nor must we judge by the standard of man's taste. Even with man, we should remember what discordant noises, the beating of tom-toms and the shrill notes of reeds, please the ears of savages. Sir S. Baker remarks,[649] that "as the stomach of the Arab prefers the raw meat and reeking liver taken hot from the animal, so does his ear prefer his equally coarse and discordant music to all other."

Love-Antics and Dances.—The curious love-gestures of various birds, especially of the Gallinaceæ, have already been incidentally noticed; so that little need here be added. In Northern America, large numbers of a grouse, the *Tetrao phasianellus*, meet every morning during the breeding-season on a selected level spot, and here they run round and round in a circle of about fifteen or twenty feet in diameter, so that the ground is worn quite bare, like a fairy-ring. In these Partridge-dances, as they are called by the hunters, the birds assume the strangest attitudes, and run round, some to the left and some to the right. Audubon describes the males of a heron *(Ardea herodias)* as walking about on their long legs with great dignity before the females, bidding defiance to their rivals. With one of the disgusting carrion-vultures *(Cathartes jota)* the same naturalist states that "the gesticulations and parade of the males at the beginning of the love-season are extremely ludicrous." Certain birds perform their love-antics on the wing, as we have seen with the black African weaver, instead of on the ground. During the spring our little white-throat *(Sylvia cinerea)* often rises a few feet or yards in the air above some bush, and "flutters with a fitful and fantastic motion, singing all the while, and then drops to its perch." The great English bustard throws himself into indescribably odd attitudes whilst courting the female, as has been figured by Wolf. An allied Indian bustard *(Otis bengalensis)* at such times "rises perpendicularly into the air with a hurried flapping of his wings, raising his crest and puffing out the feathers of his neck and breast, and then drops to the ground;" he repeats this manœuvre several times successively, at the same time humming in a peculiar tone. Such females as happen to be near "obey this saltatory summons," and when they approach he trails his wings and spreads his tail like a turkey-cock.[650]

But the most curious case is afforded by three allied genera of Australian birds, the famous Bower-birds,—no doubt the co-descendants of some ancient species which first acquired the strange instinct of constructing bowers for performing their love-antics. The bowers (fig. 46), which, as we shall hereafter see, are highly decorated with feathers, shells, bones and leaves, are built on the ground for the sole purpose of courtship, for their nests are formed in trees. Both sexes assist in the erection of the bowers, but the male is the principal workman. So strong is this instinct that it is practised under confinement, and Mr. Strange has described[651] the habits of some Satin Bower-birds, which he kept in his aviary in New South Wales. "At times the male will chase the female all over the aviary, then go to the bower, pick up a gay feather or a large leaf, utter a curious kind of note, set all his feathers erect, run round the bower and become so excited that his eyes appear ready to start from his head; he continues opening first one wing, and then the other, uttering a low, whistling note, and, like the domestic cock, seems to be picking up something from the ground, until at last the female goes gently towards him." Captain Stokes has described the habits and "play-houses" of another species, the Great Bower-bird, which was seen "amusing itself by flying backwards and forwards, taking a shell

Fig. 46. *Bower-bird, Chlamydera maculata, with bower (from Brehm).*

alternately from each side, and carrying it through the archway in its mouth." These curious structures, formed solely as halls of assemblages, where both sexes amuse themselves and pay their court, must cost the birds much labour. The bower, for instance, of the fawn-breasted species, is nearly four feet in length, eighteen inches in height, and is raised on a thick platform of sticks.

Decoration.—I will first discuss the cases in which the males are ornamented either exclusively or in a much higher degree than the females; and in a succeeding chapter those in which both sexes are equally ornamented, and finally the rare cases in which the female is somewhat more brightly-coloured than the male. As with the artificial ornaments used by savage and civilised men, so with the natural ornaments of birds, the head is the chief seat of decoration.[652] The ornaments, as mentioned at the commencement of this chapter, are wonderfully diversified. The plumes on the front or back of the head consist of variously-shaped feathers, sometimes capable of erection or expansion, by which their beautiful colours are fully displayed. Elegant ear-tufts (see fig. 39 ante) are occasionally present. The head is sometimes covered with velvety down like that of the pheasant; or is naked and vividly coloured; or supports fleshy appendages, filaments, and solid protuberances. The throat, also, is sometimes ornamented with a beard, or with wattles or caruncles. Such appendages are generally brightly coloured, and no doubt serve as ornaments, though not always ornamental in our eyes; for whilst the male is in the act of courting the female, they often swell and assume more vivid tints, as in the case of the male turkey. At such times the fleshy appendages about the head of the male Tragopan pheasant *(Ceriornis temminckii)* swell into a large lappet on the throat and into two horns, one on each side of the splendid top-knot; and these are then coloured of the most intense blue which I have ever beheld. The African hornbill *(Bucorax abyssinicus)* inflates the scarlet bladder-like wattle on its neck, and with its wings drooping and tail expanded "makes quite a grand appearance."[653] Even the iris of the eye is sometimes more brightly coloured in the male than in the female; and this is frequently the case with the beak, for instance, in our common blackbird. In *Buceros corrugatus,* the whole beak and immense casque are coloured more conspicuously in the male than in the female; and "the oblique grooves upon the sides of the lower mandible are peculiar to the male sex."[654]

The males are often ornamented with elongated feathers or plumes springing from almost every part of the body. The feathers on the throat and breast are sometimes developed into beautiful ruffs and collars. The tail-feathers are frequently increased in length; as we see in the tail-coverts of the peacock, and in the tail of the Argus pheasant. The body of this latter bird is not larger than that of a fowl; yet the length from the end of the beak to the extremity of the tail is no less than five feet three inches.[655] The wing-feathers are not elongated nearly so often as the tail-feathers; for their elongation would impede the act of flight. Yet the beautifully ocellated secondary wing-feathers of the male Argus pheasant are nearly three feet in length; and in a small African night-jar *(Cosmetornis vexillarius)* one of the primary wing-feathers, during the breeding-season, attains a length of twenty-six inches, whilst the bird itself is only ten inches in length. In another closely-allied genus of night-jars, the shafts of the elongated wing-feathers are naked, except at the extremity, where there is a disc.[656] Again, in another genus of night-jars, the tail-feathers are even still more prodigiously developed; so that we see the same kind of ornament gained by the males of closely-allied birds, through the development of widely different feathers.

It is a curious fact that the feathers of birds belonging to distinct groups have been modified in almost exactly the same peculiar manner. Thus the wing-feathers in one of the above-mentioned night-jars are bare along the shaft and terminate in a disc; or are, as they are sometimes called, spoon or racket-shaped. Feathers of this kind occur in the tail of a motmot *(Eumomota superciliaris),* of a king-fisher, finch, humming-bird, parrot, several Indian drongos *(Dicrurus* and *Edolius,* in one of which the disc stands vertically), and in the tail of certain Birds of Paradise. In these latter birds, similar feathers, beautifully ocellated, ornament the head, as is likewise the case with some gallinaceous birds. In an Indian bustard *(Sypheotides auritus)* the feathers forming the ear-tufts, which are about four inches in length, also terminate in discs.[657] The barbs of the feathers in various widely-distinct birds are filamentous or plumose, as with some Herons, Ibises, Birds of Paradise and Gallinaceæ. In other cases the barbs disappear, leaving the shafts bare; and these in the tail of the *Paradisea apoda* attain a length of thirty-four inches.[658] Smaller feathers when thus denuded appear like bristles, as on the breast of the turkey-cock. As any fleeting fashion in dress comes to be admired by man, so with birds a change of almost any kind in the structure or colouring of the feathers in the male appears to have been admired by the female. The fact of the feathers in widely distinct groups, having been modified in an analogous manner, no doubt depends primarily on all the feathers having nearly the same structure and manner of development, and consequently tending to vary in the same manner. We often see a tendency to analogous variability in the plumage of our domestic breeds belonging to distinct species. Thus top-knots have appeared in several species. In an extinct variety of the turkey, the top-knot consisted of bare quills surmounted with plumes of down, so that they resembled, to a certain extent, the racket-shaped feathers above described. In certain breeds of the pigeon and fowl the feathers are plumose, with some tendency in the shafts to be naked. In the Sebastopol goose the scapular feathers are greatly elongated, curled, or even spirally twisted, with the margins plumose.[659]

In regard to colour hardly anything need here be said; for every one knows how splendid are the tints of birds, and how harmoniously they are combined. The colours are often metallic and iridescent. Circular spots are sometimes surrounded by one or more differently shaded zones, and are thus converted into ocelli. Nor need much be said on the wonderful differences between the sexes, or of the extreme beauty of the males of many birds. The common peacock offers a striking instance. Female Birds of Paradise are obscurely coloured and destitute of all ornaments, whilst the males are probably the most highly decorated of all birds, and in so many ways, that they must be seen to be appreciated. The elongated and golden-orange plumes which spring from beneath the wings of the *Paradisea apoda* (see fig. 47 of *P. rubra,* a much less beautiful species), when vertically erected and made to vibrate, are described as forming a sort of halo, in the centre of which the head "looks like a little emerald sun with its rays formed by the two plumes."[660] In another most beautiful species the head is bald, "and of a rich cobalt blue, crossed by several lines of black velvety feathers."[661]

Fig. 47. *Paradisea rubra, male (from Brehm).*

Fig. 48. *Lophornis ornatus, male and female (from Brehm).*

Male humming-birds (figs. 48 and 49) almost vie with Birds of Paradise in their beauty, as every one will admit who has seen Mr. Gould's splendid volumes or his rich collection. It is very remarkable in how many different ways these birds are ornamented. Almost every part of the plumage has been taken advantage of and modified; and the modifications have been carried, as Mr. Gould shewed me, to a wonderful extreme in some species belonging to nearly every sub-group. Such cases are curiously like those which we see in our fancy breeds, reared by man for the sake of ornament: certain individuals originally varied in one character, and other individuals belonging to the same species in other characters; and these have been seized on by man and augmented to an extreme point—as the tail of the fantail-pigeon, the hood of the jacobin, the beak and wattle of the carrier, and so forth. The sole difference between these cases is that in the one the result is due to man's selection, whilst in the other, as with Humming-birds, Birds of Paradise, &c., it is due to sexual selection,—that is to the selection by the females of the more beautiful males.

I will mention only one other bird, remarkable from the extreme contrast in colour between the sexes, namely the famous Bell-bird *(Chasmorhynchus niveus)* of S. America, the note of which can be distinguished at the distance of nearly three miles, and astonishes every one who first hears it.

Fig. 49. *Spathura underwoodi, male and female (from Brehm).*

The male is pure white, whilst the female is dusky-green; and the former colour with terrestrial species of moderate size and inoffensive habits is very rare. The male, also, as described by Waterton, has a spiral tube, nearly three inches in length, which rises from the base of the beak. It is jet-black, dotted over with minute downy feathers. This tube can be inflated with air, through a communication with the palate; and when not inflated hangs down on one side. The genus consists of four species, the males of which are very distinct, whilst the females, as described by Mr. Sclater in a most interesting paper, closely resemble each other, thus offering an excellent instance of the common rule that within the same group the males differ much more from each other than do the females. In a second species *(C. nudicollis)* the male is likewise snow-white, with the exception of a large space of naked skin on the throat and round the eyes, which during the breeding-season is of a fine green colour. In a third species *(C. tricarunculatus)* the head and neck alone of the male are white, the rest of the body being chestnut-brown, and the male of this species is provided with three filamentous projections half as long as the body—one rising from the base of the beak and the two others from the corners of the mouth.[662]

The coloured plumage and certain other ornaments of the males when adult are either retained for life or are periodically renewed during the summer and breeding-season. At this season the beak and naked skin about the head frequently change colour, as with some herons, ibises, gulls, one of the bell-birds just noticed, &c. In the white ibis, the cheeks, the inflatable skin of the throat, and the basal portion of the beak, then become crimson.[663] In one of the rails, *Gallicrex cristatus* a large red caruncle is developed during this same period on the head of the male. So it is with a thin horny crest on the beak of one of the pelicans, *P. erythrorhynchus;* for after the breeding-season, these horny crests are shed, like horns from the heads of stags, and the shore of an island in a lake in Nevada was found covered with these curious exuviæ.[664]

Changes of colour in the plumage according to the season depend firstly on a double annual moult, secondly on an actual change of colour in the feathers themselves, and thirdly on their dull-coloured margins being periodically shed, or on these three processes more or less combined. The shedding of the deciduary margins may be compared with the shedding by very young birds of their down; for the down in most cases arises from the summits of the first true feathers.[665]

With respect to the birds which annually undergo a double moult, there are, firstly, some kinds, for instance snipes, swallow-plovers (Glareolæ), and curlews, in which the two sexes resemble each other and do not change colour at any season. I do not know whether the winter-plumage is thicker and warmer than the summer-plumage, which seems, when there is no change of colour, the most probable cause of a double moult. Secondly, there are birds, for instance certain species of Totanus and other grallatores, the sexes of which resemble each other, but have a slightly different summer and winter plumage. The difference, however, in colour in these cases is so slight that it can hardly be an advantage to them; and it may, perhaps, be attributed to the direct action of the different conditions to which the birds are exposed during the two seasons. Thirdly, there are many other birds the sexes of which are alike, but which are widely different in their summer and winter plumage. Fourthly, there are birds, the sexes of which differ from each other in colour; but the females, though moulting twice, retain the same colours throughout the year, whilst the males undergo a change, sometimes, as with certain bustards, a great change of colour. Fifthly and lastly, there are birds the sexes of which differ from each other in both their summer and winter plumage, but the male undergoes a greater amount of change at each recurrent season than the female—of which the Ruff *(Machetes pugnax)* offers a good instance.

With respect to the cause or purpose of the differences in colour between the summer and winter plumage, this may in some instances, as with the ptarmigan,[666] serve during both seasons as a protection. When the difference between the two plumages is slight it may perhaps be attributed, as already remarked, to the direct action of the conditions of life. But with many birds there can hardly be a doubt that the summer plumage is ornamental, even when both sexes are alike. We may conclude that this is the case with many herons, egrets, &c., for they acquire their beautiful plumes only during the breeding-season. Moreover, such plumes, top-knots, &c., though possessed by both sexes, are occasionally a little more highly developed in the male than in the female; and they resemble the plumes and ornaments possessed by the males alone of other birds. It is also known that confinement, by affecting the reproductive system of male birds, frequently checks the development of their secondary sexual characters, but has no immediate influence on any other characters; and I am informed by Mr. Bartlett that eight or nine specimens of the Knot *(Tringa canutus)* retained their unadorned winter plumage in the Zoological Gardens throughout the year, from which fact we may infer that the summer plumage though common to both sexes partakes of the nature of the exclusively masculine plumage of many other birds.[667]

From the foregoing facts, more especially from neither sex of certain birds changing colour during either annual moult, or changing so slightly that the change can hardly be of any service to them, and from the females of other species moulting twice yet retaining the same colours throughout the year, we may conclude that the habit of moulting twice in the year has not been acquired in order that the male should assume during the breeding-season an ornamental character; but that the double moult, having been originally acquired for some distinct purpose, has subsequently been taken advantage of in certain cases for gaining a nuptial plumage.

It appears at first sight a surprising circumstance that with closely-allied birds, some species should regularly undergo a double annual moult, and others only a single one. The ptarmigan, for instance, moults twice or even thrice in the year, and the black-cock only once: some of the splendidly-coloured honey-suckers (Nectariniæ) of India and some sub-genera of obscurely-coloured pipits (Anthus) have a double, whilst others have only a single annual moult.[668] But the gradations in the manner of moulting, which are known to occur with various birds, shew us how species, or whole groups of species, might have originally acquired their double annual moult, or having once gained the habit, have again lost it. With certain bustards and plovers the vernal moult is far from complete, some feathers being renewed, and some changed in colour. There is also reason to believe that with certain bustards and rail-like birds, which properly undergo a double moult, some of the older males retain their nuptial plumage throughout the year. A few highly modified feathers may alone be added during the spring to the plumage, as occurs with the disc-formed tail-feathers of certain drongos *(Bhringa)* in India, and with the elongated feathers on the back, neck, and crest of certain herons. By such steps as these, the vernal moult might be rendered more and more complete, until a perfect double moult was acquired. A gradation can also be shewn to exist in the length of time during which either annual plumage is retained; so that the one might come to be retained for the whole year, the other being completely lost. Thus the *Machetes pugnax* retains his ruff in the spring for barely two months. The male widow-bird *(Chera progne)* acquires in Natal his fine plumage and long tail-feathers in December or January and loses them in March; so that they are retained during only about three months. Most species which undergo a double moult keep their ornamental feathers for about six months. The male, however, of the wild *Gallus bankiva* retains his neck-hackles for nine or ten months; and when these are cast off, the underlying black feathers on the neck are fully exposed to view. But with the domesticated descendant of this species, the neck-hackles of the male are immediately replaced by new ones; so that we here see, with respect to part of the plumage, a double moult changed under domestication into a single moult.[669]

The common drake *(Anas boschas)* is well known after the breeding-season to lose his male plumage for a period of three months, during which time he assumes that of the female. The male pintail-duck *(Anas acuta)* loses his plumage for the shorter period of six weeks or two months; and Montagu remarks that "this double moult within so short a time is a most extraordinary circumstance, that seems to bid defiance to all human reasoning." But he who believes in the gradual modification of species will be far from feeling surprise at finding gradations of all kinds. If the male pintail were to acquire his new plumage within a still shorter period, the new male feathers would almost necessarily be mingled with the old, and both with some proper to the female; and this apparently is the case with the male of a not distantly-allied bird, namely the *Merganser serrator,* for the males are said to "undergo a change of plumage, which assimilates them in some measure to the female." By a little further acceleration in the process, the double moult would be completely lost.[670]

Some male birds, as before stated, become more brightly coloured in the spring, not by a vernal moult, but either by an actual change of colour in the feathers, or by their obscurely-coloured deciduary margins being shed. Changes of colour thus caused may last for a longer or shorter time. With the *Pelecanus onocrotalus* a beautiful rosy tint, with lemon-coloured marks on the breast, overspreads the whole plumage in the spring; but these tints, as Mr. Sclater states, "do not last long, disappearing generally in about six weeks or two months after they have been attained." Certain finches shed the margins of their feathers in the spring, and then become brighter-coloured, while other finches undergo no such change. Thus the

Fringilla tristis of the United States (as well as many other American species), exhibits its bright colours only when the winter is past, whilst our goldfinch, which exactly represents this bird in habits, and our siskin, which represents it still more closely in structure, undergo no such annual change. But a difference of this kind in the plumage of allied species is not surprising, for with the common linnet, which belongs to the same family, the crimson forehead and breast are displayed only during the summer in England, whilst in Madeira these colours are retained throughout the year.[671]

Display by Male Birds of their Plumage.—Ornaments of all kinds, whether permanently or temporarily gained, are sedulously displayed by the males, and apparently serve to excite, or attract, or charm the females. But the males will sometimes display their ornaments, when not in the presence of the females, as occasionally occurs with grouse at their balz-places, and as may be noticed with the peacock; this latter bird, however, evidently wishes for a spectator of some kind, and will shew off his finery, as I have often seen, before poultry or even pigs.[672] All naturalists who have closely attended to the habits of birds, whether in a state of nature or under confinement, are unanimously of opinion that the males delight to display their beauty. Audubon frequently speaks of the male as endeavouring in various ways to charm the female. Mr. Gould, after describing some peculiarities in a male humming-bird, says he has no doubt that it has the power of displaying them to the greatest advantage before the female. Dr. Jerdon[673] insists that the beautiful plumage of the male serves "to fascinate and attract the female." Mr. Bartlett, at the Zoological Gardens, expressed himself to me in the strongest terms to the same effect.

It must be a grand sight in the forests of India "to come suddenly on twenty or thirty pea-fowl, the males displaying their gorgeous trains, and strutting about in all the pomp of pride before the gratified females." The wild turkey-cock erects his glittering plumage, expands his finely-zoned tail and barred wing-feathers, and altogether, with his gorged crimson and blue wattles, makes a superb, though, to our eyes, grotesque appearance. Similar facts have already been given with respect to grouse of various kinds. Turning to another Order. The male *Rupicola crocea* (fig. 50) is one of the most beautiful birds in the world, being of a splendid orange, with some of the feathers curiously truncated and plumose. The female is brownish-green, shaded with red, and has a much smaller crest. Sir R. Schomburgk has described their courtship; he found one of their meeting-places where ten males and two females were present. The space was from four to five feet in diameter, and appeared to have been

Fig. 50. *Rupicola crocea, male (from Brehm).*

cleared of every blade of grass and smoothed as if by human hands. A male "was capering to the apparent delight of several others. Now spreading its wings, throwing up its head, or opening its tail like a fan; now strutting about with a hopping gait until tired, when it gabbled some kind of note, and was relieved by another. Thus three of them successively took the field, and then, with self-approbation, withdrew to rest." The Indians, in order to obtain their skins, wait at one of the meeting-places till the birds are eagerly engaged in dancing, and then are able to kill, with their poisoned arrows, four or five males, one after the other.[674] With Birds of Paradise a dozen or more full-plumaged males congregate in a tree to hold a dancing-party, as it is called by the natives; and here flying about, raising their wings, elevating their exquisite plumes, and making them vibrate, the whole tree seems, as Mr. Wallace remarks, to be filled with waving plumes. When thus engaged, they become so absorbed that a skilful archer may shoot nearly the whole party. These birds, when kept in confinement in the Malay Archipelago, are said to take much care in keeping their feathers clean; often spreading them out, examining them, and removing every speck of dirt. One observer, who kept several pairs alive, did not doubt that the display of the male was intended to please the female.[675]

The gold-pheasant *(Thaumalea picta)* during his courtship not only expands and raises his splendid frill, but turns it, as I have myself seen, obliquely towards the female on whichever side she may be standing, obviously in order that a large surface may be displayed before her.[676] Mr. Bartlett has observed a male Polyplectron (fig. 51) in the act of courtship, and has shewn me a specimen stuffed in the attitude then assumed. The tail and wing-feathers of this bird are ornamented with beautiful ocelli, like those on the peacock's train. Now when the peacock displays himself, he expands and erects his tail transversely to his body, for he stands in front of the female, and has to shew off, at the same time, his rich blue throat and breast. But the breast of the Polyplectron is obscurely coloured, and the ocelli are not confined to the tail-feathers. Consequently the Polyplectron does not stand in front of the female; but he erects and expands his tail-feathers a little obliquely, lowering the expanded wing on the same side, and raising that on the opposite side. In this attitude the ocelli over the whole body are exposed before the eyes of the admiring female in one grand bespangled expanse. To whichever side she may turn, the expanded wings and the obliquely-held tail are turned towards her. The male Tragopan pheasant acts in nearly the same manner, for he raises the feathers of the body, though not the wing itself, on the side which is opposite to the female, and which would otherwise be concealed, so that nearly all the beautifully-spotted feathers are exhibited at the same time.

The case of the Argus pheasant is still more striking. The immensely developed secondary wing-feathers, which are confined to the male, are ornamented with a row of from twenty to twenty-three ocelli, each above an

Fig. 51. *Polyplectron chinquis, male (from Brehm).*

inch in diameter. The feathers are also elegantly marked with oblique dark stripes and rows of spots, like those on the skin of a tiger and leopard combined. The ocelli are so beautifully shaded that, as the Duke of Argyll remarks,[677] they stand out like a ball lying loosely within a socket. But when I looked at the specimen in the British Museum, which is mounted with the wings expanded and trailing downwards, I was greatly disappointed, for the ocelli appeared flat or even concave. Mr. Gould, however, soon made the case clear to me, for he had made a drawing of a male whilst he was displaying himself. At such times the long secondary feathers in both wings are vertically erected and expanded; and these, together with the enormously elongated tail-feathers, make a grand semicircular upright fan. Now as soon as the wing-feathers are held in this position, and the light shines on them from above, the full effect of the shading comes out, and each ocellus at once resembles the ornament called a ball and socket. These feathers have been shewn to several artists, and all have expressed their admiration at the perfect shading. It may well be asked, could such artistically-shaded ornaments have been formed by means of sexual selection? But it will be convenient to defer giving an answer to this question until we treat in the next chapter of the principle of gradation.

The primary wing-feathers, which in most gallinaceous birds are uniformly coloured, are in the Argus pheasant not less wonderful objects than the secondary wing-feathers. They are of a soft brown tint with numerous dark spots, each of which consists of two or three black dots with a surrounding dark zone. But the chief ornament is a space parallel to the dark-blue shaft, which in outline forms a perfect second feather lying within the true feather. This inner part is coloured of a lighter chesnut, and is thickly dotted with minute white points. I have shewn this feather to several persons, and many have admired it even more than the ball-and-socket feathers, and have declared that it was more like a work of art than of nature. Now these feathers are quite hidden on all ordinary occasions, but are fully displayed when the long secondary feathers are erected, though in a widely different manner; for they are expanded in front like two little fans or shields, one on each side of the breast near the ground.

The case of the male Argus pheasant is eminently interesting, because it affords good evidence that the most refined beauty may serve as a charm for the female, and for no other purpose. We must conclude that this is the case, as the primary wing-feathers are never displayed, and the ball-and-socket ornaments are not exhibited in full perfection, except when the male assumes the attitude of courtship. The Argus pheasant does not possess brilliant colours, so that his success in courtship appears to have depended on the great size of his plumes, and on the elaboration of the most elegant patterns. Many will declare that it is utterly incredible that a female bird should be able to appreciate fine shading and exquisite patterns. It is undoubtedly a marvellous fact that she should possess this almost human degree of taste, though perhaps she admires the general effect rather than each separate detail. He who thinks that he can safely gauge the discrimination and taste of the lower animals, may deny that the female Argus pheasant can appreciate such refined beauty; but he will then be compelled to admit that the extraordinary attitudes assumed by the male during the act of courtship, by which the wonderful beauty of his plumage is fully displayed, are purposeless; and this is a conclusion which I for one will never admit.

Although so many pheasants and allied gallinaceous birds carefully display their beautiful plumage before the females, it is remarkable, as Mr. Bartlett informs me, that this is not the case with the dull-coloured Eared and Cheer pheasants *(Crossoptilon auritum* and *Phasianus Wallichii);* so that these birds seem conscious that they have little beauty to display. Mr. Bartlett has never seen the males of either of these species fighting together, though he has not had such good opportunities for observing the Cheer as the Eared pheasant. Mr. Jenner Weir, also, finds that all male birds with rich or strongly-characterised plumage are more quarrelsome than the dull-coloured species belonging to the same groups. The goldfinch, for instance, is far more pugnacious than the linnet, and the blackbird than the thrush. Those birds which undergo a seasonal change of plumage likewise become much more pugnacious at the period when they are most gaily ornamented. No doubt the males of some obscurely-coloured birds fight desperately together, but it appears that when sexual selection has been highly influential, and has given bright colours to the males of any species, it has also very often given a strong tendency to pugnacity. We shall meet with nearly analogous cases when we treat of mammals. On the other hand, with birds the power of song and brilliant colours have rarely been both acquired by the males of the same species; but in this case, the advantage gained would have been identically the same, namely success in charming the female. Nevertheless it must be owned that the males of several brilliantly-coloured birds have had their feathers specially modified for the sake of producing instrumental music, though the beauty of this cannot be compared, at least according to our taste, with that of the vocal music of many songsters.

We will now turn to male birds which are not ornamented in any very high degree, but which nevertheless display, during their courtship, whatever attractions they may possess. These cases are in some respects more curious than the foregoing, and have been but little noticed. I owe the following facts, selected from a large body of valuable notes, sent to me by Mr. Jenner Weir, who has long kept birds of many kinds, including all the British

Fringillidæ and Emberizidæ. The bullfinch makes his advances in front of the female, and then puffs out his breast, so that many more of the crimson feathers are seen at once than otherwise would be the case. At the same time he twists and bows his black tail from side to side in a ludicrous manner. The male chaffinch also stands in front of the female, thus shewing his red breast, and "blue bell," as the fanciers call his head; the wings at the same time being slightly expanded, with the pure white bands on the shoulders thus rendered conspicuous. The common linnet distends his rosy breast, slightly expands his brown wings and tail, so as to make the best of them by exhibiting their white edgings. We must, however, be cautious in concluding that the wings are spread out solely for display, as some birds act thus whose wings are not beautiful. This is the case with the domestic cock, but it is always the wing on the side opposite to the female which is expanded, and at the same time scraped on the ground. The male goldfinch behaves differently from all other finches: his wings are beautiful, the shoulders being black, with the dark-tipped wing-feathers spotted with white and edged with golden yellow. When he courts the female, he sways his body from side to side, and quickly turns his slightly expanded wings first to one side then to the other, with a golden flashing effect. No other British finch, as Mr. Weir informs me, turns during his courtship from side to side in this manner; not even the closely-allied male siskin, for he would not thus add to his beauty.

Most of the British Buntings are plain-coloured birds; but in the spring the feathers on the head of the male reed-bunting *(Emberiza schœniculus)* acquire a fine black colour by the abrasion of the dusky tips; and these are erected during the act of courtship. Mr. Weir has kept two species of Amadina from Australia: the *A. castanotis* is a very small and chastely-coloured finch, with a dark tail, white rump, and jet-black upper tail-coverts, each of the latter being marked with three large conspicuous oval spots of white.[678] This species, when courting the female, slightly spreads out and vibrates these parti-coloured tail-coverts in a very peculiar manner. The male *Amadina Lathami* behaves very differently, exhibiting before the female his brilliantly-spotted breast and scarlet rump and scarlet upper tail-coverts. I may here add from Dr. Jerdon, that the Indian Bulbul *(Pycnonotus hæmorrhous)* has crimson *under* tail-coverts, and the beauty of these feathers, it might be thought, could never be well exhibited; but the bird "when excited often spreads them out laterally, so that they can be seen even from above."[679] The common pigeon has iridescent feathers on the breast, and every one must have seen how the male inflates his breast whilst courting the female, thus showing off these feathers to the best advantage. One of the beautiful bronze-winged pigeons of Australia *(Ocyphaps lophotes)* behaves, as described to me by Mr. Weir, very differently: the male, whilst standing before the female, lowers his head almost to the ground, spreads out and raises perpendicularly his tail, and half expands his wings. He then alternately and slowly raises and depresses his body, so that the iridescent metallic feathers are all seen at once, and glitter in the sun.

Sufficient facts have now been given to shew with what care male birds display their various charms, and this they do with the utmost skill. Whilst preening their feathers, they have frequent opportunities for admiring themselves and of studying how best to exhibit their beauty. But as all the males of the same species display themselves in exactly the same manner, it appears that actions, at first perhaps intentional, have become instinctive. If so, we ought not to accuse birds of conscious vanity; yet when we see a peacock strutting about, with expanded and quivering tail-feathers, he seems the very emblem of pride and vanity.

The various ornaments possessed by the males are certainly of the highest importance to them, for they have been acquired in some cases at the expense of greatly impeded powers of flight or of running. The African night-jar *(Cosmetornis),* which during the pairing-season has one of its primary wing-feathers developed into a streamer of extreme length, is thus much retarded in its flight, although at other times remarkable for its swiftness. The "unwieldy size" of the secondary wing-feathers of the male Argus pheasant are said "almost entirely to deprive the bird of flight." The fine plumes of male Birds of Paradise trouble them during a high wind. The extremely long tail-feathers of the male widow-birds (Vidua) of Southern Africa render "their flight heavy;" but as soon as these are cast off they fly as well as the females. As birds always breed when food is abundant, the males probably do not suffer much inconvenience in searching for food from their impeded powers of movement; but there can hardly be a doubt that they must be much more liable to be struck down by birds of prey. Nor can we doubt that the long train of the peacock and the long tail and wing-feathers of the Argus pheasant must render them a more easy prey to any prowling tiger-cat than would otherwise be the case. Even the bright colours of many male birds cannot fail to make them conspicuous to their enemies of all kinds. Hence it probably is, as Mr. Gould has remarked, that such birds are generally of a shy disposition, as if conscious that their beauty was a source of danger, and are much more difficult to discover or approach, than the sombre-coloured and comparatively tame females, or than the young and as yet unadorned males.[680]

It is a more curious fact that the males of some birds which are provided with special weapons for battle, and which in a state of nature are so pugnacious that they often kill each other, suffer from possessing certain ornaments. Cock-fighters trim the hackles and cut off the comb and gills of their cocks; and the birds are then said to be dubbed. An undubbed bird, as Mr. Tegetmeier insists, "is at a fearful disadvantage: the comb and gills offer an easy hold to his adversary's beak, and as a cock always strikes where he holds, when once he has seized his foe, he has him entirely in his power. Even supposing that the bird is not killed, the loss of blood suffered by an undubbed cock is much greater than that sustained by one that has been trimmed."[681] Young turkey-cocks in fighting always seize hold of each other's wattles; and I presume that the old birds fight in the same manner. It may perhaps be objected that the comb and wattles are not ornamental, and cannot be of service to the birds in this way; but even to our eyes, the beauty of the glossy black Spanish cock is much enhanced by his white face and crimson comb; and no one who has ever seen the splendid blue wattles of the male Tragopan pheasant, when distended during the act of courtship, can for a moment doubt that beauty is the object gained. From the foregoing facts we clearly see that the plumes and other ornaments of the male must be of the highest importance to him; and we further see that beauty in some cases is even more important than success in battle.

CHAPTER XIV

BIRDS—CONTINUED

CHOICE EXERTED BY THE FEMALE • LENGTH OF COURTSHIP • UNPAIRED BIRDS • MENTAL QUALITIES AND TASTE FOR THE BEAUTIFUL • PREFERENCE OR ANTIPATHY SHEWN BY THE FEMALE FOR PARTICULAR MALES • VARIABILITY OF BIRDS • VARIATIONS SOMETIMES ABRUPT • LAWS OF VARIATION • FORMATION OF OCELLI • GRADATIONS OF CHARACTER • CASE OF PEACOCK, ARGUS PHEASANT, AND UROSTICTE

WHEN THE SEXES DIFFER in beauty, in the power of singing, or in producing what I have called instrumental music, it is almost invariably the male which excels the female. These qualities, as we have just seen, are evidently of high importance to the male. When they are gained for only a part of the year, this is always shortly before the breeding-season. It is the male alone who elaborately displays his varied attractions, and often performs strange antics on the ground or in the air, in the presence of the female. Each male drives away or, if he can, kills all his rivals. Hence we may conclude, that it is the object of the male to induce the female to pair with him, and for this purpose he tries to excite or charm her in various ways; and this is the opinion of all those who have carefully studied the habits of living birds. But there remains a question which has an all important bearing on sexual selection, namely, does every male of the same species equally excite and attract the female? or does she exert a choice, and prefer certain males? This question can be answered in the affirmative by much direct and indirect evidence. It is much more difficult to decide what qualities determine the choice of the females; but here again we have some direct and indirect evidence that it is to a large extent the external attractions of the male, though no doubt his vigour, courage, and other mental qualities come into play. We will begin with the indirect evidence.

Length of Courtship.—The lengthened period during which both sexes of certain birds meet day after day at an appointed place, probably depends partly on the courtship being a prolonged affair, and partly on the reiteration of the act of pairing. Thus in Germany and Scandinavia the balzens or leks of the Black-cocks, last from the middle of March, all through April into May. As many as forty or fifty, or even more birds congregate at the leks; and the same place is often frequented during successive years. The lek of the Capercailzie lasts from the end of March to the middle or even end of May. In North America "the partridge dances" of the *Tetrao phasianellus* "last for a month or more." Other kinds of grouse both in North America and Eastern Siberia[682] follow nearly the same habits. The fowlers discover the hillocks where the Ruffs congregate by the grass being trampled bare, and this shews that the same spot is long frequented. The Indians of Guiana are well acquainted with the cleared arenas, where they expect to find the beautiful Cocks of the Rock; and the natives of New Guinea know the trees where from ten to twenty full-plumaged male Birds of Paradise congregate. In this latter case it is not expressly stated that the females meet on the same trees, but the hunters, if not specially asked, would not probably mention their presence, as their skins are valueless. Small parties of an African weaver *(Ploceus)* congregate, during the breeding-season, and perform for hours their graceful evolutions. Large numbers of the Solitary snipe *(Scolopax major)* assemble during the dusk in a morass; and the same place is frequented for the same purpose during successive years; here they may be seen running about "like so many large rats," puffing out their feathers, flapping their wings, and uttering the strangest cries.[683]

Some of the above-mentioned birds, namely, the black-cock, capercailzie, pheasant-grouse, the ruff, the Solitary snipe, and perhaps some others, are, as it is believed, polygamists. With such birds it might have been thought that the stronger males would simply have driven away the weaker, and then at once have taken possession of as many females as possible; but if it be indispensable for the male to excite or please the female, we can understand the length of the courtship and the congregation of so many individuals of both sexes at the same spot. Certain species which are strictly monogamous likewise hold nuptial assemblages; this seems to be the case in Scandinavia with one of the ptarmigans, and their leks last from the middle of March to the middle of May. In Australia the lyre-bird or *Menura superba* forms "small round hillocks," and the *M. Alberti* scratches for itself shallow holes, or, as they are called by the natives, *corroborying places,* where it is believed both sexes assemble. The meetings of the *M. superba* are sometimes very large; and an account has lately been published[684] by a traveller, who heard in a valley beneath him, thickly covered with scrub, "a din which completely astonished" him; on crawling onwards he beheld to his amazement about one hundred and fifty of the magnificent lyre-cocks, "ranged in order of battle, and fighting with indescribable fury." The bowers of the Bower-birds are the resort of both sexes during the breeding-season; and "here the males meet and contend with each other for the favours of the female, and here the latter assemble and coquet with the males." With two of the genera, the same bower is resorted to during many years.[685]

The common magpie *(Corvus pica,* Linn.), as I have been informed by the Rev. W. Darwin Fox, used to assemble from all parts of Delamere Forest, in order to celebrate the "great magpie marriage." Some years ago these birds abounded in extraordinary numbers, so that a gamekeeper killed in one morning nineteen males, and another killed by a single shot seven birds at roost together. Whilst they were so numerous, they had the habit very early in the spring of assembling at particular spots, where they could be seen in flocks, chattering, sometimes fighting, bustling and flying about the trees. The whole affair was evidently considered by the birds as of the highest importance. Shortly after the meeting they all separated, and were then observed by Mr. Fox and others to be paired for the season. In any district in which a species does not exist in large numbers, great assemblages cannot, of course, be held, and the same species may have different habits in different countries. For instance, I have never met with any account of regular assemblages of black game in Scotland, yet these assemblages are so well known in Germany and Scandinavia that they have special names.

Unpaired Birds.—From the facts now given, we may conclude that with birds belonging to widely-different groups their courtship is often a prolonged, delicate, and troublesome affair. There is even reason to suspect, improbable as this will at first appear, that some males and females of the same species, inhabiting the same district, do not always please each other and in consequence do not pair. Many accounts have been published of either the male or female of a pair having been shot, and quickly replaced by another. This has been observed more frequently with the magpie than with any other bird, owing perhaps to its conspicuous appearance and nest. The illustrious Jenner states that in Wiltshire one of a pair was daily shot no less than seven times successively, "but all to no purpose, for the remaining magpie soon found another mate;" and the last pair reared their young. A new partner is generally found on the succeeding day; but Mr. Thompson gives the case of one being replaced on the evening of the same day. Even after the eggs are hatched, if one of the old birds is destroyed a mate will often be found; this occurred after an interval of two days, in a case recently observed by one of Sir J. Lubbock's keepers.[686] The first and most obvious conjecture is that male magpies must be much more numerous than the females; and that in the above cases, as well in many others which could be given, the males alone had been killed. This apparently holds good in some instances, for the gamekeepers in Delamere Forest assured Mr. Fox that the magpies and carrion-crows which they formerly killed in succession in large numbers near their nests were all males; and they accounted for this fact by the males being easily killed whilst bringing food to the sitting females. Macgillivray, however, gives, on the authority of an excellent observer, an instance of three magpies successively killed on the same nest which were all females; and another case of six magpies successively killed whilst sitting on the same eggs, which renders it probable that most of them were females, though the male will sit on the eggs, as I hear from Mr. Fox, when the female is killed.

Sir J. Lubbock's gamekeeper has repeatedly shot, but how many times he could not say, one of a pair of jays *(Garrulus glandarius),* and has never failed shortly afterwards to find the survivor rematched. The Rev. W. D. Fox, Mr. F. Bond, and others, have shot one of a pair of carrion-crows *(Corvus corone),* but the nest was soon again tenanted by a pair. These birds are rather common; but the peregrine falcon *(Falco peregrinus)* is rare, yet Mr. Thompson states that in Ireland "if either an old male or female be killed in the breeding-season (not an uncommon circumstance), another mate is found within a very few days, so that the eyries, notwithstanding such casualties, are sure to turn out their complement of young." Mr. Jenner Weir has known the same thing to occur with the peregrine falcons at Beachy Head. The same observer informs me that three kestrels, all males *(Falco tinnunculus),* were killed one after the other whilst attending the same nest; two of these were in mature plumage, and the third in the plumage of the previous year. Even with the rare golden eagle *(Aquila chrysaëtos),* Mr. Birkbeck was assured by a trustworthy gamekeeper in Scotland, that if one is killed, another is soon found. So with the white owl *(Strix flammea),* it has been observed that "the survivor readily found a mate, and the mischief went on."

White of Selborne, who gives the case of the owl, adds that he knew a man, who from believing that partridges when paired were disturbed by the males fighting, used to shoot them; and though he had widowed the same female several times she was always soon provided with a fresh partner. This same naturalist ordered the sparrows, which deprived the house-martins of their nests, to be shot: but the one which was left, "be it cock or hen, presently procured a mate, and so for several times following." I could add analogous cases relating to the chaffinch, nightingale, and redstart. With respect to the latter bird *(Phœnicura ruticilla),* the writer remarks that it was by no means common in the neighbourhood, and he expresses much surprise how the sitting female could so soon give effectual notice that she was a widow. Mr. Jenner Weir has mentioned to me a nearly similar case: at Blackheath he never sees or hears the note of the wild bullfinch, yet when one of his caged males has died, a wild one in the course of a few days has generally come and perched near the widowed female, whose call-note is far from loud. I will give only one other fact, on the authority of this same observer; one of a pair of starlings *(Sturnus vulgaris)* was shot in the morning; by noon a new mate was found; this was again shot, but before night the pair was complete; so that the disconsolate widow or widower was thrice consoled during the same day. Mr. Engleheart also informs me that he used during several years to shoot one of a pair of starlings which built in a hole in a house at Blackheath; but the loss was always immediately repaired. During one season he kept an account and found that he had shot thirty-five birds from the same nest; these consisted of both males and females, but in what proportion he could not say: nevertheless after all this destruction, a brood was reared.[687]

These facts are certainly remarkable. How is it that so many birds are ready immediately to replace a lost mate? Magpies, jays, carrion-crows, partridges, and some other birds, are never seen during the spring by themselves, and these offer at first sight the most perplexing case. But birds of the same sex, although of course not truly paired, sometimes live in pairs or in small parties, as is known to be the case with pigeons and partridges. Birds also sometimes live in triplets, as has been observed with starlings, carrion-crows, parrots, and partridges. With partridges two females have been known to live with one male, and two males with one female. In all such cases it is probable that the union would be easily broken. The males of certain birds may occasionally be heard pouring forth their love-song long after the proper time, shewing that they have either lost or never gained a mate. Death from accident or disease of either one of a pair, would leave the other bird free and single; and there is reason to believe that female birds during the breeding-season are especially liable to premature death. Again, birds which have had their nests destroyed, or barren pairs, or retarded individuals, would easily be induced to desert their mates, and would probably be glad to take what share they could of the pleasures and duties of rearing offspring, although not their own.[688] Such contingencies as these probably explain most of the foregoing cases.[689]Nevertheless it is a strange fact that within the same district, during the height of the breeding-season, there should be so many males and females always ready to repair the loss of a mated bird. Why do not such spare birds immediately pair together? Have we not some reason to suspect, and the suspicion has occurred to Mr. Jenner Weir, that inasmuch as the act of courtship appears to be with many birds a prolonged and tedious affair, so it occasionally happens that certain males and females do not succeed during the proper season, in exciting each other's love, and consequently do not pair? This suspicion will appear somewhat less improbable after we have seen what strong antipathies and preferences female birds occasionally evince towards particular males.

Mental Qualities of Birds, and their taste for the beautiful.—Before we discuss any further the question whether the females select the more attractive males or accept the first whom they may encounter, it will be advisable briefly to consider the mental powers of birds. Their reason is generally, and perhaps justly, ranked as low; yet some facts could be given[690] leading to an opposite conclusion. Low powers of reasoning, however, are compatible, as we see with mankind, with strong affections, acute perception, and a taste for the beautiful; and it is with these latter qualities that we are here concerned. It has often been said that parrots become so deeply attached to each other that when one dies the other for a long time pines; but Mr. Jenner Weir thinks that with most birds the strength of their affection has been much exaggerated. Nevertheless when one of a pair in a state of nature has been shot, the survivor has been heard for days afterwards uttering a plaintive call; and Mr. St. John gives[691] various facts proving the attachment of mated birds. Starlings, however, as we have seen, may be consoled thrice in the same day for the loss of their mates. In the Zoological Gardens parrots have clearly recognised their former masters after an interval of some months. Pigeons have such excellent local memories that they have been known to return to their former homes after an interval of nine months, yet, as I hear from Mr. Harrison Weir, if a pair which would naturally remain mated for life be separated for a few weeks during the winter and matched with other birds, the two, when brought together again, rarely, if ever, recognise each other.

Birds sometimes exhibit benevolent feelings; they will feed the deserted young even of distinct species, but this perhaps ought to be considered as a mistaken instinct. They will also feed, as shewn in an earlier part of this work, adult birds of their own species which have become blind. Mr. Buxton gives a curious account of a parrot which took care of a frost-bitten and crippled bird of a distinct species, cleansed her feathers and defended her from the attacks of the other parrots which roamed freely about his garden. It is a still more curious fact that these birds apparently evince some sympathy for the pleasures of their fellows. When a pair of cockatoos made a nest in an acacia tree, "it was ridiculous to see the extravagant interest taken in the matter by the others of the same species." These parrots, also, evinced unbounded curiosity, and clearly had "the idea of property and possession."[692]

Birds possess acute powers of observation. Every mated bird, of course, recognises its fellow. Audubon states that with the mocking-thrushes of the United States *(Mimus polyglottus)* a certain number remain all the year round in Louisiana, whilst the others migrate to the Eastern States; these latter, on their return, are instantly recognised, and always attacked, by their Southern brethren. Birds under confinement distinguish different persons, as is proved by the strong and permanent antipathy or affection which they shew, without any apparent cause, towards certain individuals. I have heard of numerous instances with jays, partridges, canaries, and especially bullfinches. Mr. Hussey has described in how extraordinary a manner a tamed partridge recognised everybody; and its likes and dislikes were very strong. This bird seemed "fond of gay colours, and no new gown or cap could be put on without catching his attention."[693] Mr. Hewitt has carefully described the habits of some ducks (recently descended from wild birds), which, at the approach of a strange dog or cat, would rush headlong into the water, and exhaust themselves in their attempts to escape; but they knew so well Mr. Hewitt's own dogs and cats that they would lie down and bask in the sun close to them. They always moved away from a strange man, and so they would from the lady who attended them, if she made any great change in her dress. Audubon relates that he reared and tamed a wild turkey which always ran away from any strange dog; this bird escaped into the woods, and some days afterwards Audubon saw, as he thought, a wild turkey, and made his dog chase it; but to his astonishment, the bird did not run away, and the dog, when he came up, did not attack the bird, for they mutually recognised each other as old friends.[694]

Mr. Jenner Weir is convinced that birds pay particular attention to the colours of other birds, sometimes out of jealousy, and sometimes as a sign of kinship. Thus he turned a reed-bunting *(Emberiza schœniculus)*, which had acquired its black head, into his aviary, and the new-comer was not noticed by any bird, except by a bullfinch, which is likewise black-headed. This bullfinch was a very quiet bird, and had never before quarrelled with any of its comrades, including another reed-bunting, which had not as yet become black-headed: but the reed-bunting with a black head was so unmercifully treated, that it had to be removed. Mr. Weir was also obliged to turn out a robin, as it fiercely attacked all birds with any red in their plumage, but no other kinds; it actually killed a red-breasted crossbill, and nearly killed a goldfinch. On the other hand, he has observed that some birds, when first introduced into his aviary, fly towards the species which resemble them most in colour, and settle by their sides.

As male birds display with so much care their fine plumage and other ornaments in the presence of the females, it is obviously probable that these appreciate the beauty of their suitors. It is, however, difficult to obtain direct evidence of their capacity to appreciate beauty. When birds gaze at themselves in a looking-glass (of which many instances have been recorded) we cannot feel sure that it is not from jealousy at a supposed rival, though this is not the conclusion of some observers. In other cases it is difficult to distinguish between mere curiosity and admiration. It is perhaps the former feeling which, as stated by Lord Lilford,[695] attracts the Ruff strongly towards any bright object, so that, in the Ionian Islands, it "will dart down to a bright-coloured handkerchief, regardless of repeated shots." The common lark is drawn down from the sky, and is caught in large numbers, by a small mirror made to move and glitter in the sun. Is it admiration or curiosity which leads the magpie, raven, and some other birds to steal and secrete bright objects, such as silver articles or jewels?

Mr. Gould states that certain humming-birds decorate the outside of their nests, "with the utmost taste; they instinctively fasten thereon beautiful pieces of flat lichen, the larger pieces in the middle, and the smaller on the part attached to the branch. Now and then a pretty feather is intertwined or fastened to the outer sides, the stem being always so placed, that the feather stands out beyond the surface." The best evidence, however, of a taste for the beautiful is afforded by the three genera of Australian bower-birds already mentioned. Their bowers (see fig. 46, p. 1060), where the sexes congregate and play strange antics, are differently constructed, but what most concerns us is, that they are decorated in a different manner by the several species. The Satin bower-bird collects gaily-coloured articles, such as the blue tail-feathers of parrakeets, bleached bones and shells, which it sticks between the twigs, or arranges at the entrance. Mr. Gould found in one bower a neatly-worked stone tomahawk and a slip of blue cotton, evidently procured from a native encampment. These objects are continually rearranged, and carried about by the birds whilst at play. The bower of the Spotted bower-bird "is beautifully lined with tall grasses, so disposed that the heads nearly meet, and the decorations are very profuse." Round stones are used to keep the grass-stems in their proper places, and to make divergent paths leading to the bower. The stones and shells are often brought from a great distance. The Regent bird, as described by Mr. Ramsay, ornaments its short bower with bleached land-shells belonging to five or six species, and with "berries of various colours, blue, red, and black, which give it when fresh a very pretty appearance. Besides these there were several newly-picked leaves and young shoots of a pinkish colour, the whole shewing a decided taste for the beautiful." Well may Mr. Gould say "these highly decorated halls of assembly must be regarded as the most wonderful instances of bird-architecture yet discovered;" and the taste, as we see, of the several species certainly differs.[696]

Preference for particular Males by the Females.—Having made these preliminary remarks on the discrimination and taste of birds, I will give all the facts known to me, which bear on the preference shewn by the female for particular males. It is certain that distinct species of birds occasionally pair in a state of nature and produce hybrids. Many instances could be given: thus Macgillivray relates how a male blackbird and female thrush "fell in love with each other," and produced offspring.[697] Several years ago eighteen cases had been recorded of the occurrence in Great Britain of hybrids between the black grouse and pheasant;[698] but most of these cases may perhaps be accounted for by solitary birds not finding one of their own species to pair with. With other birds, as Mr. Jenner Weir has reason to believe, hybrids are sometimes the result of the casual intercourse of birds building in close proximity. But these remarks do not apply to the many recorded instances of tamed or domestic birds, belonging to distinct species, which have become absolutely fascinated with each other, although living with their own species. Thus Waterton[699] states that out of a flock of twenty-three Canada geese, a female paired with a solitary Bernicle gander, although so different in appearance and size; and they produced hybrid offspring. A male Wigeon *(Mareca penelope)*, living with females of the same species, has been known to pair with a Pintail duck, *Querquedula acuta*. Lloyd describes the remarkable attachment between a shield-drake *(Tadorna vulpanser)* and a common duck. Many additional instances could be given; and the Rev. E. S. Dixon remarks that "Those who have kept many different species of geese together, well know what unaccountable attachments they are frequently forming, and that they are quite as likely to pair and rear young with individuals of a race (species) apparently the most alien to themselves, as with their own stock."

The Rev. W. D. Fox informs me that he possessed at the same time a pair of Chinese geese *(Anser cygnoides)*, and a common gander with three geese. The two lots kept quite separate, until the Chinese gander seduced one of the common geese to live with him. Moreover, of the young birds hatched from the eggs of the common geese, only four were pure, the other eighteen proving hybrids; so that the Chinese gander seems to have had prepotent charms over the common gander. I will give only one other case; Mr. Hewitt states that a wild duck, reared in captivity, "after breeding a couple of seasons with her own mallard, at once shook him off on my placing a male Pintail on the water. It was evidently a case of love at first sight, for she swam about the new-comer caressingly, though he appeared evidently alarmed and averse to her overtures of affection. From that hour she forgot her old partner. Winter passed by, and the next spring the Pintail seemed to have become a convert to her blandishments, for they nested and produced seven or eight young ones."

What the charm may have been in these several cases, beyond mere novelty, we cannot even conjecture. Colour, however, sometimes comes into play; for in order to raise hybrids from the siskin *(Fringilla spinus)* and the canary, it is much the best plan, according to Bechstein, to place birds of the same tint together. Mr. Jenner Weir turned a female canary into his aviary, where there were male linnets, goldfinches, siskins, greenfinches, chaffinches, and other birds, in order to see which she would choose; but there never was any doubt, and the greenfinch carried the day. They paired and produced hybrid offspring.

With the members of the same species the fact of the female preferring to pair with one male rather than with another is not so likely to excite attention, as when this occurs between distinct species. Such cases can best be observed with domesticated or confined birds; but these are often pampered by high feeding, and sometimes have their instincts vitiated to an extreme degree. Of this latter fact I could give sufficient proofs with pigeons, and especially with fowls, but they cannot be here related. Vitiated instincts may also account for some of the hybrid unions above referred to; but in many of these cases the birds were allowed to range freely over large ponds, and there is no reason to suppose that they were unnaturally stimulated by high feeding.

With respect to birds in a state of nature, the first and most obvious supposition which will occur to every one is that the female at the proper season accepts the first male whom she may encounter; but she has at least the opportunity for exerting a choice, as she is almost invariably pursued by many males. Audubon—and we must remember that he spent a long life in prowling about the forests of the United States and observing the birds—does not doubt that the female deliberately chooses her mate; thus, speaking of a woodpecker, he says the hen is followed by half-a-dozen gay suitors, who continue performing strange antics, "until a marked preference is shewn for one." The female of the red-winged starling *(Agelæus phœniceus)* is likewise pursued by several males, "until, becoming fatigued, she alights, receives their addresses, and soon makes a choice." He describes also how several male night-jars repeatedly plunge through the air with astonishing rapidity, suddenly turning, and thus making a singular noise; "but no sooner has the female made her choice, than the other males are driven away." With one of the vultures *(Cathartes aura)* of the United States, parties of eight or ten or more males and females assemble on fallen logs, "exhibiting the strongest desire to please mutually," and after many caresses, each male leads off his partner on the wing. Audubon likewise carefully observed the wild flocks of Canada geese *(Anser Canadensis)*, and gives a graphic description of their love-antics; he says that the birds which had been previously mated "renewed their courtship as early as the month of January, while the others would be contending or coquetting for hours every day, until all seemed satisfied with the choice they had made, after which, although they remained together, any person could easily perceive that they were careful to keep in pairs. I have observed also that the older the birds, the shorter were the preliminaries of their courtship. The bachelors and old maids, whether in regret, or not caring to be disturbed by the bustle, quietly moved aside and lay down at some distance from the rest."[700] Many similar statements with respect to other birds could be cited from this same observer.

Turning now to domesticated and confined birds, I will commence by giving what little I have learnt respecting the courtship of fowls. I have received long letters on this subject from Messrs. Hewitt and Tegetmeier, and almost an essay from the late Mr. Brent. It will be admitted by every one that these gentlemen, so well known from their published works, are careful and experienced observers. They do not believe that the females prefer certain males on account of the beauty of their plumage; but some allowance must be made for the artificial state under which they have long been kept. Mr. Tegetmeier is convinced that a game-cock, though disfigured by being dubbed with his hackles trimmed, would be accepted as readily as a male retaining all his natural ornaments. Mr. Brent, however, admits that the beauty of the male probably aids in exciting the female; and her acquiescence is necessary. Mr. Hewitt is convinced that the union is by no means left to mere chance, for the female almost invariably prefers the most vigorous, defiant, and mettlesome male; hence it is almost useless, as he remarks, "to attempt true breeding if a game-cock in good health and condition runs the locality, for almost every hen on leaving the roosting-place will resort to the game-cock, even though that bird may not actually drive away the male of her own variety." Under ordinary circumstances the males and females of the fowl seem to come to a mutual understanding by means of certain gestures, described to me by Mr. Brent. But hens will often avoid the officious attentions of young males. Old hens, and hens of a pugnacious disposition, as the same writer informs me, dislike strange males, and will not yield until well beaten into compliance. Ferguson, however, describes how a quarrelsome hen was subdued by the gentle courtship of a Shanghai cock.[701]

There is reason to believe that pigeons of both sexes prefer pairing with birds of the same breed; and dovecot-pigeons dislike all the highly improved breeds.[702] Mr. Harrison Weir has lately heard from a trustworthy observer, who keeps blue pigeons, that these drive away all other coloured varieties, such as white, red, and yellow; and from another observer, that a female dun carrier could not be matched, after repeated trials, with a black male, but immediately paired with a dun. Generally colour alone appears to have little influence on the pairing of pigeons. Mr. Tegetmeier, at my request, stained some of his birds with magenta, but they were not much noticed by the others.

Female pigeons occasionally feel a strong antipathy towards certain males, without any assignable cause. Thus MM. Boitard and Corbié, whose experience extended over forty-five years, state: "Quand une femelle éprouve de l'antipathie pour un mâle avec lequel on veut l'accoupler, malgré tous les feux de l'amour, malgré l'alpiste et le chènevis dont on la nourrit pour augmenter son ardeur, malgré un emprisonnement de six mois et même d'un an, elle refuse constamment ses caresses; les avances empressées, les agaceries, les tournoiemens, les tendres roucoulemens, rien ne peut lui plaire ni l'émouvoir; gonflée, boudeuse, blottie dans un coin de sa prison, elle n'en sort que pour boire et manger, ou pour repousser avec une espèce de rage des caresses devenues trop pressantes."[703] On the other hand, Mr. Harrison Weir has himself observed, and has heard from several breeders, that a female pigeon will occasionally take a strong fancy for a particular male, and will desert her own mate for him. Some females, according to another experienced observer, Riedel,[704] are of a profligate disposition, and prefer almost any stranger to their own mate. Some amorous males, called by our English fanciers "gay birds," are so successful in their gallantries, that, as Mr. H. Weir informs me, they must be shut up, on account of the mischief which they cause.

Wild turkeys in the United States, according to Audubon, "sometimes pay their addresses to the domesticated females, and are generally received by them with great pleasure." So that these females apparently prefer the wild to their own males.[705]

Here is a more curious case. Sir R. Heron during many years kept an account of the habits of the peafowl, which he bred in large numbers. He states that "the hens have frequently great preference to a particular peacock. They were all so fond of an old pied cock, that one year, when he was confined though still in view, they were constantly assembled close to the trellice-walls of his prison, and would not suffer a japanned peacock to touch them. On his being let out in the autumn, the oldest of the hens instantly courted him, and was successful in her courtship. The next year he was shut up in a stable, and then the hens all courted his rival."[706] This rival was a japanned or black-winged peacock, which to our eyes is a more beautiful bird than the common kind.

Lichtenstein, who was a good observer and had excellent opportunities of observation at the Cape of Good Hope, assured Rudolphi that the female widow-bird *(Chera progne)* disowns the male, when robbed of the long tail-feathers with which he is ornamented during the breeding-season. I presume that this observation must have been made on birds under confinement.[707] Here is another striking case; Dr. Jaeger,[708] director of the Zoological Gardens of Vienna, states that a male silver-pheasant, who had been triumphant over the other males and was the accepted lover of the females, had his ornamental plumage spoiled. He was then immediately superseded by a rival, who got the upper hand and afterwards led the flock.

Not only does the female exert a choice, but in some few cases she courts the male, or even fights for his possession. Sir R. Heron states that with peafowl, the first advances are always made by the female; something of the same kind takes place, according to Audubon, with the older females of the wild turkey. With the capercailzie, the females flit round the male, whilst he is parading at one of the places of assemblage, and solicit his attention.[709] We have seen that a tame wild-duck seduced after a long courtship an unwilling Pintail drake. Mr. Bartlett believes that the Lophophorus, like many other gallinaceous birds, is naturally polygamous, but two females cannot be placed in the same cage with a male, as they fight so much together. The following instance of rivalry is more surprising as it relates to bullfinches, which usually pair for life. Mr. Jenner Weir introduced a dull-coloured and ugly female into his aviary, and she immediately attacked another mated female so unmercifully that the latter had to be separated. The new female did all the courtship, and was at last successful, for she paired with the male; but after a time she met with a just retribution, for, ceasing to be pugnacious, Mr. Weir replaced the old female, and the male then deserted his new and returned to his old love.

In all ordinary cases the male is so eager that he will accept any female, and does not, as far as we can judge, prefer one to the other; but exceptions to this rule, as we shall hereafter see, apparently occur in some few groups. With domesticated birds, I have heard of only one case in which the males shew any preference for particular females, namely, that of the domestic cock, who, according to the high authority of Mr. Hewitt, prefers the younger to the older hens. On the other hand, in effecting hybrid unions between the male pheasant and common hens, Mr. Hewitt is convinced that the pheasant invariably prefers the older birds. He does not appear to be in the least influenced by their colour, but "is most capricious in his attachments."[710] From some inexplicable cause he shews the most determined aversion to certain hens, which no care on the part of the breeder can overcome. Some hens, as Mr. Hewitt informs me, are quite unattractive even to the males of their own species, so that they may be kept with several cocks during a whole season, and not one egg out of forty or fifty will prove fertile. On the other hand with the Long-tailed duck *(Harelda glacialis)*, "it has been remarked," says M. Ekström, "that certain females are much more courted than the rest. Frequently, indeed, one sees an individual surrounded by six or eight amorous males." Whether this statement is credible, I know not; but the native sportsmen shoot these females in order to stuff them as decoys.[711]

With respect to female birds feeling a preference for particular males, we must bear in mind that we can judge of choice being exerted, only by placing ourselves in imagination in the same position. If an inhabitant of another planet were to behold a number of young rustics at a fair, courting and quarrelling over a pretty girl, like birds at one of their places of assemblage, he would be able to infer that she had the power of choice only by observing the eagerness of the wooers to please her, and to display their finery. Now with birds, the evidence stands thus; they have acute powers of observation, and they seem to have some taste for the beautiful both in colour and sound. It is certain that the females occasionally exhibit, from unknown causes, the strongest antipathies and preferences for particular males. When the sexes differ in colour or in other ornaments, the males with rare exceptions are the most highly decorated, either permanently or temporarily during the breeding-season. They sedulously display their various ornaments, exert their voices, and perform strange antics in the presence of the females. Even well-armed males, who, it might have been thought, would have altogether depended for success on the law of battle, are in most cases highly ornamented; and their ornaments have been acquired at the expense of some loss of power. In other cases ornaments have been acquired, at the cost of increased risk from birds and beasts of prey. With various species many individuals of both sexes congregate at the same spot, and their courtship is a prolonged affair. There is even reason to suspect that the males and females within the same district do not always succeed in pleasing each other and pairing.

What then are we to conclude from these facts and considerations? Does the male parade his charms with so much pomp and rivalry for no purpose? Are we not justified in believing that the female exerts a choice, and that she receives the addresses of the male who pleases her most? It is not probable that she consciously deliberates; but she is most excited or attracted by the most beautiful, or melodious, or gallant males. Nor need it be supposed that the female studies each stripe or spot of colour; that the peahen, for instance, admires each detail in the gorgeous train of the peacock—she is probably struck only by the general effect. Nevertheless after hearing how carefully the male Argus pheasant displays his elegant primary wing-feathers, and erects his ocellated plumes in the right position for their full effect; or again, how the male goldfinch alternately displays his gold-bespangled wings, we ought not to feel too sure that the female does not attend to each detail of beauty. We can judge, as already remarked, of choice being exerted, only from the analogy of our own minds; and the mental powers of birds, if reason be excluded, do not fundamentally differ from ours. From these various considerations we may conclude that the pairing of birds is not left to chance; but that those males, which are best able by their various charms to please or excite the female, are under ordinary circumstances accepted. If this be admitted, there is not much difficulty in understanding how male birds have gradually acquired their ornamental characters. All animals present individual differences, and as man can modify his domesticated birds by selecting the individuals which appear to him the most beautiful, so the habitual or even occasional preference by the female of the more attractive males would almost certainly lead to their modification; and such modifications might in the course of time be augmented to almost any extent, compatible with the existence of the species.

Variability of Birds, and especially of their secondary Sexual Characters.—Variability and inheritance are the foundations for the work of selection. That domesticated birds have varied greatly, their variations being inherited, is certain. That birds in a state of nature present individual differences is admitted by every one; and that they have sometimes been modified into distinct races, is generally admitted.[712] Variations are of two kinds, which insensibly graduate into each other, namely, slight differences between all the members of the same species, and more strongly-marked deviations which occur only occasionally. These latter are rare with birds in a state of nature, and it is very doubtful whether they have often been preserved through selection, and then transmitted to succeeding generations.[713] Nevertheless, it may be worth while to give the few cases relating chiefly to colour (simple albinism and melanism being excluded), which I have been able to collect.

Mr. Gould is well known rarely to admit the existence of varieties, for he esteems very slight differences as specific; now he states[714] that near Bogota certain humming-birds belonging to the genus Cynanthus are divided into two or three races or varieties, which differ from each other in the colouring of the tail,—"some having the whole of the feathers blue, while others have the eight central ones tipped with beautiful green." It does not appear that intermediate gradations have been observed in this or the following cases. In the males alone of one of the Australian parrakeets "the thighs in some are scarlet, in others grass-green." In another parrakeet of the same country "some individuals have the band across the wing-coverts bright-yellow, while in others the same part is tinged with red."[715] In the United States some few of the males of the Scarlet Tanager *(Tanagra rubra)* have "a beautiful transverse band of glowing red on the smaller wing-coverts;"[716] but this variation seems to be somewhat rare, so that its preservation through sexual selection would follow only under unusually favourable circumstances. In Bengal the Honey buzzard *(Pernis cristata)* has either a small rudimental crest on its head, or none at all; so slight a difference however would not have been worth notice, had not this same species possessed in Southern India "a well-marked occipital crest formed of several graduated feathers."[717]

The following case is in some respects more interesting. A pied variety of the raven, with the head, breast, abdomen, and parts of the wings and tail-feathers white, is confined to the Feroe Islands. It is not very rare there, for Graba saw during his visit from eight to ten living specimens. Although the characters of this variety are not quite constant, yet it has been named by several distinguished ornithologists as a distinct species. The fact of the pied birds being pursued and persecuted with much clamour by the other ravens of the island was the chief cause which led Brünnich to conclude that it was specifically distinct; but this is now known to be an error.[718]

In various parts of the northern seas a remarkable variety of the common Guillemot *(Uria troile)* is found; and in Feroe, one out of every five birds, according to Graba's estimation, consists of this variety. It is characterised[719] by a pure white ring round the eye, with a curved narrow white line, an inch and a half in length, extending back from the ring. This conspicuous character has caused the bird to be ranked by several ornithologists as a distinct species under the name of *U. lacrymans,* but it is now known to be merely a variety. It often pairs with the common kind, yet intermediate gradations have never been seen; nor is this surprising, for variations which appear suddenly are often, as I have elsewhere shewn,[720] transmitted either unaltered or not at all. We thus see that two distinct forms of the same species may co-exist in the same district, and we cannot doubt that if the one had possessed any great advantage over the other, it would soon have been multiplied to the exclusion of the latter. If, for instance, the male pied ravens, instead of being persecuted and driven away by their comrades, had been highly attractive, like the pied peacock before mentioned, to the common black females, their numbers would have rapidly increased. And this would have been a case of sexual selection.

With respect to the slight individual differences which are common, in a greater or less degree, to all the members of the same species, we have every reason to believe that they are by far the most important for the work of selection. Secondary sexual characters are eminently liable to vary, both with animals in a state of nature and under domestication.[721] There is also reason to believe, as we have seen in our eighth chapter, that variations are more apt to occur in the male than in the female sex. All these contingencies are highly favourable for sexual selection. Whether characters thus acquired are transmitted to one sex or to both sexes, depends exclusively in most cases, as I hope to shew in the following chapter, on the form of inheritance which prevails in the groups in question.

It is sometimes difficult to form any opinion whether certain slight differences between the sexes of birds are simply the result of variability with sexually-limited inheritance, without the aid of sexual selection, or whether they have been augmented through this latter process. I do not here refer to the innumerable instances in which the male displays splendid colours or other ornaments, of which the female partakes only to a slight degree; for these cases are almost certainly due to characters primarily acquired by the male, having been transferred, in a greater or less degree, to the female. But what are we to conclude with respect to certain birds in which, for instance, the eyes differ slightly in colour in the two sexes?[722] In some cases the eyes differ conspicuously; thus with the storks of the genus *Xenorhynchus* those of the male are blackish-hazel, whilst those of the females are gamboge-yellow; with many hornbills (Buceros), as I hear from Mr. Blyth,[723] the males have intense crimson, and the females white eyes. In the *Buceros bicornis,* the hind margin of the casque and a stripe on the crest of the beak are black in the male, but not so in the female. Are we to suppose that these black marks and the crimson colour of the eyes have been preserved or augmented through sexual selection in the males? This is very doubtful; for Mr. Bartlett shewed me in the Zoological Gardens that the inside of the mouth of this Buceros is black in the male and flesh-coloured in the female; and their external appearance or beauty would not be thus affected. I observed in Chili[724] that the iris in the condor, when about a year old, is dark-brown, but changes at maturity into yellowish-brown in the male, and into bright red in the female. The male has also a small, longitudinal, leaden-coloured, fleshy crest or comb. With many gallinaceous birds the comb is highly ornamental, and assumes vivid colours during the act of courtship; but what are we to think of the dull-coloured comb of the condor, which does not appear to us in the least ornamental? The same question may be asked in regard to various other characters, such as the knob on the base of the beak of the Chinese goose *(Anser cygnoides),* which is much larger in the male than in the female. No certain answer can be given to these questions; but we ought to be cautious in assuming that knobs and various fleshy appendages cannot be attractive to the female, when we remember that with savage races of man various hideous deformities—deep scars on the face with the flesh raised into protuberances, the septum of the nose pierced by sticks or bones, holes in the ears and lips stretched widely open—are all admired as ornamental.

Whether or not unimportant differences between the sexes, such as those just specified, have been preserved through sexual selection, these differences, as well as all others, must primarily depend on the laws of variation. On the principle of correlated development, the plumage often varies on different parts of the body, or over the whole body, in the same manner. We see this well illustrated in certain breeds of the fowl. In all the breeds the feathers on the neck and loins of the males are elongated, and are called hackles; now when both sexes acquire a top-knot, which is a new character in the genus, the feathers on the head of the male become hackle-shaped, evidently on the principle of correlation; whilst those on the head of the female are of the ordinary shape. The colour also of the hackles forming the top-knot of the male, is often correlated with that of the hackles on the neck and loins, as may be seen by comparing these feathers in the Golden and Silver-spangled Polish, the Houdans, and Crève-cœur breeds. In some natural species we may observe exactly the same correlation in the colours of these same feathers, as in the males of the splendid Golden and Amherst pheasants.

The structure of each individual feather generally causes any change in its colouring to be symmetrical; we see this in the various laced, spangled, and pencilled breeds of the fowl; and on the principle of correlation the feathers over the whole body are often modified in the same manner. We are thus enabled without much trouble to rear breeds with their plumage marked and coloured almost as symmetrically as in natural species. In laced and spangled fowls the coloured margins of the feathers are abruptly defined; but in a mongrel raised by me from a black Spanish cock glossed with green and a white game hen, all the feathers were greenish-black, excepting towards their extremities, which were yellowish-white; but between the white extremities and the black bases, there was on each feather a symmetrical, curved zone of dark-brown. In some instances the shaft of the feather determines the distribution of the tints; thus with the body-feathers of a mongrel from the same black Spanish cock and a silver-spangled Polish hen, the shaft, together with a narrow space on each side, was greenish-black, and this was surrounded by a regular zone of dark-brown, edged with brownish-white. In these cases we see feathers becoming symmetrically shaded, like those which give so much elegance to the plumage of many natural species. I have also noticed a variety of the common pigeon with the wing-bars symmetrically zoned with three bright shades, instead of being simply black on a slaty-blue ground, as in the parent-species.

In many large groups of birds it may be observed that the plumage is differently coloured in each species, yet that certain spots, marks, or stripes, though likewise differently coloured, are retained by all the species. Analogous cases occur with the breeds of the pigeon, which usually retain the two wing-bars, though they may be coloured red, yellow, white, black, or blue, the rest of the plumage being of some wholly different tint. Here is a more curious case, in which certain marks are retained, though coloured in almost an exactly reversed manner to what is natural; the aboriginal pigeon has a blue tail, with the terminal halves of the outer webs of the two outer tail-feathers white; now there is a sub-variety having a white instead of a blue tail, with precisely that small part black which is white in the parent-species.[725]

Formation and variability of the Ocelli or eye-like Spots on the Plumage of Birds.— As no ornaments are more beautiful than the ocelli on the feathers of various birds, on the hairy coats of some mammals, on the scales of reptiles and fishes, on the skin of amphibians, on the wings of many Lepidoptera and other insects, they deserve to be especially noticed. An ocellus consists of a spot within a ring of another colour, like the pupil within the iris, but the central spot is often surrounded by additional concentric zones. The ocelli on the tail-coverts of the peacock offer a familiar example, as well as those on the wings of the peacock-butterfly (Vanessa). Mr. Trimen has given me a description of a S. African moth *(Gynanisa Isis)*, allied to our Emperor moth, in which a magnificent ocellus occupies nearly the whole surface of each hinder wing; it consists of a black centre, including a semi-transparent crescent-shaped mark, surrounded by successive ochre-yellow, black, ochre-yellow, pink, white, pink, brown, and whitish zones. Although we do not know the steps by which these wonderfully-beautiful and complex ornaments have been developed, the process at least with insects has probably been a simple one; for, as Mr. Trimen writes to me, "no characters of mere marking or coloration are so unstable in the Lepidoptera as the ocelli, both in number and size." Mr. Wallace, who first called my attention to this subject, shewed me a series of specimens of our common meadow-brown butterfly *(Hipparchia Janira)* exhibiting numerous gradations from a simple minute black spot to an elegantly-shaded ocellus. In a S. African butterfly *(Cyllo Leda,* Linn.) belonging to the same family, the ocelli are even still more variable. In some specimens (A, fig. 52) large spaces on the upper surface of the wings are coloured

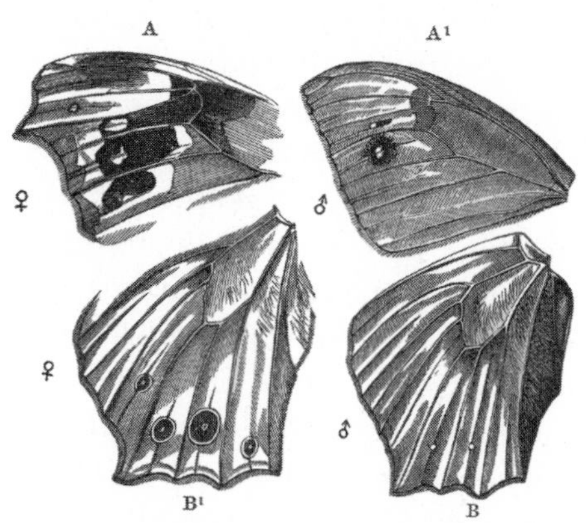

Fig. 52. *Cyllo leda, Linn., from a drawing by Mr. Trimen, shewing the extreme range of variation in the ocelli.*
A. *Specimen, from Mauritius, upper surface of fore-wing.*
A¹. *Specimen, from Natal, ditto.*
B. *Specimen, from Java, upper surface of hind-wing.*
B¹. *Specimen, from Mauritius, ditto.*

black, and include irregular white marks; and from this state a complete gradation can be traced into a tolerably perfect (A1) ocellus, and this results from the contraction of the irregular blotches of colour. In another series of specimens a gradation can be followed from excessively minute white dots, surrounded by a scarcely visible black line (B), into perfectly symmetrical and large ocelli (B1).[726] In cases like these, the development of a perfect ocellus does not require a long course of variation and selection.

With birds and many other animals it seems, from the comparison of allied species, to follow, that circular spots are often generated by the breaking up and contraction of stripes. In the Tragopan pheasant faint white lines in the female represent the beautiful white spots in the male;[727] and something of the same kind may be observed in the two sexes of the Argus pheasant. However this may be, appearances strongly favour the belief that, on the one hand, a dark spot is often formed by the colouring-matter being drawn towards a central point from a surrounding zone, which is thus rendered lighter. And, on the other hand, that a white spot is often formed by the colour being driven away from a central point, so that it accumulates in a surrounding darker zone. In either case an ocellus is the result. The colouring matter seems to be a nearly constant quantity, but is redistributed, either centripetally or centrifugally. The feathers of the common guinea-fowl offer a good instance of white spots surrounded by darker zones; and wherever the white spots are large and stand near each other, the surrounding dark zones become confluent. In the same wing-feather of the Argus pheasant dark spots may be seen surrounded by a pale zone, and white spots by a dark zone. Thus the formation of an ocellus in its simplest state appears to be a simple affair. By what further steps the more complex ocelli, which are surrounded by many successive zones of colour, have been generated, I will not pretend to say. But bearing in mind the zoned feathers of the mongrel offspring from differently-coloured fowls, and the extraordinary variability of the ocelli in many Lepidoptera, the formation of these beautiful ornaments can hardly be a highly complex process, and probably depends on some slight and graduated change in the nature of the tissues.

Gradation of Secondary Sexual Characters.—Cases of gradation are important for us, as they shew that it is at least possible that highly complex ornaments may have been acquired by small successive steps. In order to discover the actual steps by which the male of any existing bird has acquired his magnificent colours or other ornaments, we ought to behold the long line of his ancient and extinct progenitors; but this is obviously impossible. We may, however, generally gain a clue by comparing all the species of a group, if it be a large one; for some of them will probably retain, at least in a partial manner, traces of their former characters. Instead of entering on tedious details respecting various groups, in which striking instances of gradation could be given, it seems the best plan to take some one or two strongly-characterised cases, for instance that of the peacock, in order to discover if any light can thus be thrown on the steps by which this bird has become so splendidly decorated. The peacock is chiefly remarkable from the extraordinary length of his tail-coverts; the tail itself not being much elongated. The barbs along nearly the whole length of these feathers stand separate or are decomposed; but this is the case with the feathers of many species, and with some varieties of the domestic fowl and pigeon. The barbs coalesce towards the extremity of the shaft to form the oval disc or ocellus, which is certainly one of the most beautiful objects in the world. This consists of an iridescent, intensely blue, indented centre, surrounded by a rich green zone, and this by a broad coppery-brown zone, and this by five other narrow zones of slightly-different iridescent shades. A trifling character in the disc perhaps deserves notice; the barbs, for a space along one of the concentric zones are destitute, to a greater or less degree, of their barbules, so that a part of the disc is surrounded by an almost transparent zone, which gives to it a highly-finished aspect. But I have elsewhere described[728] an exactly analogous variation in the hackles of a sub-variety of the game-cock, in which the tips, having a metallic lustre, "are separated from the lower part of the feather by a symmetrically-shaped transparent zone, composed of the naked portions of the barbs." The lower margin or base of the dark-blue centre of the ocellus is deeply indented on the line of the shaft. The surrounding zones likewise shew traces, as may be seen in the drawing (fig. 53), of indentations, or rather breaks. These indentations are common to the Indian and Javan peacocks *(Pavo cristatus* and *P. muticus)*; and they seemed to me to deserve particular atten-

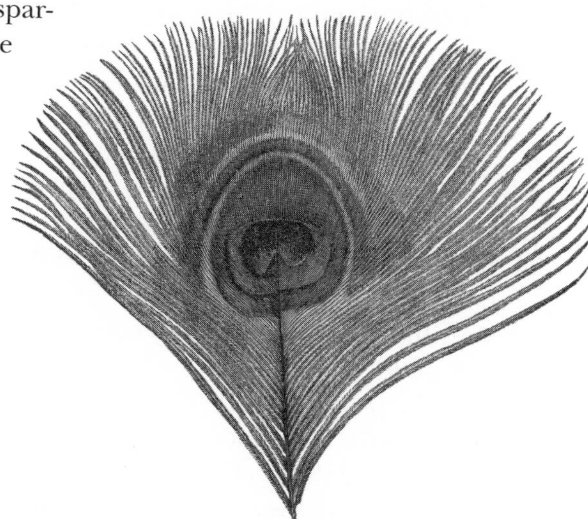

Fig. 53. *Feather of Peacock, about two-thirds of natural size, carefully drawn by Mr. Ford. The transparent zone is represented by the outermost white zone, confined to the upper end of the disc.*

tion, as probably connected with the development of the ocellus; but for a long time I could not conjecture their meaning.

If we admit the principle of gradual evolution, there must formerly have existed many species which presented every successive step between the wonderfully elongated tail-coverts of the peacock and the short tail-coverts of all ordinary birds; and again between the magnificent ocelli of the former, and the simpler ocelli or mere coloured spots of other birds; and so with all the other characters of the peacock. Let us look to the allied Gallinaceæ for any still-existing gradations. The species and sub-species of Polyplectron inhabit countries adjacent to the native land of the peacock; and they so far resemble this bird that they are sometimes called peacock-pheasants. I am also informed by Mr. Bartlett that they resemble the peacock in their voice and in some of their habits. During the spring the males, as previously described, strut about before the comparatively plain-coloured females, expanding and erecting their tail and wing-feathers, which are ornamented with numerous ocelli. I request the reader to turn back to the drawing (fig. 51, p. 1071) of a Polyplectron. In *P. Napoleonis* the ocelli are confined to the tail, and the back is of a rich metallic blue, in which respects this species approaches the Java peacock. *P. Hardwickii* possesses a peculiar top-knot, somewhat like that of this same kind of peacock. The ocelli on the wings and tail of the several species of Polyplectron are either circular or oval, and consist of a beautiful, iridescent, greenish-blue or greenish-purple disc, with a black border. This border in *P. chinquis* shades into brown which is edged with cream-colour, so that the ocellus is here surrounded with differently, though not brightly, shaded concentric zones. The unusual length of the tail-coverts is another highly remarkable character in Polyplectron; for in some of the species they are half as long, and in others two-thirds of the length of the true tail-feathers. The tail-coverts are ocellated, as in the peacock. Thus the several species of Polyplectron manifestly make a graduated approach in the length of their tail-coverts, in the zoning of the ocelli, and in some other characters, to the peacock.

Notwithstanding this approach, the first species of Polyplectron which I happened to examine almost made me give up the search; for I found not only that the true tail-feathers, which in the peacock are quite plain, were ornamented with ocelli, but that the ocelli on all the feathers differed fundamentally from those of the peacock, in there being two on the same feather, (fig. 54), one on each side of the shaft. Hence I concluded that the early progenitors of the peacock could not have resembled in any degree a Polyplectron. But on continuing my search, I observed that in some of the species the two ocelli stood very near each other; that in the tail-feathers of *P. Hardwickii* they touched each other; and, finally, that in the tail-coverts of this same species as well as of *P. malaccense* (fig. 55) they were actually confluent. As the central part alone is confluent, an indentation is left at both the upper and lower ends; and the surrounding coloured zones are likewise indented. A single ocellus is thus formed on each tail-covert, though still plainly betraying its double origin. These confluent ocelli differ from the single ocelli of the peacock in having an indentation at both ends, instead of at the lower or basal end alone. The explanation, however, of this difference is not difficult; in some species of Polyplectron the two oval ocelli on the same feather stand parallel to each other; in other species (as in *P. chinquis*) they converge towards one end; now the partial confluence of two convergent ocelli would manifestly leave a much deeper indentation at the divergent than at the convergent end. It is also manifest that if the convergence were strongly pronounced and the confluence complete, the indentation at the convergent end would tend to be quite obliterated.

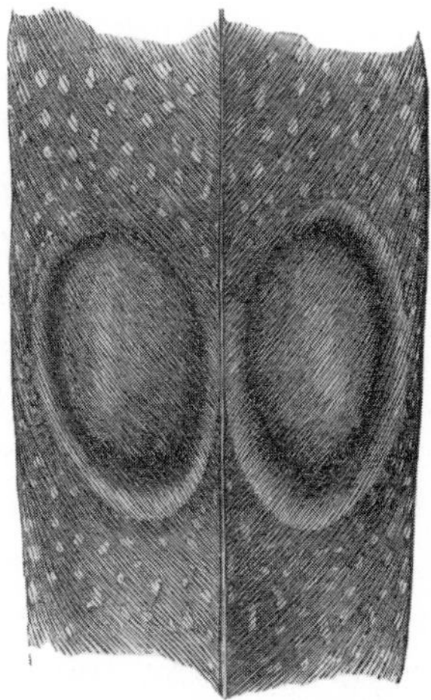

Fig. 54. *Part of a tail-covert of Polyplectron chinquis, with the two ocelli of nat. size.*

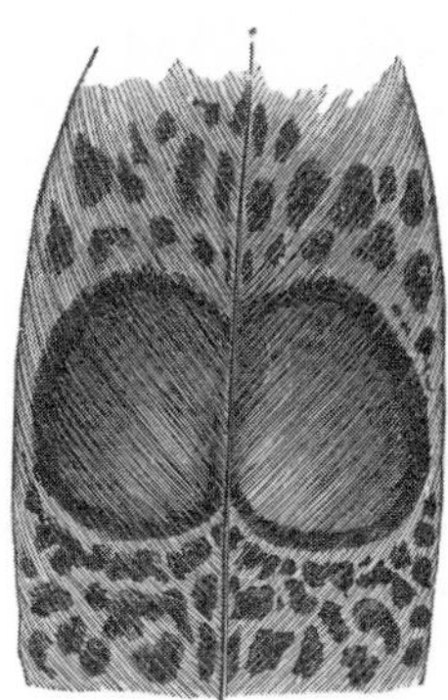

Fig. 55. *Part of a tail-covert of Polyplectron malaccense, with the two ocelli, partially confluent, of nat. size.*

The tail-feathers in both species of peacock are entirely destitute of ocelli, and this apparently is related to their being covered up and concealed by the long tail-coverts. In this respect they differ remarkably from the tail-feathers of Polyplectron, which in most of the species are ornamented with larger ocelli than those on the tail-coverts. Hence I was led carefully to examine the tail-feathers of the several species of Polyplectron in order to discover whether the ocelli in any of them shewed any tendency to disappear, and, to my great satisfaction, I was successful. The central tail-feathers of *P. Napoleonis* have the two ocelli on each side of the shaft perfectly developed; but the inner ocellus becomes less and less conspicuous on the more exterior tail-feathers, until a mere shadow or rudimentary vestige is left on the inner side of the outermost feather. Again, in *P. malaccense,* the ocelli on the tail-coverts are, as we have seen, confluent; and these feathers are of unusual length, being two-thirds of the length of the tail-feathers, so that in both these respects they resemble the tail-coverts of the peacock. Now in this species the two central tail-feathers alone are ornamented, each with two brightly-coloured ocelli, the ocelli having completely disappeared from the inner sides of all the other tail-feathers. Consequently the tail-coverts and tail-feathers of this species of Polyplectron make a near approach in structure and ornamentation to the corresponding feathers of the peacock.

As far, then, as the principle of gradation throws light on the steps by which the magnificent train of the peacock has been acquired, hardly anything more is needed. We may picture to ourselves a progenitor of the peacock in an almost exactly intermediate condition between the existing peacock, with his enormously elongated tail-coverts, ornamented with single ocelli, and an ordinary gallinaceous bird with short tail-coverts, merely spotted with some colour; and we shall then see in our mind's eye, a bird possessing tail-coverts, capable of erection and expansion, ornamented with two partially confluent ocelli, and long enough almost to conceal the tail-feathers,—the latter having already partially lost their ocelli; we shall see in short, a Polyplectron. The indentation of the central disc and surrounding zones of the ocellus in both species of peacock, seems to me to speak plainly in favour of this view; and this structure is otherwise inexplicable. The males of Polyplectron are no doubt very beautiful birds, but their beauty, when viewed from a little distance, cannot be compared, as I formerly saw in the Zoological Gardens, with that of the peacock. Many female progenitors of the peacock must, during a long line of descent, have appreciated this superiority; for they have unconsciously, by the continued preference of the most beautiful males, rendered the peacock the most splendid of living birds.

Argus pheasant.—Another excellent case for investigation is offered by the ocelli on the wing-feathers of the Argus pheasant, which are shaded in so wonderful a manner as to resemble balls lying within sockets, and which consequently differ from ordinary ocelli. No one, I presume, will attribute the shading, which has excited the admiration of many experienced artists, to chance—to the fortuitous concourse of atoms of colouring matter. That these ornaments should have been formed through the selection of many successive variations, not one of which was originally intended to produce the ball-and-socket effect, seems as incredible, as that one of Raphael's Madonnas should have been formed by the selection of chance daubs of paint made by a long succession of young artists, not one of whom intended at first to draw the human figure. In order to discover how the ocelli have been developed, we cannot look to a long line of progenitors, nor to various closely-allied forms, for such do not now exist. But fortunately the several feathers on the wing suffice to give us a clue to the problem, and they prove to demonstration that a gradation is at least possible from a mere spot to a finished ball-and-socket ocellus.

The wing-feathers, bearing the ocelli, are covered with dark stripes or rows of dark spots, each stripe or row running obliquely down the outer side of the shaft to an ocellus. The spots are generally elongated in a transverse line to the row in which they stand. They often become confluent, either in the line of the row—and then they form a longitudinal stripe—or transversely, that is, with the spots in the adjoining rows, and then they form transverse stripes. A spot sometimes breaks up into smaller spots, which still stand in their proper places.

It will be convenient first to describe a perfect ball-and-socket ocellus. This consists of an intensely black circular ring, surrounding a space shaded so as exactly to resemble a ball. The figure here given has been admirably drawn by Mr. Ford, and engraved, but a woodcut cannot exhibit the exquisite shading of the original. The ring is almost always slightly broken or interrupted (see fig. 56) at a point in the upper half, a little to the right of and above the white shade on the enclosed ball; it is also sometimes broken towards the base on the right hand. These little breaks have an important meaning. The ring is always much thickened, with the edges ill-defined towards the left-hand upper corner the feather being held erect, in the position in which it is here drawn. Beneath this thickened part there is on the surface of the ball an oblique almost pure-white mark, which shades off downwards into a pale-leaden hue, and this into yellowish and brown tints, which insensibly become darker and darker towards the lower part of the ball. It is this shading which gives so admirably the effect of light shining on a convex surface. If one of the balls be examined, it will be seen that the lower part is of a browner tint and is indistinctly separated by a curved oblique line from the upper part, which is yellower and more leaden; this oblique line runs at right angles to the longer axis of the white patch of light, and indeed of all the shading; but this difference

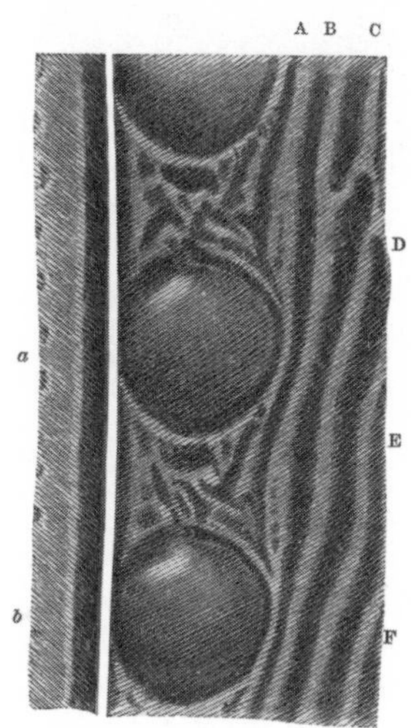

Fig. 56. *Part of Secondary wing-feather of Argus pheasant, shewing two,* a *and* b, *perfect ocelli.*

A, B, C, &c., dark stripes running obliquely down, each to an ocellus.

[Much of the web on both sides, especially to the left of the shaft, has been cut off.]

in the tints, which cannot of course be shewn in the woodcut, does not in the least interfere with the perfect shading of the ball.[729] It should be particularly observed that each ocellus stands in obvious connection with a dark stripe, or row of dark spots, for both occur indifferently on the same feather. Thus in fig. 56 stripe A runs to ocellus *a;* B runs to ocellus *b;* stripe C is broken in the upper part, and runs down to the next succeeding ocellus, not represented in the woodcut; D to the next lower one, and so with the stripes E and F. Lastly, the several ocelli arc separated from each other by a pale surface bearing irregular black marks.

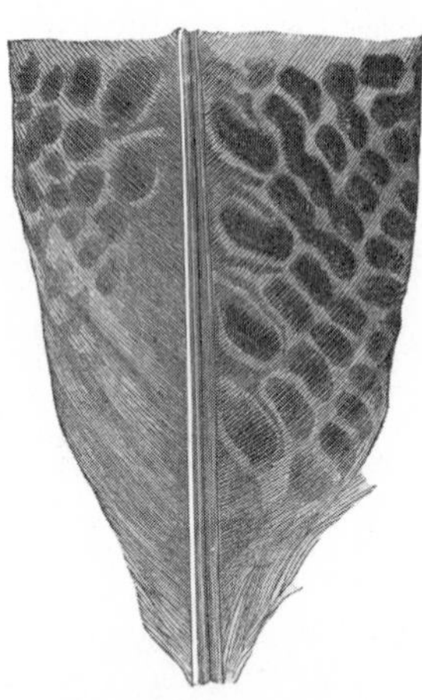

Fig. 57. *Basal part of the Secondary wing-feather, nearest to the body.*

I will next describe the other extreme of the series, namely the first trace of an ocellus. The short secondary wing-feather (fig. 57), nearest to the body, is marked like the other feathers, with oblique, longitudinal, rather irregular, rows of spots. The lowest spot, or that nearest the shaft, in the five lower rows (excluding the basal row) is a little larger than the other spots in the same row, and a little more elongated in a transverse direction. It differs also from the other spots by being bordered on its upper side with some dull fulvous shading. But this spot is not in any way more remarkable than those on the plumage of many birds, and might easily be quite overlooked. The next higher spot in each row does not differ at all from the upper ones in the same row, although in the following series it becomes, as we shall see, greatly modified. The larger spots occupy exactly the same relative position on this feather as those occupied by the perfect ocelli on the longer wing-feathers.

By looking to the next two or three succeeding secondary wing-feathers, an absolutely insensible gradation can be traced from one of the above-described lower spots, together with the next higher one in the same row, to a curious ornament, which cannot be called an ocellus, and which I will name, from the want of a better term, an "elliptic ornament." These are shewn in the accompanying figure (fig. 58). We here see several oblique rows, A, B, C, D (see the lettered diagram), &c., of dark spots of the usual character. Each row of spots runs down to and is connected with one of the elliptic ornaments, in exactly the same manner as each stripe in fig. 56 runs down to, and is connected with, one of the ball-and-socket ocelli. Looking to any one row, for instance, B, the lowest spot or mark *(b)* is thicker and considerably longer than the upper spots, and has its left extremity pointed

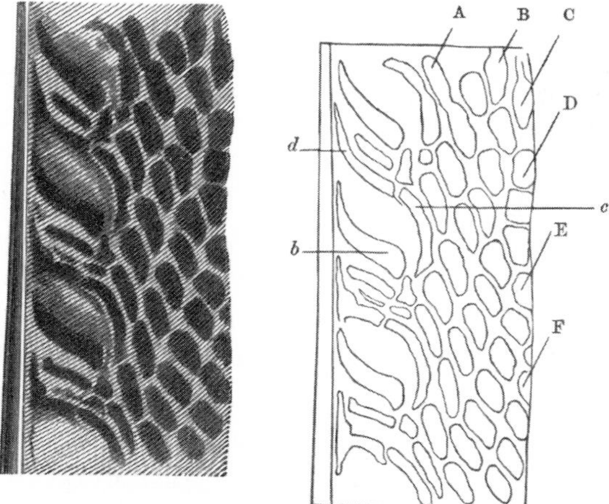

Fig. 58. *Portion of one of the Secondary wing-feathers near to the body; shewing the so-called elliptic ornaments. The right-hand figure is given merely as a diagram for the sake of the letters of reference.*

A, B, C, *&c. Rows of spots running down to and forming the elliptic ornaments.*

b. *Lowest spot or mark in row B.*

c. *The next succeeding spot or mark in the same row.*

d. *Apparently a broken prolongation of the spot c in the same row B.*

and curved upwards. This black mark is abruptly bordered on its upper side by a rather broad space of richly-shaded tints, beginning with a narrow brown zone, which passes into orange, and this into a pale leaden tint, with the end towards the shaft much paler. This mark corresponds in every respect with the larger, shaded spot, described in the last paragraph (fig. 57), but is more highly developed and more brightly coloured. To the right and above this spot *(b)*, with its bright shading, there is a long, narrow, black mark *(c)*, belonging to the same row, and which is arched a little downwards so as to face *(b)*. It is also narrowly edged on the lower side with a fulvous tint. To the left of and above *c*, in the same oblique direction, but always more or less distinct from it, there is another black mark *(d)*. This mark is generally sub-triangular and irregular in shape, but in the one lettered in the diagram is unusually narrow, elongated, and regular. It apparently consists of a lateral and broken prolongation of the mark *(c)*, as I infer from traces of similar prolongations from the succeeding upper spots; but I do not feel sure of this. These three marks, *b, c,* and *d*, with the intervening bright shades, form together the so-called elliptic ornament. These ornaments stand in a line parallel to the shaft, and manifestly correspond in position with the ball-and-socket ocelli. Their extremely elegant appearance cannot be appreciated in the drawing, as the orange and leaden tints, contrasting so well with the black marks, cannot be shewn.

Between one of the elliptic ornaments and a perfect ball-and-socket ocellus, the gradation is so perfect that it is scarcely possible to decide when the latter term ought to be used. I regret that I have not given an additional drawing, besides fig. 58, which stands about half-way in the series between one of the simple spots and a perfect ocellus. The passage from the elliptic ornament into an ocellus is effected by the elongation and greater curvature in opposed directions of the lower black mark *(b)*, and more especially of the upper one *(c)*, together with the contraction of the irregular sub-triangular or narrow mark *(d)*, so that at last these three marks become confluent, forming an irregular elliptic ring. This ring is gradually rendered more and more circular and regular, at the same time increasing in diameter. Traces of the junction of all three elongated spots or marks, especially of the two upper ones, can still be observed in many of the most perfect ocelli. The broken state of the black ring on the upper side of the ocellus in fig. 56 was pointed out. The irregular sub-triangular or narrow mark *(d)* manifestly forms, by its contraction and equalisation, the thickened portion of the ring on the left upper side of the perfect ball-and-socket ocellus. The lower part of the ring is invariably a little thicker than the other parts (see fig. 56), and this follows from the lower black mark of the elliptic ornament *(b)* having been originally thicker than the upper mark *(c)*. Every step can be followed in the process of confluence and modification; and the black ring which surrounds the ball of the ocellus is unquestionably formed by the union and modification of the three black marks, *b, c, d,* of the elliptic ornament. The irregular zigzag black marks between the successive ocelli (see again fig. 56) are plainly due to the breaking up of the somewhat more regular but similar marks between the elliptic ornaments.

The successive steps in the shading of the ball-and-socket ocelli can be followed out with equal clearness. The brown, orange, and pale-leaden narrow zones which border the lower black mark of the elliptic ornament can be seen gradually to become more and more softened and shaded into each other, with the upper lighter part towards the left-hand corner rendered still lighter, so as to become almost white. But even in the most perfect ball-and-socket ocelli a slight difference in the tints, though not in the shading, between the upper and lower parts of the ball can be perceived (as was before especially noticed), the line of separation being oblique, in the same direction with the bright coloured shades of the elliptic ornaments. Thus almost every minute detail in the shape and colouring of the ball-and-socket ocelli can be shewn to follow from gradual changes in the elliptic ornaments; and the development of the latter can be traced by equally small steps from the union of two almost simple spots, the lower one (fig. 57) having some dull fulvous shading on the upper side.

The extremities of the longer secondary feathers which bear the perfect ball-and-socket ocelli are peculiarly ornamented. (Fig. 59.) The oblique longitudinal stripes suddenly cease upwards and become confused, and above this limit the whole upper end of the feather *(a)* is covered with white dots, surrounded by little black rings, standing on a dark ground. Even the oblique stripe belonging to the uppermost ocellus *(b)* is represented only by a very short irregular black mark with the usual, curved, transverse base. As this stripe is thus abruptly cut off above, we can understand, from what has gone before, how it is that the upper thickened part of the ring is absent in the uppermost ocellus; for, as before stated, this thickened part is apparently formed by a broken prolongation of the next higher spot in the same row. From the absence of the upper and thickened part of the ring, the uppermost ocellus, though perfect in all other respects, appears as if its top had been obliquely sliced off. It would, I think, perplex any one, who believes that the plumage of the Argus-pheasant was created as we now see it, to account for the imperfect condition of the uppermost ocelli. I should add that in the secondary wing-feather farthest from the body all the ocelli are smaller and less perfect than on the other feathers, with the upper parts of the external black rings deficient, as in the case just mentioned. The imperfection here seems to be connected with the fact that the spots on this feather shew less tendency than usual to become confluent into stripes; on the contrary, they are often broken up into smaller spots, so that two or three rows run down to each ocellus.

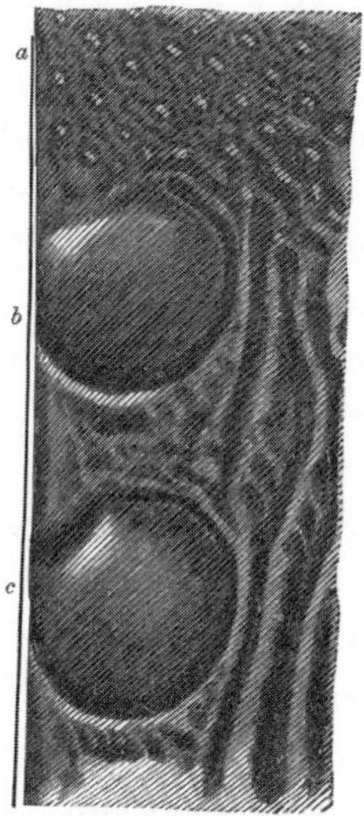

Fig. 59. *Portion near summit of one of the Secondary wing-feathers, bearing perfect ball-and-socket ocelli.*

a. *Ornamented upper part.*

b. *Uppermost, imperfect ball-and-socket ocellus. (The shading above the white mark on the summit of the ocellus is here a little too dark.)*

c. *Perfect ocellus.*

We have now seen that a perfect series can be followed, from two almost simple spots, at first quite distinct from each other, to one of the wonderful ball-and-socket ornaments. Mr. Gould, who kindly gave me some of these feathers, fully agrees with me in the completeness of the gradation. It is obvious that the stages in development exhibited by the feathers on the same bird do not at all necessarily shew us the steps which have been passed through by the extinct progenitors of the species; but they probably give us the clue to the actual steps, and they at least prove to demonstration that a gradation is possible. Bearing in mind how carefully the male Argus pheasant displays his plumes before the female, as well as the many facts rendering it probable that female birds prefer the more attractive males, no one who admits the agency of sexual selection, will deny that a simple dark spot with some fulvous shading might be converted, through the approximation and modification of the adjoining spots, together with some slight increase of colour, into one of the so-called elliptic ornaments. These latter ornaments have been shewn to many persons, and all have admitted that they are extremely pretty, some thinking them even more beautiful than the ball-and-socket ocelli. As the secondary plumes became lengthened through sexual selection, and as the elliptic ornaments increased in diameter, their colours apparently became less bright; and then the ornamentation of the plumes had to be gained by improvements in the pattern and shading; and this process has been carried on until the wonderful ball-and-socket ocelli have been finally developed. Thus we can understand—and in no other way as it seems to me—the present condition and origin of the ornaments on the wing-feathers of the Argus pheasant.

From the light reflected by the principle of gradation; from what we know of the laws of variation; from the changes which have taken place in many of our domesticated birds; and, lastly, from the character (as we shall hereafter more clearly see) of the immature plumage of young birds—we can sometimes indicate with a certain amount of confidence, the probable steps by which the males have acquired their brilliant plumage and various ornaments; yet in many cases we are involved in darkness. Mr. Gould several years ago pointed out to me a humming-bird, the *Urosticte benjamini,* remarkable from the curious differences presented by the two sexes. The male, besides a splendid gorget, has greenish-black tail-feathers, with the four *central* ones tipped with white; in the female, as with most of the allied species, the three *outer* tail-feathers on each side are tipped with white, so that the male has the four central, whilst the female has the six exterior feathers ornamented with white tips. What makes the case curious is that, although the colouring of the tail differs remarkably in both sexes of many kinds of humming-birds, Mr. Gould does not know a single species, besides the Urosticte, in which the male has the four central feathers tipped with white.

The Duke of Argyll, in commenting on this case,[730] passes over sexual selection, and asks, "What explanation does the law of natural selection give of such specific varieties as these?" He answers "none whatever;" and I quite agree with him. But can this be so confidently said of sexual selection? Seeing in how many ways the tail-feathers of humming-birds differ, why should not the four central feathers have varied in this one species alone, so as to have acquired white tips? The variations may have been gradual, or somewhat abrupt as in the case recently given of the humming-birds near Bogota, in which certain individuals alone have the "central tail-feathers tipped with beautiful green." In the female of the Urosticte I noticed extremely minute or rudimental white tips to the two outer of the four central black tail-feathers; so that here we have an indication of change of some kind in the plumage of this species. If we grant the possibility of the central tail-feathers of the male varying in whiteness, there is nothing strange in such variations having been sexually selected. The white tips, together with the small white ear-tufts, certainly add, as the Duke of Argyll admits, to the beauty of the male; and whiteness is apparently appreciated by other birds, as may be inferred from such cases as the snow-white male of the Bell-bird. The statement made by Sir R. Heron should not be forgotten, namely that his peahens, when debarred from access to the pied peacock, would not unite with any other male, and during that season produced no offspring. Nor is it strange that variations in the tail-feathers of the Urosticte should have been specially selected for the sake of ornament, for the next succeeding genus in the family takes its name of Metallura from the splendour of these feathers. Mr. Gould, after describing the peculiar plumage of the Urosticte, adds, "that ornament and variety is the sole object, I have myself but little doubt."[731] If this be admitted, we can perceive that the males which were decked in the most elegant and novel manner would have gained an advantage, not in the ordinary struggle for life, but in rivalry with other males, and would consequently have left a larger number of offspring to inherit their newly-acquired beauty.

CHAPTER XV

BIRDS—CONTINUED

DISCUSSION WHY THE MALES ALONE OF SOME SPECIES, AND BOTH SEXES OF OTHER SPECIES, ARE BRIGHTLY COLOURED • ON SEXUALLY-LIMITED INHERITANCE, AS APPLIED TO VARIOUS STRUCTURES AND TO BRIGHTLY-COLOURED PLUMAGE • NIDIFICATION IN RELATION TO COLOUR • LOSS OF NUPTIAL PLUMAGE DURING THE WINTER

WE HAVE IN THIS CHAPTER to consider, why with many kinds of birds the female has not received the same ornaments as the male; and why with many others, both sexes are equally, or almost equally, ornamented? In the following chapter we shall consider why in some few rare cases the female is more conspicuously coloured than the male.

In my 'Origin of Species'[732] I briefly suggested that the long tail of the peacock would be inconvenient, and the conspicuous black colour of the male capercailzie dangerous, to the female during the period of incubation; and consequently that the transmission of these characters from the male to the female offspring had been checked through natural selection. I still think that this may have occurred in some few instances: but after mature reflection on all the facts which I have been able to collect, I am now inclined to believe that when the sexes differ, the successive variations have generally been from the first limited in their transmission to the same sex in which they first appeared. Since my remarks appeared, the subject of sexual coloration has been discussed in some very interesting papers by Mr. Wallace,[733] who believes that in almost all cases the successive variations tended at first to be transmitted equally to both sexes; but that the female was saved, through natural selection, from acquiring the conspicuous colours of the male, owing to the danger which she would thus have incurred during incubation.

This view necessitates a tedious discussion on a difficult point, namely whether the transmission of a character, which is at first inherited by both sexes, can be subsequently limited in its transmission, by means of selection, to one sex alone. We must bear in mind, as shewn in the preliminary chapter on sexual selection, that characters which are limited in their development to one sex are always latent in the other. An imaginary illustration will best aid us in seeing the difficulty of the case: we may suppose that a fancier wished to make a breed of pigeons, in which the males alone should be coloured of a pale blue, whilst the females retained their former slaty tint. As with pigeons characters of all kinds are usually transmitted to both sexes equally, the fancier would have to try to convert this latter form of inheritance into sexually-limited transmission. All that he could do would be to persevere in selecting every male pigeon which was in the least degree of a paler blue; and the natural result of this process, if steadily carried on for a long time, and if the pale variations were strongly inherited or often recurred, would be to make his whole stock of a lighter blue. But our fancier would be compelled to match, generation after generation, his pale blue males with slaty females, for he wishes to keep the latter of this colour. The result would generally be the production either of a mongrel piebald lot, or more probably the speedy and complete loss of the pale-blue colour, for the primordial slaty tint would be transmitted with prepotent force. Supposing, however, that some pale-blue males and slaty females were produced during each successive generation, and were always crossed together; then the slaty females would have, if I may use the expression, much blue blood in their veins, for their fathers, grandfathers, &c., will all have been blue birds. Under these circumstances it is conceivable (though I know of no distinct facts rendering it probable) that the slaty females might acquire so strong a latent tendency to pale-blueness, that they would not destroy this colour in their male offspring, their female offspring still inheriting the slaty tint. If so, the desired end of making a breed with the two sexes permanently different in colour might be gained.

The extreme importance, or rather necessity, of the desired character in the above case, namely, pale-blueness, being present though in a latent state in the female, so that the male offspring should not be deteriorated, will be best appreciated as follows: the male of Sœmmerring's pheasant has a tail thirty-seven inches in length, whilst that of the female is only eight inches; the tail of the male common pheasant is about twenty inches, and that of the female twelve inches long. Now if the female Sœmmerring pheasant with her *short* tail were crossed with the male common pheasant, there can be no doubt that the male hybrid offspring would have a much *longer* tail than that of the pure offspring of the common pheasant. On the other hand, if the female common pheasant, with her tail nearly *twice as long* as that of the female Sœmmerring pheasant, were crossed with the male of the latter, the male hybrid offspring would have a much *shorter* tail than that of the pure offspring of Sœmmerring's pheasant.[734]

Our fancier, in order to make his new breed with the males of a decided pale-blue tint, and the females unchanged, would have to continue selecting the males during many generations; and each stage of paleness would have to be fixed in the males, and rendered latent in the females. The task would be an extremely difficult one, and has never been tried, but might possibly succeed. The chief obstacle would be the early and complete loss of the pale-blue tint, from the necessity of reiterated crosses with the slaty female, the latter not having at first any *latent* tendency to produce pale-blue offspring.

On the other hand, if one or two males were to vary ever so slightly in paleness, and the variations were from the first limited in their transmission to the male sex, the task of making a new breed of the desired kind would be easy, for such males would simply have to be selected and matched with ordinary females. An analogous case has actually occurred, for there are breeds of the pigeon in Belgium[735] in which the males alone are marked with black striæ. In the case of the fowl, variations of colour limited in their transmission to the male sex habitually occur. Even when this form of inheritance prevails, it might well happen that some of the successive steps in the process of variation might be transferred to the female, who would then come to resemble in a slight degree the male, as occurs in some breeds of the fowl. Or again, the greater number, but not all, of the successive steps might be transferred to both sexes, and the female would then closely resemble the male. There can hardly be a doubt that this is the cause of the male pouter pigeon having a somewhat larger crop, and of the male carrier pigeon having somewhat larger wattles, than their respective females; for fanciers have not selected one sex more than the other, and have had no wish that these characters should be more strongly displayed in the male than in the female, yet this is the case with both breeds.

The same process would have to be followed, and the same difficulties would be encountered, if it were desired to make a breed with the females alone of some new colour.

Lastly, our fancier might wish to make a breed with the two sexes differing from each other, and both from the parent-species. Here the difficulty would be extreme, unless the successive variations were from the first sexually limited on both sides, and then there would be no difficulty. We see this with the fowl; thus the two sexes of the pencilled Hamburghs differ greatly from each other, and from the two sexes of the aboriginal *Gallus bankiva;* and both are now kept constant to their standard of excellence by continued selection, which would be impossible unless the distinctive characters of both were limited in their transmission. The Spanish fowl offers a more curious case; the male has an immense comb, but some of the successive variations, by the accumulation of which it was acquired, appear to have been transferred to the female; for she has a comb many times larger than that of the females of the parent-species. But the comb of the female differs in one respect from that of the male, for it is apt to lop over; and within a recent period it has been ordered by the fancy that this should always be the case, and success has quickly followed the order. Now the lopping of the comb must be sexually limited in its transmission, otherwise it would prevent the comb of the male from being perfectly upright, which would be abhorrent to every fancier. On the other hand the uprightness of the comb in the male must likewise be a sexually-limited character, otherwise it would prevent the comb of the female from lopping over.

From the foregoing illustrations, we see that even with almost unlimited time at command, it would be an extremely difficult and complex process, though perhaps not impossible, to change through selection one form of transmission into the other. Therefore, without distinct evidence in each case, I am unwilling to admit that this has often been effected with natural species. On the other hand by means of successive variations, which were from the first sexually limited in their transmission, there would not be the least difficulty in rendering a male bird widely different in colour or in any other character from the female; the latter being left unaltered, or slightly altered, or specially modified for the sake of protection.

As bright colours are of service to the males in their rivalry with other males, such colours would be selected, whether or not they were transmitted exclusively to the same sex. Consequently the females might be expected often to partake of the brightness of the males to a greater or less degree; and this occurs with a host of species. If all the successive variations were transmitted equally to both sexes, the females would be undistinguishable from the males; and this likewise occurs with many birds. If, however, dull colours were of high importance for the safety of the female during incubation, as with many ground birds, the females which varied in brightness, or which received through inheritance from the males any marked accession of brightness, would sooner or later be destroyed. But the tendency in the males to continue for an indefinite period transmitting to their female offspring their own brightness, would have to be eliminated by a change in the form of inheritance; and this, as shewn by our previous illustration, would be extremely difficult. The more probable result of the long-continued destruction of the more brightly-coloured females, supposing the equal form of transmission to prevail, would be the lessening or annihilation of the bright colours of the males, owing to their continually crossing with the duller females. It would be tedious to follow out all the other possible results; but I may remind the reader, as shewn in the eighth chapter, that if sexually-limited variations in brightness occurred in the females, even if they were not in the least injurious to them and consequently were not eliminated, yet they would not be favoured or selected, for the male usually accepts any female, and does not select the more attractive individuals; consequently these variations would be liable to be lost, and would have little influence on the character of the race; and this will aid in accounting for the females being commonly less brightly-coloured than the males.

In the chapter just referred to, instances were given, and any number might have been added, of variations occurring at different ages, and inherited at the same age. It was also shewn that variations which occur late in life are commonly transmitted to the same sex in which they first appeared; whilst variations occurring early in life are apt to be transmitted to both sexes; not that all the cases of sexually-limited transmission can thus be accounted for. It was further shewn that if a male bird varied by becoming brighter whilst young, such variations would be of no service until the age for reproduction had arrived, and there was competition between rival males. If we suppose that three-fourths of the young males of any species are on an average destroyed by various enemies; then the chances would be as three to one against any one individual more brightly-coloured than usual surviving to propagate its kind. But in the case of birds which live on the ground and which commonly need the protection of dull colours, bright tints would be far more dangerous to the young and inexperienced than to the adult males. Consequently the males which varied in brightness whilst young would suffer much destruction and be eliminated through natural selection; on the other hand the males which varied in this manner when nearly mature, notwithstanding that they were exposed to some additional danger, might survive, and from being favoured through sexual selection, would procreate their kind. The brightly-coloured young males being destroyed and the mature ones being successful in their courtship, may account, on the principle of a relation existing between the period of variation and the form of transmission, for the males alone of many birds, having acquired and transmitted brilliant colours to their male offspring alone. But I by no means wish to maintain that the influence of age on the form of transmission is indirectly the sole cause of the great difference in brilliancy between the sexes of many birds.

As with all birds in which the sexes differ in colour, it is an interesting question whether the males alone have been modified through sexual selection, the females being left, as far as this agency is concerned, unchanged or only partially changed; or whether the females have been specially modified through natural selection for the sake of protection, I will discuss this question at considerable length, even at greater length than its intrinsic importance deserves; for various curious collateral points may thus be conveniently considered.

Before we enter on the subject of colour, more especially in reference to Mr. Wallace's conclusions, it may be useful to discuss under a similar point of view some other differences between the sexes. A breed of fowls formerly existed in Germany[736] in which the hens were furnished with spurs; they were good layers, but they so greatly disturbed their nests with their spurs that they could not be allowed to sit on their own eggs. Hence at one time it appeared to me probable that with the females of the wild Gallinaceæ the development of spurs had been checked through natural selection, from the injury thus caused to their nests. This seemed all the more probable as the wing-spurs, which could not be injurious during nidification, are often as well developed in the female as in the male; though in not a few cases they are rather larger in the male. When the male is furnished with leg-spurs the female almost always exhibits rudiments of them,—the rudiment sometimes consisting of a mere scale, as with the species of Gallus. Hence it might be argued that the females had aboriginally been furnished with well-developed spurs, but that these had subsequently been lost either through disuse or natural selection. But if this view be admitted, it would have to be extended to innumerable other cases; and it implies that the female progenitors of the existing spur-bearing species were once encumbered with an injurious appendage.

In some few genera and species, as in Galloperdix, Acomus, and the Javan peacock *(Pavo muticus),* the females, as well as the males, possess well-developed spurs. Are we to infer from this fact that they construct a different sort of nest, not liable to be injured by their spurs, from that made by their nearest allies, so that there has been no need for the removal of their spurs? Or are we to suppose that these females especially require spurs for their defence? It is a more probable conclusion that both the presence and absence of spurs in the females result from different laws of inheritance having prevailed, independently of natural selection. With the many females in which spurs appear as rudiments, we may conclude that some few of the successive variations, through which they were developed in the males, occurred very early in life, and were as a consequence transferred to the females. In the other and much rarer cases, in which the females possess fully developed spurs, we may conclude that all the successive variations were transferred to them; and that they gradually acquired the inherited habit of not disturbing their nests.

The vocal organs and the variously-modified feathers for producing sound, as well as the proper instincts for using them, often differ in the two sexes, but are sometimes the same in both. Can such differences be accounted for by the males having acquired these organs and instincts, whilst the females have been saved from inheriting them, on account of the danger to which they would have been exposed by attracting the attention of birds or beasts of prey? This does not seem to me probable, when we think of the multitude of birds which with impunity gladden the country with their voices during the spring.[737] It is a safer conclusion that as vocal and instrumental organs are of special service only to the males during their courtship, these organs were developed through sexual selection and continued use in this sex alone—the successive variations and the effects of use having been from the first limited in their transmission in a greater or less degree to the male offspring.

Many analogous cases could be advanced; for instance the plumes on the head, which are generally longer in the male than in the female, sometimes of equal length in both sexes, and occasionally absent in the female,—these several cases sometimes occurring in the same group of birds. It would be difficult to account for a difference of this kind between the sexes on the principle of the female having been benefited by possessing a slightly shorter crest than the male, and its consequent diminution or complete suppression through natural selection. But I will take a more favourable case, namely, the length of the tail. The long train of the peacock would have been not only inconvenient but dangerous to the peahen during the period of incubation and whilst accompanying her young. Hence there is not the least *à priori* improbability in the development of her tail having been checked through natural selection. But the females of various pheasants, which apparently are exposed on their open nests to as much danger as the peahen, have tails of considerable length. The females as well as the males of the *Menura superba* have long tails, and they build a domed nest, which is a great anomaly in so large a bird. Naturalists have wondered how the female Menura could manage her tail during incubation; but it is now known[738] that she "enters the nest head first, and then turns round with her tail sometimes over her back, but more often bent round by her side. Thus in time the tail becomes quite askew, and is a tolerable guide to the length of time the bird has been sitting." Both sexes of an Australian kingfisher *(Tanysiptera sylvia)* have the middle tail-feathers greatly lengthened; and as the female makes her nest in a hole, these feathers become, as I am informed by Mr. R. B. Sharpe, much crumpled during nidification.

In these two cases the great length of the tail-feathers must be in some degree inconvenient to the female; and as in both species the tail-feathers of the female are somewhat shorter than those of the male, it might be argued that their full development had been prevented through natural selection. Judging from these cases, if with the peahen, the development of the tail had been checked only when it became inconveniently or dangerously long, she would have acquired a much longer tail than she actually possesses; for her tail is not nearly so long, relatively to the size of her body, as that of many female pheasants, nor longer than that of the female turkey. It must also be borne in mind, that in accordance with this view as soon as the tail of the peahen became dangerously long, and its development was consequently checked, she would have continually reacted on her male progeny, and thus have prevented the peacock from acquiring his present magnificent train. We may therefore infer that the length of the tail in the peacock and its shortness in the peahen are the result of the requisite variations in the male having been from the first transmitted to the male offspring alone.

We are led to a nearly similar conclusion with respect to the length of the tail in the various species of pheasants. In the Eared pheasant *(Crossoptilon auritum)* the tail is of equal length in both sexes, namely, sixteen or seventeen inches; in the common pheasant it is about twenty inches long in the male, and twelve in the female; in Sœmmerring's pheasant, thirty-seven inches in the male, and only eight in the female; and lastly in Reeve's pheasant it is sometimes actually seventy-two inches long in the male and sixteen in the female. Thus in the several species, the tail of the female differs much in length, irrespectively of that of the male; and this can be accounted for as it seems to me, with much more probability, by the laws of inheritance,—that is by the successive variations having been from the first more or less closely limited in their transmission to the male sex,—than by the agency of natural selection, owing to the length of tail having been injurious in a greater or less degree to the females of the several species.

We may now consider Mr. Wallace's arguments in regard to the sexual coloration of birds. He believes that the bright tints originally acquired through sexual selection by the males, would in all or almost all cases have been transmitted to the females, unless the transference had been checked through natural selection. I may here remind the reader that various facts bearing on this view have already been given under reptiles, amphibians, fishes, and lepidoptera. Mr. Wallace rests his belief chiefly, but not exclusively, as we shall see in the next chapter, on the following statement,[739] that when both sexes are coloured in a strikingly-conspicuous manner the nest is of such a nature as to conceal the sitting bird; but when there is a marked contrast of colour between the sexes, the male being gay and the female dull-coloured, the nest is open and exposes the sitting bird to view. This coincidence, as far as it goes, certainly supports the belief that the females which sit on open nests have been specially modified for the sake of protection. Mr. Wallace admits that there are, as might have been expected, some exceptions to his two rules, but it is a question whether the exceptions are not so numerous as seriously to invalidate them.

There is in the first place much truth in the Duke of Argyll's remark[740] that a large domed nest is more conspicuous to an enemy, especially to all tree-haunting carnivorous animals, than a smaller open nest. Nor must we forget that with many birds which build open nests the males sit on the eggs and aid in feeding the young as well as the females: this is the case, for instance, with *Pyranga æstiva*,[741] one of the most splendid birds in the United States, the male being vermilion, and the female light brownish-green. Now if brilliant colours had been extremely dangerous to birds whilst sitting on their open nests, the males in these cases would have suffered greatly. It might, however, be of such paramount importance to the male to be brilliantly coloured, in order to beat his rivals, that this would more than compensate for some additional danger.

Mr. Wallace admits that with the King-crows (Dicrurus), Orioles, and Pittidæ, the females are conspicuously coloured, yet they build open nests; but he urges that the birds of the first group are highly pugnacious and could defend themselves; that those of the second group take extreme care in concealing their open nests, but this does not invariably hold good;[742] and that with the birds of the third group the females are brightly coloured chiefly on the under surface. Besides these cases the whole great family of pigeons, which are sometimes brightly, and almost always conspicuously coloured, and which are notoriously liable to the attacks of birds of prey, offers a serious exception to the rule, for pigeons almost always build open and exposed nests. In another large family, that of the Humming-birds, all the species build open nests, yet with some of the most gorgeous species the sexes are alike; and in the majority, the females, though less brilliant than the males, are very brightly coloured. Nor can it be maintained that all female humming-birds, which are brightly coloured, escape detection by their tints being green, for some display on their upper surfaces red, blue, and other colours.[743]

In regard to birds which build in holes or construct domed nests, other advantages, as Mr. Wallace remarks, besides concealment are gained, such as shelter from the rain, greater warmth, and in hot countries protection from the rays of the sun;[744] so that it is no valid objection to his view that many birds having both sexes obscurely coloured build concealed nests.[745] The female Horn-bills *(Buceros)*, for instance, of India and Africa are protected, during nidification, with extraordinary care, for the male plaisters up the hole in which the female sits on her eggs, and leaves only a small orifice through which he feeds her; she is thus kept a close prisoner during the whole period of incubation;[746] yet female hornbills are not more conspicuously coloured than many other birds of equal size which build open nests.

It is a more serious objection to Mr. Wallace's view, as is admitted by him, that in some few groups the males are brilliantly coloured and the females obscure, and yet the latter hatch their eggs in domed nests. This is the case with the Grallinæ of Australia, the Superb Warblers (Maluridæ) of the same country, the Sun-birds (Nectariniæ), and with several of the Australian Honey-suckers or Meliphagidæ.[747]

If we look to the birds of England we shall see that there is no close and general relation between the colours of the female and the nature of the nest constructed by her. About forty of our British birds (excluding those of large size which could defend themselves) build in holes in banks, rocks, or trees, or construct domed nests. If we take the colours of the female goldfinch, bullfinch, or blackbird, as a standard of the degree of conspicuousness, which is not highly dangerous to the sitting female, then out of the above forty birds, the females of only twelve can be considered as conspicuous to a dangerous degree, the remaining twenty-eight being inconspicuous.[748] Nor is there any close relation between a well-pronounced difference in colour between the two sexes, and the nature of the nest constructed. Thus the male house-sparrow *(Passer domesticus)* differs much from the female, the male tree-sparrow *(P. montanus)* differs hardly at all, and yet both build well-concealed nests. The two sexes of the common fly-catcher *(Muscicapa grisola)* can hardly be distinguished, whilst the sexes of the pied fly-catcher *(M. luctuosa)* differ considerably, and both build in holes. The female blackbird *(Turdus merula)* differs much, the female ring-ouzel *(T. torquatus)* differs less, and the female common thrush *(T. musicus)* hardly at all from their respective males; yet all build open nests. On the other hand, the not very distantly-allied water-ouzel *(Cinclus aquaticus)* builds a domed nest, and the sexes differ about as much as in the case of the ring-ouzel. The black and red grouse *(Tetrao tetrix* and *T. Scoticus)* build open nests, in equally well-concealed spots, but in the one species the sexes differ greatly, and in the other very little.

Notwithstanding the foregoing objections, I cannot doubt, after reading Mr. Wallace's excellent essay, that looking to the birds of the world, a large majority of the species in which the females are conspicuously coloured (and in this case the males with rare exceptions are equally conspicuous), build concealed nests for the sake of protection. Mr. Wallace enumerates[749] a long series of groups in which this rule holds good; but it will suffice here to give, as instances, the more familiar groups of kingfishers, toucans, trogons, puff-birds (Capitonidæ), plaintain-eaters (Musophagæ), woodpeckers, and parrots. Mr. Wallace believes that in these groups, as the males gradually acquired through sexual selection their brilliant colours, these were transferred to the females and were not eliminated by natural selection, owing to the protection which they already enjoyed from their manner of nidification. According to this view, their present manner of nesting was acquired before their present colours. But it seems to me much more probable that in most cases as the females were gradually rendered more and more brilliant from partaking of the colours of the male, they were gradually led to change their instincts (supposing that they originally built open nests), and to seek protection by building domed or concealed nests. No one who studies, for instance, Audubon's account of the differences in the nests of the same species in the Northern and Southern United States,[750] will feel any great difficulty in admitting that birds, either by a change (in the strict sense of the word) of their habits, or through the natural selection of so-called spontaneous variations of instinct, might readily be led to modify their manner of nesting.

This way of viewing the relation, as far as it holds good, between the bright colours of female birds and their manner of nesting, receives some support from certain analogous cases occurring in the Sahara Desert. Here, as in most other deserts, various birds, and many other animals, have had their colours adapted in a wonderful manner to the tints of the surrounding surface. Nevertheless there are, as I am informed by the Rev. Mr. Tristram, some curious exceptions to the rule; thus the male of the *Monticola cyanea* is conspicuous from his bright blue colour, and the female almost equally conspicuous from her mottled brown and white plumage; both sexes of two species of Dromolæa are of a lustrous black; so that these three birds are far from receiving protection from their colours, yet they are able to survive, for they have acquired the habit, when in danger, of taking refuge in holes or crevices in the rocks.

With respect to the above-specified groups of birds, in which the females are conspicuously coloured and build concealed nests, it is not necessary to suppose that each separate species had its nidifying instinct specially modified; but only that the early progenitors of each group were gradually led to build domed or concealed nests; and afterwards transmitted this instinct, together with their bright colours, to their modified descendants. This conclusion, as far as it can be trusted, is interesting, namely, that sexual selection, together with equal or nearly equal inheritance by both sexes, have indirectly determined the manner of nidification of whole groups of birds.

Even in the groups in which, according to Mr. Wallace, the females from being protected during nidification, have not had their bright colours eliminated through natural selection, the males often differ in a slight, and occasionally in a considerable degree, from the females. This is a significant fact, for such differences in colour must be accounted for on the principle of some of the variations in the males having been from the first limited in their transmission to the same sex; as it can hardly be maintained that these differences, especially when very slight, serve as a protection to the female. Thus all the species in the splendid group of the Trogons build in holes; and Mr. Gould gives figures[751] of both sexes of twenty-five species, in all of which, with one partial exception, the sexes differ sometimes slightly, sometimes conspicuously, in colour,—the males being always more beautiful than the females, though the latter are likewise beautiful. All the species of kingfisher build in holes, and with most of the species the sexes are equally brilliant, and thus far Mr. Wallace's rule holds good; but in some of the Australian species the colours of the females are rather less vivid than those of the male; and in one splendidly-coloured species, the sexes differ so much that they were at first thought to be specifically distinct.[752] Mr. R. B. Sharpe, who has especially studied this group, has shewn me some American species (Ceryle) in which the breast of the male is belted with black. Again, in Carcineutes, the difference between the sexes is conspicuous: in the male the upper surface is dull-blue banded with black, the lower surface being partly fawn-coloured, and there is much red about the head; in the female the upper surface is reddish-brown banded with black, and the lower surface white with black markings. It is an interesting fact, as shewing how the same peculiar style of sexual colouring often characterises allied forms, that in three species of Dacelo the male differs from the female only in the tail being dull-blue banded with black, whilst that of the female is brown with blackish bars; so that here the tail differs in colour in the two sexes in exactly the same manner as the whole upper surface in the sexes of Carcineutes.

With parrots, which likewise build in holes, we find analogous cases: in most of the species both sexes are brilliantly coloured and undistinguishable, but in not a few species the males are coloured rather more vividly than the females, or even very differently from them. Thus, besides other strongly-marked differences, the whole under surface of the male King Lory *(Aprosmictus scapulatus)* is scarlet, whilst the throat and chest of the female is green tinged with red: in the *Euphema splendida* there is a similar difference, the face and wing-coverts moreover of the female being of a paler blue than in the male.[753] In the family of the tits *(Parinæ)*, which build concealed nests, the female of our common blue tomtit *(Parus cæruleus)* is "much less brightly coloured" than the male; and in the magnificent Sultan yellow tit of India the difference is greater.[754]

Again in the great group of the woodpeckers,[755] the sexes are generally nearly alike, but in the *Megapicus validus* all those parts of the head, neck, and breast, which are crimson in the male are pale brown in the female. As in several woodpeckers the head of the male is bright crimson, whilst that of the female is plain, it occurred to me that this colour might possibly make the female dangerously conspicuous, whenever she put her head out of the hole containing her nest, and consequently that this colour, in accordance with Mr. Wallace's belief, had been eliminated. This view is strengthened by what Malherbe states with respect to *Indopicus carlotta;* namely, that the young females, like the young males, have some crimson about their heads, but that this colour disappears in the adult female, whilst it is intensified in the adult male. Nevertheless the following considerations render this view extremely doubtful: the male takes a fair share in incubation,[756] and would be thus far almost equally exposed to danger; both sexes of many species have their heads of an equally bright crimson; in other species the difference between the sexes in the amount of scarlet is so slight that it can hardly make any appreciable difference in the danger incurred; and lastly, the colouring of the head in the two sexes often differs slightly in other ways.

The cases, as yet given, of slight and graduated differences in colour between the males and females in the groups, in which as a general rule the sexes resemble each other, all relate to species which build domed or concealed nests. But similar gradations may likewise be observed in groups in which the sexes as a general rule resemble each other, but which build open nests. As I have before instanced the Australian parrots, so I may here instance, without giving any details, the Australian pigeons.[757] It deserves especial notice that in all these cases the slight differences in plumage between the sexes are of the same general nature as the occasionally greater differences. A good illustration of this fact has already been afforded by those kingfishers in which either the tail alone or the whole upper surface of the plumage differs in the same manner in the two sexes. Similar cases may be observed with parrots and pigeons. The differences in colour between the sexes of the same species are, also, of the same general nature as the differences in colour between the distinct species of the same group. For when in a group in which the sexes are usually alike, the male differs considerably from the female, he is not coloured in a quite new style. Hence we may infer that within the same group the special colours of both sexes when they are alike, and the colours of the male, when he differs slightly or even considerably from the female, have in most cases been determined by the same general cause; this being sexual selection.

It is not probable, as has already been remarked, that differences in colour between the sexes, when very slight, can be of service to the female as a protection. Assuming, however, that they are of service, they might be thought to be cases of transition; but we have no reason to believe that many species at any one time are undergoing change. Therefore we can hardly admit that the numerous females which differ very slightly in colour from their males are now all commencing to become obscure for the sake of protection. Even if we consider somewhat more marked sexual differences, is it probable, for instance, that the head of the female chaffinch, the crimson on the breast of the female bullfinch,—the green of the female greenfinch,—the crest of the female golden-crested wren, have all been rendered less bright by the slow process of selection for the sake of protection? I cannot think so; and still less with the slight differences between the sexes of those birds which build concealed nests. On the other hand, the differences in colour between the sexes, whether great or small, may to a large extent be explained on the principle of the successive variations, acquired by the males through sexual selection, having been from the first more or less limited in their transmission to the females. That the degree of limitation should differ in different species of the same group will not surprise any one who has studied the laws of inheritance, for they are so complex that they appear to us in our ignorance to be capricious in their action.[758]

As far as I can discover there are very few groups of birds containing a considerable number of species, in which all have both sexes brilliantly coloured and alike; but this appears to be the case, as I hear from Mr. Sclater, with the Musophagæ or plaintain-eaters. Nor do I believe that any large group exists in which the sexes of all the species are widely dissimilar in colour: Mr. Wallace informs me that the chatterers of S. America *(Cotingidæ)* offer one of the best instances; but with some of the species, in which the male has a splendid red breast, the female exhibits some red on her breast; and the females of other species shew traces of the green and other colours of the males. Nevertheless we have a near approach to close sexual similarity or dissimilarity throughout several groups: and this, from what has just been said of the fluctuating nature of inheritance, is a somewhat surprising circumstance. But that the same laws should largely prevail with allied animals is not surprising. The domestic fowl has produced a great number of breeds and sub-breeds, and in these the sexes generally differ in plumage; so that it has been noticed as a remarkable circumstance when in certain sub-breeds they resemble each other. On the other hand, the domestic pigeon has likewise produced a vast number of distinct breeds and sub-breeds, and in these, with rare exceptions, the two sexes are identically alike. Therefore if other species of Gallus and Columba were domesticated and varied, it would not be rash to predict that the same general rules of sexual similarity and dissimilarity, depending on the form of transmission, would, in both cases, hold good. In a similar manner the same form of transmission has generally prevailed throughout the same natural groups, although marked exceptions to this rule occur. Within the same family or even genus, the sexes may be identically alike or very different in colour. Instances have already been given relating to the same genus, as with sparrows, fly-catchers, thrushes and grouse. In the family of pheasants the males and females of almost all the species are wonderfully dissimilar, but are quite similar in the eared pheasant or *Crossoptilon auritum.* In two species of Chloephaga, a genus of geese, the males cannot be distinguished from the females, except by size; whilst in two others, the sexes are so unlike that they might easily be mistaken for distinct species.[759]

The laws of inheritance can alone account for the following cases, in which the female by acquiring at a late period of life certain characters proper to the male, ultimately comes to resemble him in a more or less complete manner. Here protection can hardly have come into play. Mr. Blyth informs me that the females of *Oriolus melanocephalus* and of some allied species, when sufficiently mature to breed, differ considerably in plumage from the adult males; but after the second or third moults they differ only in their beaks having a slight greenish tinge. In the dwarf bitterns (Ardetta), according to the same authority, "the male acquires his final livery at the first moult, the female not before the third or fourth moult; in the meanwhile she presents an intermediate garb, which is ultimately exchanged for the same livery as that of the male." So again the female *Falco peregrinus* acquires her blue plumage more slowly than the male. Mr. Swinhoe states that with one of the Drongo shrikes *(Dicrurus macrocercus)* the male whilst almost a nestling, moults his soft brown plumage and becomes of a uniform glossy greenish-black; but the female retains for a long time the white striæ and spots on the axillary feathers; and does not completely assume the uniform black colour of the male for the first three years. The same excellent observer remarks that in the spring of the second year the female spoonbill (Platalea) of China resembles the male of the first year, and that apparently it is not until the third spring that she acquires the same adult plumage as that possessed by the male at a much earlier age. The female *Bombycilla carolinensis* differs very little from the male, but the appendages, which like beads of red sealing-wax ornament the wing-feathers, are not developed in her so early in life as in the male. The upper mandible in the male of an Indian parrakeet *(Palæornis Javanicus)* is coral-red from his earliest youth, but in the female, as Mr. Blyth has observed with caged and wild birds, it is at first black and does not become red until the bird is at least a year old, at which age the sexes resemble each other in all respects. Both sexes of the wild turkey are ultimately furnished with a tuft of bristles on the breast, but in two-year-old birds the tuft is about four inches long in the male and hardly apparent in the female; when, however, the latter has reached her fourth year, it is from four to five inches in length.[760]

In these cases, the females follow a normal course of development in ultimately becoming like the males; and such cases must not be confounded with those in which diseased or old females assume masculine characters, or with those in which perfectly fertile females, whilst young, acquire through variation or some unknown cause the characters of the male.[761] But all these cases have so much in common that they depend, according to the hypothesis of pangenesis, on gemmules derived from each part of the male being present, though latent, in the female; their development following on some slight change in the elective affinities of her constituent tissues.

A FEW WORDS must be added on changes of plumage in relation to the season of the year. From reasons formerly assigned there can be little doubt that the elegant plumes, long pendant feathers, crests, &c., of egrets, herons, and many other birds, which are developed and retained only during the summer, serve exclusively for ornamental or nuptial purposes, though common to both sexes. The female is thus rendered more conspicuous during the period of incubation than during the winter; but such birds as herons and egrets would be able to defend themselves. As, however, plumes would probably be inconvenient and certainly of no use during the winter, it is possible that the habit of moulting twice in the year may have been gradually acquired through natural selection for the sake of casting off inconvenient ornaments during the winter. But this view cannot be extended to the many waders, in which the summer and winter plumages differ very little in colour. With defenceless species, in which either both sexes or the males alone become extremely conspicuous during the breeding-season,—or when the males acquire at this season such long wing or tail-feathers as to impede their flight, as with Cosmetornis and Vidua,—it certainly at first appears highly probable that the second moult has been gained for the special purpose of throwing off these ornaments. We must, however, remember that many birds, such as Birds of Paradise, the Argus pheasant and peacock, do not cast their plumes during the winter; and it can hardly be maintained that there is something in the constitution of these birds, at least of the Gallinaceæ, rendering a double moult impossible, for the ptarmigan moults thrice in the year.[762] Hence it must be considered as doubtful whether the many species which moult their ornamental plumes or lose their bright colours during the winter, have acquired this habit on account of the inconvenience or danger which they would otherwise have suffered.

I conclude, therefore, that the habit of moulting twice in the year was in most or all cases first acquired for some distinct purpose, perhaps for gaining a warmer winter covering; and that variations in the plumage occurring during the summer were accumulated through sexual selection, and transmitted to the offspring at the same season of the year. Such variations being inherited either by both sexes or by the males alone, according to the form of inheritance which prevailed. This appears more probable than that these species in all cases originally tended to retain their ornamental plumage during the winter, but were saved from this through natural selection, owing to the inconvenience or danger thus caused.

I HAVE ENDEAVOURED in this chapter to shew that the arguments are not trustworthy in favour of the view that weapons, bright colours, and various ornaments, are now confined to the males owing to the conversion, by means of natural selection, of a tendency to the equal transmission of characters to both sexes into transmission to the male sex alone. It is also doubtful whether the colours of many female birds are due to the preservation, for the sake of protection, of variations which were from the first limited in their transmission to the female sex. But it will be convenient to defer any further discussion on this subject until I treat, in the following chapter, on the differences in plumage between the young and old.

CHAPTER XVI

BIRDS—CONCLUDED

THE IMMATURE PLUMAGE IN RELATION TO THE CHARACTER OF THE PLUMAGE IN BOTH SEXES WHEN ADULT • SIX CLASSES OF CASES • SEXUAL DIFFERENCES BETWEEN THE MALES OF CLOSELY-ALLIED OR REPRESENTATIVE SPECIES • THE FEMALE ASSUMING THE CHARACTERS OF THE MALE • PLUMAGE OF THE YOUNG IN RELATION TO THE SUMMER AND WINTER PLUMAGE OF THE ADULTS • ON THE INCREASE OF BEAUTY IN THE BIRDS OF THE WORLD • PROTECTIVE COLOURING • CONSPICUOUSLY-COLOURED BIRDS • NOVELTY APPRECIATED • SUMMARY OF THE FOUR CHAPTERS ON BIRDS

WE MUST NOW CONSIDER the transmission of characters as limited by age in reference to sexual selection. The truth and importance of the principle of inheritance at corresponding ages need not here be discussed, as enough has already been said on the subject. Before giving the several rather complex rules or classes of cases, under which all the differences in plumage between the young and the old, as far as known to me, may be included, it will be well to make a few preliminary remarks.

With animals of all kinds when the young differ in colour from the adults, and the colours of the former are not, as far as we can see, of any special service, they may generally be attributed, like various embryological structures, to the retention by the young of the character of an early progenitor. But this view can be maintained with confidence, only when the young of several species closely resemble each other, and likewise resemble other adult species belonging to the same group; for the latter are the living proofs that such a state of things was formerly possible. Young lions and pumas are marked with feeble stripes or rows of spots, and as many allied species both young and old are similarly marked, no naturalist, who believes in the gradual evolution of species, will doubt that the progenitor of the lion and puma was a striped animal, the young having retained vestiges of the stripes, like the kittens of black cats, which when grown up are not in the least striped. Many species of deer, which when mature are not spotted, are whilst young covered with white spots, as are likewise some few species in their adult state. So again the young in the whole family of pigs (Suidæ), and in certain rather distantly-allied animals, such as the tapir, are marked with dark longitudinal stripes; but here we have a character apparently derived from an extinct progenitor, and now preserved by the young alone. In all such cases the old have had their colours changed in the course of time, whilst the young have remained but little altered, and this has been effected through the principle of inheritance at corresponding ages.

This same principle applies to many birds belonging to various groups, in which the young closely resemble each other, and differ much from their respective adult parents. The young of almost all the Gallinaceæ, and of some distantly-allied birds such as ostriches, are whilst covered with down longitudinally striped; but this character points back to a state of things so remote that it hardly concerns us. Young cross-bills (Loxia) have at first straight beaks like those of other finches, and in their immature striated plumage they resemble the mature redpole and female siskin, as well as the young of the goldfinch, greenfinch, and some other allied species. The young of many kinds of buntings (Emberiza) resemble each other, and likewise the adult state of the common bunting, *E. miliaria.* In almost the whole large group of thrushes the young have their breasts spotted—a character which is retained by many species throughout life, but is quite lost by others, as by the *Turdus migratorius.* So again with many thrushes, the feathers on the back are mottled before they are moulted for the first time, and this character is retained for life by certain eastern species. The young of many species of shrikes (Lanius), of some woodpeckers, and of an Indian pigeon *(Chalcophaps Indicus),* are transversely striped on the under surface; and certain allied species or genera when adult are similarly marked. In some closely-allied and resplendent Indian cuckoos (Chrysococcyx), the species when mature differ considerably from each other in colour, but the young cannot be distinguished. The young of an Indian goose *(Sarkidiornis melanonotus)* closely resemble in plumage an allied genus, Dendrocygna, when mature.[763] Similar facts will hereafter be given in regard to certain herons. Young black grouse *(Tetrao tetrix)* resemble the young as well as the old of certain other species, for instance the red grouse or *T. scoticus.* Finally, as Mr. Blyth, who has attended closely to this subject, has well remarked, the natural affinities of many species are best exhibited in their immature plumage; and as the true affinities of all organic beings depend on their descent from a common progenitor, this remark strongly confirms the belief that the immature plumage approximately shews us the former or ancestral condition of the species.

Although many young birds belonging to various orders thus give us a glimpse of the plumage of their remote progenitors, yet there are many other birds, both dull-coloured and bright-coloured, in which the young closely resemble their parents. With such species the young of the different species cannot resemble each other more closely than do the parents; nor can they present striking resemblances to allied forms in their adult state. They give us but little insight into the plumage of their progenitors, excepting in so far that when the young and the old are coloured in the same general manner throughout a whole group of species, it is probable that their progenitors were similarly coloured.

We may now consider the classes of cases or rules under which the differences and resemblances, between the plumage of the young and the old, of both sexes or of one sex alone, may be grouped. Rules of this kind were first enounced by Cuvier; but with the progress of knowledge they require some modification and amplification. This I have attempted to do, as far as the extreme complexity of the subject permits, from information derived from various sources; but a full essay on this subject by some competent ornithologist is much needed. In order to ascertain to what extent each rule prevails, I have tabulated the facts given in four great works, namely, by Macgillivray on the birds of Britain, Audubon on those of North America, Jerdon on those of India, and Gould on those of Australia. I may here premise, firstly, that the several cases or rules graduate into each other; and secondly, that when the young are said to resemble their parents, it is not meant that they are identically alike, for their colours are almost always rather less vivid, and the feathers are softer and often of a different shape.

RULES OR CLASSES OF CASES

I. When the adult male is more beautiful or conspicuous than the adult female, the young of both sexes in their first plumage closely resemble the adult female, as with the common fowl and peacock; or, as occasionally occurs, they resemble her much more closely than they do the adult male.

II. When the adult female is more conspicuous than the adult male, as sometimes though rarely occurs, the young of both sexes in their first plumage resemble the adult male.

III. When the adult male resembles the adult female, the young of both sexes have a peculiar first plumage of their own, as with the robin.

IV. When the adult male resembles the adult female, the young of both sexes in their first plumage resemble the adults, as with the kingfisher, many parrots, crows, hedge-warblers.

V. When the adults of both sexes have a distinct winter and summer plumage, whether or not the male differs from the female, the young resemble the adults of both sexes in their winter dress, or much more rarely in their summer dress, or they resemble the females alone; or the young may have an intermediate character; or again they may differ greatly from the adults in both their seasonal plumages.

VI. In some few cases the young in their first plumage differ from each other according to sex; the young males resembling more or less closely the adult males, and the young females more or less closely the adult females.

Class I.—In this class, the young of both sexes resemble, more or less closely, the adult female, whilst the adult male differs, often in the most conspicuous manner, from the adult female. Innumerable instances in all Orders could be given; it will suffice to call to mind the common pheasant, duck, and house-sparrow. The cases under this class graduate into others. Thus the two sexes when adult may differ so slightly, and the young so slightly from the adults, that it is doubtful whether such cases ought to come under the present, or under the third or fourth classes. So again the young of both sexes, instead of being quite alike, may differ in a slight degree from each other, as in our sixth class. These transitional cases, however, are few in number, or at least are not strongly pronounced, in comparison with those which come strictly under the present class.

The force of the present law is well shewn in those groups, in which, as a general rule, the two sexes and the young are all alike; for when the male in these groups does differ from the female, as with certain parrots, kingfishers, pigeons, &c., the young of both sexes resemble the adult female.[764] We see the same fact exhibited still more clearly in certain anomalous cases; thus the male of *Heliothrix auriculata* (one of the humming-birds) differs conspicuously from the female in having a splendid gorget and fine ear-tufts, but the female is remarkable from having a much longer tail than that of the male; now the young of both sexes resemble (with the exception of the breast being spotted with bronze) the adult female in all respects including the length of her tail, so that the tail of the male actually becomes shorter as he reaches maturity, which is a most unusual circumstance.[765] Again, the plumage of the male goosander *(Mergus merganser)* is more conspicuously coloured, with the scapular and secondary wing-feathers much longer than in the female, but differently from what occurs, as far as I know, in any other bird, the crest of the adult male, though broader than that of the female, is considerably shorter, being only a little above an inch in length; the crest of the female being two and a half inches long. Now the young of both sexes resemble in all respects the adult female, so that their crests are actually of greater length though narrower than in the adult male.[766]

When the young and the females closely resemble each other and both differ from the male, the most obvious conclusion is that the male alone has been modified. Even in the anomalous cases of the Heliothrix and Mergus, it is probable that originally both adult sexes were furnished, the one species with a much elongated tail, and the other with a much elongated crest, these characters having since been partially lost by the adult males from some unexplained cause, and transmitted in their diminished state to their male offspring alone, when arrived at the corresponding age of maturity. The belief that in the present class the male alone has been modified, as far as the differences between the male and the female together with her young are concerned, is strongly supported by some remarkable facts recorded by Mr. Blyth,[767] with respect to closely-allied species which represent each other in distinct countries. For with several of these representative species the adult males have undergone a certain amount of change and can be distinguished; the females and the young being undistinguishable, and therefore absolutely unchanged. This is the case with certain Indian chats (Thamnobia), with certain honey-suckers (Nectarinia), shrikes (Tephrodornis), certain kingfishers (Tanysiptera), Kallij pheasants (Gallophasis), and tree-partridges (Arboricola).

In some analogous cases, namely with birds having a distinct summer and winter plumage, but with the two sexes nearly alike, certain closely-allied species can easily be distinguished in their summer or nuptial plumage, yet are undistinguishable in their winter as well as in their immature plumage. This is the case with some of the closely-allied Indian wagtails or Motacillæ. Mr. Swinhoe[768] informs me that three species of Ardeola, a genus of herons, which represent each other on separate continents, are "most strikingly different" when ornamented with their summer plumes, but are hardly, if at all, distinguishable during the winter. The young also of these three species in their immature plumage closely resemble the adults in their winter dress. This case is all the more interesting because with two other species of Ardeola both sexes retain, during the winter and summer, nearly the same plumage as that possessed by the three first species during the winter and in their immature state; and this plumage, which is common to several distinct species at different ages and seasons, probably shews us how the progenitor of the genus was coloured. In all these cases, the nuptial plumage which we may assume was originally acquired by the adult males during the breeding-season, and transmitted to the adults of both sexes at the corresponding season, has been modified, whilst the winter and immature plumages have been left unchanged.

The question naturally arises, how is it that in these latter cases the winter plumage of both sexes, and in the former cases the plumage of the adult females, as well as the immature plumage of the young, have not been at all affected? The species which represent each other in distinct countries will almost always have been exposed to somewhat different conditions, but we can hardly attribute the modification of the plumage in the males alone to this action, seeing that the females and the young, though similarly exposed, have not been affected. Hardly any fact in nature shews us more clearly how subordinate in importance is the direct action of the conditions of life, in comparison with the accumulation through selection of indefinite variations, than the surprising difference between the sexes of many birds; for both sexes must have consumed the same food and have been exposed to the same climate. Nevertheless we are not precluded from believing that in the course of time new conditions may produce some direct effect; we see only that this is subordinate in importance to the accumulated results of selection. When, however, a species migrates into a new country, and this must precede the formation of representative species, the changed conditions to which they will almost always have been exposed will cause them to undergo, judging from a widely-spread analogy, a certain amount of fluctuating variability. In this case sexual selection, which depends on an element eminently liable to change—namely the taste or admiration of the female—will have had new shades of colour or other differences to act on and accumulate; and as sexual selection is always at work, it would (judging from what we know of the results on domestic animals of man's unintentional selection), be a surprising fact if animals inhabiting separate districts, which can never cross and thus blend their newly-acquired characters, were not, after a sufficient lapse of time, differently modified. These remarks likewise apply to the nuptial or summer plumage, whether confined to the males or common to both sexes.

Although the females of the above closely-allied species, together with their young, differ hardly at all from each other, so that the males alone can be distinguished, yet in most cases the females of the species within the same genus obviously differ from each other. The differences, however, are rarely as great as between the males. We see this clearly in the whole family of the Gallinaceæ: the females, for instance, of the common and Japan pheasant, and especially of the gold and Amherst pheasant—of the silver pheasant and the wild fowl—resemble each other very closely in colour, whilst the males differ to an extraordinary degree. So it is with the females of most of the Cotingidæ, Fringillidæ, and many other families. There can indeed be no doubt that, as a general rule, the females have been modified to a less extent than the males. Some few birds, however, offer a singular and inexplicable exception; thus the females of *Paradisea apoda* and *P. papuana* differ from each other more than do their respective males;[769] the female of the latter species having the under surface pure white, whilst the female *P. apoda* is deep brown beneath. So, again, as I hear from Professor Newton, the males of two species of Oxynotus (shrikes), which represent each other in the islands of Mauritius and Bourbon,[770] differ but little in colour, whilst the females differ much. In the Bourbon species the female appears to have partially retained an immature condition of plumage, for at first sight she "might be taken for the young of the Mauritian species." These differences may be compared with those which occur, independently of selection by man, and which we cannot explain, in certain sub-breeds of the game-fowl, in which the females are very different, whilst the males can hardly be distinguished.[771]

As I account so largely by sexual selection for the differences between the males of allied species, how can the differences between the females be accounted for in all ordinary cases? We need not here consider the species which belong to distinct genera; for with these, adaptation to different habits of life, and other agencies, will have come into play. In regard to the differences between the females within the same genus, it appears to me almost certain, after looking through various large groups, that the chief agent has been the transference, in a greater or less degree, to the female of the characters acquired by the males through sexual selection. In the several British finches, the two sexes differ either very slightly or considerably; and if we compare the females of the greenfinch, chaffinch, goldfinch, bullfinch, crossbill, sparrow, &c., we shall see that they differ from each other chiefly in the points in which they partially resemble their respective males; and the colours of the males may safely be attributed to sexual selection. With many gallinaceous species the sexes differ to an extreme degree, as with the peacock, pheasant, and fowl, whilst with other species there has been a partial or even complete transference of character from the male to the female. The females of the several species of Polyplectron exhibit in a dim condition, and chiefly on the tail, the splendid ocelli of their males. The female partridge differs from the male only in the red mark on her breast being smaller; and the female wild turkey only in her colours being much duller. In the guinea-fowl the two sexes are undistinguishable. There is no improbability in the plain, though peculiar spotted plumage of this latter bird having been acquired through sexual selection by the males, and then transmitted to both sexes; for it is not essentially different from the much more beautifully-spotted plumage, characteristic of the males alone of the Tragopan pheasants.

It should be observed that, in some instances, the transference of characters from the male to the female has been effected apparently at a remote period, the male having subsequently undergone great changes, without transferring to the female any of his later-gained characters. For instance, the female and the young of the black-grouse *(Tetrao tetrix)* resemble pretty closely both sexes and the young of the red-grouse *T. Scoticus*; and we may consequently infer that the black-grouse is descended from some ancient species, of which both sexes were coloured in nearly the same manner as the red-grouse. As both sexes of this latter species are more plainly barred during the breeding-season than at any other time, and as the male differs slightly from the female in his more strongly-pronounced red and brown tints,[772] we may conclude that his plumage has been, at least to a certain extent, influenced by sexual selection. If so, we may further infer that the nearly similar plumage of the female black-grouse was similarly produced at some former period. But since this period the male black-grouse has acquired his fine black plumage, with his forked and outwardly-curled tail-feathers; but of these characters there has hardly been any transference to the female, excepting that she shews in her tail a trace of the curved fork.

We may therefore conclude that the females of distinct though allied species have often had their plumage rendered more or less different by the transference in various degrees, of characters acquired, both during former and recent times, by the males through sexual selection. But it deserves especial attention that brilliant colours have been transferred much more rarely than other tints. For instance, the male of the red-throated bluebreast *(Cyanecula suecica)* has a rich blue breast, including a sub-triangular red mark; now marks of approximately the same shape have been transferred to the female, but the central space is fulvous instead of red, and is surrounded by mottled instead of blue feathers. The Gallinaceæ offer many analogous cases; for none of the species, such as partridges, quails, guinea-fowls, &c., in which the colours of the plumage have been largely transferred from the male to the female, are brilliantly coloured. This is well exemplified with the pheasants, in which the male is generally so much more brilliant than the female; but with the Eared and Cheer pheasants *(Crossoptilon auritum* and *Phasianus Wallichii)* the two sexes closely resemble each other and their colours are dull. We may go so far as to believe that if any part of the plumage in the males of these two pheasants had been brilliantly coloured, this would not have been transferred to the females. These facts strongly support Mr. Wallace's view that with birds which are exposed to much danger during nidification, the transference of bright colours from the male to the female has been checked through natural selection. We must not, however, forget that another explanation, before given, is possible; namely, that the males which varied and became bright, whilst they were young and inexperienced, would have been exposed to much danger, and would generally have been destroyed; the older and more cautious males, on the other hand, if they varied in a like manner, would not only have been able to survive, but would have been favoured in their rivalry with other males. Now variations occurring late in life tend to be transmitted exclusively to the same sex, so that in this case extremely bright tints would not have been transmitted to the females. On the other hand, ornaments of a less conspicuous kind, such as those possessed by the Eared and Cheer pheasants, would not have been dangerous, and if they appeared during early youth, would generally have been transmitted to both sexes.

In addition to the effects of the partial transference of characters from the males to the females, some of the differences between the females of closely-allied species may be attributed to the direct or definite action of the conditions of life.[773] With the males any such action would generally have been masked by the brilliant colours gained through sexual selection; but not so with the females. Each of the endless diversities in plumage, which we see in our domesticated birds is, of course, the result of some definite cause; and under natural and more uniform conditions, some one tint, assuming that it was in no way injurious, would almost certainly sooner or later prevail. The free intercrossing of the many individuals belonging to the same species would ultimately tend to make any change of colour, thus induced, uniform in character.

No one doubts that both sexes of many birds have had their colours adapted for the sake of protection; and it is possible that the females alone of some species may have been thus modified. Although it would be a difficult, perhaps an impossible process, as shewn in the last chapter, to convert through selection one form of transmission into another, there would not be the least difficulty in adapting the colours of the female, independently of those of the male, to surrounding objects, through the accumulation of variations which were from the first limited in their transmission to the female sex. If the variations were not thus limited, the bright tints of the male would be deteriorated or destroyed. Whether the females alone of many species have been thus specially modified, is at present very doubtful. I wish I could follow Mr. Wallace to the full extent; for the admission would remove some difficulties. Any variations which were of no service to the female as a protection would be at once obliterated, instead of being lost simply by not being selected, or from free intercrossing, or from being eliminated when transferred to the male and in any way injurious to him. Thus the plumage of the female would be kept constant in character. It would also be a relief if we could admit that the obscure tints of both sexes of many birds had been acquired and preserved for the sake of protection,—for example, of the hedge-warbler or kitty-wren *(Accentor modularis* and *Troglodytes vulgaris),* with respect to which we have no sufficient evidence of the action of sexual selection. We ought, however, to be cautious in concluding that colours which appear to us dull, are not attractive to the females of certain species; we should bear in mind such cases as that of the common house-sparrow, in which the male differs much from the female, but does not exhibit any bright tints. No one probably will dispute that many gallinaceous birds which live on the open ground have acquired their present colours, at least in part, for the sake of protection. We know how well they are thus concealed; we know that ptarmigans, whilst changing from their winter to their summer plumage, both of which are protective, suffer greatly from birds of prey. But can we believe that the very slight differences in tints and markings between, for instance, the female black and red-grouse serve as a protection? Are partridges, as they are now coloured, better protected than if they had resembled quails? Do the slight differences between the females of the common pheasant, the Japan and golden pheasants, serve as a protection, or might not their plumages have been interchanged with impunity? From what Mr. Wallace has observed of the habits of certain gallinaceous birds in the East he thinks that such slight differences are beneficial. For myself, I will only say that I am not convinced.

Formerly when I was inclined to lay much stress on the principle of protection, as accounting for the less bright colours of female birds, it occurred to me that possibly both sexes and the young might aboriginally have been brightly coloured in an equal degree; but that subsequently, the females from the danger incurred during incubation, and the young from being inexperienced, had been rendered dull as a protection. But this view is not supported by any evidence, and is not probable; for we thus in imagination expose during past times the females and the young to danger, from which it has subsequently been necessary to shield their modified descendants. We have, also, to reduce, through a gradual process of selection, the females and the young to almost exactly the same tints and markings, and to transmit them to the corresponding sex and period of life. It is also a somewhat strange fact, on the supposition that the females and the young have partaken during each stage of the process of modification of a tendency to be as brightly coloured as the males, that the females have never been rendered dull-coloured without the young participating in the same change; for there are no instances, as far as I can discover, of species with the females dull-coloured and the young bright-coloured. A partial exception, however, is offered by the young of certain woodpeckers, for they have "the whole upper part of the head tinged with red," which afterwards either decreases into a mere circular red line in the adults of both sexes, or quite disappears in the adult females.[774]

Finally, with respect to our present class of cases, the most probable view appears to be that successive variations in brightness or in other ornamental characters, occurring in the males at a rather late period of life have alone been preserved; and that most or all of these variations owing to the late period of life at which they appeared, have been from the first transmitted only to the adult male offspring. Any variations in brightness which occurred in the females or in the young would have been of no service to them, and would not have been selected; moreover, if dangerous, would have been eliminated. Thus the females and the young will either have been left unmodified, or, and this has much more commonly occurred, will have been partially modified by receiving through transference from the males some of the successive variations. Both sexes have perhaps been directly acted on by the conditions of life to which they have long been exposed; but the females from not being otherwise much modified will best exhibit any such effects. These changes and all others will have been kept uniform by the free intercrossing of many individuals. In some cases, especially with ground birds, the females and the young may possibly have been modified, independently of the males, for the sake of protection, so as to have acquired the same dull-coloured plumage.

Class II. When the adult female is more conspicuous than the adult male, the young of both sexes in their first plumage resemble the adult male.—This class is exactly the reverse of the last, for the females are here more brightly coloured or more conspicuous than the males; and the young, as far as they are known, resemble the adult males instead of the adult females. But the difference between the sexes is never nearly so great as occurs with many birds in the first class, and the cases are comparatively rare. Mr. Wallace who first called attention to the singular relation which exists between the less bright colours of the males and their performing the duties of incubation, lays great stress on this point,[775] as a crucial test that obscure colours have been acquired for the sake of protection during the period of nesting. A different view seems to me more probable. As the cases are curious and not numerous, I will briefly give all that I have been able to find.

In one section of the genus Turnix, quail-like birds, the female is invariably larger than the male (being nearly twice as large in one of the Australian species) and this is an unusual circumstance with the Gallinaceæ. In most of the species the female is more distinctly coloured and brighter than the male,[776] but in some few species the sexes are alike. In *Turnix taigoor* of India the male "wants the black on the throat and neck, and the whole tone of the plumage is lighter and less pronounced than that of the female." The female appears to be more vociferous, and is certainly much more pugnacious than the male; so that the females and not the males are often kept by the natives for fighting, like game-cocks. As male birds are exposed by the English bird-catchers for a decoy near a trap, in order to catch other males by exciting their rivalry, so the females of this Turnix are employed in India. When thus exposed the females soon begin their "loud purring call, which can be heard a long way off, and any females within ear-shot run rapidly to the spot, and commence fighting with the caged bird." In this way from twelve to twenty birds, all breeding-females, may be caught in the course of a single day. The natives assert that the females after laying their eggs associate in flocks, and leave the males to sit on them. There is no reason to doubt the truth of this assertion, which is supported by some observations made in China by Mr. Swinhoe.[777] Mr. Blyth believes, that the young of both sexes resemble the adult male.

The females of the three species of Painted Snipes (Rhynchæa) "are not only larger, but much more richly coloured than the males."[778] With all other birds, in which the trachea differs in structure in the two sexes it is more developed and complex in the male than in the female; but in the *Rhynchæa Australis* it is simple in the male, whilst in the female it makes four distinct convolutions before entering the lungs.[779] The female therefore of this species has acquired an eminently masculine character. Mr. Blyth ascertained, by examining many specimens, that the trachea is not convoluted in either sex of *R. Bengalensis*, which species so closely resembles *R. Australis* that it

Fig. 60. *Rhynchæa capensis (from Brehm).*

can hardly be distinguished except by its shorter toes. This fact is another striking instance of the law that secondary sexual characters are often widely different in closely-allied forms; though it is a very rare circumstance when such differences relate to the female sex. The young of both sexes of *R. Bengalensis* in their first plumage are said to resemble the mature male.[780] There is also reason to believe that the male undertakes the duty of incubation, for Mr. Swinhoe[781] found the females before the close of the summer associated in flocks, as occurs with the females of the Turnix.

The females of *Phalaropus fulicarius* and *P. hyperboreus* are larger, and in their summer plumage "more gaily attired than the males." But the difference in colour between the sexes is far from conspicuous. The male alone of *P. fulicarius* undertakes, according to Professor Steenstrup, the duty of incubation, as is likewise shewn by the state of his breast-feathers during the breeding-season. The female of the dotterel plover *(Eudromias morinellus)* is larger than the male, and has the red and black tints on the lower surface, the white crescent on the breast, and the stripes over the eyes, more strongly pronounced. The male also takes at least a share in hatching the eggs; but the female likewise attends to the young.[782] I have not been able to discover whether with these species the young resemble the adult males more closely than the adult females; for the comparison is somewhat difficult to make on account of the double moult.

Turning now to the Ostrich order: the male of the common cassowary *(Casuarius galeatus)* would be thought by any one to be the female, from his smaller size and from the appendages and naked skin about his head being much less brightly coloured; and I am informed by Mr. Bartlett that in the Zoological Gardens it is certainly the male alone who sits on the eggs and takes care of the young.[783] The female is said by Mr. T. W. Wood[784] to exhibit during the breeding-season a most pugnacious disposition; and her wattles then become enlarged and more brilliantly coloured. So again the female of one of the emus *(Dromœus irroratus)* is considerably larger than the male, and she possesses a slight top-knot, but is otherwise undistinguishable in plumage. She appears, however, "to have greater power, when angry or otherwise excited, of erecting, like a turkey-cock, the feathers of her neck and breast. She is usually the more courageous and pugilistic. She makes a deep hollow guttural boom, especially at night, sounding like a small gong. The male has a slenderer frame and is more docile, with no voice beyond a suppressed hiss when angry, or a croak." He not only performs the whole duty of incubation, but has to defend the young from their mother; "for as soon as she catches sight of her progeny she becomes violently agitated, and notwithstanding the resistance of the father appears to use her utmost endeavours to destroy them. For months afterwards it is unsafe to put the parents together, violent quarrels being the inevitable result, in which the female generally comes off conqueror."[785] So that with this emu we have a complete reversal not only of the parental and incubating instincts, but of the usual moral qualities of the two sexes; the females being savage, quarrelsome and noisy, the males gentle and good. The case is very different with the African ostrich, for the male is somewhat larger than the female and has finer plumes with more strongly contrasted colours; nevertheless he undertakes the whole duty of incubation.[786]

I will specify the few other cases known to me, in which the female is more conspicuously coloured than the male, although nothing is known about their manner of incubation. With the carrion-hawk of the Falkland Islands *(Milvago leucurus)* I was much surprised to find by dissection that the individuals, which had all their tints strongly pronounced, with the cere and legs orange-coloured, were the adult females; whilst those with duller plumage and grey legs were the males or the young. In an Australian tree-creeper *(Climacteris erythrops)* the female differs from the male in "being adorned with beautiful, radiated, rufous markings on the throat, the male having this part quite plain." Lastly in an Australian night-jar "the female always exceeds the male in size and in the brilliance of her tints; the males, on the other hand, have two white spots on the primaries more conspicuous than in the female."[787]

We thus see that the cases in which female birds are more conspicuously coloured than the males, with the young in their immature plumage resembling the adult males instead of the adult females, as in the previous class, are not numerous, though they are distributed in various Orders. The amount of difference, also, between the sexes is incomparably less than that which frequently occurs in the last class; so that the cause of the difference, whatever it may have been, has acted on the females in the present class either less energetically or less persistently than on the males in the last class. Mr. Wallace believes that the males have had their colours rendered less conspicuous for the sake of protection during the period of incubation; but the difference between the sexes in hardly any of the foregoing cases appears sufficiently great for this view to be safely accepted. In some of the cases the brighter tints of the female are almost confined to the lower surface, and the males, if thus coloured, would not have been exposed to danger whilst sitting on the eggs. It should also be borne in mind that the males are not only in a slight degree less conspicuously coloured than the females, but are of less size, and have less strength. They have, moreover, not only acquired the maternal instinct of incubation, but are less pugnacious and vociferous than the females, and in one instance have simpler vocal organs. Thus an almost complete transposition of the instincts, habits, disposition, colour, size, and of some points of structure, has been effected between the two sexes.

Now if we might assume that the males in the present class have lost some of that ardour which is usual to their sex, so that they no longer search eagerly for the females; or, if we might assume that the females have become much more numerous than the males—and in the case of one Indian Turnix the females are said to be much more commonly met with than the males"[788]—then it is not improbable that the females would have been led to court the males, instead of being courted by them. This indeed is the case to a certain extent, with some birds, as we have seen with the peahen, wild turkey, and certain kinds of grouse. Taking as our guide the habits of most male birds, the greater size and strength and the extraordinary pugnacity of the females of the Turnix and Emu, must mean that they endeavour to drive away rival females, in order to gain possession of the male; and on this view, all the facts become clear; for the males would probably be most charmed or excited by the females which were the most attractive to them by their brighter colours, other ornaments, or vocal powers. Sexual selection would then soon do its work, steadily adding to the attractions of the females; the males and the young being left not at all, or but little modified.

Class III. When the adult male resembles the adult female, the young of both sexes have a peculiar first plumage of their own.—In this class both sexes when adult resemble each other, and differ from the young. This occurs with many birds of many kinds. The male robin can hardly be distinguished from the female, but the young are widely different with their mottled dusky-olive and brown plumage. The male and female of the splendid scarlet Ibis are alike, whilst the young are brown; and the scarlet-colour, though common to both sexes, is apparently a sexual character, for it is not well developed with birds under confinement, in the same manner as often occurs in the case of brilliantly coloured male birds. With many species of herons the young differ greatly from the adults, and their summer plumage, though common to both sexes, clearly has a nuptial character. Young swans are slate-coloured, whilst the mature birds are pure white; but it would be superfluous to give additional instances. These differences between the young and the old apparently depend, as in the two last classes, on the young having retained a former or ancient state of plumage, which has been exchanged for a new plumage by the old of both sexes. When the adults are brightly coloured, we may conclude from the remarks just made in relation to the scarlet ibis and to many herons, and from the analogy of the species in the first class, that such colours have been acquired through sexual selection by the nearly mature males; but that, differently from what occurs in the two first classes, the transmission, though limited to the same age, has not been limited to the same sex. Consequently both sexes when mature resemble each other and differ from the young.

Class IV. When the adult male resembles the adult female, the young of both sexes in their first plumage resemble the adults.—In this class the young and the adults of both sexes, whether brilliantly or obscurely coloured, resemble each other. Such cases are, I think, more common than those in the last class. We have in England instances in the kingfisher, some wood-peckers, the jay, magpie, crow, and many small dull-coloured birds, such as the hedge-warbler or kitty-wren. But the similarity in plumage between the young and the old is never absolutely complete, and graduates away into dissimilarity. Thus the young of some members of the kingfisher family are not only less vividly coloured than the adults, but many of the feathers on the lower surface are edged with brown,[789]—a vestige probably of a former state of the plumage. Frequently in the same group of birds, even within the same genus, for instance in an Australian genus of parrakeets (Platycercus), the young of some species closely resemble, whilst the young of other species differ considerably from their parents of both sexes, which are alike.[790] Both sexes and the young of the common jay are closely similar; but in the Canada jay *(Perisoreus canadensis)* the young differ so much from their parents that they were formerly described as distinct species.[791]

Before proceeding, I may remark that under the present and two next classes of cases the facts are so complex, and the conclusions so doubtful, that any one who feels no especial interest in the subject had better pass them over.

The brilliant or conspicuous colours which characterise many birds in the present class, can rarely or never be of service to them as a protection; so that they have probably been gained by the males through sexual selection, and then transferred to the females and the young. It is, however, possible that the males may have selected the more attractive females; and if these transmitted their characters to their offspring of both sexes, the same results would follow as from the selection of the more attractive males by the females. But there is some evidence that this contingency has rarely, if ever, occurred in any of those groups of birds, in which the sexes are generally alike; for if even a few of the successive variations had failed to be transmitted to both sexes, the females would have exceeded to a slight degree the males in beauty. Exactly the reverse occurs under nature; for in almost every large group, in which the sexes generally resemble each other, the males of some few species are in a slight degree more brightly coloured than the females. It is again possible that the females may have selected the more beautiful males, these males having reciprocally selected the more beautiful females; but it is doubtful whether this double process of selection would be likely to occur, owing to the greater eagerness of one sex than the other, and whether it would be more efficient than selection on one side alone. It is, therefore, the most probable view that sexual selection has acted, in the present class, as far as ornamental characters are concerned, in accordance with the general rule throughout the animal kingdom, that is, on the males; and that these have transmitted their gradually-acquired colours, either equally or almost equally, to their offspring of both sexes.

Another point is more doubtful, namely, whether the successive variations first appeared in the males after they had become nearly mature, or whilst quite young. In either case sexual selection must have acted on the male when he had to compete with rivals for the possession of the female; and in both cases the characters thus acquired have been transmitted to both sexes and all ages. But these characters, if acquired by the males when adult, may have been transmitted at first to the adults alone, and at some subsequent period transferred to the young. For it is known that when the law of inheritance at corresponding ages fails, the offspring often inherit characters at an earlier age than that at which they first appeared in their parents.[792] Cases apparently of this kind have been observed with birds in a state of nature. For instance Mr. Blyth has seen specimens of *Lanius rufus* and of *Colymbus glacialis* which had assumed whilst young, in a quite anomalous manner, the adult plumage of their parents.[793] Again, the young of the common swan *(Cygnus olor)* do not cast off their dark feathers and become white until eighteen months or two years old; but Dr. F. Forel has described the case of three vigorous young birds, out of a brood of four, which were born pure white. These young birds were not albinoes, as shewn by the colour of their beaks and legs, which nearly resembled the same parts in the adults.[794]

It may be worth while to illustrate the above three modes by which, in the present class, the two sexes and the young may have come to resemble each other, by the curious case of the genus Passer.[795] In the house-sparrow *(P. domesticus)* the male differs much from the female and from the young. These resemble each other, and likewise to a large extent both sexes and the young of the sparrow of Palestine *(P. brachydactylus)*, as well as of some allied species. We may therefore assume that the female and young of the house-sparrow approximately shew us the plumage of the progenitor of the genus. Now with the tree-sparrow *(P. montanus)* both sexes and the young closely resemble the male of the house-sparrow; so that they have all been modified in the same manner, and all depart from the typical colouring of their early progenitor. This may have been effected by a male ancestor of the tree-sparrow having varied, firstly, when nearly mature, or, secondly, whilst quite young, having in either case transmitted his modified plumage to the females and the young; or, thirdly, he may have varied when adult and transmitted his plumage to both adult sexes, and, owing to the failure of the law of inheritance at corresponding ages, at some subsequent period to his young.

It is impossible to decide which of these three modes has generally prevailed throughout the present class of cases. The belief that the males varied whilst young, and transmitted their variations to their offspring of both sexes is perhaps the most probable. I may here add that I have endeavoured, with little success, by consulting various works, to decide how far with birds the period of variation has generally determined the transmission of characters to one sex or to both. The two rules, often referred to (namely, that variations occurring late in life are transmitted to one and the same sex, whilst those which occur early in life are transmitted to both sexes), apparently hold good in the first,[796] second, and fourth classes of cases; but they fail in an equal number, namely, in the third, often in the fifth,[797] and in the sixth small class. They hold good, however, as far as I can judge, with a considerable majority of the species of birds. Whether or not this be so, we may conclude from the facts given in the eighth chapter that the period of variation has been one important element in determining the form of transmission.

With birds it is difficult to decide by what standard we ought to judge of the earliness or lateness of the period of variation, whether by the age in reference to the duration of life, or to the power of reproduction, or to the number of moults through which the species passes. The moulting of birds, even within the same family, sometimes differs much without any assignable cause. Some birds moult so early, that nearly all the body-feathers are cast off before the first wing-feathers are fully grown; and we cannot believe that this was the primordial state of things. When the period of moulting has been accelerated, the age at which the colours of the adult plumage were first developed would falsely appear to us to have been earlier than it really was. This may be illustrated by the practice followed by some bird-fanciers, who pull out a few feathers from the breast of nestling bullfinches, and from the head or neck of young gold-pheasants, in order to ascertain their sex; for in the males these feathers are immediately replaced by coloured ones.[798] The actual duration of life is known in but few birds, so that we can hardly judge by this standard. And with reference to the period at which the powers of reproduction are gained, it is a remarkable fact that various birds occasionally breed whilst retaining their immature plumage.[799]

The fact of birds breeding in their immature plumage seems opposed to the belief that sexual selection has played as important a part, as I believe it has, in giving ornamental colours, plumes, &c., to the males, and, by means of equal transmission, to the females of many species. The objection would be a valid one, if the younger and less ornamented males were as successful in winning females and propagating their kind, as the older and more beautiful males. But we have no reason to suppose that this is the case. Audubon speaks of the breeding of the immature males of *Ibis tanta-*

lus as a rare event, as does Mr. Swinhoe, in regard to the immature males of Oriolus.[800] If the young of any species in their immature plumage were more successful in winning partners than the adults, the adult plumage would probably soon be lost, as the males which retained their immature dress for the longest period would prevail, and thus the character of the species would ultimately be modified.[801] If, on the other hand, the young never succeeded in obtaining a female, the habit of early reproduction would perhaps be sooner or later quite eliminated, from being superfluous and entailing waste of power.

The plumage of certain birds goes on increasing in beauty during many years after they are fully mature; this is the case with the train of the peacock, and with the crest and plumes of certain herons; for instance, the *Ardea Ludovicana*;[802] but it is very doubtful whether the continued development of such feathers is the result of the selection of successive beneficial variations, or merely of continuous growth. Most fishes continue increasing in size, as long as they are in good health and have plenty of food; and a somewhat similar law may prevail with the plumes of birds.

Class V. When the adults of both sexes have a distinct winter and summer plumage, whether or not the male differs from the female, the young resemble the adults of both sexes in their winter dress, or much more rarely in their summer dress, or they resemble the females alone; or the young may have an intermediate character; or again, they may differ greatly from the adults in both their seasonal plumages.—The cases in this class are singularly complex; nor is this surprising, as they depend on inheritance, limited in a greater or less degree in three different ways, namely by sex, age, and the season of the year. In some cases the individuals of the same species pass through at least five distinct states of plumage. With the species, in which the male differs from the female during the summer season alone, or, which is rarer, during both seasons,[803] the young generally resemble the females,—as with the so-called goldfinch of North America, and apparently with the splendid Maluri of Australia.[804] With the species, the sexes of which are alike during both the summer and winter, the young may resemble the adults, firstly, in their winter dress; secondly, which occurs much more rarely, in their summer dress; thirdly, they may be intermediate between these two states; and, fourthly, they may differ greatly from the adults at all seasons. We have an instance of the first of these four cases in one of the egrets of India *(Buphus coromandus)*, in which the young and the adults of both sexes are white during the winter, the adults becoming golden-buff during the summer. With the Gaper *(Anastomus oscitans)* of India we have a similar case, but the colours are reversed; for the young and the adults of both sexes are grey and black during the winter, the adults becoming white during the summer.[805] As an instance of the second case, the young of the razor-bill *(Alca torda,* Linn.), in an early state of plumage, are coloured like the adults during the summer; and the young of the white-crowned sparrow of North America *(Fringilla leucophrys),* as soon as fledged, have elegant white stripes on their heads, which are lost by the young and the old during the winter.[806] With respect to the third case, namely, that of the young having an intermediate character between the summer and winter adult plumages, Yarrell[807] insists that this occurs with many waders. Lastly, in regard to the young differing greatly from both sexes in their adult summer and winter plumages, this occurs with some herons and egrets of North America and India,—the young alone being white.

I will make only a few remarks on these complicated cases. When the young resemble the female in her summer dress, or the adults of both sexes in their winter dress, the cases differ from those given under Classes I. and III. only in the characters originally acquired by the males during the breeding-season, having been limited in their transmission to the corresponding season. When the adults have a distinct summer and winter plumage, and the young differ from both, the case is more difficult to understand. We may admit as probable that the young have retained an ancient state of plumage; we can account through sexual selection for the summer or nuptial plumage of the adults, but how are we to account for their distinct winter plumage? If we could admit that this plumage serves in all cases as a protection, its acquirement would be a simple affair; but there seems no good reason for this admission. It may be suggested that the widely different conditions of life during the winter and summer have acted in a direct manner on the plumage; this may have had some effect, but I have not much confidence in so great a difference, as we sometimes see, between the two plumages having been thus caused. A more probable explanation is, that an ancient style of plumage, partially modified through the transference of some characters from the summer plumage, has been retained by the adults during the winter. Finally, all the cases in our present class apparently depend on characters acquired by the adult males, having been variously limited in their transmission according to age, season, and sex; but it would not be worth while to attempt to follow out these complex relations.

Class VI. The young in their first plumage differ from each other according to sex; the young males resembling more or less closely the adult males, and the young females more or less closely the adult females.—The cases in the present class, though occurring in various groups, are not numerous; yet, if experience had not taught us to the contrary, it seems the most natural thing that the young should at first always resemble to a certain extent, and gradually become more and more like, the adults of the same sex. The adult male blackcap *(Sylvia atricapilla)* has a black head, that of the female being reddish-brown; and I am informed by Mr. Blyth, that the young of both sexes can be distinguished by this character even as nestlings. In the family of thrushes an unusual number of similar cases have been noticed; the male blackbird *(Turdus merula)* can be distinguished in the nest from the female, as the main wing-feathers, which are not moulted so soon as the body-feathers, retain a brownish tint until the second general moult.[808] The two sexes of the mocking bird *(Turdus polyglottus,* Linn.) differ very little from each other, yet the males can easily be distinguished at a very early age from the females by shewing more pure white.[809] The males of a forest-thrush and of a rock-thrush (viz. *Orocetes erythrogastra* and *Petrocincla cyanea)* have much of their plumage of a fine blue, whilst the females are brown; and the nestling males of both species have their main wing and tail-feathers edged with blue, whilst those of the female are edged with brown.[810] So that the very same feathers which in the young blackbird assume their mature character and become black after the others, in these two species assume this character and become blue before the others. The most probable view with reference to these cases is that the males, differently from what occurs in Class I., have transmitted their colours to their male offspring at an earlier age than that at which they themselves first acquired them; for if they had varied whilst quite young, they would probably have transmitted all their characters to their offspring of both sexes.[811]

In *Aïthurus polytmus* (one of the humming-birds) the male is splendidly coloured black and green, and two of the tail-feathers are immensely lengthened; the female has an ordinary tail and inconspicuous colours; now the young males, instead of resembling the adult female, in accordance with the common rule, begin from the first to assume the colours proper to their sex, and their tail-feathers soon become elongated. I owe this information to Mr. Gould, who has given me the following more striking and as yet unpublished case. Two humming-birds belonging to the genus Eustephanus, both beautifully coloured, inhabit the small island of Juan Fernandez, and have always been ranked as specifically distinct. But it has lately been ascertained that the one, which is of a rich chesnut-brown colour with a golden-red head, is the male, whilst the other, which is elegantly variegated with green and white with a metallic-green head, is the female. Now the young from the first resemble to a certain extent the adults of the corresponding sex, the resemblance gradually becoming more and more complete.

In considering this last case, if as before we take the plumage of the young as our guide, it would appear that both sexes have been independently rendered beautiful; and not that the one sex has partially transferred its beauty to the other. The male apparently has acquired his bright colours through sexual selection in the same manner as, for instance, the peacock or pheasant in our first class of cases; and the female in the same manner as the female Rhynchæa or Turnix in our second class of cases. But there is much difficulty in understanding how this could have been effected at the same time with the two sexes of the same species. Mr. Salvin states, as we have seen in the eighth chapter, that with certain humming-birds the males greatly exceed in number the females, whilst with other species inhabiting the same country the females greatly exceed the males. If, then, we might assume that during some former lengthened period the males of the Juan Fernandez species had greatly exceeded the females in number, but that during another lengthened period the females had greatly exceeded the males, we could understand how the males at one time, and the females at another time, might have been rendered beautiful by the selection of the brighter-coloured individuals of either sex; both sexes transmitting their characters to their young at a rather earlier age than usual. Whether this is the true explanation I will not pretend to say; but the case is too remarkable to be passed over without notice.

We have now seen in numerous instances under all six classes, that an intimate relation exists between the plumage of the young and that of the adults, either of one sex or both sexes. These relations are fairly well explained on the principle that one sex—this being in the great majority of cases the male—first acquired through variation and sexual selection bright colours or other ornaments, and transmitted them in various ways, in accordance with the recognised laws of inheritance. Why variations have occurred at different periods of life, even sometimes with the species of the same group, we do not know; but with respect to the form of transmission, one important determining cause seems to have been the age at which the variations first appeared.

From the principle of inheritance at corresponding ages, and from any variations in colour which occurred in the males at an early age not being then selected, on the contrary being often eliminated as dangerous, whilst similar variations occurring at or near the period of reproduction have been preserved, it follows that the plumage of the young will often have been left unmodified, or but little modified. We thus get some insight into the colouring of the progenitors of our existing species. In a vast number of species in five out of our six classes of cases, the adults of one sex or both are brightly coloured, at least during the breeding-season, whilst the young are invariably less brightly coloured than the adults, or are quite dull-coloured; for no instance is known, as far as I can discover, of the young of dull-coloured species displaying bright colours, or of the young of brightly-coloured species being more brilliantly coloured than their parents. In the fourth class, however, in which the young and the old resemble each other, there are many species (though by no means all) brightly-coloured, and as these form whole groups, we may infer that their early progenitors were likewise brightly-coloured. With this exception, if we look to the birds of the world, it appears that their beauty has been greatly increased since that period, of which we have a partial record in their immature plumage.

On the Colour of the Plumage in relation to Protection.—It will have been seen that I cannot follow Mr. Wallace in the belief that dull colours when confined to the females have been in most cases specially gained for the sake of protection. There can, however, be no doubt, as formerly remarked, that both sexes of many birds have had their colours modified for this purpose, so as to escape the notice of their enemies; or, in some instances, so as to approach their prey unobserved, in the same manner as owls have had their plumage rendered soft, that their flight may not be overheard. Mr. Wallace remarks[812] that "it is only in the tropics, among forests which never lose their foliage, that we find whole groups of birds, whose chief colour is green." It will be admitted by every one, who has ever tried, how difficult it is to distinguish parrots in a leaf-covered tree. Nevertheless, we must remember that many parrots are ornamented with crimson, blue, and orange tints, which can hardly be protective. Woodpeckers are eminently arboreal, but, besides green species, there are many black, and black-and-white kinds—all the species being apparently exposed to nearly the same dangers. It is therefore probable that strongly-pronounced colours have been acquired by tree-haunting birds through sexual selection, but that green tints have had an advantage through natural selection over other colours for the sake of protection.

In regard to birds which live on the ground, every one admits that they are coloured so as to imitate the surrounding surface. How difficult it is to see a partridge, snipe, woodcock, certain plovers, larks, and night-jars when crouched on the ground. Animals inhabiting deserts offer the most striking instances, for the bare surface affords no concealment, and all the smaller quadrupeds, reptiles, and birds depend for safety on their colours. As Mr. Tristram has remarked,[813] in regard to the inhabitants of the Sahara, all are protected by their "isabelline or sand-colour." Calling to my recollection the desert-birds which I had seen in South America, as well as most of the ground-birds in Great Britain, it appeared to me that both sexes in such cases are generally coloured nearly alike. Accordingly I applied to Mr. Tristram, with respect to the birds of the Sahara, and he has kindly given me the following information. There are twenty-six species, belonging to fifteen genera, which manifestly have had their plumage coloured in a protective manner; and this colouring is all the more striking, as with most of these birds it is different from that of their congeners. Both sexes of thirteen out of the twenty-six species are coloured in the same manner; but these belong to genera in which this rule commonly prevails, so that they tell us nothing about the protective colours being the same in both sexes of desert-birds. Of the other thirteen species, three belong to genera in which the sexes usually differ from each other, yet they have the sexes alike. In the remaining ten species, the male differs from the female; but the difference is confined chiefly to the under surface of the plumage, which is concealed when the bird crouches on the ground; the head and back being of the same sand-coloured hue in both sexes. So that in these ten species the upper surfaces of both sexes have been acted on and rendered alike, through natural selection, for the sake of protection; whilst the lower surfaces of the males alone have been diversified through sexual selection, for the sake of ornament. Here, as both sexes are equally well protected, we clearly see that the females have not been prevented through natural selection from inheriting the colours of their male parents: we must look to the law of sexually limited transmission, as before explained.

In all parts of the world both sexes of many soft-billed birds, especially those which frequent reeds or sedges, are obscurely coloured. No doubt if their colours had been brilliant, they would have been much more conspicuous to their enemies; but whether their dull tints have been specially gained for the sake of protection seems, as far as I can judge, rather doubtful. It is still more doubtful whether such dull tints can have been gained for the sake of ornament. We must, however, bear in mind that male birds, though dull-coloured, often differ much from their females, as with the common sparrow, and this leads to the belief that such colours have been gained through sexual selection, from being attractive. Many of the soft-billed birds are songsters; and a discussion in a former chapter should not be forgotten, in which it was shewn that the best songsters are rarely ornamented with bright tints. It would appear that female birds, as a general rule, have selected their mates either for their sweet voices or gay colours, but not for both charms combined. Some species which are manifestly coloured for the sake of protection, such as the jack-snipe, woodcock, and night-jar, are likewise marked and shaded, according to our standard of taste, with extreme elegance. In such cases we may conclude that both natural and sexual selection have acted conjointly for protection and ornament. Whether any bird exists which does not possess some special attraction, by which to charm the opposite sex, may be doubted. When both sexes are so obscurely coloured, that it would be rash to assume the agency of sexual selection, and when no direct evidence can be advanced shewing that such colours serve as a protection, it is best to own complete ignorance of the cause, or, which comes to nearly the same thing, to attribute the result to the direct action of the conditions of life.

There are many birds both sexes of which are conspicuously, though not brilliantly coloured, such as the numerous black, white, or piebald species; and these colours, are probably the result of sexual selection. With the common blackbird, capercailzie, black-cock, black Scoter-duck (Oidemia), and even with one of the Birds of Paradise *(Lophorina atra),* the males alone are black, whilst the females are brown or mottled; and there can hardly be a doubt that blackness in these cases has been a sexually selected character. Therefore it is in some degree probable that the complete or partial blackness of both sexes in such birds as crows, certain cockatoos, storks, and swans, and many marine birds, is likewise the result of sexual selection, accompanied by equal transmission to both sexes; for blackness can hardly serve in any case as a protection. With several birds, in which the male alone is black, and in others in which both sexes are black, the beak or skin about the head is brightly coloured, and the contrast thus afforded adds greatly to their beauty; we see this in the bright yellow beak of the male blackbird, in the crimson skin over the eyes of the black-cock and capercailzie, in the variously and brightly-coloured beak of the Scoter-drake (Oidemia), in the red beak of the chough *(Corvus graculus,* Linn.), of the black swan, and black stork. This leads me to remark that it is not at all incredible that toucans may owe the enormous size of their beaks to sexual selection, for the sake of displaying the diversified and vivid stripes of colour, with which these organs are ornamented.[814] The naked skin at the base of the beak and round the eyes is likewise often brilliantly coloured; and Mr. Gould, in speaking of one species,[815] says that the colours of the beak "are doubtless in the finest and most brilliant state during the time of pairing." There is no greater improbability in toucans being encumbered with immense beaks, though rendered as light as possible by their cancellated structure, for an object falsely appearing to us unimportant, namely, the display of fine colours, than that the male Argus pheasant and some other birds should be encumbered with plumes so long as to impede their flight.

In the same manner, as the males alone of various species are black, the females being dull-coloured; so in a few cases the males alone are either wholly or partially white, as with the several Bell-birds of South America (Chasmorhynchus), the Antarctic goose *(Bernicla antarctica),* the silver-pheasant, &c., whilst the females are brown or obscurely mottled. Therefore, on the same principle as before, it is probable that both sexes of many birds, such as white cockatoos, several egrets with their beautiful plumes, certain ibises, gulls, terns, &c., have acquired their more or less completely white plumage through sexual selection. The species which inhabit snowy regions of course come under a different head. The white plumage of some of the above-named birds appears in both sexes only when they are mature. This is likewise the case with certain gannets, tropic-birds, &c., and with the snow-goose *(Anser hyperboreus)*. As the latter breeds on the "barren grounds," when not covered with snow, and as it migrates southward during the winter, there is no reason to suppose that its snow-white adult plumage serves as a protection. In the case of the *Anastomus oscitans* previously alluded to, we have still better evidence that the white plumage is a nuptial character, for it is developed only during the summer; the young in their immature state, and the adults in their winter dress, being grey and black. With many kinds of gulls (Larus), the head and neck become pure white during the summer, being grey or mottled during the winter and in the young state. On the other hand, with the smaller gulls, or sea-mews (Gavia), and with some terns (Sterna), exactly the reverse occurs; for the heads of the young birds during the first year, and of the adults during the winter, are either pure white, or much paler-coloured than during the breeding-season. These latter cases offer another instance of the capricious manner in which sexual selection appears often to have acted.[816]

The cause of aquatic birds having acquired a white plumage so much more frequently than terrestrial birds, probably depends on their large size and strong powers of flight, so that they can easily defend themselves or escape from birds of prey, to which moreover they are not much exposed. Consequently sexual selection has not here been interfered with or guided for the sake of protection. No doubt, with birds which roam over the open ocean, the males and females could find each other much more easily when made conspicuous either by being perfectly white, or intensely black; so that these colours may possibly serve the same end as the call-notes of many land-birds. A white or black bird, when it discovers and flies down to a carcase floating on the sea or cast up on the beach, will be seen from a great distance, and will guide other birds of the same and of distinct species, to the prey; but as this would be a disadvantage to the first finders, the individuals which were the whitest or blackest would not thus have procured more food than the less strongly coloured individuals. Hence conspicuous colours cannot have been gradually acquired for this purpose through natural selection.[817]

As sexual selection depends on so fluctuating an element as taste, we can understand how it is that within the same group of birds, with habits of life nearly the same, there should exist white or nearly white, as well as black, or nearly black species,—for instance, white and black cockatoos, storks, ibises, swans, terns, and petrels. Piebald birds likewise sometimes occur in the same groups, for instance, the black-necked swan, certain terns, and the common magpie. That a strong contrast in colour is agreeable to birds, we may conclude, by looking through any large collection of specimens or series of coloured plates, for the sexes frequently differ from each other in the male having the pale parts of a purer white, and the variously coloured dark parts of still darker tints than in the female.

It would even appear that mere novelty, or change for the sake of change, has sometimes acted like a charm on female birds, in the same manner as changes of fashion with us. The Duke of Argyll says,[818]—and I am glad to have the unusual satisfaction of following for even a short distance in his footsteps—"I am more and more convinced that variety, mere variety, must be admitted to be an object and an aim in Nature." I wish the Duke had explained what he here means by Nature. Is it meant that the Creator of the universe ordained diversified results for His own satisfaction, or for that of man? The former notion seems to me as much wanting in due reverence as the latter in probability. Capriciousness of taste in the birds themselves appears a more fitting explanation. For example; the males of some parrots can hardly be said to be more beautiful, at least according to our taste, than the females, but they differ from them in such points, as the male having a rose-coloured collar instead of, as in the female, "a bright emeraldine narrow green collar;" or in the male having a black collar instead of "a yellow demi-collar in front," with a pale roseate instead of a plum-blue head.[819] As so many male birds have for their chief ornament elongated tail-feathers or elongated crests, the shortened tail, formerly described in the male of a humming-bird, and the shortened crest of the male goosander almost seem like one of the many opposite changes of fashion which we admire in our own dresses.

Some members of the heron family offer a still more curious case of novelty in colouring having apparently been appreciated for the sake of novelty. The young of the *Ardea asha* are white, the adults being dark slate-coloured; and not only the young, but the adults of the allied *Buphus coromandus* in their winter plumage are white, this colour changing into a rich golden-buff during the breeding-season. It is incredible that the young of these two species, as well as of some other members of the same family,[820] should have been specially rendered pure white and thus made conspicuous to their enemies; or that the adults of one of these two species should have been specially rendered white during the winter in a country which is never covered with snow. On the other hand we have reason to believe that whiteness has been gained by many birds as a sexual ornament. We may therefore conclude that an early progenitor of the *Ardea asha* and the *Buphus* acquired a white plumage for nuptial purposes, and transmitted this colour to their young; so that the young and the old became white like certain existing egrets; the whiteness having afterwards been retained by the young whilst exchanged by the adults for more strongly pronounced tints. But if we could look still further backwards in time to the still earlier progenitors of these two species, we should probably see the adults dark-coloured. I infer that this would be the case, from the analogy of many other birds, which are dark whilst young, and when adult are white; and more especially from the case of the *Ardea gularis,* the colours of which are the reverse of those of *A. asha,* for the young are dark-coloured and the adults white, the young having retained a former state of plumage. It appears therefore that the progenitors in their adult condition of the *Ardea asha,* the *Buphus,* and of some allies, have undergone, during a long line of descent, the following changes of colour: firstly a dark shade, secondly pure white, and thirdly, owing to another change of fashion (if I may so express myself), their present slaty, reddish, or golden-buff tints. These successive changes are intelligible only on the principle of novelty having been admired by birds for the sake of novelty.

Summary of the Four Chapters on Birds.—Most male birds are highly pugnacious during the breeding-season, and some possess weapons especially adapted for fighting with their rivals. But the most pugnacious and the best-armed males rarely or never depend for success solely on their power to drive away or kill their rivals, but have special means for charming the female. With some it is the power of song, or of emitting strange cries, or of producing instrumental music, and the males in consequence differ from the females in their vocal organs, or in the structure of certain feathers. From the curiously diversified means for producing various sounds we gain a high idea of the importance of this means of courtship. Many birds endeavour to charm the females by love-dances or antics, performed on the ground or in the air, and sometimes at prepared places. But ornaments of many kinds, the most brilliant tints, combs and wattles, beautiful plumes, elongated feathers, top-knots, and so forth, are by far the commonest means. In some cases mere novelty appears to have acted as a charm. The ornaments of the males must be highly important to them, for they have been acquired in not a few cases at the cost of increased danger from enemies, and even at some loss of power in fighting with their rivals. The males of very many species do not assume their ornamental dress until they arrive at maturity, or they assume it only during the breeding-season, or the tints then become more vivid. Certain ornamental appendages become enlarged, turgid, and brightly-coloured during the very act of courtship. The males display their charms with elaborate care and to the best effect; and this is done in the presence of the females. The courtship is sometimes a prolonged affair, and many males and females congregate at an appointed place. To suppose that the females do not appreciate the beauty of the males is to admit that their splendid decorations, all their pomp and display, are useless; and this is incredible. Birds have fine powers of discrimination, and in some few instances it can be shewn that they have a taste for the beautiful. The females, moreover, are known occasionally to exhibit a marked preference or antipathy for certain individual males.

If it be admitted that the females prefer, or are unconsciously excited by the more beautiful males, then the males would slowly but surely be rendered more and more attractive through sexual selection. That it is this sex which has been chiefly modified we may infer from the fact that in almost every genus in which the sexes differ, the males differ much more from each other than do the females; this is well shewn in certain closely-allied representative species in which the females can hardly be distinguished, whilst the males are quite distinct. Birds in a state of nature offer individual differences which would amply suffice for the work of sexual selection; but we have seen that they occasionally present more strongly-marked variations which recur so frequently that they would immediately be fixed, if they served to allure the female. The laws of variation will have determined the nature of the initial changes, and largely influenced the final result. The gradations, which may be observed between the males of allied species, indicate the nature of the steps which have been passed through, and explain in the most interesting manner certain characters, such as the indented ocelli of the tail-feathers of the peacock, and the wonderfully-shaded ocelli of the wing-feathers of the Argus pheasant. It is evident that the brilliant colours, top-knots, fine plumes, &c., of many male birds cannot have been acquired as a protection; indeed they sometimes lead to danger. That they are not due to the direct and definite action of the conditions of life, we may feel assured, because the females have been exposed to the same conditions, and yet often differ from the males to an extreme degree. Although it is probable that changed conditions acting during a lengthened period have produced some definite effect on both sexes, the more important result will have been an increased tendency to fluctuating variability or to augmented individual differences; and such differences will have afforded an excellent groundwork for the action of sexual selection.

The laws of inheritance, irrespectively of selection, appear to have determined whether the characters acquired by the males for the sake of ornament, for producing various sounds, and for fighting together, have been transmitted to the males alone or to both sexes, either permanently or periodically during certain seasons of the year. Why various characters should sometimes have been transmitted in one way and sometimes in another is, in most cases, not known; but the period of variability seems often to have been the determining cause. When the two sexes have inherited all characters in common they necessarily resemble each other; but as the successive variations may be differently transmitted, every possible gradation may be found, even within the same genus, from the closest similarity to the widest dissimilarity between the sexes. With many closely-allied species, following nearly the same habits of life, the males have come to differ from each other chiefly through the action of sexual selection; whilst the females have come to differ chiefly from partaking in a greater or lesser degree of the characters thus acquired by the males. The effects, moreover, of the definite action of the conditions of life, will not have been masked in the females, as in the case of the males, by the accumulation through sexual selection of strongly-pronounced colours and other ornaments. The individuals of both sexes, however affected, will have been kept at each successive period nearly uniform by the free intercrossing of many individuals.

With the species, in which the sexes differ in colour, it is possible that at first there existed a tendency to transmit the successive variations equally to both sexes; and that the females were prevented from acquiring the bright colours of the males, on account of the danger to which they would have been exposed during incubation. But it would be, as far as I can see, an extremely difficult process to convert, by means of natural selection, one form of transmission into another. On the other hand there would not be the least difficulty in rendering a female dull-coloured, the male being still kept bright-coloured, by the selection of successive variations, which were from the first limited in their transmission to the same sex. Whether the females of many species have actually been thus modified, must at present remain doubtful. When, through the law of the equal transmission of characters to both sexes, the females have been rendered as conspicuously coloured as the males, their instincts have often been modified, and they have been led to build domed or concealed nests.

In one small and curious class of cases the characters and habits of the two sexes have been completely transposed, for the females are larger, stronger, more vociferous and brightly-coloured than their males. They have, also, become so quarrelsome that they often fight together like the males of the most pugnacious species. If, as seems probable, they habitually drive away rival females, and by the display of their bright colours or other charms endeavour to attract the males, we can understand how it is that they have gradually been rendered, by means of sexual selection and sexually-limited transmission, more beautiful than the males—the latter being left unmodified or only slightly modified.

Whenever the law of inheritance at corresponding ages prevails, but not that of sexually-limited transmission, then if the parents vary late in life—and we know that this constantly occurs with our poultry, and occasionally with other birds—the young will be left unaffected, whilst the adults of both sexes will be modified. If both these laws of inheritance prevail and either sex varies late in life, that sex alone will be modified, the other sex and the young being left unaffected. When variations in brightness or in other conspicuous characters occur early in life, as no doubt often happens, they will not be acted on through sexual selection until the period of reproduction arrives; consequently they will be liable to be lost by the accidental deaths of the young, and if dangerous will be eliminated through natural selection. Thus we can understand how it is that variations arising late in life have chiefly been preserved for the ornamentation and arming of the males, the females and the young being left almost unaffected, and therefore like each other. With species having a distinct summer and winter plumage, the males of which either resemble or differ from the females during both seasons or during the summer alone, the degrees and kinds of resemblance between the young and the old are exceedingly complex; and this complexity apparently depends on characters, first acquired by the males, being transmitted in various ways and degrees, as limited by age, sex, and season.

As the young of so many species have been but little modified in colour and in other ornaments, we are enabled to form some judgment with respect to the plumage of their early progenitors; and we may infer that the beauty of our existing species, if we look to the whole class, has been largely increased since that period of which the immature plumage gives us an indirect record. Many birds, especially those which live much on the ground, have undoubtedly been obscurely coloured for the sake of protection. In some instances the upper exposed surface of the plumage has been thus coloured in both sexes, whilst the lower surface in the males alone has been variously ornamented through sexual selection. Finally, from the facts given in these four chapters, we may conclude that weapons for battle, organs for producing sound, ornaments of many kinds, bright and conspicuous colours, have generally been acquired by the males through variation and sexual selection, and have been transmitted in various ways according to the several laws of inheritance—the females and the young being left comparatively but little modified.[821]

Secondary Sexual Characters of Mammals

THE LAW OF BATTLE • SPECIAL WEAPONS, CONFINED TO THE MALES • CAUSE OF ABSENCE OF WEAPONS IN THE FEMALE • WEAPONS COMMON TO BOTH SEXES, YET PRIMARILY ACQUIRED BY THE MALE • OTHER USES OF SUCH WEAPONS • THEIR HIGH IMPORTANCE • GREATER SIZE OF THE MALE • MEANS OF DEFENCE • ON THE PREFERENCE SHEWN BY EITHER SEX IN THE PAIRING OF QUADRUPEDS

With mammals the male appears to win the female much more through the law of battle than through the display of his charms. The most timid animals, not provided with any special weapons for fighting, engage in desperate conflicts during the season of love. Two male hares have been seen to fight together until one was killed; male moles often fight, and sometimes with fatal results; male squirrels "engage in frequent contests, and often wound each other severely;" as do male beavers, so that "hardly a skin is without scars."[822] I observed the same fact with the hides of the guanacoes in Patagonia; and on one occasion several were so absorbed in fighting that they fearlessly rushed close by me. Livingstone speaks of the males of the many animals in Southern Africa as almost invariably shewing the scars received in former contests.

The law of battle prevails with aquatic as with terrestrial mammals. It is notorious how desperately male seals fight, both with their teeth and claws, during the breeding-season; and their hides are likewise often covered with scars. Male sperm-whales are very jealous at this season; and in their battles "they often lock their jaws together, and turn on their sides and twist about;"

so that it is believed by some naturalists that the frequently deformed state of their lower jaws is caused, by these struggles.[823]

All male animals which are furnished with special weapons for fighting, are well known to engage in fierce battles. The courage and the desperate conflicts of stags have often been described; their skeletons have been found in various parts of the world, with the horns inextricably locked together, shewing how miserably the victor and vanquished had perished.[824] No animal in the world is so dangerous as an elephant in must. Lord Tankerville has given me a graphic description of the battles between the wild bulls in Chillingham Park, the descendants, degenerated in size but not in courage, of the gigantic *Bos primigenius*. In 1861 several contended for mastery; and it was observed that two of the younger bulls attacked in concert the old leader of the herd, overthrew and disabled him, so that he was believed by the keepers to be lying mortally wounded in a neighbouring wood. But a few days afterwards one of the young bulls singly approached the wood; and then the "monarch of the chase," who had been lashing himself up for vengeance, came out and, in a short time killed his antagonist. He then quietly joined the herd, and long held undisputed sway. Admiral Sir B. J. Sulivan informs me that when he resided in the Falkland Islands he imported a young English stallion, which, with eight mares, frequented the hills near Port William. On these hills there were two wild stallions, each with a small troop of mares; "and it is certain that these stallions would never have approached each other without fighting. Both had tried singly to fight the English horse and drive away his mares, but had failed. One day they came in *together* and attacked him. This was seen by the capitan who had charge of the horses, and who, on riding to the spot, found one of the two stallions engaged with the English horse, whilst the other was driving away the mares, and had already separated four from the rest. The capitan settled the matter by driving the whole party into the coral, for the wild stallions would not leave the mares."

Male animals already provided with efficient cutting or tearing teeth for the ordinary purposes of life, as in the carnivora, insectivora, and rodents, are seldom furnished with weapons especially adapted for fighting with their rivals. The case is very different with the males of many other animals. We see this in the horns of stags and of certain kinds of antelopes in which the females are hornless. With many animals the canine teeth in the upper or lower jaw, or in both, are much larger in the males than in the females; or are absent in the latter, with the exception sometimes of a hidden rudiment. Certain antelopes, the musk-deer, camel, horse, boar, various apes, seals, and the walrus, offer instances of these several cases. In the females of the walrus the tusks are sometimes quite absent.[825] In the male elephant of India and in the male dugong[826] the upper incisors form offensive weapons. In the male narwhal one alone of the upper teeth is developed into the well-known, spirally-twisted, so called horn, which is sometimes from nine to ten feet in length. It is believed that the males use these horns for fighting together; for "an unbroken one can rarely be got, and occasionally one may be found with the point of another jammed into the broken place."[827] The tooth on the opposite side of the head in the male consists of a rudiment about ten inches in length, which is embedded in the jaw. It is not, however, very uncommon to find double-horned male narwhals in which both teeth are well developed. In the females both teeth are rudimentary. The male cachalot has a larger head than that of the female, and it no doubt aids these animals in their aquatic battles. Lastly, the adult male ornithorhynchus is provided with a remarkable apparatus, namely a spur on the fore-leg, closely resembling the poison-fang of a venomous snake; its use is not known, but we may suspect that it serves as a weapon of offence.[828] It is represented by a mere rudiment in the female.

When the males are provided with weapons which the females do not possess, there can hardly be a doubt that they are used for fighting with other males, and that they have been acquired through sexual selection. It is not probable, at least in most cases, that the females have actually been saved from acquiring such weapons, owing to their being useless and superfluous, or in some way injurious. On the contrary, as they are often used by the males of many animals for various purposes, more especially as a defence against their enemies, it is a surprising fact that they are so poorly developed or quite absent in the females. No doubt with female deer the development during each recurrent season of great branching horns, and with female elephants the development of immense tusks, would have been a great waste of vital power, on the admission that they were of no use to the females. Consequently variations in the size of these organs, leading to their suppression, would have come under the control of natural selection, and if limited in their transmission to the female offspring would not have interfered with their development through sexual selection in the males. But how on this view can we explain the presence of horns in the females of certain antelopes, and of tusks in the females of many animals, which are only of slightly less size than in the males? The explanation in almost all cases must, I believe, be sought in the laws of transmission.

As the reindeer is the single species in the whole family of Deer in which the female is furnished with horns, though somewhat smaller, thinner, and less branched than in the male, it might naturally be thought that they must be of some special use to her. There is, however, some evidence opposed to this view. The female retains her horns from the time when they are fully developed, namely in September, throughout the winter, until

May, when she brings forth her young; whilst the male casts his horns much earlier, towards the end of November. As both sexes have the same requirements and follow the same habits of life, and as the male sheds his horns during the winter, it is very improbable that they can be of any special service to the female at this season, which includes the larger proportion of the time during which she bears horns. Nor is it probable that she can have inherited horns from some ancient progenitor of the whole family of deer, for, from the fact of the males alone of so many species in all quarters of the globe possessing horns, we may conclude that this was the primordial character of the group. Hence it appears that horns must have been transferred from the male to the female at a period subsequent to the divergence of the various species from a common stock; but that this was not effected for the sake of giving her any special advantage.[829]

We know that the horns are developed at a most unusually early age in the reindeer; but what the cause of this may have been is not known. The effect, however, has apparently been the transference of the horns to both sexes. It is intelligible on the hypothesis of pangenesis, that a very slight change in the constitution of the male, either in the tissues of the forehead or in the gemmules of the horns, might lead to their early development; and as the young of both sexes have nearly the same constitution before the period of reproduction, the horns, if developed at an early age in the male, would tend to be developed equally in both sexes. In support of this view, we should bear in mind that the horns are always transmitted through the female, and that she has a latent capacity for their development, as we see in old or diseased females.[830] Moreover the females of some other species of deer either normally or occasionally exhibit rudiments of horns; thus the female of *Cervulus moschatus* has "bristly tufts, ending in a knob, instead of a horn;" and "in most specimens of the female Wapiti *(Cervus Canadensis)* there is a sharp bony protuberance in the place of the horn."[831] From these several considerations we may conclude that the possession of fairly well-developed horns by the female reindeer, is clue to the males having first acquired them as weapons for fighting with other males; and secondarily to their development from some unknown cause at an unusually early age in the males, and their consequent transmission to both sexes.

Turning to the sheath-horned ruminants: with antelopes a graduated series can be formed, beginning with the species, the females of which are completely destitute of horns—passing to those which have horns so small as to be almost rudimentary, as in *Antilocapra Americana*—to those which have fairly well-developed horns, but manifestly smaller and thinner than in the male, and sometimes of a different shape,[832] and ending with those in which both sexes have horns of equal size. As with the reindeer, so with antelopes there exists a relation between the period of the development of the horns and their transmission to one or both sexes; it is therefore probable that their presence or absence in the females of some species, and their more or less perfect condition in the females of other species, depend, not on their being of some special use, but simply on the form of inheritance which has prevailed. It accords with this view that even in the same restricted genus both sexes of some species, and the males alone of other species, are thus provided. It is a remarkable fact that, although the females of *Antilope bezoartica* are normally destitute of horns, Mr. Blyth has seen no less than three females thus furnished; and there was no reason to suppose that they were old or diseased. The males of this species have long straight spirated horns, nearly parallel to each other, and directed backwards. Those of the female, when present, are very different in shape, for they are not spirated, and spreading widely bend round, so that their points are directed forwards. It is a still more remarkable fact that in the castrated male, as Mr. Blyth informs me, the horns are of the same peculiar shape as in the female, but longer and thicker. In all cases the differences between the horns of the males and females, and of castrated and entire males, probably depend on various causes,—on the more or less complete transference of male characters to the females,—on the former state of the progenitors of the species,—and partly perhaps on the horns being differently nourished, in nearly the same manner as the spurs of the domestic cock, when inserted into the comb or other parts of the body, assume various abnormal forms from being differently nourished.

In all the wild species of goats and sheep the horns are larger in the male than in the female, and are sometimes quite absent in the latter.[833] In several domestic breeds of the sheep and goat, the males alone are furnished with horns; and it is a significant fact, that in one such breed of sheep on the Guinea coast, the horns are not developed, as Mr. Winwood Reade informs me, in the castrated male; so that they are affected in this respect like the horns of stags. In some breeds, as in that of N. Wales, in which both sexes are properly horned, the ewes are very liable to be hornless. In these same sheep, as I have been informed by a trustworthy witness who purposely inspected a flock during the lambing-season, the horns at birth are generally more fully developed in the male than in the female. With the adult musk-ox *(Ovibos moschatus)* the horns of the male are larger than those of the female, and in the latter the bases do not touch.[834] In regard to ordinary cattle Mr. Blyth remarks: "In most of the wild bovine animals the horns are both longer and thicker in the bull than in the cow, and in the cow-banteng *(Bos sondaicus)* the horns are remarkably small, and inclined much backwards. In the domestic races of cattle, both of the humped and humpless types, the horns are short and thick in the bull, longer and more slender in the cow and ox; and in the Indian buffalo, they are shorter and thicker in the bull, longer and more slender in the cow. In the wild gaour *(B. gaurus)* the horns are mostly both longer and thicker in the bull than in the cow."[835] Hence with most sheath-horned ruminants the horns of the male are either longer or stronger than those of the female. With the *Rhinoceros simus*, as I may here add, the horns of the female are generally longer but less powerful than in the male; and in some other species of rhinoceros they are said to be shorter in the female.[836] From these various facts we may conclude that horns of all kinds, even when they are equally developed in both sexes, were primarily acquired by the males in order to conquer other males, and have been transferred more or less completely to the female, in relation to the force of the equal form of inheritance.

The tusks of the elephant, in the different species or races, differ according to sex, in nearly the same manner as the horns of ruminants. In India and Malacca the males alone are provided with well-developed tusks. The elephant of Ceylon is considered by most naturalists as a distinct race, but by some as a distinct species, and here "not one in a hundred is found with tusks, the few that possess them being exclusively males."[837] The African elephant is undoubtedly distinct, and the female has large, well-developed tusks, though not so large as those of the male. These differences in the tusks of the several races and species of elephants—the great variability of the horns of deer, as notably in the wild reindeer—the occasional presence of horns in the female *Antilope bezoartica*—the presence of two tusks in some few male narwhals—the complete absence of tusks in some female walruses;—are all instances of the extreme variability of secondary sexual characters, and of their extreme liability to differ in closely-allied forms.

Although tusks and horns appear in all cases to have been primarily developed as sexual weapons, they often serve for other purposes. The elephant uses his tusks in attacking the tiger; according to Bruce, he scores the trunks of trees until they can be easily thrown down, and he likewise thus extracts the farinaceous cores of palms; in Africa he often uses one tusk, this being always the same, to probe the ground and thus to ascertain whether it will bear his weight. The common bull defends the herd with his horns; and the elk in Sweden has been known, according to Lloyd, to strike a wolf dead with a single blow of his great horns. Many similar facts could be given. One of the most curious secondary uses to which the horns of any animal are occasionally put, is that observed by Captain Hutton [838] with the wild goat *(Capra ægagrus)* of the Himalayas, and as it is said with the ibex, namely, that when the male accidentally falls from a height he bends inwards his head, and, by alighting on his massive horns, breaks the shock. The female cannot thus use her horns, which are smaller, but from her more quiet disposition she does not so much need this strange kind of shield.

Each male animal uses his weapons in his own peculiar fashion. The common ram makes a charge and butts with such force with the bases of his horns, that I have seen a powerful man knocked over as easily as a child. Goats and certain species of sheep, for instance the *Ovis cycloceros* of Afghanistan,[839] rear on their hind legs, and then not only butt, but "make a cut down and a jerk up, with the ribbed front of their scimitar-shaped horn, as with a sabre. When the *O. cycloceros* attacked a large domestic ram, who was a noted bruiser, he conquered him by the sheer novelty of his mode of fighting, always closing at once with his adversary, and catching him across the face and nose with a sharp drawing jerk of his head, and then bounding out of the way before the blow could be returned." In Pembrokeshire a male goat, the master of a flock which during several generations had run wild, was known to have killed several other males in single combat; this goat possessed enormous horns, measuring 39 inches in a straight line from tip to tip. The common bull, as every one knows, gores and tosses his opponent; but the Italian buffalo is said never to use his horns, he gives a tremendous blow with his convex forehead, and then tramples on his fallen enemy with his knees—an instinct which the common bull does not possess.[840] Hence a dog who pins a buffalo by the nose is immediately crushed. We must, however, remember that the Italian buffalo has long been domesticated, and it is by no means certain that the wild parent-form had similarly shaped horns. Mr. Bartlett informs me that when a female Cape buffalo *(Bubalus caffer)* was turned into an enclosure with a bull of the same species, she attacked him, and he in return pushed her about with great violence. But it was manifest to Mr. Bartlett that had not the bull shewn dignified forbearance, he could easily have killed her by a single lateral thrust with his immense horns. The giraffe uses his short hair-covered horns, which are rather longer in the male than in the female, in a curious manner; for with his long neck he swings his head to either side, almost upside down, with such force, that I have seen a hard plank deeply indented by a single blow.

With antelopes it is sometimes difficult to imagine how they can possibly use their curiously-shaped horns; thus the spring-boc *(Ant. euchore)* has rather short upright horns, with the sharp points bent inwards almost at a right angle, so as to face each other; Mr. Bartlett does not know how they are used, but suggests that they would inflict a fearful wound down each side of the face of an antagonist. The slightly-curved horns of the *Oryx leucoryx* (fig. 61) are directed backwards, and are of such length that their points reach beyond the middle of the back, over which they stand in an almost parallel line. Thus they seem singularly ill-fitted for fighting; but Mr. Bartlett informs me that when two of these animals prepare for battle, they kneel down, with their heads between their front legs, and in this attitude the horns stand nearly parallel and close to the ground, with the points directed for-

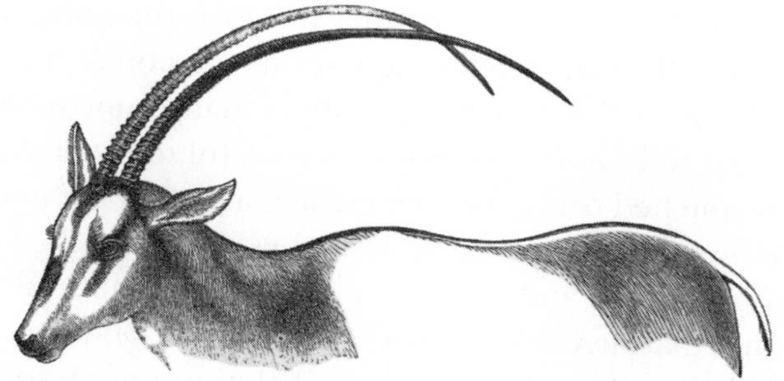

Fig. 61. *Oryx leucoryx, male (from the Knowsley Menagerie).*

wards and a little upwards. The combatants then gradually approach each other and endeavour to get the upturned points under each other's bodies; if one succeeds in doing this, he suddenly springs up, throwing up his head at the same time, and can thus wound or perhaps even transfix his antagonist. Both animals always kneel down so as to guard as far as possible against this manœuvre. It has been recorded that one of these antelopes has used his horns with effect even against a lion; yet from being forced to place his head between the fore-legs in order to bring the points of the horns forward, he would generally be under a great disadvantage when attacked by any other animal. It is, therefore, not probable that the horns have been modified into their present great length and peculiar position, as a protection against beasts of prey. We can, however, see that as soon as some ancient male progenitor of the Oryx acquired moderately long horns, directed a little backwards, he would be compelled in his battles with rival males to bend his head somewhat inwards or downwards, as is now done by certain stags; and it is not improbable that he might have acquired the habit of at first occasionally and afterwards of regularly kneeling down. In this case it is almost certain that the males which possessed the longest horns would have had a great advantage over others with shorter horns; and then the horns would gradually have been rendered longer and longer, through sexual selection, until they acquired their present extraordinary length and position.

With stags of many kinds the branching of the horns offers a curious case of difficulty; for certainly a single straight point would inflict a much more serious wound than several diverging points. In Sir Philip Egerton's museum there is a horn of the red-deer *(Cervus elaphus)* thirty inches in length, with "not fewer than fifteen snags or branches;" and at Moritzburg there is still preserved a pair of antlers of a red-deer, shot in 1699 by Frederick I., each of which bears the astonishing number of thirty-three branches. Richardson figures a pair of antlers of the wild reindeer with twenty-nine points.[841] From the manner in which the horns are branched, and more especially from deer being known occasionally to fight together by kicking with their fore-feet,[842] M. Bailly actually came to the conclusion that their horns were more injurious than useful to them! But this author overlooks the pitched battles between rival males. As I felt much perplexed about the use or advantage of the branches, I applied to Mr. McNeill of Colinsay, who has long and carefully observed the habits of red-deer, and he informs me that he has never seen some of the branches brought into action, but that the brow-antlers, from inclining downwards, are a great protection to the forehead, and their points are likewise used in attack. Sir Philip Egerton also informs me in regard both to red-deer and fallow-deer, that when they fight they suddenly dash together, and getting their horns fixed against each other's bodies a desperate struggle ensues. When one is at last forced to yield and turn round, the victor endeavours to plunge his brow-antlers into his defeated foe. It thus appears that the upper branches are used chiefly or exclusively for pushing and fencing. Nevertheless with some species the upper branches are used as weapons of offence; when a man was attacked by a Wapiti deer *(Cervus Canadensis)* in Judge Caton's park in Ottawa, and several men tried to rescue him, the stag "never raised his head from the ground; in fact he kept his face almost flat on the ground, with his nose nearly between his fore-feet, except when he rolled his head to one side to take a new observation preparatory to a plunge." In this position the terminal points of the horns were directed against his adversaries. "In rolling his head he necessarily raised it somewhat, because his antlers were so long that he could not roll his head without raising them on one side, while on the other side they touched the ground." The stag by this procedure gradually drove the party of rescuers backwards, to a distance of 150 or 200 feet; and the attacked man was killed.[843]

Although the horns of stags are efficient weapons, there can, I think, be no doubt that a single point would have been much more dangerous than a branched antler; and Judge Caton, who has had large experience with deer, fully concurs in this conclusion. Nor do the branching horns, though highly important as a means of defence against rival stags, appear perfectly well adapted for this purpose, as they are liable to become interlocked. The suspicion has therefore crossed my mind that they may serve partly as ornaments. That the branched antlers of stags, as well as the elegant lyrated horns of certain antelopes, with their graceful double curvature, (fig. 62), are ornamental in our eyes, no one will dispute. If, then, the horns, like the splendid accoutrements of the knights of old, add to the noble appearance of stags and antelopes, they may have been partly modified for this purpose, though mainly for actual service in battle; but I have no evidence in favour of this belief.

Fig. 62. *Strepsiceros Kudu (from Andrew Smith's 'Zoology of South Africa').*

An interesting case has lately been published, from which it appears that the horns of a deer in one district in the United States are now being modified through sexual and natural selection. A writer in an excellent American Journal[844] says, that he has hunted for the last twenty-one years in the Adirondacks, where the *Cervus Virginianus* abounds. About fourteen years ago he first heard of *spike-horn bucks.* These became from year to year more common; about five years ago he shot one, and subsequently another, and now they are frequently killed. "The spike-horn differs greatly from the common antler of the *C. Virginianus.* It consists of a single spike, more slender than the antler, and scarcely half so long, projecting forward from the brow, and terminating in a very sharp point. It gives a considerable advantage to its possessor over the common buck. Besides enabling him to run more swiftly through the thick woods and underbrush (every hunter knows that does and yearling bucks run much more rapidly than the large bucks when armed with their cumbrous antlers), the spike-horn is a more effective weapon than the common antler. With this advantage the spike-horn bucks are gaining upon the common bucks, and may, in time, entirely supersede them in the Adirondacks. Undoubtedly the first spike-horn buck was merely an accidental freak of nature. But his spike-horns gave him an advantage, and enabled him to propagate his peculiarity. His descendants, having a like advantage, have propagated the peculiarity in a constantly increasing ratio, till they are slowly crowding the antlered deer from the region they inhabit."

Male quadrupeds which are furnished with tusks use them in various ways, as in the case of horns. The boar strikes laterally and upwards; the musk-deer with serious effect downwards.[845] The walrus, though having so short a neck and so unwieldy a body, "can strike either upwards, or downwards, or sideways, with equal dexterity."[846] The Indian elephant fights, as I was informed by the late Dr. Falconer, in a different manner according to the position and curvature of his tusks. When they are directed forwards and upwards he is able to fling a tiger to a great distance—it is said to even thirty feet; when they are short and turned downwards he endeavours suddenly to pin the tiger to the ground, and in consequence is dangerous to the rider, who is liable to be jerked off the hoodah.[847]

Very few male quadrupeds possess weapons of two distinct kinds specially adapted for fighting with rival males. The male muntjac-deer *(Cervulus),* however, offers an exception, as he is provided with horns and exserted canine teeth. But one form of weapon, has often been replaced in the course of ages by another form, as we may infer from what follows. With ruminants the development of horns generally stands in an inverse relation with that of even moderately well-developed canine teeth. Thus camels, guanacoes, chevrotains and musk-deer, are hornless, and they have efficient canines; these teeth being "always of smaller size in the females than in the males." The Camelidæ have in their upper jaws, in addition to their true canines, a pair of canine-shaped incisors.[848] Male deer and antelopes, on the other hand, possess horns, and they rarely have canine teeth; and these when present are always of small size, so that it is doubtful whether they are of any service in their battles. With *Antilope montana* they exist only as rudiments in the young male, disappearing as he grows old; and they are absent in the female at all ages; but the females of certain other antelopes and deer have been known occasionally to exhibit rudiments of these teeth.[849] Stallions have small canine teeth, which are either quite absent or rudimentary in the mare; but they do not appear to be used in fighting, for stallions bite with their incisors, and do not open their mouths widely like camels and guanacoes. Whenever the adult male possesses canines now in an inefficient state, whilst the female has either none or mere rudiments, we may conclude that the early male progenitor of the species was provided with efficient canines, which had been partially transferred to the females. The reduction of these teeth in the males seems to have followed from some change in their manner of fighting, often caused (but not in the case of the horse) by the development of new weapons.

Tusks and horns are manifestly of high importance to their possessors, for their development consumes much organised matter. A single tusk of the Asiatic elephant,—one of the extinct woolly species,—and of the African elephant, have been known to weigh respectively 150,160, and 180 pounds; and even greater weights have been assigned by some authors.[850] With deer, in which the horns are periodically renewed, the drain on the constitution must be greater; the horns, for instance, of the moose weigh from fifty to sixty pounds, and those of the extinct Irish elk from sixty to seventy pounds,—the skull of the latter weighing on an average only five and a quarter pounds. With sheep, although the horns are not periodically renewed, yet their development, in the opinion of many agriculturists, entails a sensible loss to the breeder. Stags, moreover, in escaping from beasts of prey are loaded with an additional weight for the race, and are greatly retarded in passing through a woody country. The moose, for instance, with horns extending five and a half feet from tip to tip, although so skilful in their use that he will not touch or break a dead twig when walking quietly, cannot act so dexterously whilst rushing away from a pack of wolves. "During his progress he holds his nose up, so as to lay the horns horizontally back; and in this attitude cannot see the ground distinctly."[851] The tips of the horns of the great Irish elk were actually eight feet apart! Whilst the horns are covered with velvet, which lasts with the red-deer for about twelve weeks, they are extremely sensitive to a blow; so that in Germany the stags at this time change their habits to a certain extent, and avoid dense forests, frequenting young woods and low thickets.[852] These facts remind us, that male birds have acquired ornamental plumes at the cost of retarded flight, and other ornaments at the cost of some loss of power in their battles with rival males.

With quadrupeds, when, as is often the case, the sexes differ in size, the males are, I believe, always larger and stronger. This holds good in a marked manner, as I am informed by Mr. Gould, with the marsupials of Australia, the males of which appear to continue growing until an unusually late age. But the most extraordinary case is that of one of the seals *(Callorhinus ursinus),* a full-grown female weighing less than one-sixth of a full-grown male.[853] The greater strength of the male is invariably displayed, as Hunter long ago remarked,[854] in those parts of the body which are brought into action in fighting with rival males,—for instance, in the massive neck of the bull. Male quadrupeds are also more courageous and pugnacious than the females. There can be little doubt that these characters have been gained, partly through sexual selection, owing to a long series of victories by the stronger and more courageous males over the weaker, and partly through the inherited effects of use. It is probable that the successive variations in strength, size, and courage, whether due to so-called spontaneous variability or to the effects of use, by the accumulation of which male quadrupeds have acquired these characteristic qualities, occurred rather late in life, and were consequently to a large extent limited in their transmission to the same sex.

Under this point of view I was anxious to obtain information in regard to the Scotch deer-hound, the sexes of which differ more in size than those of any other breed (though blood-hounds differ considerably), or than in any wild canine species known to me. Accordingly, I applied to Mr. Cupples, a well-known breeder of these dogs, who has weighed and measured many of his own dogs, and who, with great kindness, has collected for me the following facts from various sources. Superior male dogs, measured at the shoulder, range from twenty-eight inches, which is low, to thirty-three, or even thirty-four inches in height; and in weight from eighty pounds, which is low, to 120, or even more pounds. The females range in height from twenty-three to twenty-seven, or even to twenty-eight inches; and in weight from fifty to seventy, or even eighty pounds.[855] Mr. Cupples concludes that from ninety-five to one hundred pounds for the male, and seventy for the female, would be a safe average; but there is reason to believe that formerly both sexes attained a greater weight. Mr. Cupples has weighed puppies when a fortnight old; in one litter the average weight of four males exceeded that of two females by six and a half ounces; in another litter the average weight of four males exceeded that of one female by less than one ounce; the same males, when three weeks old, exceeded the female by seven and a half ounces, and at the age of six weeks by nearly fourteen ounces. Mr. Wright of Yeldersley House, in a letter to Mr.Cupples, says: "I have taken notes on the sizes and weights of puppies of many litters, and as far as my experience goes, dog-puppies as a rule differ very little from bitches till they arrive at about five or six months old; and then the dogs begin to increase, gaining upon the bitches both in weight and size. At birth, and for several weeks afterwards, a bitch-puppy will occasionally be larger than any of the dogs, but they are invariably beaten by them later." Mr. McNeill, of Colinsay, concludes that "the males do not attain their full growth till over two years old, though the females attain it sooner." According to Mr. Cupples' experience; male dogs go on growing in stature till they are from twelve to eighteen months old, and in weight till from eighteen to twenty-four months old; whilst the females cease increasing in stature at the age of from nine to fourteen or fifteen months, and in weight at the age of from twelve to fifteen months. From these various statements it is clear that the full difference in size between the male and female Scotch deer-hound is not acquired until rather late in life. The males are almost exclusively used for coursing, for, as Mr. McNeill informs me, the females have not sufficient strength and weight to pull down a full-grown deer. From the names used in old legends, it appears, as I hear from Mr. Cupples, that at a very ancient period the males were the most celebrated, the females being mentioned only as the mothers of famous dogs. Hence during many generations, it is the male which has been chiefly tested for strength, size, speed, and courage, and the best will have been bred from. As, however, the males do not attain their full dimensions until a rather late period in life, they will have tended, in accordance with the law often indicated, to transmit their characters to their male offspring alone; and thus the great inequality in size between the sexes of the Scotch deer-hound may probably be accounted for.

The males of some few quadrupeds possess organs or parts developed solely as a means of defence against the attacks of other males. Some kinds of deer use, as we have seen, the upper branches of their horns chiefly or exclusively for defending themselves; and the Oryx antelope, as I am informed by Mr. Bartlett, fences most skilfully with his long, gently curved horns; but these are likewise used as organs of offence. Rhinoceroses, as the same observer remarks, in fighting parry each other's sidelong blows with their horns, which loudly clatter together, as do the tusks of boars. Although wild boars fight desperately together, they seldom, according to Brehm, receive fatal blows, as these fall on each other's tusks, or on the layer of gristly skin covering the shoulder, which the German hunters call the shield; and here we have a part specially modified for defence. With boars in the prime of life (see fig. 63) the tusks in the lower jaw are used for fighting but they become in old age, as Brehm states, so much curved inwards and upwards, over the snout, that they can no longer be thus used. They may, however, still continue to serve, and even in a still more effective manner, as a means of defence. In compensation for the loss of the lower tusks as weapons of offence, those in the upper jaw, which always project a little laterally, increase so much in length during old age, and curve so much upwards, that they can be used as a means of attack. Nevertheless an old boar is not so dangerous to man as one at the age of six or seven years.[856]

Fig. 63. *Head of common wild boar, in prime of life (from Brehm).*

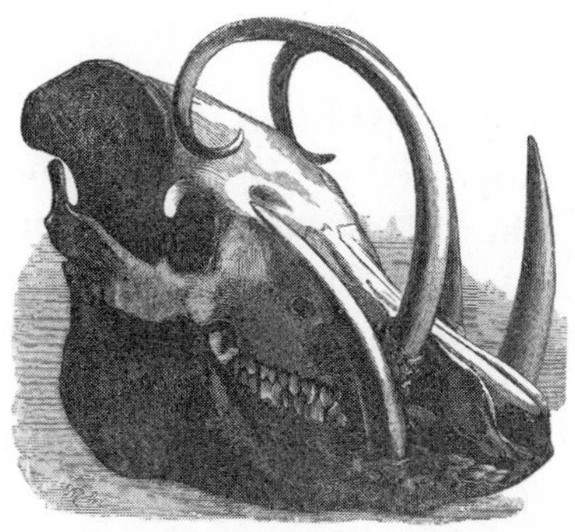

Fig. 64. *Skull of the Babirusa Pig (from Wallace's 'Malay Archipelago').*

In the full-grown male Babirusa pig of Celebes (fig. 64), the lower tusks are formidable weapons, like those of the European boar in the prime of life, whilst the upper tusks are so long and have their points so much curled inwards, sometimes even touching the forehead, that they are utterly useless as weapons of attack. They more nearly resemble horns than teeth, and are so manifestly useless as teeth that the animal was formerly supposed to rest his head by hooking them on to a branch. Their convex surfaces would, however, if the head were held a little laterally, serve as an excellent guard; and hence, perhaps it is that in old animals they "are generally broken off, as if by fighting."[857] Here, then, we have the curious case of the upper tusks of the Babirusa regularly assuming during the prime of life, a structure which apparently renders them fitted only for defence; whilst in the European boar the lower and opposite tusks assume in a less degree and only during old age nearly the same form, and then serve in like manner solely for defence.

In the wart-hog *(Phacochoerus æthiopicus,* fig. 65) the tusks in the upper jaw of the male curve upwards during the prime of life, and from being pointed, serve as formidable weapons. The tusks in the lower jaw are sharper than those in the upper, but from their shortness it seems hardly possible that they can be used as weapons of attack. They must, however, greatly strengthen those in the upper jaw, from being ground so as to fit closely against their bases. Neither the upper nor the lower tusks appear to have been specially modified to act as guards, though, no doubt, they are thus used to a certain extent. But the wart-hog is not destitute of other special means of protection, for there exists, on each side of the face, beneath the eyes, a rather stiff, yet flexible, cartilaginous, oblong pad (fig. 65), which projects two or three inches outwards; and it appeared to Mr. Bartlett and myself, when viewing the living animal, that these pads, when struck from beneath by the tusks of an opponent, would be turned upwards, and would thus protect in an admirable manner the somewhat prominent eyes. These boars, as I may add on the authority of Mr. Bartlett, when fighting together, stand directly face to face.

Fig. 65. *Head of Æthiopian Wart-hog, from 'Proc. Zool. Soc.' 1869. (I now find that this drawing represents the head of a female, but it serves to shew, on a reduced scale, the characters of the male.)*

Lastly, the African river-hog *(Potamochoerus penicillatus)* has a hard cartilaginous knob on each side of the face beneath the eyes, which answers to the flexible pad of the wart-hog; it has also two bony prominences on the upper jaw above the nostrils. A boar of this species in the Zoological Gardens recently broke into the cage of the wart-hog. They fought all night-long, and were found in the morning much exhausted, but not seriously wounded. It is a significant fact, as shewing the purpose of the above-described projections and excrescences, that these were covered with blood, and were scored and abraded in an extraordinary manner.

The mane of the lion forms a good defence against the one danger to which he is liable, namely the attacks of rival lions: for the males, as Sir A. Smith informs me, engage in terrible battles, and a young lion dares not approach an old one. In 1857 a tiger at Bromwich broke into the cage of a lion, and a fearful scene ensued; "the lion's mane saved his neck and head from being much injured, but the tiger at last succeeded in ripping up his belly, and in a few minutes he was dead."[858] The broad ruff round the throat and chin of the Canadian lynx *(Felis Canadensis)* is much longer in the male than in the female; but whether it serves as a defence I do not know. Male seals are well known to fight desperately together, and the males of certain kinds *(Otaria jubata)*[859] have great manes, whilst the females have small ones or none. The male baboon of the Cape of Good Hope *(Cynocephalus porcarius)* has a much longer mane and larger canine teeth than the female; and the mane probably serves as a protection, for on asking the keepers in the Zoological Gardens, without giving them any clue to my object, whether any of the monkeys especially attacked each other by the nape of the neck, I was answered that this was not the case, excepting with the above baboon. In the Hamadryas baboon, Ehrenberg compares the mane of the adult male to that of a young lion, whilst in the young of both sexes and in the female the mane is almost absent.

It appeared to me probable that the immense woolly mane of the male American bison, which reaches almost to the ground, and is much more developed in the males than in the females, served as a protection to them in their terrible battles; but an experienced hunter told Judge Caton that he had never observed anything which favoured this belief. The stallion has a thicker and fuller mane than the mare; and I have made particular inquiries of two great trainers and breeders who have had charge of many entire horses, and am assured that they "invariably endeavour to seize one another by the neck." It does not, however, follow from the foregoing statements, that when the hair on the neck serves as a defence, that it was originally developed for this purpose, though this is probable in some cases, as in that of the lion. I am informed by Mr. McNeill that the long hairs on the throat of the stag *(Cervus elephas)* serve as a great protection to him when hunted, for the dogs generally endeavour to seize him by the throat; but it is not probable that these hairs were specially developed for this purpose; otherwise the young and the females would, as we may feel assured, have been equally protected.

On Preference or Choice in Pairing, as shewn by either sex of Quadrupeds.—Before describing, in the next chapter, the differences between the sexes in voice, odour emitted, and ornamentation, it will be convenient here to consider whether the sexes exert any choice in their unions. Does the female prefer any particular male, either before or after the males may have fought together for supremacy; or does the male, when not a polygamist, select any particular female? The general impression amongst breeders seems to be that the male accepts any female; and this, owing to his eagerness, is, in most cases, probably the truth. Whether the female as a general rule indifferently accepts any male is much more doubtful. In the fourteenth chapter, on Birds, a considerable body of direct and indirect evidence was advanced, shewing that the female selects her partner; and it would be a strange anomaly if female quadrupeds, which stand higher in the scale of organisation and have higher mental powers, did not generally, or at least often, exert some choice. The female could in most cases escape, if wooed by a male that did not please or excite her; and when pursued, as so incessantly occurs, by several males, she would often have the opportunity, whilst they were fighting together, of escaping with, or at least of temporarily pairing with, some one male. This latter contingency has often been observed in Scotland with female red-deer, as I have been informed by Sir Philip Egerton.[860]

It is scarcely possible that much should be known about female quadrupeds exerting in a state of nature any choice in their marriage unions. The following very curious details on the courtship of one of the eared seals, *Callorhinus ursinus*, are given[861] on the authority of Capt. Bryant, who had ample opportunities for observation. He says, "Many of the females on their arrival at the island where they breed appear desirous of returning to some particular male, and frequently climb the outlying rocks to overlook the rookeries, calling out and listening as if for a familiar voice. Then changing to another place they do the same again . . . . As soon as a female reaches the shore, the nearest male goes down to meet her, making meanwhile a noise like the clucking of a hen to her chickens. He bows to her and coaxes her until he gets between her and the water so that she cannot escape him. Then his manner changes, and with a harsh growl he drives her to a place in his harem. This continues until the lower row of harems is nearly full. Then the males higher up select the time when their more fortunate neighbours are off their guard to steal their wives. This they do by taking them in their mouths and lifting them over the heads of the other females, and carefully placing them in their own harem, carrying them as cats do their kittens. Those still higher up pursue the same method until the whole space is occupied. Frequently a struggle ensues between two males for the possession of the same female, and both seizing her at once pull her in two or terribly lacerate her with their teeth. When the space is all filled, the old male walks around complacently reviewing his family, scolding those who crowd or disturb the others, and fiercely driving off all intruders. This surveillance always keeps him actively occupied."

As so little is known about the courtship of animals in a state of nature, I have endeavoured to discover how far our domesticated quadrupeds evince any choice in their unions. Dogs offer the best opportunity for observation, as they are carefully attended to and well understood. Many breeders have expressed a strong opinion on this head. Thus Mr. Mayhew remarks, "The females are able to bestow their affections; and tender recollections are as potent over them as they are known to be in other cases, where higher animals are concerned. Bitches are not always prudent in their loves, but are apt to fling themselves away on curs of low degree. If reared with a companion of vulgar appearance, there often springs up between the pair a devotion which no time can afterwards subdue. The passion, for such it really is, becomes of a more than romantic endurance." Mr. Mayhew, who attended chiefly to the smaller breeds, is convinced that the females are strongly attracted by males of large size.[862] The well-known veterinary Blaine states[863] that his own female pug became so attached to a spaniel, and a female setter to a cur, that in neither case would they pair with a dog of their own breed until several weeks had elapsed. Two similar and trustworthy accounts have been given me in regard to a female retriever and a spaniel, both of which became enamoured with terrier-dogs.

Mr. Cupples informs me that he can personally vouch for the accuracy of the following more remarkable case, in which a valuable and wonderfully-intelligent female terrier loved a retriever, belonging to a neighbour, to such a degree that she had often to be dragged away from him. After their permanent separation, although repeatedly shewing milk in her teats, she would never acknowledge the courtship of any other dog, and to the regret of her owner, never bore puppies. Mr. Cupples also states that a female deer-hound now (1868) in his kennel has thrice produced puppies, and on each occasion shewed a marked preference for one of the largest and handsomest, but not the most eager, of four deer-hounds living with her, all in the prime of life. Mr. Cupples has observed that the female generally favours a dog whom she has associated with and knows; her shyness and timidity at first incline her against a strange dog. The male, on the contrary, seems rather inclined towards strange females. It appears to be rare when the male refuses any particular female, but Mr. Wright, of Yeldersley House, a great breeder of dogs, informs me that he has known some instances; he cites the case of one of his own deer-hounds, who would not take any notice of a particular female mastiff, so that another deer-hound had to be employed. It would be superfluous to give other cases, and I will only add that Mr. Barr, who has carefully bred many blood-hounds, states that in almost every instance particular individuals of the opposite sex shew a decided preference for each other. Finally Mr. Cupples, after attending to this subject for another year, has recently written to me, "I have had full confirmation of my former statement, that dogs in breeding form decided preferences for each other, being often influenced by size, bright colour, and individual character, as well as by the degree of their previous familiarity."

In regard to horses, Mr. Blenkiron, the greatest breeder of race-horses in the world, informs me that stallions are so frequently capricious in their choice, rejecting one mare and without any apparent cause taking to another, that various artifices have to be habitually used. The famous Monarque, for instance, would never consciously look at the dam of Gladiateur, and a trick had to be practised. We can partly see the reason why valuable race-horse stallions, which are in such demand, should be so particular in their choice. Mr. Blenkiron has never known a mare to reject a horse; but this has occurred in Mr. Wright's stable, so that the mare had to be cheated. Prosper Lucas[864] quotes various statements from French authorities, and remarks, "On voit des étalons qui s'éprennent d'une jument, et négligent toutes les autres." He gives, on the authority of Baëlen, similar facts in regard to bulls. Hoffberg, in describing the domesticated reindeer of Lapland, says, "Fœmina majores et fortiores mares præ cæteris admittunt, ad eos confugiunt, a junioribus agitatæ, qui hos in fugam conjiciunt."[865] A clergyman, who has bred many pigs, assures me that sows often reject one boar and immediately accept another.

From these facts there can be no doubt that with most of our domesticated quadrupeds strong individual antipathies and preferences are frequently exhibited, and much more commonly by the female than by the male. This being the case, it is improbable that the unions of quadrupeds in a state of nature should be left to mere chance. It is much more probable that the females are allured or excited by particular males, who possess certain characters in a higher degree than other males; but what these characters are, we can seldom or never discover with certainty.

Secondary Sexual Characters of Mammals—continued

VOICE • REMARKABLE SEXUAL PECULIARITIES IN SEALS • ODOUR • DEVELOPMENT OF THE HAIR • COLOUR OF THE HAIR AND SKIN • ANOMALOUS CASE OF THE FEMALE BEING MORE ORNAMENTED THAN THE MALE • COLOUR AND ORNAMENTS DUE TO SEXUAL SELECTION • COLOUR ACQUIRED, FOR THE SAKE OF PROTECTION • COLOUR, THOUGH COMMON TO BOTH SEXES, OFTEN DUE TO SEXUAL SELECTION • ON THE DISAPPEARANCE OF SPOTS AND STRIPES IN ADULT QUADRUPEDS • ON THE COLOURS AND ORNAMENTS OF THE QUADRUMANA • SUMMARY

Quadrupeds use their voices for various purposes, as a signal of danger, as a call from one member of a troop to another, or from the mother to her lost offspring, or from the latter for protection to their mother; but such uses need not here be considered. We are concerned only with the difference between the voices of the two sexes, for instance between that of the lion and lioness, or of the bull and cow. Almost all male animals use their voices much more during the rutting-season than at any other time; and some, as the giraffe and porcupine,[866] are said to be completely mute excepting at this season. As the throats *(i.e.* the larnyx and thyroid bodies[867]) of stags become periodically enlarged at the commencement of the breeding-season, it might be thought that their powerful voices must be then in some way of high importance to them; but this is very doubtful. From information given to me by two experienced observers, Mr. McNeill and Sir P. Egerton, it seems that young stags under three years old do not roar or bellow; and that the old ones begin bellowing at the commencement of the breeding-season, at first only occasionally and moderately, whilst they restlessly wander about in search of the females. Their battles are prefaced by loud and prolonged bellowing, but during the actual conflict they are silent. Animals of all kinds which habitually use their voices, utter various noises under any strong emotion, as when enraged and preparing to fight; but this may merely be the result of their nervous excitement, which leads to the spasmodic contraction of almost all the muscles of the body, as when a man grinds his teeth and clenches his hands in rage or agony. No doubt stags challenge each other to mortal combat by bellowing; but it is not likely that this habit could have led through sexual selection, that is by the loudest-voiced males having been the most successful in their conflicts, to the periodical enlargement of the vocal organs; for the stags with the most powerful voices, unless at the same time the strongest, best-armed, and most courageous, would not have gained any advantage over their rivals with weaker voices. The stags, moreover, which had weaker voices, though not so well able to challenge other stags, would have been drawn to the place of combat as certainly as those with stronger voices.

It is possible that the roaring of the lion may be of some actual service to him in striking terror into his adversary; for when enraged he likewise erects his mane and thus instinctively tries to make himself appear as terrible as possible. But it can hardly be supposed that the bellowing of the stag, even if it be of any service to him in this way, can have been important enough to have led to the periodical enlargement of the throat. Some writers suggest that the bellowing serves as a call to the female; but the experienced observers above quoted inform me that female deer do not search for the male, though the males search eagerly for the females, as indeed might be expected from what we know of the habits of other male quadrupeds. The voice of the female, on the other hand, quickly brings to her one or more stags,[868] as is well known to the hunters who in wild countries imitate her cry. If we could believe that the male had the power to excite or allure the female by his voice, the periodical enlargement of his vocal organs would be intelligible on the principle of sexual selection, together with inheritance limited to the same sex and season of the year; but we have no evidence in favour of this view. As the case stands, the loud voice of the stag during the breeding season does not seem to be of any special service to him, either during his courtship or battles, or in any other way. But may we not believe that the frequent use of the voice, under the strong excitement of love, jealousy, and rage, continued during many generations, may at last have produced an inherited effect on the vocal organs of the stag, as well as of other male animals? This appears to me, with our present state of knowledge, the most probable view.

The male gorilla has a tremendous voice, and when adult is furnished with a laryngeal sack, as is likewise the adult male orang.[869] The gibbons rank amongst the noisiest of monkeys, and the Sumatra species *(Hylobates syndactylus)* is also furnished with a laryngeal sack; but Mr. Blyth, who has had opportunities for observation, does not believe that the male is more noisy than the female. Hence, these latter monkeys probably use their voices as a mutual call; and this is certainly the case with some quadrupeds, for instance with the beaver.[870] Another gibbon, the *H. agilis,* is highly remarkable, from having the power of emitting a complete and correct octave of musical notes,[871] which we may reasonably suspect serves as a sexual charm; but I shall have to recur to this subject in the next chapter. The vocal organs of the American *Mycetes caraya* are one-third larger in the male than in the female, and are wonderfully powerful. These monkeys, when the weather is warm, make the forests resound during the morning and evening with their overwhelming voices. The males begin the dreadful concert, in which the females, with their less powerful voices, sometimes join, and which is often continued during many hours. An excellent observer, Rengger,[872] could not perceive that they were excited to begin their concert by any special cause; he thinks that like many birds, they delight in their own music, and try to excel each other. Whether most of the foregoing monkeys have acquired their powerful voices in order to beat their rivals and to charm the females—or whether the vocal organs have been strengthened and enlarged through the inherited effects of long-continued use without any particular good being gained—I will not pretend to say; but the former view, at least in the case of the *Hylobates agilis,* seems the most probable.

I may here mention two very curious sexual peculiarities occurring in seals, because they have been supposed by some writers to affect the voice. The nose of the male sea-elephant *(Macrorhinus proboscideus),* when about three years old, is greatly elongated during the breeding-season, and can then be erected. In this state it is sometimes a foot in length. The female at no period of life is thus provided, and her voice is different. That of the male consists of a wild, hoarse, gurgling noise, which is audible at a great distance, and is believed to be strengthened by the proboscis. Lesson compares the erection of the proboscis, to the swelling of the wattles of male gallinaceous birds, whilst they court the females. In another allied kind of seal, namely, the bladder-nose *(Cystophora cristata),* the head is covered by a great hood or bladder. This is internally supported by the septum of the nose, which is produced far backwards and rises into a crest seven inches in height. The hood is clothed with short hair, and is muscular; it can be inflated until it more than equals the whole head in size! The males when rutting fight furiously on the ice, and their roaring "is said to be sometimes so loud as to be heard four miles off." When attacked by man they likewise roar or bellow; and whenever irritated the bladder is inflated. Some naturalists believe that the voice is thus strengthened, but various other uses have been assigned to this extraordinary structure. Mr. R. Brown thinks that it serves as a protection against accidents of all kinds. This latter view is not probable, if what the sealers have long maintained is correct, namely, that the hood or bladder is very poorly developed in the females and in the males whilst young.[873]

Odour.—With some animals, as with the notorious skunk of America, the overwhelming odour which they emit appears to serve exclusively as a means of defence. With shrew-mice (Sorex) both sexes possess abdominal scent-glands, and there can be little doubt, from the manner in which their bodies are rejected by birds and beasts of prey, that their odour is protective; nevertheless the glands become enlarged in the males during the breeding-season. In many quadrupeds the glands are of the same size in both sexes;[874] but their use is not known. In other species the glands are confined to the males, or are more developed in them than in the females; and they almost always become more active during the rutting-season. At this period the glands on the sides of the face of the male elephant enlarge and emit a secretion having a strong musky odour.

The rank effluvium of the male goat is well known, and that of certain male deer is wonderfully strong and persistent. On the banks of the Plata I have perceived the whole air tainted with the odour of the male *Cervus campestris,* at the distance of half a mile to leeward of a herd; and a silk handkerchief, in which I carried home a skin, though repeatedly used and washed, retained, when first unfolded, traces of the odour for one year and seven months. This animal does not emit its strong odour until more than a year old, and if castrated whilst young never emits it.[875] Besides the general odour, with which the whole body of certain ruminants seems to be permeated during the breeding-season, many deer, antelopes, sheep, and goats, possess odoriferous glands in various situations, more especially on their faces. The so-called tear-sacks or suborbital pits come under this head. These glands secrete a semi-fluid fetid matter, which is sometimes so copious as to stain the whole face, as I have seen in the case of an antelope. They are "usually larger in the male than in the female, and their development is checked by castration."[876] According to Desmarest they are altogether absent in the female of *Antilope subgutturosa.* Hence, there can be no doubt that they stand in some close relation with the reproductive functions. They are also sometimes present, and sometimes absent, in nearly-allied forms. In the adult male musk-deer *(Moschus moschiferus),* a naked space round the tail is bedewed with an odoriferous fluid, whilst in the adult female, and in the male, until two years old, this space is covered with hair and is not odoriferous. The proper musk-sack, from its position, is necessarily confined to the male, and forms an additional scent-organ. It is a singular fact that the matter secreted by this latter gland does not, according to Pallas, change in consistence, or increase in quantity, during the rutting-season; nevertheless this naturalist admits that its presence is in some way connected with the act of reproduction. He gives, however, only a conjectural and unsatisfactory explanation of its use.[877]

In most cases, when during the breeding-season the male alone emits a strong odour, this probably serves to excite or allure the female. We must not judge on this head by our own taste, for it is well known that rats are enticed by certain essential oils, and cats by valerian, substances which are far from agreeable to us; and that dogs, though they will not eat carrion, sniff and roll in it. From the reasons given when discussing the voice of the stag, we may reject the idea that the odour serves to bring the females from a distance to the males. Active and long-continued use cannot here have come into play, as in the case of the vocal organs. The odour emitted must be of considerable importance to the male, inasmuch as large and complex glands, furnished with muscles for everting the sack, and for closing or opening the orifice, have in some cases been developed. The development of these organs is intelligible through sexual selection, if the more odoriferous males are the most successful in winning the females, and in leaving offspring to inherit their gradually-perfected glands and odours.

Development of the Hair.—We have seen that male quadrupeds often have the hair on their necks and shoulders much more developed than in the females; and many additional instances could be given. This sometimes serves as a defence to the male during his battles; but whether the hair in most cases has been specially developed for this purpose is very doubtful. We may feel almost certain that this is not the case, when a thin and narrow crest runs along the whole length of the back; for a crest of this kind would afford scarcely any protection, and the ridge of the back is not a likely place to be injured; nevertheless such crests are sometimes confined to the males, or are much more developed in them than in the females. Two antelopes, the *Tragelaphus scriptus*[878] (see fig. 68, p. 1189) and *Portax picta*, may be given as instances. The crests of certain stags and of the male wild goat stand erect, when these animals are enraged or terrified;[879] but it can hardly be supposed that they have been acquired for the sake of exciting fear in their enemies. One of the above-named antelopes, the *Portax picta*, has a large well-defined brush of black hair on the throat, and this is much larger in the male than in the female. In the

Fig. 66. *Pithecia Satanas, male (from Brehm).*

Ammotragus tragelaphus of North Africa, a member of the sheep-family, the front-legs are almost concealed by an extraordinary growth of hair, which depends from the neck and upper halves of the legs; but Mr. Bartlett does not believe that this mantle is of the least use to the male, in whom it is much more developed than in the female.

Male quadrupeds of many kinds differ from the females in having more hair, or hair of a different character, on certain parts of their faces. The bull alone has curled hair on the forehead.[880] In three closely-allied sub-genera of the goat family, the males alone possess beards, sometimes of large size; in two other sub-genera both sexes have a beard, but this disappears in some of the domestic breeds of the common goat; and neither sex of the Hemitragus has a beard. In the ibex the beard is not developed during the summer, and is so small at other seasons that it may be called rudimentary.[881] With some monkeys the beard is confined to the male, as in the Orang, or is much larger in the male than in the female, as in the *Mycetes caraya* and *Pithecia satanas* (fig. 66). So it is with the whiskers of some species of Macacus,[882] and, as we have seen, with the manes of some species of baboons. But with most kinds of monkeys the various tufts of hair about the face and head are alike in both sexes.

The males of various members of the Ox family (Bovidæ), and of certain antelopes, are furnished with a dewlap, or great fold of skin on the neck, which is much less developed in the female.

Now, what must we conclude with respect to such sexual differences as these? No one will pretend that the beards of certain male-goats, or the dewlap of the bull, or the crests of hair along the backs of certain male antelopes, are of any direct or ordinary use to them. It is possible that the immense beard of the male Pithecia, and the large beard of the male Orang, may protect their throats when fighting; for the keepers in the Zoological Gardens inform me that many monkeys attack each other by the throat: but it is not probable that the beard has been developed for a distinct purpose from that which the whiskers, moustache, and other tufts of hair on the face serve; and no one will suppose that these are useful as a protection. Must we attribute to mere purposeless variability in the male all these appendages of hair or skin? It cannot be denied that this is possible; for with many domesticated quadrupeds, certain characters, apparently not derived through reversion from any wild parent-form, have appeared in, and are confined to, the males, or are more largely developed in them than in the females,—for instance the hump in the male, zebu-cattle of India, the tail in fat-tailed rams, the arched outline of the forehead in the males of several breeds of sheep, the mane in the ram of an African breed, and, lastly, the mane, long hairs on the hinder legs, and the dewlap in the male alone of the Berbura goat.[883] The mane which occurs in the rams alone of the above-mentioned African breed of sheep, is a true secondary sexual character, for it is not developed, as I hear from Mr. Winwood Reade, if the animal be castrated. Although we ought to be extremely cautious, as shewn in my work on 'Variation under Domestication,' in concluding that any character, even with animals kept by semi-civilised people, has not been subjected to selection by man, and thus augmented; yet in the cases just specified this is improbable, more especially as the characters are confined to the males, or are more strongly developed in them than in the females. If it were positively known that the African ram with a mane was descended from the same primitive stock with the other breeds of sheep, or the Berbura male-goat with his mane, dewlap, &c., from the same stock with other goats; and if selection has not been applied to these characters, then they must be due to simple variability, together with sexually-limited inheritance.

In this case it would appear reasonable to extend the same view to the many analogous characters occurring in animals under a state of nature. Nevertheless I cannot persuade myself that this view is applicable in many cases, as in that of the extraordinary development of hair on the throat and fore-legs of the male Ammotragus, or of the immense beard of the male Pithecia. With those antelopes in which the male when adult is more strongly-coloured than the female, and with those monkeys in which this is likewise the case, and in which the hair on the face is of a different colour from that on the rest of the head, being arranged in the most diversified and elegant manner, it seems probable that the crests and tufts of hair have been acquired as ornaments; and this I know is the opinion of some naturalists. If this view be correct, there can be little doubt that they have been acquired, or at least modified, through sexual selection.

Colour of the Hair and of the Naked Skin.—I will first give briefly all the cases known to me, of male quadrupeds differing in colour from the females. With Marsupials, as I am informed by Mr. Gould, the sexes rarely differ in this respect; but the great red kangaroo offers a striking exception, "delicate blue being the prevailing tint in those parts of the female, which in the male are red."[884] In the *Didelphis opossum* of Cayenne the female is said to be a little more red than the male. With Rodents Dr. Gray remarks: "African squirrels, especially those found in the tropical regions, have the fur much brighter and more vivid at some seasons of the year than at others, and the fur of the male is generally brighter than that of the female."[885] Dr. Gray informs me that he specified the African squirrels, because, from their unusually bright colours, they best exhibit this difference. The female of the *Mus minutus* of Russia is of a paler and dirtier tint than the male. In some few bats the fur of the male is lighter and brighter than in the female.[886]

The terrestrial Carnivora and Insectivora rarely exhibit sexual differences of any kind, and their colours are almost always exactly the same in both sexes. The ocelot *(Felis pardalis),* however, offers an exception, for the colours of the female, compared with those of the male, are "moins apparentes, le fauve étant plus terne, le blanc moins pur, les raies ayant moins de largeur et les taches moins de diamètre."[887] The sexes of the allied *Felis mitis* also differ, but even in a less degree, the general hues of the female being rather paler than in the male, with the spots less black. The marine Carnivora or Seals, on the other hand, sometimes differ considerably in colour, and they present, as we have already seen, other remarkable sexual differences. Thus the male of the *Otaria nigrescens* of the southern hemisphere is of a rich brown shade above; whilst the female, who acquires her adult tints earlier in life than the male, is dark-grey above, the young of both sexes being of a very deep chocolate colour. The male of the northern *Phoca groenlandica* is tawny grey, with a curious saddle-shaped dark mark on the back; the female is much smaller, and has a very different appearance, being "dull white or yellowish straw-colour, with a tawny hue on the back;" the young at first are pure white, and can "hardly be distinguished among the icy hummocks and snow, their colour thus acting as a protection."[888]

With Ruminants sexual differences of colour occur more commonly than in any other order. A difference of this kind is general with the Strepsicerene antelopes; thus the male nilghau *(Portax picta)* is bluish-grey and much darker than the female, with the square white patch on the throat, the white marks on the fetlocks, and the black spots on the ears, all much more distinct. We have seen that in this species the crests and tufts of hair are likewise more developed in the male than in the hornless female. The male, as I am informed by Mr. Blyth, without shedding his hair, periodically becomes darker during the breeding-season. Young males cannot be distinguished from young females until above twelve months old; and if the male is emasculated before this period, he never, according to the same authority, changes colour. The importance of this latter fact, as distinctive of sexual colouring, becomes obvious, when we hear[889] that neither the red summer-coat nor the blue winter-coat of the Virginian deer is at all affected by emasculation. With most or all of the highly-ornamented species of Tragelaphus the males are darker than the hornless females, and their crests of hair are more fully developed. In the male of that magnificent antelope, the *Derbyan Eland,* the body is redder, the whole neck much blacker, and the white band which separates these colours, broader, than in the female. In the Cape Eland also, the male is slightly darker than the female.[890]

In the Indian Black-buck *(A. bezoartica),* which belongs to another tribe of antelopes, the male is very dark, almost black; whilst the hornless female is fawn-coloured. We have in this species, as Mr. Blyth informs me, an exactly parallel series of facts, as with the *Portax picta,* namely in the male periodically changing colour during the breeding season, in the effects of emasculation on this change, and in the young of both sexes being undistinguishable from each other. In the *Antilope niger* the male is black, the female as well as the young being brown; in *A. sing-sing* the male is much brighter coloured than the hornless female, and his chest and belly are blacker; in the male *A. caama,* the marks and lines which occur on various parts of the body are black instead of as in the female brown; in the brindled gnu *(A. gorgon)* "the colours of the male are nearly the same as those of the female, only deeper and of a brighter hue."[891] Other analogous cases could be added.

The Banteng bull *(Bos sondaicus)* of the Malayan archipelago is almost black, with white legs and buttocks; the cow is of a bright dun, as are the young males until about the age of three years, when they rapidly change colour. The emasculated bull reverts to the colour of the female. The female Kemas goat is paler, and the female *Capra ægagrus* is said to be more uniformly tinted than their respective males. Deer rarely present any sexual differences in colour. Judge Caton, however, informs me that with the males of the Wapiti deer *(Cervus Canadensis)* the neck, belly, and legs are much darker than the same parts in the female; but during the winter the darker tints gradually fade away and disappear. I may here mention that

Judge Caton has in his park three races of the Virginian deer, which differ slightly in colour, but the differences are almost exclusively confined to the blue winter or breeding coat; so that this case may be compared with those given in a previous chapter of closely-allied or representative species of birds which differ from each other only in their nuptial plumage.[892] The females of *Cervus paludosus* of S. America, as well as the young of both sexes, do not possess the black stripes on the nose, and the blackish-brown line on the breast which characterise the adult males.[893] Lastly, the mature male of the beautifully coloured and spotted Axis deer is considerably darker, as I am informed by Mr. Blyth, than the female; and this hue the castrated male never acquires.

The last Order which we have to consider—for I am not aware that sexual differences in colour occur in the other mammalian groups—is that of the Primates. The male of the *Lemur macaco* is coal-black, whilst the female is reddish-yellow, but highly variable in colour.[894] Of the Quadrumana of the New World, the females and young of *Mycetes caraya* are greyish-yellow and alike; in the second year the young male becomes reddish-brown, in the third year black, excepting the stomach, which, however, becomes quite black in the fourth or fifth year. There is also a strongly-marked difference in colour between the sexes in *Mycetes seniculus* and *Cebus capucinus;* the young of the former and I believe of the latter species resembling the females. With *Pithecia leucocephala* the young likewise resemble the females, which are brownish-black above and light rusty-red beneath, the adult males being black. The ruff of hair round the face of *Ateles marginatus* is tinted yellow in the male and white in the female. Turning to the Old World, the males of *Hylobates hoolock* are always black, with the exception of a white band over the brows; the females vary from whity-brown to a dark tint mixed with black, but are never wholly black.[895] In the beautiful *Cercopithecus Diana* the head of the adult male is of an intense black, whilst that of the female is dark grey; in the former the fur between the thighs is of an elegant fawn-colour, in the latter it is paler. In the equally beautiful and curious moustache monkey *(Cercopithecus cephus)* the only difference between the sexes is that the tail of the male is chesnut and that of the female grey; but Mr. Bartlett informs me that all the hues become more strongly pronounced in the male when adult, whilst in the female they remain as they were during youth. According to the coloured figures given by Solomon Müller, the male of *Semnopithecus chrysomelas* is nearly black, the female being pale brown. In the *Cercopithecus cynosurus* and *griseo-viridis* one part of the body which is confined to the male sex is of the most brilliant blue or green, and contrasts strikingly with the naked skin on the hinder part of the body, which is vivid red.

Lastly, in the Baboon family, the adult male of *Cynocephalus hamadryas* differs from the female not only by his immense mane, but slightly in the colour of the hair and of the naked callosities. In the drill *(Cynocephalus leucophœus)* the females and young are much paler-coloured, with less green, than the adult males. No other member of the whole class of mammals is coloured in so extraordinary a manner as the adult male mandrill *(Cynocephalus mormon)*. The face at this age becomes of a fine blue, with the ridge and tip of the nose of the most brilliant red. According to some authors the face is also marked with whitish stripes, and is shaded in parts with black, but the colours appear to be variable. On the forehead there is a crest of hair, and on the chin a yellow beard. "Toutes les parties supérieures de leurs cuisses et le grand espace nu de leurs fesses sont également colorés du rouge le plus vif, avec un mélange de bleu qui ne manque réellement pas d'élégance."[896] When the animal is excited all the naked parts become much more vividly tinted. Several authors have used the strongest expressions in describing these resplendent colours, which they compare with those of the most brilliant birds. Another most remarkable peculiarity is that when the great canine teeth are fully developed, immense protuberances of bone are formed on each cheek, which are deeply furrowed longitudinally, and the naked skin over them is brilliantly-coloured, as just described. (Fig. 67.) In the adult females and in the young of both sexes these protuberances are scarcely perceptible; and the naked parts are much less brightly coloured, the face being almost black, tinged with blue. In the adult female, however, the nose at certain regular intervals of time becomes tinted with red.

Fig. 67. *Head of male Mandrill (from Gervais, 'Hist. Nat. des Mammifèrces').*

In all the cases hitherto given the male is more strongly or brightly coloured than the female, and differs in a greater degree from the young of both sexes. But as a reversed style of colouring is characteristic of the two sexes with some few birds, so with the Rhesus monkey *(Macacus rhesus)* the female has a large surface of naked skin round the tail, of a brilliant carmine red, which periodically becomes, as I was assured by the keepers in the Zoological Gardens, even more vivid, and her face is also pale red. On the other hand with the adult male and with the young of both sexes, as I saw in the Gardens, neither the naked skin at the posterior end of the body, nor the face, shew a trace of red. It appears, however, from some published accounts, that the male does occasionally, or during certain seasons, exhibit some traces of the red. Although he is thus less ornamented than the female, yet in the larger size of his body, larger canine teeth, more developed whiskers, more prominent superciliary ridges, he follows the common rule of the male excelling the female.

I HAVE NOW GIVEN all the cases known to me of a difference in colour between the sexes of mammals. The colours of the female either do not differ in a sufficient degree from those of the male, or are not of a suitable nature, to afford her protection, and therefore cannot be explained on this principle. In some, perhaps in many cases, the differences may be the result of variations confined to one sex and transmitted to the same sex, without any good having been thus gained, and therefore without the aid of selection. We have instances of this kind with our domesticated animals, as in the males of certain cats being rusty-red, whilst the females are tortoise-shell coloured. Analogous cases occur under nature; Mr. Bartlett has seen many black varieties of the jaguar, leopard, vulpine phalanger and wombat; and he is certain that all, or nearly all, were males. On the other hand, both sexes of wolves, foxes, and apparently of American squirrels, are occasionally born black. Hence it is quite possible that with some mammals the blackness of the males, especially when this colour is congenital, may simply be the result, without the aid of selection, of one or more variations having occurred, which from the first were sexually limited in their transmission. Nevertheless it can hardly be admitted that the diversified, vivid, and contrasted colours of certain quadrupeds, for instance of the above-mentioned monkeys and antelopes, can thus be accounted for. We should bear in mind that these colours do not appear in the male at birth, as in the case of most ordinary variations, but only at or near maturity; and that unlike ordinary variations, if the male be emasculated, they never appear or subsequently disappear. It is on the whole a much more probable conclusion that the strongly-marked colours and other ornamental characters of male quadrupeds are beneficial to them in their rivalry with other males, and have consequently been acquired through sexual selection. The probability of this view is strengthened by the differences in colour between the sexes occurring almost exclusively, as may be observed by going through the previous details, in those groups and sub-groups of mammals, which present other and distinct secondary sexual characters; these being likewise due to the action of sexual selection.

Quadrupeds manifestly take notice of colour. Sir S. Baker repeatedly observed that the African elephant and rhinoceros attacked with special fury white or grey horses. I have elsewhere shewn[897] that half-wild horses apparently prefer pairing with those of the same colour, and that herds of fallow-deer of a different colour, though living together, have long kept distinct. It is a more significant fact that a female zebra would not admit the addresses of a male ass until he was painted so as to resemble a zebra, and then, as John Hunter remarks, "she received him very readily. In this curious fact, we have instinct excited by mere colour, which had so strong an effect as to get the better of everything else. But the male did not require this, the female being an animal somewhat similar to himself, was sufficient to rouse him."[898]

In an early chapter we have seen that the mental powers of the higher animals do not differ in kind, though so greatly in degree, from the corresponding powers of man, especially of the lower and barbarous races; and it would appear that even their taste for the beautiful is not widely different from that of the Quadrumana. As the negro of Africa raises the flesh on his face into parallel ridges "or cicatrices, high above the natural surface, which unsightly deformities, are considered great personal attractions;"[899]—as negroes, as well as savages in many parts of the world, paint their faces with red, blue, white, or black bars,—so the male mandrill of Africa appears to have acquired his deeply-furrowed and gaudily-coloured face from having been thus rendered attractive to the female. No doubt it is to us a most grotesque notion that the posterior end of the body should have been coloured for the sake of ornament even more brilliantly than the face; but this is really not more strange than that the tails of many birds should have been especially decorated.

With mammals we do not at present possess any evidence that the males take pains to display their charms before the female; and the elaborate manner in which this is performed by male birds, is the strongest argument in favour of the belief that the females admire, or are excited by, the ornaments and colours displayed before them. There is, however, a striking parallelism between mammals and birds in all their secondary sexual characters, namely in their weapons for fighting with rival males, in their ornamental appendages, and in their colours. In both classes, when the male differs from the female, the young of both sexes almost always resemble each other, and in a large majority of cases resemble the adult female. In both classes the male assumes the characters proper to his sex shortly before the age for reproduction; if emasculated he either never acquires such characters or subsequently loses them. In both classes the change of colour is sometimes seasonal, and the tints of the naked parts sometimes become more vivid during the act of courtship. In both classes the male is almost always more vividly or strongly coloured than the female, and is ornamented with larger crests either of hair or feathers, or other appendages. In a few exceptional cases the female in both classes is more highly ornamented than the male. With many mammals, and at least in the case of one bird, the male is more odoriferous than the female. In both classes the voice of the male is more powerful than that of the female. Considering this parallelism there can be little doubt that the same cause, whatever it may be, has acted on mammals and birds; and the result, as far as ornamental characters are concerned, may safely be attributed, as it appears to me, to the long-continued preference of the individuals of one sex for certain individuals of the opposite sex, combined with their success in leaving a larger number of offspring to inherit their superior attractions.

Equal transmission of ornamental characters to both sexes.—With many birds, ornaments, which analogy leads us to believe were primarily acquired by the males, have been transmitted equally, or almost equally, to both sexes; and we may now enquire how far this view may be extended to mammals. With a considerable number of species, especially the smaller kinds, both sexes have been coloured, independently of sexual selection, for the sake of protection; but not, as far as I can judge, in so many cases, nor in nearly so striking a manner as in most of the lower classes. Audubon remarks that he often mistook the musk-rat,[900] whilst sitting on the banks of a muddy stream, for a clod of earth, so complete was the resemblance. The hare on her form is a familiar instance of concealment through colour; yet this principle partly fails in a closely-allied species, namely the rabbit, for as this animal runs to its burrow, it is made conspicuous to the sportsman and no doubt to all beasts of prey, by its upturned pure-white tail. No one has ever doubted that the quadrupeds which inhabit snow-clad regions, have been rendered white to protect them from their enemies, or to favour their stealing on their prey. In regions where snow never lies long on the ground a white coat would be injurious; consequently species thus coloured are extremely rare in the hotter parts of the world. It deserves notice that many quadrupeds, inhabiting moderately cold regions, although they do not assume a white winter dress, become paler during this season; and this apparently is the direct result of the conditions to which they have long been exposed. Pallas[901] states that in Siberia a change of this nature occurs with the wolf, two species of Mustela, the domestic horse, the *Equus hemionus*, the domestic cow, two species of antelopes, the musk-deer, the roe, the elk, and reindeer. The roe, for instance, has a red summer and a greyish-white winter coat; and the latter may perhaps serve as a protection to the animal whilst wandering through the leafless thickets, sprinkled with snow and hoar-frost. If the above named animals were gradually to extend their range into regions perpetually covered with snow, their pale winter-coats would probably be rendered, through natural selection, whiter and whiter by degrees, until they became as white as snow.

Although we must admit that many quadrupeds have received their present tints as a protection, yet with a host of species, the colours are far too conspicuous and too singularly arranged to allow us to suppose that they serve for this purpose. We may take as an illustration certain antelopes: when we see that the square white patch on the throat, the white marks on the fetlocks, and the round black spots on the ears, are all more distinct in the male of the *Portax picta*, than in the female;—when we see that the colours are more vivid, that the narrow white lines on the flank and the broad white bar on the shoulder are more distinct in the male *Oreas Derbyanus* than in the female;—when we see a similar difference between the sexes of the curiously-ornamented *Tragelaphus scriptus* (fig. 68),—we may conclude that these colours and various marks have been at least intensified through sexual selection. It is inconceivable that such colours and marks can be of any direct or ordinary service to these animals; and as they have almost certainly been intensified through sexual selection, it is probable that they were originally gained through this same process, and then partially transferred to the females. If this view be admitted, there can be little doubt that the equally singular colours and marks of many other antelopes, though common to both sexes, have been gained and transmitted in a like manner. Both sexes,

Fig. 68. *Tragelaphus scriptus, male (from the Knowsley Menagerie).*

for instance, of the Koodoo *(Strepsiceros Kudu,* fig. 62) have narrow white vertical lines on their hinder flanks, and an elegant angular white mark on their foreheads. Both sexes in the genus Damalis are very oddly coloured; in *D. pygarga* the back and neck are purplish-red, shading on the flanks into black, and abruptly separated from the white belly and a large white space on the buttocks; the head is still more oddly coloured, a large oblong white mask, narrowly-edged with black, covers the face up to the eyes (fig. 69); there are three white stripes on the forehead, and the ears are marked with white. The fawns of this species are of a uniform pale yellowish-brown. In *Damalis albifrons* the colouring of the head differs from that in the last species in a single white stripe replacing the three stripes, and in the ears being almost wholly white.[902] After having studied to the best of my ability the sexual differences of animals belonging to all classes, I cannot avoid the conclusion that the curiously-arranged colours of many antelopes, though common to both sexes, are the result of sexual selection primarily applied to the male.

Fig. 69. *Damalis pygarga, male (from the Knowsley Menagerie).*

The same conclusion may perhaps be extended to the tiger, one of the most beautiful animals in the world, the sexes of which cannot be distinguished by colour, even by the dealers in wild beasts. Mr. Wallace believes[903] that the striped coat of the tiger "so assimilates with the vertical stems of the bamboo, as to assist greatly in concealing him from his approaching prey." But this view does not appear to me satisfactory. We have some slight evidence that his beauty may be due to sexual selection, for in two species of Felis analogous marks and colours are rather brighter in the male than in the female. The zebra is conspicuously striped, and stripes on the open plains of South Africa cannot afford any protection. Burchell[904] in describing a herd says, "their sleek ribs glistened in the sun, and the brightness and regularity of their striped coats presented a picture of extraordinary beauty, in which probably they are not surpassed by any other quadruped." Here we have no evidence of sexual selection, as throughout the whole group of the Equidæ the sexes are identical in colour. Nevertheless he who attributes the white and dark vertical stripes on the flanks of various antelopes to sexual selection, will probably extend the same view to the Royal Tiger and beautiful Zebra.

We have seen in a former chapter that when young animals belonging to any class follow nearly the same habits of life with their parents, and yet are coloured in a different manner, it may be inferred that they have retained the colouring of some ancient and extinct progenitor. In the family of pigs, and in the genus Tapir, the young are marked with longitudinal stripes, and thus differ from every existing adult species in these two groups. With many kinds of deer the young are marked with elegant white spots, of which their parents exhibit not a trace. A graduated series can be followed from the Axis deer, both sexes of which at all ages and during all seasons are beautifully spotted (the male being rather more strongly coloured than the female)—to species in which neither the old nor the young are spotted. I will specify some of the steps in this series. The Mantchurian deer *(Cervus Mantchuricus)* is spotted during the whole year, but the spots are much plainer, as I have seen in the Zoological Gardens, during the summer, when the general colour of the coat is lighter, than during the winter, when the general colour is darker and the horns are fully developed. In the hog-deer *(Hyelaphus porcinus)* the spots are extremely conspicuous during the summer when the coat is reddish-brown, but quite disappear during the winter when the coat is brown.[905] In both these species the young are spotted. In the Virginian deer the young are likewise spotted, and about five per cent of the adult animals living in Judge Caton's park, as I am informed by him, temporarily exhibit at the period when the red summer coat is being replaced by the bluish winter coat, a row of spots on each flank, which are always the same in number, though very variable in distinctness. From this condition there is but a very small step to the complete absence of spots at all seasons in the adults; and lastly, to their absence at all ages, as occurs with certain species. From the existence of this perfect series, and more especially from the fawns of so many species being spotted, we may conclude that the now living members of the deer family are the descendants of some ancient species which, like the Axis deer, was spotted at all ages and seasons. A still more ancient progenitor probably resembled to a certain extent the *Hyomoschus aquaticus*—for this animal is spotted, and the hornless males have large exserted canine teeth, of which some few true deer still retain rudiments. It offers, also, one of those interesting cases of a form linking together two groups, as it is intermediate in certain osteological characters between the pachyderms and ruminants, which were formerly thought to be quite distinct.[906]

A curious difficulty here arises. If we admit that coloured spots and stripes have been acquired as ornaments, how comes it that so many existing deer, the descendants of an aboriginally spotted animal, and all the species of pigs and tapirs, the descendants of an aboriginally striped animal, have lost in their adult state their former ornaments? I cannot satisfactorily answer this question. We may feel nearly sure that the spots and stripes disappeared in the progenitors of our existing species at or near maturity, so that they were retained by the young and, owing to the law of inheritance at corresponding ages, by the young of all succeeding generations. It may have been a great advantage to the lion and puma from the open nature of the localities which they commonly haunt, to have lost their stripes, and to have been thus rendered less conspicuous to their prey; and if the successive variations, by which this end was gained, occurred rather late in life, the young would have retained their stripes, as we know to be the case. In regard to deer, pigs, and tapirs, Fritz Müller has suggested to me that these animals by the removal through natural selection of their spots or stripes would have been less easily seen by their enemies; and they would have especially required this protection, as soon as the carnivora increased in size and number during the Tertiary periods. This may be the true explanation, but it is rather strange that the young should not have been equally well protected, and still more strange that with some species the adults should have retained their spots, either partially or completely, during part of the year. We know, though we cannot explain the cause, that when the domestic ass varies and becomes reddish-brown, grey or black, the stripes on the shoulders and even on the spine frequently disappear. Very few horses, except dun-coloured kinds, exhibit stripes on any part of their bodies, yet we have good reason to believe that the aboriginal horse was striped on the legs and spine, and probably on the shoulders.[907] Hence the disappearance of the spots and stripes in our adult existing deer, pigs, and tapirs, may be due to a change in the general colour of their coats; but whether this change was effected through sexual or natural selection, or was due to the direct action of the conditions of life, or some other unknown cause, it is impossible to decide. An observation made by Mr. Sclater well illustrates our ignorance of the laws which regulate the appearance and disappearance of stripes; the species of Asinus which inhabit the Asiatic continent are destitute of stripes, not having even the cross shoulder-stripe, whilst those which inhabit Africa are conspicuously striped, with the partial exception of *A. tæniopus*, which has only the cross shoulder-stripe and generally some faint bars on the legs; and this species inhabits the almost intermediate region of Upper Egypt and Abyssinia.[908]

Quadrumana.—Before we conclude, it will be advisable to add a few remarks to those already given on the ornamental characters of monkeys. In most of the species the sexes resemble each other in colour, but in some, as we have seen, the males differ from the females, especially in the colour of the naked parts of the skin, in the development of the beard,

Fig. 70. *Head of Semnopithecus rubicundus.* *This and the following figures (from Prof. Gervais) are given to shew the odd arrangement and development of the hair on the head.*

whiskers, and mane. Many species are coloured either in so extraordinary or beautiful a manner, and are furnished with such curious and elegant crests of hair, that we can hardly avoid looking at these characters as having been gained for the sake of ornament. The accompanying figures (figs. 70 to 74) serve to shew the arrangement of the hair on the face and head in several species. It is scarcely conceivable that these crests of hair and the strongly-contrasted colours of the fur and skin can be the result of mere variability without the aid of selection; and it is inconceivable that they can be of any ordinary use to these animals. If so, they have probably been gained through sexual selection, though transmitted equally, or almost equally, to both sexes. With many of the Quadrumana, we have additional evidence of

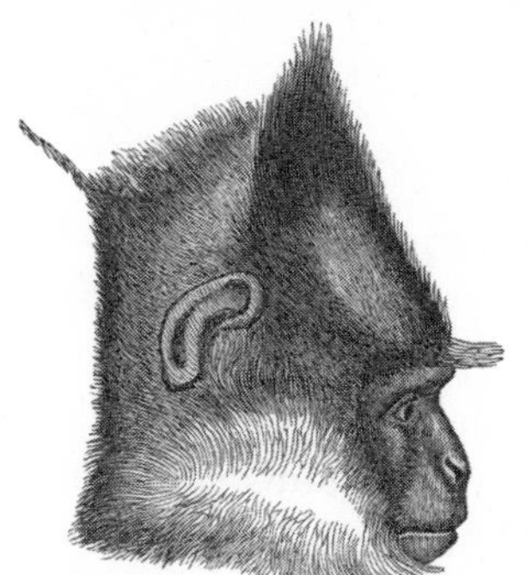

Fig. 71. *Head of Semnopithecus comatus.*

Fig. 72. *Head of Cebus capucinus.*

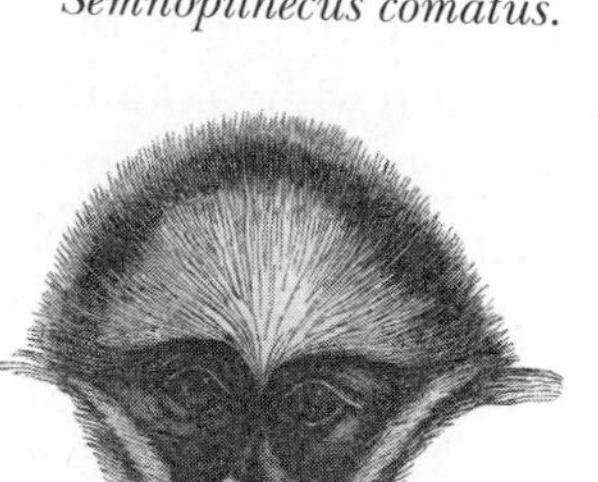

Fig. 73. *Head of Ateles marginatus.*

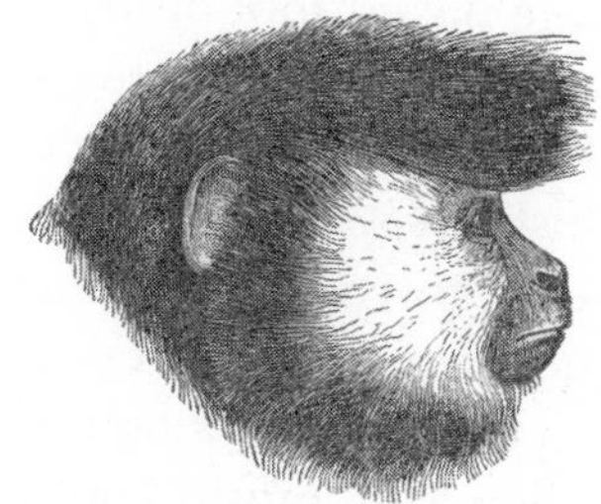

Fig. 74. *Head of Cebus vellerosus.*

Fig. 75. *Cercopithecus petaurista (from Brehm).*

the action of sexual selection in the greater size and strength of the males, and in the greater development of their canine teeth, in comparison with the females.

With respect to the strange manner in which both sexes of some species are coloured, and of the beauty of others, a few instances will suffice. The face of the *Cercopithecus petaurista* (fig. 75) is black, the whiskers and beard being white, with a defined, round, white spot on the nose, covered with short white hair, which gives to the animal an almost ludicrous aspect. The *Semnopithecus frontatus* likewise, has a blackish face with a long black beard, and a large naked spot on the forehead of a bluish-white colour. The face of *Macacus lasiotus* is dirty flesh-coloured, with a defined red spot on each cheek. The appearance of *Cercocebus æthiops* is grotesque, with its black face, white whiskers and collar, chesnut head, and a large naked white spot over each eyelid. In very many species, the beard, whiskers, and crests of hair round the face are of a different colour from the rest of the head, and when different, are always of a lighter tint,[909] being often pure white, sometimes bright yellow, or reddish. The whole face of the South American *Brachyurus calvus* is of a "glowing scarlet hue;" but this colour does not appear until the animal is nearly mature.[910] The naked skin of the face differs wonderfully in colour in the various species. It is often brown or flesh-colour, with parts perfectly white, and often as black as that of the most sooty negro. In the Brachyurus the scarlet tint is brighter than that of the most blushing Caucasian damsel. It is sometimes more distinctly orange than in any Mongolian, and in several species it is blue, passing into violet or grey. In all the species known to Mr. Bartlett, in which the adults of both sexes have strongly-coloured faces, the colours are dull or absent during early youth. This likewise holds good with the Mandrill and Rhesus, in which the face and the posterior parts of the body are brilliantly coloured in one sex alone. In these latter cases we have every reason to believe that the colours were acquired through sexual selection; and we are naturally led to extend the same view to the foregoing species, though both sexes when adult have their faces coloured in the same manner.

Although, according to our taste, many kinds of monkeys are far from beautiful, other species are universally admired for their elegant appearance and bright colours. The *Semnopithecus nemæus*, though peculiarly coloured, is described as extremely pretty; the orange-tinted face is surrounded by long whiskers of glossy whiteness, with a line of chesnut-red over the eyebrows; the fur on the back is of a delicate grey, with a square patch on the loins, the tail and the fore-arms all of a pure white; a gorget of chesnut surmounts the chest; the hind thighs are black, with the legs chesnut-red. I will mention only two other monkeys on account of their beauty; and I have selected these as they present slight sexual differences in colour, which renders it in some degree probable that both sexes owe their elegant appearance to sexual selection. In the moustache-monkey *(Cercopithecus cephus)* the general colour of the fur is mottled-greenish, with the throat white; in the male the end of the tail is chesnut; but the face is the most ornamented part, the skin being chiefly bluish-grey, shading into a blackish tint beneath the eyes, with the upper lip of a delicate blue, clothed on the lower edge with a thin black moustache; the whiskers are orange-coloured, with the upper part black, forming a band which extends backwards to the ears, the latter being clothed with whitish hairs. In the Zoological Society's Gardens I have often overheard visitors admiring the beauty of another monkey, deservedly called *Cercopithecus Diana* (fig. 76); the general colour of the fur is grey; the chest and inner surface of the fore-legs are white; a large triangular defined space on the hinder part of the back is rich chesnut; in the male the inner sides of the thighs and the abdomen are delicate fawn-coloured, and the top of the head is black; the face and ears are intensely black, finely contrasted with a white transverse crest over the eye-brows and with a long white peaked beard, of which the basal portion is black.[911]

In these and many other monkeys, the beauty and singular arrangement of their colours, and still more the diversified and elegant arrange-

Fig. 76. *Cercopithecus Diana (from Brehm).*

ment of the crests and tufts of hair on their heads, force the conviction on my mind that these characters have been acquired through sexual selection exclusively as ornaments.

Summary.—The law of battle for the possession of the female appears to prevail throughout the whole great class of mammals. Most naturalists will admit that the greater size, strength, courage, and pugnacity of the male, his special weapons of offence, as well as his special means of offence, have all been acquired or modified through that form of selection which I have called sexual selection. This does not depend on any superiority in the general struggle for life, but on certain individuals of one sex, generally the male sex, having been successful in conquering other males, and on their having left a larger number of offspring to inherit their superiority, than the less successful males.

There is another and more peaceful kind of contest, in which the males endeavour to excite or allure the females by various charms. This may be effected by the powerful odours emitted by the males during the breeding-season; the odoriferous glands having been acquired through sexual selection. Whether the same view can be extended to the voice is doubtful, for the vocal organs of the males may have been strengthened by use during maturity, under the powerful excitements of love, jealousy, or rage, and transmitted to the same sex. Various crests, tufts, and mantles of hair, which are either confined to the male, or are more developed in this sex than in the females, seem in most cases to be merely ornamental, though they sometimes serve as a defence against rival males. There is even reason to suspect that the branching horns of stags, and the elegant horns of certain antelopes, though properly serving as weapons of offence or of defence, have been partly modified for the sake of ornament.

When the male differs in colour from the female he generally exhibits darker and more strongly-contrasted tints. We do not in this class meet with the splendid red, blue, yellow, and green colours, so common with male birds and many other animals. The naked parts, however, of certain Quadrumana must be excepted; for such parts, often oddly situated, are coloured in some species in the most brilliant manner. The colours of the male in other cases may be due to simple variation, without the aid of selection. But when the colours are diversified and strongly pronounced, when they are not developed until near maturity, and when they are lost after emasculation, we can hardly avoid the conclusion that they have been acquired through sexual selection for the sake of ornament, and have been transmitted exclusively, or almost exclusively, to the same sex. When both sexes are coloured in the same manner, and the colours are conspicuous or curiously arranged, without being of the least apparent use as a protection, and especially when they are associated with various other ornamental appendages, we are led by analogy to the same conclusion, namely, that they have been acquired through sexual selection, although transmitted to both sexes. That conspicuous and diversified colours, whether confined to the males or common to both sexes, are as a general rule associated in the same groups and sub-groups with other secondary sexual characters, serving for war or for ornament, will be found to hold good if we look back to the various cases given in this and the last chapter.

The law of the equal transmission of characters to both sexes, as far as colour and other ornaments are concerned, has prevailed far more extensively with mammals than with birds; but in regard to weapons, such as horns and tusks, these have often been transmitted either exclusively, or in a much higher degree to the males than to the females. This is a surprising circumstance, for as the males generally use their weapons as a defence against enemies of all kinds, these weapons would have been of service to the female. Their absence in this sex can be accounted for, as far as we can see, only by the form of inheritance which has prevailed. Finally with quadrupeds the contest between the individuals of the same sex, whether peaceful or bloody, has with the rarest exceptions been confined to the males; so that these have been modified through sexual selection, either for fighting with each other or for alluring the opposite sex, far more commonly than the females.

CHAPTER XIX

Secondary Sexual Characters of Man

DIFFERENCES BETWEEN MAN AND WOMAN • CAUSES OF SUCH DIFFERENCES AND OF CERTAIN CHARACTERS COMMON TO BOTH SEXES • LAW OF BATTLE • DIFFERENCES IN MENTAL POWERS • AND VOICE • ON THE INFLUENCE OF BEAUTY IN DETERMINING THE MARRIAGES OF MANKIND • ATTENTION PAID BY SAVAGES TO ORNAMENTS • THEIR IDEAS OF BEAUTY IN WOMAN • THE TENDENCY TO EXAGGERATE EACH NATURAL PECULIARITY

WITH MANKIND THE DIFFERENCES between the sexes are greater than in most species of Quadrumana, but not so great as in some, for instance, the mandrill. Man on an average is considerably taller, heavier, and stronger than woman, with squarer shoulders and more plainly-pronounced muscles. Owing to the relation which exists between muscular development and the projection of the brows,[912] the superciliary ridge is generally more strongly marked in man than in woman. His body, and especially his face, is more hairy, and his voice has a different and more powerful tone. In certain tribes the women are said, whether truly I know not, to differ slightly in tint from the men; and with Europeans, the women are perhaps the more brightly coloured of the two, as may be seen when both sexes have been equally exposed to the weather.

Man is more courageous, pugnacious, and energetic than woman, and has a more inventive genius. His brain is absolutely larger, but whether relatively to the larger size of his body, in comparison with that of woman, has not, I believe been fully ascertained. In woman the face is rounder; the jaws and the base of the skull smaller; the outlines of her body rounder, in parts more prominent; and her pelvis is broader than in man;[913] but this latter character may perhaps be considered rather as a primary than a secondary sexual character. She comes to maturity at an earlier age than man.

As with animals of all classes, so with man, the distinctive characters of the male sex are not fully developed until he is nearly mature; and if emasculated they never appear. The beard, for instance, is a secondary sexual character, and male children are beardless, though at an early age they have abundant hair on their heads. It is probably due to the rather late appearance in life of the successive variations, by which man acquired his masculine characters, that they are transmitted to the male sex alone. Male and female children resemble each other closely, like the young of so many other animals in which the adult sexes differ; they likewise resemble the mature female much more closely, than the mature male. The female, however, ultimately assumes certain distinctive characters, and in the formation of her skull, is said to be intermediate between the child and the man.[914] Again, as the young of closely allied though distinct species do not differ nearly so much from each other as do the adults, so it is with the children of the different races of man. Some have even maintained that race-differences cannot be detected in the infantile skull.[915] In regard to colour, the new-born negro child is reddish nut-brown, which soon becomes slaty-grey; the black colour being fully developed within a year in the Sudan, but not until three years in Egypt. The eyes of the negro are at first blue, and the hair chesnut-brown rather than black, being curled only at the ends. The children of the Australians immediately after birth are yellowish-brown, and become dark at a later age. Those of the Guaranys of Paraguay are whitish-yellow, but they acquire in the course of a few weeks the yellowish-brown tint of their parents. Similar observations have been made in other parts of America.[916]

I have specified the foregoing familiar differences between the male and female sex in mankind, because they are curiously the same as in the Quadrumana. With these animals the female is mature at an earlier age than the male; at least this is certainly the case with the *Cebus azaræ*.[917] With most of the species the males are larger and much stronger than the females, of which fact the gorilla offers a well-known instance. Even in so trifling a character as the greater prominence of the superciliary ridge, the males of certain monkeys differ from the females,[918] and agree in this respect with mankind. In the gorilla and certain other monkeys, the cranium of the adult male presents a strongly-marked sagittal crest, which is absent in the female; and Ecker found a trace of a similar difference between the two sexes in the Australians.[919] With monkeys when there is any difference in the voice, that of the male is the more powerful. We have seen that certain male monkeys, have a well-developed beard, which is quite deficient, or much less developed in the female. No instance is known of the beard, whiskers, or moustache being larger in a female than in the male monkey. Even in the colour of the beard there is a curious parallelism between man and the Quadrumana, for when in man the beard differs in colour from the hair of the head, as is often the case, it is, I believe, invariably of a lighter tint, being often reddish. I have observed this fact in England, and Dr. Hooker, who attended to this little point for me in Russia, found no exception to the rule. In Calcutta, Mr. J. Scott, of the Botanic Gardens, was so kind as to observe with care the many races of men to be seen there, as well as in some other parts of India, namely, two races in Sikhim, the Bhoteas, Hindoos, Burmese, and Chinese. Although most of these races have very little hair on the face, yet he always found that when there was any difference in colour between the hair of the head and the beard, the latter was invariably of a lighter tint. Now with monkeys, as has already been stated, the beard frequently differs in a striking manner in colour from the hair of the head, and in such cases it is invariably of a lighter hue, being often pure white, sometimes yellow or reddish.[920]

In regard to the general hairyness of the body, the women in all races are less hairy than the men, and in some few Quadrumana the under side of the body of the female is less hairy than that of the male.[921] Lastly, male monkeys, like men, are bolder and fiercer than the females. They lead the troop, and when there is danger, come to the front. We thus see how close is the parallelism between the sexual differences of man and the Quadrumana. With some few species, however, as with certain baboons, the gorilla and orang, there is a considerably greater difference between the sexes, in the size of the canine teeth, in the development and colour of the hair, and especially in the colour of the naked parts of the skin, than in the case of mankind.

The secondary sexual characters of man are all highly variable, even within the limits of the same race or sub-species; and they differ much in the several races. These two rules generally hold good throughout the animal kingdom. In the excellent observations made on board the *Novara*,[922] the male Australians were found to exceed the females by only 65 millim. in height, whilst with the Javanese the average excess was 218 millim., so that in this latter race the difference in height between the sexes is more than thrice as great as with the Australians. The numerous measurements of various other races, with respect to stature, the circumference of the neck and chest, and the length of the back-bone and arms, which were carefully made, nearly all shewed that the males differed much more from each other than did the females. This fact indicates that, as far as these characters are concerned, it is the male which has been chiefly modified, since the races diverged from their common and primeval source.

The development of the beard and the hairiness of the body differ remarkably in the men belonging to distinct races, and even to different families in the same race. We Europeans see this amongst ourselves. In the island of St. Kilda, according to Martin,[923] the men do not acquire beards, which are very thin, until the age of thirty or upwards. On the Europæo-Asiatic continent, beards prevail until we pass beyond India, though with the natives of Ceylon they are frequently absent, as was noticed in ancient times by Diodorus.[924] Beyond India beards disappear, as with the Siamese, Malays, Kalmucks, Chinese, and Japanese; nevertheless the Ainos,[925] who inhabit the northernmost islands of the Japan archipelago, are the most hairy men in the world. With negroes the beard is scanty or absent, and they have no whiskers; in both sexes the body is almost destitute of fine down.[926] On the other hand, the Papuans of the Malay archipelago, who are nearly as black as negroes, possess well-developed beards.[927] In the Pacific Ocean the inhabitants of the Fiji archipelago have large bushy beards, whilst those of the not-distant archipelagoes of Tonga and Samoa are beardless; but these men belong to distinct races. In the Ellice group all the inhabitants belong to the same race; yet on one island alone, namely Nunemaya, "the men have splendid beards;" whilst on the other islands "they have, as a rule, a dozen straggling hairs for a beard."[928]

Throughout the great American continent the men may be said to be beardless; but in almost all the tribes a few short hairs are apt to appear on the face, especially during old age. With the tribes of North America, Catlin estimates that eighteen out of twenty men are completely destitute by nature of a beard; but occasionally there may be seen a man, who has neglected to pluck out the hairs at puberty, with a soft beard an inch or two in length. The Guaranys of Paraguay differ from all the surrounding tribes in having a small beard, and even some hair on the body, but no whiskers.[929] I am informed by Mr. D. Forbes, who particularly attended to this subject, that the Aymaras and Quichuas of the Cordillera are remarkably hairless, yet in old age a few straggling hairs occasionally appear on the chin. The men of these two tribes have very little hair on the various parts of the body where hair grows abundantly in Europeans, and the women have none on the corresponding parts. The hair on the head, however, attains an extraordinary length in both sexes, often reaching almost to the ground; and this is likewise the case with some of the N. American tribes. In the amount of hair, and in the general shape of the body, the sexes of the American aborigines do not differ from each other so much as with most other races of mankind.[930] This fact is analagous with what occurs with some allied monkeys; thus the sexes of the chimpanzee are not as different as those of the gorilla or orang.[931]

In the previous chapters we have seen that with mammals, birds, fishes, insects, &c., many characters, which there is every reason to believe were primarily gained through sexual selection by one sex alone, have been transferred to both sexes. As this same form of transmission has apparently prevailed to a large extent with mankind, it will save much useless repetition if we consider the characters peculiar to the male sex together with certain other characters common to both sexes.

Law of Battle.—With barbarous nations, for instance with the Australians, the women are the constant cause of war both between the individuals of the same tribe and between distinct tribes. So no doubt it was in ancient times; "nam fuit ante Helenam mulier teterrima belli causa." With the North American Indians, the contest is reduced to a system. That excellent observer, Hearne,[932] says:—"It has ever been the custom among these people for the men to wrestle for any woman to whom they are attached; and, of course, the strongest party always carries off the prize. A weak man, unless he be a good hunter, and well-beloved, is seldom permitted to keep a wife that a stronger man thinks worth his notice. This custom prevails throughout all the tribes, and causes a great spirit of emulation among their youth, who are upon all occasions, from their childhood, trying their strength and skill in wrestling." With the Guanas of South America, Azara states that the men rarely marry till twenty or more years old, as before that age they cannot conquer their rivals.

Other similar facts could be given; but even if we had no evidence on this head, we might feel almost sure, from the analogy of the higher Quadrumana,[933] that the law of battle had prevailed with man during the early stages of his development. The occasional appearance at the present day of canine teeth which project above the others, with traces of a diastema or open space for the reception of the opposite canines, is in all probability a case of reversion to a former state, when the progenitors of man were provided with these weapons, like so many existing male Quadrumana. It was remarked in a former chapter that as man gradually became erect, and continually used his hands and arms for fighting with sticks and stones, as well as for the other purposes of life, he would have used his jaws and teeth less and less. The jaws, together with their muscles, would then have become reduced through disuse, as would the teeth through the not well understood principles of correlation and the economy of growth; for we everywhere see that parts which are no longer of service are reduced in size. By such steps the original inequality between the jaws and teeth in the two sexes of mankind would ultimately have been quite obliterated. The case is almost parallel with that of many male Ruminants, in which the canine teeth have been reduced to mere rudiments, or have disappeared, apparently in consequence of the development of horns. As the prodigious difference between the skulls of the two sexes in the Gorilla and Orang, stands in close relation with the development of the immense canine teeth in the males, we may infer that the reduction of the jaws and teeth in the early male progenitors of man led to a most striking and favourable change in his appearance.

There can be little doubt that the greater size and strength of man, in comparison with woman, together with his broader shoulders, more developed muscles, rugged outline of body, his greater courage and pugnacity, are all due in chief part to inheritance from some early male progenitor, who, like the existing anthropoid apes, was thus characterised. These characters will, however, have been preserved or even augmented during the long ages whilst man was still in a barbarous condition, by the strongest and boldest men having succeeded best in the general struggle for life, as well as in securing wives, and thus having left a large number of offspring. It is not probable that the greater strength of man was primarily acquired through the inherited effects of his having worked harder than woman for his own subsistence and that of his family; for the women in all barbarous nations are compelled to work at least as hard as the men. With civilised people the arbitrament of battle for the possession of the women has long ceased; on the other hand, the men, as a general rule, have to work harder than the women for their mutual subsistence; and thus their greater strength will have been kept up.

Difference in the Mental Powers of the two Sexes.—With respect to differences of this nature between man and woman, it is probable that sexual selection has played a very important part. I am aware that some writers doubt whether there is any inherent difference; but this is at least probable from the analogy of the lower animals which present other secondary sexual characters. No one will dispute that the bull differs in disposition from the cow, the wild-boar from the sow, the stallion from the mare, and, as is well known to the keepers of menageries, the males of the larger apes from the females. Woman seems to differ from man in mental disposition, chiefly in her greater tenderness and less selfishness; and this holds good even with savages, as shewn by a well-known passage in Mungo Park's Travels, and by statements made by many other travellers. Woman, owing to her maternal instincts, displays these qualities towards her infants in an eminent degree; therefore it is likely that she should often extend them towards her fellow-creatures. Man is the rival of other men; he delights in competition, and this leads to ambition which passes too easily into selfishness. These latter qualities seem to be his natural and unfortunate birthright. It is generally admitted that with woman the powers of intuition, of rapid perception, and perhaps of imitation, are more strongly marked than in man; but some, at least, of these faculties are characteristic of the lower races, and therefore of a past and lower state of civilisation.

The chief distinction in the intellectual powers of the two sexes is shewn by man attaining to a higher eminence, in whatever he takes up, than woman can attain—whether requiring deep thought, reason, or imagination, or merely the use of the senses and hands. If two lists were made of the most eminent men and women in poetry, painting, sculpture, music,—comprising composition and performance, history, science, and philosophy, with half-a-dozen names under each subject, the two lists would not bear comparison. We may also infer, from the law of the deviation of averages, so well illustrated by Mr. Galton, in his work on 'Hereditary Genius,' that if men are capable of decided eminence over women in many subjects, the average standard of mental power in man must be above that of woman.

The half-human male progenitors of man, and men in a savage state, have struggled together during many generations for the possession of the females. But mere bodily strength and size would do little for victory, unless associated with courage, perseverance, and determined energy. With social animals, the young males have to pass through many a contest before they win a female, and the older males have to retain their females by renewed battles. They have, also, in the case of man, to defend their females, as well as their young, from enemies of all kinds, and to hunt for their joint subsistence. But to avoid enemies, or to attack them with success, to capture wild animals, and to invent and fashion weapons, requires the aid of the higher mental faculties, namely, observation, reason, invention, or imagination. These various faculties will thus have been continually put to the test, and selected during manhood; they will, moreover, have been strengthened by use during this same period of life. Consequently, in accordance with the principle often alluded to, we might expect that they would at least tend to be transmitted chiefly to the male offspring at the corresponding period of manhood.

Now, when two men are put into competition, or a man with a woman, who possess every mental quality in the same perfection, with the exception that the one has higher energy, perseverance, and courage, this one will generally become more eminent, whatever the object may be, and will gain the victory.[934] He may be said to possess genius—for genius has been declared by a great authority to be patience; and patience, in this sense, means unflinching, undaunted perseverance. But this view of genius is perhaps deficient; for without the higher powers of the imagination and reason, no eminent success in many subjects can be gained. But these latter as well as the former faculties will have been developed in man, partly through sexual selection,—that is, through the contest of rival males, and partly through natural selection,—that is, from success in the general struggle for life; and as in both cases the struggle will have been during maturity, the characters thus gained will have been transmitted more fully to the male than to the female offspring. Thus man has ultimately become superior to woman. It is, indeed, fortunate that the law of the equal transmission of characters to both sexes has commonly prevailed throughout the whole class of mammals; otherwise it is probable that man would have become as superior in mental endowment to woman, as the peacock is in ornamental plumage to the peahen.

It must be borne in mind that the tendency in characters acquired at a late period of life by either sex, to be transmitted to the same sex at the same age, and of characters acquired at an early age to be transmitted to both sexes, are rules which, though general, do not always hold good. If they always held good, we might conclude (but I am here wandering beyond my proper bounds) that the inherited effects of the early education of boys and girls would be transmitted equally to both sexes; so that the present inequality between the sexes in mental power could not be effaced by a similar course of early training; nor can it have been caused by their dissimilar early training. In order that woman should reach the same standard as man, she ought, when nearly adult, to be trained to energy and perseverance, and to have her reason and imagination exercised to the highest point; and then she would probably transmit these qualities chiefly to her adult daughters. The whole body of women, however, could not be thus raised, unless during many generations the women who excelled in the above robust virtues were married, and produced offspring in larger numbers than other women. As before remarked with respect to bodily strength, although men do not now fight for the sake of obtaining wives, and this form of selection has passed away, yet they generally have to undergo, during manhood, a severe struggle in order to maintain themselves and their families; and this will tend to keep up or even increase their mental powers, and, as a consequence, the present inequality between the sexes.[935]

Voice and Musical Powers.—In some species of Quadrumana there is a great difference between the adult sexes, in the power of the voice and in the development of the vocal organs; and man appears to have inherited this difference from his early progenitors. His vocal cords are about one-third longer than in woman, or than in boys; and emasculation produces the same effect on him as on the lower animals, for it "arrests that prominent growth of the thyroid, &c., which accompanies the elongation of the cords."[936] With respect to the cause of this difference between the sexes, I have nothing to add to the remarks made in the last chapter on the probable effects of the long-continued use of the vocal organs by the male under the excitement of love, rage, and jealousy. According to Sir Duncan Gibb,[937] the voice differs in the different races of mankind; and with the natives of Tartary, China, &c., the voice of the male is said not to differ so much from that of the female, as in most other races.

The capacity and love for singing or music, though not a sexual character in man, must not here be passed over. Although the sounds emitted by animals of all kinds serve many purposes, a strong case can be made out, that the vocal organs were primarily used and perfected in relation to the propagation of the species. Insects and some few spiders are the lowest animals which voluntarily produce any sound; and this is generally effected by the aid of beautifully constructed stridulating organs, which are often confined to the males alone. The sounds thus produced consist, I believe in all cases, of the same note, repeated rhythmically;[938] and this is sometimes pleasing even to the ears of man. Their chief, and in some cases exclusive use appears to be either to call or to charm the opposite sex.

The sounds produced by fishes are said in some cases to be made only by the males during the breeding-season. All the air-breathing Vertebrata necessarily possess an apparatus for inhaling and expelling air, with a pipe capable of being closed at one end. Hence when the primeval members of this class were strongly excited and their muscles violently contracted, purposeless sounds would almost certainly have been produced; and these, if they proved in any way serviceable, might readily have been modified or intensified by the preservation of properly adapted variations. The Amphibians are the lowest Vertebrates which breathe air; and many of these animals, namely, frogs and toads, possess vocal organs, which are incessantly used during the breeding-season, and which are often more highly developed in the male than in the female. The male alone of the tortoise utters a noise, and this only during the season of love. Male alligators roar or bellow during the same season. Every one knows how largely birds use their vocal organs as a means of courtship; and some species likewise perform what may be called instrumental music.

In the class of Mammals, with which we are here more particularly concerned, the males of almost all the species use their voices during the breeding-season much more than at any other time; and some are absolutely mute excepting at this season. Both sexes of other species, or the females alone, use their voices as a love-call. Considering these facts, and that the vocal organs of some quadrupeds are much more largely developed in the male than in the female, either permanently or temporarily during the breeding-season; and considering that in most of the lower classes the sounds produced by the males, serve not only to call but to excite or allure the female, it is a surprising fact that we have not as yet any good evidence that these organs are used by male mammals to charm the females. The American *Mycetes caraya* perhaps forms an exception, as does more probably one of those apes which come nearer to man, namely, the *Hylobates agilis*. This gibbon has an extremely loud but musical voice. Mr. Waterhouse states,[939] "It appeared to me that in ascending and descending the scale, the intervals were always exactly half-tones; and I am sure that the highest note was the exact octave to the lowest. The quality of the notes is very musical; and I do not doubt that a good violinist would be able to give a correct idea of the gibbon's composition, excepting as regards its loudness." Mr. Waterhouse then gives the notes. Professor Owen, who is likewise a musician, confirms the foregoing statement, and remarks that this gibbon "alone of brute mammals may be said to sing." It appears to be much excited after its performance. Unfortunately its habits have never been closely observed in a state of nature; but from the analogy of almost all other animals, it is highly probable that it utters its musical notes especially during the season of courtship.

The perception, if not the enjoyment, of musical cadences and of rhythm is probably common to all animals, and no doubt depends on the common physiological nature of their nervous systems. Even Crustaceans, which are not capable of producing any voluntary sound, possess certain auditory hairs, which have been seen to vibrate when the proper musical notes are struck.[940] It is well known that some dogs howl when hearing particular tones. Seals apparently appreciate music, and their fondness for it "was well known to the ancients, and is often taken advantage of by the hunters at the present day."[941] With all those animals, namely insects, amphibians, and birds, the males of which during the season of courtship incessantly produce musical notes or mere rhythmical sounds, we must believe that the females are able to appreciate them, and are thus excited or charmed; otherwise the incessant efforts of the males and the complex structures often possessed exclusively by them would be useless.

With man song is generally admitted to be the basis or origin of instrumental music. As neither the enjoyment nor the capacity of producing musical notes are faculties of the least direct use to man in reference to his ordinary habits of life, they must be ranked amongst the most mysterious with which he is endowed. They are present, though in a very rude and as it appears almost latent condition, in men of all races, even the most savage; but so different is the taste of the different races, that our music gives not the least pleasure to savages, and their music is to us hideous and unmeaning. Dr. Seemann, in some interesting remarks on this subject,[942] "doubts whether even amongst the nations of Western Europe, intimately connected as they are by close and frequent intercourse, the music of the one is interpreted in the same sense by the others. By travelling eastwards we find that there is certainly a different language of music. Songs of joy and dance-accompaniments are no longer, as with us, in the major keys, but always in the minor." Whether or not the half-human progenitors of man possessed, like the before-mentioned gibbon, the capacity of producing, and no doubt of appreciating, musical notes, we have every reason to believe that man possessed these faculties at a very remote period, for singing and music are extremely ancient arts. Poetry, which may be considered as the offspring of song, is likewise so ancient that many persons have felt astonishment that it should have arisen during the earliest ages of which we have any record.

The musical faculties, which are not wholly deficient in any race, are capable of prompt and high development, as we see with Hottentots and Negroes, who have readily become excellent musicians, although they do not practise in their native countries anything that we should esteem as music. But there is nothing anomalous in this circumstance: some species of birds which never naturally sing, can without much difficulty be taught to perform; thus the house-sparrow has learnt the song of a linnet. As these two species are closely allied, and belong to the order of Insessores, which includes nearly all the singing-birds in the world, it is quite possible or probable that a progenitor of the sparrow may have been a songster. It is a much more remarkable fact that parrots, which belong to a group distinct from the Insessores, and have differently-constructed vocal organs, can be taught not only to speak, but to pipe or whistle tunes invented by man, so that they must have some musical capacity. Nevertheless it would be extremely rash to assume that parrots are descended from some ancient progenitor which was a songster. Many analogous cases could be advanced of organs and instincts originally adapted for one purpose, having been utilised for some quite distinct purpose.[943] Hence the capacity for high musical development, which the savage races of man possess, may be due either to our semi-human progenitors having practised some rude form of music, or simply to their having acquired for some distinct purposes the proper vocal organs. But in this latter case we must assume that they already possessed, as in the above instance of the parrots, and as seems to occur with many animals, some sense of melody.

Music affects every emotion, but does not by itself excite in us the more terrible emotions of horror, rage, &c. It awakens the gentler feelings of tenderness and love, which readily pass into devotion. It likewise stirs up in us the sensation of triumph and the glorious ardour for war. These powerful and mingled feelings may well give rise to the sense of sublimity. We can concentrate, as Dr. Seemann observes, greater intensity of feeling in a single musical note than in pages of writing. Nearly the same emotions, but much weaker and less complex, are probably felt by birds when the male pours forth his full volume of song, in rivalry with other males, for the sake of captivating the female. Love is still the commonest theme of our own songs. As Herbert Spencer remarks, music "arouses dormant sentiments of which we had not conceived the possibility, and do not know the meaning; or, as Richter says, tells us of things we have not seen and shall not see."[944] Conversely, when vivid emotions are felt and expressed by the orator or even in common speech, musical cadences and rhythm are instinctively used. Monkeys also express strong feelings in different tones—anger and impatience by low,—fear and pain by high notes.[945] The sensations and ideas excited in us by music, or by the cadences of impassioned oratory, appear from their vagueness, yet depth, like mental reversions to the emotions and thoughts of a long-past age.

All these facts with respect to music become to a certain extent intelligible if we may assume that musical tones and rhythm were used by the half-human progenitors of man, during the season of courtship, when animals of all kinds are excited by the strongest passions. In this case, from the deeply-laid principle of inherited associations, musical tones would be likely to excite in us, in a vague and indefinite manner, the strong emotions of a long-past age. Bearing in mind that the males of some quadrumanous animals have their vocal organs much more developed than in the females, and that one anthropomorphous species pours forth a whole octave of musical notes and may be said to sing, the suspicion does not appear improbable that the progenitors of man, either the males or females, or both sexes, before they had acquired the power of expressing their mutual love in articulate language, endeavoured to charm each other with musical notes and rhythm. So little is known about the use of the voice by the Quadrumana during the season of love, that we have hardly any means of judging whether the habit of singing was first acquired by the male or female progenitors of mankind. Women are generally thought to possess sweeter voices than men, and as far as this serves as any guide we may infer that they first acquired musical powers in order to attract the other sex.[946] But if so, this must have occurred long ago, before the progenitors of man had become sufficiently human to treat and value their women merely as useful slaves. The impassioned orator, bard, or musician, when with his varied tones and cadences he excites the strongest emotions in his hearers, little suspects that he uses the same means by which, at an extremely remote period, his half-human ancestors aroused each other's ardent passions, during their mutual courtship and rivalry.

On the influence of beauty in determining the marriages of mankind.—In civilised life man is largely, but by no means exclusively, influenced in the choice of his wife by external appearance; but we are chiefly concerned with primeval times, and our only means of forming a judgment on this subject is to study the habits of existing semi-civilised and savage nations. If it can be shewn that the men of different races prefer women having certain characteristics, or conversely that the women prefer certain men, we have then to enquire whether such choice, continued during many generations, would produce any sensible effect on the race, either on one sex or both sexes; this latter circumstance depending on the form of inheritance which prevails.

It will be well first to shew in some detail that savages pay the greatest attention to their personal appearance.[947] That they have a passion for ornament is notorious; and an English philosopher goes so far as to maintain that clothes were first made for ornament and not for warmth. As Professor Waitz remarks, "however poor and miserable man is, he finds a pleasure in adorning himself." The extravagance of the naked Indians of South America in decorating themselves is shewn "by a man of large stature gaining with difficulty enough by the labour of a fortnight to procure in exchange the *chica n*ecessary to paint himself red."[948] The ancient barbarians of Europe during the Reindeer period brought to their caves any brilliant or singular objects which they happened to find. Savages at the present day everywhere deck themselves with plumes, necklaces, armlets, earrings, &c. They paint themselves in the most diversified manner. "If painted nations," as Humboldt observes, "had been examined with the same attention as clothed nations, it would have been perceived that the most fertile imagination and the most mutable caprice have created the fashions of painting, as well as those of garments."

In one part of Africa the eyelids are coloured black; in another the nails are coloured yellow or purple. In many places the hair is dyed of various tints. In different countries the teeth are stained black, red, blue, &c., and in the Malay Archipelago it is thought shameful to have white teeth like those of a dog. Not one great country can be named, from the Polar regions in the north to New Zealand in the south, in which the aborigines do not tattoo themselves. This practice was followed by the Jews of old and by the ancient Britons. In Africa some of the natives tattoo themselves, but it is much more common to raise protuberances by rubbing salt into incisions made in various parts of the body; and these are considered by the inhabitants of Kordofan and Darfur "to be great personal attractions." In the Arab countries no beauty can be perfect until the cheeks "or temples have been gashed."[949] In South America, as Humboldt remarks, "a mother would be accused of culpable indifference towards her children, if she did not employ artificial means to shape the calf of the leg after the fashion of the country." In the Old and New World the shape of the skull was formerly modified during infancy in the most extraordinary manner, as is still the case in many places, and such deformities are considered ornamental. For instance, the savages of Colombia[950] deem a much flattened head "an essential point of beauty."

The hair is treated with especial care in various countries; it is allowed to grow to full length, so as to reach to the ground, or is combed into "a compact frizzled mop, which is the Papuan's pride and glory."[951] In Northern Africa "a man requires a period of from eight to ten years to perfect his coiffure." With other nations the head is shaved, and in parts of South America and Africa even the eyebrows are eradicated. The natives of the Upper Nile knock out the four front teeth, saying that they do not wish to resemble brutes. Further south, the Batokas knock out the two upper incisors, which, as Livingstone[952] remarks, gives the face a hideous appearance, owing to the growth of the lower jaw; but these people think the presence of the incisors most unsightly, and on beholding some Europeans, cried out, "Look at the great teeth!" The great chief Sebituani tried in vain to alter this fashion. In various parts of Africa and in the Malay Archipelago the natives file the incisor teeth into points like those of a saw, or pierce them with holes, into which they insert studs.

As the face with us is chiefly admired for its beauty, so with savages it is the chief seat of mutilation. In all quarters of the world the septum, and more rarely the wings of the nose are pierced, with rings, sticks, feathers, and other ornaments inserted into the holes. The ears are everywhere pierced and similarly ornamented, and with the Botucudos and Lenguas of South America the hole is gradually so much enlarged that the lower edge touches the shoulder. In North and South America and in Africa either the upper or lower lip is pierced; and with the Botocudos the hole in the lower lip is so large that a disc of wood four inches in diameter is placed in it. Mantegazza gives a curious account of the shame felt by a South American native, and of the ridicule which he excited, when he sold his *tembeta,*—the large coloured piece of wood which is passed through the hole. In central Africa the women perforate the lower lip and wear a crystal, which, from the movement of the tongue, has "a wriggling motion indescribably ludicrous during conversation." The wife of the chief of Latooka told Sir S. Baker[953] that his "wife would be much improved if she would extract her four front teeth from the lower jaw, and wear the long pointed polished crystal in her under lip." Further south with the Makalolo, the upper lip is perforated, and a large metal and bamboo ring, called a *pelelé,* is worn in the hole. "This caused the lip in one case to project two inches beyond the tip of the nose; and when the lady smiled the contraction of the muscles elevated it over the eyes. 'Why do the women wear these things?' the venerable chief, Chinsurdi, was asked. Evidently surprised at such a stupid question, he replied, 'For beauty! They are the only beautiful things women have; men have beards, women have none. What kind of a person would she be without the pelelé? She would not be a woman at all with a 'mouth like a man, but no beard.'"[954]

Hardly any part of the body, which can be unnaturally modified, has escaped. The amount of suffering thus caused must have been wonderfully great, for many of the operations require several years for their completion, so that the idea of their necessity must be imperative. The motives are various; the men paint their bodies to make themselves appear terrible in battle; certain mutilations are connected with religious rites; or they mark the age of puberty, or the rank of the man, or they serve to distinguish the tribes. As with savages the same fashions prevail for long periods,[955] mutilations, from whatever cause first made, soon come to be valued as distinctive marks. But self-adornment, vanity, and the admiration of others, seem to be the commonest motives. In regard to tattooing, I was told by the missionaries in New Zealand, that when they tried to persuade some girls to give up the practice, they answered, "We must just have a few lines on our lips; else when we grow old we shall be so very ugly." With the men of New Zealand, a most capable judge[956] says, "to have fine tattooed faces was the great ambition of the young, both to render themselves attractive to the ladies, and conspicuous in war." A star tattooed on the forehead and a spot on the chin are thought by the women in one part of Africa to be irresistible attractions.[957] In most, but not all parts of the world, the men are more highly ornamented than the women, and often in a different manner; sometimes, though rarely, the women are hardly at all ornamented. As the women are made by savages to perform the greatest share of the work, and as they are not allowed to eat the best kinds of food, so it accords with the characteristic selfishness of man that they should not be allowed to obtain, or to use, the finest ornaments. Lastly it is a remarkable fact, as proved by the foregoing quotations, that the same fashions in modifying the shape of the head, in ornamenting the hair, in painting, tattooing, perforating the nose, lips, or ears, in removing or filing the teeth, &c., now prevail and have long prevailed in the most distant quarters of the world. It is extremely improbable that these practices which are followed by so many distinct nations are due to tradition from any common source. They rather indicate the close similarity of the mind of man, to whatever race he may belong, in the same manner as the almost universal habits of dancing, masquerading, and making rude pictures.

HAVING MADE THESE preliminary remarks on the admiration felt by savages for various ornaments, and for deformities most unsightly in our eyes, let us see how far the men are attracted by the appearance of their women, and what are their ideas of beauty. As I have heard it maintained that savages are quite indifferent about the beauty of their women, valuing them solely as slaves, it may be well to observe that this conclusion does not at all agree with the care which the women take in ornamenting themselves, or with their vanity. Burchell[958] gives an amusing account of a Bush-woman, who used so much grease, red ochre, and shining powder, "as would have ruined any but a very rich husband." She displayed also "much vanity and too evident a consciousness of her superiority." Mr. Winwood Reade informs me that the negroes of the West Coast often discuss the beauty of their women. Some competent observers have attributed the fearfully common practice of infanticide partly to the desire felt by the women to retain their good looks.[959] In several regions the women wear charms and love-philters to gain the affections of the men; and Mr. Brown enumerates four plants used for this purpose by the women of North-Western America.[960]

Hearne,[961] who lived many years with the American Indians, and who was an excellent observer, says, in speaking of the women, "Ask a Northern Indian what is beauty, and he will answer, a broad flat face, small eyes, high cheek-bones, three or four broad black lines across each cheek, a low forehead, a large broad chin, a clumsy hook nose, a tawny hide, and breasts hanging down to the belt." Pallas, who visited the northern parts of the Chinese empire, says "those women are preferred who have the Mandschú form; that is to say, a broad face, high cheek-bones, very broad noses, and enormous ears;"[962] and Vogt remarks that the obliquity of the eye, which is proper to the Chinese and Japanese, is exaggerated in their pictures for the purpose, as it "it seems, of exhibiting its beauty, as contrasted with the eye of the red-haired barbarians." It is well known, as Huc repeatedly remarks, that the Chinese of the interior think Europeans hideous with their white skins and prominent noses. The nose is far from being too prominent, according to our ideas, in the natives of Ceylon; yet "the Chinese in the seventh century, accustomed to the flat features of the Mogul races, were surprised at the prominent noses of the Cingalese; and Thsang described them as having 'the beak of a bird, with the body of a man.'"

Finlayson, after minutely describing the people of Cochin China, says that their rounded heads and faces are their chief characteristics; and he adds, "the roundness of the whole countenance is more striking in the women, who are reckoned beautiful in proportion as they display this form of face." The Siamese have small noses with divergent nostrils, a wide mouth, rather thick lips, a remarkably large face, with very high and broad cheek-bones. It is, therefore, not wonderful that "beauty, according to our notion is a stranger to them. Yet they consider their own females to be much more beautiful than those of Europe."[963]

It is well known that with many Hottentot women the posterior part of the body projects in a wonderful manner; they are steatopygous; and Sir Andrew Smith is certain that this peculiarity is greatly admired by the men.[964] He once saw a woman who was considered a beauty, and she was so immensely developed behind, that when seated on level ground she could not rise, and had to push herself along until she came to a slope. Some of the women in various negro tribes are similarly characterised; and, according to Burton, the Somal men "are said to choose their wives by ranging them in a line, and by picking her out who projects farthest *a tergo*. Nothing can be more hateful to a negro than the opposite form."[965]

With respect to colour, the negroes rallied Mungo Park on the whiteness of his skin and the prominence of his nose, both of which they considered as "unsightly and unnatural conformations." He in return praised the glossy jet of their skins and the lovely depression of their noses; this they said was "honey-mouth," nevertheless they gave him food. The African Moors, also, "knitted their brows and seemed to shudder" at the whiteness of his skin. On the eastern coast, the negro boys when they saw Burton, cried out "Look at the white man; does he not look like a white ape?" On the western coast, as Mr. Winwood Reade informs me, the negroes admire a very black skin more than one of a lighter tint. But their horror of whiteness may be partly attributed, according to this same traveller, to the belief held by most negroes that demons and spirits are white.

The Banyai of the more southern part of the continent are negroes, but "a great many of them are of a light coffee-and-milk colour, and, indeed, this colour is considered handsome throughout the whole country;" so that here we have a different standard of taste. With the Kafirs, who differ much from negroes, "the skin, except among the tribes near Delagoa Bay, is not usually black, the prevailing colour being a mixture of black and red, the most common shade being chocolate. Dark complexions, as being most common are naturally held in the highest esteem. To be told that he is light-coloured, or like a white man, would be deemed a very poor compliment by a Kafir. I have heard of one unfortunate man who was so very fair that no girl would marry him." One of the titles of the Zulu king is "You who are black."[966] Mr. Galton, in speaking to me about the natives of S. Africa, remarked that their ideas of beauty seem very different from ours; for in one tribe two slim, slight, and pretty girls were not admired by the natives.

Turning to other quarters of the world; in Java, a yellow, not a white girl, is considered, according to Madame Pfeiffer, a beauty. A man of

Cochin-China "spoke with contempt of the wife of the English Ambassador, that she had white teeth like a dog, and a rosy colour like that of potato-flowers." We have seen that the Chinese dislike our white skin, and that the N. Americans admire "a tawny hide." In S. America, the Yura-caras, who inhabit the wooded, damp slopes of the eastern Cordillera, are remarkably pale-coloured, as their name in their own language expresses; nevertheless they consider European women as very inferior to their own.[967]

In several of the tribes of North America the hair on the head grows to a wonderful length; and Catlin gives a curious proof how much this is esteemed, for the chief of the Crows was elected to this office from having the longest hair of any man in the tribe, namely ten feet and seven inches. The Aymaras and Quichuas of S. America, likewise have very long hair; and this, as Mr. D. Forbes informs me, is so much valued for the sake of beauty, that cutting it off was the severest punishment which he could inflict on them. In both halves of the continent the natives sometimes increase the apparent length of their hair by weaving into it fibrous substances. Although the hair on the head is thus cherished, that on the face is considered by the North American Indians "as very vulgar," and every hair is carefully eradicated. This practice prevails throughout the American continent from Vancouver's Island in the north to Tierra del Fuego in the south. When York Minster, a Fuegian on board the "Beagle" was taken back to his country, the natives told him he ought to pull out the few short hairs on his face. They also threatened a young missionary, who was left for a time with them, to strip him naked, and pluck the hairs from his face and body, yet he was far from a hairy man. This fashion is carried to such an extreme that the Indians of Paraguay eradicate their eyebrows and eyelashes, saying that they do not wish to be like horses.[968]

It is remarkable that throughout the world the races which are almost completely destitute of a beard dislike hairs on the face and body, and take pains to eradicate them. The Kalmucks are beardless, and they are well known, like the Americans, to pluck out all straggling hairs; and so it is with the Polynesians, some of the Malays, and the Siamese. Mr. Veitch states that the Japanese ladies "all objected to our whiskers, considering them very ugly, and told us to cut them off, and be like Japanese men." The New Zealanders are beardless; they carefully pluck out the hairs on the face, and have a saying that "There is no woman for a hairy man."[969]

On the other hand, bearded races admire and greatly value their beards; among the Anglo-Saxons every part of the body, according to their laws, had a recognised value; "the loss of the beard being estimated at twenty shillings, while the breaking of a thigh was fixed at only twelve."[970] In the East men swear solemnly by their beards. We have seen that Chinsurdi, the chief of the Makalolo in Africa, evidently thought that beards were a great ornament. With the Fijians in the Pacific the beard is "profuse and bushy, and is his greatest pride;" whilst the inhabitants of the adjacent archipelagoes of Tonga and Samoa are "beardless, and abhor a rough chin." In one island alone of the Ellice group "the men are heavily bearded, and not a little proud thereof."[971]

We thus see how widely the different races of man differ in their taste for the beautiful. In every nation sufficiently advanced to have made effigies of their gods or of their deified rulers, the sculptors no doubt have endeavoured to express their highest ideal of beauty and grandeur.[972] Under this point of view it is well to compare in our mind the Jupiter or Apollo of the Greeks with the Egyptian or Assyrian statues; and these with the hideous bas-reliefs on the ruined buildings of Central America.

I have met with very few statements opposed to the above conclusion. Mr. Winwood Reade, however, who has had ample opportunities for observation, not only with the negroes of the West Coast of Africa, but with those of the interior who have never associated with Europeans, is convinced that their ideas of beauty are *on the whole* the same as ours. He has repeatedly found that he agreed with negroes in their estimation of the beauty of the native girls; and that their appreciation of the beauty of European women corresponded with ours. They admire long hair, and use artificial means to make it appear abundant; they admire also a beard, though themselves very scantily provided. Mr. Reade feels doubtful what kind of nose is most appreciated: a girl has been heard to say, "I do not want to marry him, he has got no nose;" and this shews that a very flat nose is not an object of admiration. We should, however, bear in mind that the depressed and very broad noses and projecting jaws of the negroes of the West Coast are exceptional types with the inhabitants of Africa. Notwithstanding the foregoing statements, Mr. Reade does not think it probable that negroes would ever prefer the "most beautiful European woman, on the mere grounds of physical admiration, to a good-looking negress."[973]

The truth of the principle, long ago insisted on by Humboldt,[974] that man admires and often tries to exaggerate whatever characters nature may have given him, is shewn in many ways. The practice of beardless races extirpating every trace of a beard, and generally all the hairs on the body, offers one illustration. The skull has been greatly modified during ancient and modern times by many nations; and there can be little doubt that this has been practised, especially in N. and S. America, in order to exaggerate some natural and admired peculiarity. Many American Indians are known to admire a head flattened to such an extreme degree as to appear to us like that of an idiot. The natives on the north-western coast compress the head into a pointed cone; and it is their constant practice to gather the hair into a knot on the top of the head, for the sake, as Dr. Wilson remarks, "of increasing the apparent elevation of the favourite conoid form." The inhabitants of Arakhan "admire a broad, smooth forehead, and in order to produce it, they fasten a plate of lead on the heads of the new-born children." On the other hand, a broad, well-rounded occiput is considered a great beauty" by the natives of the Fiji Islands.[975]

As with the skull, so with the nose; the ancient Huns during the age of Attila were accustomed to flatten the noses of their infants with bandages, "for the sake of exaggerating a natural conformation." With the Tahitians, to be called *long-nose* is considered as an insult, and they compress the noses and foreheads of their children for the sake of beauty. So it is with the Malays of Sumatra, the Hottentots, certain Negroes, and the natives of Brazil.[976] The Chinese have by nature unusually small feet;[977] and it is well known that the women of the upper classes distort their feet to make them still smaller. Lastly, Humboldt thinks that the American Indians prefer colouring their bodies with red paint in order to exaggerate their natural tint; and until recently European women added to their naturally bright colours by rouge and white cosmetics; but I doubt whether many barbarous nations have had any such intention in painting themselves.

In the fashions of our own dress we see exactly the same principle and the same desire to carry every point to an extreme; we exhibit, also, the same spirit of emulation. But the fashions of savages are far more permanent than ours; and whenever their bodies are artificially modified this is necessarily the case. The Arab women of the Upper Nile occupy about three days in dressing their hair; they never imitate other tribes, "but simply vie with each other in the superlativeness of their own style." Dr. Wilson, in speaking of the compressed skulls of various American races, adds, "such usages are among the least eradicable, and long survive the shock of revolutions that change dynasties and efface more important national peculiarities."[978] The same principle comes largely into play in the art of selection; and we can thus understand, as I have elsewhere explained,[979] the wonderful development of all the races of animals and plants which are kept merely for ornament. Fanciers always wish each character to be somewhat increased; they do not admire a medium standard; they certainly do not desire any great and abrupt change in the character of their breeds; they admire solely what they are accustomed to behold, but they ardently desire to see each characteristic feature a little more developed.

No doubt the perceptive powers of man and the lower animals are so constituted that brilliant colours and certain forms, as well as harmonious and rhythmical sounds, give pleasure and are called beautiful; but why this should be so, we know no more than why certain bodily sensations are agreeable and others disagreeable. It is certainly not true that there is in the mind of man any universal standard of beauty with respect to the human body. It is, however, possible that certain tastes may in the course of time become inherited, though I know of no evidence in favour of this belief; and if so, each race would possess its own innate ideal standard of beauty. It has been argued[980] that ugliness consists in an approach to the structure of the lower animals, and this no doubt is true with the more civilised nations, in which intellect is highly appreciated; but a nose twice as prominent, or eyes twice as large as usual, would not be an approach in structure to any of the lower animals, and yet would be utterly hideous. The men of each race prefer what they are accustomed to behold; they cannot endure any great change; but they like variety, and admire each characteristic point carried to a moderate extreme.[981] Men accustomed to a nearly oval face, to straight and regular features, and to bright colours, admire, as we Europeans know, these points when strongly developed. On the other hand, men accustomed to a broad face, with high cheek-bones, a depressed nose, and a black skin, admire these points strongly developed. No doubt characters of all kinds may easily be too much developed for beauty. Hence a perfect beauty, which implies many characters modified in a particular manner, will in every race be a prodigy. As the great anatomist Bichat long ago said, if every one were cast in the same mould, there would be no such thing as beauty. If all our women were to become as beautiful as the Venus de Medici, we should for a time be charmed; but we should soon wish for variety; and as soon as we had obtained variety, we should wish to see certain characters in our women a little exaggerated beyond the then existing common standard.

CHAPTER XX

Secondary Sexual Characters of Man—continued

On the effects of the continued selection of women according to a different standard of beauty in each race • On the causes which interfere with sexual selection in civilised and savage nations • Conditions favourable to sexual selection during primeval times • On the manner of action of sexual selection with mankind • On the women in savage tribes having some power to choose their husbands • Absence of hair on the body, and development of the beard • Colour of the skin • Summary

We have seen in the last chapter that with all barbarous races ornaments, dress, and external appearance are highly valued; and that the men judge of the beauty of their women by widely different standards. We must next inquire whether this preference and the consequent selection during many generations of those women, which appear to the men of each race the most attractive, has altered the character either of the females alone or of both sexes. With mammals the general rule appears to be that characters of all kinds are inherited equally by the males and females; we might therefore expect that with mankind any characters gained through sexual selection by the females would commonly be transferred to the offspring of both sexes. If any change has thus been effected it is almost certain that the different races will have been differently modified, as each has its own standard of beauty.

With mankind, especially with savages, many causes interfere with the action of sexual selection as far as the bodily frame is concerned. Civilised men are largely attracted by the mental charms of women, by their wealth, and especially by their social position; for men rarely marry into a much lower rank of life. The men who succeed in obtaining the more beautiful women, will not have a better chance of leaving a long line of descendants than other men with plainer wives, with the exception of the few who bequeath their fortunes according to primogeniture. With respect to the opposite form of selection, namely of the more attractive men by the women, although in civilised nations women have free or almost free choice, which is not the case with barbarous races, yet their choice is largely influenced by the social position and wealth of the men; and the success of the latter in life largely depends on their intellectual powers and energy, or on the fruits of these same powers in their forefathers.

There is, however, reason to believe that sexual selection has effected something in certain civilised and semi-civilised nations. Many persons are convinced, as it appears to me with justice, that the members of our aristocracy, including under this term all wealthy families in which primogeniture has long prevailed, from having chosen during many generations from all classes the more beautiful women as their wives, have become handsomer, according to the European standard of beauty, than the middle classes; yet the middle classes are placed under equally favourable conditions of life for the perfect development of the body. Cook remarks that the superiority in personal appearance "which is observable in the erees or nobles in all the other islands (of the Pacific) is found in the Sandwich islands;" but this may be chiefly due to their better food and manner of life.

The old traveller Chardin, in describing the Persians, says their "blood is now highly refined by frequent intermixtures with the Georgians and Circassians, two nations which surpass all the world in personal beauty. There is hardly a man of rank in Persia who is not born of a Georgian or Circassian mother." He adds that they inherit their beauty, "not from their ancestors, for without the above mixture, the men of rank in Persia, who are descendants of the Tartars, would be extremely ugly."[982] Here is a more curious case: the priestesses who attended the temple of Venus Erycina at San-Giuliano in Sicily, were selected for their beauty out of the whole of Greece; they were not vestal virgins, and Quatrefages,[983] who makes this statement, says that the women of San-Giuliano are famous at the present day as the most beautiful in the island, and are sought by artists as models. But it is obvious that the evidence in the above cases is doubtful.

The following case, though relating to savages, is well worth giving from its curiosity. Mr. Winwood Reade informs me that the Jollofs, a tribe of negroes on the west coast of Africa, "are remarkable for their uniformly fine appearance." A friend of his asked one of these men, "How is it that every one whom I meet is so fine-looking, not only your men, but your women?" The Jollof answered, "It is very easily explained: it has always been our custom to pick out our worse-looking slaves and to sell them." It need hardly be added that with all savages female slaves serve as concubines. That this negro should have attributed, whether rightly or wrongly, the fine appearance of his tribe, to the long-continued elimination of the ugly women, is not so surprising as it may at first appear; for I have elsewhere shewn[984] that negroes fully appreciate the importance of selection in the breeding of their domestic animals, and I could give from Mr. Reade additional evidence on this head.

On the Causes which prevent or check the Action of Sexual Selection with Savages.—The chief causes are, firstly, so-called communal marriages or promiscuous intercourse; secondly, infanticide, especially of female infants; thirdly, early betrothals; and lastly, the low estimation in which women are held, as mere slaves. These four points must be considered in some detail.

It is obvious that as long as the pairing of man, or of any other animal, is left to chance, with no choice exerted by either sex, there can be no sexual selection; and no effect will be produced on the offspring by certain individuals having had an advantage over others in their courtship. Now it is asserted that there exist at the present day tribes which practise what Sir J. Lubbock by courtesy calls communal marriages; that is, all the men and women in the tribe are husbands and wives to each other. The licentiousness of many savages is no doubt astonishingly great, but it seems to me that more evidence is requisite before we fully admit that their existing intercourse is absolutely promiscuous. Nevertheless all those who have most closely studied the subject,[985] and whose judgment is worth much more than mine, believe that communal marriage was the original and universal form throughout the world, including the intermarriage of brothers and sisters. The indirect evidence in favour of this belief is extremely strong, and rests chiefly on the terms of relationship which are employed between the members of the same tribe, implying a connection with the tribe alone, and not with either parent. But the subject is too large and complex for even an abstract to be here given, and I will confine myself to a few remarks. It is evident in the case of communal marriages, or where the marriage-tie is very loose, that the relationship of the child to its father cannot be known. But it seems almost incredible that the relationship of the child to its mother should ever have been completely ignored, especially as the women in most savage tribes nurse their infants for a long time. Accordingly in many cases the lines of descent are traced through the mother alone, to the exclusion of the father. But in many other cases the terms employed express a connection with the tribe alone, to the exclusion even of the mother. It seems possible that the connection between the related members of the same barbarous tribe, exposed to all sorts of danger, might be so much more important, owing to the need of mutual protection and aid, than that between the mother and her child, as to lead to the sole use of terms expressive of the former relationships; but Mr. Morgan is convinced that this view of the case is by no means sufficient.

The terms of relationship used in different parts of the world may be divided, according to the author just quoted, into two great classes, the classificatory and descriptive,—the latter being employed by us. It is the classificatory system which so strongly leads to the belief that communal and other extremely loose forms of marriage were originally universal. But as far as I can see, there is no necessity on this ground for believing in absolutely promiscuous intercourse. Men and women, like many of the lower animals, might formerly have entered into strict though temporary unions for each birth, and in this case nearly as much confusion would have arisen in the terms of relationship as in the case of promiscuous intercourse. As far as sexual selection is concerned, all that is required is that choice should be exerted before the parents unite, and it signifies little whether the unions last for life or only for a season.

Besides the evidence derived from the terms of relationship, other lines of reasoning indicate the former wide prevalence of communal marriage. Sir J. Lubbock ingeniously accounts[986] for the strange and widely-extended habit of exogamy,—that is, the men of one tribe always taking wives from a distinct tribe,—by communism having been the original form of marriage; so that a man never obtained a wife for himself unless he captured her from a neighbouring and hostile tribe, and then she would naturally have become his sole and valuable property. Thus the practice of capturing wives might have arisen; and from the honour so gained might ultimately have become the universal habit. We can also, according to Sir J. Lubbock,[986] thus understand "the necessity of expiation for marriage as an infringement of tribal rites, since, according to old ideas, a man had no right to appropriateto himself that which belonged to the whole tribe." Sir J. Lubbock further gives a most curious body of facts shewing that in old times high honour was bestowed on women who were utterly licentious; and this, as he explains, is intelligible, if we admit that promiscuous intercourse was the aboriginal and therefore long revered custom of the tribe.[987]

Although the manner of development of the marriage-tie is an obscure subject, as we may infer from the divergent opinions on several points between the three authors who have studied it most closely, namely, Mr. Morgan, Mr. M'Lennan, and Sir J. Lubbock, yet from the foregoing and several other lines of evidence it seems certain that the habit of marriage has been gradually developed, and that almost promiscuous intercourse was once extremely common throughout the world. Nevertheless from the analogy of the lower animals, more particularly of those which come nearest to man in the series, I cannot believe that this habit prevailed at an extremely remote period, when man had hardly attained to his present rank in the zoological scale. Man, as I have attempted to shew, is certainly descended from some ape-like creature. With the existing Quadrumana, as far as their habits are known, the males of some species are monogamous, but live during only a part of the year with the females, as seems to be the case with the Orang. Several kinds, as some of the Indian and American monkeys, are strictly monogamous, and associate all the year round with their wives. Others are polygamous, as the Gorilla and several American species, and each family lives separate. Even when this occurs, the families inhabiting the same district are probably to a certain extent social: the Chimpanzee, for instance, is occasionally met with in large bands. Again, other species are polygamous, but several males, each with their own females, live associated in a body, as with several species of Baboons.[988] We may indeed conclude from what we know of the jealousy of all male quadrupeds, armed, as many of them are, with special weapons for battling with their rivals, that promiscuous intercourse in a state of nature is extremely improbable. The pairing may not last for life, but only for each birth; yet if the males which are the strongest and best able to defend or otherwise assist their females and young offspring, were to select the more attractive females, this would suffice for the work of sexual selection.

Therefore, if we look far enough back in the stream of time, it is extremely improbable that primeval men and women lived promiscuously together. Judging from the social habits of man as he now exists, and from most savages being polygamists, the most probable view is that primeval man aboriginally lived in small communities, each with as many wives as he could support and obtain, whom he would have jealously guarded against all other men. Or he may have lived with several wives by himself, like the Gorilla; for all the natives "agree that but one adult male is seen in a band; when the young male grows up, a contest takes place for mastery, and the strongest, by killing and driving out the others, establishes himself as the head of the community."[989] The younger males, being thus expelled and wandering about, would, when at last successful in finding a partner, prevent too close interbreeding within the limits of the same family.

Although savages are now extremely licentious, and although communal marriages may formerly have largely prevailed, yet many tribes practise some form of marriage, but of a far more lax nature than with civilised nations. Polygamy, as just stated, is almost universally followed by the leading men in every tribe. Nevertheless there are tribes, standing almost at the bottom of the scale, which are strictly monogamous. This is the case with the Veddahs of Ceylon: they have a saying, according to Sir J. Lubbock,[990] "that death alone can separate husband and wife." An intelligent Kandyan chief, of course a polygamist, "was perfectly scandalized at the utter barbarism of living with only one wife, and never parting until separated by death." It was, he said, "just like the Wanderoo monkeys." Whether savages who now enter into some form of marriage, either polygamous or monogamous, have retained this habit from primeval times, or whether they have returned to some form of marriage, after passing through a stage of promiscuous intercourse, I will not pretend to conjecture.

Infanticide.—This practice is now very common throughout the world, and there is reason to believe that it prevailed much more extensively during former times.[991] Barbarians find it difficult to support themselves and their children, and it is a simple plan to kill their infants. In South America some tribes, as Azara states, formerly destroyed so many infants of both sexes, that they were on the point of extinction. In the Polynesian Islands women have been known to kill from four or five to even ten of their children; and Ellis could not find a single woman who had not killed at least one. Wherever infanticide prevails the struggle for existence will be in so far less severe, and all the members of the tribe will have an almost equally good chance of rearing their few surviving children. In most cases a larger number of female than of male infants are destroyed, for it is obvious that the latter are of most value to the tribe, as they will when grown up aid in defending it, and can support themselves. But the trouble experienced by the women in rearing children, their consequent loss of beauty, the higher estimation set on them and their happier fate, when few in number, are assigned by the women themselves, and by various observers, as additional motives for infanticide. In Australia, where female infanticide is still common, Sir G. Grey estimated the proportion of native women to men as one to three; but others say as two to three. In a village on the eastern frontier of India, Colonel Macculloch found not a single female child.[992]

When, owing to female infanticide, the women of a tribe are few in number, the habit of capturing wives from neighbouring tribes would naturally arise. Sir J. Lubbock, however, as we have seen, attributes the practice in chief part, to the former existence of communal marriage, and to the men having consequently captured women from other tribes to hold as their sole property. Additional causes might be assigned, such as the communities being very small, in which case, marriageable women would often be deficient. That the habit of capture was most extensively practised during former times, even by the ancestors of civilised nations, is clearly shewn by the preservation of many curious customs and ceremonies, of which Mr. M'Lennan has given a most interesting account. In our own marriages the "best man" seems originally to have been the chief abettor of the bridegroom in the act of capture. Now as long as men habitually procured their wives through violence and craft, it is not probable that they would have selected the more attractive women; they would have been too glad to have seized on any woman. But as soon as the practice of procuring wives from a distinct tribe was effected through barter, as now occurs in many places, the more attractive women would generally have been purchased. The incessant crossing, however, between tribe and tribe, which necessarily follows from any form of this habit would have tended to keep all the people inhabiting the same country nearly uniform in character; and this would have greatly interfered with the power of sexual selection in differentiating the tribes.

The scarcity of women, consequent on female infanticide, leads, also, to another practice, namely polyandry, which is still common in several parts of the world, and which formerly, as Mr. M'Lennan believes, prevailed almost universally; but this latter conclusion is doubted by Mr. Morgan and Sir J. Lubbock.[993] Whenever two or more men are compelled to marry one woman, it is certain that all the women of the tribe will get married, and there will be no selection by the men of the more attractive women. But under these circumstances the women no doubt will have the power of choice, and will prefer the more attractive men. Azara, for instance, describes how carefully a Guana woman bargains for all sorts of privileges, before accepting some one or more husbands; and the men in consequence take unusual care of their personal appearance.[994] The very ugly men would perhaps altogether fail in getting a wife, or get one later in life, but the handsomer men, although the most successful in obtaining wives, would not, as far as we can see, leave more offspring to inherit their beauty than the less handsome husbands of the same women.

Early Betrothals and Slavery of Women.—With many savages it is the custom to betroth the females whilst mere infants; and this would effectually prevent preference being exerted on either side according to personal appearance. But it would not prevent the more attractive women from being afterwards stolen or taken by force from their husbands by the more powerful men; and this often happens in Australia, America, and other parts of the world. The same consequences with reference to sexual selection would to a certain extent follow when women are valued almost exclusively as slaves or beasts of burden, as is the case with most savages. The men, however, at all times would prefer the handsomest slaves according to their standard of beauty.

We thus see that several customs prevail with savages which would greatly interfere with, or completely stop, the action of sexual selection. On the other hand, the conditions of life to which savages are exposed, and some of their habits, are favourable to natural selection; and this always comes into play together with sexual selection. Savages are known to suffer severely from recurrent famines; they do not increase their food by artificial means; they rarely refrain from marriage,[995] and generally marry young. Consequently they must be subjected to occasional hard struggles for existence, and the favoured individuals will alone survive.

Turning to primeval times when men had only doubtfully attained the rank of manhood, they would probably have lived, as already stated, either as polygamists or temporarily as monogamists. Their intercourse, judging from analogy, would not then have been promiscuous. They would, no doubt, have defended their females to the best of their power from enemies of all kinds, and would probably have hunted for their subsistence, as well as for that of their offspring. The most powerful and able males would have succeeded best in the struggle for life and in obtaining attractive females. At this early period the progenitors of man, from having only feeble powers of reason, would not have looked forward to distant contingencies. They would have been governed more by their instincts and even less by their reason than are savages at the present day. They would not at that period have partially lost one of the strongest of all instincts, common to all the lower animals, namely the love of their young offspring; and consequently they would not have practised infanticide. There would have been no artificial scarcity of women, and polyandry would not have been followed; there would have been no early betrothals; women would not have been valued as mere slaves; both sexes, if the females as well as the males were permitted to exert any choice, would have chosen their partners, not for mental charms, or property, or social position, but almost solely from external appearance. All the adults would have married or paired, and all the offspring, as far as that was possible, would have been reared; so that the struggle for existence would have been periodically severe to an extreme degree. Thus during these primordial times all the conditions for sexual selection would have been much more favourable than at a later period, when man had advanced in his intellectual powers, but had retrograded in his instincts. Therefore, whatever influence sexual selection may have had in producing the differences between the races of man, and between man and the higher Quadrumana, this influence would have been much more powerful at a very remote period than at the present day.

On the Manner of Action of Sexual Selection with mankind.—With primeval men under the favourable conditions just stated, and with those savages who at the present time enter into any marriage tie (but subject to greater or less interference according as the habits of female infanticide, early betrothals, &c., are more or less practised), sexual selection will probably have acted in the following manner. The strongest and most vigorous men,—those who could best defend and hunt for their families, and during later times the chiefs or head-men,—those who were provided with the best weapons and who possessed the most property, such as a larger number of dogs or other animals, would have succeeded in rearing a greater average number of offspring, than would the weaker, poorer and lower members of the same tribes. There can, also, be no doubt that such men would generally have been able to select the more attractive women. At present the chiefs of nearly every tribe throughout the world succeed in obtaining more than one wife. Until recently, as I hear from Mr. Mantell, almost every girl in New Zealand, who was pretty, or promised to be pretty, was *tapu* to some chief. With the Kafirs, as Mr. C. Hamilton states,[996] "the chiefs generally have the pick of the women for many miles round, and are most persevering in establishing or confirming their privilege." We have seen that each race has its own style of beauty, and we know that it is natural to man to admire each characteristic point in his domestic animals, dress, ornaments, and personal appearance, when carried a little beyond the common standard. If then the several foregoing propositions be admitted, and I cannot see that they are doubtful, it would be an inexplicable circumstance, if the selection of the more attractive women by the more powerful men of each tribe, who would rear on an average a greater number of children, did not after the lapse of many generations modify to a certain extent the character of the tribe.

With our domestic animals, when a foreign breed is introduced into a new country, or when a native breed is long and carefully attended to, either for use or ornament, it is found after several generations to have undergone, whenever the means of comparison exist, a greater or less amount of change. This follows from unconscious selection during a long series of generations—that is, the preservation of the most approved individuals—without any wish or expectation of such a result on the part of the breeder. So again, if two careful breeders rear during many years animals of the same family, and do not compare them together or with a common standard, the animals are found after a time to have become to the surprise of their owners slightly different.[997] Each breeder has impressed, as Von Nathusius well expresses it, the character of his own mind—his own taste and judgment—on his animals. What reason, then, can be assigned why similar results should not follow from the long-continued selection of the most admired women by those men of each tribe, who were able to rear to maturity the greater number of children? This would be unconscious selection, for an effect would be produced, independently of any wish or expectation on the part of the men who preferred certain women to others.

Let us suppose the members of a tribe, in which some form of marriage was practised, to spread over an unoccupied continent; they would soon split up into distinct hordes, which would be separated from each other by various barriers, and still more effectually by the incessant wars between all barbarous nations. The hordes would thus be exposed to slightly different conditions and habits of life, and would sooner or later come to differ in some small degree. As soon as this occurred, each isolated tribe would form for itself a slightly different standard of beauty;[998] and then unconscious selection would come into action through the more powerful and leading savages preferring certain women to others. Thus the differences between the tribes, at first very slight, would gradually and inevitably be increased to a greater and greater degree.

WITH ANIMALS in a state of nature, many characters proper to the males, such as size, strength, special weapons, courage and pugnacity, have been acquired through the law of battle. The semi-human progenitors of man, like their allies the Quadrumana, will almost certainly have been thus modified; and, as savages still fight for the possession of their women, a similar process of selection has probably gone on in a greater or less degree to the present day. Other characters proper to the males of the lower animals, such as bright colours and various ornaments, have been acquired by the more attractive males having been preferred by the females. There are, however, exceptional cases in which the males, instead of having been the selected, have been the selectors. We recognise such cases by the females having been rendered more highly ornamented than the males,—their ornamental characters having been transmitted exclusively or chiefly to their female offspring. One such case has been described in the order to which man belongs, namely, with the Rhesus monkey.

Man is more powerful in body and mind than woman, and in the savage state he keeps her in a far more abject state of bondage than does the male of any other animal; therefore it is not surprising that he should have gained the power of selection. Women are everywhere conscious of the value of their beauty; and when they have the means, they take more delight in decorating themselves with all sorts of ornaments than do men. They borrow the plumes of male birds, with which nature decked this sex in order to charm the females. As women have long been selected for beauty, it is not surprising that some of the successive variations should have been transmitted in a limited manner; and consequently that women should have transmitted their beauty in a somewhat higher degree to their female than to their male offspring. Hence women have become more beautiful, as most persons will admit, than men. Women, however, certainly transmit most of their characters, including beauty, to their offspring of both sexes; so that the continued preference by the men of each race of the more attractive women, according to their standard of taste, would tend to modify in the same manner all the individuals of both sexes belonging to the race.

With respect to the other form of sexual selection (which with the lower animals is much the most common), namely, when the females are the selectors, and accept only those males which excite or charm them most, we have reason to believe that it formerly acted on the progenitors of man. Man in all probability owes his beard, and perhaps some other characters, to inheritance from an ancient progenitor who gained in this manner his ornaments. But this form of selection may have occasionally acted during later times; for in utterly barbarous tribes the women have more power in choosing, rejecting, and tempting their lovers, or of afterwards changing their husbands, than might have been expected. As this is a point of some importance, I will give in detail such evidence as I have been able to collect.

Hearne describes how a woman in one of the tribes of Arctic America repeatedly ran away from her husband and joined a beloved man; and with the Charruas of S. America, as Azara states, the power of divorce is perfectly free. With the Abipones, when a man chooses a wife he bargains with the parents about the price. But "it frequently happens that the girl rescinds what has been agreed upon between the parents and the bridegroom, obstinately rejecting the very mention of marriage." She often runs away, hides herself, and thus eludes the bridegroom. In the Fiji Islands the man seizes on the woman whom he wishes for his wife by actual or pretended force; but "on reaching the home of her abductor, should she not approve of the match, she runs to some one who can protect her; if, however, she is satisfied, the matter is settled forthwith." In Tierra del Fuego a young man first obtains the consent of the parents by doing them some service, and then he attempts to carry off the girl; "but if she is unwilling, she hides herself in the woods until her admirer is heartily tired of looking for her, and gives up the pursuit; but this seldom happens." With the Kalmucks there is a regular race between the bride and bridegroom, the former having a fair start; and Clarke "was assured that no instance occurs of a girl being caught, unless she has a partiality to the pursuer." So with the wild tribes of the Malay archipelago there is a similar racing match; and it appears from M.

Bourien's account, as Sir J. Lubbock remarks, that "the race 'is not to the swift, nor the battle to the strong,' but to the young man who has the good fortune to please his intended bride."

Turning to Africa: the Kafirs buy their wives, and girls are severely beaten by their fathers if they will not accept a chosen husband; yet it is manifest from many facts given by the Rev. Mr. Shooter, that they have considerable power of choice. Thus very ugly, though rich men, have been known to fail in getting wives. The girls, before consenting to be betrothed, compel the men to shew themselves off, first in front and then behind, and "exhibit their paces." They have been known to propose to a man, and they not rarely run away with a favoured lover. With the degraded bush-women of S. Africa, "when a girl has grown up to womanhood without having been betrothed, which, however, does not often happen, her lover must gain her approbation, as well as that of the parents."[999] Mr. Winwood Reade made inquiries for me with respect to the negroes of Western Africa, and he informs me that "the women, at least among the more intelligent Pagan tribes, have no difficulty in getting the husbands whom they may desire, although it is considered unwomanly to ask a man to marry them. They are quite capable of falling in love, and of forming tender, passionate, and faithful attachments."

We thus see that with savages the women are not in quite so abject a state in relation to marriage as has often been supposed. They can tempt the men whom they prefer, and can sometimes reject those whom they dislike, either before or after marriage. Preference on the part of the women, steadily acting in any one direction, would ultimately affect the character of the tribe; for the women would generally choose not merely the handsomer men, according to their standard of taste, but those who were at the same time best able to defend and support them. Such well-endowed pairs would commonly rear a larger number of offspring than the less well endowed. The same result would obviously follow in a still more marked manner if there was selection on both sides; that is if the more attractive, and at the same time more powerful men were to prefer, and were preferred by, the more attractive women. And these two forms of selection seem actually to have occurred, whether or not simultaneously, with mankind, especially during the earlier periods of our long history. We will now consider in a little more detail, relatively to sexual selection, some of the characters which distinguish the several races of man from each other and from the lower animals, namely, the more or less complete absence of hair from the body and the colour of the skin. We need say nothing about the great diversity in the shape of the features and of the skull between the different races, as we have seen in the last chapter how different is the standard of beauty in these respects. These characters will therefore probably have been acted on through sexual selection; but we have no means of judging, as far as I can see, whether they have been acted on chiefly through the male or female side. The musical faculties of man have likewise been already discussed.

Absence of Hair on the Body, and its Development on the Face and Head.—From the presence of the woolly hair or lanugo on the human fœtus, and of rudimentary hairs scattered over the body during maturity, we may infer that man is descended from some animal which was born hairy and remained so during life. The loss of hair is an inconvenience and probably an injury to man even under a hot climate, for he is thus exposed to sudden chills, especially during wet weather. As Mr. Wallace remarks, the natives in all countries are glad to protect their naked backs and shoulders with some slight covering. No one supposes that the nakedness of the skin is any direct advantage to man, so that his body cannot have been divested of hair through natural selection.[1000] Nor have we any grounds for believing, as shewn in a former chapter, that this can be due to the direct action of the conditions to which man has long been exposed, or that it is the result of correlated development.

The absence of hair on the body is to a certain extent a secondary sexual character; for in all parts of the world women are less hairy than men. Therefore we may reasonably suspect that this is a character which has been gained through sexual selection. We know that the faces of several species of monkeys, and large surfaces at the posterior end of the body in other species, have been denuded of hair; and this we may safely attribute to sexual selection, for these surfaces are not only vividly coloured, but sometimes, as with the male mandrill and female rhesus, much more vividly in the one sex than in the other. As these animals gradually reach maturity the naked surfaces, as I am informed by Mr. Bartlett, grow larger, relatively to the size of their bodies. The hair, however, appears to have been removed in these cases, not for the sake of nudity, but that the colour of the skin should be more fully displayed. So again with many birds the head and neck have been divested of feathers through sexual selection, for the sake of exhibiting the brightly-coloured skin.

As woman has a less hairy body than man, and as this character is common to all races, we may conclude that our female semi-human progenitors were probably first partially divested of hair; and that this occurred at an extremely remote period before the several races had diverged from a common stock. As our female progenitors gradually acquired this new character of nudity, they must have transmitted it in an almost equal degree to their young offspring of both sexes; so that its transmission, as in the case of many ornaments with mammals and birds, has not been limited either by age or sex. There is nothing surprising in a partial loss of hair having been esteemed as ornamental by the ape-like progenitors of man, for we have seen that with animals of all kinds innumerable strange characters have been thus esteemed, and have consequently been modified through sexual selection. Nor is it surprising that a character in a slight degree injurious should have been thus acquired; for we know that this is the case with the plumes of some birds, and with the horns of some stags.

The females of certain anthropoid apes, as stated in a former chapter, are somewhat less hairy on the under surface than are the males; and here we have what might have afforded a commencement for the process of denudation. With respect to the completion of the process through sexual selection, it is well to bear in mind the New Zealand proverb, "there is no woman for a hairy man." All who have seen photographs of the Siamese hairy family will admit how ludicrously hideous is the opposite extreme of excessive hairiness. Hence the king of Siam had to bribe a man to marry the first hairy woman in the family, who transmitted this character to her young offspring of both sexes.[1001]

Some races are much more hairy than others, especially on the male side; but it must not be assumed that the more hairy races, for instance Europeans, have retained a primordial condition more completely than have the naked races, such as the Kalmucks or Americans. It is a more probable view that the hairiness of the former is due to partial reversion, for characters which have long been inherited are always apt to return. It does not appear that a cold climate has been influential in leading to this kind of reversion; excepting perhaps with the negroes, who have been reared during several generations, in the United States,[1002] and possibly with the Ainos, who inhabit the northern islands of the Japan archipelago. But the laws of inheritance are so complex that we can seldom understand their action. If the greater hairiness of certain races be the result of reversion, unchecked by any form of selection, the extreme variability of this character, even within the limits of the same race, ceases to be remarkable.

With respect to the beard, if we turn to our best guide, namely the Quadrumana, we find beards equally well developed in both sexes of many species, but in others, either confined to the males, or more developed in them than in the females. From this fact, and from the curious arrangement, as well as the bright colours, of the hair about the heads of many monkeys, it is highly probable, as before explained, that the males first acquired their beards as an ornament through sexual selection, transmitting them in most cases, in an equal or nearly equal degree, to their offspring of both sexes. We know from Eschricht[1003] that with mankind, the female as well as the male fœtus is furnished with much hair on the face, especially round the mouth; and this indicates that we are descended from a progenitor, of which both sexes were bearded. It appears therefore at first sight probable that man has retained his beard from a very early period, whilst woman lost her beard at the same time when her body became almost completely divested of hair. Even the colour of the beard with mankind seems to have been inherited from an ape-like progenitor; for when there is any difference in tint between the hair of the head and the beard, the latter is lighter coloured in all monkeys and in man. There is less improbability in the men of the bearded races having retained their beards from primordial times, than in the case of the hair on the body; for with those Quadrumana, in which the male has a larger beard than that of the female, it is fully developed only at maturity, and the later stages of development may have been exclusively transmitted to mankind. We should then see what is actually the case, namely, our male children, before they arrive at maturity, as destitute of beards as are our female children. On the other hand the great variability of the beard within the limits of the same race and in different races indicates that reversion has come into action. However this may be, we must not overlook the part which sexual selection may have played even during later times; for we know that with savages, the men of the beardless races take infinite pains in eradicating every hair from their faces, as something odious, whilst the men of the bearded races feel the greatest pride in their beards. The women, no doubt, participate in these feelings, and if so sexual selection can hardly have failed to have effected something in the course of later times.[1004]

It is rather difficult to form a judgment how the long hair on our heads became developed. Eschricht[1005] states that in the human fœtus the hair on the face during the fifth month is longer than that on the head; and this indicates that our semi-human progenitors were not furnished with long tresses, which consequently must have been a late acquisition. This is likewise indicated by the extraordinary difference in the length of the hair in the different races; in the negro the hair forms a mere curly mat; with us it is of great length, and with the American natives it not rarely reaches to the ground. Some species of Semnopithecus have their heads covered with moderately long hair, and this probably serves as an ornament and was acquired through sexual selection. The same view may be extended to mankind, for we know that long tresses are now and were formerly much admired, as may be observed in the works of almost every poet; St. Paul says, "if a woman have long hair, it is a glory to her;" and we have seen that in North America a chief was elected solely from the length of his hair.

Colour of the Skin.—The best kind of evidence that the colour of the skin has been modified through sexual selection is wanting in the case of mankind; for the sexes do not differ in this respect, or only slightly and doubtfully. On the other hand we know from many facts already given that the colour of the skin is regarded by the men of all races as a highly important element in their beauty; so that it is a character which would be likely to be modified through selection, as has occurred in innumerable instances with the lower animals. It seems at first sight a monstrous supposition that the jet blackness of the negro has been gained through sexual selection; but this view is supported by various analogies, and we know that negroes admire their own blackness. With mammals, when the sexes differ in colour, the male is often black or much darker than the female; and it depends merely on the form of inheritance whether this or any other tint shall be transmitted to both sexes or to one alone. The resemblance of *Pithecia satanas* with his jet black skin, white rolling eyeballs, and hair parted on the top of the head, to a negro in miniature, is almost ludicrous.

The colour of the face differs much more widely in the various kinds of monkeys than it does in the races of man; and we have good reason to believe that the red, blue, orange, almost white and black tints of their skin, even when common to both sexes, and the bright colours of their fur, as well as the ornamental tufts of hair about the head, have all been acquired through sexual selection. As the newly-born infants of the most distinct races do not differ nearly as much in colour as do the adults, although their bodies are completely destitute of hair, we have some slight indication that the tints of the different races were acquired subsequently to the removal of the hair, which, as before stated, must have occurred at a very early period.

Summary.—We may conclude that the greater size, strength, courage, pugnacity, and even energy of man, in comparison with the same qualities in woman, were acquired during primeval times, and have subsequently been augmented, chiefly through the contests of rival males for the possession of the females. The greater intellectual vigour and power of invention in man is probably due to natural selection combined with the inherited effects of habit, for the most able men will have succeeded best in defending and providing for themselves, their wives and offspring. As far as the extreme intricacy of the subject permits us to judge, it appears that our male ape-like progenitors acquired their beards as an ornament to charm or excite the opposite sex, and transmitted them to man as he now exists. The females apparently were first denuded of hair in like manner as a sexual ornament; but they transmitted this character almost equally to both sexes. It is not improbable that the females were modified in other respects for the same purpose and through the same means; so that women have acquired sweeter voices and become more beautiful than men.

It deserves particular attention that with mankind all the conditions for sexual selection were much more favourable, during a very early period, when man had only just attained to the rank of manhood, than during later times. For he would then, as we may safely conclude, have been guided more by his instinctive passions, and less by foresight or reason. He would not then have been so utterly licentious as many savages now are; and each male would have jealously guarded his wife or wives. He would not then have practised infanticide; nor valued his wives merely as useful slaves; nor have been betrothed to them during infancy. Hence we may infer that the races of men were differentiated, as far as sexual selection is concerned, in chief part during a very remote epoch; and this conclusion throws light on the remarkable fact that at the most ancient period, of which we have as yet obtained any record, the races of man had already come to differ nearly or quite as much as they do at the present day.

The views here advanced, on the part which sexual selection has played in the history of man, want scientific precision. He who does not admit this agency in the case of the lower animals, will properly disregard all that I have written in the later chapters on man. We cannot positively say that this character, but not that, has been thus modified; it has, however, been shewn that the races of man differ from each other and from their nearest allies amongst the lower animals, in certain characters which are of no service to them in their ordinary habits of life, and which it is extremely probable would have been modified through sexual selection. We have seen that with the lowest savages the people of each tribe admire their own characteristic qualities,—the shape of the head and face, the squareness of the cheek-bones, the prominence or depression of the nose, the colour of the skin, the length of the hair on the head, the absence of hair on the face and body, or the presence of a great beard, and so forth. Hence these and other such points could hardly fail to have been slowly and gradually exaggerated from the more powerful and able men in each tribe, who would succeed in rearing the largest number of offspring, having selected during many generations as their wives the most strongly characterised and therefore most attractive women. For my own part I conclude that of all the causes which have led to the differences in external appearance between the races of man, and to a certain extent between man and the lower animals, sexual selection has been by far the most efficient.

CHAPTER XXI

General Summary and Conclusion

MAIN CONCLUSION THAT MAN IS DESCENDED FROM SOME LOWER FORM • MANNER OF DEVELOPMENT • GENEALOGY OF MAN • INTELLECTUAL AND MORAL FACULTIES • SEXUAL SELECTION • CONCLUDING REMARKS

A BRIEF SUMMARY will here be sufficient to recall to the reader's mind the more salient points in this work. Many of the views which have been advanced are highly speculative, and some no doubt will prove erroneous; but I have in every case given the reasons which have led me to one view rather than to another. It seemed worth while to try how far the principle of evolution would throw light on some of the more complex problems in the natural history of man. False facts are highly injurious to the progress of science, for they often long endure; but false views, if supported by some evidence, do little harm, as every one takes a salutary pleasure in proving their falseness; and when this is done, one path towards error is closed and the road to truth is often at the same time opened.

The main conclusion arrived at in this work, and now held by many naturalists who are well competent to form a sound judgment, is that man is descended from some less highly organised form. The grounds upon which this conclusion rests will never be shaken, for the close similarity between man and the lower animals in embryonic development, as well as in innumerable points of structure and constitution, both of high and of the most trifling importance,—the rudiments which he retains, and the abnormal reversions to which he is occasionally liable,—are facts which cannot be disputed. They have long been known, but until recently they told us nothing with respect to the origin of man. Now when viewed by the light of our knowledge of the whole organic world, their meaning is unmistakeable. The great principle of evolution stands up clear and firm, when these groups of facts are considered in connection with others, such as the mutual affinities of the members of the same group, their geographical distribution in past and present times, and their geological succession. It is incredible that all these facts should speak falsely. He who is not content to look, like a savage, at the phenomena of nature as disconnected, cannot any longer believe that man is the work of a separate act of creation. He will be forced to admit that the close resemblance of the embryo of man to that, for instance, of a dog—the construction of his skull, limbs, and whole frame, independently of the uses to which the parts may be put, on the same plan with that of other mammals—the occasional reappearance of various structures, for instance of several distinct muscles, which man does not normally possess, but which are common to the Quadrumana—and a crowd of analogous facts—all point in the plainest manner to the conclusion that man is the co-descendant with other mammals of a common progenitor.

We have seen that man incessantly presents individual differences in all parts of his body and in his mental faculties. These differences or variations seem to be induced by the same general causes, and to obey the same laws as with the lower animals. In both cases similar laws of inheritance prevail. Man tends to increase at a greater rate than his means of subsistence; consequently he is occasionally subjected to a severe struggle for existence, and natural selection will have effected whatever lies within its scope. A succession of strongly-marked variations of a similar nature are by no means requisite; slight fluctuating differences in the individual suffice for the work of natural selection. We may feel assured that the inherited effects of the long-continued use or disuse of parts will have done much in the same direction with natural selection. Modifications formerly of importance, though no longer of any special use, will be long inherited. When one part is modified, other parts will change through the principle of correlation, of which we have instances in many curious cases of correlated monstrosities. Something may be attributed to the direct and definite action of the surrounding conditions of life, such as abundant food, heat, or moisture; and lastly, many characters of slight physiological importance, some indeed of considerable importance, have been gained through sexual selection.

No doubt man, as well as every other animal, presents structures, which as far as we can judge with our little knowledge, are not now of any service to him, nor have been so during any former period of his existence, either in relation to his general conditions of life, or of one sex to the other. Such structures cannot be accounted for by any form of selection, or by the inherited effects of the use and disuse of parts. We know, however, that many strange and strongly-marked peculiarities of structure occasionally appear in our domesticated productions, and if the unknown causes which produce them were to act more uniformly, they would probably become common to all the individuals of the species. We may hope hereafter to understand something about the causes of such occasional modifications, especially through the study of monstrosities: hence the labours of experimentalists, such as those of M. Camille Dareste, are full of promise for the future. In the greater number of cases we can only say that the cause of each slight variation and of each monstrosity lies much more in the nature or constitution of the organism, than in the nature of the surrounding conditions; though new and changed conditions certainly play an important part in exciting organic changes of all kinds.

Through the means just specified, aided perhaps by others as yet undiscovered, man has been raised to his present state. But since he attained to the rank of manhood, he has diverged into distinct races, or as they may be more appropriately called sub-species. Some of these, for instance the Negro and European, are so distinct that, if specimens had been brought to a naturalist without any further information, they would undoubtedly have been considered by him as good and true species. Nevertheless all the races agree in so many unimportant details of structure and in so many mental peculiarities, that these can be accounted for only through inheritance from a common progenitor; and a progenitor thus characterised would probably have deserved to rank as man.

It must not be supposed that the divergence of each race from the other races, and of all the races from a common stock, can be traced back to any one pair of progenitors. On the contrary, at every stage in the process of modification, all the individuals which were in any way best fitted for their conditions of life, though in different degrees, would have survived in greater numbers than the less well fitted. The process would have been like that followed by man, when he does not intentionally select particular individuals, but breeds from all the superior and neglects all the inferior individuals. He thus slowly but surely modifies his stock, and unconsciously forms a new strain. So with respect to modifications, acquired independently of selection, and due to variations arising from the nature of the organism and the action of the surrounding conditions, or from changed habits of life, no single pair will have been modified in a much greater degree than the other pairs which inhabit the same country, for all will have been continually blended through free intercrossing.

By considering the embryological structure of man,—the homologies which he presents with the lower animals,—the rudiments which he retains,—and the reversions to which he is liable, we can partly recall in imagination the former condition of our early progenitors; and can approximately place them in their proper position in the zoological series.

We thus learn that man is descended from a hairy quadruped, furnished with a tail and pointed ears, probably arboreal in its habits, and an inhabitant of the Old World. This creature, if its whole structure had been examined by a naturalist, would have been classed amongst the Quadrumana, as surely as would the common and still more ancient progenitor of the Old and New World monkeys. The Quadrumana and all the higher mammals are probably derived from an ancient marsupial animal, and this through a long line of diversified forms, either from some reptile-like or some amphibian-like creature, and this again from some fish-like animal. In the dim obscurity of the past we can see that the early progenitor of all the Vertebrata must have been an aquatic animal, provided with branchiæ, with the two sexes united in the same individual, and with the most important organs of the body (such as the brain and heart) imperfectly developed. This animal seems to have been more like the larvæ of our existing marine Ascidians than any other known form.

THE GREATEST DIFFICULTY which presents itself, when we are driven to the above conclusion on the origin of man, is the high standard of intellectual power and of moral disposition which he has attained. But every one who admits the general principle of evolution, must see that the mental powers of the higher animals, which are the same in kind with those of mankind, though so different in degree, are capable of advancement. Thus the interval between the mental powers of one of the higher apes and of a fish, or between those of an ant and scale-insect, is immense. The development of these powers in animals does not offer any special difficulty; for with our domesticated animals, the mental faculties are certainly variable, and the variations are inherited. No one doubts that these faculties are of the utmost importance to animals in a state of nature. Therefore the conditions are favourable for their development through natural selection. The same conclusion may be extended to man; the intellect must have been all-important to him, even at a very remote period, enabling him to use language, to invent and make weapons, tools, traps, &c.; by which means, in combination with his social habits, he long ago became the most dominant of all living creatures.

A great stride in the development of the intellect will have followed, as soon as, through a previous considerable advance, the half-art and half-instinct of language came into use; for the continued use of language will have reacted on the brain, and produced an inherited effect; and this again will have reacted on the improvement of language. The large size of the brain in man, in comparison with that of the lower animals, relatively to the size of their bodies, may be attributed in chief part, as Mr. Chauncey Wright has well remarked,[1006] to the early use of some simple form of language,—

that wonderful engine which affixes signs to all sorts of objects and qualities, and excites trains of thought which would never arise from the mere impression of the senses, and if they did arise could not be followed out. The higher intellectual powers of man, such as those of ratiocination, abstraction, self-consciousness, &c., will have followed from the continued improvement of other mental faculties; but without considerable culture of the mind, both in the race and in the individual, it is doubtful whether these high powers would be exercised, and thus fully attained.

The development of the moral qualities is a more interesting and difficult problem. Their foundation lies in the social instincts, including in this term the family ties. These instincts are of a highly complex nature, and in the case of the lower animals give special tendencies towards certain definite actions; but the more important elements for us are love, and the distinct emotion of sympathy. Animals endowed with the social instincts take pleasure in each other's company, warn each other of danger, defend and aid each other in many ways. These instincts are not extended to all the individuals of the species, but only to those of the same community. As they are highly beneficial to the species, they have in all probability been acquired through natural selection.

A moral being is one who is capable of comparing his past and future actions and motives,—of approving of some and disapproving of others; and the fact that man is the one being who with certainty can be thus designated makes the greatest of all distinctions between him and the lower animals. But in our third chapter I have endeavoured to shew that the moral sense follows, firstly, from the enduring and always present nature of the social instincts, in which respect man agrees with the lower animals; and secondly, from his mental faculties being highly active and his impressions of past events extremely vivid, in which respects he differs from the lower animals. Owing to this condition of mind, man cannot avoid looking backwards and comparing the impressions of past events and actions. He also continually looks forward. Hence after some temporary desire or passion has mastered his social instincts, he will reflect and compare the now weakened impression of such past impulses, with the ever present social instinct; and he will then feel that sense of dissatisfaction which all unsatisfied instincts leave behind them. Consequently he resolves to act differently for the future—and this is conscience. Any instinct which is permanently stronger or more enduring than another, gives rise to a feeling which we express by saying that it ought to be obeyed. A pointer dog, if able to reflect on his past conduct, would say to himself, I ought (as indeed we say of him) to have pointed at that hare and not have yielded to the passing temptation of hunting it.

Social animals are partly impelled by a wish to aid the members of the same community in a general manner, but more commonly to perform certain definite actions. Man is impelled by the same general wish to aid his fellows, but has few or no special instincts. He differs also from the lower animals in being able to express his desires by words, which thus become the guide to the aid required and bestowed. The motive to give aid is likewise somewhat modified in man: it no longer consists solely of a blind instinctive impulse, but is largely influenced by the praise or blame of his fellow men. Both the appreciation and the bestowal of praise and blame rest on sympathy; and this emotion, as we have seen, is one of the most important elements of the social instincts. Sympathy, though gained as an instinct, is also much strengthened by exercise or habit. As all men desire their own happiness, praise or blame is bestowed on actions and motives, according as they lead to this end; and as happiness is an essential part of the general good, the greatest-happiness principle indirectly serves as a nearly safe standard of right and wrong. As the reasoning powers advance and experience is gained, the more remote effects of certain lines of conduct on the character of the individual, and on the general good, are perceived; and then the self-regarding virtues, from coming within the scope of public opinion, receive praise, and their opposites receive blame. But with the less civilised nations reason often errs, and many bad customs and base superstitions come within the same scope, and consequently are esteemed as high virtues, and their breach as heavy crimes.

The moral faculties are generally esteemed, and with justice, as of higher value than the intellectual powers. But we should always bear in mind that the activity of the mind in vividly recalling past impressions is one of the fundamental though secondary bases of conscience. This fact affords the strongest argument for educating and stimulating in all possible ways the intellectual faculties of every human being. No doubt a man with a torpid mind, if his social affections and sympathies are well developed, will be led to good actions, and may have a fairly sensitive conscience. But whatever renders the imagination of men more vivid and strengthens the habit of recalling and comparing past impressions, will make the conscience more sensitive, and may even compensate to a certain extent for weak social affections and sympathies.

The moral nature of man has reached the highest standard as yet attained, partly through the advancement of the reasoning powers and consequently of a just public opinion, but especially through the sympathies being rendered more tender and widely diffused through the effects of habit, example, instruction, and reflection. It is not improbable that virtuous tendencies may through long practice be inherited. With the more civilised races, the conviction of the existence of an all-seeing Deity has had a potent influence on the advancement of morality. Ultimately man no longer accepts the praise or blame of his fellows as his chief guide, though few escape this influence, but his habitual convictions controlled by reason afford him the safest rule. His conscience then becomes his supreme judge and monitor. Nevertheless the first foundation or origin of the moral sense lies in the social instincts, including sympathy; and these instincts no doubt were primarily gained, as in the case of the lower animals, through natural selection.

THE BELIEF IN GOD has often been advanced as not only the greatest, but the most complete of all the distinctions between man and the lower animals. It is however impossible, as we have seen, to maintain that this belief is innate or instinctive in man. On the other hand a belief in all-pervading spiritual agencies seems to be universal; and apparently follows from a considerable advance in the reasoning powers of man, and from a still greater advance in his faculties of imagination, curiosity and wonder. I am aware that the assumed instinctive belief in God has been used by many persons as an argument for His existence. But this is a rash argument, as we should thus be compelled to believe in the existence of many cruel and malignant spirits, possessing only a little more power than man; for the belief in them is far more general than of a beneficent Deity. The idea of a universal and beneficent Creator of the universe does not seem to arise in the mind of man, until he has been elevated by long-continued culture.

He who believes in the advancement of man from some lowly-organised form, will naturally ask how does this bear on the belief in the immortality of the soul. The barbarous races of man, as Sir J. Lubbock has shewn, possess no clear belief of this kind; but arguments derived from the primeval beliefs of savages are, as we have just seen, of little or no avail. Few persons feel any anxiety from the impossibility of determining at what precise period in the development of the individual, from the first trace of the minute germinal vesicle to the child either before or after birth, man becomes an immortal being; and there is no greater cause for anxiety because the period in the gradually ascending organic scale cannot possibly be determined.[1007]

I am aware that the conclusions arrived at in this work will be denounced by some as highly irreligious; but he who thus denounces them is bound to shew why it is more irreligious to explain the origin of man as a distinct species by descent from some lower form, through the laws of variation and natural selection, than to explain the birth of the individual through the laws of ordinary reproduction. The birth both of the species and of the individual are equally parts of that grand sequence of events, which our minds refuse to accept as the result of blind chance. The understanding revolts at such a conclusion, whether or not we are able to believe that every slight variation of structure,—the union of each pair in marriage,—the dissemination of each seed,—and other such events, have all been ordained for some special purpose.

Sexual selection has been treated at great length in these volumes; for, as I have attempted to shew, it has played an important part in the history of the organic world. As summaries have been given to each chapter, it would be superfluous here to add a detailed summary. I am aware that much remains doubtful, but I have endeavoured to give a fair view of the whole case. In the lower divisions of the animal kingdom, sexual selection seems to have done nothing: such animals are often affixed for life to the same spot, or have the two sexes combined in the same individual, or what is still more important, their perceptive and intellectual faculties are not sufficiently advanced to allow of the feelings of love and jealousy, or of the exertion of choice. When, however, we come to the Arthropoda and Vertebrata, even to the lowest classes in these two great Sub-Kingdoms, sexual selection has effected much; and it deserves notice that we here find the intellectual faculties developed, but in two very distinct lines, to the highest standard, namely in the Hymenoptera (ants, bees, &c.) amongst the Arthropoda, and in the Mammalia, including man, amongst the Vertebrata.

In the most distinct classes of the animal kingdom, with mammals, birds, reptiles, fishes, insects, and even crustaceans, the differences between the sexes follow almost exactly the same rules. The males are almost always the wooers; and they alone are armed with special weapons for fighting with their rivals. They are generally stronger and larger than the females, and are endowed with the requisite qualities of courage and pugnacity. They are provided, either exclusively or in a much higher degree than the females, with organs for producing vocal or instrumental music, and with odoriferous glands. They are ornamented with infinitely diversified appendages, and with the most brilliant or conspicuous colours, often arranged in elegant patterns, whilst the females are left unadorned. When the sexes differ in more important structures, it is the male which is provided with special sense-organs for discovering the female, with locomotive organs for reaching her, and often with prehensile organs for holding her. These various structures for securing or charming the female are often developed in the male during only part of the year, namely the breeding-season. They have in many cases been transferred in a greater or less degree to the females; and in the latter case they appear in her as mere rudiments. They are lost by the males after emasculation. Generally they are not developed in the male during early youth, but appear a short time before the age for reproduction. Hence in most cases the young of both sexes resemble each other; and the female resembles her young offspring throughout life. In almost every great class a few anomalous cases occur in which there has been an almost complete transposition of the characters proper to the two sexes; the females assuming characters which properly belong to the males. This surprising uniformity in the laws regulating the differences between the sexes in so many and such widely separated classes, is intelligible if we admit the action throughout all the higher divisions of the animal kingdom of one common cause, namely sexual selection.

Sexual selection depends on the success of certain individuals over others of the same sex in relation to the propagation of the species; whilst natural selection depends on the success of both sexes, at all ages, in relation to the general conditions of life. The sexual struggle is of two kinds; in the one it is between the individuals of the same sex, generally the male sex, in order to drive away or kill their rivals, the females remaining passive; whilst in the other, the struggle is likewise between the individuals of the same sex, in order to excite or charm those of the opposite sex, generally the females, which no longer remain passive, but select the more agreeable partners. This latter kind of selection is closely analogous to that which man unintentionally, yet effectually, brings to bear on his domesticated productions, when he continues for a long time choosing the most pleasing or useful individuals, without any wish to modify the breed.

The laws of inheritance determine whether characters gained through sexual selection by either sex shall be transmitted to the same sex, or to both sexes; as well as the age at which they shall be developed. It appears that variations which arise late in life are commonly transmitted to one and the same sex. Variability is the necessary basis for the action of selection, and is wholly independent of it. It follows from this, that variations of the same general nature have often been taken advantage of and accumulated through sexual selection in relation to the propagation of the species, and through natural selection in relation to the general purposes of life. Hence secondary sexual characters, when equally transmitted to both sexes can be distinguished from ordinary specific characters only by the light of analogy. The modifications acquired through sexual selection are often so strongly pronounced that the two sexes have frequently been ranked as distinct species, or even as distinct genera. Such strongly-marked differences must be in some manner highly important; and we know that they have been acquired in some instances at the cost not only of inconvenience, but of exposure to actual danger.

The belief in the power of sexual selection rests chiefly on the following considerations. The characters which we have the best reason for supposing to have been thus acquired are confined to one sex; and this alone renders it probable that they are in some way connected with the act of reproduction. These characters in innumerable instances are fully developed only at maturity; and often during only a part of the year, which is always the breeding-season. The males (passing over a few exceptional cases) are the most active in courtship; they are the best armed, and are rendered the most attractive in various ways. It is to be especially observed that the males display their attractions with elaborate care in the presence of the females; and that they rarely or never display them excepting during the season of love. It is incredible that all this display should be purposeless. Lastly we have distinct evidence with some quadrupeds and birds that the individuals of the one sex are capable of feeling a strong antipathy or preference for certain individuals of the opposite sex.

Bearing these facts in mind, and not forgetting the marked results of man's unconscious selection, it seems to me almost certain that if the individuals of one sex were during a long series of generations to prefer pairing with certain individuals of the other sex, characterised in some peculiar manner, the offspring would slowly but surely become modified in this same manner. I have not attempted to conceal that, excepting when the males are more numerous than the females, or when polygamy prevails, it is doubtful how the more attractive males succeed in leaving a larger number of offspring to inherit their superiority in ornaments or other charms than the less attractive males; but I have shewn that this would probably follow from the females,—especially the more vigorous females which would be the first to breed, preferring not only the more attractive but at the same time the more vigorous and victorious males.

Although we have some positive evidence that birds appreciate bright and beautiful objects, as with the Bower-birds of Australia, and although they certainly appreciate the power of song, yet I fully admit that it is an astonishing fact that the females of many birds and some mammals should be endowed with sufficient taste for what has apparently been effected through sexual selection; and this is even more astonishing in the case of reptiles, fish, and insects. But we really know very little about the minds of the lower animals. It cannot be supposed that male Birds of Paradise or Peacocks, for instance, should take so much pains in erecting, spreading, and vibrating their beautiful plumes before the females for no purpose. We should remember the fact given on excellent authority in a former chapter, namely that several peahens, when debarred from an admired male, remained widows during a whole season rather than pair with another bird.

Nevertheless I know of no fact in natural history more wonderful than that the female Argus pheasant should be able to appreciate the exquisite shading of the ball-and-socket ornaments and the elegant patterns on the wing-feathers of the male. He who thinks that the male was created as he now exists must admit that the great plumes, which prevent the wings from being used for flight, and which, as well as the primary feathers, are displayed in a manner quite peculiar to this one species during the act of courtship, and at no other time, were given to him as an ornament. If so, he must likewise admit that the female was created and endowed with the capacity of appreciating such ornaments. I differ only in the conviction that the male Argus pheasant acquired his beauty gradually, through the females having preferred during many generations the more highly ornamented males; the æsthetic capacity of the females having been advanced through exercise or habit in the same manner as our own taste is gradually improved. In the male, through the fortunate chance of a few feathers not having been modified, we can distinctly see how simple spots with a little fulvous shading on one side might have been developed by small and graduated steps into the wonderful ball-and-socket ornaments; and it is probable that they were actually thus developed.

Every one who admits the principle of evolution, and yet feels great difficulty in admitting that female mammals, birds, reptiles, and fish, could have acquired the high standard of taste which is implied by the beauty of the males, and which generally coincides with our own standard, should reflect that in each member of the vertebrate series the nerve-cells of the brain are the direct offshoots of those possessed by the common progenitor of the whole group. It thus becomes intelligible that the brain and mental faculties should be capable under similar conditions of nearly the same course of development, and consequently of performing nearly the same functions.

The reader who has taken the trouble to go through the several chapters devoted to sexual selection, will be able to judge how far the conclusions at which I have arrived are supported by sufficient evidence. If he accepts these conclusions, he may, I think, safely extend them to mankind; but it would be superfluous here to repeat what I have so lately said on the manner in which sexual selection has apparently acted on both the male and female side, causing the two sexes of man to differ in body and mind, and the several races to differ from each other in various characters, as well as from their ancient and lowly-organised progenitors.

He who admits the principle of sexual selection will be led to the remarkable conclusion that the cerebral system not only regulates most of the existing functions of the body, but has indirectly influenced the progressive development of various bodily structures and of certain mental qualities. Courage, pugnacity, perseverance, strength and size of body, weapons of all kinds, musical organs, both vocal and instrumental, bright colours, stripes and marks, and ornamental appendages, have all been indirectly gained by the one sex or the other, through the influence of love and jealousy, through the appreciation of the beautiful in sound, colour or form, and through the exertion of a choice; and these powers of the mind manifestly depend on the development of the cerebral system.

MAN SCANS with scrupulous care the character and pedigree of his horses, cattle, and dogs before he matches them; but when he comes to his own marriage he rarely, or never, takes any such care. He is impelled by nearly the same motives as are the lower animals when left to their own free choice, though he is in so far superior to them that he highly values mental charms and virtues. On the other hand he is strongly attracted by mere wealth or rank. Yet he might by selection do something not only for the bodily constitution and frame of his offspring, but for their intellectual and moral qualities. Both sexes ought to refrain from marriage if in any marked degree inferior in body or mind; but such hopes are Utopian and will never be even partially realised until the laws of inheritance are thoroughly known. All do good service who aid towards this end. When the principles of breeding and of inheritance are better understood, we shall not hear ignorant members of our legislature rejecting with scorn a plan for ascertaining by an easy method whether or not consanguineous marriages are injurious to man.

The advancement of the welfare of mankind is a most intricate problem: all ought to refrain from marriage who cannot avoid abject poverty for their children; for poverty is not only a great evil, but tends to its own increase by leading to recklessness in marriage. On the other hand, as Mr. Galton has remarked, if the prudent avoid marriage, whilst the reckless marry, the inferior members will tend to supplant the better members of society. Man, like every other animal, has no doubt advanced to his present high condition through a struggle for existence consequent on his rapid multiplication; and if he is to advance still higher he must remain subject to a severe struggle. Otherwise he would soon sink into indolence, and the more highly-gifted men would not be more successful in the battle of life than the less gifted. Hence our natural rate of increase, though leading to many and obvious evils, must not be greatly diminished by any means. There should be open competition for all men; and the most able should not be prevented by laws or customs from succeeding best and rearing the largest number of offspring. Important as the struggle for existence has been and even still is, yet as far as the highest part of man's nature is concerned there are other agencies more important. For the moral qualities are advanced, either directly or indirectly, much more through the effects of habit, the reasoning powers, instruction, religion, &c., than through natural selection; though to this latter agency the social instincts, which afforded the basis for the development of the moral sense, may be safely attributed.

The main conclusion arrived at in this work, namely that man is descended from some lowly-organised form, will, I regret to think, be highly distasteful to many persons. But there can hardly be a doubt that we are descended from barbarians. The astonishment which I felt on first seeing a party of Fuegians on a wild and broken shore will never be forgotten by me, for the reflection at once rushed into my mind—such were our ancestors. These men were absolutely naked and bedaubed with paint, their long hair was tangled, their mouths frothed with excitement, and their expression was wild, startled, and distrustful. They possessed hardly any arts, and like wild animals lived on what they could catch; they had no government, and were merciless to every one not of their own small tribe. He who has seen a savage in his native land will not feel much shame, if forced to acknowledge that the blood of some more humble creature flows in his veins. For my own part I would as soon be descended from that heroic little monkey, who braved his dreaded enemy in order to save the life of his keeper; or from that old baboon, who, descending from the mountains, carried away in triumph his young comrade from a crowd of astonished dogs—as from a savage who delights to torture his enemies, offers up bloody sacrifices, practises infanticide without remorse, treats his wives like slaves, knows no decency, and is haunted by the grossest superstitions.

Man may be excused for feeling some pride at having risen, though not through his own exertions, to the very summit of the organic scale; and the fact of his having thus risen, instead of having been aboriginally placed there, may give him hopes for a still higher destiny in the distant future. But we are not here concerned with hopes or fears, only with the truth as far as our reason allows us to discover it. I have given the evidence to the best of my ability; and we must acknowledge, as it seems to me, that man with all his noble qualities, with sympathy which feels for the most debased, with benevolence which extends not only to other men but to the humblest living creature, with his god-like intellect which has penetrated into the movements and constitution of the solar system—with all these exalted powers—Man still bears in his bodily frame the indelible stamp of his lowly origin.

Charles Darwin age 72 (1881)

The Expression of the Emotions in Man and Animals

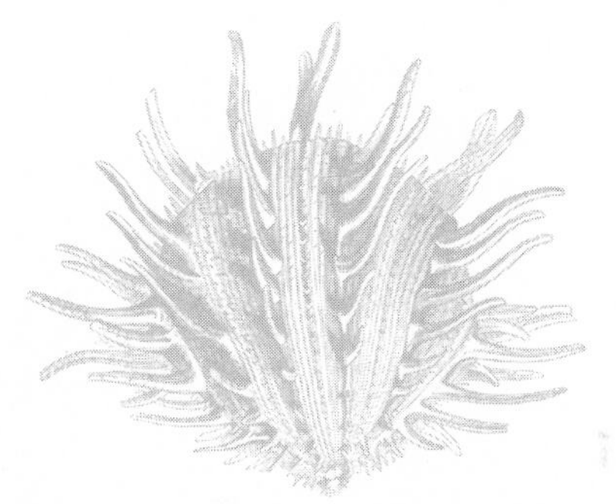

Introduction to *The Expression of the Emotions in Man and Animals*

The Expression of the Emotions in Man and Animals is both an old-fashioned descriptive treatise and the most modern of Darwin's major works. On the one hand it provides a straightforward catalog of human postures and facial expressions that might have been written in past centuries. In fact, Darwin draws effective examples from Homer and Shakespeare. This strongly empirical approach has considerable merit. So methodical and detailed are its descriptions, and so searching are the mental states Darwin imputes to them, that *The Expression of the Emotions* could serve as a guidebook for novelists—and might well have done so across the years.

On the other hand, this book is rich and accurate enough in interpretation to have served as part of the foundation of modern psychology. Darwin is, in Konrad Lorenz's words, the "patron saint" of that discipline. His pioneering contribution was to treat emotions and their manner of expression as products of evolution. Facial expressions and body postures, he argues in *The Expression of the Emotions*, are communication signals that adapt individuals for life within the complexity of social existence. The expressions of emotion, in short, are instincts, and as such they have evolved by natural selection in essentially the same manner as traits of anatomy and physiology have evolved.

The modernity of Darwin's approach goes beyond simple observation and phylogenetic deduction. What other approaches, he asks, might be taken to understand the human repertory? One is the emotional behavior of children. The appearance of particular expressions at a very early age suggests the unfolding of instinct. The insane are another source of information. Released from ordinary rational restraints, Darwin reasons, the emotions of psychotics are more openly expressed, extreme in form, and hence easily read.

Finally, these are inferences based on what today is called ethnography: if expressions and their emotional content are truly inborn, they should display consistency around the world, no matter how remote and divergent are societies in other respects. To test this hypothesis, Darwin consulted travelers who knew distant cultures well, from the Fuegians, whom he himself had observed during his visit to the tip of South America, to other aboriginal tribes in southern Africa, India, and Australia. The data he acquired from these sources, while inexact to say the least, provided favorable evidence for his thesis and established a method that is prominent in anthropology today.

Nothing if not exhaustive in his scholarship, Darwin delved yet deeper into his subject with a review of the muscles of the face that create expressions. He also conducted experiments, by presenting mirrors to monkeys to observe their startled response, and by directing his own children to raise their faces to the sky so he could record the muscles activated.

Through it all, Darwin remained the uncompromising evolutionist. It was a radical and risky position in his day, but he was right to adopt it. If natural selection is the dominant force of evolution, then biologists must consider its consequences for all things biological. Present-day biologists, as noted in my introductory essay, use two very different but complementary explanations of causation. The first is *how* a phenomenon is produced, that is, its "proximate" causes. In the case of communicative postures and facial expressions, the proximate causes are collectively the stimuli that trigger them, the sensory and nervous circuits that mediate them, and the muscles that are activated to form them. The other explanation addresses the "ultimate" causes of *why,* as a product of evolution, the behavior exists in the first place: which genes mutated to make the phenomenon possible, the social or physical environment that rendered its use profitable in survival and reproduction over many generations, and the consequent natural selection that allowed the prescribing genes to prevail in the population. No phenomenon in biology, from the genetic code on upward to the assembly of ecosystems, makes complete sense until both causes, proximate and ultimate, are identified and fitted together.

This conception of duality in biology, born of Darwin's thought from *On the Origin of Species* and continued in *The Expression of the Emotions,* led in the twentieth century to the development of new subdisciplines in the behavioral sciences. They include ethology, the study of the whole patterns of behavior under natural conditions that allow animals to express their complete hereditary range of adaptive behavior; sociobiology, the analysis of the biological basis of all forms of social behavior; and behavioral ecology, the study of the means by which behavior adapts animals to the demands and opportunities of their particular environments.

The Expression of the Emotions in Man and Animals

Charles Darwin

· 1872 ·

MANY WORKS HAVE BEEN WRITTEN on Expression, but a greater number on Physiognomy,—that is, on the recognition of character through the study of the permanent form of the features. With this latter subject I am not here concerned. The older treatises,[1] which I have consulted, have been of little or no service to me. The famous 'Conférences'[2] of the painter Le Brun, published in 1667, is the best known ancient work, and contains some good remarks. Another somewhat old essay, namely, the 'Discours,' delivered 1774–1782, by the well-known Dutch anatomist Camper,[3] can hardly be considered as having made any marked advance in the subject. The following works, on the contrary, deserve the fullest consideration.

Sir Charles Bell, so illustrious for his discoveries in physiology, published in 1806 the first edition, and in 1844 the third edition of his 'Anatomy and Philosophy of Expression.'[4] He may with justice be said, not only to have laid the foundations of the subject as a branch of science, but to have built up a noble structure. His work is in every way deeply interesting; it includes graphic descriptions of the various emotions, and is admirably illustrated. It is generally admitted that his service consists chiefly in having shown the intimate relation which exists between the movements of expression and those of respiration. One of the most important points, small as it may at first appear, is that the muscles round the eyes are involuntarily contracted during violent expiratory efforts, in order to protect these delicate organs from the pressure of the blood. This fact, which has been fully investigated for me with the greatest kindness by Professor Donders of Utrecht, throws, as we shall hereafter see, a flood of light on several of the most important expressions of the human countenance. The merits of Sir C. Bell's work have been undervalued or quite ignored by several foreign writers, but have been fully admitted by some, for instance by M. Lemoine,[5] who with great justice says:—"Le livre de Ch. Bell devrait être médité par quiconque essaye de faire parler le visage de l'homme, par les philosophes aussi bien que par les artistes, car, sous une apparence plus légère et sous le prétexte de l'esthétique, c'est un des plus beaux monuments de la science des rapports du physique et du moral."

From reasons which will presently be assigned, Sir C. Bell did not attempt to follow out his views as far as they might have been carried. He does not try to explain why different muscles are brought into action under different emotions; why, for instance, the inner ends of the eyebrows are raised, and the corners of the mouth depressed, by a person suffering from grief or anxiety.

In 1807 M. Moreau edited an edition of Lavater on Physiognomy,[6] in which he incorporated several of his own essays, containing excellent descriptions of the movements of the facial muscles, together with many valuable remarks. He throws, however, very little light on the philosophy of the subject. For instance, M. Moreau, in speaking of the act of frowning, that is, of the contraction of the muscle called by Freneh writers the *sourcilier (corrugator supercilii),* remarks with truth:—"Cette action des sourciliers est un des symptômes les plus tranchés de l'expression des affections pénibles ou concentrées." He then adds that these muscles, from their attachment and position, are fitted "à resserrer, à concentrer les principaux traits de la *face,* comme il convient dans toutes ces passions vraiment oppressives ou profondes, dans ces affections dont le sentiment semble porter l'organisation à revenir sur elle-même, à se contracter et à *s'amoindrir,* comme pour offrir moins de prise et de surface à des impressions redoutables ou importunes." He who thinks that remarks of this kind throw any light on the meaning or origin of the different expressions, takes a very different view of the subject to what I do.

In the above passage there is but a slight, if any, advance in the philosophy of the subject, beyond that reached by the painter Le Brun, who, in 1667, in describing the expression of fright, says:—"Le sourcil qui est abaissé d'un côté et élevé de l'autre, fait voir que la partie élevée semble le vouloir joindre au cerveau pour le garantir du mal que l'âme aperçoit, et le côté qui est abaissé et qui paraît enflé, nous fait trouver dans cet état par les esprits qui viennent du cerveau en abondance, comme pour couvrir l'âme et la défendre du mal qu'elle craint; la bouche fort ouverte fait voir le saisissement du cœur, par le sang qui se retire vers lui, ce qui l'oblige, voulant respirer, à faire un effort qui est cause que la bouche s'ouvre extrêmement, et qui, lorsqu'il passe par les organes de la voix, forme un son qui n'est point articulé; que si les muscles et les veines paraissent enflés, ce n'est que par les esprits que le cerveau envoie en ces parties-là." I have thought the foregoing sentences worth quoting, as specimens of the surprising nonsense which has been written on the subject.

'The Physiology or Mechanism of Blushing,' by Dr. Burgess, appeared in 1839, and to this work I shall frequently refer in my thirteenth Chapter.

In 1862 Dr. Duchenne published two editions, in folio and octavo, of his 'Mécanisme de la Physionomie Humaine,' in which he analyses by means of electricity, and illustrates by magnificent photographs, the movements of the facial muscles. He has generously permitted me to copy as many of his photographs as I desired. His works have been spoken lightly of, or quite passed over, by some of his countrymen. It is possible that Dr. Duchenne may have exaggerated the importance of the contraction of single muscles in giving expression; for, owing to the intimate manner in which the muscles are connected, as may be seen in Henle's anatomical drawings[7]—the best I believe ever published—it is difficult to believe in their separate action. Nevertheless, it is manifest that Dr. Duchenne clearly apprehended this and other sources of error, and as it is known that he was eminently successful in elucidating the physiology of the muscles of the hand by the aid of electricity, it is probable that he is generally in the right about the muscles of the face. In my opinion, Dr. Duchenne has greatly advanced the subject by his treatment of it. No one has more carefully studied the contraction of each separate muscle, and the consequent furrows produced on the skin. He has also, and this is a very important service, shown which muscles are least under the separate control of the will. He enters very little into theoretical considerations, and seldom attempts to explain why certain muscles and not others contract under the influence of certain emotions.

A distinguished French anatomist, Pierre Gratiolet, gave a course of lectures on Expression at the Sorbonne, and his notes were published (1865) after his death, under the title of 'De la Physionomie et des Mouvements d'Expression.' This is a very interesting work, full of valuable observations. His theory is rather complex, and, as far as it can be given in a single sentence (p. 65), is as follows:—"Il résulte, de tous les faits que j'ai rappelés, que les sens, l'imagination et la pensée elle-même, si élevée, si abstraite qu'on la suppose, ne peuvent s'exercer sans éveiller un sentiment corrélatif, et que ce sentiment se traduit directement, sympathiquement, symboliquement ou métaphoriquement, dans toutes les sphères des organes extérieurs, qui le racontent tous, suivant leur mode d'action propre, comme si chacun d'eux avait été directement affecté."

Gratiolet appears to overlook inherited habit, and even to some extent habit in the individual; and therefore he fails, as it seems to me, to give the right explanation, or any explanation at all, of many gestures and expressions. As an illustration of what he calls symbolic movements, I will quote his remarks (p. 37), taken from M. Chevreul, on a man playing at billiards. "Si une bille dévie légèrement de la direction que le joueur prétend lui imprimer, ne l'avez-vous pas vu cent fois la pousser du regard, de la tête et même des épaules, comme si ces mouvements, purement symboliques, pouvaient rectifier son trajet? Des mouvements non moins significatifs se produisent quand la bille manque d'une impulsion suffisante. Et, chez les joueurs novices, ils sont quelquefois accusés au point d'éveiller le sourire sur les lèvres des spectateurs." Such movements, as it appears to me, may be attributed simply to habit. As often as a man has wished to move an object to one side, he has always pushed it to that side; when forwards, he has pushed it forwards; and if he has wished to arrest it, he has pulled backwards. Therefore, when a man sees his ball travelling in a wrong direction, and he intensely wishes it to go in another direction, he cannot avoid, from long habit, unconsciously performing movements which in other cases he has found effectual.

As an instance of sympathetic movements Gratiolet gives (p. 212) the following case:— "un jeune chien à oreilles droites, auquel son maître présente de loin quelque viande appétissante, fixe avec ardeur ses yeux sur cet objet dont il suit tous les mouvements, et pendant que les yeux regardent, les deux oreilles se portent en avant comme si cet objet pouvait être entendu." Here, instead of speaking of sympathy between the ears and eyes, it appears to me more simple to believe, that as dogs during many generations have, whilst intently looking at any object, pricked their ears in order to perceive any sound; and conversely have looked intently in the direction of a sound to which they may have listened, the movements of these organs have become firmly associated together through long-continued habit.

Dr. Piderit published in 1859 an essay on Expression, which I have not seen, but in which, as he states, he forestalled Gratiolet in many of his views. In 1867 he published his 'Wissenschaftliches System der Mimik und Physiognomik.' It is hardly possible to give in a few sentences a fair notion of his views; perhaps the two following sentences will tell as much as can be briefly told: "the muscular movements of expression are in part related to imaginary objects, and in part to imaginary sensorial impressions. In this proposition lies the key to the comprehension of all expressive muscular movements." (s. 25.) Again, "Expressive movements manifest themselves chiefly in the numerous and mobile muscles of the face, partly because the nerves by which they are set into motion originate in the most immediate vicinity of the mind-organ, but partly also because these muscles serve to support the organs of sense." (s. 26.) If Dr. Piderit had studied Sir C. Bell's work, he would probably not have said (s. 101) that violent laughter causes a frown from partaking of the nature of pain; or that with infants (s. 103) the tears irritate the eyes, and thus excite the contraction of the surrounding muscles. Many good remarks are scattered throughout this volume, to which I shall hereafter refer.

Short discussions on Expression may be found in various works, which need not here be particularised. Mr. Bain, however, in two of his works has treated the subject at some length. He says,[8] "I look upon the expression so-called as part and parcel of the feeling. I believe it to be a general law of the mind that, along with the fact of inward feeling or consciousness, there is a diffusive action or excitement over the bodily members." In another place he adds, "A very considerable number of the facts may be brought under the following principle: namely, that states of pleasure are connected with an increase, and states of pain with an abatement, of some, or all, of the vital functions." But the above law of the diffusive action of feelings seems too general to throw much light on special expressions.

Mr. Herbert Spencer, in treating of the Feelings in his 'Principles of Psychology' (1855), makes the following remarks:—"Fear, when strong, expresses itself in cries, in efforts to hide or escape, in palpitations and tremblings; and these are just the manifestations that would accompany an actual experience of the evil feared. The destructive passions are shown in a general tension of the muscular system, in gnashing of the teeth and protrusion of the claws, in dilated eyes and nostrils, in growls; and these are weaker forms of the actions that accompany the killing of prey." Here we have, as I believe, the true theory of a large number of expressions; but the chief interest and difficulty of the subject lies in following out the wonderfully complex results. I infer that some one (but who he is I have not been able to ascertain) formerly advanced a nearly similar view, for Sir C. Bell says,[9] "It has been maintained that what are called the external signs of passion, are only the concomitants of those voluntary movements which the structure renders necessary." Mr. Spencer has also published[10] a valuable essay on the physiology of Laughter, in which he insists on "the general law that feeling passing a certain pitch, habitually vents itself in bodily action;" and that "an overflow of nerve-force undirected by any motive, will manifestly take first the most habitual routes; and if these do not suffice, will next overflow into the less habitual ones." This law I believe to be of the highest importance in throwing light on our subject.[11]

All the authors who have written on Expression, with the exception of Mr. Spencer—the great expounder of the principle of Evolution—appear to have been firmly convinced that species, man of course included, came into existence in their present condition. Sir C. Bell, being thus convinced, maintains that many of our facial muscles are "purely instrumental in expression;" or are "a special provision" for this sole object.[12] But the simple fact that the anthropoid apes possess the same facial muscles as we do,[13] renders it very improbable that these muscles in our case serve exclusively for expression; for no one, I presume, would be inclined to admit that monkeys have been endowed with special muscles solely for exhibiting their hideous grimaces. Distinct uses, independently of expression, can indeed be assigned with much probability for almost all the facial muscles.

Sir C. Bell evidently wished to draw as broad a distinction as possible between man and the lower animals; and he consequently asserts that with "the lower creatures there is no expression but what may be referred, more or less plainly, to their acts of volition or necessary instincts." He further maintains that their faces "seem chiefly capable of expressing rage and fear."[14] But man himself cannot express love and humility by external signs, so plainly as does a dog, when with drooping ears, hanging lips, flexuous body, and wagging tail, he meets his beloved master. Nor can these movements in the dog be explained by acts of volition or necessary instincts, any more than the beaming eyes and smiling cheeks of a man when he meets an old friend. If Sir C. Bell had been questioned about the expression of affection in the dog, he would no doubt have answered that this animal had been created with special instincts, adapting him for association with man, and that all further enquiry on the subject was superfluous.

Although Gratiolet emphatically denies[15] that any muscle has been developed solely for the sake of expression, he seems never to have reflected on the principle of evolution. He apparently looks at each species as a separate creation. So it is with the other writers on Expression. For instance, Dr. Duchenne, after speaking of the movements of the limbs, refers to those which give expression to the face, and remarks:[16] "Le créateur n'a donc pas eu à se préoccuper ici des besoins de la mécanique; il a pu, selon sa sagesse, ou—que l'on me pardonne cette manière de parler—par une divine fantaisie, mettre en action tel ou tel muscle, un seul ou plusieurs muscles à la fois, lorsqu'il a voulu que les signes caractéristiques des passions, même les plus fugaces, fussent écrits passagèrement sur la face de l'homme. Ce langage de la physionomie une fois créé, il lui a suffi, pour le rendre universel et immuable, de donner à tout être humain la faculté instinctive d'exprimer toujours ses sentiments par la contraction des mêmes muscles."

Many writers consider the whole subject of Expression as inexplicable. Thus the illustrious physiologist Müller says,[17] "The completely different expression of the features in different passions shows that, according to the kind of feeling excited, entirely different groups of the fibres of the facial nerve are acted on. Of the cause of this we are quite ignorant."

No doubt as long as man and all other animals are viewed as independent creations, an effectual stop is put to our natural desire to investigate as far as possible the causes of Expression. By this doctrine, anything and everything can be equally well explained; and it has proved as pernicious with respect to Expression as to every other branch of natural history. With mankind some expressions, such as the bristling of the hair under the influence of extreme terror, or the uncovering of the teeth under that of furious rage, can hardly be understood, except on the belief that man once existed in a much lower and animal-like condition. The community of certain expressions in distinct though allied species, as in the movements of the same facial muscles during laughter by man and by various monkeys, is rendered somewhat more intelligible, if we believe in their descent from a common progenitor. He who admits on general grounds that the structure and habits of all animals have been gradually evolved, will look at the whole subject of Expression in a new and interesting light.

The study of Expression is difficult, owing to the movements being often extremely slight, and of a fleeting nature. A difference may be clearly perceived, and yet it may be impossible, at least I have found it so, to state in what the difference consists. When we witness any deep emotion, our sympathy is so strongly excited, that close observation is forgotten or rendered almost impossible; of which fact I have had many curious proofs. Our imagination is another and still more serious source of error; for if from the nature of the circumstances we expect to see any expression, we readily imagine its presence. Notwithstanding Dr. Duchenne's great experience, he for a long time fancied, as he states, that several muscles contracted under certain emotions, whereas he ultimately convinced himself that the movement was confined to a single muscle.

In order to acquire as good a foundation as possible, and to ascertain, independently of common opinion, how far particular movements of the features and gestures are really expressive of certain states of the mind, I have found the following means the most serviceable. In the first place, to observe infants; for they exhibit many emotions, as Sir C. Bell remarks, "with extraordinary force;" whereas, in after life, some of our expressions "cease to have the pure and simple source from which they spring in infancy."[18]

In the second place, it occurred to me that the insane ought to be studied, as they are liable to the strongest passions, and give uncontrolled vent to them. I had, myself, no opportunity of doing this, so I applied to Dr. Maudsley, and received from him an introduction to Dr. J. Crichton Browne, who has charge of an immense asylum near Wakefield, and who, as I found, had already attended to the subject. This excellent observer has with unwearied kindness sent me copious notes and descriptions, with valuable suggestions on many points; and I can hardly over-estimate the value of his assistance. I owe also, to the kindness of Mr. Patrick Nicol, of the Sussex Lunatic Asylum, interesting statements on two or three points.

Thirdly, Dr. Duchenne galvanized, as we have already seen, certain muscles in the face of an old man, whose skin was little sensitive, and thus produced various expressions which were photographed on a large scale. It fortunately occurred to me to show several of the best plates, without a word of explanation, to above twenty educated persons of various ages and both sexes, asking them, in each case, by what emotion or feeling the old man was supposed to be agitated; and I recorded their answers in the words which they used. Several of the expressions were instantly recognised by almost everyone, though described in not exactly the same terms; and these may, I think, be relied on as truthful, and will hereafter be specified. On the other hand, the most widely different judgments were pronounced in regard to some of them. This exhibition was of use in another way, by convincing me how easily we may be misguided by our imagination; for when I first looked through Dr. Duchenne's photographs, reading at the same time the text, and thus learning what was intended, I was struck with admiration at the truthfulness of all, with only a few exceptions. Nevertheless, if I had examined them without any explanation, no doubt I should have been as much perplexed, in some cases, as other persons have been.

Fourthly, I had hoped to derive much aid from the great masters in painting and sculpture, who are such close observers. Accordingly, I have looked at photographs and engravings of many well-known works; but, with a few exceptions, have not thus profited. The reason no doubt is, that in works of art, beauty is the chief object; and strongly contracted facial muscles destroy beauty.[19] The story of the composition is generally told with wonderful force and truth by skilfully given accessories.

Fifthly, it seemed to me highly important to ascertain whether the same expressions and gestures prevail, as has often been asserted without much evidence, with all the races of mankind, especially with those who have associated but little with Europeans. Whenever the same movements of the features or body express the same emotions in several distinct races of man, we may infer with much probability, that such expressions are true ones,—that is, are innate or instinctive. Conventional expressions or gestures, acquired by the individual during early life, would probably have differed in the different races, in the same manner as do their languages. Accordingly I circulated, early in the year 1867, the following printed queries with a request, which has been fully responded to, that actual observations, and not memory, might be trusted. These queries were written after a considerable interval of time, during which my attention had been otherwise directed, and I can now see that they might have been greatly improved. To some of the later copies, I appended, in manuscript, a few additional remarks:—

(1.) Is astonishment expressed by the eyes and mouth being opened wide, and by the eyebrows being raised?

(2.) Does shame excite a blush when the colour of the skin allows it to be visible? and especially how low down the body does the blush extend?

(3.) When a man is indignant or defiant does he frown, hold his body and head erect, square his shoulders and clench his fists?

(4.) When considering deeply on any subject, or trying to understand any puzzle, does he frown, or wrinkle the skin beneath the lower eyelids?

(5.) When in low spirits, are the corners of the mouth depressed, and the inner corner of the eyebrows raised by that muscle which the French call the "Grief muscle"? The eyebrow in this state becomes slightly oblique, with a little swelling at the inner end; and the forehead is transversely wrinkled in the middle part, but not across the whole breadth, as when the eyebrows are raised in surprise.

(6.) When in good spirits do the eyes sparkle, with the skin a little wrinkled round and under them, and with the mouth a little drawn back at the corners?

(7.) When a man sneers or snarls at another, is the corner of the upper lip over the canine or eye tooth raised on the side facing the man whom he addresses?

(8.) Can a dogged or obstinate expression be recognized, which is chiefly shown by the mouth being firmly closed, a lowering brow and a slight frown?

(9.) Is contempt expressed by a slight protrusion of the lips and by turning up the nose, with a slight expiration?

(10.) Is disgust shown by the lower lip being turned down, the upper lip slightly raised, with a sudden expiration, something like incipient vomiting, or like something spit out of the mouth?

(11.) Is extreme fear expressed in the same general manner as with Europeans?

(12.) Is laughter ever carried to such an extreme as to bring tears into the eyes?

(13.) When a man wishes to show that he cannot prevent something being done, or cannot himself do something, does he shrug his shoulders, turn inwards his elbows, extend outwards his hands and open the palms; with the eyebrows raised?

(14.) Do the children when sulky, pout or greatly protrude the lips?

(15.) Can guilty, or sly, or jealous expressions be recognized? though I know not how these can be defined.

(16.) Is the head nodded vertically in affirmation, and shaken laterally in negation?

Observations on natives who have had little communication with Europeans would be of course the most valuable, though those made on any natives would be of much interest to me. General remarks on expression are of comparatively little value; and memory is so deceptive that I earnestly beg it may not be trusted. A definite description of the countenance under any emotion or frame of mind, with a statement of the circumstances under which it occurred, would possess much value.

To these queries I have received thirty-six answers from different observers, several of them missionaries or protectors of the aborigines, to all of whom I am deeply indebted for the great trouble which they have taken, and for the valuable aid thus received. I will specify their names, &c., towards the close of this chapter, so as not to interrupt my present remarks. The answers relate to several of the most distinct and savage races of man. In many instances, the circumstances have been recorded under which each expression was observed, and the expression itself described. In such cases, much confidence may be placed in the answers. When the answers have been simply yes or no, I have always received them with caution. It follows, from the information thus acquired, that the same state of mind is expressed throughout the world with remarkable uniformity; and this fact is in itself interesting, as evidence of the close similarity in bodily structure and mental disposition of all the races of mankind.

Sixthly, and lastly, I have attended, as closely as I could, to the expression of the several passions in some of the commoner animals; and this I believe to be of paramount importance, not of course for deciding how far in man certain expressions are characteristic of certain states of mind, but as affording the safest basis for generalisation on the causes, or origin, of the various movements of Expression. In observing animals, we are not so likely to be biassed by our imagination; and we may feel safe that their expressions are not conventional.

From the reasons above assigned, namely, the fleeting nature of some expressions (the changes in the features being often extremely slight); our sympathy being easily aroused when we behold any strong emotion, and our attention thus distracted; our imagination deceiving us, from knowing in a vague manner what to expect, though certainly few of us know what the exact changes in the countenance are; and lastly, even our long familiarity with the subject,—from all these causes combined, the observation of Expression is by no means easy, as many persons, whom I have asked to observe certain points, have soon discovered. Hence it is difficult to determine, with certainty, what are the movements of the features and of the body, which commonly characterize certain states of the mind. Nevertheless, some of the doubts and difficulties have, as I hope, been cleared away by the observation of infants,—of the insane,—of the different races of man,—of works of art,—and lastly, of the facial muscles under the action of galvanism, as effected by Dr. Duchenne.

But there remains the much greater difficulty of understanding the cause or origin of the several expressions, and of judging whether any theoretical explanation is trustworthy. Besides, judging as well as we can by our reason, without the aid of any rules, which of two or more explanations is the most satisfactory, or are quite unsatisfactory, I see only one way of testing our conclusions. This is to observe whether the same principle by which one expression can, as it appears, be explained, is applicable in other allied cases; and especially, whether the same general principles can be applied with satisfactory results, both to man and the lower animals. This latter method, I am inclined to think, is the most serviceable of all. The difficulty of judging of the truth of any theoretical explanation, and of testing it by some distinct line of investigation, is the great drawback to that interest which the study seems well fitted to excite.

Finally, with respect to my own observations, I may state that they were commenced in the year 1838; and, from that time to the present day, I have occasionally attended to the subject. At the above date, I was already inclined to believe in the principle of evolution, or of the derivation of species from other and lower forms. Consequently, when I read Sir C. Bell's great work, his view, that man had been created with certain muscles specially adapted for the expression of his feelings, struck me as unsatisfactory. It seemed probable that the habit of expressing our feelings by certain movements, though now rendered innate, had been in some manner gradually acquired. But to discover how such habits had been acquired was perplexing in no small degree. The whole subject had to be viewed under a new aspect, and each expression demanded a rational explanation. This belief led me to attempt the present work, however imperfectly it may have been executed.

I WILL NOW GIVE the names of the gentlemen to whom, as I have said, I am deeply indebted for information in regard to the expressions exhibited by various races of man, and I will specify some of the circumstances under which the observations were in each case made. Owing to the great kindness and powerful influence of Mr. Wilson, of Hayes Place, Kent, I have received from Australia no less than thirteen sets of answers to my queries. This has been particularly fortunate, as the Australian aborigines rank amongst the most distinct of all the races of man. It will be seen that the observations have been chiefly made in the south, in the outlying parts of the colony of Victoria; but some excellent answers have been received from the north.

Mr. Dyson Lacy has given me in detail some valuable observations, made several hundred miles in the interior of Queensland. To Mr. R. Brough Smyth, of Melbourne, I am much indebted for observations made by himself, and for sending me several of the following letters, namely:—From the Rev. Mr. Hagenauer, of Lake Wellington, a missionary in Gippsland, Victoria, who has had much experience with the natives. From Mr. Samuel Wilson, a landowner, residing at Langerenong, Wimmera, Victoria. From the Rev. George Taplin, superintendent of the native Industrial Settlement at Port

Macleay. From Mr. Archibald G. Lang, of Coranderik, Victoria, a teacher at a school where aborigines, old and young, are collected from all parts of the colony. From Mr. H. B. Lane, of Belfast, Victoria, a police magistrate and warden, whose observations, as I am assured, are highly trustworthy. From Mr. Templeton Bunnett, of Echuca, whose station is on the borders of the colony of Victoria, and who has thus been able to observe many aborigines who have had little intercourse with white men. He compared his observations with those made by two other gentlemen long resident in the neighbourhood. Also from Mr. J. Bulmer, a missionary in a remote part of Gippsland, Victoria.

I am also indebted to the distinguished botanist, Dr. Ferdinand Müller, of Victoria, for some observations made by himself, and for sending me others made by Mrs. Green, as well as for some of the foregoing letters.

In regard to the Maoris of New Zealand, the Rev. J. W. Stack has answered only a few of my queries; but the answers have been remarkably full, clear, and distinct, with the circumstances recorded under which the observations were made.

The Rajah Brooke has given me some information with respect to the Dyaks of Borneo.

Respecting the Malays, I have been highly successful; for Mr. F. Geach (to whom I was introduced by Mr. Wallace), during his residence as a mining engineer in the interior of Malacca, observed many natives, who had never before associated with white men. He wrote me two long letters with admirable and detailed observations on their expression. He likewise observed the Chinese immigrants in the Malay archipelago.

The well-known naturalist, H.M. Consul, Mr. Swinhoe, also observed for me the Chinese in their native country; and he made inquiries from others whom he could trust.

In India Mr. H. Erskine, whilst residing in his official capacity in the Ahmednugur District in the Bombay Presidency, attended to the expression of the inhabitants, but found much difficulty in arriving at any safe conclusions, owing to their habitual concealment of all emotions in the presence of Europeans. He also obtained information for me from Mr. West, the Judge in Canara, and he consulted some intelligent native gentlemen on certain points. In Calcutta Mr. J. Scott, curator of the Botanic Gardens, carefully observed the various tribes of men therein employed during a considerable period, and no one has sent me such full and valuable details. The habit of accurate observation, gained by his botanical studies, has been brought to bear on our present subject. For Ceylon I am much indebted to the Rev. S. O. Glenie for answers to some of my queries.

Turning to Africa, I have been unfortunate with respect to the negroes, though Mr. Winwood Reade aided me as far as lay in his power. It would have been comparatively easy to have obtained information in regard to the negro slaves in America; but as they have long associated with white men, such observations would have possessed little value. In the southern parts of the continent Mrs. Barber observed the Kafirs and Fingoes, and sent me many distinct answers. Mr. J. P. Mansel Weale also made some observations on the natives, and procured for me a curious document, namely, the opinion, written in English, of Christian Gaika, brother of the Chief Sandilli, on the expressions of his fellow-countrymen. In the northern regions of Africa Captain Speedy, who long resided with the Abyssinians, answered my queries partly from memory and partly from observations made on the son of King Theodore, who was then under his charge. Professor and Mrs. Asa Gray attended to some points in the expressions of the natives, as observed by them whilst ascending the Nile.

On the great American continent Mr. Bridges, a catechist residing with the Fuegians, answered some few questions about their expression, addressed to him many years ago. In the northern half of the continent Dr. Rothrock attended to the expressions of the wild Atnah and Espyox tribes on the Nasse River, in North-Western America. Mr. Washington Matthews, Assistant-Surgeon in the United States Army, also observed with special care (after having seen my queries, as printed in the 'Smithsonian Report') some of the wildest tribes in the Western parts of the United States, namely, the Tetons, Grosventres, Mandans, and Assinaboines; and his answers have proved of the highest value.

Lastly, besides these special sources of information, I have collected some few facts incidentally given in books of travels.

As I shall often have to refer, more especially in the latter part of this volume, to the muscles of the human face, I have had a diagram (fig. 1) copied and reduced from Sir C. Bell's work, and two others, with more accurate details (figs. 2 and 3), from Henle's well-known 'Handbuch der systematischen Anatomie des Menschen.' The same letters refer to the same muscles in all three figures, but the names are given of only the more important ones to which I shall have to allude. The facial muscles blend much together, and, as I am informed, hardly appear on a dissected face so distinct as they are here represented. Some writers consider that these muscles consist of nineteen pairs, with one unpaired;[20] but others make the number much larger, amounting even to fifty-five, according to Moreau. They are, as is admitted by everyone who has written on the subject, very variable in structure; and Moreau remarks that they are hardly alike in half-a-dozen subjects.[21] They are also variable in function. Thus the power of uncovering the canine tooth on one side differs much in different persons. The power of raising the wings of the nostrils is also, according to Dr. Piderit,[22] variable in a remarkable degree; and other such cases could be given.

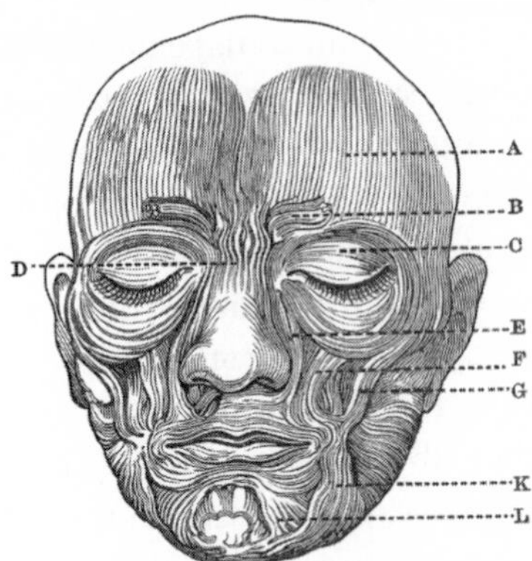

Fig. 1. *Diagram of the muscles of the face, from Sir C. Bell.*

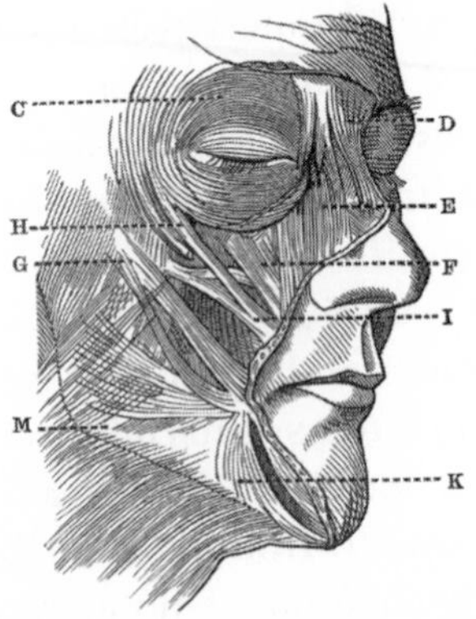

Fig. 2. *Diagram from Henle.*

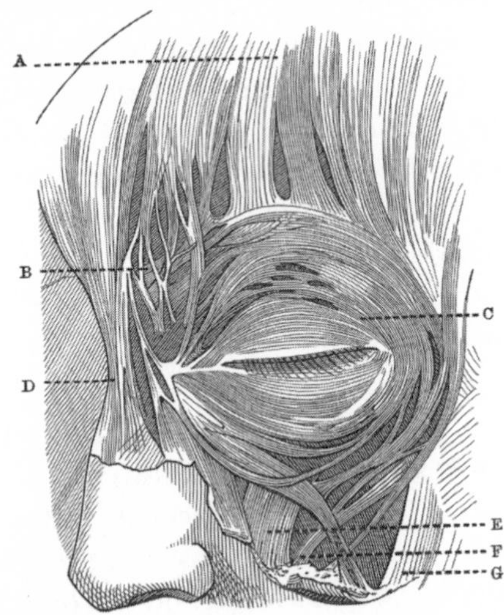

Fig. 3. *Diagram from Henle.*

A. *Occipito-frontalis, or frontal muscle.*
B. *Corrugator supercilii, or corrugator muscle.*
C. *Orbicularis palpebrarum, or orbicular muscles of the eyes.*
D. *Pyramidalis nasi, or pyramidal muscle of the nose.*
E. *Levator labii superioris alæque nasi.*
F. *Levator labii proprius.*
G. *Zygomatic.*
H. *Malaris.*
I. *Little zygomatic.*
K. *Triangularis oris, or depressor anguli oris.*
L. *Quadratus menti.*
M. *Risorius, part of the Platysma myoides.*

Finally, I must have the pleasure of expressing my obligations to Mr. Rejlander for the trouble which he has taken in photographing for me various expressions and gestures. I am also indebted to Herr Kindermann, of Hamburg, for the loan of some excellent negatives of crying infants; and to Dr. Wallich for a charming one of a smiling girl. I have already expressed my obligations to Dr. Duchenne for generously permitting me to have some of his large photographs copied and reduced. All these photographs have been printed by the Heliotype process, and the accuracy of the copy is thus guaranteed. These plates are referred to by Roman numerals.

I am also greatly indebted to Mr. T. W. Wood for the extreme pains which he has taken in drawing from life the expressions of various animals. A distinguished artist, Mr. Riviere, has had the kindness to give me two drawings of dogs—one in a hostile and the other in a humble and caressing frame of mind. Mr. A. May has also given me two similar sketches of dogs. Mr. Cooper has taken much care in cutting the blocks. Some of the photographs and drawings, namely, those by Mr. May, and those by Mr. Wolf of the Cynopithecus, were first reproduced by Mr. Cooper on wood by means of photography, and then engraved: by this means almost complete fidelity is ensured.

CHAPTER I

General Principles of Expression

THE THREE CHIEF PRINCIPLES STATED • THE FIRST PRINCIPLE • SERVICEABLE ACTIONS BECOME HABITUAL IN ASSOCIATION WITH CERTAIN STATES OF THE MIND, AND ARE PERFORMED WHETHER OR NOT OF SERVICE IN EACH PARTICULAR CASE • THE FORCE OF HABIT • INHERITANCE • ASSOCIATED HABITUAL MOVEMENTS IN MAN • REFLEX ACTIONS • PASSAGE OF HABITS INTO REFLEX ACTIONS • ASSOCIATED HABITUAL MOVEMENTS IN THE LOWER ANIMALS • CONCLUDING REMARKS

I WILL BEGIN by giving the three Principles, which appear to me to account for most of the expressions and gestures involuntarily used by man and the lower animals, under the influence of various emotions and sensations.[23] I arrived, however, at these three Principles only at the close of my observations. They will be discussed in the present and two following chapters in a general manner. Facts observed both with man and the lower animals will here be made use of; but the latter facts are preferable, as less likely to deceive us. In the fourth and fifth chapters, I will describe the special expressions of some of the lower animals; and in the succeeding chapters those of man. Everyone will thus be able to judge for himself, how far my three principles throw light on the theory of the subject. It appears to me that so many expressions are thus explained in a fairly satisfactory manner, that probably all will hereafter be found to come under the same or closely analogous heads. I need hardly premise that movements or changes in any part of the body,—as the wagging of a dog's tail, the drawing back of a horse's ears, the shrugging of a man's shoulders, or the dilatation of the capillary vessels of the skin,—may all equally well serve for expression. The three Principles are as follows.

I. The principle of serviceable associated Habits.—Certain complex actions are of direct or indirect service under certain states of the mind, in order to relieve or gratify certain sensations, desires, &c.; and whenever the same state of mind is induced, however feebly, there is a tendency through the force of habit and association for the same movements to be performed, though they may not then be of the least use. Some actions ordinarily associated through habit with certain states of the mind may be partially repressed through the will, and in such cases the muscles which are least under the separate control of the will are the most liable still to act, causing movements which we recognise as expressive. In certain other cases the checking of one habitual movement requires other slight movements; and these are likewise expressive.

II. The principle of Antithesis.—Certain states of the mind lead to certain habitual actions, which are of service, as under our first principle. Now when a directly opposite state of mind is induced, there is a strong and involuntary tendency to the performance of movements of a directly opposite nature, though these are of no use; and such movements are in some cases highly expressive.

III. The principle of actions due to the constitution of the Nervous System, independently from the first of the Will, and independently to a certain extent of Habit.—When the sensorium is strongly excited, nerve-force is generated in excess, and is transmitted in certain definite directions, depending on the connection of the nerve-cells, and partly on habit: or the supply of nerve-force may, as it appears, be interrupted. Effects are thus produced which we recognise as expressive. This third principle may, for the sake of brevity, be called that of the direct action of the nervous system.

WITH RESPECT TO our *first Principle,* it is notorious how powerful is the force of habit. The most complex and difficult movements can in time be performed without the least effort or consciousness. It is not positively known how it comes that habit is so efficient in facilitating complex movements; but physiologists admit[24] "that the conducting power of the nervous fibres increases with the frequency of their excitement." This applies to the nerves of motion and sensation, as well as to those connected with the act of thinking. That some physical change is produced in the nerve-cells or nerves which are habitually used can hardly be doubted, for otherwise it is impossible to understand how the tendency to certain acquired movements is inherited. That they are inherited we see with horses in certain transmitted paces, such as cantering and ambling, which are not natural to them,—in the pointing of young pointers and the setting of young setters—in the peculiar manner of flight of certain breeds of the pigeon, &c. We have analogous cases with mankind in the inheritance of tricks or unusual gestures, to which we shall presently recur. To those who admit the gradual evolution of species, a most striking instance of the perfection with which the most difficult consensual movements can be transmitted, is afforded by the humming-bird Sphinx-moth (*Macroglossa*); for this moth, shortly after its emergence from the cocoon, as shown by the bloom on its unruffled scales, may be seen poised stationary in the air, with its long hair-like proboscis uncurled and inserted into the minute orifices of flowers; and no one, I believe, has ever seen this moth learning to perform its difficult task, which requires such unerring aim.

When there exists an inherited or instinctive tendency to the performance of an action, or an inherited taste for certain kinds of food, some degree of habit in the individual is often or generally requisite. We find this in the paces of the horse, and to a certain extent in the pointing of dogs; although some young dogs point excellently the first time they are taken out, yet they often associate the proper inherited attitude with a wrong odour, and even with eyesight. I have heard it asserted that if a calf be allowed to suck its mother only once, it is much more difficult afterwards to rear it by hand.[25] Caterpillars which have been fed on the leaves of one kind of tree, have been known to perish from hunger rather than to eat the leaves of another tree, although this afforded them their proper food, under a state of nature;[26] and so it is in many other cases.

The power of Association is admitted by everyone. Mr. Bain remarks, that "actions, sensations, and states of feeling, occurring together or in close succession, tend to grow together, or cohere, in such a way that when any one of them is afterwards presented to the mind, the others are apt to be brought up in idea."[27] It is so important for our purpose fully to recognise that actions readily become associated with other actions and with various states of the mind, that I will give a good many instances, in the first place relating to man, and afterwards to the lower animals. Some of the instances are of a very trifling nature, but they are as good for our purpose as more important habits. It is known to everyone how difficult, or even impossible it is, without repeated trials, to move the limbs in certain opposed directions which have never been practised. Analogous cases occur with sensations, as in the common experiment of rolling a marble beneath the tips of two crossed fingers, when it feels exactly like two marbles. Everyone protects himself when falling to the ground by extending his arms, and as Professor Alison has remarked, few can resist acting thus, when voluntarily falling on a soft bed. A man when going out of doors puts on his gloves quite unconsciously; and this may seem an extremely simple operation, but he who has taught a child to put on gloves, knows that this is by no means the case.

When our minds are much affected, so are the movements of our bodies; but here another principle besides habit, namely the undirected overflow of nerve-force, partially comes into play. Norfolk, in speaking of Cardinal Wolsey, says—

> "Some strange commotion
> Is in his brain: he bites his lip and starts;
> Stops on a sudden, looks upon the ground,
> Then, lays his finger on his temple; straight,
> Springs out into fast gait; then, stops again,
> Strikes his breast hard; and anon, he casts
> His eye against the moon: in most strange postures
> We have seen him set himself."

Henry 111., act iii. sc. 2.

A vulgar man often scratches his head when perplexed in mind; and I believe that he acts thus from habit, as if he experienced a slightly uncomfortable bodily sensation, namely, the itching of his head, to which he is particularly liable, and which he thus relieves. Another man rubs his eyes when perplexed, or gives a little cough when embarrassed, acting in either case as if he felt a slightly uncomfortable sensation in his eyes or windpipe.[28]

From the continued use of the eyes, these organs are especially liable to be acted on through association under various states of the mind, although there is manifestly nothing to be seen. A man, as Gratiolet remarks, who vehemently rejects a proposition, will almost certainly shut his eyes or turn away his face; but if he accepts the proposition, he will nod his head in affirmation and open his eyes widely. The man acts in this latter case as if he clearly saw the thing, and in the former case as if he did not or would not see it. I have noticed that persons in describing a horrid sight often shut their eyes momentarily and firmly, or shake their heads, as if not to see or to drive away something disagreeable; and I have caught myself, when thinking in the dark of a horrid spectacle, closing my eyes firmly. In looking suddenly at any object, or in looking all around, everyone raises his eyebrows, so that the eyes may be quickly and widely opened; and Duchenne remarks that[29] a person in trying to remember something often raises his eyebrows, as if to see it. A Hindoo gentleman made exactly the same remark to Mr. Erskine in regard to his countrymen. I noticed a young lady earnestly trying to recollect a painter's name, and she first looked to one corner of the ceiling and then to the opposite corner, arching the one eyebrow on that side; although, of course, there was nothing to be seen there.

In most of the foregoing cases, we can understand how the associated movements were acquired through habit; but with some individuals, certain strange gestures or tricks have arisen in association with certain states of the mind, owing to wholly inexplicable causes, and are undoubtedly inherited. I have elsewhere given one instance from my own observation of an extraordinary and complex gesture, associated with pleasurable feelings, which was transmitted from a father to his daughter, as well as some other analogous facts.[30] Another curious instance of an odd inherited movement, associated with the wish to obtain an object, will be given in the course of this volume.

There are other actions which are commonly performed under certain circumstances, independently of habit, and which seem to be due to imitation or some sort of sympathy. Thus persons cutting anything with a pair of scissors may be seen to move their jaws simultaneously with the blades of the scissors. Children learning to write often twist about their tongues as their fingers move, in a ridiculous fashion. When a public singer suddenly becomes a little hoarse, many of those present may be heard, as I have been assured by a gentleman on whom I can rely, to clear their throats; but here habit probably comes into play, as we clear our own throats under similar circumstances. I have also been told that at leaping matches, as the performer makes his spring, many of the spectators, generally men and boys, move their feet; but here again habit probably comes into play, for it is very doubtful whether women would thus act.

Reflex actions.—Reflex actions, in the strict sense of the term, are due to the excitement of a peripheral nerve, which transmits its influence to certain nerve-cells, and these in their turn excite certain muscles or glands into action; and all this may take place without any sensation or consciousness on our part, though often thus accompanied. As many reflex actions are highly expressive, the subject must here be noticed at some little length. We shall also see that some of them graduate into, and can hardly be distinguished from actions which have arisen through habit.[31] Coughing and sneezing are familiar instances of reflex actions. With infants the first act of respiration is often a sneeze, although this requires the co-ordinated movement of numerous muscles. Respiration is partly voluntary, but mainly reflex, and is performed in the most natural and best manner without the interference of the will. A vast number of complex movements are reflex. As good an instance as can be given is the often-quoted one of a decapitated frog, which cannot of course feel, and cannot consciously perform, any movement. Yet if a drop of acid be placed on the lower surface of the thigh of a frog in this state, it will rub off the drop with the upper surface of the foot of the same leg. If this foot be cut off, it cannot thus act. "After some fruitless efforts, therefore, it gives up trying in that way, seems restless, as though," says Pflüger, "it was seeking some other way, and at last it makes use of the foot of the other leg and succeeds in rubbing off the acid. Notably we have here not merely contractions of muscles, but combined and harmonized contractions in due sequence for a special purpose. These are actions that have all the appearance of being guided by intelligence and instigated by will in an animal, the recognized organ of whose intelligence and will has been removed."[32]

We see the difference between reflex and voluntary movements in very young children not being able to perform, as I am informed by Sir Henry Holland, certain acts somewhat analogous to those of sneezing and coughing, namely, in their not being able to blow their noses *(i. e.* to compress the nose and blow violently through the passage), and in their not being able to clear their throats of phlegm. They have to learn to perform these acts, yet they are performed by us, when a little older, almost as easily as reflex actions. Sneezing and coughing, however, can be controlled by the will only partially or not at all; whilst the clearing the throat and blowing the nose are completely under our command.

When we are conscious of the presence of an irritating particle in our nostrils or windpipe—that is, when the same sensory nerve-cells are excited, as in the case of sneezing and coughing—we can voluntarily expel the particle by forcibly driving air through these passages; but we cannot do this with nearly the same force, rapidity, and precision, as by a reflex action. In this latter case the sensory nerve-cells apparently excite the motor nerve-cells without any waste of power by first communicating with the cerebral hemispheres—the seat of our consciousness and volition. In all cases there seems to exist a profound antagonism between the same movements, as directed by the will and by a reflex stimulant, in the force with which they are performed and in the facility with which they are excited. As Claude Bernard asserts, "L'influence du cerveau tend donc à entraver les mouvements réflexes, à limiter leur force et leur étendue."[33]

The conscious wish to perform a reflex action sometimes stops or interrupts its performance, though the proper sensory nerves may be stimulated. For instance, many years ago I laid a small wager with a dozen young men that they would not sneeze if they took snuff, although they all declared that they invariably did so; accordingly they all took a pinch, but from wishing much to succeed, not one sneezed, though their eyes watered, and all, without exception, had to pay me the wager. Sir H. Holland remarks[34] that attention paid to the act of swallowing interferes with the proper movements; from which it probably follows, at least in part, that some persons find it so difficult to swallow a pill.

Another familiar instance of a reflex action is the involuntary closing of the eyelids when the surface of the eye is touched. A similar winking movement is caused when a blow is directed towards the face; but this is an habitual and not a strictly reflex action, as the stimulus is conveyed through the mind and not by the excitement of a peripheral nerve. The whole body and head are generally at the same time drawn suddenly backwards. These latter movements, however, can be prevented, if the danger does not appear to the imagination imminent; but our reason telling us that there is no danger does not suffice. I may mention a trifling fact, illustrating this point, and which at the time amused me. I put my face close to the thick glass-plate in front of a puff-adder in the Zoological Gardens, with the firm determination of not starting back if the snake struck at me; but, as soon as the blow was struck, my resolution went for nothing, and I jumped a yard or two backwards with astonishing rapidity. My will and reason were powerless against the imagination of a danger which had never been experienced.

The violence of a start seems to depend partly on the vividness of the imagination, and partly on the condition, either habitual or temporary, of the nervous system. He who will attend to the starting of his horse, when tired and fresh, will perceive how perfect is the gradation from a mere glance at some unexpected object, with a momentary doubt whether it is dangerous, to a jump so rapid and violent, that the animal probably could not voluntarily whirl round in so rapid a manner. The nervous system of a fresh and highly-fed horse sends its order to the motory system so quickly, that no time is allowed for him to consider whether or not the danger is real. After one violent start, when he is excited and the blood flows freely through his brain, he is very apt to start again; and so it is, as I have noticed, with young infants.

A start from a sudden noise, when the stimulus is conveyed through the auditory nerves, is always accompanied in grown-up persons by the winking of the eye-lids.[35] I observed, however, that though my infants started at sudden sounds, when under a fortnight old, they certainly did not always wink their eyes, and I believe never did so. The start of an older infant apparently represents a vague catching hold of something to prevent falling. I shook a pasteboard box close before the eyes of one of my infants, when 114 days old, and it did not in the least wink; but when I put a few comfits into the box, holding it in the same position as before, and rattled them, the child blinked its eyes violently every time, and started a little. It was obviously impossible that a carefully-guarded infant could have learnt by experience that a rattling sound near its eyes indicated danger to them. But such experience will have been slowly gained at a later age during a long series of generations; and from what we know of inheritance, there is nothing improbable in the transmission of a habit to the offspring at an earlier age than that at which it was first acquired by the parents.

From the foregoing remarks it seems probable that some actions, which were at first performed consciously, have become through habit and association converted into reflex actions, and are now so firmly fixed and inherited, that they are performed, even when not of the least use,[36] as often as the same causes arise, which originally excited them in us through the volition. In such cases the sensory nerve-cells excite the motor cells, without first communicating with those cells on which our consciousness and volition depend. It is probable that sneezing and coughing were originally acquired by the habit of expelling, as violently as possible, any irritating particle from the sensitive air-passages. As far as time is concerned, there has been more than enough for these habits to have become innate or converted into reflex actions; for they are common to most or all of the higher quadrupeds, and must therefore have been first acquired at a very remote period. Why the act of clearing the throat is not a reflex action, and has to be learnt by our children, I cannot pretend to say; but we can see why blowing the nose on a handkerchief has to be learnt.

It is scarcely credible that the movements of a headless frog, when it wipes off a drop of acid or other object from its thigh, and which movements are so well co-ordinated for a special purpose, were not at first performed voluntarily, being afterwards rendered easy through long-continued habit so as at last to be performed unconsciously, or independently of the cerebral hemispheres.

So again it appears probable that starting was originally acquired by the habit of jumping away as quickly as possible from danger, whenever any of our senses gave us warning. Starting, as we have seen, is accompanied by the blinking of the eyelids so as to protect the eyes, the most tender and sensitive organs of the body; and it is, I believe, always accompanied by a sudden and forcible inspiration, which is the natural preparation for any violent effort. But when a man or horse starts, his heart beats wildly against his ribs, and here it may be truly said we have an organ which has never been under the control of the will, partaking in the general reflex movements of the body. To this point, however, I shall return in a future chapter.

The contraction of the iris, when the retina is stimulated by a bright light, is another instance of a movement, which it appears cannot possibly have been at first voluntarily performed and then fixed by habit; for the iris is not known to be under the conscious control of the will in any animal. In such cases some explanation, quite distinct from habit, will have to be discovered. The radiation of nerve-force from strongly-excited nerve-cells to other connected cells, as in the case of a bright light on the retina causing a sneeze, may perhaps aid us in understanding how some reflex actions originated. A radiation of nerve-force of this kind, if it caused a movement tending to lessen the primary irritation, as in the case of the contraction of the iris preventing too much light from falling on the retina, might afterwards have been taken advantage of and modified for this special purpose.

It further deserves notice that reflex actions are in all probability liable to slight variations, as are all corporeal structures and instincts; and any variations which were beneficial and of sufficient importance, would tend to be preserved and inherited. Thus reflex actions, when once gained for one purpose, might afterwards be modified independently of the will or habit, so as to serve for some distinct purpose. Such cases would be parallel with those which, as we have every reason to believe, have occurred with many instincts; for although some instincts have been developed simply through long-continued and inherited habit, other highly complex ones have been developed through the preservation of variations of pre-existing instincts—that is, through natural selection.

I have discussed at some little length, though as I am well aware, in a very imperfect manner, the acquirement of reflex actions, because they are often brought into play in connection with movements expressive of our emotions; and it was necessary to show that at least some of them might have been first acquired through the will in order to satisfy a desire, or to relieve a disagreeable sensation.

Associated habitual movements in the lower animals.—I have already given in the case of Man several instances of movements, associated with various states of the mind or body, which are now purposeless, but which were originally of use, and are still of use under certain circumstances. As this subject is very important for us, I will here give a considerable number of analogous facts, with reference to animals; although many of them are of a very trifling nature. My object is to show that certain movements were originally performed for a definite end, and that, under nearly the same circumstances, they are still pertinaciously performed through habit when not of the least use. That the tendency in most of the following cases is inherited, we may infer from such actions being performed in the same manner by all the individuals, young and old, of the same species. We shall also see that they are excited by the most diversified, often circuitous, and sometimes mistaken associations.

Dogs, when they wish to go to sleep on a carpet or other hard surface, generally turn round and round and scratch the ground with their forepaws in a senseless manner, as if they intended to trample down the grass and scoop out a hollow, as no doubt their wild parents did, when they lived on open grassy plains or in the woods. Jackals, fennecs, and other allied animals in the Zoological Gardens, treat their straw in this manner; but it is a rather odd circumstance that the keepers, after observing for some months, have never seen the wolves thus behave. A semi-idiotic dog—and an animal in this condition would be particularly liable to follow a senseless habit—was observed by a friend to turn completely round on a carpet thirteen times before going to sleep.

Many carnivorous animals, as they crawl towards their prey and prepare to rush or spring on it, lower their heads and crouch, partly, as it would appear, to hide themselves, and partly to get ready for their rush; and this habit in an exaggerated form has become hereditary in our pointers and setters. Now I have noticed scores of times that when two strange dogs meet on an open road, the one which first sees the other, though at the distance of one or two hundred yards, after the first glance always lowers its head, generally crouches a little, or even lies down; that is, he takes the proper attitude for concealing himself and for making a rush or spring, although the road is quite open and the distance great. Again, dogs of all kinds when intently watching and slowly approaching their prey, frequently keep one of their fore-legs doubled up for a long time, ready for the next cautious step; and this is eminently characteristic of the pointer. But from habit they behave in exactly the same manner whenever their attention is aroused (fig. 4). I have seen a dog at the foot of a high wall, listening attentively to a sound on the opposite side, with one leg doubled up; and in this case there could have been no intention of making a cautious approach.

Fig. 4. *Small dog watching a cat on a table . From a photograph taken by Mr. Rejlander.*

Dogs after voiding their excrement often make with all four feet a few scratches backwards, even on a bare stone pavement, as if for the purpose of covering up their excrement with earth, in nearly the same manner as do cats. Wolves and jackals behave in the Zoological Gardens in exactly the same manner, yet, as I am assured by the keepers, neither wolves, jackals, nor foxes, when they have the means of doing so, ever cover up their excrement, any more than do dogs. All these animals, however, bury superfluous food. Hence, if we rightly understand the meaning of the above cat-like habit, of which there can be little doubt, we have a purposeless remnant of an habitual movement, which was originally followed by some remote progenitor of the dog-genus for a definite purpose, and which has been retained for a prodigious length of time.

Dogs and jackals[37] take much pleasure in rolling and rubbing their necks and backs on carrion. The odour seems delightful to them, though dogs at least do not eat carrion. Mr. Bartlett has observed wolves for me, and has given them carrion, but has never seen them roll on it. I have heard it remarked, and I believe it to be true, that the larger dogs, which are probably descended from wolves, do not so often roll in carrion as do smaller dogs, which are probably descended from jackals. When a piece of brown biscuit is offered to a terrier of mine and she is not hungry (and I have heard of similar instances), she first tosses it about and worries it, as if it were a rat or other prey; she then repeatedly rolls on it precisely as if it were a piece of carrion, and at last eats it. It would appear that an imaginary relish has to be given to the distasteful morsel; and to effect this the dog acts in his habitual manner, as if the biscuit was a live animal or smelt like carrion, though he knows better than we do that this is not the case. I have seen this same terrier act in the same manner after killing a little bird or mouse.

Dogs scratch themselves by a rapid movement of one of their hind feet; and when their backs are rubbed with a stick, so strong is the habit, that they cannot help rapidly scratching the air or the ground in a useless and ludicrous manner. The terrier just alluded to, when thus scratched with a stick, will sometimes show her delight by another habitual movement, namely, by licking the air as if it were my hand.

Horses scratch themselves by nibbling those parts of their bodies which they can reach with their teeth; but more commonly one horse shows another where he wants to be scratched, and they then nibble each other. A friend whose attention I had called to the subject, observed that when he rubbed his horse's neck, the animal protruded his head, uncovered his teeth, and moved his jaws, exactly as if nibbling another horse's neck, for he could never have nibbled his own neck. If a horse is much tickled, as when curry-combed, his wish to bite something becomes so intolerably strong, that he will clatter his teeth together, and though not vicious, bite his groom. At the same time from habit he closely depresses his ears, so as to protect them from being bitten, as if he were fighting with another horse.

A horse when eager to start on a journey makes the nearest approach which he can to the habitual movement of progression by pawing the ground. Now when horses in their stalls are about to be fed and are eager for their corn, they paw the pavement or the straw. Two of my horses thus behave when they see or hear the corn given to their neighbours. But here we have what may almost be called a true expression, as pawing the ground is universally recognized as a sign of eagerness.

Cats cover up their excrements of both kinds with earth; and my grandfather[38] saw a kitten scraping ashes over a spoonful of pure water spilt on the hearth; so that here an habitual or instinctive action was falsely excited, not by a previous act or by odour, but by eyesight. It is well known that cats dislike wetting their feet, owing, it is probable, to their having aboriginally inhabited the dry country of Egypt; and when they wet their feet they shake them violently. My daughter poured some water into a glass close to the head of a kitten; and it immediately shook its feet in the usual manner; so that here we have an habitual movement falsely excited by an associated sound instead of by the sense of touch.

Kittens, puppies, young pigs and probably many other young animals, alternately push with their fore-feet against the mammary glands of their mothers, to excite a freer secretion of milk, or to make it flow. Now it is very common with young cats, and not at all rare with old cats of the common and Persian breeds (believed by some naturalists to be specifically distinct), when comfortably lying on a warm shawl or other soft substance, to pound it quietly and alternately with their fore-feet; their toes being spread out and claws slightly protruded, precisely as when sucking their mother. That it is the same movement is clearly shown by their often at the same time taking a bit of the shawl into their mouths and sucking it; generally closing their eyes and purring from delight. This curious movement is commonly excited only in association with the sensation of a warm soft surface; but I have seen an old cat, when pleased by having its back scratched, pounding the air with its feet in the same manner; so that this action has almost become the expression of a pleasurable sensation.

Having referred to the act of sucking, I may add that this complex movement, as well as the alternate protrusion of the fore-feet, are reflex actions; for they are performed if a finger moistened with milk is placed in the mouth of a puppy, the front part of whose brain has been removed.[39] It has recently been stated in France, that the action of sucking is excited solely through the sense of smell, so that if the olfactory nerves of a puppy are destroyed, it never sucks. In like manner the wonderful power which a chicken possesses only a few hours after being hatched, of picking up small particles of food, seems to be started into action through the sense of hearing; for with chickens hatched by artificial heat, a good observer found that "making a noise with the finger-nail against a board, in imitation of the hen-mother, first taught them to peck at their meat."[40]

I will give only one other instance of an habitual and purposeless movement. The Sheldrake *(Tadorna)* feeds on the sands left uncovered by the tide, and when a worm-cast is discovered, "it begins patting the ground with its feet, dancing as it were, over the hole;" and this makes the worm come to the surface. Now Mr. St. John says, that when his tame Sheldrakes "came to ask for food, they patted the ground in an impatient and rapid manner."[41] This therefore may almost be considered as their expression of hunger. Mr. Bartlett informs me that the Flamingo and the Kagu *(Rhinochetus jubatus)* when anxious to be fed, beat the ground with their feet in the same odd manner. So again Kingfishers, when they catch a fish, always beat it until it is killed; and in the Zoological Gardens they always beat the raw meat, with which they are sometimes fed, before devouring it.

WE HAVE NOW, I think, sufficiently shown the truth of our first Principle, namely, that when any sensation, desire, dislike, &c., has led during a long series of generations to some voluntary movement, then a tendency to the performance of a similar movement will almost certainly be excited, whenever the same, or any analogous or associated sensation, &c., although very weak, is experienced; notwithstanding that the movement in this case may not be of the least use. Such habitual movements are often, or generally inherited; and they then differ but little from reflex actions. When we treat of the special expressions of man, the latter part of our first Principle, as given at the commencement of this chapter, will be seen to hold good; namely, that when movements, associated through habit with certain states of the mind, are partially repressed by the will, the strictly involuntary muscles, as well as those which are least under the separate control of the will, are liable still to act; and their action is often highly expressive. Conversely, when the will is temporarily or permanently weakened, the voluntary muscles fail before the involuntary. It is a fact familiar to pathologists, as Sir C. Bell remarks,[42] "that when debility arises from affection of the brain, the influence is greatest on those muscles which are, in their natural condition, most under the command of the will." We shall, also, in our future chapters, consider another proposition included in our first Principle; namely, that the checking of one habitual movement sometimes requires other slight movements; these latter serving as a means of expression.

CHAPTER II

General Principles of Expression—continued

THE PRINCIPLE OF ANTITHESIS • INSTANCES IN THE DOG AND CAT • ORIGIN OF THE PRINCIPLE • CONVENTIONAL SIGNS • THE PRINCIPLE OF ANTITHESIS HAS NOT ARISEN FROM OPPOSITE ACTIONS BEING CONSCIOUSLY PERFORMED UNDER OPPOSITE IMPULSES

We will now consider our second Principle, that of Antithesis. Certain states of the mind lead, as we have seen in the last chapter, to certain habitual movements which were primarily, or may still be, of service; and we shall find that when a directly opposite state of mind is induced, there is a strong and involuntary tendency to the performance of movements of a directly opposite nature, though these have never been of any service. A few striking instances of antithesis will be given, when we treat of the special expressions of man; but as, in these cases, we are particularly liable to confound conventional or artificial gestures and expressions with those which are innate or universal, and which alone deserve to rank as true expressions, I will in the present chapter almost confine myself to the lower animals.

When a dog approaches a strange dog or man in a savage or hostile frame of mind he walks upright and very stiffly; his head is slightly raised, or not much lowered; the tail is held erect and quite rigid; the hairs bristle, especially along the neck and back; the pricked ears are directed forwards, and the eyes have a fixed stare: (see figs. 5 and 7). These actions, as will hereafter be explained, follow from the dog's intention to attack his enemy, and are thus to a large extent intelligible. As he prepares to spring with a savage growl on his enemy, the canine teeth are uncovered, and the ears are pressed close backwards on the head; but with these latter actions, we are not here concerned. Let us now suppose that the dog suddenly discovers that the man whom he is approaching, is not a stranger, but his master; and let it be observed how completely and instantaneously his whole bearing is reversed. Instead of walking upright, the body sinks downwards or even crouches, and is thrown into flexuous movements; his tail, instead of being held stiff and upright, is lowered and wagged from side to side; his hair instantly becomes smooth; his ears are depressed and drawn backwards, but not closely to the head; and his lips hang loosely. From the drawing back of the ears, the eyelids become elongated, and the eyes no longer

Fig. 5. *Dog approaching another dog with hostile intentions. By Mr. Riviere.*

Fig. 6. *The same in a humble and affectionate frame of mind. By Mr. Riviere.*

Fig. 7. *Half-bred Shepherd Dog in the same state as in Fig. 5. By Mr. A. May.*

Fig. 8. *The same caressing his master. By Mr. A. May.*

appear round and staring. It should be added that the animal is at such times in an excited condition from joy; and nerve-force will be generated in excess, which naturally leads to action of some kind. Not one of the above movements, so clearly expressive of affection, are of the least direct service to the animal. They are explicable, as far as I can see, solely from being in complete opposition or antithesis to the attitude and movements which, from intelligible causes, are assumed when a dog intends to fight, and which consequently are expressive of anger. I request the reader to look at the four accompanying sketches, which have been given in order to recall vividly the appearance of a dog under these two states of mind. It is, however, not a little difficult to represent affection in a dog, whilst caressing his master and wagging his tail, as the essence of the expression lies in the continuous flexuous movements.

We will now turn to the cat. When this animal is threatened by a dog, it arches its back in a surprising manner, erects its hair, opens its mouth and spits. But we are not here concerned with this well-known attitude, expressive of terror combined with anger; we are concerned only with that of rage or anger. This is not often seen, but may be observed when two cats are fighting together; and I have seen it well exhibited by a savage cat whilst plagued by a boy. The attitude is almost exactly the same as that of a tiger disturbed and growling over its food, which every one must have beheld in menageries. The animal assumes a crouching position, with the body extended; and the whole tail, or the tip alone, is lashed or curled from side to side. The hair is not in the least erect. Thus far, the attitude and movements are nearly the same as when the animal is prepared to spring on its prey, and when, no doubt, it feels savage. But when preparing to fight, there is this difference, that the ears are closely pressed backwards; the mouth is partially opened, showing the teeth; the fore feet are occasionally struck out with protruded claws; and the animal occasionally utters a fierce growl. (See figs. 9 and 10.) All, or almost all, these actions naturally follow (as hereafter to be explained), from the cat's manner and intention of attacking its enemy.

Let us now look at a cat in a directly opposite frame of mind, whilst feeling affectionate and caressing her master; and mark how opposite is her attitude in every respect. She now stands upright with her back slightly arched, which makes the hair appear rather rough, but it does not bristle; her tail, instead of being extended and lashed from side to side, is held quite stiff and perpendicularly upwards; her ears are erect and pointed; her mouth is closed; and she rubs against her master with a purr instead of a growl. Let it further be observed how widely different is the whole bearing of an affectionate cat from that of a dog, when with his body crouching and flexuous, his tail lowered and wagging, and ears depressed, he caresses his

Fig. 9. *Cat, savage, and prepared to fight. Drawn from life by Mr. Wood.*

Fig. 10. *Cat in an affectionate frame of mind. By Mr. Wood.*

master. This contrast in the attitudes and movements of these two carnivorous animals, under the same pleased and affectionate frame of mind, can be explained, as it appears to me, solely by their movements standing in complete antithesis to those which are naturally assumed, when these animals feel savage and are prepared either to fight or to seize their prey.

In these cases of the dog and cat, there is every reason to believe that the gestures both of hostility and affection are innate or inherited; for they are almost identically the same in the different races of the species, and in all the individuals of the same race, both young and old.

I will here give one other instance of antithesis in expression. I formerly possessed a large dog, who, like every other dog, was much pleased to go out walking. He showed his pleasure by trotting gravely before me with high steps, head much raised, moderately erected ears, and tail carried aloft but not stiffly. Not far from my house a path branches off to the right, leading to the hot-house, which I used often to visit for a few moments, to look at my experimental plants. This was always a great disappointment to the dog, as he did not know whether I should continue my walk; and the instantaneous and complete change of expression which came over him, as soon as my body swerved in the least towards the path (and I sometimes tried this as an experiment) was laughable. His look of dejection was known to every member of the family, and was called his *hot-house face.* This consisted in the head drooping much, the whole body sinking a little and remaining motionless; the ears and tail falling suddenly down, but the tail was by no means wagged. With the falling of the ears and of his great chaps, the eyes became much changed in appearance, and I fancied that they looked less bright. His aspect was that of piteous, hopeless dejection; and it was, as I have said, laughable, as the cause was so slight. Every detail in his attitude was in complete opposition to his former joyful yet dignified bearing; and can be explained, as it appears to me, in no other way, except through the principle of antithesis. Had not the change been so instantaneous, I should have attributed it to his lowered spirits affecting, as in the case of man, the nervous system and circulation, and consequently the tone of his whole muscular frame; and this may have been in part the cause.

WE WILL NOW CONSIDER how the principle of antithesis in expression has arisen. With social animals, the power of intercommunication between the members of the same community,—and with other species, between the opposite sexes, as well as between the young and the old,—is of the highest importance to them. This is generally effected by means of the voice, but it is certain that gestures and expressions are to a certain extent mutually intelligible. Man not only uses inarticulate cries, gestures, and expressions, but has invented articulate language; if, indeed, the word *invented* can be applied to a process, completed by innumerable steps, half-consciously made. Any one who has watched monkeys will not doubt that they perfectly understand each other's gestures and expression, and to a large extent, as Rengger asserts,[43] those of man. An animal when going to attack another, or when afraid of another, often makes itself appear terrible, by erecting its hair, thus increasing the apparent bulk of its body, by showing its teeth, or brandishing its horns, or by uttering fierce sounds.

As the power of intercommunication is certainly of high service to many animals, there is no *à priori* improbability in the supposition, that gestures manifestly of an opposite nature to those by which certain feelings are already expressed, should at first have been voluntarily employed under the influence of an opposite state of feeling. The fact of the gestures being now innate, would be no valid objection to the belief that they were at first intentional; for if practised during many generations, they would probably at last be inherited. Nevertheless it is more than doubtful, as we shall immediately see, whether any of the cases which come under our present head of antithesis, have thus originated.

With conventional signs which are not innate, such as those used by the deaf and dumb and by savages, the principle of opposition or antithesis has been partially brought into play. The Cistercian monks thought it sinful to speak, and as they could not avoid holding some communication, they invented a gesture language, in which the principle of opposition seems to have been employed.[44] Dr. Scott, of the Exeter Deaf and Dumb Institution, writes to me that "opposites are greatly used in teaching the deaf and dumb, who have a lively sense of them." Nevertheless I have been surprised how few unequivocal instances can be adduced. This depends partly on all the signs having commonly had some natural origin; and partly on the practice of the deaf and dumb and of savages to contract their signs as much as possible for the sake of rapidity.[45] Hence their natural source or origin often becomes doubtful or is completely lost; as is likewise the case with articulate language.

Many signs, moreover, which plainly stand in opposition to each other, appear to have had on both sides a significant origin. This seems to hold good with the signs used by the deaf and dumb for light and darkness, for strength and weakness, &c. In a future chapter I shall endeavour to show that the opposite gestures of affirmation and negation, namely, vertically nodding and laterally shaking the head, have both probably had a natural beginning. The waving of the hand from right to left, which is used as a negative by some savages, may have been invented in imitation of shaking the head; but whether the opposite movement of waving the hand in a straight line from the face, which is used in affirmation, has arisen through antithesis or in some quite distinct manner, is doubtful.

If we now turn to the gestures which are innate or common to all the individuals of the same species, and which come under the present head of antithesis, it is extremely doubtful, whether any of them were at first deliberately invented and consciously performed. With mankind the best instance of a gesture standing in direct opposition to other movements, naturally assumed under an opposite frame of mind, is that of shrugging the shoulders. This expresses impotence or an apology,— something which cannot be done, or cannot be avoided. The gesture is sometimes used consciously and voluntarily, but it is extremely improbable that it was at first deliberately invented, and afterwards fixed by habit; for not only do young children sometimes shrug their shoulders under the above states of mind, but the movement is accompanied, as will be shown in a future chapter, by various subordinate movements, which not one man in a thousand is aware of, unless he has specially attended to the subject.

Dogs when approaching a strange dog, may find it useful to show by their movements that they are friendly, and do not wish to fight. When two young dogs in play are growling and biting each other's faces and legs, it is obvious that they mutually understand each other's gestures and manners. There seems, indeed, some degree of instinctive knowledge in puppies and kittens, that they must not use their sharp little teeth or claws too freely in their play, though this sometimes happens and a squeal is the result; otherwise they would often injure each other's eyes. When my terrier bites my hand in play, often snarling at the same time, if he bites too hard and I say *gently, gently,* he goes on biting, but answers me by a few wags of the tail, which seems to say "Never mind, it is all fun." Although dogs do thus express, and may wish to express, to other dogs and to man, that they are in a friendly state of mind, it is incredible that they could ever have deliberately thought of drawing back and depressing their ears, instead of holding them erect,— of lowering and wagging their tails, instead of keeping them stiff and upright, &c., because they knew that these movements stood in direct opposition to those assumed under an opposite and savage frame of mind.

Again, when a cat, or rather when some early progenitor of the species, from feeling affectionate first slightly arched its back, held its tail perpendicularly upwards and pricked its ears, can it be believed that the animal consciously wished thus to show that its frame of mind was directly the reverse of that, when from being ready to fight or to spring on its prey, it assumed a crouching attitude, curled its tail from side to side and depressed its ears? Even still less can I believe that my dog voluntarily put on his dejected attitude and *"hot-house face"* which formed so complete a contrast to his previous cheerful attitude and whole bearing. It cannot be supposed that he knew that I should understand his expression, and that he could thus soften my heart and make me give up visiting the hot-house.

Hence for the development of the movements which come under the present head, some other principle, distinct from the will and consciousness, must have intervened. This principle appears to be that every movement which we have voluntarily performed throughout our lives has required the action of certain muscles; and when we have performed a directly opposite movement, an opposite set of muscles has been habitually brought into play,—as in turning to the right or to the left, in pushing away or pulling an object towards us, and in lifting or lowering a weight. So strongly are our intentions and movements associated together, that if we eagerly wish an object to move in any direction, we can hardly avoid moving our bodies in the same direction, although we may be perfectly aware that this can have no influence. A good illustration of this fact has already been given in the Introduction, namely, in the grotesque movements of a young and eager billiard-player, whilst watching the course of his ball. A man or child in a passion, if he tells any one in a loud voice to begone, generally moves his arm as if to push him away, although the offender may not be standing near, and although there may be not the least need to explain by a gesture what is meant. On the other hand, if we eagerly desire some one to approach us closely, we act as if pulling him towards us; and so in innumerable other instances.

As the performance of ordinary movements of an opposite kind, under opposite impulses of the will, has become habitual in us and in the lower animals, so when actions of one kind have become firmly associated with any sensation or emotion, it appears natural that actions of a directly opposite kind, though of no use, should be unconsciously performed through habit and association, under the influence of a directly opposite sensation or emotion. On this principle alone can I understand how the gestures and expressions which come under the present head of antithesis have originated. If indeed they are serviceable to man or to any other animal, in aid of inarticulate cries or language, they will likewise be voluntarily employed, and the habit will thus be strengthened. But whether or not of service as a means of communication, the tendency to perform opposite movements under opposite sensations or emotions would, if we may judge by analogy, become hereditary through long practice; and there cannot be a doubt that several expressive movements due to the principle of antithesis are inherited.

CHAPTER III

General Principles of Expression—concluded

THE PRINCIPLE OF THE DIRECT ACTION OF THE EXCITED NERVOUS SYSTEM ON THE BODY, INDEPENDENTLY OF THE WILL AND IN PART OF HABIT • CHANGE OF COLOUR IN THE HAIR • TREMBLING OF THE MUSCLES • MODIFIED SECRETIONS • PERSPIRATION • EXPRESSION OF EXTREME PAIN • OF RAGE, GREAT JOY, AND TERROR • CONTRAST BETWEEN THE EMOTIONS WHICH CAUSE AND DO NOT CAUSE EXPRESSIVE MOVEMENTS • EXCITING AND DEPRESSING STATES OF THE MIND • SUMMARY

We now come to our third Principle, namely, that certain actions, which we recognise as expressive of certain states of the mind, are the direct result of the constitution of the nervous system, and have been from the first independent of the will, and, to a large extent, of habit. When the sensorium is strongly excited nerve-force is generated in excess, and is transmitted in certain directions, dependent on the connection of the nerve-cells, and, as far as the muscular system is concerned, on the nature of the movements which have been habitually practised. Or the supply of nerve-force may, as it appears, be interrupted. Of course every movement which we make is determined by the constitution of the nervous system; but actions performed in obedience to the will, or through habit, or through the principle of antithesis, are here as far as possible excluded. Our present subject is very obscure, but, from its importance, must be discussed at some little length; and it is always advisable to perceive clearly our ignorance.

The most striking case, though a rare and abnormal one, which can be adduced of the direct influence of the nervous system, when strongly affected, on the body, is the loss of colour in the hair, which has occasionally been observed after extreme terror or grief. One authentic instance has been recorded, in the case of a man brought out for execution in India, in which the change of colour was so rapid that it was perceptible to the eye.[46]

Another good case is that of the trembling of the muscles, which is common to man and to many, or most, of the lower animals. Trembling is of no service, often of much disservice, and cannot have been at first acquired through the will, and then rendered habitual in association with any emotion. I am assured by an eminent authority that young children do not tremble, but go into convulsions under the circumstances which would induce excessive trembling in adults. Trembling is excited in different individuals in very different degrees, and by the most diversified causes,—by cold to the surface, before fever-fits, although the temperature of the body is then above the normal standard; in blood-poisoning, delirium tremens, and other diseases; by general failure of power in old age; by exhaustion after excessive fatigue; locally from severe injuries, such as burns; and, in an especial manner, by the passage of a catheter. Of all emotions, fear notoriously is the most apt to induce trembling; but so do occasionally great anger and joy. I remember once seeing a boy who had just shot his first snipe on the wing, and his hands trembled to such a degree from delight, that he could not for some time reload his gun; and I have heard of an exactly similar case with an Australian savage, to whom a gun had been lent. Fine music, from the vague emotions thus excited, causes a shiver to run down the backs of some persons. There seems to be very little in common in the above several physical causes and emotions to account for trembling; and Sir J. Paget, to whom I am indebted for several of the above statements, informs me that the subject is a very obscure one. As trembling is sometimes caused by rage, long before exhaustion can have set in, and as it sometimes accompanies great joy, it would appear that any strong excitement of the nervous system interrupts the steady flow of nerve-force to the muscles.[47]

The manner in which the secretions of the alimentary canal and of certain glands—as the liver, kidneys, or mammæ—are affected by strong emotions, is another excellent instance of the direct action of the sensorium on these organs, independently of the will or of any serviceable associated habit. There is the greatest difference in different persons in the parts which are thus affected, and in the degree of their affection.

The heart, which goes on uninterruptedly beating night and day in so wonderful a manner, is extremely sensitive to external stimulants. The great physiologist, Claude Bernard,[48] has shown how the least excitement of a sensitive nerve reacts on the heart; even when a nerve is touched so slightly that no pain can possibly be felt by the animal under experiment. Hence when the mind is strongly excited, we might expect that it would instantly affect in a direct manner the heart; and this is universally acknowledged and felt to be the case. Claude Bernard also repeatedly insists, and this deserves especial notice, that when the heart is affected it reacts on the brain; and the state of the brain again reacts through the pneumo-gastric nerve on the heart; so that under any excitement there will be much mutual action and reaction between these, the two most important organs of the body.

The vaso-motor system, which regulates the diameter of the small arteries, is directly acted on by the sensorium, as we see when a man blushes from shame; but in this latter case the checked transmission of nerve-force to the vessels of the face can, I think, be partly explained in a curious manner through habit. We shall also be able to throw some light, though very little, on the involuntary erection of the hair under the emotions of terror and rage. The secretion of tears depends, no doubt, on the connection of certain nerve-cells; but here again we can trace some few of the steps by which the flow of nerve-force through the requisite channels has become habitual under certain emotions.

A BRIEF CONSIDERATION of the outward signs of some of the stronger sensations and emotions will best serve to show us, although vaguely, in how complex a manner the principle under consideration of the direct action of the excited nervous system on the body, is combined with the principle of habitually associated, serviceable movements.

When animals suffer from an agony of pain, they generally writhe about with frightful contortions; and those which habitually use their voices utter piercing cries or groans. Almost every muscle of the body is brought into strong action. With man the mouth may be closely compressed, or more commonly the lips are retracted, with the teeth clenched or ground together. There is said to be "gnashing of teeth" in hell; and I have plainly heard the grinding of the molar teeth of a cow which was suffering acutely from inflammation of the bowels. The female hippopotamus in the Zoological Gardens, when she produced her young, suffered greatly; she incessantly walked about, or rolled on her sides, opening and closing her jaws, and clattering her teeth together.[49] With man the eyes stare wildly as in horrified astonishment, or the brows are heavily contracted. Perspiration bathes the body, and drops trickle down the face. The circulation and respiration are much affected. Hence the nostrils are generally dilated and often quiver; or the breath may be held until the blood stagnates in the purple face. If the agony be severe and prolonged, these signs all change; utter prostration follows, with fainting or convulsions.

A sensitive nerve when irritated transmits some influence to the nerve-cell, whence it proceeds; and this transmits its influence, first to the corresponding nerve-cell on the opposite side of the body, and then upwards and downwards along the cerebro-spinal column to other nerve-cells, to a greater or less extent, according to the strength of the excitement; so that; ultimately, the whole nervous system may be affected.[50] This involuntary transmission of nerve-force may or may not be accompanied by consciousness. Why the irritation of a nerve-cell should generate or liberate nerve-force is not known; but that this is the case seems to be the conclusion arrived at by all the greatest physiologists, such as Müller, Virchow, Bernard, &c.[51] As Mr. Herbert Spencer remarks, it may be received as an "unquestionable truth that, at any moment, the existing quantity of liberated nerve-force, which in an inscrutable way produces in us the state we call feeling, *must* expend itself in some direction—*must* generate an equivalent manifestation of force somewhere;" so that, when the cerebro-spinal system is highly excited and nerve-force is liberated in excess, it may be expended in intense sensations, active thought, violent movements, or increased activity of the glands.[52] Mr. Spencer further maintains that an "overflow of nerve-force, undirected by any motive, will manifestly take the most habitual routes; and, if these do not suffice, will next overflow into the less habitual ones." Consequently the facial and respiratory muscles, which are the most used, will be apt to be first brought into action; then those of the upper extremities, next those of the lower, and finally those of the whole body.[53]

An emotion may be very strong, but it will have little tendency to induce movements of any kind, if it has not commonly led to voluntary action for its relief or gratification; and when movements are excited, their nature is, to a large extent, determined by those which have often and voluntarily been performed for some definite end under the same emotion. Great pain urges all animals, and has urged them during endless generations, to make the most violent and diversified efforts to escape from the cause of suffering. Even when a limb or other separate part of the body is hurt, we often see a tendency to shake it, as if to shake off the cause, though this may obviously be impossible. Thus a habit of exerting with the utmost force all the muscles will have been established, whenever great suffering is experienced. As the muscles of the chest and vocal organs are habitually used, these will be particularly liable to be acted on, and loud, harsh screams or cries will be uttered. But the advantage derived from outcries has here probably come into play in an important manner; for the young of most animals, when in distress or danger, call loudly to their parents for aid, as do the members of the same community for mutual aid.

Another principle, namely, the internal consciousness that the power or capacity of the nervous system is limited, will have strengthened, though in a subordinate degree, the tendency to violent action under extreme suffering. A man cannot think deeply and exert his utmost muscular force. As Hippocrates long ago observed, if two pains are felt at the same time, the severer one dulls the other. Martyrs, in the ecstasy of their religious fervour have often, as it would appear, been insensible to the most horrid tortures. Sailors who are going to be flogged sometimes take a piece of lead into their mouths, in order to bite it with their utmost force, and thus to bear the pain. Parturient women prepare to exert their muscles to the utmost in order to relieve their sufferings.

We thus see that the undirected radiation of nerve-force from the nerve-cells which are first affected—the long-continued habit of attempting by straggling to escape from the cause of suffering—and the consciousness that voluntary muscular exertion relieves pain, have all probably concurred in giving a tendency to the most violent, almost convulsive, movements under extreme suffering; and such movements, including those of the vocal organs, are universally recognised as highly expressive of this condition.

As the mere touching of a sensitive nerve reacts in a direct manner on the heart, severe pain will obviously react on it in like manner, but far more energetically. Nevertheless, even in this case, we must not overlook the indirect effects of habit on the heart, as we shall see when we consider the signs of rage.

When a man suffers from an agony of pain, the perspiration often trickles down his face; and I have been assured by a veterinary surgeon that he has frequently seen drops falling from the belly and running down the inside of the thighs of horses, and from the bodies of cattle, when thus suffering. He has observed this, when there has been no struggling which would account for the perspiration. The whole body of the female hippopotamus, before alluded to, was covered with red-coloured perspiration whilst giving birth to her young. So it is with extreme fear; the same veterinary has often seen horses sweating from this cause; as has Mr. Bartlett with the rhinoceros; and with man it is a well-known symptom. The cause of perspiration bursting forth in these cases is quite obscure; but it is thought by some physiologists to be connected with the failing power of the capillary circulation; and we know that the vaso-motor system which regulates the capillary circulation, is much influenced by the mind. With respect to the movements of certain muscles of the face under great suffering, as well as from other emotions, these will be best considered when we treat of the special expressions of man and of the lower animals.

We will now turn to the characteristic symptoms of Rage. Under this powerful emotion the action of the heart is much accelerated,[54] or it may be much disturbed. The face reddens, or it becomes purple from the impeded return of the blood, or may turn deadly pale. The respiration is laboured, the chest heaves, and the dilated nostrils quiver. The whole body often trembles. The voice is affected. The teeth are clenched or ground together, and the muscular system is commonly stimulated to violent, almost frantic action. But the gestures of a man in this state usually differ from the purposeless writhings and struggles of one suffering from an agony of pain; for they represent more or less plainly the act of striking or fighting with an enemy.

All these signs of rage are probably in large part, and some of them appear to be wholly, due to the direct action of the excited sensorium. But animals of all kinds, and their progenitors before them, when attacked or threatened by an enemy, have exerted their utmost powers in fighting and in defending themselves. Unless an animal does thus act, or has the intention, or at least the desire, to attack its enemy, it cannot properly be said to be enraged. An inherited habit of muscular exertion will thus have been gained in association with rage; and this will directly or indirectly affect various organs, in nearly the same manner as does great bodily suffering.

The heart no doubt will likewise be affected in a direct manner; but it will also in all probability be affected through habit; and all the more so from not being under the control of the will. We know that any great exertion which we voluntarily make, affects the heart, through mechanical and other principles which need not here be considered; and it was shown in the first chapter that nerve-force flows readily through habitually used channels,—through the nerves of voluntary or involuntary movement, and through those of sensation. Thus even a moderate amount of exertion will tend to act on the heart; and on the principle of association, of which so many instances have been given, we may feel nearly sure that any sensation or emotion, as great pain or rage, which has habitually led to much muscular action, will immediately influence the flow of nerve-force to the heart, although there may not be at the time any muscular exertion.

The heart, as I have said, will be all the more readily affected through habitual associations, as it is not under the control of the will. A man when moderately angry, or even when enraged, may command the movements of his body, but he cannot prevent his heart from beating rapidly. His chest will perhaps give a few heaves, and his nostrils just quiver, for the movements of respiration are only in part voluntary. In like manner those muscles of the face which are least obedient to the will, will sometimes alone betray a slight and passing emotion. The glands again are wholly independent of the will, and a man suffering from grief may command his features, but cannot always prevent the tears from coming into his eyes. A hungry man, if tempting food is placed before him, may not show his hunger by any outward gesture, but he cannot check the secretion of saliva.

Under a transport of Joy or of vivid Pleasure, there is a strong tendency to various purposeless movements, and to the utterance of various sounds. We see this in our young children, in their loud laughter, clapping of hands, and jumping for joy; in the bounding and barking of a dog when going out to walk with his master; and in the frisking of a horse when turned out into an open field. Joy quickens the circulation, and this stimulates the brain, which again reacts on the whole body. The above purposeless movements and increased heart-action may be attributed in chief part to the excited state of the sensorium,[55] and to the consequent undirected overflow, as Mr. Herbert Spencer insists, of nerve-force. It deserves notice, that it is chiefly the anticipation of a pleasure, and not its actual enjoyment, which leads to purposeless and extravagant movements of the body, and to the utterance of various sounds. We see this in our children when they expect any great pleasure or treat; and dogs, which have been bounding about at the sight of a plate of food, when they get it do not show their delight by any outward sign, not even by wagging their tails. Now with animals of all kinds, the acquirement of almost all their pleasures, with the exception of those of warmth and rest, are associated, and have long been associated with active movements, as in the hunting or search for food, and in their courtship. Moreover, the mere exertion of the muscles after long rest or confinement is in itself a pleasure, as we ourselves feel, and as we see in the play of young animals. Therefore on this latter principle alone we might perhaps expect, that vivid pleasure would be apt to show itself conversely in muscular movements.

With all or almost all animals, even with birds, Terror causes the body to tremble. The skin becomes pale, sweat breaks out, and the hair bristles. The secretions of the alimentary canal and of the kidneys are increased, and they are involuntarily voided, owing to the relaxation of the sphincter muscles, as is known to be the case with man, and as I have seen with cattle, dogs, cats, and monkeys. The breathing is hurried. The heart beats quickly, wildly, and violently; but whether it pumps the blood more efficiently through the body may be doubted, for the surface seems bloodless and the strength of the muscles soon fails. In a frightened horse I have felt through the saddle the beating of the heart so plainly that I could have counted the beats. The mental faculties are much disturbed. Utter prostration soon follows, and even fainting. A terrified canary-bird has been seen not only to tremble and to turn white about the base of the bill, but to faint;[56] and I once caught a robin in a room, which fainted so completely, that for a time I thought it dead.

Most of these symptoms are probably the direct result, independently of habit, of the disturbed state of the sensorium; but it is doubtful whether they ought to be wholly thus accounted for. When an animal is alarmed it almost always stands motionless for a moment, in order to collect its senses and to ascertain the source of danger, and sometimes for the sake of escaping detection. But headlong flight soon follows, with no husbanding of the strength as in fighting, and the animal continues to fly as long as the danger lasts until utter prostration, with failing respiration and circulation, with all the muscles quivering and profuse sweating, renders further flight impossible. Hence it does not seem improbable that the principle of associated habit may in part account for, or at least augment, some of the above-named characteristic symptoms of extreme terror.

THAT THE PRINCIPLE of associated habit has played an important part in causing the movements expressive of the foregoing several strong emotions and sensations, we may, I think, conclude from considering firstly, some other strong emotions which do not ordinarily require for their relief or gratification any voluntary movement; and secondly the contrast in nature between the so-called exciting and depressing states of the mind. No emotion is stronger than maternal love; but a mother may feel the deepest love for her helpless infant, and yet not show it by any outward sign; or only by slight caressing movements, with a gentle smile and tender eyes. But let any one intentionally injure her infant, and see what a change! how she starts up with threatening aspect, how her eyes sparkle and her face reddens, how her bosom heaves, nostrils dilate, and heart beats; for anger, and not maternal love, has habitually led to action. The love between the opposite sexes is widely different from maternal love; and when lovers meet, we know that their hearts beat quickly, their breathing is hurried, and their faces flush; for this love is not inactive like that of a mother for her infant.

A man may have his mind filled with the blackest hatred or suspicion, or be corroded with envy or jealousy; but as these feelings do not at once lead to action, and as they commonly last for some time, they are not shown by any outward sign, excepting that a man in this state assuredly does not appear cheerful or good-tempered. If indeed these feelings break out into overt acts, rage takes their place, and will be plainly exhibited. Painters can hardly portray suspicion, jealousy, envy, &c., except by the aid of accessories which tell the tale; and poets use such vague and fanciful expressions as "green-eyed jealousy." Spenser describes suspicion as "Foul, ill-favoured, and grim, under his eyebrows looking still askance," &c.; Shakespeare speaks of envy "as lean-faced in her loathsome case;" and in another place he says, "no black envy shall make my grave;" and again as "above pale envy's threatening reach."

Emotions and sensations have often been classed as exciting or depressing. When all the organs of the body and mind,—those of voluntary and involuntary movement, of perception, sensation, thought, &c.,—perform their functions more energetically and rapidly than usual, a man or animal may be said to be excited, and, under an opposite state, to be depressed. Anger and joy are from the first exciting emotions, and they naturally lead, more especially the former, to energetic movements, which react on the heart and this again on the brain. A physician once remarked to me as a proof of the exciting nature of anger, that a man when excessively jaded will sometimes invent imaginary offences and put himself into a passion, unconsciously for the sake of reinvigorating himself; and since hearing this remark, I have occasionally recognized its full truth.

Several other states of mind appear to be at first exciting, but soon become depressing to an extreme degree. When a mother suddenly loses her child, sometimes she is frantic with grief, and must be considered to be in an excited state; she walks wildly about, tears her hair or clothes, and wrings her hands. This latter action is perhaps due to the principle of antithesis, betraying an inward sense of helplessness and that nothing can be done. The other wild and violent movements may be in part explained by the relief experienced through muscular exertion, and in part by the undirected overflow of nerve-force from the excited sensorium. But under the sudden loss of a beloved person, one of the first and commonest thoughts which occurs, is that something more might have been done to save the lost one. An excellent observer,[57] in describing the behaviour of a girl at the sudden death of her father, says she "went about the house wringing her hands like a creature demented, saying 'It was her fault;' 'I should never have left him;' 'If I had only sat up with him.'" &c. With such ideas vividly present before the mind, there would arise, through the principle of associated habit, the strongest tendency to energetic action of some kind.

As soon as the sufferer is fully conscious that nothing can be done, despair or deep sorrow takes the place of frantic grief. The sufferer sits motionless, or gently rocks to and fro; the circulation becomes languid; respiration is almost forgotten, and deep sighs are drawn. All this reacts on the brain, and prostration soon follows with collapsed muscles and dulled eyes. As associated habit no longer prompts the sufferer to action, he is urged by his friends to voluntary exertion, and not to give way to silent, motionless grief. Exertion stimulates the heart, and this reacts on the brain, and aids the mind to bear its heavy load.

Pain, if severe, soon induces extreme depression or prostration; but it is at first a stimulant and excites to action, as we see when we whip a horse, and as is shown by the horrid tortures inflicted in foreign lands on exhausted dray-bullocks, to rouse them to renewed exertion. Fear again is the most depressing of all the emotions; and it soon induces utter, helpless prostration, as if in consequence of, or in association with, the most violent and prolonged attempts to escape from the danger, though no such attempts have actually been made. Nevertheless, even extreme fear often acts at first as a powerful stimulant. A man or animal driven through terror to desperation, is endowed with wonderful strength, and is notoriously dangerous in the highest degree.

On the whole we may conclude that the principle of the direct action of the sensorium on the body, due to the constitution of the nervous system, and from the first independent of the will, has been highly influential in determining many expressions. Good instances are afforded by the trembling of the muscles, the sweating of the skin, the modified secretions of the alimentary canal and glands, under various emotions and sensations. But actions of this kind are often combined with others, which follow from our first principle, namely, that actions which have often been of direct or indirect service, under certain states of the mind, in order to gratify or relieve certain sensations, desires, &c., are still performed under analogous circumstances through mere habit, although of no service. We have combinations of this kind, at least in part, in the frantic gestures of rage and in the writhings of extreme pain; and, perhaps, in the increased action of the heart and of the respiratory organs. Even when these and other emotions or sensations are aroused in a very feeble manner, there will still be a tendency to similar actions, owing to the force of long-associated habit; and those actions which are least under voluntary control will generally be longest retained. Our second principle of antithesis has likewise occasionally come into play.

Finally, so many expressive movements can be explained, as I trust will be seen in the course of this volume, through the three principles which have now been discussed, that we may hope hereafter to see all thus explained, or by closely analogous principles. It is, however, often impossible to decide how much weight ought to be attributed, in each particular case, to one of our principles, and how much to another; and very many points in the theory of Expression remain inexplicable.

CHAPTER IV

Means of Expression in Animals

THE EMISSION OF SOUNDS • VOCAL SOUNDS • SOUNDS OTHERWISE PRODUCED • ERECTION OF THE DERMAL APPENDAGES, HAIRS, FEATHERS, &C., UNDER THE EMOTIONS OF ANGER AND TERROR • THE DRAWING BACK OF THE EARS AS A PREPARATION FOR FIGHTING, AND AS AN EXPRESSION OF ANGER • ERECTION OF THE EARS AND RAISING THE HEAD, A SIGN OF ATTENTION

IN THIS AND THE FOLLOWING chapter I will describe, but only in sufficient detail to illustrate my subject, the expressive movements, under different states of the mind, of some few well-known animals. But before considering them in due succession, it will save much useless repetition to discuss certain means of expression common to most of them.

The emission of Sounds.—With many kinds of animals, man included, the vocal organs are efficient in the highest degree as a means of expression. We have seen, in the last chapter, that when the sensorium is strongly excited, the muscles of the body are generally thrown into violent action; and as a consequence, loud sounds are uttered, however silent the animal may generally be, and although the sounds may be of no use. Hares and rabbits for instance, never, I believe, use their vocal organs except in the extremity of suffering; as, when a wounded hare is killed by the sportsman, or when a young rabbit is caught by a stoat. Cattle and horses suffer great pain in silence; but when this is excessive, and especially when associated with terror, they utter fearful sounds. I have often recognized, from a distance on the Pampas, the agonized death-bellow of the cattle, when caught by the lasso and hamstrung. It is said that horses, when attacked by wolves, utter loud and peculiar screams of distress.

Involuntary and purposeless contractions of the muscles of the chest and glottis, excited in the above manner, may have first given rise to the emission of vocal sounds. But the voice is now largely used by many animals for various purposes; and habit seems to have played an important part in its employment under other circumstances. Naturalists have remarked, I believe with truth, that social animals, from habitually using their vocal organs as a means of intercommunication, use them on other occasions much more freely than other animals. But there are marked exceptions to this rule, for instance, with the rabbit. The principle, also, of association, which is so widely extended in its power, has likewise played its part. Hence it follows that the voice, from having been habitually employed as a serviceable aid under certain conditions, inducing pleasure, pain, rage, &c., is commonly used whenever the same sensations or emotions are excited, under quite different conditions, or in a lesser degree.

The sexes of many animals incessantly call for each other during the breeding-season; and in not a few cases, the male endeavours thus to charm or excite the female. This, indeed, seems to have been the primeval use and means of development of the voice, as I have attempted to show in my 'Descent of Man.' Thus the use of the vocal organs will have become associated with the anticipation of the strongest pleasure which animals are capable of feeling. Animals which live in society often call to each other when separated, and evidently feel much joy at meeting; as we see with a horse, on the return of his companion, for whom he has been neighing. The mother calls incessantly for her lost young ones; for instance, a cow for her calf; and the young of many animals call for their mothers. When a flock of sheep is scattered, the ewes bleat incessantly for their lambs, and their mutual pleasure at coming together is manifest. Woe betide the man who meddles with the young of the larger and fiercer quadrupeds, if they hear the cry of distress from their young. Rage leads to the violent exertion of all the muscles, including those of the voice; and some animals, when enraged, endeavour to strike terror into their enemies by its power and harshness, as the lion does by roaring, and the dog by growling. I infer that their object is to strike terror, because the lion at the same time erects the hair of its mane, and the dog the hair along its back, and thus they make themselves appear as large and terrible as possible. Rival males try to excel and challenge each other by their voices, and this leads to deadly contests. Thus the use of the voice will have become associated with the emotion of anger, however it may be aroused. We have also seen that intense pain, like rage, leads to violent outcries, and the exertion of screaming by itself gives some relief; and thus the use of the voice will have become associated with suffering of any kind.

The cause of widely different sounds being uttered under different emotions and sensations is a very obscure subject. Nor does the rule always hold good that there is any marked difference. For instance with the dog, the bark of anger and that of joy do not differ much, though they can be distinguished. It is not probable that any precise explanation of the cause or source of each particular sound, under different states of the mind, will ever be given. We know that some animals, after being domesticated, have acquired the habit of uttering sounds which were not natural to them.[58] Thus domestic dogs, and even tamed jackals, have learnt to bark, which is a noise not proper to any species of the genus, with the exception of the *Canis latrans* of North America, which is said to bark. Some breeds, also, of the domestic pigeon have learnt to coo in a new and quite peculiar manner.

The character of the human voice, under the influence of various emotions, has been discussed by Mr. Herbert Spencer[59] in his interesting essay on Music. He clearly shows that the voice alters much under different conditions, in loudness and in quality, that is, in resonance and *timbre,* in pitch and intervals. No one can listen to an eloquent orator or preacher, or to a man calling angrily to another, or to one expressing astonishment, without being struck with the truth of Mr. Spencer's remarks. It is curious how early in life the modulation of the voice becomes expressive. With one of my children, under the age of two years, I clearly perceived that his humph of assent was rendered by a slight modulation strongly emphatic; and that by a peculiar whine his negative expressed obstinate determination. Mr. Spencer further shows that emotional speech, in all the above respects is intimately related to vocal music, and consequently to instrumental music; and he attempts to explain the characteristic qualities of both on physiological grounds,—namely, on "the general law that a feeling is a stimulus to muscular action." It may be admitted that the voice is affected through this law; but the explanation appears to me too general and vague to throw much light on the various differences, with the exception of that of loudness, between ordinary speech and emotional speech, or singing.

This remark holds good, whether we believe that the various qualities of the voice originated in speaking under the excitement of strong feelings, and that these qualities have subsequently been transferred to vocal music; or whether we believe, as I maintain, that the habit of uttering musical sounds was first developed, as a means of courtship, in the early progenitors of man, and thus became associated with the strongest emotions of which they were capable,—namely, ardent love, rivalry and triumph. That animals utter musical notes is familiar to every one, as we may daily hear in the singing of birds. It is a more remarkable fact that an ape, one of the Gibbons, produces an exact octave of musical sounds, ascending and descending the scale by half-tones; so that this monkey "alone of brute mammals may be said to sing."[60] From this fact, and from the analogy of other animals, I have been led to infer that the progenitors of man probably uttered musical tones, before they had acquired the power of articulate speech; and that consequently, when the voice is used under any strong emotion, it tends to assume, through the principle of association, a musical character. We can plainly perceive, with some of the lower animals, that the males employ their voices to please the females, and that they themselves take pleasure in their own vocal utterances; but why particular sounds are uttered, and why these give pleasure cannot at present be explained.

That the pitch of the voice bears some relation to certain states of feeling is tolerably clear. A person gently complaining of ill-treatment, or slightly suffering, almost always speaks in a high-pitched voice. Dogs, when a little impatient, often make a high piping note through their noses, which at once strikes us as plaintive;[61] but how difficult it is to know whether the sound is essentially plaintive, or only appears so in this particular case, from our having learnt by experience what it means! Rengger, states[62] that the monkeys *(Cebus azaræ)*, which he kept in Paraguay, expressed astonishment by a half-piping, half-snarling noise; anger or impatience, by repeating the sound *hu hu* in a deeper, grunting voice; and fright or pain, by shrill screams. On the other hand, with mankind, deep groans and high piercing screams equally express an agony of pain. Laughter may be either high or low; so that, with adult men, as Haller long ago remarked,[63] the sound partakes of the character of the vowels (as pronounced in German) *O* and *A*; whilst with children and women, it has more of the character of *E* and *I*; and these latter vowel-sounds naturally have, as Helmholtz has shown, a higher pitch than the former; yet both tones of laughter equally express enjoyment or amusement.

In considering the mode in which vocal utterances express emotion, we are naturally led to inquire into the cause of what is called "expression" in music. Upon this point Mr. Litchfield, who has long attended to the subject of music, has been so kind as to give me the following remarks:—"The question, what is the essence of musical 'expression' involves a number of obscure points, which, so far as I am aware, are as yet unsolved enigmas. Up to a certain point, however, any law which is found to hold as to the expression of the emotions by simple sounds must apply to the more developed mode of expression in song, which may be taken as the primary type of all music. A great part of the emotional effect of a song depends on the character of the action by which the sounds are produced. In songs, for instance, which express great vehemence of passion, the effect often chiefly depends on the forcible utterance of some one or two characteristic passages which demand great exertion of vocal force; and it will be frequently noticed that a song of this character fails of its proper effect when sung by a voice of sufficient power and range to give the characteristic passages without much exertion. This is, no doubt, the secret of the loss of effect so often produced by the transposition of a song from one key to another. The effect is thus seen to depend not merely on the actual sounds, but also in part on the nature of the action which produces the sounds. Indeed it is obvious that whenever we feel the 'expression' of a song to be due to its quickness or slowness of movement—to smoothness of flow, loudness of utterance, and so on—we are, in fact, interpreting the muscular actions which produce sound, in the same way in which we interpret muscular action generally. But this leaves unexplained the more subtle and more specific effect which we call the *musical* expression of the song—the delight given by its melody, or even by the separate sounds which make up the melody. This is an effect indefinable in language—one which, so far as I am aware, no one has been able to analyse, and which the ingenious speculation of Mr. Herbert Spencer as to the origin of music leaves quite unexplained. For it is certain that the *melodic* effect of a series of sounds does not depend in the least on their loudness or softness, or on their *absolute* pitch. A tune is always the same tune, whether it is sung loudly or softly, by a child or a man; whether it is played on a flute or on a trombone. The purely musical effect of any sound depends on its place in what is technically called a 'scale;' the same sound producing absolutely different effects on the ear, according as it is heard in connection with one or another series of sounds.

It is on this *relative* association of the sounds that all the essentially characteristic effects which are summed up in the phrase 'musical expression,' depend. But why certain associations of sounds have such-and-such effects, is a problem which yet remains to be solved. These effects must indeed, in some way or other, be connected with the well-known arithmetical relations between the rates of vibration of the sounds which form a musical scale. And it is possible—but this is merely a suggestion—that the greater or less mechanical facility with which the vibrating apparatus of the human larynx passes from one state of vibration to another, may have been a primary cause of the greater or less pleasure produced by various sequences of sounds."

But leaving aside these complex questions and confining ourselves to the simpler sounds, we can, at least, see some reasons for the association of certain kinds of sounds with certain states of mind. A scream, for instance, uttered by a young animal, or by one of the members of a community, as a call for assistance, will naturally be loud, prolonged, and high, so as to penetrate to a distance. For Helmholtz has shown[64] that, owing to the shape of the internal cavity of the human ear and its consequent power of resonance, high notes produce a particularly strong impression. When male animals utter sounds in order to please the females, they would naturally employ those which are sweet to the ears of the species; and it appears that the same sounds are often pleasing to widely different animals, owing to the similarity of their nervous systems, as we ourselves perceive in the singing of birds and even in the chirping of certain tree-frogs giving us pleasure. On the other hand, sounds produced in order to strike terror into an enemy, would naturally be harsh or displeasing.

Whether the principle of antithesis has come into play with sounds, as might perhaps have been expected, is doubtful. The interrupted, laughing or tittering sounds made by man and by various kinds of monkeys when pleased, are as different as possible from the prolonged screams of these animals when distressed. The deep grunt of satisfaction uttered by a pig, when pleased with its food, is widely different from its harsh scream of pain or terror. But with the dog, as lately remarked, the bark of anger and that of joy are sounds which by no means stand in opposition to each other; and so it is in some other cases.

There is another obscure point, namely, whether the sounds which are produced under various states of the mind determine the shape of the mouth, or whether its shape is not determined by independent causes, and the sound thus modified. When young infants cry they open their mouths widely, and this, no doubt, is necessary for pouring forth a full volume of sound; but the mouth then assumes, from a quite distinct cause, an almost quadrangular shape, depending, as will hereafter be explained, on the firm closing of the eyelids, and consequent drawing up of the upper lip. How far this square shape of the mouth modifies the wailing or crying sound, I am not prepared to say; but we know from the researches of Helmholtz and others that the form of the cavity of the mouth and lips determines the nature and pitch of the vowel sounds which are produced.

It will also be shown in a future chapter that, under the feeling of contempt or disgust, there is a tendency, from intelligible causes, to blow out of the mouth or nostrils, and this produces sounds like *pooh* or *pish*. When any one is startled or suddenly astonished, there is an instantaneous tendency, likewise from an intelligible cause, namely, to be ready for prolonged exertion, to open the mouth widely, so as to draw a deep and rapid inspiration. When the next full expiration follows, the mouth is slightly closed, and the lips, from causes hereafter to be discussed, are somewhat protruded; and this form of the mouth, if the voice be at all exerted, produces, according to Helmholtz, the sound of the vowel *O*. Certainly a deep sound of a prolonged *Oh!* may be heard from a whole crowd of people immediately after witnessing any astonishing spectacle. If, together with surprise, pain be felt, there is a tendency to contract all the muscles of the body, including those of the face, and the lips will then be drawn back; and this will perhaps account for the sound becoming higher and assuming the character of *Ah!* or *Ach!*

As fear causes all the muscles of the body to tremble, the voice naturally becomes tremulous, and at the same time husky from the dryness of the mouth, owing to the salivary glands failing to act. Why the laughter of man and the tittering of monkeys should be a rapidly reiterated sound, cannot be explained. During the utterance of these sounds, the mouth is transversely elongated by the corners being drawn backwards and upwards; and of this fact an explanation will be attempted in a future chapter. But the whole subject of the differences of the sounds produced under different states of the mind is so obscure, that I have succeeded in throwing hardly any light on it; and the remarks which I have made, have but little significance.

ALL THE SOUNDS hitherto noticed depend on the respiratory organs; but sounds produced by wholly different means are likewise expressive. Rabbits stamp loudly on the ground as a signal to their comrades; and if a man knows how to do so properly, he may on a quiet evening hear the rabbits answering him all around. These animals, as well as some others, also stamp on the ground when made angry. Porcupines rattle their quills and vibrate their tails when angered; and one behaved in this manner when a live snake was placed in its compartment. The quills on the tail are very different from those on the body: they are short, hollow, thin like a goose-quill, with their ends transversely truncated, so that they are open; they are supported, on long, thin, elastic foot-stalks. Now, when the tail is rapidly shaken, these hollow quills strike against each other and produce, as I heard in the presence of Mr. Bartlett, a peculiar continuous sound. We can, I think, understand why porcupines have been provided, through the modification of their protective spines, with this special sound-producing instrument. They are nocturnal animals, and if they scented or heard a prowling beast of prey, it would be a great advantage to them in the dark to give warning to their enemy what they were, and that they were furnished with dangerous spines. They would thus escape being attacked. They are, as I may add, so fully conscious of the power of their weapons, that when enraged they will charge backwards with their spines erected, yet still inclined backwards.

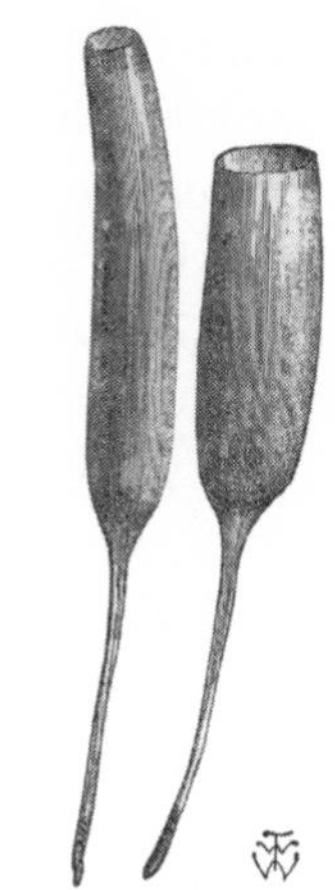

Fig. 11. *Sound-producing quills from the tail of the Porcupine.*

Many birds during their courtship produce diversified sounds by means of specially adapted feathers. Storks, when excited, make a loud clattering noise with their beaks. Some snakes produce a grating or rattling noise. Many insects stridulate by rubbing together specially modified parts of their hard integuments. This stridulation generally serves as a sexual charm or call; but it is likewise used to express different emotions.[65] Every one who has attended to bees knows that their humming changes when they are angry; and this serves as a warning that there is danger of being stung. I have made these few remarks because some writers have laid so much stress on the vocal and respiratory organs as having been specially adapted for expression, that it was advisable to show that sounds otherwise produced serve equally well for the same purpose.

Erection of the dermal appendages.—Hardly any expressive movement is so general as the involuntary erection of the hairs, feathers and other dermal appendages; for it is common throughout three of the great vertebrate classes. These appendages are erected under the excitement of anger or terror; more especially when these emotions are combined, or quickly succeed each other. The action serves to make the animal appear larger and more frightful to its enemies or rivals, and is generally accompanied by various voluntary movements adapted for the same purpose, and by the utterance of savage sounds. Mr. Bartlett, who has had such wide experience with animals of all kinds, does not doubt that this is the case; but it is a different question whether the power of erection was primarily acquired for this special purpose.

I will first give a considerable body of facts showing how general this action is with mammals, birds and reptiles; retaining what I have to say in regard to man for a future chapter. Mr. Sutton, the intelligent keeper in the Zoological Gardens, carefully observed for me the Chimpanzee and Orang; and he states that when they are suddenly frightened, as by a thunderstorm, or when they are made angry, as by being teased, their hair becomes erect. I saw a chimpanzee who was alarmed at the sight of a black coalheaver, and the hair rose all over his body; he made little starts forward as if to attack the man, without any real intention of doing so, but with the hope, as the keeper remarked, of frightening him. The Gorilla, when enraged, is described by Mr. Ford[66] as having his crest of hair "erect and projecting forward, his nostrils dilated, and his under lip thrown down; at the same time uttering his characteristic yell, designed, it would seem, to terrify his antagonists." I saw the hair on the Anubis baboon, when angered bristling along the back, from the neck to the loins, but not on the rump or other parts of the body. I took a stuffed snake into the monkey-house, and the hair on several of the species instantly became erect; especially on their tails, as I particularly noticed with the *Cercopithecus nictitans*. Brehm states[67] that the *Midas œdipus* (belonging to the American division) when excited erects its mane, in order, as he adds, to make itself as frightful as possible.

With the Carnivora the erection of the hair seems to be almost universal, often accompanied by threatening movements, the uncovering of the teeth and the utterance of savage growls. In the Herpestes, I have seen the hair on end over nearly the whole body, including the tail; and the dorsal crest is erected in a conspicuous manner by the Hyæna and Proteles. The enraged lion erects his mane. The bristling of the hair along the neck and back of the dog, and over the whole body of the cat, especially on the tail, is familiar to every one. With the cat it apparently occurs only under fear; with the dog, under anger and fear; but not, as far as I have observed, under abject fear, as when a dog is going to be flogged by a severe gamekeeper. If, however, the dog shows fight, as sometimes happens, up goes his hair. I have often noticed that the hair of a dog is particularly liable to rise, if he is half angry and half afraid, as on beholding some object only indistinctly seen in the dusk.

I have been assured by a veterinary surgeon that he has often seen the hair erected on horses and cattle, on which he had operated and was again going to operate. When I showed a stuffed snake to a Peccary, the hair rose in a wonderful manner along its back; and so it does with the boar when enraged. An Elk which gored a man to death in the United States, is described as first brandishing his antlers, squealing with rage and stamping on the ground; "at length his hair was seen to rise and stand on end," and then he plunged forward to the attack.[68] The hair likewise becomes erect on goats, and, as I hear from Mr. Blyth, on some Indian antelopes. I have seen it erected on the hairy Ant-eater; and on the Agouti, one of the Rodents. A female Bat,[69] which reared her young under confinement, when any one looked into the cage "erected the fur on her back, and bit viciously at intruding fingers."

Birds belonging to all the chief Orders ruffle their feathers when angry or frightened. Every one must have seen two cocks, even quite young birds, preparing to fight with erected neck-hackles; nor can these feathers when erected serve as a means of defence, for cock-fighters have found by experience that it is advantageous to trim them. The male Ruff *(Machetes pugnax)* likewise erects its collar of feathers when fighting. When a dog approaches a common hen with her chickens, she spreads out her wings, raises her tail, ruffles all her feathers, and looking as ferocious as possible, dashes at the intruder. The tail is not always held in exactly the same position; it is sometimes so much erected, that the central feathers, as in the accompanying drawing, almost touch the back. Swans, when angered, likewise raise their wings and tail, and erect their feathers. They open their beaks, and make by paddling little rapid starts forwards, against any one who approaches the water's edge too closely. Tropic birds[70] when disturbed on their nests are said not to fly away, but "merely to stick out their feathers and scream." The

Fig. 12. *Hen driving away a dog from her chickens. Drawn from life by Mr. Wood.*

Barn-owl, when approached "instantly swells out its plumage, extends its wings and tail, hisses and clacks its mandibles with force and rapidity."[71] So do other kinds of owls. Hawks, as I am informed by Mr. Jenner Weir, likewise ruffle their feathers, and spread out their wings and tail under similar circumstances. Some kinds of parrots erect their feathers; and I have seen this action in the Cassowary, when angered at the sight of an Anteater. Young cuckoos in the nest, raise their feathers, open their mouths widely, and make themselves as frightful as possible.

Fig. 13. *Swan driving away an Intruder. Drawn from life by Mr. Wood.*

Small birds, also, as I hear from Mr. Weir, such as various finches, buntings and warblers, when angry, ruffle all their feathers, or only those round the neck; or they spread out their wings and tail-feathers. With their plumage in this state, they rush at each other with open beaks and threatening gestures. Mr. Weir concludes from his large experience that the erection of the feathers is caused much more by anger than by fear. He gives as an instance a hybrid gold-finch of a most irascible disposition, which when approached too closely by a servant, instantly assumes the appearance of a ball of ruffled feathers. He believes that birds when frightened, as a general rule, closely adpress all their feathers, and their consequently diminished size is often astonishing. As soon as they recover from their fear or surprise, the first thing which they do is to shake out their feathers. The best instances of this adpression of the feathers and apparent shrinking of the body from fear, which Mr. Weir has noticed, has been in the quail and grass-parrakeet.[72] The habit is intelligible in these birds from their being accustomed, when in danger, either to squat on the ground or to sit motionless on a branch, so as to escape detection. Though, with birds, anger may be the chief and commonest cause of the erection of the feathers, it is probable that young cuckoos when looked at in the nest, and a hen with her chickens when approached by a dog, feel at least some terror. Mr. Tegetmeier informs me that with game-cocks, the erection of the feathers on the head has long been recognized in the cock-pit as a sign of cowardice.

The males of some lizards, when fighting together during their courtship, expand their throat pouches or frills, and erect their dorsal crests.[73] But Dr. Günther does not believe that they can erect their separate spines or scales.

We thus see how generally throughout the two higher vertebrate classes, and with some reptiles, the dermal appendages are erected under the influence of anger and fear. The movement is effected, as we know from Kölliker's interesting discovery, by the contraction of minute, unstriped, involuntary muscles,[74] often called *arrectores pili,* which are attached to the capsules of the separate hairs, feathers, &c. By the contraction of these muscles the hairs can be instantly erected, as we see in a dog, being at the same time drawn a little out of their sockets; they are afterwards quickly depressed. The vast number of these minute muscles over the whole body of a hairy quadruped is astonishing. The erection of the hair is, however, aided in some cases, as with that on the head of a man, by the striped and voluntary muscles of the underlying *panniculus carnosus.* It is by the action of these latter muscles, that the hedgehog erects its spines. It appears, also, from the researches of Leydig[75] and others, that striped fibres extend from the panniculus to some of the larger hairs, such as the vibrissæ of certain quadrupeds. The *arrectores pili* contract not only under the above emotions, but from the application of cold to the surface. I remember that my mules and dogs, brought from a lower and warmer country, after spending a night on the bleak Cordillera, had the hair all over their bodies as erect as under the greatest terror. We see the same action in our own *goose-skin* during the chill before a fever-fit. Mr. Lister has also found,[76] that tickling a neighbouring part of the skin causes the erection and protrusion of the hairs.

From these facts it is manifest that the erection of the dermal appendages is a reflex action, independent of the will; and this action must be looked at, when, occurring under the influence of anger or fear, not as a power acquired for the sake of some advantage, but as an incidental result, at least to a large extent, of the sensorium being affected. The result, in as far as it is incidental, may be compared with the profuse sweating from an agony of pain or terror. Nevertheless, it is remarkable how slight an excitement often suffices to cause the hair to become erect; as when two dogs pretend to fight together in play. We have, also, seen in a large number of animals, belonging to widely distinct classes, that the erection of the hair or feathers is almost always accompanied by various voluntary movements—by threatening gestures, opening the mouth, uncovering the teeth, spreading out of the wings and tail by birds, and by the utterance of harsh sounds; and the purpose of these voluntary movements is unmistakable. Therefore it seems hardly credible that the co-ordinated erection of the dermal appendages, by which the animal is made to appear larger and more terrible to its enemies or rivals, should be altogether an incidental and purposeless result of the disturbance of the sensorium. This seems almost as incredible as that the erection by the hedgehog of its spines, or of the quills by the porcupine, or of the ornamental plumes by many birds during their courtship, should all be purposeless actions.

We here encounter a great difficulty. How can the contraction of the unstriped and involuntary *arrectores pili* have been co-ordinated with that of various voluntary muscles for the same special purpose? If we could believe that the arrectores primordially had been voluntary muscles, and had since lost their stripes and become involuntary, the case would be comparatively simple. I am not, however, aware that there is any evidence in favour of this view; although the reversed transition would not have presented any great difficulty, as the voluntary muscles are in an unstriped condition in the embryos of the higher animals, and in the larvæ of some crustaceans. Moreover in the deeper layers of the skin of adult birds, the muscular network is, according to Leydig,[77] in a transitional condition; the fibres exhibiting only indications of transverse striation.

Another explanation seems possible. We may admit that originally the *arrectores pili* were slightly acted on in a direct manner, under the influence of rage and terror, by the disturbance of the nervous system; as is undoubtedly the case with our so-called *goose-skin* before a fever-fit. Animals have been repeatedly excited by rage and terror during many generations; and consequently the direct effects of the disturbed nervous system on the dermal appendages will almost certainly have been increased through habit and through the tendency of nerve-force to pass readily along accustomed channels. We shall find this view of the force of habit strikingly confirmed in a future chapter, where it will be shown that the hair of the insane is affected in an extraordinary manner, owing to their repeated accesses of fury and terror. As soon as with animals the power of erection had thus been strengthened or increased, they must often have seen the hairs or feathers erected in rival and enraged males, and the bulk of their bodies thus increased. In this case it appears possible that they might have wished to make themselves appear larger and more terrible to their enemies, by voluntarily assuming a threatening attitude and uttering harsh cries; such attitudes and utterances after a time becoming through habit instinctive. In this manner actions performed by the contraction of voluntary muscles might have been combined for the same special purpose with those effected by involuntary muscles. It is even possible that animals, when excited and dimly conscious of some change in the state of their hair, might act on it by repeated exertions of their attention and will; for we have reason to believe that the will is able to influence in an obscure manner the action of some unstriped or involuntary muscles, as in the period of the peristaltic movements of the intestines, and in the contraction of the bladder. Nor must we overlook the part which variation and natural selection may have played; for the males which succeeded in making themselves appear the most terrible to their rivals, or to their other enemies, if not of overwhelming power, will on an average have left more offspring to inherit their characteristic qualities, whatever these may be and however first acquired, than have other males.

The inflation of the body, and other means of exciting fear in an enemy.—Certain Amphibians and Reptiles, which either have no spines to erect, or no muscles by which they can be erected, enlarge themselves when alarmed or angry by inhaling air. This is well known to be the case with toads and frogs. The latter animal is made, in Æsop's fable of the 'Ox and the Frog,' to blow itself up from vanity and envy until it burst. This action must have been observed during the most ancient times, as, according to Mr. Hensleigh Wedgwood,[78] the word *toad* expresses in several of the languages of Europe the habit of swelling. It has been observed with some of the exotic species in the Zoological Gardens; and Dr. Günther believes that it is general throughout the group. Judging from analogy, the primary purpose probably was to make the body appear as large and frightful as possible to an enemy; but another, and perhaps more important secondary advantage is thus gained. When frogs are seized by snakes, which are their chief enemies, they enlarge themselves wonderfully; so that if the snake be of small size, as Dr. Günther informs me, it cannot swallow the frog, which thus escapes being devoured.

Chameleons and some other lizards inflate themselves when angry. Thus a species inhabiting Oregon, the *Tapaya Douglasii,* is slow in its movements and does not bite, but has a ferocious aspect; "when irritated it springs in a most threatening manner at anything pointed at it, at the same time opening its mouth wide and hissing audibly, after which it inflates its body, and shows other marks of anger."[79]

Several kinds of snakes likewise inflate themselves when irritated. The puff-adder *(Clotho arietans)* is remarkable in this respect; but I believe, after carefully watching these animals, that they do not act thus for the sake of increasing their apparent bulk, but simply for inhaling a large supply of air, so as to produce their surprisingly loud, harsh, and prolonged hissing sound. The Cobras-de-capello, when irritated, enlarge themselves a little, and hiss moderately; but, at the same time they lift their heads aloft, and dilate by means of their elongated anterior ribs, the skin on each side of the neck into a large flat disk,—the so-called hood. With their widely opened mouths, they then assume a terrific aspect. The benefit thus derived ought to be considerable, in order to compensate for the somewhat lessened rapidity (though this is still great) with which, when dilated, they can strike at their enemies or prey; on the same principle that a broad, thin piece of wood cannot be moved through the air so quickly as a small round stick. An innocuous snake, the *Tropidonotus macrophthalmus,* an inhabitant of India, likewise dilates its neck when irritated; and consequently is often mistaken for its compatriot, the deadly Cobra.[80] This resemblance perhaps serves as some protection to the Tropidonotus. Another innocuous species, the Dasypeltis of South Africa, blows itself out, distends its neck, hisses and darts at an intruder.[81] Many other snakes hiss under similar circumstances. They also rapidly vibrate their protruded tongues; and this may aid in increasing their terrific appearance.

Snakes possess other means of producing sounds besides hissing. Many years ago I observed in South America that a venomous Trigonocephalus, when disturbed, rapidly vibrated the end of its tail, which striking against the dry grass and twigs produced a rattling noise that could be distinctly heard at the distance of six feet.[82] The deadly and fierce *Echis carinata* of India produces "a curious prolonged, almost hissing sound" in a very different manner, namely by rubbing "the sides of the folds of its body against each other," whilst the head remains in almost the same position. The scales on the sides, and not on other parts of the body, are strongly keeled, with the keels toothed like a saw; and as the coiled-up animal rubs its sides together, these grate against each other.[83] Lastly, we have the well-known case of the Rattle-snake. He who has merely shaken the rattle of a dead snake, can form no just idea of the sound produced by the living animal. Professor Shaler states that it is indisguishable from that made by the male of a large Cicada (an Homopterous insect), which inhabits the same district.[84] In the Zoological Gardens, when the rattle-snakes and puff-adders were greatly excited at the same time, I was much struck at the similarity of the sound produced by them; and although that made by the rattle-snake is louder and shriller than the hissing of the puff-adder, yet when standing at some yards distance I could scarcely distinguish the two. For whatever purpose the sound is produced by the one species, I can hardly doubt that it serves for the same purpose in the other species; and I conclude from the threatening gestures made at the same time by many snakes, that their hissing,—the rattling of the rattle-snake and of the tail of the Trigonocephalus,—the grating of the scales of the Echis,—and the dilatation of the hood of the Cobra,—all subserve the same end, namely, to make them appear terrible to their enemies.[85]

It seems at first a probable conclusion that venomous snakes, such as the foregoing, from being already so well defended by their poison-fangs, would never be attacked by any enemy; and consequently would have no need to excite additional terror. But this is far from being the case, for they are largely preyed on in all quarters of the world by many animals. It is well known that pigs are employed in the United States to clear districts infested with rattle-snakes, which they do most effectually.[86] In England the hedgehog attacks and devours the viper. In India, as I hear from Dr. Jerdon, several kinds of hawks, and at least one mammal, the Herpestes, kill cobras and other venomous species;[87] and so it is in South Africa. Therefore it is by no means improbable that any sounds or signs by which the venomous species could instantly make themselves recognised as dangerous, would be of more service to them than to the innocuous species which would not be able, if attacked, to inflict any real injury.

Having said thus much about snakes, I am tempted to add a few remarks on the means by which the rattle of the rattle-snake was probably developed. Various animals, including some lizards, either curl or vibrate their tails when excited. This is the case with many kinds of snakes.[88] In the Zoological Gardens, an innocuous species, the *Coronella Sayi,* vibrates its tail so rapidly that it becomes almost invisible. The Trigonocephalus, before alluded to, has the same habit; and the extremity of its tail is a little enlarged, or ends in a bead. In the Lachesis, which is so closely allied to the rattle-snake that it was placed by Linnæus in the same genus, the tail ends in a single, large, lancet-shaped point or scale. With some snakes the skin, as Professor Shaler remarks, "is more imperfectly detached from the region about the tail than at other parts of the body." Now if we suppose that the end of the tail of some ancient American species was enlarged, and was covered by a single large scale, this could hardly have been cast off at the successive moults. In this case it would have been permanently retained, and at each period of growth, as the snake grew larger, a new scale, larger than the last, would have been formed above it, and would likewise have been retained. The foundation for the development of a rattle would thus have been laid; and it would have been habitually used, if the species, like so many others, vibrated its tail whenever it was irritated. That the rattle has since been specially developed to serve as an efficient sound-producing instrument, there can hardly be a doubt; for even the vertebræ included within the extremity of the tail have been altered in shape and cohere. But there is no greater improbability in various structures, such as the rattle of the rattle-snake,—the lateral scales of the Echis,—the neck with the included ribs of the Cobra,—and the whole body of the puff-adder,—having been modified for the sake of warning and frightening away their enemies, than in a bird, namely, the wonderful Secretary-hawk *(Gypogeranus)* having had its whole frame modified for the sake of killing snakes with impunity. It is highly probable, judging from what we have before seen, that this bird would ruffle its feathers whenever it attacked a snake; and it is certain that the Herpestes, when it eagerly rushes to attack a snake, erects the hair all over its body, and especially that on its tail.[89] We have also seen that some porcupines, when angered or alarmed at the sight of a snake, rapidly vibrate their tails, thus producing a peculiar sound by the striking together of the hollow quills. So that here both the attackers and the attacked endeavour to make themselves as dreadful as possible to each other; and both possess for this purpose specialised means, which, oddly enough, are nearly the same in some of these cases. Finally we can see that if, on the one hand, those individual snakes, which were best able to frighten away their enemies, escaped best from being devoured; and if, on the other hand, those individuals of the attacking enemy survived in larger numbers which were the best fitted for the dangerous task of killing and devouring venomous snakes;—then in the one case as in the other, beneficial variations, supposing the characters in question to vary, would commonly have been preserved through the survival of the fittest.

The Drawing back and pressure of the Ears to the Head.—The ears through their movements are highly expressive in many animals; but in some, such as man, the higher apes, and many ruminants, they fail in this respect. A slight difference in position serves to express in the plainest manner a different state of mind, as we may daily see in the dog; but we are here concerned only with the ears being drawn closely backwards and pressed to the head. A savage frame of mind is thus shown, but only in the case of those animals which fight with their teeth; and the care which they take to prevent their ears being seized by their antagonists, accounts for this position. Consequently, through habit and association, whenever they feel slightly savage, or pretend in their play to be savage, their ears are drawn back. That this is the true explanation may be inferred from the relation which exists in very many animals between their manner of fighting and the retraction of their ears.

All the Carnivora fight with their canine teeth, and all, as far as I have observed, draw their ears back when feeling savage. This may be continually seen with dogs when fighting in earnest, and with puppies fighting in play. The movement is different from the falling down and slight drawing back of the ears, when a dog feels pleased and is caressed by his master. The retraction of the ears may likewise be seen in kittens fighting together in their play, and in full-grown cats when really savage, as before illustrated in fig. 9 (p. 1292). Although their ears are thus to a large extent protected, yet they often get much torn in old male cats during their mutual battles. The same movement is very striking in tigers, leopards, &c., whilst growling over their food in menageries. The lynx has remarkably long ears; and their retraction, when one of these animals is approached in its cage, is very conspicuous, and is eminently expressive of its savage disposition. Even one of the Eared Seals, the *Otaria pusilla*, which has very small ears, draws them backwards, when it makes a savage rush at the legs of its keeper.

When horses fight together they use their incisors for biting, and their fore-legs for striking, much more than they do their hind-legs for kicking backwards. This has been observed when stallions have broken loose and have fought together, and may likewise be inferred from the kind of wounds which they inflict on each other. Every one recognises the vicious appearance which the drawing back of the ears gives to a horse. This movement is very different from that of listening to a sound behind. If an ill-tempered horse in a stall is inclined to kick backwards, his ears are retracted from habit, though he has no intention or power to bite. But when a horse throws up both hind-legs in play, as when entering an open field, or when just touched by the whip, he does not generally depress his ears, for he does not then feel vicious. Guanacoes fight savagely with their teeth; and they must do so frequently, for I found the hides of several which I shot in Patagonia deeply scored. So do camels; and both these animals, when savage, draw their ears closely backwards. Guanacoes, as I have noticed, when not intending to bite, but merely to spit their offensive saliva from a distance at an intruder, retract their ears. Even the hippopotamus, when threatening with its widely-open enormous mouth a comrade, draws back its small ears, just like a horse.

Now what a contrast is presented between the foregoing animals and cattle, sheep, or goats, which never use their teeth in fighting, and never draw back their ears when enraged! Although sheep and goats appear such placid animals, the males often join in furious contests. As deer form a closely related family, and as I did not know that they ever fought with their teeth, I was much surprised at the account given by Major Ross King of the Moose-deer in Canada. He says, when "two males chance to meet, laying back their ears and gnashing their teeth together, they rush at each other with appalling fury."[90] But Mr. Bartlett informs me that some species of deer fight savagely with their teeth, so that the drawing back of the ears by the moose accords with our rule. Several kinds of kangaroos, kept in the Zoological Gardens, fight by scratching with their fore-feet and by kicking with their hind-legs; but they never bite each other, and the keepers have never seen them draw back their ears when angered. Rabbits fight chiefly by kicking and scratching, but they likewise bite each other; and I have known one to bite off half the tail of its antagonist. At the commencement of their battles they lay back their ears, but afterwards, as they bound over and kick each other, they keep their ears erect, or move them much about.

Mr. Bartlett watched a wild boar quarrelling rather savagely with his sow; and both had their mouths open and their ears drawn backwards. But this does not appear to be a common action with domestic pigs when quarrelling. Boars fight together by striking upwards with their tusks; and Mr. Bartlett doubts whether they then draw back their ears. Elephants, which in like manner fight with their tusks, do not retract their ears, but, on the contrary, erect them when rushing at each other or at an enemy.

The rhinoceroses in the Zoological Gardens fight with their nasal horns, and have never been seen to attempt biting each other except in play; and the keepers are convinced that they do not draw back their ears, like horses and dogs, when feeling savage. The following statement, therefore, by Sir S. Baker[91] is inexplicable, namely, that a rhinoceros, which he shot in North Africa, "had no ears; they had been bitten off close to the head by another of the same species while fighting; and this mutilation is by no means uncommon."

Lastly, with respect to monkeys. Some kinds, which have moveable ears, and which fight with their teeth—for instance the *Cercopithecus ruber*—draw back their ears when irritated just like dogs; and they then have a very spiteful appearance. Other kinds, as the *Inuus ecaudatus*, apparently do not thus act. Again, other kinds—and this is a great anomaly in comparison with most other animals—retract their ears, show their teeth, and jabber, when they are pleased by being caressed. I observed this in two or three species of Macacus, and in the *Cynopithecus niger.* This expression, owing to our familiarity with dogs, would never be recognised as one of joy or pleasure by those unacquainted with monkeys.

Erection of the Ears.—This movement requires hardly any notice. All animals which have the power of freely moving their ears, when they are startled, or when they closely observe any object, direct their ears to the point towards which they are looking, in order to hear any sound from this quarter. At the same time they generally raise their heads, as all their organs of sense are there situated, and some of the smaller animals rise on their hind legs. Even those kinds which squat on the ground or instantly flee away to avoid danger, generally act momentarily in this manner, in order to ascertain the source and nature of the danger. The head being raised, with erected ears and eyes directed forwards, gives an unmistakable expression of close attention to any animal.

CHAPTER V

SPECIAL EXPRESSIONS OF ANIMALS

THE DOG, VARIOUS EXPRESSIVE MOVEMENTS OF • CATS • HORSES • RUMINANTS • MONKEYS, THEIR EXPRESSION OF JOY AND AFFECTION • OF PAIN • ANGER • ASTONISHMENT AND TERROR

The Dog.—I have already described (figs. 5 and 7) the appearance of a dog approaching another dog with hostile intentions, namely, with erected ears, eyes intently directed forwards, hair on the neck and back bristling, gait remarkably stiff, with the tail upright and rigid. So familiar is this appearance to us, that an angry man is sometimes said "to have his back up." Of the above points, the stiff gait and upright tail alone require further discussion. Sir C. Bell remarks[92] that, when a tiger or wolf is struck by its keeper and is suddenly roused to ferocity, "every muscle is in tension, and the limbs are in an attitude of strained exertion, prepared to spring." This tension of the muscles and consequent stiff gait may be accounted for on the principle of associated habit, for anger has continually led to fierce struggles, and consequently to all the muscles of the body having been violently exerted. There is also reason to suspect that the muscular system requires some short preparation, or some degree of innervation, before being brought into strong action. My own sensations lead me to this inference; but I cannot discover that it is a conclusion admitted by physiologists. Sir J. Paget, however, informs me that when muscles are suddenly contracted with the greatest force, without any preparation, they are liable to be ruptured, as when a man slips unexpectedly; but that this rarely occurs when an action, however violent, is deliberately performed.

With respect to the upright position of the tail, it seems to depend (but whether this is really the case I know not) on the elevator muscles being more powerful than the depressors, so that when all the muscles of the hinder part of the body are in a state of tension, the tail is raised. A dog in cheerful spirits, and trotting before his master with high, elastic steps, generally carries his tail aloft, though it is not held nearly so stiffly as when he is angered. A horse when first turned out into an open field, may be seen to trot with long elastic strides, the head and tail being held high aloft. Even cows when they frisk about from pleasure, throw up their tails in a ridiculous fashion. So it is with various animals in the Zoological Gardens. The position of the tail, however, in certain cases, is determined by special circumstances; thus as soon as a horse breaks into a gallop, at full speed, he always lowers his tail, so that as little resistance as possible may be offered to the air.

When a dog is on the point of springing on his antagonist, he utters a savage growl; the ears are pressed closely backwards, and the upper lip (fig. 14) is retracted out of the way of his teeth, especially of his canines. These movements may be observed with dogs and puppies in their play. But if a dog gets really savage in his play, his expression immediately changes. This, however, is simply due to the lips and ears being drawn back with much greater energy. If a dog only snarls at another, the lip is generally retracted on one side alone, namely towards his enemy.

Fig. 14. *Head of snarling Dog. From life, by Mr. Wood.*

The movements of a dog whilst exhibiting affection towards his master were described (figs. 6 and 8) in our second chapter. These consist in the head and whole body being lowered and thrown into flexuous movements, with the tail extended and wagged from side to side. The ears fall down and are drawn somewhat backwards, which causes the eyelids to be elongated, and alters the whole appearance of the face. The lips hang loosely, and the hair remains smooth. All these movements or gestures are explicable, as I believe, from their standing in complete antithesis to those naturally assumed by a savage dog under a directly opposite state of mind. When a man merely speaks to, or just notices, his dog, we see the last vestige of these movements in a slight wag of the tail, without any other movement of the body, and without even the ears being lowered. Dogs also exhibit their affection by desiring to rub against their masters, and to be rubbed or patted by them.

Gratiolet explains the above gestures of affection in the following manner: and the reader can judge whether the explanation appears satisfactory. Speaking of animals in general, including the dog, he says,[93] "C'est toujours la partie la plus sensible de leurs corps qui recherche les caresses ou les donne. Lorsque toute la longueur des flancs et du corps est sensible, l'animal serpente et rampe sous les caresses; et ces ondulations se propageant le long des muscles analogues des segments jusqu'aux extrémités de la colonne vertébrale, la queue se ploie et s'agite." Further on, he adds, that dogs, when feeling affectionate, lower their ears in order to exclude all sounds, so that their whole attention may be concentrated on the caresses of their master!

Dogs have another and striking way of exhibiting their affection, namely, by licking the hands or faces of their masters. They sometimes lick other dogs, and then it is always their chops. I have also seen dogs licking cats with whom they were friends. This habit probably originated in the females carefully licking their puppies—the dearest object of their love—for the sake of cleansing them. They also often give their puppies, after a short absence, a few cursory licks, apparently from affection. Thus the habit will have become associated with the emotion of love, however it may afterwards be aroused. It is now so firmly inherited or innate, that it is transmitted equally to both sexes. A female terrier of mine lately had her puppies destroyed, and though at all times a very affectionate creature, I was much struck with the manner in which she then tried to satisfy her instinctive maternal love by expending it on me; and her desire to lick my hands rose to an insatiable passion.

The same principle probably explains why dogs, when feeling affectionate, like rubbing against their masters and being rubbed or patted by them, for from the nursing of their puppies, contact with a beloved object has become firmly associated in their minds with the emotion of love.

The feeling of affection of a dog towards his master is combined with a strong sense of submission, which is akin to fear. Hence dogs not only lower their bodies and crouch a little as they approach their masters, but sometimes throw themselves on the ground with their bellies upwards. This is a movement as completely opposite as is possible to any show of resistance. I formerly possessed a large dog who was not at all afraid to fight with other dogs; but a wolf-like shepherd-dog in the neighbourhood, though not ferocious and not so powerful as my dog, had a strange influence over him. When they met on the road, my dog used to run to meet him, with his tail partly tucked in between his legs and hair not erected; and then he would throw himself on the ground, belly upwards. By this action he seemed to say more plainly than by words, "Behold, I am your slave."

A pleasurable and excited state of mind, associated with affection, is exhibited by some dogs in a very peculiar manner; namely, by grinning. This was noticed long ago by Somerville, who says,

"And with a courtly grin, the fawning hound
Salutes thee cow'ring, his wide op'ning nose
Upward he curls, and his large sloe-black eyes
Melt in soft blandishments, and humble joy."

The Chase, book i.

Sir W. Scott's famous Scotch greyhound, Maida, had this habit, and it is common with terriers. I have also seen it in a Spitz and in a sheep-dog. Mr. Riviere, who has particularly attended to this expression, informs me that it is rarely displayed in a perfect manner, but is quite common in a lesser degree. The upper lip during the act of grinning is retracted, as in snarling, so that the canines are exposed, and the ears are drawn backwards; but the general appearance of the animal clearly shows that anger is not felt. Sir C. Bell[94] remarks "Dogs, in their expression of fondness, have a slight eversion of the lips, and grin and sniff amidst their gambols, in a way that resembles laughter." Some persons speak of the grin as a smile, but if it had been really a smile, we should see a similar, though more pronounced, movement of the lips and ears, when dogs utter their bark of joy; but this is not the case, although a bark of joy often follows a grin. On the other hand, dogs, when playing with their comrades or masters, almost always pretend to bite each other; and they then retract, though not energetically, their lips and ears. Hence I suspect that there is a tendency in some dogs, whenever they feel lively pleasure combined with affection, to act through habit and association on the same muscles, as in playfully biting each other, or their masters' hands.

I have described, in the second chapter, the gait and appearance of a dog when cheerful, and the marked antithesis presented by the same animal when dejected and disappointed, with his head, ears, body, tail, and chops drooping, and eyes dull. Under the expectation of any great pleasure, dogs bound and jump about in an extravagant manner, and bark for joy. The tendency to bark under this state of mind is inherited, or runs in the breed: greyhounds rarely bark, whilst the Spitz-dog barks so incessantly on starting for a walk with his master that he becomes a nuisance.

An agony of pain is expressed by dogs in nearly the same way as by many other animals, namely, by howling, writhing, and contortions of the whole body.

Attention is shown by the head being raised, with the ears erected, and eyes intently directed towards the object or quarter under observation. If it be a sound and the source is not known, the head is often turned obliquely from side to side in a most significant manner, apparently in order to judge with more exactness from what point the sound proceeds. But I have seen a dog greatly surprised at a new noise, turning his head to one side through habit, though he clearly perceived the source of the noise. Dogs, as formerly remarked, when their attention is in any way aroused, whilst watching some object, or attending to some sound, often lift up one paw (fig. 4) and keep it doubled up, as if to make a slow and stealthy approach.

A dog under extreme terror will throw himself down, howl, and void his excretions; but the hair, I believe, does not become erect unless some anger is felt. I have seen a dog much terrified at a band of musicians who were playing loudly outside the house, with every muscle of his body trembling, with his heart palpitating so quickly that the beats could hardly be counted, and panting for breath with widely open mouth, in the same manner as a terrified man does. Yet this dog had not exerted himself; he had only wandered slowly and restlessly about the room, and the day was cold.

Even a very slight degree of fear is invariably shown by the tail being tucked in between the legs. This tucking in of the tail is accompanied by the ears being drawn backwards; but they are not pressed closely to the head, as in snarling, and they are not lowered, as when a dog is pleased or affectionate. When two young dogs chase each other in play, the one that runs away always keeps his tail tucked inwards. So it is when a dog, in the highest spirits, careers like a mad creature round and round his master in circles, or in figures of eight. He then acts as if another dog were chasing him. This curious kind of play, which must be familiar to every one who has attended to dogs, is particularly apt to be excited, after the animal has been a little startled or frightened, as by his master suddenly jumping out on him in the dusk. In this case, as well as when two young dogs are chasing each other in play, it appears as if the one that runs away was afraid of the other catching him by the tail; but as far as I can find out, dogs very rarely catch each other in this manner. I asked a gentleman, who had kept foxhounds all his life, and he applied to other experienced sportsmen, whether they had ever seen hounds thus seize a fox; but they never had. It appears that when a dog is chased, or when in danger of being struck behind, or of anything falling on him, in all these cases he wishes to withdraw as quickly as possible his whole hind-quarters, and that from some sympathy or connection between the muscles, the tail is then drawn closely inwards.

A similarly connected movement between the hind-quarters and the tail may be observed in the hyæna. Mr. Bartlett informs me that when two of these animals fight together, they are mutually conscious of the wonderful power of each other's jaws, and are extremely cautious. They well know that if one of their legs were seized, the bone would instantly be crushed into atoms; hence they approach each other kneeling, with their legs turned as much as possible inwards, and with their whole bodies bowed, so as not to present any salient point; the tail at the same time being closely tucked in between the legs. In this attitude they approach each other sideways, or even partly backwards. So again with deer, several of the species, when savage and fighting, tuck in their tails. When one horse in a field tries to bite the hind-quarters of another in play, or when a rough boy strikes a donkey from behind, the hind-quarters and the tail are drawn in, though it does not appear as if this were done merely to save the tail from being injured. We have also seen the reverse of these movements; for when an animal trots with high elastic steps, the tail is almost always carried aloft.

As I have said, when a dog is chased and runs away, he keeps his ears directed backwards but still open; and this is clearly done for the sake of hearing the footsteps of his pursuer. From habit the ears are often held in this same position, and the tail tucked in, when the danger is obviously in front. I have repeatedly noticed, with a timid terrier of mine, that when she is afraid of some object in front, the nature of which she perfectly knows and does not need to reconnoitre, yet she will for a long time hold her ears and tail in this position, looking the image of discomfort. Discomfort, without any fear, is similarly expressed: thus, one day I went out of doors, just at the time when this same dog knew that her dinner would be brought. I did not call her, but she wished much to accompany me, and at the same time she wished much for her dinner; and there she stood, first looking one way and then the other, with her tail tucked in and ears drawn back, presenting an unmistakable appearance of perplexed discomfort.

Almost all the expressive movements now described, with the exception of the grinning from joy, are innate or instinctive, for they are common to all the individuals young and old, of all the breeds. Most of them are likewise common to the aboriginal parents of the dog, namely the wolf and jackal; and some of them to other species of the same group. Tamed wolves and jackals, when caressed by their masters, jump about for joy, wag their tails, lower their ears, lick their master's hands, crouch down, and even throw themselves on the ground belly upwards.[95] I have seen a rather fox-like African jackal, from the Gaboon, depress its ears when caressed. Wolves and jackals, when frightened, certainly tuck in their tails; and a tamed jackal has been described as careering round his master in circles and figures of eight, like a dog, with his tail between his legs.

It has been stated[96] that foxes, however tame, never display any of the above expressive movements; but this is not strictly accurate. Many years ago I observed in the Zoological Gardens, and recorded the fact at the time, that a very tame English fox, when caressed by the keeper, wagged its tail, depressed its ears, and then threw itself on the ground, belly upwards.

The black fox of North America likewise depressed its ears in a slight degree. But I believe that foxes never lick the hands of their masters, and I have been assured that when frightened they never tuck in their tails. If the explanation which I have given of the expression of affection in dogs be admitted, then it would appear that animals which have never been domesticated—namely wolves, jackals, and even foxes—have nevertheless acquired, through the principle of antithesis, certain expressive gestures; for it is not probable that these animals, confined in cages, should have learnt them by imitating dogs.

Cats.—I have already described the actions of a cat (fig. 9), when feeling savage and not terrified. She assumes a crouching attitude and occasionally protrudes her fore-feet, with the claws exserted ready for striking. The tail is extended, being curled or lashed from side to side. The hair is not erected—at least it was not so in the few cases observed by me. The ears are drawn closely backwards and the teeth are shown. Low savage growls are uttered. We can understand why the attitude assumed by a cat when preparing to fight with another cat, or in any way greatly irritated, is so widely different from that of a dog approaching another dog with hostile intentions; for the cat uses her fore-feet for striking, and this renders a crouching position convenient or necessary. She is also much more accustomed than a dog to lie concealed and suddenly spring on her prey. No cause can be assigned with certainty for the tail being lashed or curled from side to side. This habit is common to many other animals—for instance, to the puma, when prepared to spring;[97] but it is not common to dogs, or to foxes, as I infer from Mr. St. John's account of a fox lying in wait and seizing a hare. We have already seen that some kinds of lizards and various snakes, when excited, rapidly vibrate the tips of their tails. It would appear as if, under strong excitement, there existed an uncontrollable desire for movement of some kind, owing to nerve-force being freely liberated from the excited sensorium; and that as the tail is left free, and as its movement does not disturb the general position of the body, it is curled or lashed about.

All the movements of a cat, when feeling affectionate, are in complete antithesis to those just described. She now stands upright, with slightly arched back, tail perpendicularly raised, and ears erected; and she rubs her cheeks and flanks against her master or mistress. The desire to rub something is so strong in cats under this state of mind, that they may often be seen rubbing themselves against the legs of chairs or tables, or against door-posts. This manner of expressing affection probably originated through association, as in the case of dogs, from the mother nursing and fondling her young; and perhaps from the young themselves loving each other and playing together.

Another and very different gesture, expressive of pleasure, has already been described, namely, the curious manner in which young and even old cats, when pleased, alternately protrude their fore-feet, with separated toes, as if pushing against and sucking their mother's teats. This habit is so far analogous to that of rubbing against something, that both apparently are derived from actions performed during the nursing period. Why cats should show affection by rubbing so much more than do dogs, though the latter delight in contact with their masters, and why cats only occasionally lick the hands of their friends, whilst dogs always do so, I cannot say. Cats cleanse themselves by licking their own coats more regularly than do dogs. On the other hand, their tongues seem less well fitted for the work than the longer and more flexible tongues of dogs.

Cats, when terrified, stand at full height, and arch their backs in a well-known and ridiculous fashion. They spit, hiss, or growl. The hair over the whole body, and especially on the tail, becomes erect. In the instances observed by me the basal part of the tail was held upright, the terminal part being thrown on one side; but sometimes the tail (see fig. 15) is only a little raised, and is bent almost from the base to one side. The ears are drawn back, and the teeth exposed. When two kittens are playing together, the one often thus tries to frighten the other. From what we have seen in former chapters, all the above points of expression are intelligible, except the extreme arching of the back. I am inclined to believe that, in the same manner as many birds, whilst they ruffle their feathers, spread out their wings

Fig. 15. *Cat terrified at a dog. From life, by Mr. Wood.*

and tail, to make themselves look as big as possible, so cats stand upright at their full height, arch their backs, often raise the basal part of the tail, and erect their hair, for the same purpose. The lynx, when attacked, is said to arch its back, and is thus figured by Brehm. But the keepers in the Zoological Gardens have never seen any tendency to this action in the larger feline animals, such as tigers, lions, &c.; and these have little cause to be afraid of any other animal.

Cats use their voices much as a means of expression, and they utter, under various emotions and desires, at least six or seven different sounds. The purr of satisfaction, which is made during both inspiration and expiration, is one of the most curious. The puma, cheetah, and ocelot likewise purr; but the tiger, when pleased, "emits a peculiar short snuffle, accompanied by the closure of the eyelids."[98] It is said that the lion, jaguar, and leopard, do not purr.

Horses.—Horses when savage draw their ears closely back, protrude their heads, and partially uncover their incisor teeth, ready for biting. When inclined to kick behind, they generally, through habit, draw back their ears; and their eyes are turned backwards in a peculiar manner.[99] When pleased, as when some coveted food is brought to them in the stable, they raise and draw in their heads, prick their ears, and looking intently towards their friend, often whinny. Impatience is expressed by pawing the ground.

The actions of a horse when much startled are highly expressive. One day my horse was much frightened at a drilling machine, covered by a tarpaulin, and lying on an open field. He raised his head so high, that his neck became almost perpendicular; and this he did from habit, for the machine lay on a slope below, and could not have been seen with more distinctness through the raising of the head; nor if any sound had proceeded from it, could the sound have been more distinctly heard. His eyes and ears were directed intently forwards; and I could feel through the saddle the palpitations of his heart. With red dilated nostrils he snorted violently, and whirling round, would have dashed off at full speed, had I not prevented him. The distension of the nostrils is not for the sake of scenting the source of danger, for when a horse smells carefully at any object and is not alarmed, he does not dilate his nostrils. Owing to the presence of a valve in the throat, a horse when panting does not breathe through his open mouth, but through his nostrils; and these consequently have become endowed with great powers of expansion. This expansion of the nostrils as well as the snorting, and the palpitations of the heart, are actions which have become firmly associated during a long series of generations with the emotion of terror; for terror has habitually led the horse to the most violent exertion in dashing away at full speed from the cause of danger.

Ruminants.—Cattle and sheep are remarkable from displaying in so slight a degree their emotions or sensations, excepting that of extreme pain. A bull when enraged exhibits his rage only by the manner in which he holds his lowered head, with distended nostrils, and by bellowing. He also often paws the ground; but this pawing seems quite different from that of an impatient horse, for when the soil is loose, he throws up clouds of dust. I believe that bulls act in this manner when irritated by flies, for the sake of driving them away. The wilder breeds of sheep and the chamois when startled stamp on the ground, and whistle through their noses; and this serves as a danger-signal to their comrades. The musk-ox of the Arctic regions, when encountered, likewise stamps on the ground.[100] How this stamping action arose I cannot conjecture; for from, inquiries which I have made it does not appear that any of these animals fight with their fore-legs.

Some species of deer, when savage, display far more expression than do cattle, sheep, or goats, for, as has already been stated, they draw back their ears, grind their teeth, erect their hair, squeal, stamp on the ground, and brandish their horns. One day in the Zoological Gardens, the Formosan deer *(Cervus pseudaxis)* approached me in a curious attitude, with his muzzle raised high up, so that the horns were pressed back on his neck; the head being held rather obliquely. From the expression of his eye I felt sure that he was savage; he approached slowly, and as soon as he came close to the iron bars, he did not lower his head to butt at me, but suddenly bent it inwards, and struck his horns with great force against the railings. Mr. Bartlett informs me that some other species of deer place themselves in the same attitude when enraged.

Monkeys.—The various species and genera of monkeys express their feelings in many different ways; and this fact is interesting, as in some degree bearing on the question, whether the so-called races of man should be ranked as distinct species or varieties; for, as we shall see in the following chapters, the different races of man express their emotions and sensations with remarkable uniformity throughout the world. Some of the expressive actions of monkeys are interesting in another way, namely from being closely analogous to those of man. As I have had no opportunity of observing any one species of the group under all circumstances, my miscellaneous remarks will be best arranged under different states of the mind.

Pleasure, joy, affection.—It is not possible to distinguish in monkeys, at least without more experience than I have had, the expression of pleasure or joy from that of affection. Young chimpanzees make a kind of barking noise, when pleased by the return of any one to whom they are attached. When this noise, which the keepers call a laugh, is uttered, the lips are protruded; but so they are under various other emotions. Nevertheless I could perceive that when they were pleased the form of the lips differed a little from that assumed when they were angered. If a young chimpanzee be tickled—and the armpits are particularly sensitive to tickling, as in the case of our children,—a more decided chuckling or laughing sound is uttered; though the laughter is sometimes noiseless. The corners of the mouth are then drawn backwards; and this sometimes causes the lower eyelids to be slightly wrinkled. But this wrinkling, which is so characteristic of our own laughter, is more plainly seen in some other monkeys. The teeth in the upper jaw in the chimpanzee are not exposed when they utter their laughing noise, in which respect they differ from us. But their eyes sparkle and grow brighter, as Mr. W. L. Martin,[101] who has particularly attended to their expression, states.

Young Orangs, when tickled, likewise grin and make a chuckling sound; and Mr. Martin says that their eyes grow brighter. As soon as their laughter ceases, an expression may be detected passing over their faces, which, as Mr. Wallace remarked to me, may be called a smile. I have also noticed something of the same kind with the chimpanzee. Dr. Duchenne—and I cannot quote a better authority—informs me that he kept a very tame monkey in his house for a year; and when he gave it during meal-times some choice delicacy, he observed that the corners of its mouth were slightly raised; thus an expression of satisfaction, partaking of the nature of an incipient smile, and resembling that often seen on the face of man, could be plainly perceived in this animal.

The *Cebus azaræ,*[102] when rejoiced at again seeing a beloved person, utters a peculiar tittering *(kichernden)* sound. It also expresses agreeable sensations, by drawing back the corners of its mouth, without producing any sound. Rengger calls this movement laughter, but it would be more appropriately called a smile. The form of the mouth is different when either pain or terror is expressed, and high shrieks are uttered. Another species of *Cebus* in the Zoological Gardens *(C. hypoleucus),* when pleased, makes a reiterated shrill note, and likewise draws back the corners of its mouth, apparently through the contraction of the same muscles as with us. So does the Barbary ape *(Inuus ecaudatus)* to an extraordinary degree; and I observed in this monkey that the skin of the lower eyelids then became much wrinkled. At the same time it rapidly moved its lower jaw or lips in a spasmodic manner, the teeth being exposed; but the noise produced was hardly more distinct than that which we sometimes call silent laughter. Two of the keepers affirmed that this slight sound was the animal's laughter, and when I expressed some doubt on this head (being at the time quite inexperienced), they made it attack or rather threaten a hated Entellus monkey, living in the same compartment. Instantly the whole expression of the face of the Inuus changed; the mouth was opened much more widely, the canine teeth were more fully exposed, and a hoarse barking noise was uttered.

The Anubis baboon *(Cynocephalus anubis)* was first insulted and put into a furious rage, as was easily done, by his keeper, who then made friends with him and shook hands. As the reconciliation was effected the baboon rapidly moved up and down his jaws and lips, and looked pleased. When we laugh heartily, a similar movement, or quiver, may be observed more or less distinctly in our jaws; but with man the muscles of the chest are more particularly acted on, whilst with this baboon, and with some other monkeys, it is the muscles of the jaws and lips which are spasmodically affected.

I have already had occasion to remark on the curious manner in which two or three species of Macacus and the *Cynopithecus niger* draw back their ears and utter a slight jabbering noise, when they are pleased by being caressed. With the Cynopithecus (fig. 17), the corners of the mouth are at the same time drawn backwards and upwards, so that the teeth are exposed. Hence this expression would never be recognised by a stranger as one of pleasure. The crest of long hairs on the forehead is depressed, and apparently the whole skin of the head drawn backwards. The eyebrows are thus raised a little, and the eyes assume a staring appearance. The lower eyelids also become slightly wrinkled; but this wrinkling is not conspicuous, owing to the permanent transverse furrows on the face.

Painful emotions and sensations.—With monkeys the expression of slight pain, or of any painful emotion, such as grief, vexation, jealousy, &c., is not

Fig. 16. *Cynopithecus niger, in a placid condition. Drawn from life by Mr. Wolf.*

Fig. 17. *The same, when pleased by being caressed.*

easily distinguished from that of moderate anger; and these states of mind readily and quickly pass into each other. Grief, however, with some species is certainly exhibited by weeping. A woman, who sold a monkey to the Zoological Society, believed to have come from Borneo *(Macacus maurus* or *M. inornatus* of Gray), said that it often cried; and Mr. Bartlett, as well as the keeper Mr. Sutton, have repeatedly seen it, when grieved, or even when much pitied, weeping so copiously that the tears rolled down its cheeks. There is, however, something strange about this case, for two specimens subsequently kept in the Gardens, and believed to be the same species, have never been seen to weep, though they were carefully observed by the keeper and myself when much distressed and loudly screaming. Rengger states[103] that the eyes of the *Cebus azaræ* fill with tears, but not sufficiently to overflow, when it is prevented getting some much desired object, or is much frightened. Humboldt also asserts that the eyes of the *Callithrix sciureus* "instantly fill with tears when it is seized with fear;" but when this pretty little monkey in the Zoological Gardens was teased, so as to cry out loudly, this did not occur. I do not, however, wish to throw the least doubt on the accuracy of Humboldt's statement.

The appearance of dejection in young orangs and chimpanzees, when out of health, is as plain and almost as pathetic as in the case of our children. This state of mind and body is shown by their listless movements, fallen countenances, dull eyes, and changed complexion.

Anger.—This emotion is often exhibited by many kinds of monkeys, and is expressed, as Mr. Martin remarks,[104] in many different ways. "Some species, when irritated, pout the lips, gaze with a fixed and savage glare on their foe, and make repeated short starts as if about to spring forward, uttering at the same time inward guttural sounds. Many display their anger by suddenly advancing, making abrupt starts, at the same time opening the mouth and pursing up the lips, so as to conceal the teeth, while the eyes are daringly fixed on the enemy, as if in savage defiance. Some again, and principally the long-tailed monkeys, or Guenons, display their teeth, and accompany their malicious grins with a sharp, abrupt, reiterated cry." Mr. Sutton confirms the statement that some species uncover their teeth when enraged, whilst others conceal them by the protrusion of their lips; and some kinds draw back their ears. The *Cynopithecus niger,* lately referred to, acts in this manner, at the same time depressing the crest of hair on its forehead, and showing its teeth; so that the movements of the features from anger are nearly the same as those from pleasure; and the two expressions can be distinguished only by those familiar with the animal.

Baboons often show their passion and threaten their enemies in a very odd manner, namely, by opening their mouths widely as in the act of yawning. Mr. Bartlett has often seen two baboons, when first placed in the same compartment, sitting opposite to each other and thus alternately opening their mouths; and this action seems frequently to end in a real yawn. Mr. Bartlett believes that both animals wish to show to each other that they are provided with a formidable set of teeth, as is undoubtedly the case. As I could hardly credit the reality of this yawning gesture, Mr. Bartlett insulted an old baboon and put him into a violent passion; and he almost immediately thus acted. Some species of Macacus and of Cercopithecus[105] behave in the same manner. Baboons likewise show their anger, as was observed by Brehm with those which he kept alive in Abyssinia, in another manner, namely, by striking the ground with one hand, "like an angry man striking the table with his fist." I have seen this movement with the baboons in the Zoological Gardens; but sometimes the action seems rather to represent the searching for a stone or other object in their beds of straw.

Mr. Sutton has often observed the face of the *Macacus rhesus,* when much enraged, growing red. As he was mentioning this to me, another monkey attacked a *rhesus,* and I saw its face redden as plainly as that of a man in a violent passion. In the course of a few minutes, after the battle, the face of this monkey recovered its natural tint. At the same time that the face reddened, the naked posterior part of the body, which is always red, seemed to grow still redder; but I cannot positively assert that this was the case. When the Mandrill is in any way excited, the brilliantly coloured, naked parts of the skin are said to become still more vividly coloured.

With several species of baboons the ridge of the forehead projects much over the eyes, and is studded with a few long hairs, representing our eyebrows. These animals are always looking about them, and in order to look upwards they raise their eyebrows. They have thus, as it would appear, acquired the habit of frequently moving their eyebrows. However this may be, many kinds of monkeys, especially the baboons, when angered or in any way excited, rapidly and incessantly move their eyebrows up and down, as well as the hairy skin of their foreheads.[106] As we associate in the case of man the raising and lowering of the eyebrows with definite states of the mind, the almost incessant movement of the eyebrows by monkeys gives them a senseless expression. I once observed a man who had a trick of continually raising his eyebrows without any corresponding emotion, and this gave to him a foolish appearance; so it is with some persons who keep the corners of their mouths a little drawn backwards and upwards, as if by an incipient smile, though at the time they are not amused or pleased.

A young orang, made jealous by her keeper attending to another monkey, slightly uncovered her teeth, and, uttering a peevish noise like *tish-shist,* turned her back on him. Both orangs and chimpanzees, when a little more angered, protrude their lips greatly, and make a harsh barking noise. A young female chimpanzee, in a violent passion, presented a curious resemblance to a child in the same state. She screamed loudly with widely open mouth, the lips being retracted so that the teeth were fully exposed. She threw her arms wildly about, sometimes clasping them over her head. She rolled on the ground, sometimes on her back, sometimes on her belly, and bit everything within reach. A young gibbon *(Hylobates syndactylus)* in a passion has been described[107] as behaving in almost exactly the same manner.

The lips of young orangs and chimpanzees are protruded, sometimes to a wonderful degree, under various circumstances. They act thus, not only when slightly angered, sulky, or disappointed, but when alarmed at anything—in one instance, at the sight of a turtle,[108]—and likewise when pleased. But neither the degree of protrusion nor the shape of the mouth is exactly the same, as I believe, in all cases; and the sounds which are then uttered are different. The accompanying drawing represents a chimpanzee made sulky by an orange having been offered him, and then taken away. A similar protrusion or pouting of the lips, though to a much slighter degree, may be seen in sulky children.

Many years ago, in the Zoological Gardens, I placed a looking-glass on the floor before two young orangs, who, as far as it was known, had never before seen one. At first they gazed at their own images with the most steady surprise, and often changed their point of view. They then approached close and protruded their lips towards the image, as if to kiss it, in exactly the same manner as they had previously done towards each other, when first placed, a few days before, in the same room. They next made all sorts of grimaces, and put themselves in various attitudes before the mirror; they pressed and rubbed the surface; they placed their hands at different distances behind it; looked behind it; and finally seemed almost frightened, started a little, became cross, and refused to look any longer.

Fig. 18. *Chimpanzee disappointed and sulky. Drawn from life by Mr. Wood.*

When we try to perform some little action which is difficult and requires precision, for instance, to thread a needle, we generally close our lips firmly, for the sake, I presume, of not disturbing our movements by breathing; and I noticed the same action in a young Orang. The poor little creature was sick, and was amusing itself by trying to kill the flies on the window-panes with its knuckles; this was difficult as the flies buzzed about, and at each attempt the lips were firmly compressed, and at the same time slightly protruded.

Although the countenances, and more especially the gestures, of orangs and chimpanzees are in some respects highly expressive, I doubt whether on the whole they are so expressive as those of some other kinds of monkeys. This may be attributed in part to their ears being immoveable, and in part to the nakedness of their eyebrows, of which the movements are thus rendered less conspicuous. When, however, they raise their eyebrows their foreheads become, as with us, transversely wrinkled. In comparison with man, their faces are inexpressive, chiefly owing to their not frowning under any emotion of the mind—that is, as far as I have been able to observe, and I carefully attended to this point. Frowning, which is one of the most important of all the expressions in man, is due to the contraction of the corrugators by which the eyebrows are lowered and brought together, so that vertical furrows are formed on the forehead. Both the orang and chimpanzee are said[109] to possess this muscle, but it seems rarely brought into action, at least in a conspicuous manner. I made my hands into a sort of cage, and placing some tempting fruit within, allowed both a young orang and chimpanzee to try their utmost to get it out; but although they grew rather cross, they showed not a trace of a frown. Nor was there any frown when they were enraged. Twice I took two chimpanzees from their rather dark room suddenly into bright sunshine, which would certainly have caused us to frown; they blinked and winked their eyes, but only once did I see a very slight frown. On another occasion, I tickled the nose of a chimpanzee with a straw, and as it crumpled up its face, slight vertical furrows appeared between the eyebrows. I have never seen a frown on the forehead of the orang.

The gorilla, when enraged, is described as erecting its crest of hair, throwing down its under lip, dilating its nostrils, and uttering terrific yells. Messrs. Savage and Wyman[110] state that the scalp can be freely moved backwards and forwards, and that when the animal is excited it is strongly contracted; but I presume that they mean by this latter expression that the scalp is lowered; for they likewise speak of the young chimpanzee, when crying out, as "having the eyebrows strongly contracted." The great power of movement in the scalp of the gorilla, of many baboons and other monkeys, deserves notice in relation to the power possessed by some few men, either through reversion or persistence, of voluntarily moving their scalps.[111]

Astonishment, Terror.—A living fresh-water turtle was placed at my request in the same compartment in the Zoological Gardens with many monkeys; and they showed unbounded astonishment, as well as some fear. This was displayed by their remaining motionless, staring intently with widely opened eyes, their eyebrows being often moved up and down. Their faces seemed somewhat lengthened. They occasionally raised themselves on their hind-legs to get a better view. They often retreated a few feet, and then turning their heads over one shoulder, again stared intently. It was curious to observe how much less afraid they were of the turtle than of a living snake which I had formerly placed in their compartment;[112] for in the course of a few minutes some of the monkeys ventured to approach and touch the turtle. On the other hand, some of the larger baboons were greatly terrified, and grinned as if on the point of screaming out. When I showed a little dressed-up doll to the *Cynopithecus niger,* it stood motionless, stared intently with widely opened eyes, and advanced its ears a little forwards. But when the turtle was placed in its compartment, this monkey also moved its lips in an odd, rapid, jabbering manner, which the keeper declared was meant to conciliate or please the turtle.

I was never able clearly to perceive that the eyebrows of astonished monkeys were kept permanently raised, though they were frequently moved up and down. Attention, which precedes astonishment, is expressed by man by a slight raising of the eyebrows; and Dr. Duchenne informs me that when he gave to the monkey formerly mentioned some quite new article of food, it elevated its eyebrows a little, thus assuming an appearance of close attention. It then took the food in its fingers, and, with lowered or rectilinear eyebrows, scratched, smelt, and examined it,—an expression of reflection being thus exhibited. Sometimes it would throw back its head a little, and again with suddenly raised eyebrows re-examine and finally taste the food.

In no case did any monkey keep its mouth open when it was astonished. Mr. Sutton observed for me a young orang and chimpanzee during a considerable length of time; and however much they were astonished, or whilst listening intently to some strange sound, they did not keep their mouths open. This fact is surprising, as with mankind hardly any expression is more general than a widely open mouth under the sense of astonishment. As far as I have been able to observe, monkeys breathe more freely through their nostrils than men do; and this may account for their not opening their mouths when they are astonished; for, as we shall see in a future chapter, man apparently acts in this manner when startled, at first for the sake of quickly drawing a full inspiration, and afterwards for the sake of breathing as quietly as possible.

Terror is expressed by many kinds of monkeys by the utterance of shrill screams; the lips being drawn back, so that the teeth are exposed. The hair becomes erect, especially when some anger is likewise felt. Mr. Sutton has distinctly seen the face of the *Macacus rhesus* grow pale from fear. Monkeys also tremble from fear; and sometimes they void their excretions. I have seen one which, when caught, almost fainted from an excess of terror.

SUFFICIENT FACTS have now been given with respect to the expressions of various animals. It is impossible to agree with Sir C. Bell when he says[113] that "the faces of animals seem chiefly capable of expressing rage and fear;" and again, when he says that all their expressions "may be referred, more or less plainly, to their acts of volition or necessary instincts." He who will look at a dog preparing to attack another dog or a man, and at the same animal when caressing his master, or will watch the countenance of a monkey when insulted, and when fondled by his keeper, will be forced to admit that the movements of their features and their gestures are almost as expressive as those of man. Although no explanation can be given of some of the expressions in the lower animals, the greater number are explicable in accordance with the three principles given at the commencement of the first chapter.

CHAPTER VI

Special Expressions of Man: Suffering and Weeping

THE SCREAMING AND WEEPING OF INFANTS • FORM OF FEATURES • AGE AT WHICH WEEPING COMMENCES • THE EFFECTS OF HABITUAL RESTRAINT ON WEEPING • SOBBING • CAUSE OF THE CONTRACTION OF THE MUSCLES ROUND THE EYES DURING SCREAMING • CAUSE OF THE SECRETION OF TEARS

In this and the following chapters the expressions exhibited by Man under various states of the mind will be described and explained, as far as lies in my power. My observations will be arranged according to the order which I have found the most convenient; and this will generally lead to opposite emotions and sensations succeeding each other.

Suffering of the body and mind: weeping.—I have already described in sufficient detail, in the third chapter, the signs of extreme pain, as shown by screams or groans, with the writhing of the whole body and the teeth clenched or ground together. These signs are often accompanied or followed by profuse sweating, pallor, trembling, utter prostration, or faintness. No suffering is greater than that from extreme fear or horror, but here a distinct emotion comes into play, and will be elsewhere considered. Prolonged suffering, especially of the mind, passes into low spirits, grief, dejection, and despair, and these states will be the subject of the following chapter. Here I shall almost confine myself to weeping or crying, more especially in children.

Infants, when suffering even slight pain, moderate hunger, or discomfort, utter violent and prolonged screams. Whilst thus screaming their eyes are firmly closed, so that the skin round them is wrinkled, and the forehead contracted into a frown. The mouth is widely opened with the lips retracted in a peculiar manner, which causes it to assume a squarish form; the gums or teeth being more or less exposed. The breath is inhaled almost spasmodically. It is easy to observe infants whilst screaming; but I have found photographs made by the instantaneous process the best means for observation, as allowing more deliberation. I have collected twelve, most of them made purposely for me; and they all exhibit the same general characteristics. I have, therefore, had six of them[114] (Plate I) reproduced by the heliotype process.

The firm closing of the eyelids and consequent compression of the eyeball,—and this is a most important element in various expressions,—serves to protect the eyes from becoming too much gorged with blood, as will presently be explained in detail. With respect to the order in which the several muscles contract in firmly compressing the eyes, I am indebted to Dr. Langstaff, of Southampton, for some observations, which I have since repeated. The best plan for observing the order is to make a person first raise his eyebrows, and this produces transverse wrinkles across the forehead; and then very gradually to contract all the muscles round the eyes

Plate I

with as much force as possible. The reader who is unacquainted with the anatomy of the face, ought to refer to p. 1274, and look at the woodcuts 1 to 3. The corrugators of the brow *(corrugator supercilii)* seem to be the first muscles to contract; and these draw the eyebrows downwards and inwards towards the base of the nose, causing vertical furrows, that is a frown, to appear between the eyebrows; at the same time they cause the disappearance of the transverse wrinkles across the forehead. The orbicular muscles contract almost simultaneously with the corrugators, and produce wrinkles all round the eyes; they appear, however, to be enabled to contract with greater force, as soon as the contraction of the corrugators has given them some support. Lastly, the pyramidal muscles of the nose contract; and these draw the eyebrows and the skin of the forehead still lower down, producing short transverse wrinkles across the base of the nose.[115] For the sake of brevity these muscles will generally be spoken of as the orbiculars, or as those surrounding the eyes.

When these muscles are strongly contracted, those running to the upper lip[116] likewise contract and raise the upper lip. This might have been expected from the manner in which at least one of them, the *malaris,* is connected with the orbiculars. Any one who will gradually contract the muscles round his eyes, will feel, as he increases the force, that his upper lip and the wings of his nose (which are partly acted on by one of the same muscles) are almost always a little drawn up. If he keeps his mouth firmly shut whilst contracting the muscles round the eyes, and then suddenly relaxes his lips, he will feel that the pressure on his eyes immediately increases. So again when a person on a bright, glaring day wishes to look at a distant object, but is compelled partially to close his eyelids, the upper lip may almost always be observed to be somewhat raised. The mouths of some very short-sighted persons, who are forced habitually to reduce the aperture of their eyes, wear from this same reason a grinning expression.

The raising of the upper lip draws upwards the flesh of the upper parts of the cheeks, and produces a strongly marked fold on each cheek,—the naso-labial fold,—which runs from near the wings of the nostrils to the corners of the mouth and below them. This fold or furrow may be seen in all the photographs, and is very characteristic of the expression of a crying child; though a nearly similar fold is produced in the act of laughing or smiling.[117]

As the upper lip is much drawn up during the act of screaming, in the manner just explained, the depressor muscles of the angles of the mouth (see K in wood-cuts 1 and 2) are strongly contracted in order to keep the mouth widely open, so that a full volume of sound may be poured forth. The action of these opposed muscles, above and below, tends to give to the mouth an oblong, almost squarish outline, as may be seen in the accompanying photographs. An excellent observer,[118] in describing a baby crying whilst being fed, says, "it made its mouth like a square, and let the porridge run out at all four corners." I believe, but we shall return to this point in a future chapter, that the depressor muscles of the angles of the mouth are less under the separate control of the will than the adjoining muscles; so that if a young child is only doubtfully inclined to cry, this muscle is generally the first to contract, and is the last to cease contracting. When older children commence crying, the muscles which run to the upper lip are often the first to contract; and this may perhaps be due to older children not having so strong a tendency to scream loudly, and consequently to keep their mouths widely open; so that the above-named depressor muscles are not brought into such strong action.

With one of my own infants, from his eighth day and for some time afterwards, I often observed that the first sign of a screaming-fit, when it could be observed coming on gradually, was a little frown, owing to the contraction of the corrugators of the brows; the capillaries of the naked head and face becoming at the same time reddened with blood. As soon as the screaming-fit actually began, all the muscles round the eyes were strongly contracted, and the mouth widely opened in the manner above described; so that at this early period the features assumed the same form as at a more advanced age.

Dr. Piderit[119] lays great stress on the contraction of certain muscles which draw down the nose and narrow the nostrils, as eminently characteristic of a crying expression. The *depressores anguli oris*, as we have just seen, are usually contracted at the same time, and they indirectly tend, according to Dr. Duchenne, to act in this same manner on the nose. With children having bad colds a similar pinched appearance of the nose may be noticed, which is at least partly due, as remarked to me by Dr. Langstaff, to their constant snuffling, and the consequent pressure of the atmosphere on the two sides. The purpose of this contraction of the nostrils by children having bad colds, or whilst crying, seems to be to check the downward flow of the mucus and tears, and to prevent these fluids spreading over the upper lip.

After a prolonged and severe screaming-fit, the scalp, face, and eyes are reddened, owing to the return of the blood from the head having been impeded by the violent expiratory efforts; but the redness of the stimulated eyes is chiefly due to the copious effusion of tears. The various muscles of the face which have been strongly contracted, still twitch a little, and the upper lip is still slightly drawn up or everted,[120] with the corners of the mouth still a little drawn downwards. I have myself felt, and have observed in other grown-up persons, that when tears are restrained with difficulty, as in reading a pathetic story, it is almost impossible to prevent the various muscles, which with young children are brought into strong action during their screaming-fits, from slightly twitching or trembling.

Infants whilst young do not shed tears or weep, as is well known to nurses and medical men. This circumstance is not exclusively due to the lacrymal glands being as yet incapable of secreting tears. I first noticed this fact from having accidentally brushed with the cuff of my coat the open eye of one of my infants, when seventy-seven days old, causing this eye to water freely; and though the child screamed violently, the other eye remained dry, or was only slightly suffused with tears. A similar slight effusion occurred ten days previously in both eyes during a screaming-fit. The tears did not run over the eyelids and roll down the cheeks of this child, whilst screaming badly, when 122 days old. This first happened 17 days later, at the age of 139 days. A few other children have been observed for me, and the period of free weeping appears to be very variable. In one case, the eyes became slightly suffused at the age of only 20 days; in another, at 62 days. With two other children, the tears did *not* run down the face at the ages of 84 and 110 days; but in a third child they did run down at the age of 104 days. In one instance, as I was positively assured, tears ran down at the unusually early age of 42 days. It would appear as if the lacrymal glands required some practice in the individual before they are easily excited into action, in somewhat the same manner as various inherited consensual movements and tastes require some exercise before they are fixed and perfected. This is all the more likely with a habit like weeping, which must have been acquired since the period when man branched off from the common progenitor of the genus Homo and of the non-weeping anthropomorphous apes.

The fact of tears not being shed at a very early age from pain or any mental emotion is remarkable, as, later in life, no expression is more general or more strongly marked than weeping. When the habit has once been acquired by an infant, it expresses in the clearest manner suffering of all kinds, both bodily pain and mental distress, even though accompanied by other emotions, such as fear or rage. The character of the crying, however, changes at a very early age, as I noticed in my own infants,—the passionate cry differing from that of grief. A lady informs me that her child, nine months old, when in a passion screams loudly, but does not weep; tears, however, are shed when she is punished by her chair being turned with its back to the table. This difference may perhaps be attributed to weeping being restrained, as we shall immediately see, at a more advanced age, under most circumstances excepting grief; and to the influence of such restraint being transmitted to an earlier period of life, than that at which it was first practised.

With adults, especially of the male sex, weeping soon ceases to be caused by, or to express, bodily pain. This may be accounted for by its being thought weak and unmanly by men, both of civilized and barbarous races, to exhibit bodily pain by any outward sign. With this exception, savages weep copiously from very slight causes, of which fact Sir J. Lubbock[121] has collected instances. A New Zealand chief "cried like a child because the sailors spoilt his favourite cloak by powdering it with flour." I saw in Tierra del Fuego a native who had lately lost a brother, and who alternately cried with hysterical violence, and laughed heartily at anything which amused him. With the civilized nations of Europe there is also much difference in the frequency of weeping. Englishmen rarely cry, except under the pressure of the acutest grief; whereas in some parts of the Continent the men shed tears much more readily and freely.

The insane notoriously give way to all their emotions with little or no restraint; and I am informed by Dr. J. Crichton Browne, that nothing is more characteristic of simple melancholia, even in the male sex, than a tendency to weep on the slightest occasions, or from no cause. They also weep disproportionately on the occurrence of any real cause of grief. The length of time during which some patients weep is astonishing, as well as the amount of tears which they shed. One melancholic girl wept for a whole day, and afterwards confessed to Dr. Browne, that it was because she remembered that she had once shaved off her eyebrows to promote their growth. Many patients in the asylum sit for a long time rocking themselves backwards and forwards; "and if spoken to, they stop their movements, purse up their eyes, depress the corners of the mouth, and burst out crying." In some of these cases, the being spoken to or kindly greeted appears to suggest some fanciful and sorrowful notion; but in other cases an effort of any kind excites weeping, independently of any sorrowful idea. Patients suffering from acute mania likewise have paroxysms of violent crying or blubbering, in the midst of their incoherent ravings. We must not, however, lay too much stress on the copious shedding of tears by the insane, as being due to the lack of all restraint; for certain brain-diseases, as hemiplegia, brain-wasting, and senile decay, have a special tendency to induce weeping. Weeping is common in the insane, even after a complete state of fatuity has been reached and the power of speech lost. Persons born idiotic likewise weep;[122] but it is said that this is not the case with cretins.

Weeping seems to be the primary and natural expression, as we see in children, of suffering of any kind, whether bodily pain short of extreme agony, or mental distress. But the foregoing facts and common experience show us that a frequently repeated effort to restrain weeping, in association with certain states of the mind, does much in checking the habit. On the other hand, it appears that the power of weeping can be increased through habit; thus the Rev. R. Taylor,[123] who long resided in New Zealand, asserts that the women can voluntarily shed tears in abundance; they meet for this purpose to mourn for the dead, and they take pride in crying "in the most affecting manner."

A single effort of repression brought to bear on the lacrymal glands does little, and indeed seems often to lead to an opposite result. An old and experienced physician told me that he had always found that the only means to check the occasional bitter weeping of ladies who consulted him, and who themselves wished to desist, was earnestly to beg them not to try, and to assure them that nothing would relieve them so much as prolonged and copious crying.

The screaming of infants consists of prolonged expirations, with short and rapid, almost spasmodic inspirations, followed at a somewhat more advanced age by sobbing. According to Gratiolet,[124] the glottis is chiefly affected during the act of sobbing. This sound is heard "at the moment when the inspiration conquers the resistance of the glottis, and the air rushes into the chest." But the whole act of respiration is likewise spasmodic and violent. The shoulders are at the same time generally raised, as by this movement respiration is rendered easier. With one of my infants, when seventy-seven days old, the inspirations were so rapid and strong that they approached in character to sobbing; when 138 days old I first noticed distinct sobbing, which subsequently followed every bad crying-fit. The respiratory movements are partly voluntary and partly involuntary, and I apprehend that sobbing is at least in part due to children having some power to command after early infancy their vocal organs and to stop their screams, but from having less power over their respiratory muscles, these continue for a time to act in an involuntary or spasmodic manner, after having been brought into violent action. Sobbing seems to be peculiar to the human species; for the keepers in the Zoological Gardens assure me that they have never heard a sob from any kind of monkey; though monkeys often scream loudly whilst being chased and caught, and then pant for a long time. We thus see that there is a close analogy between sobbing and the free shedding of tears; for with children, sobbing does not commence during early infancy, but afterwards comes on rather suddenly and then follows every bad crying-fit, until the habit is checked with advancing years.

On the cause of the contraction of the muscles round the eyes during screaming.—We have seen that infants and young children, whilst screaming, invariably close their eyes firmly, by the contraction of the surrounding muscles, so that the skin becomes wrinkled all around. With older children, and even with adults, whenever there is violent and unrestrained crying, a tendency to the contraction of these same muscles may be observed; though this is often checked in order not to interfere with vision.

Sir C. Bell explains[125] this action in the following manner:—"During every violent act of expiration, whether in hearty laughter, weeping, coughing, or sneezing, the eyeball is firmly compressed by the fibres of the orbicularis; and this is a provision for supporting and defending the vascular system of the interior of the eye from a retrograde impulse communicated to the blood in the veins at that time. When we contract the chest and expel the air, there is a retardation of the blood in the veins of the neck and head; and in the more powerful acts of expulsion, the blood not only distends the vessels, but is even regurgitated into the minute branches. Were the eye not properly compressed at that time, and a resistance given to the shock, irreparable injury might be inflicted on the delicate textures of the interior of the eye." He further adds, "If we separate the eyelids of a child to examine the eye, while it cries and struggles with passion, by taking off the natural support to the vascular system of the eye, and means of guarding it against the rush of blood then occurring, the conjunctiva becomes suddenly filled with blood, and the eyelids everted."

Not only are the muscles round the eyes strongly contracted, as Sir C. Bell states and as I have often observed, during screaming, loud laughter, coughing, and sneezing, but during several other analogous actions. A man contracts these muscles when he violently blows his nose. I asked one of my boys to shout as loudly as he possibly could, and as soon as he began, he firmly contracted his orbicular muscles; I observed this repeatedly, and on asking him why he had every time so firmly closed his eyes, I found that he was quite unaware of the fact: he had acted instinctively or unconsciously.

It is not necessary, in order to lead to the contraction of these muscles, that air should actually be expelled from the chest; it suffices that the muscles of the chest and abdomen should contract with great force, whilst by the closure of the glottis no air escapes. In violent vomiting or retching the diaphragm is made to descend by the chest being filled with air; it is then held in this position by the closure of the glottis, "as well as by the contraction of its own fibres."[126] The abdominal muscles now contract strongly upon the stomach, its proper muscles likewise contracting, and the contents are thus ejected. During each effort of vomiting "the head becomes greatly congested, so that the features are red and swollen, and the large veins of the face and temples visibly dilated." At the same time, as I know from observation, the muscles round the eyes are strongly contracted. This is likewise the case when the abdominal muscles act downwards with *unusual* force in expelling the contents of the intestinal canal.

The greatest exertion of the muscles of the body, if those of the chest are not brought into strong action in expelling or compressing the air within the lungs, does not lead to the contraction of the muscles round the eyes. I have observed my sons using great force in gymnastic exercises, as in repeatedly raising their suspended bodies by their arms alone, and in lifting heavy weights from the ground, but there was hardly any trace of contraction in the muscles round the eyes.

As the contraction of these muscles for the protection of the eyes during violent expiration is indirectly, as we shall hereafter see, a fundamental element in several of our most important expressions, I was extremely anxious to ascertain how far Sir C. Bell's view could be substantiated. Professor Donders, of Utrecht,[127] well known as one of the highest authorities in Europe on vision and on the structure of the eye, has most kindly undertaken for me this investigation with the aid of the many ingenious mechanisms of modern science, and has published the results.[128] He shows that during violent expiration the external, the intra-ocular, and the retro-ocular vessels of the eye are all affected in two ways, namely by the increased pressure of the blood in the arteries, and by the return of the blood in the veins being impeded. It is, therefore, certain that both the arteries and the veins of the eye are more or less distended during violent expiration. The evidence in detail may be found in Professor Donders' valuable memoir. We see the effects on the veins of the head, in their prominence, and in the purple colour of the face of a man who coughs violently from being half choked. I may mention, on the same authority, that the whole eye certainly advances a little during each violent expiration. This is due to the dilatation of the retro-ocular vessels, and might have been expected from the intimate connection of the eye and brain; the brain being known to rise and fall with each respiration, when a portion of the skull has been removed; and as may be seen along the unclosed sutures of infants' heads. This also, I presume, is the reason that the eyes of a strangled man appear as if they were starting from their sockets.

With respect to the protection of the eye during violent expiratory efforts by the pressure of the eyelids, Professor Donders concludes from his various observations that this action certainly limits or entirely removes the dilatation of the vessels.[129] At such times, he adds, we not unfrequently see the hand involuntarily laid upon the eyelids, as if the better to support and defend the eyeball.

Nevertheless much evidence cannot at present be advanced to prove that the eye actually suffers injury from the want of support during violent expiration; but there is some. It is "a fact that forcible expiratory efforts in violent coughing or vomiting, and especially in sneezing, sometimes give rise to ruptures of the little (external) vessels" of the eye.[130] With respect to the internal vessels, Dr. Gunning has lately recorded a case of exophthalmos in consequence of whooping-cough, which in his opinion depended on the rupture of the deeper vessels; and another analogous case has been recorded. But a mere sense of discomfort would probably suffice to lead to the associated habit of protecting the eyeball by the contraction of the surrounding muscles. Even the expectation or chance of injury would probably be sufficient, in the same manner as an object moving too near the eye induces involuntary winking of the eyelids. We may, therefore, safely conclude from Sir C. Bell's observations, and more especially from the more careful investigations by Professor Donders, that the firm closure of the eyelids during the screaming of children is an action full of meaning and of real service.

We have already seen that the contraction of the orbicular muscles leads to the drawing up of the upper lip, and consequently, if the mouth is kept widely open, to the drawing down of the corners by the contraction of the depressor muscles. The formation of the naso-labial fold on the cheeks likewise follows from the drawing up of the upper lip. Thus all the chief expressive movements of the face during crying apparently result from the contraction of the muscles round the eyes. We shall also find that the shedding of tears depends on, or at least stands in some connection with, the contraction of these same muscles.

In some of the foregoing cases, especially in those of sneezing and coughing, it is possible that the contraction of the orbicular muscles may serve in addition to protect the eyes from too severe a jar or vibration. I think so, because dogs and cats, in crunching hard bones, always close their eyelids, and at least sometimes in sneezing; though dogs do not do so whilst barking loudly. Mr. Sutton carefully observed for me a young orang and chimpanzee, and he found that both always closed their eyes in sneezing and coughing, but not whilst screaming violently. I gave a small pinch of snuff to a monkey of the American division, namely, a Cebus, and it closed its eyelids whilst sneezing; but not on a subsequent occasion whilst uttering loud cries.

Cause of the secretion of tears.—It is an important fact which must be considered in any theory of the secretion of tears from the mind being affected, that whenever the muscles round the eyes are strongly and involuntarily contracted in order to compress the blood-vessels and thus to protect the eyes, tears are secreted, often in sufficient abundance to roll down the cheeks. This occurs under the most opposite emotions, and under no emotion at all. The sole exception, and this is only a partial one, to the existence of a relation between the involuntary and strong contraction of these muscles and the secretion of tears, is that of young infants, who, whilst screaming violently with their eyelids firmly closed, do not commonly weep until they have attained the age of from two to three or four months. Their eyes, however, become suffused with tears at a much earlier age. It would appear, as already remarked, that the lacrymal glands do not, from the want of practice or some other cause, come to full functional activity at a very early period of life. With children at a somewhat later age, crying out or wailing from any distress is so regularly accompanied by the shedding of tears, that weeping and crying are synonymous terms.[131]

Under the opposite emotion of great joy or amusement, as long as laughter is moderate there is hardly any contraction of the muscles round the eyes, so that there is no frowning; but when peals of loud laughter are uttered, with rapid and violent spasmodic expirations, tears stream down the face. I have more than once noticed the face of a person, after a paroxysm of violent laughter, and I could see that the orbicular muscles and those running to the upper lip were still partially contracted, which together with the tear-stained cheeks gave to the upper half of the face an expression not to be distinguished from that of a child still blubbering from grief. The fact of tears streaming down the face during violent laughter is common to all the races of mankind, as we shall see in a future chapter.

In violent coughing, especially when a person is half-choked, the face becomes purple, the veins distended, the orbicular muscles strongly contracted, and tears run down the cheeks. Even after a fit of ordinary coughing, almost every one has to wipe his eyes. In violent vomiting or retching, as I have myself experienced and seen in others, the orbicular muscles are strongly contracted, and tears sometimes flow freely down the cheeks. It has been suggested to me that this may be due to irritating matter being injected into the nostrils, and causing by reflex action the secretion of tears. Accordingly I asked one of my informants, a surgeon, to attend to the effects of retching when nothing was thrown up from the stomach; and, by an odd coincidence, he himself suffered the next morning from an attack of retching, and three days subsequently observed a lady under a similar attack; and he is certain that in neither case an atom of matter was ejected from the stomach; yet the orbicular muscles were strongly contracted, and tears freely secreted. I can also speak positively to the energetic contraction of these same muscles round the eyes, and to the coincident free secretion of tears, when the abdominal muscles act with unusual force in a downward direction on the intestinal canal.

Yawning commences with a deep inspiration, followed by a long and forcible expiration; and at the same time almost all the muscles of the body are strongly contracted, including those round the eyes. During this act tears are often secreted, and I have seen them even rolling down the cheeks.

I have frequently observed that when persons scratch some point which itches intolerably, they forcibly close their eyelids; but they do not, as I believe, first draw a deep breath and then expel it with force; and I have never noticed that the eyes then become filled with tears; but I am not prepared to assert that this does not occur. The forcible closure of the eyelids is, perhaps, merely a part of that general action by which almost all the muscles of the body are at the same time rendered rigid. It is quite different from the gentle closure of the eyes which often accompanies, as Gratiolet remarks,[132] the smelling a delicious odour, or the tasting a delicious morsel, and which probably originates in the desire to shut out any disturbing impression through the eyes.

Professor Donders writes to me to the following effect: "I have observed some cases of a very curious affection when, after a slight rub *(attouchement)*, for example, from the friction of a coat, which caused neither a wound nor a contusion, spasms of the orbicular muscles occurred, with a very profuse flow of tears, lasting about an hour. Subsequently, sometimes after an interval of several weeks, violent spasms of the same muscles reoccurred, accompanied by the secretion of tears, together with primary or secondary redness of the eye." Mr. Bowman informs me that he has occasionally observed closely analogous cases, and that, in some of these, there was no redness or inflammation of the eyes.

I was anxious to ascertain whether there existed in any of the lower animals a similar relation between the contraction of the orbicular muscles during violent expiration and the secretion of tears; but there are very few animals which contract these muscles in a prolonged manner, or which shed tears. The *Macacus maurus*, which formerly wept so copiously in the Zoological Gardens, would have been a fine case for observation; but the two monkeys now there, and which are believed to belong to the same species, do not weep. Nevertheless they were carefully observed by Mr. Bartlett and myself, whilst screaming loudly, and they seemed to contract these muscles; but they moved about their cages so rapidly, that it was difficult to observe with certainty. No other monkey, as far as I have been able to ascertain, contracts its orbicular muscles whilst screaming.

The Indian elephant is known sometimes to weep. Sir E. Tennent, in describing those which he saw captured and bound in Ceylon, says, some "lay motionless on the ground, with no other indication of suffering than the tears which suffused their eyes and flowed incessantly." Speaking of another elephant he says, "When overpowered and made fast, his grief was most affecting; his violence sank to utter prostration, and he lay on the ground, uttering choking cries, with tears trickling down his cheeks."[133] In the Zoological Gardens the keeper of the Indian elephants positively asserts that he has several times seen tears rolling down the face of the old female, when distressed by the removal of the young one. Hence I was extremely anxious to ascertain, as an extension of the relation between the contraction of the orbicular muscles and the shedding of tears in man, whether elephants when screaming or trumpeting loudly contract these muscles. At Mr. Bartlett's desire the keeper ordered the old and the young elephant to trumpet; and we repeatedly saw in both animals that, just as the trumpeting began, the orbicular muscles, especially the lower ones, were distinctly contracted. On a subsequent occasion the keeper made the old elephant trumpet much more loudly, and invariably both the upper and lower orbicular muscles were strongly contracted, and now in an equal degree. It is a singular fact that the African elephant, which, however, is so different from the Indian species that it is placed by some naturalists in a distinct sub-genus, when made on two occasions to trumpet loudly, exhibited no trace of the contraction of the orbicular muscles.

From the several foregoing cases with respect to Man, there can, I think, be no doubt that the contraction of the muscles round the eyes, during violent expiration or when the expanded chest is forcibly compressed, is, in some manner, intimately connected with the secretion of tears. This holds good under widely different emotions, and independently of any emotion. It is not, of course, meant that tears cannot be secreted without the contraction of these muscles; for it is notorious that they are often freely shed with the eyelids not closed, and with the brows unwrinkled. The contraction must be both involuntary and prolonged, as during a choking fit, or energetic, as during a sneeze. The mere involuntary winking of the eyelids, though often repeated, does not bring tears into the eyes. Nor does the voluntary and prolonged contraction of the several surrounding muscles suffice. As the lacrymal glands of children are easily excited, I persuaded my own and several other children of different ages to contract these muscles repeatedly with their utmost force, and to continue doing so as long as they possibly could; but this produced hardly any effect. There was sometimes a little moisture in the eyes, but not more than apparently could be accounted for by the squeezing out of the already secreted tears within the glands.

The nature of the relation between the involuntary and energetic contraction of the muscles round the eyes, and the secretion of tears, cannot be positively ascertained, but a probable view may be suggested. The primary function of the secretion of tears, together with some mucus, is to lubricate the surface of the eye; and a secondary one, as some believe, is to keep the nostrils damp, so that the inhaled air may be moist,[134] and likewise to favour the power of smelling. But another, and at least equally important function of tears, is to wash out particles of dust or other minute objects which may get into the eyes. That this is of great importance is clear from the cases in which the cornea has been rendered opaque through inflammation, caused by particles of dust not being removed, in consequence of the eye and eyelid becoming immovable.[135] The secretion of tears from the irritation of any foreign body in the eye is a reflex action;—that is, the body irritates a peripheral nerve which sends an impression to certain sensory nerve-cells; these transmit an influence to other cells, and these again to the lacrymal glands. The influence transmitted to these glands causes, as there is good reason to believe, the relaxation of the muscular coats of the smaller arteries; this allows more blood to permeate the glandular tissue, and this induces a free secretion of tears. When the small arteries of the face, including those of the retina, are relaxed under very different circumstances, namely, during an intense blush, the lacrymal glands are sometimes affected in a like manner, for the eyes become suffused with tears.

It is difficult to conjecture how many reflex actions have originated, but, in relation to the present case of the affection of the lacrymal glands through irritation of the surface of the eye, it may be worth remarking that, as soon as some primordial form became semi-terrestrial in its habits, and was liable to get particles of dust into its eyes, if these were not washed out they would cause much irritation; and on the principle of the radiation of nerve-force to adjoining nerve-cells, the lacrymal glands would be stimulated to secretion. As this would often recur, and as nerve-force readily passes along accustomed channels, a slight irritation would ultimately suffice to cause a free secretion of tears.

As soon as by this, or by some other means, a reflex action of this nature had been established and rendered easy, other stimulants applied to the surface of the eye—such as a cold wind, slow inflammatory action, or a blow on the eyelids—would cause a copious secretion of tears, as we know to be the case. The glands are also excited into action through the irritation of adjoining parts. Thus when the nostrils are irritated by pungent vapours, though the eyelids may be kept firmly closed, tears are copiously secreted; and this likewise follows from a blow on the nose, for instance from a boxing-glove. A stinging switch on the face produces, as I have seen, the same effect. In these latter cases the secretion of tears is an incidental result, and of no direct service. As all these parts of the face, including the lacrymal glands, are supplied with branches of the same nerve, namely, the fifth, it is in some degree intelligible that the effects of the excitement of any one branch should spread to the nerve-cells or roots of the other branches.

The internal parts of the eye likewise act, under certain conditions, in a reflex manner on the lacrymal glands. The following statements have been kindly communicated to me by Mr. Bowman; but the subject is a very intricate one, as all the parts of the eye are so intimately related together, and are so sensitive to various stimulants. A strong light acting on the retina, when in a normal condition, has very little tendency to cause lacrymation; but with unhealthy children having small, old-standing ulcers on the cornea, the retina becomes excessively sensitive to light, and exposure even to common daylight causes forcible and sustained closure of the lids, and a profuse flow of tears. When persons who ought to begin the use of convex glasses habitually strain the waning power of accommodation, an undue secretion of tears very often follows, and the retina is liable to become unduly sensitive to light. In general, morbid affections of the surface of the eye, and of the ciliary structures concerned in the accommodative act, are prone to be accompanied with excessive secretion of tears. Hardness of the eyeball, not rising to inflammation, but implying a want of balance between the fluids poured out and again taken up by the intraocular vessels, is not usually attended with any lacrymation. When the balance is on the other side, and the eye becomes too soft, there is a greater tendency to lacrymation. Finally, there are numerous morbid states and structural alterations of the eyes, and even terrible inflammations, which may be attended with little or no secretion of tears.

It also deserves notice, as indirectly bearing on our subject, that the eye and adjoining parts are subject to an extraordinary number of reflex and associated movements, sensations, and actions, besides those relating to the lacrymal glands. When a bright light strikes the retina of one eye alone, the iris contracts, but the iris of the other eye moves after a measurable interval of time. The iris likewise moves in accommodation to near or distant vision, and when the two eyes are made to converge.[136] Every one knows how irresistibly the eyebrows are drawn down under an intensely bright light. The eyelids also involuntarily wink when an object is moved near the eyes, or a sound is suddenly heard. The well-known case of a bright light causing some persons to sneeze is even more curious; for nerve-force here radiates from certain nerve-cells in connection with the retina, to the sensory nerve-cells of the nose, causing it to tickle; and from these, to the cells which command the various respiratory muscles (the orbiculars included) which expel the air in so peculiar a manner that it rushes through the nostrils alone.

To return to our point: why are tears secreted during a screaming-fit or other violent expiratory efforts? As a slight blow on the eyelids causes a copious secretion of tears, it is at least possible that the spasmodic contraction of the eyelids, by pressing strongly on the eyeball, should in a similar manner cause some secretion. This seems possible, although the voluntary contraction of the same muscles does not produce any such effect. We know that a man cannot voluntarily sneeze or cough with nearly the same force as he does automatically; and so it is with the contraction of the orbicular muscles: Sir C. Bell experimented on them, and found that by suddenly and forcibly closing the eyelids in the dark, sparks of light are seen, like those caused by tapping the eyelids with the fingers; "but in sneezing the compression is both more rapid and more forcible, and the sparks are more brilliant" That these sparks are due to the contraction of the eyelids is clear, because if they "are held open during the act of sneezing, no sensation of light will be experienced." In the peculiar cases referred to by Professor Donders and Mr. Bowman, we have seen that some weeks after the eye has been very slightly injured, spasmodic contractions of the eyelids ensue, and these are accompanied by a profuse flow of tears. In the act of yawning, the tears are apparently due solely to the spasmodic contraction of the muscles round the eyes. Notwithstanding these latter cases, it seems hardly credible that the pressure of the eyelids on the surface of the eye, although effected spasmodically and therefore with much greater force than can be done voluntarily, should be sufficient to cause by reflex action the secretion of tears in the many cases in which this occurs during violent expiratory efforts.

Another cause may come conjointly into play. We have seen that the internal parts of the eye, under certain conditions, act in a reflex manner on the lacrymal glands. We know that during violent expiratory efforts the pressure of the arterial blood within the vessels of the eye is increased, and that the return of the venous blood is impeded. It seems, therefore, not improbable that the distension of the ocular vessels, thus induced, might act by reflection on the lacrymal glands—the effects due to the spasmodic pressure of the eyelids on the surface of the eye being thus increased.

In considering how far this view is probable, we should bear in mind that the eyes of infants have been acted on in this double manner during numberless generations, whenever they have screamed; and on the principle of nerve-force readily passing along accustomed channels, even a moderate compression of the eyeballs and a moderate distension of the ocular vessels would ultimately come, through habit, to act on the glands. We have an analogous case in the orbicular muscles being almost always contracted in some slight degree, even during a gentle crying-fit, when there can be no distension of the vessels and no uncomfortable sensation excited within the eyes.

Moreover, when complex actions or movements have long been performed in strict association together, and these are from any cause at first voluntarily and afterwards habitually checked, then if the proper exciting conditions occur, any part of the action or movement which is least under the control of the will, will often still be involuntarily performed. The secretion by a gland is remarkably free from the influence of the will; therefore, when with the advancing age of the individual, or with the advancing culture of the race, the habit of crying out or screaming is restrained, and there is consequently no distension of the blood-vessels of the eye, it may nevertheless well happen that tears should still be secreted. We may see, as lately remarked, the muscles round the eyes of a person who reads a pathetic story, twitching or trembling in so slight a degree as hardly to be detected. In this case there has been no screaming and no distension of the blood-vessels, yet through habit certain nerve-cells send a small amount of nerve-force to the cells commanding the muscles round the eyes; and they likewise send some to the cells commanding the lacrymal glands, for the eyes often become at the same time just moistened with tears. If the twitching of the muscles round the eyes and the secretion of tears had been completely prevented, nevertheless it is almost certain that there would have been some tendency to transmit nerve-force in these same directions; and as the lacrymal glands are remarkably free from the control of the will, they would be eminently liable still to act, thus betraying, though there were no other outward signs, the pathetic thoughts which were passing through the person's mind.

As a further illustration of the view here advanced, I may remark that if, during an early period of life, when habits of all kinds are readily established, our infants, when pleased, had been accustomed to utter loud peals of laughter (during which the vessels of their eyes are distended) as often and as continuously as they have yielded when distressed to screaming-fits, then it is probable that in after life tears would have been as copiously and as regularly secreted under the one state of mind as under the other. Gentle laughter, or a smile, or even a pleasing thought, would have sufficed to cause a moderate secretion of tears. There does indeed exist an evident tendency in this direction, as will be seen in a future chapter, when we treat of the tender feelings. With the Sandwich Islanders, according to Freycinet,[137] tears are actually recognized as a sign of happiness; but we should require better evidence on this head than that of a passing voyager. So again if our infants, during many generations, and each of them during several years, had almost daily suffered from prolonged choking-fits, during which the vessels of the eye are distended and tears copiously secreted, then it is probable, such is the force of associated habit, that during after life the mere thought of a choke, without any distress of mind, would have sufficed to bring tears into our eyes.

To sum up this chapter, weeping is probably the result of some such chain of events as follows. Children, when wanting food or suffering in any way, cry out loudly, like the young of most other animals, partly as a call to their parents for aid, and partly from any great exertion serving as a relief. Prolonged screaming inevitably leads to the gorging of the blood-vessels of the eye; and this will have led, at first consciously and at last habitually, to the contraction of the muscles round the eyes in order to protect them. At the same time the spasmodic pressure on the surface of the eye, and the distension of the vessels within the eye, without necessarily entailing any conscious sensation, will have affected, through reflex action, the lacrymal glands. Finally, through the three principles of nerve-force readily passing along accustomed channels—of association, which is so widely extended in its power—and of certain actions, being more under the control of the will than others—it has come to pass that suffering readily causes the secretion of tears, without being necessarily accompanied by any other action.

Although in accordance with this view we must look at weeping as an incidental result, as purposeless as the secretion of tears from a blow outside the eye, or as a sneeze from the retina being affected by a bright light, yet this does not present any difficulty in our understanding how the secretion of tears serves as a relief to suffering. And by as much as the weeping is more violent or hysterical, by so much will the relief be greater,—on the same principle that the writhing of the whole body, the grinding of the teeth, and the uttering of piercing shrieks, all give relief under an agony of pain.

CHAPTER VII

Low Spirits, Anxiety, Grief, Dejection, Despair

GENERAL EFFECT OF GRIEF ON THE SYSTEM • OBLIQUITY OF THE EYEBROWS UNDER SUFFERING • ON THE CAUSE OF THE OBLIQUITY OF THE EYEBROWS • ON THE DEPRESSION OF THE CORNERS OF THE MOUTH

After the mind has suffered from an acute paroxysm of grief, and the cause still continues, we fall into a state of low spirits; or we may be utterly cast down and dejected. Prolonged bodily pain, if not amounting to an agony, generally leads to the same state of mind. If we expect to suffer, we are anxious; if we have no hope of relief, we despair.

Persons suffering from excessive grief often seek relief by violent and almost frantic movements, as described in a former chapter; but when their suffering is somewhat mitigated, yet prolonged, they no longer wish for action, but remain motionless and passive, or may occasionally rock themselves to and fro. The circulation becomes languid; the face pale; the muscles flaccid; the eyelids droop; the head hangs on the contracted chest; the lips, cheeks, and lower jaw all sink downwards from their own weight. Hence all the features are lengthened; and the face of a person who hears bad news is said to fall. A party of natives in Tierra del Fuego endeavoured to explain to us that their friend, the captain of a sealing vessel, was out of spirits, by pulling down their cheeks with both hands, so as to make their faces as long as possible. Mr. Bunnet informs me that the Australian aborigines when out of spirits have a chop-fallen appearance. After prolonged suffering the eyes become dull and lack expression, and are often slightly suffused with tears. The eyebrows not rarely are rendered oblique, which is due to their inner ends being raised. This produces peculiarly-formed wrinkles on the forehead, which are very different from those of a simple frown; though in some cases a frown alone may be present. The corners of the mouth are drawn downwards, which is so universally recognised as a sign of being out of spirits, that it is almost proverbial.

The breathing becomes slow and feeble, and is often interrupted by deep sighs. As Gratiolet remarks, whenever our attention is long concentrated on any subject, we forget to breathe, and then relieve ourselves by a deep inspiration; but the sighs of a sorrowful person, owing to his slow respiration and languid circulation, are eminently characteristic.[138] As the grief of a person in this state occasionally recurs and increases into a paroxysm, spasms affect the respiratory muscles, and he feels as if something, the so-called *globus hystericus*, was rising in his throat. These spasmodic movements are clearly allied to the sobbing of children, and are remnants of those severer spasms which occur when a person is said to choke from excessive grief.[139]

Obliquity of the eyebrows.—Two points alone in the above description require further elucidation, and these are very curious ones; namely, the raising of the inner ends of the eyebrows, and the drawing down of the corners of the mouth. With respect to the eyebrows, they may occasionally be seen to assume an oblique position in persons suffering from deep dejection or anxiety; for instance, I have observed this movement in a mother whilst speaking about her sick son; and it is sometimes excited by quite trifling or momentary causes of real or pretended distress. The eyebrows assume this position owing to the contraction of certain muscles (namely, the orbiculars, corrugators, and pyramidals of the nose, which together tend to lower and contract the eyebrows) being partially checked by the more powerful action of the central fasciæ of the frontal muscle. These latter fasciæ by their contraction raise the inner ends alone of the eyebrows; and as the corrugators at the same time draw the eyebrows together, their inner ends become puckered into a fold or lump. This fold is a highly characteristic point in the appearance of the eyebrows when rendered oblique, as may be seen in figs. 2 and 5, Plate II. The eyebrows are at the same time somewhat roughened, owing to the hairs being made to project. Dr. J. Crichton Browne has also often noticed in melancholic patients who keep their eyebrows persistently oblique, "a peculiar acute arching of the upper eyelid." A trace of this may be observed by comparing the right and left eyelids of the young man in the photograph (fig. 2, Plate II); for he was not able to act equally on both eyebrows. This is also shown by the unequal furrows on the two sides of his forehead. The acute arching of the eyelids depends, I believe, on the inner end alone of the eyebrows being raised; for when the whole

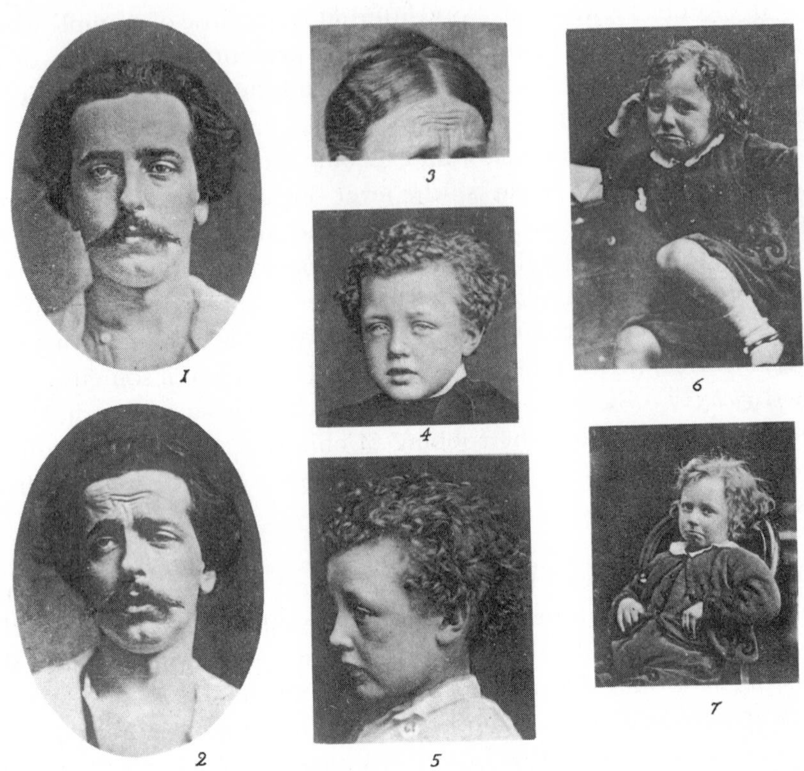

Plate II

eyebrow is elevated and arched, the upper eyelid follows in a slight degree the same movement.

But the most conspicuous result of the opposed contraction of the above-named muscles, is exhibited by the peculiar furrows formed on the forehead. These muscles, when thus in conjoint yet opposed action, may be called, for the sake of brevity, the grief-muscles. When a person elevates his eyebrows by the contraction of the whole frontal muscle, transverse wrinkles extend across the whole breadth of the forehead; but in the present case the middle fasciæ alone are contracted; consequently, transverse furrows are formed across the middle part alone of the forehead. The skin over the exterior parts of both eyebrows is at the same time drawn downwards and smoothed, by the contraction of the outer portions of the orbicular muscles. The eyebrows are likewise brought together through the simultaneous contraction of the corrugators;[140] and this latter action generates vertical furrows, separating the exterior and lowered part of the skin of the forehead from the central and raised part. The union of these vertical furrows with the central and transverse furrows (see figs. 2 and 3) produces a mark on the forehead which has been compared to a horse-shoe; but the furrows more strictly form three sides of a quadrangle. They are often conspicuous on the foreheads of adult or nearly adult persons, when their eyebrows are made oblique; but with young children, owing to their skin not easily wrinkling, they are rarely seen, or mere traces of them can be detected.

These peculiar furrows are best represented in fig. 3, Plate II, on the forehead of a young lady who has the power in an unusual degree of voluntarily acting on the requisite muscles. As she was absorbed in the attempt, whilst being photographed, her expression was not at all one of grief; I have therefore given the forehead alone. Fig. 1 on the same plate, copied from Dr. Duchenne's work,[141] represents, on a reduced scale, the face, in its natural state, of a young man who was a good actor. In fig. 2 he is shown simulating grief, but the two eyebrows, as before remarked, are not equally acted on. That the expression is true, may be inferred from the fact that out of fifteen persons, to whom the original photograph was shown, without any clue to what was intended being given them, fourteen immediately answered, "despairing sorrow," "suffering endurance," "melancholy," and so forth. The history of fig. 5 is rather curious: I saw the photograph in a shop-window, and took it to Mr. Rejlander for the sake of finding out by whom it had been made; remarking to him how pathetic the expression was. He answered, "I made it, and it was likely to be pathetic, for the boy in a few minutes burst out crying." He then showed me a photograph of the same boy in a placid state, which I have had (fig. 4) reproduced. In fig. 6, a trace of obliquity in the eyebrows may be detected; but this figure, as well as fig. 7, is given to show the depression of the corners of the mouth, to which subject I shall presently refer.

Few persons, without some practice, can voluntarily act on their grief-muscles; but after repeated trials a considerable number succeed, whilst others never can. The degree of obliquity in the eyebrows, whether assumed voluntarily or unconsciously, differs much in different persons. With some who apparently have unusually strong pyramidal muscles, the contraction of the central fasciæ of the frontal muscle, although it may be energetic, as shown by the quadrangular furrows on the forehead, does not raise the inner ends of the eyebrows, but only prevents their being so much lowered as they otherwise would have been. As far as I have been able to observe, the grief-muscles are brought into action much more frequently by children and women than by men. They are rarely acted on, at least with grown-up persons, from bodily pain, but almost exclusively from mental distress. Two persons who, after some practice, succeeded in acting on their grief-muscles, found by looking at a mirror that when they made their eyebrows oblique, they unintentionally at the same time depressed the corners of their mouths; and this is often the case when the expression is naturally assumed.

The power to bring the grief-muscles freely into play appears to be hereditary, like almost every other human faculty. A lady belonging to a family famous for having produced an extraordinary number of great actors and actresses, and who can herself give this expression "with singular precision," told Dr. Crichton Browne that all her family had possessed the power in a remarkable degree. The same hereditary tendency is said to have extended, as I likewise hear from Dr. Browne, to the last descendant of the family, which gave rise to Sir Walter Scott's novel of 'Red Gauntlet;' but the hero is described as contracting his forehead into a horse-shoe mark from any strong emotion. I have also seen a young woman whose forehead seemed almost habitually thus contracted, independently of any emotion being at the time felt.

The grief-muscles are not very frequently brought into play; and as the action is often momentary, it easily escapes observation. Although the expression, when observed, is universally and instantly recognized as that of grief or anxiety, yet not one person out of a thousand who has never studied the subject, is able to say precisely what change passes over the sufferer's face. Hence probably it is that this expression is not even alluded to, as far as I have noticed, in any work of fiction, with the exception of 'Red Gauntlet' and of one other novel; and the authoress of the latter, as I am informed, belongs to the famous family of actors just alluded to; so that her attention may have been specially called to the subject.

The ancient Greek sculptors were familiar with the expression, as shown in the statues of the Laocoon and Arrotino; but, as Duchenne remarks, they carried the transverse furrows across the whole breadth of the forehead, and thus committed a great anatomical mistake: this is likewise the case in some modern statues. It is, however, more probable that these wonderfully accurate observers intentionally sacrificed truth for the sake of beauty, than that they made a mistake; for rectangular furrows on the forehead would not have had a grand appearance on the marble. The expression, in its fully developed condition, is, as far as I can discover, not often represented in pictures by the old masters, no doubt owing to the same cause; but a lady who is perfectly familiar with this expression, informs me that in Fra Angelico's 'Descent from the Cross,' in Florence, it is clearly exhibited in one of the figures on the right-hand; and I could add a few other instances.

Dr. Crichton Browne, at my request, closely attended to this expression in the numerous insane patients under his care in the West Riding Asylum;

and he is familiar with Duchenne's photographs of the action of the grief-muscles. He informs me that they may constantly be seen in energetic action in cases of melancholia, and especially of hypochondria; and that the persistent lines or furrows, due to their habitual contraction, are characteristic of the physiognomy of the insane belonging to these two classes. Dr. Browne carefully observed for me during a considerable period three cases of hypochondria, in which the grief-muscles were persistently contracted. In one of these, a widow, aged 51, fancied that she had lost all her viscera, and that her whole body was empty. She wore an expression of great distress, and beat her semi-closed hands rhythmically together for hours. The grief-muscles were permanently contracted, and the upper eyelids arched. This condition lasted for months; she then recovered, and her countenance resumed its natural expression. A second case presented nearly the same peculiarities, with the addition that the corners of the mouth were depressed.

Mr. Patrick Nicol has also kindly observed for me several cases in the Sussex Lunatic Asylum, and has communicated to me full details with respect to three of them; but they need not here be given. From his observations on melancholic patients, Mr. Nicol concludes that the inner ends of the eyebrows are almost always more or less raised, with the wrinkles on the forehead more or less plainly marked. In the case of one young woman, these wrinkles were observed to be in constant slight play or movement. In some cases the corners of the mouth are depressed, but often only in a slight degree. Some amount of difference in the expression of the several melancholic patients could almost always be observed. The eyelids generally droop; and the skin near their outer corners and beneath them is wrinkled. The naso-labial fold, which runs from the wings of the nostrils to the corners of the mouth, and which is so conspicuous in blubbering children, is often plainly marked in these patients.

Although with the insane the grief-muscles often act persistently; yet in ordinary cases they are sometimes brought unconsciously into momentary action by ludicrously slight causes. A gentleman rewarded a young lady by an absurdly small present; she pretended to be offended, and as she upbraided him, her eyebrows became extremely oblique, with the forehead properly wrinkled. Another young lady and a youth, both in the highest spirits, were eagerly talking together with extraordinary rapidity; and I noticed that, as often as the young lady was beaten, and could not get out her words fast enough, her eyebrows went obliquely upwards, and rectangular furrows were formed on her forehead. She thus each time hoisted a flag of distress; and this she did half-a-dozen times in the course of a few minutes. I made no remark on the subject, but on a subsequent occasion I asked her to act on her grief-muscles; another girl who was present, and who could do so voluntarily, showing her what was intended. She tried repeatedly, but utterly failed; yet so slight a cause of distress as not being able to talk quickly enough, sufficed to bring these muscles over and over again into energetic action.

The expression of grief, due to the contraction of the grief-muscles, is by no means confined to Europeans, but appears to be common to all the races of mankind. I have, at least, received trustworthy accounts in regard to Hindoos, Dhangars (one of the aboriginal hill-tribes of India, and therefore belonging to a quite distinct race from the Hindoos), Malays, Negroes and Australians. With respect to the latter, two observers answer my query in the affirmative, but enter into no details. Mr. Taplin, however, appends to my descriptive remarks the words "this is exact." With respect to negroes, the lady who told me of Fra Angelico's picture, saw a negro towing a boat on the Nile, and as he encountered an obstruction, she observed his grief-muscles in strong action, with the middle of the forehead well wrinkled. Mr. Geach watched a Malay man in Malacca, with the corners of his mouth much depressed, the eyebrows oblique, with deep short grooves on the forehead. This expression lasted for a very short time; and Mr. Geach remarks it "was a strange one, very much like a person about to cry at some great loss."

In India Mr. H. Erskine found that the natives were familiar with this expression; and Mr. J. Scott, of the Botanic Gardens, Calcutta, has obligingly sent me a full description of two cases. He observed during some time, himself unseen, a very young Dhangar woman from Nagpore, the wife of one of the gardeners, nursing her baby who was at the point of death; and he distinctly saw the eyebrows raised at the inner corners, the eyelids drooping, the forehead wrinkled in the middle, the mouth slightly open, with the corners much depressed. He then came from behind a screen of plants and spoke to the poor woman, who started, burst into a bitter flood of tears, and besought him to cure her baby. The second case was that of a Hindustani man, who from illness and poverty was compelled to sell his favourite goat. After receiving the money, he repeatedly looked at the money in his hand and then at the goat, as if doubting whether he would not return it. He went to the goat, which was tied up ready to be led away, and the animal reared up and licked his hands. His eyes then wavered from side to side; his "mouth was partially closed, with the corners very decidedly depressed." At last the poor man seemed to make up his mind that he must part with his goat, and then, as Mr. Scott saw, the eyebrows became slightly oblique, with the characteristic puckering or swelling at the inner ends, but the wrinkles on the forehead were not present. The man stood thus for a minute, then heaving a deep sigh, burst into tears, raised up his two hands, blessed the goat, turned round, and without looking again, went away.

On the cause of the obliquity of the eyebrows under suffering.—During several years no expression seemed to me so utterly perplexing as this which we are here considering. Why should grief or anxiety cause the central fasciæ alone of the frontal muscle together with those round the eyes, to contract? Here we seem to have a complex movement for the sole purpose of expressing grief; and yet it is a comparatively rare expression, and often overlooked. I believe the explanation is not so difficult as it at first appears. Dr. Duchenne gives a photograph of the young man before referred to, who, when looking upwards at a strongly illuminated surface, involuntarily contracted his grief-muscles in an exaggerated manner. I had entirely forgotten this photograph, when on a very bright day with the sun behind me, I met, whilst on horseback, a girl whose eyebrows, as she looked up at me, became extremely oblique, with the proper furrows on her forehead. I have observed the same movement under similar circumstances on several subsequent occasions. On my return home I made three of my children, without giving them any clue to my object, look as long and as attentively as they could, at the summit of a tall tree standing against an extremely bright sky. With all three, the orbicular, corrugator, and pyramidal muscles were energetically contracted, through reflex action, from the excitement of the retina, so that their eyes might be protected from the bright light. But they tried their utmost to look upwards; and now a curious struggle, with spasmodic twitchings, could be observed between the whole or only the central portion of the frontal muscle, and the several muscles which serve to lower the eyebrows and close the eyelids. The involuntary contraction of the pyramidal caused the basal part of their noses to be transversely and deeply wrinkled. In one of the three children, the whole eyebrows were momentarily raised and lowered by the alternate contraction of the whole frontal muscle and of the muscles surrounding the eyes, so that the whole breadth of the forehead was alternately wrinkled and smoothed. In the other two children the forehead became wrinkled in the middle part alone, rectangular furrows being thus produced; and the eyebrows were rendered oblique, with their inner extremities puckered and swollen;—in the one child in a slight degree, in the other in a strongly marked manner. This difference in the obliquity of the eyebrows apparently depended on a difference in their general mobility, and in the strength of the pyramidal muscles. In both these cases the eyebrows and forehead were acted on under the influence of a strong light, in precisely the same manner, in every characteristic detail, as under the influence of grief or anxiety.

Duchenne states that the pyramidal muscle of the nose is less under the control of the will than are the other muscles round the eyes. He remarks that the young man who could so well act on his grief-muscles, as well as on most of his other facial muscles, could not contract the pyramidals.[142] This power, however, no doubt differs in different persons. The pyramidal muscle serves to draw down the skin of the forehead between the eyebrows, together with their inner extremities. The central fasciæ of the frontal are the antagonists of the pyramidal; and if the action of the latter is to be specially checked, these central fasciæ must be contracted. So that with persons having powerful pyramidal muscles, if there is under the influence of a bright light an unconscious desire to prevent the lowering of the eyebrows, the central fasciæ of the frontal muscle must be brought into play; and their contraction, if sufficiently strong to over-master the pyramidals, together with the contraction of the corrugator and orbicular muscles, will act in the manner just described on the eyebrows and forehead.

When children scream or cry out, they contract, as we know, the orbicular, corrugator, and pyramidal muscles, primarily for the sake of compressing their eyes, and thus protecting them from being gorged with blood, and secondarily through habit. I therefore expected to find with children, that when they endeavoured either to prevent a crying-fit from coming on, or to stop crying, they would check the contraction of the above-named muscles, in the same manner as when looking upwards at a bright light; and consequently that the central fasciæ of the frontal muscle would often be brought into play. Accordingly, I began myself to observe children at such times, and asked others, including some medical men, to do the same. It is necessary to observe carefully, as the peculiar opposed action of these muscles is not nearly so plain in children, owing to their foreheads not easily wrinkling, as in adults. But I soon found that the grief-muscles were very frequently brought into distinct action on these occasions. It would be superfluous to give all the cases which have been observed; and I will specify only a few. A little girl, a year and a half old, was teased by some other children, and before bursting into tears her eyebrows became decidedly oblique. With an older girl the same obliquity was observed, with the inner ends of the eyebrows plainly puckered; and at the same time the corners of the mouth were drawn downwards. As soon as she burst into tears, the features all changed and this peculiar expression vanished. Again, after a little boy had been vaccinated, which made him scream and cry violently, the surgeon gave him an orange brought for the purpose, and this pleased the child much; as he stopped crying all the characteristic movements were observed, including the formation of rectangular wrinkles in the middle of the forehead. Lastly, I met on the road a little girl three or four years old, who had been frightened by a dog, and when I asked her what was the matter, she stopped whimpering, and her eyebrows instantly became oblique to an extraordinary degree.

Here then, as I cannot doubt, we have the key to the problem why the central fasciæ of the frontal muscle and the muscles round the eyes contract in opposition to each other under the influence of grief;—whether their contraction be prolonged, as with the melancholic insane, or momentary, from some trifling cause of distress. We have all of us, as infants, repeatedly contracted our orbicular, corrugator, and pyramidal muscles, in order to protect our eyes whilst screaming; our progenitors before us have done the same during many generations; and though with advancing years we easily prevent, when feeling distressed, the utterance of screams, we cannot from long habit always prevent a slight contraction of the above-named muscles; nor indeed do we observe their contraction in ourselves, or attempt to stop it, if slight. But the pyramidal muscles seem to be less under the command of the will than the other related muscles; and if they be well developed, their contraction can be checked only by the antagonistic contraction of the central fasciæ of the frontal muscle. The result which necessarily follows, if these fasciæ contract energetically, is the oblique drawing up of the eyebrows, the puckering of their inner ends, and the formation of rectangular furrows on the middle of the forehead. As children and women cry much more freely than men, and as grown-up persons of both sexes rarely weep except from mental distress, we can understand why the grief-muscles are more frequently seen in action, as I believe to be the case, with children and women than with men; and with adults of both sexes from mental distress alone. In some of the cases before recorded, as in that of the poor Dhangar woman and of the Hindustani man, the action of the grief-muscles was quickly followed by bitter weeping. In all cases of distress, whether great or small, our brains tend through long habit to send an order to certain muscles to contract, as if we were still infants on the point of screaming out; but this order we, by the wondrous power of the will, and through habit, are able partially to counteract; although this is effected unconsciously, as far as the means of counteraction are concerned.

On the depression of the corners of the mouth.—This action is effected by the *depressores anguli oris* (see letter K in figs. 1 and 2). The fibres of this muscle diverge downwards, with the upper convergent ends attached round the angles of the mouth, and to the lower lip a little way within the angles.[143] Some of the fibres appear to be antagonistic to the great zygomatic muscle, and others to the several muscles running to the outer part of the upper lip. The contraction of this muscle draws downwards and outwards the corners of the mouth, including the outer part of the upper lip, and even in a slight degree the wings of the nostrils. When the mouth is closed and this muscle acts, the commissure or line of junction of the two lips forms a curved line with the concavity downwards,[144] and the lips themselves are generally somewhat protruded, especially the lower one. The mouth in this state is well represented in the two photographs (Plate II, figs. 6 and 7) by Mr. Rejlander. The upper boy (fig. 6) had just stopped crying, after receiving a slap on the face from another boy; and the right moment was seized for photographing him.

The expression of low spirits, grief or dejection, due to the contraction of this muscle has been noticed by every one who has written on the subject. To say that a person "is down in the mouth," is synonymous with saying that he is out of spirits. The depression of the corners may often be seen, as already stated on the authority of Dr. Crichton Browne and Mr. Nicol, with the melancholic insane, and was well exhibited in some photographs sent to me by the former gentleman, of patients with a strong tendency to suicide. It has been observed with men belonging to various races, namely with Hindoos, the dark hill-tribes of India, Malays, and, as the Rev. Mr. Hagenauer informs me, with the aborigines of Australia.

When infants scream they firmly contract the muscles round their eyes, and this draws up the upper lip; and as they have to keep their mouths widely open, the depressor muscles running to the corners are likewise brought into strong action. This generally, but not invariably, causes a slight angular bend in the lower lip on both sides, near the corners of the mouth. The result of the upper and lower lip being thus acted on, is that the mouth assumes a squarish outline. The contraction of the depressor muscle is best seen in infants when not screaming violently, and especially just before they begin, or when they cease to scream. Their little faces then acquire an extremely piteous expression, as I continually observed with my own infants between the ages of about six weeks and two or three months. Sometimes, when they are struggling against a crying-fit, the outline of the mouth is curved in so exaggerated a manner as to be like a horseshoe; and the expression of misery then becomes a ludicrous caricature.

The explanation of the contraction of this muscle, under the influence of low spirits or dejection, apparently follows from the same general principles as in the case of the obliquity of the eyebrows. Dr. Duchenne informs me that he concludes from his observations, now prolonged during many years, that this is one of the facial muscles which is least under the control of the will. This fact may indeed be inferred from what has just been stated with respect to infants when doubtfully beginning to cry, or endeavouring to stop crying; for they then generally command all the other facial muscles more effectually than they do the depressors of the corners of the mouth. Two excellent observers who had no theory on the subject, one of them a surgeon, carefully watched for me some older children and women as with some opposed struggling they very gradually approached the point of bursting out into tears; and both observers felt sure that the depressors began to act before any of the other muscles. Now as the depressors have been repeatedly brought into strong action during infancy in many generations, nerve-force will tend to flow, on the principle of long associated habit, to these muscles as well as to various other facial muscles, whenever in after life even a slight feeling of distress is experienced. But as the depressors are somewhat less under the control of the will than most of the other muscles, we might expect that they would often slightly contract, whilst the others remained passive. It is remarkable how small a depression of the corners of the mouth gives to the countenance an expression of low spirits or dejection, so that an extremely slight contraction of these muscles would be sufficient to betray this state of mind.

I MAY HERE MENTION a trifling observation, as it will serve to sum up our present subject. An old lady with a comfortable but absorbed expression sat nearly opposite to me in a railway carriage. Whilst I was looking at her, I saw that her *depressores anguli oris* became very slightly, yet decidedly, contracted; but as her countenance remained as placid as ever, I reflected how meaningless was this contraction, and how easily one might be deceived. The thought had hardly occurred to me when I saw that her eyes suddenly became suffused with tears almost to overflowing, and her whole countenance fell. There could now be no doubt that some painful recollection, perhaps that of a long-lost child, was passing through her mind. As soon as her sensorium was thus affected, certain nerve-cells from long habit instantly transmitted an order to all the respiratory muscles, and to those round the mouth, to prepare for a fit of crying. But the order was countermanded by the will, or rather by a later acquired habit, and all the muscles were obedient, excepting in a slight degree the *depressores anguli oris*. The mouth was not even opened; the respiration was not hurried; and no muscle was affected except those which draw down the corners of the mouth.

As soon as the mouth of this lady began, involuntarily and unconsciously on her part, to assume the proper form for a crying-fit, we may feel almost sure that some nerve-influence would have been transmitted through the long accustomed channels to the various respiratory muscles, as well as to those round the eyes, and to the vaso-motor centre which governs the supply of blood sent to the lacrymal glands. Of this latter fact we have indeed clear evidence in her eyes becoming slightly suffused with tears; and we can understand this, as the lacrymal glands are less under the control of the will than the facial muscles. No doubt there existed at the same time some tendency in the muscles round the eyes to contract, as if for the sake of protecting them from being gorged with blood, but this contraction was completely overmastered, and her brow remained unruffled. Had the pyramidal, corrugator, and orbicular muscles been as little obedient to the will, as they are in many persons, they would have been slightly acted on; and then the central fasciæ of the frontal muscle would have contracted in antagonism, and her eyebrows would have become oblique, with rectangular furrows on her forehead. Her countenance would then have expressed still more plainly than it did a state of dejection, or rather one of grief.

Through steps such as these we can understand how it is, that as soon as some melancholy thought passes through the brain, there occurs a just perceptible drawing down of the corners of the mouth, or a slight raising up of the inner ends of the eyebrows, or both movements combined, and immediately afterwards a slight suffusion of tears. A thrill of nerve-force is transmitted along several habitual channels, and produces an effect on any point where the will has not acquired through long habit much power of interference. The above actions may be considered as rudimental vestiges of the screaming-fits, which are so frequent and prolonged during infancy. In this case, as well as in many others, the links are indeed wonderful which connect cause and effect in giving rise to various expressions on the human countenance; and they explain to us the meaning of certain movements, which we involuntarily and unconsciously perform, whenever certain transitory emotions pass through our minds.

CHAPTER VIII

JOY, HIGH SPIRITS, LOVE, TENDER FEELINGS, DEVOTION

LAUGHTER PRIMARILY THE EXPRESSION OF JOY • LUDICROUS IDEAS • MOVEMENTS OF THE FEATURES DURING LAUGHTER • NATURE OF THE SOUND PRODUCED • THE SECRETION OF TEARS DURING LOUD LAUGHTER • GRADATION FROM LOUD LAUGHTER TO GENTLE SMILING • HIGH SPIRITS • THE EXPRESSION OF LOVE • TENDER FEELINGS • DEVOTION

JOY, WHEN INTENSE, leads to various purposeless movements—to dancing about, clapping the hands, stamping, &c., and to loud laughter. Laughter seems primarily to be the expression of mere joy or happiness. We clearly see this in children at play, who are almost incessantly laughing. With young persons past childhood, when they are in high spirits, there is always much meaningless laughter. The laughter of the gods is described by Homer as "the exuberance of their celestial joy after their daily banquet." A man smiles—and smiling, as we shall see, graduates into laughter—at meeting an old friend in the street, as he does at any trifling pleasure, such as smelling a sweet perfume.[145] Laura Bridgman, from her blindness and deafness, could not have acquired any expression through imitation, yet when a letter from a beloved friend was communicated to her by gesture-language, she "laughed and clapped her hands, and the colour mounted to her cheeks." On other occasions she has been seen to stamp for joy.[146]

Idiots and imbecile persons likewise afford good evidence that laughter or smiling primarily expresses mere happiness or joy. Dr. Crichton Browne, to whom, as on so many other occasions, I am indebted for the results of his wide experience, informs me that with idiots laughter is the most prevalent and frequent of all the emotional expressions. Many idiots are morose, passionate, restless, in a painful state of mind, or utterly stolid, and these never laugh. Others frequently laugh in a quite senseless manner. Thus an idiot boy, incapable of speech, complained to Dr. Browne, by the aid of signs, that another boy in the asylum had given him a black eye; and this was accompanied by "explosions of laughter and with his face covered with the broadest smiles." There is another large class of idiots who are persistently joyous and benign, and who are constantly laughing or smiling.[147] Their countenances often exhibit a stereotyped smile; their joyousness is increased, and they grin, chuckle, or giggle, whenever food is placed before them, or when they are caressed, are shown bright colours, or hear music. Some of them laugh more than usual when they walk about, or attempt any muscular exertion. The joyousness of most of these idiots cannot possibly be associated, as Dr. Browne remarks, with any distinct ideas: they simply feel pleasure, and express it by laughter or smiles. With imbeciles rather higher in the scale, personal vanity seems to be the commonest cause of laughter, and next to this, pleasure arising from the approbation of their conduct.

With grown-up persons laughter is excited by causes considerably different from those which suffice during childhood; but this remark hardly applies to smiling. Laughter in this respect is analogous with weeping, which with adults is almost confined to mental distress, whilst with children it is excited by bodily pain or any suffering, as well as by fear or rage. Many curious discussions have been written on the causes of laughter with grown-up persons. The subject is extremely complex. Something incongruous or unaccountable, exciting surprise and some sense of superiority in the laugher, who must be in a happy frame of mind, seems to be the commonest cause.[148] The circumstances must not be of a momentous nature: no poor man would laugh or smile on suddenly hearing that a large fortune had been bequeathed to him. If the mind is strongly excited by pleasurable feelings, and any little unexpected event or thought occurs, then, as Mr. Herbert Spencer remarks,[149] "a large amount of nervous energy, instead of being allowed to expend itself in producing an equivalent amount of the new thoughts and emotion which were nascent, is suddenly checked in its flow." . . . "The excess must discharge itself in some other direction, and there results an efflux through the motor nerves to various classes of the muscles, producing the half-convulsive actions we term laughter." An observation, bearing on this point, was made by a correspondent during the recent siege of Paris, namely, that the German soldiers, after strong excitement from exposure to extreme danger, were particularly apt to burst out into loud laughter at the smallest joke. So again when young children are just beginning to cry, an unexpected event will sometimes suddenly turn their crying into laughter, which apparently serves equally well to expend their superfluous nervous energy.

The imagination is sometimes said to be tickled by a ludicrous idea; and this so-called tickling of the mind is curiously analogous with that of the body. Every one knows how immoderately children laugh, and how their whole bodies are convulsed when they are tickled. The anthropoid apes, as we have seen, likewise utter a reiterated sound, corresponding with our laughter, when they are tickled, especially under the armpits. I touched with a bit of paper the sole of the foot of one of my infants, when only seven days old, and it was suddenly jerked away and the toes curled about, as in an older child. Such movements, as well as laughter from being tickled, are manifestly reflex actions; and this is likewise shown by the minute unstriped muscles, which serve to erect the separate hairs on the body, contracting near a tickled surface.[150] Yet laughter from a ludicrous idea, though involuntary, cannot be called a strictly reflex action. In this case, and in that of laughter from being tickled, the mind must be in a pleasurable condition; a young child, if tickled by a strange man, would scream from fear. The touch must be light, and an idea or event, to be ludicrous, must not be of grave import. The parts of the body which are most easily tickled are those which are not commonly touched, such as the armpits or between the toes, or parts such as the soles of the feet, which are habitually touched by a broad surface; but the surface on which we sit offers a marked exception to this rule. According to Gratiolet,[151] certain nerves are much more sensitive to tickling than others. From the fact that a child can hardly tickle itself, or in a much less degree than when tickled by another person, it seems that the precise point to be touched must not be known; so with the mind, something unexpected—a novel or incongruous idea which breaks through an habitual train of thought—appears to be a strong element in the ludicrous.

The sound of laughter is produced by a deep inspiration followed by short, interrupted, spasmodic contractions of the chest, and especially of the diaphragm.[152] Hence we hear of "laughter holding both his sides." From the shaking of the body, the head nods to and fro. The lower jaw often quivers up and down, as is likewise the case with some species of baboons, when they are much pleased.

During laughter the mouth is opened more or less widely, with the corners drawn much backwards, as well as a little upwards; and the upper lip is somewhat raised. The drawing back of the corners is best seen in moderate laughter, and especially in a broad smile—the latter epithet showing how the mouth is widened. In the accompanying figs. 1–3, Plate III, different degrees of moderate laughter and smiling have been photographed. The figure of the little girl, with the hat, is by Dr. Wallich, and the expression was a genuine one; the other two are by Mr. Rejlander. Dr. Duchenne repeatedly insists[153] that, under the emotion of joy, the mouth is acted on exclusively by the great zygomatic muscles, which serve to draw the corners

Plate III

backwards and upwards; but judging from the manner in which the upper teeth are always exposed during laughter and broad smiling, as well as from my own sensations, I cannot doubt that some of the muscles running to the upper lip are likewise brought into moderate action. The upper and lower orbicular muscles of the eyes are at the same time more or less contracted; and there is an intimate connection, as explained in the chapter on weeping, between the orbiculars, especially the lower ones, and some of the muscles running to the upper lip. Henle remarks[154] on this head, that when a man closely shuts one eye he cannot avoid retracting the upper lip on the same side; conversely, if any one will place his finger on his lower eyelid, and then uncover his upper incisors as much as possible, he will feel, as his upper lip is drawn strongly upwards, that the muscles of the lower eyelid contract. In Henle's drawing, given in woodcut, fig. 2, the *musculus malaris* (H) which runs to the upper lip may be seen to form an almost integral part of the lower orbicular muscle.

Dr. Duchenne has given a large photograph of an old man (reduced on Plate III, fig. 4), in his usual passive condition, and another of the same man (fig. 5), naturally smiling. The latter was instantly recognised by every one to whom it was shown as true to nature. He has also given, as an example of an unnatural or false smile, another photograph (fig. 6) of the same old man, with the corners of his mouth strongly retracted by the galvanization of the great zygomatic muscles. That the expression is not natural is clear, for I showed this photograph to twenty-four persons, of whom three could not in the least tell what was meant, whilst the others, though they perceived that the expression was of the nature of a smile, answered in such words as "a wicked joke," "trying to laugh," "grinning laughter," "half-amazed laughter," &c. Dr. Duchenne attributes the falseness of the expression altogether to the orbicular muscles of the lower eyelids not being sufficiently contracted; for he justly lays great stress on their contraction in the expression of joy. No doubt there is much truth in this view, but not, as it appears to me, the whole truth. The contraction of the lower orbiculars is always accompanied, as we have seen, by the drawing up of the upper lip. Had the upper lip, in fig. 6, been thus acted on to a slight extent, its curvature would have been less rigid, the naso-labial furrow would have been slightly different, and the whole expression would, as I believe, have been more natural, independently of the more conspicuous effect from the stronger contraction of the lower eyelids. The corrugator muscle, moreover, in fig. 6, is too much contracted, causing a frown; and this muscle never acts under the influence of joy except during strongly pronounced or violent laughter.

By the drawing backwards and upwards of the corners of the mouth, through the contraction of the great zygomatic muscles, and by the raising of the upper lip, the cheeks are drawn upwards. Wrinkles are thus formed under the eyes, and, with old people, at their outer ends; and these are highly characteristic of laughter or smiling. As a gentle smile increases into a strong one, or into a laugh, every one may feel and see, if he will attend to his own sensations and look at himself in a mirror, that as the upper lip is drawn up and the lower orbiculars contract, the wrinkles in the lower eyelids and those beneath the eyes are much strengthened or increased. At the same time, as I have repeatedly observed, the eyebrows are slightly lowered, which shows that the upper as well as the lower orbiculars contract at least to some degree, though this passes unperceived, as far as our sensations are concerned. If the original photograph of the old man, with his countenance in its usual placid state (fig. 4), be compared with that (fig. 5) in which he is naturally smiling, it may be seen that the eyebrows in the latter are a little lowered. I presume that this is owing to the upper orbiculars being impelled, through the force of long-associated habit, to act to a certain extent in concert with the lower orbiculars, which themselves contract in connection with the drawing up of the upper lip.

The tendency in the zygomatic muscles to contract under pleasurable emotions is shown by a curious fact, communicated to me by Dr. Browne, with respect to patients suffering from *general paralysis of the insane.*[155] "In this malady there is almost invariably optimism—delusions as to wealth, rank, grandeur—insane joyousness, benevolence, and profusion, while its very earliest physical symptom is trembling at the corners of the mouth and at the outer corners of the eyes. This is a well-recognized fact. Constant tremulous agitation of the inferior palpebral and great zygomatic muscles is pathognomic of the earlier stages of general paralysis. The countenance has a pleased and benevolent expression. As the disease advances other muscles become involved, but until complete fatuity is reached, the prevailing expression is that of feeble benevolence."

As in laughing and broadly smiling the cheeks and upper lip are much raised, the nose appears to be shortened, and the skin on the bridge becomes finely wrinkled in transverse lines, with other oblique longitudinal lines on the sides. The upper front teeth are commonly exposed. A well-marked naso-labial fold is formed, which runs from the wing of each nostril to the corner of the mouth; and this fold is often double in old persons.

A bright and sparkling eye is as characteristic of a pleased or amused state of mind, as is the retraction of the corners of the mouth and upper lip with the wrinkles thus produced. Even the eyes of microcephalous idiots, who are so degraded that they never learn to speak, brighten slightly when they are pleased.[156] Under extreme laughter the eyes are too much suffused with tears to sparkle; but the moisture squeezed out of the glands during moderate laughter or smiling may aid in giving them lustre; though this must be of altogether subordinate importance, as they become dull from grief, though they are then often moist. Their brightness seems to be chiefly due to their tenseness,[157] owing to the contraction of the orbicular muscles and to the pressure of the raised cheeks. But, according to Dr. Piderit, who has discussed this point more fully than any other writer,[158] the tenseness may be largely attributed to the eyeballs becoming filled with blood and other fluids, from the acceleration of the circulation, consequent on the excitement of pleasure. He remarks on the contrast in the appearance of the eyes of a hectic patient with a rapid circulation, and of a man suffering from cholera with almost all the fluids of his body drained from him. Any cause which lowers the circulation deadens the eye. I remember seeing a man utterly prostrated by prolonged and severe exertion during a very hot day, and a bystander compared his eyes to those of a boiled codfish.

To return to the sounds produced during laughter. We can see in a vague manner how the utterance of sounds of some kind would naturally become associated with a pleasurable state of mind; for throughout a large part of the animal kingdom vocal or instrumental sounds are employed either as a call or as a charm by one sex for the other. They are also employed as the means for a joyful meeting between the parents and their offspring, and between the attached members of the same social community. But why the sounds which man utters when he is pleased have the peculiar reiterated character of laughter we do not know. Nevertheless we can see that they would naturally be as different as possible from the screams or cries of distress; and as in the production of the latter, the expirations are prolonged and continuous, with the inspirations short and interrupted, so it might perhaps have been expected with the sounds uttered from joy, that the expirations would have been short and broken with the inspirations prolonged; and this is the case.

It is an equally obscure point why the corners of the mouth are retracted and the upper lip raised during ordinary laughter. The mouth must not be opened to its utmost extent, for when this occurs during a paroxysm of excessive laughter hardly any sound is emitted; or it changes its tone and seems to come from deep down in the throat. The respiratory muscles, and even those of the limbs, are at the same time thrown into rapid vibratory movements. The lower jaw often partakes of this movement, and this would tend to prevent the mouth from being widely opened. But as a full volume of sound has to be poured forth, the orifice of the mouth must be large; and it is perhaps to gain this end that the corners are retracted and the upper lip raised. Although we can hardly account for the shape of the mouth during laughter, which leads to wrinkles being formed beneath the eyes, nor for the peculiar reiterated sound of laughter, nor for the quivering of the jaws, nevertheless we may infer that all these effects are due to some common cause. For they are all characteristic and expressive of a pleased state of mind in various kinds of monkeys.

A graduated series can be followed from violent to moderate laughter, to a broad smile, to a gentle smile, and to the expression of mere cheerfulness. During excessive laughter the whole body is often thrown backward and shakes, or is almost convulsed; the respiration is much disturbed; the head and face become gorged with blood, with the veins distended; and the orbicular muscles are spasmodically contracted in order to protect the eyes. Tears are freely shed. Hence, as formerly remarked, it is scarcely possible to point out any difference between the tear-stained face of a person after a paroxysm of excessive laughter and after a bitter crying-fit.[159] It is probably due to the close similarity of the spasmodic movements caused by these widely different emotions that hysteric patients alternately cry and laugh with violence, and that young children sometimes pass suddenly from the one to the other state. Mr. Swinhoe informs me that he has often seen the Chinese, when suffering from deep grief, burst out into hysterical fits of laughter.

I was anxious to know whether tears are freely shed during excessive laughter by most of the races of men, and I hear from my correspondents that this is the case. One instance was observed with the Hindoos, and they themselves said that it often occurred. So it is with the Chinese. The women of a wild tribe of Malays in the Malacca peninsula, sometimes shed tears when they laugh heartily, though this seldom occurs. With the Dyaks of Borneo it must frequently be the case, at least with the women, for I hear from the Rajah C. Brooke that it is a common expression with them to say "we nearly made tears from laughter." The aborigines of Australia express their emotions freely, and they are described by my correspondents as jumping about and clapping their hands for joy, and as often roaring with laughter. No less than four observers have seen their eyes freely watering on such occasions; and in one instance the tears rolled down their cheeks. Mr. Bulmer, a missionary in a remote part of Victoria, remarks, "that they have a keen sense of the ridiculous; they are excellent mimics, and when one of them is able to imitate the peculiarities of some absent member of the tribe, it is very common to hear all in the camp convulsed with laughter." With Europeans hardly anything excites laughter so easily as mimicry; and it is rather curious to find the same fact with the savages of Australia, who constitute one of the most distinct races in the world.

In Southern Africa with two tribes of Kafirs, especially with the women, their eyes often fill with tears during laughter. Gaika, the brother of the chief Sandilli, answers my query on this head, with the words, "Yes, that is their common practice." Sir Andrew Smith has seen the painted face of a Hottentot woman all furrowed with tears after a fit of laughter. In Northern Africa, with the Abyssinians, tears are secreted under the same circumstances. Lastly, in North America, the same fact has been observed in a remarkably savage and isolated tribe, but chiefly with the women; in another tribe it was observed only on a single occasion.

Excessive laughter, as before remarked, graduates into moderate laughter. In this latter case the muscles round the eyes are much less contracted, and there is little or no frowning. Between a gentle laugh and a broad smile there is hardly any difference, excepting that in smiling no reiterated sound is uttered, though a single rather strong expiration, or slight noise—a rudiment of a laugh—may often be heard at the commencement of a smile. On a moderately smiling countenance the contraction of the upper orbicular muscles can still just be traced by a slight lowering of the eyebrows. The contraction of the lower orbicular and palpebral muscles is much plainer, and is shown by the wrinkling of the lower eyelids and of the skin beneath them, together with a slight drawing up of the upper lip. From the broadest smile we pass by the finest steps into the gentlest one. In this latter case the features are moved in a much less degree, and much more slowly, and the mouth is kept closed. The curvature of the naso-labial furrow is also slightly different in the two cases. We thus see that no abrupt line of demarcation can be drawn between the movement of the features during the most violent laughter and a very faint smile.[160]

A smile, therefore, may be said to be the first stage in the development of a laugh. But a different and more probable view may be suggested; namely, that the habit of uttering loud reiterated sounds from a sense of pleasure, first led to the retraction of the corners of the mouth and of the upper lip, and to the contraction of the orbicular muscles; and that now, through association and long-continued habit, the same muscles are brought into slight play whenever any cause excites in us a feeling which, if stronger, would have led to laughter; and the result is a smile.

Whether we look at laughter as the full development of a smile, or, as is more probable, at a gentle smile as the last trace of a habit, firmly fixed during many generations, of laughing whenever we are joyful, we can follow in our infants the gradual passage of the one into the other. It is well known to those who have the charge of young infants, that it is difficult to feel sure when certain movements about their mouths are really expressive; that is, when they really smile. Hence I carefully watched my own infants. One of them at the age of forty-five days, and being at the time in a happy frame of mind, smiled; that is, the corners of the mouth were retracted, and simultaneously the eyes became decidedly bright. I observed the same thing on the following day; but on the third day the child was not quite well and there was no trace of a smile, and this renders it probable that the previous smiles were real. Eight days subsequently and during the next succeeding week, it was remarkable how his eyes brightened whenever he smiled, and his nose became at the same time transversely wrinkled. This was now accompanied by a little bleating noise, which perhaps represented a laugh. At the age of 113 days these little noises, which were always made during expiration, assumed a slightly different character, and were more broken or interrupted, as in sobbing; and this was certainly incipient laughter. The change in tone seemed to me at the time to be connected with the greater lateral extension of the mouth as the smiles became broader.

In a second infant the first real smile was observed at about the same age, viz. forty-five days; and in a third, at a somewhat earlier age. The second infant, when sixty-five days old, smiled much more broadly and plainly than did the one first mentioned at the same age; and even at this early age uttered noises very like laughter. In this gradual acquirement, by infants, of the habit of laughing, we have a case in some degree analogous to that of weeping. As practice is requisite with the ordinary movements of the body, such as walking, so it seems to be with laughing and weeping. The art of screaming, on the other hand, from being of service to infants, has become finely developed from the earliest days.

High spirits, cheerfulness.—A man in high spirits, though he may not actually smile, commonly exhibits some tendency to the retraction of the corners of his mouth. From the excitement of pleasure, the circulation becomes more rapid; the eyes are bright, and the colour of the face rises. The brain, being stimulated by the increased flow of blood, reacts on the mental powers; lively ideas pass still more rapidly through the mind, and the affections are warmed. I heard a child, a little under four years old, when asked what was meant by being in good spirits, answer, "It is laughing, talking, and kissing." It would be difficult to give a truer and more practical definition. A man in this state holds his body erect, his head upright, and his eyes open. There is no drooping of the features, and no contraction of the eyebrows. On the contrary, the frontal muscle, as Moreau observes,[161] tends to contract slightly; and this smooths the brow, removes every trace of a frown, arches the eyebrows a little, and raises the eyelids. Hence the Latin phrase, *exporrigere frontem*—to unwrinkle the brow—means, to be cheerful or merry. The whole expression of a man in good spirits is exactly the opposite of that of one suffering from sorrow. According to Sir C. Bell, "In all the exhilarating emotions the eyebrows, eyelids, the nostrils, and the angles of the mouth are raised. In the depressing passions it is the reverse." Under the influence of the latter the brow is heavy, the eyelids, cheeks, mouth, and whole head droop; the eyes are dull; the countenance pallid, and the respiration slow. In joy the face expands, in grief it lengthens. Whether the principle of antithesis has here come into play in producing these opposite expressions, in aid of the direct causes which have been specified and which are sufficiently plain, I will not pretend to say.

With all the races of man the expression of good spirits appears to be the same, and is easily recognised. My informants, from various parts of the Old and New Worlds, answer in the affirmative to my queries on this head, and they give some particulars with respect to Hindoos, Malays, and New Zealanders. The brightness of the eyes of the Australians has struck four observers, and the same fact has been noticed with Hindoos, New Zealanders, and the Dyaks of Borneo.

Savages sometimes express their satisfaction not only by smiling, but by gestures derived from the pleasure of eating. Thus Mr. Wedgwood[162] quotes Petherick that the negroes on the Upper Nile began a general rubbing of their bellies when he displayed his beads; and Leichhardt says that the Australians smacked and clacked their mouths at the sight of his horses and bullocks, and more especially of his kangaroo dogs. The Greenlanders, "when they affirm anything with pleasure, suck down air with a certain sound;"[163] and this may be an imitation of the act of swallowing savoury food.

Laughter is suppressed by the firm contraction of the orbicular muscles of the mouth, which prevents the great zygomatic and other muscles from drawing the lips backwards and upwards. The lower lip is also sometimes held by the teeth, and this gives a roguish expression to the face, as was observed with the blind and deaf Laura Bridgman.[164] The great zygomatic muscle is sometimes variable in its course, and I have seen a young woman in whom the *depressores anguli oris* were brought into strong action in suppressing a smile; but this by no means gave to her countenance a melancholy expression, owing to the brightness of her eyes.

Laughter is frequently employed in a forced manner to conceal or mask some other state of mind, even anger. We often see persons laughing in order to conceal their shame or shyness. When a person purses up his mouth, as if to prevent the possibility of a smile, though there is nothing to excite one, or nothing to prevent its free indulgence, an affected, solemn, or pedantic expression is given; but of such hybrid expressions nothing more need here be said. In the case of derision, a real or pretended smile or laugh is often blended with the expression proper to contempt, and this may pass into angry contempt or scorn. In such cases the meaning of the laugh or smile is to show the offending person that he excites only amusement.

Love, tender feelings, &c.—Although the emotion of love, for instance that of a mother for her infant, is one of the strongest of which the mind is capable, it can hardly be said to have any proper or peculiar means of expression; and this is intelligible, as it has not habitually led to any special line of action. No doubt, as affection is a pleasurable sensation, it generally causes a gentle smile and some brightening of the eyes. A strong desire to touch the beloved person is commonly felt; and love is expressed by this means more plainly than by any other.[165] Hence we long to clasp in our arms those whom we tenderly love. We probably owe this desire to inherited habit, in association with the nursing and tending of our children, and with the mutual caresses of lovers.

With the lower animals we see the same principle of pleasure derived from contact in association with love. Dogs and cats manifestly take pleasure in rubbing against their masters and mistresses, and in being rubbed or patted by them. Many kinds of monkeys, as I am assured by the keepers in the Zoological Gardens, delight in fondling and being fondled by each other, and by persons to whom they are attached. Mr. Bartlett has described to me the behaviour of two chimpanzees, rather older animals than those generally imported into this country, when they were first brought together. They sat opposite, touching each other with their much protruded lips; and the one put his hand on the shoulder of the other. They then mutually folded each other in their arms. Afterwards they stood up, each with one arm on the shoulder of the other, lifted up their heads, opened their mouths, and yelled with delight.

We Europeans are so accustomed to kissing as a mark of affection, that it might be thought to be innate in mankind; but this is not the case. Steele was mistaken when he said "Nature was its author, and it began with the first courtship." Jemmy Button, the Fuegian, told me that this practice was unknown in his land. It is equally unknown with the New Zealanders, Tahitians, Papuans, Australians, Somals of Africa, and the Esquimaux.[166] But it is so far innate or natural that it apparently depends on pleasure from close contact with a beloved person; and it is replaced in various parts of the world, by the rubbing of noses, as with the New Zealanders and Laplanders, by the rubbing or patting of the arms, breasts, or stomachs, or by one man striking his own face with the hands or feet of another. Perhaps the practice of blowing, as a mark of affection, on various parts of the body may depend on the same principle.[167]

The feelings which are called tender are difficult to analyse; they seem to be compounded of affection, joy, and especially of sympathy. These feelings are in themselves of a pleasurable nature, excepting when pity is too deep, or horror is aroused, as in hearing of a tortured man or animal. They are remarkable under our present point of view from so readily exciting the secretion of tears. Many a father and son have wept on meeting after a long separation, especially if the meeting has been unexpected. No doubt extreme joy by itself tends to act on the lacrymal glands; but on such occasions as the foregoing vague thoughts of the grief which would have been felt had the father and son never met, will probably have passed through their minds; and grief naturally leads to the secretion of tears. Thus on the return of Ulysses :—

> "Telemachus
> Rose, and clung weeping round his father's breast.
> There the pent grief rained o'er them, yearning thus.
>
> . . . . .

Thus piteously they wailed in sore unrest,
And on their weepings had gone down the day,
But that at last Telemachus found words to say."
Worsley's Translation of the Odyssey,

Book xvi. st. 27.

So again when Penelope at last recognised her husband :—

"Then from her eyelids the quick tears did start
And she ran to him from her place, and threw
Her arms about his neck, and a warm dew
Of kisses poured upon him, and thus spake:"

Book xxiii. st. 27.

The vivid recollection of our former home, or of long-past happy days, readily causes the eyes to be suffused with tears; but here, again, the thought naturally occurs that these days will never return. In such cases we may be said to sympathize with ourselves in our present, in comparison with our former, state. Sympathy with the distresses of others, even with the imaginary distresses of a heroine in a pathetic story, for whom we feel no affection, readily excites tears. So does sympathy with the happiness of others, as with that of a lover, at last successful after many hard trials in a well-told tale.

Sympathy appears to constitute a separate or distinct emotion; and it is especially apt to excite the lacrymal glands. This holds good whether we give or receive sympathy. Every one must have noticed how readily children burst out crying if we pity them for some small hurt. With the melancholic insane, as Dr. Crichton Browne informs me, a kind word will often plunge them into unrestrained weeping. As soon as we express our pity for the grief of a friend, tears often come into our own eyes. The feeling of sympathy is commonly explained by assuming that, when we see or hear of suffering in another, the idea of suffering is called up so vividly in our own minds that we ourselves suffer. But this explanation is hardly sufficient, for it does not account for the intimate alliance between sympathy and affection. We undoubtedly sympathize far more deeply with a beloved than with an indifferent person; and the sympathy of the one gives us far more relief than that of the other. Yet assuredly we can sympathize with those for whom we feel no affection.

Why suffering, when actually experienced by ourselves, excites weeping, has been discussed in a former chapter. With respect to joy, its natural and universal expression is laughter; and with all the races of man loud laughter leads to the secretion of tears more freely than does any other cause excepting distress. The suffusion of the eyes with tears, which undoubtedly occurs under great joy, though there is no laughter, can, as it seems to me, be explained through habit and association on the same principles as the effusion of tears from grief, although there is no screaming. Nevertheless it is not a little remarkable that sympathy with the distresses of others should excite tears more freely than our own distress; and this certainly is the case. Many a man, from whose eyes no suffering of his own could wring a tear, has shed tears at the sufferings of a beloved friend. It is still more remarkable that sympathy with the happiness or good fortune of those whom we tenderly love should lead to the same result, whilst a similar happiness felt by ourselves would leave our eyes dry. We should, however, bear in mind that the long-continued habit of restraint which is so powerful in checking the free flow of tears from bodily pain, has not been brought into play in preventing a moderate effusion of tears in sympathy with the sufferings or happiness of others.

Music has a wonderful power, as I have elsewhere attempted to show,[168] of recalling in a vague and indefinite manner, those strong emotions which were felt during long-past ages, when, as is probable, our early progenitors courted each other by the aid of vocal tones. And as several of our strongest emotions—grief, great joy, love, and sympathy—lead to the free secretion of tears, it is not surprising that music should be apt to cause our eyes to become suffused with tears, especially when we are already softened by any of the tenderer feelings. Music often produces another peculiar effect. We know that every strong sensation, emotion, or excitement—extreme pain, rage, terror, joy, or the passion of love—all have a special tendency to cause the muscles to tremble; and the thrill or slight shiver which runs down the backbone and limbs of many persons when they are powerfully affected by music, seems to bear the same relation to the above trembling of the body, as a slight suffusion of tears from the power of music does to weeping from any strong and real emotion.

Devotion.—As devotion is, in some degree, related to affection, though mainly consisting of reverence, often combined with fear, the expression of this state of mind may here be briefly noticed. With some sects, both past and present, religion and love have been strangely combined; and it has even been maintained, lamentable as the fact may be, that the holy kiss of love differs but little from that which a man bestows on a woman, or a woman on a man.[169] Devotion is chiefly expressed by the face being directed towards the heavens, with the eyeballs upturned. Sir C. Bell remarks that, at the approach of sleep, or of a fainting-fit, or of death, the pupils are drawn upwards and inwards; and he believes that "when we are wrapt in devotional feelings, and outward impressions are unheeded, the eyes are raised by an action neither taught nor acquired;" and that this is due to the same cause as in the above cases.[170] That the eyes are upturned during sleep is, as I hear from Professor Donders, certain. With babies, whilst sucking their mother's breast, this movement of the eyeballs often gives to them an absurd appearance of ecstatic delight; and here it may be clearly perceived that a struggle is going on against the position naturally assumed during sleep. But Sir C. Bell's explanation of the fact, which rests on the assumption that certain muscles are more under the control of the will than others is, as I hear from Professor Donders, incorrect. As the eyes are often turned up in prayer, without the mind being so much absorbed in thought as to approach to the unconsciousness of sleep, the movement is probably a conventional one—the result of the common belief that Heaven, the source of Divine power to which we pray, is seated above us.

A humble kneeling posture, with the hands upturned and palms joined, appears to us, from long habit, a gesture so appropriate to devotion, that it might be thought to be innate; but I have not met with any evidence to this effect with the various extra-European races of mankind. During the classical period of Roman history it does not appear, as I hear from an excellent classic, that the hands were thus joined during prayer. Mr. Hensleigh Wedgwood has apparently given[171] the true explanation, though this implies that the attitude is one of slavish subjection. "When the suppliant kneels and holds up his hands with the palms joined, he represents a captive who proves the completeness of his submission by offering up his hands to be bound by the victor. It is the pictorial representation of the Latin *dare manus*, to signify submission." Hence it is not probable that either the uplifting of the eyes or the joining of the open hands, under the influence of devotional feelings, are innate or truly expressive actions; and this could hardly have been expected, for it is very doubtful whether feelings, such as we should now rank as devotional, affected the hearts of men, whilst they remained during past ages in an uncivilized condition.

CHAPTER IX

Reflection—Meditation—Ill-temper—Sulkiness—Determination

THE ACT OF FROWNING • REFLECTION WITH AN EFFORT, OR WITH THE PERCEPTION OF SOMETHING DIFFICULT OR DISAGREEABLE • ABSTRACTED MEDITATION • ILL-TEMPER • MOROSENESS • OBSTINACY • SULKINESS AND POUTING • DECISION OR DETERMINATION • THE FIRM CLOSURE OF THE MOUTH

THE CORRUGATORS, BY THEIR CONTRACTION, lower the eyebrows and bring them together, producing vertical furrows on the forehead—that is, a frown. Sir C. Bell, who erroneously thought that the corrugator was peculiar to man, ranks it as "the most remarkable muscle of the human face. It knits the eyebrows with an energetic effort, which unaccountably, but irresistibly, conveys the idea of mind." Or, as he elsewhere says, "when the eyebrows are knit, energy of mind is apparent, and there is the mingling of thought and emotion with the savage and brutal rage of the mere animal."[172] There is much truth in these remarks, but hardly the whole truth. Dr. Duchenne has called the corrugator the muscle of reflection;[173] but this name, without some limitation, cannot be considered as quite correct.

A man may be absorbed in the deepest thought, and his brow will remain smooth until he encounters some obstacle in his train of reasoning, or is interrupted by some disturbance, and then a frown passes like a shadow over his brow. A half-starved man may think intently how to obtain food, but he probably will not frown unless he encounters either in thought or action some difficulty, or finds the food when obtained nauseous. I have noticed that almost everyone instantly frowns if he perceives a strange or bad taste in what he is eating. I asked several persons, without explaining my object, to listen intently to a very gentle tapping sound, the nature and source of which they all perfectly knew, and not one frowned; but a man who joined us, and who could not conceive what we were all doing in profound silence, when asked to listen, frowned much, though not in an ill-temper, and said he could not in the least understand what we all wanted. Dr. Piderit,[174] who has published remarks to the same effect, adds that stammerers generally frown in speaking; and that a man in doing even so trifling a thing as pulling on a boot, frowns if he finds it too tight. Some persons are such habitual frowners, that the mere effort of speaking almost always causes their brows to contract.

Men of all races frown when they are in any way perplexed in thought, as I infer from the answers which I have received to my queries; but I framed them badly, confounding absorbed meditation with perplexed reflection. Nevertheless, it is clear that the Australians, Malays, Hindoos, and Kafirs of South Africa frown, when they are puzzled. Dobritzhoffer remarks that the Guaranies of South America on like occasions knit their brows.[175]

From these considerations, we may conclude that frowning is not the expression of simple reflection, however profound, or of attention, however close, but of something difficult or displeasing encountered in a train of thought or in action. Deep reflection can, however, seldom be long carried on without some difficulty, so that it will generally be accompanied by a frown. Hence it is that frowning commonly gives to the countenance, as Sir C. Bell remarks, an aspect of intellectual energy. But in order that this effect may be produced, the eyes must be clear and steady, or they may be cast downwards, as often occurs in deep thought. The countenance must not be otherwise disturbed, as in the case of an ill-tempered or peevish man, or of one who shows the effects of prolonged suffering, with dulled eyes and drooping jaw, or who perceives a bad taste in his food, or who finds it difficult to perform some trifling act, such as threading a needle. In these cases a frown may often be seen, but it will be accompanied by some other expression, which will entirely prevent the countenance having an appearance of intellectual energy or of profound thought.

We may now inquire how it is that a frown should express the perception of something difficult or disagreeable, either in thought or action. In the same way as naturalists find it advisable to trace the embryological development of an organ in order fully to understand its structure, so with the movements of expression it is advisable to follow as nearly as possible the same plan. The earliest and almost sole expression seen during the first days of infancy, and then often exhibited, is that displayed during the act of screaming; and screaming is excited, both at first and for some time afterwards, by every distressing or displeasing sensation and emotion,—by hunger, pain, anger, jealousy, fear, &c. At such times the muscles round the eyes are strongly contracted; and this, as I believe, explains to a large extent the act of frowning during the remainder of our lives. I repeatedly observed my own infants, from under the age of one week to that of two or three months, and found that when a screaming-fit came on gradually, the first sign was the contraction of the corrugators, which produced a slight frown, quickly followed by the contraction of the other muscles round the eyes. When an infant is uncomfortable or unwell, little frowns—as I record in my notes—may be seen incessantly passing like shadows over its face; these being generally, but not always, followed sooner or later by a crying-fit. For instance, I watched for some time a baby, between seven and eight weeks old, sucking some milk which was cold, and therefore displeasing to him; and a steady little frown was maintained all the time. This was never developed into an actual crying-fit, though occasionally every stage of close approach could be observed.

As the habit of contracting the brows has been followed by infants during innumerable generations, at the commencement of every crying or screaming fit, it has become firmly associated with the incipient sense of something distressing or disagreeable. Hence under similar circumstances it would be apt to be continued during maturity, although never then developed into a crying-fit. Screaming or weeping begins to be voluntarily restrained at an early period of life, whereas frowning is hardly ever restrained at any age. It is perhaps worth notice that with children much given to weeping, anything which perplexes their minds, and which would cause most other children merely to frown, readily makes them weep. So with certain classes of the insane, any effort of mind, however slight, which with an habitual frowner would cause a slight frown, leads to their weeping in an unrestrained manner. It is not more surprising that the habit of contracting the brows at the first perception of something distressing, although gained during infancy, should be retained during the rest of our lives, than that many other associated habits acquired at an early age should be permanently retained both by man and the lower animals. For instance, full-grown cats, when feeling warm and comfortable, often retain the habit of alternately protruding their forefeet with extended toes, which habit they practised for a definite purpose whilst sucking their mothers.

Another and distinct cause has probably strengthened the habit of frowning, whenever the mind is intent on any subject and encounters some difficulty. Vision is the most important of all the senses, and during primeval times the closest attention must have been incessantly directed towards distant objects for the sake of obtaining prey and avoiding danger. I remember being struck, whilst travelling in parts of South America, which were dangerous from the presence of Indians, how incessantly, yet as it appeared unconsciously, the half-wild Gauchos closely scanned the whole horizon. Now, when any one with no covering on his head (as must have been aboriginally the case with mankind), strives to the utmost to distinguish in broad daylight, and especially if the sky is bright, a distant object, he almost invariably contracts his brows to prevent the entrance of too much light; the lower eyelids, cheeks, and upper lip being at the same time raised, so as to lessen the orifice of the eyes. I have purposely asked several persons, young and old, to look, under the above circumstances, at distant objects, making them believe that I only wished to test the power of their vision; and they all behaved in the manner just described. Some of them, also, put their open, flat hands over their eyes to keep out the excess of light. Gratiolet, after making some remarks to nearly the same effect,[176] says, "Ce sont là des attitudes de vision difficile." He concludes that the muscles round the eyes contract partly for the sake of excluding too much light (which appears to me the more important end), and partly to prevent all rays striking the retina, except those which come direct from the object that is scrutinized. Mr. Bowman, whom I consulted on this point, thinks that the contraction of the surrounding muscles may, in addition, "partly sustain the consensual movements of the two eyes, by giving a firmer support while the globes are brought to binocular vision by their own proper muscles."

As the effort of viewing with care under a bright light a distant object is both difficult and irksome, and as this effort has been habitually accompanied, during numberless generations, by the contraction of the eyebrows, the habit of frowning will thus have been much strengthened; although it was originally practised during infancy from a quite independent cause, namely, as the first step in the protection of the eyes during screaming. There is, indeed, much analogy, as far as the state of the mind is concerned, between intently scrutinizing a distant object, and following out an obscure train of thought, or performing some little and troublesome mechanical work. The belief that the habit of contracting the brows is continued when there is no need whatever to exclude too much light, receives support from the cases formerly alluded to, in which the eyebrows or eyelids are acted on under certain circumstances in a useless manner, from having been similarly used, under analogous circumstances, for a serviceable purpose. For instance, we voluntarily close our eyes when we do not wish to see any object, and we are apt to close them, when we reject a proposition, as if we could not or would not see it; or when we think about something horrible. We raise our eyebrows when we wish to see quickly all round us, and we often do the same, when we earnestly desire to remember something; acting as if we endeavoured to see it.

Abstraction. Meditation.—When a person is lost in thought with his mind absent, or, as it is sometimes said, "when he is in a brown study," he does not frown, but his eyes appear vacant. The lower eyelids are generally raised and wrinkled, in the same manner as when a short-sighted person tries to distinguish a distant object; and the upper orbicular muscles are at the same time slightly contracted. The wrinkling of the lower eyelids under these circumstances has been observed with some savages, as by Mr. Dyson Lacy with the Australians of Queensland, and several times by Mr. Geach with the Malays of the interior of Malacca. What the meaning or cause of this action may be, cannot at present be explained; but here we have another instance of movement round the eyes in relation to the state of the mind.

The vacant expression of the eyes is very peculiar, and at once shows when a man is completely lost in thought. Professor Donders has, with his usual kindness, investigated this subject for me. He has observed others in this condition, and has been himself observed by Professor Engelmann. The eyes are not then fixed on any object, and therefore not, as I had imagined, on some distant object. The lines of vision of the two eyes even often become slightly divergent; the divergence, if the head be held vertically, with the plane of vision horizontal, amounting to an angle of 2° as a maximum. This was ascertained by observing the crossed double image of a distant object. When the head droops forward, as often occurs with a man absorbed in thought, owing to the general relaxation of his muscles, if the plane of vision be still horizontal, the eyes are necessarily a little turned upwards, and then the divergence is as much as 3°, or 3° 5′: if the eyes are turned still more upwards, it amounts to between 6° and 7°. Professor Donders attributes this divergence to the almost complete relaxation of certain muscles of the eyes, which would be apt to follow from the mind being wholly absorbed.[177] The active condition of the muscles of the eyes is that of convergence; and Professor Donders remarks, as bearing on their divergence during a period of complete abstraction, that when one eye becomes blind, it almost always, after a short lapse of time, deviates outwards; for its muscles are no longer used in moving the eyeball inwards for the sake of binocular vision.

Perplexed reflection is often accompanied by certain movements or gestures. At such times we commonly raise our hands to our foreheads, mouths, or chins; but we do not act thus, as far as I have seen, when we are quite lost in meditation, and no difficulty is encountered. Plautus, describing in one of his plays[178] a puzzled man, says, "Now look, he has pillared his chin upon his hand." Even so trifling and apparently unmeaning a gesture as the raising of the hand to the face has been observed with some savages.

Mr. J. Mansel Weale has seen it with the Kafirs of South Africa; and the native chief Gaika adds, that men then "sometimes pull their beards." Mr. Washington Matthews, who attended to some of the wildest tribes of Indians in the western regions of the United States, remarks that he has seen them when concentrating their thoughts, bring their "hands, usually the thumb and index "finger, in contact with some part of the face, commonly the upper lip." We can understand why the forehead should be pressed or rubbed, as deep thought tries the brain; but why the hand should be raised to the mouth or face is far from clear.

Ill-temper.—We have seen that frowning is the natural expression of some difficulty encountered, or of something disagreeable experienced either in thought or action, and he whose mind is often and readily affected in this way, will be apt to be ill-tempered, or slightly angry, or peevish, and will commonly show it by frowning. But a cross expression, due to a frown, may be counteracted, if the mouth appears sweet, from being habitually drawn into a smile, and the eyes are bright and cheerful. So it will be if the eye is clear and steady, and there is the appearance of earnest reflection. Frowning, with some depression of the corners of the mouth, which is a sign of grief, gives an air of peevishness. If a child (see Plate IV, fig. 2, p. 1406)[179] frowns much whilst crying, but does not strongly contract in the usual manner the orbicular muscles, a well-marked expression of anger or even of rage, together with misery, is displayed.

If the whole frowning brow be drawn much downward by the contraction of the pyramidal muscles of the nose, which produces transverse wrinkles or folds across the base of the nose, the expression becomes one of moroseness. Duchenne believes that the contraction of this muscle, without any frowning, gives the appearance of extreme and aggressive hardness.[180] But I much doubt whether this is a true or natural expression. I have shown Duchenne's photograph of a young man, with this muscle strongly contracted by means of galvanism, to eleven persons, including some artists, and none of them could form an idea what was intended, except one, a girl, who answered correctly, "surly reserve." When I first looked at this photograph, knowing what was intended, my imagination added, as I believe, what was necessary, namely, a frowning brow; and consequently the expression appeared to me true and extremely morose.

A firmly closed mouth, in addition to a lowered and frowning brow, gives determination to the expression, or may make it obstinate and sullen. How it comes that the firm closure of the mouth gives the appearance of determination will presently be discussed. An expression of sullen obstinacy has been clearly recognized by my informants, in the natives of six different regions of Australia. It is well marked, according to Mr. Scott, with the Hindoos. It has been recognized with the Malays, Chinese, Kafirs, Abyssinians, and in a conspicuous degree, according to Dr. Rothrock, with the wild Indians of North America, and according to Mr. D. Forbes, with the Aymaras of Bolivia. I have also observed it with the Araucanos of southern Chili. Mr. Dyson Lacy remarks that the natives of Australia, when in this frame of mind, sometimes fold their arms across their breasts, an attitude which may be seen with us. A firm determination, amounting to obstinacy, is, also, sometimes expressed by both shoulders being kept raised, the meaning of which gesture will be explained in the following chapter.

With young children sulkiness is shown by pouting, or, as it is sometimes called, "making a snout."[181] When the corners of the mouth are much depressed, the lower lip is a little everted and protruded; and this is likewise called a pout. But the pouting here referred to, consists of the protrusion of both lips into a tubular form, sometimes to such an extent as to project as far as the end of the nose, if this be short. Pouting is generally accompanied by frowning, and sometimes by the utterance of a booing or whooing noise. This expression is remarkable, as almost the sole one, as far as I know, which is exhibited much more plainly during childhood, at least with Europeans, than during maturity. There is, however, some tendency to the protrusion of the lips with the adults of all races under the influence of great rage. Some children pout when they are shy, and they can then hardly be called sulky.

From inquiries which I have made in several large families, pouting does not seem very common with European children; but it prevails throughout the world, and must be both common and strongly marked with most savage races, as it has caught the attention of many observers. It has been noticed in eight different districts of Australia; and one of my informants remarks how greatly the lips of the children are then protruded. Two observers have seen pouting with the children of Hindoos; three, with those of the Kafirs and Fingoes of South Africa, and with the Hottentots; and two, with the children of the wild Indians of North America. Pouting has also been observed with the Chinese, Abyssinians, Malays of Malacca, Dyaks of Borneo, and often with the New Zealanders. Mr. Mansel Weale informs me that he has seen the lips much protruded, not only with the children of the Kafirs, but with the adults of both sexes when sulky; and Mr. Stack has sometimes observed the same thing with the men, and very frequently with the women of New Zealand. A trace of the same expression may occasionally be detected even with adult Europeans.

We thus see that the protrusion of the lips, especially with young children, is characteristic of sulkiness throughout the greater part of the world. This movement apparently results from the retention, chiefly during youth, of a primordial habit, or from an occasional reversion to it. Young orangs and chimpanzees protrude their lips to an extraordinary degree, as described in a former chapter, when they are discontented, somewhat angry, or sulky; also when they are surprised, a little frightened, and even when slightly pleased. Their mouths are protruded apparently for the sake of making the various noises proper to these several states of mind; and its shape, as I observed with the chimpanzee, differed slightly when the cry of pleasure and that of anger were uttered. As soon as these animals become enraged, the shape of the mouth wholly changes, and the teeth are exposed. The adult orang when wounded is said to emit "a singular cry, consisting at first of high notes, which at length deepen into a low roar. While giving out the high notes he thrusts out his lips into a funnel shape, but in uttering the low notes he holds his mouth wide open."[182] With the gorilla, the lower lip is said to be capable of great elongation. If then our semi-human progenitors protruded their lips when sulky or a little angered, in the same manner as do the existing anthropoid apes, it is not an anomalous, though a curious fact, that our children should exhibit, when similarly affected, a trace of the same expression, together with some tendency to utter a noise. For it is not at all unusual for animals to retain, more or less perfectly, during early youth, and subsequently to lose, characters which were aboriginally possessed by their adult progenitors, and which are still retained by distinct species, their near relations.

Nor is it an anomalous fact that the children of savages should exhibit a stronger tendency to protrude their lips, when sulky, than the children of civilized Europeans; for the essence of savagery seems to consist in the retention of a primordial condition, and this occasionally holds good even with bodily peculiarities.[183] It may be objected to this view of the origin of pouting, that the anthropoid apes likewise protrude their lips when astonished and even when a little pleased; whilst with us this expression is generally confined to a sulky frame of mind. But we shall see in a future chapter that with men of various races surprise does sometimes lead to a slight protrusion of the lips, though great surprise or astonishment is more commonly shown by the mouth being widely opened. As when we smile or laugh we draw back the corners of the mouth, we have lost any tendency to protrude the lips, when pleased, if indeed our early progenitors thus expressed pleasure.

A little gesture made by sulky children may here be noticed, namely, their "showing a cold shoulder." This has a different meaning, as, I believe, from the keeping both shoulders raised. A cross child, sitting on its parent's knee, will lift up the near shoulder, then jerk it away as if from a caress, and afterwards give a backward push with it, as if to push away the offender. I have seen a child, standing at some distance from any one, clearly express its feelings by raising one shoulder, giving it a little backward movement, and then turning away its whole body.

Decision or determination.—The firm closure of the mouth tends to give an expression of determination or decision to the countenance. No determined man probably ever had an habitually gaping mouth. Hence, also, a small and weak lower jaw, which seems to indicate that the mouth is not habitually and firmly closed, is commonly thought to be characteristic of feebleness of character. A prolonged effort of any kind, whether of body or mind, implies previous determination; and if it can be shown that the mouth is generally closed with firmness before and during a great and continued exertion of the muscular system, then, through the principle of association, the mouth would almost certainly be closed as soon as any determined resolution was taken. Now several observers have noticed that a man, in commencing any violent muscular effort, invariably first distends his lungs with air, and then compresses it by the strong contraction of the muscles of the chest; and to effect this the mouth must be firmly closed. Moreover, as soon as the man is compelled to draw breath, he still keeps his chest as much distended as possible.

Various causes have been assigned for this manner of acting. Sir C. Bell maintains[184] that the chest is distended with air, and is kept distended at such times, in order to give a fixed support to the muscles which are thereto attached. Hence, as he remarks, when two men are engaged in a deadly contest, a terrible silence prevails, broken only by hard stifled breathing. There is silence, because to expel the air in the utterance of any sound would be to relax the support for the muscles of the arms. If an outcry is heard, supposing the struggle to take place in the dark, we at once know that one of the two has given up in despair.

Gratiolet admits[185] that when a man has to struggle with another to his utmost, or has to support a great weight, or to keep for a long time the same forced attitude, it is necessary for him first to make a deep inspiration, and then to cease breathing; but he thinks that Sir C. Bell's explanation is erroneous. He maintains that arrested respiration retards the circulation of the blood, of which I believe there is no doubt, and he adduces some curious evidence from the structure of the lower animals, showing, on the one hand, that a retarded circulation is necessary for prolonged muscular exertion, and, on the other hand, that a rapid circulation is necessary for rapid movements. According to this view, when we commence any great exertion, we close our mouths and stop breathing, in order to retard the circulation of the blood. Gratiolet sums up the subject by saying, "C'est là la vraie théorie de l'effort continu;" but how far this theory is admitted by other physiologists I do not know.

Dr. Piderit accounts[186] for the firm closure of the mouth during strong muscular exertion, on the principle that the influence of the will spreads to other muscles besides those necessarily brought into action in making any particular exertion; and it is natural that the muscles of respiration and of the mouth, from being so habitually used, should be especially liable to be thus acted on. It appears to me that there probably is some truth in this view, for we are apt to press the teeth hard together during violent exertion, and this is not requisite to prevent expiration, whilst the muscles of the chest are strongly contracted.

Lastly, when a man has to perform some delicate and difficult operation, not requiring the exertion of any strength, he nevertheless generally closes his mouth and ceases for a time to breathe; but he acts thus in order that the movements of his chest may not disturb those of his arms. A person, for instance, whilst threading a needle, may be seen to compress his lips and either to stop breathing, or to breathe as quietly as possible. So it was, as formerly stated, with a young and sick chimpanzee, whilst it amused itself by killing flies with its knuckles, as they buzzed about on the window-panes. To perform an action, however trifling, if difficult, implies some amount of previous determination.

There appears nothing improbable in all the above assigned causes having come into play in different degrees, either conjointly or separately, on various occasions. The result would be a well-established habit, now perhaps inherited, of firmly closing the mouth at the commencement of and during any violent and prolonged exertion, or any delicate operation. Through the principle of association there would also be a strong tendency towards this same habit, as soon as the mind had resolved on any particular action or line of conduct, even before there was any bodily exertion, or if none were requisite. The habitual and firm closure of the mouth would thus come to show decision of character; and decision readily passes into obstinacy.

CHAPTER X

Hatred and Anger

HATRED • RAGE, EFFECTS OF ON THE SYSTEM • UNCOVERING OF THE TEETH • RAGE IN THE INSANE • ANGER AND INDIGNATION • AS EXPRESSED BY THE VARIOUS RACES OF MAN • SNEERING AND DEFIANCE • THE UNCOVERING OF THE CANINE TOOTH ON ONE SIDE OF THE FACE

If we have suffered or expect to suffer some wilful injury from a man, or if he is in any way offensive to us, we dislike him; and dislike easily rises into hatred. Such feelings, if experienced in a moderate degree, are not clearly expressed by any movement of the body or features, excepting perhaps by a certain gravity of behaviour, or by some ill-temper. Few individuals, however, can long reflect about a hated person, without feeling and exhibiting signs of indignation or rage. But if the offending person be quite insignificant, we experience merely disdain or contempt. If, on the other hand, he is all-powerful, then hatred passes into terror, as when a slave thinks about a cruel master, or a savage about a bloodthirsty malignant deity.[187] Most of our emotions are so closely connected with their expression, that they hardly exist if the body remains passive—the nature of the expression depending in chief part on the nature of the actions which have been habitually performed under this particular state of the mind. A man, for instance, may know that his life is in the extremest peril, and may strongly desire to save it; yet, as Louis XVI. said, when surrounded by a fierce mob, "Am I afraid? feel my pulse." So a man may intensely hate another, but until his bodily frame is affected, he cannot be said to be enraged.

Rage.—I have already had occasion to treat of this emotion in the third chapter, when discussing the direct influence of the excited sensorium on the body, in combination with the effects of habitually associated actions. Rage exhibits itself in the most diversified manner. The heart and circulation are always affected; the face reddens or becomes purple, with the veins on the forehead and neck distended. The reddening of the skin has been observed with the copper-coloured Indians of South America,[188] and even, as it is said, on the white cicatrices left by old wounds on negroes.[189] Monkeys also redden from passion. With one of my own infants, under four months old, I repeatedly observed that the first symptom of an approaching passion was the rushing of the blood into his bare scalp. On the other hand, the action of the heart is sometimes so much impeded by great rage, that the countenance becomes pallid or livid,[190] and not a few men with heart-disease have dropped down dead under this powerful emotion.

The respiration is likewise affected; the chest heaves, and the dilated nostrils quiver.[191] As Tennyson writes, "sharp breaths of anger puffed her fairy nostrils out." Hence we have such expressions as "breathing out vengeance," and "fuming with anger."[192]

The excited brain gives strength to the muscles, and at the same time energy to the will. The body is commonly held erect ready for instant action, but sometimes it is bent forward towards the offending person, with the limbs more or less rigid. The mouth is generally closed with firmness, showing fixed determination, and the teeth are clenched or ground together. Such gestures as the raising of the arms, with the fists clenched, as if to strike the offender, are common. Few men in a great passion, and telling some one to begone, can resist acting as if they intended to strike or push the man violently away. The desire, indeed, to strike often becomes so intolerably strong, that inanimate objects are struck or dashed to the ground; but the gestures frequently become altogether purposeless or frantic. Young children, when in a violent rage roll on the ground on their backs or bellies, screaming, kicking, scratching, or biting everything within reach. So it is, as I hear from Mr. Scott, with Hindoo children; and, as we have seen, with the young of the anthropomorphous apes.

But the muscular system is often affected in a wholly different way; for trembling is a frequent consequence of extreme rage. The paralysed lips then refuse to obey the will, "and the voice sticks in the throat;"[193] or it is rendered loud, harsh, and discordant. If there be much and rapid speaking, the mouth froths. The hair sometimes bristles; but I shall return to this subject in another chapter, when I treat of the mingled emotions of rage and terror. There is in most cases a strongly-marked frown on the forehead; for this follows from the sense of anything displeasing or difficult, together with concentration of mind. But sometimes the brow, instead of being much contracted and lowered, remains smooth, with the glaring eyes kept widely open. The eyes are always bright, or may, as Homer expresses it, glisten with fire. They are sometimes bloodshot, and are said to protrude from their sockets—the result, no doubt, of the head being gorged with blood, as shown by the veins being distended. According to Gratiolet,[194] the pupils are always contracted in rage, and I hear from Dr. Crichton Browne that this is the case in the fierce delirium of meningitis; but the movements of the iris under the influence of the different emotions is a very obscure subject.

Shakespeare sums up the chief characteristics of rage as follows:—

"In peace there's nothing so becomes a man,
As modest stillness and humility;
But when the blast of war blows in our ears,
Then imitate the action of the tiger:
Stiffen the sinews, summon up the blood,
Then lend the eye a terrible aspect;
Now set the teeth, and stretch the nostril wide,
Hold hard the breath, and bend up every spirit
To his full height! On, on, you noblest English."

Henry V., act iii. sc. 1.

The lips are sometimes protruded during rage in a manner, the meaning of which I do not understand, unless it depends on our descent from some ape-like animal. Instances have been observed, not only with Europeans, but with the Australians and Hindoos. The lips, however, are much more commonly retracted, the grinning or clenched teeth being thus exposed. This has been noticed by almost every one who has written on expression.[195] The appearance is as if the teeth were uncovered, ready for seizing or tearing an enemy, though there may be no intention of acting in this manner. Mr. Dyson Lacy has seen this grinning expression with the Australians, when quarrelling, and so has Gaika with the Kafirs of South Africa. Dickens,[196] in speaking of an atrocious murderer who had just been caught, and was surrounded by a furious mob, describes "the people as jumping up one behind another, snarling with their teeth, and making at him like wild beasts." Every one who has had much to do with young children must have seen how naturally they take to biting, when in a passion. It seems as instinctive in them as in young crocodiles, who snap their little jaws as soon as they emerge from the egg.

A grinning expression and the protrusion of the lips appear sometimes to go together. A close observer says that he has seen many instances of intense hatred (which can hardly be distinguished from rage, more or less suppressed) in Orientals, and once in an elderly English woman. In all these cases there "was a grin, not a scowl—the lips lengthening, the cheeks settling downwards, the eyes half-closed, whilst the brow remained perfectly calm."[197]

This retraction of the lips and uncovering of the teeth during paroxysms of rage, as if to bite the offender, is so remarkable, considering how seldom the teeth are used by men in fighting, that I inquired from Dr. J. Crichton Browne whether the habit was common in the insane whose passions are unbridled. He informs me that he has repeatedly observed it both with the insane and idiotic, and has given me the following illustrations:—

Shortly before receiving my letter, he witnessed an uncontrollable outbreak of anger and delusive jealousy in an insane lady. At first she vituperated her husband, and whilst doing so foamed at the mouth. Next she approached close to him with compressed lips, and a virulent set frown. Then she drew back her lips, especially the corners of the upper lip, and showed her teeth, at the same time aiming a vicious blow at him. A second case is that of an old soldier, who, when he is requested to conform to the rules of the establishment, gives way to discontent, terminating in fury. He commonly begins by asking Dr. Browne whether he is not ashamed to treat him in such a manner. He then swears and blasphemes, paces up and down, tosses his arms wildly about, and menaces any one near him. At last, as his exasperation culminates, he rushes up towards Dr. Browne with a peculiar sidelong movement, shaking his doubled fist, and threatening destruction. Then his upper lip may be seen to be raised, especially at the corners, so that his huge canine teeth are exhibited. He hisses forth his curses through his set teeth, and his whole expression assumes the character of extreme ferocity. A similar description is applicable to another man, excepting that he generally foams at the mouth and spits, dancing and jumping about in a strange rapid manner, shrieking out his maledictions in a shrill falsetto voice.

Dr. Browne also informs me of the case of an epileptic idiot, incapable of independent movements, and who spends the whole day in playing with some toys; but his temper is morose and easily roused into fierceness. When any one touches his toys, he slowly raises his head from its habitual downward position, and fixes his eyes on the offender, with a tardy yet angry scowl. If the annoyance be repeated, he draws back his thick lips and reveals a prominent row of hideous fangs (large canines being especially noticeable), and then makes a quick and cruel clutch with his open hand at the offending person. The rapidity of this clutch, as Dr. Browne remarks, is marvellous in a being ordinarily so torpid that he takes about fifteen seconds, when attracted by any noise, to turn his head from one side to the other. If, when thus incensed, a handkerchief, book, or other article, be placed into his hands, he drags it to his mouth and bites it. Mr. Nicol has likewise described to me two cases of insane patients, whose lips are retracted during paroxysms of rage.

Dr. Maudsley, after detailing various strange animal-like traits in idiots, asks whether these are not due to the reappearance of primitive instincts— "a faint echo from a far-distant past, testifying to a kinship which man has almost outgrown." He adds, that as every human brain passes, in the course of its development, through the same stages as those occurring in the lower vertebrate animals, and as the brain of an idiot is in an arrested condition, we may presume that it "will manifest its most primitive functions, and no higher functions." Dr. Maudsley thinks that the same view may be extended to the brain in its degenerated condition in some insane patients; and asks, whence come "the savage snarl, the destructive disposition, the obscene language, the wild howl, the offensive habits, displayed by some of the insane? Why should a human being, deprived of his reason, ever become so brutal in character, as some do, unless he has the brute nature within him?"[198] This question must, as it would appear, be answered in the affirmative.

Anger, indignation.—These states of the mind differ from rage only in degree, and there is no marked distinction in their characteristic signs. Under moderate anger the action of the heart is a little increased, the colour heightened, and the eyes become bright. The respiration is likewise a little hurried; and as all the muscles serving for this function act in association, the wings of the nostrils are somewhat raised to allow of a free indraught of air; and this is a highly characteristic sign of indignation. The mouth is commonly compressed, and there is almost always a frown on the brow. Instead of the frantic gestures of extreme rage, an indignant man unconsciously throws himself into an attitude ready for attacking or striking his enemy, whom he will perhaps scan from head to foot in defiance. He carries his head erect, with his chest well expanded, and the feet planted firmly on the ground. He holds his arms in various positions, with one or both elbows squared, or with the arms rigidly suspended by his sides. With Europeans the fists are commonly clenched.[199] The figures 1 and 2 in Plate VI are fairly good representations of men simulating indignation. Any one may see in a mirror, if he will vividly imagine that he has been insulted and demands an explanation in an angry tone of voice, that he suddenly and unconsciously throws himself into some such attitude.

Rage, anger, and indignation are exhibited in nearly the same manner throughout the world; and the following descriptions may be worth giving as evidence of this, and as illustrations of some of the foregoing remarks. There is, however, an exception with respect to clenching the fists, which seems confined chiefly to the men who fight with their fists. With the Australians only one of my informants has seen the fists clenched. All agree about the body being held erect; and all, with two exceptions, state that the brows are heavily contracted. Some of them allude to the firmly-compressed mouth, the distended nostrils, and flashing eyes. According to the Rev. Mr. Taplin, rage, with the Australians, is expressed by the lips being protruded, the eyes being widely open; and in the case of the women by their dancing about and casting dust into the air. Another observer speaks of the native men, when enraged, throwing their arms wildly about.

I have received similar accounts, except as to the clenching of the fists, in regard to the Malays of the Malacca peninsula, the Abyssinians, and the natives of South Africa. So it is with the Dakota Indians of North America; and, according to Mr. Matthews, they then hold their heads erect, frown, and often stalk away with long strides. Mr. Bridges states that the Fuegians, when enraged, frequently stamp on the ground, walk distractedly about, sometimes cry and grow pale. The Rev. Mr. Stack watched a New Zealand man and woman quarrelling, and made the following entry in his note-book: "Eyes dilated, body swayed violently backwards and forwards, head inclined forwards, fists clenched, now thrown behind the body, now directed towards each other's faces." Mr. Swinhoe says that my description agrees with what he has seen of the Chinese, excepting that an angry man generally inclines his body towards his antagonist, and pointing at him, pours forth a volley of abuse.

Lastly, with respect to the natives of India, Mr. J. Scott has sent me a full description of their gestures and expression when enraged. Two low-caste Bengalees disputed about a loan. At first they were calm, but soon grew furious and poured forth the grossest abuse on each other's relations and progenitors for many generations past. Their gestures were very different from those of Europeans; for though their chests were expanded and shoulders squared, their arms remained rigidly suspended, with the elbows turned inwards and the hands alternately clenched and opened. Their shoulders were often raised high, and then again lowered. They looked fiercely at each other from under their lowered and strongly wrinkled brows, and their protruded lips were firmly closed. They approached each other, with heads and necks stretched forwards, and pushed, scratched, and grasped at each other. This protrusion of the head and body seems a common gesture with the enraged; and I have noticed it with degraded English women whilst quarrelling violently in the streets. In such cases it may be presumed that neither party expects to receive a blow from the other.

A Bengalee employed in the Botanic Gardens was accused, in the presence of Mr. Scott, by the native overseer of having stolen a valuable plant. He listened silently and scornfully to the accusation; his attitude erect, chest expanded, mouth closed, lips protruding, eyes firmly set and penetrating. He then defiantly maintained his innocence, with upraised and clenched hands, his head being now pushed forwards, with the eyes widely open and eyebrows raised. Mr. Scott also watched two Mechis, in Sikhim, quarrelling about their share of payment. They soon got into a furious passion, and then their bodies became less erect, with their heads pushed forwards; they made grimaces at each other; their shoulders were raised; their arms rigidly bent inwards at the elbows, and their hands spasmodically closed, but not properly clenched. They continually approached and retreated from each other, and often raised their arms as if to strike, but their hands were open, and no blow was given. Mr. Scott made similar observations on the Lepchas whom he often saw quarrelling, and he noticed that they kept their arms rigid and almost parallel to their bodies, with the hands pushed somewhat backwards and partially closed, but not clenched.

Sneering, Defiance: Uncovering the canine tooth on one side.—The expression which I wish here to consider differs but little from that already described, when the lips are retracted and the grinning teeth exposed. The difference consists solely in the upper lip being retracted in such a manner that the canine tooth on one side of the face alone is shown; the face itself being generally a little upturned and half averted from the person causing offence. The other signs of rage are not necessarily present. This expression may occasionally be observed in a person who sneers at or defies another, though there may be no real anger; as when any one is playfully accused of some fault, and answers, "I scorn the imputation." The expression is not a common one, but I have seen it exhibited with perfect distinctness by a lady who was being quizzed by another person. It was described by Parsons as long ago as 1746, with an engraving, showing the uncovered canine on one side.[200] Mr. Rejlander, without my having made any allusion to the subject, asked me whether I had ever noticed this expression, as he had been much struck by it. He has photographed for me (Plate IV, fig. 1) a lady, who sometimes unintentionally displays the canine on one side, and who can do so voluntarily with unusual distinctness.

1

2

Plate IV

The expression of a half-playful sneer graduates into one of great ferocity when, together with a heavily frowning brow and fierce eye, the canine tooth is exposed. A Bengalee boy was accused before Mr. Scott of some misdeed. The delinquent did not dare to give vent to his wrath in words, but it was plainly shown on his countenance, sometimes by a defiant frown, and sometimes "by a thoroughly canine snarl." When this was exhibited, "the corner of the lip over the eye-tooth, which happened in this case to be large and projecting, was raised on the side of his accuser, a strong frown being still retained on the brow." Sir C. Bell states[201] that the actor Cooke could express the most determined hate "when with the oblique cast of his eyes he drew up the outer part of the upper lip, and discovered a sharp angular tooth."

The uncovering of the canine tooth is the result of a double movement. The angle or corner of the mouth is drawn a little backwards, and at the same time a muscle which runs parallel to and near the nose draws up the outer part of the upper lip, and exposes the canine on this side of the face. The contraction of this muscle makes a distinct furrow on the cheek, and produces strong wrinkles under the eye, especially at its inner corner. The action is the same as that of a snarling dog; and a dog when pretending to fight often draws up the lip on one side alone, namely that facing his antagonist. Our word *sneer* is in fact the same as *snarl,* which was originally *snar,* the *l* "being merely an element implying continuance of action."[202]

I suspect that we see a trace of this same expression in what is called a derisive or sardonic smile. The lips are then kept joined or almost joined, but one corner of the mouth is retracted on the side towards the derided person; and this drawing back of the corner is part of a true sneer. Although some persons smile more on one side of their face than on the other, it is not easy to understand why in cases of derision the smile, if a real one, should so commonly be confined to one side. I have also on these occasions noticed a slight twitching of the muscle which draws up the outer part of the upper lip; and this movement, if fully carried out, would have uncovered the canine, and would have produced a true sneer.

Mr. Bulmer, an Australian missionary in a remote part of Gipps' Land, says, in answer to my query about the uncovering of the canine on one side, "I find that the natives in snarling at each other speak with the teeth closed, the upper lip drawn to one side, and a general angry expression of face; but they look direct at the person addressed." Three other observers in Australia, one in Abyssinia, and one in China, answer my query on this head in the affirmative; but as the expression is rare, and as they enter into no details, I am afraid of implicitly trusting them. It is, however, by no means improbable that this animal-like expression may be more common with savages than with civilized races. Mr. Geach is an observer who may be fully trusted, and he has observed it on one occasion in a Malay in the interior of Malacca. The Rev. S. O. Glenie answers, "We have observed this expression with the natives of Ceylon, but not often." Lastly, in North America, Dr. Rothrock has seen it with some wild Indians, and often in a tribe adjoining the Atnahs.

Although the upper lip is certainly sometimes raised on one side alone in sneering at or defying any one, I do not know that this is always the case, for the face is commonly half averted, and the expression is often momentary. The movement being confined to one side may not be an essential part of the expression, but may depend on the proper muscles being incapable of movement excepting on one side. I asked four persons to endeavour to act voluntarily in this manner; two could expose the canine only on the left side, one only on the right side, and the fourth on neither side. Nevertheless it is by no means certain that these same persons, if defying any one in earnest, would not unconsciously have uncovered their canine tooth on the side, whichever it might be, towards the offender. For we have seen that some persons cannot voluntarily make their eyebrows oblique, yet instantly act in this manner when affected by any real, although most trifling, cause of distress. The power of voluntarily uncovering the canine on one side of the face being thus often wholly lost, indicates that it is a rarely used and almost abortive action. It is indeed a surprising fact that man should possess the power, or should exhibit any tendency to its use; for Mr. Sutton has never noticed a snarling action in our nearest allies, namely, the monkeys in the Zoological Gardens, and he is positive that the baboons, though furnished with great canines, never act thus, but uncover all their teeth when feeling savage and ready for an attack. Whether the adult anthropomorphous apes, in the males of whom the canines are much larger than in the females, uncover them when prepared to fight, is not known.

The expression here considered, whether that of a playful sneer or ferocious snarl, is one of the most curious which occurs in man. It reveals his animal descent; for no one, even if rolling on the ground in a deadly grapple with an enemy, and attempting to bite him, would try to use his canine teeth more than his other teeth. We may readily believe from our affinity to the anthropomorphous apes that our male semi-human progenitors possessed great canine teeth, and men are now occasionally born having them of unusually large size, with interspaces in the opposite jaw for their reception.[203] We may further suspect, notwithstanding that we have no support from analogy, that our semi-human progenitors uncovered their canine teeth when prepared for battle, as we still do when feeling ferocious, or when merely sneering at or defying some one, without any intention of making a real attack with our teeth.

CHAPTER XI

Disdain—Contempt—Disgust—Guilt—Pride, &c.—Helplessness—Patience—Affirmation and Negation

CONTEMPT, SCORN AND DISDAIN, VARIOUSLY EXPRESSED • DERISIVE SMILE • GESTURES EXPRESSIVE OF CONTEMPT • DISGUST • GUILT, DECEIT, PRIDE, &C. • HELPLESSNESS OR IMPOTENCE • PATIENCE • OBSTINACY • SHRUGGING THE SHOULDERS COMMON TO MOST OF THE RACES OF MAN • SIGNS OF AFFIRMATION AND NEGATION

Scorn and disdain can hardly be distinguished from contempt, excepting that they imply a rather more angry frame of mind. Nor can they be clearly distinguished from the feelings discussed in the last chapter under the terms of sneering and defiance. Disgust is a sensation rather more distinct in its nature, and refers to something revolting, primarily in relation to the sense of taste, as actually perceived or vividly imagined; and secondarily to anything which causes a similar feeling, through the sense of smell, touch, and even of eyesight. Nevertheless, extreme contempt, or as it is often called loathing contempt, hardly differs from disgust. These several conditions of the mind are, therefore, nearly related; and each of them may be exhibited in many different ways. Some writers have insisted chiefly on one mode of expression, and others on a different mode. From this circumstance M. Lemoine has argued[204] that their descriptions are not trustworthy. But we shall immediately see that it is natural that the feelings which we have here to consider should be expressed in many different ways, inasmuch as various habitual actions serve equally well, through the principle of association, for their expression.

Scorn and disdain, as well as sneering and defiance, may be displayed by a slight uncovering of the canine tooth on one side of the face; and this movement appears to graduate into one closely like a smile. Or the smile or laugh may be real, although one of derision; and this implies that the offender is so insignificant that he excites only amusement; but the amusement is generally a pretence. Gaika in his answers to my queries remarks, that contempt is commonly shown by his countrymen, the Kafirs, by smiling; and the Rajah Brooke makes the same observation with respect to the Dyaks of Borneo. As laughter is primarily the expression of simple joy, very young children do not, I believe, ever laugh in derision.

The partial closure of the eyelids, as Duchenne[205] insists, or the turning away of the eyes or of the whole body, are likewise highly expressive of disdain. These actions seem to declare that the despised person is not worth looking at, or is disagreeable to behold. The accompanying photograph (Plate V, fig. 1) by Mr. Rejlander, shows this form of disdain. It represents a young lady, who is supposed to be tearing up the photograph of a despised lover.

The most common method of expressing contempt is by movements about the nose, or round the mouth; but the latter movements, when strongly pronounced, indicate disgust. The nose may be slightly turned up, which apparently follows from the turning up of the upper lip; or the movement may be abbreviated into the mere wrinkling of the nose. The nose is often slightly contracted, so as partly to close the passage;[206] and this is commonly accompanied by a slight snort or expiration. All these actions are the same with those which we employ when we perceive an offensive odour, and wish to exclude or expel it. In extreme cases, as Dr. Piderit remarks,[207] we protrude and raise both lips, or the upper lip alone, so as to close the nostrils as by a valve, the nose being thus turned up. We seem thus to say to the despised person that he smells offensively,[208] in nearly the same manner as we express to him by half-closing our eyelids, or turning away our faces, that he is not worth looking at. It must not, however, be supposed that such ideas actually pass through the mind when we exhibit our contempt; but as whenever we have perceived a disagreeable odour or seen a disagreeable sight, actions of this kind have been performed, they have become habitual or fixed, and are now employed under any analogous state of mind.

Various odd little gestures likewise indicate contempt; for instance, *snapping one's fingers*. This, as Mr. Tylor remarks,[209] "is not very intelligible as we generally see it; but when we notice that the same sign made quite gently, as if rolling some tiny object away between the finger and thumb, or the sign of flipping it away with the thumb-nail and forefinger, are usual and well-understood deaf-and-dumb gestures, denoting anything tiny, insignifi-

1

2 3

Plate V

cant, contemptible, it seems as though we had exaggerated and conventionalized a perfectly natural action, so as to lose sight of its original meaning. There is a curious mention of this gesture by Strabo." Mr. Washington Matthews informs me that, with the Dakota Indians of North America, contempt is shown not only by movements of the face, such as those above described, but "conventionally, by the hand being closed and held near the breast, then, as the forearm is suddenly extended, the hand is opened and the fingers separated from each other. If the person at whose expense the sign is made is present, the hand is moved towards him, and the head sometimes averted from him." This sudden extension and opening of the hand perhaps indicates the dropping or throwing away a valueless object.

The term 'disgust,' in its simplest sense, means something offensive to the taste. It is curious how readily this feeling is excited by anything unusual in the appearance, odour, or nature of our food. In Tierra del Fuego a native touched with his finger some cold preserved meat which I was eating at our bivouac, and plainly showed utter disgust at its softness; whilst I felt utter disgust at my food being touched by a naked savage, though his hands did not appear dirty. A smear of soup on a man's beard looks disgusting, though there is of course nothing disgusting in the soup itself. I presume that this follows from the strong association in our minds between the sight of food, however circumstanced, and the idea of eating it.

As the sensation of disgust primarily arises in connection with the act of eating or tasting, it is natural that its expression should consist chiefly in movements round the mouth. But as disgust also causes annoyance, it is generally accompanied by a frown, and often by gestures as if to push away or to guard oneself against the offensive object. In the two photographs (figs. 2 and 3, on Plate V) Mr. Rejlander has simulated this expression with some success. With respect to the face, moderate disgust is exhibited in various ways; by the mouth being widely opened, as if to let an offensive morsel drop out; by spitting; by blowing out of the protruded lips; or by a sound as of clearing the throat. Such guttural sounds are written *ach* or *ugh;* and their utterance is sometimes accompanied by a shudder, the arms being pressed close to the sides and the shoulders raised in the same manner as when horror is experienced.[210] Extreme disgust is expressed by movements round the mouth identical with those preparatory to the act of vomiting. The mouth is opened widely, with the upper lip strongly retracted, which wrinkles the sides of the nose, and with the lower lip protruded and everted as much as possible. This latter movement requires the contraction of the muscles which draw downwards the corners of the mouth.[211]

It is remarkable how readily and instantly retching or actual vomiting is induced in some persons by the mere idea of having partaken of any unusual food, as of an animal which is not commonly eaten; although there is nothing in such food to cause the stomach to reject it. When vomiting results, as a reflex action, from some real cause—as from too rich food, or tainted meat, or from an emetic—it does not ensue immediately, but generally after a considerable interval of time. Therefore, to account for retching or vomiting being so quickly and easily excited by a mere idea, the suspicion arises that our progenitors must formerly have had the power (like that possessed by ruminants and some other animals) of voluntarily rejecting food which disagreed with them, or which they thought would disagree with them; and now, though this power has been lost, as far as the will is concerned, it is called into involuntary action, through the force of a formerly well-established habit, whenever the mind revolts at the idea of having partaken of any kind of food, or at anything disgusting. This suspicion receives support from the fact, of which I am assured by Mr. Sutton, that the monkeys in the Zoological Gardens often vomit whilst in perfect health, which looks as if the act were voluntary. We can see that as man is able to communicate by language to his children and others, the knowledge of the kinds of food to be avoided, he would have little occasion to use the faculty of voluntary rejection; so that this power would tend to be lost through disuse.

As the sense of smell is so intimately connected with that of taste, it is not surprising that an excessively bad odour should excite retching or vomiting in some persons, quite as readily as the thought of revolting food does; and that, as a further consequence, a moderately offensive odour should cause the various expressive movements of disgust. The tendency to retch from a fetid odour is immediately strengthened in a curious manner by some degree of habit, though soon lost by longer familiarity with the cause of offence and by voluntary restraint. For instance, I wished to clean the skeleton of a bird, which had not been sufficiently macerated, and the smell made my servant and myself (we not having had much experience in such work) retch so violently, that we were compelled to desist. During the previous days I had examined some other skeletons, which smelt slightly; yet the odour did not in the least affect me, but, subsequently for several days, whenever I handled these same skeletons, they made me retch.

From the answers received from my correspondents it appears that the various movements, which have now been described as expressing contempt and disgust, prevail throughout a large part of the world. Dr. Rothrock, for instance, answers with a decided affirmative with respect to certain wild Indian tribes of North America. Crantz says that when a Greenlander denies anything with contempt or horror he turns up his nose, and gives a slight sound through it.[212] Mr. Scott has sent me a graphic description of the face of a young Hindoo at the sight of castor-oil, which he was compelled occasionally to take. Mr. Scott has also seen the same expression on the faces of high-caste natives who have approached close to some defiling object. Mr. Bridges says that the Fuegians "express contempt by shooting out the lips and hissing through them, and by turning up the nose." The tendency either to snort through the nose, or to make a noise expressed by *ugh* or *ach*, is noticed by several of my correspondents.

Spitting seems an almost universal sign of contempt or disgust; and spitting obviously represents the rejection of anything offensive from the mouth. Shakspeare makes the Duke of Norfolk say, "I spit at him—call him a slanderous coward and a villain." So, again, Falstaff says, "Tell thee what, Hal,—if I tell thee a lie, spit in my face." Leichhardt remarks that the Australians "interrupted their speeches by spitting, and uttering a noise like pooh! pooh! apparently expressive of their disgust." And Captain Burton speaks of certain negroes "spitting with disgust upon the ground."[213] Captain Speedy informs me that this is likewise the case with the Abyssinians. Mr. Geach says that with the Malays of Malacca the expression of disgust "answers to spitting from the mouth;" and with the Fuegians, according to Mr. Bridges "to spit at one is the highest mark of contempt."

I never saw disgust more plainly expressed than on the face of one of my infants at the age of five months, when, for the first time, some cold water, and again a month afterwards, when a piece of ripe cherry was put into his mouth. This was shown by the lips and whole mouth assuming a shape which allowed the contents to run or fall quickly out; the tongue being likewise protruded. These movements were accompanied by a little shudder. It was all the more comical, as I doubt whether the child felt real disgust—the eyes and forehead expressing much surprise and consideration. The protrusion of the tongue in letting a nasty object fall out of the mouth, may explain how it is that lolling out the tongue universally serves as a sign of contempt and hatred.[214]

We have now seen that scorn, disdain, contempt, and disgust are expressed in many different ways, by movements of the features, and by various gestures; and that these are the same throughout the world. They all consist of actions representing the rejection or exclusion of some real object which we dislike or abhor, but which does not excite in us certain other strong emotions, such as rage or terror; and through the force of habit and association similar actions are performed, whenever any analogous sensation arises in our minds.

Jealousy, Envy, Avarice, Revenge, Suspicion, Deceit, Slyness, Guilt, Vanity, Conceit, Ambition, Pride, Humility, &c.—It is doubtful whether the greater number of the above complex states of mind are revealed by any fixed expression, sufficiently distinct to be described or delineated. When Shakespeare speaks of Envy as *lean-faced*, or *black*, or *pale*, and Jealousy as "*the green-eyed monster*;" and when Spenser describes Suspicion as "*foul, ill-favoured, and grim*," they must have felt this difficulty. Nevertheless, the above feelings—at least many of them—can be detected by the eye; for instance, conceit; but we are often guided in a much greater degree than we suppose by our previous knowledge of the persons or circumstances.

My correspondents almost unanimously answer in the affirmative to my query, whether the expression of guilt and deceit can be recognised amongst the various races of man; and I have confidence in their answers, as they generally deny that jealousy can thus be recognised. In the cases in which details are given, the eyes are almost always referred to. The guilty man is said to avoid looking at his accuser, or to give him stolen looks. The eyes are said "to be turned askant," or "to waver from side to side," or "the eyelids to be lowered and partly closed." This latter remark is made by Mr. Hagenauer with respect to the Australians, and by Gaika with respect to the Kafirs. The restless movements of the eyes apparently follow, as will be explained when we treat of blushing, from the guilty man not enduring to meet the gaze of his accuser. I may add, that I have observed a guilty expression, without a shade of fear, in some of my own children at a very early age. In one instance the expression was unmistakably clear in a child two years and seven months old, and led to the detection of his little crime. It was shown, as I record in my notes made at the time, by an unnatural brightness in the eyes, and by an odd, affected manner, impossible to describe.

Slyness is also, I believe, exhibited chiefly by movements about the eyes; for these are less under the control of the will, owing to the force of long-continued habit, than are the movements of the body. Mr. Herbert Spencer remarks,[215] "When there is a desire to see something on one side of the visual field without being supposed to see it, the tendency is to check the conspicuous movement of the head, and to make the required adjustment entirely with the eyes; which are, therefore, drawn very much to one side. Hence, when the eyes are turned to one side, while the face is not turned to the same side, we get the natural language of what is called slyness."

Of all the above-named complex emotions, Pride, perhaps, is the most plainly expressed. A proud man exhibits his sense of superiority over others by holding his head and body erect. He is haughty *(haut),* or high, and makes himself appear as large as possible; so that metaphorically he is said to be swollen or puffed up with pride. A peacock or a turkey-cock strutting about with puffed-up feathers, is sometimes said to be an emblem of pride.[216] The arrogant man looks down on others, and with lowered eyelids hardly condescends to see them; or he may show his contempt by slight movements, such as those before described, about the nostrils or lips. Hence the muscle which everts the lower lip has been called the *musculus superbus.* In some photographs of patients affected by a monomania of pride, sent me by Dr. Crichton Browne, the head and body were held erect, and the mouth firmly closed. This latter action, expressive of decision, follows, I presume, from the proud man feeling perfect self-confidence in himself. The whole expression of pride stands in direct antithesis to that of humility; so that nothing need here be said of the latter state of mind.

Helplessness, Impotence: Shrugging the shoulders.—When a man wishes to show that he cannot do something, or prevent something being done, he often raises with a quick movement both shoulders. At the same time, if the whole gesture is completed, he bends his elbows closely inwards, raises his open hands, turning them outwards, with the fingers separated. The head is often thrown a little on one side; the eyebrows are elevated, and this causes wrinkles across the forehead. The mouth is generally opened. I may mention, in order to show how unconsciously the features are thus acted on, that though I had often intentionally shrugged my shoulders to observe how my arms were placed, I was not at all aware that my eyebrows were raised and mouth opened, until I looked at myself in a glass; and since then I have noticed the same movements in the faces of others. In the accompanying Plate VI, figs. 3 and 4, Mr. Rejlander has successfully acted the gesture of shrugging the shoulders.

Englishmen are much less demonstrative than the men of most other European nations, and they shrug their shoulders far less frequently and energetically than Frenchmen or Italians do. The gesture varies in all degrees from the complex movement, just described, to only a momentary and scarcely perceptible raising of both shoulders; or, as I have noticed in a lady sitting in an arm-chair, to the mere turning slightly outwards of the open hands with separated fingers. I have never seen very young English children shrug their shoulders, but the following case was observed with care by a medical professor and excellent observer, and has been communicated to me by him. The father of this gentleman was a Parisian, and his mother a Scotch lady. His wife is of British extraction on both sides, and my

1

2

Heliotype

Plate VI

informant does not believe that she ever shrugged her shoulders in her life. His children have been reared in England, and the nursemaid is a thorough Englishwoman, who has never been seen to shrug her shoulders. Now, his eldest daughter was observed to shrug her shoulders at the age of between sixteen and eighteen months; her mother exclaiming at the time, "Look at the little French girl shrugging her shoulders!" At first she often acted thus, sometimes throwing her head a little backwards and on one side, but she did not, as far as was observed, move her elbows and hands in the usual manner. The habit gradually wore away, and now, when she is a little over four years old, she is never seen to act thus. The father is told that he sometimes shrugs his shoulders, especially when arguing with any one; but it is extremely improbable that his daughter should have imitated him at so early an age; for, as he remarks, she could not possibly have often seen this gesture in him. Moreover, if the habit had been acquired through imitation, it is not probable that it would so soon have been spontaneously discontinued by this child, and, as we shall immediately see, by a second child, though the father still lived with his family. This little girl, it may be added, resembles her Parisian grandfather in countenance to an almost absurd degree. She also presents another and very curious resemblance to him, namely, by practising a singular trick. When she impatiently wants something, she holds out her little hand, and rapidly rubs the thumb against the

3

4

Plate VI

index and middle finger: now this same trick was frequently performed under the same circumstances by her grandfather.

This gentleman's second daughter also shrugged her shoulders before the age of eighteen months, and afterwards discontinued the habit. It is of course possible that she may have imitated her elder sister; but she continued it after her sister had lost the habit. She at first resembled her Parisian grandfather in a less degree than did her sister at the same age, but now in a greater degree. She likewise practises to the present time the peculiar habit of rubbing together, when impatient, her thumb and two of her fore-fingers.

In this latter case we have a good instance, like those given in a former chapter, of the inheritance of a trick or gesture; for no one, I presume, will attribute to mere coincidence so peculiar a habit as this, which was common to the grandfather and his two grandchildren who had never seen him.

Considering all the circumstances with reference to these children shrugging their shoulders, it can hardly be doubted that they have inherited the habit from their French progenitors, although they have only one quarter French blood in their veins, and although their grandfather did not often shrug his shoulders. There is nothing very unusual, though the fact is interesting, in these children having gained by inheritance a habit during early youth, and then discontinuing it; for it is of frequent occurrence with many kinds of animals that certain characters are retained for a period by the young, and are then lost.

As it appeared to me at one time improbable in a high degree that so complex a gesture as shrugging the shoulders, together with the accompanying movements, should be innate, I was anxious to ascertain whether the blind and deaf Laura Bridgman, who could not have learnt the habit by imitation, practised it. And I have heard, through Dr. Innes, from a lady who has lately had charge of her, that she does shrug her shoulders, turn in her elbows, and raise her eyebrows in the same manner as other people, and under the same circumstances. I was also anxious to learn whether this gesture was practised by the various races of man, especially by those who never have had much intercourse with Europeans. We shall see that they act in this manner; but it appears that the gesture is sometimes confined to merely raising or shrugging the shoulders, without the other movements.

Mr. Scott has frequently seen this gesture in the Bengalees and Dhangars (the latter constituting a distinct race) who are employed in the Botanic Garden at Calcutta; when, for instance, they have declared that they could not do some work, such as lifting a heavy weight. He ordered a Bengalee to climb a lofty tree; but the man, with a shrug of his shoulders and a lateral shake of his head, said he could not. Mr. Scott knowing that the man was lazy, thought he could, and insisted on his trying. His face now became pale, his arms dropped to his sides, his mouth and eyes were widely opened, and again surveying the tree, he looked askant at Mr. Scott, shrugged his shoulders, inverted his elbows, extended his open hands, and with a few quick lateral shakes of the head declared his inability. Mr. H. Erskine has likewise seen the natives of India shrugging their shoulders; but he has never seen the elbows turned so much inwards as with us; and whilst shrugging their shoulders they sometimes lay their uncrossed hands on their breasts.

With the wild Malays of the interior of Malacca, and with the Bugis (true Malays, though speaking a different language), Mr. Geach has often seen this gesture. I presume that it is complete, as, in answer to my query descriptive of the movements of the shoulders, arms, hands, and face, Mr. Geach remarks, "it is performed in a beautiful style." I have lost an extract from a scientific voyage, in which shrugging the shoulders by some natives (Micronesians) of the Caroline Archipelago in the Pacific Ocean, was well described. Capt. Speedy informs me that the Abyssinians shrug their shoulders, but enters into no details. Mrs. Asa Gray saw an Arab dragoman in Alexandria acting exactly as described in my query, when an old gentleman, on whom he attended, would not go in the proper direction which had been pointed out to him.

Mr. Washington Matthews says, in reference to the wild Indian tribes of the western parts of the United States, "I have on a few occasions detected men using a slight apologetic shrug, but the rest of the demonstration which you describe I have not witnessed." Fritz Müller informs me that he has seen the negroes in Brazil shrugging their shoulders; but it is of course possible that they may have learnt to do so by imitating the Portuguese. Mrs. Barber has never seen this gesture with the Kafirs of South Africa; and Gaika, judging from his answer, did not even understand what was meant by my description. Mr. Swinhoe is also doubtful about the Chinese; but he has seen them, under the circumstances which would make us shrug our shoulders, press their right elbow against their side, raise their eyebrows, lift up their hand with the palm directed towards the person addressed, and shake it from right to left. Lastly, with respect to the Australians, four of my informants answer by a simple negative, and one by a simple affirmative. Mr. Bunnett, who has had excellent opportunities for observation on the borders of the Colony of Victoria, also answers by a "yes," adding that the gesture is performed "in a more subdued and less demonstrative manner than is the case with civilized nations." This circumstance may account for its not having been noticed by four of my informants.

These statements, relating to Europeans, Hindoos, the hill-tribes of India, Malays, Micronesians, Abyssinians, Arabs, Negroes, Indians of North America, and apparently to the Australians—many of these natives having had scarcely any intercourse with Europeans—are sufficient to show that shrugging the shoulders, accompanied in some cases by the other proper movements, is a gesture natural to mankind.

This gesture implies an unintentional or unavoidable action on our own part, or one that we cannot perform; or an action performed by another person which we cannot prevent. It accompanies such speeches as, "It was not my fault;" "It is impossible for me to grant this favour;" "He must follow his own course, I cannot stop him." Shrugging the shoulders likewise expresses patience, or the absence of any intention to resist. Hence the muscles which raise the shoulders are sometimes called, as I have been informed by an artist, "the patience muscles." Shylock the Jew, says,

> "Signor Antonio, many a time and oft
> In the Rialto have you rated me
> About my monies and usances;
> Still have I borne it with a patient shrug."
>
> *Merchant of Venice*, act i. sc. 3.

Sir C. Bell has given[217] a life-like figure of a man, who is shrinking back from some terrible danger, and is on the point of screaming out in abject terror. He is represented with his shoulders lifted up almost to his ears; and this at once declares that there is no thought of resistance.

As shrugging the shoulders generally implies "I cannot do this or that," so by a slight change, it sometimes implies "I won't do it." The movement then expresses a dogged determination not to act. Olmsted describes[218] an Indian in Texas as giving a great shrug to his shoulders, when he was informed that a party of men were Germans and not Americans, thus expressing that he would have nothing to do with them. Sulky and obstinate children may be seen with both their shoulders raised high up; but this movement is not associated with the others which generally accompany a true shrug. An excellent observer[219] in describing a young man who was determined not to yield to his father's desire, says, "He thrust his hands deep down into his pockets, and set up his shoulders to his ears, which was a good warning that, come right or wrong, this rock should fly from its firm base as soon as Jack would; and that any remonstrance on the subject was purely futile." As soon as the son got his own way, he "put his shoulders into their natural position."

Resignation is sometimes shown by the open hands being placed, one over the other, on the lower part of the body. I should not have thought this little gesture worth even a passing notice, had not Dr. W. Ogle remarked to me that he had two or three times observed it in patients who were preparing for operations under chloroform. They exhibited no great fear, but seemed to declare by this posture of their hands, that they had made up their minds, and were resigned to the inevitable.

We may now inquire why men in all parts of the world when they feel,—whether or not they wish to show this feeling,—that they cannot or will not do something, or will not resist something if done by another, shrug their shoulders, at the same time often bending in their elbows, showing the palms of their hands with extended fingers, often throwing their heads a little on one side, raising their eyebrows, and opening their mouths. These states of the mind are either simply passive, or show a determination not to act. None of the above movements are of the least service. The explanation lies, I cannot doubt, in the principle of unconscious antithesis. This principle here seems to come into play as clearly as in the case of a dog, who, when feeling savage, puts himself in the proper attitude for attacking and for making himself appear terrible to his enemy; but as soon as he feels affectionate, throws his whole body into a directly opposite attitude, though this is of no direct use to him.

Let it be observed how an indignant man, who resents, and will not submit to some injury, holds his head erect, squares his shoulders, and expands his chest. He often clenches his fists, and puts one or both arms in the proper position for attack or defence, with the muscles of his limbs rigid. He frowns,—that is, he contracts and lowers his brows,—and, being determined, closes his mouth. The actions and attitude of a helpless man are, in every one of these respects, exactly the reverse. In Plate VI we may imagine one of the figures on the left side to have just said, "What do you mean by insulting me?" and one of the figures on the right side to answer, "I really could not help it." The helpless man unconsciously contracts the muscles of his forehead which are antagonistic to those that cause a frown, and thus raises his eyebrows; at the same time he relaxes the muscles about the mouth, so that the lower jaw drops. The antithesis is complete in every detail, not only in the movements of the features, but in the position of the limbs and in the attitude of the whole body, as may be seen in the accompanying plate. As the helpless or apologetic man often wishes to show his state of mind, he then acts in a conspicuous or demonstrative manner.

In accordance with the fact that squaring the elbows and clenching the fists are gestures by no means universal with the men of all races, when they feel indignant and are prepared to attack their enemy, so it appears that a helpless or apologetic frame of mind is expressed in many parts of the world by merely shrugging the shoulders, without turning inwards the elbows and opening the hands. The man or child who is obstinate, or one who is resigned to some great misfortune, has in neither case any idea of resistance by active means; and he expresses this state of mind, by simply keeping his shoulders raised; or he may possibly fold his arms across his breast.

Signs of affirmation or approval, and of negation or disapproval: nodding and shaking the head.—I was curious to ascertain how far the common signs used by us in affirmation and negation were general throughout the world. These signs are indeed to a certain extent expressive of our feelings, as we give a vertical nod of approval with a smile to our children, when we approve of their conduct; and shake our heads laterally with a frown, when we disapprove. With infants, the first act of denial consists in refusing food; and I repeatedly noticed with my own infants, that they did so by withdrawing their heads laterally from the breast, or from anything offered them in a spoon. In accepting food and taking it into their mouths, they incline their heads forwards. Since making these observations I have been informed that the same idea had occurred to Charma.[220] It deserves notice that in accepting or taking food, there is only a single movement forward, and a single nod implies an affirmation. On the other hand, in refusing food, especially if it be pressed on them, children frequently move their heads several times from side to side, as we do in shaking our heads in negation. Moreover, in the case of refusal, the head is not rarely thrown backwards, or the mouth is closed, so that these movements might likewise come to serve as signs of negation. Mr. Wedgwood remarks on this subject,[221] that "when the voice is exerted with closed teeth or lips, it produces the sound of the letter *n* or *m*. Hence we may account for the use of the particle *ne* to signify negation, and possibly also of the Greek *μή* in the same sense."

That these signs are innate or instinctive, at least with Anglo-Saxons, is rendered highly probable by the blind and deaf Laura Bridgman "constantly accompanying her *yes* with the common affirmative nod, and her *no* with our negative shake of the head." Had not Mr. Lieber stated to the contrary,[222] I should have imagined that these gestures might have been acquired or learnt by her, considering her wonderful sense of touch and appreciation of the movements of others. With microcephalous idiots, who are so degraded that they never learn to speak, one of them is described by Vogt,[223] as answering, when asked whether he wished for more food or drink, by inclining or shaking his head. Schmalz, in his remarkable dissertation on the education of the deaf and dumb, as well as of children raised only one degree above idiotcy, assumes that they can always both make and understand the common signs of affirmation and negation.[224]

Nevertheless if we look to the various races of man, these signs are not so universally employed as I should have expected; yet they seem too general to be ranked as altogether conventional or artificial. My informants assert that both signs are used by the Malays, by the natives of Ceylon, the Chinese, the negroes of the Guinea coast, and, according to Gaika, by the Kafirs of South Africa, though with these latter people Mrs. Barber has never seen a lateral shake used as a negative. With respect to the Australians, seven observers agree that a nod is given in affirmation; five agree about a lateral shake in negation, accompanied or not by some word; but Mr. Dyson Lacy has never seen this latter sign in Queensland, and Mr. Bulmer says that in Gipps' Land a negative is expressed by throwing the head a little backwards and putting out the tongue. At the northern extremity of the continent, near Torres Straits, the natives when uttering a negative "don't shake the head with it, but holding up the right hand, shake it by turning it half round and back again two or three times."[225] The throwing back of the head with a cluck of the tongue is said to be used as a negative by the modern Greeks and Turks, the latter people expressing *yes* by a movement like that made by us when we shake our heads.[226] The Abyssinians, as I am informed by Captain Speedy, express a negative by jerking the head to the right shoulder, together with a slight cluck, the mouth being closed; an affirmation is expressed by the head being thrown backwards and the eyebrows raised for an instant. The Tagals of Luzon, in the Philippine Archipelago, as I hear from Dr. Adolf Meyer, when they say "yes," also throw the head backwards. According to the Rajah Brooke, the Dyaks of Borneo express an affirmation by raising the eyebrows, and a negation by slightly contracting them, together with a peculiar look from the eyes. With the Arabs on the Nile, Professor and Mrs. Asa Gray concluded that nodding in affirmation was rare, whilst shaking the head in negation was never used, and was not even understood by them. With the Esquimaux[227] a nod means *yes* and a wink *no*. The New Zealanders "elevate the head and chin in place of nodding acquiescence."[228]

With the Hindoos Mr. H. Erskine concludes from inquiries made from experienced Europeans, and from native gentlemen, that the signs of affirmation and negation vary—a nod and a lateral shake being sometimes used as we do; but a negative is more commonly expressed by the head being thrown suddenly backwards and a little to one side, with a cluck of the tongue. What the meaning may be of this cluck of the tongue, which has been observed with various people, I cannot imagine. A native gentleman stated that affirmation is frequently shown by the head being thrown to the left. I asked Mr. Scott to attend particularly to this point, and, after repeated observations, he believes that a vertical nod is not commonly used by the natives in affirmation, but that the head is first thrown backwards either to the left or right, and then jerked obliquely forwards only once. This movement would perhaps have been described by a less careful observer as a lateral shake. He also states that in negation the head is usually held nearly upright, and shaken several times.

Mr. Bridges informs me that the Fuegians nod their heads vertically in affirmation, and shake them laterally in denial. With the wild Indians of

North America, according to Mr. Washington Matthews, nodding and shaking the head have been learnt from Europeans, and are not naturally employed. They express affirmation "by describing with the hand (all the fingers except the index being flexed) a curve downwards and outwards from the body, whilst negation is expressed by moving the open hand outwards, with the palm facing inwards." Other observers state that the sign of affirmation with these Indians is the forefinger being raised, and then lowered and pointed to the ground, or the hand is waved straight forward from the face; and that the sign of negation is the finger or whole hand shaken from side to side.[229] This latter movement probably represents in all cases the lateral shaking of the head. The Italians are said in like manner to move the lifted finger from right to left in negation, as indeed we English sometimes do.

On the whole we find considerable diversity in the signs of affirmation and negation in the different races of man. With respect to negation, if we admit that the shaking of the finger or hand from side to side is symbolic of the lateral movement of the head; and if we admit that the sudden backward movement of the head represents one of the actions often practised by young children in refusing food, then there is much uniformity throughout the world in the signs of negation, and we can see how they originated. The most marked exceptions are presented by the Arabs, Esquimaux, some Australian tribes, and Dyaks. With the latter a frown is the sign of negation, and with us frowning often accompanies a lateral shake of the head.

With respect to nodding in affirmation, the exceptions are rather more numerous, namely with some of the Hindoos, with the Turks, Abyssinians, Dyaks, Tagals, and New Zealanders. The eyebrows are sometimes raised in affirmation, and as a person in bending his head forwards and downwards naturally looks up to the person whom he addresses, he will be apt to raise his eyebrows, and this sign may thus have arisen as an abbreviation. So again with the New Zealanders, the lifting up the chin and head in affirmation may perhaps represent in an abbreviated form the upward movement of the head after it has been nodded forwards and downwards.

CHAPTER XII

Surprise—Astonishment—Fear—Horror

SURPRISE, ASTONISHMENT • ELEVATION OF THE EYEBROWS • OPENING THE MOUTH • PROTRUSION OF THE LIPS • GESTURES ACCOMPANYING SURPRISE • ADMIRATION • FEAR • TERROR • ERECTION OF THE HAIR • CONTRACTION OF THE PLATYSMA MUSCLE • DILATATION OF THE PUPILS • HORROR • CONCLUSION

Attention, if sudden and close, graduates into surprise; and this into astonishment; and this into stupefied amazement. The latter frame of mind is closely akin to terror. Attention is shown by the eyebrows being slightly raised; and as this state increases into surprise, they are raised to a much greater extent, with the eyes and mouth widely open. The raising of the eyebrows is necessary in order that the eyes should be opened quickly and widely; and this movement produces transverse wrinkles across the forehead. The degree to which the eyes and mouth are opened corresponds with the degree of surprise felt; but these movements must be coordinated; for a widely opened mouth with eyebrows only slightly raised results in a meaningless grimace, as Dr. Duchenne has shown in one of his photographs.[230] On the other hand, a person may often be seen to pretend surprise by merely raising his eyebrows.

Dr. Duchenne has given a photograph of an old man with his eyebrows well elevated and arched by the galvanization of the frontal muscle; and with his mouth voluntarily opened. This figure expresses surprise with much truth. I showed it to twenty-four persons without a word of explanation, and one alone did not at all understand what was intended. A second person answered terror, which is not far wrong; some of the others, however, added to the words surprise or astonishment, the epithets horrified, woful, painful, or disgusted.

The eyes and mouth being widely open is an expression universally recognised as one of surprise or astonishment. Thus Shakespeare says, "I saw a smith stand with open mouth swallowing a tailor's news." ('King John,' act iv. scene ii.) And again, "They seemed almost, with staring on one another, to tear the cases of their eyes; there was speech in their dumbness, language in their very gesture; they looked as they had heard of a world destroyed." ('Winter's Tale,' act v. scene ii.)

My informants answer with remarkable uniformity to the same effect, with respect to the various races of man; the above movements of the features being often accompanied by certain gestures and sounds, presently to be described. Twelve observers in different parts of Australia agree on this head. Mr. Winwood Reade has observed this expression with the negroes on the Guinea coast. The chief Gaika and others answer *yes* to my query with respect to the Kafirs of South Africa; and so do others emphatically with reference to the Abyssinians, Ceylonese, Chinese, Fuegians, various tribes of North America, and New Zealanders. With the latter, Mr. Stack states that the expression is more plainly shown by certain individuals than by others, though all endeavour as much as possible to conceal their feelings. The Dyaks of Borneo are said by the Rajah Brooke to open their eyes widely, when astonished, often swinging their heads to and fro, and beating their breasts. Mr. Scott informs me that the workmen in the Botanic Gardens at Calcutta are strictly ordered not to smoke; but they often disobey this order, and when suddenly surprised in the act, they first open their eyes and mouths widely. They then often slightly shrug their shoulders, as they perceive that discovery is inevitable, or frown and stamp on the ground from vexation. Soon they recover from their surprise, and abject fear is exhibited by the relaxation of all their muscles; their heads seem to sink between their shoulders; their fallen eyes wander to and fro; and they supplicate forgiveness.

The well-known Australian explorer, Mr. Stuart, has given[231] a striking account of stupefied amazement together with terror in a native who had never before seen a man on horseback. Mr. Stuart approached unseen and called to him from a little distance. "He turned round and saw me. What he imagined I was I do not know; but a finer picture of fear and astonishment I never saw. He stood incapable of moving a limb, riveted to the spot, mouth open and eyes staring . . . . He remained motionless until our black got within a few yards of him, when suddenly throwing down his waddies, he jumped into a mulga bush as high as he could get." He could not speak, and answered not a word to the inquiries made by the black, but, trembling from head to foot, "waved with his hand for us to be off."

That the eyebrows are raised by an innate or instinctive impulse may be inferred from the fact that Laura Bridgman invariably acts thus when astonished, as I have been assured by the lady who has lately had charge of her. As surprise is excited by something unexpected or unknown, we naturally desire, when startled, to perceive the cause as quickly as possible; and we consequently open our eyes fully, so that the field of vision may be increased, and the eyeballs moved easily in any direction. But this hardly accounts for the eyebrows being so greatly raised as is the case, and for the wild staring of the open eyes. The explanation lies, I believe, in the impossibility of opening the eyes with great rapidity by merely raising the upper lids. To effect this the eyebrows must be lifted energetically. Any one who will try to open his eyes as quickly as possible before a mirror will find that he acts thus; and the energetic lifting up of the eyebrows opens the eyes so widely that they stare, the white being exposed all round the iris. Moreover, the elevation of the eyebrows is an advantage in looking upwards; for as long as they are lowered they impede our vision in this direction. Sir C. Bell gives[232] a curious little proof of the part which the eyebrows play in opening the eyelids. In a stupidly drunken man all the muscles are relaxed, and the eyelids consequently droop, in the same manner as when we are falling asleep. To counteract this tendency the drunkard raises his eyebrows; and this gives to him a puzzled, foolish look, as is well represented in one of Hogarth's drawings. The habit of raising the eyebrows having once been gained in order to see as quickly as possible all around us, the movement would follow from the force of association whenever astonishment was felt from any cause, even from a sudden sound or an idea.

With adult persons, when the eyebrows are raised, the whole forehead becomes much wrinkled in transverse lines; but with children this occurs only to a slight degree. The wrinkles run in lines concentric with each eyebrow, and are partially confluent in the middle. They are highly characteristic of the expression of surprise or astonishment. Each eyebrow, when raised, becomes also, as Duchenne remarks,[233] more arched than it was before.

The cause of the mouth being opened when astonishment is felt, is a much more complex affair; and several causes apparently concur in leading to this movement. It has often been supposed[234] that the sense of hearing is thus rendered more acute; but I have watched persons listening intently to a slight noise, the nature and source of which they knew perfectly, and they did not open their mouths. Therefore I at one time imagined that the open mouth might aid in distinguishing the direction whence a sound proceeded, by giving another channel for its entrance into the ear through the eustachian tube. But Dr. W. Ogle[235] has been so kind as to search the best recent authorities on the functions of the eustachian tube;

and he informs me that it is almost conclusively proved that it remains closed except during the act of deglutition; and that in persons in whom the tube remains abnormally open, the sense of hearing, as far as external sounds are concerned, is by no means improved; on the contrary, it is impaired by the respiratory sounds being rendered more distinct. If a watch be placed within the mouth, but not allowed to touch the sides, the ticking is heard much less plainly than when held outside. In persons in whom from disease or a cold the eustachian tube is permanently or temporarily closed, the sense of hearing is injured; but this may be accounted for by mucus accumulating within the tube, and the consequent exclusion of air. We may therefore infer that the mouth is not kept open under the sense of astonishment for the sake of hearing sounds more distinctly; notwithstanding that most deaf people keep their mouths open.

Every sudden emotion, including astonishment, quickens the action of the heart, and with it the respiration. Now we can breathe, as Gratiolet remarks[236] and as appears to me to be the case, much more quietly through the open mouth than through the nostrils. Therefore, when we wish to listen intently to any sound, we either stop breathing, or breathe as quietly as possible, by opening our mouths, at the same time keeping our bodies motionless. One of my sons was awakened in the night by a noise under circumstances which naturally led to great care, and after a few minutes he perceived that his mouth was widely open. He then became conscious that he had opened it for the sake of breathing as quietly as possible. This view receives support from the reversed case which occurs with dogs. A dog when panting after exercise, or on a hot day, breathes loudly; but if his attention be suddenly aroused, he instantly pricks his ears to listen, shuts his mouth, and breathes quietly, as he is enabled to do, through his nostrils.

When the attention is concentrated for a length of time with fixed earnestness on any object or subject, all the organs of the body are forgotten and neglected;[237] and as the nervous energy of each individual is limited in amount, little is transmitted to any part of the system, excepting that which is at the time brought into energetic action. Therefore many of the muscles tend to become relaxed, and the jaw drops from its own weight. This will account for the dropping of the jaw and open mouth of a man stupefied with amazement, and perhaps when less strongly affected. I have noticed this appearance, as I find recorded in my notes, in very young children when they were only moderately surprised.

There is still another and highly effective cause, leading to the mouth being opened, when we are astonished, and more especially when we are suddenly startled. We can draw a full and deep inspiration much more easily through the widely open mouth than through the nostrils. Now when we start at any sudden sound or sight, almost all the muscles of the body are involuntarily and momentarily thrown into strong action, for the sake of guarding ourselves against or jumping away from the danger, which we habitually associate with anything unexpected. But we always unconsciously prepare ourselves for any great exertion, as formerly explained, by first taking a deep and full inspiration, and we consequently open our mouths. If no exertion follows, and we still remain astonished, we cease for a time to breathe, or breathe as quietly as possible, in order that every sound may be distinctly heard. Or again, if our attention continues long and earnestly absorbed, all our muscles become relaxed, and the jaw, which was at first suddenly opened, remains dropped. Thus several causes concur towards this same movement, whenever surprise, astonishment, or amazement is felt.

Although when thus affected, our mouths are generally opened, yet the lips are often a little protruded. This fact reminds us of the same movement, though in a much more strongly marked degree, in the chimpanzee and orang when astonished. As a strong expiration naturally follows the deep inspiration which accompanies the first sense of startled surprise, and as the lips are often protruded, the various sounds which are then commonly uttered can apparently be accounted for. But sometimes a strong expiration alone is heard; thus Laura Bridgman, when amazed, rounds and protrudes her lips, opens them, and breathes strongly.[238] One of the commonest sounds is a deep *Oh;* and this would naturally follow, as explained by Helmholtz, from the mouth being moderately opened and the lips protruded. On a quiet night some rockets were fired from the 'Beagle,' in a little creek at Tahiti, to amuse the natives; and as each rocket was let off there was absolute silence, but this was invariably followed by a deep groaning *Oh,* resounding all round the bay. Mr. Washington Matthews says that the North American Indians express astonishment by a groan; and the negroes on the West Coast of Africa, according to Mr. Winwood Reade, protrude their lips, and make a sound like *heigh, heigh.* If the mouth is not much opened, whilst the lips are considerably protruded, a blowing, hissing, or whistling noise is produced. Mr. R. Brough Smyth informs me that an Australian from the interior was taken to the theatre to see an acrobat rapidly turning head over heels: "he was greatly astonished, and protruded his lips, making a noise with his mouth as if blowing out a match." According to Mr. Bulmer the Australians, when surprised, utter the exclamation *korki,* "and to do this the mouth is drawn out as if going to whistle." We Europeans often whistle as a sign of surprise; thus, in a recent novel[239] it is said, "here the man expressed his astonishment and disapprobation by a prolonged whistle." A Kafir girl, as Mr. J. Mansel Weale informs me, "on hearing of the high price of an article, raised her eyebrows and whistled just as a European would." Mr. Wedgwood remarks that such sounds are written down as *whew,* and they serve as interjections for surprise.

According to three other observers, the Australians often evince astonishment by a clucking noise. Europeans also sometimes express gentle surprise by a little clicking noise of nearly the same kind. We have seen that when we are startled, the mouth is suddenly opened; and if the tongue happens to be then pressed closely against the palate, its sudden withdrawal will produce a sound of this kind, which might thus come to express surprise.

Turning to gestures of the body. A surprised person often raises his opened hands high above his head, or by bending his arms only to the level of his face. The flat palms are directed towards the person who causes this feeling, and the straightened fingers are separated. This gesture is represented by Mr. Rejlander in Plate VII, fig.1. In the 'Last Supper,' by Leonardo da Vinci, two of the Apostles have their hands half uplifted, clearly expressive of their astonishment. A trustworthy observer told me that he had lately met his wife under most unexpected circumstances: "She started, opened her mouth and eyes very widely, and threw up both her arms above her head." Several years ago I was surprised by seeing several of my young children earnestly doing something together on the ground; but the distance was too great for me to ask what they were about. Therefore I threw up my open hands with extended fingers above my head; and as soon as I had done this, I became conscious of the action. I then waited, without saying a word, to see if my children had understood this gesture; and as they came running to me they cried out, "We saw that you were astonished at us." I do not know whether this gesture is common to the various races of man, as I neglected to make inquiries on this head. That it is innate or natural may be inferred from the fact that Laura Bridgman, when amazed, "spreads her arms and turns her hands with extended fingers upwards;"[240] nor is it likely, considering that the feeling of surprise is generally a brief one, that she should have learnt this gesture through her keen sense of touch.

Huschke describes[241] a somewhat different yet allied gesture, which he says is exhibited by persons when astonished. They hold themselves erect, with the features as before described, but with the straightened arms extended backwards—the stretched fingers being separated from each other. I have never myself seen this gesture; but Huschke is probably correct; for a friend asked another man how he would express great astonishment, and he at once threw himself into this attitude.

These gestures are, I believe, explicable on the principle of antithesis. We have seen that an indignant man holds his head erect, squares his shoulders, turns out his elbows, often clenches his fist, frowns, and closes his mouth; whilst the attitude of a helpless man is in every one of these details the reverse. Now, a man in an ordinary frame of mind, doing nothing and thinking of nothing in particular, usually keeps his two arms suspended laxly by his sides, with his hands somewhat flexed, and the fingers near together. Therefore, to raise the arms suddenly, either the whole arms or the fore-arms, to open the palms flat, and to separate the fingers,—or, again, to straighten the arms, extending them backwards with separated fingers,—are movements in complete antithesis to those preserved under an indifferent frame of mind, and they are, in consequence, unconsciously assumed by an astonished man. There is, also, often a desire to display surprise in a conspicuous manner, and the above attitudes are well fitted for this purpose. It may be asked why should surprise, and only a few other states of the mind, be exhibited by movements in antithesis to others. But this principle will not be brought into play in the case of those emotions, such as terror, great joy, suffering, or rage, which naturally lead to certain lines of action, and produce certain effects on the body, for the whole system is thus preoccupied; and these emotions are already thus expressed with the greatest plainness.

There is another little gesture, expressive of astonishment, of which I can offer no explanation; namely, the hand being placed over the mouth or on some part of the head. This has been observed with so many races of man, that it must have some natural origin. A wild Australian was taken into a large room full of official papers, which surprised him greatly, and he cried out, *cluck, cluck, cluck,* putting the back of his hand towards his lips. Mrs. Barber says that the Kafirs and Fingoes express astonishment by a serious look and by placing the right hand upon the mouth, uttering the word *mawo,* which means 'wonderful.' The Bushmen are said[242] to put their right hands to their necks bending their heads backwards. Mr. Winwood Reade has observed that the negroes on the West Coast of Africa when surprised, clap their hands to their mouths, saying at the same time, "My mouth cleaves to me," *i.e.* to my hands; and he has heard that this is their usual gesture on such occasions. Captain Speedy informs me that the Abyssinians place their right hand to the forehead, with the palm outside. Lastly, Mr. Washington Matthews states that the conventional sign of astonishment with the wild tribes of the western parts of the United States "is made by placing the half-closed hand over the mouth; in doing this, the head is often bent forwards, and words or low groans are sometimes uttered." Catlin[243] makes the same remark about the hand being pressed over the mouth by the Mandans and other Indian tribes.

Admiration.—Little need be said on this head. Admiration apparently consists of surprise associated with some pleasure and a sense of approval. When vividly felt, the eyes are opened and the eyebrows raised; the eyes become bright, instead of remaining blank, as under simple astonishment; and the mouth, instead of gaping open, expands into a smile.

Fear, Terror.—The word 'fear' seems to be derived from what is sudden and dangerous;[244] and that of terror from the trembling of the vocal organs and body. I use the word 'terror' for extreme fear; but some writers think it ought to be confined to cases in which the imagination is more particularly concerned. Fear is often preceded by astonishment, and is so far akin to it, that both lead to the senses of sight and hearing being instantly aroused. In both cases the eyes and mouth are widely opened, and the eyebrows raised. The frightened man at first stands like a statue motionless and breathless, or crouches down as if instinctively to escape observation.

The heart beats quickly and violently, so that it palpitates or knocks against the ribs; but it is very doubtful whether it then works more efficiently than usual, so as to send a greater supply of blood to all parts of the body; for the skin instantly becomes pale, as during incipient faintness. This paleness of the surface, however, is probably in large part, or exclusively, due to the vaso-motor centre being affected in such a manner as to cause the contraction of the small arteries of the skin. That the skin is much affected under the sense of great fear, we see in the marvellous and inexplicable manner in which perspiration immediately exudes from it. This exudation is all the more remarkable, as the surface is then cold, and hence the term a cold sweat; whereas, the sudorific glands are properly excited into action when the surface is heated. The hairs also on the skin stand erect; and the superficial muscles shiver. In connection with the disturbed action of the heart, the breathing is hurried. The salivary glands act imperfectly; the mouth becomes dry,[245] and is often opened and shut. I have also noticed that under slight fear there is a strong tendency to yawn. One of the best-marked symptoms is the trembling of all the muscles of the body; and this is often first seen in the lips. From this cause, and from the dryness of the mouth, the voice becomes husky or indistinct, or may altogether fail. "Obstupui, steteruntque comæ, et vox faucibus hæsit."

Of vague fear there is a well-known and grand description in Job:—"In thoughts from the visions of the night, when deep sleep falleth on men, fear came upon me, and trembling, which made all my bones to shake. Then a spirit passed before my face; the hair of my flesh stood up. It stood still, but I could not discern the form thereof: an image was before my eyes, there was silence, and I heard a voice, saying, Shall mortal man be more just than God? Shall a man be more pure than his Maker?" (Job iv. 13.)

As fear increases into an agony of terror, we behold, as under all violent emotions, diversified results. The heart beats wildly, or may fail to act and faintness ensue; there is a death-like pallor; the breathing is laboured; the wings of the nostrils are widely dilated; "there is a gasping and convulsive motion of the lips, a tremor on the hollow cheek, a gulping and catching of the throat;"[246] the uncovered and protruding eyeballs are fixed on the object of terror; or they may roll restlessly from side to side, *huc illuc volvens oculos totumque pererrat.*[247] The pupils are said to be enormously dilated. All the muscles of the body may become rigid, or may be thrown into convulsive movements. The hands are alternately clenched and opened, often with a twitching movement. The arms may be protruded, as if to avert some dreadful danger, or may be thrown wildly over the head. The Rev. Mr. Hagenauer has seen this latter action in a terrified Australian. In other cases there is a sudden and uncontrollable tendency to headlong flight; and so strong is this, that the boldest soldiers may be seized with a sudden panic.

As fear rises to an extreme pitch, the dreadful scream of terror is heard. Great beads of sweat stand on the skin. All the muscles of the body are relaxed. Utter prostration soon follows, and the mental powers fail. The intestines are affected. The sphincter muscles cease to act, and no longer retain the contents of the body.

Dr. J. Crichton Browne has given me so striking an account of intense fear in an insane woman, aged thirty-five, that the description though painful ought not to be omitted. When a paroxysm seizes her, she screams out, "This is hell!" "There is a black woman!" "I can't get out!"—and other such exclamations. When thus screaming, her movements are those of alternate tension and tremor. For one instant she clenches her hands, holds her arms out before her in a stiff semi-flexed position; then suddenly bends her body forwards, sways rapidly to and fro, draws her fingers through her hair, clutches at her neck, and tries to tear off her clothes. The sterno-cleido-mastoid muscles (which serve to bend the head on the chest) stand out prominently, as if swollen, and the skin in front of them is much wrinkled. Her hair, which is cut short at the back of her head, and is smooth when she is calm, now stands on end; that in front being dishevelled by the movements of her hands. The countenance expresses great mental agony. The skin is flushed over the face and neck, down to the clavicles, and the veins of the forehead and neck stand out like thick cords. The lower lip drops, and is somewhat everted. The mouth is kept half open, with the lower jaw projecting. The cheeks are hollow and deeply furrowed in curved lines running from the wings of the nostrils to the corners of the mouth. The nostrils themselves are raised and extended. The eyes are widely opened, and beneath them the skin appears swollen; the pupils are large. The forehead is wrinkled transversely in many folds, and at the inner extremities of the eyebrows it is strongly furrowed in diverging lines, produced by the powerful and persistent contraction of the corrugators.

Mr. Bell has also described[248] an agony of terror and of despair, which he witnessed in a murderer, whilst carried to the place of execution in Turin. "On each side of the car the officiating priests were seated; and in the centre sat the criminal himself. It was impossible to witness the condition of this unhappy wretch without terror; and yet, as if impelled by some strange infatuation, it was equally impossible not to gaze upon an object so wild, so full of horror. He seemed about thirty-five years of age; of large and muscular form; his countenance marked by strong and savage features; half naked, pale as death, agonized with terror, every limb strained in anguish, his hands clenched convulsively, the sweat breaking out on his bent and contracted brow, he kissed incessantly the figure of our Saviour, painted on the flag which was suspended before him; but with an agony of wildness and despair, of which nothing ever exhibited on the stage can give the slightest conception."

I will add only one other case, illustrative of a man utterly prostrated by terror. An atrocious murderer of two persons was brought into a hospital, under the mistaken impression that he had poisoned himself; and Dr. W. Ogle carefully watched him the next morning, while he was being handcuffed and taken away by the police. His pallor was extreme, and his prostration so great that he was hardly able to dress himself. His skin perspired: and his eyelids and head drooped so much that it was impossible to catch even a glimpse of his eyes. His lower jaw hung down. There was no contraction of any facial muscle, and Dr. Ogle is almost certain that the hair did not stand on end, for he observed it narrowly, as it had been dyed for the sake of concealment.

With respect to fear, as exhibited by the various races of man, my informants agree that the signs are the same as with Europeans. They are displayed in an exaggerated degree with the Hindoos and natives of Ceylon. Mr. Geach has seen Malays when terrified turn pale and shake; and Mr. Brough Smyth states that a native Australian "being on one occasion much frightened, showed a complexion as nearly approaching to what we call paleness, as can well be conceived in the case of a very black man." Mr. Dyson Lacy has seen extreme fear shown in an Australian, by a nervous twitching of the hands, feet, and lips; and by the perspiration standing on the skin. Many savages do not repress the signs of fear so much as Europeans; and they often tremble greatly. With the Kafir, Gaika says, in his rather quaint English, the shaking "of the body is much experienced, and the eyes are widely open." With savages, the sphincter muscles are often relaxed, just as may be observed in much frightened dogs, and as I have seen with monkeys when terrified by being caught.

The erection of the hair.—Some of the signs of fear deserve a little further consideration. Poets continually speak of the hair standing on end; Brutus says to the ghost of Cæsar, "that mak'st my blood cold, and my hair to stare." And Cardinal Beaufort, after the murder of Gloucester exclaims, "Comb down his hair; look, look, it stands upright." As I did not feel sure whether writers of fiction might not have applied to man what they had often observed in animals, I begged for information from Dr. Crichton Browne with respect to the insane. He states in answer that he has repeatedly seen their hair erected under the influence of sudden and extreme terror. For instance, it is occasionally necessary to inject morphia under the skin of an insane woman, who dreads the operation extremely, though it causes very little pain; for she believes that poison is being introduced into her system, and that her bones will be softened, and her flesh turned into dust. She becomes deadly pale; her limbs are stiffened by a sort of tetanic spasm, and her hair is partially erected on the front of the head.

Dr. Browne further remarks that the bristling of the hair which is so common in the insane, is not always associated with terror. It is perhaps most frequently seen in chronic maniacs, who rave incoherently and have destructive impulses; but it is during their paroxysms of violence that the bristling is most observable. The fact of the hair becoming erect under the influence both of rage and fear agrees perfectly with what we have seen in the lower animals. Dr. Browne adduces several cases in evidence. Thus with a man now in the Asylum, before the recurrence of each maniacal paroxysm, "the hair rises up from his forehead like the mane of a Shetland pony." He has sent me photographs of two women, taken in the intervals between their paroxysms, and he adds with respect to one of these women, "that the state of her hair is a sure and convenient criterion of her mental condition." I have had one of these photographs copied, and the engraving gives, if viewed from a little distance, a faithful representation of the original, with the exception that the hair appears rather too coarse and too much curled. The extraordinary condition of the hair in the insane is due, not only to its erection, but to its dryness and harshness, consequent on the subcutaneous glands failing to act. Dr. Bucknill has said[249] that a lunatic "is a lunatic to his finger's ends;" he might have added, and often to the extremity of each particular hair.

Fig. 19. *From a photograph of an insane woman, to show the condition of her hair.*

Dr. Browne mentions as an

empirical confirmation of the relation which exists in the insane between the state of their hair and minds, that the wife of a medical man, who has charge of a lady suffering from acute melancholia, with a strong fear of death, for herself, her husband and children, reported verbally to him the day before receiving my letter as follows, "I think Mrs. —— will soon improve, for her hair is getting smooth; and I always notice that our patients get better whenever their hair ceases to be rough and unmanageable."

Dr. Browne attributes the persistently rough condition of the hair in many insane patients, in part to their minds being always somewhat disturbed, and in part to the effects of habit,—that is, to the hair being frequently and strongly erected during their many recurrent paroxysms. In patients in whom the bristling of the hair is extreme, the disease is generally permanent and mortal; but in others, in whom the bristling is moderate, as soon as they recover their health of mind the hair recovers its smoothness.

In a previous chapter we have seen that with animals the hairs are erected by the contraction of minute, unstriped, and involuntary muscles, which run to each separate follicle. In addition to this action, Mr. J. Wood has clearly ascertained by experiment, as he informs me, that with man the hairs on the front of the head which slope forwards, and those on the back which slope backwards, are raised in opposite directions by the contraction of the occipito-frontalis or scalp muscle. So that this muscle seems to aid in the erection of the hairs on the head of man, in the same manner as the homologous *panniculus carnosus* aids, or takes the greater part, in the erection of the spines on the backs of some of the lower animals.

1

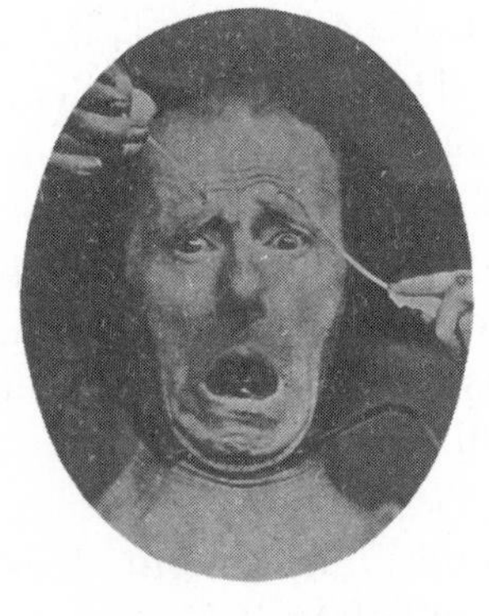

2

Plate VII

Contraction of the platysma myoides muscle.—This muscle is spread over the sides of the neck, extending downwards to a little beneath the collar-bones, and upwards to the lower part of the cheeks. A portion, called the risorius, is represented in the woodcut (M) fig. 2. The contraction of this muscle draws the corners of the mouth and the lower parts of the cheeks downwards and backwards. It produces at the same time divergent, longitudinal, prominent ridges on the sides of the neck in the young; and, in old thin persons, fine transverse wrinkles. This muscle is sometimes said not to be under the control of the will; but almost every one, if told to draw the corners of his mouth backwards and downwards with great force, brings it into action. I have, however, heard of a man who can voluntarily act on it only on one side of his neck.

Fig. 20. *Terror. From a photograph by Dr. Duchenne.*

Sir C. Bell[250] and others have stated that this muscle is strongly contracted under the influence of fear; and Duchenne insists so strongly on its importance in the expression of this emotion, that he calls it the *muscle of fright.*[251] He admits, however, that its contraction is quite inexpressive unless associated with widely open eyes and mouth. He has given a photograph (copied and reduced in the accompanying woodcut) of the same old man as on former occasions, with his eyebrows strongly raised, his mouth opened, and the platysma contracted, all by means of galvanism. The original photograph was shown to twenty-four persons, and they were separately asked, without any explanation being given, what expression was intended: twenty instantly answered, "intense fright" or "horror;" three said pain, and one extreme discomfort. Dr. Duchenne has given another photograph of the same old man, with the platysma contracted, the eyes and mouth opened, and the eye-brows rendered oblique, by means of galvanism. The expression thus induced is very striking (see Plate VII, fig. 2); the obliquity of the eyebrows adding the appearance of great mental distress. The original was shown to fifteen persons; twelve answered terror or horror, and three agony or great suffering. From these cases, and from an examination of the other photographs given by Dr. Duchenne, together with his remarks thereon, I think there can be little doubt that the contraction of the platysma does add greatly to the expression of fear. Nevertheless this muscle ought hardly to be called that of fright, for its contraction is certainly not a necessary concomitant of this state of mind.

A man may exhibit extreme terror in the plainest manner by death-like pallor, by drops of perspiration on his skin, and by utter prostration, with all the muscles of his body, including the platysma, completely relaxed. Although Dr. Browne has often seen this muscle quivering and contracting in the insane, he has not been able to connect its action with any emotional condition in them, though he carefully attended to patients suffering from great fear. Mr. Nicol, on the other hand, has observed three cases in which this muscle appeared to be more or less permanently contracted under the influence of melancholia, associated with much dread; but in one of these cases, various other muscles about the neck and head were subject to spasmodic contractions.

Dr. W. Ogle observed for me in one of the London hospitals about twenty patients, just before they were put under the influence of chloroform for operations. They exhibited some trepidation, but no great terror. In only four of the cases was the platysma visibly contracted; and it did not begin to contract until the patients began to cry. The muscle seemed to contract at the moment of each deep-drawn inspiration; so that it is very doubtful whether the contraction depended at all on the emotion of fear. In a fifth case, the patient, who was not chloroformed, was much terrified; and his platysma was more forcibly and persistently contracted than in the other cases. But even here there is room for doubt, for the muscle which appeared to be unusually developed, was seen by Dr. Ogle to contract as the man moved his head from the pillow, after the operation was over.

As I felt much perplexed why, in any case, a superficial muscle on the neck should be especially affected by fear, I applied to my many obliging correspondents for information about the contraction of this muscle under other circumstances. It would be superfluous to give all the answers which I have received. They show that this muscle acts, often in a variable manner and degree, under many different conditions. It is violently contracted in hydrophobia, and in a somewhat less degree in lockjaw; sometimes in a marked manner during the insensibility from chloroform. Dr. W. Ogle observed two male patients, suffering from such difficulty in breathing, that the trachea had to be opened, and in both the platysma was strongly contracted. One of these men overheard the conversation of the surgeons surrounding him, and when he was able to speak, declared that he had not been frightened. In some other cases of extreme difficulty of respiration, though not requiring tracheotomy, observed by Drs. Ogle and Langstaff, the platysma was not contracted.

Mr. J. Wood, who has studied with such care the muscles of the human body, as shown by his various publications, has often seen the platysma contracted in vomiting, nausea, and disgust; also in children and adults under the influence of rage,—for instance, in Irishwomen, quarrelling and brawling together with angry gesticulations. This may possibly have been due to their high and angry tones; for I know a lady, an excellent musician, who, in singing certain high notes, always contracts her platysma. So does a young man, as I have observed, in sounding certain notes on the flute. Mr. J. Wood informs me that he has found the platysma best developed in persons with thick necks and broad shoulders; and that in families inheriting these peculiarities, its development is usually associated with much voluntary power over the homologous occipito-frontalis muscle, by which the scalp can be moved.

None of the foregoing cases appear to throw any light on the contraction of the platysma from fear; but it is different, I think, with the following cases. The gentleman before referred to, who can voluntarily act on this muscle only on one side of his neck, is positive that it contracts on both sides whenever he is startled. Evidence has already been given showing that this muscle sometimes contracts, perhaps for the sake of opening the mouth widely, when the breathing is rendered difficult by disease, and during the deep inspirations of crying-fits before an operation. Now, whenever a person starts at any sudden sight or sound, he instantaneously draws a deep breath; and thus the contraction of the platysma may possibly have become associated with the sense of fear. But there is, I believe, a more efficient relation. The first sensation of fear, or the imagination of something dreadful, commonly excites a shudder. I have caught myself giving a little involuntary shudder at a painful thought, and I distinctly perceived that my platysma contracted; so it does if I simulate a shudder. I have asked others to act in this manner; and in some the muscle contracted, but not in others. One of my sons, whilst getting out of bed, shuddered from the cold, and, as he happened to have his hand on his neck, he plainly felt that this muscle strongly contracted. He then voluntarily shuddered, as he had done on former occasions, but the platysma was not then affected. Mr. J. Wood has also several times observed this muscle contracting in patients, when stripped for examination, and who were not frightened, but shivered slightly from the cold. Unfortunately I have not been able to ascertain whether, when the whole body shakes, as in the cold stage of an ague fit, the platysma contracts. But as it certainly often contracts during a shudder; and as a shudder or shiver often accompanies the first sensation of fear, we have, I think, a clue to its action in this latter case.[252] Its contraction, however, is not an invariable concomitant of fear; for it probably never acts under the influence of extreme, prostrating terror.

Dilatation of the Pupils.—Gratiolet repeatedly insists[253] that the pupils are enormously dilated whenever terror is felt. I have no reason to doubt the accuracy of this statement, but have failed to obtain confirmatory evidence, excepting in the one instance before given of an insane woman suffering from great fear. When writers of fiction speak of the eyes being widely dilated, I presume that they refer to the eyelids. Munro's statement,[254] that with parrots the iris is affected by the passions, independently of the amount of light, seems to bear on this question; but Professor Donders informs me, that he has often seen movements in the pupils of these birds which he thinks may be related to their power of accommodation to distance, in nearly the same manner as our own pupils contract when our eyes converge for near vision. Gratiolet remarks that the dilated pupils appear as if they were gazing into profound darkness. No doubt the fears of man have often been excited in the dark; but hardly so often or so exclusively, as to account for a fixed and associated habit having thus arisen. It seems more probable, assuming that Gratiolet's statement is correct, that the brain is directly affected by the powerful emotion of fear and reacts on the pupils; but Professor Donders informs me that this is an extremely complicated subject. I may add, as possibly throwing light on the subject, that Dr. Fyffe, of Netley Hospital, has observed in two patients that the pupils were distinctly dilated during the cold stage of an ague fit. Professor Donders has also often seen dilatation of the pupils in incipient faintness.

Horror.—The state of mind expressed by this term implies terror, and is in some cases almost synonymous with it. Many a man must have felt, before the blessed discovery of chloroform, great horror at the thought of an impending surgical operation. He who dreads, as well as hates a man, will feel, as Milton uses the word, a horror of him. We feel horror if we see any one, for instance a child, exposed to some instant and crushing danger. Almost every one would experience the same feeling in the highest degree in witnessing a man being tortured or going to be tortured. In these cases there is no danger to ourselves; but from the power of the imagination and of sympathy we put ourselves in the position of the sufferer, and feel something akin to fear.

Sir C. Bell remarks,[255] that "horror is full of energy; the body is in the utmost tension, not unnerved by fear." It is, therefore, probable that horror would generally be accompanied by the strong contraction of the brows; but as fear is one of the elements, the eyes and mouth would be opened, and the eyebrows would be raised, as far as the antagonistic action of the corrugators permitted this movement. Duchenne has given a photograph[256] (fig.

21) of the same old man as before, with his eyes somewhat staring, the eyebrows partially raised, and at the same time strongly contracted, the mouth opened, and the platysma in action, all effected by the means of galvanism. He considers that the expression thus produced shows extreme terror with horrible pain or torture. A tortured man, as long as his sufferings allowed him to feel any dread for the future, would probably exhibit horror in an extreme degree. I have shown the original of this photograph to twenty-three persons of both sexes and various ages; and thirteen immediately answered horror, great pain, torture, or agony; three answered extreme fright; so that sixteen answered nearly in accordance with Duchenne's belief. Six, however, said anger, guided no doubt, by the strongly contracted brows, and overlooking the peculiarly opened mouth. One said disgust. On the whole, the evidence indicates that we have here a fairly good representation of horror and agony. The photograph before referred to (Plate VII, fig. 2) likewise exhibits horror; but in this the oblique eyebrows indicate great mental distress in place of energy.

Fig. 21. *Horror and Agony. Copied from a photograph by Dr. Duchenne.*

Horror is generally accompanied by various gestures, which differ in different individuals. Judging from pictures, the whole body is often turned away or shrinks; or the arms are violently protruded as if to push away some dreadful object. The most frequent gesture, as far as can be inferred from the acting of persons who endeavour to express a vividly-imagined scene of horror, is the raising of both shoulders, with the bent arms pressed closely against the sides or chest. These movements are nearly the same with those commonly made when we feel very cold; and they are generally accompanied by a shudder, as well as by a deep expiration or inspiration, according as the chest happens at the time to be expanded or contracted. The sounds thus made are expressed by words like *uh* or *ugh.*[257] It is not, however, obvious why, when we feel cold or express a sense of horror, we press our bent arms against our bodies, raise our shoulders, and shudder.

Conclusion.—I have now endeavoured to describe the diversified expressions of fear, in its gradations from mere attention to a start of surprise, into extreme terror and horror. Some of the signs may be accounted for through the principles of habit, association, and inheritance,—such as the wide opening of the mouth and eyes, with upraised eyebrows, so as to see as quickly as possible all around us, and to hear distinctly whatever sound may reach our ears. For we have thus habitually prepared ourselves to discover and encounter any danger. Some of the other signs of fear may likewise be accounted for, at least in part, through these same principles. Men, during numberless generations, have endeavoured to escape from their enemies or danger by headlong flight, or by violently struggling with them; and such great exertions will have caused the heart to beat rapidly, the breathing to be hurried, the chest to heave, and the nostrils to be dilated. As these exertions have often been prolonged to the last extremity, the final result will have been utter prostration, pallor, perspiration, trembling of all the muscles, or their complete relaxation. And now, whenever the emotion of fear is strongly felt, though it may not lead to any exertion, the same results tend to reappear, through the force of inheritance and association.

Nevertheless, it is probable that many or most of the above symptoms of terror, such as the beating of the heart, the trembling of the muscles, cold perspiration, &c., are in large part directly due to the disturbed or interrupted transmission of nerve-force from the cerebro-spinal system to various parts of the body, owing to the mind being so powerfully affected. We may confidently look to this cause, independently of habit and association, in such cases as the modified secretions of the intestinal canal, and the failure of certain glands to act. With respect to the involuntary bristling of the hair, we have good reason to believe that in the case of animals this action, however it may have originated, serves, together with certain voluntary movements, to make them appear terrible to their enemies; and as the same involuntary and voluntary actions are performed by animals nearly related to man, we are led to believe that man has retained through inheritance a relic of them, now become useless. It is certainly a remarkable fact, that the minute unstriped muscles, by which the hairs thinly scattered over man's almost naked body are erected, should have been preserved to the present day; and that they should still contract under the same emotions, namely, terror and rage, which cause the hairs to stand on end in the lower members of the Order to which man belongs.

SELF-ATTENTION—SHAME—SHYNESS—MODESTY: BLUSHING

NATURE OF A BLUSH • INHERITANCE • THE PARTS OF THE BODY MOST AFFECTED • BLUSHING IN THE VARIOUS RACES OF MAN • ACCOMPANYING GESTURES • CONFUSION OF MIND • CAUSES OF BLUSHING • SELF-ATTENTION, THE FUNDAMENTAL ELEMENT • SHYNESS • SHAME, FROM BROKEN MORAL LAWS AND CONVENTIONAL RULES • MODESTY • THEORY OF BLUSHING • RECAPITULATION

BLUSHING IS THE MOST PECULIAR and the most human of all expressions. Monkeys redden from passion, but it would require an overwhelming amount of evidence to make us believe that any animal could blush. The reddening of the face from a blush is due to the relaxation of the muscular coats of the small arteries, by which the capillaries become filled with blood; and this depends on the proper vaso-motor centre being affected. No doubt if there be at the same time much mental agitation, the general circulation will be affected; but it is not due to the action of the heart that the network of minute vessels covering the face becomes under a sense of shame gorged with blood. We can cause laughing by tickling the skin, weeping or frowning by a blow, trembling from the fear of pain, and so forth; but we cannot cause a blush, as Dr. Burgess remarks,[258] by any physical means,—that is by any action on the body. It is the mind which must be affected. Blushing is not only involuntary; but the wish to restrain it, by leading to self-attention, actually increases the tendency.

The young blush much more freely than the old, but not during infancy,[259] which is remarkable, as we know that infants at a very early age redden from passion. I have received authentic accounts of two little girls blushing at the ages of between two and three years; and of another sensitive child, a year older, blushing, when reproved for a fault. Many children, at a somewhat more advanced age blush in a strongly marked manner. It appears that the mental powers of infants are not as yet sufficiently developed to allow of their blushing. Hence, also, it is that idiots rarely blush. Dr. Crichton Browne observed for me those under his care, but never saw a genuine blush, though he has seen their faces flush, apparently from joy, when food was placed before them, and from anger. Nevertheless some, if not utterly degraded, are capable of blushing. A microcephalous idiot, for instance, thirteen years old, whose eyes brightened a little when he was pleased or amused, has been described by Dr. Behn,[260] as blushing and turning to one side, when undressed for medical examination.

Women blush much more than men. It is rare to see an old man, but not nearly so rare to see an old woman blushing. The blind do not escape. Laura Bridgman, born in this condition, as well as completely deaf, blushes.[261] The Rev. R. H. Blair, Principal of the Worcester College, informs me that three children born blind, out of seven or eight then in the Asylum, are great blushers. The blind are not at first conscious that they are observed, and it is a most important part of their education, as Mr. Blair informs me, to impress this knowledge on their minds; and the impression thus gained would greatly strengthen the tendency to blush, by increasing the habit of self-attention.

The tendency to blush is inherited. Dr. Burgess gives the case[262] of a family consisting of a father, mother, and ten children, all of whom, without exception, were prone to blush to a most painful degree. The children were grown up; "and some of them were sent to travel in order to wear away this diseased sensibility, but nothing was of the slightest avail." Even peculiarities in blushing seem to be inherited. Sir James Paget, whilst examining the spine of a girl, was struck at her singular manner of blushing; a big splash of red appeared first on one cheek, and then other splashes, variously scattered over the face and neck. He subsequently asked the mother whether her daughter always blushed in this peculiar manner; and was answered, "Yes, she takes after me." Sir J. Paget then perceived that by asking this question he had caused the mother to blush; and she exhibited the same peculiarity as her daughter.

In most cases the face, ears and neck are the sole parts which redden; but many persons, whilst blushing intensely, feel that their whole bodies grow hot and tingle; and this shows that the entire surface must be in some manner affected. Blushes are said sometimes to commence on the forehead, but more commonly on the cheeks, afterwards spreading to the ears and neck.[263] In two Albinos examined by Dr. Burgess, the blushes commenced by a small circumscribed spot on the cheeks, over the parotidean plexus of nerves, and then increased into a circle; between this blushing circle and the blush on the neck there was an evident line of demarcation; although both arose simultaneously. The retina, which is naturally red in the Albino, invariably increased at the same time in redness.[264] Every one must have noticed how easily after one blush fresh blushes chase each other over the face. Blushing is preceded by a peculiar sensation in the skin. According to Dr. Burgess the reddening of the skin is generally succeeded by a slight pallor, which shows that the capillary vessels contract after dilating. In some rare cases paleness instead of redness is caused under conditions which would naturally induce a blush. For instance, a young lady told me that in a large and crowded party she caught her hair so firmly on the button of a passing servant, that it took some time before she could be extricated; from her sensations she imagined that she had blushed crimson; but was assured by a friend that she had turned extremely pale.

I was desirous to learn how far down the body blushes extend; and Sir J. Paget, who necessarily has frequent opportunities for observation, has kindly attended to this point for me during two or three years. He finds that with women who blush intensely on the face, ears, and nape of neck, the blush does not commonly extend any lower down the body. It is rare to see it as low down as the collar-bones and shoulder-blades; and he has never himself seen a single instance in which it extended below the upper part of the chest. He has also noticed that blushes sometimes die away downwards, not gradually and insensibly, but by irregular ruddy blotches. Dr. Langstaff has likewise observed for me several women whose bodies did not in the least redden while their faces were crimsoned with blushes. With the insane, some of whom appear to be particularly liable to blushing, Dr. J. Crichton Browne has several times seen the blush extend as far down as the collar-bones, and in two instances to the breasts. He gives me the case of a married woman, aged twenty-seven, who suffered from epilepsy. On the morning after her arrival in the Asylum, Dr. Browne, together with his assistants, visited her whilst she was in bed. The moment that he approached, she blushed deeply over her cheeks and temples; and the blush spread quickly to her ears. She was much agitated and tremulous. He unfastened the collar of her chemise in order to examine the state of her lungs; and then a brilliant blush rushed over her chest, in an arched line over the upper third of each breast, and extended downwards between the breasts nearly to the ensiform cartilage of the sternum. This case is interesting, as the blush did not thus extend downwards until it became intense by her attention being drawn to this part of her person. As the examination proceeded she became composed, and the blush disappeared; but on several subsequent occasions the same phenomena were observed.

The foregoing facts show that, as a general rule, with English women, blushing does not extend beneath the neck and upper part of the chest. Nevertheless Sir J. Paget informs me that he has lately heard of a case, on which he can fully rely, in which a little girl, shocked by what she imagined to be an act of indelicacy, blushed all over her abdomen and the upper parts of her legs. Moreau also[265] relates, on the authority of a celebrated painter, that the chest, shoulders, arms, and whole body of a girl, who unwillingly consented to serve as a model, reddened when she was first divested of her clothes.

It is a rather curious question why, in most cases the face, ears, and neck alone redden, inasmuch as the whole surface of the body often tingles and grows hot. This seems to depend, chiefly, on the face and adjoining parts of the skin having been habitually exposed to the air, light, and alternations of temperature, by which the small arteries not only have acquired the habit of readily dilating and contracting, but appear to have become unusually developed in comparison with other parts of the surface.[266] It is probably owing to this same cause, as M. Moreau and Dr. Burgess have remarked, that the face is so liable to redden under various circumstances, such as a fever-fit, ordinary heat, violent exertion, anger, a slight blow, &c.; and on the other hand that it is liable to grow pale from cold and fear, and to be discoloured during pregnancy. The face is also particularly liable to be affected by cutaneous complaints, by small-pox, erysipelas, &c. This view is likewise supported by the fact that the men of certain races, who habitually go nearly naked, often blush over their arms and chests and even down to their waists. A lady, who is a great blusher, informs Dr. Crichton Browne, that when she feels ashamed or is agitated, she blushes over her face, neck, wrists, and hands,—that is, over all the exposed portions of her skin. Nevertheless it may be doubted whether the habitual exposure of the skin of the face and neck, and its consequent power of reaction under stimulants of all kinds, is by itself sufficient to account for the much greater tendency in English women of these parts than of others to blush; for the hands are well supplied with nerves and small vessels, and have been as much exposed to the air as the face or neck, and yet the hands rarely blush. We shall presently see that the attention of the mind having been directed much more frequently and earnestly to the face than to any other part of the body, probably affords a sufficient explanation.

Blushing in the various races of man.—The small vessels of the face become filled with blood, from the emotion of shame, in almost all the races of man, though in the very dark races no distinct change of colour can be perceived. Blushing is evident in all the Aryan nations of Europe, and to a certain extent with those of India. But Mr. Erskine has never noticed that the necks of the Hindoos are decidedly affected. With the Lepchas of Sikhim, Mr. Scott has often observed a faint blush on the cheeks, base of the ears, and sides of the neck, accompanied by sunken eyes and lowered head. This has occurred when he has detected them in a falsehood, or has accused them of ingratitude. The pale, sallow complexions of these men render a blush much more conspicuous than in most of the other natives of India. With the latter, shame, or it may be in part fear, is expressed, according to Mr. Scott, much more plainly by the head being averted or bent down, with the eyes wavering or turned askant, than by any change of colour in the skin.

The Semitic races blush freely, as might have been expected, from their general similitude to the Aryans. Thus with the Jews, it is said in the Book of Jeremiah (chap. vi. 15), "Nay, they were not at all ashamed, neither could they blush." Mrs. Asa Gray saw an Arab managing his boat clumsily on the Nile, and when laughed at by his companions, "he blushed quite to the back of his neck." Lady Duff Gordon remarks that a young Arab blushed on coming into her presence.[267]

Mr. Swinhoe has seen the Chinese blushing, but he thinks it is rare; yet they have the expression "to redden with shame." Mr. Geach informs me that the Chinese settled in Malacca and the native Malays of the interior both blush. Some of these people go nearly naked, and he particularly attended to the downward extension of the blush. Omitting the cases in which the face alone was seen to blush, Mr. Geach observed that the face, arms, and breast of a Chinaman, aged 24 years, reddened from shame; and with another Chinese, when asked why he had not done his work in better style, the whole body was similarly affected. In two Malays[268] he saw the face, neck, breast, and arms blushing; and in a third Malay (a Bugis) the blush extended down to the waist.

The Polynesians blush freely. The Rev. Mr. Stack has seen hundreds of instances with the New Zealanders. The following case is worth giving, as it relates to an old man who was unusually dark-coloured and partly tattooed. After having let his land to an Englishman for a small yearly rental, a strong passion seized him to buy a gig, which had lately become the fashion with the Maoris. He consequently wished to draw all the rent for four years from his tenant, and consulted Mr. Stack whether he could do so. The man was old, clumsy, poor, and ragged, and the idea of his driving himself about in his carriage for display amused Mr. Stack so much that he could not help bursting out into a laugh; and then "the old man blushed up to the roots of his hair." Forster says that "you may easily distinguish a spreading blush" on the cheeks of the fairest women in Tahiti.[269] The natives also of several of the other archipelagoes in the Pacific have been seen to blush.

Mr. Washington Matthews has often seen a blush on the faces of the young squaws belonging to various wild Indian tribes of North America. At the opposite extremity of the continent in Tierra del Fuego, the natives, according to Mr. Bridges, "blush much, but chiefly in regard to women; but they certainly blush also at their own personal appearance." This latter statement agrees with what I remember of the Fuegian, Jemmy Button, who blushed when he was quizzed about the care which he took in polishing his shoes, and in otherwise adorning himself. With respect to the Aymara Indians on the lofty plateaus of Bolivia, Mr. Forbes says,[270] that from the colour of their skins it is impossible that their blushes should be as clearly visible as in the white races; still under such circumstances as would raise a blush in us, "there can always be seen the same expression of modesty or confusion; and even in the dark, a rise of temperature of the skin of the face can be felt, exactly as occurs in the European." With the Indians who inhabit the hot, equable, and damp parts of South America, the skin apparently does not answer to mental excitement so readily as with the natives of the northern and southern parts of the continent, who have long been exposed to great vicissitudes of climate; for Humboldt quotes without a protest the sneer of the Spaniard, "How can those be trusted, who know not how to blush?"[271] Von Spix and Martius, in speaking of the aborigines of Brazil, assert that they cannot properly be said to blush; "it was only after long intercourse with the whites, and after receiving some education, that we perceived in the Indians a change of colour expressive of the emotions of their minds."[272] It is, however, incredible that the power of blushing could have thus originated; but the habit of self-attention, consequent on their education and new course of life, would have much increased any innate tendency to blush.

Several trustworthy observers have assured me that they have seen on the faces of negroes an appearance resembling a blush, under circumstances which would have excited one in us, though their skins were of an ebony-black tint. Some describe it as blushing brown, but most say that the blackness becomes more intense. An increased supply of blood in the skin seems in some manner to increase its blackness; thus certain exanthematous diseases cause the affected places in the negro to appear blacker, instead of, as with us, redder.[273] The skin, perhaps, from being rendered more tense by the filling of the capillaries, would reflect a somewhat different tint to what it did before. That the capillaries of the face in the negro become filled with blood, under the emotion of shame, we may feel confident; because a perfectly characterized albino negress, described by Buffon,[274] showed a faint tinge of crimson on her cheeks when she exhibited herself naked. Cicatrices of the skin remain for a long time white in the negro, and Dr. Burgess, who had frequent opportunities of observing a scar of this kind on the face of a negress, distinctly saw that it "invariably became red whenever she was abruptly spoken to, or charged with any trivial offence."[275] The blush could be seen proceeding from the circumference of the scar towards the middle, but it did not reach the centre. Mulattoes are often great blushers, blush succeeding blush over their faces. From these facts there can be no doubt that negroes blush, although no redness is visible on the skin.

I am assured by Gaika and by Mrs. Barber that the Kafirs of South Africa never blush; but this may only mean that no change of colour is distinguishable. Gaika adds that under the circumstances which would make a European blush, his countrymen "look ashamed to keep their heads up."

It is asserted by four of my informants that the Australians, who are almost as black as negroes, never blush. A fifth answers doubtfully, remarking that only a very strong blush could be seen, on account of the dirty state of their skins. Three observers state that they do blush;[276] Mr. S. Wilson adding that this is noticeable only under a strong emotion, and when the skin is not too dark from long exposure and want of cleanliness. Mr. Lang answers, "I have noticed that shame almost always excites a blush, which frequently extends as low as the neck." Shame is also shown, as he adds, "by the eyes being turned from side to side." As Mr. Lang was a teacher in a native school, it is probable that he chiefly observed children; and we know that they blush more than adults. Mr. G. Taplin has seen half-castes blushing, and he says that the aborigines have a word expressive of shame. Mr. Hagenauer, who is one of those who has never observed the Australians to blush, says that he has "seen them looking down to the ground on account of shame;" and the missionary, Mr. Bulmer, remarks that though "I have not been able to detect anything like shame in the adult aborigines, I have noticed that the eyes of the children, when ashamed, present a restless, watery appearance, as if they did not know where to look."

The facts now given are sufficient to show that blushing, whether or not there is any change of colour, is common to most, probably to all, of the races of man.

Movements and gestures which accompany Blushing.—Under a keen sense of shame there is a strong desire for concealment.[277] We turn away the whole body, more especially the face, which we endeavour in some manner to hide. An ashamed person can hardly endure to meet the gaze of those present, so that he almost invariably casts down his eyes or looks askant. As there generally exists at the same time a strong wish to avoid the appearance of shame, a vain attempt is made to look direct at the person who causes this feeling; and the antagonism between these opposite tendencies leads to various restless movements in the eyes. I have noticed two ladies who, whilst blushing, to which they are very liable, have thus acquired, as it appears, the oddest trick of incessantly blinking their eyelids with extraordinary rapidity. An intense blush is sometimes accompanied by a slight effusion of tears;[278] and this, I presume, is due to the lacrymal glands partaking of the increased supply of blood, which we know rushes into the capillaries of the adjoining parts, including the retina.

Many writers, ancient and modern, have noticed the foregoing movements; and it has already been shown that the aborigines in various parts of the world often exhibit their shame by looking downwards or askant, or by restless movements of their eyes. Ezra cries out (ch. ix. 6), "O, my God! I am ashamed, and blush to lift up my head to thee, my God." In Isaiah (ch. l. 6) we meet with the words, "I hid not my face from shame." Seneca remarks (Epist. xi. 5) "that the Roman players hang down their heads, fix their eyes on the ground and keep them lowered, but are unable to blush in acting shame." According to Macrobius, who lived in the fifth century ('Saturnalia,' B. vii. c. 11), "Natural philosophers assert that nature being moved by shame spreads the blood before herself as a veil, as we see any one blushing often puts his hands before his face." Shakspeare makes Marcus ('Titus Andronicus,' act ii. sc. 5) say to his niece, "Ah! now thou turn'st away thy face for shame." A lady informs me that she found in the Lock Hospital a girl whom she had formerly known, and who had become a wretched castaway, and the poor creature, when approached, hid her face under the bedclothes, and could not be persuaded to uncover it. We often see little children, when shy or ashamed, turn away, and still standing up, bury their faces in their mother's gown; or they throw themselves face downwards on her lap.

Confusion of mind.—Most persons, whilst blushing intensely, have their mental powers confused. This is recognized in such common expressions as "she was covered with confusion." Persons in this condition lose their presence of mind, and utter singularly inappropriate remarks. They are often much distressed, stammer, and make awkward movements or strange grimaces. In certain cases involuntary twitchings of some of the facial muscles may be observed. I have been informed by a young lady, who blushes excessively, that at such times she does not even know what she is saying. When it was suggested to her that this might be due to her distress from the consciousness that her blushing was noticed, she answered that this could not be the case, "as she had sometimes felt quite as stupid when blushing at a thought in her own room."

I will give an instance of the extreme disturbance of mind to which some sensitive men are liable. A gentleman, on whom I can rely, assured me that he had been an eye-witness of the following scene:—A small dinner-party was given in honour of an extremely shy man, who, when he rose to return thanks, rehearsed the speech, which he had evidently learnt by heart, in absolute silence, and did not utter a single word; but he acted as if he were speaking with much emphasis. His friends, perceiving how the case stood, loudly applauded the imaginary bursts of eloquence, whenever his gestures indicated a pause, and the man never discovered that he had remained the whole time completely silent. On the contrary, he afterwards remarked to my friend, with much satisfaction, that he thought he had succeeded uncommonly well.

When a person is much ashamed or very shy, and blushes intensely, his heart beats rapidly and his breathing is disturbed. This can hardly fail to affect the circulation of the blood within the brain, and perhaps the mental powers. It seems however doubtful, judging from the still more powerful influence of anger and fear on the circulation, whether we can thus satisfactorily account for the confused state of mind in persons whilst blushing intensely.

The true explanation apparently lies in the intimate sympathy which exists between the capillary circulation of the surface of the head and face, and that of the brain. On applying to Dr. J. Crichton Browne for information, he has given me various facts bearing on this subject. When the sympathetic nerve is divided on one side of the head, the capillaries on this side are relaxed and become filled with blood, causing the skin to redden and to grow hot, and at the same time the temperature within the cranium on the same side rises. Inflammation of the membranes of the brain leads to the engorgement of the face, ears, and eyes with blood. The first stage of an epileptic fit appears to be the contraction of the vessels of the brain, and the first outward manifestation is an extreme pallor of countenance. Erysipelas of the head commonly induces delirium. Even the relief given to a severe headache by burning the skin with strong lotion, depends, I presume, on the same principle.

Dr. Browne has often administered to his patients the vapour of the nitrite of amyl,[279] which has the singular property of causing vivid redness of the face in from thirty to sixty seconds. This flushing resembles blushing in almost every detail: it begins at several distinct points on the face, and spreads till it involves the whole surface of the head, neck, and front of the chest; but has been observed to extend only in one case to the abdomen. The arteries in the retina become enlarged; the eyes glisten, and in one instance there was a slight effusion of tears. The patients are at first pleasantly stimulated, but, as the flushing increases, they become confused and bewildered. One woman to whom the vapour had often been administered asserted that, as soon as she grew hot, she grew *muddled.* With persons just commencing to blush it appears, judging from their bright eyes and lively behaviour, that their mental powers are somewhat stimulated. It is only when the blushing is excessive that the mind grows confused. Therefore it would seem that the capillaries of the face are affected, both during the inhalation of the nitrite of amyl and during blushing, before that part of the brain is affected on which the mental powers depend.

Conversely when the brain is primarily affected, the circulation of the skin is so in a secondary manner. Dr. Browne has frequently observed, as he informs me, scattered red blotches and mottlings on the chests of epileptic patients. In these cases, when the skin on the thorax or abdomen is gently rubbed with a pencil or other object, or, in strongly-marked cases, is merely touched by the finger, the surface becomes suffused in less than half a minute with bright red marks, which spread to some distance on each side of the touched point, and persist for several minutes. These are the *cerebral maculæ* of Trousseau; and they indicate, as Dr. Browne remarks, a highly modified condition of the cutaneous vascular system. If, then, there exists, as cannot be doubted, an intimate sympathy between the capillary circulation in that part of the brain on which our mental powers depend, and in the skin of the face, it is not surprising that the moral causes which induce intense blushing should likewise induce, independently of their own disturbing influence, much confusion of mind.

The Nature of the Mental States which induce Blushing.—These consist of shyness, shame, and modesty; the essential element in all being self-attention. Many reasons can be assigned for believing that originally self-attention directed to personal appearance, in relation to the opinion of others, was the exciting cause; the same effect being subsequently produced, through the force of association, by self-attention in relation to moral conduct. It is not the simple act of reflecting on our own appearance, but the thinking what others think of us, which excites a blush. In absolute solitude the most sensitive person would be quite indifferent about his appearance. We feel blame or disapprobation more acutely than approbation; and consequently depreciatory remarks or ridicule, whether of our appearance or conduct, causes us to blush much more readily than does praise. But undoubtedly praise and admiration are highly efficient: a pretty girl blushes when a man gazes intently at her, though she may know perfectly well that he is not depreciating her. Many children, as well as old and sensitive persons blush, when they are much praised. Hereafter the question will be discussed, how it has arisen that the consciousness that others are attending to our personal appearance should have led to the capillaries, especially those of the face, instantly becoming filled with blood.

My reasons for believing that attention directed to personal appearance, and not to moral conduct, has been the fundamental element in the acquirement of the habit of blushing, will now be given. They are separately light, but combined possess, as it appears to me, considerable weight. It is notorious that nothing makes a shy person blush so much as any remark, however slight, on his personal appearance. One cannot notice even the dress of a woman much given to blushing, without causing her face to crimson. It is sufficient to stare hard at some persons to make them, as Coleridge remarks, blush,—"account for that he who can."[280]

With the two albinos observed by Dr. Burgess,[281] "the slightest attempt to examine their peculiarities invariably" caused them to blush deeply. Women are much more sensitive about their personal appearance than men are, especially elderly women in comparison with elderly men, and they blush much more freely. The young of both sexes are much more sensitive on this same head than the old, and they also blush much more freely than the old. Children at a very early age do not blush; nor do they show those other signs of self-consciousness which generally accompany blushing; and it is one of their chief charms that they think nothing about what others think of them. At this early age they will stare at a stranger with a fixed gaze and unblinking eyes, as on an inanimate object, in a manner which we elders cannot imitate.

It is plain to every one that young men and women are highly sensitive to the opinion of each other with reference to their personal appearance; and they blush incomparably more in the presence of the opposite sex than in that of their own.[282] A young man, not very liable to blush, will blush intensely at any slight ridicule of his appearance from a girl whose judgment on any important subject he would disregard. No happy pair of young lovers, valuing each other's admiration and love more than anything else in the world, probably ever courted each other without many a blush. Even the barbarians of Tierra del Fuego, according to Mr. Bridges, blush "chiefly in regard to women, but certainly also at their own personal appearance."

Of all parts of the body, the face is most considered and regarded, as is natural from its being the chief seat of expression and the source of the voice. It is also the chief seat of beauty and of ugliness, and throughout the world is the most ornamented.[283] The face, therefore, will have been subjected during many generations to much closer and more earnest self-attention than any other part of the body; and in accordance with the principle here advanced we can understand why it should be the most liable to blush. Although exposure to alternations of temperature, &c., has probably much increased the power of dilatation and contraction in the capillaries of the face and adjoining parts, yet this by itself will hardly account for these parts blushing much more than the rest of the body; for it does not explain the fact of the hands rarely blushing. With Europeans the whole body tingles slightly when the face blushes intensely; and with the races of men who habitually go nearly naked, the blushes extend over a much larger surface than with us. These facts are, to a certain extent, intelligible, as the self-attention of primeval man, as well as of the existing races which still go naked, will not have been so exclusively confined to their faces, as is the case with the people who now go clothed.

We have seen that in all parts of the world persons who feel shame for some moral delinquency, are apt to avert, bend down, or hide their faces, independently of any thought about their personal appearance. The object can hardly be to conceal their blushes, for the face is thus averted or hidden under circumstances which exclude any desire to conceal shame, as when guilt is fully confessed and repented of. It is, however, probable that primeval man before he had acquired much moral sensitiveness would have been highly sensitive about his personal appearance, at least in reference to the other sex, and he would consequently have felt distress at any depreciatory remarks about his appearance; and this is one form of shame. And as the face is the part of the body which is most regarded, it is intelligible that any one ashamed of his personal appearance would desire to conceal this part of his body. The habit having been thus acquired, would naturally be carried on when shame from strictly moral causes was felt; and it is not easy otherwise to see why under these circumstances there should be a desire to hide the face more than any other part of the body.

The habit, so general with every one who feels ashamed, of turning away, or lowering his eyes, or restlessly moving them from side to side, probably follows from each glance directed towards those present, bringing home the conviction that he is intently regarded; and he endeavours, by not looking at those present, and especially not at their eyes, momentarily to escape from this painful conviction.

Shyness.—This odd state of mind, often called shamefacedness, or false shame, or *mauvaise honte*, appears to be one of the most efficient of all the causes of blushing. Shyness is, indeed, chiefly recognized by the face reddening, by the eyes being averted or cast down, and by awkward, nervous movements of the body. Many a woman blushes from this cause, a hundred, perhaps a thousand times, to once that she blushes from having done anything deserving blame, and of which she is truly ashamed. Shyness seems to depend on sensitiveness to the opinion, whether good or bad, of others, more especially with respect to external appearance. Strangers neither know nor care anything about our conduct or character, but they may, and often do, criticize our appearance: hence shy persons are particularly apt to be shy and to blush in the presence of strangers. The consciousness of anything peculiar, or even new, in the dress, or any slight blemish on the person, and more especially on the face—points which are likely to attract the attention of strangers—makes the shy intolerably shy. On the other hand, in those cases in which conduct and not personal appearance is concerned, we are much more apt to be shy in the presence of acquaintances, whose judgment we in some degree value, than in that of strangers. A physician told me that a young man, a wealthy duke, with whom he had travelled as medical attendant, blushed like a girl, when he paid him his fee; yet this young man probably would not have blushed and been shy, had he been paying a bill to a tradesman. Some persons, however, are so sensitive, that the mere act of speaking to almost any one is sufficient to rouse their self-consciousness, and a slight blush is the result.

Disapprobation or ridicule, from our sensitiveness on this head, causes shyness and blushing much more readily than does approbation; though the latter with some persons is highly efficient. The conceited are rarely shy; for they value themselves much too highly to expect depreciation. Why a proud man is often shy, as appears to be the case, is not so obvious, unless it be that, with all his self-reliance, he really thinks much about the opinion of others, although in a disdainful spirit. Persons who are exceedingly shy are rarely shy in the presence of those with whom they are quite familiar, and of whose good opinion and sympathy they are perfectly assured;—for instance, a girl in the presence of her mother. I neglected to inquire in my printed paper whether shyness can be detected in the different races of man; but a Hindoo gentleman assured Mr. Erskine that it is recognizable in his countrymen.

Shyness, as the derivation of the word indicates in several languages,[284] is closely related to fear; yet it is distinct from fear in the ordinary sense. A shy man no doubt dreads the notice of strangers, but can hardly be said to be afraid of them; he may be as bold as a hero in battle, and yet have no self-confidence about trifles in the presence of strangers. Almost every one is extremely nervous when first addressing a public assembly, and most men remain so throughout their lives; but this appears to depend on the consciousness of a great coming exertion, with its associated effects on the system, rather than on shyness;[285] although a timid or shy man no doubt suffers on such occasions infinitely more than another. With very young children it is difficult to distinguish between fear and shyness; but this latter feeling with them has often seemed to me to partake of the character of the wildness of an untamed animal. Shyness comes on at a very early age. In one of my own children, when two years and three months old, I saw a trace of what certainly appeared to be shyness, directed towards myself after an absence from home of only a week. This was shown not by a blush, but by the eyes being for a few minutes slightly averted from me. I have noticed on other occasions that shyness or shamefacedness and real shame are exhibited in the eyes of young children before they have acquired the power of blushing.

As shyness apparently depends on self-attention, we can perceive how right are those who maintain that reprehending children for shyness, instead of doing them any good, does much harm, as it calls their attention still more closely to themselves. It has been well urged that "nothing hurts young people more than to be watched continually about their feelings, to have their countenances scrutinized, and the degrees of their sensibility measured by the surveying eye of the unmerciful spectator. Under the constraint of such examinations they can think of nothing but that they are looked at, and feel nothing but shame or apprehension."[286]

Moral causes: guilt.—With respect to blushing from strictly moral causes, we meet with the same fundamental principle as before, namely, regard for the opinion of others. It is not the conscience which raises a blush, for a man may sincerely regret some slight fault committed in solitude, or he may suffer the deepest remorse for an undetected crime, but he will not blush. "I blush," says Dr. Burgess,[287] "in the presence of my accusers." It is not the sense of guilt, but the thought that others think or know us to be guilty which crimsons the face. A man may feel thoroughly ashamed at having told a small falsehood, without blushing; but if he even suspects that he is detected he will instantly blush, especially if detected by one whom he reveres.

On the other hand, a man may be convinced that God witnesses all his actions, and he may feel deeply conscious of some fault and pray for forgiveness; but this will not, as a lady who is a great blusher believes, ever excite a blush. The explanation of this difference between the knowledge by God and man of our actions lies, I presume, in man's disapprobation of immoral conduct being somewhat akin in nature to his depreciation of our personal appearance, so that through association both lead to similar results; whereas the disapprobation of God brings up no such association.

Many a person has blushed intensely when accused of some crime, though completely innocent of it. Even the thought, as the lady before referred to has observed to me, that others think that we have made an unkind or stupid remark, is amply sufficient to cause a blush, although we know all the time that we have been completely misunderstood. An action may be meritorious or of an indifferent nature, but a sensitive person, if he suspects that others take a different view of it, will blush. For instance, a lady by herself may give money to a beggar without a trace of a blush, but if others are present, and she doubts whether they approve, or suspects that they think her influenced by display, she will blush. So it will be, if she offers to relieve the distress of a decayed gentlewoman, more particularly of one whom she had previously known under better circumstances, as she cannot then feel sure how her conduct will be viewed. But such cases as these blend into shyness.

Breaches of etiquette.—The rules of *etiquette* always refer to conduct in the presence of, or towards others. They have no necessary connection with the moral sense, and are often meaningless. Nevertheless as they depend on the fixed custom of our equals and superiors, whose opinion we highly regard, they are considered almost as binding as are the laws of honour to a gentleman. Consequently the breach of the laws of etiquette, that is, any impoliteness or *gaucherie,* any impropriety, or an inappropriate remark, though quite accidental, will cause the most intense blushing of which a man is capable. Even the recollection of such an act, after an interval of many years, will make the whole body to tingle. So strong, also, is the power of sympathy that a sensitive person, as a lady has assured me, will sometimes blush at a flagrant breach of etiquette by a perfect stranger, though the act may in no way concern her.

Modesty.—This is another powerful agent in exciting blushes; but the word modesty includes very different states of the mind. It implies humility, and we often judge of this by persons being greatly pleased and blushing at slight praise, or by being annoyed at praise which seems to them too high according to their own humble standard of themselves. Blushing here has the usual signification of regard for the opinion of others. But modesty frequently relates to acts of indelicacy; and indelicacy is an affair of etiquette, as we clearly see with the nations that go altogether or nearly naked. He who is modest, and blushes easily at acts of this nature, does so because they are breaches of a firmly and wisely established etiquette. This is indeed shown by the derivation of the word *modest* from *modus,* a measure or standard of behaviour. A blush due to this form of modesty is, moreover, apt to be intense, because it generally relates to the opposite sex; and we have seen how in all cases our liability to blush is thus increased. We apply the term "modest," as it would appear, to those who have an humble opinion of themselves, and to those who are extremely sensitive about an indelicate word or deed, simply because in both cases blushes are readily excited, for these two frames of mind have nothing else in common. Shyness also, from this same cause, is often mistaken for modesty in the sense of humility.

Some persons flush up, as I have observed and have been assured, at any sudden and disagreeable recollection. The commonest cause seems to be the sudden remembrance of not having done something for another person which had been promised. In this case it may be that the thought passes half unconsciously through the mind, "What will he think of me?" and then the flush would partake of the nature of a true blush. But whether such flushes are in most cases due to the capillary circulation being affected, is very doubtful; for we must remember that almost every strong emotion, such as anger or great joy, acts on the heart, and causes the face to redden.

THE FACT THAT BLUSHES may be excited in absolute solitude seems opposed to the view here taken, namely that the habit originally arose from thinking about what others think of us. Several ladies, who are great blushers, are unanimous in regard to solitude; and some of them believe that they have blushed in the dark. From what Mr. Forbes has stated with respect to the Aymaras, and from my own sensations, I have no doubt that this latter statement is correct. Shakespeare, therefore, erred when he made Juliet, who was not even by herself, say to Romeo (act ii. sc. 2):—

> "Thou know'st the mask of night is on my face;
> Else would a maiden blush bepaint my cheek,
> For that which thou hast heard me speak to-night."

But when a blush is excited in solitude, the cause almost always relates to the thoughts of others about us—to acts done in their presence, or suspected by them; or again when we reflect what others would have thought of us had they known of the act. Nevertheless one or two of my informants believe that they have blushed from shame at acts in no way relating to others. If this be so, we must attribute the result to the force of inveterate habit and association, under a state of mind closely analogous to that which ordinarily excites a blush; nor need we feel surprise at this, as even sympathy with another person who commits a flagrant breach of etiquette is believed, as we have just seen, sometimes to cause a blush.

Finally, then, I conclude that blushing,—whether due to shyness—to shame for a real crime—to shame from a breach of the laws of etiquette—to modesty from humility—to modesty from an indelicacy—depends in all cases on the same principle; this principle being a sensitive regard for the opinion, more particularly for the depreciation of others, primarily in relation to our personal appearance, especially of our faces; and secondarily, through the force of association and habit, in relation to the opinion of others on our conduct.

Theory of Blushing.—We have now to consider, why should the thought that others are thinking about us affect our capillary circulation? Sir C. Bell insists[288] that blushing "is a provision for expression, as may be inferred from the colour extending only to the surface of the face, neck, and breast, the parts most exposed. It is not acquired; it is from the beginning." Dr. Burgess believes that it was designed by the Creator in "order that the soul might have sovereign power of displaying in the cheeks the various internal emotions of the moral feelings;" so as to serve as a check on ourselves, and as a sign to others, that we were violating rules which ought to be held sacred. Gratiolet merely remarks,—"Or, comme il est dans l'ordre de la nature que l'être social le plus intelligent soit aussi le plus intelligible, cette faculté de rougeur et de pâleur qui distingue l'homme, est un signe naturel de sa haute perfection."

The belief that blushing was *specially* designed by the Creator is opposed to the general theory of evolution, which is now so largely accepted; but it forms no part of my duty here to argue on the general question. Those who believe in design, will find it difficult to account for shyness being the most frequent and efficient of all the causes of blushing, as it makes the blusher to suffer and the beholder uncomfortable, without being of the least service to either of them. They will also find it difficult to account for negroes and other dark-coloured races blushing, in whom a change of colour in the skin is scarcely or not at all visible.

No doubt a slight blush adds to the beauty of a maiden's face; and the Circassian women who are capable of blushing, invariably fetch a higher price in the seraglio of the Sultan than less susceptible women.[289] But the firmest believer in the efficacy of sexual selection will hardly suppose that blushing was acquired as a sexual ornament. This view would also be opposed to what has just been said about the dark-coloured races blushing in an invisible manner.

The hypothesis which appears to me the most probable, though it may at first seem rash, is that attention closely directed to any part of the body tends to interfere with the ordinary and tonic contraction of the small arteries of that part. These vessels, in consequence, become at such times more or less relaxed, and are instantly filled with arterial blood. This tendency will have been much strengthened, if frequent attention has been paid during many generations to the same part, owing to nerve-force readily flowing along accustomed channels, and by the power of inheritance. Whenever we believe that others are depreciating or even considering our personal appearance, our attention is vividly directed to the outer and visible parts of our bodies; and of all such parts we are most sensitive about our faces, as no doubt has been the case during many past generations. Therefore, assuming for the moment that the capillary vessels can be acted on by close attention, those of the face will have become eminently susceptible. Through the force of association, the same effects will tend to follow whenever we think that others are considering or censuring our actions or character.

As the basis of this theory rests on mental attention having some power to influence the capillary circulation, it will be necessary to give a considerable body of details, bearing more or less directly on this subject. Several observers,[290] who from their wide experience and knowledge are eminently capable of forming a sound judgment, are convinced that attention or consciousness (which latter term Sir H. Holland thinks the more explicit) concentrated on almost any part of the body produces some direct physical effect on it. This applies to the movements of the involuntary muscles, and of the voluntary muscles when acting involuntarily,—to the secretion of the glands,—to the activity of the senses and sensations,—and even to the nutrition of parts.

It is known that the involuntary movements of the heart are affected if close attention be paid to them. Gratiolet[291] gives the case of a man, who by continually watching and counting his own pulse, at last caused one beat out of every six to intermit. On the other hand, my father told me of a careful observer, who certainly had heart-disease and died from it, and who positively stated that his pulse was habitually irregular to an extreme degree; yet to his great disappointment it invariably became regular as soon as my father entered the room. Sir H. Holland remarks,[292] that "the effect upon the circulation of a part from the consciousness suddenly directed and fixed upon it, is often obvious and immediate." Professor Laycock, who has particularly attended to phenomena of this nature,[293] insists that "when the attention is directed to any portion of the body, innervation and circulation are excited locally, and the functional activity of that portion developed."

It is generally believed that the peristaltic movements of the intestines are influenced by attention being paid to them at fixed recurrent periods; and these movements depend on the contraction of unstriped and involuntary muscles. The abnormal action of the voluntary muscles in epilepsy, chorea, and hysteria is known to be influenced by the expectation of an attack, and by the sight of other patients similarly affected.[294] So it is with the involuntary acts of yawning and laughing.

Certain glands are much influenced by thinking of them, or of the conditions under which they have been habitually excited. This is familiar to every one in the increased flow of saliva, when the thought, for instance, of intensely acid fruit is kept before the mind.[295] It was shown in our sixth chapter, that an earnest and long-continued desire either to repress, or to increase, the action of the lacrymal glands is effectual. Some curious cases have been recorded in the case of women, of the power of the mind on the mammary glands; and still more remarkable ones in relation to the uterine functions.[296]

When we direct our whole attention to any one sense, its acuteness is increased;[297] and the continued habit of close attention, as with blind people to that of hearing, and with the blind and deaf to that of touch, appears to improve the sense in question permanently. There is, also, some reason to believe, judging from the capacities of different races of man, that the effects are inherited. Turning to ordinary sensations, it is well known that pain is increased by attending to it; and Sir B. Brodie goes so far as to believe that pain may be felt in any part of the body to which attention is closely drawn.[298] Sir H. Holland also remarks that we become not only conscious of the existence of a part subjected to concentrated attention, but we experience in it various odd sensations, as of weight, heat, cold, tingling, or itching.[299]

Lastly, some physiologists maintain that the mind can influence the nutrition of parts. Sir J. Paget has given a curious instance of the power, not indeed of the mind, but of the nervous system, on the hair. A lady "who is subject to attacks of what is called nervous headache, always finds in the morning after such an one, that some patches of her hair are white, as if powdered with starch. The change is effected in a night, and in a few days after, the hairs gradually regain their dark brownish colour."[300]

We thus see that close attention certainly affects various parts and organs, which are not properly under the control of the will. By what means attention—perhaps the most wonderful of all the wondrous powers of the mind—is effected, is an extremely obscure subject. According to Müller,[301] the process by which the sensory cells of the brain are rendered, through the will, susceptible of receiving more intense and distinct impressions, is closely analogous to that by which the motor cells are excited to send nerve-force to the voluntary muscles. There are many points of analogy in the action of the sensory and motor nerve-cells; for instance, the familiar fact that close attention to any one sense causes fatigue, like the prolonged exertion of any one muscle.[302] When therefore we voluntarily concentrate our attention on any part of the body, the cells of the brain which receive impressions or sensations from that part are, it is probable, in some unknown manner stimulated into activity. This may account, without any local change in the part to which our attention is earnestly directed, for pain or odd sensations being there felt or increased.

If, however, the part is furnished with muscles, we cannot feel sure, as

Mr. Michael Foster has remarked to me, that some slight impulse may not be unconsciously sent to such muscles; and this would probably cause an obscure sensation in the part.

In a large number of cases, as with the salivary and lacrymal glands, intestinal canal, &c., the power of attention seems to rest, either chiefly, or as some physiologists think, exclusively, on the vaso-motor system being affected in such a manner that more blood is allowed to flow into the capillaries of the part in question. This increased action of the capillaries may in some cases be combined with the simultaneously increased activity of the sensorium.

The manner in which the mind affects the vaso-motor system may be conceived in the following manner. When we actually taste sour fruit, an impression is sent through the gustatory nerves to a certain part of the sensorium; this transmits nerve-force to the vaso-motor centre, which consequently allows the muscular coats of the small arteries that permeate the salivary glands to relax. Hence more blood flows into these glands, and they secrete a copious supply of saliva. Now it does not seem an improbable assumption, that, when we reflect intently on a sensation, the same part of the sensorium, or a closely connected part of it, is brought into a state of activity, in the same manner as when we actually perceive the sensation. If so, the same cells in the brain will be excited, though, perhaps, in a less degree, by vividly thinking about a sour taste, as by perceiving it; and they will transmit in the one case, as in the other, nerve-force to the vaso-motor centre with the same results.

To give another, and, in some respects, more appropriate illustration. If a man stands before a hot fire, his face reddens. This appears to be due, as Mr. Michael Foster informs me, in part to the local action of the heat, and in part to a reflex action from the vaso-motor centres.[303] In this latter case, the heat affects the nerves of the face; these transmit an impression to the sensory cells of the brain, which act on the vaso-motor centre, and this reacts on the small arteries of the face, relaxing them and allowing them to become filled with blood. Here, again, it seems not improbable that if we were repeatedly to concentrate with great earnestness our attention on the recollection of our heated faces, the same part of the sensorium which gives us the consciousness of actual heat would be in some slight degree stimulated, and would in consequence tend to transmit some nerve-force to the vaso-motor centres, so as to relax the capillaries of the face. Now as men during endless generations have had their attention often and earnestly directed to their personal appearance, and especially to their faces, any incipient tendency in the facial capillaries to be thus affected will have become in the course of time greatly strengthened through the principles just referred to, namely, nerve-force passing readily along accustomed channels, and inherited habit. Thus, as it appears to me, a plausible explanation is afforded of the leading phenomena connected with the act of blushing.

Recapitulation.—Men and women, and especially the young, have always valued, in a high degree, their personal appearance; and have likewise regarded the appearance of others. The face has been the chief object of attention, though, when man aboriginally went naked, the whole surface of his body would have been attended to. Our self-attention is excited almost exclusively by the opinion of others, for no person living in absolute solitude would care about his appearance. Every one feels blame more acutely than praise. Now, whenever we know, or suppose, that others are depreciating our personal appearance, our attention is strongly drawn towards ourselves, more especially to our faces. The probable effect of this will be, as has just been explained, to excite into activity that part of the sensorium which receives the sensory nerves of the face; and this will react through the vaso-motor system on the facial capillaries. By frequent reiteration during numberless generations, the process will have become so habitual, in association with the belief that others are thinking of us, that even a suspicion of their depreciation suffices to relax the capillaries, without any conscious thought about our faces. With some sensitive persons it is enough even to notice their dress to produce the same effect. Through the force, also, of association and inheritance our capillaries are relaxed, whenever we know, or imagine, that any one is blaming, though in silence, our actions, thoughts, or character; and, again, when we are highly praised.

On this hypothesis we can understand how it is that the face blushes much more than any other part of the body, though the whole surface is somewhat affected, more especially with the races which still go nearly naked. It is not at all surprising that the dark-coloured races should blush, though no change of colour is visible in their skins. From the principle of inheritance it is not surprising that persons born blind should blush. We can understand why the young are much more affected than the old, and women more than men; and why the opposite sexes especially excite each other's blushes. It becomes obvious why personal remarks should be particularly liable to cause blushing, and why the most powerful of all the causes is shyness; for shyness relates to the presence and opinion of others, and the shy are always more or less self-conscious. With respect to real shame from moral delinquencies, we can perceive why it is not guilt, but the thought that others think us guilty, which raises a blush. A man reflecting on a crime committed in solitude, and stung by his conscience, does not blush; yet he will blush under the vivid recollection of a detected fault, or of one committed in the presence of others, the degree of blushing being closely related to the feeling of regard for those who have detected, witnessed, or suspected his fault. Breaches of conventional rules of conduct, if they are rigidly insisted on by our equals or superiors, often cause more intense blushes even than a detected crime; and an act which is really criminal, if not blamed by our equals, hardly raises a tinge of colour on our cheeks. Modesty from humility, or from an indelicacy, excites a vivid blush, as both relate to the judgment or fixed customs of others.

From the intimate sympathy which exists between the capillary circulation of the surface of the head and of the brain, whenever there is intense blushing, there will be some, and often great, confusion of mind. This is frequently accompanied by awkward movements, and sometimes by the involuntary twitching of certain muscles.

As blushing, according to this hypothesis, is an indirect result of attention, originally directed to our personal appearance, that is to the surface of the body, and more especially to the face, we can understand the meaning of the gestures which accompany blushing throughout the world. These consist in hiding the face, or turning it towards the ground, or to one side. The eyes are generally averted or are restless, for to look at the man who causes us to feel shame or shyness, immediately brings home in an intolerable manner the consciousness that his gaze is directed on us. Through the principle of associated habit, the same movements of the face and eyes are practised, and can, indeed, hardly be avoided, whenever we know or believe that others are blaming, or too strongly praising, our moral conduct.

CHAPTER XIV

Concluding Remarks and Summary

THE THREE LEADING PRINCIPLES WHICH HAVE DETERMINED THE CHIEF MOVEMENTS OF EXPRESSION • THEIR INHERITANCE • ON THE PART WHICH THE WILL AND INTENTION HAVE PLAYED IN THE ACQUIREMENT OF VARIOUS EXPRESSIONS • THE INSTINCTIVE RECOGNITION OF EXPRESSION • THE BEARING OF OUR SUBJECT ON THE SPECIFIC UNITY OF THE RACES OF MAN • ON THE SUCCESSIVE ACQUIREMENT OF VARIOUS EXPRESSIONS BY THE PROGENITORS OF MAN • THE IMPORTANCE OF EXPRESSION • CONCLUSION

I HAVE NOW DESCRIBED, to the best of my ability, the chief expressive actions in man, and in some few of the lower animals. I have also attempted to explain the origin or development of these actions through the three principles given in the first chapter. The first of these principles is, that movements which are serviceable in gratifying some desire, or in relieving some sensation, if often repeated, become so habitual that they are performed, whether or not of any service, whenever the same desire or sensation is felt, even in a very weak degree.

Our second principle is that of antithesis. The habit of voluntarily performing opposite movements under opposite impulses has become firmly established in us by the practice of our whole lives. Hence, if certain actions have been regularly performed, in accordance with our first principle, under a certain frame of mind, there will be a strong and involuntary tendency to the performance of directly opposite actions, whether or not these are of any use, under the excitement of an opposite frame of mind.

Our third principle is the direct action of the excited nervous system on the body, independently of the will, and independently, in large part, of habit. Experience shows that nerve-force is generated and set free whenever the cerebro-spinal system is excited. The direction which this nerve-force follows is necessarily determined by the lines of connection between the nerve-cells, with each other and with various parts of the body. But the direction is likewise much influenced by habit; inasmuch as nerve-force passes readily along accustomed channels.

The frantic and senseless actions of an enraged man may be attributed in part to the undirected flow of nerve-force, and in part to the effects of habit, for these actions often vaguely represent the act of striking. They thus pass into gestures included under our first principle; as when an indignant man unconsciously throws himself into a fitting attitude for attacking his opponent, though without any intention of making an actual attack. We see also the influence of habit in all the emotions and sensations which are called exciting; for they have assumed this character from having habitually led to energetic action; and action affects, in an indirect manner, the respiratory and circulatory system; and the latter reacts on the brain. Whenever these emotions or sensations are even slightly felt by us, though they may not at the time lead to any exertion, our whole system is nevertheless disturbed through the force of habit and association. Other emotions and sensations are called depressing, because they have not habitually led to energetic action, excepting just at first, as in the case of extreme pain, fear, and grief, and they have ultimately caused complete exhaustion; they are consequently expressed chiefly by negative signs and by prostration. Again, there are other emotions, such as that of affection, which do not commonly lead to action of any kind, and consequently are not exhibited by any strongly marked outward signs. Affection indeed, in as far as it is a pleasurable sensation, excites the ordinary signs of pleasure.

On the other hand, many of the effects due to the excitement of the nervous system seem to be quite independent of the flow of nerve-force along the channels which have been rendered habitual by former exertions of the will. Such effects, which often reveal the state of mind of the person thus affected, cannot at present be explained; for instance, the change of colour in the hair from extreme terror or grief,—the cold sweat and the trembling of the muscles from fear, —the modified secretions of the intestinal canal,—and the failure of certain glands to act.

Notwithstanding that much remains unintelligible in our present subject, so many expressive movements and actions can be explained to a certain extent through the above three principles, that we may hope hereafter to see all explained by these or by closely analogous principles.

Actions of all kinds, if regularly accompanying any state of the mind, are at once recognised as expressive. These may consist of movements of any part of the body, as the wagging of a dog's tail, the shrugging of a man's shoulders, the erection of the hair, the exudation of perspiration, the state of the capillary circulation, laboured breathing, and the use of the vocal or other sound-producing instruments. Even insects express anger, terror, jealousy, and love by their stridulation. With man the respiratory organs are of especial importance in expression, not only in a direct, but in a still higher degree in an indirect manner.

Few points are more interesting in our present subject than the extraordinarily complex chain of events which lead to certain expressive movements. Take, for instance, the oblique eyebrows of a man suffering from grief or anxiety. When infants scream loudly from hunger or pain, the circulation is affected, and the eyes tend to become gorged with blood: consequently the muscles surrounding the eyes are strongly contracted as a protection: this action, in the course of many generations, has become firmly fixed and inherited: but when, with advancing years and culture, the habit of screaming is partially repressed, the muscles round the eyes still tend to contract, whenever even slight distress is felt: of these muscles, the pyramidals of the nose are less under the control of the will than are the others, and their contraction can be checked only by that of the central fasciæ of the frontal muscle: these latter fasciæ draw up the inner ends of the eyebrows, and wrinkle the forehead in a peculiar manner, which we instantly recognise as the expression of grief or anxiety. Slight movements, such as these just described, or the scarcely perceptible drawing down of the corners of the mouth, are the last remnants or rudiments of strongly marked and intelligible movements. They are as full of significance to us in regard to expression, as are ordinary rudiments to the naturalist in the classification and genealogy of organic beings.

That the chief expressive actions, exhibited by man and by the lower animals, are now innate or inherited,—that is, have not been learnt by the individual,—is admitted by every one. So little has learning or imitation to do with several of them that they are from the earliest days and throughout life quite beyond our control; for instance, the relaxation of the arteries of the skin in blushing, and the increased action of the heart in anger. We may see children, only two or three years old, and even those born blind, blushing from shame; and the naked scalp of a very young infant reddens from passion. Infants scream from pain directly after birth, and all their features then assume the same form as during subsequent years. These facts alone suffice to show that many of our most important expressions have not been learnt; but it is remarkable that some, which are certainly innate, require practice in the individual, before they are performed in a full and perfect manner; for instance, weeping and laughing. The inheritance of most of our expressive actions explains the fact that those born blind display them, as I hear from the Rev. R. H. Blair, equally well with those gifted with eyesight. We can thus also understand the fact that the young and the old of widely different races, both with man and animals, express the same state of mind by the same movements.

We are so familiar with the fact of young and old animals displaying their feelings in the same manner, that we hardly perceive how remarkable it is that a young puppy should wag its tail when pleased, depress its ears and uncover its canine teeth when pretending to be savage, just like an old dog; or that a kitten should arch its little back and erect its hair when frightened and angry, like an old cat. When, however, we turn to less common gestures in ourselves, which we are accustomed to look at as artificial or conventional,—such as shrugging the shoulders, as a sign of impotence, or the raising the arms with open hands and extended fingers, as a sign of wonder,—we feel perhaps too much surprise at finding that they are innate. That these and some other gestures are inherited, we may infer from their being performed by very young children, by those born blind, and by the most widely distinct races of man. We should also bear in mind that new and highly peculiar tricks, in association with certain states of the mind, are known to have arisen in certain individuals, and to have been afterwards transmitted to their offspring, in some cases, for more than one generation.

Certain other gestures, which seem to us so natural that we might easily imagine that they were innate, apparently have been learnt like the words of a language. This seems to be the case with the joining of the uplifted hands, and the turning up of the eyes, in prayer. So it is with kissing as a mark of affection; but this is innate, in so far as it depends on the pleasure derived from contact with a beloved person. The evidence with respect to the inheritance of nodding and shaking the head, as signs of affirmation and negation, is doubtful; for they are not universal, yet seem too general to have been independently acquired by all the individuals of so many races.

WE WILL NOW CONSIDER how far the will and consciousness have come into play in the development of the various movements of expression. As far as we can judge, only a few expressive movements, such as those just referred to, are learnt by each individual; that is, were consciously and voluntarily performed during the early years of life for some definite object, or in imitation of others, and then became habitual. The far greater number of the movements of expression, and all the more important ones, are, as we have seen, innate or inherited; and such cannot be said to depend on the will of the individual. Nevertheless, all those included under our first principle were at first voluntarily performed for a definite object,—namely, to escape some danger, to relieve some distress, or to gratify some desire. For instance, there can hardly be a doubt that the animals which fight with their teeth, have acquired the habit of drawing back their ears closely to their heads, when feeling savage, from their progenitors having voluntarily acted in this manner in order to protect their ears from being torn by their antagonists; for those animals which do not fight with their teeth do not thus express a savage state of mind. We may infer as highly probable that we ourselves have acquired the habit of contracting the muscles round the eyes, whilst crying gently, that is, without the utterance of any loud sound, from our progenitors, especially during infancy, having experienced, during the act of screaming, an uncomfortable sensation in their eyeballs. Again, some highly expressive movements result from the endeavour to check or prevent other expressive movements; thus the obliquity of the eyebrows and the drawing down of the corners of the mouth follow from the endeavour to prevent a screaming-fit from coming on, or to check it after it has come on. Here it is obvious that the consciousness and will must at first have come into play; not that we are conscious in these or in other such cases what muscles are brought into action, any more than when we perform the most ordinary voluntary movements.

With respect to the expressive movements due to the principle of antithesis, it is clear that the will has intervened, though in a remote and indirect manner. So again with the movements coming under our third principle; these, in as far as they are influenced by nerve-force readily passing along habitual channels, have been determined by former and repeated exertions of the will. The effects indirectly due to this latter agency are often combined in a complex manner, through the force of habit and association, with those directly resulting from the excitement of the cerebro-spinal system. This seems to be the case with the increased action of the heart under the influence of any strong emotion. When an animal erects its hair, assumes a threatening attitude, and utters fierce sounds, in order to terrify an enemy, we see a curious combination of movements which were originally voluntary with those that are involuntary. It is, however, possible that even strictly involuntary actions, such as the erection of the hair, may have been affected by the mysterious power of the will.

Some expressive movements may have arisen spontaneously, in association with certain states of the mind, like the tricks lately referred to, and afterwards been inherited. But I know of no evidence rendering this view probable.

The power of communication between the members of the same tribe by means of language has been of paramount importance in the development of man; and the force of language is much aided by the expressive movements of the face and body. We perceive this at once when we converse on an important subject with any person whose face is concealed. Nevertheless there are no grounds, as far as I can discover, for believing that any muscle has been developed or even modified exclusively for the sake of expression. The vocal and other sound-producing organs, by which various expressive noises are produced, seem to form a partial exception; but I have elsewhere attempted to show that these organs were first developed for sexual purposes, in order that one sex might call or charm the other. Nor can I discover grounds for believing that any inherited movement, which now serves as a means of expression, was at first voluntarily and consciously performed for this special purpose,—like some of the gestures and the finger-language used by the deaf and dumb. On the contrary, every true or inherited movement of expression seems to have had some natural and independent origin. But when once acquired, such movements may be voluntarily and consciously employed as a means of communication. Even infants, if carefully attended to, find out at a very early age that their screaming brings relief, and they soon voluntarily practise it. We may frequently see a person voluntarily raising his eyebrows to express surprise, or smiling to express pretended satisfaction and acquiescence. A man often wishes to make certain gestures conspicuous or demonstrative, and will raise his extended arms with widely opened fingers above his head, to show astonishment, or lift his shoulders to his ears, to show that he cannot or will not do something. The tendency to such movements will be strengthened or increased by their being thus voluntarily and repeatedly performed; and the effects may be inherited.

It is perhaps worth consideration whether movements at first used only by one or a few individuals to express a certain state of mind may not sometimes have spread to others, and ultimately have become universal, through the power of conscious and unconscious imitation. That there exists in man a strong tendency to imitation, independently of the conscious will, is certain. This is exhibited in the most extraordinary manner in certain brain diseases, especially at the commencement of inflammatory softening of the brain, and has been called the "echo sign." Patients thus affected imitate, without understanding, every absurd gesture which is made, and every word which is uttered near them, even in a foreign language.[304] In the case of animals, the jackal and wolf have learnt under confinement to imitate the barking of the dog. How the barking of the dog, which serves to express various emotions and desires, and which is so remarkable from having been acquired since the animal was domesticated, and from being inherited in different degrees by different breeds, was first learnt, we do not know; but may we not suspect that imitation has had something to do with its acquisition, owing to dogs having long lived in strict association with so loquacious an animal as man?

In the course of the foregoing remarks and throughout this volume, I have often felt much difficulty about the proper application of the terms, will, consciousness, and intention. Actions, which were at first voluntary, soon become habitual, and at last hereditary, and may then be performed even in opposition to the will. Although they often reveal the state of the mind, this result was not at first either intended or expected. Even such words as that "certain movements serve as a means of expression" are apt to mislead, as they imply that this was their primary purpose or object. This, however, seems rarely or never to have been the case; the movements having been at first either of some direct use, or the indirect effect of the excited state of the sensorium. An infant may scream either intentionally or instinctively to show that it wants food; but it has no wish or intention to draw its features into the peculiar form which so plainly indicates misery; yet some of the most characteristic expressions exhibited by man are derived from the act of screaming, as has been explained.

Although most of our expressive actions are innate or instinctive, as is admitted by everyone, it is a different question whether we have any instinctive power of recognising them. This has generally been assumed to be the case; but the assumption has been strongly controverted by M. Lemoine.[305] Monkeys soon learn to distinguish, not only the tones of voice of their masters, but the expression of their faces, as is asserted by a careful observer.[306] Dogs well know the difference between caressing and threatening gestures or tones; and they seem to recognise a compassionate tone. But as far as I can make out, after repeated trials, they do not understand any movement confined to the features, excepting a smile or laugh; and this they appear, at least in some cases, to recognise. This limited amount of knowledge has probably been gained, both by monkeys and dogs, through their associating harsh or kind treatment with our actions; and the knowledge certainly is not instinctive. Children, no doubt, would soon learn the movements of expression in their elders in the same manner as animals learn those of man. Moreover, when a child cries or laughs, he knows in a general manner what he is doing and what he feels; so that a very small exertion of reason would tell him what crying or laughing meant in others. But the question is, do our children acquire their knowledge of expression solely by experience through the power of association and reason?

As most of the movements of expression must have been gradually acquired, afterwards becoming instinctive, there seems to be some degree of *à priori* probability that their recognition would likewise have become instinctive. There is, at least, no greater difficulty in believing this than in admitting that, when a female quadruped first bears young, she knows the cry of distress of her offspring, or than in admitting that many animals instinctively recognise and fear their enemies; and of both these statements there can be no reasonable doubt. It is however extremely difficult to prove that our children instinctively recognise any expression. I attended to this point in my first-born infant, who could not have learnt anything by associating with other children, and I was convinced that he understood a smile and received pleasure from seeing one, answering it by another, at much too early an age to have learnt anything by experience. When this child was about four months old, I made in his presence many odd noises and strange grimaces, and tried to look savage; but the noises, if not too loud, as well as the grimaces, were all taken as good jokes; and I attributed this at the time to their being preceded or accompanied by smiles. When five months old, he seemed to understand a compassionate expression and tone of voice. When a few days over six months old, his nurse pretended to cry, and I saw that his face instantly assumed a melancholy expression, with the corners of the mouth strongly depressed; now this child could rarely have seen any other child crying, and never a grown-up person crying, and I should doubt whether at so early an age he could have reasoned on the subject. Therefore it seems to me that an innate feeling must have told him that the pretended crying of his nurse expressed grief; and this through the instinct of sympathy excited grief in him.

M. Lemoine argues that, if man possessed an innate knowledge of expression, authors and artists would not have found it so difficult, as is notoriously the case, to describe and depict the characteristic signs of each particular state of mind. But this does not seem to me a valid argument. We may actually behold the expression changing in an unmistakable manner in a man or animal, and yet be quite unable, as I know from experience, to analyse the nature of the change. In the two photographs given by Duchenne of the same old man (Plate III, figs. 5 and 6), almost every one recognised that the one represented a true, and the other a false smile; but I have found it very difficult to decide in what the whole amount of difference consists. It has often struck me as a curious fact that so many shades of expression are instantly recognised without any conscious process of analysis on our part. No one, I believe, can clearly describe a sullen or sly expression; yet many observers are unanimous that these expressions can be recognised in the various races of man. Almost everyone to whom I showed Duchenne's photograph of the young man with oblique eyebrows (Plate II, fig. 2) at once declared that it expressed grief or some such feeling; yet probably not one of these persons, or one out of a thousand persons, could beforehand have told anything precise about the obliquity of the eyebrows with their inner ends puckered, or about the rectangular furrows on the forehead. So it is with many other expressions, of which I have had practical experience in the trouble requisite in instructing others what points to observe. If, then, great ignorance of details does not prevent our recognising with certainty and promptitude various expressions, I do not see how this ignorance can be advanced as an argument that our knowledge, though vague and general, is not innate.

I have endeavoured to show in considerable detail that all the chief expressions exhibited by man are the same throughout the world. This fact is interesting, as it affords a new argument in favour of the several races being descended from a single parent-stock, which must have been almost completely human in structure, and to a large extent in mind, before the period at which the races diverged from each other. No doubt similar structures, adapted for the same purpose, have often been independently acquired through variation and natural selection by distinct species; but this view will not explain close similarity between distinct species in a multitude of unimportant details. Now if we bear in mind the numerous points of structure having no relation to expression, in which all the races of man closely agree, and then add to them the numerous points, some of the highest importance and many of the most trifling value, on which the movements of expression directly or indirectly depend, it seems to me improbable in the highest degree that so much similarity, or rather identity of structure, could have been acquired by independent means. Yet this must have been the case if the races of man are descended from several aboriginally distinct species. It is far more probable that the many points of close similarity in the various races are due to inheritance from a single parent-form, which had already assumed a human character.

It is a curious, though perhaps an idle speculation, how early in the long line of our progenitors the various expressive movements, now exhibited by man, were successively acquired. The following remarks will at least serve to recall some of the chief points discussed in this volume. We may confidently believe that laughter, as a sign of pleasure or enjoyment, was practised by our progenitors long before they deserved to be called human; for very many kinds of monkeys, when pleased, utter a reiterated sound, clearly analogous to our laughter, often accompanied by vibratory movements of their jaws or lips, with the corners of the mouth drawn backwards and upwards, by the wrinkling of the cheeks, and even by the brightening of the eyes.

We may likewise infer that fear was expressed from an extremely remote period, in almost the same manner as it now is by man; namely, by trembling, the erection of the hair, cold perspiration, pallor, widely opened eyes, the relaxation of most of the muscles, and by the whole body cowering downwards or held motionless.

Suffering, if great, will from the first have caused screams or groans to be uttered, the body to be contorted, and the teeth to be ground together. But our progenitors will not have exhibited those highly expressive movements of the features which accompany screaming and crying until their circulatory and respiratory organs, and the muscles surrounding the eyes, had acquired their present structure. The shedding of tears appears to have originated through reflex action from the spasmodic contraction of the eyelids, together perhaps with the eyeballs becoming gorged with blood during the act of screaming. Therefore weeping probably came on rather late in the line of our descent; and this conclusion agrees with the fact that our nearest allies, the anthropomorphous apes, do not weep. But we must here exercise some caution, for as certain monkeys, which are not closely related to man, weep, this habit might have been developed long ago in a sub-branch of the group from which man is derived. Our early progenitors, when suffering from grief or anxiety, would not have made their eyebrows oblique, or have drawn down the corners of their mouth, until they had acquired the habit of endeavouring to restrain their screams. The expression, therefore, of grief and anxiety is eminently human.

Rage will have been expressed at a very early period by threatening or frantic gestures, by the reddening of the skin, and by glaring eyes, but not by frowning. For the habit of frowning seems to have been acquired chiefly from the corrugators being the first muscles to contract round the eyes, whenever during infancy pain, anger, or distress is felt, and there consequently is a near approach to screaming; and partly from a frown serving as a shade in difficult and intent vision. It seems probable that this shading action would not have become habitual until man had assumed a completely upright position, for monkeys do not frown when exposed to a glaring light. Our early progenitors, when enraged, would probably have exposed their teeth more freely than does man, even when giving full vent to his rage, as with the insane. We may, also, feel almost certain that they would have protruded their lips, when sulky or disappointed, in a greater degree than is the case with our own children, or even with the children of existing savage races.

Our early progenitors, when indignant or moderately angry, would not have held their heads erect, opened their chests, squared their shoulders, and clenched their fists, until they had acquired the ordinary carriage and upright attitude of man, and had learnt to fight with their fists or clubs. Until this period had arrived the antithetical gesture of shrugging the shoulders, as a sign of impotence or of patience, would not have been developed. From the same reason astonishment would not then have been expressed by raising the arms with open hands and extended fingers. Nor, judging from the actions of monkeys, would astonishment have been exhibited by a widely open mouth; but the eyes would have been opened and the eyebrows arched. Disgust would have been shown at a very early period by movements round the mouth, like those of vomiting,—that is, if the view which I have suggested respecting the source of the expression is correct, namely, that our progenitors had the power, and used it, of voluntarily and quickly rejecting any food from their stomachs which they disliked. But the more refined manner of showing contempt or disdain, by lowering the eyelids, or turning away the eyes and face, as if the despised person were not worth looking at, would not probably have been acquired until a much later period.

Of all expressions, blushing seems to be the most strictly human; yet it is common to all or nearly all the races of man, whether or not any change of colour is visible in their skin. The relaxation of the small arteries of the surface, on which blushing depends, seems to have primarily resulted from earnest attention directed to the appearance of our own persons, especially of our faces, aided by habit, inheritance, and the ready flow of nerve-force along accustomed channels; and afterwards to have been extended by the power of association to self-attention directed to moral conduct. It can hardly be doubted that many animals are capable of appreciating beautiful colours and even forms, as is shown by the pains which the individuals of one sex take in displaying their beauty before those of the opposite sex. But it does not seem possible that any animal, until its mental powers had been developed to an equal or nearly equal degree with those of man, would have closely considered and been sensitive about its own personal appearance. Therefore we may conclude that blushing originated at a very late period in the long line of our descent.

From the various facts just alluded to, and given in the course of this volume, it follows that, if the structure of our organs of respiration and circulation had differed in only a slight degree from the state in which they now exist, most of our expressions would have been wonderfully different. A very slight change in the course of the arteries and veins which run to the head, would probably have prevented the blood from accumulating in our eyeballs during violent expiration; for this occurs in extremely few quadrupeds. In this case we should not have displayed some of our most characteristic expressions. If man had breathed water by the aid of external branchiæ (though the idea is hardly conceivable), instead of air through his mouth and nostrils, his features would not have expressed his feelings much more efficiently than now do his hands or limbs. Rage and disgust, however, would still have been shown by movements about the lips and mouth, and the eyes would have become brighter or duller according to the state of the circulation. If our ears had remained movable, their movements would have been highly expressive, as is the case with all the animals which fight with their teeth; and we may infer that our early progenitors thus fought, as we still uncover the canine tooth on one side when we sneer at or defy any one, and we uncover all our teeth when furiously enraged.

THE MOVEMENTS OF EXPRESSION in the face and body, whatever their origin may have been, are in themselves of much importance for our welfare. They serve as the first means of communication between the mother and her infant; she smiles approval, and thus encourages her child on the right path, or frowns disapproval. We readily perceive sympathy in others by their expression; our sufferings are thus mitigated and our pleasures increased; and mutual good feeling is thus strengthened. The movements of expression give vividness and energy to our spoken words. They reveal the thoughts and intentions of others more truly than do words, which may be falsified. Whatever amount of truth the so-called science of physiognomy may contain, appears to depend, as Haller long ago remarked,[307] on different persons bringing into frequent use different facial muscles, according to their dispositions; the development of these muscles being perhaps thus increased, and the lines or furrows on the face, due to their habitual contraction, being thus rendered deeper and more conspicuous. The free expression by outward signs of an emotion intensifies it. On the other hand, the repression, as far as this is possible, of all outward signs softens our emotions.[308] He who gives way to violent gestures will increase his rage; he who does not control the signs of fear will experience fear in a greater degree; and he who remains passive when overwhelmed with grief loses his best chance of recovering elasticity of mind. These results follow partly from the intimate relation which exists between almost all the emotions and their outward manifestations; and partly from the direct influence of exertion on the heart, and consequently on the brain. Even the simulation of an emotion tends to arouse it in our minds. Shakespeare, who from his wonderful knowledge of the human mind ought to be an excellent judge, says:—

"Is it not monstrous that this player here,
But in a fiction, in a dream of passion,
Could force his soul so to his own conceit,
That, from her working, all his visage wann'd;
Tears in his eyes, distraction in 's aspect,
A broken voice, and his whole function suiting
With forms to his conceit? And all for nothing!"

Hamlet, act ii. sc. 2.

We have seen that the study of the theory of expression confirms to a certain limited extent the conclusion that man is derived from some lower animal form, and supports the belief of the specific or sub-specific unity of the several races; but as far as my judgment serves, such confirmation was hardly needed. We have also seen that expression in itself, or the language of the emotions, as it has sometimes been called, is certainly of importance for the welfare of mankind. To understand, as far as is possible, the source or origin of the various expressions which may be hourly seen on the faces of the men around us, not to mention our domesticated animals, ought to possess much interest for us. From these several causes, we may conclude that the philosophy of our subject has well deserved the attention which it has already received from several excellent observers, and that it deserves still further attention, especially from any able physiologist.

Afterword

EVOLUTION AND RELIGION

> If I lived twenty more years and was able to work, how I should have to modify the *Origin,* and how much the views on all points will have to be modified! Well, it is a beginning, and that is something.
>
> —Charles Darwin, letter to J. D. Hooker, 1869

Darwin lived thirteen more years after writing this letter to Joseph Hooker, and he did manage to modify the theory of evolution by natural selection, expanding it in *The Descent of Man* (1871) to include human origins and in *The Expression of the Emotions in Man and Animals* (1872) to address the evolution of instinct. The ensuing 130 years have seen an enormous growth of the Darwinian heritage. Joined with molecular and cellular biology, that accumulated knowledge is today a large part of modern biology. Its centrality justifies the famous remark made by the evolutionary geneticist Theodosius Dobzhansky in 1973 that "nothing in biology makes sense except in the light of evolution." In fact, nothing in science as a whole has been more firmly established by interwoven factual documentation, or more illuminating, than the universal occurrence of biological evolution. Further, few natural processes have been more convincingly explained than evolution by the theory of natural selection or, as it is popularly called, Darwinism.

Thus it is surpassingly strange that half of Americans recently polled (2004) not only do not believe in evolution by natural selection but do not believe in evolution at all. Americans are certainly capable of belief, and with rocklike conviction if it originates in religious dogma. In evidence is the 60 percent that accept the prophecies of the Book of Revelation as truth, and yet in more evidence is the weight that faith-based positions hold in political life. Most of the religious Right opposes the teaching of evolution in public schools, either by an outright ban on the subject or, at the least, by insisting that it be treated as "only a theory" rather than a "fact."

Yet biologists, particularly those statured by the peer review and publication of substantial personal research on the subject in leading journals of science, are unanimous in concluding that evolution is a *fact*. The evidence they and thousands of others have adduced over 150 years falls together in intricate and interlocking detail. The multitudinous examples range from the small changes in DNA sequences observed as they occur in real time to finely graded sequences within larger evolutionary changes in the fossil record. Further, on the basis of comparably firm evidence, natural selection grows ever stronger as the prevailing explanation of evolution.

Many who accept the fact of evolution cannot, however, on religious grounds, accept the operation of blind chance and the absence of divine purpose implicit in natural selection. They support the alternative explanation of intelligent design. The reasoning they offer is not based on evidence but on the lack of it. The formulation of intelligent design is a default argument advanced in support of a non sequitur. It is in essence the following: There are some phenomena that have not yet been explained and that (and most importantly) the critics personally cannot imagine being explained; therefore there must be a supernatural designer at work. The designer is seldom specified, but in the canon of intelligent design it is most certainly not Satan and his angels, nor any god or gods conspicuously different from those accepted in the believer's faith.

Flipping the scientific argument upside down, the intelligent designers join the strict creationists (who insist that no evolution ever occurred in the first place) by arguing that scientists resist the supernatural theory because it is counter to their own personal secular beliefs. This may have a kernel of truth; everybody suffers from some amount of bias. But in this case bias is easily overcome. The critics forget how the reward system in science works. Any researcher who can prove the existence of intelligent design within the accepted framework of science will make history and achieve eternal fame. He will prove at last that *science and religious dogma are compatible!* Even a combined Nobel Prize and Templeton Prize (the latter designed to encourage search for just such harmony) would fall short as proper recognition. Every scientist would like to accomplish such an epoch-making advance. But no one has even come close, because unfortunately there is no evidence, no theory, and no criteria for proof that even marginally might pass for science. There is only the residue of hoped-for default, which steadily shrinks as the science of biology expands.

In all of the history of science only one other disparity of comparable magnitude to evolution has occurred between a scientific event and the impact it has had on the public mind. This was the discovery by Copernicus that Earth and therefore humanity are not the center of the universe, and the universe is not a closed spherical bubble. Copernicus delayed publication of his masterwork *On the Revolutions of the Heavenly Spheres* until the year of his death (1543). For his extension of the idea subsequently, Bruno was burned at the stake, and for its documentation Galileo was shown the instruments of torture at Rome and remained under house arrest for the remainder of his life.

Today we live in a less barbaric age, but an otherwise comparable disjunction between science and religion, the one born of Darwinism, still roils the public mind. Why does such intense and pervasive resistance to evolution continue 150 years after the publication of *The Origin of Species*, and in the teeth of the overwhelming accumulated evidence favoring it? The answer is simply that the Darwinian revolution, even more than the Copernican revolution, challenges the prehistoric and still-regnant self-image of humanity. Evolution by natural selection, to be as concise as possible, has changed everything.

In the more than slightly schizophrenic circumstances of the present era, global culture is divided into three opposing images of the human condition, each logically consistent within its own, independent premises. The dominant of these hypotheses, exemplified by the creation myths of the Abrahamic monotheistic religions (Judaism, Christianity, and Islam), sees humanity as a creation of God. He brought us into being and He guides us still as father, judge, and friend. We interpret his will from sacred scriptures and the wisdom of ecclesiastical authorities.

The second worldview is that of political behaviorism. Still beloved by the now rapidly fading Marxist-Leninist states, it says that the brain is largely a blank state devoid of any inborn inscription beyond reflexes and primitive bodily urges. As a consequence the mind originates almost wholly as a product of learning, and it is the product of a culture that itself evolves by historical contingency. Because there is no biologically based "human nature," people can be molded to the best possible political and economic system, namely, as urged upon the world through most of the twentieth century, communism. In practical politics, this belief has been repeatedly tested and, after economic collapses and tens of millions of deaths in a dozen dysfunctional states, is generally deemed a failure.

Both of these worldviews, God-centered religion and atheistic communism, are opposed by a third and in some ways more radical worldview, scientific humanism. Still held by only a tiny minority of the world's population, it considers humanity to be a biological species that evolved over millions of years in a biological world, acquiring unprecedented intelligence yet still guided by complex inherited emotions and biased channels of learning.

Human nature exists, and it was self-assembled. It is the commonality of the hereditary responses and propensities that define our species. Having arisen by evolution during the far simpler conditions in which humanity lived during more than 99 percent of its existence, it forms the behavioral part of what, in *The Descent of Man,* Darwin called the indelible stamp of our lowly origin.

To understand biological human nature in depth is to drain the fever swamps of religious and blank-slate dogma. But it also imposes the heavy burden of individual choice that goes with intellectual freedom.

Such was the long journey for Darwin, the architect of the naturalistic worldview. He began his voyage on the *Beagle* as a devout Christian who trained for the ministry. "Whilst on board the *Beagle* I was quite orthodox," he wrote much later in his autobiography, "and I remember being heartily laughed at by several of the officers (though themselves orthodox) for quoting the Bible as an unanswerable authority on some point of morality." His later drift from the religion of his birth was stepwise and slow. Still on H.M.S. *Beagle* during its circumnavigation of the globe (1831–1836) he came to believe that the "false history" and reports of God's vengeful feelings made the Old Testament "no more to be trusted than the sacred books of the Hindoos, or the beliefs of any barbarian." The miracles of Jesus seemed to him to suggest that people living at the time of the Gospels were "ignorant and credulous to a degree almost incomprehensible by us." The growth of disbelief was so slow that Darwin felt no distress. In a striking passage of his autobiography he expressed his final and complete rejection of Christian dogma based solely on blind faith:

> I can indeed hardly see how anyone ought to wish Christianity to be true; for if so the plain language of the text seems to show that the men who do not believe, and this would include my Father, Brother and almost all my best friends, will be everlastingly punished.
>
> And that is a damnable doctrine.

Did Charles Darwin recant in his last days, as some religious critics have hopefully suggested? There is not a shred of evidence that he did or that he was presented with any reason to do so. Further, it would have been wholly contrary to the deliberate, careful manner with which he approached every subject.

The great naturalist did not abandon Abrahamic and other religious dogmas because of his discovery of evolution by natural selection, as one might reasonably suppose. The reverse occurred. The shedding of blind faith gave him the intellectual fearlessness to explore human evolution wherever logic and evidence took him. And so he set forth boldly, in *The Descent of*

Man to track the origin of humanity, and in *The Expression of the Emotions in Man and Animals* to address the evolution of instinct. Thus was born scientific humanism, the only worldview compatible with science's growing knowledge of the real world and the laws of nature.

So, will science and religion find common ground, or at least agree to divide the fundamentals into mutually exclusive domains? A great many well-meaning scholars believe that such rapprochement is both possible and desirable. A few disagree, and I am one of them. I think Darwin would have held to the same position. The battle line is, as it has ever been, in biology. The inexorable growth of this science continues to widen, not to close, the tectonic gap between science and faith-based religion.

Rapprochement may be neither possible nor desirable. There is something deep in religious belief that divides people and amplifies societal conflict. In the early part of this century, the toxic mix of religion and tribalism has become so dangerous as to justify taking seriously the alternative view, that humanism based on science is the effective antidote, the light and the way at last placed before us.

In any case, the dilemma to be solved is truly profound. On the one side the input of religion on human history has been beneficent in many ways. It has generated much of which is best in culture, including the ideals of altruism and public service. From the beginning of history it has inspired the arts. Creation myths were in a sense the beginning of science itself. Fabricating them was the best the early scribes could do to explain the universe and human existence.

Yet the high risk is the ease with which alliances between religions and tribalism are made. Then comes bigotry and the dehumanization of infidels. Our gods, the true believer asserts, stand against your false idols, our spiritual purity against your corruption, our divinely sanctioned knowledge against your errancy. In past ages the posture provided an advantage. It united each tribe during life-and-death struggles with other tribes. It buoyed the devotees with a sense of superiority. It sacralized tribal laws and mores, and encouraged altruistic behaviors. Through sacred rites it lent solemnity to the passages of life. And it comforted the anxious and afflicted. For all this and more it gave people an identity and purpose, and vouchsafed tribal fitness—yet, unfortunately, at the expense of less united or otherwise less fortunate tribes.

Religions continue both to render their special services and to exact their heavy costs. Can scientific humanism do as well or better, at a lower cost? Surely that ranks as one of the great unanswered questions of philosophy. It is the noble yet troubling legacy that Charles Darwin left us.

FOOTNOTES TO *THE VOYAGE OF THE BEAGLE*

Darwin's original footnotes were numbered independently by chapter. To improve ease of access they have been renumbered here to run continuously throughout each book in turn. (*On the Origin of Species*, alone among the four books, lacks footnotes.)—EOW

PREFACE

[PAGE 29]

1 I must take this opportunity of returning my sincere thanks to Mr. Bynoe, the surgeon of the Beagle, for his very kind attention to me when I was ill at Valparaiso.

CHAPTER I

[PAGE 32]

2 I state this on the authority of Dr. E. Dieffenbach, in his German translation of the first edition of this Journal.

3 The Cape de Verd Islands were discovered in 1449. There was a tombstone of a bishop with the date of 1571; and a crest of a hand and dagger, dated 1497.

[PAGE 34]

4 I must take this opportunity of acknowledging the great kindness with which this illustrious naturalist has examined many of my specimens. I have sent (June, 1845) a full account of the falling of this dust to the Geological Society.

[PAGE 35]

5 So named according to Patrick Symes's nomenclature.

[PAGE 36]

6 See Encyclop. of Anat. and Physiol., article *Cephalopoda.*

[PAGE 37]

7 Mr. Horner and Sir David Brewster have described (Philosophical Transactions, 1836, p. 65) a singular "artificial substance resembling shell." It is deposited in fine, transparent, highly polished, brown-coloured laminæ, possessing peculiar optical properties, on the inside of a vessel, in which cloth, first prepared with glue and then with lime, is made to revolve rapidly in water. It is much softer, more transparent, and contains more animal matter, than the natural incrustation at Ascension; but we here again see the strong tendency which carbonate of lime and animal matter evince to form a solid substance allied to shell.

[PAGE 40]

8 Pers. Narr., vol. v., pt. I., p. 18.

[PAGE 41]

9 M. Montagne, in Comptes Rendus, etc., Juillet, 1844; and Annal. des Scienc. Nat., Dec. 1844.

[PAGE 43]

10 M. Lesson (Voyage de la Coquille, tom. i., p. 255) mentions red water off Lima, apparently produced by the same cause. Peron, the distinguished naturalist, in the Voyage aux Terres Australes, gives no less than twelve references to voyagers who have alluded to the discoloured waters of the sea (vol. ii. p. 239). To the references given by Peron may be added, Humboldt's Pers. Narr., vol. vi. p. 804; Flinders' Voyage, vol. i. p. 92; Labillardière, vol. i. p. 287; Ulloa's Voyage; Voyage of the Astrolabe and of the Coquille; Captain King's Survey of Australia, etc.

CHAPTER II

[PAGE 46]

11 Vênda, the Portuguese name for an inn.

[PAGE 47]

12 Annales des Sciences Naturelles for 1833.

[PAGE 51]

13 I have described and named these species in the Annals of Nat. Hist., vol. xiv. p. 241.

[PAGE 54]

14 I am greatly indebted to Mr. Waterhouse for his kindness in naming for me this and many other insects, and in giving me much valuable assistance.

15 Kirby's Entomology, vol. ii. p. 317.

[PAGE 56]

16 Mr. Doubleday has lately described (before the Entomological Society, March 3rd, 1845) a peculiar structure in the wings of this butterfly, which seems to be the means of its making its noise. He says, "It is remarkable for having a sort of drum at the base of the fore wings, between the costal nervure and the subcostal. These two nervures, moreover, have a peculiar screw-like diaphragm or vessel in the interior." I find in Langsdorff's travels (in the years 1803–7, p. 74) it is said, that in the island of St. Catherine's on the coast of Brazil, a butterfly called Februa Hoffmanseggi, makes a noise, when flying away, like a rattle.

17 I may mention, as a common instance of one day's (June 23rd) collecting, when I was not attending particularly to the Coleoptera, that I caught sixty-eight species of that order. Among these, there were only two of the Carabidæ, four Brachelytra, fifteen Rhyncophora, and fourteen of the Chrysomelidæ. Thirty-seven species of Arachnidæ, which I brought home, will be sufficient to prove that I was not paying overmuch attention to the generally favoured order of Coleoptera.

[PAGE 57]

18 In a MS. in the British Museum by Mr. Abbott, who made his observations in Georgia; see Mr. A. White's paper in the "Annals of Nat. Hist.," vol. vii. p. 472. Lieut. Hutton has described a sphex with similar habits in India, in the "Journal of the Asiatic Society," vol. i. p. 555.

[PAGE 58]

19 Don Felix Azara (vol. i. p. 175), mentioning a hymenopterous insect, probably of the same genus, says he saw it dragging a dead spider through tall grass, in a straight line to its nest, which was one hundred and sixty-three paces distant. He adds that the wasp, in order to find the road every now and then made "demi-tours d'environ trois palmes."

[PAGE 59]

20 Azara's Voyage, vol. i. p. 213.

CHAPTER III

[PAGE 65]

21 Hearne's Journey, p. 383.

[PAGE 66]

22 Maclaren, art. "America," Encyclop. Britann.

23 Azara says, "Je crois que la quantité annuelle des pluies est, dans toutes ces contrées, plus considérable qu'en Espagne."—Vol. i. p. 36.

[PAGE 68]

24 In South America I collected altogether twenty-seven species of mice, and thirteen more are known from the works of Azara and other authors. Those collected by myself have been named and described by Mr. Waterhouse at the meetings of the Zoological Society. I must be allowed to take this opportunity of returning my cordial thanks to Mr. Waterhouse, and to the other gentlemen attached to that Society, for their kind and most liberal assistance on all occasions.

25 In the stomach and duodenum of a capybara which I opened I found a very large quantity of a thin yellowish fluid, in which scarcely a fibre could be distinguished. Mr. Owen informs me that a part of the œsophagus is so constructed that nothing much larger than a crowquill can be passed down. Certainly the broad teeth and strong jaws of this animal are well fitted to grind into pulp the aquatic plants on which it feeds.

[PAGE 69]

26 At the R. Negro, in Northern Patagonia, there is an animal of the same habits, and probably a closely allied species, but which I never saw. Its noise is different from that of the Maldonado kind; it is repeated only twice instead of three or four times, and is more distinct and sonorous; when heard from a distance it so closely resembles the sound made in cutting down a small tree with an axe, that I have sometimes remained in doubt concerning it.

[PAGE 70]

27 Philosoph. Zoolog., tom. i. p. 242.

28 Magazine of Zoology and Botany, vol. i. p. 217.

[PAGE 71]

29 Read before the Academy of Sciences in Paris. L'Institut, 1834, p. 418.

[PAGE 76]

30 Geolog. Transact., vol. ii. p. 528. In the Philosoph. Transact. (1790, p. 294) Dr. Priestley has described some imperfect siliceous tubes and a melted pebble of quartz, found in digging into the ground, under a tree, where a man had been killed by lightning.

31 Annals de Chimie et de Physique, tom, xxxvii. p. 319.

[PAGE 77]

32 Azara's Voyage, vol. i. p. 36.

CHAPTER IV

[PAGE 80]

33 The corral is an enclosure made of tall and strong stakes. Every estancia, or farming estate, has one attached to it.

34 The hovels of the Indians are thus called.

[PAGE 81]

35 Report of the Agricult. Chem. Assoc. in the Agricult. Gazette, 1845, p. 93.

[PAGE 82]

36 Linnæan Trans., vol. xi. p. 205. It is remarkable how all the circumstances connected with the salt-lakes in Siberia and Patagonia are similar. Siberia, like Patagonia, appears to have been recently elevated above the waters of the sea. In both countries the salt-lakes occupy shallow depressions in the plains; in both the mud on the borders is black and fetid; beneath the crust of common salt, sulphate of soda or of magnesia occurs, imperfectly crystallized; and in both, the muddy sand is mixed with lentils of gypsum. The Siberian salt-lakes are inhabited by small crustaceous animals; and flamingoes (Edin. New Philos. Jour., Jan. 1830) likewise frequent them. As these circumstances, apparently so trifling, occur in two distant continents, we may feel sure that they are the necessary results of common causes.—See *Pallas's Travels,* 1793 to 1794, pp. 129–134.

[PAGE 85]

37 I am bound to express, in the strongest terms, my obligation to the government of Buenos Ayres for the obliging manner in which passports to all parts of the country were given me, as naturalist of the Beagle.

[PAGE 86]

38 This prophecy has turned out entirely and miserably wrong. 1845.

[PAGE 91]

39 Voyage dans l'Amérique Mérid par M. A. d'Orbigny. Part. Hist. tom. i. p. 664.

CHAPTER V

[PAGE 94]

40 Since this was written, M. Alcide d'Orbigny has examined these shells, and pronounces them all to be recent.

[PAGE 95]

41 M. Aug. Bravard has described, in a Spanish work ('Observaciones Geologicas,' 1857), this district, and he believes that the bones of the extinct mammals were washed out of the underlying Pampean deposit, and subsequently became embedded with the still existing shells, but I am not convinced by his remarks. M. Bravard believes that the whole enormous Pampean deposits is a sub-aërial formation, like sand-dunes: this seems to me to be an untenable doctrine.

42 Principles of Geology, vol. iv. p. 40.

43 This theory was first developed in the Zoology of the Voyage of the Beagle, and subsequently in Professor Owen's Memoir on Mylodon robustus.

[PAGE 96]

44 I mean by this to exclude the total amount which may have been successively produced and consumed during a given period.

[PAGE 98]

45 Travels in the Interior of South Africa, vol. ii. p. 207.

46 The elephant which was killed at Exeter Change was estimated (being partly weighed) at five tons and a half. The elephant actress, as I was informed, weighed one ton less; so that we may take five as the average of a full-grown elephant. I was told at the Surrey Gardens, that a hippopotamus which was sent to England cut up into pieces was estimated at three tons and a half; we will

call it three. From these premises we may give three tons and a half to each of the five rhinoceroses; perhaps a ton to the giraffe, and half to the bos caffer as well as to the elan (a large ox weighs from 1200 to 1500 pounds). This will give an average (from the above estimates) of 2.7 of a ton for the ten largest herbivorous animals of Southern Africa. In South America, allowing 1200 pounds for the two tapirs together, 550 for the guanaco and vicuna, 500 for three deer, 300 for the capybara, peccari, and a monkey, we shall have an average of 250 pounds, which I believe is overstating the result. The ratio will therefore be as 6048 to 250, or 24 to 1, for the ten largest animals from the two continents.

47 If we suppose the case of the discovery of a skeleton of a Greenland whale in a fossil state, not a single cetaceous animal being known to exist, what naturalist would have ventured conjecture on the possibility of a carcass so gigantic being supported on the minute crustacea and mollusca living in the frozen seas of the extreme North?

48 See Zoological Remarks to Capt. Back's Expedition, by Dr. Richardson. He says, "The subsoil north of latitude 56° is perpetually frozen, the thaw on the coast not penetrating above three feet, and at Bear Lake, in latitude 64°, not more than twenty inches. The frozen substratum does not of itself destroy vegetation, for forests flourish on the surface, at a distance from the coast."

49 See Humboldt, Fragments Asiatiques, p. 386: Barton's Geography of Plants: and Malte Brun. In the latter work it is said that the limit of the growth of trees in Siberia may be drawn under the parallel of 70°.

[PAGE 99]

50 Sturt's Travels, vol. ii. p. 74.

[PAGE 100]

51 A Gaucho assured me that he had once seen a snow-white or Albino variety, and that it was a most beautiful bird.

52 Burchell's Travels, vol. i. p. 280.

53 Azara, vol. iv. p. 173.

[PAGE 101]

54 Lichtenstein, however, asserts (Travels, vol. ii. p. 25) that the hens begin sitting when they have laid ten or twelve eggs; and that they continue laying, I presume, in another nest. This appears to me very improbable. He asserts that four or five hens associate for incubation with one cock, who sits only at night.

[PAGE 102]

55 When at the Rio Negro, we heard much of the indefatigable labours of this naturalist. M. Alcide d'Orbigny, during the years 1825 to 1833, traversed several large portions of SouthAmerica, and has made a collection, and is now publishing the results on a scale of magnificence, which at once places himself in the list of American travellers second only to Humboldt.

56 Account of the Abipones, A.D. 1749, vol. i. (English translation) p. 314.

[PAGE 104]

57 M. Bibron calls it T. crepitans.

[PAGE 107]

58 The cavities leading from the fleshy compartments of the extremity, were filled with a yellow pulpy matter, which, examined under a microscope, presented an extraordinary appearance. The mass consisted of rounded, semi-transparent, irregular grains, aggregated together into particles of various sizes. All such particles, and the separate grains, possessed the power of rapid movement; generally revolving around different axes, but sometimes progressive. The movement was visible with a very weak power, but even with the highest its cause could not be perceived. It was very different from the circulation of the fluid in the elastic bag, containing the thin extremity of the axis. On other occasions, when dissecting small marine animals beneath the microscope, I have seen particles of pulpy matter, some of large size, as soon as they were disengaged, commence revolving. I have imagined, I know not with how much truth, that this granulo-pulpy matter was in process of being converted into ova. Certainly in this zoophyte such appeared to be the case.

59 Kerr's Collection of Voyages, vol. viii. p. 119.

[PAGE 110]

60 Purchas's Collection of Voyages. I believe the date was really 1537.

[PAGE 111]

61 Azara has even doubted whether the Pampas Indians ever used bows.

CHAPTER VI

[PAGE 113]

62 I call these thistle-stalks for the want of a more correct name. I believe it is a species of Eryngium.

[PAGE 116]

63 Travels in Africa, p. 233.

[PAGE 117]

64 Two species of Tinamus, and *Eudromia elegans* of A. d'Orbigny, which can only be called a partridge with regard to its habits.

[PAGE 120]

65 History of the Abipones, vol. ii. p. 6.
66 Falconer's Patagonia, p.70.

[PAGE 121]

67 Fauna Boreali-Americana, vol. i. p. 35.

[PAGE 122]

68 See Mr. Atwater's account of the Prairies, in Silliman's N. A. Journal, vol. i. p. 117.
69 Azara's Voyages, vol. i. p. 373.
70 M. A. d'Orbigny (vol. i. p. 474) says that the cardoon and artichoke are both found wild. Dr. Hooker (Botanical Magazine, vol. lv. p. 2862), has described a variety of the Cynara from this part of South America under the name of *inermis.* He states that botanists are now generally agreed that the cardoon and the artichoke are varieties of one plant. I may add, that an intelligent farmer assured me that he had observed in a deserted garden some artichokes changing into the common cardoon. Dr. Hooker believes that Head's vivid description of the thistle of the Pampas applies to the cardoon; but this is a mistake. Captain Head referred to the plant, which I have mentioned a few lines lower down, under the title of giant thistle. Whether it is a true thistle I do not know; but it is quite different from the cardoon; and more like a thistle properly so called.

[PAGE 123]

71 It is said to contain 60,000 inhabitants. Monte Video, the second town of importance on the banks of the Plata, has 15,000.

CHAPTER VII

[PAGE 126]

72 The bizcacha (Lagostomus trichodactylus) somewhat resembles a large rabbit, but with bigger gnawing teeth and a long tail; it has, however, only three toes behind, like the agouti. During the last three or four years the skins of these animals have been sent to England for the sake of the fur.

[PAGE 127]

73 Journal of Asiatic Soc., vol. v. p. 363.

[PAGE 131]

74 I need hardly state here that there is good evidence against any horse living in America at the time of Columbus.

75 Cuvier. Ossemens Fossiles, tom. i. p. 158.

76 This is the geographical division followed by Lichtenstein, Swainson, Erichson, and Richardson. The section from Vera Cruz to Acapulco, given by Humboldt in the Polit. Essay on Kingdom of N. Spain will show how immense a barrier the Mexican table-land forms. Dr. Richardson, in his admirable Report on the Zoology of N. America read before the Brit. Assoc. 1836 (p. 157), talking of the identification of a Mexican animal with the *Synetheres prehensilis,* says, "We do not know with what propriety, but if correct, it is, if not a solitary instance, at least very nearly so, of a rodent animal being common to North and South America."

[PAGE 132]

77 See Dr. Richardson's Report, p. 157; also L'lnstitut, 1837, p. 253. Cuvier says the kinkajou is found in the larger Antilles, but this is doubtful. M. Gervais states that the Didelphis crancrivora is found there. It is certain that the West Indies possess some mammifers peculiar to themselves. A tooth of a mastodon has been brought from Bahama; Edin. New Phil. Journ., 1826, p. 395.

78 See the admirable Appendix by Dr. Buckland to Beechey's Voyage; also the writings of Chamisso in Kotzebue's Voyage.

[PAGE 133]

79 In Captain Owen's Surveying Voyage (vol. ii. p. 274) there is a curious account of the effects of a drought on the elephants, at Benguela (west coast of Africa). "A number of these animals had some time

since entered the town, in a body, to possess themselves of the wells, not being able to procure any water in the country. The inhabitants mustered, when a desperate conflict ensued, which terminated in the ultimate discomfiture of the invaders, but not until they had killed one man, and wounded several others." The town is said to have a population of nearly three thousand! Dr. Malcolmson informs me that, during a great drought in India, the wild animals entered the tents of some troops at Ellore, and that a hare drank out of a vessel held by the adjutant of the regiment.

80 Travels, vol. i. p. 374.

[PAGE 134]

81 These droughts to a certain degree seem to be almost periodical; I was told the dates of several others, and the intervals were about fifteen years.

CHAPTER VIII

[PAGE 143]

82 Mr. Waterhouse has drawn up a detailed description of this head, which I hope he will publish in some Journal.

83 A nearly similar abnormal, but I do not know whether hereditary, structure has been observed in the carp, and likewise in the crocodile of the Ganges: Histoire des Anomalies par M. Isid. Geoffroy St. Hilaire, tom. i. p. 244.

[PAGE 146]

84 M. A. d'Orbigny has given nearly a similar account of these dogs, tom. i. p. 175.

[PAGE 150]

85 I must express my obligation to Mr. Keane, at whose house I was staying on the Berquelo, and to Mr. Lumb at Buenos Ayres, for without their assistance these valuable remains would never have reached England.

[PAGE 153]

86 Lyell's Principles of Geology, vol. iii. p. 63.

[PAGE 154]

87 The flies which frequently accompany a ship for some days on its pas-

sage from harbour to harbour, wandering from the vessel, are soon lost, and all disappear.

[PAGE 155]

88 Mr. Blackwall, in his Researches in Zoology, has many excellent observations on the habits of spiders.

[PAGE 156]

89 An abstract is given in No. IV. of the Magazine of Zoology and Botany.

[PAGE 158]

90 I found here a species of cactus, described by Professor Henslow, under the name of *Opuntia Darwinii* (Magazine of Zoology and Botany, vol. i. p. 466), which was remarkable for the irritability of the stamens, when I inserted either a piece of stick or the end of my finger in the flower. The segments of the perianth also closed on the pistil, but more slowly than the stamens. Plants of this family, generally considered as tropical, occur in North America (Lewis and Clarke's Travels, p. 221), in the same high latitude as here, namely, in both cases, in 47°.

91 These insects were not uncommon beneath stones. I found one cannibal scorpion quietly devouring another.

[PAGE 161]

92 Shelley, Lines on Mt. Blanc.

[PAGE 164]

93 I have lately heard that Capt. Sulivan, R.N., has found numerous fossil bones, embedded in regular strata, on the banks of the R. Gallegos, in lat. 51° 4'. Some of the bones are large; others are small, and appear to have belonged to an armadillo. This is a most interesting and important discovery.

[PAGE 166]

94 See the excellent remarks on this subject by Mr. Lyell, in his Principles of Geology.

CHAPTER IX

[PAGE 170]

95 The deserts of Syria are characterized, according to Volney (tom. i.

p. 351), by woody bushes, numerous rats, gazelles and hares. In the landscape of Patagonia, the guanaco replaces the gazelle, and the agouti the hare.

[PAGE 174]

96 I noticed that several hours before any one of the condors died, all the lice, with which it was infested, crawled to the outside feathers. I was assured that this always happened.

[PAGE 175]

97 Loudon's Magazine of Nat. Hist., vol. vii.

[PAGE 177]

98 From accounts published since our voyage, and more especially from several interesting letters from Capt. Sulivan, R.N., employed on the survey, it appears that we took an exaggerated view of the badness of the climate of these islands. But when I reflect on the almost universal covering of peat, and on the fact of wheat seldom ripening here, I can hardly believe that the climate in summer is so fine and dry as it has lately been represented.

[PAGE 181]

99 Lesson's Zoology of the Voyage of the Coquille, tom. i. p. 168. All the early voyagers, and especially Bougainville, distinctly state that the wolf-like fox was the only native animal on the island. The distinction of the rabbit as a species, is taken from peculiarities in the fur, from the shape of the head, and from the shortness of the ears. I may here observe that the difference between the Irish and English hare rests upon nearly similar characters, only more strongly marked.

100 I have reason, however, to suspect that there is a field-mouse. The common European rat and mouse have roamed far from the habitations of the settlers. The common hog has also run wild on one islet; all are of a black colour: the boars are very fierce, and have great tusks.

101 The "culpeu" is the Canis Magellanicus brought home by Captain King from the Strait of Magellan. It is common in Chile.

[PAGE 183]

102 Pernety, Voyage aux Isles Malouines, p. 526.

[PAGE 184]

103 "Nous n'avons pas été moins saisis d'étonnement à la vûe de l'innombrable quantité de pierres de toutes grandeurs, bouleversées les unes sur les autres, et cependant rangées, comme si elles avoient été amoncelées négligemment pour remplir des ravins. On ne se lassoit pas d'admirer les effets prodigieux de la nature."—Pernety, p. 526.

[PAGE 185]

104 An inhabitant of Mendoza, and hence well capable of judging, assured me that, during the several years he had resided on these islands, he had never felt the slightest shock of an earthquake.

[PAGE 186]

105 I was surprised to find, on counting the eggs of a large white Doris (this sea-slug was three and a half inches long), how extraordinarily numerous they were. From two to five eggs (each three-thousandths of an inch in diameter) were contained in a spherical little case. These were arranged two deep in transverse rows forming a ribbon. The ribbon adhered by its edge to the rock in an oval spire. One which I found, measured nearly twenty inches in length and half in breadth. By counting how many balls were contained in a tenth of an inch in the row, and how many rows in an equal length of the ribbon, on the most moderate computation there were six hundred thousand eggs. Yet this Doris was certainly not very common; although I was often searching under the stones, I saw only seven individuals. *No fallacy is more common with naturalists, than that the numbers of an individual species depend on its powers of propagation.*

CHAPTER X

[PAGE 203]

106 This substance, when dry, is tolerably compact, and of little specific gravity: Professor Ehrenberg has examined it: he states (König Akad. der Wissen: Berlin, Feb. 1845) that it is composed of infusoria, including fourteen polygastrica, and four phytolitharia. He says that they are all inhabitants of fresh-water; this is a beautiful example of the results obtainable through Professor Ehrenberg's microscopic researches; for Jemmy Button told me that it is always collected at the bottoms of mountain-brooks. It is, moreover, a striking fact in the geographical distribution of the infusoria, which are well known to have very wide ranges, that all the species in this substance, although

brought from the extreme southern point of Tierra del Fuego, are old, known forms.

[PAGE 204]

107 One day, off the east coast of Tierra del Fuego, we saw a grand sight in several spermaceti whales jumping upright quite out of the water, with the exception of their tail-fins. As they fell down sideways, they splashed the water high up, and the sound reverberated like a distant broadside.

[PAGE 208]

108 Captain Sulivan, who, since his voyage in the Beagle, has been employed on the survey of the Falkland Islands, heard from a sealer in (1842?), that when in the western part of the Strait of Magellan, he was astonished by a native woman coming on board, who could talk some English. Without doubt this was Fuegia Basket. She lived (I fear the term probably bears a double interpretation) some days on board.

Chapter XI

[PAGE 211]

109 The south-westerly breezes are generally very dry. January 29th, being at anchor under Cape Gregory: a very hard gale from W. by S., clear sky with few cumuli; temperature 57°, dew-point 36°,—difference 21°. On January 15th, at Port St. Julian: in the morning, light winds with much rain, followed by a very heavy squall with rain,—settled into heavy gale with large cumuli,—cleared up, blowing very strong from S.S.W. Temperature 60°, dew-point 42°,—difference 18°.

[PAGE 212]

110 Rengger, Natur. der Saeugethiere von Paraguay. S. 334.

[PAGE 214]

111 Captain Fitz Roy informs me that in April (our October), the leaves of those trees which grow near the base of the mountains change colour, but not those on the more elevated parts. I remember having read some observations, showing that in England the leaves fall earlier in a warm and fine autumn than in a late and cold one. The change in the colour being here retarded in the more elevated, and therefore colder situations, must be owing to the same general law of veg-

etation. The trees of Tierra del Fuego during no part of the year entirely shed their leaves.

[PAGE 215]

112 Described from my specimens and notes by the Rev. J. M. Berkeley, in the Linnean Transactions (vol. xix. p. 37), under the name of Cyttaria Darwinii; the Chilian species is the C. Berteroii. This genus is allied to Bulgaria.

[PAGE 216]

113 I believe I must except one alpine Haltica, and a single specimen of a Melasoma. Mr. Waterhouse informs me, that of the Harpalidæ there are eight or nine species—the forms of the greater number being very peculiar; of Heteromera, four or five species; of Rhyncophora, six or seven; and of the following families one species in each: Staphylinidæ, Elateridæ, Cebrionidæ, Melolonthidæ. The species in the other orders are even fewer. In all the orders, the scarcity of the individuals is even more remarkable than that of the species. Most of the Coleoptera have been carefully described by Mr. Waterhouse in the Annals of Nat. Hist.

[PAGE 217]

114 Its geographical range is remarkably wide; it is found from the extreme southern islets near Cape Horn, as far north on the eastern coast (according to information given me by Mr. Stokes) as lat. 43°,—but on the western coast, as Dr. Hooker tells me, it extends to the R. San Francisco in California, and perhaps even to Kamtschatka. We thus have an immense range in latitude; and as Cook, who must have been well acquainted with the species, found it at Kerguelen Land, no less than 140° in longitude.

115 Voyages of the Adventure and Beagle, vol. i. p. 363.—It appears that sea-weed grows extremely quick.—Mr. Stephenson found (Wilson's Voyage round Scotland, vol. ii. p. 228) that a rock uncovered only at spring-tides, which had been chiselled smooth in November, on the following May, that is, within six months afterwards, was thickly covered with Fucus digitatus two feet, and F. esculentus six feet, in length.

[PAGE 220]

116 With respect to Tierra del Fuego, the results are deduced from the observations by Capt. King (Geographical Journal, 1830), and those taken on board the Beagle. For the Falkland Islands, I am indebted

to Capt. Sulivan for the mean of the mean temperature (reduced from careful observation at midnight, 8 A.M., noon, and 8 P.M.) of the three hottest months, viz., December, January, and February. The temperature of Dublin is taken from Barton.

117 Agüeros, Descrip. Hist. de la Prov. de Chiloé, 1791, p. 94.

[PAGE 221]

118 See the German Translation of this Journal; and for the other facts, Mr. Brown's Appendix to Flinders's Voyage.

[PAGE 222]

119 On the Cordillera of central Chile, I believe the snow-line varies exceedingly in height in different summers. I was assured that during one very dry and long summer, all the snow disappeared from Aconcagua, although it attains the prodigious height of 23,000 feet. It is probable that much of the snow at these great heights is evaporated rather than thawed.

120 Miers's Chile, vol. i. p. 415. It is said that the sugar-cane grew at Ingenio, lat. 32° to 33°, but not in sufficient quantity to make the manufacture profitable. In the valley of Quillota, south of Ingenio, I saw some large date palm-trees.

121 Bulkeley's and Cummin's Faithful Narrative of the Loss of the Wager. The earthquake happened August 25, 1741.

[PAGE 223]

122 Agüeros, Desc. Hist. de Chiloé, p. 227.

123 Geological Transactions, vol. vi. p. 415.

[PAGE 224]

124 I have given details (the first, I believe, published) on this subject in the first edition, and in the Appendix to it. I have there shown that the apparent exceptions to the absence of erratic boulders in certain hot countries, are due to erroneous observations; several statements there given I have since found confirmed by various authors.

125 Geographical Journal, 1830, pp. 65, 66.

126 Richardson's Append. to Back's Exped., and Humboldt's Fragm. Asiat., tom. ii. p. 386.

[PAGE 225]

127 Messrs. Dease and Simpson, in Geograph. Journ., vol. viii. pp. 218 and 220.

128 Cuvier (Ossemens Fossiles, tom. i. p. 151), from Billing's Voyage.

[PAGE 226]

129 In the former edition and Appendix, I have given some facts on the transportal of erratic boulders and icebergs in the Antarctic Ocean. This subject has lately been treated excellently by Mr. Hayes, in the Boston Journal (vol. iv. p. 426). The author does not appear aware of a case published by me (Geographical Journal, vol. ix. p. 528) of a gigantic boulder embedded in an iceberg in the Antarctic Ocean, almost certainly one hundred miles distant from any land, and perhaps much more distant. In the Appendix I have discussed at length the probability (at that time hardly thought of) of icebergs, when stranded, grooving and polishing rocks, like glaciers. This is now a very commonly received opinion; and I cannot still avoid the suspicion that it is applicable even to such cases as that of the Jura. Dr. Richardson has assured me that the icebergs off North America push before them pebbles and sand, and leave the submarine rocky flats quite bare; it is hardly possible to doubt that such ledges must be polished and scored in the direction of the set of the prevailing currents. Since writing that Appendix, I have seen in North Wales (London Phil. Mag., vol. xxi. p. 180) the adjoining action of glaciers and floating icebergs.

CHAPTER XII

[PAGE 236]

130 Caldcleugh, in Philosoph. Transact. for 1836.

[PAGE 237]

131 Annales des Sciences Naturelles, March, 1833. M. Gay, a zealous and able naturalist, was then occupied in studying every branch of natural history throughout the kingdom of Chile.

[PAGE 239]

132 Burchell's Travels, vol. ii. p. 45.

[PAGE 242]

133 It is a remarkable fact, that Molina, though describing in detail all the birds and animals of Chile, never once mentions this genus, the species of which are so common, and so remarkable in their habits.

Was he at a loss how to classify them, and did he consequently think that silence was the more prudent course? It is one more instance of the frequency of omissions by authors, on those very subjects where it might have been least expected.

CHAPTER XIII

[PAGE 254]

134 Horticultural Transact., vol. v. p. 249. Mr. Caldcleugh sent home two tubers, which, being well manured, even the first season produced numerous potatoes and an abundance of leaves. See Humboldt's interesting discussion on this plant, which it appears was unknown in Mexico,—in Polit. Essay on New Spain, book iv. chap. ix.

[PAGE 255]

135 By sweeping with my insect-net, I procured from these situations a considerable number of minute insects, of the family of Staphylinidæ, and others allied to Pselaphus, and minute Hymenoptera. But the most characteristic family in number, both of individuals and species, throughout the more open parts of Chiloe and Chonos is that of the Telephoridæ.

[PAGE 256]

136 It is said that some rapacious birds bring their prey alive to their nests. If so, in the course of centuries, every now and then, one might escape from the young birds. Some such agency is necessary, to account for the distribution of the smaller gnawing animals on islands not very near each other.

[PAGE 257]

137 I may mention, as a proof of how great a difference there is between the seasons of the wooded and the open parts of this coast, that on September 20th, in lat. 34°, these birds had young ones in the nest, while among the Chonos Islands, three months later in the summer, they were only laying, the difference in latitude between these two places being about 700 miles.

CHAPTER XIV

[PAGE 273]

138 M. Arago in L'Institut, 1839, p. 337. See also Miers's Chile, vol. i. p. 392; also Lyell's Principles of Geology, chap. xv., book ii.

[PAGE 276]

139 For a full account of the volcanic phenomena which accompanied the earthquake of the 20th, and for the conclusions deducible from them, I must refer to Volume V. of the Geological Transactions.

CHAPTER XV

[PAGE 281]

140 Scoresby's Arctic Regions, vol. i. p. 122.

[PAGE 282]

141 I have heard it remarked in Shropshire that the water, when the Severn is flooded from long-continued rain, is much more turbid than when it proceeds from the snow melting in the Welsh mountains. D'Orbigny (tom. 1. p. 184), in explaining the cause of the various colours of the rivers in South America, remarks that those with blue or clear water have their source in the Cordillera, where the snow melts.

[PAGE 285]

142 Dr. Gillies in Journ. of Nat. and Geograph. Science, Aug., 1830. This author gives the heights of the Passes.

[PAGE 286]

143 This structure in frozen snow was long since observed by Scoresby in the icebergs near Spitzbergen, and, lately, with more care, by Colonel Jackson (Journ. of Geograph. Soc. vol. v. p. 12) on the Neva. Mr. Lyell (Principles, vol. iv. p. 360) has compared the fissures by which the columnar structure seems to be determined, to the joints that traverse nearly all rocks, but which are best seen in the non-stratified masses. I may observe, that in the case of the frozen snow, the columnar structure must be owing to a "metamorphic" action, and not to a process during *deposition.*

[PAGE 288]

144 This is merely an illustration of the admirable laws, first laid down by Mr. Lyell, on the geographical distribution of animals, as influenced by geological changes. The whole reasoning, of course, is founded on the assumption of the immutability of species; otherwise the difference in the species in the two regions might be considered as superinduced during a length of time.

CHAPTER XVI

[PAGE 308]

145 Vol. iv. p. 11, and vol. ii. p. 217. For the remarks on Guayaquil, see Silliman's Journ., vol. xxiv. p. 384. For those on Tacna by Mr. Hamilton, see Trans. of British Association, 1840. For those on Coseguina see Mr. Caldcleugh in Phil. Trans., 1835. In the former edition I collected several references on the coincidences between sudden falls in the barometer and earthquakes; and between earthquakes and meteors.

[PAGE 310]

146 Observa. sobre el Clima de Lima, p. 67.—Azara's Travels, vol. i. p. 381.—Ulloa's Voyage, vol. ii. p. 28.—Burchell's Travels, vol. ii. p. 524.—Webster's Description of the Azores, p. 124.—Voyage à l'Isle de France par un Officier du Roi, tom. i. p. 248.—Description of St. Helena, p. 123.

[PAGE 313]

147 Temple, in his travels through Upper Peru, or Bolivia, in going from Potosi to Oruro, says, "I saw many Indian villages or dwellings in ruins, up even to the very tops of the mountains, attesting a former population where now all is desolate." He makes similar remarks in another place; but I cannot tell whether this desolation has been caused by a want of population, or by an altered condition of the land.

[PAGE 315]

148 Edinburgh Phil. Journ., Jan., 1830, p. 74; and April, 1830, p. 258—also Daubeny on Volcanoes, p. 438; and Bengal Journ., vol. vii. p. 324.

[PAGE 319]

149 Political Essay on the Kingdom of New Spain, vol. iv. p. 199.

150 A similar interesting case is recorded in the Madras Medical Quart. Journ., 1839, p. 340. Dr. Ferguson, in his admirable Paper (see 9th vol. of Edinburgh Royal Trans.), shows clearly that the poison is generated in the drying process; and hence that dry hot countries are often the most unhealthy.

CHAPTER XVII

[PAGE 331]

151 The progress of research has shown that some of these birds, which were then thought to be confined to the islands, occur on the American continent. The eminent ornithologist, Mr. Sclater, informs me that this is the case with the Strix punctatissima and Pyrocephalus nanus; and probably with the Otus Galapagoensis and Zenaida Galapagoensis: so that the number of endemic birds is reduced to twenty-three, or probably to twenty-one. Mr. Sclater thinks that one or two of these endemic forms should be ranked rather as varieties than species, which always seemed to me probable.

[PAGE 332]

152 This is stated by Dr. Günther (Zoolog. Soc., Jan 24th, 1859) to be a peculiar species, not known to inhabit any other country.

153 Voyage aux Quatre Iles d'Afrique. With respect to the Sandwich Islands, see Tyerman and Bennett's Journal, vol. i. p. 434. For Mauritius, see Voyage par un Officier, etc., part i. p. 170. There are no frogs in the Canary Islands (Webb et Berthelot, Hist. Nat. des Iles Canaries). I saw none at St. Jago in the Cape de Verds. There are none at St. Helena.

[PAGE 340]

154 Ann. and Mag. of Nat. Hist., vol. xvi. p. 19.

[PAGE 342]

155 Voyage in the U.S. ship Essex, vol. i. p. 215.

[PAGE 347]

156 Linn. Trans., vol. xii. p. 496. The most anomalous fact on this subject which I have met with is the wildness of the small birds in the Arctic parts of North America (as described by Richardson, Fauna Bor., vol. ii. p. 332), where they are said never to be persecuted. This case is

the more strange, because it is asserted that some of the same species in their winter-quarters in the United States are tame. There is much, as Dr. Richardson well remarks, utterly inexplicable connected with the different degrees of shyness and care with which birds conceal their nests. How strange it is that the English wood-pigeon, generally so wild a bird, should very frequently rear its young in shrubberies close to houses!

CHAPTER XVIII

None

CHAPTER XIX

[PAGE 375]

157 It is remarkable how the same disease is modified in different climates. At the little island of St. Helena the introduction of scarlet fever is dreaded as a plague. In some countries, foreigners and natives are as differently affected by certain contagious disorders as if they had been different animals; of which fact some instances have occurred in Chile; and, according to Humboldt, in Mexico (Polit. Essay, New Spain, vol. iv.).

158 Narrative of Missionary Enterprise, p. 282.

159 Captain Beechey (chap. iv., vol. i.) states that the inhabitants of Pitcairn Island are firmly convinced that after the arrival of every ship they suffer cutaneous and other disorders. Captain Beechey attributes this to the change of diet during the time of the visit. Dr. Macculloch (Western Isles, vol. ii. p. 32) says: "It is asserted, that on the arrival of a stranger (at St. Kilda) all the inhabitants, in the common phraseology, catch a cold." Dr. Macculloch considers the whole case, although often previously affirmed, as ludicrous. He adds, however, that "the question was put by us to the inhabitants who unanimously agreed in the story." In Vancouver's Voyage, there is a somewhat similar statement with respect to Otaheite. Dr. Dieffenbach, in a note to his translation of this Journal, states that the same fact is universally believed by the inhabitants of the Chatham Islands, and in parts of New Zealand. It is impossible that such a belief should have become universal in the northern hemisphere, at the Antipodes, and in the Pacific, without some good foundation. Humboldt (Polit. Essay on King of New Spain, vol. iv.) says, that the great epidemics of Panama and Callao are "marked" by the arrival of ships from Chile, because the people from that temperate region,

first experience the fatal effects of the torrid zones. I may add, that I have heard it stated in Shropshire, that sheep, which have been imported from vessels, although themselves in a healthy condition, if placed in the same fold with others, frequently produce sickness in the flock.

[PAGE 377]

160 Travels in Australia, vol. i. p. 154. I must express my obligation to Sir T. Mitchell, for several interesting personal communications on the subject of these great valleys of New South Wales.

[PAGE 380]

161 I was interested by finding here the hollow conical pitfall of the lion-ant, or some other insect; first a fly fell down the treacherous slope and immediately disappeared; then came a large but unwary ant; its struggles to escape being very violent, those curious little jets of sand, described by Kirby and Spence (Entomol., vol. i. p. 425) as being flirted by the insect's tail, were promptly directed against the expected victim. But the ant enjoyed a better fate than the fly, and escaped the fatal jaws which lay concealed at the base of the conical hollow. This Australian pitfall was only about half the size of that made by the European lion-ant.

[PAGE 385]

162 Physical Description of New South Wales and Van Diemen's Land, p. 354.

CHAPTER XX

[PAGE 391]

163 These plants are described in the Annals of Nat. Hist., vol. i., 1838, p. 337.

164 Holman's Travels, vol. iv. p. 378.

[PAGE 392]

165 Kotzebue's First Voyage, vol. iii. p. 155.

166 The thirteen species belong to the following orders:—In the *Coleoptera*, a minute Elater; *Orthoptera*, a Gryllus and a Blatta; *Hemiptera*, one species; *Homoptera*, two; *Neuroptera* a Chrysopa; *Hymenoptera*, two ants; *Lepidoptera nocturna*, a Diopæa, and a *Pterophorus* (?); *Diptera*, two species.

167 Kotzebue's First Voyage, vol. iii. p. 222.

168 The large claws or pincers of some of these crabs are most beautifully adapted, when drawn back, to form an operculum to the shell, nearly as perfect as the proper one originally belonging to the molluscous animal. I was assured, and as far as my observation went I found it so, that certain species of the hermit-crab always use certain species of shells.

[PAGE 397]

169 Some natives carried by Kotzebue to Kamtschatka collected stones to take back to their country.

[PAGE 398]

170 See Proceedings of Zoological Society, 1832, p. 17.

171 Tyerman and Bennett. Voyage, etc., vol. ii. p. 33.

[PAGE 399]

172 I exclude, of course, some soil which has been imported here in vessels from Malacca and Java, and likewise some small fragments of pumice, drifted here by the waves. The one block of greenstone, moreover, on the northern island must be excepted.

173 These were first read before the Geological Society in May, 1837, and have since been developed in a separate volume on the "Structure and Distribution of Coral Reefs."

[PAGE 402]

174 It is remarkable that Mr. Lyell, even in the first edition of his "Principles of Geology," inferred that the amount of subsidence in the Pacific must have exceeded that of elevation, from the area of land being very small relatively to the agents there tending to form it, namely, the growth of coral and volcanic action.

[PAGE 407]

175 It has been highly satisfactory to me to find the following passage in a pamphlet by Mr. Couthouy, one of the naturalists in the great Antarctic Expedition of the United States:—"Having personally examined a large number of coral-islands and resided eight months among the volcanic class having shore and partially encircling reefs. I may be permitted to state that my own observations have impressed a conviction of the correctness of the theory of Mr. Darwin."—The naturalists, however, of this expedition differ with me on some points respecting coral formations.

CHAPTER XXI

[PAGE 417]

176 After the volumes of eloquence which have poured forth on this subject, it is dangerous even to mention the tomb. A modern traveller, in twelve lines, burdens the poor little island with the following titles,—it is a grave, tomb, pyramid, cemetery, sepulchre, catacomb, sarcophagus, minaret, and mausoleum!

[PAGE 418]

177 It deserves notice, that all the many specimens of this shell found by me in one spot, differ as a marked variety, from another set of specimens procured from a different spot.

[PAGE 419]

178 Beatson's St. Helena. Introductory chapter, p. 4.

179 Among these few insects, I was surprised to find a small Aphodius *(nov. spec.)* and an Oryctes, both extremely numerous under dung. When the island was discovered it certainly possessed no quadruped, excepting *perhaps* a mouse: it becomes, therefore, a difficult point to ascertain, whether these stercovorous insects have since been imported by accident, or if aborigines, on what food they formerly subsisted. On the banks of the Plata, where, from the vast number of cattle and horses, the fine plains of turf are richly manured, it is vain to seek the many kinds of dung-feeding beetles, which occur so abundantly in Europe. I observed only an Oryctes (the insects of this genus in Europe generally feed on decayed vegetable matter) and two species of Phanæus, common in such situations. On the opposite side of the Cordillera in Chiloe, another species of Phanæus is exceedingly abundant, and it buries the dung of the cattle in large earthen balls beneath the ground. There is reason to believe that the genus Phanæus, before the introduction of cattle, acted as scavengers to man. In Europe, beetles, which find support in the matter which has already contributed towards the life of other and larger animals, are so numerous that there must be considerably more than one hundred different species. Considering this, and observing what a quantity of food of this kind is lost on the plains of La Plata, I imagined I saw an instance where man had disturbed that chain, by which so many animals are linked together in their native country. In Van Diemen's Land, however, I found four species of Onthophagus, two of Aphodius, and one of a third genus, very abundantly under the dung of cows; yet these latter animals had been then introduced only thirty-three years. Previously to that time the kangaroo and some other small

animals were the only quadrupeds; and their dung is of a very different quality from that of their successors introduced by man. In England the greater number of stercovorous beetles are confined in their appetites; that is, they do not depend indifferently on any quadruped for the means of subsistence. The change, therefore, in habits which must have taken place in Van Diemen's Land is highly remarkable. I am indebted to the Rev. F. W. Hope, who, I hope, will permit me to call him my master in Entomology, for giving me the names of the foregoing insects.

[PAGE 422]

180 Monats. der König. Akad. d. Wiss. zu Berlin. Vom April, 1845.

[PAGE 425]

181 I have described this Bar in detail, in the Lond. and Edin. Phil. Mag., vol. xix. (1841), p. 257.

INTRODUCTION

[PAGE 778]

1 As the works of the first-named authors are so well known, I need not give the titles; but as those of the latter are less well known in England, I will give them:—'Sechs Vorlesungen über die Darwin'sche Theorie:' zweite Auflage, 1868, von Dr. L. Büchner; translated into French under the title 'Conférences sur la Théorie Darwinienne,' 1869. 'Der Mensch, im Lichte der Darwin'sche Lehre,' 1865, von Dr. F. Rolle. I will not attempt to give references to all the authors who have taken the same side of the question. Thus G. Canestrini has published ('Annuario della Soc. d. Nat.,' Modena, 1867, p. 81) a very curious paper on rudimentary characters, as bearing on the origin of man. Another work has (1869) been published by Dr. Barrago Francesco, bearing in Italian the title of "Man, made in the image of God, was also made in the image of the ape."

[PAGE 779]

2 Prof. Häckel is the sole author who, since the publication of the 'Origin,' has discussed, in his various works, in a very able manner, the subject of sexual selection, and has seen its full importance.

CHAPTER I

[PAGE 784]

3 'Grosshirnwindungen des Menschen,' 1868, s. 96.

4 'Leç. sur la Phys.' 1866, p. 890, as quoted by M. Dally, 'L'Ordre des Primates et le Transformisme,' 1868, p. 29.

[PAGE 785]

5 'Naturgeschichte der Säugethiere von Paraguay,' 1830, s. 50.

6 Brehm, 'Thierleben,' B. i. 1864, s. 75, 86. On the Ateles, s. 105. For other analogous statements, see s. 25, 107.

7 With respect to insects see Dr. Laycock 'On a General Law of Vital Periodicity,' British Association, 1842. Dr. Macculloch, 'Silliman's North American Journal of Science,' vol. xvii. p. 305, has seen a dog suffering from tertian ague.

8 I have given the evidence on this head in my 'Variation of Animals and Plants under Domestication,' vol. ii. p. 15.

9 "Mares e diversis generibus Quadrumanorum sine dubio dignoscunt feminas humanas a maribus. Primum, credo, odoratu, postea aspectu. Mr. Youatt, qui diu in Hortis Zoologicis (Bestiariis) medicus animalium erat, vir in rebus observandis cautus et sagax, hoc mihi certissime probavit, et curatores ejusdem loci et alii e ministris confirmaverunt. Sir Andrew Smith et Brehm notabant idem in Cynocephalo. Illustrissimus Cuvier etiam narrat multa de hac re quâ ut opinor nihil turpius potest indicari inter omnia hominibus et Quadrumanis communia. Narrat enim Cynocephalum quendam in furorem incidere aspectu feminarum aliquarum, sed nequaquam accendi tanto furore ab omnibus. Semper eligebat juniores, et dignoscebat in turba, et advocabat voce gestuque."

10 This remark is made with respect to Cynocephalus and the anthropomorphous apes by Geoffroy Saint-Hilaire and F. Cuvier, 'Hist. Nat. des Mammifères,' tom. i. 1824.

11 Huxley, 'Man's Place in Nature,' 1863, p. 34.

[PAGE 786]

12 'Man's Place in Nature,' 1863, p. 67.

[PAGE 787]

13 The human embryo (upper fig.) is from Ecker, 'Icones Phys.,' 1851–1859, tab. xxx. fig. 2. This embryo was ten lines in length, so that the drawing is much magnified. The embryo of the dog is from Bischoff, 'Entwicklungsgeschichte des Hunde-Eies,' 1845, tab. xi. fig. 42 B. This drawing is five times magnified, the embryo being 25 days old. The internal viscera have been omitted, and the uterine appendages in both drawings removed. I was directed to these figures by Prof. Huxley, from whose work, 'Man's Place in Nature,' the idea of giving them was taken. Häckel has also given analogous drawings in his 'Schöpfungsgeschichte.'

14 Prof. Wyman in 'Proc. of American Acad. of Sciences,' vol. iv. 1860, p. 17.

15 Owen, 'Anatomy of Vertebrates,' vol. i. p. 533.

16 'Die Grosshirnwindungen des Menschen,' 1868, s. 95.

17 'Anatomy of Vertebrates,' vol. ii. p. 553.

18 'Proc. Soc. Nat. Hist.' Boston, 1863, vol. ix. p. 185.

19 'Man's Place in Nature,' p. 65.

20 I had written a rough copy of this chapter before reading a valuable paper, "Caratteri rudimentali in ordine all' origine del uomo" ('Annuario della Soc. d. Nat.,' Modena, 1867, p. 81), by G. Canestrini, to which paper I am considerably indebted. Häckel has given admirable discussions on this whole subject, under the title of Dysteleology, in his 'Generelle Morphologie' and 'Schöpfungsgeschichte.'

[PAGE 788]

21 Some good criticisms on this subject have been given by Messrs. Murie and Mivart, in 'Transact. Zoolog. Soc' 1869, vol. vii. p. 92.

22 'Variation of Animals and Plants under Domestication,' vol. ii. pp. 317 and 397. See also 'Origin of Species,' 5th edit. p. 535.

23 For instance M. Richard ('Annales des Sciences Nat.' 3rd series, Zoolog. 1852, tom, xviii. p. 13) describes and figures rudiments of what he calls the " muscle pédieux de la main," which he says is sometimes "infiniment petit." Another muscle, called "le tibial postérieur," is generally quite absent in the hand, but appears from time to time in a more or less rudimentary condition.

24 Prof. W. Turner, 'Proc. Royal Soc. Edinburgh,' 1866–67, p. 65.

[PAGE 789]

25 Canestrini quotes Hyrt. ('Annuario della Soc. dei Naturalisti,' Modena, 1867, p. 97) to the same effect.

26 'The Diseases of the Ear,' by J. Toynbee, F.R.S., 1860, p. 12.

[PAGE 790]

27 See also some remarks, and the drawings of the ears of the Lemuroidea, in Messrs. Murie and Mivart's excellent paper in 'Transact. Zoolog. Soc.' vol. vii. 1869, pp. 6 and 90.

[PAGE 791]

28 Müller's 'Elements of Physiology,' Eng. translat., 1842, vol. ii. p. 1117. Owen, 'Anatomy of Vertebrates,' vol. iii. p. 260; ibid. on the Walrus, 'Proc. Zoolog. Soc.' November 8th, 1854. See also R. Knox, 'Great Artists and Anatomists,' p. 106. This rudiment apparently is somewhat larger in Negroes and Australians than in Europeans, see Carl Vogt, 'Lectures on Man,' Eng. translat. p. 129.

29 'The Physiology and Pathology of Mind,' 2nd edit. 1868, p. 134.

30 Eschricht, Ueber die Richtung der Haare am menschlichen Körp er 'Müllers Archiv für Anat. und Phys.' 1837, s. 47. I shall often have to refer to this very curious paper.

31 Paget, 'Lectures on Surgical Pathology,' 1853, vol. i. p. 71.

[PAGE 792]

32 Eschricht, ibid. s. 40, 47.

33 Dr. Webb, 'Teeth in Man and the Anthropoid Apes,' as quoted by Dr. C. Carter Blake in 'Anthropological Review,' July, 1867, p. 299.

34 Owen, 'Anatomy of Vertebrates,' vol. iii. pp. 320, 321, and 325.

35 'On the Primitive Form of the Skull,' Eng. translat. in 'Anthropological Review,' Oct. 1868, p. 426.

[PAGE 793]

36 Owen, 'Anatomy of Vertebrates,' vol. iii. pp. 416, 434, 441.

37 'Annuario della Soc. d. Nat.' Modena, 1867, p. 94.

38 M. C. Martins ("De l'Unité Organique," in 'Revue des Deux Mondes,' June 15, 1862, p. 16), and Häckel ('Generelle Morphologie,' B. ii. s. 278), have both remarked on the singular fact of this rudiment sometimes causing death.

39 'The Lancet,' Jan. 24, 1863, p. 83. Dr. Knox, 'Great Artists and Anatomists,' p. 63. See also an important memoir on this process by Dr. Grube, in the 'Bulletin de l'Acad. Imp. de St. Pétersbourg,' tom. xii. 1867, p. 448.

40 "On the Caves of Gibraltar," 'Transact. Internat. Congress of Prehist. Arch.' Third Session, 1869, p. 54.

[PAGE 794]

41 Quatrefages has lately collected the evidence on this subject. 'Revue des Cours Scientifiques,' 1867–1868, p. 625.

42 Owen, 'On the Nature of Limbs,' 1849, p. 114.

[PAGE 795]

43 Leuckart, in Todd's 'Cyclop. of Anat.' 1849–52, vol. iv. p. 1415. In man this organ is only from three to six lines in length, but, like so many other rudimentary parts, it is variable in development as well as in other characters.

44 See, on this subject, Owen, 'Anatomy of Vertebrates,' vol. iii. pp. 675, 676, 706.

Chapter II

[page 797]

45 See the evidence on these points, as given by Lubbock, 'Prehistoric Times,' p. 354, &c.

[page 799]

46 'L'Instinct chez les Insectes.' 'Revue des Deux Mondes,' Feb. 1870. p. 690.

47 'The American Beaver and his Works,' 1868.

48 'The Principles of Psychology,' 2nd edit. 1870, pp. 418–443

[page 800]

49 'Contributions to the Theory of Natural Selection,' 1870, p. 212.

50 'Recherches sur les Mœurs des Fourmis,' 1810, p. 173.

[page 801]

51 All the following statements, given on the authority of these two naturalists, are taken from Rengger's 'Naturges. der Säugethiere von Paraguay,' 1830, s. 41–57, and from Brehm's 'Thierleben,' B. i. s. 10–87.

52 'Bridgewater Treatise,' p. 263.

[page 803]

53 W. C. L. Martin, 'Nat. Hist. of Mammalia,' 1841, p. 405.

54 Quoted by Vogt, 'Mémoire sur les Microcéphales,' 1867, p. 168.

55 'The Variation of Animals and Plants under Domestication,' vol. i. p. 27.

[page 804]

56 'Les Mœurs des Fourmis,' 1810, p. 150.

57 Quoted in Dr. Maudsley's 'Physiology and Pathology of Mind,' 1868, pp. 19, 220.

58 Dr. Jerdon, 'Birds of India,' vol. i. 1862, p. xxi.

59 Mr. L. H. Morgan's work on 'The American Beaver,' 1868, offers a good illustration of this remark. I cannot, however, avoid thinking that he goes too far in underrating the power of Instinct.

[page 805]

60 'The Moor and the Loch,' p. 45. Col. Hutchinson on 'Dog Breaking,' 1850, p. 46.

61 'Personal Narrative,' Eng. translat., vol. iii. p. 106.

[PAGE 806]

62 Quoted by Sir C. Lyell, 'Antiquity of Man,' p. 497.

63 'Journal of Researches during the Voyage of the "Beagle,"' 1845, p. 383. 'Origin of Species,' 5th edit. p. 260.

64 'Lettres Phil. sur l'Intelligence des Animaux,' nouvelle edit. 1802, p. 86.

65 See the evidence on this head in chap. i. vol. i. 'On the Variation of Animals and Plants under Domestication.'

[PAGE 807]

66 'Proc. Zoolog. Soc.' 1864, p. 186.

67 Savage and Wyman in 'Boston Journal of Nat. Hist.' vol. iv. 1843–44, p. 383.

68 'Säugethiere von Paraguay,' 1830, s. 51–56.

69 'Thierleben,' B. i. s. 79, 82.

70 'The Malay Archipelago,' vol. i. 1869, p. 87.

[PAGE 808]

71 'Primeval Man,' 1869, pp. 145, 147.

72 'Prehistoric Times,' 1865, p. 473, &c.

73 Quoted in 'Anthropological Review,' 1864, p. 158.

74 Rengger, ibid. s. 45.

75 See my 'Variation of Animals and Plants under Domestication,' vol. i. p. 27.

[PAGE 809]

76 See a discussion on this subject in Mr. E. B. Tylor's very interesting work, 'Researches into the Early History of Mankind,' 1865, chaps. ii. to iv.

77 Hon. Daines Barrington in 'Philosoph. Transactions,' 1773, p. 262. See also Dureau de la Malle, in 'Ann. des Sc. Nat.' 3rd series, Zoolog. tom. x. p. 119.

[PAGE 810]

78 'On the Origin of Language,' by H. Wedgwood, 1866. 'Chapters on Language,' by the Rev. F. W. Farrar, 1865. These works are most interesting. See also 'De la Phys. et de Parole,' par Albert Lemoine, 1865, p. 190. The work on this subject, by the late Prof. Aug. Schleicher, has

been translated by Dr. Bikkers into English, under the title of 'Darwinism tested by the Science of Language,' 1869.

79 Vogt, 'Mémoire sur les Microcéphales,' 1867, p. 169. With respect to savages, I have given some facts in my 'Journal of Researches,' &c, 1845, p. 209.

80 See clear evidence on this head in the two works so often quoted, by Brehm and Rengger.

81 See remarks on this head by Dr. Maudsley, 'The Physiology and Pathology of Mind,' 2nd edit. 1868, p. 199.

[PAGE 811]

82 Many curious cases have been recorded. See, for instance, 'Inquiries Concerning the Intellectual Powers,' by Dr. Abercrombie, 1838, p. 150.

83 'The Variation of Animals and Plants under Domestication,' vol. ii. p. 6.

84 See some good remarks to this effect by Dr. Maudsley, 'The Physiology and Pathology of Mind,' 1868, p. 199.

85 Macgillivray, 'Hist. of British Birds,' vol. ii. 1839, p. 29. An excellent observer, Mr. Blackwall, remarks that the magpie learns to pronounce single words, and even short sentences, more readily than almost any other British bird; yet, as he adds, after long and closely investigating its habits, he has never known it, in a state of nature, display any unusual capacity for imitation. 'Researches in Zoology,' 1834, p. 158.

86 See the very interesting parallelism between the development of species and languages, given by Sir C. Lyell in 'The Geolog. Evidences of the Antiquity of Man,' 1863, chap. xxiii.

[PAGE 812]

87 See remarks to this effect by the Rev. F. W. Farrar, in an interesting article, entitled "Philology and Darwinism" in 'Nature,' March 24th, 1870, p. 528.

88 'Nature,' Jan. 6th, 1870, p. 257.

89 Quoted by C. S. Wake, 'Chapters on Man,' 1868, p. 101.

90 Buckland, 'Bridgewater Treatise,' p. 411.

[PAGE 813]

91 See some good remarks on the simplification of languages, by Sir J. Lubbock, 'Origin of Civilisation,' 1870, p. 278.

92 'Conférences sur la Théorie Darwinienne,' French translat., 1869, p. 132.

93 The Rev. Dr. J. M'Cann, 'Anti-Darwinism,' 1869, p. 13.

[PAGE 814]

94 'The Spectator,' Dec. 4th, 1869, p. 1430.

[PAGE 815]

95 See an excellent article on this subject by the Rev. F. W. Farrar, in the 'Anthropological Review,' Aug. 1864, p. ccxvii. For further facts see Sir J. Lubbock, 'Prehistoric Times,' 2nd edit. 1869, p. 564; and especially the chapters on Religion in his 'Origin of Civilisation,' 1870.

96 The Worship of Animals and Plants, in the 'Fortnightly Review,' Oct. 1, 1869, p. 422.

97 Tylor, 'Early History of Mankind,' 1865, p. 6. See also the three striking chapters on the Development of Religion, in Lubbock's 'Origin of Civilisation,' 1870. In a like manner Mr. Herbert Spencer, in his ingenious essay in the 'Fortnightly Review' (May 1st, 1870, p. 535), accounts for the earliest forms of religious belief throughout the world, by man being led through dreams, shadows, and other causes, to look at himself as a double essence, corporeal and spiritual. As the spiritual being is supposed to exist after death and to be powerful, it is propitiated by various gifts and ceremonies, and its aid invoked. He then further shews that names or nicknames given from some animal or other object to the early progenitors or founders of a tribe, are supposed after a long interval to represent the real progenitor of the tribe; and such animal or object is then naturally believed still to exist as a spirit, is held sacred, and worshipped as a god. Nevertheless I cannot but suspect that there is a still earlier and ruder stage, when anything which manifests power or movement is thought to be endowed with some form of life, and with mental faculties analogous to our own.

[PAGE 816]

98 See an able article on the Psychical Elements of Religion, by Mr. L. Owen Pike, in 'Anthropolog. Review,' April, 1870, p. lxiii.

99 'Religion, Moral, &c, der Darwin'schen Art-Lehre,' 1869, s. 53.

100 'Prehistoric Times,' 2nd edit. p. 571. In this work (at p. 553) there will be found an excellent account of the many strange and capricious customs of savages.

CHAPTER III

[PAGE 817]

101 See, for instance, on this subject, Quatrefages, 'Unité de l'Espèce Humaine,' 1861, p. 21, &c.

102 'Dissertation on Ethical Philosophy,' 1837, p. 231, &c.

103 'Metaphysics of Ethics,' translated by J. W. Semple, Edinburgh, 1836, p. 136.

[PAGE 818]

104 Mr. Bain gives a list ('Mental and Moral Science,' 1868, p. 543–725) of twenty-six British authors who have written on this subject, and whose names are familiar to every reader; to these, Mr. Bain's own name, and those of Mr. Lecky, Mr. Shadworth Hodgson, and Sir J. Lubbock, as well as of others, may be added.

105 Sir B. Brodie, after observing that man is a social animal ('Psychological Enquiries,' 1854, p. 192), asks the pregnant question, "ought not this to settle the disputed question as to the existence of a moral sense?" Similar ideas have probably occurred to many persons, as they did long ago to Marcus Aurelius. Mr. J. S. Mill speaks, in his celebrated work, 'Utilitarianism,' (1864, p. 46), of the social feelings as a "powerful natural sentiment," and as "the natural basis of sentiment for utilitarian morality;" but on the previous page he says, "if, as is my own belief, the moral feelings are not innate, but acquired, they are not for that reason less natural." It is with hesitation that I venture to differ from so profound a thinker, but it can hardly be disputed that the social feelings are instinctive or innate in the lower animals; and why should they not be so in man? Mr. Bain (see, for instance, 'The Emotions and the Will,' 1865, p. 481) and others believe that the moral sense is acquired by each individual during his lifetime. On the general theory of evolution this is at least extremely improbable.

[PAGE 819]

106 'Die Darwin'sche Theorie,' s. 101.

107 Mr. R. Brown in 'Proc. Zoolog. Soc.' 1868, p. 409.

[PAGE 820]

108 Brehm, 'Thierleben,' B. i. 1864, s. 52, 79. For the case of the monkeys extracting thorns from each other, see s. 54. With respect to the Hamadryas turning over stones, the fact is given (s. 76) on the evi-

dence of Alvarez, whose observations Brehm thinks quite trustworthy. For the cases of the old male baboons attacking the dogs, see s. 79; and with respect to the eagle, s. 56.

109 'Annals and Mag. of Nat. Hist.' November, 1868, p. 382.

[PAGE 821]

110 Sir J. Lubbock, 'Prehistoric Times,' 2nd edit. p. 446.

111 As quoted by Mr. L. H. Morgan, 'The American Beaver,' 1868, p. 272. Capt. Stansbury also gives an interesting account of the manner in which a very young pelican, carried away by a strong stream, was guided and encouraged in its attempts to reach the shore by half a dozen old birds.

112 As Mr. Bain states, "effective aid to a sufferer springs from sympathy proper:" 'Mental and Moral Science,' 1868, p. 245.

113 'Thierleben,' B. i. s. 85.

114 'De l'Espèce et de la Class,' 1869, p. 97.

[PAGE 822]

115 'Der Darwin'schen Art-Lehre,' 1869, s. 54.

116 Brehm, 'Thierleben,' B. i. s. 76.

[PAGE 823]

117 See the first and striking chapter in Adam Smith's 'Theory of Moral Sentiments.' Also Mr. Bain's 'Mental and Moral Science,' 1868, p. 244, and 275–282. Mr. Bain states, that "sympathy is, indirectly, a source of pleasure to the sympathiser;" and he accounts for this through reciprocity. He remarks that "the person benefited, or others in his stead, may make up, by sympathy and good offices returned, for all the sacrifice." But if, as appears to be the case, sympathy is strictly an instinct, its exercise would give direct pleasure, in the same manner as the exercise, as before remarked, of almost every other instinct.

[PAGE 824]

118 This fact, the Rev. L. Jenyns states (see his edition of 'White's Nat. Hist. of Selborne,' 1853, p. 204) was first recorded by the illustrious Jenner, in 'Phil. Transact.' 1824, and has since been confirmed by several observers, especially by Mr. Blackwall. This latter careful observer examined, late in the autumn, during two years, thirty-six nests; he found that twelve contained young dead birds, five contained eggs on the point of being hatched, and three eggs not nearly hatched. Many birds not yet old enough for a prolonged flight are likewise

deserted and left behind. See Blackwall, 'Researches in Zoology,' 1834, pp. 108, 118. For some additional evidence, although this is not wanted, see Leroy, 'Lettres Phil.' 1802, p. 217.

[PAGE 825]

119 Hume remarks ('An Enquiry Concerning the Principles of Morals,' edit. of 1751, p. 132), "there seems a necessity for confessing that the happiness and misery of others are not spectacles altogether indifferent to us, but that the view of the former . . . communicates a secret joy; the appearance of the latter . . . throws a melancholy damp over the imagination."

[PAGE 826]

120 'Mental and Moral Science,' 1868, p. 254.

[PAGE 827]

121 I have given one such case, namely of three Patagonian Indians who preferred being shot, one after the other, to betraying the plans of their companions in war ('Journal of Researches,' 1845, p. 103).

[PAGE 830]

122 Dr. Prosper Despine, in his 'Psychologie Naturelle,' 1868 (tom. i. p. 243; tom ii. p. 169) gives many curious cases of the worst criminals, who apparently have been entirely destitute of conscience.

123 See an able article in the 'North British Review,' 1867, p. 395. See also Mr. W. Bagehot's articles on the Importance of Obedience and Coherence to Primitive Man, in the 'Fortnightly Review,' 1867, p. 529, and 1868, p. 457, &c.

124 The fullest account which I have met with is by Dr. Gerland, in his 'Ueber das Aussterben der Naturvölker,' 1868; but I shall have to recur to the subject of infanticide in a future chapter.

125 See the very interesting discussion on Suicide in Lecky's 'History of European Morals,' vol. i. 1869, p. 223.

[PAGE 831]

126 See, for instance, Mr. Hamilton's account of the Kaffirs, 'Anthropological Review,' 1870, p. xv.

[PAGE 832]

127 Mr. M'Lennan has given ('Primitive Marriage,' 1865, p. 176) a good collection of facts on this head.

128 Lecky, 'History of European Morals,' vol. i. 1869, p. 109.
129 'Embassy to China,' vol. ii. p. 348.
130 See on this subject copious evidence in Chap. vii. of Sir J. Lubbock, 'Origin of Civilisation,' 1870.
131 For instance Lecky, 'Hist. European Morals,' vol. i. p. 124.
132 This term is used in an able article in the 'Westminister Review,' Oct. 1869, p. 498. For the Greatest Happiness principle, see J. S. Mill, 'Utilitarianism,' p. 17.

[PAGE 834]

133 Good instances are given by Mr. Wallace in 'Scientific Opinion,' Sept. 15, 1869; and more fully in his 'Contributions to the Theory of Natural Selection,' 1870, p. 353.

[PAGE 835]

134 Tennyson, 'Idylls of the King,' p. 244.
135 'The Thoughts of the Emperor M. Aurelius Antoninus,' Eng. translat., 2nd edit., 1869, p. 112. Marcus Aurelius was born A.D. 121.
136 Letter to Mr. Mill in Bain's 'Mental and Moral Science,' 1868, p. 722.

[PAGE 836]

137 A writer in the 'North British Review' (July, 1869, p. 531), well capable of forming a sound judgment, expresses himself strongly to this effect. Mr. Lecky ('Hist. of Morals,' vol. i. p. 143) seems to a certain extent to coincide.
138 See his remarkable work on 'Hereditary Genius,' 1869, p. 349. The Duke of Argyll ('Primeval Man,' 1869, p. 188) has some good remarks on the contest in man's nature between right and wrong.

[PAGE 837]

139 'The Thoughts of Marcus Aurelius,' &c., p. 139.

CHAPTER IV

[PAGE 840]

140 'Investigations in Military and Anthropolog. Statistics of American Soldiers,' by B. A. Gould, 1869, p. 256.
141 With respect to the "Cranial forms of the American aborigines," see Dr. Aitken Meigs in 'Proc. Acad. Nat. Sci.' Philadelphia, May, 1866. On the Australians, see Huxley, in Lyell's 'Antiquity of Man,' 1863,

p. 87. On the Sandwich Islanders, Prof. J. Wyman, 'Observations on Crania,' Boston, 1868, p. 18.

142 'Anatomy of the Arteries,' by R. Quain.

143 'Transact. Royal Soc.' Edinburgh, vol. xxiv. p. 175, 189.

144 'Proc. Royal Soc.' 1867, p. 544; also 1868, p. 483, 524. There is a previous paper, 1866, p. 229.

145 'Proc. R. Irish Academy,' vol. x. 1868, p. 141.

146 'Act. Acad.,' St. Petersburg, 1778, part ii. p. 217.

[PAGE 841]

147 Brehm, 'Thierleben,' B. i. s. 58, 87. Rengger, 'Säugethiere von Paraguay,' s. 57.

148 'Variation of Animals and Plants under Domestication,' vol. ii. Chap. xii.

149 'Hereditary Genius: an Inquiry into its Laws and Consequences,' 1869.

150 Mr. Bates remarks ('The Naturalist on the Amazons,' 1863, vol. ii. p. 159), with respect to the Indians of the same S. American tribe, no two of them were at all similar in the shape of the head; one man had an oval visage with fine features, and another was quite Mongolian in breadth and prominence of cheek, spread of nostrils, and obliquity of eyes."

[PAGE 842]

151 Blumenbach, 'Treatises on Anthropolog.' Eng. translat., 1865, p. 205.

152 Godron, 'De l'Espèce,' 1859, tom. ii. livre 3. Quatrefages, 'Unité de l'Espèce Humaine,' 1861. Also Lectures on Anthropology, given in the 'Revue des Cours Scientifiques,' 1866–1868.

153 'Hist. Gen. et Part. des Anomalies de l'Organisation,' in three volumes, tom. i. 1832.

154 I have fully discussed these laws in my 'Variation of Animals and Plants under Domestication,' vol. ii. chap. xxii. and xxiii. M. J. P. Durand has lately (1868) published a valuable essay 'De l'Influence des Milieux, &c.' He lays much stress on the nature of the soil.

[PAGE 843]

155 'Investigations in Military and Anthrop. Statistics,' &c. 1869, by B. A. Gould, p. 93, 107, 126, 131, 134.

156 For the Polynesians, see Prichard's 'Physical Hist, of Mankind,' vol. v. 1847, p. 145, 283. Also Godron, 'De l'Espèce,' tom. ii. p. 289. There is also a remarkable difference in appearance between the closely-

allied Hindoos inhabiting the Upper Ganges and Bengal; see Elphinstone's 'History of India,' vol. i. p. 324.

157 'Memoirs, Anthropolog. Soc.' vol. iii. 1867–69, p. 561, 565, 567.

[PAGE 844]

158 Dr. Brakenridge, 'Theory of Diathesis,' 'Medical Times,' June 19 and July 17, 1869.

159 I have given authorities for these several statements in my 'Variation of Animals under Domestication,' vol. ii. p. 297–300. Dr. Jaeger, "Ueber das Längenwachsthum der Knochen," 'Jenaischen Zeitschrift,' B. v. Heft i.

160 'Investigations,' &c. By B. A. Gould, 1869, p. 288.

161 'Säugethiere von Paraguay,' 1830, s. 4.

162 'History of Greenland,' Eng. translat. 1767, vol. i. p. 230.

[PAGE 845]

163 'Intermarriage.' By Alex. Walker, 1838, p. 377.

164 'The Variation of Animals under Domestication,' vol. i. p. 173.

165 'Principles of Biology,' vol. i. p. 455.

166 Paget, 'Lectures on Surgical Pathology,' vol. i. 1853, p. 209.

167 'The Variation of Animals under Domestication,' vol. ii. p. 8.

168 'Säugethiere von Paraguay,' s. 8, 10. I have had good opportunities for observing the extraordinary power of eyesight in the Fuegians. See also Lawrence ('Lectures on Physiology,' &c., 1822, p. 404) on this same subject. M. Giraud-Teulon has recently collected ('Revue des Cours Scientifiques,' 1870, p. 625) a large and valuable body of evidence proving that the cause of short-sight, *"C'est le travail assidu, de près."*

169 Prichard, 'Phys. Hist. of Mankind,' on the authority of Blumenbach, vol. i. 1851, p. 311; for the statement by Pallas, vol. iv. 1844, p. 407.

170 Quoted by Prichard, 'Researches into the Phys. Hist. of Mankind,' vol. v. p. 463.

171 Mr. Forbes' valuable paper is now published in the 'Journal of the Ethnological Soc. of London,' new series, vol. ii. 1870, p. 193.

[PAGE 846]

172 Dr. Wilckens ('Landwirthschaft. Wochenblatt,' No. 10, 1869) has lately published an interesting essay shewing how domestic animals, which live in mountainous regions, have their frames modified.

[PAGE 847]

173 'Mémoire sur les Microcéphales,' 1867, p. 50, 125, 169, 171, 184–198.

[PAGE 848]

174 See Dr. A. Farre's well-known article in the 'Cyclop. of Anat. and Phys.' vol. v. 1859, p. 642. Owen 'Anatomy of Vertebrates,' vol. iii. 1868 p. 687. Prof. Turner in 'Edinburgh Medical Journal,' Feb. 1865.

175 'Annuario della Soc. dei Naturalisti in Modena,' 1867, p. 83. Prof. Canestrini gives extracts on this subject from various authorities. Laurillard remarks, that as he has found a complete similarity in the form, proportions, and connexion of the two malar bones in several human subjects and in certain apes, he cannot consider this disposition of the parts as simply accidental.

[PAGE 849]

176 A whole series of cases is given by Isid. Geoffroy St.-Hilaire, 'Hist. des Anomalies,' tom. iii. p. 437.

177 In my 'Variation of Animals under Domestication' (vol. ii. p. 57) I attributed the not very rare cases of supernumerary mammæ in women to reversion. I was led to this as a *probable* conclusion, by the additional mammæ being generally placed symmetrically on the breast, and more especially from one case, in which a single efficient mamma occurred in the inguinal region of a woman, the daughter of another woman with supernumerary mammæ. But Prof. Preyer ('Der Kampf um das Dasein,' 1869, s. 45) states that *mammæ erraticæ* have been known to occur in other situations, even on the back; so that the force of my argument is greatly weakened or perhaps quite destroyed.

With much hesitation I, in the same work (vol. ii. p. 12), attributed the frequent cases of polydactylism in men to reversion. I was partly led to this through Prof. Owen's statement, that some of the Ichthyopterygia possess more than five digits, and therefore, as I supposed, had retained a primordial condition; but after reading Prof. Gegenbaur's paper ('Jenaischen Zeitschrift,' B. v. Heft 3, s. 341), who is the highest authority in Europe on such a point, and who disputes Owen's conclusion, I see that it is extremely doubtful whether supernumerary digits can thus be accounted for. It was the fact that such digits not only frequently occur and are strongly inherited, but have the power of regrowth after amputation, like the normal digits of the lower vertebrata, that chiefly led me to the above conclusion. This extraordinary fact of their regrowth remains inexplicable, if the belief in reversion to some extremely remote progenitor must be rejected. I cannot, however, follow Prof. Gegenbaur in supposing that additional digits could not reappear through reversion, without at the same time other parts of the skeleton being simultaneously and similarly modified; for single characters often reappear through reversion.

178 'Anatomy of Vertebrates,' vol. iii. 1868, p. 323.

179 'Generelle Morphologie,' 1866, B. ii. s. clv.

180 Carl Vogt's 'Lectures on Man,' Eng. translat. 1864, p. 151.

181 C. Carter Blake, on a jaw from La Naulette, 'Anthropolog. Review,' 1867, p. 295. Schaaffhausen, ibid. 1868, p. 426.

182 'The Anatomy of Expression,' 1844, p. 110, 131.

[PAGE 850]

183 Quoted by Prof. Canestrini in the 'Annuario,' &c., 1867, p. 90.

184 These papers deserve careful study by any one who desires to learn how frequently our muscles vary, and in varying come to resemble those of the Quadrumana. The following references relate to the few points touched on in my text: 'Proc. Royal Soc.' vol. xiv. 1865, p. 379–384; vol. xv. 1866, p. 241, 242; vol. xv. 1867, p. 544; vol. xvi. 1868, p. 524. I may here add that Dr. Murie and Mr. St. George Mivart have shewn in their Memoir on the Lemuroidea ('Transact. Zoolog. Soc.' vol. vii. 1869, p. 96), how extraordinarily variable some of the muscles are in these animals, the lowest members of the Primates. Gradations, also, in the muscles leading to structures found in animals still lower in the scale, are numerous in the Lemuroidea.

185 Prof. Macalister in 'Proc. R. Irish Academy,' vol. x. 1868, p. 124.

186 Prof. Macalister (ibid. p. 121) has tabulated his observations, and finds that muscular abnormalities are most frequent in the forearms, secondly in the face, thirdly in the foot, &c.

187 The Rev. Dr. Haughton, after giving ('Proc. R. Irish Academy,' June 27, 1864, p. 715) a remarkable case of variation in the human *flexor pollicis longus,* adds, "This remarkable example shews that man may sometimes possess the arrangement of tendons of thumb and fingers characteristic of the macaque; but whether such a case should be regarded as a macaque passing upwards into a man, or a man passing downwards into a macaque, or as a congenital freak of nature, I cannot undertake to say." It is satisfactory to hear so capable an anatomist, and so embittered an opponent of evolutionism, admitting even the possibility of either of his first propositions. Prof. Macalister has also described ('Proc. R. Irish Acad.' vol. x. 1864, p.

138) variations in the *flexor pollicis longus,* remarkable from their relations to the same muscle in the Quadrumana.

[PAGE 851]

188 The authorities for these several statements are given in my 'Variation of Animals under Domestication,' vol. ii. p. 320–335.

189 This whole subject has been discussed in chap. xxiii. vol. ii. of my 'Variation of Animals and Plants under Domestication.'

190 See the ever memorable 'Essay on the Principle of Population,' by the Rev. T. Malthus, vol. i. 1826, p. 6, 517.

[PAGE 852]

191 'Variation of Animals and Plants under Domestication,' vol. ii. p. 111–113, 163.

192 Mr. Sedgwick, 'British and Foreign Medico-Chirurg. Review,' July, 1863, p. 170.

193 'The Annals of Rural Bengal,' by W. W. Hunter, 1868, p. 259.

[PAGE 853]

194 'Primitive Marriage,' 1865.

[PAGE 854]

195 See some good remarks to this effect by W. Stanley Jevons, "A Deduction from Darwin's Theory," 'Nature,' 1869, p. 231.

196 Latham, 'Man and his Migrations,' 1851, p. 135.

197 Messrs. Murie and Mivart in their "Anatomy of the Lemuroidea" ('Transact. Zoolog. Soc.' vol. vii. 1869, p. 96–98) say, "some muscles are so irregular in their distribution that they cannot be well classed in any of the above groups." These muscles differ even on the opposite sides of the same individual.

[PAGE 855]

198 'Quarterly Review,' April, 1869, p. 392. This subject is more fully discussed in Mr. Wallace's 'Contributions to the Theory of Natural Selection,' 1870, in which all the essays referred to in this work are republished. The 'Essay on Man' has been ably criticised by Prof. Claparède, one of the most distinguished zoologists in Europe, in an article published in the 'Bibliothèque Universelle,' June, 1870. The remark quoted in my text will surprise every one who has read Mr. Wallace's celebrated paper on 'The Origin of Human Races

deduced from the Theory of Natural Selection,' originally published in the 'Anthropological Review,' May, 1864, p. clviii. I cannot here resist quoting a most just remark by Sir J. Lubbock ('Prehistoric Times,' 1865, p. 479) in reference to this paper, namely, that Mr. Wallace, "with characteristic unselfishness, ascribes it *(i.e.* the idea of natural selection) unreservedly to Mr. Darwin, although, as is well known, he struck out the idea independently, and published it, though not with the same elaboration, at the same time."

199 Quoted by Mr. Lawson Tait in his "Law of Natural Selection," — 'Dublin Quarterly Journal of Medical Science,' Feb. 1869. Dr. Keller is likewise quoted to the same effect.

[PAGE 856]

200 Owen, 'Anatomy of Vertebrates,' vol. iii. p. 71

201 'Quarterly Review,' April, 1869, p. 392.

202 In *Hylobates syndactylus,* as the name expresses, two of the digits regularly cohere; and this, as Mr. Blyth informs me, is occasionally the case with the digits of *H. agilis, lar,* and *leuciscus.*

[PAGE 857]

203 Brehm, 'Thierleben,' B. i. s. 80.

204 "The Hand, its mechanism," &c. 'Bridgewater Treatise,' 1833, p. 38.

205 Häckel has an excellent discussion on the steps by which man became a biped: 'Natürliche Schöpfungsgeschichte,' 1868, s. 507. Dr. Büchner ('Conférences sur la Théorie Darwinienne,' 1869, p. 135) has given good cases of the use of the foot as a prehensile organ by man; also on the manner of progression of the higher apes to which I allude in the following paragraph: see also Owen ('Anatomy of Vertebrates,' vol. iii. p. 71) on this latter subject.

[PAGE 858]

206 "On the Primitive Form of the Skull," translated in 'Anthropological Review,' Oct. 1868, p. 428. Owen ('Anatomy of Vertebrates,' vol. ii. 1866, p. 551) on the mastoid processes in the higher apes.

[PAGE 859]

207 'Die Grenzen der Thierwelt, eine Betrachtung zu Darwin's Lehre,' 1868, s. 51.

208 Dujardin, 'Annales des Sc. Nat.' 3rd series, Zoolog. tom. xiv. 1850, p. 203. See also Mr. Lowne, 'Anatomy and Phys. of the *Musca vomi-*

toria,' 1870, p. 14. My son, Mr. F. Darwin, dissected for me the cerebral ganglia of the *Formica rufa.*

209 'Philosophical Transactions,' 1869, p. 513.

210 Quoted in C. Vogt's 'Lectures on Man,' Eng. translat. 1864, p. 88, 90. Prichard, 'Phys. Hist. of Mankind,' vol. i. 1838, p. 305.

211 'Comptes Rendus des Séances,' &c. June 1, 1868.

[PAGE 860]

212 'The Variation of Animals and Plants under Domestication,' vol. i. p. 124–129.

213 Schaaffhausen gives from Blumenbach and Busch, the cases of the spasms and cicatrix, in 'Anthropolog. Review,' Oct. 1868, p. 420. Dr. Jarrold ('Anthropologia,' 1808, p. 115, 116) adduces from Camper and from his own observations, cases of the modification of the skull from the head being fixed in an unnatural position. He believes that certain trades, such as that of a shoemaker, by causing the head to be habitually held forward, makes the forehead more rounded and prominent.

214 'Variation of Animals,' &c., vol. i. p. 117 on the elongation of the skull; p. 119, on the effect of the lopping of one ear.

215 Quoted by Schaaffhausen, in 'Anthropolog. Review,' Oct. 1868, p. 419.

[PAGE 861]

216 Owen, 'Anatomy of Vertebrates,' vol. iii. p. 619.

217 Isidore Geoffroy St.-Hilaire remarks ('Hist. Nat. Générale,' tom. ii. 1859, p. 215–217) on the head of man being covered with long hair; also on the upper surfaces of monkeys and of other mammals being more thickly clothed than the lower surfaces. This has likewise been observed by various authors. Prof. P. Gervais ('Hist. Nat. des Mammifères,' tom. i. 1854, p. 28), however, states that in the Gorilla the hair is thinner on the back, where it is partly rubbed off, than on the lower surface.

[PAGE 862]

218 Mr. St. George Mivart, 'Proc. Zoolog. Soc.' 1865, p. 562, 583. Dr. J. E. Gray, 'Cat. Brit. Mus.: Skeletons.' Owen, 'Anatomy of Vertebrates,' vol. ii. p. 517. Isidore Geoffroy, 'Hist. Nat. Gén.' tom. ii. p. 244.

[PAGE 863]

219 'The Variation of Animals and Plants under Domestication,' vol. ii. p. 280, 282.

[PAGE 865]

220 'Primeval Man,' 1869, p. 66.

CHAPTER V

[PAGE 867]

221 'Anthropological Review,' May, 1864, p. clviii.

[PAGE 868]

222 After a time the members or tribes which are absorbed into another tribe assume, as Mr. Maine remarks ('Ancient Law,' 1861, p. 131), that they are the co-descendants of the same ancestors.

223 Morlot, 'Soc. Vaud. Sc. Nat.' 1860, p. 294.

[PAGE 869]

224 I have given instances in my 'Variation of Animals under Domestication,' vol. ii. p. 196.

[PAGE 870]

225 See a remarkable series of articles on Physics and Politics in the 'Fortnightly Review,' Nov. 1867; April 1, 1868; July 1, 1869.

[PAGE 871]

226 'Origin of Civilisation,' 1870, p. 265.

227 Mr. Wallace gives cases in his 'Contributions to the Theory of Natural Selection,' 1870, p. 354.

[PAGE 872]

228 'Ancient Law,' 1861, p. 22. For Mr. Bagehot's remarks, 'Fortnightly Review,' April 1, 1868, p. 452.

229 'The Variation of Animals and Plants under Domestication,' vol. i. p. 309.

[PAGE 873]

230 'Fraser's Magazine,' Sept. 1868, p. 353. This article seems to have struck many persons, and has given rise to two remarkable essays and a rejoinder in the 'Spectator,' Oct. 3rd and 17th 1868. It has also been discussed in the 'Q. Journal of Science,' 1869, p. 152, and by Mr. Lawson Tait in the 'Dublin Q. Journal of Medical Science,' Feb. 1869,

and by Mr. E. Ray Lankester in his 'Comparative Longevity,' 1870, p. 128. Similar views appeared previously in the 'Australasian,' July 13, 1867. I have borrowed ideas from several of these writers.

231 For Mr. Wallace, see 'Anthropolog. Review,' as before cited. Mr. Galton in 'Macmillan's Magazine,' Aug. 1865, p. 318; also his great work, 'Hereditary Genius,' 1870.

[PAGE 874]

232 'Hereditary Genius,' 1870, p. 132–140.

233 See the fifth and sixth columns, compiled from good authorities, in the table given in Mr. E. R. Lankester's 'Comparative Longevity,' 1870, p.115.

[PAGE 875]

234 'Hereditary Genius,' 1870, p. 330.

235 'Origin of Species' (fifth edition, 1869), p. 104.

236 'Hereditary Genius,' 1870, p. 347.

237 E. Ray Lankester, 'Comparative Longevity,' 1870, p. 115. The table of the intemperate is from Neison's 'Vital Statistics.' In regard to profligacy, see Dr. Farr, "Influence of Marriage on Mortality," 'Nat. Assoc. for the Promotion of Social Science,' 1858.

[PAGE 876]

238 'Fraser's Magazine,' Sept. 1868, p. 353. 'Macmillan's Magazine,' Aug. 1865, p. 318. The Rev. F. W. Farrar ('Fraser's Mag.,' Aug. 1870, p. 264) takes a different view.

239 "On the Laws of the Fertility of Women," in 'Transact. Royal Soc.' Edinburgh, vol. xxiv. p. 287. See, also, Mr. Galton, 'Hereditary Genius,' p. 352–357, for observations to the above effect.

[PAGE 877]

240 'Tenth Annual Report of Births, Deaths, &c., in Scotland,' 1867, p. xxix.

241 These quotations are taken from our highest authority on such questions, namely, Dr. Farr, in his paper "On the Influence of Marriage on the Mortality of the French People," read before the Nat. Assoc. for the Promotion of Social Science, 1858.

242 Dr. Farr, ibid. The quotations given below are extracted from the same striking paper.

243 I have taken the mean of the quinquennial means, given in 'The Tenth Annual Report of Births, Deaths, &c., in Scotland,' 1867. The quo-

tation from Dr. Stark is copied from an article in the 'Daily News,' Oct. 17th, 1868, which Dr. Farr considers very carefully written.

[PAGE 878]

244 See the ingenious and original argument on this subject by Mr. Galton, 'Hereditary Genius,' p. 340–342.

245 Mr. Greg, 'Fraser's Magazine,' Sept. 1868, p. 357.

246 'Hereditary Genius,' 1870, p. 357–359. The Rev. F. H. Farrar ('Fraser's Mag.', Aug. 1870, p. 257) advances arguments on the other side. Sir C. Lyell had already ('Principles of Geology,' vol. ii. 1868, p. 489) called attention, in a striking passage, to the evil influence of the Holy Inquisition in having lowered, through selection, the general standard of intelligence in Europe.

[PAGE 879]

247 Mr. Galton, 'Macmillan's Magazine,' August, 1865, p. 325. See, also, 'Nature,' "On Darwinism and National Life," Dec. 1869, p. 184.

248 'Last Winter in the United States,' 1868, p. 29.

[PAGE 880]

249 'On the Origin of Civilisation,' 'Proc. Ethnological Soc.' Nov. 26, 1867.

250 'Primeval Man,' 1869.

251 'Royal Institution of Great Britain,' March 15, 1867. Also, 'Researches into the Early History of Mankind,' 1865.

[PAGE 881]

252 'Primitive Marriage,' 1865. See, likewise, an excellent article, evidently by the same author, in the 'North British Review,' July, 1869. Also, Mr. L. H. Morgan, "A Conjectural Solution of the Origin of the Class. System of Relationship," in 'Proc. American Acad. of Sciences,' vol. vii. Feb. 1868. Prof. Schaaffhausen ('Anthropolog. Review,' Oct. 1869, p. 373) remarks on "the vestiges of human sacrifices found both in Homer and the Old Testament."

253 Sir J. Lubbock, 'Prehistoric Times,' 2nd edit. 1869, chap. xv. and xvi. *et passim.*

254 Dr. F. Müller has made some good remarks to this effect in the 'Reise der Novara: Anthropolog. Theil,' Abtheil. iii. 1868, s. 127.

CHAPTER VI

[PAGE 883]

255 Isidore Geoffroy St.-Hilaire gives a detailed account of the position assigned to man by various naturalists in their classifications: 'Hist. Nat. Gén.' tom. ii. 1859, p. 170–189.

256 See the very interesting article, "L'Instinct chez les Insectes," by M. George Pouchet, 'Revue des Deux Mondes,' Feb. 1870, p. 682.

[PAGE 884]

257 Westwood, 'Modern Class. of Insects,' vol. ii. 1840, p. 87.

[PAGE 885]

258 'Proc. Zoolog. Soc' 1869, p. 4.

[PAGE 886]

259 'Evidence as to Man's Place in Nature,' 1863, p. 70, *et passim.*

260 Isid. Geoffroy, 'Hist. Nat. Gén.' tom. ii. 1859, p. 217.

261 "Ueber die Richtung der Haare," &c., Müller's 'Archiv für Anat. und Phys.' 1837, s. 51.

[PAGE 887]

262 On the hair in Hylobates, see 'Nat. Hist. of Mammals,' by C. L. Martin, 1841, p. 415. Also, Isid. Geoffroy on the American monkeys and other kinds, 'Hist. Nat. Gén.' vol. ii. 1859, p. 216, 243. Eschricht, ibid, s. 46, 55, 61. Owen, 'Anat. of Vertebrates,' vol. iii. p. 619. Wallace, 'Contributions to the Theory of Natural Selection,' 1870, p. 344.

263 'Origin of Species,' 5th edit. 1869, p. 194. 'The Variation of Animals and Plants under Domestication,' vol. ii. 1868, p. 348.

[PAGE 888]

264 'An Introduction to the Classification of Animals,' 1869, p. 99.

[PAGE 889]

265 This is nearly tbe same classification as that provisionally adopted by Mr. St. George Mivart ('Transact. Philosoph. Soc.' 1867, p. 300), who, after separating the Lemuridæ, divides the remainder of the Primates into the Hominidæ, the Simiadæ answering to the Catarhines, the Cebidæ, and the Hapalidæ,—these two latter groups answering to the Platyrhines.

266 'Transact. Zoolog. Soc' vol. vi. 1867, p. 214.

267 Mr. St. G. Mivart, 'Transact. Phil. Soc.' 1867, p. 410.

[PAGE 890]

268 Messrs. Murie and Mivart on the Lemuroidea, 'Transact. Zoolog. Soc.' vol. vii. 1869, p. 5.

269 Häckel has come to this same conclusion. See 'Ueber die Entstehung des Menschengeschlechts,' in Virchow's 'Sammlung. gemein. wissen. Vorträge,' 1868, s. 61. Also his 'Natürliche Schöpfungsgeschichte,' 1868, in which he gives in detail his views on the genealogy of man.

[PAGE 891]

270 'Anthropological Review,' April, 1867, p. 236.

271 'Elements of Geology,' 1865, p. 583–585. 'Antiquity of Man, 1863, p. 145.

[PAGE 892]

272 'Man's Place in Nature,' p. 105.

273 Elaborate tables are given in his 'Generelle Morphologie' (B. ii. s. cliii. and s. 425); and with more especial reference to man in his 'Natürliche Schöpfungsgeschichte,' 1868. Prof. Huxley, in reviewing this latter work ('The Academy,' 1869, p. 42) says, that he considers the phylum or lines of descent of the Vertebrata to be admirably discussed by Häckel, although he differs on some points. He expresses, also, his high estimate of the value of the general tenor and spirit of the whole work.

[PAGE 893]

274 'Palæontology,' 1860, p. 199.

275 I had the satisfaction of seeing, at the Falkland Islands, in April, 1833, and therefore some years before any other naturalist, the locomotive larvæ of a compound Ascidian, closely allied to, but apparently generically distinct from, Synoicum. The tail was about five times as long as the oblong head, and terminated in a very fine filament. It was plainly divided, as sketched by me under a simple microscope, by transverse opaque partitions, which I presume represent the great cells figured by Kowalevsky. At an early stage of development the tail was closely coiled round the head of the larva.

276 'Mémoires de l'Acad. des Sciences de St. Pétersbourg,' tom. x. No. 15, 1866.

[PAGE 895]

277 This is the conclusion of one of the highest authorities in comparative anatomy, namely, Prof. Gegenbaur: 'Grundzüge der vergleich. Anat.' 1870, s. 876. The result has been arrived at chiefly from the study of the Amphibia; but it appears from the researches of Waldeyer (as quoted in Humphry's 'Journal of Anat. and Phys.' 1869, p. 161), that the sexual organs of even "the higher vertebrata are, in their early condition, hermaphrodite." Similar views have long been held by some authors, though until recently not well based.

278 The male Thylacinus offers the best instance. Owen, 'Anatomy of Vertebrates,' vol. iii. p. 771.

279 Serranus is well known often to be in an hermaphrodite condition; but Dr. Günther informs me that he is convinced that this is not its normal state. Descent from an ancient androgynous prototype would, however, naturally favour and explain, to a certain extent, the recurrence of this condition in these fishes.

[PAGE 896]

280 Mr. Lockwood believes (as quoted in 'Quart. Journal of Science,' April, 1868, p. 269), from what he has observed of the development of Hippocampus, that the walls of the abdominal pouch of the male in some way afford nourishment. On male fishes hatching the ova in their mouths, see a very interesting paper by Prof. Wyman, in 'Proc. Boston Soc. of Nat. Hist.' Sept. 15, 1857; also Prof. Turner, in 'Journal of Anat. and Phys.' Nov. 1, 1866, p. 78. Dr. Günther has likewise described similar cases.

[PAGE 897]

281 All vital functions tend to run their course in fixed and recurrent periods, and with tidal animals the periods would probably be lunar; for such animals must have been left dry or covered deep with water,—supplied with copious food or stinted,—during endless generations, at regular lunar intervals. If then the Vertebrata are descended from an animal allied to the existing tidal Ascidians, the mysterious fact, that with the higher and now terrestrial Vertebrata, not to mention other classes, many normal and abnormal vital processes run their course according to lunar periods, is rendered intelligible. A recurrent period, if approximately of the right duration, when once gained, would not, as far as we can judge, be liable to be changed; consequently it might be thus transmitted during almost any number of generations. This conclusion, if it could be proved sound, would

be curious; for we should then see that the period of gestation in each mammal, and the hatching of each bird's eggs, and many other vital processes, still betrayed the primordial birthplace of these animals.

CHAPTER VII

[PAGE 900]

282 'History of India,' 1841, vol. i. p. 323. Father Ripa makes exactly the same remark with respect to the Chinese.

283 A vast number of measurements of Whites, Blacks, and Indians, are given in the 'Investigations in the Military and Anthropolog. Statistics of American Soldiers,' by B. A. Gould, 1869, p. 298–358; on the capacity of the lungs, p. 471. See also the numerous and valuable tables, by Dr. Weisbach, from the observations of Dr. Scherzer and Dr. Schwarz, in the 'Reise der Novara: Anthropolog. Theil,' 1867.

284 See, for instance, Mr. Marshall's account of the brain of a Bush-woman, in 'Phil. Transact.' 1864, p. 519.

[PAGE 901]

285 Wallace, 'The Malay Archipelago,' vol. ii. 1869, p. 178.

286 With respect to the figures in the famous Egyptian caves of Abou-Simbel, M. Pouchet says ('The Plurality of the Human Races,' Eng. translat. 1864, p. 50), that he was far from finding recognisable representations of the dozen or more nations which some authors believe that they can recognise. Even some of the most strongly-marked races cannot be identified with that degree of unanimity which might have been expected from what has been written on the subject. Thus Messrs. Nott and Gliddon ('Types of Mankind,' p. 148) state that Rameses II., or the Great, has features superbly European; whereas Knox, another firm believer in the specific distinction of the races of man ('Races of Man,' 1850, p. 201), speaking of young Memnon (the same person with Rameses II., as I am informed by Mr. Birch) insists in the strongest manner that he is identical in character with the Jews of Antwerp. Again, whilst looking in the British Museum with two competent judges, officers of the establishment, at the statue of Amunoph III., we agreed that he had a strongly negro cast of features; but Messrs. Nott and Gliddon (ibid. p. 146, fig. 53) describe him as "a hybrid, but not of negro intermixture."

287 As quoted by Nott and Gliddon, 'Types of Mankind,' 1854, p. 439. They give also corroborative evidence; but C. Vogt thinks that the subject requires further investigation.

288 "Diversity of Origin of the Human Races," in the 'Christian Examiner,' July, 1850.

[PAGE 902]

289 'Transact. R. Soc. of Edinburgh,' vol. xxii. 1861, p. 567.

290 'On the Phenomena of Hybridity in the Genus Homo,' Eng. translat. 1861.

[PAGE 903]

291 See the interesting letter by Mr. T. A. Murray, in the 'Anthropolog. Review,' April, 1868, p. liii. In this letter Count Strzelecki's statement, that Australian women who have borne children to a white man are afterwards sterile with their own race, is disproved. M. A. de Quatrefages has also collected ('Revue des Cours Scientifiques,' March, 1869, p. 239) much evidence that Australians and Europeans are not sterile when crossed.

292 'An Examination of Prof. Agassiz's Sketch of the Nat. Provinces of the Animal World,' Charleston, 1855, p. 44.

293 'Military and Anthropolog. Statistics of American Soldiers,' by B. A. Gould, 1869, p. 319.

294 'The Variation of Animals and Plants under Domestication,' vol. ii. p. 109. I may here remind the reader that the sterility of species when crossed is not a specially-acquired quality; but, like the incapacity of certain trees to be grafted together, is incidental on other acquired differences. The nature of these differences is unknown, but they relate more especially to the reproductive system, and much less to external structure or to ordinary differences in constitution. One important element in the sterility of crossed species apparently lies in one or both having been long habituated to fixed conditions; for we know that changed conditions have a special influence on the reproductive system, and we have good reason to believe (as before remarked) that the fluctuating conditions of domestication tend to eliminate that sterility which is so general with species in a natural state when crossed. It has elsewhere been shewn by me (ibid. vol. ii. p. 185, and 'Origin of Species,' 5th edit. p. 317) that the sterility of crossed species has not been acquired through natural selection: we can see that when two forms have already been rendered very sterile, it is scarcely possible that their sterility should be augmented by the preservation or survival of the more and more sterile individuals; for as the sterility increases fewer and fewer offspring will be produced from which to breed, and at last only single individuals will be produced, at the rarest intervals. But there is even a higher grade of steril-

ity than this. Both Gärtner and Kölreuter have proved that in genera of plants including numerous species, a series can be formed from species which when crossed yield fewer and fewer seeds, to species which never produce a single seed, but yet are affected by the pollen of the other species, for the germen swells. It is here manifestly impossible to select the more sterile individuals, which have already ceased to yield seeds; so that the acme of sterility, when the germen alone is affected, cannot be gained through selection. This acme, and no doubt the other grades of sterility, are the incidental results of certain unknown differences in the constitution of the reproductive system of the species which are crossed.

[PAGE 904]

295 'The Variation of Animals,' &c., vol. ii. p. 92.

296 M. de Quatrefages has given ('Anthropolog. Review,' Jan. 1869, p. 22) an interesting account of the success and energy of the Paulistas in Brazil, who are a much crossed race of Portuguese and Indians, with a mixture of the blood of other races.

[PAGE 905]

297 For instance with the aborigines of America and Australia. Prof. Huxley says ('Transact. Internat. Congress of Prehist. Arch.' 1868, p. 105) that the skulls of many South Germans and Swiss are "as short and as broad as those of the Tartars," &c.

298 See a good discussion on this subject in Waitz, 'Introduct. to Anthropology,' Eng. translat. 1863, p. 198–208, 227. I have taken some of the above statements from H. Tuttle's 'Origin and Antiquity of Physical Man,' Boston, 1866, p. 35.

[PAGE 906]

299 Prof. Nägeli has carefully described several striking cases in his 'Botanische Mittheilungen,' B. ii. 1866, s. 294–369. Prof. Asa Gray has made analogous remarks on some intermediate forms in the Compositæ of N. America.

300 'Origin of Species,' 5th edit. p. 68.

[PAGE 907]

301 See Prof. Huxley to this effect in the 'Fortnightly Review,' 1865, p. 275.

302 'Lectures on Man,' Eng. translat. 1864, p. 468.

303 'Die Racen des Schweines,' 1860, s. 46. 'Vorstudien für Geschiehte,

&c., Schweineschädel,' 1864, s. 104. With respect to cattle, see M. de Quatrefages, 'Unité' de l'Espèce Humaine,' 1861, p. 119.

[PAGE 909]

304 Tylor's 'Early History of Mankind,' 1865; for the evidence with respect to gesture-language, see p. 54. Lubbock's 'Prehistoric Times,' 2nd edit. 1869.

305 'The Primitive Inhabitants of Scandinavia,' Eng. translat. edited by Sir J. Lubbock, 1868, p. 104.

306 Hodder M. Westropp, on Cromlechs, &c., 'Journal of Ethnological Soc.' as given in 'Scientific Opinion,' June 2nd, 1869, p. 3.

307 'Journal of Researches: Voyage of the "Beagle,"' p. 83.

[PAGE 910]

308 'Prehistoric Times,' 1869, p. 574.

[PAGE 911]

309 Translation in 'Anthropological Review,' Oct. 1868, p. 431.

310 'Transact. Internat. Congress of Prehistoric Arch.' 1868, p. 172–175. See also Broca (translation) in 'Anthropological Review,' Oct. 1868, p. 410.

[PAGE 912]

311 Dr. Gerland, 'Ueber das Aussterben der Naturvölker,' 1868, s. 82.

312 Gerland (ibid. s. 12) gives facts in support of this statement.

313 See remarks to this effect in Sir H. Holland's 'Medical Notes and Reflections,' 1839, p. 390.

314 I have collected ('Journal of Researches, Voyage of the "Beagle,"' p. 375) a good many cases bearing on this subject: see also Gerland, ibid. s. 8. Poeppig speaks of the "breath of civilisation as poisonous to savages."

315 Sproat, 'Scenes and Studies of Savage Life,' 1868, p. 284.

[PAGE 913]

316 Bagehot, "Physics and Politics," 'Fortnightly Review,' April 1, 1868, p. 455.

317 "On Anthropology," translation, 'Anthropolog. Review,' Jan. 1868, p. 38.

[PAGE 914]

318 'The Annals of Rural Bengal,' 1868, p. 134.

319 'The Variation of Animals and Plants under Domestication,' vol. ii. p. 95.

320 Pallas, 'Act. Acad. St. Petersburgh,' 1780, part ii. p. 69. He was followed by Rudolphi, in his 'Beyträge zur Anthropologie,' 1812. An excellent summary of the evidence is given by Godron, 'De l'Espèce,' 1859, vol. ii. p. 246, &c.

321 Sir Andrew Smith, as quoted by Knox, 'Races of Man,' 1850, p. 473.

322 See De Quatrefages on this head, 'Revue des Cours Scientifiques,' Oct. 17, 1868, p. 731.

323 Livingstone's 'Travels and Researches in S. Africa,' 1857, p. 338, 339. D'Orbigny, as quoted by Godron, 'De l'Espèce,' vol. ii. p. 266.

324 See a paper read before the Royal Soc. in 1813, and published in his Essays in 1818. I have given an account of Dr. Wells' views in the Historical Sketch (p. xvi) to my 'Origin of Species.' Various cases of colour correlated with constitutional peculiarities are given in my 'Variation of Animals under Domestication,' vol. ii. p. 227, 335.

325 See, for instance, Nott and Gliddon. 'Types of Mankind,' p. 68.

[PAGE 915]

326 Major Tulloch, in a paper read before the Statistical Society, April 20th, 1840, and given in the 'Athenæum,' 1840, p. 353.

327 'The Plurality of the Human Race' (translat.), 1864, p. 60.

328 Quatrefages, 'Unité de l'Espèce Humaine,' 1861, p. 205. Waitz, 'Introduct. to Anthropology,' translat. vol. i. 1863, p. 124. Livingstone gives analogous cases in his 'Travels.'

329 In the spring of 1862 I obtained permission from the Director-General of the Medical department of the Army, to transmit to the surgeons of the various regiments on foreign service a blank table, with the following appended remarks, but I have received no returns. "As several well-marked cases have been recorded with our domestic animals of a relation between the colour of the dermal appendages and the constitution; and it being notorious that there is some limited degree of relation between the colour of the races of man and the climate inhabited by them; the following investigation seems worth consideration. Namely, whether there is any relation in Europeans between the colour of their hair, and their liability to the diseases of tropical countries. If the surgeons of the several regiments, when stationed in unhealthy tropical districts, would be so good as first to count, as a standard of comparison, how many men, in the force whence the sick are drawn, have dark and light-coloured hair, and hair of intermediate or doubtful tints; and if a similar account were kept by the same medical gentlemen, of all the men who suffered

from malarious and yellow fevers, or from dysentery, it would soon be apparent, after some thousand cases had been tabulated, whether there exists any relation between the colour of the hair and constitutional liability to tropical diseases. Perhaps no such relation would be discovered, but the investigation is well worth making. In case any positive result were obtained, it might be of some practical use in selecting men for any particular service. Theoretically the result would be of high interest, as indicating one means by which a race of men inhabiting from a remote period an unhealthy tropical climate, might have become dark-coloured by the better preservation of dark- haired or dark-complexioned individuals during a long succession of generations."

330 'Anthropological Review,' Jan. 1866, p. xxi.

[PAGE 916]

331 See, for instance, Quatrefages ('Revue des Cours Scientifiques,' Oct. 10, 1868, p. 724) on the effects of residence in Abyssinia and Arabia, and other analogous cases. Dr. Rolle ('Der Mensch, seine Abstammung,' &c, 1865, s. 99) states, on the authority of Khanikof, that the greater number of German families settled in Georgia, have acquired in the course of two generations dark hair and eyes. Mr. D. Forbes informs me that the Quichuas in the Andes vary greatly in colour, according to the position of the valleys inhabited by them.

332 Harlan, 'Medical Researches,' p. 532. Quatrefages ('Unité de l'Espèce Humaine,' 1861, p. 128) has collected much evidence on this head.

333 See Prof. Schaaffhausen, translat. in 'Anthropological Review,' Oct. 1868, p. 429.

[PAGE 917]

334 Mr. Catlin states ('N. American Indians,' 3rd edit. 1842,vol. i. p. 49) that in the whole tribe of the Mandans, about one in ten or twelve of the members of all ages and both sexes have bright silvery grey hair, which is hereditary. Now this hair is as coarse and harsh as that of a horse's mane, whilst the hair of other colours is fine and soft.

335 On the odour of the skin, Godron, 'Sur l'Espèce,' tom. ii. p. 217. On the pores in the skin, Dr. Wilckens, 'Die Aufgaben der landwirth. Zootechuik,' 1869, s. 7.

CHAPTER VIII

[PAGE 922]

336 Westwood, 'Modern Class. of Insects,' vol. ii. 1840, p. 541. In regard to the statement about Tanais, mentioned below, I am indebted to Fritz Müller.

337 Kirby and Spence, 'Introduction to Entomology,' vol. iii. 1826, p. 309.

[PAGE 925]

338 Even with those of plants in which the sexes are separate, the male flowers are generally mature before the female. Many hermaphrodite plants are, as first shewn by C. K. Sprengel, dichogamous; that is, their male and female organs are not ready at the same time, so that they cannot be self-fertilised. Now with such plants the pollen is generally mature in the same flower before the stigma, though there are some exceptional species in which the female organs are mature before the male.

[PAGE 927]

339 I have received information, hereafter to be given, to this effect with respect to poultry. Even with birds, such as pigeons, which pair for life, the female, as I hear from Mr. Jenner Weir, will desert her mate if he is injured or grows weak.

[PAGE 929]

340 On the Gorilla, Savage and Wyman, 'Boston Journal of Nat. Hist.' vol. v. 1845–47, p. 423. On Cynocephalus, Brehm, 'Illust. Thierleben,' B. i. 1864, s. 77. On Mycetes, Rengger, 'Naturgesch.: Säugethiere von Paraguay,' 1830, s. 14, 20. On Cebus, Brehm, ibid. s. 108.

341 Pallas, 'Spicilegia Zoolog.' Fasc. xii. 1777, p. 29. Sir Andrew Smith, 'Illustrations of the Zoology of S. Africa,' 1849, pl. 29, on the Kobus. Owen, in his 'Anatomy of Vertebrates' (vol. iii. 1868, p. 633) gives a table incidentally showing which species of Antelopes pair and which are gregarious.

[PAGE 930]

342 Dr. Campbell, in 'Proc. Zoolog. Soc.' 1869, p. 138. See also an interesting paper, by Lieut. Johnstone, in 'Proc. Asiatic Soc. of Bengal,' May, 1868.

343 'The Ibis,' vol. iii. 1861, p. 133, on the Progne Widow-bird. See also on the Vidua axillaris, ibid. vol. ii. 1860, p. 211. On the

polygamy of the Capercailzie and Great Bustard, see L. Lloyd, 'Game Birds of Sweden,' 1867, p. 19, and 182. Montagu and Selby speak of the Black Grouse as polygamous and of the Red Grouse as monogamous.

[PAGE 931]

344 The Rev. E. S. Dixon, however, speaks positively ('Ornamental Poultry,' 1848, p. 76) about the eggs of the guinea-fowl being infertile when more than one female is kept with the same male.

345 Noel Humphreys, 'River Gardens,' 1857.

[PAGE 932]

346 Kirby and Spence, 'Introduction to Entomology,' vol. iii. 1826, p. 342.

347 One parasitic Hymenopterous insect (Westwood, 'Modern Class. of Insects,' vol. ii. p. 160) forms an exception to the rule, as the male has rudimentary wings, and never quits the cell in which it is born, whilst the female has well-developed wings. Audouin believes that the females are impregnated by the males which are born in the same cells with them; but it is much more probable that the females visit other cells, and thus avoid close interbreeding. We shall hereafter meet with a few exceptional cases, in various classes, in which the female, instead of the male, is the seeker and wooer.

[PAGE 933]

348 'Essays and Observations,' edited by Owen, vol. i. 1861, p. 194.

349 Prof. Sachs ('Lehrbuch der Botanik,' 1870, s. 633) in speaking of the male and female reproductive cells, remarks, "verhält sich die eine bei der Vereinigung activ, . . . die andere erscheint bei der Vereinigung passiv."

[PAGE 934]

350 'Reise der Novara: Anthropolog. Theil,' 1867, s. 216–269. The results were calculated by Dr. Weisbach from measurements made by Drs. K. Scherzer and Schwarz. On the greater variability of the males of domesticated animals, see my 'Variation of Animals and Plants under Domestication,' vol. ii. 1868, p. 75.

351 'Proceedings Royal Soc.' vol. xvi. July, 1868, p. 519 and 524.

352 'Proc. Royal Irish Academy,' vol. x. 1868, p.123.

353 'Massachusetts Medical Soc.' vol. ii. No. 3, 1868, p. 9.

[PAGE 937]

354 'The Variation of Animals and Plants under Domestication,' vol. ii. 1868, p. 75. In the last chapter but one, the provisional hypothesis of pangenesis, above alluded to, is fully explained.

355 These facts are given on the high authority of a great breeder, Mr. Teebay, in Tegetmeier's 'Poultry Book,' 1868, p. 158. On the characters of chickens of different breeds, and on the breeds of the pigeon, alluded to in the above paragraph, see 'Variation of Animals,' &c., vol. i. p. 160, 249; vol. ii. p. 77.

[PAGE 938]

356 'Novæ species Quadrupedum e Glirium ordine,' 1778, p. 7. On the transmission of colour by the horse, see 'Variation of Animals, &c. under Domestication,' vol. i. p. 21. Also vol. ii. p. 71, for a general discussion on Inheritance as limited by Sex.

357 Dr. Chapuis, 'Le Pigeon Voyageur Belge,' 1865, p. 87. Boitard et Corbié, 'Les Pigeons de Volière,' &c., 1824, p. 173.

[PAGE 940]

358 References are given in my 'Variation of Animals under Domestication,' vol. ii. p. 72.

[PAGE 942]

359 I am much obliged to Mr. Cupples for having made enquiries for me in regard to the Roebuck and Red Deer of Scotland from Mr. Robertson, the experienced head-forester to the Marquis of Breadalbane. In regard to Fallow-deer, I am obliged to Mr. Eyton and others for information. For the *Cervus alces* of N. America, see 'Land and Water,' 1868, p. 221 and 254; and for the *C. Virginianus* and *strongyloceros* of the same continent, see J. D. Caton, in 'Ottawa Acad. of Nat. Sc.' 1868, p. 13. For *Cervus Eldi* of Pegu, see Lieut. Beavan, 'Proc. Zoolog. Soc.' 1867, p. 762.

360 *Antilocapra Americana.* Owen, 'Anatomy of Vertebrates,' vol. iii. p. 627.

361 I have been assured that the horns of the sheep in North Wales can always be felt, and are sometimes even an inch in length, at birth. With cattle Youatt says ('Cattle,' 1834, p. 277) that the prominence of the frontal bone penetrates the cutis at birth, and that the horny matter is soon formed over it.

362 I am greatly indebted to Prof. Victor Carus for having made inquiries for me, from the highest authorities, with respect to the merino sheep

of Saxony. On the Guinea coast of Africa there is a breed of sheep in which, as with merinos, the rams alone bear horns; and Mr. Winwood Reade informs me that in the one case observed, a young ram born on Feb. 10th first showed horns on March 6th, so that in this instance the development of the horns occurred at a later period of life, conformably with our rule, than in the Welch sheep, in which both sexes are horned.

363 In the common peacock *(Pavo cristatus)* the male alone possesses spurs, whilst both sexes of the Java peacock *(P. muticus)* offer the unusual case of being furnished with spurs. Hence I fully expected that in the latter species they would have been developed earlier in life than in the common peacock; but M. Hegt of Amsterdam informs me, that with young birds of the previous year, belonging to both species, compared on April 23rd, 1869, there was no difference in the development of the spurs. The spurs, however, were as yet represented merely by slight knobs or elevations. I presume that I should have been informed if any difference in the rate of development had subsequently been observed.

[PAGE 943]

364 In some other species of the Duck Family the speculum in the two sexes differs in a greater degree; but I have not been able to discover whether its full development occurs later in life in the males of such species, than in the male of the common duck, as ought to be the case according to our rule. With the allied *Mergus cucullatus* we have, however, a case of this kind: the two sexes differ conspicuously in general plumage, and to a considerable degree in the speculum, which is pure white in the male and greyish-white in the female. Now the young males at first resemble, in all respects, the female, and have a greyish-white speculum, but this becomes pure white at an earlier age than that at which the adult male acquires his other more strongly-marked sexual differences in plumage: see Audubon, 'Ornithological Biography,' vol. iii. 1835, p. 249–250.

[PAGE 944]

365 'Das Ganze der Taubenzucht,' 1837, s. 21, 24. For the case of the streaked pigeons, see Dr. Chapuis, 'Le Pigeon Voyageur Belge,' 1865, p. 87.

[PAGE 945]

366 For full particulars and references on all these points respecting

the several breeds of the Fowl, see 'Variation of Animals and Plants under Domestication,' vol. i. p. 250, 256. In regard to the higher animals, the sexual differences which have arisen under domestication are described in the same work under the head of each species.

[PAGE 949]

367 'Twenty-ninth Annual Report of the Registrar-General for 1866.' In this report (p. xii) a special decennial table is given.

368 For Norway and Russia, see abstract of Prof. Faye's researches, in 'British and Foreign Medico-Chirurg. Review,' April, 1867, p. 343, 345. For France, the 'Annuaire pour l'An 1867,' p. 213.

369 In regard to the Jews, see M. Thury, 'La Loi de Production des Sexes,' 1863, p. 25.

370 Babbage, 'Edinburgh Journal of Science,' 1829, vol. i. p. 88; also p. 90, on still-born children. On illegitimate children in England, see 'Report of Registrar-General for 1866,' p. xv.

371 'British and Foreign Medico-Chirurg. Review,' April, 1867, p. 343. Dr. Stark also remarks ('Tenth Annual Report of Births, Deaths, &c., in Scotland,' 1867, p. xxviii) that "These examples may suffice to shew that, at almost every stage of life, the males in Scotland have a greater liability to death and a higher death-rate than the females. The fact, however, of this peculiarity being most strongly developed at that infantile period of life when the dress, food, and general treatment of both sexes are alike, seems to prove that the higher male death-rate is an impressed, natural, and constitutional peculiarity due to sex alone."

372 With the savage Guaranys of Paraguay, according to the accurate Azara ('Voyages dans l'Amérique mérid.' tom. ii. 1809, p. 60, 179), the women in proportion to the men are as 14 to 13.

373 Leuckart (in Wagner, 'Handwörterbuch der Phys.' B. iv. 1853, s. 774.

374 'Anthropological Review,' April, 1870, p. cviii.

[PAGE 950]

375 During the last eleven years a record has been kept of the number of mares which have proved barren or prematurely slipped their foals; and it deserves notice, as shewing how infertile these highly-nurtured and rather closely-interbred animals have become, that not far from one-third of the mares failed to produce living foals. Thus during 1866, 809 male colts and 816 female colts were born, and 743 mares failed to produce offspring. During 1867, 836 males and 902 females were born, and 794 mares failed.

[PAGE 951]

376 I am much indebted to Mr. Cupples for having procured for me the above returns from Scotland, as well as some of the following returns on cattle. Mr. R. Elliot, of Laighwood, first called my attention to the premature deaths of the males,—a statement subsequently confirmed by Mr. Aitchison and others. To this latter gentleman, and to Mr. Payan, I owe my thanks for the larger returns on sheep.

377 Bell, 'History of British Quadrupeds,' p. 100.

378 'Illustrations of the Zoology of S. Africa,' 1849, pl. 29.

[PAGE 952]

379 Brehm ('Illust. Thierleben,' B. iv. s. 990) comes to the same conclusion.

380 On the authority of L. Lloyd, 'Game Birds of Sweden,' 1867, p. 12, 132.

381 'Nat. Hist. of Selbourne,' letter xxix. edit. of 1825, vol. i. p. 139.

382 Mr. Jenner Weir received similar information, on making enquiries during the following year. To shew the number of chaffinches caught, I may mention that in 1869 there was a match between two experts; and one man caught in a day 62, and another 40, male chaffinches. The greatest number ever caught by one man in a single day was 70.

[PAGE 953]

383 'Ibis,' vol. ii. p. 260, as quoted in Gould's 'Trochilidæ,' 1861, p. 52. For the foregoing proportions, I am indebted to Mr. Salvin for a table of his results.

384 'Ibis,' 1860, p. 137; and 1867, p. 369.

385 'Ibis,' 1862, p. 137.

386 Leuckart quotes Bloch (Wagner, 'Handwörterbuch der Phys.' B. iv. 1853, s. 775), that with fish there are twice as many males as females.

387 Quoted in the 'Farmer,' March 18, 1869, p. 369.

[PAGE 954]

388 'The Stormontfield Piscicultural Experiments,' 1866, p. 23. The 'Field' newspaper, June 29th, 1867.

389 'Land and Water,' 1868, p. 41.

390 Yarrell, 'Hist. British Fishes,' vol. i. 1836, p. 307; on the *Cyprinus carpio,* p. 331; on the *Tinca vulgaris,* p. 331; on the *Abramis brama,* p. 336. See, for the minnow *(Leuciscus phoxinus),* 'Loudon's Mag. of Nat. Hist.' vol. v. 1832, p. 682.

[PAGE 955]

391 Leuckart quotes Meinecke (Wagner, 'Handwörterbuch der Phys.' B. iv. 1853, s. 775) that with Butterflies the males are three or four times as numerous as the females.

392 'The Naturalist on the Amazons,' vol. ii. 1863, p. 228, 347.

393 Four of these cases are given by Mr. Trimen in his 'Rhopalocera Africæ Australis.'

394 Quoted by Trimen, 'Transact. Ent. Soc.' vol. v. part iv. 1866, p. 330.

395 'Transact. Linn. Soc.' vol. xxv. p. 37.

396 'Proc. Entomolog. Soc.' Feb. 17th, 1868.

397 Quoted by Dr. Wallace in 'Proc. Ent. Soc.' 3rd series, vol. v. 1867, p. 487.

[PAGE 956]

398 Blanchard, 'Metamorphoses, Mœurs des Insectes,' 1868, p. 225–226.

[PAGE 957]

399 'Lepidopteren-Doubblettren Liste,' Berlin, No. x. 1866.

400 This naturalist has been so kind as to send me some results from former years, in which the females seemed to preponderate; but so many of the figures were estimates, that I found it impossible to tabulate them.

[PAGE 958]

401 Günther's 'Record of Zoological Literature,' 1867, p. 260. On the excess of female Lucanus, ibid. p. 250. On the males of Lucanus in England, Westwood, 'Modern Class. of Insects,' vol. i. p. 187. On the Siagonium, ibid. p. 172.

402 Walsh, in 'The American Entomologist,' vol. i. 1869, p. 103. F. Smith, 'Record of Zoological Literature,' 1867, p. 328.

403 'Farm Insects,' p. 45–46.

[PAGE 959]

404 'Observations on N. American Neuroptera,' by H. Hagen and B. D. Walsh, 'Proc. Ent. Soc. Philadelphia,' Oct. 1863, p. 168, 223, 239.

405 'Proc. Ent. Soc. London,' Feb. 17, 1868.

406 Another great authority in this class, Prof. Thorell of Upsala ('On European Spiders,' 1869–70, part i. p. 205) speaks as if female spiders were generally commoner than the males.

407 See, on this subject, Mr. Pickard-Cambridge, as quoted in 'Quarterly Journal of Science,' 1868, p. 429.

[PAGE 961]

408 I have often been struck with the fact, that in several species of Primula the seeds in the capsules which contained only a few were very much larger than the numerous seeds in the more productive capsules.

409 'Principles of Biology,' vol. ii. 1867, chaps. ii.–xi.

Chapter IX

[PAGE 965]

410 'De l'Espèce et de la Class.' &c., 1869, p. 106.

[PAGE 966]

411 See, for instance, the account which I have given in my 'Journal of Researches,' 1845, p. 7.

412 I have given ('Geolog. Observations on Volcanic Islands,' 1844, p. 53) a curious instance of the influence of light on the colours of a frondescent incrustation, deposited by the surf on the coast-rocks of Ascension, and formed by the solution of triturated sea-shells.

[PAGE 968]

413 'Facts and Arguments for Darwin,' English translat. 1869, p. 20. See the previous discussion on the olfactory threads. Sars has described a somewhat analogous case (as quoted in 'Nature,' 1870, p. 455) in a Norwegian crustacean, the *Pontoporeia affinis*.

414 See Sir J. Lubbock in 'Annals. and Mag. of Nat. Hist.' vol. xi. 1853, pl. i. and x.; and vol. xii. (1853) pl. vii. See also Lubbock in 'Transact. Ent. Soc.' vol. iv. new series, 1856–1858, p. 8. With respect to the zigzagged antennæ mentioned below, see Fritz Müller, 'Facts and Arguments for Darwin,' 1869, p. 40, foot-note.

[PAGE 969]

415 See a paper by Mr. C. Spence Bate, with figures, in 'Proc. Zoolog. Soc.' 1868, p. 363; and on the nomenclature of the genus, ibid. p. 585. I am greatly indebted to Mr. Spence Bate for nearly all the above statements with respect to the chelæ of the higher crustaceans.

416 'Hist. Nat. des Crust.' tom. ii. 1837, p. 50.

[PAGE 970]

417 Fritz Müller, 'Facts and Arguments for Darwin,' 1869, p. 25–28.

[PAGE 971]

418 'Travels in the Interior of Brazil,' 1846, p. 111. I have given, in my 'Journal of Researches,' p. 397, an account of the habits of the Birgus.

[PAGE 972]

419 Mr. Ch. Fraser, in 'Proc. Zoolog. Soc.' 1869, p. 3. I am indebted to Mr. Bate for the statement from Dr. Power.

420 Claus, 'Die freilebenden Copepoden,' 1863, s. 35.

421 'Facts and Arguments,' &c, p. 79.

[PAGE 973]

422 'A History of the Spiders of Great Britain,' 1861–64. For the following facts, see p. 102, 77, 88.

423 Aug. Vinson ('Aranéides des Iles de la Réunion,' pl. vi. figs. 1 and 2) gives a good instance of the small size of the male in *Epeira nigra*. In this species, as I may add, the male is testaceous and the female black with legs banded with red. Other even more striking cases of inequality in size between the sexes have been recorded ('Quarterly Journal of Science,' 1868, July, p. 429); but I have not seen the original accounts.

[PAGE 974]

424 Kirby and Spence, 'Introduction to Entomology,' vol. i. 1818, p. 280.

425 Theridion (Asagena, Sund.) serratipes, 4-punctatum et guttatum; see Westring, in Kroyer, 'Naturhist. Tidskrift,' vol. iv. 1842–1843, p. 349; and vol. ii. 1846–1849, p. 342. See, also, for other species, 'Araneæ Svecicæ,' p. 184.

426 Walckenaer et P. Gervais, 'Hist. Nat. des Insectes: Aptères," tom. iv. 1847, p. 17, 19, 68.

CHAPTER X

[PAGE 975]

427 Sir J. Lubbock, 'Transact. Linnean Soc.' vol. xxv. 1866, p. 484. With respect to the Mutillidæ see Westwood, 'Modern Class. of Insects,'vol. ii. p. 213.

428 These organs in the male often differ in closely-allied species, and

afford excellent specific characters. But their importance, under a functional point of view, as Mr. R. MacLachlan has remarked to me, has probably been overrated. It has been suggested, that slight differences in these organs would suffice to prevent the intercrossing of well-marked varieties or incipient species, and would thus aid in their development. That this can hardly be the case, we may infer from the many recorded cases (see for instance, Bronn, 'Geschichte der Natur,' B. ii. 1843, s. 164; and Westwood, 'Transact. Ent. Soc' vol. iii. 1842, p. 195) of distinct species having been observed in union. Mr. MacLachlan informs me (vide 'Stett. Ent. Zeitung,' 1867, s. 155) that when several species of Phryganidæ, which present strongly-pronounced differences of this kind, were confined together by Dr. Aug. Meyer, *they coupled,* and one pair produced fertile ova.

429 'The Practical Entomologist,' Philadelphia, vol. ii. May, 1867, p. 88.

[PAGE 976]

430 Mr. Walsh, ibid. p. 107.

431 'Modern Classification of Insects,' vol. ii. 1840, p. 206, 205. Mr. Walsh, who called my attention to this double use of the jaws, says that he has repeatedly observed this fact.

432 We have here a curious and inexplicable case of dimorphism, for some of the females of four European species of Dytiscus, and of certain species of Hydroporus, have their elytra smooth; and no intermediate gradations between sulcated or punctured and quite smooth elytra have been observed. See Dr. H. Schaum, as quoted in the 'Zoologist,' vol. v.–vi. 1847–48, p. 1896. Also Kirby and Spence, 'Introduction to Entomology,' vol. iii. 1826, p. 305.

433 Westwood, 'Modern Class.' vol. ii. p. 193. The following statement about Penthe, and others in inverted commas, are taken from Mr. Walsh, 'Practical Entomologist,' Philadelphia, vol. ii. p. 88.

434 Kirby and Spence, 'Introduct.' &c., vol. iii. p. 332–336.

[PAGE 977]

435 'Insecta Maderensia,' 1854, p. 20.

436 E. Doubleday, 'Annals and Mag. of Nat. Hist.' vol. i. 1848, p. 379. I may add that the wings in certain Hymenoptera (see Shuckard, 'Fossorial Hymenop.' 1837, p. 39–43) differ in neuration according to sex.

437 H. W. Bates, in 'Journal of Proc. Linn. Soc.' vol. vi. 1862, p. 74. Mr. Wonfor's observations are quoted in 'Popular Science Review,' 1868, p. 343.

438 Kirby and Spence, 'Introduction to Entomology,' vol. iii. p. 299.

439 Robinet, 'Vers à Soie,' 1848, p. 207.

[PAGE 978]

440 'Transact. Ent. Soc.' 3rd series, vol. v. p. 486.

441 'Journal of Proc. Ent. Soc.' Feb. 4th, 1867, p. lxxi.

442 For this and other statements on the size of the sexes, see Kirby and Spence, ibid. vol. iii. p. 300; on the duration of life in insects, see p. 344.

[PAGE 979]

443 'Transact. Linnean Soc.' vol. xxvi. 1868, p. 296.

444 'The Malay Archipelago,' vol. ii. 1869, p. 313.

445 'Modern Classification of Insects,' vol. ii. 1840, p. 526.

446 See Mr. B. T. Lowne's very interesting work, 'On the Anatomy of the Blow-Fly, Musca vomitoria,' 1870, p. 14.

[PAGE 980]

447 Westwood, 'Modern Class. of Insects,' vol. ii. p. 473.

448 These particulars are taken from Westwood's 'Modern Class. of Insects,' vol. ii. 1840, p. 422. See, also, on the Fulgoridæ, Kirby and Spence, 'Introduct.' vol. ii. p. 401.

449 'Zeitschrift fur wissenschaft. Zoolog.' B. xvii. 1867, s. 152–158.

450 I am indebted to Mr. Walsh for having sent me this extract from a 'Journal of the Doings of Cicada septemdecim,' by Dr. Hartman.

[PAGE 981]

451 L. Guilding, 'Transact. Linn. Soc.' vol. xv. p. 154.

452 Köppen, as quoted in the 'Zoological Record,' for 1867, p. 460.

453 Gilbert White, 'Nat. Hist. of Selborne,' vol. ii. 1825, p. 262.

454 Harris, 'Insects of New England,' 1842, p. 128.

455 'The Naturalist on the Amazons,' vol. i. 1863, p. 252. Mr. Bates gives a very interesting discussion on the gradations in the musical apparatus of the three families. See also Westwood, 'Modern Class.' vol. ii. p. 445 and 453.

456 'Proc. Boston Soc. of Nat. Hist.' vol. xi. April, 1868.

457 'Nouveau Manuel d'Anat. Comp.' (French translat.), tom. i. 1850, p. 567.

[PAGE 982]

458 'Zeitschrift für wissenschaft. Zoolog.' B. xvii. 1867, s. 117.

459 Westwood, 'Modern Class. of Insects,' vol. i. p. 440.

460 Westwood, 'Modern Class. of Insects,' vol. i. p. 453.

[PAGE 983]

461 Landois, ibid. s. 121, 122.

462 Mr. Walsh also informs me that he has noticed that the female of the *Platyphyllum concavum,* "when captured makes a feeble grating noise by shuffling her wing-covers together."

[PAGE 984]

463 Landois, ibid. s. 113.

464 'Insects of New England,' 1842, p. 133.

465 Westwood, 'Modern Classification,' vol. i. p. 462.

466 Westwood, ibid. vol. i. p. 453.

467 Landois, ibid. s. 115, 116, 120, 122.

[PAGE 985]

468 'Transact. Ent. Soc.' 3rd series, vol. ii. ('Journal of Proceedings, p. 117.)

469 Westwood, 'Modern Class. of Insects,' vol. i. p. 427; for crickets, p. 445.

[PAGE 986]

470 Mr. Ch. Horne, in 'Proc. Ent. Soc.' May 3, 1869, p. xii.

471 The *Oecanthus nivalis.* Harris, 'Insects of New England,' 1842, p. 124.

472 Platyblemnus: Westwood, 'Modern. Class.' vol. i. p. 447.

473 B. D. Walsh, the Pseudo-neuroptera of Illinois, in 'Proc. Ent. Soc. of Philadelphia,' 1862, p. 361.

474 'Modern Class.' vol. ii. p. 37.

475 Walsh, ibid. p. 381. I am indebted to this naturalist for the following facts on Hetærina, Anax, and Gomphus.

476 'Transact. Ent. Soc.' vol. i. 1836, p. lxxxi.

[PAGE 987]

477 See abstract in the 'Zoological Record' for 1867, p. 450.

478 Kirby and Spence, 'Introduct. to Entomology,' vol. ii. 1818, p. 35.

479 See an interesting article, "The Writings of Fabre," in 'Nat. Hist. Review,' April, 1862, p. 122.

[PAGE 988]

480 'Journal of Proc. of Entomolog. Soc.' Sept. 7th, 1863, p. 169.

481 P. Huber, 'Recherches sur les Mœurs des Fourmis,' 1810, p. 150, 165.

482 'Proc. Entomolog. Soc. of Philadelphia,' 1866, p. 238–239.

483 Quoted by Westwood, 'Modern Class. of Insects,' vol. ii. p. 214.

[PAGE 989]

484 Pyrodes pulcherrimus, in which the sexes differ conspicuously, has been described by Mr. Bates in 'Transact. Ent. Soc.' 1869, p. 50. I will specify the few other cases in which I have heard of a difference in colour between the sexes of beetles. Kirby and Spence ('Introduct. to Entomology,' vol. iii. p. 301) mention a Cantharis, Meloe, Rhagium, and the *Leptura testacea*; the male of the latter being testaceous, with a black thorax, and the female of a dull red all over. These two latter beetles belong to the Order of Longicorns. Messrs. B. Trimen and Waterhouse, junr., inform me of two Lamellicorns, viz., a Peritrichia and Trichius, the male of the latter being more obscurely coloured than the female. In *Tillus elongatus* the male is black, and the female always, as it is believed, of a dark blue colour with a red thorax. The male, also, of *Orsodacna atra,* as I hear from Mr. Walsh, is black, the female (the so-called *O. ruficollis)* having a rufous thorax.

[PAGE 990]

485 'Proc. Entomolog. Soc. of Philadelphia,' 1864, p. 228.

486 Kirby and Spence, 'Introduct. Entomolog.' vol. iii. p. 300.

[PAGE 993]

487 Kirby and Spence, ibid. vol. iii. p. 329.

[PAGE 994]

488 'Modern Classification of Insects,' vol. i. p. 172. On the same page there is an account of Siagonium. In the British Museum I noticed one male specimen of Siagonium in an intermediate condition, so that the dimorphism is not strict.

489 'The Malay Archipelago,' vol. ii. 1869, p. 276.

490 'Entomological Magazine,' vol. i. 1833, p. 82. See also on the conflicts of this species, Kirby and Spence, ibid. vol. iii. p. 314; and Westwood, ibid. vol. i. p. 187.

491 Quoted from Fischer, in 'Dict. Class. d'Hist. Nat.' tom. x. p. 324.

[PAGE 995]

492 'Ann. Soc. Entomolog. France,' 1866, as quoted in 'Journal of Travel,'

by A. Murray, 1868, p. 135.

493 Westwood, 'Modern Class.' vol. i. p. 184.

[PAGE 996]

494 Wollaston, On certain musical Curculionidæ, 'Annals and Mag. Of Nat. Hist.' vol. vi. 1860, p. 14.

495 'Zeitschrift für wiss. Zoolog.' B. xvii. 1867, s. 127.

496 I am greatly indebted to Mr. G. R. Crotch for having sent me numerous prepared specimens of various beetles belonging to these three families and others, as well as for valuable information of all kinds. He believes that the power of stridulation in the Clythra has not been previously observed. I am also much indebted to Mr. E. W. Janson, for information and specimens. I may add that my son, Mr. F. Darwin, finds that *Dermestes murinus* stridulates, but he searched in vain for the apparatus. Scolytus has lately been described by Mr. Algen as a stridulator, in the 'Edinburgh Monthly Magazine,' 1869, Nov., p. 130.

497 Schiödte, translated in 'Annals and Mag. of Nat. Hist.' vol. xx. 1867, p. 37.

498 Westring has described (Kroyer, 'Naturhist. Tidskrift,' B. ii. 1848–49, p. 334) the stridulating organs in these two, as well as in other families. In the Carabidæ I have examined *Elaphrus uliginosus* and *Blethisa multipunctata,* sent to me by Mr. Crotch. In Blethisa the transverse ridges on the furrowed border of the abdominal segment do not come into play, as far as I could judge, in scraping the rasps on the elytra.

[PAGE 997]

499 I am indebted to Mr. Walsh, of Illinois, for having sent me extracts from Leconte's 'Introduction to Entomology,' p. 101, 143.

[PAGE 999]

500 M. P. de la Brulerie, as quoted in 'Journal of Travel,' A. Murray, vol. i. 1868, p. 135.

501 Mr. Doubleday informs me that "the noise is produced by the insect raising itself on its legs as high as it can, and then striking its thorax five or six times, in rapid succession, against the substance upon which it is sitting." For references on this subject see Landois, 'Zeitschrift für wissen. Zoolog.' B. xvii. s. 131. Olivier says (as quoted by Kirby and Spence, 'Introduct.' vol. ii. p. 395) that the female of *Pimelia striata* produces a rather loud sound by striking her abdomen against any hard substance, "and that the male, obedient to this call, soon attends her and they pair."

CHAPTER XI

[PAGE 1000]

502 Apatura Iris: 'The Entomologist's Weekly Intelligencer,' 1859, p. 139. For the Bornean Butterflies see C. Collingwood, 'Rambles of a Naturalist,' 1868, p. 183.

[PAGE 1001]

503 See my 'Journal of Researches,' 1845, p. 56. Mr. Doubleday has detected ('Proc. Ent. Soc.' March 3rd, 1845, p. 123) a peculiar membranous sac at the base of the front wings, which is probably connected with the production of the sound.

504 See also Mr. Bates' paper in 'Proc. Ent. Soc. of Philadelphia,' 1865, p. 206. Also Mr. Wallace on the same subject, in regard to Diadema, in 'Transact. Entomolog. Soc. of London,' 1869, p. 278.

[PAGE 1003]

505 'The Naturalist on the Amazons,' vol. i. 1863, p. 19.

[PAGE 1004]

506 See the interesting article in the 'Westminster Review,' July, 1867, p. 10. A woodcut of the Kallima is given by Mr. Wallace in 'Hardwicke's Science Gossip,' Sept. 1867, p. 196.

[PAGE 1005]

507 See the interesting observations by Mr. T. W. Wood, 'The Student,' Sept. 1868, p. 81.

508 Mr. Wallace in 'Hardwicke's Science Gossip,' Sept. 1867, p. 193.

509 See also, on this subject, Mr. Weir's paper in 'Transact. Ent. Soc.' 1869, p. 23.

510 'Westminster Review,' July, 1867, p. 16.

[PAGE 1006]

511 For instance, Lithosia; but Prof. Westwood ('Modern Class. of Insects,' vol. ii. p. 390) seems surprised at this case. On the relative colours of diurnal and nocturnal Lepidoptera, see ibid. p. 333 and 392; also Harris, 'Treatise on the Insects of New England,' 1842, p. 315.

512 Such differences between the upper and lower surfaces of the wings of several species of Papilio, may be seen in the beautiful plates to

Mr. Wallace's Memoir on the Papilionidæ of the Malayan Region, in 'Transact. Linn. Soc.' vol. xxv. part i. 1865.

513 'Proc. Ent. Soc.' March 2nd, 1868.

514 See also an account of the S. American genus Erateina (one of the Geometræ) in 'Transact. Ent. Soc.' new series, vol. v. pl. xv. and xvi.

[PAGE 1007]

515 'Proc. Ent. Soc. of London,' July 6, 1868, p. xxvii.

516 Harris, 'Treatise,' &c., edited by Flint, 1862, p. 395.

517 For instance, I observe in my son's cabinet that the males are darker than the females in the *Lasiocampa quercus, Odonestis potatoria, Hypogymna dispar, Dasychira pudibunda,* and *Cycnia mendica.* In this latter species the difference in colour between the two sexes is strongly marked; and Mr. Wallace informs me that we here have, as he believes, an instance of protective mimickry confined to one sex, as will hereafter be more fully explained. The white female of the Cycnia resembles the very common *Spilosoma menthrasti,* both sexes of which are white; and Mr. Stainton observed that this latter moth was rejected with utter disgust by a whole brood of young turkeys, which were fond of eating other moths; so that if the Cycnia was commonly mistaken by British birds for the Spilosoma, it would escape being devoured, and its white deceptive colour would thus be highly beneficial.

[PAGE 1008]

518 'Rambles of a Naturalist in the Chinese Seas,' 1868, p. 182.

[PAGE 1009]

519 Wallace on the Papilionidæ of the Malayan Region, in 'Transact. Linn. Soc.' vol. xxv. 1865, p. 8, 36. A striking case of a rare variety, strictly intermediate between two other well-marked female varieties, is given by Mr. Wallace. See also Mr. Bates, in 'Proc. Entomolog. Soc.' Nov. 19th, 1866, p. xl.

520 Mr. R. MacLachlan, 'Transact. Ent. Soc.' vol. ii. part 6th, 3rd series, 1866, p. 459.

[PAGE 1010]

521 H. W. Bates, 'The Naturalist on the Amazons,' vol. ii. 1863, p. 228. A. R. Wallace, in 'Transact. Linn. Soc." vol. xxv. 1865, p. 10.

522 On this whole subject see 'The Variation of Animals and Plants under Domestication,' vol. ii. 1868, chap. xxiii.

[PAGE 1011]

523 A. R. Wallace, in 'The Journal of Travel,' vol. i. 1868, p. 88. 'Westminster Review,' July, 1867, p. 37. See also Messrs. Wallace and Bates in 'Proc. Ent. Soc.' Nov. 19th, 1866, p. xxxix.

[PAGE 1013]

524 'The Variation of Animals and Plants under Domestication,' vol. ii. chap. xii. p. 17.

[PAGE 1014]

525 'Transact. Linn. Soc.' vol. xxiii. 1862, p. 495.

526 'Proc. Ent. Soc.' Dec. 3rd, 1866, p. xlv.

[PAGE 1015]

527 'Transact. Linn. Soc.' vol. xxv. 1865, p. 1; also 'Transact. Ent. Soc.' vol. iv. (3rd series), 1867, p. 301.

528 See an ingenious article entitled, "Difficulties of the Theory of Natural Selection," in the 'Month,' 1869. The writer strangely supposes that I attribute the variations in colour of the Lepidoptera, by which certain species belonging to distinct families have come to resemble others, to reversion to a common progenitor; but there is no more reason to attribute these variations to reversion than in the case of any ordinary variation.

529 Wallace, "Notes on Eastern Butterflies," 'Transact. Ent. Soc.' 1869, p. 287.

530 Wallace, in 'Westminster Review,' July, 1867, p. 37; and in 'Journal of Travel and Nat. Hist.' vol. i. 1868, p. 88.

531 See remarks by Messrs. Bates and Wallace, in 'Proc. Ent. Soc.' Nov. 19, 1866, p. xxxix.

[PAGE 1016]

532 See Mr. Wallace in 'Westminster Review,' July, 1867, p. 11 and 37. The male of no butterfly, as Mr. Wallace informs me, is known to differ in colour, as a protection, from the female; and he asks me how I can explain this fact on the principle that one sex alone has varied and has transmitted its variations exclusively to the same sex, without the aid of selection to check the variations being inherited by the other sex. No doubt if it could be shewn that the females of very many species had been rendered beautiful through protective mimickry, but that this has never occurred with the males, it would be a serious difficulty.

But the number of cases as yet known hardly suffices for a fair judgment. We can see that the males, from having the power of flying more swiftly, and thus escaping danger, would not be so likely as the females to have had their colours modified for the sake of protection; but this would not in the least have interfered with their receiving protective colours through inheritance from the females. In the second place, it is probable that sexual selection would actually tend to prevent a beautiful male from becoming obscure, for the less brilliant individuals would be less attractive to the females. Supposing that the beauty of the male of any species had been mainly acquired through sexual selection, yet if this beauty likewise served as a protection, the acquisition would have been aided by natural selection. But it would be quite beyond our power to distinguish between the two processes of sexual and ordinary selection. Hence it is not likely that we should be able to adduce cases of the males having been rendered brilliant exclusively through protective mimickry, though this is comparatively easy with the females, which have rarely or never been rendered beautiful, as far as we can judge, for the sake of sexual attraction, although they have often received beauty through inheritance from their male parents.

[PAGE 1017]

533 'Proc. Entomolog. Soc.' Dec. 3rd, 1866, p. xlv., and March 4th, 1867, p. lxxx.

534 See Mr. J. Jenner Weir's paper on insects and insectivorous birds, in 'Transact. Ent. Soc.' 1869, p. 21; also Mr. Butler's paper, ibid. p. 27.

CHAPTER XII

[PAGE 1023]

535 Yarrell's 'Hist. of British Fishes,' vol. ii. 1836, p. 417, 425, 436. Dr. Günther informs me that the spines in *R. clavata* are peculiar to the female.

536 See Mr. R. Warington's interesting articles in 'Annals and Mag. of Nat. Hist.' Oct. 1852 and Nov. 1855.

537 Noel Humphreys, 'River Gardens,' 1857.

538 Loudon's 'Mag. of Natural History,' vol. iii. 1830, p. 331.

539 'The Field,' June 29th, 1867. For Mr. Shaw's statement, see 'Edinburgh Review,' 1843. Another experienced observer (Scrope's

'Days of Salmon Fishing,' p. 60) remarks that the male would, if he could, keep, like the stag, all other males away.

[PAGE 1024]

540 Yarrell, 'History of British Fishes,' vol. ii. 1836, p. 10.
541 'The Naturalist in Vancouver's Island,' vol. i. 1866, p. 54.
542 'Scandinavian Adventures,' vol. i. 1854, p. 100, 104.

[PAGE 1025]

543 See Yarrell's account of the Rays in his 'Hist. of British Fishes,' vol. ii. 1836, p. 416, with an excellent figure, and p. 422, 432.
544 As quoted in 'The Farmer,' 1868, p. 369.
545 I have drawn up this description from Yarrell's 'British Fishes,' vol. i. 1836, p. 261 and 266.

[PAGE 1026]

546 'Catalogue of Acanth. Fishes in the British Museum,' by Dr. Günther, 1861, p. 138–151.
547 'Game Birds of Sweden,' &c., 1867, p. 466.
548 With respect to this and the following species I am indebted to Dr. Günther for information: see also his paper on the Fishes of Central America, in 'Transact. Zoolog. Soc.' vol. vi. 1868, p. 485.

[PAGE 1027]

549 Dr. Günther makes this remark; 'Catalogue of Fishes in the British Museum,' vol. iii. 1861, p. 141.
550 See Dr. Günther on this genus, in 'Proc. Zoolog. Soc.' 1868, p. 232.

[PAGE 1028]

551 F. Buckland, in 'Land and Water,' July, 1868, p. 377, with a figure.
552 Dr. Günther, 'Catalogue of Fishes,' vol. iii. p. 221 and 240.
553 See also 'A Journey in Brazil,' by Prof. and Mrs. Agassiz, 1868, p. 220.

[PAGE 1029]

554 Yarrell, 'British Fishes,' vol. ii. 1836, p. 10, 12, 35.
555 W. Thompson, in 'Annals and Mag. of Nat. History,' vol. vi. 1841, p. 440.
556 'The American Agriculturist,' 1868, p. 100.
557 'Annals and Mag. of Nat. Hist.' Oct. 1852.

558 Loudon's 'Mag. of Nat. Hist.' vol. v. 1832, p. 681.

[PAGE 1030]

559 Bory de Saint Vincent, in 'Dict. Class. d'Hist. Nat.' tom. ix. 1826, p. 151.

560 Owing to some remarks on this subject, made in my work 'On the Variation of Animals under Domestication,' Mr. W. F. Mayers ('Chinese Notes and Queries,' Aug. 1868, p. 123) has searched the ancient Chinese encyclopedias. He finds that gold-fish were first reared in confinement during the Sung Dynasty, which commenced A.D. 960. In the year 1129 these fishes abounded. In another place it is said that since the year 1548 there has been "produced at Hangchow a variety called the fire-fish, from its intensely red colour. It is universally admired, and there is not a household where it is not cultivated, *in rivalry as to its colour,* and as a source of profit."

561 'Westminster Review,' July, 1867, p. 7.

562 "Indian Cyprinidæ," by Mr. J. M'Clelland, 'Asiatic Researches,' vol. xix. part ii. 1839, p. 230.

[PAGE 1031]

563 'Proc. Zoolog. Soc.' 1865, p. 327, pl. xiv. and xv.

564 Yarrell, 'British Fishes,' vol. ii. p. 11.

[PAGE 1032]

565 According to the observations of M. Gerbe; see Günther's 'Record of Zoolog. Literature,' 1865, p. 194.

566 Cuvier, 'Règne Animal,' vol. ii. 1829, p. 242.

567 See Mr. Warington's most interesting description of the habits of the *Gasterosteus leiurus,* in 'Annals and Mag. of Nat. Hist.' November, 1855.

568 Prof. Wyman, in 'Proc. Boston Soc. of Nat. Hist.' Sept. 15, 1857. Also W. Turner, in 'Journal of Anatomy and Phys.' Nov. 1, 1866, p. 78. Dr. Günther has likewise described other cases.

[PAGE 1033]

569 Yarrell, 'Hist. of British Fishes,' vol. ii. 1836, p. 329, 338.

570 Dr. Günther, since publishing an account of this species in 'The Fishes of Zanzibar,' by Col. Playfair, 1866, p. 137, has re-examined the specimens, and has given me the above information.

[PAGE 1034]

571 The Rev. C. Kingsley, in 'Nature,' May, 1870, p. 40.
572 Bell, 'History of British Reptiles,' 2nd edit. 1849, p. 156–159.

[PAGE 1035]

573 Bell, ibid. p. 146, 151.
574 'Zoology of the Voyage of the "Beagle,"' 1843. "Reptiles," by Mr. Bell, p. 49.
575 'The Reptiles of India,' by Dr. A. Günther, Ray Soc. 1864, p. 413.

[PAGE 1036]

576 Bell, 'History of British Reptiles,' 1849, p. 93.
577 J. Bishop, in 'Todd's Cyclop. of Anat. and Phys.' vol. iv. p. 1503.
578 Bell, ibid. p. 112–114.

[PAGE 1037]

579 Mr. C. J. Maynard, 'The American Naturalist,' Dec. 1869, p. 555.
580 See my 'Journal of Researches during the Voyage of the "Beagle,"' 1845, p. 384.
581 'Travels through Carolina,' &c., 1791, p. 128.
582 Owen, 'Anatomy of Vertebrates,' vol. i. 1866, p. 615.
583 Sir Andrew Smith, 'Zoolog. of S. Africa: Reptilia,' 1849, pl. x.
584 Dr. A. Günther, 'Reptiles of British India,' Ray Soc. 1864, p. 304, 308.

[PAGE 1038]

585 Owen, 'Anatomy of Vertebrates,' vol. i. 1866, p. 615.
586 The celebrated botanist Schleiden incidently remarks ('Ueber den Darwinismus: Unsere Zeit,' 1869, s. 269), that Rattle-snakes use their rattles as a sexual call, by which the two sexes find each other. I do not know whether this suggestion rests on any direct observations. These snakes pair in the Zoological Gardens, but the keepers have never observed that they use their rattles at this season more than at any other.
587 "Rambles in Ceylon," 'Annals and Mag. of Nat. Hist.,' 2nd series, vol. ix. 1852, p. 333.
588 'Westminster Review,' July 1st, 1867, p. 32.

[PAGE 1039]

589 Mr. N. L. Austen kept these animals alive for a considerable time; see 'Land and Water,' July, 1867, p. 9.

[PAGE 1040]

590 All these statements and quotations, in regard to Cophotis, Sitana and Draco, as well as the following facts in regard to Ceratophora, are taken from Dr. Günther's magnificent work on the 'Reptiles of British India,' Ray Soc. 1864, p. 122, 130, 135.

[PAGE 1041]

591 Bell, 'History of British Reptiles,' 2nd edit. 1849, p. 40.

592 For Proctotretus see 'Zoology of the Voyage of the "Beagle:" Reptiles,' by Mr. Bell, p. 8. For the Lizards of S. Africa, see 'Zoology of S. Africa: Reptiles,' by Sir Andrew Smith, pl. 25 and 39. For the Indian Calotes, see 'Reptiles of British India,' by Dr. Günther, p. 143.

CHAPTER XIII

[PAGE 1043]

593 'Ibis,' vol. iii. (new series) 1867, p. 414.

594 Gould, 'Handbook to the Birds of Australia,' 1865, vol. ii. p. 383.

[PAGE 1044]

595 Quoted by Mr. Gould, 'Introduction to the Trochilidæ,' 1861, p. 29.

[PAGE 1045]

596 Gould, ibid. p. 52.

597 W. Thompson, 'Nat. Hist. of Ireland: Birds,' vol. ii. 1850, p. 327.

598 Jerdon, 'Birds of India,' 1863, vol. ii. p. 96.

[PAGE 1046]

599 Macgillivray, 'Hist. Brit. Birds,' vol. iv. 1852, p. 177–181.

600 Sir R. Schomburgk, in 'Journal of R. Geograph. Soc.' vol. xiii. 1843, p. 31.

601 'Ornithological Biography,' vol. i. p. 191. For pelicans and snipes, see vol. iii. p. 381, 477.

602 Gould, 'Handbook of Birds of Australia,' vol. i. p. 395; vol. ii. p. 383.

603 Mr. Hewitt in the 'Poultry Book by Tegetmeier,' 1866, p. 137.

[PAGE 1047]

604 Layard, 'Annals and Mag. of Nat. Hist.' vol. xiv. 1854, p. 63.

605 Jerdon, 'Birds of India,' vol. iii. p. 574.

606 Brehm, 'Illust. Thierleben,' 1867, B. iv. s. 351. Some of the foregoing statements are taken from L. Lloyd, 'The Game Birds of Sweden,' &c., 1867, p. 79.

[PAGE 1048]

607 Jerdon, 'Birds of India:' on Ithaginis, vol. iii. p. 523; on Galloperdix, p. 541.

608 For the Egyptian goose, see Macgillivray, 'British Birds,' vol. iv. p. 639. For Plectropterus, 'Livingstone's Travels,' p. 254. For Palamedea, Brehm's 'Thierleben,' B, iv. s. 740. See also on this bird Azara, 'Voyages dans l'Amérique mérid.' tom. iv. 1809, p. 179, 253.

[PAGE 1049]

609 See, on our peewit, Mr. R. Carr in 'Land and Water,' Aug. 8th, 1868, p. 46. In regard to Lobivanellus, see Jerdon's 'Birds of India,' vol. iii. p. 647, and Gould's 'Handbook of Birds of Australia,' vol. ii. p. 220. For the Holopterus, see Mr. Allen in the 'Ibis,' vol. v. 1863, p. 156.

610 Audubon, 'Ornith. Biography,' vol. ii. p. 492; vol. i. p. 4–13.

611 Mr. Blyth, 'Land and Water,' 1867, p. 212.

612 Richardson, on Tetrao umbellus, 'Fauna Bor. Amer.: Birds,' 1831, p. 343. L. Lloyd, 'Game Birds of Sweden,' 1867, p. 22, 79, on the capercailzie and black-cock. Brehm, however, asserts ('Thierleben,' &c., B. iv. s. 352) that in Germany the grey-hens do not generally attend the Balzen of the black-cocks, but this is an exception to the common rule; possibly the hens may lie hidden in the surrounding bushes, as is known to be the case with the grey-hens in Scandinavia, and with other species in N. America.

613 'Ornithological Biography,' vol. ii. p. 275.

614 Brehm, 'Thierleben,' &c., B. iv. 1867, p. 990. Audubon, 'Ornith. Biography,' vol. ii. p. 492.

[PAGE 1050]

615 'Land and Water,' July 25th, 1868, p. 14.

616 Audubon's 'Ornitholog. Biography;' on Tetrao cupido, vol. ii. p. 492; on the Sturnus, vol. ii. p. 219.

617 'Ornithological Biograph.' vol. v. p. 601.

618 The Hon. Daines Barrington, 'Philosoph. Transact.' 1773, p. 252.

[PAGE 1051]

619 'Ornithological Dictionary,' 1833, p. 475.

620 'Naturgeschichte der Stubenvögel,' 1840, s. 4. Mr. Harrison Weir

likewise writes to me:—"I am informed that the best singing males generally get a mate first when they are bred in the same room."

621 'Philosophical Transactions,' 1773, p. 263. White's 'Natural History of Selborne,' vol. i. 1825, p. 246.

622 'Naturges. der Stubenvögel,' 1840, s. 252.

623 Mr. Bold, 'Zoologist,' 1843–44, p. 659.

[PAGE 1052]

624 D. Barrington, 'Phil. Transact.' 1773, p. 262. Bechstein, 'Stubenvögel,' 1840, s. 4.

625 This is likewise the case with the water-ouzel, see Mr. Hepburn in the 'Zoologist,' 1845–1846, p. 1068.

626 L. Lloyd, 'Game Birds of Sweden,' 1867, p. 25.

627 Barrington, ibid. p. 264. Bechstein, ibid. s. 5.

628 Dureau de la Malle gives a curious instance ('Annales des Sc. Nat.' 3rd series, Zoolog. tom. x. p. 118) of some wild blackbirds in his garden in Paris which naturally learnt from a caged bird a republican air.

629 Bishop, in 'Todd's Cyclop. of Anat. and Phys.' vol. iv. p. 1496.

630 As stated by Barrington in 'Philosoph. Transact.' 1773, p. 262.

631 Gould, 'Handbook to the Birds of Australia,' vol. i. 1865, p. 308–310. See also Mr. T. W. Wood in the 'Student,' April, 1870, p. 125.

[PAGE 1053]

632 See remarks to this effect in Gould's 'Introduction to the Trochilidæ,' 1861, p. 22.

633 'The Sportsman and Naturalist in Canada,' by Major W. Ross King, 1866, p. 144–146. Mr. T. W. Wood gives in the 'Student' (April, 1870, p. 116) an excellent account of the attitude and habits of this bird during its courtship. He states that the ear-tufts or neck-plumes are erected, so that they meet over the crown of the head.

634 Richardson, 'Fauna Bor. Americana: Birds,' 1831, p. 359. Audubon, ibid. vol. iv. p. 507.

[PAGE 1054]

635 The following papers have been lately written on this subject:—Prof. A. Newton, in the 'Ibis,' 1862, p. 107; Dr. Cullen, ibid. 1865, p. 145; Mr. Flower, in 'Proc. Zool. Soc.' 1865, p. 747; and Dr. Murie, in 'Proc. Zool. Soc.' 1868, p. 471. In this latter paper an excellent figure is given of the male Australian Bustard in full display with the sack distended.

636 Bates, 'The Naturalist on the Amazons,' 1863, vol. ii. p. 284; Wallace, in 'Proc. Zool. Soc.' 1850, p. 206. A new species, with a still larger neck-appendage *(C. penduliger),* has lately been discovered, see 'Ibis,' vol. i. p. 457.

[PAGE 1055]

637 Bishop, in Todd's 'Cyclop. of Anat. and Phys.' vol. iv. p. 1499.

638 The spoonbill (Platalea) has its trachea convoluted into a figure of eight, and yet this bird (Jerdon, 'Birds of India,' vol. iii. p. 763) is mute; but Mr. Blyth informs me that the convolutions are not constantly present, so that perhaps they are now tending towards abortion.

639 'Elements of Comp. Anat.' by R. Wagner, Eng. translat. 1845, p. 111. With respect to the swan, as given above, Yarrell's 'Hist. of British Birds,' 2nd edit. 1845, vol. iii. p. 193.

640 C. L. Bonaparte, quoted in the 'Naturalist Library: Birds,' vol. xiv. p. 126.

641 L. Lloyd, 'The Game Birds of Sweden,' &c., 1867, p. 22, 81.

642 Jenner, 'Philosoph. Transactions,' 1824, p. 20.

[PAGE 1056]

643 For the foregoing several facts see, on Birds of Paradise, Brehm, 'Thierleben,' Band iii. s. 325. On Grouse, Richardson, 'Fauna Bor. Americ.: Birds,' p. 343 and 359; Major W. Ross King, 'The Sportsman in Canada,' 1866, p. 156; Audubon, 'American Ornitholog. Biograph.' vol. i. p. 216. On the Kalij-pheasant, Jerdon, 'Birds of India,' vol. iii. p. 533. On the Weavers, 'Livingstone's Expedition to the Zambesi,' 1865, p. 425. On Woodpeckers, Macgillivray, 'Hist. of British Birds,' vol. iii. 1840, p. 84, 88, 89, and 95. On the Hoopoe, Mr. Swinhoe, in 'Proc. Zoolog. Soc.' June 23, 1863. On the Night-Jar, Audubon, ibid. vol. ii. p. 255. The English Night-Jar likewise makes in the spring a curious noise during its rapid flight.

[PAGE 1057]

644 See M. Meves' interesting paper in 'Proc. Zool. Soc.' 1858, p. 199. For the habits of the snipe, Macgillivray, 'Hist. British Birds,' vol. iv. p. 371. For the American snipe, Capt. Blakiston, 'Ibis,' vol. v. 1863, p. 131.

645 Mr. Salvin, in 'Proc. Zool. Soc.' 1867, p. 160. I am much indebted to this distinguished ornithologist for sketches of the feathers of the Chamæpetes, and for other information.

646 Jerdon, 'Birds of India,' vol. iii. p. 618, 621.

647 Gould, 'Introduction to the Trochilidæ,' 1861, p. 49. Salvin, 'Proc. Zoolog. Soc.' 1867, p. 160.

[PAGE 1058]

648 Sclater, in 'Proc. Zool. Soc.' 1860, p. 90, and in 'Ibis,' vol. iv. 1862, p. 175. Also Salvin, in 'Ibis,' 1860, p. 37.

[PAGE 1059]

649 'The Nile Tributaries of Abyssinia,' 1867, p. 203.

[PAGE 1060]

650 For Tetrao phasianellus, see Richardson, 'Fauna Bor. America,' p. 361, and for further particulars Capt. Blakiston, 'Ibis,' 1863, p. 125. For the Cathartes and Ardea, Audubon, 'Ornith. Biography,' vol. ii. p. 51, and vol. iii. p. 89. On the White-throat, Macgillivray, 'Hist. British Birds,' vol. ii. p. 354. On the Indian Bustard, Jerdon, 'Birds of India,' vol. iii. p. 618.

651 Gould, 'Handbook to the Birds of Australia,' vol. i. p. 444, 449, 455. The bower of the Satin Bower-bird may always be seen in the Zoological Society's Gardens, Regent's Park.

[PAGE 1061]

652 See remarks to this effect on the "Feeling of Beauty among Animals," by Mr. J. Shaw, in the 'Athenæum,' Nov. 24th, 1866, p. 681.

653 Mr. Monteiro, 'Ibis,' vol. iv. 1862, p. 339.

654 'Land and Water,' 1868, p. 217.

[PAGE 1062]

655 Jardine's 'Naturalist Library: Birds,' vol. xiv. p. 166.

656 Sclater, in the 'Ibis,' vol. vi. 1864, p. 114. Livingstone, 'Expedition to the Zambesi,' 1865, p. 66.

657 Jerdon, 'Birds of India,' vol. iii. p. 620.

658 Wallace, in 'Annals and Mag. of Nat. Hist.' vol. xx. 1857, p. 416; and in his 'Malay Archipelago,' vol. ii. 1869, p. 390.

[PAGE 1063]

659 See my work on 'The Variation of Animals and Plants under Domestication,' vol. i. p. 289, 293.

660 Quoted from M. de Lafresnaye, in 'Annals and Mag. of Nat. Hist.'

vol. xiii. 1854, p. 157: see also Mr. Wallace's much fuller account in vol. xx. 1857, p. 412, and in his Malay Archipelago.

661 Wallace, 'The Malay Archipelago,' vol. ii. 1869, p. 405.

[PAGE 1065]

662 Mr. Sclater, 'Intellectual Observer,' Jan. 1867. Waterton's 'Wanderings,' p. 118. See also Mr. Salvin's interesting paper, with a plate, in the 'Ibis,' 1865, p. 90.

[PAGE 1066]

663 'Land and Water,' 1867, p. 394.

664 Mr. D. G. Elliot, in 'Proc. Zool. Soc.' 1869, p. 589.

665 'Nitzsch's Pterylography,' edited by P. L. Sclater. Ray Soc. 1867, p. 14.

666 The brown mottled summer plumage of the ptarmigan is of as much importance to it, as a protection, as the white winter plumage; for in Scandinavia, during the spring, when the snow has disappeared, this bird is known to suffer greatly from birds of prey, before it has acquired its summer dress: see Wilhelm von Wright, in Lloyd, 'Game Birds of Sweden,' 1867, p. 125.

[PAGE 1067]

667 In regard to the previous statements on moulting, see, on snipes, &c., Macgillivray, 'Hist. Brit. Birds,' vol. iv. p. 371; on Glareolæ, curlews, and bustards, Jerdon, 'Birds of India,' vol. iii. p. 615, 630, 683; on Totanus, ibid, p. 700; on the plumes of herons, ibid, p. 738, and Macgillivray, vol. iv. p. 435 and 444, and Mr. Stafford Allen, in the 'Ibis,' vol. v. 1863, p. 33.

668 On the moulting of the ptarmigan, see Gould's 'Birds of Great Britain.' On the honey-suckers, Jerdon, 'Birds of India,' vol. i. p. 359, 365, 369. On the moulting of Anthus, see Blyth, in 'Ibis,' 1867, p. 32.

[PAGE 1068]

669 For the foregoing statements in regard to partial moults, and on old males retaining their nuptial plumage, see Jerdon, on bustards and plovers, in 'Birds of India,' vol. iii. p. 617, 637, 709, 711. Also Blyth in 'Land and Water,' 1867, p. 84. On the Vidua, 'Ibis,' vol. iii. 1861, p. 133. On the Drongo shrikes, Jerdon, ibid. vol. i. p. 435. On the vernal moult of the *Herodias bubulcus,* Mr. S. S. Allen, in 'Ibis,' 1863, p. 33. On *Gallus bankiva,* Blyth, in 'Annals and Mag. of Nat. Hist.' vol. i. 1848, p. 455; see, also, on this subject, my 'Variation of Animals under Domestication,' vol. i. p. 236.

670 See Macgillivray, 'Hist. British Birds' (vol. v. p. 34, 70, and 223), on the moulting of the Anatidæ, with quotations from Waterton and Montagu. Also Yarrell, 'Hist. of British Birds,' vol. iii. p. 243.

[PAGE 1069]

671 On the pelican, see Sclater, in 'Proc. Zool. Soc.' 1868, p. 265. On the American finches, see Audubon, 'Ornith. Biography,' vol. i. p. 174, 221, and Jerdon, 'Birds of India,' vol. ii. p. 383. On the *Fringilla cannabina* of Madeira, Mr. E. Vernon Harcourt, 'Ibis,' vol. v., 1863, p. 230.

672 See also 'Ornamental Poultry,' by Rev. E. S. Dixon, 1848, p. 8.

673 'Birds of India,' introduct. vol. i. p. xxiv.; on the peacock, vol. iii. p. 507. See Gould's 'Introduction to the Trochilidæ,' 1861, p. 15 and 111.

[PAGE 1070]

674 'Journal of R. Geograph, Soc.' vol. x. 1840, p. 236.

675 'Annals and Mag. of Nat. Hist.' vol. xiii. 1854, p. 157; also Wallace, ibid. vol. xx. 1857, p. 412, and 'The Malay Archipelago,' vol. ii. 1869, p. 252. Also Dr. Bennett, as quoted by Brehm, 'Thierleben,' B. iii. s. 326.

[PAGE 1071]

676 Mr. T. W. Wood has given ('The Student,' April, 1870, p. 115) a full account of this manner of display, which he calls the lateral or one-sided, by the gold pheasant and by the Japanese pheasant, *Ph. versicolor*.

[PAGE 1072]

677 'The Reign of Law,' 1867, p. 203.

[PAGE 1074]

678 For the description of these birds, see Gould's 'Handbook to the Birds of Australia,' vol. i. 1865, p. 417.

679 'Birds of India,' vol. ii. p. 96.

[PAGE 1075]

680 On the Cosmetornis, see Livingstone's 'Expedition to the Zambesi,' 1865, p. 66. On the Argus pheasant, Jardine's 'Nat. Hist. Lib.: Birds,' vol. xiv. p. 167. On Birds of Paradise, Lesson, quoted by Brehm, 'Thierleben,' B. iii. s. 325. On the widow-bird, Barrow's 'Travels in

Africa,' vol. i. p. 243, and 'Ibis,' vol. iii. 1861, p. 133. Mr. Gould, on the shyness of male birds, 'Handbook to Birds of Australia,' vol. i. 1865, p. 210, 457.

[PAGE 1076]

681 Tegetmeier, 'The Poultry Book,' 1866, p. 139.

CHAPTER XIV

[PAGE 1078]

682 Nordmann describes ('Bull. Soc. Imp. des Nat. Moscow,' 1861, tom. xxxiv. p. 264) the balzen of *Tetrao urogalloides* in Amur Land. He estimated the number of assembled males at above a hundred, the females, which lie hid in the surrounding bushes, not being counted. The noises uttered differ from those of the *T. urogallus* or the capercailzie.

683 With respect to the assemblages of the above named grouse see Brehm, 'Thierleben,' B. iv. s. 350; also L. Lloyd, 'Game Birds of Sweden,' 1867, p. 19, 78. Richardson, 'Fauna Bor. Americana,' Birds, p. 362. References in regard to the assemblages of other birds have previously been given. On Paradisea see Wallace, in 'Annals and Mag. of Nat. Hist.' vol. xx. 1857, p. 412. On the snipe, Lloyd, ibid. p. 221.

684 Quoted by Mr. T. W. Wood in the 'Student,' April, 1870, p. 125.

[PAGE 1079]

685 Gould, 'Handbook of Birds of Australia,' vol. i. p. 300, 308, 448, 451. On the ptarmigan, above alluded to, see Lloyd, ibid. p. 129.

[PAGE 1080]

686 On magpies, Jenner, in 'Phil. Transact.' 1824, p. 21. Macgillivray, 'Hist. British Birds,' vol. i. p. 570. Thompson, in 'Annals and Mag. of Nat. Hist.' vol. viii. 1842, p. 494.

[PAGE 1081]

687 On the peregrine falcon see Thompson, 'Nat. Hist. of Ireland: Birds,' vol. i. 1849, p. 39. On owls, sparrows, and partridges, see White, 'Nat. Hist. of Selborne,' edit. of 1825, vol. i. p. 139. On the Phœnicura, see Loudon's 'Mag. of Nat. Hist.' vol. vii. 1834, p. 245. Brehm, ('Thierleben,' B. iv. s. 991) also alludes to cases of birds thrice mated during same day.

688 See White ('Nat. Hist. of Selborne,' 1825, vol. i. p. 140) on the existence, early in the season, of small coveys of male partridges, of which fact I have heard other instances. See Jenner, on the retarded state of the generative organs in certain birds, in 'Phil. Transact.' 1824. In regard to birds living in triplets, I owe to Mr. Jenner Weir the cases of the starling and parrots, and to Mr. Fox, of partridges; on carrion-crows, see the 'Field,' 1868, p. 415. On various male birds singing after the proper period, see Rev. L. Jenyns, 'Observations in Natural History,' 1846, p. 87.

689 The following case has been given ('The Times,' Aug. 6th, 1868) by the Rev. F. O. Morris, on the authority of the Hon. and Rev. O. W. Forester. "The gamekeeper here found a hawk's nest this year, with five young ones in it. He took four and killed them, but left one with its wings clipped as a decoy to destroy the old ones by. They were both shot next day, in the act of feeding the young one, and the keeper thought it was done with. The next day he came again and found two other charitable hawks, who had come with an adopted feeling to succour the orphan. These two he killed, and then left the nest. On returning afterwards he found two more charitable individuals on the same errand of mercy. One of these he killed; the other he also shot, but could not find. No more came on the like fruitless errand."

[PAGE 1082]

690 For instance, Mr. Yarrell states ('Hist. British Birds,' vol. iii. 1845, p. 585) that a gull was not able to swallow a small bird which had been given to it. The gull "paused for a moment, and then, as if suddenly recollecting himself, ran off at full speed to a pan of water, shook the bird about in it until well soaked, and immediately gulped it down. Since that time he invariably has had recourse to the same expedient in similar cases."

691 'A Tour in Sutherlandshire,' vol. i. 1849, p. 185.

[PAGE 1083]

692 'Acclimatization of Parrots,' by C. Buxton, M.P. 'Annals and Mag. of Nat. Hist.' Nov. 1868, p. 381.

693 'The Zoologist,' 1847–1848, p. 1602.

694 Hewitt on wild ducks, 'Journal of Horticulture,' Jan. 13, 1863, p. 39. Audubon on the wild turkey, 'Ornith. Biography,' vol. i. p. 14. On the mocking thrush, ibid. vol. i. p. 110.

[PAGE 1084]

695 The 'Ibis,' vol. ii. 1860, p. 344.

696 On the ornamented nests of humming-birds, Gould, 'Introduction to the Trochilidæ,' 1861, p. 19. On the bower-birds, Gould, 'Handbook to the Birds of Australia,' 1865, vol. i. p. 444–461. Mr. Ramsay in the 'Ibis,' 1867, p. 456.

[PAGE 1085]

697 'Hist. of British Birds,' vol. ii. p. 92.

698 'Zoologist,' 1853–1854, p. 3946.

699 Waterton, 'Essays on Nat. Hist.' 2nd series, p. 42, 117. For the following statements, see on the wigeon, Loudon's 'Mag. of Nat. Hist.' vol. ix. p. 616; L. Lloyd, 'Scandinavian Adventures,' vol. i. 1854, p. 452; Dixon, 'Ornamental and Domestic Poultry,' p. 137; Hewitt, in 'Journal of Horticulture,' Jan. 13, 1863, p. 40; Bechstein, 'Stubenvögel,' 1840, s. 230.

[PAGE 1087]

700 Audubon, 'Ornitholog. Biography,' vol. i. p. 191, 349; vol. ii. p. 42, 275; vol. iii. p. 2.

701 'Rare and Prize Poultry,' 1854, p. 27.

702 'The Variation of Animals and Plants under Domestication,' vol. ii. p. 103.

[PAGE 1088]

703 Boitard and Corbié, 'Les Pigeons,' 1824, p. 12. Prosper Lucas ('Traité de l'Héréd. Nat.' tom. ii. 1850, p. 296) has himself observed nearly similar facts with pigeons.

704 'Die Taubenzucht,' 1824, s. 86.

705 'Ornithological Biography,' vol. i. p. 13.

706 'Proc. Zool. Soc.' 1835, p. 54. The japanned peacock is considered by Mr. Sclater as a distinct species, and has been named *Pavo nigripennis.*

[PAGE 1089]

707 Rudolphi, 'Beyträge zur Anthropologie,' 1812, s. 184.

708 'Die Darwin'sche Theorie, und ihre Stellung zu Moral und Religion,' 1869, s. 59.

709 In regard to peafowl, see Sir R. Heron, 'Proc. Zoolog. Soc.' 1835, p. 54, and the Rev. E. S. Dixon, 'Ornamental Poultry,' 1848, p. 8. For the turkey, Audubon, ibid. p. 4. For the capercailzie, Lloyd, 'Game Birds of Sweden,' 1867, p. 23.

710 Mr. Hewitt, quoted in 'Tegetmeier's Poultry Book,' 1866, p. 165.

[PAGE 1090]

711 Quoted in Lloyd's 'Game Birds of Sweden,' p. 345.

[PAGE 1091]

712 According to Dr. Blasius ('Ibis,' vol. ii. 1860, p. 297), there are 425 indubitable species of birds which breed in Europe, besides 60 forms, which are frequently regarded as distinct species. Of the latter, Blasius thinks that only ten are really doubtful, and that the other fifty ought to be united with their nearest allies; but this shews that there must be a considerable amount of variation with some of our European birds. It is also an unsettled point with naturalists, whether several North American birds ought to be ranked as specifically distinct from the corresponding European species.

713 'Origin of Species,' fifth edit. 1869, p. 104. I had always perceived, that rare and strongly-marked deviations of structure, deserving to be called monstrosities, could seldom be preserved through natural selection, and that the preservation of even highly-beneficial variations would depend to a certain extent on chance. I had also fully appreciated the importance of mere individual differences, and this led me to insist so strongly on the importance of that unconscious form of selection by man, which follows from the preservation of the most valued individuals of each breed, without any intention on his part to modify the characters of the breed. But until I read an able article in the 'North British Review' (March, 1867, p. 289, *et seq.*) which has been of more use to me than any other Review, I did not see how great the chances were against the preservation of variations, whether slight or strongly pronounced, occurring only in single individuals.

714 'Introduct. to the Trochilidæ,' p. 102.

715 Gould, 'Handbook of Birds of Australia,' vol. ii. p. 32 and 68.

716 Audubon, 'Ornitholog. Biography,' 1838, vol. iv. p. 389.

[PAGE 1092]

717 Jerdon, 'Birds of India,' vol. i. p. 108; and Mr. Blyth, in 'Land and Water,' 1868, p. 381.

718 Graba, 'Tagebuch, Reise nach Färo,' 1830, s. 51–54. Macgillivray, 'Hist. British Birds,' vol. iii. p. 745. 'Ibis,' vol. v. 1863, p. 469.

719 Graba, ibid. s. 54. Macgillivray, ibid. vol. v. p. 327.

720 'Variation of Animals and Plants under Domestication,' vol. ii. p. 92

721 On these points see also 'Variation of Animals and Plants under Domestication,' vol. i. p. 253; vol. ii. p. 73, 75.

[PAGE 1093]

722 See, for instance, on the irides of a Podica and Gallicrex in 'Ibis,' vol. ii. 1860, p. 206; and vol. v. 1863, p. 426.

723 See also Jerdon, 'Birds of India,' vol. i. p. 243–245.

724 'Zoology of the Voyage of H.M.S. Beagle,' 1841, p. 6.

[PAGE 1095]

725 Bechstein, 'Naturgeschichte Deutschlands,' B. iv. 1795, s. 31, on a sub-variety of the Monck pigeon.

[PAGE 1096]

726 This woodcut has been engraved from a beautiful drawing, most kindly made for me by Mr. Trimen; see also his description of the wonderful amount of variation in the coloration and shape of the wings of this butterfly, in his 'Rhopalocera Africæ Australis,' p. 186. See also an interesting paper by the Rev. H. H. Higgins, on the origin of the ocelli in the Lepidoptera in the 'Quarterly Journal of Science,' July, 1868, p. 325.

727 Jerdon, 'Birds of India,' vol. iii. p. 517.

[PAGE 1097]

728 'Variation of Animals and Plants under Domestication,' vol. i. p. 254.

[PAGE 1102]

729 When the Argus pheasant displays his wing-feathers like a great fan, those nearest to the body stand more upright than the outer ones, so that the shading of the ball-and-socket ocelli ought to be slightly different on the different feathers, in order to bring out their full effect, relatively to the incidence of the light. Mr. T. W. Wood, who has the experienced eye of an artist, asserts ('Field,' Newspaper, May 28, 1870, p. 457) that this is the case; but after carefully examining two mounted specimens (the proper feathers from one having been given to me by Mr. Gould for more accurate comparison) I cannot perceive that this acme of perfection in the shading has been attained; nor can others to whom I have shewn these feathers recognise the fact.

[PAGE 1106]

730 'The Reign of Law,' 1867, p. 247.

[PAGE 1107]

731 'Introduction to the Trochilidæ,' 1861, p. 110.

CHAPTER XV

[PAGE 1108]

732 Fourth edition, 1866, p. 241.

733 'Westminster Review,' July, 1867. 'Journal of Travel,' vol. i. 1868, p. 73.

[PAGE 1110]

734 Temminck says that the tail of the female *Phasianus Sœmmerringii* is only six inches long, 'Planches coloriées,' vol. v. 1838, p. 487 and 488: the measurements above given were made for me by Mr. Sclater. For the common pheasant, see Macgillivray, 'Hist. Brit. Birds,' vol. i. p. 118–121.

735 Dr. Chapuis, 'Le Pigeon Voyageur Belge,' 1865, p. 87.

[PAGE 1113]

736 Bechstein, 'Naturgesch. Deutschlands,' 1793, B. iii. s. 339.

[PAGE 1114]

737 Daines Barrington, however, thought it probable ('Phil. Transact.' 1773, p. 164) that few female birds sing, because the talent would have been dangerous to them during incubation. He adds, that a similar view may possibly account for the inferiority of the female to the male in plumage.

738 Mr. Ramsay, in 'Proc. Zoolog. Soc.' 1868, p. 50.

[PAGE 1115]

739 'Journal of Travel,' edited by A. Murray, vol. i. 1868, p. 78.

[PAGE 1116]

740 'Journal of Travel,' edited by A. Murray, vol. i. 1868, p. 281.

741 Audubon, 'Ornithological Biography,' vol. i. p. 233.

742 Jerdon, 'Birds of India,' vol. ii. p. 108. Gould's 'Handbook of the Birds of Australia,' vol. i. p. 463.

743 For instance, the female *Eupetomena macroura* has the head and tail dark blue with reddish loins; the female *Lampornis porphyrurus* is blackish-green on the upper surface, with the lores and sides of the

throat crimson; the female *Eulampis jugularis* has the top of the head and back green, but the loins and the tail are crimson. Many other instances of highly conspicuous females could be given. See Mr. Gould's magnificent work on this family.

744 Mr. Salvin noticed in Guatemala ('Ibis,' 1864, p. 375) that humming-birds were much more unwilling to leave their nests during very hot weather, when the sun was shining brightly, than during cool, cloudy, or rainy weather.

745 I may specify, as instances of obscurely-coloured birds building concealed nests, the species belonging to eight Australian genera, described in Gould's 'Handbook of the Birds of Australia,' vol. i. p. 340, 362, 365, 383, 387, 389, 391, 414.

746 Jerdon, 'Birds of India,' vol. i. p. 244.

[PAGE 1117]

747 On the nidification and colours of these latter species, see Gould's 'Handbook,' &c., vol. i. p. 504, 527.

748 I have consulted, on this subject, Macgillivray's 'British Birds,' and though doubts may be entertained in some cases in regard to the degree of concealment of the nest, and of the degree of conspicuousness of the female, yet the following birds, which all lay their eggs in holes or in domed nests, can hardly be considered, according to the above standard, as conspicuous: Passer, 2 species; Sturnus, of which the female is considerably less brilliant than the male; Cinclus; Motacilla boarula (?); Erithacus (?); Fruticola, 2 sp.; Saxicola; Ruticilla, 2 sp.; Sylvia, 3 sp.; Parus, 3 sp.; Mecistura; Anorthura; Certhia; Sitta; Yunx; Muscicapa, 2 sp.; Hirundo, 3 sp.; and Cypselus. The females of the following 12 birds may be considered as conspicuous according to the same standard, viz., Pastor, Motacilla alba, Parus major and p. cæruleus, Upupa, Picus, 4 sp., Coracias, Alcedo, and Merops.

749 'Journal of Travel,' edited by A. Murray, vol. i. p. 78.

[PAGE 1118]

750 See many statements in the 'Ornithological Biography.' See, also, some curious observations on the nests of Italian birds by Eugenio Bettoni, in the 'Atti della Società Italiana,' vol. xi. 1869, p. 487.

[PAGE 1119]

751 See his 'Monograph of the Trogonidæ,' first edition.

752 Namely Cyanalcyon. Gould's 'Handbook of the Birds of Australia,' vol. i. p. 133; see, also, p. 130, 136.

753 Every gradation of difference between the sexes may be followed in the parrots of Australia. See Gould's 'Handbook,' &c., vol. ii. p. 14–102.

754 Macgillivray's 'British Birds,' vol. ii. p. 433. Jerdon, 'Birds of India,' vol. ii. p. 282.

755 All the following facts are taken from M. Malherbe's magnificent 'Monographie des Picidées,' 1861.

[PAGE 1120]

756 Audubon's 'Ornithological Biography,' vol. ii. p. 75; see also the 'Ibis,' vol. i. p. 268.

757 Gould's 'Handbook of the Birds of Australia,' vol. ii. p. 109–149.

[PAGE 1121]

758 See remarks to this effect in my work on 'Variation under Domestication,' vol. ii. chap. xii.

[PAGE 1122]

759 The 'Ibis,' vol. vi. 1864, p. 122.

760 On Ardetta, Translation of Cuvier's 'Règne Animal,' by Mr. Blyth, footnote, p. 159. On the Peregrine Falcon, Mr. Blyth, in Charlesworth's 'Mag. of Nat. Hist.' vol. i. 1837, p. 304. On Dicrurus, 'Ibis,' 1863, p. 44. On the Platalea, 'Ibis,' vol. vi. 1864, p. 366. On the Bombycilla, Audubon's 'Ornitholog. Biography,' vol. i. p. 229. On the Palæornis, see, also, Jerdon, 'Birds of India,' vol. i. p. 263. On the wild turkey, Audubon, ibid. vol. i. p. 15: I hear from Judge Caton that in Illinois the female very rarely acquires a tuft.

[PAGE 1123]

761 Mr. Blyth has recorded (Translation of Cuvier's 'Règne Animal,' p. 158) various instances with Lanius, Ruticilla, Linaria, and Anas. Audubon has also recorded a similar case ('Ornith. Biog.' vol. v. p. 519) with *Tyranga æstiva*.

762 See Gould's 'Birds of Great Britain.'

CHAPTER XVI

[PAGE 1126]

763 In regard to thrushes, shrikes, and woodpeckers, see Mr. Blyth, in Charlesworth's 'Mag. of Nat. Hist.' vol. i. 1837, p. 304; also footnote to his translation of Cuvier's 'Règne Animal,' p. 159. I give the case

of Loxia from Mr. Blyth's information. On thrushes, see also Audubon, 'Ornith. Biography,' vol. ii. p. 195. On Chrysococcyx and Chalcophaps, Blyth, as quoted in Jerdon's 'Birds of India,' vol. iii. p. 485. On Sarkidiornis, Blyth, in 'Ibis,' 1867, p. 175.

[PAGE 1128]

764 See, for instance, Mr. Gould's account ('Handbook of the Birds of Australia,' vol. i. p. 133) of Cyanalcyon (one of the Kingfishers) in which, however, the young male, though resembling the adult female, is less brilliantly coloured. In some species of Dacelo the males have blue tails, and the females brown ones; and Mr. R. B. Sharpe informs me that the tail of the young male of *D. Gaudichaudi* is at first brown. Mr. Gould has described (ibid. vol. ii. p. 14, 20, 37) the sexes and the young of certain Black Cockatoos and of the King Lory, with which the same rule prevails. Also Jerdon ('Birds of India,' vol. i. p. 260) on the *Palæornis rosa*, in which the young are more like the female than the male. See Audubon ('Ornith. Biograph.' vol. ii. p. 475) on the two sexes and the young of *Columba passerina.*

765 I owe this information to Mr. Gould who shewed me the specimens; see also his 'Introduction to the Trochilidæ,' 1861, p. 120.

[PAGE 1129]

766 Macgillivray, 'Hist. Brit. Birds,' vol. v. p. 207–214.

767 See his admirable paper in the 'Journal of the Asiatic Soc. of Bengal,' vol. xix. 1850, p. 223; see also Jerdon, 'Birds of India,' vol. i. introduction, p. xxix. In regard to Tanysiptera, Prof. Schlegel told Mr. Blyth that he could distinguish several distinct races, solely by comparing the adult males.

768 See also Mr. Swinhoe, in 'Ibis,' July, 1863, p. 131; and a previous paper, with an extract from a note by Mr, Blyth, in 'Ibis,' Jan. 1861, p. 52.

[PAGE 1131]

769 Wallace, 'The Malay Archipelago,' vol. ii. 1869, p. 394.

770 These species are described, with coloured figures, by M. F. Pollen, in 'Ibis,' 1866, p. 275.

771 'Variation of Animals, &c., under Domestication,' vol. i. p. 251.

[PAGE 1132]

772 Macgillivray, 'Hist. British Birds,' vol. i. p. 172–174.

[PAGE 1133]

773 See, on this subject, chap. xxiii. in the 'Variation of Animals and Plants under Domestication.'

[PAGE 1134]

774 Audubon, 'Ornith. Biography,' vol. i. p. 193. Macgillivray, 'Hist. Brit. Birds,' vol. iii. p. 85. See also the case before given of *Indopicus carlotta.*

[PAGE 1135]

775 'Westminster Review,' July, 1867, and A. Murray, 'Journal of Travel,' 1868, p. 83.

776 For the Australian species, see Gould's 'Handbook,' &c., vol. ii. p. 178, 180, 186, and 188. In the British Museum specimens of the Australian Plain-wanderer *(Pedionomus torquatus)* may be seen, shewing similar sexual differences.

[PAGE 1136]

777 Jerdon, 'Birds of India,' vol. iii. p. 596. Mr. Swinhoe, in 'Ibis,' 1865, p. 542; 1866, p. 131, 405.

778 Jerdon, 'Birds of India,' vol. iii. p. 677.

779 Gould's 'Handbook of the Birds of Australia,' vol. ii. p. 275.

[PAGE 1137]

780 'The Indian Field,' Sept. 1858, p. 3.

781 'Ibis,' 1866, p. 298.

782 For these several statements, see Mr. Gould's 'Birds of Great Britain.' Prof. Newton informs me that he has long been convinced, from his own observations and from those of others, that the males of the above-named species take either the whole or a large share of the duties of incubation, and that they "shew much greater devotion towards their young, when in danger, than do the females." So it is, as he informs me, with *Limosa lapponica* and some few other Waders, in which the females are larger and have more strongly contrasted colours than the males.

783 The natives of Ceram (Wallace, 'Malay Archipelago,' vol. ii. p. 150) assert that the male and female sit alternately on the eggs; but this assertion, as Mr. Bartlett thinks, may be accounted for by the female visiting the nest to lay her eggs.

784 'The Student,' April, 1870, p. 124.

[PAGE 1138]

785 See the excellent account of the habits of this bird under confinement, by Mr. A. W. Bennett, in 'Land and Water,' May, 1868, p. 233.

786 Mr. Sclater, on the incubation of the Struthiones, 'Proc. Zoo. Soc.,' June 9, 1863.

787 For the Milvago, see 'Zoology of the Voyage of the "Beagle,"' Birds, 1841, p. 16. For the Climacteris and night-jar (Eurostopodus), see Gould's 'Handbook of the Birds of Australia,' vol. i. p. 602 and 97. The New Zealand shieldrake *(Tadorna variegata)* offers a quite anomalous case: the head of the female is pure white, and her back is redder than that of the male; the head of the male is of a rich dark bronzed colour, and his back is clothed with finely pencilled slate-coloured feathers, so that he may altogether be considered as the more beautiful of the two. He is larger and more pugnacious than the female, and does not sit on the eggs. So that in all these respects this species comes under our first class of cases; but Mr. Sclater ('Proc. Zool. Soc.' 1866, p. 150) was much surprised to observe that the young of both sexes, when about three months old, resembled in their dark heads and necks the adult males, instead of the adult females; so that it would appear in this case that the females have been modified, whilst the males and the young have retained a former state of plumage.

[PAGE 1139]

788 Jerdon, 'Birds of India,' vol. iii. p. 598.

[PAGE 1140]

789 Jerdon, 'Birds of India,' vol. i. p. 222, 228. Gould's 'Handbook of the Birds of Australia,' vol. i. 124, 130.

790 Gould, Ibid. vol. ii. p. 37, 46, 56.

791 Audubon, 'Ornith. Biography,' vol. ii. p. 55.

[PAGE 1141]

792 'Variation of Animals and Plants under Domestication,' vol. ii. p. 79.

793 Charlesworth, 'Mag. of Nat. Hist.' vol. i. 1837, p. 305, 306.

794 'Bulletin de la Soc. Vaudoise des Sc. Nat.' vol. x. 1869, p. 132. The young of the Polish swan, *Cygnus immutabilis* of Yarrell, are always white; but this species, as Mr. Sclater informs me, is believed to be nothing more than a variety of the Domestic Swan *(Cygnus olor)*.

795 I am indebted to Mr. Blyth for information in regard to this genus. The sparrow of Palestine belongs to the sub-genus Petronia.

[PAGE 1142]

796 For instance, the males of *Tanagra æstiva* and *Fringilla cyanea* require three years, the male of *Fringilla ciris* four years, to complete their beautiful plumage. (See Audubon, 'Ornith. Biography,' vol. i. p. 233, 280, 378.) The Harlequin duck takes three years (ibid. vol. iii. p. 614). The male of the Gold pheasant, as I hear from Mr. J. Jenner Weir, can be distinguished from the female when about three months old, but he does not acquire his full splendour until the end of the September in the following year.

797 Thus the *Ibis tantalus* and *Grus Americanus* take four years, the Flamingo several years, and the *Ardea Ludovicana* two years, before they acquire their perfect plumage. See Audubon, ibid. vol. i. p. 221; vol. iii. p. 133, 139, 211.

798 Mr. Blyth, in Charlesworth's 'Mag. of Nat. Hist.' vol. i. 1837, p. 300. Mr. Bartlett has informed me in regard to gold-pheasants.

799 I have noticed the following cases in Audubon's 'Ornith. Biography,' The Redstart of America *(Muscicapa ruticilla,* vol. i. p. 203). The *Ibis tantalus* takes four years to come to full maturity, but sometimes breeds in the second year (vol. iii. p. 133). The *Grus Americanus* takes the same time, but breeds before acquiring its full plumage (vol. iii. p. 211). The adults of *Ardea cærulea* are blue and the young white; and white, mottled, and mature blue birds may all be seen breeding together (vol. iv. p. 58): but Mr. Blyth informs me that certain herons apparently are dimorphic, for white and coloured individuals of the same age may be observed. The Harlequin duck *(Anas histrionica,* Linn.) takes three years to acquire its full plumage, though many birds breed in the second year (vol. iii. p. 614). The White-headed Eagle *(Falco leucocephalus,* vol. iii. p. 210) is likewise known to breed in its immature state. Some species of Oriolus (according to Mr. Blyth and Mr. Swinhoe, in 'Ibis,' July, 1863, p. 68) likewise breed before they attain their full plumage.

[PAGE 1143]

800 See the last foot-note.

801 Other animals, belonging to quite distinct classes, are either habitually or occasionally capable of breeding before they have fully acquired their adult characters. This is the case with the young males of the salmon. Several amphibians have been known to breed whilst retaining their larval structure. Fritz Müller has shewn ('Facts and Arguments for Darwin,' Eng. trans. 1869, p. 79) that the males of several amphipod crustaceans become sexually mature whilst young; and I infer that this is a case of premature breeding, because they have not

as yet acquired their fully-developed claspers. All such facts are highly interesting, as bearing on one means by which species may undergo great modifications of character, in accordance with Mr. Cope's views, expressed under the terms of the "retardation and acceleration of generic characters;" but I cannot follow the views of this eminent naturalist to their full extent. See Mr. Cope, "On the Origin of Genera," from the 'Proc. of Acad. Nat. Sc. of Philadelphia,' Oct. 1868.

802 Jerdon, 'Birds of India,' vol. iii. p. 507, on the peacock. Audubon, ibid. vol. iii. p. 139, on the Ardea.

803 For illustrative cases see vol. iv. of Macgillivray's 'Hist. Brit. Birds;' on Tringa, &c., p. 229, 271; on the Machetes, p. 172; on the *Charadrius hiaticula,* p. 118; on the *Charadrius pluvialis,* p. 94.

804 For the goldfinch of N. America, *Fringilla tristis,* Linn., see Audubon, 'Ornith. Biography,' vol. i. p. 172. For the Maluri, Gould's 'Handbook of the Birds of Australia,' vol. i. p. 318.

[PAGE 1144]

805 I am indebted to Mr. Blyth for information in regard to the Buphus; see also Jerdon, 'Birds of India,' vol. iii. p. 749. On the Anastomus, see Blyth, in 'Ibis,' 1867, p. 173.

806 On the Alca, see Macgillivray, 'Hist. Brit. Birds,' vol. v. p. 347. On the *Fringilla leucophrys,* Audubon, ibid. vol. ii. p. 89. I shall have hereafter to refer to the young of certain herons and egrets being white.

807 'History of British Birds,' vol. i. 1839, p. 159.

[PAGE 1145]

808 Blyth, in Charlesworth's 'Mag. of Nat. Hist.' vol. i. 1837, p. 362; and from information given to me by him.

809 Audubon, 'Ornith. Biography,' vol. i. p. 113.

810 Mr. C. A. Wright, in 'Ibis,' vol. vi. 1864, p. 65. Jerdon, 'Birds of India,' vol. i. p. 515.

811 The following additional cases may be mentioned: the young males of *Tanagra rubra* can be distinguished from the young females (Audubon, 'Ornith. Biography,' vol. iv. p. 392), and so it is with the nestlings of a blue nuthatch, *Dendrophila frontalis* of India (Jerdon, 'Birds of India,' vol. i. p. 389). Mr. Blyth also informs me that the sexes of the stonechat, *Saxicola rubicola,* are distinguishable at a very early age.

[PAGE 1147]

812 'Westminster Review,' July, 1867, p. 5.

813 'Ibis,' 1859, vol. i. p. 429, *et seq.*

[PAGE 1149]

814 No satisfactory explanation has ever been offered of the immense size, and still less of the bright colours, of the toucan's beak. Mr. Bates ('The Naturalist on the Amazons,' vol. ii. 1863, p. 341) states that they use their beak for reaching fruit at the extreme tips of the branches; and likewise, as stated by other authors, for extracting eggs and young birds from the nests of other birds. But as Mr. Bates admits, the beak "can scarcely be considered a very perfectly-formed instrument for the end to which it is applied." The great bulk of the beak, as shewn by its breadth, depth, as well as length, is not intelligible on the view, that it serves merely as an organ of prehension.

815 Ramphastos carinatus, Gould's 'Monograph of Ramphastidæ.'

[PAGE 1150]

816 On Larus, Gavia, and Sterna, see Macgillivray, 'Hist. Brit. Birds,' vol. v. p. 515, 584, 626. On the Anser hyperboreus, Audubon, 'Ornith. Biography,' vol. iv. p. 562. On the Anastomus, Mr. Blyth, in 'Ibis,' 1867, p. 173.

817 It may be noticed that with vultures, which roam far and wide through the higher regions of the atmosphere, like marine birds over the ocean, three or four species are almost wholly or largely white, and many other species are black. This fact supports the conjecture that these conspicuous colours may aid the sexes in finding each other during the breeding-season.

[PAGE 1151]

818 'The Journal of Travel,' edited by A. Murray, vol. i. 1868, p. 286.

819 See Jerdon on the genus Palæornis, 'Birds of India,' vol. i. p. 258–260.

820 The young of *Ardea rufescens* and *A. cærulea* of the U. States are likewise white, the adults being coloured in accordance with their specific names. Audubon ('Ornith. Biography,' vol. iii. p. 416; vol. iv. p. 58) seems rather pleased at the thought that this remarkable change of plumage will greatly "disconcert the systematists."

[PAGE 1155]

821 I am greatly indebted to the kindness of Mr. Sclater for having looked over these four chapters on birds, and the two following ones on mammals. By this means I have been saved from making mistakes about the names of the species, and from giving any facts which are actually known to this distinguished naturalist to be erroneous. But of course he is not at all answerable for the accuracy of the statements, quoted by me from various authorities.

CHAPTER XVII

[PAGE 1156]

822 See Waterton's account of two hares fighting, 'Zoologist,' vol. i. 1843, p. 211. On moles, Bell, 'Hist. of British Quadrupeds,' 1st edit. p. 100. On squirrels, Audubon and Bachman, 'Viviparous Quadrupeds of N. America,' 1846, p. 269. On beavers, Mr. A. H. Green, in 'Journal of Lin. Soc. Zoolog.' vol. x. 1869, p. 362.

[PAGE 1157]

823 On the battles of seals, see Capt. C. Abbott in 'Proc. Zool. Soc.' 1868, p. 191; also Mr. R. Brown, ibid. 1869, p. 436; also L. Lloyd, 'Game Birds of Sweden,' 1867, p. 412; also Pennant. On the sperm-whale, see Mr. J. H. Thompson, in 'Proc. Zool. Soc.' 1867, p. 246.

824 See Scrope ('Art of Deer-stalking,' p. 17) on the locking of the horns with the Cervus elephas. Richardson, in 'Fauna Bor. Americana,' 1829, p. 252, says that the wapiti, moose, and reindeer have been found thus locked together. Sir A. Smith found at the Cape of Good Hope the skeletons of two gnus in the same condition.

825 Mr. Lamont ('Seasons with the Sea-Horses,' 1861, p. 143) says that a good tusk of the male walrus weighs 4 pounds, and is longer than that of the female, which weighs about 3 pounds. The males are described as fighting ferociously. On the occasional absence of the tusks in the female, see Mr. R. Brown, 'Proc. Zool. Soc.' 1868, p. 429.

[PAGE 1158]

826 Owen, 'Anatomy of Vertebrates,' vol. iii. p. 283.

827 Mr. R. Brown, in 'Proc. Zool. Soc.' 1869, p. 553.

828 Owen on the Cachalot and Ornithorhynchus, ibid. vol. iii. p. 638, 641.

[PAGE 1159]

829 On the structure and shedding of the horns of the reindeer, Hoffberg, 'Amœnitates Acad.' vol. iv. 1788, p. 149. See Richardson, 'Fauna Bor. Americana,' p. 241, in regard to the American variety or species; also Major W. Boss King, 'The Sportsman in Canada,' 1866, p. 80.

830 Isidore Geoffrey St.-Hilaire, 'Essais de Zoolog. Générale,' 1841, p. 513. Other masculine characters, besides the horns, are sometimes similarly transferred to the female; thus Mr. Boner, in speaking of an old female chamois ('Chamois Hunting in the Mountains of Bavaria,' 1860, 2nd edit. p. 363), says, "not only was the head very

male-looking, but along the back there was a ridge of long hair, usually to be found only in bucks."

831 On the Cervulus, Dr. Gray, 'Catalogue of the Mammalia in British Museum,' part iii. p. 220. On the *Cervus Canadensis* or Wapiti see Hon. J. D. Caton, 'Ottawa Acad. of Nat. Sciences,' May, 1868, p. 9.

832 For instance the horns of the female *Ant. Euchore* resemble those of a distinct species, viz. the *Ant. Dorcas* var. *Corine*, see Desmarest, 'Mammalogie,' p. 455.

[PAGE 1160]

833 Gray, 'Catalogue Mamm. Brit. Mus.' part iii. 1852, p. 160.

834 Richardson, 'Fauna Bor. Americana,' p. 278.

[PAGE 1161]

835 'Land and Water' 1867, p. 346.

836 Sir Andrew Smith, 'Zoology of S. Africa,' pl. xix. Owen, 'Anatomy of Vertebrates,' vol. iii. p. 624.

837 Sir J. Emerson Tennent, 'Ceylon,' 1859, vol. ii. p. 274. For Malacca, 'Journal of Indian Archipelago,' vol. iv. p. 357.

838 'Calcutta Journal of Nat. Hist.' vol. ii. 1843, p. 526.

[PAGE 1162]

839 Mr. Blyth, in 'Land and Water,' March, 1867, p. 134, on the authority of Capt. Hutton and others. For the wild Pembrokeshire goats see the 'Field,' 1869, p. 150.

840 M. E. M. Bailly, "sur l'usage des Cornes," &c., 'Annal. des Sc. Nat.' tom. ii. 1824, p. 369.

[PAGE 1164]

841 Owen, on the Horns of Red-deer, 'British Fossil Mammals,' 1846, p. 478; 'Forest Creatures,' by Charles Boner, 1861, p. 76, 62. Richardson on the Horns of the Reindeer, 'Fauna Bor. Americana,' 1829, p. 240.

842 Hon. J. D. Caton ('Ottawa Acad. of Nat. Science,' May, 1868, p. 9), says that the American deer fight with their fore-feet, after "the question of superiority has been once settled and acknowledged in the herd." Bailly, "Sur l'usage des Cornes," 'Annales des Sc. Nat.' tom. ii. 1824, p. 371.

843 See a most interesting account in the Appendix to Hon. J. D. Caton's paper, as above quoted.

[PAGE 1165]

844 'The American Naturalist,' Dec. 1869, p. 552.

845 Pallas, 'Spicilegia Zoologica,' fasc. xiii. 1779, p. 18.

846 Lamont, 'Seasons with the Sea-Horses,' 1861, p. 141.

847 See also Corse ('Philosoph. Transact.' 1799, p. 212) on the manner in which the short-tusked Mooknah variety of the elephant attacks other elephants.

[PAGE 1166]

848 Owen, 'Anatomy of Vertebrates,' vol. iii. p. 349.

849 See Rüppell (in 'Proc. Zoolog. Soc.' Jan. 12, 1836, p. 3) on the canines in deer and antelopes, with a note by Mr. Martin on a female American deer. See also Falconer ('Palæont. Memoirs and Notes,' vol. i. 1868, p. 576) on canines in an adult female deer. In old males of the musk-deer the canines (Pallas, 'Spic. Zoolog.' fasc. xiii. 1779, p. 18) sometimes grow to the length of three inches, whilst in old females a rudiment projects scarcely half an inch above the gums.

850 Emerson Tennent, 'Ceylon,' 1859, vol. ii. p. 275; Owen, 'British Fossil Mammals,' 1846, p. 245.

[PAGE 1167]

851 Richardson, 'Fauna Bor. Americana,' on the moose, *Alces palmata,* p. 236, 237; also on the expanse of the horns 'Land and Water,' 1869, p. 143. See also Owen, 'British Fossil Mammals,' on the Irish elk, p. 447, 455.

852 'Forest Creatures,' by C. Boner, 1861, p. 60.

853 See the very interesting paper by Mr. J. A. Allen in 'Bull. Mus. Comp. Zoolog. of Cambridge, United States,' vol. ii. No. 1, p. 82. The weights were ascertained by a careful observer, Capt. Bryant.

854 'Animal Economy,' p. 45.

[PAGE 1168]

855 See also Richardson's 'Manual on the Dog,' p. 59. Much valuable information on the Scottish deer-hound is given by Mr. McNeill, who first called attention to the inequality in size between the sexes, in Scrope's 'Art of Deer Stalking.' I hope that Mr. Cupples will keep to his intention of publishing a full account and history of this famous breed.

[PAGE 1169]

856 Brehm, 'Thierleben,' B. ii. s. 729–732.

857 See Mr. Wallace's interesting account of this animal, 'The Malay Archipelago,' 1869, vol. i. p. 435.

[PAGE 1171]

858 'The Times,' Nov. 10th, 1857. In regard to the Canada lynx, see Audubon and Bachman, 'Quadrupeds of N. America,' 1846, p. 139.

859 Dr. Murie, on Otaria, 'Proc. Zoolog. Soc.' 1869, p. 109. Mr. J. A. Allen, in the paper above quoted (p. 75), doubts whether the hair, which is longer on the neck in the male than in the female, deserves to be called a mane.

[PAGE 1172]

860 Mr. Boner in his excellent description of the habits of the red-deer in Germany ('Forest Creatures,' 1861, p. 81) says, "while the stag is defending his rights against one intruder, another invades the sanctuary of his harem, and carries off trophy after trophy." Exactly the same thing occurs with seals, see Mr. J. A. Allen, ibid. p. 100.

861 Mr. J. A. Allen in 'Bull. Mus. Comp. Zoolog. of Cambridge, United States,' vol. ii. No. 1, p. 99.

[PAGE 1173]

862 'Dogs: their Management,' by E. Mayhew, M.R.C.V.S., 2nd edit. 1864, p. 187–192.

863 Quoted by Alex. Walker 'On Intermarriage,' 1838, p. 276, see also p. 244.

[PAGE 1174]

864 'Traité de l'Héréd. Nat.' tom. ii. 1850, p. 296.

865 'Amœnitates Acad.' vol. iv. 1788, p. 160.

CHAPTER XVIII

[PAGE 1175]

866 Owen, 'Anatomy of Vertebrates,' vol. iii. p. 585.

867 Ibid. p. 595.

[PAGE 1176]

868 See, for instance, Major W. Ross King ('The Sportsman in Canada,' 1866, p. 53, 131) on the habits of the moose and wild reindeer.

[PAGE 1177]

869 Owen, 'Anatomy of Vertebrates,' vol. iii. p. 600.

870 Mr. Green, in 'Journal of Linn. Soc.' vol. x. Zoology, 1869, p. 362.

871 C. L. Martin, 'General Introduction to the Nat. Hist. of Mamm. Animals,' 1841, p. 431.

872 'Naturgeschichte der Säugethiere von Paraguay,' 1830, s. 15, 21.

[PAGE 1178]

873 On the sea-elephant, see an article by Lesson, in 'Dict. Class. Hist. Nat.' tom. xiii. p. 418. For the Cystophora or Stemmatopus, see Dr. Dekay, 'Annals of Lyceum of Nat. Hist. New York,' vol. i. 1824, p. 94. Pennant has also collected information from the sealers on this animal. The fullest account is given by Mr. Brown, who doubts about the rudimentary condition of the bladder in the female, in 'Proc. Zoolog. Soc.' 1868, p. 435.

874 As with the castoreum of the beaver, see Mr. L. H. Morgan's most interesting work, 'The American Beaver,' 1868, p. 300. Pallas ('Spic. Zoolog.' fasc. viii. 1779, p. 23) has well discussed the odoriferous glands of mammals. Owen ('Anat. of Vertebrates,' vol. iii. p. 634) also gives an account of these glands, including those of the elephant, and (p. 763) those of shrew-mice.

875 Rengger, 'Naturgeschichte der Säugethiere von Paraguay,' 1830, s. 355. This observer also gives some curious particulars in regard to the odour emitted.

876 Owen, 'Anatomy of Vertebrates,' vol. iii. p. 632. See, also, Dr. Murie's observations on their glands in 'Proc. Zoolog. Soc.' 1870, p. 340. Desmarest, On the *Antilope subgutturosa*, 'Mammalogie,' 1820, p. 455.

[PAGE 1179]

877 Pallas, 'Spicilegia Zoolog.' fasc. xiii. 1799, p. 24; Desmoulins, 'Dict. Class. d'Hist. Nat.' tom. iii. p. 586.

878 Dr. Gray, 'Gleanings from the Menagerie at Knowsley,' pl. 28.

879 Judge Caton on the wapiti, 'Transact. Ottawa Acad. Nat. Sciences,' 1868, p. 36, 40; Blyth, 'Land and Water,' on *Capra ægagrus*, 1867, p. 87.

[PAGE 1180]

880 'Hunter's Essays and Observations,' edited by Owen, 1861, vol. i. p. 236.

881 See Dr. Gray's 'Cat. of Mammalia in British Museum,' part iii. 1852, p. 144.

882 Rengger, 'Säugethiere,' &c., s. 14; Desmarest, 'Mammalogie,' p. 66.

[PAGE 1181]

883 See the chapters on these several animals in vol. i. of my 'Variation of Animals under Domestication;' also vol. ii. p. 73; also chap. xx. on the practice of selection by semi-civilised people. For the Berbura goat, see Dr. Gray, 'Catalogue,' ibid. p. 157.

[PAGE 1182]

884 *Osphranter rufus,* Gould, 'Mammals of Australia.' vol. ii. 1863. On the Didelphis, Desmarest, 'Mammalogie,' p. 256.

885 'Annals and Mag. of Nat. Hist.' Nov. 1867, p. 325. On the *Mus minutus,* Desmarest, 'Mammalogie,' p. 304.

886 J. A. Allen, in 'Bulletin of Mus. Comp. Zoolog. of Cambridge, United States,' 1869, p. 207.

887 Desmarest, 'Mammalogie,' 1820, p. 223. On *Felis mitis,* Rengger, ibid. s. 194.

888 Dr. Murie on the Otaria, 'Proc. Zool. Soc.' 1869, p. 108. Mr. R. Brown, on the *P. groenlandica,* ibid. 1868, p. 417. See also on the colours of seals, Desmarest, ibid. p. 243, 249.

[PAGE 1183]

889 Judge Caton, in 'Trans. Ottawa Acad. of Nat. Sciences,' 1868, p. 4.

890 Dr. Gray, 'Cat. of Mamm. in Brit. Mus.' part iii. 1852, p. 134–142; also Dr. Gray, 'Gleanings from the Menagerie of Knowsley,' in which there is a splendid drawing of the Oreas derbianus: see the text on Tragelaphus. For the Cape Eland *(Oreas canna),* see Andrew Smith, 'Zoology of S. Africa,' pl. 41 and 42. There are also many of these antelopes in the Zoological Society's Gardens.

891 On the *Ant. niger,* see 'Proc. Zool. Soc.' 1850, p. 133. With respect to an allied species, in which there is an equal sexual difference in colour, see Sir S. Baker, 'The Albert Nyanza,' 1866, vol. ii. p. 327. For the *A. sing-sing,* Gray, 'Cat. B. Mus.' p. 100. Desmarest, 'Mammalogie,' p. 468, on the *A. caama.* Andrew Smith, 'Zoology of S. Africa,' on the Gnu.

[PAGE 1184]

892 'Ottawa Academy of Sciences,' May 21, 1868, p. 3, 5.

893 S. Müller, on the Banteng, 'Zoog. Indischen Archipel.' 1839–1844, tab. 35: see also Baffles, as quoted by Mr. Blyth, in 'Land and Water,' 1867, p. 476. On goats, Dr. Gray, 'Cat. Brit. Mus.' p. 146; Desmarest, 'Mammalogie,' p. 482. On the *Cervus paludosus,* Rengger, ibid. s. 345.

894 Sclater, 'Proc. Zool. Soc.' 1866, p. 1. The same fact has also been fully ascertained by MM. Pollen and van Dam.

895 On Mycetes, Rengger, ibid. s. 14; and Brehm, 'Illustrirtes Thierleben,' B. i. s. 96, 107. On Ateles, Desmarest, 'Mammalogie,' p. 75. On Hylobates, Blyth, 'Land and Water,' 1867, p. 135. On the Semnopithecus, S. Müller, 'Zoog. Indischen Archipel.' tab. x.

[PAGE 1185]

896 Gervais, 'Hist. Nat. des Mammifères,' 1854, p. 103. Figures are given of the skull of the male. Desmarest, 'Mammalogie,' p. 70. Geoffroy St.-Hilaire and F. Cuvier, 'Hist. Nat. des Mamm.' 1824, tom. i.

[PAGE 1187]

897 'The Variation of Animals and Plants under Domestication,' 1868, vol. ii. p. 102, 103.

898 'Essays and Observations by J. Hunter,' edited by Owen, 1861, vol. i. p. 194.

899 Sir S. Baker, 'The Nile Tributaries of Abyssinia,' 1867.

[PAGE 1188]

900 *Fiber zibethicus,* Audubon and Bachman, 'The Quadrupeds of N. America,' 1846, p. 109.

901 'Novæ species Quadrupedum e Glirium ordine,' 1778, p. 7. What I have called the roe is the *Capreolus Sibiricus subecaudatus* of Pallas.

[PAGE 1190]

902 See the fine plates in A. Smith's 'Zoology of S. Africa,' and Dr. Gray's 'Gleanings from the Menagerie of Knowsley.'

903 'Westminster Review,' July 1, 1867, p. 5.

904 'Travels in South Africa,' 1824, vol. ii. p. 315.

[PAGE 1191]

905 Dr. Gray, 'Gleanings from the Menagerie of Knowsley,' p. 64. Mr. Blyth, in speaking ('Land and Water,' 1869, p. 42) of the hog-deer of Ceylon, says it is more brightly spotted with white than the common hog-deer, at the season when it renews its horns.

906 Falconer and Cautley, 'Proc. Geolog. Soc.' 1843; and Falconer's 'Pal. Memoirs,' vol. i. p. 196.

[PAGE 1192]

907 'The Variation of Animals and Plants under Domestication,' 1868, vol. i. p. 61–64.

908 'Proc. Zool. Soc.' 1862, p. 164. See, also, Dr. Hartmann, 'Ann. d. Landw.' Bd. xliii. s. 222.

[PAGE 1194]

909 I observed this fact in the Zoological Gardens; and numerous cases may be seen in the coloured plates in Geoffrey St.-Hilaire and F. Cuvier, 'Hist. Nat. des Mammifères,' tom. i. 1824.

910 Bates, 'The Naturalist on the Amazons,' 1863, vol. ii. p. 310.

[PAGE 1195]

911 I have seen most of the above-named monkeys in the Zoological Society's Gardens. The description of the *Semnopithecus nemæus* is taken from Mr. W. C. Martin's 'Nat. Hist. of Mammalia,' 1841, p. 460; see also p. 475, 523.

CHAPTER XIX

[PAGE 1198]

912 Schaaffhausen, translation in 'Anthropological Review,' Oct. 1868, p. 419, 420, 427.

[PAGE 1199]

913 Ecker, translation in 'Anthropological Review,' Oct. 1868, p. 351–356. The comparison of the form of the skull in men and women has been followed out with much care by Welcker.

914 Ecker and Welcker, ibid. p. 352, 355; Vogt, 'Lectures on Man,' Eng. translat. p. 81.

915 Schaaffhausen, 'Anthropolog. Review,' ibid. p. 429.

916 Pruner-Bey, on negro infants, as quoted by Vogt, 'Lectures on Man,'

Eng. translat. 1864, p. 189: for further facts on negro infants, as quoted from Winterbottom and Camper, see Lawrence, 'Lectures on Physiology,' &c. 1822, p. 451. For the infants of the Guaranys, see Rengger, 'Säugethiere,' &c. s. 3. See also Godron, 'De l'Espèce,' tom. ii. 1859, p. 253. For the Australians, Waitz, 'Introduct. to Anthropology,' Eng. translat. 1863, p. 99.

917 Rengger, 'Säugethiere,' &c. 1830, s. 49.

918 As in *Macacus cynomolgus* (Desmarest, 'Mammalogie,' p. 65) and in *Hylobates agilis* (Geoffroy St.-Hilaire and F. Cuvier, 'Hist. Nat. des Mamm.' 1824, tom. i. p. 2).

919 'Anthropological Review,' Oct. 1868, p. 353.

[PAGE 1200]

920 Mr. Blyth informs me that he has never seen more than one instance of the beard, whiskers, &c., in a monkey becoming white with old age, as is so commonly the case with us. This, however, occurred in an aged and confined *Macacus cynomolgus,* whose moustaches were "remarkably long and human-like." Altogether this old monkey presented a ludicrous resemblance to one of the reigning monarchs of Europe, after whom he was universally nick-named. In certain races of man the hair on the head hardly ever becomes grey; thus Mr. D. Forbes has never seen, as he informs me, an instance with the Aymaras and Quichuas of S. America.

921 This is the case with the females of several species of Hylobates, see Geoffrey St.-Hilaire and F. Cuvier, 'Hist. Nat. des Mamm.' tom. i. See, also, on *H. lar.* 'Penny Encyclopedia,' vol. ii. p. 149, 150.

922 The results were deduced by Dr. Weisbach from the measurements made by Drs. K. Scherzer and Schwarz, see 'Reise der *Novara*: Anthropolog. Theil,' 1867, s. 216, 231, 234, 236, 239, 269.

[PAGE 1201]

923 'Voyage to St. Kilda' (3rd edit. 1753) p. 37.

924 Sir J. E. Tennent, 'Ceylon,' vol. ii. 1859, p. 107.

925 Quatrefages, 'Revue des Cours Scientifiques,' Aug. 29, 1868, p. 630; Vogt, 'Lectures on Man,' Eng. translat. p. 127.

926 On the beards of negroes, Vogt, 'Lectures,' &c. ibid. p. 127; Waitz, 'Introduct. to Anthropology,' Engl. translat. 1863, vol. i. p. 96. It is remarkable that in the United States ('Investigations in Military and Anthropological Statistics of American Soldiers,' 1869, p. 569) the pure negroes and their crossed offspring seem to have bodies almost as hairy as those of Europeans.

927 Wallace, 'The Malay Arch.' vol. ii. 1869, p. 178.

928 Dr. J. Barnard Davis on Oceanic Races, in 'Anthropolog. Review,' April, 1870, p. 185, 191.

929 Catlin, 'North American Indians,' 3rd edit. 1842, vol. ii. p. 227. On the Guaranys, see Azara, 'Voyages dans l'Amérique Mérid.' tom. ii 1809, p. 58; also Rengger, 'Säugethiere von Paraguay,' s. 3.

930 Prof. and Mrs. Agassiz ('Journey in Brazil,' p. 530) remark that the sexes of the American Indians differ less than those of the negroes and of the higher races. See also Rengger, ibid. p. 3, on the Guaranys.

931 Rütimeyer, 'Die Grenzen der Thierwelt; eine Betrachtung zu Darwin's Lehre,' 1868, s. 54.

[PAGE 1202]

932 'A Journey from Prince of Wales Fort.' 8vo. edit. Dublin, 1796, p. 104. Sir J. Lubbock ('Origin of Civilisation,' 1870, p. 69) gives other and similar cases in North America. For the Guanas of S. America see Azara, 'Voyages,' &c. tom. ii. p. 94.

933 On the fighting of the male gorillas, see Dr. Savage, in 'Boston Journal of Nat. Hist.' vol. v. 1847, p. 423. On *Presbytis entellus*, see the 'Indian Field,' 1859, p. 146.

[PAGE 1204]

934 J. Stuart Mill remarks ('The Subjection of Women,' 1869, p. 122), "the things in which man most excels woman are those which require most plodding, and long hammering at single thoughts." What is this but energy and perseverance?

[PAGE 1205]

935 An observation by Vogt bears on this subject: he says, it is a "remarkable circumstance, that the difference between the sexes, as regards the cranial cavity, increases with the development of the race, so that the male European excels much more the female, than the negro the negress. Welcker confirms this statement of Huschke from his measurements of negro and German skulls." But Vogt admits ('Lectures on Man,' Eng. translat. 1864, p. 81) that more observations are requisite on this point.

936 Owen, 'Anatomy of Vertebrates,' vol. iii. p. 603.

[PAGE 1206]

937 'Journal of the Anthropolog. Soc.' April, 1869, p. lvii. and lxvi.

938 Dr. Scudder, "Notes on Stridulation," in 'Proc. Boston Soc. of Nat. Hist.' vol. xi. April, 1868.

[PAGE 1207]

939 Given in W. C. L. Martin's 'General Introduct. to Nat. Hist. of Mamm. Animals,' 1841, p. 432; Owen, 'Anatomy of Vertebrates,' vol. iii. p. 600.

940 Helmholtz, 'Théorie Phys. de la Musique,' 1868, p. 187.

941 Mr. R. Brown, in 'Proc. Zoo. Soc.' 1868, p. 410.

942 'Journal of Anthropolog. Soc.' Oct. 1870, p. clv. See also the several later chapters in Sir John Lubbock's 'Prehistoric Times,' second edition, 1869, which contain an admirable account of the habits of savages.

[PAGE 1208]

943 Since this chapter has been printed I have seen a valuable article by Mr. Chauncey Wright ('North Amer. Review,' Oct. 1870, page 293), who, in discussing the above subject, remarks, "There are many consequences of the ultimate laws or uniformities of nature through which the acquisition of one useful power will bring with it many resulting advantages as well as limiting disadvantages, actual or possible, which the principle of utility may not have comprehended in its action." This principle has an important bearing, as I have attempted to shew in the second chapter of this work, on the acquisition by man of some of his mental characteristics.

[PAGE 1209]

944 See the very interesting discussion on the Origin and Function of Music, by Mr. Herbert Spencer, in his collected 'Essays,' 1858, p. 359. Mr. Spencer comes to an exactly opposite conclusion to that at which I have arrived. He concludes that the cadences used in emotional speech afford the foundation from which music has been developed; whilst I conclude that musical notes and rhythm were first acquired by the male or female progenitors of mankind for the sake of charming the opposite sex. Thus musical tones became firmly associated with some of the strongest passions an animal is capable of feeling, and are consequently used instinctively, or through association, when strong emotions are expressed in speech. Mr. Spencer does not offer any satisfactory explanation, nor can I, why high or deep notes should be expressive, both with man and the lower animals, of certain emotions. Mr. Spencer gives also an interesting discussion on the relations between poetry, recitative, and song.

945 Rengger, 'Säugethiere von Paraguay,' s. 49.

946 See an interesting discussion on this subject by Häckel, 'Generelle Morph.' B. ii. 1866, s. 246.

[PAGE 1210]

947 A full and excellent account of the manner in which savages in all parts of the world ornament themselves is given by the Italian traveller, Prof. Mantegazza, 'Rio de la Plata, Viaggi e Studi,' 1867, p. 525–545; all the following statements, when other references are not given, are taken from this work. See, also, Waitz, 'Introduct. to Anthropolog.' Eng. transl. vol. i. 1863, p. 275, *et passim.* Lawrence also gives very full details in his 'Lectures on Physiology,' 1822. Since this chapter was written Sir J. Lubbock has published his 'Origin of Civilisation,' 1870, in which there is an interesting chapter on the present subject, and from which (p. 42, 48) I have taken some facts about savages dyeing their teeth and hair, and piercing their teeth.

948 Humboldt, 'Personal Narrative,' Eng. translat. vol. iv. p. 515; on the imagination shewn in painting the body, p. 522; on modifying the form of the calf of the leg, p. 466.

949 'The Nile Tributaries,' 1867; 'The Albert N'yanza,' 1866, vol. i. p. 218.

[PAGE 1211]

950 Quoted by Prichard, 'Phys. Hist. of Mankind,' 4th edit. vol. i. 1851, p. 321.

951 On the Papuans, Wallace, 'The Malay Archipelago,' vol. ii. p. 445. On the coiffure of the Africans, Sir S. Baker, 'The Albert N'yanza.' vol. i. p. 210.

952 'Travels,' p. 533.

953 'The Albert N'yanza,' 1866, vol. i. p. 217.

[PAGE 1212]

954 Livingstone, 'British Association,' 1860; report given in the 'Athenæum,' July 7, 1860, p. 29.

955 Sir S. Baker (ibid. vol. i. p. 210) speaking of the natives of Central Africa says, "every tribe has a distinct and unchanging fashion for dressing the hair." See Agassiz ('Journey in Brazil,' 1868, p. 318) on the invariability of the tattooing of the Amazonian Indians.

956 Rev. R. Taylor, 'New Zealand and its Inhabitants,' 1855, p. 152.

957 Mantegazza, 'Viaggi e Studi,' p. 542.

[PAGE 1213]

958 'Travels in S. Africa,' 1824, vol. i. p. 414.

959 See, for references, 'Gerland über das Aussterben der Naturvölker,' 1868, s. 51, 53, 55; also Azara, 'Voyages,' &c. tom. ii. p. 116.

960 On the vegetable productions used by the North-Western American Indians, 'Pharmaceutical Journal,' vol. x.

961 'A Journey from Prince of Wales Fort,' 8vo. edit. 1796, p. 89.

962 Quoted by Prichard, 'Phys. Hist. of Mankind,' 3rd edit. vol. iv. 1844, p. 519; Vogt, 'Lectures on Man,' Eng. translat. p. 129. On the opinion of the Chinese on the Cingalese, E. Tennent, 'Ceylon,' vol. ii. 1859, p. 107.

[PAGE 1214]

963 Prichard, as taken from Crawfurd and Finlayson, 'Phys. Hist. of Mankind,' vol. iv. p. 534, 535.

964 Idem illustrissimus viator dixit mihi præcinctorium vel tabula fæminæ, quod nobis teterrimum est, quondam permagno æstimari ab hominibus in hac gente. Nunc res mutata est, et censet talem conformationem minime optandam est."

965 'The Anthropological Review,' November, 1864, p. 237. For additional references, see Waitz, 'Introduct. to Anthropology,' Eng. translat. 1863, vol. i. p. 105.

966 'Mungo Park's Travels in Africa,' 4to. 1816, p. 53, 131. Burton's statement is quoted by Schaaffhausen, 'Archiv für Anthropolog.' 1866, s. 163. On the Banyai, Livingstone, 'Travels,' p. 64. On the Kafirs, the Rev. J. Shooter, 'The Kafirs of Natal and the Zulu Country,' 1857, p. 1.

[PAGE 1215]

967 For the Javanese and Cochin-Chinese, see Waitz, 'Introduct. to Anthropology,' Eng. translat. vol. i. p. 305. On the Yura-caras, A. d'Orbigny, as quoted in Prichard, 'Phys. Hist. of Mankind,' vol. v. 3rd edit. p. 476.

968 'North American Indians,' by G. Catlin, 3rd edit. 1842, vol. i. p. 49; vol. ii. p. 227. On the natives of Vancouver Island, see Sproat, 'Scenes and Studies of Savage Life,' 1868, p. 25. On the Indians of Paraguay, Azara, 'Voyages,' tom. ii. p. 105.

969 On the Siamese, Prichard, ibid. vol. iv. p. 533. On the Japanese, Veitch in 'Gardeners' Chronicle,' 1860, p. 1104. On the New Zealanders, Mantegazza, 'Viaggi e Studi,' 1867, p. 526. For the other nations mentioned, see references in Lawrence, 'Lectures on Physiology,' &c. 1822, p. 272.

970 Lubbock, 'Origin of Civilisation,' 1870, p. 321.

[PAGE 1216]

971 Dr. Barnard Davis quotes Mr. Pritchard and others for these facts in regard to the Polynesians, in 'Anthropological Review,' April, 1870, p. 185, 191.

972 Ch. Comte has remarks to this effect in his 'Traité de Legislation,' 3rd edit. 1837, p. 136.

973 The Fuegians, as I have been informed by a missionary who long resided with them, consider European women as extremely beautiful; but from what we have seen of the judgment of the other aborigines of America, I cannot but think that this must be a mistake, unless indeed the statement refers to the few Fuegians who have lived for some time with Europeans, and who must consider us as superior beings. I should add that a most experienced observer, Capt. Burton, believes that a woman whom we consider beautiful is admired throughout the world, 'Anthropological Review,' March, 1864, p. 245.

974 'Personal Narrative,' Eng. translat. vol. iv. p. 518, and elsewhere. Mantegazza, in his 'Viaggi e Studi,' 1867, strongly insists on this same principle.

[PAGE 1217]

975 On the skulls of the American tribes, see Nott and Gliddon, 'Types of Mankind,' 1854, p. 440; Prichard, 'Phys. Hist. of Mankind,' vol. i. 3rd edit. p. 321; on the natives of Arakhan, ibid. vol. iv. p. 537. Wilson, 'Physical Ethnology,' Smithsonian Institution, 1863, p. 288; on the Fijians, p. 290. Sir J. Lubbock ('Prehistoric Times,' 2nd edit. 1869, p. 506) gives an excellent résumé on this subject.

976 On the Huns, Godron, 'De l'Espèce,' tom. ii. 1859, p. 300. On the Tahitians, Waitz, 'Anthropolog.' Eng. translat. vol. i. p. 305. Marsden, quoted by Prichard, 'Phys. Hist. of Mankind,' 3rd edit, vol. v. p. 67. Lawrence, 'Lectures on Physiology,' p. 337.

977 This fact was ascertained in the 'Reise der *Novara:* Anthropolog. Theil,' Dr. Weisbach, 1867, s. 265.

978 'Smithsonian Institution,' 1863, p. 289. On the fashions of Arab women, Sir S. Baker, 'The Nile Tributaries,' 1867, p. 121.

979 'The Variation of Animals and Plants under Domestication,' vol. i. p. 214; vol. ii. p. 240.

[PAGE 1218]

980 Schaaffhausen, 'Archiv für Anthropologie,' 1866, s. 164.

981 Mr. Bain has collected ('Mental and Moral Science,' 1868, p. 304–314) about a dozen more or less different theories of the idea of beauty; but none are quite the same with that here given.

CHAPTER XX

[PAGE 1220]

982 These quotations are taken from Lawrence ('Lectures on Physiology,' &c. 1822, p. 393), who attributes the beauty of the upper classes in England to the men having long selected the more beautiful women.
983 "Anthropologie," 'Revue des Cours Scientifiques,' Oct. 1868, p. 721.

[PAGE 1221]

984 'The Variation of Animals and Plants under Domestication,' vol. i. p. 207.
985 Sir J. Lubbock, 'The Origin of Civilisation,' 1870, chap. iii. especially p. 60–67. Mr. M'Lennan, in his extremely valuable work on 'Primitive Marriage,' 1865, p. 163, speaks of the union of the sexes "in the earliest times as loose, transitory, and in some degree promiscuous." Mr. M'Lennan and Sir J. Lubbock have collected much evidence on the extreme licentiousness of savages at the present time. Mr. L. H. Morgan, in his interesting memoir on the classificatory system of relationship ('Proc. American Acad. of Sciences,' vol. vii. Feb. 1868, p. 475) concludes that polygamy and all forms of marriage during primeval times were essentially unknown. It appears, also, from Sir J. Lubbock's work, that Bachofen likewise believes that communal intercourse originally prevailed.

[PAGE 1222]

986 Address to British Association 'On the Social and Religious Condition of the Lower Races of Man,' 1870, p. 20.
987 'Origin of Civilisation,' 1870, p. 86. In the several works above quoted there will be found copious evidence on relationship through the females alone, or with the tribe alone.

[PAGE 1223]

988 Brehm ('Illust. Thierleben,' B. i. p. 77) says *Cynocephalus hamadryas* lives in great troops containing twice as many adult females as adult males. See Rengger on American polygamous species, and Owen ('Anat. of Vertebrates,' vol. iii. p. 746) on American monogamous species. Other references might be added.
989 Dr. Savage, in 'Boston Journal of Nat. Hist.' vol. v. 1845–47, p. 423.

[PAGE 1224]

990 'Prehistoric Times,' 1869, p. 424.

991 Mr. M'Lennan, 'Primitive Marriage,' 1865. See especially on exogamy and infanticide, p. 130, 138, 165.

992 Dr. Gerland ('Ueber das Aussterben der Naturvölker,' 1868) has collected much information on infanticide, see especially s. 27, 51, 54. Azara ('Voyages,' &c. tom. ii. p. 94, 116) enters in detail on the motives. See also M'Lennan (ibid. p. 139) for cases in India.

[PAGE 1225]

993 'Primitive Marriage,' p. 208; Sir J. Lubbock, 'Origin of Civilisation,' p. 100. See also Mr. Morgan, loc. cit., on former prevalence of polyandry.

994 'Voyages,' &c. tom. ii. p. 92–95.

[PAGE 1226]

995 Burchell says ('Travels in S. Africa,' vol. ii. 1824, p. 58), that among the wild nations of Southern Africa, neither men nor women ever pass their lives in a state of celibacy. Azara ('Voyages dans l'Amérique Merid.,' tom. ii. 1809, p. 21) makes precisely the same remark in regard to the wild Indians of South America.

[PAGE 1227]

996 'Anthropological Review,' Jan. 1870, p. xvi.

997 'The Variation of Animals and Plants under Domestication,' vol. ii. p. 210–217.

[PAGE 1228]

998 An ingenious writer argues, from a comparison of the pictures of Raphael, Rubens, and modern French artists, that the idea of beauty is not absolutely the same even throughout Europe; see the 'Lives of Haydn and Mozart,' by M. Bombet, English translat. p. 278.

[PAGE 1230]

999 Azara, 'Voyages,' &c. tom. ii. p. 23. Dobrizhoffer, 'An Account of the Abipones,' vol. ii. 1822, p. 207. Williams on the Fiji Islanders, as quoted by Lubbock, 'Origin of Civilisation,' 1870, p. 79. On the Fuegians, King and FitzRoy, 'Voyages of the *Adventure* and *Beagle*,' vol. ii. 1839, p. 182. On the Kalmucks, quoted by M'Lennan, 'Primitive Marriage,' 1865, p. 32. On the Malays, Lubbock, ibid. p. 76. The Rev. J. Shooter, 'On the Kafirs of Natal,' 1857, p. 52–60. On the Bush-women, Burchell, 'Travels in S. Africa,' vol. ii. 1824, p. 59.

[PAGE 1231]

1000 'Contributions to the Theory of Natural Selection,' 1870, p. 346. Mr. Wallace believes (p. 350) "that some intelligent power has guided or determined the development of man;" and he considers the hairless condition of the skin as coming under this head. The Rev. T. R. Stebbing, in commenting on this view ('Transactions of Devonshire Assoc. for Science,' 1870) remarks, that had Mr. Wallace "employed his usual ingenuity on the question of man's hairless skin, he might have seen the possibility of its selection through its superior beauty or the health attaching to superior cleanliness. At any rate it is surprising that he should picture to himself a superior intelligence plucking the hair from the backs of savage men (to whom, according to his own account it would have been useful and beneficial), in order that the descendants of the poor shorn wretches might after many deaths from cold and damp in the course of many generations," have been forced to raise themselves in the scale of civilisation through the practice of various arts, in the manner indicated by Mr. Wallace.

[PAGE 1232]

1001 'The Variation of Animals and Plants under Domestication,' vol. ii. 1868, p. 327.

1002 'Investigations into Military and Anthropological Statistics of American Soldiers,' by B. A. Gould, 1869, p. 568:—Observations were carefully made on the pilosity of 2129 black and coloured soldiers, whilst they were bathing; and by looking to the published table, "it is manifest at a glance that there is but little, if any, difference between the white and the black races in this respect." It is, however, certain that negroes in their native and much hotter land of Africa, have remarkably smooth bodies. It should be particularly observed, that pure blacks and mulattoes were included in the above enumeration; and this is an unfortunate circumstance, as in accordance with the principle, the truth of which I have elsewhere proved, crossed races would be eminently liable to revert to the primordial hairy character of their early ape-like progenitors.

1003 "Ueber die Richtung der Haare am Menschlichen Körper," in Müller's 'Archiv für Anat. und Phys.' 1837, s. 40.

[PAGE 1233]

1004 Mr. Sproat ('Scenes and Studies of Savage Life,' 1868, p. 25) suggests, with reference to the beardless natives of Vancouver's Island, that the custom of plucking out the hairs on the face, "continued from one generation to another, would perhaps at last produce a race distinguishable by a thin and straggling growth of beard." But the custom would not have arisen until the beard had already become, from some independent cause, greatly reduced. Nor have we any direct evidence that the continued eradication of the hair would lead to any inherited effect. Owing to this cause of doubt, I have not hitherto alluded to the belief held by some distinguished ethnologists, for instance M. Gosse of Geneva, that artificial modifications of the skull tend to be inherited. I have no wish to dispute this conclusion; and we now know from Dr. Brown-Séquard's remarkable observations, especially those recently communicated (1870) to the British Association, that with guinea-pigs the effects of operations are inherited.

1005 'Ueber die Richtung,' ibid. s. 40.

CHAPTER XXI

[PAGE 1239]

1006 On the "Limits of Natural Selection," in the 'North American Review,' Oct. 1870, p. 295.

[PAGE 1242]

1007 The Rev. J. A. Picton gives a discussion to this effect in his 'New Theories and the Old Faith,' 1870.

FOOTNOTES TO *THE EXPRESSION OF THE EMOTIONS IN MAN AND ANIMALS*

INTRODUCTION

[PAGE 1261]

1 J. Parsons, in his paper in the Appendix to the 'Philosophical Transactions' for 1746, p. 41, gives a list of forty-one old authors who have written on Expression.

2 'Conférences sur l'expression des différents Caractères des Passions,' Paris, 4to, 1667. I always quote from the republication of the 'Conférences' in the edition of Lavater, by Moreau, which appeared in 1820, as given in vol. ix. p. 257.

3 'Discours par Pierre Camper sur le moyen de représenter les diverses passions,' &c. 1792.

4 I always quote from the third edition, 1844, which was published after the death of Sir C. Bell, and contains his latest corrections. The first edition of 1806 is much inferior in merit, and does not include some of his more important views.

5 'De la Physionomie et de la Parole,' par Albert Lemoine, 1865, p. 101.

[PAGE 1262]

6 'L'Art de connaître les Hommes,' &c., par G. Lavater. The earliest edition of this work, referred to in the preface to the edition of 1820 in ten volumes, as containing the observations of M. Moreau, is said to have been published in 1807; and I have no doubt that this is correct, because the 'Notice sur Lavater' at the commencement of volume i. is dated April 13, 1806. In some bibliographical works, however, the date of 1805–1809 is given; but it seems impossible that 1805 can be correct. Dr. Duchenne remarks ('Mécanisme de la Physionomie Humaine,' 8vo edit. 1862, p. 5, and 'Archives Générales de Médecine, Jan. et Fév. 1862) that M. Moreau *"a composé pour son ouvrage un article important,"* &c., in the year 1805; and I find in volume i. of the edition of 1820 passages bearing the dates of December 12, 1805, and another January 5,1806, besides that of April 13, 1806,

above referred to. In consequence of some of these passages having thus been *composed* in 1805, Dr. Duchenne assigns to M. Moreau the priority over Sir C. Bell, whose work, as we have seen, was published in 1806. This is a very unusual manner of determining the priority of scientific works; but such questions are of extremely little importance in comparison with their relative merits. The passages above quoted from M. Moreau and from Le Brun are taken in this and all other cases from the edition of 1820 of Lavater, tom. iv. p. 228, and tom. ix. p. 279.

[PAGE 1263]

7 'Handbuch der systematischen Anatomie des Menschen,' Band I., dritte Abtheilung, 1858.

[PAGE 1264]

8 'The Senses and the Intellect,' 2nd edit. 1864, pp. 96 and 288. The preface to the first edition of this work is dated June, 1855. See also the 2nd edition of Mr. Bain's work on the 'Emotions and Will.'

[PAGE 1265]

9 'The Anatomy of Expression,' 3rd edit. p. 121.

10 'Essays, Scientific, Political, and Speculative,' Second Series, 1863, p. 111. There is a discussion on Laughter in the First Series of Essays, which discussion seems to me of very inferior value.

11 Since the publication of the essay just referred to, Mr. Spencer has written another, on "Morals and Moral Sentiments," in the 'Fortnightly Review,' April 1,1871, p. 426. He has, also, now published his final conclusions in vol. ii. of the second edit. of the 'Principles of Psychology,' 1872, p. 539. I may state, in order that I may not be accused of trespassing on Mr. Spencer's domain, that I announced in my 'Descent of Man,' that I had then written a part of the present volume: my first MS. notes on the subject of expression bear the date of the year 1838.

12 'Anatomy of Expression,' 3rd edit. pp. 98, 121, 131.

13 Professor Owen expressly states ('Proc. Zoolog. Soc.' 1830, p. 28) that this is the case with respect to the Orang, and specifies all the more important muscles which are well known to serve with man for the expression of his feelings. See, also, a description of several of the facial muscles in the Chimpanzee, by Prof. Macalister, in 'Annals and Magazine of Natural History,' vol. vii. May, 1871, p. 342.

[PAGE 1266]

14 'Anatomy of Expression,' pp. 121, 138.
15 'De la Physionomie,' pp. 12, 73.
16 'Mécanisme de la Physionomie Humaine,' 8vo edit. p. 31.
17 'Elements of Physiology,' English translation, vol. ii. p. 934.

[PAGE 1267]

18 'Anatomy of Expression,' 3rd edit. p. 198.

[PAGE 1268]

19 See remarks to this effect in Lessing's 'Laocoon,' translated by W. Ross, 1836, p. 19.

[PAGE 1273]

20 Mr. Partridge in Todd's 'Cyclopædia of Anatomy and Physiology,' vol. ii. p. 227.
21 'La Physionomie,' par G. Lavater, tom. iv. 1820, p. 274. On the number of the facial muscles, see vol. iv. pp. 209–211.
22 'Mimik und Physiognomik,' 1867, s. 91.

CHAPTER I

[PAGE 1276]

23 Mr. Herbert Spencer ('Essays,' Second Series, 1863, p. 138) has drawn a clear distinction between emotions and sensations, the latter being "generated in our corporeal framework." He classes as Feelings both emotions and sensations.

[PAGE 1277]

24 Müller, 'Elements of Physiology,' Eng. translat. vol. ii. p. 939. See also Mr. H. Spencer's interesting speculations on the same subject, and on the genesis of nerves, in his 'Principles of Biology,' vol. ii. p. 346; and in his 'Principles of Psychology,' 2nd edit. pp. 511–557.

[PAGE 1278]

25 A remark to much the same effect was made long ago by Hippocrates and by the illustrious Harvey; for both assert that a young animal forgets in the course of a few days the art of sucking, and cannot with-

out some difficulty again acquire it. I give these assertions on the authority of Dr. Darwin, 'Zoonomia,' 1794, vol. i. p. 140.

26 See for my authorities, and for various analogous facts, 'The Variation of Animals and Plants under Domestication,' 1868, vol. ii. p. 304.

27 'The Senses and the Intellect,' 2nd edit. 1864, p. 332. Prof. Huxley remarks ('Elementary Lessons in Physiology,' 5th edit. 1872, p. 306), "It may be laid down as a rule, that, if any two mental states be called up together, or in succession, with due frequency and vividness, the subsequent production of the one of them will suffice to call up the other, and that whether we desire it or not."

[PAGE 1279]

28 Gratiolet ('De la Physionomie,' p. 324), in his discussion on this subject, gives many analogous instances. See p. 42, on the opening and shutting of the eyes. Engel is quoted (p. 323) on the changed paces of a man, as his thoughts change.

29 'Mécanisme de la Physionomie Humaine,' 1862, p. 17.

[PAGE 1280]

30 'The Variation of Animals and Plants under Domestication,' vol. ii. p. 6. The inheritance of habitual gestures is so important for us, that I gladly avail myself of Mr. F. Galton's permission to give in his own words the following remarkable case:—"The following account of a habit occurring in individuals of three consecutive generations is of peculiar interest, because it occurs only during sound sleep, and therefore cannot be due to imitation, but must be altogether natural. The particulars are perfectly trustworthy, for I have enquired fully into them and speak from abundant and independent evidence. A gentleman of considerable position was found by his wife to have the curious trick, when he lay fast asleep on his back in bed, of raising his right arm slowly in front of his face, up to his forehead, and then dropping it with a jerk, so that the wrist fell heavily on the bridge of his nose. The trick did not occur every night, but occasionally, and was independent of any ascertained cause. Sometimes it was repeated incessantly for an hour or more. The gentleman's nose was prominent, and its bridge often became sore from the blows which it received. At one time an awkward sore was produced, that was long in healing, on account of the recurrence, night after night, of the blows which first caused it. His wife had to remove the button from the wrist of his night-gown as it made severe scratches, and some means were attempted of tying his arm.

Many years after his death, his son married a lady who had never heard of the family incident. She, however, observed precisely the same peculiarity in her husband; but his nose, from not being particularly prominent, has never as yet suffered from the blows. The trick does not occur when he is half-asleep, as, for example, when dozing in his arm-chair, but the moment he is fast asleep it is apt to begin. It is, as with his father, intermittent; sometimes ceasing for many nights, and sometimes almost incessant during a part of every night. It is performed, as it was by his father, with his right hand.

One of his children, a girl, has inherited the same trick. She performs it, likewise, with the right hand, but in a slightly modified form; for, after raising the arm, she does not allow the wrist to drop upon the bridge of the nose, but the palm of the half-closed hand falls over and down the nose, striking it rather rapidly. It is also very intermittent with this child, not occurring for periods of some months, but sometimes occurring almost incessantly."

31 Prof. Huxley remarks ('Elementary Physiology,' 5th edit. p. 305) that reflex actions proper to the spinal cord are *natural;* but, by the help of the brain, that is through habit, an infinity of *artificial* reflex actions may be acquired. Virchow admits ('Sammlung wissenschaft. Vorträge,' &c., "Ueber das Rückenmark," 1871, ss. 24, 31) that some reflex actions can hardly be distinguished from instincts; and, of the latter, it may be added, some cannot be distinguished from inherited habits.

[PAGE 1281]

32 Dr. Maudsley, 'Body and Mind,' 1870, p. 8.

33 See the very interesting discussion on the whole subject by Claude Bernard, 'Tissus Vivants,' 1866, pp. 353–356.

34 'Chapters on Mental Physiology,' 1858, p. 85.

[PAGE 1282]

35 Müller remarks ('Elements of Physiology,' Eng. tr. vol. ii. p. 1311) on starting being always accompanied by the closure of the eyelids.

[PAGE 1283]

36 Dr. Maudsley remarks ('Body and Mind,' p. 10) that "reflex movements which commonly effect a useful end may, under the changed circumstances of disease, do great mischief, becoming even the occasion of violent suffering and of a most painful death."

[PAGE 1285]

37 See Mr. F. H. Salvin's account of a tame jackal in 'Land and Water,' October, 1869.

[PAGE 1286]

38 Dr. Darwin, 'Zoonomia,' 1794, vol. i. p. 160. I find that the fact of cats protruding their feet when pleased is also noticed (p. 151) in this work.

[PAGE 1287]

39 Carpenter, 'Principles of Comparative Physiology,' 1854, p. 690; and Müller's 'Elements of Physiology,' Eng. translat. vol. ii. p. 936.

40 Mowbray on 'Poultry,' 6th edit. 1830, p. 54.

41 See the account given by this excellent observer in 'Wild Sports of the Highlands,' 1846, p. 142.

[PAGE 1288]

42 'Philosophical Transactions,' 1823, p. 182.

CHAPTER II

[PAGE 1293]

43 'Naturgeschichte der Säugethiere von Paraguay,' 1830, s. 55.

[PAGE 1294]

44 Mr. Tylor gives an account of the Cistercian gesture-language in his 'Early History of Mankind' (2nd edit. 1870, p. 40), and makes some remarks on the principle of opposition in gestures.

45 See on this subject Dr. W. R. Scott's interesting work, 'The Deaf and Dumb,' 2nd edit. 1870, p. 12. He says, "This contracting of natural gestures into much shorter gestures than the natural expression requires, is very common amongst the deaf and dumb. This contracted gesture is frequently so shortened as nearly to lose all semblance of the natural one, but to the deaf and dumb who use it, it still has the force of the original expression."

CHAPTER III

[PAGE 1298]

46 See the interesting cases collected by M. G. Pouchet in the 'Revue des

Deux Mondes,' January 1, 1872, p. 79. An instance was also brought some years ago before the British Association at Belfast.

47 Müller remarks ('Elements of Physiology,' Eng. translat. vol. ii. p. 934) that when the feelings are very intense, "all the spinal nerves become affected to the extent of imperfect paralysis, or the excitement of trembling of the whole body."

48 'Leçons sur les Prop. des Tissus Vivants,' 1866, pp. 457–466.

[PAGE 1299]

49 Mr. Bartlett, "Notes on the Birth of a Hippopotamus," Proc. Zoolog. Soc. 1871, p. 255.

[PAGE 1300]

50 See, on this subject, Claude Bernard, 'Tissus Vivants,' 1866, pp. 316, 337, 358. Virchow expresses himself to almost exactly the same effect in his essay "Ueber das Rückenmark" (Sammlung wissenschaft, Vorträge, 1871, s. 28).

51 Müller ('Elements of Physiology,' Eng. translat. vol. ii. p. 932) in speaking of the nerves, says, "any sudden change of condition of whatever kind sets the nervous principle into action." See Virchow and Bernard on the same subject in passages in the two works referred to in my last foot-note.

52 H. Spencer, 'Essays, Scientific, Political,' &c., Second Series, 1863, pp. 109, 111.

53 Sir H. Holland, in speaking ('Medical Notes and Reflexions,' 1839, p. 328) of that curious state of body called the *fidgets*, remarks that it seems due to "an accumulation of some cause of irritation which requires muscular action for its relief."

[PAGE 1301]

54 I am much indebted to Mr. A. H. Garrod for having informed me of M. Lorain's work on the pulse, in which a sphygmogram of a woman in a rage is given; and this shows much difference in the rate and other characters from that of the same woman in her ordinary state.

[PAGE 1303]

55 How powerfully intense joy excites the brain, and how the brain reacts on the body, is well shown in the rare cases of Psychical Intoxication. Dr. J. Crichton Browne ('Medical Mirror,' 1865) records the case of a young man of strongly nervous temperament, who, on hearing by a telegram that a fortune had been bequeathed him, first became pale

then exhilarated, and soon in the highest spirits, but flushed and very restless. He then took a walk with a friend for the sake of tranquillising himself, but returned staggering in his gait, uproariously laughing, yet irritable in temper, incessantly talking, and singing loudly in the public streets. It was positively ascertained that he had not touched any spirituous liquor, though every one thought that he was intoxicated. Vomiting after a time came on, and the half-digested contents of his stomach were examined, but no odour of alcohol could be detected. He then slept heavily, and on awaking was well, except that he suffered from headache, nausea, and prostration of strength.

56 Dr. Darwin, 'Zoonomia,' 1794, vol, i. p. 148.

[PAGE 1305]

57 Mrs. Oliphant, in her novel of 'Miss Majoribanks,' p. 362.

CHAPTER IV

[PAGE 1309]

58 See the evidence on this head in my 'Variation of Animals and Plants under Domestication,' vol. i. p. 27. On the cooing of pigeous, vol. i. pp. 154, 155.

59 'Essays, Scientific, Political, and Speculative,' 1858. 'The Origin and Function of Music,' p. 359.

[PAGE 1310]

60 'The Descent of Man,' 1870, p. 1207. The words quoted are from Professor Owen. It has lately been shown that some quadrupeds much lower in the scale than monkeys, namely Rodents, are able to produce correct musical tones: see the account of a singing Hesperomys, by the Rev. S. Lockwood, in the 'American Naturalist,' vol. v. December, 1871, p. 761.

61 Mr. Tylor ('Primitive Culture,' 1871, vol. i. p. 166), in his discussion on this subject, alludes to the whining of the dog.

62 'Naturgeschichte der Säugethiere von Paraguay,' 1830, s. 46.

63 Quoted by Gratiolet, 'De la Physionomie,' 1865, p. 115.

[PAGE 1311]

64 'Theorie Physiologique de la Musique,' Paris, 1868, p. 146. Helmholtz has also fully discussed in this profound work the relation of the form of the cavity of the mouth to the production of vowel-sounds.

[PAGE 1314]

65 I have given some details on this subject in my 'Descent of Man,' pp. 981, 999.

66 As quoted in Huxley's 'Evidence as to Man's Place in Nature,' 1863, p. 52.

67 Illust. Thierleben, 1864, B. i. s. 130.

[PAGE 1315]

68 The Hon. J. Caton, Ottawa Acad. of Nat. Sciences, May, 1868, pp. 36, 40. For the *Capra Ægagrus,* 'Land and Water,' 1867, p. 37.

69 'Land and Water,' July 20, 1867, p. 659.

70 *Phaeton rubricauda:* 'Ibis,' vol. iii. 1861, p. 180.

[PAGE 1316]

71 On the *Strix flammea,* Audubon, 'Ornithological Biography,' 1864, vol. ii. p. 407. I have observed other cases in the Zoological Gardens.

[PAGE 1317]

72 *Melopsittacus undulatus.* See an account of its habits by Gould, 'Handbook of Birds of Australia,' 1865, vol. ii. p. 82.

73 See, for instance, the account which I have given ('Descent of Man,' p. 1039) of an Anolis and Draco.

74 These muscles are described in his well-known works. I am greatly indebted to this distinguished observer for having given me in a letter information on this same subject.

75 'Lehrbuch der Histologie des Menschen,' 1857, s. 82. I owe to Prof. W. Turner's kindness an extract from this work.

[PAGE 1318]

76 'Quarterly Journal of Microscopical Science,' 1853, vol. i. p. 262.

77 'Lehrbuch der Histologie,' 1857, s. 82.

[PAGE 1319]

78 'Dictionary of English Etymology,' p. 403.

[PAGE 1320]

79 See the account of the habits of this animal by Dr. Cooper, as quoted in 'Nature,' April 27, 1871, p. 512.

80 Dr. Günther, 'Reptiles of British India,' p. 262.

81 Mr. J. Mansel Weale, 'Nature,' April 27, 1871, p. 508.

82 'Journal of Researches during the Voyage of the "Beagle,"' 1845, p. 104. I here compared the rattling thus produced with that of the Rattle-snake.

[PAGE 1321]

83 See the account by Dr. Anderson, Proc. Zool. Soc. 1871, p. 196.

84 The 'American Naturalist,' Jan. 1872, p. 32. I regret that I cannot follow Prof. Shaler in believing that the rattle has been developed, by the aid of natural selection, for the sake of producing sounds which deceive and attract birds, so that they may serve as prey to the snake. I do not, however, wish to doubt that the sounds may occasionally subserve this end. But the conclusion at which I have arrived, viz. that the rattling serves as a warning to would-be devourers, appears to me much more probable, as it connects together various classes of facts. If this snake had acquired its rattle and the habit of rattling, for the sake of attracting prey, it does not seem probable that it would have invariably used its instrument when angered or disturbed. Prof. Shaler takes nearly the same view as I do of the manner of development of the rattle; and I have always held this opinion since observing the Trigonocephalus in South America.

85 From the accounts lately collected, and given in the 'Journal of the Linnean Society,' by Mrs. Barber, on the habits of the snakes of South Africa; and from the accounts published by several writers, for instance by Lawson, of the rattle-snake in North America,—it does not seem improbable that the terrific appearance of snakes and the sounds produced by them, may likewise serve in procuring prey, by paralysing, or as it is sometimes called fascinating, the smaller animals.

86 See the account by Dr. R. Brown, in Proc. Zool. Soc. 1871, p. 39. He says that as soon as a pig sees a snake it rushes upon it; and a snake makes off immediately on the appearance of a pig.

87 Dr. Günther remarks ('Reptiles of British India,' p. 340) on the destruction of cobras by the ichneumon or herpestes, and whilst the cobras are young by the jungle-fowl. It is well known that the peacock also eagerly kills snakes.

88 Prof. Cope enumerates a number of kinds in his 'Method of Creation of Organic Types,' read before the American Phil. Soc., December 15th, 1871, p. 20. Prof. Cope takes the same view as I do of the use of the gestures and sounds made by snakes. I briefly alluded to this subject in the last edition of my 'Origin of Species.' Since the passages in the text above have been printed, I have been pleased to find that

Mr. Henderson ('The American Naturalist,' May, 1872, p. 260) also takes a similar view of the use of the rattle, namely "in preventing an attack from being made."

[PAGE 1322]

89 Mr. des Vœux, in Proc. Zool. Soc. 1871, p. 3.

[PAGE 1324]

90 'The Sportsman and Naturalist in Canada,' 1866, p. 53.
91 'The Nile Tributaries of Abyssinia,' 1867, p. 443.

CHAPTER V

[PAGE 1326]

92 'The Anatomy of Expression,' 1844, p. 190.

[PAGE 1328]

93 'De la Physionomie,' 1865, pp. 187, 218.

[PAGE 1329]

94 'The Anatomy of Expression,' 1844, p. 140.

[PAGE 1331]

95 Many particulars are given by Gueldenstädt in his account of the jackal in Nov. Comm. Acad. Sc. Imp. Petrop. 1775, tom. xx. p. 449. See also another excellent account of the manners of this animal and of its play, in 'Land and Water,' October, 1869. Lieut. Annesley, R.A., has also communicated to me some particulars with respect to the jackal. I have made many inquiries about wolves and jackals in the Zoological Gardens, and have observed them for myself.
96 'Land and Water,' November 6, 1869.

[PAGE 1332]

97 Azara, 'Quadrupèdes du Paraguay,' 1801, tom. i. p. 136.

[PAGE 1334]

98 'Land and Water,' 1867, p. 657. See also Azara on the Puma in the work above quoted.

99 Sir C. Bell, 'Anatomy of Expression,' 3rd edit. p. 123. See also p. 126, on horses not breathing through their mouths, with reference to their distended nostrils.

[PAGE 1335]

100 'Land and Water,' 1869, p. 152.

[PAGE 1336]

101 'Natural History of Mammalia,' 1841, vol. i. pp. 383, 410.

102 Rengger ('Säugethiere von Paraguay,' 1830, s. 46) kept these monkeys in confinement for seven years in their native country of Paraguay.

[PAGE 1338]

103 Rengger, ibid. s. 46. Humboldt, 'Personal Narrative,' Eng. translat. vol. iv. p. 527.

104 Nat. Hist. of Mammalia, 1841, p. 351.

[PAGE 1339]

105 Brehm, 'Thierleben,' B. i. s. 84. On baboons striking the ground, s. 61.

106 Brehm remarks ('Thierleben,' s. 68) that the eyebrows of the *Inuus ecaudatus* are frequently moved up and down when the animal is angered.

[PAGE 1340]

107 G. Bennett, 'Wanderings in New South Wales,' &c., vol. ii. 1834, p. 153.

108 W. C. Martin, Nat. Hist. of Mamm. Animals, 1841, p. 405.

[PAGE 1341]

109 Prof. Owen on the Orang, Proc. Zool. Soc. 1830, p.28. On the Chimpanzee, see Prof. Macalister, in Annals and Mag. of Nat. Hist. vol. vii. 1871, p. 342, who states that the *corrugator supercilii* is inseparable from the *orbicularis palpebrarum.*

110 Boston Journal of Nat. Hist. 1845–47, vol. v. p. 423. On the Chimpanzee, ibid. 1843–44, vol. iv. p. 365.

111 See on this subject, 'Descent of Man,' p. 789.

[PAGE 1342]

112 'Descent of Man,' p. 802.

[PAGE 1343]

113 'Anatomy of Expression,' 3rd edit. 1844, pp. 138, 121.

CHAPTER VI

[PAGE 1345]

114 The best photographs in my collection are by Mr. Rejlander, of Victoria Street, London, and by Herr Kindermann, of Hamburg. Figs. 1, 3, 4, and 6 are by the former; and figs. 2 and 5, by the latter gentleman. Fig. 6 is given to show moderate crying in an older child.

[PAGE 1346]

115 Henle (Handbuch d. Syst. Anat. 1858, B. i. s. 139) agrees with Duchenne that this is the effect of the contraction of the *pyramidalis nasi.*

116 These consist of the *levator labii superioris alæque nasi,* the *levator labii proprius,* the *malaris,* and the *zygomaticus minor,* or little zygomatic. This latter muscle runs parallel to and above the great zygomatic, and is attached to the outer part of the upper lip. It is represented in fig. 2 (I. p. 1274), but not in figs. 1 and 3. Dr. Duchenne first showed ('Mécanisme de la Physionomie Humaine,' Album, 1862, p. 39) the importance of the contraction of this muscle in the shape assumed by the features in crying. Henle considers the above-named muscles (excepting the *malaris*) as subdivisions of the *quadratus labii superioris.*

117 Although Dr. Duchenne has so carefully studied the contraction of the different muscles during the act of crying, and the furrows on the face thus produced, there seems to be something incomplete in his account; but what this is I cannot say. He has given a figure (Album, fig. 48) in which one half of the face is made, by galvanizing the proper muscles, to smile; whilst the other half is similarly made to begin crying. Almost all those (viz. nineteen out of twenty-one persons) to whom I showed the smiling half of the face instantly recognized the expression; but, with respect to the other half, only six persons out of twenty-one recognized it,—that is, if we accept such terms as "grief," "misery," "annoyance," as correct;—whereas, fifteen persons were ludicrously mistaken; some of them saying the face expressed "fun," "satisfaction," "cunning," "disgust," &c. We may infer from this that there is something wrong in the expression. Some of the fifteen persons may, however, have been partly misled by not expecting to see an old man crying, and by tears not being secreted. With respect to another figure by Dr. Duchenne (fig. 49), in which the muscles of

half the face are galvanized in order to represent a man beginning to cry, with the eyebrow on the same side rendered oblique, which is characteristic of misery, the expression was recognized by a greater proportional number of persons. Out of twenty-three persons, fourteen answered correctly, "sorrow," "distress," "grief," "just going to cry," "endurance of pain," &c. On the other hand, nine persons either could form no opinion or were entirely wrong, answering, " cunning leer," "jocund," "looking at an intense light," "looking at a distant object," &c.

[PAGE 1347]

118 Mrs. Gaskell, 'Mary Barton,' new edit. p. 84.

119 'Mimik und Physiognomik,' 1867, s. 102. Duchenne, Mécanisme de la Phys. Humaine, Album, p. 34.

120 Dr. Duchenne makes this remark, ibid. p. 39.

[PAGE 1349]

121 'The Origin of Civilization,' 1870, p. 355.

122 See, for instance, Mr. Marshall's account of an idiot in Philosoph. Transact. 1864, p. 526. With respect to cretins, see Dr. Piderit, 'Mimik und Physiognomik,' 1867, s. 61.

123 'New Zealand and its Inhabitants,' 1855, p. 175.

[PAGE 1350]

124 'De la Physionomie,' 1865, p. 126.

[PAGE 1351]

125 'The Anatomy of Expression,' 1844, p. 106. See also his paper in the 'Philosophical Transactions,' 1822, p. 284, ibid. 1823, pp. 166 and 289. Also 'The Nervous System of the Human Body,' 3rd edit. 1836, p. 175.

126 See Dr. Brinton's account of the act of vomiting, in Todd's Cyclop. of Anatomy and Physiology, 1859, vol. v. Supplement, p. 318.

[PAGE 1352]

127 I am greatly indebted to Mr. Bowman for having introduced me to Prof. Donders, and for his aid in persuading this great physiologist to undertake the investigation of the present subject. I am likewise much indebted to Mr. Bowman for having given me, with the utmost kindness, information on many points.

128 This memoir first appeared in the 'Nederlandsch Archief voor

Genees en Natuurkunde,' Deel 5, 1870. It has been translated by Dr. W. D. Moore, under the title of " On the Action of the Eyelids in determination of Blood from expiratory effort," in 'Archives of Medicine,' edited by Dr. L. S. Beale, 1870, vol. v. p. 20.

129 Prof. Donders remarks (ibid. p. 28), that, "After injury to the eye, after operations, and in some forms of internal inflammation, we attach great value to the uniform support of the closed eyelids, and we increase this in many instances by the application of a bandage. In both cases we carefully endeavour to avoid great expiratory pressure, the disadvantage of which is well known." Mr. Bowman informs me that in the excessive photophobia, accompanying what is called scrofulous ophthalmia in children, when the light is so very painful that during weeks or months it is constantly excluded by the most forcible closure of the lids, he has often been struck on opening the lids by the paleness of the eye,—not an unnatural paleness, but an absence of the redness that might have been expected when the surface is somewhat inflamed, as is then usually the case; and this paleness he is inclined to attribute to the forcible closure of the eyelids.

130 Donders, ibid. p. 36.

[PAGE 1354]

131 Mr. Hensleigh Wedgwood (Dict. of English Etymology, 1859, vol i. p. 410) says, "the verb to weep comes from Anglo-Saxon *wop*, the primary meaning of which is simply outcry."

[PAGE 1355]

132 'De la Physionomie,' 1865, p. 217.

133 'Ceylon,' 3rd edit. 1859, vol. ii. pp. 364, 376. I applied to Mr. Thwaites, in Ceylon, for further information with respect to the weeping of the elephant; and in consequence received a letter from the Rev. Mr. Glenie, who, with others, kindly observed for me a herd of recently captured elephants. These, when irritated, screamed violently; but it is remarkable that they never when thus screaming contracted the muscles round the eyes. Nor did they shed tears; and the native hunters asserted that they had never observed elephants weeping. Nevertheless, it appears to me impossible to doubt Sir E. Tennent's distinct details about their weeping, supported as they are by the positive assertion of the keeper in the Zoological Gardens. It is certain that the two elephants in the Gardens, when they began to trumpet loudly, invariably contracted their orbicular muscles. I can reconcile these conflicting statements only by supposing that the recently captured elephants in Ceylon, from being enraged or frightened,

desired to observe their persecutors, and consequently did not contract their orbicular muscles, so that their vision might not be impeded. Those seen weeping by Sir E. Tennent were prostrate, and had given up the contest in despair. The elephants which trumpeted in the Zoological Gardens at the word of command, were, of course, neither alarmed nor enraged.

[PAGE 1356]

134 Bergeon, as quoted in the 'Journal of Anatomy and Physiology,' Nov. 1871, p. 235.

[PAGE 1357]

135 See, for instance, a case given by Sir Charles Bell, 'Philosophical Transactions,' 1823, p. 177.

[PAGE 1358]

136 See, on these several points, Prof. Donders 'On the Anomalies of Accommodation and Refraction of the Eye,' 1864, p. 573.

[PAGE 1360]

137 Quoted by Sir J. Lubbock, 'Prehistoric Times,' 1865, p. 458.

CHAPTER VII

[PAGE 1363]

138 The above descriptive remarks are taken in part from my own observations, but chiefly from Gratiolet ('De la Physionomie,' pp. 53, 337; on Sighing, 232), who has well treated this whole subject. See, also, Huschke, 'Mimices et Physiognomices, Fragmentum Physiologicum,' 1821, p. 21. On the dulness of the eyes, Dr. Piderit, 'Mimik und Physiognomik,' 1867, s. 65.

139 On the action of grief on the organs of respiration, see more especially Sir C. Bell, 'Anatomy of Expression,' 3rd edit. 1844, p. 151.

[PAGE 1364]

140 In the foregoing remarks on the manner in which the eyebrows are made oblique, I have followed what seems to be the universal opinion of all the anatomists, whose works I have consulted on the action of the above-named muscles, or with whom I have conversed. Hence throughout this work I shall take a similar view of the action of the

corrugator supercilii, orbicularis, pyramidalis nasi, and *frontalis* muscles. Dr. Duchenne, however, believes, and every conclusion at which he arrives deserves serious consideration, that it is the corrugator, called by him the *sourcilier,* which raises the inner corner of the eyebrows and is antagonistic to the upper and inner part of the orbicular muscle, as well as to the *pyramidalis nasi* (see Mécanisme de la Phys. Humaine, 1862, folio, art. v., text and figures 19 to 29: octavo edit. 1862, p. 43 text). He admits, however, that the corrugator draws together the eyebrows, causing vertical furrows above the base of the nose, or a frown. He further believes that towards the outer two-thirds of the eyebrow the corrugator acts in conjunction with the upper orbicular muscle; both here standing in antagonism to the frontal muscle. I am unable to understand, judging from Henle's drawings (woodcut, fig. 3), how the corrugator can act in the manner described by Duchenne. See, also, on this subject, Prof. Donders' remarks in the 'Archives of Medicine,' 1870, vol. v. p. 34. Mr. J. Wood, who is so well known for his careful study of the muscles of the human frame, informs me that he believes the account which I have given of the action of the corrugator to be correct. But this is not a point of any importance with respect to the expression which is caused by the obliquity of the eyebrows, nor of much importance to the theory of its origin.

[PAGE 1365]

141 I am greatly indebted to Dr. Duchenne for permission to have these two photographs (figs. 1 and 2) reproduced by the heliotype process from his work in folio. Many of the foregoing remarks on the furrowing of the skin, when the eyebrows are rendered oblique, are taken from his excellent discussion on this subject.

[PAGE 1370]

142 Mécanisme de la Phys. Humaine, Album, p. 15.

[PAGE 1371]

143 Henle, Handbuch der Anat. des Menschen, 1858, B. i. s. 148, figs. 68 and 69.

[PAGE 1372]

144 See the account of the action of this muscle by Dr. Duchenne, 'Mécanisme de la Physionomie Humaine,' Album (1862), viii. p. 34.

CHAPTER VIII

[PAGE 1375]

145 Herbert Spencer, 'Essays Scientific,' &c., 1858, p. 360.

146 F. Lieber on the vocal sounds of L. Bridgman, 'Smithsonian Contributions,' 1851, vol. ii. p. 6.

[PAGE 1376]

147 See, also, Mr. Marshall, in Phil. Transact. 1864, p. 526.

148 Mr. Bain ('The Emotions and the Will,' 1865, p. 247) has a long and interesting discussion on the Ludicrous. The quotation above given about the laughter of the gods is taken from this work. See, also, Mandeville, 'The Fable of the Bees,' vol. ii. p. 168.

149 'The Physiology of Laughter,' Essays, Second Series, 1863, p. 114.

[PAGE 1377]

150 J. Lister in 'Quarterly Journal of Microscopical Science,' 1853, vol. i, p. 266.

151 'De la Physionomie,' p. 186.

152 Sir C. Bell (Anat. of Expression, p. 147) makes some remarks on the movement of the diaphragm during laughter.

153 'Mécanisme de la Physionomie Humaine,' Album, Légende vi.

[PAGE 1378]

154 Handbuch der System. Anat. des Menschen, 1858, B. i. s. 144. See my woodcut (H. fig. 2).

[PAGE 1380]

155 See, also, remarks to the same effect by Dr. J. Crichton Browne in 'Journal of Mental Science,' April, 1871, p. 149.

156 C. Vogt, 'Mémoire sur les Microcéphales,' 1867, p. 21.

157 Sir C. Bell, 'Anatomy of Expression,' p. 133.

158 'Mimik und Physiognomik,' 1867, s. 63–67.

[PAGE 1381]

159 Sir J. Reynolds remarks ('Discourses,' xii. p. 100), "It is curious to observe, and it is certainly true, that the extremes of contrary passions are, with very little variation, expressed by the same action." He gives as an instance the frantic joy of a Bacchante and the grief of a Mary Magdalen.

[PAGE 1383]

160 Dr. Piderit has come to the same conclusion, ibid. s. 99.

[PAGE 1384]

161 'La Physionomie,' par G. Lavater, edit. of 1820, vol. iv. p. 224. See, also, Sir C. Bell, 'Anatomy of Expression,' p. 172, for the quotation given below.

[PAGE 1385]

162 A 'Dictionary of English Etymology,' 2nd edit. 1872, Introduction, p. xliv.

163 Crantz, quoted by Tylor, 'Primitive Culture,' 1871. vol. i. p. 169.

164 F. Lieber, 'Smithsonian Contributions,' 1851, vol. ii. p. 7.

165 Mr. Bain remarks ('Mental and Moral Science,' 1868, p. 239), "Tenderness is a pleasurable emotion, variously stimulated, whose effort is to draw human beings into mutual embrace."

[PAGE 1386]

166 Sir J. Lubbock, 'Prehistoric Times,' 2nd edit. 1869, p. 552, gives full authorities for these statements. The quotation from Stecle is taken from this work.

167 See a full account, with references, by E. B. Tylor, 'Researches into the Early History of Mankind,' 2nd edit. 1870, p. 51.

[PAGE 1388]

168 'The Descent of Man,' p. 1209.

169 Dr. Maudsley has a discussion to this effect in his 'Body and Mind,' 1870, p. 85.

[PAGE 1389]

170 'The Anatomy of Expression,' p, 103, and 'Philosophical Transactions,' 1823, p. 182.

171 'The Origin of Language,' 1866, p. 146. Mr. Tylor ('Early History of Mankind,' 2nd edit. 1870, p. 48) gives a more complex origin to the position of the hands during prayer.

CHAPTER IX

[PAGE 1390]

172 'Anatomy of Expression,' pp. 137, 139. It is not surprising that the cor-

rugators should have become much more developed in man than in the anthropoid apes; for they are brought into incessant action by him under various circumstances, and will have been strengthened and modified by the inherited effects of use. We have seen how important a part they play, together with the orbiculares, in protecting the eyes from being too much gorged with blood during violent expiratory movements. When the eyes are closed as quickly and as forcibly as possible, to save them from being injured by a blow, the corrugators contract. With savages or other men whose heads are uncovered, the eyebrows are continually lowered and contracted to serve as a shade against a too strong light; and this is effected partly by the corrugators. This movement would have been more especially serviceable to man, as soon as his early progenitors held their heads erect. Lastly, Prof. Donders believes ('Archives of Medicine,' ed. by L. Beale, 1870, vol. v. p. 34), that the corrugators are brought into action in causing the eyeball to advance in accommodation for proximity in vision.

173 'Mécanisme de la Physionomie Humaine,' Album, Légende iii.

[PAGE 1391]

174 'Mimik und Physiognomik,' s. 46.

175 'History of the Abipones,' Eng. translat. vol. ii. p. 59, as quoted by Lubbock, 'Origin of Civilisation,' 1870, p. 355.

[PAGE 1393]

176 'De la Physionomie,' pp. 15, 144, 146. Mr. Herbert Spencer accounts for frowning exclusively by the habit of contracting the brows as a shade to the eyes in a bright light: see 'Principles of Psychology,' 2nd edit. 1872, p. 546.

[PAGE 1394]

177 Gratiolet remarks (De la Phys. p. 35), "Quand l'attention est fixée sur quelque image intérieure, l'œil regarde dans le vide et s'associe automatiquement à la contemplation de l'esprit." But this view hardly deserves to be called an explanation.

178 'Miles Gloriosus,' act ii. sc. 2.

[PAGE 1395]

179 The original photograph by Herr Kindermann is much more expressive than this copy, as it shows the frown on the brow more plainly.

180 'Mécanisme de la Physionomie Humaine,' Album, Légende iv. figs. 16–18.

[PAGE 1396]

181 Hensleigh Wedgwood on 'The Origin of Language,' 1866, p. 78.

[PAGE 1397]

182 Müller, as quoted by Huxley, 'Man's Place in Nature,' 1863, p. 38.
183 I have given several instances in my 'Descent of Man,' vol. i. chap. iv.

[PAGE 1398]

184 Anatomy of Expression,' p. 190.
185 'De la Physionomie,' pp. 118–121.

[PAGE 1399]

186 'Mimik und Physiognomik,' s. 79.

CHAPTER X

[PAGE 1400]

187 See some remarks to this effect by Mr. Bain, 'The Emotions and the Will,' 2nd edit. 1865, p. 127.

[PAGE 1401]

188 Rengger, Naturgesch. der Säugethiere von Paraguay, 1830, s. 3.
189 Sir C. Bell, 'Anatomy of Expression,' p. 96. On the other hand, Dr. Burgess ('Physiology of Blushing,' 1839, p. 31) speaks of the reddening of a cicatrix in a negress as of the nature of a blush.
190 Moreau and Gratiolet have discussed the colour of the face under the influence of intense passion: see the edit. of 1820 of Lavater, vol. iv. pp. 282 and 300; and Gratiolet, 'De la Physionomie,' p. 345.
191 Sir C. Bell ('Anatomy of Expression,' pp. 91, 107) has fully discussed this subject. Moreau remarks (in the edit. of 1820 of 'La Physionomie, par G. Lavater,' vol. iv. p. 237), and quotes Portal in confirmation, that asthmatic patients acquire permanently expanded nostrils, owing to the habitual contraction of the elevatory muscles of the wings of the nose. The explanation by Dr. Piderit ('Mimik und Physiognomik,' s. 82) of the distension of the nostrils, namely, to allow free breathing whilst the mouth is closed and the teeth clenched, does not appear to be nearly so correct as that by Sir C. Bell, who attributes it to the sympathy (*i.e.* habitual co-action) of all the respiratory muscles. The nostrils of an angry man may be seen to become dilated, although his mouth is open.

192 Mr. Wedgwood, 'On the Origin of Language,' 1866, p. 76. He also observes that the sound of hard breathing "is represented by the syllables *puff, huff, whiff,* whence a *huff* is a fit of ill-temper."

193 Sir C. Bell ('Anatomy of Expression,' p. 95) has some excellent remarks on the expression of rage.

[PAGE 1402]

194 'De la Physionomie,' 1865, p. 346.

195 Sir C. Bell, 'Anatomy of Expression,' p. 177. Gratiolet (De la Phys. p. 369) says, "les dents se découvrent, et imitent symboliquement l'action de déchirer et de mordre." If, instead of using the vague term *symboliquement,* Gratiolet had said that the action was a remnant of a habit acquired during primeval times when our semi-human progenitors fought together with their teeth, like gorillas and orangs at the present day, he would have been more intelligible. Dr. Piderit ('Mimik,' &c., s. 82) also speaks of the retraction of the upper lip during rage. In an engraving of one of Hogarth's wonderful pictures, passion is represented in the plainest manner by the open glaring eyes, frowning forehead, and exposed grinning teeth.

196 'Oliver Twist,' vol. iii. p. 245.

[PAGE 1403]

197 'The Spectator,' July 11, 1868, p. 819.

[PAGE 1404]

198 'Body and Mind,' 1870, pp. 51–53.

199 Le Brun, in his well-known 'Conférence sur l'Expression' ('La Physionomie, par Lavater,' edit. of 1820, vol. ix. p. 268), remarks that anger is expressed by the clenching of the fists. See, to the same effect, Huschke, 'Mimices et Physiognomices, Fragmentum Physiologicum, 1824, p. 20. Also Sir C. Bell, 'Anatomy of Expres-sion,' p. 219.

[PAGE 1406]

200 Transact. Philosoph. Soc., Appendix, 1746, p. 65.

[PAGE 1407]

201 'Anatomy of Expression,' p. 136. Sir C. Bell calls (p. 131) the muscles which uncover the canines the *snarling muscles.*

202 Hensleigh Wedgwood, 'Dictionary of English Etymology,' 1865, vol. iii. pp. 240, 243.

[PAGE 1408]

203 'The Descent of Man,' p. 849.

CHAPTER XI

[PAGE 1409]

204 'De la Physionomie et la Parole,' 1865, p. 89.

[PAGE 1410]

205 'Physionomie Humaine,' Album, Légende viii. p. 35. Gratiolet also speaks (De la Phys. 1865, p. 52) of the turning away of the eyes and body.

206 Dr. W. Ogle, in an interesting paper on the Sense of Smell ('Medico-Chirurgical Transactions,' vol. liii. p. 268), shows that when we wish to smell carefully, instead of taking one deep nasal inspiration, we draw in the air by a succession of rapid short sniffs. If "the nostrils be watched during this process, it will be seen that, so far from dilating, they actually contract at each sniff. The contraction does not include the whole anterior opening, but only the posterior portion." He then explains the cause of this movement. When, on the other hand, we wish to exclude any odour, the contraction, I presume, affects only the anterior part of the nostrils.

207 'Mimik und Physiognomik,' ss. 84, 93. Gratiolet (ibid. p. 155) takes nearly the same view with Dr. Piderit respecting the expression of contempt and disgust.

208 Scorn implies a strong form of contempt; and one of the roots of the word 'scorn' means, according to Mr. Wedgwood (Dict. of English Etymology, vol. iii. p. 125), ordure or dirt. A person who is scorned is treated like dirt.

209 'Early History of Mankind,' 2nd edit. 1870, p. 45.

[PAGE 1412]

210 See, to this effect, Mr. Hensleigh Wedgwood's Introduction to the 'Dictionary of English Etymology,' 2nd edit. 1872, p. xxxvii.

211 Duchenne believes that in the eversion of the lower lip, the corners are drawn downwards by the *depressores anguli oris.* Henle (Handbuch d. Anat. des Menschen, 1858, B. i. s. 151) concludes that this is effected by the *musculus quadratus menti.*

[PAGE 1413]

212 As quoted by Tylor, 'Primitive Culture,' 1871, vol. i. p. 169.

[PAGE 1414]

213 Both these quotations are given by Mr. H. Wedgwood, 'On the Origin of Language,' 1866, p. 75.

214 This is stated to be the case by Mr. Tylor (Early Hist. of Mankind, 2nd edit. 1870, p. 52); and he adds, "it is not clear why this should be so."

[PAGE 1415]

215 'Principles of Psychology,' 2nd edit. 1872, p. 552.

216 Gratiolet (De la Phys. p. 351) makes this remark, and has some good observations on the expression of pride. See Sir C. Bell ('Anatomy of Expression,' p. 111) on the action of the *musculus superbus.*

[PAGE 1420]

217 'Anatomy of Expression,' p. 166.

218 'Journey through Texas,' p. 352.

219 Mrs. Oliphant,' The Brownlows,' vol. ii. p. 206.

[PAGE 1422]

220 'Essai sur le Langage,' 2nd edit. 1846. I am much indebted to Miss Wedgwood, for having given me this information, with an extract from the work.

221 'On the Origin of Language,' 1866, p. 91.

222 'On the Vocal Sounds of L. Bridgman;' Smithsonian Contributions, 1851, vol. ii. p. 11.

223 'Mémoire sur les Microcéphales,' 1867, p. 27.

224 Quoted by Tylor, 'Early History of Mankind,' 2nd edit. 1870, p. 38.

[PAGE 1423]

225 Mr. J. B. Jukes, 'Letters and Extracts,' &c., 1871, p. 248.

226 F. Lieber, 'On the Vocal Sounds,' &c., p. 11. Tylor, ibid. p. 53.

227 Dr. King, Edinburgh Phil. Journal, 1845, p. 313.

228 Tylor, 'Early History of Mankind,' 2nd edit. 1870, p. 53.

[PAGE 1424]

229 Lubbock, 'The Origin of Civilization,' 1870, p. 277. Tylor, ibid. p. 38. Lieber (ibid. p. 11) remarks on the negative of the Italians.

CHAPTER XII

[PAGE 1425]

230 'Mécanisme de la Physionomie,' Album, 1862, p. 42.

[PAGE 1426]

231 'The Polyglot News Letter,' Melbourne, Dec. 1858, p. 2.

[PAGE 1427]

232 'The Anatomy of Expression,' p. 106.

233 'Mécanisme de la Physionomie,' Album, p. 6.

234 See, for instance, Dr. Piderit ('Mimik und Physiognomik,' s. 88), who has a good discussion on the expression of surprise.

235 Dr. Murie has also given me information leading to the same conclusion, derived in part from comparative anatomy.

[PAGE 1428]

236 'De la Physionomie,' 1865, p. 234.

237 See, on this subject, Gratiolet, ibid. p. 254.

[PAGE 1429]

238 Lieber, 'On the Vocal Sounds of Laura Bridgman,' Smithsonian Contributions, 1851, vol. ii. p. 7.

239 'Wenderholme,' vol. ii. p. 91.

[PAGE 1430]

240 Lieber, 'On the Vocal Sounds,' &c., ibid. p. 7.

241 Huschke, 'Mimices et Physiognomices,' 1821, p. 18. Gratiolet (De la Phys. p. 255) gives a figure of a man in this attitude, which, however, seems to me expressive of fear combined with astonishment. Le Brun also refers (Lavater, vol. ix. p. 299) to the hands of an astonished man being opened.

[PAGE 1431]

242 Huschke, ibid. p. 18.

243 'North American Indians,' 3rd edit. 1842, vol. i. p. 105.

[PAGE 1432]

244 H. Wedgwood, Dict. of English Etymology, vol. ii. 1862, p. 35. See,

also, Gratiolet ('De la Physionomie,' p. 135) on the sources of such words as 'terror, horror, rigidus, frigidus,' &c.

245 Mr. Bain ('The Emotions and the Will,' 1865, p. 54) explains in the following manner the origin of the custom "of subjecting criminals in India to the ordeal of the morsel of rice. The accused is made to take a mouthful of rice, and after a little time to throw it out. If the morsel is quite dry, the party is believed to be guilty,—his own evil conscience operating to paralyse the salivating organs."

[PAGE 1433]

246 Sir C. Bell, Transactions of Royal Phil. Soc. 1822, p. 308. 'Anatomy of Expression,' p. 88 and pp. 164–169.

247 See Moreau on the rolling of the eyes, in the edit. of 1820 of Lavater, tome iv. p. 263. Also, Gratiolet, De la Phys. p. 17.

248 'Observations on Italy,' 1825, p. 48, as quoted in 'The Anatomy of Expression,' p. 168.

[PAGE 1435]

249 Quoted by Dr. Maudsley,' Body and Mind,' 1870, p. 41.

[PAGE 1437]

250 'Anatomy of Expression,' p. 168.

251 Mécanisme de la Phys. Humaine, Album, Légende xi.

[PAGE 1439]

252 Duchenne takes, in fact, this view (ibid. p. 45), as he attributes the contraction of the platysma to the shivering of fear *(frisson de la peur);* but he elsewhere compares the action with that which causes the hair of frightened quadrupeds to stand erect; and this can hardly be considered as quite correct.

[PAGE 1440]

253 'De la Physionomie,' pp. 51, 256, 346.

254 As quoted in White's 'Gradation in Man,' p. 57.

255 'Anatomy of Expression,' p. 169.

256 'Mécanisme de la Physionomie,' Album, pl. 65, pp. 44, 45.

[PAGE 1441]

257 See remarks to this effect by Mr. Wedgwood, in the Introduction to his 'Dictionary of English Etymology,' 2nd edit. 1872, p. xxxvii.

CHAPTER XIII

[PAGE 1443]

258 'The Physiology or Mechanism of Blushing,' 1839, p. 156. I shall have occasion often to quote this work in the present chapter.

259 Dr. Burgess, ibid. p. 56. At p. 33 he also remarks on women blushing more freely than men; as stated below.

[PAGE 1444]

260 Quoted by Vogt, 'Mémoire sur les Microcéphales,' 1867, p. 20. Dr. Burgess (ibid. p. 56) doubts whether idiots ever blush.

261 Lieber 'On the Vocal Sounds,' &c.; Smithsonian Contributions, 1851, vol. ii. p. 6.

262 Ibid. p. 182.

263 Moreau, in edit. of 1820 of Lavater, vol. iv. p. 303.

[PAGE 1445]

264 Burgess, ibid. p. 38, on paleness after blushing, p. 177.

[PAGE 1446]

265 See Lavater, edit. of 1820, vol. iv. p. 303.

266 Burgess, ibid. pp. 114, 122. Moreau in Lavater, ibid. vol. iv. p. 293.

[PAGE 1447]

267 'Letters from Egypt,' 1865, p. 66. Lady Gordon is mistaken when she says Malays and Mulattoes never blush.

268 Capt. Osborn ('Quedah,' p. 199), in speaking of a Malay, whom he reproached for cruelty, says he was glad to see that the man blushed.

269 J. R. Forster, 'Observations during a Voyage round the World,' 4to, 1778, p. 229. Waitz gives ('Introduction to Anthropology,' Eng. translat. 1863, vol. i. p. 135) references for other islands in the Pacific. See, also, Dampier 'On the Blushing of the Tunquinese' (vol. ii. p. 40); but I have not consulted this work. Waitz quotes Bergmann, that the Kalmucks do not blush, but this may be doubted after what we have seen with respect to the Chinese. He also quotes Roth, who denies that the Abyssinians are capable of blushing. Unfortunately, Capt. Speedy, who lived so long with the Abyssinians, has not answered my inquiry on this head. Lastly, I must add that the Rajah Brooke has never observed the least sign of a blush with the Dyaks of Borneo; on the contrary, under circumstances which would excite a blush in us, they assert "that they feel the blood drawn from their faces."

[PAGE 1448]

270 Transact. of the Ethnological Soc. 1870, vol. ii. p. 16.

271 Humboldt, 'Personal Narrative,' Eng. translat. vol. iii. p. 229.

272 Quoted by Prichard, Phys. Hist. of Mankind, 4th edit. 1851, vol. i. p. 271.

273 See, on this head, Burgess, ibid. p. 32. Also Waitz, 'Introduction to Anthropology,' Eng. edit. vol. i. p. 135. Moreau gives a detailed account ('Lavater,' 1820, tom. iv. p. 302) of the blushing of a Madagascar negress-slave when forced by her brutal master to exhibit her naked bosom.

274 Quoted by Prichard, Phys. Hist. of Mankind, 4th edit. 1851, vol. i. p. 225.

[PAGE 1449]

275 Burgess, ibid. p. 31. On mulattoes blushing, see p. 33. I have received similar accounts with respect to mulattoes.

276 Barrington also says that the Australians of New South Wales blush, as quoted by Waitz, ibid. p. 135.

277 Mr. Wedgwood says (Dict. of English Etymology, vol. iii. 1865, p.155) that the word shame "may well originate in the idea of shade or concealment, and may be illustrated by the Low German *scheme*, shade or shadow." Gratiolet (De la Phys. pp. 357–362) has a good discussion on the gestures accompanying shame; but some of his remarks seem to me rather fanciful. See, also, Burgess (ibid. pp. 69, 134) on the same subject.

[PAGE 1450]

278 Burgess, ibid. pp. 181, 182. Boerhaave also noticed (as quoted by Gratiolet, ibid. p. 361) the tendency to the secretion of tears during intense blushing. Mr. Bulmer, as we have seen, speaks of the "watery eyes" of the children of the Australian aborigines when ashamed.

[PAGE 1451]

279 See also Dr. J. Crichton Browne's 'Memoir' on this subject in the West Riding Lunatic Asylum Medical Report, 1871, pp. 95–98.

[PAGE 1453]

280 In a discussion on so-called animal magnetism in 'Table Talk,' vol. i.

281 Ibid. p. 40.

282 Mr. Bain ('The Emotions and the Will,' 1865, p. 65) remarks on "the

shyness of manners which is induced between the sexes . . . . from the influence of mutual regard, by the apprehension on either side of not standing well with the other."

283 See, for evidence on this subject, 'The Descent of Man,' &c., pp. 1061, 1211.

[PAGE 1455]

284 H. Wedgwood, Dict. English Etymology, vol. iii. 1865, p. 184. So with the Latin word *verecundus*.

285 Mr. Bain ('The Emotions and the Will,' p. 64) has discussed the "abashed" feelings experienced on these occasions, as well as the *stage-fright* of actors unused to the stage. Mr. Bain apparently attributes these feelings to simple apprehension or dread.

[PAGE 1456]

286 'Essays on Practical Education,' by Maria and R. L. Edgeworth, new edit. vol. ii. 1822, p. 38. Dr. Burgess (ibid. p. 187) insists strongly to the same effect.

287 Ibid. p. 50.

[PAGE 1459]

288 Bell, 'Anatomy of Expression,' p. 95. Burgess, as quoted below, ibid. p. 49. Gratiolet, De la Phys. p. 94.

289 On the authority of Lady Mary Wortley Montague; see Burgess, ibid. p. 43.

[PAGE 1460]

290 In England, Sir H. Holland was, I believe, the first to consider the influence of mental attention on various parts of the body, in his 'Medical Notes and Reflections,' 1839, p. 64. This essay, much enlarged, was reprinted by Sir H. Holland in his 'Chapters on Mental Physiology,' 1858, p. 79, from which work I always quote. At nearly the same time, as well as subsequently, Prof. Laycock discussed the same subject: see 'Edinburgh Medical and Surgical Journal,' 1839, July, pp. 17–22. Also his 'Treatise on the Nervous Diseases of Women,' 1840, p. 110; and 'Mind and Brain,' vol. ii. 1860, p. 327. Dr. Carpenter's views on mesmerism have a nearly similar bearing. The great physiologist Müller treated ('Elements of Physiology,' Eng. translat, vol. ii. pp. 937, 1085) of the influence of the attention on the senses. Sir J. Paget discusses the influence of the mind on the nutrition of parts, in his 'Lectures on Surgical Pathology,' 1853, vol. i. p.

39: I quote from the 3rd edit. revised by Prof. Turner, 1870, p. 28. See, also, Gratiolet, De la Phys. pp. 283–287.

291 De la Phys. p. 283.

292 'Chapters on Mental Physiology,' 1858, p. 111.

293 'Mind and Brain,' vol ii. 1860, p. 327.

294 'Chapters on Mental Physiology,' pp. 104–106.

[PAGE 1461]

295 See Gratiolet on this subject, De la Phys. p. 287.

296 Dr. J. Crichton Browne, from his observations on the insane, is convinced that attention directed for a prolonged period on any part or organ may ultimately influence its capillary circulation and nutrition. He has given me some extraordinary cases; one of these, which cannot here be related in full, refers to a married woman fifty years of age, who laboured under the firm and long-continued delusion that she was pregnant. When the expected period arrived, she acted precisely as if she had been really delivered of a child, and seemed to suffer extreme pain, so that the perspiration broke out on her forehead. The result was that a state of things returned, continuing for three days, which had ceased during the six previous years. Mr. Braid gives, in his 'Magic, Hypnotism,' &c., 1852, p. 95, and in his other works, analogous cases, as well as other facts showing the great influence of the will on the mammary glands, even on one breast alone.

297 Dr. Maudsley has given ('The Physiology and Pathology of Mind,' 2nd edit. 1868, p. 105), on good authority, some curious statements with respect to the improvement of the sense of touch by practice and attention. It is remarkable that when this sense has thus been rendered more acute at any point of the body, for instance, in a finger, it is likewise improved at the corresponding point on the opposite side of the body.

298 'The Lancet,' 1838, pp. 39–40, as quoted by Prof. Laycock, 'Nervous Diseases of Women,' 1840, p. 110.

299 'Chapters on Mental Physiology,' 1858, pp. 91–93.

300 'Lectures on Surgical Pathology,' 3rd edit, revised by Prof. Turner, 1870, pp. 28, 31.

301 'Elements of Physiology,' Eng. translat. vol. ii. p. 938.

302 Prof. Laycock has discussed this point in a very interesting manner. See his 'Nervous Diseases of Women,' 1840, p. 110.

[PAGE 1462]

303 See, also, Mr. Michael Foster, on the action of the vaso-motor system, in his interesting Lecture before the Royal Institution, as translated in the 'Revue des Cours Scientifiques,' Sept. 25, 1869, p. 683.

CHAPTER XIV

[PAGE 1470]

304 See the interesting facts given by Dr. Bateman on 'Aphasia,' 1870, p. 110.

[PAGE 1471]

305 'La Physionomie et la Parole,' 1865, pp. 103, 118.

306 Rengger, 'Naturgeschichte der Säugethiere von Paraguay,' 1830, s. 55.

[PAGE 1476]

307 Quoted by Moreau, in his edition of Lavater, 1820, tom. iv. p. 211.

308 Gratiolet ('De la Physionomie,' 1865, p. 66) insists on the truth of this conclusion.

Index to *The Voyage of the Beagle*

Index to *On the Origin of Species*

Index to *The Descent of Man*

INDEX TO *THE EXPRESSION OF THE EMOTIONS IN MAN AND ANIMALS*

General Index

This combined index covers all four of Darwin's books transcribed in the present collection. It contains key names and concepts used by Darwin himself, plus modern terms for ideas either discussed explicitly or foreshadowed in the Darwin texts. Fine details, in the language of 1845–1872, are preserved in the original Darwin indexes, also reproduced in this collection.

—E. O. Wilson

A

B

C

D

E

F

G

J

K

L

M

N

O

P

R

W

Y

Z

Acknowledgments

THE SOURCES FOR this collection of the four major works by Charles Darwin are as follows: *Journal of researches into the natural history and geology of the countries visited during the voyage of H.M.S.* Beagle *round the world,* Charles Darwin, the 1845 first edition republished as *The Voyage of the Beagle,* with an introduction by Steve Jones (New York: Modern Library, 2001); *On the Origin of Species,* Charles Darwin, with an introduction by Ernst Mayr (Cambridge, MA: Harvard University Press, 1964; 18th printing, 2003), a facsimile of the first edition, published in 1859; *The Descent of Man, and Selection in Relation to Sex,* Charles Darwin, with an introduction by John Tyler Bonner and Robert M. May (Princeton, NJ: Princeton University Press, 1981), a photoreproduction of the 1871 first edition published by John Murray, London; *The Expression of the Emotions in Man and Animals,* Charles Darwin, with an introduction by S. J. Rachman (New York: St. Martin's Press, 1979), the text being that of the 1872 first edition.

The map of the route of H.M.S. *Beagle* is taken from *The Origin of Species,* revised edition, by Charles Darwin, introduced and abridged by Philip Appleman (New York: W. W. Norton, 2002). The other illustrations were taken from: *Journal of Researches into the Natural History and Geology of the Countries Visited during the Voyage of H.M.S.* Beagle *Round the World,* by Charles Darwin, in 2 volumes (New York: Harper & Brothers, 1846; original publication in 1845); *On the Origin of Species,* by Charles Darwin, a facsimile of the first edition, with an introduction by Ernst Mayr (Cambridge, MA: Harvard University Press, 1964); *The Descent of Man, and Selection in Relation to Sex,* by Charles Darwin, in 2 volumes (New York: D. Appleton and Company, 1871); and *The Expression of the Emotions in Man and Animals,* by Charles Darwin (London: John Murray, 1872). Charles Darwin age 31 (1840) is a portrait by George Richmond reproduced in Gavin de Beer, *Charles Darwin, Evolution by Natural Selection* (London: Thomas Nelson & Sons Ltd., 1963). Charles Darwin age 51 (1860) is a photograph by Maull &

Fox reproduced in Nora Barlow, *The Autobiographpy of Charles Darwin 1809–1882* (New York: Harcourt, Brace & Co., 1958). Charles Darwin age 65 (1874) © The Natural History Museum, London. Charles Darwin age 72 (1881) is a photograph by Elliott & Fray, London, reproduced in Francis Darwin, ed., *More Letters of Charles Darwin; A record of his work in a series of hitherto unpublished letters,* Vol. II (New York: Appleton and Company, 1903). The contrast of each figure was adjusted in a way to maximize clarity and definition, as must have been intended for the original drawings and photographs, and expressed in the fresh first editions.

I am grateful to Ms. Kathleen Horton for her meticulous work on the illustrations, as well as in proofing and organizing the transcription of the Darwin books.

—E. O. Wilson